W9-CCI-616

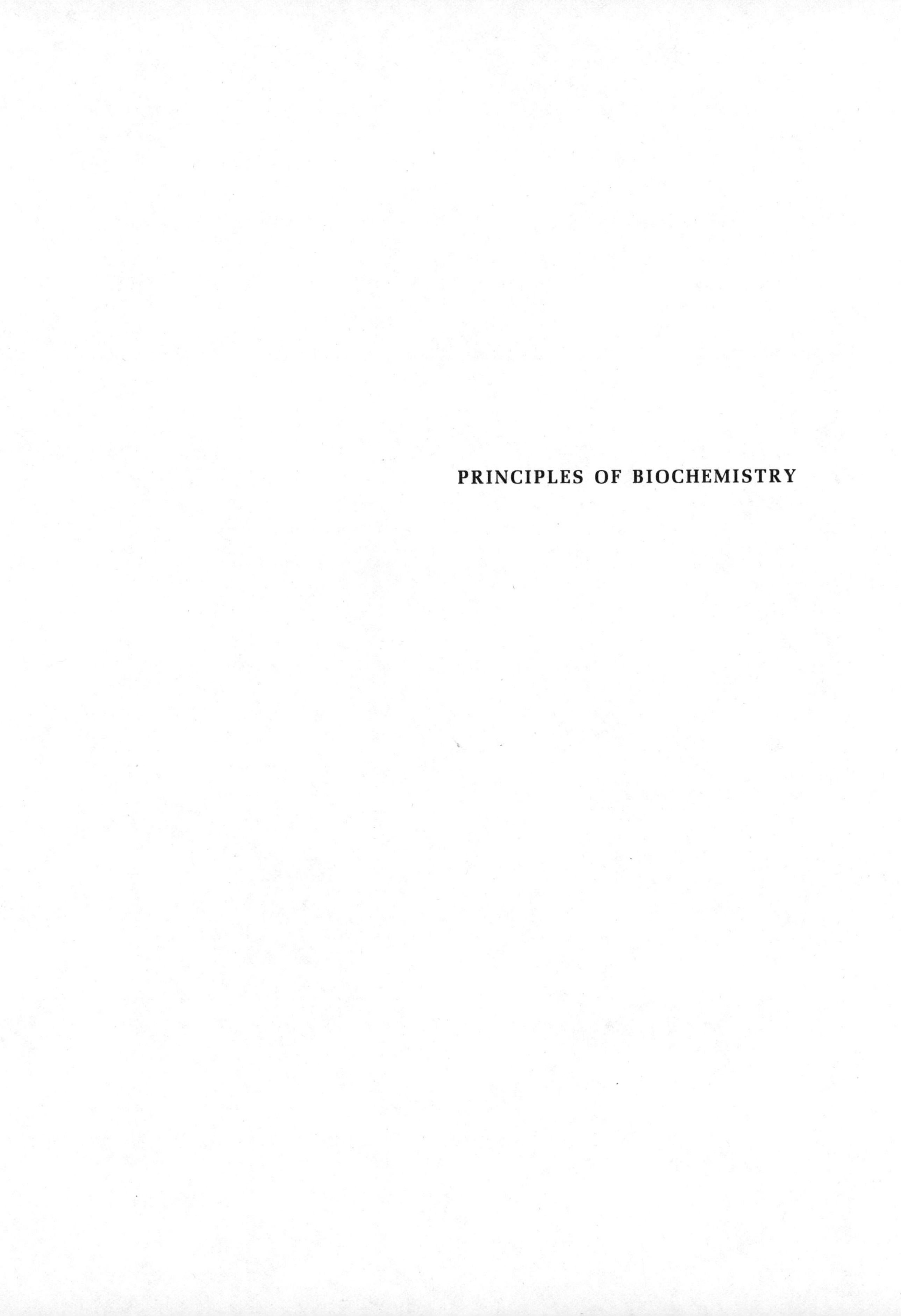

PRINCIPLES OF BIOCHEMISTRY

Principles of Biochemistry

ALBERT L. LEHNINGER

THE JOHNS HOPKINS UNIVERSITY

SCHOOL OF MEDICINE

WORTH PUBLISHERS, INC.

FOR JAN

Principles of Biochemistry

Printed in the United States of America

Library of Congress Catalog Card No. 82-70015

ISBN: 0-87901-136-X

Third Printing, December 1984

Editors: Sally Anderson, June Fox

Production: Kenneth Ekkens

Illustrator: Shirley Baty

Picture Editor: Anne Feldman

Design: Malcolm Grear Designers

Typographer: Progressive Typographers, Inc.

Printing and Binding: Rand McNally and Company

Cover: Computer graphics representation of bovine trypsin inhibitor–trypsin complex (inhibitor surface in red, trypsin surface in green), courtesy of Robert Langridge

Worth Publishers, Inc.

444 Park Avenue South

New York, New York 10016

Preface

Principles of Biochemistry is intended primarily for students taking their first course in biochemistry. It is a new book, not simply an updating of my earlier books, *Biochemistry* (1970, 1975) and *Short Course in Biochemistry* (1973). In setting out to prepare new editions of those books, I grew increasingly uneasy about my objectives. The first edition of *Biochemistry*, published in 1970, was primarily for an undergraduate audience, for those taking their first, perhaps their only, course in biochemistry. The second edition in 1975 grew by more than 20 percent. A third edition, with a proportional effort to include new biochemical advances, would become a volume of 1,500 pages. Such a book could of course play an important role in biochemical education, but it would not properly serve the undergraduate audience for whom I had written *Biochemistry* in the first place. *Principles of Biochemistry* is, then, a return to my original objective; it is, in a manner of speaking, the 1982 rebirth of the first edition of *Biochemistry*.

Size was not the only consideration. The time has come when a single biochemistry textbook cannot be all things to all students. A comprehensive book that describes the full panoply of today's biochemistry at a level that would satisfy the needs of graduate students would surely be found intimidating by most undergraduates in their first encounter with the field. Textbooks have a tendency, moreover, to acquire more complex structure and become more densely written in successive editions, with the result that they often lose the very clarity of exposition and organization that made their first editions successful. A fresh start seemed called for.

I once felt that biochemistry should be primarily a graduate subject, to be approached only after a thorough grounding in chemistry and biology. Today I have quite a different view. Biochemistry should be taught much earlier, since it has become the *lingua franca* of the life sciences and greatly illuminates subsequent study in any area of biology. And not only biology: An early course in biochemistry for students of chemistry or physics provides challenging glimpses of how living organisms solve some of the most fundamental chemical and physical problems.

Considered more broadly, an undergraduate course in biochemistry also has a place in educating young people for a future in which there will be ever-greater concern for the health and well-being of mankind. The extraordinary advances in biochemical genetics and genetic engineering, together with their social implications, are already matters of wide public interest. The growing world population, with its increasing demands for food, raw materials, and energy, can even now be seen to impinge on the delicate ecological balances within the biosphere. Increasingly, society must make important decisions involving conflicts between biological principles and political, industrial, or ethical concerns. It can therefore be argued that a knowledge of biochemistry is useful for all well-informed citizens, whatever their calling—quite apart from the special intellectual excitement it offers to those who wish to explore and understand the molecular interactions that take place in living organisms.

Principles of Biochemistry is made up of four parts: biomolecules, bioenergetics and metabolism, aspects of human biochemistry, and the fundamentals of molecular genetics; it is written in the same style and language I used in *Biochemistry*. Throughout the book I have tried to emphasize the framework and the molecular logic of biochemistry rather than encyclopedic detail, always with full explanations and descriptions of fundamental processes.

The book opens with chapters on cell structure and on some rudiments of organic chemistry relevant to biomolecules; thus it can be useful to those with a minimal background in biology and organic chemistry. After considering the properties of water, the structure and biological functions of proteins are described in depth. Hemoglobin is examined in detail to show how amino acid sequence and primary structure determine conformation, and how conformation can influence cell structure and function. Enzymes and the regulation of enzyme activity are then treated in depth, with repeated emphasis on conformation illustrated by a "gallery" of enzyme structures. Chapters on vitamins and coenzymes, on carbohydrates, and on lipids and membranes complete Part I.

Part II deals with bioenergetics and cell metabolism—the "meat and potatoes" of biochemistry. A thorough grounding in cell bioenergetics is followed by detailed discussions of glycolysis, the citric acid cycle, electron transport, and oxidative phosphorylation. Chapters on the catabolism of fatty acids and amino acids follow, succeeded by chapters on biosynthetic pathways and photosynthesis. Regulation of metabolic pathways is discussed in detail.

Part III is devoted to human biochemistry. It includes chapters on organ relationships in metabolism, endocrine regulation, and human nutrition. To me, nutrition is not simply a matter of knowing that a given vitamin serves as part of a given coenzyme. The science of nutrition is one of biochemistry's greatest contributions to human welfare, and I believe it deserves a more holistic treatment than it usually receives.

In Part IV I have provided an especially full treatment of the "cutting edges" of molecular genetics; these chapters take into account the rapid pace of new developments (through 1981), including the techniques of DNA cloning.

Throughout the book there are many interest-provoking items of related information, some historical in nature, many dealing with medicine and human health, still others touching on zoology and animal physiology, agriculture and food, environmental issues, and world nutritional problems. Occasionally, there are brief sections covering more difficult, quantitative, or interesting but optional information. This material may not be covered in all courses, so it is boxed for easy identification. Examples include the derivation of the Henderson-Hasselbalch equation, the *RS* system, how to tell a person's age through amino acid chemistry, and the sequencing of DNA.

There are almost 850 illustrations, tables, charts, and photographs in the book. Each chapter has a summary as well as a useful list of readings and references. There is a comprehensive glossary of over 400 biochemical terms at the end of the book.

Particularly noteworthy are the problems at the end of each chapter, over 350 altogether, most of them written by Paul van Eikeren of Harvey Mudd College. The problems are not merely numerical; they focus on biochemical reasoning and require thoughtful analysis. All of them, together with their answers in an appendix, have been thoroughly reviewed by experienced teachers of undergraduate biochemistry.

In presenting this new book, I again welcome suggestions and criticisms from teachers and students alike.

Acknowledgments

I am very grateful to those who have helped me prepare this book. First, I must thank Paul van Eikeren for writing most of the problems in the book and the answers in the appendix. Carl Shonk of Central Michigan State University went over each problem and its solution with a fine-toothed comb and made many valuable suggestions to enhance their didactic value. I want also to thank Barbara Sollner-Webb of Johns Hopkins School of Medicine for providing the problems for the chapters on genetic biochemistry.

The entire text, both in drafts and final version, was reviewed in detail by Edward Harris of Texas A & M University, James Hageman of New Mexico State University, and Carl Shonk. Specific sections of the manuscript were also closely scrutinized by Norman Sansing, University of Georgia; James Bamburg, Colorado State University; Michael Dahmus, University of California, Davis; and Paul Englund and Barbara Sollner-Webb of the Johns Hopkins School of Medicine. Keith

Roberts of the John Innes Institute provided useful suggestions for illustrations in the early chapters. Geoffrey Martin carefully checked the accuracy of all the equations and structural formulas in the book, and Linda Hansford proofread the entire book and prepared the index. I must, however, take sole responsibility for errors of fact or emphasis.

I am especially grateful to Peggy Jane Ford, who not only typed the manuscript, several times over, but also marshaled my time and attention to the competing demands of teaching, research, administration, and book writing. I also wish to thank June Fox and particularly Sally Anderson of Worth Publishers, who edited and guided the book through production. Indeed, I wish to thank the entire staff of Worth Publishers for their understanding, encouragement, and practical help. An author could not ask for better cooperation in seeing his brainchild into print.

Finally, I must acknowledge with deep appreciation the indispensable aid and encouragement of my wife, who not only tolerated the occupational agonies that beset the long-distance writer, but also served as my keenest critic of style and language.

ALBERT L. LEHNINGER

Sparks, Maryland
January 1982

Contents in Brief

PART I

Biomolecules 1

1. Biochemistry: The Molecular Logic of Living Organisms 3
2. Cells 15
3. The Composition of Living Matter: Biomolecules 45
4. Water 67
5. Amino Acids and Peptides 95
6. Proteins: Covalent Structure and Biological Function 121
7. Fibrous Proteins 147
8. Globular Proteins: The Structure and Function of Hemoglobin 169
9. Enzymes 207
10. Vitamins and Trace Elements in the Function of Enzymes 249
11. Carbohydrates: Structure and Biological Function 277
12. Lipids and Membranes 303

PART II

Bioenergetics and Metabolism 331

13. A Survey of Metabolism 333
14. The ATP Cycle and Cell Bioenergetics 361
15. Glycolysis: A Central Pathway of Glucose Catabolism 397
16. The Citric Acid Cycle 435
17. Electron Transport, Oxidative Phosphorylation, and Regulation of ATP Production 467
18. The Oxidation of Fatty Acids in Animal Tissues 511
19. Oxidative Degradation of Amino Acids: The Urea Cycle 531
20. Biosynthesis of Carbohydrates in Animal Tissues 561
21. The Biosynthesis of Lipids 583
22. Biosynthesis of Amino Acids and Nucleotides 615
23. Photosynthesis 645

PART III

Some Aspects of Human Biochemistry 681

24. Digestion, Transport, and the Integration of Metabolism 683
25. Hormones 721
26. Human Nutrition 753

PART IV

Molecular Transmission of Genetic Information 791

27. DNA: The Structure of Chromosomes and Genes 793
28. Replication and Transcription of DNA 837
29. Protein Synthesis and Its Regulation 871
30. More about Genes: Repair, Mutation, Recombination, and Cloning 913

APPENDIX A: Common Abbreviations in Biochemical Research Literature 946

APPENDIX B: Unit Abbreviations, Prefixes, Constants, and Conversion Factors 948

APPENDIX C: International Atomic Weights 949

APPENDIX D: Logarithms 950

APPENDIX E: Answers to Problems 952

APPENDIX F: Glossary 969

Illustration Acknowledgments 981

Index 983

Contents

PART I

Biomolecules 1

CHAPTER 1

Biochemistry: The Molecular Logic of Living Organisms 3

Living Matter Has Several Identifying Characteristics 3

Biochemistry Seeks to Understand the Living State 5

All Living Organisms Contain Organic Macromolecules Built According to a Common Plan 5

Living Organisms Exchange Energy and Matter 7

Enzymes, the Catalysts of Living Cells, Promote Sequences of Organized Chemical Reactions 8

Cells Transmit Energy in a Chemical Form 9

Cell Metabolism Is Constantly Regulated 10

Living Organisms Replicate Themselves Accurately 10

CHAPTER 2

Cells 15

All Cells Share Some Structural Features 15

Cells Must Have Very Small Dimensions 16

There Are Two Great Classes of Cells: Prokaryotes and Eukaryotes 17

The Prokaryotes Are the Simplest and Smallest Cells 18

Escherichia coli Is the Best-Known Prokaryotic Cell 20

Eukaryotic Cells Are Larger and More Complex than Prokaryotes 22

The Nucleus of Eukaryotes Is a Very Complex Structure 24

Mitochondria Are the Power Plants of Eukaryotic Cells 25

The Endoplasmic Reticulum Forms Channels through the Cytoplasm 27

Golgi Bodies Are Secretory Organelles 28

Lysosomes Are Packets of Hydrolyzing Enzymes 28

Peroxisomes Are Peroxide-Destroying Vesicles 29

Microfilaments Function in Contractile Processes of Cells 29

Microtubules Also Function in Cell Movements 30

Microfilaments, Microtubules, and the Microtrabecular Network Constitute the Cytoskeleton 30

Cilia and Flagella Give Cells Propulsive Power 31

The Cytoplasm Also Contains Granular Bodies 32

The Cytosol Is the Continuous Aqueous Phase of the Cytoplasm 33

The Cell Membrane Presents a Large Surface Area 33

The Surface of Many Animal Cells Also Contains "Antennae" 34

Eukaryotic Plant Cells Have Some Special Features 35

Viruses Are Supramolecular Parasites 37

Summary 39

References 40

Problems 41

CHAPTER 3

The Composition of Living Matter: Biomolecules 45

The Chemical Composition of Living Matter Differs from That of the Earth's Crust 45

Most Biomolecules Are Compounds of Carbon 46

Organic Biomolecules Have Specific Shapes and Dimensions 47

Functional Groups of Organic Biomolecules Determine Their Chemical Properties 50

Many Biomolecules Are Asymmetric 51

The Major Classes of Biomolecules in Cells Are Very Large Molecules 53

Macromolecules Are Constructed from Small Building-Block Molecules 55

The Building-Block Molecules Have Simple Structures 55

There Is a Hierarchy in Cell Structure 58

Biomolecules First Arose by Chemical Evolution 59

Chemical Evolution Can Be Simulated 61

Summary 62

References 63

Problems 64

CHAPTER 4

Water 67

The Unusual Physical Properties of Water Are Due to Hydrogen Bonding 67

Hydrogen Bonds Are Common in Biological Systems 69

Water Has Unusual Solvent Properties 70

Solutes Change the Properties of Water 71

The Equilibrium Point of Reversible Reactions Is Expressed by an Equilibrium Constant 73

The Ionization of Water Is Expressed by an Equilibrium Constant 74

The pH Scale Designates the H^+ and OH^- Concentrations 76

Box 4-1 The ion product of water 77

Acids and Bases Reflect the Properties of Water 78

Weak Acids Have Characteristic Titration Curves 79

Buffers Are Mixtures of Weak Acids and Their Conjugate Bases 81

Phosphate and Bicarbonate Are Important Biological Buffers 83

Box 4-2 The Henderson-Hasselbalch equation 84

Box 4-3 How the bicarbonate buffer system of blood works 86

The Fitness of the Aqueous Environment for Living Organisms 87

Acid Rain Is Polluting Our Lakes and Streams 88

Summary 89

References 90

Problems 90

CHAPTER 5

Amino Acids and Peptides 95

Amino Acids Have Common Structural Features 95

Nearly All Amino Acids Have an Asymmetric Carbon Atom 96

Stereoisomers Are Named on the Basis of Their Absolute Configuration 97

The Optically Active Amino Acids of Proteins Are L Stereoisomers 98

Box 5-1 The RS system of designating optical isomers 99

Box 5-2 How to tell a person's age through amino acid chemistry 100

Amino Acids Can Be Classified on the Basis of Their R Groups 100

Eight Amino Acids Have Nonpolar R Groups 102

Seven Amino Acids Have Uncharged Polar R Groups 102

Two Amino Acids Have Negatively Charged (Acidic) R Groups 102

Three Amino Acids Have Positively Charged (Basic) R Groups 103

Some Proteins Also Contain "Special" Amino Acids 103

Amino Acids Are Ionized in Water Solutions 103

Amino Acids Can Act as Acids and as Bases 104

Amino Acids Have Characteristic Titration Curves 104

The Titration Curve Predicts the Electric Charge of Amino Acids 106

Amino Acids Differ in Their Acid-Base Properties 107

Their Acid-Base Properties Are the Basis for the Analysis of Amino Acids 108

Paper Electrophoresis Separates Amino Acids According to Electric Charge 108

Ion-Exchange Chromatography Is a More Useful Separation Process 109

Amino Acids Have Characteristic Chemical Reactions 110

Peptides Are Chains of Amino Acids 111

Peptides Can Be Separated on the Basis of Their Ionization Behavior 112

Peptides Have Characteristic Chemical Reactions 113

Some Peptides Have Intense Biological Activity 114

Summary 115

References 115

Problems 116

CHAPTER 6

Proteins: Covalent Structure and Biological Function 121

Box 6-1 How many amino acid sequences are possible? 122

Proteins Have Many Different Biological Functions 122

Proteins Can Also Be Classified According to Shape 124

Proteins Yield Their Amino Acids on Hydrolysis 124

Some Proteins Contain Chemical Groups Other Than Amino Acids 125

Proteins Are Very Large Molecules 126

Proteins Can Be Separated and Purified 127

The Amino Acid Sequence of Polypeptide Chains Can Be Determined 129

Insulin Was the First Protein to Be Sequenced 134

Many Other Proteins Have Been Sequenced Since 135

Homologous Proteins from Different Species Have Homologous Sequences 137

The Immune Response Can Detect Differences between Homologous Proteins 138

Proteins Undergo a Structural Change Called Denaturation 140

Summary 141

References 142

Problems 142

CHAPTER 7

Fibrous Proteins 147

Configuration and Conformation Have Different Meanings 147

Paradoxically, Native Proteins Appear to Have Only One or a Few Conformations 148

α-Keratins Are Fibrous Proteins Made by Epidermal Cells 149

X-Ray Analysis of Keratins Shows That They Have Repeating Structural Units 150

X-Ray Studies of Peptides Show the Peptide Bond to Be Rigid and Planar 150

In α-Keratin the Polypeptide Chains Form an α Helix 151

Some Amino Acids Are Not Compatible with the α Helix 152

The α-Keratins Are Rich in Amino Acids Compatible with α-Helical Structures 153

In Native α-Keratins α-Helical Polypeptide Chains Are Supercoiled into Ropes 154

The Insolubility of α-Keratins Is a Reflection of Their Nonpolar R Groups 154

β-Keratins Contain a Different Conformation of Their Polypeptide Chain: β Structure 155

Permanent Waving Is Biochemical Engineering 156

Collagen and Elastin Are the Major Fibrous Proteins of Connective Tissues 157

Collagen Is the Most Abundant Protein in the Body 157

Collagen Has Both Familiar and Unusual Properties 158

The Polypeptides in Collagen Are Three-Stranded Helical Structures 159

The Structure of Elastin Confers Distinctive Properties on Elastic Tissue 160

What Fibrous Proteins Tell Us About Protein Structure 162

Other Types of Fibrillar or Filamentous Proteins Occur in Cells 162

Summary 163

References 164

Problems 165

CHAPTER 8

Globular Proteins: The Structure and Function of Hemoglobin 169

The Polypeptide Chain(s) of Globular Proteins Are Tightly Folded 169

X-Ray Analysis of Myoglobin Was the Breakthrough 170

Myoglobins from Different Species Have Similar Conformations 173

The Tertiary Structure of Each Type of Globular Protein Is Distinctive 173

Amino Acid Sequence Determines Tertiary Structure 177

Four Different Forces Stabilize the Tertiary Structure of Globular Proteins 178

The Rate of Folding of Polypeptide Chains Is Critical 179

Oligomeric Proteins Have Both Tertiary and Quaternary Structure 180

X-Ray Analysis Has Revealed the Complete Structure of Hemoglobin 181

Myoglobin and the α and β Chains of Hemoglobin Have Nearly the Same Tertiary Structure 183

Quaternary Structures of Other Oligomeric Proteins Have Been Determined 184

Red Blood Cells Are Specialized to Carry Oxygen 185

Myoglobin and Hemoglobin Differ in Their Oxygen-Binding Curves 186

The Cooperative Binding of Oxygen Enhances the Efficiency of Hemoglobin as an Oxygen Carrier 187

Hemoglobin Also Transports H^+ and CO_2 188

Oxygenation of Hemoglobin Causes a Change in Its Three-Dimensional Conformation 190

Box 8-1 *Diphosphoglycerate and the oxygen affinity of hemoglobin* 192

Sickle-Cell Anemia Is a Molecular Disease of Hemoglobin 194

Sickle-Cell Hemoglobin Has an Altered Amino Acid Sequence 196

Sickling Is Caused by the Tendency of Hemoglobin S Molecules to Stick Together 198

Proteins Containing "Wrong" Amino Acids Are the Result of Gene Mutations 198

Can a Molecular Cure for Sickle Hemoglobin Be Found? 199

Summary 200

References 201

Problems 202

CHAPTER 9

Enzymes 207

Much of the History of Biochemistry Is the History of Enzyme Research 208

Enzymes Show All the Properties of Proteins 209

Enzymes Are Classified on the Basis of the Reactions They Catalyze 210

Enzymes Enhance the Rate of Chemical Reactions by Lowering Their Activation Energy 211

The Substrate Concentration Has a Profound Effect on the Rate of Enzyme-Catalyzed Reactions 212

There Is a Quantitative Relationship between the Substrate Concentration and the Rate of an Enzymatic Reaction 213

Box 9-1 *The Michaelis-Menten equation* 214

Each Enzyme Has a Characteristic K_M for a Given Substrate 216

Box 9-2 *Transformations of the Michaelis-Menten equation: The double-reciprocal plot* 217

Many Enzymes Catalyze Reactions in Which There Are Two Substrates 217

Enzymes Have an Optimum pH 218

Enzymes Can Be Assayed Quantitatively 218

Enzymes Show Specificity toward Their Substrates 220

Enzymes Can Be Inhibited by Specific Chemical Agents 221

There Are Two Kinds of Reversible Inhibitors: Competitive and Noncompetitive 223

Noncompetitive Inhibition Is Also Reversible but Not by the Substrate 224

Box 9-3 *Kinetic tests for distinguishing between competitive and noncompetitive inhibition* 225

Several Factors Contribute to the Catalytic Efficiency of Enzymes 225

X-Ray Analysis Has Revealed Important Structural Features of Enzymes 226

Box 9-4 *A gallery of enzyme structures revealed by x-ray diffraction analysis* 228

Enzyme Systems Have a Pacemaker or Regulatory Enzyme 233

Allosteric Enzymes Are Regulated by Noncovalent Binding of Modulator Molecules 233

Allosteric Enzymes May Be Inhibited or Stimulated by Their Modulators 235

Allosteric Enzymes Deviate from Michaelis-Menten Behavior 235

Allosteric Enzymes Show Communication between Subunits 237

Some Enzymes Are Regulated by Reversible Covalent Modification 237

Box 9-5 *The three-dimensional structure of the regulatory enzyme aspartate transcarbamoylase* 238

Many Enzymes Occur in Multiple Forms 239

Enzymes May Be Catalytically Defective Due to Genetic Mutation 241

Summary 242

References 243

Problems 244

CHAPTER 10

Vitamins and Trace Elements in the Function of Enzymes 249

Vitamins Are Essential Organic Micronutrients 250

Vitamins Are Essential Components of Coenzymes and Enzyme Prosthetic Groups 250

Vitamins Can Be Grouped into Two Classes 251

Thiamine (Vitamin B_1) Functions in the Form of Thiamine Pyrophosphate 252

Riboflavin (Vitamin B_2) Is a Component of the Flavin Nucleotides 254

Nicotinamide Is the Active Group of the Coenzymes NAD and NADP 255

Pantothenic Acid Is a Component of Coenzyme A 256

Pyridoxine (Vitamin B_6) Is Important in Amino Acid Metabolism 258

Biotin Is the Active Component of Biocytin, the Prosthetic Group of Some Carboxylating Enzymes 259

Folic Acid Is the Precursor of the Coenzyme Tetrahydrofolic Acid 260

Vitamin B_{12} Is the Precursor of Coenzyme B_{12} 262

The Biochemical Function of Vitamin C (Ascorbic Acid) Is Not Known 264

The Fat-Soluble Vitamins Are Derivatives of Isoprene 264

Vitamin A Probably Has Several Functions 265

Vitamin D Is the Precursor of a Hormone 267

Vitamin E Protects Cell Membranes against Oxygen 268

Vitamin K Is a Component of a Carboxylating Enzyme 269

Many Inorganic Elements Are Required in Animal Nutrition 269

There Are Many Iron-Requiring Enzymes 270

Copper Also Functions in Some Oxidative Enzymes 271

Zinc Is Essential in the Action of Many Enzymes 271

Manganese Ions Are Required by Several Enzymes 271

Cobalt Is Part of Vitamin B_{12} 272

Selenium Is Both an Essential Trace Element and a Poison 272

Other Trace Elements Are Known to Be Required by Some Enzymes 272

Summary 273

References 274

Problems 274

CHAPTER 11

Carbohydrates: Structure and Biological Function 277

There Are Three Classes of Carbohydrates, Based on the Number of Sugar Units 277

There Are Two Families of Monosaccharides: Aldoses and Ketoses 278

The Common Monosaccharides Have Several Asymmetric Centers 279

The Common Monosaccharides Occur in Ring Forms 281

Simple Monosaccharides Are Reducing Agents 284

Disaccharides Contain Two Monosaccharide Units 284

Polysaccharides Contain Many Monosaccharide Units 287

Some Polysaccharides Serve as a Storage Form of Cell Fuel 287

Cellulose Is the Most Abundant Structural Polysaccharide 289

Cell Walls Are Rich in Structural and Protective Polysaccharides 292

Glycoproteins Are Hybrid Molecules 294

Animal Cell Surfaces Contain Glycoproteins 295

Acid Mucopolysaccharides and Proteoglycans Are Important Components of Connective Tissue 296

Summary 297

References 298

Problems 299

CHAPTER 12

Lipids and Membranes 303

Fatty Acids Are Building-Block Components of Most Lipids 303

Triacylglycerols Are Fatty Acid Esters of Glycerol 306

Triacylglycerols Are Storage Lipids 308

Waxes Are Fatty Acid Esters of Long-Chain Alcohols 309

Phospholipids Are Major Components of Membrane Lipids 310

Sphingolipids Are Also Important Components of Membranes 312

Steroids Are Nonsaponifiable Lipids with Specialized Functions 315

Lipoproteins Blend the Properties of Lipids and Proteins 315

Polar Lipids Form Micelles, Monolayers, and Bilayers 317

The Major Components of Membranes Are Polar Lipids and Proteins 318

Box 12-1 Electron microscopy of membranes 320

Membranes Have a Fluid-Mosaic Structure 321

Membranes Have a Specific Sidedness or Asymmetry 322

Red-Blood-Cell Membranes Have Been Studied in Detail 322

Lectins Are Specific Proteins Capable of Binding to or Agglutinating Certain Cells 324

Membranes Have Very Complex Functions 325

Summary 326

References 327

Problems 328

PART II

Bioenergetics and Metabolism 331

CHAPTER 13

A Survey of Metabolism 333

Living Organisms Participate in the Cycling of Carbon and Oxygen 333

Nitrogen Is Cycled in the Biosphere 335

Metabolic Pathways Are Promoted by Sequential Enzyme Systems 337

Metabolism Consists of Catabolic (Degradative) Pathways and Anabolic (Biosynthetic) Pathways 337

Catabolic Pathways Converge to a Few End Products 338

Biosynthetic (Anabolic) Pathways Diverge to Yield Many Products 340

There Are Important Differences between Corresponding Catabolic and Anabolic Pathways 341

ATP Carries Energy from Catabolic to Anabolic Reactions 343

NADPH Carries Energy in the Form of Reducing Power 344

Cell Metabolism Is an Economical, Tightly Regulated Process 345

Metabolic Pathways Are Regulated at Three Levels 346

Secondary Metabolism 347

There Are Three Main Approaches to Identification of a Metabolic Sequence 348

Mutants of Organisms Allow Identification of Intermediate Steps in Metabolism 349

Isotopic Tracers Provide a Powerful Method of Studying Metabolism 351

Metabolic Pathways Are Compartmented in Cells 352

Summary 356

References 356

Problems 357

CHAPTER 14

The ATP Cycle and Cell Bioenergetics 361

The First and Second Laws of Thermodynamics 361

Box 14-1 The concept of entropy 364

Cells Require Free Energy 366

The Standard-Free-Energy Change of a Chemical Reaction Can Be Calculated 366

$\Delta G^{\circ\prime}$ Has Characteristic Values for Different Chemical Reactions 368

There Is an Important Difference between $\Delta G^{\circ\prime}$ and ΔG 369

Standard-Free-Energy Values of Chemical Reactions Are Additive 370

ATP Is the Major Chemical Link between Energy-Yielding and Energy-Requiring Cell Activities 371

The Chemistry of ATP Is Well Known 373

ATP Has a Characteristic Standard Free Energy of Hydrolysis 374

Why Does ATP Have a Relatively High Standard Free Energy of Hydrolysis? 374

ATP Acts as a Common Intermediate in Phosphate-Transfer Reactions 376

Box 14-2 The free energy of hydrolysis of ATP in intact cells 377

Two Super High-Energy Phosphate Compounds Are Generated by Breakdown of Glucose to Lactate 378

Transfer of a Phosphate Group from ATP to an Acceptor Molecule Can Energize It 379

ATP Is Used to Energize Muscle Contraction 380

Phosphocreatine Is a Temporary Storage Form of High-Energy Phosphate Groups in Muscles 383

ATP Also Energizes Active Transport across Membranes 384

ATP Can Also Be Broken Down to AMP and Pyrophosphate 386

Box 14-3 ATP provides the energy for firefly bioluminescence 388

There Are Other Energy-Rich Nucleoside 5′-Triphosphates besides ATP 389

The ATP System Functions in a Dynamic Steady State 391

Summary 392

References 393

Problems 394

CHAPTER 15

Glycolysis: A Central Pathway of Glucose Catabolism 397

Glycolysis Is a Central Pathway in Most Organisms 397

ATP Formation Is Coupled to Glycolysis 399

Much Free Energy Remains in the Products of Glycolysis 400

Glycolysis Has Two Phases 400

Box 15-1 Anaerobic glycolysis, oxygen debt, alligators, and coelacanths 401

Glycolysis Takes Place via Phosphorylated Intermediates 403

The First Phase of Glycolysis Results in Cleavage of the Hexose Chain 403

The Second Phase of Glycolysis Is Energy-Conserving 408

"Feeder" Pathways Lead from Glycogen and Other Carbohydrates into the Central Glycolytic Pathway 414

Other Monosaccharides Can Enter the Glycolytic Sequence 417

Disaccharides Must First Be Hydrolyzed to Monosaccharides 419

The Entry of Glucose Residues into the Glycolytic Sequence Is Regulated 420

Hormones Ultimately Regulate the Interconversion of Phosphorylase *a* and *b* 422

The Glycolytic Sequence Itself Is Regulated at Two Major Points 423

How Are the Regulated Steps of Glycolysis Identified in Intact Cells? 425

Alcoholic Fermentation Differs from Glycolysis Only in Its Terminal Steps 426

Box 15-2 Brewing beer 428

Summary 428

References 429

Problems 430

CHAPTER 16

The Citric Acid Cycle 435

Oxidation of Glucose to CO_2 and H_2O Releases Much More Energy than Glycolysis 437

Pyruvate Must First Be Oxidized to Acetyl-CoA and CO_2 437

The Citric Acid Cycle Is a Circular Rather than a Linear Enzyme System 441

How Did the Idea of the Citric Acid Cycle Arise? 441

The Citric Acid Cycle Has Eight Steps 444

Summary of the Cycle 448

Why a Citric Acid Cycle? 448

Isotopic Tests of the Citric Acid Cycle 449

The Conversion of Pyruvate to Acetyl-CoA Is Regulated 449

Box 16-1 Is citric acid the first tricarboxylic acid formed in the cycle? 450

The Citric Acid Cycle Is Regulated 452

Citric Acid Cycle Intermediates Are Used for Other Metabolic Purposes and Can Be Replenished 453

The Glyoxylate Cycle Is a Modification of the Citric Acid Cycle 455

There Are Secondary Pathways of Glucose Catabolism: The Pentose Phosphate Pathway 456

The Secondary Pathway from Glucose to Glucuronic Acid and Ascorbic Acid 457

Summary 459

References 460

Problems 461

CHAPTER 17

Electron Transport, Oxidative Phosphorylation, and Regulation of ATP Production 467

Electron Flow from Substrates to Oxygen Is the Source of ATP Energy 467

Electron Transport and Oxidative Phosphorylation Take Place in the Inner Mitochondrial Membrane 469

Electron-Transferring Reactions Are Oxidation-Reduction Reactions 470

Each Conjugate Redox Couple Has a Characteristic Standard Potential 472

Free-Energy Changes Accompany Electron Transfers 474

There Are Many Electron Carriers in the Electron-Transport Chain 476

The Pyridine Nucleotides Have a Collecting Function 476

NADH Dehydrogenase Accepts Electrons from NADH 478

Ubiquinone Is a Lipid-Soluble Quinone 479

The Cytochromes Are Electron-Carrying Heme Proteins 480

Incomplete Reduction of Oxygen Causes Cell Injury 481

The Electron Carriers Always Function in a Specific Sequence 482

Electron-Transport Energy Is Conserved by Oxidative Phosphorylation 484

The ATP-Synthesizing Enzyme Has Been Isolated and Reconstituted 484

How Is the Redox Energy of Electron Transport Delivered to ATP Synthetase? 487

The Chemiosmotic Hypothesis Postulates That a Proton Gradient Carries Energy from Electron Transport to ATP Synthesis 489

Electron-Transport Energy Is Useful for Other Purposes 491

Box 17-1 Many questions on the mechanism of oxidative phosphorylation remain to be answered 492

Bacteria and Chloroplasts Also Contain H^+-Transporting Electron-Transport Chains 493

The Inner Mitochondrial Membrane Contains Specific Transport Systems 495

Shuttle Systems Are Required for Oxidation of Extramitochondrial NADH 496

The Complete Oxidation of Glucose Leads to Synthesis of 38 ATPs 497

ATP Formation by Oxidative Phosphorylation Is Regulated by the Cell's Energy Needs 498

The Energy Charge Is Another Index of Cellular Energy Status 500

Glycolysis, the Citric Acid Cycle, and Oxidative Phosphorylation Have Interlocking and Concerted Regulatory Mechanisms 500

Cells Contain Other Oxygen-Using Enzymes 502

Summary 504

References 505

Problems 506

CHAPTER 18

The Oxidation of Fatty Acids in Animal Tissues 511

Fatty Acids Are Activated and Oxidized in Mitochondria 511

Fatty Acids Enter Mitochondria by a Three-Step Transport Process 512

Fatty Acids Are Oxidized in Two Stages 514

The First Stage in the Oxidation of Saturated Fatty Acids Has Four Steps 515

The First Stage of Fatty Acid Oxidation Yields Acetyl-CoA and ATP 518

In the Second Stage of Fatty Acid Oxidation Acetyl-CoA Is Oxidized via the Citric Acid Cycle 519

The Oxidation of Unsaturated Fatty Acids Requires Two Additional Enzymatic Steps 520

Oxidation of Fatty Acids with an Odd Number of Carbons 521

Hypoglycin, a Toxic Agent of Some Plants, Inhibits Fatty Acid Oxidation 523

Formation of Ketone Bodies in the Liver and Their Oxidation in Other Organs 524

Regulation of Fatty Acid Oxidation and Ketone-Body Formation 526

Summary 527

References 527

Problems 528

CHAPTER 19

Oxidative Degradation of Amino Acids: The Urea Cycle 531

Transfer of α-Amino Groups Is Catalyzed by Transaminases 531

Ammonia Is Formed from Glutamate 534

Box 19-1 Transaminases and other enzymes in the blood are useful in medical diagnosis 535

The Carbon Skeletons of Amino Acids Are Degraded by 20 Different Pathways 536

Ten Amino Acids Yield Acetyl-CoA during Their Degradation 537

Phenylalanine Catabolism Is Genetically Defective in Some People 540

Box 19-2 The human, social, and economic costs of some genetic diseases 542

Five Amino Acids Are Converted into α-Ketoglutarate 543

Three Amino Acids Are Converted into Succinyl-CoA 544

Phenylalanine and Tyrosine Yield Fumarate 544

The Oxaloacetate Pathway 544

Some Amino Acids Can Be Converted into Glucose and Some into Ketone Bodies 545

Ammonia Is Toxic to Animals 545

Glutamine Carries Ammonia from Many Peripheral Tissues to the Liver 546

Alanine Carries Ammonia from Muscles to the Liver 546

Excretion of Amino Nitrogen Is Another Biochemical Problem 548

Glutaminase Participates in Excretion of Ammonia 549

Urea Is Formed by the Urea Cycle 549

The Urea Cycle Has Several Complex Steps 550

The Energy Cost of Urea Synthesis 554

Genetic Defects in the Urea Cycle Lead to Excess Ammonia in the Blood 554

Birds, Snakes, and Lizards Excrete Uric Acid 555

Summary 556

References 557

Problems 557

CHAPTER 20

Biosynthesis of Carbohydrates in Animal Tissues 561

The Pathway of Gluconeogenesis Shares Seven Steps with the Pathway of Glycolysis 562

Conversion of Pyruvate into Phosphoenolpyruvate Requires a Bypass 564

The Second Bypass Reaction in Gluconeogenesis Is the Conversion of Fructose 1,6-Diphosphate into Fructose 6-Phosphate 565

Conversion of Glucose 6-Phosphate into Free Glucose Is the Third Bypass Reaction 566

Gluconeogenesis Is Costly 566

Gluconeogenesis and Glycolysis Are Regulated Reciprocally 567
Citric Acid Cycle Intermediates Are Also Precursors of Glucose 568
Most Amino Acids Are Glucogenic 568
Gluconeogenesis Takes Place during Recovery from Muscular Exercise 569
Gluconeogenesis Is an Especially Active Process in Ruminant Animals 569
Alcohol Consumption Inhibits Gluconeogenesis 570
"Futile Cycles" in Carbohydrate Metabolism 571
Biosynthesis of Glycogen Proceeds by a Pathway Different from That of Glycogen Breakdown 572
Glycogen Synthase and Glycogen Phosphorylase Are Reciprocally Regulated 574
Glycogen Metabolism Is Subject to Genetic Defects 576
Lactose Synthesis Is Regulated in a Unique Way 576
Summary 577
References 578
Problems 579

CHAPTER 21

The Biosynthesis of Lipids 583

The Biosynthesis of Fatty Acids Proceeds by a Distinctive Pathway 583
Malonyl-CoA Is Formed from Acetyl-CoA 585
The Fatty Acid Synthase System Has Seven Active Sites 587
The Sulfhydryl Groups of Fatty Acid Synthase Are First Charged with Acyl Groups 588
Addition of Each 2-Carbon Unit Requires Four Steps 589
Palmitic Acid Is the Precursor of Other Long-Chain Fatty Acids 594
Regulation of Fatty Acid Biosynthesis 595
The Biosynthesis of Triacylglycerols and Glycerol Phosphatides Begins with Common Precursors 595
Triacylglycerol Biosynthesis Is Regulated by Hormones 597
Triacylglycerols: Energy Sources in Some Hibernating Animals 598
Box 21-1 Another biological function of triacylglycerols 599
Biosynthesis of Phosphoglycerides Requires a Head Group 600
Phosphatidylcholine Is Made by Two Different Pathways 602
Polar Lipids Are Inserted into Cell Membranes 603
Lipid Metabolism Is Subject to Genetic Defects 604
There Are Many Lysosomal Diseases 606
Cholesterol and Other Steroids Are Also Made from 2-Carbon Precursors 607
Isopentenyl Pyrophosphate Is the Precursor of Many Other Lipid-Soluble Biomolecules 610
Summary 611
References 611
Problems 612

CHAPTER 22

Biosynthesis of Amino Acids and Nucleotides 615

Some Amino Acids Must Be Obtained from the Diet 615
Glutamate, Glutamine, and Proline Share a Common Biosynthetic Pathway 616
Alanine, Aspartate, and Asparagine Also Arise from Central Metabolites 618
Tyrosine Is Made from an Essential Amino Acid, Phenylalanine 618
Cysteine Is Made from Two Other Amino Acids, Methionine and Serine 618
Serine Is a Precursor of Glycine 620
Biosynthesis of the Essential Amino Acids 621
Amino Acid Biosynthesis Is under Allosteric Regulation 622
Amino Acid Biosynthesis Is Also Regulated by Changes in Enzyme Concentration 624
Glycine Is a Precursor of Porphyrins 625
Porphyrin Derivatives Accumulate in Some Genetic Disorders 626
Degradation of Heme Groups Yields Bile Pigments 627
Purine Nucleotides Are Made by a Complex Pathway 627
Purine Nucleotide Biosynthesis Is Regulated by Feedback Control 630
Pyrimidine Nucleotides Are Made from Aspartate and Ribose Phosphate 631
Regulation of Pyrimidine Nucleotide Biosynthesis 632
Ribonucleotides Are the Precursors of the Deoxyribonucleotides 632
Degradation of Purines Leads to Uric Acid in Human Beings 634
Purine Bases Are Recycled by a Salvage Pathway 634
Overproduction of Uric Acid Causes Gout 636
The Nitrogen Cycle 636
Not Many Organisms Can Fix Nitrogen 637

Fixation of Nitrogen Is a Complex Enzymatic Process 638

Summary 640

References 641

Problems 642

CHAPTER 23

Photosynthesis 645

How the Photosynthetic Equation Was Discovered 646

Photosynthetic Organisms Are Very Diverse 646

Photosynthetic Organisms Rely on Different Hydrogen Donors 648

Photosynthesis Has Light and Dark Phases 649

Plant Photosynthesis Takes Place in Chloroplasts 649

Absorption of Light Excites Molecules 651

Chlorophylls Are the Major Light-Absorbing Pigments 653

Thylakoids Also Contain Accessory Pigments 654

The Thylakoid Membrane Contains Two Kinds of Photochemical Reaction Systems 654

Electron Flow Is Induced by Illumination of Chloroplasts 655

Captured Light Energy Causes Uphill Flow of Electrons 656

Photosystems I and II Cooperate to Carry Electrons from H_2O to $NADP^+$ 658

The Z Scheme Shows the Energy Profile of Photosynthetic Electron Transport 658

Several Electron Carriers Participate in Photosynthetic Electron Transport 659

Phosphorylation of ADP Is Coupled to Photosynthetic Electron Transport 660

Chloroplasts Also Promote Cyclic Electron Flow and Cyclic Photophosphorylation 661

Photosynthetic Phosphorylation Resembles Oxidative Phosphorylation 661

The Overall Equation of Plant Photosynthesis 662

Photosynthetic Formation of Hexoses Involves Net Reduction of Carbon Dioxide 663

Carbon Dioxide Is Fixed into Phosphoglycerate 663

Glucose Is Made from CO_2 by the Calvin Cycle 664

Glucose Is the Precursor of the Plant Carbohydrates Sucrose, Starch, and Cellulose 668

Regulation of the Dark Reactions 669

Tropical Plants Employ the C_4 or Hatch-Slack Pathway 669

The C_4 Pathway Serves to Concentrate CO_2 672

Photorespiration Limits the Efficiency of C_3 Plants 672

Photorespiration Is an Important Problem in Temperate-Zone Agriculture 673

Salt-Loving Bacteria Use Light Energy to Make ATP 674

Photosynthetic Organisms Are Models for the Design of Solar Energy Cells 675

Summary 676

References 676

Problems 678

PART III

Some Aspects of Human Biochemistry 681

CHAPTER 24

Digestion, Transport, and the Integration of Metabolism 683

Foods Are Enzymatically Digested to Prepare Them for Absorption 683

The Liver Processes and Distributes Nutrients 692

In the Liver Sugars Follow Five Metabolic Routes 692

Amino Acids Also Have Five Metabolic Routes 693

Lipids Follow Five Pathways 695

Each Organ Has Specialized Metabolic Functions 696

Skeletal Muscle Uses ATP for Intermittent Mechanical Work 696

Heart Muscle Must Work Constantly and Rhythmically 698

The Brain Uses Energy for Transmission of Impulses 699

Adipose Tissue Has an Active Metabolism 701

The Kidneys Use ATP to Do Osmotic Work 703

Blood Is a Very Complex Fluid 705

Large Volumes of Oxygen Are Transported by the Blood 707

Hemoglobin Is the Carrier of Oxygen 708

Red Blood Cells Also Transport CO_2 709

Diagnosing and Treating Diabetes Mellitus Relies on Biochemical Measurements 712

Ketosis Occurs in Diabetes 714

Urea Excretion Is Elevated in Diabetics 714

Acidosis Accompanies Severe Diabetes 715

Summary 715

References 716

Problems 717

CHAPTER 25

Hormones 721

Hormones Function in a Complex Interrelated Hierarchy 721

Some General Characteristics of Hormones 723

The Hormones of the Hypothalamus and Pituitary Gland Are Peptides 725

Box 25-1 Radioimmunoassay of polypeptide hormones 727

The Adrenal Medulla Secretes the Amine Hormones Adrenaline and Noradrenaline 727

Adrenaline Stimulates the Formation of Cyclic Adenylate 729

Cyclic AMP Stimulates Protein Kinase Activity 731

The Stimulation of Glycogen Breakdown by Adrenaline Occurs by an Amplification Cascade 732

Adrenaline Also Inhibits Glycogen Synthesis 733

Phosphodiesterase Inactivates Cyclic Adenylate 734

The Pancreas Secretes Several Hormones Regulating Metabolic Pathways 735

Insulin Is the Hypoglycemic Hormone 736

Insulin Secretion Is Regulated Primarily by Blood Glucose 737

The Second Messenger of Insulin Is Still Unknown 738

Insulin Influences Many Other Aspects of Metabolism 738

Glucagon Is the Hyperglycemic Pancreatic Hormone 739

Somatostatin Inhibits Secretion of Insulin and Glucagon 739

Somatotropin Also Affects the Action of Insulin 740

Adrenal Cortical Hormones Are Steroids 740

The Thyroid Hormones Control Metabolic Rate 742

The Sex Hormones Are Steroids 743

Estrogen Action on Target Cells Is Becoming Understood 744

Many Other Hormones Are Known 745

Prostaglandins and Thromboxanes Modulate the Action of Some Hormones 746

Summary 747

References 748

Problems 749

CHAPTER 26

Human Nutrition 753

There Are Five Basic Components in an Adequate Diet 753

Energy Is Provided by Oxidation of the Bulk Organic Nutrients 755

Alcohol Also Contributes Calories 761

Obesity Is the Result of Caloric Overnutrition 762

Proteins Are Required for Their Amino Acid Content 763

Certain Plant Proteins Can Supplement Each Other in the Diet 765

Marasmus and Kwashiorkor Are World Health Problems 765

Deficiencies of Some Vitamins Can Be Life-Threatening 767

Thiamine Deficiency Is Still a Nutritional Problem 767

Nicotinamide and Tryptophan Nutrition Are Related 769

Many Foods Are Low in Ascorbic Acid 770

Box 26-1 One of the first recorded accounts of the cure of scurvy, among the men of Jacques Cartier's Expedition to Newfoundland, 1535 771

Marginal Riboflavin Deficiency Is Also Common 772

The Most Common Vitamin Deficiency Is of Folic Acid 772

Human Deficiencies of Pyridoxine, Biotin, and Pantothenic Acid Are Rare 773

True Dietary Deficiency of Vitamin B_{12} Is Very Rare 773

Vitamin A Deficiency Has Multiple Effects 774

Vitamin D Deficiency Causes Rickets and Osteomalacia 776

Human Deficiencies of Vitamin E or K Are Very Rare 776

Many Elements Are Required in Human Nutrition 777

Calcium and Phosphorus Are Essential for Development of Bones and Teeth 778

Marginal Magnesium Deficiency Is Relatively Common 779

Sodium and Potassium Intake Are Important in Prevention and Treatment of Hypertension 779

Iron and Copper Are Needed for Heme Protein Synthesis 780

Goiter Is the Result of Iodine Deficiency 781

Dental Caries Is an Important Nutritional Problem 782

Zinc and Several Other Microelements Are Essential in the Diet 783

A Balanced Diet Must Have Variety 784

Labeling Food Products Protects Consumers 784

References 786

Problems 787

PART IV

Molecular Transmission of Genetic Information 791

CHAPTER 27

DNA: The Structure of Chromosomes and Genes 793

DNA and RNA Perform Different Functions 794

The Nucleotide Units of DNA and RNA Have Characteristic Bases and Pentoses 796

Phosphodiester Bonds Link the Successive Nucleotides of Nucleic Acids 797

DNA Stores Genetic Information 800

DNAs of Different Species Have Distinctive Base Compositions 803

Watson and Crick Postulated a Double-Helix Model of DNA Structure 804

The Base Sequence of DNA Constitutes a Template 807

Double-Helical DNA May Undergo Denaturation or Unwinding 808

DNA Strands from Two Different Species Can Form DNA-DNA Hybrids 809

Some Physical Properties of Double-Helical DNAs Reflect the Proportion of G≡C versus A=T Pairs 810

Native DNA Molecules Are Very Fragile 810

Viral DNA Molecules Are Relatively Small 811

The Chromosomes of Prokaryotic Cells Are Single, Very Large DNA Molecules 812

Circular DNAs Are Supercoiled 813

Some Bacteria Also Contain DNA in the Form of Plasmids 814

Eukaryotic Cells Contain Much More DNA than Prokaryotes 815

Eukaryotic Chromosomes Consist of Chromatin Fibers 816

Histones Are Small Basic Proteins 816

DNA-Histone Complexes Form Beadlike Nucleosomes 818

Eukaryotic Cells Also Contain Cytoplasmic DNA 818

Genes Are Segments of DNA That Code for Polypeptide Chains and RNAs 820

There Are Many Genes in a Single Chromosome 821

How Large Are Genes? 822

Bacterial DNA Is Protected by Restriction-Modification Systems 823

Eukaryotic DNA Contains Base Sequences That Are Repeated Many Times 825

A Few Eukaryotic Genes Occur in Multiple Copies 826

Eukaryotic DNA Has Many Palindromes 826

Many Eukaryotic Genes Contain Intervening Nontranscribed Sequences (Introns) 827

The Base Sequences of Some DNAs Have Been Determined 827

Box 27-1 The sequencing of a short fragment of DNA by the chemical method of Maxam and Gilbert 830

Summary 832

References 833

Problems 834

CHAPTER 28

Replication and Transcription of DNA 837

DNA Undergoes Semiconservative Replication 837

Circular DNA Is Replicated Bidirectionally 840

Eukaryotic DNAs Have Many Origins of Replication 841

Sometimes DNA Is Replicated by a Rolling-Circle Process 842

Bacterial Extracts Contain DNA Polymerase 843

Preformed DNA Is Necessary for the Action of DNA Polymerase 844

DNA Replication Requires Many Enzymes and Protein Factors 845

There Are Three DNA Polymerases in *E. coli* 845

Simultaneous Replication of Both Strands of DNA Poses a Problem 846

The Discovery of Okazaki Fragments Solved the Problem 847

Synthesis of Okazaki Fragments Requires an RNA Primer 847

The Okazaki Fragments Are Spliced by DNA Ligase 848

Replication Requires Physical Separation of the Parent Duplex DNA 848

DNA Polymerases Can Proofread and Correct Errors 850

Replication in Eukaryotic Cells Is Very Complex 852

Genes Are Transcribed to Yield RNAs 852

Messenger RNAs Code for Polypeptide Chains 853

Messenger RNA Is Made by a DNA-Directed RNA Polymerase 854

Eukaryotic Cell Nuclei Have Three RNA Polymerases 856

DNA-Directed RNA Polymerase Can Be Selectively Inhibited 856

RNA Transcripts Are Processed Further 857

Heterogeneous Nuclear RNAs Are Precursors of Eukaryotic Messenger RNAs 859

Intron RNA Must Be Removed from mRNA Precursors 859

Small Nuclear RNAs Aid in Removing Intron RNA 861

The Transcription Process Can Be Visualized 862

DNA Is Transcribed from Some Viral RNAs by Reverse Transcriptase 863

Some Viral RNAs Are Replicated by an RNA-Directed RNA Polymerase 864

Polynucleotide Phosphorylase Makes Random RNA-Like Polymers 865

Summary 866

References 867

Problems 868

CHAPTER 29

Protein Synthesis and Its Regulation 871

Early Discoveries Set the Stage 872

Protein Synthesis Occurs in Five Major Stages 873

Transfer RNAs Are Required for the Activation of Amino Acids 874

Aminoacyl-tRNA Synthetases Attach the Correct Amino Acids to Their tRNAs 877

Transfer RNA Is an Adapter 879

Polypeptide Chains Originate at the Amino-Terminal End 879

N-Formylmethionine Is the Initiating Amino Acid in Prokaryotes and Methionine in Eukaryotes 880

Ribosomes Are Molecular Machines for Making Polypeptide Chains 881

The Cytoplasmic Ribosomes of Eukaryotes Are Larger and More Complex 882

Initiation of a Polypeptide Has Several Steps 883

Elongation of the Polypeptide Chain Is a Repetitive Process 885

Termination of Polypeptide Synthesis Requires a Special Signal 888

Energy Is Required to Guarantee Fidelity of Protein Synthesis 888

Polyribosomes Make Possible Rapid Translation of a Single Messenger 890

Polypeptide Chains Undergo Folding and Processing 891

Newly Made Proteins Are Often Directed to Their Destinations 893

Protein Synthesis Is Inhibited by Many Different Antibiotics 894

The Genetic Code Has Been Solved 895

The Genetic Code Has Some Interesting Characteristics 896

Wobble Allows Some tRNAs to Recognize More than One Codon 898

Viral DNAs Sometimes Contain Genes within Genes or Overlapping Genes 900

Protein Synthesis Is Regulated 901

Bacteria Contain Constitutive and Induced Enzymes 902

Prokaryotes Also Can Repress Protein Synthesis 903

The Operon Hypothesis 903

Repressor Molecules Have Been Isolated 906

Operons Also Have a Promoter Site 906

Summary 908

References 909

Problems 910

CHAPTER 30

More about Genes: Repair, Mutation, Recombination, and Cloning 913

DNA Is Constantly Subject to Damage 914

Ultraviolet Damage Can Be Excised and Repaired 914

Spontaneous Deamination of Cytosine to Uracil Can Be Repaired 915

Damage by External Chemical Agents Can Also Be Repaired 916

Alteration of a Single Base Pair Causes Point Mutations 918

Insertion or Deletion of Nucleotides Causes Frame-Shift Mutations 919

Mutations Are Random, Rare Events in Individuals 921

Many Mutagenic Agents Are Also Carcinogenic 922

Genes Often Undergo Recombination 923

Segments of Chromosomes Often Undergo Transposition 926

Antibody Variety Is the Result of Transposition and Recombination Events 926

Genes from Different Organisms Can Be Artificially Recombined 929

Plasmids and Phage Lambda Are Vectors for Introducing Foreign Genes into Bacteria 930

Isolation of Genes and the Preparation of cDNAs 932

Construction of the Gene-Bearing Vector 934

Insertion of the "Loaded" Plasmids into the *E. coli* Chromosome 934

Cloned cDNAs Can Be Used to Recover the Corresponding Natural Gene 935

Expression of Cloned Genes Is Enhanced by a Promoter 936

Many Genes Have Been Cloned in Different Host Cells 937

Recombinant DNA and Gene Cloning Open New Lines of Genetic Research 938

Recombinant DNA Research May Have Many Practical Applications 938

Interferon Genes Have Been Cloned 939

Summary 940

References 941

Problems 942

APPENDIX A

Common Abbreviations in Biochemical Research Literature 946

APPENDIX B

Unit Abbreviations, Prefixes, Constants, and Conversion Factors 948

APPENDIX C

International Atomic Weights 949

APPENDIX D

Logarithms 950

APPENDIX E

Answers to Problems 952

APPENDIX F

Glossary 969

Illustration Acknowledgments 981

Index 983

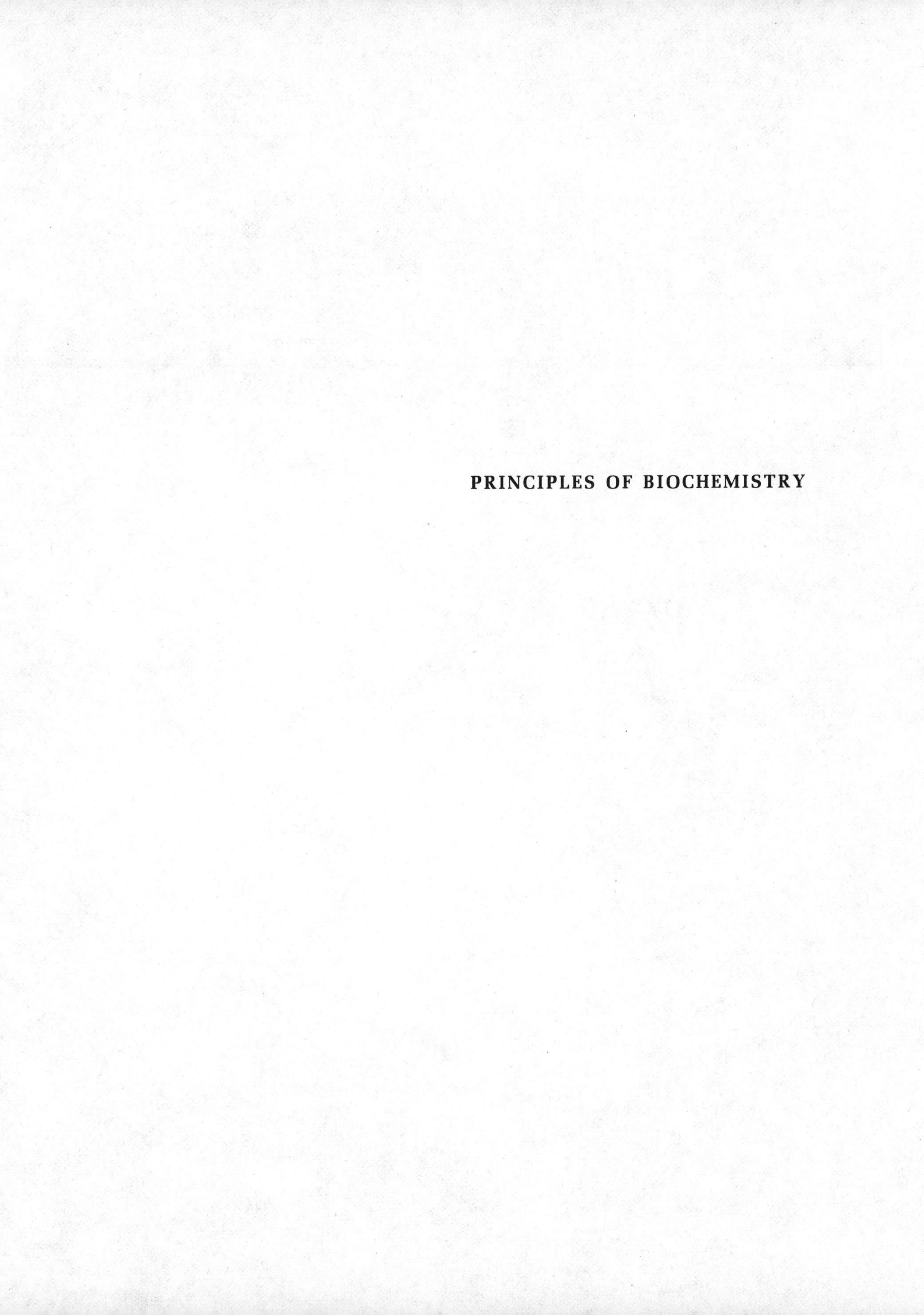

PRINCIPLES OF BIOCHEMISTRY

PART I

Biomolecules

Some 20 billion years ago the universe arose, with a cataclysmic explosion that hurled hot, energy-rich subatomic particles into all space. Gradually, as the universe cooled, these elementary particles combined to form positively charged nuclei to which negatively charged electrons were attracted. Thus were created the hundred or more chemical elements. Every single atom in the universe today, including those of all living organisms, was born of the "big bang." Thus we humans and, indeed, all living things are made of stardust.

The simple organic compounds from which living organisms are constructed are unique to life and do not otherwise occur on the earth today, except as products of biological activity. These building-block compounds, called biomolecules, were selected during the course of biological evolution for their fitness in performing specific cell functions. They are identical in all organisms. Biomolecules are related to each other and interact in a kind of molecular "game" or logic. The size, shape, and chemical reactivity of biomolecules enable them not only to serve as building blocks of the intricate structure of cells, but also to participate in their dynamic, self-sustaining transformations of energy and matter. Biomolecules must therefore be examined from two viewpoints, that of the chemist and that of the biologist. Biochemistry is superchemistry.

Interstellar gas and dust in the constellation Sagittarius. *Also present in interstellar space are many simple organic compounds of the type thought to be precursors of biomolecules on the earth and, who knows, elsewhere in the universe.*

CHAPTER 1

Biochemistry: The Molecular Logic of Living Organisms

Living things are composed of lifeless molecules. When these molecules are isolated and examined individually, they conform to all the physical and chemical laws that describe the behavior of inanimate matter. Yet living organisms possess extraordinary attributes not shown by collections of inanimate molecules. If we examine some of these special properties, we can approach the study of biochemistry with a better understanding of the fundamental questions it seeks to answer.

Living Matter Has Several Identifying Characteristics

One of the most conspicuous attributes of living organisms is that they are complicated and highly organized. They possess intricate internal structures and contain many kinds of complex molecules. Furthermore, living organisms occur in millions of different species. In contrast, the inanimate matter in our environment, as represented by clay, sand, rocks, and seawater, usually consists of random mixtures of relatively simple chemical compounds.

Second, each component part of a living organism appears to have a specific purpose or function. This is true not only of macroscopic structures, such as the heart, the lungs, and the brain, but also of microscopic intracellular structures, such as the nucleus. Even individual chemical compounds in cells, such as their proteins and lipids, have specific functions. Thus, it is quite legitimate to ask what the purpose of a given molecule or chemical reaction in a living organism may be. But it is meaningless to ask about the function of different chemical compounds in inanimate matter; they just happen to be there.

Third, and now we come closer to the essence of the life process, living organisms have the ability to extract, transform, and use energy from their environment, either in the form of organic nutrients or the radiant energy of sunlight. Such energy enables living organisms to build and maintain their own intricate energy-rich structures, to do the mechanical work of

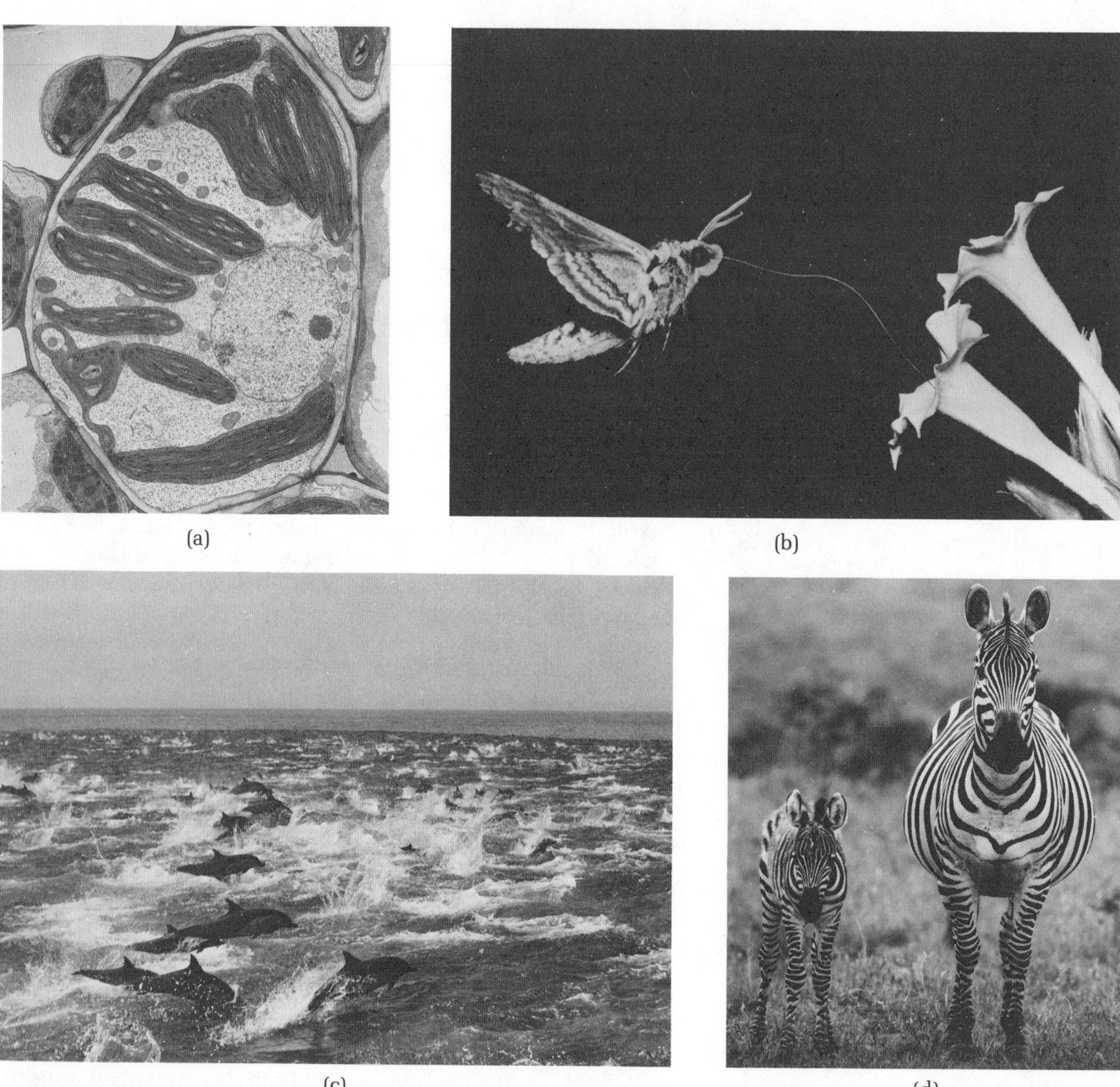

(a) (b) (c) (d)

Figure 1-1
Some characteristics of living matter: the "signs of life." (a) *Cross section of a photosynthetic cell showing its intricate and complex internal structure. The dark objects are chloroplasts, which contain thousands of chlorophyll molecules oriented to trap solar energy.* (b) *The sphinx moth's long tongue was adapted over the course of biological evolution for a specific purpose: to obtain nectar from deep-belled flowers.* (c) *Dolphins, which feed on small fish, convert the chemical energy of food into powerful bursts of muscular energy.* (d) *Biological reproduction occurs with near-perfect fidelity.*

locomotion, and to transport materials across membranes. Living organisms are never at equilibrium within themselves or with their surroundings. On the other hand, inanimate matter does not use energy in a purposeful way to maintain its structure and to do work. Rather, when left to itself it tends to decay toward a more random state with time, to come to equilibrium with its surroundings.

But the most extraordinary attribute of living organisms is their capacity for precise self-replication, a property which can be regarded as the very quintessence of the living state. Mixtures of inanimate matter with which we are familiar show no apparent capacity to grow and reproduce in forms identical in mass, shape, and internal structure, generation after generation.

Biochemistry Seeks to Understand the Living State

We may now ask: If living organisms are composed of molecules that are intrinsically inanimate, why is it that living matter differs so radically from nonliving matter, which also consists of inanimate molecules? Why does the living organism appear to be more than the sum of its inanimate parts? Philosophers once answered that living organisms are endowed with a mysterious and divine life-force. But this doctrine, called *vitalism*, has been rejected by modern science, which seeks rational and, above all, testable explanations of natural phenomena. The basic goal of the science of biochemistry is to determine how the collections of inanimate molecules that constitute living organisms interact with each other to maintain and perpetuate the living state. To be sure, biochemistry also yields important insights and practical applications in medicine, agriculture, nutrition, and industry, but its ultimate concern is with the wonder of life and living organisms.

The molecules of which living organisms are composed conform to all the familiar laws of chemistry, but they also interact with each other in accordance with another set of principles, which we shall refer to collectively as the *molecular logic of the living state*. These principles do not necessarily involve new or as yet undiscovered physical laws or forces. Instead, they are a unique set of relationships characterizing the nature, function, and interactions of *biomolecules*, i.e., the kinds of molecules found in living organisms.

Now let us see if we can identify some of the important axioms in the molecular logic of the living state.

All Living Organisms Contain Organic Macromolecules Built According to a Common Plan

Most of the chemical components of living organisms are organic compounds of carbon, in which carbon atoms are covalently joined with other carbon atoms and with hydrogen, oxygen, or nitrogen. Organic compounds in living matter occur in extraordinary variety, and many of them are extremely large and complex. For example, even the simplest and smallest cells, the bacteria, contain a very large number of different organic molecules. A single cell of the common bacterium *Escherichia coli* contains about 5000 different kinds of organic compounds, including as many as 3000 different kinds of proteins and 1000 different kinds of nucleic acids. Moreover, proteins and nucleic acids are very large and complex molecules (*macromolecules*); the precise structures of only a few of them are known. In the much more complex human organism there may be well over 50,000 different kinds of proteins. It is unlikely that any of the protein molecules of *E. coli* is identical with any of the proteins found in human beings, although many function in quite similar ways. In fact, each species of organism has its own set of protein and nucleic acid molecules, nearly all of which are distinctively different from those of other species. Since there are probably close to 10 million species of living organisms, we can see that all species together must contain, at the very least, 10^{11} different kinds of protein molecules and almost as many different kinds of nucleic acids.

For biochemists to attempt to isolate, identify, and synthesize all the different organic molecules present in living matter would appear to be a hopeless undertaking. Paradoxically, however, the immense diversity of organic molecules in living organisms is reducible to a basic simplicity. This is because all the macromolecules of cells are composed of simple, small building-block molecules of a few different kinds, strung into long chains containing from 50 to many thousands of units. The long, chainlike molecules of deoxyribonucleic acids (DNA) are constructed from only four different kinds of building blocks, the deoxyribonucleotides, arranged in a characteristic sequence. Proteins consist of covalently linked chains of 20 different kinds of amino acids, small organic compounds of known structure. The 20 different amino acids can be arranged in many different sequences to form many different kinds of proteins, just as the 26 letters of the English alphabet can be arranged into an almost limitless number of words, sentences, or, indeed, books. Moreover, the few kinds of nucleotides from which all nucleic acids are built and the 20 different kinds of amino acids from which all proteins are built are identical in all species, whether microbial, plant, or animal. This fact strongly suggests that all living organisms may be descended from a common ancestor.

Figure 1-2
(a) Segment of a DNA molecule, showing the sequence of its four different nucleotide building blocks. (b) Segment of a protein molecule, showing the sequence of its amino acid units.

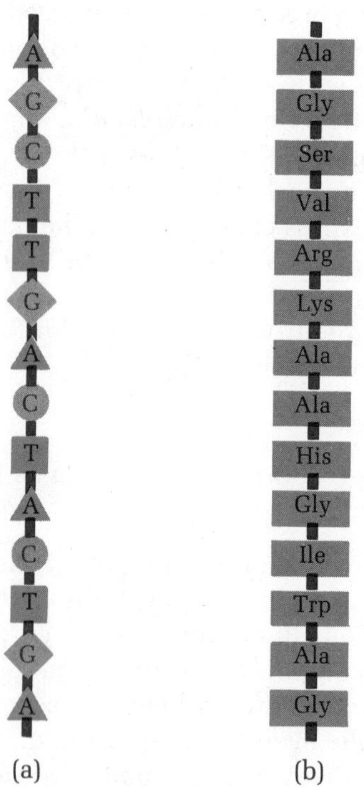

The simple building-block molecules from which all macromolecules are constructed have another striking characteristic. Each of them serves more than one function in living cells. The different amino acids serve not only as building blocks of protein molecules but also as precursors of hormones, alkaloids, pigments, and many other kinds of biomolecules. The nucleotides serve not only as building blocks of nucleic acids but also as coenzymes and as energy-carrying molecules. So far as we know, living organisms normally contain no functionless compounds, although there are some biomolecules whose functions we do not yet understand.

From these considerations we can now set out some of the axioms or principles in the molecular logic of the living state:

There is a basic simplicity in the structure of biological macromolecules.

All living organisms use the same kinds of building-block molecules and thus appear to have a common ancestry.

The identity of each species or organism is preserved by its possession of distinctive sets of nucleic acids and of proteins.

All biomolecules have specific functions in cells.

Living Organisms Exchange Energy and Matter

Living organisms are not exceptions to the physical laws governing energy exchanges. Their growth and maintenance require energy, which must be paid for in some way. Living organisms absorb from their environment forms of energy that are useful to them under the special conditions of temperature and pressure in which they live. They then return to the environment an equivalent amount of energy in some other, less useful, form. The useful kind of energy that cells require is *free energy*, which may be simply defined as that type of energy which can do work at constant temperature and pressure. The less useful type of energy cells return to their environment consists largely of heat, which is dissipated into the surroundings and becomes randomized. Here we have another axiom in the molecular logic of the living state:

> *Living organisms create and maintain their complex, orderly, purposeful structures at the expense of free energy from their environment, to which they return energy in less useful forms.*

Although living organisms are energy-transforming systems, they are very different from familiar man-made machines. The energy-transforming systems of living cells are built entirely of relatively fragile and unstable organic molecules, unable to withstand high temperatures, strong electric currents, or extremely acid or basic conditions. All parts of a living cell have essentially the same temperature; furthermore, there are

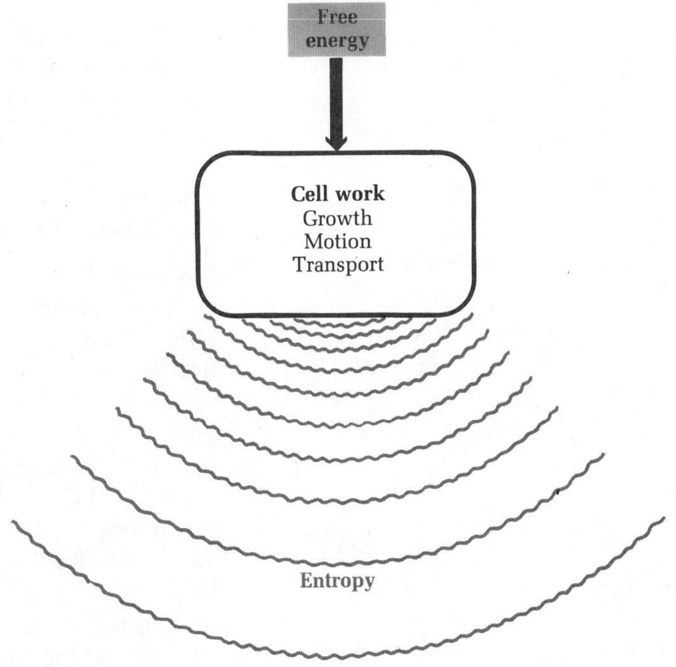

Figure 1-3
Living organisms carry out various forms of work at the expense of free energy from the environment. They return to their surroundings an equal amount of heat and other forms of energy that become randomized and useless. Such dissipated energy is called entropy.

Figure 1-4
Sunlight is the ultimate source of all biological energy.

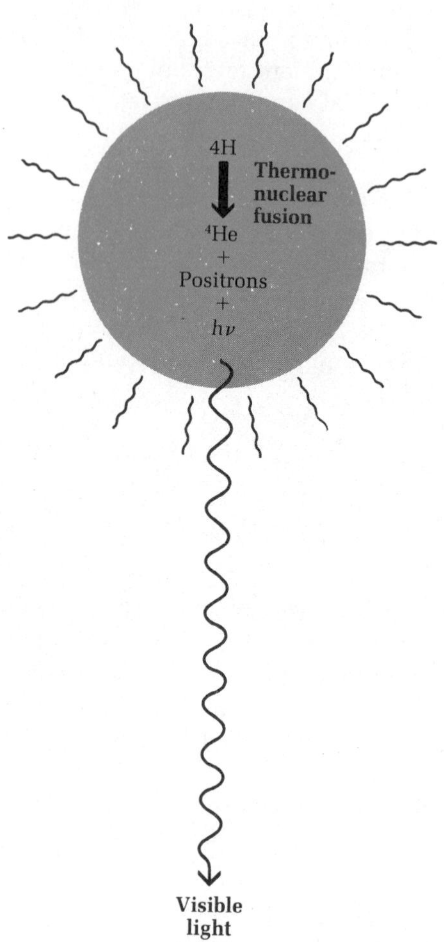

no significant differences in pressure from one part of a cell to another. Thus we can conclude that cells are unable to use heat as a source of energy, since heat can do work only if it passes from one body to another at a lower temperature. Nor do living cells resemble steam engines or electric motors, the types of engines most familiar to us. Instead, and this is another important axiom in the molecular logic of the living state:

> *Living cells are chemical engines that function at constant temperature.*

Cells use chemical energy to carry out the chemical work of cell repair and growth, the osmotic work required to transport nutrients into cells, and the mechanical work of contraction and motion.

Ultimately, all living organisms in the earth's biosphere derive their energy from sunlight, which arises from the nuclear fusion of hydrogen atoms to form helium in the immensely high temperature of the sun. Photosynthetic cells of the plant world trap the radiant energy of sunlight and use it to convert carbon dioxide and water into various energy-rich plant products, e.g., starch and cellulose. In doing so they set molecular oxygen free into the atmosphere. Ultimately, nonphotosynthetic organisms obtain energy for their needs from energy-rich plant products by oxidizing them at the expense of atmospheric oxygen to form carbon dioxide and other end products, which return to the environment and are recycled by the plant world. Thus other axioms of the living state emerge:

> *The energy needs of all organisms are provided, directly or indirectly, by solar energy.*
>
> *The plant and animal worlds—indeed, all living organisms—are dependent on each other through exchanges of energy and matter via the environment.*

Enzymes, the Catalysts of Living Cells, Promote Sequences of Organized Chemical Reactions

Figure 1-5
Enzymes greatly increase the rate of specific chemical reactions.

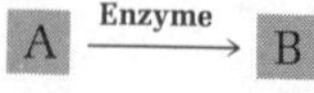

Cells can function as chemical engines because they possess enzymes, catalysts capable of greatly enhancing the rate of specific chemical reactions without being consumed in the process. Enzymes are highly specialized protein molecules, made by cells from simple amino acid building blocks. Each type of enzyme can catalyze only one specific type of chemical reaction; thus, hundreds of different types of enzymes are required in the metabolism of any given kind of cell. Enzymes are far superior to catalysts made by chemists since they are much more specific, have higher catalytic efficiency, and can function under mild conditions of temperature and hydrogen-ion concentration. Enzymes can catalyze in seconds complex sequences of reactions that would require days, weeks, or months

Figure 1-6
Enzymes often function in sequences.

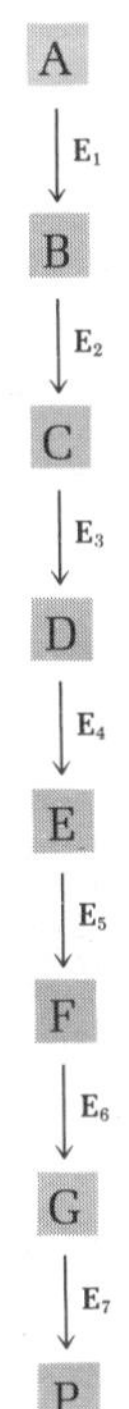

of work in the chemical laboratory. Moreover, enzyme-catalyzed reactions proceed with 100 percent yields; there are no by-products. In contrast, when the organic chemist carries out reactions in the laboratory, they are nearly always accompanied by the formation of one or more by-products. Because enzymes can enhance a single reaction pathway of a given molecule without enhancing its other possible reactions, living cells can carry out many different chemical reactions simultaneously without bogging down in a morass of useless by-products.

The hundreds of enzyme-catalyzed chemical reactions in cells are organized into many different sequences of consecutive reactions. Such sequences may have anywhere from 2 to 20 or more reaction steps. Some of these sequences of enzyme-catalyzed reactions degrade organic nutrients into simple end products, in order to extract chemical energy. Others start from small precursor molecules and build large and complex macromolecules, in a step-by-step fashion, for which energy must be provided. These enzyme-catalyzed pathways, which together constitute the metabolism of cells, have many interconnections.

Cells Transmit Energy in a Chemical Form

Living cells capture, store, and transport energy in a chemical form, largely as *adenosine triphosphate* (ATP). ATP functions as the major carrier of chemical energy in the cells of all living species. ATP can transfer its energy to certain other biomolecules, and in so doing, its terminal phosphate group is detached; the energy-rich ATP molecule thus becomes the energy-depleted *adenosine diphosphate* (ADP), which can in turn regain a phosphate group to become ATP, at the expense of either solar energy in photosynthetic cells or chemical energy in animal cells. ATP is the major connecting link between two large networks of enzyme-catalyzed reactions in the cell. One of these networks conserves chemical energy derived from the environment, largely by causing the phosphorylation of the energy-poor ADP to the energy-rich ATP. The other network utilizes the energy of ATP to carry out the biosynthesis of cell components from simple precursors, to do the mechanical

Figure 1-7
(a) Structural formula and (b) space-filling model for adenosine triphosphate (ATP).

(a)

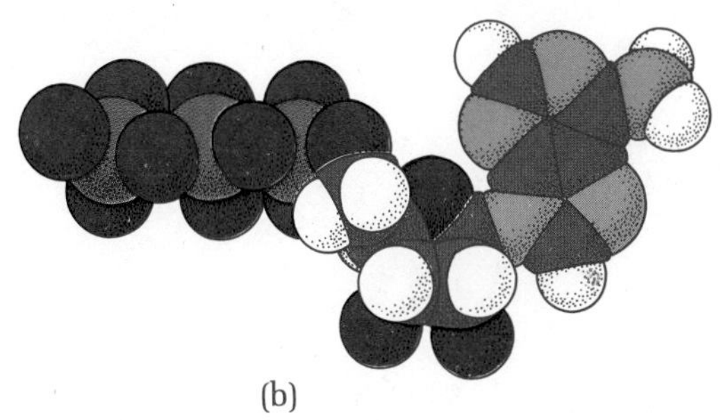

(b)

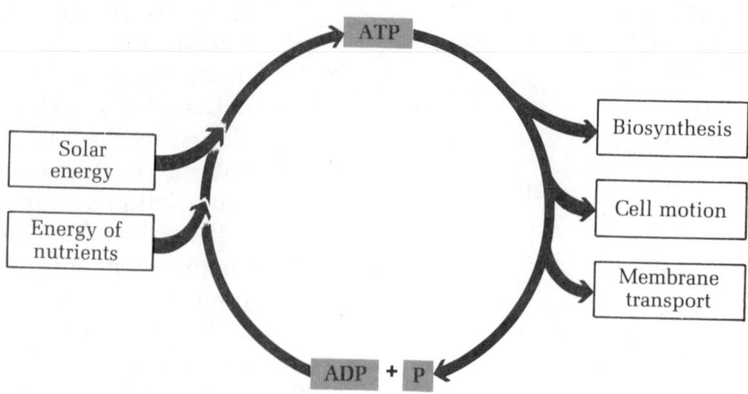

Figure 1-8
ATP is a chemical transmitter linking energy sources to energy-requiring cell processes.

work needed for contraction and motion, and to carry out the osmotic work of membrane transport. Like the building-block biomolecules, these consecutively linked networks of enzyme-catalyzed reactions are virtually identical in most living species.

Cell Metabolism Is Constantly Regulated

Growing cells can simultaneously synthesize thousands of different kinds of protein and nucleic acid molecules in the precise proportions required to constitute the living, functional protoplasm characteristic of their species. The enzyme-catalyzed reactions of metabolism are thus intricately regulated to make only the requisite number of each type of building-block molecule and to assemble them into a certain number of molecules of each type of nucleic acid, each type of protein, and each type of lipid or polysaccharide. Moreover, living cells also possess the power to regulate the synthesis of their own catalysts, the enzymes. Thus the cell can "turn off" the synthesis of an enzyme required to make a given product from its precursors whenever that product is available, ready-made, from the environment. Such self-adjusting and self-regulating properties allow living cells to maintain themselves in a steady state, even when the external environment fluctuates. We can then define another axiom in the molecular logic of the living state:

> *Living cells are self-regulating chemical engines, tuned to operate on the principle of maximum economy.*

Living Organisms Replicate Themselves Accurately

The most remarkable of all the properties of living cells is their ability to reproduce themselves with nearly perfect fidelity for hundreds and thousands of generations. Three features immediately stand out. First, living organisms are so immensely complex that the amount of genetic information transmitted

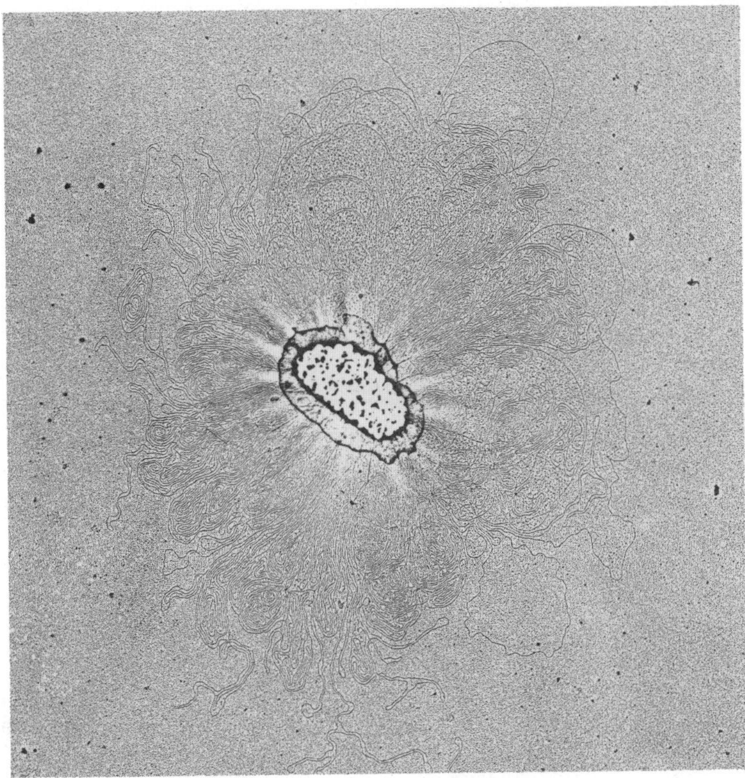

Figure 1-9
The single DNA molecule seeping out of a disrupted cell of the bacterium Hemophilus influenzae. *The DNA molecule is hundreds of times longer than the cell itself.*

seems out of all proportion to the minute size of the cell nucleus, which is the repository of genetic information. We know today that all the genetic information of a bacterial cell is contained in a single large molecule of deoxyribonucleic acid (DNA). And the immensely greater amount of genetic information in a single human germ cell is encoded in a set of DNA molecules weighing altogether only 6×10^{-12} g. We therefore come to another axiom in the molecular logic of life:

> *Genetic information is encoded in units that are submolecular in dimensions; these units are the four kinds of nucleotides of which DNA is composed.*

A second remarkable characteristic of the self-replicating property of living organisms is the extraordinary stability of the genetic information stored in DNA. Very few early historical records have survived for long, even though they have been etched in copper or carved in stone. The Dead Sea scrolls and the Rosetta stone, which gave the key for deciphering ancient Egyptian hieroglyphics, are only a few thousand years old. But there is good reason to believe that many bacteria have nearly the same size, shape, and internal structure and contain the same kinds of building-block molecules and the same kinds of enzymes as those which lived billions of years ago. And this pertains despite the fact that bacteria, like all organisms, have

Figure 1-10
Complementarity between the coding elements of the two strands of DNA.

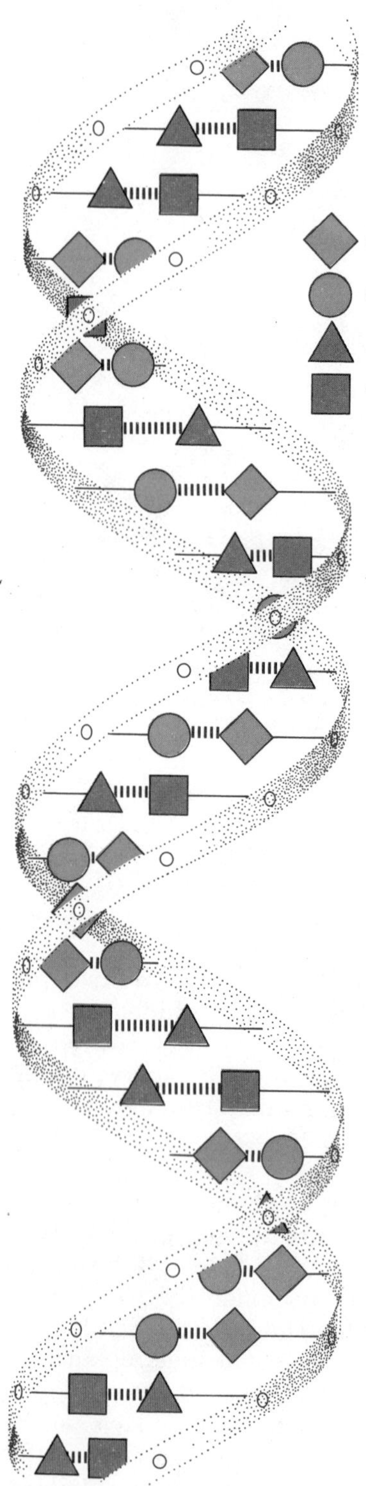

been undergoing constant evolutionary change. Genetic information is preserved, not on copper or stone, but in the form of DNA, an organic molecule so fragile that it will break into many pieces if a solution containing it is merely stirred or pipetted. Even in the intact cell DNA strands may break frequently, but they are quickly and automatically repaired. The remarkable capacity of living cells to preserve their genetic material is the result of *structural complementarity*. One DNA strand serves as the template, or pattern, for the enzymatic replication or repair of a structurally complementary DNA strand. Yet, despite the near-perfect fidelity of genetic replication, DNA occasionally undergoes a very small change, a *mutation*, which sometimes results in better or more fit progeny and sometimes in progeny less able to survive. In this way living organisms can continually improve their ability to survive and thus allow differentiation and further evolution of new species to occur as the environment changes with time.

There is a third remarkable characteristic of genetic information transfer in living organisms. The genetic information is encoded in the form of a linear one-dimensional sequence of the nucleotide building blocks of DNA. But living cells are three-dimensional in structure and have three-dimensional parts or components. The one-dimensional information of DNA is converted into the three-dimensional information inherent in living organisms by translation of DNA structure into protein structure, a process in which ribonucleic acid (RNA) is a participant. Unlike DNA molecules, which all have a basically identical shape, different protein molecules spontaneously curl up and fold into an enormous variety of specific three-dimensional structures, each necessary in performing a specific function. The precise geometry of each type of protein is determined by its amino acid sequence, and this in turn is determined by the nucleotide sequence in DNA.

We can now summarize the various axioms of the molecular logic of cells:

A living cell is a self-assembling, self-adjusting, self-perpetuating isothermal system of organic molecules which extracts free energy and raw materials from its environment.

It carries out many consecutive organic reactions promoted by organic catalysts, which it produces itself.

It maintains itself in a dynamic steady state, far from equilibrium with its surroundings. It functions on the principle of maximum economy of parts and processes.

Its nearly precise self-replication through many generations is ensured by a self-repairing linear coding system.

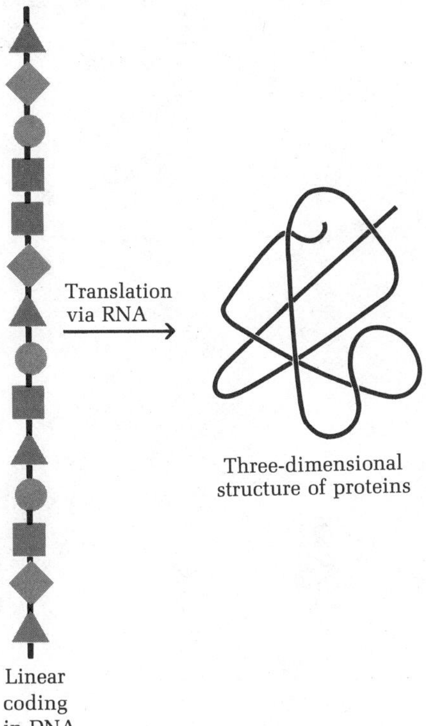

Figure 1-11
The linear nucleotide sequence of DNA *is translated into the three-dimensional structure of proteins.*

It is the goal of biochemistry to understand how the interactions of biomolecules with each other bring about these characteristics of the living state. At no point in our examination of the molecular logic of living cells have we encountered any violation of known physical laws, nor have we needed to define new ones. The "soft," organic machinery of living cells functions within the same set of laws that governs the operation of man-made machines, but the chemical reactions and regulatory processes of cells have been refined far beyond the present capabilities of the chemical engineer.

This set of principles, which summarizes the molecular logic of the living state, appears to be applicable to all cells, even though we may perceive these statements to be simplistic and rather mechanical. Can such a molecular logic of the living state apply to the much greater complexity of multicellular organisms, particularly the higher forms of life? Can it apply to men and women as organisms, with their extraordinary and unique capacity for thought, language, and creativity? We cannot yet begin to answer these questions, although we know today that the development and the behavior of higher organisms depend upon, and can be modified by, molecular agents and thus may have a biochemical basis. But we shall have to wait before we can answer these larger questions, because today the science of biochemistry knows only a very small fraction of what there is to know about living organisms.

In this orienting survey we have seen that biochemistry has an underlying system, a set of organizing principles. It is not merely a collection of unrelated chemical facts about living matter. As we now begin to study biochemistry, these organizing principles should serve as our frame of reference. In this book we first describe the various classes of biomolecules. We then proceed to analyze the isothermal, consecutively linked, self-adjusting, enzyme-catalyzed reactions which make possible metabolism, the flow of matter and energy between the organism and the environment. Finally, we shall consider the molecular basis of the self-replication of cells and the translation of the one-dimensional information of DNA into three-dimensional proteins. Along the way we shall see how biochemistry also provides new and important insights into human physiology, nutrition, and medicine, into plant biology and agriculture, into evolution and ecology, and into the great cycles of energy and matter between the sun, the earth, and the plant and animal worlds.

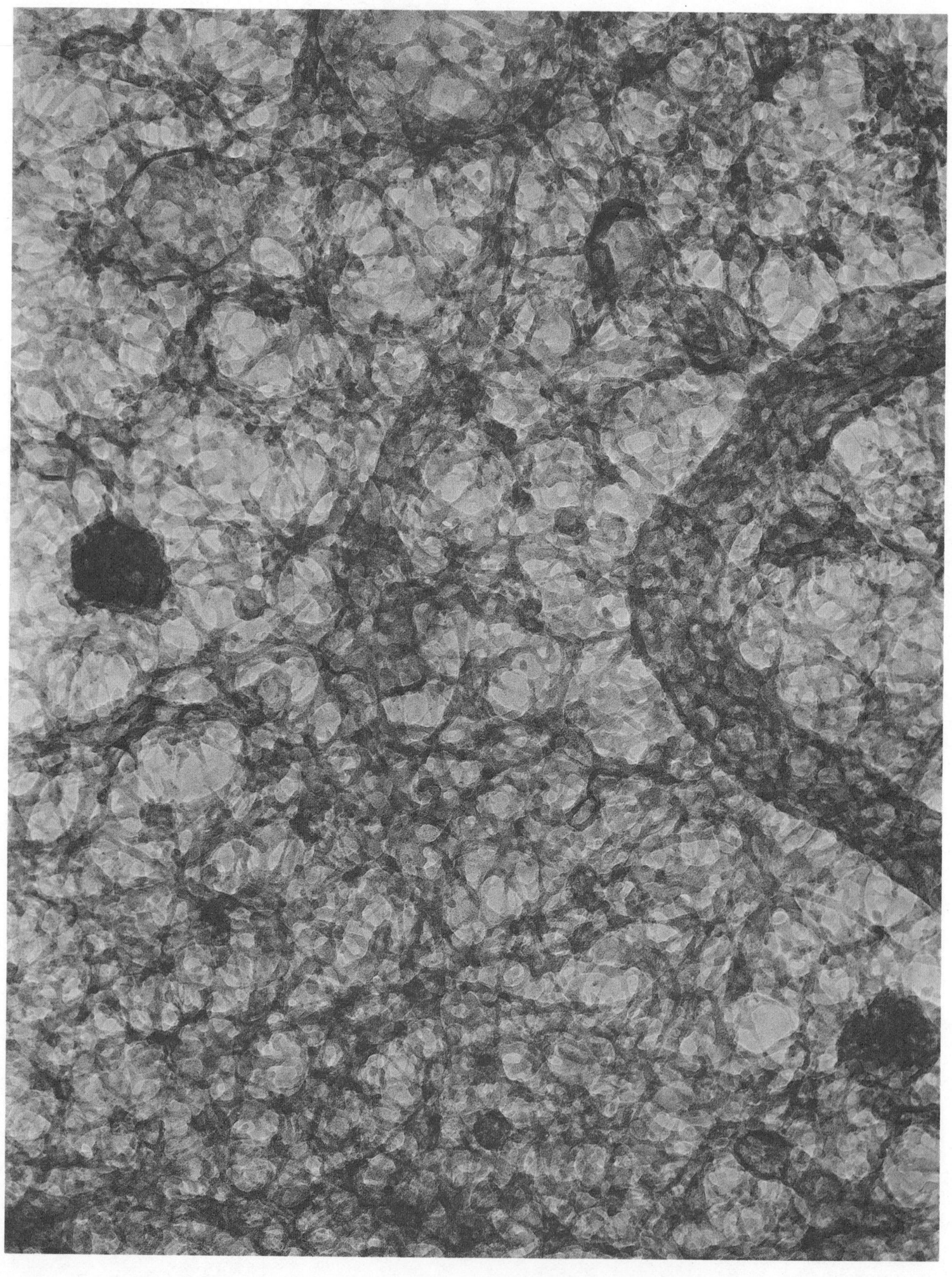

CHAPTER 2

Cells

Lost in the cell cytoplasm! Extremely high magnification of the complex meshwork or lattice of very fine filaments that form the "ground substance" of animal cells. This network, which cannot be visualized by standard electron microscopy, was recently discovered using a new high-voltage electron microscope capable of very high resolution. This picture shows only a very tiny portion of the cell cytoplasm, which is entirely pervaded by this complex three-dimensional network.

Readers may wonder why a discussion of biochemistry begins with cells. Indeed, some will already have studied cell biology. Nevertheless, it is important to begin in this way because cells, not test tubes or beakers, are the natural setting of most biochemical reactions. A major difference between biochemistry and "ordinary" chemistry is that biochemical reactions take place within the constraints imposed by the size of cells and their internal compartments, as well as the physical and chemical conditions that are compatible with cellular life.

The ordinary chemical reactions with which we are familiar proceed in reaction vessels that are made of nonbiological materials and are enormous in size compared to the molecules undergoing transformation. Moreover, ordinary chemical reactions may require high temperatures or pressures, strong reagents, or the application of electric energy; often they are carried out in organic solvents. In contrast, the biochemical reactions of living cells take place within the extremely small dimensions of living cells or their subcompartments, whose walls are fragile and only a few molecules thick. Moreover, biochemical reactions take place in an aqueous medium at relatively low and constant temperatures. Cells simply cannot tolerate extremes of temperature, pressure, or acidity or the presence of powerful reagents. From here on we must think of the chemistry of life processes in the context of cell dimensions, structure, and activities. We must also bring together two points of view, that of the chemist and that of the cell biologist.

In this chapter we shall orient ourselves to the structure and biochemical activities of representative cells. Later chapters will consider some of these relationships in greater detail.

All Cells Share Some Structural Features

Cells are the structural and functional units of living organisms. The smallest organisms consist of single cells; in contrast, the human body is believed to contain at least 10^{14} cells.

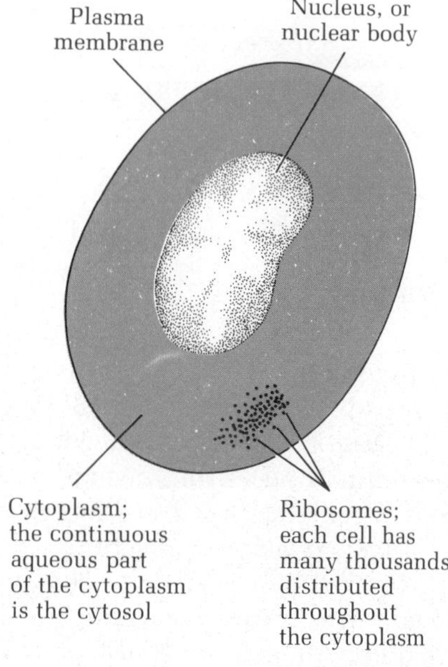

Figure 2-1
All cells possess a surrounding membrane, either a nucleus or nuclear body, and ribosomes. Later we shall see that there are two great classes of cells, which differ in other structural respects.

There are many different types of cells, which vary enormously in size, shape, and specialized functions. In a handful of farm soil or a cup of pond water there may be dozens of different types of unicellular organisms. And in each higher multicellular organism, whether it is the human body or a corn plant, there are dozens or hundreds of different types of cells, all highly specialized to function together in the form of tissues and organs. Yet no matter how large and complex the organism, each of its cell types retains some individuality and independence.

Despite their many outward differences, various kinds of cells are remarkably similar in their basic structural features (Figure 2-1). Every cell has a very thin surrounding membrane, which makes it self-contained and to some extent self-sufficient. The cell membrane, also called the *plasma membrane* or *cytoplasmic membrane*, is selectively permeable. It allows necessary nutrients and salts to pass into the cell and waste products to leave but usually excludes the entry of unneeded substances from the environment. In all cells the molecular architecture of the plasma membrane is basically the same, consisting of two layers of lipid molecules in which a variety of specialized proteins are embedded. Some of the membrane proteins are enzymes, whereas others can bind nutrients from the environment and transport them into the cell. Within each cell there is the *cytoplasm*, in which most of the enzyme-catalyzed reactions of cell metabolism occur. Here cells use chemical energy to do the work of building and maintaining their structure and of bringing about cell motion or contractility. Also present in the cytoplasm of all cells are *ribosomes*, small granules from 18 to 22 nm in diameter, which function to synthesize proteins. And all living cells have either a *nucleus* or a *nuclear body*, in which the genetic material is replicated and stored in the form of deoxyribonucleic acid (DNA).

Cells Must Have Very Small Dimensions

Cells share another important structural feature. They are all relatively small (they have to be). In the laboratory ordinary chemical reactions take place in vessels containing many milliliters or even liters. The contents of such reaction vessels must be stirred to mix the reagents thoroughly, so that the reaction rate is not limited by the rate of diffusion of the reacting molecules. In living cells, however, biochemical reactions occur in compartments of microscopic volume; for example, the volume of a single cell of the bacterium *Escherichia coli* is only about 2×10^{-12} milliliters (mL). To appreciate fully some of the implications of cell size for the chemical aspects of cell activities, we must first become familiar with the dimensions of biomolecules and cells. Table 2-1 shows the important units of length now used in expressing cell dimensions, the *nanometer* (nm) and *micrometer* (μm); they are based on the metric system.

Table 2-1 The International System of Units

Basic units	
Length	Meter (m)
Mass	Kilogram (kg)
Time	Second (s)
Prefixes	
10^3, kilo (k)	10^{-3}, milli (m)
10^6, mega (M)	10^{-6}, micro (μ)
10^9, giga (G)	10^{-9}, nano (n)
Units of length common in cell biology and biochemistry	
Nanometer (nm)	$= 10^{-9}$ m
	$= 10^{-6}$ mm
	$= 10^{-3}$ μm
Micrometer (μm)	$= 10^{-6}$ m
	$= 10^{-3}$ mm
	= 1000 nm
Older units, superseded but still often used	
1 micron (μ)	= 1 micrometer (μm)
1 millimicron (mμ)	= 1 nanometer (nm)
1 angstrom (Å)	= 0.1 nanometer (nm)

Table 2-2 Dimensions of Some Biological Structures

Structure	Long dimension, nm
Alanine, an amino acid	0.5
Glucose, a sugar	0.7
Phosphatidylcholine, a membrane lipid	3.5
Myoglobin, a small protein	3.6
Hemoglobin, a medium-sized protein	6.8
Ribosome of *E. coli*	18
Poliomyelitis virus	30
Myosin, a long, rod-shaped protein	160
Tobacco mosaic virus	300
Mitochondrion of a liver cell	1,500
E. coli cell	2,000
Chloroplast of a spinach leaf cell	8,000
Liver cell	20,000

Although older units, such as the *angstrom*, or older names, such as the *micron*, are used less and less, we should also be familiar with them. For some orientation with respect to the dimensions of cells Table 2-2 shows the size of some important biological structures, including small biomolecules (alanine and glucose); macromolecules (three proteins and a lipid); supramolecular systems (ribosomes and viruses); cell organelles (mitochondria and chloroplasts); a bacterium; and a liver cell. Many bacterial cells are about 2 μm in length, and most cells in higher animals are 20 or 30 μm.

One may well ask why the dimensions of living cells are what they are. Why aren't cells much smaller or much larger than we know them to be? There are important reasons. The smallest complete cells, microorganisms known as *Mycoplasma*, cannot be much smaller than they are, simply because the building-block molecules of which their organic matter is composed have fixed sizes, set by the dimensions of carbon, hydrogen, oxygen, and nitrogen atoms. Since a certain minimum number of different biomolecules is necessary for life, if cells were to be much smaller, they would have to be built of smaller atoms or molecules.

On the other hand, cells probably cannot be much larger than they are, simply because their rates of metabolism might then become limited by the rate of diffusion of nutrient molecules to all their internal parts, thus limiting the power of cells to regulate their metabolism. The upper limit to the size of cells is therefore set by fundamental laws of physics, which describe the rate of diffusion of solute molecules in aqueous systems. Indeed, in the largest types of cells, the cytoplasm is compartmented into structures of smaller dimensions, the cell *organelles*, in part to facilitate rapid interactions between specific molecules by shortening the distance they must move in order to collide and react. Quite simply, one reason cells are small is because they must get along without mechanical or electrical stirring devices. Another reason is that there is an optimum ratio of the surface area of cells to their volume. A large surface area in relation to cell volume allows many molecules of nutrients to pass into the cell per unit time. The ratio of surface area to volume decreases sharply with an increase in the diameter of a sphere, as some calculations will show. (Try this. Calculate the surface-to-volume ratios for spheres of 1, 10, and 100 μm diameter. The surface area of a sphere is $4\pi r^2$, and its volume is $\frac{4}{3}\pi r^3$, where r is the radius and π is 3.14.)

There Are Two Great Classes of Cells: Prokaryotes and Eukaryotes

The simplest and smallest cells, those with the most ancient origins, are known as *prokaryotes;* they consist of different families of unicellular microorganisms generally called *bacteria*. Prokaryotic cells were the first to arise in biological evolution; fossil remains of such cells, dated as being well over 3 bil-

Figure 2-2
This bacterium-like microfossil, found in South Africa in a deposit of a flintlike rock called black chert, is about 3.4 billion years old. One of the oldest fossils now known, it has been given the name Eobacterium isolatum, *meaning "single bacterium at the dawn (of life)."*

The short, straight line at the bottom of this micrograph and those that follow is a reference marker for size; 1 μm is 1/10,000 cm.

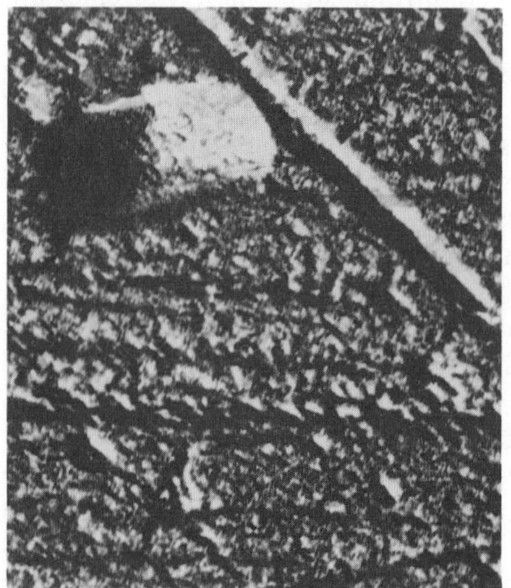

lion (3×10^9) years old, have been found in ancient shales in Africa (Figure 2-2), as well as in Australia. *Eukaryotic cells,* which arose perhaps a billion years after the prokaryotes, are much larger, more complex, and show a wider range of diversity and differentiation. They are the type of cells found in all multicellular animals, plants, and fungi.

The terms prokaryotic and eukaryotic derive from the Greek *karyon,* "nut" or "kernel," i.e., nucleus. Prokaryotic means "before the nucleus," and eukaryotic means with a "well-formed nucleus." In prokaryotes the genetic material is localized in a rather irregular *nuclear body* or *nucleoid,* which has no surrounding membrane. Eukaryotes, on the other hand, have a highly developed, very complex nucleus, surrounded by a *nuclear envelope* consisting of two membranes. We shall now examine prokaryotic and eukaryotic cells in more detail.

The Prokaryotes Are the Simplest and Smallest Cells

The prokaryotes consist of about 3000 species of bacteria, including the organisms commonly called the *blue-green algae.* The blue-green algae are a special family of bacteria; their modern and preferred name is *cyanobacteria* (*cyano-*, "blue"). The cyanobacteria are distinctive and often set apart because they have an oxygen-evolving photosynthetic system much like that in higher green plants. Although certain other classes of bacteria can engage in photosynthesis, they do not evolve oxygen. In fact, most species of bacteria are nonphotosynthetic and obtain their energy from the breakdown of nutrients from the environment. There are some 20 different families of prokaryotes, which are classified or named according to their shape (see Figure 2-3), their capacity for motion, their staining characteristics, the nutrients they favor, or the products they make. Some bacteria are *pathogenic* (disease-causing), but many are highly beneficial. Among the prokaryotes are families of very small cells that normally live as parasites within other cells.

Prokaryotic cells, although invisible to the naked eye and not as familiar to us as the higher animals and plants, form a very important part of the total biomass on the earth. Perhaps three-fourths of all the living matter on the earth consists of microscopic organisms, most of them prokaryotes. Moreover, prokaryotes play an important role in the biological exchanges of matter and energy on the earth. Photosynthetic bacteria in both fresh and marine waters trap solar energy and use it to generate carbohydrates and other cell materials, which are used in turn as food by other forms of life. Some bacteria can fix molecular nitrogen (N_2) of the atmosphere to form biologically useful nitrogenous compounds. Prokaryotes thus form the starting point of many food chains in the biosphere. Moreover, prokaryotes also participate as ultimate consumers, since various bacteria degrade the organic structures of dead plants and animals, returning the end products to the atmosphere, soil, and seas. There they are used over again in the biological cycling of the elements carbon, nitrogen, and oxygen.

Prokaryotic cells are also particularly important in the study of biochemistry and molecular biology because of their simple structure, the speed and ease with which large numbers of such cells can be grown, and their relatively simple mechanisms for reproduction and transmission of genetic information. Under optimal conditions the bacterium *E. coli* will divide every 20 to 30 min at 37°C in a simple nutrient medium of glucose, ammonium salts, and minerals. Another important feature of prokaryotic cells is that they reproduce in a very simple asexual manner. They grow until they double in size and then divide into two identical daughter cells, each receiving one copy of the genetic material (DNA) of the parent cell. Prokaryotic cells have only one chromosome, consisting of a double-stranded DNA molecule. Moreover, genetic mutants of prokaryotes can be readily induced and cultured. Because of these properties bacteria have yielded much of our knowledge of the basic molecular processes involved in the transmission of genetic information.

Figure 2-3
Some prokaryotes are classified and named according to their shapes.

(a) The bacilli, *which are rod-shaped, include* Escherichia coli *as well as pathogenic organisms causing diphtheria, tetanus, and tuberculosis.*

(b) *The cocci are spherical bacteria. Sometimes they associate into pairs (diplococci), clusters (staphylococci), or strands (streptococci, shown here). The bacteria causing scarlet fever, some forms of pneumonia, and wound infections are cocci.*

(c) The spirilla, *helical in shape, are often very long, up to 500 μm. Those shown here are 10 to 15 μm long. The causative organism of syphilis is a member of this class.*

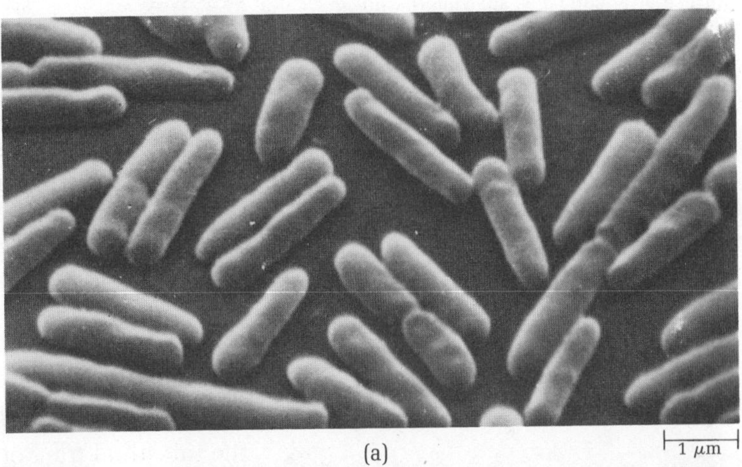

(a)

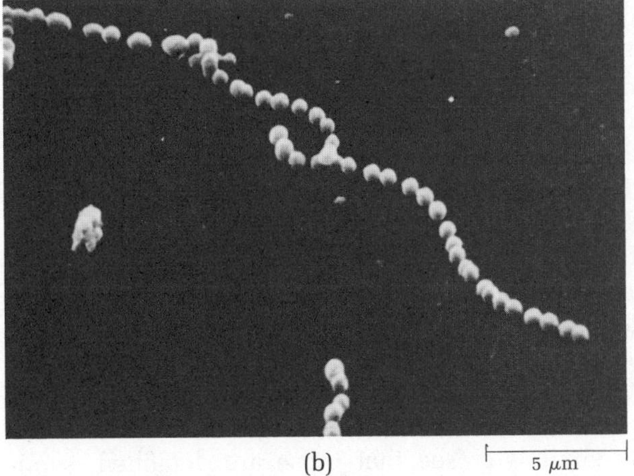

(b)

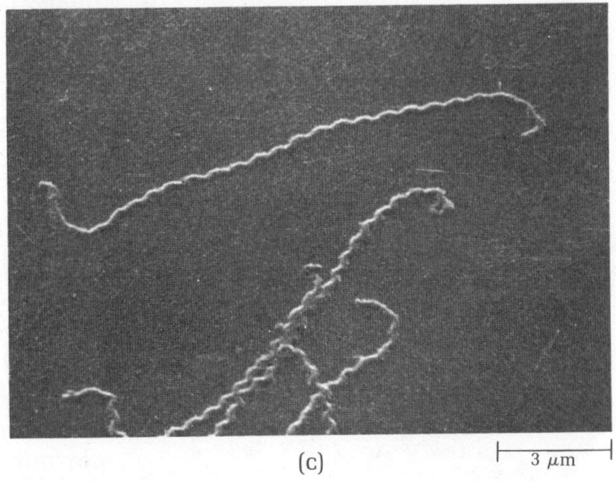

(c)

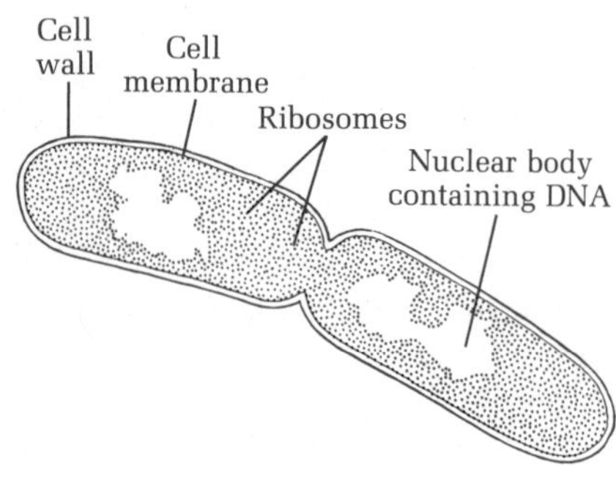

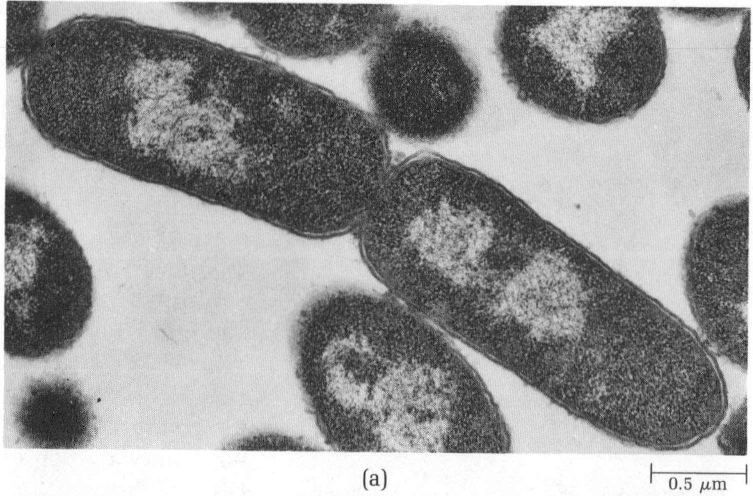

(a)

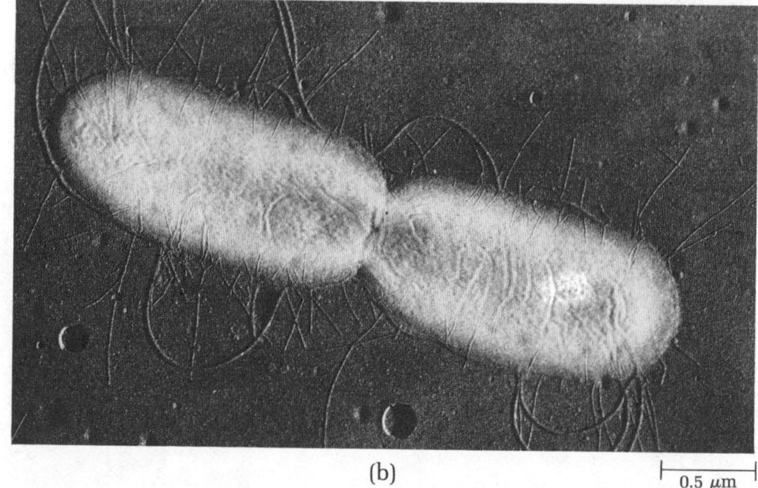

(b)

Figure 2-4
Two views of E. coli *cells.* (a) *An electron micrograph of a thin section. In the center are two cells that have just finished dividing but have not yet separated. The less dense central areas in each cell are the nuclear bodies or nucleoids, containing DNA. The very dense granules in the cytoplasm are ribosomes.* (b) *Electron micrograph of the surface of* E. coli *cells, showing the pili and flagella.*

Escherichia coli Is the Best-Known Prokaryotic Cell

Escherichia coli (Figure 2-4), a usually harmless inhabitant of the intestinal tract of human beings and many other higher animals, is the most intensively studied of the prokaryotes and probably the best-understood cell of any kind. *E. coli* cells are about 2 μm long and a little less than 1 μm in diameter. They have a protective cell wall, a delicate cell membrane inside the wall, a cytoplasm enclosed within the membrane, and a nuclear body containing a single molecule of double-stranded DNA in the form of a very long, endless loop, often called a *circle*. The DNA molecule of an *E. coli* cell is almost 1000 times longer than the cell itself and thus must be very tightly folded to fit into the nuclear body, which is usually less than 1 μm in its longest dimension. As with all prokaryotes, no membrane surrounds the genetic material in *E. coli*. In addition to the main body of DNA in the nucleoid, the cytoplasm of most bacteria contains very small circular segments of DNA called *plasmids*. We shall later see that these are detached, semi-independent genetic elements, which are leading today to important new developments in genetic biochemistry and engineering.

The outer cell wall of *E. coli* cells is covered with a sheath or capsule of a slimy substance through which protrude short, hairlike structures called *pili,* whose function is not entirely clear. Strains of *E. coli* and other bacteria that are motile also possess one or more long *flagella,* which can propel the bacterium through its aqueous surroundings. Bacterial flagella are thin, rigid, curved rods, about 10 to 20 nm across. They are attached to a structure inside the membrane resembling an auto transmission, which rotates the flagellum. The cell membrane consists of a very thin bilayer of lipid molecules penetrated by proteins. It is selectively permeable and contains proteins capable of transporting certain nutrients into the cell and waste products out. The cell membrane of most prokaryotes also contains important electron-carrying proteins that can convert oxidative energy into the chemical energy of ATP. In the photosynthetic bacteria internal membranes derived from the plasma membrane contain chlorophyll and other light-trapping pigments (Figure 2-5).

In the cytoplasm of *E. coli* are a number of granular elements. Most conspicuous are the densely staining *ribosomes,* which in prokaryotes are about 18 nm in diameter. Ribosomes, which contain ribonucleic acid and a number of protein molecules, carry out the synthesis of cell proteins. They often occur in clusters called *polyribosomes* or *polysomes.* Also present in

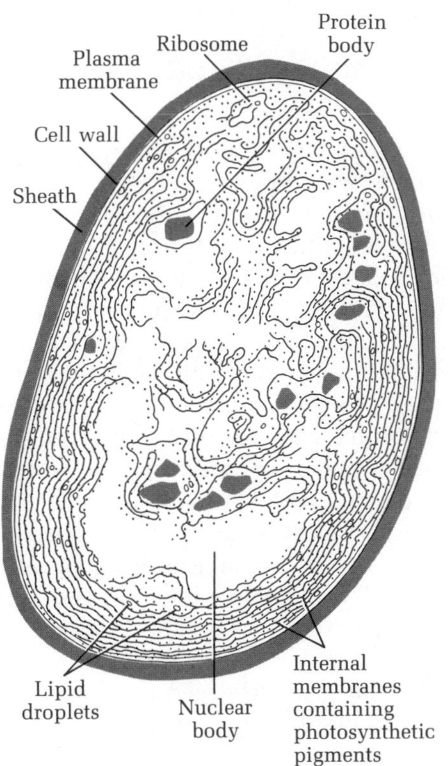

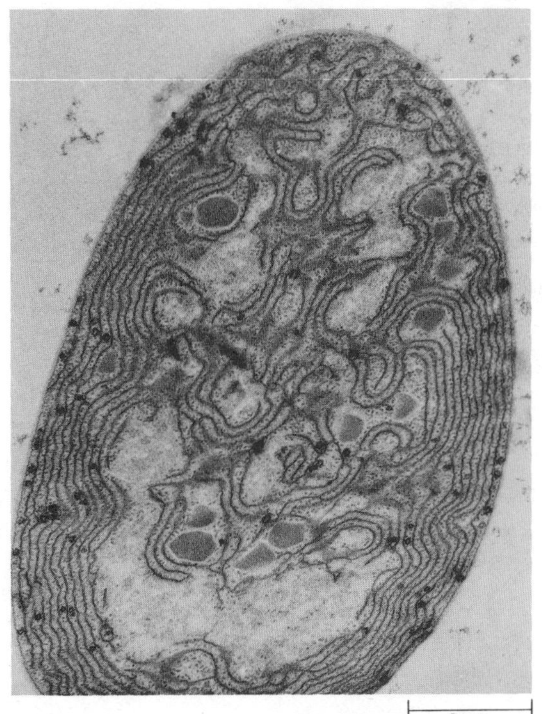

Figure 2-5
Electron micrograph of the blue-green alga, or cyanobacterium, Anabaena azollae. *The abundant internal membranes are derived from the plasma membrane; they contain chlorophyll and other photosynthetic pigments. The cyanobacteria often associate into long strings or filaments.*

the cytoplasm of many bacteria are granules containing stored nutrients; some consist of starch, and others contain fat. The cytosol, the aqueous phase of the cytoplasm, also contains in dissolved form many enzymes, many building-block molecules functioning as precursors of cell macromolecules, and a number of inorganic salts.

Even in the simple bacteria we note a primitive division of labor within the cell. The cell wall is the outer frontier, protective in nature; the cell membrane transports nutrients inward and wastes outward and also generates chemical energy as ATP. The cytoplasm is the site of many enzyme-catalyzed reactions leading to the synthesis of many cell components; the ribosomes manufacture proteins; and the nuclear body participates in the storage and transmission of genetic information.

Although prokaryotes are relatively simple and small compared with eukaryotic cells, some of them are capable of surprisingly complex activities. For example, many bacteria show the phenomenon of chemotaxis. They are attracted to and can move toward certain chemical substances, particularly nutrients, and are repelled by and move away from toxic substances. Thus they have a primitive sensory system that can communicate signals to their flagella, which propel the cells toward or away from a given attractant or repellant (Figure 2-6). They also have a primitive memory.

The cells of some species of prokaryotes tend to associate together into clusters or filaments, which give the impression of being primitive multicellular organisms, but true multicellular organisms contain only eukaryotic cells.

Figure 2-6
Chemotaxis in bacteria. Motile bacteria can sense small concentration gradients and will "swim" toward attractants, such as nutrients. The cell, propelled by one or more flagella, moves through a series of straight runs interrupted by "tumbles." Movement toward the attractant (or away from the repellant) is not direct but proceeds by a devious pathway.

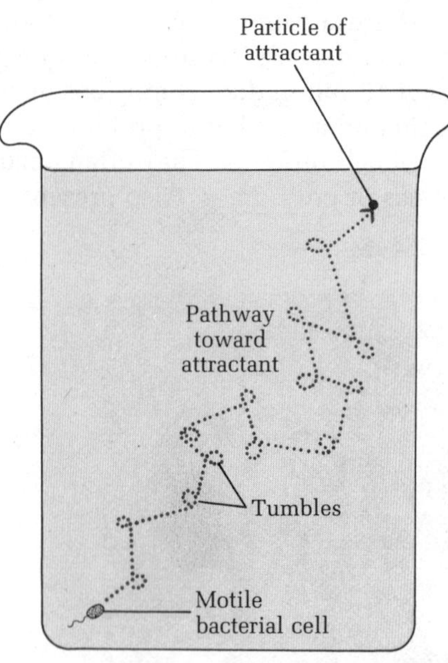

Eukaryotic Cells Are Larger and More Complex than Prokaryotes

Typical eukaryotic cells are much larger than prokaryotes. For example, hepatocytes, the major cell type in the liver of higher animals, have a diameter of some 20 to 30 μm, compared to 1 to 2 μm for bacteria. But more significant is the cell volume, which in most eukaryotic cells is from 1000 to 10,000 times larger than in bacteria. The relative volumes of eukaryotic and prokaryotic cells can be approximated from the formula for the volume of a sphere (page 17). Some eukaryotic cells are very much larger than this range, e.g., the unfertilized egg of a chicken, but its enormous volume is almost entirely taken up by stored nutrients required for the growing embryo. Some eukaryotic cells are extremely long; certain motor cells of the human nervous system, for example, may exceed 1 m in length. Most characteristic, however, is that eukaryotic cells have a well-formed nucleus, surrounded by paired membranes and with a complex internal structure. Like prokaryotes, eukaryotic cells also undergo asexual division, but this occurs by a much more complex process, called mitosis. The germ cells of eukaryotic organisms may also undergo complex sexual conjugation, with an exchange of genes.

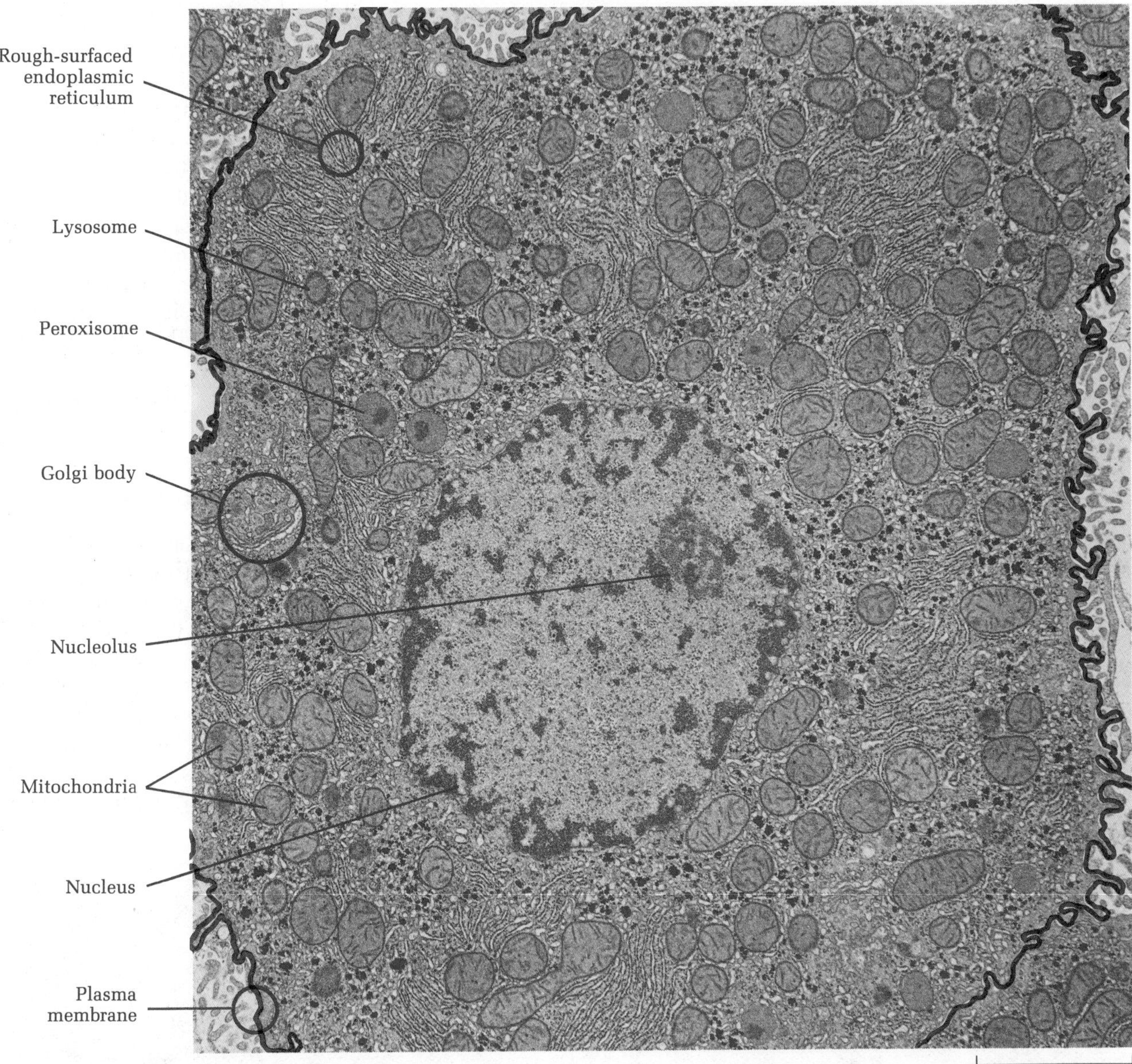

Figure 2-7
Transmission electron micrograph of a rat hepatocyte, the major type of cell in the liver. The plasma membrane, which is extensively folded and has a large surface area, is outlined in color. The many finger-like projections of the cell membrane, which greatly increase its surface area, are more strikingly shown in the scanning electron micrograph of an isolated hepatocyte in Figure 2-20.

The other striking difference between eukaryotes and prokaryotes is that eukaryotes contain, in addition to a well-formed nucleus, a number of other internal membrane-surrounded organelles, such as the *mitochondria*, the *endoplasmic reticulum*, and the *Golgi bodies*, each having a specific role in the metabolism and housekeeping activities of the cell. Figure 2-7 shows a typical eukaryotic cell, a rat liver cell, which has an extremely intricate and highly compartmented internal structure. As we shall see, eukaryotic cells have a more sophisticated division of labor among their many internal cell structures, each of which plays a specific role in cell activity.

The cells of all higher animals, plants, and fungi are eukaryotic. There are also many unicellular eukaryotes, including diverse species of protozoa, diatoms, euglenoids, yeasts, and slime molds. Because they have a much larger amount of genetic material, and because they often undergo sexual conjugation in which genes can be exchanged, eukaryotic forms of life are capable of a broader range of differentiation and specialization than prokaryotes. Thus eukaryotic organisms occur in millions of different species, compared with only a few thousand prokaryotic species. On the other hand, prokaryotic organisms are better able to tolerate changes in their environment and can reproduce at much higher rates, giving them the ability to survive under very unfavorable conditions.

Let us now examine the structure and role of the component parts of eukaryotic cells in more detail.

The Nucleus of Eukaryotes Is a Very Complex Structure

The nucleus contains nearly all the DNA of eukaryotic cells. In both animal cells (Figure 2-7) and plant cells (Figure 2-8) the nucleus is surrounded by a nuclear envelope, composed of two closely paired membranes separated by a narrow space. At intervals the two membranes of the nuclear envelope are fused together around openings called nuclear pores, which have a diameter of about 90 nm. Through these windowlike openings various substances can pass between the nucleus and cytoplasm. Inside the nucleus is the nucleolus (Figure 2-8), which stains more densely because of its rich content of ribonucleic

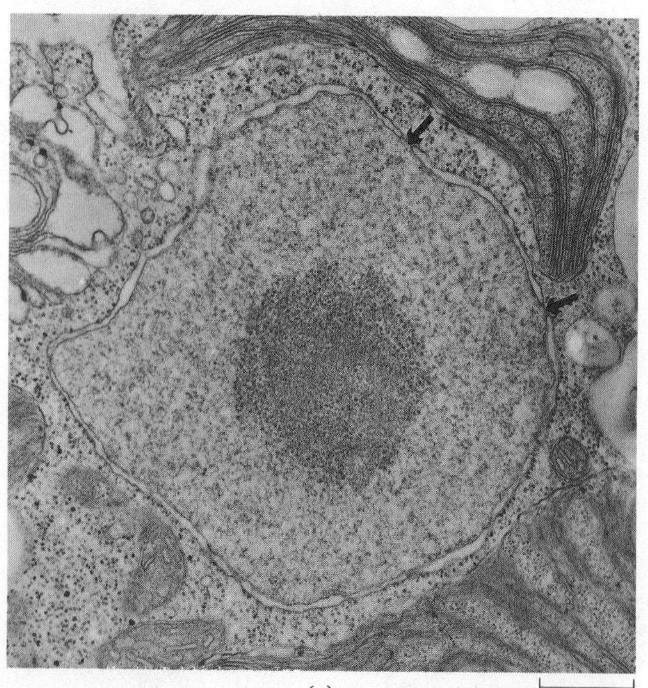

(a)

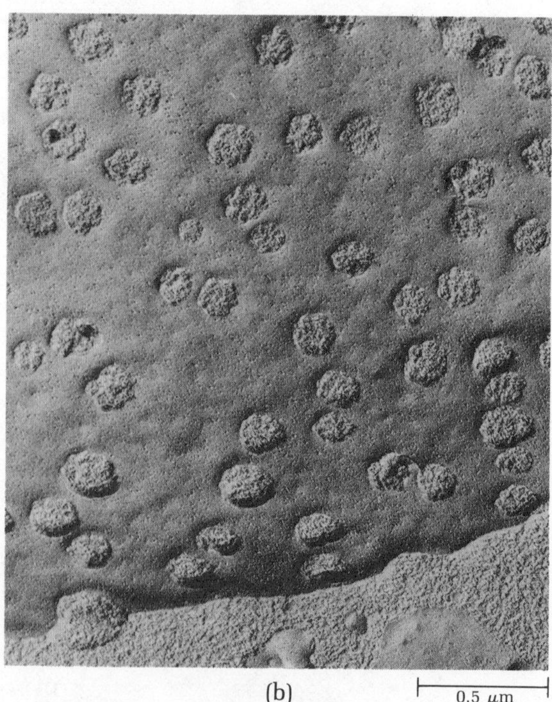

(b)

Figure 2-8
(a) Electron micrograph of the well-formed nucleus of the eukaryotic alga Chlamydomonas. *The dark body in the center of the nucleus is the nucleolus, the site of production of the major components of ribosomes. Partially formed ribosomes can be seen around its periphery. The nuclear envelope has paired membranes with* nuclear pores; *two are shown by arrows.*

(b) The surface of the nuclear envelope showing the numerous nuclear pores. The interior of the nucleus can be seen through the pores. This electron micrograph was made by the freeze-fracture method.

acid (RNA). The nucleolus is an RNA "factory"; here also the first stages in the synthesis of ribosomes take place.

The rest of the nucleus contains chromatin, so called because it stains in a characteristic way (Figures 2-7 and 2-8). Chromatin consists of DNA, RNA, and a number of specialized proteins. Between cell divisions chromatin is dispersed in a rather random fashion throughout the nucleus, but just before cell division the chromatin becomes organized into discrete granular bodies, the chromosomes. Each species of eukaryotic cell has a characteristic number of chromosomes; in human somatic cells there are 46. After the chromosomes undergo replication, the daughter chromosomes are separated and delivered into the daughter cells in the process of mitosis, a very complex sequence of events, of which some important stages are shown in Figure 2-9. After mitosis is complete the chromatin disperses again. Thus the well-formed nucleus of eukaryotes is very complex in both its structure and biological activity, compared with the relatively simple nuclear body of prokaryotes.

Mitochondria Are the Power Plants of Eukaryotic Cells

Very conspicuous in the cytoplasm of eukaryotic cells are the mitochondria (singular, mitochondrion), shown in Figure 2-10. These structures vary widely in size, shape, number, and location, depending on the species of cell. In each liver cell of the rat there are perhaps 1000 mitochondria. They have a diameter of about 1 μm, close to the size of bacterial cells. Some types of eukaryotic cells, e.g., sperm cells or yeast cells, contain only a few very large mitochondria, whereas others, e.g., egg cells, contain many thousands. Sometimes mitochondria are highly branched and ramify through a large volume of cytoplasm.

Each mitochondrion has two membrane systems. The outer membrane is smooth, surrounding the mitochondrion completely. The inner membrane has infoldings called cristae. In liver mitochondria there are relatively few cristae, but in the mitochondria of heart cells the cristae are very numerous and parallel. The inner compartment of mitochondria is filled with the gel-like matrix.

Figure 2-9
Major steps in mitosis of a eukaryotic cell.

(a) The period between cell divisions. Chromatin is dispersed throughout the nucleus.

(b) The beginning of cell division. The chromatin condenses to form chromosomes and is replicated. The nuclear envelope begins to break down. The spindle apparatus develops at the poles of the cell and the nucleolus disperses.

(c) The chromosomes are pulled apart to opposite poles. Each daughter receives a complete set of chromosomes.

(d) Two daughter nuclei. Their nuclear envelopes and nucleoli are formed, the chromatin disperses, and the parent cell begins to cleave to form the daughter cells.

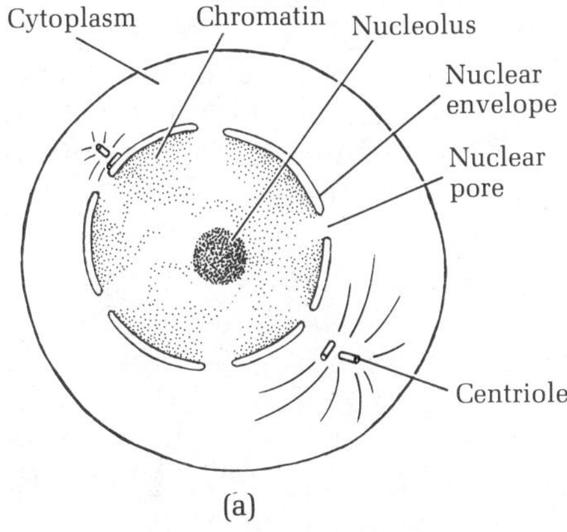

(a)

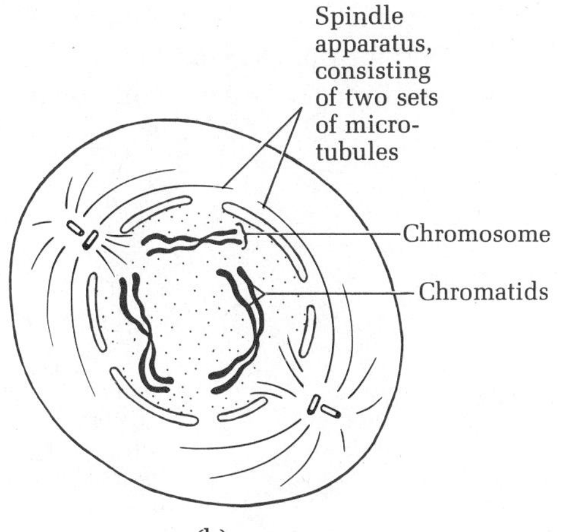

(b)

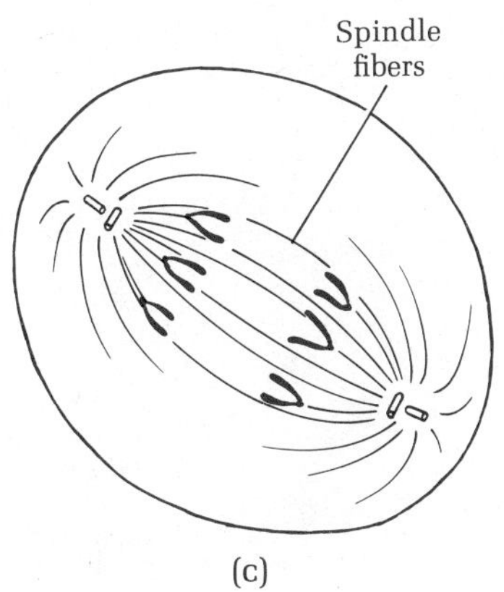

(c)

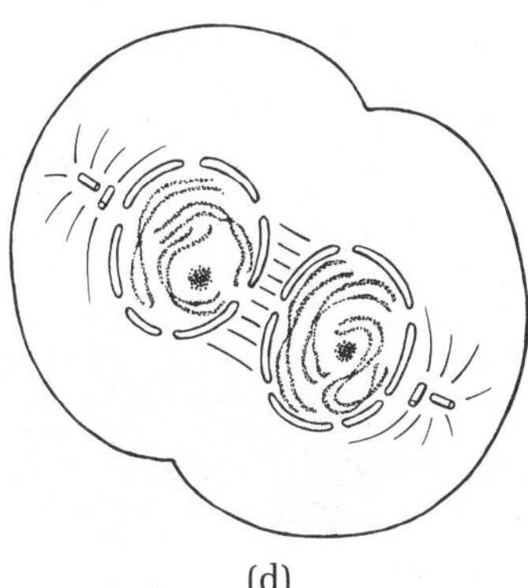

(d)

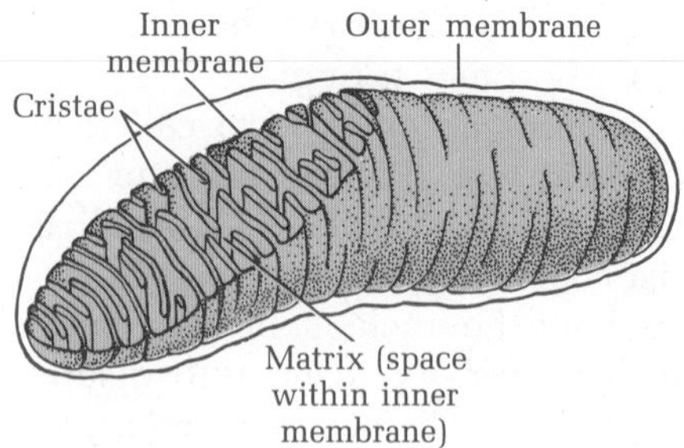

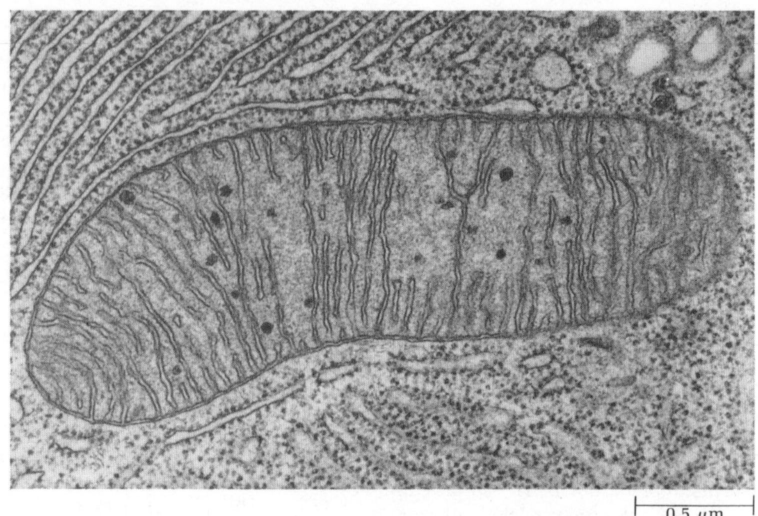

Figure 2-10
Structure of mitochondria. Their name is derived from Greek mitos, *"thread," and* chondros, *"grain" or "seed." In some cells mitochondria are long, almost filamentous, but in most others they are elliptical or spherical. This electron micrograph of a mitochondrion in a pancreatic cell shows the smooth outer membrane and the numerous infoldings of the inner membrane, called* cristae.

The mitochondria are the power plants of the cell. They contain many enzymes that together catalyze the oxidation of organic cell nutrients by molecular oxygen to yield carbon dioxide and water. Some of these enzymes are located in the matrix and some in the inner membrane. Much chemical energy is released during these oxidations, which is used to generate *adenosine triphosphate* (ATP), the major energy-carrying molecule of cells. ATP formed by the mitochondria diffuses to all parts of the cell, where it is used to carry out cellular work.

Mitochondria also contain small amounts of DNA, as well as RNA and ribosomes. Mitochondrial DNA codes for the synthesis of certain specific proteins of the inner membrane. One may well ask why mitochondria contain DNA. This question has led to the interesting concept that mitochondria originally arose during biological evolution by the invasion of the cytoplasm of large anaerobic prokaryotic cells by other, smaller prokaryotes capable of using molecular oxygen to oxidize their nutrients (Figure 2-11). The invading bacteria thus became parasites within the host cells. With time and further evolution this relationship became symbiotic, of benefit to both the host and parasite. We know today that mitochondria actually divide during the process of cell division. Mitochondrial DNA and mitochondrial ribosomes thus may be evolutionary descendants of the DNA and ribosomes of these small invading bacteria.

Figure 2-11
A plausible theory for the evolutionary origin of mitochondria. It is based on a number of striking biochemical and genetic similarities between bacteria and the mitochondria of eukaryotic cells. During the course of evolution of eukaryotic cells, the invading bacteria became symbiotic with the host cell, to their mutual advantage. Ultimately the cytoplasmic bacteria became the mitochondria.

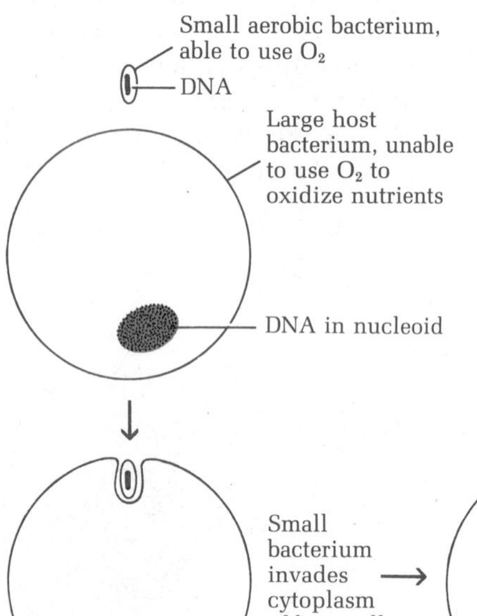

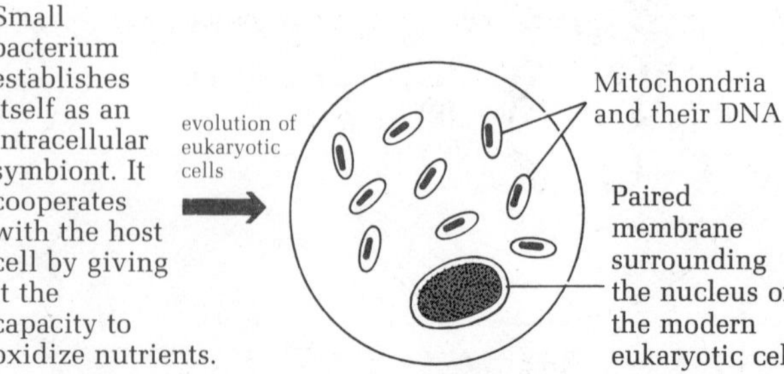

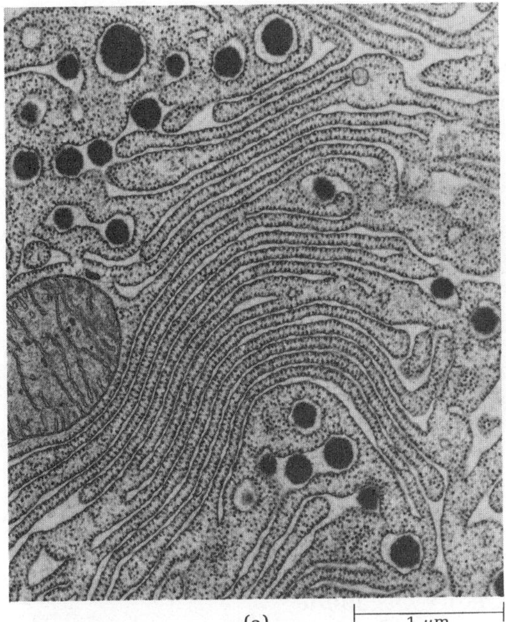

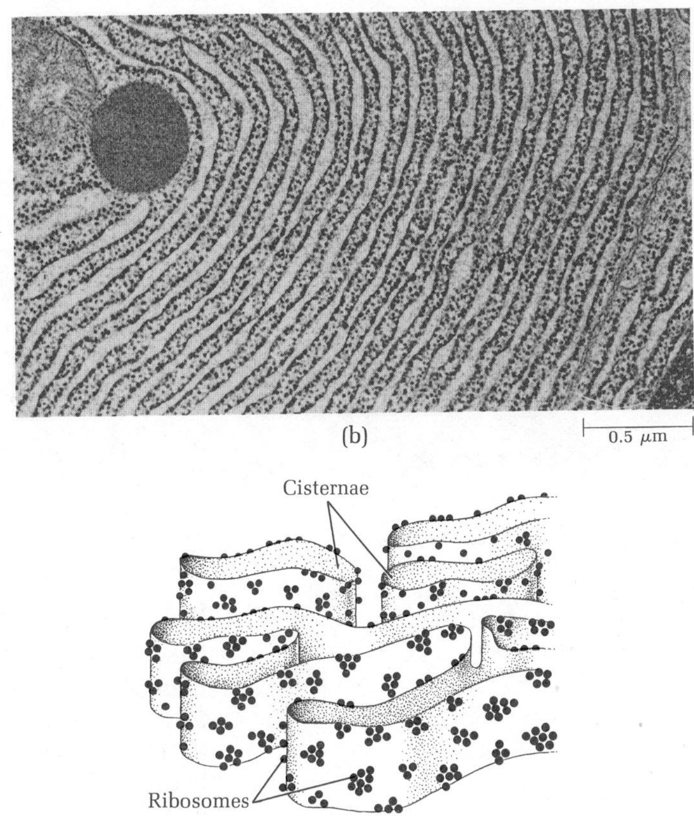

Figure 2-12
(a) *Endoplasmic reticulum (rough-surfaced) of a pancreatic cell. These cells are very active in synthesizing proteins, which are made by the ribosomes adhering to the outer surface of the membrane and are then secreted into the cisternae. A portion of a mitochondrion is seen at the left.*
(b) *A higher magnification, showing individual ribosomes and the cisternae.*
(c) *A three-dimensional interpretation of the endoplasmic reticulum, showing how the narrow cisternae make continuous, maze-like channels throughout much of the cytoplasm.*

The Endoplasmic Reticulum Forms Channels through the Cytoplasm

In the cytoplasm of nearly all eukaryotic cells there is a very complex three-dimensional maze of membrane channels, the *endoplasmic reticulum,* which forms many folds and convolutions throughout the cytoplasmic space (Figure 2-12). The spaces enclosed within the endoplasmic reticulum, called *cisternae,* serve as channels for the transport of various products through the cell, usually to its exterior. In some cells, however, the cisternae serve as storage spaces. There are two types of endoplasmic reticulum, *rough* and *smooth*. In the former type the outer surface of the membrane is studded with ribosomes; smooth endoplasmic reticulum is devoid of ribosomes. The ribosomes that are attached to the rough endoplasmic reticulum are engaged in the biosynthesis of proteins that are to be temporarily stored or transported out of the cell. Proteins synthesized by membrane-bound ribosomes are extruded through the membrane into the cisternal space and ultimately pass to the cell exterior. The endoplasmic reticulum also plays a role in the biosynthesis of lipids. In different types of eukaryotic cells the endoplasmic reticulum has different forms and functions. In skeletal-muscle cells, whose contraction is stimulated by Ca^{2+}, the reticulum participates in the relaxation process by reabsorbing Ca^{2+} ions.

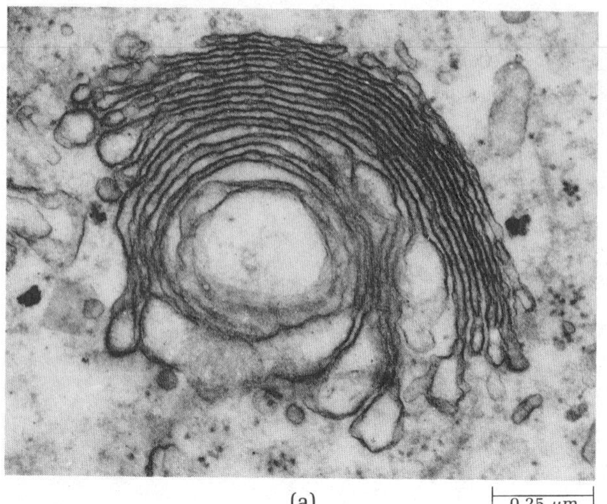

(b)

Figure 2-13
(a) *Golgi body in an amoeba. Small spherical vesicles are being pinched off from the edges of the large flattened vesicles.* (b) *A drawing to show a Golgi body in a three-dimensional projection.*

Golgi Bodies Are Secretory Organelles

Nearly all eukaryotic cells have characteristic clusters of membrane-surrounded vesicles called *Golgi bodies* (Figure 2-13), after Camillo Golgi, an Italian cytologist who first described them at the end of the nineteenth century. Golgi bodies take different forms in different types of eukaryotic cells, but the most characteristic arrangement is that of a stack of flattened vesicles, each surrounded by a single membrane. Near the ends of the Golgi vesicles are much smaller spherical vesicles that are pinched off from the edges of the large ones.

The Golgi bodies receive certain cell products from the endoplasmic reticulum and "package" them into secretory vesicles, which find their way to the outer plasma membrane of the cell and fuse to it. The fused portion may then open to discharge the vesicle contents to the exterior, a process called *exocytosis*. In some plant cells this process is used to carry prefabricated sections of the outer cell wall from the inside of the cell, where they are made, to the outside of the cell, there to be attached to the growing cell wall.

Lysosomes Are Packets of Hydrolyzing Enzymes

Lysosomes are membrane-surrounded spherical vesicles in the cytoplasm (Figure 2-14). They vary in size but are usually no larger than mitochondria. Lysosomes contain many different enzymes capable of digesting, i.e., destroying by hydrolysis, cell proteins, polysaccharides, and lipids that are no longer needed. Since such enzymes are harmful to the rest of the cell, they are segregated in the lysosomes. The proteins and other components to be degraded are selectively brought into the lysosomes and hydrolyzed there into their simple building-block components, which are then discharged back into the cytoplasm. In the human genetic disorder *Tay-Sachs disease*,

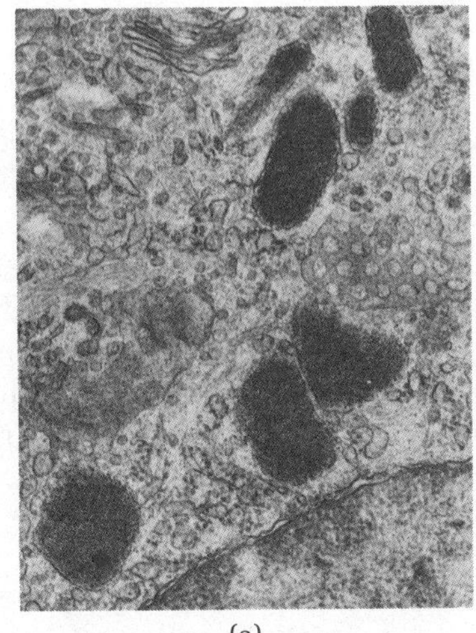

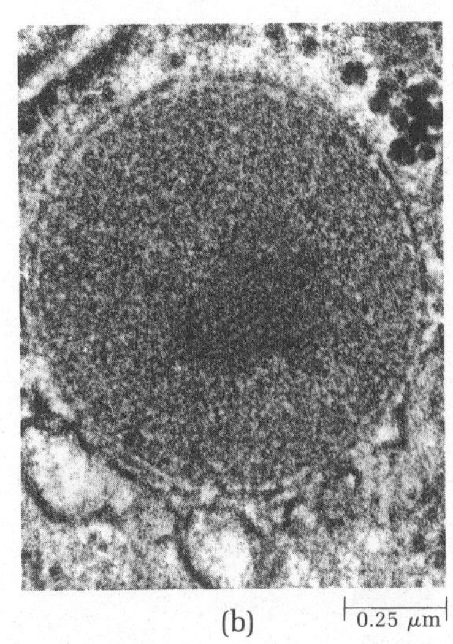

Figure 2-14
Lysosomes and peroxisomes are "bags" of enzymes.

(a) *Portion of a cell from the adrenal cortex. The very dense oval or irregularly shaped bodies are lysosomes. They are smaller than mitochondria. In addition to 40 or more hydrolytic enzymes, lysosomes sometimes contain whorls of "extra" membrane used to envelop proteins and other cell components.*

(b) *A peroxisome. The crystalline material is urate oxidase, one of several peroxide-forming enzymes present in peroxisomes.*

the lysosomes are defective in their content of certain lipid-hydrolyzing enzymes, causing some lipids to accumulate in the brain and other tissues and leading to mental retardation.

Peroxisomes Are Peroxide-Destroying Vesicles

Another type of membrane-surrounded organelle in the cytoplasm is the peroxisome (Figure 2-14). These structures, also known as microbodies, are somewhat larger than lysosomes, have only a single outer membrane, and contain much protein, often in crystalline form. Segregated within these structures are enzymes that form and use hydrogen peroxide, hence the name peroxisomes. Hydrogen peroxide (H_2O_2), which is extremely toxic to cellular life, is decomposed into water and oxygen by another enzyme in the peroxisomes called catalase. By keeping the hydrogen peroxide-forming enzymes and catalase inside the peroxisome the rest of the cell is protected from the destructive action of peroxides.

Microfilaments Function in Contractile Processes of Cells

High-resolution electron microscopy has shown that the cytoplasm of most eukaryotic cells contains many filaments, consisting of strings of protein molecules. Such filaments occur in three classes differing in diameter, composition, and function. Smallest are the microfilaments, which are about 5 nm in diameter (Figure 2-15). Often they form a loose web directly beneath the cell membrane. Microfilaments have been found to be identical with the thin actin filaments of the contractile system of skeletal muscle. Microfilaments participate in the generation of tension: in the contraction of muscle, the folding or extension of the cell membrane, and in movements of structures within cells.

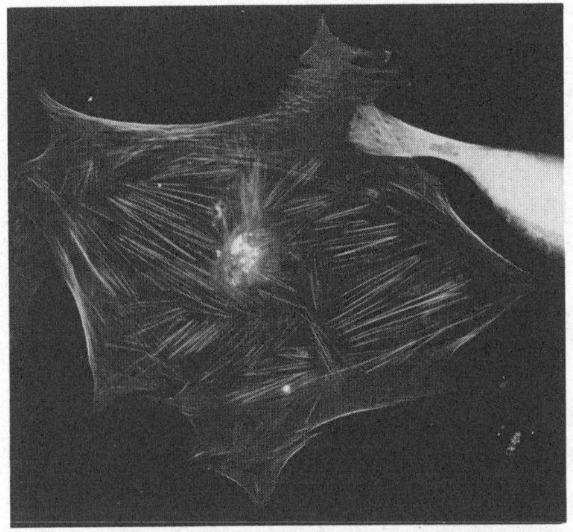

(b) 25 μm

Figure 2-15
Actin and myosin filaments in fibroblasts, specialized cells of connective tissue.
(a) *Thin actin filaments visualized by a fluorescent marker.* (b) *Myosin filaments near the nucleus.*

Figure 2-16
Microtubules. These long, hollow structures play many roles in cell biology. They give shape to cells, participate in cell division (Figure 2-9), convey materials in cells, function as structural and motile units in eukaryotic cilia and flagella (Figure 2-18), and form part of the cytoskeleton (Figure 2-17).

(a) *Structure of a microtubule. It is assembled from complexes of two proteins, α- and β-tubulin, arranged in a helix to form 13 vertical filaments around a hollow core. The diameter and pitch vary somewhat in different cells.*

(b) *Cross section of a microtubule, showing the profiles of the 13 vertical strands.*

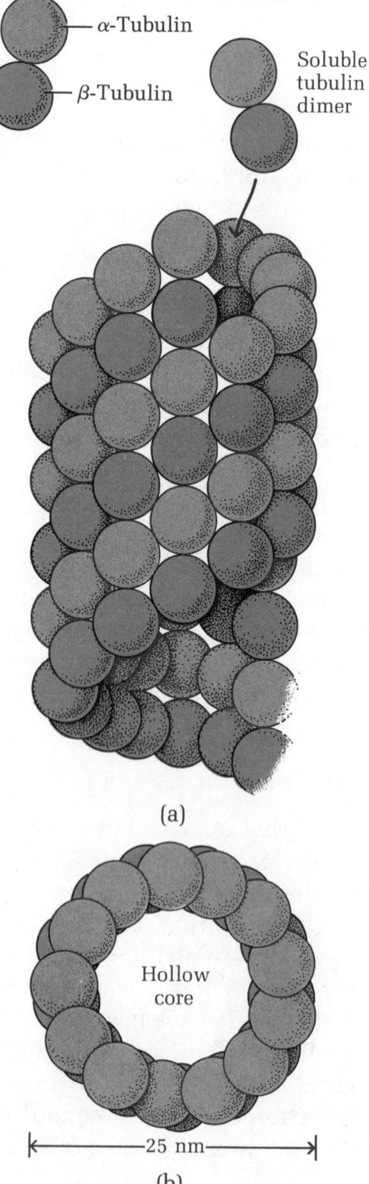

The second type of filament in eukaryotic cells is the *myosin filament*, which is much thicker than the actin filaments (Figure 2-15). Myosin filaments are major components of the contractile system of skeletal muscle, but they are also found in nonmuscle cells, often in association with thin or actin filaments. In some kinds of cells myosin filaments are attached to the cell membrane. Actin and myosin filaments are involved in various types of cellular or intracellular movement.

A third type of filament is still thicker, with a diameter of about 10 nm. The 10-nm filaments are found in many cells and have been given different names.

Microtubules also Function in Cell Movements

Many eukaryotic cells, particularly the long cells of the nervous system of animals, contain *microtubules*, about 25 nm in diameter (Figure 2-16). Each microtubule consists of 13 strands of protein molecules closely arranged around a hollow central space. In nerve cells bundles of microtubules participate in the movement of materials from the cell body toward the end of the cell extensions, or *axons*. Microtubules have many functions. They participate in the action of the mitotic spindle during cell division, and they also serve as the motile units in eukaryotic cilia and flagella.

Microfilaments, Microtubules, and the Microtrabecular Network Constitute the Cytoskeleton

In many eukaryotic cells the different kinds of microfilaments and the microtubules together form a flexible framework, which has been called the *cytoskeleton*. Recently a third element of the cytoskeleton has been discovered, the *microtrabecular network*, shown in the frontispiece to this chapter. This structure was visualized by high-voltage electron microscopy, which has very high resolving power. Until the application of this method, the microtrabecular network was seen only as an amorphous background called the *ground substance* of the cytoplasm. The microtrabecular network consists of very thin, interlacing filaments; their chemical composition is not yet known, but they almost certainly contain proteins. The precise three-dimensional arrangement and function of the microtrabecular network in different kinds of cells is under active study.

The cytoskeleton gives cells their characteristic shape and form, provides attachment points for organelles and formed bodies, fixing their location in cells, and also makes communication between parts of the cell possible. The cytoskeleton is not to be regarded as a rigid, permanent framework of the cell but as a dynamic, changing structure. Microtubules, for example, undergo constant assembly from their building blocks and disassembly. An interpretation of the structure of the cytoskeleton of a human fibroblast cell is shown in Figure 2-17.

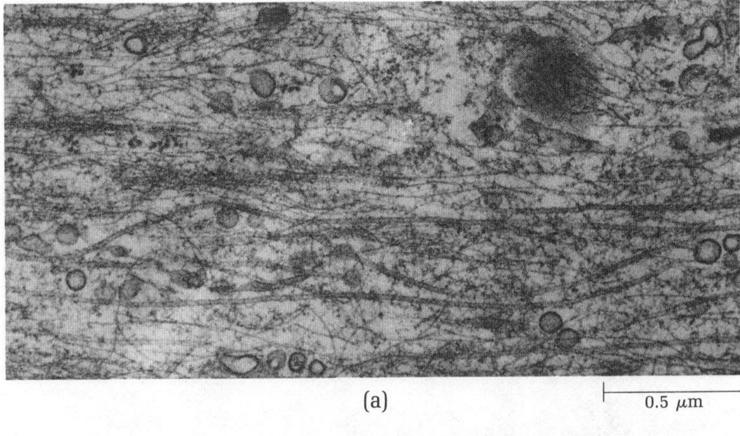

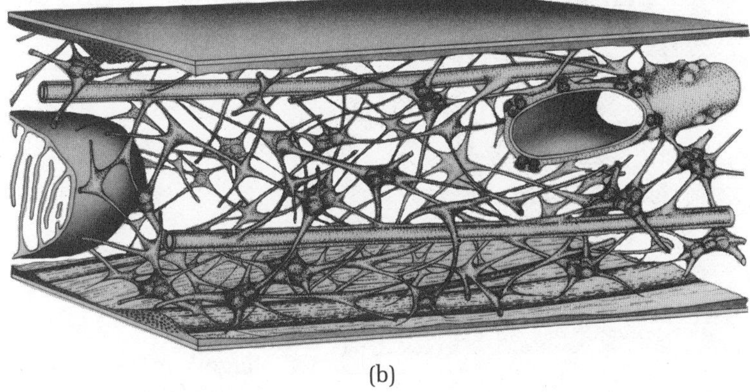

Figure 2-17
(a) Electron micrograph of the filamentous cytoskeleton of a fibroblast. (b) A schematic representation of the cytoskeleton of a fibroblast. The long tubes are microtubules; the thinner elements are different kinds of microfilaments.

Cilia and Flagella Give Cells Propulsive Power

Cilia and flagella, motile structures or appendages extending from the surface of many unicellular eukaryotes and also certain cells in animal tissues (but not those of plants), are constructed on the same basic architectural plan (Figure 2-18). But at this point it must be emphasized that eukaryotic flagella are very different from the flagella of prokaryotes. Prokaryotic flagella are much thinner (10 to 20 nm) and consist of single protein strands. They are rigid, curved rods whose rotating motion is imparted entirely by their "motors" in the cell membrane. Eukaryotic flagella are much thicker (200 nm), have a much more complex structure, and are capable of generating motion along their entire length. Eukaryotic cilia and flagella, which are sheathed by an extension of the cell membrane, contain 9 pairs of microtubules arranged around 2 central tubules, in what is often called the *9 + 2 arrangement* (Figure 2-18). Cilia and flagella have the same diameter, but cilia are short, less than 10 μm long, whereas flagella are much longer, up to 200 μm. In most cases cilia serve to move materials past cells, through a wavelike, sweeping motion, whereas flagella function to propel cells. Sperm cells of animals have a single long

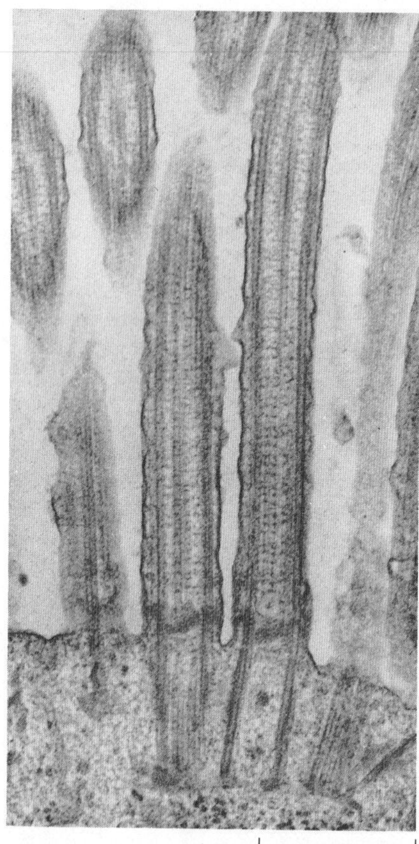

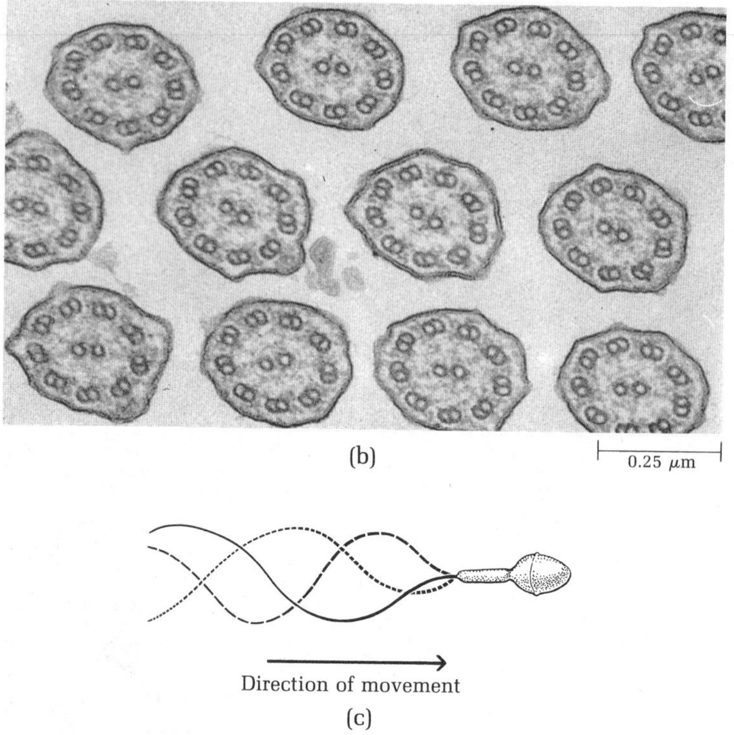

Figure 2-18
Cilia and flagella have the same architecture, although cilia are much shorter. The paired microtubules are surrounded by an extension of the cell cytoplasm and of the cell membrane. Sliding and twisting of the microtubules along each other, promoted by ATP, cause the flagella to undergo undulating movements.

(a) *Longitudinal section, showing the parallel microtubules of the cilia.*

(b) *Cross section of the cilia, showing the 9 + 2 arrangement of 9 paired or double microtubules spaced around 2 central tubules.*

(c) *Propulsive action of the single flagellum of a sperm cell.*

flagellum (Figure 2-18). Movements of cilia and flagella are brought about by complex sliding motions of individual microtubules along each other within the 9 + 2 structure.

A sliding movement of filaments or microtubules along each other, which requires the energy of ATP, is the basic process making possible the contraction of skeletal muscle, the propelling action of cilia and flagella, and the characteristic engulfing, ruffling, and locomotor activities of the cell membrane, as in amoebas.

The Cytoplasm also Contains Granular Bodies

The cytoplasm of eukaryotes also contains granular elements that are not surrounded by a membrane. Chief among these are *ribosomes* (Figure 2-19), which occur in free form in the cytoplasm as well as attached to the endoplasmic reticulum. The ribosomes of eukaryotes are larger than those of prokaryotes, but they perform the same basic function, namely, the biosynthesis of proteins from amino acids.

Another type of granular element in the cytoplasm of eukaryotic cells, especially in liver cells, is *glycogen granules* (Figure 2-19). Glycogen is a macromolecule made up of highly branched, chainlike strings of glucose molecules. Glycogen granules function as a reserve fuel source, especially in liver and muscle cells. Some eukaryotic cells contain *fat droplets,* which also store energy-rich fuel.

The Cytosol Is the Continuous Aqueous Phase of the Cytoplasm

Bathing the cytoplasmic organelles, ribosomes, and granular elements is the continuous pervading aqueous phase, called the *cytosol*. The cytosol is not simply a dilute aqueous solution; it is quite complex in composition and almost gel-like in consistency. The cytosol contains in dissolved form many enzymes and enzyme systems, as well as other proteins that bind, store, or transport nutrients, trace minerals, and oxygen. It also contains as solutes a great many different kinds of small biomolecules, not only the building blocks, such as the amino acids and nucleotides, but also hundreds of small organic molecules called *metabolites*, which represent intermediates in the biosynthesis or degradation of building-block molecules and macromolecules. For example, the conversion of blood glucose into lactic acid by working skeletal muscle takes place by sequential formation of 10 successive intermediates, the last being directly converted into lactic acid.

A third class of solutes in the cytosol consists of various *coenzymes*, as well as ATP and ADP, central components of the cellular energy-transfer system. Finally, the cytosol also contains various *mineral ions*, such as K^+, Mg^{2+}, Ca^{2+}, Cl^-, HCO_3^-, and HPO_4^{2-}.

All the components of the cytosol are maintained in balanced, constant concentrations and proportions, through the activity of various transport processes operating across the plasma membrane.

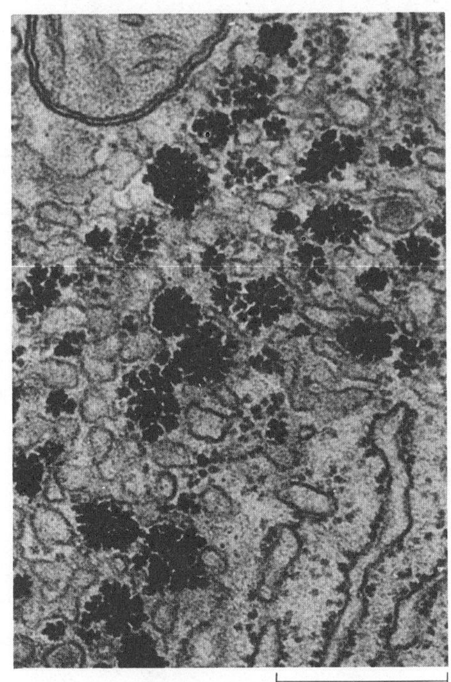

Figure 2-19
Glycogen granules (deeply stained) in the cytoplasm of a hamster liver cell. They consist of large clusters of glycogen molecules. They are much larger than the ribosomes, visible on the surface of the endoplasmic reticulum in the lower right corner.

The Cell Membrane Presents a Large Surface Area

We have seen that it is generally advantageous for cells to have a large surface area in relation to their volume, so that the rate of diffusion of nutrients and oxygen into cells does not limit the rate of their metabolism. This problem is solved in eukaryotes by a combination of two factors. First, most eukaryotic cells have an intrinsically lower rate of metabolism than prokaryotic cells, whose principal purpose in life is to grow and multiply as fast as possible so that the species can survive. Since cell growth requires energy, which is supplied by external nutrients or absorbed light energy, prokaryotes are dependent on a large ratio of membrane surface to cell volume for maximal rates of nutrient intake or trapping of light energy. Eukaryotic cells are not under such pressure to grow and divide and generally have a much lower rate of energy flow.

Eukaryotic cells do, however, utilize special structural features to ensure a maximal surface-to-volume ratio. Nerve cells, which have a relatively high rate of metabolism, are long and thin and thus present a large surface area. Other cells may be highly branched or star-shaped, but a more general device for maximizing the surface area is the occurrence of very numerous ruffles or fingerlike projections (called *microvilli*) of

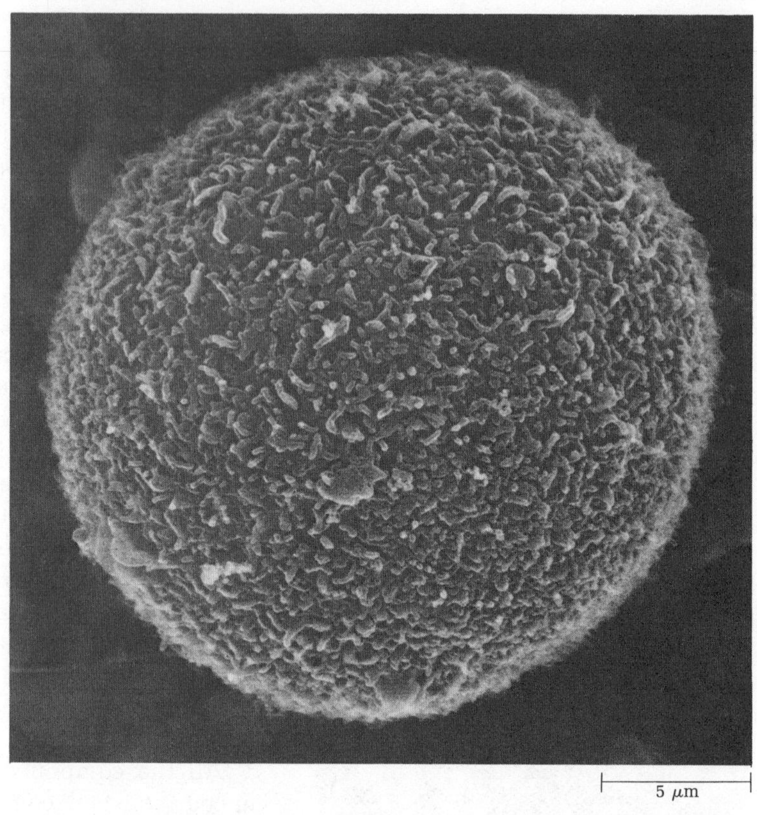

Figure 2-20
A scanning electron micrograph of the surface of an isolated rat liver cell. It shows how the surface area of the cell is greatly increased by the protruding microvilli, which constantly change shape and position.

the cell membrane. As is seen in the transmission electron micrograph of a liver cell in Figure 2-7, the cell membrane is not smooth but highly convoluted. This feature is much more strikingly shown by the scanning electron micrograph of the surface of a liver cell in Figure 2-20. Many animal cells have such microvilli, particularly cells of the lining of the small intestine, through which nutrient molecules must pass at a high rate during absorption of digested food.

The Surface of Many Animal Cells also Contains "Antennae"

Outside the plasma membrane, many cells of animal tissues have a thin, flexible cell coat. It consists of a number of different polysaccharide, lipid, and protein groups on the outer surface of the plasma membrane. The cell surface contains a number of different molecular structures that sense or recognize external signals. Among these are *cell-recognition sites,* by which cells recognize other cells of the same kind and which provide points of adhesion to maintain the structure of specific tissues. The surface of many animal cells also contains various kinds of *hormone receptor sites*. Hormones are chemical messengers secreted by certain cells into the blood, which can regulate the activity of other kinds of cells elsewhere in the body. When hormone molecules bind to their receptor sites on the surface of their target cells, they stimulate some special aspect of cell activity. Other specific sites on the surface of animal cells can recognize and bind certain proteins that are foreign to the cells. Binding of foreign proteins to such sites causes cellular responses that result in allergy. Such cell-specific sites are also responsible for the rejection of surgically transplanted tissues or organs that are not compatible with those of the recipient. Thus the surface of many animal cells is really a complex mosaic of different kinds of sensitive molecular "antennae," by which cells recognize the outside world and are stimulated to react to specific agents in their environment.

Eukaryotic Plant Cells Have Some Special Features

Although fundamentally similar, the eukaryotic cells of higher plants (see Figure 2-21) differ in some details from those of higher animals. Perhaps the most obvious difference is that most plant cells contain *plastids*. Plastids are specialized organelles in the cytoplasm; they have two surrounding membranes. Most conspicuous of the plastids and characteristically present in all green plant cells are the *chloroplasts* (Figure 2-22). Like mitochondria, the chloroplasts may be considered to be power plants, the important difference being that chloroplasts are *solar* power plants, using light energy, whereas mitochondria are *chemical* power plants, using the chemical energy of nutrient molecules. Chloroplasts absorb light energy and use it to reduce carbon dioxide to form carbohydrates, such as starch,

Figure 2-21
Electron micrograph of a cell from the leaf of a corn plant. Note that chloroplasts are much larger than mitochondria but are fewer in number. The vacuole usually becomes larger with the age of the cell. The cell wall is relatively thick and rigid.

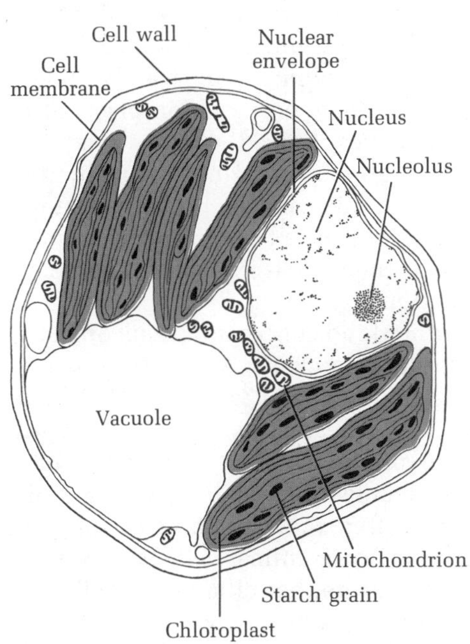

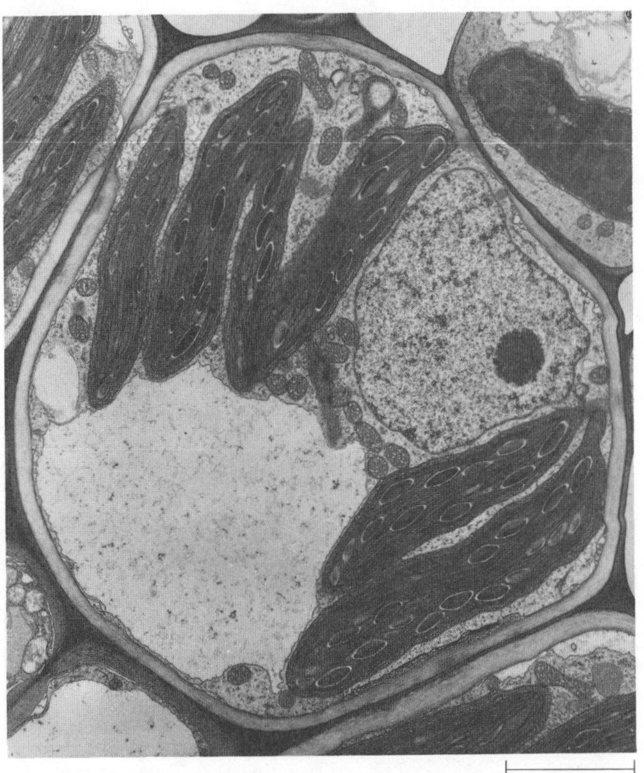

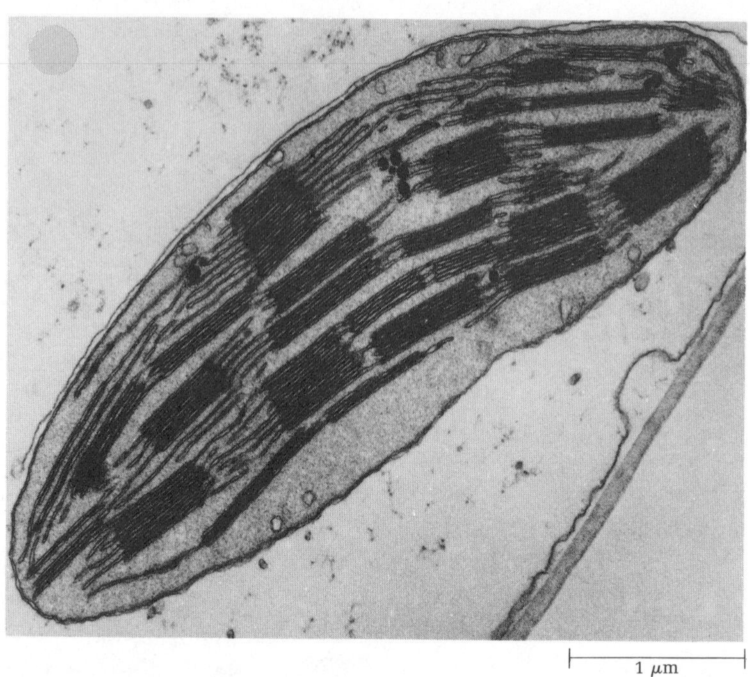

Figure 2-22
A chloroplast in a photosynthetic cell of a lettuce leaf.

with the liberation of molecular oxygen (O_2). Photosynthetic plant cells contain both chloroplasts and mitochondria, chloroplasts serving as power plants in the light and the mitochondria in the dark, when they oxidize carbohydrate generated by photosynthesis during daylight hours.

Chloroplasts are considerably larger than mitochondria and occur in many different shapes. Because they contain large amounts of the pigment *chlorophyll*, photosynthetic cells are generally green, but they may have other colors, depending on the relative amounts of other pigments in the chloroplasts. These pigment molecules, which together can absorb light energy over much of the visible spectrum, are localized in the inner membrane of the chloroplast, which is intricately folded to form *thylakoid disks*. Like mitochondria, chloroplasts also contain DNA, RNA, and ribosomes. Indeed, chloroplasts also appear to have had their evolutionary origin in parasitizing prokaryotes (Figure 2-11), but in this case the prokaryotes that gained entrance into host cells may have been primitive cyanobacteria, which brought with them their capacity for photosynthesis and oxygen formation.

Plant cells also contain other kinds of plastids. *Leucoplasts*, which are colorless, serve to store starch and oils. Also conspicuous in many plant cells are large *vacuoles* surrounded by a single membrane (Figure 2-21). These vesicles are filled with cell sap and cellular waste products, often aggregated into crystalline deposits. Such vacuoles are small in young cells but become larger with age; often they fill most of the cell volume. Vacuoles also occur in some animal cells but are very much smaller. Plant cells possess neither cilia nor flagella.

Most cells of higher plants are totally enclosed by *cell walls*,

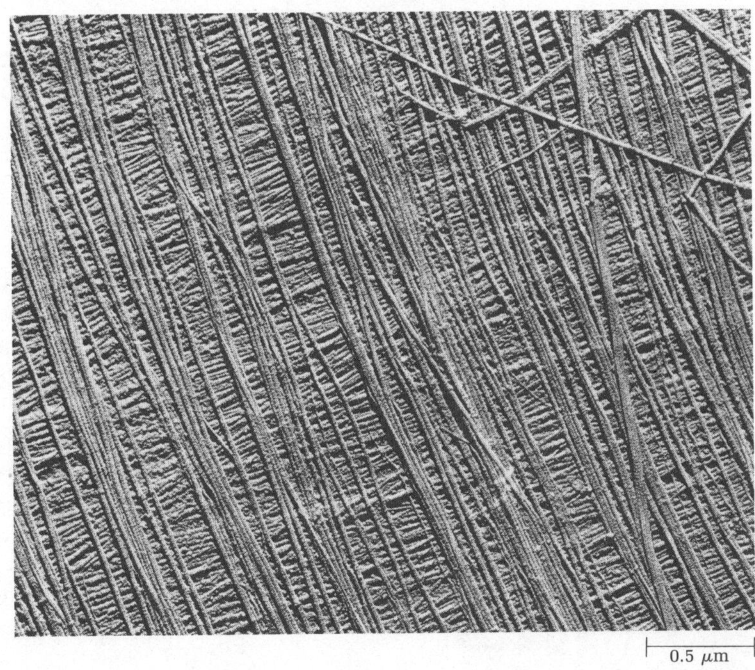

Figure 2-23
Electron micrograph of a plant cell wall. It consists of crisscross layers of cellulose fibrils embedded in an organic "glue." Plant cell walls are very strong and are analogous in structure to concrete reinforced with steel rods.

which serve largely as rigid, protective shells. They are relatively thick, porous, and very strong (Figures 2-21 and 2-23). Plant cell walls consist of fibrils of cellulose "glued" together by complex polymeric cement substances. Plant cell walls allow water and small molecules to pass through their pores readily, but they resist swelling or expansion of the enclosed cell. In the woody parts of plants and in the trunks of trees the primary cell walls are surrounded by thick, strong *outer* or *secondary walls*, which collectively can support large weights.

Viruses Are Supramolecular Parasites

Our survey of cells as units of life cannot be complete without consideration of viruses. Viruses are nonliving but biologically formed supramolecular assemblies that can replicate themselves in appropriate host cells. They consist of a nucleic acid molecule surrounded by a protective shell, or *capsid*, made up of protein molecules. Viruses exist in two states. Outside the cells that formed them viruses are simply nonliving particles, called *virions*, which have a regular size, shape, and composition. Some viruses may even be crystallized and thus behave like extremely large molecules. However, once a viral particle or its nucleic acid component gains entry into its specific host cell, it assumes a different kind of existence, that of an intracellular parasite. The viral nucleic acid carries the genetic message for specifying the entire structure of the intact virion. It takes over and alters the host cell's normal activities and directs the biochemical parasitization of the cell by diverting its enzymes and ribosomes from their normal cellular roles into

Figure 2-24
Replication of a bacteriophage in its host cell.

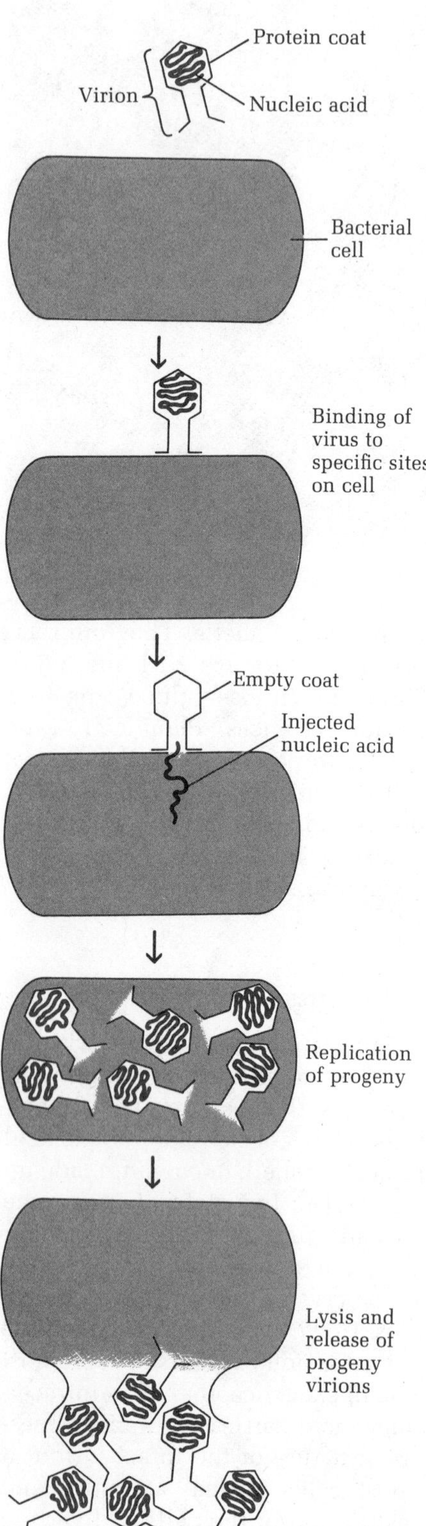

the manufacture of many new daughter viral particles. As a result, dozens or hundreds of progeny viral particles may arise from the single virion that infected the host cell (Figure 2-24). In some host-virus systems, the progeny virions are released from the host cell, which subsequently dies and undergoes lysis. In other host-virus systems, however, the newly manufactured nucleic acid of the virus remains inside the host cell, sometimes having little effect on its survival but often causing profound changes in its appearance and activity. Some viruses contain DNA; others contain RNA.

Hundreds of different viruses are known, each specific for a certain type of host cell. The host may be an animal cell, a plant cell, or a bacterial cell (Table 2-3). Viruses specific for bacteria are known as bacteriophages or simply phages (the word "phage" means to eat or consume). The virus may contain only one kind of protein in its capsid, like the plant virus tobacco mosaic virus, one of the simplest viruses and the first to be crystallized (Figure 2-25). Other viruses may contain dozens or hundreds of different kinds of proteins. Viruses differ greatly in size. Bacteriophage ϕX174, one of the smallest, has a diameter of 18 nm. Vaccinia virus is one of the largest; its virions are almost as large as the smallest bacteria. Viruses also differ in shape and complexity of structure. Among the more complex is bacteriophage T4 (Figure 2-25), whose host cell is *E. coli*. T4 has a head, a tail, and a complex set of tail fibers, which together act as a "stinger" or hypodermic syringe for injection of the viral DNA into the host cell. Figure 2-25 and Table 2-3 summarize the size, shape, and particle weights of a number of

Table 2-3 Properties of Some Viruses

Virus	Nucleic acid	Particle weight, 10^6 daltons	Long dimension, nm	Shape
E. coli bacteriophages				
ϕX174	DNA	6	18	Polyhedral
T4	DNA	220	200	Tadpole
λ (lambda)	DNA	50	120	Tadpole
MS2	RNA	3.6	20	Polyhedral
Plant viruses				
Tobacco mosaic	RNA	40	300	Rod
Tomato bushy stunt	RNA	10.6	28	Polyhedral
Animal viruses				
Poliomyelitis	RNA	6.7	30	Polyhedral
Simian virus 40 (SV40), cancer-causing in newborn animals	DNA	28	45	Spherical
Adenovirus (common colds)	DNA	200	70	Polyhedral
Smallpox	DNA	4000	250	Spherical

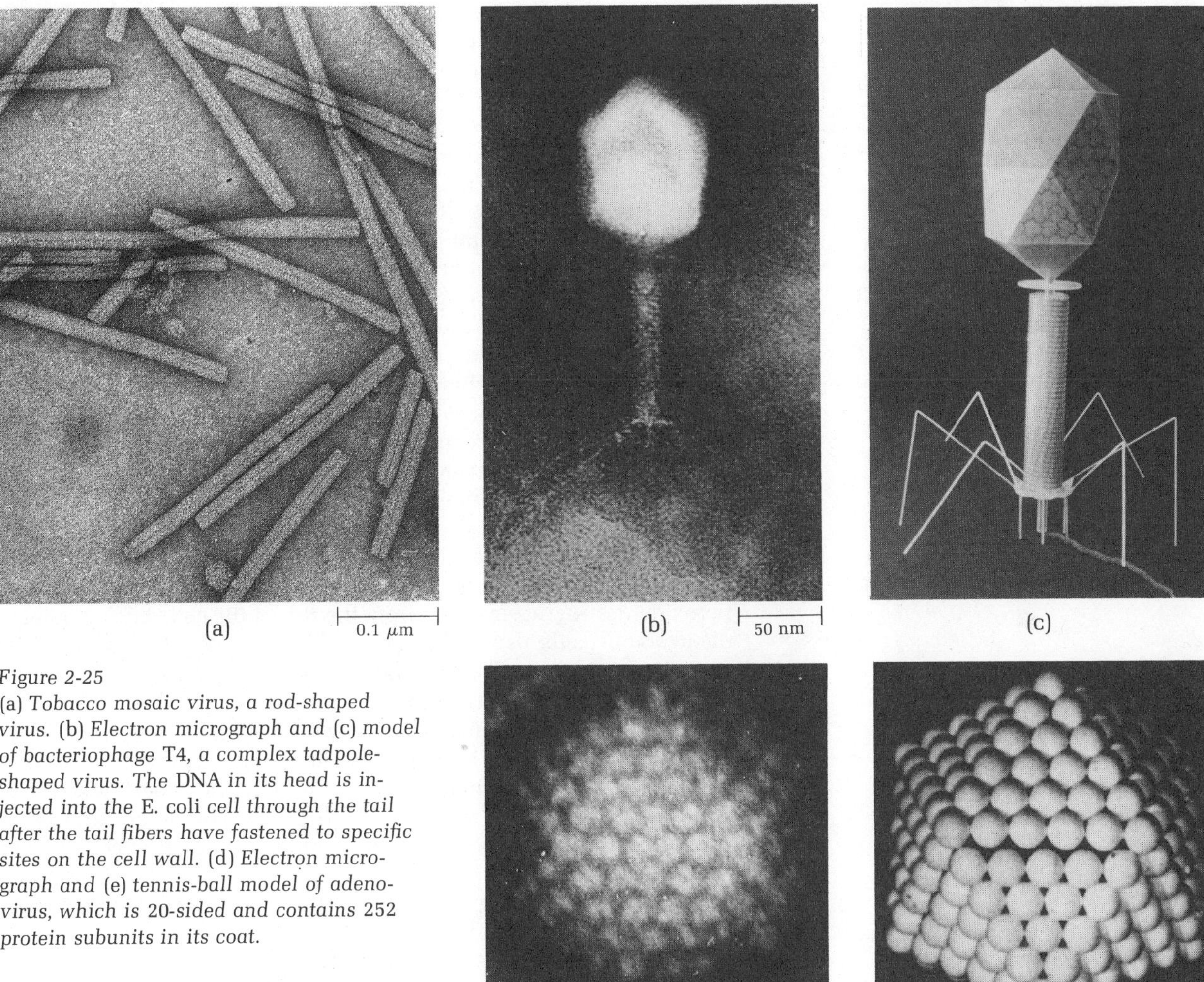

Figure 2-25
(a) Tobacco mosaic virus, a rod-shaped virus. (b) Electron micrograph and (c) model of bacteriophage T4, a complex tadpole-shaped virus. The DNA in its head is injected into the E. coli cell through the tail after the tail fibers have fastened to specific sites on the cell wall. (d) Electron micrograph and (e) tennis-ball model of adenovirus, which is 20-sided and contains 252 protein subunits in its coat.

viruses, as well as the type and size of their nucleic acid components. Some viruses are highly pathogenic for man, e.g., those causing smallpox, poliomyelitis, influenza, the common cold, infectious mononucleosis, and shingles. Latent viruses are also believed to be the cause of cancer in animals. Viruses play an exceedingly important role in the study of biochemistry, since they have given us enormously useful information about the structure of chromosomes, the enzymatic mechanisms of nucleic acid synthesis, and the regulation of the flow of genetic information.

Summary

All cells have a surrounding plasma membrane, cytoplasm, ribosomes, and a nuclear zone or nucleus. Cell size and shape are affected by the rates of physical diffusion of nutrient molecules and oxygen and by the ratio of the surface area to cell volume. There are two great classes of cells: prokaryotic and eukaryotic. The prokaryotes, which consist of the bacteria and blue-green algae, are simple, small cells, which characteristically lack a membrane surrounding their genetic material. They have a cell wall and a plasma membrane, and some have flagella for propulsion. The cytoplasm of prokaryotic cells contains no membrane-surrounded organelles but does contain ribosomes and nutrient granules. Prokaryotic cells grow and divide very rapidly. *Escherichia coli* is the best understood of all prokaryotes, extremely useful in biochemical and genetic studies.

Eukaryotic cells are much larger than prokaryotes; their volume is 1000 to 10,000 times greater. In addition to a well-formed membrane-surrounded nucleus with multiple chromosomes, eukaryotic cells also contain membrane-surrounded organelles. Among them are the mitochondria, which function to oxidize cell fuels and generate ATP, and chloroplasts (in photosynthetic cells), which capture light energy for the conversion of CO_2 into glucose. Mitochondria and chloroplasts are believed to have a bacterial origin. Other eukaryotic organelles include the endoplasmic reticulum, which functions to guide and convey secretory products to the Golgi bodies, there to be packaged for delivery outside the cell. The lysosomes contain degradative enzymes, and peroxisomes segregate peroxide-forming and peroxide-destroying enzymes from the rest of the cell. The cytoplasm of eukaryotic cells also contains microfilaments of at least three kinds, as well as microtubules. Together the microfilaments, microtubules, and microtrabecular network form a flexible internal framework, the cytoskeleton. Many animal cells have cilia and flagella, whose propelling activity is due to their content of paired microtubules. Eukaryotic cells also contain ribosomes, some free and some bound to the surface of the rough endoplasmic reticulum. The outer surfaces of animal cells contain specific recognition and binding sites for other cells and for hormones.

Viruses are nonliving supramolecular structures consisting of a nucleic acid molecule surrounded by a protein coat. They are capable of infecting specific host cells, causing them to replicate the viral particles according to genetic instructions provided by the viral nucleic acid. Viruses have yielded much valuable information on the biochemical aspects of genetic information flow.

References

Textbooks

CURTIS, H.: *Biology,* 3d ed., Worth, New York, 1979. A beautifully written and illustrated general biology textbook.

DYSON, R. D.: *Cell Biology: A Molecular Approach,* 2d ed., Allyn and Bacon, Boston, 1978. A biochemically oriented textbook.

FAWCETT, D. W.: *The Cell,* 2d ed., Saunders. Philadelphia, 1981. Electron micrographs of different cells and tissues.

KARP, G.: *Cell Biology,* McGraw-Hill, New York, 1979. Molecular aspects of genetic information transfer are emphasized.

LEDBETTER, M. C., and K. R. PORTER: *Introduction to the Fine Structure of Plant Cells,* Springer-Verlag, New York, 1970.

LOEWY, A. G., and P. SIEKEVITZ: *Cell Structure and Function,* 3d ed., Holt, New York, 1979.

Interesting Short Books

LURIA, S.: *Life: The Unfinished Experiment,* Scribner's, New York, 1973. A Nobel prize molecular biologist looks at modern biology and its implications.

MARGULIES, L.: *Origin of Eukaryotic Cells,* Yale University Press, New Haven, Conn., 1970. Interesting development of a theory.

THOMAS, L.: *The Lives of a Cell: Notes of a Biology Watcher,* Viking, New York, 1974. Engaging personal essays on cells by a noted leader in biomedical research.

Some Articles on Special Topics in Cell Biology

BERG, H.: "How Bacteria Survive," *Sci. Am.,* **233**:36–44, August (1975).

CAPALDI, R. A.: "A Dynamic Model of Cell Membranes," *Sci. Am.,* **230**:26–33, March (1974).

EVERHART, T. E., and T. L. HAYES: "The Scanning Electron Microscope," *Sci. Am.,* **226**:54–69, January (1972).

MAZIA, D.: "The Cell Cycle," *Sci. Am.,* **230**:54–64, January (1974).

PORTER, K. R., and J. B. TUCKER: "The Ground Substance of the Living Cell," *Sci. Am.,* **244**:56–67, March (1981).

SATIR, P.: "How Cilia Move," *Sci. Am.,* **231**:44–63, October (1974).

SLOBODA, R. D.: "The Role of Microtubules in Cell Structure and Cell Division," *Amer. Sci.,* **68**:290–298 (1980).

WESSELLS, N. K.: "How Living Cells Change Shape," *Sci. Am.,* **225**: 76–85, October (1971).

Problems

To understand the molecular logic of cells we must be able to consider biomolecules and their interactions in both qualitative and quantitative terms. We must also be able to analyze complex cell phenomena in terms of the simplest components and processes involved. For this reason each chapter concludes with a series of problems illustrating important biochemical principles. Some are numerical problems that reveal important dimensions of molecules and cells or the rates of biochemical processes. Others require application of basic principles and some thought to analyze a given biochemical structure or cell process. Some of the problems are relatively easy and straightforward, but others are more challenging. Problem solving is the best way to consolidate your command of biochemical principles.

Useful books that provide guidance and practice in solving biochemical problems include:

MONTGOMERY, R., and C. A. SWENSON: *Quantitative Problems in Biochemical Sciences,* 2d ed., Freeman, San Francisco, 1976.

SEGEL, I.: *Biochemical Calculations,* 2d ed., Wiley, New York, 1976. Especially good on enzyme kinetics.

WOOD, W. B., J. H. WILSON, R. M. BENBOW, and L. E. HOOD: *Biochemistry: A Problems Approach*, 2d ed., Benjamin, Menlo Park, Calif., 1981. This unique, comprehensive book reviews the major topics of biochemistry and provides true-false and multiple-choice drill questions and many numerical problems. The sequence of topics is about the same as in this volume.

Some problems on the contents of Chapter 2 follow. They involve simple geometrical and numerical relationships concerning cell structure and activities. Each has a title for easy reference and discussion.

1. *The Smallness of Cells and Their Components* From the data in Table 2-2 calculate the approximate number of (a) liver cells, (b) mitochondria, and (c) myoglobin molecules that can be placed in a single layer on the head of a pin (0.5 mm in diameter). Assume each structure is spherical. The area of a circle is given by πr^2, where $\pi = 3.14$.

2. *Number of Solute Molecules in the Smallest Known Cells* Mycoplasmas are the smallest known cells. They are spherical and have a diameter of about 0.33 μm. Because of their small size they readily pass through filters designed to trap larger bacteria. One species, *Mycoplasma pneumoniae*, is the causative organism of the disease primary atypical pneumonia.
   (a) D-Glucose is the major energy-yielding nutrient of mycoplasma cells. Its concentration within such cells is about 1.0 m*M*. Calculate the number of glucose molecules in a single mycoplasma cell. Avogadro's number, the number of molecules in 1 mol of a nonionized substance, is 6.02×10^{23}. The volume of a sphere is $\frac{4}{3}\pi r^3$.
   (b) The intracellular fluid of mycoplasma cells contains 10 g of hexokinase (MW 100,000) per liter. Calculate the molar concentration of hexokinase, the first enzyme required for the energy-yielding metabolism of glucose.

3. *Components of E. coli* *E. coli* cells are rod-shaped, about 2 μm long and 0.8 μm in diameter. The volume of a cylinder is given by $\pi r^2 h$, where h is its length.
   (a) If the average density of *E. coli* (mostly water) is 1.1 g/cm³, what is the weight of a single cell?
   (b) The protective cell wall of *E. coli* is 10 nm thick. What percentage of the total volume of the bacterium does the wall occupy?
   (c) *E. coli* grows and multiplies rapidly because of the inclusion of some 15,000 spherical ribosomes (diameter, 18 nm), which carry out protein synthesis. What percentage of the total cell volume do the ribosomes occupy?

4. *Genetic Information in E. coli DNA* The genetic information contained in DNA consists of a linear sequence of successive code words, known as codons. Each codon is a specific sequence of three nucleotides (three nucleotide pairs in double-stranded DNA), and each codes for a single amino acid unit in a protein. The molecular weight of an *E. coli* DNA molecule is very great, about 2.5×10^9. The average molecular weight of a nucleotide pair is 660, and each nucleotide pair contributes 0.34 nm to the length of DNA.
   (a) Given this information, calculate the length of an *E. coli* DNA molecule. Compare the length of the DNA molecule with the actual cell dimensions. How does it fit in?

(b) Assume that the average protein in *E. coli* consists of a chain of 400 amino acids. What is the maximum number of proteins that can be coded by an *E. coli* DNA molecule?

5. *The High Rate of Bacterial Metabolism* Bacterial cells have a much higher rate of metabolism than animal cells. Under ideal conditions a bacterium will double in size and divide in 20 min, while it may require animal cells some 24 h. The high rate of bacterial metabolism requires a high ratio of surface area to cell volume.
(a) Why would the surface-to-volume ratio have an effect on the maximum rate of metabolism?
(b) Calculate the surface-to-volume ratio for the spherical bacterium *Neisseria gonorrhoeae* (diameter 0.5 μm), responsible for the disease gonorrhea. Compare it with the surface-to-volume ratio for a globular ameba, a large eukaryotic cell of diameter 150 μm.
(c) Estimate the surface-to-volume ratio for a typical 70-kg human being. (*Hint:* Treat the person as a sphere plus a set of cylinders.) Compare it with the surface-to-volume ratio for a bacterium.

Problem 6

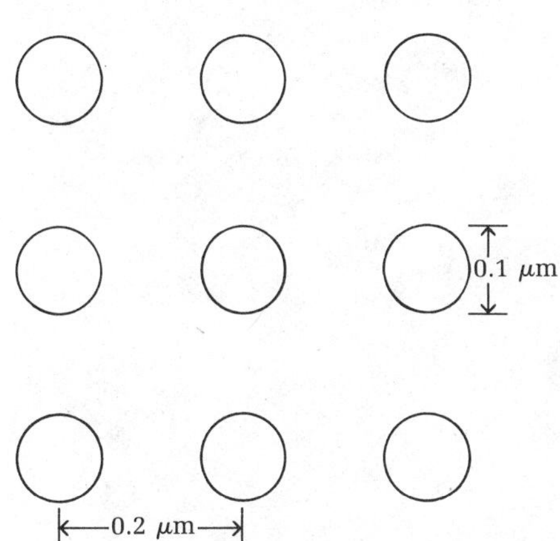

Arrangement of microvilli on the "patch"

6. *A Strategy to Increase the Surface Area of Cells* Certain cells whose function is to absorb nutrients from the surroundings, e.g., the cells lining the small intestine or the root hair cells of a plant, are optimally adapted to their role because their surface area exposed to nutrients is increased by microvilli. Consider a spherical epithelial cell (20 μm in diameter) lining the small intestine. Since only a part of the surface faces the interior of the intestine, assume that a "patch" corresponding to 25 percent of the cell area is covered with microvilli. Furthermore, assume that the microvilli are cylinders 0.1 μm in diameter, 1.0 μm long, and spaced in a regular grid 0.2 μm on center. The surface area of a sphere is $4\pi r^2$.
(a) Calculate the number of microvilli on the patch.
(b) Calculate the surface area of the patch, assuming it has no microvilli.
(c) Calculate the surface area of the patch, assuming it does have microvilli.
(d) What percentage improvement of the absorptive capacity (reflected by the surface-to-volume ratio) does the presence of microvilli provide?

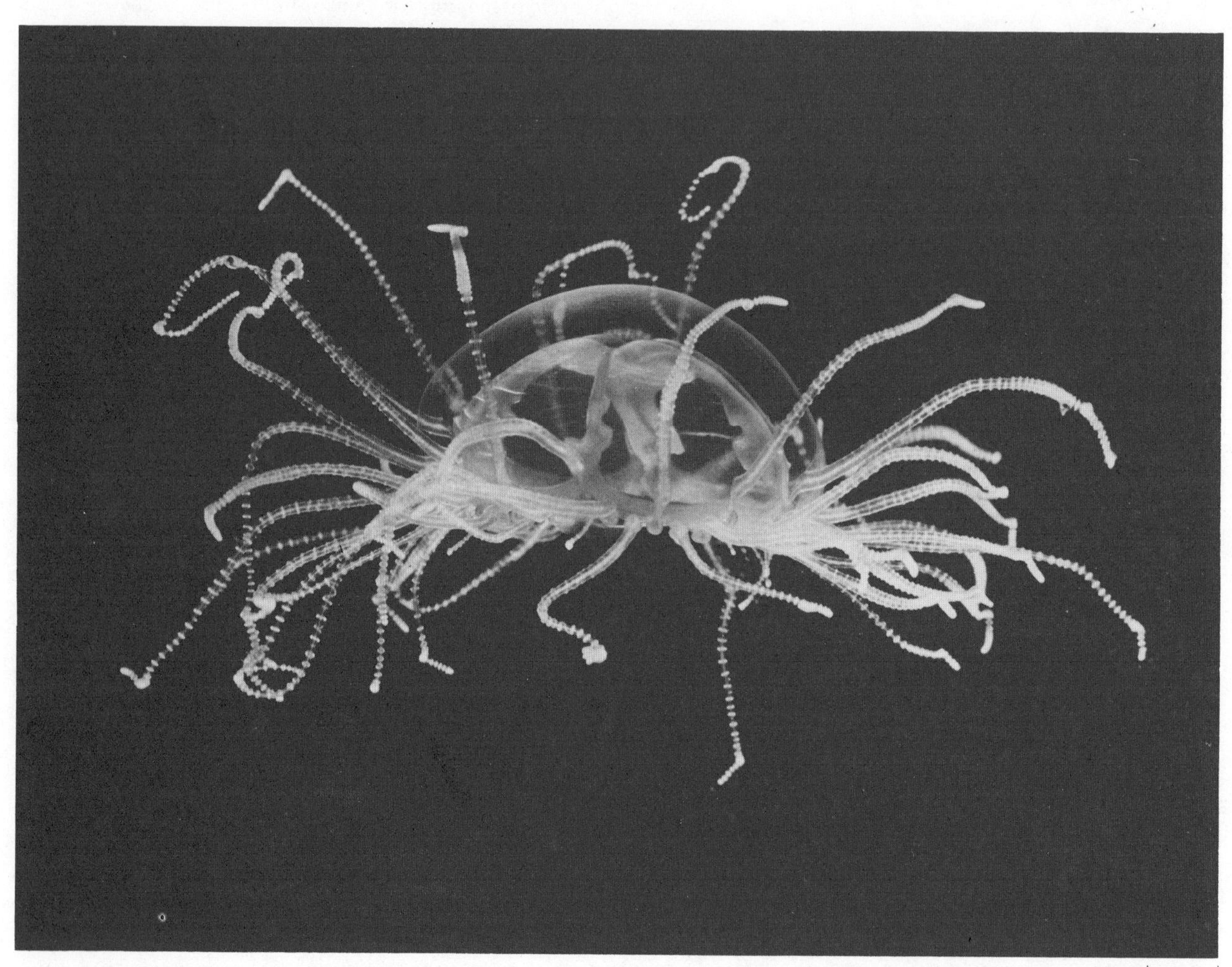

The chemical composition of living matter —such as the jellyfish Gonionemus murbachii*—differs from that of its physical environment, which, for this organism, is salt water.*

CHAPTER 3

The Composition of Living Matter: Biomolecules

Table 3-1 The Bioelements
The following elements have been found to be essential components of living matter; however, not all the trace elements listed are necessarily required by every species

Bulk elements of organic matter	
Carbon	C
Hydrogen	H
Oxygen	O
Nitrogen	N
Phosphorus	P
Sulfur	S
Elements occurring as ions	
Sodium	Na^+
Potassium	K^+
Magnesium	Mg^{2+}
Calcium	Ca^{2+}
Chlorine	Cl^-
Trace elements	
Iron	Fe
Copper	Cu
Zinc	Zn
Manganese	Mn
Cobalt	Co
Iodine	I
Molybdenum	Mo
Vanadium	V
Nickel	Ni
Chromium	Cr
Fluorine	F
Selenium	Se
Silicon	Si
Tin	Sn
Boron	B
Arsenic	As

We have seen that there are many similarities in the chemical composition of different animal and plant species. For example, all protein molecules in all living species are made from the same set of 20 amino acids. Similarly, all nucleic acids in all species are made from the same sets of nucleotides. Yet we shall now see that the chemical composition of living matter is very different from that of the inanimate matter of the earth's crust. We may therefore ask some basic questions as we begin the study of biomolecules and their interactions. What chemical elements are found in cells? What kinds of molecules are present in living matter? In what proportions? How did they come to be there? In what ways are the kinds of molecules found in living cells especially suited for their roles?

As we seek answers to these questions, we must examine biomolecules as we would nonbiological molecules, by the same principles and approaches used in classical chemistry. But we must also consider them from the biological viewpoint, in the light of the concept that the different kinds of molecules in living matter are related to each other and interact together in what we have called the molecular logic of the living state.

The Chemical Composition of Living Matter Differs from That of the Earth's Crust

Only 27 of the 92 natural chemical elements are essential for different forms of life; they are listed in Table 3-1. Most of the elements in living matter have relatively low atomic numbers; only three have atomic numbers above 34. Moreover, the distribution of the elements found in living organisms is not in proportion to their occurrence in the earth's crust (Table 3-2). The four most abundant elements in living organisms, in terms of the percentage of the total number of atoms, are hydrogen, oxygen, carbon, and nitrogen; together they make up over 99 percent of the mass of most cells. Three of these elements, hydrogen, nitrogen, and carbon, occur in much greater abun-

Table 3-2 The Eight Most Abundant Elements in the Earth's Crust and in the Human Body, as a Percentage of the Total Number of Atoms

Earth's crust		Human body	
Element	%	Element	%
O	47	H	63
Si	28	O	25.5
Al	7.9	C	9.5
Fe	4.5	N	1.4
Ca	3.5	Ca	0.31
Na	2.5	P	0.22
K	2.5	Cl	0.08
Mg	2.2	K	0.06

dance in living matter than in the earth's crust. The difference in the elementary composition of the earth's crust and living matter is even more striking when we consider the composition by weight of the dry or solid portion of living matter, excluding its water content, which amounts to over 75 percent. Carbon makes up from 50 to 60 percent by weight of the solid matter of living cells, nitrogen almost 8 to 10 percent, oxygen about 25 to 30 percent, and hydrogen almost 3 to 4 percent. In contrast, carbon, hydrogen, and nitrogen together make up much less than 1 percent of the mass of the earth's crust. But on the other hand, eight of the ten most abundant elements in the human body are also among the ten most abundant elements in seawater.

Two tentative conclusions follow from these data. The first is that chemical compounds containing carbon, hydrogen, oxygen, and nitrogen, the elements predominating in living organisms, may have been selected during evolution because they possess unique fitness for life processes. The second is that seawater may have been the liquid medium in which living organisms first arose in the early history of the earth.

Most Biomolecules Are Compounds of Carbon

The chemistry of living organisms is organized around the element carbon, which accounts for over one-half of their dry weight. Carbon, like hydrogen, oxygen, and nitrogen, is capable of forming covalent bonds, i.e., bonds formed by sharing electron pairs (Figure 3-1). The hydrogen atom needs one electron, oxygen needs two, nitrogen three, and the carbon atom four to fill their respective outer shells. Thus a carbon atom can share four electron pairs with four hydrogen atoms to form the compound *methane* (CH_4), in which each of the shared electron pairs is a *single bond*. Carbon can also form single bonds to oxygen and nitrogen atoms. But most significant in biology is the ability of carbon atoms to share electron pairs with each other

Figure 3-1
Covalent bonding. Two atoms with unpaired electrons in their outer shells can form covalent bonds with each other by sharing electron pairs. Atoms participating in covalent bonding tend to fill their outer shells.

Atom	Number of unpaired electrons (in color)	Number of electrons in complete outer shell
H·	1	2
:Ȯ·	2	8
:N̈·	3	8
·Ċ·	4	8

H· + H· ⟶ H:H = H—H
Dihydrogen

2H· + ·Ö: ⟶ H:Ö: (H) = H—O—H
Water

:N· + 3H· ⟶ H:N:H (H) = N bonded to three H
Ammonia

·C· + 4H· ⟶ H:C:H (H, H) = H—C—H (H, H)
Methane

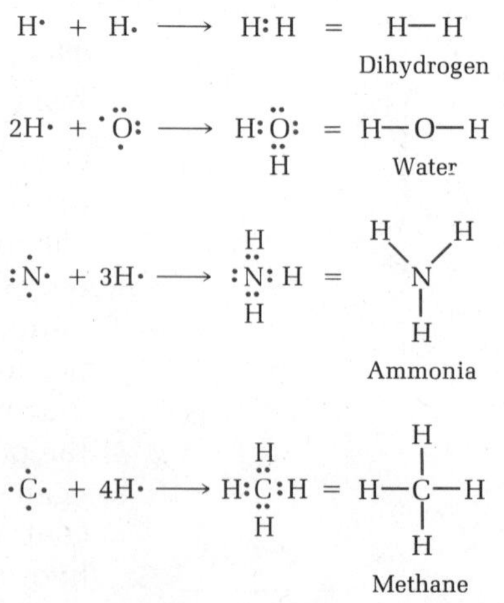

Figure 3-2
Versatility of carbon in forming covalent single and double bonds, particularly between carbon atoms. Triple bonds occur only rarely in organic biomolecules.

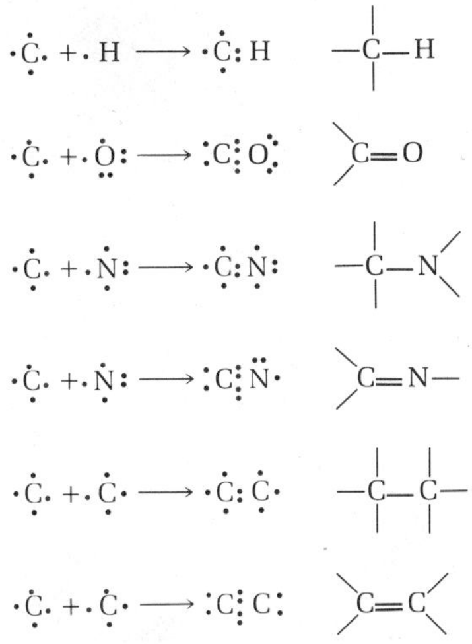

to form very stable carbon-carbon single bonds. Each carbon atom can form single bonds with one, two, three, or four other carbon atoms. Moreover, two carbon atoms also can share two electron pairs with each other, thus forming carbon-carbon double bonds (Figure 3-2). Thanks to these properties, covalently linked carbon atoms can be joined to form many kinds of structures—linear chains, branched chains, cyclic and cage-like structures, and combinations thereof—to form the skeletons of a great many different kinds of organic molecules (Figure 3-3). To such carbon backbones other kinds of atomic groups can be attached, since carbon can also form covalent bonds with oxygen, hydrogen, and nitrogen, as well as with sulfur. Molecules containing such covalently bonded carbon backbones we recognize as organic compounds, which can occur in an almost limitless variety. Because most biomolecules are organic compounds of carbon, we can infer that the bonding versatility of the element carbon may have been a major factor in the selection of carbon compounds for the molecular machinery of cells during the origin and evolution of living organisms.

Organic Biomolecules Have Specific Shapes and Dimensions

The four covalent single bonds of a carbon atom are spaced in a tetrahedral arrangement with an angle of about 109.5° between any two of them (Figure 3-4). This angle varies little from one carbon atom to another in different organic molecules. Because of this characteristic, different organic compounds of carbon can have many different kinds of three-dimensional structures. No other chemical element can form molecules of such widely different sizes and shapes or with such a variety of side chains and functional groups. Much of the intricate complexity of internal cell structures is a reflection of the different sizes and shapes of the organic molecules of which they are composed.

Figure 3-3
Carbon-carbon bonds form the backbones of many types of organic molecules.

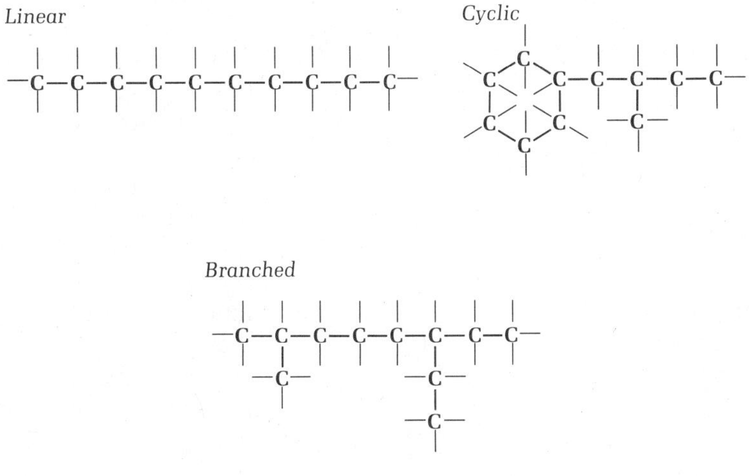

Linear

Cyclic

Branched

Figure 3-4
(a) *Carbon atoms have a characteristic tetrahedral arrangement of their four single bonds, which are about 0.154 nm long and are spaced 109.5° apart.*
(b) *Carbon-carbon single bonds have complete freedom of rotation, shown by different representations of the compound ethane* ($H_3C—CH_3$).
(c) *Carbon-carbon double bonds are shorter and do not allow free rotation. The single bonds on each doubly bonded carbon make an angle of 120° with each other. The two doubly bonded carbons and the atoms designated A, B, X, and Y all lie in the same rigid plane.*

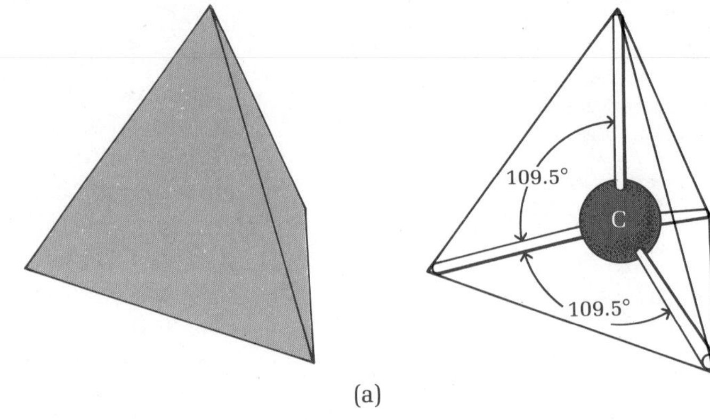

Overhead views

End views

(b)

(c)

The second important property of organic compounds is that there is complete freedom of rotation around each carbon-carbon single bond unless there are very large or highly charged groups attached to both carbon atoms, in which case rotation may be restricted. Thus organic molecules with many single bonds can assume a number of different shapes, called *conformations*, depending on the degree to which each single bond is rotated (Figure 3-4).

The third important feature of the covalent bonds of carbon is that they have characteristic bond lengths. Carbon-carbon single bonds have an average length of 0.154 nm (or 1.54 Å, the older unit used in structural chemistry), whereas carbon-carbon double bonds are shorter, about 0.134 nm long. In contrast to single bonds, carbon-carbon double bonds are rigid and do not allow complete rotation. They also make a larger angle with the remaining single bonds around a carbon atom (Figure 3-4). Bond angles and center-to-center bond lengths in organic molecules are best shown with *ball-and-stick models*, whereas the outer contours of organic molecules are best shown by *space-filling models* (Figure 3-5), which are proportional in dimensions to the radius of different atoms (Table 3-3). From these considerations we see that organic biomolecules have characteristic sizes and three-dimensional space-filling properties, depending on their backbone structures and their substituent groups.

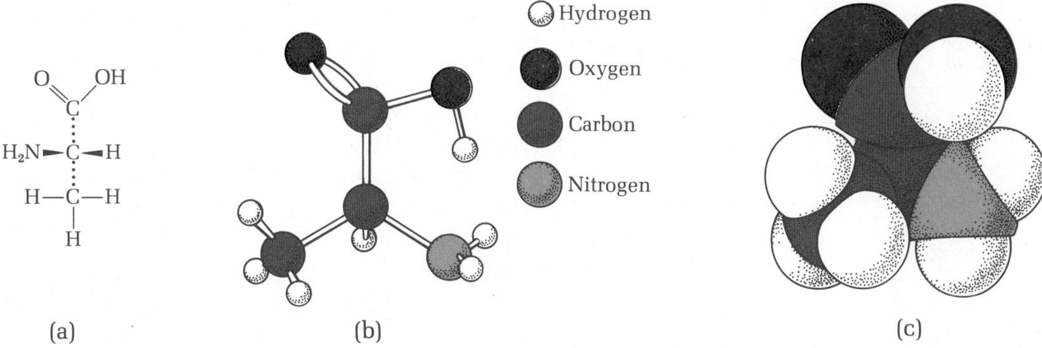

Figure 3-5
Models of the structure of the amino acid alanine. (a) Structural formula in perspective form. (b) Ball-and-stick model, showing relative bond lengths and the bond angles. The balls indicate the approximate size of the atomic nuclei. (c) Space-filling model, in which each atom is shown in its correct relative diameter (see also Table 3-3).

Table 3-3 Radii of Some Atoms

The data are for van der Waals radii, which indicate space-filling dimensions of atoms; however, when these atoms are joined in covalent linkage, the atomic radius at the point of bonding is less because the joined atoms are pulled together by the shared electron pair

Element	Radius, nm	Space-filling model
		0.1 nm
Carbon	0.077	
Hydrogen	0.037	
Oxygen	0.066	
Nitrogen	0.070	
Phosphorus	0.110	
Sulfur	0.104	
Some covalent bonds		
	—C—H	
	—C—N	
	H—O—H	

Figure 3-6
Complementary fit of a substrate molecule to the active or catalytic site on an enzyme molecule. The catalytic site recognizes and acts on only those molecules whose space-filling characteristics and dimensions fit it precisely.

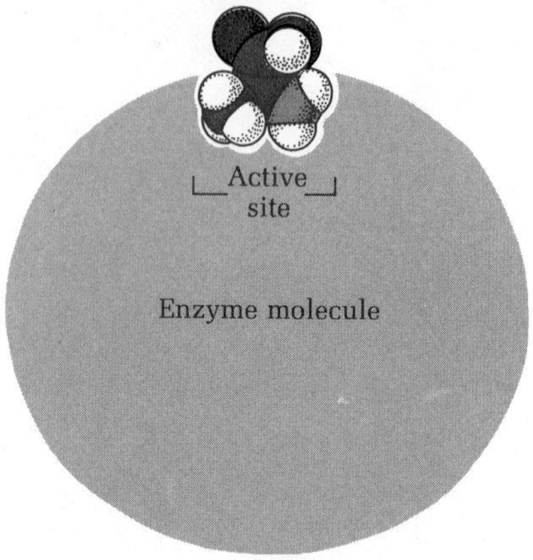

Table 3-4 Functional Groups Characterizing Families of Organic Compounds

R_1 and R_2 designate hydrocarbons into which the functional group is substituted

Functional group	Structure	Family
Hydroxyl	R_1—O—H	Alcohols
Aldehyde	R_1—C(=O)—H	Aldehydes
Carbonyl	R_1—C(=O)—R_2	Ketones
Carboxyl	R_1—C(=O)—OH	Acids
Amino	R_1—N(H)—H	Amines
Amido	R_1—C(=O)—N(H)—H	Amides
Thiol	R_1—S—H	Thiols
Ester	R_1—C(=O)—O—R_2	Esters
Ether	R_1—O—R_2	Ethers

The three-dimensional conformation of organic biomolecules is of the utmost importance in many aspects of biochemistry, e.g., in the reaction between the catalytic site of an enzyme and the substrate it acts upon (Figure 3-6). The two molecules must fit each other exactly, in a *complementary* fashion, for proper biological function. Such precise complementarity also is required in the binding of a hormone molecule to its receptor on a cell surface and in the replication of DNA and many other cell activities. For this reason the study of the three-dimensional structure of biomolecules with precise physical methods is an important part of modern research on cell structure and biochemical function.

Functional Groups of Organic Biomolecules Determine Their Chemical Properties

Nearly all organic biomolecules can be regarded as derivatives of *hydrocarbons*, compounds of carbon and hydrogen in which the backbone consists of carbon atoms joined by covalent bonds and the other bonds of the carbons are shared with hydrogen atoms. The backbones of such hydrocarbons are very stable, because carbon-carbon single and double bonds share their electron pairs equally.

One or more hydrogen atoms of hydrocarbons may be replaced by different kinds of *functional groups* to yield different families of organic compounds. Typical families of organic compounds and their characteristic functional groups are the *alcohols*, which have one or more *hydroxyl groups; amines*, which have *amino* groups; *ketones*, which have *carbonyl* groups; and *acids*, which have *carboxyl* groups (Table 3-4). Several other common functional groups are also important in biomolecules (Table 3-5).

Such functional groups of organic biomolecules are chemically much more reactive than the saturated hydrocarbon backbones, which are not readily attacked by most chemical reagents. Functional groups can alter the electron distribution and the geometry of neighboring atoms and thus affect the chemical reactivity of an organic molecule as a whole. From the functional groups present in organic biomolecules it is possible to analyze and predict their chemical behavior and reactions. As we shall see, enzymes, the catalysts of living cells, function by recognizing a specific functional group in a biomolecule and catalyzing a characteristic chemical change in its structure.

Most of the biomolecules we shall examine are *polyfunctional*, containing two or more different kinds of functional groups. In such molecules each type of functional group has its own chemical characteristics and reactions. For illustration we may use the *amino acids*, an important family of biomolecules that serve primarily as building blocks of proteins. All amino acids contain at least two different kinds of functional groups, an amino group and a carboxyl group, as shown for *alanine*

Table 3-5 Some Other Functional Groups Found in Biomolecules

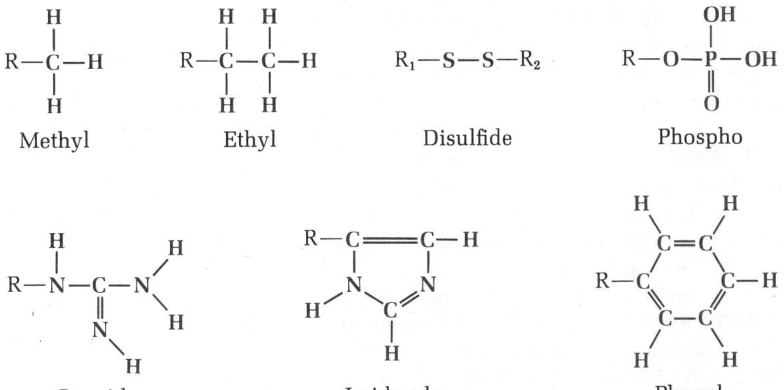

R—C(H)(H)—H	Methyl
R—C(H)(H)—C(H)(H)—H	Ethyl
R_1—S—S—R_2	Disulfide
R—O—P(OH)(=O)—OH	Phospho
R—N(H)—C(=N—H)—N(H)(H)	Guanido
R—C═C—H, N(H)—C(H)═N ring	Imidazole
R—C ring: C(H)=C(H), C—H, C(H)—C(H)	Phenyl

Figure 3-7
Biomolecules with multiple functional groups.

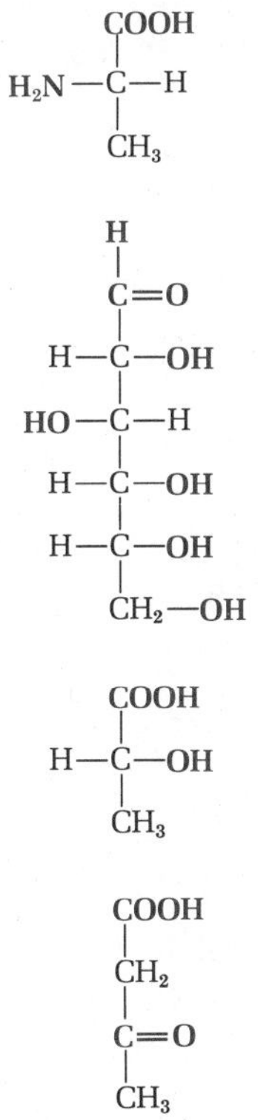

H_2N—C(COOH)(CH$_3$)—H — Alanine, an amino acid building block of proteins

H—C=O; H—C—OH; HO—C—H; H—C—OH; H—C—OH; CH_2—OH — Glucose, a sugar

COOH; H—C—OH; CH_3 — Lactic acid, a product of glucose metabolism

COOH; CH_2; C=O; CH_3 — Acetoacetic acid, a metabolic product of fat oxidation

(Figure 3-7). The chemical properties of this amino acid are profoundly dependent on the chemical properties of the amino group and the carboxyl group. Among other common examples of polyfunctional biomolecules is the simple sugar glucose, which contains two kinds of functional groups, hydroxyl groups and an aldehyde group (Figure 3-7). Over and over again we shall see that the functional groups of biomolecules play very important roles in their biological activities.

Many Biomolecules Are Asymmetric

The tetrahedral arrangement of single bonds around a carbon atom confers another very striking property on some organic molecules which is of the utmost importance in biology. Whenever there are four *different* atoms or functional groups singly bonded to a given carbon atom in an organic molecule, the carbon atom is said to be asymmetric since it can exist in two isomeric forms, called *enantiomers,* which have different configurations in space. As can be seen in Figure 3-8, enantiomers are nonsuperimposable mirror images of each other. Enantiomers, also called optical isomers or stereoisomers, are chemically identical in their reactions, but they differ in a very characteristic physical property, the ability to rotate the plane of plane-polarized light. A solution of one of the two enantiomers rotates the plane of such light to the right, and a solution of the other rotates it to the left, a property that can be measured in a polarimeter. Compounds not having an asymmetric carbon atom do not rotate the plane of plane-polarized light.

Figure 3-8 shows that the amino acid alanine is an asymmetric molecule, since its central carbon atom contains four different substituent groups: a methyl group, an amino group, a carboxyl group, and a hydrogen atom. We see further that the two different enantiomers of alanine are nonsuperimposable mirror images of each other. The two forms of alanine bear the same relationship to each other as the right and left hands; we

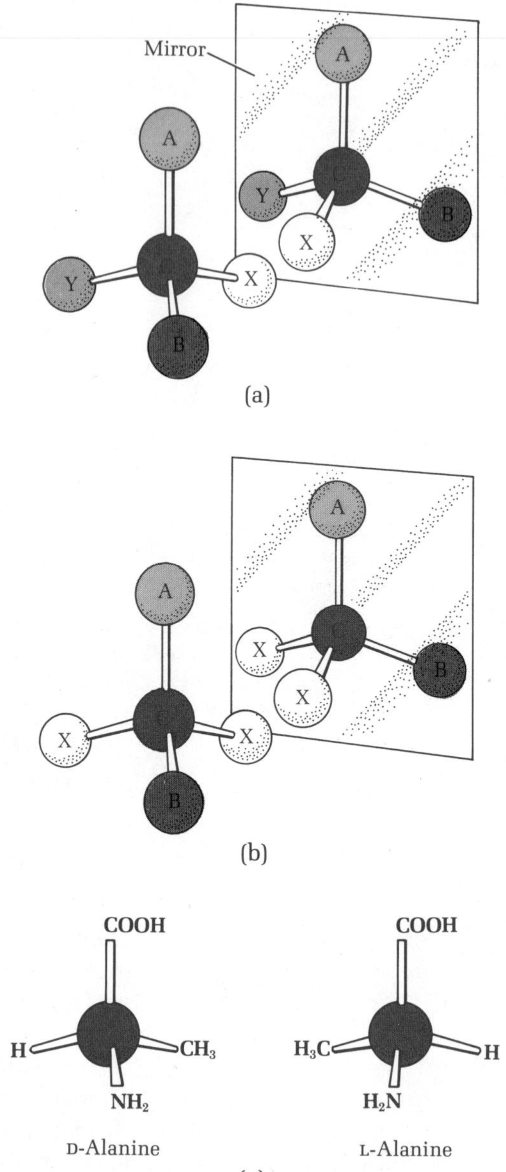

Figure 3-8
Chiral molecules. (a) When a carbon atom has four different substituent groups (A, B, X, Y), they can be arranged in two ways that represent nonsuperimposable mirror images of each other. Such a carbon atom is asymmetric and is called a chiral atom *or* center.

(b) When there are only three dissimilar groups around the carbon atom, only one configuration in space is possible and the molecule is symmetrical, or achiral. *In this case the molecule can be shown as two mirror images, but they are superimposable. If the molecule on the left is rotated counterclockwise (when looking down its vertical bond to A), it will be superimposable on the molecule on the right.*

(c) Alanine is a chiral molecule, since it has four different groups around a central carbon atom, to form nonsuperimposable mirror images of each other. The two different chiral forms of alanine are designated D*-alanine and* L*-alanine. The basis for the designations* D *and* L *for chiral forms of molecules will be discussed in Chapter 5.*

know from experience that the right hand will not fit into a left-handed glove. Because compounds with asymmetric carbon atoms can be regarded as occurring in left- and right-handed forms, they are called *chiral compounds* (Greek, *chiros*, "hand"). Correspondingly, the asymmetric atom or center of chiral compounds is called the *chiral atom* or *center*.

Many organic biomolecules besides amino acids are chiral and contain one or more asymmetric carbon atoms. The common sugar glucose is another example; it has no less than five asymmetric carbons. Moreover, the chiral molecules in living organisms are usually present in only one of their possible chiral forms. For example, the amino acids present in protein molecules, of which alanine is an example, occur in only one of their possible chiral forms. Similarly, glucose, the building-block unit of starch, occurs biologically in only one of *its* numerous chiral forms. In contrast, when the organic chemist in the laboratory synthesizes an organic compound in which there is an asymmetric carbon atom, the ordinary, nonbiological reactions used form the two possible chiral forms at an equal rate, leading to an equimolar mixture. The two chiral forms in such a mixture can be separated from each other only by painstaking physical methods. In living cells chiral biomolecules are made by enzymes in such a way that only one of the two enantiomers is formed. This occurs because the enzyme molecules themselves are chiral structures.

The stereospecificity of many biomolecules is a characteristic feature of the molecular logic of living cells, and it also argues strongly that the three-dimensional shape of biomolecules is extremely important in their biological function. We shall examine chiral molecules and stereoisomerism in more detail later, when we survey the amino acids (Chapter 5) and the sugars (Chapter 11).

The Major Classes of Biomolecules in Cells Are Very Large Molecules

Table 3-6 shows the major classes of biomolecules found in the bacterium *Escherichia coli*, the contribution of each class to the total weight of the cell, and an approximation of the different kinds of biomolecules in each class. Water is the most abundant single compound in *E. coli*, as it is in all types of cells and organisms. Inorganic salts and mineral elements, on the other hand, constitute only a very small fraction of the total solids, but many of them are in approximate proportion to their distribution in seawater. Nearly all the solid matter in *E. coli* cells, and for that matter all kinds of cells, is organic and is present in

Table 3-6 Molecular Components of an *E. coli* Cell

	Percentage of total weight	Approximate number of molecular species
Water	70	1
Proteins	15	3000
Nucleic acids		
DNA	1	1
RNA	6	>3000
Polysaccharides	3	5
Lipids	2	20
Building-block molecules and intermediates	2	500
Inorganic ions	1	20

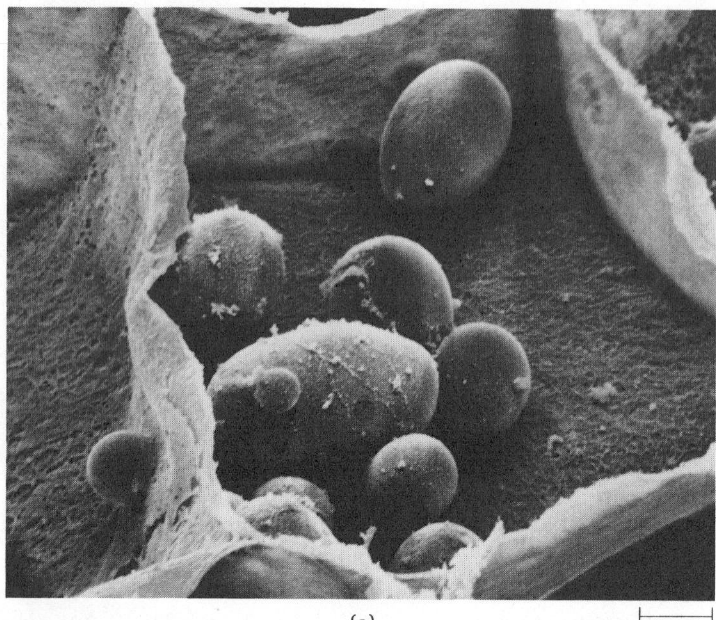

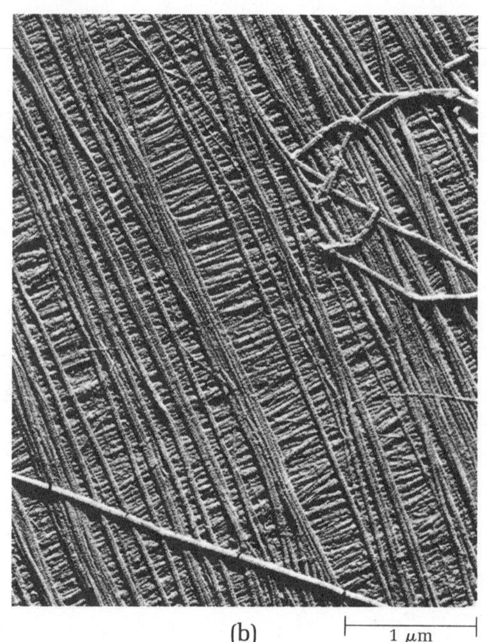

Figure 3-9
(a) Scanning electron micrograph of starch granules in a potato cell. (b) Electron micrograph showing how strands of cellulose in layers form the structural framework of plant cell walls.

four forms: *proteins*, *nucleic acids*, *polysaccharides*, and *lipids*. Proteins constitute the largest fraction of living matter, not only in *E. coli* but in all types of cells. Indeed, the name protein (Greek, *proteios*) means "first" or "foremost." The proteins are direct products and effectors of gene action in all forms of life. Many proteins have specific catalytic activity and function as enzymes. Other proteins serve as structural elements in cells and tissues. Still others are present in cell membranes and promote the transport of certain substances into or out of cells. Many other biological functions are served by proteins, perhaps the most versatile of all biomolecules. The *nucleic acids*, DNA and RNA, have the same universal functions in all cells, to participate in the storage, transmission, and translation of genetic information. DNA serves as the repository of genetic information, whereas different kinds of RNAs help translate this information into protein structure.

The *polysaccharides* have two major functions. Some, like *starch*, are storage forms of energy-yielding fuels, and others, like *cellulose*, function as extracellular structural elements (Figure 3-9). The *lipids*, greasy or oily substances, serve two chief roles: as major structural components of membranes and as a storage form of energy-rich fuel.

These four great classes of biomolecules have a common characteristic: they are all relatively large structures with high molecular weights and are therefore called *macromolecules*. Proteins have molecular weights in the range from 5000 to over 1 million; the various kinds of nucleic acids have molecular weights ranging up to several billion; and polysaccharides, such as starch, also have molecular weights into the millions. Individual lipid molecules are much smaller (MW 750 to 1500), but since lipid molecules usually associate together in thousands to form very large structures that function like macromolecular systems, in particular the "core" of cell membranes, we can include such lipid structures among the macromolecules.

Macromolecules Are Constructed from Small Building-Block Molecules

Although living organisms contain a very large number of different proteins and different nucleic acids, we have seen (Chapter 1) that there is a fundamental simplicity underlying their structure. The simple building-block molecules from which all proteins and all nucleic acids are constructed are few in number and identical in all living species. The proteins of all species are made up of only 20 different amino acids, which are arranged in different linear sequences to form long chains. Similarly, the long, chainlike nucleic acids of all organisms are built from a small number of nucleotides, also arranged in many different sequences. Proteins and nucleic acids are *informational macromolecules:* each protein and each nucleic acid has a characteristic information-rich building-block sequence.

Polysaccharides also are constructed from many building-block units. Starch and cellulose, for example, consist of long chains of only a single type of building block, namely, the sugar glucose. Because polysaccharides are built from only a single kind of unit, or from two different alternating units, they cannot carry encoded genetic information (Figure 3-10).

Figure 3-10
Building-block sequences in informational and noninformational macromolecules. A, T, G, and C represent the four bases of DNA, an informational molecule. Glc represents glucose, the repeating unit in cellulose, which does not function to carry information.

Segment of a strand of DNA, an informational molecule	Segment of cellulose
A	Glc
T	Glc
G	Glc
C	Glc
C	Glc
T	Glc
A	Glc
G	Glc
G	Glc
T	Glc
A	Glc
C	Glc
A	Glc
T	Glc
G	Glc

Well over 90 percent of the solid organic matter of living organisms, containing many thousands of different macromolecules, is constructed from only about three dozen different kinds of simple, small organic molecules. Thus we need to know the structure and properties of only a relatively few organic compounds in order to comprehend the structure of biological macromolecules and some of the organizing principles of biochemistry.

The Building-Block Molecules Have Simple Structures

Figure 3-11 shows the structures of the building-block biomolecules, arranged in families. The building blocks of proteins are 20 different amino acids; all have an amino group and a carboxyl group attached to the same carbon atom. These amino acids differ from each other only in the part of the molecule called the *R group,* shown in Figure 3-11a.

The recurring structural units of all nucleic acids are eight different *nucleotides;* four kinds of nucleotides are the building blocks of DNA, and four others are the structural units of RNA. Each nucleotide in turn contains three smaller units: (1) a nitrogenous organic base, (2) a 5-carbon sugar, and (3) phosphoric acid. The five different organic bases and the two different sugar components of nucleotides are shown in Figure 3-11b.

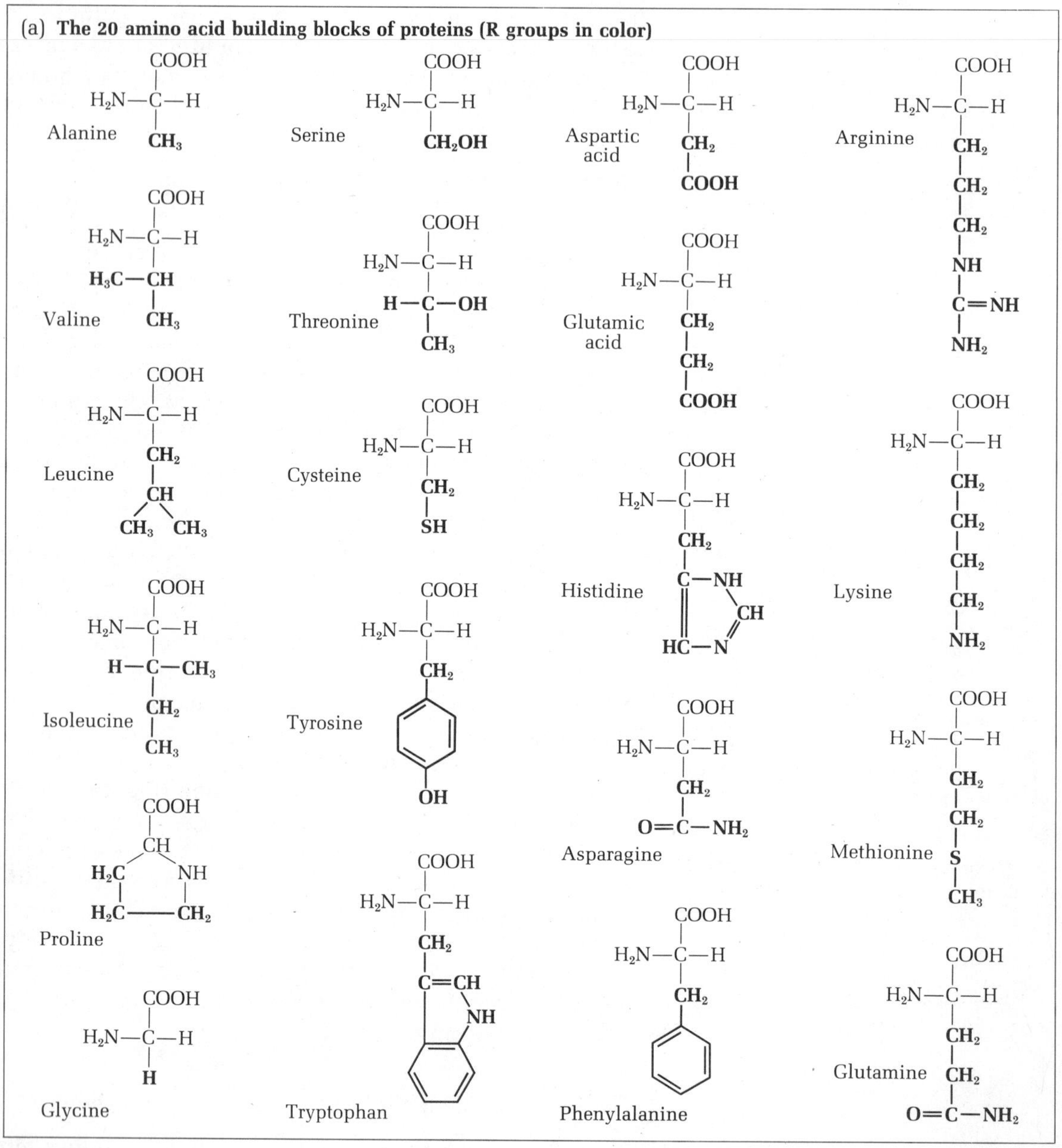

(a) The 20 amino acid building blocks of proteins (R groups in color)

Alanine: COOH—C(H)(H₂N)—**CH₃**

Valine: COOH—C(H)(H₂N)—**CH(CH₃)—H₃C** (**H₃C—CH—CH₃**)

Leucine: COOH—C(H)(H₂N)—**CH₂—CH(CH₃)CH₃**

Isoleucine: COOH—C(H)(H₂N)—**H—C—CH₃, CH₂, CH₃**

Proline: COOH—CH, NH, H₂C, H₂C—CH₂ (ring)

Glycine: COOH—C(H)(H₂N)—**H**

Serine: COOH—C(H)(H₂N)—**CH₂OH**

Threonine: COOH—C(H)(H₂N)—**H—C—OH, CH₃**

Cysteine: COOH—C(H)(H₂N)—**CH₂—SH**

Tyrosine: COOH—C(H)(H₂N)—**CH₂—(benzene ring)—OH**

Tryptophan: COOH—C(H)(H₂N)—**CH₂—C=CH—NH (indole ring)**

Aspartic acid: COOH—C(H)(H₂N)—**CH₂—COOH**

Glutamic acid: COOH—C(H)(H₂N)—**CH₂—CH₂—COOH**

Histidine: COOH—C(H)(H₂N)—**CH₂—C—NH, CH, HC—N (ring)**

Asparagine: COOH—C(H)(H₂N)—**CH₂—O=C—NH₂**

Phenylalanine: COOH—C(H)(H₂N)—**CH₂—(benzene ring)**

Arginine: COOH—C(H)(H₂N)—**CH₂—CH₂—CH₂—NH—C=NH—NH₂**

Lysine: COOH—C(H)(H₂N)—**CH₂—CH₂—CH₂—CH₂—NH₂**

Methionine: COOH—C(H)(H₂N)—**CH₂—CH₂—S—CH₃**

Glutamine: COOH—C(H)(H₂N)—**CH₂—CH₂—O=C—NH₂**

Figure 3-11
The primordial building-block biomolecules, the ABC's of biochemistry. Shown on these two pages are (a) the 20 amino acids from which the proteins of all organisms are built, (b) the five nitrogenous bases and two 5-carbon sugars from which all nucleic acids are built, (c) the major building blocks of lipids, and (d) α-D-glucose, the parent sugar from which most carbohydrates are derived.

We have already seen that the most abundant polysaccharides in nature, starch and cellulose, are constructed of repeating units of D-glucose. Lipids also are constructed from a relatively few kinds of organic building-block molecules. Most lipid molecules contain one or more long-chain fatty acids, of which *palmitic acid* and *oleic acid* (Figure 3-11c) are parent compounds. Many lipids also contain an alcohol, e.g., *glycerol*, and some contain phosphoric acid. Thus only about three dozen different organic compounds, summarized in Figure 3-11, are the parents of most biomolecules.

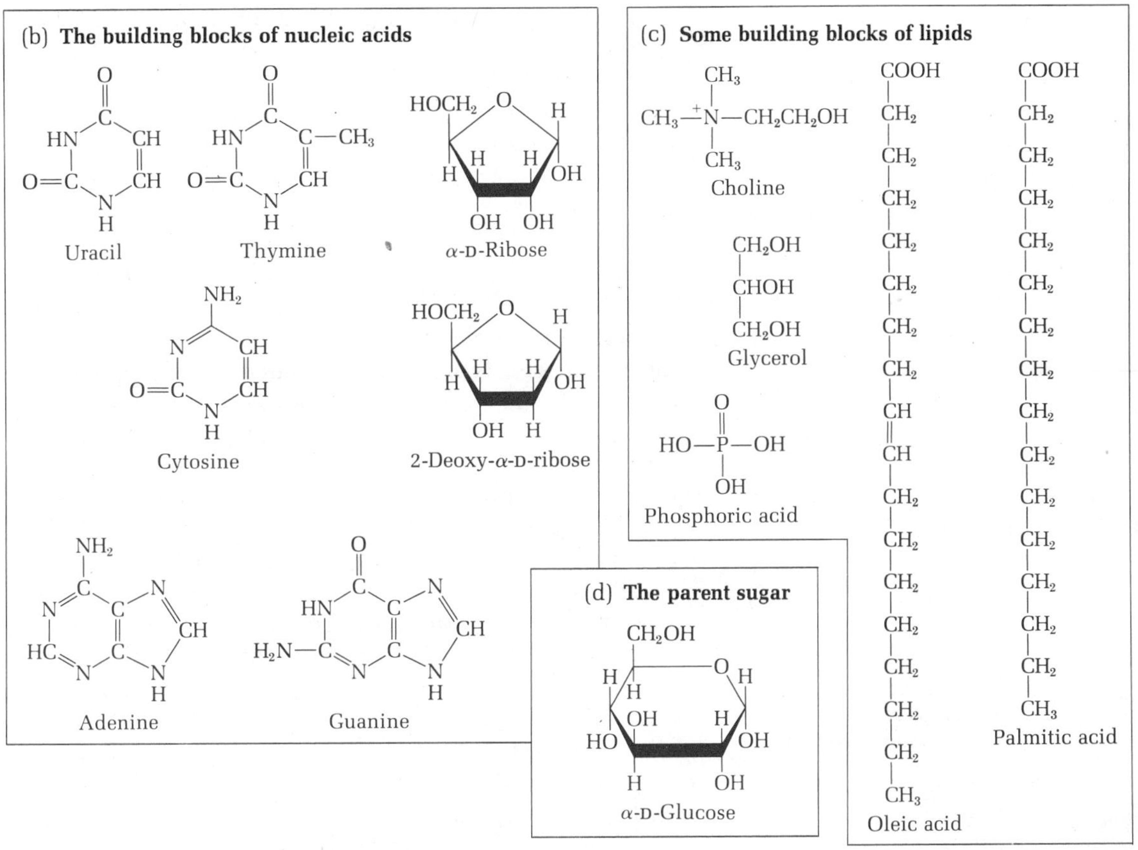

(b) **The building blocks of nucleic acids**

Uracil

Thymine

α-D-Ribose

Cytosine

2-Deoxy-α-D-ribose

Adenine

Guanine

(c) **Some building blocks of lipids**

Choline

Glycerol

Phosphoric acid

Oleic acid

Palmitic acid

(d) **The parent sugar**

α-D-Glucose

Figure 3-12
Each primordial building-block molecule is a precursor of many other kinds of biomolecules.

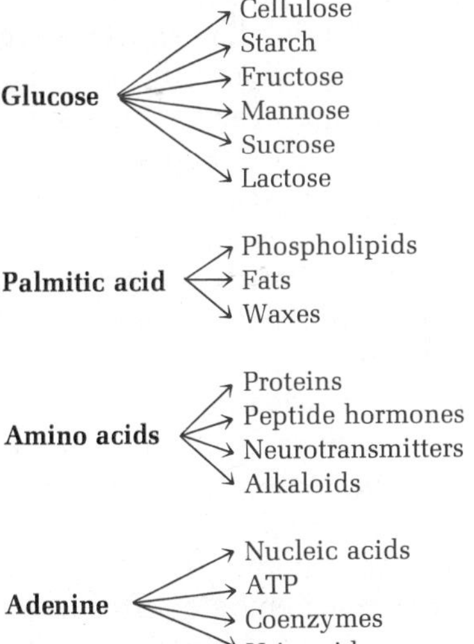

The building-block biomolecules in Figure 3-11 have multiple functions in living organisms. For example, Figure 3-12 shows that D-glucose serves not only as the building-block unit of the storage carbohydrate starch and of the structural carbohydrate cellulose but also is the precursor of other sugars such as D-*fructose*, D-*mannose*, and *sucrose*, or cane sugar. Fatty acids serve not only as components of the complex lipids in the structure of cell membranes but also as components of the fats, energy-rich molecules that serve as fuel storage depots. Fatty acids also are components of the protective waxy coats on the leaves and fruits of plants, and they serve as precursors of other specialized molecules. Amino acids are not only the building blocks of proteins, but some also act as neurotransmitter substances and as precursors of some hormones and of toxic alkaloids in some plants. Adenine serves as a building block in the structure of nucleic acids, of some coenzymes, and of ATP, the cellular energy carrier.

The building-block biomolecules shown in Figure 3-11 are thus the ancestors or parents of most other biomolecules. We may look upon them as the molecular ABC's of living matter.

We should regard this group of simple organic substances with some awe and wonder, since they were selected during the course of evolution and became partners in the extraordinary and unique set of relationships that we have called the molecular logic of living organisms.

There Is a Hierarchy in Cell Structure

The building-block biomolecules we have examined are very small compared with biological macromolecules. For example, an amino acid molecule, such as alanine, is less than 0.7 nm long, whereas a typical protein, such as *hemoglobin,* the red oxygen-carrying protein of red blood cells, consists of some 600 amino acid units strung into long chains and folded into a globular shape. Protein molecules in turn are small compared, for example, with *ribosomes* of animal tissues, which contain some 70 different proteins and 4 nucleic acids. And ribosomes in turn are small compared with organelles such as mitochondria. Thus it is a far jump from the simple biomolecules to the larger macroscopic aspects of cell structure.

Figure 3-13 shows that there is a structural *hierarchy* in cellular organization. *Organelles,* the largest subdivisions of eukaryotic cells, consist of smaller substructures called *supramolecular assemblies,* and these in turn consist of *macromolecules.* For example, the cell nucleus, an organelle, contains several kinds of supramolecular assemblies, such as *membranes, chromatin,* and *ribosomes.* Each such supramolecular assembly consists of macromolecules; e.g., chromatin consists of DNA, RNA, and many different proteins. Each macromolecule is in turn built of small building blocks.

Figure 3-13
The structural hierarchy in the molecular organization of cells.

Cell

Organelles	Nucleus Mitochondria Golgi body Endoplasmic reticulum
Supramolecular assemblies	Membranes Ribosomes Chromatin Microtubules
Macromolecules	Proteins DNA RNA Polysaccharides
Building blocks	Amino acids Glucose Adenine and other bases Palmitic acid, etc.

In proteins and nucleic acids and polysaccharides the individual building-block units are joined to each other by covalent bonds, but in supramolecular cell assemblies, e.g., ribosomes, membranes, or chromatin, the macromolecules are joined together by much weaker interactions. Among these are *hydrogen bonds,* which have a bond energy of only a few kilocalories compared with covalent bonds, which have bond energies of 80 to 100 kcal/mol. The many proteins and the RNA molecules of ribosomes are also held together in a characteristic and specific three-dimensional arrangement by their precisely fitting structures and many individually weak but collectively strong forces of attraction, such as hydrogen bonds (Figure 3-14).

Even though the building-block molecules are very small in relation to the size of cells and their organelles, they can influence the shape and function of these much larger structures. For example, in the human genetic disease *sickle-cell anemia,* the oxygen-carrying hemoglobin molecule of the red blood cells is defective because in its synthesis 2 of the nearly 600 amino acid units of which hemoglobin is composed are the wrong kind. This very small difference in the structure of a tiny portion of the molecule prevents the hemoglobin of affected people from functioning normally: their red blood cells are de-

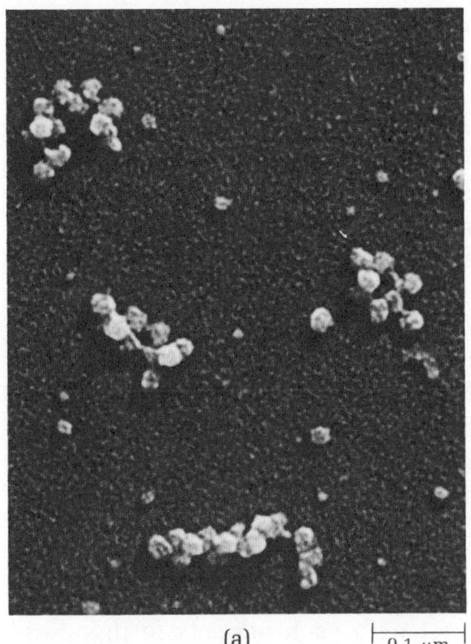

Figure 3-14
Ribosomes. (a) *Electron micrograph of groups of ribosomes of yeast cells.* (b) *Structural organization of a ribosome of* E. coli. *The two subunits of* E. coli *ribosomes are actually irregular in shape, as will be seen in Chapter 29.*

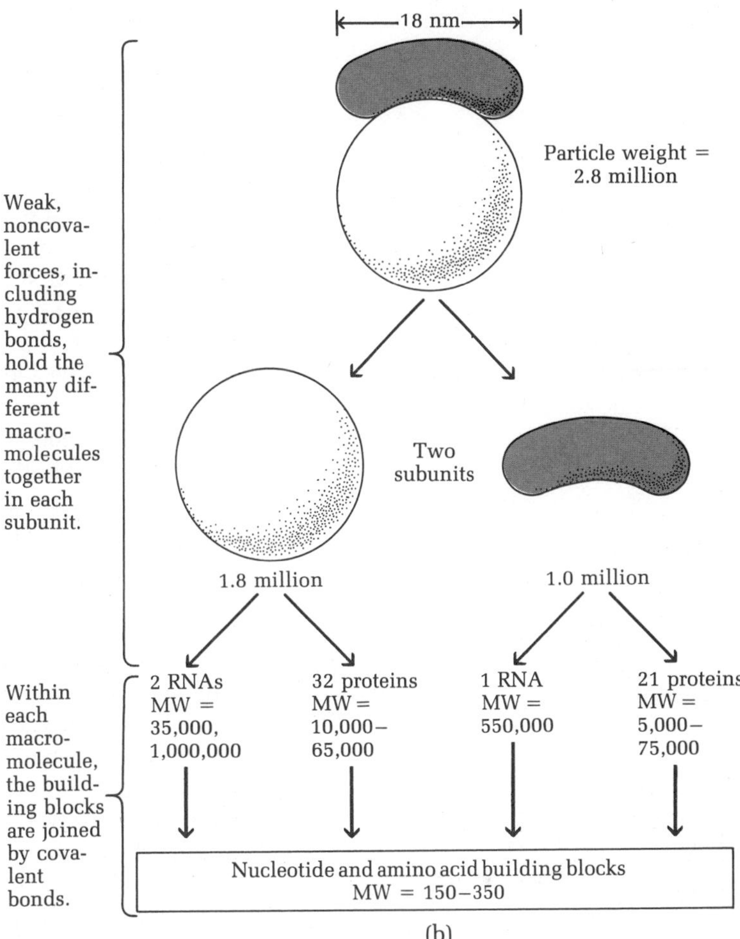

formed and abnormal in function. Thus we see that the size, shape, and biological function not only of macromolecules but also of entire cells may depend on the size and shape of their simple building-block components.

Biomolecules First Arose by Chemical Evolution

Since the macromolecules in all species of living organism are universally made from only a few dozen building-block molecules, it has been suggested that all living organisms may have descended from a single primordial cell line. Thus the first cells to have arisen on earth and survived may have been built from only a few dozen different organic molecules which happened to have, singly and collectively, the most appropriate combination of chemical and physical properties for their function as building blocks of macromolecules and for carrying out the basic energy-transforming and self-replicating features of a living cell. This set of primordial biomolecules may have been retained during biological evolution over billions of years because of their unique "fitness."

But here we come to a dilemma. Apart from their occurrence in living organisms, organic compounds, including the basic biomolecules, occur only in traces in the earth's crust today. How, then, did the first living organisms acquire their characteristic organic building blocks? In the 1920s A. I. Oparin proposed that early in the history of the earth, many different organic compounds were present in its surface waters, possibly in a rather high concentration. From this warm "soup" of organic compounds the first primitive living cells somehow arose, over 3 billion (3000 million) years ago. Oparin suggested that natural chemical and physical processes occurring on the primitive earth could have led to the spontaneous formation of simple organic compounds, such as amino acids and sugars, from components of the primitive atmosphere, which had a composition very different from our air today. According to his theory, the electric energy of lightning discharges or the heat formed during volcanic activity (Figure 3-15) could have activated methane, ammonia, water vapor, and other components of the primitive atmosphere to react with each other to form simple organic compounds. These compounds are believed to have condensed and dissolved in the primitive sea, which over many centuries gradually became enriched with a large variety of simple organic compounds. In this warm solution some organic molecules had a greater tendency than others to associate together into larger complexes and structures. These in turn, very slowly and step by step over millions of years, ultimately assembled spontaneously to form membranes, proteins, and catalysts, which came together to become the precursors of the first primitive cells to survive. For many years Oparin's views remained speculative and appeared untestable.

Figure 3-15
Lightning evoked by a volcanic eruption that resulted in the appearance of the island of Surtsey off the coast of Iceland in 1963. The intense fields of electric, thermal, and shock-wave energy generated by such cataclysms, which were frequent on the primitive earth, could have been a major factor in the origin of organic compounds.

Chemical Evolution Can Be Simulated

Today this concept of the origin of biomolecules is supported by laboratory studies. A classical experiment on the abiotic (nonbiological) origin of organic biomolecules was carried out in 1953 by Stanley Miller. He subjected gaseous mixtures of methane, ammonia, water vapor, and hydrogen to electric sparking across a pair of electrodes, to simulate lightning, for periods of a week or more (Figure 3-16). Then he chilled the contents of the closed vessel to condense its water-soluble components and analyzed the products. He found that the gas phase contained carbon monoxide, carbon dioxide, and nitrogen, which were evidently formed from the gases introduced initially. In the dark-colored condensate he found significant amounts of water-soluble organic substances. Among the compounds Miller identified in this mixture were α-amino acids, including some known to be present in proteins. He also found several of the simple organic acids known to occur in living organisms, such as acetic acid.

Figure 3-16
Spark-discharge apparatus for demonstrating abiotic formation of organic compounds under primitive-atmosphere conditions.

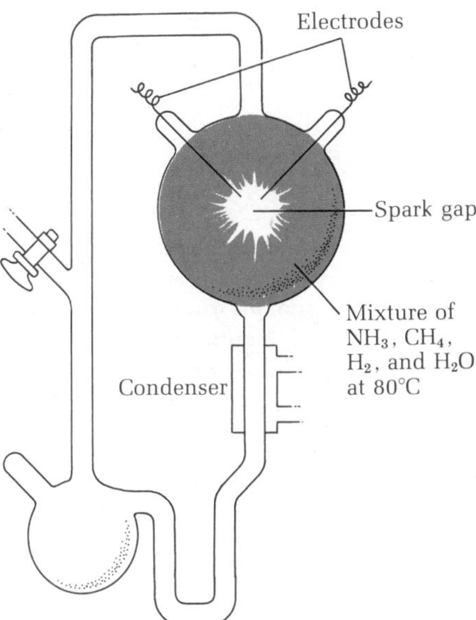

Miller proposed that hydrogen cyanide (HCN), a very reactive substance, was formed from methane and ammonia and then reacted with other components of the gas mixture to yield certain amino acids. Other investigators have since carried out many other such simulation experiments with various mixtures of gases, including also nitrogen, hydrogen, carbon monoxide, and carbon dioxide, again finding that amino acids and other organic biomolecules are readily formed when a source of energy is applied. The many different forms of energy or radiation tested and found to activate the formation of simple organic molecules include heat, visible light, ultraviolet light, x-rays, γ radiation, sparking and silent electric discharges, ultrasonic waves, shock waves, and α- and β-particles. Several hundred different organic compounds are readily formed in such primitive-earth-simulation experiments, including representatives of all the important types of molecules found in cells as well as many not found in cells. Among them are many of the common amino acids present in proteins, the nitrogenous bases that serve as the building blocks of nucleic acids, and many biologically occurring organic acids and sugars. Thus it appears quite possible that the primitive sea became enriched in dissolved organic compounds, which may have included many or all of the basic building-block molecules we recognize in living cells today.

That simple organic molecules can be formed nonbiologically has received important support from the discovery that hundreds of different organic molecules have been detected in interstellar space by spectroscopic methods. Such observations have raised the possibility that life may have arisen in other parts of the universe. The term <u>chemical evolution</u> is used to refer to the origin and development of organic molecules from inorganic precursors in the presence of energy. We now know that the earth was first formed about 4800 million years ago. It is believed that chemical evolution then took place on the earth for at least the first 1000 million years of its life. Then the first living cells arose, perhaps about 3500 million years ago. Then began the process of biological evolution, which still continues.

The oceans today no longer contain rich concentrations of organic compounds; indeed, biomolecules are found only in traces outside of living organisms. What happened to this rich organic soup? It is believed that the earliest living cells eventually used up the organic compounds of the seas, not only as building blocks for their own structures but also as nutrients or fuel, to provide themselves with the energy required for growth. Gradually, through the ages, the organic compounds of the primitive sea were consumed, faster than they were created by natural forces. This idea, and indeed the whole concept of chemical evolution, was considered over 100 years ago by Charles Darwin, as indicated in the following excerpt of a letter he wrote in 1871 to Sir Joseph Hooker:†

> It is often said that all the conditions for the first production of a living organism are now present, which could ever have been present. But if (and oh what a big if) we could conceive in some warm little pond, with all sorts of ammonia and phosphoric acid salts, light, heat, electricity, etc., present, that a protein compound was chemically formed, ready to undergo still more complex changes, at the present day such matter would be instantly devoured or absorbed, which would not have been the case before living creatures were formed.

As organic molecules disappeared from the seas, living organisms began to "learn" how to make their own organic biomolecules. They learned to use the energy of sunlight through photosynthesis to make sugars and other organic molecules from carbon dioxide; they learned to fix atmospheric nitrogen and convert it into nitrogenous biomolecules, such as amino acids. As evolution progressed further, different kinds of organisms ultimately began to interact with each other, to exchange nutrients and energy, thus forming increasingly complex ecological systems.

With these orienting chapters on cells and the interacting biomolecules of which they consist, we are now ready to examine the molecular components of cells in some detail, always with the outlook that they participate in an underlying logic. We shall begin with water, the liquid matrix of all living organisms.

Summary

Most of the solid matter of living organisms consists of organic compounds of carbon covalently bonded to carbon, hydrogen, oxygen, and nitrogen atoms. Carbon appears to have been selected in the course of biological evolution because of a number of desirable properties.

† Quoted in Melvin Calvin, *Chemical Evolution*, Oxford University Press, London, 1969.

Among them is the ability of carbon atoms to form single and double bonds with each other, making possible formation of linear, branched, and cyclic backbone structures in great variety, to which are attached different kinds of functional groups. Organic biomolecules also have characteristic shapes or conformations in three dimensions. Many biomolecules occur in asymmetric or chiral forms called enantiomers.

Most of the organic matter in living cells consists of four major types of macromolecules, i.e., nucleic acids, proteins, polysaccharides, and assemblies of lipid molecules. Each type of macromolecular system is composed of small, covalently linked building-block molecules of relatively few kinds. Proteins consist of chains made from twenty different kinds of amino acids, nucleic acids are chains of four different nucleotide units, and polysaccharides are chains of simple, recurring sugar units. Nucleic acids and proteins are called informational macromolecules because the characteristic sequences of their building-block units reflect the genetic individuality of a species. Polysaccharides, on the other hand, are not informational molecules, since they consist of single repeating units.

There is a structural hierarchy in the molecular organization of cells. Cells contain organelles, such as nuclei and mitochondria, which in turn contain supramolecular structures, such as membranes and ribosomes, and these consist in turn of clusters of macromolecules that associate together, bound by many relatively weak forces of attraction. In macromolecules, however, the building-block units are linked to each other by covalent bonds.

The common building-block biomolecules probably first arose spontaneously from atmospheric gases and water under the influence of energy during the early history of the earth. Such processes, called chemical evolution, can be simulated in the laboratory. The building-block biomolecules of today appear to have been selected during early biological evolution as being the most fit for their biological functions. The building-block biomolecules are relatively few in number, but they are very versatile. Each can perform different functions in cells.

References

BAKER, J. J., and G. E. ALLEN: *Matter, Energy, and Life: An Introduction for Biology Students*, 4th ed., Addison-Wesley, Reading, Mass., 1981.

CALLEWAERT, D. M., and J. GENYEA: *Basic Chemistry: General, Organic, Biological*, Worth, New York, 1980.

CALVIN, M.: *Chemical Evolution*, Oxford University Press, London, 1969.

DICKERSON, R. E., and I. GEIS: *Chemistry, Matter, and the Universe*, Benjamin, Menlo Park, Calif., 1976.

FRIEDEN, E.: "The Chemical Elements of Life," *Sci. Am.*, **227**:52–64, July (1972).

LEHNINGER, A. L.: *Biochemistry*, 2d ed., Worth, New York, 1975. Chapter 37 gives a more detailed treatment of the origin of biomolecules and of cells.

MORRISON, R. T., and R. N. BOYD: *Organic Chemistry*, 3d ed., Allyn and Bacon, Boston, 1973. Excellent textbook of organic chemistry; it includes many biomolecules.

ORGEL, L.: *The Origins of Life: Molecules and Natural Selection*, Wiley, New York, 1973. Paperback; very interesting.

WEINBERG, S.: *The First Three Minutes: A Modern View of the Origin of the Universe*, Basic Books, New York, 1977.

WHITE, E. H.: *Chemical Background for the Biological Sciences*, 2d ed., Prentice-Hall, Englewood Cliffs, N.J., 1970.

Problems

1. *Vitamin C: Is the Synthetic Vitamin as Good as the Natural One?* One of the claims put forth by purveyors of health foods is that vitamins obtained from natural sources are more healthful than those obtained by chemical synthesis. For example, it is claimed that pure L-ascorbic acid (vitamin C) obtained from rose hips is better for you than pure L-ascorbic acid manufactured in a chemical plant. Are the vitamins from the two sources different? Can the body distinguish the source of the vitamin?

2. *Identification of Functional Groups* Tables 3-4 and 3-5 show the common functional groups. Since the properties and biological activities of biomolecules are largely determined by their functional groups, it is important to be able to identify them. In each of the biomolecules below identify and name the constituent functional groups.

(a)
```
     H   H
     |   |
H2N—C—C—OH
     |   |
     H   H
```
Ethanolamine

(b)
```
   H
   |
H—C—OH
   |
H—C—OH
   |
H—C—OH
   |
   H
```
Glycerol

(c)
```
          O
          ‖
      ⁻O—P—O⁻
          |
 H        O
  \       |
   C=C—COO⁻
  /
 H
```
Phosphoenolpyruvate, an intermediate in glucose metabolism

(d)
```
      COOH
       |
  H2N—C—H
       |
   H—C—OH
       |
      CH3
```
Threonine, an amino acid

(e)
```
  ⁻O   O
    \ //
     C
     |
    CH2
     |
    CH2
     |
    NH
     |
     C=O
     |
    HC—OH
     |
CH3—C—CH3
     |
    CH2OH
```
Pantothenate, a vitamin

(f)
```
  H   O
   \ //
    C
    |
 H—C—NH2
    |
HO—C—H
    |
 H—C—OH
    |
 H—C—OH
    |
   CH2OH
```
D-Glucosamine

3. *Drug Activity and Stereochemistry* In some cases the quantitative differences in biological activity between the two enantiomers of a compound are quite large. For example the D isomer of the drug isoproterenol, used to treat mild asthma, is 50 to 80 times more effec-

Problem 3

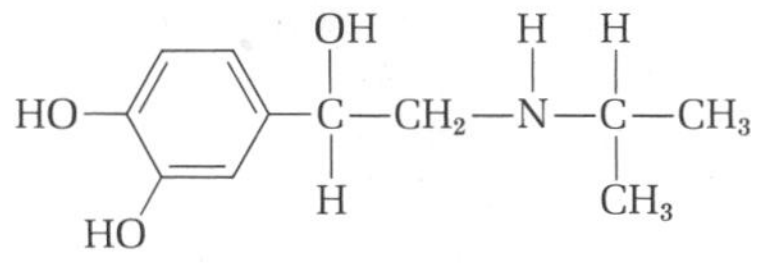

HO, HO, OH, H H, $-C-CH_2-N-C-CH_3$, H, CH_3

Isoproterenol

Problem 4

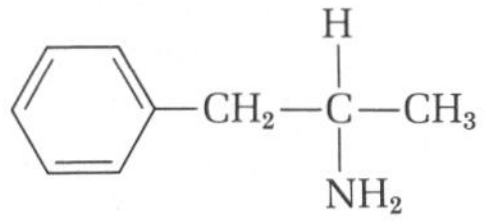

H, $-CH_2-C-CH_3$, NH_2

tive as a bronchodilator than the L isomer. Identify the chiral center in isoproterenol. Why would the two enantiomers have such radically different bioactivity?

4. *Drug Action and Shape of Molecules* Two drug companies market an antidepressant drug under the trade names Dexedrine and Benzedrine. The structure of the drug is shown at the left. The physical properties (C, H, and N analysis, melting point, solubility, etc.) of Dexedrine and Benzedrine are identical. The recommended oral dosage of Dexedrine is 5 mg/d, but the recommended dosage of Benzedrine is significantly higher. Apparently it requires considerably more Benzedrine than Dexedrine to yield the same physiological response. Explain this apparent contradiction.

5. *Building Blocks of Complex Biomolecules* Although the number and complexity of naturally occurring biomolecules is enormous, their structures contain an inherent simplicity because they are constructed from a limited set of building blocks. Figure 3-11 shows the structures of the major building blocks of complex biomolecules. For each of the three important biomolecules below, identify the constituent building blocks.

   (a) Adenosine triphosphate (ATP), an energy-carrying biomolecule

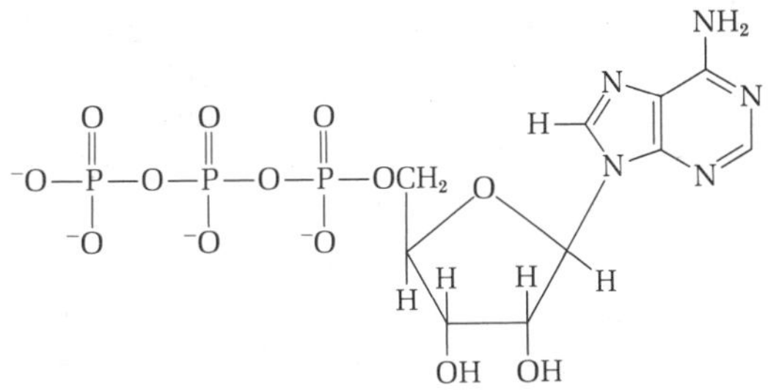

   NH_2, N, N, H, N, N, $^-O-P-O-P-O-P-OCH_2$, O, O, O, ^-O, ^-O, ^-O, O, H, H, H, H, OH, OH

   (b) Phosphatidylcholine, a major membrane component in cells of higher organisms

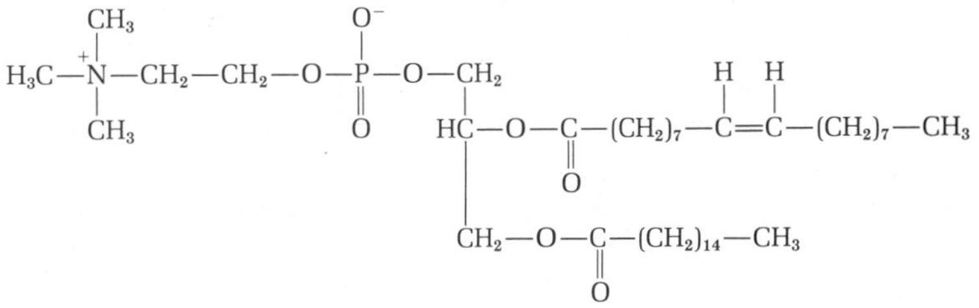

   CH_3, O^-, $H_3C-\overset{+}{N}-CH_2-CH_2-O-P-O-CH_2$, CH_3, O, H H, $HC-O-C-(CH_2)_7-C=C-(CH_2)_7-CH_3$, O, $CH_2-O-C-(CH_2)_{14}-CH_3$, O

   (c) Methionine enkephalin, the brain's own opiate

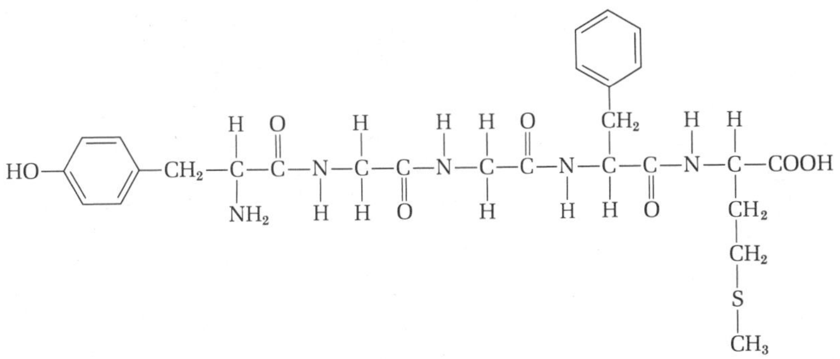

   $HO-C_6H_4-CH_2-C(H)(NH_2)-C(=O)-N(H)-C(H)(H)-C(=O)-N(H)-C(H)(H)-C(=O)-N(H)-C(CH_2C_6H_5)(H)-C(=O)-N(H)-C(H)(CH_2CH_2SCH_3)-COOH$

6. *Determination of the Structure of a Biomolecule* An unknown substance, X, was isolated from rabbit muscle. The structure of X was determined from the following observations and experiments. Quantitative analysis showed that X was composed entirely of C, H, and O. A weighed sample of X was completely oxidized, and the amount of H_2O and CO_2 produced was measured. From this analysis it was concluded that X contains 40.00% C, 6.71% H, and 53.29% O by weight. The molecular weight of X was determined by a mass spectrometer and found to be 90.0. An infrared spectrum of X showed that it contained one double bond. X dissolved readily in water to give an acidic solution. A solution of X was tested in a polarimeter, which demonstrated that X was optically active with a specific rotation $[\alpha]_D = +2.6°$.
   (a) Determine the empirical and molecular formula of X.
   (b) Draw the possible structures for X that fit the molecular formula and contain one double bond. Consider *only* linear or branched structures and disregard cyclic structures. Note that oxygen makes very poor bonds to itself.
   (c) What is the structural significance of the observed optical activity? Which structures in (b) does this observation eliminate? Which structures are consistent with the observation?
   (d) What is the structural significance of the observation that a solution of X was acidic? Which structures in (b) are now eliminated? Which structures are consistent with the observation?
   (e) What is the structure of X? Is more than one structure consistent with all the data?

CHAPTER 4

Water

Water is the most abundant substance in living systems and makes up 70 percent or more of the weight of most forms of life. Moreover, we have seen that the first living organisms probably arose in the primeval oceans; water is thus the great mother of us all. Because it pervades all portions of every cell, water is the medium in which the transport of nutrients, the enzyme-catalyzed reactions of metabolism, and the transfer of chemical energy occur. Therefore all aspects of cell structure and function are necessarily adapted to the physical and chemical properties of water. Indeed, we shall also see how cells have learned to capitalize on its properties to carry out some of their activities.

We often take water for granted as a bland, inert liquid, convenient for many practical purposes. Although it is chemically stable, water is a substance with rather unusual properties. Indeed, water and its ionization products, H^+ and OH^- ions, profoundly influence the properties of many important components of cells, such as enzymes, proteins, nucleic acids, and lipids. For example, the catalytic activity of enzymes is sharply dependent upon the concentration of H^+ and OH^- ions.

Table 4-1 Melting Point, Boiling Point, and Heat of Vaporization of Some Common Liquids

	mp, °C	bp, °C	Heat of vaporization, cal/g†
Water	0	100	**540**
Methanol	− 98	65	**263**
Ethanol	−117	78	**204**
Propanol	−127	97	**164**
Acetone	− 95	56	**125**
Hexane	− 98	69	**101**
Benzene	6	80	**94**
Chloroform	− 63	61	**59**

† The number of calories of heat energy required to convert 1.0 g of a liquid at its boiling point (atmospheric pressure) into its gaseous state at the same temperature.

The Unusual Physical Properties of Water Are Due to Hydrogen Bonding

Water has a higher melting point, boiling point, and heat of vaporization than most common liquids (see Table 4-1). This fact indicates that there are strong forces of attraction between adjacent water molecules, which give liquid water great internal cohesion. For example, the heat of vaporization is a direct measure of the amount of energy required to overcome the attractive forces between adjacent molecules in a liquid so that they can escape from each other and enter the gaseous state.

Why does liquid water show such strong intermolecular attraction? The answer lies in the structure of the water mole-

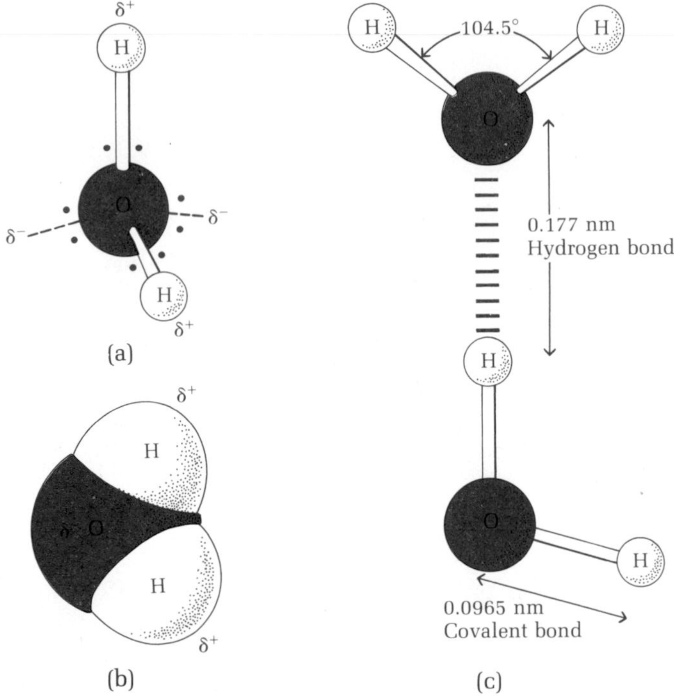

Figure 4-1
The dipolar nature of the H_2O molecule, shown by ball-and-stick (a) and space-filling (b) models. Because of the nearly tetrahedral arrangement of the valence electron pairs around the oxygen atom, the two hydrogen atoms have localized partial positive charges and the oxygen atom has two localized partial negative charges.

(c) Two H_2O molecules joined by a hydrogen bond (designated by colored dashes) between the oxygen atom of the upper molecule and a hydrogen atom of the lower one. Each H_2O molecule can theoretically hydrogen-bond with as many as 4 other H_2O molecules, as in ice (see Figure 4-2).

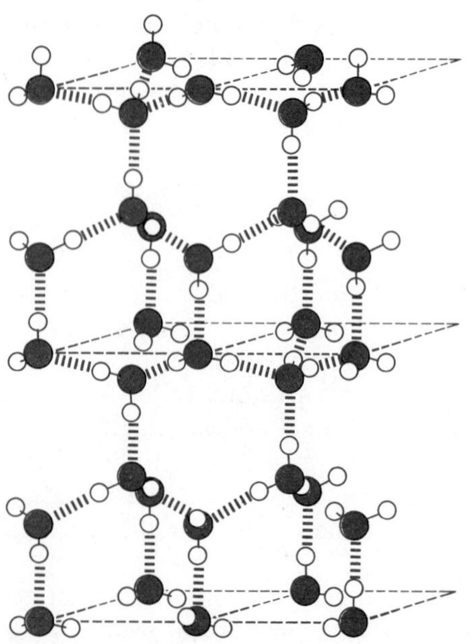

Figure 4-2
Each water molecule is maximally hydrogen-bonded to 4 other water molecules to form a regular crystal lattice. In liquid water at room temperature each water molecule engages in hydrogen-bonding with about 3.4 other water molecules. The crystal lattice of ice is more "open" than the arrangement of H_2O molecules in liquid water, accounting for the fact that ice is less dense than liquid water and therefore floats.

cule. Each of its two hydrogen atoms shares an electron pair with the oxygen atom. The geometry of the shared electron pairs causes the molecule to be V-shaped (Figure 4-1). The two unshared electron pairs of the oxygen atom give it a localized partial negative charge at the apex of the V, and the strong electron-withdrawing tendency of oxygen gives the two hydrogen nuclei partial positive charges. Although the water molecule is electrically neutral, its partial positive and partial negative charges are separated, with the result that the molecule is an electric dipole. Because of this separation of charges two water molecules can be attracted to each other by the electrostatic force between the partial negative charge on the oxygen atom of one water molecule and the partial positive charge on a hydrogen atom of the other (Figure 4-1). This type of electrostatic attraction is called a hydrogen bond.

Because of the nearly tetrahedral arrangement of the electrons about the oxygen atom (Figure 4-1), each water molecule theoretically can form hydrogen bonds with as many as 4 neighboring water molecules. At any given instant in liquid water at room temperature, each water molecule is believed to form hydrogen bonds with an average of 3.4 other water molecules. Since water molecules are in continuous motion in the liquid state these hydrogen bonds are constantly and rapidly being broken and reformed. But in ice each water molecule is fixed in space and hydrogen-bonds with the maximum of 4 other water molecules, to yield a regular lattice structure (Figure 4-2). This accounts for the relatively high melting point of ice (Table 4-1). Water is an example of a polar liquid. In con-

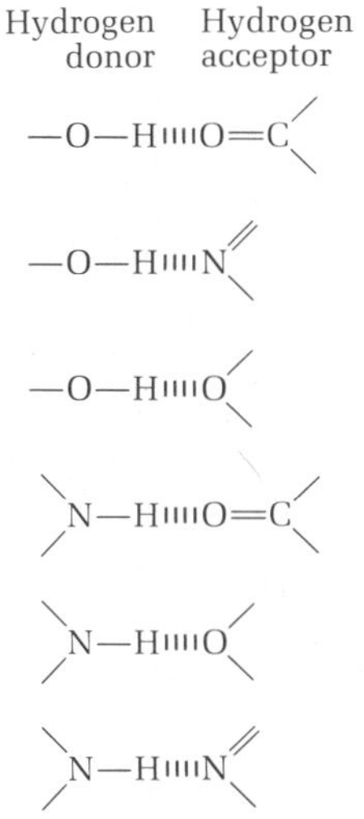

Figure 4-3
Hydrogen bonds. In this type of bond a hydrogen atom is shared unequally between two electronegative atoms. The atom to which the H *is covalently bound is the* hydrogen donor; *the other electronegative atom is the* acceptor. *In biological systems the electronegative atoms participating in hydrogen bonding are oxygen and nitrogen; only rarely can carbon participate in hydrogen bonds. The distance between two hydrogen-bonded atoms varies from* 0.26 *to* 0.31 nm. *The common types of hydrogen bonds are shown below.*

Hydrogen donor — Hydrogen acceptor

—O—H⋯O=C<

—O—H⋯N<

—O—H⋯O<

>N—H⋯O=C<

>N—H⋯O<

>N—H⋯N<

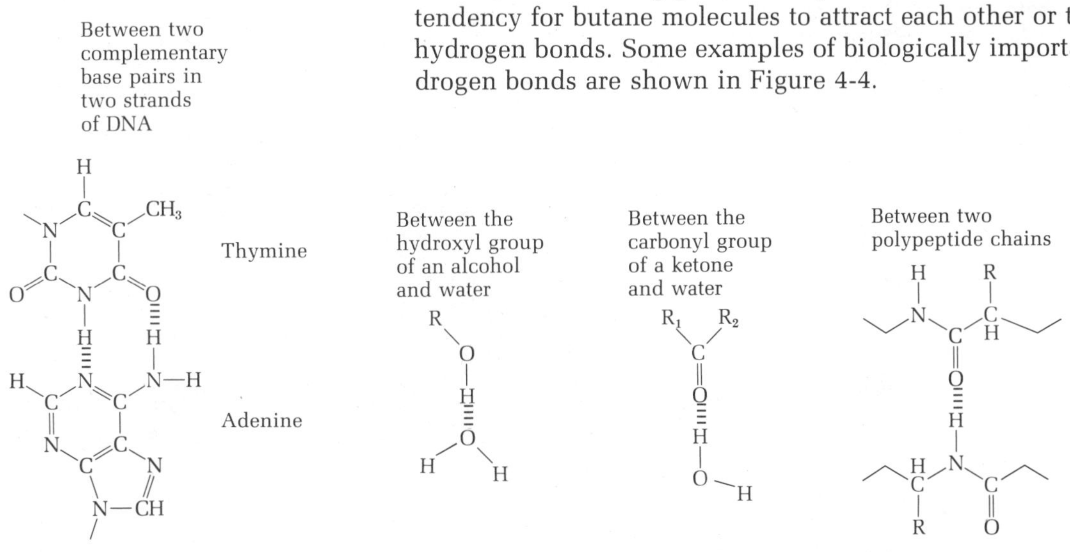

Figure 4-4
Some hydrogen bonds of biological importance.

Between two complementary base pairs in two strands of DNA: Thymine, Adenine

Between the hydroxyl group of an alcohol and water

Between the carbonyl group of a ketone and water

Between two polypeptide chains

trast, the molecules of nonpolar liquids, like benzene or hexane, show relatively little tendency to attract each other electrostatically. Much less energy is required to separate molecules of these liquids; for this reason the heats of vaporization of hexane and benzene are much lower than those for water (Table 4-1).

Hydrogen bonds are weak compared with covalent bonds. The hydrogen bonds in liquid water are estimated to have a bond energy (the energy required to break a bond) of only about 4.5 kcal/mol, compared with 110 kcal/mol for the covalent H—O bonds in water molecules. Nevertheless, because of their large number, hydrogen bonds confer great internal cohesion on liquid water. Although at any given time most of the molecules in liquid water are hydrogen-bonded, the half-life of each hydrogen bond is less than 1×10^{-9} s. Consequently, liquid water is not viscous but very fluid. The apt term *flickering clusters* has been applied to the short-lived groups of hydrogen-bonded molecules in liquid water.

Hydrogen Bonds Are Common in Biological Systems

Hydrogen bonds are not unique to water. They readily form between an electronegative atom (usually oxygen or nitrogen) and a hydrogen atom covalently bonded to another electronegative atom in the same or another molecule (Figure 4-3). Hydrogen atoms bonded to strongly electronegative atoms such as oxygen tend to have strong partial positive charges. However, hydrogen atoms covalently bonded to carbon atoms, which are not electronegative, do not participate in hydrogen bonding. This difference is responsible for the fact that butanol ($CH_3CH_2CH_2CH_2OH$), which has one of its hydrogens bonded to oxygen and thus can form hydrogen bonds with other butanol molecules, has a relatively high boiling point of 117°C. On the other hand, butane ($CH_3CH_2CH_2CH_3$), which cannot form hydrogen bonds because all its hydrogen atoms are bonded to carbon, has a boiling point of only −0.5°C since there is little tendency for butane molecules to attract each other or to form hydrogen bonds. Some examples of biologically important hydrogen bonds are shown in Figure 4-4.

Figure 4-5
Directionality of the hydrogen bond. The upper structure has a stronger hydrogen bond because the attraction between the partial electric charges is greatest in this orientation.

Strong hydrogen bond

Weak hydrogen bond

Hydrogen bonds have another characteristic: they are strongest when the bonded molecules are oriented to allow maximum electrostatic interaction (Figure 4-5). Hydrogen bonds are thus directional and capable of holding two hydrogen-bonded molecules or groups in a specific geometrical arrangement. We shall see later that this property of hydrogen bonds confers very precise three-dimensional structures upon protein and nucleic acid molecules, in which there are many intramolecular hydrogen bonds (Chapters 7, 8, and 27).

Water Has Unusual Solvent Properties

Water is a much better solvent than most common liquids. Most crystalline salts, e.g., sodium chloride, readily dissolve in water but are nearly insoluble in nonpolar liquids like chloroform or benzene. This property is a reflection of the dipolar character of the water molecule. The crystal lattice of a salt is held together by very strong electrostatic attractions between alternating positive and negative ions. When crystalline NaCl is exposed to water, the dipolar water molecules are very strongly attracted to Na^+ and Cl^- ions and pull them away from the lattice to form the hydrated Na^+ and Cl^- ions in solution (Figure 4-6). Water also dissolves many simple organic compounds having carboxyl or amino groups, which tend to ionize by interaction with water.

A second class of substances readily dissolved by water includes many neutral organic compounds with polar functional groups, such as sugars, alcohols, aldehydes, and ketones. Their solubility is due to the propensity of water molecules to form hydrogen bonds with the hydroxyl groups of sugars and alcohols or the carbonyl groups of aldehydes and ketones, as shown in Figure 4-4.

Figure 4-6
Water dissolves many crystalline salts by causing hydration of their component ions.

The crystal lattice of NaCl is held together by electrostatic attraction between Na^+ and Cl^- ions.

Water dissolves the crystal by hydrating the Na^+ and Cl^- ions, thus pulling them away from the lattice.

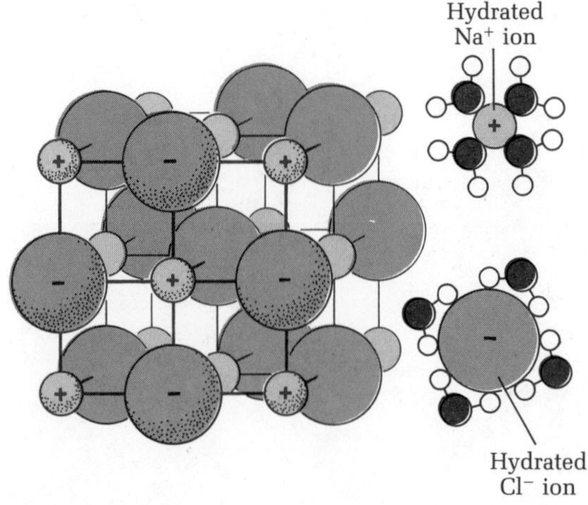

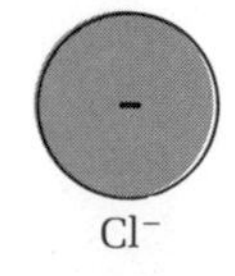

Figure 4-7
Formation of a soap micelle in water. The nonpolar tails of the sodium oleate molecules are inside the micelle, hidden from the water, whereas the negatively charged carboxyl groups are exposed on the micelle surface. The number of Na^+ ions in the water phase equals the number of negative charges on the micelle, so that the solution is electroneutral.

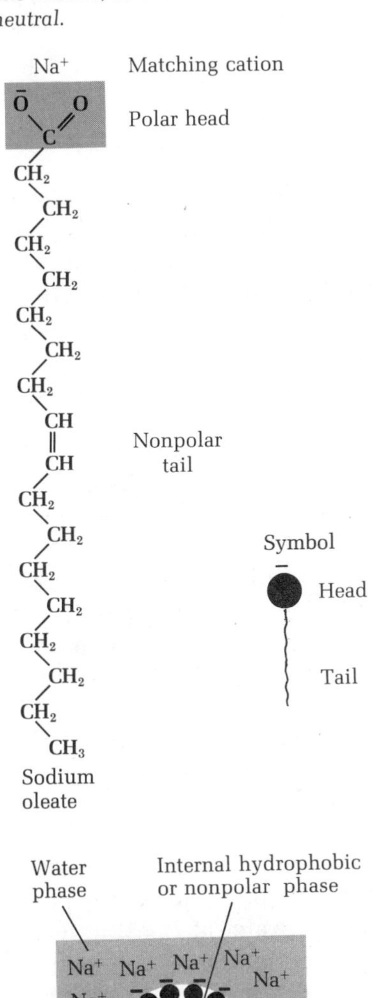

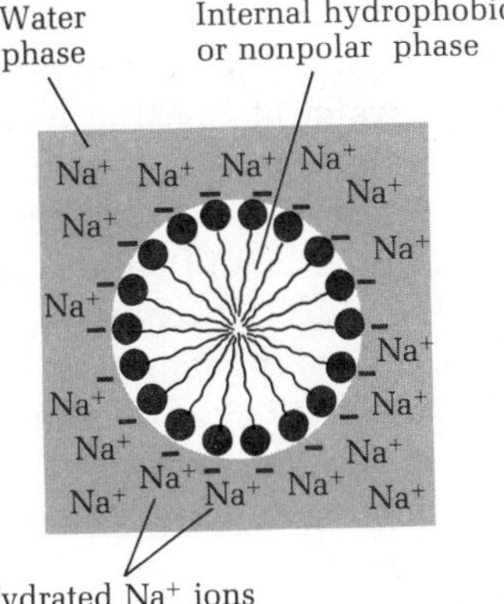

The third class of substances dispersed by water includes those with both hydrophobic (water-hating) and hydrophilic (water-loving) groups, often called amphipathic compounds. A simple example is the sodium salt of the long-chain fatty acid oleic acid. Because its long hydrocarbon chain is hydrophobic and intrinsically insoluble in water, there is very little tendency for sodium oleate (a soap) to *dissolve* in water in the form of a truly molecular solution. However, it readily *disperses* in water to form aggregates called micelles, in which the hydrophilic, negatively charged carboxyl groups of oleate are exposed to, and interact with, the water phase and the hydrophobic, nonpolar hydrocarbon chains are hidden within the structure (Figure 4-7). Micelles may contain hundreds or thousands of soap molecules. Such soap micelles remain evenly suspended in water because they are all negatively charged and thus tend to repel each other. Soapy water tends to be turbid because the micelles are relatively large and scatter light.

The characteristic internal location of the nonpolar groups in such micelles is the result of the tendency of the surrounding water molecules to hydrogen-bond with each other and to associate with the hydrophilic carboxyl groups, thus forcing the hydrocarbon chains into the interior of the micelle, where they have no contact with water. Water "likes" water and carboxyl groups more than it "likes" hydrocarbon chains, which cannot form hydrogen bonds. We use the term hydrophobic interaction to refer to the association of the hydrophobic portions of amphipathic molecules inside such micelles, but it is the tendency of the surrounding water molecules to form hydrogen bonds with each other that provides the driving force for the formation and stability of micelles. The phospholipids (page 317), the proteins (page 179), and the nucleic acids (Chapter 27) are among the many cell components that are amphipathic and tend to form structures in which the nonpolar, hydrophobic parts are hidden from water. Moreover, as we shall see later (Chapter 12), micellar arrangements of amphipathic lipid molecules form the "core" of biological membranes.

Solutes Change the Properties of Water

There are four particular properties of liquid water, designated colligative properties, which are profoundly modified by dissolved solutes, namely, (1) the freezing point, (2) the boiling point, (3) the vapor pressure, and (4) the osmotic pressure. The term "colligative" means "bound together" and refers to the fact that these four properties have a common basis and are all modified predictably by dissolved solutes.

A solution of 1.00 gram molecular weight (1.00 mol) of any ideal nonvolatile solute in 1000 g of water is 1.00 molal (*m*). In such a solution at a pressure of 760 atm the presence of the solute depresses the freezing point of the water, normally 0°C, to −1.86°C, elevates its boiling point, normally 100°C, to

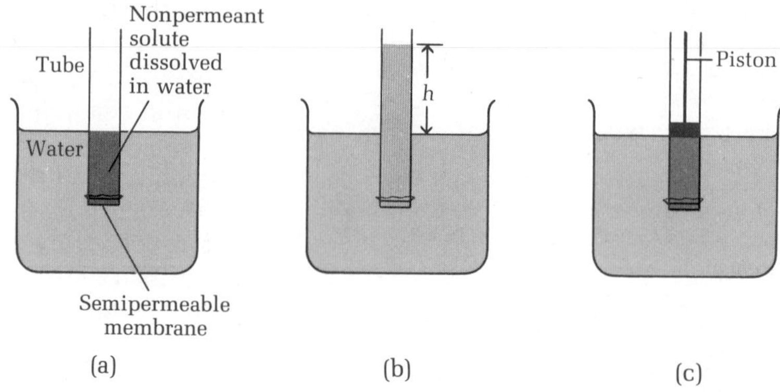

Figure 4-8
Osmosis and osmotic pressure.
(a) Initial state. *Water will flow from the outside compartment into the solution inside the membrane, as it seeks to equalize its concentration across the membrane.*
(b) Final state. *Water has moved into the solution of the nonpermeant compound, causing it to become diluted. At equilibrium, the height h of the column just counterbalances the osmotic pressure, the tendency of water to flow toward a zone where its concentration is less.*
(c) *Osmotic pressure is the force that must be applied to exactly oppose osmotic flow. It is equal to the hydrostatic head h.*

100.543°C, and yields an osmotic pressure of 22.4 atm in an appropriate apparatus (Figure 4-8). An ideal solute is one that neither dissociates into two or more components nor associates to reduce the total number of solute particles. The colligative properties depend only on the *number* of solute molecules per unit volume of solvent and are independent of their chemical structure. This is because 1.00 mol of any nonionizing compound contains 6.02×10^{23} molecules (Avogadro's number). Thus 1.00 *m* aqueous solutions of glycerol (MW 92) or glucose (MW 180) can be expected to have the same freezing point (−1.86°C), boiling point (100.543°C), or osmotic pressure (22.4 atm), because both contain the same number of molecules per liter of water. But a 0.100 *m* solution of glucose would have a freezing-point depression only one-tenth as great; it would freeze at −0.186°C because it has only one-tenth the number of molecules per liter as a 1.0 *m* solution. A 0.100 *m* solution of NaCl, which is completely dissociated into Na^+ and Cl^- ions, would have a freezing point of −0.372°C since it contains twice as many solute particles per liter as a 0.100 *m* solution of glucose. The colligative laws and these constants hold accurately only in dilute aqueous solutions.

These effects of solutes on the properties of water have considerable biological importance. For one thing, they permit a freshwater fish to remain active in water at freezing temperature because the total solute concentration in the blood of the fish is high enough to depress the freezing point of the blood below that of water. Moreover, the presence of solutes in the blood that are incapable of passing through capillary membranes, particularly the proteins, gives the blood a higher osmotic pressure than the extracellular fluid. As a consequence, water tends to diffuse into the blood capillaries from the extracellular fluid compartment, thus keeping the vascular system "full" and preventing it from collapse.

Another way in which dissolved solutes influence the properties of water is by disturbing the hydrogen-bonding tendency between water molecules. The presence of an ionic solute such as NaCl causes a distinct change in the structure of liquid water, since each Na^+ and Cl^- ion is surrounded by a hydration shell of water dipoles. The hydrated Na^+ and Cl^- ions have a geometry and properties somewhat different from the clusters of hydrogen-bonded water molecules; they are more highly ordered and more regular in structure. Dissolved salts thus tend to "break" the normal structure of liquid water and change its solvent properties. We shall see later that the solubility of proteins is profoundly decreased by high concentrations of neutral salts, such as NaCl, Na_2SO_4, and $(NH_4)_2SO_4$, which change the properties of water and reduce its effectiveness as a solvent for proteins. This effect of dissolved neutral salts can be used to separate different proteins from each other, since they vary in their tendency to precipitate from salt solutions.

The Equilibrium Point of Reversible Reactions Is Expressed by an Equilibrium Constant

Water molecules have a slight tendency to undergo reversible ionization to yield a hydrogen ion (H^+) and a hydroxide ion (OH^-), so that we have the equilibrium

$$H_2O \rightleftharpoons H^+ + OH^- \tag{1}$$

Because the reversible ionization of water is very important in its properties and in its effects on cell function, we must have a means of expressing the *extent* of ionization of water in quantitative terms. To this end we shall digress briefly to review some properties of reversible chemical reactions.

The position of equilibrium of any given chemical reaction is given by the *equilibrium constant*. For the generalized reaction

$$A + B \rightleftharpoons C + D \tag{2}$$

an expression for the equilibrium constant is easily derived by making use of the *law of mass action*. It states that a chemical reaction like (2) will proceed to the right to a new equilibrium position if we increase the concentration of either A or B or both. Conversely, the reaction proceeds to the left to a new equilibrium point on an increase in the concentration of C and/or D. This principle can be stated in more quantitative terms. The rate of the forward reaction v_F, proceeding from left to right, is proportional to the product of the active concentrations of the reactants A and B

$$v_F = k_F [A] [B]$$

where k_F is a proportionality constant and the square brackets indicate molar concentration. The velocity v_R of the reverse reaction, from right to left, is similarly

$$v_R = k_R [C] [D]$$

Since equilibrium is defined as the condition in which the rates of the forward and reverse reactions are equal, at equilibrium we have the equality

$$v_F = v_R$$

and therefore

$$k_F\,[\mathrm{A}][\mathrm{B}] = k_R\,[\mathrm{C}][\mathrm{D}]$$

Rearranging gives

$$\frac{k_F}{k_R} = \frac{[\mathrm{C}][\mathrm{D}]}{[\mathrm{A}][\mathrm{B}]}$$

The ratio of the two constants k_F/k_R can be replaced by a single new constant K'_{eq}, the *equilibrium constant*

$$K'_{\mathrm{eq}} = \frac{[\mathrm{C}][\mathrm{D}]}{[\mathrm{A}][\mathrm{B}]}$$

(The prime sign in K'_{eq} indicates that the equilibrium constant is based on the measured molar concentrations of the reactants and products.)

The equilibrium constant is fixed and characteristic for any given chemical reaction at a specified temperature. It defines the composition of the final equilibrium mixture of any given reaction, regardless of the starting amounts of reactants and products. Conversely, we can calculate the equilibrium constant for a given reaction at a given temperature if we know the concentrations of all its reactants and products at the point of equilibrium.

The Ionization of Water Is Expressed by an Equilibrium Constant

Now let us return from our digression to consider the ionization of water in more quantitative terms. This reversible process results in the formation of hydrogen and hydroxide ions. However, when we use the term "hydrogen ion" and the symbol H^+ it must be understood that "bare" hydrogen ions, i.e., protons, do not exist in water; hydrogen ions, like most other ions, are always hydrated. The hydrated form of the H^+ ion is called the *hydronium ion*. This is often designated H_3O^+, but actually each H^+ is closely surrounded by several H_2O molecules, the number depending on the temperature.

The ionization of water according to the equation

$$H_2O \rightleftharpoons H^+ + OH^- \quad (3)$$

proceeds to only a very slight extent; at 25°C only about 1 out of every 10 million molecules in pure water is ionized at any instant. Although water has only a very slight tendency to ionize, the products H^+ and OH^- have very profound biological effects. For this reason we must be able to express the extent of ionization of water quantitatively.

We can do this by writing the expression for the equilibrium constant for the reversible reaction (3)

$$K'_{eq} = \frac{[H^+][OH^-]}{[H_2O]}$$

We can now simplify this expression since the concentration of H_2O is relatively very high (it is equal to the number of grams of H_2O in 1 L divided by the gram molecular weight, or $1000/18 = 55.5\ M$) and thus is essentially constant in relation to the very low concentrations of H^+ and OH^- ions in pure water at 25°C, namely, $1 \times 10^{-7}\ M$. Accordingly we can substitute 55.5 in the equilibrium constant expression to yield

$$K'_{eq} = \frac{[H^+][OH^-]}{55.5}$$

which on rearranging becomes

$$55.5K'_{eq} = [H^+][OH^-]$$

The value for K'_{eq} has been carefully estimated from electrical-conductivity measurements of water (only the ions arising from the dissociation of H_2O can carry current in pure water) and found to be 1.8×10^{-16} at 25°C. Substituting this value for K'_{eq} in the above equation gives

$$(55.5)(1.8 \times 10^{-16}) = [H^+][OH^-]$$

$$99.9 \times 10^{-16} = [H^+][OH^-]$$

$$1.0 \times 10^{-14} = [H^+][OH^-]$$

The symbol K_w is used to designate the product $55.5K'_{eq}$, and we have the relationship

$$K_w = 1.0 \times 10^{-14} = [H^+][OH^-]$$

K_w, called the *ion product* of water, has the value 1.0×10^{-14} at 25°C. What this means is that the product $[H^+][OH^-]$ in aqueous solutions at 25°C always equals the fixed number 1×10^{-14}. When there are exactly equal concentrations of both H^+ and OH^-, as in pure water, the solution is said to be *neutral*. Under these conditions the concentration of H^+ and OH^- can be calculated from the ion product of water as follows:

$$K_w = [H^+][OH^-] = [H^+]^2$$

Solving for H^+ gives

$$[H^+] = \sqrt{K_w} = \sqrt{1 \times 10^{-14}}$$

$$[H^+] = [OH^-] = 10^{-7}\,M$$

Furthermore, the ion product of water says that whenever the concentration of H^+ ions is greater than $1 \times 10^{-7}\,M$, the concentration of OH^- must become less than $1 \times 10^{-7}\,M$, and vice versa. Thus, when the concentration of H^+ is very high, as in a solution of hydrochloric acid, the OH^- concentration must be very low, since the product of their concentrations must be 1×10^{-14}. Conversely, when the concentration of OH^- is very high, as in a solution of sodium hydroxide, the concentration of H^+ must be very low. Thus, from the ion product of water we can calculate the H^+ concentration if we know the OH^- concentration, or vice versa (Box 4-1).

The pH Scale Designates the H^+ and OH^- Concentrations

K_w, the ion product of water, is the basis for the *pH scale* (Table 4-2), a convenient means of designating the actual concentration of H^+ (and thus of OH^-) in any aqueous solution in the range of acidity between 1.0 *M* H^+ and 1.0 *M* OH^-. The term pH is defined by the expression

$$\text{pH} = \log \frac{1}{[H^+]} = -\log [H^+]$$

Table 4-2 The pH Scale

$[H^+]$, *M*	pH	$[OH^-]$, *M*	pOH
1.0	0	10^{-14}	14
0.1	1	10^{-13}	13
0.01	2	10^{-12}	12
0.001	3	10^{-11}	11
10^{-4}	4	10^{-10}	10
10^{-5}	5	10^{-9}	9
10^{-6}	6	10^{-8}	8
10^{-7}	7	10^{-7}	7
10^{-8}	8	10^{-6}	6
10^{-9}	9	10^{-5}	5
10^{-10}	10	10^{-4}	4
10^{-11}	11	0.001	3
10^{-12}	12	0.01	2
10^{-13}	13	0.1	1
10^{-14}	14	1.0	0

In a precisely neutral solution at 25°C, where the hydrogen-ion concentration is $1.0 \times 10^{-7}\,M$, the pH would be given by

$$\text{pH} = \log \frac{1}{1 \times 10^{-7}} = \log (1 \times 10^{7}) = \log 1.0 + \log 10^{7}$$

$$= 0 + 7$$

$$\text{pH} = 7$$

The value of 7.0 for the pH of a precisely neutral solution is not an arbitrarily chosen figure; it is derived from the absolute value of the ion product of water at 25°C. Solutions having a pH greater than 7 are alkaline because the concentration of OH^- is greater than that of H^+. Conversely, solutions having a pH less than 7 are acid (Table 4-2).

It is especially important to note that the pH scale is logarithmic, not arithmetic. To say that two solutions differ in pH by 1 pH unit means that one solution has 10 times the H^+ concentration of the other but does not tell us the absolute magnitude of the difference. Figure 4-9 gives the pH of some common aqueous fluids. Note that a cola drink (pH 3.0) or red wine (pH 3.7) has an H^+ concentration approximately 10,000 times greater than that of blood.

Box 4-1 The ion product of water

The ion product of water makes it possible to calculate the concentration of H^+, given the concentration of OH^- and vice versa, as the following problems show.

1. What is the concentration of H^+ in a solution of 0.1 *N* NaOH?

$$K_w = [H^+][OH^-]$$

Solving for $[H^+]$ gives

$$[H^+] = \frac{K_w}{[OH^-]} = \frac{1 \times 10^{-14}}{0.1}$$
$$= \frac{10^{-14}}{10^{-1}} = 10^{-13}\ M \qquad ans.$$

2. What is the concentration of OH^- in a solution in which H^+ concentration is 0.00013 *M*?

$$K_w = [H^+][OH^-]$$

Solving for $[OH^-]$ gives

$$[OH^-] = \frac{K_w}{[H^+]} = \frac{1 \times 10^{-14}}{0.00013}$$
$$= \frac{1 \times 10^{-14}}{1.3 \times 10^{-4}}$$
$$= 7.7 \times 10^{-11}\ M \qquad ans.$$

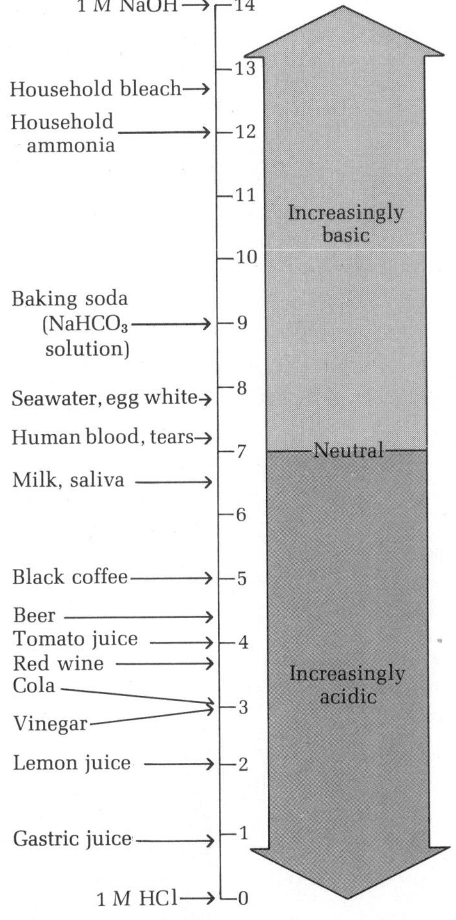

Figure 4-9
pH of some fluids.

Sometimes the expression pOH is used to denote the basicity, or OH^- concentration, of a solution; pOH is defined by the expression

$$\text{pOH} = \log \frac{1}{[OH^-]} = -\log\,[OH^-]$$

which is analogous to the expression for pH. Thus the pOH of 0.1 *M* OH^- is 1 and the pOH of a 10^{-7} *M* OH^- solution is 7.0. It is useful to remember that pH and pOH are related to each other in a very simple way:

$$\text{pH} + \text{pOH} = 14$$

Table 4-2 shows the inverse relationships between pH and pOH.

The pH of an aqueous solution can be approximated by using various indicator dyes, including litmus, phenolphthalein, and phenol red, but accurate measurements of pH in the chemical or clinical laboratory are made with a special glass electrode that is selectively sensitive to H^+ concentration but insensitive to Na^+, K^+, and other cations. In an instrument called a *pH meter* the signal from this electrode is amplified and compared with the signal generated by a solution having an accurately known pH.

Measurement of pH is one of the most important and frequently used procedures in biochemistry since the pH determines many important features of the structure and activity of biological macromolecules, such as the catalytic activity of enzymes. Moreover, measurements of the pH of the blood and urine are commonly used in diagnosing disease. For example, the pH of the blood plasma of severely diabetic persons is often lower than the normal value of 7.4; this condition is called *acidosis*. In contrast, in certain other disease states the pH of the blood is higher than normal, the condition of *alkalosis*.

Acids and Bases Reflect the Properties of Water

Hydrochloric, sulfuric, and nitric acids, commonly called *strong acids*, are completely ionized in dilute aqueous solutions; similarly, the *strong bases* NaOH and KOH are also completely ionized.

In biology we are more concerned with the behavior of *weak acids* and *weak bases*, which are not completely ionized when dissolved in water. An example of a weak acid is *acetic acid* (CH_3COOH), which gives vinegar its acid taste; an example of a weak base is *ammonia* (NH_3), used as a household cleaner. Weak acids and bases are common in biological systems and play important roles in metabolism and its regulation. The behavior of aqueous solutions of weak acids and bases is best understood if we first define some terms.

First, acids are defined as *proton donors* and bases as *proton acceptors*. A proton donor and its corresponding proton acceptor make up a *conjugate acid-base pair* (Table 4-3). For example, acetic acid (CH_3COOH), a proton donor, and the acetate anion (CH_3COO^-), the corresponding proton acceptor, constitute a conjugate acid-base pair, related to each other by the reversible reaction

Table 4-3 Some Conjugate Acid-Base Pairs
Each pair consists of a proton donor and a proton acceptor

Proton donor	Proton acceptor
CH_3COOH	CH_3COO^-
$H_2PO_4^-$	HPO_4^{2-}
NH_4^+	NH_3

$$\underset{\text{Proton donor, conjugate acid}}{CH_3COOH} \rightleftharpoons \underset{\text{Proton}}{H^+} + \underset{\text{Proton acceptor, conjugate base}}{CH_3COO^-}$$

Each acid has a characteristic tendency to lose its proton in an aqueous solution. The stronger the acid the greater its tendency to lose its proton. The tendency of any acid HA to lose a proton and form its conjugate base A^- is defined by the equilibrium constant K' for the reversible reaction

$$HA \rightleftharpoons H^+ + A^-$$

which is

$$K' = \frac{[H^+][A^-]}{[HA]}$$

Equilibrium constants for ionization reactions like these are more usually called *ionization* or *dissociation constants*. The dissociation constants, often designated K_a' (*a* for acid), of some acids are given in Table 4-4. Note that acids differ in their tendency to lose a proton. The stronger acids in Table 4-4, such as formic and lactic acids, have higher dissociation constants, whereas the weaker acids, such as the ion $H_2PO_4^-$, have lower dissociation constants. Among the weakest acids in Table 4-4 is the NH_4^+ ion, which has only a very slight tendency to

Table 4-4 Dissociation Constant and pK′ of Some Common Acids at 25°C
Formic acid is present in ant sting, acetic acid gives vinegar its acidity, propionic acid is formed in the rumen of cattle, and lactic acid is a metabolic breakdown product of glucose in working skeletal muscle

Acid (proton donor)	K′, M	pK′
HCOOH (formic acid)	1.78×10^{-4}	3.75
CH_3COOH (acetic acid)	1.74×10^{-5}	4.76
CH_3CH_2COOH (propionic acid)	1.35×10^{-5}	4.87
$CH_3CHOHCOOH$ (lactic acid)	1.38×10^{-4}	3.86
H_3PO_4 (phosphoric acid)	7.25×10^{-3}	2.14
$H_2PO_4^-$ (dihydrogen phosphate ion)	1.38×10^{-7}	6.86
HPO_4^{2-} (monohydrogen phosphate ion)	3.98×10^{-13}	12.4
H_2CO_3 (carbonic acid)	1.70×10^{-4}	3.77
HCO_3^- (bicarbonate ion)	6.31×10^{-11}	10.2
NH_4^+ (ammonium ion)	5.62×10^{-10}	9.25

donate a proton, as shown by its very low dissociation constant. Its conjugate base is ammonia (NH_3), which attracts protons very strongly.

Table 4-4 also gives values for the expression pK′, which is defined by the equation

$$pK' = \log \frac{1}{K'} = -\log K'$$

The symbol p denotes "negative logarithm of" in both pH and pK′. The more strongly dissociated the acid the lower its pK′. As we shall now see, the pK′ of any weak acid can be determined quite easily.

Weak Acids Have Characteristic Titration Curves

Titration is used to determine the amount of an acid in a given solution. In this procedure a measured volume of the acid is titrated with a solution of a base, usually sodium hydroxide (NaOH), the concentration of which is accurately known. The NaOH is added in small increments until the acid is exactly neutralized, as determined with an indicator or a pH meter. From the volume of NaOH added and its concentration, the concentration of the acid in the solution being titrated can be calculated.

From such a titration of a weak acid we can gain much additional information if we carefully measure the pH of the acid being titrated after each increment of NaOH has been added, until the point of neutralization has been reached. A plot of the pH of the solution vs. the amount of NaOH added to this point is called a *titration curve*. The titration curve of acetic acid, a typical weak acid, is shown in Figure 4-10. Let us trace the course of the titration of a 0.1 *M* solution of acetic acid with

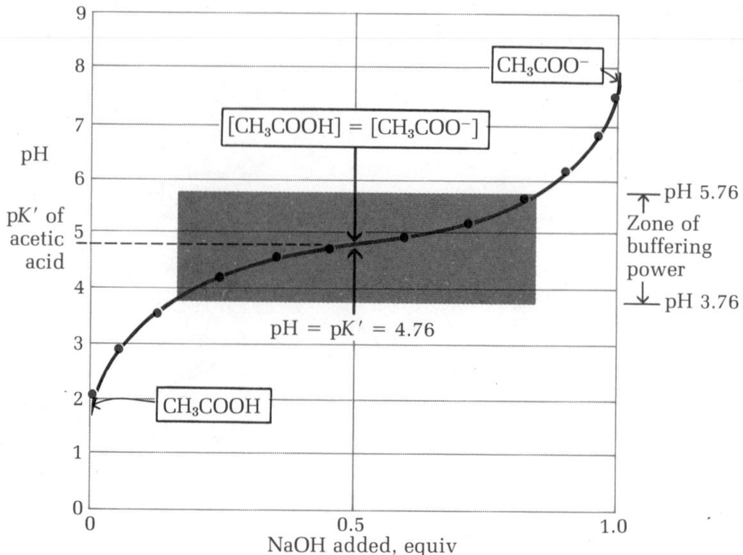

Figure 4-10
The titration curve of acetic acid (see text for details). After each addition of standard NaOH to the solution of acetic acid the pH of the mixture is measured. This value is plotted against the fraction of the total amount of NaOH required to neutralize the acetic acid, i.e., bring it to pH ≈ 7. The points so obtained yield the titration curve. Shown in the boxes are the predominant ionic forms of acetic acid at the points designated. At the midpoint of the titration the concentrations of the proton donor and proton acceptor are equal. The pH at this point is numerically equal to the pK′ of acetic acid. The shaded zone is the region of buffering power.

0.1 *M* NaOH at 25°C, keeping in mind that two reversible equilibria are involved in the process

$$H_2O \rightleftharpoons H^+ + OH^- \tag{4}$$

$$HAc \rightleftharpoons H^+ + Ac^- \tag{5}$$

which must conform to their characteristic equilibrium constants, which are, respectively,

$$K_w = [H^+][OH^-] = 1 \times 10^{-14} \tag{6}$$

$$K' = \frac{[H^+][Ac^-]}{[HAc]} = 1.74 \times 10^{-5}\,M \tag{7}$$

At the beginning of the titration, before any NaOH is added, the acetic acid is already slightly ionized, to an extent that can be calculated from its dissociation constant (7). Try this calculation. To make it simple, assume that the ionization of acetic acid is so small that the concentration of undissociated acetic acid is not significantly less than its total concentration of 0.1 *M*.

When we add successive increments of NaOH, the added OH^- will combine with the free H^+ in the solution to form H_2O, to an extent that will satisfy the equilibrium relationship $K_w = [H^+][OH^-] = 1 \times 10^{-14}$. But as soon as free H^+ is removed in this way, some of the undissociated HAc immediately dissociates further to satisfy its own equilibrium constant (7). The net result as we proceed with the titration is that the HAc undergoes increased ionization with each increment of NaOH added. Consequently [HAc] decreases and $[Ac^-]$ increases, as equilibria (4) and (5) satisfy their equilibrium constants (6) and (7) at each step. At the midpoint of the titration (Figure 4-10), at which exactly 0.5 equiv of NaOH has been added, one-half of the original acetic acid has undergone dissociation, so that the proton-donor concentration [HAc] now equals the proton-acceptor concentration [Ac^-]. At this midpoint a very important relationship holds: the pH of the equimolar solution of acetic acid and acetate, namely, pH 4.76, is exactly equal to the pK' of acetic acid, as can be seen on comparison of the pK' values in Table 4-4 and the titration curve in Figure 4-10. We shall soon see the basis for this important relationship, which holds for all weak acids.

As we continue the titration by adding further increments of NaOH, the remaining undissociated acetic acid is gradually converted into acetate (CH_3COO^-) as H^+ is removed by reaction with the added OH^- to form H_2O. Finally, we shall reach the endpoint of the titration, at about pH 7.0, at which all the acetic acid will have lost its protons to OH^-, to form H_2O and acetate. Throughout the titration the two equilibria (4) and (5) will coexist, each always conforming to its equilibrium constant. These two ionization reactions are reversible and, being ionic, are established virtually instantaneously. We can therefore reverse the titration procedure. Starting at the neutralization point, we can add H^+ back to the system and titrate the acetate back to the initial state. We shall get exactly the same curve as in Figure 4-10, after adjusting for volume changes as the titration proceeds. During this back-titration the added H^+ will react with Ac^- to make HAc, and the ratio [Ac^-]/[HAc] will thus decline as more H^+ is added, until it approaches the initial state, where we began the titration with NaOH.

Figure 4-11 compares the titration curves of three weak acids with greatly different dissociation constants, namely, acetic acid (pK' = 4.76); dihydrogen phosphate, or $H_2PO_4^-$ (pK' = 6.86); and ammonium ion, or NH_4^+ (pK' = 9.25). Although the titration curves of these acids have the same shape, they are displaced along the pH axis, simply because these acids differ in their strength. Acetic acid is the strongest and loses its proton to OH^- most readily, since its K' is highest (pK' lowest) of the three. Acetic acid is already half-dissociated at pH 4.76. $H_2PO_4^-$ loses a proton to OH^- less readily than acetic acid, being half-dissociated at pH 6.86. NH_4^+ is the weakest acid of the three and does not become half-dissociated until pH 9.25.

Now we come to a most important point about the titration curve of a weak acid: it shows graphically that a weak acid and its anion can act as a buffer.

Buffers Are Mixtures of Weak Acids and Their Conjugate Bases

Buffers are aqueous systems that tend to resist changes in their pH when small amounts of acid (H^+) or base (OH^-) are added. A buffer system consists of a weak acid (the proton donor) and its conjugate base (the proton acceptor). As an example, a mixture of equal concentrations of acetic acid and acetate ion, such

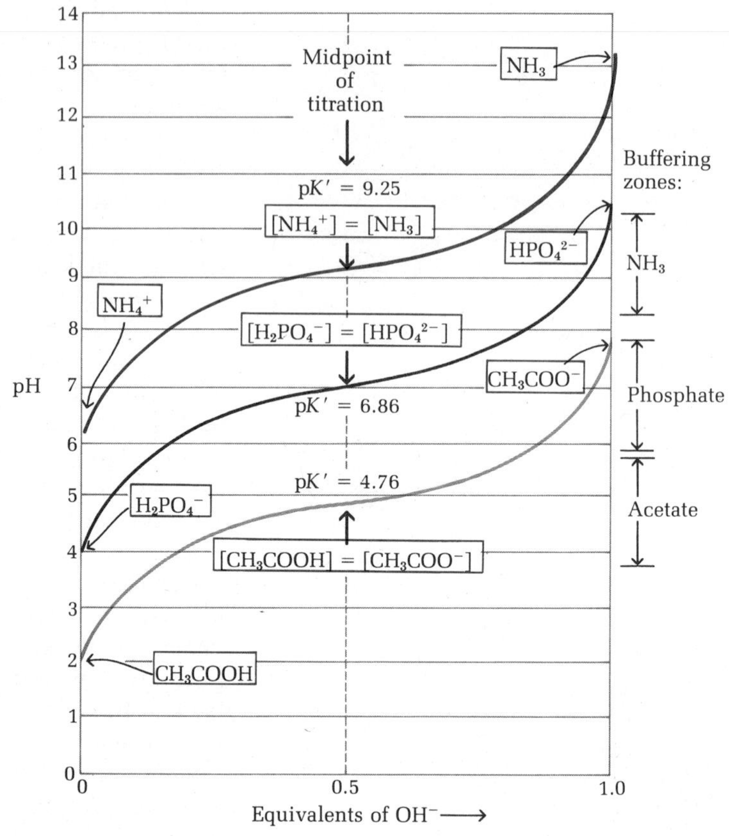

Figure 4-11
Comparison of the titration curves of three weak acids, acetic acid, $H_2PO_4^-$, and NH_4^+. The predominant ionic forms at the pH points shown are given in boxes. The zones of buffering capacity are indicated at the right. Conjugate acid-base pairs are effective buffers between 25 and 75 percent neutralization of the proton-donor species.

as occurs at the midpoint of the titration curve in Figure 4-10, is a buffer system. As we can see, the titration curve of acetic acid has a relatively flat zone extending about 1.0 pH unit on either side of its midpoint pH of 4.76. In this zone there is only a small change in pH when increments of OH^- (or H^+) are added to the system. This relatively flat zone is the buffering region of the acetic acid–acetate buffer pair. At the midpoint of the buffering region, where the concentration of the proton donor acetic acid exactly equals that of its conjugate base acetate, the buffering power of the system is maximal; i.e., its pH will change least on addition of an increment of H^+ or OH^-. Moreover, the pH of this important point in the titration curve of acetic acid is equal to its pK'. It is very important to note that the acetate buffer system *does* change in pH slightly when a small amount of H^+ or OH^- is added, but this change is very small compared with the pH change that would result if the same amount of H^+ (or OH^-) were added to pure water or to a solution of a salt of a strong acid and strong base, such as NaCl, neither of which has buffering power.

Buffering power is not a matter of black magic; it is simply an expression of two fundamental reversible reaction equilibria that occur in a solution of a proton donor and its conjugate proton acceptor when they are present in nearly equal concentrations. Let us see how a buffer system works, with the help of

Figure 4-12
Capacity of the acetic acid–acetate couple to act as a buffer system, capable of absorbing either H^+ or OH^- through the reversibility of the dissociation of acetic acid (see text).

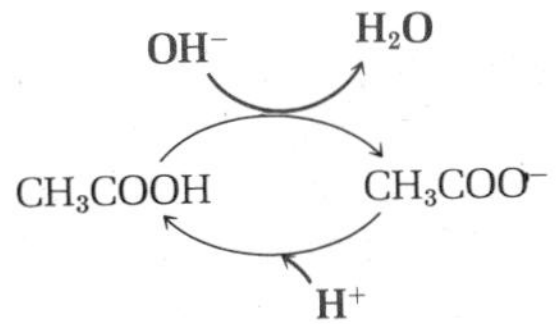

Figure 4-12. The proton donor or weak acid component (HA) of a buffer pair contains reserve bound H^+, which can be released to neutralize an addition of OH^- to the system with the formation of H_2O. This will happen because the term $[H^+][OH^-]$ will momentarily exceed 1×10^{-14}. The equilibrium quickly adjusts itself so that the product $[H^+][OH^-]$ is equal to 1×10^{-14} (at 25°C), thus momentarily reducing the concentration of H^+. But now the quotient $[H^+][A^-]/[HA]$ is less than the value of K', so that the ionization of the acid HA adjusts itself to restore equilibrium by further dissociation. Similarly, the conjugate base component of the buffer, the anion A^-, is capable of reacting with H^+ ions added to the buffer solution and is thus converted into HA. Again, the two ionization reactions adjust themselves and come to equilibrium. We can now see how a conjugate acid-base pair tends to resist a change in pH when small amounts of acid or base are added. Buffer action is simply the automatic consequence of two reversible reactions taking place and seeking their respective points of equilibrium defined by their equilibrium constants K_w and K'. Whenever we add H^+ or OH^- to a buffer, the result is a small change in the ratio of the relative concentrations of the weak acid and its anion and thus a small change in pH. The decrease in concentration of one component of a buffer system on adding a small amount of acid or base is balanced exactly by an increase in the other. The *sum* of the buffer components does not change, only their ratio.

The next important point is this. The acetic acid–acetate couple, as we have seen, is an effective buffer near pH 4.76, the automatic consequence of the fact that the pK′ of acetic acid is 4.76. Obviously this system cannot act as a buffer at the pH of blood, which is about 7.4. If we now return to Figure 4-11, we see that each conjugate acid-base couple has a characteristic pH zone in which it is an effective buffer. We see that the $H_2PO_4^-$–HPO_4^{2-} couple has a pK′ of 6.86 and thus can serve as a buffer system near pH 6.86, while the NH_4^+–NH_3 couple, with a pK′ of 9.25, can act as a buffer near pH 9.25. Of these, only the $H_2PO_4^-$–HPO_4^{2-} couple can be an effective buffer at the pH of blood (pH 7.4).

The quantitative relationship between pH, the buffering action of a mixture of weak acid with its conjugate base, and the pK′ of the weak acid is given by the *Henderson-Hasselbalch equation*. This simple equation and its usefulness in solving buffer problems are shown in Box 4-2.

Phosphate and Bicarbonate Are Important Biological Buffers

The intracellular and extracellular fluids of all organisms tend to have a characteristic and constant pH, which is regulated by various biological activities. However, the first line of defense of living organisms against changes in their internal pH is provided by buffer systems. The two most important buffer systems in mammals are the phosphate and bicarbonate sys-

Box 4-2 The Henderson-Hasselbalch equation

It will be noted that the titration curves of acetic acid, $H_2PO_4^-$, and NH_4^+ (Figure 4-11) are nearly identical in shape, suggesting that they are all reflections of a fundamental law or relationship. This is indeed the case. The shape of the titration curve of any weak acid is expressed by the Henderson-Hasselbalch equation, which is important for understanding buffer action and acid-base balance in the blood and tissues of the mammalian organism. Its simple derivation is shown below, together with some problems that this equation can solve.

The Henderson-Hasselbalch equation is simply another way of stating the expression for the dissociation constant of an acid,

$$K' = \frac{[H^+][A^-]}{[HA]}$$

First we solve for $[H^+]$

$$[H^+] = K' \frac{[HA]}{[A^-]}$$

We then take the negative logarithm of both sides

$$-\log [H^+] = -\log K' - \log \frac{[HA]}{[A^-]}$$

Substituting pH for $-\log [H^+]$ and pK′ for $-\log K$; we get

$$\text{pH} = \text{p}K' - \log \frac{[HA]}{[A^-]}$$

If we now invert $-\log [HA]/[A^-]$, which involves changing its sign, we obtain the Henderson-Hasselbalch equation

$$\text{pH} = \text{p}K' + \log \frac{[A^-]}{[HA]}$$

which in more general form is stated as

$$\text{pH} = \text{p}K' + \log \frac{[\text{proton acceptor}]}{[\text{proton donor}]}$$

The Henderson-Hasselbalch equation fits the titration curve of all weak acids and enables us to deduce a number of important quantitative relationships. For example, from it we can now see why the pK′ of a weak acid is equal to the pH of the solution at the midpoint of the titration. At this point $[HA] = [A^-]$, and we have

$$\begin{aligned} \text{pH} &= \text{p}K' + \log 1.0 \\ &= \text{p}K' + 0 \\ \text{pH} &= \text{p}K' \end{aligned}$$

The Henderson-Hasselbalch equation also makes it possible to calculate the pK′ of any acid from the molar ratio of proton-donor and proton-acceptor species at any given pH, to calculate the pH of a conjugate acid-base pair of a given pK′ and a given molar ratio, and to calculate the molar ratio of proton donor and proton acceptor at any pH given the pK′ of the weak acid.

Samples of these three types of problems are shown below, together with their solutions

1. Calculate the pK′ of lactic acid, given the fact that when the concentration of free lactic acid is 0.010 *M* and the concentration of lactate ion is 0.087 *M*, the pH is 4.80.

$$\text{pH} = \text{p}K' + \log \frac{[\text{lactate}]}{[\text{lactic acid}]}$$

$$\begin{aligned} \text{p}K' &= \text{pH} - \log \frac{[\text{lactate}]}{[\text{lactic acid}]} \\ &= 4.80 - \log \frac{0.087}{0.010} = 4.80 - \log 8.7 \\ &= 4.80 - 0.94 = 3.86 \quad \textit{ans.} \end{aligned}$$

2. Calculate the pH of a mixture of 0.1 *M* acetic acid and 0.2 *M* sodium acetate. The pK′ of acetic acid is 4.76.

$$\begin{aligned} \text{pH} &= \text{p}K' + \log \frac{[\text{acetate}]}{[\text{acetic acid}]} \\ &= 4.76 + \log \frac{0.2}{0.1} = 4.76 + 0.301 \\ &= 5.06 \quad \textit{ans.} \end{aligned}$$

3. Calculate the ratio of the concentrations of acetate and acetic acid required in a buffer system of pH 5.30.

$$\text{pH} = \text{p}K' + \log \frac{[\text{acetate}]}{[\text{acetic acid}]}$$

$$\begin{aligned} \log \frac{[\text{acetate}]}{[\text{acetic acid}]} &= \text{pH} - \text{p}K' \\ &= 5.30 - 4.76 = 0.54 \end{aligned}$$

$$\frac{[\text{acetate}]}{[\text{acetic acid}]} = \text{antilog } 0.54 = 3.47 \quad \textit{ans.}$$

tems. The phosphate buffer system, important in intracellular fluid, consists of the conjugate acid-base pair $H_2PO_4^-$, as proton donor, and HPO_4^{2-}, as proton acceptor,

$$\underset{\text{Proton donor}}{H_2PO_4^-} \rightleftharpoons H^+ + \underset{\text{Proton acceptor}}{HPO_4^{2-}}$$

The phosphate buffer system works exactly like the acetate buffer system except for the pH range in which it functions. The phosphate buffer system is maximally effective near pH 6.86, since the pK′ of $H_2PO_4^-$ is 6.86 (Table 4-4; Figure 4-11). Thus the phosphate buffer pair $H_2PO_4^-$–HPO_4^{2-} tends to resist changes in pH in the range between about 6.1 and 7.7 and would therefore be effective in providing buffering power in intracellular fluid, whose pH is in the range 6.9–7.4.

The major buffer system in blood plasma is the bicarbonate buffer system, consisting of carbonic acid (H_2CO_3) as the proton donor and bicarbonate (HCO_3^-) as the proton acceptor:

$$H_2CO_3 \rightleftharpoons H^+ + HCO_3^-$$

This system, which has its own equilibrium constant

$$K_1' = \frac{[H^+][HCO_3^-]}{[H_2CO_3]}$$

functions as a buffer in the same way as other conjugate acid-base pairs, but it is unique in that one of its components, carbonic acid (H_2CO_3), is formed from dissolved (*d*) carbon dioxide and water, according to the reversible reaction

$$CO_2(d) + H_2O \rightleftharpoons H_2CO_3$$

which has an equilibrium constant given by the expression

$$K_2' = \frac{[H_2CO_3]}{[CO_2(d)][H_2O]}$$

Since carbon dioxide is a gas under normal conditions, the concentration of the dissolved CO_2 is in turn the result of equilibration with the CO_2 of the gas phase

$$CO_2(g) \rightleftharpoons CO_2(d)$$

This process has an equilibrium constant given by

$$K_3' = \frac{[CO_2(d)]}{[CO_2(g)]}$$

The pH of a bicarbonate buffer system depends on the concentration of the dissolved H_2CO_3 and HCO_3^-, the proton donor

Box 4-3 How the bicarbonate buffer system of blood works

This buffer system involves a series of three reversible equilibria between the gaseous CO_2 in the lungs and the bicarbonate (HCO_3^-) in the blood plasma (Figure 1). When H^+ is added to blood as it passes through the tissues, it momentarily increases the concentration of H^+. This increase causes reaction 3 to proceed toward a new equilibrium, in which the concentration of H_2CO_3 increases, causing an increase in the concentration of $CO_2(d)$ in the blood. This results in an increase in the pressure of CO_2 in the gas phase in the lungs, and the extra CO_2 is exhaled.

Conversely, when some OH^- is added to the blood plasma, the opposite sequence of events occurs: the H^+ concentration is lowered, causing more H_2CO_3 to dissociate into $H^+ + HCO_3^-$. This in turn causes more $CO_2(g)$ from the lungs to dissolve in the blood plasma. The rate of breathing, i.e., rate of intake and exhaling of CO_2, thus can quickly adjust these equilibria to keep the blood pH nearly constant.

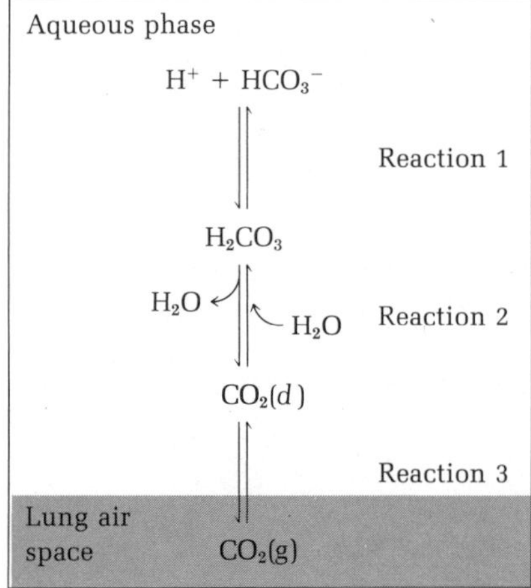

Figure 1
The CO_2 in the air space of the lungs is in equilibrium with the bicarbonate buffer in the blood plasma passing through capillaries in the lungs. Because the concentration of dissolved CO_2 can be adjusted rapidly through changes in the rate of breathing, the bicarbonate buffer system of the blood is in near-equilibrium with a large potential reservoir of CO_2.

and acceptor components. But since the concentration of H_2CO_3 in turn depends on the concentration of dissolved CO_2 and this in turn on the partial pressure of CO_2 in the gas phase, the pH of a bicarbonate buffer exposed to a gas phase is ultimately determined by the concentration of HCO_3^- in the aqueous phase and the partial pressure of CO_2 in the gas phase (Box 4-3).

The bicarbonate buffer system is an effective physiological buffer near pH 7.4 because the proton donor H_2CO_3 in the blood plasma is in labile equilibrium with a large reserve capacity of gaseous CO_2 in the air space of the lungs. Under any conditions in which the blood must absorb excess OH^-, the H_2CO_3 of the blood that is converted into HCO_3^- by reaction with OH^- is quickly restored from the large pool of gaseous CO_2 in the lungs. It dissolves in the blood to make $CO_2(d)$, which in turn combines with water to form H_2CO_3 (Box 4-3). Conversely, whenever the pH of the blood falls, some of the HCO_3^- of the buffer combines with the excess H^+ to form H_2CO_3. This decomposes to yield dissolved CO_2, which in turn is released as CO_2

in the gas phase in the lungs and then ultimately exhaled. As the blood passes through the many fine capillaries in the lungs, its bicarbonate buffer system quickly comes into near-equilibrium with the CO_2 in the gas space of the lungs. The cooperation between the bicarbonate buffer system and the action of the lungs provides a very responsive mechanism for maintaining the blood pH constant.

The pH of blood plasma is held at remarkably constant values. Human blood plasma normally has a pH close to 7.40. Should the pH-regulating mechanisms fail, as may happen in severe uncontrolled diabetes, because of the acidosis caused by overproduction of metabolic acids, the pH of the blood can fall to 6.8 or below and can lead to irreparable damage and death. And in other diseases the pH may rise to the point of no return. We may ask: What molecular mechanisms in cells are so extraordinarily sensitive that an increase in H^+ concentration of as little as 5×10^{-8} *M* (approximately the difference between blood at pH 7.4 and blood at pH 6.8) can be life-threatening? Although many aspects of cell structure and function are influenced by pH, it is the catalytic activity of enzymes that is especially sensitive. The typical pH-activity curves of some enzymes in Figure 4-13 show that they have maximal activity at a characteristic pH, called the *optimum pH*. On either side of the optimum pH their activity often declines sharply. Thus a small change in pH can make a large difference in the rate of some enzyme-catalyzed reaction crucial to the organism, for example, in skeletal muscles or in the brain. Biological control of the pH of cells and body fluids thus is of central importance in all aspects of metabolism and cellular activities.

Figure 4-13
The effect of pH *on the activity of some enzymes. Each enzyme has a characteristic* pH-activity profile.

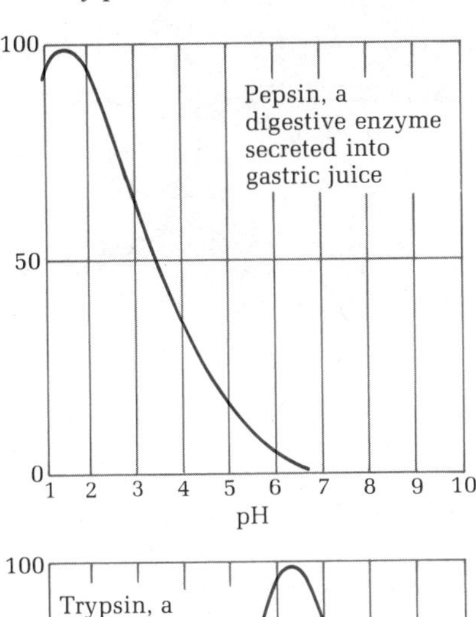

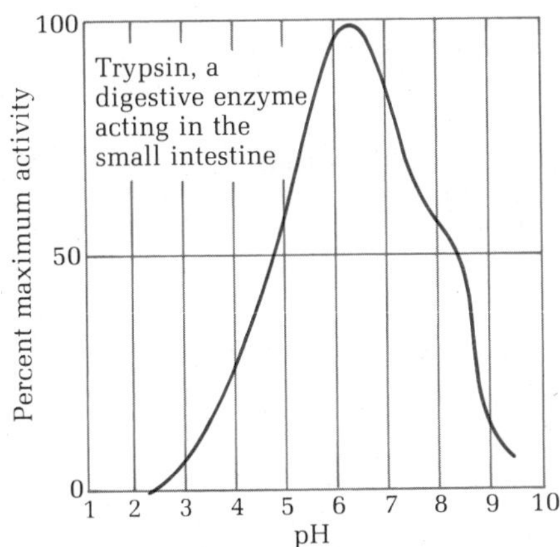

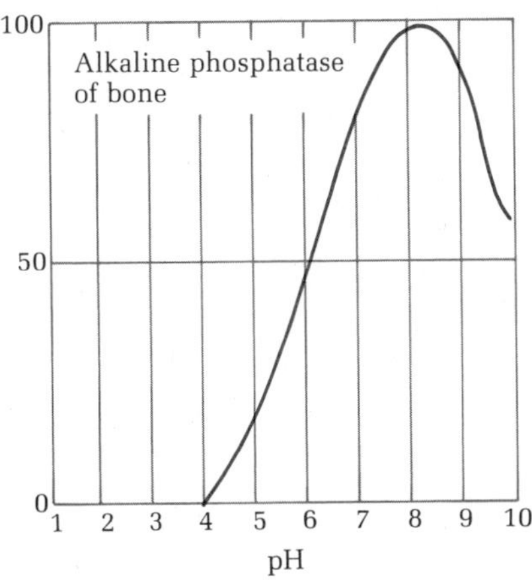

The Fitness of the Aqueous Environment for Living Organisms

Living organisms have effectively adapted to their aqueous environment and have even evolved means of exploiting the unusual properties of water. The high specific heat of water is useful to the cell since it allows water to act as a "heat buffer," permitting the temperature of an organism to remain relatively constant as the temperature of the air fluctuates. Furthermore the high heat of vaporization of water is exploited by some vertebrates as a means of losing excess body heat by evaporation of sweat. The high degree of internal cohesion of liquid water, due to hydrogen bonding, is exploited by plants as a means of transporting dissolved nutrients from the roots up to the leaves during the process of transpiration. Even the fact that ice has a lower density than liquid water and therefore floats has important biological consequences in the life cycles of aquatic organisms. But most fundamental to all living organisms is the fact that many important biological properties of cell macromolecules, particularly the proteins and nucleic acids, derive from

Figure 4-14
The water strider exploits the high surface tension of water. This insect, which lives on the surface of ponds, has specialized hairs on its first and third pairs of legs that enable it to rest upon the surface film, depressing it but not penetrating. The middle pair of legs, which penetrate the film, function like a pair of propelling oars.

their interactions with water molecules of the surrounding medium. We shall soon see that the characteristic three-dimensional structures of proteins, which determine their biological activities, are maintained by the properties of water. Even the precision of replication of the double-helical structure of DNA depends on the properties of water.

Acid Rain Is Polluting Our Lakes and Streams

Pure water exposed to "normal" air has a pH of about 5.6, rather than the theoretical value 7.0. This is because air contains a very small amount of gaseous CO_2 (about 0.04 percent or a partial pressure of 0.3 mmHg). When pure water at pH 7.0 comes into equilibrium with the CO_2 of air, the following reversible reactions, which lead to the formation of H^+ and HCO_3^- (see Box 4-3), change the pH of the water to about 5.6:

$$CO_2(g) \rightleftharpoons CO_2(d)$$

$$CO_2(d) + H_2O \rightleftharpoons H_2CO_3$$

$$H_2CO_3 \rightleftharpoons H^+ + HCO_3^-$$

Over the past hundred or more years the acidity of rain or snow in the eastern United States and in northern Europe has steadily increased, up to thirtyfold, with the result that the pH of the lakes and rivers in these areas has declined from about 5.6 to values significantly below pH 5.0.

Acid rain is formed by the reaction of rainwater with sulfur dioxide and nitrogen oxides in the atmosphere, emanating from the combustion of coal and oil, which contain small amounts of sulfur and nitrogen compounds. As a result rainwater becomes a dilute solution of sulfuric and nitric acids. Because the smoke from coal- and oil-burning power plants and steel mills is usually vented into the atmosphere by tall smokestacks to avoid local air pollution, the upper air in whole regions of the globe has become polluted with these acids, which are washed to the ground by rain. Sometimes local rains can be extremely acid; during a rainstorm in Scotland in 1974 the pH of the rain was found to be 2.4, lower than the pH of vinegar!

As a consequence of acid rain, many lakes in Scandinavian countries and in eastern Canada, northern New England, the Adirondacks, and Florida have undergone severe acidification and with it severe depletion or complete loss of their fish life, since many fish species cannot tolerate acidity significantly below pH 5.0. Moreover, the increased acidity has disturbed the delicate balance between the animal and plant life in some freshwater ecosystems. As it becomes necessary to burn more coal in the future, we can expect more severe contamination of our freshwater resources unless industrial power plants are equipped with efficient devices for removal of these contaminants from their smoke emissions.

Summary

Water is the most abundant compound in living organisms. Its relatively high freezing point, boiling point, and heat of vaporization are the result of strong intermolecular attractions in the form of hydrogen bonding between adjacent water molecules. Liquid water has considerable short-range order and consists of hydrogen-bonded clusters of very short half-life. The polarity and hydrogen-bonding properties of the water molecule make it a potent solvent for many ionic compounds and other polar molecules. Water also disperses amphipathic molecules, such as soaps, to form micelles, clusters of molecules in which the hydrophobic groups are hidden from exposure to water and the charged groups are located on the external surface.

Water ionizes very slightly to form H^+ and OH^- ions. In dilute aqueous solutions, the concentrations of H^+ and OH^- ions are inversely related by the expression $K_w = [H^+][OH^-] = 1 \times 10^{-14}$ (at 25°C). The hydrogen-ion concentration of biological systems is usually expressed in terms of pH, defined as $\text{pH} = -\log [H^+]$. The pH of aqueous solutions is measured by means of glass electrodes sensitive to H^+ concentration.

Acids are defined as proton donors and bases as proton acceptors. A conjugate acid-base pair consists of a proton donor HA and its corresponding proton acceptor A^-. The tendency of an acid HA to donate protons is expressed by its dissociation constant ($K' = [H^+][A^-]/[HA]$) or by the function pK', defined as $-\log K'$. The pH of a solution of a weak acid is quantitatively related to its pK' and to the ratio of the concentrations of its proton-donor and proton-acceptor species.

A conjugate acid-base pair can act as a buffer and resist changes in pH; its capacity to do so is greatest at a pH equal to its pK'. The most important biological buffer pairs are H_2CO_3–HCO_3^- and $H_2PO_4^-$–HPO_4^{2-}. The catalytic activity of enzymes is strongly influenced by pH.

References

CALLEWAERT, D. M., and J. GENYEA: *Basic Chemistry: General, Organic, Biological,* Worth, New York, 1980. This and other general chemistry textbooks will be useful for review.

DICK, D. A. T.: *Cell Water,* Butterworths, Washington, 1966. Properties and functions of water in living organisms.

EISENBERG, D., and W. KAUZMANN: *The Structure and Properties of Water,* Oxford University Press, Fair Lawn, N.J., 1969. Advanced treatment of the physical chemistry of water.

LIKENS, G. E., R. F. WRIGHT, J. N. GALLOWAY, and T. J. BUTLER: "Acid Rain," *Sci. Am.,* **241**:43–51, October (1979).

MONTGOMERY, R., and C. A. SWENSON: *Quantitative Problems in Biochemical Sciences,* 2d ed., Freeman, San Francisco, 1976.

SEGEL, I. H.: *Biochemical Calculations,* 2d ed., Wiley, New York, 1976.

SOLOMON, A. K.: "The State of Water in Red Cells," *Sci. Am.,* **244**:88–96, February (1971). An article describing research on the structure of water within cells.

Problems

1. *Radiator Antifreeze* Ethylene glycol, an alcohol with two hydroxyl groups ($CH_2OH—CH_2OH$) is commonly used as a solute to depress the freezing point of water in making automobile radiator fluid. Calculate the approximate concentration of ethylene glycol in moles per liter that will just suffice to protect radiator fluid from freezing at 0°F. [To convert Fahrenheit degrees into Celsius use the relationship $°C = \frac{5}{9}(°F - 32)$.]

2. *Simulating Vinegar* One way to make vinegar (*not* the preferred way) is to prepare a solution of acetic acid, the sole acid component of vinegar, to the proper pH (see Table 4-4) and add appropriate flavoring agents. Acetic acid is a liquid at 25°C with a density of 1.049 g/mL. Calculate the amount that must be added to distilled water to make 1 L of simulated vinegar.

3. *Acidity of Gastric HCl* In a hospital laboratory a 10.0-mL sample of gastric juice, obtained several hours after a meal, was titrated with 0.1 *N* NaOH to neutrality; 7.2 mL was required. Since the stomach contained no ingested food or drink, it can be assumed that no buffers are present. What is the pH of the gastric juice?

4. *Measurement of Acetylcholine Levels by pH Changes* The concentration of the neurotransmitter molecule acetylcholine can be determined from the pH changes that accompany its hydrolysis. When incubated with a catalytic amount of the enzyme acetylcholinesterase, acetylcholine is quantitatively converted into choline and acetic acid, which dissociates to yield acetate and a hydrogen ion:

$$\underset{\text{Acetylcholine}}{CH_3—\overset{\overset{\displaystyle O}{\|}}{C}—O—CH_2—CH_2—\overset{+}{N}(CH_3)_3} \xrightarrow{H_2O} \underset{\text{Choline}}{HO—CH_2—CH_2—\overset{+}{N}(CH_3)_3} + \underset{\text{Acetate}}{CH_3—\underset{\underset{\displaystyle O}{\|}}{C}—O^-} + H^+$$

In a typical analysis, the pH of 15 mL of an aqueous solution containing an unknown amount of acetylcholine was 7.65. When incubated with acetylcholinesterase, the pH of the solution decreased to a final value of 6.87. Assuming that there was no buffer in the assay mixture, determine the number of moles of acetylcholine in the 15 mL of unknown.

5. *Significance of the pK′ of an Acid* One common description of the pK′ of an acid is that it represents the pH at which the acid is half-ionized, i.e., the pH at which the acid exists as a 50:50 mixture of the acid and the conjugate base. Demonstrate that this relationship is correct, starting from the equilibrium-constant expression.

6. *Properties of a Buffer* The amino acid glycine is often used as the main ingredient of a buffer in biochemical experiments. The amino group of glycine, which has a pK′ of 9.3, can exist in either the protonated form ($—NH_3^+$) or the free base ($—NH_2$) because of the reversible equilibrium

$$—\overset{+}{N}H_3 \rightleftharpoons —NH_2 + H^+$$

   (a) In what pH zone can glycine be used as an effective buffer due to its amino group?
   (b) In a 0.1 *M* solution of glycine at pH 9.0, what fraction of glycine has its amino group in the $—NH_3^+$ form?
   (c) How much 5 *M* KOH must be added to 1.0 L of 0.1 *M* glycine at pH 9.0 to bring its pH back to exactly 10.0?
   (d) In order to have 99 percent of the glycine in a form in which its amino group is protonated, what must the numerical relation be between the pH of the solution and the pK′ of the amino group of glycine?

7. *The Effect of pH on Solubility* The strongly polar hydrogen-bonding nature of water makes it an excellent solvent for dissolving ionic (charged) species. By contrast, un-ionized, nonpolar organic molecules, such as benzene, are relatively insoluble in water. In principle, the solubility of all organic acids or bases in water can be increased by deprotonation or protonation, respectively, to form a charged species. For example, the solubility of benzoic acid in water is low. The addition of sodium bicarbonate raises the pH of the solution and deprotonates the benzoic acid to form benzoate ion, which is quite soluble in water.

   Are the molecules in (a) to (c) more soluble in water with 0.1 *M* NaOH or 0.1 *M* HCl?

Problem 7

Benzoic acid (C—OH with C=O on benzene ring)
pK′ ~ 5
insoluble in water

Benzoate ion (C—O⁻ with C=O on benzene ring)
(charged)
soluble in water

(a) Pyridine ion (N^+—H)
pK′ ~ 5

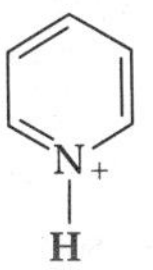

(b) β-Naphthol (OH)
pK′ ~ 10

(c) N-Acetyltyrosine methyl ester (H_3C—C(=O)—N(H)—C(H)(—CH_2—C₆H₄—OH)—C(=O)—O—CH_3)
pK′ ~ 10

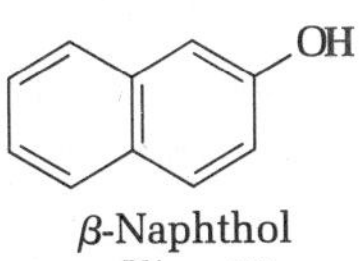

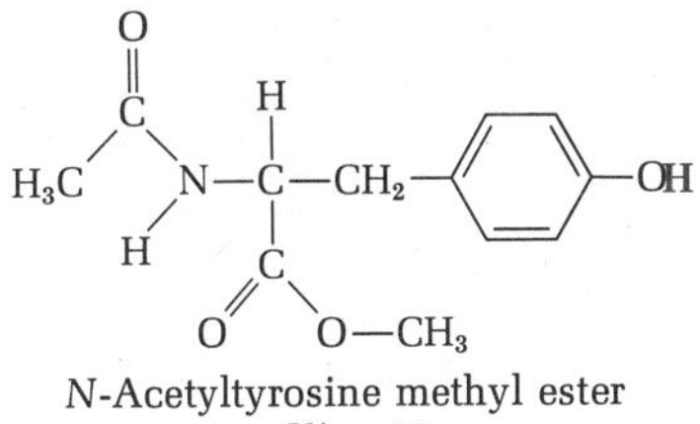

8. *Treatment of Poison Ivy Rash* Catechols substituted with long-chain alkyl groups are the compounds in poison ivy and poison oak that produce the characteristic itchy rash.

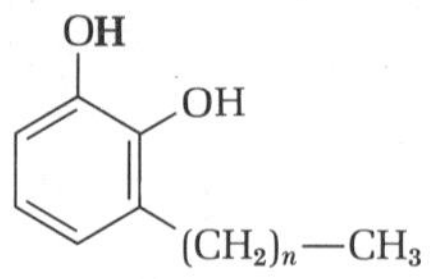

OH
OH
$(CH_2)_n—CH_3$
pK′ = 8

If you were accidentally exposed to poison ivy, which of the treatments below would you apply to the affected area? Justify your choice.
(a) Wash the area with cold water.
(b) Wash the area with dilute vinegar or lemon juice.
(c) Wash the area with soap and water.
(d) Wash the area with soap, water, and baking soda (sodium bicarbonate).

9. *pH and Drug Absorption* The widely prescribed drug aspirin is a weak acid with a pK′ of 3.5.

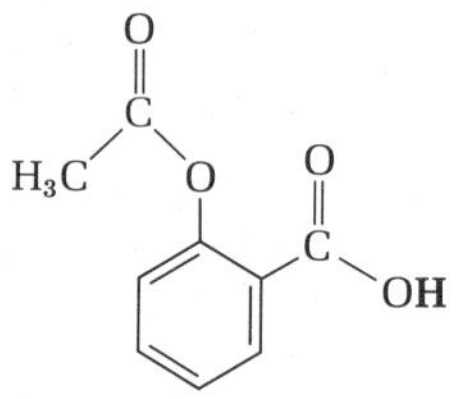

O
‖
C
H_3C O O
‖
C
OH

pK′ = 3.5

Aspirin is absorbed into the blood through the cells lining the stomach and the small intestine. For a substance to be absorbed it must readily pass through the cell membrane. Passage through the cell membrane is determined by the polarity of the molecule: ionic (charged) and highly polar molecules pass slowly, while neutral hydrophobic ones pass through rapidly. Since the pH of the gastric juice in the stomach is about 1 and pH in the small intestine is about 6, where is more aspirin absorbed into the bloodstream, stomach or small intestine? Clearly justify your choice.

10. *Preparation of Standard Buffer for Calibration of a pH Meter* The glass electrode used in commercial pH meters gives an electrical response proportional to the hydrogen-ion concentration. To convert these responses into pH correctly, glass electrodes must be calibrated against standard solutions of known hydrogen-ion concentration. Determine the weight in grams of sodium dihydrogen phosphate ($NaH_2PO_4 \cdot H_2O$; FW 138.01) and disodium hydrogen phosphate (Na_2HPO_4; FW 141.98), where FW stands for formula weight, needed to prepare 1 L of a standard buffer at pH 7.00 whose total phosphate concentration is 0.100 *M*. The pK′ of dihydrogen phosphate is 6.86 at 25°C.

11. *Control of Blood pH by the Rate of Respiration*
(a) The partial pressure of CO_2 in the lungs can be varied rapidly by the rate and depth of breathing. For example, a common remedy to alleviate hiccups is to increase the concentration

of CO_2 in the lungs. This can be achieved by holding one's breath, by very slow and shallow breathing (hypoventilation), or by breathing in and out of a paper bag. Under such conditions, the partial pressure of CO_2 in the lung airspace rises above normal. Qualitatively explain the effect of these procedures on the blood pH.

(b) A common practice of competitive short-distance runners is to breathe rapidly and deeply (hyperventilation) for about $\frac{1}{2}$ min to remove CO_2 from their lungs just before running in, say, a 100-m dash. Their blood pH may rise to 7.60. Explain why the blood pH goes up.

(c) During a short-distance run the muscles produce a large amount of lactic acid from their glucose stores. In view of this fact why is hyperventilation before a dash useful?

CHAPTER 5

Amino Acids and Peptides

Proteins are the most abundant macromolecules in living cells and constitute 50 percent or more of their dry weight. They are found in all cells and all parts of cells. Proteins also occur in great variety; hundreds of different kinds may be found in a single cell. Moreover, proteins have many different biological roles since they are the molecular instruments through which genetic information is expressed. It is therefore appropriate to begin the study of biological macromolecules with the proteins, whose name means "first" or "foremost."

The key to the structure of the thousands of different proteins is the group of relatively simple building-block molecules from which proteins are built. All proteins, whether from the most ancient lines of bacteria or from the highest forms of life, are constructed from the same basic set of 20 amino acids, covalently linked in characteristic sequences. Because each of these amino acids has a distinctive side chain which lends it chemical individuality, this group of 20 building-block molecules may be regarded as the alphabet of protein structure.

In this chapter we shall also examine *peptides*, short chains of two or more amino acids joined by covalent bonds. What is most remarkable is that cells can join the 20 amino acids in many different combinations and sequences, yielding peptides and proteins having strikingly different properties and activities. From these building blocks different organisms can make such widely diverse products as enzymes, hormones, the lens protein of the eye, feathers, spider webs, tortoise shell (Figure 5-1), nutritive milk proteins, enkephalins (the body's own opiates), antibiotics, mushroom poisons, and many other substances having specific biological activity.

Figure 5-1
The protein keratin is formed by all vertebrates. It is the chief structural component of hair, scales, horns, wool, nails, and feathers. Keratin is also the principal component of the tough armor of the tortoise.

Amino Acids Have Common Structural Features

When proteins are boiled with strong acid or base, their amino acid building blocks are released from the covalent linkages that join them into chains. The free amino acids so formed are relatively small molecules, and their structures are all known.

Figure 5-2
General structure of the amino acids found in proteins, shown in nonionic form. The portion in black is common to all the α-amino acids of proteins (except proline). R, in color, represents the side chain, or R group, which is different in each amino acid. In all the amino acids except glycine the α carbon atom has four different substituent groups and is thus an asymmetric or chiral carbon.

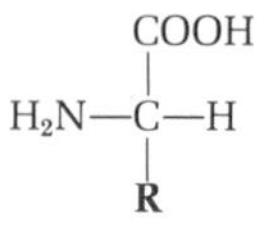

$$H_2N-\overset{\overset{\displaystyle COOH}{|}}{\underset{\underset{\displaystyle \mathbf{R}}{|}}{C}}-H$$

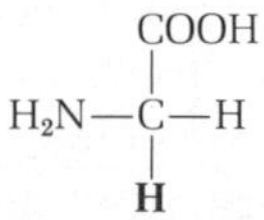

The first amino acid to be discovered was asparagine, in 1806. The last of the 20 to be found, threonine, was not identified until 1938. All the amino acids have trivial or common names, sometimes derived from the source from which they were first isolated. Asparagine was first found in asparagus, as one might guess; glutamic acid was found in wheat gluten; and glycine (Greek, *glykos*, "sweet") was so named because of its sweet taste.

All of the 20 amino acids found in proteins have as common denominators a carboxyl group and an amino group bonded to the same carbon atom (Figure 5-2). They differ from each other in their side chains, or R groups, which vary in structure, size, electric charge, and solubility in water. The 20 amino acids of proteins are often referred to as the *standard*, *primary*, or *normal* amino acids, to distinguish them from other kinds of amino acids present in living organisms but not in proteins. The standard amino acids have been assigned three-letter abbreviations and one-letter symbols (Table 5-1), which are used as shorthand to indicate the composition and sequence of amino acids in polypeptide chains.

Table 5-1 Amino Acid Shorthand

Amino acid	Three-letter abbreviation	One-letter symbol
Alanine	Ala	A
Arginine	Arg	R
Asparagine	Asn	N
Aspartic acid	Asp	D
Cysteine	Cys	C
Glutamine	Gln	Q
Glutamic acid	Glu	E
Glycine	Gly	G
Histidine	His	H
Isoleucine	Ile	I
Leucine	Leu	L
Lysine	Lys	K
Methionine	Met	M
Phenylalanine	Phe	F
Proline	Pro	P
Serine	Ser	S
Threonine	Thr	T
Tryptophan	Trp	W
Tyrosine	Tyr	Y
Valine	Val	V

Nearly All Amino Acids Have an Asymmetric Carbon Atom

We note from Figure 5-2 that all the standard amino acids except one have an *asymmetric* carbon atom, the α carbon, to which are bonded four different substituent groups, i.e., a carboxyl group, an amino group, an R group, and a hydrogen atom. The asymmetric α carbon atom is thus a *chiral center* (page 52). As we have seen, compounds with a chiral center occur in two different isomeric forms, which are identical in all chemical and physical properties except one, the direction in which they can cause the rotation of plane-polarized light in a *polarimeter* (page 51). With the single exception of *glycine*, which has no asymmetric carbon atom (Figure 5-3), all of the 20 amino acids obtained from the hydrolysis of proteins under sufficiently mild conditions are *optically active*; i.e., they can rotate the plane-polarized light in one direction or the other. Because of the tetrahedral arrangement of the valence bonds around the α carbon atom of amino acids the four different substituent groups can occupy two different arrangements in space, which are nonsuperimposable mirror images of each other (Figure 5-4). These two forms are called *optical isomers*, *enantiomers*, or *stereoisomers*. A solution of one stereoisomer of a given amino acid will rotate plane-polarized light to the left (counterclockwise) and is called the *levorotatory* isomer [designated (−)]; the other stereoisomer will rotate plane-polarized light to the same extent but to the right (clockwise) and is called the *dextrorotatory* isomer [designated (+)]. An equimolar mixture of the (+) and (−) forms will not rotate plane-polarized light. Because all the amino acids (except glycine) when carefully isolated from proteins do rotate plane-polarized light, they evidently occur in only one of their stereoisomeric forms in protein molecules.

Figure 5-3
Glycine, the only amino acid without an asymmetric carbon atom. Its R group, a hydrogen atom, is in color.

$$H_2N-\overset{\overset{\displaystyle COOH}{|}}{\underset{\underset{\displaystyle \mathbf{H}}{|}}{C}}-H$$

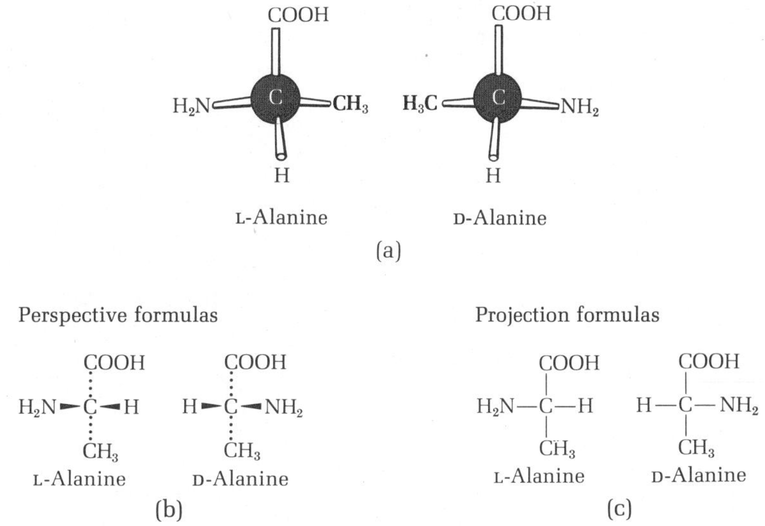

Figure 5-4
(a) The two optical isomers of alanine, with their reference groups (the carboxyl groups) on the vertical bond to the chiral center. L- *and* D-*alanine are nonsuperimposable mirror images of each other. (b, c) Two different conventions for showing the configurations in space of optical isomers. In perspective formulas the wedge-shaped bonds project above the plane of the paper, the dotted bonds behind it. In projection formulas the horizontal bonds are assumed to project above the plane of the paper, the vertical bonds behind. However, projection formulas are sometimes loosely used without reference to stereochemical configuration.*

Optical activity of a stereoisomer is expressed quantitatively by its specific rotation, determined from measurements of the degree of rotation of a solution of the pure stereoisomer at a given concentration in a tube of a given length in a polarimeter:

$$[\alpha]_D^{25^\circ C} = \frac{\text{observed rotation, deg}}{\text{length of tube, dm} \times \text{concentration, g/mL}}$$

where the abbreviation dm stands for decimeters (0.1 m). The temperature and the wavelength of the light employed (usually the D line of sodium, 589 nm) must be specified. Table 5-2 shows the specific rotation of several amino acids; note that some are levorotatory and others dextrorotatory.

Table 5-2 Specific Rotation of Some Amino Acids Isolated from Proteins

All are of the L configuration; note that some are dextrorotatory and some levorotatory

Amino acid	Specific rotation $[\alpha]_D^{25^\circ}$
L-Alanine	+ 1.8
L-Arginine	+12.5
L-Isoleucine	+12.4
L-Phenylalanine	−34.5
L-Glutamic acid	+12.0
L-Histidine	−38.5
L-Lysine	+13.5
L-Serine	− 7.5
L-Proline	−86.2
L-Threonine	−28.5

Stereoisomers Are Named on the Basis of Their Absolute Configuration

A more systematic basis for classifying and naming stereoisomers than the direction of rotation of plane-polarized light is the *absolute configuration* of the four different substituents in the tetrahedron around the asymmetric carbon atom. For this purpose a reference compound has been chosen, to which all other optically active compounds are compared. This reference compound is the 3-carbon sugar *glyceraldehyde* (Figure 5-5), the smallest sugar to have an asymmetric carbon atom (the structure of sugars is described in Chapter 11). By convention the two stereoisomers of glyceraldehyde are designated L and D (note the use of small capital letters). They have the configurations shown in Figure 5-5, which have been established by x-ray analysis. Directly underneath the two stereoisomers of glyceraldehyde are shown the two corresponding stereoisomers of the amino acid *alanine*. The *reference atom* is the most oxidized carbon atom attached to the asymmetric carbon.

Figure 5-5
Steric relationship of the enantiomers of alanine to the absolute configuration of L- and D-glyceraldehyde.

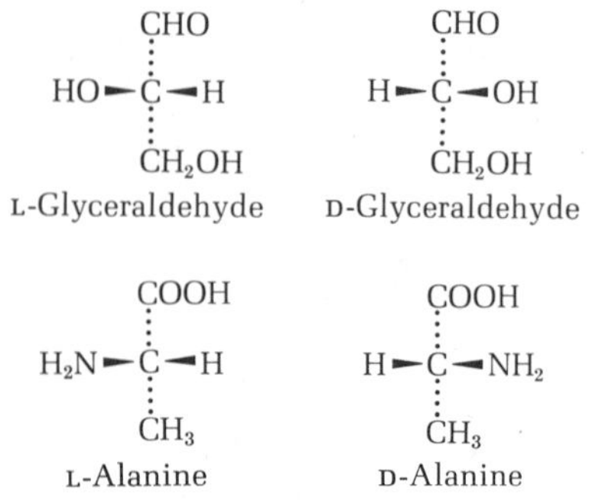

L-Glyceraldehyde	D-Glyceraldehyde
CHO / HO—C—H / CH_2OH	CHO / H—C—OH / CH_2OH

L-Alanine	D-Alanine
COOH / H_2N—C—H / CH_3	COOH / H—C—NH_2 / CH_3

Thus the carboxyl carbon of L-alanine and the aldehyde carbon of L-glyceraldehyde are the reference atoms, which are taken to occupy the same position in space. Once the reference atoms are set, we can see that the amino group of L-alanine is related to the hydroxyl group of L-glyceraldehyde and the methyl, or R group, of L-alanine is related to the —CH_2OH group of L-glyceraldehyde. Similarly, the absolute configuration of the substituent groups in D-alanine is related to those of the D stereoisomer of glyceraldehyde. *The stereoisomers of all chiral compounds having a configuration related to that of L-glyceraldehyde are designated L, and the stereoisomers related to D-glyceraldehyde are designated D, regardless of the direction in which they rotate the plane-polarized light.* The symbols L and D thus refer to the *absolute configuration* of the four substituents around the chiral carbon, *not* the direction of rotation of plane-polarized light. Table 5-2 shows the specific rotation of some L-amino acids.

When a compound has two or more chiral centers, it has 2^n possible stereoisomers, where n is the number of chiral centers. Glycine has no asymmetric carbon atom (Figure 5-3) and therefore cannot exist in stereoisomeric forms. All the other amino acids commonly found in proteins have one asymmetric carbon, except *threonine* and *isoleucine*, which have two and thus have $2^n = 2^2 = 4$ stereoisomers; however, only one of the four possible isomers of these amino acids occurs in protein molecules.

Under the above rules, naming the stereoisomers of compounds having two or more chiral centers sometimes becomes difficult or leads to ambiguities. A new naming system, called the *RS system,* is being used more and more to designate stereoisomers, particularly those with two or more chiral centers (Box 5-1).

The Optically Active Amino Acids of Proteins Are L Stereoisomers

Nearly all biologically occurring compounds containing a chiral center are found in nature in only one stereoisomeric form, either D or L. Except for glycine, which has no asymmetric carbon atom, the amino acids present in protein molecules are the L stereoisomers. This conclusion is the result of a great deal of painstaking chemical research on the optical properties and the organic reactions of the amino acids. We shall see later (page 115) that some D-amino acids do occur in living matter, but they have never been found in proteins.

That the amino acids of proteins are L stereoisomers is remarkable, since ordinary, nonbiological chemical reactions used to synthesize a compound having an asymmetric carbon atom always yield optically inactive products. This is because ordinary chemical reactions form both D and L stereoisomers at the same rate. As a result the product is a *racemic mixture* or *racemate,* an equimolar mixture of the D and L isomers, which

Box 5-1 The *RS* system of designating optical isomers

When a compound has two or more chiral centers, ambiguities may arise in designating its isomers by the DL system. The *RS* system for naming optical isomers has been invented to avoid this and other possible ambiguities. To show how it works let us consider an example. First, we must inspect and rank the four different substituent atoms around each asymmetric carbon atom in order of decreasing atomic number or in order of decreasing valence density when the *smallest* or lowest-rank group is pointing directly away from the viewer. It is as if the viewer were looking down on a steering wheel with three spokes, with the steering column representing the bond to the lowest-priority substituent. If the decreasing rank order then is seen to be clockwise, the configuration around this chiral center is *R* (Latin *rectus,* "right"); if the decreasing rank order is counterclockwise, the configuration is *S* (*sinister,* "left"). Each chiral center is named similarly.

Some common functional groups of biomolecules are ranked in order of decreasing priority as follows:

$$SH > OR > OH > NHCOR > NH_2 > COOH > CHO > CH_2OH > C_6H_5 > CH_3 > H$$

Although most of the amino acids found in proteins are unambiguously named by the DL system, threonine and isoleucine have two asymmetric carbon atoms and thus have $2^2 = 4$ stereoisomers. The configurations around the two chiral centers of these isomers in relationship to L- and D-glyceraldehyde have been established, and the isomers have been given appropriate names. The prefix *allo* derives from the Greek *allos,* "other."

```
       COOH                 COOH
        |                    |
  H2N—C—H            H2N—C—H
        |                    |
    H—C—OH             HO—C—H
        |                    |
       CH3                  CH3
  L-Threonine       L-allo-Threonine

       COOH                 COOH
        |                    |
    H—C—NH2             H—C—NH2
        |                    |
   HO—C—H               H—C—OH
        |                    |
       CH3                  CH3
  D-Threonine       D-allo-Threonine
```

In amino acids the designations L and D refer to the configuration of the α, or 2, carbon atom. The *allo* designations refer to the configuration at carbon 3. The isomers of compounds having two chiral centers are known as *diastereoisomers.*

Let us now see how L-threonine, the normal isomer found in proteins, is named by the *RS* system. We begin with carbon atom 2. Its four substituent groups, ranked in order of decreasing atomic number or valence density of the atom bound to carbon atom 2, are $—NH_2$, $—COOH$, $—CHOHCH_3$, $—H$. If we now change the position of the entire L-threonine molecule so that the bond from carbon 2 to the H atom becomes the steering column, pointing down behind the plane of the page, we shall see the other three groups as

```
          NH2
           |
          (C2)
         /    \
     HOOC      CHOH
                   \
                    CH3
```

We note that the order of *decreasing* priority is counterclockwise (see arrows). Therefore the configuration of carbon atom 2 is *S*. In the same way we can show a comparable steric diagram for carbon atom 3, with the bond to the lowest-priority group, the H atom, pointing behind the page.

```
          OH
           |
          (C3)     NH2
         /    \   /
     H3C        C
               / \
              H   COOH
```

The priority sequence at carbon 3 runs clockwise, and the configuration is therefore *R*. Consequently we designate L-threonine as *(2S,3R)-threonine* in the *RS* system.

The traditional designations L-, D-, L-*allo,* and D-*allo* for threonine have been so well known for so long that they continue to be in common use. This is also true of the names and stereodesignations of simple sugars (Chapter 11); but the *RS* system is the ultimate unambiguous method for the designation of optical isomers of molecules in which the relationship to the glyceraldehyde isomers may not have been established or the DL system is ambiguous. This often happens in the case of complex, naturally occurring substances having two or more chiral centers.

Box 5-2 How to tell a person's age through amino acid chemistry

The optical isomers of amino acids very slowly undergo spontaneous, nonenzymatic racemization, so that over a very long period of time an equimolar mixture of the D and L isomers will be formed from the pure L or the pure D isomer. Each L-amino acid racemizes at a known rate at a given temperature. This fact can be used to determine the age of living people and animals or the age of fossil bones. For example, the L-aspartate of the protein *dentine*, present in the outer hard enamel of the teeth, spontaneously racemizes at the rate of 0.10 percent per year at body temperature. Dentine contains only L-aspartate at the time the tooth is formed in childhood. The dentine can be isolated from a single tooth of a person and its content of the D-aspartate determined. Such analyses have been made on the dentine of inhabitants of mountain villages in Ecuador, where individuals have claimed to be exceptionally long-lived. While some of these claims have been discounted, the racemization test has been found to be rather accurate. For example, the test yielded an age of 99 for a woman who was 97 years old, according to verified records.

Tests made on prehistoric fossil skeletons of elephants, porpoises, and bears have shown that this method agrees very well with the results of dating based on the rate of decay of radioisotopes.

Table 5-3 Classification of the Amino Acids According to the Polarity of Their R Groups (at pH 7)

Nonpolar R groups
Alanine
Isoleucine
Leucine
Methionine
Phenylalanine
Proline
Tryptophan
Valine
Polar but uncharged R groups
Asparagine
Cysteine
Glutamine
Glycine
Serine
Threonine
Tyrosine
Negatively charged R groups
Aspartic acid
Glutamic acid
Positively charged R groups
Arginine
Histidine
Lysine

does not rotate plane-polarized light in either direction. Racemic mixtures can be resolved, or separated, into their D and L isomers only by very laborious physical fractionation procedures. Pure D or L isomers will eventually revert to a racemic mixture with time (Box 5-2).

Living cells have the unique ability to synthesize their L-amino acids by *stereospecific* enzymes. The stereospecificity of these enzymes is made possible by the asymmetric nature of their active sites. Later we shall see that the characteristic three-dimensional structures of proteins, which are responsible for their diverse biological activities, require that *all* their building-block amino acids be of one stereochemical series.

Amino Acids Can Be Classified on the Basis of Their R Groups

Let us now examine the structures of the 20 amino acids found in proteins. This guided tour will be made easy by the fact that the amino acids can be grouped into families based on the properties of their R groups, in particular, their *polarity* (Table 5-3), i.e., their tendency to interact with water at biological pH (near pH 7.0). The R groups of the amino acids vary widely with respect to their polarity, from totally nonpolar or hydrophobic (water-hating) R groups to highly polar or hydrophilic (water-loving) R groups.

The structures of the 20 standard amino acids are shown in Figure 5-6. There are four main families of amino acids, those with (1) nonpolar or hydrophobic R groups, (2) polar but uncharged R groups, (3) negatively charged R groups, and (4) positively charged R groups. Within each family there are gradations of polarity, size, and shape of the R groups.

Figure 5-6
The 20 amino acids common in proteins. They are shown with their amino and carboxyl groups ionized, as they would occur at pH 7.0. The shaded portions are those common to all the amino acids. The unshaded portions are their R groups.

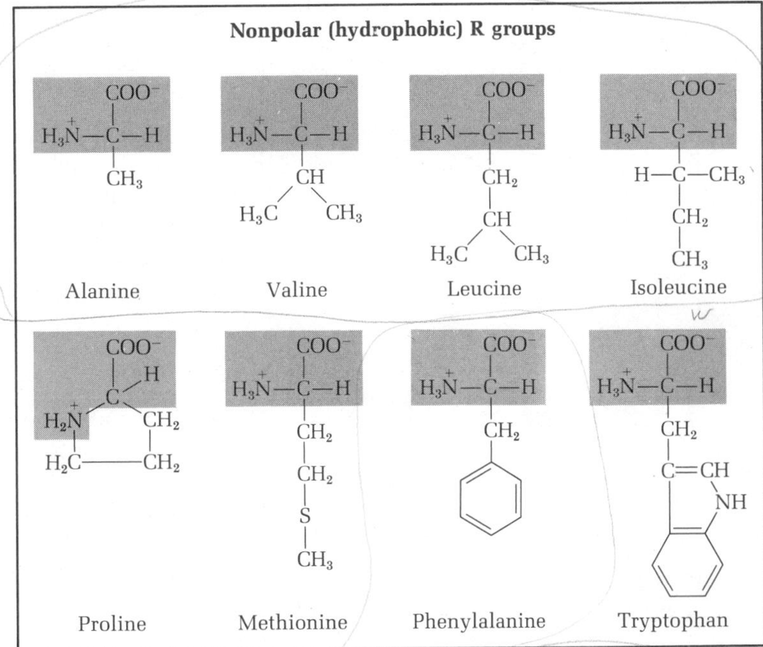

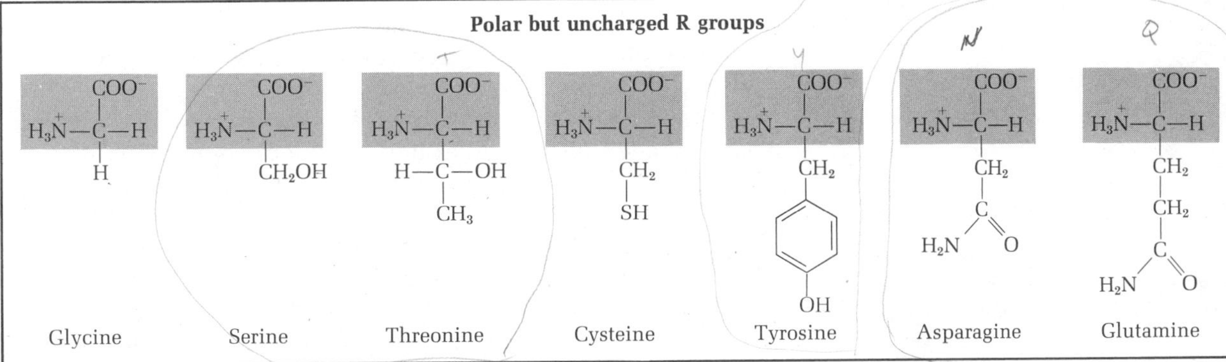

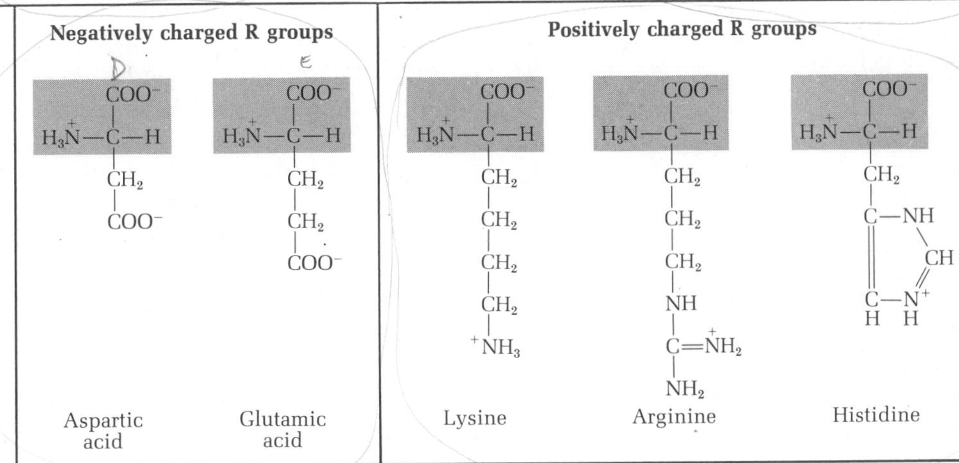

Eight Amino Acids Have Nonpolar R Groups

The R groups in this class of amino acids are hydrocarbon in nature and thus hydrophobic (Figure 5-6). This group includes five amino acids with aliphatic R groups (*alanine, valine, leucine, isoleucine,* and *proline*), two with aromatic rings (*phenylalanine* and *tryptophan*), and one containing sulfur (*methionine*). Of this group, proline requires special mention because its α-amino group is not free but is substituted by a portion of its R group to yield a cyclic structure (Figure 5-6).

Seven Amino Acids Have Uncharged Polar R Groups

The R groups of these amino acids (Figure 5-6) are more soluble in water, i.e., more hydrophilic, than those of the nonpolar amino acids, because they contain functional groups that form hydrogen bonds with water. This class of amino acids includes *glycine, serine, threonine, cysteine, tyrosine, asparagine,* and *glutamine.* The polarity of serine, threonine, and tyrosine is contributed by their *hydroxyl groups;* that of asparagine and glutamine by their *amide groups;* and that of cysteine by its *sulfhydryl,* or *thiol group.* The R group of glycine, a single hydrogen atom, is too small to influence the high degree of polarity of the α-amino and α-carboxyl groups.

Asparagine and glutamine are the amides of two other amino acids also found as building blocks of proteins, *aspartic acid* and *glutamic acid,* respectively, to which asparagine and glutamine are easily hydrolyzed by acid or base. Cysteine and tyrosine have R groups that tend to dissociate H^+ ions, but the thiol group of cysteine and the phenolic hydroxyl group of tyrosine are only slightly ionized at pH 7.0.

Cysteine requires special mention for another reason. It may occur in proteins in two forms, either as cysteine itself or as *cystine,* in which two cysteine molecules are covalently joined by a disulfide bridge formed by oxidation of their thiol groups (Figure 5-7). Cystine plays a special role in the structure of some proteins, e.g., the hormone *insulin* and the *immunoglobulins,* or *antibodies.* In such proteins the two halves of the cystine molecule serve as building blocks in two different polypeptide chains, which are thus cross-linked by the disulfide bond (page 134). Such cystine cross-linkages are not usually found in intracellular proteins but are common in proteins secreted into, and functioning in, extracellular fluids.

Figure 5-7
Cysteine and cystine. The thiol (—SH) groups of two molecules of cysteine are readily oxidized to yield the disulfide group of cystine. Both cysteine and cystine may be present in proteins.

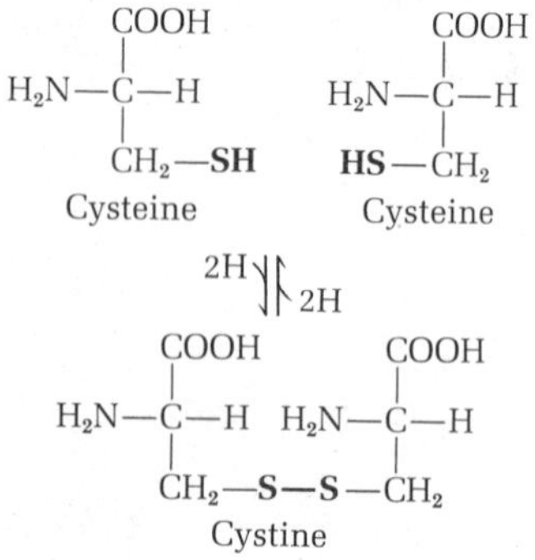

```
     COOH                 COOH
      |                    |
H2N—C—H              H2N—C—H
      |                    |
     CH2—SH           HS—CH2
   Cysteine             Cysteine

          2H ↓↑ 2H

     COOH            COOH
      |               |
H2N—C—H        H2N—C—H
      |               |
     CH2—S—S—CH2
          Cystine
```

Two Amino Acids Have Negatively Charged (Acidic) R Groups

The two amino acids whose R groups have a net negative charge at pH 7.0 are *aspartic acid* and *glutamic acid,* each with a second carboxyl group (Figure 5-6). These amino acids are the parent compounds of asparagine and glutamine, respectively (see above).

Figure 5-8
(a) Some special amino acids found in proteins. The groups in color are the extra functional groups added to the standard amino acids proline and lysine. (b) Desmosine, found in the protein elastin, is formed from four molecules of lysine, whose carbon backbones are shown in boldface and shaded in black.

$HO-\overset{H}{\underset{4}{C}}-{}_3CH_2$, ${}_5CH_2$, ${}^2CH-COOH$ (1), N–H (ring)

4-Hydroxyproline

$\overset{6}{H_2N-CH_2}-\underset{OH}{\overset{5}{CH}}-\overset{4}{CH_2}-\overset{3}{CH_2}-\underset{NH_2}{\overset{2}{CH}}-\overset{1}{COOH}$

5-Hydroxylysine

$CH_3-NH-\overset{6}{CH_2}-\overset{5}{CH_2}-\overset{4}{CH_2}-\overset{3}{CH_2}-\underset{NH_2}{\overset{2}{CH}}-\overset{1}{COOH}$

6-N-Methyllysine

(a)

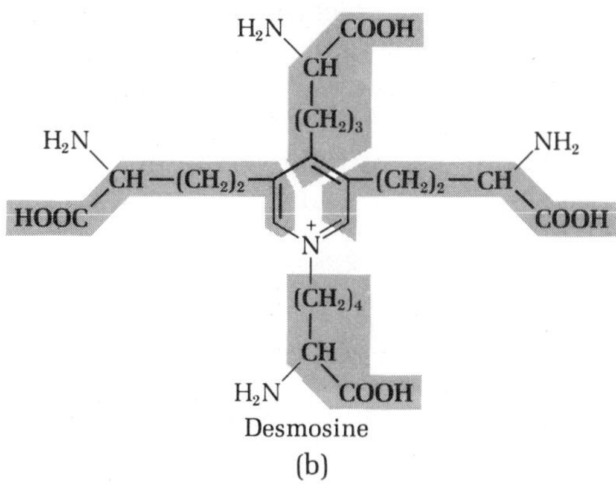

Desmosine

(b)

Three Amino Acids Have Positively Charged (Basic) R Groups

The amino acids in which the R groups have a net positive charge at pH 7.0 are lysine, which has a second amino group at the ϵ position on its aliphatic chain; arginine, which has a positively charged guanidino group; and histidine, which contains the weakly ionized imidazole group (Figure 5-6).

Some Proteins Also Contain "Special" Amino Acids

In addition to the 20 *standard* amino acids that are common in all proteins, other *special* amino acids have been found as components of only certain types of proteins (Figure 5-8). Each of these special amino acids is derived from one of the 20 common amino acids. Among the special amino acids are *4-hydroxyproline*, a derivative of proline, and *5-hydroxylysine*; both are found in the fibrous protein *collagen* of connective tissues (Chapter 7). *N-Methyllysine* is found in myosin, a muscle protein functioning in contraction (page 380). Another important special amino acid is *γ-carboxyglutamic acid*, found in the blood-clotting protein *prothrombin* (Chapter 25), as well as in certain other proteins that bind Ca^{2+} ions in their biological function. More complicated is the special amino acid *desmosine*, a derivative of lysine, found only in the fibrous protein *elastin* (page 161).

Amino Acids Are Ionized in Water Solutions

Amino acids in aqueous solution are ionized and can act as acids or bases. Knowledge of the acid-base properties of amino acids is extremely important in understanding many properties of proteins. Moreover, the art of separating, identifying, and quantitating the different amino acids, which are necessary steps in determining the amino acid composition and sequence of protein molecules, is based on their characteristic acid-base behavior.

Those α-amino acids having a single amino group and a single carboxyl group crystallize from neutral aqueous solutions in a fully ionized species called a *dipolar ion* or *zwitterion* (German for "hybrid ion") (Figure 5-9). Although such dipolar ions are electrically neutral and do not move in an electric field, they have opposite electric charges at their two "poles." The dipolar nature of amino acids was first suggested by the fact that crystalline amino acids have melting points that are much higher than those of other organic molecules of similar size. The crystal lattice of amino acids is held together by strong electrostatic forces between positively and negatively charged functional groups of neighboring molecules, resembling the stable ionic crystal lattice of NaCl (page 70). Very high temperatures must be applied to such an ionic lattice to separate the strongly interacting positive and negative charges from each other and thus cause the crystal to melt. In

Figure 5-9
Nonionic and dipolar (zwitterion) forms of amino acids. Note the separation of the + and − charges in the dipolar form, which makes it an electric dipole.

Nonionic form	Dipolar form (also called zwitterion)
$H_2N-\overset{COOH}{\underset{R}{C}}-H$	$H_3\overset{+}{N}-\overset{COO^-}{\underset{R}{C}}-H$

contrast, most simple nonionic organic compounds of similar molecular weights have relatively low melting points, consonant with their relatively "soft" and unstable nonionic crystal lattices.

Amino Acids Can Act as Acids and as Bases

When a crystalline amino acid, e.g., alanine, is dissolved in water, it occurs as the dipolar ion, which can act either as an acid (proton donor)

$$R-\underset{\overset{|}{\overset{+}{N}H_3}}{\overset{\overset{H}{|}}{C}}-COO^- \rightleftharpoons R-\underset{\overset{|}{NH_2}}{\overset{\overset{H}{|}}{C}}-COO^- + H^+$$

or as a base (proton acceptor)

$$R-\underset{\overset{|}{\overset{+}{N}H_3}}{\overset{\overset{H}{|}}{C}}-COO^- + H^+ \rightleftharpoons R-\underset{\overset{|}{\overset{+}{N}H_3}}{\overset{\overset{H}{|}}{C}}-COOH$$

Substances having this two-way property are *amphoteric* (Greek *amphi*, "both") and are often called *ampholytes*, from "amphoteric electrolytes." A simple monoamino monocarboxylic α-amino acid, such as alanine, is actually a *diprotic* acid when it is fully protonated, i.e., when both its carboxyl group and amino group have accepted protons. In this form it has two groups that can ionize to yield protons, as indicated in the following equations:

$$R-\underset{\overset{|}{\overset{+}{N}H_3}}{\overset{\overset{H}{|}}{C}}-COOH \xrightarrow{H^+\uparrow} R-\underset{\overset{|}{\overset{+}{N}H_3}}{\overset{\overset{H}{|}}{C}}-COO^- \xrightarrow{H^+\uparrow} R-\underset{\overset{|}{NH_2}}{\overset{\overset{H}{|}}{C}}-COO^-$$

Amino Acids Have Characteristic Titration Curves

Figure 5-10 shows the titration curve of the diprotic form of alanine. The plot has two distinct stages, each corresponding to the removal of one proton. Each of the two stages resembles in shape the titration curve of a monoprotic acid, such as acetic acid (Figure 4-10, page 80), and can be analyzed in the same way. At the beginning of the titration of alanine the predominant species is $^+NH_3-CHR-COOH$, the fully protonated form (in this formula R represents the methyl group of alanine). At the midpoint in the first stage of the titration, in which the carboxyl group of alanine loses its proton, equimolar concen-

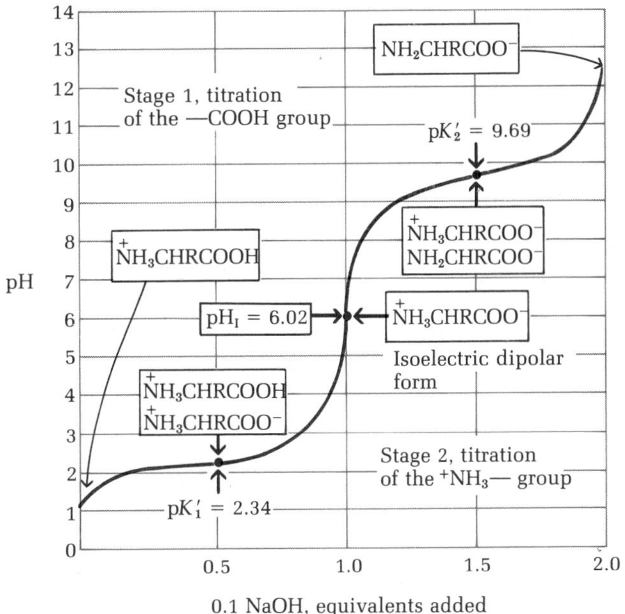

Figure 5-10
The titration curve of 0.1 M alanine. The ionic species predominating at various pH values are shown in boxes. R designates the methyl group of alanine. The flat portions of the titration curve, centered about $pK'_1 = 2.34$ and $pK'_2 = 9.69$, are the zones having buffering power.

trations of proton-donor ($^+NH_3$—CHR—COOH) and proton-acceptor ($^+NH_3$—CHR—COO^-) species are present. At the midpoint of a titration, we recall (page 82), the pH is equal to the pK′ of the protonated group being titrated. Since the midpoint pH is 2.34, the carboxyl group of alanine has a pK′ of 2.34. If we now proceed further with the titration, we come to another important point, at pH 6.02. Here there is a point of inflection, at which we have completed the removal of the first proton and begin to remove the second. At this pH the alanine is present largely as the dipolar ion $^+NH_3$—CHR—COO^-. We shall return to the significance of this point in the titration curve shortly.

The second stage of the titration corresponds to the removal of a proton from the $^+NH_3$— group of alanine. At the midpoint we have equimolar concentrations of $^+NH_3$—CHR—COO^- and NH_2—CHR—COO^-. The pH at this point is 9.69, equal to the pK′ for the $^+NH_3$— group. The titration is complete at pH of about 12, at which point the predominant form of alanine is NH_2—CHR—COO^-.

From the titration curve of alanine we can derive several important pieces of information. First, it gives us a quantitative measure of the pK′ of each of the two ionizing groups, that of the carboxyl group being 2.34 and that of the substituted ammonium group 9.69. Note that the carboxyl group of alanine is over 100 times more strongly dissociated than the carboxyl group of acetic acid, which has a pK′ of 4.76. This may seem unexpected, since the pK′ values of other simple monocarboxylic acids such as formic acid or propionic acid are similar to that of acetic acid (see Table 4-4, page 79). The enhanced tendency of the carboxyl group of alanine to ionize is due to electrostatic repulsion of the carboxyl proton by the nearby

Figure 5-11
The amino group of α-amino acids increases the tendency of the carboxyl group to ionize because of the mutual repulsion between the positive charge on the $^+NH_3$— group and the positively charged H^+ ion (color). Thus the carboxyl group of α-amino acids tends to be more completely ionized than in acetic acid.

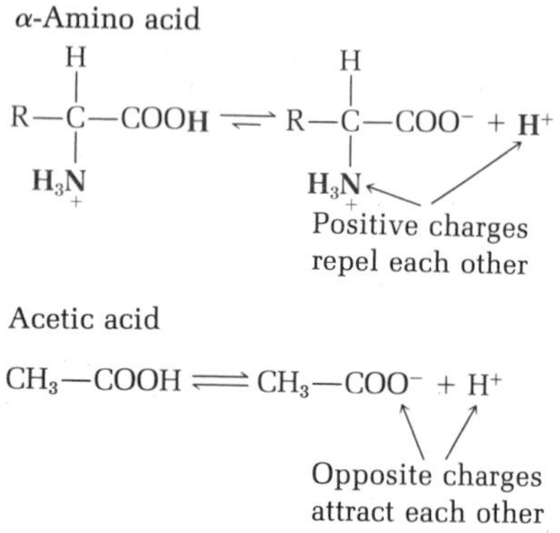

α-Amino acid

$$R\text{—}\underset{H_3\overset{+}{N}}{\overset{H}{C}}\text{—COOH} \rightleftharpoons R\text{—}\underset{H_3\overset{+}{N}}{\overset{H}{C}}\text{—COO}^- + H^+$$

Positive charges repel each other

Acetic acid

$$CH_3\text{—COOH} \rightleftharpoons CH_3\text{—COO}^- + H^+$$

Opposite charges attract each other

positively charged $^+NH_3$— group on the α carbon atom (Figure 5-11). For this reason the ionization equilibrium of the carboxyl group of alanine is such as to favor dissociation of H^+. No such repelling force influences the ionization of the —COOH group of acetic acid (Figure 5-11).

The second piece of information given by the titration curve of alanine (Figure 5-10) is that this amino acid has *two* regions of buffering power (see page 80). One of these is the relatively flat portion of the curve centered about the first pK′ of 2.34, indicating that alanine is a good buffer in the pH zone near 2.34. The other buffering zone is between pH 8.7 and pH 10.7. Note also that alanine is not a good buffer at the pH of intracellular fluid or blood, about 7.4.

With the Henderson-Hasselbalch equation (page 84) we can calculate the proportions of proton-donor and proton-acceptor species of alanine required to make a buffer at a given pH within the buffering ranges of alanine; it also makes it possible to solve other kinds of buffer problems involving amino acids.

The Titration Curve Predicts the Electric Charge of Amino Acids

A third important piece of information derived from the titration curve of an amino acid is the relationship between its net electric charge and the pH of the solution. At pH 6.02, the point of inflection between the two stages in its titration curve, alanine is present as its dipolar or zwitterion form, which is fully ionized but has no *net* electric charge (Figure 5-10). The alanine molecule at this pH is electrically neutral and will not move in an electric field. This characteristic pH is called the isoelectric pH or isoelectric point (pH_I or pI). The isoelectric point is the arithmetic mean of the two pK′ values:

$$pH_I = \tfrac{1}{2}(pK_1' + pK_2')$$

which in the case of alanine is

$$pH_I = \tfrac{1}{2}(2.34 + 9.69) = 6.02$$

At any pH above the isoelectric point, alanine has a net negative charge and will thus move toward the positive electrode (the anode) when placed in an electric field. At any pH below its isoelectric pH, alanine has a net positive charge, as can be seen in Figure 5-10, and will move toward the negative electrode, the cathode. The farther the pH of an alanine solution is from the isoelectric point the greater the net electric charge of the population of alanine molecules. At pH 1.0, for example, alanine exists entirely as the form $^+NH_3$—CHR—COOH, with a net positive charge of 1.0. At pH 2.34, where we have an equal mixture of $^+NH_3$—CHR—COOH and $^+NH_3$—CHR—COO^-, the average or net positive charge is 0.5. In the same way we can predict the sign and the magnitude of the net charge at other pH values of any amino acid at any given pH.

This information has practical importance, as we shall soon see, since the different amino acids can be separated from each other on the basis of the direction and relative rate of migration of each when a mixture of them is placed in an electric field at a known pH.

Table 5-4 pK′ Values for the Ionizing Groups of Some Amino Acids at 25°C

Amino acid	pK'_1 —COOH	pK'_2 $—NH_3^+$	pK'_R R group
Glycine	2.34	9.6	
Alanine	2.34	9.69	
Leucine	2.36	9.60	
Serine	2.21	9.15	
Threonine	2.63	10.43	
Glutamine	2.17	9.13	
Aspartic acid	2.09	9.82	3.86
Glutamic acid	2.19	9.67	4.25
Histidine	1.82	9.17	6.0
Cysteine	1.71	10.78	8.33
Tyrosine	2.20	9.11	10.07
Lysine	2.18	8.95	10.53
Arginine	2.17	9.04	12.48

Amino Acids Differ in Their Acid-Base Properties

We have analyzed the titration curve of alanine in Figure 5-10 in some detail. What about the other 19 amino acids? Fortunately, we can make some simplifying generalizations about the acid-base behavior of different classes of amino acids.

All amino acids having a single α-amino group, a single carboxyl group, and an R group that does not ionize have titration curves resembling that of alanine. This group of amino acids (Table 5-3) is characterized by having very similar, although not identical, values for pK'_1 in the range 2.0 to 3.0 and for pK'_2 in the range 9.0 to 10.0, as seen from some examples in Table 5-4. Therefore, all the amino acids in this group behave like alanine and have titration curves resembling that in Figure 5-10.

Amino acids with an ionizable R group (Table 5-3) have more complex titration curves with *three* stages corresponding to the three possible ionization steps: thus they have three pK′ values. The third stage for the titration of the ionizable R group merges to some extent with the others. The titration curves of two representatives of this group, glutamic acid and histidine, are shown in Figure 5-12. The isoelectric points of amino acids in this class reflect the type of ionizing R groups present. For example, glutamic acid, with two carboxyl groups and one amino group, has an isoelectric point of 3.22 (the average of the pK′ values of the two carboxyl groups), considerably lower than that of alanine. Similarly, the isoelectric point of lysine with two amino groups is 9.74, much higher than that of alanine.

Figure 5-12
The titration curves of glutamic acid and histidine. The pK′ of the R group is designated pK'_R.

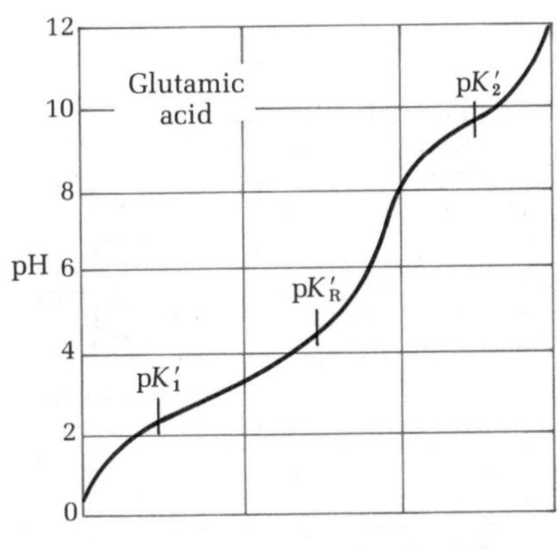

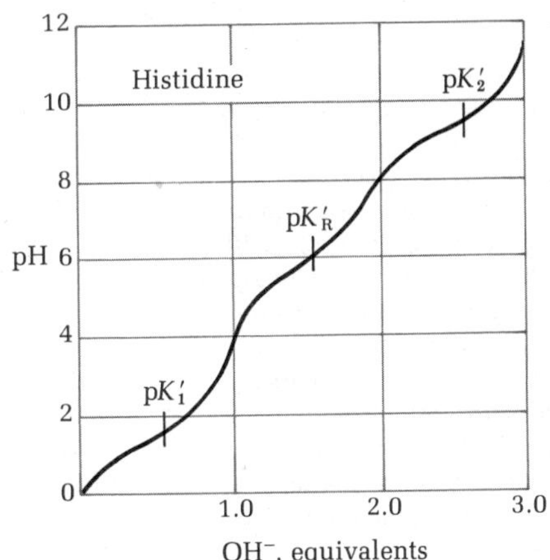

Another important generalization can be made about the acid-base behavior of the 20 standard amino acids. For all practical purposes only one, histidine, has significant buffering power near the pH of intracellular fluids and blood. As can be seen in Table 5-4 and Figure 5-12, the R group of histidine has a pK′ of 6.0, which gives it significant buffering capacity at pH 7. All the other amino acids have pK′ values too far away from pH 7 to be effective buffers at this pH. The oxygen-carrying protein hemoglobin of the red blood cell is unique in having a high content of histidine units in its structure. They give hemoglobin considerable buffering power near pH 7, which is important to the role of red blood cells in the transport of oxygen and carbon dioxide by the blood (Chapter 25).

Their Acid-Base Properties Are the Basis for the Analysis of Amino Acids

As we shall see in Chapter 6, the first step in establishing the structure of a given protein is to hydrolyze it into its component amino acids and then determine how many of each kind are present. To separate, identify, and measure quantitatively the amounts of each of the 20 amino acids in a mixture of all may seem tedious and difficult, but very sensitive and powerful methods are available to perform this task quickly, in particular, *electrophoresis* and *ion-exchange chromatography*. Both these methods exploit differences in the acid-base behavior of the different amino acids, i.e., differences in the sign and magnitude of their net electric charges at a given pH, which are predictable from their pK′ values and their titration curves.

Paper Electrophoresis Separates Amino Acids according to Electric Charge

The simplest method for separating amino acids is paper electrophoresis (Figure 5-13). A drop of an aqueous solution of the amino acid mixture is placed on a filter-paper strip moistened with a buffer at a given pH. A high-voltage electric field is applied to the strip. Because of their different pK′ values, the

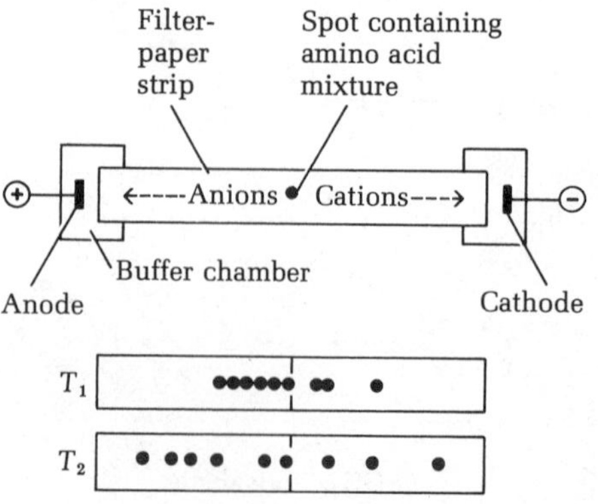

Figure 5-13
Separation of amino acids by paper electrophoresis. A drop of a solution of the amino acid mixture is dried on the paper. The paper strip is moistened with a buffer at a given pH and placed between cooling plates. The ends of the strip are immersed in the electrode compartments. Application of a direct-current electric field separates the amino acids according to their net electric charge at the pH used. Amino acids that are cations at the pH used will migrate toward the cathode or negative pole; anionic amino acids will move to the anode or positive pole, as indicated at time T_1. At the end of the process, T_2, the paper is dried, sprayed with ninhydrin, and heated, revealing the location of the amino acids, which are identified by comparison with the position of authentic amino acids as markers.

amino acids migrate in different directions and at different rates along the strip, depending on the pH of the buffer system and the voltage applied. For example, at pH 1.0 histidine, arginine, and lysine have a charge of +2 and move more rapidly to the negatively charged cathode than the other amino acids, which have a charge of +1. At pH 6.0, on the other hand, the positively charged amino acids (lysine, arginine, histidine) move to the cathode and the negatively charged amino acids (aspartic acid and glutamic acid) to the anode. All the other amino acids will remain at or near the origin, since they have no ionizing groups other than their α-amino and α-carboxyl groups and thus have about the same isoelectric point, as can be determined from the pK_1' and pK_2' values in Table 5-4. To locate the amino acids on the paper, it is dried, sprayed with ninhydrin (page 110), and heated. Blue or purple spots, each indicating the presence of an amino acid, will appear on the paper. Known specimens of the different amino acids are run under the same conditions as "markers," to establish their characteristic locations (Figure 5-13).

Ion-Exchange Chromatography Is a More Useful Separation Process

Ion-exchange chromatography is the most widely used method for separating, identifying, and quantitating the amounts of each amino acid in a mixture. It also exploits differences in the acid-base behavior of amino acids, but additional factors make major contributions to the effectiveness of this procedure. The chromatographic column consists of a long tube filled with granules of a synthetic resin containing fixed charged groups. Resins with fixed anionic groups are called cation-exchange resins; those with fixed cationic groups are anion exchangers. In the simplest form of ion-exchange chromatography, amino acids can be separated on columns of cation-exchange resins in which the fixed anionic groups, e.g., sulfonic acid groups ($—SO_3^-$), are first "charged" with Na^+ (Figure 5-14). An acid solution (pH 3.0) of the amino acid mixture to be analyzed is then poured on the column and allowed to percolate through slowly. At pH 3.0 the amino acids are largely cations with net positive charges, but they differ in the extent to which they are ionized. As the mixture passes down the column, the positively charged amino acids will displace the bound Na^+ ions from the fixed $—SO_3^-$ groups of the resin particles. At pH 3.0 the amino acids with the largest positive charge (lysine, arginine, and histidine) will displace Na^+ from the resin first and will be bound

Figure 5-14
Different ionic forms of a cation-exchange resin. The negatively charged sulfonate ($—SO_3^-$) groups attract and bind cations, such as H^+, Na^+, or cationic forms of amino acids (below). At pH 3 most amino acids are cations but differ in the net strength of their positive charges and thus in the degree to which they can displace Na^+ from the fixed anionic groups. Lysine would be bound most tightly because of its two $—^+NH_3$ groups, whereas glutamic acid and aspartic acid would be bound least tightly, since they would have the least amount of positive charge at pH 3. The binding of amino acids to ion-exchange resins is also affected by their degree of adsorption or their solubility in the resin particles.

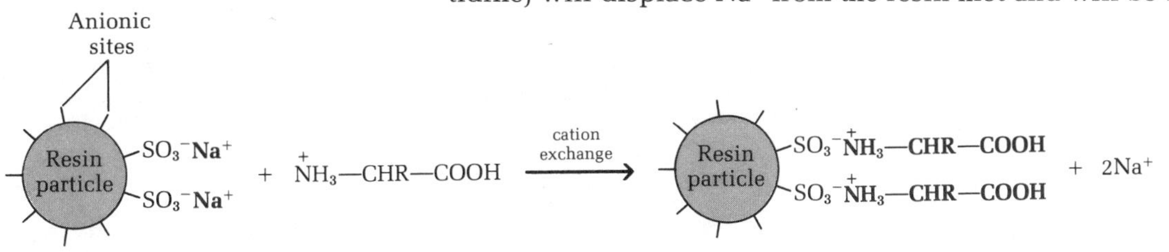

Anionic sites

Resin particle $(SO_3^-Na^+)_2$ + $\overset{+}{N}H_3—CHR—COOH$ $\xrightarrow{\text{cation exchange}}$ Resin particle $(SO_3^-\overset{+}{N}H_3—CHR—COOH)_2$ + $2Na^+$

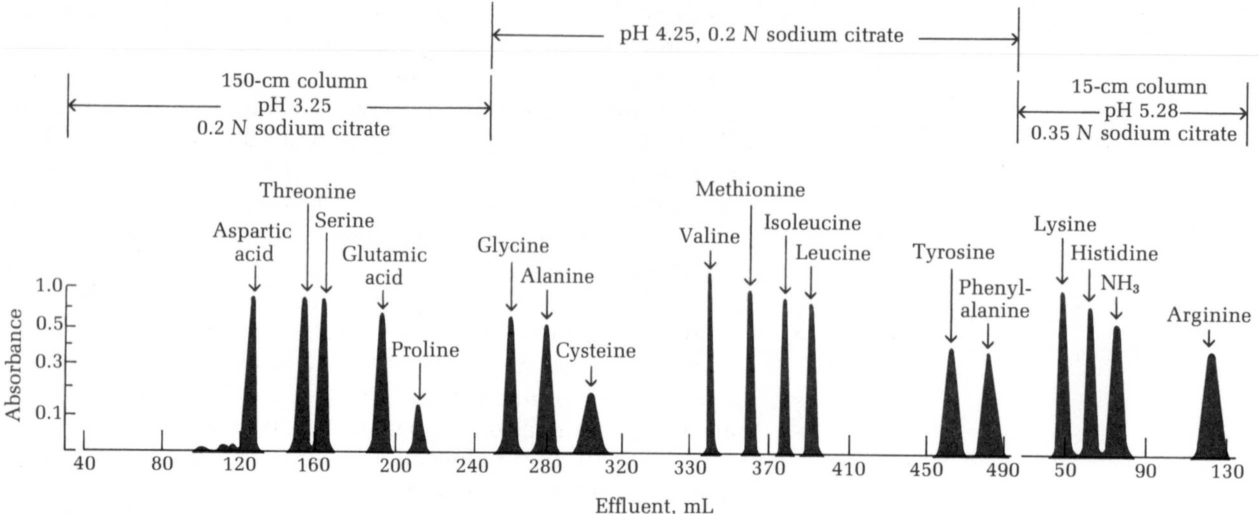

Figure 5-15
Automatically recorded chromatographic analysis of amino acids on a cation-exchange resin. The elution is carried out with different buffers of successively higher pH. The effluent is caught in small volumes, and the amino acid content of each tube is automatically analyzed. The area under each peak is proportional to the amount of each amino acid in the mixture.

to the resin most tightly. The amino acids with the least amount of positive charge at pH 3.0 (glutamic acid and aspartic acid) will be bound least. All the other amino acids will have intermediate amounts of positive charge. The different amino acids will therefore move down the resin column at different rates, which depend largely on their pK' values but also in part on their adsorption or solubility in the resin particle. Glutamic and aspartic acids will move down the column at the highest rates, since they will be bound least at pH 3.0, whereas lysine, arginine, and histidine will move most slowly. Small fractions of a few milliliters each are collected from the bottom of the column and analyzed quantitatively. The entire procedure has been automated, so that elution, collection of fractions, analysis of each fraction, and recording of data are performed automatically in an *amino acid analyzer*. Figure 5-15 shows a chromatogram of an amino acid mixture analyzed in this way.

Amino Acids Have Characteristic Chemical Reactions

As for all organic compounds, the chemical reactions of amino acids are those characteristic of their functional groups (Chapter 3). Since all amino acids contain amino and carboxyl groups, all will give chemical reactions characteristic for these groups. For example, their amino groups can be acetylated, and their carboxyl groups can be esterified. Although we shall not examine all such organic reactions of amino acids, there are two important reactions that are widely used for the detection, measurement, and identification of amino acids.

The first is the *ninhydrin reaction* (Figure 5-16), which is used to detect and estimate amino acids quantitatively in small amounts. Heating with excess ninhydrin yields a purple product with all amino acids having a free α-amino group, whereas a yellow product is formed from *proline*, in which the α-amino group is substituted. Under appropriate conditions the intensity

Figure 5-16
The ninhydrin reaction for detection and measurement of α-amino acids. The fate of the amino acid atoms is traced in color. Two molecules of ninhydrin and the nitrogen atom of the amino acid end up in the purple pigment.

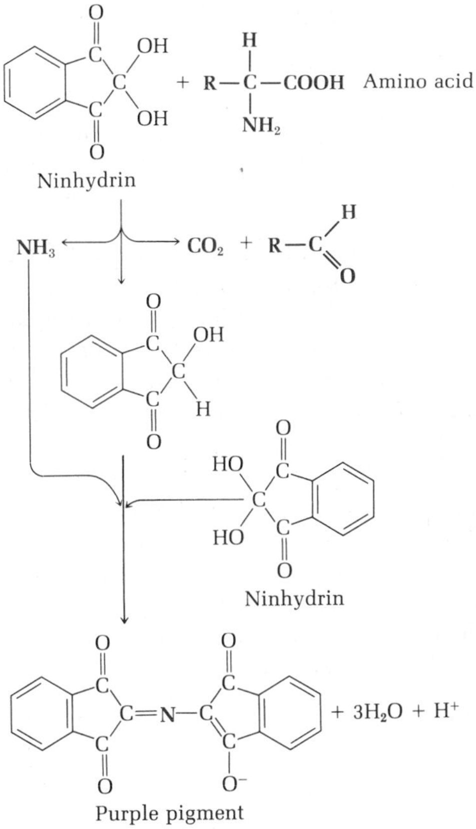

Figure 5-17
Formation of 2,4-dinitrophenyl derivatives of amino acids.

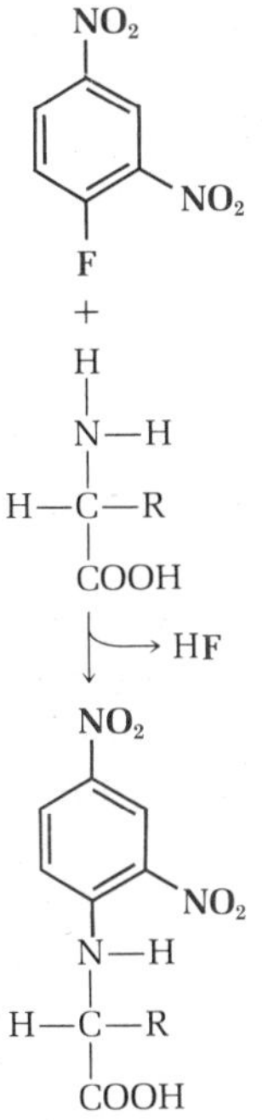

of color produced can be used to measure amino acid concentration colorimetrically. This is a very sensitive method for measuring amino acid concentration.

A second important reaction of amino acids is with the reagent *1-fluoro-2,4-dinitrobenzene* (FDNB). In mildly alkaline solution FDNB reacts with α-amino acids to yield *2,4-dinitrophenyl* derivatives (Figure 5-17), useful in the identification of individual amino acids. Later we shall see the importance of this reaction in determining the amino acid sequence of peptides.

Peptides Are Chains of Amino Acids

Two amino acid molecules can be covalently joined through a substituted amide linkage (page 50), termed a *peptide bond*, to yield a *dipeptide*. Such a linkage is formed by removal of the elements of H_2O from the carboxyl group of one amino acid and the α-amino group of the other by the action of strong condensing agents (Figure 5-18). Three amino acids can be joined by two peptide bonds in a similar manner to form a *tripeptide;*

Figure 5-18
Formation of a dipeptide.

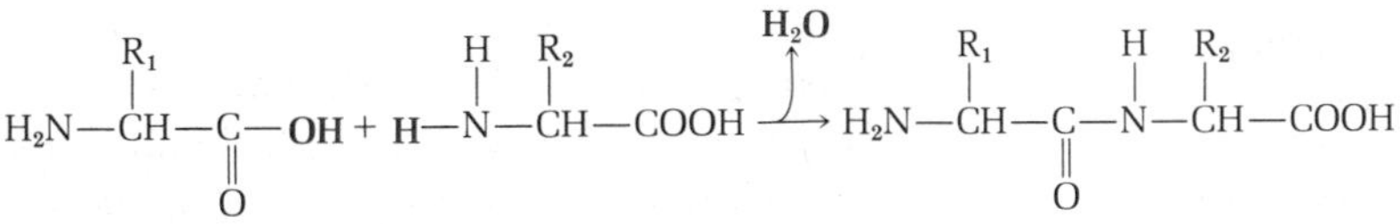

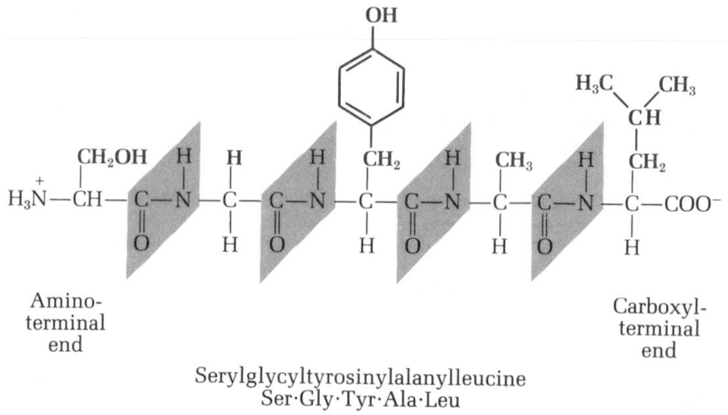

Figure 5-19
Structure of serylglycyltyrosinylalanylleucine, a pentapeptide. Peptides are named beginning with the amino-terminal residue. The peptide bonds are shown with gray shading, the R groups with color.

Amino-terminal end

Carboxyl-terminal end

Serylglycyltyrosinylalanylleucine
Ser·Gly·Tyr·Ala·Leu

similarly, we have *tetrapeptides* and *pentapeptides*. When there are many amino acids joined in this fashion, the structure is called a *polypeptide*. Peptides of varying length are formed by the partial hydrolysis of the very long polypeptide chains of proteins, which may have hundreds of amino acid units.

Figure 5-19 shows the structure of a pentapeptide. The amino acid units in a peptide are usually called *residues* (they are no longer amino acids since they have lost a hydrogen atom from their amine groups and a portion of their carboxyl groups). The amino acid residue at that end of a peptide having a free α-amino group is the *amino-terminal* residue (also *N-terminal residue*); the residue at the opposite end, which has a free carboxyl group, is the *carboxyl-terminal* or *C-terminal residue*. Peptides are named from the sequence of their constituent amino acids beginning from the N-terminal residue, as shown in Figure 5-19.

Figure 5-20
The 39 possible dipeptides containing glycine. The three-letter abbreviations for amino acids are used to show the large number of possible dipeptides containing glycine. The amino-terminal residue is at the left.

Gly·Gly	
Gly·Ala	Ala·Gly
Gly·Val	Val·Gly
Gly·Leu	Leu·Gly
Gly·Ile	Ile·Gly
Gly·Pro	Pro·Gly
Gly·Met	Met·Gly
Gly·Phe	Phe·Gly
Gly·Trp	Trp·Gly
Gly·Ser	Ser·Gly
Gly·Thr	Thr·Gly
Gly·Cys	Cys·Gly
Gly·Tyr	Tyr·Gly
Gly·Asn	Asn·Gly
Gly·Gln	Gln·Gly
Gly·Asp	Asp·Gly
Gly·Glu	Glu·Gly
Gly·Lys	Lys·Gly
Gly·Arg	Arg·Gly
Gly·His	His·Gly

Peptides Can Be Separated on the Basis of Their Ionization Behavior

Peptides formed by partial hydrolysis of proteins occur in enormous variety. For example, a given amino acid can form a dipeptide with each of the 20 standard amino acids in two different sequences to give a total of 39 dipeptides (Figure 5-20). Each of the other 19 amino acids can also form a similar series of dipeptides. If we count up all the possible dipeptides that can be made from the 20 amino acids, we will have $20 \times 20 = 400$ dipeptides all together. The number of possible tripeptides and tetrapeptides of different composition and sequences is very much greater. The quantitative separation of short peptides is therefore a much more difficult task than the separation of the 20 amino acids. Nevertheless, it is possible to separate complex mixtures of peptides by exploiting differences in their acid-base behavior and their polarity.

Peptides contain only one free α-amino group and one free α-carboxyl group at their terminal residues (Figure 5-21). These groups ionize as they do in simple amino acids. All the other α-amino and α-carboxyl groups of the constituent amino acids

Figure 5-21
Ionization and electric charge of peptides. The groups ionized at pH 6.0 are in color. The cationic, isoelectric, and anionic forms of the dipeptide alanylalanine are shown above. Below is shown a tetrapeptide.

Cationic form (below pH 3)

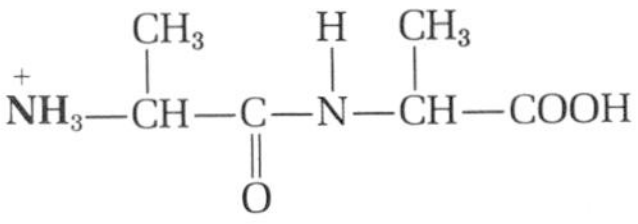

$$\overset{+}{N}H_3-\underset{|}{\overset{CH_3}{|}}{CH}-\underset{\|}{C}(=O)-\overset{H}{\overset{|}{N}}-\overset{CH_3}{\overset{|}{CH}}-COOH$$

Isoelectric form

$$\overset{+}{N}H_3-\overset{CH_3}{\overset{|}{CH}}-C(=O)-\overset{H}{\overset{|}{N}}-\overset{CH_3}{\overset{|}{CH}}-COO^-$$

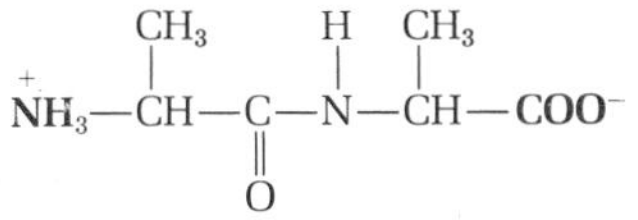

Anionic form (above pH 10)

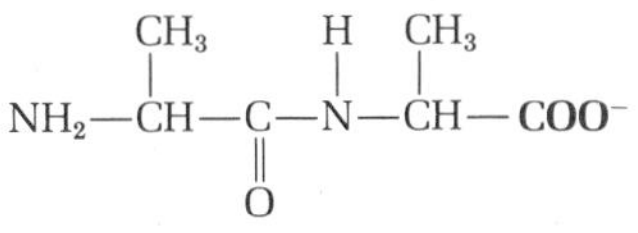

$$NH_2-\overset{CH_3}{\overset{|}{CH}}-C(=O)-\overset{H}{\overset{|}{N}}-\overset{CH_3}{\overset{|}{CH}}-COO^-$$

Alanylglutamylglycyllysine, a tetrapeptide with two residues having ionizing R groups

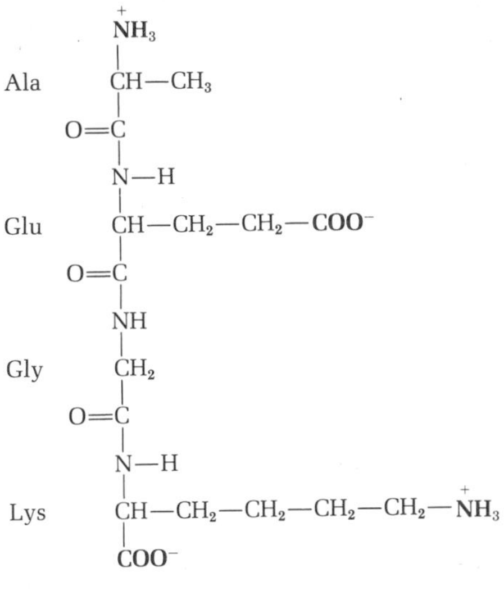

Ala $\overset{+}{N}H_3-CH(CH_3)-C(=O)-$
$N-H$
Glu $CH(-CH_2-CH_2-COO^-)-C(=O)-$
NH
Gly $CH_2-C(=O)-$
$N-H$
Lys $CH(-CH_2-CH_2-CH_2-CH_2-\overset{+}{N}H_3)-COO^-$

are covalently joined in the form of peptide bonds, which do not ionize and thus do not contribute to the total acid-base behavior of peptides. However, the R groups of some amino acids can ionize (Table 5-4). When such amino acids are present in a peptide, their R groups contribute to its acid-base properties (Figure 5-21). Thus the total acid-base behavior of a peptide can be predicted from its single free α-amino and α-carboxyl groups at the ends of the chain and the nature and number of its ionizing R groups. Like free amino acids, peptides have characteristic titration curves and a characteristic isoelectric pH at which they do not move in an electric field.

Despite the very large number of different peptides that can be formed in the incomplete hydrolysis of proteins, complex mixtures of peptides can be separated from each other by ion-exchange chromatography or electrophoresis or a combination of both, on the basis of the differences in their acid-base behavior and polarity at different pH values.

Peptides Have Characteristic Chemical Reactions

Like other organic molecules, peptides undergo chemical reactions that are characteristic not only of their functional groups, e.g., the free amino and carboxyl groups, but also of their R groups.

There are two very useful reactions of peptides. Their peptide bonds can be hydrolyzed by boiling with either strong acid or base to yield the constituent amino acids in free form

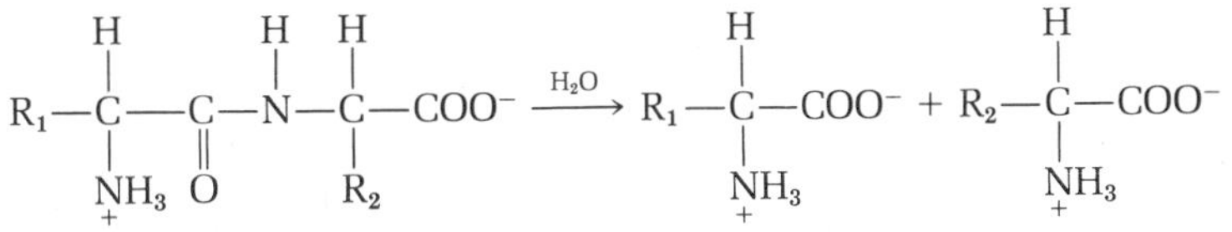

$$R_1-\underset{\underset{+}{N}H_3}{\overset{H}{C}}-\underset{O}{\overset{\|}{C}}-\overset{H}{N}-\underset{R_2}{\overset{H}{C}}-COO^- \xrightarrow{H_2O} R_1-\underset{\underset{+}{N}H_3}{\overset{H}{C}}-COO^- + R_2-\underset{\underset{+}{N}H_3}{\overset{H}{C}}-COO^-$$

Hydrolysis of peptide bonds in this manner is a necessary step in determining the amino acid composition and sequence of proteins. Peptide bonds can also be hydrolyzed by certain enzymes, such as *trypsin* and *chymotrypsin*, proteolytic (protein-dissolving) enzymes secreted into the intestine to aid in digesting, i.e., hydrolyzing, food proteins. Although boiling with acid or base hydrolyzes all peptide bonds, regardless of the nature and sequence of the amino acid units they join, trypsin and chymotrypsin are selective in their catalysis of peptide hydrolysis. Trypsin hydrolyzes only those peptide bonds in which the carboxyl group has been contributed by a lysine or arginine residue. Chymotrypsin, on the other hand, attacks only those peptide bonds in which the carboxyl group is contributed by phenylalanine, tryptophan, and tyrosine. As we shall see, such selective enzymatic hydrolysis is very useful in analysis of amino acid sequence.

The other important chemical reaction of peptides, also used in determining their amino acid sequence, is with *1-fluoro-2,4-dinitrobenzene*. We have seen (Figure 5-17) that this

Figure 5-22
Labeling the amino-terminal residue of a tetrapeptide with 1-fluoro-2,4-dinitrobenzene (FDNB).

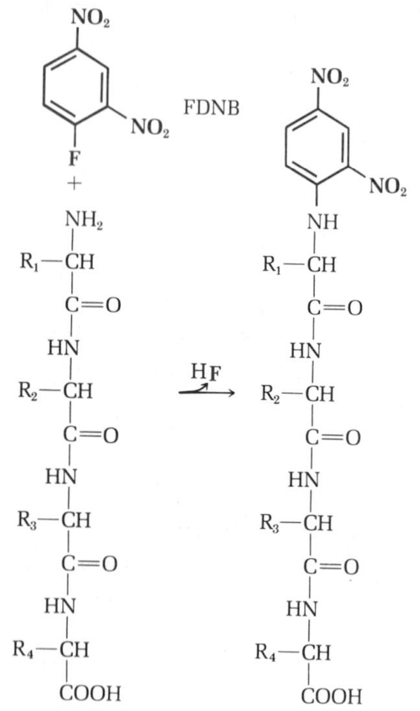

Tetrapeptide

2,4-Dinitrophenyl-tetrapeptide

reagent reacts with the α-amino group of a free amino acid to yield a 2,4-dinitrophenylamino acid. It also reacts with the α-amino group of the amino-terminal residue of peptides, whether short or long, to form a dinitrophenylpeptide. By this reaction the amino-terminal residue of a peptide is labeled (Figure 5-22). In Chapter 6 we shall see how this and other reactions for labeling the amino-terminal residues of peptides are used in determining the amino acid sequence of polypeptide chains.

Some Peptides Have Intense Biological Activity

In addition to the peptides formed as products of the partial hydrolysis of protein molecules, many peptides occur in free form in living matter, not associated with protein structure. Most interesting is the fact that many such free peptides have intense biological activity. For example, a number of hormones are known to be peptides or polypeptides. Hormones are chemical messengers, secreted by specific cells of endocrine glands, such as the pancreas, pituitary, or adrenal cortex, and sent via the blood to stimulate specific functions of other tissues or organs. The hormone *insulin*, secreted by the B cells of the pancreas, is a chemical messenger carried by the blood to other organs, especially the liver and muscles, where it becomes bound to receptors on cell surfaces and stimulates the capacity of these cells to use glucose as a metabolic fuel. Insulin contains two polypeptide chains, one having 30 amino acid residues and the other 21. Other polypeptide hormones include *glucagon*, a pancreatic hormone that opposes the action of insulin, and *corticotropin*, a hormone of the anterior pituitary gland that stimulates the adrenal cortex. Corticotropin has 39 amino acid residues.

Some hormones have much shorter peptide chains. Among them are *oxytocin* (nine amino acid residues), a hormone secreted by the posterior pituitary that stimulates uterine contractions; *bradykinin* (nine residues), a hormone that inhibits inflammation of tissues; and *thyrotropin-releasing factor* (three residues), which is formed in the hypothalamus and stimulates the release of another hormone, *thyrotropin*, from the anterior pituitary gland (Figure 5-23). Especially noteworthy among short peptides are the *enkephalins*, formed in the central nervous system. When enkephalins bind to specific receptors in certain cells of the brain, they induce analgesia, deadening of pain sensations. Enkephalins (the word means "in the head") are the body's own opiates, since they become bound to sites in the brain that also bind morphine, heroin, and other addicting opiate drugs. Some extremely toxic mushroom poisons, such as *amanitin*, are also peptides, as are many *antibiotics*, which may be regarded as "chemical warfare" agents, made by some species of microorganisms but toxic to others.

Figure 5-23
Some naturally occurring peptides with intense biological activity. The amino-terminal residues are at the left end.

Bradykinin, *a hormonelike peptide that inhibits inflammatory reactions.*

Arg·Pro·Pro·Gly·Phe·Ser·Pro·Phe·Arg

Oxytocin, *formed by the posterior pituitary gland. The shaded portion is a residue of glycinamide* ($NH_2CH_2CONH_2$).

S—S (disulfide bridge between the two Cys residues)
Cys·Tyr·Ile·Gln·Asn·Cys·Pro·Leu·Gly·NH_2

Thyrotropin-releasing factor, *formed by the hypothalamus.*

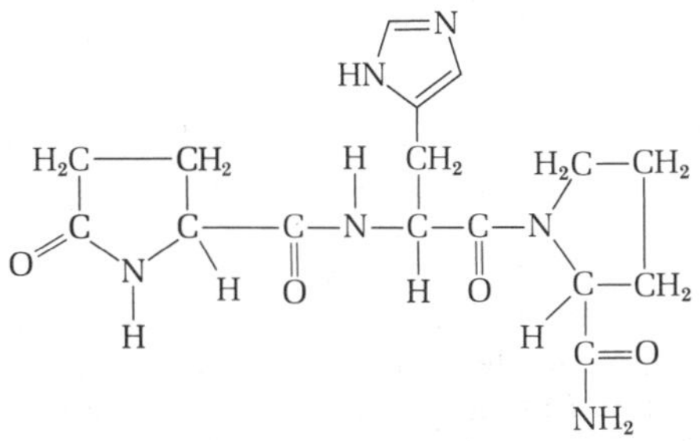

Pyroglutamic acid — Histidine — Prolinamide

Enkephalins, *brain peptides with opiatelike activity.*

Tyr·Gly·Gly·Phe·Met
Tyr·Gly·Gly·Phe·Leu

Gramicidin S, *an antibiotic. The arrows lead from the amino-terminal toward the carboxyl-terminal residues.* Orn *is the symbol for ornithine, an amino acid that does not occur in proteins. Note that gramicidin S contains two residues of a* D*-amino acid.*

D-Phe→ L-Leu→ L-Orn→ L-Val→ L-Pro
↑ ↓
L-Pro ←L-Val ←L-Orn ←L-Leu ←D-Phe

It is most remarkable that these peptides have such potent biological effects, despite the fact that the amino acids of which they are composed are harmless, nontoxic substances. Clearly, it is the sequence of amino acids in peptides and polypeptides that gives them their striking biological effects and specificity.

Summary

The 20 amino acids commonly found as hydrolysis products of proteins contain an α-carboxyl group, an α-amino group, and a distinctive R group substituted on the α carbon atom. The α carbon atom of the amino acids (except glycine) is asymmetric and thus can exist in at least two stereoisomeric forms. Only the L stereoisomers, which are related to L-glyceraldehyde, are found in proteins. The amino acids are classified on the basis of the polarity of their R groups. The nonpolar class includes alanine, leucine, isoleucine, valine, proline, phenylalanine, tryptophan, and methionine. The polar neutral class includes glycine, serine, threonine, cysteine, tyrosine, asparagine, and glutamine. The negatively charged (acidic) class contains aspartic acid and glutamic acid, and the positively charged (basic) class contains arginine, lysine, and histidine.

Monoamino monocarboxylic amino acids are diprotic acids ($^+NH_3CHRCOOH$) at low pH. As the pH is raised to about 6, the isoelectric point, the proton is lost from the carboxyl group to form the dipolar or zwitterion species $^+NH_3CHRCOO^-$, which is electrically neutral. Further increase in pH causes loss of the second proton, to yield the ionic species $NH_2CHRCOO^-$. Amino acids with ionizable R groups may exist in additional ionic species, depending on the pH. Amino acids form colored derivatives with ninhydrin. Complex mixtures of amino acids can be separated, identified, and estimated by electrophoresis or ion-exchange chromatography.

Amino acids can be joined covalently through peptide bonds to form peptides, which are also formed on incomplete hydrolysis of polypeptides. The acid-base behavior of a peptide is a function of its amino-terminal amino group, its carboxyl-terminal carboxyl group, and those R groups which ionize. Peptides can be hydrolyzed to yield free amino acids. The amino-terminal residue of a peptide can react with 1-fluoro-2,4-dinitrobenzene to yield a characteristic yellow derivative. Some peptides occur free in cells and tissues and have specific biological functions. They include hormones, antibiotics, and other agents having intense biological activity.

References

CANTOR, C. R., and P. R. SCHIMMEL: *Biophysical Chemistry*, pt. I, *The Conformation of Biological Macromolecules*, Freeman, San Francisco, 1980. Excellent textbook outlining properties of biological macromolecules and their building blocks.

COOPER, T. G.: *The Tools of Biochemistry*, Wiley, New York, 1977. Theory and practical instructions for carrying out chromatography and electrophoresis of amino acids.

CORRIGAN, J. T.: "D-Amino Acids in Animals," *Science*, **164**:142–148 (1969).

DICKERSON, R. E., and I. GEIS: *Proteins: Structure, Function, and Evolution*, 2d ed., Benjamin/Cummings, Menlo Park, Calif., 1983. Beautifully illustrated and interesting account.

HASCHEMEYER, R., and A. H. HASCHEMEYER: *Proteins: A Guide to Study by Physical and Chemical Methods*, Wiley, New York, 1973.

LEHNINGER, A. L.: *Biochemistry*, 2d ed., Worth, New York, 1975. Chapters 4 and 5 contain more detailed treatment of amino acids and peptides.

MEISTER, A.: *Biochemistry of the Amino Acids*, 2d ed., 2 vols., Academic, New York, 1965. Encyclopedic treatment of the properties, occurrence, and metabolism of amino acids.

SEGEL, I. H.: *Biochemical Calculations*, 2d ed., Wiley, New York, 1976.

Problems

1. *Specific Rotation of an Amino Acid from Watermelons* The amino acid citrulline was first isolated from the watermelon, *Citrullus vulgaris*, but it is found in most animal tissues. Although citrulline does not occur in proteins, it is the metabolic precursor of arginine and of urea, the excreted end product of amino-group metabolism. Its structure has been established as

$$\begin{array}{l} \quad CH_2CH_2CH_2\text{—}NH\text{—}\underset{\underset{\displaystyle O}{\|}}{C}\text{—}NH_2 \\ \quad\ \vdots \\ H \blacktriangleright C \blacktriangleleft NH_2 \\ \quad\ \vdots \\ \quad COOH \end{array}$$

At 25°C a 20-cm tube filled with a 5.0% solution of citrulline in 0.3 N HCl rotates the plane of plane-polarized light 1.79° to the right. What is the *specific rotation* of citrulline? Can the specific rotation of citrulline indicate whether citrulline is a D- or L-amino acid?

2. *Absolute Configuration of Citrulline* Is citrulline from watermelons (see the structure given in problem 1) a D- or L-amino acid? Explain.

3. *Relation between the Structures and Chemical Properties of the Amino Acids* Since the amino acids serve as the building blocks of proteins, understanding of their structures and chemical properties is crucial to understanding how proteins carry out their biological functions. The structures of the side chains or R groups of 16 of the amino acids (Ala, Arg, Asn, Asp, Cys, Glu, Gly, His, Lys, Met, Phe, Pro, Ser, Trp, Tyr, and Val) are given in the following list. Match each structure with the name of the amino acid and the most appropriate description of its properties from the list in the right-hand column. Some of the descriptions may be used more than once.

R-group structures

(1) —H

(2) $—CH_3$

(3) $—C(H)(CH_3)—CH_3$

(4) $—CH_2OH$

(5) $—CH_2—CH_2—CH_2—$

(6) $—CH_2—C_6H_5$ (benzene ring)

(7) $—CH_2$— (indole ring, N—H)

(8) $—CH_2—C_6H_4—OH$

(9) $—CH_2—C(=O)—O^-$

(10) $—CH_2—CH_2—C(=O)—O^-$

(11) $—CH_2—CH_2—S—CH_3$

(12) $—CH_2—SH$

(13) $—CH_2$— (imidazole ring, N, N—H)

(14) $—CH_2—CH_2—CH_2—N(H)—C(—NH_2)={}^{+}N(H)—H$

(15) $—CH_2—CH_2—CH_2—CH_2—\overset{+}{N}H_3$

(16) $—CH_2—C(=O)—NH_2$

Properties of R groups

(a) Small polar R group containing a hydroxyl group. This amino acid is important in the active site of some enzymes.
(b) Provides the least amount of steric hindrance.
(c) R group has $pK' \approx 10.5$, making it positively charged at cell pH.
(d) Sulfur-containing R group; neutral at all pHs.
(e) Aromatic R group, hydrophobic in nature and neutral at all pHs.
(f) Saturated hydrocarbon, important in hydrophobic interactions.
(g) The only amino acid having an ionizing R group with a pK' near 7. It is an important group in the active site of some enzymes.
(h) The only amino acid having a substituted α-amino group. Influences protein folding by forcing a bend in the chain.
(i) R group has a pK' near 4 and thus is negatively charged at pH 7.
(j) An aromatic R group capable of forming hydrogen bonds; it has a pK' near 10.
(k) Forms disulfide cross-links between polypeptide chains; the pK' of its functional group is about 8.
(l) R group with $pK' \approx 12$, making it positively charged at all physiological pHs. Its positive charge is important in some proteins for binding negatively charged phosphate groups.
(m) When this polar but uncharged R group is hydrolyzed, this amino acid is converted into another amino acid having a negatively charged R group at pH near 7.

Problem 4

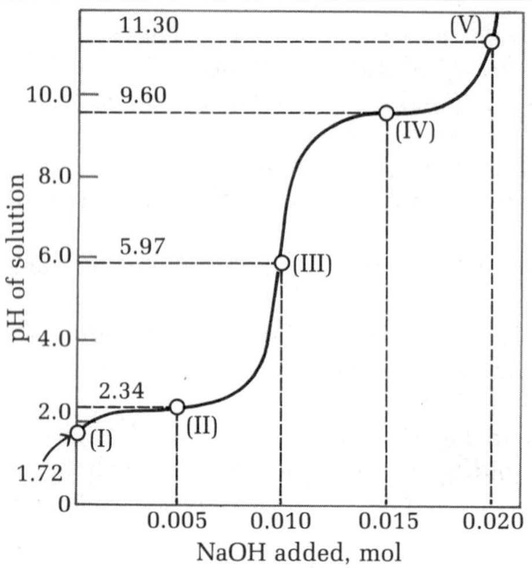

4. *Relationship between the Titration Curve and the Acid-Base Properties of Glycine* A 100-mL solution of 0.1 *M* glycine at pH 1.72 was titrated with 2 *M* NaOH solution. During the titration, the pH was monitored and the results were plotted in the graph shown. The key points in the titration are designated I to V on the graph. For each of the statements below, *identify* the appropriate key point in the titration and *justify* your choice.
   (a) What point gives us the pH of a 0.1 *M* solution of glycine as the species $H_3N^+—CH_2—COOH$?
   (b) At what point is the *average* net charge on glycine $+\frac{1}{2}$?
   (c) At what point is the amino group of half of the glycine molecules ionized?
   (d) At what point is the pH equal to the pK′ of the ionization of the carboxyl group of glycine?
   (e) At what point is the pH equal to the pK′ of the ionization of the protonated amino group ($—\overset{+}{N}H_3$) of glycine?
   (f) At what point does glycine have its maximum buffering capacity?
   (g) At what point is the *average* net charge of glycine zero?
   (h) At what point has the carboxyl group of glycine been completely titrated (first equivalence point)?
   (i) At what point is half of the carboxyl group ionized?
   (j) At what point is glycine completely titrated (second equivalence point)?
   (k) At what point is the structure of the predominant species $H_3\overset{+}{N}—CH_2—COO^-$?
   (l) At what point do the structures of the predominant species correspond to a 50-50 mixture of $H_3\overset{+}{N}—CH_2—COO^-$ and $H_2N—CH_2—COO^-$?
   (m) At what point is the average net charge of glycine -1?
   (n) At what point do the structures of the predominant species consist of a 50-50 mixture of $H_3\overset{+}{N}—CH_2—COOH$ and $H_3\overset{+}{N}—CH_2—COO^-$?
   (o) What point corresponds to the isoelectric point of glycine?
   (p) At what point is the *average* net charge on glycine $-\frac{1}{2}$?
   (q) What point represents the end of the titration?
   (r) If one wanted to use glycine as an efficient buffer, which points would represent the *worst* pHs for buffering power?
   (s) At what point in the titration is the predominant species $H_2N—CH_2—COO^-$?

Problem 5

$H_3\overset{+}{N}—CH_2—C(=O)O^-$

Dipolar (zwitterion)

$H_2N—CH_2—C(=O)OH$

Uncharged

5. *How Much Glycine Is Present as the Completely Uncharged Species?* At a pH equal to the isoelectric point, the *net* charge on glycine is zero. Although two structures of glycine can be drawn which have a net charge of zero (dipolar and uncharged forms), the predominant form of glycine at the isoelectric point is dipolar.
   (a) Explain why the form of glycine at the isoelectric point is dipolar rather than completely uncharged.
   (b) Estimate the fraction of glycine present as the completely uncharged form at the isoelectric point. Justify your assumptions.

6. *Ionization State of Amino Acids* Each ionizable group of an amino acid can exist in one of two states, charged or neutral. The electric charge on the functional group is determined by the relationship between its pK′ and the pH of the solution. This relationship is described by the Henderson-Hasselbalch equation.

(a) Histidine has three ionizable functional groups. Write the relevant equilibrium equations for its three ionizations and assign the proper equilibrium constant (pK′) for each ionization. Draw the structure of histidine in each ionization state. What is the net charge on the histidine molecule in each ionization state?
(b) Draw the structures of the predominant ionization state of histidine at pH 1, 4, 8, and 12. Note that the ionization state can be determined by treating each ionizable group independently.
(c) What is the net charge of histidine at pH 1, 4, 8, and 12? For each pH will histidine migrate toward the anode (+) or cathode (−) during electrophoresis?

7. *Preparation of a Glycine Buffer* Glycine is commonly used as a buffer. Preparation of a 0.1 *M* glycine buffer starts with 0.1 *M* solutions of glycine hydrochloride ($^+NH_3$—CH_2—COOH Cl^-) and glycine ($^+NH_3$—CH_2—COO^-), two forms of glycine that are commercially available. What volumes of these two solutions must be mixed to prepare 1 L of 0.1 *M* glycine buffer having a pH of 3.2?

8. *Paper Electrophoresis of Amino Acids* A drop of a solution containing a mixture of glycine, alanine, glutamic acid, lysine, arginine, and histidine was placed in the center of a paper strip and dried. The paper was moistened with a buffer of pH 6.0 and an electric current was applied to the ends of the strip.
(a) Which amino acid(s) moved toward the anode?
(b) Which amino acid(s) moved toward the cathode?
(c) Which remained at or near the origin?

9. *Separation of Amino Acids by Ion-Exchange Chromatography* Mixtures of amino acids are analyzed by first separating the mixture into its components through ion-exchange chromatography. A small amount of the mixture is placed on top of a column of polystyrene beads containing sulfonic acid residues (see Figure 5-14). The column is then eluted with a buffer solution. The amino acids flow down the column at different rates because of two factors that retard their movement: (1) the ionic attraction between the negatively charged sulfonic acid residues on the column and positively charged functional groups on the amino acid and (2) the hydrophobic interaction between the amino acid side chains and the strongly hydrophobic backbone of the polystyrene resin. For each pair of amino acids listed, determine which member of the pair will be eluted first from an ion-exchange column (retained the least) by a pH 7.0 buffer.
(a) Asp and Lys
(b) Arg and Met
(c) Glu and Val
(d) Gly and Leu
(e) Ser and Ala

10. *Variety of Tripeptides* Suppose you want to synthesize tripeptides using glycine, alanine, and serine as building blocks.
(a) How many different tripeptides can you prepare using any of the three amino acids in any of the three positions if you can use an amino acid more than once?

Problem 11

$$\begin{array}{c} COOH \\ | \\ H_2N-C-H \\ | \\ H-C-CH_3 \\ | \\ CH_2 \\ | \\ CH_3 \end{array}$$

Isoleucine

(b) How many different tripeptides can you prepare using each amino acid only once?

11. *Naming the Optical Isomers of Isoleucine* The structure of the amino acid isoleucine is shown at the left.
(a) How many chiral centers does it have?
(b) How many optical isomers?
(c) Draw perspective formulas for all the optical isomers of isoleucine.
(d) How would you name each of these by the *RS* system? (*Hint:* The $CH_3CH_2^-$ group has a valence density between that of the $C_6H_5^-$ and CH_3^- groups.)

12. *Comparison of the pK's of an Amino Acid and Its Peptides* The titration curve of the amino acid alanine shows the ionization of two functional groups with pK's 2.34 and 9.69, corresponding to the ionization of the carboxylic acid and the protonated amine, respectively. The titration of di-, tri-, and oligopeptides of alanine having more than four alanine residues also shows the ionization of only two functional groups, although the values of the experimental pK's are different. The trend in pK′ values is summarized in the table.
(a) Draw the structure of Ala-Ala-Ala. Identify the functional groups associated with pK_1' and pK_2'.
(b) The value of pK_1' *increases* in going from Ala to an Ala oligopeptide. Provide an explanation for this trend.
(c) The value of pK_2' *decreases* in going from Ala to an Ala oligopeptide. Provide an explanation for this trend.

Amino acid or peptide	pK_1'	pK_2'
Ala	2.34	9.69
Ala-Ala	3.12	8.30
Ala-Ala-Ala	3.39	8.03
Ala-(Ala)$_n$-Ala, $n \geq 4$	3.42	7.94

CHAPTER 6

Proteins: Covalent Structure and Biological Function

The proteins, whose name means "first" or "foremost," are the most abundant macromolecules in cells and constitute over half the dry weight of most organisms. In the preceding chapter we examined amino acids, the units of protein structure, and simple peptides, which consist of a few amino acids joined by peptide bonds. Now we shall examine the structure of proteins, which consist of polypeptides, extremely long chains of many amino acid units.

Proteins are the instruments by which genetic information is expressed. Just as there are thousands of genes in the cell nucleus, each specifying one distinctive characteristic of the organism, there are, correspondingly, thousands of different kinds of proteins in the cell, each carrying out a specific function determined by its gene. Proteins thus are not only the most abundant macromolecules but also are extremely versatile in their functions.

It is extraordinary that all proteins in all species, regardless of their function or biological activity, are built from the same basic set of 20 standard amino acids, which by themselves have no intrinsic biological activity. What is it, then, that gives one protein enzymatic activity, another protein hormone activity, and still others antibody activity? How do they differ chemically? Quite simply, proteins differ from each other because each has a distinctive *sequence* of its amino acid units. The amino acids are the alphabet of protein structure, since they can be arranged in an almost infinite number of sequences to make an almost infinite number of different proteins (Box 6-1).

It is our task in this chapter to examine the *primary structure* of protein molecules, by which we mean their covalent backbone structure and the sequence of amino acid residues. We shall also examine some relationships between amino acid sequence and biological function.

Box 6-1 How many amino acid sequences are possible?

There are thousands of different proteins in each species of organism, and there are perhaps 10 million different species. Can only 20 amino acids really be assembled into, say, 10^{11} or more different sequences?

Mathematics alone can tell us. In a dipeptide containing two different amino acids there can be two sequence isomers; in a tripeptide with three different amino acids, A, B, and C, six sequential arrangements are possible, ABC, ACB, BAC, BCA, CAB, and CBA. The general expression for calculating the number of possible sequences of a set of objects is $n!$ (read "n factorial"), where n is the number of objects. For a tetrapeptide of 4 different amino acids we would have $4! = 4 \cdot 3 \cdot 2 \cdot 1 = 24$ possible sequences. For a polypeptide of 20 different amino acids, each occurring once, the number of sequences is $20! = 20 \cdot 19 \cdot 18 \cdot \cdots$, which comes out to the startling figure of about 2×10^{18}. But this is only a very small polypeptide, with 20 residues and a molecular weight of about 2600. For a protein of MW 34,000 containing 12 different amino acids in *equal* numbers, over 10^{300} sequences are possible. If we further assume that this protein is made from 20 amino acids occurring in equal numbers, the number of possible sequences would be far greater. If there were only one molecule of each possible sequence isomer of such a protein, their total weight would far exceed the weight of the earth!

Twenty amino acids can thus be arranged in enough sequences to account not only for the thousands of proteins in each of all living species of organism but also for all species of organisms that ever existed in the past or that will exist in the future. Living species of organisms are believed to represent only about one-thousandth of all that ever existed on the earth. The molecular logic of amino acids and proteins amply allows for the ever diverging nature of biological evolution.

Proteins Have Many Different Biological Functions

Let us first examine the range and scope of the biological functions of proteins, since it is an important goal of biochemistry today to find out how the amino acid sequence of different proteins enables them to serve their different functions. We can identify several major classes of proteins according to their biological roles (Table 6-1).

Enzymes

The most varied and most highly specialized proteins are those with catalytic activity, the enzymes. Nearly all the chemical reactions of organic biomolecules in cells are catalyzed by enzymes. To date over 2000 different enzymes, each capable of catalyzing a different kind of chemical reaction, have been discovered in different forms of life.

Transport Proteins

Transport proteins in blood plasma bind and carry specific molecules or ions from one organ to another. Hemoglobin of red blood cells binds oxygen as the blood passes through the lungs and carries it to the peripheral tissues, where the oxygen is released to carry out the energy-yielding oxidation of nutrients. The blood plasma contains *lipoproteins*, which carry lipids from the liver to other organs. Other kinds of transport proteins are present in cell membranes and are adapted to bind and transport glucose, amino acids, and other nutrients across the membrane into cells.

Nutrient and Storage Proteins

The seeds of many plants store nutrient proteins required for the growth of the embryonic plant. Particularly familiar examples are the seed proteins of wheat, corn, and rice. *Ovalbumin*, the major protein of egg white, and *casein*, the major protein of milk, are other examples of nutrient proteins. *Ferritin* of animal tissues stores iron.

Contractile or Motile Proteins

Some proteins endow cells and organisms with the ability to contract, to change shape, or to move about. *Actin* and *myosin* are filamentous proteins functioning in the contractile system of skeletal muscle and also in many nonmuscle cells (page 29). Another example is *tubulin*, the protein from which microtubules are built. Microtubules are important components of flagella and cilia (page 30), which can propel cells.

Structural Proteins

Many proteins serve as supporting filaments, cables, or sheets, to give biological structures strength or protection. The major component of tendons and cartilage is the fibrous protein *collagen*, which has very high tensile strength. Leather is almost pure collagen. Ligaments contain *elastin*, a structural protein capable of stretching in two dimensions. Hair, fingernails, and feathers consist largely of the tough, insoluble protein *keratin*. The major component of silk fibers and spider webs is the protein *fibroin*.

Defense Proteins

Many proteins defend organisms against invasion by other species or protect them from injury. The *immunoglobulins* or *antibodies* of vertebrates are specialized proteins made by lymphocytes which can recognize and precipitate or neutralize invading bacteria, viruses, or foreign proteins from another species. *Fibrinogen* and *thrombin* are blood-clotting proteins that prevent loss of blood when the vascular system is injured. *Snake venoms*, *bacterial toxins*, and toxic plant proteins, such as *ricin*, also appear to function in defense.

Regulatory Proteins

Some proteins help regulate cellular or physiological activity. Among them are many *hormones*, such as *insulin*, which regulates sugar metabolism and whose deficiency is a cause of dia-

Table 6-1 Classification of Proteins According to Biological Function

Class	Example
Enzymes	Ribonuclease Trypsin
Transport proteins	Hemoglobin Serum albumin Myoglobin β_1-Lipoprotein
Nutrient and storage proteins	Gliadin (wheat) Ovalbumin (egg) Casein (milk) Ferritin
Contractile or motile proteins	Actin Myosin Tubulin Dynein
Structural proteins	Keratin Fibroin Collagen Elastin Proteoglycans
Defense proteins	Antibodies Fibrinogen Thrombin Botulinus toxin Diphtheria toxin Snake venoms Ricin
Regulatory proteins	Insulin Growth hormone Corticotropin Repressors

betes, growth hormone of the pituitary, and parathyroid hormone, which regulates Ca^{2+} and phosphate transport. Other regulatory proteins, called repressors, regulate the biosynthesis of enzymes by bacterial cells.

Other Proteins

There are numerous other proteins whose functions are rather exotic and not easily classified. Monellin, a protein of an African plant, has an intensely sweet taste. It is being studied as a nonfattening, nontoxic food sweetener for human use. The blood plasma of some Antarctic fish contains "antifreeze" proteins, which protect their blood from freezing. The wing hinges of some insects are made of the protein resilin, which has nearly perfect elastic properties.

It is extraordinary that all these proteins, with their very different properties and functions, are made from the same basic group of 20 amino acids.

(a)

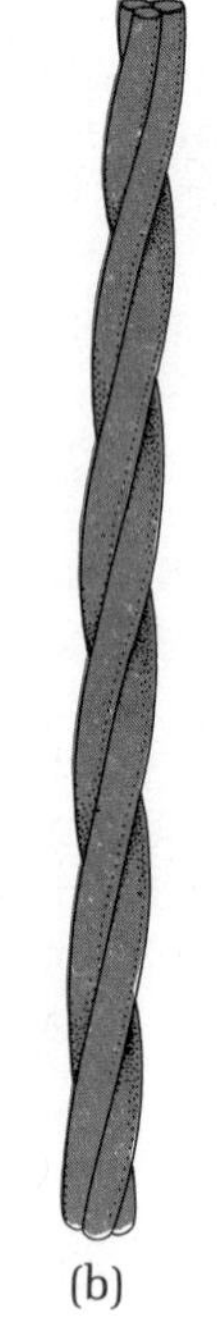

(b)

Figure 6-1
Globular and fibrous proteins. (a) In globular proteins the polypeptide chain is tightly folded. They are usually soluble in aqueous media. (b) In keratin, the fibrous protein of hair, the polypeptide chains are arranged along one axis. This drawing shows three keratin molecules twisted around each other to make a ropelike structure. Fibrous proteins are insoluble in water.

Proteins Can Also Be Classified According to Shape

Proteins can also be divided into two great classes on the basis of their shape and certain physical characteristics: globular and fibrous proteins (Figure 6-1). In globular proteins the polypeptide chain or chains are tightly folded into compact spherical or globular shapes. Globular proteins are usually soluble in aqueous systems and diffuse readily; most have a mobile or dynamic function. Nearly all enzymes are globular proteins, as are blood transport proteins, antibodies, and nutrient storage proteins. Fibrous proteins are water-insoluble, long, stringy molecules, with the polypeptide chains extended along one axis rather than folded into a globular shape. Most fibrous proteins serve in a structural or protective role. Typical fibrous proteins are α-keratin of hair and wool, fibroin of silk, and collagen of tendons.

We can also include in this class filamentous proteins that participate in contractile events in both muscle and nonmuscle cells, such as actin and myosin, as well as the protofilaments from which microtubules are constructed.

Proteins Yield Their Amino Acids on Hydrolysis

Many proteins have been isolated in pure crystalline form. Crystals of pure trypsin, a digestive enzyme secreted into the intestine, and cytochrome c, an electron-carrying protein of mitochondria, are shown in Figure 6-2. As is true for simple peptides (page 113), hydrolysis of proteins with acid or base yields a mixture of free α-amino acids, their building-block units. Each type of protein on complete hydrolysis yields a characteristic proportion or mixture of the different amino acids. Table 6-2 shows the composition of the amino acid mix-

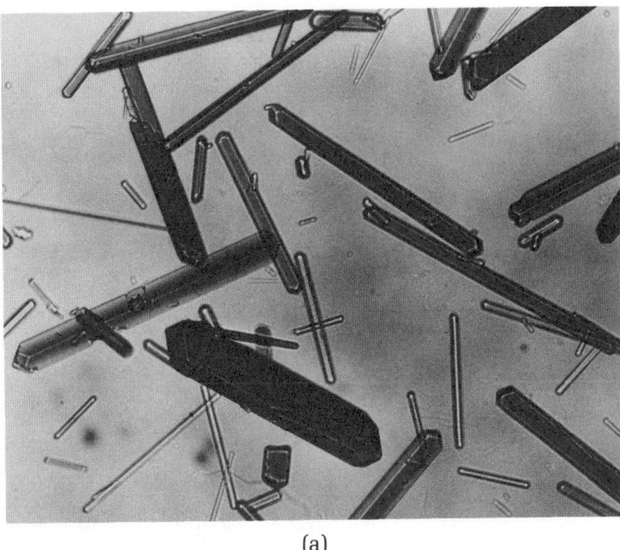

(a)

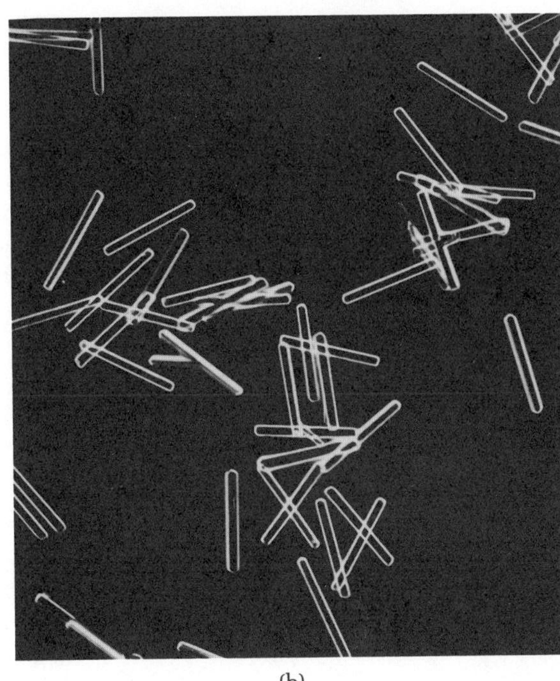

(b)

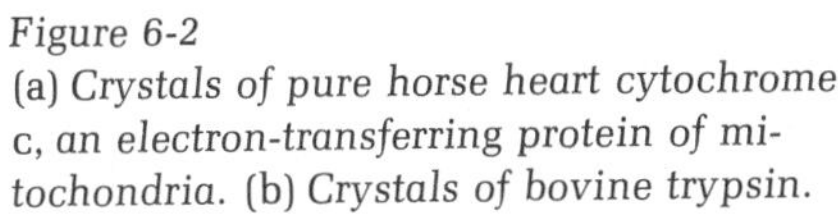

Figure 6-2
(a) Crystals of pure horse heart cytochrome c, an electron-transferring protein of mitochondria. (b) Crystals of bovine trypsin.

Table 6-2 Amino Acid Composition of Two Proteins

Data are in residues of each amino acid per molecule of the protein

Amino acid	Human cytochrome *c*	Bovine chymotrypsinogen
Ala	6	22
Arg	2	4
Asn	5	15
Asp	3	8
Cys	2	10
Gln	2	10
Glu	8	5
Gly	13	23
His	3	2
Ile	8	10
Leu	6	19
Lys	18	14
Met	3	2
Phe	3	6
Pro	4	9
Ser	2	28
Thr	7	23
Trp	1	8
Tyr	5	4
Val	3	23
Total	104	245

tures obtained on complete hydrolysis of *cytochrome c* and *bovine chymotrypsinogen*, the inactive precursor of the digestive enzyme *chymotrypsin*. We note that these two proteins, with very different functions, also differ greatly in the relative numbers of each kind of amino acid unit they contain. The 20 amino acids never occur in equal amounts in proteins. Some amino acids may occur only once per molecule in a given type of protein; others may occur in large numbers. Moreover, not all proteins contain all the 20 common amino acids. Each type of protein contains distinctively different proportions of the amino acid building blocks.

Some Proteins Contain Chemical Groups Other Than Amino Acids

Many proteins, such as the enzymes ribonuclease and chymotrypsinogen, contain only amino acids and no other chemical groups; they are called *simple proteins*. However, other kinds of proteins yield on hydrolysis some other chemical component in addition to amino acids; these are called *conjugated proteins*. The nonamino acid part of a conjugated protein is usually called its *prosthetic group*. Conjugated proteins are classified on the basis of the chemical nature of their prosthetic groups (Table 6-3). *Lipoproteins* contain lipids, *glycoproteins* contain sugar groups (Greek *glykos*, "sweet"), and *metalloproteins* contain one or another specific metal, such as iron, copper, or zinc. Usually the prosthetic group of a protein plays an important role in its biological function.

Table 6-3 Conjugated Proteins

Class	Prosthetic group	Example
Lipoproteins	Lipids	β_1-Lipoprotein of blood
Glycoproteins	Carbohydrates	γ-Globulin of blood
Phosphoproteins	Phosphate groups	Casein of milk
Hemoproteins	Heme (iron porphyrin)	Hemoglobin
Flavoproteins	Flavin nucleotides	Succinate dehydrogenase
Metalloproteins	Iron	Ferritin
	Zinc	Alcohol dehydrogenase

Proteins Are Very Large Molecules

How long are the polypeptide chains in proteins? Table 6-2 shows that human cytochrome *c* has 104 amino acid units linked in a single chain; bovine chymotrypsinogen has 245 amino acid units. The polypeptides of different proteins may have anywhere from about 100 to as many as 1800 or more amino acid residues. Proteins are not merely mixtures of a number of polypeptides of differing length, composition, or sequence. *All the molecules of any given type of protein are identical in amino acid composition, sequence, and length of the polypeptide chain.*

Some proteins contain only a single polypeptide chain, but others, called *oligomeric* proteins, have two or more (Table 6-4). For example, the enzyme ribonuclease has one polypeptide chain, whereas hemoglobin has four.

The molecular weights of proteins, which can be determined by various physicochemical methods, may range from about 12,000 for small proteins like cytochrome *c*, which has only 104 residues, to molecular weights as high as 10^6 or greater, in the case of proteins with very long polypeptide chains or those

Table 6-4 Molecular Data on Some Proteins

	MW	No. of residues	No. of chains
Insulin (bovine)	5,733	51	2
Ribonuclease (bovine pancreas)	12,640	124	1
Lysozyme (egg white)	13,930	129	1
Myoglobin (horse heart)	16,890	153	1
Chymotrypsin (bovine pancreas)	22,600	241	3
Hemoglobin (human)	64,500	574	4
Serum albumin (human)	68,500	~550	1
Hexokinase (yeast)	102,000	~800	2
γ-Globulin (horse)	149,900	~1250	4
Glutamate dehydrogenase (bovine liver)	1,000,000	~8300	~40

with several polypeptide chains. The molecular weights of some typical proteins are given in Table 6-4. No simple generalizations can be made about the molecular weights of proteins in relation to their function. Different enzymes, for example, may have molecular weights that vary over a very wide range.

We can calculate the approximate number of amino acid residues in a simple protein containing no prosthetic group by dividing its molecular weight by 110. Although the average molecular weight of the 20 different amino acids in proteins is about 138, the smaller amino acids predominate in most proteins, so that the average molecular weight is nearer 128. Since a molecule of water (MW 18.0) is removed to create each peptide bond, the average amino acid residue weight is about $128 - 18 = 110$. Table 6-4 gives the number of amino acid residues present in different proteins.

Proteins Can Be Separated and Purified

Cells contain hundreds if not thousands of different kinds of proteins. Clearly it is necessary to have a pure preparation of a given protein before we can determine its amino acid composition and sequence. How, then, can one protein, e.g., an enzyme, be separated from the hundreds of other kinds of proteins in a given cell or tissue extract and brought to purity?

First of all, proteins can be separated from low-molecular-weight substances present in a cell or tissue extract by the process of *dialysis* (Figure 6-3). Large molecules, such as proteins, are retained inside a bag of a material with ultramicroscopic pores, like cellophane. Thus when such a bag containing a cell or tissue extract is suspended in water, the small molecules in the tissue extract, such as salts, will pass out through the pores, but the high-molecular-weight proteins will stay behind.

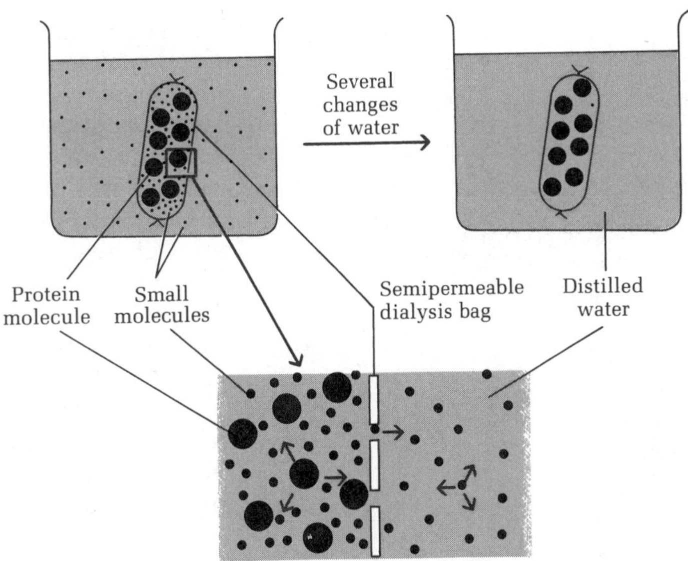

Figure 6-3
Dialysis. The membrane enclosing the protein solution allows water and small solutes, such as NaCl or glucose, to pass through freely but does not permit passage of large solutes, such as proteins. Small molecules will diffuse out of the dialysis bag into the outer chamber, since diffusing molecules tend to move toward a zone in which their concentration is lower. By replacing the outer aqueous phase with distilled H_2O several times, the concentration of small solute molecules in the protein solution can be decreased to a vanishingly small amount.

Dextran bead

Large protein molecule cannot penetrate into bead

Small protein molecule can penetrate into pores of bead

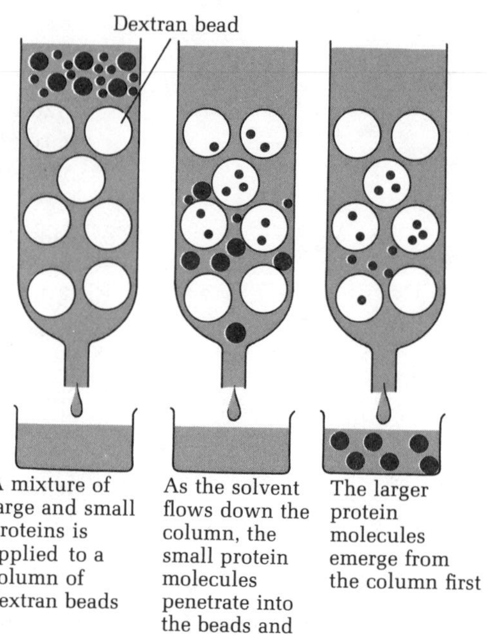

Figure 6-4
Separation of proteins on the basis of size by gel filtration. The mixture of proteins in solution is passed down a column of very small, porous beads of a hydrophilic polymer; a dextran derivative is widely used. Small protein molecules can penetrate into the beads, but large ones cannot. The molecular weight of a given protein can be determined by comparing its rate of passage down the column with the rate of flow of other proteins of known molecular weight.

Once a mixture of proteins has been freed of small molecules by dialysis, the proteins can be sorted out on the basis of size by *gel filtration*. In this procedure, which is a form of chromatography, the solution containing the mixture of proteins is passed down a column containing very small porous beads of a highly hydrated polymer. The smaller protein molecules can penetrate into the pores in the beads and thus are retarded in their flow down the column (Figure 6-4), but large protein molecules cannot penetrate into the beads and pass down the column more rapidly. Proteins of intermediate size will pass down the column at intermediate rates, depending on the degree to which they can penetrate into the beads. Such a gel-filtration column is also called a *molecular sieve*.

Proteins can also be separated from each other by *electrophoresis* (page 108) on the basis of the sign and number of the electric charges contributed by their R groups and by the charged amino-terminal and carboxyl-terminal groups. Like simple peptides, the polypeptide chains of proteins have characteristic *isoelectric points*, which will reflect the relative number of acidic and basic R groups (Table 6-5). At any given pH a mixture of proteins will contain some that have a net negative charge, some that have a net positive charge, and some with no net charge. When such a mixture is placed in an electric field, the proteins with a positive charge will move toward the negatively charged electrode, those with a net negative charge will move toward the positive electrode, and those with no net charge will remain stationary. Furthermore, those protein molecules with a relatively high density of charges will move toward the electrode faster than a protein with a relatively low charge density. Often electrophoresis is carried out in a matrix consisting of a strip of paper or cellulose acetate or a slab of a hydrophilic gel to keep the separated protein

Table 6-5 The Isoelectric Points of Some Proteins

	Isoelectric pH
Pepsin	<1.0
Egg albumin	4.6
Serum albumin	4.9
Urease	5.0
β-Lactoglobulin	5.2
γ_1-Globulin	6.6
Hemoglobin	6.8
Myoglobin	7.0
Chymotrypsinogen	9.5
Cytochrome c	10.7
Lysozyme	11.0

molecules from rapidly diffusing throughout the aqueous phase (Figure 6-5).

Yet another powerful method for separating proteins from each other is by <u>ion-exchange chromatography</u>, based in large part on the differences between proteins with respect to the density and sign of their electric charges at a given pH. Thus ion-exchange chromatography can be used to separate amino acids (page 109), peptides (page 113), or proteins.

To separate one specific protein from many others requires a distinctive means of measuring it, so that the progress of the purification steps can be monitored. If we are purifying an enzyme, for example, we can measure its catalytic activity, which will distinguish that protein from all others.

The Amino Acid Sequence of Polypeptide Chains Can Be Determined

Two major discoveries in 1953 ushered in the modern era of biochemistry. In that year James D. Watson and Francis Crick at Cambridge University in England deduced the double-helical structure of DNA and proposed a structural basis for the precise replication of DNA. Implicit in their proposal was the idea that the sequence of nucleotide units in DNA bears encoded genetic information. In that same year Frederick Sanger, working at the same university, worked out the sequence of amino acids in the polypeptide chains of the hormone insulin. This was a most important achievement in itself, since it had long been thought that elucidation of the amino acid sequence of a polypeptide would be a hopelessly difficult task. Sanger's achievement, coming at the same time as the proposal of Watson and Crick, also suggested that the nucleotide sequence of DNA and the amino acid sequence of proteins are somehow related. This idea, within a decade, led to the identification of the nucleotide code words of DNA and RNA that determine the amino acid sequence of protein molecules.

Until Sanger's achievement, which required several years of work, it was not even certain that all molecules of any given protein are precisely identical in molecular weight and amino acid composition. Today the amino acid sequences of hundreds of different proteins from many different species are known. The amino acid sequence of a polypeptide chain is determined by principles first developed by Sanger. They are still used today, although with many variations and great improvements in detail. There are six basic steps in solving the amino acid sequence of any polypeptide.

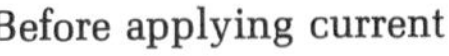

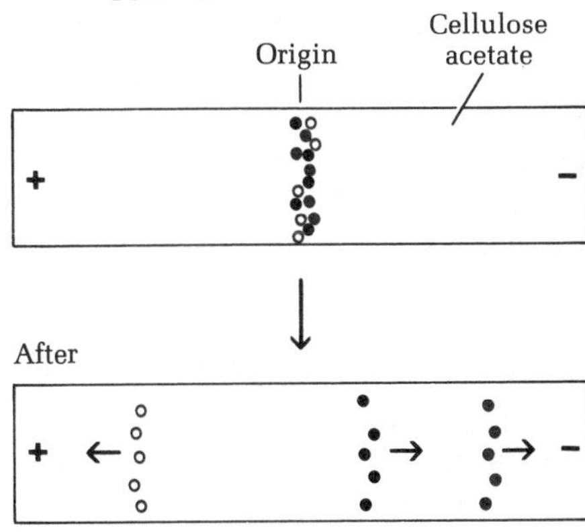

Figure 6-5
Electrophoresis of a mixture of three proteins in a strip of cellulose acetate soaked with a buffer of fixed pH. The ends of the strip are placed in electrode chambers. The cellulose acetate serves as a matrix to prevent the protein molecules from diffusing randomly. The protein mixture is applied to the center of the strip and subjected to direct current applied to the electrodes. The three proteins migrate to either the positive or negative pole and at different rates, which depend on the pH of the medium and the acid-base properties of each protein. At the end of the process the positions of the proteins are made visible with stains that bind to proteins.

Step 1: Determining the Amino Acid Composition

The first step is to hydrolyze all the peptide bonds of the pure polypeptide. The amino acid mixture so formed is then analyzed by ion-exchange chromatography (page 109) to determine which amino acids are present and the relative numbers of each.

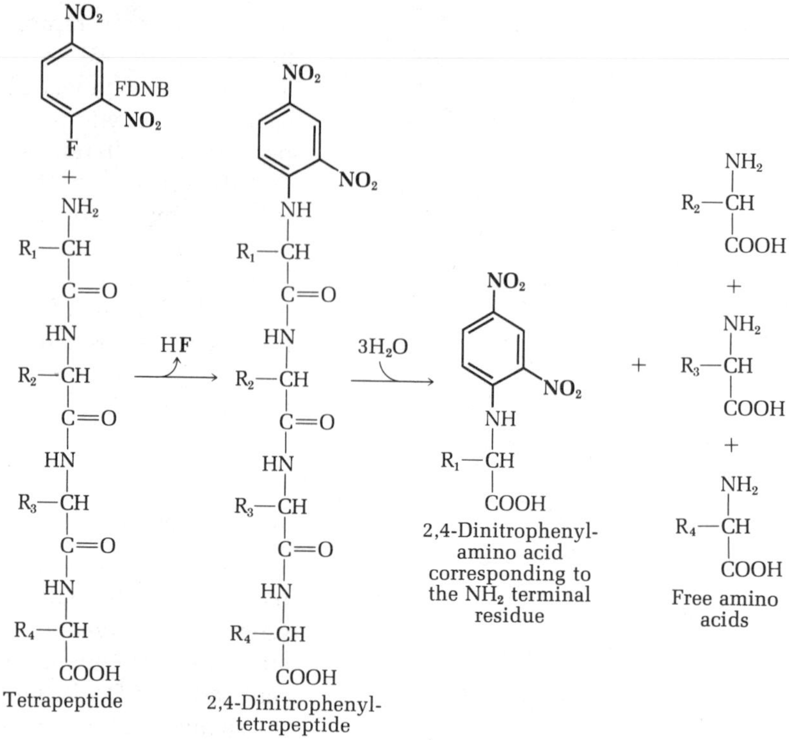

Figure 6-6
Identification of the amino-terminal residue of a tetrapeptide as its 2,4-dinitrophenyl derivative. The tetrapeptide is allowed to react with 1-fluoro-2,4-dinitrobenzene (FDNB) to yield its 2,4-dinitrophenyl derivative. The latter is then boiled with 6 N HCl to hydrolyze all the peptide bonds, yielding the amino-terminal amino acid as its 2,4-dinitrophenyl derivative.

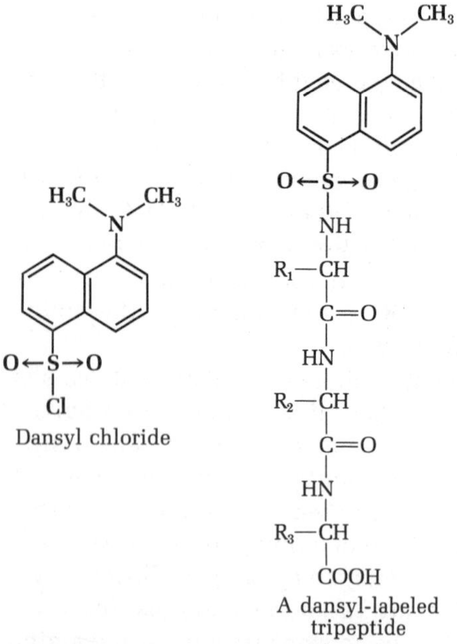

Figure 6-7
Labeling the amino-terminal residue of a tripeptide with dansyl chloride. The dansyl derivative of the amino-terminal residue can be separated and identified after hydrolysis of all the peptide bonds. Because the fluorescent dansyl group can be detected in much smaller amounts than the dinitrophenyl group, the dansyl method is far more sensitive than the fluorodinitrobenzene method.

Step 2: Identifying the Amino- and Carboxyl-Terminal Residues

The next step is to identify the amino acid residue at that end of the polypeptide chain bearing the free α-amino group, the amino-terminal end. For this purpose Sanger developed the reagent 1-fluoro-2,4-dinitrobenzene (page 113), which can label the amino-terminal residue of the chain as the yellow 2,4-dinitrophenyl (DNP) derivative. When such a DNP derivative of a polypeptide is subjected to hydrolysis with acid, all the peptide bonds in the chain are hydrolyzed. However, the covalent bond between the 2,4-dinitrophenyl group and the α-amino group of the amino-terminal residue is resistant to this treatment. Consequently, the amino-terminal residue will be present in the hydrolyzate as its 2,4-dinitrophenyl derivative (Figure 6-6). This derivative can easily be separated from the unsubstituted free amino acids and identified by chromatographic comparison with authentic dinitrophenyl derivatives of the different amino acids.

Another reagent used to label the amino-terminal residue is dansyl chloride (Figure 6-7), which reacts with the free α-amino group to yield a dansyl derivative. This is highly fluorescent and can thus be detected and measured in much lower concentrations than dinitrophenyl derivatives.

The carboxyl-terminal amino acid residue of a polypeptide can also be identified. In one procedure the polypeptide is incubated with the enzyme carboxypeptidase, which hydrolyzes only the peptide bond at the carboxyl-terminal end of the

Table 6-6 The Specificity of Four Important Methods for Fragmenting Polypeptide Chains

Treatment	Cleavage points (residues furnishing the carbonyl group of the peptide bond cleaved)
Trypsin	Lysine Arginine
Chymotrypsin	Phenylalanine Tryptophan Tyrosine
Pepsin	Phenylalanine Tryptophan Tyrosine Several others
Cyanogen bromide	Methionine

chain. By determining which amino acid is released first by the action of carboxypeptidase on the polypeptide, the carboxyl-terminal residue can be identified.

With the amino-terminal and carboxyl-terminal residues of the polypeptide identified, we shall have established two important points of reference in its amino acid sequence.

Step 3: Fragmenting the Polypeptide Chain

Another sample of the intact polypeptide chain is now fragmented into smaller pieces, short peptides having on the average 10 to 15 amino acid residues. The objective is to separate these fragments and determine the amino acid sequence of each.

Several methods can be used for fragmenting the polypeptide chain. A common procedure is partial enzymatic hydrolysis of the polypeptide by the digestive enzyme *trypsin*. This enzyme is very specific in its catalytic action. It catalyzes the hydrolysis of only those peptide bonds in which the carboxyl group is contributed by either a lysine or an arginine residue, regardless of the length or amino acid sequence of the chain (Table 6-6). The number of smaller peptides produced by trypsin cleavage can thus be predicted from the total number of lysine or arginine residues in the original polypeptide. A polypeptide with five lysine and/or arginine residues will usually yield six smaller peptides on cleavage with trypsin. Moreover, all except one of these will have a lysine or arginine residue at the carboxyl-terminal position. The fragments produced by trypsin action are then separated from each other by ion-exchange chromatography on a column, by paper electrophoresis, or by chromatography, often carried out in two dimensions to yield a *peptide map* (Figure 6-8).

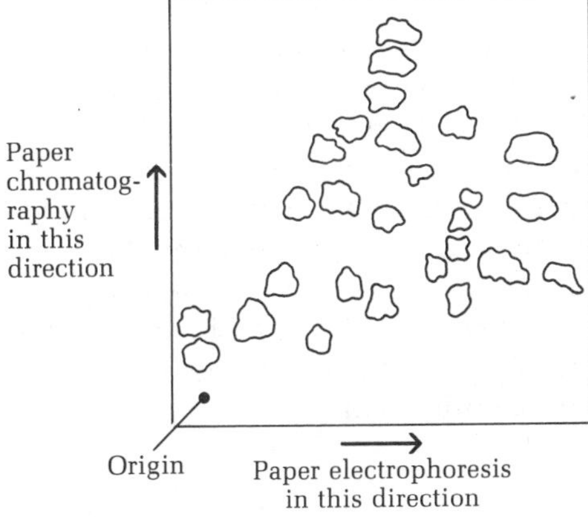

Figure 6-8
A map of the peptides obtained when normal human hemoglobin is cleaved with trypsin. Each spot contains one of the peptides. Such a two-dimensional map is obtained by carrying out electrophoresis of the peptide mixture along one direction of the paper square, drying the paper, and then carrying out chromatographic separation of the peptides in the other direction. Neither process alone will separate the peptides completely, but the sequential combination of the two is very effective in separating complex peptide mixtures.

Step 4: Identifying the Sequences of the Peptide Fragments

The amino acid sequence of each peptide fragment resulting from step 3 is then determined. A chemical method devised by Pehr Edman is usually employed for this purpose. The *Edman degradation* procedure labels and removes only the amino-terminal residue from the peptides, while leaving all the other peptide bonds intact (Figure 6-9). After removal and identification of the amino-terminal residue by this method, the new amino-terminal residue so exposed can now be labeled and removed by repeating the same series of reactions. In a residue-by-residue manner the Edman degradation can yield the entire amino acid sequence of a peptide using only a single sample. Figure 6-9 shows how the Edman degradation works. The peptide is first reacted with *phenylisothiocyanate*, which combines with the free α-amino group of the amino-terminal residue. Treatment of the peptide with cold dilute acid removes the amino-terminal residue and converts it into a *phenylthiohydantoin* derivative, which can now be identified by chro-

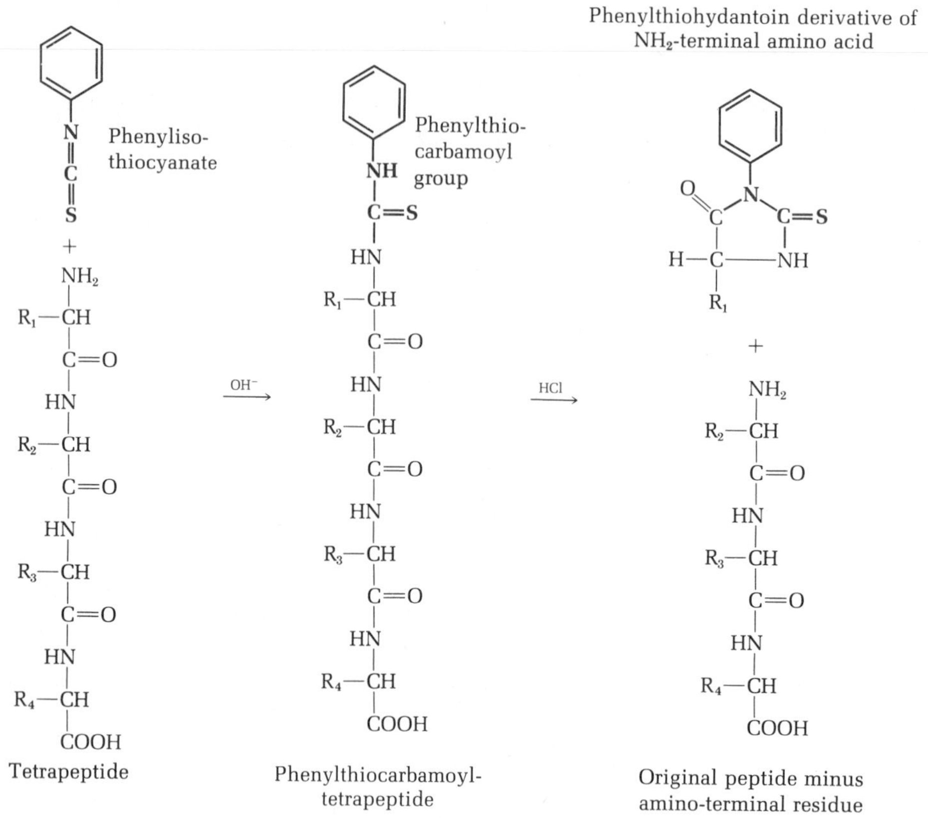

Phenyliso-thiocyanate $\xrightarrow{OH^-}$ Phenylthio-carbamoyl group $\xrightarrow{HCl}$ Phenylthiohydantoin derivative of NH_2-terminal amino acid

Tetrapeptide

Phenylthiocarbamoyl-tetrapeptide

Original peptide minus amino-terminal residue

Figure 6-9
Steps in the Edman degradation procedure for sequencing a peptide. The starting tetrapeptide is reacted with phenylisothiocyanate to yield the phenylthiocarbamoyl derivative of the amino-terminal residue. The latter is cleaved from the rest of the peptide without splitting any of the other peptide bonds and is recovered as the phenylthiohydantoin derivative, which can be identified chromatographically. The tripeptide remaining is now treated by the same cycle of reactions, leading to the identification of the second residue, and so on until each residue has been identified.

matographic procedures. The rest of the peptide chain remains intact. The shortened peptide can now be subjected to another round of these reactions, which permits the identification of the new amino-terminal residue. Thus, through repetitive removal of successive amino-terminal residues of a peptide by this "subtractive" method, the amino acid sequence of peptides having up to 10 or 20 residues can easily be determined.

All the peptide fragments resulting from the action of trypsin are sequenced in this manner. We are then left with the problem of determining how these trypsin fragments were arranged in the original polypeptide chain.

Step 5: Cleavage of the Original Polypeptide Chain by a Second Procedure

In order to establish the order of the peptide fragments formed by trypsin, another sample of the intact polypeptide is cleaved into small fragments by a different method, one that cleaves peptide bonds at points other than those cleaved by trypsin. Here it is often advantageous to use a chemical rather than an enzymatic method. The reagent cyanogen bromide is particularly useful, since it cleaves only those peptide bonds in which the carbonyl group is contributed by methionine residues (Table 6-6). Thus, if the polypeptide contains eight methionine residues, cleavage with cyanogen bromide will usually yield nine peptide fragments. The fragments resulting from this procedure are now separated from each other by electrophoresis or chromatography. Each of these small peptides is now subjected to the repetitive Edman degradation, as in step 4 above, to identify its amino acid sequence.

We now have obtained two different sets of peptide fragments, one by action of trypsin and the other set by chemical cleavage with cyanogen bromide. We also know the amino acid sequence in each of the two sets of peptide fragments.

Step 6: Ordering Peptide Fragments by Establishing Overlaps

The sequences of amino acids in each fragment obtained from the original polypeptides by the two cleavage procedures are now examined, with the objective of finding peptides from the second procedure whose sequences establish continuity or overlaps between the fragments obtained by the first cleavage procedure. The principle is shown in Figure 6-10. Overlapping peptides obtained from the second fragmentation yield the correct order of the peptide fragments produced in the first cleavage. Moreover, the two sets of fragments check each other out for possible errors in determining the amino acid sequence of each fragment.

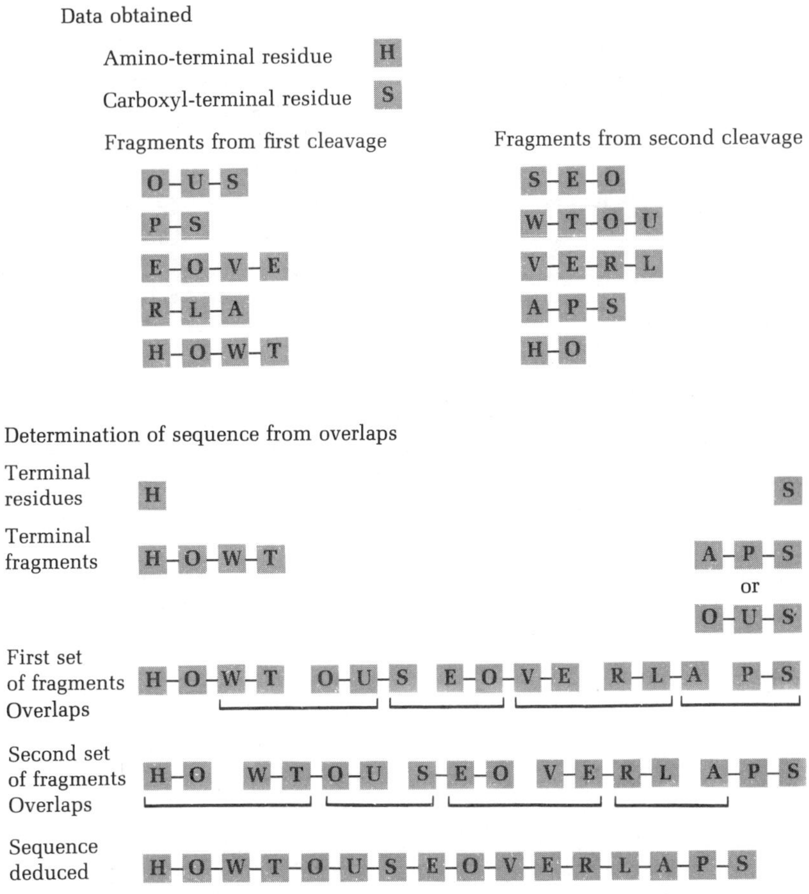

Figure 6-10
Placing peptide fragments in their proper order with overlaps. In this example a polypeptide having 16 amino acid residues is subjected to determination of the amino- and carboxyl-terminal residues and to two different types of fragmentation. The data obtained are shown above, and the determination of the sequence from the overlaps is shown below.

Data obtained

Amino-terminal residue H

Carboxyl-terminal residue S

Fragments from first cleavage

O–U–S
P–S
E–O–V–E
R–L–A
H–O–W–T

Fragments from second cleavage

S–E–O
W–T–O–U
V–E–R–L
A–P–S
H–O

Determination of sequence from overlaps

Terminal residues: H ... S

Terminal fragments: H–O–W–T ... A–P–S or O–U–S

First set of fragments: H–O–W–T O–U–S E–O–V–E R–L–A P–S
Overlaps

Second set of fragments: H–O W–T–O–U S–E–O V–E–R–L A–P–S
Overlaps

Sequence deduced: H–O–W–T–O–U–S–E–O–V–E–R–L–A–P–S

Figure 6-11
The structure of bovine insulin, showing the amino acid sequence of the two chains and the cross-linkages. The A chain of the insulins of man, pig, dog, rabbit, and sperm whale are identical. The B chains of the cow, pig, dog, goat, and horse are identical. The amino acid replacements in the A chain usually occur in positions 8, 9, and 10, shown in color.

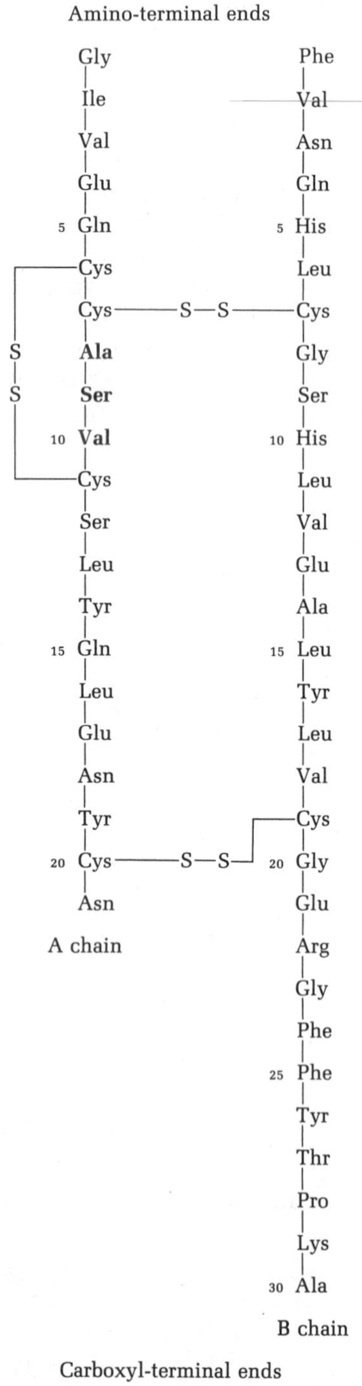

Amino-terminal ends

A chain: Gly—Ile—Val—Glu—5 Gln—Cys—Cys—**Ala**—**Ser**—10 **Val**—Cys—Ser—Leu—Tyr—15 Gln—Leu—Glu—Asn—Tyr—20 Cys—Asn

B chain: Phe—Val—Asn—Gln—5 His—Leu—Cys—Gly—Ser—10 His—Leu—Val—Glu—Ala—15 Leu—Tyr—Leu—Val—Cys—20 Gly—Glu—Arg—Gly—Phe—25 Phe—Tyr—Thr—Pro—Lys—30 Ala

S—S (A6 Cys—A11 Cys); Cys (A7)—S—S—Cys (B7); Cys (A20)—S—S—Cys (B19)

A chain

B chain

Carboxyl-terminal ends

Sometimes the second cleavage procedure fails to establish overlaps for two or more peptides from the first cleavage. In this case a third or even a fourth cleavage method must be used to obtain a set of peptides that can provide the necessary overlap(s) to complete the sequence. Cleavage of the polypeptide with other proteolytic enzymes, such as chymotrypsin or pepsin, may be used, although these enzymes are less specific in their action on peptide bonds than trypsin (Table 6-6).

Insulin Was the First Protein to Be Sequenced

We have now seen the logic and approach used to determine the amino acid sequence of polypeptide chains. Let us examine the results of Sanger's determination of the amino acid sequence of bovine insulin, shown in Figure 6-11. Bovine insulin has a molecular weight of about 5700. It has two polypeptide chains, the A chain with 21 and the B chain with 30 amino acid residues. The two chains are connected by two disulfide (—S—S—) cross-linkages, and one of the chains has an internal disulfide linkage. The two polypeptide chains were first separated by cleaving the disulfide cross-linkages. For this purpose Sanger employed the oxidizing agent performic acid, which cleaves each cystine residue into two cysteic acid residues (Figure 6-12), one in each of the chains. The chains were then separated and the sequences of each determined. Examination of the amino acid sequences of the two chains shows no obvious pattern or periodicity in the occurrence of any of the amino acids; moreover, the sequences of the two chains are quite different.

The successful determination of the amino acid sequence of the insulin chains has prompted intensive study of the relationships between the structure of insulins isolated from different species and their biological activity in promoting glucose metabolism. Both the A and B chains of insulin are required for its biological activity; moreover, the disulfide cross-linkages

Figure 6-12
Cleavage of cystine cross-linkages by performic acid.

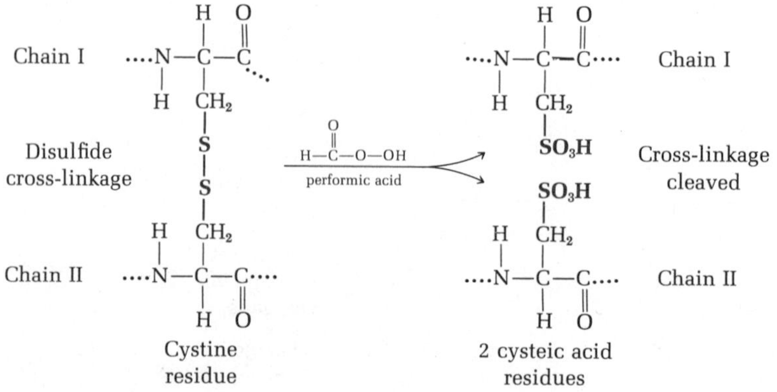

Chain I

Disulfide cross-linkage

Chain II

Cystine residue

H—C(=O)—O—OH
performic acid

Chain I

Cross-linkage cleaved

Chain II

2 cysteic acid residues

$\cdots N(H)-C(H)(CH_2-S-S-CH_2)-C(=O)\cdots$ → $\cdots N(H)-C(H)(CH_2-SO_3H)-C(=O)\cdots$ + $HO_3S-CH_2-\cdots$

must be intact. Removal of parts of either chain by selective cleavage results in loss of some or all of the activity. Although insulins isolated from the pancreas of several species, e.g., cow, pig, sheep, and whale, are hormonally active in human beings and are used in treatment of diabetic patients, they are not identical with human insulin. What is particularly significant is that at certain positions in each chain of insulin the amino acids found are always the same, regardless of the species from which the insulin was obtained. However, at other positions the amino acids may differ from one species to another. Such observations strongly indicate that the biological activity of insulin depends on the amino acid sequences of its chains, as well as cross-linking of the chains at specific points.

Many Other Proteins Have Been Sequenced Since

Soon after Sanger's success other investigators were emboldened to determine the amino acid sequence of even longer polypeptide chains. In short order the sequence of amino acids in corticotropin, the hormone of the anterior pituitary gland that stimulates the adrenal cortex, was worked out. This hormone has a single chain of 39 residues with a molecular weight of about 4600. In the late 1950s the first sequence analysis of an enzyme protein, ribonuclease, was achieved by Stanford Moore and William Stein at the Rockefeller Institute, and by Christian Anfinsen and a group at the National Institutes of Health. Bovine ribonuclease has 124 amino acid residues in a single chain, which contains four intrachain —S—S— cross-linkages (Figure 6-13).

Human corticotropin

Ser·Tyr·Ser·Met·Glu·His·Phe·Arg·Trp·Gly·$_{10}$
Lys·Pro·Val·Gly·Lys·Lys·Arg·Arg·Pro·Val·$_{20}$
Lys·Val·Tyr·Pro·Asp·Ala·Gly·Glu·Asp·Gln·$_{30}$
Ser·Ala·Glu·Ala·Phe·Pro·Leu·Glu·Phe$_{39}$

Bovine ribonuclease

Lys·Glu·Thr·Ala·Ala·Ala·Lys·Phe·Glu·Arg·$_{10}$
Gln·His·Met·Asp·Ser·Ser·Thr·Ser·Ala·Ala·$_{20}$
Ser·Ser·Ser·Asn·Tyr·Cys·Asn·Gln·Met·Met·$_{30}$
Lys·Ser·Arg·Asn·Leu·Thr·Lys·Asp·Arg·Cys·$_{40}$
Lys·Pro·Val·Asn·Thr·Phe·Val·His·Glu·Ser·$_{50}$
Leu·Ala·Asp·Val·Gln·Ala·Val·Cys·Ser·Gln·$_{60}$
Lys·Asn·Val·Ala·Cys·Lys·Asn·Gly·Gln·Thr·$_{70}$
Asn·Cys·Tyr·Gln·Ser·Tyr·Ser·Thr·Met·Ser·$_{80}$
Ile·Thr·Asp·Cys·Arg·Glu·Thr·Gly·Ser·Ser·$_{90}$
Lys·Tyr·Pro·Asn·Cys·Ala·Tyr·Lys·Thr·Thr·$_{100}$
Gln·Ala·Asn·Lys·His·Ile·Ile·Val·Ala·Cys·$_{110}$
Glu·Gly·Asn·Pro·Tyr·Val·Pro·Val·His·Phe·$_{120}$
Asp·Ala·Ser·Val$_{124}$

The drawing of the ribonuclease molecule shows the positions of the four disulfide cross-linkages contributed by the cystine residues.

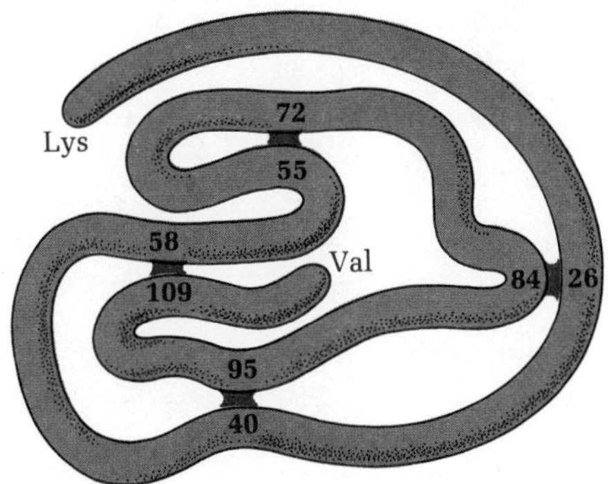

Figure 6-13
The amino acid sequences of the pituitary hormone corticotropin (39 residues) and bovine ribonuclease (124 residues).

The next important landmark was the identification of the amino acid sequences of the two types of polypeptide chains from crystalline hemoglobin. This was the first sequence analysis of an oligomeric protein. Hemoglobin contains four polypeptide chains, called globins, two identical α-globins (141 residues) and two identical β-globins (146 residues). These chains are not covalently attached to each other in the hemoglobin molecule but are tightly associated through hydrogen bonding and by hydrophobic interactions (Chapter 7). The α- and β-globins were separated and the sequence of each determined by two groups in the United States and another in Germany. Among the longer polypeptide chains for which complete amino acid sequences have since been deduced are bovine trypsinogen (229 residues), bovine chymotrypsinogen (245 residues), glyceraldehyde-3-phosphate dehydrogenase (333 residues), and the single chain of human serum albumin (582 residues).

The many individual steps and the careful bookkeeping required in the determination of the amino acid sequence of long polypeptide chains can now be carried out by programmed and automated analyzers. Even the Edman degradation is carried out on a programmed machine, called a sequenator, which mixes reagents in the proper proportions, carries out separa-

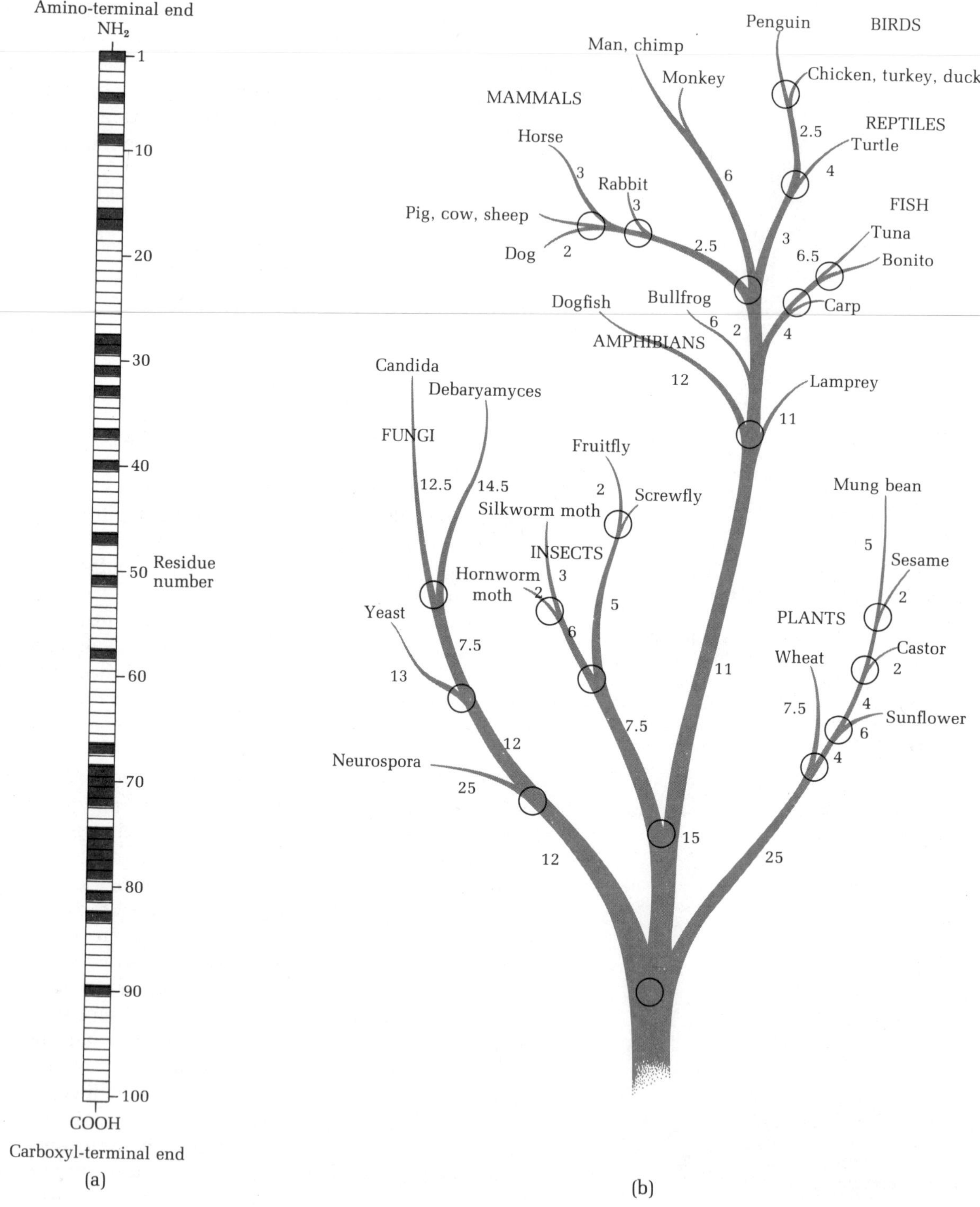
Amino-terminal end
NH_2
1
10
20
30
40
50
Residue number
60
70
80
90
100
COOH
Carboxyl-terminal end
(a)
Penguin
BIRDS
Man, chimp
Monkey
Chicken, turkey, duck
MAMMALS
2.5
REPTILES
Horse
Turtle
6
4
3
Rabbit
3
FISH
Pig, cow, sheep
Tuna
Dog
2
2.5
3
6.5
Bonito
Dogfish
Bullfrog
Carp
6
2
4
AMPHIBIANS
Candida
12
Lamprey
Debaryamyces
11
FUNGI
Fruitfly
12.5
14.5
2
Screwfly
Mung bean
Silkworm moth
INSECTS
5
Sesame
Hornworm moth
3
2
2
5
Yeast
PLANTS
6
7.5
Castor
Wheat
11
2
13
4
7.5
Sunflower
7.5
6
12
Neurospora
4
25
15
12
25
(b)

Figure 6-14
(a) The location of the 27 invariant residues in cytochrome c from over 60 species, including mammals, fishes, reptiles, amphibians, birds, insects and other invertebrates, fungi, and plants. As more species of cytochrome c are examined, the number of invariant residues may become smaller. (b) Main branches of the evolutionary tree constructed from the number of amino acid differences between cytochrome c molecules isolated from different species. The numbers in black give the number of residues by which the cytochrome c of a given line of organism differs from its ancestors. The circles represent points of divergence in evolution.

tions of the products, identifies them, and records the results on a strip chart. Such instruments have greatly reduced the time and labor required to determine the amino acid sequence of polypeptides. Moreover, these new methods are extremely sensitive. Amino acid analyzers can quickly determine how much of each amino acid is present in a single human thumbprint! Moreover, only a milligram of protein is required to determine its complete amino acid sequence.

Homologous Proteins from Different Species Have Homologous Sequences

Several important conclusions have come from study of the amino acid sequence of *homologous* proteins from different species. Homologous proteins are those which perform the same function in different species; an example is hemoglobin, which has the same oxygen-transport function in different vertebrates. Homologous proteins from different species usually have polypeptide chains that are identical or nearly identical in length. Moreover many positions in the amino acid sequence of homologous proteins are occupied by the same amino acid in all species and are thus called *invariant residues*. But in other positions there may be considerable variation in the amino acid from one species to another; these are called *variable residues*. Such a set of similarities in the amino acid sequence of homologous proteins is called *sequence homology*, and it suggests that the animals from which such homologous proteins were obtained have had a common evolutionary origin.

The biological significance of sequence homology is best illustrated by *cytochrome c*, an iron-containing mitochondrial protein that transfers electrons during biological oxidations in eukaryotic cells. The polypeptide chain of this protein has a molecular weight of about 12,500 and has about 100 amino acid residues in most species. The amino acid sequences of cytochrome *c*'s from over 60 different species have been determined, and 27 positions in the chain of amino acid residues are identical in all species tested (Figure 6-14), suggesting that they are the important residues that specify the biological activity of cytochrome *c*. The amino acid residues in other positions in the chain may vary from one species to another. Another important conclusion has been reached from examination of cytochrome *c* sequences: the number of residues that differ in cytochrome *c* sequences from any two species is in proportion to the phylogenetic difference between the species. For example, 48 amino acid residues are different in the cytochrome *c* molecules of the horse and of yeast, which are very widely separated species, whereas only 2 residues differ in the cytochrome *c*'s of the much more closely related duck and chicken. In fact, the cytochrome *c* molecule has identical amino acid sequences in the chicken and the turkey. The sequences of cytochrome *c* are also identical in the pig, cow, and sheep. Information regarding the number of residue differences between homologous proteins of different species allows the construction of evolutionary maps that show the origin and sequence of development of different animals and plants during the evolution of species (Figure 6-14).

The Immune Response Can Detect Differences between Homologous Proteins

That homologous proteins from different species are usually not identical has also come from study of a class of proteins called *antibodies* or *immunoglobulins*. Antibody molecules appear in the blood serum and certain tissues of a given vertebrate species in response to injection of an *antigen*, a protein or other macromolecule foreign to that species. Each foreign protein elicits the formation of a different antibody. This response of the body, which is highly specific for the injected protein, is called the *immune response*, and it is the basis of the whole field of immunology. Antibody molecules, which are formed by specialized cells called *lymphocytes*, can combine with the antigen which elicited their formation to form an *antigen-antibody complex*. Immunity to infectious diseases can often be conferred by injecting very small amounts of certain macromolecular components, i.e., antigens, of the causative microorganism or virus. Antibodies to these foreign antigens are then formed by the host lymphocytes. If the microorganism which furnished the antigen happens to gain access to the blood or lymph of the immunized animal at some later time, the antibodies formed by the animal neutralize or inactivate the invading microorganism or virus by combining with its antigenic components. The immune response is given only by vertebrates and is thus a rather recent product of biological evolution.

The formation of antibodies can be quantitatively studied by means of the *precipitin reaction*. A recipient vertebrate, often a rabbit, is immunized against a specific foreign protein, such as ovalbumin from a hen's egg. The blood serum of the immunized rabbit (the *antiserum*), which contains the antibody, is then mixed with a small amount of antigen, i.e., hen ovalbumin. The mixture becomes turbid, and a precipitate is formed, which contains the antigen-antibody complex. No such precipitate is formed when blood serum of an unimmunized animal is mixed with the antigen.

Antibodies are Y-shaped protein molecules containing four polypeptide chains. They possess binding sites that are complementary to specific structural features of the antigen molecule. The antibody molecule has two such binding sites, making possible the formation of a three-dimensional lattice of alternating antigen and antibody molecules (Figure 6-15).

Antibodies are highly specific for the foreign proteins that evoke their formation. A rabbit antibody formed to horse serum albumin, for example, will combine with the latter but will not usually combine with other horse proteins, such as horse hemoglobin. While the antibodies generated by the rabbit to horse serum albumin will react best with the latter, they will also react to a significant extent with the serum albumin of those species most closely related to the horse, i.e., zebra, cow, pig, and other ungulates, whereas serum albumins from rodents, birds, and amphibians will be far less reactive. These observations are thus fully consistent with studies of amino acid sequences of homologous proteins. They also indicate that the

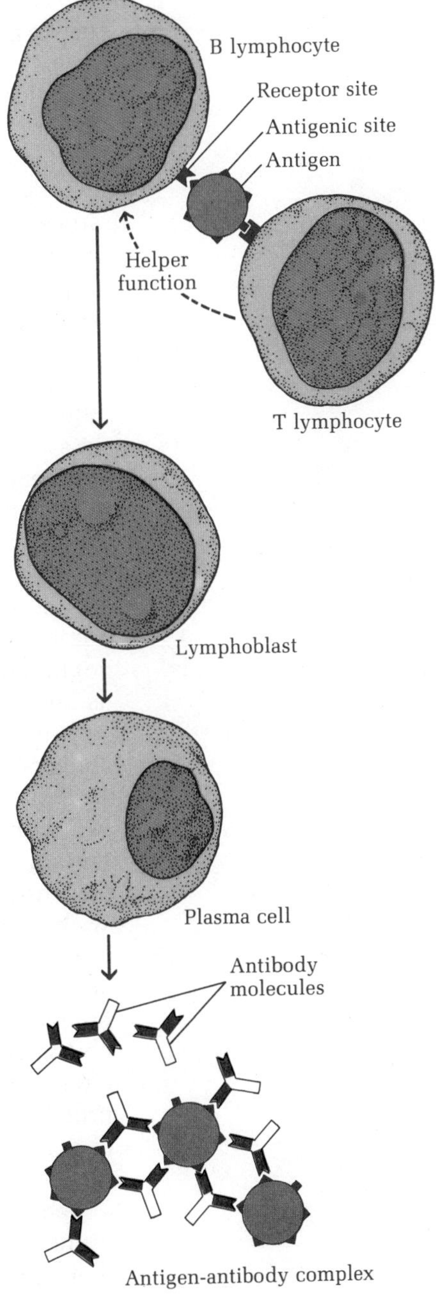

Figure 6-15
The immune response and the action of antibodies. When a macromolecule foreign to the body, particularly a protein from some other species, gains entrance into the blood or tissues, it evokes a defensive reaction called the immune response. The foreign macromolecule, called the antigen, binds to a special type of white blood cell, a B lymphocyte, causing it to undergo development into a plasma cell, which produces a large amount of an antibody against that antigen. The formation of specific antibodies by the B lymphocytes is helped by other cells called T lymphocytes. Antibodies, or immunoglobulins, are complex high-molecular-weight proteins containing four polypeptide chains and several carbohydrate groups. They can combine specifically with the antigen that evoked their formation but not with other proteins. Immunoglobulins are Y-shaped and have two binding sites for the antigen. They can precipitate the antigen by forming an insoluble, latticelike aggregate. Each antigen can evoke its own specific type of antibody, which will recognize and combine with only the antigen that evoked it or closely related molecules.

specificity of the immune response is related to the amino acid sequences of proteins. We shall examine other aspects of antibodies later, since they are of extreme importance in medicine and also tell much about the structure of proteins and the action of genes (Chapter 30).

Proteins Undergo a Structural Change Called Denaturation

Throughout this chapter we have emphasized the relationship between the amino acid sequence of proteins, their biological activity, and their species specificity. But there is much more to the story of proteins than their *primary structure*, the term we use to refer to their covalent backbone structure and amino acid sequence. This is most readily shown by a long-known and familiar property of proteins that we have not yet considered. When a solution of a protein, such as egg albumin, is slowly heated to about 60 or 70°C, the solution gradually becomes cloudy and a stringy coagulum forms. This is a familiar process since it occurs when an egg is poached. The white of the egg, which contains the albumin, coagulates to a white solid on heating. After the egg white has been coagulated by heat in this way, it will not redissolve on cooling to yield a clear solution as in the original unheated egg white. Heating of egg albumin has therefore changed it, seemingly in an irreversible manner. This effect of heat occurs with virtually all globular proteins, regardless of their size or biological function, although the precise temperature at which it occurs may vary. The change in a protein that is produced by heat is known as *denaturation*. Proteins in their natural state are called *native proteins;* after the change they are *denatured proteins*.

There is a second important consequence of protein denaturation: the protein nearly always loses its characteristic biological activity. Thus, when an aqueous solution of an enzyme is heated to the boiling point for a few minutes and cooled, it will usually become insoluble and, most important, it will no longer be catalytically active. Denaturation of proteins can be produced not only by heat, but also by extremes of pH; by certain miscible organic solvents such as alcohol or acetone; by certain solutes such as urea; by exposure of the protein to detergents; or simply by vigorous shaking of a protein solution with an air phase until it foams. Each of these agencies causing denaturation is a relatively mild treatment. In fact, direct tests show that when proteins undergo denaturation, no covalent bonds in the polypeptide chain backbone are broken. Thus the characteristic amino acid sequence of the protein is still intact after denaturation; nevertheless, the biological activity of most proteins is lost. We must conclude that the biological activity of proteins depends on something more than amino acid sequence alone.

What is the answer to this mystery? Quite simply it is that proteins have higher orders of structure, above and beyond their primary backbone structure. In brief, the covalently

Figure 6-16
Heat and various other treatments can denature or unfold native globular proteins without breaking the covalent backbone. The denatured protein can assume many different random forms, which usually are biologically inactive.

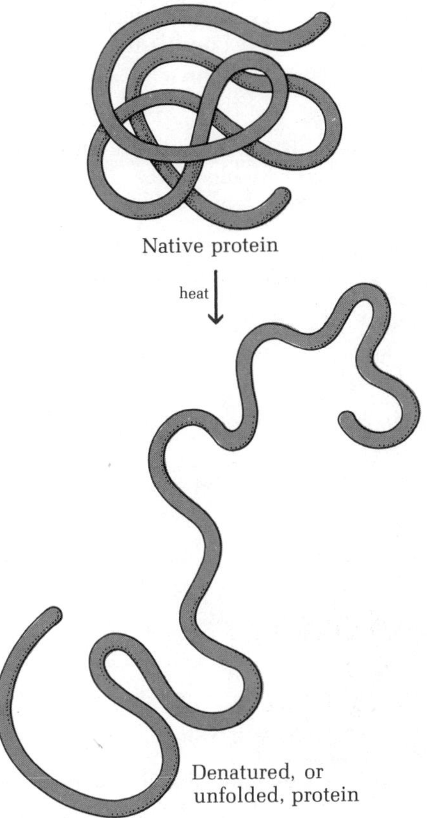

linked polypeptide chains of native proteins are folded in three dimensions in a pattern that is characteristic for each type of protein. The specific manner in which the chain is folded gives each protein its characteristic biological activity (Figure 6-16). When a protein is denatured, the characteristic three-dimensional arrangement of its polypeptide chain(s) is disrupted and it unfolds into random structures, without damage to the covalent backbone structure. Native protein molecules are therefore very fragile and readily disrupted by heat and other seemingly mild treatments. When we attempt to isolate and purify proteins and study their biological behavior, we must treat them gently to avoid denaturing them.

The next two chapters will show how the polypeptide chains of native proteins are folded into characteristic and specific conformations and how these conformations depend on amino acid sequence.

Summary

Proteins are the most abundant macromolecules in cells, making up over half their dry weight. They consist of very long polypeptide chains having from 100 to over 1000 amino acid units joined by peptide linkages. Simple proteins yield only amino acids on hydrolysis; conjugated proteins contain in addition some other component, either a metal ion or an organic prosthetic group. Some proteins are fibrous and insoluble; others are globular, with highly folded polypeptide chains.

Cells contain hundreds or thousands of different proteins, each with a different function or biological activity. Nevertheless, all are made from the same set of 20 amino acids, but they differ in the sequence of their amino acid units. The amino acid sequence of polypeptide chains can be established by fragmenting them into smaller pieces and determining the amino acid sequence of each fragment by the Edman degradation procedure. The peptide fragments are then placed in the correct order by finding sequence overlaps between them. To do this the original polypeptide is fragmented by a second procedure that cleaves it at points other than those cleaved in the first procedure. From the amino acid sequences of the second set of fragments the required sequence overlaps are obtained.

Homologous proteins from different species of organisms show sequence homology; i.e., certain critical positions in the polypeptide chains of homologous proteins contain the same amino acids, regardless of the species. In other positions of homologous proteins the amino acids may differ. The more closely related the species the more nearly identical the amino acid sequences of their homologous proteins. Thus the sequences of homologous proteins indicate that organisms containing them arose from a common ancestor but underwent changes as different species diverged during evolution. Similar conclusions have been drawn from the results of studies of the specificity of antibodies to antigens of homologous species.

Globular proteins on heating, exposure to extremes of pH, or treatment with certain reagents usually become insoluble and lose their biological activity without damage to the polypeptide backbone. This process, called denaturation, is due to unfolding of the polypeptide chain.

References

Some references in Chapter 5 will also be useful.

Books

COOPER, T. G.: *The Tools of Biochemistry*, Wiley, New York, 1977. Guide to experimental methods in protein biochemistry.

DAYHOFF, M. O.: *Atlas of Protein Sequence and Structure*, vol. 5, suppl. 1–3. National Biomedical Research Foundation, Washington, D.C., 1972–1979. An encyclopedia of amino acid sequences of proteins.

DICKERSON, R. E., and I. GEIS: *Proteins: Structure, Function, and Evolution*, 2d ed., Benjamin/Cummings, Menlo Park, Calif., 1983. Beautifully illustrated introduction to proteins.

HASCHEMEYER, R., and A. H. HASCHEMEYER: *Proteins: A Guide to Study by Physical and Chemical Methods*, Wiley, New York, 1973.

NEURATH, H., and R. L. HILL (eds.): *The Proteins*, 3d ed., Academic, New York, 1977. Comprehensive, detailed articles on proteins, in a series of volumes.

SCHULTZ, G. E., and R. H. SCHIRMER: *Principles of Protein Structure*, Springer, New York, 1979.

Articles

DICKERSON, R. E.: "Structure and History of an Ancient Protein," *Sci. Am.*, **226**:58–72, April (1972).

EDELMAN, G. M.: "The Structure and Function of Antibodies," *Sci. Am.*, **223**:34–42, August (1970).

MOORE, S., and W. H. STEIN: "Chemical Structures of Pancreatic Ribonuclease and Deoxyribonuclease," *Science*, **180**:458–464 (1973).

O'FARRELL, P. H.: "High Resolution Two-Dimensional Analysis of Proteins," *J. Biol. Chem.*, **250**:4007–4021 (1975). Interesting attempt to count all the proteins in the *E. coli* cell.

SRINAVASAN, P. R., J. S. FRUTON, and J. T. EDSALL (eds.): "The Origins of Modern Biochemistry. A Retrospective on Proteins," *Ann. N. Y. Acad. Sci.*, **325** (1979). A collection of very interesting articles on the history of protein research.

Problems

1. *How Many β-Galactosidase Molecules Are Present in an E. coli Cell? E. coli* is a rod-shaped bacterium 2 μm long and 1 μm in diameter. When they are grown on lactose (a sugar found in milk), the bacteria synthesize the enzyme β-galactosidase (MW 450,000), which catalyzes the breakdown of lactose. The average density of the bacterial cell is 1.2 g/mL, and 14 percent of its total mass is soluble protein, of which 1.0 percent is β-galactosidase. Calculate the number of β-galactosidase molecules in an *E. coli* cell grown on lactose.

2. *Number of Tryptophan Residues in Bovine Serum Albumin* A quantitative amino acid analysis reveals that bovine serum albumin contains 0.58 percent by weight of tryptophan, which has a molecular weight of 204.

(a) Calculate the minimum molecular weight of bovine serum albumin.
(b) Gel filtration of bovine serum albumin gives a molecular weight estimate of about 70,000. How many tryptophan residues are present in a molecule of serum albumin?

3. *The Molecular Weight of Ribonuclease* Lysine makes up 10.5 percent of the weight of ribonuclease. Calculate the minimum molecular weight of ribonuclease. The ribonuclease molecule contains 10 lysine residues. Calculate the molecular weight of ribonuclease.

4. *Net Electric Charge of Polypeptides* A polypeptide isolated from the brain has the sequence

   Glu–His–Trp–Ser–Tyr–Gly–Leu–Arg–Pro–Gly

   Determine the net charge on the molecule at pH 3. What is the net charge at pH 5.5? At pH 8? At pH 11? The pK′ values of the R groups of Glu, His, Ser, Tyr, and Arg are 4.3, 6.0, 13.6, 10, and 12.48, respectively. Calculate the isoelectric pH for this polypeptide.

5. *The Isoelectric Point of Pepsin* Pepsin of gastric juice (pH $\approx$ 1.5) has an isoelectric point of about 1, much lower than that of other proteins (see Table 6-5). What functional groups must be present in relatively large numbers to give pepsin such a low isoelectric point? What amino acids can contribute such groups?

6. *The Isoelectric Point of Histones* Histones are proteins of eukaryotic cell nuclei. They are tightly bound to deoxyribonucleic acid (DNA), which has many phosphate groups. The isoelectric point of histones is very high, about 10.8. What amino acids must be present in relatively large numbers in histones? In what way do these residues contribute to the strong binding of histones to DNA?

7. *Solubility of Polypeptides* One method for separating polypeptides makes use of their differential solubilities. As pointed out in the text, the solubility of large polypeptides in water depends upon the relative polarity of their R groups, particularly on the number of ionized groups: the more ionized groups there are, the more soluble the polypeptide. Which of each pair of polypeptides below is more soluble at the indicated conditions?
   (a) $(Gly)_{20}$ or $(Glu)_{20}$ at pH 7.0
   (b) $(Lys–Ala)_3$ or $(Phe–Met)_3$ at pH 7.0
   (c) $(Ala–Ser–Gly)_5$ or $(Asn–Ser–His)_5$ at pH 9.0
   (d) $(Ala–Asp–Gly)_5$ or $(Asn–Ser–His)_5$ at pH 3.0

8. *Fragmentation of a Polypeptide Chain by Proteolytic Enzymes* Trypsin and chymotrypsin are specific enzymes that catalyze the hydrolysis of polypeptides at specific locations (Table 6-6). The sequence of the B chain of the polypeptide hormone insulin is shown below. Note that the cystine cross-linkage between the A and B chain has been cleaved through the action of performic acid (see Figure 6-12).

   Phe–Val–Asn–Gln–His–Leu–$CySO_3H$–Gly–Ser–His–
   Leu–Val–Glu–Ala–Leu–Tyr–Leu–Val–$CySO_3H$–Gly–Glu–
   Arg–Gly–Phe–Phe–Tyr–Thr–Pro–Lys–Ala

   Indicate the points in the B chain that are cleaved by (a) trypsin and (b) chymotrypsin.

9. *Sequence Determination of the Brain Peptide Leucine Enkephalin* A group of polypeptides that influence nerve transmission in certain parts of the brain has been isolated from normal brain tissue. These polypeptides are known as *opioids* because they bind to specific receptors that bind opiate drugs, such as morphine and naloxone. Opioids thus mimic some of the properties of opiates. Some researchers consider these polypeptides to be the brain's own "pain killers." Using the information below, determine the amino acid sequence of the opioid leucine enkephalin. Explain how your structure is consistent with each piece of information.
   (a) Complete hydrolysis by 1 *M* HCl at 110°C followed by amino acid analysis indicated the presence of Gly, Leu, Phe, and Tyr, in a 2:1:1:1 molar ratio.
   (b) Treatment of the polypeptide with 2,4-dinitrofluorobenzene followed by complete hydrolysis and chromatography indicated the presence of the 2,4-dinitrophenyl derivative of tyrosine. No free tyrosine could be found.
   (c) Partial hydrolysis of the polypeptide by chymotrypsin followed by chromatography yielded Leu, Tyr, and a shorter peptide. Complete hydrolysis of the latter followed by amino acid analysis indicated the presence of Gly and Phe in a 2:1 ratio.

10. *Electrophoresis of Peptides* When placed in an electric field, ionized amino acids and peptides will migrate either toward the anode or cathode depending upon the pH (see Figure 6-5). This technique has been used extensively to separate peptides according to their net charge. It is particularly versatile because the net charge on a peptide can be varied by changing the pH of the medium.
    (a) For each of the amino acids and peptides below, determine the direction of migration (anode or cathode) at the indicated pH
        (1) Glu (pH 7)
        (2) Glu (pH 1)
        (3) Asp–His (pH 1)
        (4) Asp–His (pH 10)
    (b) Determine the pH at which the three dipeptides Gly–Lys, Asp–Val, and Ala–His can be readily separated by electrophoresis.

11. *Solubility: Salting-Out of Proteins*
    (a) Most pure proteins are insoluble in pure distilled water but dissolve in dilute salt solutions. However, the addition of high concentrations of neutral salts to an aqueous solution of protein causes it to precipitate. This phenomenon is referred to as salting out. For example, most proteins are soluble in 0.1 *M* $(NH_4)_2SO_4$, but when the concentration of $(NH_4)_2SO_4$ is raised to 3 *M* they will precipitate. When the excess $(NH_4)_2SO_4$ is removed by dialysis, the proteins will redissolve. Suggest a molecular explanation for the observation that high concentrations of added salts decrease the solubility of proteins.
    (b) The solubilities of two proteins as a function of the concentration of $(NH_4)_2SO_4$ are shown in the diagram at the left. How can this observation be used to separate proteins *A* and *B*?

Problem 11

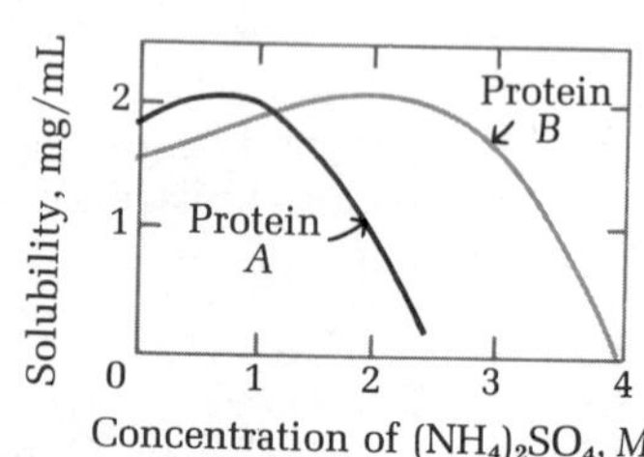

12. *Affinity Chromatography: A Highly Specific and Efficient Method of Isolating Specific Proteins* Because of the sensitive nature of most proteins, biochemists had to develop special methods for

isolating and purifying proteins because many of the conventional methods available for organic molecules, e.g., distillation and solvent extractions, were not suitable. It is not uncommon to have to isolate a single protein present at a concentration of 10^{-3} to 10^{-6} M in a mixture with several thousand other biomolecules. A technique known as affinity chromatography has made a major impact on the isolation and purification of certain enzymes, immunoglobulins, and receptor proteins. The technique makes use of the well-known fact that such a protein in its normal biological action will *reversibly* bind another specific molecule, called a ligand, very tightly to form a *noncovalent* protein-ligand complex.

Protein + Ligand ⇌ Protein-ligand complex

In this technique the specific ligand for the protein to be isolated is *covalently* attached to insoluble 10- to 50-μm polymer beads.

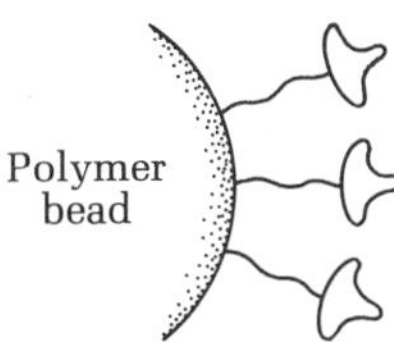

To isolate this protein from a cell extract a sample of the extract is placed on a column composed of the ligand-attached polymer beads and the column is washed repeatedly with buffer. The only proteins that are retained on the column are those with a high affinity for the ligand anchored to the beads; the rest of the proteins are simply washed through. Since the affinity and specificity of the protein for its ligand is very high, one can often isolate and purify a miniscule amount of a specific protein from a cell extract containing hundreds of other different proteins in a single step.

How does one get the retained protein *off* the affinity column in pure form? Explain the basis of your proposed procedure.

13. *Structure of an Antibiotic Polypeptide from Bacillus brevis* Extracts from the bacterium *Bacillus brevis* contain a peptide with antibiotic properties. Such peptide antibiotics form complexes with metal ions and apparently disrupt ion transport across the cell membrane, killing certain bacterial species. The structure of the polypeptide has been determined from the following observations:
   (a) Complete acid hydrolysis of the peptide followed by amino acid analysis yielded equimolar amounts of Leu, Orn, Phe, Pro, and Val. Orn is ornithine, an amino acid not present in proteins but present in some peptides. It has the structure

$$\overset{+}{N}H_3—CH_2—CH_2—CH_2—\underset{\underset{\overset{+}{N}H_3}{|}}{\overset{\overset{H}{|}}{C}}—CO_2^-$$

(b) Molecular-weight measurements gave an approximate value of 1200.

(c) When treated with the enzyme carboxypeptidase, the peptide failed to undergo hydrolysis.

(d) Treatment of the intact polypeptide with fluorodinitrobenzene, followed by complete hydrolysis and chromatography, yielded only free amino acids and the following derivative:

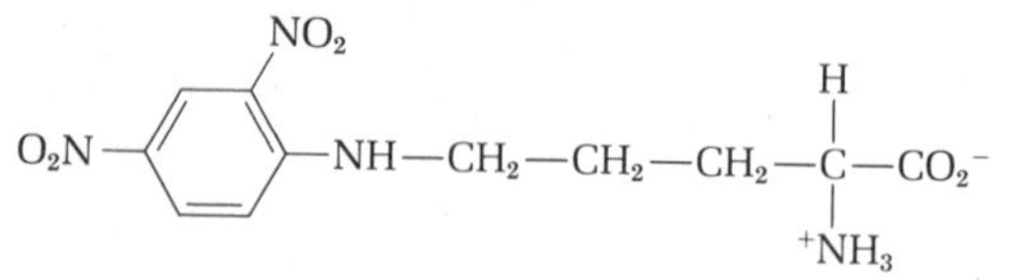

$$O_2N\text{—}C_6H_3(NO_2)\text{—}NH\text{—}CH_2\text{—}CH_2\text{—}CH_2\text{—}\underset{^+NH_3}{\overset{H}{C}}\text{—}CO_2^-$$

(*Hint:* Note that the 2,4-dinitrophenyl derivative involves the amino group of the side chain rather than the usual α nitrogen.)

(e) Partial hydrolysis of the polypeptide followed by chromatographic separation and sequence analysis yielded the di- and tripeptides below (the amino-terminal amino acid is always at the left)

Leu–Phe Phe–Pro Phe–Pro–Val
Val–Orn–Leu Orn–Leu Val–Orn Pro–Val–Orn

Given the above information, deduce the amino acid sequence of the antibiotic polypeptide. Show your reasoning. When you have arrived at a structure, go back and demonstrate that your proposed structure is consistent with *each* experimental observation above.

CHAPTER 7

Fibrous Proteins

Proteins can be classified into two major sorts: *fibrous proteins*, whose polypeptide chains are arranged in long strands or sheets, and *globular proteins*, whose polypeptide chains are tightly folded into a spherical or globular shape. In this chapter we shall examine the three-dimensional structure of the fibrous proteins. From the biological point of view fibrous proteins play very important roles in the anatomy and physiology of animals. They may constitute one-half or more of the total body protein in larger vertebrates. They provide external protection, since they are the major components of the outer layer of skin, hair, feathers, nails, and horns. Fibrous proteins also provide support, shape, and form, since they are major organic components of connective tissues, including tendons, cartilage, bone, and the deeper layers of skin.

There is another reason to examine the fibrous proteins first: They have simpler structures than globular proteins. This fact allowed determination of the three-dimensional structure of fibrous proteins by x-ray analysis. This was a pioneering advance, not only for the new insights into the structure and function of fibrous proteins, but also because it was the key to x-ray analysis of the structure and function of globular proteins.

Configuration and Conformation Have Different Meanings

In this and the following chapter we shall discuss the three-dimensional arrangement of polypeptide chains. We therefore must first define two frequently used terms that refer to the structure of molecules in space: *configuration* and *conformation*. These terms are *not* synonyms. Configuration denotes the arrangement in space of an organic molecule that is conferred by the presence of either (1) *double bonds*, around which there is no freedom of rotation, or (2) *chiral centers*, around which substituent groups are arranged in a specific sequence. Figure 7-1 shows the configurations of *fumaric acid*, an intermediate

Figure 7-1
Configurations of stereoisomers. Such isomers cannot be interconverted without breaking covalent bonds.

Geometrical (cis-trans) isomers

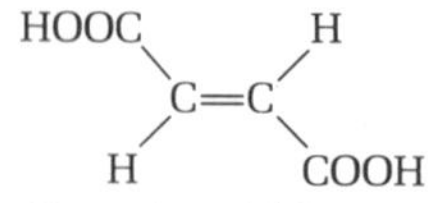

HOOC—C(H)=C(H)—COOH
Fumaric acid (trans)

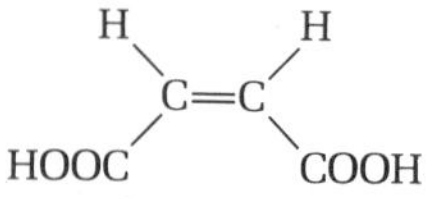

H, H / C=C / HOOC, COOH
Maleic acid (cis)

Optical isomers (enantiomers)

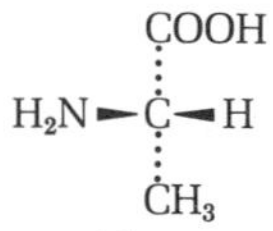

COOH / H_2N—C—H / CH_3
L-Alanine

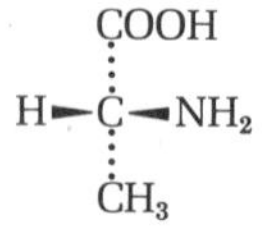

COOH / H—C—NH_2 / CH_3
D-Alanine

in sugar metabolism, and its isomer *maleic acid,* which occurs in some plants. These compounds are *geometrical* or *cis-trans* isomers; they differ in the arrangement of their substituent groups with respect to their nonrotating double bonds. Fumaric acid is the trans isomer and maleic acid the cis isomer, each of which is a well-defined compound that can be isolated in pure form. Figure 7-1 also shows the L and D isomers of alanine (pages 52 and 97), in which the substituent groups have two different configurations around a chiral center. *The identifying characteristic of configurational isomers is that they cannot be interconverted without breaking one or more covalent bonds.*

The term *conformation,* on the other hand, refers to the spatial arrangement of substituent groups of organic molecules that *are* free to assume different positions in space without breaking any bonds, because of the freedom of rotation about their carbon-carbon single bonds. In the simple hydrocarbon *ethane,* for example, there is complete freedom of rotation around the C—C single bond. Many different conformations of the ethane molecule are therefore possible, depending upon the degree of rotation of one of the carbon atoms in relation to the other, but they are all freely interconvertible through rotation. The *staggered* conformation of ethane (Figure 7-2) is more stable than all others and thus predominates, whereas the *eclipsed* form is least stable. It is not possible to isolate either one of these two conformational forms free of the other, since they are freely interconvertible and in equilibrium with each other. However, as we can predict from the models in Figure 7-2, if one or more hydrogen atoms in the two carbons of ethane were replaced by functional groups that are either very large or electrically charged, freedom of rotation around the carbon-carbon single bond might be hindered, thus limiting the number of possible conformations of ethane.

Paradoxically, Native Proteins Appear to Have Only One or a Few Conformations

Since the covalent backbone of polypeptide chains contains only single bonds, we would expect a polypeptide to be able to assume an infinite number of conformations in space. Moreover, we might expect the conformation of a polypeptide to undergo constant change because of thermal motion and random rotation of segments of the chain around each of the single bonds in the backbone. Thus it may seem paradoxical to learn that the polypeptide chain of a native protein has only one conformation (or a very few) in space under normal biological conditions of temperature and pH. This *native conformation* is sufficiently stable for the protein to be isolated with its biological activity intact *if* we handle the protein carefully and do not allow it to become denatured or unfolded. This fact therefore implies that the single bonds in the backbone of native proteins do not rotate freely. We shall now see that this is so. But first let

Figure 7-2
Two extreme conformations of the ethane molecule. Many conformations are possible because of freedom of rotation around the C—C bond. The different conformational forms are rapidly interconvertible and cannot be separated from each other. The staggered form is more stable and predominates.

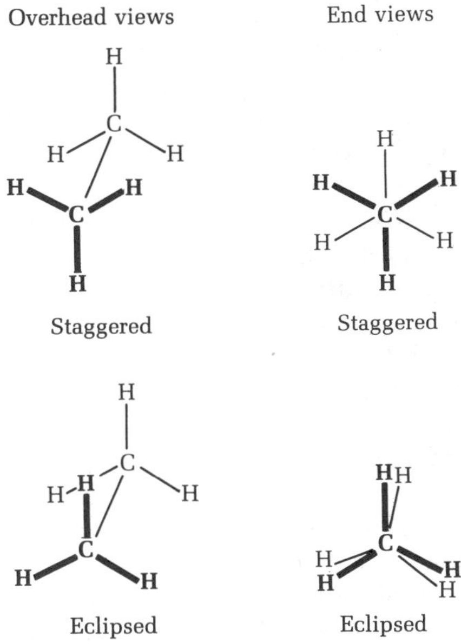

us review something of the biology of the keratins, the class of fibrous proteins whose structure became the key to the study of protein conformation.

α-Keratins Are Fibrous Proteins Made by Epidermal Cells

The α-keratins are the major fibrous proteins that provide external protection to vertebrates. They constitute almost the entire dry weight of hair, wool, feathers, nails, claws, quills, scales, horns, hooves, tortoise shell (page 95), and much of the outer layer of skin. The α-keratins are a family of proteins. They are generally similar in their amino acid composition, they characteristically contain cystine residues, and they all have a similar spatial arrangement of their polypeptide chains.

While we might think of hair, feathers, nails, and other such external structures of this sort as being extracellular in location, actually the keratins are made within epidermal cells of the skin, from which hair, feathers, and nails are derived. The polypeptide chains of keratin first become organized into filaments in hair cells. These filaments assemble into ropelike and cablelike structures, which ultimately fill the hair cell. The hair cells then flatten and die, the cell walls forming a tubelike covering around the hair called the *cuticle*. Nails, feathers, and the scales of reptiles develop in a similar manner but in much more complex patterns (Figure 7-3).

α-Keratin, particularly the form occurring in hair and wool, has played a very important role in the development of our present knowledge of protein structure. Its polypeptide chains

Figure 7-3
Feather structure. Although feathers are almost pure α-keratin, their polypeptide chains are organized into extraordinarily complex structures during their development in epidermal cells. Feathers are strong and flexible; they provide insulation and, through their pigmentation, some protection from enemies. Above all, they must be very light. They also show microscopic adaptations to their biological function. Shown here is a scanning electron micrograph of the iridescent feathers of Anna's hummingbird. The feather shaft has thousands of barbs and each barb has many barbules. At the outer ends of the feathers the barbules have tiny hooks that catch on the barbules of adjacent feathers, like the two parts of Velcro tape. When a bird preens, it is "zipping" its feathers back together into a streamlined form.

occur in long, coiled arrays which lend themselves readily to analysis by x-ray methods and thus became a stepping-stone toward understanding the much more complex globular proteins.

X-Ray Analysis of Keratins Shows That They Have Repeating Structural Units

The spacing of atoms in crystals can be determined by measuring the angles and the intensities at which a beam of x-rays of a given wavelength is diffracted by the electron shells around the atoms in the lattice. For example, x-ray analysis of sodium chloride crystals shows that Na^+ and Cl^- ions are arranged in a simple cubic lattice. The spacing of the different kinds of atoms in complex organic molecules, even very large ones such as proteins, can also be analyzed by x-ray diffraction methods. However this is far more difficult than for simple salt crystals because the very large number of atoms in a protein molecule yields thousands of diffraction spots that must be analyzed by computer methods.

In the early 1930s William Astbury in England carried out the first pioneering x-ray studies of proteins. He found that an x-ray beam directed at a hair or wool fiber, which consists largely of α-keratin, gave a characteristic diffraction pattern. From this pattern he deduced that hair and wool possess a periodicity or repeating structural unit of about 0.54 nm along the long axis of the hair. His observations suggested further that the polypeptide chains in this family of fibrous proteins are not fully extended but twisted or coiled in some regular way.

On the other hand, Astbury found that *fibroin*, the protein of silk fibers, a β-keratin, has a somewhat different x-ray diffraction pattern, suggesting a repeat unit of 0.70 nm. Significantly, when hair or wool was steamed and stretched, it yielded an x-ray pattern resembling that of silk, with a periodicity approaching 0.70 nm. For some years these observations could not be analyzed further, for lack of knowledge regarding the three-dimensional structure of polypeptides.

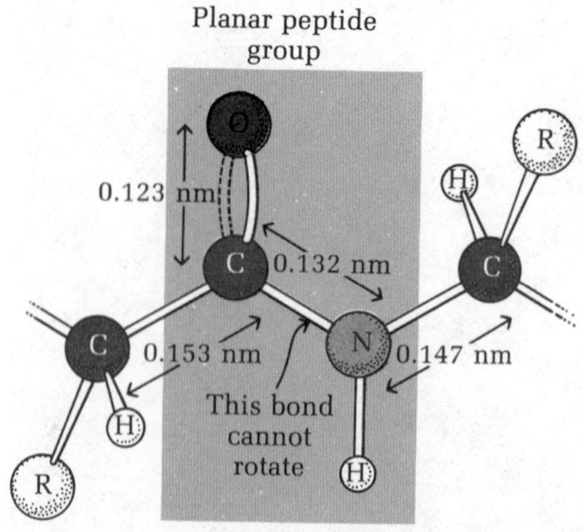

Figure 7-4
The planar peptide group. Its C—N bond has some double-bond character and cannot rotate. Note that the oxygen and hydrogen atoms in the plane are on opposite sides of the C—N bond. This is the trans configuration.

X-Ray Studies of Peptides Show the Peptide Bond to Be Rigid and Planar

The next stage in the development of our knowledge of the spatial structure of proteins came from the work of Linus Pauling and Robert Corey in the United States in the 1940s and early 1950s. They recorded the x-ray diffraction patterns of crystals of amino acids and of simple dipeptides and tripeptides and from them deduced the precise structure of the peptide bond. They found that the C—N bond joining two amino acids in a peptide linkage (Figure 7-4) is shorter than most other C—N bonds, such as that in a simple amine. Moreover, the C—N bond of the peptide linkage has some double-bond character

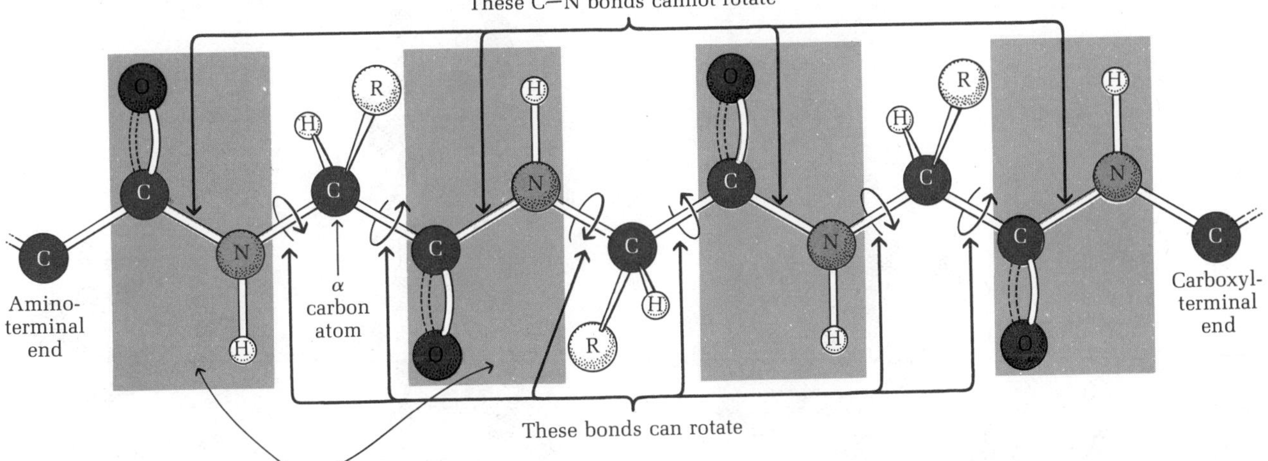

Figure 7-5
Limited rotation around the single bonds of a polypeptide chain. The C—N bonds in the planar peptide groups (shaded in color), which make up one-third of all the backbone bonds, are not free to rotate. Other single bonds in the backbone may also be hindered, depending on the size and charge of the R groups.

and thus cannot rotate freely. They further deduced that the four atoms of the peptide group lie in a single plane, in such a way that the oxygen atom of the carbonyl group and the hydrogen atom of the —NH— group are trans to each other (Figure 7-4). From these studies they concluded that the C—N bonds, which make up one-third of all the bonds in the backbone of polypeptides, are unable to rotate freely because of their partial double-bond character. The backbone of a polypeptide chain can thus be pictured as a series of rigid planes separated by substituted methylene groups (—CHR—), as shown in Figure 7-5. Now we can see how the rigid peptide bonds impose some constraints upon the number of conformations in space that can be assumed by a polypeptide chain.

In α-Keratin the Polypeptide Chains Form an α Helix

With the help of precisely constructed models, Pauling and Corey then studied how a polypeptide chain can twist or fold in view of the constraints imposed by the rigid, planar peptide bonds. Particularly they sought conformations that could account for the repeat unit of about 0.54 nm in the α-keratin of hair. The simplest arrangement the polypeptide chain could assume with its rigid peptide bonds (but with the other single bonds free to rotate) is the helical structure shown in Figure 7-6, which they called the *α helix*. In this structure the polypeptide backbone is tightly wound around the long axis of the molecule, while the R groups of the amino acid residues protrude outward from the helical backbone. The repeating unit is a single turn of the helix, which extends about 0.54 nm along the long axis, corresponding closely to the periodicity observed on x-ray analysis of hair keratin.

The question then was: What forces hold the α helix in this conformation? Why should such a helix form when other conformations are possible? The answer is that the α helix is a stable, preferred conformation of the polypeptide chain of α-keratin because it permits a hydrogen bond to form between

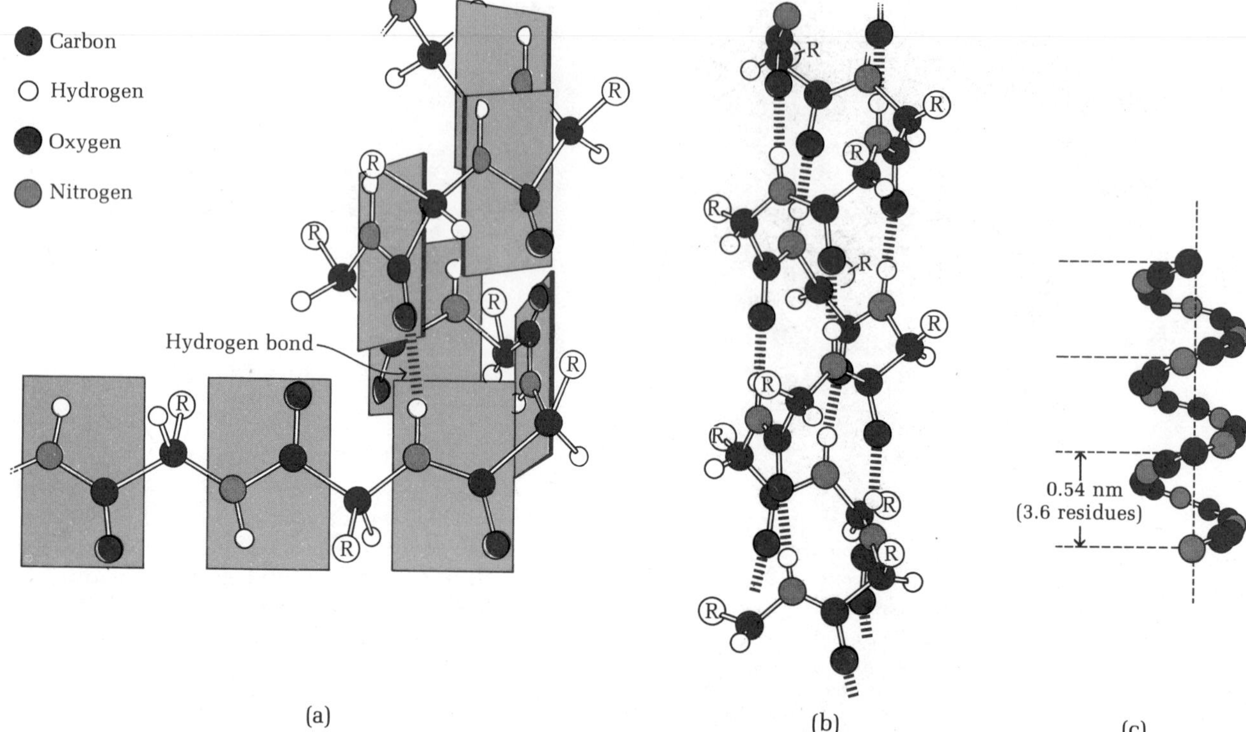

Figure 7-6
Three models of the α helix, showing different aspects of its structure. (a) *Formation of a right-handed α helix. The planes of the rigid peptide bonds are parallel to the long axis of the helix. One of the hydrogen bonds is shown in color.* (b) *Ball-and-stick model of an α helix showing the intrachain hydrogen bonds.* (c) *The repeat unit is a single turn of the helix.*

every H atom attached to the electronegative nitrogen atom of the peptide linkage and the electronegative carbonyl oxygen atom of the fourth amino acid behind it in the helix (Figure 7-6). *Every* peptide bond of the chain participates in such hydrogen bonding. Thus each successive coil of the α helix is held to the adjacent coils by several hydrogen bonds, giving the entire structure considerable stability. Thus we see that the α helix is a stable conformation of the polypeptide chain because of two types of constraints to free rotation around single bonds, i.e., the nonrotating planar peptide bonds and the formation of many intrachain hydrogen bonds. Many lines of evidence now make it certain that the polypeptide chains of α-keratins have this characteristic α-helical conformation.

Further model-building experiments have shown that an α helix can form with either L- or D-amino acids. However, the amino acids must be all one or the other stereoisomer, since a helix cannot form from a peptide chain containing a mixture of L and D residues. Furthermore, starting from the naturally occurring L-amino acids, either right- or left-handed helical coils can be built; the right-handed helix is the form present in most fibrous proteins.

Some Amino Acids Are Not Compatible with the α Helix

Although the α helix is the conformation of the polypeptide chain in α-keratins, not all polypeptides can fold to form a stable α helix. For example, if a polypeptide chain has many glutamic acid residues in a long block, this segment of the

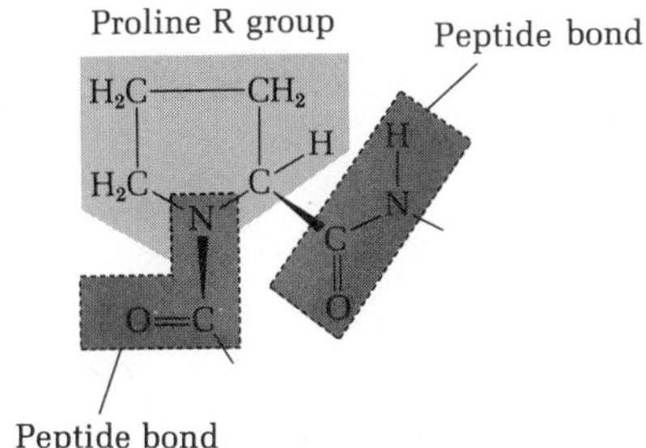

Figure 7-7
A proline residue in a polypeptide chain introduces a bend. The areas shaded in gray show the rigid peptide linkages. The rigid proline R group is shaded in color.

chain will not form an α helix at pH 7.0. The reason is that the negatively charged carboxyl groups of adjacent residues of glutamic acid repel each other so strongly that they overcome the stabilizing influence of hydrogen bonds on the α helix. For the same reason, if there are many closely adjacent lysine and/or arginine residues, whose R groups have a net positive charge at pH 7.0, they will also repel each other and prevent formation of the α helix. Moreover, certain other amino acids, such as asparagine, serine, threonine, and leucine, also tend to prevent formation of the α helix if they occur close together in the chain, because of the bulk and shape of their R groups.

Yet another constraint to the formation of the α helix occurs if one or more proline residues are present in the polypeptide chain. In proline the nitrogen atom is part of a rigid ring, and no rotation of the ring N—C bond is ever possible (Figure 7-7). Moreover, there is no substituent hydrogen on the nitrogen atom of a proline residue in peptide linkage. Thus no intrachain hydrogen bond can form with the proline peptide bond as partner. The consequence is that wherever a proline residue occurs in a polypeptide chain, it causes a kink or bend and thus interrupts the α-helical structure.

Thus we have four different kinds of constraints on the conformation in space of a polypeptide chain: (1) the rigidity and trans configuration of the peptide bonds, (2) the electrostatic repulsion (or attraction) between amino acid residues with charged R groups, (3) the bulkiness of adjacent R groups, and (4) the occurrence of proline residues.

Here we introduce an important new term. We have seen that the covalent peptide bonds and the amino acid sequence of polypeptide chains are referred to as their *primary structure.* Now we introduce the term *secondary structure* to refer to the arrangement in space, i.e., the *conformation,* of successive adjacent amino acid residues in polypeptide chains. In the case of α-keratin the secondary structure of the polypeptide chain is the α helix.

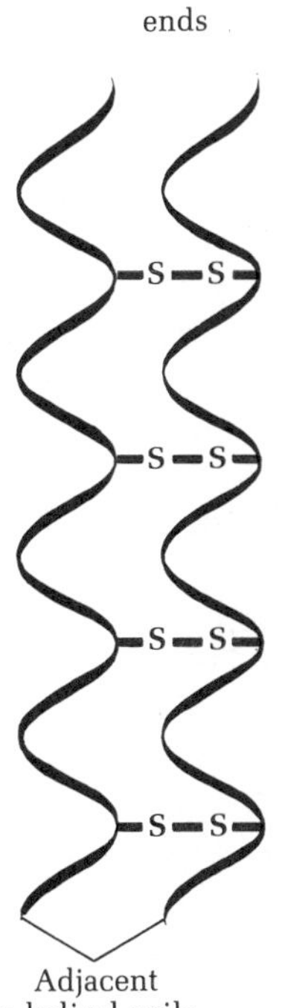

Figure 7-8
Cystine cross-links between adjacent α-helical coils in α-keratin. In "hard" keratins, such as that of tortoise shell, the cross-linking cystine residues are very numerous.

The α-Keratins Are Rich in Amino Acids Compatible with α-Helical Structures

The backbone of polypeptides will automatically tend to assume that conformation in space which is allowed or permitted by the particular set of constraints afforded by its amino acid content and sequence. In the polypeptide chains of native α-keratins the amino acid composition and sequence are such as to allow the chain to form the α helix spontaneously and to give it considerable stability through the many intrachain hydrogen bonds. The α-keratins are rich in amino acids that favor α-helix formation and contain few of those amino acids that are incompatible with it, such as proline. α-Keratins are also particularly rich in cystine residues (Figure 7-8), which we recall (page 102) can provide disulfide (—S—S—) cross-links between adjacent polypeptide chains. Such cross-links are co-

Figure 7-9
The structure of hair and hair α-keratin. The basic unit of structure is an α-keratin polypeptide chain in its native α-helix form. Three α-helical chains form a supercoiled three-strand "rope" and 11 such three-strand ropes constitute a hair microfibril.

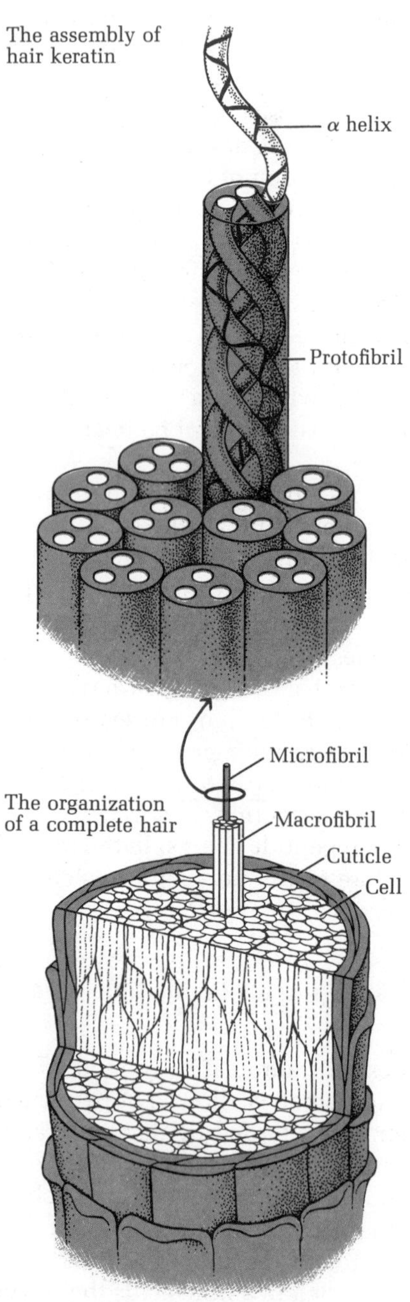

valent and thus are very strong. The covalent cross-links provided by the many cystine residues bind adjacent α helixes together and give fibers of α-keratin great cohesive strength.

In Native α-Keratins α-Helical Polypeptide Chains Are Supercoiled into Ropes

When native hair or wool is examined under the electron microscope, particularly at frayed ends where the hair has been broken, it can be seen that the hair is made up of many fibrils and that each fibril in turn is made up of sets of even smaller fibers twisted around each other in ropelike arrays. From more detailed x-ray evidence it has been concluded that the α-helical polypeptide chains of hair are twisted around each other, to form three-strand supercoiled structures resembling ropes (Figure 7-9), each strand being an α-helical coil. In such structures the component α-helical polypeptides all run in the same direction, so that all the amino-terminal residues are at the same end. The three-strand arrangement of the polypeptide chains of hair is held together tightly because of covalent cross-links between the adjacent polypeptides provided by cystine residues. α-Keratins from different sources differ in their cystine content. The hardest and toughest α-keratins, e.g., those of tortoise shell, have a very high content of cystine, up to 18 percent.

The Insolubility of α-Keratins Is a Reflection of Their Nonpolar R Groups

α-Keratins are remarkable not only for their physical strength but also for their complete insolubility in water at pH 7.0 at body temperatures. This is in striking contrast to a globular protein such as serum albumin, which is so soluble in water that 60 percent solutions can be made. How can these differences between two proteins made from the same set of 20 amino acids be explained?

Part of the answer lies in the types of amino acids that predominate in these two proteins. α-Keratins are especially rich in amino acids that have hydrophobic, or water-insoluble, R groups, such as phenylalanine, isoleucine, valine, methionine, and alanine. Moreover, we have seen that in the polypeptide chains of α-keratins the R groups of the amino acid residues are on the outside of the rodlike helix. Because of the many fixed hydrophobic R groups that are exposed to water on the outside of α-keratin fibrils, they are essentially insoluble. Although globular proteins may also contain many hydrophobic R groups, the polypeptide chains of these proteins are folded in such a way that the hydrophobic R groups are inside the globular conformation, hidden from exposure to water, whereas the hydrophilic, or polar, R groups are exposed on the outer surface. Globular proteins, such as serum albumin, are therefore usually soluble in aqueous systems.

β-Keratins Contain a Different Conformation of Their Polypeptide Chain: β Structure

β-Keratins, such as *fibroin*, the protein of silk and spider webs, are also stringy, insoluble proteins, but they are flexible and supple; moreover, they do not stretch. They differ from α-keratins in having a different periodicity of structure, which repeats at 0.70-nm intervals. Another difference is an important clue to establishing the structure of β-keratins. α-Keratin in the form of hair can be stretched to nearly double its normal length when it is steamed. In this stretched condition the x-ray pattern of stretched hair resembles that of silk fibroin. These observations led to the conclusion that when the α-keratin of hair is stretched, its α-helical conformation is disrupted and the polypeptide chain becomes extended in a conformation almost twice the length of the α helix. This happens because the moist heat disrupts the hydrogen bonds holding the α helix together, allowing the coils to be stretched into a longer conformation. If the steamed α-keratin is now allowed to cool and the stress removed, it will spontaneously revert to its original α-helical conformation.

The more extended conformation of the polypeptide chains in silk and in stretched hair and wool has been identified by x-ray analysis. It is called the *β conformation*. Figure 7-10 shows that in the β conformation the backbone of the polypeptide chain is extended into a zigzag rather than helical structure. In fibroin the zigzag polypeptide chains are arranged side by side to form a structure resembling a series of pleats; in fact, such a structure is called a *pleated sheet* (Figure 7-10). In the β conformation there are no *intrachain* hydrogen bonds. Instead, there are *interchain* hydrogen bonds between the peptide linkages of *adjacent* polypeptide chains in their more extended conformation. All the peptide linkages of β-keratin participate in such interchain hydrogen bonding. The R groups of the amino acids protrude out from the zigzag structure, as seen in the edge view.

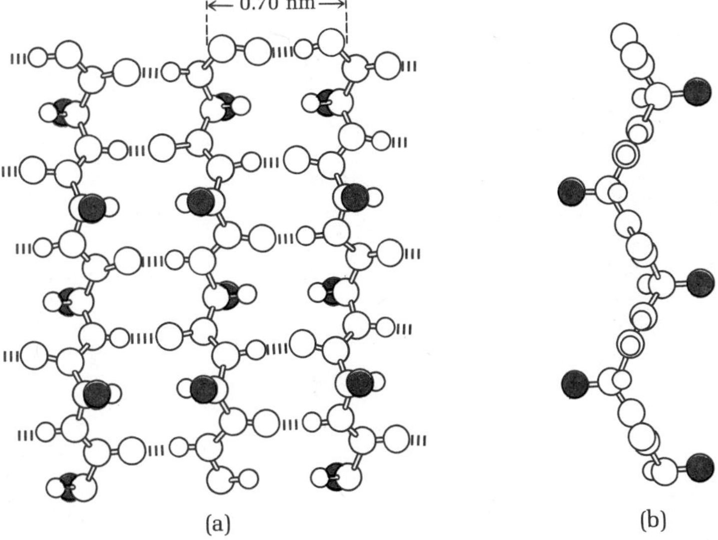

Figure 7-10
The β conformation of the polypeptide chains of β-keratins. (a) Top view of three chains arranged in a pleated sheet, showing the hydrogen-bond cross-links between adjacent chains. The R groups are indicated in color. (b) An edge view showing the R groups extending out from the pleated sheet.

There are two other important differences between α- and β-keratins. There are no cystine cross-linkages between the side-by-side chains of β-keratins; moreover, in β-keratins adjacent polypeptide chains are usually oriented in opposite or *antiparallel* directions, rather than in the *parallel* fashion typical of the chains in α-keratin (Figure 7-8).

The β structure can form only in polypeptides having the appropriate kinds of amino acids in the appropriate sequence. It can form only if the R groups of the amino acid residues are relatively small. Thus we find that silk fibroin and other β-keratins, such as the protein of spider webs, have a very high content of glycine and alanine, the amino acids with the smallest R groups. Indeed, in silk fibroin every other amino acid is glycine.

Permanent Waving Is Biochemical Engineering

We have seen above that α-keratins exposed to moist heat can be stretched into the β conformation but that on cooling the α-keratin reverts to the α-helical conformation spontaneously. This is because the R groups of α-keratins are larger on the average than in β-keratins and thus are not compatible with a stable β conformation. This characteristic of α-keratins, as well as their content of disulfide cross-linkages, is the basis for the permanent-wave industry (Figure 7-11). The hair to be waved is first bent around a form of appropriate shape. A solution of a reducing agent, usually a compound containing a thiol or sulfhydryl group (—SH), is then applied with heat. The reducing agent cleaves the disulfide cross-linkages by reducing cystine to cysteine residues, one in each chain. The moist heat breaks hydrogen bonds and causes the α-helical structure of the polypeptide chains in the hair keratin to uncoil and stretch. After a time the reducing solution is removed, and an oxidizing agent is added to establish *new* disulfide bonds between pairs of cysteine residues of adjacent polypeptide chains but not the same pairs that existed before the treatment. On washing and cooling

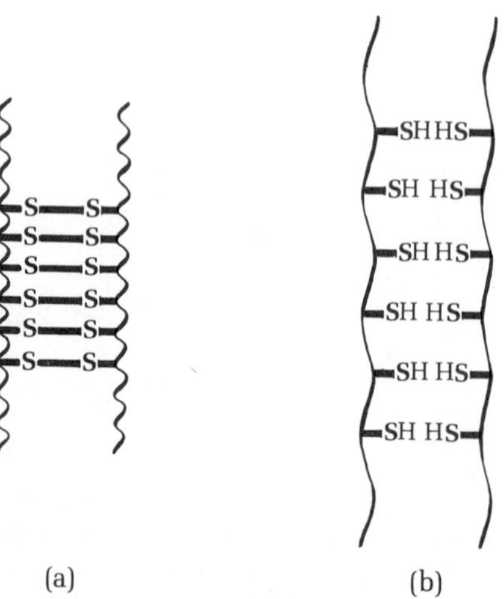

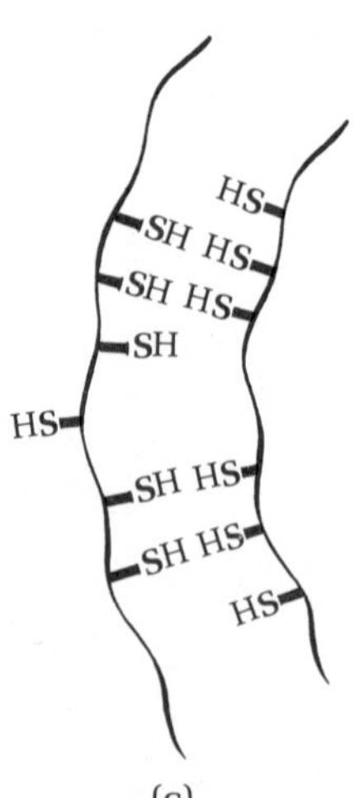

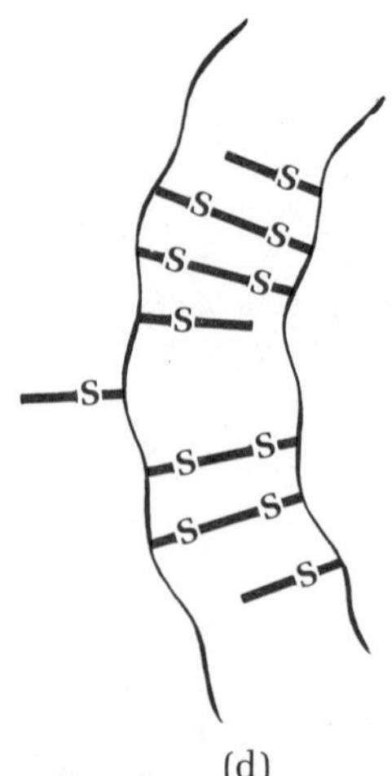

Figure 7-11
Steps in permanent waving. (a) In straight hair the α-helical coils in the keratin are held in a straight position by disulfide cross-links. (b) To impart curls the cross-links are broken by reducing agents, which convert the disulfide bonds of cystine to the thiol groups of cysteine residues, one in each chain. (c) The hair is formed into curls on a form. As the polypeptide coils are bent, the corresponding thiol groups are displaced from each other. (d) New cystine cross-links are established by oxidation of the —SH groups. The new cross-links make the curls "permanent."

the hair, the polypeptide chains revert to their α-helical conformation. The hair fibers now curl in the desired fashion because the new disulfide cross-linkages were formed where they would exert some torsion or twist on the bundles of α-helical coils in the hair fibers.

Collagen and Elastin Are the Major Fibrous Proteins of Connective Tissues

Connective tissue consists of extracellular structural and supportive elements of the body, which make up a large part of the total organic matter of higher animals. Tendons, ligaments, cartilage, and the organic matrix of bones are its most familiar elements. Connective tissue also envelops blood vessels, forms an important structural layer under the skin, helps bind cells together into tissues, and forms the extracellular ground substance between cells. There are three major molecular components of connective tissue: the two fibrous proteins *collagen* and *elastin*, which occur together in most connective tissues but vary in their proportions, and *proteoglycans*, a family of hybrid molecules consisting of proteins covalently joined to polysaccharides.

Collagen fibrils do not stretch, whereas elastin fibrils are elastic. Tendons, which transmit muscle tension to the skeleton, are largely composed of collagen. Ligaments, which are rich in elastin, bind and hold the bones composing a skeletal joint in place. They are necessarily flexible and elastic. When the ligaments around the knee or shoulder are stretched too far, these joints become dislocated but can be put back together. The proteoglycans differ from glycoproteins (page 126) in that they contain much polysaccharide and relatively little protein, whereas glycoproteins contain much more protein and less carbohydrate. Proteoglycans function as the *ground substance*, in which the fibrous elements of connective tissues are embedded or coated. They also serve to cushion tissues and to provide lubrication for the joints.

Collagen, elastin, and the proteoglycans are made by *fibroblasts* and *chondrocytes*, specialized cells in connective tissues. These proteins are then extruded from the cells and become organized into various connective-tissue structures.

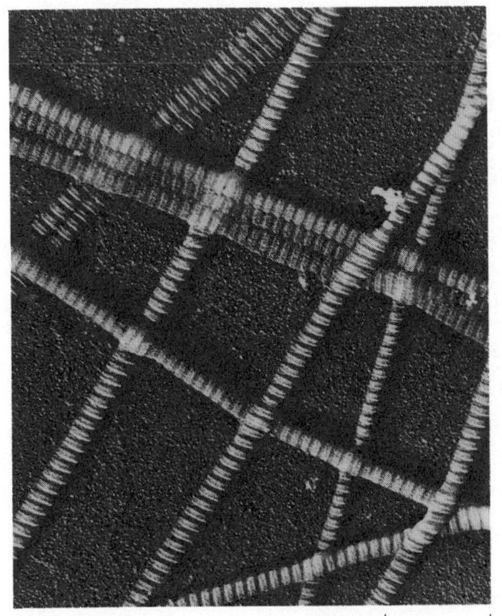

Figure 7-12
Electron micrograph of collagen fibrils of connective tissue. Note the periodicity of the cross-striations, which have a repeat distance of about 64 nm in collagens from many tissues.

Collagen Is the Most Abundant Protein in the Body

Collagen makes up almost a third of the total protein mass of vertebrates; it is the most abundant protein in the body. The larger the animal the greater the fraction of its total proteins contributed by collagen. A cow weighing 1000 lb is largely supported and held together by the strong, tough collagen fibers in its hide, tendons, cartilage, and bones.

Collagenous connective tissue consists of fibers; these in turn are made up of collagen *fibrils*, which have a cross-striated appearance (Figure 7-12). These fibrils are organized in different ways, depending on the biological function of the connective tissue. In tendons the collagen fibrils are arranged in cross-linked parallel bundles to yield structures of very great tensile strength with no capacity for stretch. Collagen fibrils can support at least 10,000 times their own weight and are said to have greater tensile strength than an equal cross section of steel wire. In the hide of the cow the collagen fibrils form an irregular, interlacing, and very tough network; recall that leather is almost pure collagen. The cornea of the eye has sheets of collagen fibrils laid down in a crisscross array. Whatever their arrangement in different connective tissues, collagen fibrils under the electron microscope always show characteristic cross-striations, which are repeated at intervals of 60 to 70 nm, depending on their source. The spacing varies somewhat, because collagens are a family of very similar proteins with some variations in structure, depending on their anatomical function and the species. The most common collagens have repeat units of 64 nm.

Collagen Has Both Familiar and Unusual Properties

Boiling in water transforms collagen, which is stringy, insoluble, and indigestible, into *gelatin*, a soluble mixture of polypeptides that is the basis of gelatin desserts. This conversion involves hydrolysis of some of the covalent bonds of collagen and is a major reason for cooking meat, since it is the collagen of the connective tissues and blood vessels that makes meat tough. Meat-tenderizing preparations available in food stores contain enzymes from plants that can hydrolyze some of the peptide bonds of collagen to yield soluble polypeptides, which are readily digested.

Collagens contain about 35 percent *glycine* and about 11 percent *alanine* residues, unusually high proportions of these amino acids. Even more distinctive is their high content of *proline* and the unusual *4-hydroxyproline* (Figure 7-13), an amino acid rarely found in proteins other than collagen and elastin. Together proline and hydroxyproline amount to about 21 percent of the amino acid residues in collagens. The unusual amino acid composition of collagen, very rich in four amino acids but low in nearly all the others, is the basis for the relatively poor nutritional quality of gelatin as a food protein. The best food proteins contain all 20 amino acids, particularly a group of 10, called *essential amino acids*, which are required in the nutrition of most animals.

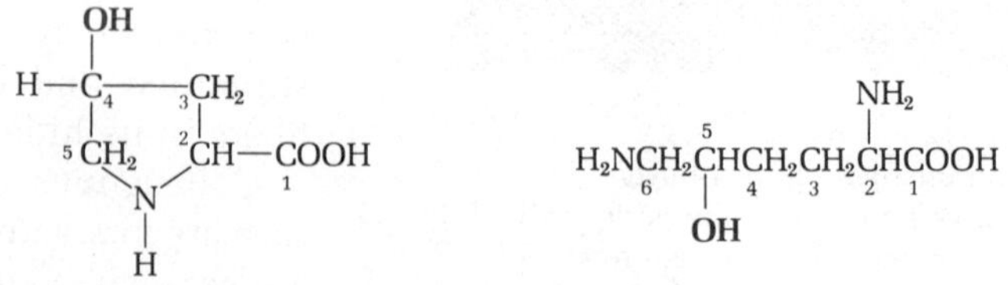

Figure 7-13
Two special amino acids found in collagen. They are derived from the standard amino acids proline and lysine by addition of hydroxyl groups (color).

4-Hydroxyproline

$H_2NCH_2CH(OH)CH_2CH_2CH(NH_2)COOH$

5-Hydroxylysine

Remember that proline and hydroxyproline form bends in polypeptide chains and that they are not compatible with α-helix structure.

The Polypeptides in Collagen Are Three-Stranded Helical Structures

Collagen fibrils consist of recurring polypeptide subunits called tropocollagen, arranged head to tail in parallel bundles (Figure 7-14). The heads of the tropocollagen molecules are staggered along the length of the fibers, accounting for the characteristic 64-nm spacing of the cross-striations in most collagens.

Collagen fibers also have a distinctive x-ray diffraction pattern, different from those of α- and β-keratins. From x-ray analysis it has been deduced that the tropocollagen subunits consist of three polypeptide chains tightly coiled into a three-strand rope.

Tropocollagen is a rod-shaped molecule, about 300 nm long and only 1.5 nm thick. It has a molecular weight of about 300,000. The three helically intertwined polypeptides are of equal length, each having about 1000 amino acid residues. In some collagens all three chains are identical in amino acid sequence, but in others two chains are identical and the third differs. Much progress is being made to determine the amino acid sequence of the major types of collagen chains, which are among the longest found in any protein.

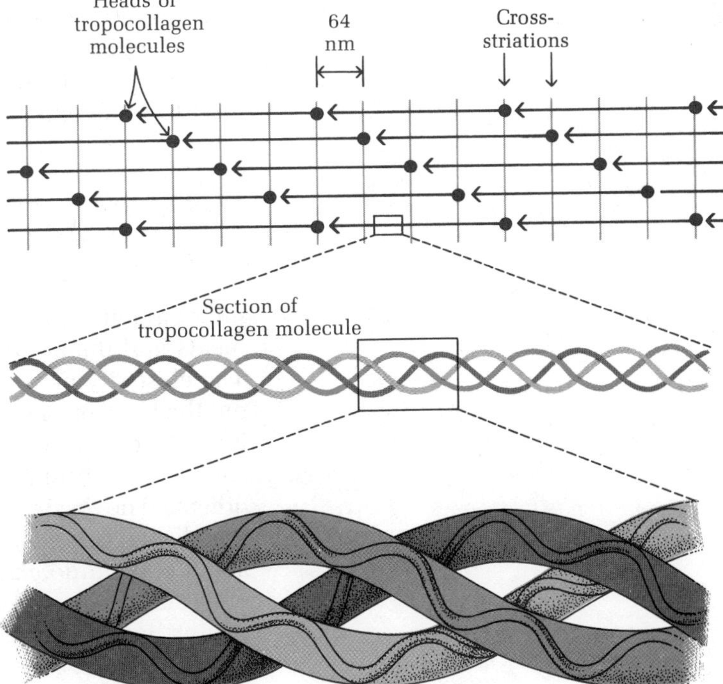

Figure 7-14
The arrangement of tropocollagen molecules in collagen fibrils. Each tropocollagen molecule extends through four of the 64-nm cross-striations. The "heads" of the tropocollagen molecules are arranged so that they are in register at 64-nm intervals. Below the schematic diagram of the fibril is a representation of a section of a tropocollagen molecule, showing the backbone of the tropocollagen triple helix. The further enlargement at the bottom shows that each of the three polypeptide chains of tropocollagen is itself a helix, whose pitch and spacing is determined by the rigid R groups of the numerous proline and hydroxyproline residues.

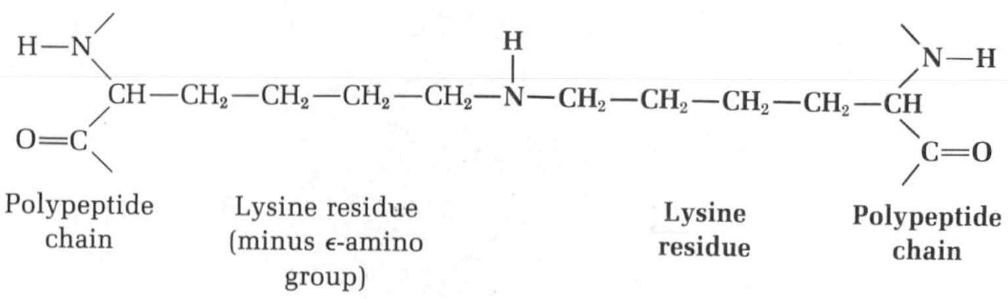

Figure 7-15
One type of cross-link between parallel collagen chains. It is formed from two lysine residues, one from each chain, by enzymatic action.

$$\begin{array}{l} H-N \\ \quad CH-CH_2-CH_2-CH_2-CH_2-\overset{H}{\overset{|}{N}}-CH_2-CH_2-CH_2-CH_2-CH \\ O=C \end{array} \qquad \begin{array}{l} N-H \\ \\ C=O \end{array}$$

Polypeptide chain | Lysine residue (minus ε-amino group) | **Lysine residue** | **Polypeptide chain**

X-ray studies have shown that each polypeptide chain of tropocollagen is itself a helix, with a periodicity and dimensions very different from those of the α helix. The helix is left-handed and has only three residues per turn. Because of its many proline and hydroxyproline residues, which impart a rigid, kinked conformation, the three helical polypeptide chains are tightly twisted about each other. They are also cross-linked to each other by hydrogen bonds and by an unusual type of covalent cross-link found only in collagen, formed between lysine residues of two chains (Figure 7-15). Adjacent tropocollagen triple helixes are also cross-linked to each other. Because of the tightness of coiling of the triple helix of tropocollagen and its cross-linkages, it has no capacity to stretch. Tropocollagen also contains some carbohydrate side chains attached to the hydroxyl groups of hydroxylysine.

As we become older, more and more covalent cross-links form in and between tropocollagen units, rendering the collagen fibrils in our connective tissues more rigid and brittle. Since collagen is present in so many structures, its increasing brittleness and rigidity on aging alters the mechanical properties of tendons and cartilage, makes bones more brittle, and causes the cornea of the eye to be less transparent.

The collagen helix is unique and found in no other protein, in contrast to the α helix and β conformation, which occur, at least to some extent, in many globular proteins.

The Structure of Elastin Confers Distinctive Properties on Elastic Tissue

The major connective tissues that are rich in elastin but contain minor amounts of collagen are the yellow elastic tissue of ligaments and the elastic connective tissue layer of large arteries. The elasticity of the arterial wall helps distribute and smooth out the force of the pulsatile pumping of the blood by the heart. Elastic connective tissue contains a fibrous protein which shares some properties with collagen but is very different in others. The basic subunit of elastin fibrils is *tropoelastin*, which has a molecular weight of about 72,000 and contains about 800 amino acid residues. Like collagen, it is rich in glycine and alanine. Tropoelastin differs from tropocollagen in having many lysine residues but few prolines. Tropoelastin forms a special type of helix, different from the α helix and col-

Figure 7-16
A residue of desmosine, a special amino acid found only in elastin. Its central ring structure is derived from the R groups (in color) of four molecules of lysine. It cross-links polypeptide chains of elastin. See Figure 7-17.

H_2N COOH
CH
$(CH_2)_3$
HOOC NH_2
CH—$(CH_2)_2$— —$(CH_2)_2$—CH
H_2N COOH
+N
$(CH_2)_4$
CH
H_2N COOH

lagen helix. It consists of lengths of helix rich in glycine residues separated by short regions containing lysine and alanine residues. The helical portions stretch on applying tension but revert to their original length when tension is released. The regions containing lysine residues engage in the formation of covalent cross-linkages. Four lysine R groups come together and are converted by enzymes into desmosine (page 103), shown in Figure 7-16, and a similar compound isodesmosine. They are capable of joining tropoelastin chains into arrays that can be stretched reversibly in all directions (Figure 7-17).

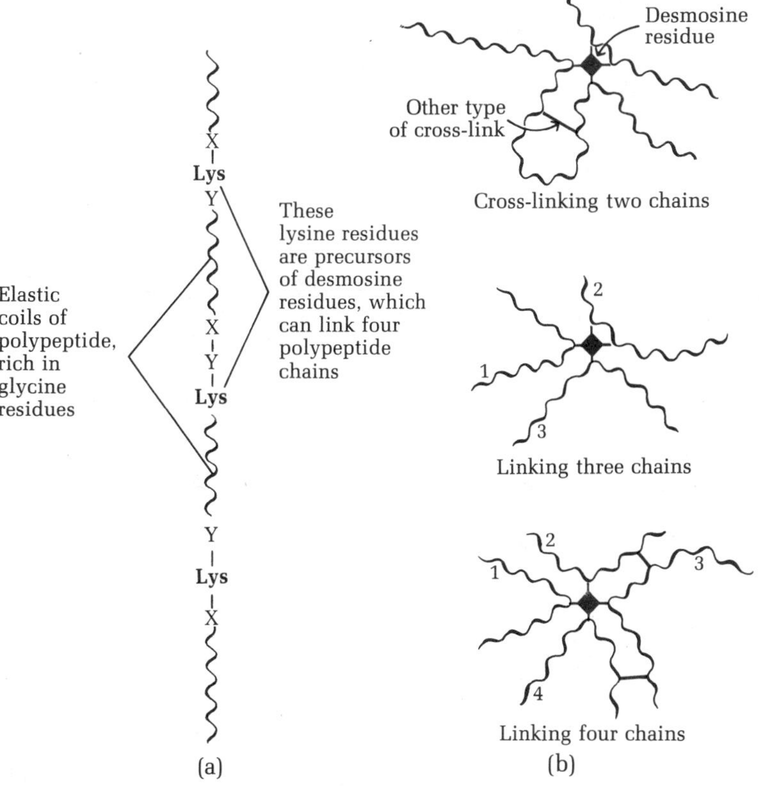

Figure 7-17
Tropoelastin molecules and their linkage to form a network of polypeptide chains in elastin. (a) Segment of a tropoelastin molecule. (b) The precise structure of elastin is not known. It consists of tropoelastin molecules cross-linked in such a way as to give two-dimensional or three-dimensional elasticity. In addition to desmosine residues (color), which can link 2, 3, or 4 tropoelastin molecules, as shown, elastin contains other kinds of cross-linkages, also designated in color.

What Fibrous Proteins Tell Us About Protein Structure

The fibrous proteins we have considered teach three major lessons about protein structure. First, we have seen that proteins possess not only primary structure, i.e., a covalent backbone, but also a characteristic secondary structure, the manner in which successive amino acid residues are arranged in space.

The second important lesson is that the secondary structure of polypeptides, such as the α helix and the β conformation, is the spontaneous and automatic result of its amino acid content and sequence. The characteristic secondary structure of a protein is its most stable form under a given set of biological conditions. The α helix and β structures are stable because they possess many hydrogen bonds, *intrachain* in the case of the α helix and *interchain* in the case of β structure. Although hydrogen bonds are individually relatively weak, collectively they give the α helix and β structure considerable stability.

The third lesson is that the three-dimensional conformations of fibrous proteins are adapted to serve specific biological functions. Thus the α-helical structure of α-keratin is well adapted to provide external protection, in the form of hair, feathers, horns, and scales in vertebrates; the β conformation is adapted to provide the supple, nonstretching characteristics of the silk fibers of cocoons and spider webs; and the collagen conformation provides the high tensile strength required of tendons (Table 7-1). Ultimately these specific biological properties are conferred by the specific amino acid sequence of fibrous proteins. The characteristic secondary structures of fibrous proteins were important stepping-stones to understanding the much more complex three-dimensional structure of globular proteins. As we shall see in the next chapter, segments of α helix and of β structure are often found in globular proteins.

Table 7-1 Secondary Structures and Properties of Fibrous Proteins

Structure	Characteristics	Examples
α helix, cross-linked by cystine	Tough, insoluble protective structures of varying hardness and flexibility	Hair, feathers, nails
β conformation	Soft, flexible filaments	Silk
Collagen triple helix	High tensile strength, without stretch	Tendons, bone matrix
Elastin chains cross-linked by desmosine	Two-way stretch with elasticity	Ligaments

Other Types of Fibrillar or Filamentous Proteins Occur in Cells

Skeletal muscles, as well as many nonmuscle cells, contain two proteins that form characteristic fibrillar or filamentous structures, *myosin* and *actin*. They are not primarily structural in their biological function; instead they participate in energy-dependent contractile activity.

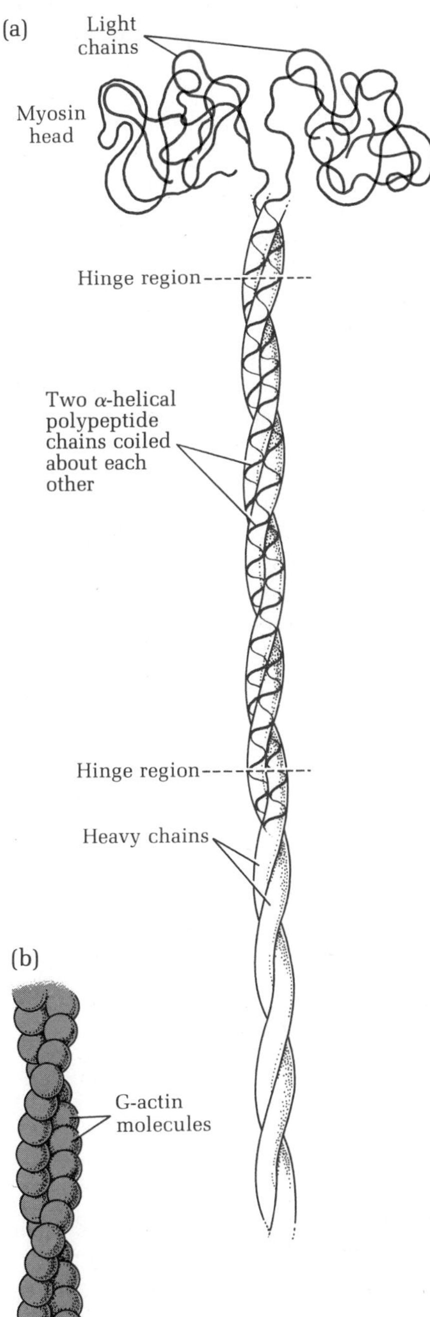

Figure 7-18
Myosin and actin, the two filamentous proteins of contractile systems. (a) *Myosin contains a long tail consisting of two supercoiled α-helical polypeptide chains (heavy chains). The head of the molecule, which contains four light chains, is an enzyme capable of hydrolyzing ATP.* (b) *A representation of F-actin, which consists of two chains of G-actin units coiled about each other.*

Myosin is a very long, rodlike molecule with a tail of two α-helical polypeptides coiled about each other; it also has a complex "head," which possesses enzymatic activity (Figure 7-18). It has a total molecular weight of 450,000, is about 160 nm long, and actually contains six polypeptide chains. The long tail consists of two chains of molecular weight 200,000 each; these are the *heavy chains*. It has hingelike, flexible joints. The head is globular and contains the ends of the heavy chains and also four light chains, each of about 18,000 molecular weight, folded into a globular conformation. The head of the myosin molecule has enzyme activity; it catalyzes hydrolysis of ATP to yield ADP and phosphate. Many myosin molecules assemble together to form the *thick filaments* of skeletal muscles. Myosin also occurs in nonmuscle cells (page 29).

Closely associated with the thick filaments in skeletal muscle are the *thin filaments*, which consist of the protein actin. Actin occurs in two forms, *globular actin* (G-actin) and *fibrous actin* (F-actin). Fibrous actin is actually a long string of G-actin molecules (MW 46,000) associated to form a filament. Two F-actin filaments coil about each other to form a two-stranded ropelike structure (Figure 7-18).

The thick filaments (formed from myosin molecules) and the thin filaments (formed from G-actin) are arranged in parallel arrays in the contractile system of muscle. As we shall see later (Chapters 14 and 25), the contraction of skeletal muscle is due to the sliding of thin filaments along the thick filaments, induced in the presence of certain other muscle proteins and Ca^{2+}. ATP is required for this sliding action, which is responsible for the shortening of skeletal muscle during contraction.

Another type of long, filamentous protein system is represented by *microtubules*, described earlier (page 30). Microtubules are long, hollow tubes composed of 13 protein filaments arranged in a parallel fashion about a hollow core. Each filament consists of two alternating globular proteins, *α-tubulin* and *β-tubulin*. Microtubules are present in eukaryotic cilia and flagella, and their sliding or twisting along each other endows them with their characteristic corkscrew, rotary, or whiplash motions used for propulsion. Microtubules participate in many other cell activities, such as cell division, and also give shape and form to some cells (page 29). The movement of microtubules in flagella is also dependent upon ATP hydrolysis (Chapters 2 and 14).

Summary

There are four kinds of fibrous proteins that have protective or structural roles in animal organisms: α-keratin, β-keratin, collagen, and elastin. They have provided important insights into the relationship between structure and function in protein molecules. The α-keratins are the characteristic insoluble, tough proteins found in hair, wool, feathers, scales, horns, hooves, and tortoise shells. X-ray analysis shows that α-keratin fibrils have repeat units about 0.54 nm long and that their polypeptide chains are coiled. X-ray analysis also shows that peptide bonds are rigid, planar structures, due to the partial double-bond character of the C—N bonds in the backbone of polypeptides. These observations led to the conclusion that the polypeptide chains of α-keratins consist of right-handed α-helical coils, in which each loop or coil has 3.6 amino acid residues and extends 0.54 nm along the long axis of the helix. All the peptide bonds participate in intrachain hydrogen bonding, which makes the α helix stable. The α helix is destabilized by adjacent residues having R groups of the same electric charge, having large, bulky R groups, and by proline residues, which introduce bends into the α helix. Hair fibers consist of multistranded ropes or cables in which each strand is an α-helical polypeptide chain and the strands are supercoiled around each other. There are many cystine cross-linkages in α-keratins.

The β-keratins, of which silk fibroin is the best example, have a periodicity of about 0.70 nm, which is also given by steamed and stretched α-keratins. In β-keratins the polypeptide chain is extended along one axis in a zigzag fashion. The adjacent polypeptide chains in β-keratins are hydrogen-bonded to each other, antiparallel in direction, and arranged into a pleated sheet, with the R groups protruding above and below the sheet structure. β-Keratins contain many glycine and alanine residues.

Collagen is the most abundant protein in vertebrates. It is found in tendons, skin fibers, blood vessels, bone, and cartilage. Collagen fibrils consist of three coiled polypeptide chains, each arranged in a special type of kinked helix, which contains about 21 percent proline and hydroxyproline residues. Collagen fibrils do not stretch, and they have great tensile strength. Collagen is converted into gelatin, a soluble, digestible mixture of polypeptides, by partial hydrolysis. Elastin, the characteristic protein of elastic connective tissue, consists of a network of polypeptide chains cross-linked by desmosine. It has elastic properties. Myosin, actin, and tubulin are units of intracellular filamentous proteins participating in contractile and motile activities dependent upon ATP energy.

References

Books

See also the references in Chapters 6 and 8.

CANTOR, C. R., and P. R. SCHIMMEL: *Biophysical Chemistry*, pt. I, *The Conformation of Biological Macromolecules*, Freeman, San Francisco, 1980.

DICKERSON, R. E., and I. GEIS: *Proteins: Structure, Function, and Evolution*, Benjamin/Cummings, Menlo Park, Calif., 1983.

SCHULTZ, G. E., and R. H. SCHIRMER: *Principles of Protein Structure*, Springer-Verlag, New York, 1979.

Some Interesting Articles

EYRE, DAVID R.: "Collagen: Molecular Diversity in the Body's Protein Scaffold," *Science*, **207**:1315, March (1980).

FRASER, R. D. B.: "Keratins," *Sci. Am.*, **221**:86–96, August (1969).

GROSS, J.: "Collagen," *Sci. Am.*, **204**:120–130, May (1961).

Problems

1. *Properties of the Peptide Bond* In x-ray studies of crystalline peptides Linus Pauling and Robert Corey found that the C—N bond in the peptide link is intermediate in length (0.132 nm) between a typical C—N single bond (0.149 nm) and a C=N double bond (0.127 nm). They also found that the peptide linkage is flat, i.e., all four atoms attached to the C—N group are located in the same plane, and that the two α carbon atoms attached to the C—N are always trans, i.e., on opposite sides of the peptide bond:

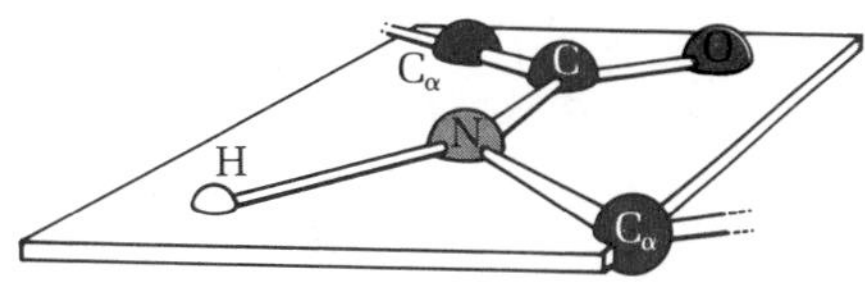

   (a) What does the length of the C—N bond in the peptide linkage indicate about its strength and its bond order, i.e., whether it is single, double, or triple?
   (b) In light of the answer in part (a), provide an explanation for the observation that such a C—N bond is intermediate in length between a double and single bond.
   (c) What do the observations of Pauling and Corey tell us about the ease of rotation about the C—N peptide bond?

2. *Early Observations on the Structure of Wool* William Astbury discovered that the x-ray pattern of wool shows a repeating structural unit spaced about 0.54 nm along the direction of the wool fiber. When he steamed and stretched the wool, the x-ray pattern showed a new repeating structural unit at a spacing of 0.70 nm. Steaming the wool and letting it shrink gave an x-ray pattern consistent with the original spacing of about 0.54 nm. Although these observations provided important clues to the molecular structure of wool, Astbury was unable to interpret them at the time. Given the current understanding of the structure of wool, interpret Astbury's observations.

3. *Rate of Synthesis of Hair α-Keratin* In human dimensions, the growth of hair is a relatively slow process, occurring at a rate of 6 to 8 in/year. All this growth is concentrated at the base of the hair fiber, where α-keratin filaments are synthesized inside living epidermal cells and assembled into ropelike structures (Figure 7-9). The basic structural element of α-keratin is the α helix, which has 3.6 amino acids per turn and a rise of 0.54 nm per turn (see Figure 7-6). Assuming that the biosynthesis of α-helical keratin chains is the rate-limiting factor in the growth of hair, calculate the rate at which peptide bonds of α-keratin chains must be synthesized (peptide bonds per second) to account for the observed yearly growth of hair.

4. *The Effect of pH on the Conformations of Polyglutamic Acid and Polylysine* The unfolding of the α-helical conformation of a polypeptide to a randomly coiled conformation is accompanied by a large decrease in its specific rotation. Polyglutamic acid, a polypeptide made up of only glutamic acid residues, has the α-helical conformation at pH 3. However, when the pH is raised to 7, there

is a large decrease in the specific rotation of the solution. Similarly, polylysine is an α helix at pH 10, but when the pH is lowered to 7, the specific rotation also decreases, as shown by the graph.

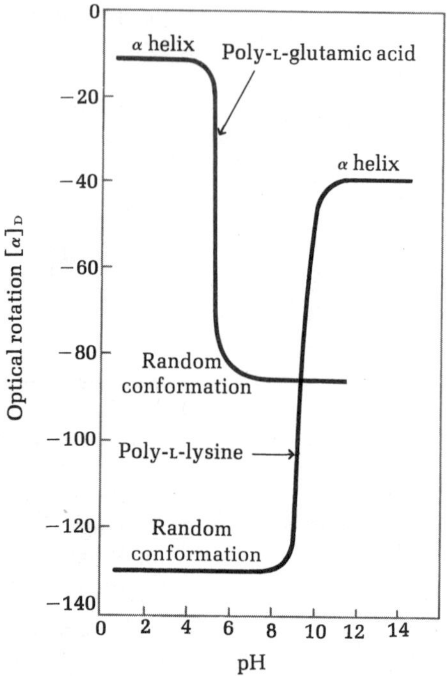

What is the explanation for the effect of the pH changes on the conformations of polyglutamic acid and polylysine? Why does the transition occur over such a narrow range of pH?

5. *The Cystine Content Determines the Mechanical Properties of Many Proteins* A number of natural proteins are very rich in cystine, and their mechanical properties (tensile strength, viscosity, hardness, etc.) are correlated with the cystine content. For example, glutenin, a cystine-rich protein found in wheat, is responsible for the cohesive and elastic character of dough made from wheat flour. Similarly, the hard, tough nature of tortoise shell is due to the high cystine content of its α-keratin. What is the molecular basis for the correlation between cystine content and mechanical properties of the protein?

6. *Why Does Wool Shrink?* When wool sweaters or socks are washed in hot water and/or dried in an electric dryer, they shrink. From what you know of α-keratin structure, how can you account for this? Silk, on the other hand, does not shrink under the same conditions. Explain.

7. *Stability of Cystine-Containing Proteins to Heat* Most globular proteins are denatured (undergo unfolding of their structures) with loss of their activity when briefly heated to 65°C. But globular proteins that contain multiple cystine residues often must be heated longer at higher temperatures to denature them. One such is ribonuclease, which has 104 amino acid residues in a single chain

and contains four disulfide cross-links contributed by cystine residues. Ribonuclease must be heated to higher temperatures to unfold it; when such a solution is quickly cooled, it is enzymatically active. Can you suggest a molecular basis for this property?

8. *Cleavage of Cystine Cross-Links* The —S—S— cross-links of cystine in proteins are formed when two cysteine residues within one chain or in two chains undergo oxidation in the presence of an oxidizing agent to make an —S—S— bond. In the determination of the amino acid sequence of a protein, for a variety of practical reasons one must first cleave all the —S—S— cross-links. Since this process is the reverse of oxidation, it can be carried out by a reducing agent:

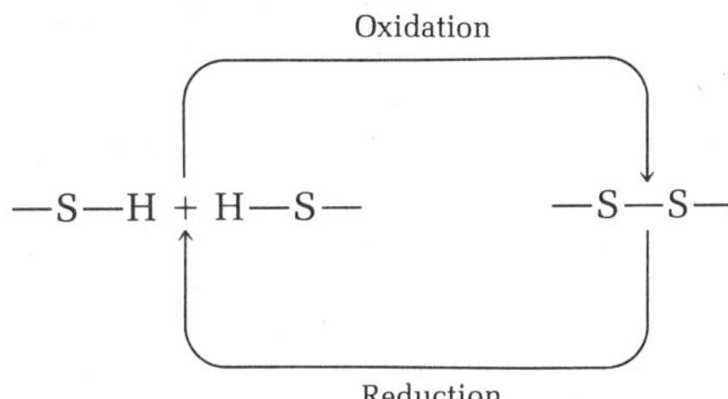

(a) One of the standard techniques for cleaving disulfide bridges in a protein is to treat it with an excess of 2-mercaptoethanol ($HSCH_2CH_2OH$). Explain the rationale for this practice.
(b) One of the drawbacks of this procedure is that the cystine cross-links, once cleaved, reform spontaneously. Why does this occur?

9. *Spacing between β Sheets of Silk* Chemical studies of digested silk from the silk fibroin of *Bombyx mori* show that a basic six-residue unit is repeated many times along its polypeptide chains:

$$(\text{–Gly–Ser–Gly–Ala–Gly–Ala–})_n$$

Furthermore, x-ray crystallographic studies have shown that the major repeating unit has a spacing of 0.70 nm, as described in the text. However, there are also two other kinds of repeating units of 0.35 and 0.57 nm, which represent the distances between β sheets. Suggest how the six-residue units that constitute the polypeptide chains give rise to the distances between β sheets in silk.

10. *Bacteriorhodopsin in Purple Membrane Proteins* Under the proper environmental conditions, the salt-loving bacterium *Halobacterium halobium* synthesizes a membrane protein (MW 26,000) known as bacteriorhodopsin. Molecules of this protein, which is purple because it contains retinal, aggregate into "purple patches" in the cell membrane. Bacteriorhodopsin acts as a light-activated proton pump and serves to provide energy for cell functions. X-ray analysis of this protein reveals that it consists of seven parallel α-helical segments, each of which traverses the 4.5-nm-thick bacterial cell membrane. Calculate the minimal number of amino acids necessary for one segment of α helix to traverse the membrane completely. Estimate the fraction of the bacteriorhodopsin protein that occurs in α-helical form. Justify all your assumptions.

11. *Biosynthesis of Collagen* Collagen, the most abundant protein in mammals, has an unusual amino acid composition. Unlike most other proteins, collagen is very rich in proline and hydroxyproline (Figure 7-13). Since hydroxyproline is not one of the 20 amino acids commonly found in proteins, its incorporation in collagen can be explained by two routes: (1) proline is hydroxylated by enzymes *before* incorporation into collagen; (2) proline is hydroxylated *after* incorporation into collagen. To differentiate between these two hypotheses, the following experiments were performed. When ^{14}C-proline is administered to a rat and the collagen from the tail isolated, the newly synthesized tail collagen is found to be radioactive. If, however, ^{14}C-hydroxyproline is administered to a rat, no radioactivity is observed in the newly synthesized collagen. How do these experiments differentiate between the two hypotheses?

12. *Pathogenic Action of Bacteria That Cause Gas Gangrene* The highly pathogenic anaerobic bacterium *Clostridium perfringens*, responsible for gas gangrene, a condition in which tissue structure is destroyed, secretes an enzyme that efficiently catalyzes the hydrolysis of the indicated peptide bond in the sequence

$$-\text{X}—\text{Gly}-\text{Pro}-\text{Y}- \xrightarrow[\text{enz}]{H_2O} -\text{X}—CO_2^- + H_3\overset{+}{\text{N}}—\text{Gly}-\text{Pro}-\text{Y}-$$

where X and Y are any of the 20 amino acids. How does the secretion of this enzyme contribute to the invasiveness of this bacterium in human tissues? Why does this enzyme not affect the bacterium itself?

CHAPTER 8

Globular Proteins: The Structure and Function of Hemoglobin

In globular proteins the polypeptide chain is folded into a compact, globular shape. As a class they are more complex in conformation than fibrous proteins, have a far greater variety of biological functions, and are dynamic rather than static in their activities. Nearly all of the 2000 or more enzymes are globular proteins. Other globular proteins function to transport oxygen, nutrients, and inorganic ions in the blood; some serve as antibodies, some as hormones, and still others as components of membranes and ribosomes.

In this chapter we shall see how the polypeptide chains of some globular proteins are folded and how their amino acid sequences determine their three-dimensional structures. We shall also see how the native folded conformation of globular proteins is necessary for their biological activity. Then we shall examine the chemistry, biology, and some medical implications of the oxygen-transporting protein hemoglobin of red blood cells, to illustrate how the three-dimensional structure of globular proteins is adapted to carrying out their important biological functions.

The Polypeptide Chain(s) of Globular Proteins Are Tightly Folded

Two important pieces of evidence indicate that the polypeptide chains of globular proteins are tightly folded and that their folded conformations are important to their biological functions. The first is that native globular proteins undergo *denaturation* when they are heated, exposed to extremes of pH, or treated with urea (page 140). When a globular protein is denatured, its covalent backbone structure remains intact but the polypeptide chain is unfolded into random, irregular, and changing conformations in space. A denatured globular protein generally becomes insoluble in aqueous systems at pH near 7 and usually loses its biological activity.

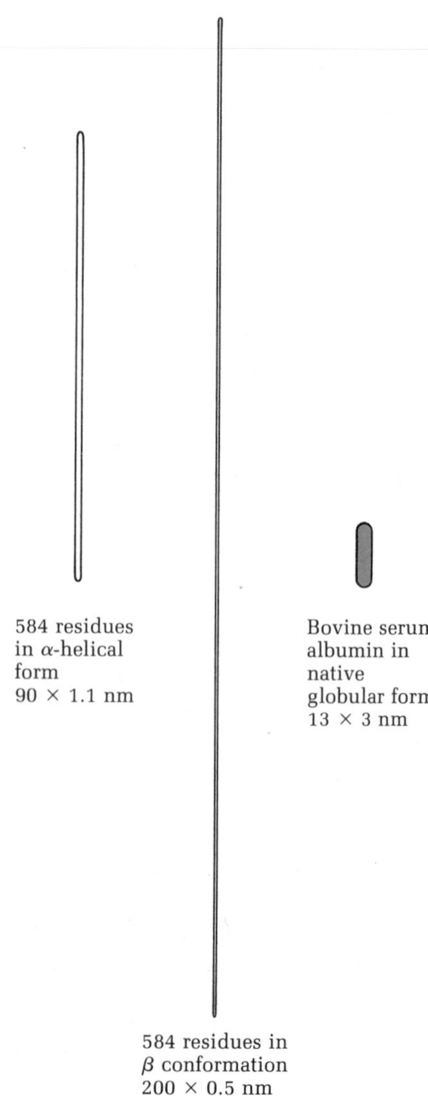

Figure 8-1
Dimensions of bovine serum albumin in its native globular conformation. Serum albumin has 584 residues in a single chain. Shown at the left are the approximate dimensions of its single polypeptide chain if it occurred as an α helix or in extended β conformation. The actual dimensions of native serum albumin are shown at the right.

The second piece of evidence for the folding of globular proteins comes from comparisons of the length of their polypeptide chains with their actual molecular dimensions as revealed by physicochemical measurements. For example, serum albumin (MW 64,500) has a single polypeptide chain with 584 amino acid residues. If this chain were in a fully extended β conformation, it would be almost 200 nm long and about 0.5 nm thick; if it occurred as an α helix, it would be about 90 nm long and 1.1 nm thick (Figure 8-1). However, physicochemical measurements on native serum albumin show that its long dimension is only about 13 nm and that it is about 3 nm wide (Figure 8-1). Clearly the polypeptide chain of serum albumin must be very tightly folded to fit into these dimensions. It is now certain that *all* globular proteins are compactly folded in special ways that are important for their biological activity. *We use the term tertiary structure to designate how the polypeptide chains of globular proteins are folded into compact spherical or globular shapes.*

Obvious questions now arise. How can the folding pattern of globular proteins be determined? Is the polypeptide chain folded identically in all globular proteins? What factors keep the polypeptide chain in its folded conformation?

X-Ray Analysis of Myoglobin Was the Breakthrough

The answer to these questions came from the powerful method of x-ray analysis, which, as we have seen, revealed the structures of different fibrous proteins. However, x-ray analysis of globular protein structure is much more difficult than for fibrous proteins, which are extended along one dimension and usually have periodic structures. To deduce the three-dimensional structure of globular proteins from x-ray diffraction patterns requires many complex computer calculations.

The breakthrough came from x-ray studies of the globular protein *myoglobin* carried out by John Kendrew and his colleagues in England in the 1950s. Myoglobin is a relatively small, oxygen-binding protein (MW 16,700) found in muscle cells. It functions there to store bound oxygen and to enhance its transport to the mitochondria, which consume oxygen during oxidation of cell nutrients. Myoglobin contains a single polypeptide chain of 153 amino acid residues of known sequence and a single iron-porphyrin, or *heme*, group (Figure 8-2), identical with that of hemoglobin, the oxygen-binding protein of red blood cells. The heme group is responsible for the deep red-brown color of both myoglobin and hemoglobin. Myoglobin is particularly abundant in the muscles of diving mammals such as the whale, seal, and porpoise, whose muscles are so rich in this protein that they are brown. Storage of oxygen by muscle myoglobin permits these animals to remain submerged for long periods.

Structure of heme

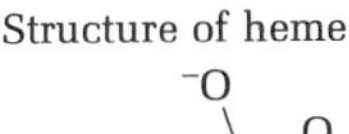

Space-filling model of heme

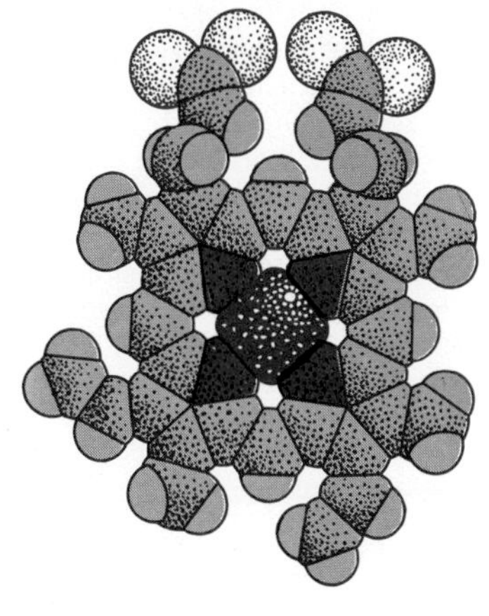

The coordination bonds of iron

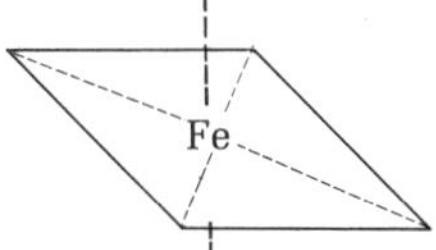

Edge view

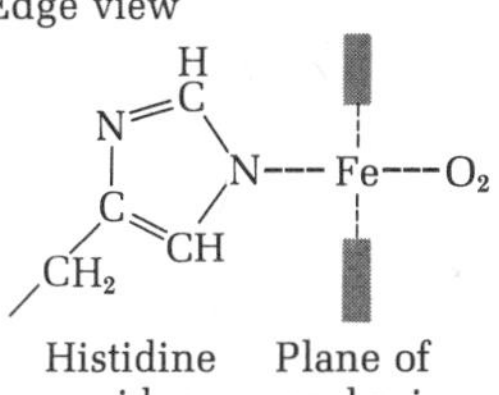

Figure 8-2
The heme group, present in myoglobin, hemoglobin, and many other heme proteins, consists of a complex organic ring structure, protoporphyrin, to which an iron atom in its ferrous Fe(II) state is bound. The iron atom has six coordination bonds, four to the flat porphyrin molecule and two perpendicular to it. In myoglobin and hemoglobin one of the latter is filled by a nitrogen atom of a histidine residue. The other is "open" and serves as the binding site for an oxygen molecule, as shown in the edge view at the lower right. In myoglobin and hemoglobin, carbon monoxide (CO) can compete with O_2 for binding to the open bond; CO is bound about 200 times more strongly than O_2. In carbon monoxide poisoning, much of the hemoglobin is present as carbon monoxide hemoglobin, thus causing inhibition of O_2 transport to the tissues.

Figure 8-3
Photograph of the x-ray diffraction pattern of crystalline sperm whale myoglobin. From the positions and intensities of the spots, which represent reflections of the x-ray beam by specific atoms in the myoglobin molecule, the precise three-dimensional structure of myoglobin was calculated.

Kendrew found that the x-ray diffraction pattern (Figure 8-3) of crystalline myoglobin from muscles of the sperm whale is very complex, with nearly 25,000 reflections. Computer analysis of these reflections took place in stages. In the first stage, completed in 1957, the three-dimensional structure of myoglobin was calculated to 0.6 nm resolution. This degree of resolution, although insufficient to reveal the exact positions of individual atoms, indicated how the polypeptide chain backbone is folded in the myoglobin molecule. The sausagelike outline of the tertiary structure of the myoglobin chain, which is folded into an odd, irregular form, is shown in Figure 8-4. Because the R groups are missing, the structure appears much more open than it actually is. The drawing also shows the flat heme group, which is tightly but noncovalently bound to the polypeptide chain. In the second stage the x-ray analysis of myoglobin was carried out to 0.2 nm resolution, which was high enough to identify most of the R groups. In the third stage, at a resolution of 0.14 nm, all the amino acid residues could be identified. Their sequence agreed with that obtained by chemical analysis.

Figure 8-4 shows the detailed, residue-by-residue *secondary structure* of the polypeptide backbone of myoglobin within the sausagelike outline of the polypeptide chain and how the

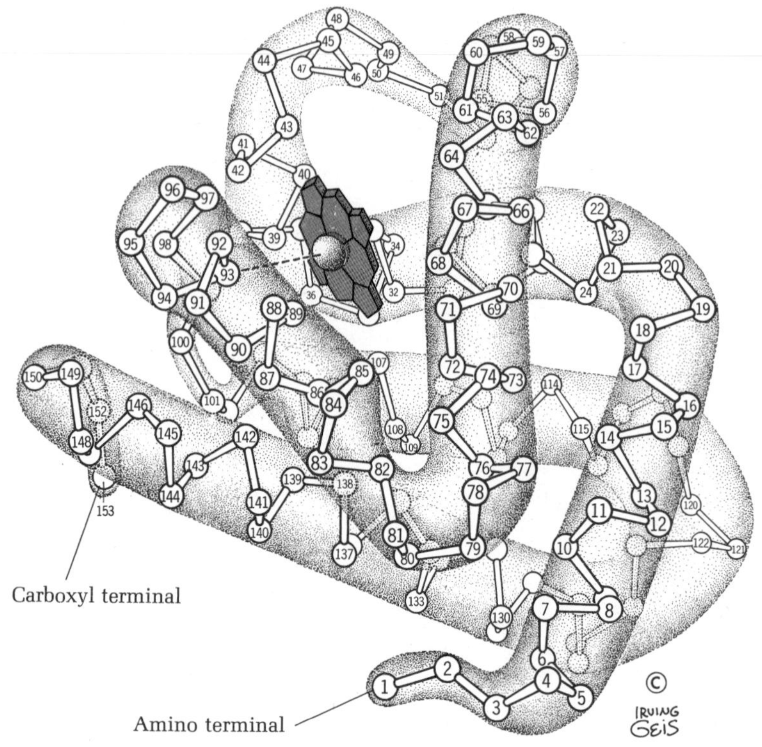

Figure 8-4
Tertiary structure of sperm whale myoglobin from x-ray analysis. The drawing shows the backbone structure deduced from the data to 0.2 nm resolution. Only the backbone atoms are shown within the sausagelike outline of the chain. The spaces between loops of the chain are not empty but are filled with the R groups, omitted in this drawing. The myoglobin molecule has eight α-helical segments. The heme group is shown in color.

whole chain is folded in three dimensions, i.e., its *tertiary structure*. The backbone of the myoglobin molecule is made up of eight relatively straight segments interrupted by bends. Each straight segment consists of a length of α helix, the longest having 23 amino acid residues and the shortest only 7. All the α-helical segments have been identified as right-handed. Almost 80 percent of the amino acids in the myoglobin molecule are in these α-helical regions. X-ray analysis also revealed the precise position of each of the R groups, which extend out from the sausagelike outline of the backbone and occupy nearly all the open space between the folded loops.

From precisely constructed models of the myoglobin molecule, patterned on the x-ray data, other important conclusions were drawn:

1. The myoglobin molecule is so compact that in its interior there is room for only four molecules of water.

2. All except two of the polar R groups of the myoglobin chain are located on the outer surface of the molecule, and all of them are hydrated.

3. Most of the hydrophobic R groups are in the interior of the myoglobin molecule, hidden from exposure to water (for reference, Table 8-1 lists the amino acids with the most hydrophilic and most hydrophobic R groups, and those with R groups that are in between in their properties).

Table 8-1 Classification of Amino Acids According to Polarity and Location in the Structure of Globular Proteins

Highly hydrophilic amino acids, nearly always found on the outer surface of globular proteins
Aspartic acid
Glutamic acid
Asparagine
Glutamine
Lysine
Arginine
Histidine
Highly hydrophobic amino acids, found largely in the interior of globular proteins
Phenylalanine
Leucine
Isoleucine
Methionine
Valine
Tryptophan
Amino acids of intermediate polarity, found in both the interior and exterior of globular proteins
Proline
Threonine
Serine
Cysteine
Alanine
Glycine
Tyrosine

4. Each of the four proline residues of myoglobin occurs at a bend (recall that the rigid R group of proline is incompatible with α-helical structure) (page 153). Other bends or turns contain serine, threonine, and asparagine residues, which are among the amino acids that tend to be incompatible with α-helical structure if they are in close juxtaposition (page 153).
5. All the peptide bonds are in the planar trans configuration (page 151).
6. The flat heme group rests in a crevice, or pocket, in the myoglobin molecule. The iron atom in the center of the heme group has two coordination bonds that are perpendicular to the plane of the heme. One of these is bound to the R group of the histidine residue at position 93, whereas the other coordination bond is the site to which an O_2 molecule is bound.

Myoglobins from Different Species Have Similar Conformations

We have seen that in a series of homologous proteins from different species, such as cytochrome *c*'s, the amino acid residues at certain positions in the sequence are *invariant*, always the same, whereas at other positions the amino acid residues may vary (page 136). This is also true for myoglobins isolated from different species of whales, from the seal, and from some terrestrial vertebrates. The strong implication that all myoglobins have a common ancestor and some common features in the way their polypeptide chains are folded has been completely substantiated by x-ray analysis of myoglobins from several other species. Since their tertiary structures are very similar to that of sperm whale myoglobin, the amino acid sequence homologies and the similar tertiary structures of different myoglobins lead to the conclusion that the amino acid sequence of myoglobin somehow must determine its three-dimensional folding pattern. Other sets of homologous proteins also show this relationship: in each case there are sequence homologies as well as similar tertiary structures.

The Tertiary Structure of Each Type of Globular Protein Is Distinctive

Are all globular proteins folded exactly like myoglobin? This question can now be answered, since the tertiary structures of other small, single-chain globular proteins have been analyzed by x-ray methods. Especially interesting is the electron-carrying mitochondrial protein *cytochrome c*, whose amino acid sequence has been studied in well over 60 species (page 137). Like myoglobin, cytochrome *c* is a small heme protein (MW 12,400) containing a single polypeptide chain of about

100 residues and a single heme group, which in this case is covalently attached to the polypeptide. Like myoglobin, cytochrome *c* is also compactly folded, with most of its hydrophilic R groups outside and most of its hydrophobic R groups inside its globular structure. Because cytochrome *c* and myoglobin are heme proteins, one might have expected them to have similar tertiary structures. Not so. X-ray analysis of cytochrome *c* shows it to have an entirely different three-dimensional structure (Figure 8-5 and Table 8-2). Whereas al-

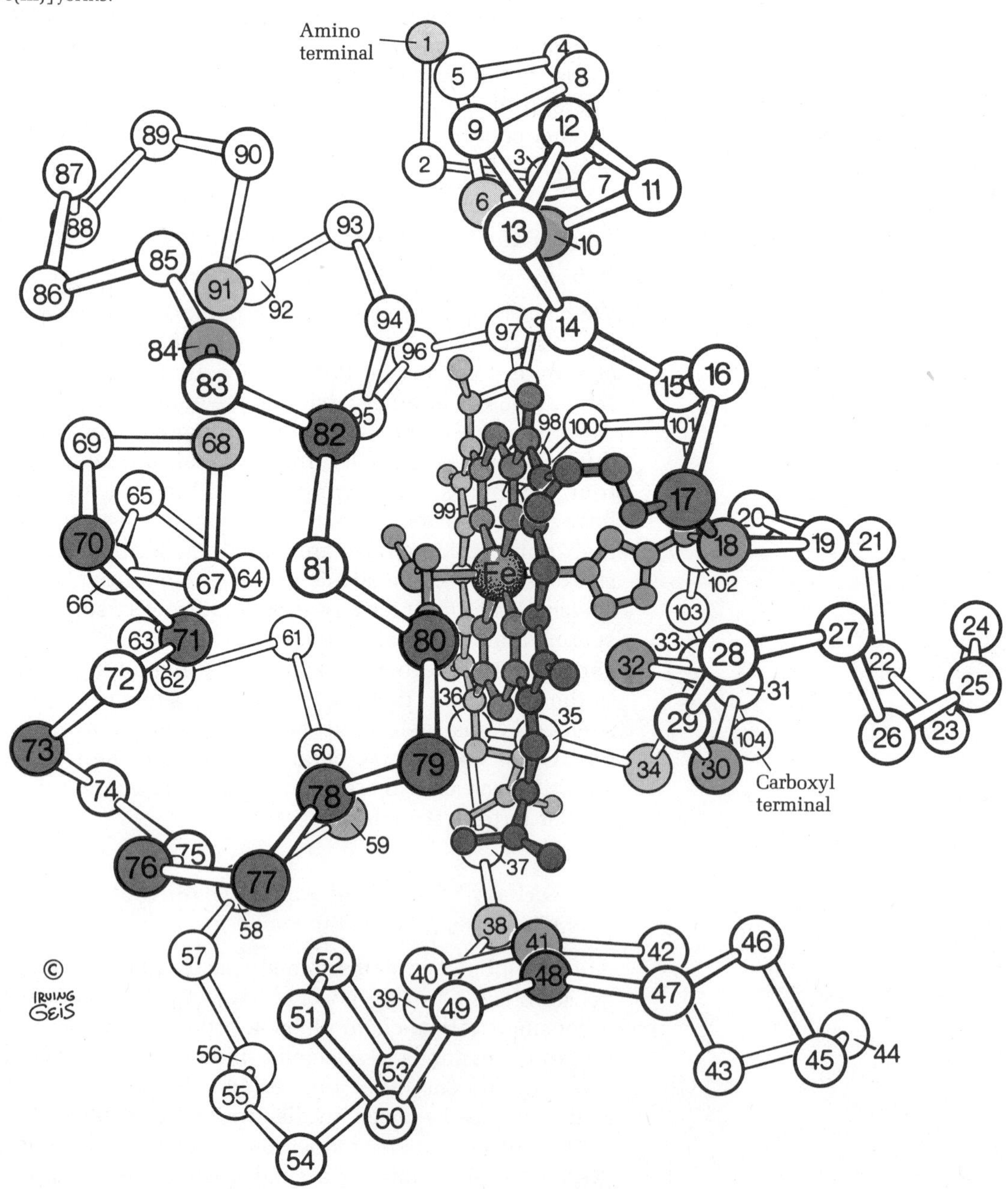

Figure 8-5
Backbone of the cytochrome c *molecule. The heme group, covalently bound in a crevice in the molecule, is shown in gray. The invariant amino acid residues are in color. In cytochrome* c *the fifth and sixth bonds of iron are occupied by the R groups of residues 18 and 80. In its normal electron-transferring function the iron of cytochrome* c *oscillates between ferrous [Fe(II)] and ferric [Fe(III)] forms.*

Table 8-2 Approximate Amounts of α Helix and β Structure in Some Single-Chain Proteins†

Protein	Total residues	Residues, %	
		α helix	β structure
Myoglobin	153	78	0
Cytochrome *c*	104	39	0
Lysozyme	129	40	12
Ribonuclease	124	26	35
Chymotrypsin	247	14	45
Carboxypeptidase	307	38	17

† Portions of the polypeptide chains that are not accounted for by α-helical or β regions consist of bends, reverse turns, and irregularly coiled or extended stretches. Segments of α helix and β structure sometimes deviate slightly from their normal dimensions and geometry. Data from C. R. Cantor and P. R. Schimmel, *Biophysical Chemistry*, pt. I, *The Conformation of Biological Macromolecules*, p. 100, Freeman, San Francisco, 1980.

most 80 percent of the myoglobin chain is α-helical, only about 40 percent of cytochrome *c* is in α-helical segments. The rest of the cytochrome *c* chain contains various bends, turns, and irregularly coiled and extended segments. Thus, although cytochrome *c* and myoglobin are both small heme proteins, they differ markedly in secondary and tertiary structure, and in amino acid sequence, in keeping with their very different biological functions.

The tertiary structure of two other small globular proteins may be compared. Lysozyme is an enzyme in egg white and human tears that catalyzes the hydrolytic cleavage of complex polysaccharides in the protective cell walls of some families of bacteria. Lysozyme is so named because it can lyse, or dissolve, bacterial cell walls and thus serve as a bactericidal agent. Like myoglobin and cytochrome *c*, lysozyme has a compactly folded conformation and has most of its hydrophobic R groups inside the globular structure, shielded from water, and its hydrophilic R groups outside, facing the aqueous medium. About 40 percent of its 129 amino acid residues are in α-helical segments (Table 8-2), which line a long crevice in the side of the molecule. This crevice is the catalytic site of the enzyme. As we shall see (Chapter 9), each enzyme has a catalytic site which binds the substrate, the molecule it acts upon. The bacterial polysaccharide on which lysozyme acts fits into this crevice during the catalytic action of the enzyme. Figure 8-6 shows a space-filling model of lysozyme, which illustrates the compactness of the structure of globular proteins.

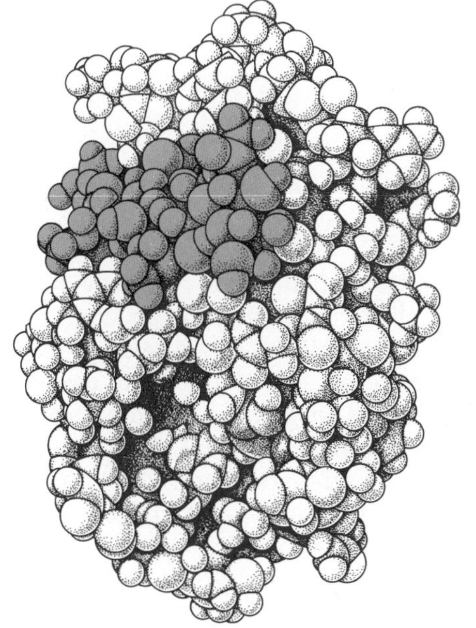

Figure 8-6
Space-filling model of the lysozyme molecule with its tightly bound polysaccharide substrate (color). Note that the lysozyme molecule is very compact, with little free space inside. Another view of lysozyme is shown in Chapter 9 on page 228.

Ribonuclease, another small globular protein, is an enzyme secreted by the pancreas into the small intestine, where it catalyzes the hydrolysis of certain bonds in the ribonucleic acids present in ingested food. Its tertiary structure, determined by

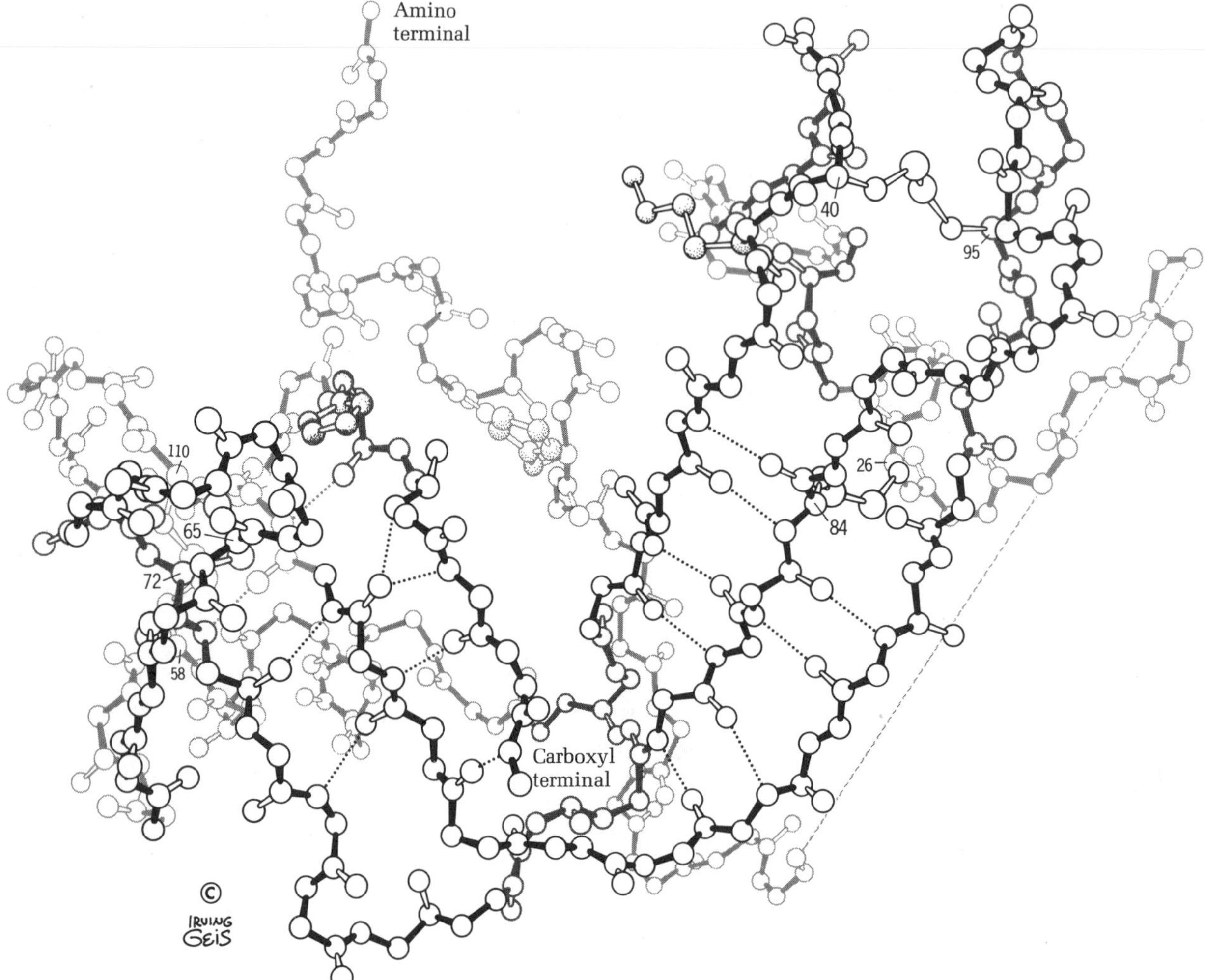

Figure 8-7
The conformation of the ribonuclease molecule as deduced from x-ray analysis. The dotted lines represent hydrogen bonds between loops of the polypeptide chain that are arranged in a β-pleated sheet. The crevice in the upper center of the molecule is the binding site for the substrate. Figure 8-8 shows the position of the disulfide intrachain cross-linkages.

x-ray analysis (Figure 8-7), shows that although very little of its polypeptide chain is in α-helical conformation, it contains many segments in the β conformation. Ribonuclease also differs from myoglobin, cytochrome *c*, and lysozyme in having four cystine residues that form covalent disulfide cross-linkages between loops of the polypeptide chain. These serve to hold the tertiary structure firmly in place (Figure 8-7). Many proteins, particularly those which function outside of cells, contain such intrachain disulfide cross-links.

Clearly, these four small, single-chain globular proteins differ from each other substantially (Table 8-2). They have widely varying amounts of α helix and β conformation and diverse folding patterns. They also have different amino acid sequences and very different biological functions. From x-ray and sequence data on many kinds of globular proteins, it is now well established that each type has a distinctive three-dimensional conformation, specialized for its particular biological function.

Amino Acid Sequence Determines Tertiary Structure

We have seen from consideration of the α helix and the β conformation that the residue-by-residue secondary structure of the polypeptide chain is determined by its amino acid sequence. Only when sets of neighboring amino acid residues along the chain have the appropriate sequences of R groups will the α helix or β conformation form spontaneously and remain stable. The tertiary structure of globular proteins is also determined by their amino acid sequence. Whereas secondary structure of polypeptide chains is determined by *short-range* sequences of R groups, tertiary structure is conferred by *longer-range* aspects of amino acid sequence. The formation of bends in the polypeptide chain and the direction and angle of these bends are determined by the precise location of specific bend-producing amino acid residues, such as proline, threonine, and serine. Moreover, loops of the highly folded polypeptide chain are held in their characteristic tertiary positions by different kinds of interactions between R groups of adjacent loops, as we shall see.

Many of the invariant amino acid residues of homologous proteins, those which are always present at specific positions, regardless of species, appear to occur at critical points along the polypeptide chain. Some occur at or near bends in the chain; others, such as cystine residues, at cross-linking points between loops in the tertiary structure. Still others occur in such specific locations as the catalytic sites of enzymes or sites in the molecule where a prosthetic group is bound, e.g., the heme group of cytochrome *c*.

But the most important proof that the tertiary structure of a globular protein is determined by its amino acid sequence comes from experiments showing that denaturation of some proteins is reversible. Heat or extremes of pH cause most globular proteins to unfold and lose their biological activity without breaking covalent bonds in the polypeptide backbone. For many years the denaturation of proteins was thought to be irreversible; e.g., egg white protein coagulated by cooking does not spontaneously go back into solution again on cooling. However, it has been found that some globular proteins denatured by heat or extremes of pH will in fact regain their native structure and their biological activity if they are slowly cooled or slowly returned to their normal pH, a process called *renaturation*.

A classical case of renaturation is given by ribonuclease, a single-chain protein with four intrachain disulfide bonds. Crystalline ribonuclease can be denatured by exposure to a concentrated urea solution in the presence of a reducing agent, which cleaves the disulfide bonds of the four cystine residues to yield eight cysteine residues, thus freeing the entire polypeptide from its folded conformation. Under these conditions the enzyme loses its catalytic activity and undergoes complete unfolding into a randomly looped form (Figure 8-8). If we now

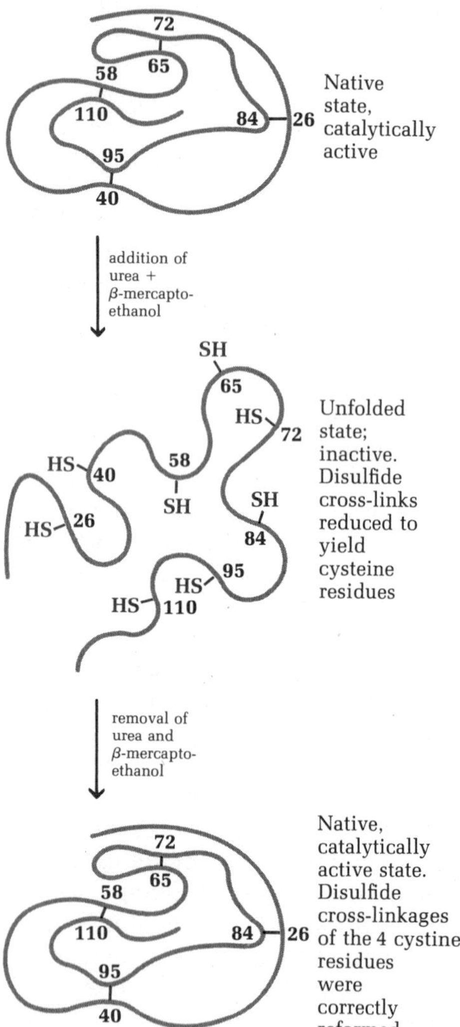

Figure 8-8
Renaturation of unfolded, denatured ribonuclease, with reestablishment of correct disulfide cross-links. The urea was added to dissociate the hydrogen bonds of ribonuclease, and the mercaptoethanol (CH_3CH_2SH) *to reduce and thus cleave the disulfide bonds of the four cystine residues to yield eight cysteine residues.*

place the solution of denatured ribonuclease into a dialysis bag (page 127) and suspend the bag in water, the small-molecular-weight solutes (urea and the reducing agent) will diffuse out of the denatured ribonuclease solution. As these agents are gradually removed, the randomly coiled, denatured ribonuclease slowly and spontaneously refolds itself into its correct three-dimensional tertiary structure, with full restoration of its catalytic activity (Figure 8-8). This experiment proves that the information required for the polypeptide chain of ribonuclease to refold itself properly is inherent in the primary structure of its polypeptide chain, i.e., its amino acid sequence.

The accuracy of the refolding of ribonuclease is further verified by another part of this experiment. The eight cysteine residues of the completely reduced and unfolded ribonuclease were reoxidized by atmospheric oxygen to form four intrachain cystine residues, in exactly the same positions in the molecule as in the original native ribonuclease. This finding is remarkable. Although the eight cysteine residues could theoretically have recombined at random to form cystine residues in 105 different ways, only *one* specific set of disulfide cross-linkages, that present in native ribonuclease, was formed during renaturation (Figure 8-8). Thus the polypeptide chain of denatured ribonuclease will refold very precisely to yield its unique, biologically active conformation rather than some "wrong" conformation. This classical experiment, carried out by Christian Anfinsen in the 1950s, proved that the amino acid sequence of the polypeptide chain of proteins contains all the information required to fold the chain into its native three-dimensional structure.

Four Different Forces Stabilize the Tertiary Structure of Globular Proteins

We have seen how secondary structure of polypeptides is formed and maintained. *Intrachain* hydrogen bonds hold the α helix together, and *interchain* hydrogen bonds hold the β conformation together in pleated sheets (page 155). Given the proper amino acid sequences, the α helix and the β conformation are the most stable structures these chains can assume; i.e., they are the structures having the least free energy. What forces stabilize the tertiary structure of globular proteins? Four kinds of interactions cooperate to hold the looped polypeptide chains of globular proteins in position (Figure 8-9) under biological conditions of temperature, pH, and ionic concentration.

1. *Hydrogen bonding between R groups of residues in adjacent loops of the chain.* For example, the hydroxyl group of a serine residue in one segment of a polypeptide chain may form a hydrogen bond with a ring nitrogen atom of a histidine residue in an adjacent loop of the same chain.

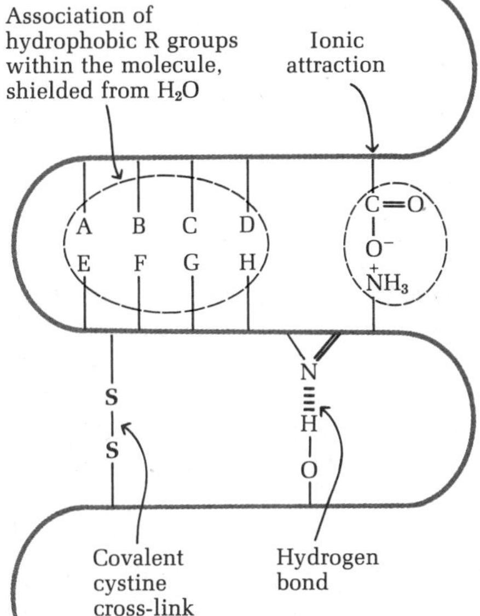

Figure 8-9
Factors maintaining the tertiary structure of globular proteins.

2. *Ionic attractions between oppositely charged R groups.* As an example, the negatively charged carboxylate group ($—COO^-$) of a glutamic acid residue may be attracted to the positively charged ε-amino group ($—\overset{+}{N}H_3$) of a lysine residue in an adjacent loop.

3. *Hydrophobic interactions.* The hydrophobic R groups of some amino acid residues (see Table 8-1) shun an aqueous environment and tend to associate together in the interior of the globular structure, shielded from water.

4. *Covalent cross-linkages.* Adjacent loops of the polypeptide chain of some proteins, such as ribonuclease, contain intrachain cystine residues, which form covalent cross-links between adjacent loops. Such covalent cross-links are of course much stronger than the noncovalent cross-linking forces listed above, but not all proteins contain them. Proteins lacking disulfide cross-linkages therefore maintain their characteristic tertiary structures through many noncovalent interactions or contacts, which are individually weak but collectively strong.

Although the native tertiary conformations of globular proteins are the stablest forms their polypeptide chains can assume, i.e., their minimum-energy forms, the tertiary conformation of globular proteins must not be regarded as absolutely rigid and fixed. Many globular proteins normally undergo changes in their conformation in the course of their biological function. For example, the hemoglobin molecule, which we shall examine later, changes its conformation when it binds oxygen but reverts to its original conformation when the oxygen is released. Moreover, many enzyme molecules also change their conformation when they bind their substrates, a process that is part of their catalytic action. Globular proteins also have a certain amount of flexibility in their backbones and undergo short-range internal fluctuations; i.e., they "breathe."

The Rate of Folding of Polypeptide Chains Is Critical

In living cells proteins are made from amino acids at a very high rate. For example, *E. coli* cells can make a complete, biologically active protein molecule containing 100 amino acid residues in about 5 s at 37°C. Yet calculations show that at least 10^{50} years would be required for a polypeptide chain having 100 amino acid residues to fold itself spontaneously by a random process in which the polypeptide tries out all possible conformations around every single bond in its backbone until it finds its native, biologically active conformation!

Thus proteins cannot fold into their correct conformations by a completely random trial-and-error process. There simply must be shortcuts. We do not know exactly how or by what pathway any protein undergoes spontaneous folding, whether it starts at one end of the chain, in the middle, or at several points simultaneously. But it appears that the spontaneous folding of polypeptide chains into their correct tertiary structures is highly *cooperative*. That is, once a certain minimum length of the chain has folded properly, the probability that the rest of the chain will fold correctly is greatly increased.

Oligomeric Proteins Have Both Tertiary and Quaternary Structure

Oligomeric proteins are those with two or more separate polypeptide chains. Some have only two, some have several, and some have dozens of chains (Table 8-3). The polypeptides in oligomeric proteins may be identical or different. The number of polypeptide chains in an oligomeric protein can be found by determining the number of amino-terminal residues per molecule of protein, using an appropriate labeling reagent, such as 2,4-dinitrofluorobenzene (page 130). An oligomeric protein with four separate polypeptide chains, such as hemoglobin, will have four amino-terminal residues, one for each chain. Although insulin has two chains, they are covalently cross-linked.

Oligomeric proteins have higher molecular weights and more complex functions than single-chain proteins. The best-known example of an oligomeric protein is hemoglobin. Its function as an oxygen-carrying protein in red blood cells is regulated by blood pH and CO_2 concentration, as we shall see. Among the larger, more complex oligomeric proteins are the enzyme *RNA polymerase* of *E. coli* (five subunit chains),

Table 8-3 Some Oligomeric Proteins†
Most have an even number of chains

Protein	MW	Chains per molecule
Hemoglobin (mammals)	64,500	4
Adenylate kinase (rat liver)	18,000	3
Hexokinase (yeast)	102,000	2
Lactate dehydrogenase (beef heart)	140,000	4
Cytochrome oxidase	200,000	7
Glutamate dehydrogenase (beef liver)	320,000	6
F_1ATPase	380,000	9 or 10
RNA polymerase (*E. coli*)	400,000	5
Aspartate transcarbamoylase (*E. coli*)	310,000	12
Isocitrate dehydrogenase (beef heart)	1,000,000	10
Glutamine synthetase (*E. coli*)	600,000	12
Pyruvate dehydrogenase complex	4,600,000	72

† Data from D. W. Darnall and I. M. Klotz, "Subunit Constitution of Proteins: A Table," *Arch. Biochem. Biophys.*, **166**:651–682 (1975).

responsible for initiation and synthesis of RNA chains; the enzyme *aspartate transcarbamoylase* (twelve chains), important in the synthesis of nucleotides; and, as an extreme case, the enormous *pyruvate dehydrogenase complex* of mitochondria, which is a cluster of three enzymes containing 72 polypeptide chains altogether (Table 8-3).

In oligomeric proteins each of the subunit polypeptide chains has its own characteristic secondary and tertiary conformation in space. However, such proteins have another level of conformation, called *quaternary structure. This term denotes the arrangement of the subunit chains in relationship to each other, i.e., how they fit or pack together in the native conformation of oligomeric proteins.*

X-ray analysis is much more difficult for oligomeric than for single-chain proteins. Nevertheless, sufficient information is already available to yield some very important clues to the more complex biological activities of oligomeric proteins.

X-Ray Analysis Has Revealed the Complete Structure of Hemoglobin

The first oligomeric protein to be subjected to x-ray analysis was hemoglobin (MW 64,500), which contains four polypeptide chains and four heme prosthetic groups, in which the iron atoms are in the ferrous [Fe(II)] state. The protein portion, called *globin*, consists of two α chains (141 residues each) and two β chains (146 residues each). Since hemoglobin is 4 times as large as myoglobin, much more time and effort was required to solve its three-dimensional structure, finally achieved by Max Perutz and his colleagues at Cambridge. X-ray analysis has revealed that the hemoglobin molecule is roughly spherical, with a diameter of about 5.5 nm. Each of the four chains has a characteristic *tertiary structure*, in which the chain is folded. As in myoglobin, the α and β chains of hemoglobin contain several segments of α helix separated by bends. The four polypeptide chains fit together in an approximately tetrahedral arrangement, to constitute the characteristic *quaternary structure* of hemoglobin (Figure 8-10). There is one heme group bound to each chain. The hemes are rather far apart, about 2.5 nm from each other, and tilted at different angles. Each heme is partially buried in a pocket lined with hydrophobic R groups. It is bound to its polypeptide chain through a coordination bond of the iron atom to the R group of a histidine residue, as in Figure 8-2. The sixth coordination bond of the iron atom of each heme is available to bind a molecule of O_2.

Closer examination of the quaternary structure of hemoglobin, with the help of models, shows that although there is little direct contact between the two α chains or between the two β chains, there are many contact points between the α and β chains of the dissimilar chain pairs $\alpha_1\beta_1$ and $\alpha_2\beta_2$. These contact points consist largely of hydrophobic R groups of amino

Figure 8-10
The three-dimensional structure of oxyhemoglobin and deoxyhemoglobin revealed by x-ray diffraction analysis. The quaternary structure, i.e., how the four subunits are packed together, is shown. The subunits occur in pairs, $\alpha_1\beta_1$ and $\alpha_2\beta_2$. There are few contact points between α_1 and α_2 or between β_1 and β_2, but there are many holding the $\alpha_1\beta_1$ and $\alpha_2\beta_2$ pairs together. Although irregular in shape, the molecule has twofold symmetry, since 180° rotation of the molecule around its central axis, which is perpendicular to the page, would superimpose α_1 on α_2 and β_2 on β_1. The residue numbers are indicated in each chain. The central cavity plays an important role, discussed in Box 8-1, on page 193. Note that the heme groups are relatively far apart. The differences between oxyhemoglobin and deoxyhemoglobin are not large, but they are very important in the function of hemoglobin. They are discussed later in this chapter.

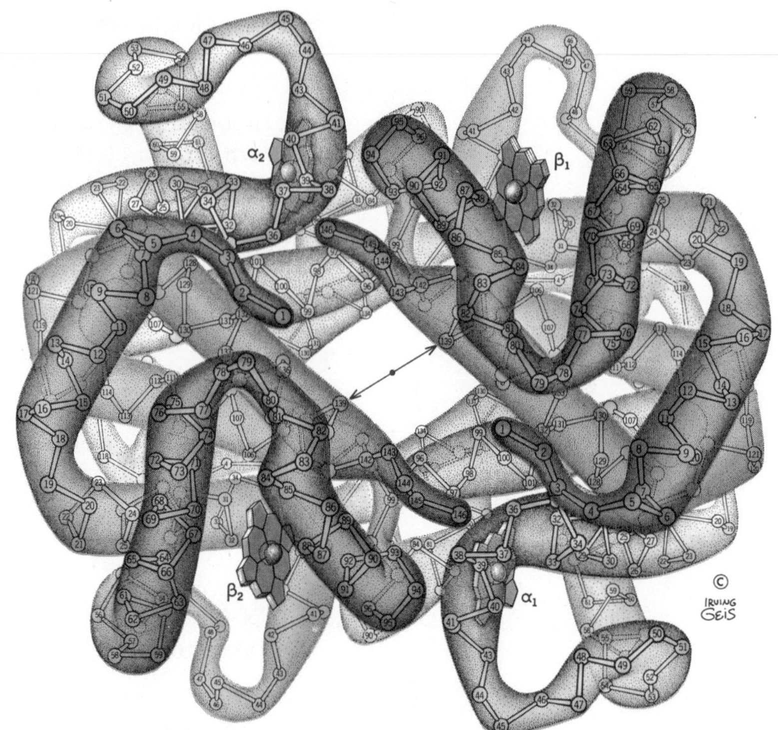

Oxyhemoglobin

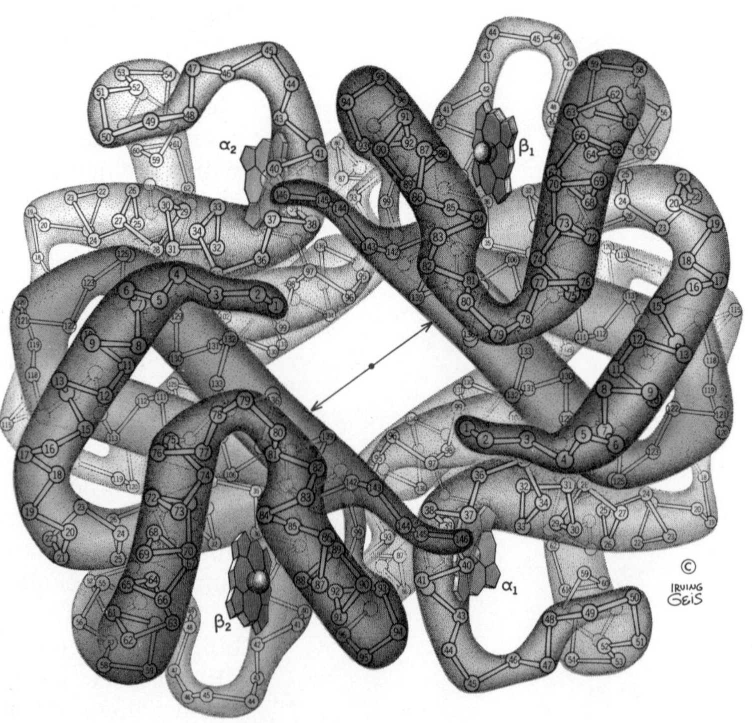

Deoxyhemoglobin

acid residues. The $\alpha_1\beta_1$ and $\alpha_2\beta_2$ pairs, being made up of irregularly shaped polypeptide chains, do not fit each other precisely. In fact, there is a central open channel, or cavity, running right through the hemoglobin molecule, which can be seen in the top view of the molecule (Figure 8-10). We shall refer to this channel again later.

Myoglobin and the α and β Chains of Hemoglobin Have Nearly the Same Tertiary Structure

X-ray diffraction and chemical analysis of hemoglobin structure have revealed an important set of relationships. First, it has been found that the α and β chains of hemoglobin have nearly identical tertiary structures. Both have well over 70 percent α-helical character, both have similar lengths of α-helical segments, and the bends or turns have about the same angles.

Second, the hemoglobins of many different vertebrate species have approximately the same tertiary structure of their polypeptide chains. Moreover, the quaternary structures of different hemoglobins closely resemble each other.

The third important point is that the tertiary structures of the α and β chains of hemoglobin are very similar to the tertiary structure of myoglobins. Thus the similar tertiary structures of myoglobin and the α and β chains of hemoglobin can be related to the capacity of both proteins to bind oxygen in their biological functions.

The family relationship between myoglobin and the hemoglobin chains is further shown by a comparison of the amino acid sequences of sperm whale myoglobin and the α and β chains of horse hemoglobin. Figure 8-11 shows that they share 27 identical residues in comparable positions and have very closely related residues in 40 other positions, e.g., the similar residues aspartic and glutamic acid or valine and isoleucine. Here again we see that amino acid sequences of homologous proteins share a number of corresponding invariant residues and that homologous proteins tend to have similar three-dimensional structures.

Figure 8-11

(a) Positions (color) of the invariant amino acid residues shared by the α and β chains of horse hemoglobin and sperm whale myoglobin. The black lines show the positions that have identical amino acids in the α and β chains.

(b) The similar tertiary structure of the β chain of horse hemoglobin and sperm whale myoglobin.

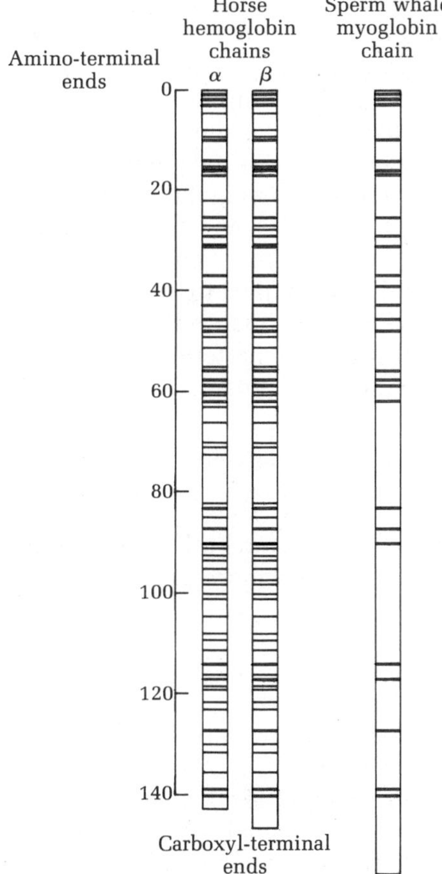

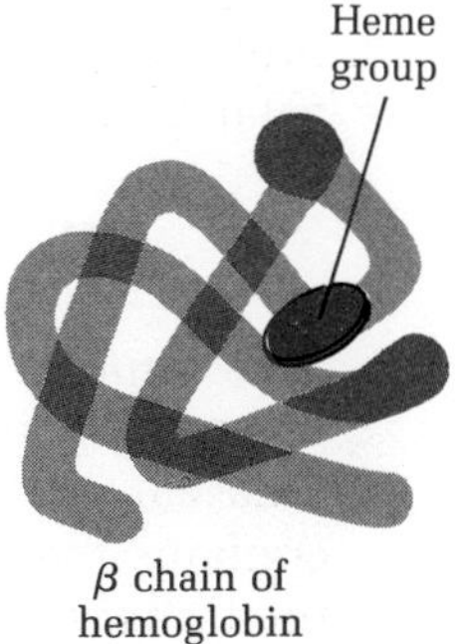

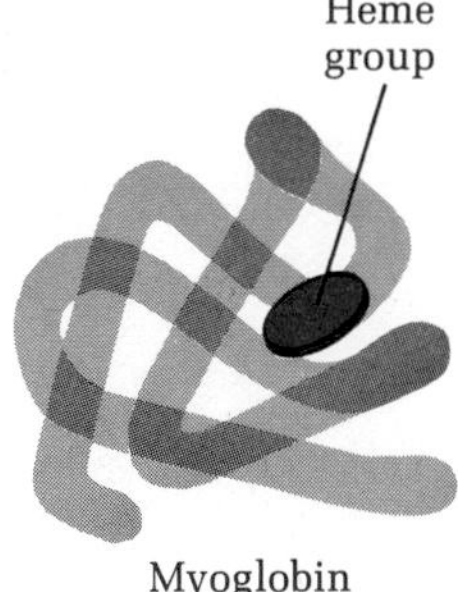

(b)

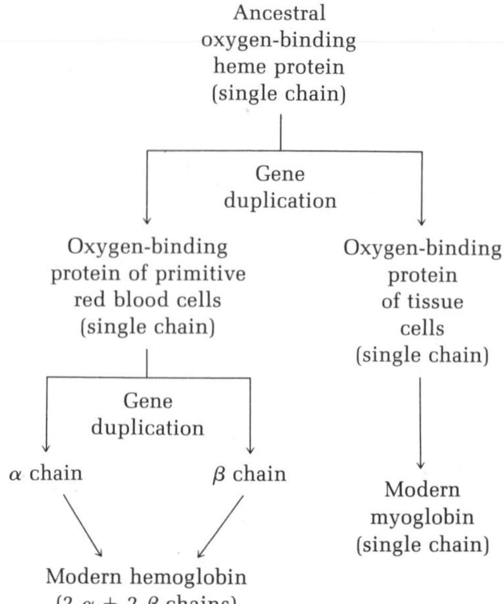

Figure 8-12
Evolution of myoglobin and hemoglobin from an ancestral oxygen-binding heme protein. All myoglobins and the α and β chains of all modern hemoglobins (altogether 145 sequences have been examined) share six invariant residues and have many other positions in which closely related amino acids are found. Presumably the single gene coding for the ancestral single-chain heme protein became duplicated, one becoming the myoglobin gene and the other the original hemoglobin gene. The two genes subsequently underwent independent mutation. The original hemoglobin gene may have undergone another duplication to yield the modern α and β genes.

In addition to the α and β chains of normal adult hemoglobin, human beings also possess a gene for coding a γ chain present in fetal hemoglobin, which has the composition $\alpha_2\gamma_2$. Fetal hemoglobin has a higher affinity for oxygen than adult hemoglobin. Adult red blood cells also contain a minor hemoglobin containing δ chains; its composition is $\alpha_2\delta_2$.

Yet another principle emerges from the structural data on myoglobin and hemoglobin. It appears very probable that myoglobin and hemoglobin are descended from a common ancestral oxygen-binding heme protein (Figure 8-12), which may have had but a single polypeptide chain. At some point in the further evolution of species the gene coding for the ancestral oxygen-binding protein may have become duplicated. These two gene copies then underwent mutations independently, so that one of them gradually coded for the myoglobin type of protein, adapted to storage of oxygen in cells, and the other gene underwent a different pathway of mutation to code ultimately for the α and β chains of hemoglobin, adapted to transportation of oxygen in red blood cells. We shall see many other instances of functionally and structurally related proteins which may have originated from common ancestral proteins.

Quaternary Structures of Other Oligomeric Proteins Have Been Determined

X-ray analysis has been carried out on several other oligomeric proteins. An example is the enzyme *hexokinase* of yeast, which catalyzes the reaction

$$\text{ATP} + \text{D-glucose} \longrightarrow \text{ADP} + \text{D-glucose 6-phosphate}$$

This important reaction takes place in virtually all organisms since it is a necessary step in the metabolism of glucose. Hexokinase from yeast has a molecular weight of about 108,000 and contains two polypeptide chains. X-ray analysis has revealed the tertiary structure of the two polypeptide chains, as well as the quaternary structure of hexokinase, in which the two polypeptides fit together closely (Figure 8-13). Of special inter-

Figure 8-13
The structure of the oligomeric protein yeast hexokinase. Its two subunits are related to each other by a screw axis. When the molecule is rotated 180° about the vertical axis shown and is simultaneously moved upward, the lower subunit will be superimposed on the upper one. Hexokinase is a regulatory enzyme that controls the rate of entry of glucose into cell metabolism. The subunit on the right contains a bound glucose molecule at its catalytic site. Another view of hexokinase is shown in Chapter 9 on page 232.

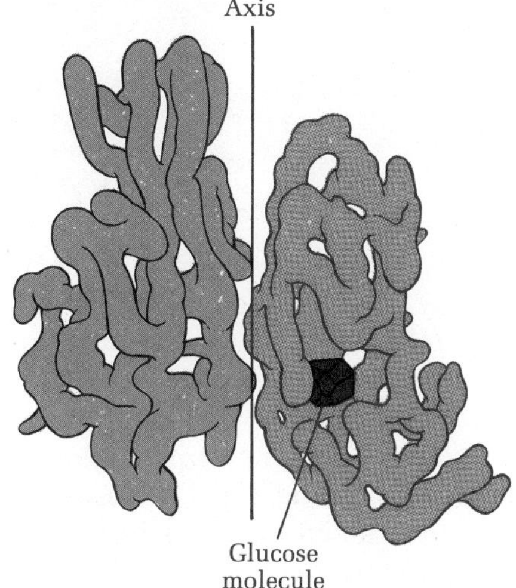

Figure 8-14
The subunit structure of glutamine synthetase of E. coli. This regulatory enzyme has 12 subunits arranged as shown.

Side view

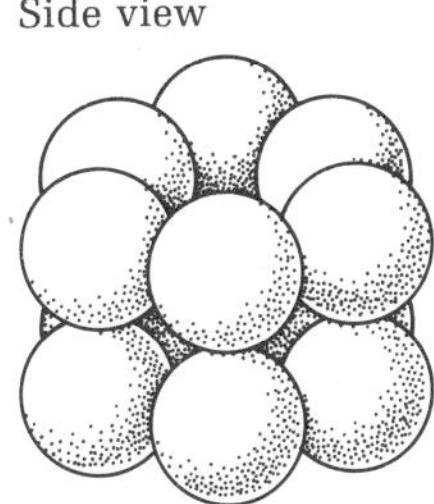

Top view

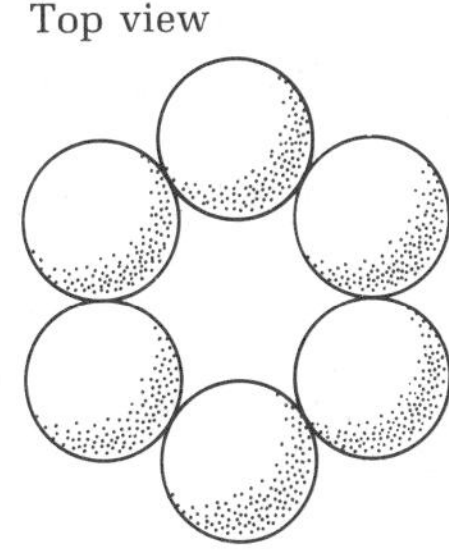

est is the structure of the catalytic site on the enzyme molecule, the site to which ATP and glucose must bind in order to undergo the catalytic process. As we shall see in Chapter 9, the conformation of the hexokinase molecule changes its geometry during the catalytic cycle.

Another oligomeric protein whose structure has been determined is the enzyme *lactate dehydrogenase* of skeletal muscle, which catalyzes the last step in the metabolic conversion of glucose into lactate. Lactate dehydrogenase has a molecular weight of 140,000 and contains four polypeptide chains. The tertiary structures of these chains are very different from those of hemoglobin.

Yet another protein whose quaternary structure has been examined is the enzyme *glutamine synthetase* of *E. coli,* which catalyzes the formation of glutamine from glutamate and ammonia at the expense of energy provided by ATP (Chapter 19). It is a much more complex oligomeric protein than hemoglobin or hexokinase. Figure 8-14 shows the arrangement of its 12 subunits.

These three oligomeric proteins—all enzymes—share a common feature with hemoglobin. They all participate in some aspect of biological regulation as an intrinsic part of their function. Hexokinase, lactate dehydrogenase, and *E. coli* glutamine synthetase are members of a special class of enzymes called *regulatory enzymes,* which, as we shall see in the next chapter, not only catalyze specific reactions but also help regulate the rate of the metabolic pathway in which they participate as catalysts. Hemoglobin also plays a regulatory role. It not only carries oxygen from the lungs to the peripheral tissues but also *regulates* the binding of oxygen in the lungs and the release of oxygen in the tissues in response to certain signals, particularly the pH and the CO_2 concentration. Many oligomeric proteins appear to function in such a regulatory capacity.

We shall now examine hemoglobin more closely, to see how it is adapted to carrying oxygen from the lungs to the tissues and H^+ and CO_2 from the tissues to the lungs and how its quaternary structure enables it to help regulate these important transport activities. Hemoglobin is an instructive prototype or model for the function of many regulatory oligomeric proteins.

Red Blood Cells Are Specialized to Carry Oxygen

There are between 5 and 6 L of blood in the adult human body. Between one-third and one-half of the blood volume is contributed by the erythrocytes, which are suspended in the protein-rich *blood plasma.* The blood must carry about 600 L of oxygen from the lungs to the tissues every day, but very little of this is carried by the blood plasma, since oxygen is only sparingly soluble in aqueous solutions. Nearly all the oxygen carried by whole blood is bound and transported by the hemoglobin of the erythrocytes. The hemoglobin in 100 mL of whole blood binds about 20 mL of gaseous oxygen.

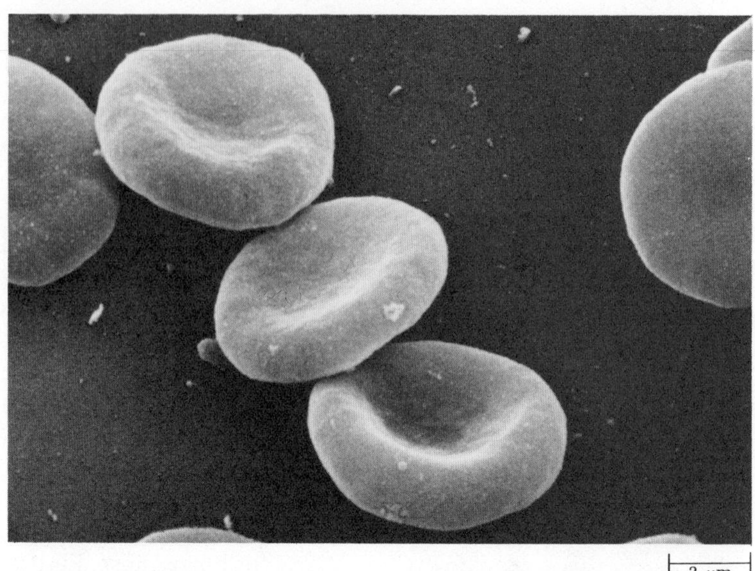

Figure 8-15
Scanning electron micrograph of normal human red blood cells.

Normal human erythrocytes are rather small (6 to 9 μm) biconcave disks (Figure 8-15). They have no nucleus, mitochondria, endoplasmic reticulum, or other organelles. Erythrocytes are formed from precursor cells, called *reticulocytes*. In this maturation process the reticulocytes lose their normal intracellular organelles and form large amounts of hemoglobin. Red blood cells are thus incomplete, vestigial cells, unable to reproduce and destined to last only about 120 days in human beings. Their main function is to carry hemoglobin, which is dissolved in the aqueous cytosol in a very concentrated solution of about 34 percent hemoglobin.

The hemoglobin of the red blood cells in arterial blood passing from the lungs to the peripheral tissues is about 96 percent saturated with oxygen. In the venous blood returning to the heart the hemoglobin is only about 64 percent saturated. Thus each 100 mL of blood passing through a tissue releases about one-third of the oxygen it carries, equal to about 6.5 mL of oxygen gas at atmospheric pressure and body temperature.

Myoglobin and Hemoglobin Differ in Their Oxygen-Binding Curves

The special properties of the hemoglobin molecule that make it such an effective oxygen carrier in the blood are best understood by comparing the oxygen-binding affinity of myoglobin and hemoglobin. Figure 8-16 shows their *oxygen saturation curves*, plots of the percentage of the total oxygen-binding sites of hemoglobin or myoglobin that are occupied by oxygen molecules when solutions of these proteins are in equilibrium with different partial pressures of oxygen in the gas phase.

We note first of all that myoglobin has a very high affinity for oxygen; it is already 50 percent saturated with oxygen at only 1

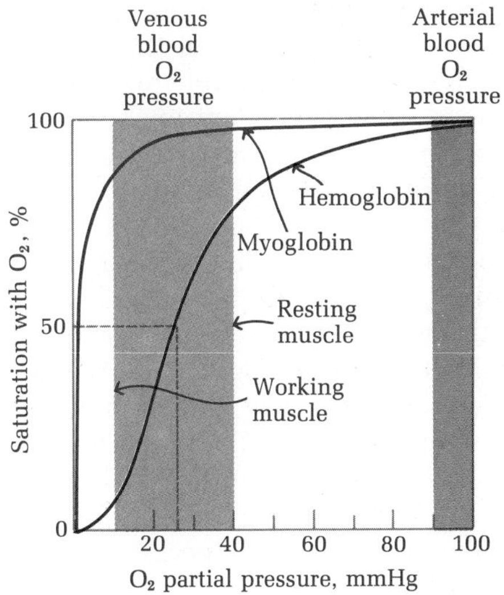

Figure 8-16
The oxygen saturation curves of myoglobin and hemoglobin. Myoglobin has a much higher affinity for oxygen than hemoglobin. It is 50 percent saturated at only 1 to 2 mmHg O_2 pressure, whereas hemoglobin requires about 26 mmHg O_2 pressure for 50 percent saturation. Note that although both hemoglobin and myoglobin are over 95 percent saturated at the partial pressure of oxygen in arterial blood leaving the lungs (~100 mmHg), hemoglobin is only about 75 percent saturated in resting muscle, where the O_2 pressure is 40 mmHg, and only 10 percent saturated in working muscle, where the O_2 pressure is only 10 mmHg. Thus hemoglobin can release its oxygen very effectively in muscle and other peripheral tissues. Myoglobin, on the other hand, is still about 90 percent saturated at 10 mmHg O_2 pressure and therefore unloads very little oxygen even at very low partial pressures of oxygen. Thus the sigmoid saturation curve of hemoglobin is a molecular adaptation for its transport function in red blood cells.

to 2 mmHg pressure. Second, we see that the oxygen saturation curve of myoglobin is a simple *hyperbolic curve,* as might be expected from the mass action of oxygen on the equilibrium

$$\text{Myoglobin} + O_2 \rightleftharpoons \text{oxymyoglobin}$$

Myoglobin is already over 95 percent saturated at only about 20 mmHg. In contrast, the affinity of hemoglobin for oxygen is much lower; moreover, the oxygen saturation curve of hemoglobin is *sigmoid* (S-shaped) (Figure 8-16). This shape indicates that the affinity of hemoglobin for binding the first molecule of oxygen is relatively very low (shown by the bottom part of the S at less than 10 mmHg pressure), but subsequent oxygen molecules are bound with a very much higher affinity, accounting for the steeply rising portion of the S-shaped curve. In fact, the increase in the affinity of hemoglobin for oxygen, after the first oxygen is bound, is almost 500-fold. Thus the four heme-polypeptide subunits of hemoglobin are not identical and independent in their oxygen affinity.

We see that once the first heme-polypeptide subunit of a hemoglobin molecule binds an oxygen molecule, it communicates this information to the remaining subunits, which respond by greatly increasing their oxygen affinity. Such communication between the four heme-polypeptide subunits of hemoglobin is the result of *cooperative interactions* between the subunits. Because binding of one molecule of oxygen increases the probability that further molecules of oxygen will be bound by the remaining subunits, we speak of hemoglobin's having *positive cooperativity.* Sigmoid binding curves, like that of hemoglobin for oxygen; are characteristic of positive cooperative binding. Cooperative oxygen binding does not occur with myoglobin, which has only one heme group and thus can bind only one oxygen atom; its saturation curve is therefore hyperbolic. We can now see why there is a fundamental difference between the oxygen-binding actions of myoglobin and hemoglobin.

We use the term *ligand* to refer to a specific molecule, such as oxygen in the case of hemoglobin, that is bound by a protein (the word ligand comes from Latin, to tie or bind, and means "bound entity"). Many other oligomeric proteins have multiple binding sites for their ligands and, like hemoglobin, show positive cooperativity. However, some oligomeric proteins show *negative cooperativity:* binding of one ligand molecule *decreases* the probability that further ligand molecules will be bound.

The Cooperative Binding of Oxygen Enhances the Efficiency of Hemoglobin as an Oxygen Carrier

In the lungs the partial pressure of oxygen in the air spaces is about 100 mmHg; at this pressure hemoglobin is about 96 percent saturated with oxygen. However, in the cells of a working muscle the partial pressure of oxygen is only about 26 mmHg, because muscle cells use oxygen at a high rate and thus lower its local concentration. As the blood passes through the muscle capillaries, oxygen will be released from the nearly saturated hemoglobin in the red blood cells into the blood plasma and thence into the muscle cells. We see from the oxygen saturation curve of hemoglobin in Figure 8-16 that it will release about a third of its bound oxygen as it passes through the muscle capillaries, so that when it leaves the muscle, it will be only about 64 percent saturated. When the blood returns to the lungs, where the partial pressure of oxygen is much higher at 100 mmHg, the hemoglobin quickly binds more oxygen until it is 96 percent saturated again.

Now suppose that the hemoglobin in the red blood cell is replaced by myoglobin. We see from the hyperbolic oxygen saturation curve of myoglobin (Figure 8-16) that only 1 or 2 percent of the bound oxygen can be released from myoglobin as the oxygen partial pressure is lowered from 100 mmHg, the oxygen pressure in the lungs, to 26 mmHg, its pressure in the muscle. Myoglobin therefore is not very well adapted for carrying oxygen from the lungs to the tissues, since it has a much higher affinity for oxygen and releases very little of it at the oxygen pressures that exist in muscles and other peripheral tissues. Hemoglobin, on the other hand, is very efficient in this function, because its sigmoid oxygen saturation curve allows it to release a large part of its oxygen at the oxygen partial pressure existing in the tissues.

We must not conclude from this comparison that myoglobin is an inefficient or badly designed protein. In its true biological function *within* muscle cells, which is to store oxygen and make it available to the mitochondria, myoglobin is in fact much better suited than hemoglobin because its very high affinity for oxygen at low oxygen pressures enables it to bind and store oxygen effectively. Thus hemoglobin and myoglobin are specialized and adapted for different kinds of oxygen-binding functions. We shall now see that hemoglobin is specialized still further.

Hemoglobin Also Transports H^+ and CO_2

In addition to carrying oxygen from the lungs to the tissues, hemoglobin also carries two end products of tissue respiration, namely, H^+ and CO_2, from the tissues to the lungs and the kidneys, the two organs concerned in the excretion of these products. In the cells of the peripheral tissues organic fuels are oxidized by the mitochondria, using oxygen brought from the lungs by hemoglobin, with the formation of carbon dioxide, water, and other products. The formation of CO_2 causes an increase in H^+ concentration (i.e., a *decrease* in pH) in the tissues, since the hydration of CO_2 yields H_2CO_3, a weak acid, which dissociates to form H^+ and bicarbonate

$$H_2CO_3 \rightleftharpoons H^+ + HCO_3^-$$

Besides carrying nearly all the required oxygen from the lungs to the tissues, hemoglobin transports a significant portion, about 20 percent of the total CO_2 and H^+ formed in the tissues, to the lungs and the kidneys.

Long ago it was found that the binding of oxygen by hemoglobin is profoundly influenced by pH and CO_2 concentration and, further, that the binding of H^+ and CO_2 is inversely related to the binding of O_2. At the relatively low pH and high CO_2 concentration in the peripheral tissues, the affinity of hemoglobin for oxygen is decreased as H^+ and CO_2 are bound. Conversely, in the lung capillaries, as CO_2 is excreted and the blood pH consequently rises, the affinity of hemoglobin for oxygen is increased. This effect of pH and CO_2 concentration on the binding and release of oxygen by hemoglobin is called the *Bohr effect,* after Christian Bohr, the Danish physiologist who discovered it.

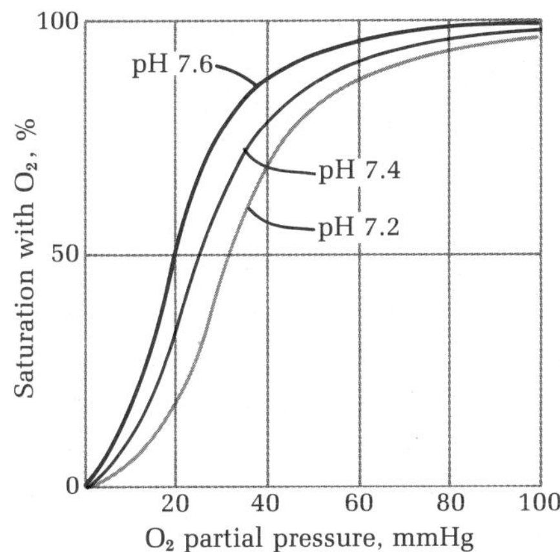

Figure 8-17
Effect of pH on the oxygen saturation curve of hemoglobin. At the low pH of the tissues (pH 7.2) oxygen is more readily released, whereas at the higher pH in the lungs (pH 7.6) oxygen is more readily taken up.

The Bohr effect is the result of an equilibrium involving not only the ligand oxygen but also two other ligands that can be bound by hemoglobin, H^+ and CO_2. The binding equilibrium for oxygen, which we have designated so far by the reaction

$$Hb + O_2 \rightleftharpoons HbO_2$$

is really an incomplete statement. To account for the effect of H^+ concentration on the binding of oxygen by hemoglobin we can rewrite this reaction as

$$HHb^+ + O_2 \rightleftharpoons HbO_2 + H^+$$

where HHb^+ denotes a protonated form of hemoglobin. This equation tells us that the oxygen saturation curve of hemoglobin is influenced by the H^+ concentration (Figure 8-17). Both oxygen and H^+ are bound by hemoglobin but in an inverse manner. When the oxygen concentration is high, as in the lungs, it will be bound by hemoglobin and H^+ ions will be released. When the oxygen concentration is low, as it is in the tissues, H^+ will be bound. However, the oxygen and the H^+ are *not* bound at the same sites in hemoglobin. Oxygen is bound to the iron atoms of the hemes, whereas the H^+ is bound to the R groups of histidine residues 146 in the β globin chains and to two other residues in the α chains. Thus we see that the four polypeptide chains of hemoglobin communicate with each other not only with regard to binding oxygen by their heme groups but also with regard to binding H^+ by specific amino acid residues.

But there is more to the story, since hemoglobin also binds carbon dioxide, again in a manner inversely related to the binding of oxygen. CO_2 is bound by the α-amino group at the amino-terminal end of each of the four polypeptide chains of hemoglobin to form *carbaminohemoglobin*

$$O{=}C{=}O + H_2N{-}\underset{R}{\overset{H}{C}}{-}\underset{}{C}({=}O){-} \xrightarrow{\;\;} {}^{-}O{-}C({=}O){-}\overset{H}{N}{-}\underset{R}{\overset{H}{C}}{-}C({=}O){-} + H^+$$

Amino-terminal residue

Carbamino-terminal residue

At a high concentration of CO_2, as occurs in tissues, some CO_2 will be bound to hemoglobin, and the affinity for O_2 will decrease, causing its release. Conversely, when O_2 is bound in the lungs, the affinity of hemoglobin for CO_2 decreases. Thus the oxygen saturation curve of hemoglobin is influenced by both the pH and the CO_2 concentration. This inverse relationship between the binding of O_2 on one hand and the binding of H^+ and CO_2 on the other is highly advantageous to the organism. In the tissues the low pH and high CO_2 concentration tend to cause release of O_2 from hemoglobin, and in the lungs the high concentration of O_2 tends to promote release of H^+ and CO_2. Thus its capacity to communicate ligand binding information from one polypeptide subunit to the others means that the hemoglobin molecule is beautifully adapted to integrating the transport of O_2, CO_2, and H^+ by the red blood cells.

But new questions arise. What is there about the structure of hemoglobin that allows it to show these mutually advantageous inverse changes in its affinity for O_2 vs. its affinities for H^+ and CO_2? How is binding information transmitted from one polypeptide subunit of hemoglobin to the others? Why does hemoglobin have these capabilities and myoglobin not?

Oxygenation of Hemoglobin Causes a Change in Its Three-Dimensional Conformation

The answer to these questions comes from evidence that the deoxyhemoglobin molecule undergoes a change in its conformation when it binds oxygen. Such a change first came to light when it was found that crystals of deoxyhemoglobin prepared in the absence of oxygen shattered when they were exposed to an atmosphere of oxygen. This observation suggested that binding of oxygen caused the dimensions of the hemoglobin molecules to change, thus becoming incompatible with the precise crystal lattice of deoxyhemoglobin. This evidence has been verified by comparative x-ray analysis, which showed that deoxyhemoglobin and oxyhemoglobin differ in their three-dimensional conformations (Figures 8-10 and 8-18). There is no change in the *tertiary structure* of the α and β chains when deoxyhemoglobin is oxygenated or in the way the α and β chains fit together to form the $\alpha_1\beta_1$ and $\alpha_2\beta_2$ pairs. But when oxygen is bound to the heme groups of deoxyhemoglobin, the $\alpha_1\beta_1$ and $\alpha_2\beta_2$ halves of the molecule, which remain rigid, change their positions slightly with respect to each other and

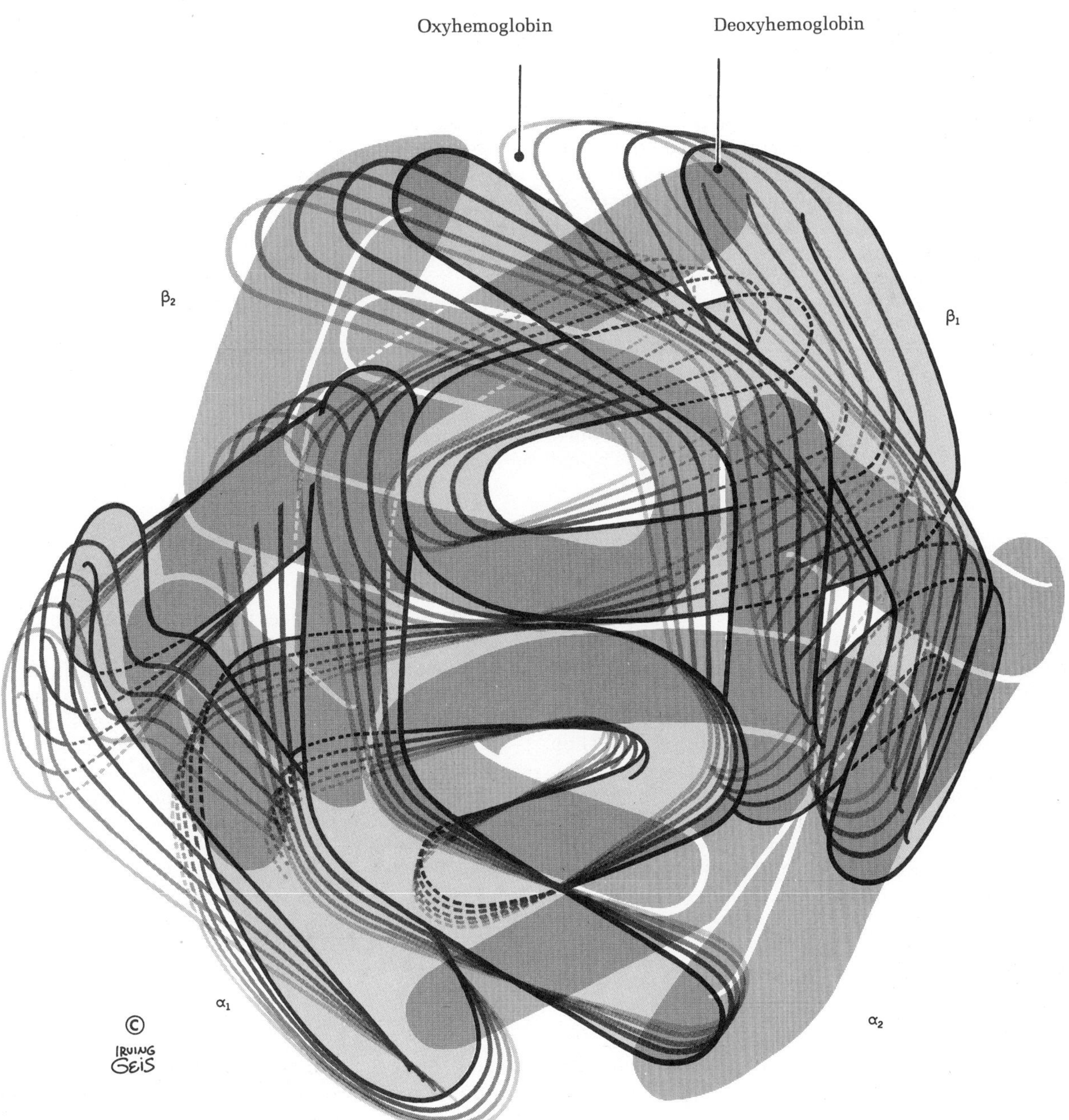

Figure 8-18
Scale drawing showing in "slow motion" a schematic representation of the change in quaternary structure as the $\alpha_1\beta_1$ subunit pair changes its position in relation to the $\alpha_2\beta_2$ pair (fixed in position) when oxyhemoglobin loses its oxygen to become deoxyhemoglobin.

come closer together. That is, there is a change in the *quaternary structure*, or packing arrangement, of the subunits. As a result the oxyhemoglobin molecule has a somewhat more compact structure than deoxyhemoglobin, and the central cavity becomes smaller. The two β hemes come closer together, and the two α hemes move away from each other, which may be related to the sigmoid oxygen-saturation relationship. Along with these changes, the H^+-binding residues in the α and β

Box 8-1 Diphosphoglycerate and the oxygen affinity of hemoglobin

It has long been known that *2,3-diphosphoglycerate* (Figure 1) is present in relatively high concentrations in red blood cells, but its function remained quite mysterious until it was found to have a profound effect on the affinity of hemoglobin for oxygen. When 2,3-diphosphoglycerate (DPG) is added to a solution of pure hemoglobin, it greatly reduces its affinity for oxygen. DPG produces this effect because it is itself bound by deoxyhemoglobin. We can therefore write an equation for a fourth binding equilibrium of hemoglobin

$$\text{HbDPG} + O_2 \rightleftharpoons \text{HbO}_2 + \text{DPG}$$

Thus we have an inverse relationship between the binding of oxygen and the binding of DPG, which are bound at different sites in the hemoglobin molecule.

DPG regulates the oxygen-binding affinity of hemoglobin in the red blood cell in relationship to the oxygen partial pressure in the lungs. Within a few hours after a normal human subject moves from sea level to an altitude of 12,000 ft, the DPG level in the red blood cells increases, causing more DPG to be bound and the oxygen affinity of the hemoglobin to decrease. At high altitudes the partial pressure of oxygen is significantly less than at sea level. Correspondingly, the pressure of oxygen in the tissues is less. The increase in DPG on ascending to high altitudes allows hemoglobin to release oxygen more readily in the tissues. The reverse change occurs when people acclimated to high altitudes, like those of the Himalayas or Andes, are brought to sea level. Increases in the DPG level of red cells also occur in patients suffering from *hypoxia,* lowered oxygenation of their tissues due to inadequate function of the lungs or the circulatory system.

Where on the hemoglobin molecule is DPG bound? Hemoglobin has a central open cavity, or channel, as we have seen in Figure 8-10. This cavity, lined with many positively charged R groups, is the binding site for DPG, which is bound to deoxyhemoglobin so that it forms a cross-link or bridge between the two β subunits. When oxygen is bound, the DPG is displaced from the cavity. Hemoglobin binds only one molecule of DPG, in contrast to four O_2, four CO_2, and close to four H^+ (Figure 1).

Figure 1
(a) Structure of 2,3-diphosphoglycerate (DPG). Its charged groups, which bind to the two β chains, are in color.
(b) Location of the DPG molecule (color) in the central cavity of hemoglobin. The negatively charged groups of DPG are attracted to neighboring, positively charged R groups (black) of the β chains.

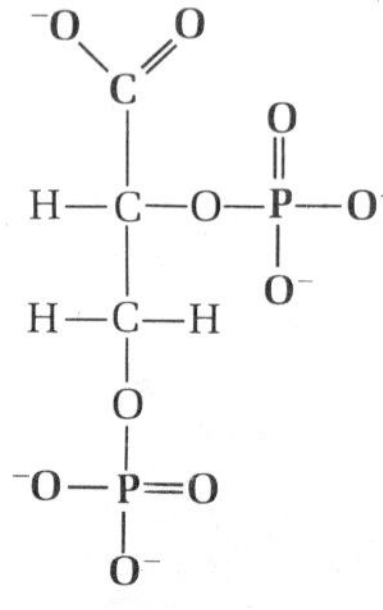

(a)

chains undergo a change from relatively hydrophilic to relatively hydrophobic environments. This shift increases the tendency of these protonated groups to lose H^+; that is, they become stronger acids as the hemoglobin is oxygenated, a change that can account for the Bohr effect. Thus these changes in the quaternary structure of hemoglobin when it is oxygenated appear to be related to the inverse binding affinities of hemoglobin for oxygens vs. those for H^+ and CO_2.

Yet another facet of the regulatory action of hemoglobin has been revealed with the discovery by Reinhold and Ruth Benesch of a *fourth* ligand for this protein, *2,3-diphosphoglycerate.* Box 8-1 outlines this interesting story.

Hemoglobin may be regarded as a molecular computer, which can sense changes in the concentration of any of its four ligands,

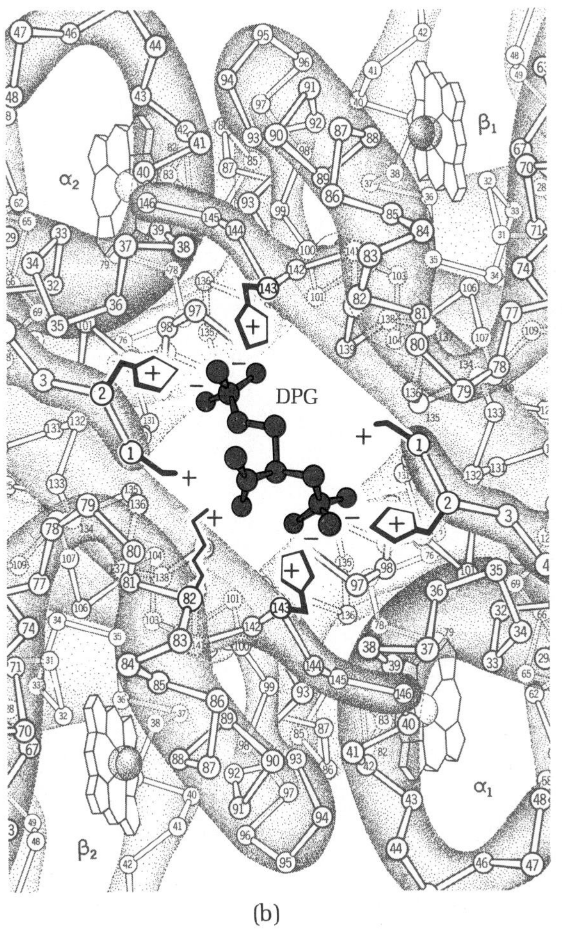

(b)

As it is normally isolated from blood, hemoglobin already contains considerable DPG, which is difficult to remove completely. When hemoglobin is "stripped" of its DPG, its sigmoid oxygen-binding relationship is largely lost and it shows a much higher affinity for oxygen. When excess DPG is added, the oxygen binding affinity decreases (Figure 2). DPG is therefore essential for the normal release of oxygen from hemoglobin in the tissues. The erythrocytes of some birds do not contain DPG, but another phosphate compound, inositol hexaphosphate, which is even more potent than DPG in reducing the oxygen affinity of hemoglobin.

Figure 2
Effect of DPG *on the oxygen saturation curve of hemoglobin.*

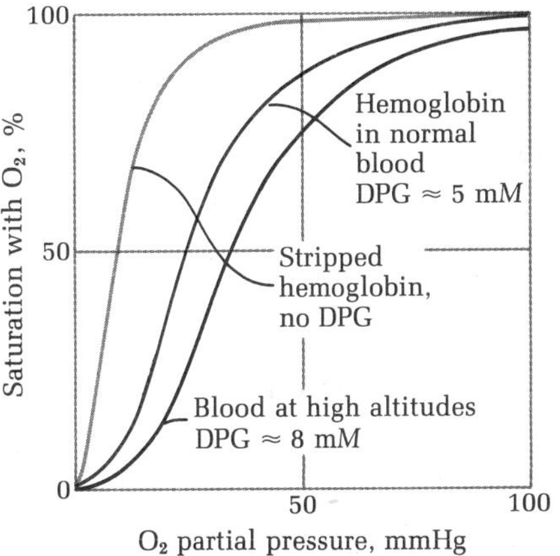

transmit this information through conformational changes of the molecule, and thus bring about adjustments in the binding affinities for the other ligands. It is believed that when one or two of the subunits bind oxygen, the small conformational changes induced exert leverage on the empty subunits, so that the quaternary structure of the entire molecule undergoes a change to a form having a high affinity for oxygen and a low affinity for H^+ and CO_2. Conversely, when O_2 is released, the quaternary structure reverts to its original form and H^+ and CO_2 become bound. Various theories have been proposed to account for the detailed structural and binding changes occurring in the oxygenation-deoxygenation cycle (Figure 8-19). Whatever the details, hemoglobin is a useful model of the computerlike regulatory activity of many other oligomeric proteins, particularly enzymes

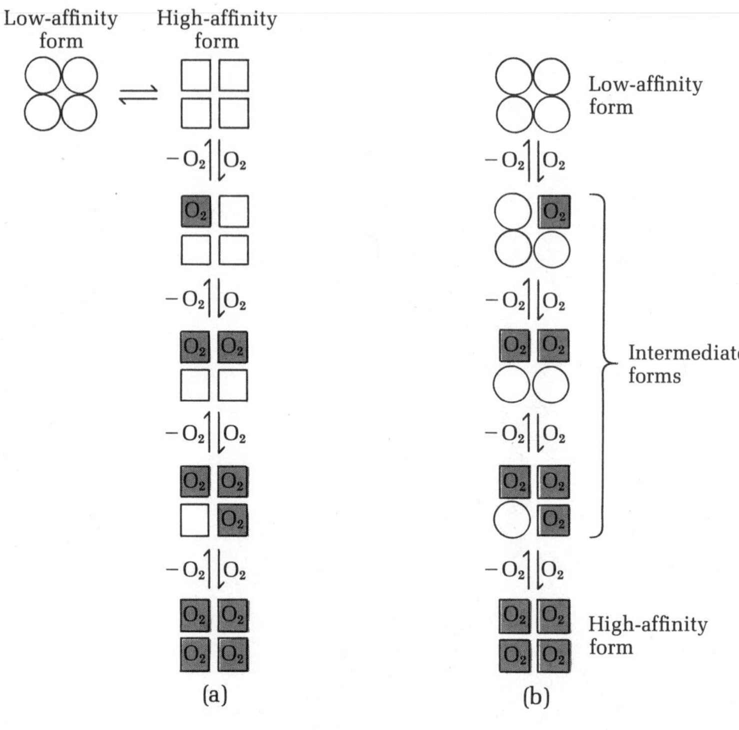

Figure 8-19
The symmetry (all-or-none) and sequential (induced fit) models for the cooperative binding of oxygen by hemoglobin. In both models the subunits have two different states. The circles represent the subunit state having low affinity for oxygen and the squares the state with high affinity for oxygen.

(a) *In the symmetry, or all-or-none, model it is assumed that hemoglobin occurs in only two forms, one with all the subunits in the low-affinity state and the other with all the subunits in the high-affinity state. In the absence of oxygen the two forms of hemoglobin are in equilibrium, but the low-affinity form predominates. When oxygen is present, it binds preferentially to the high-affinity form, pulling the equilibrium to the right, and increasing the binding of the remaining oxygen. In the symmetry model there are no intermediate forms between the low- and high-affinity states.*

(b) *In the sequential, or induced-fit, model, there are several intermediate conformations before the high-affinity form is reached. When one oxygen molecule is bound to a low-affinity subunit, its binding induces a change to form a high-affinity subunit. Once formed, it increases the probability that the other subunits will also undergo an induced fit and shift into their high-affinity forms as they bind oxygen molecules in sequence.*

having both catalytic and regulatory activity. Most proteins possessing built-in regulatory features have two or more polypeptide chains that fit together in a characteristic quaternary conformation, which appears to change as the protein molecule switches from one activity state to another.

Sickle-Cell Anemia Is a Molecular Disease of Hemoglobin

The great importance of the amino acid sequence in determining the secondary, tertiary, and quaternary structure of globular proteins and thus their biological function is strikingly shown in the hereditary human disease *sickle-cell anemia*, in which hemoglobin molecules are genetically altered. Patients with sickle-cell anemia suffer from repeated crises brought on by physical exertion, in which they become weak, dizzy, and short of breath and experience heart murmurs and an increased pulse rate. The hemoglobin content of the blood of such patients is only about half the normal value of 15 to 16 g per 100 mL; thus they are *anemic* (the word means "lacking blood," i.e., red cells). Moreover, microscopic examination shows that the red blood cells are not only fewer in number but also abnormal. In addition to an unusually large number of immature red cells, there are many long, thin, crescentlike red blood cells, which look like a blade of a sickle (Figure 8-20).

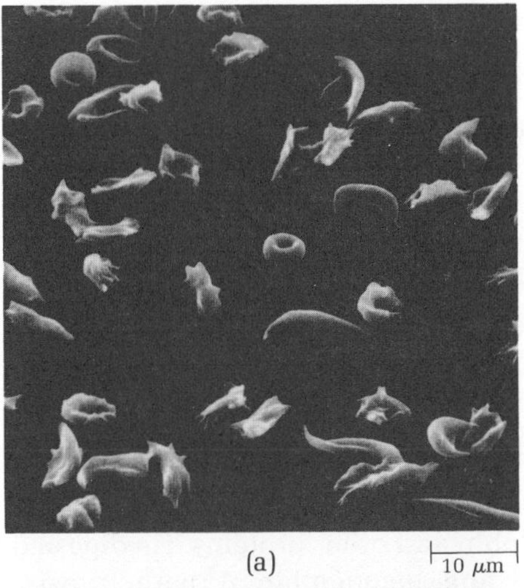

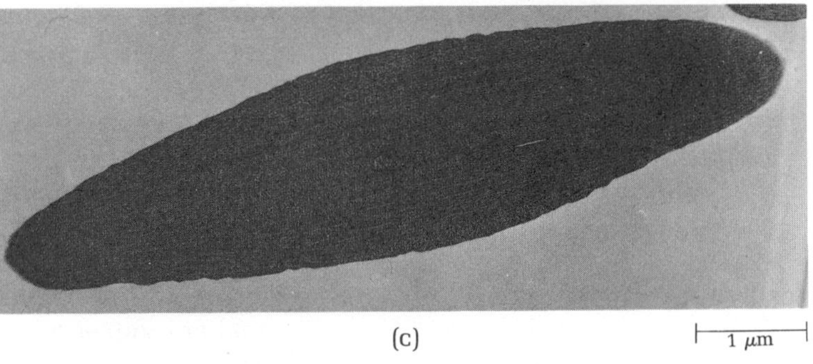

Figure 8-20
(a) Scanning electron micrograph of sickled red blood cells. In the center is a normal cell for comparison (see also Figure 8-15). (b) Some sickle cells greatly enlarged to show the bizarre shapes they may assume. (c) Transmission electron micrograph of a sickle cell showing the longitudinal arrays of sickle hemoglobin polymers that give these cells their bizarre shapes.

The number of such sickled cells is greatly increased when the blood is deoxygenated. One consequence is that the sickled cells are very fragile and undergo rupture, accounting for the low hemoglobin level in the blood. The more serious consequence is that small blood capillaries in different organs become blocked by the long, abnormally shaped red cells, a major factor in the early death of many patients with the disease.

Sickle-cell anemia is a genetic disease in which the individual has inherited a mutant hemoglobin gene from *both* parents. Sickle-cell anemia must be distinguished from a milder disease called *sickle-cell trait*, in which individuals have received the abnormal hemoglobin gene from only *one* parent. In people with the sickle-cell trait, which affects about 8 percent of the black population in the United States, only about 1 percent of the red cells become sickled. People with the sickle-cell trait may live completely normal lives if they avoid vigorous exercise or other stresses on the circulatory system.

The different shape of the red cells in sickle-cell anemia is caused by the abnormal form of hemoglobin they contain. Hemoglobin from sickle cells is called *hemoglobin S*; that from normal human adults is *hemoglobin A*. When hemoglobin S is deoxygenated, it becomes insoluble and forms bundles of tubular fibers (Figure 8-21), whereas hemoglobin A remains soluble on deoxygenation. The insoluble fibers of deoxygenated hemoglobin S are responsible for deforming the red blood cells into the sickle shape.

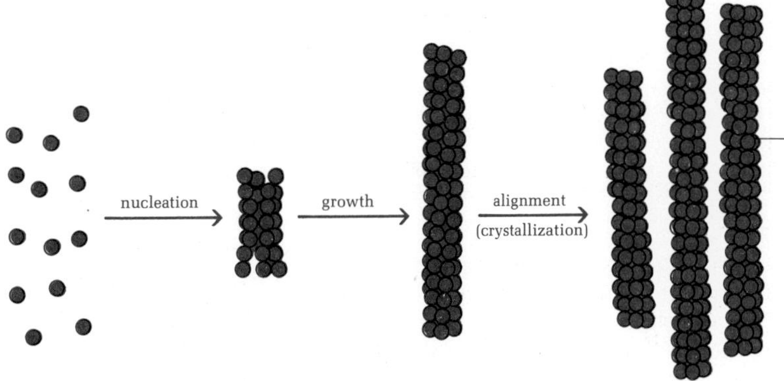

Figure 8-21
Association of deoxyhemoglobin S molecules to form bundles of insoluble tubular fibers, which distort the erythrocytes into a sickle shape.

Sickle-Cell Hemoglobin Has an Altered Amino Acid Sequence

In the late 1940s Linus Pauling and Harvey Itano showed that sickle-cell hemoglobin migrates toward the positively charged electrode at a slightly lower rate than normal hemoglobin A when both are placed in an electric field and subjected to electrophoresis (page 128). They concluded that the polypeptide chains of the hemoglobin S molecule must have a slightly smaller number of negatively charged R groups than hemoglobin A. Later Vernon Ingram devised a simple experimental procedure that pinpointed the difference between hemoglobin S and hemoglobin A more precisely. His technique, called *peptide fingerprinting* or *mapping*, is still widely used to detect genetic variants of hemoglobin and other proteins. He digested samples of hemoglobin S and hemoglobin A with trypsin, which cleaves only those peptide bonds which involve the carboxyl group of lysine and arginine residues (page 131). This treatment resulted in the formation of 28 kinds of peptide fragments, since there is a total of 27 lysine and arginine residues in the combined α and β chains. The mixture of peptides from each type of hemoglobin was spotted on filter paper moistened with a buffer solution. The peptide mixtures were then subjected to electrophoresis along one direction of the paper, causing the peptide fragments to separate, although incompletely, into different zones. The paper was then dried and placed in a fresh buffer of a different pH and subjected to paper chromatography at right angles to the direction of electrophoresis (page 113). When the paper was dried and heated with ninhydrin (page 111), a characteristic two-dimensional fingerprint, or map, of all the peptide fragments was obtained. As seen in Figure 8-22, hemoglobin S differs from hemoglobin A in only one peptide spot, which occupies a different position in the fingerprints of the two proteins, consistent with the view that only this peptide differs in its electric charge in the two forms of hemoglobin. When these spots were cut out of the two papers and the peptides eluted and analyzed, Ingram found that they differed in only one respect: *the displaced peptide from hemoglobin S contained a valine residue where the same peptide from normal hemoglobin A has a glutamic acid residue* (Figure 8-22). All the other amino acid residues in the two chains of hemoglobin S are identical with those in the hemoglobin A chains. The location of the aberrant residue was found to be position 6 in the β chains of hemoglobin. Since the R group of valine has no electric charge whereas glutamic acid has a negative charge at pH 8 (Figure 8-23), it follows that sickle-cell hemoglobin must have two fewer negative charges than hemoglobin A, one for each of the two β chains in the hemoglobin molecule. This difference causes hemoglobin S to move at a slightly different rate than hemoglobin A in an electric field.

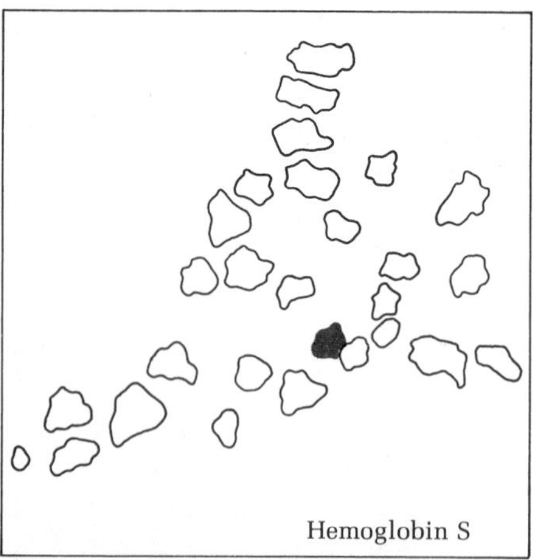

Figure 8-22
Fingerprints or peptide maps of trypsin peptides of hemoglobin A and hemoglobin S. Only a single peptide (color) is displaced; it contains the genetically replaced amino acid.

Figure 8-23
(a) The genetic defect in sickle-cell hemoglobin. As the result of a mutation in the β-chain gene the glutamic acid residue normally present in the 6 position of the β chain of hemoglobin A is replaced by a valine residue. This replacement results in the loss of one negative charge in each of the two β chains.

(b) Positions of 163 mutations (black circles) in human hemoglobin observed as of 1979. There are 105 mutations in the β chains and 58 in the α chains. The invariant positions in hemoglobin are shown in color. Mutations occurring near the heme groups are most likely to lead to serious defects in hemoglobin function.

Glutamic acid residue

COO^-
|
CH_2
|
H CH_2
| |
—N—C—C—
| ‖
H O

↑

1 2 3 4 5 6 7 8
Val·His·Leu·Thr·Pro·**Glu**·Glu·Lys · (hemoglobin A)

Val·His·Leu·Thr·Pro·**Val**·Glu·Lys · (hemoglobin S)

↓

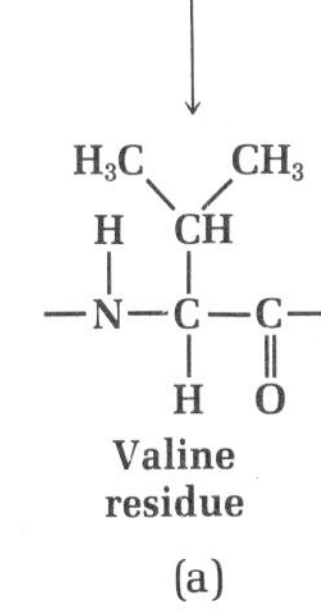

(a)

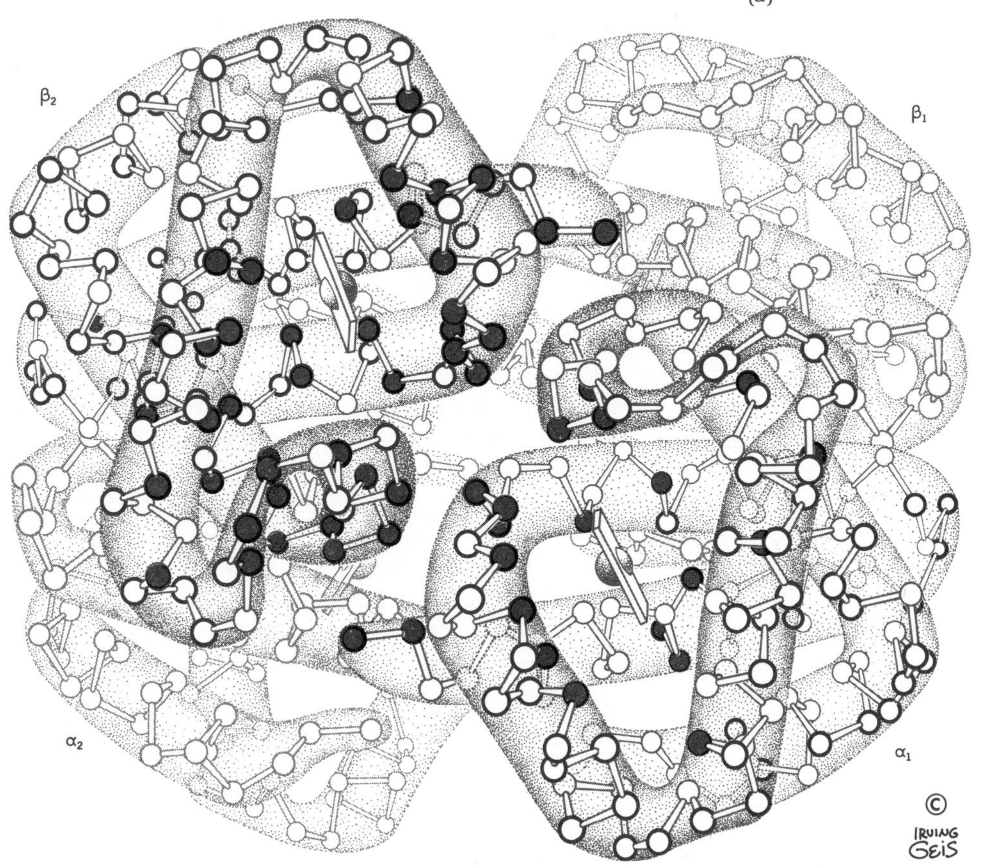

(b)

Sickling Is Caused by the Tendency of Hemoglobin S Molecules to Stick Together

The substitution of two valine residues for two glutamic acid residues in a protein molecule having 574 amino acid residues altogether may not appear to be a very significant alteration, but position 6 of the β chain is a critical point in the quaternary structure of hemoglobin. Replacement of the normal glutamic acid residue by valine creates a "sticky" hydrophobic contact point at position 6 of the β chain, which is on the outer surface of the molecule. These sticky spots cause deoxyhemoglobin S molecules to associate abnormally with each other, to form the long, fibrous aggregates responsible for sickling of the red cells (Figure 8-21).

Proteins Containing "Wrong" Amino Acids Are the Result of Gene Mutations

In people with sickle-cell anemia the gene coding for the β chain of hemoglobin has undergone an irreversible mutation so that it codes for a valine residue where a glutamic acid residue should occur; all the other amino acids of the β chain are normal. Sickle-cell hemoglobin is the result of only 1 of over 300 different mutations of the hemoglobin genes that have been discovered in human subjects, most of which represent a single amino acid replacement in either the α or β chains (Figure 8-23; Table 8-4). Many of these mutations have been detected by electrophoretic tests and/or fingerprinting of the hemoglobin of patients with disorders of the red blood cells.

Hemoglobin is not the only protein in the human body that may be altered genetically by mutation. *All* the proteins in the body, whether globular or fibrous, are subject to mutation. It just happens that hemoglobin has been studied most in human beings because aberrations in its molecular structure tend to produce conspicuous circulatory or respiratory symptoms; moreover, hemoglobin is easy to isolate from small samples of human blood. Mutations leading to alterations in amino acid sequence have been found in many other human proteins, including various enzymes concerned in metabolism, as well as in fibrous proteins such as collagen.

Although it is common to think of mutations resulting in altered protein molecules as genetic "defects," mutation of genes specifying protein structure can also result in "improvement" of a protein molecule, so that it is *better* able to function and enhances the ability of an organism bearing it to survive in its natural habitat. This is in fact true of the gene mutation yielding sickle-cell hemoglobin. In some parts of Africa up to 40 percent of the black population carries the sickle gene, suggesting that this mutation must have had some survival value. Native Africans with sickle-cell genes show a much lower susceptibility to malaria, caused by a mosquito-borne parasite that invades red

Table 8-4 Some Mutated Human Hemoglobins

These mutations are often designated by the city or region of discovery. Only hemoglobin S causes sickling. Other hemoglobin mutants show other functional changes in their properties.

Abnormal hemoglobin	Position and normal residue	Replacement
	α chain	
I	16 Lys	Glu
$G_{Honolulu}$	30 Glu	Gln
Norfolk	57 Gly	Asp
M_{Boston}	58 His	Tyr
$G_{Philadelphia}$	68 Asn	Lys
$O_{Indonesia}$	116 Glu	Lys
	β chain	
C	6 Glu	Lys
S	6 Glu	Val
$G_{San\ Jose}$	7 Glu	Gly
E	26 Glu	Lys
$M_{Saskatoon}$	63 His	Tyr
Zürich	63 His	Arg
$M_{Milwaukee}$	67 Val	Glu
D_{Punjab}	121 Glu	Gln

blood cells. Sickled red blood cells are less favorable for the growth of the malaria parasite than normal red blood cells. The geographic areas of Africa having the highest incidence of the sickle gene are also the areas that once had the highest prevalence of the malaria parasite (Figure 8-24). Thus possession of the sickle gene conferred significant survival value on individuals in those parts of Africa where malaria was once endemic and often fatal in childhood.

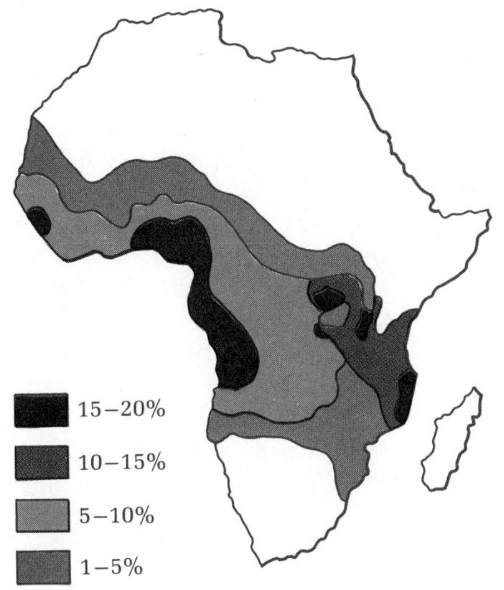

Figure 8-24
Relative frequency of the sickle-cell gene in different areas of Africa. The zones of high frequency lie in an area where malaria was a major cause of death.

Can a Molecular Cure for Sickle Hemoglobin Be Found?

Although the sickle-cell gene may have been advantageous in parts of Africa where malaria was endemic, it is certainly disadvantageous today, particularly to black Africans living in other parts of the world where malaria is an uncommon disease and the normal life span is much longer. With our advanced knowledge of the structure and conformation of hemoglobin, attempts are being made to develop a rational molecular treatment of sickle-cell anemia. A search is under way for an otherwise harmless drug or chemical agent that can react with one or more specific functional groups of the hemoglobin S molecule to eliminate the deleterious sickling effect of the

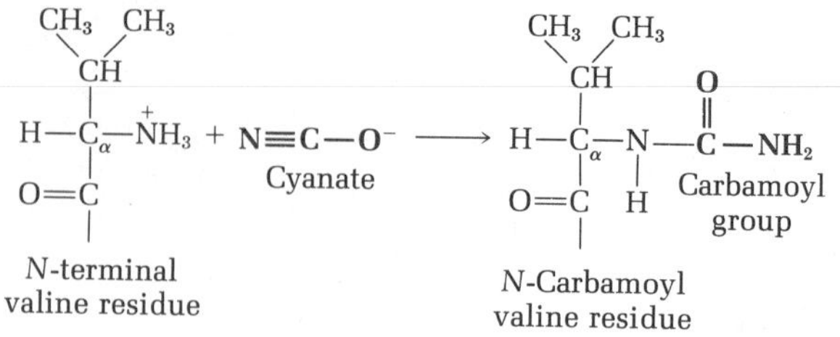

Figure 8-25
Formation of the N-carbamoyl derivative of the amino-terminal valine residue of the β chain of hemoglobin S on treatment in vitro with potassium cyanate. This single chemical change prevents erythrocytes from a sickle-cell anemia patient from sickling when they are deoxygenated.

$$(CH_3)_2CH{-}C_\alpha H(\overset{+}{N}H_3){-}C{=}O \;+\; N{\equiv}C{-}O^- \longrightarrow (CH_3)_2CH{-}C_\alpha H(NH{-}C({=}O){-}NH_2){-}C{=}O$$

N-terminal valine residue + Cyanate ⟶ N-Carbamoyl valine residue (Carbamoyl group)

substitution of valine for glutamic acid in the β chains. Several chemical agents have in fact been found that will largely prevent the sickling of red cells in the test tube. *Potassium cyanate*, for example, combines with certain amino groups of hemoglobin S to form carbamoyl derivatives (Figure 8-25), which do not sickle on deoxygenation. However, administration of potassium cyanate to sickle-cell patients is unsafe because of toxic side effects. Nevertheless, current research on antisickling agents gives some hope that by appropriate biochemical engineering a reagent may be devised that not only prevents sickling of hemoglobin S but also is completely safe, without deleterious effects on other body proteins.

Summary

The polypeptide chains of globular proteins are tightly folded, as revealed by studies of their size and shape. X-ray diffraction analysis of myoglobin and other small single-chain proteins, such as cytochrome *c*, lysozyme, and ribonuclease, shows each of them to have its own characteristically different tertiary folding pattern. In all globular proteins the chain is very compactly folded, with little or no interior space for water molecules. Nearly all the hydrophobic R groups are inside the molecule, shielded from water; and most of the ionic R groups are hydrated and on the outside, facing the aqueous medium. The folded tertiary structure is held together by a combination of noncovalent interactions (particularly hydrophobic interactions between hydrophobic R groups), electrostatic attractions between oppositely charged R groups, and hydrogen bonding. These interactions are individually weak but collectively strong. In some globular proteins tertiary structure is also determined and fixed by covalent cystine cross-linkages. The information determining tertiary structure is inherent in the amino acid sequence of polypeptides, as suggested by the fact that homologous proteins of different species share not only many invariant residues but also identical conformations. Proof is given by the discovery that many denatured globular proteins, devoid of their normal biological activity, will spontaneously renature again with restoration of their biological activity.

Oligomeric globular proteins, those with two or more polypeptide chains, are larger and more complex in structure and often participate in regulatory activities. How their subunit chains fit together is termed their quaternary structure. X-ray analysis of hemoglobin and other oligomeric proteins shows them also to be very compact in structure, most hydrophobic R groups being inside and most hydrophilic groups outside. In hemoglobin, which has two α and two β chains, there are few contacts between like chains but many contacts holding the $\alpha_1\beta_1$ and $\alpha_2\beta_2$ pairs together. Hemoglobin has a sigmoid oxygenation curve, which is well adapted to binding oxygen in the lungs and releasing it in the peripheral tissues. Myoglobin, in contrast, has a much higher affinity for oxygen and a hyperbolic saturation curve, which endow it with properties desirable for storage of oxygen in muscles. The binding of oxygen by hemoglobin is enhanced by increased pH and low CO_2 concentration, whereas release of oxygen is enhanced by decreased pH and high CO_2 concentration. These relationships, as well as the regulation of O_2 affinity by the binding of 2,3-diphosphoglycerate, are mediated by specific binding sites for O_2, H^+, CO_2, and DPG and by changes in the quaternary structure of hemoglobin during the oxygenation-deoxygenation cycle. Thus the subunits of hemoglobin, like those of other oligomeric proteins, are capable of communicating regulatory interactions through conformational interactions. Changes in the amino acid sequence of globular proteins through gene mutations, as in sickle-cell anemia, can cause significant changes in their conformations and thus their biological functions.

References

Books

CANTOR, C. R., and P. R. SCHIMMEL: *Biophysical Chemistry*, pt. I, *The Conformation of Biological Macromolecules*, Freeman, San Francisco, 1980.

DICKERSON, R. E., and I. GEIS: *Hemoglobin: Structure, Function, Evolution, and Pathology*, Benjamin/Cummings, Menlo Park, Calif., 1982.

FERMI, G., and M. PERUTZ: *Atlas of Molecular Structures in Biology*, vol. 2, *Hemoglobin and Myoglobin*, Oxford University Press, New York, 1981.

GLUSKER, J. P., and K. N. TRUEBLOOD: *Crystal Structure Analysis: A Primer*, Oxford University Press, New York, 1972.

HASCHEMEYER, R., and A. H. HASCHEMEYER: *Proteins: A Guide to Study by Physical and Chemical Methods*, Wiley, New York, 1973.

SCHULTZ, G. E., and R. H. SCHIRMER: *Principles of Protein Structure*, Springer-Verlag, New York, 1979.

Articles

ANFINSEN, C. B.: "Principles That Govern the Folding of Polypeptide Chains," *Science*, **181**:223–230 (1973).

CERAMI, A., and C. M. PETERSON: "Cyanate and Sickle Cell Disease," *Sci. Am.*, **232**:44, April (1975) (offprint 1319).

DICKERSON, R. E.: "Cytochrome C and the Evolution of Energy Metabolism," *Sci. Am.*, **242**:236, March (1980).

INGRAM, V. M.: "Gene Mutation in Hb: The Chemical Difference Between Normal and Sickle Cell Hemoglobin," *Nature*, **180**:326–328 (1957). Discovery of the amino acid replacement in hemoglobin S.

KENDREW, J. C.: "The Three-Dimensional Structure of a Protein Molecule," *Sci. Am.*, **205**:96–111, December (1961) (offprint 121).

KOSHLAND, D. E., JR.: "Protein Shape and Biological Control," *Sci. Am.*, **229**:52, October (1973).

PAULING, L., H. ITANO, S. J. SINGER, and I. C. WELLS: "Sickle Cell Anemia: A Molecular Disease," *Science*, **110**:543–548 (1949). Classical description of the electrophoretic difference between hemoglobins A and S.

PERUTZ, M. F.: "Hemoglobin Structure and Respiratory Transport," *Sci. Am.*, **239**:92, December (1978).

PERUTZ, M. F., and H. LEHMANN: "Molecular Pathology of Human Hemoglobin," *Nature*, **219**:902–909 (1968).

Problems

1. *Formation of Bends and Intrachain Cross-Linkages in Polypeptide Chains* In the following polypeptide where might bends or turns occur? Where might intrachain disulfide cross-linkages be formed?

1	2	3	4	5	6	7	8	9	10	11
Ile ·	Ala ·	His ·	Thr ·	Tyr ·	Gly ·	Pro ·	Phe ·	Glu ·	Ala ·	Ala ·
12	13	14	15	16	17	18	19	20	21	22
Met ·	Cys ·	Lys ·	Trp ·	Glu ·	Ala ·	Gln ·	Pro ·	Asp ·	Gly ·	Met ·
23	24	25	26	27	28					
Glu ·	Cys ·	Ala ·	Phe ·	His ·	Arg					

2. *Location of Specific Amino Acids in Globular Proteins* X-ray analysis of the tertiary structure of myoglobin and other small single-chain globular proteins has led to some generalizations about how the polypeptide chains of soluble proteins fold. With these in mind, indicate the probable location, whether in the interior or on the external surface, of the following amino acid residues in native globular proteins: aspartate, leucine, serine, valine, glutamine, and lysine. Explain your reasoning.

3. *Functional Proteins from Linear Polymers* A protein is biologically active only if it has its correct three-dimensional structure. Proteins are synthesized from information contained in a linear, i.e., one-dimensional, coding sequence in DNA. Furthermore, ribosomes assemble linear, one-dimensional sequences of amino acids. In view of these facts explain how biologically active proteins having specific three-dimensional structures can be formed in cells. Provide some experimental verification for your explanation.

4. *Disulfide Cross-Links and Protein Folding* The hypothesis that the linear amino acid sequence of a protein determines its folding pattern (secondary and tertiary structure) can be tested by unfolding a protein and then allowing it to refold spontaneously. A comparison of the biological activity of the protein before unfolding and after refolding provides a measure of the fraction of the protein that has returned to its native state. For example, ribonuclease is

completely unfolded by cleaving the four disulfide cross-links followed by treatment with 8 *M* urea. If the urea is now removed by dialysis and the disulfide cross-links allowed to reform under controlled conditions, 95 to 100 percent of the activity is recovered. This experiment is illustrated in Figure 8-8. The results of similar experiments on other proteins are tabulated.

Protein	No. of disulfide bonds	Recovered activity, %	
		Observed	Predicted†
Ribonuclease	4	95–100	~1
Lysozyme	4	50–80	~1
Alkaline phosphatase	2	80	~33
Insulin (bovine)	3	5–10	~6–7

† If disulfide bonds form randomly

(a) If the four disulfide cross-links in unfolded ribonuclease were formed by completely random pairing of the cysteine residues, the predicted activity would only be about 1 percent of the original activity. Why would you expect such a low recovery?

(b) The observed recovery of the activities of ribonuclease, lysozyme, and alkaline phosphatase after unfolding and refolding is considerably higher than the activities predicted if the disulfide cross-links occurred by random pairing. Provide an explanation for this observation.

(c) One of the examples in the table clearly stands apart from the rest, namely, insulin. The observed recovery of activity is very low and practically the same as that predicted by random disulfide-link formation. What does this observation indicate about the native structure of insulin? Speculate on how insulin achieves its native structure.

5. *The Number of Polypeptide Chains in an Oligomeric Protein* A sample (660 mg) of an oligomeric protein having a molecular weight of 132,000 was treated with an excess of 2,4-dinitrofluorobenzene under slightly alkaline conditions until the chemical reaction was complete. The peptide bonds of the protein were then hydrolyzed completely by heating it with concentrated HCl. The hydrolyzate was found to contain 5.5 mg of the following compound:

CH_3 CH_3
NO_2 CH
O_2N—⟨ring⟩—NH—C—COOH
H

However, 2,4-dinitrophenyl derivatives of the α-amino groups of other amino acids could not be found.

(a) Explain why this information can be used to determine the number of polypeptide chains in an oligomeric protein.

(b) Calculate the number of polypeptide chains in this protein.

6. *Molecular Weight of Hemoglobin* The first indication that proteins have molecular weights greatly exceeding those of the organic compounds then known was obtained over 100 years ago. For example, it was known at that time that hemoglobin contains 0.34 percent by weight of iron.
   (a) From this information determine the minimal molecular weight of hemoglobin.
   (b) Subsequent experiments indicated that the true molecular weight of hemoglobin is 64,500. What information did this provide about the number of iron atoms in hemoglobin?

7. *The Packing of Hemoglobin in the Human Red Blood Cell* Human blood contains 160 g of hemoglobin per liter of blood. Blood contains some 5.0×10^9 red blood cells per milliliter. Although each red blood cell is a biconcave disk, to simplify calculations we may regard the cells as simple cylinders, with the following dimensions:

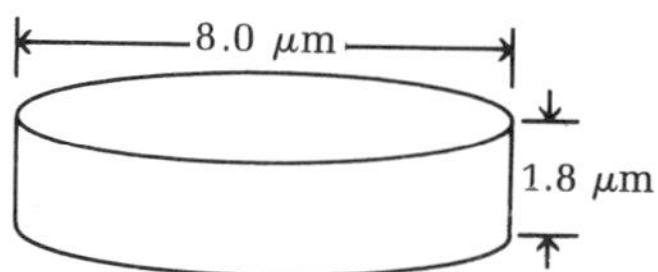

   (a) Calculate the weight of hemoglobin in one red blood cell.
   (b) Calculate the number of molecules of hemoglobin in one red blood cell.
   (c) Calculate the volume of the red blood cell.
   (d) Hemoglobin is a globular protein 6.8 nm in diameter. What fraction of the total volume of the red blood cell is occupied by hemoglobin?
   (e) The ratio of the total volume of hemoglobin to the total volume of the red blood cell [part (d) above] gives a deceptive picture of how tightly hemoglobin is packed in the red blood cell. One must remember that when spheres are packed, the voids between the spheres occupy a substantial portion of the total volume. Assuming that hemoglobin is packed in the red cell in a cubic array, as indicated in the drawing,

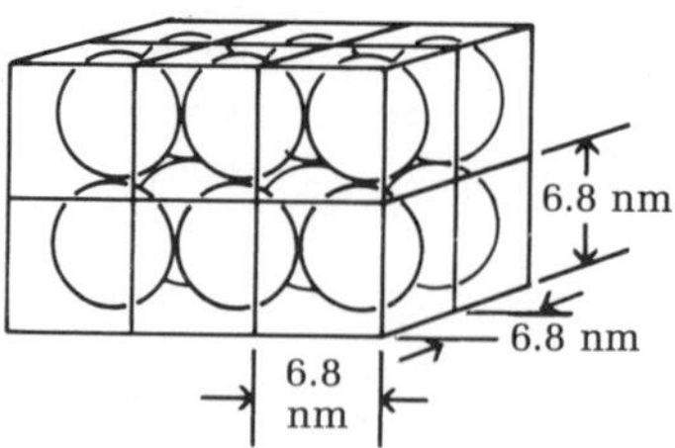

   calculate the total volume of the array of hemoglobin molecules present in a single red blood cell and compare it with the cell volume. How tightly are hemoglobin molecules packed in the red blood cell?
   (f) In light of your observations in part (e), are the hemoglobin molecules close enough to interact with each other? If so, is it possible that interaction of hemoglobin S molecules in sickled red cells could affect the shape of the red blood cell?

8. *The Effect of Myoglobin on the Oxygen Storage Ability of Tissues*
   (a) Animal tissues contain about 70 percent by weight of water. The concentration of oxygen in tissue water is normally about $3.5 \times 10^{-5} M$ in living animals. Calculate the weight of oxygen per kg of tissue that can be stored in the form of oxygen dissolved in tissue water.
   (b) Most mammalian tissues contain myoglobin to store oxygen. In humans, myoglobin is most abundant in heart tissue, where it constitutes 0.7 percent by weight of the total tissue mass. Calculate the weight of oxygen that can be stored in 1 kg of human heart tissue. Compare with your answer to (a).
   (c) The skeletal muscle of deep-diving mammals contains much more myoglobin, and its concentration is in proportion to the duration of their dives. Fresh seal muscle exposed to air was found to contain 0.15 g of oxygen per kilogram wet weight. Calculate the percent of myoglobin in fresh seal muscle.

9. *Comparison of Fetal and Maternal Hemoglobin* Studies of oxygen transport in pregnant females have shown that the oxygen saturation curves of fetal and maternal blood are markedly different when measured at the same conditions. This phenomenon arises from the fact that fetal erythrocytes contain a structural variant of hemoglobin (hemoglobin F, $\alpha_2\gamma_2$) while maternal erythrocytes contain the usual hemoglobin A ($\alpha_2\beta_2$).

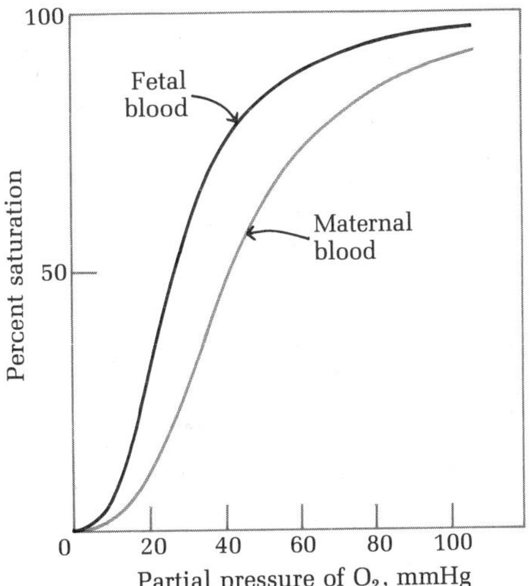

   (a) Which hemoglobin has a higher affinity for oxygen under physiological conditions, hemoglobin A or hemoglobin F? Explain.
   (b) What is the physiological significance of the different oxygen affinities? Explain.
   (c) When all the 2,3-diphosphoglycerate (DPG) is carefully removed from samples of hemoglobin A and F, the measured oxygen saturation curves (and consequently the oxygen affinities) are displaced to the left. However, hemoglobin A now has a greater affinity for oxygen than hemoglobin F. When DPG is returned, the oxygen saturation curves return to normal as shown in the figure. What is the effect of DPG on the oxygen affinity of hemoglobin? How can the above information be used to explain the different oxygen affinities of fetal and maternal hemoglobin?

10. *Identification of Mutant Hemoglobin* A sample of a mutant adult hemoglobin is subjected to trypsin hydrolysis, and a peptide map or "fingerprint" is prepared. It differs from that of normal hemoglobin A in having a lysine residue instead of an asparagine residue in one of the peptides.
    (a) What is the purpose of the trypsin hydrolysis procedure?
    (b) Which mutant of hemoglobin could the unknown be (see Table 8-4)?
    (c) How could you screen for it by a quicker and simpler procedure?

11. *Distinguishing Hemoglobin C from Hemoglobin S* Two samples of blood, one containing hemoglobin C and the other hemoglobin S, lost their labels on storage in a refrigerator. How would you determine which was which? (See Table 8-4.)

CHAPTER 9

Enzymes

We now come to the most remarkable and highly specialized proteins, those with catalytic activity. Enzymes have extraordinary catalytic power, which is generally far greater than that of synthetic catalysts. They have a high degree of specificity for their substrates, they accelerate specific chemical reactions without formation of by-products, and they function in dilute aqueous solutions under very mild conditions of temperature and pH. Few nonbiological catalysts show all these properties.

Enzymes are the functional units of cell metabolism. Acting in organized sequences, they catalyze the hundreds of stepwise reactions by which nutrient molecules are degraded, chemical energy is conserved and transformed, and cell macromolecules are made from simple precursors. Among the many enzymes participating in metabolism there is a special class, called *regulatory enzymes*, which can sense various metabolic signals and change their catalytic rates accordingly. Through their action enzyme systems are highly coordinated to yield a harmonious interplay between the many different metabolic activities necessary to sustain the living state.

In some human diseases, especially heritable genetic disorders, there may be a deficiency or even a total absence of one or more enzymes in the tissues. In other abnormal conditions the excessive activity of a specific enzyme can sometimes be controlled by a drug designed to inhibit its catalytic activity. Moreover, measurements of the activity of certain enzymes in the blood plasma, red blood cells, or tissue samples are important in diagnosing disease. Enzymes have become important practical tools, not only in medicine but also in the chemical industry, in food processing, and in agriculture. Even in everyday activities in the home enzymes play a part.

In this chapter we shall not try to catalog the nearly 2000 different known enzymes; instead we shall survey the properties and characteristics common to most enzymes. In succeeding chapters, as we examine the pathways of cell metabolism, we shall describe many different enzymes and their action in more detail and show some examples of the practical use of enzymes.

Much of the History of Biochemistry Is the History of Enzyme Research

That some form of catalysis takes place in biological systems was first recognized in the early 1800s from studies of the digestion of meat by secretions of the stomach and the conversion of starch into sugar by saliva and various plant extracts. Subsequently a great many instances of biological catalysis, now known to be enzymatic, were recorded. In the 1850s Louis Pasteur concluded that fermentation of sugar into alcohol by yeast is catalyzed by "ferments." He postulated that these ferments, later named *enzymes* ("in yeast"), are inseparable from the structure of living yeast cells, a view that prevailed for many years. It was therefore a major landmark in the history of biochemistry when in 1897 Eduard Buchner succeeded in extracting in soluble active form from yeast cells the set of enzymes that catalyzes the fermentation of sugar to alcohol. This discovery proved that these important enzymes, which catalyze a major energy-yielding metabolic pathway, can still function when removed from the structure of living cells. It also encouraged biochemists to attempt the isolation of many different enzymes and to examine their catalytic properties.

At the turn of the twentieth century Emil Fischer carried out the first systematic studies of enzyme specificity. Others studied the kinetics of enzyme activity and formulated theories of enzyme action. But it was not until 1926 that an enzyme was isolated in pure crystalline form for the first time. This was *urease*, obtained from extracts of the jack bean by James Sumner at Cornell University. Sumner found that the urease crystals consisted entirely of protein. He was therefore led to postulate that all enzymes are proteins, but his conclusions were vigorously opposed by the renowned German biochemist Richard Willstätter, an authoritative figure, who insisted that enzymes were low-molecular-weight compounds and that the protein found in urease crystals was merely a contaminant. Only in the 1930s, after John Northrop and his colleagues crystallized pepsin and trypsin and found them also to be proteins, was the protein nature of enzymes widely accepted. Then followed a period of intensive research on the enzymes catalyzing the reactions of cell metabolism. Today some 2000 different enzymes have been identified, each catalyzing a different chemical reaction. Hundreds have been obtained in pure crystalline form (Figure 9-1).

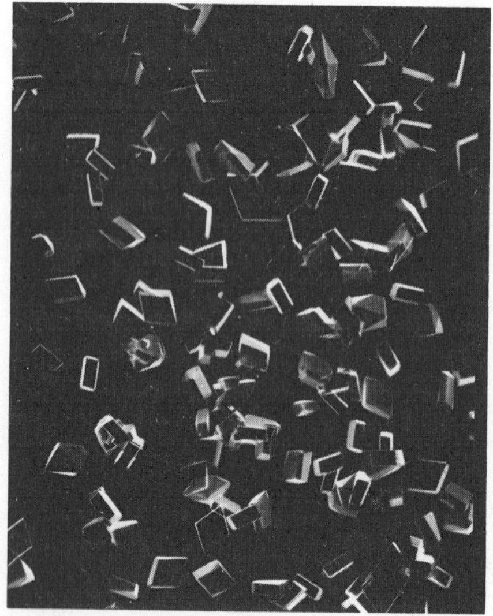

Figure 9-1
Crystals of bovine chymotrypsin.

But even today many questions about enzymes still remain to be fully answered. Why were proteins selected to become cell catalysts? Why are enzyme molecules so large compared with the substrate structures they act upon? How can amino acids, which cannot themselves accelerate chemical reactions, produce such powerful catalytic activity when linked into specific sequences? How is enzyme action regulated?

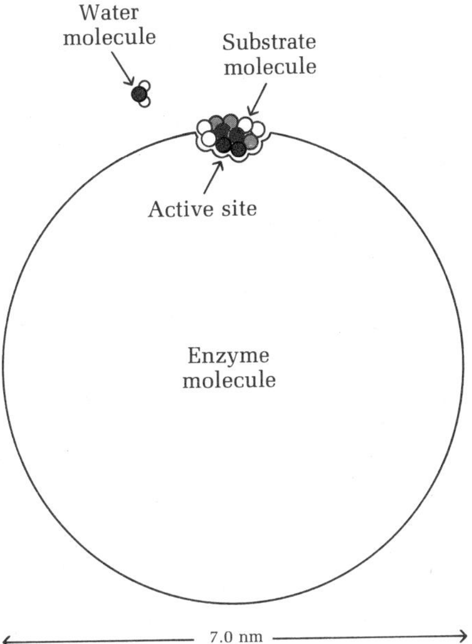

Figure 9-2
Relative dimensions of a medium-sized enzyme molecule (MW 100,000; diameter 7 nm) and a typical substrate molecule (MW 250; length 0.8 nm). The active site occupies only a small fraction of the surface area of the enzyme molecule. Also shown for comparison is a water molecule.

Enzymes Show All the Properties of Proteins

All pure enzymes examined to date are proteins; moreover, their catalytic activity depends upon the integrity of their structure as proteins. For example, if an enzyme is boiled with strong acid or incubated with trypsin, treatments that cleave polypeptide chains, its catalytic activity will usually be destroyed, showing that the primary backbone structure of the enzyme protein is required for activity. Moreover, if we disrupt the characteristic folding of the polypeptide chain(s) of a native enzyme protein by heat, by exposure to extremes of pH, or by treatment with other denaturing agents, catalytic activity will also be lost. Thus the primary, secondary, and tertiary structures of enzyme proteins are important in their catalytic activity.

Enzymes, like other proteins, have molecular weights ranging from about 12,000 to over 1 million. They are therefore very large compared with the substrates or functional groups they act upon (Figure 9-2). Some enzymes consist only of polypeptides and contain no chemical groups other than amino acid residues; an example is pancreatic ribonuclease. Other enzymes, however, require for activity an additional chemical component called a cofactor. The cofactor may be either inorganic, such as Fe^{2+}, Mn^{2+}, or Zn^{2+} ions (Table 9-1), or it may be a complex organic molecule called a coenzyme (Table 9-2). Some enzymes require *both* a coenzyme and one or more metal ions for activity. In some enzymes the coenzyme or metal ion is only loosely and transiently bound to the protein, but in others it is tightly and permanently bound, in which case it is called a prosthetic group. A complete, catalytically active enzyme together with its coenzyme or metal is called a holoenzyme. Coenzymes and metal ions are stable on heating, whereas the protein part of an enzyme, called the apoenzyme, is denatured by heat. Coenzymes, which we shall examine in Chapter 10, function as transient carriers of specific functional groups (Table 9-2).

Table 9-1 Some Enzymes Containing or Requiring Essential Inorganic Elements as Cofactors

Fe^{2+} or Fe^{3+}	Cytochrome oxidase Catalase Peroxidase
Cu^{2+}	Cytochrome oxidase
Zn^{2+}	DNA polymerase Carbonic anhydrase Alcohol dehydrogenase
Mg^{2+}	Hexokinase Glucose 6-phosphatase
Mn^{2+}	Arginase
K^+	Pyruvate kinase (also requires Mg^{2+})
Ni^{2+}	Urease
Mo	Nitrate reductase
Se	Glutathione peroxidase

Table 9-2 Coenzymes Serving as Transient Carriers of Specific Atoms or Functional Groups†

Coenzyme	Entity transferred
Thiamine pyrophosphate	Aldehydes
Flavin adenine dinucleotide	Hydrogen atoms
Nicotinamide adenine dinucleotide	Hydride ion (H^-)
Coenzyme A	Acyl groups
Pyridoxal phosphate	Amino groups
5′-Deoxyadenosylcobalamine (coenzyme B_{12})	H atoms and alkyl groups
Biocytin	CO_2
Tetrahydrofolate	Other one-carbon groups

† Their structures and mode of action are described in Chapter 10.

Enzymes Are Classified on the Basis of the Reactions They Catalyze

Many enzymes have been named by adding the suffix *-ase* to the name of their substrate. Thus *urease* catalyzes hydrolysis of urea and *arginase* catalyzes hydrolysis of arginine. But many enzymes have been given names that do not denote their substrates, e.g., *pepsin* and *trypsin*. It has also happened that one and the same enzyme is known by two or more names or that two different enzymes have been given the same name. Because of these and other ambiguities, as well as the ever-increasing number of newly discovered enzymes, a systematic basis for naming and classifying enzymes has been adopted by international agreement. This system places all enzymes in six major classes, each with subclasses, based on the type of reaction catalyzed (Table 9-3). Each enzyme is assigned a four-digit classification number and a systematic name, which identifies the reaction catalyzed. An example is the naming of the enzyme catalyzing the reaction

$$\text{ATP} + \text{D-glucose} \longrightarrow \text{ADP} + \text{D-glucose 6-phosphate}$$

The formal systematic name of this enzyme is *ATP:glucose phosphotransferase*, which indicates that it catalyzes the transfer of a phosphate group from ATP to glucose. It is placed in class 2 of Table 9-3, and its classification number is 2.7.1.1, where the first digit (2) stands for the class name (transferase), the second digit (7) for the subclass (phosphotransferases), the third digit (1) for the sub-subclass (phosphotransferases with a hydroxyl group as acceptor), and the fourth digit (1) for D-glucose as the phosphate-group acceptor. When the systematic name of an enzyme is long or cumbersome, a trivial name may be used; in this case the trivial name is *hexokinase*.

Table 9-3 International Classification of Enzymes, Based on the Reactions They Catalyze

Most enzymes catalyze the transfer of electrons, atoms, or functional groups. They are therefore classified, given code numbers, and assigned names according to the type of transfer reaction, the group donor, and the group acceptor. There are six major classes.

No.	Class	Type of reaction catalyzed
1	Oxidoreductases	Transfer of electrons
2	Transferases	Group-transfer reactions
3	Hydrolases	Hydrolysis reactions (transfer of functional groups to water)
4	Lyases	Addition of groups to double bonds or the reverse
5	Isomerases	Transfer of groups within molecules to yield isomeric forms
6	Ligases	Formation of C—C, C—S, C—O, and C—N bonds by condensation reactions coupled to ATP cleavage

Enzymes Enhance the Rate of Chemical Reactions by Lowering Their Activation Energy

Enzymes are true catalysts. They greatly enhance the rate of specific chemical reactions that would otherwise occur only very slowly. They cannot change the equilibrium point of the reactions they promote, nor are they used up or permanently changed by these reactions.

How do catalysts, including enzymes, enhance the rate of chemical reactions? First we must recall that the energy content of individual molecules in a population at constant temperature varies greatly and may be represented by a bell-shaped curve. Some molecules are very rich in energy, some are very poor, but most have an energy content near a mean. A chemical reaction, such as A → P, takes place because a certain fraction of A molecules at any given instant possesses more internal energy than the rest of the population, sufficient to bring them to the top of the energy "hill" (Figure 9-3), to a reactive form called the *transition state*. The *activation energy* of a reaction is the amount of energy in calories required to bring all the molecules in 1 mol of a substance at a given temperature to the transition state at the top of the energy barrier. At this point there is equal probability for them to undergo reaction to form the products or to fall back into the pool of unreacted A molecules (Figure 9-3). The rate of any chemical reaction is proportional to the concentration of the transition-state species. Thus the rate of a chemical reaction will be very high if a large fraction of A molecules is in the energy-rich transition state but very low if only a small fraction of A is in the transition state.

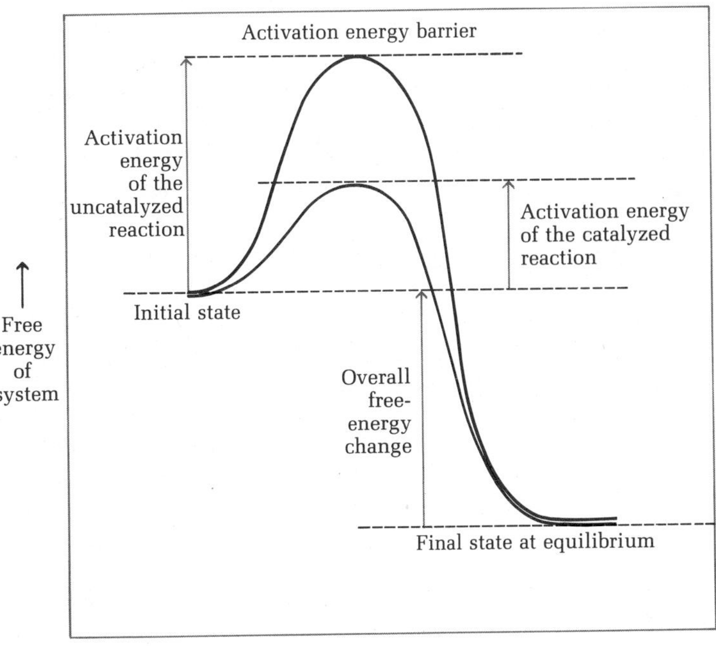

Figure 9-3
Catalysts lower the activation energy barrier of chemical reactions without altering the overall free energy change of the reaction or the final equilibrium position. At the peak of the activation energy barrier, there is a transition state.

There are two general ways in which the rate of a chemical reaction can be increased. One is to increase the temperature, which increases the thermal motion of the molecules and thus increases the fraction having sufficient internal energy to enter the transition state. Usually the rate of a chemical reaction is approximately doubled by a 10°C rise in temperature.

The second way to accelerate a chemical reaction is to add a catalyst. Catalysts accelerate chemical reactions by finding a lower "pass" over the energy barrier. The catalyst, designated C, combines transiently with the reactant A to produce a new complex or compound CA whose transition state has a much lower activation energy than the transition state of A in the uncatalyzed reaction (Figure 9-3). The catalyst-reactant complex CA then reacts to form the product P, releasing the free catalyst, which can then combine with another molecule of A and repeat the cycle. In this way catalysts lower the activation energy of chemical reactions, allowing a much larger fraction of the molecules in a given population to react per unit time than in the absence of the catalyst. Much evidence shows that enzymes, like other catalysts, also combine with their substrates during the course of their catalytic cycles.

The Substrate Concentration Has a Profound Effect on the Rate of Enzyme-Catalyzed Reactions

Figure 9-4
Effect of substrate concentration on the initial rate of an enzyme-catalyzed reaction. V_{max} must be approximated from such a plot, since the initial rate will approach but never quite reach V_{max}. The substrate concentration at which the rate is one-half maximal is K_M, the Michaelis-Menten constant.

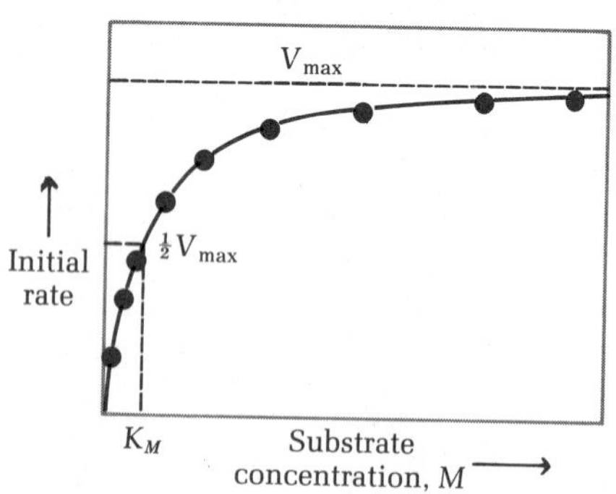

Let us examine the effect of varying the substrate concentration on the initial rate of an enzyme-catalyzed reaction when the enzyme concentration is held constant (Figure 9-4). At very low concentrations of substrate the rate of the reaction is very low, but it will increase with an increase in the substrate concentration. If we test the effect of higher and higher substrate concentrations, each time measuring the initial rate of the catalyzed reaction, we shall find that the rate increases by smaller and smaller amounts. Finally, a point will be reached beyond which there are only vanishingly small increases in the reaction rate with increasing substrate concentration (Figure 9-4). No matter how high the substrate concentration is raised beyond this point, the reaction rate will approach but never quite reach a plateau. At this plateau, called the *maximum rate* (V_{max}), the enzyme is "saturated" with its substrate and can function no faster.

This saturation effect is exhibited by nearly all enzymes. It led Victor Henri to the conclusion in 1903 that an enzyme combines with its substrate molecule to form an enzyme-substrate complex as a necessary step in enzyme catalysis. This idea was expanded into a general theory of enzyme action, particularly by Leonor Michaelis and Maud Menten in 1913. They postulated that the enzyme E first combines reversibly with its substrate S to form an enzyme-substrate complex ES in a relatively fast reversible reaction

$$E + S \rightleftharpoons ES$$

The ES complex then breaks down in a second reversible reaction, which is slower, to yield the reaction product P and the free enzyme E

$$ES \rightleftharpoons P + E$$

Since the second reaction is the rate-limiting step, the overall rate of the enzyme-catalyzed reaction must be proportional to the concentration of the enzyme-substrate complex ES. At any given instant in an enzyme-catalyzed reaction, the enzyme exists in two forms, the free or uncombined form E and the combined form ES. The rate of the catalyzed reaction will obviously be at a maximum when virtually all of the enzyme is present as the ES complex and the concentration of free enzyme E is vanishingly small. This condition will exist at a very high concentration of the substrate, since, by the law of mass action, the equilibrium of the first reaction will be pushed to the right when we increase the concentration of S

$$E + S \rightleftharpoons ES$$

If we increase S to high enough levels, essentially all the free enzyme E will have been converted into the ES form. In the second reaction of the catalytic cycle, the ES complex continuously and rapidly breaks down, to yield the product P and the free enzyme. But if the concentration of S is high enough, the free enzyme E will immediately combine with another molecule of S. Under these conditions a steady state is achieved in which the enzyme is always saturated with its substrate and the reaction rate is maximum.

There Is a Quantitative Relationship between the Substrate Concentration and the Rate of an Enzymatic Reaction

If we examine Figure 9-4, which shows the relationship between the substrate concentration and the rate of an enzymatic reaction, we shall see that it is difficult to say from the closer and closer approach of the reaction rate to the maximum velocity V_{max} exactly what substrate concentration is required to attain V_{max}. However, because the curve expressing this relationship has the same general shape for most enzymes (it is a rectangular hyperbola), Michaelis and Menten defined a constant, today designated K_M, that is useful in establishing the precise relationship between the substrate concentration and the velocity of the enzyme-catalyzed reaction. K_M, the *Michaelis-Menten constant,* can be most simply defined as the *concentration of the specific substrate at which a given enzyme yields one-half its maximum velocity* (shown in Figure 9-4).

The characteristic shape of the substrate-saturation curve for an enzyme (Figure 9-4) can be expressed mathematically by the

Box 9-1 The Michaelis-Menten equation

Many enzymes show the typical hyperbolic curve (Figure 9-4) relating the reaction velocity to the substrate concentration, with a gradual approach to saturation of the enzyme with substrate. We have seen that there are two cardinal points in such plots: (1) K_M, the substrate concentration giving half-maximal velocity, and (2) V_{max}, or the maximum velocity toward which the rate approaches at infinitely high substrate concentration. Michaelis and Menten showed that much additional useful information can be derived from the hyperbolic saturation curves of enzymes when they are translated into a simple mathematical form. The *Michaelis-Menten equation* is an algebraic expression of the hyperbolic shape of such curves, in which the important terms are *substrate concentration* ([S]), *initial velocity* (v_0), V_{max}, and K_M. This equation is fundamental to all studies of enzyme kinetics because it makes possible the quantitative calculation of enzyme characteristics and the analysis of enzyme inhibition.

Here we shall develop the basic logic and the algebraic steps in a modern derivation of the Michaelis-Menten equation. The derivation first starts with the two basic reactions involved in the formation and breakdown of the enzyme-substrate complex

$$E + S \underset{k_{-1}}{\overset{k_1}{\rightleftharpoons}} ES \qquad (a)$$

$$ES \underset{k_{-2}}{\overset{k_2}{\rightleftharpoons}} E + P \qquad (b)$$

Then $[E_t]$ represents the total enzyme concentration (the sum of the free and combined enzyme), [ES] is the concentration of the enzyme-substrate complex, and $[E_t] - [ES]$ represents the concentration of free or uncombined enzyme. [S], the substrate concentration, is ordinarily far greater than $[E_t]$, so that the amount of S bound by E at any given time is negligible compared with the total concentration of S. The derivation begins by considering the rates of formation and breakdown of ES.

1. *Rate of formation of ES*. The rate of formation of ES in reaction (*a*) is

$$\text{Rate of formation} = k_1([E_t] - [ES])[S] \qquad (c)$$

where k_1 is the rate constant of reaction (*a*). The rate of formation of ES from E + P by reversal of reaction (*b*) is very small and may thus be neglected.

2. *Rate of breakdown of ES*. The rate of breakdown of ES is

$$\text{Rate of breakdown} = k_{-1}[ES] + k_2[ES]$$

in which k_{-1} and k_2 are the rate constants for the reverse of reaction (*a*) and the forward direction of reaction (*b*), respectively.

3. *The steady state*. When the rate of formation of ES is equal to its rate of breakdown, the ES concentration will be constant and the reaction system is in a steady state:

$$\text{Rate of formation of ES} = \text{rate of breakdown of ES}$$
$$k_1([E_t] - [ES])[S] = k_{-1}[ES] + k_2[ES] \qquad (d)$$

4. *Separation of the rate constants*. The left side of equation (*d*) is multiplied out to give

$$k_1[E_t][S] - k_1[ES][S]$$

and the right side is simplified to give $(k_{-1} + k_2)[ES]$. We then have

$$k_1[E_t][S] - k_1[ES][S] = (k_{-1} + k_2)[ES]$$

Michaelis-Menten equation

$$v_0 = \frac{V_{max}[S]}{K_M + [S]} \qquad (1)$$

where v_0 = initial rate at substrate concentration [S]
V_{max} = maximum rate
K_M = Michaelis-Menten constant of enzyme for particular substrate

This equation was derived by Michaelis and Menten starting from the basic hypothesis that the rate-limiting step in enzymatic reactions is the breakdown of the ES complex to form the product and the free enzyme. Box 9-1 gives a modern derivation of the Michaelis-Menten equation.

When we transpose and change the sign of the term $-k_1[ES][S]$, we get

$$k_1[E_t][S] = k_1[ES][S] + (k_{-1} + k_2)[ES]$$

This obviously can be simplified further to

$$k_1[E_t][S] = (k_1[S] + k_{-1} + k_2)[ES]$$

Now we can solve this equation for [ES]

$$[ES] = \frac{k_1[E_t][S]}{k_1[S] + k_{-1} + k_2}$$

This can now be simplified further, in such a way as to combine the rate constants into one expression

$$[ES] = \frac{[E_t][S]}{[S] + (k_2 + k_{-1})/k_1} \qquad (e)$$

5. *Definition of initial velocity v_0 in terms of [ES].* The initial velocity, according to the Michaelis and Menten theory, is determined by the rate of breakdown of [ES] in reaction (*b*), whose rate constant is k_2. Thus we have

$$v_0 = k_2[ES]$$

But since [ES] is the entire right side of equation (*e*), we get

$$v_0 = \frac{k_2[E_t][S]}{[S] + (k_2 + k_{-1})/k_1} \qquad (f)$$

Now let us simplify further by defining K_M (the Michaelis-Menten constant) as $(k_2 + k_{-1})/k_1$ and by defining V_{max} as $k_2[E_t]$, that is, the rate when all the available E is present as ES. We substitute these terms in equation (*f*) and get

$$\boxed{v_0 = \frac{V_{max}[S]}{[S] + K_M}}$$

This is the Michaelis-Menten equation, the *rate equation* for a one-substrate enzyme-catalyzed reaction. It is a statement of the quantitative relationship between the initial velocity v_0, the maximum velocity V_{max}, and the initial substrate concentration, all related through the Michaelis-Menten constant K_M.

An important numerical relationship emerges from the Michaelis-Menten equation in the special case when the initial reaction rate is exactly one-half the maximum velocity, i.e., when $v_0 = \frac{1}{2}V_{max}$ (Figure 9-4). Then

$$\frac{V_{max}}{2} = \frac{V_{max}[S]}{K_M + [S]}$$

If we divide by V_{max}, we obtain

$$\frac{1}{2} = \frac{[S]}{K_M + [S]}$$

Solving for K_M, we get

$$K_M + [S] = 2[S]$$

$$K_M = [S] \qquad \text{when } v_0 \text{ is exactly } \tfrac{1}{2}V_{max}$$

The Michaelis-Menten equation can be algebraically transformed into equivalent equations that are useful in the practical determination of K_M and V_{max} and in the analysis of inhibitor action (Box 9-2).

The Michaelis-Menten equation is basic to all aspects of the kinetics of enzyme action. If we know K_M and V_{max}, we can calculate the reaction rate of an enzyme at any given concentration of its substrate. Most enzymatic reactions, including those with two or more participating substrates (see below), can be quantitatively analyzed by the Michaelis-Menten theory. This fact has constituted strong evidence that enzymes catalyze reactions by combining transiently with their substrates, thus lowering the activation energy of the overall reaction. The formation of enzyme-substrate complexes can often be detected directly by physicochemical methods, e.g., by characteristic changes in the absorption spectrum of the enzyme when its substrate is added.

Each Enzyme Has a Characteristic K_M for a Given Substrate

The key element in the Michaelis-Menten equation is K_M, which is characteristic for any given enzyme for a specific substrate under defined conditions of pH and temperature. An *approximate* value of K_M can be obtained by a simple graphical procedure, as shown in Figure 9-4. However, it is difficult to determine V_{max} accurately from the type of plot in Figure 9-4 because V_{max} is only approached and never actually attained. A more accurate value of K_M can be obtained by plotting the same data in a different manner, called a *double-reciprocal plot,* using an algebraic transformation of the Michaelis-Menten equation described in Box 9-2.

Table 9-4 K_M for Some Enzymes

Enzyme	Substrate	K_M, mM
Catalase	H_2O_2	25
Hexokinase (brain)	ATP	0.4
	D-Glucose	0.05
	D-Fructose	1.5
Carbonic anhydrase	HCO_3^-	9
Chymotrypsin	Glycyltyrosinylglycine	108
	N-Benzoyltyrosinamide	2.5
β-Galactosidase	D-Lactose	4.0
Threonine dehydratase	L-Threonine	5.0

Table 9-4 shows the K_M values for a number of enzymes. We note that some enzymes, such as carbonic anhydrase and catalase, require a relatively high concentration of the substrate to attain their half-maximal rates of catalysis. Other enzymes, such as brain hexokinase, which catalyzes the transfer of a phosphate group from ATP to glucose, show a half-maximal rate at a very low substrate concentration. Enzymes that have two or more substrates, such as hexokinase or *aspartate aminotransferase,* which catalyzes the reversible reaction

$$\text{Aspartate} + \alpha\text{-ketoglutarate} \rightleftharpoons \text{oxaloacetate} + \text{glutamate}$$

may have a different K_M for each substrate. When an enzyme can act on many different substrates sharing a structural feature, such as *chymotrypsin,* it may have widely different K_M values for different substrates (Table 9-4).

Under intracellular conditions enzymes are not usually saturated with their substrates and thus are not usually functioning at their highest possible rates. Regulation of the reaction rate of enzymes in cells may in part be brought about by changes in the intracellular concentrations of their substrates.

Box 9-2 Transformations of the Michaelis-Menten equation: The double-reciprocal plot

The Michaelis-Menten equation

$$v_0 = \frac{V_{max}[S]}{[S] + K_M} \qquad (a)$$

can be algebraically transformed into other forms that are more useful in plotting experimental data. One common transformation is derived simply by taking the reciprocal of both sides of the Michaelis-Menten equation (*a*) to give

$$\frac{1}{v_0} = \frac{K_M + [S]}{V_{max}[S]}$$

Separating the components of the numerator on the right side of the equation gives

$$\frac{1}{v_0} = \frac{K_M}{V_{max}[S]} + \frac{[S]}{V_{max}[S]}$$

which simplifies to

$$\frac{1}{v_0} = \frac{K_M}{V_{max}} \frac{1}{[S]} + \frac{1}{V_{max}} \qquad (b)$$

Equation (*b*) is a transformation of the Michaelis-Menten equation called the *Lineweaver-Burk equation.* For enzymes obeying the Michaelis-Menten relationship exactly a plot of $1/v_0$ against $1/[S]$ yields a straight line (Figure 1). This line will have a slope of K_M/V_{max}, an intercept of $1/V_{max}$ on the $1/v_0$ axis, and an intercept of $-1/K_M$ on the $1/[S]$ axis. The *double-reciprocal* or *Lineweaver-Burk plot* has the great advantage of allowing a more accurate determination of V_{max}, which can only be approximated from a simple plot of v_0 vs. [S], as shown in Figure 2.

Other transformations of the Michaelis-Menten equation have been derived and used. Each has some particular advantage in analyzing enzyme kinetic data.

The double-reciprocal plot of enzyme rate data is very useful in analyzing enzyme inhibition, as we shall see.

Figure 1

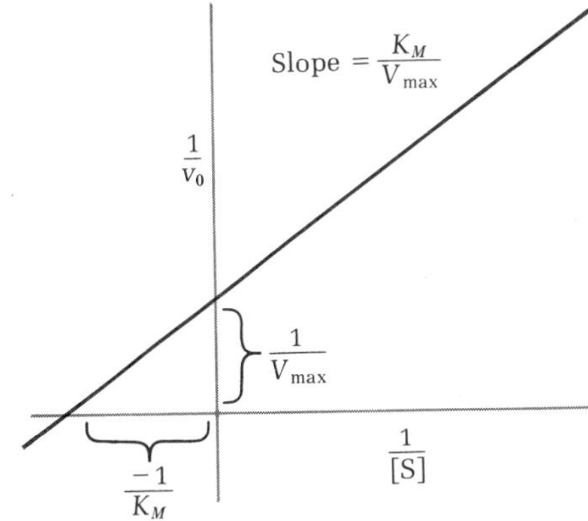

Figure 2

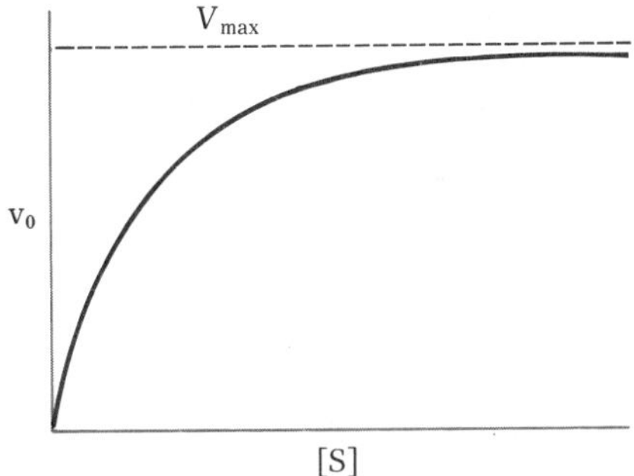

Many Enzymes Catalyze Reactions in Which There Are Two Substrates

We have seen how the substrate concentration affects the rate of a simple enzyme reaction in which there is only one substrate molecule, in reactions having the form A → P. In many enzymatic reactions of metabolism, however, there are actually two (and sometimes even three) different substrate molecules that bind to the enzyme and participate in the reaction. For example, in the reaction catalyzed by hexokinase, ATP and glucose are the substrate molecules, and ADP and glucose 6-phosphate are the products:

$$\text{ATP + glucose} \longrightarrow \text{ADP + glucose 6-phosphate}$$

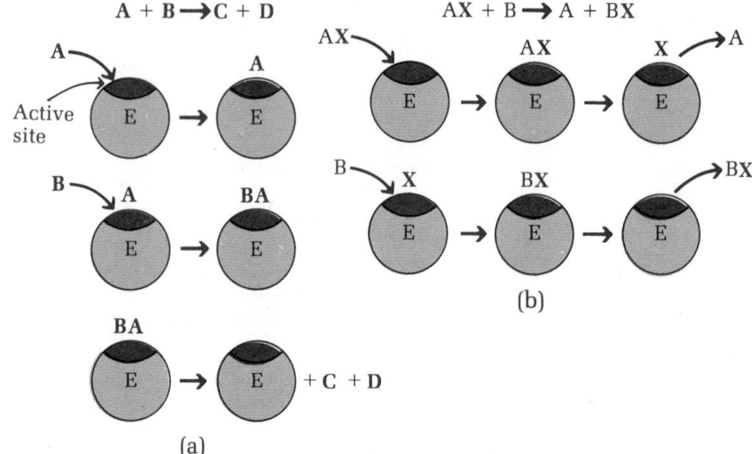

Figure 9-5
Schematic representations of two classes of bisubstrate reactions.

(a) Single-displacement reaction. In some single-displacement reactions, substrates A and B may combine with the enzyme in either sequence. In others there is a specific order in which A and B bind.

(b) Double-displacement (ping-pong) reaction. The first substrate AX donates group X to the enzyme. A must then leave before substrate B, the acceptor of X, is bound.

The rates of such bisubstrate reactions can also be analyzed by the Michaelis-Menten approach. As we have seen, hexokinase has a characteristic K_M for each of its two substrates (Table 9-4).

Enzymatic reactions in which there are two substrates usually involve transfer of an atom or a functional group from one of the substrates to the other. Such reactions proceed by two different types of pathways (Figure 9-5). In the first class, called single-displacement reactions, the two substrates A and B are bound to the enzyme E in either a specific or a random order to form an EAB complex, which then reacts to form the products C + D. In the other class of bisubstrate reactions, called double-displacement or ping-pong reactions, only one of the two substrates is bound to the catalytic site at any given time. Binding of the first substrate is followed by transfer of its functional group to the enzyme molecule. Only after the product of the first substrate leaves can the second substrate bind to the enzyme and accept the functional group.

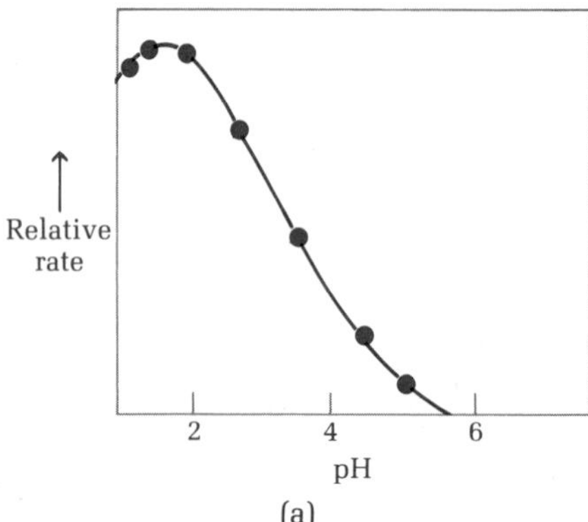

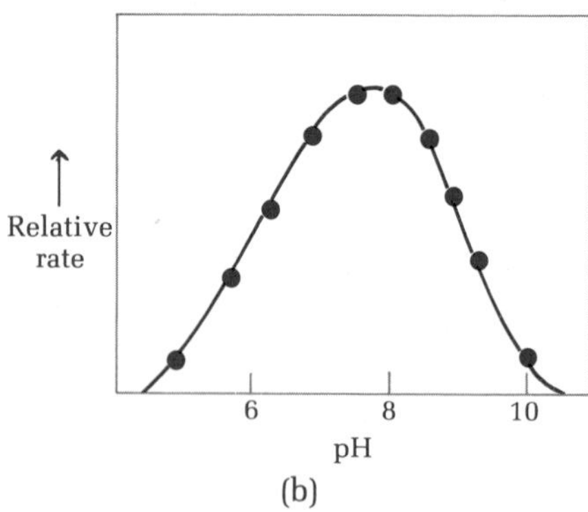

Figure 9-6
pH-activity profiles of two enzymes. Such curves are constructed from measurements of the initial rates when the reaction is carried out in buffers of different pH values.

(a) The pH-activity profile of pepsin, which hydrolyzes certain peptide bonds of proteins during digestion in the stomach. The pH of gastric juice is between 1 and 2.

(b) The pH-activity profile of glucose 6-phosphatase of liver cells, responsible for releasing glucose into the blood. The normal pH of the cytosol of liver cells is about 7.2.

Enzymes Have an Optimum pH

Enzymes have a characteristic optimum pH at which their activity is maximal (Figure 9-6 and Table 9-5). The pH-activity profiles of enzymes reflect the pH at which important proton-donating or proton-accepting groups in the enzyme catalytic site are in their required state of ionization. The optimum pH of an enzyme is not necessarily identical with the pH of its normal surroundings, which may be just above or below the optimum pH. The catalytic activity of enzymes in cells may thus be regulated in part by changes in the pH of the surrounding medium.

Enzymes Can Be Assayed Quantitatively

The amount of an enzyme in a given solution or tissue extract can be assayed quantitatively in terms of the catalytic effect it

Table 9-5 Optimum pH of Some Enzymes

Enzyme	Optimum pH
Pepsin	1.5
Trypsin	7.7
Catalase	7.6
Arginase	9.7
Fumarase	7.8
Ribonuclease	7.8

produces. For this purpose it is necessary to know (1) the overall equation of the reaction catalyzed, (2) an analytical procedure for determining the disappearance of the substrate or the appearance of the reaction products, (3) whether the enzyme requires cofactors such as metal ions or coenzymes, (4) the dependence of the enzyme activity on substrate concentration, i.e., the K_M for the substrate, (5) the optimum pH, and (6) a temperature zone in which the enzyme is stable and has high activity. Ordinarily enzymes are assayed at their optimum pH, at some convenient temperature, usually within the range 25 to 38°C, and with a near-saturating concentration of substrate. Under these conditions the initial reaction rate is usually proportional to enzyme concentration, at least over a given range of enzyme concentration (Figure 9-7).

By international agreement, *1.0 unit of enzyme activity is defined as that amount causing transformation of 1.0 micromole (μmol = 10^{-6} mol) of substrate per minute at 25°C under optimal conditions of measurement*. The *specific activity* is the number of enzyme units per milligram of protein. The specific activity is a measure of enzyme purity: it increases during purification of an enzyme and becomes maximal and constant when the enzyme is in the pure state.

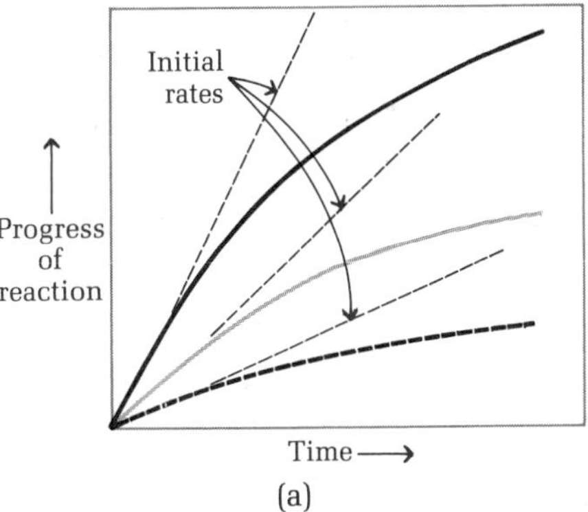

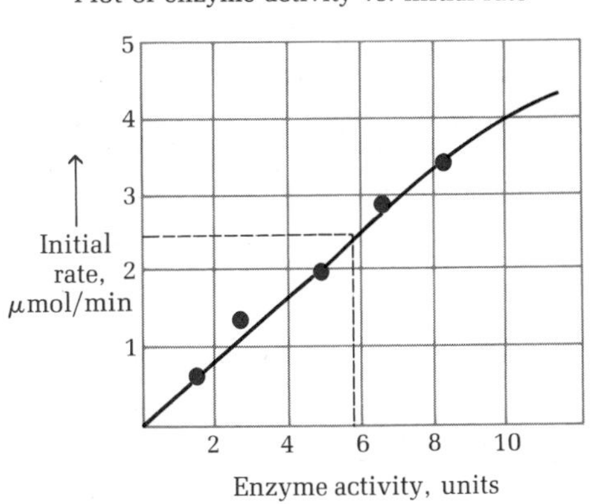

Figure 9-7
Quantitative assay of enzyme activity. There are three steps: (1) Determination of K_M (Figure 9-4 and Box 9-2), (2) measurement of the initial rates at different concentrations of the enzyme, as in (a), with substrate concentration near saturation, for example, at 10 × K_M, and (3) plotting the initial rates against enzyme concentration, as in (b). The linear portion of the plot can be used to determine the number of enzyme units in an unknown sample yielding a given initial rate. In this example the unknown, which gave an initial rate of 2.5 μmol/min, contains 5.8 units of enzyme activity.

Table 9-6 Turnover Numbers of Some Enzymes (Substrate Molecules Transformed per Minute at 20–38°C)

Carbonic anhydrase	36,000,000
β-Amylase	1,100,000
β-Galactosidase	12,500
Phosphoglucomutase	1,240

The *turnover number* of an enzyme is the number of substrate molecules transformed per unit time by a single enzyme molecule (or by a single catalytic site) when the enzyme concentration alone is the rate-limiting factor (Table 9-6). The enzyme *carbonic anhydrase*, an important enzyme found in high concentrations in the red blood cell, is among the most active of all enzymes, with a turnover number of 36,000,000 per minute per enzyme molecule. It catalyzes the reversible hydration of dissolved carbon dioxide to form carbonic acid, an otherwise slow reaction:

$$CO_2 + H_2O \rightleftharpoons H_2CO_3$$

The hydration of CO_2 in red blood cells is an important step in the transport of CO_2 from the tissues, where it is formed, to the lungs, where it is released and exhaled (page 85; Chapter 24).

Enzymes Show Specificity toward Their Substrates

Some enzymes have nearly absolute specificity for a given substrate and will not attack even very closely related molecules. A good example is the enzyme *aspartase*, found in many plants and bacteria. It catalyzes the reversible addition of ammonia to the double bond of fumaric acid to form L-aspartate (Figure 9-8). However, aspartase does not promote addition of ammonia to any other unsaturated acid. Aspartase also has rigid optical and geometrical specificity: it will not act on D-aspartate, nor will it add ammonia to maleate, the cis geometrical isomer of fumarate.

At the other extreme are enzymes which have relatively broad specificity and act on many compounds with a common structural feature. For example, chymotrypsin catalyzes hydrolysis of many different peptides or polypeptides but cleaves only those peptide bonds in which the carbonyl group is contributed by phenylalanine, tyrosine, or tryptophan (page 131). A somewhat different example is intestinal phosphatase, which catalyzes hydrolysis of many different esters of phos-

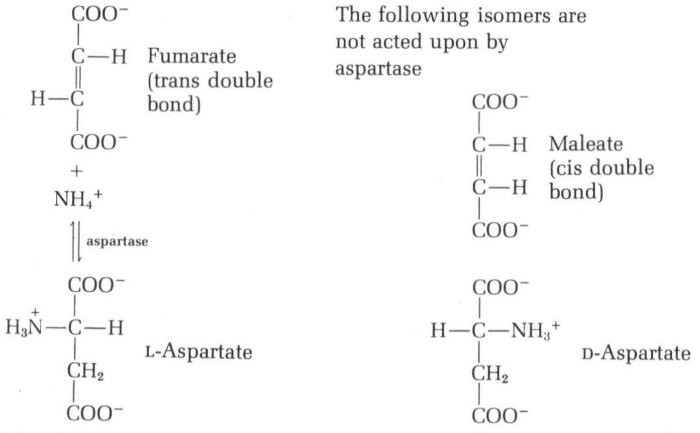

COO^-
|
C—H
‖
H—C
|
COO^-

Fumarate (trans double bond)

+

NH_4^+

⇅ aspartase

COO^-
|
$H_3\overset{+}{N}$—C—H
|
CH_2
|
COO^-

L-Aspartate

The following isomers are not acted upon by aspartase

COO^-
|
C—H
‖
C—H
|
COO^-

Maleate (cis double bond)

COO^-
|
H—C—NH_3^+
|
CH_2
|
COO^-

D-Aspartate

Figure 9-8
The aspartase reaction and its substrate specificity. Aspartase is absolutely specific for fumarate in the forward direction and for L-aspartate in the reverse direction. It will not attack maleate, the cis isomer of fumarate, or D-aspartate.

Figure 9-9
Substrate specificity of chymotrypsin. Although chymotrypsin functions biologically as a peptidase, it also can hydrolyze amides and esters, as well as some synthetic nonbiological compounds having the required susceptible bond and a hydrophobic positioning group.

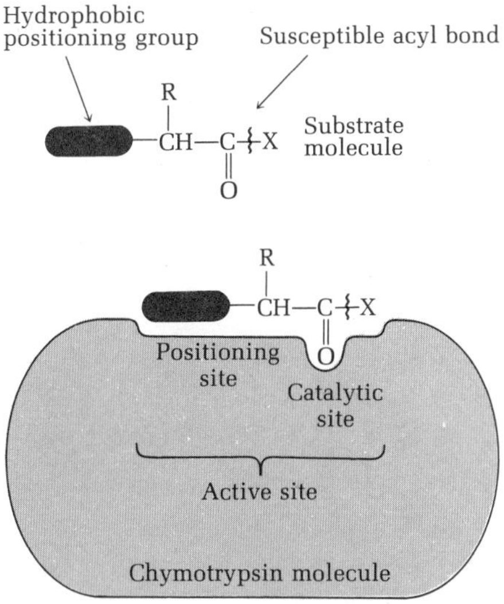

Some synthetic compounds hydrolyzed by chymotrypsin. Each has a hydrophobic positioning group and a susceptible acyl bond, both in color.

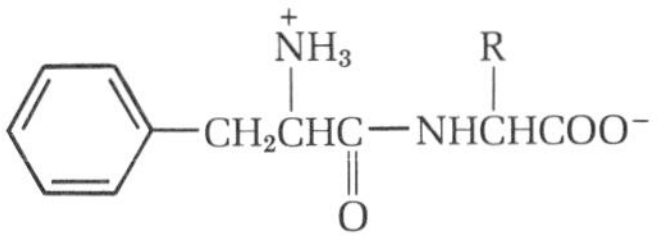

phoric acid but at rates that vary considerably. Studies of the substrate specificity of enzymes have led to the concept of a complementary "lock-and-key" relationship between the substrate molecule and a specific area on the surface of the enzyme molecule, the active site or catalytic site, to which the substrate molecule is bound as it undergoes the catalytic reaction.

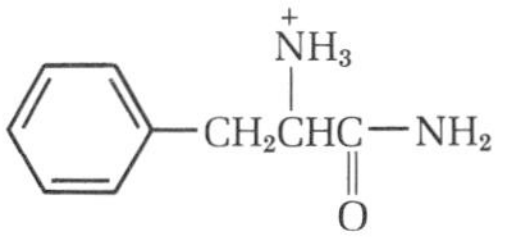

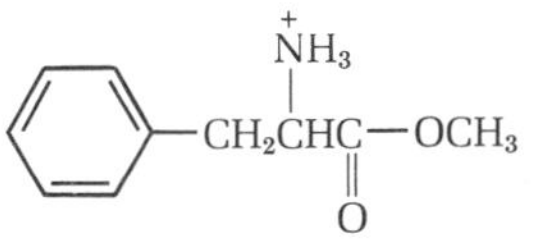

Work on the specificity of enzymes shows that a substrate molecule must have two distinct structural features: (1) the specific chemical bond or linkage that can be attacked by the enzyme and (2) usually some other functional group, a binding group, which binds to the enzyme and positions the substrate molecule properly on the active site so that the susceptible bond is precisely located in relation to the catalytic group of the enzyme. Figure 9-9 shows the substrate specificity of chymotrypsin, which normally hydrolyzes those peptide bonds in proteins and simple peptides in which the carbonyl group is contributed by amino acids having an aromatic ring, i.e., by tyrosine, tryptophan, and phenylalanine residues. However, tests of dozens of different possible synthetic substrate molecules have shown that chymotrypsin can also cleave simple amides as well as ester linkages. Moreover, the aromatic R groups of tyrosine, tryptophan, and phenylalanine, for which chymotrypsin is specific in polypeptides, appear to serve only as hydrophobic binding groups. The evidence for this is that chymotrypsin will also accept as substrates synthetic peptides in which large hydrophobic alkyl groups replace the aromatic rings of the natural amino acids.

Such substrate-specificity studies, together with studies of enzyme inhibition, permit us to map the active sites of enzymes.

Enzymes Can Be Inhibited by Specific Chemical Agents

Most enzymes can be poisoned or inhibited by certain chemical reagents. From the study of enzyme inhibitors valuable information has been obtained about the substrate specificity of enzymes, the nature of the functional groups at the active site, and the mechanism of the catalytic activity. Enzyme inhibitors also are very useful in elucidating metabolic pathways in cells. Moreover, some drugs useful in medicine appear to function because they can inhibit certain enzymes in malfunctioning cells.

There are two major types of enzyme inhibitors: irreversible and reversible. Irreversible inhibitors are those which combine with or destroy a functional group on the enzyme molecule that is necessary for its catalytic activity. An example of an irreversible inhibitor is the compound diisopropylfluorophosphate (DFP), which inhibits the enzyme acetylcholinesterase, important in the transmission of nerve impulses. Acetylcholinesterase catalyzes the hydrolysis of acetylcholine (Figure 9-10), a neurotransmitter substance functioning in certain portions of the nervous system. Acetylcholine is released by a stimulated

Figure 9-10
Irreversible inhibition of acetylcholinesterase by diisopropylfluorophosphate. (a) *The reaction catalyzed by acetylcholinesterase.* (b) *Reaction of diisopropylfluorophosphate with the essential serine hydroxyl group.*

$$CH_3C(=O)-O-CH_2CH_2-N^+(CH_3)_2-CH_3 \quad \text{Acetylcholine}$$

$$+$$

$$H_2O$$

$$\downarrow \text{acetylcholinesterase}$$

$$CH_3COO^- \quad \text{Acetate}$$

$$+$$

$$HOCH_2CH_2-N^+(CH_3)_2-CH_3 \quad \text{Choline}$$

(a)

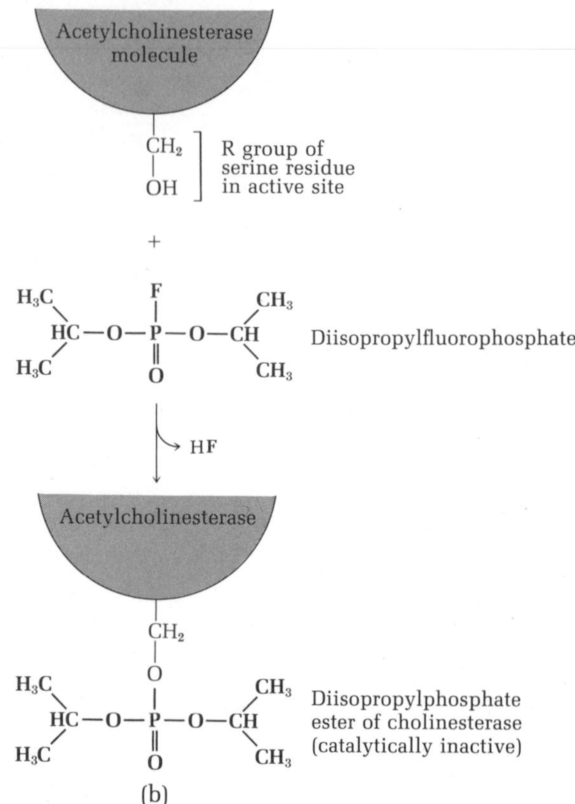

nerve cell into the <u>synapse</u>, or junction, with another nerve cell. Once acetylcholine has been secreted into the synapse, it binds to receptor sites on the next nerve cell, causing the latter to propagate the nerve impulse. Before a second impulse can be transmitted through the synapse, however, the acetylcholine secreted after the first impulse must be hydrolyzed by the acetylcholinesterase in the junction. The products of its action, acetate and choline (Figure 9-10), have no transmitter activity. The irreversible inhibitor DFP is very reactive and combines with the hydroxyl group of an essential serine residue at the active site of acetylcholinesterase to form a catalytically inactive derivative. Once this derivative has been formed, the enzyme molecule can no longer function. Animals treated with DFP, one of the first nerve gases to be discovered, become paralyzed in certain functions because of the failure of nerve impulses to be transmitted properly. But there is a useful side to DFP. It has led to the development of *malathion* and other insecticides that are relatively nontoxic for people and animals. Malathion is inactive by itself and is degraded by higher animals into products that are believed to be harmless to them, but it is converted by enzymes of insects into an active inhibitor of their acetylcholinesterase.

DFP has been found to inhibit a whole class of enzymes, many of them capable of catalyzing hydrolysis of peptide or ester linkages. They include not only *acetylcholinesterase* but also *trypsin*, *chymotrypsin*, *elastase*, *phosphoglucomutase*,

Figure 9-11
Irreversible inhibition of an —SH enzyme by iodoacetamide.

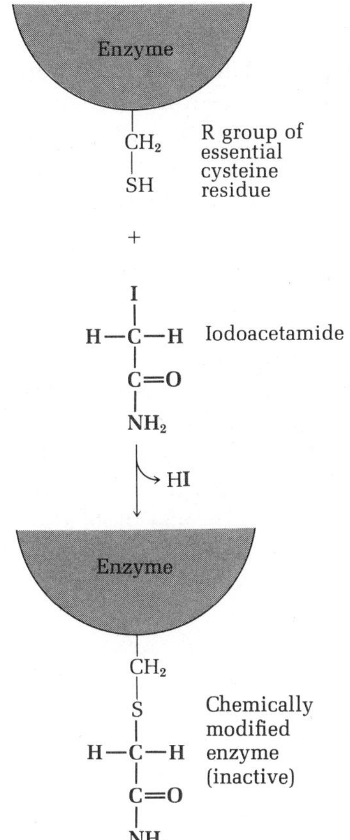

and *cocoonase*, an enzyme secreted by the silkworm larva to hydrolyze the silk fibers of the cocoon, allowing the larva to escape. All the DFP-inhibited enzymes have an essential serine residue in their active site, which participates in their catalytic activity (Figure 9-10).

Another irreversible inhibitor of some enzymes is *iodoacetamide* (Figure 9-11), which can react with sulfhydryl (—SH) groups of essential cysteine residues or with the imidazole group of essential histidine residues. With the help of such inhibitors the hydroxyl group of serine, the thiol group of cysteine, and the imidazole group of histidine have been identified as participating in the catalytic activity of different classes of enzymes.

There Are Two Kinds of Reversible Inhibitors: Competitive and Noncompetitive

Reversible inhibitors of enzymes have also given much important information on the structure of the active sites of different enzymes. *A competitive inhibitor competes with the substrate for binding to the active site but, once bound, cannot be transformed by the enzyme.* It is the hallmark of competitive inhibition that it can be reversed or relieved simply by increasing the substrate concentration. For example, if an enzyme is 50 percent inhibited at given concentrations of the substrate and competitive inhibitor, we can diminish the percentage inhibition by raising the substrate concentration.

Competitive inhibitors usually resemble the normal substrate in three-dimensional structure. Because of this resemblance the competitive inhibitor "tricks" the enzyme into binding it. In fact, competitive inhibition can be quantitatively analyzed by the Michaelis-Menten theory. The competitive inhibitor I simply combines reversibly with the enzyme to form an EI complex

$$E + I \rightleftharpoons EI$$

However, the inhibitor I cannot be attacked by the enzyme to form new reaction products.

The classical example is the competitive inhibition of succinate dehydrogenase by the malonate anion (Figure 9-12). Succinate dehydrogenase is a member of the group of enzymes catalyzing the citric acid cycle, the final metabolic pathway for the oxidative degradation of carbohydrates and fats in the mitochondria. This enzyme catalyzes the removal of two hydrogen atoms from succinate, one from each of the two methylene (—CH_2—) groups. Succinate dehydrogenase is inhibited by malonate, which resembles succinate in having two ionized carboxyl groups at pH 7.0 but differs in having only three carbon atoms. However, malonate is not dehydrogenated by succinate dehydrogenase; it simply occupies the active site, keeping it from acting on its normal substrate. The reversibility

Figure 9-12
The succinate dehydrogenase reaction and its competitive inhibition. Note that the competitive inhibitors resemble succinate in having two appropriately spaced negatively charged groups, which can fit the active site.

COO^-—CH_2—CH_2—COO^- Succinate

$\xrightarrow{\text{succinate dehydrogenase}}$ 2H

COO^-—CH=HC—COO^- Fumarate

Active site of succinate dehydrogenase

Succinate, Active site, Enzyme molecule

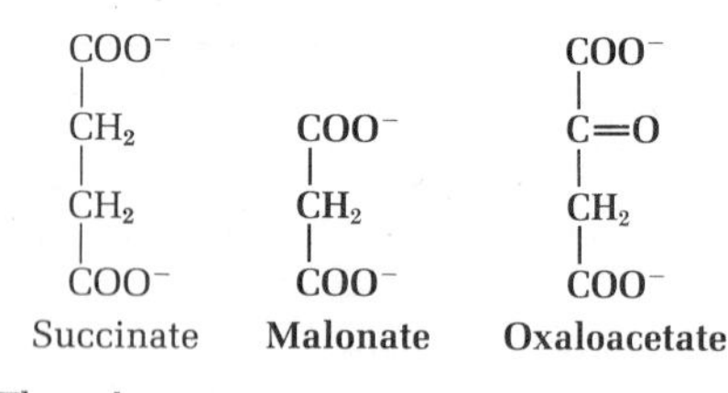

of the inhibition by malonate is shown by the fact that increasing the succinate concentration will reduce the extent of inhibition by a given concentration of malonate.

Other compounds with the proper distance between two anionic groups may act as competitive inhibitors of succinate dehydrogenase; among them is *oxaloacetate,* a normal intermediate in the citric acid cycle (Figure 9-12). From these structural relationships it has been concluded that the catalytic site of succinate dehydrogenase has two appropriately spaced positively charged groups capable of attracting the two negatively charged carboxylate groups of the succinate anion. The catalytic site of succinate dehydrogenase thus shows complementarity to the structure of its substrate (Figure 9-12).

Competitive inhibition is most easily recognized experimentally by determining the effect of the inhibitor concentration on the relationship between substrate concentration and initial rate. The double-reciprocal transformation of the Michaelis-Menten equation (see Box 9-2) is very useful in determining whether reversible enzyme inhibition is competitive or noncompetitive (Box 9-3). Double-reciprocal plots also yield the dissociation constant K_I of the enzyme-inhibitor complex. For the dissociation reaction

$$\text{EI} \rightleftharpoons \text{E} + \text{I}$$

the dissociation constant is

$$K_I = \frac{[\text{E}][\text{I}]}{[\text{EI}]}$$

Noncompetitive Inhibition Is Also Reversible but Not by the Substrate

In noncompetitive inhibition the inhibitor binds at a site on the enzyme other than the substrate binding site, altering the conformation of the enzyme molecule so that reversible inactivation of the catalytic site results. Noncompetitive inhibitors bind reversibly to both the free enzyme and the ES complex to form the inactive complexes EI and ESI:

$$\text{E} + \text{I} \rightleftharpoons \text{EI}$$

$$\text{ES} + \text{I} \rightleftharpoons \text{ESI}$$

Noncompetitive inhibition of enzymes is differentiated from competitive inhibition by double-reciprocal plots of rate data, as shown in Box 9-3.

The most important noncompetitive inhibitors are naturally occurring metabolic intermediates that can combine *reversibly* with specific sites on certain regulatory enzymes, thus changing the activity of their catalytic sites. An example is the inhibition of L-threonine dehydratase by L-isoleucine, discussed on page 234.

Box 9-3 Kinetic tests for distinguishing between competitive and noncompetitive inhibition

The double-reciprocal plot of enzyme rate data offers an easy way of determining whether an enzyme inhibitor is competitive or noncompetitive. Two sets of rate experiments are carried out; the enzyme concentration is held constant in both sets. In one set the substrate concentration is held constant and the effect of increasing the inhibitor concentration on the initial rate v_0 is determined by appropriate measurements. In the other set the inhibitor concentration is held constant and the substrate concentration is varied. The reciprocals $1/v_0$ of the initial rates v_0 are plotted against the reciprocal $1/[S]$ of the enzyme concentration [S].

Figure 1 shows a set of double-reciprocal plots obtained in the absence of the inhibitor and with two different concentrations of a competitive inhibitor. Competitive inhibitors yield a family of lines with a common intercept on the $1/v_0$ axis but with different slopes. Since the intercept on the $1/v_0$ axis is equal to $1/V_{max}$, we can see that V_{max} is unchanged by the presence of a competitive inhibitor. That is, regardless of the concentration of a competitive inhibitor, there is always some high substrate concentration that will displace the competitive inhibitor from the active site.

In noncompetitive inhibition similar plots of the rate data give the family of lines shown in Figure 2, having a common intercept on the $1/[S]$ axis, indicating that K_M for the substrate is not altered by a noncompetitive inhibitor but V_{max} decreases.

Figure 1

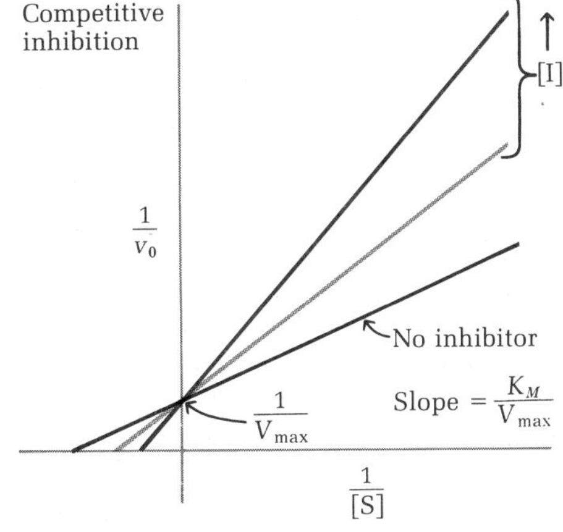

Figure 2

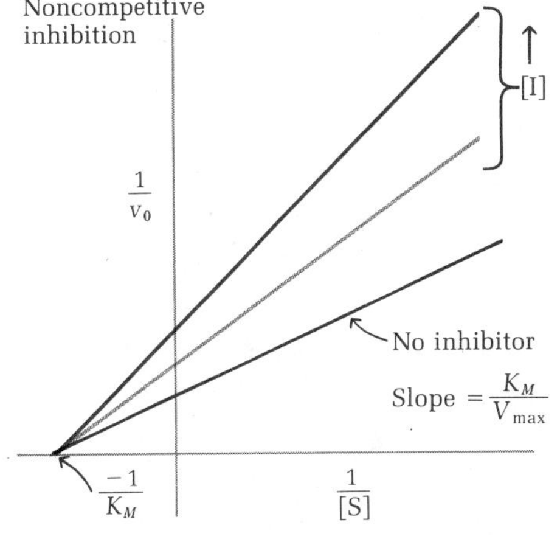

Several Factors Contribute to the Catalytic Efficiency of Enzymes

Enzymes enhance the rates of the reactions they catalyze by a factor of between 10^8 and 10^{20}. As an example, at pH 8 and 20°C urease accelerates the rate of hydrolysis of urea 10^{14} times. How do enzymes make such enormous catalytic power possible under such mild conditions of temperature and pH?

There appear to be four major factors (Table 9-7) by which enzymes accelerate the rates of chemical reactions.

Table 9-7 Factors That Contribute to the Catalytic Efficiency of Enzymes

1. Proximity and orientation of the substrate in relation to the catalytic group
2. Strain and distortion of the susceptible bond by induced fit of the enzyme
3. General acid-base catalysis
4. Covalent catalysis

Proximity and Orientation. The enzyme may bind the substrate molecule in such a way that the susceptible bond is not only in close proximity to the catalytic group but also is precisely

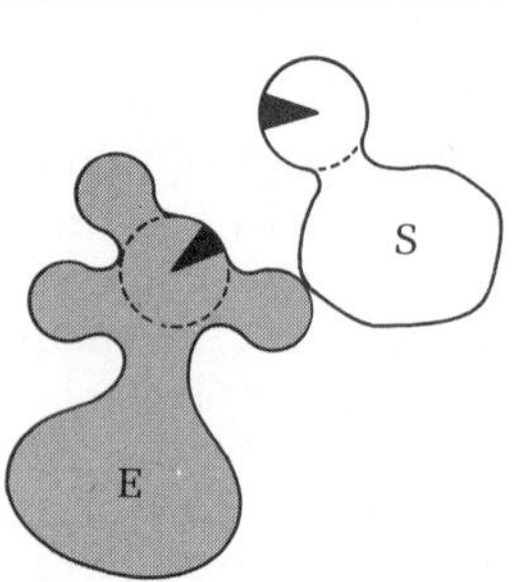

Unfavorable orientation, unfavorable proximity

Favorable proximity, unfavorable orientation

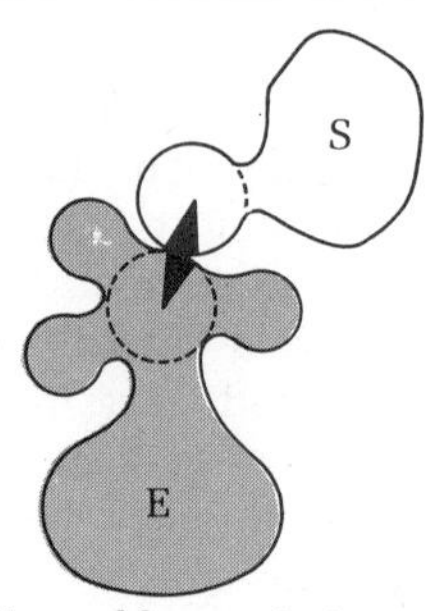

Favorable proximity, favorable orientation

Figure 9-13
Schematic representation of the factors of proximity and orientation in the interaction of the substrate molecule S with a catalytic group on the enzyme active site E.

oriented to it, thus greatly increasing the probability that the ES complex will enter the transition state (Figure 9-13).

Strain and Distortion: Induced Fit. Binding of the substrate may induce a conformational change in the enzyme molecule, which strains the structure of the active site and also distorts the bound substrate, thus helping to bring the ES complex into its transition state. Such a change is called the *induced fit* of the enzyme to the substrate (Figure 9-14). Changes in the tertiary or quaternary structure of the relatively large enzyme molecule thus can exert mechanical leverage on the substrate. This concept may explain why enzymes are proteins and thus much larger than most substrate molecules.

Figure 9-14
Induced fit of the active site of the enzyme to a strained form of the substrate molecule.

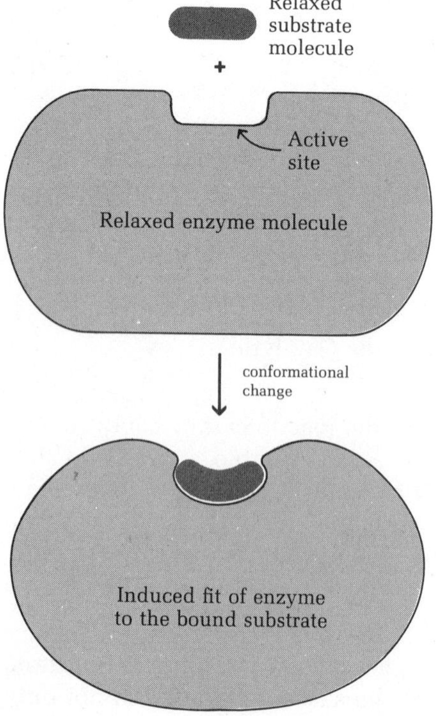

General Acid-Base Catalysis. The active site of the enzyme may furnish R groups of specific amino acid residues that are good proton donors or proton acceptors (Figure 9-15). Such general-acid or general-base groups are powerful catalysts for many organic reactions in aqueous systems.

Covalent Catalysis. Some enzymes react with their substrates to form very unstable, covalently joined enzyme-substrate complexes, which undergo further reaction to form the products much more readily than the uncatalyzed reaction (Figure 9-16).

Although these four factors (Table 9-7) are believed to provide contributions of different magnitudes to the rate accelerations produced by different kinds of enzymes, in no case do we yet have exact knowledge of the mechanism by which an enzyme produces its characteristic rate acceleration.

X-Ray Analysis Has Revealed Important Structural Features of Enzymes

Much important information on the structure and catalytic mechanisms of enzymes has come from x-ray diffraction analy-

Figure 9-15
Many organic reactions are promoted by proton donors or proton acceptors, i.e., general acids or general bases. The active sites of some enzymes contain functional groups of amino acid residues, such as those shown, which participate in the catalytic process as proton donors or proton acceptors. The —SH group is contributed by cysteine and the imidazole group by histidine.

Some proton-donating groups

—COOH

$—\overset{+}{N}H_3$

—SH

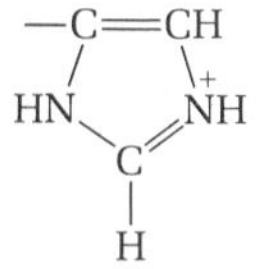

$-C{=}CH$ / HN—C(H)=NH$^+$ (imidazolium)

Some proton-accepting groups

$—COO^-$

$—NH_2$

$—S^-$

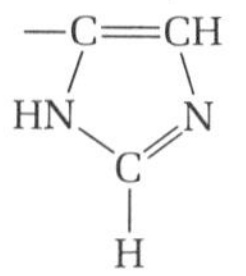

$-C{=}CH$ / HN—C(H)=N (imidazole)

Figure 9-16
A model of covalent catalysis. In some enzymatic reactions the enzyme displaces a functional group R of the substrate RX to form a covalent EX complex, which is unstable and is hydrolyzed far more rapidly than RX. Chymotrypsin is an example (Box 9-4B on page 229).

Uncatalyzed reaction

$$RX + H_2O \longrightarrow ROH + HX$$

Catalyzed reaction

$$RX + \mathbf{E{-}OH} \longrightarrow ROH + \mathbf{E}X$$
$$\mathbf{E}X + H_2O \longrightarrow \mathbf{E{-}OH} + HX$$

Sum: $RX + H_2O \longrightarrow ROH + HX$

sis (page 173). Many crystalline enzymes have been examined by this method, which has supplemented the information gained from chemical studies of enzymes. Some of the important advances in our knowledge of enzymes made possible by x-ray analysis are shown in the next few pages, which present a "gallery" of enzyme structures (Box 9-4).

First, x-ray analysis reveals the secondary, tertiary, and quaternary structures of different enzyme molecules, enabling them to be compared with noncatalytic globular proteins. From such comparisons, it is not possible to identify any particular three-dimensional conformation of their polypeptide chain(s) that is characteristic of all enzymes and different from those of noncatalytic proteins. Members of a given class of enzymes, such as those that catalyze transfer of phosphate groups from ATP to phosphate-acceptor molecules, may, however, share some common structural features.

Second, x-ray analysis has permitted identification of the catalytic or active sites of many enzymes. Often the active site consists of a crevice or pocket in the enzyme molecule into which the substrate molecule fits in a complementary fashion. The active site of some enzymes is lined with back-and-forth loops of polypeptide chains in the β conformation, and in others the active site has the form of a pocket lined with amino acid residues having charged polar groups. In some cases it has been found possible to determine the structure of an enzyme-substrate complex by x-ray methods. An example is the enzyme *lysozyme* (Box 9-4A), which cleaves certain bonds in the backbone of a bacterial polysaccharide.

Combined x-ray and chemical studies have revealed a detailed picture of the active site of chymotrypsin, which consists of three polypeptide chains connected by cystine cross linkages (Box 9-4B). Chemical studies have shown that when chymotrypsin is inactivated by diisopropylfluorophosphate, a covalent derivative of a serine residue at position 195 of the polypeptide chain is formed, thus identifying this residue as part of the active catalytic site. Other chemical studies showed that a histidine residue at position 57 and an aspartic acid residue at position 102 are also involved in catalysis by this enzyme. Although these residues are far apart with respect to their position in the polypeptide backbone and one is on a different chain from the others, x-ray analysis shows them to be very close together in the folded three-dimensional structure of native chymotrypsin (Box 9-4B). This precise structural information, taken together with chemical studies of the enzyme, has made it possible to propose mechanisms for the catalytic action of chymotrypsin, such as that shown in Box 9-4C, and to exclude certain other mechanisms that are inconsistent with the active-site structure. Although we do not yet know in detail exactly how chymotrypsin works, we are closer to understanding its action than that of any other enzyme.

Box 9-4 A gallery of enzyme structures revealed by x-ray diffraction analysis

Many crystalline enzymes have been subjected to x-ray structural analysis. These studies have often been correlated with chemical studies of (1) amino acid sequence, (2) substrate specificity, (3) the action of specific inhibitors, and (4) the identification of specific functional groups in the catalytic site. Representatives of most of the major classes of enzymes (see Table 9-3) have been examined in efforts to determine possible relationships between the catalytic action of enzymes and their three-dimensional structure. In this gallery are shown scale drawings illustrating different features of enzyme structure and action deduced from x-ray analysis of three crystalline enzymes.

A. The lysozyme-substrate complex

Although enzyme-substrate complexes normally undergo very rapid catalytic decomposition, it is sometimes possible to prepare or modify a substrate molecule chemically so that it is accepted and bound at the enzyme active site but is not acted upon by the enzyme. Such a substrate molecule has been found for lysozyme, which normally hydrolyzes certain bonds in the backbone of bacterial polysaccharides. Figure 1 is a scale drawing of the proposed structure of the normal enzyme-substrate complex of lysozyme. It was deduced from x-ray diffraction patterns of a crystalline complex of lysozyme and a "false" or unreactive substrate, an analog of the enzyme's normal substrate. These measurements were carried out by David C. Phillips and his colleagues at Oxford University. The polypeptide chain, including the R groups and the H atoms, is shown in color. A segment of the substrate molecule, outlined in black, lies in a channel or cleft in the lysozyme molecule and is held in position by specific hydrogen bonds (shown in solid color) between the enzyme and the substrate. The substrate molecule is a polymer containing alternating units of N-acetylglucosamine and N-acetylmuramic acid, each in cyclic form, joined by glycosidic bonds designated A–F (Chapter 11). The point at which the substrate molecule is cleaved is shown by a dashed line.

Figure 1

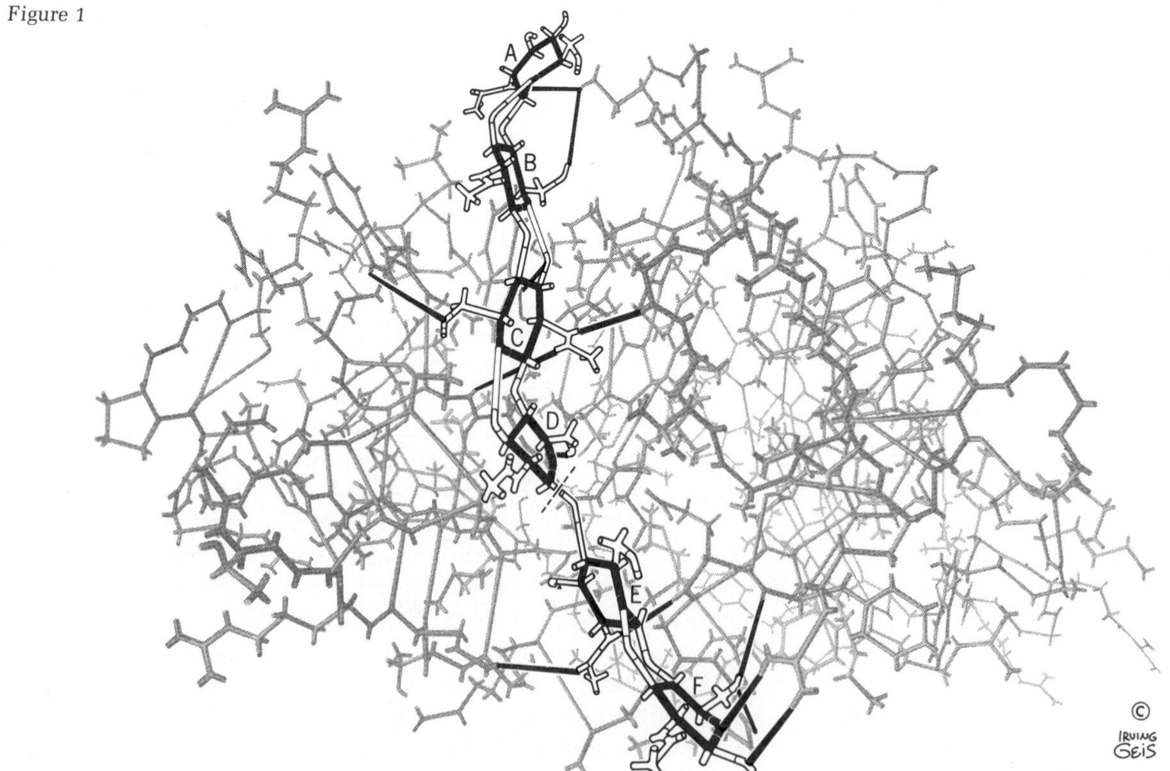

B. The active site of chymotrypsin

Chymotrypsin is a proteolytic enzyme secreted into the small intestine by the pancreas in the form of an inactive precursor or zymogen called chymotrypsinogen. Chymotrypsinogen, which has a single polypeptide chain of 245 residues and five intrachain disulfide cross-links contributed by five cystine residues, is activated by the action of trypsin, another proteolytic enzyme in the intestine. Trypsin removes two dipeptides from positions 14–15 and 147–148 of chymotrypsinogen by hydrolysis to yield active chymotrypsin, which thus has three polypeptide chains, covalently connected by two disulfide cross-links, one between chains A and B and the other between chains B and C, as shown in Figure 1. Chymotrypsin requires for activity histidine residue 57 and aspartic acid residue 102 in chain B, as well as serine residue 195 in chain C. Although they are far apart in the sequence, and one is actually in a different chain from the others, these three residues lie very close together in the three-dimensional structure of the enzyme molecule. This is shown in the scale drawing of the backbone of the chymotrypsin molecule (Figure 2), as deduced from the x-ray diffraction pattern of crystalline chymotrypsin by David M. Blow and his colleagues of the University of Cambridge. In this drawing the R groups of only the three specific residues at the active site are shown.

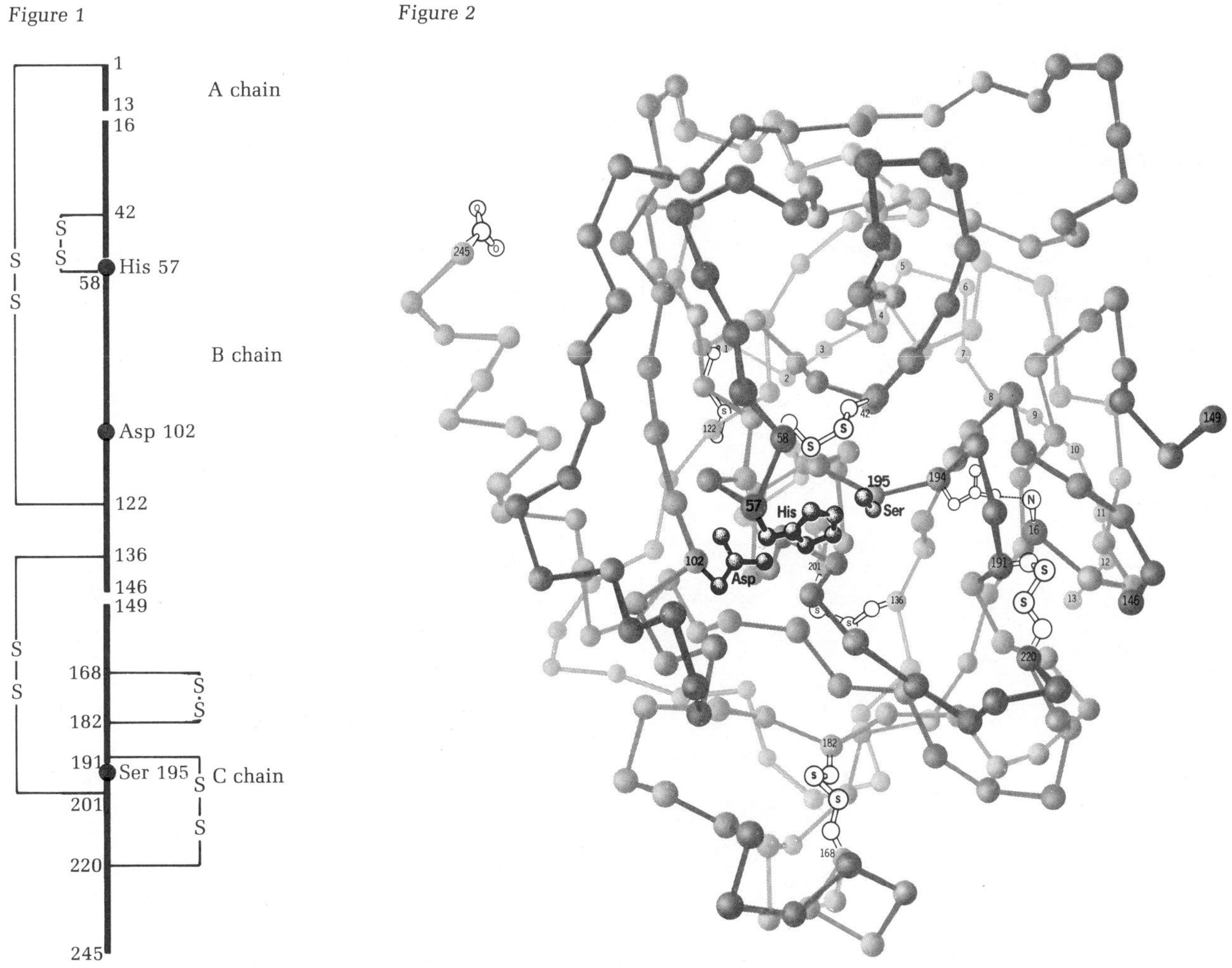

Figure 1

Figure 2

Box 9-4 A gallery of enzyme structures revealed by x-ray diffraction analysis (*Continued*)

C. A possible mechanism for the hydrolysis of specific peptides by chymotrypsin

The close proximity of amino acid residues histidine 57, aspartic acid 102, and serine 195 in the chymotrypsin molecule, as revealed by x-ray analysis (see Part B), has suggested a possible mechanism for the participation of these residues in the catalytic cycle of chymotrypsin. The substrate molecule has been proposed to bind to the active site in such a way that its hydrophobic positioning group (page 221) fits into a hydrophobic pocket at the active site (Figure 1). The peptide bond that is cleaved is thus positioned near the hydroxyl group of the serine 195 residue (Figure 2), which is known to form a covalent compound with the acyl group of the substrate (page 226). The serine hydroxyl group readily undergoes loss of its H atom, which is strongly attracted by a hydrogen bond to the electronegative nitrogen atom of the imidazole R group of histidine 57. Concerted formation of a covalent ester linkage between the serine oxygen and the acyl carbon atom of the substrate, with cleavage of the peptide bond, then takes place through a transient intermediate (Figure 3). As a result the first product leaves the active site, and

Figure 1
Substrate diffuses to enzyme.

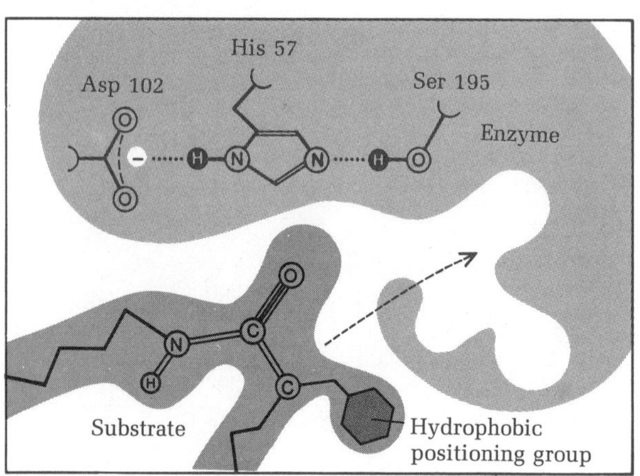

Figure 2
Substrate binds to enzyme.

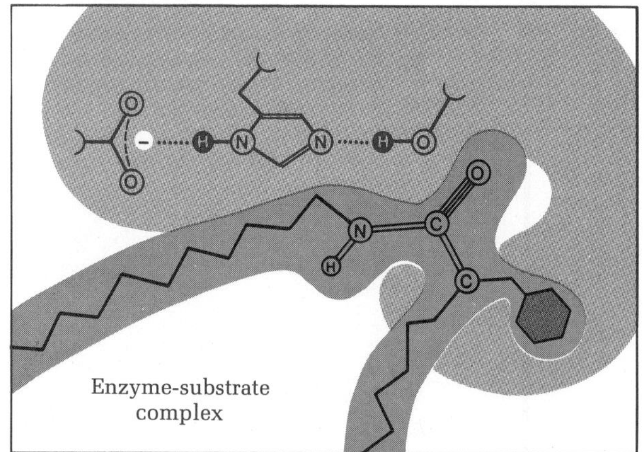

Figure 5
Water molecule approaches.

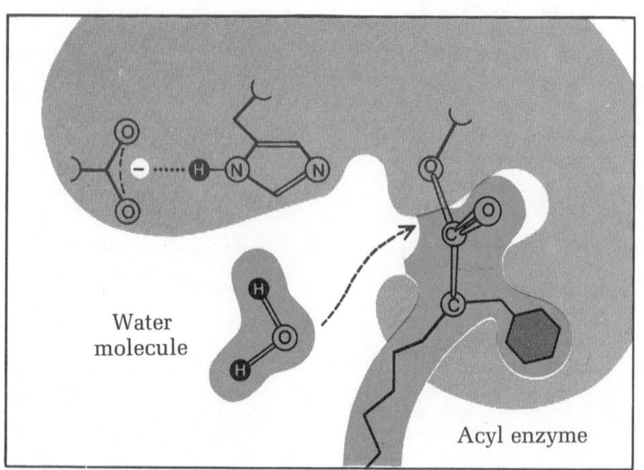

Figure 6
Water reacts with acyl enzyme.

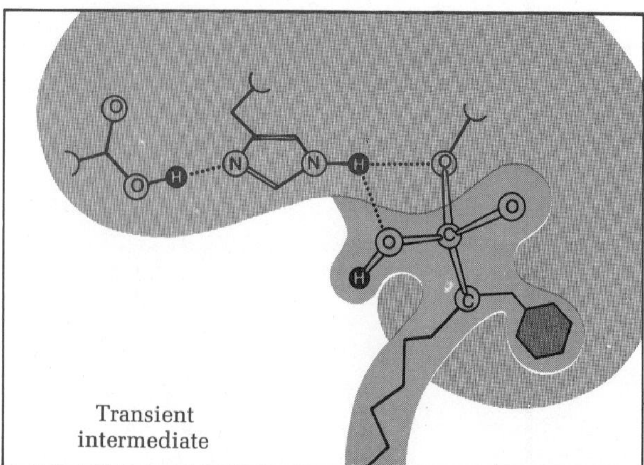

the acyl group of the substrate is covalently bound to the serine hydroxyl group of the enzyme; this derivative is called the *acyl enzyme* (Figure 4). Its ester linkage, which is very unstable compared with the original peptide bond of the substrate, undergoes hydrolysis to yield the second product, the carboxylic portion of the substrate, with restoration of the hydrogen atom on the serine hydroxyl group (Figures 5 and 6), to form the enzyme-product complex (Figure 7). The second product then leaves the active site, thus completing the catalytic cycle (Figure 8). The acyl enzyme is the crucial intermediate in this case of covalent catalysis. The imidazole group of histidine 57 participates in the associated proton movements by general-acid–general-base catalysis.

It has been suggested that the function of aspartic acid residue 102, with its strong negative charge, is to help labilize the imidazole group of histidine 57 in order to attract the hydrogen atom from serine 195. However, doubts have been raised about whether such a "relay" of electric charges takes place because of the distance between aspartic acid 102 and histidine 57. Whatever the mechanism of its action, aspartic acid 102 is a necessary residue for catalysis.

Figure 3
Substrate bond is cleaved.

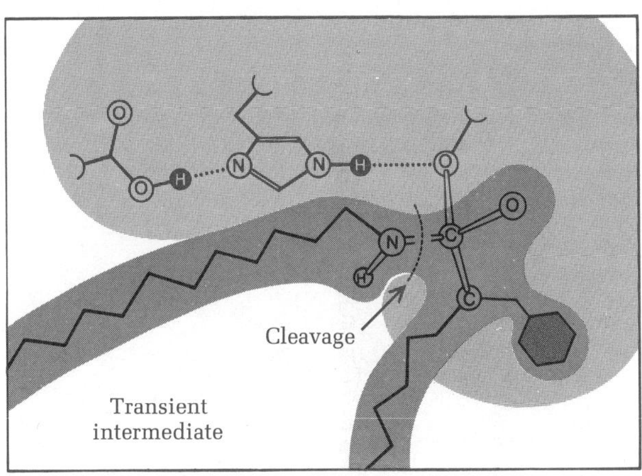

Figure 4
First product leaves.

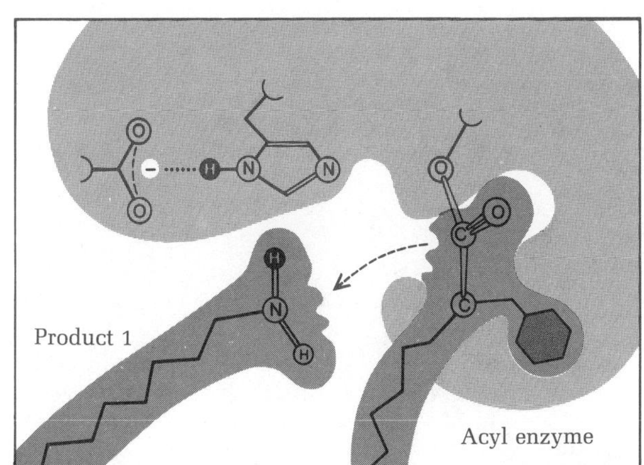

Figure 7
Acyl bond to enzyme is broken.

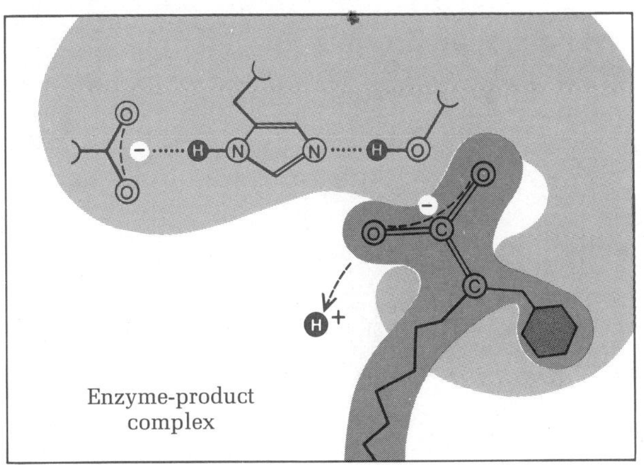

Figure 8
Second product leaves.

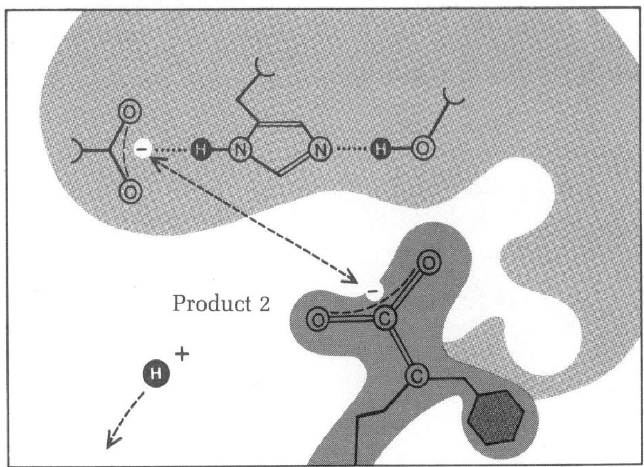

Box 9-4 A gallery of enzyme structures revealed by x-ray diffraction analysis (*Continued*)

D. Induced fit of the hexokinase molecule to D*-glucose, one of its substrates*

Hexokinase, which catalyzes the phosphorylation of D-glucose and other hexoses by ATP

$$\text{ATP} + \text{D-glucose} \longrightarrow \text{ADP} + \text{D-glucose 6-phosphate}$$

has two polypeptide-chain subunits, as differentiated in the scale drawing of a space-filling model (Figure 1). The free D-glucose molecule (solid color) is also shown, together with the "empty" hexokinase molecule. When D-glucose is bound to the active site of the enzyme, in the absence of the second substrate ATP, the two subunits move together to enclose the glucose molecule in a pocket (Figure 2). Thus a rather large change in the quaternary structure of hexokinase takes place as the enzyme undergoes an induced fit to form the hexokinase-glucose complex in the absence of ATP. This complex is stable enough to be crystallized. The x-ray measurements of these structures were made by Thomas A. Steitz of Yale University.

When both ATP and glucose are present, they are bound to their specific sites and the catalytic cycle then takes place rapidly, with formation and release of ADP and D-glucose 6-phosphate from the enzyme, which then resumes its empty conformation to start another cycle.

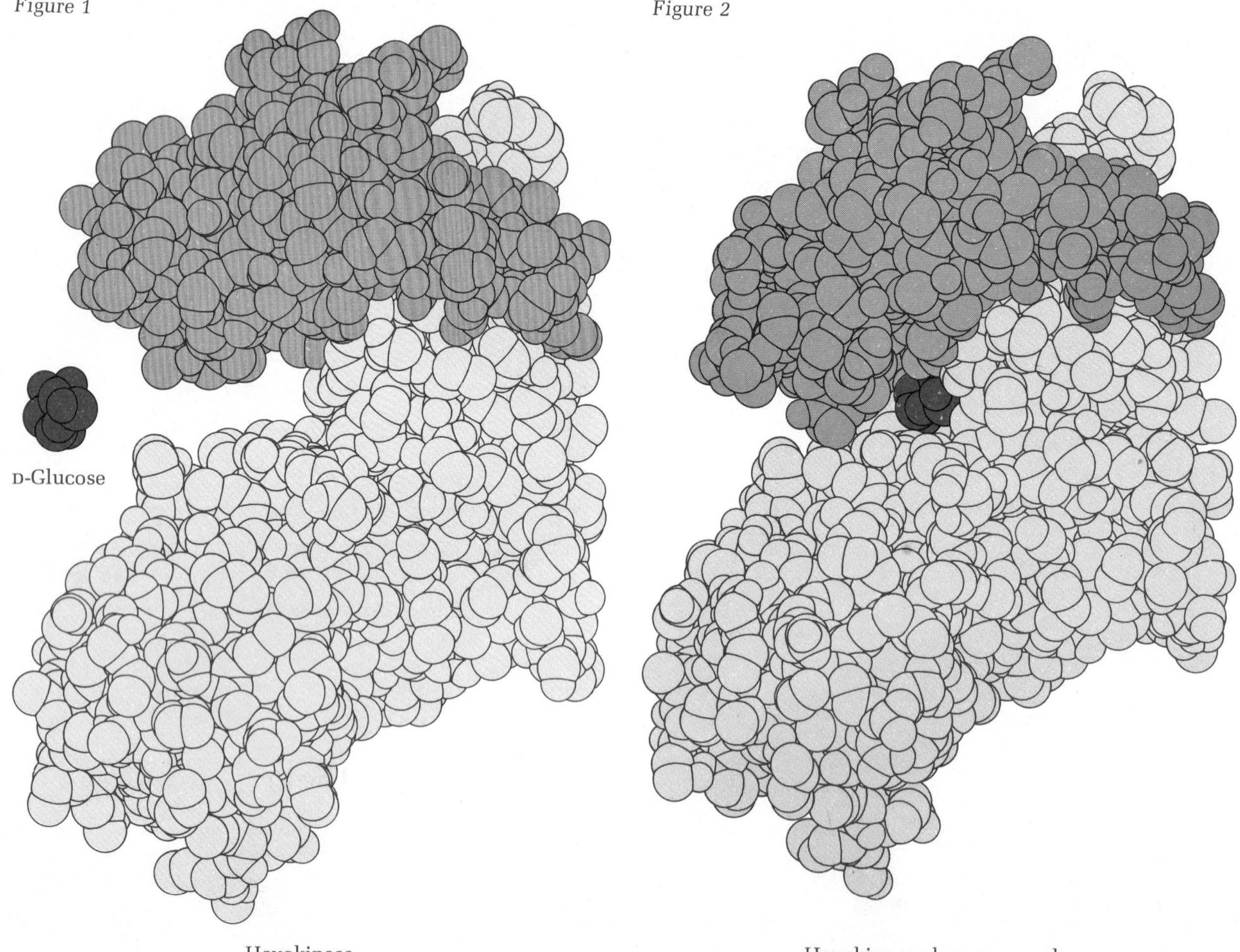

Figure 1 Hexokinase

Figure 2 Hexokinase-glucose complex

Yet another aspect of enzyme action has been revealed by x-ray studies, namely, the participation of conformational changes in the enzyme molecule as it binds and acts on its substrate. An outstanding example is given by hexokinase (page 210), which catalyzes the phosphorylation of D-glucose by ATP. As shown in Box 9-4D, binding of the relatively small glucose molecule to the active site of hexokinase causes its two polypeptide-chain subunits to move together like the jaws of a trap, enclosing the glucose molecule and making it ready for attack by the ATP molecule. Presumably this induced fit of the enzyme distorts the glucose molecule, bringing it more readily into the transition state.

Enzyme Systems Have a Pacemaker or Regulatory Enzyme

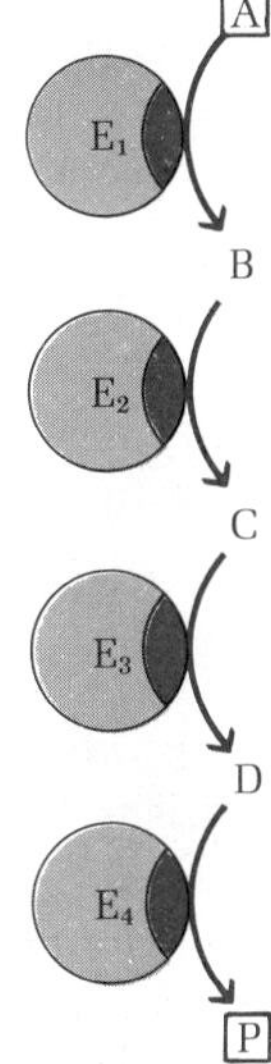

Figure 9-17
Schematic drawing of a multienzyme system responsible for converting A into P via four sequential enzyme-catalyzed steps.

In cell metabolism groups of enzymes work together in sequential chains or systems to carry out a given metabolic process, such as the conversion of glucose into lactic acid in skeletal muscle or the synthesis of an amino acid from simpler precursors. In such enzyme systems the reaction product of the first enzyme becomes the substrate of the next, and so on (Figure 9-17). Multienzyme systems may have as many as 15 or more enzymes acting in a specific sequence.

In each enzyme system there is at least one enzyme, the "pacemaker," that sets the rate of the overall sequence because it catalyzes the slowest or rate-limiting step. Such pacemaker enzymes not only have a catalytic function but are also capable of increasing or decreasing their catalytic activity in response to certain signals. By the action of such pacemaker enzymes the rate of each metabolic sequence is constantly adjusted, on a minute-by-minute basis, to changes in the cell's demands for energy and for building-block molecules required in cell growth and repair. In most multienzyme systems the *first* enzyme of the sequence is the pacemaker enzyme. The other enzymes in the sequence, which are usually present in amounts providing a large excess of catalytic activity, simply follow the pacemaker; they can promote their reactions only as fast as their substrates are made available from preceding steps.

Such pacemaker enzymes, whose activity is modulated through various types of molecular signals, are called *regulatory enzymes*. There are two major classes of regulatory enzymes: *allosteric* or *noncovalently regulated enzymes* and *covalently regulated enzymes*.

Allosteric Enzymes Are Regulated by Noncovalent Binding of Modulator Molecules

In some multienzyme systems the first, or regulatory, enzyme has a distinctive characteristic: it is inhibited by the end product of the multienzyme system. Whenever the end product of such a metabolic sequence increases above its usual steady-state concentration, indicating that it is being produced in

Figure 9-18
Feedback inhibition of the conversion of L-*threonine into* L-*isoleucine, which is catalyzed by a sequence of five enzymes* (E_1 *to* E_5) *via four intermediates* A, B, C, *and* D. *The first enzyme, threonine dehydratase* (E_1), *is specifically inhibited by* L-*isoleucine, the end product of the sequence, but not by intermediates* A, B, C, *or* D. *Such inhibition is indicated by the dashed feedback line and the colored bar across the reaction arrow for threonine dehydratase.*

$$\begin{array}{c} COO^- \\ | \\ H_3\overset{+}{N}-C-H \\ | \\ H-C-OH \\ | \\ CH_3 \end{array} \quad \text{L-Threonine}$$

$\downarrow E_1$ threonine dehydratase

A

$\downarrow E_2$

B

$\downarrow E_3$

C

$\downarrow E_4$

D

$\downarrow E_5$

$$\begin{array}{c} COO^- \\ | \\ H_3\overset{+}{N}-C-H \\ | \\ H-C-CH_3 \\ | \\ CH_2 \\ | \\ CH_3 \end{array} \quad \text{L-Isoleucine}$$

excess of the cell's needs, the end product of the sequence acts as a specific inhibitor of the first, or regulatory, enzyme in the sequence. The whole enzyme system thus slows down to bring the rate of production of its end product back into balance with the cell's needs. This type of regulation is called *feedback inhibition.* A classical example of such allosteric feedback inhibition, one of the first discovered, is the bacterial enzyme system that catalyzes the conversion of L-threonine into L-isoleucine (Figure 9-18). In this sequence of five enzymes, the first, *threonine dehydratase,* is inhibited by isoleucine, the product of the last enzyme of the sequence. Isoleucine is quite specific as an inhibitor. No other intermediate in this sequence of reactions is inhibitory to threonine dehydratase, nor is any other enzyme in the sequence inhibited by isoleucine. Feedback inhibition is one of several types of allosteric regulation.

The inhibition of threonine dehydratase by isoleucine is reversible; if the isoleucine concentration decreases, the rate of threonine dehydratase activity increases. Thus threonine dehydratase activity responds very rapidly and reversibly to fluctuations in the concentration of isoleucine in the cell. Although it is a very specific inhibitor of this enzyme, isoleucine does not bind to its substrate site. Instead it binds to another specific site on the enzyme molecule, the *regulatory* site. The binding of isoleucine to the regulatory site of threonine dehydratase is noncovalent and thus is readily reversible. *Threonine dehydratase is a typical member of the class of allosteric enzymes, regulatory enzymes that function through reversible, noncovalent binding of a modulator molecule.* The term allosteric derives from Greek *allo,* "other," and *stereos,* "space" or "site." Allosteric enzymes are those having "other sites."

The properties of allosteric enzymes are significantly different from those of simple nonregulatory enzymes discussed earlier in this chapter. First, like all enzymes, allosteric enzymes have *catalytic* sites which bind the substrate and transform it, but they also have one or more *regulatory* or *allosteric* sites for binding the regulating metabolite, which is called the *effector* or *modulator* (Figure 9-19). Just as the catalytic site of

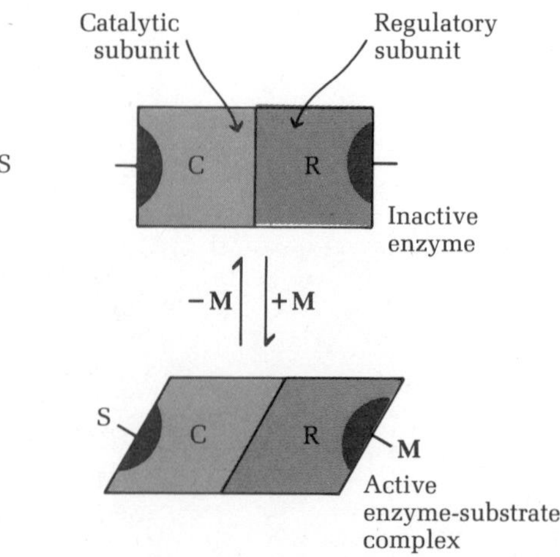

Figure 9-19
Schematic model of the subunit interactions in an allosteric enzyme. In many allosteric enzymes the substrate binding site and the modulator binding site are on different subunits, the catalytic (C) *and regulatory* (R) *subunits, respectively. Binding of the positive modulator* M *to its specific site on the regulatory subunit is communicated to the catalytic subunit through a conformational change, rendering the catalytic subunit active and capable of binding the substrate* S *with high affinity. On dissociation of the modulator* M *from the regulatory subunit the enzyme reverts to its inactive or less active form.*

an enzyme is specific for its substrate, the allosteric site is specific for its modulator. Second, allosteric enzyme molecules are generally larger and more complex than those of simple enzymes. Most of them have two or more polypeptide chains or subunits. Third, allosteric enzymes usually show significant deviations from classical Michaelis-Menten behavior, one of the features by which they were first recognized.

Allosteric Enzymes May Be Inhibited or Stimulated by Their Modulators

When the allosteric site is occupied by a specific *inhibitory* or *negative* modulator, which happens when the modulator concentration in the cell rises, the enzyme undergoes a change to a less active or inactive form; i.e., it is "turned off." When the inhibitory modulator leaves the allosteric site, which happens when the modulator concentration in the cell decreases, the enzyme switches back to its active, or "on," form.

But there are also allosteric enzymes that are *stimulated* by their modulator molecules. In this case the *stimulatory* or *positive* modulator is not the end product of the enzyme sequence but some other metabolite that serves as the molecular signal to the enzyme to hurry up (Figure 9-20). Often the stimulatory modulator of this type of allosteric enzyme is the *substrate molecule itself*. Allosteric enzymes of this class, which are called *homotropic* (because the substrate and the modulator are identical), have two or more binding sites for the substrate. These binding sites often serve a dual function, acting *both* as catalytic sites and as regulatory sites. This type of allosteric enzyme responds to situations in which the substrate accumulates in excessive amounts and must be removed by the subsequent reactions. Thus we have two types of allosteric enzymes: those which are inhibited by their modulator, usually by a molecule *other* than their substrates (they are thus called *heterotropic* enzymes), and those which are stimulated by their modulator, often the substrate itself. In many ways the on-and-off switching of an allosteric enzyme resembles the on-and-off switching of hemoglobin by diphosphoglycerate (Chapter 8).

Figure 9-20
Designation of inhibitory and stimulatory allosteric regulation.

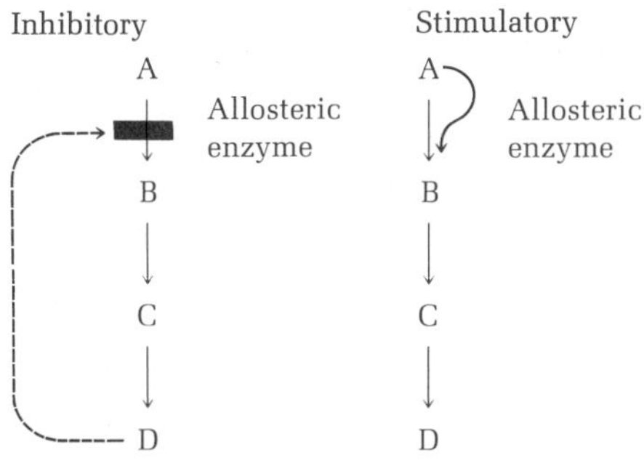

Some allosteric enzymes have two or more modulators, which may be opposite in effect, so that one or more modulators of the enzyme are stimulatory and one or more are inhibitory. In these more complex enzymes each modulator has its own specific allosteric site, which, when it is occupied, signals the enzyme either to speed up its catalytic action or slow it down.

Allosteric Enzymes Deviate from Michaelis-Menten Behavior

Allosteric enzymes show relationships between substrate concentration and rate that differ from classical Michaelis-Menten behavior in various ways, depending on whether the enzyme has an inhibitory or stimulatory modulator. Allosteric enzymes do exhibit "saturation" with the substrate when the latter is added in high enough concentrations, but when the initial rate of some allosteric enzymes is plotted against the substrate concentration (Figure 9-21), a *sigmoid* saturation curve results, rather than the classical *hyperbolic* substrate saturation curve shown by nonregulatory enzymes. Although we can find a point on the sigmoid saturation curve of allosteric enzymes at which the velocity of the reaction is half-maximal, we cannot refer to it with the designation K_M since the enzyme does not follow the hyperbolic Michaelis-Menten relationship. Instead the symbol $[S]_{0.5}$ or $K_{0.5}$ is used to symbolize the substrate concentration giving half-maximal velocity of an allosteric enzyme.

The hyperbolic saturation curve for a nonregulatory enzyme, shown earlier in Figure 9-4, closely resembles the binding curve for oxygen given by myoglobin (page 187). On the other hand, the saturation curve of the allosteric enzyme in Figure 9-21a is sigmoid, resembling the oxygen-binding curve of hemoglobin (page 187). Indeed, myoglobin and hemoglobin may be regarded as useful models for the interpretation of the behavior of ordinary vs. allosteric enzymes. Myoglobin has but one binding site for its ligand oxygen, a single polypeptide chain, and it gives a hyperbolic oxygen saturation curve. Similarly, many nonregulatory enzymes also have but a single binding site for their substrates, possess only a single polypeptide chain, and give a hyperbolic substrate saturation curve.

Hemoglobin, on the other hand, has four binding sites, one on each of its four subunits, and they act cooperatively. Recall that when one binding site of hemoglobin is occupied by an oxygen molecule, the affinity of the remaining oxygen-binding sites increases, causing the oxygen saturation curve to rise very sharply after the first oxygen is bound and giving the curve its sigmoid shape. In a similar way a *homotropic* allosteric enzyme (Figure 9-21a) has multiple binding sites for its substrate and they act cooperatively, so that the binding of one molecule of the substrate greatly enhances the binding of subsequent

Figure 9-21
Substrate-activity curves for representative allosteric enzymes.
(a) *The sigmoid curve given by a homotropic enzyme, in which the substrate also serves as a positive (stimulatory) modulator.* $K_{0.5}$ *is the substrate concentration giving half-maximal velocity. Note that a relatively small increase in substrate concentration in the steep part of the curve can cause a very large increase in the reaction rate. Note also the resemblance to the oxygen saturation curve of hemoglobin.*
(b) *The effects of a stimulatory or positive modulator (+), an inhibitory or negative modulator (−), and no modulator (0) on an allosteric enzyme in which* $K_{0.5}$ *of the enzyme is modulated without change in* V_{max}.
(c) *A less common type of modulation, in which* V_{max} *is modulated with* $K_{0.5}$ *nearly constant.*
These are examples of a variety of sometimes very complex responses given by allosteric enzymes to their modulators.

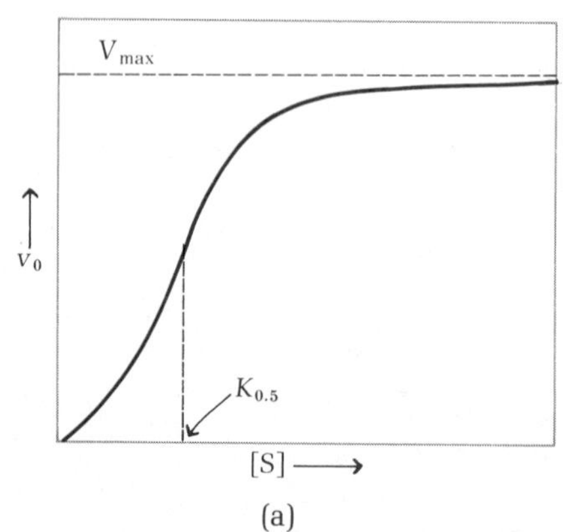

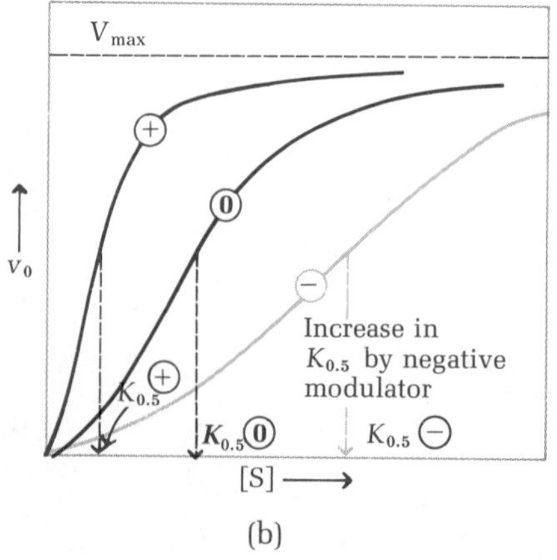

substrate molecules. This accounts for the sigmoid rather than hyperbolic increase in the rate of enzyme activity on increasing the substrate concentration.

With *heterotropic* enzymes, in which the modulator is some metabolite other than the substrate itself, it is difficult to generalize about the shape of the substrate saturation curve, which is related to whether the modulator is positive (stimulatory) or negative (inhibitory). If the modulator is stimulatory, it may cause the substrate saturation curve to become more nearly hyperbolic, with a decrease in $K_{0.5}$ but no change in V_{max}, thus resulting in an increased rate at a fixed substrate concentration (Figure 9-21b). Other allosteric enzymes respond to a stimulatory modulator by an increase in V_{max}, with little change in $K_{0.5}$ (Figure 9-21c). If the modulator is negative or inhibitory, the substrate saturation curve may become more sigmoid, with an increase in $K_{0.5}$ (Figure 9-21b). Allosteric enzymes therefore show different kinds of responses in their substrate-activity curves because some have inhibitory modulators, some have stimulatory modulators, and some have both.

Allosteric Enzymes Show Communication between Subunits

Allosteric enzymes and hemoglobin have other similarities. First, like hemoglobin, allosteric enzymes generally have multiple polypeptide-chain subunits; indeed, some allosteric enzymes may have six, eight, or a dozen or more subunits. Second, in allosteric enzymes there appears to be communication between the binding site for the modulator and the catalytic site for the substrate, similar to the communication established when one oxygen molecule is bound to one subunit of hemoglobin, signaling to the other subunits to increase their oxygen affinity. Third, allosteric enzymes undergo conformational changes on binding the modulator molecule and thus undergo switching between a relatively inactive and a relatively active state (Figure 9-19). Here again we find a resemblance to the conformational changes and switching properties of hemoglobin.

Some allosteric enzymes are extremely complex in structure and contain many polypeptide chains. An important example is *aspartate transcarbamoylase,* which has 12 polypeptide chains organized into *catalytic* and *regulatory* subunits. Box 9-5 shows the very complex quaternary structure of this enzyme, deduced from x-ray analysis. This enzyme catalyzes an important reaction in the biosynthesis of nucleotides. We shall examine its function and regulatory behavior in more detail in Chapter 22.

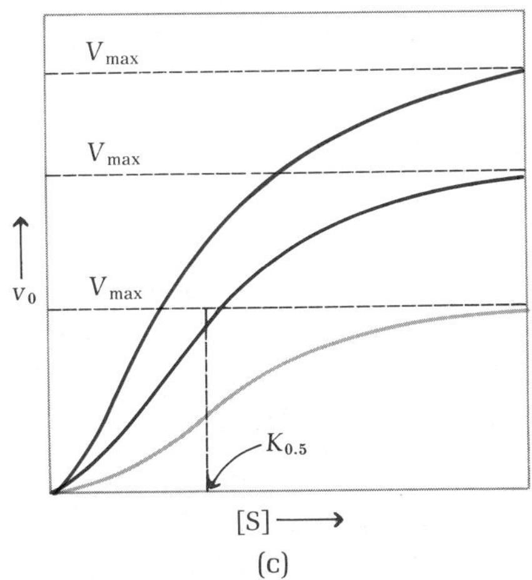

(c)

Some Enzymes Are Regulated by Reversible Covalent Modification

Another important class of regulatory enzymes is modulated through interconversion of their active and inactive forms by *covalent modification* of the enzyme molecule. An important

Box 9-5 The three-dimensional structure of the regulatory enzyme aspartate transcarbamoylase

This allosteric regulatory enzyme (Figure 1) contains two catalytic clusters, each containing three catalytic polypeptide chains in their folded tertiary structures, and three regulatory clusters (in color), each containing two regulatory polypeptide chains. Heavily outlined is a catalytic cluster, showing three catalytic polypeptides in folded conformation. Underneath is the other catalytic cluster. This structure was deduced by William Lipscomb and his associates at Harvard University.† The role of this enzyme in nucleotide synthesis and its regulation will be discussed in Chapter 22.

† See also C. R. Cantor and P. R. Schimmel, *Biophysical Chemistry*, pt. I, pp. 139–144, Freeman, San Francisco, 1980, for an excellent description.

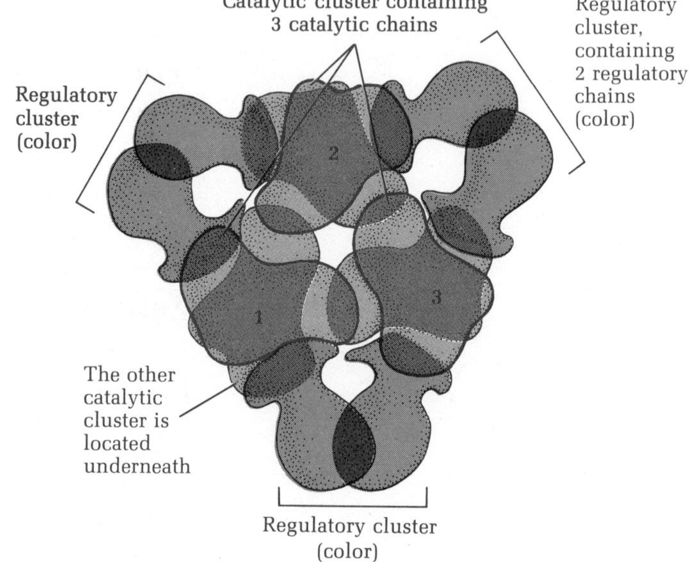

Figure 1
The three numbered, heavily outlined structures represent the three polypeptide chains of one of the catalytic clusters. The other catalytic cluster, which also contains three chains, is underneath.

example is the regulatory enzyme glycogen phosphorylase of muscle and liver, which catalyzes the reaction

$$\underset{\text{Glycogen}}{(\text{Glucose})_n} + \text{phosphate} \longrightarrow \underset{\substack{\text{Shortened}\\\text{glycogen}\\\text{chain}}}{(\text{glucose})_{n-1}} + \text{glucose 1-phosphate}$$

The glucose 1-phosphate so formed can then be broken down into lactic acid in the muscle or free glucose in the liver. Glycogen phosphorylase occurs in two forms, the active form phosphorylase *a* and the relatively inactive form phosphorylase *b* (Figure 9-22). Phosphorylase *a* has two polypeptide-chain subunits, each with one specific serine residue in its sequence that is phosphorylated at its hydroxyl group. These serine phosphate residues are required for maximum activity of the enzyme. The phosphate groups can be hydrolytically removed from phosphorylase *a* by an enzyme called phosphorylase phosphatase

$$\underset{\text{(More active)}}{\text{Phosphorylase } a} + 2H_2O \xrightarrow[\text{phosphatase}]{\text{phosphorylase}} \underset{\text{(Less active)}}{\text{phosphorylase } b} + 2P_i$$

Figure 9-22
Regulation of glycogen phosphorylase activity by covalent modification. In the active form of the enzyme, phosphorylase a, *specific serine residues, one on each subunit, are in the phosphorylated state. Phosphorylase* a *is converted into phosphorylase* b, *which is relatively inactive, by enzymatic loss of these phosphate groups, promoted by phosphorylase phosphatase. Phosphorylase* b *can be reactivated to form phosphorylase* a *by the action of phosphorylase kinase, which catalyzes the phosphorylation of the serine hydroxyl groups by ATP.*

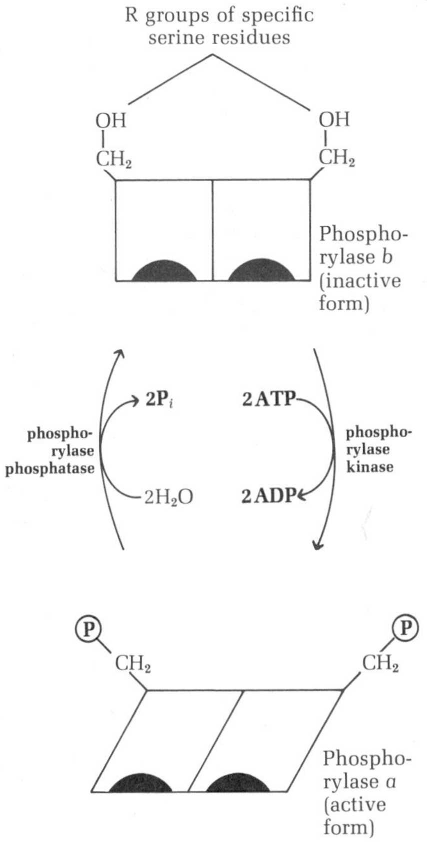

In this reaction phosphorylase *a* is converted into phosphorylase *b*, which is much less active than phosphorylase *a* in catalyzing glycogen breakdown. Thus the active form of glycogen phosphorylase is converted into the relatively inactive form by the cleavage of two covalent bonds between phosphoric acid and two specific serine residues in the enzyme.

Phosphorylase *b* can in turn be reactivated, i.e., covalently transformed back into active phosphorylase *a*, by another enzyme, phosphorylase kinase, which catalyzes transfer of phosphate groups from ATP to the hydroxyl groups of the specific serine residues in phosphorylase *b* (Figure 9-22):

$$\underset{\text{(Less active)}}{2\text{ATP} + \text{phosphorylase } b} \xrightarrow{\text{phosphorylase kinase}} \underset{\text{(More active)}}{2\text{ADP} + \text{phosphorylase } a}$$

Thus, the breakdown of glycogen in skeletal muscles and the liver is regulated through variations in the ratio of the active and inactive forms of the enzyme. These two forms differ in their quaternary structure, so that the catalytic site undergoes changes in structure and consequently changes in catalytic activity.

Although in most known cases covalent regulation of enzyme action is brought about by phosphorylation and dephosphorylation of specific serine residues, as described for glycogen phosphorylase, other kinds of covalent modulation are brought about by the methylation of specific amino acid residues or by the attachment of adenylate groups. Further examples of covalent modulation of regulatory enzymes will be described in subsequent chapters.

Some of the more complex regulatory enzymes are modulated by both noncovalent and covalent mechanisms. Such enzymes are located at particularly crucial points in metabolism, so that they respond to multiple regulatory metabolites through both allosteric and covalent modification. Actually glycogen phosphorylase, just discussed, is an example. Although its primary regulation is through covalent modification as described above, it is secondarily modulated in a noncovalent allosteric manner by adenylate, which acts as a stimulating modulator of phosphorylase *b* (Chapter 20).

Another example is glutamine synthetase of *E. coli*, one of the most complex regulatory enzymes known. It has many allosteric modulators and is also regulated by a reversible covalent modification (Chapter 23). Both enzymes will be discussed later in connection with their roles in metabolism.

Many Enzymes Occur in Multiple Forms

Many enzymes occur in more than one molecular form in the same species, in the same tissue, or even in the same cell. In such cases the different forms of the enzyme catalyze the same reaction, but since they have different kinetic properties and

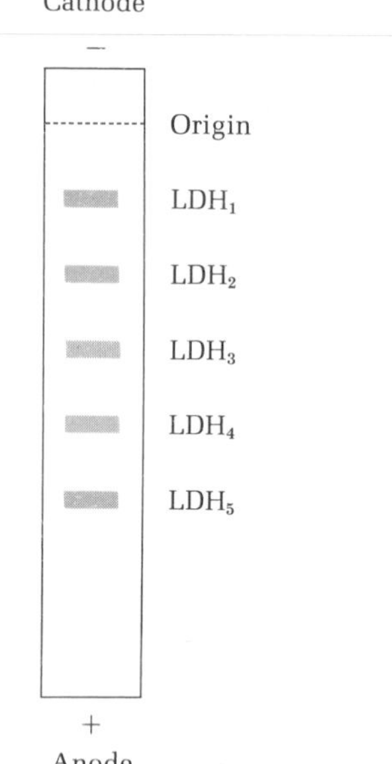

Figure 9-23
Electrophoretic separation of lactate dehydrogenase (LDH) isozymes. Isozymes are usually given numbers indicating the relative rate of electrophoretic transport in polyacrylamide or starch gels under standardized conditions. LDH enzymes are also designated by letters, as indicated in the text.

Isozyme analysis is frequently used in medical diagnosis. The LDH isozyme predominating in heart tissue, LDH_1 (also designated H_4 or B_4), greatly increases in the blood plasma after a heart attack, in which the blood circulation to a portion of the heart is limited or cut off. The membranes of the damaged heart cells no longer function properly and allow some of the cytosolic enzymes, including LDH, to leak into the blood. In some diseases of the liver, such as infectious hepatitis, the characteristic LDH isozymes of the liver, LDH_4 and LDH_5, increase in the blood.

distinctly different amino acid composition or sequence, they can be distinguished and separated by appropriate procedures. Such multiple forms of enzymes are called *isoenzymes* or *isozymes*. One of the first enzymes found to have such multiple forms is *lactate dehydrogenase*, which catalyzes the reversible oxidation-reduction reaction

$$\text{Lactate} + \text{NAD}^+ \rightleftharpoons \text{pyruvate} + \text{NADH} + \text{H}^+$$

in which lactate loses two hydrogen atoms and is thus oxidized to become pyruvate (NAD^+ and NADH are the oxidized and reduced forms of the coenzyme *nicotinamide adenine dinucleotide*, discussed in the next chapter). Lactate dehydrogenase occurs in animal tissues as five different isozymes separable by electrophoresis (Figure 9-23). All the lactate dehydrogenase isozymes contain four polypeptide chains, each of molecular weight 33,500, but the five isozymes contain varying ratios of two kinds of polypeptides which differ in composition and sequence. The A chains (also designated M for muscle) and the B chains (also designated H for heart) are coded by two different genes. In skeletal muscle the lactate dehydrogenase isozyme that predominates contains four A chains, and in heart the predominant isozyme contains four B chains. The lactate dehydrogenase isozymes in other tissues are a mixture of the five possible forms, which may be designated A_4, A_3B, A_2B_2, AB_3, and B_4. The different lactate dehydrogenase isozymes differ significantly in the maximum activities V_{max} and in the Michaelis constant K_M for their substrates, particularly for pyruvate. The properties of LDH isozyme A_4 favor rapid reduction of very low concentrations of pyruvate to lactate in skeletal muscle, whereas those of isozyme B_4 tend to favor rapid oxidation of lactate to pyruvate in the heart. The other lactate dehydrogenase isozymes have intermediate kinetic properties.

Many different enzymes participating in cell metabolism have been found to occur in multiple isozyme forms. All the isozyme forms of a given enzyme catalyze the same reaction but differ in kinetic properties and may also differ in their response to allosteric modulators. The distribution of the different isozyme forms of any given enzyme reflects at least four factors:

1. *The differing metabolic patterns in different organs*. For example, the LDH isozymes in heart and skeletal muscle reflect the metabolic differences of these organs.
2. *The different locations and metabolic roles of a given enzyme within one type of cell*. For example, the enzyme malate dehydrogenase occurs in different forms in the mitochondria and the cytosol, where they play somewhat different roles (Chapter 17).
3. *The differentiation and development of adult tissues from their embryonic or fetal forms*. For example, the fetal liver

has a characteristic isozyme distribution of lactate dehydrogenase which changes as the organ undergoes differentiation to its adult form. An interesting discovery is that some of the enzymes of glucose breakdown in malignant cancer cells occur in their fetal isozyme forms.

4. *Fine tuning of metabolic rates through the different responses of isozyme forms to allosteric modulators.* Some regulatory enzymes occur in isozyme forms that differ in their response to modulators (Chapter 22).

Enzymes May Be Catalytically Defective Due to Genetic Mutation

Many human genetic diseases are known in which one enzyme or another is either totally inactive or is otherwise defective in its catalytic or regulatory function. In such diseases the defective enzyme molecule may contain one or more "wrong" amino acids in its polypeptide chain(s) as a result of a mutation in the DNA coding for it. Catalytic activity of an enzyme depends not only on the occurrence of specific amino acid residues in the catalytic and regulatory sites but also on its overall three-dimensional conformation. Replacement of a single amino acid residue at some critical position in the chain may therefore alter or destroy its catalytic activity, just as replacement of a single amino acid residue renders sickle-cell hemoglobin defective in its function (page 196). When the genetically altered enzyme is a member of an enzyme system catalyzing a central metabolic pathway, the consequence may be a serious or even lethal disturbance in metabolism.

Some of the major human genetic diseases in which there are aberrations in one enzyme or another are given in Table 9-8. These diseases will be discussed in chapters to follow. Many efforts are being made to prevent the unfavorable consequences of such genetic defects in enzymes. One approach being investigated is the introduction of the normal, active form of the defective enzyme into the body, immobilized in a filter capsule inserted into a blood vessel. In this way it is hoped that the metabolites accumulating in the body as a result of the genetic de-

Table 9-8 Some Human Genetic Diseases in Which Specific Enzymes Are Defective

Disease	Defective enzyme
Albinism	Tyrosine 3-monooxygenase
Alkaptonuria	Homogentisate 1,2-dioxygenase
Galactosemia	Galactose 1-phosphate uridylyl transferase
Homocystinuria	Cystathionine β-synthase
Phenylketonuria	Phenylalanine 4-monooxygenase
Tay-Sachs disease	Hexosaminidase A

Figure 9-24
The characteristic color pattern of the Siamese cat is a result of genetic alteration of an enzyme involved in the synthesis of a dark hair pigment. Because of this defect the enzyme is active only in the cooler parts of the body.

fect can be transformed into their normal products as the blood passes through the capsule containing the active normal enzyme.

Genetic alterations in enzymes are not always harmful. Often they produce variations in secondary characteristics of an organism, such as changes in eye color or hair characteristics (Figure 9-24). Sometimes a genetic alteration in an enzyme may make it more efficient, giving the organism some advantage in the struggle for survival.

Summary

Enzymes are proteins that catalyze specific chemical reactions. They bind the substrate molecule to form a transient enzyme-substrate complex, which breaks down to form the free enzyme and its products. As the concentration of the substrate S is increased, the catalytic activity of a fixed concentration of an enzyme E will increase in a hyperbolic fashion to approach a characteristic maximum rate V_{max}, at which essentially all the enzyme is in the form of the ES complex and is thus saturated with S. The substrate concentration giving one-half V_{max} is the Michaelis-Menten constant K_M, which is characteristic for each enzyme acting on a given substrate. The Michaelis-Menten equation

$$v_0 = \frac{V_{max}[S]}{K_M + [S]}$$

relates the rate of an enzymatic reaction to the substrate concentration and V_{max} through the constant K_M. This equation is also applicable to bisubstrate reactions, which occur by either single-displacement or double-displacement (ping-pong) pathways. Each enzyme also has an optimum pH, as well as a characteristic specificity for the substrates on which it acts. Enzymes can be inactivated by irreversible modification of some functional group essential for catalytic activity. They can also be reversibly inhibited, competitively or noncompetitively. Competitive inhibitors, which usually resemble the substrate in structure, compete reversibly with the substrate for binding to the active site, but they are not transformed by the enzyme. Noncompetitive inhibitors bind to some other site on both the free enzyme and the enzyme-substrate complex; their action cannot be reversed by the substrate. Enzymes promote chemical reactions by orienting the substrate in close proximity to the catalytic site, by providing catalytic proton-donating or proton-accepting groups, by forming unstable covalent intermediates with the substrate, or by exerting strain or distortion on the substrate.

In addition to catalytic activity, some enzymes possess regulatory activity and serve as pacemakers of the rate of metabolism. Some regulatory enzymes, called allosteric enzymes, are modulated in rate by reversible, noncovalent binding of a specific modulator or effector molecule to a regulatory or allosteric site. Such modulators may be either the substrate itself or some other metabolic intermediate. The other class of regulatory enzymes consists of those modulated by covalent modification of some specific functional group necessary for activity. Some enzymes occur in multiple forms, called isozymes, that have different kinetic characteristics. In many human genetic diseases one or more enzymes are defective in function as a result of heritable mutations.

References

Some Reference Books

BARMAN, T.: *Enzyme Handbook*, vol. 1, Springer-Verlag, New York, 1969. Valuable summary of the major properties of the then-known enzymes, classified according to the international rules.

Enzyme Nomenclature, Academic, New York, 1979. Recommendations of an International Commission on Enzymes.

FERSHT, A.: *Enzyme Structure and Mechanism*, Freeman, San Francisco, 1977. A clearly written, concise introduction.

FRIEDMANN, HERBERT (ed.): *Benchmark Papers in Biochemistry*, vol. 1, *Enzymes*, Hutchinson Ross, Stroudsburg, Pa., 1981. A collection of classic papers in enzyme chemistry with historical commentaries by the editor. Extremely interesting.

NEWSHOLME, E. A., and C. START: *Regulation in Metabolism*, Wiley, New York, 1973. Chapters 1 and 2 review the properties of regulatory enzymes.

SEGEL, I. H.: *Enzyme Kinetics: Behavior and Analysis of Rapid Equilibrium and Steady State Enzyme Systems*, Wiley, New York, 1975. More advanced.

Some Articles

ANDERSON, C. M., F. H. ZUCKER, and T. A. STEITZ: "Space-Filling Models of Kinase Clefts and Conformational Changes," *Science*, **204**:375–380 (1979). Structure of hexokinase and other enzymes utilizing ATP.

DISCHE, Z.: "The Discovery of Feedback Inhibition," *Trends Biochem. Sci.*, **1**:269 (1976).

"Enzymes: One Hundred Years," *FEBS Lett. Suppl.*, vol. 62 (1976). A series of interesting papers presented at the hundredth anniversary of the coining of the word "enzyme."

KOSHLAND, D. E., JR., and K. E. NEET: "The Catalytic and Regulatory Properties of Enzymes," *Annu. Rev. Biochem.*, **37**:359–410 (1968).

MONOD, J., J.-P. CHANGEUX, and F. JACOB: "Allosteric Proteins and Cellular Control Systems," *J. Mol. Biol.*, **6**:306–329 (1963). A classic paper that introduced the concept of allosteric regulation.

MOSBACH, K.: "Enzymes Bound to Artificial Matrixes," *Sci. Am.*, **224**:26–33, March (1971). Besides mimicking how some enzymes act in cells, enzymes bound to matrixes are used as biocatalysts in industry and offer a new medical tool.

PHILLIPS, D. C.: "The Three-Dimensional Structure of an Enzyme Molecule," *Sci. Am.*, **215**:78–90, November (1966) (offprint 1055). A superb article on the three-dimensional structure and catalytic mechanism of lysozyme.

SCHMIDT, E., and F. W. SCHMIDT: "Clinical Enzymology," *FEBS Lett. Suppl.*, vol. 62 (1976). Use of enzyme measurements in medical diagnosis.

SEGAL, H. L.: "Enzymatic Interconversion of Active and Inactive Forms of Enzymes," *Science*, **180**:25–31 (1973).

Problems

1. *Keeping the Sweet Taste of Corn* The sweet taste of freshly picked corn is due to the high level of sugar in the kernels. Store-bought corn (several days after picking) is not as sweet, since about 50 percent of the free sugar of corn is converted into starch within one day of picking. To preserve the sweetness of fresh corn the husked ears are immersed in boiling water for a few minutes ("blanched") and then cooled in cold water. Corn processed in this way and stored in a freezer maintains its sweetness. What is the biochemical basis for this procedure?

2. *Intracellular Concentration of Enzymes* To obtain a first approximation to the actual concentration of enzymes in a bacterial cell, assume that it contains 1000 different enzymes in solution in the cytosol. We can greatly simplify the problem by further assuming that each has a molecular weight of 100,000 and that all the 1000 enzymes are present in equal concentration. The bacterial cell is a cylinder of 1 μm diameter and 2.0 μm in height. If the cytosol (specific gravity, 1.20) is 20 percent soluble protein by weight, and if the soluble protein consists entirely of different enzymes, calculate the *average* molar concentration of the enzymes in this hypothetical cell.

3. *Rate Enhancement by Urease* The enzyme urease enhances the rate of urea hydrolysis at pH 8.0 and 20°C by a factor of 10^{14}. If a given quantity of urease can completely hydrolyze a given quantity of urea in 5 min at 20°C and pH 8.0, how long will it take for this amount of urea to be hydrolyzed under the same conditions in the absence of urease? Assume that both reactions take place in sterile systems so that bacteria cannot attack the urea.

4. *Requirements of Active Sites in Enzymes* The active site of an enzyme usually consists of a pocket in the enzyme surface lined with the amino acid side chains necessary to bind the substrate and catalyze its chemical transformation. Carboxypeptidase, which sequentially removes the carboxyl-terminal amino acid residue from its peptide substrates, consists of a single chain of 307 amino acids. The three essential catalytic groups in the active site are furnished by arginine 145, tyrosine 248, and glutamic acid 270, where the numbers denote the position of the amino acid in the amino acid sequence.
   (a) If the carboxypeptidase chain were a perfect α helix, how far apart (in nanometers) would arginine 145 and tyrosine 248 be? Arginine 145 and glutamic acid 270? (*Hint:* See Figure 7-6.)
   (b) Explain how it is that these three amino acids, so distantly separated in the sequence, can catalyze a reaction occurring in the space of a few tenths of a nanometer.
   (c) If only these three catalytic groups are involved in the mechanism of hydrolysis, why is it necessary for the enzyme to contain such a large number of amino acid residues?

5. *Quantitative Assay for Lactate Dehydrogenase* The muscle enzyme lactate dehydrogenase catalyzes the reaction

$$\underset{\text{Pyruvate}}{CH_3{-}\overset{\overset{\displaystyle O}{\|}}{C}{-}COO^-} + \underset{\substack{\text{Absorbs}\\\text{light at}\\\text{340 nm}}}{NADH} + H^+ \longrightarrow \underset{\text{Lactate}}{CH_3{-}\underset{\underset{\displaystyle H}{|}}{\overset{\overset{\displaystyle OH}{|}}{C}}{-}COO^-} + \underset{\substack{\text{Does not}\\\text{absorb light}\\\text{at 340 nm}}}{NAD^+}$$

In contrast to NAD^+, solutions of NADH absorb light at 340 nm in the near-ultraviolet region. This property is used to determine the concentration of NADH in solution by measuring the amount of light absorbed at 340 nm by the solution with a spectrophotometer. Explain how these properties of NADH can be used to design a quantitative assay for lactate dehydrogenase.

6. *Estimation of V_{max} and K_M by Inspection* Although graphic methods are available for accurate determination of the values of V_{max} and K_M of an enzyme-catalyzed reaction (for example, see Box 9-2), these values can be quickly estimated from the reaction velocities at increasing substrate concentrations. Using the definition of V_{max} and K_M, estimate the approximate value of V_{max} and K_M for the enzyme-catalyzed reaction from which the following data were obtained:

[S], M	V, μmol/L·min
2.5×10^{-6}	28
4.0×10^{-6}	40
1×10^{-5}	70
2×10^{-5}	95
4×10^{-5}	112
1×10^{-4}	128
2×10^{-3}	139
1×10^{-2}	140

7. *Meaning of V_{max}* In a laboratory experiment two students independently isolated the enzyme lactate dehydrogenase, which catalyzes the reduction of pyruvate to lactate, from chicken heart. The enzyme was obtained in the form of a concentrated solution. The students then measured the activity of their own enzyme solutions as a function of substrate concentration under identical conditions and thus determined the V_{max} and K_M's of their preparations. When they compared their results, they noticed that their values of K_M were identical but their values of V_{max} were radically different. One student argued that the different V_{max} values confirmed that they had isolated *different* forms of the same enzyme. The other student argued that in spite of different V_{max} values, they had isolated the *same* form of the enzyme. Which student was correct? Explain. How can they resolve the discrepancy? Explain.

8. *Relation between Reaction Velocity and Substrate Concentration: Michaelis-Menten Equation*
   (a) At what substrate concentration will an enzyme having a maximum substrate-transformation velocity of 30 μmol/min·mg and a K_M of 0.005 M show one-quarter of its maximum rate?
   (b) Determine the fraction of V_{max} that would be found at a substrate concentration of $\frac{1}{2}K_M$, $2K_M$, and $10K_M$.

9. *Graphical Analysis of V_{max} and K_M Values* The following experimental data were collected during a study of the catalytic activity of an intestinal peptidase capable of hydrolyzing the dipeptide glycylglycine:

$$\text{Glycylglycine} + H_2O \longrightarrow 2\ \text{glycine}$$

[S], mM	1.5	2.0	3.0	4.0	8.0	16.0
Product formed, mg/min	0.21	0.24	0.28	0.33	0.40	0.45

From these data determine by graphical analysis (see Box 9-2) the values of K_M and V_{max} for this enzyme preparation.

10. *The Turnover Number of Carbonic Anhydrase* Carbonic anhydrase of red blood cells, which has a molecular weight of 30,000, is among the most active known enzymes. It catalyzes the reversible hydration of CO_2

$$H_2O + CO_2 \rightleftharpoons H_2CO_3$$

which is important in the transport of CO_2 from the tissues to the lungs. If 10 μg of pure carbonic anhydrase catalyzes the hydration of 0.30 g of CO_2 in 1 min at 37°C under optimal conditions, calculate the turnover number of carbonic anhydrase.

11. *Irreversible Inhibition of an Enzyme* Many enzymes are inhibited irreversibly by heavy-metal ions such as Hg^{2+}, Cu^{2+}, or Ag^+, which can react with essential sulfhydryl groups to form mercaptides

$$E{-}SH + Ag^+ \longrightarrow E{-}S{-}Ag + H^+$$

The affinity of Ag^+ for sulfhydryl groups is so great that Ag^+ can be used to titrate —SH groups quantitatively. To 10 mL of a solution containing 1.0 mg/mL of a pure enzyme was added just enough $AgNO_3$ to inactivate the enzyme completely. A total of 0.342 μmol of $AgNO_3$ was required. Calculate the *minimum* molecular weight of the enzyme. Why does the value obtained in this way give only the minimum molecular weight?

12. *Protection of an Enzyme against Denaturation by Heat* When enzyme solutions are heated, there is a progressive loss of catalytic activity with time. This loss is the result of the unfolding of the native enzyme molecule to a randomly coiled conformation, because of its increased thermal energy. A solution of the enzyme hexokinase incubated at 45°C lost 50 percent of its activity in 12 min, but when the hexokinase was incubated at 45°C in the presence of a very large concentration of one of its substrates, glucose, it lost only 3 percent of its activity. Explain why thermal denaturation of hexokinase was retarded in the presence of one of its substrates.

13. *Clinical Application of Differential Enzyme Inhibition* Human blood serum contains a class of enzymes known as acid phosphatases which hydrolyze biological phosphate esters under slightly acidic conditions (pH 5.0):

$$R{-}O{-}\overset{\displaystyle O^-}{\underset{\displaystyle O}{\overset{|}{\underset{\|}{P}}}}{-}O^- + H_2O \longrightarrow R{-}OH + HO{-}\overset{\displaystyle O^-}{\underset{\displaystyle O}{\overset{|}{\underset{\|}{P}}}}{-}O^-$$

Acid phosphatases are produced by erythrocytes, the liver, kidney, spleen, and prostate gland. From the clinical standpoint, the enzyme from the prostate gland is most important because an increased concentration in the blood is frequently an indication of cancer of the prostate gland. The phosphatase from the prostate gland is strongly inhibited by the tartrate ion, while acid phosphatases from other tissues are not. How can this information be used to develop a specific procedure for measuring the activity of the acid phosphatase of the prostate gland in human blood serum?

14. *Inhibition of Carbonic Anhydrase by Acetazolamide* Carbonic anhydrase is strongly inhibited by the drug acetazolamide, which is used to increase the secretion of urine and to treat glaucoma, characterized by excessively high pressure within the eyeball. In these and other secretory processes carbonic anhydrase plays an important role since it participates in regulating the pH and bicarbonate content of a number of body fluids. The experimental curve of reaction velocity versus substrate concentration of the carbonic anhydrase reaction is illustrated. When the experiment is repeated in the presence of acetazolamide, the lower curve is obtained. From an inspection of the curves and your knowledge of the kinetic properties of competitive and noncompetitive enzyme inhibitors determine the nature of the inhibition given by acetazolamide. Explain.

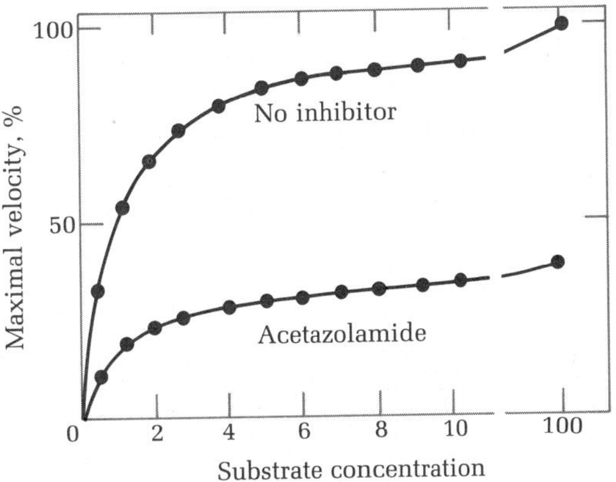

15. *Treatment of Methanol Poisoning* Methanol (wood alcohol), a commercial solvent once used as an automotive antifreeze, is very toxic and may cause death if a person swallows as little as 30 mL. The extreme toxicity is due not to the methanol itself but to its metabolic product, formaldehyde. Methanol is rapidly oxidized to formaldehyde by the action of the liver enzyme alcohol dehydrogenase:

$$NAD^+ + \underset{\text{Methanol}}{CH_3\text{—}OH} \longrightarrow NADH + H^+ + \underset{\text{Formaldehyde}}{H_2C\text{=}O}$$

Part of the medical treatment for methanol poisoning is to give the patient ethanol (ethyl alcohol), either by mouth or intravenously, in amounts that would cause intoxication in a normal individual. Explain why this treatment is effective.

16. *pH Optimum of Lysozyme* The enzymatic activity of lysozyme is optimal at pH 5.2 and decreases above and below this pH value (see figure at left). Lysozyme contains two amino acid residues in the active site essential for catalysis: glutamic acid at position 35 and aspartic acid at position 52 in the protein chain. The pK′ values of the carboxyl side chains of these two residues are 5.9 and 4.5, respectively. What is the ionization state (protonated or deprotonated) of each residue at the pH optimum of lysozyme? How can the ionization states of these two amino acid residues explain the pH-activity profile of lysozyme shown in the figure?

Problem 16

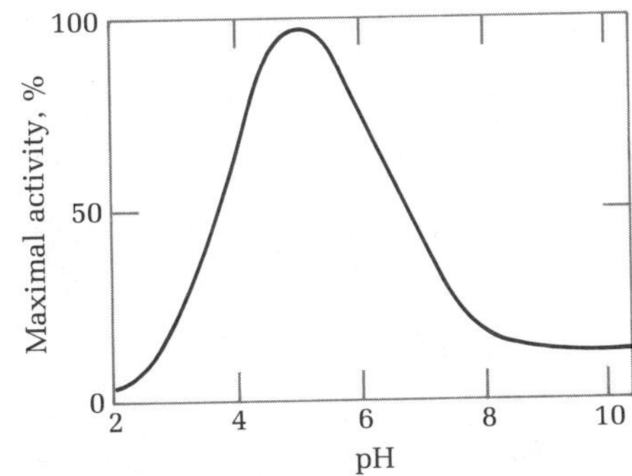

CHAPTER 10

Vitamins and Trace Elements in the Function of Enzymes

Many enzymes require a nonprotein cofactor for their catalytic function, either an organic molecule called a coenzyme or some inorganic component such as a metal ion. In some enzymes the cofactor is a direct participant in the catalytic process, and in others it serves as a transient carrier of some specific functional group derived from the substrate. Although such enzyme cofactors occur in only very small amounts in cells, they are essential for the action of many enzymes and therefore play a vital role in cell metabolism.

In this chapter we shall examine the chemical nature and function of these substances. We shall see that the vitamins, organic nutrients required in small amounts in the diet of human beings and most animals for proper growth and function, are essential precursors of various coenzymes. The discovery of the vitamins and their lifesaving value in the prevention and cure of nutritional deficiency diseases is one of the most important contributions of biochemistry to medicine and human welfare, as we shall see in Chapter 26. An equally important advance was provided by research that revealed how the vitamins and the essential minerals function in the activity of enzymes, since it has led to an understanding of the relationship between good nutrition and good health.

This chapter collects information on the structure of the vitamins and coenzymes and the essential minerals and how they function as cofactors in the action of different kinds of enzymes, as preparation for subsequent chapters on cell metabolism. In the latter we shall meet these coenzymes and minerals again and see in more detail how the deficiency of different vitamins and minerals can seriously impair specific metabolic pathways. Later, in Chapter 26, we shall examine another aspect of the vitamins and the essential elements: their role in human nutrition.

Vitamins Are Essential Organic Micronutrients

Because vitamins are required in the human diet in only milligram or microgram quantities per day, they are called *micronutrients*. This term serves to distinguish them from the *macronutrients*, namely, carbohydrates, proteins, and fats, which are required in the human diet in bulk amounts, in hundreds or at least dozens of grams per day. Macronutrients are needed in large quantities to provide energy, to yield organic precursors of many body components, and to furnish amino acids for synthesis of body proteins. The vitamins, on the other hand, are needed in only small amounts because they are catalytic in their action, making possible the numerous chemical transformations of macronutrients that together we call metabolism. Like enzymes, the active forms of the vitamins are present in only very low concentrations in the tissues.

Today 13 different vitamins are known to be required in the diet of human beings and many animal species for normal growth and function, in addition to the bulk nutrients carbohydrate, fat, and protein. The name "vitamine" was first applied to a specific organic micronutrient required to prevent the nutritional deficiency disease *beriberi*, once prevalent in rice-eating countries. Because this factor had the properties of an amine, Casimir Funk, the Polish biochemist who first purified it, called it "vitamine," denoting an amine essential for life. Later, after a number of other essential organic micronutrients were discovered, the *-e* was dropped when it was found that not all of them are amines.

Nearly all the known vitamins are present in the cells of all animals and most plants and microorganisms and perform the same important biochemical functions there. However, not all of the known vitamins are necessarily required in the diet of every animal species. For example, although vitamin C is required in the diet of people, monkeys, guinea pigs, and the Indian fruit bat, most other animals do not require vitamin C in the diet because they have enzymes to manufacture it from its simple precursor glucose. The term vitamin is thus more generally applied to a group of organic substances participating in very small amounts in the normal function of cells, which some organisms are unable to synthesize and must obtain from exogenous sources.

Vitamins Are Essential Components of Coenzymes and Enzyme Prosthetic Groups

The biochemical function of some of the vitamins first became known in the 1930s through the convergence of two lines of research, one on the chemical structure of coenzymes and the other on the structure of vitamins. In 1935 the German biochemist Otto Warburg had succeeded in isolating and identifying the structure of a coenzyme now called *nicotinamide adenine dinucleotide phosphate*, necessary in certain enzyme-catalyzed oxidation-reduction reactions in the cell. One of

Figure 10-1
Nicotinamide, the pellagra-preventive factor. It functions as a component of the coenzyme nicotinamide adenine dinucleotide (see Figure 10-6, on page 255).

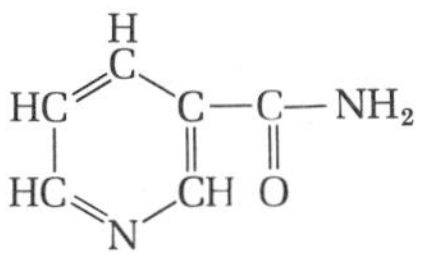

the components of this coenzyme he found to be the simple organic compound *nicotinamide* (Figure 10-1), which had first been isolated from tobacco many years earlier. Somewhat later the American biochemists D. Wayne Woolley and Conrad Elvehjem were trying to identify the chemical structure of a substance they had isolated from meat and other foods which could prevent or cure a nutritional deficiency disease of dogs called *blacktongue*, the counterpart of the human *pellagra*. Woolley and Elvehjem, struck by some chemical similarities between nicotinamide and the factor they were studying, tested pure nicotinamide in their deficient dogs and found that it would cure blacktongue. Very soon it was found that nicotinamide also cured pellagra in people. Today we know that nicotinamide is an important component of a coenzyme required in enzymatic catalysis of certain vital oxidative reactions. Although nicotinamide is a very simple molecule, most animals cannot make enough of it for their needs and must obtain it from the diet.

Soon other vitamins were found to function as components of other coenzymes or of enzyme prosthetic groups. Since coenzymes are catalytic in function and occur in very low concentrations in cells, the nutritional requirements for vitamins can be satisfied by very small amounts. For example, the human minimum daily requirement of vitamin B_6 is about 2 mg, and of vitamin B_{12} less than 3 μg.

Simultaneous with these discoveries came the recognition that a number of inorganic elements are also necessary in animal nutrition, for a similar reason: they are necessary components in the action of certain other enzymes. For example, the element zinc, essential in the diet of people and animals, is an essential component of a great many enzymes.

We shall now outline the coenzyme functions of the different vitamins. In later chapters we shall see why vitamin deficiencies, and hence deficient enzyme action, can cause serious impairment of the central pathways involved in the metabolism of carbohydrates, fats, and proteins.

Vitamins Can Be Grouped into Two Classes

Vitamins are divided into two classes, *water-soluble* and *fat-soluble* (Table 10-1). The water-soluble vitamins include thiamine (vitamin B_1), riboflavin (vitamin B_2), nicotinic acid, pantothenic acid, pyridoxine (vitamin B_6), biotin, folic acid, vitamin B_{12}, and ascorbic acid (vitamin C). The coenzyme function of nearly all of these is known. The biochemical functions of the fat-soluble vitamins A, D, E, and K, which are oily substances not readily soluble in water, are not as well understood. In addition to these well-established vitamins, there are other substances which are required by a few species but which are not generally regarded as vitamins. They include *carnitine* (page 514), *inositol*, and *lipoic acid* (Chapter 26).

Table 10-1 Vitamins and Their Role in Enzyme Function

Vitamin	Coenzyme form (or active form)	Type of reaction or process promoted
Water-soluble		
Thiamine	Thiamine pyrophosphate	Decarboxylation of α-keto acids
Riboflavin	Flavin mononucleotide, flavin adenine dinucleotide	Oxidation-reduction reactions
Nicotinic acid	Nicotinamide adenine dinucleotide, nicotinamide adenine dinucleotide phosphate	Oxidation-reduction reactions
Pantothenic acid	Coenzyme A	Acyl-group transfer
Pyridoxine	Pyridoxal phosphate	Amino-group transfer
Biotin	Biocytin	CO_2 transfer
Folic acid	Tetrahydrofolic acid	1-carbon-group transfer
Vitamin B_{12}	Deoxyadenosyl cobalamin	1,2 hydrogen shifts
Ascorbic acid	Not known	Cofactor in hydroxylation reactions
Fat-soluble		
Vitamin A	Retinal	Visual cycle
Vitamin D	1,25-Dihydroxycholecalciferol	Regulation of Ca^{2+} metabolism
Vitamin E	Not known	Protection of membrane lipids
Vitamin K	Not known	Cofactor in carboxylation reactions

We shall first examine the nature and properties of the water-soluble vitamins, their function as components of specific coenzymes, and examples of enzymatic reactions in which these coenzymes are involved (Table 10-1).

Thiamine (Vitamin B_1) Functions in the Form of Thiamine Pyrophosphate

Vitamin B_1, or thiamine, is necessary in the nutrition of most vertebrates and of some microbial species. Its deficiency in the human diet causes beriberi, a disease characterized by neurological disorders, paralysis, and loss of weight. In Asia in the 1800s and early 1900s beriberi caused the death of perhaps hundreds of thousands of people subsisting on diets in which polished or refined white rice was the major staple food. The husks, which are removed in the refining process, contain nearly all the thiamine of rice. Thiamine was first isolated in pure form in 1926. Its chemical structure was determined in the early 1930s by Robert R. Williams in the United States, and its chemical synthesis was achieved soon thereafter.

Figure 10-2
Thiamine and its active forms.

Thiamine (vitamin B_1)

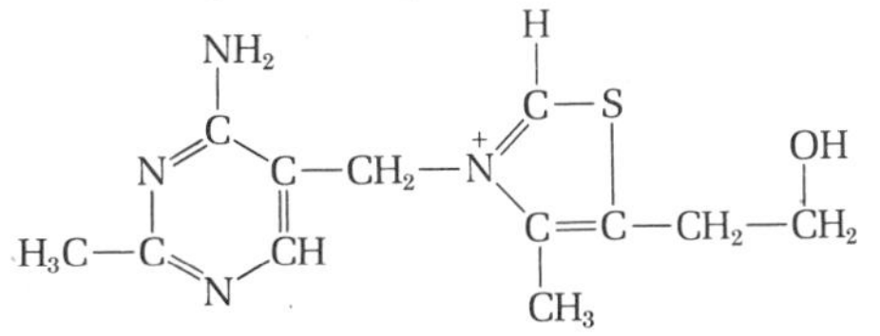

Thiamine pyrophosphate, the coenzyme form. The reactive group is in color.

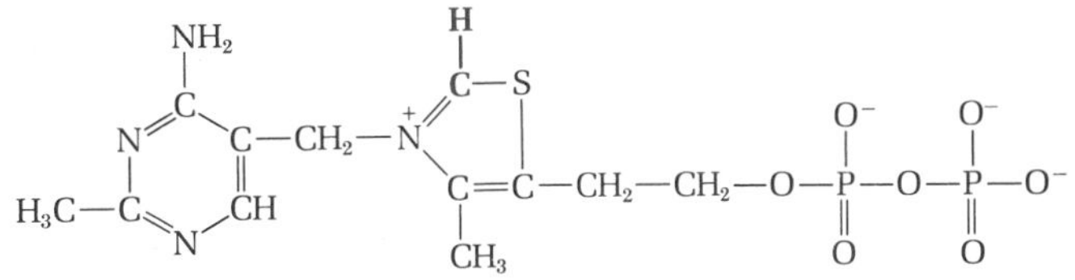

α-Hydroxyethylthiamine pyrophosphate, the transient form carrying an "active" acetaldehyde group, shown in color

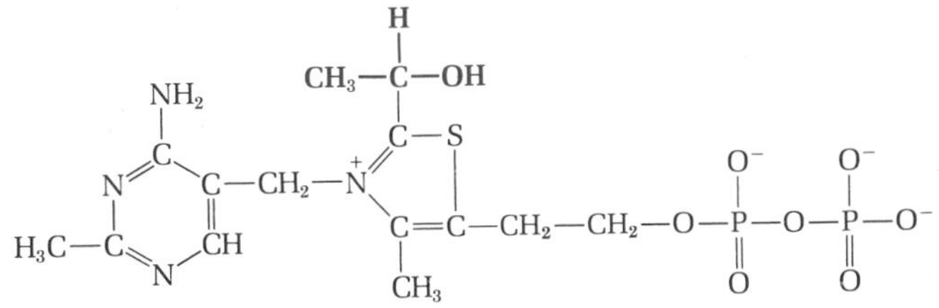

Thiamine contains two ring systems, a pyrimidine and a thiazole (Figure 10-2). In animal tissues it is largely present as *thiamine pyrophosphate*, its coenzyme form (Figure 10-2). Thiamine pyrophosphate functions as a coenzyme in several enzymatic reactions in which aldehyde groups are transferred from a donor to an acceptor molecule. In such reactions the thiamine pyrophosphate serves as a transient intermediate carrier of the aldehyde group, which is covalently attached to the thiazole ring. A simple example is the reaction catalyzed by *pyruvate decarboxylase* (Figure 10-3), an important step in the fermentation of glucose by yeast to yield alcohol. In the pyruvate decarboxylase reaction the carboxyl group of pyruvate is lost as CO_2, and the rest of the pyruvate molecule, sometimes referred to as *active acetaldehyde*, is simultaneously transferred to the 2 position of the thiazole ring of the tightly bound

Figure 10-3
The enzymatic decarboxylation of pyruvate by pyruvate decarboxylase, designated E, which requires thiamine pyrophosphate (TPP) as a tightly bound prosthetic group. The overall reaction is shown first. Below it the overall reaction is broken down into separate steps, to show the role of the thiamine pyrophosphate as intermediate carrier of the acetaldehyde.

Overall reaction

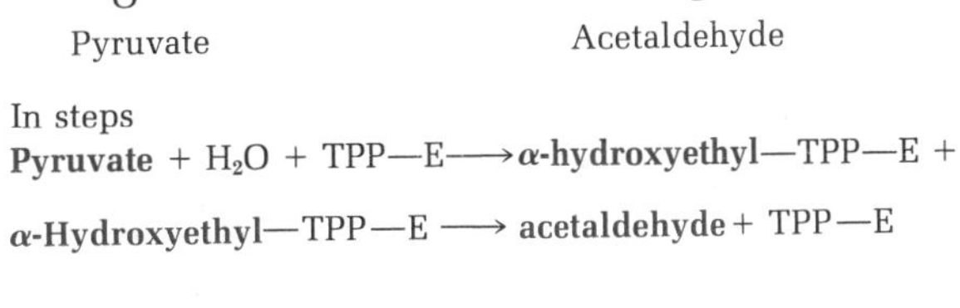

$$\underset{\text{Pyruvate}}{CH_3{-}\overset{\displaystyle O}{\overset{\|}{C}}{-}COO^-} + H_2O \xrightarrow{\text{pyruvate decarboxylase}} \underset{\text{Acetaldehyde}}{CH_3{-}\overset{\displaystyle O}{\overset{\|}{C}}{-}H} + HCO_3^-$$

In steps

Pyruvate + H_2O + TPP—E ⟶ **α-hydroxyethyl**—TPP—E + $\mathbf{HCO_3^-}$

α-Hydroxyethyl—TPP—E ⟶ **acetaldehyde** + TPP—E

thiamine pyrophosphate, to yield its hydroxyethyl derivative. This intermediate exists only transiently, since the hydroxyethyl group is quickly cleaved from the coenzyme to yield free acetaldehyde. Thiamine pyrophosphate also serves as the coenzyme in the more complex *pyruvate dehydrogenase* and *α-ketoglutarate dehydrogenase* reactions taking place in the main pathway of oxidation of carbohydrate in cells.

Riboflavin (Vitamin B_2) Is a Component of the Flavin Nucleotides

Vitamin B_2, or *riboflavin*, first isolated from milk, was identified and synthesized in 1935. Its intense yellow color is due to its complex isoalloxazine ring system (Figure 10-4). Later it was discovered that riboflavin is a component of two closely related coenzymes, *flavin mononucleotide* (FMN) and *flavin adenine dinucleotide* (FAD), shown in Figure 10-4. They function as tightly bound prosthetic groups of a class of dehydrogenases known as *flavoproteins* or *flavin dehydrogenases*. In the reactions catalyzed by these enzymes the isoalloxazine ring of the flavin nucleotides serves as a transient carrier of a pair of hydrogen atoms removed from the substrate molecule (Figure 10-5). *Succinate dehydrogenase*, discussed before (page 223), is an example of a flavin dehydrogenase. It contains

Figure 10-4
Riboflavin (vitamin B_2) and its coenzyme forms.

Riboflavin

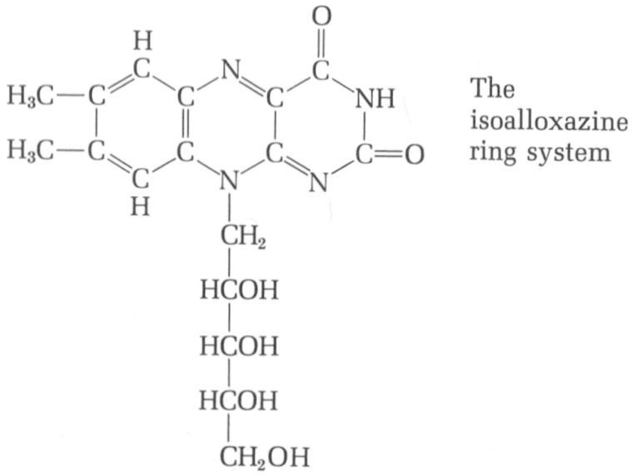

The isoalloxazine ring system

Flavin mononucleotide (FMN). The reactive group is in color.

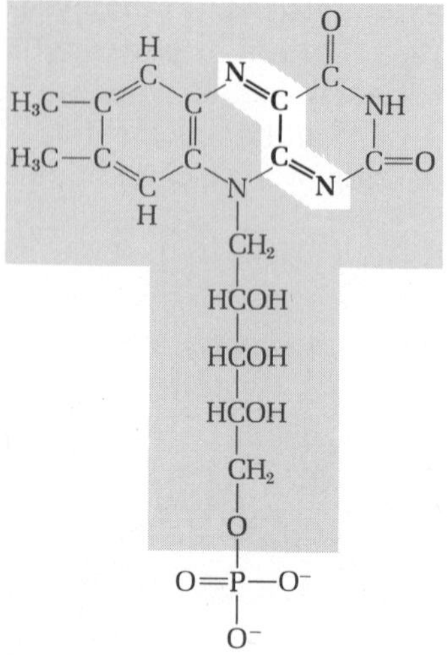

Flavin adenine dinucleotide (FAD). The reactive group is in color.

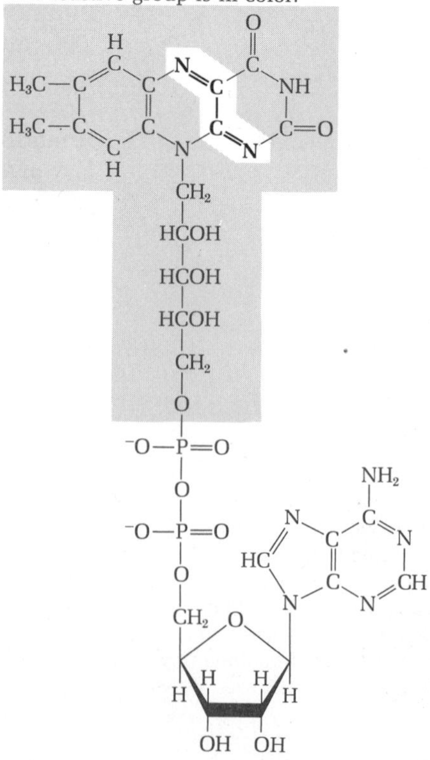

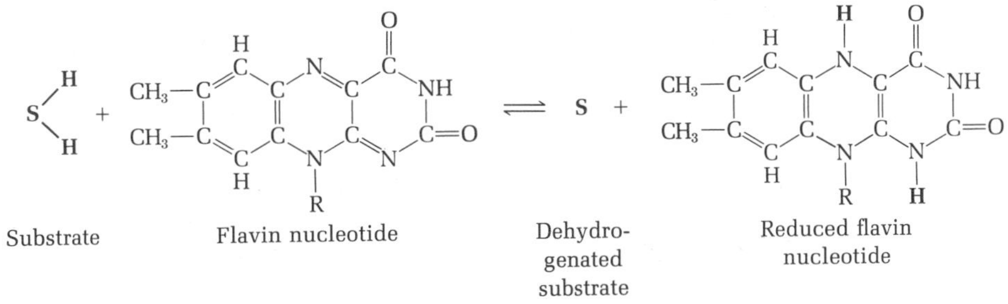

Figure 10-5
The reversible transfer of a pair of H atoms from the substrate to the isoalloxazine ring of FMN or FAD by a flavin dehydrogenase. R designates the remainder of the flavin nucleotide.

a covalently bound prosthetic group of FAD and catalyzes the reaction

$$\text{Succinate} + \text{E—FAD} \longrightarrow \text{fumarate} + \text{E—FADH}_2$$

in which E—FAD designates the succinate dehydrogenase molecule with its bound FAD. In most other flavin dehydrogenases the FMN or FAD is noncovalently bound. Some flavin dehydrogenases also contain iron or some other metal as part of their active sites.

Nicotinamide Is the Active Group of the Coenzymes NAD and NADP

Deficiency of nicotinic acid (Figure 10-6) in the diet leads to the human deficiency disease pellagra (from Italian meaning "rough skin"). Pellagra occurs in many regions of the world where the diet is poor in meat, milk, and eggs and heavily dependent on corn as a staple. Both nicotinic acid and its amide nicotinamide are active in preventing and curing pel-

Figure 10-6
(a) Two forms of the pellagra-preventive vitamin and (b) the structures of its active coenzyme forms, nicotinamide adenine dinucleotide (NAD^+) and nicotinamide adenine dinucleotide phosphate ($NADP^+$). They contain two nucleotide units, each consisting of a nitrogenous base (nicotinamide or adenine), a 5-carbon sugar (D-ribose), and a phosphate group. The structures shown are the oxidized forms of the nucleotides. The reduced form of NAD is shown in Figure 10-7.

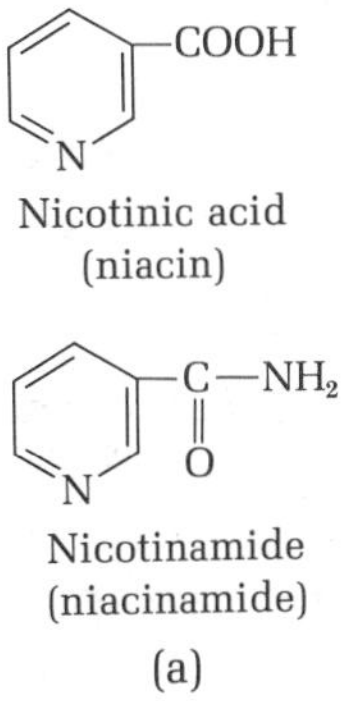

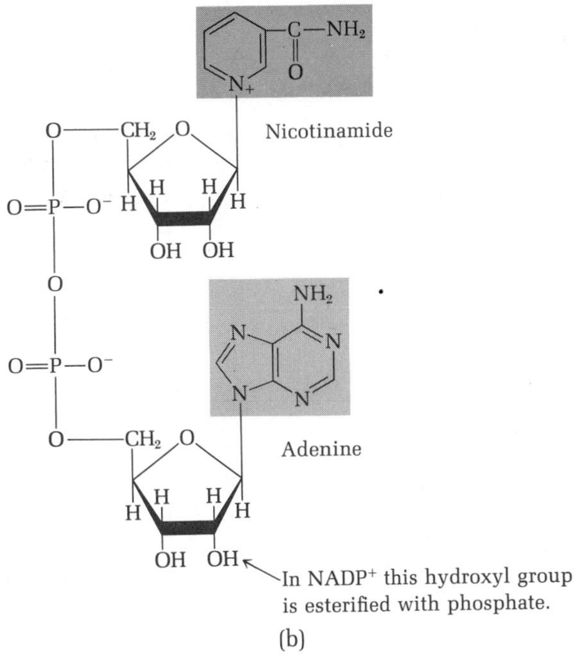

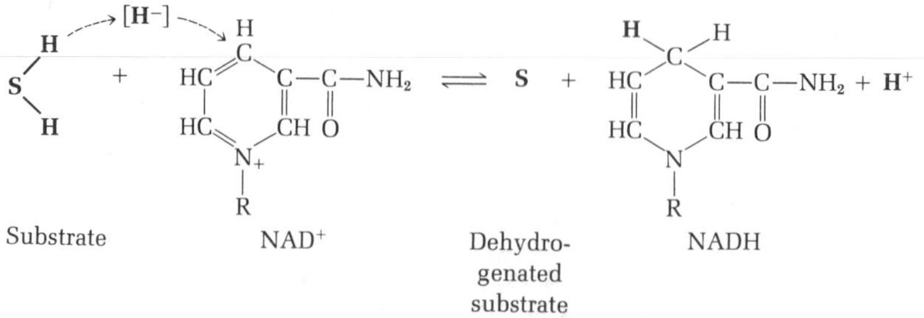

Substrate NAD$^+$ Dehydrogenated substrate NADH

Figure 10-7
General reaction showing how NAD$^+$ acts as a coenzyme in enzymatic dehydrogenation reactions. The substrate molecule and its reaction products are in color. Only the nicotinamide portion of NAD$^+$ is shown, the remainder of the NAD$^+$ molecule being denoted by R.

lagra. Because their names might mislead some people into thinking that tobacco is nutritious, nicotinic acid has been given the alternative name *niacin* for public use.

Nicotinamide is a component of the two related coenzymes *nicotinamide adenine dinucleotide* (NAD) and *nicotinamide adenine dinucleotide phosphate* (NADP), shown in Figure 10-6. NADP consists of NAD with an extra phosphate group. These coenzymes occur in oxidized (designated NAD$^+$ and NADP$^+$) and reduced (NADH and NADPH) forms. The nicotinamide component of these coenzymes serves as the transient intermediate carrier of a *hydride ion* that is enzymatically removed from a substrate molecule by the action of certain *dehydrogenases* (Figure 10-7). An example of such an enzymatic reaction is that catalyzed by *malate dehydrogenase*, which dehydrogenates malate to yield oxaloacetate, a step in the oxidation of carbohydrates and fatty acids. This enzyme catalyzes the reversible transfer of a hydride ion from malate to NAD$^+$ to form NADH; the other hydrogen atom leaves the hydroxyl group of malate to appear as a free H$^+$ ion:

$$\text{L-Malate} + \text{NAD}^+ \xrightleftharpoons{\text{malate dehydrogenase}} \text{oxaloacetate} + \text{NADH} + \text{H}^+$$

Many dehydrogenases of this type are known, each specific for a given substrate. Some can function only with NAD$^+$ as coenzyme, others require NADP$^+$, while a few are active with either coenzyme. In most dehydrogenases NAD (or NADP) is bound to the enzyme protein only transiently during the catalytic cycle; but in others, the coenzyme is very tightly bound and remains on the enzyme active site permanently.

Pantothenic Acid Is a Component of Coenzyme A

Pantothenic acid (Figure 10-8) was first isolated in 1938 from yeast and liver extracts by Roger Williams (the brother of Robert Williams, who established the structure of thiamine). Pantothenic acid (*pan-* means "everywhere") has been found in all plant and animal tissues tested and also in microorganisms, but it was not until a dozen years after its isolation that its coenzyme function was identified by Fritz Lipmann and Nathan Kaplan. They had discovered a heat-stable cofactor necessary for promoting ATP-dependent enzymatic acetylation of alcohols or amines. On purification and analysis of this factor, called

Figure 10-8
Pantothenic acid and coenzyme A.

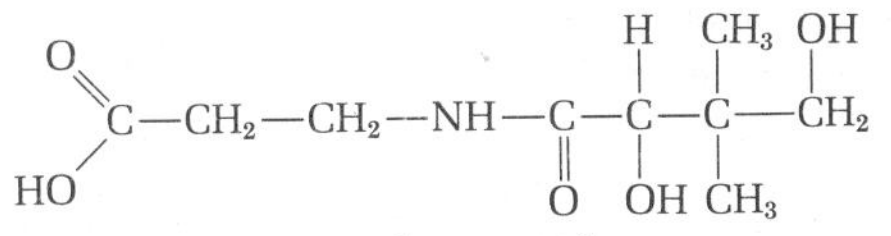

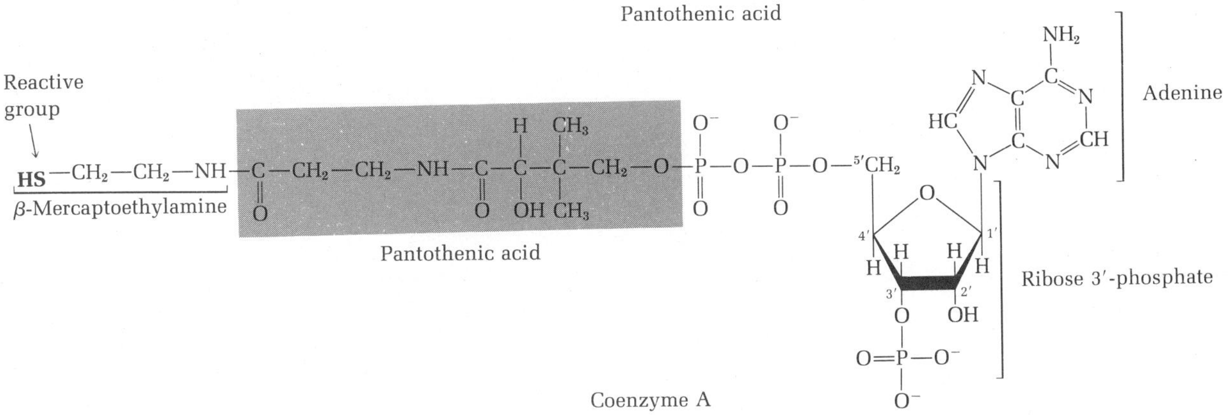

coenzyme A (for acetylation), they discovered that it contains pantothenic acid in a bound form. Today we know that coenzyme A has a much broader significance, since it is required in many different enzymatic reactions involving not only acetyl groups but acyl groups in general. Coenzyme A (abbreviated as CoA or CoA-SH) is a transient carrier of acyl groups.

The coenzyme A molecule (Figure 10-8) has a reactive thiol (—SH) group, to which acyl groups become covalently linked to form thioesters during acyl-group transfer reactions. Thioesters are esters of thiols (symbolized R—SH) rather than alcohols (R—OH). Figure 10-9 shows how the thiol group of coenzyme A functions as an acyl-group carrier. In the first reaction in Figure 10-9 acetyl-CoA is formed during the oxidative decarboxylation of pyruvate by the pyruvate dehydrogenase complex. In the second reaction, the acetyl group of acetyl-CoA is transferred to oxaloacetate to yield citrate by citrate synthase (Figure 10-9). The latter is the first reaction of the citric acid cycle, the central pathway for the oxidative degradation of carbohydrate and fatty acids in aerobic cells (Chapter 16).

Figure 10-9
The role of coenzyme A in the pyruvate dehydrogenase and citrate synthase reactions. The acetyl group being transferred is in color.

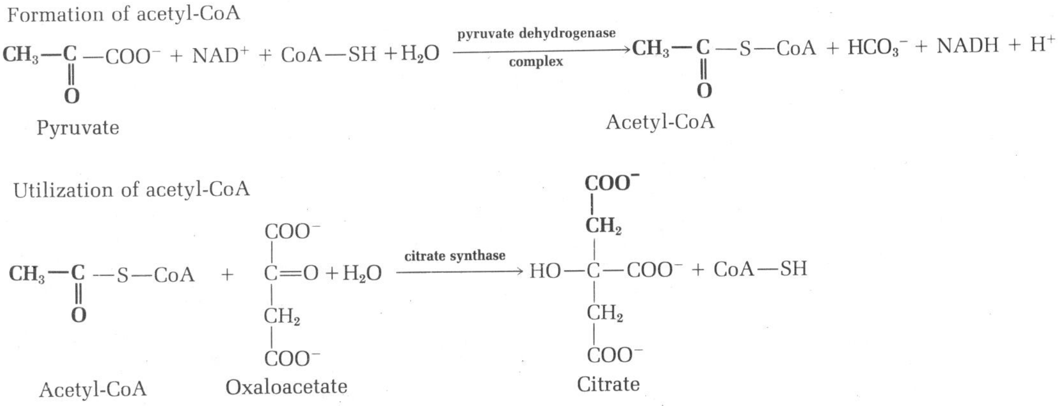

Formation of acetyl-CoA

$$\underset{\text{Pyruvate}}{CH_3-\overset{O}{\overset{\|}{C}}-COO^-} + NAD^+ + CoA-SH + H_2O \xrightarrow[\text{complex}]{\text{pyruvate dehydrogenase}} \underset{\text{Acetyl-CoA}}{CH_3-\overset{O}{\overset{\|}{C}}-S-CoA} + HCO_3^- + NADH + H^+$$

Utilization of acetyl-CoA

$$\underset{\text{Acetyl-CoA}}{CH_3-\overset{O}{\overset{\|}{C}}-S-CoA} + \underset{\text{Oxaloacetate}}{\begin{array}{c}COO^-\\|\\C{=}O\\|\\CH_2\\|\\COO^-\end{array}} + H_2O \xrightarrow{\text{citrate synthase}} \underset{\text{Citrate}}{\begin{array}{c}COO^-\\|\\CH_2\\|\\HO-C-COO^-\\|\\CH_2\\|\\COO^-\end{array}} + CoA-SH$$

Pyridoxine (Vitamin B_6) Is Important in Amino Acid Metabolism

The vitamin B_6 group consists of three closely related compounds, pyridoxine, pyridoxal, and pyridoxamine (Figure 10-10), which are readily interconvertible biologically. The active form of vitamin B_6 is pyridoxal phosphate, which also occurs in its amino form pyridoxamine phosphate. Pyridoxal phosphate serves as the tightly bound prosthetic group of a number of enzymes catalyzing reactions of amino acids. The

Figure 10-10
Active forms of vitamin B_6, its coenzyme forms, and the transaminase reaction.

Active forms of vitamin B_6

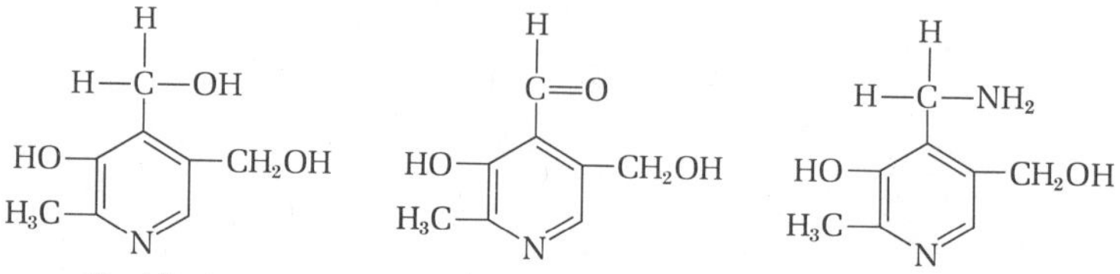

Pyridoxine — Pyridoxal — Pyridoxamine

Coenzyme forms of vitamin B_6

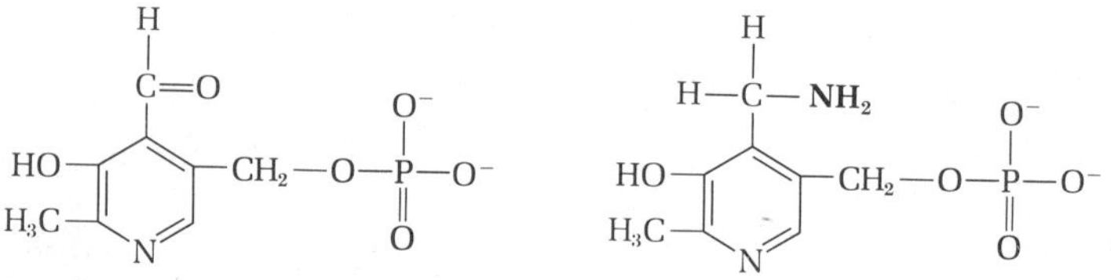

Pyridoxal phosphate, the amino-group-acceptor form — Pyridoxamine phosphate, the amino-group donor form

A transamination reaction, in which pyridoxal phosphate acts as intermediate amino-group carrier on the enzyme active site. It is written in two steps, with the transaminase and its prosthetic group designated by the two forms $E{-}\phi{-}\underset{\|}{\overset{}{C}}{-}H$ (with =O) and $E{-}\phi{-}C{-}H$ (with H and NH_2).

$$\text{Glutamate} + E{-}\phi{-}C({=}O){-}H \rightleftharpoons \alpha\text{-Ketoglutarate} + E{-}\phi{-}CH_2{-}NH_2$$

$$E{-}\phi{-}CH_2{-}NH_2 + \text{Oxaloacetate} \rightleftharpoons E{-}\phi{-}C({=}O){-}H + \text{Aspartate}$$

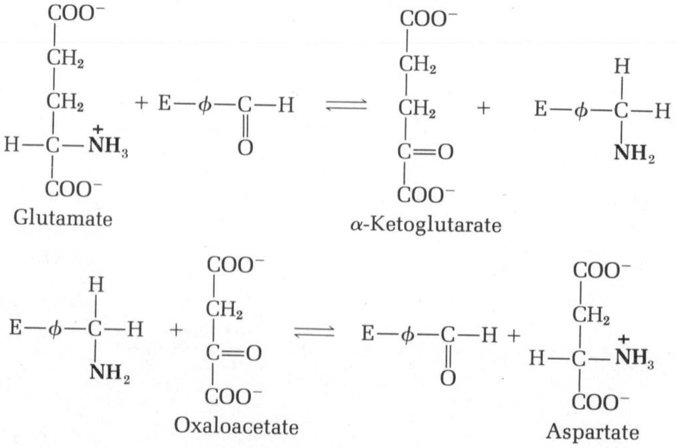

Sum: Glutamate + oxaloacetate $\rightleftharpoons$ α-ketoglutarate + aspartate

most common and best known of these are *transaminations*, reactions in which an amino group of an α-amino acid is reversibly transferred to the α carbon atom of an α-keto acid (Figure 10-10). In such reactions, catalyzed by enzymes called *transaminases* or *aminotransferases*, the tightly bound pyridoxal phosphate serves as a transient intermediate carrier of the amino group from its donor, the α-amino acid, to the amino-group acceptor, the α-keto acid. In the catalytic cycle of transaminases the amino group of the incoming α-amino acid substrate is transferred to the enzyme-bound pyridoxal phosphate. The resulting amino derivative of the coenzyme, pyridoxamine phosphate, now donates its amino group to the second substrate, the α-keto acid, and the coenzyme reverts to its pyridoxal phosphate form. Transaminations like this can occur from any one of many different amino acids to *α-ketoglutarate*, which acts as a common amino-group acceptor, to yield *glutamic acid*, a central metabolite in amino-group metabolism.

Transaminases typically catalyze *double-displacement*, or *ping-pong*, reactions (page 218). Note that the amino group is transferred from the first substrate, the amino acid, to the coenzyme, followed by departure of the resulting α-keto acid, before the second substrate, the incoming α-keto acid, is bound. The amino group is then transferred from the pyridoxamine phosphate to the second substrate.

Biotin Is the Active Component of Biocytin, the Prosthetic Group of Some Carboxylating Enzymes

In 1935 Frits Kögl, a Dutch biochemist, isolated in crystalline form from 250 kg of dried egg yolks about 1 mg of a growth factor required by yeast cells and also by rats fed large amounts of raw egg white. The new growth factor was called *biotin*. Although eggs are otherwise nutritious and do contain biotin, consumption of large amounts of uncooked egg whites can cause biotin deficiency in animals. This paradox is the result of the fact that egg white contains a protein, *avidin*, which binds biotin very tightly and does not allow the vitamin to be absorbed from the intestine.

In biotin-dependent enzymes the biotin molecule is covalently attached to the enzyme protein through an amide linkage with the ϵ-amino group of a specific lysine residue at the enzyme active site. The name *biocytin* is given to this *biotinyllysine* residue, which can be isolated from biotin-containing enzymes after acid or enzymatic hydrolysis (Figure 10-11). Biotin is a transient carrier of a carboxy ($—COO^-$) group in a number of enzymatic carboxylation reactions requiring ATP. The carboxy group is transiently attached to a nitrogen atom of the double-ring system of biotin. An example of a biotin-dependent carboxylation reaction is that catalyzed by *pyruvate carboxylase*, which carboxylates pyruvate to yield oxaloacetate (Figure 10-11).

Figure 10-11
Biotin and its active form, biocytin, the prosthetic group of some carboxylating enzymes.

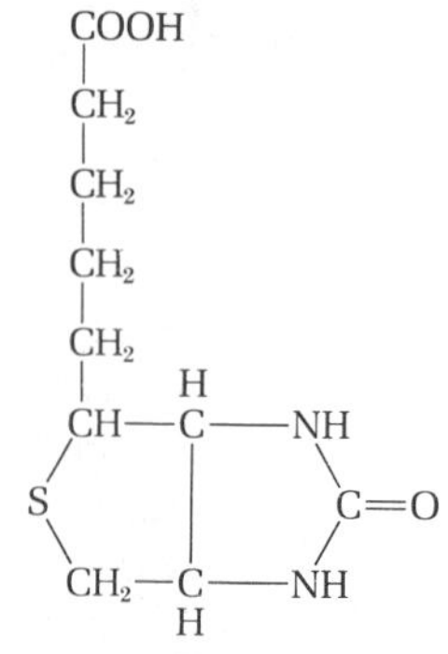

COOH
CH_2
CH_2
CH_2
CH_2
H
CH—C—NH
S C=O
CH_2—C—NH
H

Biotin

Biocytin (biotinyllysine residue). The reactive group is in color.

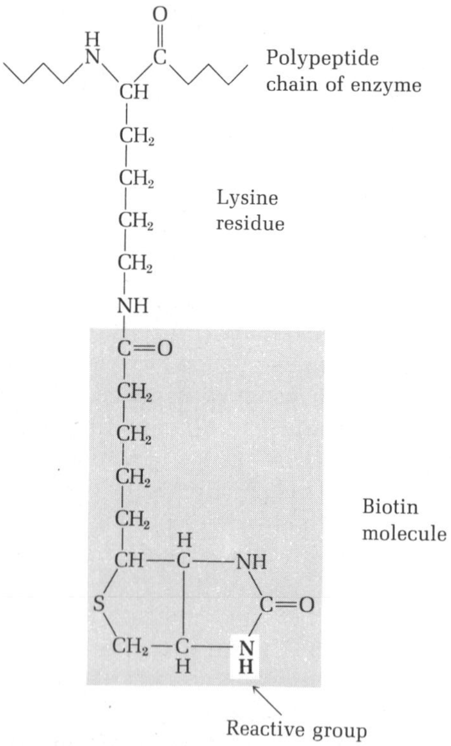

Polypeptide chain of enzyme

Lysine residue

Biotin molecule

Reactive group

The *N*-carboxy derivative of biocytin formed as transient intermediate in biotin-dependent carboxylation reactions. Only the ring system of biocytin is shown.

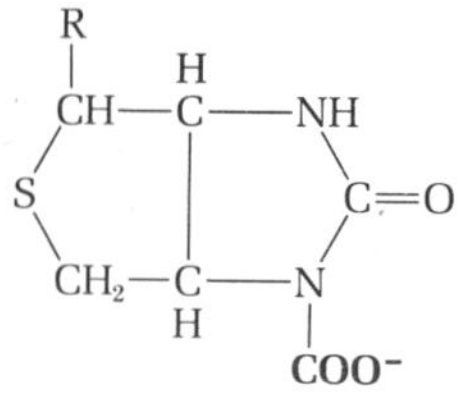

R
H
CH—C—NH
S C=O
CH_2—C—N
H
COO^-

The carboxylation of pyruvate to oxaloacetate, an important step in the biosynthesis of glucose from pyruvate. It is catalyzed by pyruvate carboxylase, a biotin-dependent enzyme.

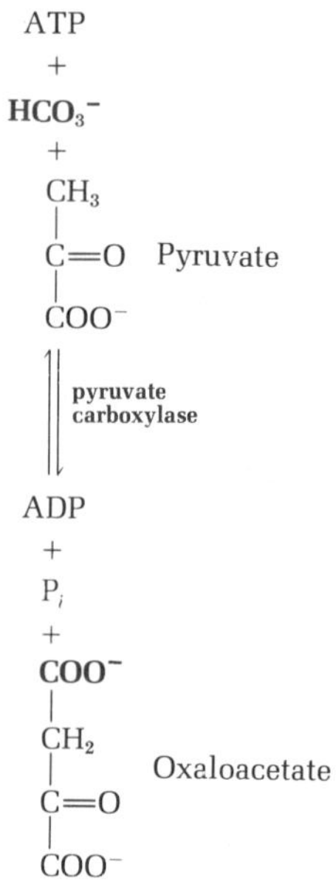

ATP + HCO_3^- + CH_3—C(=O)—COO^- (Pyruvate)

⇌ pyruvate carboxylase

ADP + P_i + COO^-—CH_2—C(=O)—COO^- (Oxaloacetate)

Folic Acid Is the Precursor of the Coenzyme Tetrahydrofolic Acid

Folic acid (Latin *folium,* "leaf") was first isolated from spinach leaves but has a very broad biological distribution. It has three major components: glutamic acid, *p*-aminobenzoic acid, and a derivative of the heterocyclic fused-ring compound pteridine (Figure 10-12). Deficiency of folic acid, also known as pteroylglutamic acid, causes a type of anemia in which red blood cells do not mature properly. Folic acid has no coenzyme activity itself, but it is enzymatically reduced in the tissues to tetrahydrofolic acid (FH_4), its active coenzyme form. Tetrahydrofolate functions as an intermediate carrier of 1-carbon groups in a number of complex enzymatic reactions in which methyl

Figure 10-12
Folic acid, its coenzyme form tetrahydrofolate, and methylenetetrahydrofolate. The methylene group is one of five different 1-carbon groups that can be transferred via tetrahydrofolate.

Folic acid

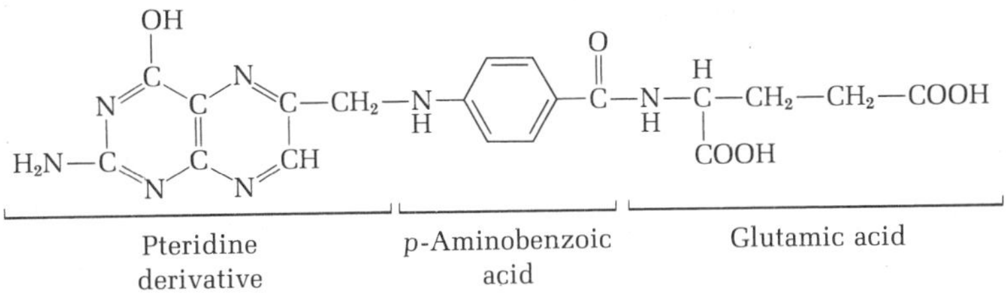

Pteridine derivative | p-Aminobenzoic acid | Glutamic acid

Tetrahydrofolate. The coenzyme form of folic acid. The four hydrogen atoms added are in color.

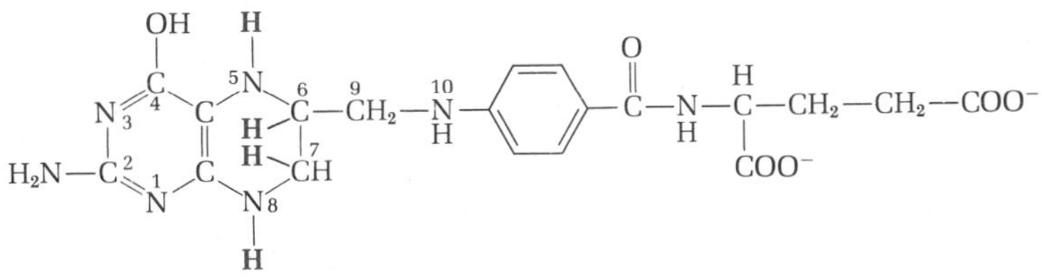

N^5,N^{10}-Methylenetetrahydrofolate. The methylene group being transferred is in color.

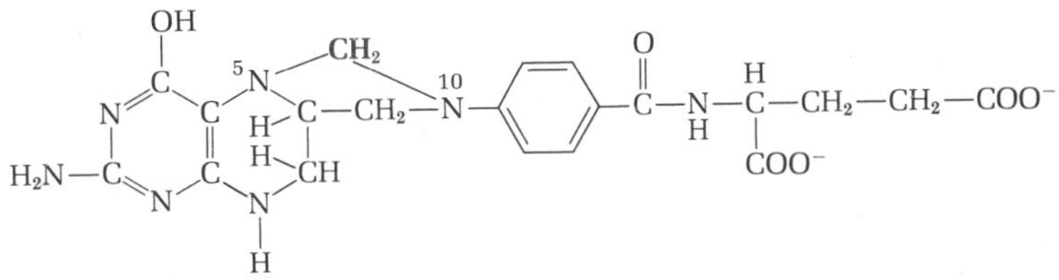

The 1-carbon groups transferred by enzymes requiring tetrahydrofolate

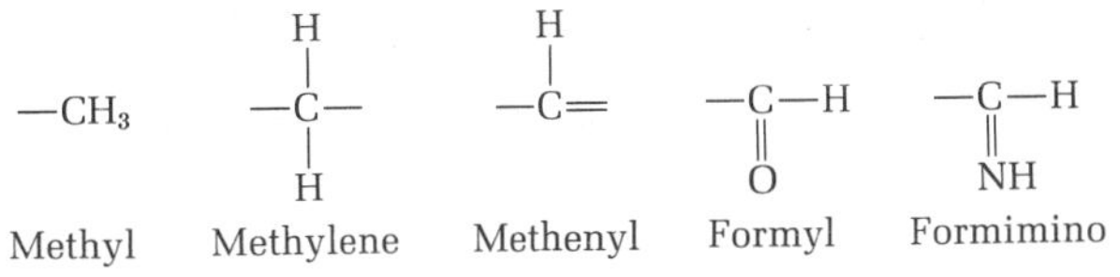

Methyl ($-CH_3$) Methylene Methenyl Formyl Formimino

($-CH_3$), methylene ($-CH_2-$), methenyl ($-CH=$), formyl ($-CHO$), or formimino ($-CH=NH$) groups are transferred from one molecule to another (Figure 10-12). An example of such a reaction is shown in Figure 10-13.

The reduction of the vitamin folic acid to its active form tetrahydrofolate occurs in two steps, in which two successive pairs of hydrogen atoms are added. The second step, catalyzed by *dihydrofolate reductase*, is severely inhibited by certain drugs useful in the treatment of some forms of cancer. Since tetrahydrofolate is an essential coenzyme in the biosynthesis of *thymidylic acid*, a nucleotide building block of DNA, these drugs inhibit the replication of DNA in susceptible cancer cells.

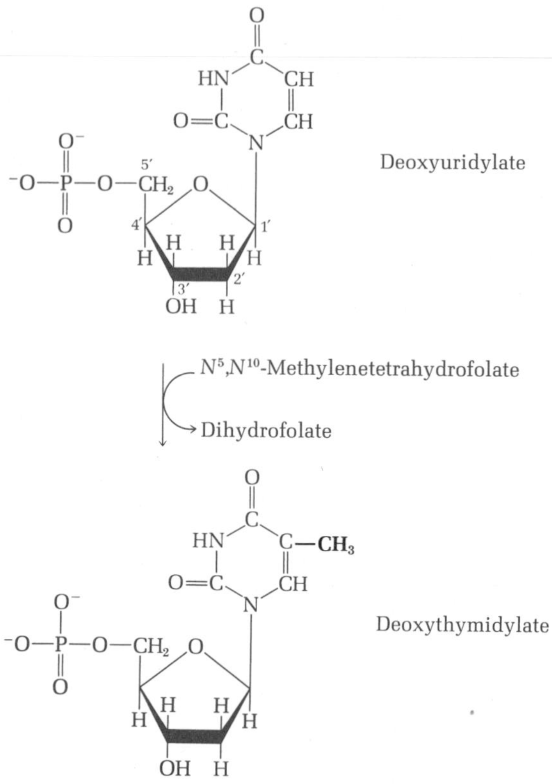

Figure 10-13
Role of N^5,N^{10}-methylenetetrahydrofolate (Figure 10-12) as a methyl-group donor in the enzymatic formation of thymidylic acid, a building block of DNA. The newly inserted methyl group is in color.

Deoxyuridylate

N^5,N^{10}-Methylenetetrahydrofolate

Dihydrofolate

Deoxythymidylate

Figure 10-14
The similarity between the structures of p-aminobenzoic acid and sulfanilamide, a competitive inhibitor of the enzyme system that incorporates p-aminobenzoate into folic acid.

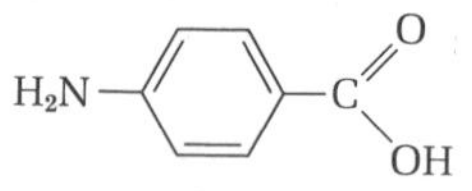

p-Aminobenzoic acid

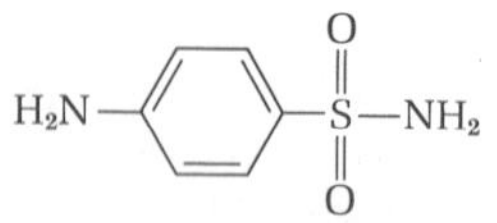

Sulfanilamide

Some bacteria do not require preformed folic acid as a growth factor because they can make it if *p-aminobenzoic acid*, one of the components of folic acid, is available. Thus p-aminobenzoic acid is a vitamin for these bacteria. This was a valuable discovery because it led to our understanding of the mode of action of *sulfanilamide*, an important drug that inhibits the growth of pathogenic bacteria requiring p-aminobenzoic acid. Figure 10-14 shows the close similarity between the structures of p-aminobenzoate and sulfanilamide, which competes with p-aminobenzoate in the enzymatic synthesis of folic acid.

Vitamin B_{12} Is the Precursor of Coenzyme B_{12}

Vitamin B_{12}, the most complex of the vitamins, has had an extraordinary history. In 1926 two American physicians, George Minot and William Murphy, discovered that large amounts of partially cooked liver in the diet cure patients of pernicious anemia, a complex and often fatal disease for which no animal counterpart is known. Little progress was made in isolating the liver anti-pernicious-anemia factor until the late 1940s, when Mary Shorb identified a bacterial species that requires it for growth. The growth response of this organism could be used for a rapid, simple assay of the factor. Vitamin B_{12} was isolated in crystalline form in 1948 by E. Lester Smith in England and by Edward Rickes and Karl Folkers in the United States, but it took another 10 years to determine its very complex structure (Figure 10-15), which required x-ray diffraction analysis.

Vitamin B_{12} is unique among all the vitamins in that it contains not only a complex organic molecule but also an essential trace element, *cobalt*. Vitamin B_{12} as usually isolated is called *cyanocobalamine* because it contains a cyano group attached to the cobalt. The complex *corrin ring* system of vitamin B_{12} (Figure 10-15), to which cobalt is coordinated, is chemically related to the porphyrin ring system of heme and heme proteins (page 171). In the coenzyme form of vitamin B_{12}, called *5′-deoxyadenosylcobalamine*, the cyano group is replaced by the *5′-deoxyadenosyl* group (Figure 10-15).

Vitamin B_{12} is not made by either plants or animals and can be synthesized by only a few species of microorganisms. It is required in only minute amounts, about 3 μg per day, by otherwise healthy people. Its complex role in the cure of pernicious anemia will be described in Chapter 26.

Figure 10-15
Vitamin B_{12} and its coenzyme form.

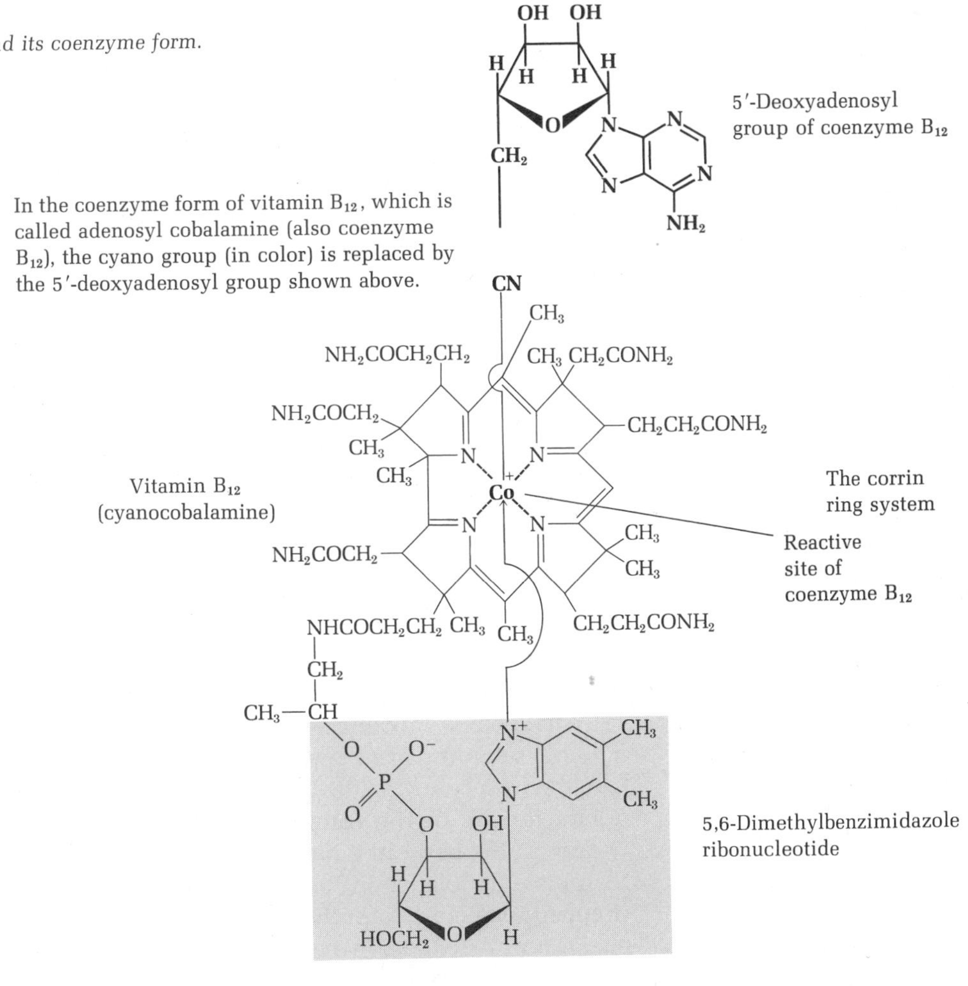

Figure 10-16
Reactions promoted by coenzyme forms of vitamin B_{12}.

The 1,2 shift of a hydrogen atom in exchange with the group X, promoted by enzymes dependent on coenzyme B_{12}.

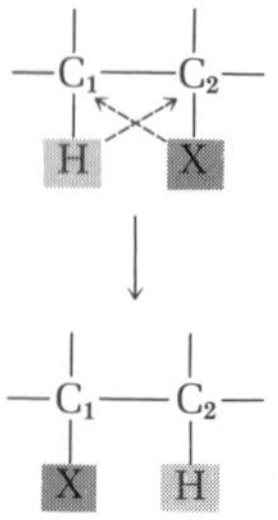

$$-\overset{|}{\underset{|}{C_1}}-\overset{|}{\underset{|}{C_2}}- \quad (H \text{ on } C_1,\ X \text{ on } C_2) \longrightarrow -\overset{|}{\underset{|}{C_1}}-\overset{|}{\underset{|}{C_2}}- \quad (X \text{ on } C_1,\ H \text{ on } C_2)$$

A coenzyme B_{12} reaction, catalyzed by methylaspartate mutase. Note the exchange of H and X groups between adjacent carbon atoms.

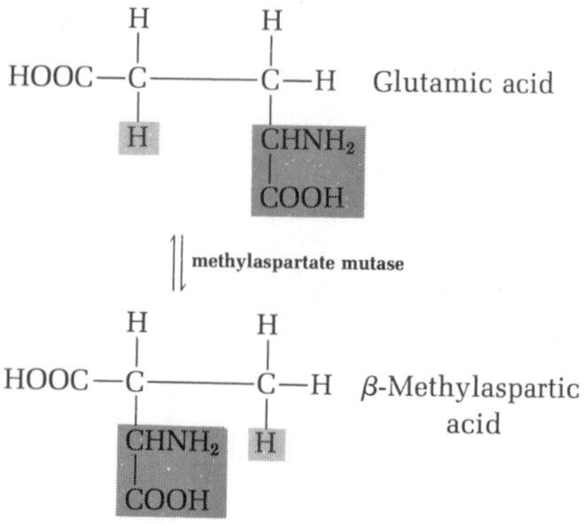

$$HOOC-CH_2-CH_2-CH(NH_2)-COOH \quad \text{Glutamic acid}$$

$$\rightleftharpoons \text{ methylaspartate mutase}$$

$$HOOC-CH(CHNH_2COOH)-CH_3 \quad \beta\text{-Methylaspartic acid}$$

Enzymes requiring coenzyme B_{12} have as a common denominator the ability to carry out the shift of a hydrogen atom from one carbon atom to an adjacent one, in exchange for an alkyl, carboxyl, hydroxyl, or amino group, as shown in Figure 10-16. Also given is one of the characteristic reactions in which coenzyme B_{12} or cobamide serves as an essential cofactor. *Methylcobalamine*, another coenzyme form of vitamin B_{12}, participates in some enzymatic reactions involving the transfer of methyl groups.

The Biochemical Function of Vitamin C (Ascorbic Acid) Is Not Known

Although it had been known since the 1790s that a factor in citrus fruit prevents scurvy, the factor was not isolated and identified until 1933, when C. Glen King and W. A. Waugh in the United States finally isolated the antiscurvy factor from lemon juice. Determination of its structure soon followed (Figure 10-17). Ascorbic acid is present in all animal and higher plant tissues. It is required in the diet of human beings and only a few other vertebrates; most animals and probably all plants can synthesize ascorbic acid from glucose. Ascorbic acid is not present in or required by microorganisms.

Ascorbic acid appears to act as a cofactor in the enzymatic hydroxylation of proline residues of the collagen of connective tissue of vertebrates, to form *4-hydroxyproline* residues (Figure 5-7). Hydroxyproline residues are found only in collagen (page 157) and no other animal proteins. Although ascorbic acid thus appears to function in the formation and maintenance of a major component of connective tissue of higher animals, it is still uncertain that this is the sole or even a major function of this vitamin.

Figure 10-17
Ascorbic acid (vitamin C) and its oxidation product, dehydroascorbic acid. The latter, although biologically active, is very unstable and decomposes readily.

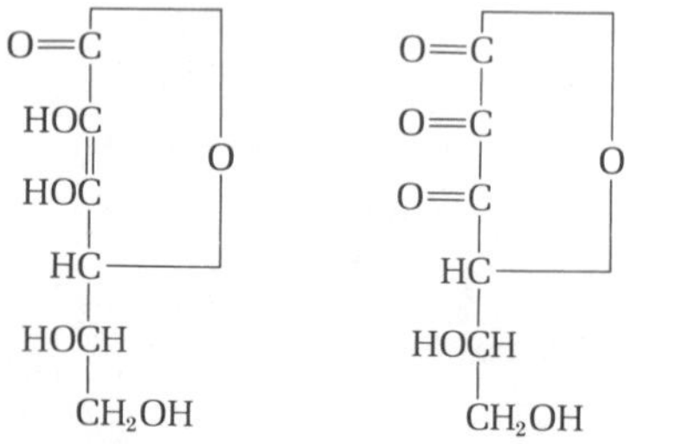

L-Ascorbic acid: O=C–HOC=HOC–HC(–O–)–HOCH–CH_2OH

L-Dehydroascorbic acid: O=C–O=C–O=C–HC(–O–)–HOCH–CH_2OH

The Fat-Soluble Vitamins Are Derivatives of Isoprene

The four fat-soluble vitamins (A, D, E, and K) are formed biologically from units of the 5-carbon hydrocarbon *isoprene*, also called *2-methylbutadiene* (Figure 10-18), a building block of many naturally occurring oily, greasy, or rubbery substances of plant origin. Natural rubber and gutta-percha, used to make golf-ball covers, are polymers of isoprene. To show the isoprenoid origin of the four fat-soluble vitamins the isoprene units are designated in their structural formulas as shown in the structure of vitamin A (Figure 10-19).

The specific biochemical or coenzyme functions of the fat-soluble vitamins have been obscure for a long time, but much progress is being made in their elucidation. One significant characteristic of the fat-soluble vitamins is that they can be stored in the body in large amounts, so that the effects of their complete deficiency in the diet may not be manifested physiologically for many months.

Vitamin A Probably Has Several Functions

Vitamin A was first recognized as an essential nutritional factor by Elmer McCollum in 1915 and later isolated from fish-liver oils. There are two natural forms, vitamin A_1, or retinol, obtained from marine fish livers, and vitamin A_2, from livers of freshwater fish (Figure 10-19). Both are 20-carbon alcohols formed of isoprene units. Vitamin A itself does not occur in plants, but many plants contain isoprenoid compounds known as carotenoids, which can be enzymatically converted into vitamin A by most animals. Figure 10-19 shows how vitamin A is formed by cleavage of *β-carotene*, which gives carrots, sweet potatoes, and other yellow vegetables their characteristic color.

Figure 10-18
Isoprene, the unit of structure of isoprenoid compounds.

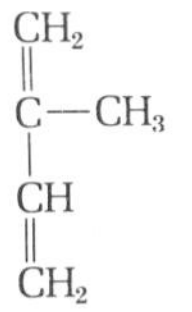

Isoprene

Head-to-tail arrangement of isoprene units (separated by the dashed color line)

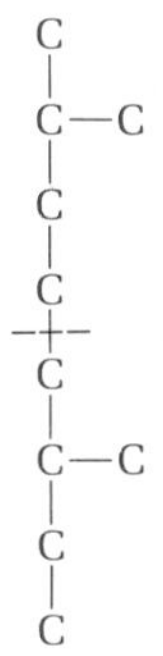

Tail-to-tail arrangement

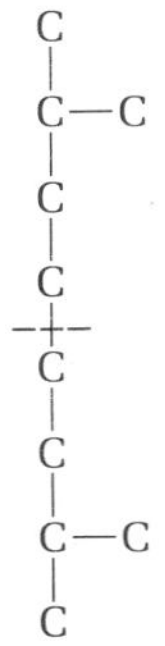

Vitamin A_1

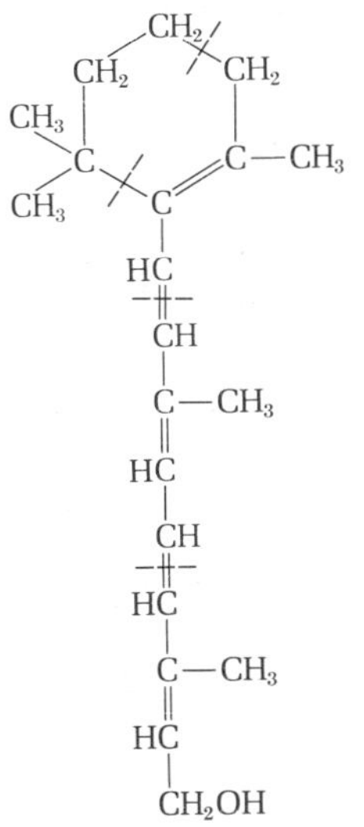

β-Carotene, a plant precursor of vitamin A

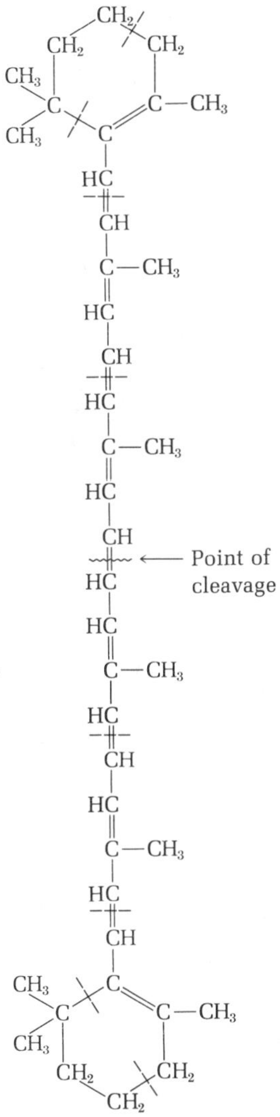

Figure 10-19
Vitamin A_1 and its precursor β-carotene. The isoprene structural units are set off by colored dashed lines. Cleavage of β-carotene yields two molecules of vitamin A_1. This reaction takes place in the small intestine.

Deficiency of vitamin A leads to a variety of characteristic symptoms in people and experimental animals, with no apparent common denominators. They include dry skin, xerophthalmia ("dry eyes"), dry mucous membranes, retarded development and growth, sterility in male animals, and night blindness, an early symptom commonly used for diagnosis of vitamin A deficiency in people (page 775).

Through intensive biochemical and biophysical studies, first pioneered by George Wald at Harvard University, we now have comprehensive information on the function of vitamin A in vision. Figure 10-20 outlines the visual cycle in rod cells of the retina. These cells function in sensing low-intensity light but do not sense colors. In the visual cycle an oxidized form of retinol, retinal or vitamin A aldehyde, is the active component, bound to a protein called opsin. The retinal-opsin complex, called rhodopsin, is present in stacked intracellular membranes in the rod cells. When rhodopsin is excited by visible light, the retinal, in which the double bond at the 11 position is in the cis form (the rest of the double bonds are trans), undergoes a number of very complex, rapid molecular changes and is finally isomerized to form all-*trans*-retinal. These alterations, which change the geometrical configuration of the retinal (Figure 10-20), are believed to accompany a change in the shape of the entire rhodopsin molecule. This event is a molecular trigger producing in the optic-nerve endings an impulse that is transmitted to the brain. The all-*trans*-retinal formed during illumination is then enzymatically converted back into 11-*cis*-retinal in "dark" reactions.

Retinal is also present in bacteriorhodopsin, a light-sensitive purple retinal-protein complex in the cell membrane of the halobacteria, salt-loving prokaryotes that obtain much of their energy from light absorbed by this pigment (Chapter 17).

Figure 10-20
The visual cycle. When a rhodopsin molecule is excited by visible light its 11-cis-retinal prosthetic group, which absorbs light energy, is isomerized in several steps to all-trans-retinal, a process that triggers the nerve impulse. The all-trans-retinal no longer fits the active site on the opsin protein and leaves it. Through two subsequent enzymatic reactions the all-trans-retinal is converted back into 11-cis-retinal, which binds to the opsin again to reconstitute rhodopsin.

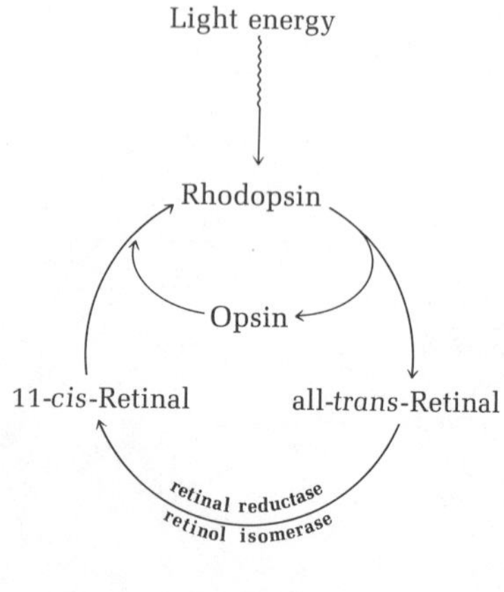

Light energy
Rhodopsin
Opsin
11-*cis*-Retinal
all-*trans*-Retinal
retinal reductase
retinol isomerase

Cis-trans isomerization of 11-*cis*-retinal

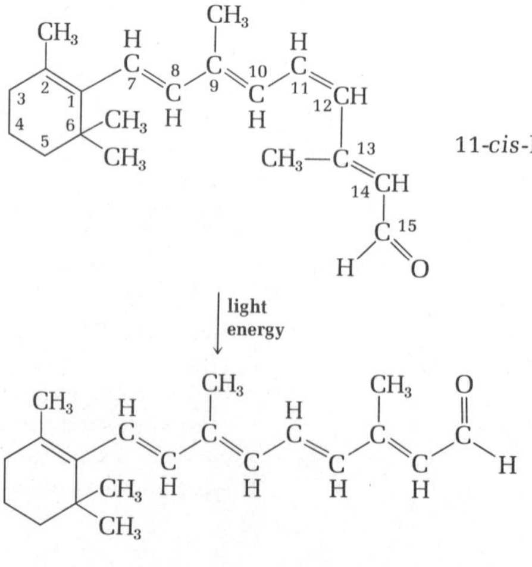

11-*cis*-Retinal

light energy

all-*trans*-Retinal

Vitamin D Is the Precursor of a Hormone

Deficiency of vitamin D leads to abnormal calcium and phosphorus metabolism and defective bone formation in the children's disease rickets, which results in bowlegs and pigeonbreast. *Vitamin D_3*, or *cholecalciferol* (Figure 10-21), is normally made in the skin of people and animals from an inactive precursor, *7-dehydrocholesterol*, by reactions that are promoted by exposure to the ultraviolet component of sunlight. It is also abundant in fish-liver oils. The other common form is *vitamin D_2*, or *ergocalciferol*, a commercial product made by ultraviolet irradiation of *ergosterol* of yeast. People do not need supplementary vitamin D as long as the skin receives sufficient exposure to sunlight.

The biochemical function of vitamin D has been intensively studied in recent years. Vitamin D_3 itself is not biologically active, but it is the precursor of *1,25-dihydroxycholecalciferol* (Figure 10-22). Vitamin D_3 is hydroxylated in two stages, first in the liver and then in the kidney. Because 1,25-dihydroxycholecalciferol is generated by the kidneys and is

Figure 10-21
Forms of vitamin D.

Formation of vitamin D_3 in animals

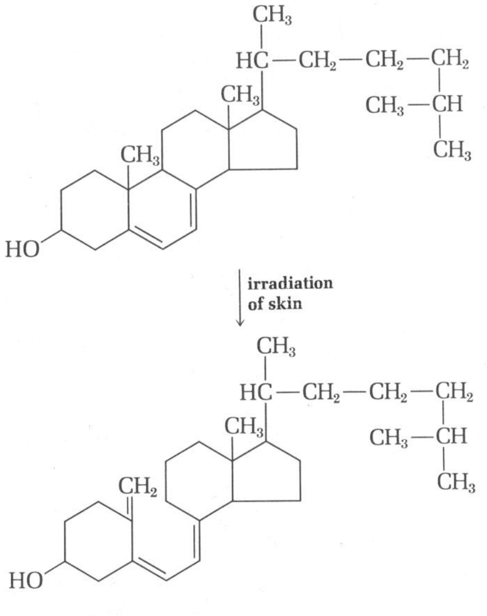

7-Dehydrocholesterol

irradiation of skin

Vitamin D_3 (cholecalciferol)

Vitamin D_2 (ergocalciferol), a commercial product formed by irradiation of the ergosterol of yeast and other fungi

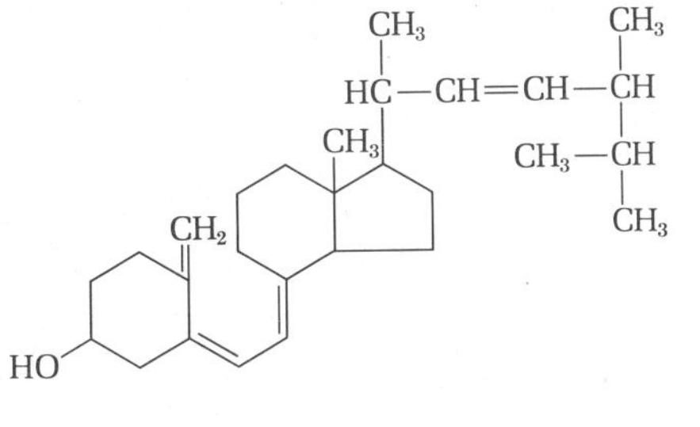

Figure 10-22
Formation and function of the active form of vitamin D_3, 1,25-dihydroxycholecalciferol.

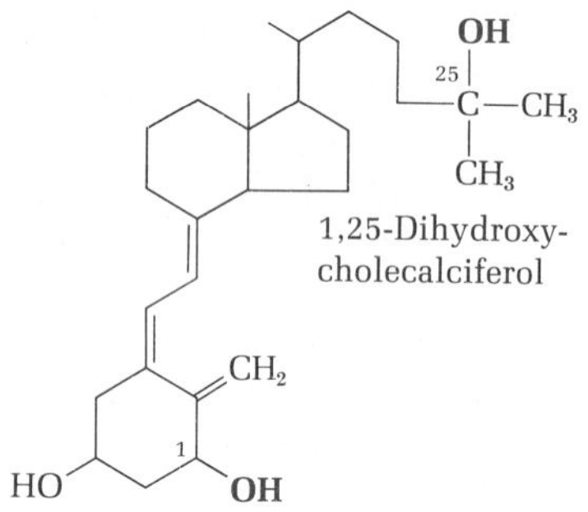

1,25-Dihydroxy-
cholecalciferol

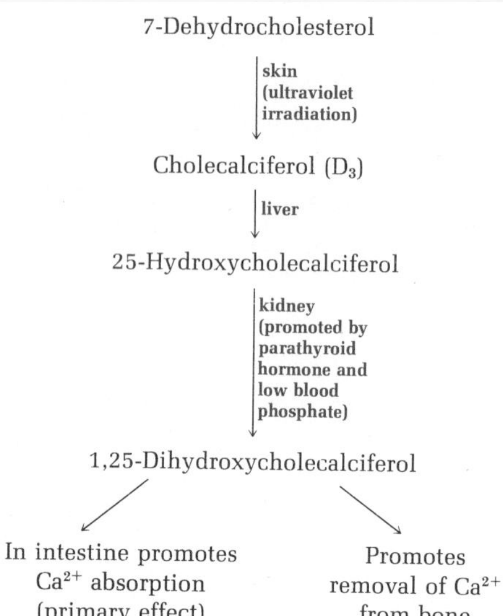

transmitted to other sites in the body, particularly the small intestine and bones, where it regulates Ca^{2+} and phosphate metabolism, it is considered to be a hormone, defined as a chemical "messenger" synthesized by one organ for the regulation of the biological activity of another tissue.

Vitamin E Protects Cell Membranes against Oxygen

Vitamin E consists of at least three molecular species, the α-, β-, and *γ-tocopherols*, of which α-tocopherol is most important (Figure 10-23). Tocopherols are found in vegetable oils and are especially abundant in wheat germ. Deficiency of vitamin E in rats and other animals leads to a scaly skin, muscular weakness, and sterility. The name derives from Greek *tokos*, "childbirth." Whether vitamin E affects human fertility is unknown. Tocopherol deficiency causes other symptoms, including degeneration of the liver and altered membrane function. The tocopherols contain a substituted aromatic ring and a long isoprenoid side chain. The precise biochemical action of vitamin E has not yet been identified, but it probably participates in preventing the destructive attack of oxygen on the lipids of cell membranes (page 308).

Figure 10-23
Vitamin E (α-tocopherol). The isoprene units of the side chain are separated by colored dashed lines.

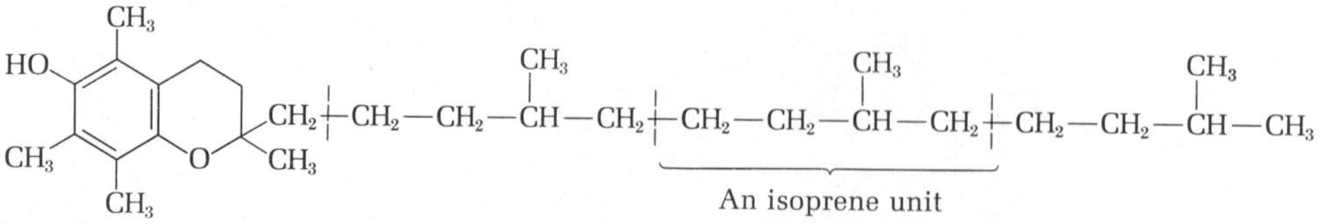

An isoprene unit

Figure 10-24
Forms of vitamin K.

Vitamin K_1 (phylloquinone). This form, found in plants, has four isoprene units in its side chain.

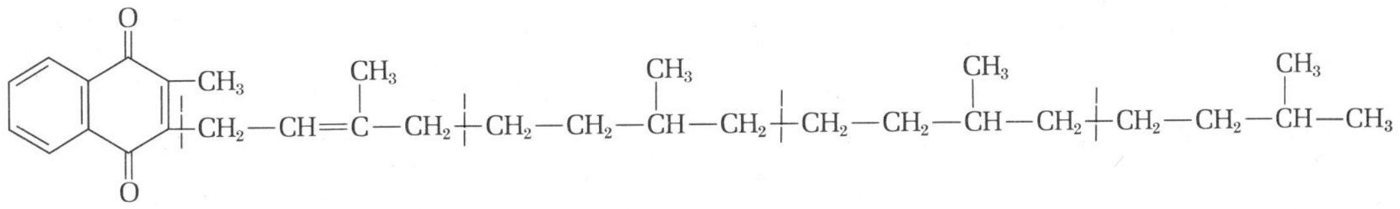

Vitamin K_2 (menaquinone). In this form, found in animals, the side chain has six isoprene units, each with a double bond.

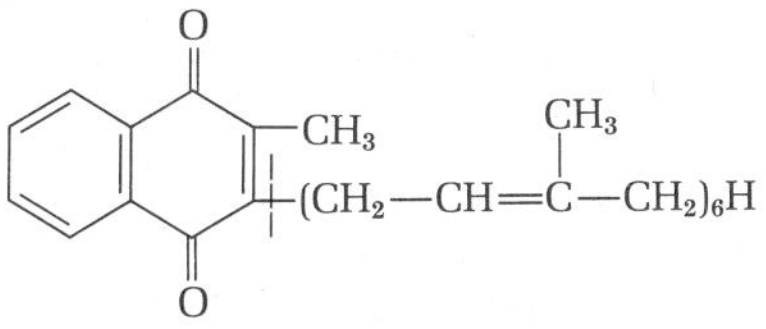

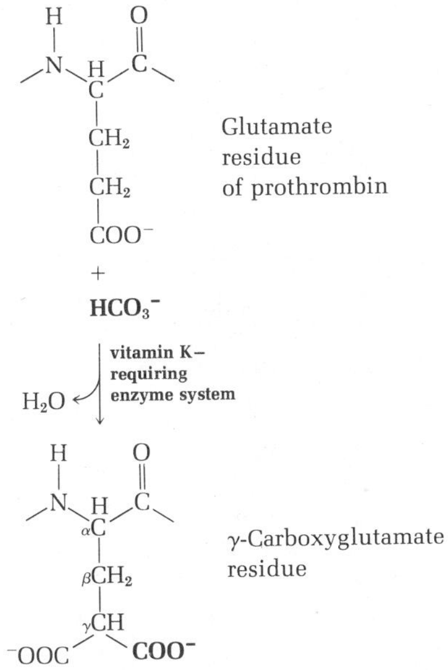

Vitamin K Is a Component of a Carboxylating Enzyme

The two major forms of vitamin K are vitamin K_1 and vitamin K_2, abundant in most higher plants. They are naphthoquinones with isoprenoid side chains of differing length (Figure 10-24). Deficiency of vitamin K in chicks and other animals results in faulty clotting of the blood.

The biochemical function of vitamin K in the blood-clotting mechanisms has recently been discovered. Vitamin K is necessary for the proper formation of the blood plasma protein prothrombin, the inactive precursor of thrombin, an enzyme that converts the protein fibrinogen of blood plasma into fibrin, the insoluble, fibrous protein that holds blood clots together. Prothrombin itself must bind Ca^{2+} before it can be activated to yield thrombin. In vitamin K–deficient animals the prothrombin molecule is defective and unable to bind Ca^{2+} properly. Normal prothrombin contains several residues of the special amino acid γ-carboxyglutamic acid, which functions in binding Ca^{2+}. In vitamin K–deficient animals the prothrombin molecule contains glutamic acid in place of the normal γ-carboxyglutamic residues. John Suttie, of the University of Wisconsin, has discovered an enzyme system that converts the glutamic acid residues of prothrombin from vitamin K–deficient animals into γ-carboxyglutamic acid residues. This enzyme requires vitamin K for activity (Figure 10-25). Several other Ca^{2+}-binding proteins in the body also contain γ-carboxyglutamic residues.

Figure 10-25
Function of vitamin K as cofactor in the formation of γ-carboxyglutamate residues of prothrombin and other proteins.

Glutamate residue of prothrombin

$+ HCO_3^-$

vitamin K–requiring enzyme system

H_2O

γ-Carboxyglutamate residue

Many Inorganic Elements Are Required in Animal Nutrition

In addition to vitamins in the diet, animals and people require a number of chemical elements in inorganic form for proper growth and biological function. These elements may be grouped in two classes: bulk elements and trace elements. The bulk elements, which include calcium, magnesium, sodium, potassium, phosphorus, sulfur, and chlorine, are required in relatively large amounts, in the neighborhood of grams per day. Often they have more than one function. For example, calcium is a structural component of bone mineral, or hydroxyapatite, whose approximate composition is given by $[Ca_3(PO_4)_2]_3 \cdot Ca(OH)_2$. Free Ca^{2+} also serves as a very important regulatory

Table 10-2 Trace Elements and Their Biological Functions

Element	Examples of biochemical function
Iron	Prosthetic group of heme enzymes (catalase, cytochrome oxidase)
Iodine	Required in structure of thyroid hormones
Copper	Prosthetic group of cytochrome oxidase
Manganese	Cofactor of arginase and other enzymes
Zinc	Cofactor in dehydrogenases, DNA polymerase, carbonic anhydrase
Cobalt	Component of vitamin B_{12}
Molybdenum	Cofactor of xanthine oxidase
Selenium	Cofactor of glutathione peroxidase and other enzymes
Vanadium	Cofactor of nitrate reductase
Nickel	Cofactor of urease

Table 10-3 Required Trace Elements Whose Precise Biochemical Role Is Still Unknown

Element	Apparent role
Chromium	Proper utilization of blood glucose
Tin	Formation of bone
Fluorine	Formation of bone
Silicon	Formation of connective tissue and bone
Arsenic	Not known

agent in the cell cytosol, where its concentration is less than 10^{-6} *M*. Phosphorus in the form of phosphate is a vital component of the intracellular energy-transferring ATP system.

Of more immediate concern with respect to enzyme action are the *essential trace elements* (Tables 10-2 and 10-3), required in only milligram or microgram amounts per day, comparable to the requirements of the vitamins. Some 15 trace elements are known to be required in animal nutrition.

Most of the essential trace elements function as enzyme cofactors or prosthetic groups. Essential elements appear to function in such enzymes in one of at least three different ways: (1) The essential element may already have inherent activity in catalyzing a chemical reaction, which is greatly enhanced by the enzyme protein. This is especially true of the metals iron and copper. (2) The essential metal ion may form a complex with both the substrate and the enzyme active site, thus bringing them together in an active form. (3) An essential metal ion may function as a potent electron-withdrawing agent at some point in the catalytic cycle. Enzymes requiring a metal ion for activity are often called *metalloenzymes.*

There Are Many Iron-Requiring Enzymes

Iron is among the best known of the trace elements with regard to biological function. Iron is a component of the heme groups of the oxygen-carrying proteins hemoglobin and myoglobin and of the electron-carrying mitochondrial protein *cytochrome c* (Chapter 8). Several important enzymes also contain heme prosthetic groups (Figure 10-26). A prime example is *cytochrome oxidase,* which catalyzes the reduction of molecular oxygen to water by electrons coming from nutrient molecules. In cytochrome oxidase the iron atoms undergo cyclic valence changes between the ferric [Fe(III)] and ferrous [Fe(II)] forms and thus function to transfer electrons from cytochrome *c* to molecular oxygen. *Cytochrome P450,* which participates in enzymatic hydroxylation reactions, also can transfer electrons to oxygen.

Other heme enzymes include *catalase,* which catalyzes the decomposition of hydrogen peroxide, and *peroxidase,* which catalyzes oxidation of various organic substances by peroxides. In catalase the iron atom itself is an active participant in the catalytic cycle. Simple iron salts, such as $FeSO_4$, already possess some catalytic activity in promoting the decomposition of hydrogen peroxide to H_2O and O_2. Presumably the porphyrin group and the protein part of catalase greatly enhance the inherent catalytic activity of iron.

The *iron-sulfur enzymes* are another important class of iron-containing enzymes, which also function in electron-transferring reactions in animal, plant, and bacterial cells. The iron-sulfur enzymes do not have heme groups but characteristically contain equal numbers of iron and sulfur atoms, in a special labile form decomposed by acid. An example is the *fer-*

redoxin of chloroplasts, which functions to carry electrons from light-excited chlorophyll to various electron acceptors (Chapter 23). Other iron-sulfur enzymes function in electron-transfer reactions of mitochondria, as we shall see (page 479).

Some flavoproteins contain iron in addition to a flavin nucleotide.

Copper Also Functions in Some Oxidative Enzymes

Copper plays an important role in the catalytic activity of *cytochrome oxidase*, which contains both iron and copper in its electron-transferring prosthetic groups. The copper atoms of cytochrome oxidase undergo cyclic Cu(II)-Cu(I) valence transitions as they participate in carrying electrons to oxygen. Copper is also present in the active group of *lysyl oxidase*, an enzyme that makes the cross-linkages between polypeptide chains in collagen and elastin (page 160). Animals that are copper-deficient develop defective collagen molecules lacking cross-links, with the result that the collagen and elastin in the walls of major arteries become weakened and the arteries tend to rupture. Copper is also required for the proper utilization of iron in the body.

Figure 10-26
The iron-porphyrin, or heme, group. It is the prosthetic group of several heme enzymes, such as cytochrome oxidase, catalase, and peroxidase. See also Figure 8-5, page 174.

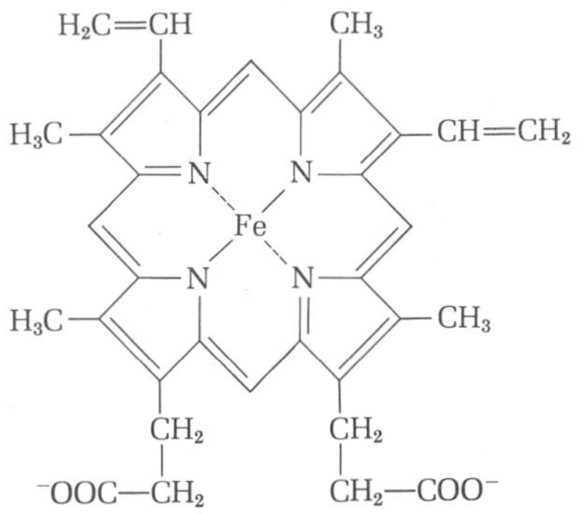

Zinc Is Essential in the Action of Many Enzymes

Zn^{2+} is an essential component of nearly a hundred different enzymes. It is present in many *NAD-* and *NADP-linked dehydrogenases*, enzymes that promote the transfer of hydride ions from substrate molecules to the coenzymes NAD^+ and $NADP^+$. For example, the NAD-requiring enzyme *alcohol dehydrogenase* of the liver, which catalyzes dehydrogenation of ethanol to yield acetaldehyde, contains two atoms of Zn^{2+}, which appear to bind the NAD^+ coenzyme to the active site of the enzyme. Zn^{2+} is also an essential component of *DNA* and *RNA polymerases* and thus participates in important enzymatic reactions involved in the replication and transcription of genetic information. Zn^{2+} is also present in *carbonic anhydrase*, which catalyzes the hydration of CO_2 to H_2CO_3, and in the proteolytic enzyme *carboxypeptidase*, secreted into the small intestine. The hormone insulin is stored as a zinc complex. Among the most interesting roles of zinc is the proper functioning of the taste and smell receptors of the tongue and nasal passages.

Manganese Ions Are Required by Several Enzymes

The enzyme *arginase*, which hydrolyzes arginine to form urea, an end product of human amino group metabolism, contains tightly bound Mn^{2+}, which is essential for its activity. Mn^{2+} also serves as a cofactor of some phosphate-transferring enzymes and in the enzymatic reaction by which oxygen is produced during plant photosynthesis in chloroplasts.

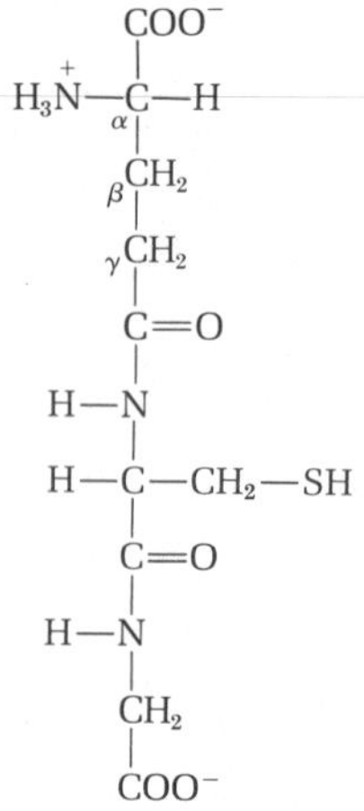

$$\begin{array}{c} COO^- \\ | \\ H_3\overset{+}{N}-\underset{\alpha}{C}-H \\ | \\ {}_{\beta}CH_2 \\ | \\ {}_{\gamma}CH_2 \\ | \\ C{=}O \\ | \\ H-N \\ | \\ H-C-CH_2-SH \\ | \\ C{=}O \\ | \\ H-N \\ | \\ CH_2 \\ | \\ COO^- \end{array}$$

Figure 10-27
Glutathione (GSH), a tripeptide containing L-glutamic acid, L-cysteine, and glycine. Note that the glutamic acid residue is attached to the cysteine residue by its γ-carboxyl group, rather than its α-carboxyl. Glutathione is present in all animal cells in high concentrations. One of its functions is to serve as the reductant of toxic peroxides by the action of glutathione peroxidase (see text). It also acts as a cofactor in a few enzymatic reactions but is not regarded as a coenzyme. Glutathione has also been postulated to function in the transport of amino acids across cell membranes.

Cobalt Is Part of Vitamin B_{12}

Vitamin B_{12} is unique in having both an organic and an inorganic component, cobalt. The microorganisms that make vitamin B_{12} clearly require intake of cobalt. Since vitamin B_{12} is found in only trace amounts in animal cells and some microorganisms, only trace amounts of cobalt are required for biosynthesis of the vitamin.

Selenium Is Both an Essential Trace Element and a Poison

Selenium, in the form of selenite and selenate, has for years been known to be very toxic to cattle grazing in parts of Montana and the Dakotas, as well as in other parts of the world where selenium salts are naturally present in large amounts in the soil. It was therefore a major surprise when it was found that selenium in much smaller amounts is actually required in the diet of rats and chicks. Recent research has shown that selenium combined in an amino acid is an essential component of the prosthetic groups of several enzymes, particularly glutathione peroxidase, which functions together with the peptide glutathione (Figure 10-27) to protect cells against the destructive effects of hydrogen peroxide. In red blood cells the iron of hemoglobin is normally in the ferrous [Fe(II)] form, but it is readily oxidized to the ferric form by hydrogen peroxide to yield methemoglobin, which is inactive in carrying oxygen. Glutathione peroxidase protects against the formation of methemoglobin by consuming hydrogen peroxide in the reaction.

$$\underset{\text{Reduced glutathione}}{2GSH} + H_2O_2 \longrightarrow \underset{\text{Oxidized glutathione}}{GSSG} + 2H_2O$$

The active site of glutathione peroxidase contains a residue of the unusual amino acid selenocysteine (Figure 10-28), in which the sulfur atom of cysteine is replaced by a selenium atom. Presumably the —SeH group of this residue has advantageous properties in the mechanism of this and other selenium enzymes.

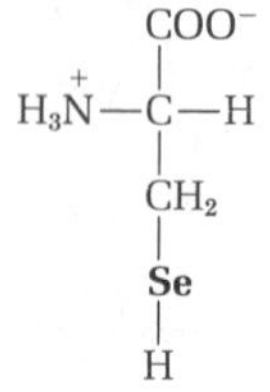

$$\begin{array}{c} COO^- \\ | \\ H_3\overset{+}{N}-C-H \\ | \\ CH_2 \\ | \\ \mathbf{Se} \\ | \\ H \end{array}$$

Figure 10-28
Selenocysteine, an analog of cysteine in which selenium replaces sulfur. A selenocysteine residue is present in the active site of glutathione peroxidase and other selenium-dependent enzymes.

Other Trace Elements Are Known to Be Required by Some Enzymes

Molybdenum and vanadium have been found to function in the active sites of certain flavin dehydrogenases. For example, the enzyme xanthine oxidase, which catalyzes the oxidation of certain purines to yield the excretory product uric acid, has a prosthetic group of FAD and also contains molybdenum and iron as its other essential components.

Silicon, ubiquitous as silica (SiO_2), the main component of sand, and as calcium aluminum silicate, the major mineral component of clay, is required by the rat, the chick, and other animals provided synthetic diets scrupulously freed of traces of silicon. The function of silicon is not yet known, but it occurs in greatest concentration in bone and connective tissues of animals, apparently in an organic form.

Nickel has recently been discovered as a component of *urease*, the first enzyme obtained in crystalline form in 1926 (page 208). *Chromium*, closely related to nickel in the periodic table, is involved in the regulation of glucose uptake by animal tissues. *Tin* is required for proper development of the skeletal system, presumably in calcification processes.

Some plants have been shown to require two additional elements, *boron* and *aluminum*.

Summary

The vitamins are trace organic substances which are essential in the function of most forms of life but which some organisms are unable to synthesize and must obtain from exogenous sources. Most of the water-soluble vitamins function as components of different coenzymes or prosthetic groups of enzymes important in cell metabolism. Thiamine (vitamin B_1) is the active component of thiamine pyrophosphate, a coenzyme required as a transient carrier of acetaldehyde in the enzymatic decarboxylation of pyruvate, a major product of glucose breakdown in cells. Riboflavin (vitamin B_2) is a component of the coenzymes flavin mononucleotide (FMN) and flavin adenine dinucleotide (FAD), which function as hydrogen-carrying prosthetic groups of certain oxidative enzymes. Nicotinic acid is a component of the nicotinamide adenine dinucleotides (NAD and NADP), which serve as transient carriers of hydride ions in the action of certain dehydrogenases. Pantothenic acid is an essential component of coenzyme A, which functions as a transient acyl-group carrier during the enzymatic oxidation of pyruvate and fatty acids. Vitamin B_6 (pyridoxine) is an essential precursor of pyridoxal phosphate, the prosthetic group of transaminases and other amino acid–transforming enzymes. Biotin functions as the prosthetic group of certain carboxylases; it serves as a carrier of carboxy groups. Folic acid is the precursor of tetrahydrofolic acid, a coenzyme functioning in the enzymatic transfer of 1-carbon compounds. Vitamin B_{12}, as its 5′-deoxyadenosyl derivative, functions in the enzymatic exchange of hydrogen atoms and certain substituent groups between adjacent carbon atoms.

The fat-soluble vitamins serve other important roles. Vitamin A is a precursor of a light-sensitive pigment in the visual cycle of rod cells in vertebrates. Vitamin D_3, or cholecalciferol, formed from 7-dehydrocholesterol by exposure of the skin to sunlight, is the major biological precursor of 1,25-dihydroxycholecalciferol, which has a hormonelike action in regulating Ca^{2+} metabolism in the small intestine and bones. Vitamin K is a cofactor in the enzymatic formation of the γ-carboxyglutamyl residues of prothrombin, a Ca^{2+}-binding plasma protein important in blood clotting. Iron, copper, zinc, manganese, cobalt, molybdenum, selenium, and nickel are essential components of a variety of different enzymes. In addition, a number of other elements, including vanadium, tin, chromium, and silicon, are essential in the diet, but their precise functions are not known.

References

Nutritional aspects of vitamins and minerals are discussed in Chapter 26.

Books

BOYER, P. D., H. LARDY, and K. MYRBACK (eds.): The *Enzymes*, 3d ed., 14 vols., Academic, New York, 1970–1981. Comprehensive reviews of the function of coenzymes.

FLORKIN, M., and E. H. STOTZ (eds.): *Comprehensive Biochemistry*, vol. 21, *Metabolism of Vitamins and Trace Elements*, American Elsevier, New York, 1971. Detailed accounts of the chemistry, metabolism, and biochemical roles of the vitamins and trace elements.

HUTCHINSON, D. W.: *Nucleotides and Coenzymes*, Wiley, New York, 1973.

Articles

HARRIS, L. J.: "The Discovery of Vitamins," in Joseph Needham (ed.), *The Chemistry of Life*, Cambridge University Press, New York, 1970, pp. 156–170.

HUBBARD, R.: "100 Years of Rhodopsin," *Trends Biochem. Sci.*, **1:** 154–158 (1976).

SIMKISS, K.: "Metal Ions in Cells," *Endeavour*, **3:**2–6 (1979).

STAUDINGER, H. A.: "Ascorbic Acid," *Trends Biochem. Sci.*, **3:** 211–212 (1978).

Problems

1. *Nutritional Requirements of Nicotinic Acid* A dietary deficiency of the vitamin nicotinic acid causes the disease pellagra.
   (a) The daily requirement of nicotinic acid for a human adult, which is 7.5 mg, is decreased when a high level of the amino acid tryptophan is present in the diet. What does this observation suggest about the metabolic relationship between nicotinic acid and tryptophan?
   (b) Pellagra was once a common disease, especially in the southern United States, before and after the turn of the century, when the diet in rural areas was low in meat and contained corn (maize) products as a staple. Can you suggest why these conditions led to nicotinic acid deficiency?

2. *Polyneuritis in Pigeons* In a classic experiment pigeons placed on an experimental diet became unable to retain their balance and coordination. The levels of pyruvate in their blood and brains were found to be much higher than in normal birds. The condition could be prevented or cured by feeding the birds meat scraps. Suggest an explanation for these observations.

3. *Bacterial Assay for Riboflavin* *Lactobacillus casei*, a member of a family of bacteria used in the preparation of such fermented products as yogurt, sauerkraut, and pickles, cannot synthesize ribo-

flavin. A characteristic feature of these bacteria is that they obtain their energy from the breakdown of glucose to lactic acid (pK′ 3.5). How would you devise a quantitative assay for riboflavin based on this information?

4. *Pyridoxine and Bacterial Amino Acid Requirement* The bacterium *Lactobacillus casei* grows on a simple culture medium containing the vitamins riboflavin and pyridoxine and four amino acids. If a complete mixture of amino acids is added to the culture medium, together with riboflavin, the amount of pyridoxine required for optimal growth is reduced by 90 percent. Explain these observations.

5. *Egg Whites Prevent Spoilage of Egg Yolk* Eggs will keep in the refrigerator for 4 to 6 weeks without spoiling. Isolated egg yolks (no egg white), on the other hand, spoil rapidly even under refrigeration.
   (a) What causes the spoiling?
   (b) How can you explain the observation that the presence of egg white helps to prevent the spoiling of egg yolks?
   (c) What is the biological advantage of this mode of protection to a bird?

6. *Nutritional Requirement for Folic Acid in Streptococcus faecalis* A bacterium present in the large intestine, *Streptococcus faecalis*, has a nutritional requirement for the vitamin folic acid. If the growth medium contains adenine and thymidine, the bacteria can be readily grown in the absence of folic acid. An examination of such cultures shows that the bacteria grown in this way are devoid of folic acid. Why do the bacteria need folic acid? Why does the addition of adenine and thymidine to the culture medium abolish this need?

7. *Cobalt Requirement in Ruminants* Much of the weight of terrestrial plants consists of insoluble polysaccharides, of which cellulose is the major component. Although most animals lack the enzymes necessary to digest cellulose, ruminants (e.g., cows, horses, sheep, and goats) make use of microbial action to predigest grasses and leafy plants. Unlike most animals, ruminants have a high nutritional requirement for cobalt. In regions where the cobalt in the soil is low, such as Australia, cobalt deficiency in cattle and sheep is a serious problem. Suggest a reason for the high nutritional requirement of cobalt in ruminants.

8. *Frequency of Vitamin Intake* Although it is possible to administer several weeks' supply of vitamins A and D in a single dose, the B-complex vitamins must be supplied more frequently. Why?

9. *Vitamin A Deficiency* Xerophthalmia, a blinding disease of the eyes characterized by a dry and lusterless condition of the eyeballs, is caused by the deficiency of vitamin A. The disease affects large numbers of children but few adults. In tropical countries the disease annually causes blindness in tens of thousands of children aged between 18 and 36 months. In contrast, adult volunteers fed a vitamin A–deficient diet for more than 2 years show only impaired night vision. This minimal damage can be readily reversed by the administration of vitamin A. Suggest an explanation for the observed differences in the effect of vitamin A deficiency on children and adults.

10. *Renal Osteodystrophy* Renal osteodystrophy, also called renal rickets, a disease associated with extensive demineralization of the bones, frequently occurs in patients with damaged kidneys, in spite of a well-balanced diet. What vitamin is involved in bone mineralization? Why does kidney damage lead to demineralization?

Problem 11

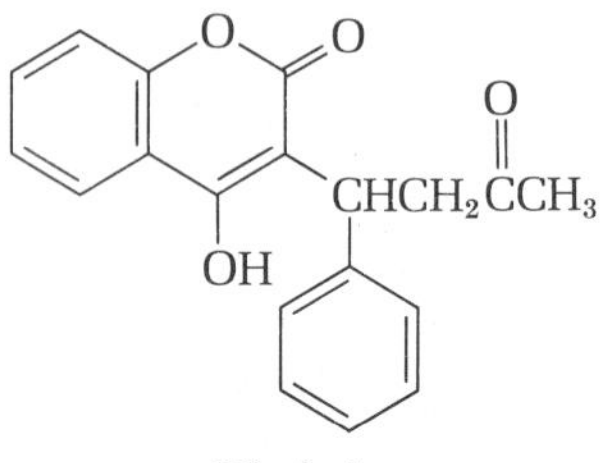

Warfarin

11. *Action of Warfarin and Dicoumarol* Warfarin, prepared commercially as a rodenticide, is a powerful antagonist of vitamin K, i.e., its administration opposes the action of vitamin K.
   (a) Provide a molecular explanation for how warfarin could act as an antagonist.
   (b) Why does the ingestion of warfarin by rodents lead to their death?
   (c) If cattle or horses are fed improperly cured sweet clover, they develop a disease characterized by abnormal internal bleeding. The agent responsible for this effect has been identified as dicoumarol, which arises from the action of microorganisms on coumarin, a natural constituent of sweet clover. Dicoumarol is used clinically to treat patients with acute thrombophlebitis (formation of obstructive blood clots). Explain the basis of this treatment.

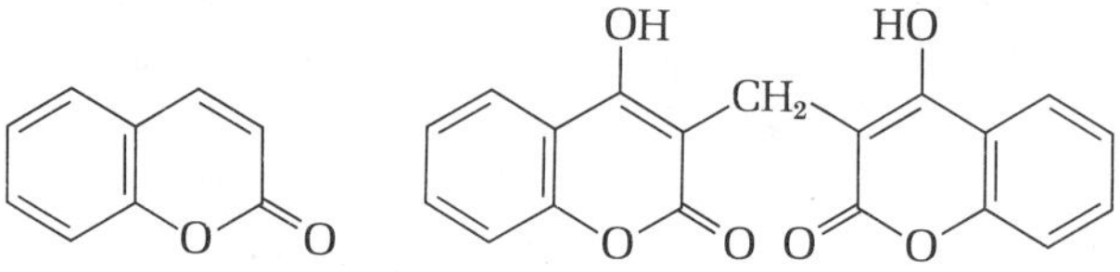

Coumarin   Dicoumarol

12. *Methylmalonic Acid in the Urine* A clinic patient was found to be suffering from a metabolic disturbance characterized by acidosis, i.e., low blood and urine pH. Chemical analysis of the patient's body fluids revealed that a large amount of methylmalonic acid was excreted in the urine. When this compound is fed to normal animals, it is known to be converted into succinic acid. Can you suggest a nutritional explanation for these observations?

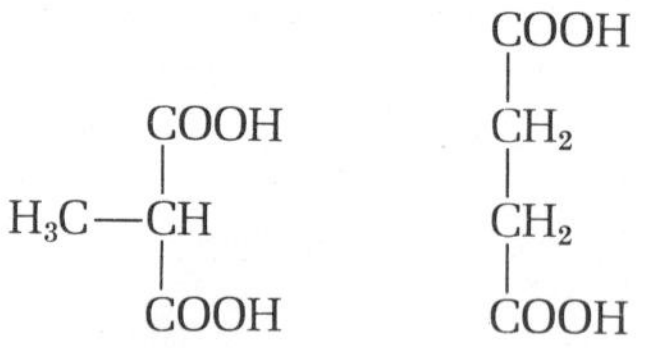

Methylmalonic acid: $H_3C—CH(COOH)_2$ Succinic acid: $HOOC—CH_2—CH_2—COOH$

Problem 13

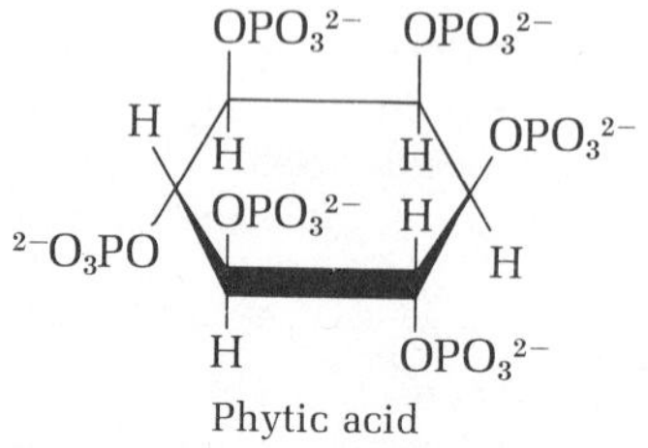

Phytic acid

13. *Human Zinc Deficiency* In certain parts of the Middle East, principally Iran and Egypt, human zinc deficiency is observed. This highly localized nutritional deficiency is ascribed to the consumption of a diet very high in cereal grains. Cereals contain large amounts of phytic acid (myoinositol hexaphosphate), a substance that binds divalent metal cations, especially Zn^{2+}, very tightly.
   (a) Why does the consumption of cereals lead to zinc deficiency?
   (b) The disease is most prevalent in rural villages where unleavened ("flat") bread, made without yeast to raise the dough, is a principal part of the diet. In larger towns, where yeast leavening is used, the incidence of zinc deficiency is much lower. Suggest an explanation for this observation.

CHAPTER 11

Carbohydrates: Structure and Biological Function

It is now appropriate, as we approach the study of cell metabolism, to examine carbohydrates, the "staff of life" for most organisms. Carbohydrates in the form of sugar and starch represent a major part of the total caloric intake for humans and for most animal life, as well as for many microorganisms. Carbohydrates are also central in the metabolism of green plants and other photosynthetic organisms that utilize solar energy to synthesize carbohydrates from CO_2 and H_2O. The vast amounts of starch and other carbohydrates made by photosynthesis become the ultimate energy and carbon sources for nonphotosynthetic cells of the animal, plant, and microbial worlds.

Carbohydrates have other important biological functions. Starch and glycogen serve as temporary stores of glucose. Insoluble carbohydrate polymers serve as structural and supportive elements in the cell walls of bacteria and plants and in the connective tissues and cell coats of animal organisms. Other carbohydrates serve to lubricate skeletal joints, to provide adhesion between cells, and to confer biological specificity on the surface of animal cells.

In this chapter we shall examine the structures, properties, and functions of the major carbohydrates. We shall also see how the properties of carbohydrates are blended with those of proteins in hybrid molecules called *glycoproteins* and *proteoglycans*, important components of cell surfaces and extracellular supportive systems in animals.

There Are Three Classes of Carbohydrates, Based on the Number of Sugar Units

Carbohydrates are polyhydroxy aldehydes or ketones or substances that yield such compounds on hydrolysis. The name carbohydrate owes its origin to the fact that most substances of this class have empirical formulas suggesting they are carbon "hydrates," in which the ratio of carbon to hydrogen to oxygen atoms is 1:2:1. For example, the empirical formula of D-glucose is $C_6H_{12}O_6$, which can also be written $(CH_2O)_6$ or $C_6(H_2O)_6$. Although many common carbohydrates conform to the empirical formula $(CH_2O)_n$, others do not show this ratio and some also contain nitrogen, phosphorus, or sulfur.

There are three major classes of carbohydrates: *monosaccharides*, *oligosaccharides*, and *polysaccharides* (the word "saccharide" is derived from a Greek word meaning sugar). *Monosaccharides*, or *simple sugars*, consist of a *single* polyhydroxy aldehyde or ketone unit. The most abundant monosaccharide in nature is the 6-carbon sugar D-glucose.

Oligosaccharides (Greek *oligos*, "few") consist of short chains of monosaccharide units joined together by covalent bonds. Of these the most abundant are the *disaccharides*, which have *two* monosaccharide units. Typical is *sucrose*, or *cane sugar*, which consists of the 6-carbon sugars D-glucose and D-fructose joined in covalent linkage. Most oligosaccharides having three or more units do not occur free but are joined as side chains to polypeptides in *glycoproteins* and *proteoglycans*, to be discussed later.

Polysaccharides consist of long chains having hundreds or thousands of monosaccharide units. Some polysaccharides, such as *cellulose*, have linear chains, whereas others, such as *glycogen*, have branched chains. The most abundant polysaccharides, *starch* and *cellulose* of the plant world, consist of recurring units of D-glucose, but they differ in how the D-glucose units are linked to each other.

The common monosaccharides and disaccharides all have names ending with the suffix *-ose*.

There Are Two Families of Monosaccharides: Aldoses and Ketoses

Monosaccharides are colorless, crystalline solids that are freely soluble in water but insoluble in nonpolar solvents. Most have a sweet taste.

As indicated above, the common monosaccharides have the empirical formula $(CH_2O)_n$, where $n = 3$ or some larger number. The backbone of monosaccharides is an unbranched, single-bonded carbon chain. One of the carbon atoms is double-bonded to an oxygen atom to form a carbonyl group; each of the other carbon atoms has a hydroxyl group. If the carbonyl group is at the end of the carbon chain, the monosaccharide is an *aldehyde* (page 50) and is called an *aldose*; if the carbonyl group is at any other position, the monosaccharide is a *ketone* (page 50) and is called a *ketose*. The simplest monosaccharides are the two 3-carbon *trioses*: *glyceraldehyde*, an aldose, and *dihydroxyacetone*, a ketose (Figure 11-1).

Monosaccharides having 4, 5, 6, and 7 carbon atoms in their backbones are called, respectively, *tetroses*, *pentoses*, *hexoses*, and *heptoses*. Each of these exists in two series: aldotetroses and ketotetroses, aldopentoses and ketopentoses, aldohexoses

Figure 11-1
The two trioses. The carbonyl groups are in color.

Glyceraldehyde, an aldose

Dihydroxyacetone, a ketose

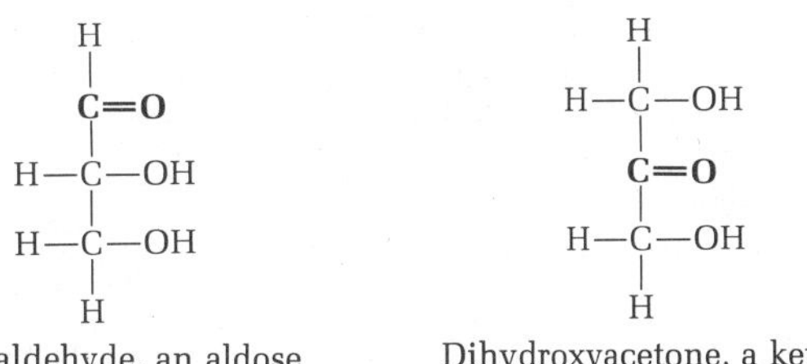

Figure 11-2
Two common hexoses.

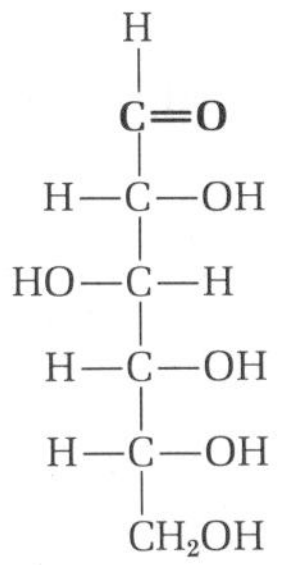

D-Glucose, an aldohexose

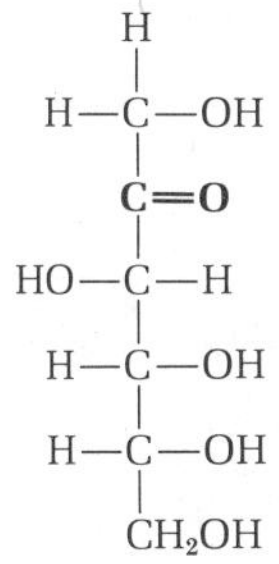

D-Fructose, a ketohexose

Figure 11-3
The pentose components of nucleic acids.

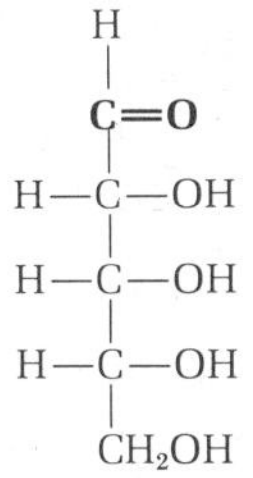

D-Ribose, the sugar component of ribonucleic acid (RNA)

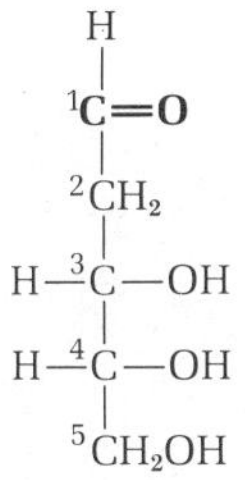

2-Deoxy-D-ribose, the sugar component of deoxyribonucleic acid (DNA)

and ketohexoses, and so on. The hexoses, which include the aldohexose D-*glucose* and the ketohexose D-*fructose* (Figure 11-2), are the most abundant monosaccharides in nature. The aldopentoses D-*ribose* and *2-deoxy-*D-*ribose* (Figure 11-3) are components of nucleic acids.

The Common Monosaccharides Have Several Asymmetric Centers

All the monosaccharides except dihydroxyacetone contain one or more asymmetric or chiral carbon atoms (page 52) and thus occur in optically active isomeric forms. The simplest aldose, glyceraldehyde, contains only one chiral center and thus is capable of existing as two different optical isomers that are nonsuperimposable mirror images of each other (page 98). The al-

Figure 11-4
The family of D-aldoses having from 3 to 6 carbon atoms, shown in straight-chain structural formulas. Those named in boxes are the most abundant in nature. The color-shaded carbon atoms are chiral centers.

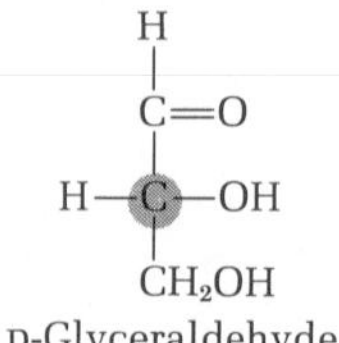

D-Erythrose	D-Threose
H	H
C=O	C=O
H—C—OH	HO—C—H
H—C—OH	H—C—OH
CH_2OH	CH_2OH

D-Ribose	D-Arabinose	D-Xylose	D-Lyxose
H	H	H	H
C=O	C=O	C=O	C=O
H—C—OH	HO—C—H	H—C—OH	HO—C—H
H—C—OH	H—C—OH	HO—C—H	HO—C—H
H—C—OH	H—C—OH	H—C—OH	H—C—OH
CH_2OH	CH_2OH	CH_2OH	CH_2OH

D-Allose	D-Altrose	D-Glucose	D-Mannose	D-Gulose	D-Idose	D-Galactose	D-Talose
H	H	H	H	H	H	H	H
C=O	C=O	C=O	C=O	C=O	C=O	C=O	C=O
H—C—OH	HO—C—H	H—C—OH	HO—C—H	H—C—OH	HO—C—H	H—C—OH	HO—C—H
H—C—OH	H—C—OH	HO—C—H	HO—C—H	H—C—OH	H—C—OH	HO—C—H	HO—C—H
H—C—OH	H—C—OH	H—C—OH	H—C—OH	HO—C—H	HO—C—H	HO—C—H	HO—C—H
H—C—OH	H—C—OH	H—C—OH	H—C—OH	H—C—OH	H—C—OH	H—C—OH	H—C—OH
CH_2OH	CH_2OH	CH_2OH	CH_2OH	CH_2OH	CH_2OH	CH_2OH	CH_2OH

Figure 11-5
The stereoisomers of glyceraldehyde.

Perspective formulas

D-Glyceraldehyde	L-Glyceraldehyde
CHO	CHO
H—C—OH	HO—C—H
CH_2OH	CH_2OH

Projection formulas; the horizontal bonds are assumed to project forward of the plane of the page (see page 97).

D-Glyceraldehyde	L-Glyceraldehyde
CHO	CHO
H—C—OH	HO—C—H
CH_2OH	CH_2OH

dohexoses have four chiral centers and can exist in the form of $2^n = 2^4 = 16$ different stereoisomers, among which is the common form of glucose, namely D-glucose. Figure 11-4 shows the structures of all the stereoisomers of aldotrioses, aldotetroses, aldopentoses, and aldohexoses of the D series. They are given as *projection formulas* (page 97), in which the horizontal bonds are assumed to extend forward from the plane of the paper and the vertical bonds to extend to the rear (Figure 11-5). Later we shall see two other ways of representing the three-dimensional structure of sugar molecules, *Haworth projections* and *conformational formulas.*

The common naturally occurring monosaccharides (with the exception of dihydroxyacetone) are optically active. D-Glucose, the common form of glucose in nature, is dextrorotatory, with a specific rotation of $[\alpha]_D^{20} = +52.7°$, while D-fructose, the common form of fructose, is levorotatory ($[\alpha]_D^{20} = -92.4°$). Like the stereoisomeric forms of the amino acids (Chapter 5), the stereo-

Figure 11-6
Three important ketoses.

```
   CH2OH              CH2OH
   |                  |
   C=O                C=O
   |                  |
H—C—OH           HO—C—H
   |                  |
H—C—OH            H—C—OH
   |                  |
   CH2OH           H—C—OH
                      |
                      CH2OH

D-Ribulose         D-Fructose

        CH2OH
        |
        C=O
        |
    HO—C—H
        |
     H—C—OH
        |
     H—C—OH
        |
     H—C—OH
        |
        CH2OH
  D-Sedoheptulose
```

isomers of monosaccharides can all be related to a standard reference compound, glyceraldehyde, which has one D form and one L form (Figure 11-5). *However, since many of the aldoses have two or more chiral centers, the prefixes* D- *and* L- *are used to refer to the configuration of the chiral carbon most distant from the carbonyl carbon atom.* When the hydroxyl group on the most distant chiral carbon projects to the *right* in the projection formula, it designates a D sugar; when it projects to the *left,* it designates an L sugar. Although most of the different possible D-aldoses (Figure 11-4) have been found in nature, the most useful to remember are the pentose D-*ribose* and the hexoses D-*glucose,* D-*mannose,* and D-*galactose.*

In a similar way one can write the structures of all the D-ketoses with up to 6 carbon atoms; they have the same configuration at the asymmetric carbon atom most distant from the carbonyl group. Ketoses are systematically designated by inserting *ul* into the name of the corresponding aldose; for example, D-*ribulose* is the ketopentose corresponding to the aldopentose D-*ribose.* However, a few ketoses have trivial names, such as fructose. Biologically the most important ketoses are the ketopentose D-*ribulose,* the ketohexose D-*fructose,* and the ketoheptose D-*sedoheptulose* (Figure 11-6). Some aldoses and ketoses of the L series occur in nature, but they are relatively uncommon.

When two sugars differ only in the configuration around *one* specific carbon atom, they are called *epimers* of each other. D-Glucose and D-mannose are epimers with respect to carbon atom 2, and D-glucose and D-galactose are epimers with respect to carbon atom 4 (Figure 11-7).

The Common Monosaccharides Occur in Ring Forms

In Figures 11-1 to 11-4 and 11-6 the structures of various aldoses and ketoses are written in straight-chain forms. Although such structures are correct for the trioses and tetroses, monosaccharides with five or more carbon atoms in the backbone usually occur in solution as cyclic or ring structures, in which the carbonyl group is not free as written but has formed a covalent bond with one of the hydroxyl groups along the chain. One

Figure 11-7
Two epimers of D-*glucose.*

```
D-Glucose and D-mannose,              D-Glucose and D-galactose,
epimers at carbon 2                   epimers at carbon 4

    CHO             CHO                   CHO     1      CHO
    |               |                     |              |
 H—C—OH         HO—C—H                 H—C—OH   2    H—C—OH
    |               |                     |              |
HO—C—H          HO—C—H                HO—C—H    3   HO—C—H
    |               |                     |              |
 H—C—OH          H—C—OH                H—C—OH   4   HO—C—H
    |               |                     |              |
 H—C—OH          H—C—OH                H—C—OH   5    H—C—OH
    |               |                     |              |
    CH2OH           CH2OH                 CH2OH   6      CH2OH
D-Glucose       D-Mannose             D-Glucose      D-Galactose
```

indication that D-glucose has a ring structure is that it has two crystalline forms of slightly different properties. If D-glucose is crystallized from water, a form called α-D-glucose results, for which the specific rotation (page 97) is $[\alpha]_D^{20} = +112.2°$. If D-glucose is crystallized from pyridine, β-D-glucose results; its $[\alpha]_D^{20}$ is +18.7°. The two forms are identical in chemical composition. From various chemical considerations it has been deduced that the α and β isomers of D-glucose are not straight-chain structures but two different six-membered ring compounds (Figure 11-8). Such cyclic forms of sugars are called *pyranoses* because they resemble the six-membered ring compound *pyran*. The systematic names for the two ring forms of D-glucose are *α-D-glucopyranose* and *β-D-glucopyranose* (Figure 11-8).

When α-D-glucose is dissolved in water, its specific rotation gradually changes with time and reaches a stable value of 52.7°; when β-D-glucose is similarly treated, its rotation also ultimately attains the same value. This change, called *mutarotation*, is due to the formation from either α- or β-D-glucose of an equilibrium mixture consisting of about one-third α-D-glucose and two-thirds β-D-glucose and a very small amount of straight-chain compound at 25°C. Thus the α and β isomers of D-glucose are interconvertible in aqueous solution.

The formation of pyranose rings in D-glucose is the result of a general reaction between aldehydes and alcohols to form derivatives called *hemiacetals* (Figure 11-9), which contain an asymmetric carbon atom and thus can exist in two stereoisomeric forms. D-Glucopyranose is an *intramolecular* hemiacetal, in which the free hydroxyl group at carbon atom 5 has reacted with the aldehydic carbon atom 1, rendering the latter asymmetric. D-Glucopyranose therefore can exist as two different stereoisomers, designated α and β (Figure 11-8). The net result is that D-glucose behaves as though it had one more asymmetric center than is given by its straight-chain formula. *Isomeric forms of monosaccharides that differ from each other only in their configuration about the hemiacetal carbon atom, such as α-D-glucose and β-D-glucose, are called anomers.* The hemiacetal or carbonyl carbon atom is called the *anomeric carbon.* Only aldoses having 5 or more carbon atoms can form stable pyranose rings. Aldohexoses also exist in cyclic forms having five-membered rings. Because such rings resemble the five-membered ring compound *furan,* they are called *furanoses* (Figure 11-10). However, the six-membered aldopyranose ring is much more stable than the aldofuranose ring and predominates in aldohexose solutions.

Figure 11-8
Formation of the two forms of D-glucopyranose. When the aldehyde group at C1 and the hydroxyl group at C5 react to form the hemiacetal linkage, two different stereoisomers may be formed at carbon 1, designated α and β.

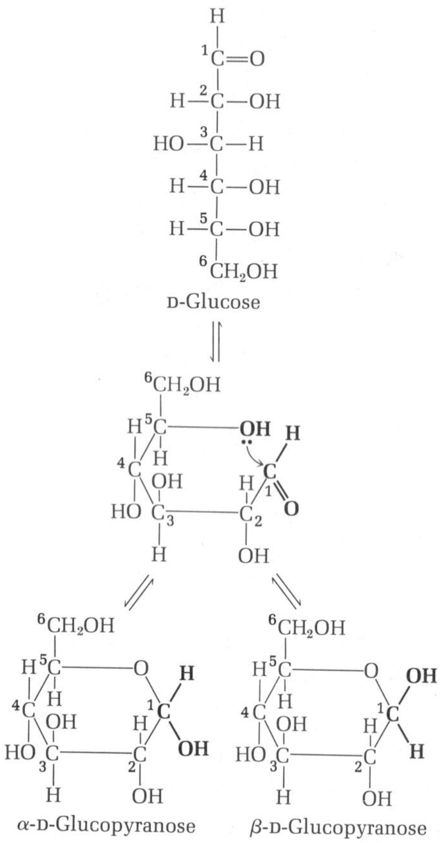

D-Glucose

α-D-Glucopyranose

β-D-Glucopyranose

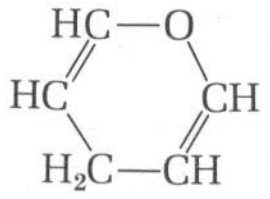

Pyran, the parent compound of pyranoses

Figure 11-9
Aldehydes and ketones can react with alcohols to yield hemiacetals and hemiketals. The carbonyl carbon becomes chiral in these reactions.

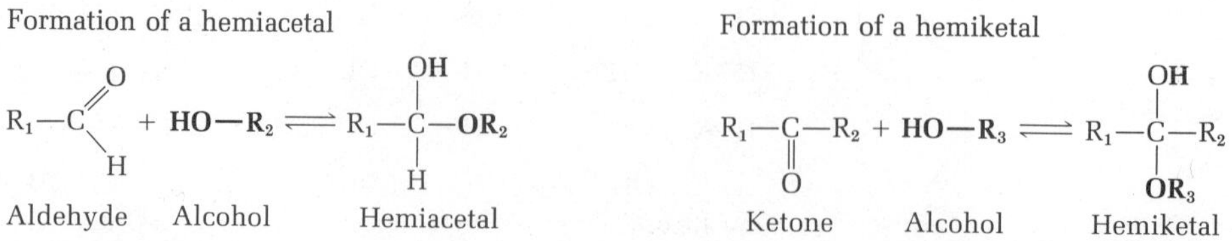

Formation of a hemiacetal

Aldehyde + Alcohol ⇌ Hemiacetal

$R_1{-}CHO + HO{-}R_2 \rightleftharpoons R_1{-}CH(OH){-}OR_2$

Formation of a hemiketal

Ketone + Alcohol ⇌ Hemiketal

$R_1{-}CO{-}R_2 + HO{-}R_3 \rightleftharpoons R_1{-}C(OH)(OR_3){-}R_2$

Figure 11-10
Haworth projection formulas of the pyranose forms of D-glucose and the furanose forms of D-fructose.

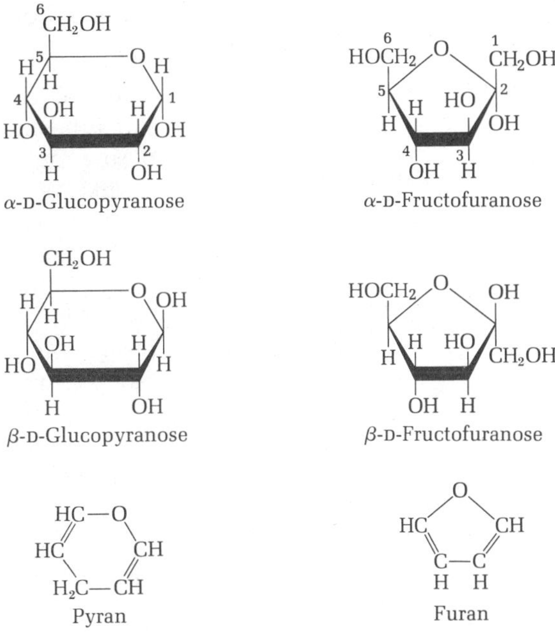

Ketohexoses also occur in α and β anomeric forms. In these compounds the hydroxyl group on carbon atom 5 reacts with the carbonyl group at carbon atom 2, forming a five-membered furanose ring containing a hemiketal linkage (Figure 11-9). D-Fructose forms two different furanoses (Figure 11-10); the more common form is *β-D-fructofuranose*.

Haworth projection formulas are commonly used to show the ring forms of monosaccharides. Although the edge of the ring nearest the reader is usually represented by bold lines (Figure 11-10), the six-membered pyranose ring is not planar, as Haworth projections suggest. In most sugars it occurs in the *chair* conformation, but in some it assumes the *boat* form; these forms are represented by *conformational formulas* (Figure 11-11). The specific three-dimensional conformations of the simple 6-carbon sugars are important in determining the biological properties and function of some polysaccharides, as we shall see.

Figure 11-11
(a) Conformational formulas of the boat and chair forms of the pyranose ring. (b) The conformational formula of α-D-glucopyranose, which has the chair conformation.

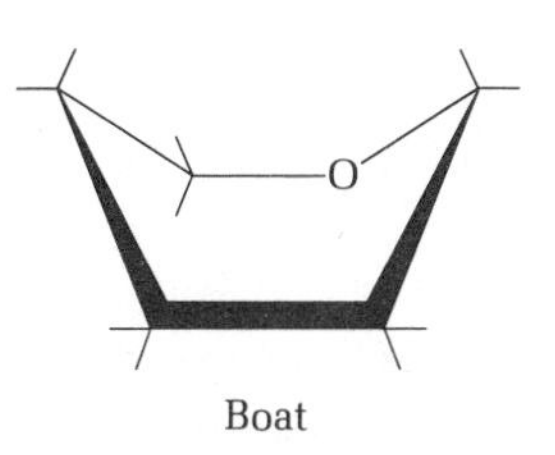

Chair

(a)

α-D-Glucopyranose

(b)

Simple Monosaccharides Are Reducing Agents

Monosaccharides readily reduce such oxidizing agents as ferricyanide, hydrogen peroxide, or cupric ion (Cu^{2+}). In such reactions the sugar is oxidized at the carbonyl group, and the oxidizing agent becomes reduced. (Remember that reducing agents are electron donors and oxidizing agents are electron acceptors.) Glucose and other sugars capable of reducing oxidizing agents are called *reducing sugars*. This property is useful in the analysis of sugars. By measuring the amount of an oxidizing agent that is reduced by a solution of a sugar, it is possible to estimate the concentration of the sugar. In this way blood and urine can be analyzed for their content of glucose in the diagnosis of *diabetes mellitus*, in which the blood glucose level is abnormally high and there is excessive urinary excretion of glucose.

Disaccharides Contain Two Monosaccharide Units

Disaccharides consist of two monosaccharides covalently bound to each other. In most disaccharides the chemical bond that joins the two monosaccharide units is called a *glycosidic bond* and is formed when the hydroxyl group on one of the sugars reacts with the anomeric carbon on the second sugar. Glycosidic bonds are readily hydrolyzed by acid but resist cleavage by base. Thus disaccharides can be hydrolyzed to yield their free monosaccharide components by boiling with dilute acid.

Disaccharides are also abundant in nature; the most common are *sucrose*, *lactose*, and *maltose* (Figure 11-12). *Maltose*, the simplest, contains two D-glucose residues joined by a glycosidic linkage between carbon atom 1 (the anomeric carbon) of the first glucose residue and carbon atom 4 of the second glucose (Figure 11-12). The configuration of the anomeric carbon atom in the glycosidic linkage between the two D-glucose residues is α, and the linkage is thus symbolized $\alpha(1 \rightarrow 4)$. The monosaccharide unit bearing the anomeric carbon is designated by the first number or *locant* in this symbol. Both the glucose residues of maltose are in pyranose form. Maltose is a reducing sugar since it has one potentially free carbonyl group that can be oxidized. The second glucose residue of maltose is capable of existing in α and β forms; the α form shown in Figure 11-12 is formed by the action of the salivary enzyme *amylase* on starch (page 686). Maltose is hydrolyzed to two molecules of D-glucose by the intestinal enzyme *maltase*, which is specific for the $\alpha(1 \rightarrow 4)$ bond. The disaccharide *cellobiose* also contains two D-glucose residues, but they are joined in a $\beta(1 \rightarrow 4)$ linkage.

The disaccharide *lactose* (Figure 11-12), which yields D-galactose and D-glucose on hydrolysis, occurs only in milk. Since it has a potentially free carbonyl group on the glucose residue, lactose is a reducing disaccharide. During digestion lactose undergoes enzymatic hydrolysis by the lactase of the intestinal mucosal cells. This enzyme is very active in suckling infants, but only northern Europeans and a few African tribes tend to retain intestinal lactase activity in adulthood. Adults of most other human groups, including Orientals, Arabs, Jews, most Africans, Indians, and Mediterranean peoples, have little intestinal lactase, and many show lactose intolerance. This difference is genetic in nature. Since lactose itself cannot be absorbed from the intestine into the bloodstream unless it is first hydrolyzed into its monosaccharide units, lactose remains unabsorbed in the intestinal tract of lactose-intolerant peoples.

Figure 11-12
Important disaccharides. The structure of maltose is shown as a Haworth projection and as a conformational formula.

Maltose (β form) [O-α-D-glucopyranosyl-(1 → 4)-β-D-glucopyranose]

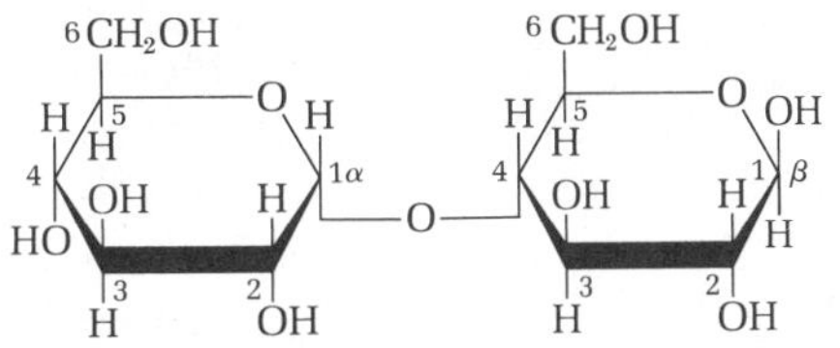

Haworth projection

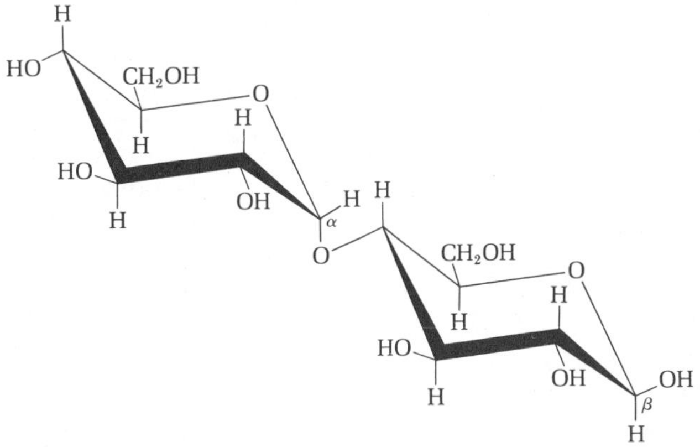

Conformational formula

Lactose (β form) [O-β-D-galactopyranosyl-(1 → 4)-β-D-glucopyranose]

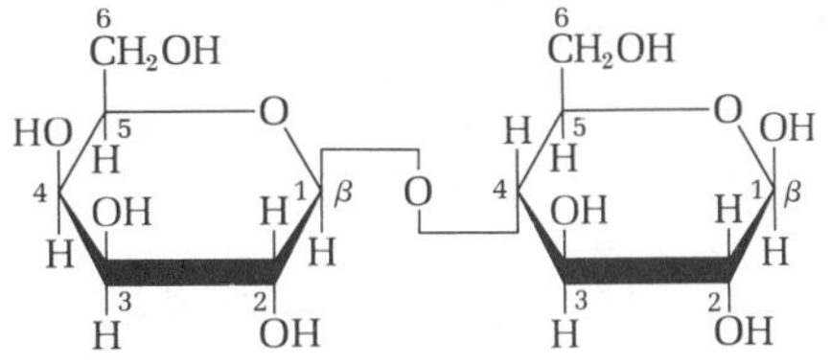

Sucrose [O-β-D-fructofuranosyl-(2 → 1)-α-D-glucopyranoside]

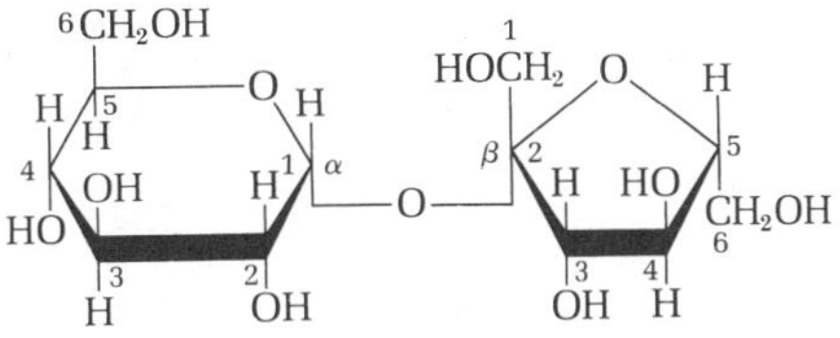

In such individuals, lactose ingested in large amounts in the form of milk causes watery diarrhea, abnormal intestinal flow, and colic pains. Lactose intolerance is not to be confused with the genetic disease *galactosemia* (page 419).

Sucrose, or cane sugar, is a disaccharide of glucose and fructose. It is formed by many plants but does not occur in higher animals. In contrast to maltose and lactose, sucrose contains no free anomeric carbon atom, since the anomeric carbons of both its component monosaccharide units are linked to each other (Figure 11-12). For this reason sucrose is not a reducing sugar. Sucrose is something of a riddle in plant biochemistry. Although D-glucose is the predominant building block of both starch and cellulose, sucrose is a major intermediate product of photosynthesis. In many plants it is the major form in which sugar is transported from the leaves to other portions of plants via their vascular systems. The advantage of sucrose over D-glucose as the transport form of sugar may be that its anomeric carbon atoms are joined, thus protecting sucrose from oxidative or hydrolytic attack by plant enzymes until it reaches its final destination in the plant.

Animals cannot absorb sucrose as such, but it is made available for absorption by the enzyme *sucrase*, also called *invertase*, in the cells lining the small intestine (page 685). This enzyme catalyzes the hydrolysis of sucrose to D-glucose and D-fructose, which are readily absorbed in the bloodstream.

Sucrose is by far the sweetest of the three common disaccharides. It is also sweeter than glucose (Table 11-1). Because of the increasing cost of imported cane sugar, made from sugar cane and sugar beets, and because of the availability in the United States of vast amounts of D-glucose obtained by hydrolysis of cornstarch, a new industrial process has been developed for making a sweeter sweetener out of D-glucose. In this process starch is first hydrolyzed to yield D-glucose in the form of corn syrup, a concentrated neutral solution of D-glucose. It is allowed to flow through a large column of an inert support material to which the enzyme *glucose isomerase*, isolated from plant sources, has been attached by covalent bonding. This enzyme, immobilized on the inert support, catalyzes the reversible reaction

$$\text{D-Glucose} \rightleftharpoons \text{D-fructose}$$

to form an equimolar mixture of D-glucose and D-fructose from corn syrup. Since D-fructose is about 2.5 times sweeter than D-glucose (Table 11-1), the sweetening power of the corn syrup is greatly increased. This new product, which is cheaper and at the same time as nutritious as sucrose, is now widely used in the food, soft drink, and ice cream industries. Recently a new product, 90 percent pure fructose, also made by the isomerase procedure, has been put on the market for use as a table sweetener, but it costs twice as much as sucrose, weight for weight. Although it is sweeter than sucrose, thus requiring consump-

Table 11-1 Sweetness of Some Sugars and Saccharin

Sugar	Relative sweetness
Sucrose	100
Glucose	70
Fructose	170
Maltose	30
Lactose	16
Saccharin	40,000

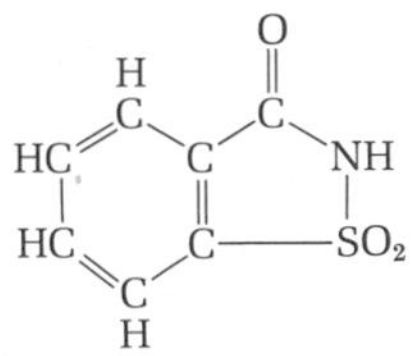

Figure 11-13
Saccharin, an artificial sweetener having no food value.

tion of fewer calories for a given amount of sweetening, it otherwise has no nutritional advantage over sucrose to justify its much higher cost.

Artificial sweeteners having no food value have been developed, particularly for obese or diabetic patients, for whom excess sugar consumption is harmful. Artificial sweeteners stimulate the same taste buds of the tongue that are stimulated by sugars, but they are not used as food by the body (see Chapter 26). The most widely used artificial sweetener is *saccharin* (Figure 11-13), which is 400 times sweeter than sucrose.

Polysaccharides Contain Many Monosaccharide Units

Most of the carbohydrates found in nature occur as polysaccharides of high molecular weight. Some polysaccharides serve biologically as storage forms of monosaccharides, whereas others serve as structural elements in cell walls and connective tissues. On complete hydrolysis with acid or by the action of specific enzymes, polysaccharides yield monosaccharides or their derivatives.

Polysaccharides, also called *glycans*, differ in the nature of their recurring monosaccharide units, in the length of their chains, and in the degree of branching. There are two kinds: *homopolysaccharides*, which contain only a single type of monomeric unit, and *heteropolysaccharides*, which contain two or more different kinds of monomeric units. An example of a homopolysaccharide is the storage carbohydrate *starch*, which contains only D-glucose units. An example of a heteropolysaccharide is the *hyaluronic acid* of connective tissue, which contains alternating residues of two different sugar units. Polysaccharides generally do not have precise molecular weights, like those of proteins; they are mixtures of molecules of high molecular weight, to which monosaccharide units may be added or from which they may be subtracted enzymatically, depending on the metabolic needs of the cells in which they are stored.

Some Polysaccharides Serve as a Storage Form of Cell Fuel

The most important storage polysaccharides in nature are *starch*, characteristic of plant cells, and *glycogen* in animal cells. Both starch and glycogen occur intracellularly in the form of large clusters or granules (Figure 11-14). Starch and glycogen molecules are highly hydrated since they have many exposed hydroxyl groups. Consequently, glycogen and starch, when extracted from the granules with hot water, will form turbid colloidal solutions or dispersions.

Starch is especially abundant in tubers, such as potatoes, and in seeds, especially corn, but the ability to form starch is present in most plant cells (Figure 11-14). Starch contains two types of glucose polymer, *α-amylose* and *amylopectin*. The former consists of long, unbranched chains of D-glucose units

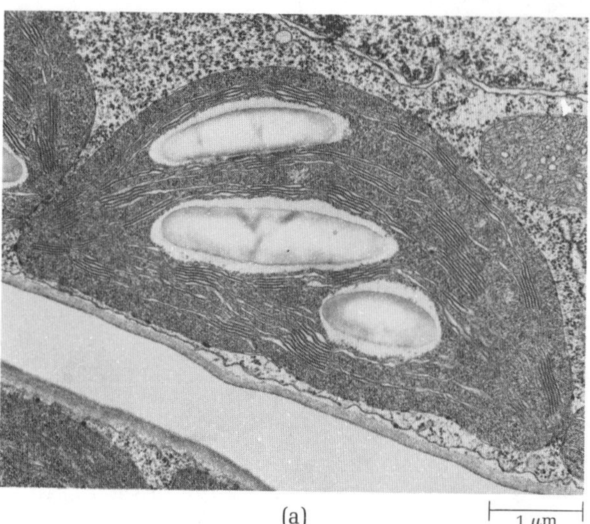

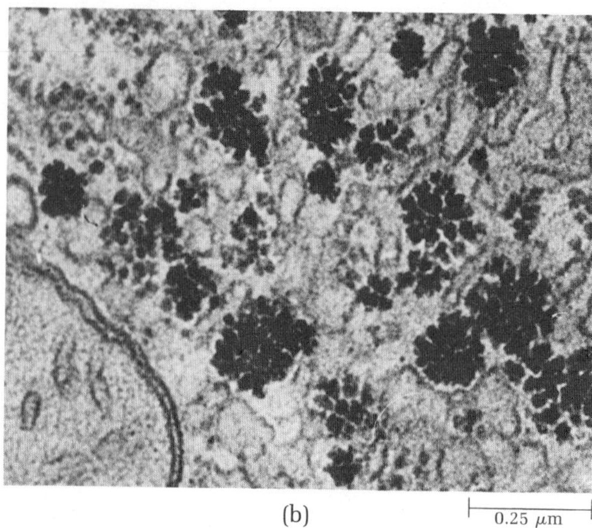

Figure 11-14
Starch and glycogen are stored as granules in plant and animal cells, respectively.
(a) *Large starch granules in a single chloroplast. In many photosynthetic leaf cells starch is made from D-glucose formed photosynthetically.*
(b) *Electron micrograph of glycogen granules in a liver cell of the hamster. These granules are much smaller than the starch granules shown at the left.*

connected by $\alpha(1 \rightarrow 4)$ linkages. Such chains vary in molecular weight from a few thousand to 500,000. Amylopectin also has a high molecular weight but is highly branched (Figure 11-15). The glycosidic linkages joining successive glucose residues in amylopectin chains are $\alpha(1 \rightarrow 4)$, but the branch points of amylopectin are $\alpha(1 \rightarrow 6)$ linkages. When potatoes are boiled, amylose is extracted by the hot water, turning it milky and opalescent. The amylopectin left behind accounts for most of the starch in boiled potatoes.

Glycogen is the main storage polysaccharide of animal cells, the counterpart of starch in plant cells. Like amylopectin, glycogen is a branched polysaccharide of D-glucose in $\alpha(1 \rightarrow 4)$ linkage (Figure 11-15), but it is more highly branched and more compact. The branch linkages are $\alpha(1 \rightarrow 6)$. Glycogen is especially abundant in the liver, where it may attain up to 7 percent of the wet weight; it is also present in skeletal muscle. In liver cells glycogen is found in large granules, which are themselves clusters of smaller granules composed of single, highly branched glycogen molecules with an average molecular weight of several million (Figure 11-14). Such liver glycogen granules also contain, in tightly bound form, the enzymes responsible for the synthesis of glycogen and its degradation.

Glycogen and starch are hydrolyzed in the digestive tract by *amylases,* which are secreted into the digestive tract. Saliva and pancreatic juice contain α-amylases, which hydrolyze the $\alpha(1 \rightarrow 4)$ linkages of the outer branches of glycogen and amylopectin to yield D-glucose, a small amount of maltose, and a resistant "core," called a *limit dextrin* (Figure 11-15). Dextrins form the base of mucilage or library paste. The limit dextrin is not hydrolyzed further by α-amylase, which cannot attack the $\alpha(1 \rightarrow 6)$ linkages at the branch points. For this purpose a debranching enzyme, *$\alpha(1 \rightarrow 6)$-glucosidase,* is required. This enzyme can hydrolyze the branch linkages, thus exposing another tier of $\alpha(1 \rightarrow 4)$-linked branches to the action of α-

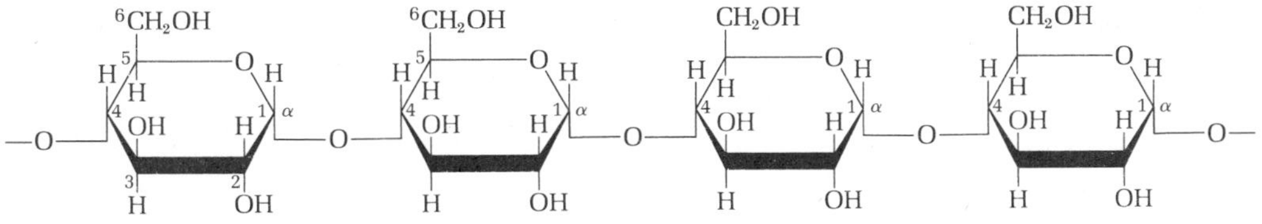

(a)

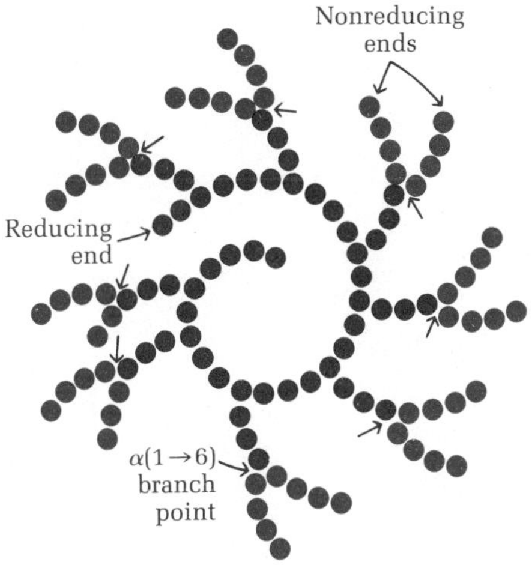

Nonreducing ends

Reducing end

$\alpha(1 \rightarrow 6)$ branch point

(b)

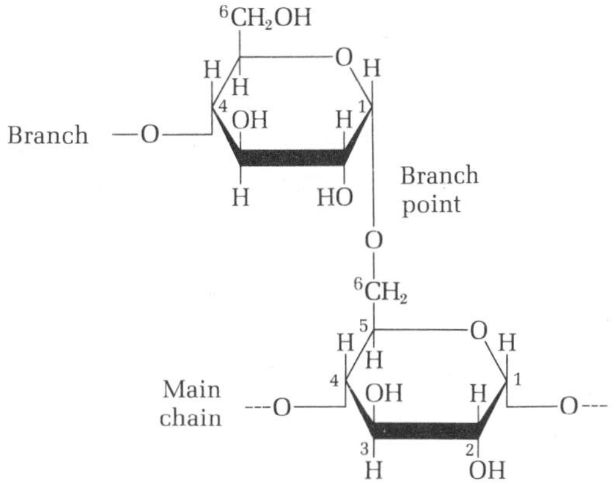

Branch

Branch point

Main chain

(c)

Figure 11-15
Amylose and amylopectin, the polysaccharides of starch.

(a) *Amylose, a linear polymer of* D*-glucose units in* $\alpha(1 \rightarrow 4)$ *linkage.*

(b) *Amylopectin. Each circle represents one glucose residue. The colored circles represent the glucose residues of the outer branches, which are removed by* α*-amylase. The black circles represent the structure of limit dextrin after* α*-amylase has removed all the exterior glucose residues. The* $\alpha(1 \rightarrow 6)$ *linkages at the branch points (indicated by small arrows) are cleaved by* $\alpha(1 \rightarrow 6)$*-glucosidase, thus opening another tier of* D*-glucose residues in* $\alpha(1 \rightarrow 4)$ *linkage to amylase action. Glycogen has a similar structure but is more highly branched and more compact.*

(c) *Structure of a branch point.*

amylase. Another set of branch points is then reached, which again is cleaved by $\alpha(1 \rightarrow 6)$-glucosidase. The combined action of α-amylase and $\alpha(1 \rightarrow 6)$-glucosidase can therefore completely degrade glycogen and amylopectin to glucose and a small amount of maltose. In animal cells, however, glycogen is degraded by a different type of enzyme, glycogen phosphorylase, which breaks down glycogen to yield glucose 1-phosphate, rather than free glucose (pages 414 and 574).

The enzyme β-amylase of malt differs from α-amylase in that it hydrolyzes *alternating* $\alpha(1 \rightarrow 4)$ linkages, thus yielding mostly maltose and little glucose. In this case the α- and β- in the names of α- and β-amylase do not refer to α- and β-glycosidic linkages but are simply an arbitrary way of differentiating between two different kinds of amylases.

Cellulose Is the Most Abundant Structural Polysaccharide

Many polysaccharides serve as extracellular structural elements in the cell walls of unicellular microorganisms and higher plants and in the outer surfaces of animal cells. Other polysaccharides are components of the connective tissue of vertebrates and the exoskeletons of arthropods. Structural polysaccharides provide protection, shape, and support to cells, tissues, or organs.

There are many different structural polysaccharides. We shall examine one of them, cellulose, to see how its specific molecular structure makes it suitable for its biological function. Cellulose, a fibrous, tough, water-insoluble substance, is found in the protective cell walls of plants, particularly in stalks, stems, trunks, and all the woody portions of plant tissues.

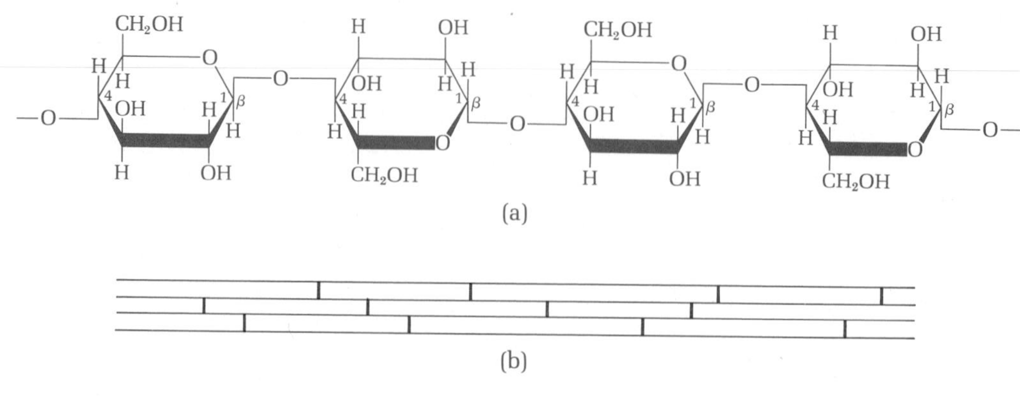

(a)

(b)

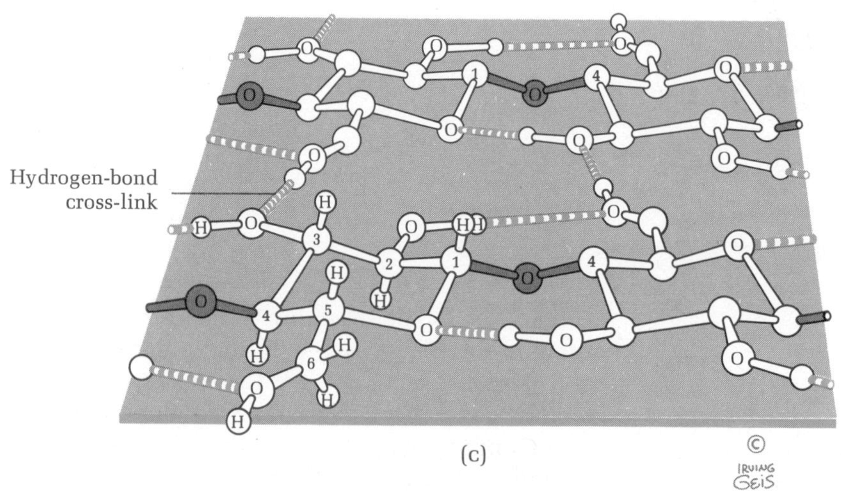

(c)

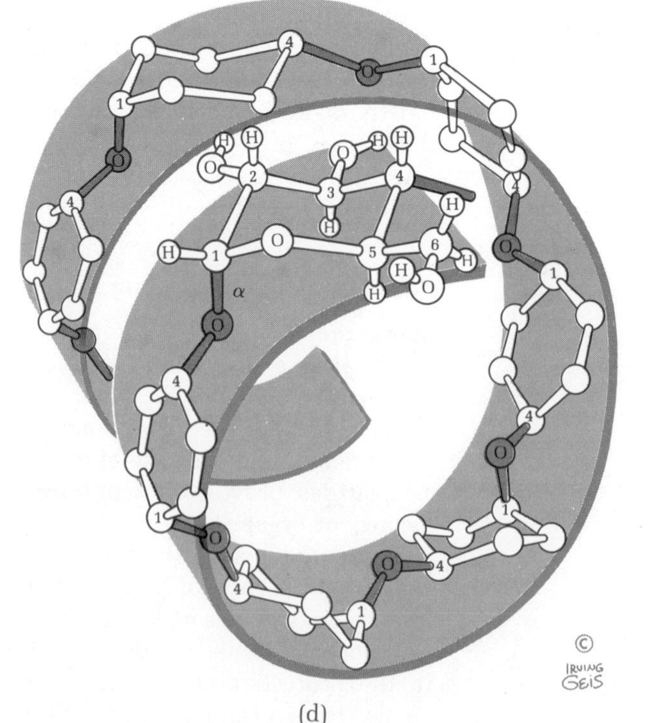

(d)

Figure 11-16
The structure of cellulose and the different conformations assumed by the $\beta(1 \rightarrow 4)$ cellulose chains and the $\alpha(1 \rightarrow 4)$ starch and glycogen chains.

(a) Cellulose chain; the D-glucose units are in $\beta(1 \rightarrow 4)$ linkage.

(b) Schematic drawing showing how parallel cellulose chains are held together by cross-links of hydrogen bonds (color).

(c) Scale drawing of segments of two parallel cellulose chains, showing the actual conformation of the D-glucose residues and the hydrogen-bond cross-links.

(d) Scale drawing of a segment of amylose. The $\alpha(1 \rightarrow 4)$ linkages of amylose, amylopectin, and glycogen cause the chain to assume a tightly coiled helical structure, in which many of the hydroxyl groups face outward.

Wood is largely made up of cellulose and other polymeric substances; cotton is almost pure cellulose. Cellulose is not only the most abundant extracellular structural polysaccharide of the plant world but also the most abundant of *all* biomolecules, plant or animal. Proteins are of course the most abundant intracellular macromolecules (page 54).

Since cellulose is a linear, unbranched homopolysaccharide of 10,000 or more D-glucose units connected by $1 \rightarrow 4$ glycosidic bonds, it would appear to resemble amylose and the main chains of glycogen. But there is a very important difference: in cellulose the $1 \rightarrow 4$ linkages are in the β configuration, whereas in amylose, amylopectin, and glycogen the $1 \rightarrow 4$ linkages are α. This seemingly trivial difference in the structure of cellulose and amylose results in polymeric structures of very different properties (Figure 11-16). Thanks to the geometry of their $\alpha(1 \rightarrow 4)$ linkages, the main chains of D-glucose units in glycogen and starch tend to assume a coiled, helical conformation, which favors formation of dense granules, observed in many animal and plant cells. The $\alpha(1 \rightarrow 4)$ linkages of glycogen and starch are readily hydrolyzed by α-amylase in the intestinal tract of vertebrates, and the D-glucose so formed is absorbed and used as an energy-rich fuel. On the other hand, because of their β linkages, D-glucose chains in cellulose assume an extended conformation and undergo side-by-side aggregation into insoluble fibrils (Figure 11-16). The $\beta(1 \rightarrow 4)$ linkages of cellulose are not hydrolyzed by α-amylases. Since no enzyme capable of hydrolyzing cellulose is secreted by the intestinal tract of vertebrates, cellulose cannot be digested; its D-glucose units are thus completely unavailable as food in most higher organisms. Termites readily digest cellulose but only because their intestinal tract harbors a parasitic microorganism, Trichonympha (Figure 11-17), which secretes cellulase, a cellulose-hydrolyzing enzyme enabling termites to digest wood. Wood-rot fungi and bacteria also produce cellulase.

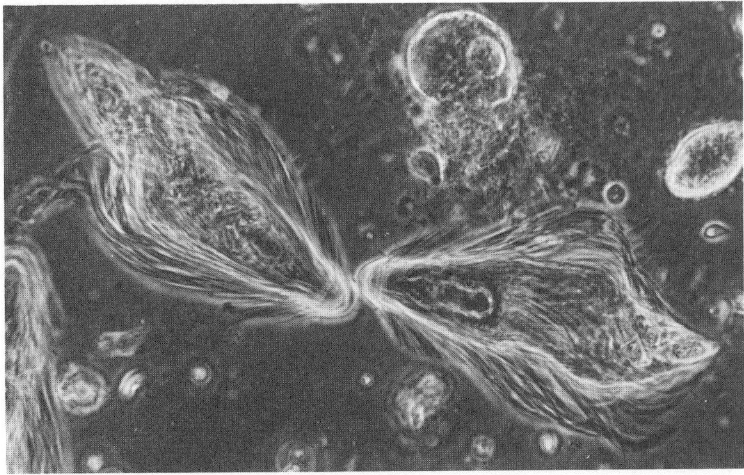

Figure 11-17
Two Trichonympha *cells. These protozoa, which inhabit the intestinal tract of termites, secrete cellulase. The host termites are unable to digest cellulose without the help of these parasites.*

The only vertebrates able to use cellulose as food are cattle and other ruminant animals (sheep, goats, camels, giraffes), which do so in a very indirect manner. Much of the abdominal volume and 15 percent of the total weight of a cow are occupied by its series of four stomachs. The first two, which constitute the rumen, teem with microorganisms that secrete cellulase and break cellulose down to yield D-glucose, which they ferment into short-chain fatty acids (Chapter 12), carbon dioxide, and methane gas (CH_4). The fatty acids produced by the rumen organisms are absorbed into the bloodstream of the cow, taken up by the tissues, and used as fuel. The CO_2 and methane, produced at a rate of 2 L/min, are released by eructation, a continuous, scarcely audible reflex process akin to belching. In the remaining two stomachs of ruminants the microorganisms, having done their duty, are digested by enzymes secreted by the stomach lining, to yield amino acids, sugars, and other hydrolysis products that are absorbed and thus become available for the nutrition of the cow. This represents a useful symbiotic relationship for both the cow and the microorganisms, which enjoy a short but happy life in a warm and favorable environment in which the cellulose of grass and clover is the major fuel source for both beast and "bugs."

Enormous amounts of cellulose are made annually by the plant world, not only by forest growth but also by cultivated crop plants. It is estimated that every day some 50 kg of cellulose is synthesized by the plant world for each human being on earth. Cellulose is also a very useful substance. Wood, cotton, paper, and cardboard are all largely cellulose. Moreover, cellulose is the base of many other products, such as rayon, insulating tiles, and other packing and building materials.

The tough insoluble shells, or exoskeletons, of lobsters, crabs, and many insects consist largely of the structural polysaccharide *chitin*, a linear polymer of *N-acetyl-D-glucosamine* connected by β linkages (Figure 11-18). The chitin framework of lobster and crab shells is impregnated and hardened with calcium carbonate.

Figure 11-18
N-Acetyl-D-glucosamine, an important building-block of chitin and many other structural polysaccharides. D-Glucosamine, an amino sugar, has an amino group (shaded in color) at carbon 2 instead of a hydroxyl group.

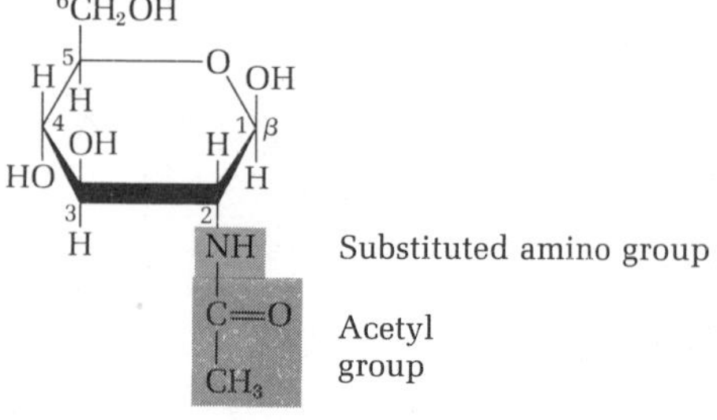

Cell Walls Are Rich in Structural and Protective Polysaccharides

Most plant cells are surrounded by rigid, very strong polysaccharide structures, which have been likened to fiberglass-reinforced plastic. The framework of plant cell walls consists of crisscross layers of long, extended cellulose fibrils, which are stronger than steel wire of the same diameter (Figure 11-19). This fibrillar framework is impregnated with a cementlike matrix consisting of other kinds of structural polysaccharides and another polymeric substance called lignin. In the woody portion of tree trunks the cell walls are very thick, capable of withstanding enormous compression forces (Figure 11-19).

The bacterial cell wall (Figure 11-20), which is located outside the cell membrane, forms a rigid, porous shell around

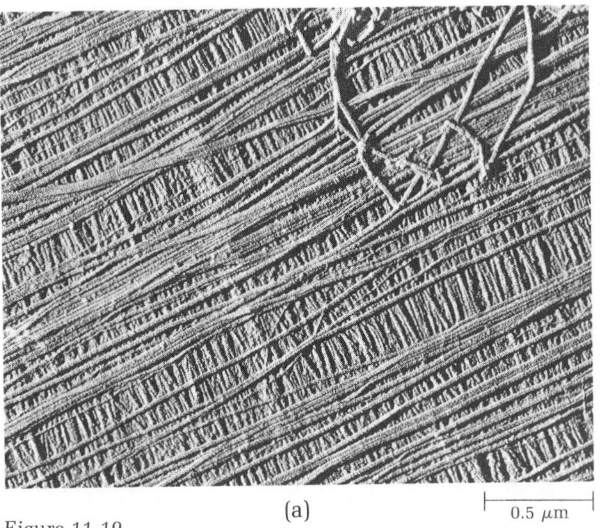

Figure 11-19
Cellulose is the major component of plant cell walls.
(a) *An electron micrograph of the cell wall of an alga* (Chaetomorpha). *It consists of crisscross layers of cellulose fibrils impregnated with cementlike polymeric substances.*
(b) *Cross section of a locust tree, showing its annual growth rings. The wood produced in the spring has large cells with thin walls; that produced late in the season has small cells with thicker walls and more layers of cellulose fibrils. The lighter wood around the heartwood is sapwood.*

the cell. It provides physical protection to the delicate cell membrane and to the cytoplasm within. The cell walls of most bacteria consist of a covalently bonded structural framework completely surrounding the cell. It is made up of long, parallel polysaccharide chains cross-linked to each other at intervals by short polypeptide chains. The polysaccharide chains consist of alternating monosaccharide units of *N-acetyl-D-glucosamine* (Figure 11-18) and *N-acetylmuramic acid,* a complex 9-carbon sugar; these are linked by $\beta(1 \rightarrow 4)$ bonds (Figure 11-20). Attached to each *N*-acetylmuramic unit is a tetrapeptide side chain. The parallel polysaccharide chains are cross-linked by short polypeptide chains, which differ somewhat in structure from species to species. In the pus bacterium *Staphylococcus*

Figure 11-20
(a) *Schematic drawing of the peptidoglycan of the cell wall of the gram-positive bacterium* Staphylococcus aureus.
(b) *Structure of the repeating disaccharide unit in the backbone of the peptidoglycan.*

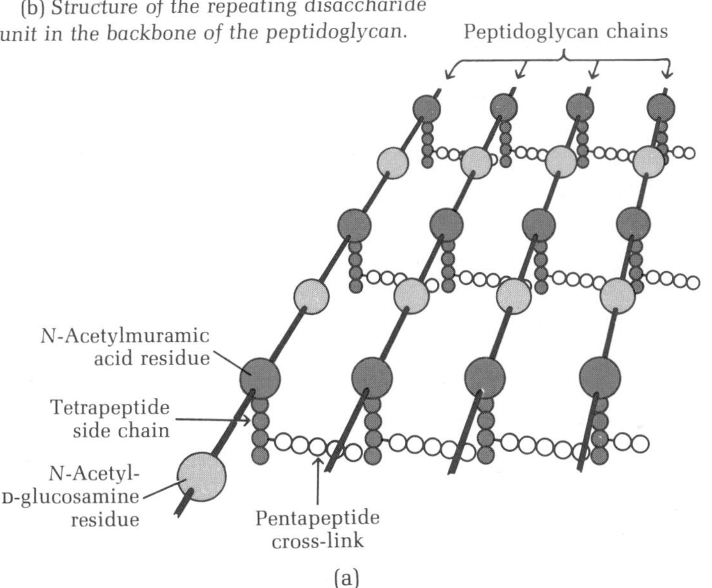

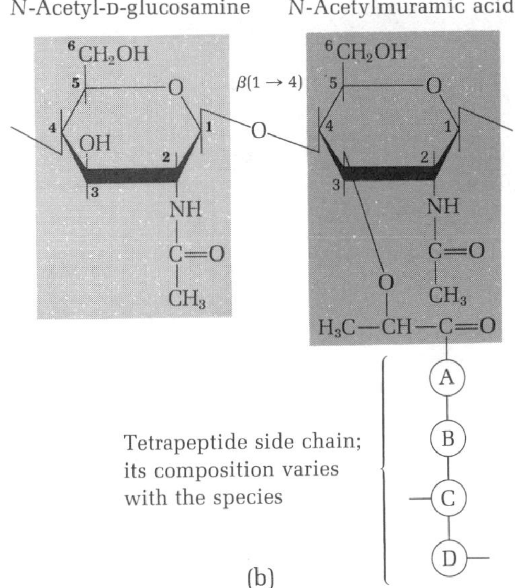

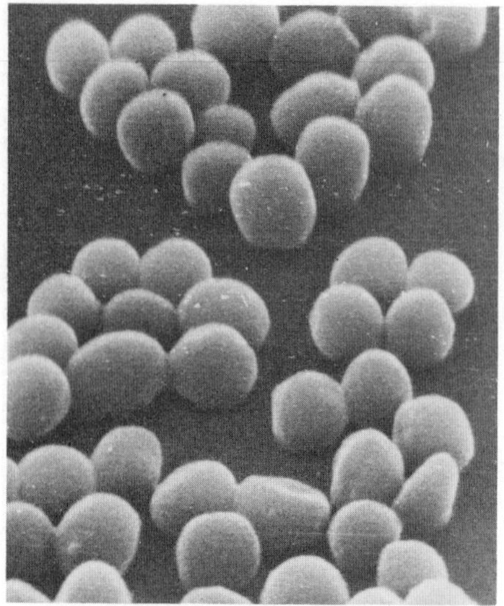

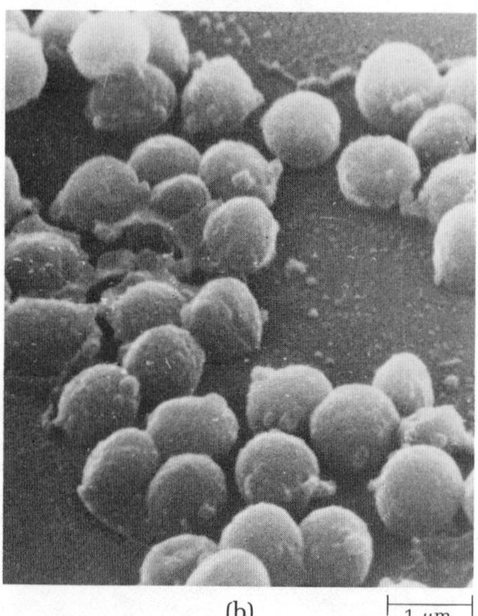

Figure 11-21
The action of penicillin on Staphylococcus aureus *cells. (a) Cells before treatment. (b) After exposure to penicillin the cell walls are defective and collapse.*

aureus, which causes boils and wound infections, the *N*-acetylmuramic acid residues in adjacent polysaccharide chains are cross-linked to each other by pentapeptides of five glycine residues. The entire cross-linked structure surrounding the cell is called a *murein* (Latin *murus,* "wall") or a *peptidoglycan,* a term indicating the hybrid nature of the structure, which combines peptide and polysaccharide elements. The peptidoglycan, which is continuous around the entire bacterial cell, may be regarded as a single, enormous cagelike molecule. In the gram-positive bacteria (which give the gram stain, a color reaction with the dye crystal violet) there are several concentric layers of peptidoglycan around the cell, interwoven with other macromolecular components. In the gram-negative species, such as *E. coli,* the peptidoglycan framework is covered with a lipid-rich outer coat containing hydrophobic proteins (Chapter 12). Intact cell walls are vital for the protection, growth, and division of bacteria. *Penicillin,* one of the most valuable antibiotics in treatment of bacterial infections, inhibits a late step in the enzymatic synthesis of peptidoglycans in penicillin-susceptible organisms, so that the cell wall is incomplete and fails to sustain the normal growth of the cells (Figure 11-21).

Glycoproteins Are Hybrid Molecules

Glycoproteins are proteins that contain covalently attached carbohydrates, either single monosaccharides or relatively short oligosaccharides. The carbohydrate portion may constitute anywhere from less than 1 percent carbohydrate up to 30 percent or more. Some glycoproteins have only one or a few carbohydrate groups; others have numerous oligosaccharide side chains, which may be linear or branched (Figure 11-22). Nearly all proteins on the outer surface of animal cells are glycoproteins. In addition, most proteins that are secreted to the cell exterior are glycoproteins, as are most of the proteins of the blood plasma. To generalize, most proteins having an extracellular exposure, location, or function are glycoproteins.

A most remarkable extracellular glycoprotein is the *antifreeze protein* in the blood of some Arctic and Antarctic fish species, as well as in winter flounder and codfish of the eastern coast of North America. The antifreeze proteins vary somewhat in structure from species to species, but the best known consists of a polypeptide backbone of the recurring tripeptide unit Ala-Ala-Thr repeated up to 50 times (Figure 11-22). To each threonine residue is attached the disaccharide *D-galactosyl-N-acetyl-D-galactosamine.* Antifreeze proteins depress the freezing point of water, apparently because they inhibit the formation of ice crystals. The combination of the antifreeze proteins and the very high concentration of NaCl in the blood [which also depresses its freezing point (page 71)] allows these fish species to tolerate the low temperatures of polar seawater, at which the blood of terrestrial vertebrates would freeze.

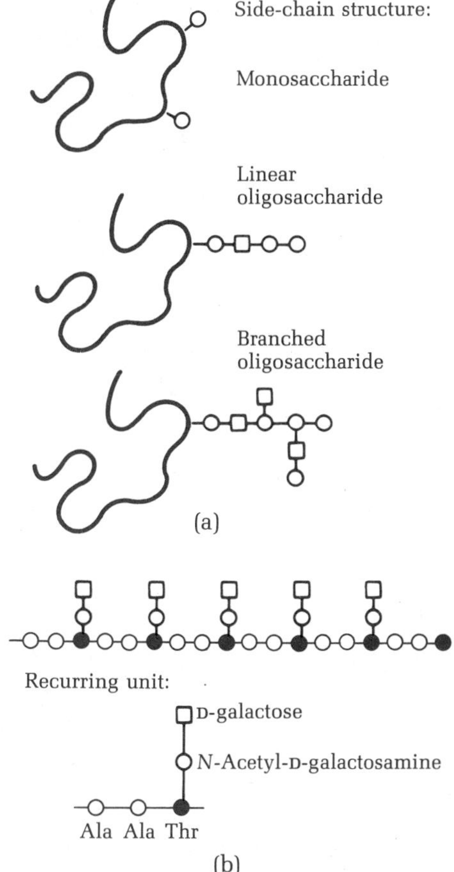

Figure 11-22
Some glycoproteins and a proteoglycan structure.
(a) Three types of glycoproteins differing in the size and composition of their carbohydrate side chains.
(b) The recurring unit of the antifreeze glycoprotein of some polar fish species. It is repeated many times.

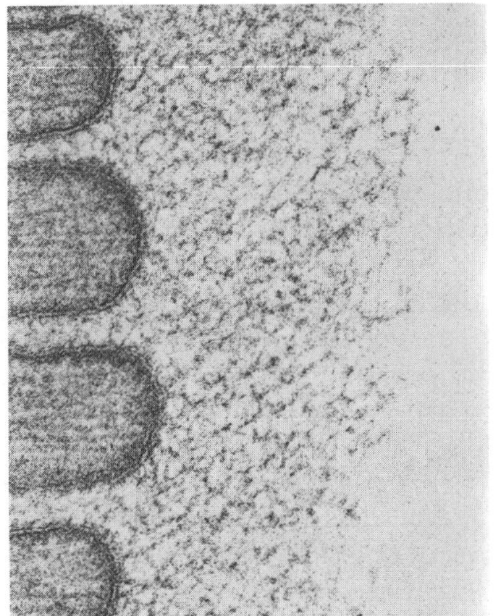

Figure 11-23
The glycocalyx, or fuzzy coat, a filamentous, meshlike oligosaccharide structure on the microvilli (at left) of epithelial cells of the intestine.

Animal Cell Surfaces Contain Glycoproteins

Instead of rigid surrounding walls the cells of animal tissues have a soft, flexible outer surface, sometimes called the *cell coat,* which contains different kinds of oligosaccharide chains. In the cells lining the intestine there is a very thick carbohydrate-rich coat called the *glycocalyx* or "fuzzy coat" (Figure 11-23). The oligosaccharides in such cell coats are contributed largely by specific glycoproteins present in the plasma membrane, which also contains another class of hybrid molecules with carbohydrate groups, the *glycolipids.*

One of the best-known membrane glycoproteins is *glycophorin* of the red blood cell membrane (Chapter 12, page 323), which contains almost 50 percent carbohydrate in the form of a long polysaccharide chain covalently attached to one end of the polypeptide chain. The polysaccharide chain extends from the outer surface of the cell membrane, whereas the polypeptide chain is buried inside the membrane. Another membrane glycoprotein is *fibronectin* (Latin, *fibra,* "fiber," and *nectere,* to "bind" or "tie"). Fibronectin appears to promote adhesion of like cells to each other (page 34). These and other glycoproteins will be considered further in the discussion of membrane structure in the following chapter.

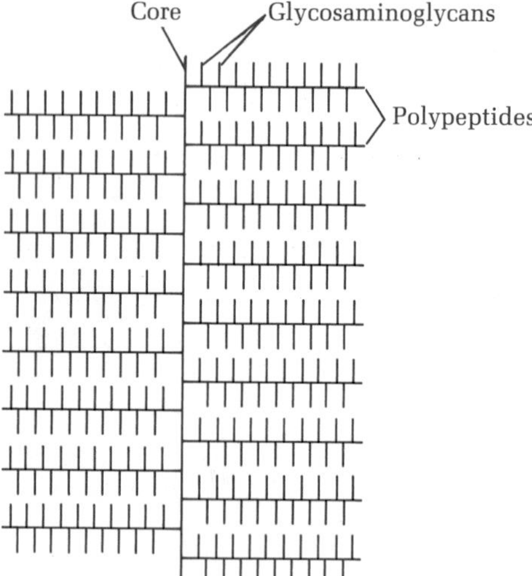

Figure 11-24
A schematic representation of a proteoglycan. It consists of a branched core or framework of protein, to which are attached many side chains of glycosaminoglycans. The entire structure is highly hydrated.

Figure 11-25
D-*Glucuronate. It has a carboxyl group at carbon* 6, *which is dissociated at* pH *near* 7, *as shown.*

Acid Mucopolysaccharides and Proteoglycans Are Important Components of Connective Tissue

Another group of structural and protective polysaccharides are the *acid mucopolysaccharides.* They are usually attached to proteins to form *proteoglycans,* a name reserved for hybrid molecules of polysaccharides and protein in which the polysaccharide makes up most of the weight, often 95 percent or more. In contrast, the glycoproteins are hybrid molecules of protein and carbohydrate in which the protein predominates.

The acid mucopolysaccharides or *glycosaminoglycans* consist of recurring disaccharide units, each of which contains a derivative of an aminohexose, usually D-glucosamine or D-galactosamine. At least one of the two sugars in the recurring disaccharide unit of acid mucopolysaccharides contains an acidic group having a negative charge at pH 7, either a carboxylate or sulfate group. An example of an acidic hexose is D-glucuronate, derived from D-glucose by oxidation of the 6-carbon atom to a carboxylate group (Figure 11-25). Acid mucopolysaccharides are thus heteropolysaccharides since they consist of two kinds of monosaccharides in alternating sequence (Table 11-2). The prefix *muco-* refers to the fact that these polysaccharides were first isolated from *mucin,* the slippery, lubricating proteoglycan of mucous secretions. Today, however, the term "acid mucopolysaccharide" is used more broadly to refer to acidic polysaccharides from various vertebrate sources. Proteoglycans are found in the jellylike *ground substance,* or *intercellular cement,* filling the space between the cells of most tissues. They are also present in cartilage, tendons, and skin, and in the *synovial fluid,* the secretion that lubricates skeletal joints.

The acid mucopolysaccharide *hyaluronic acid* of the intercellular ground substance of animal tissues contains many alternating units of *D-glucuronic acid* and *N-acetyl-D-glucosamine* (Figure 11-26). Hyaluronic acid forms highly viscous, jellylike solutions. Hyaluronic acid is often combined with other mucopolysaccharides. *Hyaluronidase,* an enzyme secreted by some pathogenic (disease-causing) bacteria, can hydrolyze the glycosidic linkages of hyaluronic acid, rendering tissues more susceptible to invasion by the bacteria. This enzyme also hydrolyzes an outer acid polysaccharide coat of the ovum of vertebrates and makes possible penetration by the sperm cell during fertilization. *Chondroitin,* a major polysac-

Table 11-2 Monosaccharide Components of Representative Acid Polysaccharides of Connective Tissue

Polysaccharide	Components
Hyaluronic acid	D-Glucuronate + N-acetyl-D-glucosamine
Chondroitin	D-Glucuronate + N-acetyl-D-galactosamine
Dermatan sulfate	D-Iduronate + N-acetyl-D-galactosamine-4-sulfate

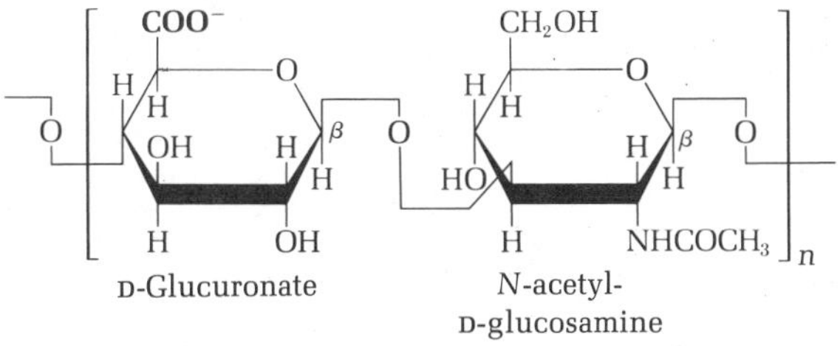

Figure 11-26
Repeating unit of hyaluronate, a linear polymer containing thousands of such disaccharide units. At pH 7 the carboxyl group of the D-glucuronate residues is fully ionized.

charide of cartilage proteoglycans, contains alternating units of *D-glucuronic acid* and *N-acetyl-D-galactosamine*. *Dermatan sulfate* of skin also contains alternating units of two different sugars (Table 11-2).

Another important acidic polysaccharide is *heparin*, generated by certain types of cells that are especially abundant in the lining of arterial blood vessels. It has repeating units of six sugar residues, each consisting of an alternating sequence of sulfate derivatives of N-acetyl-D-glucosamine and D-iduronate. Heparin is a very powerful inhibitor of blood clotting and aids in preventing the formation of clots in circulating blood. Heparin isolated from lung tissue is used medically to inhibit the clotting of blood specimens withdrawn by syringe, to prevent clotting during intravenous administration of fluids, and to inhibit clotting within blood vessels in various pathological conditions, e.g., after heart attacks.

The acid mucopolysaccharide molecules in the proteoglycans of cartilage are covalently bound to protein. A typical cartilage proteoglycan contains about 150 polysaccharide chains, each of molecular weight 20,000, covalently bound as side chains to core polypeptides. Such proteoglycans are highly hydrated structures (Figure 11-24).

Summary

Carbohydrates are polyhydroxylic aldehydes or ketones having the empirical formula $(CH_2O)_n$. They are classified as monosaccharides or sugars (one aldehyde or ketone unit); oligosaccharides (several monosaccharide units); and polysaccharides, large linear or branched molecules containing many monosaccharide units. Monosaccharides or simple sugars have a single aldehyde or ketone unit. They also have at least one asymmetric carbon atom and thus exist in stereoisomeric forms. Most common naturally occurring sugars, such as ribose, glucose, fructose, and mannose, are of the D series. Simple sugars having 5 or more carbon atoms may exist in the form of closed-ring hemiacetals, either furanoses (five-membered ring) or pyranoses (six-membered ring). Furanoses and pyranoses occur in anomeric α and β forms, which are interconverted in the process of mutarotation. Sugars able to reduce oxidizing agents are called reducing sugars.

Disaccharides consist of two monosaccharides joined by a covalent linkage. Maltose contains two D-glucose residues in $\alpha(1 \rightarrow 4)$ glycosidic linkage. Lactose contains D-galactose and D-glucose. Sucrose, a nonreducing sugar, contains D-glucose and D-fructose units joined by their anomeric carbon atoms.

Polysaccharides (glycans) contain many monosaccharide units in glycosidic linkage. Some function as storage forms of carbohydrate. The most important storage polysaccharides are starch and glycogen, high-molecular-weight, branched polymers of glucose having $\alpha(1 \rightarrow 4)$ linkages in the main chains and $\alpha(1 \rightarrow 6)$ linkages at the branch points. The former linkages are hydrolyzed by α-amylase and the latter by $\alpha(1 \rightarrow 6)$-glucosidase. Other polysaccharides play a structural role in cell walls. Cellulose, the structural polysaccharide of plants, has D-glucose units in $\beta(1 \rightarrow 4)$ linkage. Cellulose is not attacked by α- or β-amylases and is not digested by vertebrates, except for ruminant animals, in which cellulase, secreted by bacteria, degrades cellulose to D-glucose. The rigid porous walls of bacterial cells contain peptidoglycans, linear polysaccharides of alternating *N*-acetylmuramic acid and *N*-acetylglucosamine units, cross-linked by short peptide chains. Plant cell walls contain a rigid, strong framework of cellulose fibrils, impregnated with other polymeric substances. Animal cells have a soft, flexible glycocalyx or outer coat, containing oligosaccharide chains linked to lipids and proteins. Glycoproteins contain one or more sugar residues; most cell surface or extra-cellular proteins are glycoproteins. Connective tissues of animals contain several acid mucopolysaccharides, consisting of alternating sugar units, one of which has an acid group. Such structures, in which the polysaccharide predominates, are called proteoglycans.

References

Books

LEHNINGER, A. L.: *Biochemistry*, 2d ed., Worth, New York, 1975. Chapters 10 and 23 contain more detailed information on carbohydrate chemistry.

DAVIDSON, E. A.: *Carbohydrate Chemistry*, Holt, New York, 1967. An excellent survey.

FLORKIN, M., and E. H. STOTZ (eds.): *Comprehensive Biochemistry*, sec. II, vol. 5, *Carbohydrates*, Elsevier, New York, 1963. Reference work.

LENNARZ, W. J. (ed.): *The Biochemistry of Glycoproteins and Proteoglycans*, Plenum, New York, 1980. Up-to-date review of this very complex field.

Some Articles

COSTERTON, J. W., G. G. GEESEY, and K.-J. CHENG: "How Bacteria Stick," *Sci. Am.*, **238**:86–95, January (1978). Polysaccharides made by bacteria enable them to stick firmly to surfaces, such as human teeth.

JAQUES, L. B.: "Heparin: An Old Drug with a New Paradigm," *Science*, **206**:528–533 (1979).

KRETCHMER, N.: "Lactose and Lactase," *Sci. Am.*, **227**:70–78, October (1972) (offprint 1259). Discussion of lactose intolerance.

POINTS, H. G.: "On the Scent of the Riddle of Sucrose," *Trends Biochem. Sci.*, **3**:137–139 (1978). The function of sucrose in plants.

YAMADA, K. M., and K. OLDEN: "Fibronectins, Adhesive Glycoproteins of Cell Surface and Blood," *Nature*, **275**:179–184 (1978).

Problems

1. *Interconversion of D-Galactose Forms* A freshly prepared solution of the α form (1 g/mL in a 10-cm cell) shows an optical rotation of $+150.7°$. When allowed to stand for a prolonged period of time the rotation gradually decreases and reaches an equilibrium value of $+80.2°$. In contrast, a freshly prepared solution (1 g/mL) of the β form shows an optical rotation of only $+52.8°$. Moreover, when the solution is allowed to stand for several hours, the rotation increases to an equilibrium value of $+80.2°$, identical to the equilibrium value reached by α-D-galactose.
   (a) Draw the Haworth projection formulas of the α and β forms of galactose. What feature distinguishes the two forms?
   (b) Why does the rotation of a freshly prepared solution of the α form gradually decrease with time? Explain. Why do solutions of the α and β forms (at equal concentrations) reach the same optical rotation value at equilibrium? Explain.
   (c) Calculate the percent composition of the two forms of galactose at equilibrium.

2. *Invertase "Inverts" Sucrose* The hydrolysis of sucrose ($[\alpha]_D^{20} = +66.5°$) yields an equimolar mixture of D-glucose ($[\alpha]_D^{20} = +52.5°$) and D-fructose ($[\alpha]_D^{20} = -92°$).
   (a) Suggest a convenient way to determine the rate of hydrolysis of sucrose by an enzyme preparation extracted from the lining of the small intestine.
   (b) Explain why an equimolar mixture of D-glucose and D-fructose formed by hydrolysis of sucrose is called invert sugar in the food industry.
   (c) The enzyme invertase is allowed to act on a solution of sucrose until the optical rotation of the system becomes zero. What fraction of the sucrose has been hydrolyzed? (Today the preferred name for invertase is sucrase.)

3. *Manufacture of Liquid-Filled Chocolates* The manufacture of chocolates containing a liquid center is an interesting application of enzyme engineering. The flavored liquid center consists largely of an aqueous solution of sugars rich in fructose to provide sweetness. The technical dilemma is the following: the chocolate coating must be prepared by pouring hot melted chocolate over a solid core, yet the final product must have a liquid, fructose-rich center. Suggest a way to solve this problem. (*Hint:* The solubility of sucrose is much lower than the combined solubility of glucose and fructose.)

4. *Anomers of Lactose* Lactose, a disaccharide composed of galactose and glucose, exists in two anomeric forms designated α and β. The properties of the two anomers are significantly different. For example, the β anomer has a sweeter taste than the α anomer. Moreover, the β anomer is more soluble than the α anomer; consequently the α anomer may crystallize when ice cream is stored in the freezer for a long time, giving the ice cream a sandy texture.
   (a) Draw the Haworth projection formulas of the two anomeric forms of lactose.
   (b) Draw the Haworth projection formula of all the products that result when the α anomer is hydrolyzed to galactose and glucose. Do the same for the β anomer.

5. *Anomers of Sucrose?* Although the disaccharide lactose exists in two anomeric forms, no anomeric forms of the disaccharide sucrose have been reported. Why?

6. *Growth Rate of Bamboo* The stems of bamboo, a tropical grass, can grow at the phenomenal rate of 1 ft/d under optimal conditions. Given that the stems are composed almost entirely of cellulose fibers oriented in the direction of growth, calculate the number of sugar residues per second that must be added by the action of enzymes to growing cellulose chains to account for the growth rate. Each D-glucose unit in the cellulose molecule is about 0.45 nm long.

7. *Comparison of Cellulose and Glycogen* The practically pure cellulose obtained from the seed threads of the plant species *Gossypium* (cotton) is tough, fibrous, and completely insoluble in water. In contrast, glycogen obtained from muscle or liver disperses readily in hot water to make a turbid solution. Although they have markedly different physical properties, both substances are composed of 1,4-linked polymers of D-glucose of comparable molecular weight. What features of their structures cause these two polysaccharides to differ in their properties? What are the biological advantages of their respective physical properties?

8. *Glycogen as Energy Storage: How Long Can a Game Bird Fly?* Since ancient times it had been observed that certain game birds, such as grouse, quail, and pheasants, are easily fatigued. The Greek historian Xenophon (434–355 B.C.) wrote: "The bustards, on the other hand, can be caught if one is quick in starting them up, for they will fly only a short distance, like partridges, and soon tire; and their flesh was delicious." The flight muscles of game birds rely almost entirely on the metabolic breakdown of glucose 1-phosphate for the necessary energy (see glycolysis, Chapter 15). In game birds, glucose 1-phosphate is formed by the breakdown of stored muscle glycogen, catalyzed by the enzyme glycogen phosphorylase. The rate of energy production for flight (in the form of ATP) is limited by the rate at which glycogen can be broken down. During a "panic flight," the game bird's rate of glycogen breakdown is quite high, approximately 120 μmol/min of glucose 1-phosphate produced per gram of fresh tissue. Given that the flight muscles usually contain about 0.35% glycogen by weight, calculate how long a game bird can fly.

Problem 9

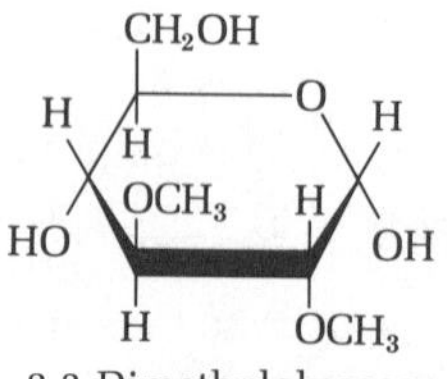

CH_2OH H O H H OCH_3 H HO OH H OCH_3

2,3-Dimethylglucose

9. *Determination of the Extent of Branching in Amylopectin* The extent of branching [number of $\alpha(1 \rightarrow 6)$ glycosidic bonds] in amylopectin can be determined by the following procedure. A weighed sample of amylopectin is exhaustively treated with a methylating agent that replaces all the hydrogens on the sugar hydroxyls with a methyl group (—O—H → —O—CH_3). Subsequently, all the glycosidic bonds in the treated sample above are hydrolyzed with aqueous acid. The amount of 2,3-dimethylglucose in the hydrolyzed sample is then determined.
   (a) Explain the basis of this procedure for determining the number of $\alpha(1 \rightarrow 6)$ branch points in amylopectin. What happens to the nonbranched glucose residues in amylopectin during the above procedure?
   (b) A 258-mg sample of amylopectin treated as described above yielded 12.4 mg of 2,3-dimethylglucose. Determine what percent of the glucose molecules in amylopectin contain an $\alpha(1 \rightarrow 6)$ branch.

10. *Structure Determination of Trehalose* Almost 30 percent of the cocoon of the parasitic beetle *Larinus maculatus* consists of the carbohydrate trehalose. Upon acid hydrolysis, trehalose gives D-glucose as the only product. If trehalose is exhaustively methylated (all sugar —OH groups converted into $—OCH_3$ groups) and subsequently hydrolyzed with aqueous acid, only one product, 2,3,4,6-tetramethylglucose, is obtained. What is the structure of trehalose? Demonstrate that your proposed structure is consistent with the data presented above.

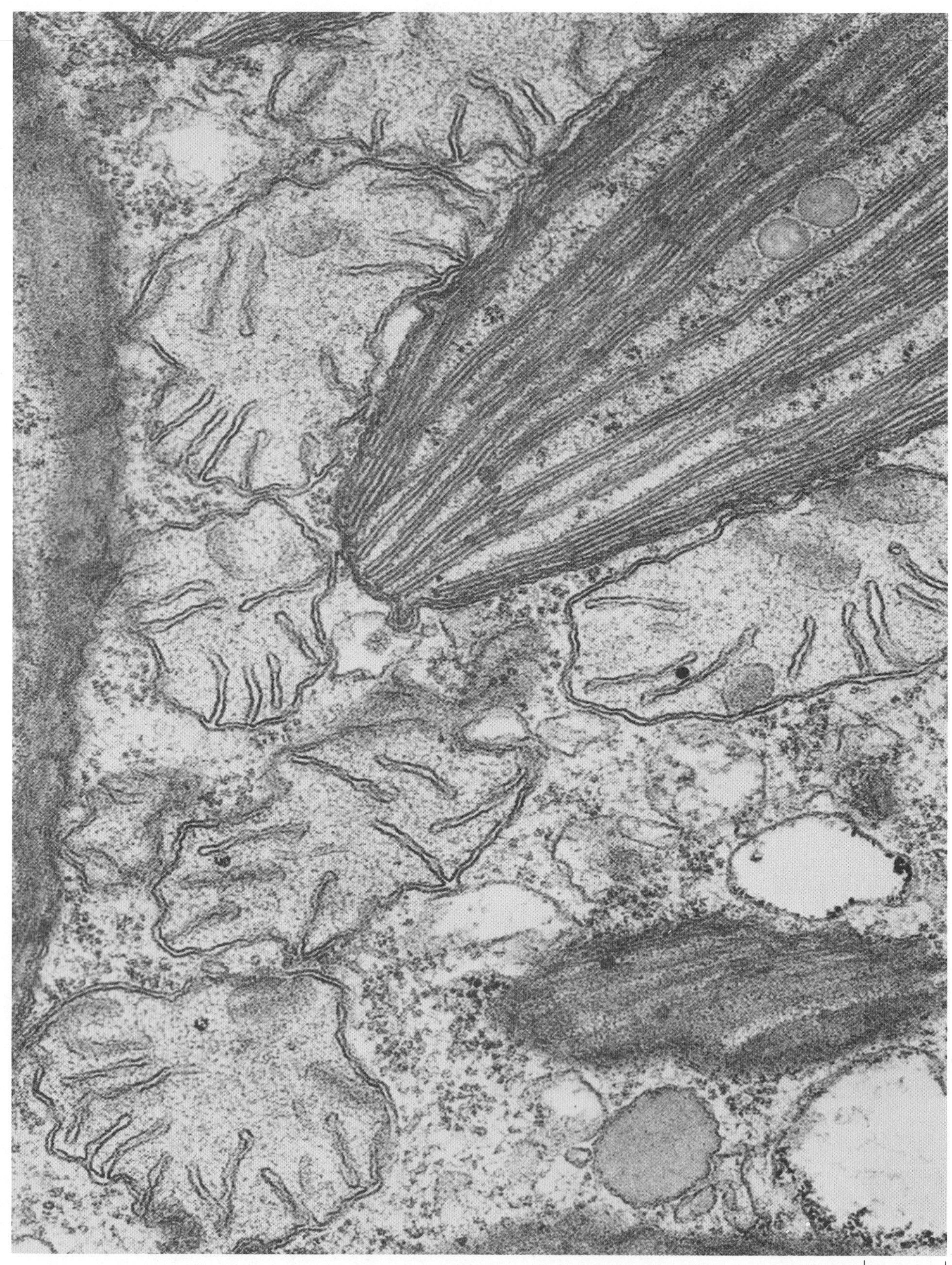
0.2 μm

CHAPTER 12

Lipids and Membranes

In the preceding chapters we have examined several important components of cells: water, proteins, enzymes, coenzymes, and carbohydrates. There is yet another family of biomolecules, the *lipids*, to be examined before we begin the study of cell metabolism. Lipids are water-insoluble, oily or greasy organic substances that are extractable from cells and tissues by nonpolar solvents, such as chloroform or ether. The most abundant kinds of lipids are the fats or *triacylglycerols*, which are major fuels for most organisms. Indeed, they are the most important storage form of chemical energy.

But there is another reason to examine the lipids at this point. *Polar lipids*, another class, are major components of cell membranes, the "containers" in which metabolic reactions occur. Membranes enclose cells from the environment and also make possible internal compartmentalization of metabolic activities within cells. But membranes are not simply inert "skins" surrounding cells: they contain many important enzymes and transport systems. Moreover, on the outer surface of cell membranes are located many different recognition or receptor sites that can recognize other cells, bind certain hormones, and sense other types of signals from the external environment. Many of the properties of cell membranes are reflections of their polar lipid content.

Lipids play an important role in cell structure and function. In this electron micrograph of the cytoplasm of the photosynthetic alga Euglena, *the lipid-containing membranes of a chloroplast (upper right) and several mitochondria (surrounding the chloroplast and lower left) are clearly visible. Two lipid droplets, which serve to store chemical energy, can be seen in the chloroplast. The gray oval structure at the lower right is a lipid-filled inclusion in the cytoplasm.*

Fatty Acids Are Building-Block Components of Most Lipids

There are several classes of lipids, and each has specific biological functions; the major types are listed in Table 12-1. We begin our discussion with the *fatty acids*, characteristic building-block components of most of the lipids. Fatty acids are long-chain organic acids having from 4 to 24 carbon atoms; they have a single carboxyl group and a long, nonpolar hydrocarbon "tail" (Figure 12-1), which gives most lipids their water-insoluble and oily or greasy nature. Fatty acids do not occur in free or uncombined form in cells or tissues but are present in

Table 12-1 Major Types of Lipids, Grouped according to Chemical Structure
Several other types of lipids are known, but they are less abundant in animal tissues

Triacylglycerols
Waxes
Phosphoglycerides
Phosphatidylethanolamine
Phosphatidylcholine
Phosphatidylserine
Phosphatidylinositol
Cardiolipin
Sphingolipids
Sphingomyelin
Cerebrosides
Gangliosides
Sterols and their fatty acid esters

covalently bound form in different classes of lipids, from which they can be released by chemical or enzymatic hydrolysis. Many different kinds of fatty acids have been isolated from the lipids of various species. They differ from each other in chain length and in the presence, number, and position of their double bonds; some fatty acids also have methyl-group branches. Figure 12-1 and Table 12-2 give the structures of some important fatty acids found in naturally occurring lipids.

Nearly all fatty acids in nature have an even number of carbon atoms; those with 16 and 18 carbons are the most abundant. The long hydrocarbon tail may be fully saturated, i.e., contain only single bonds, or it may be unsaturated, with one or more double bonds. In general, unsaturated fatty acids are twice as abundant as saturated fatty acids in both animal and plant lipids. In most of the unsaturated fatty acids there is a double bond (designated Δ^9) between carbon atoms 9 and 10. If there are additional double bonds, they usually occur between the Δ^9 double bond and the methyl-terminal end of the chain. In fatty acids containing two or more double bonds, the double bonds are never conjugated (—CH=CH—CH=CH—) but are separated by a methylene group:

$$—CH{=}CH—CH_2—CH{=}CH—$$

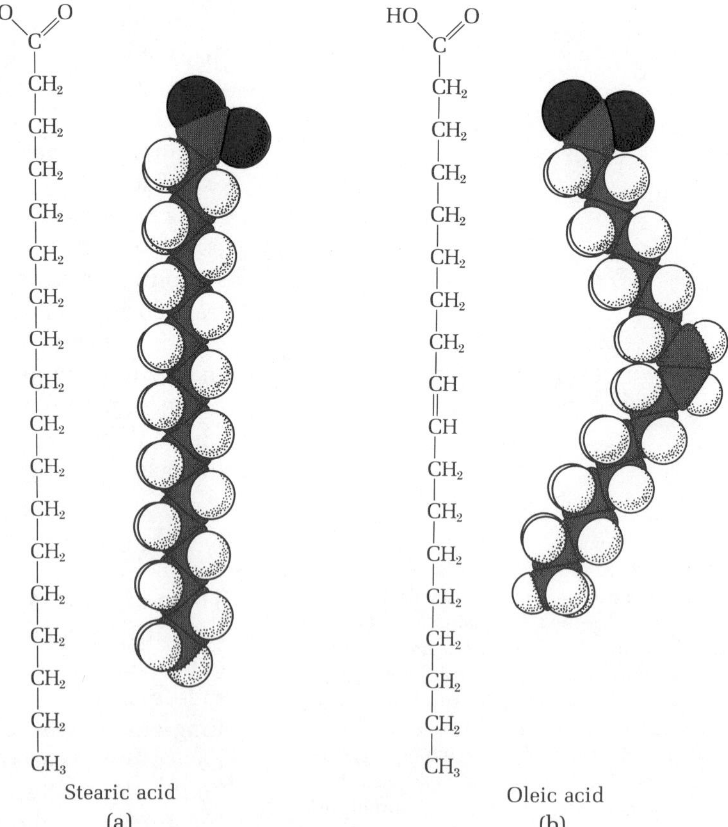

Figure 12-1
Two common fatty acids, shown as structural formulas and as space-filling models.
(a) *Although stearic acid is shown in its extended form, it is not a rigid, linear molecule. Because each single bond is free to rotate, the tail of stearic acid and all other saturated fatty acids can assume many different conformations and thus is flexible and floppy.*
(b) *In oleic acid, however, the cis double bond introduces a rigid bend in the hydrocarbon tail. All the other bonds in the tail of oleic acid are single bonds and thus are free to rotate.*

Table 12-2 Some Naturally Occurring Fatty Acids

Carbon atoms	Structure	Systematic name	Common name	mp, °C
		Saturated fatty acids		
12	$CH_3(CH_2)_{10}COOH$	*n*-Dodecanoic	Lauric acid	44.2
14	$CH_3(CH_2)_{12}COOH$	*n*-Tetradecanoic	Myristic	53.9
16	$CH_3(CH_2)_{14}COOH$	*n*-Hexadecanoic	Palmitic	63.1
18	$CH_3(CH_2)_{16}COOH$	*n*-Octadecanoic	Stearic	69.6
20	$CH_3(CH_2)_{18}COOH$	*n*-Eicosanoic	Arachidic	76.5
24	$CH_3(CH_2)_{22}COOH$	*n*-Tetracosanoic	Lignoceric	86.0
		Unsaturated fatty acids		
16	$CH_3(CH_2)_5CH{=}CH(CH_2)_7COOH$		Palmitoleic	−0.5
18	$CH_3(CH_2)_7CH{=}CH(CH_2)_7COOH$		Oleic	13.4
18	$CH_3(CH_2)_4CH{=}CHCH_2CH{=}CH(CH_2)_7COOH$		Linoleic	−5
18	$CH_3CH_2CH{=}CHCH_2CH{=}CHCH_2CH{=}CH(CH_2)_7COOH$		Linolenic	−11
20	$CH_3(CH_2)_4CH{=}CHCH_2CH{=}CHCH_2CH{=}CHCH_2CH{=}CH(CH_2)_3COOH$		Arachidonic	−49.5

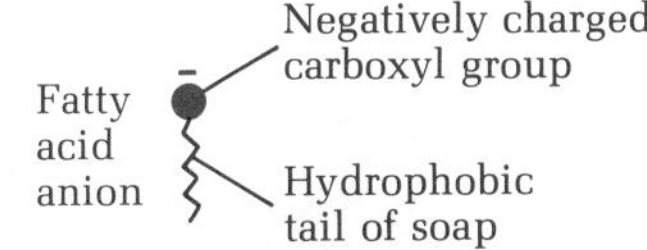

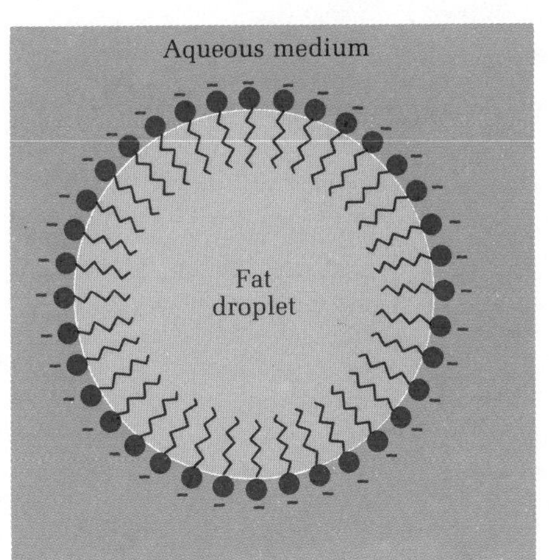

Figure 12-2
Action of a soap in emulsifying fat or grease. The fat is dispersed into droplets by the soap, which forms a shell of hydrophilic, highly polar carboxylate groups around the fat droplets, to form a stable emulsion. The negatively charged carboxylate groups are electrically balanced by an equal number of positive ions, for example, Na^+.

The double bonds of nearly all the naturally occurring unsaturated fatty acids are in the cis geometrical configuration, which produces a rigid bend in the aliphatic chain (Figure 12-1). Fatty acids with multiple double bonds, such as *arachidonic acid*, which has four, are kinked and relatively rigid compared with saturated fatty acids, which, thanks to the freedom of rotation around their single bonds, are more flexible and elongated. The saturated fatty acids from C_{12} to C_{24} are solids having a waxy consistency; unsaturated fatty acids, on the other hand, are oily liquids at body temperature.

The common fatty acids are insoluble in water but can be dispersed into *micelles* (page 71) in dilute NaOH or KOH, which converts fatty acids into *soaps*, the name given to salts of fatty acids. Bath soap is largely a mixture of potassium salts of fatty acids. K^+ or Na^+ soaps are *amphipathic* (page 71): the ionized carboxyl group constitutes the polar head and the hydrocarbon chain is the nonpolar tail. K^+ or Na^+ soaps have the property of emulsifying oily or greasy water-insoluble substances. The hydrophobic tails of the soaps extend into the grease droplets, whereas the polar heads of the soap molecules face toward the water. Thus soaps provide a hydrophilic coat around the grease droplets to form a fine dispersion or emulsion (Figure 12-2).

The Ca^{2+} and Mg^{2+} soaps of fatty acids are very insoluble and thus do not emulsify oily substances. They are precipitated as white insoluble curds when bath soap (largely K^+ soaps) is used in hard water, which contains Ca^{2+} and Mg^{2+} salts.

Triacylglycerols Are Fatty Acid Esters of Glycerol

The simplest and most abundant lipids containing fatty acids as building blocks are the *triacylglycerols,* also often referred to as *fats, neutral fats,* or *triglycerides.* Triacylglycerols are esters of the alcohol *glycerol* with three fatty acid molecules (Figure 12-3). Triacylglycerols are the major components of storage or depot fats in plant and animal cells but are not normally found in membranes. Note that triacylglycerols are nonpolar, hydrophobic molecules since they contain no electrically charged or highly polar functional groups.

Triacylglycerols occur in many different types, depending on the identity and position of the three fatty acid components esterified to glycerol. Those containing a single kind of fatty acid in all three positions are called *simple triacylglycerols;* they are named after the fatty acids they contain. Examples are tristearoylglycerol, tripalmitoylglycerol, and trioleylglycerol, which contain *stearic acid, palmitic acid,* and *oleic acid,* respectively. The trivial and more commonly used names of these simple triacylglycerols are *tristearin, tripalmitin,* and *triolein,* respectively. Triacylglycerols containing two or more *different* fatty acids are called *mixed triacylglycerols.* Most natural fats, such as those in olive oil, butter, and other food fats, are com-

Figure 12-3
Glycerol and the general structure of triacylglycerols. Note that if there are two different fatty acids in positions 1 and 3 of the glycerol, glycerol carbon atom 2 (color) becomes a chiral center. Such triacylglycerols in living matter have the L configuration. Tripalmitin, shown at right, and all other triacylglycerols having identical fatty acids in positions 1 and 3 are optically inactive.

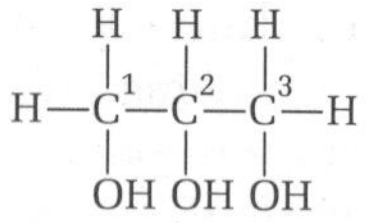

Glycerol

General structure of triacylglycerols; R_1, R_2, and R_3 are the hydrocarbon tails of three fatty acids

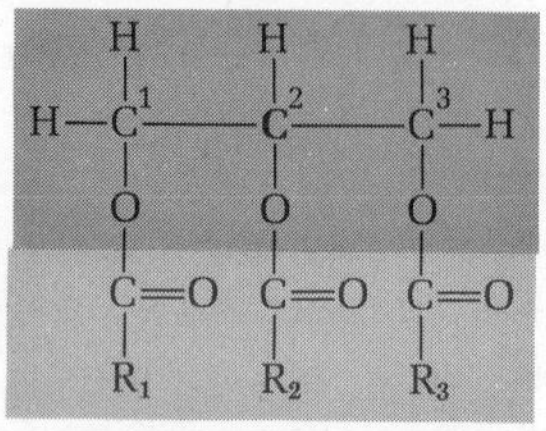

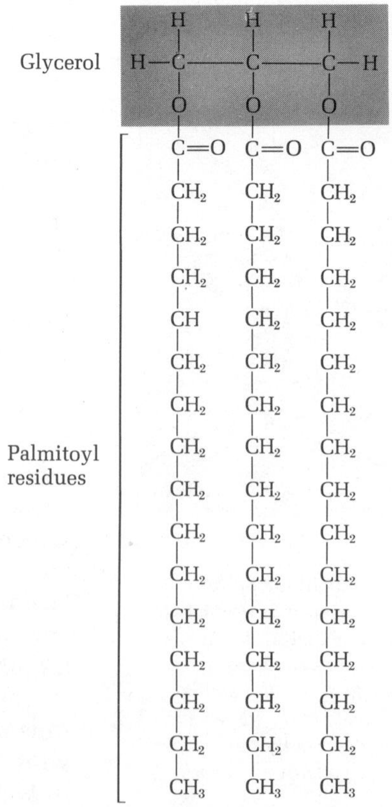

Tripalmitin

Table 12-3 Fatty Acid Composition of Three Natural Food Fats†

	Percent of total fatty acids				
	Saturated				Unsaturated
	C_4–C_{12}	C_{14}	C_{16}	C_{18}	C_{16} + C_{18}
Olive oil	<2	<2	13	3	80
Butter	11	10	26	11	40
Beef fat	<2	<2	29	21	46

† These fats consist of mixtures of triacylglycerols, differing in their fatty acid composition and thus in their melting points. Olive oil, which is liquid at room temperature, contains largely unsaturated (liquid) fatty acids. Beef fat, which is rich in long-chain saturated fatty acids, is solid at room temperature. Butter fat, which contains significant amounts of short-chain fatty acids, is soft at room temperature.

plex mixtures of simple and mixed triacylglycerols containing a variety of fatty acids differing in chain length and degree of saturation (Table 12-3).

Triacylglycerols containing only saturated fatty acids, such as tristearin, the major component of beef suet, are white greasy solids at room temperature. Triacylglycerols containing three unsaturated fatty acids, such as triolein, the major component of olive oil, are liquids. Butter is a mixture of triacylglycerols, some of which have fatty acids with relatively short chains. Since the shorter-chain fatty acids have lower melting points, they give butter its softness at room temperature (Table 12-3).

Naturally occurring triacylglycerols are insoluble in water. They have a lower specific gravity than water, which accounts for the fact that the oil forms the upper layer of oil-and-vinegar salad dressing. Triacylglycerols readily dissolve in nonpolar solvents, such as chloroform, benzene, or ether, which are often used to extract fats from tissues. Triacylglycerols undergo hydrolysis when boiled with acids or bases or when acted upon enzymatically by the enzyme lipase, which is secreted into the small intestine by the pancreas. Hydrolysis of triacylglycerols with KOH or NaOH, called saponification (meaning "soap formation"), yields a mixture of K^+ or Na^+ soaps and glycerol (Figure 12-4). This is the primary chemical reaction involved in making household soap from triacylglycerols.

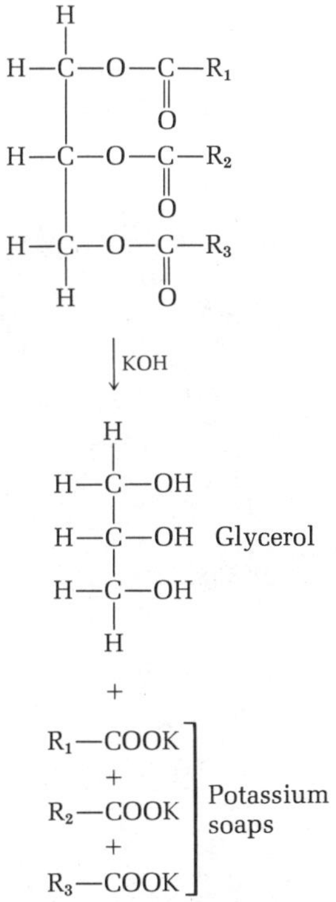

$$\begin{array}{l} \text{H} \\ \text{H—C—O—C(=O)—R}_1 \\ \text{H—C—O—C(=O)—R}_2 \\ \text{H—C—O—C(=O)—R}_3 \\ \text{H} \end{array} \xrightarrow{\text{KOH}} \begin{array}{l} \text{H} \\ \text{H—C—OH} \\ \text{H—C—OH} \\ \text{H—C—OH} \\ \text{H} \end{array} \text{Glycerol} + \left.\begin{array}{l} \text{R}_1\text{—COOK} \\ + \\ \text{R}_2\text{—COOK} \\ + \\ \text{R}_3\text{—COOK} \end{array}\right] \text{Potassium soaps}$$

Figure 12-4
Saponification (alkaline hydrolysis) of a triacylglycerol. Household soap is made by hydrolyzing a mixture of triacylglycerols with KOH. The K^+ *soaps of fatty acids are collected, washed free of KOH, and compressed into cakes.*

Triacylglycerols with largely unsaturated fatty acids, which are thus liquid at room temperature, can be converted chemically into solid fats by partial hydrogenation of their double bonds. Large amounts of liquid vegetable oils, such as corn oil, are converted into solid shortenings, e.g., Crisco, by a catalytic hydrogenation process, which causes the conversion of some of their double bonds into single bonds.

When exposed to air, triacylglycerols containing highly unsaturated fatty acids tend to undergo a complex process called autoxidation. Molecular oxygen can attack fatty acids having two or more double bonds to yield complex products that are responsible for the off-taste of rancid fats. Linseed oil, a plant oil used as a vehicle or base for paints, is very rich in highly unsaturated fatty acids. It undergoes autoxidation when exposed to air, followed by polymerization to a hard, resinous coating as it "dries" or oxidizes. Autoxidation of unsaturated fats does not normally take place in cells, since it is kept in check by the inhibitory action of vitamin E (page 268), by various enzymes, and possibly also by ascorbic acid (page 264). However, it may occur in some diseases, leading to the formation of abnormal lipid deposits in certain tissues.

Triacylglycerols Are Storage Lipids

Triacylglycerols function largely as storage lipids. In most animal and plant cells triacylglycerols occur as microscopic, oily droplets, finely dispersed and emulsified in the cytosol (Figure 12-5). In adipocytes, or fat cells, specialized cells of the connective tissue of animals, very large amounts of triacylglycerols are stored as fat droplets, which fill almost the entire cell volume (Figure 12-5a). Fat cells are found in large numbers under the skin, in the abdominal cavity, and in the mammary glands. In obese persons many kilograms of triacylglycerols are deposited in the fat cells of the body, sufficient to supply basal energy needs of the body for several months. In contrast, the body can store less than a day's energy supply in the form of glycogen (page 583). Triacylglycerols are much better adapted

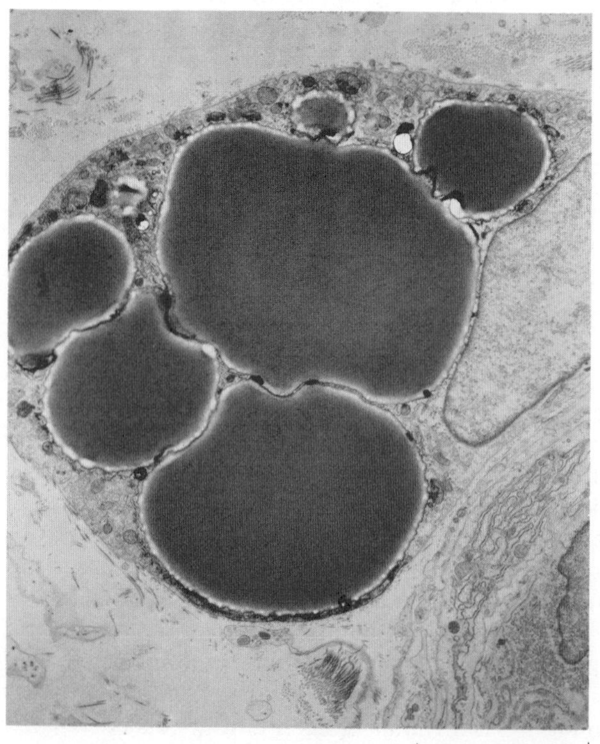

(a)

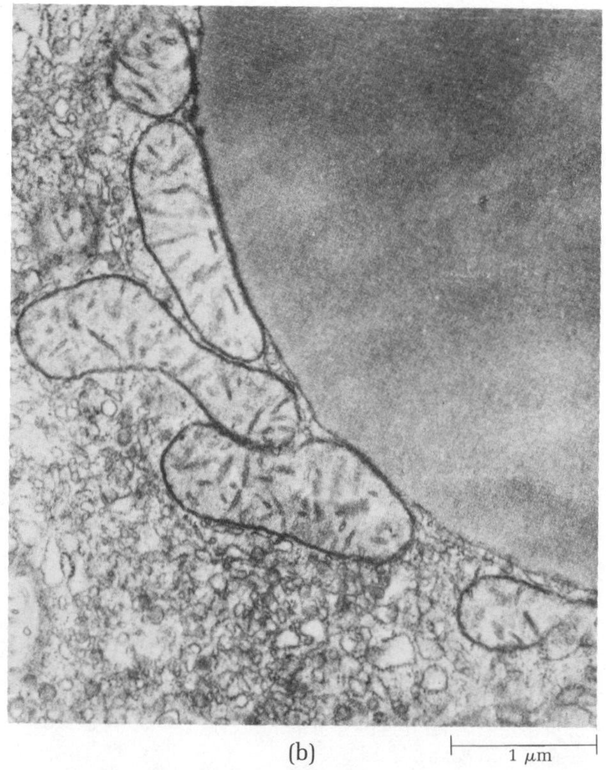

(b)

Figure 12-5
Fat stores in cells.
(a) *An adipocyte or fat cell from adipose tissue of a pig. The huge fat droplets virtually fill the cell volume. The nucleus is the pale, oblong structure at the far right.*
(b) *A portion of a large fat droplet in the liver cytoplasm of a fasted hamster. Livers of well-fed animals contain only a few small fat droplets, which increase in size and number during fasting, when fat is imported from adipose tissue to the liver to serve as the major fuel. Several mitochondria, in which fatty acid oxidation occurs, are mobilized on the surface of the droplet.*

Figure 12-6
The Weddell seal of the Antarctic. The very thick layer of fat under the skin serves not only for fat storage but also doubles as a built-in wet suit for insulation against the cold.

than glycogen to serve as a storage form of energy. Not only can they be stored in very large amounts and in an almost pure unhydrated form, but they also yield over twice as much energy, gram for gram, as carbohydrates (page 511; Chapter 26).

In some animals triacylglycerols stored under the skin serve a double purpose, both as important energy storage depots, and as insulation against very low temperatures. Seals, walruses, penguins and other warm-blooded Arctic and Antarctic animals are amply padded with triacylglycerols (Figure 12-6).

Waxes Are Fatty Acid Esters of Long-Chain Alcohols

Waxes are esters of long-chain saturated and unsaturated fatty acids (having from 14 to as many as 36 carbon atoms) with long-chain alcohols (having from 16 to 22 carbon atoms) (Figure 12-7). In vertebrates waxes are secreted by skin glands as a protective coating, to keep the skin pliable, lubricated, and waterproof. Hair, wool, and fur are also coated with waxy secretions. Birds, particularly waterfowl, secrete waxes in their preen glands to make their feathers water-repellent. The leaves of many plants are coated with a protective layer of waxes. The shiny appearance of the leaves of many tropical plants, of holly and rhododendrons, and, indeed, of poison ivy is a reflection of their waxy coats.

Waxes are formed and used in very large amounts in marine life, especially in plankton organisms, in which wax serves as the chief storage form of caloric fuel. Since some whales, herring, salmon, and many other marine species consume plankton in large amounts, waxes are major food and storage lipids in oceanic food chains.

Figure 12-7
Structure of a wax, in this case an ester of oleic acid with oleyl alcohol.

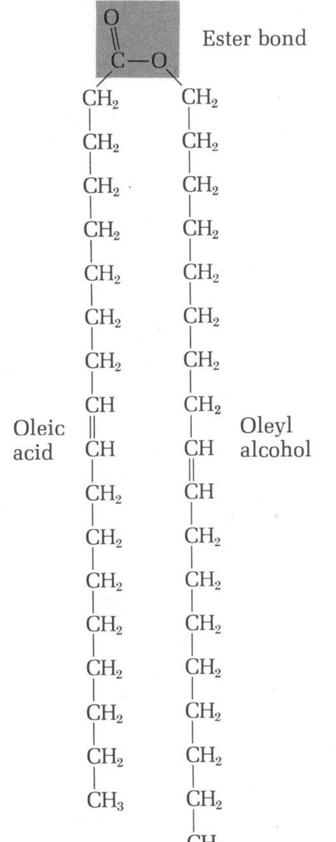

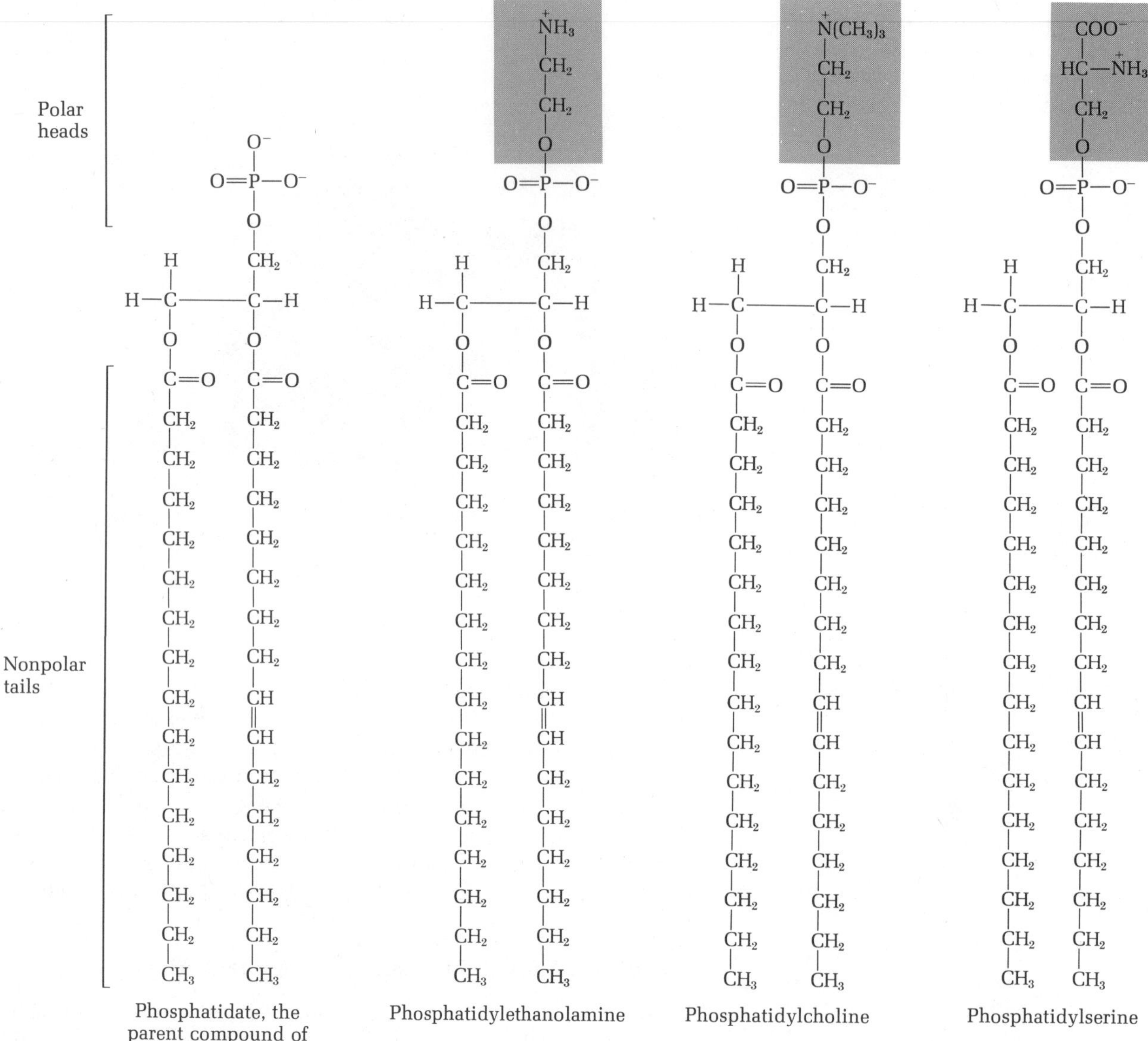

Figure 12-8
The common phosphoglycerides. Their polar heads consist of a negatively charged phosphate group (at pH 7.0) esterified to the head alcohol (color screen). *The latter may also have an electrical charge.*

Phospholipids Are Major Components of Membrane Lipids

There are several classes of membrane lipids. They differ from triacylglycerols in possessing one or more highly polar "head" groups, in addition to their hydrocarbon tails. For this reason they are often called *polar lipids*. The most abundant membrane lipids are the *phospholipids*. Phospholipids serve primarily as structural elements of membranes and are never stored in large amounts. As their name implies, this group of lipids contains phosphorus in the form of phosphoric acid groups. The major phospholipids found in membranes are the *phosphoglycerides* (Figure 12-8), which contain two fatty acid molecules esterified to the first and second hydroxyl groups of glycerol. The third hydroxyl group of glycerol forms an ester linkage with phosphoric acid. In addition phosphoglycerides contain a second alcohol, which is also esterified to the phosphoric acid. The second alcohol group is thus located on the polar head of the phosphoglyceride molecule. There are several different classes of phosphoglycerides, which differ in their head alcohol groups. However, all phosphoglycerides contain two nonpolar tails, contributed by their long-chain fatty acids. The most abundant fatty acids found in phosphoglycerides have 16 or 18 carbon atoms. Usually one of the fatty acids is saturated and the other unsaturated; the latter is always esterified to the middle or 2-hydroxyl of the glycerol. Note that carbon atom 2 of the glycerol part of phosphoglycerides is a chiral center; it has the L configuration since it is related to L-glyceraldehyde (page 97).

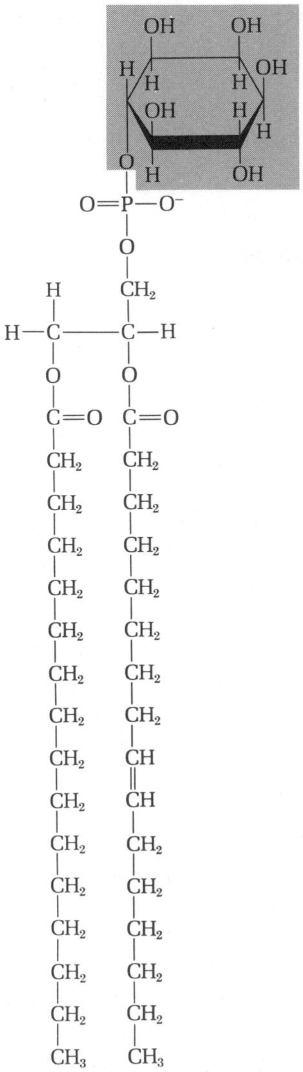

Different types of phosphoglycerides are named according to the alcohol on their polar heads (Figure 12-8). The parent compound of phosphoglycerides is *phosphatidic acid*, which has no head alcohol. It occurs in free form only in small amounts, but it is an intermediate in the biosynthesis of phosphoglycerides. The most abundant phosphoglycerides are the closely related *phosphatidylethanolamine* and *phosphatidylcholine*, which contain the alcohols *ethanolamine* and *choline*, respectively, on their polar heads. Each of these can contain different combinations of fatty acids. Other phosphoglycerides include *phosphatidylserine*, containing as head group the hydroxyamino acid *serine* (page 102), and *phosphatidylinositol*, containing the cyclic alcohol *inositol*. *Cardiolipin*, characteristically found in the inner membrane of mitochondria, differs from the rest of the phosphoglycerides; it is a "double" phosphoglyceride (Figure 12-9).

All phosphoglycerides have a negative charge on the phosphoric group at pH 7.0. In addition, the head alcohol group may also contribute one or more electric charges at pH near 7, as shown in Figure 12-8. Thus phosphoglycerides have two very different kinds of groups, a polar, hydrophilic head group and hydrophobic, nonpolar tails. They are therefore *amphipathic* (page 71). In general, membrane lipids are amphipathic, whereas the storage lipids, the triacylglycerols and waxes, are not.

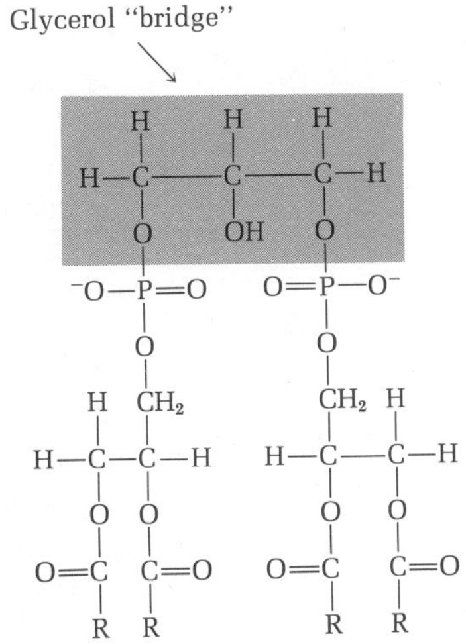

Phosphatidylinositol

Glycerol "bridge"

Figure 12-9
Cardiolipin, a "double" phosphoglyceride, is present in large amounts in mitochondrial and bacterial membranes. The R's stand for the tails of long-chain fatty acids.

Phosphoglycerides undergo hydrolysis when heated with acids or bases, to yield their building-block components: fatty acids, glycerol, phosphoric acid, and the head alcohol. They can also be hydrolyzed enzymatically by phospholipases of different types, which catalyze hydrolysis of specific linkages in the phosphoglyceride molecule.

Sphingolipids Are Also Important Components of Membranes

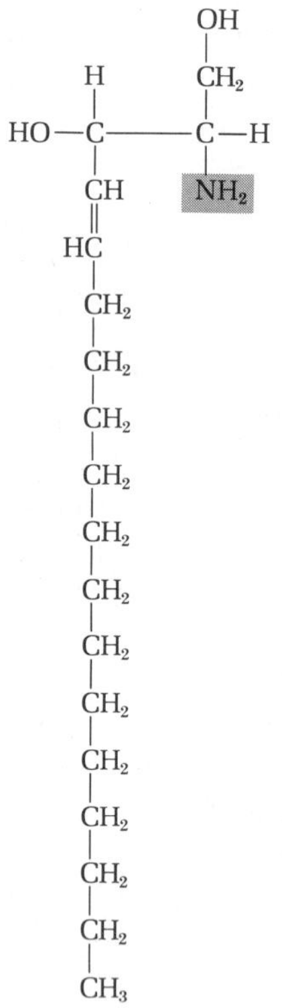

Sphingolipids, the second large class of membrane lipids, also have a polar head and two nonpolar tails, but they contain no glycerol. Sphingolipids are composed of one molecule of a long-chain fatty acid, one molecule of the long-chain amino alcohol *sphingosine* or one of its derivatives, and a polar head alcohol.

Figure 12-10
Sphingosine. In dihydrosphingosine the double bond of sphingosine is reduced. In sphingolipids a fatty acid is joined by an amide bond to the amino group (color).

```
           OH
           |
     H    CH2
     |     |
HO — C ——— C — H
     |     |
    CH    NH2
    ||
    HC
     |
    CH2
     |
    CH2
     |
    CH2
     |
    CH2
     |
    CH2
     |
    CH2
     |
    CH2
     |
    CH2
     |
    CH2
     |
    CH2
     |
    CH2
     |
    CH2
     |
    CH3
```

Sphingosine is the parent compound of a number of long-chain amino alcohols found in different sphingolipids. In mammals sphinogosine and *dihydrosphingosine* (Figure 12-10) are the most abundant. In sphingolipids the polar head group is attached to the hydroxyl group of sphingosine, and the fatty acid component forms an amide bond with the amino group. There are three subclasses of sphingolipids: *sphingomyelins*, *cerebrosides*, and *gangliosides*. Sphingomyelins contain phosphorus, but cerebrosides and gangliosides do not.

Sphingomyelins (Figure 12-11) are the simplest and most abundant sphingolipids. They characteristically contain phosphocholine or phosphoethanolamine as their polar head groups. Because sphingomyelins contain phosphorus, they may also be classed as phospholipids, together with phosphoglycerides. Indeed, sphingomyelins closely resemble the phosphoglycerides phosphatidylethanolamine and phosphatidylcholine in their general properties, and they have similar electric charges. Sphingomyelins are present in most membranes of animal cells; the *myelin sheath* surrounding certain nerve cells is very rich in sphingomyelins.

Cerebrosides contain no phosphorus and have no electric charge since their polar head groups are neutral. Since the head group characteristically consists of one or more sugar units, cerebrosides are often called *glycosphingolipids* (Greek *glykos*, "sweet"). They are *glycolipids*, a generic name for lipids having sugar groups. Figure 12-12 shows the structure of a *galactocerebroside*, which contains the sugar D-*galactose* (page 281) as its polar head group. Galactocerebrosides are characteristically found in the cell membranes of the brain. *Glucocerebrosides*, which contain D-glucose as head group, are present in the cell membranes of nonneural tissues.

In addition there are also cerebrosides containing two, three, or four sugar units, which may be D-glucose, D-galactose, or N-acetyl-D-galactosamine. These more complex cerebrosides occur largely in the outer layer of cell membranes and constitute important components of cell surfaces, as we shall see.

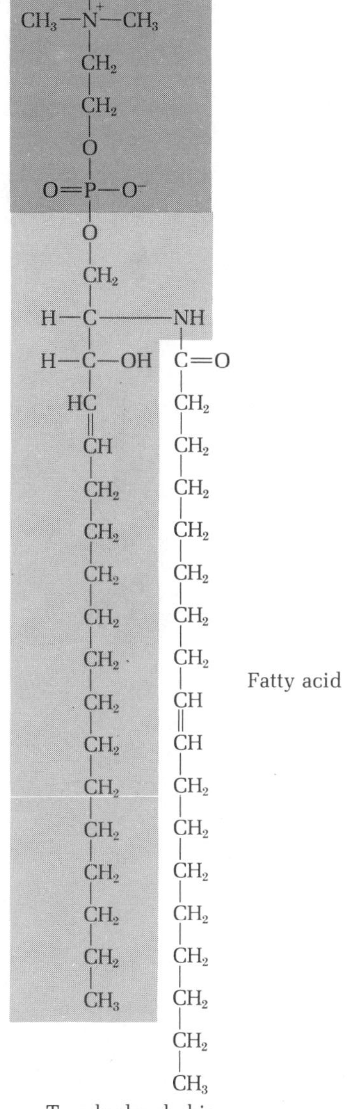

Figure 12-11
Structure of sphingomyelin. Although it was first isolated from myelin, a membranous sheath surrounding certain cells in the brain, sphingomyelin is found in membranes of many animal tissues.

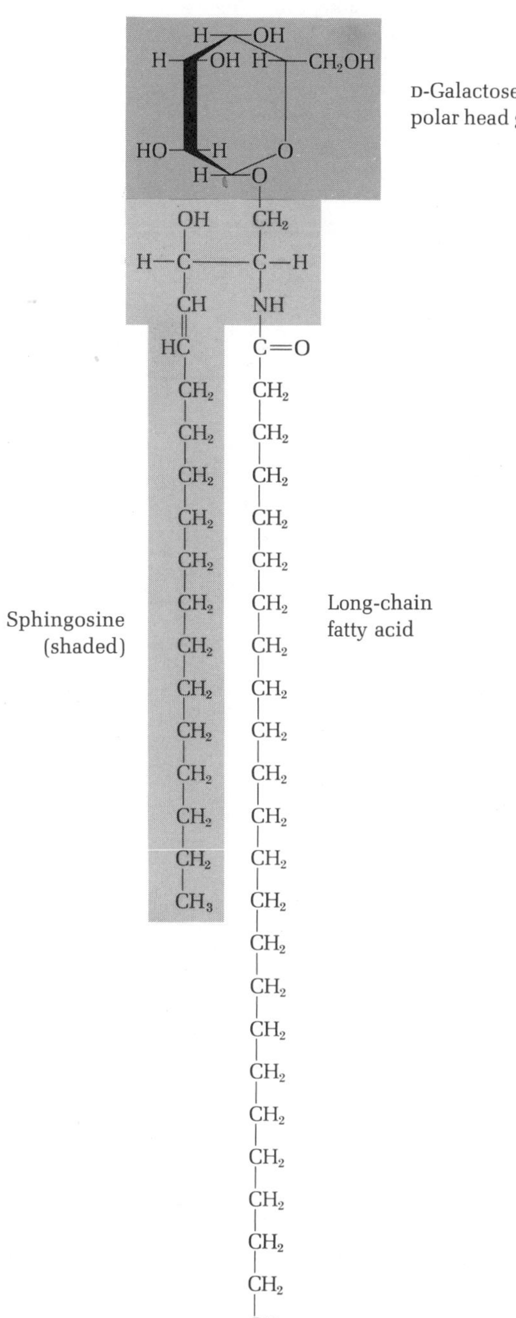

Figure 12-12
A galactocerebroside. The fatty acid component of cerebrosides usually has 24 carbon atoms.

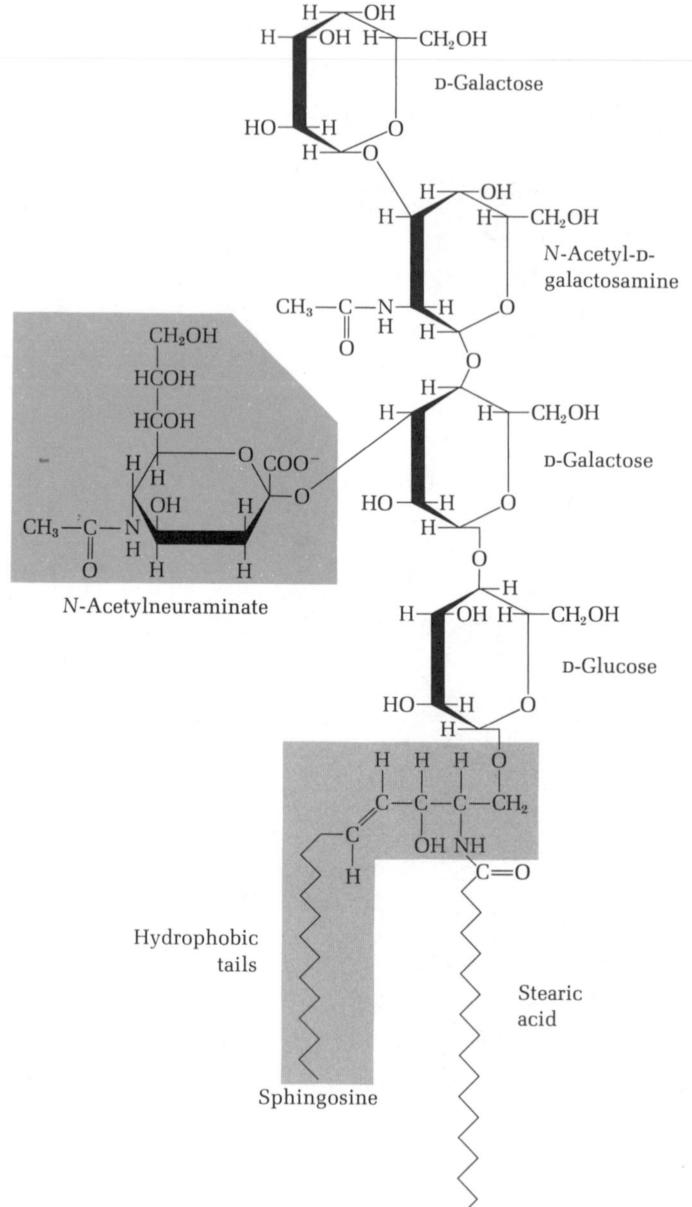

Figure 12-13
Structure of ganglioside GM_1. Gangliosides have very complex oligosaccharide head groups containing at least one residue of N-acetylneuraminic acid (sialic acid), which is ionized at pH 7.0. Gangliosides are given identifying symbols that indicate the structure of the head group. Fifteen or more different classes of gangliosides have been found in cell membranes. They are especially abundant at nerve endings and at specific hormone-receptor sites on cell surfaces. The hydrophobic tails are much longer than suggested by their backbone symbols.

Gangliosides, the most complex sphingolipids (Figure 12-13), contain very large polar heads made up of several sugar units. Characteristically, one or more of the terminal sugar units of gangliosides is *N-acetylneuraminic acid*, also called *sialic acid* (Figure 12-13), which has a negative charge at pH 7. N-Acetylneuraminate residues are also present in the oligosaccharide side chains of some membrane glycoproteins. Gangliosides make up about 6 percent of the membrane lipids in the gray matter of the brain. They are also found in lesser amounts in the membranes of most nonneural tissues. Gangliosides are important components of specific *receptor sites* on the surface of cell membranes. For example, they are found in the specific sites on nerve endings to which neurotransmitter molecules become bound during the chemical transmission of an impulse from one nerve cell to the next.

Steroids Are Nonsaponifiable Lipids with Specialized Functions

The lipids discussed to this point are saponifiable; i.e., they are hydrolyzed by heating with alkali to yield soaps of their fatty acid components. Cells also contain nonsaponifiable lipids, which contain no fatty acids and thus cannot form soaps. There are two major classes of nonsaponifiable lipids, steroids and terpenes. We shall consider here only those steroids which are important membrane components.

Steroids are complex fat-soluble molecules with four fused rings (Figure 12-14). The most abundant steroids are the sterols, which are steroid alcohols. Cholesterol is the major sterol in animal tissues. Cholesterol and its esters with long-chain fatty acids are important components of plasma lipoproteins and of the outer cell membrane. Plant cell membranes contain other kinds of sterols, particularly stigmasterol, which differs from cholesterol only in having a double bond between carbons 22 and 23. The cholesterol molecule has a polar head group, the hydroxyl group at position 3. The rest of the molecule is a relatively rigid nonpolar structure, as indicated in the space-filling model in Figure 12-14.

Lipoproteins Blend the Properties of Lipids and Proteins

Some lipids are associated with specific proteins to form lipoproteins. In blood plasma there are three major classes of plasma lipoproteins, which may contain from 50 to 90 percent lipid. In these conjugated proteins no covalent linkages exist between the tightly bound lipid molecules and the polypep-

Figure 12-14
(a) Cholesterol, a steroid alcohol. The designation of the rings of steroids and the numbering of the carbon atoms are shown. Because the four rings are rigid, the presence of cholesterol reduces the fluidity of membranes. The hydroxyl group (color) constitutes its polar head; the rest of the molecule is hydrophobic. (b) Space-filling model of cholesterol. The hydroxyl group is at the top. (c) A cholesterol ester. Like triacylglycerols, cholesterol esters are saponifiable.

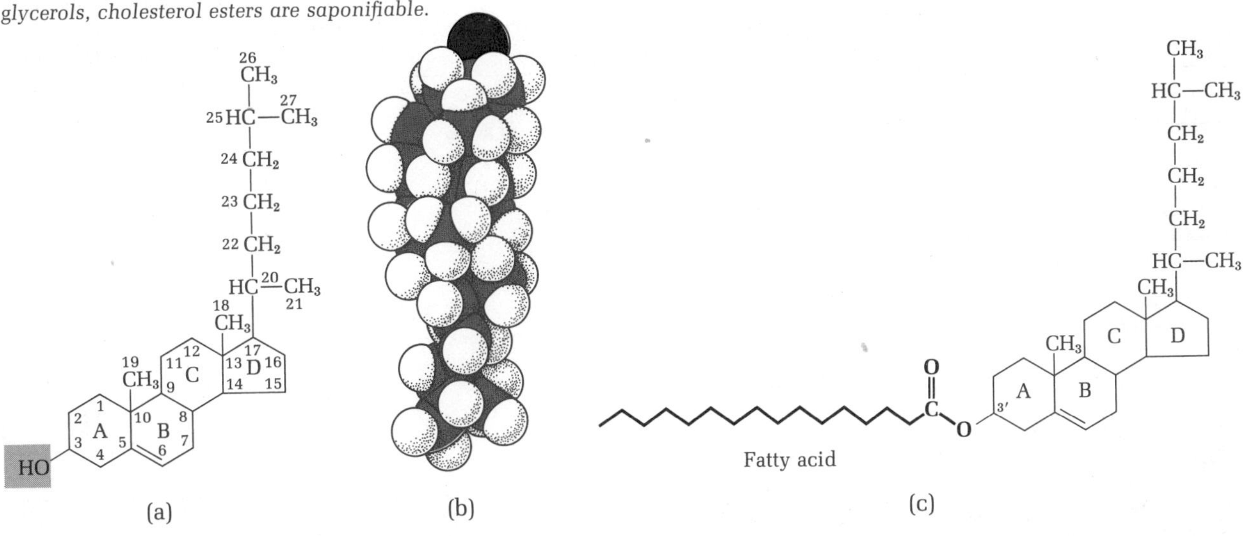

Figure 12-15
Schematic model of a plasma lipoprotein. Polar segments of the polypeptide chain, as well as polar heads of phospholipid molecules, are on the outer surface, facing the aqueous medium. The water-insoluble inner core of triacylglycerols and cholesterol is thus shielded from water.

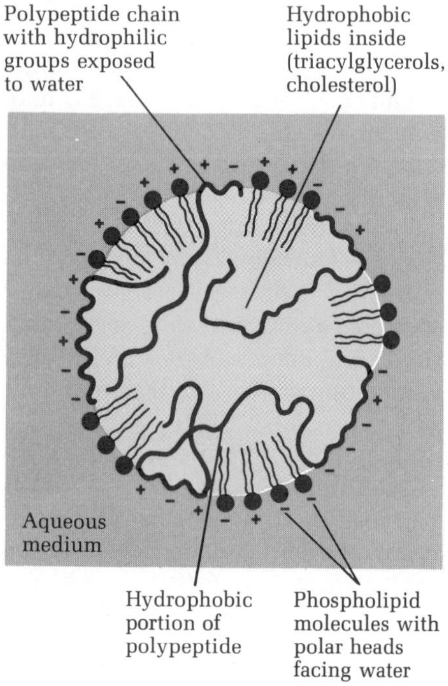

tides. Plasma lipoproteins contain both polar lipids and triacylglycerols, as well as cholesterol and its esters. The nonpolar triacylglycerols and cholesterol are hidden inside an outer coat of water-soluble, hydrophilic segments of the polypeptide chains and the hydrophilic polar heads of phosphoglyceride molecules (Figure 12-15). The outer hydrophilic coat of the lipoproteins faces the water and gives these lipid-rich structures water solubility, well adapted for the transport of lipids via the blood from the small intestine to the fat depots and tissues (Chapter 24). The lipoproteins of blood plasma are classified on the basis of their density, which in turn is a reflection of their lipid content (Table 12-4). The greater their lipid content the lower their density and the greater their tendency to move upward, i.e., to float, when blood plasma is centrifuged at very high speeds. In addition to the three classes of lipoproteins, blood plasma also contains *chylomicrons,* particularly after fat-rich meals. Chylomicrons are droplets of nearly pure triacylglycerols, coated by a very thin layer of protein (Table 12-4). They are much larger than the lipoproteins. Chylomicrons carry triacylglycerols from the small intestine, where they are absorbed during digestion, to the fat depots.

Much evidence suggests that the combination of a high plasma level of *very-low-density lipoproteins* (VLDL) with a low level of *high-density lipoproteins* (HDL) is an important factor in causing *atherosclerosis,* the formation of thick deposits of cholesterol and its esters on the inner surfaces of blood vessels. Atherosclerosis predisposes toward strokes and coronary infarction, conditions caused by the restriction of blood flow through the clogged blood vessels of the brain and heart, respectively (Chapters 24 and 26).

Table 12-4 Approximate Composition of Blood Plasma Lipoproteins†

Type	Density, g/mL	Protein, %	Triacylglycerols, %	Phospholipids, %	Cholesterol, %	Relative amounts of triacylglycerols and proteins
Chylomicrons	0.92–0.96	1.7	96	0.8	1.7	
Very low density (VLDL)	0.95–1.00	10	60	18	15	
Low density (LDL)	1.00–1.06	25	10	22	45	
High density (HDL)	1.06–1.21	50	3	30	18	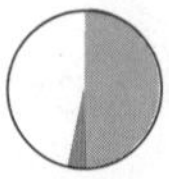

† There are three major classes of lipoproteins according to their density, i.e., lipid content. Blood plasma also contains chylomicrons, much larger structures of very low density. See also Chapter 24.

Polar Lipids Form Micelles, Monolayers, and Bilayers

Like the soaps (page 305), the polar lipids are amphipathic (page 71). In aqueous systems the polar lipids spontaneously disperse to form *micelles*, in which the hydrocarbon tails of the lipids are hidden from the aqueous environment and the electrically charged hydrophilic heads are exposed on the surface, facing the aqueous medium (Figure 12-16). Such micelles may contain thousands of lipid molecules. Polar lipids also spread spontaneously on the surface of aqueous solutions to form a layer one molecule thick, a *monolayer*. In such systems the hydrocarbon tails are exposed to air, thus avoiding the water, and the hydrophilic heads extend into the polar water phase (Figure 12-16).

Polar lipids also readily and spontaneously form very thin *bilayers* separating two aqueous compartments. In these structures the hydrocarbon tails of the lipid molecules extend inward from the two surfaces to form a continuous inner hydrocarbon core, and the hydrophilic heads face outward, extending into the aqueous phases. Phospholipid bilayers are about 6 to 7 nm thick, depending on the nature of the fatty acids in the lipids. They are nonrigid, fluid, and very flexible. In the laboratory such bilayers are readily formed by agitating phospholipids in aqueous suspensions at high frequencies, a process which results in the formation of *liposomes*, closed vesicles surrounded by a continuous lipid bilayer (Figure 12-16). Phospholipid bilayers can also be formed in small apertures separating two aqueous compartments. Such phospholipid bilayers and liposomes have been studied intensively, since their properties are very similar to those of natural membranes. For example, polar lipid bilayers and natural membranes have high electrical resistance; they allow water to pass through readily, but they do not allow cations or anions to pass.

Figure 12-16
Polar lipids, especially the phosphoglycerides, can spontaneously form micelles, monolayers, and bilayers. They also can form enclosed vesicles called liposomes, which are useful models for the behavior of the membranes surrounding cells and organelles.

Symbol of a polar lipid molecule

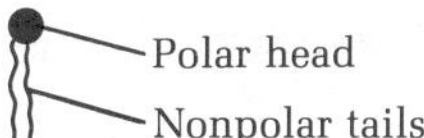

Phosphoglyceride micelles

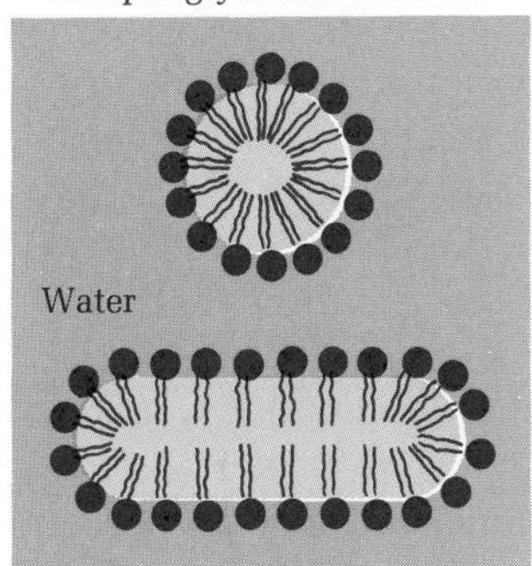

A phosphoglyceride monolayer

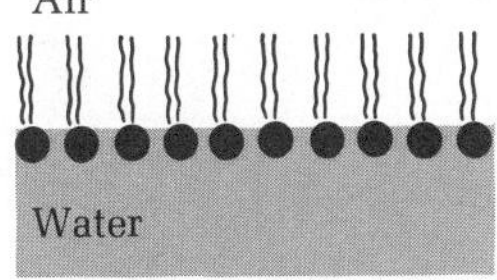

A phosphoglyceride bilayer

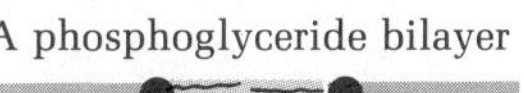

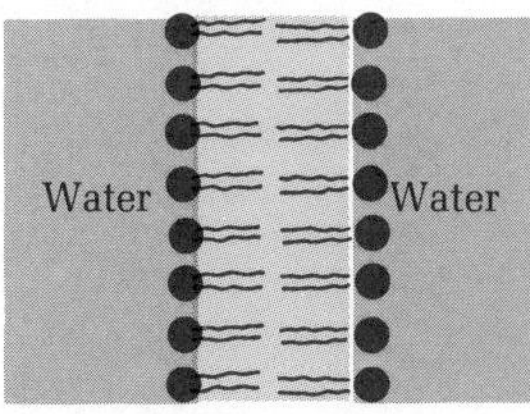

A liposome

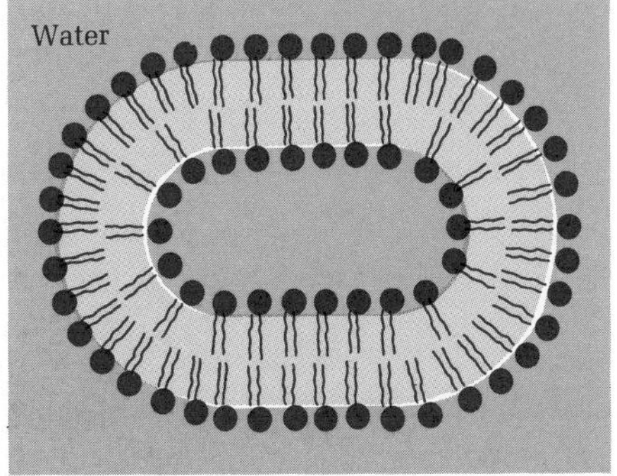

When liposomes are injected into the bloodstream, they are absorbed and their lipids are metabolized by cells of the reticuloendothelial system, which are largely located in the bone marrow and spleen. This fact has led to the use of liposomes to carry specific drugs to the reticuloendothelial system to concentrate their effects on this tissue. For this purpose liposomes are "loaded" with a solution of the drug and then injected. It has been found in experimental animals that using liposomes as a carrier greatly increases the potency and safety of drugs effective against the protozoa causing *leishmaniasis* (*kala azar*), a debilitating parasitic disease affecting millions of people in tropical countries. Liposomes can also be fused into the plasma membranes of cells under some conditions, making it possible to study experimentally the effect of altering the lipid composition of cell membranes.

Phospholipids spontaneously form bilayers and liposomes in aqueous systems through the operation of the same forces that stabilize the structure of globular proteins. Recall that a polypeptide chain in water will fold itself into such a conformation that its hydrophobic amino acid R groups will be inside the globular structure, shielded from water, and the hydrophilic polar R groups will be on the outside, facing the aqueous medium. Exactly the same process occurs in polar lipid molecules; they tend to arrange themselves into structures in which the nonpolar hydrocarbon chains are hidden and the polar heads face toward water. Triacylglycerols alone cannot form micelles, since they lack polar heads; but when they are mixed with phosphoglycerides, they form finely emulsified droplets, which contain phosphoglyceride molecules on the surface with the triacylglycerols inside. Fat droplets in cells (Figure 12-5), as well as chylomicrons, are structures of this kind.

The Major Components of Membranes Are Polar Lipids and Proteins

The outer, or plasma, membranes of many cells, as well as the membranes of intracellular organelles such as mitochondria and chloroplasts, have been isolated and their molecular composition studied. All membranes contain polar lipids, which make up from 20 to 80 percent of the membrane mass, depending on the type of membrane; the remainder is mostly protein. Although the plasma membranes of animal cells generally contain about equal amounts of lipid and protein, the inner mitochondrial membrane contains about 80 percent protein and only 20 percent lipid, and the myelin membranes of the brain contain about 80 percent lipid and only 20 percent protein. The lipid part of membranes consists of a mixture of different kinds of polar or amphipathic lipids; animal cell membranes contain largely phosphoglycerides, with lesser amounts of sphingolipids. Triacylglycerols are present in only minor amounts in membranes. Some animal cell membranes, particularly the outer plasma membrane, contain considerable

Table 12-5 Approximate Lipid Composition (Percent by Weight) of the Subcellular Membranes of Rat Liver

Note the characteristically high concentration of cholesterol and cholesterol esters in the plasma membrane, as well as the high glycolipid content, much of it gangliosides

Membrane	Phospholipids	Cholesterol	Glycolipids	Cholesterol esters and minor components
Plasma	57	15	6	22
Golgi	57	9	0	34
Endoplasmic reticulum	85	5	0	10
Inner mitochondrial	92	0	0	8
Nuclear	85	5	0	10

amounts of cholesterol and cholesterol esters. Each type of membrane of any given animal cell has a characteristic and rather constant lipid composition (Table 12-5).

Natural membranes are very thin, from 6 to 9 nm, flexible, and fluid. They are freely permeable to water but are intrinsically impermeable to electrically charged ions like Na^+, Cl^-, or H^+ and to polar but uncharged molecules, such as sugars. Natural membranes allow only those polar molecules to pass for which there are specific transport systems or carriers. On the other hand lipid-soluble molecules readily pass through natural membranes since they go into solution in the hydrocarbon core of the membrane. Moreover, both natural membranes and polar lipid bilayers have high electrical resistance and are thus good insulators. Because of such common properties it is believed that natural membranes consist of a continuous, sheetlike polar lipid bilayer containing a number of proteins.

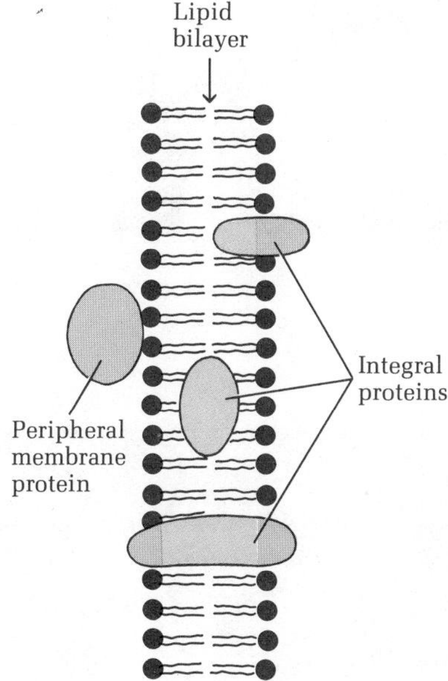

Figure 12-17
Membrane proteins. Peripheral (extrinsic) proteins are easily removed, but integral membrane proteins are difficult to extract into aqueous solution.

The proteins of different membranes make up from about 20 to 80 percent of their mass. The red-blood-cell membrane contains about 20 different proteins, whereas the inner membrane of mitochondria contains a much larger number. Some of the proteins in membranes are enzymes; others function to bind and transport polar molecules across the membrane. Membrane proteins differ in their association with the membrane structure. Some, called *extrinsic* or *peripheral proteins*, are loosely attached to the membrane surface. Others are embedded in the membrane structure and may even extend completely through the membrane; these are called *intrinsic* or *integral* proteins (Figure 12-17). The peripheral proteins are usually easy to extract from the membrane structure, whereas integral proteins can be removed only with detergents or organic solvents.

In addition to data obtained from chemical analysis of membranes, much information has come from the application of electron microscopy to the study of membrane structure (Box 12-1).

Box 12-1 Electron microscopy of membranes

Combined with different staining procedures and tissue-preparation methods, electron microscopy has revealed many important details of membrane structure. Shown here are three different aspects of the plasma membrane of the red blood cell, visualized by electron microscopy after three different ways of preparing the cells for examination.

Figure 1 is an edge view of the plasma membrane of a red blood cell showing the two dense lines ("railroad tracks") seen after osmium tetroxide fixation of cells. The two dense lines correspond to the outer and inner polar layers of the membrane lipid head groups. The clear zone between the lines is the hydrophobic portion of the lipid bilayer, which contains the nonpolar fatty acid tails. This is a *transmission electron micrograph*.

Figure 2 shows the glycocalyx (page 295) on the outer surface of the red blood cell, visualized by a special staining procedure. The "fuzzy coat," which consists of hydrophilic oligosaccharide groups of glycoproteins and glycolipids, is over 100 nm thick, some 10 times the thickness of the lipid bilayer itself.

A view of the inside of the red-blood-cell membrane visualized by the *freeze-fracture method* is shown in Figures 3 and 4. In this procedure the cells are frozen and the frozen block shattered or split. The fracture lines formed sometimes split a membrane along a plane between the two lipid layers (Figure 3). Replicas of the two facing surfaces are made and examined by electron microscopy (Figure 4). The inside surface of one lipid layer forms the smooth background; the clusters of globular bodies are molecules of integral membrane proteins. The arrow points to the outside edge of the fracture.

Other features of membrane structure have been observed by *scanning electron microscopy* [e.g., the microvilli protruding from cell surfaces (Figure 2-20, page 34)] and by *negative-contrast staining*, which shows the occurrence of large peripheral proteins [e.g., the F_1ATPase of the inner mitochondria membrane (Chapter 17)].

Figure 1

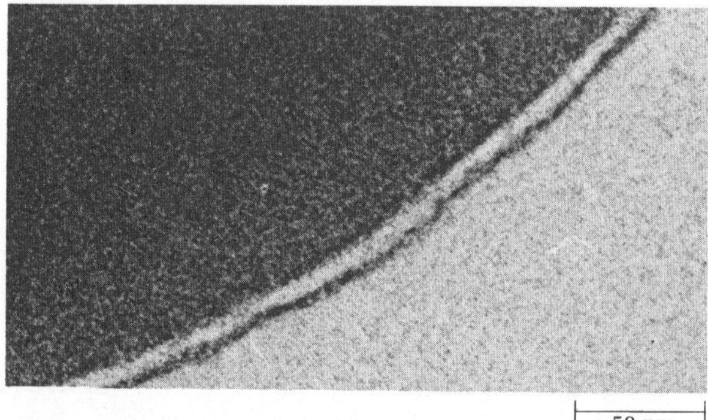

Figure 2
The erythrocyte glycocalyx. This cell has an unusually lush 140-nm-thick coat of oligosaccharide filaments 1.2 to 2.5 nm in diameter.

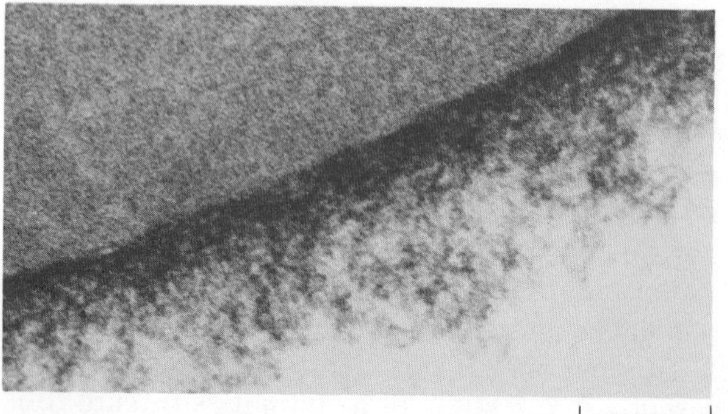

Figure 3

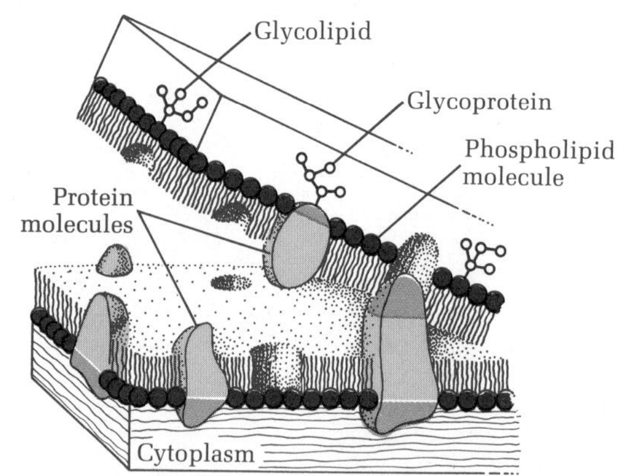

Figure 4

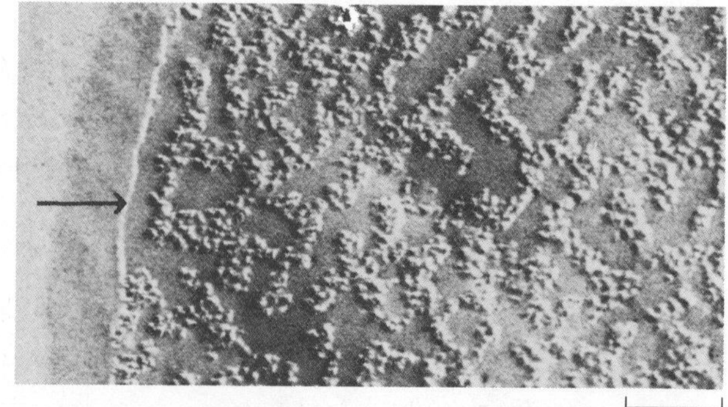

Membranes Have a Fluid-Mosaic Structure

From the results of chemical and electron-microscopic evidence and from the similarity in properties of synthetic phospholipid bilayers and the properties of natural membranes, S. Jonathan Singer and Garth Nicolson postulated in 1972 a unifying theory of membrane structure called the *fluid-mosaic model* (Figure 12-18). They proposed that the matrix, or continuous part of membrane structure, is a polar lipid bilayer. The bilayer is fluid because the hydrophobic tails of its polar lipids consist of an appropriate mixture of saturated and unsaturated fatty acids that is fluid at the normal temperature of the cell. The fluid-mosaic model proposes that the integral proteins of membranes have hydrophobic amino acid R groups on their surfaces, which would cause such proteins to "dissolve" in the central hydrophobic portion of the bilayer. On the other hand, the model proposes that peripheral or extrinsic membrane proteins have essentially hydrophilic R groups on their surfaces, which are bound by electrostatic attraction to the hydrophilic, electrically charged polar heads of the bilayer lipids. The integral membrane proteins, which include enzymes and transport systems, are inactive unless they are placed inside the hydrophobic core of the bilayer, which produces the proper three-dimensional conformation for their activity. Again, note that there are no covalent bonds between lipid molecules of the bilayer or between the protein components and the lipids.

The fluid-mosaic model further allows the membrane proteins to move around laterally. The peripheral proteins literally float on the surface of the bilayer "sea," whereas the integral proteins are like icebergs, almost completely submerged in the hydrocarbon core (Figure 12-18). The freedom of membrane proteins to move laterally in the lipid bilayer may be restricted, however, by attractions between functionally related membrane proteins to form interacting clusters. Thus there is a

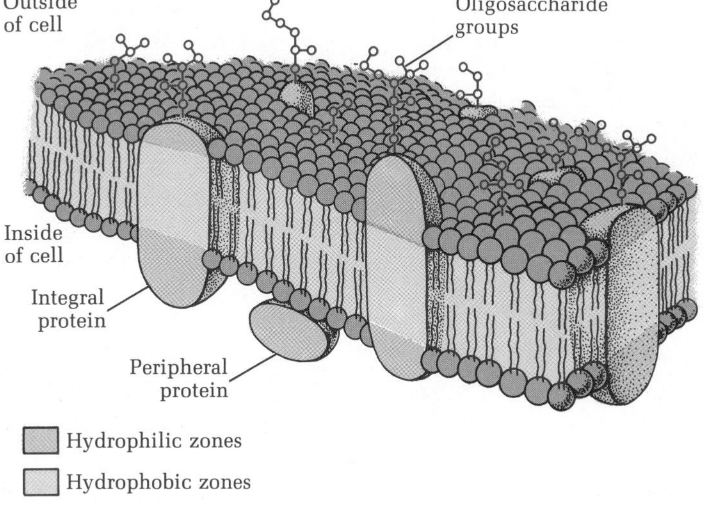

Figure 12-18
The Singer-Nicolson fluid-mosaic model of membrane structure.

mosaic, or surface pattern, of membrane proteins in the fluid lipid bilayer. It was suggested that such clusters of membrane proteins move laterally in the bilayer, to account for the fact that in the life cycle of some cell types certain membrane proteins may move to a specific site or zone of the membrane, a process called *capping*. The Singer-Nicolson model can explain many of the physical, chemical, and biological properties of membranes and has been widely accepted as the most probable molecular arrangement of lipids and proteins of membranes. However, as we shall soon see, biological membranes may have additional structural features above and beyond the elements of the fluid-mosaic model.

Membranes Have a Specific Sidedness or Asymmetry

Most membranes have a specific sidedness, or asymmetry, which also is accounted for by the fluid-mosaic model. In the first place, the polar lipids in the outer and inner faces of the plasma membrane of bacterial and animal cells differ in composition. For example, the inner lipid layer of the human erythrocyte membrane contains most of the phosphatidylethanolamine and phosphatidylserine, whereas the outer layer contains most of the phosphatidylcholine and sphingomyelin. Second, some of the transport systems of membranes function in only one direction. For example, red blood cells contain a membrane transport system or "pump," that pumps Na^+ out of the cell into the surrounding medium and K^+ in, at the expense of energy furnished by the hydrolysis of ATP within the cell. This pump, called the *Na^+K^+-transporting ATPase,* never pumps Na^+ and K^+ ions in the opposite directions. Third, the outer surface of plasma membranes is rich in oligosaccharide groups contributed by the head groups of glycolipids and the oligosaccharide side chains of membrane glycoproteins, whereas the inner surface of the plasma membrane is nearly devoid of such oligosaccharide groups.

This lipid molecule has "flipped over" from the other surface

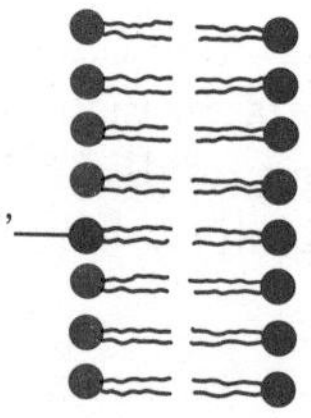

Figure 12-19
The asymmetry of lipid distribution in the two faces of a lipid bilayer in cell membranes. Only rarely does a lipid molecule "flip" from one side to the other, as shown in the diagram. On the other hand, the polar lipid molecules on each side are free to move laterally in the plane of the membrane, but not across the membrane.

The sidedness of biological membranes is in large part maintained and preserved by the resistance of the lipid bilayer to the transfer of individual phospholipid molecules from one face of the membrane to the other (Figure 12-19). This resistance is caused by the large amount of energy required to push the polar, charged phospholipid heads through the hydrocarbon core of the membrane. A polar lipid molecule on one surface of a membrane is therefore free to move in its own side of the bilayer but is restrained from flipping over to the other surface (Figure 12-19).

Red-Blood-Cell Membranes Have Been Studied in Detail

Studies of the proteins present in the plasma membrane of red blood cells have given new insights into the structure of membranes. In particular, they have suggested that at least some membranes may have a "skeleton." The membrane of human

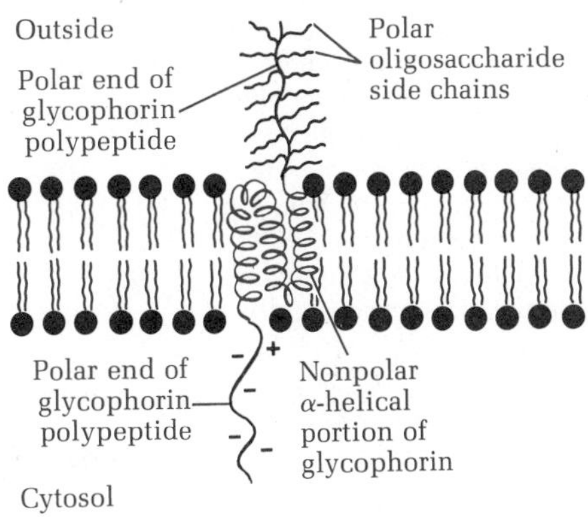

Figure 12-20
Glycophorin molecule in the red-blood-cell membrane. The bushy carbohydrate chains facing out contain characteristic blood-group recognition sites and sites where some viruses attach.

red blood cells contains five major proteins and many minor ones. Most of the membrane proteins are glycoproteins. Among the integral proteins of the red-blood-cell membrane is *glycophorin* ("sugar bearer"). It has a molecular weight of 30,000, a polypeptide chain of 130 residues, and many sugar residues, which make up about 60 percent of the glycophorin molecule. At one end of the polypeptide chain is the complex hydrophilic head, made up of perhaps 15 oligosaccharide branches, each with about 10 sugar units. At the other end of the glycophorin polypeptide are many glutamic and aspartic acid residues, which have negative charges at pH 7.0 (Figure 12-20). In the middle, between the two hydrophilic ends, is a sequence of about 30 hydrophobic amino acid residues. The sugar-bearing end of glycophorin is located on the outer surface of the red-blood-cell membrane, forming a bushlike projection. The hydrophobic middle segment is believed to run through the lipid bilayer, and the other end, with its highly polar, negatively charged amino acid residues, is believed to protrude into the cytosol. The sugar-rich head of glycophorin contains blood-group antigens, specific for the blood type (A, B, or O) of the organism. It also contains sites to which some disease-causing viruses attach.

Another major protein of the red-blood-cell membrane is *spectrin*, a peripheral or extrinsic protein of the inner surface of the membrane which makes up over 20 percent of the total membrane protein. It is easily extracted from the membrane. Spectrin consists of four polypeptide chains which together have a molecular weight of almost 1 million and which form long, flexible rods 100 to 200 nm long. Spectrin molecules bind to certain specific proteins and lipid molecules on the inner surface of the red-blood-cell membrane, to form a flexible, rod-like network on its inner surface, which presumably acts as a skeletal structure for the membrane. Also attached to spectrin are microfilaments of actin, which seem to bind spectrin rods together. Thus it appears that the red-blood-cell membrane has a skeleton or framework to which specific lipids and membrane proteins are anchored (Figure 12-21).

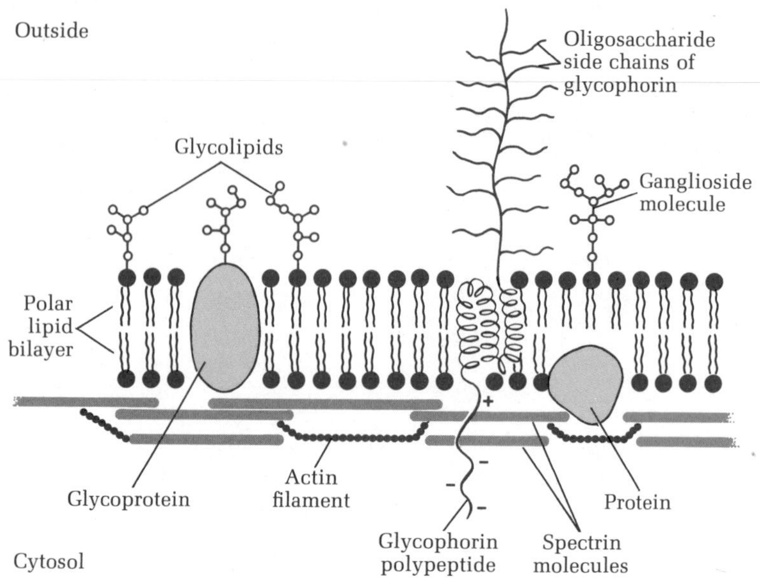

Figure 12-21
Schematic representation of a portion of the red-blood-cell membrane, showing the oligosaccharide antennae of membrane glycoproteins and glycolipids, the oligosaccharide side chains of glycophorin, and the underlying skeleton of the membrane, composed of spectrin molecules linked by short actin filaments bound to the inner surface of the membrane.

The plasma membranes of other types of cells have more complex structures. Another major membrane glycoprotein, present on the outer surface of cells in many solid tissues, is *fibronectin* (page 295), with highly adhesive properties which appear to provide adhesion between cells of the same kind in solid tissues.

Lectins Are Specific Proteins Capable of Binding to or Agglutinating Certain Cells

Many years ago it was discovered that certain plant proteins, sometimes called *phytohemagglutinins*, can bind to and agglutinate red blood cells. It has since been found that they will bind to the surfaces of many other kinds of animal cells. They are especially abundant in plants, particularly of the legume family, and are also found in many invertebrate tissues. *Concanavalin A* of the jack bean and *ricin* of the castor bean were among the first hemagglutinins discovered. These and many other plant and animal proteins capable of binding to cell surfaces are called *lectins* (Latin *legere*, "to pick or choose") because they bind to certain specific carbohydrate groups on cell surfaces. For example, concanavalin A binds D-glucose and D-mannose residues, while soybean lectin binds D-galactose and *N*-acetyl-D-galactosamine. Over 1000 lectins have been identified. Most remarkable and significant is the fact that some lectins preferentially agglutinate malignant tumor cells. Tumor cells must therefore have a different surface structure from normal cells; the specific carbohydrate residues to which lectins bind are apparently more exposed on tumor cell surfaces.

The specificity of lectins, as well as their practical usefulness, is shown by their ability to distinguish between type A, B, and O red blood cells, which differ in the structure of the oligosaccharide groups of glycophorin. Lima bean lectin will agglutinate only type A red blood cells, a lectin from a lotus species will agglutinate type B cells, and a third lectin from yet another plant will agglutinate only type O cells. Lima bean lectin binds to glycoproteins having largely N-acetyl-D-galactosamine residues, whereas lectins reactive with type O cells bind to surface glycoproteins containing the sugar *fucose*. Not only red blood cells but other cells of animal tissues have type-specific glycoproteins on their surfaces. This is why the transplantation of skin or an organ, e.g., kidney or heart, from one human being to another requires the tissue type of the donor and recipient to be identical if the transplanted tissue is to take successfully.

Lectins of plants and invertebrates appear to be defensive proteins that protect these organisms, which have no immune system and thus no antibodies, from invasion by microbial parasites. Lectins are believed to be located on the surface of plant cells.

Membranes Have Very Complex Functions

It is now clear that membranes are not simply inert skins holding the cell together, nor are they static, fixed structures, since they perform many complex dynamic functions and have rather remarkable biological properties. Most membranes contain enzymes, some acting on substrates outside the membrane and others on substrates inside the membrane-enclosed compartment. The inner mitochondrial membrane and the thylakoid membrane of chloroplasts contain complex systems of many enzymes and proteins. Most membranes also contain transport systems, which translocate certain specific organic nutrient molecules, such as glucose, or allow specific inorganic ions to enter and certain products to leave. These transport systems help preserve the steady-state constancy of the internal medium of cells by regulating the flow of materials in and out of cells. Membranes also contain electrically charged surface groups, which help support a difference of electrical potential across the membrane structure. This property is very important in the function of nerve cells, which can transmit impulses in the form of a very fast, wavelike change in the electrical properties of the membrane along the extended cell body or axon. Cell membranes also have self-sealing characteristics. If they are punctured or disrupted mechanically, they quickly and automatically reseal.

On their outer surfaces membranes also contain specific *recognition sites*, distinctive patches that function to recognize certain molecular signals. For example, the membranes of some bacteria can sense a small difference in the concentration of a nutrient, stimulating them to swim toward the source of the nutrient, a phenomenon called *chemotaxis*. The outer surface of animal cell membranes contains sites that recognize other cells of the same kind, promoting their association during the orderly development of tissue structure. Other kinds of recognition sites on cell surfaces serve as specific receptors for hormone molecules. For example, liver and muscle cell surfaces contain sites that recognize and bind the hormones *insulin*, *glucagon*, and *adrenaline*. Once they are occupied, these hormone receptor sites convey signals through the membrane to intracellular enzyme systems, regulating their activity. Other sites on cell surfaces, called *histocompatibility sites*, are characteristic for given individuals of a species.

In many specific recognition or receptor sites on animal cell membranes gangliosides appear to be important components. Although gangliosides are quantitatively minor components of membrane lipids, they appear to be concentrated at specific sites. Because there are many different kinds of gangliosides, each with a different oligosaccharide head group, it is believed that different gangliosides as well as glycoproteins are arranged in specific mosaic patterns to constitute the different kinds of receptor sites on the outer surface of cells. Their polar, negatively charged head groups may be regarded as receptors or antennae projecting from the membrane surface, sensitive to certain signaling molecules, such as hormones.

Cell membranes thus are highly complex structures, in which many different kinds of molecular assemblies are arranged in ordered two-dimensional mosaics which give surfaces their biological specificity. The molecular architecture of cell membranes is one of the most important objectives of research in cell biology and biochemistry today.

Summary

Lipids are oily or greasy water-insoluble components of cells that can be extracted by nonpolar solvents. Some lipids serve as structural components of membranes and others as storage forms of fuel. Fatty acids, which are the greasy components of lipids, usually have an even number of carbon atoms; the most abundant have 16 or 18 carbon atoms. Fatty acids may be saturated or unsaturated; unsaturated fatty acids have the cis configuration. In most unsaturated fatty acids, one double bond is at the Δ^9 position. Sodium or potassium salts of fatty acids are called soaps. Triacylglycerols contain three fatty acid molecules esterified to the three hydroxyl groups of glycerol. Simple triacylglycerols contain only one type of fatty acid; mixed triacylglycerols contain at least two different types. Triacylglycerols are primarily storage fats.

The polar lipids, which have polar heads and nonpolar tails, are major components of membranes. The most abundant are the phosphoglycerides, which contain two fatty acid molecules esterified to the two free hydroxyl groups of glycerol 3-phosphate and a second alcohol, the head group, esterified to the phosphoric acid. Phosphoglycerides differ in the structure of the head group. The most common phosphoglycerides are phosphatidylethanolamine and phosphatidylcholine. The polar heads of the phosphoglycerides carry electric charges at pH near 7. The sphingolipids, also membrane components, contain the base sphingosine but no glycerol. Sphingomyelin possesses, in addition to phosphoric acid and choline, two long hydrocarbon chains, one contributed by a fatty acid and the other by sphingosine, a long-chain aliphatic amino alcohol. Cholesterol, a sterol, is a precursor of many steroids and is also an important component of plasma membranes.

All polar lipids possess polar or charged heads and nonpolar hydrocarbon tails; they spontaneously form micelles, monolayers, and bilayers, stabilized by hydrophobic interactions. Polar lipid bilayers function as the structural core of cell membranes, which also contain many different kinds of proteins, some (extrinsic proteins) on the surface and others (intrinsic proteins) inside the membrane structure. Cell membranes possess sidedness and also contain, on their outer surface, the hydrophilic oligosaccharide groups of glycoproteins and glycolipids. Some of these oligosaccharide groups play important roles in cell-cell recognition and adhesion, tissue typing, and hormone receptor sites.

References

Books

ANSELL, G. B., J. N. HAWTHORNE, and R. M. C. DAWSON: *Form and Function of the Phospholipids*, 2d ed., Elsevier, New York, 1973. A comprehensive treatment.

GURR, A. I., and A. T. JAMES: *Lipid Biochemistry: An Introduction*, 3d ed., Methuen, New York, 1980.

HANSON, J. R.: *Introduction to Steroid Chemistry*, Pergamon, New York, 1968. A brief summary.

HARRISON, R., and G. G. LUNT: *Biological Membranes, Their Structure and Function*, 2d ed., Halsted, New York, 1980. Excellent up-to-date paperback. Many outstanding illustrations.

WEISSMANN, G., and R. CLAIBORNE (eds.): *Cell Membranes: Biochemistry, Cell Biology, and Pathology*, H. P. Publishing Co., New York, 1975. Excellent, well-illustrated survey.

Articles

BENSON, A. A., and R. F. LEE: "The Role of Wax in Oceanic Food Chains," *Sci. Am.*, **232**:77–86, March (1975).

CAPALDI, R. A.: "A Dynamic Model of Cell Membranes," *Sci. Am.*, **230**:26–33, March (1974). Refinements of the Singer-Nicolson model.

FISHMAN, P. H., and R. O. BRADY: "Biosynthesis and Function of Gangliosides," *Science*, **194**:906–915 (1976).

LODISH, H. F., and J. E. ROTHMAN: "The Assembly of Cell Membranes," *Sci. Am.*, **240**:48–63, January (1979).

LUX, S. E.: "Dissecting the Red Cell Membrane Skeleton," *Nature*, **281**:426–429 (1979).

MARCHESI, V. T.: "Spectrin: Present Status of a Putative Cyto-Skeletal Protein of the Red Cell Membrane," *J. Membrane Biol.*, **51**:101–131 (1979).

MARX, J. L.: "Liposomes: Research Applications Grow," *Science*, **199**:1056–1128 (1978).

SHARON, N.: "Lectins," *Sci. Am.*, **236**:108–119, June (1977).

SINGER, S. J., and G. L. NICOLSON: "The Fluid Mosaic Model of the Structure of Membranes," *Science*, **175**:720–731 (1972).

Problems

1. *Melting Points of Fatty Acids* The melting points of a series of 18-carbon fatty acids are stearic acid (69.6°), oleic acid (13.4°), linoleic acid (−5°), and linolenic acid (−11°). What structural aspect of these 18-carbon fatty acids can be correlated to the melting point? Provide a molecular explanation for the trend in melting points.

2. *Spoilage of Cooking Fats* Some fats used in cooking, such as olive oil, spoil rapidly upon exposure to air at room temperature, while others, such as solid shortening (Crisco) remain unchanged. Why?

3. *Preparation of Bearnaise Sauce* During the preparation of bearnaise sauce, phosphatidylcholine (lecithin) from egg yolks is incorporated into melted butter to stabilize the sauce and avoid separation. Suggest why this works.

4. *Hydrolysis of Lipids* Name the products of mild hydrolysis of the following with dilute sodium hydroxide:
   (a) 1-stearoyl-2,3-dipalmitoylglycerol
   (b) 1-palmitoyl-2-oleyl phosphatidylcholine
   What is the product of the action of hot concentrated NaOH on (b)?

5. *Net Electrical Charge of Phospholipids* At pH 7.0 what is the electrical charge of (a) phosphatidylcholine, (b) phosphatidylethanolamine, and (c) phosphatidylserine?

6. *Protection of Succulent Plants* The surfaces of succulent plants native to arid regions are generally covered by a coat of wax. How does this help the plant survive?

7. *Number of Detergent Molecules per Micelle* When a small amount of sodium dodecyl sulfate [$CH_3(CH_2)_{11}OSO_3^-Na^+$, a common commercial detergent] is dissolved in water, the detergent ions go into solution as monomeric species. As more detergent is added, a point is reached (the critical micelle concentration) at which the monomers associate to form micelles (Figure 12-16). The critical micelle concentration of sodium dodecyl sulfate is 8.2 m*M*. An examination of the micelles shows that they have an average molecular weight of 18,000. Calculate the number of detergent molecules in the average micelle.

8. *Hydrophobic and Hydrophilic Units of Membrane Lipids* A common structural feature of membrane lipid molecules is their amphipathic nature; i.e., they contain poth hydrophobic and hydrophilic groups. For example, in phosphatidylcholine, the two fatty acid chains are hydrophobic and the phosphocholine head group is the hydrophilic portion of the molecule. For each of the following membrane lipids, give the names of the building blocks that serve as the hydrophobic and hydrophilic units:
   (a) phosphatidylethanolamine
   (b) sphingomyelin
   (c) galactocerebroside
   (d) ganglioside
   (e) cholesterol

9. *Properties of Lipids and Lipid Bilayers* Lipid bilayers formed between two aqueous phases have three important properties:

they form two-dimensional sheets, the edges of the sheet close upon themselves, and the sheets undergo self-sealing to form liposomes.

(a) What properties of lipids are responsible for these properties of bilayers? Explain.
(b) What are the biological consequences of these properties with regard to the structures of biological membranes?

10. *Movement of Ions through Cell Membranes* The lipid bilayer of the cell membrane prevents the rapid escape of ions, such as K^+, Cl^-, and Mg^{2+}, from cells. Why?

11. *Extraction of Integral Membrane Proteins* In contrast to proteins found in the cytosol, many membrane-embedded proteins are virtually impossible to remove from the membrane into aqueous solution (Figure 12-17). However, removal and solubilization of such proteins can frequently be achieved if the extracting solution contains sodium dodecyl sulfate (Problem 7) or some other detergent such as sodium cholate. What is the basis for this procedure?

12. *Orientation of Membrane Glycoproteins by Sugar Residues* Examination of different membrane glycoproteins shows that the sugar residues are always located on the outside surface of the membrane (e.g., Figure 12-18). One explanation for this phenomenon is that the sugar serves to maintain the asymmetric orientation of the glycoprotein in the membrane.
(a) Why are the sugar residues located on the outer surface of the membrane rather than in the interior of the membrane?
(b) How do sugar residues on the glycoprotein maintain the asymmetric orientation in the membrane? Explain.

13. *Membrane Fluidity and Function* A central hypothesis in the field of membrane research is that the membrane lipids must be fluid (as opposed to "frozen") in order for the membrane to carry out its functions. Support for this hypothesis is provided by the observation that the fatty acid composition of bacterial membranes can be changed by the conditions under which the bacteria are grown. For example, if the bacteria are grown at lower than normal temperatures, the observed amounts of unsaturated fatty acids (relative to the saturated fatty acid content) are above normal. Conversely, if the bacteria are grown at higher than normal temperatures, the observed amounts of unsaturated fatty acids in the membrane lipids (relative to the saturated fatty acids) are below normal.
(a) Suggest reasons why the lipid in the bacterial membrane must be fluid for the intact membrane to operate properly.
(b) Explain how the observed change in unsaturated fatty acid levels relative to saturated fatty acid levels at different growth temperatures supports the membrane fluidity hypothesis.

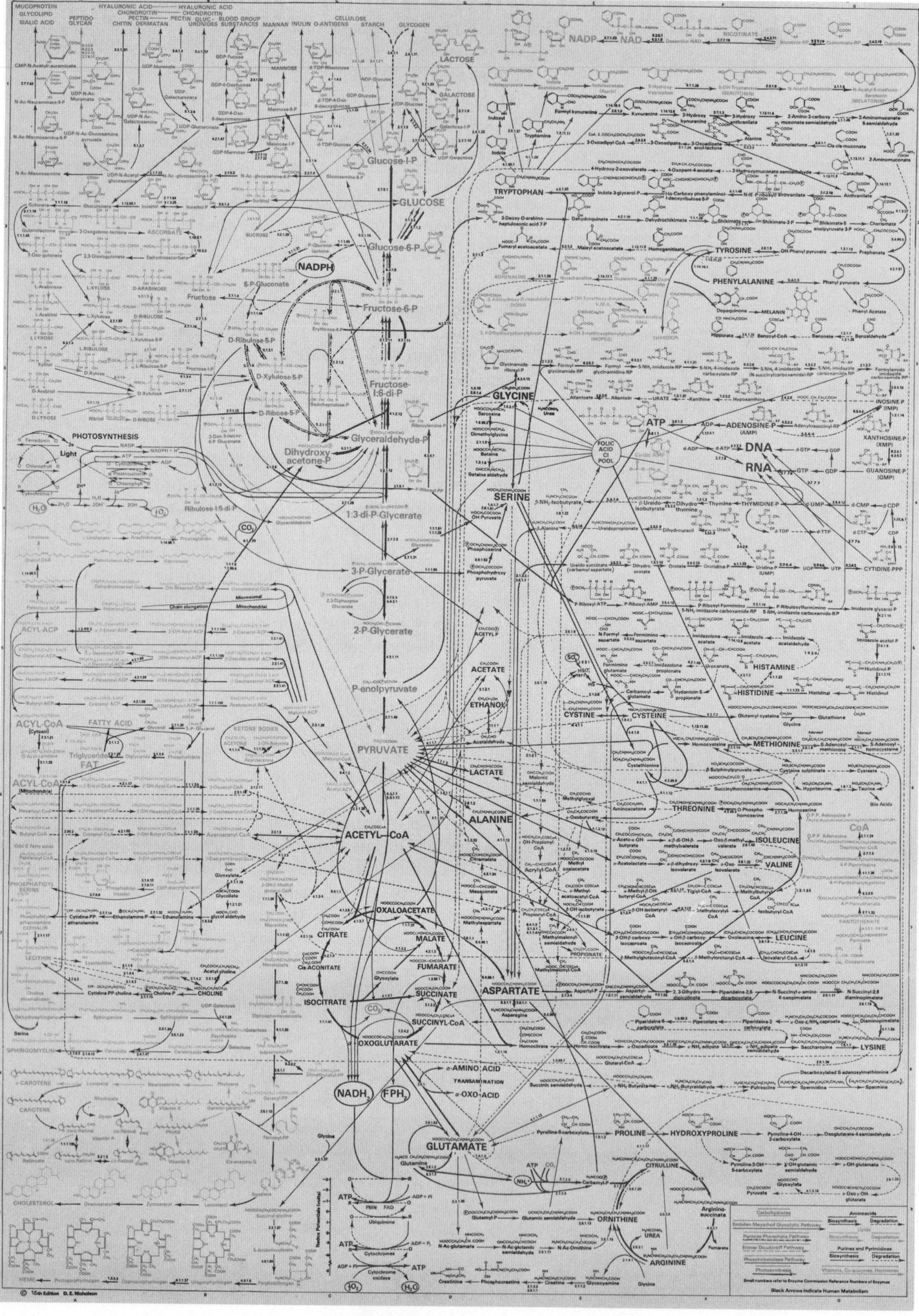

NADP
NAD
LACTOSE
GALACTOSE
Glucose-1-P
GLUCOSE
Glucose-6-P
NADPH
Fructose
Fructose-6-P
D-Ribulose-5-P
D-Xylulose-5-P
D-Ribose-5-P
Fructose-1:6-di-P
Glyceraldehyde-P
Dihydroxy acetone-P
PHOTOSYNTHESIS
Light
Ribulose 1:5-di-P
1:3-di-P-Glycerate
3-P-Glycerate
2-P-Glycerate
P-enolpyruvate
PYRUVATE
ACYL-ACP
FATTY ACID
ACYL-CoA
(Cytosol)
Triglyceride
FAT
ACYL-CoA
(Mitochondria)
KETONE BODIES
ACETYL-CoA
PHOSPHATIDYL SERINE
CHOLINE
LECITHIN
SPHINGOMYELIN
CAROTENE
CHOLESTEROL
HEME
CITRATE
OXALOACETATE
MALATE
FUMARATE
Cis-ACONITATE
ISOCITRATE
SUCCINATE
SUCCINYL-CoA
OXOGLUTARATE
NADH₂
FPH₂
α-AMINO ACID
TRANSAMINATION
α-OXO-ACID
GLUTAMATE
ATP
TRYPTOPHAN
TYROSINE
PHENYLALANINE
GLYCINE
SERINE
FOLIC ACID C1 POOL
ATP
ADENOSINE-P (AMP)
DNA
RNA
INOSINE-P (IMP)
XANTHOSINE-P (XMP)
GUANOSINE-P (GMP)
THYMIDINE-P
CYTIDINE-PPP
ACETATE
ETHANOL
LACTATE
ALANINE
CYSTINE
CYSTEINE
HISTAMINE
HISTIDINE
METHIONINE
THREONINE
ISOLEUCINE
VALINE
LEUCINE
CoA
PANTOTHENATE
ASPARTATE
LYSINE
PROLINE
HYDROXYPROLINE
CITRULLINE
ORNITHINE
UREA
ARGININE
Carbohydrates
Embden Meyerhof Glycolytic Pathway
Pentose Phosphate Pathway
Entner Doudoroff Pathway
Phospholipid Pathway
Photosynthesis
Aminoacids
Biosynthesis
Degradation
Lipids
Purines and Pyrimidines
Vitamins, Co-enzymes, Hormones
Small numbers refer to Enzyme Commission Reference Numbers of Enzymes
Black Arrows indicate Human Metabolism
© 15th Edition D. E. Nicholson

PART II

Bioenergetics and Metabolism

We humans, and all living organisms, have very complex and intricate structures. Many of us are vaguely aware that the complexity of living organisms is sustained by the consumption and transformation of our nutrients. But to the physicist, living organisms have been a very special challenge, because they seemingly contradict one of the fundamental laws of physical science. This principle, the second law of thermodynamics, says that organized collections of matter tend to undergo disorganization, to become more random and disordered as time goes on. However, we know today that living cells are not exempt from this law. They get around it by maintaining their internal order in a dynamic steady state, at the expense of nutrients and free energy obtained from the environment and transformed by metabolism.

To gain some appreciation of the pathways, energetics, and dynamics of cell metabolism, we must begin by examining the characteristic energy exchanges occurring in single chemical reactions as they are catalyzed by specific enzymes under the conditions of constant temperature and pressure that prevail in cells. We will see how several enzyme-catalyzed reactions can be linked into a chain or system by common intermediates, to make possible efficient transfer of chemical energy. Then we will examine the central metabolic pathways of cells, the sequences of consecutive enzyme-catalyzed reactions that lead to the degradation of the major foodstuff molecules—carbohydrates, fats, and amino acids—and the release of some of their free energy as ATP. Subsequently, we will also examine, step-by-step, some important central pathways by which the major cell macromolecules are built from simple precursors at the expense of chemical energy. The rate of each of these metabolic pathways, whether degradative or biosynthetic in direction, is under exquisitely sensitive control and regulation. The result of these activities, involving the coordinated activity of hundreds of enzymes, is a marvelously engineered network

A metabolic map. Over 2000 known enzymes, and many more yet to be discovered, participate in the metabolism of cells. Some catalyze the heavily traveled major routes of metabolism. Other paths lead to many specialized products needed in only small amounts. Ultimately all the pathways of metabolism are interconnected. As we might do in becoming acquainted with a metropolis, we shall study metabolism by first exploring its main avenues, from which all side streets and by-ways can be reached.

of enzymatic reactions, which functions with computerlike efficiency to preserve and maintain the internal order of cells, despite fluctuations in their environment.

CHAPTER 13

A Survey of Metabolism

Myriads of enzyme-catalyzed chemical reactions take place in living cells. Although we collectively refer to these reactions as *metabolism*, we must not think of cell metabolism in terms of a membrane-surrounded bag of randomly acting enzymes. Metabolism is a highly coordinated and purposeful cell activity, in which many multienzyme systems cooperate. Metabolism has four specific functions: (1) to obtain chemical energy from the degradation of energy-rich nutrients from the environment or from captured solar energy, (2) to convert nutrient molecules into the building-block precursors of cell macromolecules, (3) to assemble these building blocks into proteins, nucleic acids, lipids, polysaccharides, and other cell components, and (4) to form and degrade biomolecules required in specialized functions of cells.

Although metabolism involves hundreds of different enzyme-catalyzed reactions, the central metabolic pathways, our major concern, are few in number, and they are identical in most forms of life. In this chapter we shall survey the sources of matter and energy for metabolism, the central pathways by which the major cell components are synthesized and degraded, and the mechanisms by which chemical energy is transferred. We shall also see how the pathways of cell metabolism are studied experimentally.

Living Organisms Participate in the Cycling of Carbon and Oxygen

Let us begin by considering metabolism in its macroscopic aspects, in the total metabolic interplay of living organisms in our biosphere. Living organisms can be divided into two large groups according to the chemical form of carbon they require from the environment. *Autotrophic* ("self-feeding") cells can use carbon dioxide from the atmosphere as the sole source of carbon and construct all their carbon-containing biomolecules from it. Examples are photosynthetic bacteria and green leaf cells of plants. Some autotrophic organisms, such as the cyanobacteria, can also use nitrogen of the atmosphere to generate all their nitrogenous components. Heterotrophic cells ("feeding on others") cannot use atmospheric carbon dioxide and must obtain carbon from their environment in the form of relatively complex organic molecules, such as glucose. The cells of higher animals and most microorganisms are heterotrophic. Autotrophs are relatively self-sufficient cells, whereas heterotrophs, with their requirements for carbon in fancier forms, must subsist on the products formed by other cells.

There is another major difference. Many autotrophic organisms are photosynthetic and obtain their energy from sunlight, whereas heterotrophic cells obtain their energy from the degradation of organic nutrients made by autotrophs. In our biosphere autotrophs and heterotrophs live together in a vast, interdependent cycle in which autotrophic organisms use atmospheric CO_2 to build their organic biomolecules and some of them generate oxygen. Heterotrophs in turn use the organic products of autotrophs as nutrients and return CO_2 to the atmosphere. Thus carbon and oxygen are constantly cycled between the animal and plant worlds, solar energy ultimately providing the driving force for this massive process (Figure 13-1).

Autotrophic and heterotrophic organisms can be divided into subclasses. For example, there are two great classes of heterotrophic organisms, aerobic and anaerobic. Aerobes live in air and use molecular oxygen to oxidize their organic nutrient molecules. Anaerobes live in the absence of oxygen and degrade their nutrients without using oxygen. Many cells, such as yeasts, can live either aerobically or anaerobically; such orga-

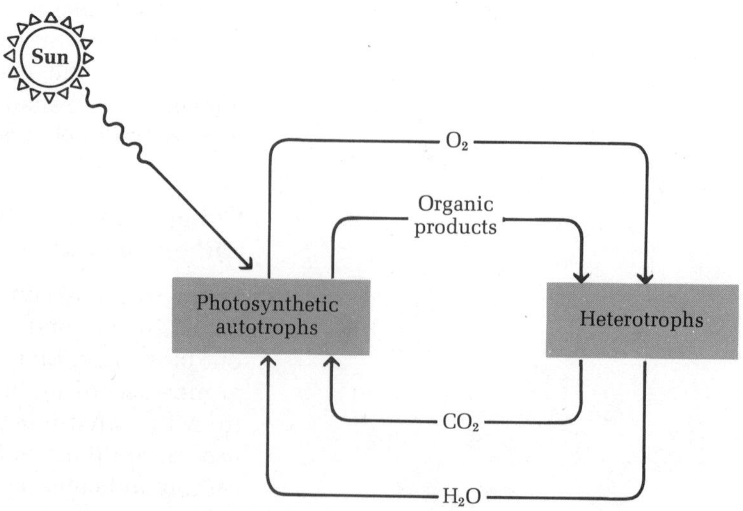

Figure 13-1
The cycling of carbon dioxide and oxygen between the photosynthetic and heterotrophic domains in the earth's biosphere. The magnitude of this cycle is enormous. Over 3.5×10^{11} tons of carbon are turned over in the biosphere annually. The balance between CO_2 production and use is believed to be an important factor in the earth's climate. Over the last 100 years the CO_2 content of the atmosphere has increased about 25 percent, due to the increasing rate of combustion of oil and coal. Some earth scientists believe that further increases in atmospheric CO_2 will cause an increase in the average temperature of the atmosphere (the "greenhouse effect"), but others disagree, because of the difficulty of accurately estimating the amounts of CO_2 produced and recycled by the biosphere and the amounts absorbed by the oceans. Almost 300 years are required for all the atmospheric CO_2 to cycle through the plant world.

nisms are called *facultative*. Anaerobes that cannot use oxygen at all, and indeed may be poisoned by it, e.g., microorganisms present in deep soils or the ocean floor, are called *strict anaerobes*. Most heterotrophic cells, particularly those of higher organisms, are facultative, but if oxygen is available to them, they will use their aerobic pathways to oxidize their nutrients.

Not all cells of a given organism are of the same class. For example, in higher plants the green chlorophyll-containing cells of leaves are photosynthetic autotrophs, but their root cells, which contain no chlorophyll, are heterotrophs. Moreover, green leaf cells are autotrophic only in daylight. In the dark they function as heterotrophs and obtain their energy by oxidizing carbohydrate manufactured in daylight.

Nitrogen Is Cycled in the Biosphere

In addition to sources of carbon, oxygen, and energy, all living organisms require a source of nitrogen. Nitrogen is necessary for the biosynthesis of amino acids and the purine and pyrimidine bases, the nitrogenous building blocks of proteins and nucleic acids. Here again living organisms differ widely in the chemical form of nitrogen they can use. In most higher animals, at least some of the nitrogen intake must be in the form of amino acids. In human beings and the albino rat, for example, 10 of the 20 standard amino acids must be obtained already formed from the diet since these organisms are unable to synthesize them from simpler precursors. Plants are generally able to use either ammonia or soluble nitrates as their sole source of nitrogen. Only a relatively few organisms are capable of using (“*fixing*”) the gaseous nitrogen (N_2) that makes up some 80 percent of our atmosphere. Since the earth's crust contains very little inorganic nitrogen in the form of soluble salts, all living organisms are ultimately dependent upon atmospheric nitrogen and nitrogen-fixing organisms. Nitrogen can be fixed by the cyanobacteria, or blue-green algae, which are totally self-sufficient cells, since they are completely autotrophic. They not only can use atmospheric nitrogen but also can carry out photosynthesis and thus obtain all their carbon requirements from atmospheric carbon dioxide. Most other species of nitrogen-fixing bacteria occur in the soil. Some of them live symbiotically in root nodules of certain plants, particularly members of the legume family, and carry out *symbiotic nitrogen fixation*.

Other microbial organisms (*nitrifying bacteria*) carry out the oxidation of ammonia to nitrites and nitrates, and still others (*denitrifying bacteria*) can convert nitrates back into ammonia. Thus, in addition to a massive carbon and oxygen cycle (Figure 13-1) we have a nitrogen cycle in our biosphere, in which huge amounts of nitrogen undergo cycling and turnover (Figure 13-2). The cycling of carbon, oxygen, and nitrogen, which involves many species of living organisms, obviously depends on

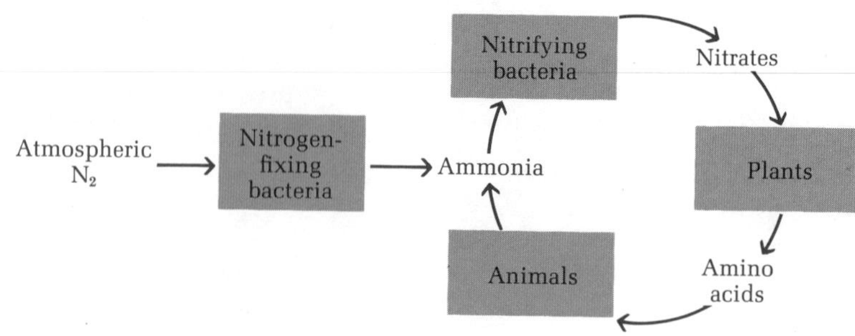

Figure 13-2
The cycling of nitrogen in the biosphere.

a proper balance between the producers and consumers in our biosphere (Figure 13-3). Moreover, these great cycles of matter in our biosphere are accompanied by an enormous flow of energy, which begins with the capture of solar energy by photosynthetic organisms and its use to generate energy-rich carbohydrates and other organic nutrients, which are then used as energy sources by heterotrophic organisms. In the metabolism of each organism participating in these metabolic cycles and in the performance of different kinds of energy-requiring activities there is a loss of *free* (i.e., useful) energy and an inevitable increase in the amount of *unavailable* (i.e., useless) energy. At almost every step through these biological cycles, heat and other forms of energy are dissipated into the surroundings and become randomized and unavailable. The flow of energy through the biosphere is thus a one-way flow, rather than a cycling process, since high-grade useful energy can never be regenerated from unavailable, dissipated energy. Thus carbon, oxygen, and nitrogen recycle continuously, but free, or useful, energy is constantly degraded to an unavailable form.

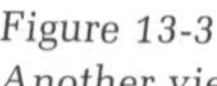

Figure 13-3
Another view of the flow of solar energy, carbon, oxygen, and nitrogen. In this isolated ecosystem photosynthesis by the grasses fixes carbon dioxide of the atmosphere and generates organic nutrients and molecular oxygen. Soil microorganisms fix atmospheric nitrogen and convert it into ammonia and nitrates, which are used as nitrogen sources by the grass for the synthesis of proteins and nucleic acids. The zebras obtain oxygen from the air, as well as organic carbon and amino acids from the oxidation of the starch, proteins, and other components of the grasses. The lions feed on the zebras and return waste materials to the soil and its microorganisms to complete the cycle.

Solar energy is the driving force for the whole cycle, but at each step in this food chain less than 10 percent of the useful energy input is realized as biomass, all the rest being dissipated in the surroundings and thus made unavailable. Less than 0.1 percent of the solar-energy input ends up in the chemical energy stored in the lions of this ecosystem. This is why a very large grazing area is required to support a herd of zebras and a large herd of zebras is required to support two lions.

As we now turn from these macroscopic aspects of metabolism to the microscopic metabolic events occurring in living cells, we must always keep in mind that each kind of cell has its own characteristic requirements for carbon, oxygen, and nitrogen sources, and also for a source of energy. Cell metabolism, then, concerns itself with the enzymatic transformations of both matter and energy, starting from raw materials and leading to the biosynthesis of living matter.

Figure 13-4
Multienzyme systems.

A linear pathway. Precursor A is converted into product E by four consecutive enzyme-catalyzed steps in such a way that a product of one step becomes the substrate of the next.

A circular or cyclic pathway. The oxidation of acetyl residues to CO_2 and H_2O proceeds by such a pathway, the citric acid cycle.

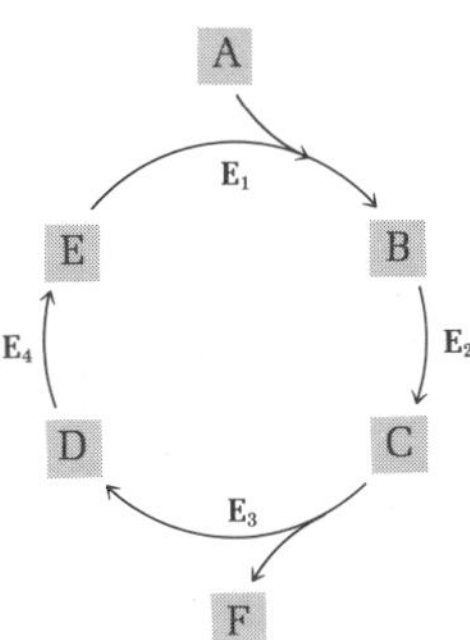

Metabolic Pathways Are Promoted by Sequential Enzyme Systems

Enzymes are of course the simplest units of metabolic activity, each catalyzing a specific chemical reaction. But metabolism is best discussed in terms of *multienzyme sequences*, each promoting the sequential catalytic steps involved in a given metabolic pathway. Such enzyme systems may have anywhere from 2 to 20 enzymes acting in a consecutive, linked fashion, so that the product of the first enzyme becomes the substrate of the second, and so on (Figure 13-4). The successive transformation products in such a pathway (B, C, D, etc., in Figure 13-4) are known as *metabolic intermediates* or *metabolites*. Each of the consecutive steps in these pathways brings about a small, specific chemical change, usually the removal, transfer, or addition of a specific atom, molecule, or functional group. Through such orderly, step-by-step changes the incoming biomolecule is transformed into its metabolic end product. Most metabolic pathways are linear, but some are circular or cyclic (Figure 13-4). Usually metabolic pathways have branches leading in or out. The term *intermediary metabolism* is often used to denote the specific sequences of intermediates involved in the pathways of cell metabolism.

Metabolism Consists of Catabolic (Degradative) Pathways and Anabolic (Biosynthetic) Pathways

Intermediary metabolism has two phases: *catabolism* and *anabolism*. Catabolism is the *degradative* phase of metabolism, in which organic nutrient molecules, e.g., carbohydrates, lipids, and proteins, coming either from the environment or from the cell's own nutrient stores, are degraded by stepwise reactions into smaller, simpler end products, e.g., lactic acid, CO_2, and ammonia. Catabolism is accompanied by release of the free energy inherent in the complex structure of large organic molecules. At certain steps in a catabolic pathway much of the free energy is conserved, by means of coupled enzymatic reactions in the form of the energy-carrying molecule *adenosine triphosphate* (ATP). Some may be conserved as energy-rich hydrogen atoms carried by the coenzyme *nicotinamide adenine dinucleotide phosphate* in its reduced form NADPH (Figure 13-5).

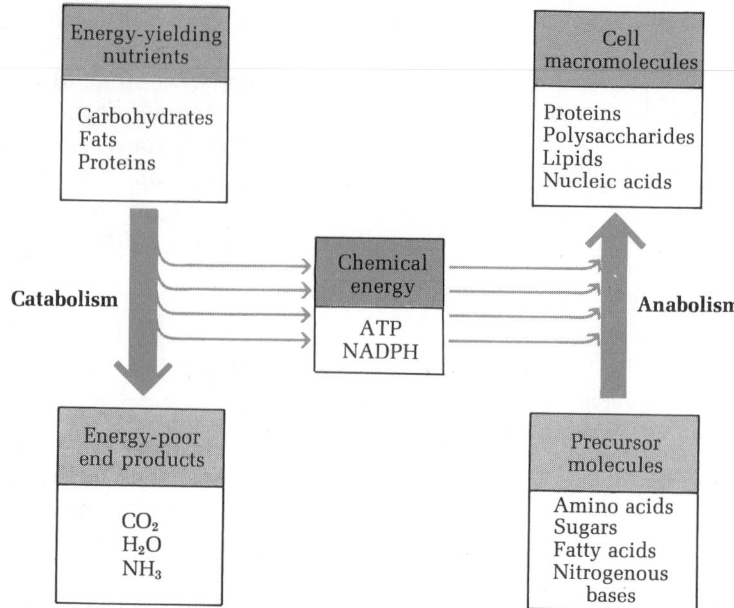

Figure 13-5
Energy relationships between catabolic and anabolic pathways. Catabolic pathways deliver chemical energy in the form of ATP and NADPH. These are used in biosynthetic pathways to convert small precursor molecules into cell macromolecules.

In *anabolism*, also called *biosynthesis*, the building-up or synthetic phase of metabolism, small precursor or building-block molecules are built up into large macromolecular components of cells, such as proteins and nucleic acids. Since biosynthesis results in increased size and complexity of structure it requires input of free energy, which is furnished by the breakdown of ATP to ADP and phosphate. Biosynthesis of some cell components also requires high-energy hydrogen atoms, which are donated by NADPH (Figure 13-5). Catabolism and anabolism take place simultaneously in cells, and their rates are regulated independently.

Catabolic Pathways Converge to a Few End Products

Now let us look at catabolism more closely. The enzymatic degradation of each of the major energy-yielding nutrients of cells (carbohydrates, lipids, and proteins) proceeds in a stepwise manner through a number of consecutive enzymatic reactions. There are three major stages in aerobic catabolism, as shown in Figure 13-6. In stage I, cell macromolecules are degraded to their major building blocks. Thus polysaccharides are degraded to hexoses or pentoses; lipids are degraded to fatty acids, glycerol, and other components; and proteins are hydrolyzed to their 20 component amino acids.

In stage II of catabolism, the various products formed in stage I are collected and converted into a smaller number of yet simpler molecules. Thus, the hexoses, pentoses, and glycerol from stage I are degraded to a single 3-carbon intermediate, *pyruvate*, which is then converted into a single 2-carbon unit, the acetyl group of *acetyl–coenzyme A*. Similarly, the fatty acids and the carbon skeletons of most of the amino acids are

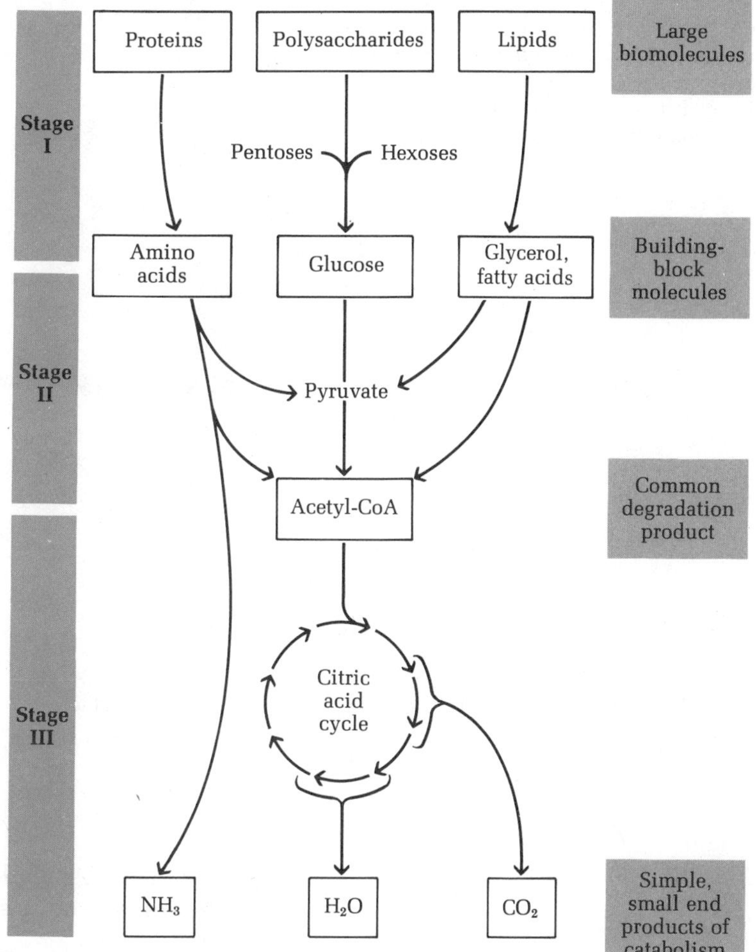

Figure 13-6
The three stages of catabolism of the major energy-yielding nutrients. In stage I hundreds of proteins and many kinds of polysaccharides and lipids are broken down into their building-block components, which are relatively few in number. In stage II the building-block molecules are further degraded into a common product, the acetyl groups of acetyl-CoA. Finally, in stage III, catabolism converges into the citric acid cycle with the formation of only three major end products. Although nucleic acids also undergo degradation in stages, this process is not shown since their breakdown does not contribute significantly to the energy needs of cells.

also broken down to form acetyl groups in the form of acetyl-CoA. Acetyl-CoA is thus the common end product of stage II of catabolism.

In stage III the acetyl group of acetyl-CoA is fed into the *citric acid cycle,* the final common pathway by which most energy-yielding nutrients are ultimately oxidized to carbon dioxide. Water and ammonia (or other nitrogenous products) are the other end products of catabolism.

It is important to note that the pathways of catabolism *converge* toward the citric acid cycle in stage III. During catabolism in stage I, dozens or even hundreds of different proteins are degraded to 20 amino acids; in stage II, the 20 amino acids are degraded, largely to acetyl-CoA and ammonia (NH_3); and in stage III, the acetyl groups of acetyl-CoA are oxidized by the citric acid cycle to only two products, CO_2 and H_2O. Similarly, in stage I many different polysaccharides and disaccharides are degraded to a few simple sugars, which are all finally converted into acetyl-CoA in stage II and into CO_2 and H_2O in stage III. The final pathway of catabolism thus resembles a widening river, fed by many tributary streams (Figure 13-7).

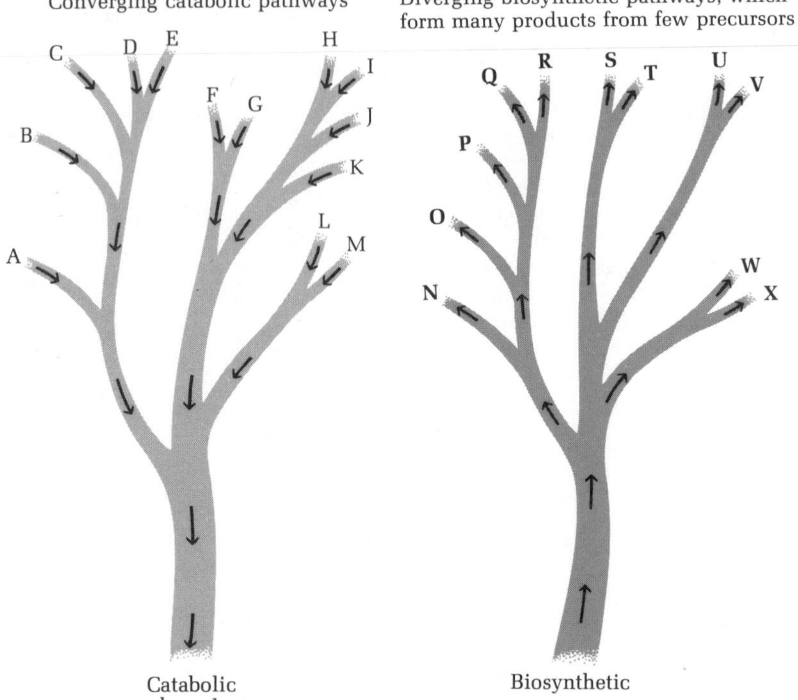

Figure 13-7
Convergence of catabolic pathways and divergence of anabolic pathways. Catabolism begins with many diverse cell components but ends in a final common pathway, with only a few end products.

Biosynthetic (Anabolic) Pathways Diverge to Yield Many Products

Anabolism, or biosynthesis, also takes place in three stages, beginning with small precursor molecules. For example, protein synthesis begins with the formation of α-keto acids and other precursors. In the next stage the α-keto acids are aminated by amino-group donors to form α-amino acids. In the final stage of anabolism the amino acids are assembled into polypeptide chains to form many different proteins. Similarly, acetyl groups are built up into fatty acids, and they in turn are assembled to form various lipids. Just as catabolism is a converging process, anabolism is a diverging one, since it begins with a few simple precursor molecules from which a large variety of different macromolecules are made (Figure 13-7). The central pathways of anabolism thus have many branches leading to hundreds of different cell components.

Each major stage in the catabolism or anabolism of a given biomolecule is catalyzed by a multienzyme system. The sequential chemical changes taking place in each of the central routes of metabolism are virtually identical in all forms of life. For example, the catabolism of D-glucose to yield pyruvic acid is accomplished through the same chemical intermediates and through the same number of reactions in most living organisms.

There Are Important Differences between Corresponding Catabolic and Anabolic Pathways

The catabolic and the corresponding but oppositely directed anabolic pathways between a given precursor and a given product are usually not identical. They may use different reaction intermediates or different enzymatic reactions in the intermediate steps. For example, a sequence of 11 specific enzymes catalyzes the successive steps in the degradation of glucose into pyruvic acid in the liver. Although it might seem logical and economical to make glucose from pyruvic acid by simple reversal of all the enzymatic steps used in glucose breakdown, the biosynthesis of glucose in the liver proceeds differently. It uses only 9 of the 11 enzymatic steps used in glucose degradation, the other 2 steps being replaced by an entirely different set of enzyme-catalyzed reactions used only in the direction of glucose synthesis. Similarly, the corresponding and opposite catabolic and anabolic pathways between proteins and amino acids, for example, or between fatty acids and acetyl-CoA, are not identical.

It may seem wasteful to have two metabolic pathways between two given points, one for catabolism and one for anabolism, but there are important reasons for corresponding catabolic and anabolic routes to be different. The first is that the pathway taken in the degradation of a biomolecule may be energetically impossible for its biosynthesis. Degradation of a complex organic molecule is usually a "downhill" process, proceeding with loss of free energy, whereas biosynthesis is an "uphill" process, requiring input of energy. To take a simple example (Figure 13-8), a boulder dislodged at the top of a hill will roll downhill, losing energy as it goes. At one or more points in its descent, the boulder may lose a large amount of energy all at once, as it falls over very steep places. A tractor may not be able to haul the boulder back up the hill by the path of its descent, but may succeed if it bypasses the steep portions by following a more gradual route (Figure 13-8). Thus biosynthetic routes usually require input of extra energy to get around steep points in the energy "hill."

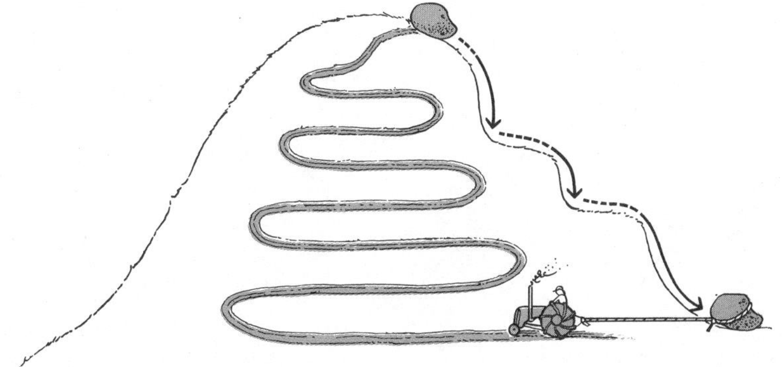

Figure 13-8
The hill-and-boulder analogy. Catabolism is a downhill process and proceeds with a loss of free energy. The energy losses are especially great at the points where the boulder falls most precipitously (solid arrows). Anabolism is an uphill process and requires input of energy, which can be provided only in small, fixed amounts. Thus the tractor must take a more gradual route to the top of the hill, to bypass the steep points where the energy requirement is particularly large.

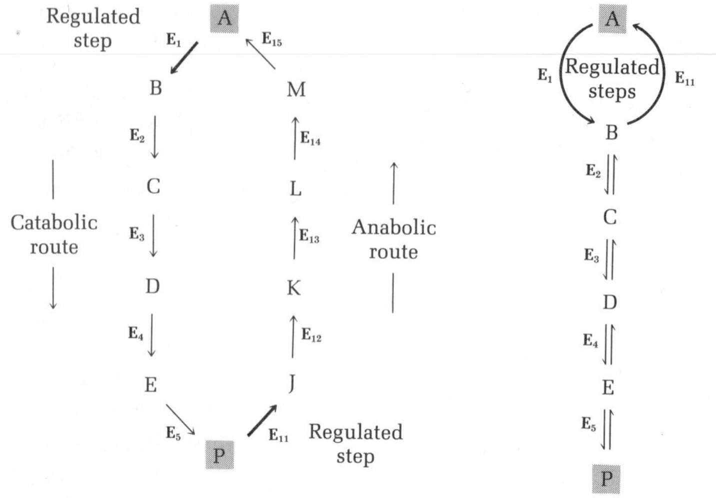

Figure 13-9
Parallel catabolic and anabolic routes must differ in at least one enzymatic step, so that they can be regulated independently. Shown are two examples of independent regulation of catabolic and anabolic routes between A and P. In (a) the parallel routes proceed via a totally different set of enzymes. In (b) only one enzyme differs between the catabolic and anabolic routes. In both cases the regulated steps are designated by colored arrows.

There is a second reason for different pathways in corresponding catabolic and anabolic routes: they must be independently regulated. If only one pathway were used reversibly in both directions, slowing down the catabolic pathway by inhibiting one of its enzymes would also slow down the corresponding biosynthetic process. Independent regulation is possible only if the opposite pathways are completely different or, in the case when they have some common enzymatic steps, the rate-controlling enzymes must occur in those portions not shared by the opposing pathways (Figure 13-9).

Sometimes opposite catabolic and anabolic routes take place in different parts of the cell. For example, the oxidation of fatty acids to the stage of acetyl-CoA in the liver takes place by the action of a set of enzymes largely localized in the mitochondria, where oxidative events are favored, whereas the biosynthesis of fatty acids from acetyl-CoA, which requires input of hydrogen atoms, i.e., reducing power, takes place by a completely different set of enzymes located in the cytosol, where reducing reactions are favored (Figure 13-10).

Figure 13-10
Separate compartmentation for chemically opposite metabolic pathways. Fatty acid oxidation takes place largely in the mitochondria, whereas fatty acid synthesis, which requires reducing power, takes place in the cytosol.

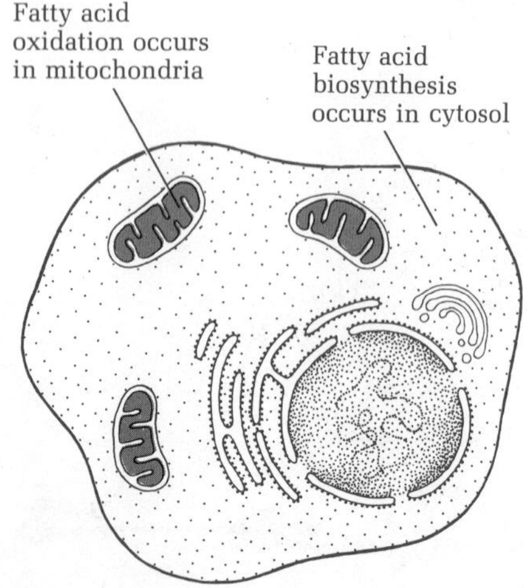

Although the corresponding pathways of catabolism and anabolism are not identical, stage III of catabolism, which consists of the citric acid cycle and some auxiliary enzymes (Figure 13-6), acts as a central meeting ground accessible to *both* catabolic and anabolic routes. This is sometimes called the *amphibolic* stage of metabolism; it has a *dual* function (Greek *amphi*, "both"). Stage III is used catabolically to complete the degradation of small molecules derived from stage II of catabolism, but it is also used anabolically, to furnish small molecules as precursors for biosynthesis of amino acids, fatty acids, and carbohydrates, as we shall see later.

Nearly all the reactions of metabolism are ultimately linked to each other because the product of one enzymatic reaction becomes the substrate of the next in consecutive sequences. Indeed, we can speak of metabolism as an extremely complex network of enzyme-catalyzed reactions. If the flow of nutrients through one part of this network is diminished or perturbed, the entire network may undergo changes to counterbalance or compensate for the change. Each of the central pathways of metabolism, whether catabolic or anabolic, can adjust its rate according to the needs of the moment in the cell economy. Moreover, it also appears that the reactions of catabolism and anabolism are adjusted to take place as economically as possible, with the least wasteful use of both matter and energy. For example, cells oxidize their nutrients at rates just sufficient to supply their energy needs at any given moment.

ATP Carries Energy from Catabolic to Anabolic Reactions

We have seen in outline how organic nutrients are transformed by enzyme action during metabolism. Now let us see how energy is transformed. Complex nutrient molecules such as glucose contain much potential energy because of their high degree of structural order. When the glucose molecule is degraded by oxidation to form the simple, small end products CO_2 and H_2O, much free energy becomes available. *Free energy* is that form of energy capable of doing work under conditions of constant temperature and pressure. However, unless there is some way of capturing or conserving the free energy released when glucose is oxidized, it will simply appear as heat. Although heat energy is useful in maintaining body temperature in higher animals, it cannot be used to do the mechanical work of muscle contraction or the chemical work of biosynthesis. Heat can do work at constant pressure only when it can flow from a warmer to a cooler body, which is impossible in living cells since they are *isothermal:* the temperature is the same in all their parts. Instead, much of the free energy released from glucose and other cellular fuels during their catabolism is conserved by the coupled synthesis of *adenosine triphosphate* (ATP) from *adenosine diphosphate* (ADP) (Figure 13-11) and *inorganic phosphate*. ATP, ADP, and phosphate are present in

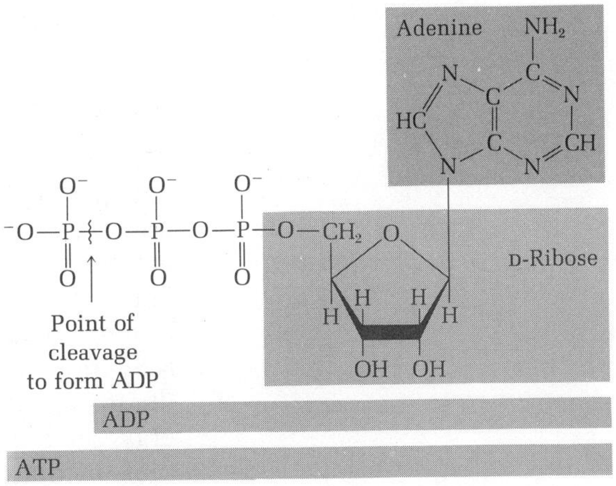

Figure 13-11
Adenosine triphosphate (ATP) in its ionized form at pH 7.0. Adenosine diphosphate (ADP) has one less phosphate group. The organic building blocks of the ATP molecule are the purine base adenine and the sugar D-ribose (discussed further in Chapter 14).

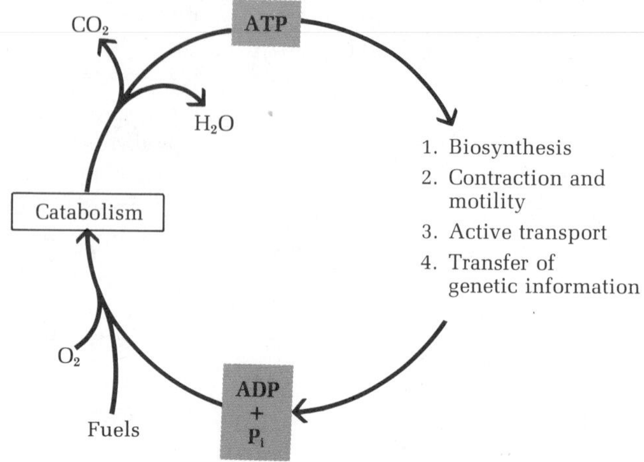

Figure 13-12
Energy-dependent activities of cells depend upon transmission of energy by ATP, which is broken down to ADP and phosphate. ATP is regenerated at the expense of energy yielded by catabolism of cell fuels.

all living cells and serve universally as an energy-transmitting system. The chemical energy so conserved in the form of ATP can do work of four different kinds (Figure 13-12). (1) It can provide the energy required for the chemical work of biosynthesis. In this process the terminal phosphate group or groups of ATP are transferred enzymatically to precursor building-block molecules which thus become "energized" and prepared for their assembly into macromolecules. (2) ATP is also the energy source for cell motility or contraction and (3) for the transport of nutrients through membranes against concentration gradients. (4) ATP energy is also used in subtle ways to ensure accurate genetic-information transfer during the biosynthesis of DNA, RNA, and proteins; indeed, information itself is a form of energy. Whenever the chemical energy of ATP is used to do cell work, its terminal phosphate group is lost (Figure 13-11) and appears as inorganic phosphate, leaving ADP, the discharged form of the energy-carrying system. The ADP can then be recharged with a phosphate group, thus regenerating ATP, in reactions that are coupled to the energy-yielding degradation of cell fuels. Thus we have an energy cycle in cells, in which ATP serves as the energy-carrying link between energy-yielding and energy-requiring cellular processes.

NADPH Carries Energy in the Form of Reducing Power

A second way of carrying chemical energy from reactions of catabolism to the energy-requiring reactions of biosynthesis is in the form of hydrogen atoms or electrons. When glucose is formed from carbon dioxide during photosynthesis, or when fatty acids are made from acetate in the liver of an animal, reducing power in the form of hydrogen atoms is required for the reduction of double bonds to single bonds. To be effective as reducing agents hydrogen atoms must have considerable free energy. Such high-energy hydrogen atoms are obtained from cell fuels by dehydrogenases, which catalyze removal of hydrogen

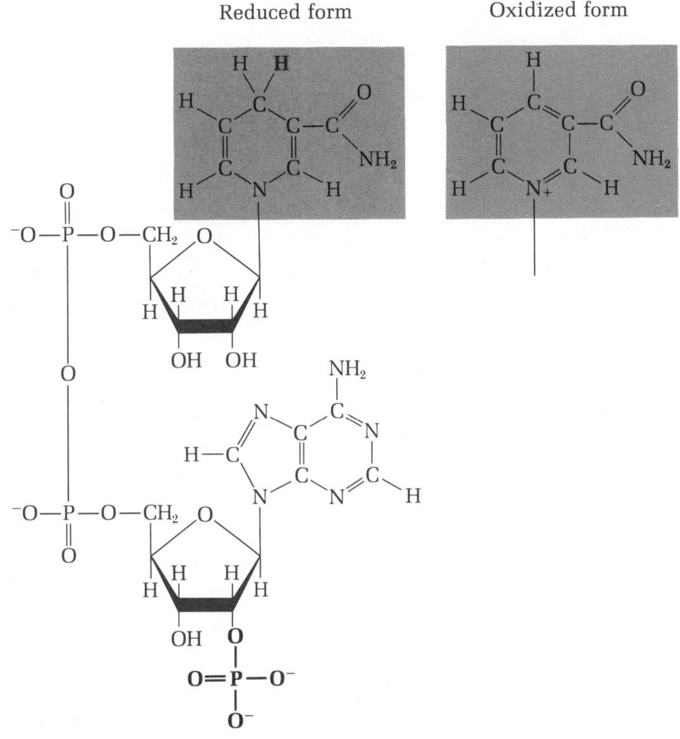

Figure 13-13
Nicotinamide adenine dinucleotide phosphate in its reduced form NADPH. The nicotinamide ring, which carries the energy-rich hydrogen atom and electrons, is shaded in color. The oxidized form of the nicotinamide portion is also shown (see also pages 255, 266).

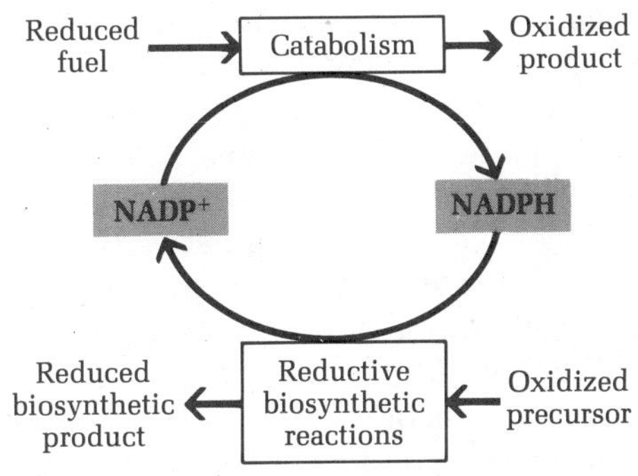

Figure 13-14
Transfer of reducing power from catabolic to biosynthetic reactions via the NADP cycle.

atoms from fuel molecules and their transfer to specific coenzymes, particularly to the oxidized form of nicotinamide adenine dinucleotide phosphate ($NADP^+$) (Figure 13-13). The reduced, or hydrogen-carrying, form of this coenzyme, designated NADPH, is a carrier of energy-rich electrons from catabolic reactions to electron-requiring biosynthetic reactions (Figure 13-14), just as ATP is a carrier of energy-rich phosphate groups.

Cell Metabolism Is an Economical, Tightly Regulated Process

Cell metabolism appears to operate according to the principle of maximum economy. The overall rate of energy-yielding catabolism is controlled by the needs of the cell for energy in the form of ATP and NADPH, not simply by the availability or concentration of cell fuels. Thus cells consume just enough nutrients to meet the rate of energy utilization at any given time. Similarly, the rate of biosynthesis of building-block molecules and of cell macromolecules is also adjusted to immediate needs. For example, growing cells synthesize each of the 20 different amino acids at just the rate and in just the proportions that suffice for the assembly of only the new proteins required at a given moment, so that none of the 20 amino acids are overproduced and left over. Many animals and plants can store energy-supplying and carbon-supplying nutrients, such as fat and carbohydrate, but they generally cannot store proteins, nucleic acids, or simple building-block biomolecules, which are made only when needed and in the amounts required. Plant seeds and egg cells are exceptions; frequently they contain large amounts of stored proteins, which are the source of amino acids required in the growth of embryos.

Catabolic pathways are very sensitive and responsive to changes in energy needs. For example, when a housefly takes off into full flight, its rate of oxygen and fuel consumption increases a hundredfold within less than a second, due to the sudden increase in ATP consumption required by the flight muscles. Thus the regulatory mechanisms of the central metabolic pathways, particularly those providing energy as ATP, are capable of responding to metabolic needs quickly and with great sensitivity.

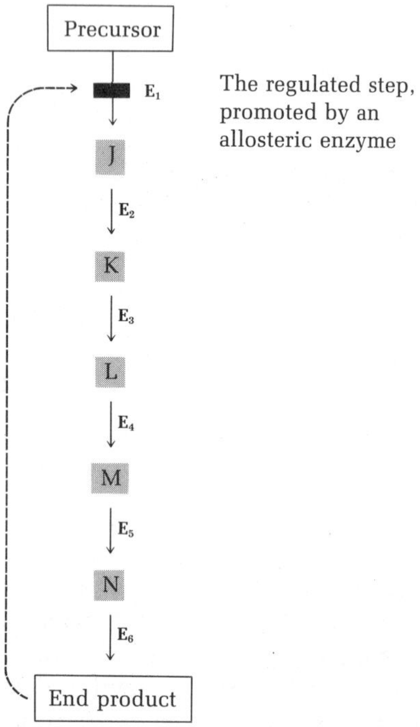

Precursor
E_1 The regulated step, promoted by an allosteric enzyme
J
E_2
K
E_3
L
E_4
M
E_5
N
E_6
End product

Figure 13-15
Regulation of a catabolic pathway by feedback, or end-product, inhibition of an allosteric enzyme. The letters J, K, L, etc. denote the chemical intermediates in the pathway, and E_1, E_2, *and* E_3, *etc. denote the enzymes acting at each step. The first enzyme in the pathway* (E_1) *is an allosteric enzyme. It is inhibited by the end product of the sequence. Allosteric inhibition is indicated, as before, by a dashed colored arrow leading from the inhibitory metabolite to the colored bar across the reaction catalyzed by the allosteric enzyme. The regulated step, catalyzed by enzyme* E_1, *is usually essentially irreversible in the cell.*

Metabolic Pathways Are Regulated at Three Levels

The regulation of metabolic pathways is brought about by three different types of mechanisms. The first and most immediately responsive form of regulation is through the action of *allosteric enzymes* (Figure 13-15), which are capable of changing their catalytic activity in response to stimulatory or inhibitory effector molecules (page 233). Allosteric enzymes are usually located at or near the beginning of a multienzyme sequence and catalyze its rate-limiting step, which is usually an essentially irreversible reaction. In catabolic pathways, which lead to the generation of ATP from ADP, the end product ATP often functions as an allosteric inhibitor of an early step in catabolism. In anabolic pathways the biosynthetic end product, such as an amino acid, often functions as the allosteric inhibitor of an early step (page 234). Some allosteric enzymes are stimulated by specific positive modulators. For example, an allosteric enzyme regulating a catabolic sequence may be stimulated by the positive modulators ADP or AMP and inhibited by the negative modulator ATP. An allosteric enzyme in a given pathway may also be specifically responsive to intermediates or products of other metabolic pathways. In this way the rates of different enzyme systems can be coordinated with each other.

Metabolic control is exerted at a second level in higher organisms by *hormonal regulation* (Figure 13-16). Hormones are chemical messengers secreted by different endocrine glands and carried by the blood to other tissues or organs, where they may stimulate or inhibit some specific metabolic activity. For example, the hormone *adrenaline*, secreted by the medulla of the adrenal gland, is carried by the blood to the liver, where it stimulates the breakdown of glycogen to glucose, thus increasing the blood sugar level. Adrenaline also stimulates the breakdown of glycogen in skeletal muscles to yield lactate and energy in the form of ATP. Adrenaline produces these effects by binding to specific *adrenaline receptor sites* on the cell surfaces in liver and muscle. Binding of adrenaline is a signal that is communicated to the interior of the cell, ultimately causing

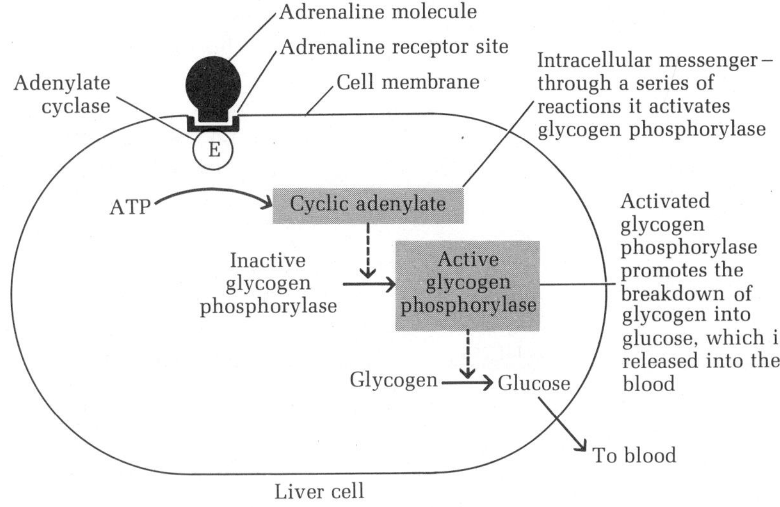

Figure 13-16
Hormonal regulation of an enzymatic reaction. Binding of the hormone adrenaline to its specific receptors on the liver cell surface promotes the formation of cyclic adenylate by membrane-bound adenylate cyclase. Cyclic adenylate is an allosteric activator, an intracellular messenger, which ultimately brings about the conversion of the inactive form of glycogen phosphorylase into its active form, thus increasing the rate of breakdown of liver glycogen into blood glucose. Details of this pathway are described in Chapter 25.

the covalent conversion of a less active to a more active form of *glycogen phosphorylase* (page 237), the first enzyme in a sequence that leads to formation of glucose and subsequent products from glycogen (Figure 13-16).

The third level at which metabolic regulation is exerted is through control of the *concentration* of a given enzyme in the cell. The concentration of an enzyme at any given time is the result of a balance between the rate of its synthesis and the rate of its degradation. The rate of synthesis of certain enzymes is greatly accelerated under some conditions so that the actual concentration of the enzyme in the cell is increased substantially. For example, when an animal is on a high-carbohydrate, low-protein diet, the enzymes of the liver that normally degrade amino acids to acetyl-CoA are present in very low concentration. Since there is little need for these enzymes as long as the animal is maintained on a low-protein diet, they are simply not made in large amounts. But when the animal is placed on a protein-rich diet, within a day its liver will show substantially increased concentrations of the enzymes required in the degradation of ingested amino acids. Thus the liver cell can turn the biosynthesis of specific enzymes on or off, depending on the nature of the incoming nutrients. This is called *enzyme induction* (Figure 13-17).

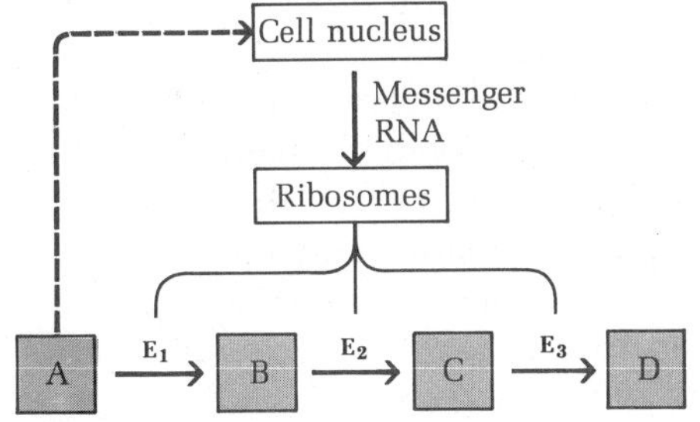

Figure 13-17
Enzyme induction. The presence of substrate A in high concentrations in a cell can induce increased biosynthesis of enzymes $\mathbf{E}_1$, $\mathbf{E}_2$, *and* $\mathbf{E}_3$, *thus increasing their concentration in the cell, allowing the reaction sequence to proceed at a higher rate to remove or transform the excess substrate A. Excess A thus serves as a signal to the nucleus to "turn on" the genes specifying enzymes* $\mathbf{E}_1$, $\mathbf{E}_2$, *and* $\mathbf{E}_3$. *The resulting messenger RNA enables the ribosomes to increase the rate of synthesis of* $\mathbf{E}_1$, $\mathbf{E}_2$, *and* $\mathbf{E}_3$.

Secondary Metabolism

Up to this point we have been speaking largely of the *central* metabolic pathways in which the *bulk* nutrients of cells, carbohydrates, fats, and proteins, are transformed. In these central pathways the flow of metabolites is relatively large. For example, several hundred grams of glucose are oxidized to CO_2 and H_2O each day by an adult human being. But there are other metabolic pathways in which the flow is much smaller, involving the formation or degradation of substances in terms of only milligrams per day. These pathways constitute the *secondary metabolism* of cells, involving specialized products that are required by cells in only small amounts. Such secondary metabolic pathways are involved in the biosynthesis of coenzymes and hormones, for example, which are made and used in only trace amounts. The secondary pathways of metabolism in different forms of life lead to hundreds of highly specialized biomolecules, e.g., nucleotides, pigments, toxins, antibiotics, and alkaloids. While such products are very important to the life of the organisms that make them and each has a specific biological purpose, they are made by specialized secondary pathways that are not always known in detail. In this book we cannot examine these specialized pathways, which can be quite complex, and shall concern ourselves mainly with the *central* or *primary* pathways of metabolism.

There Are Three Main Approaches to Identification of a Metabolic Sequence

Three major approaches are used, singly or in combination, to work out the chemical details of a metabolic pathway. The first and most direct is to study the pathway in vitro (Latin, "in glass," i.e., in the test tube) in a cell-free extract of a tissue capable of catalyzing the overall metabolic process in question. For example, it has been known since the mid-1800s that yeast converts glucose into ethanol and CO_2. The real beginning of our knowledge of the steps in this metabolic pathway, which provides anaerobic yeast cells with nearly all their energy requirement, was the discovery by Eduard Buchner in 1898 that the fermentation of glucose to yield ethanol and CO_2 can occur in a cell-free extract of yeast (page 208). Subsequently it was found that the breakdown of glucose in such extracts required the addition of inorganic phosphate, which disappeared from the extract as glucose was consumed. It was then found that a phosphorylated derivative of a hexose accumulated in the medium, which had all the expected properties of an intermediate in the conversion of glucose into ethanol and CO_2. Once this intermediate was identified, an enzyme was found in the yeast extract that acted upon it to form another product, also ultimately isolated and identified. Thus two intermediates of glucose breakdown were identified. Addition of enzyme inhibitors to yeast extracts caused other intermediates to accumulate. By a combination of such approaches the 11 metabolites that are the intermediates in the conversion of glucose into ethanol in yeast were ultimately isolated and identified. Each of the 11 enzymes involved in this sequence has been isolated and purified. Many other metabolic pathways have been worked out by this direct approach, in which the successive intermediates and the enzymes forming and degrading them are identified, one by one. When the entire sequence is known, it can be reconstituted in the test tube from purified components.

Mutants of Organisms Allow Identification of Intermediate Steps in Metabolism

Another important approach to elucidation of metabolic pathways is the study of genetic mutations of organisms in which a given enzyme fails to be synthesized in active form. Such a defect, if not lethal, may result in the accumulation and excretion of the substrate of the defective enzyme. From genetic defects in specific enzymes occurring in people it has been possible, for example, to deduce the nature of certain steps in the metabolism of amino acids (Figure 13-18). Such human genetic disorders are rare and do not lend themselves to systematic investigation. However, genetic defects in metabolic pathways can be produced at will in microorganisms by subjecting them to mutagenic agencies, e.g., irradiation with x-rays or treatment with certain chemicals, which can alter the structure of specific genes in their DNA. Such mutant microorganisms, in which one enzyme or another is defective, are powerful tools for study of metabolism.

Let us see how such mutants can be used. Normal or unmutated cells of the common mold *Neurospora crassa* (Figure 13-19) can grow on a simple medium containing glucose as the

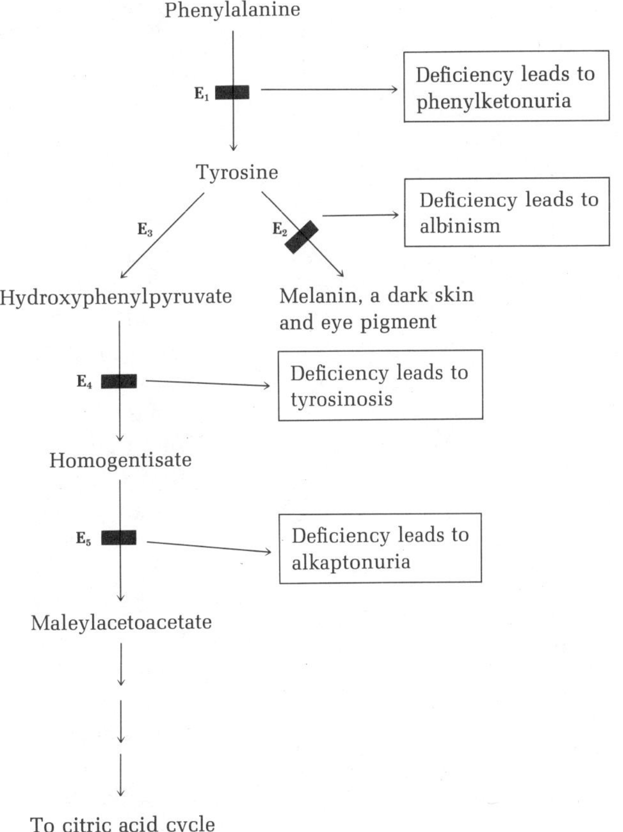

Figure 13-18
Some single-gene genetic deficiencies found in the metabolism of the amino acid phenylalanine in some human beings. Recognition of these defects provided evidence for the nature of the intermediates in phenylalanine metabolism.

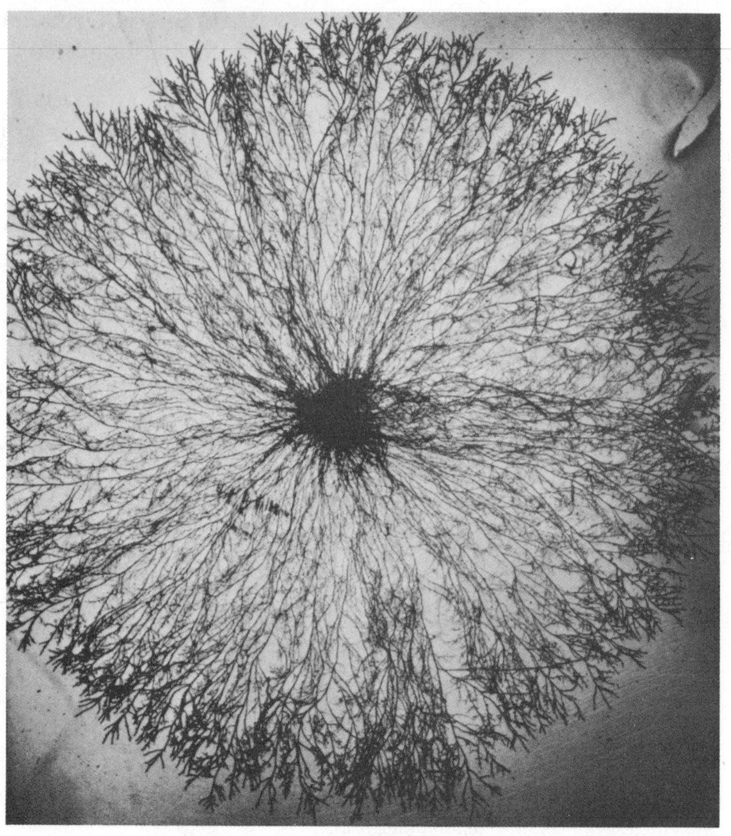

Figure 13-19
The vegetative form, or mycelium, of the bread mold Neurospora crassa. *Mutants of this organism are easily produced and have been very useful in tracing some metabolic pathways. Use of these mutants led to the one gene–one enzyme concept.*

sole carbon source and ammonia as the sole nitrogen source. However, if *Neurospora* spores are irradiated with x-rays, some mutant cells will arise that are no longer capable of growing on this simple medium but will grow if the medium is supplemented with a specific metabolite. For example, certain *Neurospora* mutants will grow normally if the basal medium is supplemented with the amino acid arginine, which is not required by unmutated cells. In such mutants one of the enzymes required in the synthesis of arginine from ammonia is evidently defective or lacking. For lack of arginine these cells cannot manufacture proteins containing arginine residues and thus they fail to grow. But when arginine is supplied in the culture medium, along with glucose and ammonia, the mutant cells grow readily. Such mutants defective in a biosynthetic pathway, whose growth can be restored by providing them with the normal product of the pathway, are called *auxotrophic mutants* (the Greek prefix *auxo-* refers to an increase, i.e., an increased requirement for growth).

Mutants of *Neurospora* defective in the ability to make arginine are not all identical; they differ with respect to the specific step in arginine biosynthesis that is defective (Figure 13-20). Such a set of "arginine-less" mutants can be used to identify intermediates in the sequence of enzymatic reactions leading to the synthesis of arginine. When mutant I (Figure

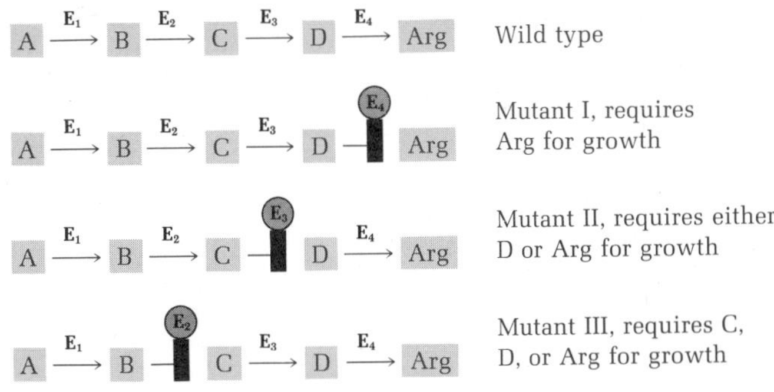

Figure 13-20
Auxotrophic mutants of Neurospora crassa *with defective enzymes (color screen) at different points in the biosynthesis of arginine (Arg) from precursor* A, B, C, *and* D *are intermediates in this conversion. Mutant I is deficient in* $\mathbf{E_4}$ *but will grow in a medium enriched in arginine. Under these conditions intermediate* D *accumulates in the culture medium. Mutant II is deficient in enzyme* $\mathbf{E_3}$ *but retains* $\mathbf{E_4}$ *activity. Thus it can grow if the medium is supplemented with either arginine or intermediate* D, *which can be made by mutant I. Similarly, mutant III, deficient in* $\mathbf{E_2}$, *will grow if supplemented with either* C, D, *or arginine, since it can make arginine if supplied with either intermediate* C *or* D.

13-20) is grown in the presence of very small, limiting amounts of arginine, the cells will grow until all the arginine is used up in the synthesis of cell protein. Simultaneously, the blocked precursor D will accumulate in the culture medium since it cannot be converted into arginine. If we now remove the cells of mutant I from their culture medium by filtration and replace them with cells of a different arginine-less mutant, designated mutant II, which cannot make intermediate D, the culture medium from mutant I will support the growth of mutant II since it supplies the precursor D. The filtered medium of mutant II, however, will not support the growth of mutant I cells. The precursor generated by mutant I can therefore be assayed by using the rate of growth of mutant II. In this way, precursors A, B, C, and D of arginine were ultimately identified. The pathway of biosynthesis of many of the amino acids was originally worked out by such cross-feeding procedures, employing auxotrophic mutants of *Neurospora crassa* or *E. coli.*

Isotopic Tracers Provide a Powerful Method of Studying Metabolism

Another powerful method for establishing the general outline of a metabolic pathway is to use an isotopic form of an element (Table 13-1) to label a given metabolite. For example, the radioactive isotope ^{14}C (the normal or average atomic weight of carbon is 12.01) is frequently used to label a specific carbon atom in an organic molecule. Such a ^{14}C-labeled molecule is chemically indistinguishable from a normal unlabeled molecule but can easily be detected and measured through its radioactivity. For example, acetic acid can be synthesized in the laboratory in such a way that its carboxyl carbon atom is enriched in ^{14}C, which otherwise occurs in only extremely small and constant amounts in the carbon compounds found in the biosphere and geosphere. When such a sample of radioactively labeled acetate is fed to an animal, its metabolic fate can readily be traced. For example, the respiratory CO_2 exhaled by the animal will be found to contain ^{14}C, indicating that some of the acetate is metabolized in such a way that its carboxyl carbon atom is converted into CO_2. Moreover, if palmitic acid is subse-

Table 13-1 Some Isotopes Used as Tracers

Element	Average atomic weight	Isotope useful as tracer	Type of isotope	Half-life
H	1.01	^{2}H	Stable	
		^{3}H	Radioactive	12.1 y
C	12.01	^{13}C	Stable	
		^{14}C	Radioactive	5700 y
N	14.01	^{15}N	Stable	
O	16.00	^{18}O	Stable	
Na	22.99	^{24}Na	Radioactive	15 h
P	30.97	^{32}P	Radioactive	14.3 d
S	32.06	^{35}S	Radioactive	87.1 d
K	39.10	^{42}K	Radioactive	12.5 h
Fe	55.85	^{59}Fe	Radioactive	45 d
I	126.90	^{131}I	Radioactive	8 d

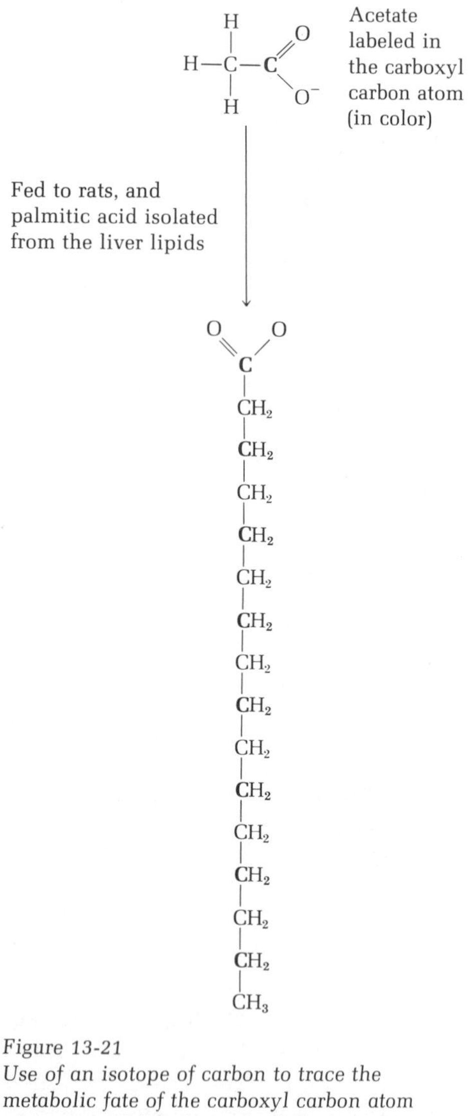

Acetate labeled in the carboxyl carbon atom (in color)

H_3C-COO^-

Fed to rats, and palmitic acid isolated from the liver lipids

$^-OOC-(CH_2)_{14}-CH_3$

Figure 13-21
Use of an isotope of carbon to trace the metabolic fate of the carboxyl carbon atom of acetate. Although much of the isotopic carbon of labeled acetate appears as exhaled CO_2, a significant amount is found in the palmitic acid of the liver lipids. Since the isotope is found only in alternating carbon atoms, as shown, this experiment tells us that palmitic acid is made by joining eight molecules of acetate in a head-to-tail fashion.

quently isolated from the liver lipids of the animal, it will also be found to contain ^{14}C, indicating that the carboxyl carbon atom of acetate is a biosynthetic precursor of palmitic acid. Furthermore, when the labeled palmitic acid molecule is chemically degraded, it will be found to have excess ^{14}C in only the alternate carbon atoms beginning with the carboxyl carbon (Figure 13-21). However, if acetate containing ^{14}C only in the methyl group is fed, again palmitic acid will be labeled, but in alternate carbon atoms starting from the α, or 2, carbon atom. These observations led to the conclusion that all the carbons of palmitic acid ultimately derive from acetate molecules, in a pathway that results in a head-to-tail linkage of the carbon skeleton of acetate molecules.

The isotope tracer method can also be used to determine the rate of metabolic processes in intact organisms. One of the most significant advances made with this powerful method is the discovery that the macromolecular components of cells and tissues undergo constant *metabolic turnover;* i.e., they exist in a dynamic steady state in the cell, in which constant biosynthesis is exactly counterbalanced by an equal rate of degradation. For example, isotopic measurements have shown that the proteins of rat liver have a half-life of about 5 to 6 d (Table 13-2). On the other hand, the proteins of skeletal-muscle tissue or the brain turn over much more slowly.

The isotope tracer technique has made possible an extraordinary range of important metabolic observations.

Metabolic Pathways Are Compartmented in Cells

We have seen in Chapter 1 that there are two great classes of cells, prokaryotic and eukaryotic, which differ greatly in size, internal structure, and genetic and metabolic organization. Pro-

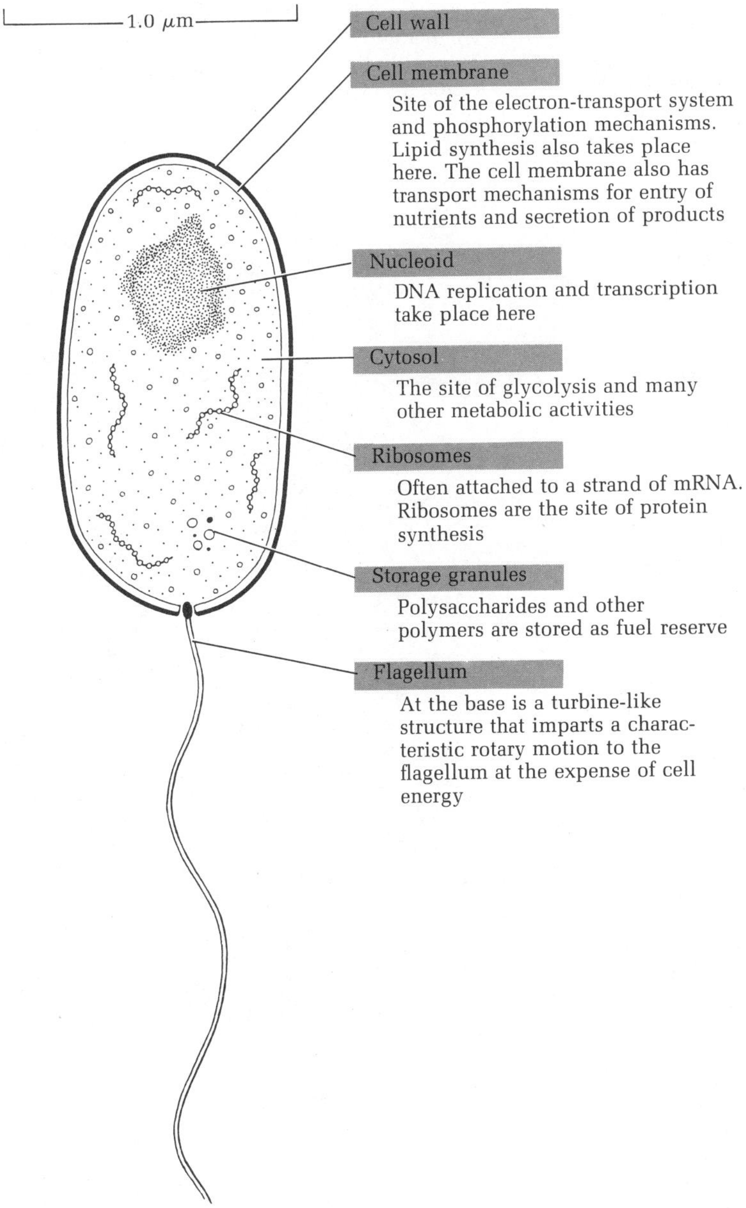

Figure 13-22
Localization of some metabolic activities in a bacterial cell.

Table 13-2 Metabolic Turnover of Some Components of Rat Tissues, from an Early Study with Isotopic Carbon

Tissue	Half-life, d
Liver	
Total protein	5.0–6.0
Glycogen	0.5–1.0
Phosphoglycerides	1–2
Triacylglycerols	1–2
Cholesterol	5–7
Mitochondrial proteins	9.7
Muscle	
Total protein	~50
Glycogen	0.5–1.0
Brain	
Triacylglycerols	10–15
Phospholipid	200
Cholesterol	>100

karyotic cells, which include the bacteria and blue-green algae, are very small, simple cells with only a single membrane system, that surrounding the cell.

Prokaryotic cells contain no compartments separated by internal membranes, yet there is some degree of segregation of certain enzyme systems in bacteria (Figure 13-22). For example, most of the enzymes participating in biosynthesis of proteins are located in the ribosomes, and some of the enzymes participating in the biosynthesis of phospholipids are located in the bacterial cell membrane.

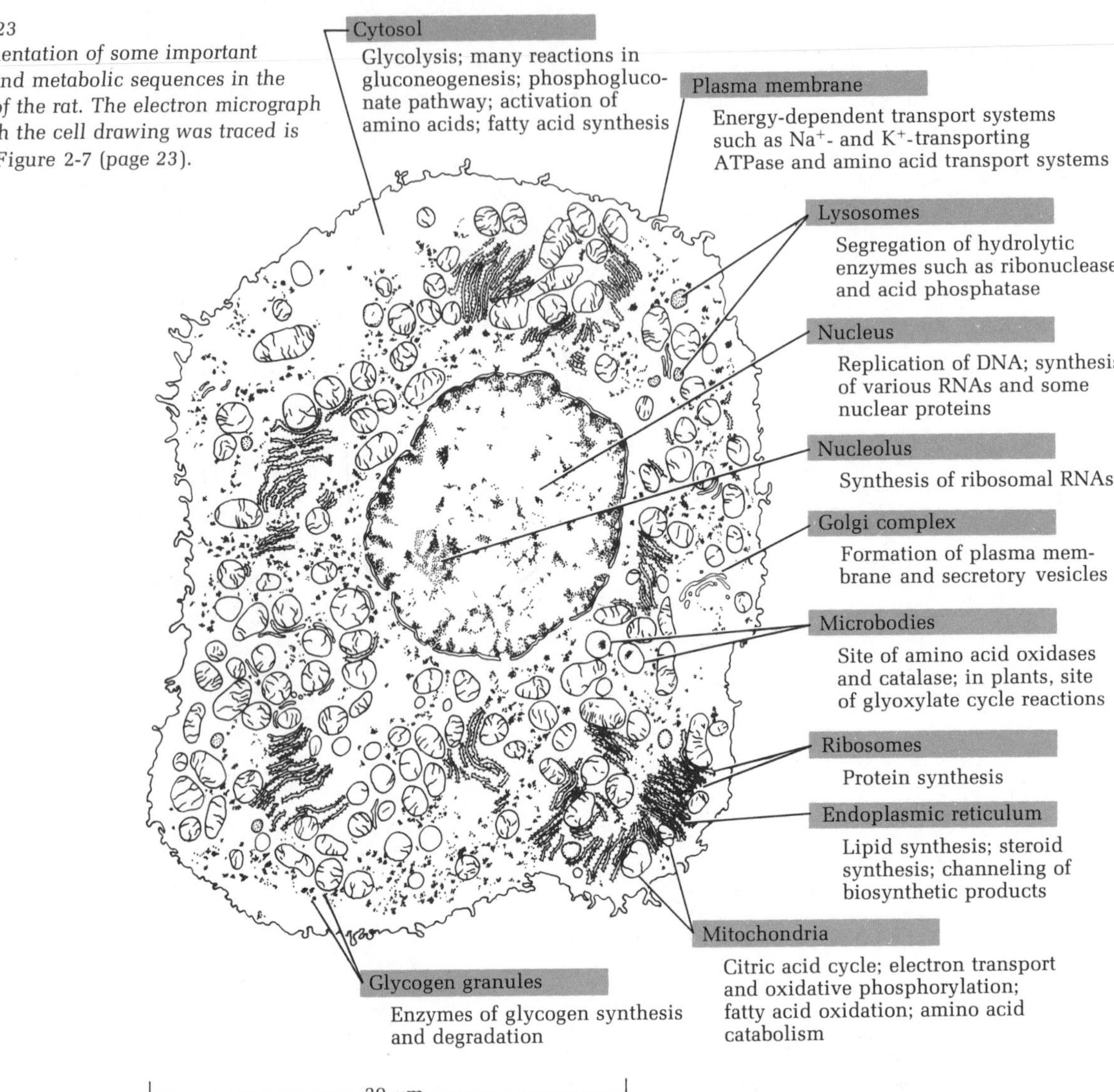

Figure 13-23
Compartmentation of some important enzymes and metabolic sequences in the liver cell of the rat. The electron micrograph from which the cell drawing was traced is shown in Figure 2-7 (page 23).

Eukaryotic cells, which include those of higher animals and plants as well as the fungi, the protozoa, and higher algae, are much larger and much more complex than prokaryotic cells (Figure 13-23). In addition to the cell envelope, eukaryotic cells have a distinct membrane-surrounded nucleus containing several or many chromosomes. They also contain other membranous internal organelles (mitochondria, endoplasmic reticulum, Golgi bodies, and in green plant cells, chloroplasts). In eukaryotic cells the enzymes catalyzing metabolic pathways are often localized in a specific organelle or compartment. How do we know? It is possible to isolate cell organelles from cells or tissues by a centrifugation procedure (Figure 13-24). Animal or plant tissues are first gently homogenized in an isotonic sucrose medium, a process that ruptures the plasma membrane

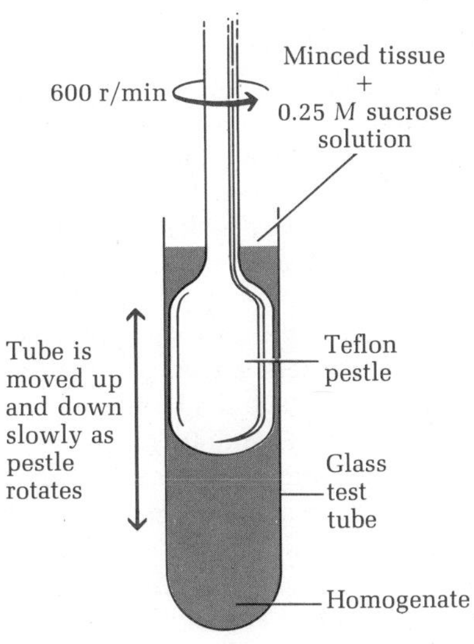

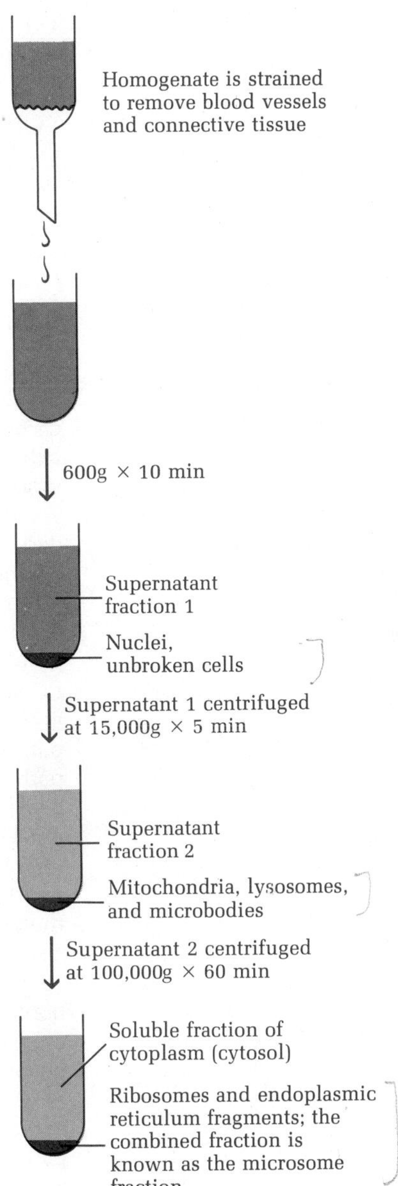

Figure 13-24
Fractionation of a cell extract by differential centrifugation. The cell membrane is ruptured by the shearing forces developed by the rotating homogenizer pestle. Following removal of connective tissue and fragments of blood vessels by a stainless-steel sieve, the cell extract is centrifuged at a series of increasing rotor speeds.

but leaves most of the internal organelles intact. Sucrose is used because it does not pass through membranes readily and thus does not cause internal organelles such as chloroplasts and mitochondria to swell. The subcellular organelles, e.g., nuclei and mitochondria, which differ in size and specific gravity and thus sediment at different rates in a centrifugal field, can then be isolated from the homogenate by *differential centrifugation* (Figure 13-24). The nuclei, mitochondria, and other fractions obtained in this way can be tested for their ability to catalyze a given metabolic sequence. From this approach it has been found that different metabolic pathways take place in different intracellular locations in eukaryotic cells (Figure 13-23). For example, the entire sequence of enzymes concerned in the conversion of glucose into lactate in some cells is located in the *cytosol*, the soluble portion of the cell cytoplasm, whereas the enzymes of the citric acid cycle are located in the *mitochondria*, as are the enzymes of electron transport and those concerned in conservation of oxidative energy as ATP.

Summary

Organisms can be classified on the basis of their carbon requirements. Autotrophs can utilize carbon dioxide, whereas heterotrophs require carbon in a reduced organic form, such as glucose. Many autotrophic cells, e.g., those of green plants, obtain their energy from sunlight; heterotrophs obtain energy from the oxidation of organic nutrients.

Intermediary metabolism can be divided into catabolism, the degradation of energy-rich nutrient molecules, and anabolism, the biosynthesis of new cellular components. Catabolism and anabolism occur in three major stages. In the first stage of catabolism polysaccharides, lipids, and proteins are enzymatically degraded to their building blocks, in the second stage the building blocks are oxidized to acetyl-CoA as major product, and in the third stage the acetyl groups of acetyl-CoA are oxidized to carbon dioxide. Catabolic pathways converge into a final common pathway, whereas anabolic pathways diverge to yield many different biosynthetic products from a few precursors. Corresponding anabolic and catabolic pathways are not enzymatically identical; they are regulated differently; and they are often located in different parts of the cell. Catabolism of nutrient molecules is accompanied by conservation of some of the energy of the nutrient in the form of adenosine triphosphate (ATP). ATP serves as a carrier of chemical energy from catabolic reactions to cellular processes requiring energy: biosynthesis, contraction or motion, membrane transport, and the transfer of genetic information. Chemical energy in the form of reducing power is also carried from catabolic to anabolic pathways as the reduced coenzyme NADPH.

Metabolism is regulated (1) by allosteric enzymes, (2) by hormonal control, and (3) by regulation of enzyme synthesis. Metabolic pathways are studied in extracts of cells or tissues, from which the component enzymes and metabolic intermediates can be isolated. Microorganisms that are genetically defective in a given pathway (auxotrophs) provide a powerful tool for analyzing metabolic pathways, as does the isotope tracer technique. In eukaryotic cells, the enzymes catalyzing various pathways of metabolism are compartmented in different organelles, e.g., nuclei, mitochondria, and endoplasmic reticulum, which can be isolated for direct study.

References

Books

COLOWICK, S. P., and N. O. KAPLAN (eds.): *Methods in Enzymology*, Academic, New York, 1955–. Continuing series of reference volumes on all aspects of enzymology and metabolism and their measurement.

DAGLEY, S., and D. E. NICHOLSON: *Introduction to Metabolic Pathways*, Wiley, New York, 1970. A compendium of metabolic "maps."

ROODYN, D. B. (ed.): *Subcellular Biochemistry*, vols. I–VII, Plenum, New York, 1972–1980. A series of reference volumes with articles on the compartmentation of biochemical activities within cells.

STANIER, R. Y., M. DOUDOROFF, and E. A. ADELBERG: *The Microbial World*, 4th ed., Prentice-Hall, Englewood Cliffs, N.J., 1976. An important textbook of general microbiology. It considers the various subdivisions of autotrophic and heterotrophic organisms and also discusses the large contribution of microorganisms to the total metabolism of the world's biomass.

Articles

The entire September 1970 issue (vol. 223, no. 3) of *Scientific American* is devoted to the biosphere. The following articles in this issue are especially recommended:

PENMAN, H. L.: "The Water Cycle," p. 98.

CLOUD, P., and A. GIBOR: "The Oxygen Cycle," p. 110.

BOLIN, B.: "The Carbon Cycle," p. 124.

DELWICHE, C. C.: "The Nitrogen Cycle," p. 136.

BROWN, L. R.: "Human Food Production as a Process in the Biosphere," p. 160.

HALL, D. O.: "Photobiological Energy Conversion," *FEBS Lett.*, **64**:6–16 (1976). A general article relating photosynthesis to food and energy needs of mankind.

WOODWELL, G. M.: "The Carbon Dioxide Question," *Sci. Am.*, **238**: 34–43, January (1978). Discusses the possible effects and fate of the increasing CO_2 content of the atmosphere.

Problems

1. *Aspects of a Metabolic Pathway* The accounting process involved in a metabolic pathway is very similar to balancing a checkbook. In the latter case, one keeps a record of the transactions (deposits and withdrawals), and at the end of the month the balance is determined. Similarly, a metabolic pathway consists of a sequence of chemical transformations (transactions) necessary to carry out a metabolic process. Like checking accounts, chemical transformations are quantitative entities, and thus every transformation must consist of a balanced chemical equation. In intact cells such an accounting is not always easy because one or more of the intermediates in the pathway may be diverted into other metabolic routes. The enzyme-catalyzed reactions necessary for the catabolism of glyceraldehyde 3-phosphate to ethanol in yeast fermentations are given below. Note that although each chemical transformation is a balanced equation, the order of reactions is not necessarily that found in the metabolic pathway:

Glyceraldehyde 3-phosphate + P_i + NAD^+ →
3-phosphoglyceroyl phosphate + NADH + H^+
Phosphoenolpyruvate + ADP → pyruvate + ATP
Ethanol + NAD^+ → acetaldehyde + NADH + H^+
3-Phosphoglyceroyl phosphate + ADP → 3-phosphoglycerate + ATP
2-Phosphoglycerate ⇌ 3-phosphoglycerate
2-Phosphoglycerate → phosphoenolpyruvate + H_2O
Pyruvate → CO_2 + acetaldehyde

(a) Use all these word equations and your knowledge of the chemical structure of each intermediate to deduce and write the sequence of chemical transformations (the metabolic pathway) involved in the catabolism of glyceraldehyde 3-phosphate to ethanol.
(b) What is the *net* equation for the overall process? That is, do the equivalent of balancing a checkbook by accounting for all inputs and outputs.
(c) Write the metabolic pathway in flow-chart form, as in Figure 13-4. Show the interrelationships between different parts of the pathway.

2. *Cyclic Metabolic Pathway* Some bacteria of the genus *Pseudomonas* can use oxalate ($^-OOC—COO^-$, a highly toxic substance to most mammals) as a fuel source. The metabolism of oxalate in such bacteria is carried out by the cyclic pathway shown schematically:

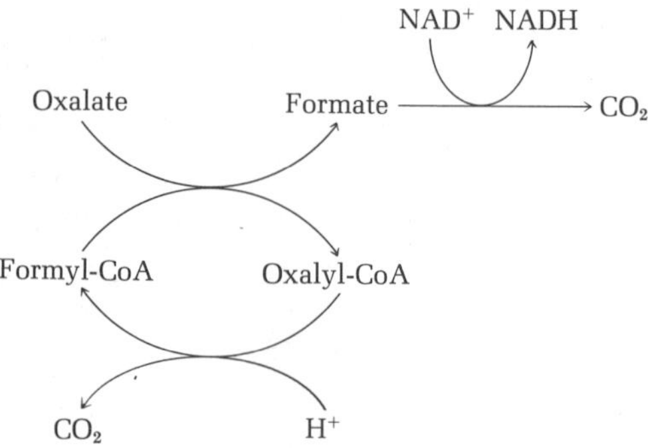

(a) Write the metabolic pathway of oxalate as a sequence of chemical transformations (balanced equations) in which the product of one transformation serves as the reactant for the next. Show structural formulas for each step.
(b) Write the balanced *net* equation for the catabolism of oxalate to CO_2, again using structural formulas.

3. *^{14}C Incorporation into Amino Acids* In 1955 R. B. Roberts and his colleagues published their studies on the growth of *E. coli* in a medium containing uniformly labeled [^{14}C]glucose as their only carbon source. The objective of their experiments was to label the cell's biomolecules with ^{14}C. They observed that growing the bacteria on [^{14}C]glucose led to the rapid incorporation of ^{14}C into all the amino acids. If the bacteria were grown on [^{14}C]glucose in a medium that included unlabeled histidine (glucose and histidine are now carbon sources), ^{14}C was incorporated into all the amino acids *except* histidine. In the absence of unlabeled histidine, why is ^{14}C incorporated into histidine? Why does the presence of unlabeled histidine prevent the formation of ^{14}C-labeled histidine? Is this a case of feedback inhibition or of enzyme repression?

4. *Regulation of Lactose Metabolism in E. coli* *E. coli* can be grown in a simple medium using lactose as the only carbon source. The hydrolysis of lactose to galactose and glucose catalyzed by the enzyme β-galactosidase is essential to the metabolism of lactose and consequently to the survival of the bacterium. When *E. coli* are grown on lactose, each bacterial cell contains several thousand molecules of β-galactosidase (see Problem 1 on page 142) to carry out this hydrolysis. In contrast, if the *E. coli* is grown using glucose or glycerol as the sole carbon source, each cell contains only five to ten β-galactosidase molecules.
   (a) How is the metabolism of lactose regulated? Explain.
   (b) If the culture medium is changed from lactose to glycerol, why does the level of β-galactosidase drop? Why doesn't the enzyme level remain constant?
   (c) When the culture medium contains methyl β-galactoside as the sole source of carbon, the cells also grow at a high rate and contain thousands of β-galactosidase molecules, but when methyl α-galactoside is the sole source of carbon, the cells grow very slowly and contain very little β-galactosidase. Explain.

5. *Comparison of Catabolic and Anabolic Pathways* The interconversion of glucose and fructose 1,6-diphosphate, a key series of steps in carbohydrate metabolism, is shown in the accompanying figure. The breakdown of glucose constitutes the catabolic route, while the biosynthesis of glucose from fructose 1,6-diphosphate corresponds to the anabolic route. Both routes use the same hexose monophosphate intermediates. Although these pathways are very similar, there are distinct differences. The purpose of this problem is to recognize them.

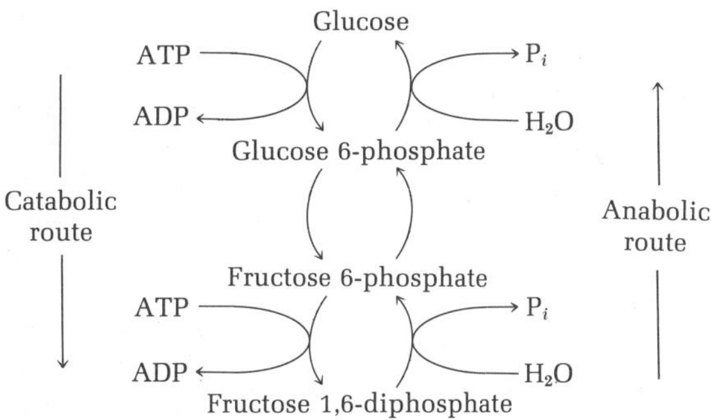

   (a) Write a balanced equation for each step in the *catabolic* route. Write the net equation that results when the individual steps are added.
   (b) Repeat part (a) for the anabolic route.
   (c) What are the distinguishing differences, based on the net reactions, between the catabolic and anabolic route? Are the two pathways simply the reverse of each other?
   (d) What assures the directionality of the catabolism of glucose, i.e., what prevents the catabolism of glucose from running backward?
   (e) Could the glucose–glucose 6-phosphate interconversion in the anabolic and catabolic routes be catalyzed by the same enzyme? Explain. Could the glucose 6-phosphate–fructose 6-phosphate interconversion in the anabolic and catabolic routes be catalyzed by the same enzyme? Explain.

6. *Measurement of Radioisotopes* Quantitative measurement of the radioisotopes commonly used in biological studies (^{3}H, ^{14}C, ^{32}P, and ^{35}S) is conveniently carried out by a technique known as liquid scintillation counting. The unit of radioactivity is the curie (Ci), the quantity of radioactivity that will produce 2.22×10^{12} radioactive disintegrations per minute (dpm). Since the measuring efficiency of liquid scintillation counters is less than 100 percent, the experimental radioactivity levels are expressed in counts per minute (cpm), and the counting efficiency becomes

$$\text{Efficiency} = \frac{\text{cpm}}{\text{dpm}} \times 100$$

Radioactive labeling is extremely useful in biochemical studies because the radioactivity levels can be used to determine the concentration of a chemical species. The specific activity of a compound is the quantity of radioactivity per unit (gram, mole, equivalent, milliliter, etc.). It is important to realize that the use of specific activity does *not* require that every molecule in the sample be labeled. All that is necessary is that the experimental counts be proportional to concentration, the specific activity and efficiency serving as the proportionality constants.

(a) A commercial supplier of radioactive compounds sells you 250 μCi of ^{14}C-glucose dissolved in 1 mL of water, with a specific activity of 500 mCi/mmol. Calculate the concentration of glucose in the supplier's 1-mL vial.

(b) If 10 μL of the supplier's sample is removed and counted with 70 percent efficiency, how many counts per minute will be observed?

7. *Measurement of the Intracellular Concentration of Methionine* When *E. coli* cells are grown on radioactive $^{35}SO_4^{2-}$ as the only sulfur source, all sulfur-containing amino acids and proteins will be labeled with ^{35}S. In a typical experiment, *E. coli* is grown in a medium containing 0.85 m*M* $^{35}SO_4^{2-}$. A 250-μL sample of this medium measured at an efficiency of 87 percent gave a total of 4.50×10^5 cpm. After maximal growth of the culture had been achieved, the cells were isolated by filtration and washed with cold water. The free (uncombined) amino acids were extracted from the cells with boiling water and separated by ion-exchange chromatography. The extraction of 1.85 g of wet cells gave a total of 3.2×10^5 cpm of L-[^{35}S]methionine measured at a counting efficiency of 82 percent. Assuming that the wet cells consist of 80 percent water and the remainder solids, calculate the intracellular concentration of free L-methionine in *E. coli*.

CHAPTER 14

The ATP Cycle and Cell Bioenergetics

Today more than ever we are aware that energy, the capacity to do work, is vital to our modern civilization. We need energy to manufacture goods, to transport materials and people, to heat homes and workplaces, and to carry out many lesser tasks. In similar ways energy is just as vital in the microcosm of a living cell. Cells constantly manufacture new substances, they perform the mechanical work of motion, they transport substances, and they generate heat. Through billions of years of evolution cells have learned to use energy more economically and efficiently than most man-made machines. Indeed, cells are now being looked to as models for the development of useful new energy-transforming devices, particularly for the capture of solar energy.

Bioenergetics is the field of biochemistry concerned with the transformation and use of energy by living cells. In this chapter we shall first consider a few basic principles of thermodynamics, the branch of physical science that deals with energy changes. Then we shall see how the ATP system operates as an efficient energy-carrying system in cells, linking energy-yielding catabolism with energy-requiring cell processes.

The First and Second Laws of Thermodynamics

Energy is familiar to us in a variety of forms, e.g., electric, mechanical, chemical, heat, and light energy. We also know that different forms of energy can be interconverted. An electric motor transforms electric into mechanical energy, a storage battery converts chemical into electric energy, and a steam engine converts heat into mechanical energy. Moreover, different forms of energy are quantitatively related to each other. For example, 1.0 calorie of heat energy is theoretically equivalent to 4.185×10^7 ergs of mechanical energy.

But we also know that when one kind of energy is converted into another, there is always some loss. When an electric motor converts electric into mechanical energy, the useful energy output is always less than the input, due to friction in the motor, which generates heat. The heat in turn is dissipated into the surrounding environment and is no longer useful. Almost always, when energy is used to do work or when one form of energy is converted into another, there is a loss of *useful* energy. Indeed, in many machines less than 25 percent of the input energy is recovered in the form of useful work. Many quantitative observations made by physicists and chemists on the interconversion of different forms of energy have led to two fundamental laws of thermodynamics, which may be stated in a very simple, approximate form as follows:

The First Law

In any physical or chemical change the total amount of energy in the universe remains constant.

The first law is the principle of the conservation of energy; paraphrased, it says that energy cannot be created or destroyed. Thus, whenever energy is used to do work or is transformed from one kind into another, the total amount of energy is unchanged.

The Second Law

All physical or chemical changes tend to proceed in such a direction that useful energy undergoes irreversible degradation into a randomized form called entropy. They come to a stop at an equilibrium point, at which the entropy formed is the maximum possible under the existing conditions.

This simple, somewhat abstract definition requires some explanation. First, we must define the terms useful energy and entropy more precisely. There are two kinds of useful energy: (1) *free energy*, the kind of energy that can do work *at constant temperature and pressure*, and (2) *heat energy*, which can do work only through a *change of temperature*. Entropy is energy in a state of randomness or disorder; it is unavailable, "useless" energy. A quantitative, rigorous definition of entropy requires mathematical statements of randomness; however, some simple examples of the formation of entropy are given in Box 14-1.

There is another aspect of the second law that is necessary for an understanding of its operation, particularly in biological systems. First, we must identify the reaction *system*, the collection of matter that is undergoing a particular chemical or physical process; this may be an animal, a cell, or two reacting compounds. Then we must define the *surroundings* with which the reacting system can exchange energy. The system plus the surroundings together constitute the *universe* (Figure 14-1), which quite literally includes the entire earth; indeed, it includes even outer space. Some chemical or physical processes can be made to take place in *isolated* systems, unable to ex-

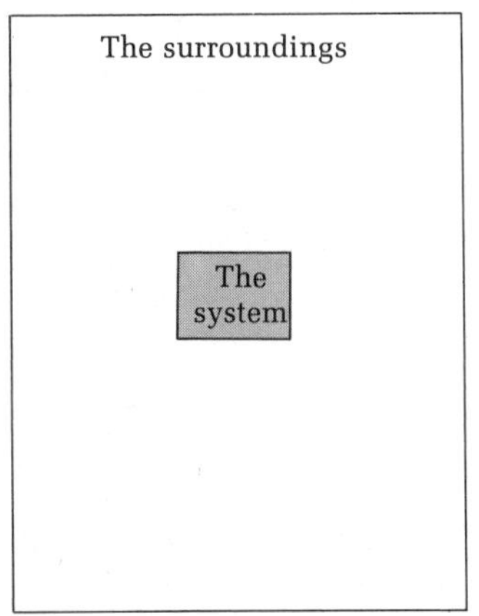

Figure 14–1
Schematic representation of a reacting system and its surroundings. Energy can be exchanged between the system and the surroundings, but such energy exchanges must conform to the laws of thermodynamics. First, the total amount of energy in the "universe," i.e., the system + surroundings, must stay constant. The second law says that when the system undergoes some physical or chemical change, the entropy of the universe tends to increase. Accompanying these changes heat may flow from the system to the surroundings or from the surroundings to the system.

In the special case where temperature and pressure of the system are constant, there is a decrease in free energy. The symbol G was chosen to represent free energy to honor Willard Gibbs, an American pioneer in thermodynamic theory.

change energy with their surroundings. But in the real world, and particularly in the biological world, reacting systems do exchange energy, and often they exchange matter, with their surroundings. We shall now see the importance of the distinction between the system and its surroundings during energy exchanges.

Changes in free energy, heat, and entropy in chemical reactions at constant temperature and pressure, the conditions existing in biological systems, are related to each other quantitatively by the equation

$$\Delta G = \Delta H - T\,\Delta S \tag{1}$$

in which ΔG is the change in free energy of the reacting *system*, ΔH is the change in its heat content or *enthalpy* (the word means "warming within"), T is the absolute temperature at which the process is taking place, and ΔS is the change in entropy *of the universe*, which includes the reacting system. When any chemical reaction proceeds to its equilibrium point, the entropy of the universe (system + surroundings) always increases and ΔS therefore always has a positive sign in the real world. Only theoretically, in an ideal system, can a reaction proceed with no increase in entropy. Corresponding to the *increase* in entropy of the *universe* during a reaction, there is a *decrease* in free energy of the reacting *system*. Therefore ΔG *of a reacting system* is negative. The enthalpy change, ΔH, is defined as the amount of heat the reacting system releases to or absorbs from its surroundings at constant temperature and pressure. When the reacting system loses (i.e., releases) heat, the sign of ΔH is negative; when the system absorbs heat from the surroundings, ΔH is positive.

There is another important characteristic of entropy changes, directly relevant to biological systems. The second law says that entropy *of the universe* increases during chemical or physical processes. It does not say that the entropy increase necessarily must take place in the reacting system itself; it may take place anywhere in the universe. Living organisms do not undergo an increase in their internal disorder or entropy when they metabolize their nutrients: we know, as a matter of observation, that a fly or an elephant (i.e., the system) retains its characteristically complex and orderly structure as it carries out its activities. Rather, it is the *surroundings* of living organisms that increase in entropy during life processes. *Living organisms preserve their internal order by taking in free energy of nutrients (or sunlight) from their surroundings and returning to their surroundings an equal amount of energy in a less useful form, largely heat, that becomes randomized throughout the rest of the universe.*

Finally, we must emphasize that the formation of entropy or disorder is not a totally useless activity. The increase in entropy of the universe during biological processes, because it is irreversible, provides their driving force and gives direction

Box 14-1 *The concept of entropy*

The term entropy, which literally means "a change within," was first used in 1851 by Rudolf Clausius of Germany, one of the promulgators of the second law. A rigorous quantitative definition of entropy involves statistical and probability considerations. However, its nature can be illustrated qualitatively by three simple examples, each of which shows one aspect of entropy. The key descriptors of entropy are randomness or disorder, manifested in different ways.

CASE 1: The Teakettle and the Randomization of Heat We know that steam generated from boiling water can do useful work. But suppose we turn off the burner under a teakettle full of water at 100°C (the "system") in the kitchen (the "surroundings") and allow it to cool. As it cools, no work will be done, but heat will pass from the teakettle to the surroundings, raising the temperature of the surroundings (the kitchen) by an infinitesimally small amount until complete equilibrium is attained. At this point all parts of the teakettle and the kitchen will be at precisely the same temperature. The free energy that was once contained in the gases used by the stove to heat the water, which was *potentially* capable of doing work, has disappeared. Its equivalent in heat energy is still present in the teakettle + kitchen (i.e., the universe) but has become completely randomized throughout its contents. This energy is no longer available and can do no work because there is no temperature differential within the kitchen. Moreover, the increase in entropy of the kitchen (the surroundings) is irreversible. We know from everyday experience that heat will *never* spontaneously pass back from the kitchen into the teakettle to raise the temperature of the water to 100°C again.

CASE 2: The Oxidation of Glucose Entropy is a state or condition not only of energy but also of matter. Aerobic organisms extract free energy from glucose obtained from their surroundings. To obtain this energy they oxidize the glucose with molecular oxygen, also obtained from the surroundings. The end products of the oxidative metabolism of glucose are CO_2 and H_2O, which are then returned to the surroundings. In this process the surroundings have undergone an increase in entropy, whereas the organism itself remains in a steady state and undergoes no change in its internal order. Although some of the entropy arises from the dissipation of heat, entropy also arises from another kind of disorder, illustrated by the equation for the oxidation of glucose by living organisms, which we can write as

$$C_6H_{12}O_6 + 6O_2 \longrightarrow 6CO_2 + 6H_2O$$

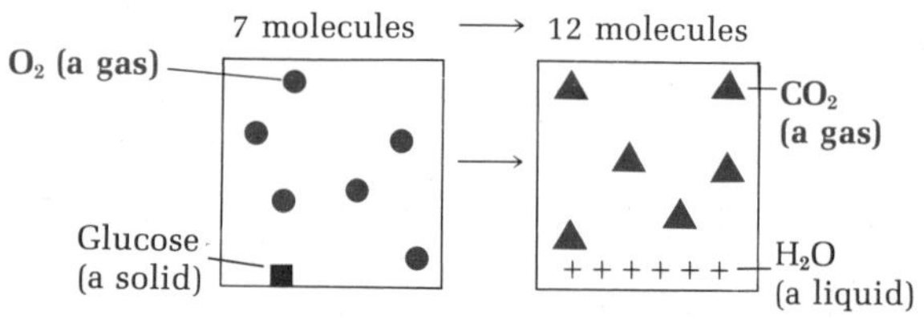

to all biological activities. Living organisms constantly produce entropy in their surroundings as a necessary cost of maintaining their own internal order.

It may be helpful at this point to examine an actual chemical reaction that takes place in cells, to get some feeling for the magnitude of the changes in these forms of energy. Aerobic cells carry out the oxidation of glucose ($C_6H_{12}O_6$) to CO_2 and H_2O at constant temperature and pressure:

$$C_6H_{12}O_6 + 6O_2 \longrightarrow 6CO_2 + 6H_2O$$

If we assume that the temperature is 25°C or 298 K and the pressure is 1.0 atm (760 mmHg), which are the standard conditions in thermodynamic calculations, the following energy changes take place per mole of glucose oxidized:

The atoms contained in 1 molecule of glucose plus 6 molecules of oxygen, a total of 7 molecules, are more randomly dispersed by the reaction, so that they are now present in a total of 12 molecules ($6CO_2$ + $6H_2O$).

Whenever a chemical reaction proceeds so that there is an increase in the number of molecules, or when a solid substance, such as glucose, is converted into liquid or gaseous products, which have more freedom to move or fill space than a solid, there is an increase in molecular disorder and thus an increase in entropy.

CASE 3: Information and Entropy The following short passage from *Julius Caesar*, Act IV, Scene 3, is spoken by Brutus, when he realizes that he must face Mark Antony's army. It is an information-rich, non-random arrangement of 125 letters from the English alphabet:

There is a tide in the affairs of men,
Which taken at the flood, leads on to fortune;
Omitted, all the voyage of their life
Is bound in shallows and in miseries.

In addition to what this quotation says overtly, it has many hidden meanings. It not only reflects a complex sequence of events in the play, but it also echoes the play's ideas on conflict, ambition, and the demands of leadership. Since it is also permeated with Shakespeare's understanding of human nature, it is very rich in information.

However, if the 125 letters making up this quotation were allowed to fall into a completely random, chaotic pattern, as shown below, they would have no meaning whatsoever. In this form the 125 letters would contain

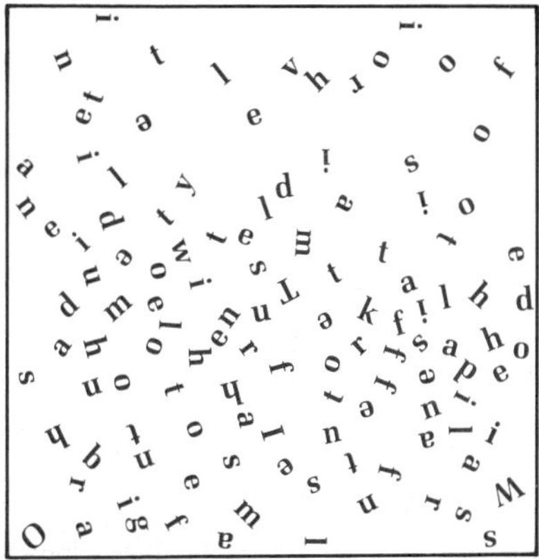

little or no information, but would be very rich in entropy. Such considerations have led to the conclusion that information is a form of energy; information has been called "negative entropy." In fact, the branch of mathematics called information theory, which is basic to the programming logic of computers, is closely related to thermodynamic theory. Living organisms are highly ordered, nonrandom structures, immensely rich in information and thus entropy-poor.

$\Delta G = -686{,}000$ cal/mol (the free energy of the system, i.e., the reacting molecules, has decreased)

$\Delta H = -673{,}000$ cal/mol (the reacting molecules have released heat)

$$\Delta S = \frac{\Delta H - \Delta G}{T}$$

$$= \frac{-673{,}000 - (-686{,}000)}{298}$$

$= +44$ cal/deg (the entropy of the universe has increased)

The increase in molecular disorder or entropy resulting from the oxidation of glucose may be visualized as shown in Box 14-1.

Cells Require Free Energy

Heat is not a significant source of energy for living cells because heat can do work only as it passes from a zone or object at one temperature to another at a lower temperature. Furthermore, the efficiency of a heat engine in doing work is related to the temperature differential through which the heat passes: the greater the temperature drop, the greater the percentage of the input heat energy that can be realized as work output. Since living cells have essentially the same temperature throughout, they cannot make significant use of heat energy to do work. Heat is useful to cells, however, to preserve an optimal working temperature.

The form of energy that cells can and must use is *free energy*, which *can* do work at constant temperature and constant pressure. Heterotrophic cells obtain their free energy from energy-rich nutrient molecules, and photosynthetic cells obtain their free energy from absorbed solar radiation. Both kinds of cells transform these free-energy inputs into common forms of chemical energy and use them to do their work by processes that involve no significant temperature differentials. Quite simply, *cells are chemical engines* that function at constant temperature and pressure.

Now we must consider how the free energy of chemical reactions is measured and expressed in quantitative terms.

The Standard-Free-Energy Change of a Chemical Reaction Can Be Calculated

Every chemical reaction has a characteristic *standard-free-energy change* ΔG°. [As we shall see, ΔG° is *not* the same as ΔG as given in equation (1).] The standard-free-energy change is a constant for any given reaction, and it can be calculated from the equilibrium constant of the reaction under *standard conditions*, i.e., a temperature of 25°C (298 K) and a pressure of 1.00 atm (760 mmHg). The equilibrium constant (page 74) for the reaction $A + B \rightleftharpoons C + D$, for example, is given by

$$K'_{eq} = \frac{[C][D]}{[A][B]}$$

in which [A], [B], [C], and [D] are the molar concentrations of the reaction components at the point of equilibrium of the reaction under the standard conditions. In reactions in which more than one molecule of any reactant or product participates, as in the general reaction

$$aA + bB \rightleftharpoons cC + dD$$

where *a*, *b*, *c*, and *d* are the number of molecules of A, B, C, and D participating, the equilibrium constant is given by

$$K'_{eq} = \frac{[C]^c[D]^d}{[A]^a[B]^b}$$

Once we have determined the equilibrium constant K'_{eq} of a chemical reaction, we can calculate its standard-free-energy change in units of calories per mole of reactant. The calorie, the most widely used unit of energy in biology, is defined as the amount of energy which in the form of heat can raise the temperature of 1.00 g of water at 15°C by 1°C. To calculate the standard-free-energy change $\Delta G°$ we use the relationship

$$\Delta G° = -2.303\ RT \log K'_{eq}$$

in which R is the gas constant (1.987 cal/mol·K) and T is the absolute temperature, in this case 298 K. If the equilibrium constant for a given chemical reaction is 1.0, the standard-free-energy change of that reaction is $\Delta G° = 0.0$, since the logarithm of 1.0 is zero (logarithm tables are in the Appendix). If the equilibrium constant of a reaction is greater than 1.0, its standard-free-energy change $\Delta G°$ is negative. If the equilibrium constant is less than 1.0, $\Delta G°$ is positive in sign.

It will be helpful to define the standard-free-energy change in another way. $\Delta G°$ is the difference between the free-energy content of the reactants and the free energy of the products under standard conditions, i.e., when the temperature is 298 K, the pressure is 1.0 atm, and the reactants and products are all present in their standard concentrations, namely 1.0 *M*. When $\Delta G°$ is negative, it means that the products contain less free energy than the reactants and therefore the reaction will proceed to form the products under standard conditions, since all chemical reactions will tend to go in that direction resulting in a decrease in the free energy of the system. When $\Delta G°$ is positive, it means that the products of the reaction contain more free energy than the reactants. The reaction will therefore tend to go in the reverse direction if we start with 1.0 *M* concentrations of all components. To be more explicit, reactions with a negative $\Delta G°$ proceed forward in the direction written when they start with all reactants and products at 1.0 *M*, until they reach equilibrium. Reactions with a positive $\Delta G°$ will proceed in the reverse of the direction written when they start with all components at 1.0 *M*. Table 14-1 summarizes these points. *Basically, the standard-free-energy change of a chemical reaction is simply a different mathematical way of expressing its equilibrium constant*. Table 14-2 shows the numerical relationship between $\Delta G°$ and the equilibrium constant K'_{eq}.

Two additional points must now be made. Because biochemical reactions take place near pH 7.0 and often involve H^+ formation or use, *pH 7.0 is by convention designated the standard pH in biochemical energetics*. The standard-free-energy change at pH 7.0 in biochemical systems is designated by $\Delta G°'$, which we shall use henceforth.

The second point concerns the units of energy. By international agreement the basic SI (Système International) unit of energy is the *joule* (J), named after James Joule (1818–1889), the British engineer who provided the experimental basis for

Table 14-1 Relationships between K_{eq}, $\Delta G°'$, and the Direction of Chemical Reactions under Standard Conditions

When K_{eq} is:	$\Delta G°'$ is:	Starting with 1 *M* components the reaction:
>1.0	Negative	Proceeds forward
1.0	Zero	Is at equilibrium
<1.0	Positive	Proceeds in reverse

Table 14-2 Relationship between the Equilibrium Constants of Chemical Reactions and Their Standard-Free-Energy Changes

When K'_{eq} is:	$\Delta G°'$ (cal/mol) is:
0.001	+4089
0.01	+2726
0.1	+1363
1.0	0
10.0	−1363
100.0	−2726
1000.0	−4089

the first law, the conservation of energy. However, in biology and medicine heat and energy measurements are commonly expressed in calories, which we shall use in this book. Calories and joules are easily interconverted: 1.000 cal = 4.184 J.

$\Delta G^{\circ\prime}$ Has Characteristic Values for Different Chemical Reactions

Now let us make a sample calculation of the standard-free-energy change of the reaction catalyzed by the enzyme phosphoglucomutase (whose role in the cell we examine in the next chapter):

$$\text{Glucose 1-phosphate} \xrightleftharpoons{\text{phosphoglucomutase}} \text{glucose 6-phosphate}$$

Chemical analysis shows that if we start with, say, 0.0200 *M* glucose 1-phosphate in the presence of excess phosphoglucomutase and allow the reaction to go in the forward direction, or if we start with 0.0200 *M* glucose 6-phosphate and allow the reaction to go in the reverse direction, the final equilibrium mixture in either case will contain 0.00100 *M* glucose 1-phosphate and 0.019 *M* glucose 6-phosphate at 25°C and pH 7.0. (Remember that enzymes do not affect the point of equilibrium of a reaction; they merely hasten its attainment.) From these data we can then calculate the equilibrium constant to three significant figures:

$$K'_{eq} = \frac{[\text{glucose 6-phosphate}]}{[\text{glucose 1-phosphate}]} = \frac{0.0190}{0.00100} = 19.0$$

From this value of K'_{eq} we can calculate the standard-free-energy change:

$$\begin{aligned}\Delta G^{\circ\prime} &= -2.303\,RT \log K'_{eq} \\ &= -2.30(1.987)(298) \log 19.0 \\ &= -1360(1.28) = -1740 \text{ cal/mol}\end{aligned}$$

Since the sign of the standard-free-energy change is negative, the conversion of glucose 1-phosphate into glucose 6-phosphate proceeds with a loss of free energy when it starts with 1.0 *M* glucose 1-phosphate and 1.0 *M* glucose 6-phosphate. In biochemical reactions it is more convenient to express $\Delta G^{\circ\prime}$ in kilocalories; thus $\Delta G^{\circ\prime}$ in this example becomes −1.74 kcal/mol.

Table 14-3 gives the standard-free-energy changes for representative chemical reactions. Note that hydrolysis of simple esters, amides, peptides, and glycosides, as well as rearrangements and eliminations, proceeds with relatively small standard-free-energy changes, whereas hydrolysis of acid anhydrides occurs with relatively large decreases in standard free energy. Oxidation of organic compounds to CO_2 and H_2O proceeds with especially large decreases in standard free energy. However, as we shall see later (Chapters 15 and 17),

Table 14-3 Standard-Free-Energy Changes of Some Chemical Reactions at pH 7.0 and 25°C

Reaction type	$\Delta G^{\circ\prime}$, kcal/mol
Hydrolysis reactions	
Acid anhydrides	
Acetic anhydride + $H_2O \rightarrow$ 2 acetate	−21.8
ATP + $H_2O \rightarrow$ ADP + phosphate	−7.3
Esters	
Ethyl acetate + $H_2O \rightarrow$ ethanol + acetate	−4.7
Glucose 6-phosphate + $H_2O \rightarrow$ glucose + phosphate	−3.3
Amides and peptides	
Glutamine + $H_2O \rightarrow$ glutamate + NH_4^+	−3.4
Glycylglycine + $H_2O \rightarrow$ 2 glycine	−2.2
Glycosides	
Maltose + $H_2O \rightarrow$ 2 glucose	−3.7
Lactose + $H_2O \rightarrow$ glucose + galactose	−3.8
Rearrangements	
Glucose 1-phosphate $\rightarrow$ glucose 6-phosphate	−1.74
Fructose 6-phosphate $\rightarrow$ glucose 6-phosphate	−0.40
Elimination of water	
Malate $\rightarrow$ fumarate + H_2O	+0.75
Oxidations with molecular oxygen	
Glucose + $6O_2 \rightarrow 6CO_2 + 6H_2O$	−686
Palmitic acid + $23O_2 \rightarrow 16CO_2 + 16H_2O$	−2338

standard-free-energy changes, such as those in Table 14-3, do not tell us how much of the free energy is biologically available.

There Is an Important Difference between $\Delta G^{\circ\prime}$ and ΔG

We must be careful to distinguish between two different quantities, the free-energy change ΔG and the standard-free-energy change $\Delta G^{\circ\prime}$. We have seen that in every spontaneous chemical or physical process the free energy of the reacting system always decreases, i.e., ΔG is negative. But we have also seen that each chemical reaction has a characteristic *standard-free-energy change* $\Delta G^{\circ\prime}$, which may be positive, negative, or zero, depending on the equilibrium constant of the reaction. The standard-free-energy change $\Delta G^{\circ\prime}$ tells us in which direction and how far a given reaction will go in order to reach equilibrium when it occurs under standard conditions, i.e., *when the initial concentration of all components is 1.0 M, the pH is 7.0, and the temperature is 25°C.* $\Delta G^{\circ\prime}$ is thus an immutable constant: it has a characteristic, unchanging value for a given reaction. But the *actual* free-energy change ΔG of a given chemical reaction is a function of the conditions of concentration, pH, and temperature actually prevailing during the reaction, which are not necessarily the standard conditions as defined above. Moreover, the ΔG of any reaction proceeding toward its equilibrium is always negative, becomes smaller (i.e., less negative) as the reaction proceeds, and is zero at the point of equilibrium, indicating that no more work can be done by the reaction.

ΔG and $\Delta G^{\circ\prime}$ for any reaction A + B → C + D are related by the equation

$$\Delta G = \Delta G^{\circ\prime} + \mathbf{2.303}\ \boldsymbol{RT}\ \mathbf{log}\ \frac{[\mathbf{C}][\mathbf{D}]}{[\mathbf{A}][\mathbf{B}]}$$

in which the terms in color are those *actually prevailing* in the system under question. As a simple example, let us suppose that the reaction A + B → C + D is taking place at the standard conditions of temperature (25°C) and pressure (1.0 atm) but that the initial concentrations of A, B, C, and D are *not* equal and that none of them are present at the standard concentration of 1.0 *M*. To determine the actual free-energy change ΔG that will occur under these nonstandard conditions of concentration, as the reaction proceeds to equilibrium, we simply put in the *actual* initial concentrations of A, B, C, and D; the values of *R*, *T*, and $\Delta G^{\circ\prime}$ will of course be the standard values. Solving this equation gives ΔG, the free-energy change of the reaction under the actual concentration conditions in which it is taking place. ΔG will be negative and will decline with time as the reaction proceeds because the actual concentrations of A and B will be getting smaller and the concentrations of C and D will be getting larger. Thus ΔG of an ongoing chemical reaction is negative and always progresses toward zero; $\Delta G^{\circ\prime}$, on the other hand, is a constant.

It is important to realize that $\Delta G^{\circ\prime}$ and ΔG give the *maximum* amount of free energy that a given reaction can *theoretically* deliver. This amount of energy can be realized only if there is some perfectly efficient device available to trap or harness it. If no such device is available, no work is done by the reaction at constant temperature and pressure.

Standard-Free-Energy Values of Chemical Reactions Are Additive

In the case of two sequential chemical reactions

$$A \longrightarrow B \qquad \Delta G_1^{\circ\prime}$$

$$B \longrightarrow C \qquad \Delta G_2^{\circ\prime}$$

each reaction has its own equilibrium constant, and each has its characteristic standard-free-energy change, $\Delta G_1^{\circ\prime}$ and $\Delta G_2^{\circ\prime}$. Since the two reactions are sequential, B cancels out and the overall reaction is

$$A \longrightarrow C$$

Reaction A → C will also have its own equilibrium constant and thus will also have its own standard-free-energy change $\Delta G_S^{\circ\prime}$. Now we come to a very important property of standard-free-energy changes: *the $\Delta G^{\circ\prime}$ values of sequential* *chemical reactions are additive*. For the overall reaction A → C, $\Delta G_s^{\circ\prime}$ is the algebraic sum of the individual standard-free-energy changes $\Delta G_1^{\circ\prime}$ and $\Delta G_2^{\circ\prime}$ of the two separate reactions:

$$\Delta G_s^{\circ\prime} = \Delta G_1^{\circ\prime} + \Delta G_2^{\circ\prime}$$

This relationship is very useful since it makes it easy to calculate the standard-free-energy change for a sequence of metabolic reactions. For example, the following sequence of reactions takes place during glycogen breakdown in muscle, as we shall see in the next chapter:

$$\text{Glucose 1-phosphate} \xrightarrow{\text{phosphoglucomutase}} \text{Glucose 6-phosphate} \qquad \Delta G^{\circ\prime} = -1.74 \text{ kcal/mol}$$

$$\text{Glucose 6-phosphate} \xrightarrow{\text{glucose phosphate isomerase}} \text{Fructose 6-phosphate} \qquad \Delta G^{\circ\prime} = +0.40 \text{ kcal/mol}$$

The sum of the two reactions

$$\text{Glucose 1-phosphate} \longrightarrow \text{fructose 6-phosphate}$$

therefore has a standard-free-energy change of

$$\Delta G_s^{\circ\prime} = -1.74 + (+0.40) = -1.34 \text{ kcal/mol}$$

It is often possible to calculate the standard-free-energy change of a given reaction even though we do not know its actual equilibrium constant. If it can be made sequential with another reaction of known $\Delta G^{\circ\prime}$, and if we can determine the equilibrium constant of the sum of the two sequential reactions, we can calculate $\Delta G^{\circ\prime}$ of the first reaction.

ATP Is the Major Chemical Link between Energy-Yielding and Energy-Requiring Cell Activities

Now that we have developed some fundamental principles of energy changes in chemical systems, we can examine the energy cycle in cells. Heterotrophic cells obtain free energy in a chemical form from the degradation (catabolism) of nutrient molecules, particularly carbohydrates and fats, and use it (1) to synthesize biomolecules from smaller precursors, (2) do mechanical work, as in muscular contraction, (3) to transport substances across membranes against gradients, and (4) to ensure fidelity of information transfer. Adenosine triphosphate (ATP) (Figure 14-2) serves as the major linking intermediate between energy-yielding and energy-requiring chemical reactions in cells. During the catabolism, or breakdown, of energy-rich cell fuels

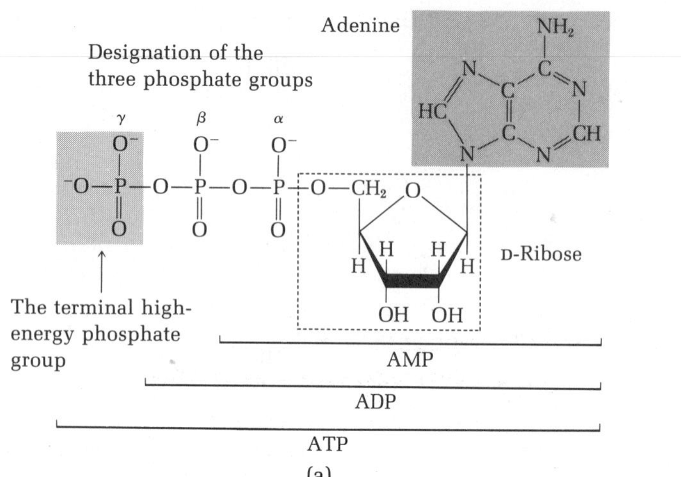

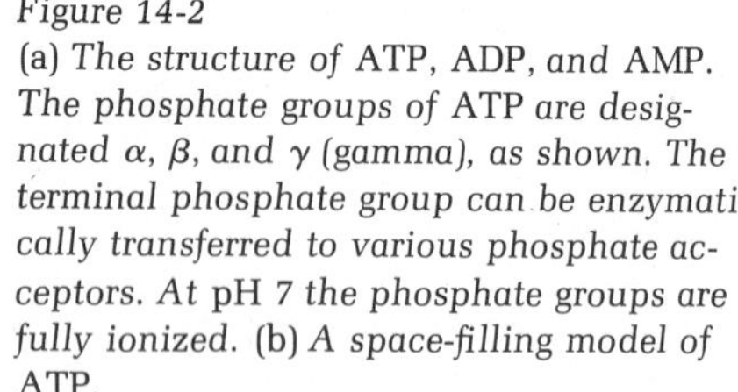

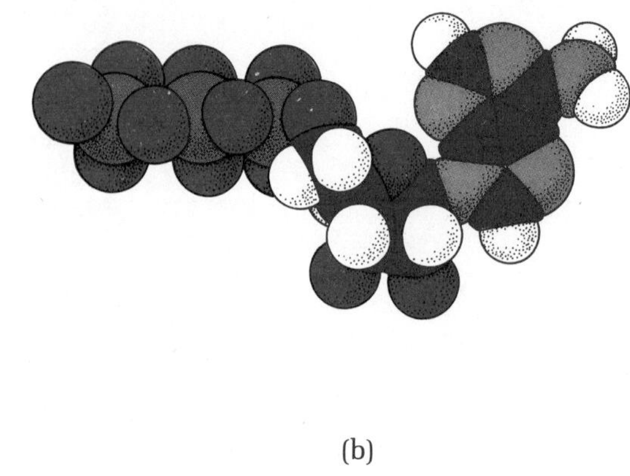

Figure 14-2
(a) The structure of ATP, ADP, and AMP. The phosphate groups of ATP are designated α, β, and γ (gamma), as shown. The terminal phosphate group can be enzymatically transferred to various phosphate acceptors. At pH 7 the phosphate groups are fully ionized. (b) A space-filling model of ATP.

some of their free energy is harnessed to make ATP from adenosine diphosphate (ADP) and phosphate (P_i), a process that requires input of free energy. ATP then donates much of its chemical energy to energy-requiring processes by undergoing breakdown to ADP and phosphate (Figure 14-3). ATP thus serves as a carrier of chemical energy from energy-yielding processes to the basic energy-requiring cell activities. ATP is also used to carry out more exotic kinds of cell work, e.g., by the firefly to generate its luminescent mating signals.

ATP was discovered in extracts of skeletal muscle by Karl Lohmann in Germany and almost simultaneously by Cyrus Fiske and Yellapragada Subbarow in the United States in 1929. At first ATP was thought to be concerned only in muscle activity, but it was subsequently found to be present in all types of cells—animal, plant, and microbial. Later ATP was also found to participate in a number of cell activities other than muscle contraction. In 1941 Fritz Lipmann saw the broad significance of these observations and postulated the unifying concept that ATP is the primary and universal carrier of chemical energy in cells. It was he who first proposed the ATP cycle shown in a present-day form in Figure 14-3.

Figure 14-3
The ATP cycle in cells.

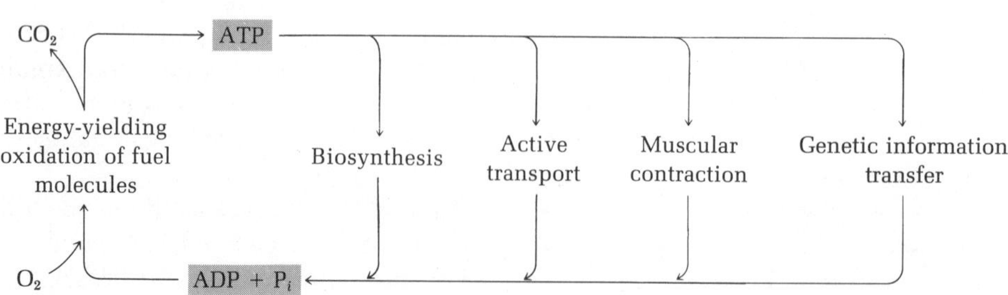

The Chemistry of ATP Is Well Known

ATP and its successive hydrolysis products, *adenosine diphosphate* (ADP) and *adenosine monophosphate* (AMP), are *nucleotides* (Figure 14-2). Recall (Chapter 3) that nucleotides consist of a heterocyclic purine or pyrimidine base, a 5-carbon sugar, and one or more phosphate groups. In ATP, ADP, and AMP the base is the purine *adenine*, and the 5-carbon sugar is D-*ribose* (Figure 14-2). Many nucleotides are known, differing in their sugars and nitrogenous bases. Nucleotides have a variety of cell functions but are particularly well known as building blocks of DNA and RNA, in which they function as coding elements. ATP, ADP, and AMP (Figure 14-2) are present in all forms of life and serve the same universal functions. They occur not only in the cell cytosol but also in mitochondria and the cell nucleus. In normal respiring cells ATP makes up 80 percent or more of the sum of all three adenine ribonucleotides (Table 14-4).

Table 14-4 Adenine Nucleotide, Phosphate, and Phosphocreatine (PCr) Concentrations in Some Cells, in m*M*†

	ATP	ADP	AMP	P_i	PCr
Rat liver	3.38	1.32	0.29	4.8	0
Rat muscle	8.05	0.93	0.04	8.05	28
Human erythrocytes	2.25	0.25	0.02	1.65	0
Rat brain	2.59	0.73	0.06	2.72	4.7
E. coli	7.90	1.04	0.82	7.9	0

† For erythrocytes the concentrations are those of the cytosol, since erythrocytes lack a nucleus and mitochondria. In the other types of cells, the data are for the entire cell contents, although it is known that the cytosol and the mitochondria have very different concentrations of ADP. Phosphocreatine is discussed later in this chapter.

At pH 7.0 both ATP and ADP occur as the multiply charged anions ATP^{4-} and ADP^{3-} because their phosphate groups are almost completely ionized at this pH. However, in the intracellular fluid, which contains high concentrations of Mg^{2+}, ATP and ADP exist largely as the $MgATP^{2-}$ and $MgADP^{-}$ complexes (Figure 14-4). In many enzymatic reactions in which ATP participates as phosphate donor, its active form is actually the $MgATP^{2-}$ complex. In cells the concentration of ATP normally remains relatively constant in a steady state, the rate of its formation equaling the rate of its breakdown. Thus the terminal phosphate group of ATP undergoes continuous removal and replacement from the pool of inorganic phosphate during cell metabolism.

ATP has been synthesized in the laboratory, and its structure and properties are well understood. Moreover, it functions to link energy-yielding reactions with energy-requiring reactions by well-known chemical principles, which we now examine.

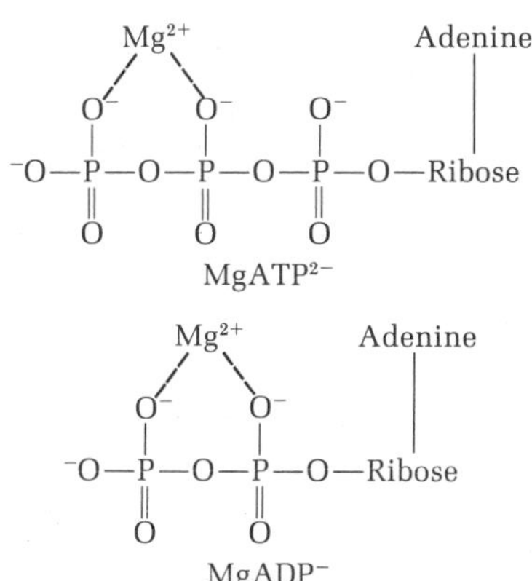

Figure 14-4
Mg^{2+} *complexes of ATP and ADP.*

ATP Has a Characteristic Standard Free Energy of Hydrolysis

When ATP undergoes loss of its terminal phosphate group by hydrolysis with the formation of ADP and inorganic phosphate

$$ATP + H_2O \longrightarrow ADP + P_i$$

the standard-free-energy change (see Table 14-3) is −7.3 kcal/mol.

Standard-free-energy changes have also been determined for the hydrolysis of other phosphorylated compounds (Table 14-5). Some phosphate compounds yield more and some yield less free energy than ATP when they undergo hydrolysis under standard conditions. For example, for the enzymatic reaction

$$\text{Glucose 6-phosphate} + H_2O \xrightarrow{\text{glucose 6-phosphatase}} \text{glucose} + \text{phosphate}$$

$\Delta G^{\circ\prime}$ is −3.3 kcal/mol, indicating that this reaction yields much less free energy than the hydrolysis of ATP ($\Delta G^{\circ\prime}$ = −7.3 kcal/mol) under standard conditions. Because early research showed ATP to yield significantly more free energy on hydrolysis than glucose 6-phosphate or a number of other phosphate esters, ATP was called a *high-energy phosphate compound* and glucose 6-phosphate was designated a *low-energy phosphate compound*. It was later found that cells also contain some phosphate compounds, e.g., *phosphoenolpyruvate* and *3-phosphoglyceroyl phosphate* (Table 14-5), with much larger standard free energies of hydrolysis than ATP. They are also high-energy phosphate compounds. However, the designations "high" and "low" do not clearly indicate that there are *three* classes of phosphate compounds. Such phosphate compounds as phosphoenolpyruvate and 3-phosphoglyceroyl phosphate, whose standard free energies of hydrolysis are much greater than that of ATP (Table 14-5), we may designate as "super" high-energy compounds. We shall soon see that the *intermediate* $\Delta G^{\circ\prime}$ value of ATP is very important to its biological function.

Table 14-5 Standard Free Energy of Hydrolysis of Some Phosphorylated Compounds

	$\Delta G^{\circ\prime}$, kcal/mol
Phosphoenolpyruvate	−14.8
3-Phosphoglyceroyl phosphate (→ 3-phosphoglycerate + P_i)	−11.8
Phosphocreatine	−10.3
ADP (→ AMP + P_i)	−7.3
ATP (→ ADP + P_i)	−7.3
AMP (→ adenosine + P_i)	−3.4
Glucose 1-phosphate	−5.0
Fructose 6-phosphate	−3.8
Glucose 6-phosphate	−3.3
Glycerol 1-phosphate	−2.2

Note in Table 14-5 that when ADP is hydrolyzed to AMP and phosphate, the standard-free-energy change is −7.3 kcal/mol, the same as $\Delta G^{\circ\prime}$ for the terminal phosphate group of ATP. Thus the two terminal phosphate groups of ATP (the β and γ groups) are both high-energy groups. $\Delta G^{\circ\prime}$ of hydrolysis of AMP to yield adenosine and phosphate, on the other hand, is much lower, only −3.4 kcal/mol; it is therefore in the low-energy class. Later we shall see that the phosphate groups can also be used in energy-requiring reactions in the cell through the action of a special enzyme (page 387).

Why Does ATP Have a Relatively High Standard Free Energy of Hydrolysis?

What structural features of the ATP molecule cause it to deliver considerably more free energy on hydrolysis of the terminal phosphate group than the hydrolysis of glucose 6-phosphate, for example? The answer to this question is to be found in the properties of *both* the substrate and the reaction products, because the standard-free-energy change is a measure of the *difference* in the free-energy content of the reactants and products. There are three major structural determinants of the standard free energy of hydrolysis of ATP. The first is the *degree of ionization* of ATP and its hydrolysis products. At pH 7.0 ATP is almost completely ionized as the ATP^{4-} ion. On hydrolysis it actually yields three products, ADP^{3-}, HPO_4^{2-}, and H^+. The overall equation of ATP hydrolysis can be given approximately by

$$ATP^{4-} + H_2O \longrightarrow ADP^{3-} + HPO_4^{2-} + H^+$$

Under standard conditions ATP^{4-}, ADP^{3-}, and HPO_4^{2-} will be present at 1.0 *M* concentrations. However, at pH 7.0 (the standard pH for $\Delta G^{\circ\prime}$ calculations) the hydrogen-ion concentration is only 10^{-7} *M*. This means that by the law of mass action the equilibrium for ATP hydrolysis tends to be pulled far to the right, since $[H^+]$ at pH 7.0 is very small compared with the standard concentration of 1.0 *M* for the other components of the reaction. On the other hand, when glucose 6-phosphate is hydrolyzed at pH 7, no significant extra H^+ is formed and therefore no such pulling force takes place:

$$\text{Glucose 6-phosphate}^{2-} + H_2O \longrightarrow \text{glucose} + HPO_4^{2-}$$

The second reason for the higher $\Delta G^{\circ\prime}$ for ATP is that at pH 7.0 ATP molecules have four closely spaced negative charges, which repel each other strongly (see Figure 14-2). When the terminal phosphate bond is hydrolyzed, some of this electrical stress within the ATP molecule is relieved by the separation of the negatively charged products ADP^{3-} and HPO_4^{2-}. These products have relatively little tendency to approach each other to react in the reverse direction to form ATP again. In contrast, when glucose 6-phosphate undergoes hydrolysis, one product, glucose, has no net charge. Thus glucose and HPO_4^{2-}, the other product, do not repel each other and tend to recombine more readily.

A third major contribution to the large negative $\Delta G^{\circ\prime}$ value of ATP hydrolysis is the fact that each of the two products ADP^{3-} and HPO_4^{2-} are *resonance hybrids*, special stable forms in which certain electrons are in a configuration possessing much less energy than in their original positions in the ATP molecule. Thus, when ATP is hydrolyzed, the electrons in the products ADP^{3-} and HPO_4^{2-} can sink to lower energy levels than in unhydrolyzed ATP. This causes the ADP^{3-} and HPO_4^{2-} anions, released from each other, to contain less free energy than when they were still combined as ATP^{4-}.

High-energy phosphate compounds, those whose hydrolysis proceeds with a large standard-free-energy decrease, are often said to contain a "high-energy phosphate bond," symbolized in structural formulas by a squiggle ~. The term "high-energy phosphate bond," although in long use among biochemists, is incorrect and misleading, since it wrongly suggests that the bond itself contains the energy. This is not the case; actually, the breaking of chemical bonds requires *input* of energy. The free energy released by hydrolysis of phosphate esters thus does *not* come from the specific bond that is broken but results from the fact that the *products of the reaction have a smaller free-energy content than the reactants*. It is however quite appropriate to use the term "high-energy phosphate compound" when referring to ATP or other phosphate compounds having a large negative $\Delta G°'$ of hydrolysis.

One last and very important point should be made about the free-energy changes of biochemical reactions. Although the $\Delta G°'$ for ATP hydrolysis is −7.3 kcal/mol under standard conditions, the *actual* free energy of hydrolysis of ATP in intact living cells is very different. This is because the concentrations of ATP, ADP, and P_i in living cells are not identical and are much lower than the standard 1.0 *M* concentrations. The actual free-energy change for ATP hydrolysis at concentrations other than the standard concentrations can be calculated. Box 14-2 shows how ΔG for ATP hydrolysis in the intact red blood cell can be calculated from the data in Table 14-4. ΔG for ATP hydrolysis in intact cells, usually designated ΔG_p, is much higher than $\Delta G°'$; in most cells ΔG_p ranges from −12 to −16 kcal/mol. ΔG_p is often called the *phosphorylation potential*; we shall refer to it again later (page 498).

ATP Acts as a Common Intermediate in Phosphate-Transfer Reactions

We have seen that ATP has an *intermediate* $\Delta G°'$ value in the thermodynamic scale of phosphate compounds. This is one of the properties making it possible for ATP to serve as an intermediate carrier of phosphate groups from super high-energy compounds, i.e., those releasing more energy on hydrolysis than ATP, to acceptor molecules whose phosphate derivatives have a low $\Delta G°'$ and thus release less free energy than ATP when hydrolyzed under standard conditions.

Exactly how can ATP act as such a mediator? We have seen that the reactions of metabolism take place via chains of consecutive or sequential enzyme-catalyzed reactions linked by *common intermediates* (page 337), in which a product of one reaction becomes the substrate of the next. Thus the reactions

$$A + B \longrightarrow C + D$$

$$D + E \longrightarrow F + G$$

are linked by the common intermediate D. The only way chemical energy can be transferred from one chemical reaction to another at constant temperature and pressure is for the two reactions to have a common intermediate. Thus the component D in

Box 14-2 The free energy of hydrolysis of ATP in intact cells

The standard free energy of hydrolysis of ATP has the value −7.3 kcal/mol. In the cell, however, the concentrations of ATP, ADP, and phosphate are not only unequal but also much lower than the standard 1 *M* concentrations (Table 14-4). Moreover, the pH inside cells may differ somewhat from the standard pH of 7.0. Thus the *actual* free energy of hydrolysis of ATP under intracellular conditions (ΔG_p) differs from the standard free energy $\Delta G^{\circ\prime}$. We can easily calculate the *actual* free-energy change ΔG_p of ATP hydrolysis in the cell. For example, in human erythrocytes the concentrations of ATP, ADP, and P_i are 2.25, 0.25, and 1.65 m*M*, respectively (Table 14-4). Let us assume for simplicity that the pH is 7.0 and the temperature is 25°C, the standard pH and temperature. The actual free energy ΔG of hydrolysis of ATP in the red blood cell under these conditions is given by the relationship

$$\Delta G = \Delta G^{\circ\prime} + 2.303\, RT \log \frac{[\mathrm{ADP}][\mathrm{P}_i]}{[\mathrm{ATP}]}$$

Substituting the appropriate values we obtain

$$\begin{aligned}\Delta G &= -7300 + 1360 \log \frac{(2.50 \times 10^{-4})(1.65 \times 10^{-3})}{2.25 \times 10^{-3}} \\ &= -7300 + 1360 \log 1.83 \times 10^{-4} \\ &= -7300 + 1360(-3.74) \\ &= -7300 - 5100 = -12{,}400 \text{ cal/mol} \\ &= -12.4 \text{ kcal/mol}\end{aligned}$$

Thus ΔG_p, the actual free-energy change for ATP hydrolysis in the intact erythrocyte (−12.4 kcal/mol), is much larger than the standard-free-energy change (−7.3 kcal/mol). By the same token, the free energy required to *synthesize* 1 mol of ATP from ADP and phosphate under the conditions prevailing in the red blood cell would be +12.4 kcal/mol.

Since the concentrations of ATP, ADP, and P_i may differ from one cell type to another (Table 14-4), ΔG_p for ATP hydrolysis may differ from one type of cell to another. Moreover, in any given cell ΔG_p can vary from time to time, depending on the metabolic conditions in the cell and how they influence the concentrations of ATP, ADP, and phosphate and the pH at any given time. We can calculate the actual free-energy change for any given metabolic reactions as it occurs in the cell providing we know the concentrations of all the reactants and products of the reaction and other factors, such as pH, temperature, and the concentration of Mg^{2+}, that may affect the equilibrium constant and thus the $\Delta G^{\circ\prime}$.

these two reactions can serve as the means of energy transfer from the first to the second reaction.

ATP functions as such an energy-carrying common intermediate in the cell, linking the reactions delivering free energy and those requiring it. During energy-yielding catabolic reactions super high-energy phosphate compounds are generated at the expense of energy released on degradation of organic cell nutrients. A specific enzyme called a *kinase* then catalyzes the transfer of a phosphate group from such a super high-energy phosphate compound, designated X—Ⓟ, to ADP to form ATP. In the second step another specific kinase catalyzes the transfer of the terminal phosphate group from ATP to an acceptor molecule, say Y, whose energy content is increased when it accepts phosphate to become Y—Ⓟ. These two reactions can be written

$$\mathrm{X{-}Ⓟ} + \mathrm{ADP} \longrightarrow \mathrm{X} + \mathrm{ATP}$$

$$\mathrm{ATP} + \mathrm{Y} \longrightarrow \mathrm{ADP} + \mathrm{Y{-}Ⓟ}$$

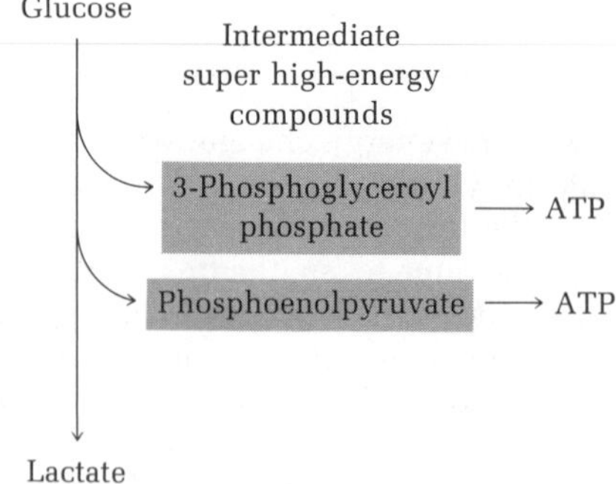

Figure 14-5
Formation of two kinds of intermediate super high-energy phosphate compounds during the energy-yielding breakdown of glucose to lactate. Each can donate a phosphate group to ADP to form ATP.

The net effect of these two reactions, coupled by the common intermediate ATP, is the transfer of chemical energy from X—Ⓟ to Y via transfer of the phosphate group. ATP is nearly always the mediator of such phosphate-group transfer reactions, since cells generally do not contain kinases that can transfer phosphate groups *directly* from super high-energy phosphate compounds to low-energy acceptors.

Two Super High-Energy Phosphate Compounds Are Generated by Breakdown of Glucose to Lactate

Two important donors of phosphate groups to ADP are the compounds *3-phosphoglyceroyl phosphate* and *phosphoenolpyruvate* (Table 14-5). Both are formed during the energy-yielding degradation of glucose to yield lactate (Figure 14-5), a process called *glycolysis*, to be examined in the next chapter. A large part of the free energy released on degradation of glucose to lactate is conserved in these two compounds. In the cell these high-energy phosphate compounds do not undergo hydrolysis; instead their phosphate groups are transferred to ADP to yield ATP by the action of specific kinases. For 3-phosphoglyceroyl phosphate (Figure 14-6) the phosphate-group transfer reaction, catalyzed by *phosphoglycerate kinase*, is

$$\text{3-Phosphoglyceroyl phosphate} + \text{ADP} \rightleftharpoons \text{ATP} + \text{3-phosphoglycerate}$$

Only one of the two phosphate groups of phosphoglyceroyl phosphate is transferred to ADP, that of its carboxyl group. (The other phosphate group, on carbon 3, is a low-energy phosphate group.) Although this kinase reaction is reversible, it proceeds far to the right under standard conditions because

Figure 14-6
Transfer of a phosphate group from 3-phosphoglyceroyl phosphate to ADP.

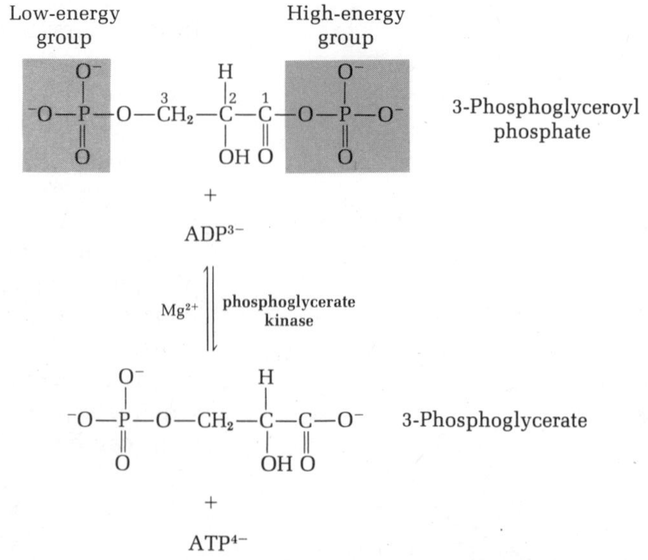

Low-energy group — High-energy group

$$^{-}O-P(O^{-})(=O)-O-\overset{3}{C}H_2-\overset{2}{C}H(OH)-\overset{1}{C}(=O)-O-P(O^{-})(=O)-O^{-}$$ 3-Phosphoglyceroyl phosphate

$+$ ADP^{3-}

Mg^{2+} ⇅ phosphoglycerate kinase

$$^{-}O-P(O^{-})(=O)-O-CH_2-CH(OH)-C(=O)-O^{-}$$ 3-Phosphoglycerate

$+$ ATP^{4-}

3-phosphoglyceroyl phosphate has a higher $\Delta G^{\circ\prime}$ value for hydrolysis (-11.8 kcal/mol) than ATP (-7.3 kcal/mol).

Phosphoenolpyruvate, the other high-energy phosphate compound formed during breakdown of glucose to pyruvate, also donates its phosphate group to ADP in a similar reaction (Figure 14-7), catalyzed by *pyruvate kinase*

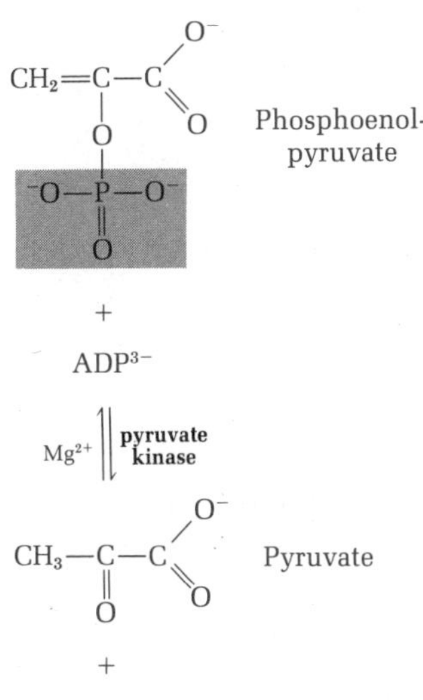

Figure 14-7
Transfer of a phosphate group from phosphoenolpyruvate to ATP.

$$\text{Phosphoenolpyruvate} + \text{ADP} \xrightarrow{Mg^{2+}} \text{pyruvate} + \text{ATP}$$

This reaction also tends to go far to the right under standard conditions because $\Delta G^{\circ\prime}$ for hydrolysis of phosphoenolpyruvate (-14.8 kcal/mol) is much larger than $\Delta G^{\circ\prime}$ for hydrolysis of ATP. This reaction is irreversible in the cell. Thus both phosphoenolpyruvate and 3-phosphoglyceroyl phosphate, which contain a large part of the chemical energy released in the anaerobic breakdown of glucose, can transfer much of their energy to ADP to yield ATP.

Transfer of a Phosphate Group from ATP to an Acceptor Molecule Can Energize It

ATP can now transfer its phosphate group to various acceptor molecules to yield low-energy phosphate compounds, most of which are phosphoric esters of alcohols (Table 14-5). Such reactions are also catalyzed by kinases. Among them is *hexokinase,* which catalyzes phosphate-group transfer from ATP to D-glucose

$$\text{ATP} + \text{D-glucose} \xrightarrow{Mg^{2+}} \text{ADP} + \text{D-glucose 6-phosphate}$$

Another is *glycerol kinase,* which catalyzes the reaction

$$\text{ATP} + \text{glycerol} \xrightarrow{Mg^{2+}} \text{ADP} + \text{glycerol 3-phosphate}$$

α-D-Glucose 6-phosphate

L-Glycerol 3-phosphate

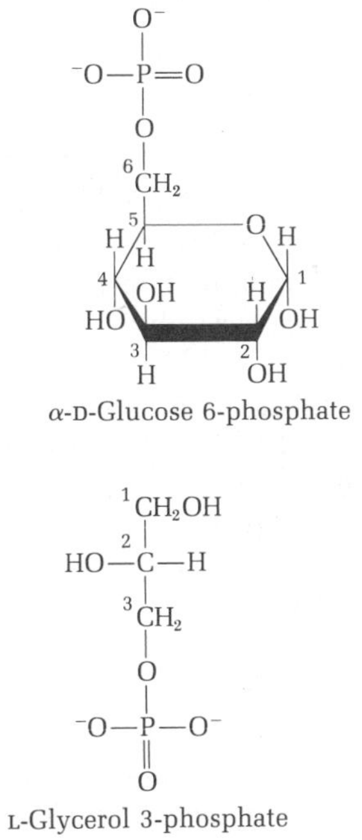

Figure 14-8
Two low-energy phosphate compounds. They are esters of phosphoric acid with hydroxyl groups.

In both cases a hydroxyl group of the acceptor molecule is phosphorylated to yield a phosphoric ester (Figure 14-8). Since the $\Delta G^{\circ\prime}$ values for the hydrolysis of glucose 6-phosphate ($\Delta G^{\circ\prime} = -3.3$ kcal/mol) and glycerol 3-phosphate ($\Delta G^{\circ\prime} = -2.2$ kcal/mol) are smaller than for ATP, the above reactions tend to go to the right as written if we begin with equal 1.0 M concentrations of the substrates and reactants.

Glucose 6-phosphate and glycerol 3-phosphate have a higher energy content than free or unphosphorylated glucose and glycerol. We may consider them as "energized" forms of glucose and glycerol, respectively, which can now undergo further enzymatic reactions in which they serve as activated building blocks for synthesis of larger molecules. For example, glucose 6-phosphate is an activated precursor that can lead to the biosynthesis of glycogen; similarly, glycerol 3-phosphate is an activated building block in lipid biosynthesis. Thus some of the free energy originally yielded by the degradation of glucose to lactate, which was conserved in the form of 3-

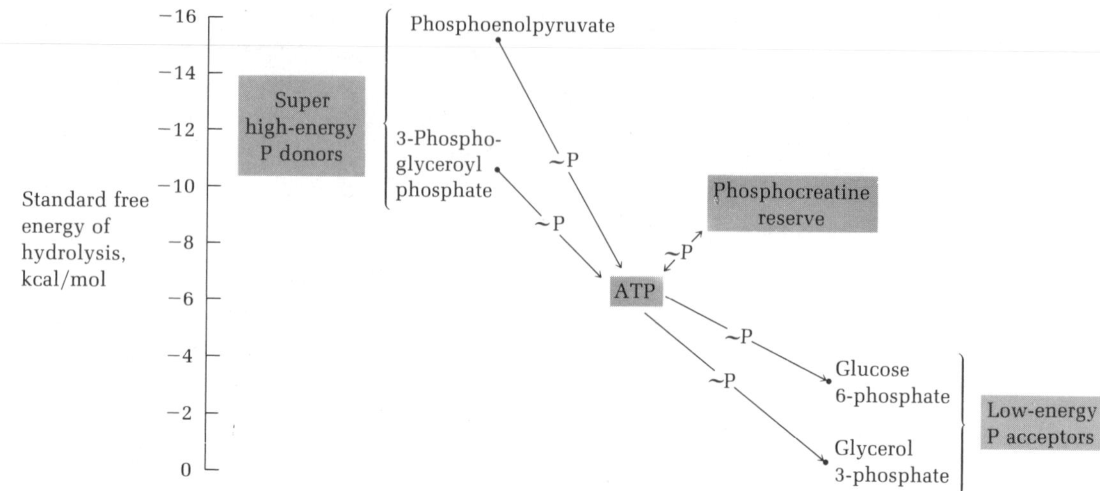

Figure 14-9
Flow of phosphate groups from super high-energy phosphate donors via ATP to various acceptor molecules to form their low-energy phosphate derivatives. This flow of phosphate groups, which is catalyzed by kinases, proceeds with an overall loss of free energy under intracellular conditions. Phosphocreatine functions as a reserve supply of high-energy phosphate groups in muscle and nerve cells.

phosphoglyceroyl phosphate and phosphoenolpyruvate, can be transmitted to glycerol, glucose, or other phosphate acceptors, with ATP serving as the intermediate carrier of chemical energy in the form of phosphate groups.

Figure 14-9 shows the flow sheet of enzymatic phosphate transfer reactions in the cell. An important feature is that virtually all super high-energy phosphate compounds must pass their phosphate groups to low-energy phosphate acceptor molecules via ATP, in two steps catalyzed by specific kinases.

ATP Is Used to Energize Muscle Contraction

In addition to energizing precursor molecules to prepare them for the biosynthesis of various cell components, ATP provides chemical energy for other major forms of cell work, such as the *mechanical work* of muscular contraction and the *osmotic work* required to transport substances against gradients of concentration.

In the contractile system of skeletal muscle cells there are two major types of filaments (Figure 14-10). The thick filaments are composed of bundles of parallel, rodlike molecules of *myosin*, and the thin filaments consist of two strands of *fibrous actin* (F actin) twisted about each other. F actin in turn is made up of globular actin (G actin) molecules arranged into strings. In muscle fibrils the thick and thin filaments are regularly arranged in parallel, interdigitated fashion, in repeating sets called *sarcomeres*. During muscle contraction the thick filaments slide into spaces between the thin filaments in each sarcomere, causing shortening of the entire muscle fiber. The hydrolysis of ATP to ADP and phosphate furnishes the chemical energy for the filaments to slide. As shown in Figure 14-10, each myosin molecule in a thick filament has a large head.

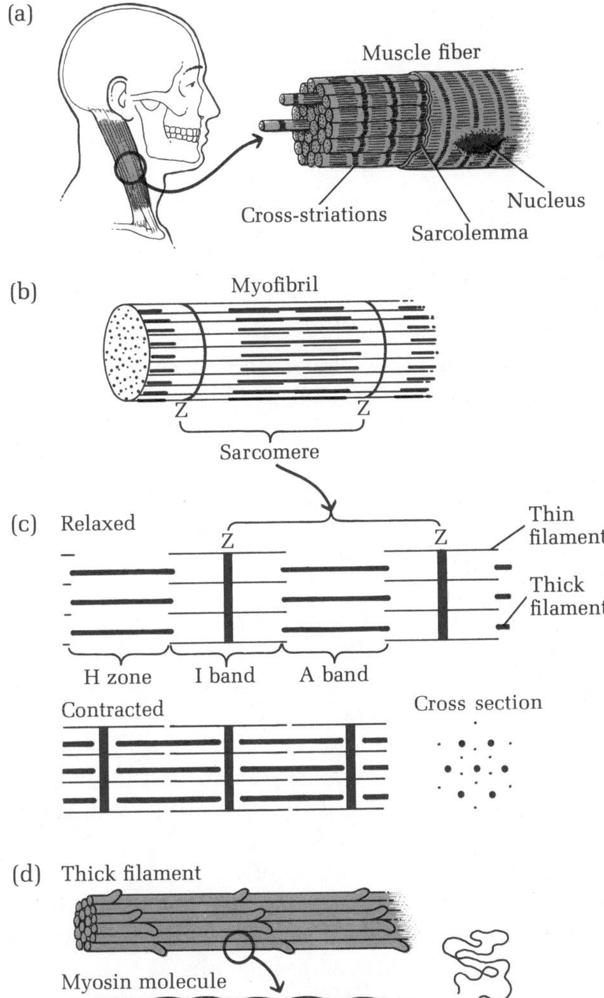

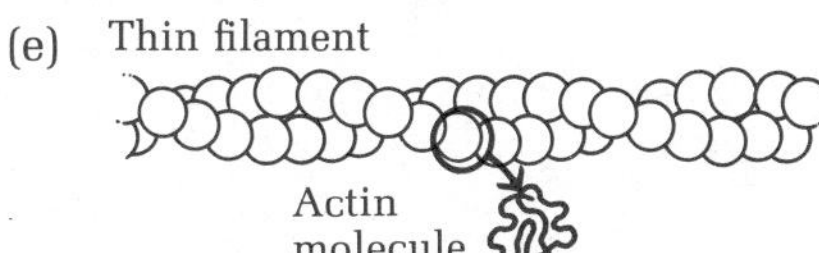

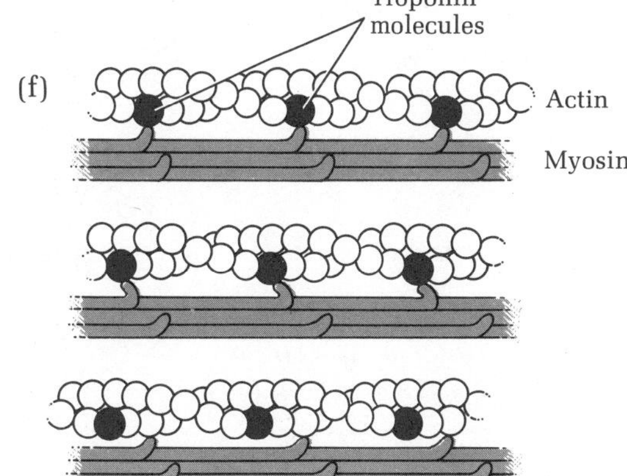

Figure 14-10
The contractile system of skeletal muscle. (a) Skeletal muscle is composed of bundles of parallel muscle fibers. These are very long cells with numerous nuclei. (b) Each muscle fiber contains many myofibrils, parallel sets of contractile filaments. Myofibrils are divided into sarcomeres by dark partitions, the Z lines. (c) Each sarcomere is made up of regularly spaced thick and thin filaments. The thick filaments can slide along the thin filaments. (d) The thick filaments consist of bundles of a long, slender protein, myosin. Each myosin molecule has two α-helical polypeptide chains twisted around each other. One end is folded into a globular structure, the head, which can hydrolyze ATP to ADP and P_i. (e) Each thin filament consists of two F-actin strands coiled about each other. Each strand is a string of globular G-actin molecules. (f) The globular heads of the myosin molecules protrude from the thick filaments. In the presence of ATP, they are believed to function like levers by attaching to the thin filaments, pulling them toward the center of the sarcomere, shortening it and contracting the myofibril. Simultaneously ATP is hydrolyzed to ADP and P_i by the myosin heads. The contraction process is controlled by a Ca^{2+}-binding protein, troponin, which is attached to the actin filaments at intervals.

These heads, which are regularly spaced along the thick filaments, are actually enzymes. They hydrolyze ATP as they perform repetitive make-and-break contacts with the thin filaments in such a way that sliding force is exerted, causing the thick filaments to move along the thin filaments toward the ends of the sarcomeres. The hydrolysis of ATP is believed to be accompanied by a change in the shape or conformation of the myosin head, producing mechanical force. In this way myosin and actin, as well as other proteins of the contractile system, are specialized to transform the chemical energy of ATP into the mechanical energy of muscular contraction.

The contraction and relaxation of skeletal muscles is controlled by the Ca^{2+} concentration in the cytosol. Normally the Ca^{2+} concentration in resting muscle is very low. When the motor nerve stimulates the muscle fiber, Ca^{2+} is released from membranous tubules that run across the muscle cell. The Ca^{2+} so released binds to a complex regulatory protein, *troponin*, located at intervals along the thin filaments. The troponin molecule serves as a trigger. It undergoes a conformational change that sets off the ATPase activity of the heads of the myosin molecule in the thick filaments, thus initiating contraction. So long as free Ca^{2+} is present in the muscle cytosol, the troponin will remain active. Relaxation of muscle comes about when the nerve impulses cease and Ca^{2+} is transported from the sarcoplasm into the cisternae of the sarcoplasmic reticulum through action of a Ca^{2+}-pumping ATPase activity in the membrane. Thus ATP energy is required not only for the contraction of muscles but also for their relaxation. As we shall see, energy released by ATP hydrolysis can be used to pump other ions through membranes.

Muscles are specialized in different ways. Smooth muscles of the intestine, for example, contract very slowly, whereas the flight muscles of insects, such as flies and mosquitos, have very high rates of contraction and relaxation. Some muscles can be locked in their contracted state, such as the adductor, or "catch," muscles of clams and oysters. Heart muscle shows rhythmic contractions. Skeletal muscles may also be specialized. White skeletal muscle is "fast" muscle, which can function without oxygen, whereas red skeletal muscle, which responds more slowly, requires oxygen. The skeletal muscles of some animals are very powerful and efficient in their action (Figure 14-11). All muscles, however specialized, use actin, myosin, and troponin as the molecular components of their contractile elements, and all use ATP energy. Other types of cell movement and propulsion (Chapter 2) are afforded by the

Figure 14-11
The powerful and highly coordinated skeletal muscles of the cheetah enable it to attain a speed of over 70 mi/h for several hundred yards.

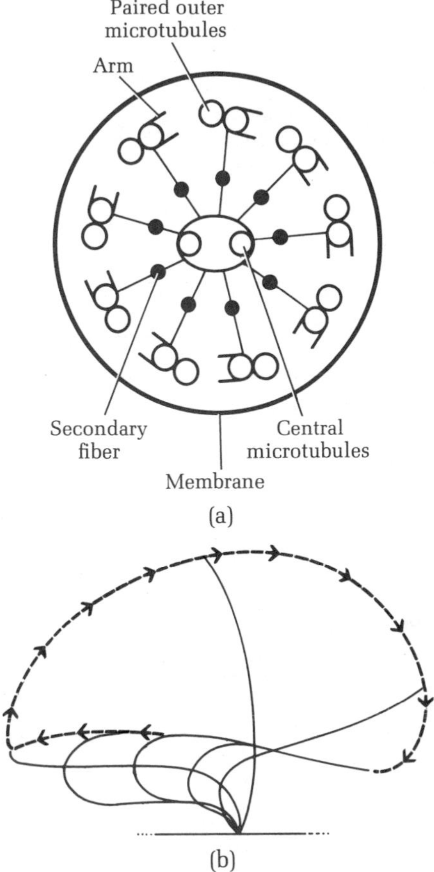

Figure 14-12
Eukaryotic cilia and flagella use ATP for generating mechanical force. (a) *Cross section of a cilium. These structures consist of an outer ring of 9 sets of paired microtubules, plus 2 unpaired central tubules, the 9 + 2 arrangement* (page 31). *Cilia are surrounded by an extension of the cell membrane. The characteristic undulating, sweeping, or rotatory motions of cilia are driven by ATP hydrolysis. These motions are imparted by sliding or twisting of the paired microtubules, in a manner resembling the ATP-dependent sliding of thick and thin filaments along each other in skeletal muscle. The outer paired microtubules contain "arms" at intervals, comparable to the myosin heads of muscle thick filaments. These arms consist of molecules of dynein, a large protein which has ATPase activity. Hydrolysis of ATP by dynein provides energy for the mechanical sliding or twisting of the microtubules. It has been suggested that the central tubules are the pacemakers in the activity of cilia.* (b) *Stages in the beat of a cilium in the gills of a marine worm, in which the cilia are about 30 μm long. These characteristic motions are imparted by the ATP-dependent sliding of tubular filaments along each other.*

cilia and flagella of eukaryotic cells (Figure 14-12). The separation of chromosomes by the contractile action of the spindle apparatus during mitosis also uses ATP as an energy source, to bring about twisting and sliding of microtubules along each other.

The actin-myosin-troponin system is a unique type of chemical engine because it converts chemical into mechanical energy *directly* at constant temperature and pressure. No familiar man-made machines are capable of this type of energy transformation. Thus living organisms have perfected a type of energy conversion that engineers have yet to achieve.

Phosphocreatine Is a Temporary Storage Form of High-Energy Phosphate Groups in Muscles

One high-energy phosphate compound plays a unique role in the energetics of muscle and other excitable tissues, such as brain and nerve, namely, *phosphocreatine* (Figure 14-13). This compound serves as a temporary storage form of high-energy phosphate groups. Phosphocreatine (also called *creatine phosphate*) has a $\Delta G^{\circ\prime}$ of hydrolysis of -10.3 kcal/mol, somewhat

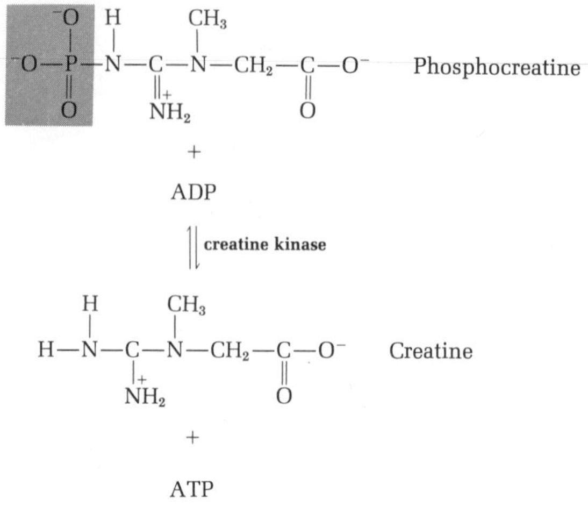

Figure 14-13
Function of phosphocreatine as a reserve donor of high-energy phosphate groups in muscles. Phosphocreatine may be regarded as a "buffer" of high-energy phosphate groups, which helps keep the concentration of ATP constant.

$^{-}O-P(=O)(O^{-})-NH-C(=\overset{+}{N}H_2)-N(CH_3)-CH_2-C(=O)-O^{-}$ Phosphocreatine

+

ADP

$\rightleftharpoons$ creatine kinase

$H-NH-C(\overset{+}{N}H_2)-N(CH_3)-CH_2-C(=O)-O^{-}$ Creatine

+

ATP

greater than the $\Delta G^{\circ\prime}$ for ATP hydrolysis. Phosphocreatine can transfer its phosphate group to ATP in a reaction catalyzed by the enzyme *creatine kinase:*

$$\text{Phosphocreatine} + \text{ADP} \rightleftharpoons \text{creatine} + \text{ATP}$$

Phosphocreatine functions to keep the ATP concentration in muscle cells at constant high levels, particularly in skeletal muscle, which must perform intermittent and sometimes strenuous work at a high rate. Whenever some of the ATP of the muscle cell is used for contraction, ADP is formed. Through the action of creatine kinase, phosphocreatine quickly donates its phosphate group to ADP to restore the normal ATP level. Since the phosphocreatine content of muscle is about 3 to 4 times greater than that of ATP (Table 14-4), it can store enough phosphate groups to keep the ATP level constant during short periods of intense activity. During the subsequent recovery period, the accumulated creatine is rephosphorylated to phosphocreatine by ATP, since the creatine kinase reaction is reversible. Because there is no other metabolic pathway for the formation and breakdown of phosphocreatine, this compound is well adapted for this temporary storage function.

In many invertebrates the muscles contain *phosphoarginine* instead of phosphocreatine as a reserve form of energy. Compounds like phosphocreatine and phosphoarginine that function as energy reserves are called *phosphagens*.

ATP Also Energizes Active Transport across Membranes

The chemical energy of ATP can also be used to bring about concentration work, i.e., the work required to transport an ion or a molecule across a membrane into another aqueous compartment where its concentration is higher. We can calculate the amount of free energy required to transport 1.0 mol of a nonionic solute from one compartment to another, e.g., from

Figure 14-14
Active transport of a solute against a concentration gradient. Starting from equilibrium, at which the solute concentration is equal in both compartments, active transport of a solute from one compartment into the other causes its movement against a concentration gradient. Free-energy input is required to generate and maintain a gradient of solute concentration across the membrane. If the free-energy input is interrupted, the solute in the enriched compartment will diffuse back until equilibrium is reached again, at which point the solute concentration is the same in both compartments.

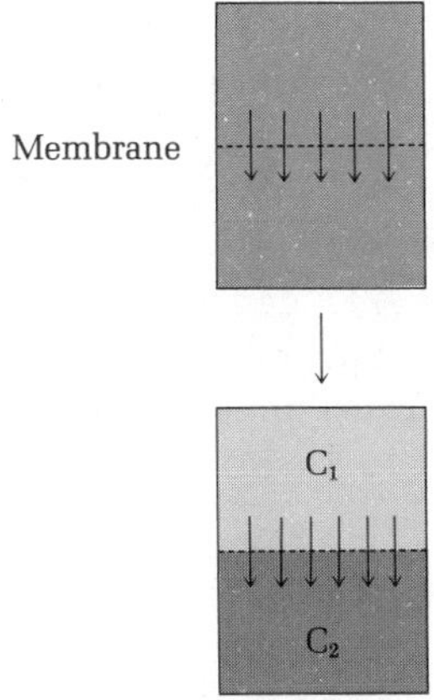

In active transport the free energy of the system increases

the surrounding medium into the cell, if we know the concentration of the free solute in each compartment (Figure 14-14). The basic relationship is given by

$$\Delta G = 2.303RT \log \frac{C_2}{C_1}$$

where C_1 is the molar concentration of the solute in the surrounding medium and C_2 is its molar concentration in the cell; R and T have their usual meanings. With this equation we can calculate the free energy required to transport 1 mol of glucose up a hundredfold gradient, from the medium in which its concentration is, say, 1.0 m*M* into a compartment in which its final concentration is 100 m*M*. We then have

$$\begin{aligned}\Delta G &= 2.30(1.98)(298) \log \frac{0.100}{0.001} \\ &= 1360(2.0) = 2720 \text{ cal/mol} \\ &= 2.72 \text{ kcal/mol}\end{aligned}$$

Since the free-energy change is positive, the transport of 1 mole of glucose (or any neutral molecule) against a hundredfold concentration gradient requires at least 2.72 kcal of free energy applied to the system through some type of energy-harnessing device.

The concentration gradients across the membranes of cells vary widely. Perhaps the largest gradient in the body occurs across the plasma membrane of the *parietal cells* of the stomach lining; these cells secrete hydrochloric acid into the gastric juice. Since the concentration of HCl in gastric juice may be as high as 0.1 *M* (pH 1.0) and the concentration of H^+ in the cells is about 10^{-7} *M* (pH 7.0), parietal cells can secrete H^+ ions against a concentration gradient of about 1 million to 1. These cells must have very active membrane "pumps" for secreting H^+ ions, since considerable energy is required to achieve this concentration gradient. The transport of solutes across membranes against concentration gradients is called *active transport*. The formation of gastric HCl is promoted by a membrane-bound enzyme called *H^+-transporting ATPase*. For each molecule of cytosolic ATP hydrolyzed to ADP and phosphate, 2 H^+ ions are transported across the plasma membrane from the cytosol to the exterior in the formation of gastric juice.

Another important example of an active-transport process is the transport of Na^+ and K^+ across the plasma membrane of all animal cells. In red blood cells, in which this process has been most extensively studied, the K^+ concentration in the cytosol is about 110 m*M* whereas the K^+ concentration in the blood plasma is only about 3 mM. On the other hand, the Na^+ concentration in the blood plasma is relatively high, about 140 m*M*, and the Na^+ concentration in the red blood cell is only about 4 m*M*. Maintenance of these gradients across the plasma mem-

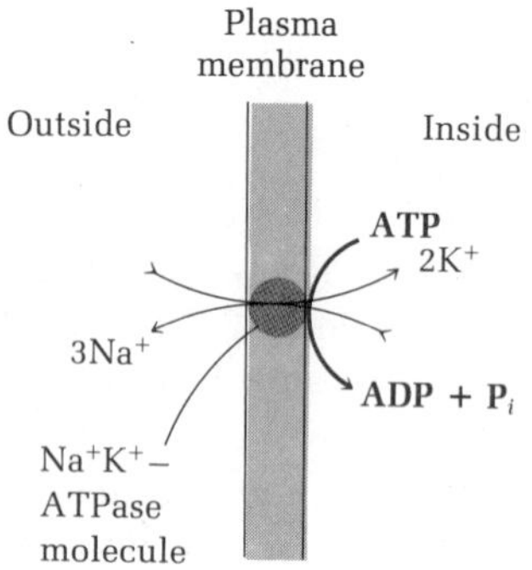

Figure 14-15
Schematic representation of the action of the Na^+K^+–ATPase. The transport of K^+ into the cell, where the K^+ concentration is high, and transport of Na^+ out of the cell into the extracellular water, where Na^+ concentration is high, requires free energy, furnished by hydrolysis of ATP. For each molecule of ATP hydrolyzed to ADP and P_i, three Na^+ ions are extruded and two K^+ ions are accumulated in the cell. The transport process takes place in two steps. In the first the ATPase molecule is phosphorylated by ATP, which enables it to bind Na^+. In the second step K^+ binds and causes the transport of Na^+ and K^+ across the membrane, accompanied by release of free phosphate into the cytosol. ATP and its hydrolysis products ADP and P_i remain in the cell.

brane depends upon the input of ATP. The red cell membrane contains a specialized enzyme, called the *Na^+K^+-transporting ATPase,* which functions both as an enzyme and as a molecular pump. It catalyzes the hydrolysis of ATP to ADP and phosphate and uses the free energy set free to pump K^+ inward and Na^+ outward (Figure 14-15). As the energizing step in this process, the terminal phosphate group of ATP is transferred to the Na^+K^+–ATPase molecule. This enzyme-bound phosphate group undergoes hydrolysis and appears as inorganic phosphate in the cytosol as it supplies energy to transport Na^+ and K^+ in opposite directions across the membrane, each against a gradient. The ADP and phosphate formed then are recharged to ATP by the energy-yielding breakdown of glucose. The Na^+K^+–ATPase does not hydrolyze ATP unless Na^+ is available from the inside of the cell and K^+ is available from the outside. The Na^+K^+–ATPase molecule, which has 2 α and 2 β subunits, spans the plasma membrane. It undergoes conformational changes during the ion-transporting step. In the kidney, which must excrete excess Na^+ in the urine and conserve K^+ in the blood, almost two-thirds of the ATP generated by respiration is used to transport Na^+ and K^+. Several different ion-transporting ATPases play important roles in different cell membranes (Table 14-6).

Table 14-6 Cation-Transporting Membrane ATPases

Name	Cell type	Location	Function
Na^+K^+–ATPase	Most animal cells	Plasma membrane	Maintains high intracellular K^+ concentration
H^+–ATPase	Parietal cells of gastric mucosa	Plasma membrane	Secretes H^+ in gastric juice
H^+–ATPase	Animal and plant cells	Mitochondrial inner membrane	Participates in oxidative and photosynthetic phosphorylation of ADP to ATP
	Plant cells	Chloroplast inner membrane	
	Bacteria	Plasma membrane	
Ca^{2+}–ATPase	Animal cells	Plasma membrane	Pumps Ca^{2+} out of cells to help maintain cytosolic Ca^{2+}
		Sarcoplasmic reticulum	Pumps Ca^{2+} into cisternae of sarcoplasmic reticulum to relax muscle

ATP Can Also Be Broken Down to AMP and Pyrophosphate

Although ADP and phosphate are the products of many ATP-using reactions and ADP is the direct phosphate acceptor in energy-yielding reactions, in some ATP-using reactions of cells the *two* terminal phosphate groups of ATP, that is, the β and γ groups (Figure 14-2), are removed *in one piece* as inorganic pyrophosphate (symbolized PP_i), leaving adenosine mono-

Figure 14-16
Palmitoyl-coenzyme A, a typical fatty acyl–CoA. The thioester or carbon-sulfur bond (color) linking the fatty acid to CoA-SH has a high $\Delta G^{\circ\prime}$ for hydrolysis, about −7.5 kcal/mol. Fatty acyl–CoA's are activated precursors in the biosynthesis of lipids.

CH_3—$(CH_2)_{14}$—C=O—S—Coenzyme A

CH_3
|
CH_2
|
CH_2
|
CH_2
|
CH_2
|
CH_2
|
CH_2
|
CH_2
|
CH_2 Palmitoyl group
|
CH_2
|
CH_2
|
CH_2
|
CH_2
|
CH_2
|
CH_2
|
C=O
|
S
|
Coenzyme A

phosphate (AMP) as the other product of ATP breakdown. An example is the enzymatic activation of a fatty acid to form its coenzyme A ester (page 513), a reaction that "energizes" the fatty acid and converts it into an activated precursor, *fatty acyl–CoA* (Figure 14-16), for the biosynthesis of lipids:

$$\text{ATP} + \underset{\text{Fatty acid}}{\text{RCOOH}} + \text{CoA—SH} \rightleftharpoons \text{AMP} + \text{PP}_i + \underset{\text{Fatty acyl–CoA}}{\text{RCO—S—CoA}}$$
$$\Delta G^{\circ\prime} = +0.2 \text{ kcal/mol}$$

This activation reaction proceeds by a *pyrophosphate cleavage* of ATP, to yield pyrophosphate and AMP, in contrast to the usual *orthophosphate cleavage*, in which ATP loses a single orthophosphate group, as in the hexokinase reaction

$$\text{ATP} + \text{D-glucose} \longrightarrow \text{ADP} + \text{D-glucose 6-phosphate}$$
$$\Delta G^{\circ\prime} = -3.3 \text{ kcal/mol}$$

The hydrolysis of ATP to yield AMP + PP_i

$$\text{ATP} + H_2O \longrightarrow \text{AMP} + \text{PP}_i$$

proceeds with $\Delta G^{\circ\prime} = -7.7$ kcal/mol, slightly greater than the $\Delta G^{\circ\prime}$ for the hydrolysis of the terminal or γ phosphate bond. The inorganic pyrophosphate is subsequently hydrolyzed by *pyrophosphatase* to yield two molecules of inorganic orthophosphate

$$\text{Pyrophosphate} + H_2O \longrightarrow 2 \text{ phosphate}$$
$$\Delta G^{\circ\prime} = -6.9 \text{ kcal/mol}$$

The overall reaction

$$\text{ATP} + 2H_2O \longrightarrow \text{AMP} + 2\text{P}_i$$
$$\Delta G^{\circ\prime} = -14.6 \text{ kcal/mol}$$

has a $\Delta G^{\circ\prime}$ that is the sum of the $\Delta G^{\circ\prime}$ values of the two sequential component reactions. We also note that the $\Delta G^{\circ\prime}$ of the overall reaction is exactly twice the $\Delta G^{\circ\prime}$ of the terminal phosphate groups of ATP and ADP.

Using *two* phosphate groups of ATP to energize one molecule of a precursor molecule may appear to be wasteful of phosphate-group energy, but we shall see later that this is an important device to ensure the completeness of certain biosynthetic reactions. An unusual use of the pyrophosphate cleavage of ATP occurs in the firefly, which uses it as a source of energy to generate light flashes (Box 14-3).

AMP can return to the ATP cycle again by the action of an enzyme present in all animal cells, *adenylate kinase*, which catalyzes the reversible phosphorylation of AMP to yield ADP:

$$\text{ATP} + \text{AMP} \xrightleftharpoons{Mg^{2+}} \text{ADP} + \text{ADP}$$

The ADP so formed can now be phosphorylated to ATP again.

Box 14-3 ATP provides the energy for firefly bioluminescence

Many fungi, marine microorganisms, jellyfish, and crustaceans as well as the firefly (Figure 1) are capable of generating bioluminescence, which requires considerable amounts of energy. In the firefly a combination of ATP and oxidative energy is used in a set of reactions that convert chemical energy into light energy. From many thousands of firefly lanterns collected by children in and around Baltimore, William McElroy and his colleagues at Johns Hopkins University isolated the principal biochemical components involved, *luciferin* (Figure 2), a complex carboxylic acid, and *luciferase,* an enzyme. The generation of a light flash requires that luciferin first be activated by an enzymatic reaction with ATP in which a pyrophosphate cleavage of ATP occurs, to form *luciferyl adenylate* (Figure 2). This compound is now acted upon by molecular oxygen and luciferase to bring about the oxidative decarboxylation of the luciferin to yield *oxyluciferin*. This reaction, which has intermediate steps, is accompanied by emission of light (Figure 3). The color of the light flash differs with firefly species and appears to be determined by differences in the structure of the luciferase. Luciferin is then regenerated from oxyluciferin in a subsequent series of reactions. Other bioluminescent organisms use other types of enzymatic reactions to generate light.

Pure firefly luciferin and luciferase are used to measure minute quantities of ATP by the intensity of the light flash produced. As little as a few picomoles (10^{-12} mol) of ATP can be measured in this way.

Figure 1
The firefly, a species of beetle.

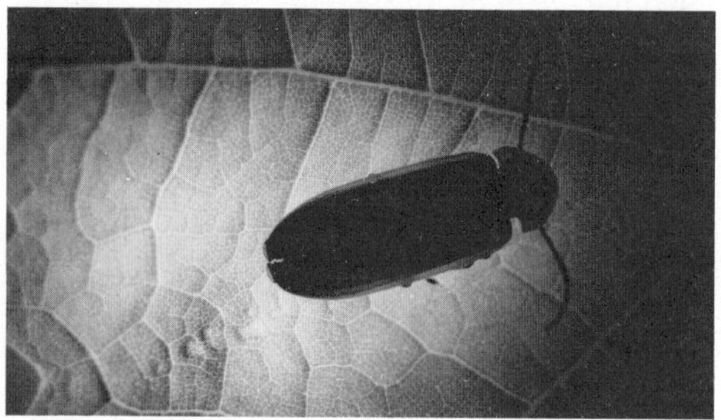

Figure 2
Important components in firefly bioluminescence.

Luciferin

Luciferyl adenylate (—O—P—O—Ribose—Adenine: AMP)

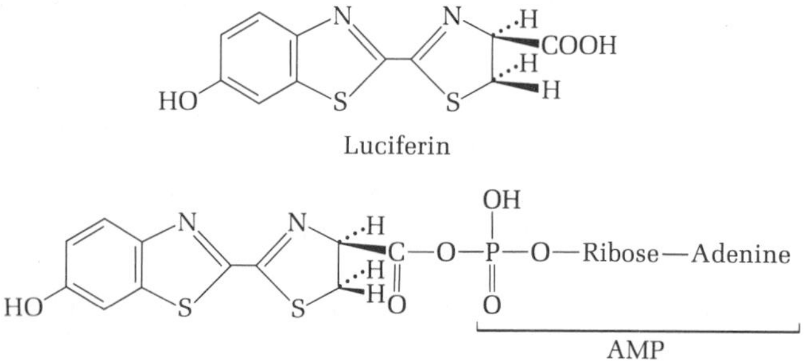

Figure 3
The firefly cycle.

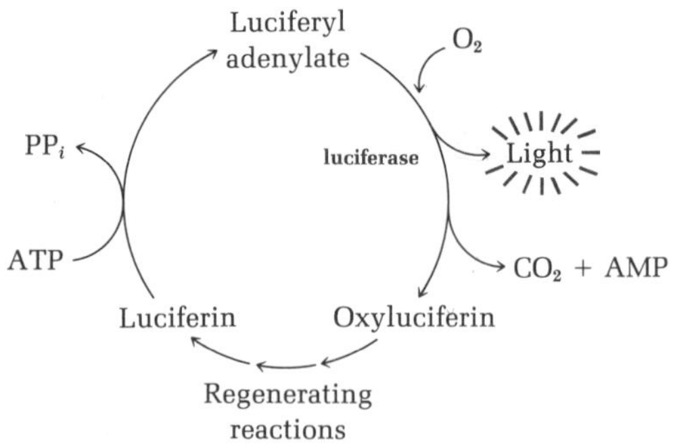

Luciferyl adenylate → (O_2, luciferase) → Light; CO_2 + AMP; Oxyluciferin → Regenerating reactions → Luciferin; ATP → PP_i

Adenylate kinase has a second important function. When it acts in the reverse direction

$$2ADP \xrightleftharpoons{Mg^{2+}} ATP + AMP$$

it helps maintain the ATP level in the cell since it promotes the transfer of the terminal phosphate group of one molecule of ADP to a second molecule, converting the latter into ATP. In contracting muscles, adenylate kinase thus allows the use of both the γ and β phosphate groups of ATP as an energy source (Figure 14-17). Consequently, adenylate kinase acting on ADP

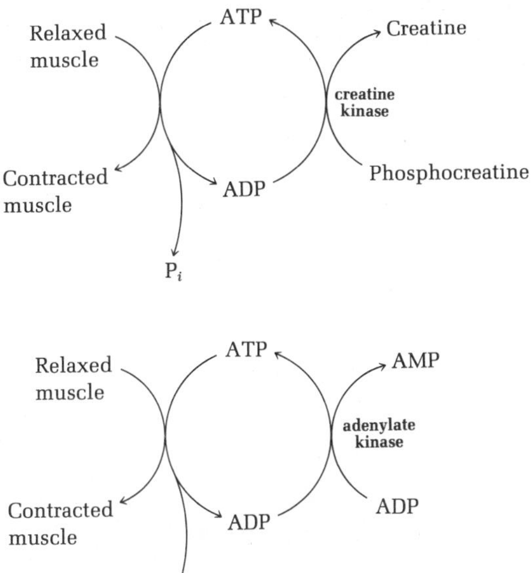

Figure 14-17
Two ways in which the ATP level is maintained in rapidly contracting anaerobic skeletal muscle.

can supplement phosphocreatine as a source of ATP during muscle contraction.

There Are Other Energy-Rich Nucleoside 5′-Triphosphates besides ATP

Uridine triphosphate (UTP), *guanosine triphosphate* (GTP), and *cytidine triphosphate* (CTP) are phosphorylated ribonucleotides analogous in structure to ATP (Figure 14-18) and with the same $\Delta G^{\circ\prime}$ for hydrolysis. They are present in all cells but in much lower concentrations than ATP. Also found in cells in low concentrations are the corresponding *deoxyribonucleoside 5′-triphosphates: 2′-deoxyadenosine 5′-triphosphate* (dATP), *2′-deoxyguanosine 5′-triphosphate* (dGTP), *2-deoxycytidine 5′-triphosphate* (dCTP), and *2′-deoxythymidine 5′-triphosphate* (dTTP). Although ATP is the mainstream carrier of phosphate groups in the cell, the other types of nucleoside 5′-triphosphates are specialized to serve in certain biosynthetic pathways. They acquire their terminal phosphate groups from ATP in reactions catalyzed by Mg^{2+}-dependent enzymes called *nucleoside diphosphokinases*, which promote the following types of reversible reactions:

$$\text{ATP} + \text{UDP} \rightleftharpoons \text{ADP} + \text{UTP}$$

$$\text{ATP} + \text{GDP} \rightleftharpoons \text{ADP} + \text{GTP}$$

$$\text{ATP} + \text{CDP} \rightleftharpoons \text{ADP} + \text{CTP}$$

$$\text{GTP} + \text{UDP} \rightleftharpoons \text{GDP} + \text{UTP}$$

$$\text{ATP} + \text{dCDP} \rightleftharpoons \text{ADP} + \text{dCTP}$$

$$\text{GTP} + \text{dADP} \rightleftharpoons \text{GDP} + \text{dATP}$$

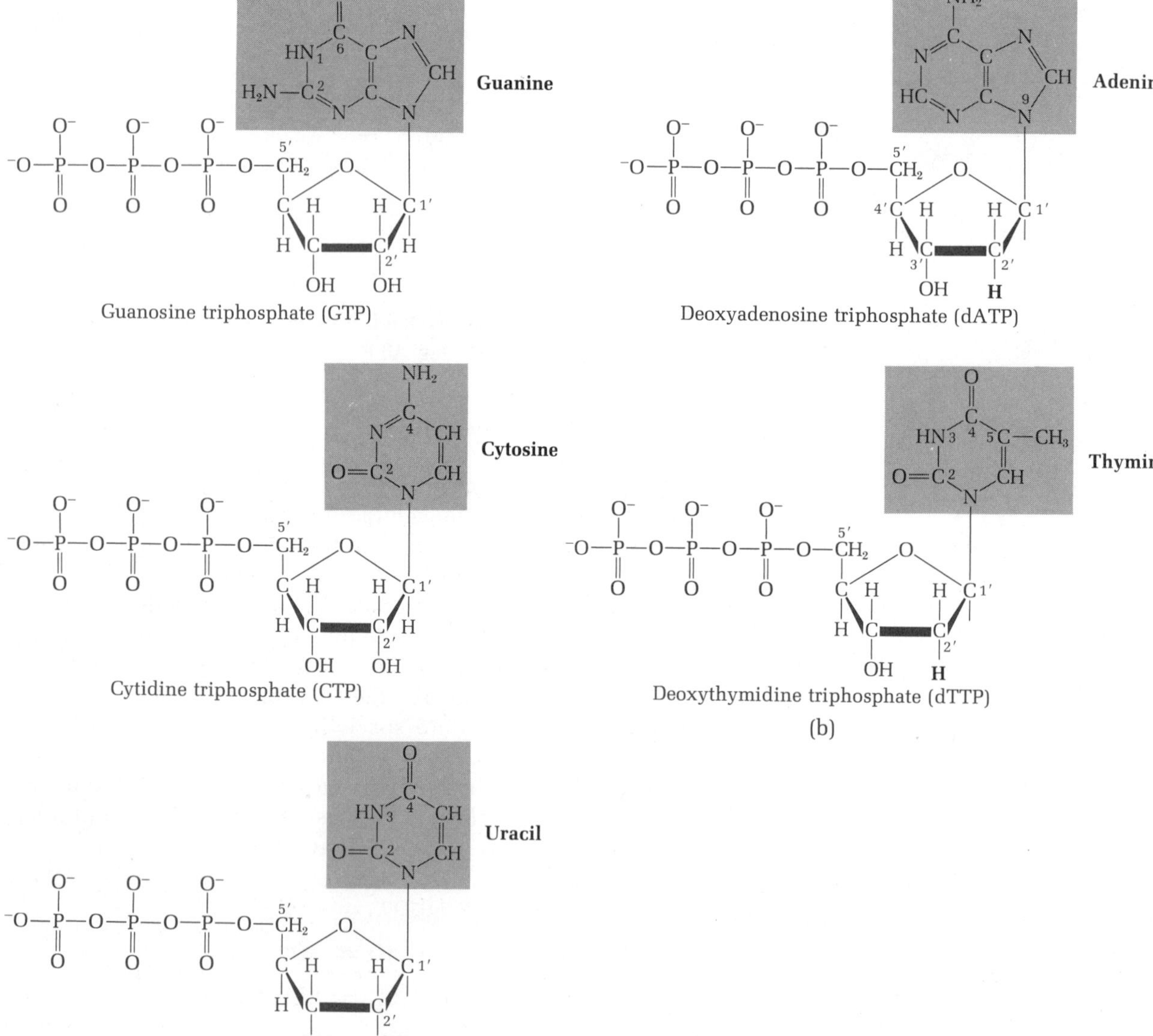

Figure 14-18
(a) The four nucleoside 5′-triphosphates. Each has a characteristic base (color). (b) Deoxynucleoside 5′-triphosphates contain a hydrogen instead of a hydroxyl at position 2′. Deoxythymidine triphosphate is the precursor of the thymidylate residues of DNA. These are not found in RNA, which contains uridylate residues instead (derived from uridine triphosphate).

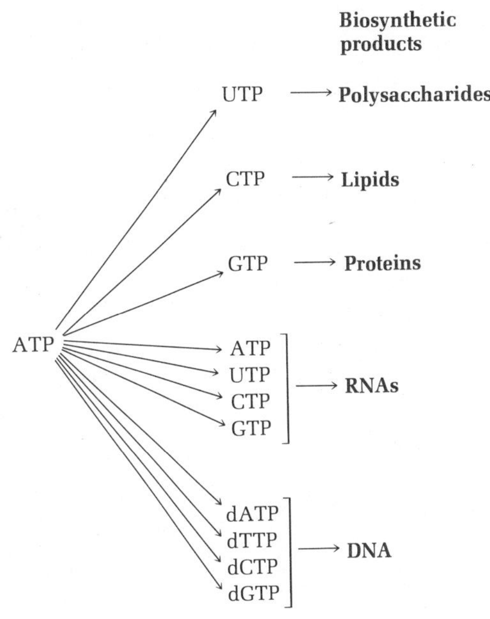

Figure 14-19
Channeling of ATP energy into different biosynthetic pathways via different nucleoside and deoxynucleoside triphosphates.

Figure 14-19 shows how the other nucleoside and deoxynucleoside triphosphates serve to channel energy and building blocks into specific metabolic pathways for the biosynthesis of lipids and proteins, and particularly for the biosynthesis of DNA and RNA.

The ATP System Functions in a Dynamic Steady State

How do we know that the ATP system actually functions to transfer phosphate-group energy in *intact* cells as indicated in Figure 14-3? This can be answered by experiments with radioactive phosphorus (^{32}P) as a tracer. One way is to determine the rate of turnover of the terminal phosphate group of the ATP present in cells or tissues with the use of inorganic phosphate labeled with ^{32}P. Labeled inorganic phosphate is introduced into respiring cells, and the ATP is isolated from samples of the cells at short time intervals thereafter. It is found that although the concentration of ATP does not change, its terminal phosphate group quickly becomes radioactive, indicating that the terminal or γ phosphate group is rapidly lost and just as rapidly replaced with radioactive phosphate from the pool of inorganic phosphate. This turnover process continues until the specific radioactivity of the terminal phosphate group of ATP is the same as that of the inorganic phosphate pool. The rate of turnover of the terminal phosphate group of ATP is very high. The half-time for replacement of the terminal phosphate group of ATP in respiring liver cells is only a minute or two; in aerobic bacterial cells, which respire at much higher rates than animal cells, it is a matter of only seconds. On the other hand the α phosphate group of ATP attached directly to the D-ribose part of the molecule turns over very slowly.

Despite the very high rate of turnover of the terminal phosphate group of ATP in living cells, the total concentration of ATP stays constant and we thus have a dynamic steady state. In this condition the rate of use of ATP to supply energy, with loss of the terminal phosphate group, is exactly counterbalanced by the rate of the rephosphorylation of ADP to form ATP, coupled to the energy-yielding oxidation of cell fuels. The rate of ATP synthesis at the expense of cell fuels is regulated, so that ATP is formed from ADP and phosphate at a rate just sufficient to provide the energy needs of the cell at any given time.

Summary

Chemical reactions proceed in such a direction that at equilibrium the entropy S of the system plus surroundings is at a maximum and the free energy G of the reacting molecules is at a minimum. Every chemical reaction has a characteristic standard-free-energy change $\Delta G^{\circ\prime}$ at the standard temperature of 25°C and the standard pressure of 1.0 atm with all reactants and products at 1 M concentration and the pH at 7.0. $\Delta G^{\circ\prime}$ can be calculated from the equilibrium constant K'_{eq} by the equation $\Delta G^{\circ\prime} = -2.30\ RT \log K'_{eq}$. The $\Delta G^{\circ\prime}$ of hydrolysis of ATP to ADP and phosphate is -7.3 kcal/mol at pH 7.0 and 25°C. Some phosphorylated compounds, such as 3-phosphoglyceroyl phosphate and phosphoenolpyruvate, which are generated during breakdown of glucose to lactate, have much larger negative $\Delta G^{\circ\prime}$ values for their hydrolysis than ATP and may be regarded as super high-energy compounds. Other phosphates, such as glucose 6-phosphate, have lower $\Delta G^{\circ\prime}$ values than ATP and are low-energy phosphate compounds. Phosphate groups can be transferred from super high-energy phosphates generated in catabolism to ADP to yield ATP by the action of specific kinases. Other specific kinases can transfer the terminal phosphate group of ATP to certain acceptor molecules to form low-energy phosphate compounds, which thus become energized for biosynthetic reactions. Consequently, ATP is the obligatory common intermediate or carrier of phosphate groups in metabolism. ATP also carries energy to the contractile filaments of actin and myosin of skeletal muscle, enabling them to slide along each other to bring about muscle shortening; simultaneously, ATP is hydrolyzed to ADP and phosphate. Phosphocreatine is a temporary storage form of high-energy phosphate groups in muscle and nerve cells. It can donate its phosphate group to ADP by the action of creatine kinase. ATP also provides chemical energy to membrane ATPases, energizing them to transport H^+ and other cations across membranes against concentration gradients.

ATP may undergo loss of either an orthophosphate or a pyrophosphate group during its use in biosynthetic reactions, to form either ADP or AMP, respectively. AMP formed in pyrophosphate cleavage reactions is rephosphorylated to ADP by the adenylate kinase reaction, ATP + AMP $\rightleftharpoons$ 2ADP. Other nucleoside 5′-triphosphates such as GTP, UTP, CTP, dATP, dTTP, etc., also participate as carriers of high-energy phosphate groups, which they channel into different biosynthetic routes; such triphosphates are also precursors in nucleic acid biosynthesis. The terminal phosphate group of ATP undergoes constant and extremely rapid replacement by inorganic phosphate in intact respiring cells, in a dynamic steady state where the rate of use of the terminal phosphate of ATP is exactly counterbalanced by the rate of resynthesis from phosphate and ADP.

References

Books

ATKINSON, D. E.: *Cellular Energy Metabolism and Its Regulation,* Academic Press, New York, 1977. A valuable book stressing the importance of the energy charge in the regulation of cell metabolism.

BECKER, W. M.: *Energy and the Living Cell,* Harper & Row, New York, 1977. Many problems and solutions.

BLUM, H. F.: *Time's Arrow and Evolution,* 3d ed., Princeton University Press, Princeton, N.J., 1968. Essays and reflections on entropy in biology.

BRODA, E.: *The Evolution of Bioenergetic Processes,* Pergamon, Oxford, 1975. Comparative and evolutionary aspects of bioenergetics.

KREBS, H. A., and H. L. KORNBERG: *Energy Transformations in Living Matter,* Springer, New York, 1967. A classical analysis of the energetics of glycolysis and respiration.

LEHNINGER, A. L.: *Bioenergetics,* 2d ed., Benjamin, Menlo Park, Calif., 1971. An introductory book on different aspects of cell bioenergetics.

LIPMANN, F.: *Wanderings of a Biochemist,* Wiley, New York, 1971. Reprints of classic papers by a pioneer in bioenergetics, as well as personal reminiscences.

MILLER, G. T., JR.: *Energetics, Kinetics and Life,* Academic, New York, 1970. Bioenergetics and ecology.

WOOD, W. B., J. H. WILSON, R. M. BENBOW, and L. E. HOOD: *Biochemistry: A Problems Approach,* 2d ed., Benjamin, Inc., Menlo Park, Calif., 1981.

Some Articles

ERECINSKA, M., and D. F. WILSON: "Homeostatic Regulation of Cellular Energy Metabolism," *Trends Biochem. Sci.,* **3**:221–223 (1978).

GATES, D. M.: "The Flow of Energy in the Biosphere," *Sci. Am.,* **224**: 88–100, September (1971).

INGRAHAM, L. L., and A. B. PARDEE: "Free Energy and Entropy in Metabolism," pp. 1–46 in D. M. Greenberg (ed.), *Metabolic Pathways,* 3d ed., vol. 1, Academic, New York, 1967.

SATIR, P.: "How Cilia Move," *Sci. Am.,* **231**:44–52, October (1974).

TRIBUS, M., and E. C. MCIRVINE: "Energy and Information," *Sci. Am.,* **225**:179–188, September (1971).

Problems

1. *Calculation of $\Delta G^{\circ\prime}$ From Equilibrium Constants* Calculate the standard-free-energy changes of the following metabolically important enzyme-catalyzed reactions at 25°C from the equilibrium constants given (pH 7.0):

(a) Glutamate + oxaloacetate $\xrightleftharpoons[]{\text{aspartate transaminase}}$ aspartate + α-ketoglutarate

$K'_{eq} = 6.8$

(b) Dihydroxyacetone phosphate $\xrightleftharpoons[]{\text{triose phosphate isomerase}}$ glyceraldehyde 3-phosphate $K'_{eq} = 0.0475$

(c) Fructose 6-phosphate + ATP $\xrightleftharpoons[]{\text{phosphofructokinase}}$ fructose 1,6-diphosphate + ADP $K'_{eq} = 254$

2. *Calculation of Equilibrium Constants from $\Delta G^{\circ\prime}$* Calculate the equilibrium constants for the following reactions at pH 7.0 and 25°C, using the $\Delta G^{\circ\prime}$ values of Table 14-3:

(a) Glucose 6-phosphate + H_2O $\xrightarrow{\text{glucose 6-phosphatase}}$ glucose + phosphate

(b) Lactose + H_2O $\xrightarrow{\beta\text{-galactosidase}}$ glucose + galactose

(c) Malate $\xrightarrow{\text{fumarase}}$ fumarate + H_2O

3. *The $\Delta G^{\circ\prime}$ for Coupled Reactions* Glucose 1-phosphate is converted into fructose 6-phosphate in two successive reactions:

Glucose 1-phosphate ⟶ glucose 6-phosphate
Glucose 6-phosphate ⟶ fructose 6-phosphate

Using the $\Delta G^{\circ\prime}$ values in Table 14-3, calculate the equilibrium constant K'_{eq} for the sum of the two reactions at 25°C

Glucose 1-phosphate ⟶ fructose 6-phosphate

4. *Strategy for Overcoming an Unfavorable Reaction: ATP-Dependent Chemical Coupling* The phosphorylation of glucose to glucose 6-phosphate is the initial step in the catabolism of glucose. The direct phosphorylation of glucose by inorganic phosphate is described by the equation

Glucose + phosphate ⟶ glucose 6-phosphate + H_2O

$\Delta G^{\circ\prime} = +3.3$ kcal/mol

(a) Calculate the equilibrium constant for the above reaction. In the rat liver cell the physiological concentrations of glucose and phosphate are maintained at approximately 4.8 m*M*. What is the equilibrium concentration of glucose 6-phosphate obtained by the direct phosphorylation of glucose by inorganic phosphate? Does this route represent a reasonable metabolic route for the catabolism of glucose? Explain.
(b) In principle, at least, one way to increase the concentration of glucose 6-phosphate is to drive the equilibrium reaction to the right by increasing the intracellular concentrations of glucose and phosphate. Assuming a fixed concentration of phosphate at 4.8 mM, how high would the intracellular concentration of glucose have to be raised to have an equilibrium concentration of glucose 6-phosphate of 250 μ*M* (normal phy-

siological concentration)? Would this route be a physiologically reasonable approach given that the maximum solubility of glucose is less than 1 *M*?

(c) As described in this chapter, the phosphorylation of glucose in the cell is coupled to the hydrolysis of ATP; that is, part of the free energy of ATP hydrolysis is utilized to effect the unfavorable phosphorylation of glucose.

$$\text{Glucose} + \text{phosphate} \longrightarrow \text{glucose 6-phosphate} + H_2O \qquad \Delta G^{\circ\prime} = +3.3 \text{ kcal/mol}$$

$$\text{ATP} + H_2O \longrightarrow \text{ADP} + \text{phosphate} \qquad \Delta G^{\circ\prime} = -7.3 \text{ kcal/mol}$$

$$\text{Net: Glucose} + \text{ATP} \longrightarrow \text{glucose 6-phosphate} + \text{ADP}$$

Calculate $\Delta G^{\circ\prime}$ and K'_{eq} for the net reaction. When the ATP-dependent phosphorylation of glucose is carried out, what concentration of glucose is needed to achieve a 250 μM intracellular concentration of glucose 6-phosphate when the concentrations of ATP and ADP are 3.38 and 1.32 m*M*, respectively? Does this coupling process provide a feasible route, at least in principle, for the phosphorylation of glucose as it occurs in the cell? Explain.

(d) Although coupling ATP hydrolysis to glucose phosphorylation makes thermodynamic sense, how this coupling is to take place has not been specified. Since coupling requires a common intermediate, one conceivable route is to use the ATP hydrolysis to raise the intracellular concentration of inorganic phosphate and thus drive the unfavorable phosphorylation of glucose by inorganic phosphate. Is this a reasonable route? Explain.

(e) The ATP-coupled phosphorylation of glucose is catalyzed in the liver cell by the enzyme glucokinase. This enzyme binds ATP and glucose to form a glucose-ATP-enzyme complex, and the phosphate is transferred directly from ATP to glucose. Explain the advantages of this route.

5. *Calculations of $\Delta G^{\circ\prime}$ for ATP-Coupled Reactions* From data in Table 14-5 calculate the $\Delta G^{\circ\prime}$ value for the reactions

(a) Phosphocreatine + ADP ⟶ creatine + ATP

(b) ATP + fructose ⟶ ADP + fructose 6-phosphate

6. *Calculations of $\Delta G'$ at Physiological Concentrations* Calculate the physiological $\Delta G'$ (not $\Delta G^{\circ\prime}$) for the reaction

$$\text{Phosphocreatine} + \text{ADP} \longrightarrow \text{creatine} + \text{ATP}$$

at 25°C as it occurs in the cytosol of brain cells, in which phosphocreatine is present at 4.7 m*M*, creatine at 1.0 m*M*, ADP at 0.20 m*M*, and ATP at 2.6 m*M*.

7. *Free Energy Required for ATP Synthesis under Physiological Conditions* In the cytosol of rat liver cells, the mass-action ratio is

$$Q = \frac{[\text{ATP}]}{[\text{ADP}][P_i]} = 5.33 \times 10^2$$

Calculate the free energy required to synthesize ATP in the rat liver cell.

8. *Daily ATP Utilization by Human Adults*
   (a) A total of 7.3 kcal of free energy is needed to synthesize ATP from ADP and inorganic phosphate (P_i) when the reactants and products are both at 1 *M* concentration (standard state). Since the actual physiological concentrations of ADP, P_i, and ATP in the cells are not 1 *M*, the free energy required to synthesize ATP under physiological conditions is different from $\Delta G^{\circ\prime}$. Calculate the free energy required to synthesize ATP in the human liver cell when the physiological concentrations of ATP, ADP, and P_i are 3.5, 1.50, and 5.0 m*M*, respectively.
   (b) A normal 68-kg (150-lb) adult requires a caloric intake of 2000 kcal of food per 24-h period. This food is metabolized and the free energy used to synthesize ATP, which is then utilized to do the body's daily chemical and mechanical work. Assuming that the efficiency of converting food energy into ATP is 50 percent, calculate the weight of ATP utilized by a human adult in a 24-h period. What percent of the body weight does this represent?
   (c) Although adults synthesize large amounts of ATP daily, their body weight, structure, and composition do not change significantly during this period. Explain this apparent contradiction.

9. *ATP Reserve in Muscle Tissue* The ATP concentration in muscle tissue (approximately 70 percent water) is about 8.0 m*M*. During strenuous activity each gram of muscle tissue uses ATP at the rate of 300 μmol/min for contraction.

   (a) How long would the reserve ATP last during a 100-meter dash?
   (b) The phosphocreatine level in muscle is about 40.0 m*M*. How does this help extend the reserve of muscle ATP?
   (c) Given the size of the reserve ATP pool, how can a person run a marathon?

10. *Cleavage of ATP to AMP and PP_i During Metabolism* The synthesis of the activated form of acetate (acetyl-CoA) is carried out in an ATP-dependent process

$$\text{Acetate} + \text{CoA} + \text{ATP} \longrightarrow \text{acetyl-CoA} + \text{AMP} + \text{PP}_i$$

   (a) The $\Delta G^{\circ\prime}$ for the hydrolysis of acetyl-CoA to acetate and CoA and of ATP to AMP and PP_i are -7.5 and -7.3 kcal, respectively. Calculate the value of $\Delta G^{\circ\prime}$ for the ATP-dependent synthesis of acetyl-CoA.
   (b) Almost all cells contain the enzyme inorganic pyrophosphatase, which catalyzes the hydrolysis of PP_i to inorganic phosphate (P_i). What effect does the presence of this enzyme have on the synthesis of acetyl-CoA? Explain.

CHAPTER 15

Glycolysis: A Central Pathway of Glucose Catabolism

Now that we have examined the organizing principles of cell metabolism and bioenergetics, we are finally ready to see how the chemical energy inherent in the structure of the glucose molecule is released in a useful form to perform various kinds of biological work in cells. Remember that glucose is the major fuel of most organisms, that it is rich in energy, and that it can quickly be mobilized from glycogen stores when sudden demands for energy are made.

In this chapter we shall consider *glycolysis*, the process by which the glucose molecule, which has 6 carbon atoms, is enzymatically degraded in a sequence of 10 enzyme-catalyzed reactions to yield two molecules of pyruvate, which has 3 carbon atoms. During the sequential reactions of glycolysis much of the free energy released from glucose is conserved in the form of ATP. Since glycolysis (from Greek words meaning "sugar" and "dissolution") is the best-understood central metabolic pathway, we shall examine it in some detail; the principles involved in its function and regulation are common to all pathways of cell metabolism. Also to be considered are the "feeder" pathways leading from glycogen, disaccharides, and monosaccharides into glycolysis.

Glycolysis Is a Central Pathway in Most Organisms

Glycolysis is an almost universal central pathway of glucose catabolism, not only in animals and plants, but also in a great many microorganisms. The glycolytic sequence of reactions differs from one species to another only in how its rate is regulated and in the subsequent metabolic fate of the pyruvate formed.

There are three important routes taken by the pyruvate after glycolysis. In aerobic organisms, glycolysis constitutes only the first stage in the complete aerobic degradation of glucose to CO_2 and H_2O (Figure 15-1). The pyruvate so formed is then oxi-

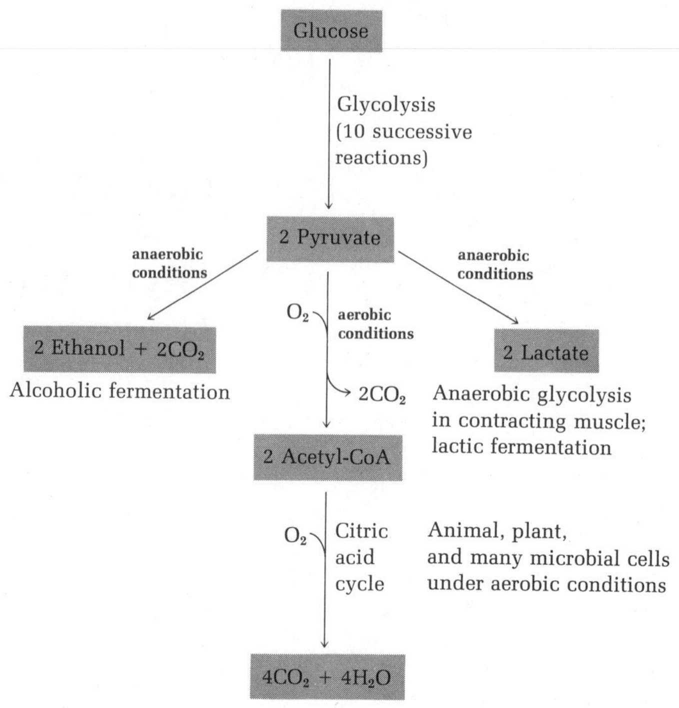

Figure 15-1
Pyruvate, the end product of glycolysis, follows different catabolic pathways depending on the organism and the metabolic conditions.

dized, with loss of its carboxyl group as CO_2, to form the acetyl group of *acetyl–coenzyme A* (page 257). Then the acetyl group is oxidized completely to CO_2 and H_2O by the *citric acid cycle,* with the intervention of molecular oxygen (Figure 15-1). This is the pathway taken by pyruvate in aerobic animal and plant cells.

The second pathway of pyruvate is its reduction to lactate. When some animal tissues must function anaerobically, particularly vigorously contracting skeletal muscle, the pyruvate formed from glucose cannot be oxidized further for lack of oxygen. Under these conditions pyruvate formed by glycolysis is reduced to *lactate.* In skeletal muscle this process, called *anaerobic glycolysis,* is an important source of ATP energy in very intense physical activity. Lactate is also the product of glycolysis in anaerobic microorganisms that carry out *lactic acid fermentation* (Figure 15-1). Production of lactic acid from sugar by lactic acid bacteria causes the souring of milk and the mildy acid taste of sauerkraut, made by the fermentation of cabbage.

The third major pathway of pyruvate leads to ethanol. In some microorganisms, e.g., brewer's yeast, the pyruvate formed from glucose by glycolysis is converted anaerobically into ethanol and CO_2, a process called *alcoholic fermentation* (Figure 15-1). *Fermentation* is a general term denoting the *anaerobic* degradation of glucose or other organic nutrients into various products (characteristic for different organisms) for the purpose of obtaining energy in the form of ATP (page 378). Since living organisms first arose in an atmosphere lacking oxygen, anaerobic breakdown of glucose is the most ancient type of biological mechanism for obtaining energy from organic fuel molecules.

ATP Formation Is Coupled to Glycolysis

During glycolysis much of the free energy of the glucose molecule is conserved in the form of ATP. This is easiest to show if we write the complete balanced equation for anaerobic glycolysis as it occurs in very active skeletal muscle:

$$\text{Glucose} + 2\text{P}_i + 2\text{ADP} \longrightarrow 2\ \text{lactate}^- + 2\text{H}^+ + 2\text{ATP} + 2\text{H}_2\text{O}$$

Thus for each molecule of glucose degraded, two molecules of ATP are generated from ADP and P_i. We can now resolve the equation of anaerobic glycolysis into two processes: (1) the conversion of glucose into lactate, which results in release of free energy,

$$\text{Glucose} \longrightarrow 2\ \text{lactate}^- + 2\text{H}^+ \qquad (1)$$
$$\Delta G_1^{\circ\prime} = -47.0\ \text{kcal/mol}$$

and (2) the formation of ATP from ADP and phosphate, which requires input of free energy:

$$2\text{P}_i + 2\text{ADP} \longrightarrow 2\text{ATP} + 2\text{H}_2\text{O} \qquad (2)$$
$$\Delta G_2^{\circ\prime} = +2(7.30) = +14.6\ \text{kcal/mol}$$

These two processes cannot proceed independently; they are obligatorily coupled. However, when the energy-yielding and energy-using reactions are written in this way, we clearly see that the breakdown of 1 mol of glucose to lactate would yield, under standard conditions, more than enough free energy (47.0 kcal) to generate 2 mol of ATP from ADP and phosphates, which requires input of 2(7.3) = +14.6 kcal. At the actual *intracellular* concentrations of ATP, ADP, and phosphate, as well as glucose and lactate, the efficiency of recovery of the energy of glycolysis in the form of ATP is over 60 percent (page 377).

If we now write the sum of equations (1) and (2), we can also determine the *overall* standard-free-energy change of glycolysis, including ATP formation, as the algebraic sum ΔG_s of $\Delta G_1^{\circ\prime}$ and $\Delta G_2^{\circ\prime}$

$$\text{Glucose} + 2\text{P}_i + 2\text{ADP} \longrightarrow 2\ \text{lactate}^- + 2\text{H}^+ + 2\text{ATP} + 2\text{H}_2\text{O} \qquad (3)$$
$$\Delta G_s^{\circ\prime} = \Delta G_1^{\circ\prime} + \Delta G_2^{\circ\prime} = -47.0 + 14.6 = -32.4\ \text{kcal/mol}$$

Thus we see that the overall coupled reaction of glycolysis proceeds with a very large decrease in free energy. Either under standard conditions or as it takes place in cells, glycolysis is an essentially irreversible reaction driven to completion by the large decrease in free energy.

Much Free Energy Remains in the Products of Glycolysis

Glycolysis releases only a small fraction of the total available energy of the glucose molecule. When glucose is oxidized completely to CO_2 and H_2O, the total standard-free-energy change is -686 kcal/mol (see page 369). The glycolytic degradation of glucose to two molecules of lactate ($\Delta G^{\circ\prime} = -47.0$ kcal/mol) therefore yields only $(47/686)100 = 6.9$ percent of the total energy that can be set free if the glucose is oxidized completely to CO_2 and H_2O. The two molecules of lactate formed by glycolysis contain most of the biologically available energy of the glucose molecule. This can be released only when the products of glycolysis are oxidized completely to CO_2 and H_2O, with molecular oxygen as acceptor, as we shall see in the next chapter. Anaerobic glycolysis to the stage of lactate is not to be considered wasteful, however. On the contrary, glycolysis is a marvelously engineered process that succeeds in extracting free energy from glucose *without* oxidizing it. In animals the lactate formed by active muscles, which diffuses into the blood, can be recycled; it is taken up by the liver and converted back into blood glucose during the recovery from severe muscle activity. In some animals anaerobic glycolysis is extremely important in muscular activity (Box 15-1).

Glycolysis Has Two Phases

Before we examine the individual enzymatic steps let us first take a birds'-eye view of glycolysis. The breakdown of the 6-carbon glucose into two molecules of the 3-carbon pyruvate is brought about by the sequential action of 10 enzymes, each of which has been isolated in pure form from several different species and studied in detail. The first five steps constitute the *preparatory phase* (Figure 15-2). In these reactions glucose is enzymatically phosphorylated by ATP, first at the 6 carbon atom and later at carbon 1, to yield *fructose 1,6-diphosphate,* which is then cleaved in half to yield two molecules of the 3-carbon *glyceraldehyde 3-phosphate,* the product of the first phase of glycolysis. Note that two molecules of ATP must be invested to activate, or prime, the glucose molecule and prepare it for its cleavage into two 3-carbon pieces; later there will be a good return on this investment. Other hexoses, particularly D-fructose, D-galactose, and D-mannose, also may enter the preparatory phase of glycolysis following their phosphorylation, as we shall see later in this chapter. To sum up: *the preparatory phase of glycolysis serves to collect the carbon chains of all the metabolized hexoses in the form of one common product, glyceraldehyde 3-phosphate.*

The second phase of glycolysis, promoted by the remaining five enzymes, represents the payoff of glycolysis, in which the energy freed when two molecules of glyceraldehyde 3-phosphate are converted into two molecules of pyruvate is conserved by the coupled phosphorylation of four molecules of ADP to ATP (Figure 15-2). Although four molecules of ATP are

Box 15-1 Anaerobic glycolysis, oxygen debt, alligators, and coelacanths

Most vertebrates are essentially aerobic organisms, which first convert glucose into pyruvate by glycolysis and then oxidize the pyruvate completely to CO_2 and H_2O using molecular oxygen. *Anaerobic* glycolysis comes into play in most vertebrates, including human beings, during short bursts of extreme muscular activity, e.g., in a 100-m sprint, during which oxygen cannot be carried to the muscles fast enough to oxidize pyruvate for generating ATP. Instead, the muscles use their stored glycogen as fuel to generate ATP by anaerobic glycolysis, with production of lactate as the end product. In a sprint the lactate in the blood thus builds up to high concentrations. It is slowly converted back into glucose by the liver in the subsequent rest or recovery period, during which oxygen is consumed at a gradually diminishing rate until the respiratory rate returns to normal. The excess oxygen consumed in the recovery period represents the repayment of the *oxygen debt*. This is the amount of oxygen required to supply sufficient ATP during recovery respiration to regenerate the liver and muscle glycogen "borrowed" to carry out intense muscular activity in the sprint.

Use of anaerobic glycolysis as an energy source for muscular contraction is particularly pronounced in white muscles. Although most skeletal muscles contain both white and red fibers, some muscles have nearly all red and some nearly all white fibers. The flight muscles of the domestic turkey are white; they are capable of only very short flights. The leg muscles of the horse, which is capable of sustained running, are largely made up of red fibers. White muscle fibers, which contain few mitochondria, have very fast rates of contraction. They obtain most of their ATP energy from anaerobic glycolysis and thus can function at maximum rates for only short periods, due to the inefficient use of their glycogen stores. In contrast, red muscles contract more slowly, are rich in mitochondria, obtain most of their energy by oxidation of their fuels with oxygen, and are capable of sustained activity over long periods.

Generally speaking, the circulatory systems of small animals can bring oxygen to their muscles fast enough to avoid having to use muscle glycogen anaerobically. For example, migrating birds often fly great distances at high speeds without rest and without incurring an oxygen debt. Many running animals of moderate size also have an essentially aerobic metabolism in their red skeletal muscle. In much larger animals, however, the circulatory system cannot completely sustain aerobic metabolism of their muscles during long bursts of muscular activity. Such animals generally are slow-moving under normal circumstances and engage in intense muscular activity only in the gravest emergencies, because such bursts of activity require long recovery periods to repay the oxygen debt.

Alligators and crocodiles, for example, are normally sluggish and torpid. Yet when provoked these animals are capable of lightning-fast charges and dangerous lashings of their powerful tails. Such intense bursts of activity are short and must be followed by long periods of recovery. The fast emergency movements require anaerobic glycolysis to generate ATP in their white skeletal muscles. Since the stores of glycogen in muscles are not large, they are rapidly expended in intense muscular activity. Moreover, in such bursts of action, lactate, the product of anaerobic glycolysis, reaches very high concentration in muscles and extracellular fluid. Whereas a trained athlete can recover from a 100-m sprint in 30 min or less, an alligator may require many hours of rest and extra oxygen consumption to clear the excess lactate from the blood and regenerate muscle glycogen.

Other large animals, e.g., the elephant and rhinoceros, have similar metabolic problems, as do diving mammals, e.g., whales and seals. Dinosaurs and other huge prehistoric animals probably had to depend on anaerobic glycolysis for muscle energy followed by very long recovery periods, during which they were vulnerable to attack by smaller predators better able to use oxygen and thus better adapted to continuous, sustained muscular activity.

Deep-sea explorations have revealed many species of marine life at great ocean depths, where the oxygen concentration is near zero. For example, the primitive coelacanth, a large fish recovered from depths of 12,000 ft or more by fishermen off the coast of South Africa, has been found to have an essentially anaerobic metabolism in virtually all its tissues, converting carbohydrates by anaerobic mechanisms into lactate and other products, most of which must be excreted. Indeed, some marine vertebrates actually ferment glucose, not to lactate but to ethanol and CO_2, in order to obtain energy in the form of ATP.

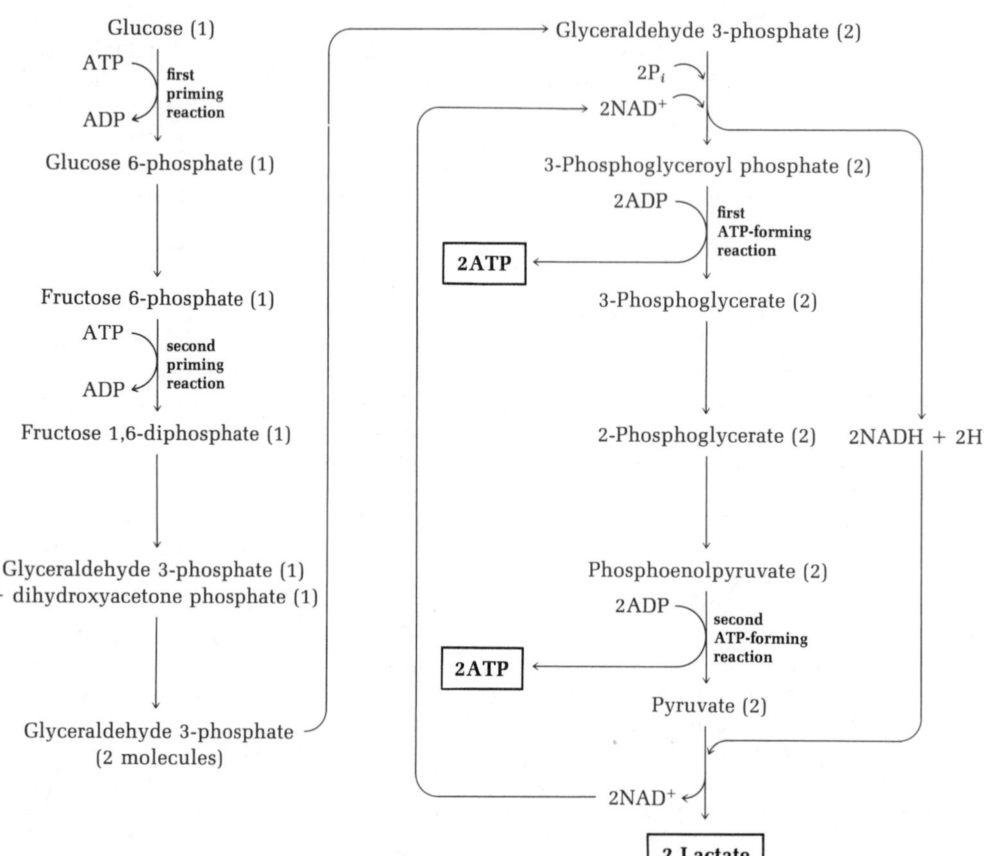

Figure 15-2
The two phases of anaerobic glycolysis. The numerals after the 3-carbon intermediates indicate that two molecules of each undergo the designated reaction.

formed in the second phase, the net overall yield is two molecules of ATP per molecule of glucose used, since two molecules of ATP were invested in the first phase of glycolysis.

Three different types of chemical transformation take place during glycolysis: (1) the degradation of the carbon skeleton of glucose to yield pyruvate, i.e., the pathway of the carbon atoms; (2) the phosphorylation of ADP to ATP by high-energy phosphate compounds formed during glycolysis, i.e., the pathway of phosphate groups, and (3) the transfer of hydrogen atoms or electrons. We shall trace these pathways as we discuss the sequential reactions of glycolysis.

One more point. In most kinds of cells the enzymes that promote glycolysis are present in dissolved form in the cell cytosol, the continuous aqueous medium of the cytoplasm (page 33). In contrast, the enzymes promoting the oxygen-requiring phase of carbohydrate oxidation are located in the mitochondrial membrane in eukaryotic cells and in the plasma membrane of prokaryotic cells.

Glycolysis Takes Place via Phosphorylated Intermediates

The first important point to note as we travel down the glycolytic sequence is that each of the nine metabolic intermediates between glucose and pyruvate is a phosphorylated compound (see Figure 15-2). There appear to be three functions of the phosphate groups.

1. The phosphate groups are completely ionized at pH 7, thus giving each of the intermediates of glycolysis a net negative charge. Since cell membranes are generally impermeable to molecules having an electric charge, the glycolytic intermediates cannot escape from the cell. Glucose can enter cells and lactate or pyruvate can leave only because cell membranes have specific transport systems that allow these particular molecules to pass.
2. The second function of the phosphate groups is obvious: they are essential components in the enzymatic conservation of metabolic energy since they are ultimately transferred to ADP to yield ATP.
3. The phosphate groups serve as recognition or binding groups required for the proper fit of the glycolytic intermediates to the active sites of their corresponding enzymes.

Nearly all the glycolytic enzymes require Mg^{2+} for activity. Since Mg^{2+} forms complexes with the phosphate groups of the glycolytic intermediates and of ADP and ATP (page 373), the substrate binding sites of many of the glycolytic enzymes appear to be specific for the Mg^{2+} complexes of the phosphorylated intermediates.

The First Phase of Glycolysis Results in Cleavage of the Hexose Chain

Figure 15-3 shows, with structural formulas, the enzymatic steps in the first phase of glycolysis. With this and Figure 15-2 to refer to, we shall now see how the 6-carbon chain of glucose is cleaved to yield two molecules of the 3-carbon glyceraldehyde phosphate.

Phosphorylation of Glucose

In the first step the D-glucose molecule is primed for subsequent reactions by its phosphorylation at the 6 position to yield glucose 6-phosphate (Figure 15-3) at the expense of ATP. This reaction, which is irreversible under intracellular conditions, is catalyzed by hexokinase, found in most animal, plant, and microbial cells

$$\text{ATP}^{4-} + \alpha\text{-D-glucose} \xrightarrow{Mg^{2+}} \text{ADP}^{3-} + \alpha\text{-D-glucose 6-phosphate}^{2-} + \text{H}^+$$
$$\Delta G^{\circ\prime} = -4.0 \text{ kcal/mol}$$

Figure 15-3
The reaction steps in the first phase of glycolysis. The enzyme names are shown in color. Numbers on the structures designate the carbon atoms. Glucose 6-phosphate, fructose 6-phosphate, and fructose 1,6-diphosphate are shown in open-chain form for simplicity, although they occur in their α-anomeric ring forms under cell conditions.

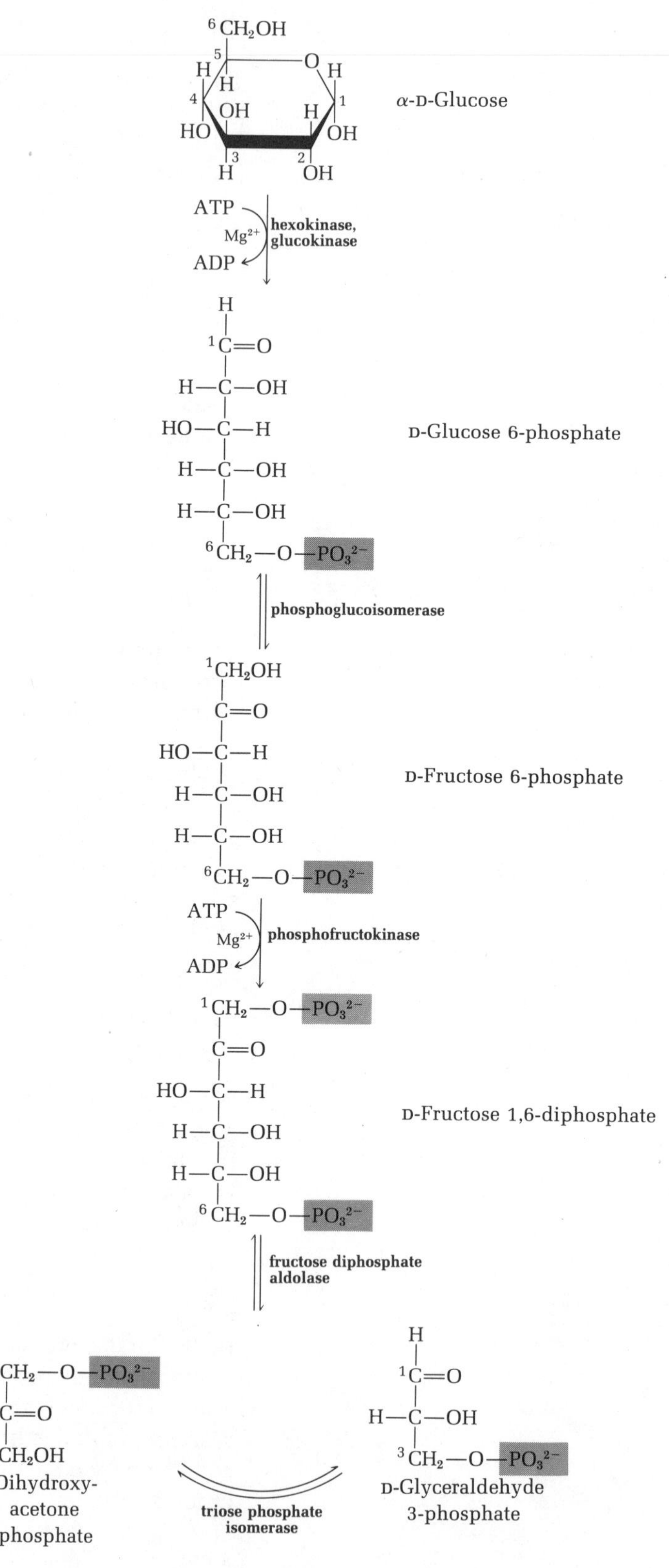

Hexokinase promotes the phosphorylation not only of D-glucose but also of certain other common hexoses, such as D-fructose and D-mannose. The hexokinase of yeast cells has been crystallized, and its three-dimensional structure has been studied in detail by x-ray analysis. The hexokinase molecule undergoes a profound change in its shape, an induced fit, when it binds the hexose molecule (page 232). Hexokinase requires Mg^{2+} because the true substrate of the enzyme is not ATP^{4-} but the $MgATP^{2-}$ complex (page 373).

Hexokinase occurs in different isozyme forms (page 239) in different organisms or tissues. Although they all catalyze the reaction in Figure 15-3, they differ with respect to their kinetic properties. In muscle cells, for example, the hexokinase has a low K_M for glucose (about 0.1 m*M*) and thus will phosphorylate blood glucose (4 to 5 m*M*) at maximal rates. Muscle hexokinase is strongly inhibited by its product, glucose 6-phosphate. This and other evidence has led to the conclusion that, in muscle, hexokinase is a regulatory enzyme in which glucose 6-phosphate is both a reaction product and an allosteric inhibitor. Whenever the concentration of glucose 6-phosphate in the cell rises above its normal level, it temporarily and reversibly inhibits hexokinase to bring the rate of its formation into balance with the rate of its utilization.

In the liver there is another form, called *glucokinase*, which is not found in other tissues. Glucokinase differs from isozymes of hexokinase in three respects: (1) it is specific for D-glucose and does not act on other hexoses; (2) it is not inhibited by glucose 6-phosphate; and (3) it has a much higher K_M for glucose (about 10 m*M*) than hexokinase. Liver glucokinase comes into play when the glucose concentration in the blood becomes high, e.g., after a meal rich in sugar. Under these conditions liver glucokinase acts on the excess blood glucose to convert it into glucose 6-phosphate for storage as liver glycogen. Glucokinase is very important in another respect: it is deficient in the disease *diabetes mellitus*. In this disease the pancreas fails to secrete insulin in normal amounts (Chapter 25), the blood glucose level is very high, and little liver glycogen is formed.

Conversion of Glucose 6-Phosphate to Fructose 6-Phosphate

Phosphoglucoisomerase, which has been isolated in highly purified form from muscle tissue, catalyzes the reversible isomerization of glucose 6-phosphate, an aldose, to yield *fructose 6-phosphate*, a ketose (Figure 15-3). This reaction involves a shift in the carbonyl oxygen from carbon atom 1 to carbon atom 2:

$$\alpha\text{-D-Glucose 6-phosphate} \xrightleftharpoons{Mg^{2+}} \alpha\text{-D-fructose 6-phosphate}$$
$$\Delta G^{\circ\prime} = +0.4 \text{ kcal/mol}$$

The reaction proceeds readily in either direction, as we might predict from the relatively small standard-free-energy change. Phosphoglucoisomerase requires Mg^{2+} and is specific for glucose 6-phosphate and fructose 6-phosphate.

Phosphorylation of Fructose 6-Phosphate to Fructose 1,6-Diphosphate

This is the second of the two priming reactions of glycolysis. *Phosphofructokinase* (Figure 15-3), which requires Mg^{2+}, catalyzes the transfer of a phosphate group from ATP to phosphorylate D-fructose 6-phosphate in the 1 position to yield *fructose 1,6-diphosphate*

$$\text{ATP} + \text{D-fructose 6-phosphate} \xrightarrow{Mg^{2+}} \text{ADP} + \text{D-fructose 1,6-diphosphate} + H^+ \qquad \Delta G^{\circ\prime} = -3.40 \text{ kcal/mol}$$

The phosphofructokinase reaction is essentially irreversible under the conditions existing in cells.

This reaction is the second important control point in glycolysis. Phosphofructokinase, like hexokinase, is a regulatory enzyme (Chapter 9), one of the most complex known. It is the major regulatory enzyme in muscle glycolysis. The activity of phosphofructokinase is accelerated whenever the cell's supply of ATP becomes depleted or there is an excess of ATP breakdown products, ADP and AMP, particularly the latter. It is inhibited whenever the cell has ample ATP and is well supplied by other fuels, such as citrate or fatty acids. The regulatory action of phosphofructokinase will be discussed further later.

Cleavage of Fructose 1,6-Diphosphate

This reaction is catalyzed by the enzyme *fructose diphosphate aldolase*, often simply called *aldolase*. This enzyme is readily isolated in crystalline form from rabbit muscle extracts. The reaction it catalyzes is a reversible aldol condensation (Figure 15-3). Fructose 1,6-diphosphate is cleaved reversibly to yield two different triose phosphates, *glyceraldehyde 3-phosphate*, an aldose, and *dihydroxyacetone phosphate*, a ketose:

$$\text{D-Fructose 1,6-diphosphate} \rightleftharpoons \text{dihydroxyacetone phosphate} + \text{D-glyceraldehyde 3-phosphate} \qquad \Delta G^{\circ\prime} = +5.73 \text{ kcal/mol}$$

The aldolase of animal tissues does not require Mg^{2+}, but in many microorganisms aldolase is a Zn^{2+}-containing enzyme. Although the aldolase reaction has a strongly positive standard-free-energy change, under the pH and concentration conditions existing in cells it can readily proceed in either direction. In the forward direction the reaction products are quickly removed by the next step.

Interconversion of the Triose Phosphates

Only one of the two triose phosphates formed by aldolase, namely, glyceraldehyde 3-phosphate, can be directly degraded in the subsequent reaction steps of glycolysis. However, dihydroxyacetone phosphate can be rapidly and reversibly converted into glyceraldehyde 3-phosphate by the fifth enzyme of the glycolytic sequence, *triose phosphate isomerase* (Figure 15-3):

$$\text{Dihydroxyacetone phosphate} \rightleftharpoons \text{D-glyceraldehyde 3-phosphate} \qquad \Delta G^{\circ\prime} = +1.83 \text{ kcal/mol}$$

Note that by this reaction carbon atoms 1, 2, and 3 of the starting glucose now become indistinguishable from carbon atoms 6, 5, and 4, respectively (Figure 15-4).

This reaction completes the first phase of glycolysis, in which the hexose molecule has been phosphorylated in the 1 and 6 positions and then cleaved to form, ultimately, two molecules of glyceraldehyde 3-phosphate. Other hexoses, such as D-fructose, D-mannose, and D-galactose, are also convertible into glyceraldehyde 3-phosphate, as we shall see later in this chapter.

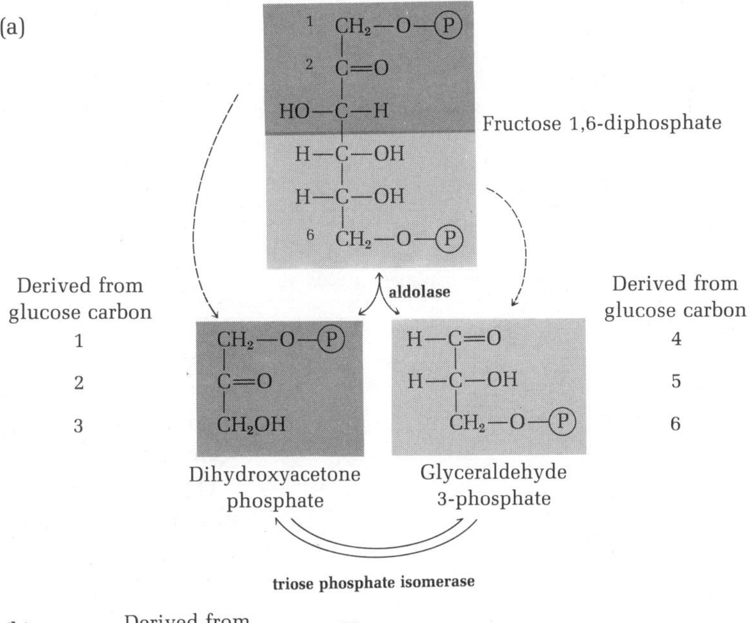

Figure 15-4
Fate of the carbon atoms of glucose in the formation of glyceraldehyde 3-phosphate. (a) *The aldolase and triose phosphate isomerase reactions, showing the origin of the two 3-carbon products.* (b) *As a result of the triose phosphate isomerase reaction, the two halves of the original glucose molecule are now two molecules of glyceraldehyde 3-phosphate. Each of the three carbon atoms of glyceraldehyde 3-phosphate are derived from either of two specific carbons of glucose, as shown. The numbering of the carbon atoms of glyceraldehyde 3-phosphate is not identical with the numbering of the carbon atoms of D-glucose. This is important for interpreting experiments with D-glucose in which a single carbon is labeled with a carbon isotope.*

Figure 15-5
The second phase of glycolysis.

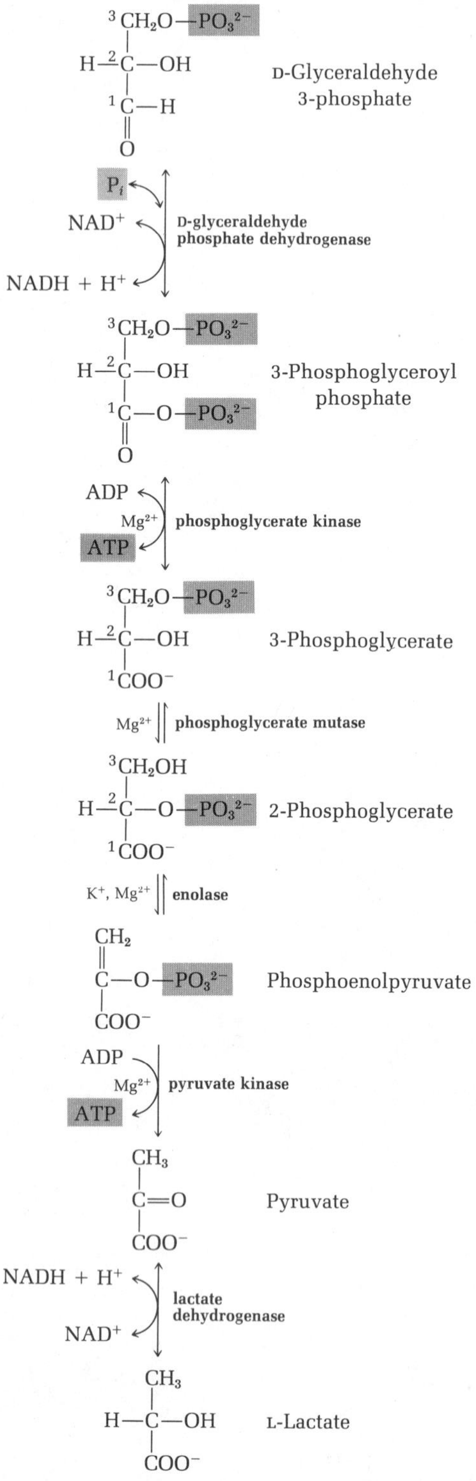

The Second Phase of Glycolysis Is Energy-Conserving

The second phase of glycolysis (see details in Figure 15-5) contains the energy-conserving phosphorylation steps, in which the free energy of the glucose molecule is conserved in the form of ATP. Since one molecule of glucose can yield two of glyceraldehyde 3-phosphate, both halves of the glucose molecule follow the same pathway in the second phase of glycolysis. The conversion of two molecules of glyceraldehyde 3-phosphate into two of pyruvate is accompanied by the formation of four molecules of ATP from ADP. However, the *net* yield of ATP per molecule of glucose degraded is only two, because two molecules of ATP were invested in the first phase of glycolysis to phosphorylate the two ends of the hexose molecule.

Oxidation of Glyceraldehyde 3-Phosphate to 3-Phosphoglyceroyl Phosphate

This is the first of the two energy-conserving reactions of glycolysis leading to the formation of ATP (Figure 15-5). The enzyme catalyzing this step is *glyceraldehyde phosphate dehydrogenase*, which promotes the reversible reaction

$$\text{D-Glyceraldehyde 3-phosphate} + NAD^+ + P_i \rightleftharpoons \text{3-phosphoglyceroyl phosphate} + NADH + H^+ \qquad \Delta G^{\circ\prime} = +1.5 \text{ kcal/mol}$$

In this complex reaction the aldehyde group of D-glyceraldehyde 3-phosphate is dehydrogenated, not to a free carboxyl group, as one might expect, but to a carboxylic anhydride with phosphoric acid, *3-phosphoglyceroyl phosphate*. This type of anhydride, called an *acyl phosphate*, has a very high standard free energy of hydrolysis ($\Delta G^{\circ\prime} = -11.8$ kcal/mol) and is thus a super high-energy phosphate (page 374). The standard free energy of hydrolysis of the 3-phosphate group at the other end of 3-phosphoglyceroyl phosphate is only about 3.2 kcal/mol. Much of the free energy of oxidation of the aldehyde group of glyceraldehyde 3-phosphate is conserved in the high-energy acyl phosphate group.

The acceptor of hydrogen in the glyceraldehyde phosphate dehydrogenase reaction is the coenzyme NAD^+ (Figure 15-6), the oxidized form of *nicotinamide adenine dinucleotide*, which contains the vitamin nicotinamide (page 255). The reduction of NAD^+ proceeds by the enzymatic transfer of a *hydride ion* ($:H^-$) from the aldehyde group of glyceraldehyde 3-phosphate to the 4 position of the nicotinamide ring of NAD^+, leading to its reduction at ring positions 1 and 4 to yield the reduced coenzyme NADH (Figure 15-6). The other hydrogen atom of the substrate molecule appears in the medium as H^+. For this reason the enzymatic reduction of NAD^+ is written to include the hydrogen ion formed:

$$\text{Substrate} + NAD^+ \rightleftharpoons \text{dehydrogenated substrate} + NADH + H^+$$

(a)

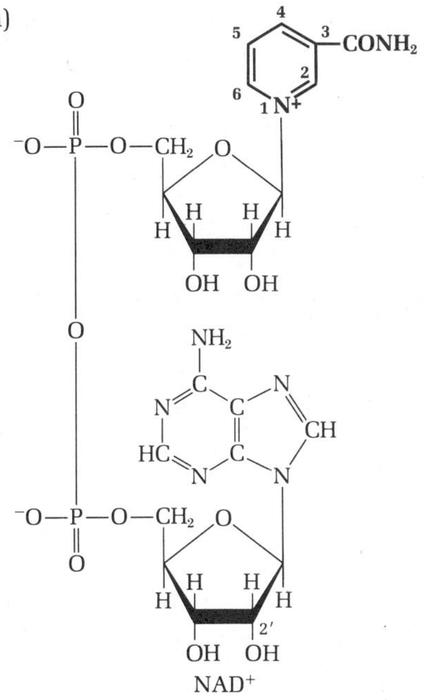

NAD$^+$

(b)

RH_2 + NAD$^+$ ⇌ R + NADH + H^+

The mechanism of action of glyceraldehyde phosphate dehydrogenase is rather complex (Figure 15-7). The substrate first reacts with an —SH group of an essential cysteine residue at the active site of the enzyme. The enzyme then promotes transfer of a hydride ion from the covalently bound substrate to NAD$^+$, also tightly bound at the active site. In this process a high-energy covalent acyl-enzyme complex is formed, which then reacts with inorganic phosphate to discharge free 3-phosphoglyceroyl phosphate, regenerating the free enzyme. The NADH formed in this reaction must be reoxidized to NAD$^+$ again, so that it can participate in the breakdown of many

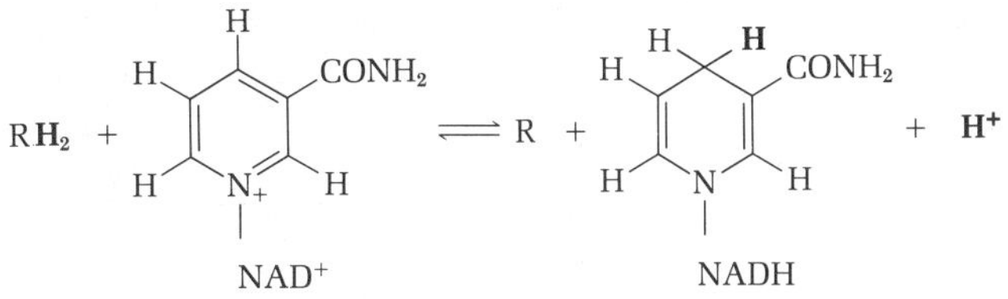

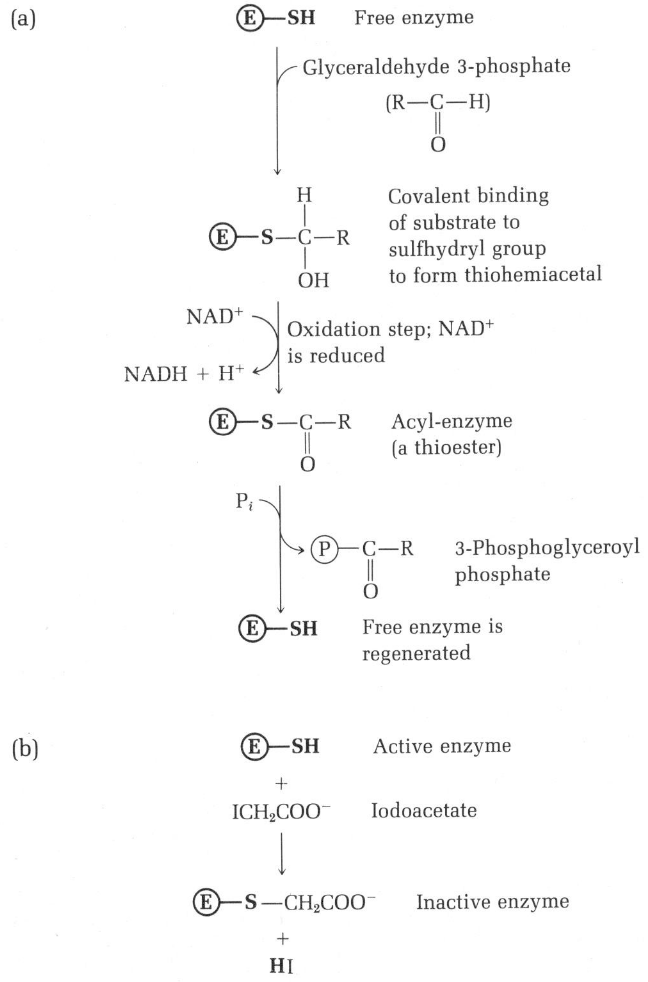

Figure 15-6
(a) Structure of nicotinamide adenine dinucleotide in its oxidized form (NAD$^+$). (b) Reduction of NAD$^+$ by transfer of a hydride ion (:H$^-$) from a substrate RH_2 to position 4 of the nicotinamide ring. See also page 256.

(a)

Ⓔ—SH Free enzyme

Glyceraldehyde 3-phosphate (R—C(=O)—H)

Ⓔ—S—CH(OH)—R Covalent binding of substrate to sulfhydryl group to form thiohemiacetal

NAD$^+$ → NADH + H^+ Oxidation step; NAD$^+$ is reduced

Ⓔ—S—C(=O)—R Acyl-enzyme (a thioester)

P_i → Ⓟ—C(=O)—R 3-Phosphoglyceroyl phosphate

Ⓔ—SH Free enzyme is regenerated

(b)

Ⓔ—SH Active enzyme

\+

ICH_2COO^- Iodoacetate

↓

Ⓔ—S—CH_2COO^- Inactive enzyme

\+

HI

Figure 15-7
(a) Reaction scheme for glyceraldehyde phosphate dehydrogenase, showing the formation of a covalent thiohemiacetal linkage between the substrate and the essential —SH group of the enzyme. This enzyme-substrate intermediate is oxidized by NAD$^+$, also bound to the active site, converting it into a covalent acyl-enzyme intermediate, a thioester. The bond between the acyl group and the thiol group of the enzyme has a very high standard free energy of hydrolysis. In the last step the thioester bond undergoes phosphorolysis to release the free enzyme and an acyl phosphate, which conserves much of the free energy liberated during oxidation of the aldehyde group.

(b) Iodoacetate is a potent inhibitor of glyceraldehyde phosphate dehydrogenase because it forms a covalent derivative of the essential —SH group of the enzyme, rendering it inactive.

molecules of glucose to pyruvate. Since NAD^+ occurs in only limited concentrations in cells, glycolysis would soon come to a halt for lack of NAD^+ unless the NADH can be reoxidized.

Glyceraldehyde phosphate dehydrogenase has been isolated in crystalline form from rabbit skeletal muscle. It has a molecular weight of 140,000 and contains four identical subunits, each consisting of a single polypeptide chain of some 330 amino acid residues. The enzyme is inhibited by iodoacetate (page 223), which combines with the essential —SH group of the enzyme, thus preventing its participation in catalysis (Figure 15-7b). The discovery that iodoacetate inhibits glycolysis was important in the history of research on enzyme systems (page 348).

Transfer of Phosphate from 3-Phosphoglyceroyl Phosphate to ADP

The enzyme phosphoglycerate kinase transfers the high-energy phosphate group from the carboxyl group of 3-phosphoglyceroyl phosphate to ADP, thus forming ATP and leaving 3-phosphoglycerate (Figure 15-5)

$$\text{3-Phosphoglyceroyl phosphate} + \text{ADP} \xrightleftharpoons{Mg^{2+}} \text{3-phosphoglycerate} + \text{ATP}$$
$$\Delta G^{\circ\prime} = -4.50 \text{ kcal/mol}$$

This and the preceding reaction of glycolysis together constitute an energy-coupling process. If we write these two reaction equations again, we shall see that 3-phosphoglyceroyl phosphate is the common intermediate. It is formed in the first reaction, and its acyl phosphate group is transferred to ADP to form ATP in the second reaction

$$\text{Glyceraldehyde 3-phosphate} + P_i + NAD^+ \rightleftharpoons \textbf{3-phosphoglyceroyl phosphate} + NADH + H^+$$

$$\textbf{3-Phosphoglyceroyl phosphate} + \text{ADP} \rightleftharpoons \text{3-phosphoglycerate} + \text{ATP}$$

The sum of these two sequential reactions, coupled to each other by the common intermediate 3-phosphoglyceroyl phosphate, is given by the combined equation

$$\text{Glyceraldehyde 3-phosphate} + P_i + ADP + NAD^+ \rightleftharpoons \text{3-phosphoglycerate} + ATP + NADH + H^+$$
$$\Delta G^{\circ\prime} = -3.0 \text{ kcal/mol}$$

The outcome of these two reactions, both reversible under cell conditions, is that the energy released on oxidation of an aldehyde to a carboxylate group has been conserved by the coupled formation of ATP from ADP and phosphate. The generation of ATP coupled to an enzyme transformation of a "substrate," i.e., a metabolic intermediate such as glyceraldehyde 3-phosphate, is spoken of as a substrate-level phosphorylation. We shall see other examples later.

Conversion of 3-Phosphoglycerate to 2-Phosphoglycerate

This reaction, in which there is a reversible shift of the phosphate group within the substrate molecule, is catalyzed by the enzyme phosphoglycerate mutase

$$\text{3-Phosphoglycerate} \xrightleftharpoons{Mg^{2+}} \text{2-phosphoglycerate} \qquad \Delta G^{\circ\prime} = +1.06 \text{ kcal/mol}$$

Mg^{2+} is essential for this reaction, which involves transfer of the phosphate group from the 3 to the 2 position of glycerate (Figure 15-5). The term mutase is often used to designate enzymes catalyzing intramolecular shifts of functional groups.

Dehydration of 2-Phosphoglycerate to Phosphoenolpyruvate

This is the second reaction of the glycolytic sequence in which a high-energy phosphate compound is generated. This reaction is catalyzed by enolase, which promotes reversible removal of a molecule of water from 2-phosphoglycerate to yield phosphoenolpyruvate (Figure 15-5)

$$\text{2-Phosphoglycerate} \xrightleftharpoons{Mg^{2+}} \text{phosphoenolpyruvate} + H_2O \qquad \Delta G^{\circ\prime} = +0.44 \text{ kcal/mol}$$

Despite the relatively small standard-free-energy change in this reaction, there is a very large difference in the standard free energy of hydrolysis of the phosphate groups of the reactant and product. That of 2-phosphoglycerate (a low-energy phosphate) is about −4.2 kcal and that of phosphoenolpyruvate (a super high-energy phosphate) about −14.8 kcal (page 374). Although 2-phosphoglycerate and phosphoenolpyruvate contain nearly the same *total* amount of energy, the loss of the water molecule from the former causes a redistribution of energy *within* the molecule, so that there is a much greater decrease in free energy when the phosphate group of phosphoenolpyruvate is hydrolyzed.

Enolase has been isolated in pure crystalline form from several sources (MW 85,000). It requires Mg^{2+}, which makes a complex with the enzyme before the substrate is bound. Enolase is characteristically inhibited by the simultaneous presence of fluoride (F^-) and phosphate; the fluorophosphate ion, which binds Mg^{2+}, is the true inhibitory agent.

Transfer of the Phosphate Group from Phosphoenolpyruvate to ADP

The last step in glycolysis is the transfer of the high-energy phosphate group from phosphoenolpyruvate to ADP (Figure

15-5), which is catalyzed by pyruvate kinase. In this reaction, which is another example of a substrate-level phosphorylation, the product pyruvate appears in its enol form.

$$\text{Phosphoenolpyruvate} + \text{ADP} \xrightarrow[K^+]{Mg^{2+}} \text{enolpyruvate} + \text{ATP}$$

However, the enol form of pyruvate rapidly and nonenzymatically rearranges to yield the keto form of pyruvate, the form that predominates at pH 7.0:

$$\underset{\text{Enolpyruvate}}{CH_2{=}C(OH){-}COO^-} \rightleftharpoons \underset{\text{Ketopyruvate}}{CH_3{-}C({=}O){-}COO^-}$$

Because the point of equilibrium of this reaction is very far to the right, it "pulls" the preceding pyruvate kinase reaction to the right by mass action. The sum of the pyruvate kinase reaction and the nonenzymatic formation of ketopyruvate is given by

$$\text{Phosphoenolpyruvate} + \text{ADP} + H^+ \xrightarrow{Mg^{2+},\,K^+} \text{ketopyruvate} + \text{ATP}$$
$$\Delta G^{\circ\prime} = -7.5 \text{ kcal/mol}$$

The overall reaction has a very large negative $\Delta G^{\circ\prime}$ value, due in large part to the spontaneous conversion of the enol form of pyruvate into the keto form. The $\Delta G^{\circ\prime}$ of phosphoenolpyruvate hydrolysis is -14.8 kcal/mol; about half of this energy is recovered as ATP ($\Delta G^{\circ\prime} = -7.3$ kcal/mol) and the rest (-7.5 kcal/mol) constitutes a large driving force pushing the reaction far to the right. The pyruvate kinase reaction is essentially irreversible under intracellular conditions.

Pyruvate kinase has been obtained in pure crystalline form (MW 250,000); it requires K^+ and either Mg^{2+} or Mn^{2+}. It is an important regulatory enzyme whose action will be described later.

Reduction of Pyruvate to Lactate

Pyruvate represents an important junction point in carbohydrate catabolism. In animal tissues under aerobic conditions pyruvate is the product of glycolysis, and the NADH formed by the dehydrogenation of glyceraldehyde 3-phosphate is then reoxidized to NAD^+ by O_2 (Chapter 17). However, under anaerobic conditions, in very active skeletal muscles or in lactic acid bacteria, NADH generated by glycolysis cannot be reoxidized by O_2 but must be reoxidized to NAD^+ by pyruvate, converting the latter into lactate. Under these conditions electrons originally donated by glyceraldehyde 3-phosphate to NAD^+ are carried in the form of NADH to pyruvate. The reduction of pyruvate is catalyzed by lactate dehydrogenase, which forms the L isomer of lactate:

$$\text{Pyruvate} + \text{NADH} + \text{H}^+ \rightleftharpoons \text{L-lactate} + \text{NAD}^+ \qquad \Delta G^{\circ\prime} = -6.0 \text{ kcal/mol}$$

The overall equilibrium of this reaction is far to the right, as shown by the large negative value of $\Delta G^{\circ\prime}$. Because dehydrogenation of the two molecules of glyceraldehyde 3-phosphate derived from each molecule of glucose requires two NAD^+ and yields two NADH, the regeneration of two NAD^+ by the reduction of two pyruvate to two lactate allows NAD to be used over and over again in the glycolytic sequence.

As we have seen (page 240), lactate dehydrogenase of most tissues occurs in five different isozyme forms, which differ in their K_M values for pyruvate, their turnover numbers or V_{max}, and the degree of their allosteric inhibition by pyruvate. The heart type of lactate dehydrogenase (designated H_4) consists of four H-type polypeptide chains. It has a low K_M for pyruvate and is strongly inhibited by pyruvate, whereas the muscle isozyme (designated M_4) has a higher K_M for pyruvate, is not inhibited by pyruvate, and is more active catalytically. Although many efforts have been made to provide a comprehensive rationale for the function and role of the lactate dehydrogenase isozymes in different tissues, especially in the heart, skeletal muscle, and liver, much controversy and little agreement surround this topic. Thus the role of the lactate dehydrogenase isozymes and the two genes responsible for their synthesis are not yet understood. It is of interest that a 64-year-old man was found to have no heart-type lactate dehydrogenase, as a result of a genetic defect; he suffered no discernible cardiac or metabolic problems, however. This observation raises the question whether *every* enzyme or protein in a cell or tissue is in fact essential; some may be vestigial and no longer used.

The Overall Balance Sheet

A balance sheet for anaerobic glycolysis can now be constructed to account for (1) the fate of the carbon skeleton of glucose, (2) the pathway of electrons in the oxidoreduction reactions, and (3) the input of phosphate and ADP and the output of ATP. The left-hand part of the following equation shows all the inputs of ATP, P_i, ADP, NAD^+, NADH, and H^+ (consult Figures 15-4 and 15-5), and the right-hand part shows all the outputs (keep in mind that each molecule of glucose yields two molecules of glyceraldehyde 3-phosphate):

$$\text{Glucose} + 2\text{ATP} + 2\text{P}_i + 2\text{NAD}^+ + 2\text{NADH} + 2\text{H}^+ + 4\text{ADP} \longrightarrow 2\text{ lactate}^- + 2\text{H}^+ + 4\text{ATP} + 2\text{H}_2\text{O} + 2\text{NADH} + 2\text{H}^+ + 2\text{NAD}^+ + 2\text{ADP}$$

If we now cancel out common terms on both sides of the equation, we get

$$\text{Glucose} + 2\text{P}_i + 2\text{ADP} \longrightarrow 2\text{ lactate}^- + 2\text{H}^+ + 2\text{ATP} + 2\text{H}_2\text{O}$$

the overall equation for anaerobic glycolysis in anaerobic skeletal muscle or in lactate fermentation.

In the overall process D-glucose is converted into two molecules of lactate (the pathway of carbon). Two molecules of ADP and two of phosphate are converted into two of ATP (the pathway of phosphate groups). Four electrons (in the form of two hydride ions) are transferred, via two molecules of NAD^+, from two molecules of glyceraldehyde 3-phosphate to two of pyruvate to yield two of lactate (the pathway of electrons). Although two oxidoreduction steps have taken place in the glycolytic sequence, there is no *net* change in the oxidation state of carbon as glucose is converted into lactate. This can be seen by comparing the empirical formula of glucose ($C_6H_{12}O_6$) with that of lactic acid ($C_3H_6O_3$). The ratio of C to H to O atoms is the same for both, showing that no *net* oxidation of carbon has occurred. Nevertheless, some of the energy of the glucose molecule has been extracted by anaerobic glycolysis, sufficient to give a net yield of two ATP per glucose consumed.

Under aerobic conditions pyruvate rather than lactate is the product of the glycolytic breakdown of glucose. Under these conditions the two NADH formed by dehydrogenation of two glyceraldehyde 3-phosphate are not reoxidized by pyruvate. Thus the overall equation for glycolysis in this case is

$$\text{Glucose} + 2P_i + 2\text{ADP} + 2\text{NAD}^+ \longrightarrow 2\text{ pyruvate}^- + 2\text{ATP} + 2\text{NADH} + 2\text{H}^+ + 2\text{H}_2\text{O}$$

Under these aerobic conditions the two molecules of NADH formed by glycolysis in the cytosol are reoxidized to NAD^+ by transfer of their electrons to the electron-transport chain, which in eukaryotic cells is located in the mitochondria. Here these electrons are ultimately passed to oxygen, reducing it to form H_2O:

$$2\text{NADH} + 2\text{H}^+ + \text{O}_2 \longrightarrow 2\text{NAD}^+ + 2\text{H}_2\text{O}$$

"Feeder" Pathways Lead from Glycogen and Other Carbohydrates into the Central Glycolytic Pathway

In addition to D-glucose, many other carbohydrates ultimately enter the glycolytic sequence to undergo energy-yielding degradation. The most important are the storage polysaccharides *glycogen* and *starch*, the disaccharides *maltose*, *lactose*, and *sucrose*, and the monosaccharides *fructose*, *mannose*, and *galactose*. We shall now consider the pathways by which these carbohydrates can enter the glycolytic sequence. They are summarized in Figure 15-8.

The D-glucose units of the outer branches of glycogen and starch gain entrance into the glycolytic pathway through the sequential action of two enzymes, *glycogen phosphorylase* (or the similar *starch phosphorylase* in plants) and *phosphogluco-*

Figure 15-8
Entry of glycogen and different hexoses into the first stage of glycolysis.

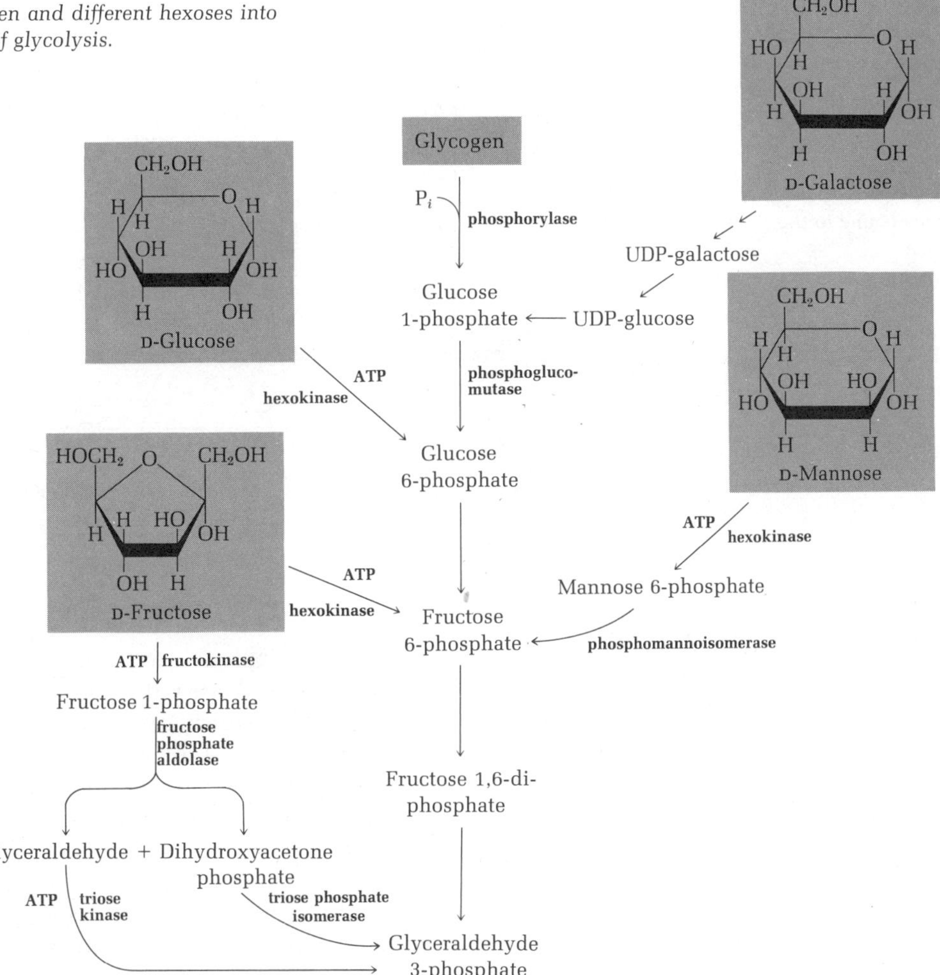

mutase. Glycogen phosphorylase, which is widely distributed in animal cells, catalyzes the general reaction below, in which $(\text{glucose})_n$ designates a glycogen (or starch) branch containing n D-glucose residues in $\alpha(1 \rightarrow 4)$ linkage, whereas $(\text{glucose})_{n-1}$ designates the glycogen branch shortened by removal of one glucose residue from the end of the branch (see page 289 for the structures of glycogen and starch):

$$(\text{Glucose})_n + P_i \longrightarrow (\text{glucose})_{n-1} + \alpha\text{-D-glucose 1-phosphate} \qquad \Delta G^{\circ\prime} = +0.73 \text{ kcal/mol}$$

Under intracellular conditions, in which the phosphate concentration is relatively high, the glycogen phosphorylase reaction proceeds only in the direction of degradation to yield glucose 1-phosphate. In this reaction the terminal $\alpha(1 \rightarrow 4)$ glycosidic linkage at the nonreducing end of the glycogen branch undergoes *phosphorolysis*, the removal of the terminal glucose

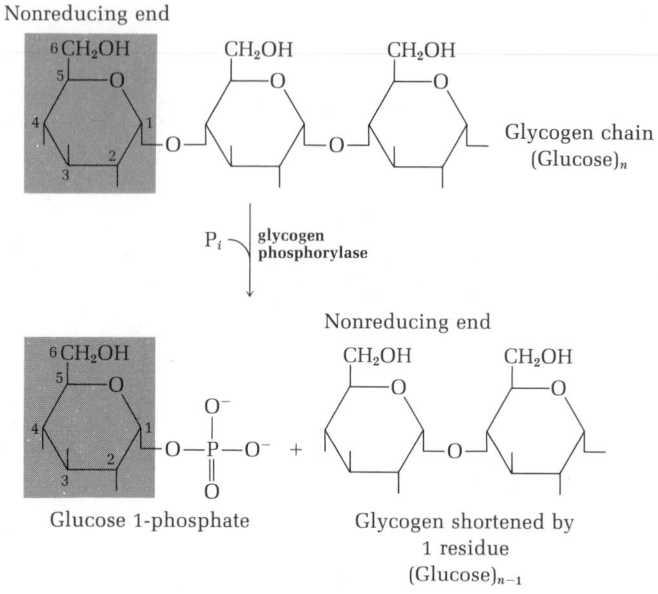

Figure 15-9
Removal of a terminal glucose residue from the nonreducing end of a glycogen chain by the action of glycogen phosphorylase. This process is repetitive and removes successive glucose residues until it reaches the fourth glucose unit from the branch point (see text). Note the shorthand designation of the hydroxyl groups of the glucose residues; the hydrogen atoms bound to the pyranose rings are not shown.

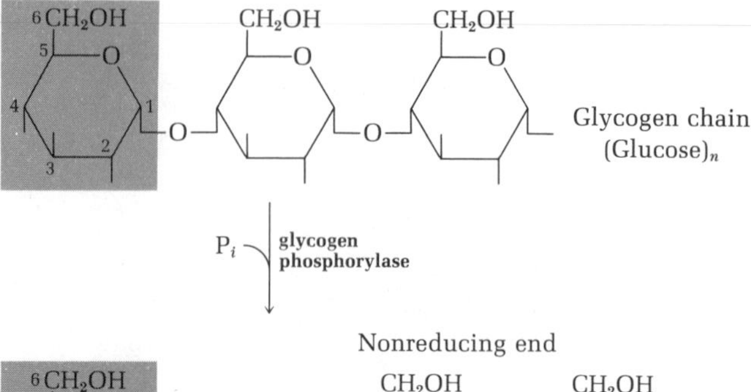

residue by attack of phosphate, to yield *α-D-glucose 1-phosphate*. Left behind is a chain with one less glucose unit (Figure 15-9). Glycogen phosphorylase acts repetitively on the nonreducing ends of glycogen branches until it reaches a point four glucose residues ahead of an $\alpha(1 \rightarrow 6)$ branch point (page 289). Here glycogen phosphorylase action stops.

Further degradation of glycogen by phosphorylase can occur only after the action of another enzyme, *$\alpha(1 \rightarrow 6)$ glucosidase,* which catalyzes two reactions. In the first the enzyme removes three of the remaining glucose residues and transfers them to the end of some other outer branch. The remaining glucose residue on the first branch, which is attached by an $\alpha(1 \rightarrow 6)$ linkage at the branch point, is removed by a second action of $\alpha(1 \rightarrow 6)$ glucosidase. It hydrolyzes the $\alpha(1 \rightarrow 6)$ linkage at the branch point, thus yielding a molecule of D-glucose and making available another length of glucose residues in $\alpha(1 \rightarrow 4)$ linkage to the action of glycogen phosphorylase.

Glucose 1-phosphate, the end product of the glycogen phosphorylase (and starch phosphorylase) reactions, is converted into glucose 6-phosphate by *phosphoglucomutase,* which has been obtained in pure form from many sources. This enzyme catalyzes the reversible reaction

$$\text{Glucose 1-phosphate} \rightleftharpoons \text{glucose 6-phosphate} \qquad \Delta G^{\circ\prime} = -1.74 \text{ kcal/mol}$$

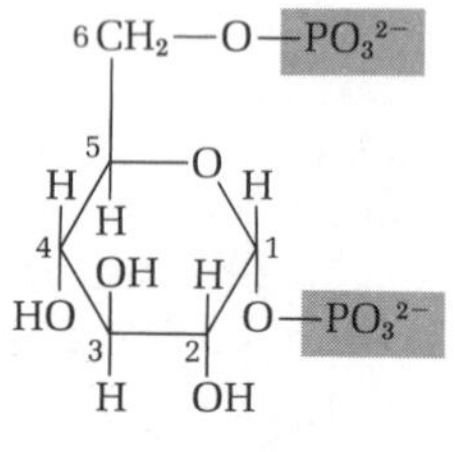

Phosphoglucomutase requires a cofactor, *glucose 1,6-diphosphate* (Figure 15-10). Its role is indicated by the following sequence of intermediate steps in the action of the enzyme, which cycles between two forms, a phosphorylated and nonphosphorylated or dephospho form:

Figure 15-10
α-D-Glucose 1,6-diphosphate, a cofactor required by phosphoglucomutase.

Phosphoenzyme + glucose 1-phosphate $\rightleftharpoons$
dephosphoenzyme + **glucose 1,6-diphosphate**

Glucose 1,6-diphosphate + dephosphoenzyme $\rightleftharpoons$
phosphoenzyme + glucose 6-phosphate

Sum: Glucose 1-phosphate $\rightleftharpoons$ glucose 6-phosphate

Phosphoglucomutase is noteworthy from another point of view. It is one of a large group of enzymes containing an essential serine residue at its active site. It is the hydroxyl group of this serine residue that is phosphorylated by glucose 1,6-diphosphate. This and other enzymes of the *serine class* (page 221) are irreversibly inhibited by certain organic phosphates, such as *diisopropylfluorophosphate*, which form catalytically inactive organophosphate esters with the essential serine hydroxyl groups (page 222).

Other Monosaccharides Can Enter the Glycolytic Sequence

In animals monosaccharides other than glucose may be transformed so that they can enter into the glycolytic sequence and yield energy during their degradation (Figure 15-8).

D-*Fructose*, present in free form in many fruits and formed by hydrolysis of sucrose or table sugar in the small intestine, can be phosphorylated by *hexokinase*, which acts on a number of different hexoses:

$$\text{D-Fructose} + \text{ATP} \xrightarrow{Mg^{2+}} \text{D-fructose 6-phosphate} + \text{ADP} + H^+$$

This is a major pathway in muscles and the kidney. In the liver, however, fructose gains entry into glycolysis by a different pathway. The liver enzyme *fructokinase* catalyzes the phosphorylation of fructose, not at carbon atom 6, but at carbon 1 (Figure 15-11).

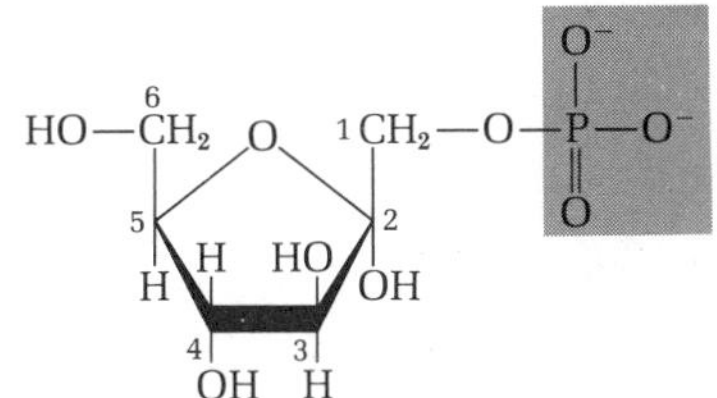

Figure 15-11
α-D-Fructose 1-phosphate, an intermediate in the conversion of fructose into glyceraldehyde phosphate.

$$\text{D-Fructose} + \text{ATP} \xrightarrow{Mg^{2+}} \text{D-fructose 1-phosphate} + \text{ADP} + H^+$$

The fructose 1-phosphate is then cleaved to form D-glyceraldehyde and dihydroxyacetone phosphate by *aldolase:*

D-Fructose 1-phosphate $\rightleftharpoons$
D-glyceraldehyde + dihydroxyacetone phosphate

Dihydroxyacetone phosphate will be recognized as an intermediate of the glycolytic sequence which can be converted into glyceraldehyde 3-phosphate. The other product, D-glyceraldehyde, is phosphorylated by ATP and *triose kinase* to D-glyceraldehyde 3-phosphate, also an intermediate in glycolysis:

D-Glyceraldehyde + ATP $\xrightarrow{Mg^{2+}}$
D-glyceraldehyde 3-phosphate + ADP + H^+

Thus the liver converts D-fructose into two molecules of glyceraldehyde 3-phosphate.

D-*Galactose*, derived by hydrolysis of the disaccharide lactose (milk sugar), is first phosphorylated at the 1 position at the expense of ATP by the enzyme *galactokinase*

$$\text{ATP} + \text{D-galactose} \xrightarrow{Mg^{2+}} \text{D-galactose 1-phosphate} + \text{ADP} + H^+$$

The resulting product, D-*galactose 1-phosphate*, is then converted into its epimer at carbon atom 4, namely, D-glucose 1-phosphate, in a set of reactions shown in Figure 15-12. In these reactions *uridine diphosphate* (UDP) (page 390) functions as a coenzyme-like carrier of hexose groups. In the human liver, galactose 1-phosphate reacts with UDP-glucose to yield UDP–D-galactose and glucose 1-phosphate, catalyzed by the enzyme *UDP-glucose:α-D-galactose 1-phosphate uridylyltransferase*. The galactose residue of UDP–D-galactose is then enzymatically epimerized at carbon 4 to yield UDP–D-glucose by the enzyme *UDP-glucose 4-epimerase* (Figure 15-12). UDP-glucose is

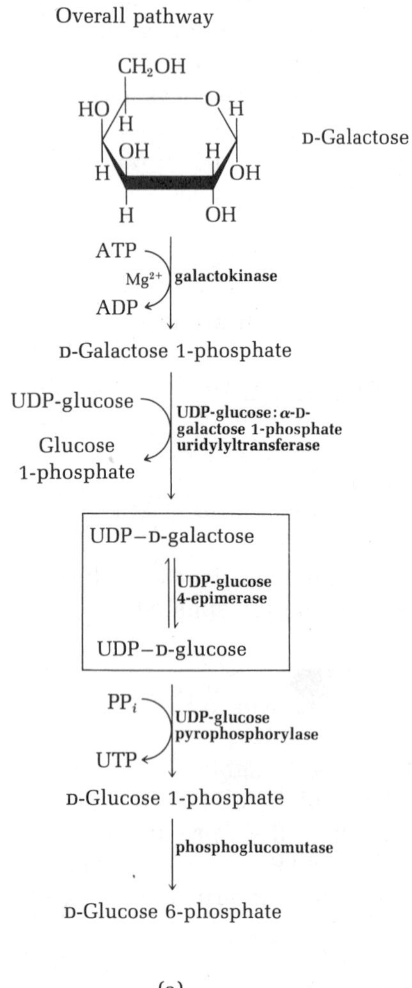

Figure 15-12
(a) Pathway of the conversion of D-galactose into D-glucose. Details of the step enclosed by the box are shown in (b). In the interconversion of UDP-glucose and UDP-galactose the NAD required by the enzyme appears to accept two hydrogens from the 4 carbon atom of the glucose residue and then return them in such a way as to yield the 4 epimer.

Galactosemia, a human genetic disease of infants and children, is usually caused by a deficiency of UDP-glucose:α-D-galactose 1-phosphate uridylyltransferase. Because of the defect in conversion of galactose into glucose, galactose and galactose 1-phosphate derived from dietary lactose accumulate in the tissues, causing damage to the brain and liver, as well as cataract formation. Free galactose also appears in the blood in substantial amounts. A milder form of galactosemia occurs in galactokinase deficiency.

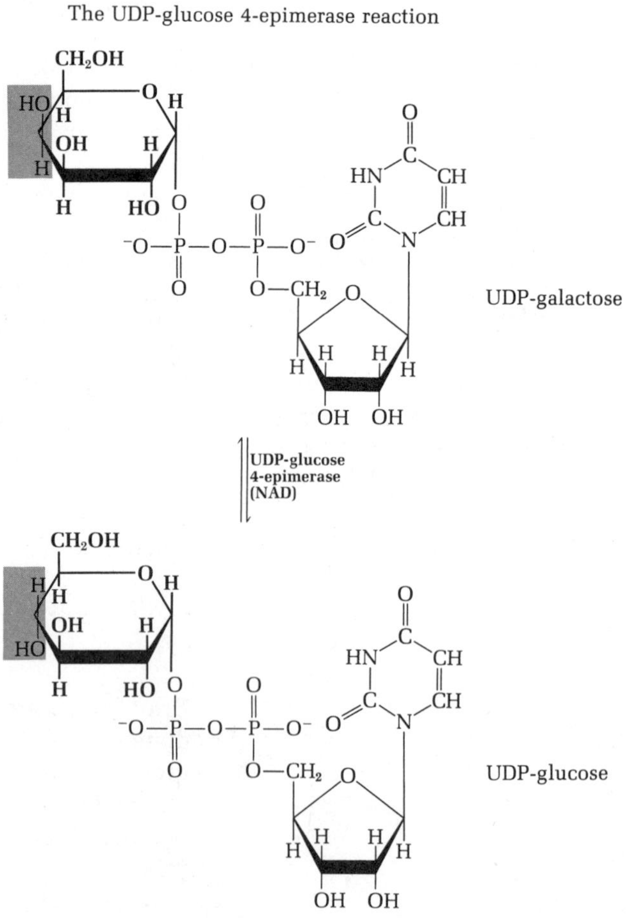

then cleaved by UDP-glucose pyrophosphorylase to D-glucose 1-phosphate, which is converted into glucose 6-phosphate by phosphoglucomutase. This sequence of reactions not only is responsible for the conversion of D-galactose into D-glucose but is also used in reverse in the mammary gland for the synthesis of D-galactose, required in the formation of lactose or milk sugar. The important role of UDP as a carrier of sugar groups in metabolism was first discovered by Luis Leloir, an Argentinian biochemist. We shall later see other pathways in which UDP derivatives of sugars act as intermediates.

In people with the most common form of the genetic deficiency disease *galactosemia*, UDP-glucose:α-D-galactose 1-phosphate uridylyltransferase (Figure 15-12) is genetically defective, preventing the overall conversion of D-galactose into D-glucose. As a result D-galactose and D-galactose 1-phosphate cannot be metabolized and accumulate in the blood and tissues. The liver and other organs become enlarged, vision becomes impaired because of the formation of cataracts, and there is mental retardation. Since the major dietary source of D-galactose is the lactose of milk, this genetic deficiency appears in infants. Galactosemia can be alleviated by withholding milk and milk products from the diet. Other forms of galactosemia result when either galactokinase or UDP-glucose 4-epimerase are genetically defective.

D-Mannose, which arises from the digestion of various polysaccharides and glycoproteins present in natural foods, can be phosphorylated at the 6 position by *hexokinase*

$$\text{D-Mannose} + \text{ATP} \xrightarrow{Mg^{2+}} \text{D-mannose 6-phosphate} + \text{ADP} + H^+$$

D-Mannose 6-phosphate is then isomerized by the action of *phosphomannoisomerase*, to yield D-fructose 6-phosphate, an intermediate of glycolysis:

$$\text{D-Mannose 6-phosphate} \rightleftharpoons \text{D-fructose 6-phosphate}$$

Figure 15-8 summarizes these feeder pathways, which bring several different sugars into the central glycolytic pathway.

Disaccharides Must First Be Hydrolyzed to Monosaccharides

Disaccharides as such cannot enter the glycolytic cycle. In fact, when they are injected into the bloodstream they are not used. Before they can be used by the body, ingested disaccharides must first be enzymatically hydrolyzed in the cells lining the small intestine, to yield their hexose units:

$$\text{Maltose} + H_2O \xrightarrow{\text{maltase}} \text{D-glucose} + \text{D-glucose}$$

$$\text{Lactose} + H_2O \xrightarrow{\text{lactase}} \text{D-galactose} + \text{D-glucose}$$

$$\text{Sucrose} + H_2O \xrightarrow{\text{sucrase}} \text{D-fructose} + \text{D-glucose}$$

The monosaccharides so formed are absorbed into the blood and pass to the liver, where they are phosphorylated and converted into intermediates of the glycolytic sequence as described above.

The *lactose intolerance* (pages 285 and 686) of most human races except Northern Europeans and some Africans is due to the disappearance of most or all of the lactase activity of the intestinal cells after childhood, so that lactose can no longer be completely digested and absorbed. In persons with lactose intolerance the lactose remaining in the intestinal tract causes discomfort and diarrhea. It is not a serious problem since in those parts of the world where lactose intolerance is general, milk is simply not used as food by adults. There is no known association or relationship between lactose intolerance and galactosemia; the former is very common, the latter very rare.

The Entry of Glucose Residues into the Glycolytic Sequence Is Regulated

The rates of the central catabolic pathways that degrade glucose to produce chemical energy in the form of ATP adjust themselves to the minute-by-minute needs of the cell for ATP, whether it is to be used for biosynthetic reactions, for active-transport processes, or for contractile or mechanical work. Since degradation products of glucose are also important precursors or intermediates in other aspects of metabolism, the regulatory enzymes in carbohydrate catabolism also recognize and respond to appropriate signals from other metabolic pathways. We shall now examine the various regulatory enzymes that control the rate of breakdown of carbohydrates via glycolysis.

First we shall look at regulation of the *entry* of glucose residues into the glycolytic sequence. The two important reactions by which glucose residues enter the glycolytic pathway are both controlled by regulatory enzymes. One is the entry of free glucose, which takes place by the phosphorylation of carbon 6 of glucose by ATP, catalyzed by *hexokinase*. In some tissues, such as skeletal muscle, hexokinase is an allosteric enzyme which is inhibited by its reaction product glucose 6-phosphate, as schematized in Figure 15-13. Whenever there is a significant rise in the concentration of glucose 6-phosphate in the cell, indicating that it is not being used as fast as it is formed, hexokinase is inhibited by glucose 6-phosphate, thus preventing any further phosphorylation of glucose until the excess glucose 6-phosphate has been used. The liver, on the other hand, contains largely glucokinase, which is not inhibited by glucose 6-phosphate (page 405). Thus the liver, which can store large amounts of glycogen, can phosphorylate excess blood glucose to form glucose 6-phosphate, which is converted via glucose 1-phosphate into glycogen for storage. The hormone insulin, secreted by the pancreas into the blood whenever the blood glucose concentration is high (Chapter 25), stimulates the

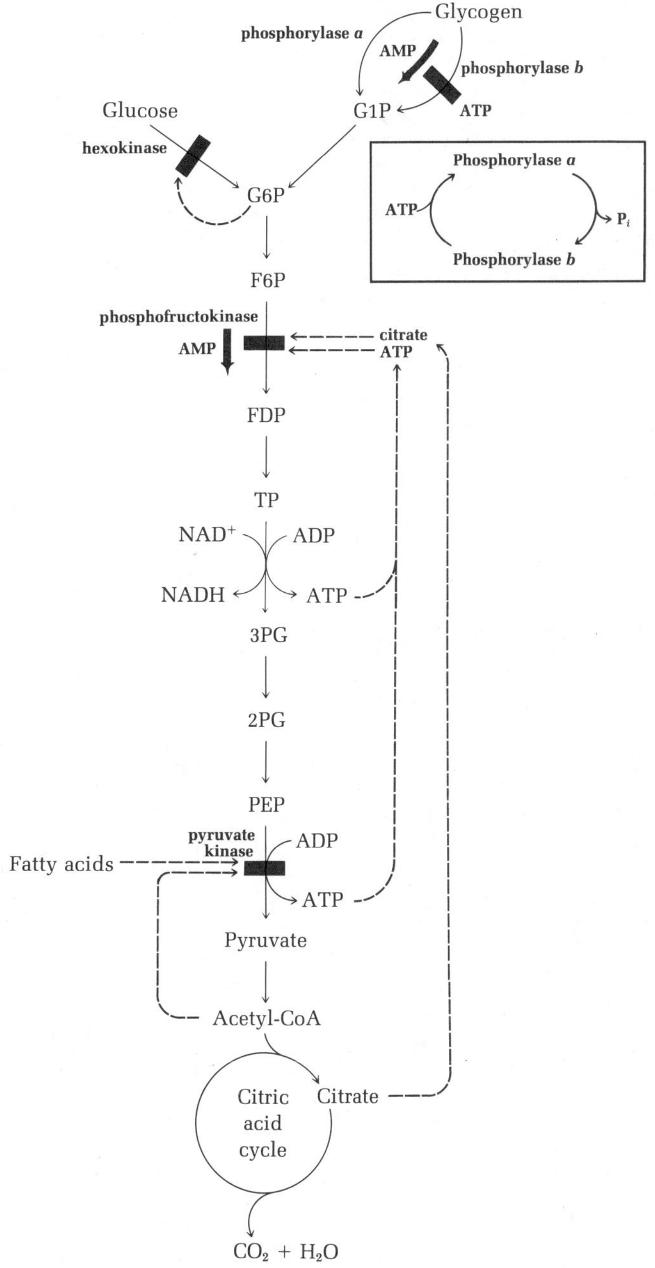

Figure 15-13
Regulation of the entry and use of glucose residues in glycolysis. Regulatory inhibition is shown by dashed feedback arrows leading to bars across the reaction arrows (color); regulatory stimulation is shown by a colored arrow parallel to the reaction arrow. G1P designates glucose 1-phosphate; G6P, glucose 6-phosphate; F6P, fructose 6-phosphate; FDP, fructose 1,6-diphosphate; TP, triose phosphates; 3PG, 3-phosphoglycerate; 2PG, 2-phosphoglycerate; PEP, phosphoenolpyruvate.

synthesis of glucokinase. In diabetes and during starvation glucokinase activity is deficient.

The second point of entry of glucose residues into glycolysis is from glycogen, through the action of *glycogen phosphorylase*, also a regulatory enzyme. In both the liver and muscles glycogen phosphorylase is situated at a strategic point between the fuel reservoir glycogen and the glycolytic system for using the fuel. In skeletal muscle the enzyme occurs in two forms, a catalytically active phosphorylated form (*phosphorylase a*) and a much less active dephosphorylated form (*phosphorylase b*). Phosphorylase *a* has been crystallized. It has a molecular

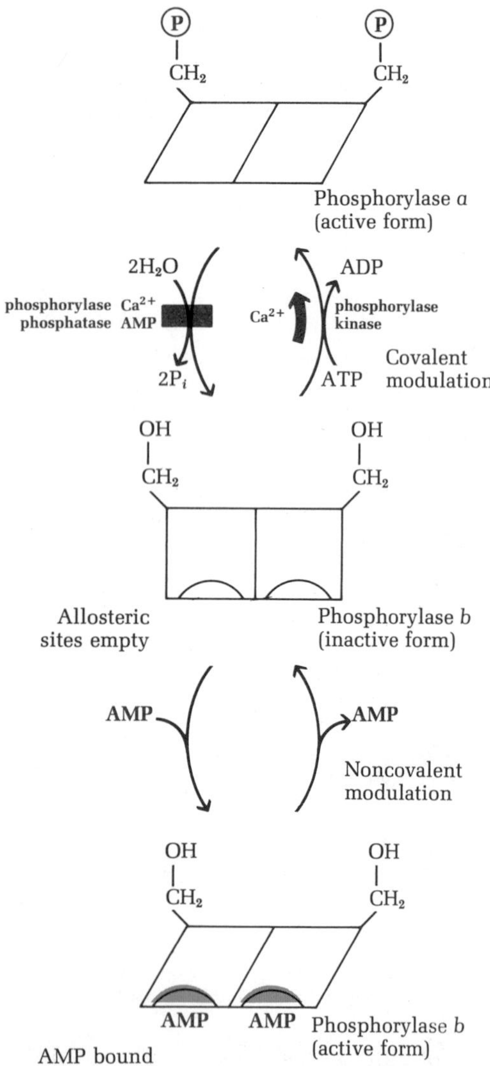

Figure 15-14
Regulation of glycogen phosphorylase. The enzyme has two subunits, each of which can be phosphorylated by phosphorylase kinase on an essential serine hydroxyl group to give phosphorylase a*, a reaction promoted by* Ca^{2+}*. Dephosphorylation of phosphorylase* a *is blocked by* Ca^{2+} *and AMP. Phosphorylase* b *can also be activated by noncovalent binding of AMP at its allosteric sites. Conformational changes of the enzyme are indicated schematically.*

weight of 190,000 and consists of two identical subunits, each containing an essential serine residue that is phosphorylated (Figure 15-14). In muscle the rate of conversion of glycogen units into glucose 1-phosphate is regulated by the *ratio* of the active phosphorylase *a* to the less active phosphorylase *b*.

The interconversion of the active and inactive forms of glycogen phosphorylase is brought about by specific enzymes that promote covalent modification (page 237) of phosphorylase. Phosphorylase *a* is converted into the less active phosphorylase *b* by an enzyme called *phosphorylase a phosphatase*, which removes by hydrolysis the essential phosphate groups in the phosphorylase *a* molecule (Figure 15-14). Phosphorylase *b* can be converted back into active phosphorylase *a* by the enzyme *phosphorylase b kinase*, which promotes a reaction in which ATP phosphorylates the essential serine residues of phosphorylase *b* to yield phosphorylase *a*. Thus, by the action of phosphorylase *a* phosphatase and phosphorylase *b* kinase the ratio of the active phosphorylase *a* to the relatively inactive phosphorylase *b* in the cell can be varied.

Glycogen phosphorylase of muscle is regulated in a second way. Phosphorylase *b*, the relatively inactive form, can be stimulated by noncovalent binding of its allosteric modulator AMP, which increases in concentration in muscle during ATP breakdown by the contractile system (Figure 15-14; see also page 389). The stimulation of phosphorylase *b* by AMP is prevented by ATP, a negative modulator. The activity of phosphorylase *b* thus reflects the ratio of AMP to ATP. On the other hand, phosphorylase *a* is not stimulated by AMP. For this reason phosphorylase *a* is sometimes referred to as the AMP-independent form and phosphorylase *b* as the AMP-dependent form. Glycogen phosphorylase of skeletal muscle therefore is regulated by two different mechanisms: (1) covalent modification, through phosphorylation and dephosphorylation of its essential serine hydroxyl groups, and (2) allosteric regulation of phosphorylase *b* by the noncovalent binding of AMP or ATP. In resting muscle nearly all the phosphorylase is in the *b* form, which is inactive because ATP is present in a much higher concentration than AMP.

Hormones Ultimately Regulate the Interconversion of Phosphorylase *a* and *b*

We have seen that the interconversion of phosphorylase *a* and *b* in muscle is promoted by two enzymes, phosphorylase *a* phosphatase and phosphorylase *b* kinase, which acting together determine the ratio of phosphorylase *a* and *b* and thus are major elements in determining the rate of glycogen breakdown to glucose 1-phosphate. But we may now well ask: How are the activities of phosphorylase phosphatase and phosphorylase kinase regulated?

We shall develop the answer to this important question in more detail later (Chapter 25), but for the present we can say that the hormone *adrenaline* secreted into the blood by the adrenal medulla when the organism is suddenly confronted by an emergency, is a molecular signal to the liver and muscles. When it receives the signal, the liver turns on its glycogen phosphorylase and increases the formation of blood glucose, thus making fuel available to the muscles. It is also a signal to skeletal muscle to turn on the breakdown of muscle glycogen to lactate by glycolysis and thus increase the production of ATP to meet the emergency. Through a cascade of sequential reactions, examined later, adrenaline ultimately stimulates the activity of phosphorylase *b* kinase, which causes a large increase in the ratio of phosphorylase *a* to phosphorylase *b*. When the secretion of adrenaline ceases after the emergency is over, the phosphorylase *b* kinase reverts to its original, slower rate and the ratio of phosphorylase *a* to phosphorylase *b* returns to normal (Chapter 25).

The liver also contains glycogen phosphorylase in *a* and *b* forms, which in principle function like muscle phosphorylase but differ somewhat in structure and regulatory properties. Glycogen breakdown has a different purpose in the liver from that in muscle, namely, to yield free blood glucose. Glucose 1-phosphate formed by liver phosphorylase is converted into glucose 6-phosphate, which is the immediate precursor of free blood glucose through the action of *glucose 6-phosphatase:*

$$\text{D-Glucose 6-phosphate} + H_2O \longrightarrow \text{D-glucose} + P_i$$

Stimulation of the formation of phosphorylase *a* from phosphorylase *b* by the action of adrenaline in the liver thus results in an increase in the concentration of glucose in the blood, which prepares the organism for emergency. Later we shall develop further details of glycogen breakdown and synthesis and their regulation (Chapters 20 and 25).

The Glycolytic Sequence Itself Is Regulated at Two Major Points

In addition to regulation of the rate of glycolysis through control of the entry of free glucose or the glucosyl residues of glycogen, as described above, the sequence of reactions from glucose 6-phosphate to pyruvate is also under biological control. There are two major regulatory points in this sequence, the steps catalyzed by *phosphofructokinase* and by *pyruvate kinase*.

Phosphofructokinase (PFK) is a complex allosteric enzyme, with many stimulatory and inhibitory modulators. Dozens of research papers have been written about its regulation, which differs with cell type. In skeletal muscle its rate is regulated by the concentration of its substrates ATP and fructose 6-phosphate and its products ADP and fructose 1,6-diphosphate, all of which may function as allosteric regulators. Also very important as modulators are AMP, citrate, Mg^{2+}, phosphate, and certain other metabolites in muscle tissue (Table 15-1). While regulation of PFK involves a complex interplay of factors, ATP

Table 15-1 Some Allosteric Activators and Inhibitors of Phosphofructokinase

Activators	Inhibitors
AMP	ATP
Fructose 1,6-diphosphate	Citrate
ADP	Mg^{2+}
Phosphate, K^+	Ca^{2+}

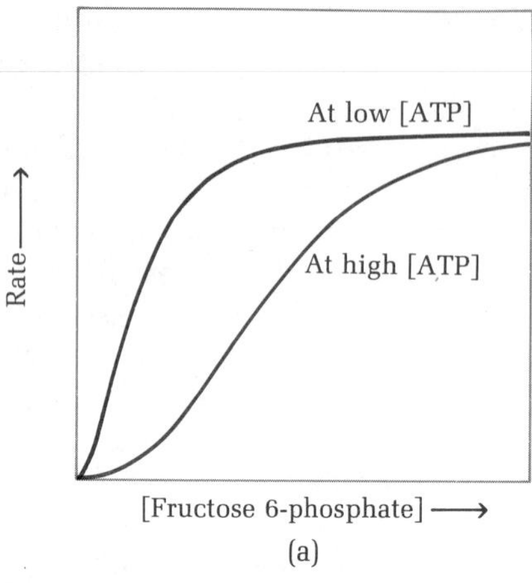

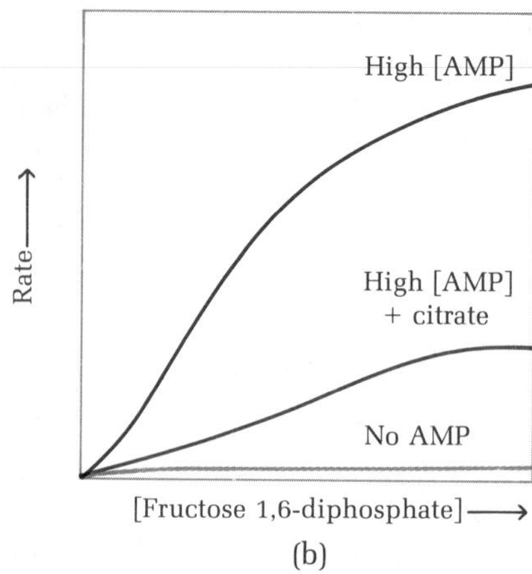

Figure 15-15
Some factors in the allosteric regulation of muscle phosphofructokinase. (a) Effect of the concentration of ATP and fructose 6-phosphate on the rate of the phosphofructokinase reaction. At low concentrations of ATP the K_M of the enzyme for fructose 6-phosphate is relatively low, enabling the enzyme to function at a high rate at relatively low concentrations of fructose 6-phosphate. At high ATP, K_M for fructose 6-phosphate is greatly increased as indicated by the sigmoid relationship.

(b) Effect of AMP, citrate, and fructose 1,6-diphosphate. Fructose 1,6-diphosphate is a strong activator but requires AMP for maximum stimulation. Citrate, on the other hand, is a strong inhibitor.

These are only a few of the complex relationships between the many allosteric modulators of phosphofructokinase.

and citrate are the most important inhibitory modulators, and AMP and fructose 1,6-diphosphate the most active stimulatory modulators. Whenever the concentration of ATP falls during very active muscular contraction and more energy is required, PFK activity is accelerated, even when the fructose 6-phosphate concentration is very low, as is shown by the hyperbolic substrate concentration curve in Figure 15-15a. However, when the cell already has a high level of ATP relative to ADP and AMP, the apparent affinity of phosphofructokinase for fructose 6-phosphate is greatly decreased, as shown by the sigmoidal curve in Figure 15-15*a*. In this case phosphofructokinase will generate its products only at rather high concentrations of fructose 6-phosphate. Citrate, an intermediate of the citric acid cycle, accentuates the inhibitory action of high concentrations of ATP on phosphofructokinase activity. On the other hand, increasing concentrations of AMP, formed in contracting muscle by the adenylate kinase reaction (page 389), is a very potent stimulating modulator, which opposes the inhibitory effect of ATP on the phosphofructokinase reaction (Figure 15-15*b*). As a result of these complex allosteric interactions, the rate of the phosphofructokinase reaction accelerates several hundredfold when resting skeletal muscle is excited to maximal activity.

The *pyruvate kinase* reaction is a secondary control point in glycolysis. It is also an allosteric enzyme. Pyruvate kinase occurs in at least three isozyme forms (page 239), which differ somewhat in their tissue distribution and in their response to modulators. At high ATP concentrations the apparent affinity of pyruvate kinase for phosphoenolpyruvate is relatively low, and the reaction rate will accordingly be low at normal phosphoenolpyruvate concentrations. Pyruvate kinase is also inhibited by acetyl-CoA and by long-chain fatty acids, both important fuels for the citric acid cycle. Thus, whenever the cell already has a high concentration of ATP, or whenever ample fuels are already available for energy-yielding respiration, glycolysis is inhibited by the action of either phospho-

(a) Active muscle

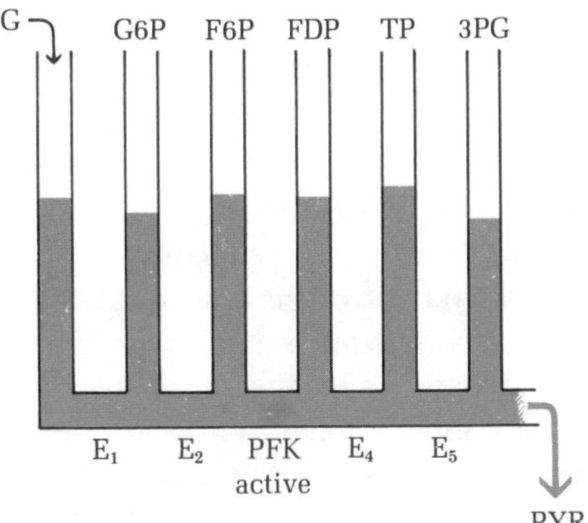

(b) Muscle at rest

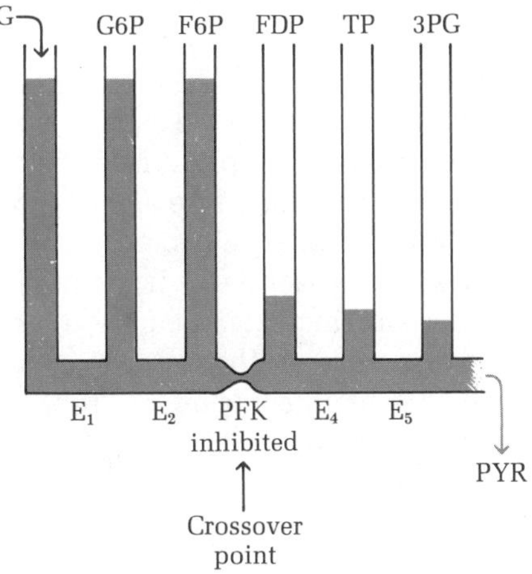

Figure 15-16
Hydraulic analogy of a crossover point in the regulation of muscle glycolysis. Measurement of the concentrations of the successive intermediates (a) in an active intact muscle and (b) in the muscle at rest can reveal the regulated reaction step. The crossover point is the reaction promoted by the enzyme whose substrate increases in concentration and whose product(s) decrease in concentration when the muscle returns from its active to its resting state. The crossover point shown is that catalyzed by phosphofructokinase (PFK), which regulates the rate at which pyruvate is formed. G designates glucose; G6P, glucose 6-phosphate; F6P, fructose 6-phosphate; FDP, fructose 1,6-diphosphate; TP, triose phosphates; 3PG, 3-phosphoglycerate; PYR, pyruvate. Not all the glycolytic intermediates are shown.

fructokinase or pyruvate kinase, depending upon conditions. On the other hand, at low ATP concentrations, the apparent affinity of pyruvate kinase for phosphoenolpyruvate increases, enabling the enzyme to transfer the phosphate group from phosphoenolpyruvate to ADP even when the concentration of phosphoenolpyruvate is relatively low. Certain amino acids also modulate pyruvate kinase activity, especially in the liver.

Glycolysis in each type of cell is regulated with computer-like efficiency, so that changes in the level of many different metabolites can influence its overall rate. Since glycolysis is the most ancient catabolic pathway and is central to so much of metabolism, it is not surprising that its regulation is complex.

How Are the Regulated Steps of Glycolysis Identified in Intact Cells?

Although regulatory effects on the action of allosteric enzymes such as phosphofructokinase are readily observed in the test tube with purified enzyme preparations, a question arises. How do we *really* know that the phosphofructokinase reaction is a major regulated step in *intact* cells? The most important test is provided by measurements of the changes in the concentration of the different intermediates of glycolysis in intact cells or tissues in response to changes in the rate of glycolysis. If we turn again to Figure 15-13 and imagine glycolysis taking place from glucose 6-phosphate to pyruvate at a steady, unchanging rate in resting muscle, the concentration of each of the intermediates will be constant in a steady state. If we now suddenly cause the phosphofructokinase reaction to be inhibited, the inhibition of this enzyme will be manifested by a sudden *increase* in the concentration of its substrate fructose 6-phosphate, which will tend to accumulate, and by a *decrease* in the concentration of the product of this enzyme, fructose 1,6-diphosphate, and all the subsequent glycolytic intermediates, since they will continue to be converted into pyruvate at the same rate as before. The step from fructose 6-phosphate to fructose 1,6-diphosphate, in which the concentration of the former increases and the latter decreases when PFK is inhibited, is called a crossover point. The crossover point can identify the place where an enzyme system is regulated in its transitions between rest and activity. Figure 15-16 shows a hydraulic analogy of the PFK crossover point. Thus the measurement of the concentrations of the different intermediates of a metabolic sequence in an intact tissue, and how these concentrations change as the overall rate of the pathway changes, enables us to determine which reaction(s) in the pathway are regulated. By this approach phosphofructokinase has been proved to be the major regulatory step in glycolysis in skeletal muscle, brain, and other tissues. The concentration of all the intermediates in resting vs. stimulated cells is determined by quick-freezing the cells or tissue in liquid nitrogen to stop further enzymatic action (this is called *freeze-clamping*), extracting the frozen tissue with an acidic reagent that denatures and inactivates enzymes, and then analyzing the concentrations of the intermediates in the tissue extracts.

At any given time under any set of conditions only one of the regulatory reactions of glycolysis determines the rate of breakdown of glucose (or glycogen) to pyruvate. Why, then, does glycolysis have more than one step that can be regulated? It is because cell metabolism is a very complex process. Under some metabolic circumstances it is strategically important for the cell to regulate the rate of glycolysis by controlling the *entry* of glucose residues, either by controlling hexokinase or glycogen phosphorylase. Under another set of conditions it may be advantageous to the cell to control the phosphofructokinase reaction, and in still other conditions it may be desirable to control the pyruvate kinase reaction. Because each tissue or cell type has somewhat different metabolic activities and functions, as well as different fuel mixes, the point at which glycolysis is regulated may differ from one cell type to another and from one set of metabolic conditions to another. Thus multiple regulatory points in the central glycolytic pathway give the cell great metabolic flexibility.

Two further points are important to remember about regulation of glycolysis—and for that matter all metabolic pathways: (1) The regulated steps in a metabolic sequence are usually irreversible under intracellular conditions. Phosphorylase, hexokinase, phosphofructokinase, and pyruvate kinase all catalyze reactions that proceed with a large free-energy decrease under intracellular conditions and are thus essentially irreversible. (2) In glycolysis most of the other enzyme-catalyzed steps, i.e., the nonregulated steps, are at or near equilibrium. Nevertheless, because it has steps that are irreversible, the overall process of glycolysis must also be irreversible. Many *individual* enzymatic reactions in intact cells may be at or near equilibrium, but considered overall, living organisms and their metabolic functions are never at equilibrium.

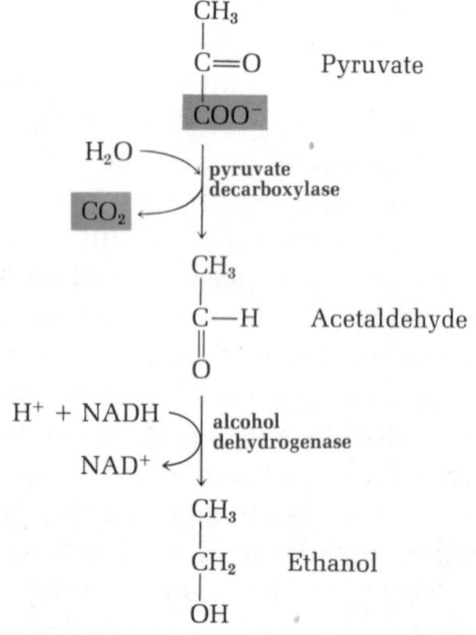

Figure 15-17
Terminal steps in alcoholic fermentation.

Alcoholic Fermentation Differs from Glycolysis Only in Its Terminal Steps

In yeast and other microorganisms that ferment glucose to ethanol and CO_2, rather than to lactate, the enzymatic pathway of glucose degradation is identical to that described for anaerobic glycolysis, except for the step catalyzed by lactate dehydrogenase. In yeast, which contains no lactate dehydrogenase like that of muscle tissue, two alternate enzymatic reactions take place (Figure 15-17). In the first, pyruvate resulting from the breakdown of glucose loses its carboxyl group by the action of pyruvate decarboxylase. This reaction is a simple decarboxylation and does not involve the net oxidation of the pyruvate:

$$\underset{\text{Pyruvate}}{CH_3-\underset{\underset{O}{\|}}{C}-COO^-} + H^+ \longrightarrow \underset{\text{Acetaldehyde}}{CH_3-\underset{\underset{O}{\|}}{C}-H} + CO_2$$

This reaction is irreversible in the cell. Pyruvate decarboxylase requires Mg^{2+} and has a tightly bound coenzyme, *thiamine pyrophosphate,* whose function as a transient carrier of the acetaldehyde group was discussed earlier (page 253).

In the final step of alcoholic fermentation, acetaldehyde is reduced to ethanol, with NADH derived from glyceraldehyde 3-phosphate dehydrogenation furnishing the reducing power, through the action of *alcohol dehydrogenase:*

$$CH_3{-}\underset{\underset{\displaystyle O}{\|}}{C}{-}H + NADH + H^+ \rightleftharpoons \underset{\text{Ethanol}}{CH_3{-}CH_2OH} + NAD^+$$

Ethanol and CO_2, instead of lactate, are thus the end products of alcoholic fermentation. The overall equation of alcoholic fermentation can therefore be written

$$\text{Glucose} + 2P_i + 2\text{ADP} \longrightarrow 2\text{ ethanol} + 2CO_2 + 2\text{ATP} + 2H_2O$$

We note that there is no net change in the ratio of hydrogen to carbon atoms when D-glucose (H/C ratio = 12/6 = 2) is fermented to two ethanol and two CO_2 (H/C ratio = 12/6 = 2). In all anaerobic fermentations the H/C ratio of the reactants and products remains the same.

Pyruvate decarboxylase is characteristically present in brewer's yeast and all other organisms that promote alcoholic fermentation, but it is absent in animal tissues and other organisms that carry out lactate fermentation, such as the lactic acid bacteria.

While the biochemistry of alcoholic fermentation can be expressed by a series of enzyme-catalyzed chemical reactions, brewing and wine making are ancient arts practiced centuries before the science of chemistry was born. Indeed, the timeworn recipes for making beer and wine provided important clues to some of the most fundamental discoveries in the early history of biology and biochemistry. For example, it was not until 1856 that it was conclusively proved by Louis Pasteur that the fermentation of sugar to yield alcohol is caused by microorganisms and not by black magic. Pasteur was employed by wine makers in France to determine why the wine produced in certain years was bad and turned into vinegar. He carried out classical experiments demonstrating that sterile glucose solutions do not undergo fermentation, whereas glucose solutions exposed to unfiltered air do, the result of contamination by airborne spores of yeasts and other microorganisms. From the bloom on freshly picked grapes, he was able to culture yeast cells and showed them to be responsible for the subsequent fermentation of the juice extracted from the crushed grapes. Formation of acetic acid from ethanol Pasteur found to be due to the action of other microorganisms, the acetic acid bacteria, which are aerobic and oxidize ethanol to acetic acid. Brewing beer, another ancient art, involves a number of enzymatic processes in addition to the reactions of alcoholic fermentation (Box 15-3).

Box 15-2 Brewing beer

Beer is made by the alcoholic fermentation of the carbohydrates present in cereal grains such as barley. But these carbohydrates, largely polysaccharides, are not available to the glycolytic enzymes in yeast cells, which can accept only disaccharides and monosaccharides. The barley must first undergo a process called *malting.* The cereal seeds are allowed to germinate until they form the appropriate enzymes required to break down the polysaccharides of the cell walls of the seeds, as well as the starch and other polysaccharide food reserves within the cells of the seeds. Germination is then stopped by controlled heating, before further growth of the seedlings occurs. The product is *malt,* which now contains enzymes such as α-amylase and maltase, capable of breaking down starch to maltose, glucose, and other simple sugars. The malt also contains enzymes specific for the β linkages of cellulose and other cell-wall polysaccharides of the barley husks, which must be broken down in order to allow α-amylase to act on the starch within the grains.

In the next step the brewer prepares the *wort,* the nutrient medium required for the subsequent fermentation by yeast cells. The malt is mixed with water and then mashed or crushed. This allows the enzymes formed in the malting process to act on the cereal polysaccharides to form maltose, glucose, and other simple sugars, which are soluble in the aqueous medium. The remaining cell matter is then separated and the liquid wort or medium is boiled with hops, to give flavor. The wort is cooled and then aerated.

Now the yeast cells are added. In the aerobic wort the yeast grows and reproduces very rapidly, using energy obtained from some of the sugars in the wort. In this phase no alcohol is formed because the yeast, being amply supplied with oxygen, oxidizes the pyruvate formed by glycolysis via the citric acid cycle to CO_2 and H_2O. The aerobic metabolism of the yeast yields very fast growth of the cells and is controlled by adding just the right amount of oxygen. When all the dissolved oxygen in the vat of wort has been consumed, the yeast cells, being facultative (page 335), switch over to anaerobic utilization of the sugar in the wort. From this point on, the yeast ferments the sugars of the wort into ethanol and carbon dioxide. The fermentation process is controlled in part by the concentration of the ethanol formed, by the pH, and by the remaining sugar. After the fermentation has been stopped, the cells are removed, and the "raw" beer is ready for final processing. Light beers, which have become very popular, contain less sugar and alcohol than normal beer but retain the flavor of standard beer.

In the final steps of brewing, the amount of foam or head on the beer, which is given by dissolved proteins, is adjusted. Normally this is controlled by the action of proteolytic enzymes that appear in the malting process. If they act on the beer proteins too long, the beer will have very little head and will be flat; if the proteolytic enzymes do not act long enough, the beer will not be clear when it is cold. Sometimes proteolytic enzymes from other sources are added to control the head. An important factor in the flavor of beer is a trace compound, *dimethyl sulfide.* This substance in a high concentration has a very disagreeable taste, but without it lager beer is insipid and tasteless. The dimethyl sulfide is formed by enzymes developed in the malting process and must be very carefully controlled.

Thus, many important ingredients in the brewing of beer are still matters of the skill of the brewer and are not yet completely known by the biochemist. Perhaps it is best that this ancient art remains an art!

Summary

Glycolysis, the conversion of D-glucose into two molecules of pyruvate, is a central pathway for recovery of chemical energy as ATP in most living organisms. Under anaerobic conditions the pyruvate is reduced to lactate in most animal and plant tissues or ethanol and CO_2 in alcoholic fermentation by yeast. The overall equation for anaerobic glycolysis in muscle and for lactate fermentation in some microorganisms is

$$\text{Glucose} + 2\text{ADP} + 2\text{P}_i \rightarrow 2\ \text{lactate}^- + 2\text{H}^+ + 2\text{ATP} + 2\text{H}_2\text{O}$$

and for alcoholic fermentation is

$$\text{Glucose} + 2\text{ADP} + 2\text{P}_i \rightarrow 2\ \text{ethanol} + 2\text{CO}_2 + 2\text{ATP} + 2\text{H}_2\text{O}$$

In aerobic cells, pyruvate is oxidized to acetyl-CoA and CO_2 instead of being reduced to lactate (or ethanol + CO_2). Glycolysis is thus the obligatory first stage in aerobic glucose catabolism in many organisms.

The conversion of glucose into pyruvate is catalyzed by 10 enzymes acting in sequence and takes place in two phases. In the first, which involves five enzyme-catalyzed steps, D-glucose is enzymatically phosphorylated by ATP and ultimately cleaved to yield two molecules of D-glyceraldehyde 3-phosphate. In the second phase of glycolysis the glyceraldehyde 3-phosphate is oxidized by NAD^+ with uptake of inorganic phosphate to form 3-phosphoglyceroyl phosphate. The latter donates its high-energy phosphate group to ADP to yield ATP plus 3-phosphoglycerate, which is then isomerized to 2-phosphoglycerate. After dehydration of the latter by enolase, the phosphoenolpyruvate formed donates its phosphate group to ADP and thus yields free pyruvate. Two molecules of ATP enter the first phase of glycolysis, and four are formed from ADP in the second phase, giving a net yield of two ATPs from one molecule of glucose. In the absence of oxygen the NADH formed by the dehydrogenation of glyceraldehyde 3-phosphate is reoxidized to NAD^+ by pyruvate and lactate dehydrogenase to form lactate in animal tissues.

The conversion of glucose residues of glycogen and starch into glucose 6-phosphate is made possible by glycogen or starch phosphorylase and phosphoglucomutase. Hexoses other than glucose, such as fructose, mannose, and galactose, are also phosphorylated and converted into intermediates of the glycolytic pathway. The entry of glucose into glycolysis, promoted by hexokinase, is regulated by glucose 6-phosphate, which is an inhibitory modulator. Glycogen phosphorylase, which catalyzes conversion of glucose units of glycogen into glucose 1-phosphate, is a regulatory enzyme existing in active (phosphorylase *a*) and less active (phosphorylase *b*) forms; the latter form is stimulated by AMP. Phosphofructokinase, which is inhibited by ATP and citrate and stimulated by AMP, is the major regulatory enzyme of the glycolytic sequence. The pyruvate kinase reaction is a second point of regulation in glycolysis. In alcoholic fermentation, the reaction sequence is identical up to the stage of pyruvate, but instead of being reduced to lactate, pyruvate is first decarboxylated to acetaldehyde, which is then reduced to ethanol by NADH and alcohol dehydrogenase.

References

Books

ATKINSON, D. E.: *Cellular Energy Metabolism and Its Regulation*, Academic, New York, 1977. Interesting treatment of the energetics and regulation of glycolysis.

DICKENS, F., P. J. RANDLE, and W. J. WHELAN: *Carbohydrate Metabolism and Its Disorders*, 2 vols., Academic, New York, 1968. A comprehensive collection of review articles.

FRUTON, J. S.: *Molecules and Life*, Wiley, New York, 1972. Includes a detailed historical account of research on glycolysis.

HOCHACHKA, P.: *Living without Oxygen*, Harvard University Press, Cambridge, Mass., 1980. Comparative biochemistry and physiology of anaerobic glycolysis in different organisms.

KALCKAR, H. M. (ed.): *Biological Phosphorylations: Development of Concepts,* Prentice-Hall, Englewood Cliffs, N.J., 1969. Contains classical papers on glycolysis.

LEHNINGER, A. L.: *Biochemistry,* 2d ed., Worth, New York, 1975. Chapter 16 has further details on glycolysis.

NEWSHOLME, E. A., and C. START: *Regulation in Metabolism,* Wiley, New York, 1973. Regulation of glycolysis is covered in Chapters 3 and 6.

Articles

COULSON, R. A.: "Anaerobic Glycolysis: The Smith and Wesson of the Heterotherms," *Perspec. Biol. Med.,* **22:**465–479 (1979). Very interesting analysis of the comparative role of anaerobic glycolysis in large animals, developed from experimental observations on alligators and other animals.

OTTAWAY, J. H., and J. MOWBRAY: "The Role of Compartmentation in the Control of Glycolysis," *Curr. Topics Cell Regulation,* **12:**108–195 (1977).

Problems

1. *Equation for the First Stage of Glycolysis* Write balanced equations for the sequence of reactions in the catabolism of D-glucose to two molecules of D-glyceraldehyde 3-phosphate (the first stage of glycolysis). For each equation write the standard-free-energy change. Then write the overall or *net* equation for the first stage of glycolysis, including the net standard-free-energy change.

2. *The Second Stage of Glycolysis in Skeletal Muscle* In working skeletal muscle under anaerobic conditions, glyceraldehyde 3-phosphate is converted into lactate (the second stage of glycolysis). Write balanced equations for the sequence of reactions in this process, with standard-free-energy changes for each. Then write the overall or net equation for the second stage of glycolysis, including the net standard-free-energy change.

3. *Metabolism of Fructose in Spermatozoa* Fructose is found in bovine and human semen at concentrations up to $\sim$12 m*M*. Spermatozoa use fructose anaerobically to produce the ATP necessary for flagellar motion (swimming). The major catabolic pathway of fructose to lactate in these cells bypasses the phosphofructokinase reaction in glycolysis by using an enzyme that cleaves fructose 1-phosphate into two 3-carbon compounds (Figure 15-8). Write equations for the sequence of chemical transformations involved. Also write the net equation for the anaerobic catabolism of fructose to lactate by spermatozoa.

4. *Pathway of Atoms in Fermentation* A pulse-chase experiment using radioactively labeled carbon sources is carried out on a yeast extract maintained under strictly anaerobic conditions to produce ethanol. The experiment consists of incubating a small amount of radioactively labeled substrate (the *pulse*) with the yeast extract just long enough for each intermediate in the pathway to become labeled. The label is then "chased" through

Problem 4

Glucose

the pathway by the addition of excess unlabeled substrate. The purpose of the chase is to prevent recycling of labeled products and their entry into other pathways.

(a) If glucose labeled in the 1 position with ^{14}C is used as a substrate, what is the location of ^{14}C in the product ethanol? Explain. The position of the label in glucose is shown in the structure at the left.

(b) Where would ^{14}C have to be located in the starting glucose molecule in order to assure that all the ^{14}C activity will be liberated as $^{14}CO_2$ during alcoholic fermentation? Explain.

5. *Relation between Enzyme Kinetic Properties and Physiological Functions* The glucose concentration inside mammalian cells is relatively low compared with its concentration in blood plasma, because glucose transport into the cell is regulated and because glucose is rapidly phosphorylated by the reaction

$$\text{Glucose} + \text{ATP} \longrightarrow \text{glucose 6-phosphate} + \text{ADP} + \text{H}^+$$

In mammals two different enzymes catalyze this reaction; they show markedly different properties. Skeletal muscle contains only one of them, hexokinase, which has a K_M value of 0.1 m*M* and is inhibited by glucose 6-phosphate. The liver, on the other hand, contains both hexokinase and glucokinase, the latter predominating. Glucokinase has a much higher K_M value, 10.0 m*M*, and is not inhibited by glucose 6-phosphate. What is the significance of the two different K_M values of muscle hexokinase and liver glucokinase? Rationalize the differences in the kinetic properties of these two enzymes (K_M and inhibition by glucose 6-phosphate) in terms of the physiological roles they play in muscle vs. liver.

6. *Role of Lactate Dehydrogenase* During strenuous activity, muscle tissue demands vast quantities of ATP compared with resting tissue. In white skeletal muscle, e.g., rabbit leg muscle or turkey flight muscle, this ATP is produced almost exclusively by anaerobic glycolysis. As shown in Figure 15-5, ATP is produced in the second stage of glycolysis by two enzymatic reactions, promoted by phosphoglycerate kinase and pyruvate kinase. Suppose white skeletal muscle were devoid of lactate dehydrogenase. Could it carry out strenuous physical activity, i.e., generate ATP at a high rate by glycolysis? Explain. Remember that the lactate dehydrogenase reaction does not involve ATP. A clear understanding of the answer to this question is essential for comprehension of the glycolytic cycle.

7. *Arsenate Poisoning* Since arsenate is structurally and chemically similar to phosphate (P_i), many enzymes that require phosphate will also utilize arsenate. Organic compounds of arsenate are less stable than analogous phosphate compounds, however. For example, acyl arsenates decompose rapidly by hydrolysis in the absence of catalysts

$$\text{R—C(=O)—O—As(=O)(O}^-\text{)—O}^- + \text{H}_2\text{O} \longrightarrow \text{R—C(=O)—O}^- + \text{HOAsO}_3^{2-} + \text{H}^+$$

On the other hand, acyl phosphates, such as 3-phosphoglyceroyl phosphate, are more stable and are transformed in cells by enzymatic action.

(a) Predict the effect on the net reaction catalyzed by glyceraldehyde 3-phosphate dehydrogenase if phosphate is replaced by arsenate.

(b) What would be the consequence to an organism if arsenate were substituted for phosphate? Arsenate is very toxic to most organisms. Explain why.

8. *Requirement of Phosphate in Alcoholic Fermentation* In 1905 Harden and Young carried out a series of classic studies on anaerobic alcoholic fermentation of D-glucose to ethanol and CO_2 by extracts of brewer's yeast and made the following observations: (1) Inorganic phosphate was essential to fermentation; when the supply of phosphate was exhausted, fermentation ceased before all the glucose was used. (2) During fermentation under these conditions ethanol, carbon dioxide, and a hexose diphosphate accumulated. (3) When arsenate was substituted for phosphate, no hexose diphosphate accumulated but the fermentation proceeded until all the glucose was converted into ethanol and carbon dioxide.

(a) Why does fermentation cease when the supply of phosphate is exhausted?

(b) Why do ethanol and carbon dioxide accumulate? Is the conversion of pyruvate into ethanol and carbon dioxide essential? Why? Identify the hexose diphosphate that accumulates. Why does it accumulate?

(c) Why does the substitution of arsenate for phosphate prevent the accumulation of hexose diphosphate yet allow the fermentation to ethanol and carbon dioxide to go to completion? (See Problem 7.)

Problem 9

$$HO{-}CH_2{-}\overset{\displaystyle OH}{\underset{\displaystyle H}{C}}{-}CH_2{-}OH$$

Glycerol

9. *Metabolism of Glycerol* Glycerol obtained from the breakdown of fat is metabolized by being converted into dihydroxyacetone phosphate, an intermediate in glycolysis, in two enzyme-catalyzed reactions. Propose a reaction sequence for the metabolism of glycerol. On which known enzyme-catalyzed reactions is your proposal based? Write the net equation for the conversion of glycerol into pyruvate based on your scheme.

10. *Measurement of Intracellular Metabolite Concentrations* Measuring the concentrations of metabolic intermediates in the living cell represents a difficult experimental problem. Since the enzymes contained in the cell rapidly catalyze metabolic interconversions, a common problem associated with perturbing the cell experimentally is that the measured concentrations of metabolites reflect not the physiological concentrations but the equilibrium concentrations. Hence, a reliable experimental technique requires all enzyme-catalyzed reactions to be instantaneously stopped in the intact tissue, so that the metabolic intermediates do not undergo change. This objective is accomplished by rapidly compressing the tissue between large aluminum plates cooled with liquid nitrogen (−190°C), a process called *freeze-clamping*. After freezing, which stops enzyme action instantly, the tissue is powdered and the enzymes inactivated by precipitation with perchloric acid. The precipitate is removed by centrifugation and the clear supernatant extract is analyzed for metabolites by specific enzymatic measurements. The actual concentration of the metabolite in the cell is determined from the total water content of the tissue and a measurement of the extracellular volume. The actual

Table 1

Metabolite	Apparent concentration, mM (μmol/mL of intracellular H_2O)
Fructose 6-phosphate	0.087
Fructose 1,6-diphosphate	0.022
ATP	11.52
ADP	1.32

From J. R. Williamson, *J. Biol. Chem.*, **240:** 2308 (1965).

cellular concentrations of the substrates and products involved in the phosphorylation of fructose 6-phosphate by the enzyme phosphofructokinase in isolated rat heart tissue are given in Table 1.

(a) Using the information in the table, calculate the mass-action ratio Q for the phosphofructokinase reaction at physiological conditions, where

$$Q = \frac{[\text{fructose 1,6-diphosphate}][\text{ADP}]}{[\text{fructose 6-phosphate}][\text{ATP}]}$$

(b) Given that $\Delta G^{\circ\prime}$ for the phosphofructokinase reaction is -3.4 kcal/mol, calculate the equilibrium constant for the phosphofructokinase reaction.

(c) Compare the values of Q and K'_{eq}. Is the physiological reaction at equilibrium? Explain. What does this experiment say about the role of phosphofructokinase as a regulatory enzyme?

11. *Regulation of Phosphofructokinase* The effect of ATP on the allosteric enzyme phosphofructokinase is shown below. For a given concentration of fructose 6-phosphate, the phosphofructokinase activity increases with increasing concentrations of ATP, but a point is reached beyond which increasing concentrations of ATP cause inhibition of phosphofructokinase.

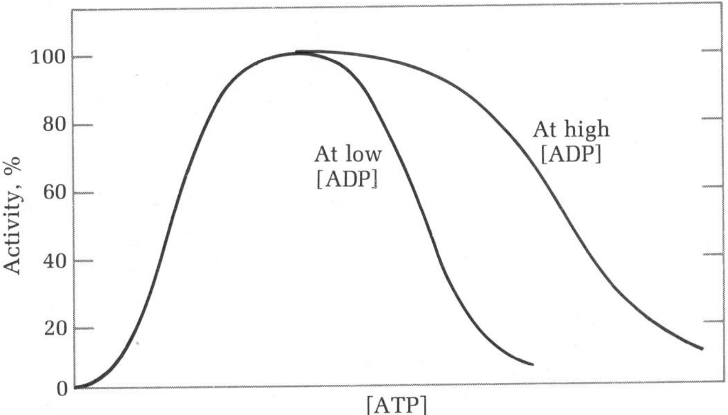

(a) Explain how ATP can be both a substrate and an inhibitor of phosphofructokinase. How is the enzyme regulated by ATP?

(b) In what ways is glycolysis regulated by ATP levels?

(c) The inhibition of phosphofructokinase by ATP is diminished when the ADP concentration is high. How can this observation be explained?

12. *Enzyme Activity and Physiological Function* The V_{max} value for the enzyme glycogen phosphorylase from skeletal muscle is much larger than the V_{max} value of the same enzyme from liver tissue.

(a) What is the physiological function of glycogen phosphorylase in skeletal muscle? In liver tissue?

(b) Why does the V_{max} of the muscle enzyme need to be larger than that of the liver enzyme?

13. *Enzyme Defects in Carbohydrate Metabolism* Summaries of four clinical case studies are given on the following page. For each case determine which enzyme is defective and designate the appropriate treatment from the lists provided. Justify your choices. Answer the questions contained in each case study.

Case A The patient develops vomiting and diarrhea shortly after milk ingestion. A lactose tolerance test is administered. (The patient ingests a standard amount of lactose, and the blood plasma glucose and galactose concentrations are measured at intervals. In normal individuals the levels increase to a maximum in about 1 h and then recede.) The patient's blood concentrations of glucose and galactose do not rise but remain constant. Explain why the blood glucose and galactose increase and then decrease in normal individuals. Why do they fail to rise in the patient?

Case B The patient, who is mentally retarded, develops vomiting and diarrhea after ingestion of milk. He has a low concentration of blood glucose but a much higher than normal blood concentration of reducing sugars. The urine gives a positive test for galactose. Why is the reducing sugar concentration in the blood high? Why does galactose appear in the urine?

Case C The patient complains of painful muscle cramps when performing strenuous physical exercise but is otherwise normal. A muscle biopsy indicates that muscle glycogen concentration is much higher than in normal individuals. Why does glycogen accumulate?

Case D The patient is lethargic, her liver is enlarged, and a biopsy of the liver shows large amounts of excess glycogen. She also has a lower-than-normal level of blood glucose. Account for the low blood glucose concentration in this patient.

Enzyme Defects

(a) Muscle phosphofructokinase
(b) Phosphomannoisomerase
(c) Galactose 1-phosphate uridylyltransferase
(d) Liver phosphorylase
(e) Triose kinase
(f) Lactase in intestinal mucosa
(g) Maltase in intestinal mucosa

Treatments

1. Jogging 3 miles per day
2. Fat-free diet
3. Low-lactose diet
4. Avoiding strenuous exercise
5. Large doses of niacin
6. Frequent and regular feedings

14. *Severity of Clinical Symptoms Due to Enzyme Deficiency* The clinical symptoms of the two forms of galactosemia involving the deficiency of galactokinase and galactose 1-phosphate uridylyltransferase show radically different severity. Although both deficiencies produce gastric discomfort upon milk ingestion, the deficiency of the latter enzyme leads to liver, kidney, spleen, and brain dysfunction and eventual death. What products accumulate in the blood and tissues with each enzyme deficiency? Estimate the relative toxicities of these products from the above information.

CHAPTER 16

The Citric Acid Cycle

In Chapter 15 we saw how cells obtain ATP energy from the breakdown of glucose in the absence of oxygen. However most animal and plant cells normally are aerobic and oxidize their organic fuels completely to carbon dioxide and water. Under these conditions the pyruvate formed in the glycolytic breakdown of glucose is not reduced to lactate or to ethanol and CO_2, as occurs in anaerobic conditions. Instead it is oxidized to CO_2 and H_2O in the aerobic phase of catabolism, which biochemists call *respiration*. We usually think of the term respiration in its physiological or macroscopic sense, to refer to the uptake of O_2 and release of CO_2 by the lungs, but the biochemist and cell biologist use the term in a microscopic sense, to refer to the molecular processes involved in O_2 consumption and CO_2 formation by cells.

Cell respiration occurs in three major stages, as shown in the flowsheet in Figure 16-1. In the first stage organic fuel molecules—carbohydrates, fatty acids, and also some amino acids—are oxidized to yield 2-carbon fragments, the acetyl groups of acetyl–coenzyme A. In the second stage these acetyl groups are fed into the citric acid cycle, which enzymatically degrades them to yield energy-rich hydrogen atoms and to set free CO_2, the final oxidation product of organic fuels. In the third stage of respiration the hydrogen atoms are separated into protons (H^+) and energy-rich electrons, which are transferred along a chain of electron-carrying molecules, the *respiratory chain*, to molecular oxygen, which they reduce to form H_2O. During this process of *electron transport* much energy is released and conserved in the form of ATP in a process called *oxidative phosphorylation*. Respiration is far more complex than glycolysis; it has been said that respiration is to glycolysis what a modern jet turbine is to a one-cylinder reciprocating engine. This is why we must devote two chapters to respiration.

In this chapter we examine the Krebs *citric acid cycle*, also called the *tricarboxylic acid cycle*, the final common pathway

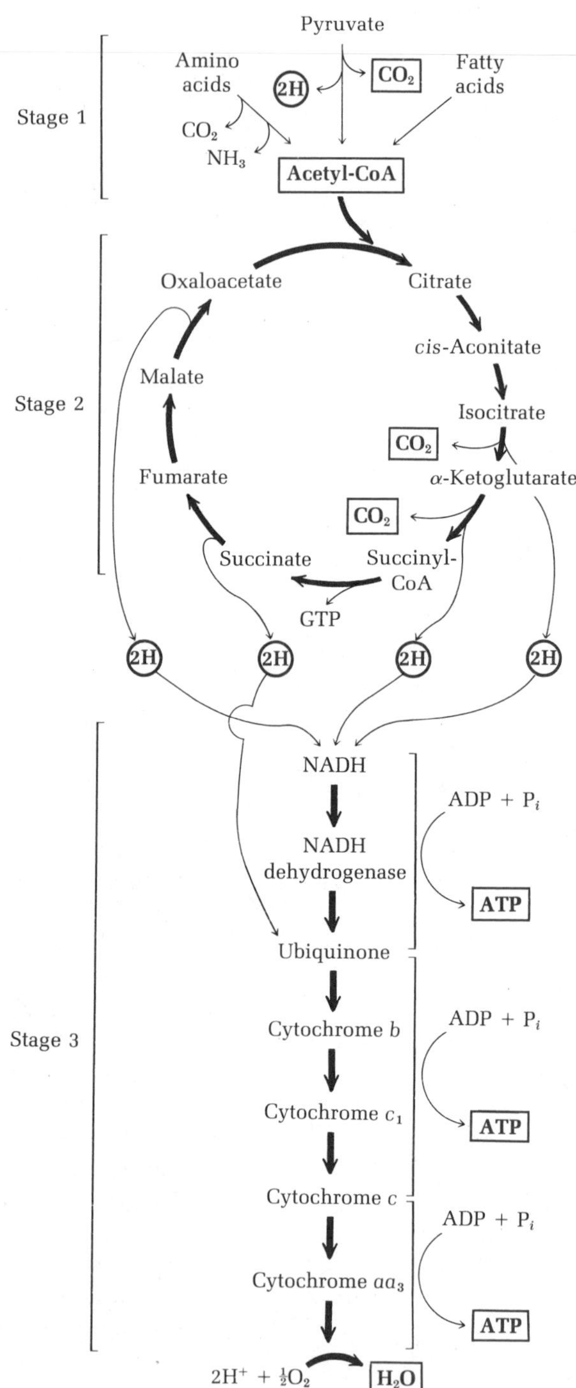

Figure 16-1
Stages in cell respiration. Stage 1: Mobilization of acetyl-CoA from glucose, fatty acids, and some amino acids. Stage 2: The citric acid cycle. Stage 3: Electron transport and oxidative phosphorylation. Each pair of H atoms entering the electron-transport chain as NADH yields 3 ATPs.

for oxidation of acetyl groups, into which the major organic fuel molecules of the cell, carbohydrates, fatty acids, and amino acids, converge during catabolism.

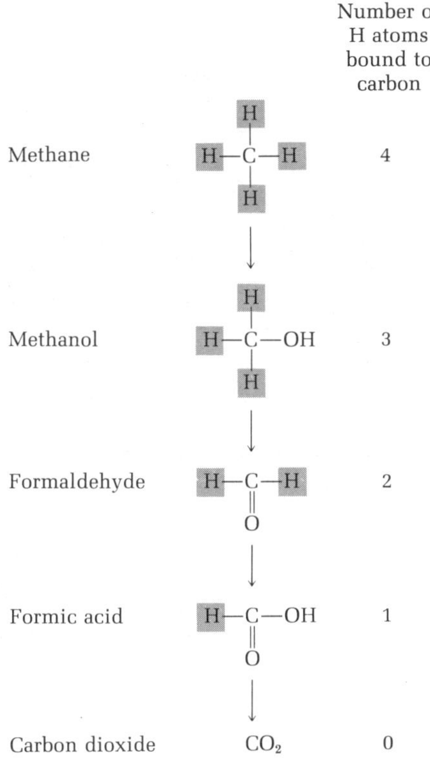

Figure 16-2
Successive steps in oxidation of methane to CO_2. As the four H atoms are removed from the carbon atom, one by one, there is a stepwise loss of available free energy. The ratio of carbon-bound H atoms to C atoms of simple organic molecules is in approximate proportion to their standard free energy of oxidation to CO_2.

Oxidation of Glucose to CO_2 and H_2O Releases Much More Energy than Glycolysis

We have seen (page 401) that the breakdown of glucose to lactate by glycolysis releases only a very small fraction of the chemical energy potentially available in the structure of the glucose molecule. Much more energy is released when the glucose molecule is oxidized completely to CO_2 and H_2O, as shown by the standard-free-energy changes of these two reactions:

$$\text{Glucose} \longrightarrow 2\ \text{lactate} + 2\text{H}^+ \qquad \Delta G^{\circ\prime} = -47.0\ \text{kcal/mol}$$

$$\text{Glucose} + 6\text{O}_2 \longrightarrow 6\text{CO}_2 + 6\text{H}_2\text{O} \qquad \Delta G^{\circ\prime} = -686\ \text{kcal/mol}$$

When cells break glucose down anaerobically, the lactate formed still contains some 93 percent of the available energy of the original glucose. This is because lactic acid is almost as complex a molecule as glucose and has undergone no net oxidation. The free energy released on complete combustion of organic molecules is in approximate proportion to the ratio of carbon-bound hydrogen atoms to the total number of carbons, as shown in a very simple case in Figure 16-2. Only by removal of all the hydrogen atoms from the carbon atoms of glucose or other organic fuels and their replacement with oxygen to yield CO_2 can all their biologically available free energy be released.

Pyruvate Must First Be Oxidized to Acetyl-CoA and CO_2

Carbohydrates, fatty acids, and most of the amino acids are ultimately oxidized to CO_2 and H_2O via the citric acid cycle. First, however, before these nutrients can enter the cycle their carbon backbones must be degraded so that they yield the acetyl groups of acetyl-CoA, the form in which the citric acid cycle accepts most of its fuel input. In Chapters 18 and 19 we shall see how fatty acids and amino acids, respectively, yield acetyl groups to the cycle. Here we consider how pyruvate, derived from glucose by glycolysis, is dehydrogenated to yield acetyl-CoA and CO_2 by a structured cluster of enzymes called the *pyruvate dehydrogenase complex*, located in the mitochondria of eukaryotic cells and in the cytoplasm of prokaryotes. The overall reaction it catalyzes is

$$\text{Pyruvate} + \text{NAD}^+ + \text{CoA-SH} \longrightarrow \text{acetyl-CoA} + \text{NADH} + \text{CO}_2 \qquad \Delta G^{\circ\prime} = -8.0\ \text{kcal/mol}$$

In this rather complicated reaction the pyruvate undergoes *oxidative decarboxylation*, a dehydrogenation process in which the carboxyl group is removed as a molecule of CO_2 and the acetyl group appears as acetyl-CoA. The two hydrogen atoms removed from pyruvate appear as NADH and H^+. The NADH so formed then gives up its electrons to the electron-transport chain (Figure 16-1), which in turn carries them to molecular oxygen.

The combined dehydrogenation and decarboxylation of pyruvate to acetyl-CoA involves the sequential action of three different enzymes, *pyruvate dehydrogenase* (E_1), *dihydrolipoyl transacetylase* (E_2), and *dihydrolipoyl dehydrogenase* (E_3), as well as five different coenzymes or prosthetic groups, *thiamine pyrophosphate* (TPP), *flavin adenine dinucleotide* (FAD), *coenzyme A* (CoA), *nicotinamide adenine dinucleotide* (NAD^+), and *lipoic acid*. All these enzymes and coenzymes are organized into a multienzyme cluster first isolated and studied in detail by Lester Reed and his colleagues at the University of Texas. Four different vitamins required in human nutrition are vital components of this system: thiamine (in TPP), riboflavin (in FAD), pantothenic acid (in CoA), and nicotinamide (in NAD^+). In addition, the reaction requires *lipoic acid* (Figure 16-3). It is an essential vitamin or growth factor for several microorganisms, but it can be made by higher animals from readily available precursors. This large multienzyme cluster has a particle weight of over 6 million as isolated from *E. coli*; it

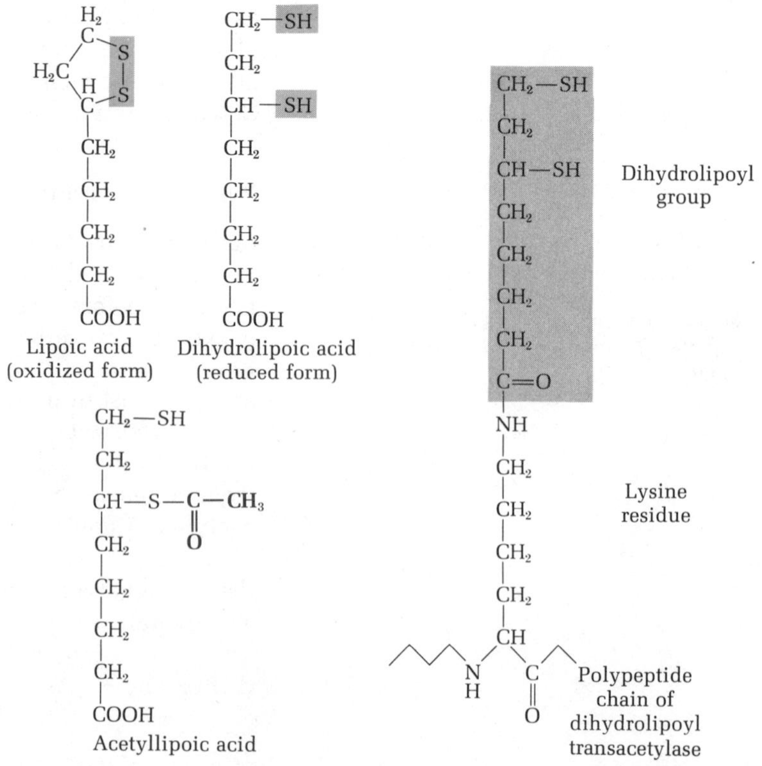

Figure 16-3
Lipoic acid and its active form, the prosthetic group of dihydrolipoyl transacetylase. Lipoic acid and the lipoyl group occur in oxidized or disulfide form, in reduced or dithiol form, and in acetylated form. The lipoyl group therefore acts as a hydrogen carrier and as an acetyl-group carrier. The lipoyllysyl group is about 1.4 nm long and acts as a swinging arm. It carries H atoms from pyruvate dehydrogenase to dihydrolipoyl dehydrogenase and also carries the acetyl group to CoA-SH.

Lipoic acid is sometimes called a pseudovitamin.

is about 45 nm in diameter and slightly larger than a ribosome. The "core" of the cluster, to which the other enzymes are attached, is dihydrolipoyl transacetylase (MW 200,000), which consists of 24 polypeptide chain subunits, each containing two lipoic acid groups in amide linkage with the ε-amino groups of specific lysine residues in the active sites of the subunits (Figure 16-3). Attached to the dihydrolipoyl transacetylase are the very large molecules pyruvate dehydrogenase and dihydrolipoyl dehydrogenase. Pyruvate dehydrogenase contains bound thiamine pyrophosphate, and dihydrolipoyl dehydrogenase contains bound FAD. The lipoyllysyl groups of the core enzyme are about 1.4 nm long and serve as "swinging arms" that can transfer hydrogen atoms and acetyl groups from one enzyme molecule to another in the pyruvate dehydrogenase complex. Also attached to the pyruvate dehydrogenase complex are two other enzymes (discussed later) that function to regulate the pyruvate dehydrogenase reaction.

Figure 16-4 shows schematically how the pyruvate dehydrogenase complex, which is localized in the mitochondria of animal cells, carries out the successive reactions involved in the decarboxylation and dehydrogenation of pyruvate in a sequence of five steps. In step 1 pyruvate loses its carboxyl group as it reacts with the bound thiamine pyrophosphate of pyruvate dehydrogenase (E_1) to form the hydroxyethyl derivative of the thiazole ring of the thiamine pyrophosphate (page 253). Pyruvate dehydrogenase also carries out step 2, the transfer of H

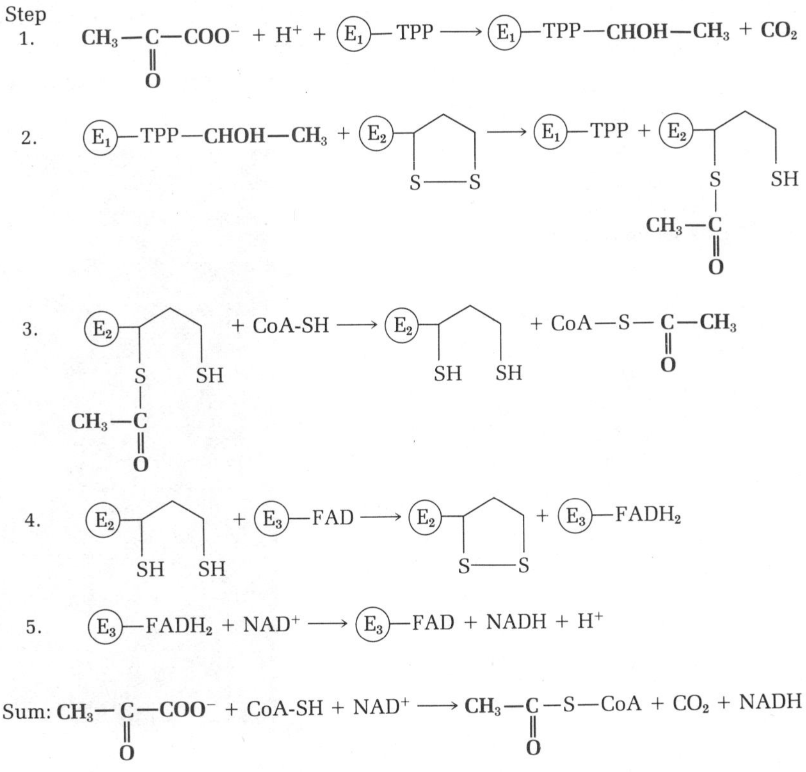

Figure 16-4
Steps in the oxidative decarboxylation of pyruvate to acetyl-CoA by the pyruvate dehydrogenase complex. The fate of pyruvate is traced in color. The structures of thiamine pyrophosphate and its α-hydroxyethyl derivative are given in Figure 10-2.

Key:
E_1 = pyruvate dehydrogenase
TPP = thiamine pyrophosphate
TPP—CHOH—CH_3 = α-hydroxyethylthiamine pyrophosphate
E_2 = dihydrolipoyl transacetylase
E_3 = dihydrolipoyl dehydrogenase

Step

1. $CH_3—C(=O)—COO^- + H^+ + (E_1)—TPP \longrightarrow (E_1)—TPP—CHOH—CH_3 + CO_2$

2. $(E_1)—TPP—CHOH—CH_3 + (E_2)$—lipoyl (S—S) $\longrightarrow (E_1)—TPP + (E_2)$—lipoyl (S—C(=O)—$CH_3$, SH)

3. (E_2)—lipoyl (S—C(=O)—CH_3, SH) + CoA-SH $\longrightarrow (E_2)$—lipoyl (SH, SH) + CoA—S—C(=O)—CH_3

4. (E_2)—lipoyl (SH, SH) + (E_3)—FAD $\longrightarrow (E_2)$—lipoyl (S—S) + (E_3)—$FADH_2$

5. $(E_3)—FADH_2 + NAD^+ \longrightarrow (E_3)—FAD + NADH + H^+$

Sum: $CH_3—C(=O)—COO^- + CoA\text{-}SH + NAD^+ \longrightarrow CH_3—C(=O)—S—CoA + CO_2 + NADH$

atoms and the acetyl group from thiamine pyrophosphate to the oxidized form of the lipoyllysyl prosthetic groups of the core enzyme dihydrolipoyl transacetylase, to form the 6-acetyl thioester of the reduced lipoyl groups. In step 3 a molecule of CoA-SH reacts with the acetyl derivative of dihydrolipoyl transacetylase to yield acetyl-S-CoA and the fully reduced, or dithiol form, of the lipoyl groups. In step 4 the reduced form of dihydrolipoyl transacetylase is acted upon by dihydrolipoyl dehydrogenase, which promotes transfer of hydrogen atoms from the reduced lipoyl groups to the FAD prosthetic group of dihydrolipoyl dehydrogenase. In step 5, the last, the reduced FAD group of dihydrolipoyl dehydrogenase transfers hydrogen to NAD^+, forming NADH. Central to the process are the lipoyllysyl swinging arms, which pass the H atoms and acetyl group from one enzyme to the next. All these enzymes and coenzymes are structured into a cluster to keep the prosthetic groups close together, allowing the reaction intermediates to react quickly with each other. If these very large enzyme molecules were separated from each other in the cytosol, they would require much more time to diffuse long distances before colliding and reacting with each other. Figure 16-5 shows an electron micrograph of the pyruvate dehydrogenase complex.

In the nutritional disease beriberi there is a deficiency of vitamin B_1 or thiamine (page 252). As one might now predict from this fact, thiamine-deficient animals are unable to oxidize pyruvate normally, particularly in the brain, which usually obtains all its energy by aerobic oxidation of glucose and for which pyruvate oxidation is therefore vital. Defective function of the pyruvate dehydrogenase is responsible for the characteristic polyneuritis or generalized malfunction of the motor nervous system in beriberi (Chapter 10).

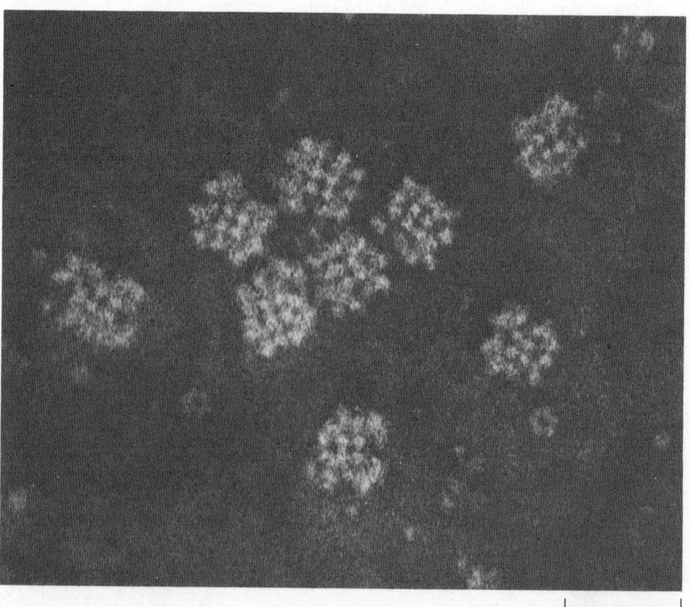

Figure 16-5
Electron micrograph of the pyruvate dehydrogenase complex isolated from E. coli. *showing its subunit structure.*

It is important to note here that the reaction catalyzed by the pyruvate dehydrogenase complex is irreversible in animal tissues. This has been proved by isotopic experiments which show that radioactive CO_2 cannot be reattached to acetyl-CoA to yield pyruvate labeled in the carboxyl group.

Regulation of the activity of the pyruvate dehydrogenase complex is an important element in the biological control of respiration, as we shall see later.

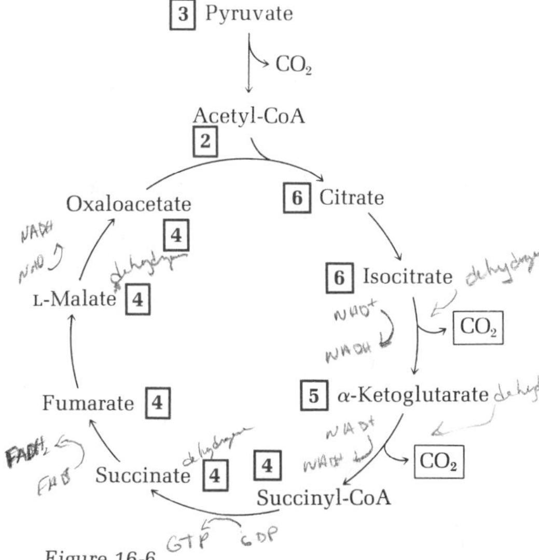

Figure 16-6
Outline of the citric acid cycle. The colored boxes give the number of carbon atoms in each intermediate of the cycle. Succinyl-CoA has four carbon atoms in its succinyl group, the portion of the molecule converted into free succinate.

The Citric Acid Cycle Is a Circular Rather than a Linear Enzyme System

Now that we have seen how acetyl-CoA is formed from pyruvate, we are ready to examine the citric acid cycle. It will be helpful to start with a bird's-eye view of how it functions. First we note a fundamental difference between glycolysis and the citric acid cycle. Glycolysis takes place by a *linear* sequence of enzyme-catalyzed steps. In contrast, the enzyme system that carries out the citric acid cycle functions in a cyclic manner. To begin a turn of the cycle (Figure 16-6), acetyl-CoA donates its acetyl group to the 4-carbon compound *oxaloacetate* to form the 6-carbon *citrate*. Citrate is then transformed into *isocitrate*, also a 6-carbon molecule, which is dehydrogenated with loss of CO_2 to yield the 5-carbon compound *α-ketoglutarate*. The latter then undergoes loss of CO_2 and ultimately yields the 4-carbon compound *succinate* and a second molecule of CO_2. Succinate is then enzymatically converted in three steps into the 4-carbon *oxaloacetate*, with which the cycle began. Thus oxaloacetate is regenerated after one turn of the cycle and is ready to react with another molecule of acetyl-CoA to start a second turn. In each turn of the cycle one acetyl group (2 carbons) enters as acetyl-CoA and two molecules of CO_2 come out. In each turn one molecule of oxaloacetate is used to form citrate, but after a series of reactions the oxaloacetate is regenerated again. Therefore no *net* removal of oxaloacetate occurs when the citric acid cycle operates; one molecule of oxaloacetate can theoretically suffice to bring about oxidation of an infinite number of acetyl groups.

How Did the Idea of the Citric Acid Cycle Arise?

This is a natural question, since such a complex cycle for oxidation of the 2-carbon acetyl groups to CO_2 via the 6-carbon citric acid may seem unnecessarily complicated and not in keeping with the principle of maximum economy in the biochemical logic of living cells.

The citric acid cycle was first postulated as the pathway of pyruvate oxidation in animal tissues in 1937 by Hans Krebs. The idea of the cycle came to him during a study of the effect of the anions of various organic acids on the rate of oxygen consumption by suspensions of minced pigeon-breast-muscle oxidizing pyruvate. This muscle, used in flight, has a very high

Figure 16-7
The naturally occurring tri- and dicarboxylic acids that promote the oxidation of pyruvate by muscle suspensions. Other naturally occurring organic acids, such as tartaric acid, oxalic acid, and ketoadipic acid, do not promote pyruvate oxidation. The active acids are shown in the sequence in which they participate in the citric acid cycle. In each step a single chemical change takes place. The point of inhibition by malonate is also shown; in its presence citrate is oxidized to succinate, which accumulates.

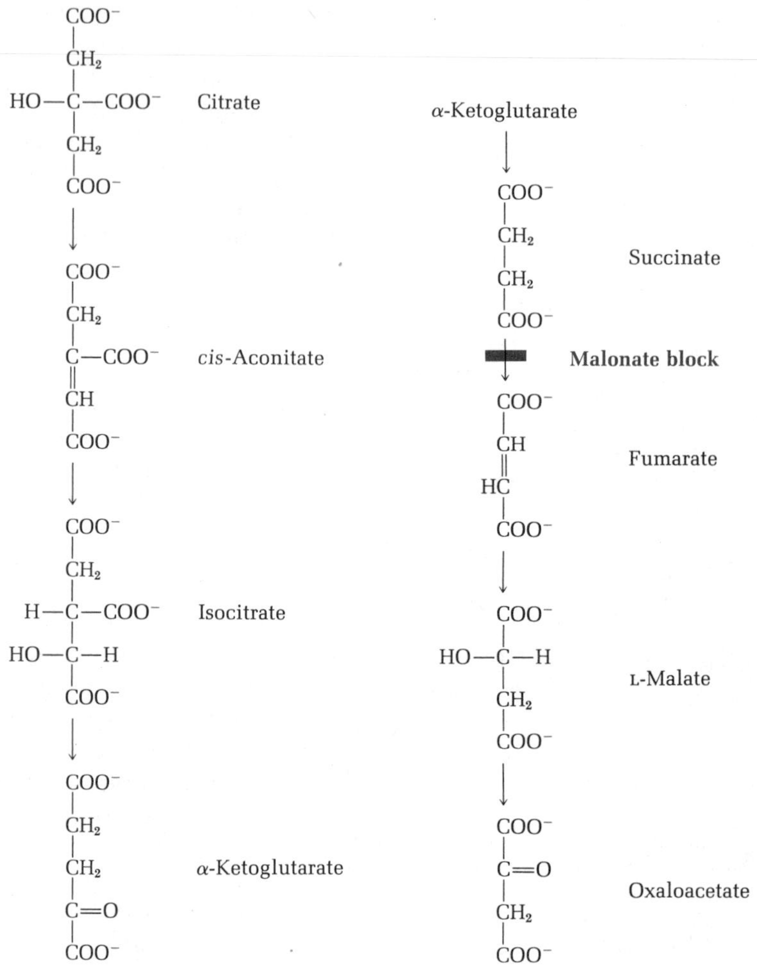

Citrate: COO^-—CH_2—HO—C—COO^-—CH_2—COO^-

↓

cis-Aconitate: COO^-—CH_2—C(—COO^-)═CH—COO^-

↓

Isocitrate: COO^-—CH_2—H—C—COO^-—HO—C—H—COO^-

↓

α-Ketoglutarate: COO^-—CH_2—CH_2—C═O—COO^-

α-Ketoglutarate

↓

Succinate: COO^-—CH_2—CH_2—COO^-

Malonate block

↓

Fumarate: COO^-—CH═HC—COO^-

↓

L-Malate: COO^-—HO—C—H—CH_2—COO^-

↓

Oxaloacetate: COO^-—C═O—CH_2—COO^-

Malonate: COO^-—CH_2—COO^- Succinate: COO^-—CH_2—CH_2—COO^-

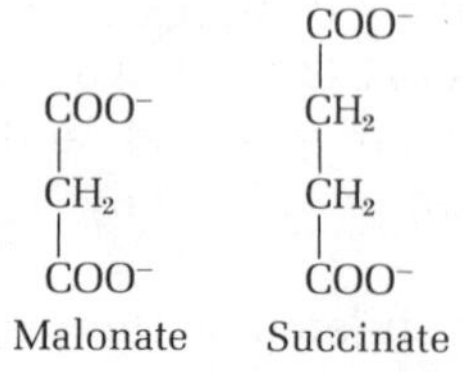

Figure 16-8
Malonate, a competitive inhibitor of succinate dehydrogenase (see also Figure 19-12 on page 224). Note the similarity in structure of malonate and succinate.

rate of respiration and was thus especially appropriate for the study of oxidative activity. Earlier investigators, particularly Albert Szent-Györgyi in Hungary, had found that certain 4-carbon dicarboxylic organic acids known to be present in animal tissues, succinic, fumaric, malic, and oxaloacetic acids, stimulate the consumption of oxygen by muscle. Krebs confirmed this observation and found that they also stimulate oxidation of pyruvate. Moreover, he found that oxidation of pyruvate by muscle is also stimulated by the 6-carbon tricarboxylic acids citric, cis-aconitic, and isocitric acids, and the 5-carbon α-ketoglutaric acid. The structures of these acids are shown in Figure 16-7. No other naturally occurring organic acids that were tested possessed such activity. The stimulatory action of the active acids was remarkable, since the addition of even a small quantity of any one of them could promote the oxidation of many times that amount of pyruvate.

The second important observation made by Krebs was that malonate (Figure 16-8), a specific competitive inhibitor of succinate dehydrogenase (page 224), inhibits the aerobic utiliza-

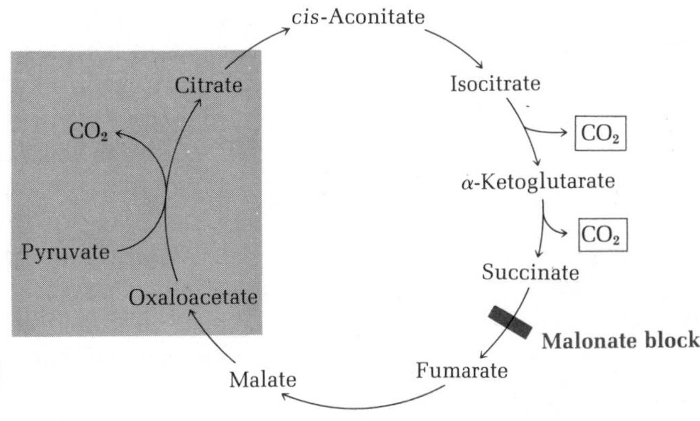

Figure 16-9
Closing the circle in the earliest formulation of the cycle. When Krebs found that pyruvate and oxaloacetate react to yield citrate (reactions in color), it proved that the reaction sequence is circular. Note that succinate will accumulate at the malonate block when pyruvate and oxaloacetate are oxidized via citrate.

Figure 16-10
Sir Hans Krebs in a photograph made on the occasion of his eightieth birthday in August 1980. Krebs was born in Germany and received his medical education there. From 1926 to 1930 he worked in Berlin with Otto Warburg, himself one of the great pioneers of modern biochemistry. In 1932, while engaged as an assistant in medicine at the University of Freiburg, Krebs worked out the outlines of the urea cycle with a medical student, Kurt Henseleit (Chapter 19). In 1933 Krebs emigrated to England to the University of Cambridge. Later he moved to the University of Sheffield, where he carried out most of the work on the citric acid cycle. From 1954 on he was head of the Biochemistry Department at Oxford. On his "retirement" from that position in 1967 he started a whole new research career examining the dynamics and regulation of metabolism in the Department of Medicine at Oxford, where he continued to work actively with several collaborators until his death in November, 1981. He was a frequent and popular lecturer in universities the world around. The citric acid cycle has been regarded as the most important single discovery in the history of metabolic biochemistry.

tion of pyruvate by muscle suspensions, regardless of which active organic acid is added. This indicated that succinate and succinate dehydrogenase must be essential components in the enzymatic reactions involved in the oxidation of pyruvate. Krebs further found that when malonate is used to inhibit the aerobic utilization of pyruvate by a suspension of muscle tissue, there is an accumulation of citrate, α-ketoglutarate, and succinate in the suspending medium, suggesting that citrate and α-ketoglutarate are normally converted into succinate when malonate is not present.

From these basic observations and other evidence Krebs reasoned that the active tri- and dicarboxylic acids listed above can be arranged in a logical chemical sequence, in which each step is a simple chemical transformation catalyzed by a specific enzyme (Figure 16-7). Further, since the incubation of pyruvate and oxaloacetate with ground muscle tissue resulted in accumulation of citrate in the medium, Krebs reasoned that this sequence functions in a circular rather than linear manner, in such a way that its beginning and end are linked together (Figure 16-9). For the missing link that closes the circle he proposed the reaction

$$\text{Pyruvate} + \text{oxaloacetate} \longrightarrow \text{citrate} + CO_2$$

From these simple experiments and logical reasoning Krebs postulated what he called the <u>citric acid cycle</u> as the main pathway for oxidation of carbohydrate in muscle. In the years since its discovery, the citric acid cycle has been found to function not only in muscles but in virtually all tissues of higher animals and plants and in many aerobic microorganisms. For this important discovery, Krebs (Figure 16-10) was awarded the Nobel prize in 1953, which he shared with Fritz Lipmann, the "father" of the ATP cycle (page 372).

The citric acid cycle is also called the <u>tricarboxylic acid cycle</u> because for some years after Krebs postulated the cycle it was uncertain whether citric acid or some other tricarboxylic acid, e.g., isocitric acid, was the first product formed by reaction of pyruvate and oxaloacetate. This uncertainty was subse-

Figure 16-11
Location of citric acid cycle enzymes in mitochondria.

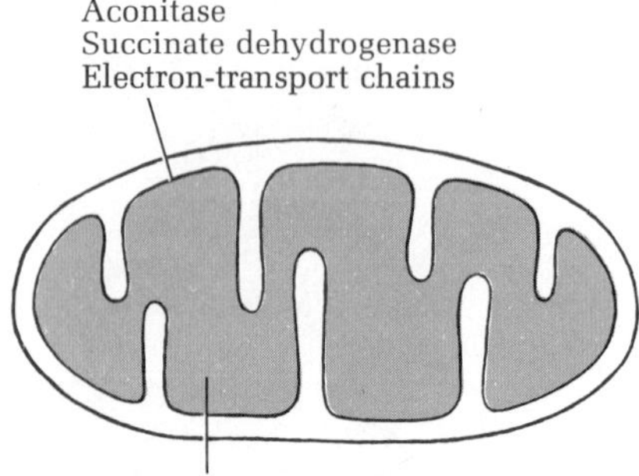

quently removed, as we shall see later. Today we know that citric acid is indeed the first tricarboxylic acid formed. Thus it is most appropriate to use the name citric acid cycle or, quite simply, the Krebs cycle.

Later it was shown by Eugene Kennedy and Albert Lehninger that the entire set of reactions of the citric acid cycle takes place in the mitochondria of animal cells. Isolated rat liver mitochondria (page 26) were found to contain not only all the enzymes and coenzymes required for the citric acid cycle but also all the enzymes and proteins necessary for the last stage of respiration, namely, electron transport and oxidative phosphorylation. Thus the mitochondria are aptly called the "power plants" of the cell. The location of the enzymes of the citric acid cycle in mitochondria is shown in Figure 16-11.

The Citric Acid Cycle Has Eight Steps

Now let us examine the eight successive reaction steps of the citric acid cycle, with special reference to the chemical transformations taking place as the acetyl group of acetyl-CoA is rearranged and ultimately dismembered to yield CO_2 and H atoms, trapped in the form of the reduced coenzymes NADH and $FADH_2$. Figures 16-12 and 16-13 show balanced equations of the cycle reactions and the structures of the intermediates.

Citrate Is Formed by Condensation of Acetyl-CoA with Oxaloacetate

The first reaction of the cycle is the condensation of acetyl-CoA with oxaloacetate to form citrate, catalyzed by citrate synthase (Figure 16-12). In this reaction the methyl carbon of the acetyl group of acetyl-CoA condenses with the carbonyl group of oxaloacetate; simultaneously, the thioester bond is cleaved to release free coenzyme A

$$\text{Acetyl-S-CoA} + \text{oxaloacetate} + H_2O \xrightarrow[\text{synthase}]{\text{citrate}} \text{citrate} + \text{CoA-SH} + H^+$$
$$\Delta G^{\circ\prime} = -7.7 \text{ kcal/mol}$$

Under most circumstances this reaction goes far to the right in the cell, as might be predicted from its large negative standard free energy of hydrolysis. The CoA-SH formed in this reaction is now free to participate in the oxidative decarboxylation of another molecule of pyruvate to yield another molecule of acetyl-CoA for entry into the cycle. Cityl-CoA is believed to be a transient intermediate in the citrate synthase reaction. It is formed on the active site of the enzyme and rapidly undergoes hydrolysis to yield free CoA-SH and citrate, which are then released from the active site.

Citrate synthase is a regulatory enzyme, and in many types of cells the reaction it catalyzes is the rate-limiting step of the citric acid cycle.

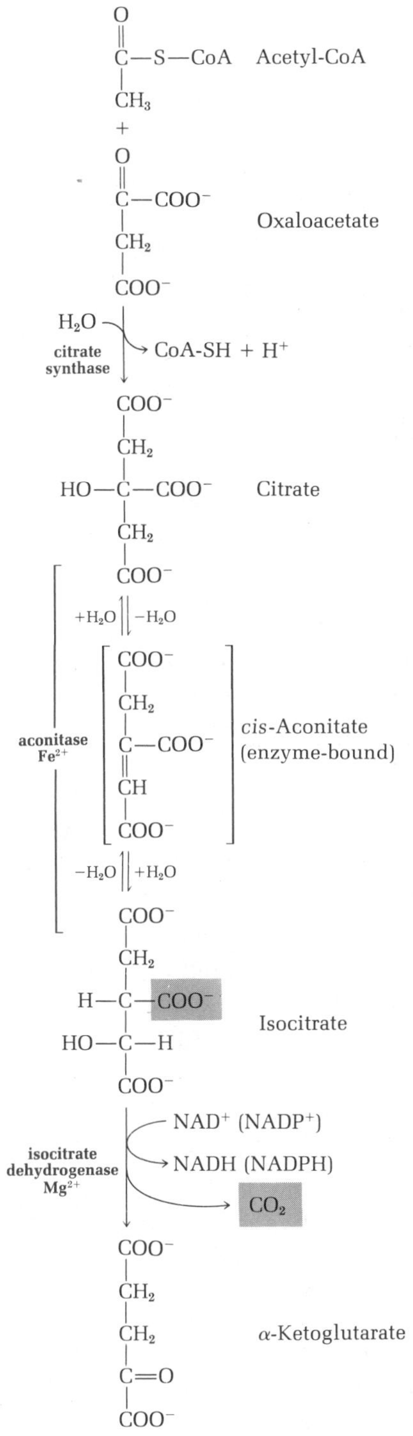

Figure 16-12
The first four reactions of the citric acid cycle in complete balanced form.

Acetyl-CoA: CH_3—C(=O)—S—CoA
+
Oxaloacetate: ^-OOC—C(=O)—CH_2—COO^-

H_2O → (citrate synthase) → CoA-SH + H^+

Citrate: ^-OOC—CH_2—C(OH)(COO^-)—CH_2—COO^-

aconitase Fe^{2+}: $+H_2O$ / $-H_2O$

cis-Aconitate (enzyme-bound): ^-OOC—CH_2—C(COO^-)=CH—COO^-

$-H_2O$ / $+H_2O$

Isocitrate: ^-OOC—CH_2—H—C—COO^- | HO—C—H—COO^-

isocitrate dehydrogenase Mg^{2+}: NAD^+ ($NADP^+$) → NADH (NADPH), CO_2

α-Ketoglutarate: ^-OOC—CH_2—CH_2—C=O—COO^-

Citrate Is Converted into Isocitrate via cis-Aconitate

The enzyme aconitase catalyzes the reversible transformation of citrate into isocitrate, through the intermediary formation of the tricarboxylic *cis-aconitate*, which normally does not dissociate from the enzyme active site (Figure 16-12). Aconitase promotes the reversible addition of H_2O to the double bond of enzyme-bound *cis*-aconitate in two different ways, one leading to citrate and the other to isocitrate

$$\text{Citrate} \underset{+H_2O}{\overset{-H_2O}{\rightleftharpoons}} [cis\text{-aconitate}] \underset{-H_2O}{\overset{+H_2O}{\rightleftharpoons}} \text{isocitrate}$$

Although the equilibrium mixture at pH 7.4 and 25°C contains less than 10 percent isocitrate, in the cell the reaction is pulled to the right because the product isocitrate is rapidly transformed in the subsequent steps of the cycle. Aconitase is a rather complex enzyme. It contains iron and acid-labile sulfur atoms arranged in a cluster called an *iron-sulfur center* (page 479), but the precise function of this cluster, which is presumably the prosthetic group of the enzyme, is still unknown.

Isocitrate Is Dehydrogenated to Yield α-Ketoglutarate and CO_2

In the next step isocitrate is dehydrogenated to α-ketoglutarate and CO_2 (Figure 16-12) by *isocitrate dehydrogenase*. There are two different isocitrate dehydrogenases, one requiring NAD^+ as electron acceptor and the other $NADP^+$. The overall reactions catalyzed by the two isocitrate dehydrogenases are otherwise identical

$$\text{Isocitrate} + NAD^+ (NADP^+) \longrightarrow \alpha\text{-ketoglutarate} + CO_2 + NADH\ (NADPH) + H^+$$
$$\Delta G^{\circ\prime} = -5.0 \text{ kcal/mol}$$

Mitochondria contain both the NAD-linked and NADP-linked isocitrate dehydrogenases; the former is found only in mitochondria, whereas the latter is found in both mitochondria and the cytosol. Both the mitochondrial enzymes appear to participate in the citric acid cycle, but the NAD-linked isocitrate dehydrogenase is predominant. It requires Mg^{2+} or Mn^{2+} and is virtually inactive in the absence of its positive modulator ADP. The occurrence of two isocitrate dehydrogenases in mitochondria may be related to the regulation of the cycle.

α-Ketoglutarate Is Oxidized to Succinate and CO_2

In the next step α-ketoglutarate undergoes oxidative decarboxylation to form *succinyl-CoA* and CO_2 (Figure 16-13) by the action of the *α-ketoglutarate dehydrogenase complex*, which catalyzes the reaction

$$\alpha\text{-Ketoglutarate} + NAD^+ + \text{CoA} \longrightarrow \text{succinyl-CoA} + CO_2 + NADH$$

Figure 16-13
The remaining reactions of the citric acid cycle. See Figure 16-12 for preceding reactions.

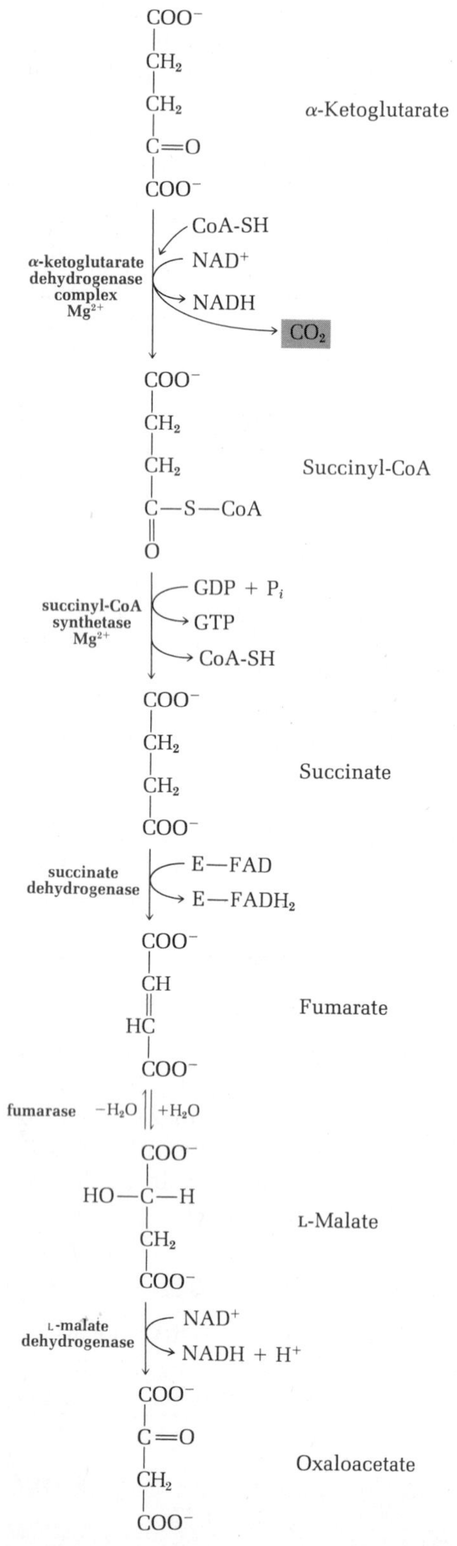

COO^- — CH_2 — CH_2 — C=O — COO^- α-Ketoglutarate

α-ketoglutarate dehydrogenase complex Mg^{2+}: CoA-SH, NAD^+ → NADH, → CO_2

COO^- — CH_2 — CH_2 — C(=O)—S—CoA Succinyl-CoA

succinyl-CoA synthetase Mg^{2+}: GDP + P_i → GTP, → CoA-SH

COO^- — CH_2 — CH_2 — COO^- Succinate

succinate dehydrogenase: E—FAD → E—$FADH_2$

COO^- — CH = HC — COO^- Fumarate

fumarase $-H_2O$ ⇅ $+H_2O$

COO^- — HO—C—H — CH_2 — COO^- L-Malate

L-malate dehydrogenase: NAD^+ → NADH + H^+

COO^- — C=O — CH_2 — COO^- Oxaloacetate

Note that this reaction is virtually identical to the pyruvate dehydrogenase reaction discussed above; both promote the oxidation of an α-keto acid with loss of the carboxyl group as CO_2. In both structure and function the α-ketoglutarate dehydrogenase complex closely resembles the pyruvate dehydrogenase complex. It contains three enzymes, analogous to those in the pyruvate system, as well as enzyme-bound thiamine pyrophosphate, Mg^{2+}, coenzyme A, NAD^+, FAD, and lipoic acid. An important difference, however, is that the α-ketoglutarate dehydrogenase system does not have quite so elaborate a regulatory mechanism as the pyruvate dehydrogenase complex.

Conversion of Succinyl-CoA into Succinate

Succinyl-CoA, the product of the preceding step, is a high-energy compound. Like acetyl-CoA, it has a strongly negative $\Delta G^{\circ\prime}$ for hydrolysis of the thioester bond:

$$\text{Succinyl-S-CoA} + H_2O \longrightarrow \text{succinate} + \text{CoA-SH} + H^+ \qquad \Delta G^{\circ\prime} = -8.0 \text{ kcal/mol}$$

In cells, however, succinyl-CoA does not lose its CoA group by a simple hydrolysis, which would waste free energy. It undergoes instead an energy-conserving coupled reaction in which cleavage of the thioester bond is accompanied by the phosphorylation of *guanosine diphosphate* (GDP) to *guanosine triphosphate* (GTP) (Figure 16-13):

$$\text{Succinyl-S-CoA} + P_i + \text{GDP} \xrightarrow{Mg^{2+}} \text{succinate} + \text{GTP} + \text{CoA-SH} \qquad \Delta G^{\circ\prime} = -0.7 \text{ kcal/mol}$$

The enzyme catalyzing this reaction, *succinyl-CoA synthetase*, yields free succinate and causes the formation of the terminal high-energy phosphate group of GTP from GDP and P_i at the expense of the free energy yielded on cleavage of succinyl-CoA. This energy-conserving reaction has an intermediate step in which the enzyme molecule itself becomes phosphorylated at a histidine residue in the active site. This phosphate group, which has high-energy characteristics, is transferred to GDP to make the GTP. The coupled formation of GTP at the expense of the energy yielded by the oxidative decarboxylation of α-ketoglutarate is another example of a substrate-level phosphorylation. We will recall an earlier example of this type of phosphorylation, the coupled synthesis of ATP at the expense of the energy generated by the dehydrogenation of glyceraldehyde 3-phosphate in glycolysis (page 410). Such reactions are called substrate-level phosphorylations because the energy required to bring them about arises from the dehydrogenation of an organic substrate molecule. This term distinguishes these phosphorylations from *oxidative phosphorylation*, which is coupled to electron transport and is thus also called *respiratory-chain phosphorylation* (Chapter 17).

The GTP formed by succinyl-CoA synthetase then may donate its terminal phosphate group to ADP to form ATP, by the reversible action of a nucleoside diphosphokinase (page 389)

$$\mathrm{GTP} + \mathrm{ADP} \xrightleftharpoons{\mathrm{Mg}^{2+}} \mathrm{GDP} + \mathrm{ATP}$$

$$\Delta G^{\circ\prime} = 0.0 \text{ kcal/mol}$$

Dehydrogenation of Succinate to Fumarate

In the next step of the cycle succinate formed from succinyl-CoA is dehydrogenated to fumarate (Figure 16-13) by the flavoprotein succinate dehydrogenase, which contains covalently bound flavin adenine dinucleotide. This reducible prosthetic group functions as the hydrogen acceptor in the following reaction, in which E designates the enzyme protein:

$$\text{Succinate} + \text{E—FAD} \longrightarrow \text{fumarate} + \text{E—FADH}_2$$

Succinate dehydrogenase is tightly bound to the inner mitochondrial membrane. As isolated from beef heart mitochondria, it has a molecular weight of about 100,000 and contains one molecule of covalently bound FAD, as well as two iron-sulfur clusters or centers, one containing two and the other four iron atoms. These iron atoms undergo Fe(II)–Fe(III) valence changes during the action of the enzyme and thus are believed to carry electrons (Chapter 17).

Succinate dehydrogenase is competitively inhibited by malonate (pages 224 and 442); as we have seen, inhibition by malonate played an important role in establishing the outlines of the citric acid cycle.

Fumarate Is Hydrated to Form Malate

The reversible hydration of fumarate to L-malate (Figure 16-13)

$$\text{Fumarate} + \mathrm{H_2O} \rightleftharpoons \text{L-malate}$$

$$\Delta G^{\circ\prime} \approx 0 \text{ kcal/mol}$$

is catalyzed by fumarate hydratase. This enzyme, more commonly called fumarase, has been obtained in crystalline form from pig heart. Fumarase is highly specific; it hydrates the trans double bond of fumarate but does not act on maleate, the cis isomer of fumarate, or monocarboxylic unsaturated acids with either cis or trans double bonds. In the reverse direction (from L-malate to fumarate) fumarase is optically specific; it does not dehydrate D-malate. Fumarase has a molecular weight of about 200,000 and contains four polypeptide chain subunits; it requires no coenzyme.

Malate Is Dehydrogenated to Form Oxaloacetate

In the last reaction of the citric acid cycle, the NAD-linked L-malate dehydrogenase, which is present in the mitochondrial matrix, catalyzes the dehydrogenation of L-malate to oxaloacetate (Figure 16-13)

$$\text{L-Malate} + NAD^+ \rightleftharpoons \text{oxaloacetate} + NADH + H^+ \qquad \Delta G^{\circ\prime} = +7.1 \text{ kcal}$$

The equilibrium of this reaction is far to the left under standard thermodynamic conditions (1 M concentration of all components and pH 7.0). In intact cells, however, the reaction proceeds to the right since oxaloacetate, the reaction product, is rapidly removed by the citrate synthase reaction, so that the concentration of oxaloacetate in the cell is extremely low, less than 10^{-6} M.

Summary of the Cycle

We have now completed one turn around the citric acid cycle. An acetyl group, containing two carbon atoms, was fed into the cycle by combining it with oxaloacetate. Two carbon atoms emerged from the cycle as carbon dioxide. At the end of the cycle a molecule of oxaloacetate was regenerated. Four pairs of hydrogen atoms were removed from four of the cycle intermediates by enzymatic dehydrogenation; three pairs were used to reduce three molecules of NAD^+ to NADH and one pair to reduce the FAD of succinate dehydrogenase to $FADH_2$. The four pairs of electrons from these hydrogen atoms pass down the electron-transport chain and ultimately reduce two molecules of O_2 to form four molecules of H_2O. Again note that the two carbon atoms appearing as CO_2 are not the same two carbons that entered in the form of the acetyl group. As can be seen from Figures 16-12 and 16-13, additional turns around the cycle are required before the carbon atoms that entered as an acetyl group finally appear as CO_2.

As a by-product of the cycle, one molecule of ATP was formed from ADP and phosphate via GTP yielded by the succinyl-CoA synthetase reaction. In the next chapter we shall see that the four pairs of electrons, derived from the four pairs of hydrogen atoms removed in the dehydrogenase reactions of the cycle, pass down the electron-transport chain to molecular oxygen, with formation of H_2O. Although the Krebs cycle itself generates only one molecule of ATP per turn, its four dehydrogenation steps provide a large flow of energy-rich electrons into the respiratory chain and thus eventually lead to formation of a large number of ATP molecules during oxidative phosphorylation (Chapter 17).

Why a Citric Acid Cycle?

We may now ask an important question. Why is such a complex cycle involving successive 6-, 5-, and 4-carbon intermediates necessary to oxidize the simple 2-carbon acetyl group? The answer is inherent in some principles of organic chemistry.

Acetic acid, although a simple, small molecule, is very resistant to chemical oxidation of its methyl carbon atom. Quite drastic conditions, incompatible with the cell milieu, are required to oxidize acetate directly to two molecules of CO_2. Living cells have "learned" through evolution to take an easier but roundabout way to oxidize acetic acid, one with a lower free energy of activation. Cells have learned to combine acetic acid with another molecule (oxaloacetate) to yield a product (citrate) that is much more amenable to dehydrogenation and decarboxylation than acetate itself. Although some metabolic reactions may appear to be much more complex than necessary, close study of such cases in the light of the basic principles of organic reaction mechanisms shows that they represent chemically easier ways of promoting a given transformation.

Isotopic Tests of the Citric Acid Cycle

The citric acid cycle was first postulated from experiments carried out on suspensions of ground muscle tissue. Subsequently its details were worked out by study of the highly purified enzymes of the cycle. One might ask whether these enzymes really function in a cycle in intact living cells and whether the rate of the cycle is high enough to account for the overall rate of glucose oxidation in animal tissues. These questions have been studied by use of isotopically labeled metabolites, such as pyruvate or acetate, in which the isotopes ^{13}C or ^{14}C were used to mark a given carbon atom. That the citric acid cycle actually takes place in intact cells at a high rate has been verified by many stringent tests with the isotope tracer technique.

Some of the earliest isotopic experiments produced an unexpected result, however, which aroused considerable controversy about the pathway and mechanism of the citric acid cycle. In fact, these experiments at first seemed to show that citric acid was *not* the first tricarboxylic acid to be made. For this reason the name citric acid cycle was superseded by the name tricarboxylic acid cycle, abbreviated TCA cycle. Ultimately it was found that citric acid is indeed the first tricarboxylic acid formed. Box 16-1 gives some details of this episode in the scientific history of the cycle.

The Conversion of Pyruvate to Acetyl-CoA Is Regulated

We have seen (Chapter 15) that the rate of glycolysis is controlled at two levels. First the amount of fuel flowing into the glycolytic sequence is regulated. Thus hexokinase, which promotes phosphorylation of D-glucose to glucose 6-phosphate, and glycogen phosphorylase, which promotes the first step in formation of glucose 6-phosphate from glycogen, are the two regulatory enzymes that control the entry of glucose into the glycolytic sequence. Similarly, the rate of the citric acid cycle is regulated in the first instance by the rate of formation of its fuel acetyl-CoA, which arises from the oxidation of pyruvate

Box 16-1 Is citric acid the first tricarboxylic acid formed in the cycle?

When the heavy-carbon isotope ^{13}C and the radioactive carbon isotopes ^{11}C and ^{14}C became available, they were very quickly put to use to trace the pathway of carbon atoms through the citric acid cycle. In one such experiment, which initiated the controversy over the role of citric acid, acetate labeled in the carboxyl group (designated $CH_3\overset{*}{C}OO^-$) was incubated aerobically with a tissue preparation. Since acetate is enzymatically converted into acetyl-CoA in animal tissues (page 513), the pathway of the carboxyl carbon atom of the acetyl group of acetyl-CoA in the cycle reactions could be traced. α-Ketoglutaric acid was isolated from the tissue after incubation. The α-ketoglutaric acid was then degraded by known chemical reactions to establish the positions of the isotopic carbon derived from carboxyl-labeled acetate. Condensation of unlabeled oxaloacetate with carboxyl-labeled acetate would be expected to produce citrate labeled in one of the two primary carboxyl groups (Figure 1). Because citric acid is a symmetrical molecule, with no asymmetric carbon, the two terminal carboxyl groups of citrate are chemically indistinguishable. Therefore, half of the labeled citrate molecules formed from labeled acetate were ex-

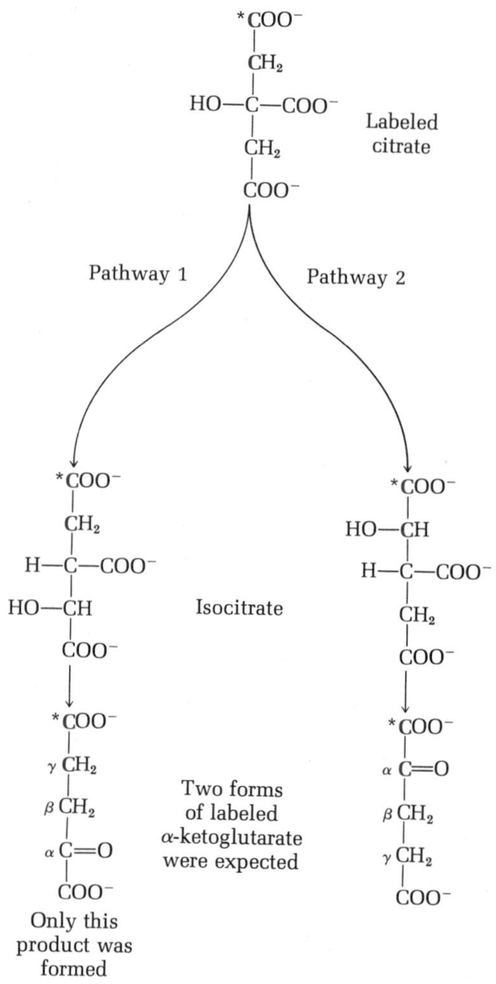

Figure 1
Incorporation of isotopic carbon of the labeled acetyl group into α-ketoglutaric acid by the citric acid cycle. Citric acid formed in the citrate synthase reaction is a symmetrical molecule with no chiral center. It therefore was expected to yield two different kinds of α-ketoglutaric acid, labeled as shown. Yet only one form resulted, that yielded in pathway 1, labeled in the γ-carboxyl group. The carbon atoms of the entering acetyl group of acetyl-CoA are shown in color.

and from the oxidation of fatty acids (Chapter 18). Figure 16-14 shows how the formation of acetyl-CoA by the pyruvate dehydrogenase complex is regulated in animal tissues through a process of covalent modification (page 237). When the ATP concentration in the mitochondrion is relatively high and ample acetyl-CoA and Krebs cycle intermediates are already available to meet the cell's energy needs, this condition constitutes a signal that further formation of acetyl-CoA should be slowed down. Under these conditions ATP serves as a stimulatory modulator to activate an auxiliary enzyme called *pyruvate dehydrogenase kinase*. This enzyme uses ATP to phosphorylate a specific serine residue in the pyruvate dehydrogenase

(a) CH_2COO^- — C — HO, CH_2COO^-, COO^- — Susceptible bond

(b) Z, C, X, Z, Y

(c) This bond cannot be positioned correctly and is not attacked → Z, C, X, Y, Z — This bond can be positioned correctly and is attacked — X′, Y′, Z′ — Active site with complementary binding points

Figure 2
The prochiral nature of citric acid. A simple way to explain this property of citric acid is to assume that its binding to the active site of an enzyme, e.g., aconitase, is through a specific three-point attachment. Although citric acid has no asymmetric carbon, its three different substituents around the central carbon can fit three complementary points of attachment in only one way. (a) Structure of citric acid. (b) Schematic representation of citric acid: X = —OH, Y = —COO^-, Z = —CH_2COO^-. (c) Correct complementary fit of citric acid to the binding site on aconitase. There is only one way in which the three specified groups of citric acid can fit on the three binding sites. Thus only one of the two —CH_2COO^- groups will be attacked by aconitase.

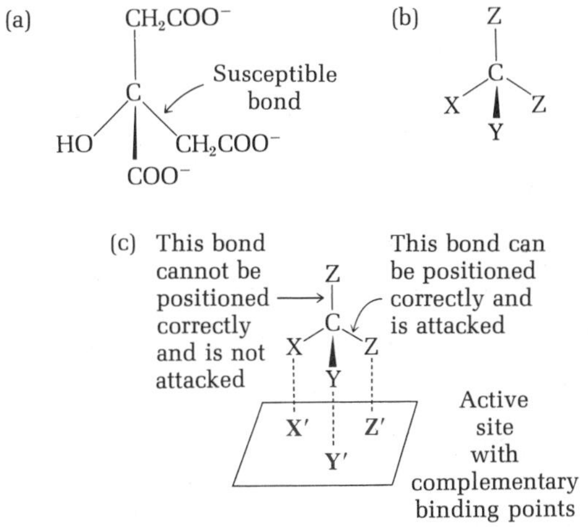

pected to yield α-ketoglutarate with the isotope in the α-carboxyl group, and the other half to yield α-ketoglutarate with the isotope in the γ-carboxyl group; i.e., the α-ketoglutarate isolated should contain isotopic carbon in both carboxyl groups. Contrary to this expectation, the labeled α-ketoglutarate isolated from the tissue suspension contained the isotope only in the γ-carboxyl group of the α-ketoglutarate, not in both carboxyl carbons (Figure 1). It was concluded that citric acid itself, or any other symmetrical molecule, could not possibly be an intermediate in the pathway from acetate to α-ketoglutarate. Hence it was postulated that an asymmetric tricarboxylic acid, presumably *cis*-aconitic acid or isocitric acid, had to be the first condensation product formed from acetate and oxaloacetate, rather than citrate. This is how the name tricarboxylic acid cycle came into use.

In 1948, however, Alexander Ogston, a biochemist at Oxford University, pointed out that although citric acid has no asymmetric carbon atom, it has the *potential* of reacting asymmetrically if the enzyme acting upon it has an active site that is asymmetric. He suggested that the active site of aconitase, the enzyme acting on the newly formed citrate, may have three points to which the citrate molecule must be bound and that the citrate molecule must undergo a specific three-point attachment on these binding points. As seen in Figure 2, the binding of citrate to the three points can happen in only one way, which would account for the fact that only one type of labeled α-ketoglutarate is formed.

Organic molecules which have no chiral center but which are *potentially* capable of reacting asymmetrically with an asymmetric active site are called prochiral molecules.

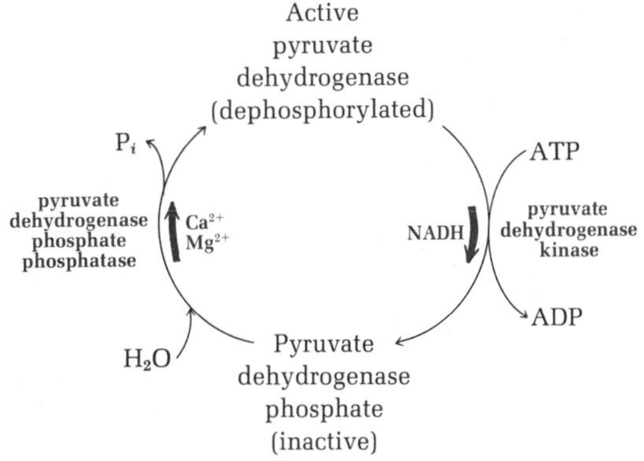

Figure 16-14
Regulation of the pyruvate dehydrogenase reaction by covalent interconversion between active and inactive forms. Ca^{2+} promotes formation of the active enzyme by stimulating pyruvate dehydrogenase phosphate phosphatase.

molecule to yield its inactive form pyruvate dehydrogenase phosphate (Figure 16-14). However, if the demand for ATP increases, causing the ATP level to decline, the inactive phosphorylated form of pyruvate dehydrogenase can be reactivated again. This occurs by hydrolytic removal of the inhibitory phosphate group from pyruvate dehydrogenase by another enzyme, pyruvate dehydrogenase phosphate phosphatase. It is stimulated by a rise in the concentration of free Ca^{2+}, an important metabolic messenger whose concentration rises at times of ATP need. Pyruvate dehydrogenase kinase and pyruvate dehydrogenase phosphate phosphatase are also present in the pyruvate dehydrogenase complex, which is therefore a very complex, self-contained, self-regulating enzyme system.

The pyruvate dehydrogenase complex is also regulated through allosteric modulation. It is strongly inhibited by ATP, as well as by acetyl-CoA and NADH, the products of the pyruvate dehydrogenase reaction, which also serve as allosteric inhibitors of the system. The allosteric inhibition of pyruvate oxidation is greatly enhanced when long-chain fatty acids are available; we shall see later (Chapter 18) that fatty acids are also sources of acetyl-CoA. Thus the active form of the pyruvate dehydrogenase complex is turned off when ample fuel is available in the form of fatty acids and acetyl-CoA and when the cell's ATP and its $NADH/NAD^+$ ratio are high.

The Citric Acid Cycle Is Regulated

Now let us examine the regulation of the citric acid cycle itself (Figure 16-15). As in most metabolic cycles, the initial steps of the citric acid cycle are believed to be rate-setting for the cycle as a whole. In many tissues the first reaction in the cycle,

$$\text{Acetyl-CoA} + \text{oxaloacetate} \longrightarrow \text{citrate} + \text{CoA}$$

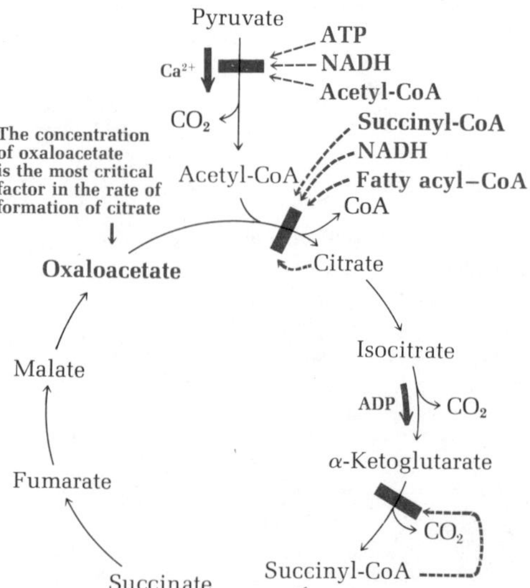

Figure 16-15
Regulation of the citric acid cycle during oxidation of pyruvate in animal cells. ATP, NADH, acetyl-CoA, and Ca^{2+} control the rate of formation of acetyl-CoA from pyruvate, whereas the concentration of oxaloacetate and the activities of citrate synthase and isocitrate dehydrogenase regulate the rate of the cycle itself.

appears to set the overall pace of the cycle. The rate of the citrate synthase reaction is of course controlled by the concentration of acetyl-CoA, one of its substrates, which is in turn controlled by the activity of the pyruvate dehydrogenase complex. It is also controlled by the concentration of oxaloacetate, which may be the most important factor since its concentration in mitochondria is very low and depends upon metabolic conditions. The activity of citrate synthase is also regulated through variations in the concentration of succinyl-CoA, a later intermediate of the cycle. Whenever succinyl-CoA rises above its normal steady-state level, it inhibits citrate synthase by decreasing its affinity for acetyl-CoA. Fatty acid precursors of acetyl-CoA also inhibit citrate synthase by allosteric effects. In some cells citrate and NADH are inhibitors of citrate synthase.

In most cells the oxidation of isocitrate to α-ketoglutarate and CO_2, which can occur by the action of two different isocitrate dehydrogenases, appears to be regulated through the allosteric stimulation of the NAD-linked enzyme by ADP. On the other hand, NADH and NADPH are inhibitory modulators of isocitrate dehydrogenase activity. The activity of the α-ketoglutarate dehydrogenase complex is inhibited by its product succinyl-CoA. Thus at least three steps in the citric acid cycle are regulated, the details differing from one cell type to another.

Under normal conditions the rate of glycolysis and the rate of the citric acid cycle are integrated with each other so that only as much glucose is broken down to form pyruvate as is needed to supply the citric acid cycle with its fuel, the acetyl groups of acetyl-CoA. Normally neither pyruvate, lactate, nor acetyl-CoA accumulates to high levels in aerobic cells, their concentrations tending to remain instead in a dynamic steady state. The rate of glycolysis is matched to the rate of the citric acid cycle not only by its inhibition by high levels of ATP and NADH, which are common components of both the glycolytic and respiratory stages of glucose oxidation, but also by the concentration of citrate. Citrate, the product of the first step of the citric acid cycle, serves as an important allosteric inhibitor of the phosphorylation of fructose 6-phosphate by phosphofructokinase in the glycolytic sequence (page 424 and Figure 15-15).

Citric Acid Cycle Intermediates Are Used for Other Metabolic Purposes and Can Be Replenished

The citric acid cycle is an *amphibolic* pathway (page 342); it functions not only in the oxidative catabolism of carbohydrates, fatty acids, and amino acids but also as the first stage in many biosynthetic pathways, for which it provides precursors. By the action of several important auxiliary enzymes, certain intermediates of the citric acid cycle, particularly α-ketoglutarate, succinate, and oxaloacetate, can be removed from the cycle to serve as precursors of amino acids (Chapter 22). When this happens, the rate of the citric acid cycle would be expected to decline because the drain on its intermediates would result in lowering their concentration in the cell. However, the intermediates of the citric acid cycle can be replenished again, by the action of another set of enzymes. Under normal circumstances the reactions by which the cycle intermediates are drained away and those by which they are replenished are in dynamic balance, so that the concentration of the citric acid cycle intermediates in mitochondria usually remains relatively constant.

The special enzymatic mechanisms by which the pool of cycle intermediates can be replenished are called *anaplerotic* ("filling up") reactions. In animal tissues the most important is the enzymatic carboxylation of pyruvate by CO_2 to form oxaloacetate (Figure 16-16), through the action of *pyruvate carboxylase*, which catalyzes the reversible reaction

$$\text{Pyruvate} + CO_2 + \text{ATP} + H_2O \xrightleftharpoons{Mg^{2+}} \text{oxaloacetate} + \text{ADP} + P_i + 2H^+ \qquad \Delta G^{\circ\prime} = -0.5 \text{ kcal}$$

Figure 16-16
The pyruvate carboxylase reaction and its stimulation by the positive modulator acetyl-CoA. The incorporated CO_2 is shown in color.

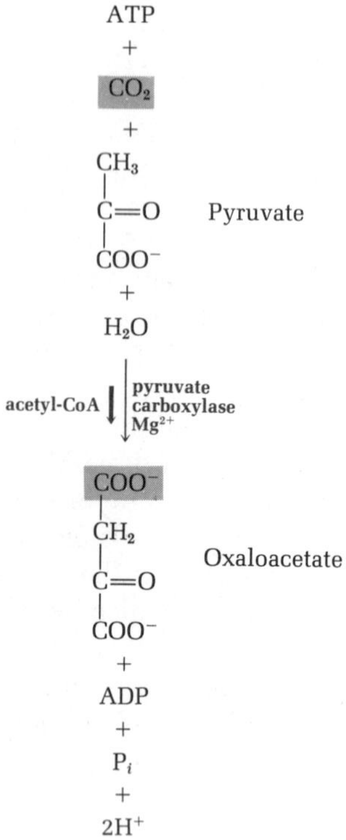

$$\text{ATP} + CO_2 + \underset{\text{Pyruvate}}{CH_3{-}C({=}O){-}COO^-} + H_2O \xrightarrow[\text{acetyl-CoA}]{\text{pyruvate carboxylase, } Mg^{2+}} \underset{\text{Oxaloacetate}}{COO^-{-}CH_2{-}C({=}O){-}COO^-} + \text{ADP} + P_i + 2H^+$$

When the citric acid cycle is deficient in oxaloacetate or any of the other intermediates, pyruvate is carboxylated to produce more oxaloacetate. The enzymatic addition of a carboxyl group to the pyruvate molecule requires energy, which is supplied by the coupled splitting of ATP to ADP and phosphate. Since the standard-free-energy change of the overall reaction is very small, we can conclude that the free energy required to attach a carboxyl group to pyruvate is about equal to the free energy delivered by cleavage of ATP.

Pyruvate carboxylase is a very complex enzyme. It has a molecular weight of about 650,000 and contains four prosthetic groups, each consisting of a molecule of the vitamin *biotin* (page 259) covalently attached through an amide linkage with the ϵ-amino group of a specific lysine residue in the active site (Figure 16-17). Free CO_2, the precursor of the new carboxyl group of oxaloacetate, is first "energized" through its covalent union with a ring nitrogen atom of the biotin prosthetic group. This occurs in the first step of the overall reaction catalyzed by pyruvate carboxylase, which requires ATP (E stands for enzyme):

$$\text{ATP} + CO_2 + \text{E—biotin} + H_2O \rightleftharpoons \text{ADP} + P_i + \text{E—biotin—COO}^- + 2H^+$$

In the second step, also occurring at the enzyme active site, the new carboxyl group covalently bound to the biotin prosthetic group is transferred to pyruvate to form oxaloacetate (Figure 16-16):

$$\text{E—biotin—COO}^- + \text{pyruvate} \rightleftharpoons \text{E—biotin} + \text{oxaloacetate}$$

Pyruvate carboxylase is a regulatory enzyme. The rate of its forward reaction leading to oxaloacetate is very low unless

Figure 16-17
Prosthetic group of pyruvate carboxylase. The biotin carboxyl group occurs in a peptide linkage with the ϵ-amino group of a lysine residue in the active site of the enzyme. CO_2 is activated to form an N-carboxyl derivative of the biotinyl prosthetic group. The carboxyl group, the immediate donor of CO_2 to pyruvate, is then transferred to pyruvate.

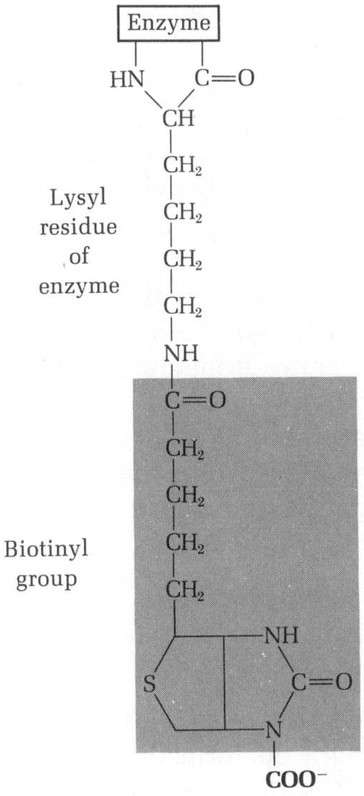

acetyl-CoA, its positive allosteric modulator, is present (Figure 16-16). Thus, whenever acetyl-CoA, which carries the fuel for the citric acid cycle, is present in excess, it stimulates the pyruvate carboxylase reaction to produce more oxaloacetate, thus enabling the cycle to use more acetyl-CoA in the citrate synthase reaction.

The pyruvate carboxylase reaction is the most important anaplerotic reaction in the liver and kidney. Other anaplerotic reactions take place in the heart and muscles. One such reaction is that catalyzed by *phosphoenolpyruvate carboxykinase* (Chapter 20).

$$\text{Phosphoenolpyruvate} + CO_2 + \text{GDP} \xrightarrow{Mn^{2+}} \text{oxaloacetate} + \text{GTP}$$

In this reaction the breakdown of phosphoenolpyruvate, a super-high-energy phosphate compound generated by glycolysis, furnishes the energy for the carboxylation to yield oxaloacetate and at the same time generates GTP.

The Glyoxylate Cycle Is a Modification of the Citric Acid Cycle

In plants and some microorganisms, such as *E. coli,* acetyl groups must often serve both as an energy-rich fuel and as a source of intermediates required to synthesize the carbon skeletons of carbohydrates. In such cells the citric acid cycle operates in two modes: (1) it can carry out the standard cycle reactions for the oxidation of acetyl-CoA to CO_2, as occurs in most tissues, and (2) it can also operate in a specialized modification called the *glyoxylate cycle* (Figure 16-18). The overall reaction equation of the glyoxylate cycle, which may also be regarded as an anaplerotic pathway, is

$$2\ \text{Acetyl-S-CoA} + NAD^+ + 2H_2O \longrightarrow \text{succinate} + 2\text{CoA-SH} + \text{NADH} + 3H^+$$

Figure 16-18
The glyoxylate cycle. The series of reactions in color are catalyzed by isocitrate lyase and malate synthase. The others are reactions of the citric acid cycle. Isocitrate lyase is also called isocitrase.

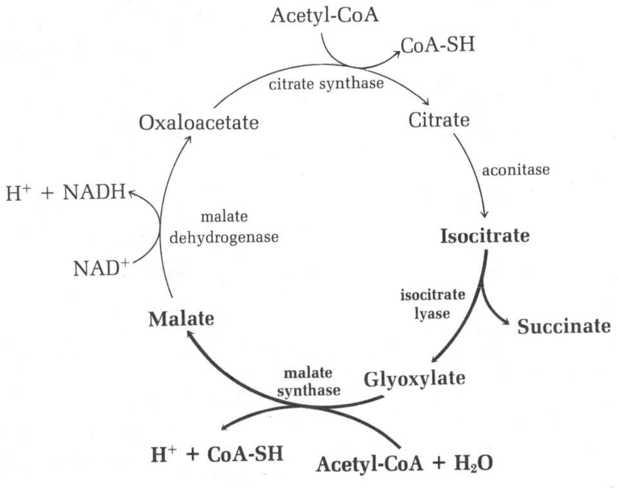

Figure 16-19
Specialized reactions of the glyoxylate cycle. (a) *Conversion of isocitrate to succinate and glyoxylate.* (b) *Condensation of glyoxylate and acetyl-CoA to give malate.*

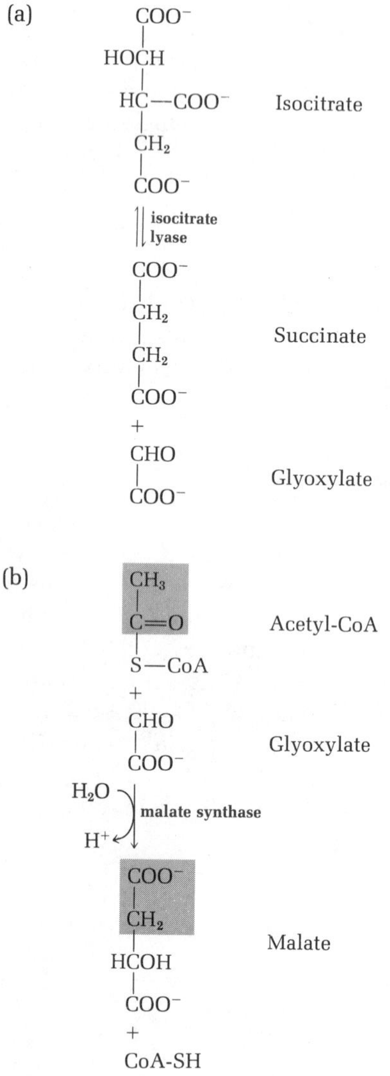

(a)

COO^- | HOCH | HC—COO^- | CH_2 | COO^- — Isocitrate

⇅ isocitrate lyase

COO^- | CH_2 | CH_2 | COO^- — Succinate

+

CHO | COO^- — Glyoxylate

(b)

CH_3 | C=O | S—CoA — Acetyl-CoA

+

CHO | COO^- — Glyoxylate

H_2O → H^+ ↓ malate synthase

COO^- | CH_2 | HCOH | COO^- — Malate

+

CoA-SH

In the glyoxylate cycle acetyl-CoA condenses with oxaloacetate to form citrate (Figure 16-18). The breakdown of isocitrate does not occur via the usual isocitrate dehydrogenase reaction of the citric acid cycle, however, but through a cleavage catalyzed by the enzyme *isocitrate lyase*, to form succinate and glyoxylate. The glyoxylate then condenses with acetyl-CoA to yield malate by the action of *malate synthase* (Figure 16-19). The malate is then dehydrogenated to oxaloacetate, which can condense with another molecule of acetyl-CoA to start another turn of the cycle. In each turn of the glyoxylate cycle, two molecules of acetyl-CoA enter and one molecule of succinate is formed, which is used for biosynthetic purposes. The succinate may be converted via fumarate and malate into oxaloacetate, which can then be converted into phosphoenolpyruvate by the reversal of the phosphoenolpyruvate carboxykinase reaction described above. Phosphoenolpyruvate is then used as the precursor for biosynthesis of glucose (Chapter 20). The glyoxylate cycle pathway does not occur in animals, which lack the enzymes isocitrate lyase and malate synthase. Animals have other means of making carbohydrates from simple precursors (Chapter 20). The glyoxylate cycle is very active in germinating plant seeds, where it converts acetyl residues derived from the fatty acids of their storage triacylglycerols into glucose. In plant cells the enzymes isocitratase and malate synthase are localized in cytoplasmic organelles called *glyoxysomes*.

There Are Secondary Pathways of Glucose Catabolism: The Pentose Phosphate Pathway

Most of the glucose catabolized by animal tissues proceeds down the glycolytic sequence with the formation of pyruvate. Most of the pyruvate in turn is oxidized via the citric acid cycle. The main function of glucose catabolism via this route is to generate ATP energy. There are, however, some minor catabolic pathways taken by glucose which are specialized for certain purposes. Such pathways constitute part of the *secondary metabolism* of glucose. These pathways are not mainline pathways but lead to specialized products needed by the cell. Two such secondary pathways can be outlined.

The *pentose phosphate pathway*, also called the *phosphogluconate pathway* (Figure 16-20), leads to two special products in animal tissues: NADPH and ribose 5-phosphate. Recall that NADPH is a carrier of chemical energy in the form of *reducing power* (page 344). This function is especially prominent in tissues actively carrying out the biosynthesis of fatty acids and steroids from small precursors, particularly the mammary gland, adipose or fat tissue, the adrenal cortex, and the liver. The biosynthesis of fatty acids requires reducing power in the form of NADPH to reduce the double bonds of intermediates in this process. However, other tissues less active in synthesizing fatty acids, such as skeletal muscle, are virtually lacking in the pentose phosphate pathway. A second function of the pentose

Figure 16-20
The pentose phosphate pathway.

Glucose 6-phosphate: HCOH, HCOH, HOCH, HCOH, HC, $CH_2OPO_3^{2-}$

$NADP^+$ → glucose 6-phosphate dehydrogenase, Mg^{2+} → H^+ + NADPH

6-Phospho-glucono-δ-lactone: C=O, HCOH, HOCH, HCOH, HC, $CH_2OPO_3^{2-}$

H_2O → lactonase, Mg^{2+} → H^+

6-Phospho-gluconate: COO^-, HCOH, HOCH, HCOH, HCOH, $CH_2OPO_3^{2-}$

$NADP^+$ → 6-phosphogluconate dehydrogenase, Mg^{2+} → H^+ + NADPH, CO_2

D-Ribulose 5-phosphate: CH_2OH, C=O, HCOH, HCOH, $CH_2OPO_3^{2-}$

phosphopentose isomerase

D-Ribose 5-phosphate: CHO, HCOH, HCOH, HCOH, $CH_2OPO_3^{2-}$

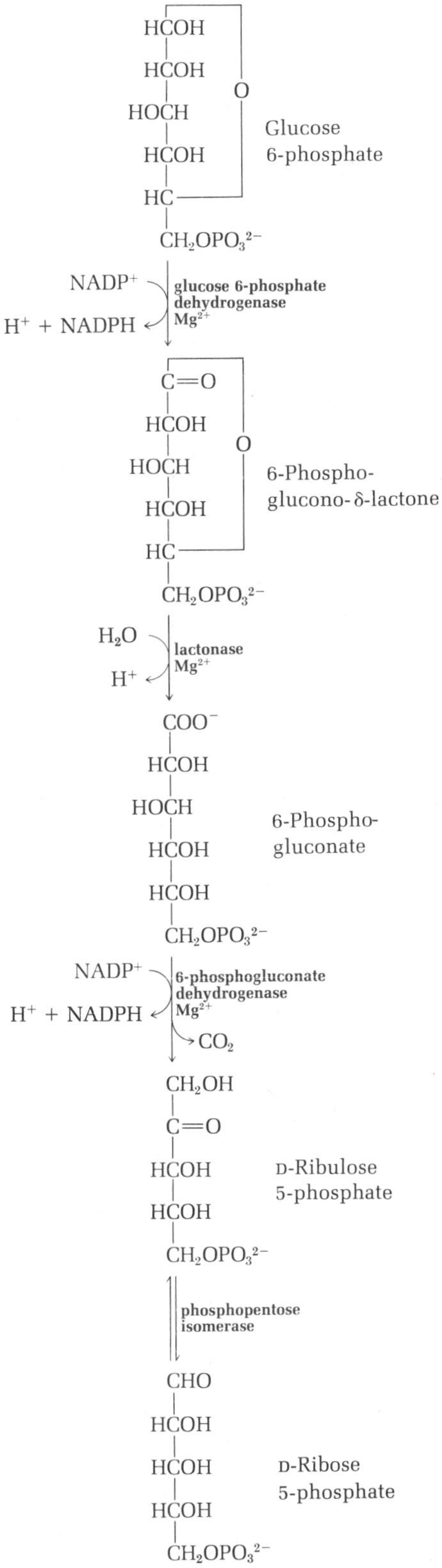

phosphate pathway is to generate pentoses, particularly D-ribose, used in the biosynthesis of nucleic acids.

The first reaction of the pentose phosphate pathway is the enzymatic dehydrogenation of glucose 6-phosphate by *glucose 6-phosphate dehydrogenase* to form *6-phosphogluconate* (Figure 16-20). $NADP^+$ is the electron acceptor. The first product is *6-phosphoglucono-δ-lactone,* which is hydrolyzed to the free acid by a specific *lactonase* (Figure 16-20). The overall equilibrium lies far in the direction of formation of NADPH. In the next step 6-phosphogluconate undergoes dehydrogenation and decarboxylation by *6-phosphogluconate dehydrogenase* to form the ketopentose *D-ribulose 5-phosphate* (Figure 16-20), a reaction that generates a second molecule of NADPH. *Phosphopentose isomerase* then converts D-ribulose 5-phosphate into its aldose isomer, *D-ribose 5-phosphate* (Figure 16-20), which can be used in the biosynthesis of ribonucleotides and deoxyribonucleotides. In some cells the pentose phosphate pathway ends at this point, and its overall equation is then written

$$\text{Glucose 6-phosphate} + 2\text{NADP}^+ + \text{H}_2\text{O} \longrightarrow \text{D-ribose 5-phosphate} + \text{CO}_2 + 2\text{NADPH} + 2\text{H}^+$$

The net result is the production of NADPH for reductive biosynthetic reactions in the extramitochondrial cytoplasm and the production of D-ribose 5-phosphate as a precursor for nucleotide synthesis.

The pentose phosphate pathway is also active in human red blood cells. The NADPH produced is required for preventing the unsaturated fatty acids in the cell membrane from undergoing abnormal reactions with oxygen and for keeping the iron atoms of hemoglobin in their normal ferrous (Fe^{2+}) valence state. There is a group of human genetic deficiency diseases in which glucose 6-phosphate dehydrogenase or some other enzymes of the pentose phosphate pathway are defective or inactive. The red blood cells of such individuals tend to undergo *hemolysis,* the loss of hemoglobin through the damaged membrane, causing anemia. This condition is greatly worsened when certain drugs are ingested, particularly the antimalarial drug *primaquine.* Many millions of people in Africa and Asia have this genetic deficiency.

Later we shall see other aspects of the pentose phosphate pathway (Chapter 23).

The Secondary Pathway from Glucose to Glucuronic Acid and Ascorbic Acid

Another secondary pathway for glucose in animal tissues leads to two specialized products, *D-glucuronate,* important in detoxification and excretion of foreign organic compounds, and *L-ascorbic acid* or *vitamin C.* In this pathway (Figure 16-21) D-glucose 1-phosphate is first converted into *UDP-glucose* by reaction with UTP. The glucose portion of UDP-glucose is then

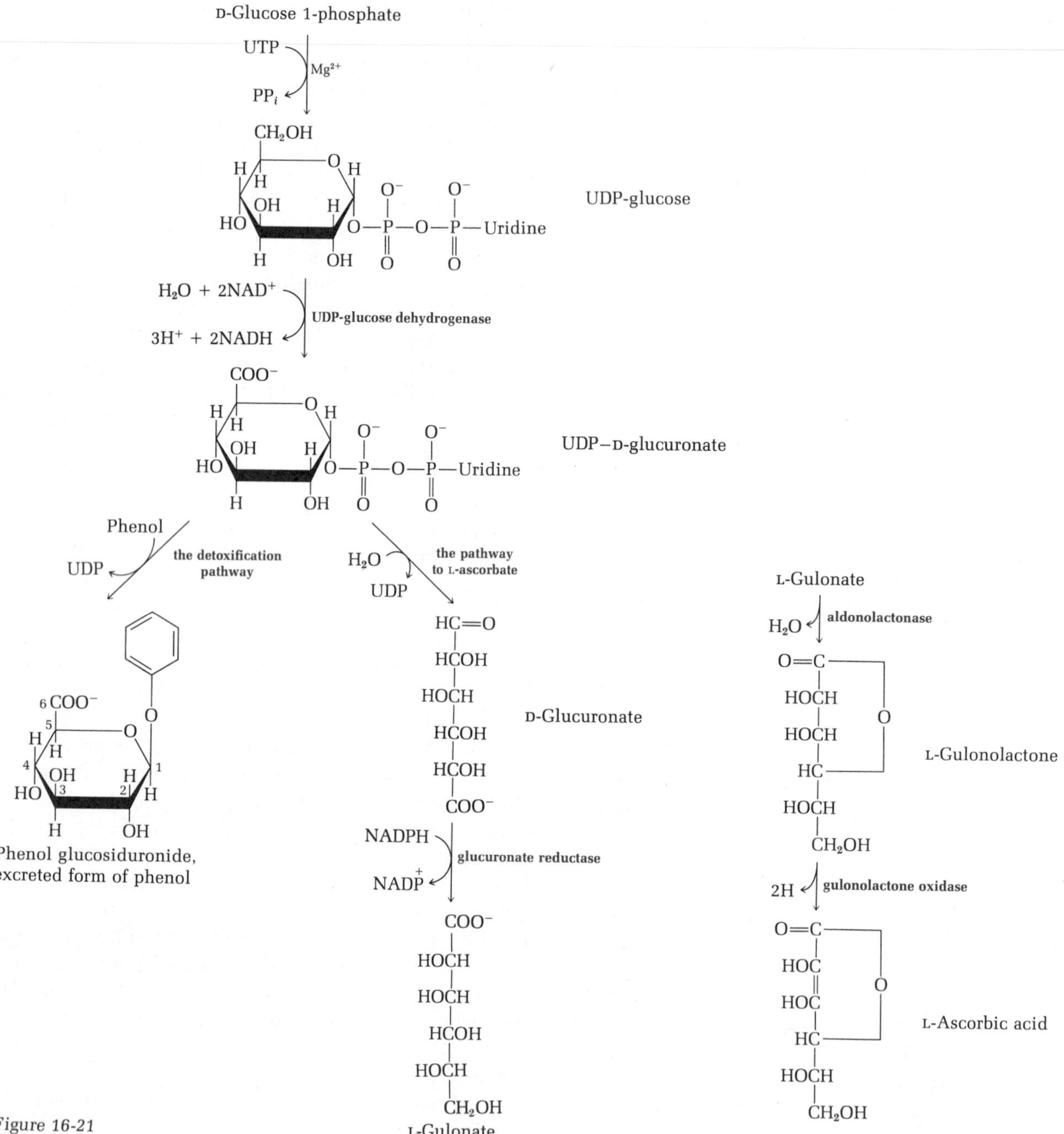

Figure 16-21
Secondary pathways for glucose via UDP-glucuronic acid.

enzymatically dehydrogenated to yield *UDP–D-glucuronate,* a reaction which provides another example of the use of UDP derivatives as intermediates in the enzymatic transformations of sugars (see page 418). UDP–D-glucuronate can then be used to detoxify some foreign compounds or drugs, e.g., phenol, thus enhancing their excretion by the kidney (Figure 16-21). UDP-glucuronate is also the precursor of D-glucuronate residues of such acid polysaccharides as hyaluronic acid and heparin (page 297).

D-Glucuronate is also an intermediate in the conversion of D-glucose into L-ascorbic acid. It is reduced by NADPH to the 6-carbon sugar acid *L-gulonate,* which is converted into its lactone. *L-Gulonolactone* then undergoes dehydrogenation by the flavoprotein *gulonolactone oxidase,* to yield L-ascorbate or vitamin C. L-Ascorbate is synthesized by this pathway by all plants and animals capable of making their own vitamin C. Some species, including human beings, guinea pigs, monkeys, some birds, the Indian fruit bat, and some fish, are unable to synthesize ascorbate and require it ready-made in the diet. In human beings, the guinea pig, and various monkey species, the enzyme gulonolactone oxidase is genetically lacking, accounting for their inability to make this vitamin. It appears possible that all organisms once had the enzymatic capacity to make ascorbate but that some species lost it by a mutation which was not lethal if they normally consumed plant foods containing ample quantities of the vitamin.

Although the amount of D-glucose diverted into this secondary pathway is very small compared with the large amounts of glucose proceeding through glycolysis and the citric acid cycle, the products of such secondary pathways are vital to the organism.

Summary

Cell respiration occurs in three stages: (1) the oxidative formation of acetyl-CoA from pyruvate, fatty acids, and amino acids, (2) the degradation of acetyl residues by the citric acid cycle to yield CO_2 and H atoms, and (3) the transport of electrons to molecular oxygen, which is accompanied by coupled oxidative phosphorylation of ADP to ATP. The oxidative catabolism of glucose yields much more energy than anaerobic glycolysis. Pyruvate, the end product of glycolysis under aerobic conditions, first undergoes dehydrogenation and decarboxylation by the pyruvate dehydrogenase complex, which contains three sequentially acting enzymes, to yield acetyl-CoA and CO_2. The citric acid cycle, which takes place in the mitochondria, begins when citrate synthase catalyzes the condensation of acetyl-CoA with oxaloacetate to form citrate. Then aconitase catalyzes the reversible formation of isocitrate from citrate. Isocitrate is then dehydrogenated to α-ketoglutarate plus CO_2 by NAD- and NADP-linked isocitrate dehydrogenases. The α-ketoglutarate is then dehydrogenated and decarboxylated to succinyl-CoA and CO_2. Succinyl-CoA reacts with GDP and phosphate by the action of succinyl-CoA synthetase to form free succinate and GTP, which transfers its terminal phosphate group to ADP. The succinate is then oxidized to fumarate by succinate dehydrogenase, a flavin-linked enzyme. Fumarate is reversibly hydrated by fumarase to L-malate, which is oxidized by NAD-linked L-malate dehydrogenase to regenerate a molecule of oxaloacetate. The latter can then combine with another molecule of acetyl-CoA and start another revolution of the cycle. Isotopic tracer tests with carbon-labeled fuel molecules or intermediates have established that the citric acid cycle is the major pathway of carbohydrate oxidation in animal cells. The overall rate of the cycle in the liver is controlled by the rate of conversion of pyruvate to acetyl-CoA and by the rate of reaction of acetyl-CoA to form citrate, as catalyzed by citrate synthase, an allosteric enzyme inhibited by succinyl-CoA and other negative modulators.

The citric acid cycle intermediates are also used as precursors in biosynthesis of amino acids and other biomolecules. The cycle intermediates are then replenished by anaplerotic reactions, the most important of which is the ATP-requiring carboxylation of pyruvate to oxaloacetate. In plants and some microorganisms living on acetate as sole carbon source for synthesis of carbohydrate, a variation of the citric acid cycle, the glyoxylate cycle, comes into play and makes possible the net formation of succinate and other cycle intermediates from acetyl-CoA.

Glucose may undergo secondary catabolic reactions leading to specialized products. In the pentose phosphate pathway glucose 6-phosphate can be dehydrogenated to yield ribose 5-phosphate and NADPH by a sequence of reactions in the soluble cytosol. The ribose phosphates are precursors in nucleotide and nucleic acid synthesis, and the NADPH is the major reductant in the biosynthesis of hydrogen-rich biomolecules, such as fatty acids and cholesterol. Glucose yields UDP–D-glucuronate, which functions to detoxify some foreign substances and acts as the precursor of L-ascorbic acid (vitamin C).

References

Books

GOODWIN, T. W. (ed.): *The Metabolic Roles of Citrate,* Academic, New York, 1968. A symposium in honor of Sir Hans Krebs.

LEHNINGER, A. L.: *Biochemistry,* 2d ed., Worth, New York, 1975. Chapter 17 has a more detailed treatment of the cycle.

LOWENSTEIN, J. M. (ed.): *Citric Acid Cycle: Control and Compartmentation,* Dekker, New York, 1969.

Articles

HANSFORD, R. G.: "Control of Mitochondrial Substrate Oxidation," *Curr. Top. Bioenerget.,* **10**:217–278 (1980). Detailed review of the regulation of the citric acid cycle.

KREBS, H. A.: "The History of the Tricarboxylic Acid Cycle," *Perspect. Biol. Med.,* **14**:154–170 (1970). An absorbing personal account of the origin of the cycle.

LOWENSTEIN, J. M.: "The Tricarboxylic Acid Cycle," pp. 146–270 in D. M. Greenberg (ed.), *Metabolic Pathways,* 3d ed., vol. 1, Academic, New York, 1967.

REED, L. J.: "Multienzyme Complexes," *Acc. Chem. Res.,* **7**:40–46 (1974).

SRERE, P. A.: "The Enzymology of the Formation and Breakdown of Citrate," *Adv. Enzymol.,* **43**:57–101 (1975).

WILLIAMSON, D. H.: "Sir Hans Krebs, the First 80 Years," *Trends Biochem. Sci.,* **5**:vi–viii (August 1980). A brief biography.

Problems

1. *Balance Sheet for the Citric Acid Cycle* The citric acid cycle uses eight enzymes to catabolize acetyl-CoA: citrate synthase, aconitase, isocitrate dehydrogenase, α-ketoglutarate dehydrogenase, succinyl-CoA synthetase, succinate dehydrogenase, fumarase, and malate dehydrogenase.
   (a) Write a balanced equation for the reaction catalyzed by each enzyme.
   (b) What cofactor(s) are required by each enzyme reaction?
   (c) For each enzyme determine which of the following describes the type of reaction catalyzed: condensation (carbon-carbon bond formation); dehydration (loss of water); hydration (addition of water); decarboxylation (loss of CO_2); oxidation-reduction; substrate-level phosphorylation; isomerization.
   (d) Write a balanced net equation for the catabolism of acetyl-CoA to carbon dioxide.

2. *Recognizing Oxidation and Reduction Reactions in Metabolism* The biochemical strategy of living organisms is the stepwise oxidation of organic compounds to carbon dioxide and water. By properly coupling these reactions a major part of the energy produced in oxidation is conserved in the form of ATP. It is important to be able to recognize oxidation-reduction processes in metabolism from an examination of the chemical transformation. The reduction of an organic molecule results from the hydrogenation (addition of hydrogen H—H) of a double bond (1) or a single bond with accompanying cleavage (2). Conversely, the oxidation of an organic molecule results from dehydrogenation (removal of hydrogen H—H). In biochemical redox reactions (see Problem 3) the coenzyme couples NAD^+–NADH and FAD–$FADH_2$ have the function of dehydrogenating-hydrogenating organic molecules in the presence of the proper enzymes.

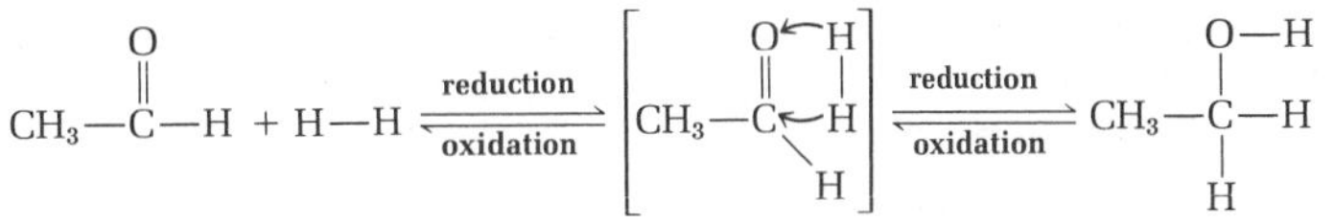

$$\underset{\text{Acetaldehyde}}{CH_3-\overset{\overset{\large O}{\|}}{C}-H} + H-H \underset{\text{oxidation}}{\overset{\text{reduction}}{\rightleftharpoons}} \left[CH_3-\overset{\overset{O-H}{\|\;\;|}}{C}-H\;\;\text{(H)}\right] \underset{\text{oxidation}}{\overset{\text{reduction}}{\rightleftharpoons}} \underset{\text{Ethanol}}{CH_3-\overset{\overset{O-H}{|}}{\underset{\underset{H}{|}}{C}}-H} \qquad (1)$$

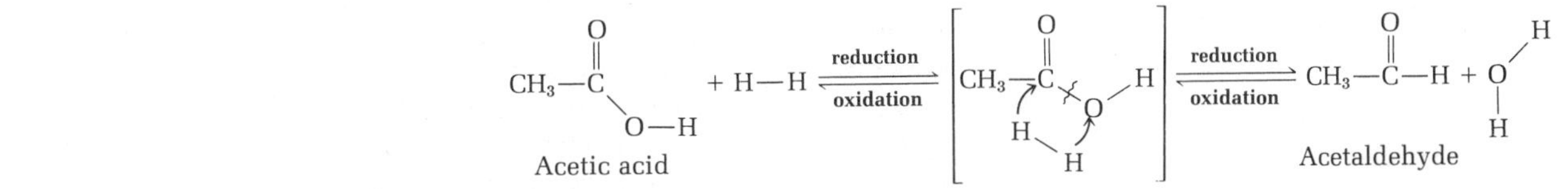

$$\underset{\text{Acetic acid}}{CH_3-\overset{\overset{O}{\|}}{C}-O-H} + H-H \underset{\text{oxidation}}{\overset{\text{reduction}}{\rightleftharpoons}} \left[CH_3-\overset{\overset{O}{\|}}{C}\cdots O(H)\cdots H-H\right] \underset{\text{oxidation}}{\overset{\text{reduction}}{\rightleftharpoons}} \underset{\text{Acetaldehyde}}{CH_3-\overset{\overset{O}{\|}}{C}-H} + H-O-H \qquad (2)$$

For each of the following metabolic transformations, determine whether oxidation or reduction has occurred. Balance each transformation by inserting H—H.

(a) $\underset{\text{Methanol}}{CH_3-OH} \longrightarrow \underset{\text{Formaldehyde}}{H-\overset{\overset{O}{\|}}{C}-H}$

(b) $\underset{\text{Formaldehyde}}{H-\overset{\overset{O}{\|}}{C}-H} \longrightarrow \underset{\text{Formic acid}}{H-\overset{\overset{O}{\|}}{C}-OH}$

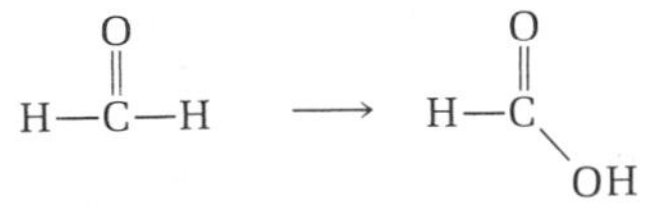

(c) O=C=O ⟶ H—C(=O)—O—H

Carbon dioxide ⟶ Formic acid

(d) $CH_2(OH)$—C(OH)(H)—C(=O)—O—H ⟶ $CH_2(OH)$—C(OH)(H)—C(=O)—H

Glyceric acid ⟶ Glyceraldehyde

(e) $CH_2(OH)$—C(OH)(H)—$CH_2(OH)$ ⟶ $CH_2(OH)$—C(=O)—$CH_2(OH)$

Glycerol ⟶ Dihydroxyacetone

(f) C_6H_5—C(H)(H)—H ⟶ C_6H_5—C(=O)—O—H

Toluene ⟶ Benzoic acid

(g) ^{-}O—C(=O)—CH_2—CH_2—C(=O)—O^{-} ⟶ ^{-}O—C(=O)—CH=CH—C(=O)—O^{-}

Succinate ⟶ Fumarate

(h) CH_3—C(=O)—C(=O)—O—H ⟶ CH_3—C(=O)—O—H + CO_2

Pyruvic acid ⟶ Acetic acid

3. *Nicotinamide Coenzymes as Reversible Redox Carriers* The nicotinamide coenzymes (Chapter 10) can undergo reversible oxidation-reduction reactions with specific substrates in the presence of the appropriate dehydrogenase. The nicotinamide ring is the portion of the coenzyme involved in the redox reaction; the remaining portion of the coenzyme serves as a binding group recognized by the dehydrogenase. Formally, reduced NADH + H^+ serves as the hydrogen source (H—H), as described in Problem 2. Whenever the coenzyme is oxidized, a substrate must be simultaneously reduced:

$$\underset{\text{Oxidized}}{\text{Substrate}} + \underset{\text{Reduced}}{\text{NADH}} + H^+ \rightleftharpoons \underset{\text{Reduced}}{\text{substrate}} + \underset{\text{Oxidized}}{\text{NAD}^+}$$

For each of the following reactions, determine whether the substrate has been oxidized, reduced, or unchanged in oxidation state (see Problem 2). For substrates that have undergone a redox change, balance the reaction with the necessary amount of NAD^+, NADH, H^+, and H_2O. The objective is to recognize when a redox coenzyme is *necessary* in a metabolic reaction.

(a) $CH_3CH_2OH \longrightarrow CH_3{-}CHO$

Ethanol → Acetaldehyde

(b) $^{2-}O_3PO{-}CH_2{-}CH(OH){-}C(=O){-}OPO_3^{2-} \longrightarrow {}^{2-}O_3PO{-}CH_2{-}CH(OH){-}CHO + HPO_4^{2-}$

3-Phosphoglyceroyl phosphate → Glyceraldehyde 3-phosphate

(c) $CH_3{-}C(=O){-}COO^- \longrightarrow CH_3{-}CHO + CO_2$

Pyruvate → Acetaldehyde

(d) $CH_3{-}C(=O){-}COO^- \longrightarrow CH_3{-}COO^- + CO_2$

Pyruvate → Acetate

(e) $^-OOC{-}CH_2{-}C(=O){-}COO^- \longrightarrow {}^-OOC{-}CH_2{-}CH(OH){-}COO^-$

Oxaloacetate → Malate

(f) $CH_3{-}C(=O){-}CH_2{-}COO^- + H^+ \longrightarrow CH_3{-}C(=O){-}CH_3 + CO_2$

Acetoacetate → Acetone

4. *Stimulation of Oxygen Consumption by Oxaloacetate and Malate* In the early 1930s, Albert Szent-Györgyi reported the interesting observation that the addition of small amounts of oxaloacetate or malate to suspensions of minced pigeon breast muscle stimulate the oxygen consumption of the preparation. Surprisingly, when the amount of oxygen consumed was measured, it was about 7 times more than the amount necessary to oxidize the added oxaloacetate or malate completely to carbon dioxide and water.
   (a) Why does the addition of oxaloacetate or malate stimulate oxygen consumption?
   (b) Why is the amount of oxygen consumed so much greater than the amount necessary to oxidize the added oxaloacetate or malate completely?

5. *The Number of Molecules of Oxaloacetate in a Mitochondrion* In the last reaction of the citric acid cycle, malate is dehydrogenated to regenerate the oxaloacetate necessary for the entry of acetyl-CoA via the citrate synthase reaction

$$\text{L-Malate} + NAD^+ \longrightarrow \text{oxaloacetate} + NADH + H^+ \qquad \Delta G^{\circ\prime} = +7.1\ \text{kcal}$$

   (a) Calculate the equilibrium constant for the reaction at 25°C.

(b) Since $\Delta G^{\circ\prime}$ assumes a standard pH of 7, the equilibrium constant obtained in (a) corresponds to

$$K'_{eq} = \frac{[\text{oxaloacetate}][\text{NADH}]}{[\text{L-malate}][\text{NAD}^+]}$$

The measured concentration of L-malate in rat liver mitochondria is about 0.20 mM with an $NAD^+/NADH$ concentration ratio of 10. Calculate the concentration of oxaloacetate at pH 7 in rat liver mitochondrion.

(c) Rat liver mitochondria are spherical and have a diameter of about 2 μm. To appreciate the magnitude of the oxaloacetate concentration in mitochondria, calculate the number of oxaloacetate molecules in a single rat liver mitochondrion.

6. *Respiration Studies in Isolated Mitochondria* Cell respiration can be studied using isolated mitochondrial preparations and measuring their oxygen consumption under different conditions. If 0.01 M sodium malonate is added to actively respiring mitochondria using pyruvate as a fuel source, respiration soon stops and a metabolic intermediate accumulates.
   (a) What is the structure of the accumulated intermediate?
   (b) Explain why it accumulates.
   (c) Explain why oxygen consumption stops.
   (d) Aside from removing malonate, how can the inhibition of respiration by malonate be overcome? Explain.

7. *Labeling Studies in Isolated Mitochondria* The metabolic pathway of organic compounds has often been delineated by using a radioactively labeled substrate and following the fate of the label.
   (a) How can you determine if glucose added to a suspension of isolated mitochondria is metabolized to CO_2 and H_2O?
   (b) Suppose you add pyruvate labeled at the methyl position with ^{14}C to the mitochondria. After one turn of the citric acid cycle what is the location of the ^{14}C in the oxaloacetate? Explain by tracing the ^{14}C label through the pathway.
   (c) How many turns of the citric acid cycle must the ^{14}C go through before all the isotope is released as $^{14}CO_2$? Explain.

8. *Glucose-1-^{14}C Catabolism* If an actively respiring bacterial culture is briefly incubated with glucose-1-^{14}C and the glycolytic and citric acid cycle intermediates are isolated, where is the ^{14}C in each of the intermediates listed below? Consider only the initial incorporation of ^{14}C into these molecules.
   (a) Fructose 1,6-diphosphate
   (b) Glyceraldehyde 3-phosphate
   (c) Phosphoenolpyruvate
   (d) Acetyl-CoA
   (e) Citrate
   (f) α-Ketoglutarate
   (g) Oxaloacetate

9. *Synthesis of Oxaloacetate by the Citric Acid Cycle* Oxaloacetate is formed in the last step of the citric acid cycle by the NAD^+-dependent oxidation of L-malate. Can a *net* synthesis of oxaloacetate take place from acetyl-CoA using only the enzymes and cofactors of the citric acid cycle, without depleting the intermediates of the cycle? Explain. How is oxaloacetate replenished?

10. *Mode of Action of the Rodenticide Fluoroacetate* Fluoroacetate, prepared commercially for rodent control, is also produced naturally by a South African plant. After entering the cells, fluoroacetate is converted into fluoroacetyl-CoA in a reaction catalyzed by the enzyme acetate thiokinase

$$\text{F—CH}_2\text{COO}^- + \text{CoA-SH} + \text{ATP} \longrightarrow \text{F—CH}_2\overset{\displaystyle\text{O}}{\underset{}{\overset{\|}{\text{C}}}}\text{—S—CoA} + \text{AMP} + \text{PP}_i$$

To study the toxic effect of fluoroacetate, a metabolic experiment on intact isolated rat heart was carried out. After the heart was perfused with 0.22 m*M* fluoroacetate, the measured rate of glucose uptake and glycolysis decreased and glucose 6-phosphate and fructose 6-phosphate accumulated. An examination of the citric acid cycle intermediates indicated that their concentrations were below normal except for citrate, whose concentration was 10 times higher than normal.

(a) Where does the block in the citric acid cycle occur? What causes citrate to accumulate and the other intermediates to be depleted?
(b) Fluoroacetyl-CoA is enzymatically transformed in the citric acid cycle. What is the structure of the metabolic end product of fluoroacetate? Why does it block the citric acid cycle? How might the inhibition be overcome?
(c) Why do glucose uptake and glycolysis decrease upon fluoroacetate perfusion? Why do hexose monophosphates accumulate?
(d) Why is fluoroacetate poisoning fatal?

11. *Net Synthesis of α-Ketoglutarate* α-Ketoglutarate plays a central role in the biosynthesis of several amino acids. Write a series of known enzymatic reactions that result in the *net* synthesis of α-ketoglutarate from pyruvate. Your proposed sequence must not involve the net consumption of other citric acid cycle intermediates. Write the net reaction for your proposed sequence and identify the source of each reactant.

12. *Glyoxylate Cycle in Plant Seeds* Animals cannot synthesize carbohydrates from fats because they cannot convert acetyl-CoA (from fatty acid breakdown) into either pyruvate or oxaloacetate (required for the biosynthesis of glucose). Some microorganisms and plants, on the other hand, have the enzymes isocitrate lyase and malate synthase (Figure 16-18), enabling them to synthesize oxaloacetate from acetyl-CoA via the glycoxylate cycle. The seeds of higher plants contain large quantities of oils, which furnish the germinating plant with fatty acids as precursors of cellulose in the early stages of development, before the photosynthetic apparatus becomes available. Write the series of known enzyme reactions that allow the *net* production of oxaloacetate used in the biosynthesis of glucose from acetyl-CoA. Your scheme must not involve the net consumption of any citric acid cycle intermediate. Write the net or overall equation for the synthesis of oxaloacetate from acetyl-CoA. Indicate the source of all cofactors.

13. *Glucose Catabolism: Glycolytic vs. Pentose Phosphate Pathway* An isotopic approach can be used to assess the fraction of glucose catabolism in a given cell or tissue proceeding via the glycolytic

versus the pentose phosphate pathway. The cells are divided into two batches: one is incubated with glucose-1-^{14}C and the other with glucose-6-^{14}C. A comparison is then made of the initial rates at which ^{14}C appears in the CO_2 formed by oxidation of glucose. Explain the chemical basis of this approach. Predict the relative initial rates of $^{14}CO_2$ formation for liver cells assuming that glucose degradation is equally distributed between the two pathways.

CHAPTER 17

Electron Transport, Oxidative Phosphorylation, and Regulation of ATP Production

Now we come to the culminating events in cell respiration, electron transport and oxidative phosphorylation. All the enzymatic steps in the oxidative degradation of carbohydrates, fats, and amino acids in aerobic cells converge into this final stage of cell respiration, in which electrons flow from organic substrates to oxygen, yielding energy for the generation of ATP from ADP and phosphate.

An approximate calculation will show the quantitative importance of oxidative phosphorylation in the human body. A normal 70-kg adult male in a business occupation requires about 2800 kcal of energy per day. This amount of energy can be yielded by the hydrolysis under standard conditions of about 2800/7.3 = 384 mol or 190 *kilograms* of ATP. However, the total amount of ATP actually present in his body is only about 50 *grams*. In order to furnish chemical energy for body needs, the 50 g of ATP must be broken down into ADP and phosphate and resynthesized again thousands of times in the course of a day. Moreover, the rate of ATP turnover in the body must also vary greatly, from its minimal rate during sleep to its maximal rate during extreme muscular activity. Oxidative phosphorylation is not only a vital, continuous process, but its rate must be regulated over a very wide range.

Electron Flow from Substrates to Oxygen Is the Source of ATP Energy

Figure 17-1 shows the overall organization of electron transport and oxidative phosphorylation. In each turn around the citric acid cycle, four pairs of hydrogen atoms are removed, from isocitrate, α-ketoglutarate, succinate, and malate, by the action of specific dehydrogenases. These hydrogen atoms at some point donate their electrons to the electron-transport chain and become H^+ ions, which escape into the aqueous medium. The electrons are transported along a chain of electron-carrying molecules until they reach *cytochrome* aa_3,

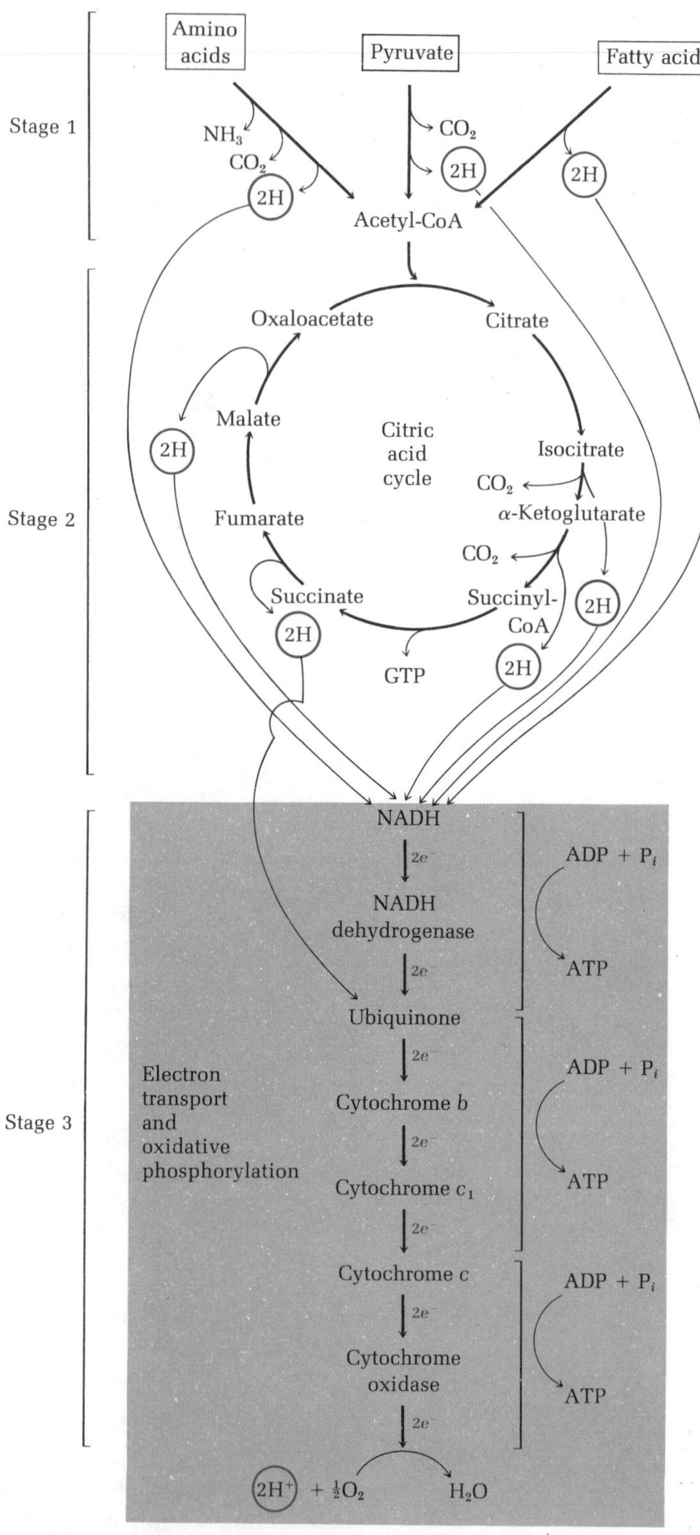

Figure 17-1
The flow sheet of respiration, showing the origin of the pairs of hydrogen atoms removed by dehydrogenases, with transfer of their electrons ($2e^-$) to the electron-transport chain, which carries them to oxygen. Reduction of each atom of oxygen requires $2e^- + 2H^+$. Energy set free during transport of a pair of electrons from NADH to oxygen is harnessed to cause the coupled synthesis of three molecules of ATP from ADP and phosphate in the process of oxidative phosphorylation. The electron-transport chain is shown here in abbreviated form.

or *cytochrome oxidase*, which promotes the transfer of the electrons to oxygen, the final electron acceptor in aerobic organisms. As each atom of oxygen accepts two electrons from the chain, two H^+, equivalent to the two H^+ lost earlier from the two hydrogen atoms removed by dehydrogenases, are taken up from the aqueous medium to form H_2O.

In addition to the four pairs of hydrogens arising from the citric acid cycle, others come from the dehydrogenases that act upon pyruvate, fatty acids, and amino acids during their degradation to acetyl-CoA and other products (Figure 17-1). Virtually all the hydrogen atoms derived by the action of dehydrogenases on fuel molecules in aerobic cells ultimately donate their electrons to the respiratory chain, the final common pathway leading to the terminal electron acceptor, oxygen.

The respiratory chain consists of a series of proteins with tightly bound prosthetic groups capable of accepting and donating electrons. Each member can accept electrons from the preceding member and transfer them to the following one, in a specific sequence. The electrons entering the electron-transport chain are energy-rich, but as they pass down the chain to oxygen in a step-by-step manner, they lose free energy. Much of this energy is conserved in the form of ATP by molecular mechanisms in the inner mitochondrial membrane. As indicated in Figure 17-1, as each pair of electrons passes down the respiratory chain from NADH to oxygen, the coupled synthesis of three molecules of ATP from ADP and phosphate takes place. The three segments of the respiratory chain that provide energy to generate ATP by oxidative phosphorylation are called the *energy-conserving sites* or *segments*.

Electron Transport and Oxidative Phosphorylation Take Place in the Inner Mitochondrial Membrane

In eukaryotic cells nearly all the specific dehydrogenases required in the oxidation of pyruvate and other fuels via the citric acid cycle are located in the internal compartment of mitochondria, the *matrix* (Figure 17-2). The electron-transferring molecules of the respiratory chain and the enzyme molecules that synthesize ATP from ADP and phosphate are embedded in the inner membrane. Fuels of the citric acid cycle, such as pyruvate, must pass from the cytosol, where they are formed, through both mitochondrial membranes into the internal matrix compartment, where they are acted upon by dehydrogenases. Similarly, the ADP formed from ATP during energy-requiring activities in the cytosol must pass into the mitochondrial matrix, to be rephosphorylated to ATP. The new ATP formed must then pass back to the cytosol. Special membrane transport systems (page 495) in the inner mitochondrial membrane promote not only the entry of pyruvate and other fuels into the mitochondria but also the entry of phosphate and ADP and the exit of ATP during oxidative phosphorylation.

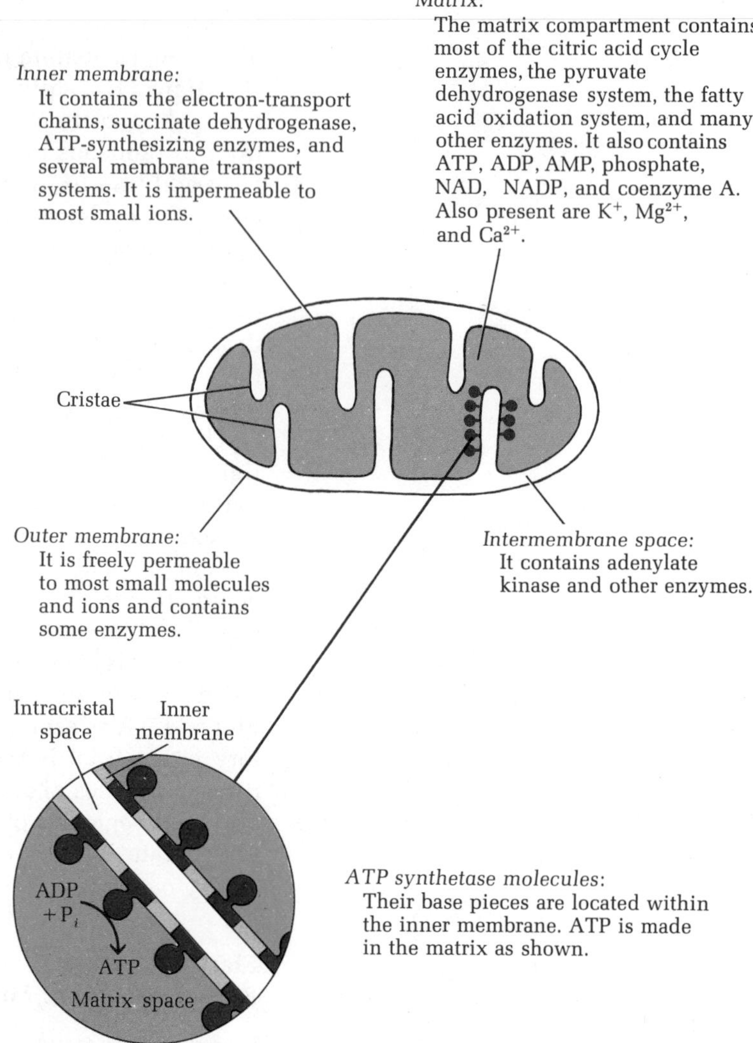

Figure 17-2
The biochemical anatomy of mitochondria, showing the location of the enzymes of the citric acid cycle, the electron-transport chains, the enzymes catalyzing oxidative phosphorylation, and the internal pool of coenzymes. The inner membrane of a single liver mitochondrion may have over 10,000 sets of electron-transport chains and ATP synthetase molecules. The number of sets is proportional to the area of the inner membrane. Heart mitochondria, which have very profuse cristae and thus a much larger area of inner membrane, contain over 3 times as many sets of electron-transport systems as liver mitochondria. The internal pool of coenzymes and intermediates is functionally separate from the cytosolic pool. See Chapter 2 for other details of mitochondrial structure.

The inner mitochondrial membrane is thus a complex structure, containing the electron-carrier molecules, a number of enzymes, and several membrane transport systems, which altogether make up 75 percent or more of the total membrane weight, the remainder being lipid. The inner membrane has an intricate, mosaic structure whose integrity is essential for its life-supporting activity in generating ATP.

Electron-Transferring Reactions Are Oxidation-Reduction Reactions

Although we have examined some enzyme-catalyzed reactions in which hydrogen atoms or electrons are transferred from one molecule to another, we must now review the properties of such reactions in a more quantitative way. Chemical reactions in which electrons are transferred from one molecule to another are called *oxidation-reduction reactions* (also *oxidoreductions* or *redox reactions*). The *electron-donating* molecule in such a reaction is called the *reducing agent* or *reductant;* the *electron-accepting molecule* is the *oxidizing agent* or *oxidant.*

Reducing and oxidizing agents function as *conjugate reductant-oxidant pairs* (*redox pairs*), just as acids and bases function as conjugate acid-base pairs (page 78). Recall that in acid-base reactions we can write the general equation

$$\text{Proton donor} \rightleftharpoons \text{H}^+ + \text{proton acceptor}$$

In redox reactions we can write a similar general equation

$$\text{Electron donor} \rightleftharpoons \text{e}^- + \text{electron acceptor}$$

A specific example is the reaction

$$Fe^{2+} \rightleftharpoons e^- + Fe^{3+}$$

in which the ferrous ion (Fe^{2+}) is the electron donor and the ferric ion (Fe^{3+}) the electron acceptor. Together Fe^{2+} and Fe^{3+} constitute a *conjugate redox pair.*

Electrons are transferred from one molecule to another in one of four different ways:

1. They may be transferred directly as electrons. For example, the Fe^{2+}–Fe^{3+} redox pair can transfer an electron to the Cu^+–Cu^{2+} redox pair:

$$Fe^{2+} + Cu^{2+} \longrightarrow Fe^{3+} + Cu^+$$

2. Electrons may be transferred in the form of hydrogen atoms. Recall that a hydrogen atom consists of a proton (H^+) and a single electron (e^-). In this case we can write the general equation

$$AH_2 \rightleftharpoons A + 2e^- + 2H^+$$

where AH_2 is the hydrogen (or electron) donor, A is the hydrogen acceptor, and AH_2 and A together constitute a conjugate redox pair, which can reduce the electron acceptor B by transfer of H atoms:

$$AH_2 + B \longrightarrow A + BH_2$$

3. Electrons may be transferred from an electron donor to an acceptor in the form of a *hydride ion* ($:H^-$), which bears *two* electrons, as in the case of NAD-linked dehydrogenases (see page 256).

4. Electron transfer also takes place when there is a direct combination of an organic reductant with oxygen, to give a product in which the oxygen is covalently incorporated, as in the oxidation of a hydrocarbon to an alcohol:

$$R{-}CH_3 + \tfrac{1}{2}O_2 \longrightarrow R{-}CH_2{-}OH$$

In this reaction the hydrocarbon is the electron donor and the oxygen atom is the electron acceptor.

All four types of electron transfer occur in cells. The neutral term *reducing equivalent* is commonly used to designate a single electron equivalent participating in an oxidoreduction, whether it is in the form of an electron per se, a hydrogen atom, a hydride ion, or takes place in a reaction with oxygen to yield an oxygenated product. As we shall see, in mitochondrial electron transport electrons are transferred in different forms, as hydride ions, as hydrogen atoms, and, in the later steps catalyzed by cytochromes, as electrons.

Because biological fuel molecules usually undergo enzymatic dehydrogenation to lose two reducing equivalents at a time, and because each oxygen atom can accept two reducing equivalents, it is customary to think of the unit of biological oxidations as a *pair* of reducing equivalents passing from substrate to oxygen.

Each Conjugate Redox Couple Has a Characteristic Standard Potential

The tendency of a conjugate acid-base pair to lose a proton reversibly is given by the dissociation constant *K* (page 78). Similarly, the tendency of a given conjugate redox pair to lose an electron can also be specified quantitatively by a constant, the *standard oxidation-reduction potential* E_0'. This is defined as the electromotive force (emf) in volts given by a responsive electrode placed in a solution containing both the electron donor and its conjugate electron acceptor at 1.0 *M* concentration, 25°C, and pH = 7.0 (Figure 17-3). The electrode must be able to accept electrons from the electron donor and donate them to the conjugate electron acceptor. Such an electrode immersed in a mixture of the conjugate redox pair constitutes a *half-cell.* In order to determine its emf it must be electrically connected to a *reference half-cell* whose emf is known (Figure 17-3). The ultimate reference half-cell is the *hydrogen electrode,* which has an emf arbitrarily set at zero when 1 *M* H^+ is in equilibrium with 1.0 atm H_2 gas at 25°C at pH 0. At pH 7, however, the hydrogen electrode has an emf of −0.41 V (Figure 17-3).

It is the convention in biochemistry to express the standard potentials of conjugate redox pairs as *reduction potentials,* which assign increasingly negative values to systems having an increasing tendency to lose electrons, and increasingly posi-

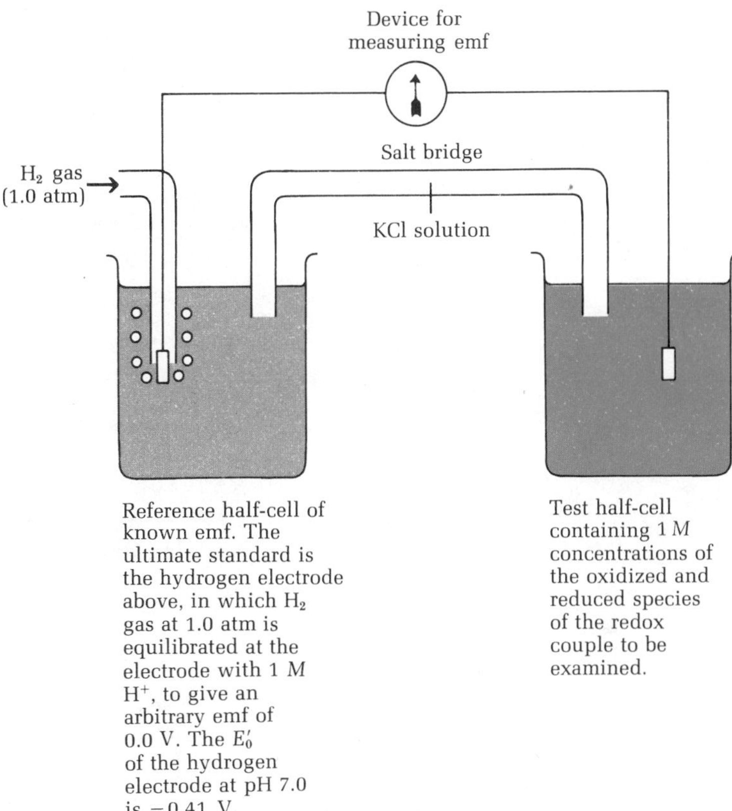

Figure 17-3
Measurement of the standard reduction potential. The solution containing the mixture of 1.0 M oxidized and reduced forms of the redox couple to be examined is placed in the right vessel. The electrode, usually platinum, is connected by an external circuit to a reference half-cell of a redox couple of known potential (at the left). The ultimate reference cell is the hydrogen electrode, a platinum electrode in contact with 1.0 M H^+ (i.e., pH 0) and saturated with H_2 gas at 1.0 atm, which is arbitrarily assigned a standard potential of 0.0. The electrodes can accept or donate electrons to the redox couple in each cell, depending upon their respective potentials. A salt bridge containing a saturated KCl solution provides an electrical connection between the test cell and the reference cell. Electrons will flow from the test electrode to the reference electrode, or vice versa, via the external circuit, in a direction depending upon the relative electron "pressure" or potential of the two cells, but always from the cell of more negative to the cell of more positive potential. From the observed emf and the known emf of the reference cell, the emf of the test cell containing the redox couple is obtained.

tive values to systems having an increasing tendency to accept electrons. The terms *standard reduction potential, standard potential,* and *standard oxidation-reduction potential* are used interchangeably.

Table 17-1 gives the standard reduction potentials of a number of conjugate redox pairs important in biological electron transport. They are listed in order of increasing potential, i.e., in the order of decreasing tendency to lose electrons. Thus conjugate redox pairs having relatively negative standard potentials tend to lose electrons to those lower in the table, which have more positive standard potentials. For example, when the isocitrate/α-ketoglutarate + CO_2 couple is present in 1.0 M concentrations it has a standard potential E_0' of −0.38 V. This redox couple tends to pass electrons to the redox couple NADH/NAD^+, which has a relatively more positive potential, in the presence of isocitrate dehydrogenase (page 445). Conversely, the strongly positive standard potential of the water-oxygen couple, 0.82 V, indicates that the water molecule has very little tendency to lose electrons to form molecular oxygen. Put another way, molecular oxygen has a very high affinity for electrons or hydrogen atoms. Although standard potentials are given in units of *volts,* they are often expressed in *millivolts* for convenience.

Table 17-1 The Standard Reduction Potentials E_0' of Some Conjugate Redox Couples Participating in Oxidative Metabolism†

Redox couple	E_0'
Some substrate couples	
Acetyl-CoA + CO_2 + $2H^+$ + $2e^-$ ⟶ pyruvate + CoA	−0.48
α-Ketoglutarate + CO_2 + $2H^+$ + $2e^-$ ⟶ isocitrate	−0.38
3-Phosphoglyceroyl phosphate + $2H^+$ + $2e^-$ ⟶ glyceraldehyde 3-phosphate + P_i	−0.29
Pyruvate + $2H^+$ + $2e^-$ ⟶ lactate	−0.19
Oxaloacetate + $2H^+$ + $2e^-$ ⟶ malate	−0.18
Fumarate + $2H^+$ + $2e^-$ ⟶ succinate	+0.03
Components of the electron-transport chain	
$2H^+$ + $2e^-$ ⟶ H_2	**−0.41**
NAD^+ + H^+ + $2e^-$ ⟶ NADH	−0.32
$NADP^+$ + H^+ + $2e^-$ ⟶ NADPH	−0.32
NADH dehydrogenase (FMN form) + $2H^+$ + $2e^-$ ⟶ NADH dehydrogenase ($FMNH_2$ form)	−0.30
Ubiquinone + $2H^+$ + $2e^-$ ⟶ ubiquinol	+0.04
Cytochrome *b* (ox) + e^- ⟶ cytochrome *b* (red)	+0.07
Cytochrome c_1 (ox) + e^- ⟶ cytochrome c_1 (red)	+0.23
Cytochrome *c* (ox) + e^- ⟶ cytochrome *c* (red)	+0.25
Cytochrome *a* (ox) + e^- ⟶ cytochrome *a* (red)	+0.29
Cytochrome a_3 (ox) + e^- ⟶ cytochrome a_3 (red)	+0.55
$\frac{1}{2}O_2$ + $2H^+$ + $2e^-$ ⟶ H_2O	**+0.82**

† Assuming 1 *M* concentrations of all components, pH = 7.0, and 25°C. Half-reactions which express the affinity of each system for electrons are shown. The more negative the E_0', the lower the affinity of the system for electrons. Conversely, the more positive the E_0' of a system, the greater its electron affinity. Thus electrons tend to flow from one redox couple to another in the direction of the more positive system. Two landmark potentials (color) are for the $H_2/2H^+$ couple and the $H_2O/\frac{1}{2}O_2$ couple.

Free-Energy Changes Accompany Electron Transfers

The E_0' values of various redox couples allow us to predict the direction of flow of electrons from one redox couple to another when both are present under standard conditions and a catalyst is available. Electrons usually will not flow from one redox couple to another unless a catalyst or enzyme is present to accelerate the process; the catalyst, however, does not alter the direction of flow or affect the final equilibrium attained. Under such conditions, electrons will tend to flow from a relatively electronegative conjugate redox pair, such as $NADH/NAD^+$ ($E_0' = -0.32$ V), to more electropositive electron acceptors, such as reduced cytochrome *c*/oxidized cytochrome *c* ($E_0' = +0.23$ V). Similarly, they will also tend to flow from the cytochrome *c* redox pair to the water-oxygen pair ($E_0' = +0.82$ V). The tendency for electrons to flow from electronegative toward electropositive systems is the result of the loss of free energy, since electrons always tend to move in such a direction that the free energy of the reacting system decreases. The greater the difference in the standard potentials between two redox pairs, the greater the free-energy loss as electrons pass from the electronegative to the electropositive couple. Therefore, when electrons flow down the complete electron-transport chain from NADH ($E'_0 = -0.32$ V) to oxygen ($E'_0 = +0.82$ V), via the several electron-carrying molecules of the electron-transport chain, they lose a large amount of free energy because the difference between the standard potentials of the redox pairs $NADH/NAD^+$ and $H_2O/\frac{1}{2}O_2$ is relatively great.

Let us now calculate exactly how much free energy will become available as a pair of electrons passes from NADH to oxygen. The standard-free-energy change of a reaction in which there is a transfer of electrons is given by

$$\Delta G^{\circ\prime} = -n\mathcal{F}\,\Delta E'_0$$

where $\Delta G^{\circ\prime}$ is the standard-free-energy change in calories, n is the number of electrons transferred, $\mathcal{F}$ is a constant called the *faraday* (23,062 cal/V·mol), and $\Delta E'_0$ is the difference between the standard potential of the electron-donor system and that of the electron-acceptor system. All components are assumed to be at 1.0 *M* concentration at 25°C and pH 7.0. The standard-free-energy change as a pair of electron equivalents passes from the $NADH/NAD^+$ pair ($E'_0 = -0.32$ V) to the $H_2O/\frac{1}{2}O_2$ pair ($E'_0 = +0.82$ V) is thus

$$\Delta G^{\circ\prime} = -2(23{,}062)[0.82 - (-0.32)] = -52.6 \text{ kcal}$$

The 52.6 kcal of free energy released as a pair of electrons passes from NADH to oxygen under standard conditions is clearly more than enough to bring about the synthesis of three molecules of ATP, which requires input of 3(7.3) = 21.9 kcal under standard conditions.

In the same way, using the expression $\Delta G^{\circ\prime} = -n\mathcal{F}\,\Delta E'_0$, we can calculate the free-energy changes for individual segments of the electron-transport chain from the differences in the standard potentials of the electron-donating redox pair and the electron-accepting pair. Figure 17-4 is an energy diagram

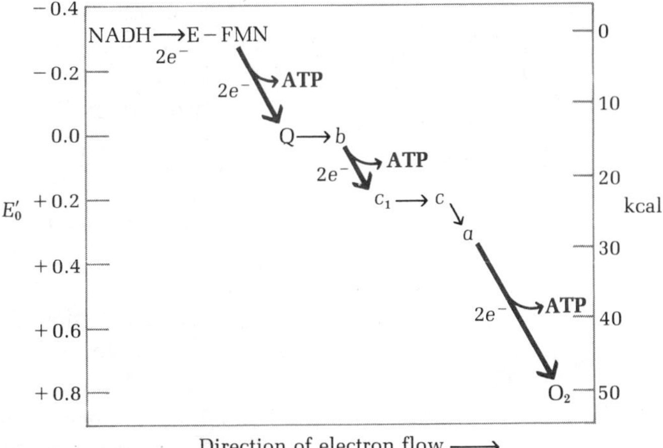

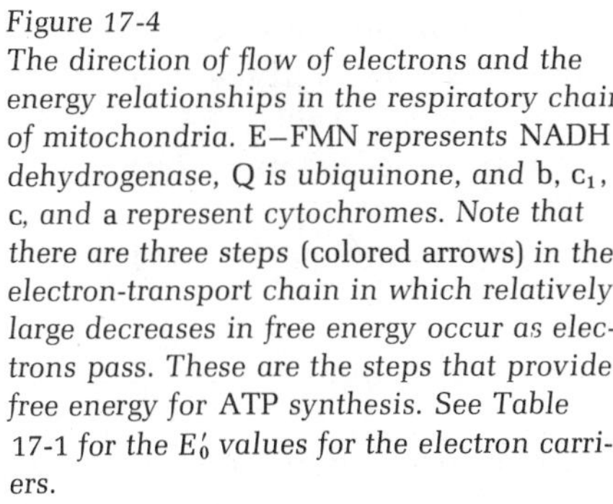
Figure 17-4
The direction of flow of electrons and the energy relationships in the respiratory chain of mitochondria. E–FMN represents NADH dehydrogenase, Q is ubiquinone, and b, c_1, c, and a represent cytochromes. Note that there are three steps (colored arrows) in the electron-transport chain in which relatively large decreases in free energy occur as electrons pass. These are the steps that provide free energy for ATP synthesis. See Table 17-1 for the E'_0 values for the electron carriers.

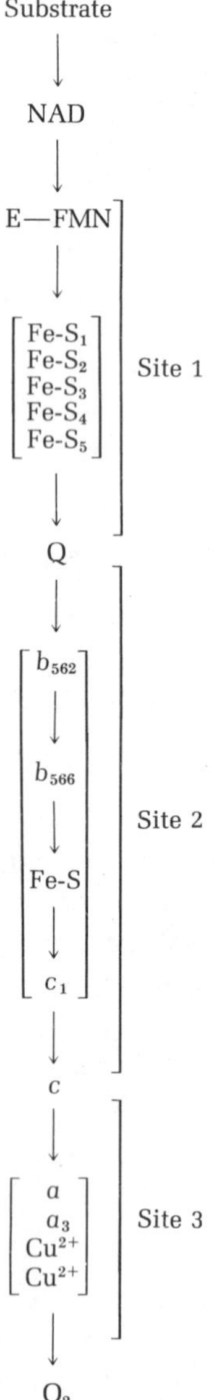

Figure 17-5
The complete set of electron carriers of the respiratory chain. In site 1 there are at least five different iron-sulfur centers. In site 2 there are two different forms of cytochrome b, with different light-absorption peaks, as well as an iron-sulfur center distinct from those in site 1. In site 3 there are two copper ions in addition to cytochromes a and a_3. The precise sequence and function of all the redox centers is not known with certainty.

showing (1) the standard potentials of some of the electron carriers of the respiratory chain, (2) the direction of electron flow, which is always "downhill" toward oxygen, and (3) the relative free-energy change at each step. Note that there are large free-energy decreases in three of the steps along the electron-transport chain. These are the energy-conserving sites that provide energy for ATP synthesis.

There Are Many Electron Carriers in the Electron-Transport Chain

The respiratory chain of mitochondria contains a large number of electron-carrying proteins that act in sequence to transfer electrons from substrates to oxygen. Although Figure 17-1 shows the respiratory chain to have seven electron carriers, this is an abbreviated representation. There are 15 or more chemical groups in the electron-transport chain that can accept and transfer reducing equivalents in sequence, summarized in Figure 17-5.

Note the several different kinds of electron-carrying groups, all associated with proteins. They include *nicotinamide adenine dinucleotide* (NAD), active with various dehydrogenases; *flavin mononucleotide* (FMN), in NADH dehydrogenase; *ubiquinone* or *coenzyme Q*, an isoprenoid lipid-soluble quinone, which functions in association with one or more proteins; two different kinds of iron-containing proteins, the *iron-sulfur centers* (Fe-S) and the *cytochromes;* and copper of *cytochrome aa_3*. A third important point is that nearly all the electron-carrying proteins of the chain are water-insoluble and are embedded in the inner mitochondrial membrane.

The Pyridine Nucleotides Have a Collecting Function

Most of the electron pairs entering the respiratory chain arise from the action of dehydrogenases that use the coenzymes NAD^+ or $NADP^+$ (Figure 17-6) as electron acceptors. As a group they are designated the *NAD(P)-linked dehydrogenases.* We have already met several such dehydrogenases in our discussions of glycolysis and the citric acid cycle, but many more are known. Some important ones are listed in Table 17-2. All catalyze reversible reactions of the following general types:

$$\text{Reduced substrate} + NAD^+ \rightleftharpoons \text{oxidized substrate} + NADH + H^+$$

$$\text{Reduced substrate} + NADP^+ \rightleftharpoons \text{oxidized substrate} + NADPH + H^+$$

The great majority of the pyridine-linked dehydrogenases are specific for NAD^+ (Table 17-2). However, certain others require $NADP^+$ as electron acceptor, such as *glucose 6-phosphate dehydrogenase* (page 457). A very few, such as *glutamate dehydrogenase,* can react with either NAD^+ or $NADP^+$ (Table 17-2).

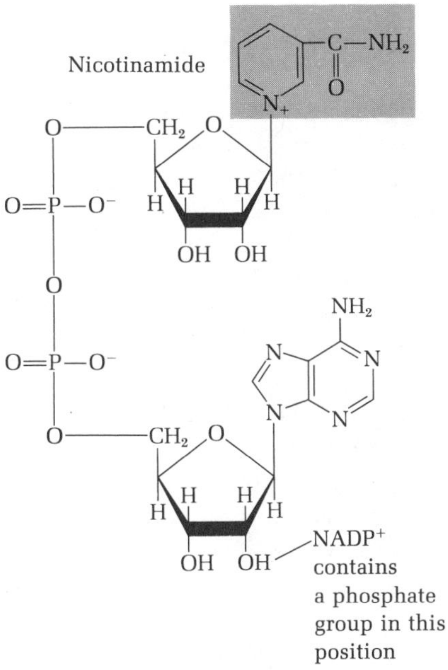

Nicotinamide

$NADP^+$ contains a phosphate group in this position

(a)

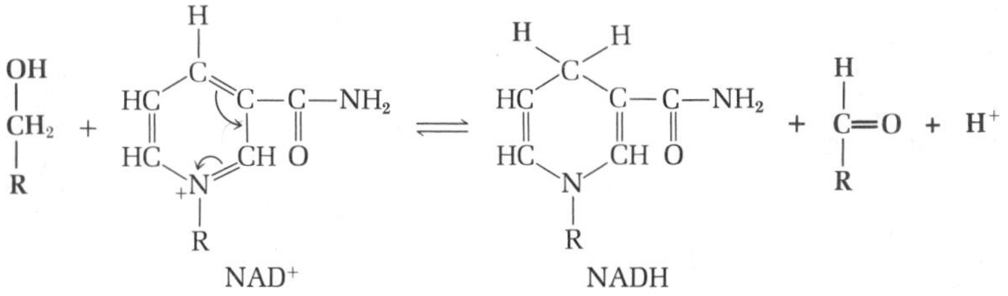

NAD^+ NADH

(b)

Figure 17-6
Nicotinamide adenine dinucleotide (NAD^+) *and nicotinamide adenine dinucleotide phosphate* ($NADP^+$). (a) *Oxidized forms* (NAD^+ *and* $NADP^+$). *Nicotinamide* (*shaded*) *is a vitamin of the B complex* (*page 255*) *and is the portion of the molecule participating in electron transfer.* (b) *Reduction of the nicotinamide ring of* NAD^+ *by substrate. The two reducing equivalents are transferred from the substrate (designated* RCH_2OH*) to* NAD^+ *in the form of a hydride ion* ($:H^-$)*. The other hydrogen removed from the substrate becomes* H^+.

Some pyridine-linked dehydrogenases are located in the cytosol, some in the mitochondria, and still others in both. Cytosolic dehydrogenases can react only with cytosolic pyridine nucleotides; similarly, mitochondrial dehydrogenases generally react only with mitochondrial pyridine nucleotides in the matrix. The cytosolic and mitochondrial pools of NAD and NADP are separated by the inner mitochondrial membrane, which is impermeable to these coenzymes. We shall come back to this point later.

The most important NAD-linked dehydrogenases functioning in carbohydrate catabolism are *glyceraldehyde phosphate dehydrogenase* and *lactate dehydrogenase* of the glycolytic system, located in the cytosol, and *pyruvate dehydrogenase*, present in the mitochondria (Table 17-2). Three NAD-linked dehydrogenases participate in the citric acid cycle in the mitochondria: *isocitrate*, *α-ketoglutarate*, and *malate dehydrogenases*. Other mitochondrial dehydrogenases of importance are *3-hydroxyacyl-CoA dehydrogenase* of the fatty acid

Table 17-2 Some Important Reactions Catalyzed by NAD(P)-Linked Dehydrogenases

Reaction	Location†
NAD-linked	
Isocitrate + NAD^+ ⇌ α-ketoglutarate + CO_2 + NADH + H^+	M
α-Ketoglutarate + CoA + NAD^+ ⇌ succinyl-CoA + CO_2 + NADH + H^+	M
L-Malate + NAD^+ ⇌ oxaloacetate + NADH + H^+	M and C
Pyruvate + CoA + NAD^+ ⇌ acetyl-CoA + CO_2 + NADH + H^+	M
Glyceraldehyde 3-phosphate + P_i + NAD ⇌ 1,3-diphosphoglycerate + NADH + H^+	C
Lactate + NAD^+ ⇌ pyruvate + NADH + H^+	C
NADP-linked	
Isocitrate + $NADP^+$ ⇌ α-ketoglutarate + CO_2 + NADPH + H^+	M and C
Glucose 6-phosphate + $NADP^+$ ⇌ 6-phosphogluconate + NADPH + H^+	C
NAD or NADP	
L-Glutamate + H_2O + NAD^+ ($NADP^+$) ⇌ α-ketoglutarate + NH_3 + NADH (NADPH) + H^+	M

† M = mitochondria and C = cytosol.

Figure 17-7
The collecting function of NAD and ubiquinone (Q). NAD collects reducing equivalents from many NAD-linked substrates and from NADPH. Ubiquinone collects reducing equivalents from NADH dehydrogenase and various substrates acted on by other flavin dehydrogenases. Pairs of reducing equivalents collected from most flavin dehydrogenases do not pass through the first phosphorylation site and thus give rise to only two ATPs.

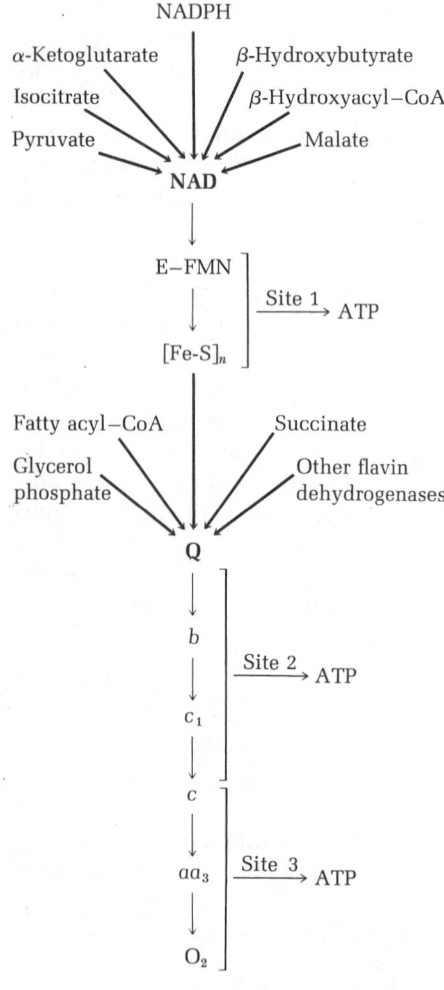

oxidation cycle, β-hydroxybutyrate dehydrogenase (Chapter 18), and glutamate dehydrogenase, which functions in amino acid catabolism (Chapter 19).

The pyridine-linked dehydrogenases remove two hydrogen atoms from their substrates. One of these is transferred as a hydride ion (:H^-) to the NAD^+ or $NADP^+$; the other appears as H^+ in the medium. Each hydride ion carries two reducing equivalents; one is transferred as a hydrogen atom to carbon 4 of the nicotinamide ring, the other as an electron to the ring nitrogen (Figure 17-6).

Since most dehydrogenases in cells transfer H atoms from their substrates to NAD^+, this coenzyme *collects* pairs of reducing equivalents from many different substrates in one molecular form, NADH (Figure 17-7). Ultimately, NAD^+ can also collect reducing equivalents from substrates acted upon by NADP-linked dehydrogenases. This is made possible by pyridine nucleotide transhydrogenase, a complex enzyme that catalyzes the reaction

$$\text{NADPH} + \text{NAD}^+ \rightleftharpoons \text{NADP}^+ + \text{NADH}$$

NADH Dehydrogenase Accepts Electrons from NADH

In the next step in the mainstream of electron transport (Figure 17-5), a pair of reducing equivalents is transferred from NADH to NADH dehydrogenase, located in the inner mitochondrial membrane. In this reaction the tightly bound prosthetic group of NADH dehydrogenase becomes reduced (Figure 17-8). This

Oxidized form (FMN)

The double-bond system that accepts two H atoms

CH_2–CHOH–CHOH–CHOH–$CH_2OPO_3^{2-}$

NADH + H^+

NAD^+

Reduced form ($FMNH_2$)

Figure 17-8
Transfer of reducing equivalents from NADH to flavin mononucleotide (FMN), the prosthetic group of NADH dehydrogenase. R represents the five-carbon phosphorylated side chain.

Figure 17-9
Postulated arrangement of iron (color) and sulfur atoms in iron-sulfur centers. The number of iron and acid-labile sulfur atoms (shaded) in these clusters is always equal, but some iron-sulfur centers have two and others four iron atoms. The iron-sulfur center shown here is a four-iron cluster. The peripheral sulfur atoms are those of four cysteine residues of the polypeptide chain of the enzyme.

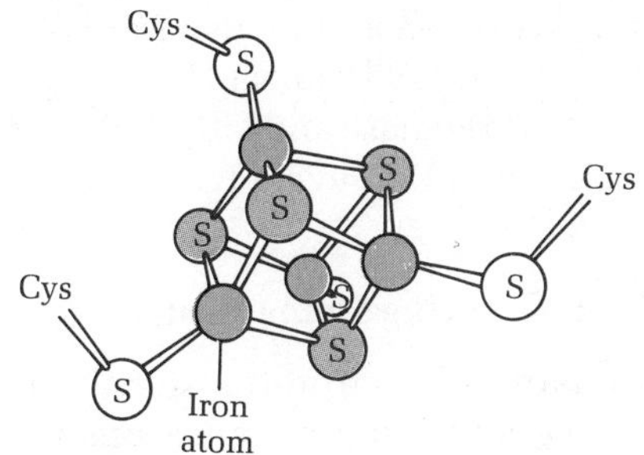

prosthetic group is *flavin mononucleotide* (FMN), which contains a molecule of vitamin B_2 or *riboflavin* (page 254). NADH dehydrogenase is a member of the class of *flavin-linked dehydrogenases* or *flavoproteins*. Transfer of two reducing equivalents from NADH to NADH dehydrogenase (here designated E–FMN) reduces the FMN to $FMNH_2$:

$$\text{NADH} + \text{H}^+ + \text{E–FMN} \rightleftharpoons \text{NAD}^+ + \text{E–FMNH}_2$$

In addition to its flavin nucleotide prosthetic group, NADH dehydrogenase also contains several *nonheme iron atoms*. These are grouped into several clusters, in which the iron atoms are paired with an equal number of acid-labile sulfur atoms. Such clusters are called *iron-sulfur centers* (Figure 17-9). Recall that iron-sulfur centers are also associated with *succinate dehydrogenase* (page 447). The iron atoms in these clusters, by undergoing Fe(II)–Fe(III) cycles, transfer reducing equivalents from the $FMNH_2$ prosthetic group of NADH dehydrogenase to the next electron carrier in the chain, *ubiquinone*. The complex of NADH dehydrogenase with the iron-sulfur proteins, which is called *NADH-ubiquinone reductase*, thus contains two kinds of electron-carrying structures: FMN and several iron-sulfur centers, which appear to function in sequence.

Ubiquinone Is a Lipid-Soluble Quinone

The next carrier of reducing equivalents in the respiratory chain is *ubiquinone* or *coenzyme Q* (for quinone). The name ubiquinone is a reflection of the ubiquity of this compound (it occurs in virtually all cells). Ubiquinone is a fat-soluble quinone with a very long isoprenoid side chain (Figure 17-10). In most mammalian tissues ubiquinone has ten 5-carbon isoprenoid units in its side chain and is thus designated Q_{10} or CoQ_{10}. In other organisms it may have only six or eight isoprene units (Q_6 or Q_8). When reduced NADH dehydrogenase

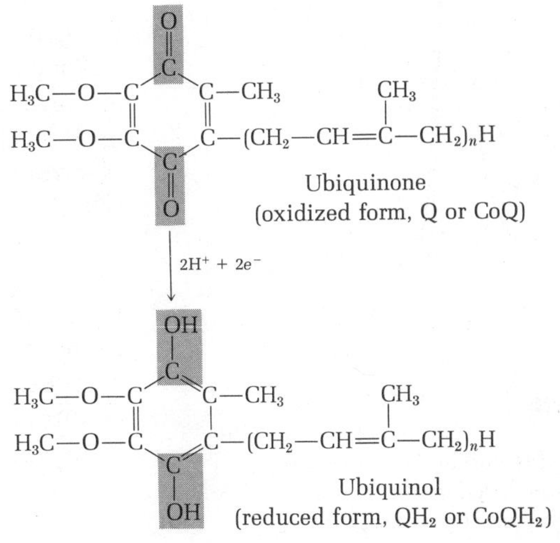

Ubiquinone
(oxidized form, Q or CoQ)

$2H^+ + 2e^-$

Ubiquinol
(reduced form, QH_2 or $CoQH_2$)

Figure 17-10
Ubiquinone, or coenzyme Q. The subscript n indicates the number of isoprenoid units in the side chain (see text). The groups in color function in carrying H atoms. Note that there is also a shift in the ring double bonds when ubiquinone is reduced to ubiquinol.

(E—$FMNH_2$) donates its reducing equivalents via the iron-sulfur centers to ubiquinone, the latter becomes reduced to *ubiquinol* or QH_2 (Figure 17-10), with regeneration of the oxidized form of NADH dehydrogenase:

$$E{-}FMNH_2 + Q \rightleftharpoons E{-}FMN + QH_2$$

Ubiquinone, a much longer molecule than the phospholipids in the inner membrane, occurs in both free and protein-bound forms. The function of ubiquinone is to *collect* reducing equivalents, not only from NADH dehydrogenase but also from other flavin-linked dehydrogenases of mitochondria (see Figure 17-7), particularly succinate dehydrogenase and fatty acyl–CoA dehydrogenase of the fatty acid oxidation cycle (Chapter 18).

The Cytochromes Are Electron-Carrying Heme Proteins

The cytochromes are iron-containing electron-transferring red or brown proteins that act in sequence to carry electrons from ubiquinone to molecular oxygen. They are *heme proteins,* in which the iron is present as an iron-porphyrin, or *heme,* prosthetic group resembling that of hemoglobin (page 271). Although discovered many years earlier and originally named *histohematins,* the cytochromes were first shown to function in biological oxidations by David Keilin beginning in 1925. With a spectroscope he observed that the flight muscles of living insects contain red-brown pigments. He found that these pigments underwent marked changes in their spectra whenever the insect, fastened to a microscope slide, struggled to free itself. He renamed these pigments *cytochromes* and postulated that they undergo oxidation and reduction as they carry electrons from nutrients to oxygen. There are three classes of cytochromes, *a, b,* and *c,* distinguished by differences in their light-absorption spectra. Each type of cytochrome in its reduced, or ferrous, state has three distinctive absorption bands in the visible range (Figure 17-11). Keilin showed that the cytochromes act in a sequence, the last one donating electrons to oxygen.

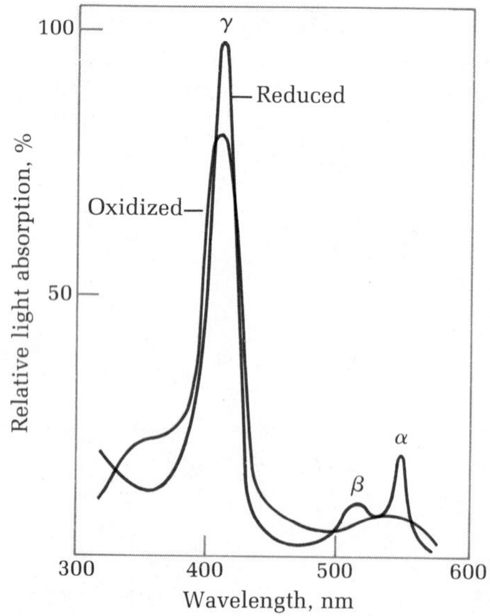

Figure 17-11
The absorption spectra of cytochrome c in its oxidized (color) and reduced (black) forms. Also labeled are the characteristic α, β, and γ bands of the reduced forms.

Today we know that the cytochromes in the respiratory chain are arranged in the sequence $b \rightarrow c_1 \rightarrow c \rightarrow aa_3$ (Figures 17-1 and 17-7). Cytochrome *b,* which occurs in two forms, accepts electrons from ubiquinone and transfers them to cytochrome c_1, which in turn donates the electrons it receives to cytochrome *c*. Each cytochrome in its ferric [Fe(III)] form accepts one electron to become the ferrous form [Fe(II)]. Also participating in the transfer of electrons from ubiquinone to cytochrome *c* is an iron-sulfur protein (Figure 17-5). The last electron carrier is *cytochrome* aa_3 or *cytochrome oxidase,* which can donate electrons directly to oxygen to complete the process of electron transport.

Cytochrome c is the best known of the cytochromes. It is a small protein (MW 12,500) with an iron-porphyrin group covalently attached to its single polypeptide chain (page 173). Its amino acid sequence (page 136) and three-dimensional structure (page 174) have been studied in detail. Cytochrome *c* is easily extracted from mitochondria and has been obtained in crystalline form from many species. As we have seen (page 136), cytochrome *c* is an ancient protein, since its amino acid sequence has many points of similarity in all eukaryotic organisms—microbial, plant, or animal.

Cytochrome aa_3 is different from the other cytochromes. It contains two molecules of tightly bound heme A, which differs from the protoheme of hemoglobin in that the porphyrin ring has a long hydrocarbon side chain. Moreover, cytochrome aa_3 also contains two essential copper atoms. After the cytochrome *a* component receives electrons from cytochrome *c* and thus becomes reduced to the Fe(II) form, it passes its electrons to cytochrome a_3. Reduced cytochrome a_3 then passes electrons to molecular oxygen (O_2). Participating with the two heme groups in this process are the two bound copper atoms, which undergo cuprous-cupric redox changes [Cu(I)–Cu(II)] in their function. This is a complex and important step in electron transport, since *four* electrons must be passed almost simultaneously to O_2 to yield two H_2O, with uptake of four H^+ from the aqueous medium. Of all the members of the electron-transport chain, only cytochrome aa_3 can react directly with oxygen.

Incomplete Reduction of Oxygen Causes Cell Injury

It is very important to the cell that the oxygen molecule be completely reduced to two H_2O by accepting four electrons. If O_2 is only partially reduced by accepting two electrons, the product is hydrogen peroxide (H_2O_2). If O_2 accepts only one electron, the product is the *superoxide radical* ($:O_2^-$). Hydrogen peroxide and superoxide are extremely toxic to cells because they attack unsaturated fatty acid components of membrane lipids, thus damaging membrane structure. Aerobic cells protect themselves against superoxide and peroxide by the action of *superoxide dismutase*, a metal-containing enzyme that converts the superoxide radical into hydrogen peroxide, and *catalase*, which converts hydrogen peroxide into H_2O and molecular oxygen:

$$2O_2^- + 2H^+ \xrightarrow[\text{dismutase}]{\text{superoxide}} H_2O_2 + O_2$$

$$2H_2O_2 \xrightarrow{\text{catalase}} 2H_2O + O_2$$

Although toxic, hydrogen peroxide can be useful. The bombardier beetle generates a concentrated solution of hydrogen peroxide in one sac of its spray gland and a solution of a hydroquinone in the other sac. When threatened, this remarkable insect startles (and poisons) its enemies by firing a hot (100°F) spray of a toxic quinone, resulting from the explosive oxidation of the hydroquinone by the hydrogen peroxide.

The Electron Carriers Always Function in a Specific Sequence

How do we know that the electron carriers in the respiratory chain function in this sequence? First, their standard redox potentials (Figure 17-4 and Table 17-1) are successively more positive going toward oxygen, as we might expect, since electrons tend to flow from electronegative to electropositive systems, causing a decrease in free energy. Second, each member of the chain is specific for a given electron donor and acceptor. For example, NADH can transfer electrons to NADH dehydrogenase but cannot transfer them directly to cytochrome *b* or to cytochrome *c*. Third, structured complexes of functionally related electron carriers have been isolated from the mitochondrial membrane (Figure 17-12). Complex I consists of NADH dehydrogenase and its iron-sulfur centers, which are closely linked in their function. Complex II consists of succinate dehydrogenase and its iron-sulfur centers. Complex III consists of cytochromes *b* and c_1 and a specific iron-sulfur center. Cytochromes *a* and a_3 together constitute complex IV. Ubiquinone

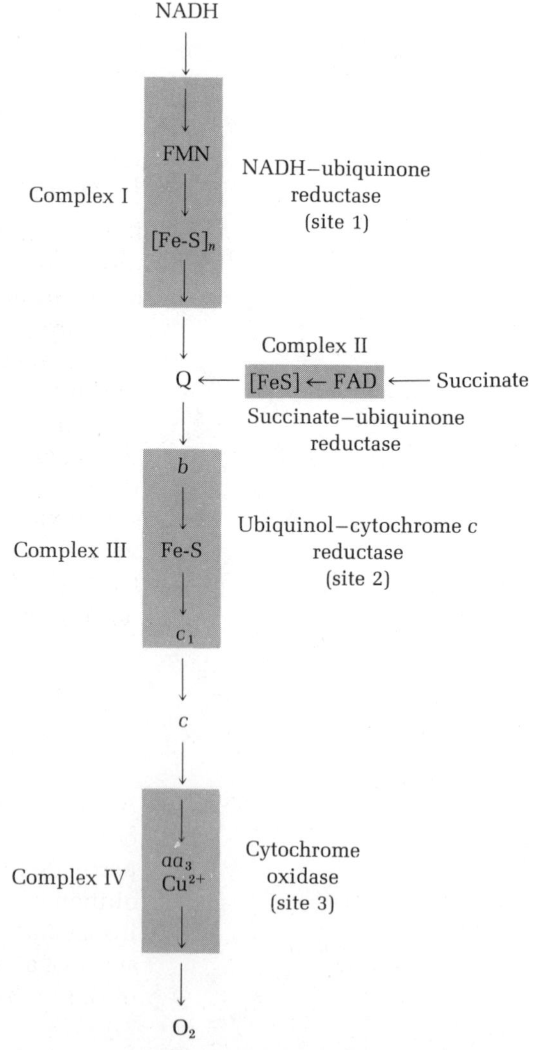

Figure 17-12
The electron-transport complexes. They can be isolated as functional assemblies.

is the connecting link between complexes I, II, and III, and cytochrome *c* is the connecting link between complex III and complex IV (Figure 17-12).

Specific inhibitors acting at certain points in the chain have also greatly helped in the study of electron transport. The most important are (1) rotenone, an extremely toxic plant product used by South American Indians as a fish poison, which blocks electron transport from NADH to ubiquinone; (2) the toxic antibiotic antimycin A, isolated from a strain of *Streptomyces*, which blocks transfer of electrons from ubiquinone to cytochrome *c*; and (3) cyanide, one of the most deadly poisons known, which blocks the reduction of oxygen catalyzed by cytochrome aa_3 (Figure 17-13). Another important inhibitor of cytochrome aa_3 is carbon monoxide. As a hydraulic analogy of the respiratory chain in Figure 17-14 shows, inhibition of electron transport at a given site creates a crossover point (page 425). At such a point the electron carriers just before the block become more reduced and those after the block become more oxidized. These changes can be detected in a spectrophotometer, since the oxidized and reduced forms of the electron carriers have characteristically different spectra.

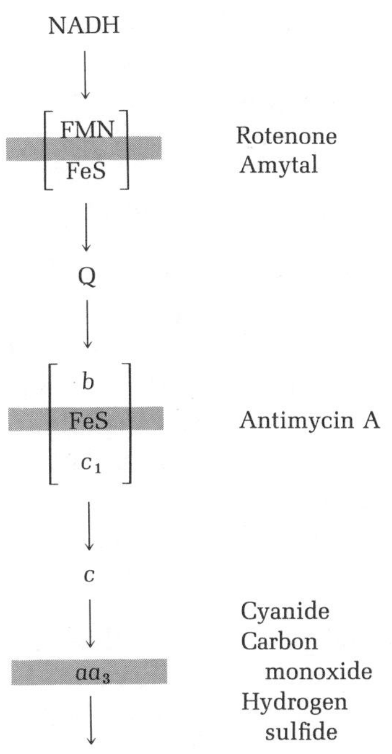

Figure 17-13
Sites of action of various inhibitors of electron transport. Amytal is a barbiturate drug used as a sedative. In addition to cyanide, hydrogen sulfide and carbon monoxide are also potent inhibitors of cytochrome oxidase.

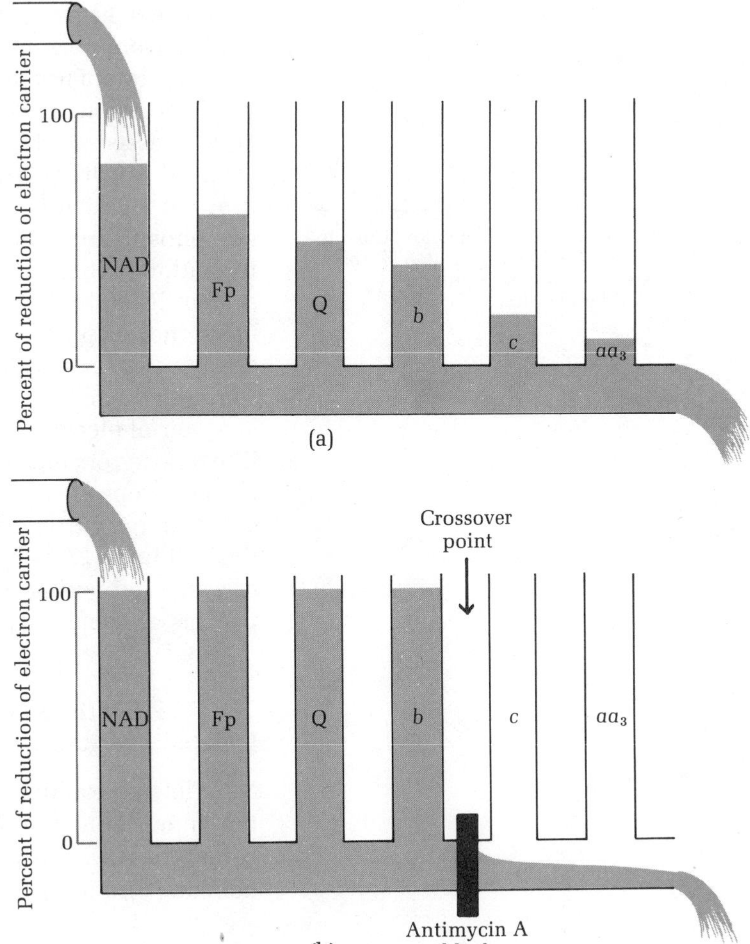

Figure 17-14
Hydraulic analogy of the respiratory chain. (a) Normal respiratory steady state. The extent of reduction of the successive electron-carrier molecules in a population of mitochondria decreases as electrons proceed from substrates to oxygen. (b) The respiratory-chain inhibitor antimycin A produces a crossover point in the redox state of the carriers.

Electron-Transport Energy Is Conserved by Oxidative Phosphorylation

We have seen that there are three energy-conserving segments of the electron-transport chain that provide the energy required to make ATP from ADP and phosphate in the process of oxidative phosphorylation (Figures 17-4, 17-7, and 17-12). Electron pairs that originate from NAD-linked dehydrogenases pass through all three energy-conserving sites and thus yield a maximum of three ATPs. The complete equation for phosphorylating electron transport from NADH to oxygen is:

$$NADH + H^+ + \tfrac{1}{2}O_2 + 3P_i + 3ADP \longrightarrow NAD^+ + 3ATP + 4H_2O$$

However, when succinate is dehydrogenated by the flavin-linked succinate dehydrogenase, only two ATPs are formed per pair of electrons flowing to oxygen (Figure 17-7 and Table 17-3). This is because the electron pair from succinate enters at ubiquinone, i.e., at a point *beyond* site 1. Electron pairs arising from other flavin-linked dehydrogenases, such as fatty acyl–CoA dehydrogenase of the fatty acid oxidation cycle (Chapter 18), also give rise to only two ATPs (Figure 17-7). Oxidative phosphorylation is not limited to the dehydrogenations of the citric acid cycle but occurs during electron transport originating from all mitochondrial dehydrogenases taking part in carbohydrate, fatty acid, and amino acid catabolism.

Table 17-3 The Number of ATP Molecules Formed at Each Oxidative Step of the Citric Acid Cycle

Step	ATPs formed
Isocitrate $\longrightarrow$ α-ketoglutarate + CO_2	3
α-Ketoglutarate $\longrightarrow$ succinate + CO_2	4†
Succinate $\longrightarrow$ fumarate	2
Malate $\longrightarrow$ oxaloacetate	3
Total	12

† Since the conversion of succinyl-CoA to succinate yields GTP and from it ATP (page 446), the steps from α-ketoglutarate to succinate yield altogether four ATPs.

The formation of three ATPs conserves much of the free energy released during electron transport. Since transfer of an electron pair from NADH to oxygen yields 52.6 kcal, and since input of 7.3 kcal is required to make 1 mol of ATP from ADP and phosphate under standard thermodynamic conditions, the formation of three molecules of ATP can theoretically conserve a significant fraction of the total free-energy decline during electron transport from NADH to oxygen. Now we can begin to see why the respiratory chain has so many electron carriers. Through them, the rather large decline in free energy occurring as a pair of electrons flows from NADH to oxygen is broken up into a series of steps that have smaller free-energy drops. Three of these steps yield an amount of energy approximately the size of the energy currency of the cell, i.e., the free energy of ATP synthesis (Figure 17-4). The respiratory chain is therefore a stepwise cascade delivering free energy derived from cell fuels in packets of useful size.

The ATP-Synthesizing Enzyme Has Been Isolated and Reconstituted

Next let us examine the ATP-synthesizing enzyme system of the inner mitochondrial membrane. This enzyme complex, called the *ATP synthetase* or *$F_0F_1ATPase$*, has two major components, F_0 and F_1 (F for factor). The F_1 component resembles a

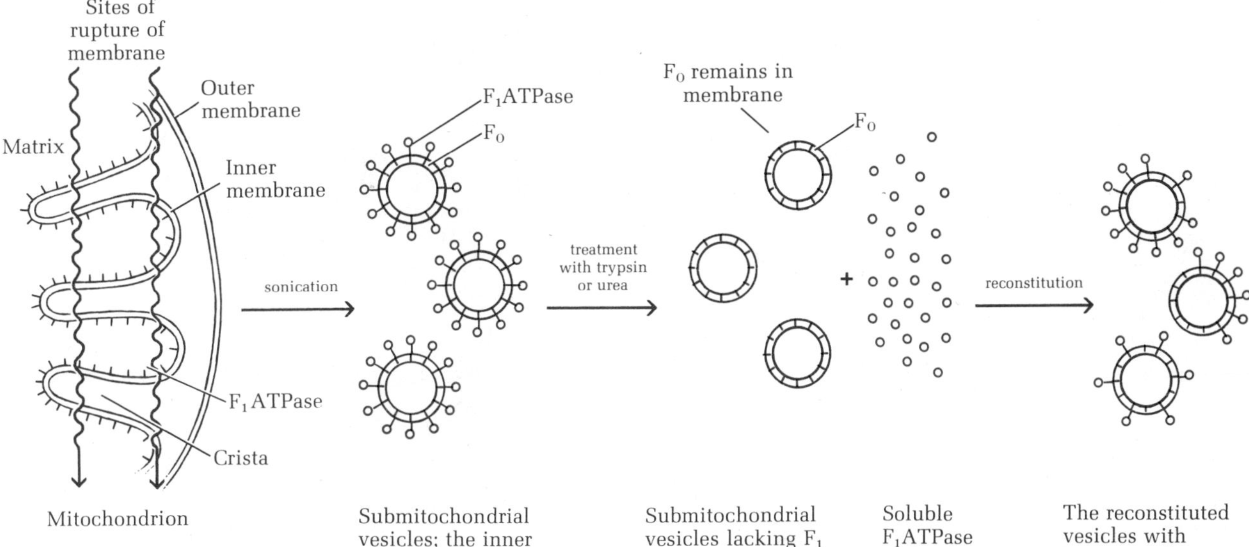

Figure 17-15
Formation of inner-membrane vesicles by sonic treatment of mitochondria and the reconstitution of oxidative phosphorylation. During the sonic treatment the cristae membranes are fragmented and reseal to form vesicles in which the F_1 spheres are on the outside, rather than the inside. When these inverted vesicles are treated with urea or trypsin, the F_1 spheres become detached; such vesicles, which still contain F_O, promote electron transport but not phosphorylation. When F_1 molecules are added back to the depleted vesicles, they become reattached to F_O units in the inner-membrane vesicles. Such reconstituted vesicles can now catalyze both electron transport and oxidative phosphorylation.

doorknob protruding into the matrix from the inner membrane (Figures 17-2 and 17-15). It is attached by a stalk to F_O, which is embedded in the inner membrane and extends across it. The subscript of F_O is not zero but the letter O, to denote that it is the portion of the ATP synthetase that binds the toxic antibiotic oligomycin, a potent inhibitor of this enzyme and thus of oxidative phosphorylation.

F_1 was first extracted from the mitochondrial inner membrane and purified by Efraim Racker and his colleagues. F_1 alone cannot make ATP from ADP and phosphate in its isolated form, but it can hydrolyze ATP to ADP and phosphate and is therefore also called F_1ATPase. When F_1 is carefully extracted from inside-out vesicles prepared from the inner mitochondrial membrane (Figure 17-15), the vesicles still contain intact respiratory chains and can catalyze electron transport. However, since they no longer contain the F_1 knobs, as confirmed by electron microscopy, they cannot make ATP. But when a preparation of isolated F_1 is added back to such depleted vesicles under appropriate conditions, the normal inner membrane structure, complete with F_1 knobs, is reconstituted and the capacity of the inner membrane vesicles to carry out energy coupling between electron transport and ATP is restored (Figure 17-15). Membrane-reconstitution experiments of this kind, pioneered by Racker, have opened the door to much important research on membrane structure and function.

F_1 has recently been isolated in pure crystalline form (Figure 17-16). It has a molecular weight of about 380,000, contains

Figure 17-16
The structure of F_0F_1ATPase (ATP synthetase). (a) F_0F_1ATPase *was first discovered as knoblike projections from the inner surface of the membrane, as shown in the electron micrograph.* (b) *A model of* F_0F_1ATPase, *showing one possible arrangement of its subunits.* (c) *Crystals of the* F_1 *portion of the complex from rat liver mitochondria.* (d) *Electron micrograph showing two* F_0F_1ATPase *molecules isolated from rat liver mitochondria.*

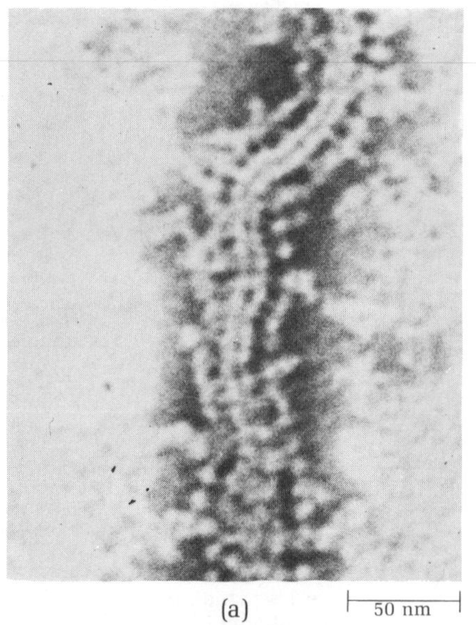

(a)

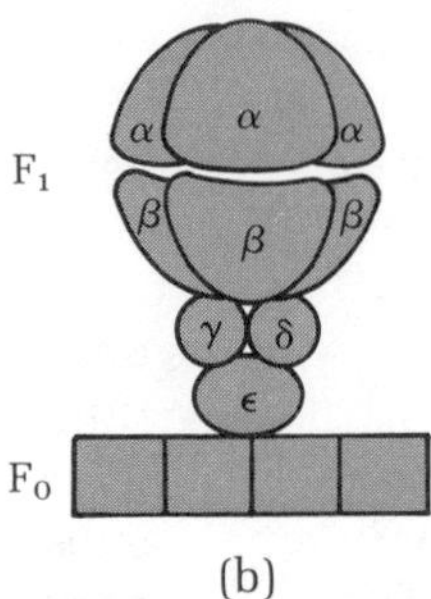

(b)

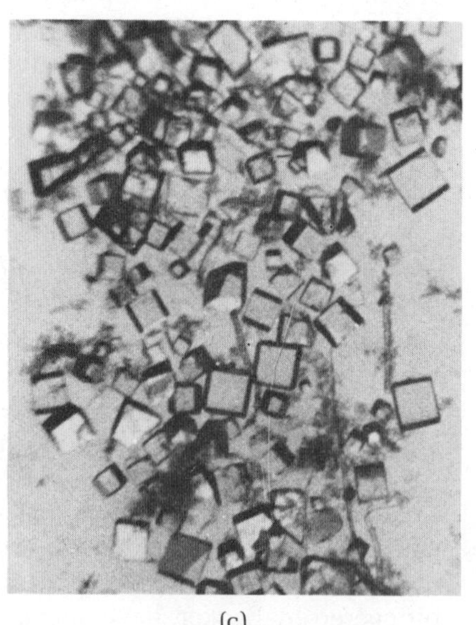

(c)

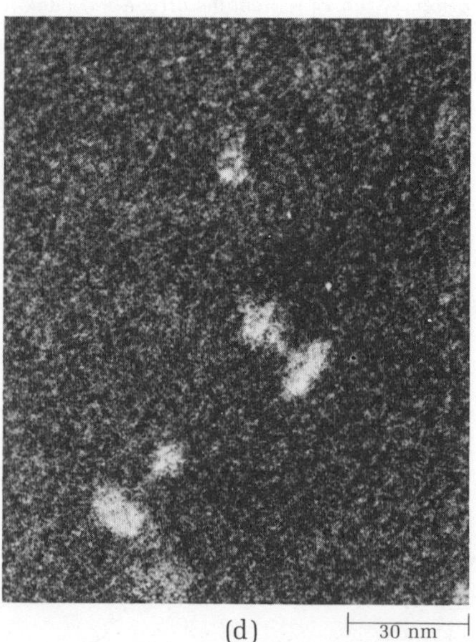

(d)

nine polypeptide chain subunits of five different kinds arranged into a cluster, and also has several binding sites for ATP and ADP. The complete F_0F_1ATPase molecule has also been obtained in highly purified form. High-resolution electron micrographs show it to have the knoblike F_1 head, a stalk, and a base piece, which normally extends across the inner membrane (Figure 17-16). F_0F_1ATPase is called an ATPase because in isolated form it hydrolyzes ATP to ADP plus P_i. However, since its major biological function in intact mitochondria is to make ATP from ADP and P_i, it is more appropriately called *ATP synthetase.*

How Is the Redox Energy of Electron Transport Delivered to ATP Synthetase?

In the preceding sections of this chapter we have examined the process of electron transport and the structure of ATP synthetase. Now we come to a central question: How does the electron-transport chain cooperate with the ATP synthetase to bring about oxidative phosphorylation of ADP to ATP? This has been one of the most challenging and difficult problems in biochemical and cell research. Although we know a great deal today about the utilization of ATP energy in biosynthetic reactions, we still do not know precisely, in molecular terms, how ATP is generated during oxidative phosphorylation. One of the reasons is that the enzymes concerned in electron transport and oxidative phosphorylation are very complex and they are embedded in the inner mitochondrial membrane, rendering detailed study of their interactions difficult. Three different mechanisms have been proposed for energy transfer between electron transport and ATP synthesis.

The *chemical coupling hypothesis* proposes that electron transport is coupled to ATP synthesis by a sequence of consecutive reactions in which a high-energy covalent intermediate is formed by electron transport and subsequently is cleaved and donates its energy to make ATP. This proposal is reminiscent of the role of the common intermediate 3-phosphoglycerol phosphate in ATP formation by the glycolytic sequence (page 410).

The *conformational-coupling hypothesis* postulated that the transfer of electrons along the respiratory chain causes a conformational change in protein components of the inner membrane to yield a high-energy form. The conformational change so produced is transmitted to the F_0F_1ATPase molecule, causing it to become energized. Relaxation of the energized F_0F_1ATPase to its normal conformation was proposed to provide energy for the synthesis of ATP and its release from the enzyme.

The *chemiosmotic hypothesis,* proposed by the British biochemist Peter Mitchell, involves a very different and novel principle. It postulated that electron transport pumps H^+ from the matrix across the inner mitochondrial membrane to the outer aqueous phase, thus generating an H^+ gradient across the inner membrane. The osmotic energy inherent in this gradient was postulated to drive the energy-requiring synthesis of ATP. This hypothesis appears to be closest to describing the organizing principle of oxidative phosphorylation. Let us now examine some characteristic properties of oxidative phosphorylation that support the chemiosmotic hypothesis.

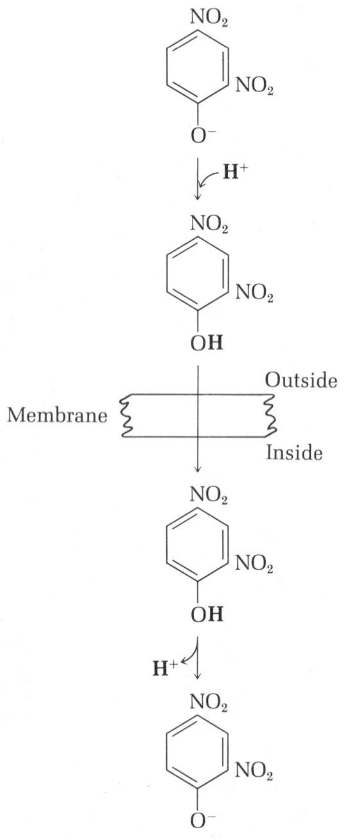

NO_2 NO_2 O^- H^+ NO_2 NO_2 OH Outside Membrane Inside NO_2 NO_2 OH H^+ NO_2 NO_2 O^-

Figure 17-17
Action of the typical uncoupling agent 2,4-dinitrophenol. At pH near 7 this agent exists largely as the anion, which is not soluble in lipids. In its protonated form it is lipid-soluble and thus can pass through the membrane, carrying a proton. The H^+ so carried is discharged on the other side of the membrane. In this way uncouplers can prevent formation of H^+ gradients across the membrane. Such H^+-conducting uncouplers are also called protonophores.

No "High-Energy" Intermediates Linking Electron Transport to ATP Synthesis Have Been Found

Intensive searches over many years have failed to detect hypothetical chemical intermediates in oxidative phosphorylation.

Oxidative Phosphorylation Requires Intact Inner-Membrane Structure

The inner membrane must be intact and form a completely enclosed vesicle. Breaks or holes in the inner membrane destroy its capacity for oxidative phosphorylation, although electron transport from substrates to oxygen may still continue.

The Inner Mitochondrial Membrane Is Impermeable to H^+, OH^-, K^+, and Cl^- Ions

This property also is related to oxidative phosphorylation. If the membrane is damaged or treated in such a way that these or certain other ions can pass through readily, oxidative phosphorylation will not occur. Such observations have suggested that a difference in ionic composition or concentration across the inner membrane of respiring mitochondria is essential for ATP synthesis.

Oxidative Phosphorylation Can Be Prevented by Uncoupling Agents

Certain chemical substances, for example 2,4-dinitrophenol (Figure 17-17), allow electron transport in mitochondria to occur but prevent the phosphorylation of ADP to ATP by uncoupling the essential linkage between electron transport and ATP synthesis. Hence they are called uncoupling agents. In their presence the free energy released by electron transport appears as heat rather than as newly made ATP. Uncoupling agents greatly increase the permeability of the inner membrane to H^+. They are lipophilic substances capable of binding H^+ from one side of the membrane and carrying it through the membrane toward the side with the lower H^+ concentration.

Certain uncoupling agents were once used as "reducing drugs" to treat obesity by lowering the efficiency of ATP production. These agents proved to be extremely toxic, however, and their use has long since been discontinued.

Figure 17-18
Valinomycin, a toxic antibiotic that promotes transport of K^+ through membranes. This K^+ ionophore consists of L- *and* D-*valine, lactate, and hydroxyisovalerate residues joined to form a ringlike compound. Valinomycin forms a specific complex with a K^+ ion* (color), *located inside its hydrophilic interior. The lipid-soluble exterior* (gray) *of the valinomycin molecule permits the valinomycin-K^+ complex to pass through the mitochondrial membrane at a high rate.*

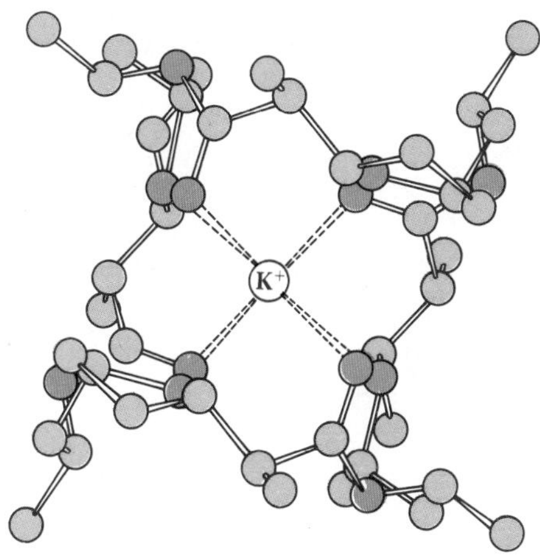

Oxidative Phosphorylation Can Also Be Prevented by Certain Ionophores

Ionophores ("ion carriers") are lipid-soluble substances capable of binding and carrying specific ions through the membrane. They differ from the uncoupling agents in that they promote the transport of cations other than H^+ through the membrane. The toxic antibiotic *valinomycin* (Figure 17-18) forms a lipid-soluble complex with K^+ which readily passes through the mitochondrial inner membrane, whereas K^+ alone in the absence of valinomycin penetrates only very slowly. The ionophore *gramicidin* induces the penetration not only of K^+ but also of Na^+ and several other monovalent cations through the inner membrane. Thus increasing the permeability of the mitochondrial membrane to H^+, K^+, or Na^+ by uncouplers or ionophores prevents oxidative phosphorylation.

Electron Flow Causes Respiring Mitochondria to Eject H^+

The energy released as electrons flow from substrates to oxygen along the respiratory chain can under certain conditions cause the transport of H^+ from the mitochondrial matrix into the medium. As a consequence the inner matrix compartment becomes more alkaline, i.e., increases in pH, and the suspending medium becomes more acid. Thus the inner membrane contains H^+ "pumps" that use the free energy of electron flow as the driving force to transport H^+ outward against a gradient of H^+. Accompanying ejection of H^+ is formation of an electric potential across the inner membrane, since the movement of H^+ from the matrix to the outside causes the outside of the membrane to become more electropositive and the inside more electronegative. Thus electron transport generates an *electrochemical H^+ gradient* which has two components; the membrane potential contributes the greater part of the energy of the gradient:

$$\Delta\tilde{\mu}_H = \Delta\Psi - Z\,\Delta pH$$

$\Delta\tilde{\mu}_H$	$\Delta\Psi$	$Z\,\Delta pH$
Electrochemical H^+ gradient	Membrane potential (negative inside)	The pH gradient (acid outside)

The term Z is a factor for converting pH units into millivolts, the units in which $\Delta\tilde{\mu}_H$ and $\Delta\Psi$ are usually expressed. The membrane potential contributes about 75 percent of the total H^+ electrochemical gradient generated by electron transport.

The Chemiosmotic Hypothesis Postulates That a Proton Gradient Carries Energy from Electron Transport to ATP Synthesis

The mitochondrial properties just described constitute the foundation stones of the chemiosmotic hypothesis (Figure 17-19). It proposes that the function of electron transport,

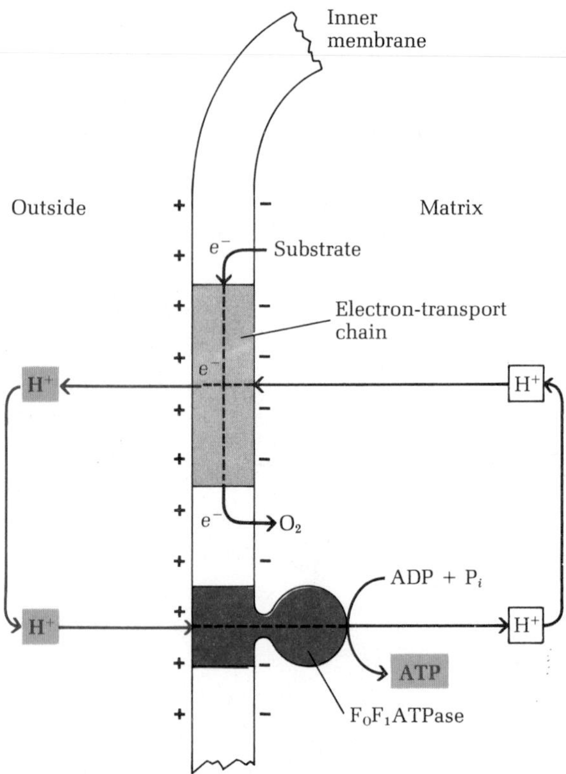

Figure 17-19
Principle of the chemiosmotic hypothesis. The electron-transport chain is proposed to be an H^+ pump; the energy released during electron transport is used to translocate H^+ from the matrix to the outside, generating an acid-outside electrochemical gradient of H^+ This process also generates an electric potential across the membrane, which is positive outside. External H^+ then flows back into the matrix down its electrochemical gradient through the F_0F_1ATPase, generating ATP at the expense of free energy released as H^+ passes into the zone of lower H^+ concentration. The hypothesis thus proposes a continuous cycling of H^+ out of and back into the mitochondria, driven by electron transport. See Box 17-1 for other details.

which takes place in the inner membrane, is to "pump" H^+ from the matrix to the outer medium, thus generating an acid-outside gradient of H^+ between two bulk aqueous phases separated by the inner membrane. Such a gradient of H^+, in which the H^+ concentration outside of the mitochondrion is greater than that inside, contains potential energy (page 385). The chemiosmotic hypothesis further proposes that the H^+ ions ejected by electron transport flow back into the mitochondrial matrix through a specific H^+ channel or "pore" in the F_0F_1ATPase molecule, driven by the concentration gradient of H^+. The free energy released as H^+ flows back through the ATPase causes the coupled synthesis of ATP from ADP and phosphate.

We note that no high-energy chemical agent is required as a common intermediate to link electron transport and ATP synthesis. Instead, it is the gradient of H^+ generated across the membrane that serves as the carrier of energy from electron transport to ATP synthesis. We may also note that the chemiosmotic hypothesis requires that the membrane must be intact in the form of a completely closed vesicle, either in intact mitochondria or in submitochondrial vesicles made from the inner membrane (Figure 17-15), since an H^+ gradient across the membrane could not otherwise exist. We can also see that if a "leak" of H^+ across the membrane is induced by uncoupling agents, as described in Figure 17-17, the H^+ gradient would be discharged and energy-coupling would fail. Finally, it has been found that expulsion of H^+ from the mitochondria during electron transport and absorption of external H^+ by the ATP synthetase occur at rates compatible with the overall rate of oxidative phosphorylation by intact mitochondria.

Although the chemiosmotic principle accounts for most properties of the oxidative phosphorylation process, there are several aspects of it that are still rather puzzling and little understood. Among these is the mechanism by which the electron-transport chain "pumps" H^+ from the matrix to the outside medium (Box 17-1).

Electron-Transport Energy Is Useful for Other Purposes

Figure 17-20
Central role of the transmembrane gradient of H^+ in providing energy for different cell activities.

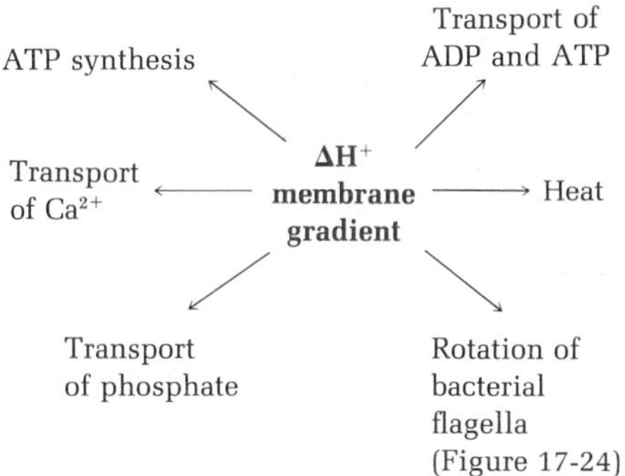

The primary role of electron transport in mitochondria is to furnish energy for the synthesis of ATP during oxidative phosphorylation. But electron-transport energy can be used for other biological purposes (Figure 17-20). For example, it can be used to generate heat. Human infants, other mammals born hairless, and some hibernating animals have a special type of fatty tissue, called *brown fat,* in the neck and upper back. Its function is to generate heat from fat oxidation. It is brown because it has profuse mitochondria, which in turn are rich in the red-brown cytochromes. The specialized mitochondria in this tissue (Figure 17-21) do not normally make ATP. Instead they dissipate the free energy of electron transport as heat in order to maintain the body temperature of young animals. Brown fat mitochondria have special H^+ pores in their inner membrane that allow the external H^+ pumped out by electron transport to flow back into the mitochondria, rather than through the F_0F_1ATPase. As a result, the free energy of electron transport is diverted from ATP synthesis into the production of heat.

Figure 17-21
Cross section of a mitochondrion of rat brown fat. It has many long, closely stacked cristae rich in cytochromes, and therefore it has a very high respiratory activity. Nearly all the energy generated by electron transport in brown fat mitochondria is converted into heat, because the H^+ pumped outward passes back into the matrix through open H^+ pores, rather than through the ATP synthetase.

Box 17-1 Many questions on the mechanism of oxidative phosphorylation remain to be answered

Although the chemiosmotic hypothesis is widely accepted as the central organizing principle for energy transfer from electron transport to ATP synthesis in mitochondria, bacteria, and chloroplasts (Chapter 23), it leaves many important problems still unsolved. Perhaps the most debated issue is the manner in which the process of electron transport in the inner membrane pumps H^+ from the mitochondrial matrix to the exterior. Mitchell has proposed a clever scheme (Figure 1). It capitalizes on the fact that reducing equivalents are transferred as H atoms by some of the electron carriers, such as ubiquinone, and as electrons by others, such as iron-sulfur centers and cytochromes. He proposed that the hydrogen-carrying and electron-carrying proteins alternate in the respiratory chain to form three "loops." In each loop two H atoms are carried outward through the membrane and deliver two H^+ to the outer medium; the corresponding pair of electrons is then carried back from the outer to the inner surface (Figure 1). Each pair of reducing equivalents passing through such a loop carries two H^+ from the matrix to the external medium. Each loop was proposed to provide the osmotic energy to make one molecule of ATP.

Attractive as this mechanism seems, it is not consistent with all experimental facts. For one thing, it requires a sequence and sidedness of the hydrogen- and electron-carrying centers of the chain that does not entirely agree with available evidence. Second, it can translocate only two H^+ per pair of electrons per loop, since each electron moving outward is accompanied by only one H^+. However, recent research indicates that at least three and possibly four H^+ are translocated per electron pair per loop, and that three or four H^+ ions pass back into the matrix to make one ATP.

Other questions have been raised. Do H^+ ions actually flow out of and back into mitochondria during normal oxidative phosphorylation? It appears that at least part of the H^+ movements may occur within or on the surface of the membrane rather than between the aqueous compartments separated by the membrane. Another question remains to be answered: Exactly how does the flow of H^+ through the complex ATP synthetase system generate the new covalent bond to the terminal phosphate group of ATP?

Thus, much more remains to be learned about the molecular components and the properties of the energy-transducing membranes of mitochondria, bacteria, and chloroplasts. Some day, after many more experiments and ideas have been tested, we shall know the answers to these questions, still largely hidden in the complex structure of the inner membrane. This is the way of research: hypotheses must be proposed starting from experimental observations and then tested, again and again, to make sure that every observed fact can be accounted for. In a sense biological research is never really finished. Often what appears to be a well-established and certain conclusion is later seen to be only an approximation, a step toward better understanding as new facts are uncovered and new insights arise. Research on the molecular logic of cells is truly an endless frontier.

The H^+ gradient generated by electron transport is also used to transport Ca^{2+} from the medium or cytosol into the inner matrix compartment of animal mitochondria (Figure 17-22). The inward transport of Ca^{2+} is counterbalanced by Ca^{2+} efflux, whose rate is regulated. In this way mitochondria help regulate the characteristically low concentrations of Ca^{2+} in cells (about 10^{-7} *M*). Free Ca^{2+} is an important intracellular messenger in controlling many cell functions. Increases in Ca^{2+} concentration initiate or promote such functions as muscle contraction (page 382), glycogen breakdown (page 732), and the oxidation of pyruvate (page 452); decreases in Ca^{2+} reverse these effects.

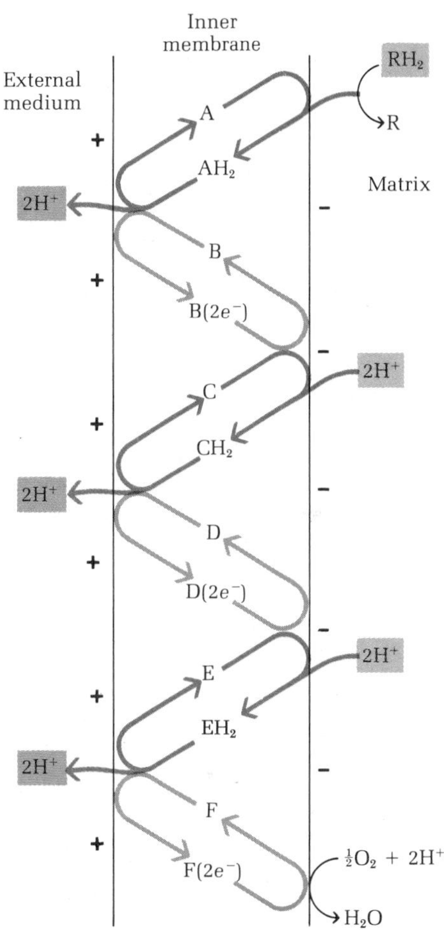

Figure 1
The loop mechanism of H^+ translocation in the chemiosmotic hypothesis. It is proposed that there are three H^+-conducting loops in the arrangement of the successive respiratory carrier molecules (A–F) in the chain, each loop translocating two H^+ from the matrix to the outside via a carrier (color) that transports reducing equivalents as H atoms. The two electrons remaining after discharge of two H^+ to the outside are then conducted back to the other side of the inner membrane by a carrier (gray) that transports reducing equivalents as electrons. Three such loops carry $3 \times 2 = 6$ H^+ from the matrix to the medium per pair of electrons passing from the substrate RH_2 to oxygen. It is assumed that the members of the respiratory chain are fixed in the membrane to give the necessary sidedness.

Bacteria and Chloroplasts Also Contain H^+-Transporting Electron-Transport Chains

Aerobic bacteria also carry out a process of electron transport from NAD-linked substrates to oxygen, to which is coupled the phosphorylation of cytosolic ADP to ATP. The dehydrogenases are located in the bacterial cytosol, but the electron carriers of the respiratory chain are located in the plasma membrane of the bacteria, as are the coupling mechanisms by which ATP is generated. Bacteria also can pump H^+ outward during electron transport. These similarities between bacteria and mitochon-

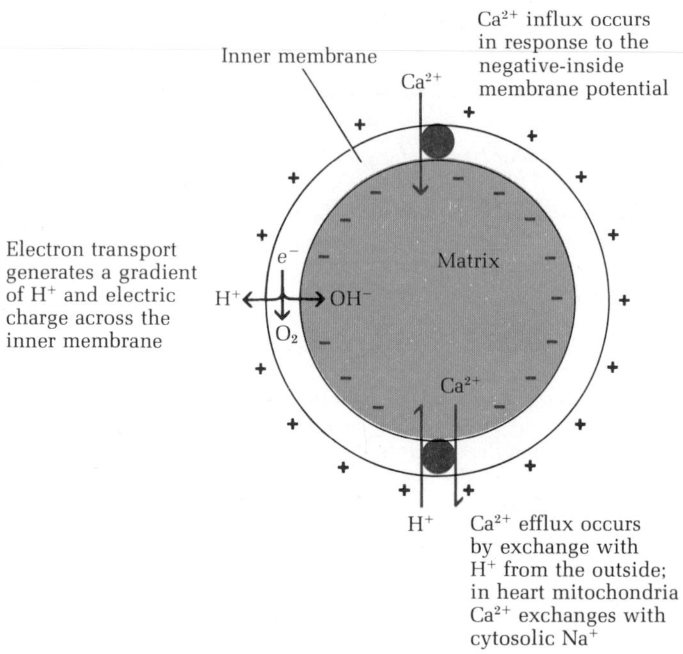

Figure 17-22
Energy-dependent Ca^{2+} influx and efflux in mitochondria. The inner membrane contains two transport systems for Ca^{2+}, which are probably proteins. One transports Ca^{2+} inward, the other transports Ca^{2+} outward. The external Ca^{2+} concentration, which is maintained at a very low level, is the result of a balance between the rates of Ca^{2+} influx and efflux.

dria with respect to the organization of their electron-transport chains (Figure 17-23) constitute additional evidence that mitochondria may have arisen from invading aerobic bacteria during the course of the evolution of eukaryotic cells (pages 26 and 880). Bacterial flagella (page 21) are given their rotary motion by "proton turbines" present in the bacterial cell membrane (Figure 17-24).

Chloroplasts of photosynthetic plant cells, which generate ATP from ADP and phosphate at the expense of captured light energy, also have a complex inner membrane containing electron-transport chains and ATP-synthesizing enzymes, as we shall see in Chapter 23. Both bacterial and chloroplast phosphorylation take place by mechanisms very similar to those occurring in mitochondria. Here we have another example of the molecular continuity of different kinds of living organisms.

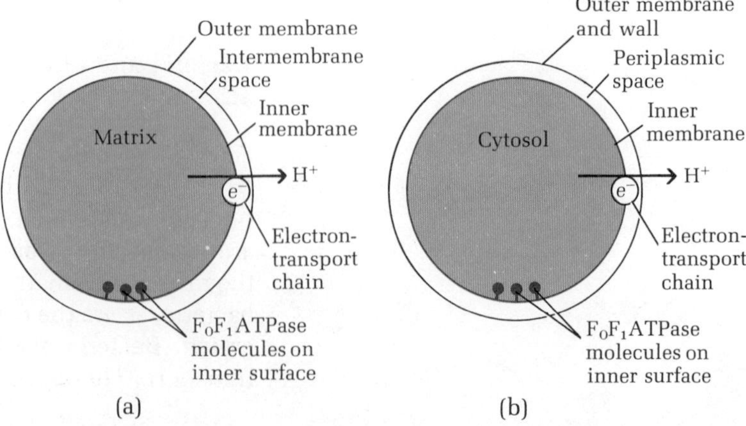

Figure 17-23
Similarities in the arrangement of electron-transport chains, H^+ pumping activity, and ATP synthetase molecules in (a) mitochondria and (b) aerobic bacteria.

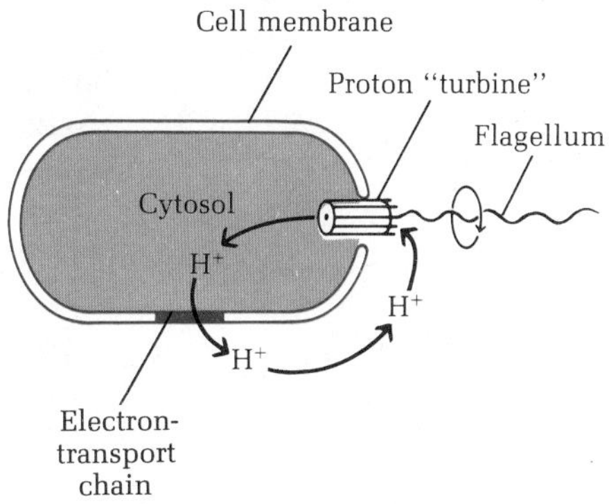

Figure 17-24
Rotation of bacterial flagella by "protonmotive force." Bacterial flagella are rigid structures, which differ from those of eukaryotic cells. They are given rotatory motion at their base in the cell membrane by a structure that has been called a "proton turbine." H^+ *ions ejected by electron transport flow back into the cell through the "turbine," causing rotation of the flagellum.*

The Inner Mitochondrial Membrane Contains Specific Transport Systems

The mitochondrial inner membrane is intrinsically impermeable not only to H^+, OH^-, and K^+ but also to many other ionic solutes. How, then, can the electrically charged ADP^{3-} and phosphate^{2-} formed in the cytosol on breakdown of cytosolic ATP enter the mitochondrial matrix, and how can the newly made ATP^{4-} leave again, since oxidative phosphorylation takes place within the inner matrix compartment?

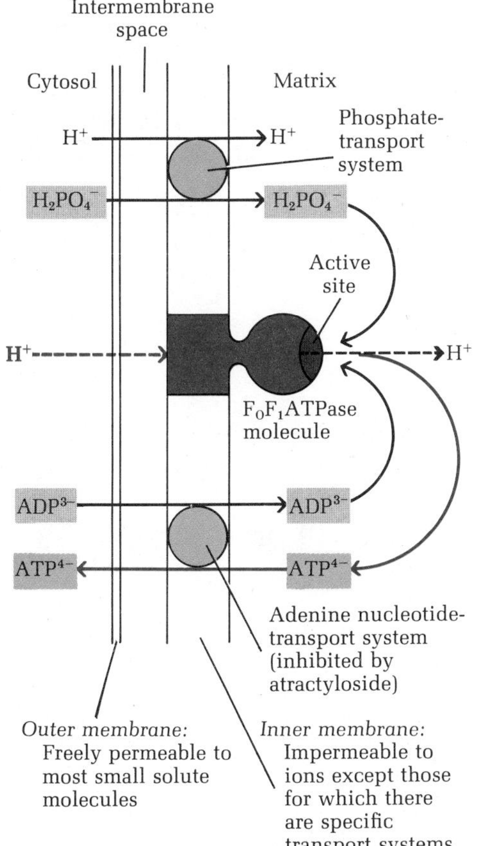

Figure 17-25
Inner-membrane transport systems bring phosphate and ADP into the matrix and allow the newly synthesized ATP to leave.

Two specific transport systems in the inner mitochondrial membrane (Figure 17-25) make these events possible. The first, called the *adenine nucleotide translocase,* transports one molecule of ADP^{3-} inward in exchange for one ATP^{4-} coming out. It consists of a specific protein that extends across the inner membrane and is capable of binding ADP^{3-} to a specific site on the outside surface. Through a conformational change in the translocase molecule the ADP^{3-} is transported inward in exchange for an ATP^{4-} molecule transported outward. The adenine nucleotide translocase system is specific. It will carry only ATP and ADP—not AMP or any other nucleotides, such as GDP or GTP. Adenine nucleotide translocase is very specifically inhibited by *atractyloside,* a toxic glycoside formed by a species of thistle that grows in some Mediterranean countries. For centuries it has been known that grazing cattle are poisoned when they ingest this plant at certain times of the year. The isolation of this toxic factor and its identification as a specific inhibitor of adenine nucleotide translocation was the result of some brilliant detective work by biochemists in Italy, France, Germany, and the United States. Obviously, if the transport of ADP into and ATP out of the mitochondria is inhibited, cytosolic ATP cannot be regenerated from ADP.

The second membrane transport system functioning in oxidative phosphorylation promotes cotransport of the $H_2PO_4^-$ ion from the outside with an H^+ ion into the matrix compartment (Figure 17-25). This carrier, called the *phosphate carrier* or *translocase,* is specific for phosphate; it is also inhibited by certain chemical agents. The combined action of the phosphate and the adenine nucleotide translocases allows external phosphate and ADP to enter the matrix and the resulting ATP to return to the cytosol, where most of the ATP-requiring cell activities take place.

The inner membrane of liver mitochondria also contains specific transport systems for pyruvate, to allow it to enter the matrix from the cytosol, where pyruvate is formed; for dicarboxylates, such as malate and succinate; and for the tricarboxylates citrate and isocitrate. Mitochondria also have transport systems specific for aspartate and glutamate, as we now see.

Shuttle Systems Are Required for Oxidation of Extramitochondrial NADH

The NADH dehydrogenase of the inner mitochondrial membrane can accept electrons only from NADH in the matrix. Since the inner membrane is not permeable to external or cytosolic NADH, how can the NADH generated by glycolysis, which takes place outside of the mitochondria, be reoxidized to NAD^+ by molecular oxygen via the respiratory chain?

Special *shuttle systems* carry reducing equivalents from cytosolic NADH into the mitochondria by an indirect route. The most active NADH shuttle, which functions in liver, kidney, and heart mitochondria, is called the *malate-aspartate shuttle*. Its principle is shown in Figure 17-26. The reducing equivalents of cytosolic NADH are first transferred to cytosolic oxaloacetate to yield malate by the action of cytosolic malate dehydrogenase. The malate formed, which carries the reducing equivalents donated by cytosolic NADH, passes through the inner membrane into the matrix on a dicarboxylate-transport

Figure 17-26
The malate-aspartate shuttle for transporting reducing equivalents from cytosolic NADH into the mitochondrial matrix. Malate, which carries the reducing equivalents, is transported across the inner membrane by a dicarboxylate-transport system (A). The reducing equivalents are then transferred to matrix NAD^+ by matrix malate dehydrogenase. The resulting matrix NADH is oxidized by the mitochondrial electron-transport chain, leading to oxidative phosphorylation. The oxaloacetate so formed cannot pass through the membrane from the matrix back into the cytosol, but is converted by transaminase into aspartate, which can pass via the amino acid–transport system C. The function of the other reactions and of transport system B is to regenerate oxaloacetate in the cytosol. Transport system B makes possible the exchange of glutamate for aspartate. The dicarboxylate-transport system A carries α-ketoglutarate out in exchange for malate passing inward. Transaminase (Chapter 19) catalyzes the reversible transfer of amino groups from glutamate to oxaloacetate:

$$\text{Glutamate} + \text{oxaloacetate} \rightleftharpoons \alpha\text{-ketoglutarate} + \text{aspartate}$$

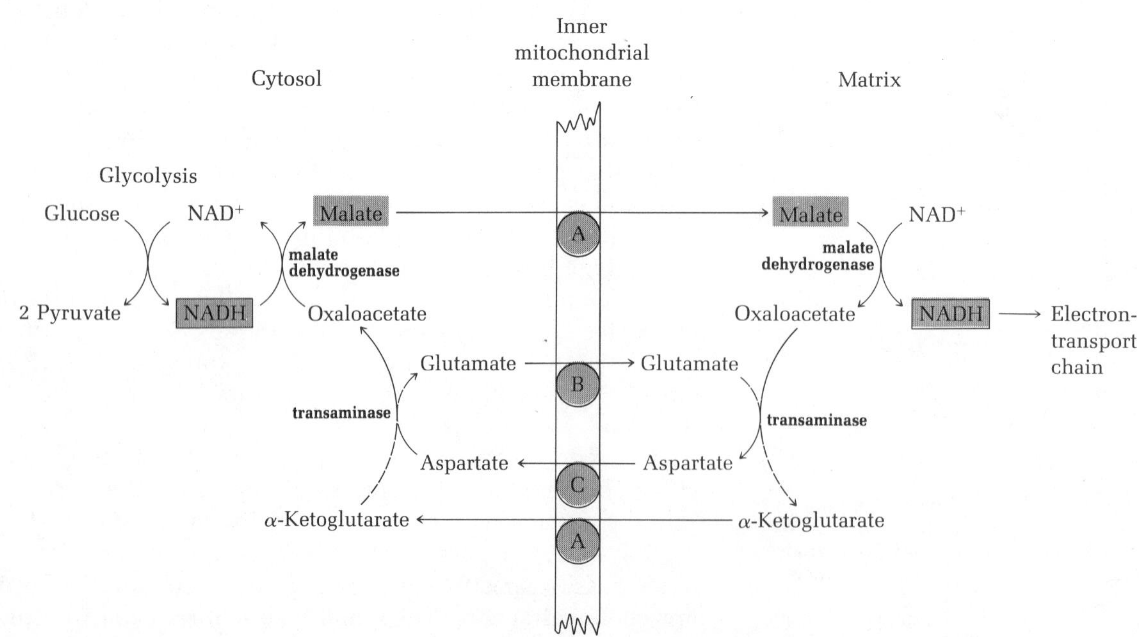

system. Once inside the mitochondrion, the reducing equivalents carried by malate are passed to matrix NAD^+ by the action of matrix malate dehydrogenase, reducing matrix NAD^+ to NADH, which can then pass its electrons directly to the respiratory chain in the inner membrane. Three molecules of ATP are generated as this pair of electrons passes to oxygen . The rest of the shuttle (see Figure 17-26) is concerned with the regeneration of cytosolic oxaloacetate to start another cycle of the shuttle.

In skeletal muscle and in the brain another type of NADH shuttle, the *glycerol phosphate shuttle,* occurs. It differs in its end result from the malate-aspartate shuttle in that it delivers the reducing equivalents from NADH into site 2, not site 1. Thus only two ATPs are formed by oxidation of NADH via the glycerol phosphate shuttle.

The Complete Oxidation of Glucose Leads to Synthesis of 38 ATPs

Now let us sum up the recovery of the chemical energy in the form of ATP as glucose is oxidized to CO_2 and H_2O in animal cells. First, glycolysis of one molecule of glucose under aerobic conditions yields two molecules of pyruvate, two of NADH, and two of ATP, the entire process taking place in the cytosol:

$$\text{Glucose} + 2P_i + 2\text{ADP} + 2\text{NAD}^+ \longrightarrow 2\text{ pyruvate} + 2\text{ATP} + 2\text{NADH} + 2\text{H}^+ + 2\text{H}_2\text{O}$$

Next, two pairs of electrons from the two cytosolic NADH yielded by glyceraldehyde phosphate dehydrogenase during glycolysis (page 408) are carried into the mitochondria by the malate-aspartate shuttle and then enter the electron-transport chain and flow to oxygen. This process will generate $2(3) = 6$ ATP, since two NADH are oxidized according to the equation

$$2\text{NADH} + 2\text{H}^+ + 6P_i + 6\text{ADP} + \text{O}_2 \longrightarrow 2\text{NAD}^+ + 6\text{ATP} + 8\text{H}_2\text{O}$$

(Of course, if the glycerol phosphate shuttle is used instead of the malate-aspartate shuttle, only two ATP will be generated per NADH.)

Next we have the complete equation for the dehydrogenation of two molecules of pyruvate to yield two acetyl-CoA and two CO_2 in the mitochondria, which results in the formation of two NADH and the subsequent transfer of its two electron pairs to oxygen via the chain, each pair furnishing three ATPs:

$$2\text{ Pyruvate} + 2\text{CoA} + 6P_i + 6\text{ADP} + \text{O}_2 \longrightarrow 2\text{ acetyl-CoA} + 2\text{CO}_2 + 6\text{ATP} + 8\text{H}_2\text{O}$$

Then we have the equation for the oxidation of two molecules of acetyl-CoA to CO_2 and H_2O via the citric acid cycle, together with the oxidative phosphorylations coupled to electron transport from isocitrate, α-ketoglutarate, and malate to oxygen, each of which yields three ATPs; the oxidation of succinate, which yields two ATPs; and the formation of two ATPs via GTP generated from succinyl-CoA (page 447):

$$2\ \text{Acetyl-CoA} + 24P_i + 24\text{ADP} + 4O_2 \longrightarrow 2\text{CoA-SH} + 4CO_2 + 24\text{ATP} + 26H_2O$$

If we now add these four equations, canceling out terms that appear on both sides, we shall get the overall equation of glycolysis plus respiration,

$$\text{Glucose} + 38P_i + 38\text{ADP} + 6O_2 \longrightarrow 6CO_2 + 38\text{ATP} + 44H_2O$$

Thus for each molecule of glucose undergoing complete oxidation to CO_2 and H_2O in the liver, kidney, and heart, in which the malate-aspartate shuttle functions, a maximum of 38 mol of ATP is generated. (When the glycerol phosphate shuttle is used, 36 ATP are formed per glucose oxidized.) The theoretical recovery of free energy in the complete oxidation of glucose is thus 38(7.3/686)(100) = 40 percent under standard (1.0 *M*) conditions. But in the intact cell the efficiency of this conversion is probably over 70 percent because the cellular concentrations of glucose, O_2, P_i, ADP, and ATP are unequal and much lower than the concentration of 1.0 *M* assumed in standard-free-energy calculations (page 377).

ATP Formation by Oxidative Phosphorylation Is Regulated by the Cell's Energy Needs

Now let us consider how ATP formation coupled to electron transport is regulated. The reaction equation for oxidation of NADH by mitochondria shows that phosphate and ADP, as well as oxygen, are required for electron transport to take place:

$$\text{NADH} + H^+ + 3P_i + 3\text{ADP} + \tfrac{1}{2}O_2 \longrightarrow \text{NAD}^+ + 3\text{ATP} + 4H_2O$$

As electron transport proceeds, phosphate and ADP will be removed from the cytosol and ATP will accumulate. The system finally reaches a point where nearly all the ADP will have been transformed into ATP by oxidative phosphorylation. Although the inorganic phosphate concentration will also be lowered, it is usually present in much higher concentrations in cells than ADP. Thus when the ADP in the cytosol is largely depleted, the rate of oxygen consumption by the mitochondria must necessarily slow down to only a fraction of its maximal rate, because it is limited by the low concentration of ADP (Figure 17-27). This is the "idling" or "resting" respiration. Respiration will increase to its maximum rate only when the concentration of ADP in the cytosol increases. This will occur when the rate of some energy-requiring process in the cell increases, thus

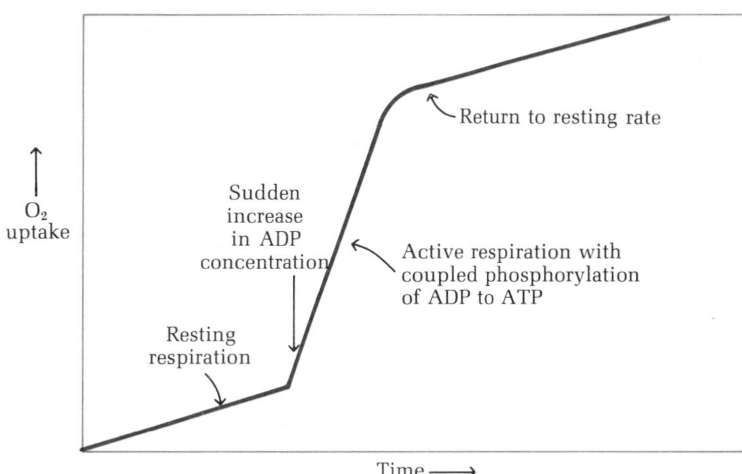

Figure 17-27
Acceptor control of respiration. During resting respiration nearly all the available ADP has been phosphorylated to ATP, and the system is idling. When ADP concentration is suddenly increased, as during abrupt initiation of muscle activity, the rate of oxygen consumption increases to the active rate. During this period the ADP is phosphorylated to ATP. When nearly all the ADP has been phosphorylated to ATP, respiration returns to the resting rate again.

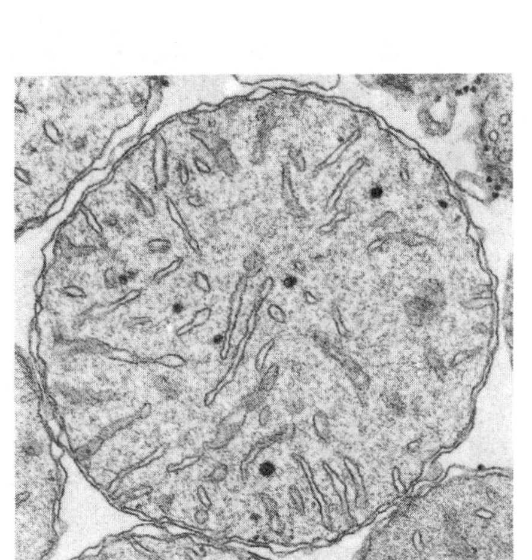

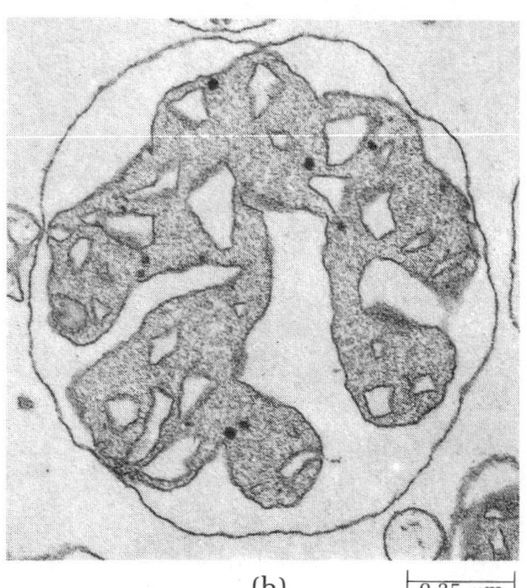

Figure 17-28
Electron micrographs of (a) *"idling" or fully energized mouse liver mitochondria and* (b) *actively respiring mitochondria generating ATP at a maximum rate. Both the inner membrane and the matrix undergo striking structural changes in the transitions between resting and active respiration.*

increasing the rate of breakdown of ATP to ADP and making ADP available to be phosphorylated during electron transport. The dependence of the rate of oxygen consumption on the concentration of ADP as the phosphate acceptor is called *acceptor control* of respiration. In various animal and human tissues the *acceptor-control ratio* of respiration, the ratio of the *maximal* rate of ADP-induced oxygen consumption to the *resting* rate is at least 10. In some humans acceptor control is faulty, probably because of a genetic defect. As a result their tissues show a high rate of oxygen consumption at all times.

One way of expressing the energy status of cells is given by the *mass-action ratio* of the ATP system, the quotient

$$\frac{[\text{ATP}]}{[\text{ADP}][\text{P}_i]}$$

where the brackets denote molar concentrations. Normally this ratio is very high, so that the ATP–ADP system is almost fully phosphorylated. Under these conditions the concentration of ADP is very low and is insufficient to give a maximal rate of respiration. In this case the rate of ATP synthesis is sufficient to meet current needs. But if some ATP-requiring cellular activity suddenly increases in rate, some of the cell ATP will be split into ADP and phosphate, lowering the $[\text{ATP}]/[\text{ADP}][\text{P}_i]$ ratio. The increased concentration of ADP will now automatically increase the rate of electron transport and oxidative phosphorylation, causing ATP to be regenerated from the ADP. This will continue until the $[\text{ATP}]/[\text{ADP}][\text{P}_i]$ ratio is brought back up to its normal high level; at this point respiration will slow down again. Normally, the rate of oxidation of cell fuels is regulated with such sensitivity and precision that the $[\text{ATP}]/[\text{ADP}][\text{P}_i]$ ratio fluctuates only slightly in most tissues, even during varying energy demands. ATP is formed only as fast as it is used in energy-requiring cell activities.

The Energy Charge Is Another Index of Cellular Energy Status

We have seen that important regulatory enzymes of glycolysis, the citric acid cycle, and oxidative phosphorylation respond to ATP and ADP as modulators; ATP is an inhibitory modulator and ADP is generally stimulatory in the catabolism of carbohydrates. Thus any change in the [ATP]/[ADP][P_i] mass-action ratio, which is normally very high, may cause a corresponding change in the activity of certain regulatory enzymes in the central catabolic pathways. However, some regulatory enzymes also respond to AMP as a stimulatory modulator. To account for the participation of AMP, as well as ATP and ADP, in metabolic regulation, an alternative expression, the *energy charge*, has been proposed by Daniel Atkinson for the energy status of cells. It is a measure of the extent to which the *total* adenine nucleotide system, the sum of ATP, ADP, and AMP, is "filled" with high-energy phosphate groups:

$$\text{Energy charge} = \frac{[\text{ATP}] + \frac{1}{2}[\text{ADP}]}{[\text{ATP}] + [\text{ADP}] + [\text{AMP}]}$$

The energy charge is 1 if the total adenine nucleotide pool is fully phosphorylated to ATP and 0 if the adenine nucleotides are "empty" and present only as AMP. Normally the energy charge of cells is about 0.9, meaning that the adenylate system is almost completely charged. Under some conditions the energy-charge quotient is better able than the mass-action ratio to account for allosteric signaling between energy-yielding and energy-requiring metabolic processes. This is a controversial question, however, and perhaps there is no single all-encompassing index of energy status that applies to the regulation of all metabolic systems.

Glycolysis, the Citric Acid Cycle, and Oxidative Phosphorylation Have Interlocking and Concerted Regulatory Mechanisms

In carbohydrate catabolism there are three energy-yielding stages: glycolysis (Chapter 15), the citric acid cycle (Chapter 16), and oxidative phosphorylation. Each is so regulated by its own set of controls that it proceeds at a rate just sufficient to satisfy the minute-to-minute needs of the cell for its products. More than this, these three stages are coordinated with each other so that they function together in an economical and self-regulating manner, like a smoothly running piece of machinery, to produce ATP, the ultimate product of energy-yielding catabolism, and specific intermediates, e.g., pyruvate and citrate, that are required as precursors in the biosynthesis of other cell components. The integration of these three stages is made possible by interlocking regulatory mechanisms. In Figure 17-29, we see that the relative concentrations of ATP and ADP (or the

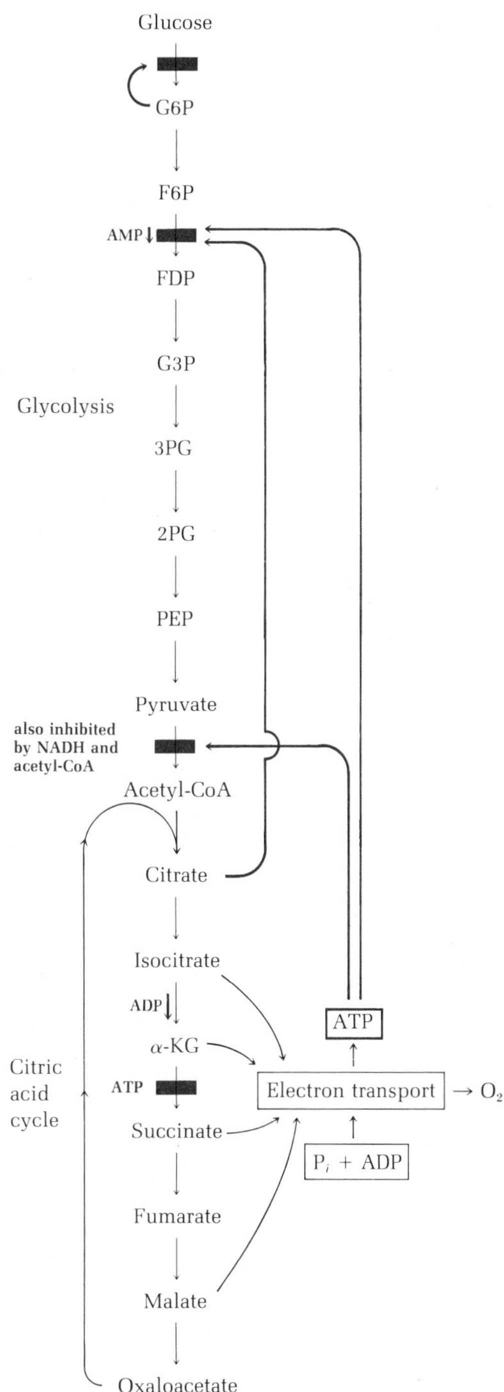

Figure 17-29
Interlocking regulation of glycolysis, pyruvate oxidation, the citric acid cycle, and oxidative phosphorylation by the relative concentrations of ATP, ADP, and AMP. The regulatory controls, both inhibitory and stimulatory, are shown by colored bars and arrows. When ATP concentration is high, and ADP and AMP correspondingly low, the rates of glycolysis, pyruvate oxidation, the citric acid cycle, and oxidative phosphorylation are at a minimum. When there is a large increase in the rate of ATP utilization by the cell, with increased formation of ADP, AMP, and P_i, all four processes are accelerated. Interlocking of glycolysis and the citric acid cycle by citrate, also shown in the diagram, supplements the action of the adenylate system. In addition, increased levels of NADH and acetyl-CoA also inhibit the oxidation of pyruvate to acetyl-CoA. G6P designates glucose 6-phosphate; F6P, fructose 6-phosphate; FDP, fructose diphosphate; G3P, glyceraldehyde 3-phosphate; 3PG, 3-phosphoglycerate; 2PG, 2-phosphoglycerate; PEP, phosphoenolpyruvate; α-KG, α-ketoglutarate.

mass-action ratio of the ATP system) control not only the rate of electron transport and oxidative phosphorylation but also the rates of the citric acid cycle, pyruvate oxidation, and glycolysis. Whenever there is an increased drain on ATP, causing ATP concentration to fall and ADP and P_i to rise, the rate of electron transport and oxidative phosphorylation will immediately increase. Simultaneously, the rate of pyruvate oxidation via the citric acid cycle will increase, thus increasing the flow of electrons into the respiratory chain. These events will in turn evoke an increase in the rate of glycolysis, thus providing an increased rate of pyruvate formation. When the $[ATP]/[ADP][P_i]$ ratio is restored to its normal high level, electron transport and oxidative phosphorylation will slow down again, because the concentration of ADP will be brought back to the low resting level. Similarly, the citric acid cycle and the glycolytic sequence will also slow down, because of the interlocking action of ATP as allosteric inhibitor of glycolysis and pyruvate oxidation.

The regulatory enzymes of glycolysis and the citric acid cycle also function in a *concerted* manner. Whenever ATP, produced by oxidative phosphorylation, and citrate, the first intermediate of the citric acid cycle, increase to higher than normal levels, they produce *concerted* allosteric inhibition of phosphofructokinase (Figure 17-29), the two together being more inhibitory than the sum of their individual effects. Thus a network of interlocking and concerted regulatory mechanisms controls glycolysis so that pyruvate is produced only as fast as it is needed by the citric acid cycle, the donor of electrons for the oxidative phosphorylation process.

In cancer cells, this interlocking coordination appears to be defective: Glycolysis proceeds at a very much higher rate than required by the citric acid cycle. As a result, aerobic cancer cells use far more blood glucose than normal cells, but cannot oxidize the pyruvate formed by glycolysis and therefore reduce most of it to lactate, which leaves the cells and is carried away by the blood.

Cells Contain Other Oxygen-Using Enzymes

Over 90 percent of the oxygen consumed by most cells is reduced by the cytochrome oxidase of mitochondria. However, some tissues contain other types of oxygen-consuming enzymes carrying out specialized oxidative reactions in which oxygen atoms are directly incorporated into the substrate molecule, forming a new hydroxyl or carboxyl group, for example. These enzymes are called *oxygenases*. Although these specialized oxidative reactions consume only a small fraction of the total oxygen uptake of cells, they are extremely important to the organism.

There are two classes of oxygenases, *dioxygenases* and *monooxygenases*. Dioxygenases catalyze reactions in which *both* the oxygen atoms of O_2 are incorporated into the organic substrate molecule. An example of a dioxygenase is *pyrocatechase,* which catalyzes the opening of the ring of *catechol* by reaction with O_2:

$$\text{Catechol (C}_6\text{H}_4\text{(OH)}_2\text{)} + O{-}O \longrightarrow HOOC{-}CH{=}CH{-}CH{=}CH{-}COOH$$

Catechol

When this reaction occurs in the presence of molecular oxygen labeled with ^{18}O isotope, the isotopic oxygen atoms (in color above) are found in the two carboxyl groups of the product.

Monooxygenases, which are more abundant and more complex in their action, catalyze reactions in which only one of the two oxygen atoms is incorporated into the organic substrate, the other being reduced to H_2O. Monooxygenases require *two* substrates to serve as reductants of the two oxygen atoms of O_2. *The main substrate* accepts one of the two oxygens, and a *cosubstrate* furnishes H atoms to reduce the other oxygen to H_2O. The general reaction equation for monooxygenases is

$$AH + BH_2 + O{-}O \longrightarrow A{-}OH + B + H_2O$$

AH is the main substrate that accepts an oxygen atom, and BH_2 provides H to reduce the other oxygen to H_2O. Because most monooxygenases catalyze reactions in which the main substrate becomes hydroxylated, they are also called *hydroxylases.* They are sometimes called *mixed-function oxidases* since they oxidize two different substrates simultaneously.

There are different classes of monooxygenases, depending upon the nature of the cosubstrate that furnishes the two H atoms for the formation of H_2O. Some use reduced flavin nucleotides ($FMNH_2$ or $FADH_2$), others use NADH or NADPH, and still others use α-ketoglutarate as the cosubstrate. An important example (Chapter 19) is the monooxygenase that hydroxylates the phenyl ring of phenylalanine to give tyrosine. This enzyme is defective in the human genetic disease *phenylketonuria.*

The most numerous and most complex monooxygenation reactions are those employing a type of heme protein called *cytochrome P-450.* This type of cytochrome is usually present in the endoplasmic reticulum rather than the mitochondria. Like mitochondrial cytochrome oxidase, cytochrome P-450 can react with oxygen and with carbon monoxide, but it can be differentiated from cytochrome oxidase because the carbon monoxide complex of its reduced form absorbs light strongly at 450 nm.

Cytochrome P-450 catalyzes hydroxylation reactions in which an organic substrate RH is hydroxylated to R—OH at the expense of one oxygen atom of O_2, the other being reduced by reducing equivalents furnished by NADH or NADPH but usually passed to P-450 by an iron-sulfur protein. Figure 17-30

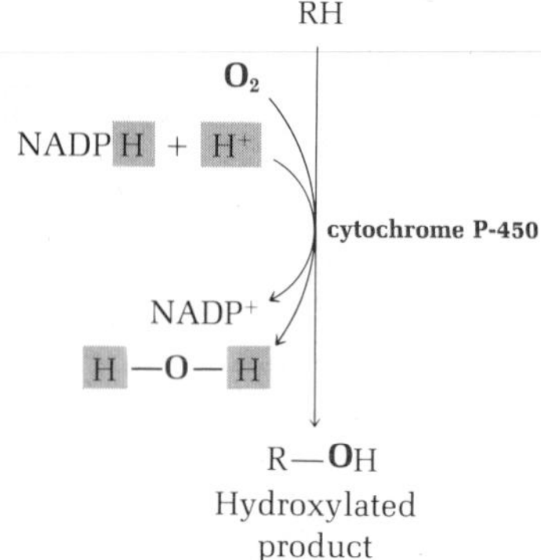

Figure 17-30
Hydroxylation of a lipid-soluble drug RH by cytochrome P-450, which functions as a monooxygenase. The product, R—OH, has increased water solubility and thus is more readily excreted. The cosubstrate is NADPH + H^+, which provides the H atoms (shaded) to reduce the other oxygen atom to H_2O.

shows a simplified outline of the action of cytochrome P-450, which has many complex intermediate steps not yet fully understood. In the adrenal cortex cytochrome P-450 participates in the hydroxylation of steroids to yield the adrenal cortical hormones, for example. Cytochrome P-450 is also important in the hydroxylation of many different drugs and other substances that are foreign to the body, particularly if they are relatively insoluble. Hydroxylation of such foreign compounds makes them more soluble in water and is an important step in their detoxification and excretion (Chapter 24). Cytochrome P-450 occurs in different forms with different specificities for their substrates.

Summary

In electron-transferring (oxidation-reduction) reactions the tendency for the electron donor (reductant) to donate electrons is given by the standard reduction potential E_0'. Redox systems with a relatively electronegative E_0' tend to donate electrons to redox systems with a more positive E_0'. The standard-free-energy change of electron-transfer reactions is calculated from the relationship $\Delta G^{\circ\prime} = -n\mathcal{F}\ \Delta E_0'$. In mitochondria H atoms removed from substrates by the action of dehydrogenases donate their electrons to the electron-transport chain, which transfers them to molecular oxygen, reducing it to H_2O. The energy released in electron transport is used for the oxidative phosphorylation of ADP to ATP, which takes place in the inner mitochondrial membrane.

Reducing equivalents from all NAD-linked dehydrogenations are transferred to mitochondrial NADH dehydrogenase, which contains FMN as its prosthetic group. They are then passed via a series of iron-sulfur centers to ubiquinone, which transfers the electrons to cytochrome *b*. The electrons are then passed to, successively, cytochrome c_1 and *c*, and then to cytochrome aa_3 (cytochrome oxidase), which contains copper. Cytochrome oxidase passes electrons to O_2, which requires four electrons and four H^+ to reduce O_2 completely to two H_2O. Electron transport can be inhibited at characteristic points by rotenone, antimycin A, and cyanide. The large decrease in free energy of electron transport is harnessed by three energy-conserving sites in the respiratory chain, each capable of generating ATP from ADP and P_i. Oxidative phosphorylation may be uncoupled from electron transport by uncoupling agents or by ionophores such as valinomycin. The inner membrane must be intact and impermeable to H^+ and other ions for oxidative phosphorylation to occur. H^+ ejection from mitochondria accompanies electron transport. The chemiosmotic hypothesis, one of three different hypotheses for the mechanism of oxidative phosphorylation, proposes that an acid-outside gradient of H^+ across the mitochondrial inner membrane is generated by electron transport. This gradient is proposed to drive the synthesis of ATP as H^+ passes through the F_0F_1ATPase molecule into the matrix. The mitochondrial inner membrane contains specific transport systems for adenine nucleotides, phosphate, and various metabolites. The rate of electron transport is low when ADP is low but increases when ADP increases in concentration during ATP-using cell processes. The rates of glycolysis, the citric acid cycle, and oxidative phosphorylation are coordinated with each other through interlocking regulatory mechanisms that respond to the [ATP]/[ADP][P_i] ratio and to other central metabolites that reflect the energy status of cells.

Cells also promote oxygenation reactions in which oxygen atoms are incorporated into organic molecules, particularly relatively hydrophobic foreign molecules and drugs, to yield hydroxylated or carboxylated products.

References

For Orientation

DICKERSON, R. E.: "Cytochrome *c* and the Evolution of Energy Metabolism," *Sci. Am.*, **242:**137–153, March (1980).

HINKLE, P., and R. E. MCCARTY: "How Cells Make ATP," *Sci. Am.*, **238:** 104–123, March (1978).

WHITTAKER, D. A., and S. M. DANKS: *Mitochondria: Structure, Function and Assembly*, Longman, London, 1978. A short, up-to-date, clearly written, and well-illustrated account. Excellent for further orientation.

History and Background

KEILIN, D.: *The History of Cell Respiration and Cytochromes*, Cambridge University Press, London, 1966.

LEHNINGER, A. L.: *The Mitochondrion: Molecular Basis of Structure and Function*, Benjamin, New York, 1965. The history and function of mitochondria as power plants.

RACKER, E.: *A New Look at Mechanisms in Bioenergetics*, Academic, New York, 1976. Personal account, with amusing asides, of one investigator's trail of research.

Mechanism of Oxidative Phosphorylation

BOYER, P. D., B. CHANCE, L. ERNSTER, P. MITCHELL, E. RACKER, and E. C. SLATER: "Oxidative Phosphorylation and Photophosphorylation," *Ann. Rev. Biochem.*, **46:**955–1026 (1977). Some leading investigators of oxidative phosphorylation present different views on its mechanism.

MITCHELL, P.: "Keilin's Respiratory Chain Concept and Its Chemiosmotic Consequences," *Science*, **206**:1148–1159 (1979). The author's Nobel lecture outlines the evolution of the chemiosmotic hypothesis.

TEDESCHI, H.: *Mitochondria: Structure, Biogenesis, and Transducing Functions*, Springer-Verlag, New York, 1976. A more detailed reference book.

Special Topics

LEHNINGER, A. L.: "Mitochondria and Biological Mineralization Processes: an Exploration," in E. Quagliariello, F. Palmieri, and T. Singer (eds.), *Horizons in Biochemistry and Biophysics*, vol. 4, pp. 1–30, Addison-Wesley, Reading, Mass., 1977. Because they can accumulate Ca^{2+} as well as phosphate, mitochondria may be the site in which the first stages of biological calcification occur.

LUFT, R., D. IRKOS, G. PALMIERI, L. ERNSTER, and B. AFZELIUS: "A Case of Severe Hypermetabolism of Non-Thyroid Origin with a Defect in Mitochondrial Respiratory Control: A Correlated Clinical, Biochemical, and Morphological Study," *J. Clin. Invest.*, **41**:1776–1804 (1962). The first case study of a human being with a defect in the regulation of electron transport.

Problems

1. *Oxidation-Reduction Reactions* The NADH dehydrogenase complex of the mitochondrial electron-transport chain promotes the following series of oxidation-reduction reactions, in which Fe^{3+} and Fe^{2+} represent the iron in iron-sulfur centers, Q is ubiquinone, QH_2 is ubiquinol, and E is the enzyme protein.

$$\text{(1) NADH} + \text{H}^+ + \text{E—FMN} \rightarrow \text{NAD}^+ + \text{E—FMNH}_2$$

$$\text{(2) E—FMNH}_2 + 2\text{Fe}^{3+} \rightarrow \text{E—FMN} + 2\text{Fe}^{2+} + 2\text{H}^+$$

$$\text{(3) } 2\text{Fe}^{2+} + 2\text{H}^+ + \text{Q} \rightarrow 2\text{Fe}^{3+} + \text{QH}_2$$

$$\textit{Net:}\ \text{NADH} + \text{H}^+ + \text{Q} \rightarrow \text{NAD}^+ + \text{QH}_2$$

For each of the three reactions catalyzed by the NADH dehydrogenase complex, identify (a) the electron donor, (b) the electron acceptor, (c) the conjugate redox pair, (d) the reducing agent, and (e) the oxidizing agent.

2. *Standard Reduction Potentials* The standard reduction potential of any redox couple is defined for the half-cell reaction:

$$\text{Oxidizing agent} + (n)\ \text{electrons} = \text{reducing agent}$$

The standard reduction potentials of the NAD^+/NADH and pyruvate/lactate conjugate pairs are −0.32 and −0.19 V, respectively.
(a) Which conjugate pair has the greater tendency to lose electrons? Explain.
(b) Which is the stronger oxidizing agent? Explain.
(c) If we begin with 1 *M* concentrations of each reactant and product at pH 7, in which direction will the following reaction proceed?

$$\text{Pyruvate} + \text{NADH} + \text{H}^+ = \text{lactate} + \text{NAD}^+$$

(d) What is the standard-free-energy change $\Delta G^{\circ\prime}$ at 25°C for this reaction?
(e) What is the equilibrium constant for this reaction at 25°C?

3. *Sequence of Electron Carriers in a Plant Electron-Transport Chain* Research on an electron-transport chain in the cells of spinach leaves has revealed the presence of a series of reversible electron-carrier molecules having the standard reduction potentials shown.

Reduced form	Oxidized form	E_0', V
Cytochrome b_6 (Fe^{2+})	Cytochrome b_6 (Fe^{3+})	−0.06
Cytochrome f (Fe^{2+})	Cytochrome f (Fe^{3+})	+0.365
Ferredoxin (reduced)	Ferredoxin (oxidized)	−0.432
Ferredoxin-reducing substrate (reduced)	FRS (oxidized)	−0.60
Plastocyanin (reduced)	Plastocyanin (oxidized)	+0.40

Predict the probable sequence of these electron carriers in this electron-transport chain, based on their standard reduction potentials. Draw an energy diagram analogous to Figure 17-4. Which steps in this chain are *unlikely* to yield sufficient free energy, under standard conditions, to generate a molecule of ATP per pair of electrons transferred?

4. *Balance Sheet for ATP Synthesis During Substrate Oxidation* Calculate the number of ATP molecules that are generated during the complete oxidative degradation of one molecule of each of the following to CO_2 and H_2O:
(a) Fructose 6-phosphate
(b) Acetyl-CoA
(c) Glyceraldehyde 3-phosphate
(d) Sucrose

5. *Energy Span of the Respiratory Chain* Electron transport in the mitochondrial respiratory chain may be represented by the net reaction equation:

$$NADH + H^+ + \tfrac{1}{2}O_2 = H_2O + NAD^+$$

(a) Calculate the value of $\Delta E_0'$ for the net reaction of mitochondrial electron transport.
(b) Calculate the standard-free-energy change $\Delta G^{\circ\prime}$ for this reaction.
(c) How many ATP molecules can *theoretically* be generated by this reaction if the standard free energy of ATP synthesis is +7.3 kcal/mol?

6. *Use of FAD Rather Than NAD⁺ in the Oxidation of Succinate* All the dehydrogenation steps in glycolysis and the citric acid cycle utilize NAD^+ ($E_0' = -0.32$ V) as the electron acceptor except succinate dehydrogenase, which utilizes covalently bound FAD ($E_0' = +0.05$ V). Why is FAD a more appropriate electron acceptor than NAD^+ in the dehydrogenation of succinate? Give a possible explanation based on a comparison of the E_0' values of the succinate-fumarate system, the NAD^+/NADH couple, and the FAD/$FADH_2$ couple.

7. *Degree of Reduction of Electron Carriers in the Respiratory Chain* The degree of reduction of each electron carrier in the respiratory chain is determined by the conditions existing in the mitochondria. For example, when the supply of NADH and O_2 is abundant, the steady-state degree of reduction of the carriers decreases as electrons pass from the substrate to oxygen. When electron transport is blocked, the carriers before the block become more reduced while those beyond the block become more oxidized, as illustrated by the hydraulic analogy of respiration in Figure 17-14. For each of the conditions below, draw the appropriate hydraulic analogy of the respiratory chain.
   (a) Abundant supply of NADH and O_2 but cyanide added
   (b) Abundant supply of NADH but O_2 exhausted
   (c) Abundant supply of O_2 but NADH exhausted
   (d) Abundant supply of NADH and O_2

8. *The Effect of Rotenone and Antimycin A on Electron Transport* Rotenone, a toxic substance obtained from a plant species, strongly inhibits NADH dehydrogenase of mitochondria. Antimycin A, a toxic antibiotic, strongly inhibits the oxidation of ubiquinol.
   (a) Explain why rotenone ingestion is lethal to some insects and fish species.
   (b) Explain why antimycin A is a poison in animal tissues.
   (c) Assuming that rotenone and antimycin A are equally effective in blocking their respective sites in the electron-transport chain, which would be a more potent poison? Explain.

9. *Uncouplers of Oxidative Phosphorylation* In normal mitochondria the rate of electron transport is tightly coupled to the demand for ATP. Thus when the rate of utilization of ATP is relatively low, the rate of electron transport is also low. Conversely, when ATP is demanded at a high rate, electron transport is rapid. Under such conditions of tight coupling, the P/O ratio, the number of ATPs produced per atom of oxygen consumed when NADH is the electron donor, is close to 3.
   (a) Predict the effect of a relatively low and a relatively high concentration of an uncoupling agent on the rate of electron transport and the P/O ratio.
   (b) The ingestion of uncouplers causes profuse sweating and an increase in body temperature. Explain this phenomenon in molecular terms. What happens to the P/O ratio in the presence of uncouplers?
   (c) The uncoupler 2,4-dinitrophenol was once prescribed as a weight-reducing drug. How can this agent, in principle, serve as a reducing aid? Such uncoupling agents were no longer used medically after some deaths occurred following their use. Why can the ingestion of uncouplers lead to death?
   (d) Some pathogenic microorganisms produce a soluble toxin that leads to a very high fever. Suggest a mode of action for the toxin.

10. *Mode of Action of Dicyclohexylcarbodiimide (DCCD)* When DCCD is added to a suspension of tightly coupled, actively respiring mitochondria, the rate of electron transport measured by O_2 consumption and the rate of ATP production measured by the formation of ATP dramatically decrease. If a solution of 2,4-dinitrophenol is now added to the inhibited mitochondrial prepa-

ration, the O_2 consumption goes back up to normal but the production of ATP remains inhibited.

(a) What process in electron transport or oxidative phosphorylation is affected by DCCD?

(b) Why does DCCD affect the O_2 consumption of mitochondria? Explain the effect of 2,4-dinitrophenol on the inhibited mitochondrial preparation.

(c) Which of the following inhibitors does DCCD resemble most in its action: antimycin A, rotenone, oligomycin, arsenate?

11. *Oxidative Phosphorylation in Inverted Submitochondrial Vesicles* The chemiosmotic hypothesis proposes that electron transport in intact mitochondria causes "pumping" of H^+ from the mitochondrial matrix compartment to the outside, to generate a pH gradient across the membrane. The H^+ gradient so formed is energy-rich and causes H^+ to be driven back into the matrix through the F_0F_1ATPase, thus generating ATP from ADP and P_i. It has been shown that inverted inner-membrane vesicles, with the F_0F_1ATPase knobs on the outside surface (see Figure 17-15), are also capable of oxidative phosphorylation.

(a) Draw a diagram showing the direction of H^+ pumping during electron transport in submitochondrial vesicles.

(b) In the diagram indicate the direction of H^+ flow through the F_0F_1ATPase during ATP synthesis.

(c) Predict the effect of oligomycin and atractyloside on electron transport and ATP synthesis in such submitochondrial vesicles.

12. *Brown Fat Mitochondria* Brown fat, a type of adipose tissue present in the neck and upper back of newborn infants but largely absent in adults, contains very numerous mitochondria, which give this tissue a brown appearance. Brown fat is also found extensively in some hibernating or cold-adapted animals. The mitochondria in brown fat are found to produce less than one ATP per atom of oxygen consumed during NADH oxidation, compared to the normal three ATPs per O atom in liver mitochondria.

(a) Suggest a physiological function for the low P/O ratio in the brown fat of newborn infants.

(b) Suggest possible mechanisms for the low P/O ratio characteristic of brown fat mitochondria.

13. *The Dicarboxylate-Transport System of Mitochondria* The dicarboxylate-transport system of the inner mitochondrial membrane, which promotes the transport of malate and α-ketoglutarate across the membrane, is inhibited by *n*-butylmalonate. Suppose *n*-butylmalonate is added to a suspension of aerobic kidney cells using glucose exclusively as fuel. Predict the effect of *n*-butylmalonate on

(a) Glycolysis

(b) Oxygen consumption

(c) Lactate formation

(d) ATP synthesis

14. *The Pasteur Effect* When oxygen was admitted to an anaerobic suspension of cells utilizing glucose at a high rate, the rate of glucose consumption declined dramatically as the added oxygen was consumed. In addition, the accumulation of lactate ceased. This effect, first observed by Louis Pasteur in the 1860s, is character-

istic of most cells capable of both aerobic and anaerobic utilization of glucose and was thus named after Pasteur.

(a) Why did the accumulation of lactate cease after oxygen was admitted?

(b) Why does the presence of oxygen decrease the rate of glucose utilization?

(c) How does the onset of oxygen consumption slow down the rate of glucose consumption? Explain in terms of specific enzymes.

15. *Changes in the Energy Charge* During a change in the physiological activity of skeletal muscle cells the energy charge of the cells suddenly decreased from its normal value of 0.89 to about 0.70, and then gradually returned to the normal level.

(a) What kind of change in activity could be responsible for the sudden decrease in energy charge? Explain.

(b) What effect did this sudden change have on the rate of glycolysis and respiration?

(c) State precisely how the energy charge can influence glycolysis and respiration.

16. *How Many H^+ Ions in a Mitochondrion?* The chemiosmotic hypothesis proposes that electron transport functions to translocate H^+ from the mitochondrial matrix to the external medium to establish a pH gradient across the inner membrane, the outside more acidic than the inside. The tendency of H^+ to diffuse from the outside into the matrix, where H^+ concentration is lower, is proposed to be the driving force for ATP synthesis via the F_0F_1ATPase. During oxidative phosphorylation by a suspension of mitochondria in a medium of pH 7.4, it has been found that the internal pH of the matrix is 7.7.

(a) Calculate the molar concentrations of H^+ in the external medium and the matrix under these conditions.

(b) What is the outside/inside ratio of H^+ concentrations? Comment on the energy inherent in this concentration. (*Hint:* See Chapter 14.)

(c) Calculate the number of H^+ ions in a respiring liver mitochondrion, assuming its inner matrix compartment is a sphere of diameter 1.5 μm.

(d) From these data would you think the pH gradient alone is sufficiently great to generate ATP?

(e) If not, can you suggest how the necessary energy for synthesis of ATP arises?

CHAPTER 18

The Oxidation of Fatty Acids in Animal Tissues

The triacylglycerols play an extremely important role in furnishing energy in animals. They have the highest energy content of the major nutrients (over 9 kcal/g), they are deposited in cells as almost pure droplets of fat, and they can be stored in very large amounts in the adipose tissues. On the average, 40 percent or more of the daily energy requirement of people in highly developed countries is met by dietary triacylglycerols. They provide more than half the energy requirements of some organs, particularly the liver, heart, and resting skeletal muscle. Moreover, stored triacylglycerols are virtually the sole source of energy in hibernating animals and migrating birds. We shall also see (Chapter 21) that carbohydrates consumed in excess of the very limited capacity of the body to store glycogen are converted into triacylglycerols for long-term storage.

About 95 percent of the biologically available energy of triacylglycerols resides in their three long-chain fatty acid components; only 5 percent is contributed by the glycerol portion. In this chapter we shall examine the metabolic pathways and energy yield when these energy-rich fatty acids are oxidized to carbon dioxide and water in animal tissues. We shall see that the oxidation of fatty acids shares a final common pathway with the oxidation of carbohydrate, namely, the citric acid cycle.

Fatty Acids Are Activated and Oxidized in Mitochondria

Because nearly all the fatty acids of animal tissues have an even number of carbon atoms, it was long suspected that fatty acids are synthesized and degraded by addition or subtraction of 2-carbon fragments. Classical experiments by Franz Knoop in Germany in the early years of this century confirmed this view and led him to conclude that fatty acids are oxidized by successive loss of 2-carbon fragments in a scheme of *β oxidation*, in which the β carbon atom of the fatty acid is oxidized to yield a β-keto acid, which then undergoes cleavage to form a 2-carbon

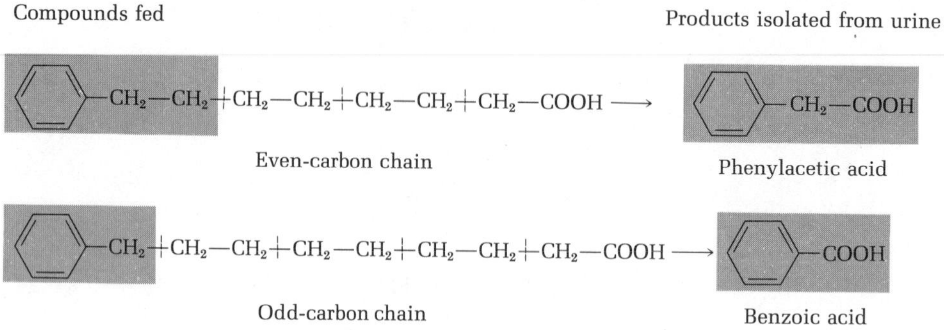

Compounds fed | Products isolated from urine

C_6H_5—CH_2—CH_2 ⫶ CH_2—CH_2 ⫶ CH_2—CH_2 ⫶ CH_2—COOH ⟶ C_6H_5—CH_2—COOH

Even-carbon chain — Phenylacetic acid

C_6H_5—CH_2 ⫶ CH_2—CH_2 ⫶ CH_2—CH_2 ⫶ CH_2—CH_2 ⫶ CH_2—COOH ⟶ C_6H_5—COOH

Odd-carbon chain — Benzoic acid

Figure 18-1
Knoop's experiments on oxidation of phenyl fatty acids fed to rabbits. The phenyl groups were substituted on the ω, i.e., methyl-terminal carbon atom. Even-carbon-chain ω-phenyl fatty acids always yielded phenylacetic acid, and odd-carbon-chain ω-phenyl acids always yielded benzoic acid as the end product excreted in the urine. From these results he concluded that oxidative attack begins at the β carbon atom, followed by cleavage of successive 2-carbon fragments from the chain at the dashed lines, presumably as acetate, which is oxidized to CO_2 and H_2O. The remaining portion of the molecule (color) is not oxidized further and is therefore excreted.

fragment, presumably acetic acid, and a fatty acid chain shorter by 2 carbon atoms (Figure 18-1). But for decades no success attended efforts to demonstrate fatty acid oxidation in cell-free extracts or homogenates of animal tissues. An important clue came when Albert Lehninger in the United States found that ATP restores the ability of liver homogenates to oxidize fatty acids. He postulated that ATP was required to "prime" or activate the fatty acid at its carboxyl group by an enzymatic reaction. He also found that the oxidation of fatty acids in liver homogenates yields active 2-carbon fragments that can enter the citric acid cycle. Later, it was found that fatty acid oxidation takes place in the mitochondria of liver cells. The next clue, which quickly led to recognition of the nature of the enzymatic steps of fatty acid oxidation, came from the work of Feodor Lynen and his colleagues in Munich. They found that the ATP-dependent activation of fatty acids involves the enzymatic esterification of the carboxyl group with the thiol group of coenzyme A and that all the subsequent intermediates in the fatty acid oxidation sequence are thioesters of coenzyme A. Now let us examine the pathway of fatty acid oxidation as we understand it today.

Fatty Acids Enter Mitochondria by a Three-Step Transport Process

Fatty acids are delivered into the cytosol from two sources. Some free fatty acids arrive via the blood, bound to serum albumin. They are released and pass through the cell membrane into the cytosol. The second source is the breakdown of cell triacylglycerols by the action of lipases. The free fatty acids in the cytosol cannot pass through the mitochondrial membranes as such. They must first undergo a series of three enzymatic reactions in order to enter the mitochondrial matrix, where fatty acid oxidation occurs. The first step is catalyzed by enzymes present in the outer mitochondrial membrane, acyl-CoA synthetases, which promote the reaction

$$\text{RCOOH} + \text{ATP} + \text{CoA-SH} \rightleftharpoons \underbrace{\text{R}-\overset{\overset{\text{O}}{\|}}{\text{C}}-\text{S}-\text{CoA}}_{\text{Fatty acyl–CoA}} + \text{AMP} + \text{PP}_i \quad (1)$$

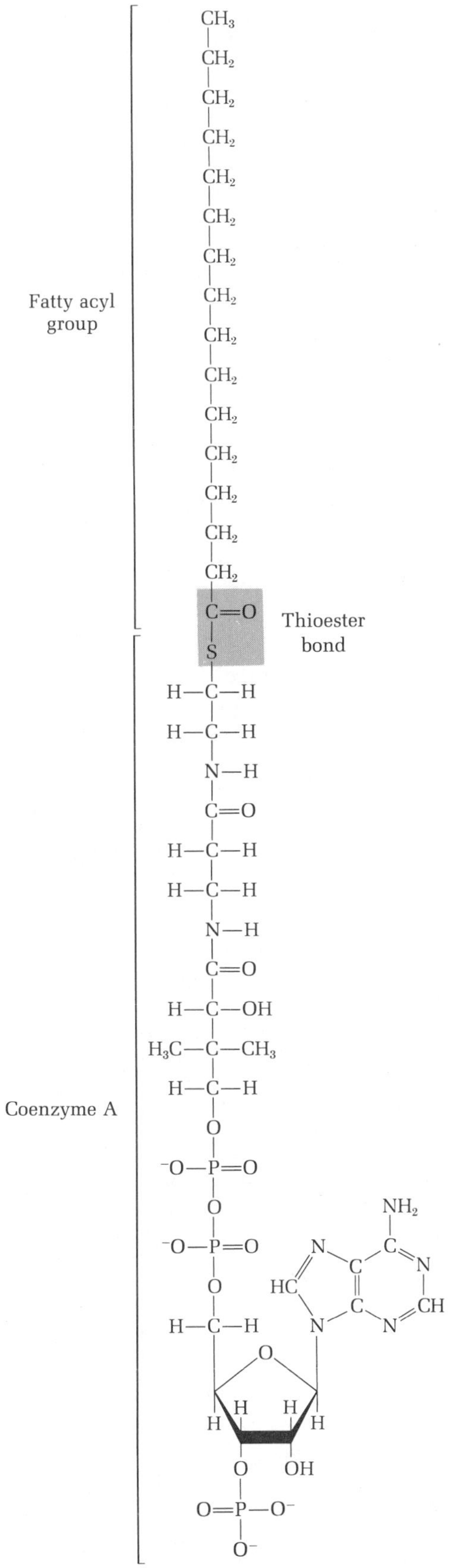

Fatty acyl group

Thioester bond

Coenzyme A

Figure 18-2
Palmitoyl-CoA. The carboxyl group of the palmitic acid (16 carbons) and the thiol group of coenzyme A are linked to form a thioester bond. Note that fatty acyl–CoAs are very large molecules.

in which RCOOH represents a long-chain fatty acid and PP_i represents inorganic pyrophosphate. In this reaction a thioester linkage is formed between the fatty acid carboxyl group and the thiol group of coenzyme A to yield a fatty acyl–CoA (Figure 18-2); simultaneously, the ATP undergoes cleavage to yield AMP and inorganic pyrophosphate. This is a coupled reaction: the energy derived from the cleavage of the ATP to AMP and pyrophosphate on the active site is used to make the new thioester bond. Fatty acyl–CoAs, like acetyl-CoA, are high-energy compounds; their hydrolysis to free fatty acid and CoA-SH has a large negative $\Delta G^{\circ\prime}$ of about -7.5 kcal/mol.

The overall reaction shown in equation (1) is readily reversible, since its $\Delta G^{\circ\prime}$ is only -0.20 kcal/mol. An enzyme-bound intermediate in this reaction has been identified. It is an anhydride of the fatty acid and the phosphate group of AMP, called a fatty acyl adenylate (Figure 18-3). It is formed on the active site of the enzyme, where it reacts with free CoA-SH to yield free fatty acyl–CoA and AMP as products.

The pyrophosphate formed in the activation reaction can then be hydrolyzed by a second enzyme, inorganic pyrophosphatase:

$$\text{Pyrophosphate} + H_2O \longrightarrow 2\ \text{phosphate} \qquad (2)$$
$$\Delta G^{\circ\prime} = -6.9\ \text{kcal/mol}$$

Because the hydrolysis of pyrophosphate in intact cells goes far toward completion, it tends to pull the preceding activation reaction (1) strongly to the right, in the direction of the formation of fatty acyl–CoA. The overall reaction, the sum of reactions (1) and (2), is

$$\text{Fatty acid} + \text{ATP} + \text{CoA-SH} \longrightarrow \text{fatty acyl–S–CoA} + \text{AMP} + 2P_i$$
$$\Delta G^{\circ\prime} = -7.1\ \text{kcal/mol}$$

We shall see other instances in which pyrophosphate cleavage of ATP (page 386), followed by hydrolysis of pyrophosphate, occurs in the activation of biomolecules.

Fatty acyl–CoA esters as such cannot cross the inner mitochondrial membrane, but the enzyme carnitine acyltransferase I, present on the outer surface of the inner membrane, catalyzes the second reaction in the entry process:

$$\text{Fatty acyl–S–CoA} + \text{carnitine} \rightleftharpoons \text{fatty acyl–carnitine} + \text{CoA-SH}$$

Figure 18-3
Structure of a fatty acyl adenylate intermediate. The acyl group is in color; again, keep in mind that the fatty acyl group (**R—CO—**) *is very long in comparison with the adenylate portion* (see Figure 18-2).

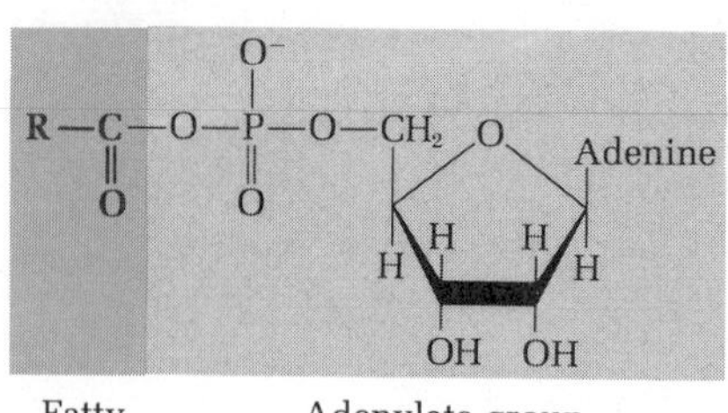

Fatty acyl group

Adenylate group (AMP)

Figure 18-4
The reversible carnitine acyltransferase reaction.

$$H_3C{-}\overset{\displaystyle CH_3}{\underset{\displaystyle CH_3}{N^+}}{-}CH_2{-}\overset{\displaystyle H}{\underset{\displaystyle OH}{C}}{-}CH_2{-}COO^-$$

Carnitine

Palmitoyl-S-CoA

carnitine acyltransferase

CoA-SH

$$H_3C{-}\overset{\displaystyle CH_3}{\underset{\displaystyle CH_3}{N^+}}{-}CH_2{-}\overset{\displaystyle H}{C}{-}CH_2{-}COO^-$$

O

C=O

Oxygen ester linkage

$(CH_2)_{14}$ … CH_3

The palmitoyl group

Palmitoyl carnitine

The fatty acyl–carnitine ester now passes through the inner membrane into the internal matrix compartment of the mitochondrion. Note that the fatty acyl–carnitine is an oxygen ester, whereas fatty acyl–CoA is a thioester. Carnitine (Figure 18-4) is present in most animal and plant tissues. Some lower organisms, e.g., the mealworm, lack the capacity to make it and thus require it ready-made in the diet. Human beings and other vertebrates manufacture carnitine from lysine.

In the third and final step of the entry process the fatty acyl group is enzymatically transferred from carnitine to *intra*mitochondrial CoA by *carnitine acyltransferase II;* this form of the enzyme is located on the inside surface of the inner membrane, where it regenerates fatty acyl–CoA and releases it into the matrix:

$$\text{Fatty acyl–carnitine} + \text{CoA-SH} \rightleftharpoons \text{fatty acyl–S–CoA} + \text{carnitine} \quad (3)$$

Although this three-step process [equations (1) to (3)] for transferring fatty acids into the mitochondrion may appear unnecessarily complex, it has the effect of keeping the cytosolic and intramitochondrial pools of CoA separated, since they have different functions. The mitochondrial pool of CoA is largely used in oxidative degradation of pyruvate, fatty acids, and some amino acids, whereas the cytosolic pool of CoA is used in the biosynthesis of fatty acids. Recall that the cytosolic and intramitochondrial pools of NAD and ATP are also kept separate by the inner mitochondrial membrane (page 470). There is a second point of importance. The first enzyme in the entry process, carnitine acyltransferase I, is a regulatory enzyme and controls the rate of fatty acyl group entry and thus the rate of fatty acid oxidation, as we shall see.

The fatty acyl–CoA is now ready for the oxidation of its fatty acid component by a set of specific enzymes in the mitochondrial matrix.

Fatty Acids Are Oxidized in Two Stages

After fatty acids have gained entrance into the mitochondria, their oxidation takes place in two major stages (Figure 18-5). In

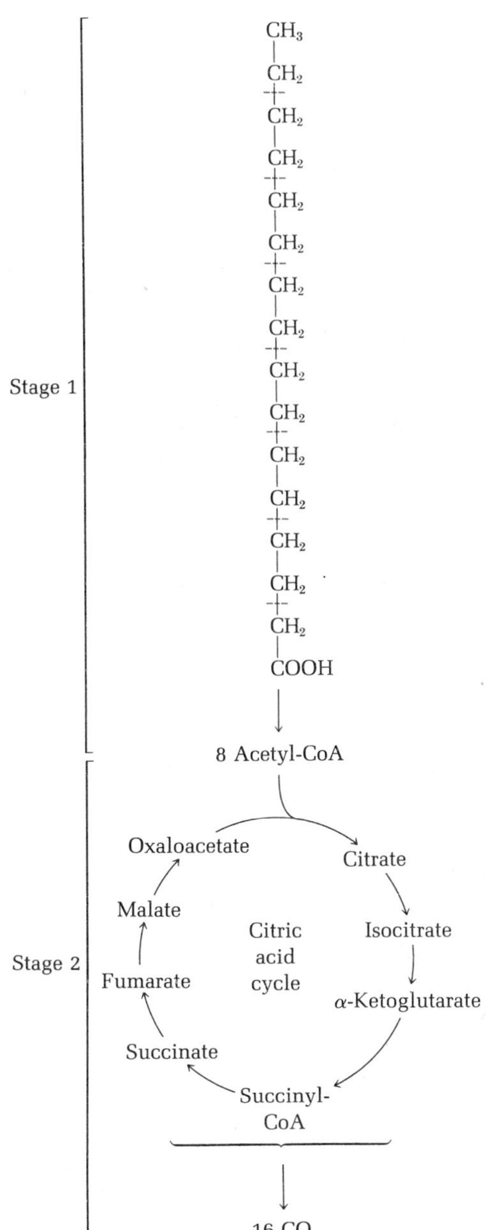

Figure 18-5
Stages of fatty acid oxidation. Stage 1: Oxidation of the long-chain fatty acid to yield acetyl residues in the form of acetyl-CoA. Stage 2: Oxidation of the acetyl residues to CO_2.

the first the fatty acids undergo oxidative removal of successive 2-carbon units, starting from the carboxyl end of the fatty acid chain, by a repeating series of passes through a set of enzymes that removes one 2-carbon acetyl unit at a time, in the form of acetyl-CoA. Thus the 16-carbon fatty acid palmitic acid undergoes seven passes through this set of enzymes, each pass chopping off a 2-carbon unit as acetyl-CoA. At the end of seven passes the last 2-carbon unit of palmitic acid also appears as acetyl-CoA. The overall result is the conversion of the 16-carbon chain of palmitic acid into eight 2-carbon fragments in the form of the acetyl groups of acetyl-CoA. Formation of each molecule of acetyl-CoA requires removal of four hydrogen atoms from the fatty acid by the action of dehydrogenases.

In the second stage of fatty acid oxidation the acetyl residues of acetyl-CoA are oxidized to CO_2 and H_2O via the citric acid cycle. This also takes place in the mitochondria. Acetyl-CoA derived from fatty acid oxidation thus enters a final common pathway of oxidation along with acetyl-CoA coming from glucose via pyruvate oxidation (page 468).

Both stages of fatty acid oxidation result in a flow of hydrogen atoms or their corresponding electrons via the mitochondrial electron-transport chain to oxygen. Coupled to the flow of electrons is the oxidative phosphorylation of ADP to ATP. Thus energy delivered in both stages of fatty acid oxidation is conserved in the form of ATP.

The First Stage in the Oxidation of Saturated Fatty Acids Has Four Steps

Four enzyme-catalyzed reactions are involved in the first stage of fatty acid oxidation.

The First Dehydrogenation Step

After entry of the saturated fatty acyl–CoA ester into the matrix, it undergoes enzymatic dehydrogenation at the α and β carbon atoms (carbon atoms 2 and 3) to form a double bond in the carbon chain, thus yielding a *trans*-Δ^2-enoyl-CoA as the product, in a step catalyzed by *acyl-CoA dehydrogenase*, an enzyme (designated E) containing FAD as its prosthetic group:

$$\text{Fatty acyl–S–CoA} + \text{E–FAD} \longrightarrow trans\text{-}\Delta^2\text{-enoyl–S–CoA} + \text{E–FADH}_2$$

The symbol Δ^2 designates the position of the double bond (Figure 18-6). It is important to note that the new double bond formed has the trans configuration; recall (page 305) that naturally occurring unsaturated fatty acids normally have their double bonds in the cis configuration. We shall come back to this apparent discrepancy later. In this reaction the hydrogen atoms removed from the fatty acyl–CoA are transferred to FAD (page 305), the tightly bound prosthetic group of acyl-CoA dehydrogenase. The reduced form of the dehydrogenase then

$$R{-}CH_2{-}CH_2{-}\mathbf{CH_2{-}C}{-}S{-}CoA \quad (\text{C}{=}\mathbf{O}) \quad \text{Palmitoyl-CoA}$$

$$R{-}CH_2{-}C(H){=}\mathbf{C}(\mathbf{H}){-}\mathbf{C}({=}\mathbf{O}){-}S{-}CoA \quad trans\text{-}\Delta^2\text{-Enoyl-CoA}$$

$$R{-}CH_2{-}C(OH)(H){-}\mathbf{CH_2{-}C}({=}\mathbf{O}){-}S{-}CoA \quad \text{L-3-Hydroxyacyl-CoA}$$

$$R{-}CH_2{-}C({=}O){-}\mathbf{CH_2{-}C}({=}\mathbf{O}){-}S{-}CoA \quad \text{3-Ketoacyl-CoA}$$

$$(C_{14})\ R{-}CH_2{-}C({=}O){-}S{-}CoA + \mathbf{CH_3{-}C}({=}\mathbf{O}){-}S{-}CoA$$

(C_{14}) Acyl-CoA (meristoyl-CoA

Acetyl-CoA

(a)

(b)

Figure 18-6
The fatty acid oxidation cycle. (a) *In one round of the cycle one acetyl residue* (color) *is removed in the form of acetyl-CoA from the carboxyl end of palmitic acid* (C_{16}), *which enters as palmitoyl-CoA.* (b) *Six more rounds of the cycle take place, yielding seven more acetyl-CoAs, the seventh arising from the last 2 carbon atoms of the 16-carbon chain.*

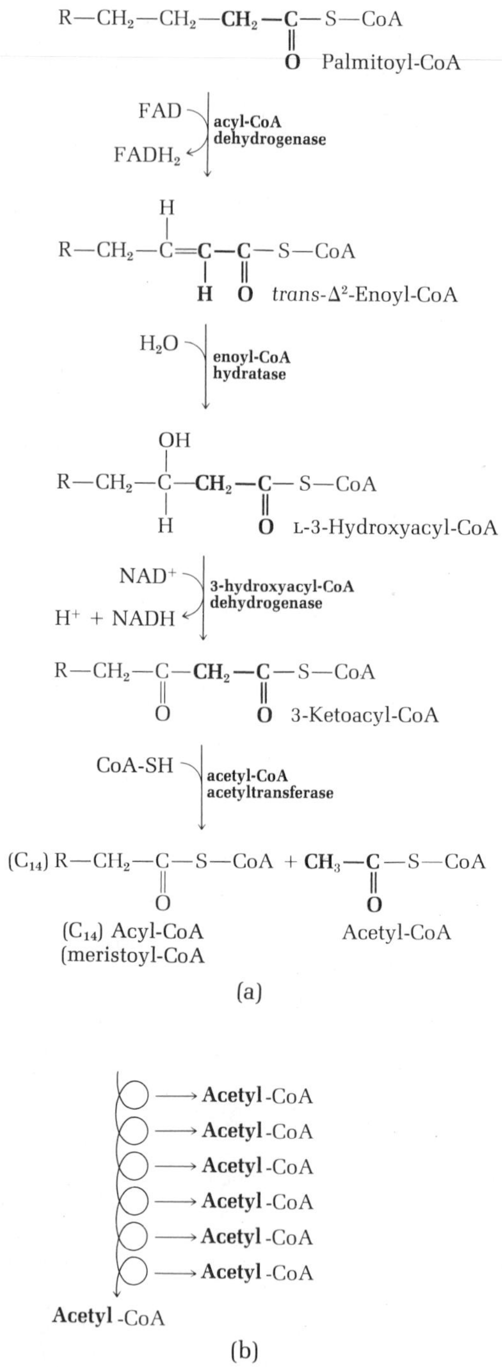

donates its electrons to a connecting electron carrier, called the *electron-transferring flavoprotein* (ETFP), which in turn transfers a pair of electrons to ubiquinone of the mitochondrial respiratory chain (Figure 18-7). During the subsequent transport of this electron pair to oxygen by the respiratory chain two ATPs are generated by oxidative phosphorylation (page 478).

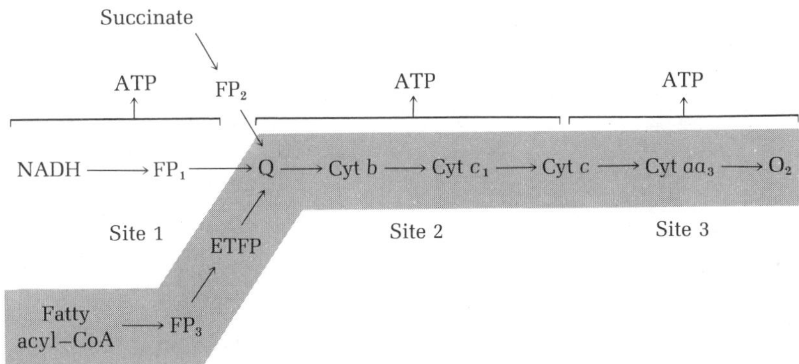

Figure 18-7
Reducing equivalents derived from fatty acyl–CoA by fatty acyl–CoA dehydrogenase (FP_3, for flavoprotein 3) are passed by electron-transferring flavoprotein (ETFP) to ubiquinone (Q) of the mitochondrial electron-transport chain. Two ATPs are generated per pair of electrons passing from ubiquinone to oxygen. Ubiquinone thus collects electrons from NADH dehydrogenase (FP_1), succinate dehydrogenase (FP_2), and fatty acyl–CoA dehydrogenase (FP_3).

The Hydration Step

In the second step of the fatty acid oxidation cycle, water adds to the double bond of the *trans*-Δ^2-enoyl-CoA to form the L stereoisomer of *β-hydroxyacyl-CoA* (also designated *3-hydroxyacyl-CoA*), catalyzed by *enoyl-CoA hydratase*, which has been isolated in crystalline form:

$$trans\text{-}\Delta^2\text{-Enoyl-S-CoA} + H_2O \rightleftharpoons \text{L-3-hydroxyacyl-S-CoA}$$

The reaction catalyzed is shown in Figure 18-6.

The Second Dehydrogenation Step

In the third step of the fatty acid oxidation cycle, the L-3-hydroxyacyl-CoA is dehydrogenated to form *3-ketoacyl-CoA* (Figure 18-6) by the action of *3-hydroxyacyl-CoA dehydrogenase;* NAD^+ is the specific electron acceptor. The reaction is

$$\text{L-3-Hydroxyacyl-S-CoA} + NAD^+ \rightleftharpoons \text{3-ketoacyl-S-CoA} + NADH + H^+$$

This enzyme is absolutely specific for the L stereoisomer. The NADH formed in this reaction then donates its reducing equivalents to the NADH dehydrogenase of the respiratory chain (Figure 18-7). As in all other NAD-linked substrate dehydrogenations in mitochondria, three ATP molecules are generated from ADP per pair of electrons passing from NADH to oxygen via the electron-transport chain (page 484).

The Cleavage Step

The fourth and last step of the fatty acid oxidation cycle is catalyzed by *acetyl-CoA acetyltransferase* (more commonly known as *thiolase*), which promotes reaction of 3-ketoacyl-CoA with a molecule of free CoA-SH to split off the carboxyl-terminal 2-carbon fragment of the original fatty acid as acetyl-CoA, leaving as the other product the CoA ester of the original fatty acid, now shortened by two carbon atoms (Figure 18-6):

$$\text{3-Ketoacyl-S-CoA} + \text{CoA-SH} \rightleftharpoons \text{shortened fatty acyl–S–CoA} + \text{acetyl-S-CoA}$$

This reaction is often called a *thiolysis*, by analogy with the process of hydrolysis, since the β-ketoacyl-CoA is cleaved by reaction with the thiol group of CoA (Figure 18-6).

The First Stage of Fatty Acid Oxidation Yields Acetyl-CoA and ATP

We have now completed one pass through the fatty acid oxidation cycle. One molecule of acetyl-CoA and two pairs of hydrogen atoms were removed from the entering long-chain fatty acyl–CoA, to shorten it by 2 carbon atoms. The equation for one pass, beginning with the CoA ester of palmitic acid (16 carbon atoms), is

$$\text{Palmitoyl-S-CoA} + \text{CoA-SH} + \text{FAD} + \text{NAD}^+ + H_2O \longrightarrow \text{myristoyl-S-CoA} + \text{acetyl-S-CoA} + \text{FADH}_2 + \text{NADH} + H^+$$

Following removal of one acetyl-CoA unit from palmitoyl-CoA, we are left with the CoA ester of the shortened fatty acid, namely, the 14-carbon *myristic acid*. The myristoyl-CoA can now enter the fatty acid oxidation cycle and go through another set of four reactions, exactly analogous to the first, to yield a second molecule of acetyl-CoA and *lauroyl-CoA*, the CoA ester of the homologous 12-carbon fatty acid, namely, *lauric acid*. Altogether seven passes through the fatty acid oxidation cycle are required to oxidize one molecule of palmitoyl-CoA to yield eight molecules of acetyl-CoA (Figure 18-6):

$$\text{Palmitoyl-S-CoA} + 7\text{CoA-SH} + 7\text{FAD} + 7\text{NAD}^+ + 7H_2O \longrightarrow 8\ \text{acetyl-S-CoA} + 7\text{FADH}_2 + 7\text{NADH} + 7H^+$$

Each molecule of FADH_2 formed during oxidation of the fatty acid donates a pair of electrons to ubiquinone of the respiratory chain; two molecules of ATP are generated from ADP and phosphate during the ensuing transport of the electron pair to oxygen and the coupled oxidative phosphorylations (Figure 18-7). Similarly, each molecule of NADH formed delivers a pair of electrons to the mitochondrial NADH dehydrogenase; the subsequent transport of each pair of electrons to oxygen results in formation of three molecules of ATP from ADP and phosphate. Thus five molecules of ATP are formed per molecule of acetyl-CoA removed in each pass through the sequence as it occurs in animal tissues, such as the liver or heart. We can therefore write the overall equation for the oxidation of palmitoyl-CoA to eight molecules of acetyl-CoA, including the electron transport and oxidative phosphorylations:

$$\text{Palmitoyl-S-CoA} + 7\text{CoA-SH} + 7O_2 + 35P_i + 35\text{ADP} \longrightarrow 8\ \text{acetyl-S-CoA} + 35\text{ATP} + 42H_2O \quad (4)$$

This is the overall equation for the *first stage* of fatty acid oxidation (see Figure 18-5).

In the Second Stage of Fatty Acid Oxidation Acetyl-CoA Is Oxidized via the Citric Acid Cycle

The acetyl-CoA produced from the oxidation of fatty acids is no different from the acetyl-CoA formed from pyruvate. Its acetyl group is ultimately oxidized to CO_2 and H_2O by the same pathway, i.e., the citric acid cycle (page 436). The following equation represents the balance sheet for the second stage in fatty acid oxidation (Figure 18-5), the oxidation of the eight acetyl-CoA molecules formed from palmitoyl-CoA, together with the coupled phosphorylations:

$$\text{8 Acetyl-S-CoA} + 16O_2 + 96P_i + 96\text{ADP} \longrightarrow 8\text{CoA-SH} + 96\text{ATP} + 104H_2O + 16CO_2 \quad (5)$$

Combining equations (4) and (5) for the first and second stages of fatty acid oxidation, we get the overall equation for the complete oxidation of palmitoyl-CoA to carbon dioxide and water

$$\text{Palmitoyl-S-CoA} + 23O_2 + 131P_i + 131\text{ADP} \longrightarrow \text{CoA-SH} + 131\text{ATP} + 16CO_2 + 146H_2O \quad (6)$$

Table 18-1 summarizes the yields of NADH, $FADH_2$, and ATP in the successive steps in fatty acid oxidation. The standard-free-energy change of the oxidation of palmitic acid to $CO_2 + H_2O$ is about 2340 kcal/mol. Under standard conditions $7.3 \times 129 = 942$ kcal are recovered as the phosphate-bond energy of ATP. However, when the free-energy changes are calculated from *actual* concentrations of reactants and products under intracellular conditions, the free-energy recovery is over 80 percent.

Table 18-1 The Yields of ATP in the Oxidative Steps during Oxidation of One Molecule of Palmitoyl-CoA to $CO_2 + H_2O$

	NAD-linked steps	FAD-linked steps	ATP
Acyl-CoA dehydrogenase		7	14
3-Hydroxyacyl-CoA dehydrogenase	7		21
Isocitrate dehydrogenase	8		24
α-Ketoglutarate dehydrogenase	8		24
Succinyl-CoA synthetase†			8
Succinate dehydrogenase		8	16
Malate dehydrogenase	8		24
Total ATP formed			131

† Assuming that the GTP formed reacts with ADP to yield ATP.

Figure 18-8
Oxidative removal of three acetyl-CoA units from oleyl-CoA to yield a 12-carbon cis-Δ^3-enoyl-CoA.

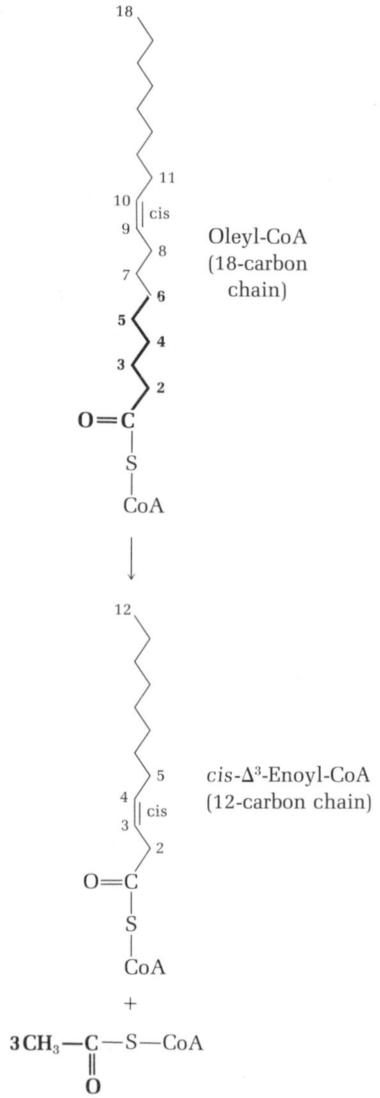

18, 11, 10, 9, cis, 8, 7, 6, 5, 4, 3, 2
O=C—S—CoA
Oleyl-CoA (18-carbon chain)

12, 5, 4, 3, cis, 2
O=C—S—CoA
cis-Δ^3-Enoyl-CoA (12-carbon chain)

+

$3CH_3—C(=O)—S—CoA$

The Oxidation of Unsaturated Fatty Acids Requires Two Additional Enzymatic Steps

The fatty acid oxidation sequence just described outlines the pathway taken when the incoming fatty acid is saturated, i.e., has only single bonds in its carbon chain. We have seen, however, that most of the fatty acids found in triacylglycerols and phospholipids of animals and plants are unsaturated and thus have one or more double bonds (page 304). These are in the cis configuration; moreover, they are not usually in the specific position in the fatty acid chain that can be acted upon by enoyl-CoA hydratase, the enzyme normally catalyzing the addition of H_2O to the double bond of the Δ^2-enoyl-CoA generated during β oxidation of fatty acids.

However, by the action of two auxiliary enzymes, the fatty acid oxidation cycle described above can also oxidize the common unsaturated fatty acids used by cells as fuel. The action of these two enzymes, one an *isomerase* and the other an *epimerase,* can be illustrated by two examples. First, let us follow the course of oxidation of *oleic acid,* an abundant 18-carbon unsaturated fatty acid, whose cis double bond is between carbons 9 and 10 (denoted Δ^9). Oleic acid is first converted into *oleyl-CoA* (Figure 18-8) which is transported through the mitochondrial membrane as *oleyl-carnitine* and then converted back into oleyl-CoA in the matrix, as already shown for palmitic acid. The oleyl-CoA then undergoes three passes through the fatty acid oxidation cycle, to yield three molecules of acetyl-CoA and the CoA ester of a 12-carbon unsaturated fatty acid, with its cis double bond between carbons 3 and 4 (Figure 18-8). This product cannot be acted upon by the next enzyme in the normal fatty acid cycle, enoyl-CoA hydratase, which acts only on trans double bonds. However, by the action of one of the two auxiliary enzymes, namely *enoyl-CoA isomerase,* the *cis*-Δ^3-enoyl-CoA is isomerized to yield *trans*-Δ^2-enoyl-CoA (Figure 18-9), a normal substrate for enoyl-CoA hydratase, which then converts it into the corresponding L-3-hydroxyacyl-CoA. This product is now acted upon by the remaining enzymes of the fatty acid cycle to yield acetyl-CoA and a 10-carbon saturated fatty acid as its CoA ester. The latter then undergoes four more passes through the normal fatty acid cycle to yield altogether nine acetyl-CoAs from one molecule of the 18-carbon oleic acid.

The other auxiliary enzyme, the epimerase, is required for oxidation of polyunsaturated fatty acids. As an example, we take the 18-carbon *linoleic acid,* which has two cis double bonds, one between carbons 9 and 10 (Δ^9) and the other between carbons 12 and 13 (Δ^{12}). Linoleyl-CoA undergoes three passes through the standard fatty acid oxidation sequence, yielding three molecules of acetyl-CoA and the CoA ester of a 12-carbon unsaturated fatty acid with a cis double bond between carbons 3 and 4, as in the case of oleyl-CoA, and another cis double bond between carbons 6 and 7. The cis double bond at Δ^3 is then isomerized by enoyl-CoA isomerase to *trans*-Δ^2-enoyl-CoA, which undergoes the subsequent reactions

Figure 18-9
Action of enoyl-CoA isomerase in converting a cis-Δ^3-enoyl-CoA into a trans-Δ^2-enoyl-CoA. The latter is then converted into a 3-hydroxyacyl-CoA.

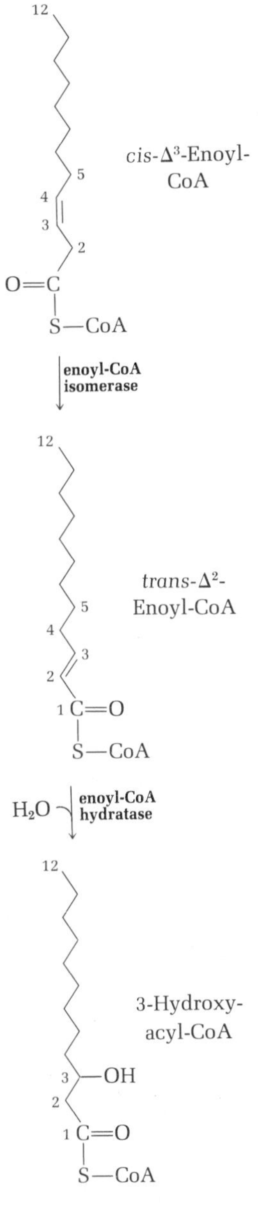

12, 5, 4, 3, 2
cis-Δ^3-Enoyl-CoA
O=C
S—CoA
enoyl-CoA isomerase
12, 5, 4, 3, 2
trans-Δ^2-Enoyl-CoA
1 C=O
S—CoA
H_2O **enoyl-CoA hydratase**
12
3-Hydroxy-acyl-CoA
3 —OH
2
1 C=O
S—CoA

of the normal oxidative sequence to yield a molecule of acetyl-CoA. Another pass yields, in addition to acetyl-CoA, an 8-carbon unsaturated fatty acyl–CoA with a cis Δ^2 double bond. This *can* be acted upon by enoyl-CoA hydratase, but the product in this case is the D stereoisomer of 3-hydroxyacyl-CoA instead of the L stereoisomer normally formed during saturated fatty acid oxidation. At this point the second auxiliary enzyme, *3-hydroxyacyl-CoA epimerase*, comes into play. It epimerizes the D- to the L-3-hydroxyacyl-CoA (Figure 18-10), which now undergoes the normal reactions to yield acetyl-CoA and a 6-carbon saturated acyl-CoA. The latter is then oxidized as usual to yield three more acetyl-CoAs. The overall result is that linoleic acid is converted into nine acetyl-CoAs, with the help of the two auxiliary enzymes.

Oxidation of Fatty Acids with an Odd Number of Carbons

Although most naturally occurring lipids contain fatty acids with an even number of carbon atoms, fatty acids with an odd number of carbons are found in significant amounts in the lipids of many plants and some marine organisms. Moreover, cattle and other ruminant animals form large amounts of the 3-carbon *propionic acid* during fermentation of carbohydrates in the rumen. The propionate so formed is absorbed into the blood and oxidized by the liver and other tissues. Long-chain odd-carbon fatty acids are oxidized by the same pathway as the even-carbon acids, beginning at the carboxyl end of the chain. However, the substrate for the last pass through the fatty acid oxidation sequence is a fatty acyl–CoA in which the fatty acid has 5 carbon atoms. When this is oxidized and ultimately cleaved, the products are *acetyl-CoA* and *propionyl-CoA*, the latter derived from the last 3 carbon atoms of the long-chain odd-carbon fatty acid. The acetyl-CoA is of course oxidized via the citric acid cycle, but propionyl-CoA derived from odd-

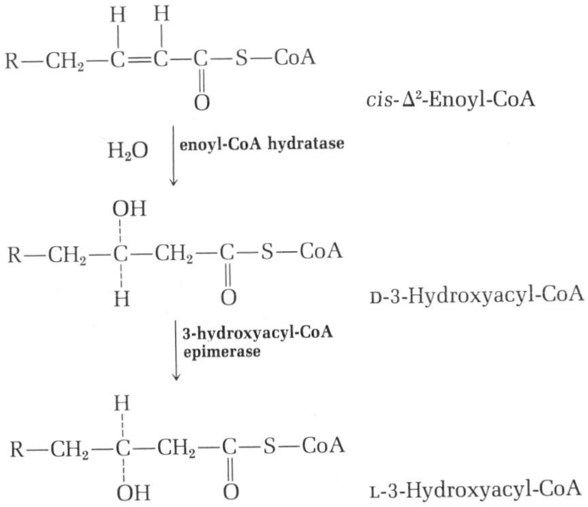

$$R{-}CH_2{-}\overset{H}{C}{=}\overset{H}{C}{-}\underset{O}{\overset{}{C}}{-}S{-}CoA$$ *cis*-Δ^2-Enoyl-CoA

H_2O **enoyl-CoA hydratase**

$$R{-}CH_2{-}\underset{H}{\overset{OH}{C}}{-}CH_2{-}\underset{O}{C}{-}S{-}CoA$$ D-3-Hydroxyacyl-CoA

3-hydroxyacyl-CoA epimerase

$$R{-}CH_2{-}\underset{OH}{\overset{H}{C}}{-}CH_2{-}\underset{O}{C}{-}S{-}CoA$$ L-3-Hydroxyacyl-CoA

Figure 18-10
Formation of a D-3-hydroxyacyl-CoA and its conversion into the L stereoisomer. The latter can then undergo the subsequent reactions of the fatty acid oxidation sequence.

Figure 18-11
Carboxylation of propionyl-CoA to D-methylmalonyl-CoA, and conversion of the latter to succinyl-CoA. See also Figure 18-12.

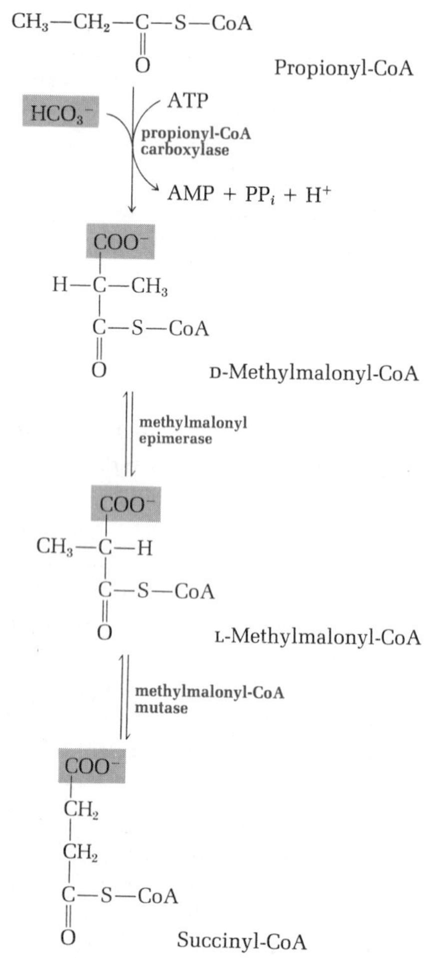

$CH_3—CH_2—C(=O)—S—CoA$ Propionyl-CoA

HCO_3^-, ATP → propionyl-CoA carboxylase → AMP + PP_i + H^+

COO^-—C(H)(CH_3)—C(=O)—S—CoA D-Methylmalonyl-CoA

methylmalonyl epimerase

COO^-—C(CH_3)(H)—C(=O)—S—CoA L-Methylmalonyl-CoA

methylmalonyl-CoA mutase

COO^-—CH_2—CH_2—C(=O)—S—CoA Succinyl-CoA

carbon fatty acids and other sources takes a rather unusual enzymatic pathway. Propionyl-CoA is carboxylated to the D stereoisomer of methylmalonyl-CoA (Figure 18-11) by a biotin-containing enzyme called propionyl-CoA carboxylase. In this reaction bicarbonate is the precursor of the new carboxyl group and ATP furnishes the energy required to form the new covalent bond through its pyrophosphate cleavage to yield AMP and pyrophosphate:

$$\text{Propionyl-CoA} + \text{ATP} + CO_2 \longrightarrow \text{D-methylmalonyl-CoA} + \text{AMP} + PP_i$$

Mg^{2+} is also required in this reaction. The D-methylmalonyl-CoA so formed is then enzymatically epimerized (page 281) to form its L stereoisomer by the action of methylmalonyl epimerase (Figure 18-11):

$$\text{D-Methylmalonyl-CoA} \rightleftharpoons \text{L-methylmalonyl-CoA}$$

The L-methylmalonyl-CoA now undergoes a most unusual intramolecular rearrangement to form succinyl-CoA (Figure 18-11), catalyzed by methylmalonyl-CoA mutase, which requires as its coenzyme deoxyadenosylcobalamin, a coenzyme form of vitamin B_{12} or cobalamin (page 263):

$$\text{L-Methylmalonyl-CoA} \rightleftharpoons \text{succinyl-CoA}$$

Succinyl-CoA is of course an intermediate of the citric acid cycle and ultimately leads to oxaloacetate.

This sequence of metabolic reactions from propionyl-CoA to succinyl-CoA may seem like the hard way to bring about this overall transformation. One might expect that succinyl-CoA could be made in one step by attaching CO_2 to carbon 3 of the propionic acid portion of propionyl-CoA. Instead, cells have chosen to put the CO_2 on carbon atom 2 and on its wrong side to boot. After the epimerase has put the CO_2 on the "right" side of carbon 2, that is, to form L-methylmalonyl-CoA, it would seem a simple matter to shift the carboxyl group from carbon 2 to carbon 3 of the propionyl group (see Figure 18-12). Instead, the very bulky —CO—S—CoA group is moved, with the help of the complex coenzyme deoxyadenosylcobalamin. Here again we may presume that cells have learned to solve a difficult chemical problem by bypassing it in a roundabout way.

The methylmalonyl-CoA mutase reaction is remarkable. It consists of an exchange of the group —CO—S—CoA (Figure 18-12) at carbon atom 2 of the original propionyl group of methylmalonyl-CoA for an H atom on carbon atom 3. This is one of the relatively few enzymatic reactions in biology in which there is an exchange of an alkyl or substituted alkyl group with a hydrogen atom on an adjacent carbon. The enzymes catalyzing these unusual reactions all contain 5′-deoxyadenosylcobalamin (page 262). Recall that deficiency in

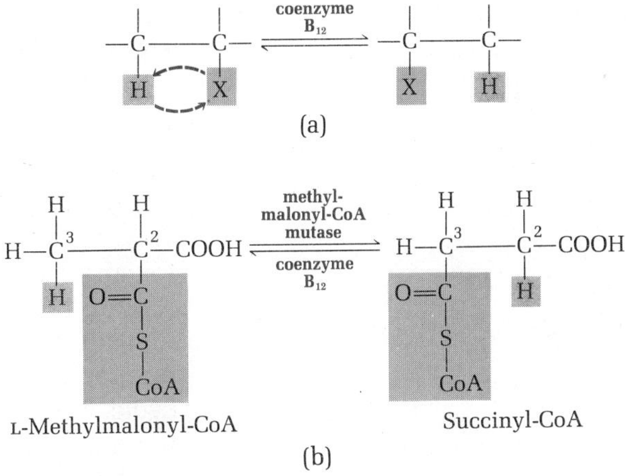

coenzyme B_{12}

(a)

methyl-malonyl-CoA mutase

coenzyme B_{12}

L-Methylmalonyl-CoA

Succinyl-CoA

(b)

Figure 18-12
The intramolecular rearrangement in the methylmalonyl-CoA mutase reaction. Coenzyme B_{12} participates in reactions in which a hydrogen atom and a group X on an adjacent carbon are exchanged. (a) Model reaction; (b) methylmalonyl-CoA mutase reaction.

the absorption of vitamin B_{12} from the intestine results in pernicious anemia. As it happens, methylmalonyl-CoA is an intermediate not only in the oxidation of odd-carbon fatty acids but also in the oxidative degradation of three amino acids (page 544); they are *methionine, valine,* and *isoleucine.* Several genetic disorders in methylmalonyl-CoA metabolism have been identified in human beings, usually in infants. There is a heritable defect in methylmalonyl-CoA mutase, resulting in the failure of methylmalonyl-CoA to be converted into succinyl-CoA. As a consequence, methylmalonic acid cannot be metabolized further and appears in large amounts in the blood and urine, causing a lowering of the blood pH. This condition, called *methylmalonic acidemia,* can be alleviated in some of these patients by injecting large amounts of vitamin B_{12}, since the genetic defect results in a lowered rate of the enzymatic reaction by which vitamin B_{12} is converted into its active coenzyme form. But in another class of patients suffering from methylmalonic acidemia the genetic defect is in the protein part of methylmalonyl-CoA mutase and cannot be alleviated by administration of vitamin B_{12}; in these cases the disease may be fatal.

Hypoglycin A

O_2 → NH_3, CO_2

Methylene cyclopropylacetate

ATP + CoA-SH → AMP + PP_i

Methylene cyclopropyl-acetyl-CoA, a potent inhibitor of butyroyl-CoA oxidation

Figure 18-13
Hypoglycin A, present in unripe ackee fruit, is enzymatically converted into an extremely potent inhibitor of the oxidation of short-chain fatty acyl–CoAs.

Hypoglycin, a Toxic Agent of Some Plants, Inhibits Fatty Acid Oxidation

Consumption of the unripe fruit of the ackee plant has long been known to cause an endemic metabolic disease in undernourished people in Jamaica. It is characterized by hypoglycemia (low blood sugar) and disturbances in fatty acid metabolism. The toxic principle of the ackee fruit is *hypoglycin,* a derivative of propionic acid (Figure 18-13). Hypoglycin is metabolically converted into a substance which, in the form of its CoA ester, is a powerful and specific inhibitor of the oxidation of short-chain acyl-CoAs, particularly of butyroyl-CoA. As a consequence butyroyl-CoA undergoes hydrolysis to yield free butyrate, which accumulates in the blood in abnormal amounts and indirectly causes lowering of the blood-sugar level.

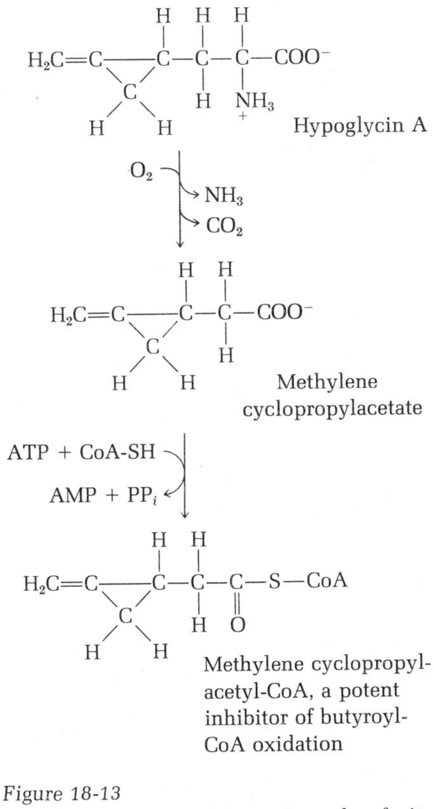

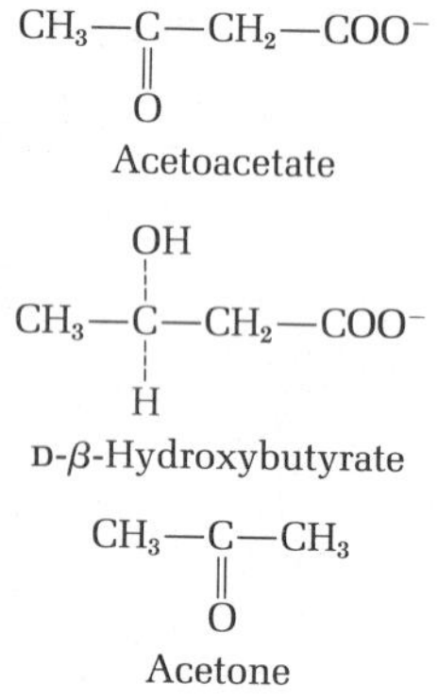

Figure 18-14
The ketone bodies.

$$CH_3-\underset{\underset{O}{\parallel}}{C}-CH_2-COO^-$$

Acetoacetate

$$CH_3-\underset{\underset{H}{|}}{\overset{\overset{OH}{|}}{C}}-CH_2-COO^-$$

D-β-Hydroxybutyrate

$$CH_3-\underset{\underset{O}{\parallel}}{C}-CH_3$$

Acetone

Formation of Ketone Bodies in the Liver and Their Oxidation in Other Organs

In human beings and most other mammals the further metabolism of the acetyl-CoA formed during oxidation of fatty acids may take place by way of two routes in the liver. One, oxidation via the citric acid cycle, has been described. The other pathway leads to *acetoacetate* and *D-β-hydroxybutyrate*, which, together with *acetone*, are collectively called *ketone bodies* (Figure 18-14). Acetoacetate and β-hydroxybutyrate are not oxidized further in the liver but are transported by the blood to the peripheral tissues, where they are oxidized via the citric acid cycle. The first step in formation of acetoacetate in the liver is the enzymatic condensation of two molecules of acetyl-CoA, catalyzed by thiolase:

$$\text{Acetyl-S-CoA} + \text{acetyl-S-CoA} \rightleftharpoons \text{acetoacetyl-S-CoA} + \text{CoA-SH}$$

The acetoacetyl-CoA then undergoes loss of CoA to become free acetoacetate in a sequence of two reactions (Figure 18-15), whose end result is given by the overall equation

$$\text{Acetoacetyl-S-CoA} + H_2O \longrightarrow \text{acetoacetate} + \text{CoA-SH}$$

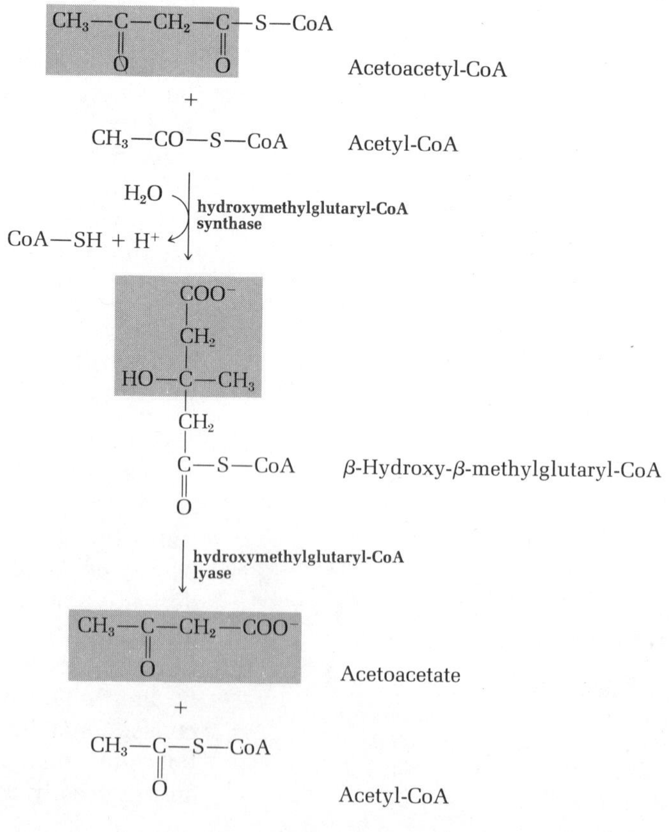

Figure 18-15
Removal of CoA from acetoacetyl-CoA. This process is called deacylation. *Hydroxymethylglutaryl-CoA is also an important intermediate in cholesterol biosynthesis (Chapter 21).*

$$CH_3-\underset{\underset{O}{\parallel}}{C}-CH_2-\underset{\underset{O}{\parallel}}{C}-S-CoA \quad \text{Acetoacetyl-CoA}$$

+

$$CH_3-CO-S-CoA \quad \text{Acetyl-CoA}$$

H_2O → ; CoA—SH + H^+ ← ; **hydroxymethylglutaryl-CoA synthase**

$$^-OOC-CH_2-\underset{\underset{CH_3}{|}}{\overset{\overset{OH}{|}}{C}}-CH_2-\underset{\underset{O}{\parallel}}{C}-S-CoA \quad \text{β-Hydroxy-β-methylglutaryl-CoA}$$

hydroxymethylglutaryl-CoA lyase

$$CH_3-\underset{\underset{O}{\parallel}}{C}-CH_2-COO^- \quad \text{Acetoacetate}$$

+

$$CH_3-\underset{\underset{O}{\parallel}}{C}-S-CoA \quad \text{Acetyl-CoA}$$

The free acetoacetate so produced is reversibly reduced by D-β-hydroxybutyrate dehydrogenase, a mitochondrial enzyme, to D-β-hydroxybutyrate:

$$\text{Acetoacetate} + \text{NADH} + \text{H}^+ \rightleftharpoons \text{D-}\beta\text{-hydroxybutyrate} + \text{NAD}^+$$

This enzyme is specific for the D stereoisomer; it does not act on L-β-hydroxyacyl-CoAs. D-β-Hydroxybutyrate dehydrogenase thus is not to be confused with L-3-hydroxyacyl-CoA dehydrogenase (page 517). Acetoacetate is also the precursor of acetone, formed in small amounts when acetoacetate, an unstable molecule, loses its carboxyl group spontaneously or by the action of *acetoacetate decarboxylase:*

$$\underset{\text{Acetoacetate}}{CH_3{-}\overset{\displaystyle O}{\overset{\|}{C}}{-}CH_2{-}COO^-} + H^+ \longrightarrow \underset{\text{Acetone}}{CH_3{-}\overset{\displaystyle O}{\overset{\|}{C}}{-}CH_3} + CO_2$$

Acetone, a volatile compound, is present in significant amounts in the blood of diabetics, in whom it imparts a characteristic sweet odor to the breath, often wrongly taken to be alcoholic. Free acetoacetate and D-β-hydroxybutyrate resulting from the above reactions then diffuse out of liver cells into the bloodstream and pass to the peripheral tissues.

The purpose of ketone-body formation is to divert some of the acetyl-CoA formed from fatty acids in the liver from further oxidation there and to send it, in the form of the ketone bodies, to other tissues for oxidation to CO_2 and H_2O. Ketone-body formation is an "overflow" pathway. This is one of the many ways the liver uses to distribute fuels to the rest of the body. Normally, the concentration of ketone bodies in the blood is very low, but in fasting or in diabetes mellitus it may reach extremely high levels. This condition, known as *ketosis*, arises when the rate of formation of the ketone bodies by the liver exceeds the capacity of the peripheral tissues to use them up. In diabetes there is a failure of the tissues to use glucose from the blood. As a consequence the liver compensates by burning more fatty acids as fuel, but this causes an overproduction of ketone bodies by the liver, beyond the capacity of the peripheral tissues to oxidize them.

In the peripheral tissues D-β-hydroxybutyrate is oxidized to acetoacetate by D-β-hydroxybutyrate dehydrogenase:

$$\text{D-}\beta\text{-Hydroxybutyrate} + \text{NAD}^+ \rightleftharpoons \text{acetoacetate} + \text{NADH} + \text{H}^+$$

The acetoacetate so formed is then activated to form its CoA ester, by transfer of CoA from succinyl-CoA, an intermediate of the citric acid cycle (page 445), in a reaction catalyzed by *3-ketoacyl-CoA transferase:*

$$\text{Succinyl-S-CoA} + \text{acetoacetate} \rightleftharpoons \text{succinate} + \text{acetoacetyl-S-CoA}$$

The acetoacetyl-CoA so formed is then cleaved by thiolase to yield acetyl-CoA:

$$\text{Acetoacetyl-S-CoA} + \text{CoA-SH} \rightleftharpoons 2\ \text{acetyl-S-CoA}$$

The resulting acetyl-CoA then enters the citric acid cycle for complete oxidation in the peripheral tissues.

Regulation of Fatty Acid Oxidation and Ketone-Body Formation

In the liver, fatty acyl–CoAs formed in the cytosol have two major pathways open to them: (1) oxidation in the mitochondria or (2) conversion into triacylglycerols and phospholipids by enzymes in the cytosol. Which pathway is taken depends upon the rate of transfer of long-chain fatty acyl–CoAs into the mitochondria. The three-step membrane transport process by which fatty acyl groups are carried from cytosolic fatty acyl–CoA into the mitochondrial matrix via carnitine is the rate-controlling step for fatty acid oxidation. Once fatty acyl groups have entered the mitochondria, they are committed to oxidation to yield acetyl-CoA.

Carnitine acyltransferase I, which catalyzes the transfer of fatty acyl groups from fatty acyl–CoAs to carnitine outside the matrix compartment, is an allosteric enzyme. It is specifically inhibited by its modulator *malonyl-CoA* (Figure 18-16), a metabolite we have not met before. It is the first intermediate in the biosynthesis of long-chain fatty acids from acetyl-CoA, which takes place in the cytosol. Malonyl-CoA increases in concentration whenever the animal is well supplied with carbohydrate, since excess glucose that cannot be oxidized or stored as glycogen is converted in the cytosol into triacylglycerols for storage. Thus the oxidation of fatty acids is turned off whenever the liver is amply supplied with glucose as fuel and is actively making triacylglycerols from excess glucose. The turnoff is accomplished by allosteric inhibition of the entry of fatty acyl groups into the mitochondria.

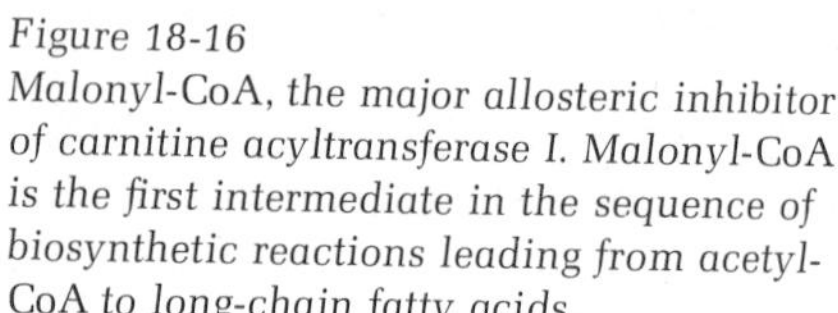

Figure 18-16
Malonyl-CoA, the major allosteric inhibitor of carnitine acyltransferase I. Malonyl-CoA is the first intermediate in the sequence of biosynthetic reactions leading from acetyl-CoA to long-chain fatty acids.

COO^-
|
CH_2
|
C—S—CoA
‖
O

Once acetyl-CoA has been formed by oxidation of fatty acids in liver mitochondria, it has two possible fates: it may be oxidized to CO_2 via the citric acid cycle or it may be converted into the ketone bodies to be circulated to the peripheral tissues. A major determinant of the pathway taken by acetyl-CoA in liver mitochondria is the availability of oxaloacetate to initiate entry of acetyl-CoA into the citric acid cycle. When oxaloacetate concentration is very low, little acetyl-CoA enters the cycle; ketone-body formation is then favored. Oxaloacetate concentration tends to be low when an animal is fasting or on a low carbohydrate diet. In this case the rate of fatty acid oxidation is increased and much of the acetyl-CoA formed is converted via hydroxymethylglutaryl-CoA into free acetoacetate and D-β-hydroxybutyrate, which are circulated to the peripheral tissues. There the ketone bodies serve as a major fuel and are oxidized to CO_2 and H_2O via the citric acid cycle.

Summary

The fatty acid components of lipids furnish a large fraction of the oxidative energy in animals. Free fatty acids are first activated by esterification with CoA to form acyl-CoA esters at the outer mitochondrial membrane and are then converted into fatty acyl–carnitine esters, which can cross the inner mitochondrial membrane into the matrix, where fatty acyl–CoA esters are formed again. All subsequent steps in the oxidation of fatty acids take place in the form of their CoA esters, within the mitochondrial matrix. Four reaction steps are required to remove each acetyl-CoA residue from the carboxyl end of saturated fatty acyl–CoAs: (1) the dehydrogenation of carbon atoms 2 and 3 by FAD-linked acyl-CoA dehydrogenases, (2) hydration of the resulting trans Δ^2 double bond by enoyl-CoA hydratase, (3) dehydrogenation of the resulting L-3-hydroxyacyl-CoA by NAD^+-linked 3-hydroxyacyl-CoA dehydrogenase, and (4) the CoA-requiring cleavage by thiolase of the resulting β-keto fatty acyl–CoA to form acetyl-CoA and the CoA ester of the original fatty acid, shortened by 2 carbons. The shortened fatty acid–CoA ester can then reenter the sequence, with ultimate loss of another acetyl-CoA. The 16-carbon palmitic acid yields altogether eight molecules of acetyl-CoA, which are then oxidized to CO_2 via the citric acid cycle. A large fraction of the standard free energy of oxidation of palmitic acid is recovered by oxidative phosphorylation as ATP energy. Oxidation of unsaturated fatty acids requires the action of two additional enzymes, enoyl-CoA isomerase and β-hydroxyacyl-CoA epimerase, needed to convert the D- into the L-stereoisomer of 3-hydroxyacyl-CoAs. Odd-carbon fatty acids are oxidized by the same basic pathway but yield one molecule of propionyl-CoA, which is carboxylated to methylmalonyl-CoA. The latter is isomerized to succinyl-CoA by a very complex reaction catalyzed by methylmalonyl-CoA mutase, which requires coenzyme B_{12}. The ketone bodies acetoacetate, D-β-hydroxybutyrate, and acetone are formed in the liver and are carried to other tissues, where they are oxidized via acetyl-CoA and the citric acid cycle. Fatty acid oxidation in the liver is regulated by the rate of entry of fatty acyl groups into the mitochondria, specifically by the allosteric inhibition of carnitine acyltransferase I by malonyl-CoA, an early intermediate of fatty acid biosynthesis in the cytosol. Whenever an animal is on a high carbohydrate intake, fatty acid oxidation is suppressed in favor of fatty acid biosynthesis.

References

Book

CUNNINGHAM, E. B.: *Biochemistry: Mechanisms of Metabolism*, McGraw-Hill, New York, 1978. Chapter 12 contains further details on the chemistry and enzymology of fatty acid oxidation pathways.

Articles

GREVILLE, D. G., and P. V. TUBBS: "Catabolism of Long-Chain Fatty Acids in Mammalian Tissues," *Essays Biochem.*, **4**:155–212 (1968). An early review, but basic to more recent developments.

MCGARRY, J. D., G. F. LEATHERMAN, and D. W. FOSTER: "Carnitine Palmitoyltransferase I. The Site of Inhibition of Hepatic Fatty Acid Oxidation by Malonyl-CoA," *J. Biol. Chem.*, **253**:4128–4136 (1978).

WILLIAMSON, D. H.: "Recent Developments in Ketone Body Metabolism," *Biochem. Soc. Proc.*, **7**:1313–1321 (1979).

"Recent Progress in β-Oxidation of Fatty Acids," *Biochem. Soc. Trans.*, **7**:68–88 (1978). A useful series of articles by different experts on various aspects of fatty acid oxidation and ketone-body metabolism.

Problems

1. *Energy in Triacylglycerols* On a per-carbon basis, where does the largest amount of biologically available energy in triacylglycerols reside: in the fatty acid portion or the glycerol portion? Indicate how knowledge of the chemical structure of triacylglycerols provides the answer.

2. *Fuel Reserves in Adipose Tissue*
   (a) If 15 percent of the body mass of a 70-kg adult consists of triacylglycerols, calculate the total available fuel reserve in kilocalories in the form of triacylglycerols.
   (b) If the basal energy requirement is approximately 2000 kcal/d, how long could this person survive if the oxidation of fatty acids stored as triacylglycerols were the only source of energy?
   (c) What would be the weight loss per day in pounds under such starvation conditions?

3. *Common Reaction Steps in the Fatty Acid Oxidation Cycle and Citric Acid Cycle* Cells often follow the same enzyme reaction pattern for bringing about analogous metabolic reactions. For example, the steps in the oxidation of pyruvate and α-ketoglutarate to acetyl-CoA and succinyl-CoA, although catalyzed by different enzymes, are very similar. The first stage in the oxidation of fatty acids follows a reaction sequence closely resembling one in the citric acid cycle. Show by equations the analogous reaction sequences in the two pathways.

4. *The Chemistry of the Acyl-CoA Synthetase Reaction* Fatty acids are converted into their coenzyme A esters by the reversible reaction

$$R{-}COOH + ATP + CoA{-}SH \rightleftharpoons R{-}\overset{\overset{\displaystyle O}{\|}}{C}{-}S{-}CoA + AMP + PP_i$$

   (a) The enzyme-bound intermediate in this reaction has been identified as the mixed anhydride of the fatty acid and adenosine monophosphate (AMP).

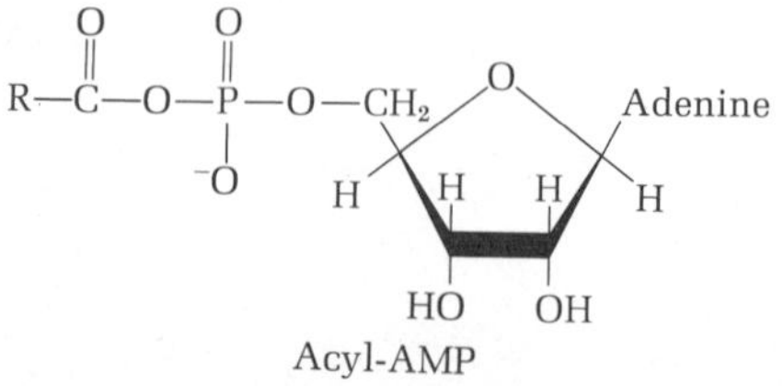

$$R{-}\overset{\overset{\displaystyle O}{\|}}{C}{-}O{-}\underset{\underset{\displaystyle ^{-}O}{|}}{\overset{\overset{\displaystyle O}{\|}}{P}}{-}O{-}CH_2\text{(ribose: H, H, H, H; HO, OH)}\text{—Adenine}$$

Acyl-AMP

   Write two equations corresponding to the sequence of two steps involved in the reaction catalyzed by acyl-CoA synthetase with acyl-AMP as the intermediate.

(b) The reaction above is readily reversible, with an equilibrium constant near 1. How can this reaction be made to favor formation of AMP? How can it be made to favor formation of

$$R{-}\overset{\displaystyle O}{\overset{\|}{C}}{-}S{-}CoA?$$

5. *Oxidation of Tritiated Palmitate* Palmitate uniformly labeled with tritium (^3H) to a specific activity of 2.48×10^8 cpm per micromole of palmitate is added to a mitochondrial preparation that oxidizes it to acetyl-CoA. The acetyl-CoA is isolated and hydrolyzed to acetate. The specific activity of the isolated acetate per micromole of acetate is 1.00×10^7 cpm. Is this result consistent with the β-oxidation pathway? Explain. What is the final fate of the removed tritium?

6. *Fatty Acids as a Source of Water* Contrary to legend, camels do not store water in their humps, which actually consist of a large fat deposit. How can these fat deposits serve as a source of water? Calculate the amount of water (gallons) that can be produced by the camel from 1 lb of fat. Assume for simplicity that the fat consists entirely of tripalmitin.

7. *Petroleum as a Microbial Food Source* Some microorganisms of the genus *Nocardia* and *Pseudomonas* can grow in an environment where hydrocarbons are the only food source. These bacteria oxidize straight-chain aliphatic hydrocarbons to their corresponding carboxylic acids, e.g.,

$$NAD^+ + \underset{\text{Octane}}{CH_3(CH_2)_6CH_3} + O_2 \longrightarrow CH_3(CH_2)_6C\begin{smallmatrix}{/\!\!/}O\\ \diagdown OH\end{smallmatrix} + H^+ + NADH$$

How can these bacteria be used to clean up oil spills?

8. *Metabolism of a Straight-Chain Phenylated Fatty Acid* A crystalline metabolite was isolated from the urine of a rabbit which had been fed a straight-chain fatty acid containing a terminal phenyl group

$$C_6H_5{-}CH_2(CH_2)_nCOOH$$

An aqueous solution of this metabolite was acidic. A 302-mg sample of the metabolite was completely neutralized by adding 22.2 mL of 0.1 M NaOH.

(a) What is the probable molecular weight and structure of the metabolite?

(b) Did the straight-chain fatty acid fed to the rabbit contain an even or an odd number of methylene ($-CH_2-$) groups, i.e., is n even or odd? Explain.

9. *Fatty Acid Oxidation in Diabetics* When the acetyl-CoA produced during β oxidation in the liver exceeds the capacity of the citric acid cycle, the excess acetyl-CoA reacts to form the ketone bodies acetoacetate, D-β-hydroxybutyrate, and acetone. This condition exists in severe diabetics because the tissues of such patients are unable to utilize glucose and oxidize large amounts

of fatty acids instead. Although acetyl-CoA is not toxic, the mitochondrion must divert the acetyl-CoA to ketone bodies. Why? How does this diversion solve the problem?

10. *Consequences of a High-Fat Diet with No Carbohydrates* Suppose you had to subsist on a diet of whale and seal blubber with little or no carbohydrate.
    (a) What would be the effect of carbohydrate deprivation on the utilization of fats for energy?
    (b) If your diet were totally devoid of carbohydrate, would it be better to consume odd- or even-numbered fatty acids? Explain.

11. *Formation of Acetyl-CoA from Fatty Acid Precursors* Write a balanced net equation for the formation of acetyl-CoA from the following substances, including all activation steps:
    (a) Myristoyl-CoA
    (b) Stearic acid
    (c) D-β-Hydroxybutyric acid

12. *Pathway of Labeled Atoms during Fatty Acid Oxidation* Palmitic acid labeled with ^{14}C in carbon atom 9 is oxidized under conditions in which the citric acid cycle is operating. What will the location of ^{14}C be in (a) acetyl-CoA, (b) citric acid, and (c) butyryl-CoA? Assume only one turn of the citric acid cycle.

13. *Net Equation for Complete Oxidation of β-Hydroxybutyric Acid* Write the net equation for the complete oxidation of β-hydroxybutyric acid in the kidney. Include any required activation steps and all oxidative phosphorylations.

CHAPTER 19

Oxidative Degradation of Amino Acids: The Urea Cycle

Most of the metabolic energy generated in the tissues comes from the oxidation of carbohydrates and triacylglycerols, which together furnish up to 90 percent of the energy requirement of the adult human male. The remainder, from 10 to 15 percent, depending on the diet, is furnished by the oxidation of amino acids.

Although amino acids function primarily as building blocks for the biosynthesis of protein, they can undergo oxidative degradation in three different metabolic circumstances. (1) During the normal dynamic turnover of body proteins the amino acids released, if not needed for synthesis of new body proteins, may undergo oxidative degradation. (2) When amino acids are ingested in excess of the body's needs for protein synthesis, the surplus may be catabolized, since amino acids cannot be stored. (3) During fasting or in diabetes mellitus, when carbohydrates are either unavailable or not properly utilized, body proteins are called upon as fuel. Under these different circumstances, amino acids undergo loss of their amino groups, and the α-keto acids so formed may undergo oxidation to carbon dioxide and water, in part via the citric acid cycle.

In this chapter we shall examine the metabolic pathways taken when the 20 amino acids occurring in proteins undergo oxidative degradation. We shall also see that the ammonia released from amino acids is excreted in different chemical forms, depending upon the species of organism.

Transfer of α-Amino Groups Is Catalyzed by Transaminases

The α-amino groups of the 20 L-amino acids commonly found in proteins are ultimately removed at some stage in their oxidative degradation. If not reused for synthesis of new amino acids or other nitrogenous products, these amino groups are collected and ultimately converted into a single excretory end

Figure 19-1
The transaminase reaction. The transferred amino group is in color. In most transaminases α-ketoglutarate is the amino-group acceptor.

$$H_3\overset{+}{N}-CH(COO^-)-R_1 \quad \text{L-Amino acid}$$

$$+$$

$$^-OOC-C(=O)-CH_2-CH_2-COO^- \quad \alpha\text{-Ketoglutarate}$$

$$\rightleftharpoons \text{ transaminase (pyridoxal phosphate)}$$

$$^-OOC-C(=O)-R_1 \quad \alpha\text{-Keto acid}$$

$$+$$

$$H_3\overset{+}{N}-CH(COO^-)-CH_2-CH_2-COO^- \quad \text{L-Glutamate}$$

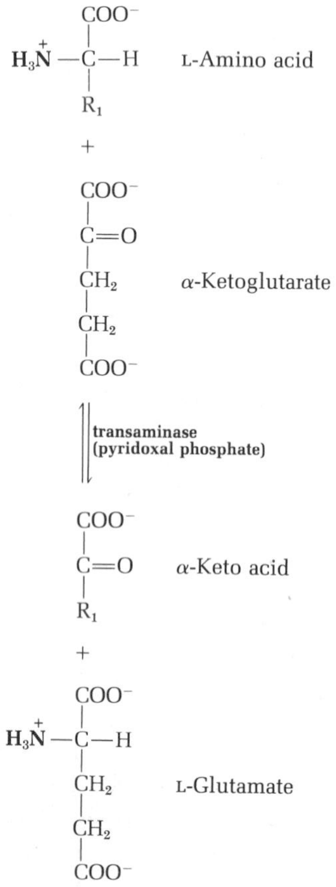

product, which in human beings and most other terrestrial vertebrates is urea. The removal of the α-amino groups of most of the L-amino acids is promoted by enzymes called transaminases or aminotransferases. In these reactions, called transaminations, the α-amino group is enzymatically transferred from the amino acid to the α carbon atom of α-ketoglutarate, leaving behind the corresponding α-keto acid analog of the incoming amino acid and causing the amination of the α-ketoglutarate to form L-glutamate (Figure 19-1):

$$\text{L-}\alpha\text{-Amino acid} + \alpha\text{-ketoglutarate} \rightleftharpoons \alpha\text{-keto acid} + \text{L-glutamate}$$

It will be noted that there is no *net* deamination, i.e., loss of amino groups, in such reactions since the α-ketoglutarate becomes aminated as the α-amino acid is deaminated. The whole point of transamination reactions is to *collect* the amino groups from many different amino acids in the form of only one, namely, L-glutamate. Thus amino-group catabolism converges into a single product.

Most transaminases are specific for α-ketoglutarate as the amino-group acceptor in the reaction as written above. However, they are less specific for the other substrate, the L-amino acid that donates the amino group. Some of the most important transaminases, which are named for the amino-group donor, are designated by the following equations:

$$\text{L-Alanine} + \alpha\text{-ketoglutarate} \xrightleftharpoons{\text{alanine transaminase}} \text{pyruvate} + \text{L-glutamate}$$

$$\text{L-Aspartate} + \alpha\text{-ketoglutarate} \xrightleftharpoons{\text{aspartate transaminase}} \text{oxaloacetate} + \text{L-glutamate}$$

$$\text{L-Leucine} + \alpha\text{-ketoglutarate} \xrightleftharpoons{\text{leucine transaminase}} \alpha\text{-ketoisocaproate} + \text{L-glutamate}$$

$$\text{L-Tyrosine} + \alpha\text{-ketoglutarate} \xrightleftharpoons{\text{tyrosine transaminase}} p\text{-hydroxyphenylpyruvate} + \text{L-glutamate}$$

Thus α-ketoglutarate is the common acceptor of amino groups from most of the other amino acids. The L-glutamate so formed serves to channel amino groups into certain biosynthetic pathways (Chapter 22) or into a final sequence of reactions by which nitrogenous waste products are formed and then excreted. The reactions catalyzed by the transaminases are freely reversible, since they have an equilibrium constant of about 1.0. $\Delta G^{\circ\prime}$ for such reactions is therefore close to zero (page 367).

All transaminases have a tightly bound prosthetic group and share a common reaction mechanism. The prosthetic group is pyridoxal phosphate, a derivative of pyridoxine or vitamin B_6 (page 258). Pyridoxal phosphate functions as an intermediate carrier of amino groups on the active site of transaminases (Figure 19-2). During the catalytic cycle it undergoes reversible

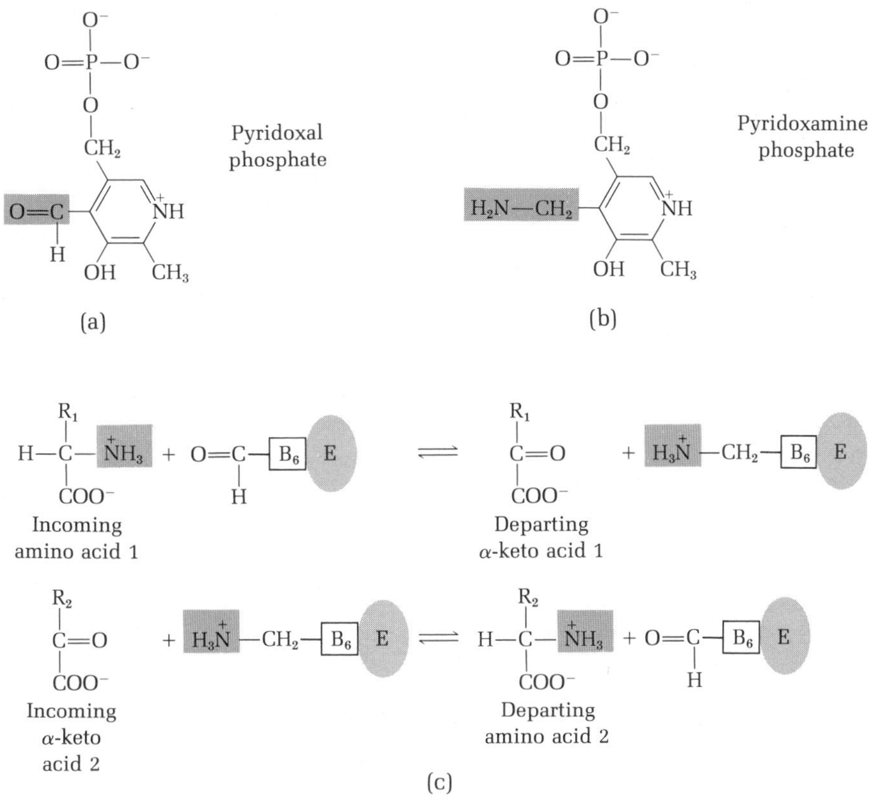

Pyridoxal phosphate

(a)

Pyridoxamine phosphate

(b)

Incoming amino acid 1

Departing α-keto acid 1

Incoming α-keto acid 2

Departing amino acid 2

(c)

Figure 19-2
The prosthetic group of transaminases. Pyridoxal phosphate (a) and its aminated form pyridoxamine phosphate (b) are the tightly bound coenzymes of transaminases. The functional groups involved in their action are shaded. (c) Pyridoxal phosphate is the intermediate carrier of an amino group in the action of transaminases. E represents the enzyme protein, and $O{=}\underset{H}{C}{-}\boxed{B_6}$ *the tightly bound pyridoxal phosphate.*

Transaminases catalyze bimolecular ping-pong reactions (page 218). The first substrate, α-amino acid 1, departs as α-keto acid 1, after losing its amino group, before the second substrate, α-keto acid 2, is bound.

transitions between its aldehyde form, pyridoxal phosphate, which can accept amino groups, and its aminated form, pyridoxamine phosphate, which can donate its amino group to α-ketoglutarate. In this way the prosthetic group acts as a reversible, transient carrier of amino groups from an α-amino acid to α-ketoglutarate (Figure 19-2). Transaminases are classical examples of enzymes catalyzing bimolecular ping-pong reactions (page 218). In such reactions the first substrate must leave the active site before the second substrate can bind. Thus the incoming amino acid binds to the active site, donates its amino group to pyridoxal phosphate, and departs in the form of an α-keto acid. Then the incoming α-keto acid is bound, accepts the amino group from pyridoxamine phosphate, and departs, now in the form of an amino acid.

Figure 19-3 shows how the carbonyl group of enzyme-bound pyridoxal phosphate condenses with the α-amino group of the incoming amino acid to form a covalently linked intermediate, a type of compound called a Schiff base. A shift of the C=N double bond follows and the carbon skeleton of the amino acid is then hydrolyzed off, leaving the amino group covalently bound to the prosthetic group in the form of pyridoxamine phosphate. The latter now forms a Schiff base with the incoming α-ketoglutarate, to which the amino acid group is transferred, essentially through reversal of the reactions by which it was formed.

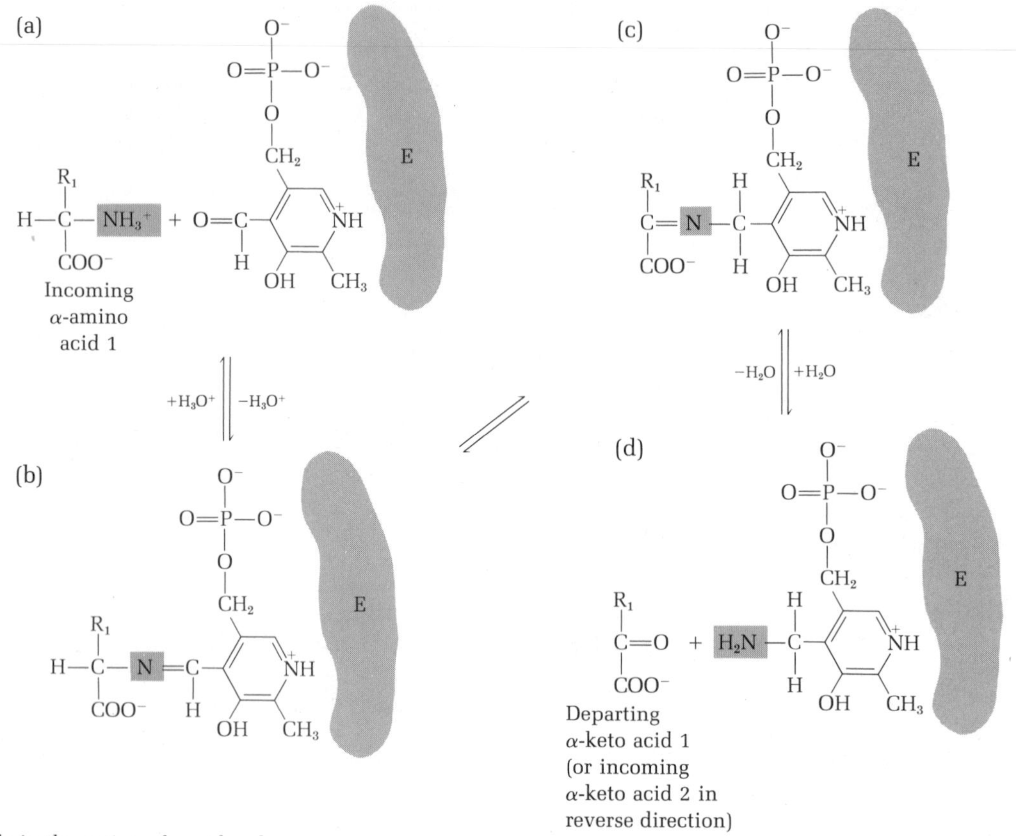

Figure 19-3
Some details in the action of pyridoxal phosphate in transaminases. The amino group of the incoming α-amino acid (a) *reacts with the carbonyl group of the pyridoxal phosphate, tightly bound to the enzyme, to yield a Schiff base intermediate* (b), *which undergoes transformation to its tautomeric form* (c). *The latter is hydrolyzed to yield the corresponding α-keto acid, which departs, leaving the transaminase with the amino group covalently bound as pyridoxamine phosphate* (d). *Since these reactions are reversible, the aminated form of the transaminase then transfers its amino group to an incoming α-keto acid 2, to yield the new amino acid.*

The measurement of alanine and aspartate transaminases in blood serum is an important diagnostic procedure in medicine, used to determine the severity of heart attacks and to monitor recovery. It is also used to detect the toxic effects of some industrial chemicals (Box 19-1).

Ammonia Is Formed from Glutamate

We have seen that amino groups are removed from nearly all the α-amino acids by transamination to α-ketoglutarate to form L-glutamate. How are amino groups removed from glutamate in order to prepare them for excretion?

Glutamate undergoes *oxidative deamination* by the action of L-*glutamate dehydrogenase,* which requires NAD^+ as the acceptor of the reducing equivalents:

$$\text{L-Glutamate}^- + NAD^+ + H_2O \rightleftharpoons \alpha\text{-ketoglutarate}^{2-} + NH_4^+ + NADH + H^+$$

This enzyme is present only in the mitochondria, where it is localized in the matrix. Glutamate dehydrogenase is responsible for most of the ammonia formed in animal tissues, since glutamate is the only amino acid whose α-amino group can be

Box 19-1 Transaminases and other enzymes in the blood are useful in medical diagnosis

Alanine transaminase (also called glutamate-pyruvate transaminase, GPT) and aspartate transaminase (also called glutamate-oxaloacetate transaminase, GOT) are important in the diagnosis of heart and liver damage. Occlusion of a coronary artery by lipid deposits can cause severe local anoxia and ultimately the degeneration of a localized portion of the heart muscle; this process is called *myocardial infarction*. Such damage causes these enzymes, among others, to leak from the injured heart cells into the bloodstream. Measurements of the concentration in the blood serum of these two transaminases by the SGPT and SGOT tests (S for serum) and of another heart enzyme, *creatine kinase* (the SCK test), can give important information regarding the severity and the stage of the damage to the heart. Creatine kinase is the first heart enzyme to appear in the blood after a coronary attack; it also disappears quickly from the blood. GOT is the next to appear, and GPT follows later. Lactate dehydrogenase also leaks from injured or anaerobic heart muscle.

SGOT and SGPT are also important in industrial medicine to determine whether persons exposed to carbon tetrachloride, chloroform, or other solvents used in chemical, dry-cleaning, and other industries have suffered liver damage. These solvents cause liver degeneration, with resulting leakage into the blood of various enzymes from the injured liver cells. Transaminases, which are very active in liver and whose activity can be detected in very small amounts, are most useful in monitoring the blood serum of persons exposed to such industrial chemicals.

Analysis of different enzyme activities in blood serum gives valuable diagnostic information for a number of disease conditions.

directly removed at a high rate in this manner. Glutamate and glutamate dehydrogenase therefore have a unique role in amino-group metabolism.

Glutamate dehydrogenase is a complex allosteric enzyme. Its molecular weight is about 300,000, and it consists of six identical subunits, each containing a single polypeptide chain of 500 residues. It is strongly activated by the positive modulator ADP but inhibited by GTP, the product of the succinyl-CoA synthetase reaction in the citric acid cycle (page 446). Whenever the liver cell needs fuel for the citric acid cycle to form more ATP, glutamate dehydrogenase activity is increased, making α-ketoglutarate available for the citric acid cycle and releasing NH_3 to be excreted. On the other hand, whenever GTP accumulates in the mitochondria as a result of high citric acid cycle activity, oxidative deamination of glutamate is inhibited.

Ammonia can be salvaged and reused in the synthesis of amino acids. In this case glutamate dehydrogenase acts in reverse, reducing ammonia and α-ketoglutarate to form glutamate. Instead of being the simple reverse of the NAD-linked reaction written above, however, this reaction is promoted by the NADP-linked reaction

$$\text{NADPH} + \text{H}^+ + \text{NH}_4^+ + \alpha\text{-ketoglutarate}^{2-} \longrightarrow \text{NADP}^+ + \text{glutamate}^- + \text{H}_2\text{O}$$

The use of two different coenzymes by glutamate dehydrogenase for the release and uptake of NH_3 makes possible independent regulation of the deamination of glutamate and the amination of α-ketoglutarate, even though both are catalyzed by the same enzyme.

We shall now consider the oxidative pathways taken by the deaminated amino acids, in accordance with the theme of the last few chapters, namely, the catabolic pathways by which the major nutrients are ultimately oxidized and used as a source of energy. Later in the chapter we shall resume discussion of the fate of the amino groups.

The Carbon Skeletons of Amino Acids Are Degraded by 20 Different Pathways

There are 20 standard amino acids in proteins, all having different carbon skeletons. Correspondingly, there are 20 different catabolic pathways for their degradation. Since these pathways taken together normally account for only 10 percent of the body's energy, each of the 20 accounts for only about 0.5 percent of the total catabolism, on the average. The amino acid pathways, taken singly, are therefore nowhere near as active as glycolysis and the citric acid cycle. For this reason we shall not examine them all in detail. As it happens, the 20 amino acid catabolic pathways converge to form only five products, all of which enter the citric acid cycle for complete oxidation to CO_2 and H_2O (Figure 19-4).

Figure 19-4 shows that the carbon skeletons of 10 of the amino acids are ultimately broken down to yield *acetyl-CoA*.

Figure 19-4
Entry of the carbon skeletons of the common amino acids into the citric acid cycle. Leucine and tryptophan are degraded to yield both acetoacetyl-CoA and acetyl-CoA.

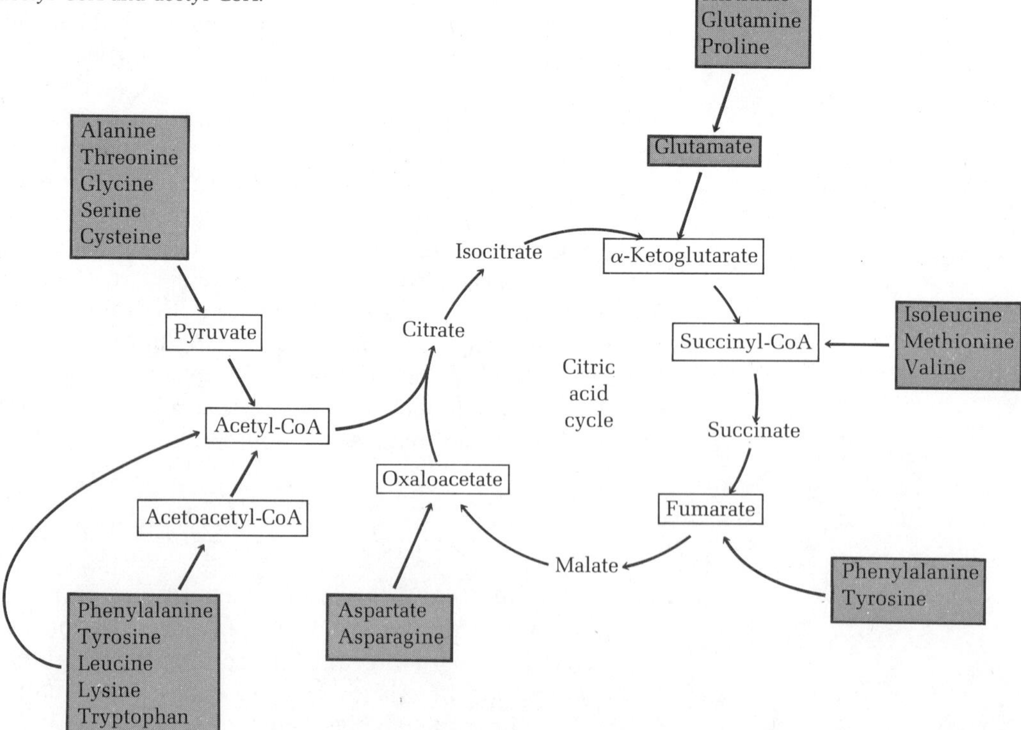

Five amino acids are converted into *α-ketoglutarate*, three into *succinyl-CoA*, two into *oxaloacetate*, and two yield *fumarate*. The individual pathways for the 20 amino acids will be summarized by means of flow sheets, each leading to a specific point of entry into the citric acid cycle. In these flow sheets the amino acid carbon atoms that enter the citric acid cycle are given in color. Some of the enzymatic reactions in these pathways that are particularly noteworthy for their mechanisms or their medical significance will be singled out for special discussion.

Ten Amino Acids Yield Acetyl-CoA during Their Degradation

The carbon skeletons of 10 amino acids yield acetyl-CoA, which enters the citric acid cycle directly. Five of the ten are degraded to acetyl-CoA via *pyruvate*. The other five are converted into *acetoacetyl-CoA*, which is then cleaved to form acetyl-CoA (Figure 19-4). The five amino acids entering via pyruvate are *alanine*, *cysteine*, *glycine*, *serine*, and *threonine* (Figure 19-5). Alanine yields pyruvate directly on transamination with α-ketoglutarate. The 4-carbon amino acid *threonine* is degraded into the 2-carbon *glycine*. Glycine has two pathways.

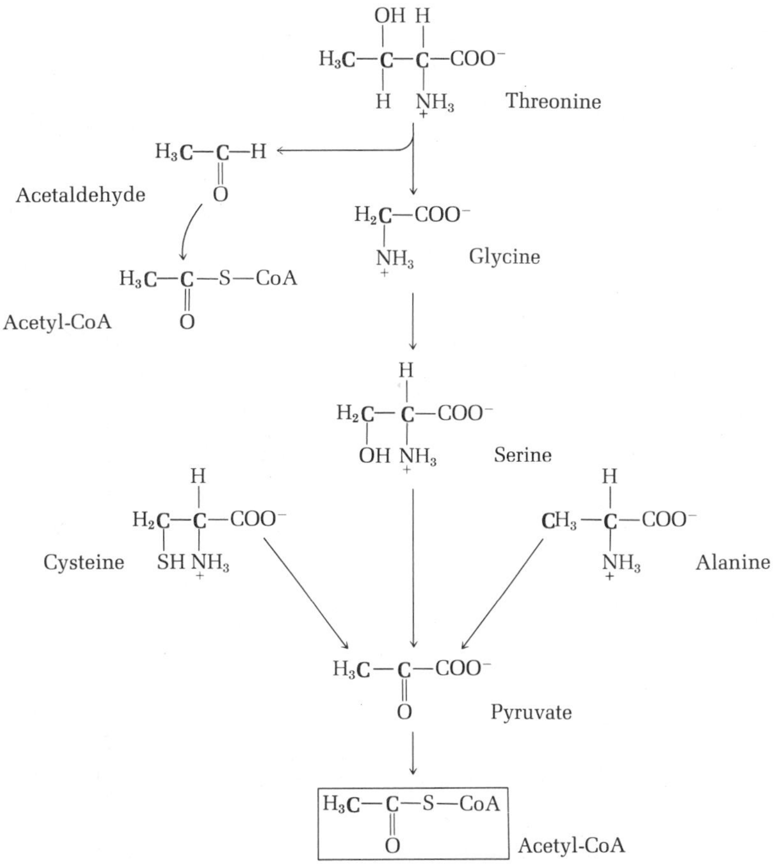

Figure 19-5
Outline of the pathways from threonine, glycine, serine, cysteine, and alanine to acetyl-CoA via pyruvate.

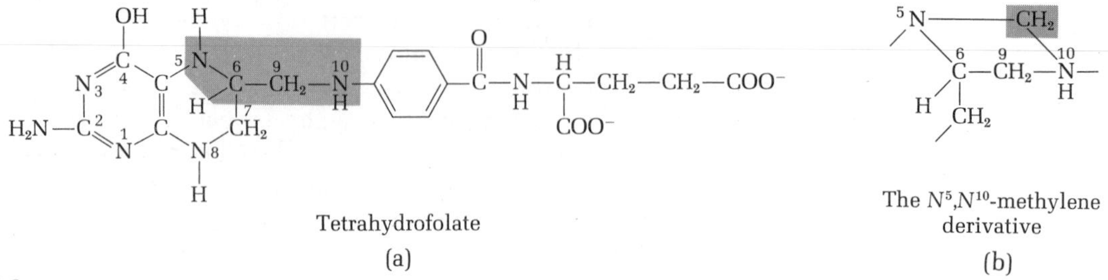

Tetrahydrofolate

(a)

The N^5,N^{10}-methylene derivative

(b)

Figure 19-6
(a) Tetrahydrofolate (FH_4). The part of the molecule carrying the 1-carbon group is shaded. (b) The N^5,N^{10}-methylene derivative of FH_4, formed in the glycine synthase reaction. The methylene group is shaded.

It can be converted into serine, a 3-carbon amino acid, by enzymatic addition of a hydroxymethyl group carried by the coenzyme *tetrahydrofolate* (Figure 19-6). As we have seen (page 260), tetrahydrofolate functions as a carrier of 1-carbon groups, such as methyl, formyl, formimino, and hydroxymethyl groups (Figure 19-6). However, the major pathway of glycine degradation is by a different tetrahydrofolate-requiring reaction, which brings about oxidative cleavage of glycine into CO_2, NH_4^+, and a methylene group ($—CH_2—$), which is accepted by tetrahydrofolate. This readily reversible reaction is catalyzed by *glycine synthase*

$$\underset{\text{Glycine}}{H_3\overset{+}{N}—CH_2—COO^-} + FH_4 + NAD^+ \rightleftharpoons N^5,N^{10}\text{-methylene } FH_4 + CO_2 + NADH + NH_4^+$$

In this pathway the two carbon atoms of glycine do not enter the citric acid cycle. One is lost as CO_2, and the other becomes the methylene group of N^5,N^{10}-methylene tetrahydrofolate (Figure 19-6), which is used as a methyl-group donor in certain biosynthetic pathways.

Portions of the carbon backbone of the amino acids *phenylalanine, tyrosine, lysine, tryptophan,* and *leucine* yield *acetoacetyl-CoA,* which is then converted into acetyl-CoA (Figure 19-7).

Two pathways in this set deserve special mention. The pathway from tryptophan to acetyl-CoA is the most complex of all the pathways of amino acid catabolism in animal tissues: it has 13 steps. Some of the intermediates in tryptophan catabolism are required precursors for biosynthesis of other important biomolecules, including *serotonin,* a neurohormone, and *nicotinic acid,* a vitamin (Figure 19-8). The tryptophan catabolic pathway thus has a number of branches that make possible the formation of several other products from the single precursor tryptophan.

The second noteworthy pathway in this group is that leading from *phenylalanine* (Figure 19-9). Phenylalanine and its oxidation product tyrosine are degraded into two fragments, both of which can enter the citric acid cycle but at different points. Four of the nine carbon atoms of phenylalanine and tyrosine yield free acetoacetate, which is then converted into

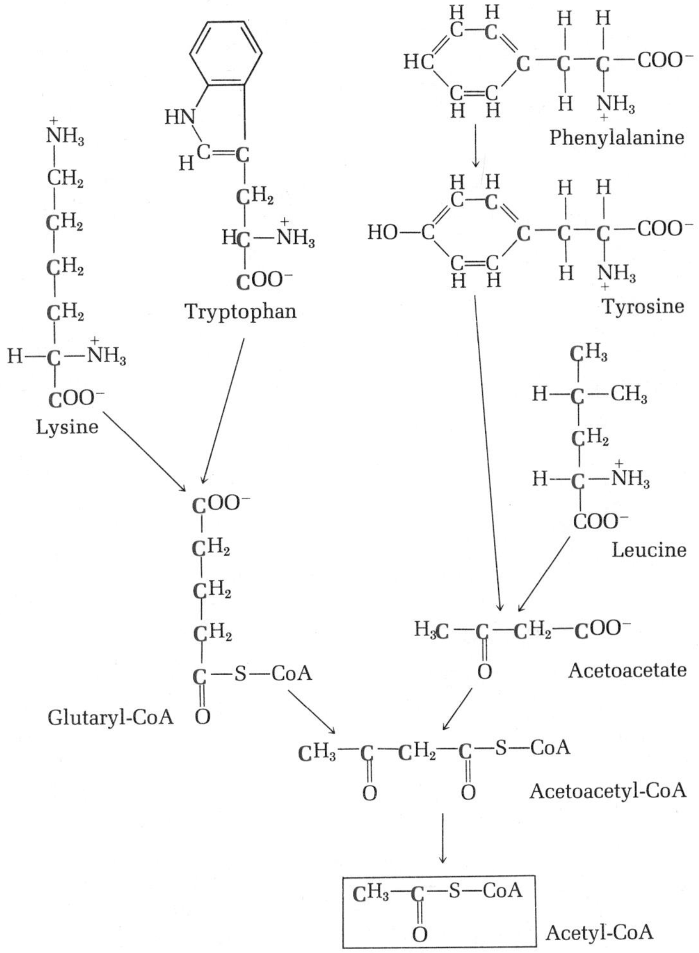

Figure 19-7
Outline of the pathways from lysine, tryptophan, phenylalanine, tyrosine, and leucine to acetyl-CoA via acetoacetyl-CoA.

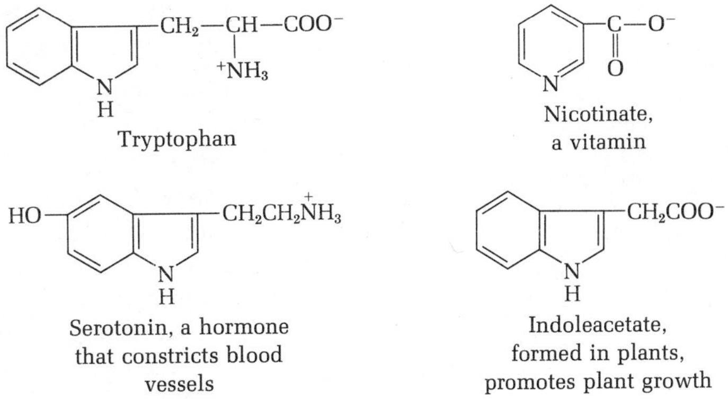

Lysine

Tryptophan

Phenylalanine

Tyrosine

Leucine

Glutaryl-CoA

Acetoacetate

Acetoacetyl-CoA

Acetyl-CoA

Figure 19-8
Tryptophan and some of its important metabolic products.

Tryptophan

Nicotinate, a vitamin

Serotonin, a hormone that constricts blood vessels

Indoleacetate, formed in plants, promotes plant growth

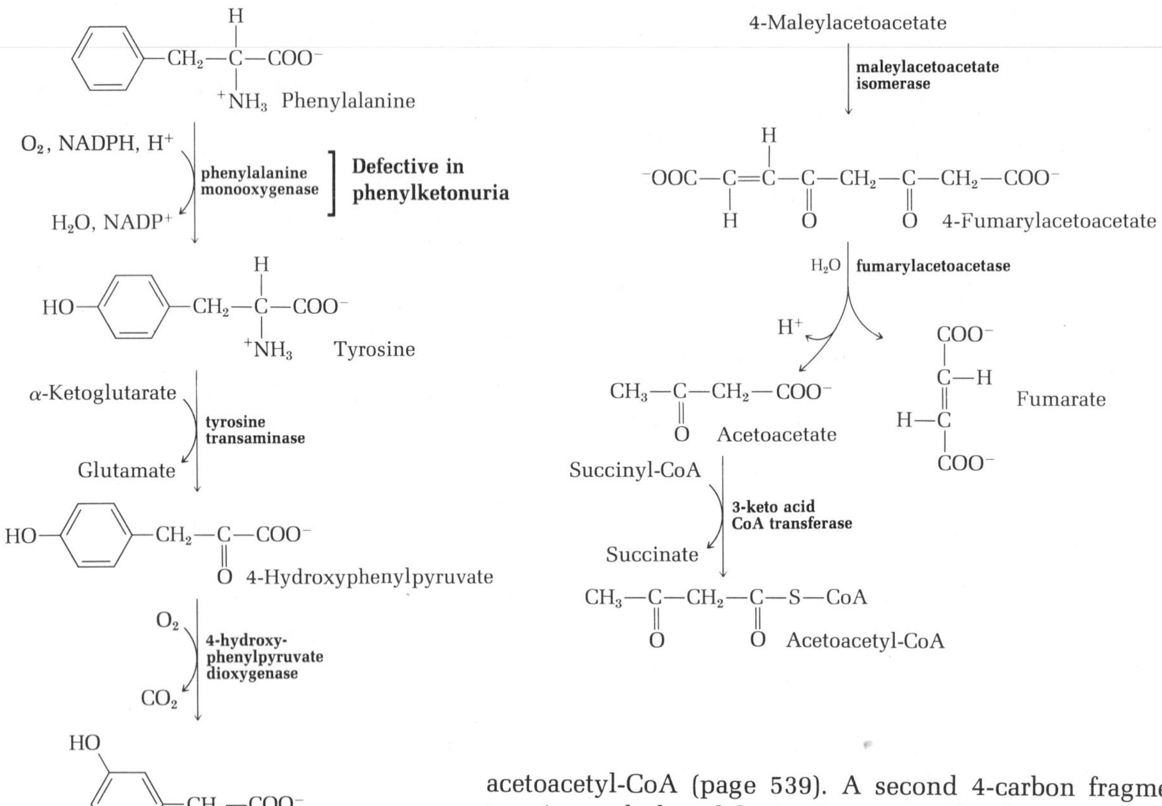

Figure 19-9
Normal pathway for the conversion of phenylalanine and tyrosine into aceto-acetyl-CoA and fumaric acid. The first enzyme in this pathway is defective in the genetic disease phenylketonuria.

acetoacetyl-CoA (page 539). A second 4-carbon fragment of tyrosine and phenylalanine is recovered as *fumarate*, an intermediate of the citric acid cycle (see below). Eight of the nine carbon atoms of these two amino acids thus enter the citric acid cycle; the remaining carbon is lost as CO_2. Phenylalanine, after its hydroxylation to yield tyrosine, is also the ultimate precursor of the thyroid hormone *thyroxine* and of the hormones *adrenaline* and *noradrenaline*, secreted by the adrenal medulla (Chapter 25).

Phenylalanine Catabolism Is Genetically Defective in Some People

Many different genetic defects in amino acid metabolism have been identified in human beings. In these conditions, most of which are rare, a specific gene coding the amino acid sequence of some enzyme in an amino acid pathway has undergone mutation. As a consequence, the enzyme synthesized in response to the altered gene is defective because it contains the wrong amino acid in some critical position in the polypeptide chain or has lost or gained amino acid residues. Such a genetically altered enzyme may be completely inactive or it may show only a fraction of the activity of the normal enzyme, resulting from an altered K_M or V_{max}. Most of the human genetic defects involving amino acid metabolism cause specific intermediates to accumulate. In some of these defects, certain nerve bundles fail to develop properly and thus lead to mental retardation.

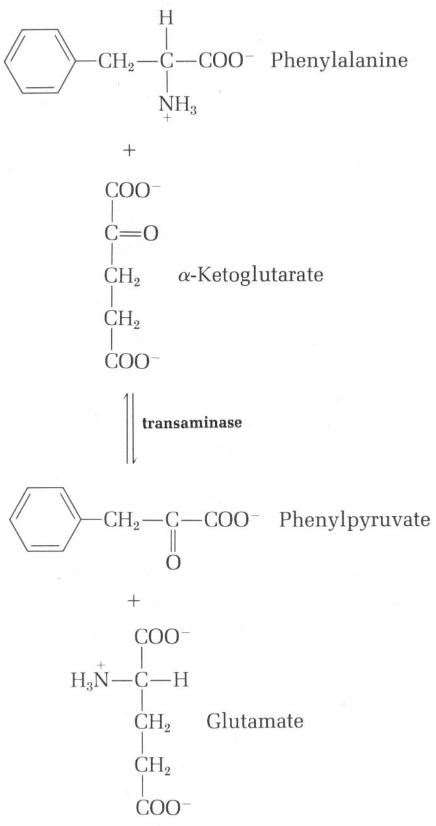

Figure 19-10
Formation of phenylpyruvate by an alternate pathway in phenylketonuria.

The phenylalanine-tyrosine pathway requires special mention in this respect, since three of its enzymatic steps are particularly prone to genetic alteration, leading to three different types of heritable human metabolic defects. The first enzyme in this pathway (Figure 19-9), *phenylalanine 4-monooxygenase* (also called *phenylalanine hydroxylase*), which catalyzes the hydroxylation of phenylalanine to tyrosine, is defective in some people. This defect is responsible for the genetic disease *phenylketonuria*. Phenylalanine monooxygenase inserts one of the two oxygen atoms of O_2 into phenylalanine to form the hydroxyl group of tyrosine; the other oxygen atom is reduced to H_2O by the NADH also required in the reaction:

$$\text{L-Phenylalanine} + \text{NADH} + \text{H}^+ + \text{O}_2 \longrightarrow \text{L-tyrosine} + \text{NAD}^+ + \text{H}_2\text{O}$$

When phenylalanine 4-monooxygenase is genetically defective, a secondary pathway of phenylalanine metabolism, normally little used, comes into play. In this minor pathway phenylalanine undergoes transamination with α-ketoglutarate to yield *phenylpyruvate* (Figure 19-10)

$$\text{Phenylalanine} + \alpha\text{-ketoglutarate} \rightleftharpoons \text{phenylpyruvate} + \text{glutamate}$$

However, phenylpyruvate is not metabolized further and thus is a dead end, since it (and also phenylalanine) accumulates in the blood and tissues and is excreted in the urine. Excess phenylpyruvate in the blood in early life impairs the normal development of the brain and causes severe mental retardation. *Phenylketonuria* (PKU) was among the first human genetic defects of metabolism discovered. When this condition is recognized early enough in infancy, mental retardation can largely be prevented by a proper choice of diet, so that foods containing proteins with a high content of phenylalanine are withheld. Since nearly all proteins contain at least some phenylalanine, and since phenylalanine is also required in small amounts for proper growth [it is an *essential* amino acid (Chapter 26)], the composition of the diet must be carefully controlled. Natural proteins, such as casein of milk, must first be hydrolyzed and the phenylalanine removed.

Phenylketonuria must be diagnosed and dietary treatment begun in the first weeks after birth; otherwise irreversible mental retardation develops. Many untreated phenylketonurics die before they reach the age of 25; others require institutional care for life, at great human and social cost (Box 19-2). Phenylketonuria is a serious public health problem. It is relatively common, since 1 out of every 10,000 infants is born with the defect. Most states now require every newborn infant to be tested for this defect, which is easily detected by measurement of phenylalanine or phenylpyruvate in the urine.

Box 19-2 The human, social, and economic costs of some genetic diseases

There are believed to be well over 2000 kinds of human genetic defects or diseases, and the number continues to increase rapidly. More than 120,000 infants with genetic diseases are born each year in the United States. In many cases the emotional costs to the human beings involved are very great and cannot be counted. The costs to society are also large and call for increased counseling of prospective parents and for the better education of the public.

Phenylketonuria is an example of a relatively common genetic disease that can be detected and treated. If it is discovered in infancy and the child is kept on a carefully controlled diet during the first 6 years of life, the child will develop into a normal adult. Detection and treatment are very costly. Nevertheless, quite apart from humanitarian considerations, the costs of *not* diagnosing and treating PKU children are much greater. In 1980 the cost of a PKU test was about $2 per infant. That comes to about $6 million to test the 3 million children born each year in the United States. Since the incidence of positive PKU tests is 1 in 10,000 births, the testing program will reveal as many as 300 afflicted children each year who must receive a special diet free of phenylalanine. The cost of administering the diet is over $1000 a year per child. The annual cost of dietary treatment is thus as high as $1.8 million. The total annual cost of this program nationwide thus approaches $8 million and can be expected to increase. This may seem to be a very large amount to spend on 300 children, but the alternative is much more costly. If no tests were carried out at birth, the 300 children might have to be maintained for life in a mental institution for an average of approximately 30 years, at a total cost of $10,000 per year per child. Such a cost-benefit analysis shows that carrying out the PKU program nationwide, while costing over $8 million annually, ultimately saves up to 10 times that amount. Thus, for PKU at least, the effort is clearly worth the cost, and the same is probably true for other genetic disorders as well.

In some genetic diseases it is possible to screen prospective parents to detect carriers of defective genes. Such screening is not infallible and is in most cases carried out on a voluntary basis. Sickle-cell anemia (page 194) and Tay-Sachs disease (Chapter 21) are among those that can be detected in carriers, but unfortunately many hereditary diseases cannot. For some genetic diseases, amniocentesis can be used to detect afflicted fetuses before birth. Among these is Tay-Sachs disease (Chapter 21). Unfortunately, the only "treatment" for some genetic diseases diagnosed by amniocentesis is abortion, a difficult choice to make.

Most tragic are those genetic diseases which cannot be predicted from tests on prospective parents, for which early detection is not possible, or for which virtually nothing can be done at this time. The costs of long-term care for victims of some of these diseases cannot be met by private means and require the patients to become wards of society. Even if biochemists could successfully trace the origins of all genetic diseases to faulty gene structure, biological science alone cannot answer the social and ethical questions that are posed.

The fourth enzyme in the phenylalanine pathway (Figure 19-9), *homogentisate dioxygenase,* also is defective in some people as the result of a heritable genetic mutation. Persons with this defect are unable to degrade homogentisate, an intermediate in phenylalanine catabolism. As a result homogentisate accumulates in the body fluids and is excreted in the urine. Allowed to stand exposed to air, the urine of such a person will turn black. As the urine becomes alkaline due to decomposition of some of the urea to form ammonia, homogentisate is spontaneously oxidized by atmospheric O_2 to a black pigment, related to that present in the skin of black people. This genetic defect is known as *alkaptonuria.* People with this condition usually suffer no significant impairment of health, apart from concern over their black urine. Historical records tell us that people who passed black urine literally worried themselves sick; black urine was regarded as a bad omen.

Many other genetic diseases involving amino acid metabolism are known (Table 19-1).

Table 19-1 Some Human Genetic Disorders Affecting Amino Acid Metabolism

Name	Defective enzyme or process
Albinism	Tyrosine 3-monooxygenase
Alkaptonuria	Homogentisate 1,2-dioxygenase
Argininosuccinic acidemia	Argininosuccinate lyase
Homocystinuria	Cystathionine β-synthase
Maple syrup urine disease	Branched-chain α-keto acid dehydrogenase
Phenylketonuria	Phenylalanine 4-monooxygenase
Hypervalinemia	Valine transaminase

Five Amino Acids Are Converted into α-Ketoglutarate

The carbon skeletons of five amino acids (arginine, histidine, glutamic acid, glutamine, and proline) enter the citric acid cycle via α-ketoglutarate (Figure 19-11).

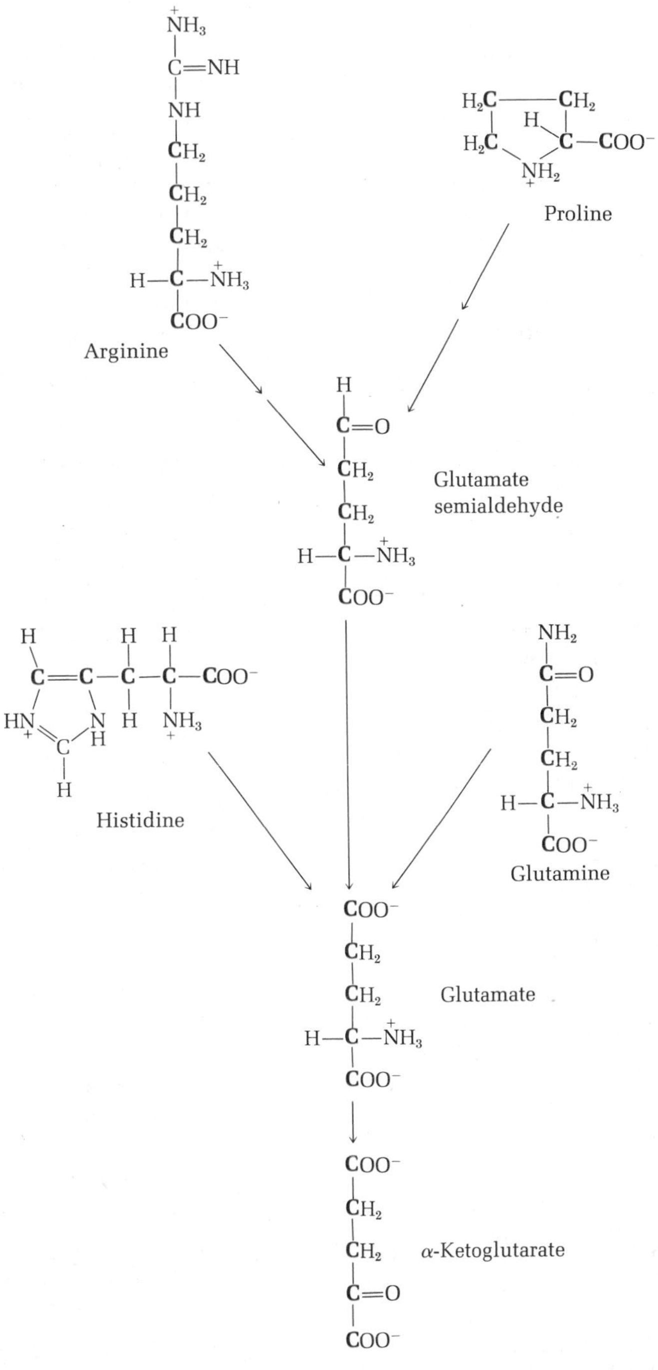

Figure 19-11
Outline of the pathways from arginine, histidine, proline, glutamine, and glutamate to α-ketoglutarate.

Arginine

Proline

Glutamate semialdehyde

Histidine

Glutamine

Glutamate

α-Ketoglutarate

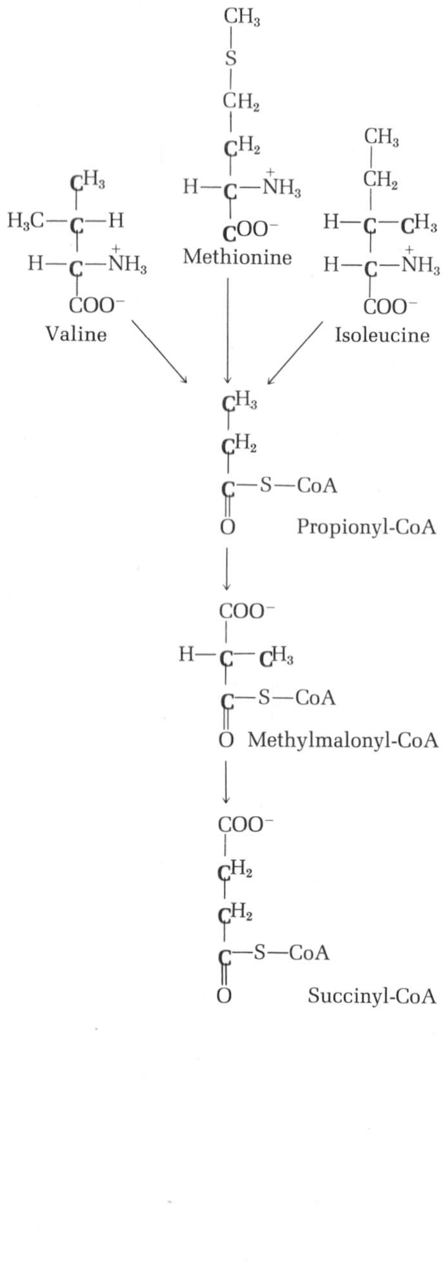

Figure 19-12
Outline of the pathways from isoleucine, methionine, and valine to succinyl-CoA.

Three Amino Acids Are Converted into Succinyl-CoA

The carbon skeletons of *methionine, isoleucine,* and *valine* are ultimately degraded by pathways which yield *succinyl-CoA,* an intermediate of the citric acid cycle (Figure 19-12). Isoleucine and valine have rather similar patterns of degradation. Both undergo transamination, followed by oxidative decarboxylation of the resulting α-keto acids. Four of the five carbon atoms of valine are converted into succinic acid, as are three of the six carbon atoms of isoleucine.

The three α-keto acids derived by deamination of valine, isoleucine, and leucine undergo oxidative decarboxylation by the same enzyme complex, the *α-keto acid dehydrogenase* system, which is genetically defective in some people. Thus these α-keto acids accumulate in the blood and "spill over" into the urine. This relatively rare condition, which results in abnormal development of the brain and early death of infants unless treated, is called *maple syrup urine disease* because of the characteristic odor imparted to the urine by the keto acids. It is treated by rigid control over the diet to exclude, as far as possible, the common amino acids valine, isoleucine, and leucine. Such treatment is extremely expensive.

Phenylalanine and Tyrosine Yield Fumarate

As was pointed out above, *phenylalanine* and *tyrosine* each yield two 4-carbon products, *acetoacetate* and *fumarate* (see Figure 19-9). The acetoacetate enters the citric acid cycle in the form of acetyl-CoA, whereas fumarate is of course an intermediate of the cycle.

The Oxaloacetate Pathway

The carbon skeletons of asparagine and aspartic acid ultimately enter the citric acid cycle via oxaloacetate (Figure 19-4). The enzyme *asparaginase* catalyzes the hydrolysis of asparagine to yield aspartate

$$\text{Asparagine} + H_2O \longrightarrow \text{aspartate}^- + NH_4^+$$

The aspartate then donates its amino group to α-ketoglutarate in a transamination reaction to yield glutamate

$$\text{Aspartate} + \alpha\text{-ketoglutarate} \rightleftharpoons \text{oxaloacetate} + \text{glutamate}$$

The remaining carbon skeleton of aspartate, namely oxaloacetate, enters the citric acid cycle.

We have now seen how the 20 different amino acids, after their deamination, are degraded by dehydrogenation, decarboxylation, and other reactions to yield portions of their carbon backbones in the form of five well-known central metabolites that can enter the citric acid cycle. Here they become oxidized

Table 19-2 Glucogenic and Ketogenic Amino Acids

Glucogenic
Alanine
Arginine
Asparagine
Aspartic acid
Cysteine
Glutamic acid
Glutamine
Glycine
Histidine
Methionine
Proline
Serine
Threonine
Tryptophan
Valine
Ketogenic
Leucine
Lysine
Tryptophan
Ketogenic and glucogenic
Phenylalanine
Tyrosine

completely to carbon dioxide and water. During electron transport, ATP is generated by oxidative phosphorylation, and in this way amino acids contribute to the total energy supply of the organism.

Some Amino Acids Can Be Converted into Glucose and Some into Ketone Bodies

We have seen that five of the amino acids on degradation are converted into acetoacetyl-CoA. These amino acids can yield ketone bodies in the liver, since acetoacetyl-CoA can be converted into acetoacetate and β-hydroxybutyrate (page 524). These five amino acids are called *ketogenic* (Table 19-2). Their ability to form ketone bodies is particularly evident in untreated diabetes mellitus, in which large amounts of ketone bodies are produced by the liver, not only from fatty acids but also from the ketogenic amino acids.

The amino acids that can be converted into pyruvate, α-ketoglutarate, succinate, and oxaloacetate, 15 altogether, can be converted into glucose and glycogen by pathways described in Chapter 20. They are called *glucogenic* amino acids (Table 19-2). The division between ketogenic and glucogenic amino acids is not sharp since two amino acids (phenylalanine and tyrosine) are both ketogenic and glucogenic. Some of the amino acids that can be converted into pyruvate, particularly alanine, cysteine, and serine, can also potentially form acetoacetate via acetyl-CoA, particularly in diabetes mellitus (pages 524, 536).

Ammonia Is Toxic to Animals

Some pages ago we put aside discussion of amino-group metabolism to consider the energy-yielding degradation of the deaminated amino acids. Now we return to where we left off, to consider the fate of the ammonia resulting from the oxidative deamination of glutamate by glutamate dehydrogenase, a process that occurs in nearly all tissues. Here we come into a serious biochemical problem, because ammonia is a very toxic substance, particularly to the brain. Ammonia is so toxic that injection of even very dilute solutions into the bloodstream can render animals comatose.

The toxicity of ammonia to the brain is not completely understood, but two major factors can be identified. (1) The pK' of ammonia is quite high, so that at the pH of the blood it occurs almost entirely as ammonium ion (NH_4^+). NH_4^+ ions are not readily permeant through the plasma membrane or mitochondrial membranes. However, free ammonia (NH_3), a neutral molecule, is freely permeant. Although only about 1 percent of the total ammonia in the blood occurs in the form of free NH_3 at pH 7.4, this small amount can penetrate membranes and gain entry into brain cells and their mitochondria. (2) The entry of ammonia into brain mitochondria leads to the formation of glutamate from ammonia and α-ketoglutarate, through the reverse action of glutamate dehydrogenase

$$NH_4^+ + \alpha\text{-ketoglutarate}^{2-} + NADPH + H^+ \longrightarrow \text{glutamate}^- + NADP^+ + H_2O$$

The net result is that α-ketoglutarate is withdrawn from the pool of citric acid cycle intermediates in brain mitochondria, lowering the rate of oxidation of glucose, the major fuel of the brain. Although these two factors are highly significant, other aspects of the sensitivity of the brain to ammonia are not fully understood.

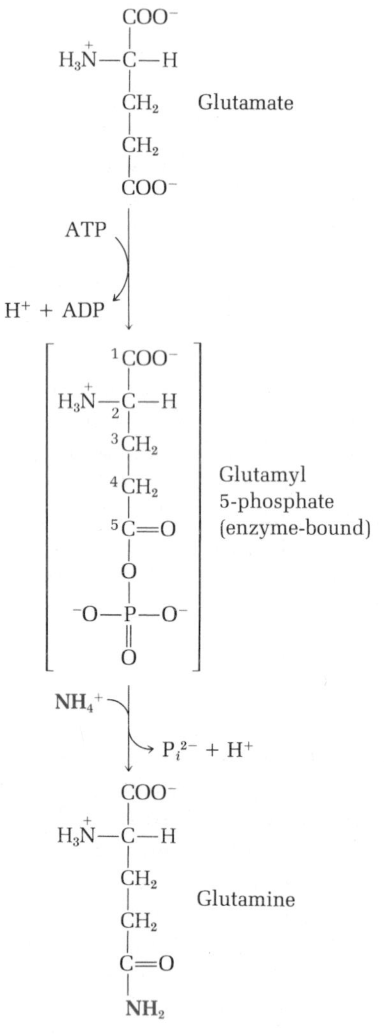

Glutamate → Glutamyl 5-phosphate (enzyme-bound) → Glutamine

(ATP → H^+ + ADP; NH_4^+ → P_i^{2-} + H^+)

Figure 19-13
Formation of glutamyl 5-phosphate as the enzyme-bound intermediate in the glutamine synthetase reaction.

Glutamine Carries Ammonia from Many Peripheral Tissues to the Liver

The question now is: How does the toxic ammonia get from the peripheral tissues to the organs that detoxify or excrete it, without putting the brain at risk?

The way taken to transport ammonia from the peripheral tissues to the liver or kidneys in most animals is to convert it into a nontoxic compound before exporting it via the blood. In many tissues, including the brain, ammonia is enzymatically combined with glutamate to yield *glutamine* by the action of *glutamine synthetase*, which promotes the reaction

$$ATP + NH_4^+ + \text{glutamate} \longrightarrow ADP + P_i + \text{glutamine} + H^+$$

In this reaction *glutamyl 5-phosphate* (Figure 19-13) is an enzyme-bound high-energy intermediate, an acyl phosphate resulting from the phosphorylation of the 5-carboxyl group of glutamate by ATP. The bound glutamyl 5-phosphate combines with ammonia on the active site to form glutamine and release phosphate. The glutamine so formed is a neutral nontoxic compound which can readily pass through cell membranes, whereas glutamate, which bears a net negative charge (page 101), cannot. The glutamine is carried via the blood to the liver in most land animals, where it can be acted upon by *glutaminase* to yield glutamate and ammonia

$$\text{Glutamine} + H_2O \longrightarrow \text{glutamate}^- + NH_4^+$$

The ammonia so formed is converted by the liver into urea. Glutamine is a major transport form of ammonia; it is present in normal blood in much higher concentrations than other amino acids.

Alanine Carries Ammonia from Muscles to the Liver

Alanine also plays a special role in transporting ammonia to the liver in a nontoxic form. Muscles, like other tissues, produce ammonia during the degradation of amino acids. In addition, ammonia also arises from the deamination of adenylate

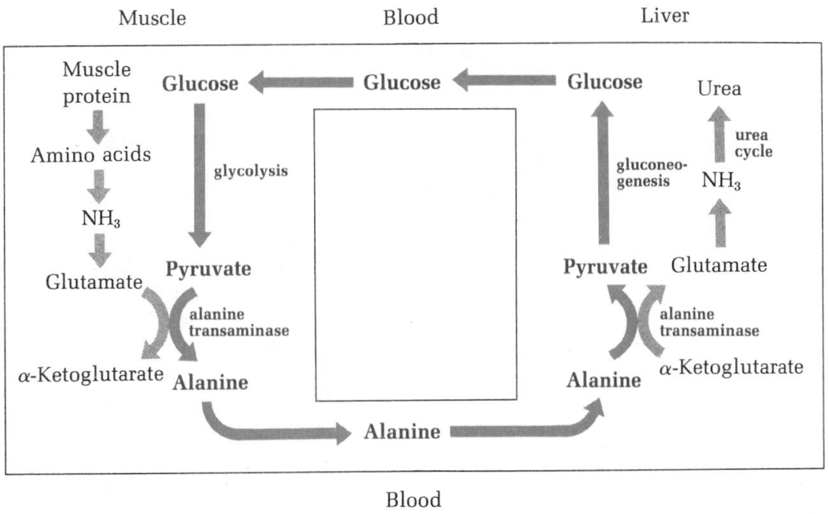

Figure 19-14
The glucose-alanine cycle. It functions in a dual manner (1) to carry amino groups from skeletal muscle to the liver, to be converted into urea, and (2) to provide the working muscle with blood glucose made by the liver from the carbon backbone of alanine.

(AMP), a rather prominent process in very active skeletal muscles. Ammonia formed from these two sources is carried from the muscles to the liver by the amino acid alanine, through the action of the *glucose-alanine cycle* (Figure 19-14). In this cycle ammonia is converted into the amino group of glutamate by the action of glutamate dehydrogenase

$$NH_4^+ + \alpha\text{-ketoglutarate}^{2-} + NADPH + H^+ \longrightarrow glutamate^- + NADP^+ + H_2O$$

The glutamate so formed now transfers its α-amino group to pyruvate, a readily available product of muscle glycolysis, by the action of *alanine transaminase*

$$\text{Glutamate} + \text{pyruvate} \rightleftharpoons \alpha\text{-ketoglutarate} + \text{alanine}$$

The alanine, a neutral amino acid with no net charge at pH near 7, escapes into the blood and is carried to the liver. Here the alanine transfers its amino group to α-ketoglutarate by the action of alanine transaminase to yield glutamate, which then undergoes deamination to yield α-ketoglutarate and ammonia through the action of glutamate dehydrogenase. The ammonia so yielded is converted by the liver into urea.

The choice of alanine to transport ammonia from hard-working skeletal muscles to the liver is another example of the intrinsic economy of living organisms. Vigorously contracting skeletal muscles not only produce ammonia but also form large amounts of pyruvate during glycolysis. Both these products must find their way to the liver, where ammonia is converted into urea for excretion and pyruvate is rebuilt into blood glucose, which returns to the muscles. Animals have learned to solve two problems with one cycle: they combine the ammonia with the pyruvate to form alanine, a nontoxic neutral amino acid, which is passed via the blood to the liver for further processing (Figure 19-14).

Excretion of Amino Nitrogen Is Another Biochemical Problem

How is excess amino nitrogen finally excreted from the body? From comparative biochemical studies of different animal species it has been found that amino nitrogen is excreted in one of three major forms, as *ammonia*, as *urea*, or as *uric acid*. Most aquatic species, e.g., the teleost or bony fishes, excrete amino nitrogen as *ammonia* and are thus called *ammonotelic* animals; most terrestrial animals excrete amino nitrogen in the form of *urea* and are thus *ureotelic;* and birds, lizards, and snakes excrete amino nitrogen as *uric acid* and are called *uricotelic* (Figure 19-15).

The basis for these differences lies in the anatomy and physiology of different organisms in relation to their usual habitat. The bony fishes transport amino nitrogen in the blood as glutamine but excrete it in the form of ammonia through their gills, which contain glutaminase and thus allow glutamine to be hydrolyzed to glutamate and ammonia. Since ammonia is freely soluble in water, it is quickly swept away and diluted by the large volume of water that passes through the gills. The bony fishes thus do not require a complex urinary system to excrete ammonia.

However, as biological evolution occurred and some aquatic species learned to live on dry land, the excretion of amino nitrogen as ammonia through gills was no longer possible. Terrestrial animals ultimately evolved different methods of excreting amino nitrogen. Such animals need kidneys and a urinary bladder to excrete water-soluble nitrogenous waste products, but since free NH_3 can penetrate membranes readily, excretion of large amounts of ammonia directly into the urine could result in its reabsorption into the blood. There was another disadvantage: since ammonia occurs in the blood largely as NH_4^+ ion, its excretion would require the excretion of an equivalent number of anions, such as chloride or phosphate. To

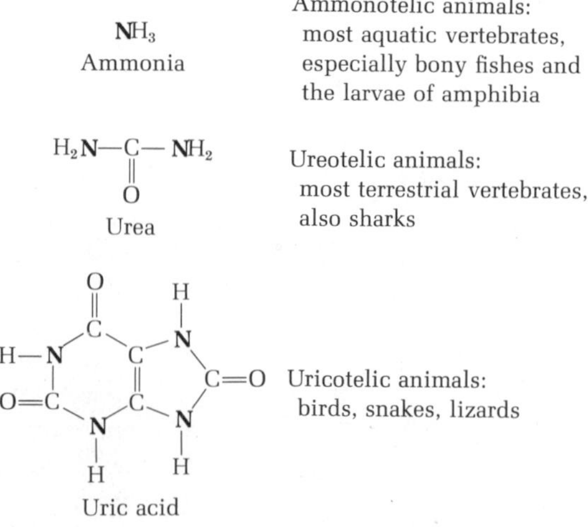

Figure 19-15
Excretory forms of amino-group nitrogen in different forms of life.

avoid such complications most terrestrial animals acquired the ability to excrete amino nitrogen as urea, a neutral, highly soluble, and nontoxic molecule. The capacity to make and excrete urea has a cost, however; as we shall see, it requires considerable ATP energy.

In birds weight is an important consideration. Since excretion of urea into urine requires that a rather large volume of water be excreted as well, birds in their evolution learned to excrete amino nitrogen in a form that does not require a large volume of water to be carried around. They convert amino nitrogen into uric acid, a relatively insoluble compound that is excreted as a semisolid mass of uric acid crystals, containing very little water (Figure 19-15). For the advantage of excreting amino nitrogen in the form of solid uric acid, birds must carry out considerable metabolic work, since the biosynthesis of uric acid is a complex energy-requiring process.

The importance of the habitat in excretion of amino nitrogen is illustrated by the change in the pathway of nitrogen excretion as the tadpole undergoes metamorphosis into the adult frog. Tadpoles are entirely aquatic and excrete amino nitrogen as ammonia through their gills. The tadpole liver lacks the necessary enzymes to make urea, but during metamorphosis it acquires them and loses the ability to excrete ammonia. In the adult frog, which is more terrestrial in habit, amino nitrogen is excreted almost entirely as urea.

Glutaminase Participates in Excretion of Ammonia

In ammonotelic animals, amino groups are transaminated from the various amino acids to α-ketoglutarate to form glutamate, which then undergoes oxidative deamination via glutamate dehydrogenase to yield free ammonia in the liver mitochondria. Because free ammonia is very toxic and cannot be transported via the blood, it is converted into the amide group of glutamine by *glutamine synthetase*. The nontoxic neutral glutamine is then transported via the blood to the gills, where it loses its amide nitrogen as ammonium ion (NH_4^+) by the action of *glutaminase*

$$\text{Glutamine} + H_2O \longrightarrow \text{glutamate}^- + NH_4^+$$

Urea Is Formed by the Urea Cycle

In ureotelic animals the ammonia resulting from the deamination of amino acids is converted into urea in the liver by a cyclic mechanism, the *urea cycle*, first discovered by Hans Krebs (page 443) and Kurt Henseleit in 1932. Krebs thus was the discoverer of two major metabolic cycles. He found the urea cycle first, in research carried out while he was a member of the medical staff of a hospital in Freiburg, Germany. He and Henseleit, a medical student, found that the rate of urea formation from ammonia by thin slices of liver suspended in a buffered aerobic

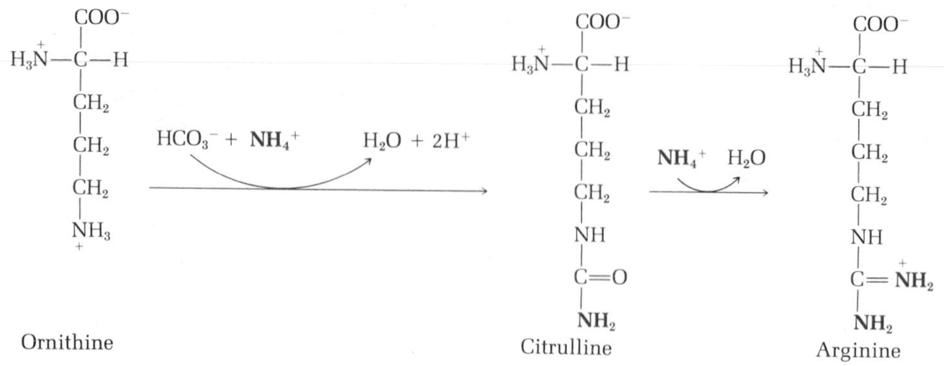

$$\underset{\text{Ornithine}}{\begin{array}{c} COO^- \\ | \\ H_3\overset{+}{N}-C-H \\ | \\ CH_2 \\ | \\ CH_2 \\ | \\ CH_2 \\ | \\ \overset{+}{N}H_3 \end{array}} \xrightarrow[\quad]{HCO_3^- + \mathbf{NH_4^+} \;\; H_2O + 2H^+} \underset{\text{Citrulline}}{\begin{array}{c} COO^- \\ | \\ H_3\overset{+}{N}-C-H \\ | \\ CH_2 \\ | \\ CH_2 \\ | \\ CH_2 \\ | \\ NH \\ | \\ C=O \\ | \\ \mathbf{NH_2} \end{array}} \xrightarrow{\mathbf{NH_4^+} \;\; H_2O} \underset{\text{Arginine}}{\begin{array}{c} COO^- \\ | \\ H_3\overset{+}{N}-C-H \\ | \\ CH_2 \\ | \\ CH_2 \\ | \\ CH_2 \\ | \\ NH \\ | \\ C=\overset{+}{\mathbf{NH_2}} \\ | \\ \mathbf{NH_2} \end{array}}$$

Figure 19-16
The three amino acids found by Krebs to stimulate urea formation from ammonia in liver slices. As shown, ornithine and citrulline can be regarded as successive precursors of arginine. The groups formed from ammonia are in color.

medium was greatly accelerated by adding any one of three specific compounds, *ornithine, citrulline,* or *arginine* (Figure 19-16). Arginine is of course one of the standard amino acids found in proteins. Although ornithine and citrulline are also α-amino acids, they do not occur as building blocks of protein molecules. These three compounds stimulated urea synthesis to a far greater extent than any other common nitrogenous compounds tested. The structures of the three active compounds suggested that they might be related in a sequence, with ornithine the precursor of citrulline and citrulline in turn the precursor of arginine (Figure 19-16). Arginine had long been known to be hydrolyzed to ornithine and urea by the action of the enzyme *arginase*

$$\text{Arginine} + H_2O \longrightarrow \text{ornithine} + \text{urea}$$

From these facts Krebs deduced that a cyclic process occurs, in which ornithine plays a role resembling that of oxaloacetate in the citric acid cycle. A molecule of ornithine combines with one molecule of NH_3 and one of CO_2 to form citrulline. A second molecule of ammonia is added to citrulline to form arginine, which is then hydrolyzed to yield urea, with regeneration of a molecule of ornithine (Figure 19-17). All organisms capable of arginine biosynthesis can catalyze these reactions to the point of arginine, but only ureotelic animals have large amounts of the enzyme *arginase,* which catalyzes the irreversible hydrolysis of arginine to form urea and ornithine. The ornithine that is regenerated is then ready for the next turn of the urea cycle. Urea, the product, is a neutral, nontoxic, water-soluble substance. It is passed via the blood to the kidneys and excreted into the urine.

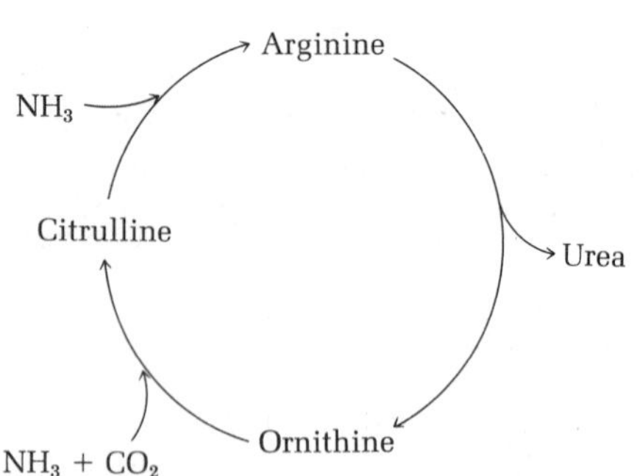

Figure 19-17
The urea cycle as originally postulated by Krebs and Henseleit.

The Urea Cycle Has Several Complex Steps

Now let us look at the successive steps of the urea cycle as we know them today (Figure 19-18). The first amino group to enter the urea cycle arises in the form of free ammonia by the oxida-

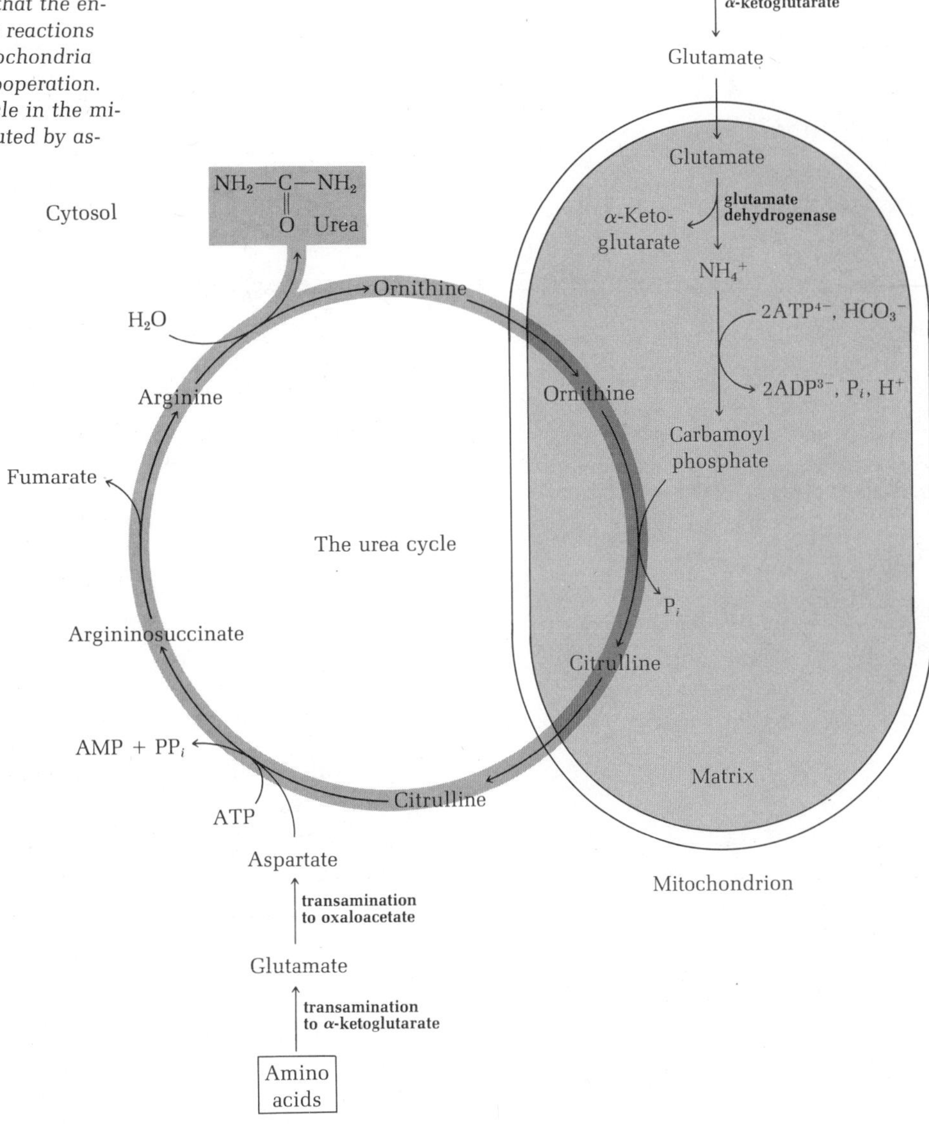

Figure 19-18
The urea cycle (shaded in color) as it is known today. Also shown are the pathways of entry of amino groups. Note that the enzymes catalyzing the urea cycle reactions are distributed between the mitochondria and the cytosol, in metabolic cooperation. One amino group enters the cycle in the mitochondria; the other is contributed by aspartate from the cytosol.

tive deamination of glutamate inside the mitochondria of the liver cell, catalyzed by glutamate dehydrogenase, which requires NAD^+

$$\text{Glutamate}^- + NAD^+ + H_2O \rightleftharpoons \alpha\text{-ketoglutarate}^{2-} + NH_4^+ + NADH + H^+$$

The free ammonia so formed is immediately used, together with carbon dioxide generated in the mitochondria by respiration, to form carbamoyl phosphate in the matrix in an ATP-dependent reaction catalyzed by the enzyme carbamoyl phosphate synthetase I. The Roman numeral designates the

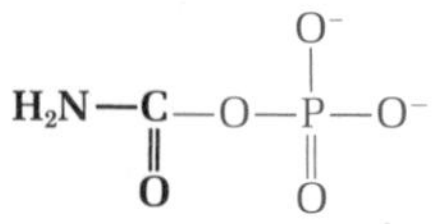

Figure 19-19
Carbamoyl phosphate. Note that it is an acyl phosphate, a mixed anhydride of a carboxylic acid and phosphoric acid. It is therefore a high-energy compound. The carbamoyl group is in color.

$$H_2N{-}C({=}O){-}O{-}P(O^-)({=}O){-}O^-$$

mitochondrial form of this enzyme to distinguish it from the cytosolic form (II). The latter has a different function, being required in nucleotide biosynthesis (Chapter 22). The mitochondrial reaction is

$$HCO_3^- + NH_4^+ + 2ATP^{4-} \longrightarrow H_2N{-}C({=}O){-}O{-}PO_3^{2-} + 2ADP^{3-} + P_i^- + H^+$$

Carbamoyl phosphate

$$\Delta G^{\circ\prime} = -3.3 \text{ kcal/mol}$$

Carbamoyl phosphate synthetase I is a regulatory enzyme; it requires N-acetylglutamate as a positive or stimulating modulator. Carbamoyl phosphate (Figure 19-19) is a high-energy compound: it may be regarded as an activated carbamoyl-group donor. Note that the terminal phosphate groups of *two* molecules of ATP are used to form one molecule of carbamoyl phosphate.

In the next step of the urea cycle carbamoyl phosphate donates its carbamoyl group to ornithine to form citrulline and release phosphate (Figure 19-20), in a reaction catalyzed by *ornithine transcarbamoylase*, a Mg^{2+}-requiring mitochondrial enzyme

$$\text{Carbamoyl phosphate} + \text{ornithine} \longrightarrow \text{citrulline} + P_i^- + H^+$$

The citrulline so formed now leaves the mitochondria and passes into the cytosol of the liver cell.

The second amino group now is introduced in the form of L-aspartate, which in turn acquired it from L-glutamate by the action of *aspartate transaminase*

$$\text{Oxaloacetate} + \text{L-glutamate} \rightleftharpoons \text{L-aspartate} + \alpha\text{-ketoglutarate}$$

L-Glutamate of course receives its amino group from most of the other common amino acids by transaminations to α-ketoglutarate. The transfer of the second amino group to citrulline occurs by a condensation reaction between the amino group of aspartate and the carbonyl carbon of citrulline in the presence of ATP to form *argininosuccinate* (Figure 19-21). This reaction is catalyzed by *argininosuccinate synthetase* of the liver cytosol, a Mg^{2+}-dependent enzyme

$$\text{Citrulline} + \text{aspartate} + \text{ATP} \longrightarrow \text{argininosuccinate} + \text{AMP} + PP_i + H^+$$

In the next step argininosuccinate is reversibly cleaved by *argininosuccinate lyase* (Figure 19-22) to form free arginine and fumarate

$$\text{Argininosuccinate} \rightleftharpoons \text{arginine} + \text{fumarate}$$

Figure 19-20
Formation of citrulline from ornithine and carbamoyl phosphate. The newly introduced carbamoyl group is in color.

Ornithine: $COO^-{-}CH(NH_3^+){-}CH_2{-}CH_2{-}CH_2{-}NH_3^+$

+

Carbamoyl phosphate: $NH_2{-}C({=}O){-}O{-}P(O^-)({=}O){-}O^-$

↓ ornithine transcarbamoylase ($\rightarrow P_i$)

Citrulline: $COO^-{-}CH(NH_3^+){-}CH_2{-}CH_2{-}CH_2{-}NH{-}C({=}O){-}NH_2$

+

H^+

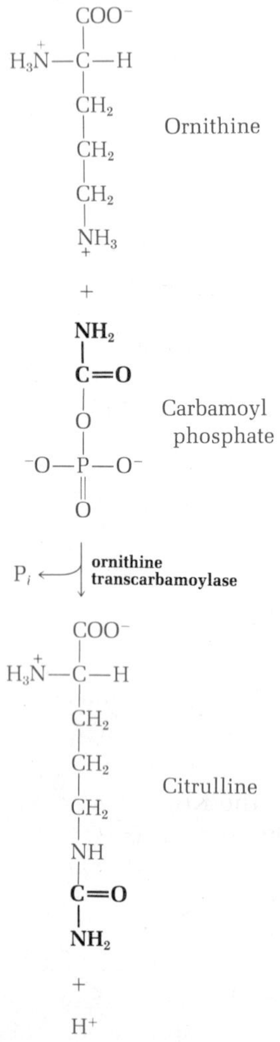

Figure 19-21
Formation of argininosuccinate. The carbamoyl group of citrulline and the amino group contributed by aspartate are in color. The remainder of the aspartate molecule is shaded.

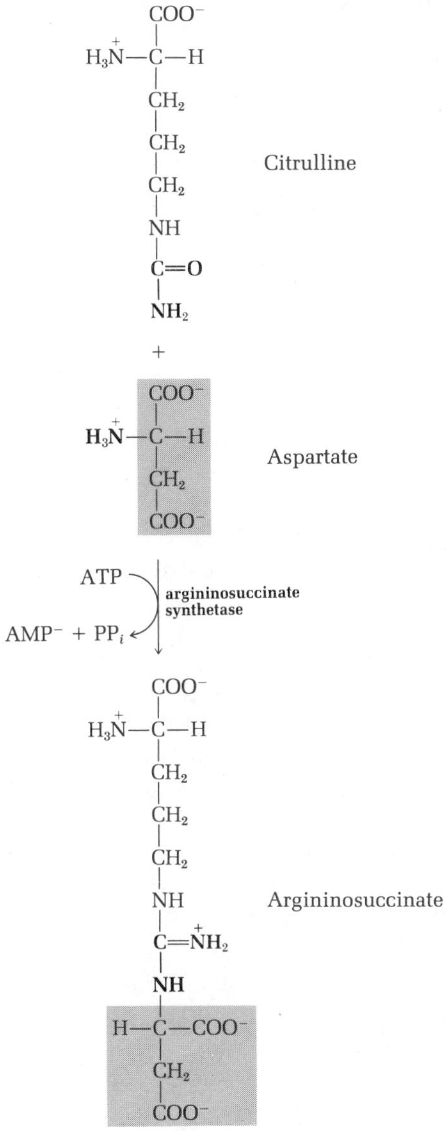

COO^- — $H_3\overset{+}{N}$—C—H — CH_2 — CH_2 — CH_2 — NH — C=O — NH_2 Citrulline

\+

COO^- — $H_3\overset{+}{N}$—C—H — CH_2 — COO^- Aspartate

ATP → $AMP^- + PP_i$ argininosuccinate synthetase

COO^- — $H_3\overset{+}{N}$—C—H — CH_2 — CH_2 — CH_2 — NH — C=$\overset{+}{N}H_2$ — NH — H—C—COO^- — CH_2 — COO^- Argininosuccinate

Figure 19-22
Formation of arginine from argininosuccinate.

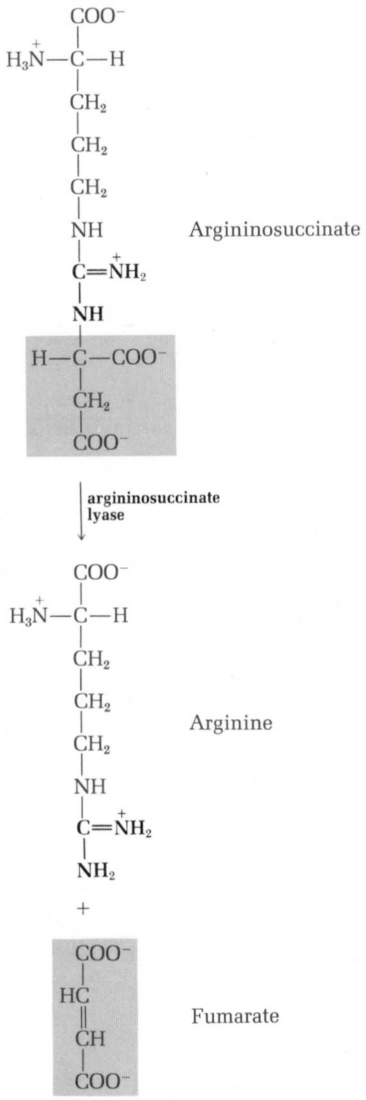

COO^- — $H_3\overset{+}{N}$—C—H — CH_2 — CH_2 — CH_2 — NH — C=$\overset{+}{N}H_2$ — NH — H—C—COO^- — CH_2 — COO^- Argininosuccinate

argininosuccinate lyase

COO^- — $H_3\overset{+}{N}$—C—H — CH_2 — CH_2 — CH_2 — NH — C=$\overset{+}{N}H_2$ — NH_2 Arginine

\+

COO^- — HC ═ CH — COO^- Fumarate

The fumarate so formed returns to the pool of citric acid cycle intermediates. Note that we have here a link between the urea cycle and the citric acid cycle (indeed, the two Krebs cycles together have been referred to as the Krebs *bi*-cycle!).

In the last reaction of the urea cycle the liver enzyme *arginase* cleaves arginine to yield urea and ornithine (Figure 19-23)

$$\text{Arginine} + H_2O \longrightarrow \text{ornithine} + \text{urea}$$

Ornithine is thus regenerated and can enter the mitochondria again to initiate another round of the urea cycle.

Figure 19-23
Formation of urea in the arginase reaction.

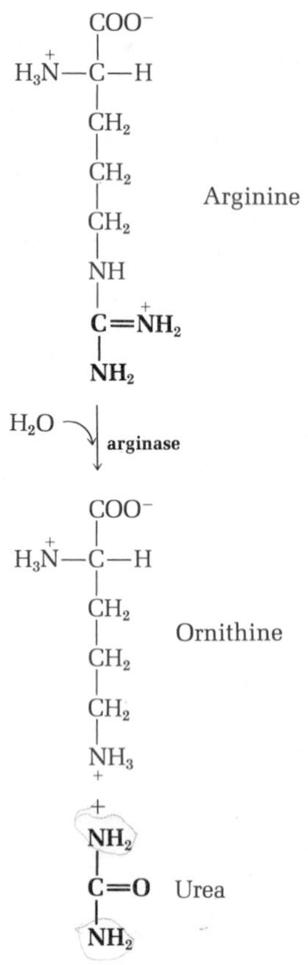

$$\begin{array}{l} COO^- \\ H_3\overset{+}{N}-C-H \\ CH_2 \\ CH_2 \\ CH_2 \\ NH \\ C=\overset{+}{N}H_2 \\ NH_2 \end{array} \quad \text{Arginine}$$

$$H_2O \xrightarrow{\text{arginase}}$$

$$\begin{array}{l} COO^- \\ H_3\overset{+}{N}-C-H \\ CH_2 \\ CH_2 \\ CH_2 \\ \underset{+}{NH_3} \end{array} \quad \text{Ornithine}$$

$$\begin{array}{l} \overset{+}{N}H_2 \\ C=O \\ NH_2 \end{array} \quad \text{Urea}$$

The overall equation of the urea cycle is

$$2NH_4^+ + HCO_3^- + 3ATP^{4-} + H_2O \longrightarrow \text{urea} + 2ADP^{3-} + 2P_i^- + AMP^- + PP_i^{3-} + H^+$$

The urea cycle brings together two amino groups and HCO_3^- to form a molecule of urea, which diffuses from the liver cells into the bloodstream, thence to be excreted into the urine by the kidneys. Thus the toxic ammonia is converted into the harmless urea in ureotelic animals.

Note that for every molecule of urea formed a molecule of HCO_3^- must be consumed. The urea cycle thus disposes of *two* waste products, ammonia and bicarbonate. Moreover, this fact suggests that the urea cycle participates in the regulation of blood pH, which depends upon the ratio of dissolved CO_2 to HCO_3^- (page 86).

The Energy Cost of Urea Synthesis

We note in the above equation that the synthesis of one molecule of urea requires four high-energy phosphate groups. Two ATPs are required to make carbamoyl phosphate, and one is required to make argininosuccinate. In the latter reaction, however, the ATP undergoes a pyrophosphate cleavage (page 387) to AMP and pyrophosphate, which may be hydrolyzed to yield two orthophosphates. The ultimate cost of making a molecule of urea is four ATPs.

It has been estimated that for the privilege of excreting urea instead of ammonia, ureotelic animals lose about 15 percent of the energy of the amino acids from which the urea was derived. This loss is made up in some ruminant animals. The cow, for example, secretes much urea from its blood into the rumen. There the rumen microorganisms use urea as a source of NH_3 to manufacture amino acids, which are then absorbed and used by the cow. By secreting urea into its gastrointestinal tract and recycling it in the same way, the camel avoids water loss connected with the urinary excretion of urea. This is one of several biochemical and physiological adaptations that enable the camel to get along with a very small water intake. Neither ruminants nor nonruminants can by themselves, without the aid of microorganisms, use urea as a source of amino groups for amino acid synthesis, since they lack the enzymes necessary to hydrolyze or use urea.

Genetic Defects in the Urea Cycle Lead to Excess Ammonia in the Blood

People with heritable genetic defects in one enzyme or another of the urea cycle have an impaired ability to form urea from ammonia. They cannot tolerate a protein-rich diet, since amino acids ingested in excess over the minimum daily requirements

Figure 19-24
Amination of the α-keto acid analogs of essential amino acids by transamination from nonessential amino acids.

$$\begin{array}{c} COO^- \\ | \\ C{=}O \\ | \\ R_E \end{array} + \begin{array}{c} COO^- \\ | \\ H_3\overset{+}{N}{-}C{-}H \\ | \\ R_N \end{array}\text{ Nonessential amino acid} \rightleftharpoons \begin{array}{c} COO^- \\ | \\ C{=}O \\ | \\ R_N \end{array} + \begin{array}{c} COO^- \\ | \\ H_3\overset{+}{N}{-}C{-}H \\ | \\ R_E \end{array}\text{ Essential amino acid}$$

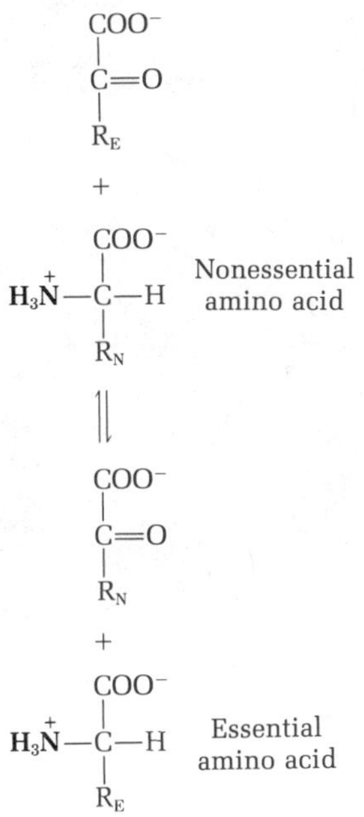

for protein synthesis would be deaminated in the liver and cause free ammonia to appear in the blood. As we have seen, ammonia is very toxic and causes mental disorders, retarded development, and, in high amounts, coma and death. Patients with defects in the urea cycle are often treated by substituting in the diet α-keto acid analogs for amino acids that are essential for growth (Chapter 26). Because the essential parts of the indispensable amino acids are their carbon skeletons, not their amino groups, the α-keto analogs of the essential amino acids can accept amino groups from excess nonessential amino acids by transaminase action (Figure 19-24). In this way the nonessential amino acids are kept from delivering their amino groups to the blood in the form of ammonia.

Birds, Snakes, and Lizards Excrete Uric Acid

In the uricotelic animals (birds, snakes, and lizards) uric acid (Figure 19-25) is the chief form in which amino nitrogen is excreted. Uric acid also happens to be the chief end product of purine metabolism in primates, birds, and reptiles. Uric acid is a complex molecule with two condensed-ring systems called the purine nucleus. The adenine and guanine portions of adenine and guanine nucleotides are also purines. The pathway of synthesis of uric acid from amino groups has many steps since the purine nucleus must be built stepwise from a number of simple precursors. Figure 19-25 shows the origin of the carbon and nitrogen atoms of uric acid, deduced from isotope tracer studies. We shall examine the complex pathway of synthesis of purines and uric acid in Chapter 22. Here it will suffice to say that this pathway has many steps and requires considerable energy. Thus uricotelic animals must pay a heavy price for the advantage of excreting amino nitrogen in a semisolid form. There is a compensation, however; uric acid is the common excretory end product not only of amino nitrogen but also of purine catabolism (Chapter 22).

Uric acid is deposited in enormous amounts on many islands off the coast of South America, which serve as immense rookeries for sea birds (Figure 19-26). These huge guano deposits are used as fertilizer, thus returning organic nitrogen to the soil, to be used again for the synthesis of amino acids by plants and soil microorganisms.

Figure 19-25
Excretion of amino nitrogen as uric acid by birds, snakes, and lizards. The nitrogen atoms (color) of uric acid are derived from α-amino groups of amino acids by very complex pathways. Uric acid is called an acid because it occurs in tautomeric forms which can ionize to form urates. Na^+ *and* K^+ *urates are only slightly more soluble than uric acid.*

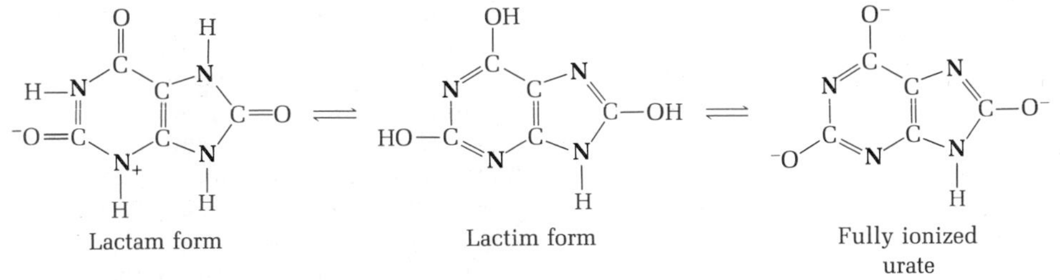

Lactam form ⇌ Lactim form ⇌ Fully ionized urate

Figure 19-26
A view on San Lorenzo Island, one of the guano islands off the coast of Peru. Hundreds of thousands of "gooney" birds nest on these islands, and, over the centuries, enormous cliff-like deposits of guano, which are largely solid uric acid, have built up. Guano is a valuable fertilizer worth well over $100 a ton. In the late 1800s fleets of sailing vessels carried on rich world-wide trade in guano.

Summary

A small fraction of oxidative energy in man comes from the oxidative catabolism of amino acids. After removal of amino groups by transamination to α-ketoglutarate, the carbon skeletons of amino acids undergo oxidative degradation to compounds that can enter the citric acid cycle for oxidation to CO_2 and H_2O. There are five pathways by which carbon skeletons of amino acids enter the citric acid cycle: (1) via acetyl-CoA, (2) via α-ketoglutarate, (3) via succinate, (4) via fumarate, and (5) via oxaloacetate. The amino acids entering via acetyl-CoA are divided into two groups. The first (alanine, cysteine, glycine, serine, and threonine) yields acetyl-CoA via pyruvate, and the second (leucine, lysine, phenylalanine, tyrosine, and tryptophan) yields acetyl-CoA via acetoacetyl-CoA. The amino acids proline, histidine, arginine, glutamine, and glutamic acid enter via α-ketoglutarate; methionine, isoleucine, and valine enter via succinate; four carbon atoms of phenylalanine and tyrosine enter via fumarate; and asparagine and aspartic acid enter via oxaloacetate. A number of human genetic defects in amino acid catabolic pathways occur. Phenylketonuria is a particularly serious and relatively common defect.

Ammonotelic animals (bony fishes, tadpoles) excrete amino nitrogen from their gills as ammonia, which derives from the hydrolysis of glutamine. Ureotelic animals (most terrestrial animals) excrete amino nitrogen as urea. Urea is formed in the liver by the urea cycle, discovered by Hans Krebs. Arginine is the immediate precursor of urea by the action of arginase, which yields urea and ornithine. Arginine is resynthesized from ornithine by carbamoylation of the latter to citrulline at the expense of carbamoyl phosphate, followed by addition of an amino group to citrulline at the expense of aspartate. Ornithine is regenerated in each turn of the cycle. Uricotelic animals (birds, snakes, and lizards) excrete amino nitrogen in semisolid form as uric acid, a derivative of purine. The formation of the nontoxic urea and of solid uric acid has a high ATP cost.

References

Books

BALDWIN, E.: *An Introduction to Comparative Biochemistry*, 4th ed., Cambridge University Press, New York, 1964.

CUNNINGHAM, E. B.: *Biochemistry: Mechanisms of Metabolism*, McGraw-Hill, New York, 1978. Chapter 14 has an excellent account of amino acid pathways and their enzymatic mechanisms.

DAGLEY, S., and D. E. NICHOLSON: *An Introduction to Metabolic Pathways*, Wiley, New York, 1970. A reference book.

GRISOLIA, S., R. BAGUENA, and F. MAYOR: *The Urea Cycle*, Wiley, New York, 1976. Interesting proceedings of a symposium on the urea cycle, which honored Hans Krebs.

HOCHACHKA, P. W., and G. N. SOMERO: *Strategies of Biochemical Adaptation*, Holt, Rinehart & Winston, New York, 1973. Comparative biochemistry.

SCRIVER, C. R., and L. E. ROSENBERG: *Amino Acid Metabolism and Its Disorders*, Saunders, Philadelphia, 1973.

Articles

HOLMES, F. L.: "Hans Krebs and the Discovery of the Ornithine Cycle," *Fed. Proc.*, **39**:216–225 (1980). A medical historian reconstructs the events leading to the discovery of the cycle.

Problems

1. *Products of Amino Acid Transamination* Draw the structure and give the name of the α-keto acid resulting when the following amino acids undergo transamination with α-ketoglutarate:
   (a) Aspartate
   (b) Glutamate
   (c) Alanine
   (d) Phenylalanine

2. *Measurement of the Alanine Transaminase Reaction Rate* The activity (reaction rate) of alanine transaminase is usually measured by including an excess of pure lactate dehydrogenase and NADH in the reaction system. The rate of alanine disappearance is equal to the rate of NADH disappearance measured spectrophotometrically. How does this work? Explain.

3. *Distribution of Amino Nitrogen* If your diet is rich in alanine but deficient in aspartate, will you show signs of aspartate deficiency? Explain.

4. *A Genetic Defect in Amino Acid Metabolism: A Case History* A two-year-old child was brought to the hospital. His mother indicated that he vomited frequently, especially after feedings. The child's weight and physical development were below normal. His hair, although dark, contained patches of white. A urine sample treated with ferric chloride ($FeCl_3$) gave a green color characteristic of the presence of phenylpyruvic acid. Quantitative analysis of urine samples gave the results shown in the table on the next page.

Substance	Patient, mmol/L	Normal, mmol/L
Phenylalanine	7.0	0.01
Phenylpyruvate	4.8	0
Phenyllactate	10.3	0

(a) Suggest which enzyme might be deficient. Propose a treatment for this condition.
(b) Why does phenylalanine appear in the urine in large amounts?
(c) What is the source of phenylpyruvate and phenyllactate? Why does this pathway (normally not functional) come into play when the concentration of phenylalanine rises?
(d) Why does the patient's hair contain patches of white?

5. *Role of Cobalamin in Amino Acid Catabolism* Pernicious anemia is caused by impaired absorption of cobalamin due to the deficiency of a glycoprotein (intrinsic factor) secreted by the stomach. What is the effect of this impairment on the catabolism of amino acids? Are all amino acids affected equally?

6. *Lactate vs. Alanine as Metabolic Fuel: The Cost of Nitrogen Removal* The three carbons in lactate and alanine have identical states of oxidation, and animals can use either carbon source as a metabolic fuel. Compare the net ATP yield (moles of ATP per mole of substrate) for the complete oxidation (to CO_2 and H_2O) of lactate vs. alanine when the cost of nitrogen excretion as urea is included.

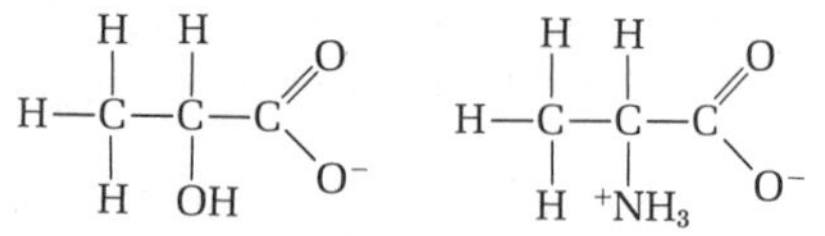

Problem 6

Lactate   Alanine

7. *Pathway of Carbon and Nitrogen in Glutamate Metabolism* When glutamic acid labeled with ^{14}C in carbon 2 and ^{15}N in the amino group undergoes oxidative degradation in the liver of a rat, in which atoms of the following metabolites will each isotope be found?
(a) Urea
(b) Succinate
(c) Arginine
(d) Citrulline
(e) Ornithine
(f) Aspartate

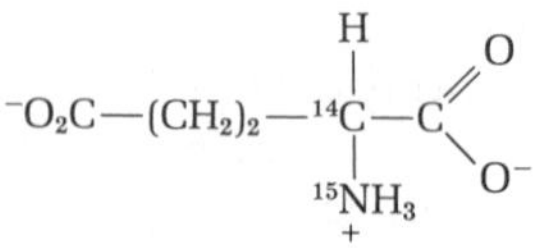

Problem 7

$^{-}O_2C{-}(CH_2)_2{-}{}^{14}CH(^{15}NH_3^+){-}COO^-$

Glutamic acid

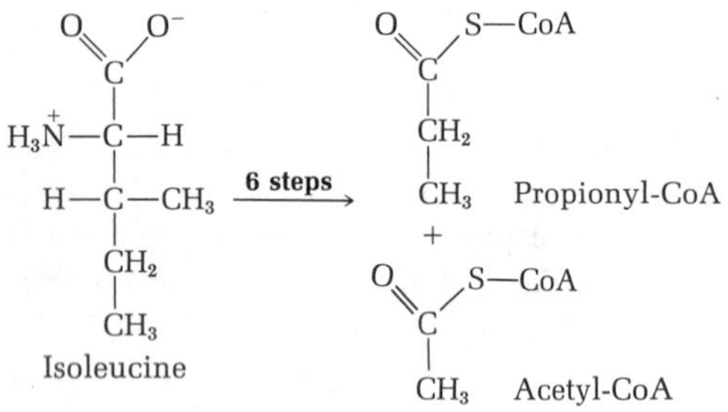

8. *Chemical Strategy of Isoleucine Catabolism* Isoleucine is degraded by a series of six steps to propionyl-CoA and acetyl-CoA, as shown at the left.
(a) The chemical process of isoleucine degradation consists of strategies analogous to those found in the citric acid cycle and the β oxidation of fatty acids. The intermediates involved in isoleucine degradation (**I** to **V**) shown below are not in the proper order. Use your knowledge and understanding of the citric acid cycle and the fatty acid oxidation pathway to arrange the intermediates into the proper metabolic sequence for isoleucine degradation.
(b) For each step proposed above, describe the chemical process, provide an analogous example from the citric acid cycle or β-oxidation pathway, and indicate any necessary cofactors.

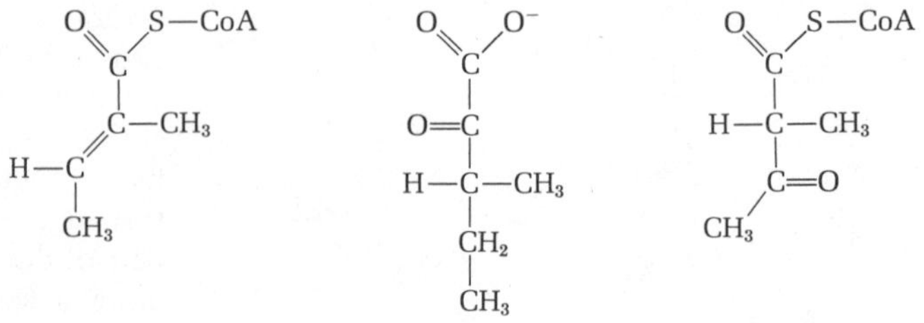

Problem 8

Isoleucine —6 steps→ Propionyl-CoA + Acetyl-CoA

I   **II**   **III**

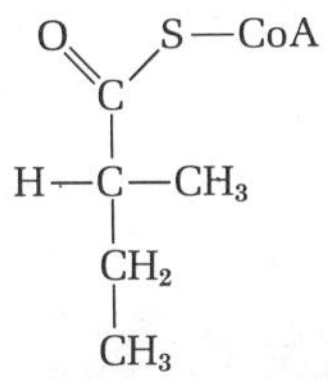

O S—CoA
C
H—C—CH_3
CH_2
CH_3
IV

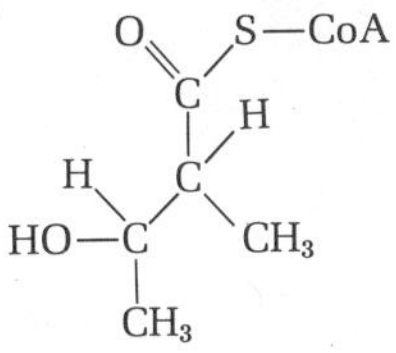

O S—CoA
C
H
H C
HO—C CH_3
CH_3
V

9. *Ammonia Intoxication Resulting from Arginine-Deficient Diet* In a recent report [J. Morris and Q. Rogers, *Science*, **199**:431 (1978)] near-adult cats after fasting overnight were given a single meal of a complete amino acid diet without arginine. Within 2 h, blood ammonia levels increased from a normal level of 18 μg/L to 140 μg/L and the cats showed the clinical symptoms of ammonia toxicity. One cat died 4.5 h after ingesting only 8 g of the diet. A control group fed a complete amino acid diet or an amino acid diet where arginine was replaced by ornithine showed no unusual clinical symptoms.
   (a) What was the role of fasting in the experiment?
   (b) What caused the ammonia levels to rise? Why did the absence of arginine lead to ammonia toxicity? Is arginine an essential amino acid in cats?
   (c) Why can ornithine be substituted for arginine?

10. *Oxidation of Glutamate* Write a series of balanced equations and the net reaction describing the oxidation of 2 mol of glutamate to 2 mol of α-ketoglutarate plus 1 mol of excreted urea.

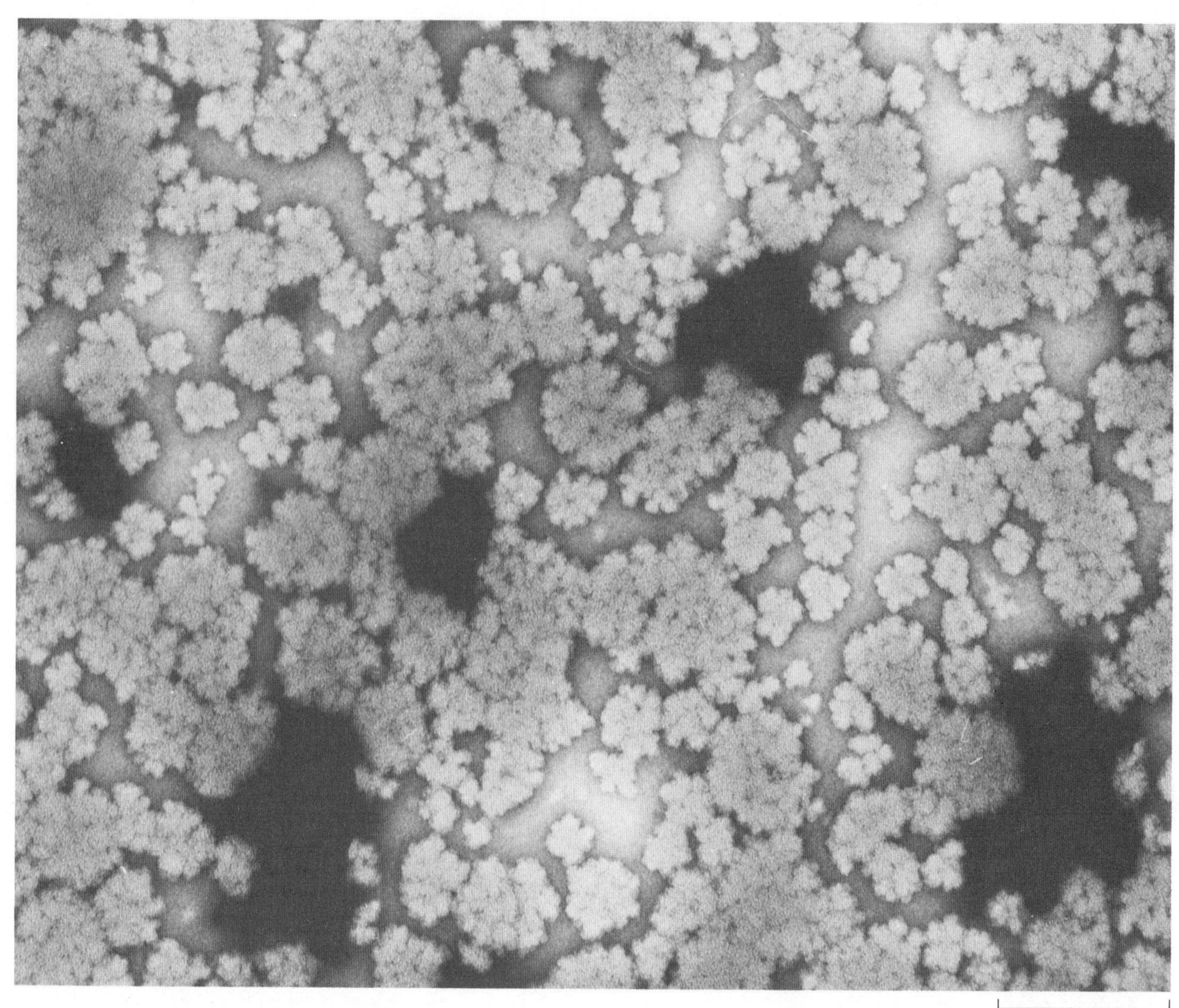

Negative contrast electron micrograph of glycogen granules isolated from rat liver. These granules, which represent the storage form of glucose fuel in the liver, are called α particles, and the smaller units of which they are composed are β particles. These granules contain not only glycogen but also the enzymes necessary for its synthesis and degradation, as well as the enzymes that regulate these processes in a reciprocal manner.

CHAPTER 20

Biosynthesis of Carbohydrates in Animal Tissues

We have now reached a turning point in the study of cell metabolism. In preceding chapters we have seen how the major foodstuffs—carbohydrates, fatty acids, and amino acids—are degraded via converging *catabolic* pathways to enter the citric acid cycle and to yield their energy-rich electrons to the respiratory chain. As the electrons flow downhill to oxygen, they provide energy to make ATP. Now we shall examine *anabolic* pathways, which use chemical energy in the form of ATP and NADPH to synthesize important cell components from simple precursor molecules. Catabolism and anabolism proceed simultaneously in a dynamic steady state, so that the energy-yielding degradation of cell components is counterbalanced by biosynthetic processes, which create and maintain the intricate orderliness of living cells.

Some of the organizing principles of biosynthesis (Chapter 10) require brief review and reemphasis at this point. *The first principle is that the pathway taken in the biosynthesis of a biomolecule is not usually identical to the pathway taken in its degradation*. The two opposing pathways may share one or even several reversible reactions, but there is always at least one enzymatic step that is dissimilar. If the reactions of catabolism and anabolism were catalyzed by the same set of enzymes acting reversibly, no stable biological structure of any complexity could exist because cell macromolecules would change in amount whenever the concentrations of their precursors fluctuated.

Second, biosynthetic pathways are controlled by different regulatory enzymes from those controlling the corresponding catabolic pathways. Usually corresponding biosynthetic and degradative pathways are regulated in a coordinated, reciprocal manner, so that stimulation of the biosynthetic pathway is accompanied by inhibition of the corresponding degradative pathway, or vice versa. Moreover, biosynthetic pathways are usually regulated at their initial steps. In this way the cell

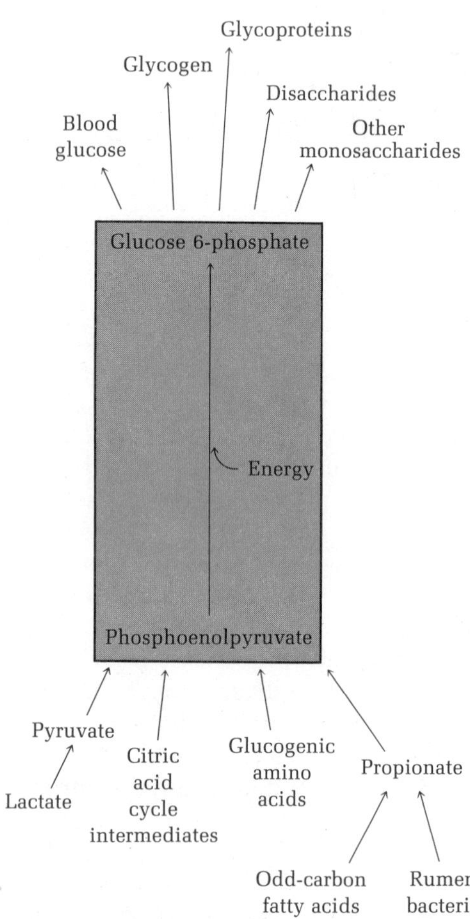

Figure 20-1
The pathway from phosphoenolpyruvate to glucose 6-phosphate is common to the biosynthetic conversion of many different precursors into various carbohydrates in animal tissues.

avoids wasting precursors to make unneeded intermediates. These relationships emphasize again that an intrinsic economy prevails in the molecular logic of living cells.

Third, energy-requiring biosynthetic processes are obligatorily coupled to the energy-yielding breakdown of ATP, in such a way that the overall process is essentially irreversible, just as the overall process of catabolism is irreversible. Thus the total amount of ATP (and NADPH) energy used in a given biosynthetic pathway always exceeds the minimum amount of free energy required to convert the precursor into the biosynthetic product.

We shall begin our survey of biosynthetic processes with the central pathway that leads to the formation of different carbohydrates from noncarbohydrate precursors in animal tissues. The biosynthesis of D-glucose is an absolute necessity in all higher animals, since the brain and nervous system, as well as the kidney medulla, the testes, erythrocytes, and embryonic tissues, require D-glucose from the blood as their sole or major fuel source. The human brain alone requires over 120 g of glucose per day. Animals constantly make D-glucose from simpler precursors, such as pyruvate and certain amino acids, in a set of carefully regulated biosynthetic reactions, and then pass the glucose into the blood. Other important carbohydrates are also made from noncarbohydrate precursors (Figure 20-1). Most important is the biosynthesis of glycogen in the liver and muscles. Liver glycogen serves as a reservoir of glucose, readily converted into blood glucose, whereas muscle glycogen is an important source of ATP energy for muscular contraction through its breakdown via glycolysis. In animals the formation of D-glucose from noncarbohydrate precursors is called *gluconeogenesis* ("formation of new sugar"). The important precursors of D-glucose in animals are lactate, pyruvate, glycerol, most of the amino acids, and the intermediates of the citric acid cycle (Figure 20-1). In animals gluconeogenesis occurs largely in the liver and to a much smaller extent in kidney cortex.

In Chapter 23 we shall see that enormous amounts of glucose and other carbohydrates are made in the plant world by the photosynthetic reduction of carbon dioxide; animals, however, cannot bring about net conversion of CO_2 into new glucose.

The Pathway of Gluconeogenesis Shares Seven Steps with the Pathway of Glycolysis

Just as the glycolytic conversion of glucose into pyruvate is a central pathway of catabolism of carbohydrates, the conversion of pyruvate into glucose is a central pathway in gluconeogenesis. These pathways are not identical, although they share several steps (Figure 20-2). Seven enzymatic reactions of glycolysis also take part in gluconeogenesis; all seven are freely reversible.

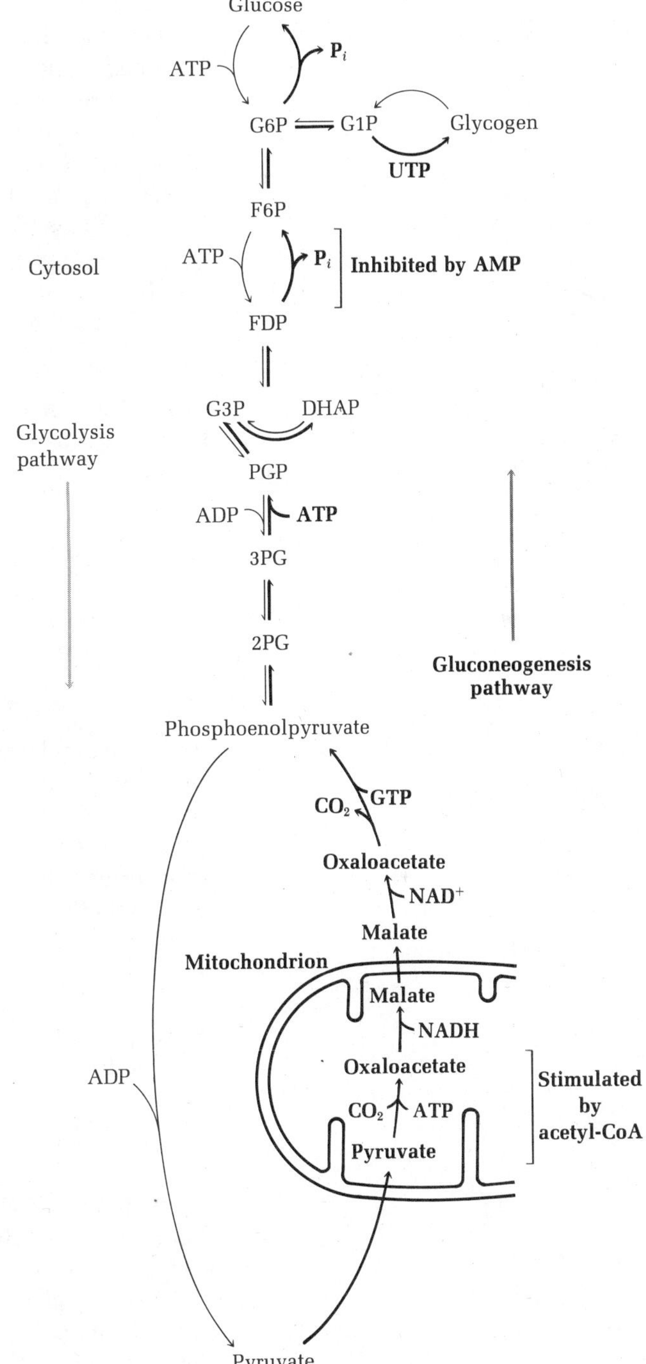

Figure 20-2
The opposing pathways of glycolysis and gluconeogenesis in rat liver. The route of gluconeogenesis is in color. In some species phosphoenolpyruvate can be made in the cytosol, without mitochondrial involvement. Two major sites of regulation of gluconeogenesis are also shown.

Key:
- G6P = glucose 6-phosphate
- G1P = glucose 1-phosphate
- F6P = fructose 6-phosphate
- FDP = fructose 1,6-diphosphate
- G3P = glyceraldehyde 3-phosphate
- DHAP = dihydroxyacetone phosphate
- PGP = 3-phosphoglyceroyl phosphate
- 3PG = 3-phosphoglycerate
- 2PG = 2-phosphoglycerate

On the other hand, three steps in glycolysis are essentially irreversible and therefore cannot be used in gluconeogenesis. They are bypassed by an alternate set of enzymes, catalyzing different reactions with different stoichiometries; they function in gluconeogenesis but not in glycolysis (Figure 20-2). These bypass reactions (each discussed below) are irreversible in the direction of glucose synthesis. Thus, both glycolysis and gluconeogenesis are irreversible processes in cells. Moreover, we shall see that gluconeogenesis and glycolysis are independently regulated through controls exerted on specific enzymatic steps that are *not* common to the two pathways.

Conversion of Pyruvate into Phosphoenolpyruvate Requires a Bypass

The first of the bypass reactions in gluconeogenesis is the conversion of pyruvate into phosphoenolpyruvate (Figure 20-2). This reaction cannot occur by reversal of the pyruvate kinase reaction (page 411)

$$\text{Phosphoenolpyruvate} + \text{ADP} \longrightarrow \text{pyruvate} + \text{ATP} \qquad \Delta G^{\circ\prime} = -7.5 \text{ kcal/mol}$$

which has a large negative standard-free-energy change and has been found to be irreversible in intact cells. Instead, the phosphorylation of pyruvate is achieved by a roundabout sequence of reactions which in some animals requires the cooperation of enzymes in both the cytosol and the mitochondria of liver cells (Figure 20-2). The first step in this bypass sequence is catalyzed by pyruvate carboxylase of mitochondria, a biotin-containing enzyme, which catalyzes formation of oxaloacetate from pyruvate (Figure 20-3), an anaplerotic reaction (page 454) capable of "filling up" the pool of citric acid cycle intermediates.

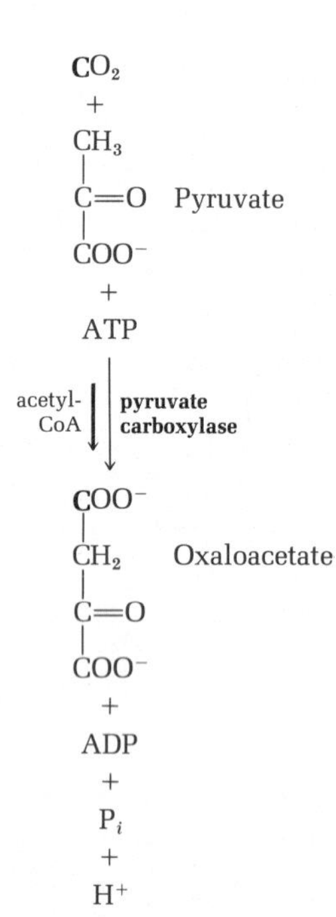

Figure 20-3
Carboxylation of pyruvate to oxaloacetate. The CO_2 incorporated into oxaloacetate will be lost again in a subsequent reaction (Figure 20-4).

$$\text{Pyruvate} + CO_2 + \text{ATP} \xrightarrow{\text{acetyl-CoA}} \text{oxaloacetate} + \text{ADP} + P_i \qquad (1)$$

Pyruvate carboxylase is a regulatory enzyme, which is almost completely inactive in the absence of its positive modulator acetyl-CoA.

The oxaloacetate formed from pyruvate in the mitochondria is then reversibly reduced to malate by mitochondrial malate dehydrogenase (page 447) at the expense of NADH:

$$\text{NADH} + \text{H}^+ + \text{oxaloacetate} \rightleftharpoons \text{NAD}^+ + \text{malate} \qquad (2)$$

The malate then leaves the mitochondria via a special dicarboxylate transport system in the inner mitochondrial membrane (page 495) to gain entry into the cytosol, where malate is reoxidized by the cytosolic form of NAD-linked malate dehydrogenase to yield extramitochondrial oxaloacetate

Figure 20-4
Conversion of oxaloacetate into phosphoenolpyruvate. The CO_2 fixed in the pyruvate carboxylase reaction (Figure 20-3) is now lost again as CO_2.

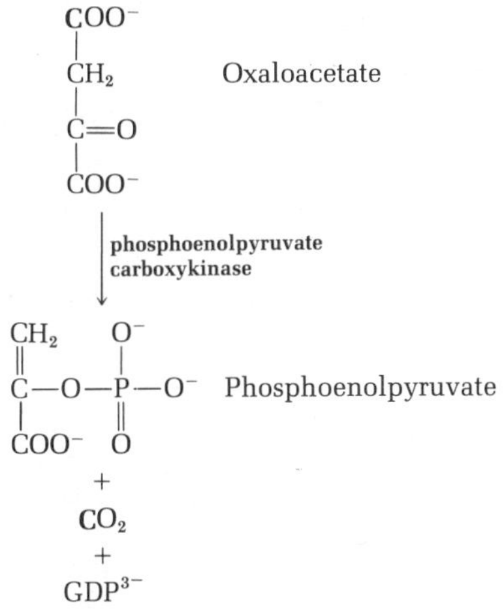

$$\begin{array}{l} COO^- \\ | \\ CH_2 \\ | \\ C{=}O \\ | \\ COO^- \end{array} \quad \text{Oxaloacetate}$$

$$\Big\downarrow \text{phosphoenolpyruvate carboxykinase}$$

$$\begin{array}{l} CH_2 \quad\;\; O^- \\ \| \qquad\quad | \\ C{-}O{-}P{-}O^- \\ | \qquad\quad \| \\ COO^- \;\; O \end{array} \quad \text{Phosphoenolpyruvate}$$

$$+ \; CO_2 \; + \; GDP^{3-}$$

$$\text{Malate} + NAD^+ \longrightarrow \text{oxaloacetate} + NADH + H^+ \qquad (3)$$

The oxaloacetate so formed is then acted upon by *phosphoenolpyruvate carboxykinase* (page 455) to yield phosphoenolpyruvate, a Mg^{2+}-dependent reaction in which *guanosine triphosphate* (GTP) serves as the phosphate donor (Figure 20-4)

$$\text{Oxaloacetate} + GTP \rightleftharpoons \text{phosphoenolpyruvate} + CO_2 + GDP \qquad (4)$$

This reaction is reversible under intracellular conditions. Phosphoenolpyruvate carboxykinase is found only in the cytosol of rat liver cells but occurs in both the mitochondria and cytosol in the liver of several other species.

We can now write the overall equation of this set of bypass reactions for the formation of phosphoenolpyruvate from pyruvate, i.e., the sum of reactions (1) to (4)

$$\text{Pyruvate} + ATP + GTP \longrightarrow \text{phosphoenolpyruvate} + ADP + GDP + P_i \qquad (5)$$
$$\Delta G^{\circ\prime} = +0.2 \text{ kcal/mol}$$

We see that *two* high-energy phosphate groups, one from ATP and one from GTP, each yielding -7.3 kcal/mol under standard conditions, must be expended to phosphorylate one molecule of pyruvate to phosphoenolpyruvate, which requires input of 14.8 kcal/mol under standard conditions. In contrast, when phosphoenolpyruvate is converted into pyruvate during glycolysis, only one ATP is generated from ADP. Although the *standard*-free-energy change $\Delta G^{\circ\prime}$ of the net reaction leading to phosphoenolpyruvate synthesis is $+0.2$ kcal/mol, the *actual* free-energy change $\Delta G'$ under intracellular conditions is very strongly negative, about -6.0 kcal; it is thus essentially irreversible.

The Second Bypass Reaction in Gluconeogenesis Is the Conversion of Fructose 1,6-Diphosphate into Fructose 6-Phosphate

The second reaction of the downhill glycolytic sequence that cannot participate in the uphill process of gluconeogenesis is the phosphorylation of fructose 6-phosphate by phosphofructokinase

$$ATP + \text{fructose 6-phosphate} \longrightarrow ADP + \text{fructose 1,6-diphosphate}$$

This reaction, which is irreversible in intact cells, is bypassed (Figure 20-2) by the enzyme *fructose diphosphatase*, which carries out the essentially irreversible hydrolysis of the 1-phosphate group to yield fructose 6-phosphate

$$\text{Fructose 1,6-diphosphate} + H_2O \xrightarrow{Mg^{2+}} \text{fructose 6-phosphate} + P_i$$
$$\Delta G^{\circ\prime} = -3.9 \text{ kcal}$$

Fructose diphosphatase has a molecular weight of 150,000 and requires Mg^{2+} for activity. It is also a regulatory enzyme. It is strongly inhibited by the negative modulator AMP but stimulated by the positive modulator ATP.

Conversion of Glucose 6-Phosphate into Free Glucose Is the Third Bypass Reaction

The third bypass reaction, the final reaction in formation of D-glucose, is the dephosphorylation of glucose 6-phosphate to yield free glucose, which is delivered from the liver into the blood (Figure 20-2). This does not occur by reversal of the hexokinase reaction (page 403), which is irreversible in the liver, but is brought about by *glucose 6-phosphatase* which catalyzes the irreversible hydrolytic reaction

$$\text{Glucose 6-phosphate} + H_2O \longrightarrow \text{glucose} + P_i \qquad \Delta G^{\circ\prime} = -2.9 \text{ kcal/mol}$$

This enzyme, which requires Mg^{2+}, is characteristically found in the endoplasmic reticulum fraction of the liver of vertebrates. Glucose 6-phosphatase is not present in muscles or in the brain, which therefore cannot furnish free glucose to the blood.

Gluconeogenesis Is Costly

Table 20-1 summarizes the biosynthetic reactions leading from pyruvate to free blood glucose. The overall sum of these reactions is

$$2 \text{ Pyruvate} + 4ATP + 2GTP + 2NADH + 2H^+ + 4H_2O \longrightarrow \text{glucose} + 2NAD^+ + 4ADP + 2GDP + 6P_i$$

For each molecule of glucose formed from pyruvate *six* high-energy phosphate groups are required, four coming from ATP and two from GTP. In addition, two molecules of NADH are required for the reduction steps. This equation is clearly not the simple reverse of the equation for the conversion of glucose into pyruvate by glycolysis, which yields only *two* molecules of ATP

$$\text{Glucose} + 2ADP + 2P_i + 2NAD^+ \longrightarrow 2 \text{ pyruvate} + 2ATP + 2NADH + 2H^+ + 2H_2O$$

Thus the synthesis of glucose from pyruvate is a relatively costly process. However, much of this high cost is necessary to ensure that gluconeogenesis is irreversible. Under intracellular conditions, in which ΔG_p for ATP may be as high as 16 kcal/mol (page 376), the overall free-energy change of glycolysis is at least -15 kcal per mole. Under the same conditions the overall free-energy change of gluconeogenesis from pyruvate is considerably higher. Thus glycolysis and gluconeogenesis are both essentially irreversible processes under normal intracellular conditions.

Table 20-1 Sequential Reactions in Gluconeogenesis Starting from Pyruvate†

Reaction	
Pyruvate + CO_2 + ATP → oxaloacetate + ADP + P_i	×2
Oxaloacetate + GTP ⇌ phosphoenolpyruvate + CO_2 + GDP	×2
Phosphoenolpyruvate + H_2O ⇌ 2-phosphoglycerate	×2
2-Phosphoglycerate ⇌ 3-phosphoglycerate	×2
3-Phosphoglycerate + ATP ⇌ 3-phosphoglyceroyl phosphate + ADP	×2
3-Phosphoglyceroyl phosphate + NADH + H^+ → glyceraldehyde 3-phosphate + NAD^+ + P_i	×2
Glyceraldehyde 3-phosphate ⇌ dihydroxyacetone phosphate	
Glyceraldehyde 3-phosphate + dihydroxyacetone phosphate ⇌ fructose 1,6-diphosphate	
Fructose 1,6-phosphate + H_2O → fructose 6-phosphate + P_i	
Fructose 6-phosphate ⇌ glucose 6-phosphate	
Glucose 6-phosphate + H_2O → glucose + P_i	
Sum: 2 Pyruvate + 4ATP + 2GTP + 2NADH + $2H^+$ + $4H_2O$ → glucose + $2NAD^+$ + 4ADP + 2GDP + $6P_i$	

† The bypass reactions are in color; all other reactions are reversible steps of the glycolytic sequence. The figures at the right indicate that the reaction is to be multiplied by 2, since two 3-carbon precursors are required to make a molecule of glucose.

Gluconeogenesis and Glycolysis Are Regulated Reciprocally

Figure 20-2 shows the points in the gluconeogenesis and glycolytic pathways where regulation occurs. The first control point in gluconeogenesis is the reaction catalyzed by the regulatory enzyme pyruvate carboxylase. This enzyme is virtually inactive in the absence of its positive allosteric modulator, acetyl-CoA. As a consequence, the biosynthesis of glucose from pyruvate is promoted whenever excess mitochondrial acetyl-CoA builds up beyond the immediate needs of the cell for fuel for the citric acid cycle. Since acetyl-CoA is a negative or inhibitory modulator of the pyruvate dehydrogenase complex, the oxidation of pyruvate to acetyl-CoA is diminished and the biosynthetic conversion of pyruvate to glucose is favored by accumulation of acetyl-CoA.

The second control point in gluconeogenesis is the reaction catalyzed by fructose diphosphatase, which is strongly inhibited by AMP. Since the corresponding glycolytic enzyme, phosphofructokinase, is stimulated by AMP and ADP but inhibited by citrate and ATP (page 423), these opposing steps in the two pathways are regulated in a coordinate or reciprocal manner. Thus, whenever the citric acid cycle is already amply supplied with fuel, either as acetyl-CoA or as its first intermediate citrate, or whenever the cell is amply supplied with ATP, the biosynthetic pathway from pyruvate to glucose is favored, thus promoting the storage of glucose in the form of glycogen.

Some degree of regulation of gluconeogenesis is exerted indirectly by controls over pyruvate kinase, one of the glycolytic enzymes that is not used in gluconeogenesis. Pyruvate kinase exists in two forms, L (for liver) and M (for muscle). The

L form, which predominates in tissues capable of gluconeogenesis, is allosterically inhibited by excess ATP and by certain amino acids, particularly alanine, which is a glucose precursor in gluconeogenesis. Through inhibition of L pyruvate kinase, glycolysis is slowed down under conditions in which ample energy and glucose precursors are available, favoring the gluconeogenesis pathway. M pyruvate kinase is not regulated in this manner.

We shall see in Chapter 25 that gluconeogenesis is also regulated by certain hormones.

Citric Acid Cycle Intermediates Are Also Precursors of Glucose

The biosynthetic pathway to glucose just described allows the net synthesis of glucose not only from pyruvate but also from various precursors of pyruvate or phosphoenolpyruvate (Figure 20-1). Chief among them are the citric acid cycle intermediates citrate, isocitrate, α-ketoglutarate, succinate, fumarate, and malate. All may undergo oxidation in the citric acid cycle to yield oxaloacetate, which is converted into phosphoenolpyruvate by phosphoenolpyruvate carboxykinase, as already shown (Figure 20-2). However, only three carbon atoms of each of the cycle intermediates can be converted into glucose.

It is important to note that normally acetyl-CoA cannot be used as a precursor of glucose by animals, since acetyl-CoA cannot be converted into pyruvate. We will recall that the pyruvate dehydrogenase reaction is irreversible under intracellular conditions (page 441). Thus, under normal conditions there is no *net* conversion of even-carbon fatty acids into glucose by animals, since such fatty acids yield only acetyl-CoA on oxidative cleavage.

Table 20-2 Glucogenic Amino Acids†

Converted into pyruvate
Alanine
Serine
Cysteine
Glycine
Converted into oxaloacetate
Asparagine
Aspartate
Converted into succinyl-CoA
Valine
Threonine
Methionine
Converted into α-ketoglutarate
Glutamate
Glutamine
Proline
Arginine
Histidine
Furnish carbon atoms for synthesis of *both* glucose and ketone bodies
Phenylalanine
Tyrosine
Isoleucine
Tryptophan
Lysine

† These amino acids are precursors of blood glucose or liver glycogen because they can be converted into pyruvate or into citric acid cycle intermediates. They are grouped into families according to their site of entry. Only leucine is totally unable to furnish carbon for net glucose synthesis.

Most Amino Acids Are Glucogenic

As shown in Chapter 19, some or all of the carbon atoms of many of the amino acids derived from proteins are ultimately converted by animals into either pyruvate or certain intermediates of the citric acid cycle. Such amino acids can therefore undergo net conversion into glucose and glycogen and are called *glucogenic* amino acids (Table 20-2). Obvious examples are alanine, glutamate, and aspartate, which on deamination yield pyruvate, α-ketoglutarate, and oxaloacetate, respectively, all of which are precursors of phosphoenolpyruvate by the reactions described above. In diabetes mellitus the net conversion of the glucogenic amino acids into glucose is a very active process and proceeds at a much higher rate than in normal people. As a secondary consequence large amounts of urea, derived from the deamination of the glucogenic amino acids, are excreted into the urine by diabetics.

Gluconeogenesis Takes Place during Recovery from Muscular Exercise

The synthesis of glucose from smaller precursors proceeds at a particularly high rate in the recovery period following all-out muscular effort, such as a 100-meter sprint (page 401). During such very intense muscular activity the circulatory system cannot bring oxygen and glucose to skeletal muscles fast enough to meet the overwhelming demand of the muscles for ATP. In this case muscle glycogen is used as a reserve fuel and is rapidly broken down by glycolysis to form lactate, thus generating ATP, which is the energy source for the contraction of the muscle. Since insufficient oxygen is available under these circumstances, the lactate cannot be metabolized further in the muscles and diffuses into the blood, where its concentration may become very high during such intense muscular effort. At the end of a 100-meter sprint a runner continues to breathe hard, but gradually the breathing rate declines to its normal level. During this recovery period the blood lactate also returns back to its normal low level. Much of the extra oxygen consumed during the recovery period, the oxygen debt, is used to generate the ATP necessary for resynthesis of blood glucose and muscle glycogen from the lactate formed anaerobically during the sprint. During full recovery, which may take as long as 30 min, the lactate is removed from the blood by the liver and converted into blood glucose via the gluconeogenesis pathway described above. The blood glucose returns to the muscles, where it is rebuilt into glycogen (Figure 20-5). Because six high-energy phosphate groups of ATP are ultimately expended to rebuild one molecule of glucose from two molecules of lactate, whereas only two molecules of ATP were originally obtained from the breakdown of a molecule of glucose in the muscles, use of muscle glycogen as an anaerobic fuel is energetically very expensive. Moreover, long recovery periods are required to rebuild glycogen stores (page 401). While this may appear to be a rather high price to pay for the advantage of being able to sprint all-out for short periods, this capacity has very significant survival value in the evolutionary struggle between prey and predator (Box 15-1, page 400).

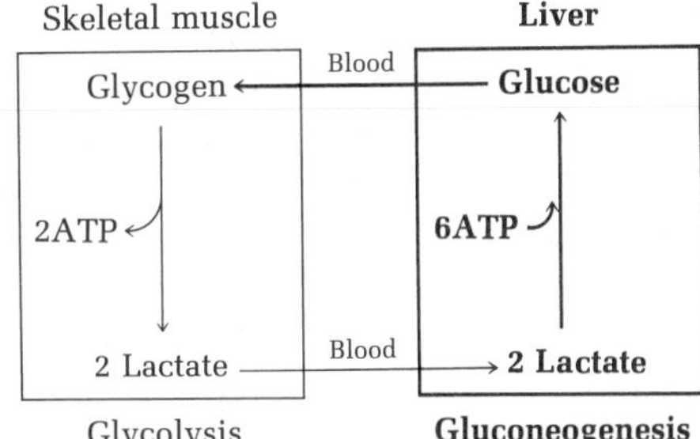

Figure 20-5
Cooperation between skeletal muscle and the liver in recovery from strenuous muscle activity, during which glycogen is broken down anaerobically to yield 2 lactate and 2ATP per glucose unit degraded. During recovery (color) blood lactate arising from the muscles is recovered and converted into blood glucose by the liver. Six ATP are required to generate a molecule of glucose from 2 lactate. Glucose returns to the muscles via the blood and is stored as glycogen.

Large marine mammals, e.g., seals and walrus, and amphibious reptiles, e.g., alligators and turtles, can remain submerged for long periods not only because they can store significant amounts of oxygen as oxymyoglobin but also because they can use glycogen via anaerobic glycolysis to generate ATP.

Gluconeogenesis Is an Especially Active Process in Ruminant Animals

Although the laboratory rat has a metabolism remarkably similar to that of human beings, this is not true of all familiar animals. In ruminant animals, such as cattle, ingested plant food is subjected to bacterial fermentation in the rumen, the first of

Figure 20-6
The rumen, which occupies a large fraction of the abdominal cavity of cattle, is a huge fermentation chamber where cellulose is enzymatically hydrolyzed by bacteria to yield glucose. The glucose is then fermented to lactate, propionate, acetate, and butyrate, which are absorbed into the blood. Lactate and propionate are converted into glucose at a high rate by the liver.

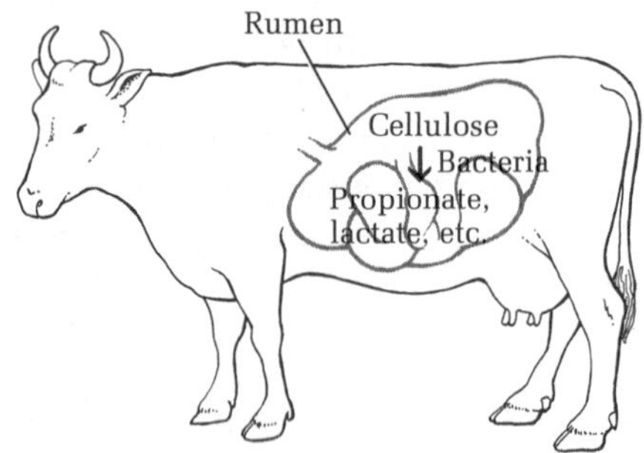

the four compartments of the stomach. In the cow the rumen has a huge capacity, some 70 L. It constitutes a large fermentation chamber (Figure 20-6) in which different bacterial species cooperate to degrade major plant components, particularly cellulose, which is not hydrolyzed by any of the normal digestive enzymes secreted by animals. The rumen bacteria hydrolyze the cellulose, which has $\beta(1 \rightarrow 4)$ linkages between its glucose residues, to yield free D-glucose. But the bacteria do not stop here. They ferment nearly all the glucose to form lactate and other products, particularly acetate, propionate, and butyrate. In the cow only a few grams of unfermented glucose pass from the intestinal tract into the bloodstream in a 24-h period. Nevertheless, cows, like rats and people, need blood glucose, not only to supply the brain and other tissues with fuel but also as a precursor of lactose (milk sugar) if they are lactating.

How does the cow obtain glucose if ingested carbohydrates are almost completely fermented to short-chain organic acids in the rumen? The answer is that cattle are constantly dependent upon gluconeogenesis, which proceeds at a very high rate in the bovine liver. Lactate formed in the rumen by bacterial fermentation is absorbed into the blood and converted into glucose by the liver via the pathway we have already described, exactly as in people and rats. Another major product of glucose fermentation in the rumen, the 3-carbon *propionate* (page 521), is converted into glucose by a pathway that occurs in both ruminant and nonruminant animals but is quantitatively much more important in the former. This pathway (Figure 20-7) is remarkable for two special reasons: (1) it includes a step in which carbon dioxide is used or "fixed" into organic form by the carboxylation of propionyl-CoA; (2) the gluconeogenic pathway from propionate to glucose includes a step promoted by an enzyme containing a tightly bound coenzyme form of vitamin B_{12}, *deoxyadenosylcobalamine* (page 262), as its prosthetic group. This enzyme is *methylmalonyl-CoA mutase* (page 522). In this reaction a complex substituted alkyl group is moved from one carbon atom to the next in exchange for an H atom, to form *succinyl-CoA* (Figure 20-7). Once succinyl-CoA has been formed, it is converted into malate, which is a precursor of phosphoenolpyruvate (Figure 20-2) and ultimately of D-glucose. Note that the CO_2 fixed in the carboxylation of propionyl-CoA is lost again later. The conversion of propionate to glucose is much slower in nonruminant animals, where propionate occurs only during the oxidation of fatty acids with an odd number of carbon atoms (page 521) and during the oxidative degradation of the amino acids methionine and valine.

Alcohol Consumption Inhibits Gluconeogenesis

Another aspect of gluconeogenesis is of special interest in human biology and medicine. Excessive consumption of ethyl alcohol is highly inhibitory to gluconeogenesis in the liver and can result in a deficiency of glucose in the blood, called *hypo-*

Figure 20-7
Conversion of propionate into succinyl-CoA, which in turn can be converted into phosphoenolpyruvate and thence into glucose. The CO_2 taken up in the formation of D-methylmalonyl-CoA is later lost when oxaloacetate is converted into phosphoenolpyruvate. Note the exchange of the substituents on two adjacent carbons of L-methylmalonyl-CoA by methylmalonyl-CoA mutase, which requires coenzyme B_{12}. See also page 522.

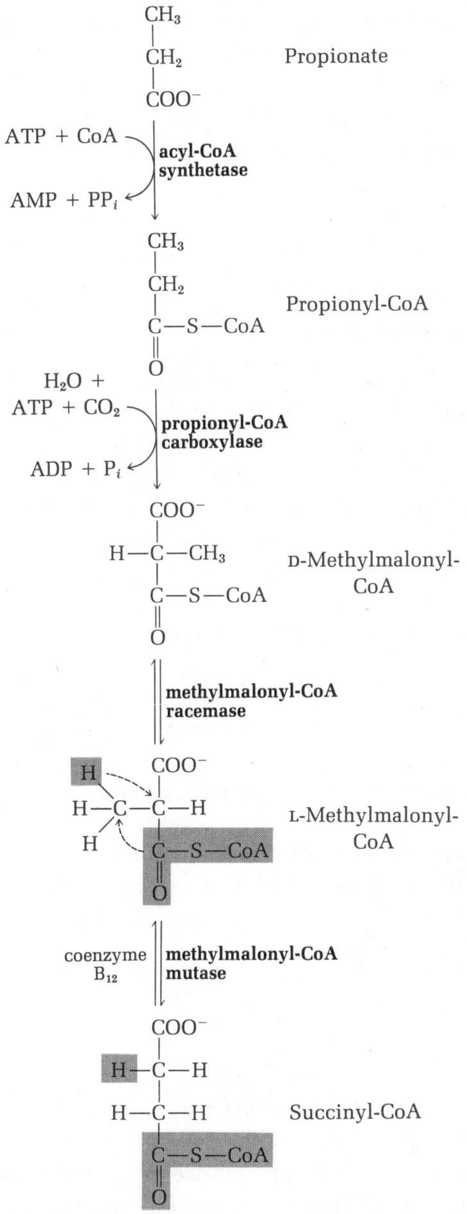

CH_3—CH_2—COO^- Propionate

ATP + CoA → AMP + PP_i (**acyl-CoA synthetase**)

CH_3—CH_2—C(=O)—S—CoA Propionyl-CoA

H_2O + ATP + CO_2 → ADP + P_i (**propionyl-CoA carboxylase**)

COO^-—H—C—CH_3—C(=O)—S—CoA D-Methylmalonyl-CoA

⇅ **methylmalonyl-CoA racemase**

COO^-—H—C(H)(H)—C—H—C(=O)—S—CoA L-Methylmalonyl-CoA

⇅ coenzyme B_{12} **methylmalonyl-CoA mutase**

COO^-—H—C—H—H—C—H—C(=O)—S—CoA Succinyl-CoA

glycemia. This effect of alcohol is particularly crucial after periods of heavy physical exertion or lowered intake of food. If ethanol is given to human subjects after a period of exhaustive physical effort, the blood glucose level can decline to only 30 or 40 percent of its normal concentration. Hypoglycemia is harmful to brain function, since it specifically affects those portions of the brain concerned in temperature regulation, with the result that the rectal temperature may drop 2°C or more under such circumstances. If glucose is given to such subjects by mouth, normal body temperature is quickly restored. The old practice of giving brandy or whiskey to exhausted, starving persons rescued from the sea or the wilderness is physiologically unsound and even dangerous; the rational treatment is administration of glucose.

"Futile Cycles" in Carbohydrate Metabolism

Sharp-eyed readers examining the pathways of glycolysis and gluconeogenesis shown in Figure 20-2 will come up with a very troublesome question. As shown, there are three points in the opposing metabolic routes between glucose and pyruvate where the enzymatic reaction in the catabolic direction is bypassed by a different type of enzymatic reaction in the anabolic pathway. For example, phosphofructokinase catalyzes phosphorylation of fructose 6-phosphate by ATP, whereas its bypass counterpart in gluconeogenesis, fructose diphosphatase, catalyzes hydrolysis of fructose 1,6-diphosphate to yield fructose 6-phosphate. The two opposing reactions are

$$\text{ATP} + \text{fructose 6-phosphate} \longrightarrow \text{ADP} + \text{fructose 1,6-diphosphate}$$
$$\text{Fructose 1,6-diphosphate} + H_2O \longrightarrow \text{fructose 6-phosphate} + P_i$$

It will be seen that the sum of these two reactions is

$$\text{ATP} + H_2O \longrightarrow \text{ADP} + P_i$$

an energy-wasting reaction resulting in the net hydrolysis of ATP without any net metabolic work being done. Clearly, if these two reactions occur simultaneously at a high rate in the same cell, they can cause loss of a large amount of energy, which would appear as heat. Such an ATP-degrading cycle is called a futile cycle. A similar futile cycle could occur with the corresponding enzyme pair hexokinase and glucose 6-phosphatase:

$$\text{ATP} + \text{glucose} \longrightarrow \text{glucose 6-phosphate} + \text{ADP}$$
$$\text{Glucose 6-phosphate} + H_2O \longrightarrow \text{glucose} + P_i$$
$$\text{Sum:} \quad \text{ATP} + H_2O \longrightarrow \text{ADP} + P_i$$

Under normal circumstances futile cycles probably do not take place, since they are prevented by reciprocal regulatory mechanisms, i.e., whenever the net flow is catabolic, in the direction of glycolysis, fructose diphosphatase activity is turned off. Conversely, when the net flow is in the direction of gluconeogenesis, phosphofructokinase is turned off.

Recent research indicates, however, that futile cycling may sometimes take place physiologically and have a real biological purpose, i.e., the production of heat. An interesting example of futile cycling occurs in some insects. In cold weather bumblebees cannot fly until they have warmed up their engines, i.e., until the temperature of their muscles has been raised to about 30°C and maintained by futile cycling of fructose 6-phosphate and fructose 1,6-diphosphate and the consequent heat-generating hydrolysis of ATP. It is also believed that futile cycling for heat production may take place in some species of animals undergoing arousal from the state of hibernation, in which the body temperature may be much below its normal level.

Biosynthesis of Glycogen Proceeds by a Pathway Different from That of Glycogen Breakdown

Now that we have seen how glucose is synthesized from simpler precursors, let us examine the biosynthetic conversion of glucose residues into glycogen. In animals glycogen synthesis occurs in virtually all the tissues but is especially prominent in the liver and skeletal muscles. The starting point for synthesis of glycogen from free glucose is the *hexokinase* reaction, which phosphorylates glucose to glucose 6-phosphate

$$\text{ATP} + \text{D-glucose} \longrightarrow \text{D-glucose 6-phosphate} + \text{ADP}$$

In the next step glucose 6-phosphate is reversibly converted into glucose 1-phosphate by *phosphoglucomutase* (page 416)

$$\text{Glucose 6-phosphate} \rightleftharpoons \text{glucose 1-phosphate}$$

Now we come to a key reaction in glycogen biosynthesis, one not participating in glycogen breakdown. This is the formation of *uridine diphosphate glucose* (UDP-glucose) (Figure 20-8) by the action of *glucose 1-phosphate uridylyltransferase*

$$\text{UTP} + \text{glucose 1-phosphate} \longrightarrow \text{UDP-glucose} + \text{PP}_i$$

This reaction is caused to proceed to the right by the action of pyrophosphatase, which hydrolyzes pyrophosphate (PP_i) to orthophosphate (P_i). UDP-glucose, we have seen earlier, is an intermediate in the conversion of D-galactose into D-glucose (page 418). UDP-glucose is the immediate donor of glucose residues in the enzymatic formation of glycogen by the action of *glycogen synthase,* which promotes the transfer of the glucosyl residue from UDP-glucose to a nonreducing end of the

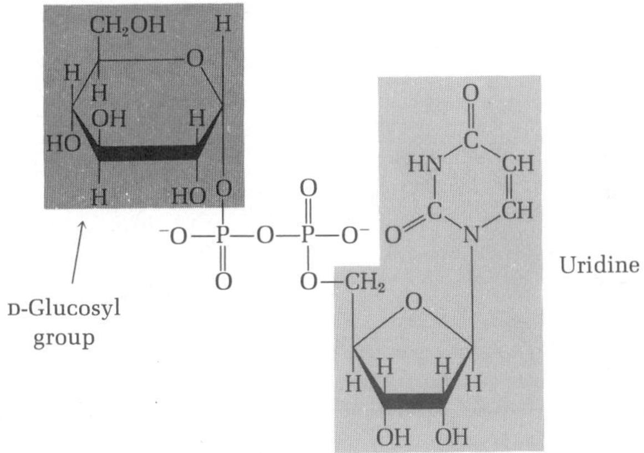

Figure 20-8
Uridine diphosphate glucose (UDP-glucose), the glucosyl donor in the glycogen synthase reaction.

branched glycogen molecule (Figure 20-9). In this reaction a new $\alpha(1 \rightarrow 4)$ linkage is established between carbon atom 1 of the incoming glucose and carbon atom 4 of the terminal glucose residue of a glycogen branch

$$\text{UDP-glucose} + \underset{\text{Glycogen branch}}{(\text{glucose})_n} \longrightarrow \text{UDP} + \underset{\text{Lengthened glycogen branch}}{(\text{glucose})_{n+1}}$$

The overall equilibrium of this set of three reactions greatly favors synthesis of glycogen. Glycogen synthase requires as a primer an $\alpha(1 \rightarrow 4)$ polyglucose chain or branch having at least four glucose residues, to which it adds successive glucosyl groups at the nonreducing end.

Figure 20-9
Elongation of a glycogen chain by glycogen synthase. The D-glucosyl residue of UDP–D-glucose is transferred to the nonreducing end of a glycogen branch to make a new $\alpha(1 \rightarrow 4)$ linkage.

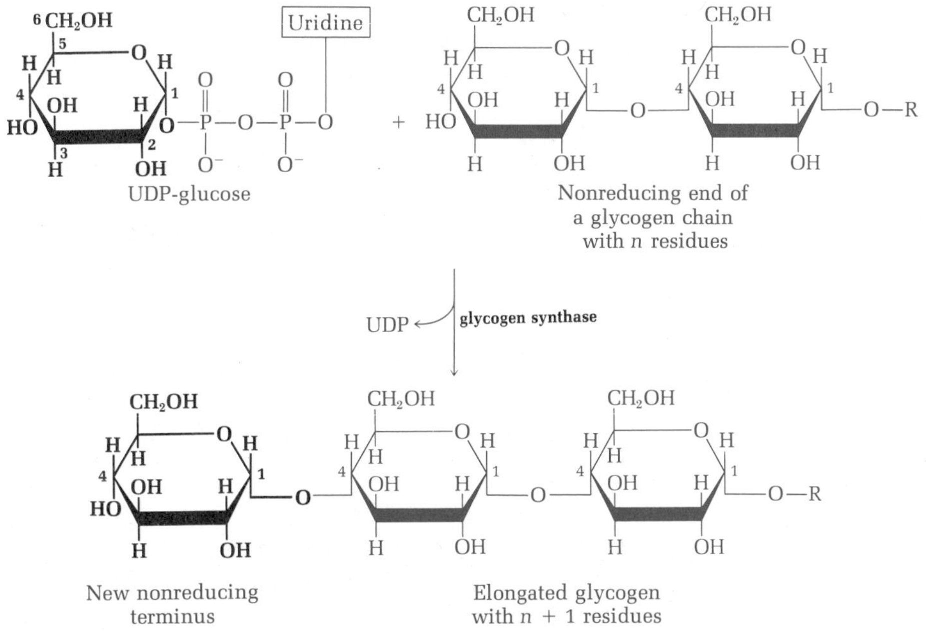

Uridine

UDP-glucose

Nonreducing end of a glycogen chain with n residues

UDP ← glycogen synthase

New nonreducing terminus

Elongated glycogen with $n + 1$ residues

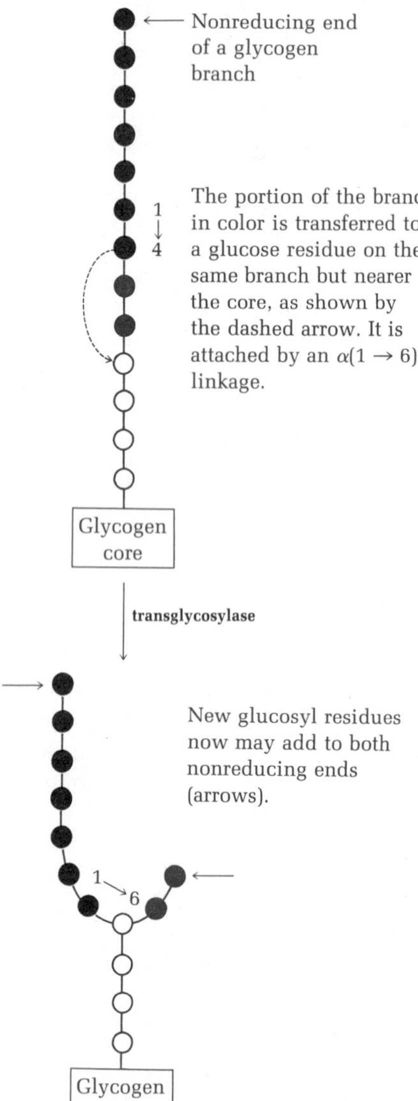

Figure 20-10
How glycogen branching enzyme [glycosyl (4 → 6) transferase] forms a new branch point during glycogen synthesis.

The role of UTP and UDP-glucose in the biosynthesis of glycogen and many other carbohydrate derivatives was discovered by the Argentine biochemist Luis Leloir, awarded the Nobel prize in 1970. We have seen other instances in which nucleoside diphosphate sugars are intermediates in the biosynthesis of other carbohydrates and their derivatives (page 418).

Glycogen synthase cannot make the α(1 → 6) bonds found at the branch points of glycogen (pages 288 and 416), but a glycogen-branching enzyme, *glycosyl-(4 → 6)-transferase*, catalyzes transfer of a terminal oligosaccharide fragment of six or seven glucosyl residues from the nonreducing end of a glycogen branch having at least eleven residues to the 6-hydroxyl group of a glucose residue of the same or another glycogen chain at a more interior point, thus creating a new branch (Figure 20-10). Further glucosyl residues may be added to the new branch by glycogen synthase. The biological effect of branching is to make the glycogen molecule more soluble and to increase the number of nonreducing ends, thus making the glycogen more reactive to glycogen phosphorylase and glycogen synthase.

Starch is synthesized in plants by a similar pathway that utilizes ADP-glucose instead of UDP-glucose as a glucosyl donor.

Glycogen Synthase and Glycogen Phosphorylase Are Reciprocally Regulated

Earlier we saw that the breakdown of glycogen is regulated by both covalent and allosteric modulation of glycogen phosphorylase (page 421). *Phosphorylase a*, the active form, which contains essential phosphorylated serine residues, is dephosphorylated by *phosphorylase phosphatase* to yield *phosphorylase b*, the relatively inactive form, which can be stimulated by AMP, its allosteric modulator. Phosphorylase kinase can convert phosphorylase *b* back into the active phosphorylase *a* at the expense of ATP, which rephosphorylates the essential serine residues.

Glycogen synthase also occurs in phosphorylated and dephosphorylated forms, but it is regulated in a reciprocal manner, opposite to that of glycogen phosphorylase (Figure 20-11). Its active form, *glycogen synthase a*, is dephosphorylated. When it is phosphorylated by ATP at two serine hydroxyl groups by protein kinase, glycogen synthase *a* is converted into its less active form *glycogen synthase b*

$$\underset{\text{Active}}{\text{Glycogen synthase } a} + 2\text{ATP} \xrightarrow[\text{kinase}]{\text{protein}} \underset{\text{Less active}}{\text{glycogen synthase } b} + 2\text{ADP}$$

The conversion of the less active glycogen synthase *b* back into the active form is promoted by *phosphoprotein phosphatase*, which removes the phosphate groups from the serine residues

$$\text{Glycogen synthase } b + 2H_2O \longrightarrow \text{glycogen synthase } a + 2P_i$$

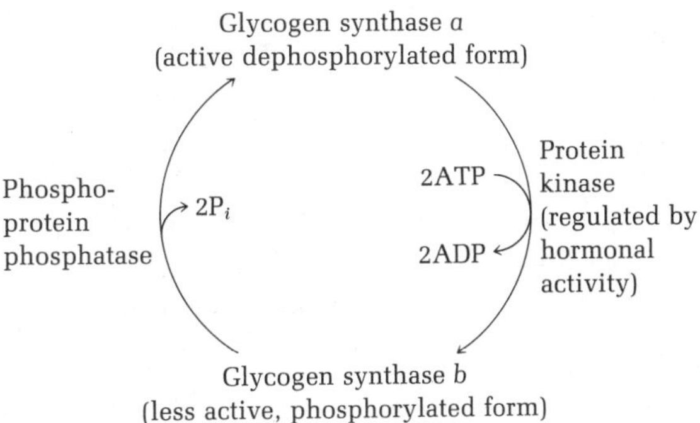

Figure 20-11
Regulation of glycogen synthase activity by its enzymatic phosphorylation and dephosphorylation. Protein kinase itself occurs in two forms, active and inactive, whose ratio is hormonally regulated (Chapter 25).

Glycogen phosphorylase and glycogen synthase are therefore reciprocally regulated; when one is stimulated, the other is inhibited (Figure 20-12). Thus it appears that these two enzymes are never fully active simultaneously.

Glycogen synthase can also be modulated allosterically. The less active form glycogen synthase *b* is stimulated by its allosteric modulator glucose 6-phosphate. Because glycogen synthase *b* *depends* upon glucose 6-phosphate for activity, it has been given the alternative name *glycogen synthase D*. Glycogen synthase *a* is not stimulated by, i.e., is *independent* of, glucose 6-phosphate and is therefore given the alternate name *glycogen synthase I*.

The balance between the rates of glycogen synthesis and breakdown in the liver is ultimately controlled by two hormones, *adrenaline*, of the adrenal medulla, and *glucagon*, secreted by the pancreas. These hormones act by regulating the ratio of the active to inactive forms of both glycogen phosphorylase and glycogen synthase. Adrenaline secretion stimulates breakdown of glycogen in both the liver and muscles by

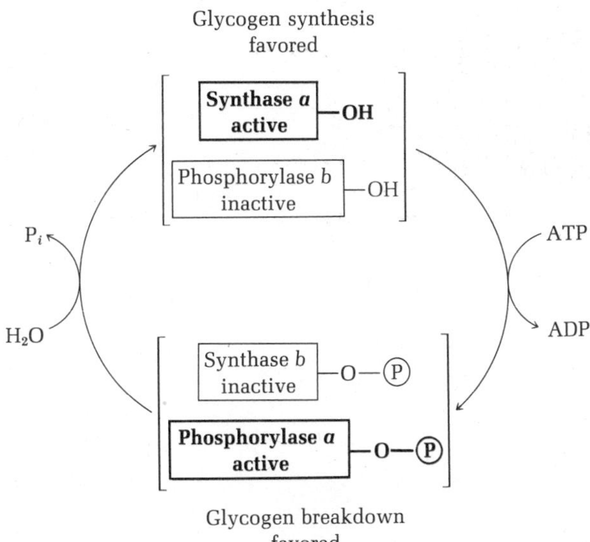

Figure 20-12
Reciprocal regulation of glycogen synthase and glycogen phosphorylase by phosphorylation and dephosphorylation. The active forms of each enzyme are shown in color, the inactive forms in black. The phosphorylated serine groups are indicated by —O—Ⓟ.

causing an increase in the ratio of phosphorylase *a* to phosphorylase *b*, but a decrease in the ratio of glycogen synthase *a* to glycogen synthase *b*. Although glucagon has a similar end effect, its action is exerted by a different pathway. We shall examine the details of the hormonal regulation of glycogen metabolism in Chapter 25.

Glycogen Metabolism Is Subject to Genetic Defects

Several different human genetic diseases affecting the synthesis or degradation of glycogen have been discovered. Among the first to be described was the case of an 8-year-old girl in Germany who had a chronically enlarged liver as well as metabolic symptoms. After the girl's death from influenza, the pathologist found her liver to be 3 times normal size; it contained a huge amount of glycogen, almost 40 percent of the liver dry weight. The glycogen isolated from the liver was chemically normal in all respects, but when a sample of the liver was ground, suspended in buffer, and incubated, the glycogen remained intact and neither lactate nor glucose was formed. When glycogen from the diseased liver was added to a suspension of normal liver, it underwent rapid breakdown to glucose. From this biochemical detective work it was concluded that the patient in question had a defect in the breakdown of glycogen, often called <u>von Gierke's disease</u> after the pathologist who discovered it. At first it was thought that glucose 6-phosphatase was the defective enzyme, since no glucose was formed by the diseased liver, but the fact that no lactate was formed suggested that either glycogen phosphorylase or debranching enzyme [$\alpha(1 \rightarrow 6)$ glucosidase, page 416] was defective. It has since been suggested that the defective enzyme in this classical case was the debranching enzyme. As a result only the outer branches of liver glycogen molecules could undergo degradation to yield glucose or lactate, resulting in accumulation of many glycogen molecules having large cores that could not be degraded further.

Today 12 or more different inborn errors of glycogen synthesis or degradation are known, each affecting a different enzyme (Table 20-3). The most serious, ultimately lethal, are deficiencies in glucose 6-phosphatase, debranching enzyme, and branching enzyme. Also lethal are genetic defects in pyruvate carboxylase and phosphoenolpyruvate carboxykinase, which catalyze early steps in gluconeogenesis.

Table 20-3 Human Genetic Errors of Glycogen Metabolism and Gluconeogenesis

Defective enzyme	Glycogen storage disease
Glucose 6-phosphatase	Type I
$\alpha(1 \rightarrow 6)$ Glucosidase	Type II
Debranching enzyme	Type III
Branching enzyme	Type IV
Muscle phosphorylase	Type V
Liver phosphorylase	
Liver phosphorylase kinase	
Muscle phosphofructokinase	
Liver glycogen synthase	
Fructose 1,6-diphosphatase	
Pyruvate carboxylase	
Phosphoenolpyruvate carboxykinase	

Lactose Synthesis Is Regulated in a Unique Way

Most tissues of vertebrates contain the enzyme <u>galactosyltransferase</u>, which promotes the transfer of a D-galactose residue to *N*-acetylglucosamine

$$\text{UDP-D-galactose} + N\text{-acetylglucosamine} \longrightarrow \text{UDP} + \text{D-galactosyl-}N\text{-acetyl-D-glucosamine}$$

This reaction is a step in the biosynthesis of the carbohydrate portion of galactose-containing glycoproteins (page 294) in animal tissues. In the lactating mammary gland, however, D-galactose is a precursor in another biosynthetic process, the synthesis of *lactose* or milk sugar, a disaccharide of D-galactose and D-glucose (page 284). In the lactating mammary gland, galactosyltransferase participates in lactose synthesis in a most unusual way. During pregnancy galactosyltransferase, which is very active with N-acetylglucosamine, but only feebly active with D-glucose as galactosyl acceptor, is present in the mammary gland, as in most tissues. However, when lactation begins after parturition, the specificity of galactosyltransferase changes: it now transfers the D-galactosyl group to D-glucose at a very high rate, thus making lactose according to the reaction

$$\text{UDP-D-galactose} + \text{D-glucose} \longrightarrow \text{UDP} + \text{D-lactose}$$

This "new" enzyme is called *lactose synthase*.

The change in specificity of galactosyltransferase is caused by the formation of *α-lactalbumin*, a protein present in milk, whose function was long unknown. α-Lactalbumin has been found to be an *enzyme modifier*; its synthesis in the mammary gland, which is regulated by the hormones promoting lactation, leads to the formation of a *lactalbumin-galactosyltransferase* complex, i.e., *lactose synthase*. Thus, by hormone action the synthesis of lactose is switched on in the mammary gland by the formation of a specificity-modulating subunit of lactose synthetase.

Summary

Gluconeogenesis is the formation of "new" sugar from nonsugar precursors, the most important being pyruvate, lactate, citric acid cycle intermediates, and many of the amino acids. As is true for all biosynthetic pathways, gluconeogenesis proceeds by an enzymatic pathway that differs from the corresponding catabolic pathway, that is independently regulated, and that requires input of chemical energy in the form of ATP. The biosynthetic pathway from pyruvate to glucose, which in vertebrates takes place largely in the liver and secondarily in the kidney, employs eight of the glycolytic enzymes, which function reversibly and are present in large excess. There are, however, three irreversible steps in the downhill glycolytic pathway that cannot be used in gluconeogenesis and are bypassed by alternative reactions catalyzed by quite different enzymes. The first bypass consists of the conversion of pyruvate into phosphoenolpyruvate via the formation of oxaloacetate; the second is the dephosphorylation of fructose 1,6-diphosphate by fructose diphosphatase; and the third is the dephosphorylation of glucose 6-phosphate by glucose 6-phosphatase. For each molecule of D-glucose made from pyruvate the terminal phosphate groups of four molecules of ATP and two of GTP must be used. Gluconeogenesis is regulated at two major points: (1) the carboxylation of pyruvate by pyruvate carboxylase, which is stimulated by the allosteric effector acetyl-CoA, and (2) the dephosphorylation of fructose 1,6-diphosphate by fructose diphosphatase, which is inhibited by

AMP but stimulated by citrate. Three carbon atoms of each of the citric acid cycle intermediates and many of the amino acids are convertible into glucose. Even-carbon fatty acids or acetyl-CoA do not lead to net glucose formation, but three carbon atoms of odd-carbon fatty acids and propionate (formed by rumen bacteria) are convertible into glucose via the intermediate formation of methylmalonyl-CoA and its conversion into succinyl-CoA, which requires participation of coenzyme B_{12}. During recovery from strenuous physical exercise gluconeogenesis is a very active process resulting in conversion of blood lactate into glycogen and glucose.

Glycogen synthesis also proceeds via a pathway different from its breakdown. It involves the conversion of glucose 1-phosphate into uridine diphosphate glucose, which then donates its glucosyl residue to the nonreducing end of glycogen branches by the action of glycogen synthase. New glycogen branches are started by $\alpha(1{,}4 \rightarrow 1{,}6)$transglycosylase. Glycogen synthesis and breakdown are independently and reciprocally regulated. The balance between the rates of synthesis and breakdown is controlled by adrenaline and glucagon. Glycogen synthesis and breakdown are subject to genetic defects.

Lactose synthesis in the mammary gland is brought about by the lactalbumin-galactosyltransferase enzyme complex, in which the lactalbumin serves as a specificity-modifying subunit, whose formation is regulated by hormones promoting lactation.

References

Books

CUNNINGHAM, E. B.: *Biochemistry: Mechanisms of Metabolism*, McGraw-Hill, New York, 1978. Chapter 9 has further details on carbohydrate biosynthesis, particularly on the enzymology.

DICKENS, F., P. J. RANDLE, and W. J. WHELAN (eds.): *Carbohydrate Metabolism and Its Disorders*, vols. I and II, Academic, New York, 1968. A collection of valuable reviews.

NEWSHOLME, E. A., and C. START: *Regulation in Metabolism*, Wiley, New York, 1973. Regulation of gluconeogenesis and glycogen synthesis is developed further in Chapters 4 to 6.

Articles

BENT, H. A.: "Energy and Exercise," *J. Chem. Ed.*, vol. 55, nos. 7–12, July–December (1978).

HOWELL, R. R.: "The Glycogen Storage Diseases," pp. 160–181 in J. B. Stanbury, J. B. Wyngaarden, and D. S. Frederickson (eds.), *The Metabolic Basis of Inherited Disease*, 4th ed., McGraw-Hill, New York, 1978.

KATZ, J., and R. ROGNSTAD: "Futile Cycles in the Metabolism of Glucose," *Curr. Top. Cell Regul.*, **10**:238–287 (1976).

KREBS, H. A.: "Some Aspects of the Regulation of Fuel Supply in Omnivorous Animals," *Adv. Enz. Regul.*, **10**:397–420 (1972).

SHARON, N.: "Carbohydrates," *Sci. Am.*, **243**:90–116, November (1980).

Problems

1. *Role of Oxidative Phosphorylation in Gluconeogenesis* Is it possible to obtain a net synthesis of glucose from pyruvate if the citric acid cycle and oxidative phosphorylation are totally inhibited?

2. *Pathway of Atoms in Gluconeogenesis* A liver extract capable of carrying out all the normal metabolic reactions is briefly incubated in separate experiments with the following ^{14}C-labeled precursors:

   (a) 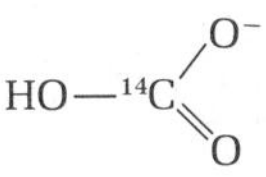 $HO{-}^{14}C(=O)O^-$

   ^{14}C-Bicarbonate

   (b) 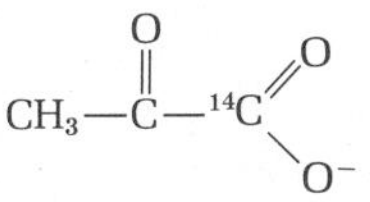 $CH_3{-}C(=O){-}^{14}C(=O)O^-$

   1-^{14}C-Pyruvate

   Trace the pathway of each precursor through gluconeogenesis. Indicate the location of ^{14}C in all intermediates of the process and in the product glucose.

3. *Pathway of CO_2 in Gluconeogenesis* In the first bypass step in gluconeogenesis, the conversion of pyruvate to phosphoenolpyruvate, pyruvate is carboxylated by pyruvate carboxylase to oxaloacetate and then subsequently decarboxylated by phosphoenolpyruvate carboxykinase to yield phosphoenolpyruvate. The observation that the addition of CO_2 is directly followed by the loss of CO_2 suggests that ^{14}C of $^{14}CO_2$ would not be incorporated into phosphoenolpyruvate, glucose, or any of the intermediates in gluconeogenesis. However, it has been found that if rat liver slices synthesize glucose in the presence of $^{14}CO_2$, ^{14}C will slowly appear in the phosphoenolpyruvate, and eventually in carbons 3 and 4 of glucose. How does the ^{14}C label get into phosphoenolpyruvate and carbons 3 and 4 of glucose? (*Hint:* During gluconeogenesis in the presence of $^{14}CO_2$, several of the 4-carbon intermediates in the citric acid cycle also become labeled.)

4. *Regulation of Fructose Diphosphatase and Phosphofructokinase* What are the effects of increasing concentrations of ATP and AMP on the catalytic activities of fructose diphosphatase and phosphofructokinase? What are the consequences of these effects of ATP and AMP on the relative flow of metabolites through gluconeogenesis and glycolysis?

5. *Glucogenic Substrates* A common procedure for the determination of the effectiveness of compounds as precursors of glucose is to fast the animal until the liver glycogen stores are depleted and then administer the substrate in question. A substrate that leads to a *net* increase in liver glycogen is termed glucogenic since it must first be converted to glucose 6-phosphate. Show by means of known enzymatic reactions which of the following substances are glucogenic.

   (a) $^-OOC{-}CH_2{-}CH_2{-}COO^-$

   Succinate

(b) $\mathrm{\underset{|}{\overset{OH}{CH_2}}—\underset{H}{\overset{OH}{C}}—\overset{OH}{CH_2}}$

Glycerol

(c) $\mathrm{CH_3—\overset{O}{\overset{\|}{C}}—S—CoA}$

Acetyl-CoA

(d) $\mathrm{CH_3—\overset{O}{\overset{\|}{C}}—C\overset{\diagup O}{\underset{\diagdown O^-}{}}}$

Pyruvate

(e) $\mathrm{CH_3—CH_2—CH_2—C\overset{\diagup O}{\underset{\diagdown O^-}{}}}$

Butyrate

Problem 6

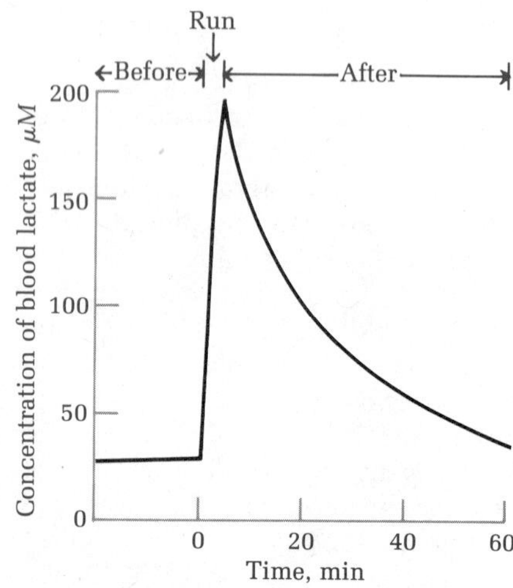

6. *Blood Lactate during Vigorous Exercise* The concentration of lactate in blood plasma before, during, and after a 400-meter sprint are shown in the accompanying figure.
   (a) What causes the rapid rise in lactate concentration?
   (b) What causes the decline in lactate level after completion of the run? Why does the decline occur more slowly than the rise?
   (c) Why is the concentration of lactate not zero during the resting state?

7. *Excess Oxygen Uptake during Gluconeogenesis* Lactate absorbed by the liver is converted to glucose. This process requires the input of 6 mol of ATP for every mole of glucose produced. The extent of this process in rat liver slices can be monitored by administering ^{14}C-lactate and measuring the amount of ^{14}C-glucose produced. Since the stoichiometry between O_2 consumption and ATP production is known (Chapter 17), we can predict the extra O_2 consumption above the normal rate when a given amount of lactate is administered. The extra amount of O_2 necessary for the synthesis of glucose from lactate, however, when actually measured is always higher than predicted by known stoichiometric relationships. Suggest a possible explanation for this observation.

8. *At What Point Is Glycogen Synthesis Regulated?* Explain how the two following observations serve to identify the point of regulation in the synthesis of glycogen in skeletal muscle.
   (a) The measured activity of glycogen synthase in resting muscle, expressed in micromoles of UDP-glucose used per gram per minute, is lower than the activity of phosphoglucomutase or UDP-glucose pyrophosphorylase, each measured in terms of micromoles of substrate transformed per gram per minute.
   (b) Stimulation of glycogen synthesis leads to a small decrease in the concentrations of glucose 6-phosphate and glucose 1-phosphate, a large decrease in the concentration of UDP-glucose, but a substantial increase in the concentration of UDP.

9. *What Is the Cost of Storing Glucose in the Form of Glycogen?* Write the sequence of the steps and the net reaction required to calculate the cost in number of ATPs of converting cytoplas-

mic glucose 6-phosphate into glycogen and back into glucose 6-phosphate. What fraction of the maximum number of ATPs that are available from complete catabolism of glucose 6-phosphate does this cost represent?

10. *Identification of a Defective Enzyme in Carbohydrate Metabolism* A sample of liver tissue was obtained post-mortem from the body of a patient believed to be genetically deficient in one of the enzymes of carbohydrate metabolism. A homogenate of the liver sample had the following characteristics: (1) it degraded glycogen to glucose 6-phosphate, (2) it was unable to make glycogen from any sugar or to utilize galactose as an energy source, and (3) it synthesized glucose 6-phosphate from lactate. Which of the following three enzymes is deficient?
(a) Glycogen phosphorylase
(b) Fructose diphosphatase
(c) UDP-glucose pyrophosphorylase
Give reasons for your choice.

11. *Ketosis in Sheep* The udder uses almost 80 percent of the total glucose synthesized by the ewe. The glucose is used for milk production, principally in the synthesis of lactose and of glycerol phosphate, used in the formation of milk triacylglycerols. During the winter when food quality is poor, milk production decreases and the ewes sometimes develop ketosis, i.e., increased levels of plasma ketone bodies. Why do these changes occur? A standard treatment for this condition is the administration of large doses of propionate. How does this work?

12. *Adaptation to Galactosemia* Galactosemia is a pathological condition in which there is deficient utilization of galactose derived from lactose in the diet. One form of this disease is due to the absence of the enzyme galactose-1-phosphate uridylyltransferase. If the patient survives the disease in early life, some capacity to metabolize ingested galactose sometimes develops in later life, because of increased production of the enzyme UDP-galactose pyrophosphorylase, which catalyzes the reaction

$$\text{UTP} + \text{galactose 1-phosphate} \longrightarrow \text{UDP-galactose} + \text{PP}_i$$

How does the presence of this enzyme increase the capacity of such patients to metabolize galactose?

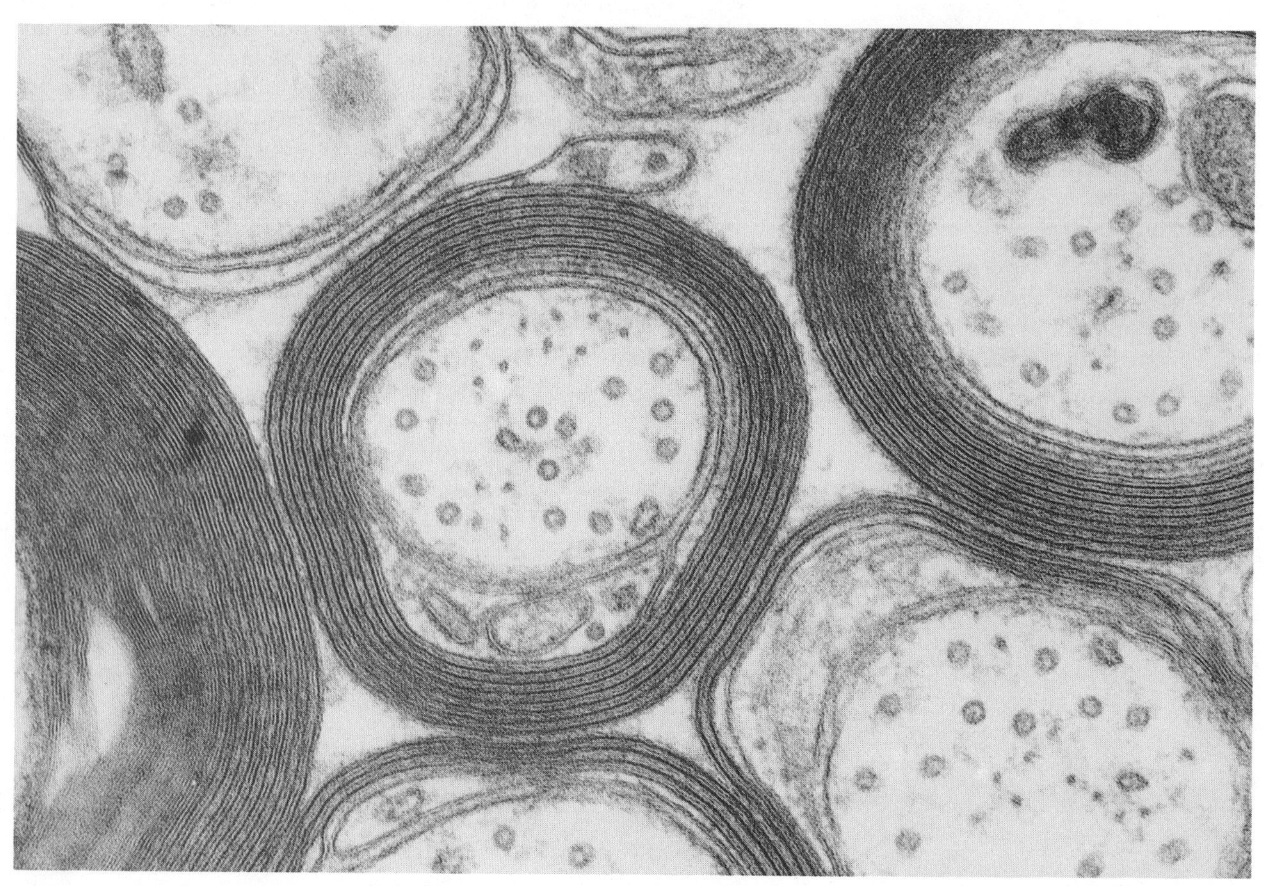

Cross section showing the myelin sheaths surrounding the axons of several nerve fibers. The sheath, which is composed largely of polar lipids and some proteins, consists of the plasma membrane of a Schwann cell. As the Schwann cell grows, it wraps itself around the axon and extrudes its cytoplasm from between the layers. The myelin sheath insulates certain nerve fibers that must transmit impulses very rapidly.

CHAPTER 21

The Biosynthesis of Lipids

The biosynthesis of triacylglycerols is a very active metabolic process in animals, particularly because their capacity to store triacylglycerols is quite large. In human beings only a few hundred grams of glycogen can be stored in the liver and muscles, barely enough to supply the body's energy needs for some 12 hours. In contrast, the total amount of stored triacylglycerols in a 70-kg human male of average build is about 12 kg, enough to supply his basal energy needs for as long as 8 weeks. Whenever carbohydrate is consumed in excess of the capacity to store glycogen, it is converted into triacylglycerols, which can be stored in large amounts in fat cells in different parts of the body, especially under the skin and in the abdominal cavity. Plants also manufacture triacylglycerols as a storage form of energy-rich fuel, especially in fruits, nuts, and seeds.

The polar lipids of membranes, such as the various phospholipids and sphingolipids, which are not stored, are also constantly synthesized by animals as membranes undergo continuous metabolic turnover. For example, the half-life of the membrane phospholipids in the liver of the rat is less than 3 days.

In this chapter we shall describe first the biosynthesis of fatty acids, the major building blocks of triacylglycerols and polar lipids. Well over 90 percent of the stored energy of triacylglycerols resides in their fatty acid components. Fatty acids also give triacylglycerols and phospholipids their hydrophobic character. Then we shall examine the synthesis of triacylglycerols, the simpler types of membrane phospholipids, and cholesterol, an important component of some membranes and the precursor of such important steroid products as the bile acids, sex hormones, and adrenal cortical hormones.

The Biosynthesis of Fatty Acids Proceeds by a Distinctive Pathway

We recall that the predominant fatty acids in the lipids of animal tissues have an even number of carbon atoms. This fact long ago suggested that both oxidation and synthesis of fatty acids proceed by loss or gain, respectively, of 2-carbon units. Accordingly, when fatty acid oxidation was found to occur by oxidative removal of successive acetyl groups in the form of acetyl-CoA, it was expected that the biosynthesis of fatty acids would prove to proceed by simple reversal of the same enzymatic steps used in their oxidation. It turned out, however, that fatty acid biosynthesis occurs by a different pathway, is catalyzed by a different set of enzymes, and takes place in a different part of the cell. Moreover, it was found that a 3-carbon intermediate participates in fatty acid biosynthesis and that CO_2 is also required.

Today we know that the *fatty acid synthase system* catalyzes the following overall reaction, in which one molecule of acetyl-CoA and seven molecules of the 3-carbon *malonic acid*, in the form of its CoA thioester *malonyl-CoA* (Figure 21-1), are assembled in succession to make a molecule of the 16-carbon palmitic acid, with release of seven molecules of CO_2

Figure 21-1
Malonyl-CoA, the immediate precursor of 2-carbon units of fatty acid chains. Malonyl-CoA is a derivative of malonate, a potent competitive inhibitor of succinate dehydrogenase (page 223). It may therefore appear surprising that malonyl-CoA is a normal biosynthetic precursor of fatty acids. However, malonyl-CoA does not inhibit succinate dehydrogenase, presumably because it lacks two free carboxylate groups, properly spaced to fit the succinate binding site. Moreover, free malonate is not a normal metabolite. Malonyl-CoA is made by the direct carboxylation of acetyl-CoA, as we shall see.

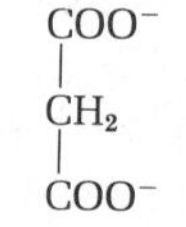

$$\begin{array}{c} COO^- \\ | \\ CH_2 \\ | \\ COO^- \end{array}$$

Malonate

$$\begin{array}{l} {}^3COO^- \\ \;\;| \\ {}^2CH_2 \\ \;\;| \\ {}^1C{-}S{-}CoA \\ \;\;\| \\ \;\;O \end{array}$$

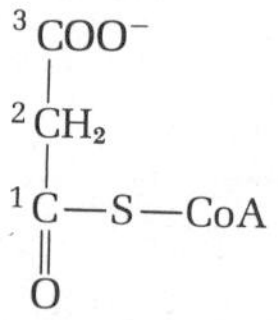

Malonyl-CoA

$$\text{Acetyl-CoA} + 7\ \text{malonyl-S-CoA} + 14\text{NADPH} + 20\text{H}^+ \longrightarrow \\ CH_3(CH_2)_{14}COO^- + 7CO_2 + 8\text{CoA-SH} + 14\text{NADP}^+ + 6H_2O$$

We also note that the reducing power required to make the singly bonded hydrocarbon backbone of fatty acids is furnished by NADPH. An unexpected feature of this pathway is that the 3-carbon malonyl group is the immediate precursor of all except one of the eight 2-carbon units from which the palmitic acid chain is built. The single molecule of acetyl-CoA required in fatty acid synthesis serves as the "starter" unit. The methyl and carboxyl carbon atoms of its acetyl group become carbon atoms 16 and 15, respectively, of the palmitic acid formed (Figure 21-2). Chain growth begins with the initiating acetyl residue and grows by successive additions of 2-carbon units, each derived from malonyl-CoA, proceeding toward the carboxyl end of palmitic acid (Figure 21-2). Each successive 2-carbon unit, after the initiating acetyl group, originates from the 2 carbon atoms of the malonyl group that are nearest the CoA; simultaneously, the third carbon atom of malonyl-CoA, that of the unesterified carboxyl group, is lost as CO_2. Ultimately, however, *all* the carbon atoms of fatty acids arise from acetyl-CoA, since malonyl-CoA is itself formed from acetyl-CoA and carbon dioxide, as we shall see.

A second distinctive feature of the mechanism of fatty acid biosynthesis is that the acyl intermediates in the process are thioesters, not of CoA, as occurs in fatty acid oxidation, but of a small-molecular-weight protein called *acyl carrier protein* (ACP), which has essential —SH groups.

The third feature is that fatty acid biosynthesis takes place in the cytosol of eukaryotic cells, whereas fatty acid oxidation takes place largely in the mitochondria. Fatty acids made in the cytosol are then used as building blocks to make either triacylglycerols or phospholipids, as we shall see later.

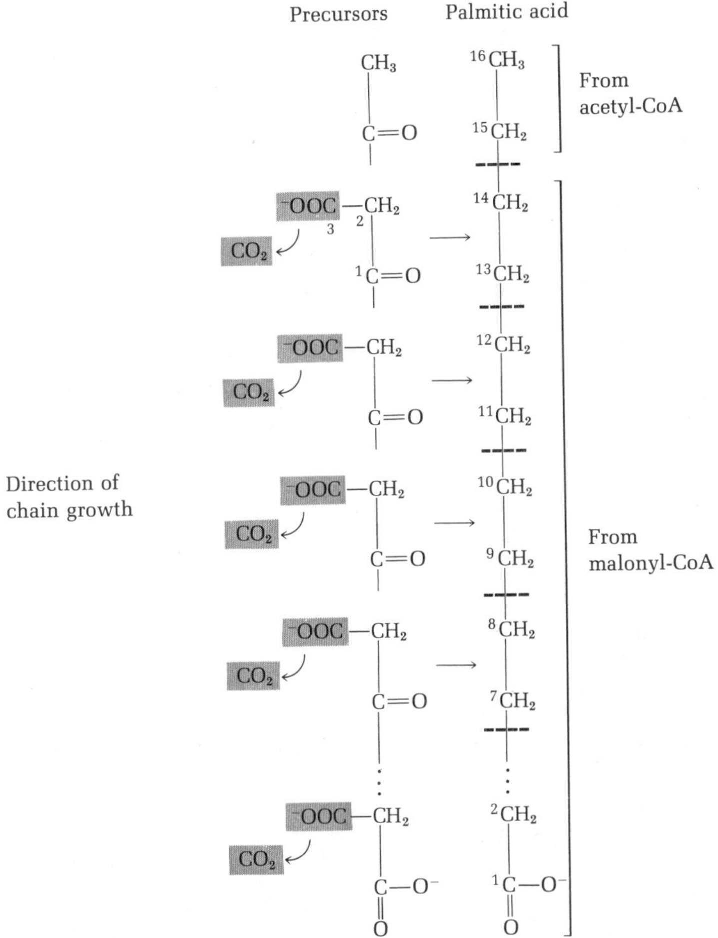

Figure 21-2
Source of carbon atoms during fatty acid biosynthesis. Carbons 1 and 2 of the precursor malonyl groups are added to the growing chain, whereas carbon atom 3 is released as CO_2.

Malonyl-CoA Is Formed from Acetyl-CoA

Although malonyl-CoA is the immediate precursor of most of the 2-carbon groups entering into the biosynthesis of fatty acids, it is first formed from acetyl-CoA in the cytosol. Cytosolic acetyl-CoA in turn arises from intramitochondrial acetyl-CoA. Let us trace the successive steps in the formation of malonyl-CoA.

Nearly all the acetyl-CoA used in metabolism is formed in the mitochondria from pyruvate oxidation (Chapter 16), from the oxidation of fatty acids (Chapter 18), and from the degradation of the carbon skeletons of amino acids (Chapter 19). Moreover, we have seen that acetyl-CoA cannot pass through the mitochondrial membrane (page 514). How, then, can cytosolic acetyl-CoA arise from mitochondrial acetyl-CoA? The answer is that an *acetyl-group shuttle* is required to transfer acetyl groups across the mitochondrial membrane (Figure 21-3). In this shuttle system intramitochondrial acetyl-CoA first reacts with oxaloacetate to form citrate, i.e., the first step in the citric acid cycle, by the action of *citrate synthase* (page 444).

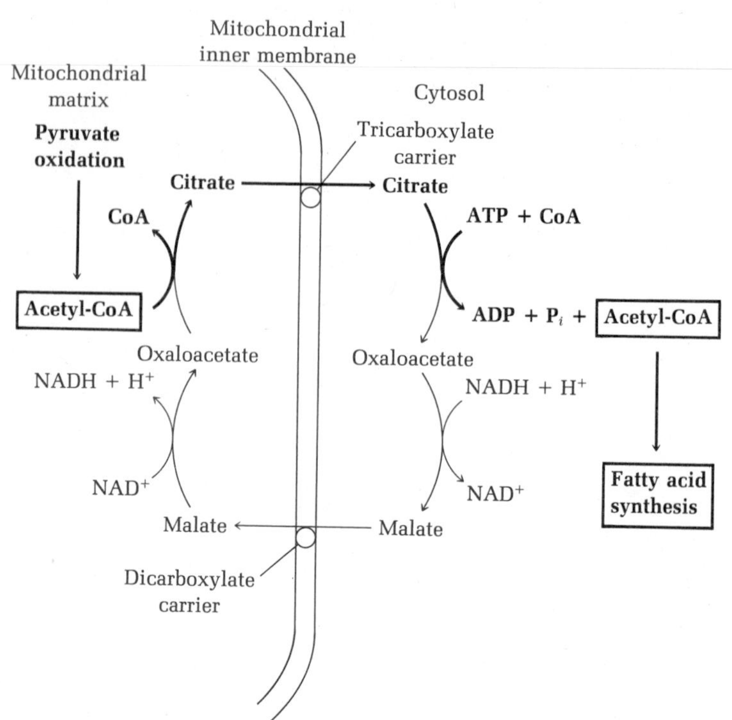

Figure 21-3
The acetyl-group shuttle for transfer of acetyl groups from mitochondrial acetyl-CoA to the cytosol for fatty acid synthesis. The colored portion shows the route by which acetyl groups pass out of the mitochondria as citrate into the cytosol, where they are delivered as acetyl-CoA for fatty acid synthesis. The arrows in black show how the shuttle cycle is completed by regenerating mitochondrial oxaloacetate.

$$\text{Acetyl-CoA} + \text{oxaloacetate} + H_2O \longrightarrow \text{citrate} + \text{CoA} + H^+$$

The citrate so formed then passes from the matrix into the cytosol through the mitochondrial inner membrane on the specific tricarboxylate-transport system (page 495). The citrate then undergoes a reaction with cytosolic CoA and ATP to yield cytosolic acetyl-CoA, catalyzed by citrate lyase, also called citrate-cleavage enzyme

$$\text{Citrate} + \text{ATP} + \text{CoA} \longrightarrow \text{acetyl-CoA} + \text{ADP} + P_i + \text{oxaloacetate}$$

The oxaloacetate formed cannot return to the mitochondria as such. It is reduced by cytosolic malate dehydrogenase to malate, which returns via the dicarboxylate-transport system to the mitochondrial matrix, where it is reoxidized to oxaloacetate to complete the shuttle.

After acetyl-CoA is formed in the cytosol it undergoes carboxylation to yield malonyl-CoA, the immediate precursor of 14 of the 16 carbon atoms of palmitic acid. This irreversible reaction (Figure 21-4) is catalyzed by acetyl-CoA carboxylase

$$\text{ATP} + \text{acetyl-CoA} + CO_2 + H_2O \longrightarrow \text{malonyl-CoA} + \text{ADP} + P_i + H^+$$

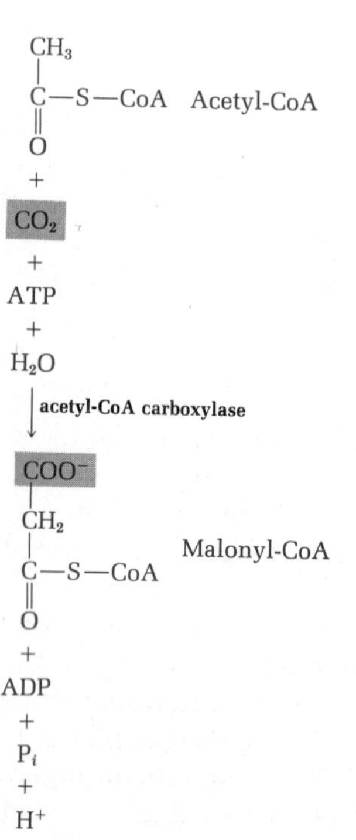

Figure 21-4
The acetyl-CoA carboxylase reaction. Biotin serves as part of the prosthetic group, which is a transient carrier of CO_2. The structures of biotin and biocytin are shown on page 260.

The CO_2 entering into the reaction becomes the free carboxylate group of malonyl-CoA. Acetyl-CoA carboxylase is a very complex enzyme. It contains biotin (page 259) as its prosthetic group, covalently bound in amide linkage to the ϵ-amino group of a lysine residue on one of the four subunits of the enzyme molecule. The biotinyl group serves as a "swinging arm" carrier of CO_2 (page 260) to acetyl-CoA:

$$CO_2 + ATP + H_2O + \text{biotin–E} \longrightarrow \text{carboxybiotin–E} + ADP + P_i + H^+$$

$$\text{Carboxybiotin–E} + \text{acetyl-CoA} \longrightarrow \text{malony-CoA} + \text{biotin–E}$$

$$\text{Sum:}\quad ATP + \text{acetyl-CoA} + CO_2 + H_2O \longrightarrow \text{malonyl-CoA} + ADP + P_i + H^+$$

ATP furnishes the driving force to attach the new carboxyl group to acetyl-CoA and to make the reaction irreverisible.

Acetyl-CoA carboxylase is a regulatory enzyme; the reaction it catalyzes is the rate-limiting step in the biosynthesis of fatty acids in animal tissues. Its major positive modulator is citrate, which converts the enzyme into a highly active filamentous polymer. Whenever the citrate level in the mitochondria rises, which happens when there is a high rate of formation of mitochondrial acetyl-CoA and of ATP, citrate moves out of the mitochondria and becomes both the precursor of cytosolic acetyl-CoA and the allosteric signal for the activation of acetyl-CoA carboxylase.

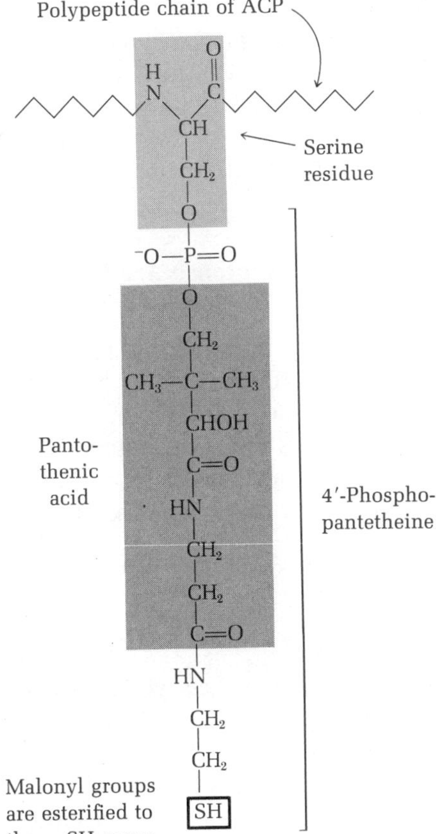

Figure 21-5
The prosthetic group of acyl carrier protein is 4′-phosphopantetheine, which is covalently attached to the hydroxyl group of a serine residue in the polypeptide chain. Phosphopantetheine, which contains the B vitamin pantothenic acid, is also a part of the CoA molecule (see page 256). Its —SH group is the site of entry of malonyl groups during fatty acid synthesis.

The Fatty Acid Synthase System Has Seven Active Sites

In animal tissues the enzymes involved in fatty acid biosynthesis, seven in number, are organized into a cluster or complex, the *fatty acid synthase system,* whose total molecular weight is about 400,000. This complex is localized in the cell cytosol. The seven enzymes are presumably organized in this way to facilitate the successive steps in the fatty acid synthesis cycle.

Central to the system is the *acyl carrier protein* (ACP), to which the acyl intermediates of fatty acid synthesis are covalently bound. ACP is a relatively small heat-stable protein of molecular weight 9000. Its prosthetic group is *4′-phosphopantetheine* (Figure 21-5), which also forms part of the structure of coenzyme A. Phosphopantetheine contains the vitamin *pantothenic acid* (page 256) as well as a sulfhydryl group. Phosphopantetheine is covalently linked via its phosphate group to the hydroxyl group of a serine residue in the ACP molecule.

The function of ACP in fatty acid biosynthesis is analogous to the role of coenzyme A in fatty acid oxidation. The acyl intermediates are esterified to ACP during the reactions in which the fatty acid chain is built up, whereas in fatty acid oxidation the fatty acyl intermediates are esterified to coenzyme A. The 4′-phosphopantetheine prosthetic group of ACP, together with the serine residue to which it is attached, has been proposed to serve as a "swinging arm" to carry the covalently bound acyl groups from one enzyme active site to the next in the proper sequence (Figure 21-6), as in the case of the pyruvate dehydrogenase complex of mitochondria (page 437). The fatty

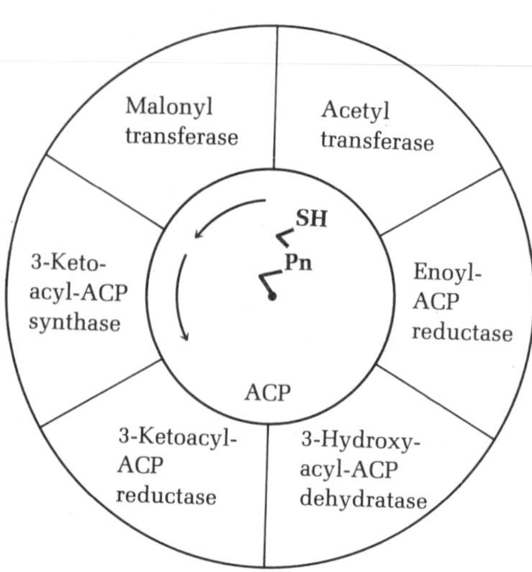

Figure 21-6
A schematic representation of the fatty acid synthase complex. In animal tissues the enzymes participating in fatty acid synthesis are clustered around the acyl carrier protein (ACP). The 4′-phosphopantetheine prosthetic group (Pn) is attached to a serine residue (see Figure 21-5). It forms a swinging arm 2.0 nm long, carrying the fatty acyl group from one enzyme active site to the next, counterclockwise in this representation. In bacteria and plant cells, the enzymes of the fatty acid synthase system occur as separate polypeptides rather than in a cluster.

acid synthase has two types of essential sulfhydryl groups (Figure 21-7). One is furnished by the single 4′-phosphopantetheine prosthetic group of ACP and the other is furnished by a specific cysteine residue of 3-ketoacyl-ACP synthase (page 589). Both —SH groups participate in fatty acid biosynthesis.

Figure 21-7
The fatty acid synthase has two essential —SH groups. One is contributed by the 4′-phosphopantetheine prosthetic group (Pn), the other by a specific cysteine (Cys) residue. Both participate in fatty acid synthesis. The phosphopantetheine —SH group of ACP is the site of entry of malonyl groups. E designates the entire fatty acid synthase complex.

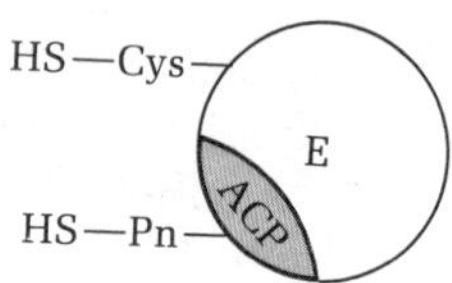

The Sulfhydryl Groups of Fatty Acid Synthase Are First Charged with Acyl Groups

Before the actual steps involved in building up the fatty acid chain can begin, the two sulfhydryl groups must be "loaded" with the correct acyl groups. This occurs in two enzyme-catalyzed steps (Figure 21-8). In the first reaction, catalyzed by ACP-*acetyltransferase*, the acetyl group of acetyl-S-CoA is transferred to the cysteine —SH group of the synthase (designated E)

Acetyl—S—CoA + (HS—Cys)(HS—ACP)E ⟶ CoA—SH + (acetyl—S—Cys)(HS—ACP)E

In the second reaction the malonyl group of malonyl-S-CoA is transferred to the phosphopantetheine sulfhydryl group of ACP, in a reaction catalyzed by *ACP-malonyltransferase*

Malonyl—S—CoA + (acetyl—S—Cys)(HS—ACP)E ⟶ CoA—SH + (acetyl—S—Cys)(malonyl—S—ACP)E

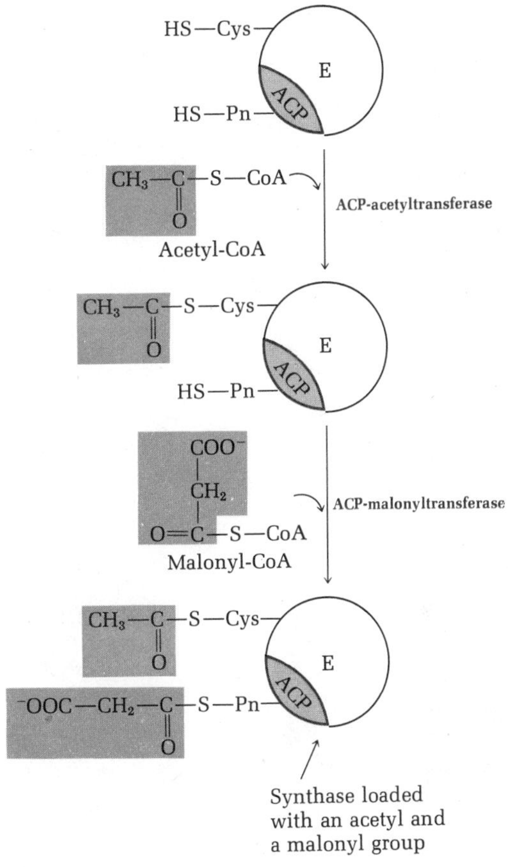

Figure 21-8
Loading the synthase with an acetyl and a malonyl group, the former on the Cys —SH, the latter on the —SH of 4′-phosphopantetheine (Pn). Incoming malonyl groups are always attached to the Pn —SH group.

The net result of these two reactions is that the synthase now has two covalently bound acyl groups, an acetyl group at the cysteine —SH group and a malonyl group at the phosphopantetheine —SH group. The two acyl groups are rather close together on the synthase, which is now ready for the actual chain-lengthening process. It is important to remember that the malonyl group binds only to the pantetheine —SH group.

Addition of Each 2-Carbon Unit Requires Four Steps

The Condensation Step

In the first of the four steps in the elongation of the fatty acid chain, the acetyl group and the malonyl group, covalently bonded to —SH groups of the synthase, undergo a condensation reaction to form an acetoacetyl group bound to the phosphopantetheine —SH group; simultaneously, a molecule of CO_2 is set free. This reaction is catalyzed by *3-ketoacyl-ACP synthase*

$$\begin{matrix}\text{Acetyl—S—Cys} \searrow & & & \text{HS—Cys} \searrow & \\ & \text{E} + \text{H}^+ \longrightarrow & & & \text{E} + \text{CO}_2 \\ \text{Malonyl—S—ACP} \nearrow & & & \text{acetoacetyl—S—ACP} \nearrow & \end{matrix}$$

Figure 21-9
The condensation reaction of fatty acid biosynthesis. The loss of CO_2 from the malonyl group (color) furnishes the driving force for the transfer of the acetyl group (gray) from Cys—SH to carbon 2 of the malonyl group. The starter acetyl group (gray) finally becomes the last 2-carbon unit at the methyl end of the completed fatty chain.

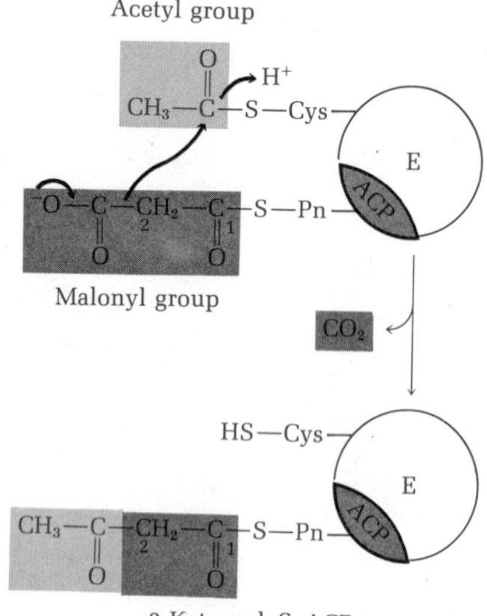

Note that the acetyl group is transferred from the cysteine —SH group to the malonyl group on the —SH of phosphopantetheine, so that it becomes the methyl-terminal 2-carbon unit of the new acetoacetyl group (Figure 21-9). In effect the acetyl group displaces the free carboxyl of the malonyl group as CO_2. The CO_2 formed in this reaction is the same CO_2 that was originally introduced into malonyl-CoA by the acetyl-CoA carboxylase reaction described above. Thus carbon dioxide is not permanently fixed in covalent linkage during fatty acid biosynthesis but plays a *catalytic* role in fatty acid synthesis, being regenerated as each 2-carbon unit is inserted.

Why do cells go to the trouble of adding CO_2 to make the malonyl group from an acetyl group, only to lose CO_2 again during the formation of acetoacetate? The answer is that the loss of CO_2 from the malonyl group momentarily makes a powerfully reactive group of the remaining 2-carbon portion, enabling it to react readily with the acetyl group (Figure 21-9).

The 3-Keto Reduction Step

The acetoacetyl-S-ACP next undergoes reduction at the carbonyl group, at the expense of NADPH as electron donor, to form *D-3-hydroxybutyryl-S-ACP* (Figure 21-10), in a reaction catalyzed by *3-ketoacyl-ACP reductase*

HS—Cys
E + NADPH + H^+ ⟶
Acetoacetyl—S—ACP

HS—Cys
E + $NADP^+$
D-3-hydroxybutyryl—S—ACP

It is noteworthy that the D-3-hydroxybutyryl group is not the same stereoisomeric form as the L-3-hydroxyacyl intermediate in fatty acid oxidation (page 517).

The Dehydration Step

In the third step of the fatty acid synthesis cycle, the D-3-hydroxybutyryl-S-ACP is dehydrated by *3-hydroxyacyl-ACP dehydratase* to yield *trans*-Δ^2-butenoyl-S-ACP (Figure 21-10)

HS—Cys
E ⟶
D-3-Hydroxybutyryl—S—ACP

HS—Cys
E + H_2O
trans-Δ^2-butenoyl—S—ACP

Figure 21-10
The three succeeding steps of the fatty acid synthesis cycle. After the last reaction the butyryl group is transferred to the Cys —SH group. Now the Pn —SH group is ready to receive the next incoming malonyl group from malonyl-CoA.

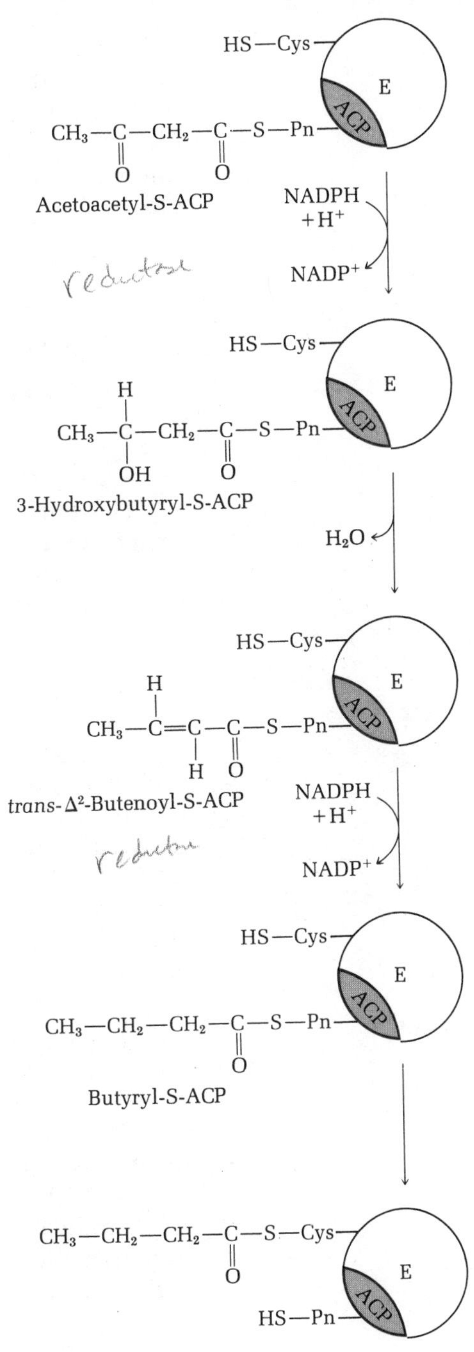

The Saturation Step

In the fourth step, which completes one pass through the fatty acid synthase complex, the double bond of *trans*-Δ^2-butenoyl-S-ACP is reduced or saturated to form *butyryl-S-ACP* by the action of *enoyl-ACP reductase* (Figure 21-10), NADPH again being the electron donor

$$\text{trans-}\Delta^2\text{-Butenoyl—S—ACP}\diagup\text{E}\diagdown\text{Cys—SH} + \text{NADPH} + \text{H}^+ \longrightarrow \text{butyryl—S—ACP}\diagup\text{E}\diagdown\text{Cys—SH} + \text{NADP}^+$$

The butyryl group is now transferred from the phosphopantetheine —SH group to the cysteine —SH group

$$\text{Butyryl—S—ACP}\diagup\text{E}\diagdown\text{Cys—SH} \longrightarrow \text{HS—ACP}\diagup\text{E}\diagdown\text{Cys—S—butyryl}$$

The newly lengthened fatty acyl group now occupies the —SH group originally occupied by the acetyl group.

To start the next round of reactions, in order to lengthen the chain by another 2-carbon unit, the next malonyl group is transferred from malonyl-CoA to the phosphopantetheine —SH group of ACP (Figure 21-11). The butyryl group then leaves the HS-Cys group and displaces CO_2 from the malonyl group on the HS-ACP group. We now have a 6-carbon acyl group, covalently bound to the phosphopantetheine —SH group. Its 3-keto group is reduced in the next three steps of the synthase cycle to yield the 6-carbon saturated acyl group, exactly as in the first round of reactions. Then the hexanoyl group is transferred from the phosphopantetheine —SH to the cysteine —SH group.

After a total of seven such cycles palmitoyl-S-ACP is the end product. The lengthening process stops at 16 carbons, and free palmitic acid is released from the ACP molecule by the action of a hydrolytic enzyme

$$\text{Palmitoyl—S—ACP}\diagup\text{E}\diagdown\text{Cys—SH} + \text{H}_2\text{O} \longrightarrow \text{palmitic acid} + \text{HS—ACP}\diagup\text{E}\diagdown\text{Cys—SH}$$

Note that the biosynthesis of palmitic acid requires input of chemical energy in two forms: as the phosphate-group energy of ATP and the reducing power of NADPH. The ATP is required to generate the thioester linkage of acetyl-CoA and to attach CO_2 to acetyl-CoA to make malonyl-CoA; the NADPH is required to reduce the double bonds.

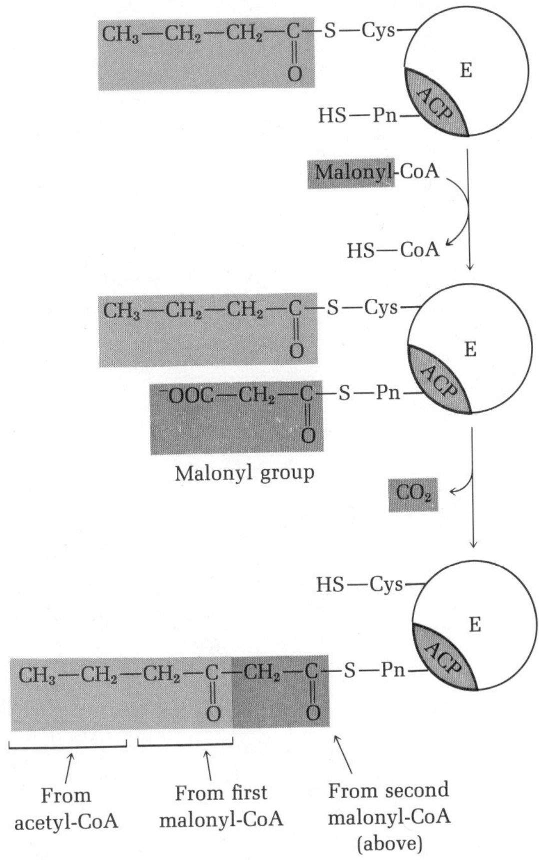

Figure 21-11
Beginning of the second round of the fatty acid synthesis cycle. The butyryl group is on the Cys —SH group. The incoming malonyl group is transferred to the Pn —SH group. In the condensation step the entire butyryl group on the Cys —SH is exchanged for the free carboxyl group of the malonyl residue, which is lost as CO_2 (color). The product, a 6-carbon 3-keto fatty acyl group, now contains 4 carbons derived from malonyl-CoA and 2 derived from the acetyl-CoA that started the reaction.

The NADPH required for the reductive steps in fatty acid biosynthesis arises from two major sources, depending upon the cell type. In the liver it is largely formed by the reactions of the pentose phosphate pathway (page 456), especially the action of glucose 6-phosphate dehydrogenase (page 457)

$$\text{Glucose 6-phosphate} + NADP^+ \longrightarrow \text{6-phosphogluconate} + NADPH + H^+$$

In fat or adipose cells NADPH is largely generated by malate enzyme

$$\text{Malate} + NADP^+ \longrightarrow \text{pyruvate} + CO_2 + NADPH + H^+$$

Both these reactions producing NADPH occur in the cytosol, in which the molar ratio $NADPH/NADP^+$ is very high, about 75, furnishing a strongly reducing environment for the reductive synthesis of fatty acids. In contrast, the molar ratio $NADH/NAD^+$ in the cytosol is much smaller, only about 8×10^{-4}. NADPH of the cytosol is clearly very well adapted for its role as the primary donor of hydrogen atoms in reductive biosynthetic reactions.

To summarize, the enzymatic biosynthesis of palmitic acid differs from the enzymatic oxidation of palmitic acid in (1) its

Table 21-1 Differences between Enzymatic Biosynthesis and Enzymatic Oxidation of Palmitic Acid

	Biosynthesis	Oxidation
Intracellular location	Cytosol	Mitochondria
Acyl-group carrier	ACP	CoA
Form in which 2-carbon units participate	Malonyl-CoA	Acetyl-CoA
Stereoisomeric form of 3-hydroxyacyl group	D	L
Electron donor or acceptor	NADPH	FAD, NAD^+
CO_2 as participant	Yes	No

intracellular location, (2) the nature of the acyl-group carrier, (3) the form in which the 2-carbon units are added or removed, (4) the stereoisomeric configuration of the β-hydroxyacyl intermediate, (5) the pyridine nucleotide employed in the redox steps, and (6) the participation of CO_2. These differences (Table 21-1) illustrate how these opposite metabolic pathways are chemically and physically distinct from each other.

Palmitic Acid Is the Precursor of Other Long-Chain Fatty Acids

Palmitic acid, the normal product of the fatty acid synthase system in animal cells, is the precursor of other long-chain fatty acids (Figure 21-12). It may be lengthened to form *stearic acid* (18 carbons) or even longer saturated fatty acids by further additions of acetyl groups, through the action of *fatty acid elongation systems,* which occur in the endoplasmic reticulum and the mitochondria. The elongation system of the endoplasmic reticulum, the more active, adds an additional 2-carbon unit, donated in the form of malonyl-CoA, to palmitoyl-S-ACP to form stearoyl-ACP by exactly the same pathway as in palmitate synthesis.

Palmitic and stearic acids in turn serve as precursors of the two most common monounsaturated fatty acids of animal tissues (Figure 21-12), namely, *palmitoleic acid* (16 carbons) and *oleic acid* (18 carbons), each of which has a single cis double bond in the Δ^9 position (page 304). The double bond is introduced into the fatty acid chain by an oxidative reaction catalyzed by *fatty acyl–CoA oxygenase*

$$\text{Palmitoyl-CoA} + \text{NADPH} + H^+ + O_2 \longrightarrow \text{palmitoleyl-CoA} + \text{NADP}^+ + 2H_2O$$

$$\text{Stearoyl-CoA} + \text{NADPH} + H^+ + O_2 \longrightarrow \text{oleyl-CoA} + \text{NADP}^+ + 2H_2O$$

These reactions are examples of *mixed-function oxidations,* since two different groups are oxidized, the single bond of the fatty acid and the cosubstrate NADPH.

Figure 21-12
Routes of synthesis of other fatty acids. Palmitic acid is the precursor of stearic acid and longer-chain saturated fatty acids, as well as the monounsaturated palmitoleic and oleic acids. Animals cannot convert oleic acid into linoleic acid, required in the diet as an essential fatty acid. Its conversion into other polyunsaturated fatty acids and prostaglandins is shown. Unsaturated fatty acids are symbolized by indicating the number of carbons and the number and position of the double bonds. Thus linoleic acid ($C_{18}\Delta^{9,12}$) has 18 carbons and two double bonds, one between carbons 9 and 10 and one between 12 and 13.

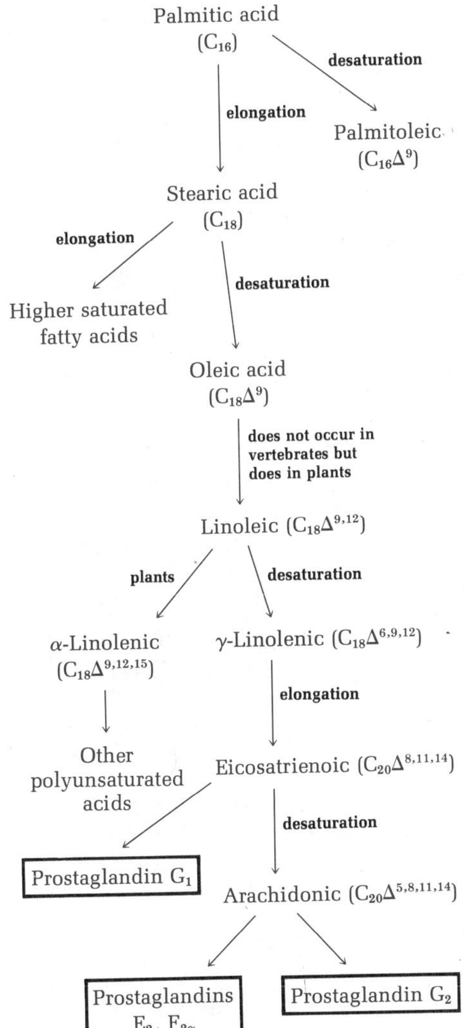

Animal tissues can readily introduce double bonds at the Δ^9 position of fatty acids but cannot introduce additional double bonds between the Δ^9 double bond and the methyl-terminal end of the fatty acid chain. *Linoleic acid*, with two double bonds at Δ^9 and Δ^{12}, and *α-linolenic acid* ($C_{18}\Delta^{9,12,15}$) cannot be synthesized by mammals. Because they are necessary precursors for synthesis of other products, these fatty acids are required in the diet and are obtained from plant sources. They are therefore called *essential fatty acids*. Lack of linoleic acid in the diet of rats causes a scaly dermatitis. Once ingested by mammals, linoleic acid may be converted into certain other polyunsaturated acids, particularly *γ-linolenic* and *arachidonic acids* (page 305), which can be made only from linoleic acid (Figure 21-12). Arachidonic acid is a 20-carbon acid with double bonds at Δ^5, Δ^8, Δ^{11}, and Δ^{14}. It is a crucially important fatty acid since it is an essential precursor of most *prostaglandins* and *thromboxanes*, hormonelike substances that regulate many different cell functions (Chapter 25).

Regulation of Fatty Acid Biosynthesis

The rate of fatty acid biosynthesis is determined primarily by the rate of the acetyl-CoA carboxylase reaction, which forms malonyl-CoA. Acetyl-CoA carboxylase is an allosteric enzyme. It is barely active in the absence of its stimulatory modulator, citrate. Whenever citrate concentration in the mitochondria rises, it escapes to the cytosol via the shuttle. In the cytosol citrate constitutes an allosteric signal that the citric acid cycle is amply supplied with fuel and that excess acetyl-CoA is to be stored as fat. Citrate becomes bound to the allosteric site of acetyl-CoA carboxylase, causing a large increase in the rate of conversion of acetyl-CoA into malonyl-CoA. The cytosolic citrate also is the source of the acetyl-CoA required in fatty acid synthesis. On the other hand, whenever there is overproduction of palmitoyl-CoA (the product of fatty acid synthesis and the immediate precursor of triacylglycerols), it serves as an allosteric signal that inhibits acetyl-CoA carboxylase. Since fatty acids are not stored as such but only as triacylglycerols, the concentration of glycerol phosphate, a required precursor in the synthesis of triacylglycerols, can control fatty acid synthesis. Other factors, discussed later, coordinate fatty acid biosynthesis with carbohydrate metabolism.

Let us now examine the assembly of fatty acids and glycerol to form triacylglycerols.

The Biosynthesis of Triacylglycerols and Glycerol Phosphatides Begins with Common Precursors

Triacylglycerols and the major phospholipids *phosphatidylethanolamine* and *phosphatidylcholine* share two common precursors and several enzymatic steps in their biosynthesis in animal tissues. The common precursors are *fatty acyl–CoAs* and *glycerol 3-phosphate*. Glycerol phosphate can be formed in two ways. It arises from dihydroxyacetone phosphate generated during glycolysis by the action of the cytosolic NAD-linked *glycerol phosphate dehydrogenase*

$$\text{Dihydroxyacetone phosphate} + \text{NADH} + \text{H}^+ \rightleftharpoons \text{L-glycerol 3-phosphate} + \text{NAD}^+$$

It can also be made from glycerol by the action of *glycerol kinase* (page 379)

$$\text{ATP} + \text{glycerol} \longrightarrow \text{glycerol 3-phosphate} + \text{ADP}$$

The other precursors of triacylglycerols are fatty acyl–CoAs, formed from fatty acids by *fatty acyl–CoA synthetases* (page 512)

$$\text{Fatty acid} + \text{ATP} + \text{CoA-SH} \longrightarrow \text{fatty acyl–S–CoA} + \text{AMP} + \text{PP}_i$$

The first stage in the biosynthesis of triacylglycerols is the acylation of the two free hydroxyl groups of glycerol phosphate by two molecules of fatty acyl–CoA to yield *diacylglycerol 3-phosphate* (Figure 21-13)

$$\text{Fatty acyl–S–CoA} + \text{glycerol} \longrightarrow \text{monoacylglycerol 3-phosphate} + \text{CoA-SH}$$

$$\text{Monoacylglycerol 3-phosphate} + \text{fatty acyl–S–CoA} \longrightarrow \text{diacylglycerol 3-phosphate} + \text{CoA-SH}$$

Diacylglycerol 3-phosphate, more commonly called *phosphatidic acid*, occurs in only trace amounts in cells, but it is an important intermediate in lipid biosynthesis. In the pathway to triacylglycerols phosphatidate is hydrolyzed by *phosphatidate phosphatase* to form a *1,2-diacylglycerol* (Figure 21-13)

$$\text{Phosphatidate} + \text{H}_2\text{O} \longrightarrow \text{1,2-diacylglycerol} + \text{P}_i$$

Diacylglycerols are then converted into triacylglycerols by reaction with a third molecule of a fatty acyl–CoA

$$\text{Fatty acyl–S–CoA} + \text{1,2-diacylglycerol} \longrightarrow \text{triacylglycerol} + \text{CoA-SH}$$

The formation of each ester linkage of triacylglycerols requires input of considerable free energy. In order to make the ester linkage the fatty acid must first be activated by forming its CoA ester, which requires two high-energy phosphate groups of ATP, since this reaction proceeds by a pyrophosphate cleavage of ATP and subsequent hydrolysis of the pyrophosphate formed.

Figure 21-13
Synthesis of diacylglycerols. The two incoming fatty acyl groups react in sequence. Usually two different long-chain fatty acyl groups R_1 and R_2 are introduced. Keep in mind that fatty acyl groups are very long in relation to the size of the glycerol phosphate molecule (page 306), as suggested by the diacylglycerol structure.

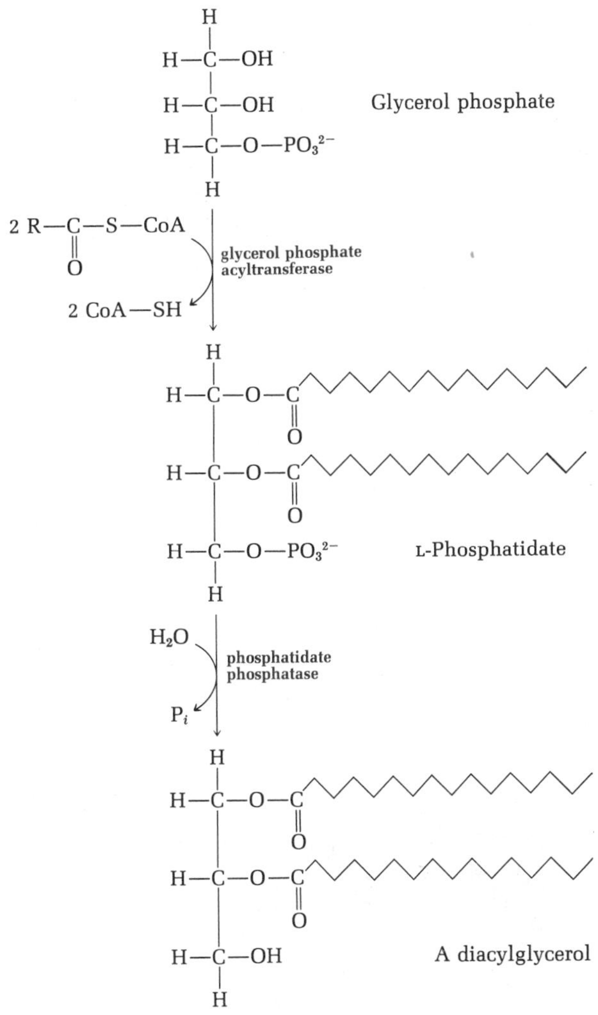

Glycerol phosphate

$2\,R{-}C(=O){-}S{-}CoA$

glycerol phosphate acyltransferase

$2\,CoA{-}SH$

L-Phosphatidate

H_2O

phosphatidate phosphatase

P_i

A diacylglycerol

Triacylglycerol Biosynthesis Is Regulated by Hormones

In normal human adults and in animals triacylglycerol biosynthesis and oxidation occur simultaneously in a steady state, so that the amount of body fat stays relatively constant over long periods, although there may be minor short-term changes as the caloric intake fluctuates. However, if carbohydrate, fat, or protein is consumed in excess above normal energy needs, the excess calories are stored in the form of triacylglycerols. Both carbohydrates (Chapter 16) and the carbon chains of amino acids (Chapter 18) can give rise to acetyl-CoA, required for *net* biosynthesis of fatty acids and triacylglycerols. The excess fat stored in this way can be drawn upon for energy and enables the body to withstand periods of fasting (Chapter 26).

The rate of triacylglycerol biosynthesis is profoundly altered by the action of several hormones. Insulin, for example, promotes the conversion of carbohydrate into triacylglycerols. In severe diabetes mellitus, due to failure of insulin secretion or action, patients not only are unable to use glucose properly but also fail to synthesize fatty acids and triacylglycerols from carbohydrates or amino acids. They show increased rates of fat oxidation and ketone-body formation. As a consequence they also lose weight. Triacylglycerol metabolism is also influenced by the secretion of pituitary growth hormone, by adrenal cortical hormones, and by glucagon (Chapter 25).

Triacylglycerols: Energy Sources in Some Hibernating Animals

Many animals depend on fat stores for energy during hibernation or dormancy, during migratory periods, and in other situations involving radical metabolic adjustments, as in the case of the camel, which can obtain its water supply from the oxidation of fat.

One of the most pronounced adjustments of fat metabolism known occurs in the hibernation of the grizzly bear. Bears go into a continuous state of dormancy for periods as long as 7 months without arousal. Unlike most other hibernating species, the bear maintains its body temperature between 32 and 35°C, nearly the normal level. Although the bear in this state expends about 6000 cal/d, it does not eat, drink, urinate, or defecate for months at a time. When accidentally aroused, the bear is almost immediately alert and ready to defend itself.

Experimental studies have shown that the bear uses body fat as its sole fuel during hibernation. The oxidation of fat yields sufficient energy for maintaining body temperature, for active synthesis of amino acids and proteins, as well as other energy-requiring activities, such as membrane transport. Fat oxidation also releases large amounts of water, as we have seen (page 519), which replenishes water loss in the breath. In addition, degradation of triacylglycerols yields glycerol, which is converted into blood glucose, following its enzymatic phosphorylation to glycerol phosphate and oxidation to dihydroxyacetone phosphate. Urea formed during the degradation of amino acids is reabsorbed and recycled by the bear, the amino groups being used to make new amino acids for maintaining body proteins.

Figure 21-14
A well-fattened dormouse just before hibernation. Hibernating animals accumulate large amounts of body fat, not only as a food reserve but also as insulation against cold. During hibernation the dormouse curls up into a ball, thus presenting the lowest possible body surface-to-volume ratio, to minimize heat loss.

Bears store an enormous amount of body fat in preparation for their long hibernation periods. Normally an adult grizzly bear consumes about 9000 kcal/d during the late spring and summer. But as winter approaches bears will feed 20 h/d and consume up to 20,000 kcal, in response to seasonal changes in hormone secretion. Large amounts of body triacylglycerols are formed from the huge amounts of carbohydrate consumed during the fattening-up period. Other hibernating species, including the tiny dormouse, also accumulate large amounts of body fat (Figure 21-14). The camel, although not a hibernator, can synthesize and store triacylglycerols in large amounts in its hump, a metabolic source of both energy and water under desert conditions.

Box 21-1 Another biological function of triacylglycerols

Recent research on the anatomy and feeding behavior of sperm whales has uncovered another way in which triacylglycerols are biologically useful. Sperm whales attain an average length of almost 60 ft. The sperm whale's head is very large, accounting for one-fourth of its total length and over one-third of its total weight (Figure 1). About 90 percent of the weight of the head is made up of the *spermaceti organ*, located above the long upper jaw. It is a blubbery mass consisting of some muscles surrounded by oily connective tissue. This organ contains up to 4 tons of spermaceti oil, a mixture of triacylglycerols containing an abundance of unsaturated fatty acids. It is liquid at the normal resting body temperature of the whale, about 37°C, but it begins to crystallize at about 31°C and becomes solid when the temperature drops several more degrees.

The biological function of spermaceti oil has long been the subject of speculation. Only recently, however, has its probable function been deduced from studies of the anatomy and feeding behavior of the sperm whale. Sperm whales feed almost exclusively on squid in very deep water. In their feeding dives they remain submerged for some 50 min; they require only 10 min on the surface to replenish their oxygen supply and blow off CO_2. Sperm whales make very deep descents of 1000 m or more in their feeding expeditions; the record dive recorded is 3000 m, almost 2 mi. At these depths the sperm whale has no competitors for the very plentiful squid.

In its deep dives after squid only about 25 percent of the time is spent making the round trip; the rest of the time the sperm whale rests quietly at great depth waiting for schools of squid to pass and be gobbled up. Now we come back to the spermaceti oil. For a marine animal to remain at a given depth it must have the same density as the water surrounding it. To make this possible some species have a swim bladder filled with air or nitrogen; others store body fat, which has a lower density than seawater. But the sperm whale is capable of changing its buoyancy, keeping it neutral not only on the tropical ocean surface but also at great depths, where the water is much colder and thus has a greater density.

Figure 1
Silhouette of a sperm whale, showing the large spermaceti organ, surrounded by its case. It is a huge enlargement of the snout and lies above the upper jaw.

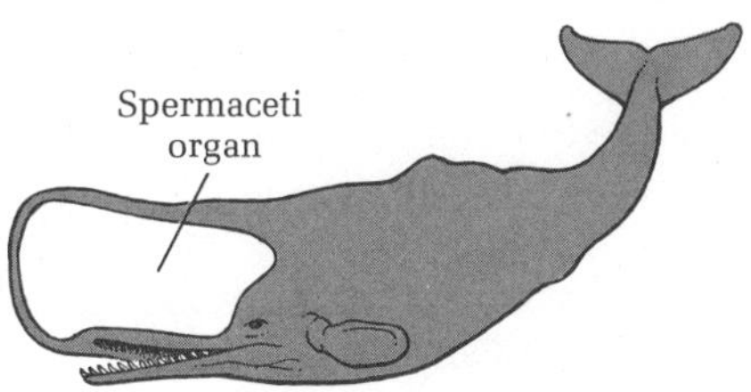

The key to the sperm whale's ability to change its buoyancy is the freezing point of spermaceti oil. When the temperature of liquid spermaceti oil is lowered several degrees during a deep dive, it congeals or crystallizes and becomes more dense, thus changing the buoyancy of the whale to match the density of seawater at great depths. To promote rapid cooling of the oil during a dive, the spermaceti organ is richly supplied with blood capillaries. The loss of body heat from the organ made possible by the rapid circulation of the blood is enhanced by a long nasal passage through the spermaceti organ, which the whale can close off and fill with colder seawater during a dive. During the return to the surface, the congealed spermaceti oil is warmed again and melted, decreasing its density to give it neutral buoyancy equal to that of the surface water.

Thus we see in the sperm whale a remarkable anatomical and biochemical adaptation, perfected by evolution. The triacylglycerols synthesized by the sperm whale contain fatty acids of the proper chain length and degree of unsaturation to give the spermaceti oil the proper melting point. Thus the animal can remain at great depths to feed with minimum expenditure of energy, without the constant swimming effort that would otherwise be required.

Stored triacylglycerols have other important biological uses. We have seen that Arctic seals and walruses have thick layers of fat under the skin as insulation (page 309). Stored triacylglycerols are used for a very different purpose in the head of the sperm whale (Box 21-1).

Biosynthesis of Phosphoglycerides Requires a Head Group

Let us now examine the synthesis of membrane lipids. The phosphoglycerides *phosphatidylethanolamine* and *phosphatidylcholine*, which serve as major components of membrane lipids, are also made from 1,2-diacylglycerols, as shown in the flow sheet in Figure 21-15. To make phosphoglycerides, their characteristic "head" groups (page 310) are made and attached by a type of intermediate and enzymatic reaction we have not yet discussed. In the synthesis of phosphatidylethanolamine the *phosphoethanolamine* head group is introduced by a reaction of diacylglycerol with *cytidine diphosphate ethanolamine*, which must first be made from three precursors, *ethanolamine*, *ATP*, and *cytidine triphosphate* (CTP; page 390). In the first step phosphoethanolamine is formed by *ethanolamine kinase*.

$$\text{ATP} + \text{ethanolamine} \xrightarrow{Mg^{2+}} \text{ADP} + \text{phosphoethanolamine}$$

The phosphoethanolamine then reacts with CTP to form *cytidine diphosphate ethanolamine* and pyrophosphate (Figure

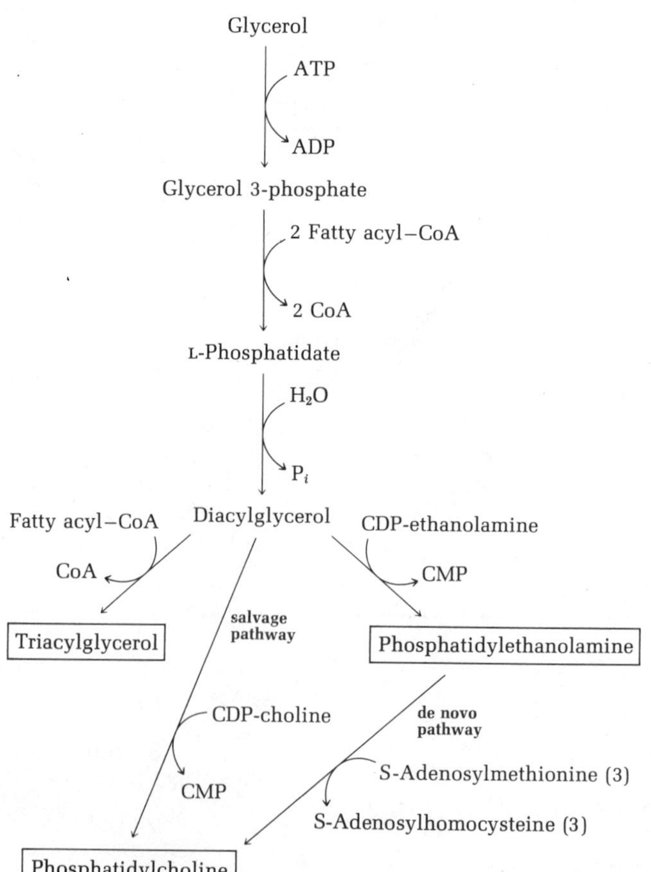

Figure 21-15
Schematic summary of glycerol lipid synthesis, showing that triacylglycerols and the two major glycerol-containing phospholipids share common precursors.

The enzymes involved in the biosynthesis of the phosphoglycerides from diacylglycerol are tightly bound to the endoplasmic reticulum.

Figure 21-16
(a) Formation of cytidine diphosphate ethanolamine. Cytidine diphosphate choline is made by a similar reaction from phosphocholine (b).

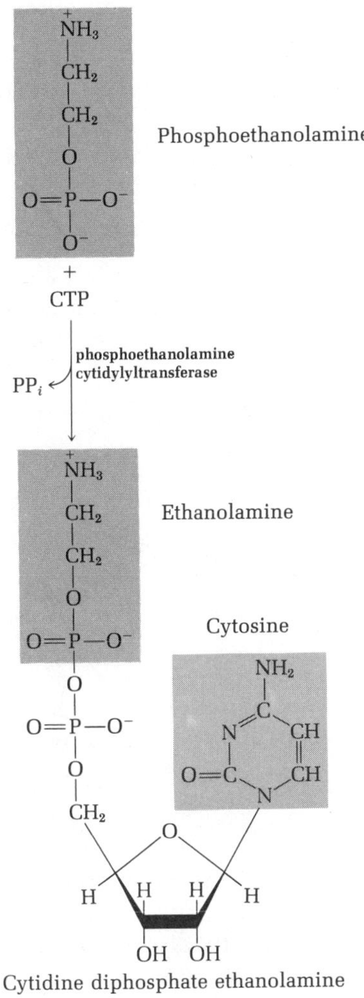

Phosphocholine
(b)

Figure 21-17
Formation of phosphatidylethanolamine from CDP-ethanolamine and diacylglycerol. The long hydrocarbon tails of the fatty acyl groups are symbolized by the zigzag bonds.

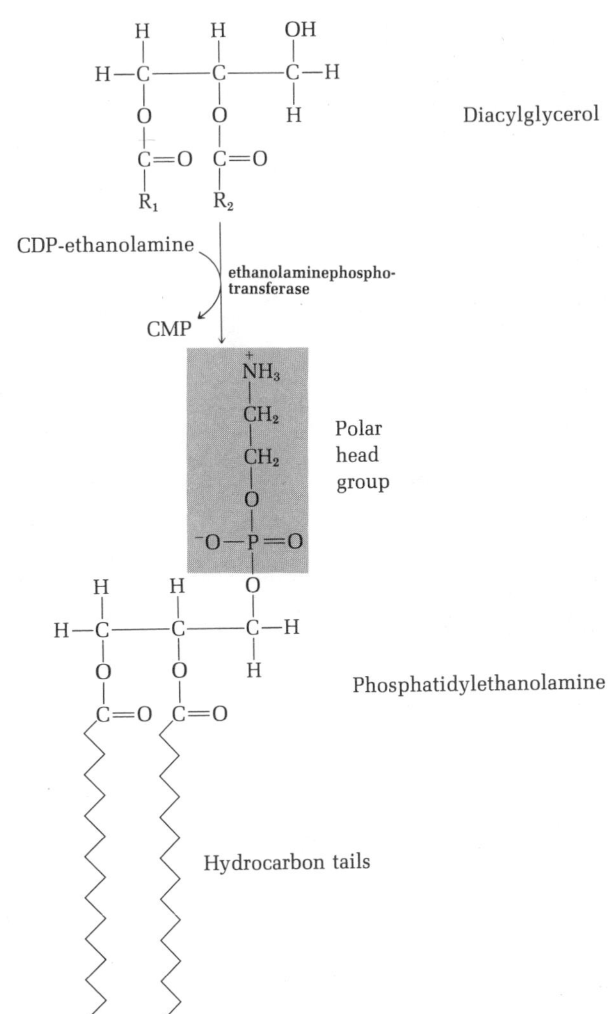

21-16), catalyzed by <u>*phosphoethanolamine cytidylyltransferase*</u>

$$\text{CTP} + \text{phosphoethanolamine} \longrightarrow \text{CDP-ethanolamine} + \text{PP}_i$$

To attach the head group to the diacylglycerol to make phosphatidylethanolamine (Figure 21-17), the enzyme <u>*ethanolaminephosphotransferase*</u> catalyzes the reaction

$$\text{CDP-ethanolamine} + \text{diacylglycerol} \longrightarrow \text{phosphatidylethanolamine} + \text{CMP}$$

Just as ATP is a carrier of activated phosphate groups and UDP-glucose is a carrier of glucosyl groups, CDP-ethanolamine carries activated phosphoethanolamine groups. Here we see another way in which nucleotides function as carriers of specific chemical groups in cell metabolism. Cytidine nucleotides are quite specific in this role; other nucleoside 5′-triphosphates cannot replace CTP in animal tissues. The central importance of cytidine nucleotides in lipid biosynthesis was discovered by Eugene P. Kennedy.

Phosphatidylcholine Is Made by Two Different Pathways

One pathway of phosphatidylcholine biosynthesis is called the *de novo* ("from the beginning") pathway since it does not require preformed choline as a precursor. In the de novo pathway (Figure 21-15) the choline portion of the phosphatidylcholine molecule is not introduced into the molecule as such but is formed from the ethanolamine part of phosphatidylethanolamine by three methylation steps. The methyl-group donor is *S-adenosylmethionine* (SAM; Figure 21-18), an activated form of methionine in which the methyl group is especially reactive. The reactions are

Figure 21-18
S-Adenosylmethionine (SAM) and its demethylated product, S-adenosylhomocysteine.

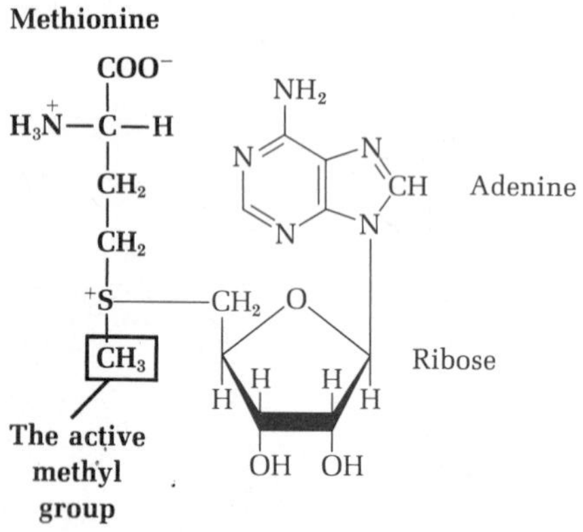

Methionine

Adenine

Ribose

The active methyl group

S-Adenosylmethionine

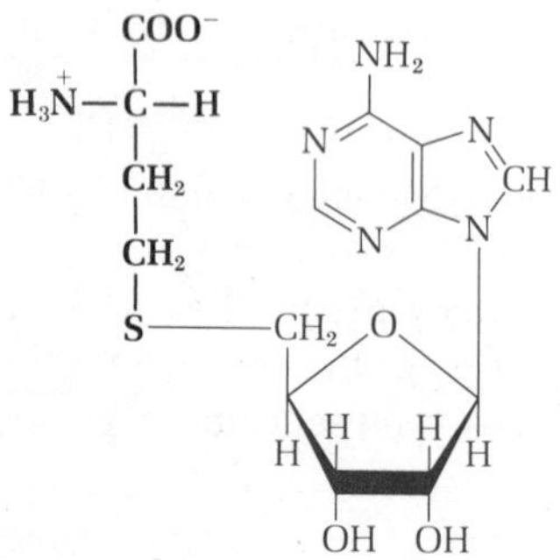

S-Adenosylhomocysteine

Phosphatidylethanolamine + S-adenosylmethionine ⟶ phosphatidylmonomethylethanolamine + S-adenosylhomocysteine

Phosphatidylmonomethylethanolamine + SAM ⟶ phosphatidyldimethylethanolamine + S-adenosylhomocysteine

Phosphatidyldimethylethanolamine + SAM ⟶ phosphatidylcholine + S-adenosylhomocysteine

The other pathway of phosphatidylcholine synthesis, the *salvage pathway*, is so named because the free choline released by the metabolic degradation of phosphatidylcholine is salvaged and used over again to rebuild phosphatidylcholine. The salvage pathway (Figure 21-15) closely resembles the pathway of phosphatidylethanolamine synthesis. The free choline is first activated by ATP through the action of *choline kinase* to form *phosphocholine*

$$\text{ATP} + \text{choline} \xrightarrow{Mg^{2+}} \text{ADP} + \text{phosphocholine}$$

The phosphocholine then reacts with CTP to form *cytidine diphosphate choline*

$$\text{CTP} + \text{phosphocholine} \longrightarrow \text{CDP-choline} + PP_i$$

The CDP-choline now reacts with 1,2-diacylglycerol to yield phosphatidylcholine

$$\text{CDP-choline} + \text{1,2-diacylglycerol} \longrightarrow \text{CMP} + \text{phosphatidylcholine}$$

Table 21-2 Major Membrane Lipids in Animal Cells (See Chapter 12)

Phosphatidylcholine
Phosphatidylethanolamine
Phosphatidylserine
Phosphatidylinositol
Cardiolipin
Sphingomyelin
Cerebrosides
Gangliosides

Many higher animals need this salvage pathway since their ability to make phosphatidylcholine by the de novo pathway is limited. This is because the methyl groups required in the form of S-adenosylmethionine derive from the nutritionally essential amino acid methionine. If the methionine intake is marginal, the capacity to methylate phosphatidylethanolamine and certain other methyl-group acceptors is limited. In this case the organism makes every effort to salvage free choline, which is already methylated. In fact, when the methionine intake of animals is deficient, it can in part be made up by preformed choline in the diet. Thus, under these special circumstances dietary choline serves as a supplementary vitamin (Chapter 26).

Other phospholipids found in membranes (Table 21-2) such as *phosphatidylserine*, *phosphatidylinositol*, and *cardiolipin* (page 310), are made from diacylglycerols and CDP derivatives in pathways resembling those described above.

Polar Lipids Are Inserted into Cell Membranes

The polar lipids, including the phosphoglycerides just described, sphingolipids, and glycolipids, are not deposited in fat cells for storage but are inserted into cell membranes in specific proportions or ratios. The phosphoglycerols, which are synthesized by enzymes on the endoplasmic reticulum, are for the most part inserted into the lipid bilayer of the reticulum, whose total area is quite large particularly in the liver and pancreas. The endoplasmic reticulum membrane is in turn the precursor of the Golgi body membranes. Membrane vesicles constantly pinch off from the Golgi body and find their way to the plasma membrane, bearing secretory products (Figure 21-19). Such vesicles often fuse with the plasma membrane. Phosphoglycerides may also be transported by carrier proteins from the endoplasmic reticulum to the mitochondria. Thus there is a directed flow of newly made polar lipids into the various types of membranes in the cell.

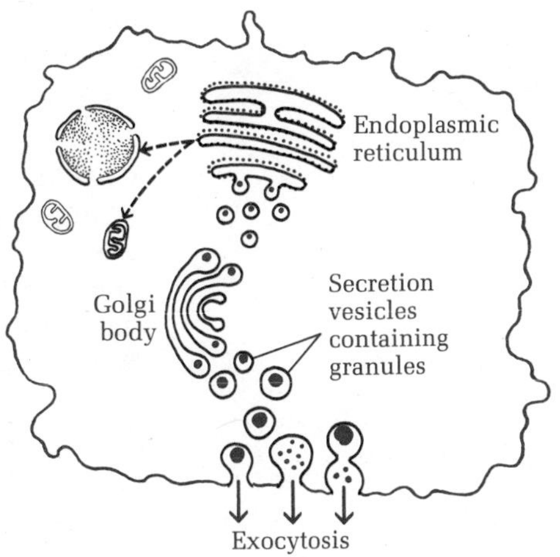

Figure 21-19
Once synthesized, different polar lipid molecules are inserted into the lipid bilayer of cell membranes in specific proportions. The major route of polar lipids is insertion into the endoplasmic reticulum bilayer. These lipids thus find their way, successively, into the membranes of Golgi bodies, secretion vesicles, and the plasma membrane. Lipids formed on the endoplasmic reticulum may also be carried by specific proteins through the cytosol to be inserted into mitochondrial membranes. The pathway of the membrane lipids is shown in color.

Figure 21-20
Sites of action of phospholipases on phosphatidylcholine. R_1 *and* R_2 *are long-chain fatty acyl groups.*

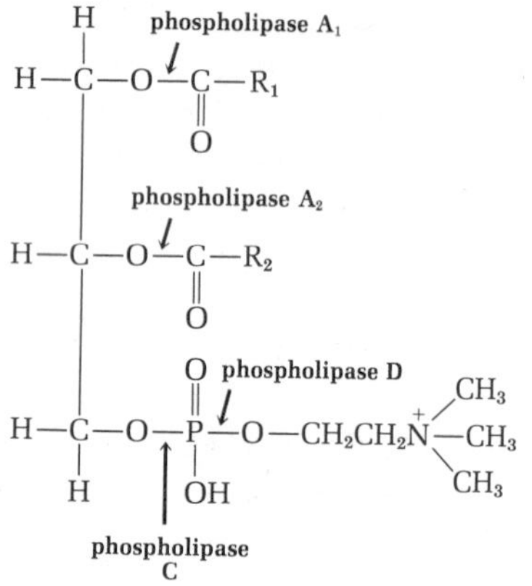

phospholipase A_1

phospholipase A_2

phospholipase D

phospholipase C

Lipid Metabolism Is Subject to Genetic Defects

All the polar lipids of membranes undergo constant metabolic turnover in a dynamic steady state, in which the rate of their synthesis is normally counterbalanced by an equal rate of breakdown. The breakdown of lipids is promoted by hydrolytic enzymes, each capable of hydrolyzing a specific covalent bond. For example, the degradation of phosphatidylcholine, a major membrane lipid, takes place by the action of several different phospholipases, whose actions are shown in Figure 21-20.

The metabolism of sphingolipids of membranes, which include sphingomyelin, cerebrosides, and gangliosides (page 312), is particularly prone to genetic defects of enzymes concerned with their degradation. As a consequence sphingolipids or their partial breakdown products accumulate in large amounts in the tissues, since their synthesis is normal in rate but their degradation is interrupted. For example, in the rare genetic abnormality called *Niemann-Pick disease*, sphingomyelin accumulates in the brain, spleen, and liver. The disease first becomes evident in infants and causes mental retardation and early death. Niemann-Pick disease is caused by a genetic defect in the degradation of sphingomyelin by the enzyme *sphingomyelinase*, which cleaves phosphocholine from sphingomyelin (structure on page 314).

Much more common is *Tay-Sachs disease*, in which a specific type of ganglioside accumulates in the brain and spleen owing to the lack of the lysosomal enzyme *N-acetylhexosaminidase*, a degradative enzyme that normally hydrolyzes a specific bond between an *N*-acetyl-D-galactosamine and a D-galactose residue in the polar head of the ganglioside (Figure 21-21). As a result, the degradation of the ganglioside is halted at an intermediate step, and large amounts of the partially degraded ganglioside accumulate, causing degeneration of the nervous system and leading to mental retardation, blindness, and early death (Figure 21-22).

Tay-Sachs disease is rare in the population at large (1 in 300,000 births) but has a very high incidence (1 in 3600 births) in Ashkenazic Jews (those of Central European extraction), who make up more than 90 percent of the Jewish population in America; 1 in 28 Ashkenazic Jews carries the defective gene in recessive form. When both parents are carriers, their children are likely to develop Tay-Sachs disease. Because the disease is irreversible and requires total institutional care, genetic counseling of parents has become important in averting its consequences. Tests have also been devised to determine the incidence of the recessive gene in prospective parents. Carriers of the gene can be detected by culturing fibroblast cells obtained from the skin of the parents and testing the cells for the presence of hexosaminidase A. Tests of the fetus can also be made during pregnancy by assay of the enzyme in cells obtained from the amniotic fluid in which the growing fetus is cushioned. The sampling of the amniotic fluid is known as *amniocentesis*.

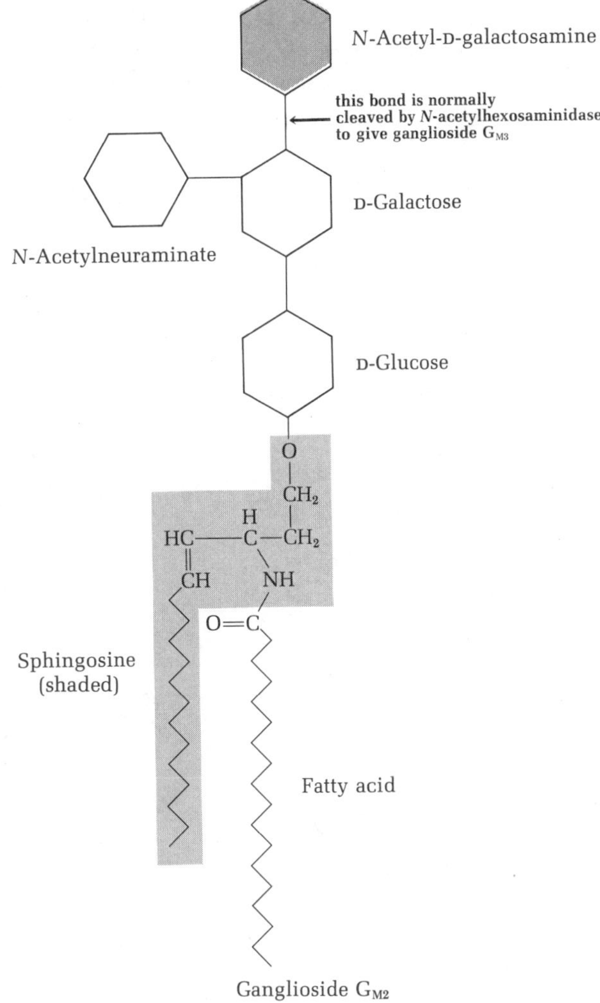

Figure 21-21
The genetic defect in Tay-Sachs disease. Ganglioside G_{M2} in the course of its normal degradation is acted upon by the lysosomal enzyme N-acetylhexosaminidase to yield N-acetyl-D-galactosamine and ganglioside G_{M3}. This enzyme is defective in Tay-Sachs disease, and as a result ganglioside G_{M2} accumulates in the lysosomes, particularly in brain cells.

Figure 21-22
(a) *A 1-year-old infant with Tay-Sachs disease. Deterioration of the brain has already begun and will be followed by blindness. Few children having the disease survive past their fifth year.* (b) *An electron micrograph of a portion of an affected brain cell, showing the abnormal ganglioside deposits in the lysosomes.*

(a)

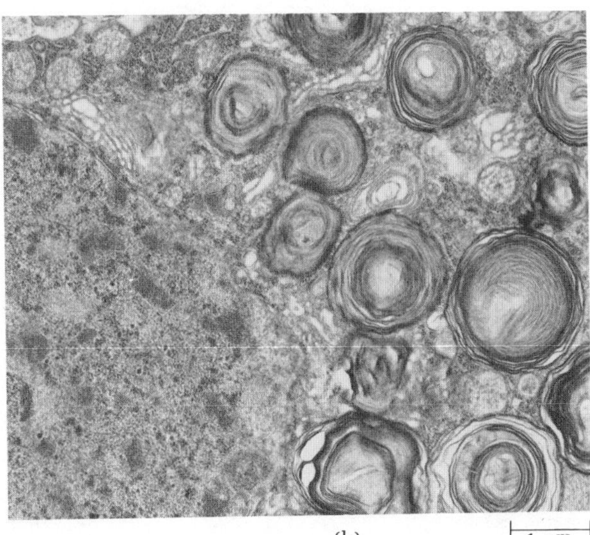

(b)

Table 21-3 Some Lysosomal Diseases

Most lysosomal diseases involve defects in enzymes that participate in the hydrolysis or degradation of complex lipids, glycogen, glycoproteins, and proteoglycans

Disease	Defective enzyme
Fabry's disease	Trihexosylceramide galactosylhydrolase
Gangliosidosis	β-Galactosidase
Hurler's syndrome	α-L-Iduronidase
Gaucher's disease	Glucocerebrosidase
Krabbe's disease	Galactosylceramide β-galactosylhydrolase
Mannosidosis	α-Mannosidase
Niemann-Pick disease	Sphingomyelinase
Tay-Sachs disease	N-acetyl-hexosaminidase
Glycogen storage disease (types I, II, and III)	See Table 20-3, page 576

There Are Many Lysosomal Diseases

Tay-Sachs disease, Niemann-Pick disease, and many other genetic diseases in which sphingolipids and proteoglycans undergo incomplete degradation are *lysosomal diseases* (Table 21-3). They are given this general term because many steps in the enzymatic degradation of sphingolipids and proteoglycans take place in the *lysosomes,* small, membrane-surrounded cytoplasmic vesicles (page 28) which function in intracellular digestion of some cell components. Lysosomes engulf cell macromolecules and insoluble cell components and degrade them by the action of hydrolytic enzymes into smaller water-soluble products, which then diffuse into the cytosol and are metabolized further. Lysosomes contain over 50 different hydrolytic enzymes active against lipids, mucopolysaccharides, glycogen, and proteins. The interior of lysosomes is somewhat acid, about pH 5.5. When a lysosomal hydrolytic enzyme is genetically faulty, the incompletely degraded macromolecule or insoluble lipid accumulates in the lysosomes, which become swollen to such an extent that they disrupt normal cell function. Lysosomal diseases in which the degradation of sphingolipids is faulty tend to cause mental retardation because the brain is exceedingly rich in sphingolipids compared with other tissues. When the degradation of sphingolipids is interrupted, the normal function of brain cells is disturbed.

Another example of a lysosomal disease is *Hurler's syndrome* or *gargoylism,* in which an enzyme participating in the normal degradation of the acid mucopolysaccharide portion of certain proteoglycans (page 296) is faulty and partial-degradation products accumulate. Infants with this disorder develop coarse facial features, thick ridges in the skin, which is rich in proteoglycans, mental retardation, and blindness. Early death is inevitable.

Much research is under way to correct genetic deficiencies in lysosomal enzymes by means of biochemical engineering. A major objective is to replace the defective enzyme with its normal, catalytically active form. For example, the genetically defective enzyme in living cells isolated from humans with Hurler's disease can be corrected *in vitro* by addition of the active enzyme obtained from normal cells. However, great difficulties are faced in introducing a normal enzyme into the body to replace a defective lysosomal enzyme. The normal enzyme must be of human origin or at least compatible with the human immune system; moreover, it must be introduced in such a way that it enters into the lysosomes of the specific cells in the body in which the genetic deficiency is manifest. Another approach, described in Chapter 30, is to introduce the normal active gene for the enzyme that is defective into the chromosomes of the cells, so that they can synthesize the normal active enzyme from the information provided by the gene and direct the entry of the new enzyme into the lysosomes.

Cholesterol and Other Steroids Are Also Made From 2-Carbon Precursors

Cholesterol is not only an important component of some cell membranes (page 319) and of plasma lipoproteins (page 316) but also the precursor of many other biologically important steroids, such as bile acids and various steroid hormones. Like the long-chain fatty acids, cholesterol is made from acetyl-CoA, but the acetyl groups are linked together in a different way. This was deduced from isotopic labeling experiments in which two kinds of isotopically labeled acetate were fed to animals. One type was labeled with ^{14}C in the methyl carbon; the other was labeled in the carboxyl carbon atom. The cholesterol was then isolated from the tissues of animals fed the two different types of labeled acetate. The labeled cholesterol obtained in each case was then degraded, step by step, by known chemical reactions to yield characteristic products. Determination of the radioactivity of these products revealed the location in the cholesterol molecule of the carbon atoms derived from the methyl carbon of acetate and those derived from its carboxyl carbon. The results of these pioneering experiments, carried out by Konrad Bloch, Robert Woodward, and others, are shown in Figure 21-23. This information became the blueprint for working

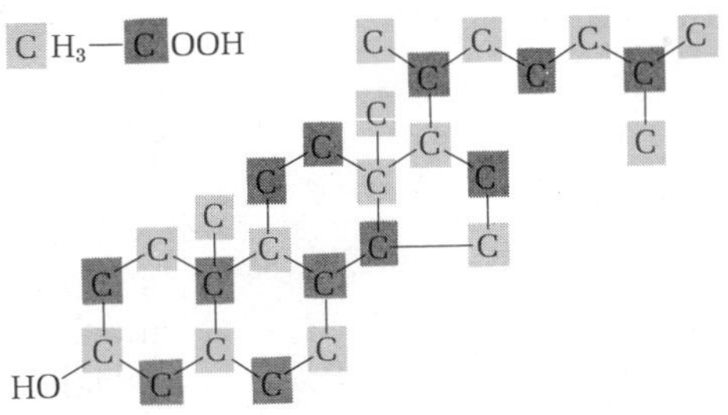

Figure 21-23
The origin of the carbon atoms of cholesterol deduced from tracer experiments with acetic acid labeled in the methyl carbon (shaded in gray) and with acetic acid labeled at the carboxyl carbon (shaded in color).

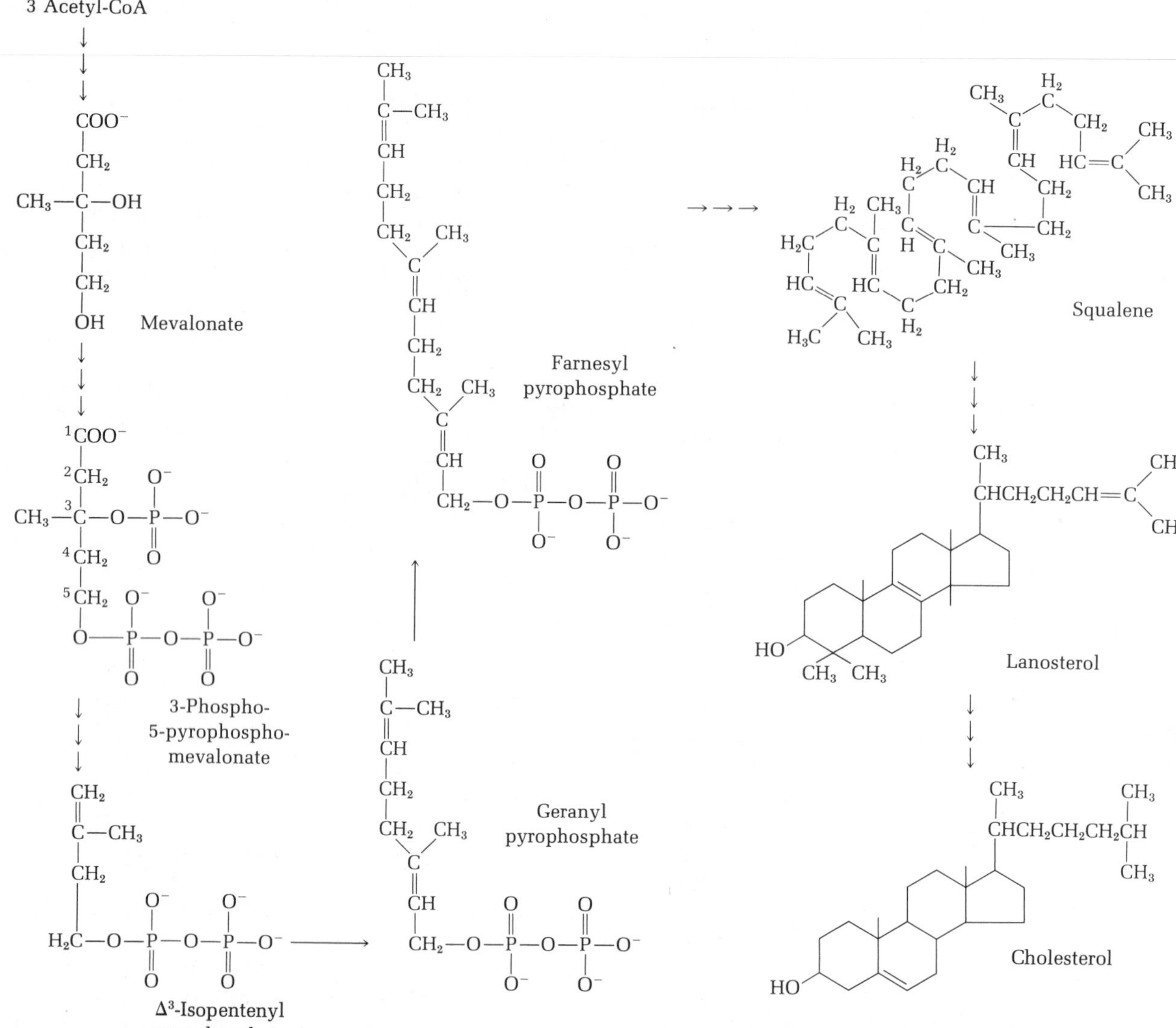

Figure 21-24
Steps in the biosynthesis of cholesterol. Three molecules of acetyl-CoA combine to yield mevalonate, which becomes phosphorylated to 3-phospho-5-pyrophosphomevalonate. On loss of CO_2 and phosphate, 3-isopentenyl pyrophosphate is formed. Stepwise assembly of six molecules of the latter ultimately yields the linear hydrocarbon squalene. Squalene then cyclizes to form lanosterol, which is then converted into cholesterol.

out the enzymatic steps in cholesterol biosynthesis, which takes place in several stages (Figure 21-24).

The first stage in cholesterol biosynthesis leads to the intermediate mevalonic acid (Figure 21-25), in the following sequence of reactions:

$$\text{Acetyl-CoA} + \text{acetyl-CoA} \xrightarrow{\text{thiolase}} \text{acetoacetyl-CoA} + \text{CoA}$$

$$\text{Acetoacetyl-CoA} + \text{acetyl-CoA} + H_2O \xrightarrow{\text{synthase}} \text{3-hydroxy-3-methylglutaryl-CoA} + \text{CoA} + H^+$$

$$\text{3-Hydroxy-3-methylglutaryl-CoA} + 2\text{NADPH} + 2H^+ \xrightarrow{\text{hydroxymethylglutaryl-CoA reductase}} \text{mevalonate} + \text{CoA} + 2\text{NADP}^+$$

In the next sequence of reactions three phosphate groups are attached to mevalonate, following which the phosphorylated mevalonate so formed loses a carboxyl group and a pair of hy-

Figure 21-25
Formation of mevalonate from acetyl-CoA. The origin of carbons 1 and 2 of mevalonate from acetyl-CoA is shown in color.

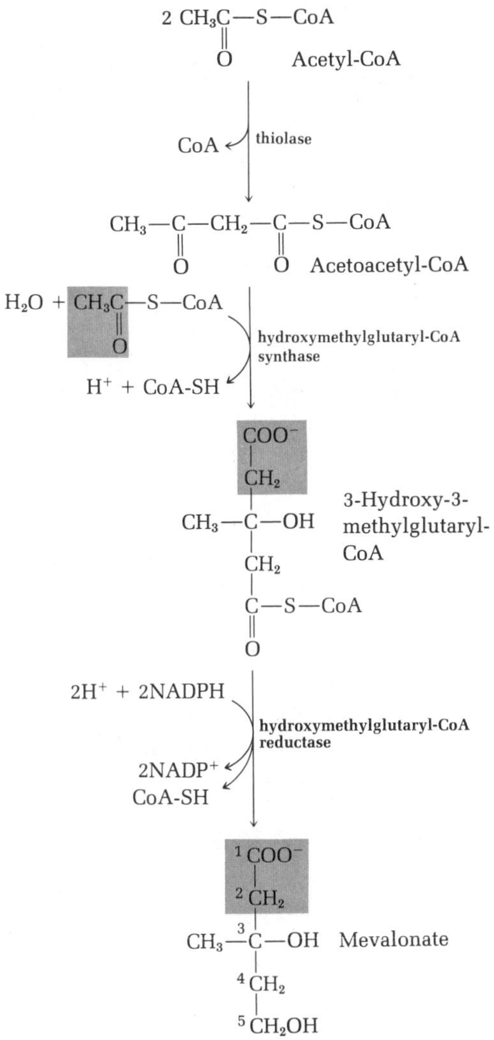

drogen atoms to yield <u>*Δ³-isopentenyl pyrophosphate*</u> (Figure 21-26), an activated form of an <u>*isoprene unit*</u> (page 264). Six isopentenyl groups are then assembled, with loss of their pyrophosphate groups, to yield the hydrocarbon <u>*squalene*</u> (Figure 21-27), which has 30 carbon atoms, 24 in the chain and 6 in the form of methyl-group branches. Squalene was first isolated from the liver of sharks (genus *Squalus*).

In the third series of reactions in cholesterol biosynthesis squalene undergoes a series of complex enzymatic reactions in which its linear structure is folded and cyclized to form <u>*lanosterol*</u>, which has the four condensed rings characteristic of steroids (Figure 21-24). Lanosterol is finally converted, after a fourth set of reactions, into cholesterol. Elucidation of this extraordinary biosynthetic pathway, one of the most complex known, was rewarded in 1961 by Nobel prizes to Konrad Bloch, an American, Feodor Lynen, a German, and John Cornforth, an Englishman.

The regulation of cholesterol biosynthesis is also complex. The rate-limiting step is early in the sequence, the reaction in which hydroxymethylglutaryl-CoA is converted into mevalonate (Figure 21-25). The enzyme catalyzing this reaction, <u>*hydroxymethylglutaryl-CoA reductase*</u>, is a complex regulatory enzyme whose activity is modulated over a hundredfold range. It is inhibited by cholesterol, the end product of the entire

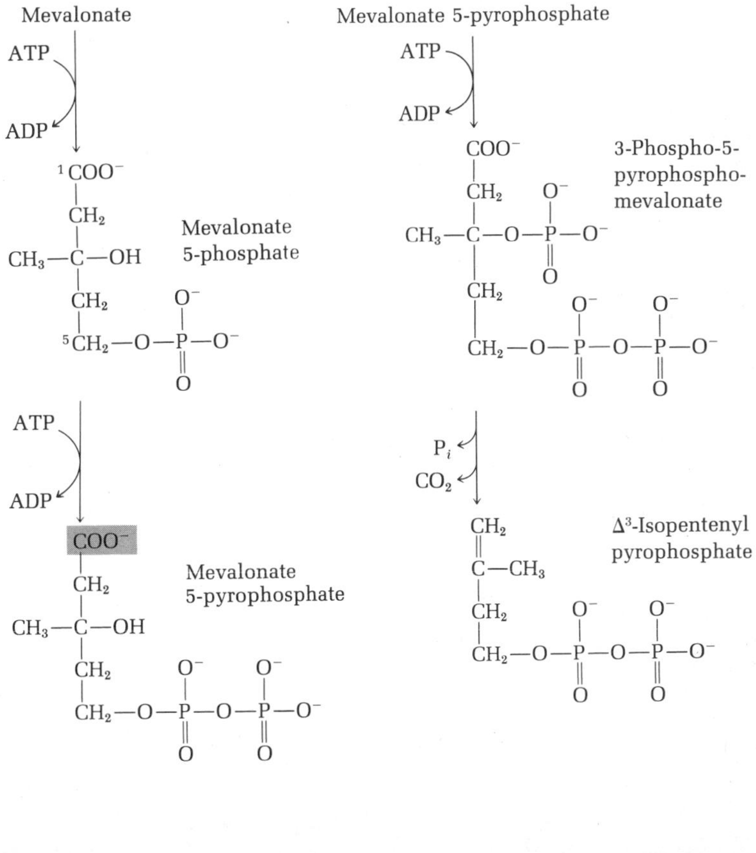

Figure 21-26
Conversion of mevalonate into Δ³-isopentenyl pyrophosphate, the activated isoprenoid unit. Altogether six of these will combine to form squalene.

Figure 21-27
Squalene, a 30-carbon isoprenoid hydrocarbon, the precursor of lanosterol and cholesterol. The isoprene units are set off by colored dashes.

CH_3
|
$C—CH_3$
‖
CH
|
CH_2
|
CH_2
|
$C—CH_3$
‖
CH
|
CH_2
|
CH_2
|
$C—CH_3$
‖
CH
|
CH_2
|
CH_2
|
CH
‖
$C—CH_3$
|
CH_2
|
CH_2
|
CH
‖
$C—CH_3$
|
CH_2
|
CH_2
|
CH
‖
$C—CH_3$
|
CH_3

pathway, as well as by mevalonate. The enzyme is present in endoplasmic reticulum and occurs in both phosphorylated (inactive) and dephosphorylated (active) forms. Cholesterol biosynthesis is also controlled by the concentration of a specific protein, *sterol carrier protein,* which binds the water-insoluble intermediates of the sequence and thus makes them more readily available for the subsequent enzymatic steps. The rate of cholesterol biosynthesis is altered not only by tissue levels of cholesterol and other steroids but also by fasting, by diurnal variations in food intake, and in cancer-bearing animals. It is also inhibited when certain cholesterol-containing plasma lipoproteins bind to specific receptors on cell surfaces.

Faulty regulation of cholesterol biosynthesis is one of the factors involved in the pathological process of *atherogenesis,* the formation of cholesterol- and lipid-rich deposits in arteries and arterioles. These deposits can limit blood flow and cause heart attacks or strokes by depriving the tissue of an adequate supply of oxygen (Chapter 26).

Isopentenyl Pyrophosphate Is the Precursor of Many Other Lipid-Soluble Biomolecules

Isopentenyl pyrophosphate, which is derived from acetyl-CoA, is the activated building-block precursor of many important biomolecules that contain isoprene units (Figure 21-28). They include vitamins A, E, and K; the carotenoids; rubber; gutta-percha; the phytol side chain of chlorophyll (Chapter 23); many essential oils, such as the fragrant principles of lemon oil, eucalyptus, and musk; as well as the hydrocarbons found in turpentine.

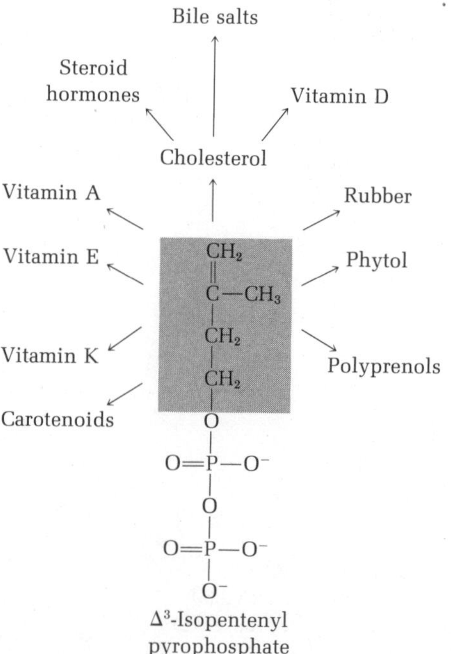

Figure 21-28
Isopentenyl pyrophosphate is the precursor of many isoprenoid compounds, in which isoprene units (color) are joined in different ways to form long chains or rings.

Summary

Long-chain saturated fatty acids are synthesized from acetyl-CoA by a cytosolic complex of enzymes including acyl carrier protein (ACP), which contains phosphopantetheine as its prosthetic group. ACP contains two types of —SH groups, one furnished by the phosphopantetheine (Pn) and the other by a cysteine (Cys) residue. ACP functions as the carrier of the fatty acyl intermediates. Acetyl-S-Cys-ACP, which is formed from acetyl-CoA and CO_2, reacts with malonyl-S-ACP, formed from malonyl-CoA, to yield acetoacetyl-S-ACP with release of CO_2. Reduction to the D-3-hydroxy derivative and its hydration to *trans*-Δ^2-unsaturated acyl-S-ACP is followed by reduction of the latter to butyryl-S-ACP at the expense of NADPH. Six more molecules of malonyl-S-ACP react successively at the carboxyl end of the growing fatty acid chain to form palmitoyl-S-Cys-ACP, the end product of the fatty acid synthase complex. Free palmitic acid is then released by hydrolysis. Palmitic acid may be elongated to yield the 18-carbon stearic acid. Palmitic and stearic acids in turn can be desaturated to yield palmitoleic and oleic acids, respectively, by the action of mixed-function oxygenases. Mammals cannot make linoleic acid and must obtain it from plant sources; they can convert exogenous linoleic acid into arachidonic acid, which is in turn the precursor of prostaglandins.

Triacylglycerols are formed by reaction of two molecules of fatty acyl–CoA with glycerol 3-phosphate to form phosphatidic acid, which is dephosphorylated to a diacylglycerol. The latter is then acylated by a third molecule of fatty acyl–CoA to yield a triacylglycerol. Diacylglycerols are also the major precursors of phosphoglycerides. The head group of phosphatidylethanolamine is first preformed by reaction of cytidine triphosphate (CTP) with phosphoethanolamine to form cytidine diphosphate ethanolamine. The phosphoethanolamine group of the latter is then transferred to diacylglycerol to yield phosphatidylethanolamine. Phosphatidylcholine is formed by methylation of phosphatidylethanolamine or by reaction of diacylglycerol with cytidine diphosphate choline. Cholesterol is also formed from acetyl-CoA in a very complex series of reactions via the important intermediates hydroxymethylglutaryl-CoA, mevalonate, and the linear hydrocarbon squalene, which is cyclized to yield the condensed steroid ring system and side chain. Cholesterol biosynthesis is inhibited by dietary cholesterol.

References

Fatty Acid Biosynthesis

CUNNINGHAM, E. B.: "Biochemistry: Mechanisms of Metabolism," McGraw-Hill, New York, 1978. Chapter 12 has many additional details on the enzymology and mechanism of fatty acid synthesis.

JEFFCOAT, R.: "The Biosynthesis of Unsaturated Fatty Acids and Its Control in Mammalian Liver," *Essays Biochem.*, **15**:1–36 (1979).

Lipid Biosynthesis

SNYDER, F. (ed.): *Lipid Metabolism in Mammals*, vols. 1 and 2, Plenum, New York, 1977.

Cholesterol Metabolism

BLOCH, K. S.: "The Biological Synthesis of Cholesterol," *Science*, **150:** 19–28 (1965).

BROWN, M. S., and J. L. GOLDSTEIN: "Receptor-Mediated Control of Cholesterol Metabolism," *Science*, **191:**150–154 (1976).

General

CLARKE, M. R.: "The Head of the Sperm Whale," *Sci. Am.*, **240:**128–141, January (1979). Further details of the function of spermaceti oil in the biology of the sperm whale.

NELSON, R. A.: "Protein and Fat Metabolism in Hibernating Bears," *Fed. Proc.*, **39:**2955–2958 (1980).

Genetic Defects in Lipid Metabolism

ALLISON, A. C.: "Lysosomes," in J. J. Head (ed.), *Carolina Biology Readers*, Carolina Biological Supply Company, Burlington, N.C., 1977.

BROWN, M. S., and J. L. GOLDSTEIN: "Familial Hypercholesterolemia: Defective Binding of Lipoproteins to Cultured Fibroblasts Associated with Impaired Regulation of 3-Hydroxy-3-methylglutaryl Coenzyme A Reductase Activity," *Proc. Natl. Acad. Sci. USA*, **71:** 788–792 (1974).

DIETSCHY, J. M., A. M. GOTTO, JR., and J. A. ONTKO (eds.): *Disturbances in Lipid and Lipoprotein Metabolism*, American Physiological Society, 1978.

STANBURY, J. B., J. B. WYNGAARDEN, and D. S. FREDRICKSON (eds.): *The Metabolic Basis of Inherited Disease*, 4th ed., McGraw-Hill, New York, 1978.

Problems

1. *Role of Carbon Dioxide in Fatty Acid Synthesis* Carbon dioxide is an indispensable participant in the biosynthesis of fatty acids. What is the specific role of CO_2? If a soluble liver fraction is incubated with $^{14}CO_2$ and other components required in fatty acid biosynthesis, does the resulting palmitate contain ^{14}C? Explain.

2. *Pathway of Carbon in Fatty Acid Synthesis* Using your knowledge of fatty acid biosynthesis, provide an explanation for the following experimental observations:
   (a) The addition of uniformly labeled ^{14}C-acetyl-CoA to a soluble liver fraction yields palmitate uniformly labeled with ^{14}C.
   (b) However, the addition of a *trace* of uniformly labeled ^{14}C-acetyl-CoA in the presence of an excess of malonyl-CoA to a soluble liver fraction yields palmitate labeled with ^{14}C only in carbons 15 and 16.

3. *Net Equation of Fatty Acid Synthesis* Write the net equation for the biosynthesis of palmitic acid in rat liver starting from mitochondrial acetyl-CoA and cytosolic NADPH, ATP, and CO_2.

4. *Pathway of Hydrogen in Fatty Acid Synthesis* Consider a preparation that contains all the enzymes and cofactors necessary for fatty acid biosynthesis from added acetyl-CoA and malonyl-CoA.
   (a) If acetyl-CoA labeled with deuterium (heavy isotope of hydrogen)

$$CD_3{-}C(=O){-}S{-}CoA$$

   and an excess of unlabeled malonyl-CoA are added as substrates, how many deuterium atoms are incorporated into every molecule of palmitate? What are their locations? Explain.
   (b) If unlabeled acetyl-CoA and deuterium-labeled malonyl-CoA

$$^{-}OOC{-}CD_2{-}C(=O){-}S{-}CoA$$

   are added as substrates, how many deuterium atoms are incorporated into every molecule of palmitate? What are their locations? Explain.

5. *Generation of NADPH for Fatty Acid Biosynthesis* Since the inner mitochondrial membrane is not permeable to acetyl-CoA, mitochondrial acetyl groups are shuttled into the cytosol by the route outlined in Figure 21-3. The cytosol contains an NADP-dependent malate enzyme which catalyzes the reaction

$$\text{Malate} + NADP^+ \longrightarrow \text{pyruvate} + CO_2 + NADPH + H^+$$

   Given that the inner mitochondrial membrane is permeable to pyruvate (in addition to citrate and malate), devise a shuttle scheme for production of cytosolic NADPH that utilizes the NADP-dependent malate enzyme and other enzymes known to be in the mitochondrial and cytosolic compartments. Write a net reaction for the transfer of acetyl groups from the mitochondrion to the cytosol.

6. *Modulation of Acetyl-CoA Carboxylase* Acetyl-CoA carboxylase is the principal regulation point in the biosynthesis of fatty acids. Some of the properties of the enzyme are described below:
   (a) The addition of citrate or isocitrate raises the V_{max} of the enzyme by as much as a factor of 10.
   (b) The enzyme exists in two interconvertible forms which differ

$$\underset{\text{Inactive}}{\text{Protomer}} \rightleftharpoons \underset{\text{Active}}{\text{filamentous polymer}}$$

   markedly in their activities. Citrate and isocitrate bind preferentially to the filamentous form while palmitoyl-CoA binds preferentially to the protomer.

   Explain how these properties are consistent with the regulatory role of acetyl-CoA carboxylase in the biosynthesis of fatty acids.

7. *Cost of Triacylglycerol Synthesis* Show by an appropriate net equation for the biosynthesis of tripalmitin from glycerol and palmitic acid how many ATPs are required per molecule of tripalmitin formed.

8. *The Energy Cost of the Synthesis of Phosphatidylcholine* Write the sequence of steps and the net reaction for the biosynthesis of phosphatidylcholine by the *salvage* pathway from oleic acid, palmitic acid, dihydroxyacetone phosphate, and choline. Starting from oleic acid, palmitic acid, and dihydroxyacetone phosphate, what is the cost in number of ATPs of the synthesis of phosphatidylcholine by the salvage pathway?

Problem 9

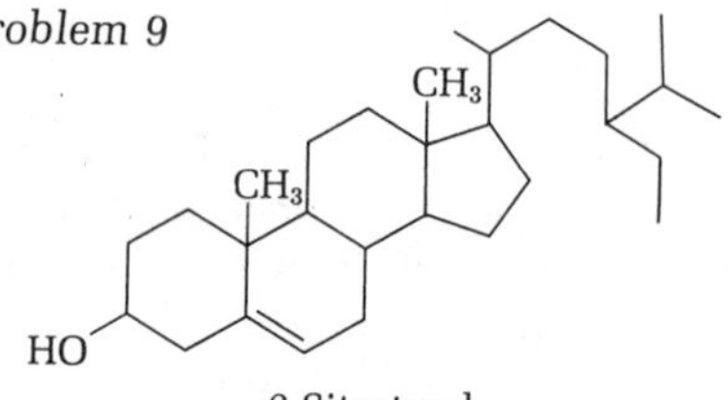

CH_3 CH_3 HO

β-Sitosterol

9. *Treatment of Hypercholesterolemia* Plants do not synthesize cholesterol but make other types of sterols known as phytosterols. Among these is β-sitosterol, whose structure is shown at the left. When people with hypercholesterolemia are fed β-sitosterol, the plasma cholesterol level is lowered, an effect that would decrease the risk of atherosclerosis. Suggest possible ways in which β-sitosterol might produce this effect.

10. *Coordination of Amino Acid and Fatty Acid Metabolism* A sample of 3-^{14}C-alanine

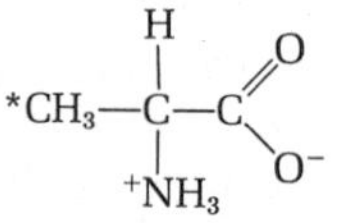

$$^{*}CH_3-\underset{^{+}NH_3}{\overset{H}{C}}-C\begin{matrix}\nearrow O \\ \searrow O^-\end{matrix}$$

is injected into a rat. After 1 h the animal is sacrificed, the liver removed, and the lipids extracted. The isolated palmitate contains ^{14}C. Explain. What is the location of ^{14}C in the isolated palmitate? Can alanine be used as a precursor for the *net* synthesis of new palmitate?

11. *Distinguishing Features of Anabolic and Catabolic Pathways of Fatty Acids* One turn of the anabolic and catabolic pathways of a short-chain fatty acid is represented by the equation

$$CH_3-CH_2-CH_2-COO^- \underset{\text{anabolic}}{\overset{\text{catabolic}}{\rightleftharpoons}} 2CH_3-COO^-$$

(a) Compare the net equations for the corresponding catabolic and anabolic routes. Are they simply the reverse of each other? How are they different?

(b) What specific factors in these pathways allow them to proceed independently of each other?

CHAPTER 22

Biosynthesis of Amino Acids and Nucleotides

For several reasons the biosynthetic pathways leading to the amino acids and nucleotides are considered together in this chapter. Both kinds of biomolecules contain nitrogen atoms, which arise from common biological sources. Moreover, amino acids are precursors in the biosynthesis of nucleotides. Amino acids and nucleotides are related in another way: they are the basic units in the biochemistry of inheritance. Nucleotides, the coding elements of nucleic acids, are essential in the preservation and transfer of genetic information. Amino acids, the building blocks of proteins, are needed for the expression of genetic information.

Not only are the biosynthetic pathways that lead to the 20 amino acids of proteins and the eight common nucleotides of nucleic acids numerous, but most are rather complex. We shall not try to examine them all in detail; instead we shall emphasize the major metabolic principles involved. Since each of the amino acids and nucleotides is required in relatively small amounts, the metabolic flow through most of these pathways is nowhere near so large as the biosynthetic flow leading to carbohydrate or fat in animal tissues. On the other hand, since the different amino acids and nucleotides must be made in the correct ratios and at the right time for protein and nucleic acid synthesis, their biosynthetic pathways must be accurately regulated and coordinated with each other.

As is true for other metabolic pathways we have examined, the routes taken in biosynthesis of the amino acids and nucleotides are not the same as those followed in their degradation. Moreover, the corresponding biosynthetic and catabolic pathways are independently regulated.

Some Amino Acids Must Be Obtained from the Diet

Living organisms differ considerably with respect to their ability to synthesize the 20 different amino acids. They also differ with respect to the forms of nitrogen they can use as precursors

Table 22-1 Nonessential and Essential Amino Acids for Humans and the Albino Rat

Nonessential	Essential
Glutamate	Isoleucine
Glutamine	Leucine
Proline	Lysine
Aspartate	Methionine
Asparagine	Phenylalanine
Alanine	Threonine
Glycine	Tryptophan
Serine	Valine
Tyrosine	Arginine†
Cysteine	Histidine

† Essential in young, growing animals but not in adults.

of amino groups. For example, the albino rat and human beings can synthesize only 10 of the 20 amino acids required as building blocks for protein biosynthesis (Table 22-1). These 10 are called the *nonessential* or *dispensable amino acids,* which are made from ammonia and various carbon sources. The other 10 must be obtained from the diet and are thus called the *nutritionally essential* or *indispensable amino acids.* Higher plants are more versatile; they can make all the amino acids required for protein synthesis. Moreover, they can use either ammonia or nitrate as precursors of their amino groups. Microorganisms differ widely in their capacity to synthesize amino acids. For example, *Escherichia coli* can make all the amino acids required for protein synthesis from simple precursors, but the lactic acid bacteria cannot and must obtain certain amino acids from their environment.

Glutamate, Glutamine, and Proline Share a Common Biosynthetic Pathway

First let us examine the biosynthesis of the nonessential amino acids, those which can be synthesized by human beings, the albino rat, and other mammals. In most cases the precursor of the backbone of the nonessential amino acids is the corresponding α-keto acid, ultimately derived from citric acid cycle intermediates. The amino groups are usually furnished by transamination reactions (page 531) from glutamate, catalyzed by the transaminases, which contain pyridoxal phosphate as their prosthetic groups (page 258).

The biosynthetic pathways to the related amino acids *glutamate, glutamine,* and *proline* (Figure 22-1) are simple and appear to be identical in all forms of life. *Glutamate* is formed from ammonia and α-ketoglutarate, an intermediate of the citric acid cycle, by the action of *L-glutamate dehydrogenase.* The required reducing power is furnished by NADPH

$$NH_4^+ + \alpha\text{-ketoglutarate} + NADPH \rightleftharpoons \text{L-glutamate} + NADP^+ + H_2O$$

This reaction is of fundamental importance in the biosynthesis of *all* amino acids, since glutamate is the amino group donor in the biosynthesis of other amino acids through transamination reactions. L-Glutamate dehydrogenase is located in the mitochondrial matrix.

Glutamine is formed from glutamate by the action of *glutamine synthetase*

$$\text{Glutamate} + NH_4^+ + ATP \longrightarrow \text{glutamine} + ADP + P_i + H^+$$

We recall that enzyme-bound *glutamyl 5-phosphate* (page 546) is an intermediate in this reaction, which takes place in two steps

Figure 22-1
Biosynthesis of the glutamate class of amino acids. Details of the conversion of glutamate into proline are shown in Figure 22-2.

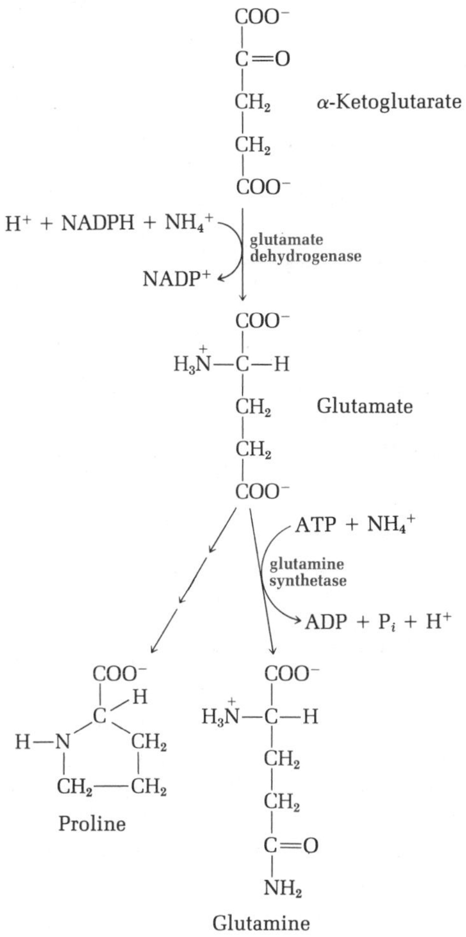

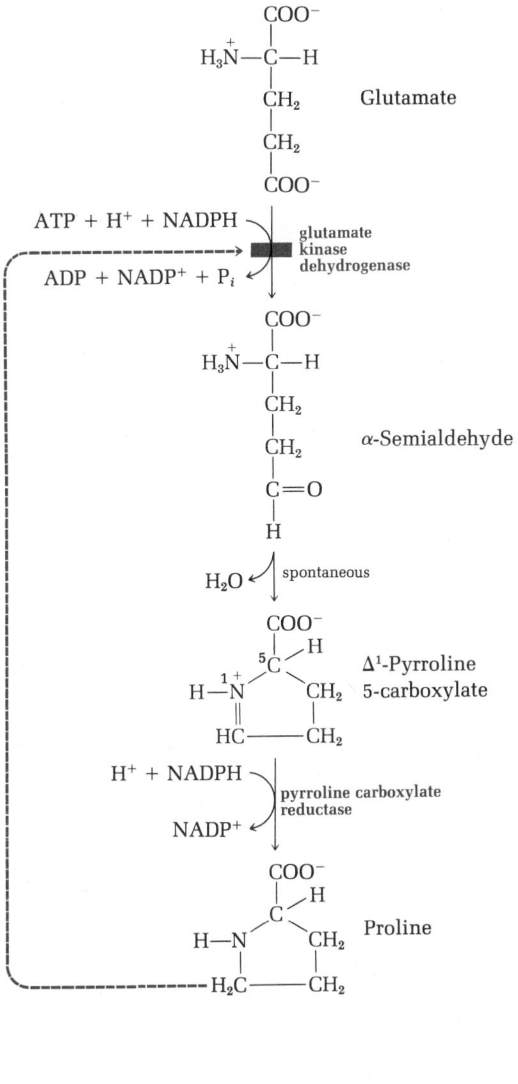

$$\text{Glutamate} + \text{ATP} \rightleftharpoons \text{glutamyl 5-phosphate} + \text{ADP}$$
$$\text{Glutamyl 5-phosphate} + NH_4^+ \rightleftharpoons \text{glutamine} + P_i + H^+$$
$$\text{Sum: Glutamate} + \text{ATP} + NH_4^+ \rightleftharpoons \text{glutamine} + \text{ADP} + P_i + H^+$$

This is also an important central reaction in amino acid metabolism, since it is the main pathway for converting free ammonia, which is toxic, into the nontoxic glutamine for transport in the blood (page 546). Glutamine synthetase is an allosteric enzyme. In *E. coli* and other prokaryotes the catalytic activity of glutamine synthetase is regulated by a number of different metabolites, as will be shown.

Proline, a cyclized derivative of glutamate, is formed by the pathway shown in Figure 22-2. Glutamate is first reduced to its γ-semialdehyde, which is then cyclized and reduced further to yield proline.

Figure 22-2
Biosynthesis of L*-proline. All five carbon atoms arise from glutamic acid. Proline is an allosteric inhibitor of the first reaction in its biosynthesis. This negative-feedback action is shown by the colored arrow and the point of inhibition by the colored bar.*

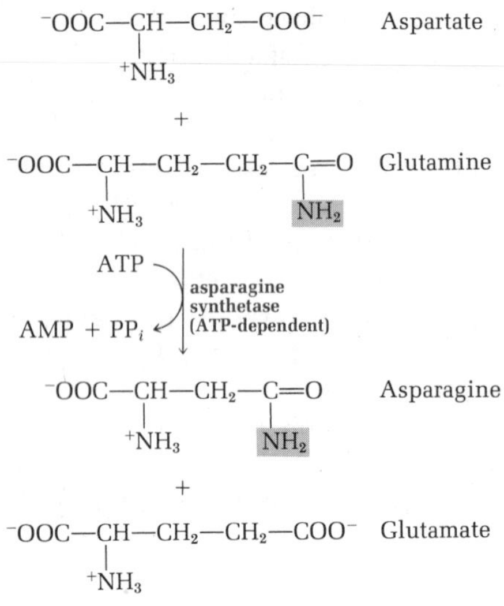

$^{-}OOC{-}CH(NH_3^+){-}CH_2{-}COO^-$ Aspartate

+

$^{-}OOC{-}CH(NH_3^+){-}CH_2{-}CH_2{-}C({=}O){-}NH_2$ Glutamine

ATP → AMP + PP_i (asparagine synthetase (ATP-dependent))

$^{-}OOC{-}CH(NH_3^+){-}CH_2{-}C({=}O){-}NH_2$ Asparagine

+

$^{-}OOC{-}CH(NH_3^+){-}CH_2{-}CH_2{-}COO^-$ Glutamate

Figure 22-3
Asparagine formation from aspartate in animal tissues. Asparagine is formed by a different reaction in many bacteria (see text).

Alanine, Aspartate, and Asparagine Also Arise from Central Metabolites

In most organisms the nonessential amino acids *alanine* and *aspartate* arise from pyruvate and oxaloacetate, respectively, by transaminations from glutamate

$$\text{Glutamate} + \text{pyruvate} \rightleftharpoons \alpha\text{-ketoglutarate} + \text{alanine}$$

$$\text{Glutamate} + \text{oxaloacetate} \rightleftharpoons \alpha\text{-ketoglutarate} + \text{aspartate}$$

In many bacteria aspartate is the direct precursor of *asparagine* in a reaction catalyzed by *asparagine synthetase,* which is analogous to the glutamine synthetase reaction

$$\text{Aspartate} + NH_4^+ + \text{ATP} \rightleftharpoons \text{asparagine} + \text{ADP} + P_i + H^+$$

In mammals, however, there is a different pathway for asparagine synthesis: the amino group is transferred from the amide group of glutamine to the β-carboxyl group of aspartate (Figure 22-3) by *asparagine synthetase (ATP-dependent)*

$$\text{Glutamine} + \text{aspartate} + \text{ATP} + H_2O \longrightarrow \text{glutamate} + \text{asparagine} + \text{AMP} + PP_i$$

Tyrosine Is Made from an Essential Amino Acid, Phenylalanine

Tyrosine is a nonessential amino acid, but it is made by animals from the essential amino acid *phenylalanine* via hydroxylation at the 4 position of the phenyl group by *phenylalanine oxygenase,* which (page 541) also participates in the degradation of phenylalanine. In this reaction NADPH is required as coreductant of a molecule of oxygen; recall that phenylalanine oxygenase is a *monooxygenase* or *mixed function oxidase* (page 503). The reaction catalyzed is

$$\text{Phenylalanine} + \text{NADPH} + H^+ + O_2 \longrightarrow \text{tyrosine} + \text{NADP}^+ + H_2O$$

COO⁻, $H_3\overset{+}{N}$—C—H, CH_2, CH_2, ^+S—CH_2, CH_3; Methionine; NH_2, N, N, CH, N, N; Adenine; O, H, H, H, H, OH OH; Ribose

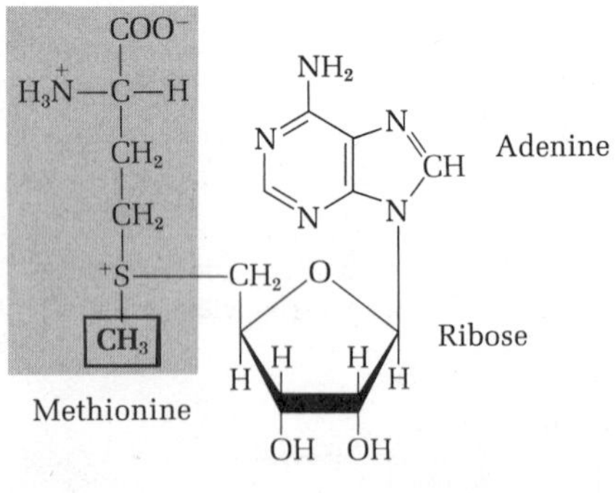

Figure 22-4
The structure of S-adenosylmethionine. In this derivative of methionine the stable sulfur atom in the thioether linkage has been converted into its very reactive sulfonium derivative, which now readily donates its methyl group to different acceptors.

Cysteine Is Made from Two Other Amino Acids, Methionine and Serine

In mammals *cysteine* is made from two other amino acids, *methionine,* which is essential in the diet, and *serine,* which is not. Methionine furnishes the sulfur atom and serine furnishes the carbon skeleton in the synthesis of cysteine. In the first reaction in this pathway methionine is converted into *S-adenosylmethionine* (Figure 22-4) by reaction with ATP

$$\text{L-Methionine} + \text{ATP} + H_2O \longrightarrow \text{S-adenosylmethionine} + PP_i + P_i$$

The adenosyl group may be regarded as a carrier of the methionine molecule. In this form the methyl group of methionine is very reactive and can be enzymatically transferred to any of a number of different methyl-group acceptors, leaving *S-adenosylhomocysteine* as the demethylated product

S-Adenosylmethionine + methyl-group acceptor ⟶
S-adenosylhomocysteine + methylated acceptor

For example, we have seen how S-adenosylmethionine serves as the methyl-group donor in the conversion of phosphatidylethanolamine into phosphatidylcholine (page 602).

After removal of the methyl group, the S-adenosylhomocysteine is ready for the subsequent reactions in the pathway to cysteine, shown in Figure 22-5. In the next step free *homocysteine* is formed by the reaction

S-Adenosylhomocysteine + H_2O ⟶ adenosine + homocysteine

Figure 22-5
Biosynthesis of cysteine from methionine, which furnishes the sulfur atom, and from serine, which furnishes the carbon chain.

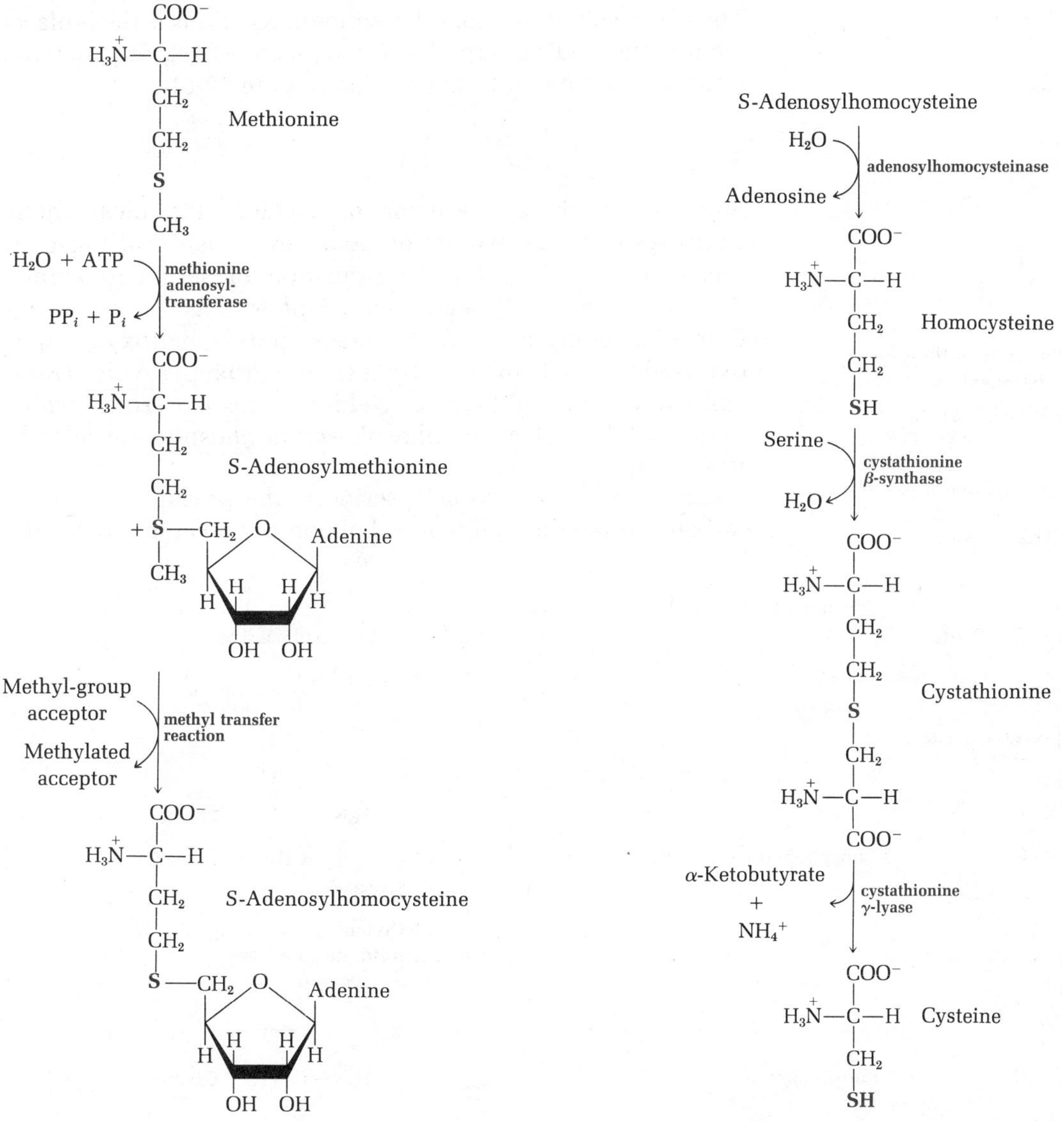

Homocysteine next reacts with serine in a reaction catalyzed by *cystathionine β-synthase* to yield *cystathionine* (Figure 22-5)

$$\text{Homocysteine} + \text{serine} \longrightarrow \text{cystathionine} + H_2O$$

In the last step *cystathionine γ-lyase,* also a pyridoxal phosphate enzyme, catalyzes the removal of ammonia and cleavage of cystathionine to yield free cysteine (Figure 22-5)

$$\text{Cystathionine} + H^+ \longrightarrow \alpha\text{-ketobutyrate} + NH_4^+ + \text{cysteine}$$

When we add all the separate steps together, the overall equation of cysteine synthesis becomes

$$\text{L-Methionine} + \text{ATP} + \text{methyl acceptor} + H_2O + H^+ + \text{serine} \longrightarrow \text{methylated acceptor} + \text{adenosine} + \alpha\text{-ketobutyrate} + NH_4^+ + \text{cysteine} + PP_i + P_i$$

The end result of this complex series of reactions is the replacement of the —OH group of serine with an —SH group, derived from methionine, to form cysteine (Figure 22-5).

Serine Is a Precursor of Glycine

Since serine is a precursor of glycine, the biosynthetic pathways of these two amino acids are considered together. The major pathway for the formation of serine in animal tissues (Figure 22-6) begins with *3-phosphoglycerate,* an intermediate of glycolysis. In the first step its α-hydroxy group is oxidized by NAD^+ to yield *3-phosphohydroxypyruvate.* Transamination from glutamate yields *3-phosphoserine,* which undergoes hydrolysis by *phosphoserine phosphatase* to yield free serine.

The 3-carbon amino acid serine is the precursor of the 2-carbon glycine through removal of one carbon atom, that at the

Figure 22-6
Biosynthesis of serine from 3-phosphoglycerate and the subsequent conversion of serine into glycine. Glycine is also made from CO_2 + NH_3 by the action of glycine synthase, which uses N^5,N^{10}-methylenetetrahydrofolate as methyl donor (see text).

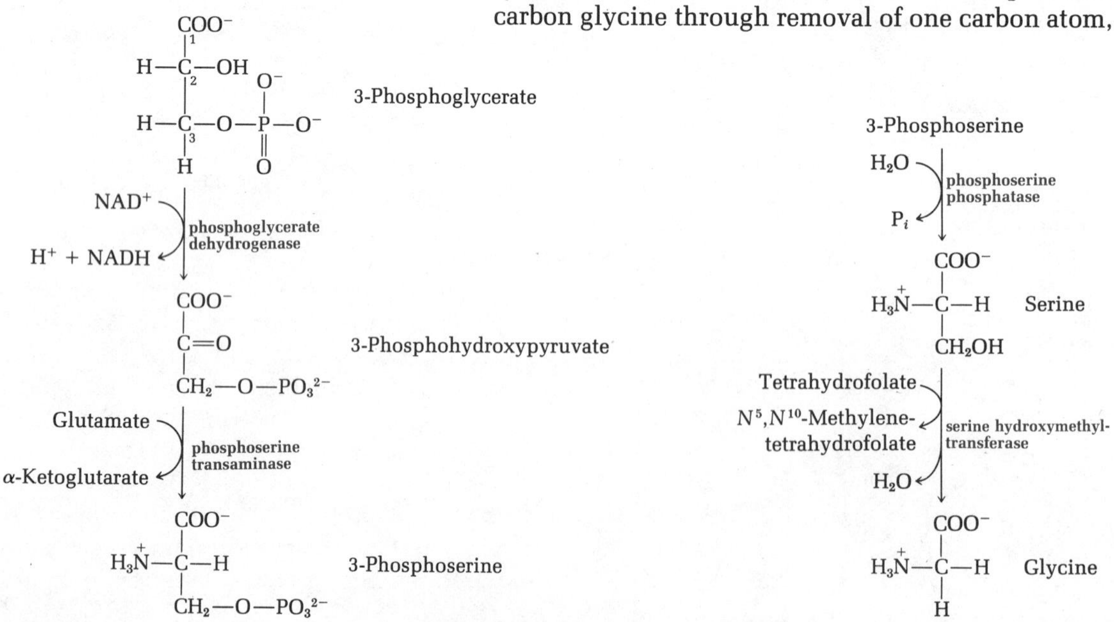

3-Phosphoglycerate: $^{-}OOC^{1}$—$C^{2}H(OH)$—$C^{3}H_2$—O—PO_3^{2-}

NAD^+ → $H^+ + NADH$ (phosphoglycerate dehydrogenase)

3-Phosphohydroxypyruvate: COO^-—C=O—CH_2—O—PO_3^{2-}

Glutamate → α-Ketoglutarate (phosphoserine transaminase)

3-Phosphoserine: COO^-—$\overset{+}{H_3N}$—C—H—CH_2—O—PO_3^{2-}

3-Phosphoserine

H_2O → P_i (phosphoserine phosphatase)

Serine: COO^-—$\overset{+}{H_3N}$—C—H—CH_2OH

Tetrahydrofolate → N^5,N^{10}-Methylenetetrahydrofolate, H_2O (serine hydroxymethyltransferase)

Glycine: COO^-—$\overset{+}{H_3N}$—C—H—H

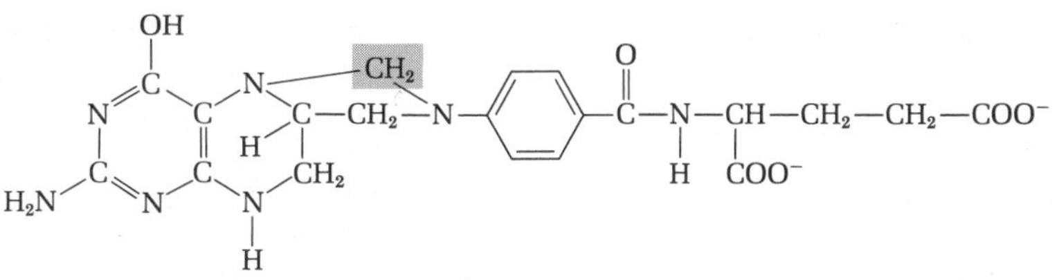

Figure 22-7
Structure of N^5,N^{10}*-methylenetetrahydrofolate. The methylene group that is transferred is in color (see page 261).*

β or 3 position (Figure 22-6). This reaction is accomplished by an enzyme that requires the coenzyme *tetrahydrofolate,* the active form of the vitamin *folic acid* (page 260). Tetrahydrofolate is the acceptor of the β carbon atom of serine during its cleavage to yield glycine. The carbon atom removed from serine forms a methylene bridge (page 538) between nitrogen atoms 5 and 10 of tetrahydrofolate to yield N^5,N^{10}-methylenetetrahydrofolate (Figure 22-7). The overall reaction, which is reversible, is

$$\text{Serine} + \text{tetrahydrofolate} \rightleftharpoons \text{glycine} + N^5,N^{10}\text{-methylenetetrahydrofolate} + H_2O$$

This reaction completes the formation of glycine from serine.

N^5,N^{10}-Methylenetetrahydrofolate is a member of the family of folic acid coenzymes, which are capable, along with S-adenosylmethionine and coenzyme B_{12}, of carrying different kinds of 1-carbon groups (pages 260 and 261). The 1-carbon unit derived from serine and carried by tetrahydrofolate can be transferred to various acceptor molecules.

In the liver of vertebrates glycine can be made by another route (page 538), catalyzed by the enzyme *glycine synthase*

$$CO_2 + NH_4^+ + \text{NADH} + H^+ + N^5,N^{10}\text{-methylenetetrahydrofolate} \rightleftharpoons \text{glycine} + \text{NAD}^+ + \text{tetrahydrofolate}$$

Biosynthesis of the Essential Amino Acids

The pathways for the synthesis of the amino acids essential in the nutrition of the albino rat and humans have been deduced from biochemical and genetic studies of those microorganisms capable of making them. We shall not examine these pathways, but a few generalizations about them can be made. The pathways leading to synthesis of the essential amino acids are generally longer (5 to 15 steps) and more complex than those leading to nonessential amino acids, most of which have fewer than 5 steps. Higher animals are unable to make several of the essential amino acids for lack of one or two enzymes in their pathways. The most complex of the pathways to the essential amino acids are those leading to *phenylalanine, tryptophan,* and *histidine,* which have benzene or heterocyclic rings. The synthesis of these rings, particularly the two condensed rings of tryptophan, requires numerous and complex enzyme-catalyzed steps.

Five of the amino acids essential for animals are synthesized by plants and microorganisms from nonessential amino acids: threonine, methionine, and lysine are formed from aspartate, and arginine and histidine are formed from glutamate. Isoleucine is formed by bacteria from the essential amino acid threonine.

Amino Acid Biosynthesis Is under Allosteric Regulation

The most responsive manner in which amino acid synthesis is controlled is through allosteric inhibition of the first reaction in the biosynthetic sequence by its final end product (pages 233 and 346). The first reaction of such a sequence, which is usually irreversible, is catalyzed by an allosteric enzyme. As an example, Figure 22-8 shows the allosteric regulation of the syn-

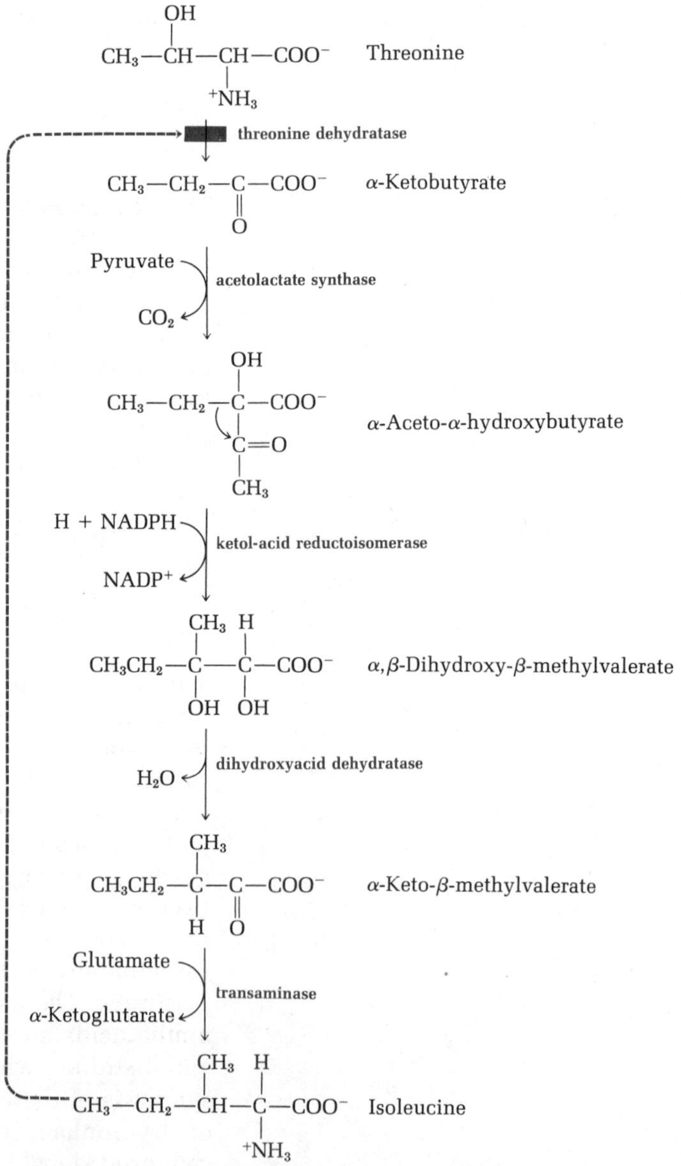

Figure 22-8
Biosynthesis of isoleucine from threonine in E. coli. The first reaction in the sequence is inhibited by the end product, isoleucine. This was one of the first examples of allosteric feedback inhibition to be discovered. Valine can reverse or prevent the inhibitory action of isoleucine.

$CH_3—CH(OH)—CH(^+NH_3)—COO^-$ Threonine

threonine dehydratase

$CH_3—CH_2—C(=O)—COO^-$ α-Ketobutyrate

Pyruvate → CO_2 acetolactate synthase

$CH_3—CH_2—C(OH)(C(=O)CH_3)—COO^-$ α-Aceto-α-hydroxybutyrate

H + NADPH → $NADP^+$ ketol-acid reductoisomerase

$CH_3CH_2—C(CH_3)(OH)—C(H)(OH)—COO^-$ α,β-Dihydroxy-β-methylvalerate

→ H_2O dihydroxyacid dehydratase

$CH_3CH_2—C(CH_3)(H)—C(=O)—COO^-$ α-Keto-β-methylvalerate

Glutamate → α-Ketoglutarate transaminase

$CH_3—CH_2—CH(CH_3)—C(H)(^+NH_3)—COO^-$ Isoleucine

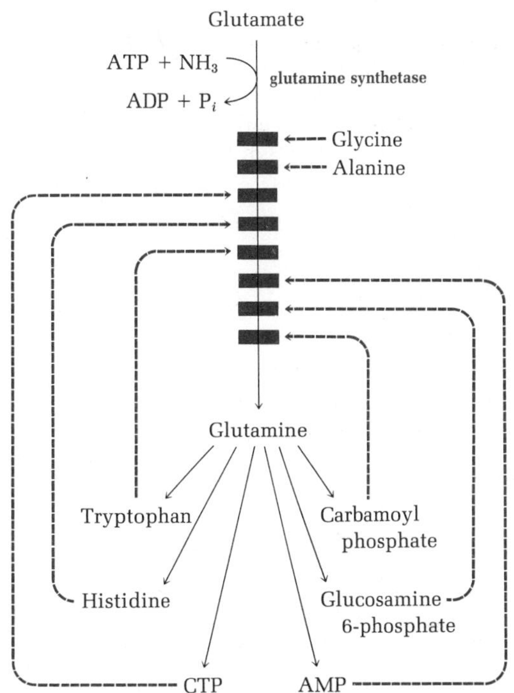

Figure 22-9
Allosteric inhibition of glutamine synthetase in E. coli. *In this organism glutamine is a precursor of the products shown, all of which can act as feedback inhibitors. Such inhibition by several negative modulators is called* concerted inhibition. *Glutamine synthetase is also strongly inhibited by excess* ATP, *which converts it into an inactive form by a covalent modification of essential tyrosine groups in its subunits. The regulation of glutamine synthetase in animal tissues is much simpler.*

thesis of isoleucine from threonine, discussed earlier (page 234). The end product isoleucine is a negative modulator of the first reaction in the sequence. Such allosteric or noncovalent modulation of amino acid synthesis is responsive on a minute-to-minute basis in bacteria.

Another noteworthy example is the remarkable set of allosteric controls exerted on the activity of glutamine synthetase of *E coli.* In these bacteria glutamine is the amino-group donor in the biosynthesis of many metabolic products (Figure 22-9). Eight products of glutamine metabolism in *E. coli* are now known to serve as negative-feedback modulators of the activity of glutamine synthetase, one of the most complex regulatory enzymes known.

Since the 20 amino acids must be made in the correct proportions for protein synthesis, cells have developed ways not only of controlling the rate of synthesis of individual amino acids but also of *coordinating* their formation. Such coordination is especially well developed in fast-growing bacterial cells. Figure 22-10 shows how *E. coli* cells coordinate the synthesis of the amino acids lysine, methionine, threonine, and isoleucine, all made from aspartate. The step from aspartate to aspartyl phosphate is catalyzed by three *isozyme* forms (page 239), each of which can be independently controlled. The steps from aspartate semialdehyde to homoserine and from threonine to α-ketobutyrate are also catalyzed by dual, independently controlled isozymes. One of the isozymes for the conversion of aspartate to aspartyl phosphate can be allosterically

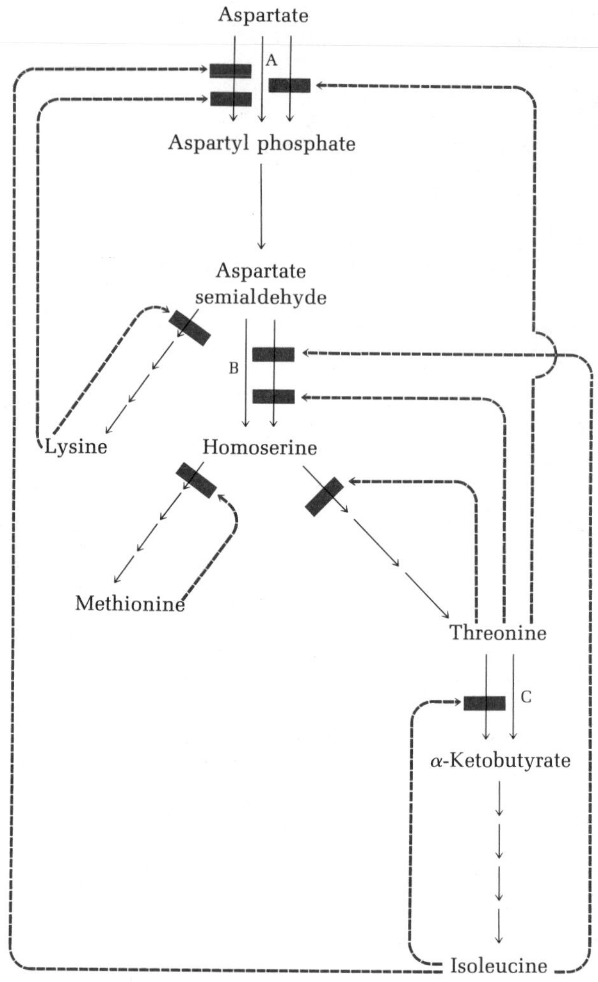

Figure 22-10
Interlocking network of regulatory mechanisms in the biosynthesis of several amino acids derived from aspartate in E. coli. *The different types of regulation involved are described in the text, as are the labels A, B, and C.*

inhibited by two different modulators, lysine and isoleucine, whose action is more than additive. This is another example of *concerted* inhibition (page 502). The sequence from aspartate to isoleucine shows multiple, overlapping negative-feedback inhibition; i.e., isoleucine inhibits the conversion of threonine into α-ketobutyrate, and threonine inhibits its own formation at three points, from homoserine, from aspartate semialdehyde, and from aspartate. This is called *sequential* feedback inhibition.

Amino Acid Biosynthesis Is Also Regulated by Changes in Enzyme Concentration

Another mechanism for regulation of the biosynthesis of amino acids is through control of the concentration of the biosynthetic enzymes. When cells do not need a given amino acid because it is already available in high concentration, the enzymes required in its synthesis are present in very low levels. However,

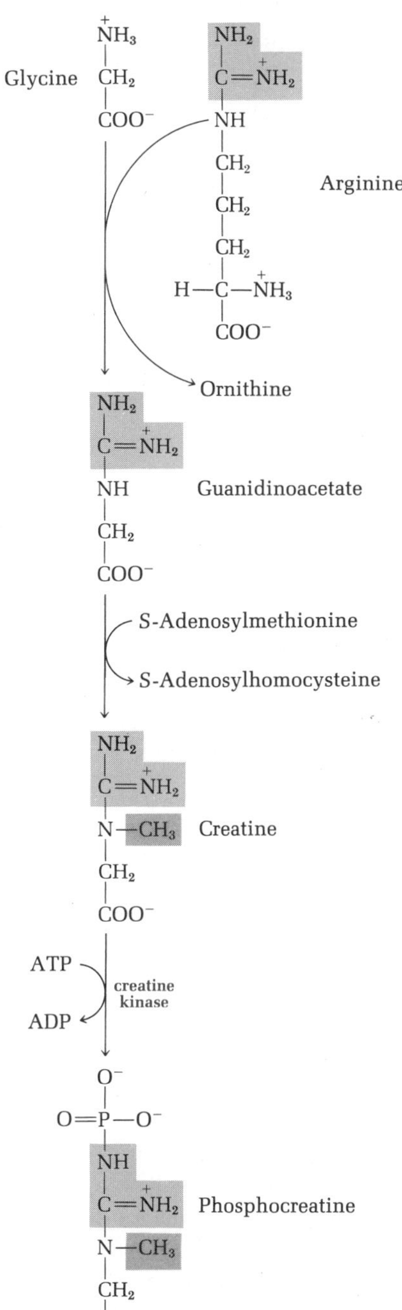

Figure 22-11
Biosynthesis of creatine and phosphocreatine. Creatine is made from three amino acids, glycine, arginine, and methionine. Arginine is the donor of the guanidino group (color) and methionine is the donor of the methyl group (gray). This pathway shows the versatility of amino acids as precursors in the biosynthesis of other nitrogenous biomolecules.

if the concentration of this amino acid falls to lower than adequate levels, the cells begin to make more of the enzymes required for manufacture of the amino acid. This type of regulation is brought about by changes in the activity of the genes coding for these enzymes. Whenever the product of an amino acid biosynthetic pathway is available in ample concentrations the genes coding for the enzymes of this pathway are inactivated or repressed. When the level of the biosynthetic product of this sequence falls, the genes for the sequence become *derepressed* and cause more of these particular enzymes to be synthesized. Later we shall see how enzyme synthesis is regulated by genetic repression mechanisms (Chapter 29).

There are three isozymes in Figure 22-10, labeled A, B, and C, which are not indicated as having allosteric modulators. These isozymes are subject to regulation by alterations in their rate of synthesis by the cell and are called *repressible enzymes.* The synthesis of both isozyme A and isozyme B is repressed in *E. coli* whenever methionine is present in ample amounts. Similarly, the synthesis of isozyme C is repressed when ample isoleucine is present in the medium. Regulation of amino acid biosynthesis by repression or derepression (Chapter 29) is generally slower in response than allosteric regulation.

Thus we have an interlocking network of different kinds of controls in the biosynthesis of this group of related amino acids. Fast-growing bacterial cells coordinate the synthesis of all the amino acids with computerlike efficiency.

Glycine Is a Precursor of Porphyrins

In addition to their role as building blocks of proteins, the amino acids are precursors of many specialized biomolecules, including various hormones, vitamins, coenzymes, alkaloids, cell-wall polymers, porphyrins, antibiotics, pigments, and neurotransmitter substances, all of which serve essential biological roles. Space will not permit full development of the many secondary biosynthetic pathways that yield these products, but two examples will be described briefly. One example is the synthesis of creatine, important in the bioenergetics of muscle and nerve in the form of phosphocreatine (page 383). It is made from three amino acids: glycine, arginine, and methionine (Figure 22-11).

The biosynthesis of porphyrins, for which glycine is again a major precursor, is another important example because of the central importance of the porphyrin nucleus in heme proteins such as hemoglobin and the cytochromes, as well as the Mg^{2+}-containing porphyrin derivative *chlorophyll.* The porphyrins are constructed from four molecules of the monopyrrole derivative *porphobilinogen,* which is synthesized in the sequence shown in Figure 22-12. This pathway was largely elucidated in isotopic tracer studies carried out by David Shemin. In the first reaction, glycine reacts with succinyl-CoA to yield *α-amino-β-ketoadipic acid,* which is then decarboxylated to

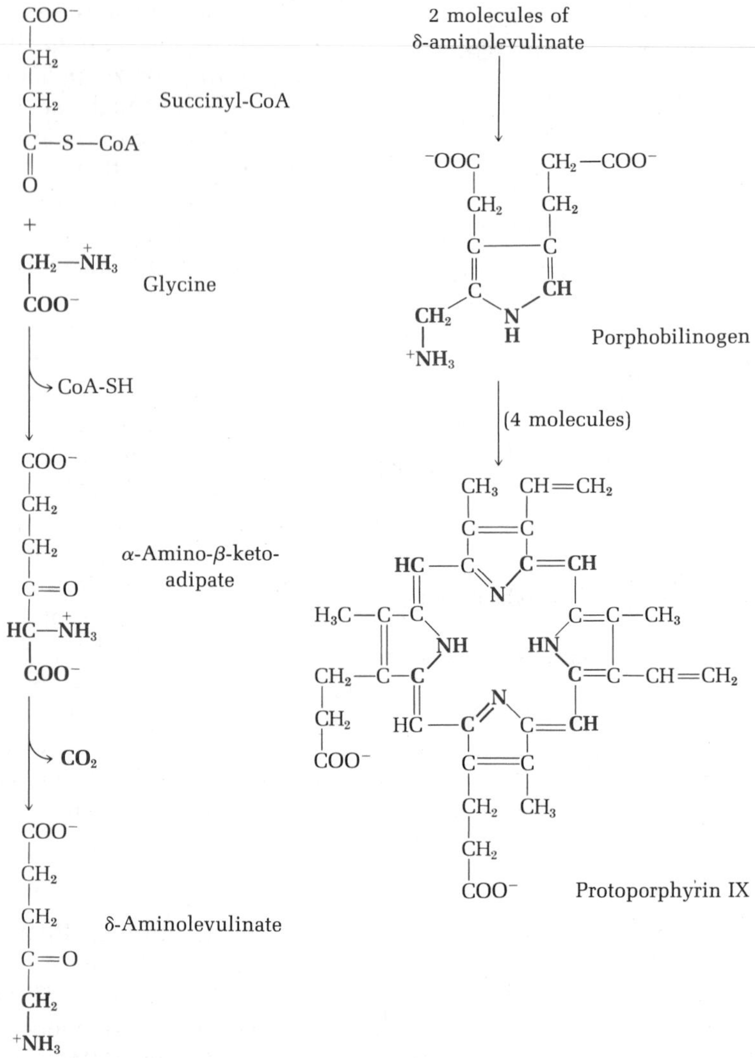

Figure 22-12
Biosynthesis of protoporphyrin IX, the porphyrin of hemoglobin and myoglobin. The carbon and nitrogen atoms furnished by glycine are shown in color. The remaining carbon atoms are derived from the succinyl group of succinyl-CoA.

Succinyl-CoA

+

Glycine

$\rightarrow$ CoA-SH

α-Amino-β-keto-adipate

$\rightarrow CO_2$

δ-Aminolevulinate

2 molecules of δ-aminolevulinate

Porphobilinogen

(4 molecules)

Protoporphyrin IX

give <u>δ-aminolevulinic acid</u> (δ is Greek delta) and carbon dioxide. Two molecules of δ-aminolevulinic acid next condense to form porphobilinogen. Four molecules of porphobilinogen come together to form <u>protoporphyrin</u>, through a series of complex enzymatic reactions. The iron atom is incorporated after the protoporphyrin has been assembled. Figure 22-12 shows the origin of the carbon and nitrogen atoms of protoporphyrin IX from glycine. Porphyrin biosynthesis is regulated by the concentration of the heme protein product, such as hemoglobin, which can serve as a feedback inhibitor of early steps in porphyrin synthesis.

Porphyrin Derivatives Accumulate in Some Genetic Disorders

Genetic defects of certain enzymes in the biosynthetic pathway from glycine to porphyrins lead to the accumulation of specific porphyrin precursors in the red blood cells, in body fluids, and

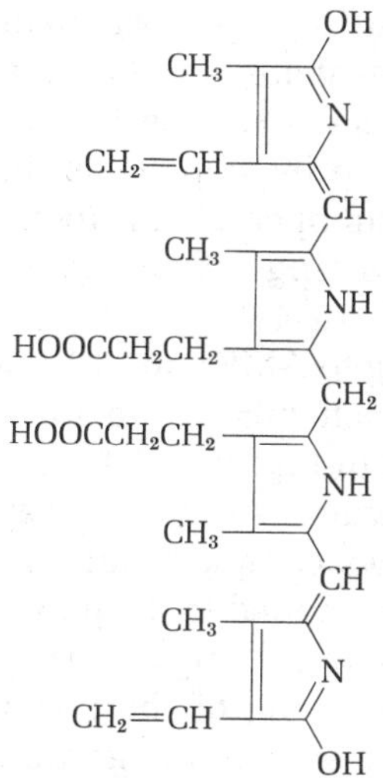

Figure 22-13
Bilirubin, a bile pigment. It is formed by the cleavage of the porphyrin ring system of heme into the linear or open tetrapyrrole shown. Measurement of its concentration in the blood is useful in the diagnosis of some liver disorders.

in the liver. These genetic diseases are known as *porphyrias*. In one of the porphyrias, which affects mainly the red blood cells, there is accumulation of *uroporphyrinogen I*, an abnormal isomer of a precursor of protoporphyrin. It stains the urine red and causes the teeth to fluoresce strongly in ultraviolet light and the skin to show abnormal sensitivity to sunlight.

Another type of porphyria causes accumulation of the porphyrin precursor *porphobilinogen* in the liver, as well as intermittent neurological and behavioral aberrations. George III, King of England at the time of the American War for Independence in 1776–1781, is believed to have suffered from this type of porphyria. Medical historians have suggested that the behavioral symptoms of this disease may have been responsible for the unreasonable insistence of George III on excessive taxation and punishment of the American colonies.

Degradation of Heme Groups Yields Bile Pigments

The iron-porphyrin or heme group of hemoglobin, released from dying red blood cells in the spleen, is degraded to yield free Fe^{3+} and ultimately *bilirubin*, a linear or open tetrapyrrole derivative (Figure 22-13). Bilirubin becomes bound to serum albumin of the blood and is transported to the liver, where it is further transformed into a water-soluble derivative excreted into the bile. Bilirubin is the pigment that is responsible for yellowing of the skin and eyeballs in *jaundice*, which results from impaired liver function. Determination of the concentration of bilirubin in the blood is useful in diagnosis of this and other kinds of liver disease.

Purine Nucleotides Are Made by a Complex Pathway

Now let us examine the biosynthesis of the nucleotides, in which amino acids are important precursors. Nearly all living organisms, except for some bacteria, appear able to synthesize the purine and pyrimidine nucleotides. The two parent purine nucleotides of nucleic acids are *adenosine 5'-monophosphate* (AMP), also called *adenylate*, and *guanosine 5'-monophosphate* (GMP), or *guanylate*. These nucleotides contain the purine bases *adenine* and *guanine*, respectively. The first important clues to the biosynthetic origin of the purine nucleotides came from experiments in which various isotopic metabolites were fed to animals and the sites of incorporation of the labeled atoms into the purine ring determined. Such experiments were carried out in birds, which excrete nitrogen largely in the form of an oxidation product of purines, *uric acid* (page 555), which is easily isolated in pure form from bird droppings. Figure 22-14 shows the origin of the carbon and nitrogen atoms of the purine ring system as revealed by tracer experiments carried out in pigeons. Note that each of the four nitrogen atoms of purines arises from an amino acid.

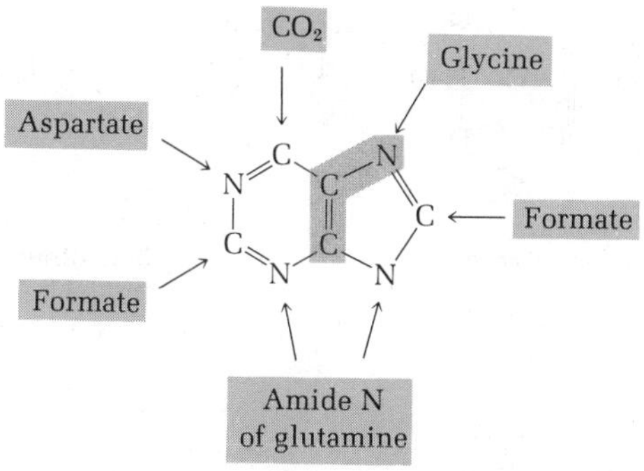

Figure 22-14
Origin of the ring atoms of purines, determined from isotopic experiments with precursors labeled with ^{14}C or ^{15}N.

Although one might expect the purine ring to be made first and then joined to the ribose phosphate portion, the biosynthesis of purine nucleotides begins with ribose 5-phosphate and the purine ring is built on it, step by step. Thus the first steps in this de novo pathway yield an open-chain ribonucleotide, which subsequently undergoes ring closure to yield a purine nucleotide.

Major steps in the pathway leading to AMP and GMP, which was elucidated by the American biochemists John Buchanan and G. Robert Greenberg, are shown in Figures 22-15 to 22-17. First, an amino group donated by glutamine is attached at position 1 of ribose 5-phosphate in two complex steps (Figure 22-15). The amino acid glycine is then added to the new amino group. Other stepwise reactions follow, and ring closure occurs to yield the five-membered imidazole ring of the purine nucleus (Figure 22-16). Aspartate donates its amino group to the imidazole ring, and a second ring closure takes place, to yield the second of the two fused rings of the purine nucleus. The first intermediate to have a complete purine ring is inosinic acid (Figures 22-16 and 17). The subsequent conversion of inosinic acid into adenylic acid (AMP) requires the insertion of an amino group derived from aspartate; this takes place by a rather

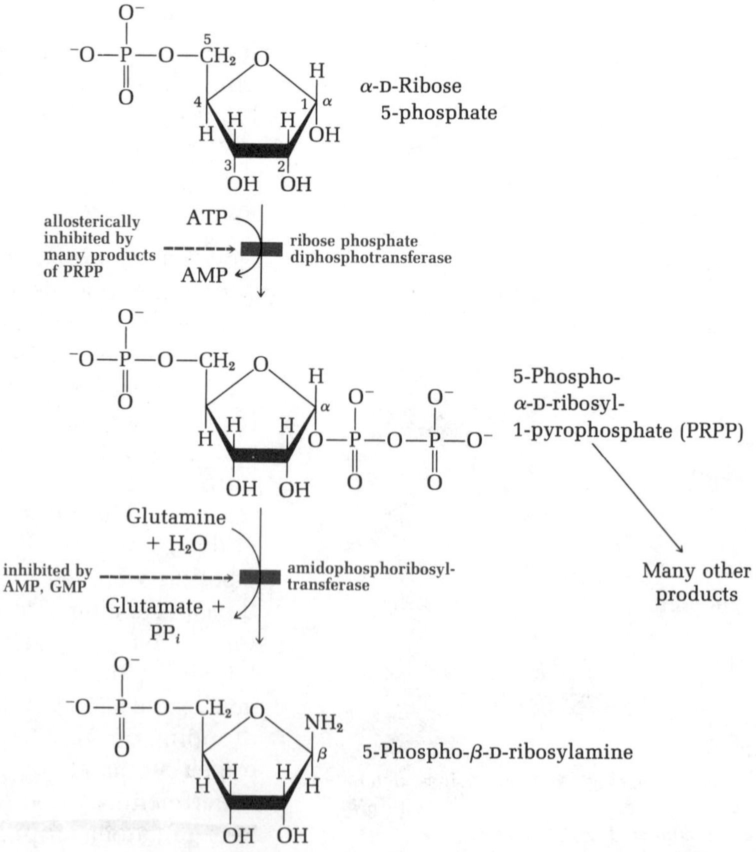

Figure 22-15
Conversion of D*-ribose 5-phosphate into 5-phospho-β-*D*-ribosylamine.*

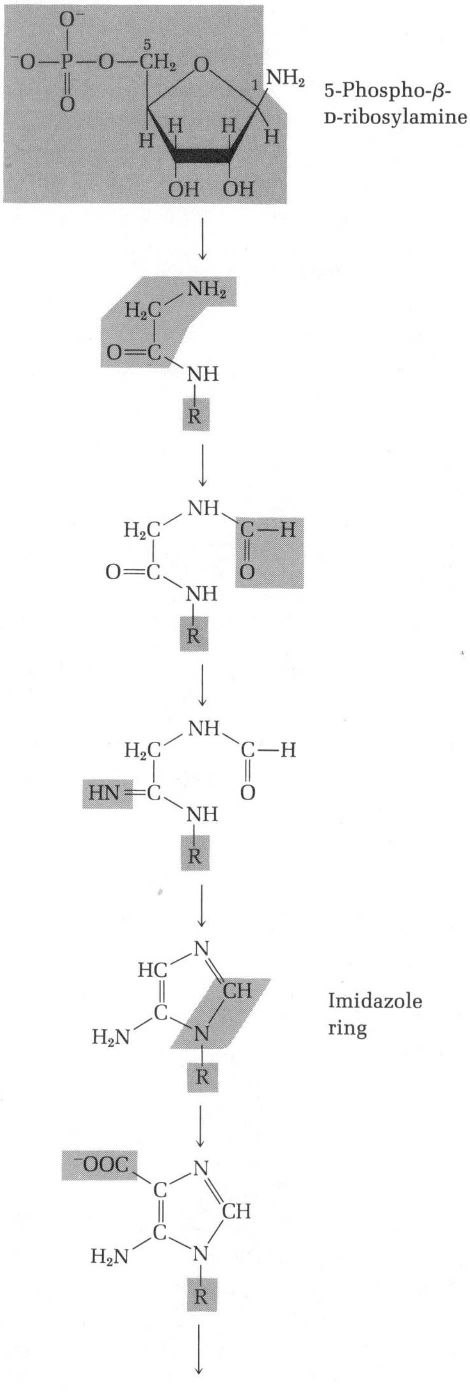

5-Phospho-β-D-ribosylamine

Imidazole ring

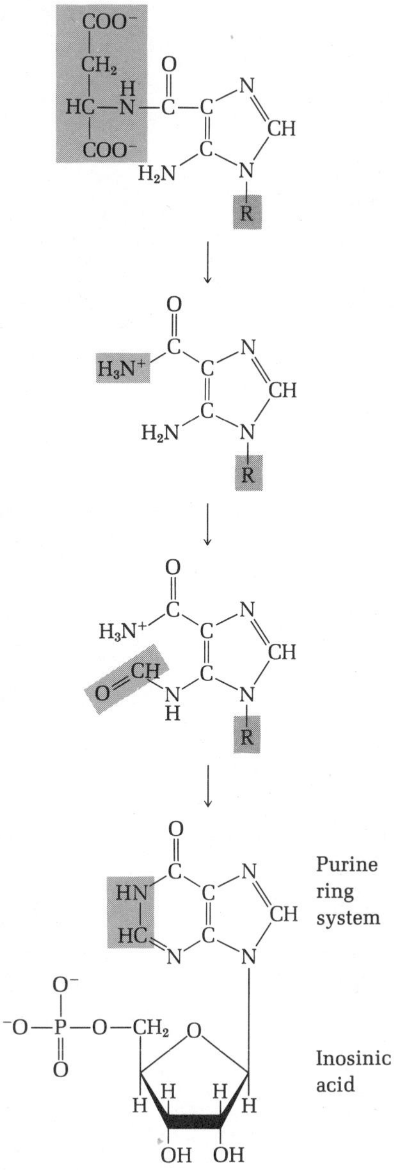

Purine ring system

Inosinic acid

Figure 22-16
Steps in the construction of the purine ring of inosinic acid. The result of each successive enzymatic step is shown in color. R symbolizes the 5-phospho-D-ribosyl group (gray) on which the purine ring is built.

complex set of reactions that we shall not detail here. Inosinic acid is converted into *guanylic acid* (GMP) via reactions that require ATP (Figure 22-17). This pathway to the purine nucleotides is called the *de novo* pathway. Later we shall see that purine nucleotides can also be made by a salvage pathway.

The phosphorylation of AMP to ADP is promoted by *adenylate kinase* (page 387)

$$\text{ATP} + \text{AMP} \rightleftharpoons 2\text{ADP}$$

The ADP so formed is then phosphorylated to ATP by the glycolytic sequence or by the respiratory chain. ATP next brings about the successive formation of GDP and GTP by the action of

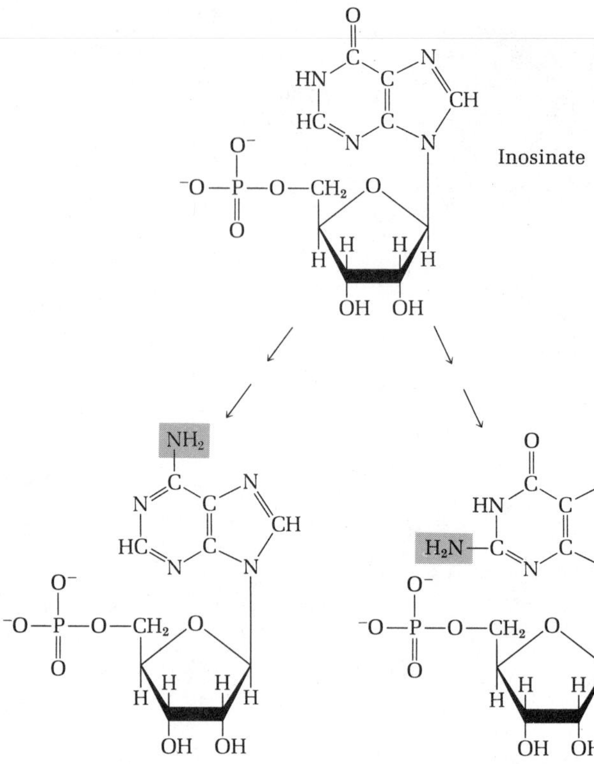

Figure 22-17
Conversion of inosinic acid into adenylic and guanylic acids. Changes in the purine ring are in color.

Inosinate

Adenylate

Guanylate

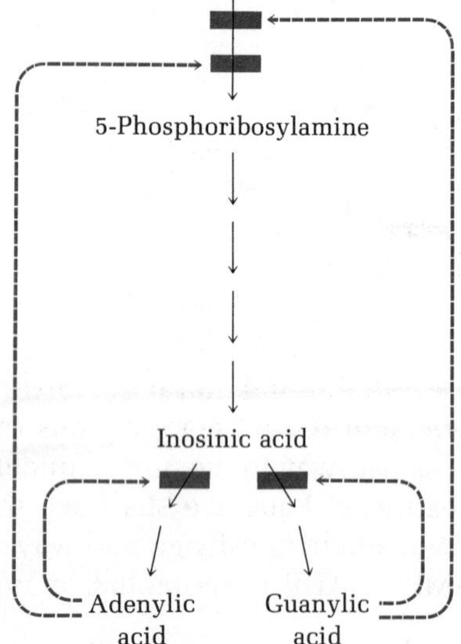

Figure 22-18
Feedback-control mechanisms in the biosynthesis of adenine and guanine nucleotides in E. coli. Regulation of these pathways in other organisms varies somewhat.

nucleoside monophosphate kinase and *nucleoside diphosphate kinase,* respectively

$$ATP + GMP \rightleftharpoons GDP + ADP$$

$$ATP + GDP \rightleftharpoons GTP + ADP$$

Purine Nucleotide Biosynthesis Is Regulated by Feedback Control

Three major feedback mechanisms cooperate in regulating the overall rate of de novo purine nucleotide synthesis and the relative rates of formation of the two end products adenylic acid and guanylic acid (Figure 22-18). The first of these control mechanisms is exerted on the first reaction that is unique to purine synthesis, i.e., the step leading to the transfer of an amino group to *5-phosphoribosyl-1-pyrophosphate* (PRPP) to form *5-phosphoribosylamine* (Figure 22-15). This reaction is catalyzed by an allosteric enzyme that is inhibited by the end products AMP and GMP. Thus, whenever either AMP or GMP accumulates to excess, the first step in its biosynthesis from PRPP is inhibited. In the second control mechanism, exerted at a later stage (Figure 22-18), an excess of GMP in the cell brings about allosteric inhibition of its formation from inosinic acid,

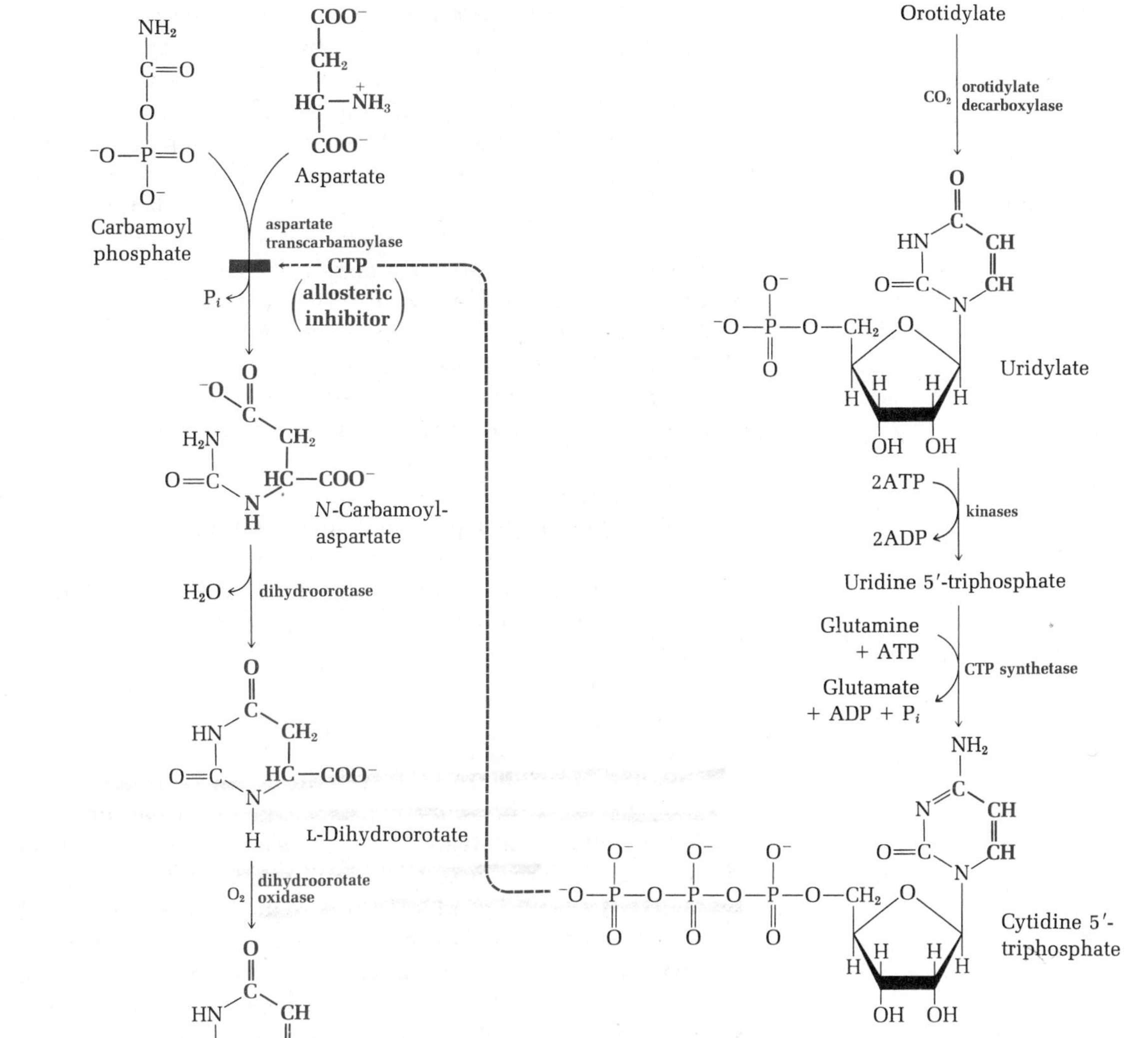

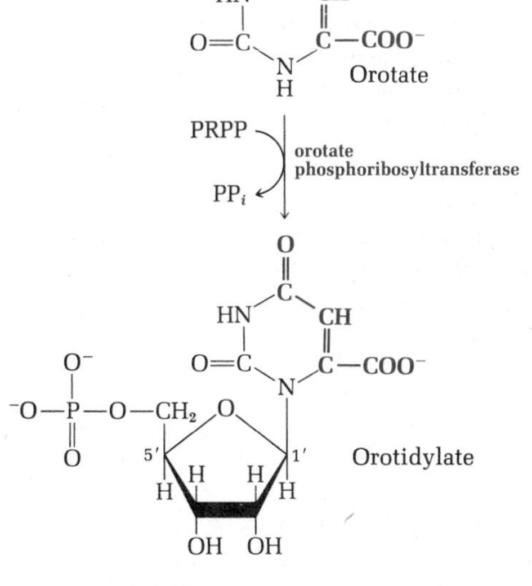

Figure 22-19
Biosynthesis of the pyrimidine nucleotides UTP and CTP via orotidylic acid.

without affecting the formation of AMP (Figure 22-18). Conversely, an accumulation of adenylic acid results in inhibition of its formation without affecting the biosynthesis of GMP.

Pyrimidine Nucleotides Are Made from Aspartate and Ribose Phosphate

The common pyrimidine nucleotides are *cytidine 5′-monophosphate* (CMP), or *cytidylate,* and *uridine 5′-monophosphate* (UMP), or *uridylate,* which contain the pyrimidines *cytosine* and *uracil,* respectively. Pyrimidine nucleotide biosynthesis (Figure 22-19) proceeds in a somewhat different manner from that of purine nucleotides; in this case the six-membered pyrimidine ring is made first and then attached to ribose phosphate. Required in this process is *carbamoyl phosphate,* also an intermediate in the urea cycle (page

551). However, the carbamoyl phosphate required in urea synthesis is made in the mitochondria by a mitochondrial enzyme, carbamoyl phosphate synthetase I, whereas the carbamoyl phosphate required in pyrimidine biosynthesis is made in the cytosol by a different form of the enzyme, carbamoyl phosphate synthetase II. The cytosolic carbamoyl phosphate reacts with aspartate to yield *N-carbamoylaspartate*. This reaction is catalyzed by *aspartate transcarbamoylase*, one of the most thoroughly studied allosteric enzymes (page 238), which we shall discuss again below. By removal of water from N-carbamoylaspartate, a reaction catalyzed by *dihydroorotase*, the pyrimidine ring is closed with formation of *L-dihydroorotic acid*. This compound is now oxidized to yield the pyrimidine derivative *orotic acid*, a reaction in which NAD^+ is the ultimate electron acceptor.

At this point, the D-ribose 5-phosphate side chain, provided by 5-phosphoribosyl-1-pyrophosphate, is attached to orotate to yield *orotidylic acid*. Orotidylate is then decarboxylated to yield *uridylate*, which is phosphorylated to UTP and then accepts an amino group from glutamine to form cytidine triphosphate, or CTP (Figure 22-19).

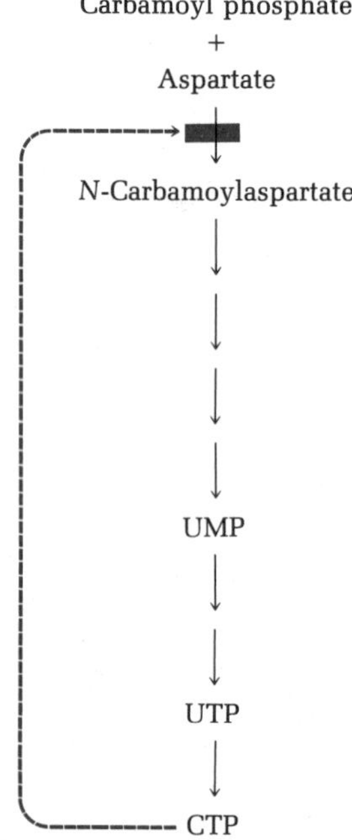

Figure 22-20
Regulation of the biosynthetic pathway to CTP by end-product inhibition of aspartate transcarbamoylase. The inhibitory effect of CTP is prevented by ATP.

Regulation of Pyrimidine Nucleotide Biosynthesis

The regulation of the rate of pyrimidine nucleotide synthesis occurs through the enzyme *aspartate transcarbamoylase* (ATCase), which catalyzes the first reaction in the sequence (Figure 22-20). This enzyme is inhibited by cytidine triphosphate (CTP), the end product of this sequence of reactions. The ATCase molecule consists of six catalytic subunits and six regulatory subunits (page 238). The catalytic subunits bind the substrate molecules, and the allosteric subunits bind the allosteric inhibitor CTP. The entire ATCase molecule, as well as its subunits, exists in two conformations, active and inactive. When the regulatory subunits are empty, the enzyme is maximally active. However, when CTP accumulates, it is bound by the regulatory subunits to cause a change in their conformation. This change is transmitted to the catalytic subunits, which then also shift to an inactive conformation. The presence of ATP prevents the changes induced by CTP. Figure 22-21 shows the effects of the allosteric regulators on the activity of ATCase.

Ribonucleotides Are the Precursors of the Deoxyribonucleotides

Deoxyribonucleotides, the building blocks of DNA, are derived from the corresponding ribonucleotides by reactions in which the 2′ carbon atom of the D-ribose portion of the ribonucleotide is directly reduced to form the 2′-deoxy derivative. In this way, for example, adenosine diphosphate (ADP) is reduced to form *2′-deoxyadenosine diphosphate* (dADP), and GDP is reduced to

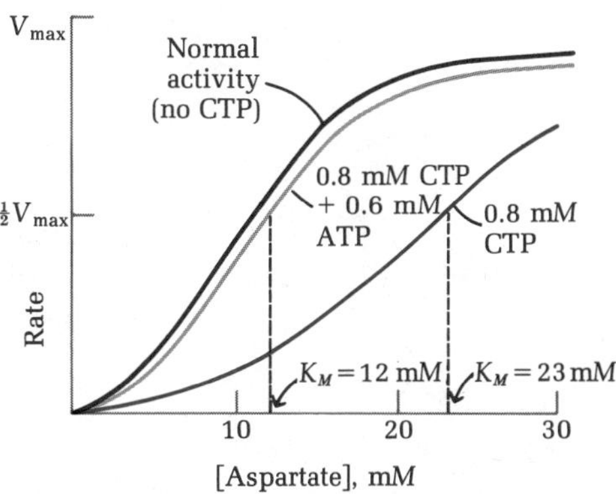

Figure 22-21
Effect of the allosteric modulators CTP and ATP on the rate of conversion of aspartate into carbamoylaspartate by aspartate transcarbamoylase. Note that the addition of CTP, its allosteric inhibitor, causes K_M for aspartate to increase. ATP fully reverses this effect.

dGDP. The reduction of the D-ribose portion of the ribonucleoside diphosphates to 2′-deoxy-D-ribose requires a pair of hydrogen atoms, which are ultimately donated by NADPH via an intermediate hydrogen-carrying protein, thioredoxin. This protein has pairs of —SH groups that serve to carry H atoms from NADPH to the ribonucleoside diphosphate. The oxidized or disulfide form of thioredoxin is reduced by NADPH in a reaction catalyzed by *thioredoxin reductase*

$$\text{NADPH} + \text{H}^+ + \text{thioredoxin}\langle{}^{S}_{S} \longrightarrow \text{NADP}^+ + \text{thioredoxin}\langle{}^{SH}_{SH}$$

The reduced thioredoxin then reduces the nucleoside diphosphates (NDPs) to deoxyribonucleoside diphosphates (dNDPs) in a reaction catalyzed by *ribonucleotide reductase* (Figure 22-22)

$$\text{Thioredoxin}\langle{}^{SH}_{SH} + \text{NDP} \longrightarrow \text{thioredoxin}\langle{}^{S}_{S} + \text{dNDP} + H_2O$$

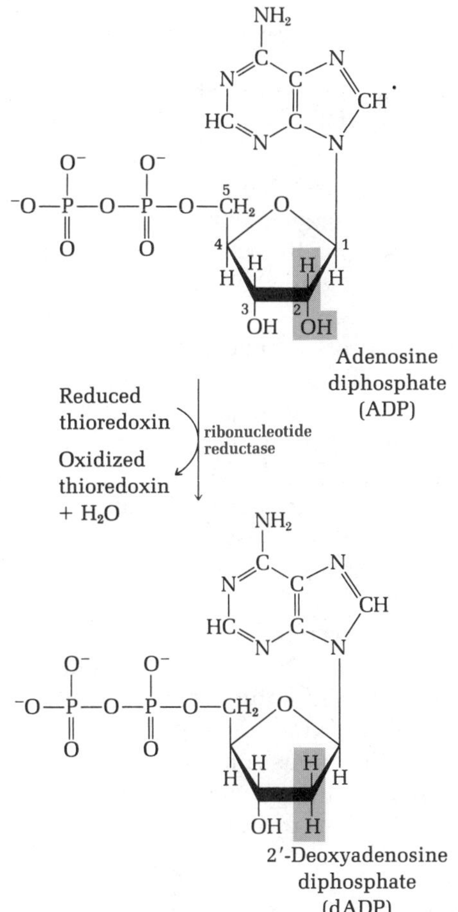

Figure 22-22
Conversion of ADP into dADP. Other ribonucleoside 5′-diphosphates (NDPs) are converted into their 2′-deoxy forms (dNDPs) by the same reaction.

DNA contains *thymidylate* or dTMP residues instead of the uridylate or UMP residues present in RNA (page 390). To make them, deoxyuridine diphosphate (dUDP), formed from uridine diphosphate (UDP), is hydrolyzed to dUMP

$$\text{dUDP} + H_2O \longrightarrow \text{dUMP} + P_i$$

The dUMP is then converted into the corresponding deoxythymidylate (dTMP) via methylation of dUMP by N^5,N^{10}-methylenetetrahydrofolate catalyzed by *thymidylate synthase*

$$\text{dUMP} + N^5,N^{10}\text{-methylenetetrahydrofolate} \longrightarrow \text{dTMP} + \text{dihydrofolate}$$

dTMP is then phosphorylated to yield dTDP

$$\text{dTMP} + \text{ATP} \longrightarrow \text{dTDP} + \text{ADP}$$

Tetrahydrofolate is regenerated from dihydrofolate by the enzyme *tetrahydrofolate dehydrogenase:*

$$\text{Dihydrofolate} + \text{NADPH} + H^+ \longrightarrow \text{tetrahydrofolate} + \text{NADP}^+$$

To complete the synthesis of the deoxyribonucleoside 5′-triphosphates, which are direct precursors in DNA biosynthesis (Chapter 28), the following kinase reactions occur

$$\text{ATP} + \text{dADP} \longrightarrow \text{ADP} + \text{dATP}$$

$$\text{ATP} + \text{dCDP} \longrightarrow \text{ADP} + \text{dCTP}$$

$$\text{ATP} + \text{dTDP} \longrightarrow \text{ADP} + \text{dTTP}$$

$$\text{ATP} + \text{dGDP} \longrightarrow \text{ADP} + \text{dGTP}$$

Degradation of Purines Leads to Uric Acid in Human Beings

Purine nucleotides are degraded by a pathway (Figure 22-23) in which the phosphate group is lost by the action of *5′-nucleotidase.* Adenylate yields *adenosine,* which is then deaminated to *inosine.* Inosine is then hydrolyzed to yield its purine base *hypoxanthine* and D-ribose. Hypoxanthine is oxidized successively to *xanthine* and then to *uric acid* by *xanthine oxidase,* a complex flavoenzyme that also contains an atom of molybdenum and four iron-sulfur centers (page 479) in its prosthetic group. Molecular oxygen is the electron acceptor in this complex reaction.

GMP catabolism also yields *urate* as end product. GMP is first hydrolyzed to yield the nucleoside *guanosine,* which is then cleaved to free guanine. Guanine undergoes hydrolytic removal of its amino group to yield *xanthine,* which is converted into urate by xanthine oxidase (Figure 22-24).

Uric acid is the excreted end product of purine catabolism in primates. However, in many other vertebrates uric acid is degraded further to the excretory product *allantoin,* by the action of *urate oxidase* (Figure 22-24).

The total amount of uric acid excreted by the normal adult human is about 0.6 g per 24 h, arising in part from ingested purines and in part from the turnover of the purine nucleotides of nucleic acids. Genetic aberrations in human purine metabolism have been found, some with serious consequences; but before we discuss them, let us examine the metabolic salvage of free purines, from which nucleotides can be resynthesized.

Figure 22-23
Pathways of purine nucleotide catabolism. Details of the final steps are shown in Figure 22-24.

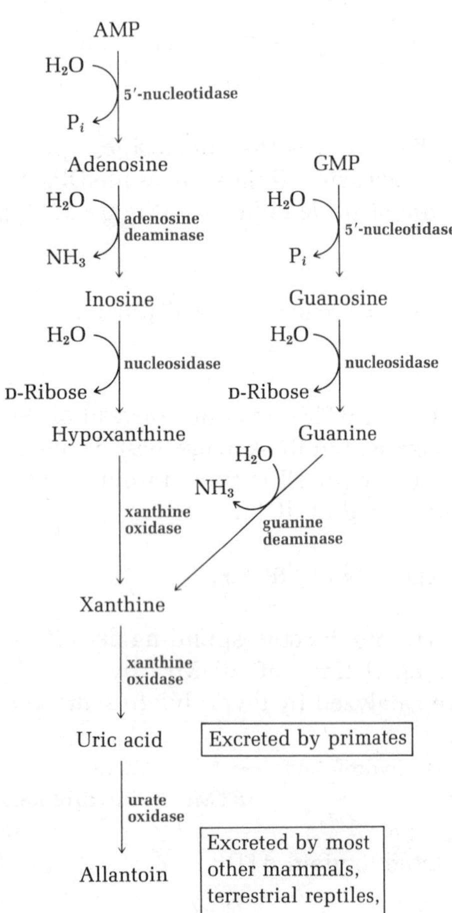

Purine Bases Are Recycled by a Salvage Pathway

Free purine and pyrimidine bases are constantly formed in cells during the metabolic degradation of nucleotides by pathways described above. However, free purines formed on degradation of purine nucleotides are in large part salvaged and used over again to remake nucleotides. This occurs by a pathway that is quite different from the de novo biosynthesis of purines described earlier, in which the purine ring system is assembled step by step on ribose 5-phosphate in a long series of reactions. The salvage pathway is much simpler. It consists of a single reaction, in which free adenine reacts with 5-phosphoribosyl-1-pyrophosphate (PRPP) to yield the corresponding adenine nucleotide

$$\text{Adenine} + \text{PRPP} \longrightarrow \text{AMP} + \text{PP}_i$$

Free guanine is salvaged in the same way by a different enzyme

$$\text{Guanine} + \text{PRPP} \longrightarrow \text{GMP} + \text{PP}_i$$

It is curious that these salvage pathways are not only simpler than the de novo pathway, but also "cheaper," since they require less ATP energy.

The same enzyme that salvages guanine can also salvage *hypoxanthine,* the deamination product of adenine, to yield inosinic acid (IMP)

$$\text{Hypoxanthine} + \text{PRPP} \longrightarrow \text{IMP} + \text{PP}_i$$

The enzyme catalyzing the latter two reactions, called *hypoxanthine-guanine phosphoribosyltransferase,* has been found to be missing in children suffering from a sex-linked recessive genetic defect. The consequences of this enzyme deficiency are both catastrophic and paradoxical. The genetic lack of hypoxanthine-guanine phosphoribosyltransferase activity, seen only in male children, results in a bizarre set of symptoms, called the *Lesch-Nyhan syndrome,* after Michael Lesch, then a medical student, and William Nyhan, a pediatrician on the faculty of the Johns Hopkins School of Medicine, who discovered and identified this defect in 1964 through some interesting biochemical detective work. Children with this genetic disorder, which becomes manifest by the age of 2 years, are mentally retarded and badly coordinated. In addition, they are extremely hostile and aggressive. Moreover, they show compulsive self-destructive tendencies: they mutilate themselves by biting their fingers, toes, and lips.

Lesch-Nyhan disease is also somewhat paradoxical, metabolically speaking. We have seen that there are two pathways by which purine nucleotides are synthesized, the de novo pathway, which builds the purine ring system in a stepwise fashion, and the salvage pathway, which merely reassembles purine nucleotides from the free purines and PRPP. In the Lesch-Nyhan disorder, the genetic defect is in the salvage pathway, but the purine nucleotides are still made by the de novo pathway. However, the problem is not a deficiency in purine nucleotide synthesis but an overproduction, due to an aberration in some still unknown regulatory mechanism. The

Figure 22-24
Final steps in catabolism of purines, showing the complex reactions catalyzed by xanthine oxidase and urate oxidase. Xanthine oxidase is inhibited by allopurinol (Figure 22-25).

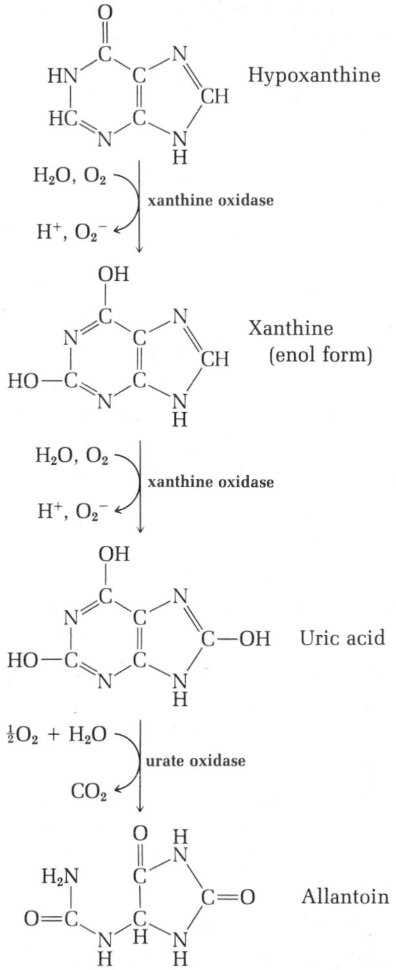

Hypoxanthine

H_2O, O_2 → H^+, O_2^- (xanthine oxidase)

Xanthine (enol form)

H_2O, O_2 → H^+, O_2^- (xanthine oxidase)

Uric acid

$\frac{1}{2}O_2 + H_2O$ → CO_2 (urate oxidase)

Allantoin

overproduction of purine nucleotides is accompanied by an overproduction of both uric acid and PRPP. Why this defect in what appears to be a minor metabolic pathway should disturb human behavior so profoundly is not known. Very little can be done for victims of Lesch-Nyhan disease, although cells from such patients can be "cured" in vitro by genetic engineering (Chapter 30).

Overproduction of Uric Acid Causes Gout

The disease gout, long erroneously thought to be due to "high living," is a disease of the joints, usually in males, caused by an elevated concentration of uric acid in the blood and tissues. The joints become inflamed, painful, and arthritic, due to the abnormal deposition of crystals of sodium urate. The kidneys are also affected, since excess urate is also deposited in the kidney tubules. The precise cause of gout is not known, but it is suspected to be due to partial genetic deficiency of one or another enzyme concerned in purine metabolism.

Gout can be effectively treated by a combination of nutritional and drug therapy. Foods especially rich in nucleotides and nucleic acids, such as liver or glandular products, are withheld from the diet, as are coffee and tea, which contain the purines caffeine and theobromine. In addition, major improvement follows the use of the drug allopurinol, an inhibitor of xanthine oxidase, the enzyme responsible for converting purines into uric acid (Figure 22-25).

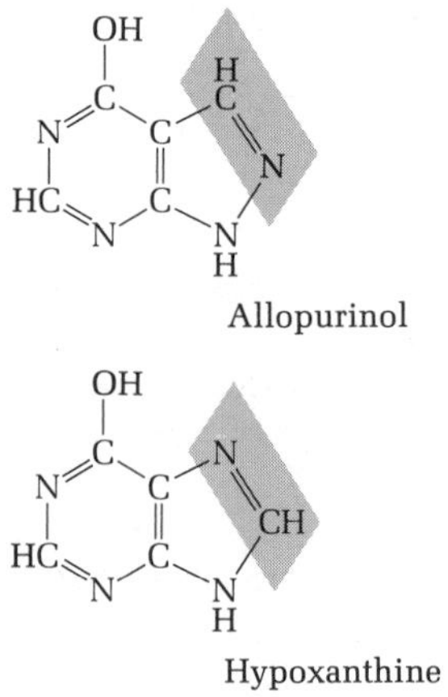

Allopurinol

Hypoxanthine

Figure 22-25
Allopurinol, an inhibitor of xanthine oxidase. Only a slight alteration (color) in the structure of the substrate hypoxanthine yields a medically effective enzyme inhibitor. Allopurinol is an example of a useful drug that was designed to be a competitive inhibitor.

The Nitrogen Cycle

The biosynthetic pathways to the amino acids and nucleotides share a requirement for nitrogen, but soluble, biologically useful nitrogen compounds are generally scarce in natural environments. For this reason ammonia, amino acids, and nucleotides are used economically by most organisms, particularly since such compounds are precursors of their vital nucleic acids and proteins. Indeed, we have seen that free amino acids, purines, and pyrimidines, formed during metabolic turnover, are often salvaged and used over again.

The most abundant form of nitrogen is present in air, which is four-fifths molecular nitrogen (N_2). However, only a relatively few species can convert atmospheric nitrogen into forms useful to living organisms, which therefore salvage and reuse biologically available nitrogen in a vast nitrogen cycle (Figure 22-26). The first step in the nitrogen cycle is the fixation of atmospheric nitrogen by nitrogen-fixing organisms to yield ammonia. Ammonia can be used by most living organisms. However, there are some important soil bacteria that derive their energy by oxidizing ammonia to form nitrite and ultimately nitrate. Because these organisms are extremely abundant and active, nearly all ammonia reaching the soil ultimately becomes oxidized to nitrate, a process known as nitrifi-

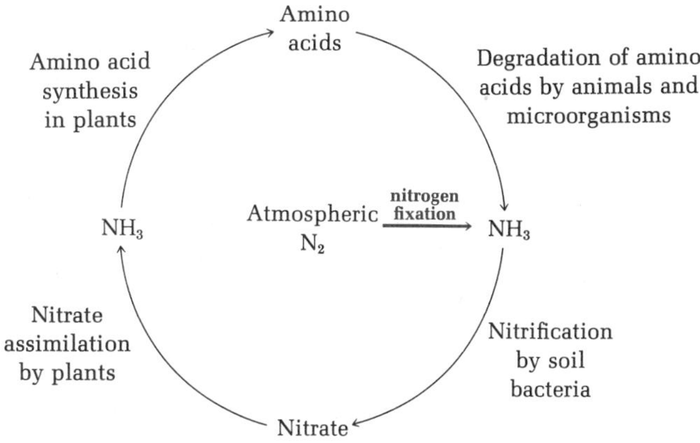

Figure 22-26
The nitrogen cycle. The total amount of nitrogen fixed annually in the biosphere exceeds) 10^{11} kg.

cation. Plants and many bacteria can readily reduce nitrate back into ammonia again by the action of *nitrate reductases*. Ammonia so formed can be built into amino acids by plants, which are then used by animals as a source of both nonessential and essential amino acids to build animal proteins. On the death of animals the microbial degradation of their proteins returns ammonia to the soil, where nitrifying bacteria convert it into nitrite (NO_2^-) and nitrate (NO_3^-) again.

Now let us examine the process of nitrogen fixation, important to every form of life.

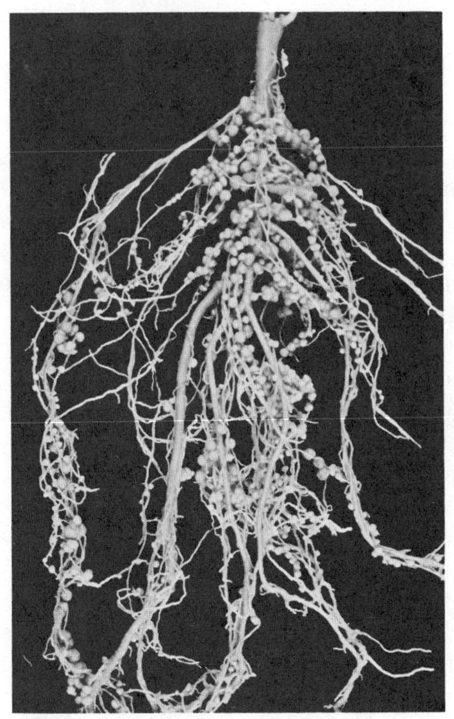

Figure 22-27
Nitrogen-fixing nodules on the roots of bird's-foot trefoil, a legume. The symbiotic nitrogen-fixing bacteria grow in large clusters closely associated with root cells. The latter provide some factors essential for nitrogen fixation, particularly leghemoglobin, which has a very high affinity for binding oxygen. Oxygen is highly inhibitory to nitrogenase.

Not Many Organisms Can Fix Nitrogen

Only a relatively few species of microorganisms and plants can fix atmospheric nitrogen. Some free-living bacteria, such as the *cyanobacteria* or *blue-green algae*, which inhabit not only fresh and salt waters but also soils, as well as other kinds of soil bacteria, such as *Azotobacter*, are capable of fixing atmospheric nitrogen. The first important product of nitrogen fixation in these organisms is ammonia (NH_3), which can be used by other forms of life, either directly or after its conversion into other soluble compounds, such as nitrites, nitrates, or amino acids.

Another type of nitrogen fixation occurs in *leguminous plants*, which include peas, beans, clover, and alfalfa, in a process that requires the cooperation of the host plant and symbiotic bacteria present in its root nodules. This type of nitrogen fixation is called *symbiotic* nitrogen fixation. The nitrogen-fixing enzymes are actually located in the root-nodule bacteria, but the plant supplies some essential components that the bacteria lack (Figure 22-27). In addition to the legumes, a few other higher plants can fix nitrogen, but the vast majority of nonleguminous plants, as well as all animal species, are unable to use atmospheric nitrogen.

The stringent requirement of nitrogen for life is shown by the fact that for optimal growth most crop plants need large amounts of soluble nitrogen, more than can be furnished by the soil. Nitrogen may be furnished by "natural" fertilizers, such as manure or guano (page 555), or by "chemical" fertilizers, such as ammonium nitrate (NH_4NO_3). Another way in which essential nitrogen is furnished for the growth of crop plants is by crop rotation. Corn can be grown in a given field only for one or two crops if no source of nitrogen is applied, but if the field is planted with peas, beans, alfalfa, or clover every second or third year, these plants, thanks to their ability to carry out symbiotic nitrogen fixation, can form enough ammonia and nitrates to enrich the soil sufficiently for a subsequent crop of corn.

Fixation of Nitrogen Is a Complex Enzymatic Process

Nitrogen fixation is catalyzed by an enzyme complex, the *nitrogenase* system, whose action is still not fully understood. Because the nitrogenase system is unstable and is quickly inactivated by exposure to atmospheric oxygen, it has been difficult to isolate in active form and purify. The first stable product of nitrogen fixation that has been recognized is ammonia (NH_3); thus the overall process is believed to consist of the reduction of molecular nitrogen (N_2) to two molecules of ammonia

$$N_2 + 3H_2 \longrightarrow 2NH_3 \qquad \Delta G^{\circ\prime} = -8.0 \text{ kcal/mol}$$

Since the standard-free-energy change is strongly negative, the reaction will proceed to the right under standard conditions. However, molecular nitrogen is a relatively unreactive gas with a very strong chemical bond between the two nitrogen atoms. Its reduction to ammonia has a very large activation-energy barrier (page 211), which the nitrogenase system must in some unknown way overcome.

NADPH is the ultimate hydrogen donor for the nitrogenase complex. The reducing equivalents of NADPH are first transferred to the iron-sulfur protein *ferredoxin* (page 479), the immediate donor of reducing equivalents for reduction of nitrogen. Ferredoxin has a molecular weight of 6000 and contains seven iron atoms as well as an equal number of acid-labile sulfur atoms. Ferredoxin then transmits its reducing equivalents to the nitrogenase complex, which consists of two metalloenzymes, the first an iron protein and the second a protein containing both iron and molybdenum (Figure 22-28). ATP is also required and is hydrolyzed to ADP and phosphate during nitrogen fixation. The precise function of ATP is not clear, since no phosphorylated intermediates have been found. Presumably the free energy of hydrolysis helps in lowering the activation energy. It appears likely that 12 molecules of ATP are hydrolyzed to ADP and P_i for each molecule of N_2 reduced to two NH_4^+. The overall equation for nitrogen fixation can therefore be written

$$N_2 + 3NADPH + 5H^+ + 12ATP + 12H_2O \longrightarrow 2NH_4^+ + 3NADP^+ + 12ADP + 12P_i$$

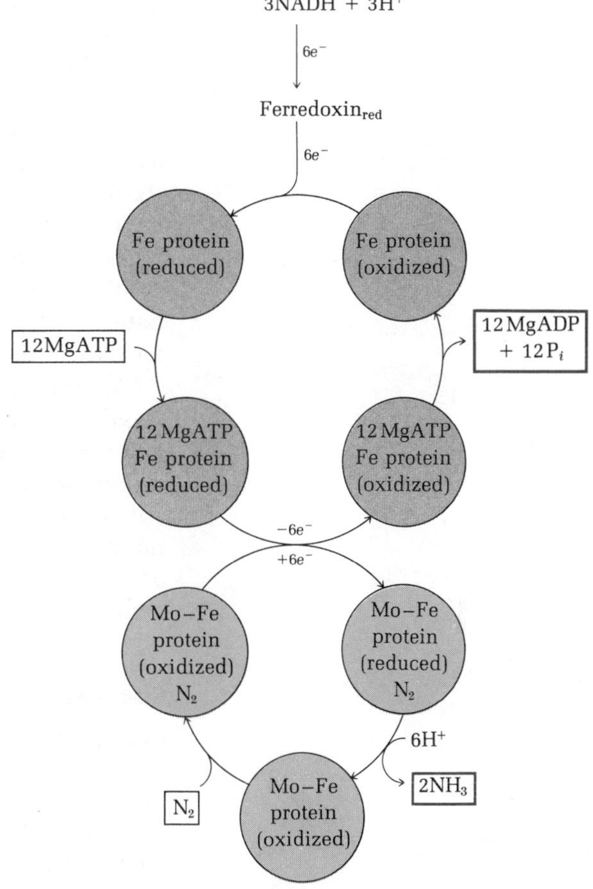

Figure 22-28
A proposed pathway of the nitrogenase reaction. It is believed that a number of intermediate steps occur on the active sites of these enzymes.

The nitrogenase system has an interesting property that permits the quantitative study of nitrogen fixation in intact plants. Nitrogenase can not only reduce N_2 but also reduces *acetylene* ($HC{\equiv}CH$) to *ethylene* ($H_2C{=}CH_2$). Since the ratio of ethylene to acetylene in the atmosphere can be measured by physical methods, it is possible to measure the activity of the nitrogenase system of plant and soil systems in an experimental greenhouse by introducing acetylene into its gas phase and measuring the rate of formation of ethylene.

The nitrogenase system is the subject of much experimental study because of its immense practical importance. Ammonia for fertilizers is produced commercially from atmospheric nitrogen by catalytic reduction in the Haber process, for which the reaction is

$$N_2 + 3H_2 \longrightarrow 2NH_3$$

Very high temperatures and pressures are required. If the nitrogenase reaction could be simulated by a simpler set of cheap catalysts, it might be possible to generate soluble ammonium salts for fertilizers inexpensively, particularly in developing countries, which cannot afford the energy-intensive Haber process.

Biological approaches to make atmospheric nitrogen more available are also being taken. One effort is to determine whether the roots of common nonleguminous crop plants, such as corn, can be infected with different species of nitrogen-fixing bacteria or mutants of them, with the aim of creating a new and fruitful association or symbiosis. Indeed, it has been found that the roots of a number of nonleguminous plants in tropical countries possess nitrogen-fixing bacteria. Unfortunately, nitrogen fixation by such plants requires very warm soil; they do not fix nitrogen when grown in temperate zones.

Another approach is to isolate the DNA coding for the nitrogenase system of bacteria and incorporate it into the genomes of other microorganisms or plants devoid of nitrogenase. Such gene transfers to nonfixing bacteria such as *E. coli* have been achieved. Permanent and productive recombination of nitrogenase DNA into the genomes of higher plants is a far more difficult problem. With increasing knowledge of genetic engineering it may some day be solved.

Summary

Human beings and the albino rat can synthesize 10 of the 20 protein amino acids. The remainder, which are required in the diet and are called essential amino acids, can be synthesized by plants and bacteria. Among the nonessential amino acids glutamate is formed by reductive amination of α-ketoglutarate. Glutamate is the precursor of glutamine and proline. Alanine and aspartate are formed by transamination to pyruvate and oxaloacetate, respectively. Tyrosine is formed by hydroxylation of phenylalanine, an essential amino acid. Cysteine is formed from methionine and serine by a complex series of reactions in which S-adenosylmethionine and cystathionine are intermediates. The carbon chain of serine is derived from 3-phosphoglycerate. Serine is a precursor of glycine; the β carbon atom of serine is transferred to tetrahydrofolate. The pathways of biosynthesis of the essential amino acids in bacteria and plants are more complex and longer. They are formed from certain of the nonessential amino acids and other metabolites. The biosynthetic pathways leading to the amino acids are subject to allosteric end-product inhibition; the regulatory enzyme is usually the first in the sequence. Amino acids are precursors of many other important biomolecules. The porphyrin ring of heme proteins is derived from glycine and succinyl-CoA.

The purine ring system of purine nucleotides is built up in a step-by-step fashion on the 1 carbon atom of 5-phosphoribosylamine. Amino acids furnish all the nitrogen atoms of purines. Two ring-closure steps ensue to form the purine nucleus. The pyrimidines are formed from aspartic acid, CO_2, and ammonia. Ribose 5-phosphate is then attached to yield the pyrimidine ribonucleotides. Free purines are salvaged and rebuilt into nucleotides by a separate pathway. Genetic deficiency in one of the salvage enzymes results in Lesch-Nyhan disease, characterized by bizarre behavioral symptoms. Gout, another genetic disease, results in the accumulation of uric acid crystals in the joints.

Fixation of atmospheric nitrogen takes place in certain soil bacteria and in the root nodules of leguminous plants by the action of the complex nitrogenase system. Formation of ammonia by fixation of molecular nitrogen in legume root nodules, nitrification of ammonia to form nitrate by soil organisms, the denitrification of nitrate by higher plants to form ammonia, and the synthesis of amino acids from ammonia by plants and animals constitute the nitrogen cycle.

References

Pathways of Amino Acid Biosynthesis

BENDER, D. A.: *Amino Acid Metabolism,* Wiley, New York, 1975.

BLAKLEY, R. L.: *The Biochemistry of Folic Acid and Related Pteridines,* North-Holland, Amsterdam, 1969.

CUNNINGHAM, E. B.: *Biochemistry: Mechanisms of Metabolism,* McGraw-Hill, New York, 1978. Excellent description of the enzymatic steps.

MEISTER, A.: *Biochemistry of Amino Acids,* 2d ed., Academic, New York, 1965. This two-volume work is a comprehensive reference book.

UMBARGER, H. E.: "Amino Acid Biosynthesis and Its Regulation," *Annu. Rev. Biochem.,* **47:**533–606 (1978). Definitive review by a pioneer in research on the regulation of these pathways.

Genetic Disorders in Amino Acid and Nucleotide Metabolism

NYHAN, W. L. (ed.): *Heritable Disorders of Amino Acid Metabolism,* Wiley, New York, 1974.

STANBURY, J. B., J. B. WYNGAARDEN, and D. S. FREDRICKSON: *The Metabolic Basis of Inherited Disease,* 4th ed., McGraw-Hill, New York, 1978. The articles on gout and Lesch-Nyhan disease are especially good.

Nitrogen Fixation

BRILL, W. J.: "Biological Nitrogen Fixation," *Sci. Am.,* **236:**68–81, March (1977).

DELWICHE, C. C.: "The Nitrogen Cycle," *Sci. Am.,* **223:**136–147, September (1970).

JONES, T.: "Nitrogen Fixation and Bioenergetics: The Role of ATP in Nitrogenase Catalysis," *FEBS Lett.,* **98:**1–8 (1979).

MORTENSON, L. E., and R. N. F. THORNELEY: "Structure and Function of Nitrogenase," *Annu. Rev. Biochem.,* **48:**387–418 (1979).

Heme Metabolism

GRANICK, S., and S. I. BEALE: "Hemes, Chlorophyll, and Related Compounds: Biosynthesis and Metabolic Regulation," *Adv. Enzymol.,* **40:**33–203 (1978).

Nucleotide Metabolism

HENDERSON, J. F., and A. R. P. PATERSON: *Nucleotide Metabolism: An Introduction*, Academic, New York, 1973.

JONES, M. E.: "Pyrimidine Nucleotide Biosynthesis in Animals," *Annu. Rev. Biochem.*, **49**:253–279 (1980).

Problems

1. *Defect in Phenylalanine Hydroxylase and Diet* In normal individuals, tyrosine is a nonessential amino acid, but patients who have a genetic defect in phenylalanine hydroxylase require tyrosine in their diet for normal growth. Explain.

2. *Equation for the Synthesis of Aspartate from Glucose* Write the net equation for the synthesis of the nonessential amino acid aspartate from glucose, carbon dioxide, and ammonia.

Problem 3

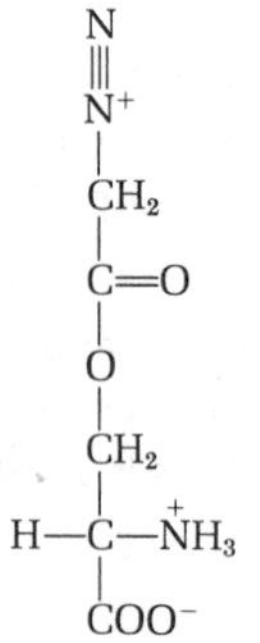

$$N{\equiv}N^+{-}CH_2{-}C({=}O){-}O{-}CH_2{-}CH(\overset{+}{N}H_3){-}COO^-$$

O-(2-Diazoacetyl)-L-serine (azaserine)

3. *Inhibition of Nucleotide Synthesis by Azaserine* The diazo compound *O*-(2-diazoacetyl)-L-serine, known also as azaserine, is a powerful inhibitor of those enzymes that transfer ammonia from glutamine to an acceptor (amidotransferases) during biosynthesis. If cells that are actively carrying out purine biosynthesis are treated with azaserine, what intermediate in the synthesis of inosinic acid from α-D-ribose 5-phosphate will accumulate? Explain.

4. *Nucleotide Biosynthesis in Amino Acid Auxotrophic Bacteria* Although normal *E. coli* cells can synthesize all the amino acids, some mutants, called amino acid auxotrophs, are unable to synthesize specific amino acids and require the addition of that amino acid to the culture medium for optimal growth. In addition to their requirement for protein synthesis, specific amino acids are also required in the biosynthesis of other nitrogenous cell products. Consider the three amino acid auxotrophs that are unable to synthesize glycine, glutamine, and aspartate. For each mutant what nitrogenous cell products other than proteins would fail to be synthesized?

5. *Anticancer Drugs: Blockage of Deoxythymidylate Synthesis*
   (a) Deoxyuridine monophosphate (dUMP) is converted into deoxythymidylate monophosphate (dTMP), which is required in DNA synthesis, via the methylation of dUMP by N^5,N^{10}-methylenetetrahydrofolate, catalyzed by the enzyme thymidylate synthase (page 633)

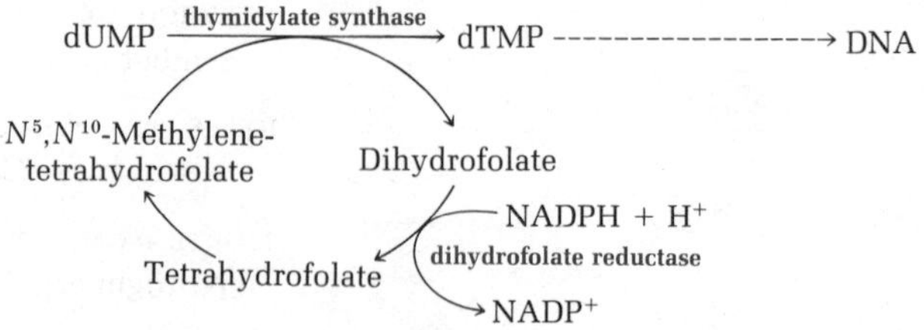

dUMP —thymidylate synthase→ dTMP - - - - → DNA

N^5,N^{10}-Methylene-tetrahydrofolate → Dihydrofolate

Dihydrofolate + NADPH + H^+ —dihydrofolate reductase→ Tetrahydrofolate + $NADP^+$

Tetrahydrofolate → N^5,N^{10}-Methylene-tetrahydrofolate

Problem 5

(Structure: 5-fluorouracil ring with H–N, C=O, F, N–H)

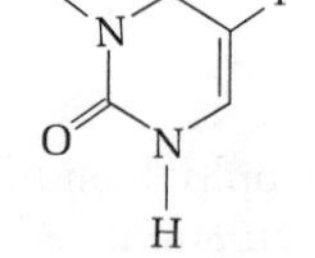

Fluorouracil

The uridine derivative, fluorouracil (see left), is transformed by the cell into fluorodeoxyuridylate (F-dUMP), a powerful irreversible inhibitor of thymidylate synthase. How can you account for the fact that fluorouracil inhibits the growth of rapidly dividing cancer cells in experimental animals?

(b) Dihydrofolate, produced in the thymidylate synthase reaction, is converted back into tetrahydrofolate by dihydrofolate reductase. How is tetrahydrofolate transformed into N^5,N^{10}-methylenetetrahydrofolate? Dihydrofolate reductase is strongly inhibited ($K_i = 10^{-9}$ *M*) by the drug methotrexate, clinically useful in treating some forms of cancer.

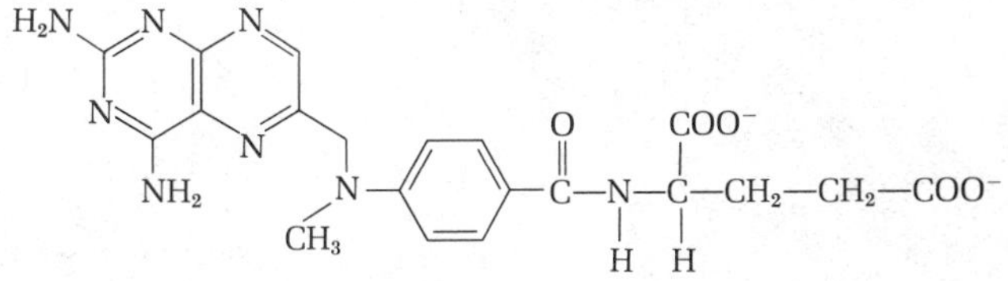

Methotrexate

How can this drug inhibit growth of cancer cells? Would you expect it to inhibit growth of normal cells?

6. *Nucleotides Are Poor Sources of Energy* In most organisms, nucleotides are not employed as energy-yielding fuels. What observations support this conclusion? Why are nucleotides relatively poor sources of energy in mammals?

7. *Mode of Action of Sulfa Drugs* Some bacteria require the inclusion of p-aminobenzoic acid in the culture medium for normal growth. The growth of such bacteria is severely inhibited by the addition of sulfanilamide, one of the earliest antibacterial sulfa drugs. Moreover, in its presence, 5′-phosphoribosyl-4-carboxamide-5-aminoimidazole accumulates in the culture medium. Both effects are reversed by the addition of excess p-aminobenzoic acid.
   (a) What is the role of p-aminobenzoic acid? (*Hint:* See Figure 22-7 and page 260).
   (b) Why does 5′-phosphoribosyl-4-carboxamide-5-aminoimidazole accumulate in the presence of sulfanilamide? (See Figure 22-16.)
   (c) Why is the inhibition and accumulation reversed by the addition of excess p-aminobenzoic acid?

Problem 7

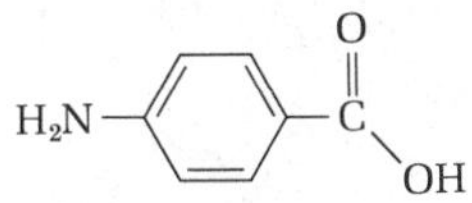

p-Aminobenzoic acid

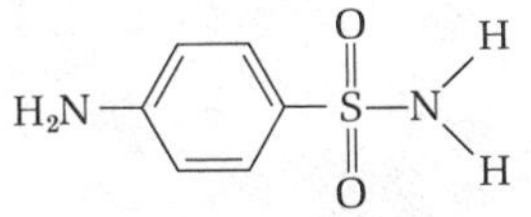

Sulfanilamide

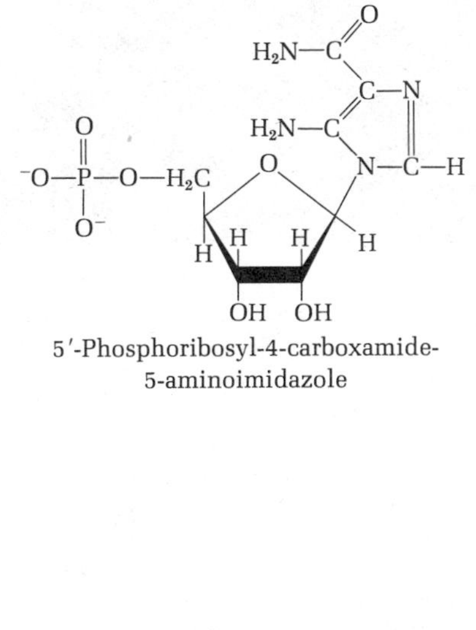

5′-Phosphoribosyl-4-carboxamide-5-aminoimidazole

8. *Treatment of Gout* Allopurinol (Figure 22-25), an inhibitor of xanthine oxidase, is used to treat chronic gout. Explain the biochemical basis for this treatment. Patients treated with allopurinol sometimes develop xanthine stones, although the incidence of kidney damage is much lower than with untreated gout. Explain this observation in light of the following solubilities in urine: uric acid (0.15 g/L), xanthine (0.05 g/L), and hypoxanthine (1.4 g/L).

9. *ATP Consumption by Root Nodules in Legumes* The bacteria residing in the root nodules of the pea plant consume over 20 percent of all the ATP produced by the pea plant. Suggest a reason why these bacteria consume so much ATP.

10. *Pathway of Carbon in Pyrimidine Biosynthesis* What are the locations of ^{14}C in orotate in cells grown on a small amount of uniformly labeled ^{14}C-succinate? Explain.

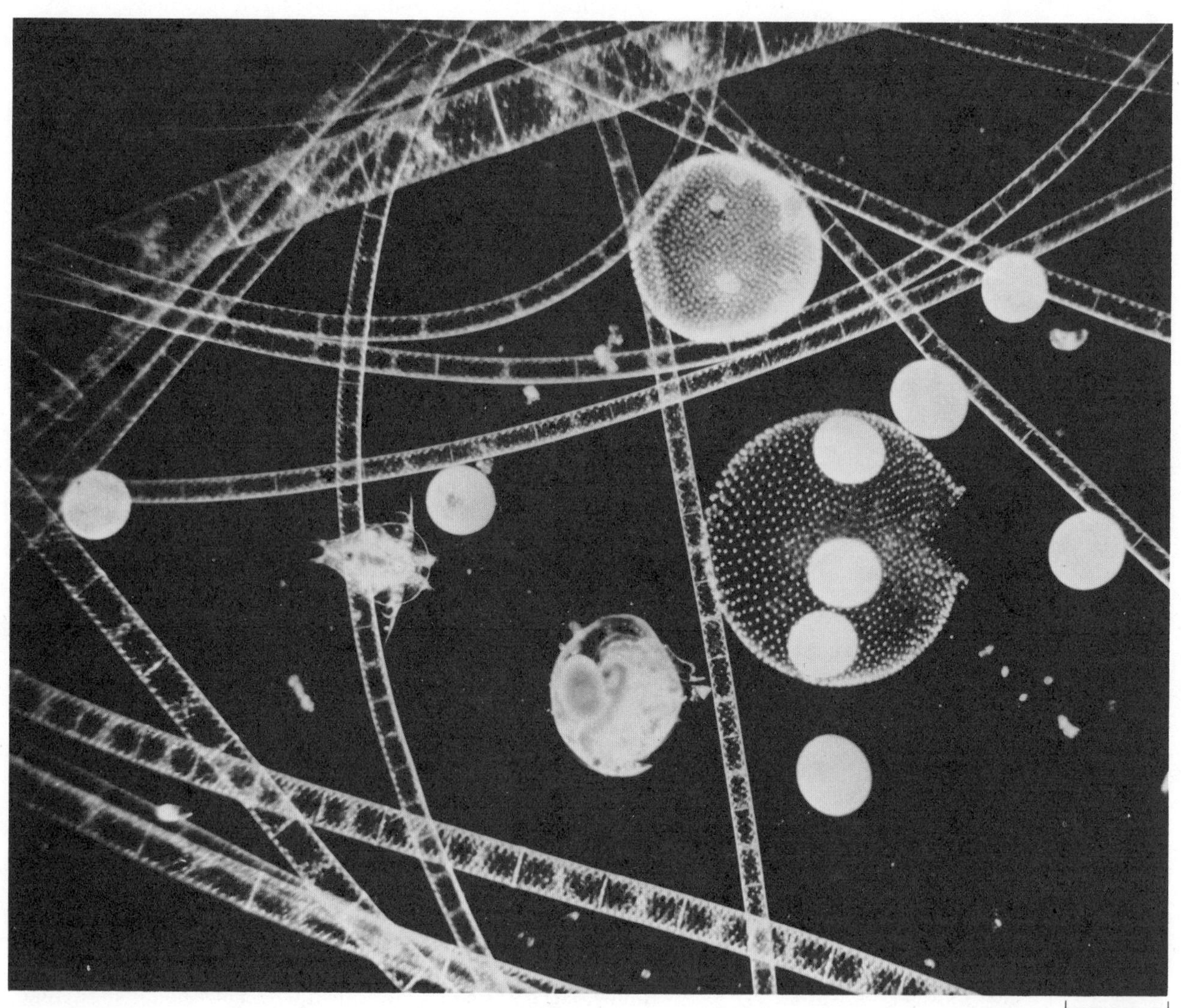

Two freshwater photosynthetic algae. The long filaments are Spirogyra, *strings of photosynthetic cells in which the chloroplasts are bandlike spirals. The large spherical structures are* Volvox, *each a colony of hundreds of cells. The large one at the right center has just burst, discharging daughter colonies. At left center is a very small crustacean, a copepod. Other photosynthetic cells are shown in Figure 23-2. Freshwater and marine organisms together account for more photosynthetic activity than terrestrial higher plants.*

CHAPTER 23

Photosynthesis

Now we come to the ultimate source of nearly all biological energy, the capture of solar energy by photosynthetic organisms and its conversion into the energy of biomass. Photosynthetic and heterotrophic organisms live in a balanced steady state in our biosphere (Figure 23-1). Photosynthetic plants trap solar energy in the form of ATP and NADPH, which they use as energy sources to make carbohydrates and other organic cell components from carbon dioxide and water; simultaneously, they release oxygen into the atmosphere. Aerobic heterotrophs, on the other hand, use the oxygen so formed to degrade the energy-rich organic products of photosynthesis to CO_2 and H_2O, in order to generate ATP for their own activities. The carbon dioxide formed by respiration in heterotrophs returns to the atmosphere, to be used again by photosynthetic organisms. Solar energy thus provides the driving force for the continuous cycling of atmospheric carbon dioxide and oxygen through our biosphere (Figure 23-1).

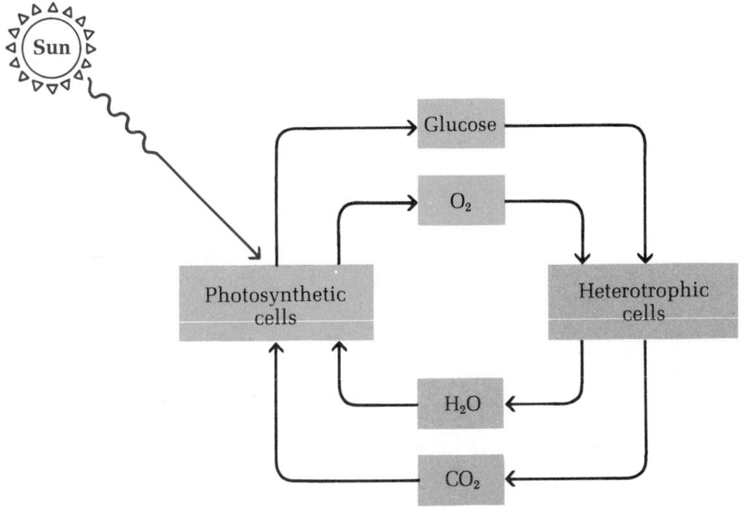

Figure 23-1
Solar energy is the ultimate source of all biological energy. Photosynthetic cells use the energy of sunlight to manufacture glucose and other organic cell products, which heterotrophic cells use as energy and carbon sources.

Enormous amounts of energy are stored as products of photosynthesis. Each year at least 10^{17} kcal of free energy is generated by the plant world at the expense of solar energy. This is over 10 times greater than all the fossil-fuel energy used per annum by mankind the world over. Even fossil fuels (coal, oil, and natural gas) are the products of photosynthesis that took place millions of years ago. Because of our global dependence upon solar energy, past and present, for both energy and food, the mechanism of photosynthesis poses the most fundamental of all biochemical problems.

How the Photosynthetic Equation Was Discovered

Joseph Priestley, one of the discoverers of oxygen, carried out some of the first important experiments on photosynthesis in 1770 to 1780. He found that a volume of air enclosed in a jar can be "depleted" by a burning candle so that it no longer supports combustion or the life of a mouse. Priestley then found that if a sprig of mint is then placed in the jar, the air is slowly "restored," so that it again allows a candle to burn and a mouse to live. He concluded that green plants evolve oxygen, a process that appeared to be the reverse of respiration in animals, which results in consumption of oxygen. It is quite astonishing that Priestley, having made these acute observations, did not realize that light was required for the "restoration" of air by the sprig of mint. The importance of light was discovered some years later by Jan Ingenhousz, a Dutch physician and gentleman-scientist, in experiments carried out in his private home laboratory. Ingenhousz further found that only the green part of a plant carries out the formation of oxygen in the light.

Later, in the early nineteenth century, the first quantitative measurements were made to account for the amounts of carbon dioxide assimilated, oxygen evolved, and the plant matter produced during photosynthesis. And in 1842 Robert Mayer, the discoverer of the first law of thermodynamics, the conservation of energy, published a paper in which he concluded that sunlight provides the energy for the formation of the photosynthetic products. Thus, in the mid-nineteenth century, it was recognized that the general equation of plant photosynthesis has the form

$$CO_2 + H_2O \xrightarrow{\text{light}} O_2 + \text{organic matter}$$

Photosynthetic Organisms Are Very Diverse

Photosynthesis occurs not only in the familiar green plants we see around us but also in lower eukaryotic organisms not visible to the naked eye, such as *algae*, *euglenids*, *dinoflagellates*, and *diatoms* (Figure 23-2). Moreover, photosynthesis also occurs in prokaryotic organisms. The photosynthetic prokaryotes include the *cyanobacteria*, the *green sulfur bacteria* found in mountain lakes, and the *purple sulfur bacteria* common in sulfur springs. The cyanobacteria or blue-green

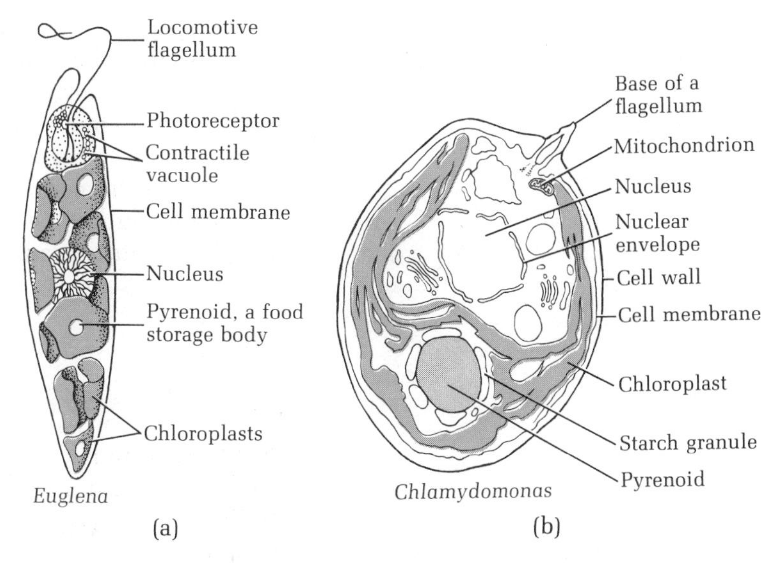

Figure 23-2
Photosynthetic organisms of plankton. (a) and (b) Two representative freshwater organisms, Euglena *and* Chlamydomonas, *which have flagella and orient themselves toward light. (c) to (e) Three diatom species. Diatoms are very abundant in ocean plankton. Their cell walls are extremely intricate double shells, distinctive for each species. The shells are largely silica* (SiO_2)*. Microscopists once used their empty shells (e) to test the resolving power of lenses. (f) and (g) Dinoflagellates, also typical photosynthetic organisms of seawater plankton. They possess flagella and often have bizarre shapes, given by their outer cellulose walls. Some species are bioluminescent, and others are toxic to fish and people. "Red tides" off the coast of the United States are caused by immense growths of red dinoflagellates, which secrete a powerful nerve toxin.*

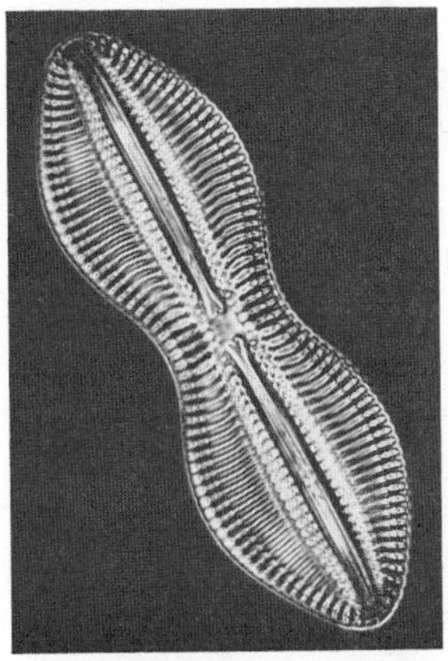

(c) 50 μm

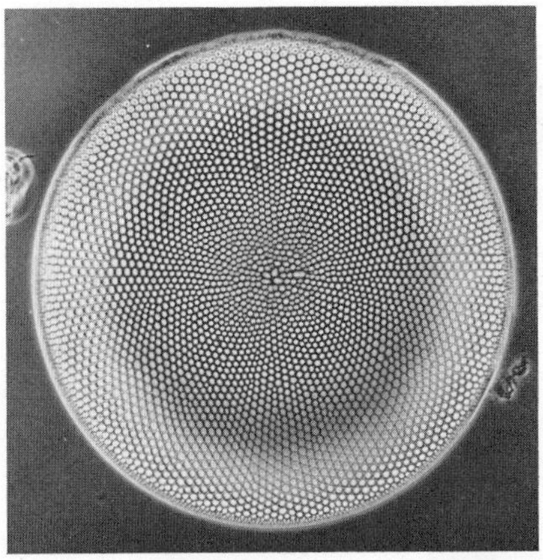

(d) 10 mm

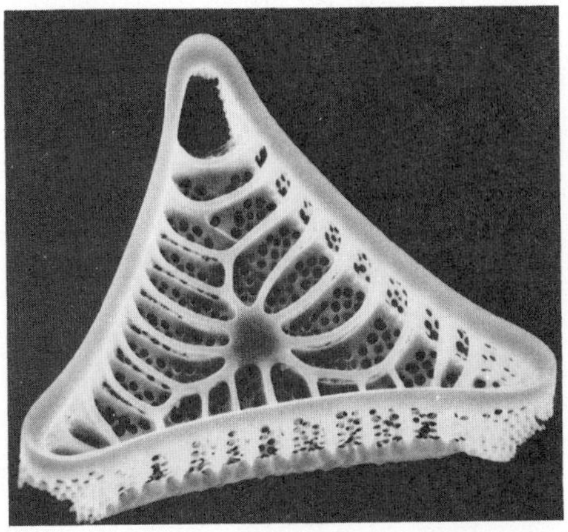

(e) 30 μm

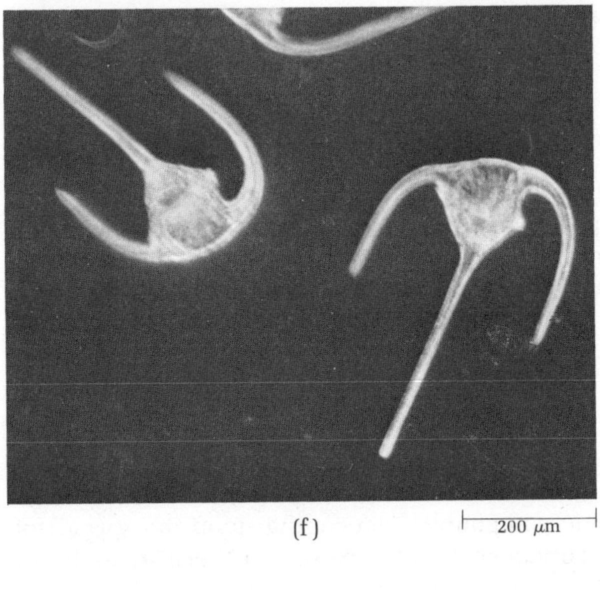

(f) 200 μm

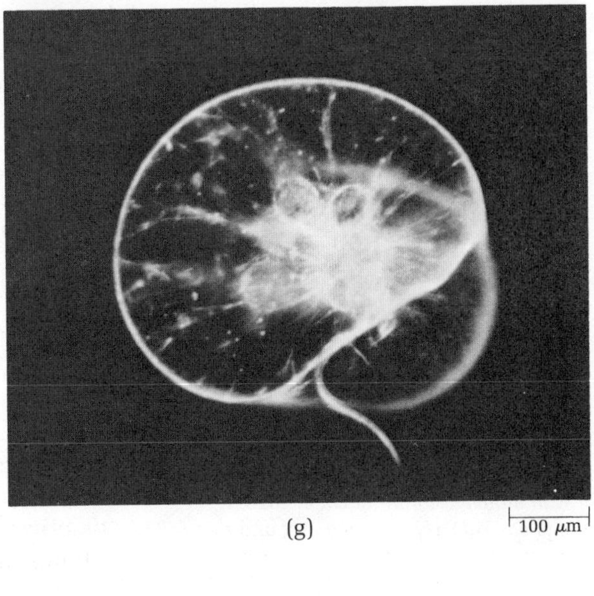

(g) 100 μm

algae (page 18), found in both fresh and salt waters, are perhaps the most versatile of photosynthetic organisms. Because they can also fix atmospheric nitrogen (page 637), cyanobacteria are among the most self-sufficient organisms in our biosphere. Altogether, at least half of all the photosynthetic activity on the earth takes place in oceans, rivers, and lakes, brought about by the many different microorganisms that constitute the phytoplankton.

Photosynthetic Organisms Rely on Different Hydrogen Donors

Photosynthetic organisms can be divided into two classes: those which produce oxygen and those which do not. Green leaf cells of higher plants are oxygen producers. They use water as the hydrogen donor for the reduction of carbon dioxide and in this process yield molecular oxygen, according to the overall equation

$$n\mathrm{H_2O} + n\mathrm{CO_2} \xrightarrow{\text{light}} (\mathrm{CH_2O})_n + n\mathrm{O_2} \qquad (1)$$

in which n is often assigned the value 6 to correspond with formation of glucose $[(CH_2O)_6 = C_6H_{12}O_6]$ as the end product of CO_2 reduction.

With the exception of the cyanobacteria, which have an oxygen-producing photosynthetic system resembling that of green plants, photosynthetic bacteria do not produce oxygen. In fact, many of them are strict anaerobes and cannot tolerate oxygen. As hydrogen donors some photosynthetic bacteria use inorganic compounds. For example, the green sulfur bacteria use hydrogen sulfide as hydrogen donor according to the equation

$$2\mathrm{H_2S} + \mathrm{CO_2} \xrightarrow{\text{light}} (\mathrm{CH_2O}) + \mathrm{H_2O} + 2\mathrm{S}$$

These bacteria, instead of giving off molecular O_2, extrude elemental sulfur as the oxidation product of H_2S. Other photosynthetic bacteria use organic compounds as hydrogen donors, for example, lactate:

$$2\ \text{Lactate} + \mathrm{CO_2} \xrightarrow{\text{light}} (\mathrm{CH_2O}) + \mathrm{H_2O} + 2\ \text{pyruvate}$$

Cornelis van Niel, a pioneer in the study of comparative metabolism, postulated that plant and bacterial photosynthesis are fundamentally similar processes despite the differences in the hydrogen donors they employ. This similarity is obvious if the equation of photosynthesis is written in a more general form,

$$2\mathrm{H_2D} + \mathrm{CO_2} \xrightarrow{\text{light}} (\mathrm{CH_2O}) + \mathrm{H_2O} + 2\mathrm{D}$$

in which H_2D symbolizes a hydrogen donor and D is its oxidized form. H_2D thus may be water, hydrogen sulfide, lactate, or other organic compounds, depending upon the species of photosynthetic organism (Figure 23-3). Van Niel also predicted

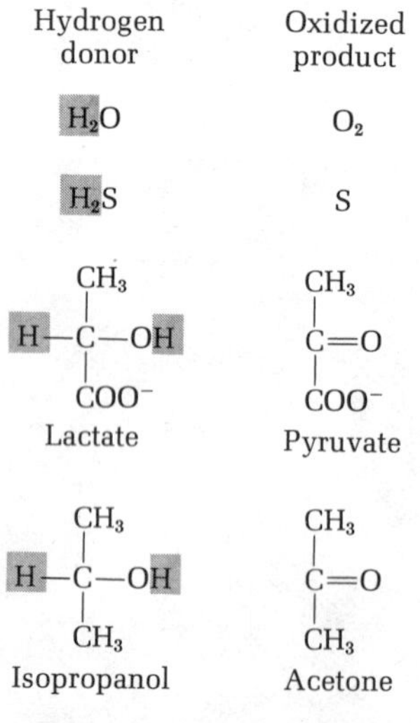

Hydrogen donor	Oxidized product
H_2O	O_2
H_2S	S
CH_3–CH(OH)–COO^- (H–C–OH) Lactate	CH_3–C(=O)–COO^- Pyruvate
CH_3–CH(OH)–CH_3 (H–C–OH) Isopropanol	CH_3–C(=O)–CH_3 Acetone

Figure 23-3
Some hydrogen (electron) donors for different photosynthetic organisms. The hydrogen atoms used are in color. Green plants use H_2O as electron donor and from it evolve O_2.

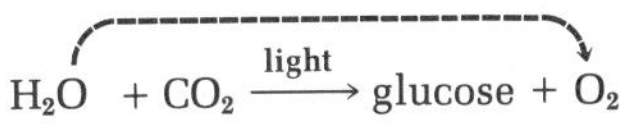

Figure 23-4
The source of the O_2 evolved in plant photosynthesis is H_2O.

that the molecular oxygen formed during plant photosynthesis is derived exclusively from water and not from the carbon dioxide. Isotope tracer experiments with ^{18}O-labeled water or carbon dioxide supported his prediction (Figure 23-4). This chapter will deal largely with oxygen-evolving photosynthesis of higher plants.

Photosynthesis Has Light and Dark Phases

Photosynthesis in green plants takes place in two stages, the *light reactions,* which occur only when plants are illuminated, and the *dark reactions,* which can occur in the absence or presence of light. In the light reactions chlorophyll and other pigments of the photosynthetic cells absorb light energy and conserve it in chemical form as the two energy-rich products ATP and NADPH; simultaneously, oxygen is evolved. In the dark reactions, the ATP and NADPH generated in the light reactions are used to reduce carbon dioxide to form glucose and other organic products (Figure 23-5). The formation of oxygen, which occurs only in the light, and the reduction of carbon dioxide, which does not require light, thus are distinct and separate processes.

But here we must introduce an important point, to be discussed later. Although the reactions involved in the reduction of CO_2 to glucose can proceed in the dark, they are *regulated* by light.

Figure 23-5
The light reactions generate energy-rich NADPH and ATP at the expense of solar energy. They are used in the dark reactions to reduce CO_2 to form glucose.

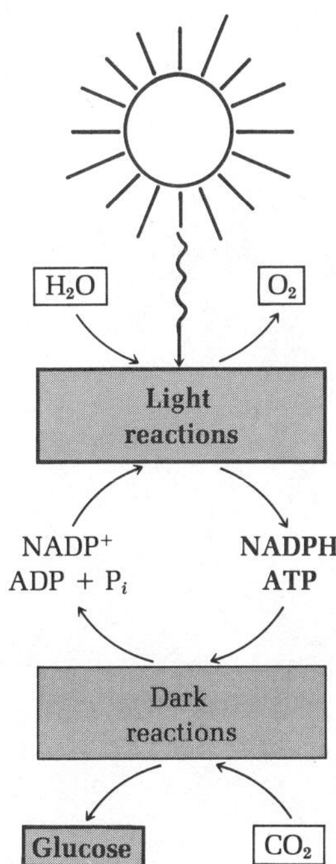

Plant Photosynthesis Takes Place in Chloroplasts

In eukaryotic photosynthetic cells both the light and dark reactions take place in the *chloroplasts,* which we may regard as the major power plants of such cells. We must recall, however, that green leaf cells also contain mitochondria (page 35). At night, when solar energy is not available, mitochondria generate ATP for the cell's needs by using oxygen to oxidize carbohydrate generated in chloroplasts during daytime hours.

Chloroplasts may assume many different shapes in different species, but they are usually much larger in volume than mitochondria (Figure 23-6). They are surrounded by a continuous outer membrane, which is rather fragile. An inner membrane system encloses the internal compartment. Inside the latter, and often connected to the inner membrane, are many flattened, membrane-surrounded vesicles or sacs, called *thylakoids* ("baggy trousers"), which are usually arranged in stacks called *grana* (Figure 23-6). The thylakoid membranes contain all the photosynthetic pigments of the chloroplast and all the

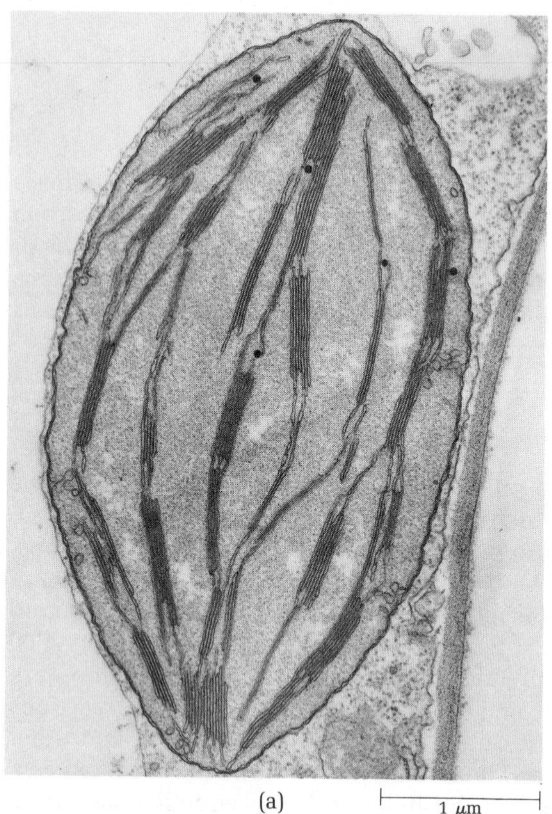

(a)

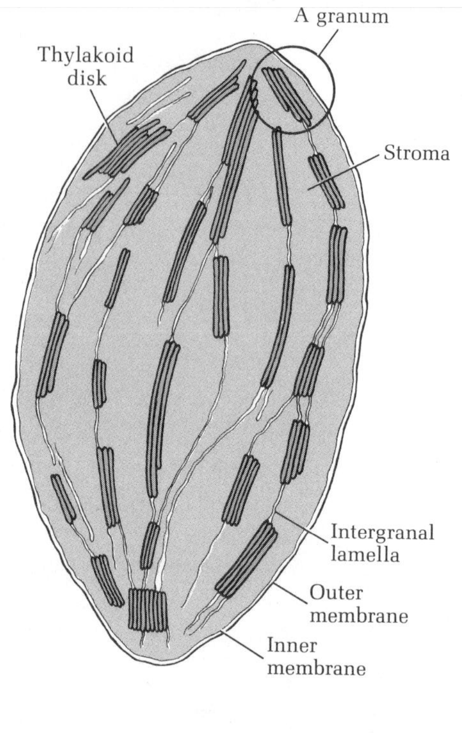

(b)

Figure 23-6
(a) An electron micrograph of a chloroplast of a spinach leaf. (b) A schematic diagram of its structure. (c) A freeze-fracture electron micrograph of the internal surface of a thylakoid membrane. The pyramidal particles are thought to be molecules of enzymes participating in photosynthesis.

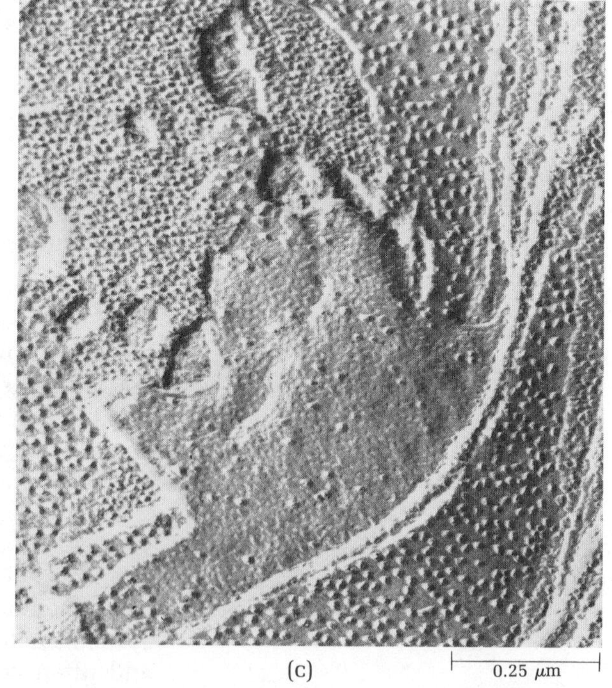

(c)

enzymes required for the primary light-dependent reactions. The fluid in the compartment surrounding the thylakoid vesicles, the *stroma*, contains most of the enzymes required for the dark reactions, in which CO_2 is reduced to form glucose. In many species the dark reactions also take place in the cell cytosol. Chloroplasts are readily isolated from extracts of ground spinach leaves by differential centrifugation (page 354).

Absorption of Light Excites Molecules

Visible light is electromagnetic radiation of wavelength 400 to 700 nm. Solar light arises from the nuclear fusion of hydrogen atoms to form helium atoms and electrons, a process that is promoted by the enormously high temperatures in the interior of the sun. The overall reaction is

$$4H \longrightarrow {}^{4}He + 2e^{-} + h\nu$$

where $h\nu$ represents a *quantum* of light energy, also called a *photon*. Remember that light has both wavelike and corpuscular characteristics. The energy carried by photons is inversely proportional to the wavelength of the light (Table 23-1). Photons of short wavelength, those at the violet end of the visible spectrum, carry the greatest amount of energy.

Table 23-1 Energy Content of Photons

Wavelength, nm	Color	kcal/einstein†
400	Violet	71.8
500	Blue	57.7
600	Yellow	47.8
700	Red	40.6

† An einstein (a "mole" of photons) contains 6.023×10^{23} photons.

The ability of a chemical compound to absorb light depends upon the arrangement of electrons around the atomic nuclei in its structure. When a photon is absorbed by a molecule, an electron is boosted to a higher energy level. This happens on an all-or-none basis: to boost an electron, a photon must have a certain minimum quantity of energy, hence the term *quantum*. A molecule that has absorbed a photon is in an energy-rich *excited state*, which is usually very unstable. When the light source is turned off, the high-energy electrons usually return rapidly to their normal low-energy orbitals as the excited molecule reverts to its original stable condition, called the *ground state*. During the return to the ground state, the packet of energy absorbed when the molecule was excited, called an *exciton*, is lost again, either as light or as heat. The light emitted on decay of excited molecules is called *fluorescence* (Figure 23-7). Excitation of molecules by light and their fluorescent decay are extremely fast processes. Only a few picoseconds (1 ps = 10^{-12} s) are required to excite a chlorophyll molecule in the test tube. Moreover, excited molecules remain excited for only a very short time: it has been calculated that a Concorde jet plane at top speed travels only 6 μm in the time a chlorophyll molecule remains excited.

Now we come to a very fundamental point. When isolated chlorophyll molecules in the test tube are excited by light, the absorbed energy is quickly released as fluorescence and heat. But when chlorophyll in intact spinach leaves is excited by vis-

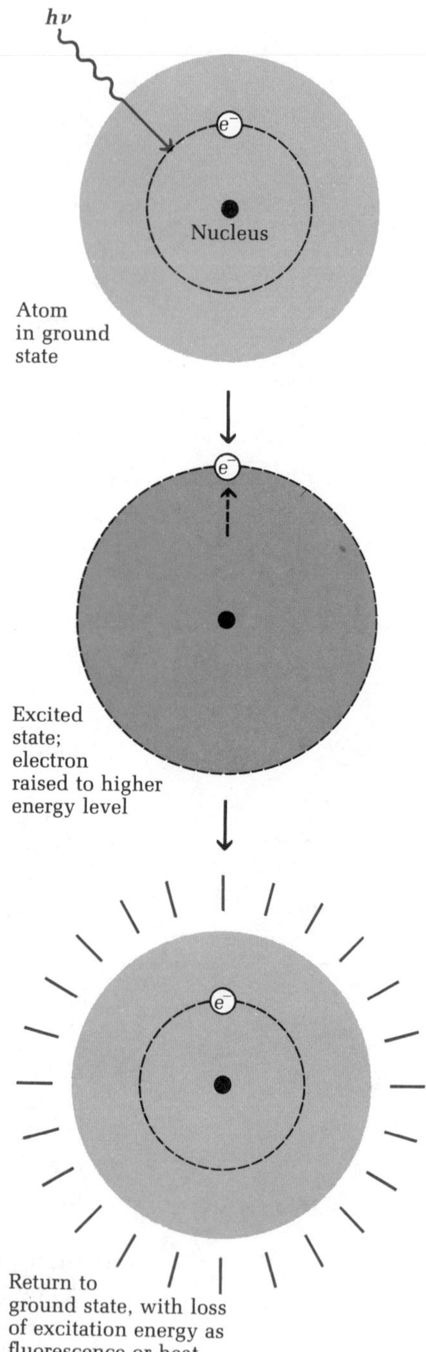

Figure 23-7
Excitation of an atom by absorption of light energy. The return of the atom to its normal ground state results in loss of the absorbed light energy as fluorescence or heat. When photosynthetic cells are excited by light, however, the absorbed energy does not appear as fluorescence but is conserved to generate NADPH and ATP.

ible light, no fluorescence is observed. Instead, the high-energy electrons are expelled from the excited chlorophyll molecule and "jump" to the first member of a chain of electron carriers. Coupled to such light-dependent electron flow along this chain are processes that generate ATP and NADPH.

Chlorophylls Are the Major Light-Absorbing Pigments

Now let us examine the light-absorbing pigments in the thylakoid membranes. First in importance are the green chlorophylls, Mg^{2+} complexes of molecules resembling the protoporphyrin of hemoglobin (page 171). Chlorophyll *a*, present in the chloroplasts of all green plant cells, contains four substituted pyrrole rings, one of which (ring IV) is reduced (Figure 23-8). Chlorophyll *a* also has a fifth ring, which is not a pyrrole. This characteristic five-ring porphyrin derivative is called *pheoporphyrin* (*pheo-*, "a plant"). Chlorophyll *a* also has a long isoprenoid side chain, consisting of the alcohol *phytol* esterified to a carboxyl-group substituent in ring IV (Figure 23-8). The four central nitrogen atoms of chlorophyll *a* are coordinated with Mg^{2+}.

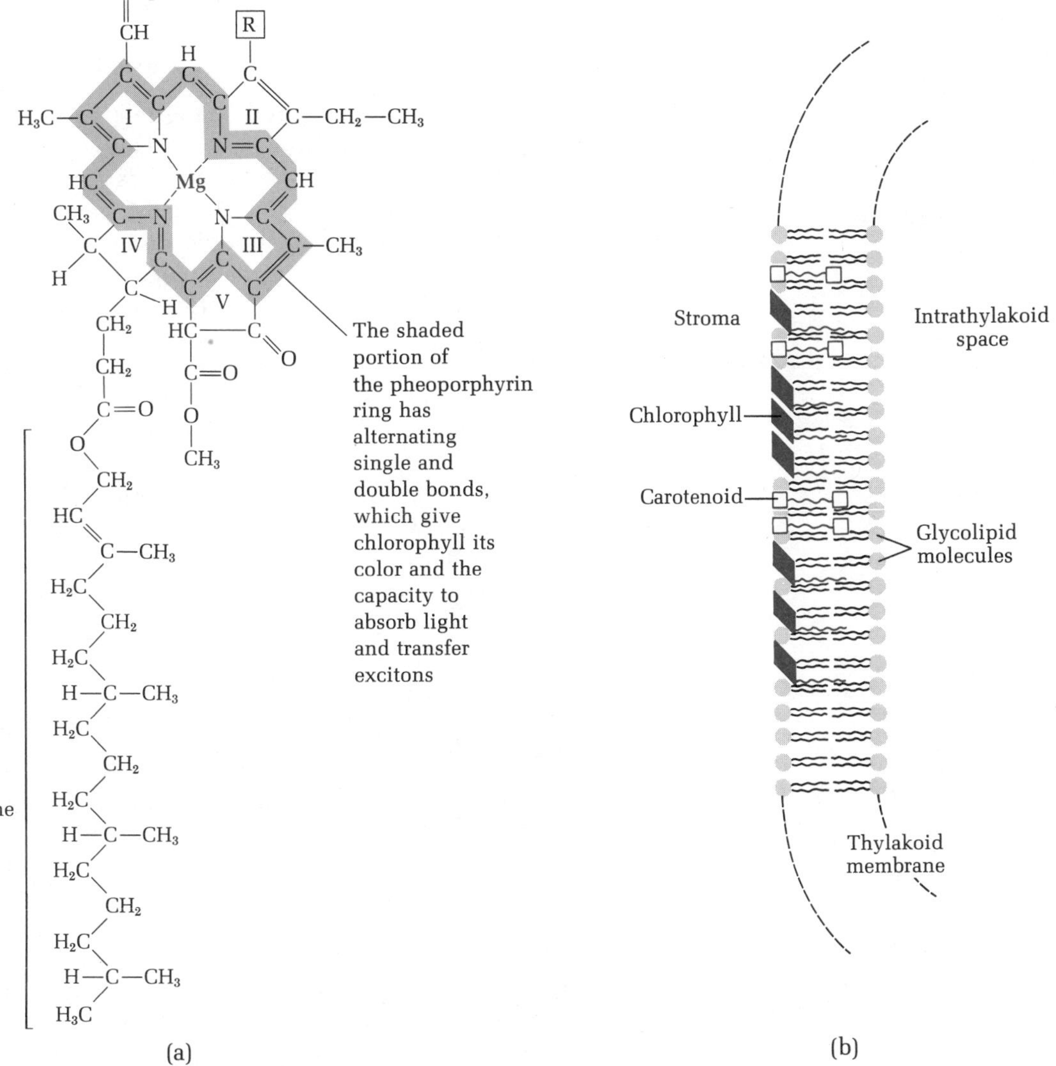

Figure 23-8
(a) *Chlorophyll* a *and* b. *Ring V is the extra ring, not found in protoporphyrin.* R = —CH_3 *in chlorophyll* a; R = CHO *in chlorophyll* b. (b) *Arrangement of light-absorbing chlorophyll and carotenoid pigments in the thylakoid membrane. They are oriented and grouped into patches or photosystems. Chlorophyll molecules are bound to specific proteins in the thylakoid membrane. For simplicity they are not shown in the illustration.*

Photosynthetic cells of higher plants always contain two types of chlorophyll. One is always chlorophyll *a*, and the second in many species is *chlorophyll b*, which has an aldehyde group instead of a methyl group attached to ring II (Figure 23-8). Pure chlorophyll *a* and *b* can be isolated from leaf extracts by chromatographic procedures. Although both are green, their absorption spectra are slightly different. Most higher plants contain about twice as much chlorophyll *a* as chlorophyll *b*.

Figure 23-8 also shows how the structure of chlorophyll *a* adapts it for its biological activity. The deeply colored five-ring system, which itself forms a larger ring around the Mg, endows the molecule with light-absorbing power; the Mg promotes formation of chlorophyll aggregates, which facilitate light trapping; and the long hydrophobic side chain not only anchors but also orients the chlorophyll molecule in the lipid bilayer of the membrane.

Figure 23-9
β-Carotene, an accessory pigment in green leaves. Many other carotenoids serve as accessory pigments in different species. Note that β-carotene, like chlorophyll, contains many conjugated double bonds, which enable the molecule to absorb light and transmit excitons.

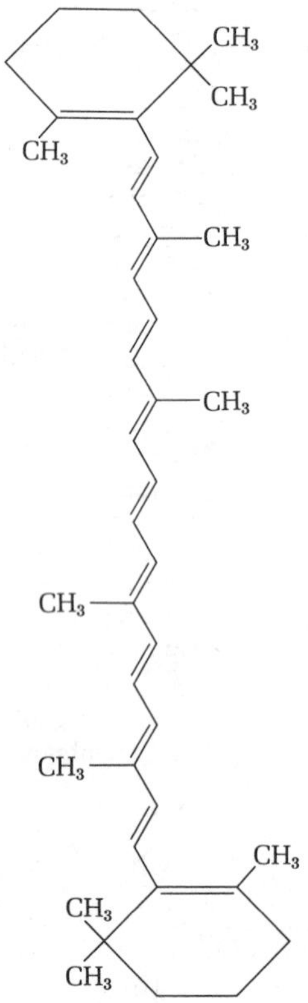

CH_3 CH_3 CH_3 CH_3 CH_3 CH_3 CH_3 CH_3 CH_3 CH_3

Thylakoids Also Contain Accessory Pigments

In addition to chlorophylls, the thylakoid membranes contain secondary light-absorbing pigments, together called the *accessory pigments*. They include various *carotenoids*, which may be yellow, red, or purple. The most important are *β-carotene* (Figure 23-9), a red isoprenoid compound which is the precursor of vitamin A in animals (page 265), and the yellow carotenoid *xanthophyll*. The carotenoid pigments absorb light at wavelengths other than those absorbed by the chlorophylls and thus are supplementary light receptors. The relative amounts of the chlorophylls and the different carotenoids vary characteristically from one species of plant to another. Indeed, variations in the proportions of these pigments are responsible for the characteristic colors of photosynthetic cells, which vary from a deep blue-green, as in spruce needles, to a greener green as in maple leaves, to the red, brown, or even purple color of different species of multicellular algae and the leaves of some decorative plants.

The Thylakoid Membrane Contains Two Kinds of Photochemical Reaction Systems

The light-absorbing pigments of thylakoid membranes are arranged in functional sets or clusters, as is shown in Figure 23-8b. In spinach chloroplasts these clusters, called *photosystems*, contain about 200 molecules of chlorophylls and about 50 molecules of carotenoids. The clusters can absorb light over the entire visible spectrum but especially well between 400 to 500 and 600 to 700 nm (Figure 23-10). All the pigment molecules in a photosystem can absorb photons, but there is only *one* molecule in each cluster that can actually con-

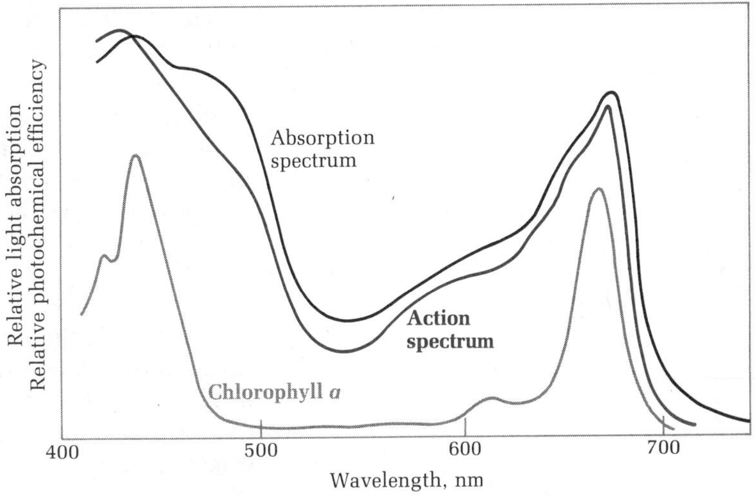

Figure 23-10
The absorption and photochemical action spectra of a green leaf. The absorption spectrum indicates the amount of light absorbed by the leaf as a function of wavelength. The photochemical action spectrum indicates the efficiency of light of different wavelengths in promoting photosynthesis. Essentially all the absorbed visible light promotes photosynthesis, but the regions 400 to 500 and 600 to 700 nm are most effective. Also shown for comparison is the spectrum of pure chlorophyll a, *which absorbs relatively weakly between 500 and 600 nm. Some photosynthetic cells contain accessory pigments effective in absorbing light in this region, which thus supplement the chlorophylls.*

vert the light energy into chemical energy. This specialized energy-converting pigment molecule consists of a chlorophyll molecule combined with a specific protein and is called the *photochemical reaction center.* All the other pigment molecules in such a photosystem or cluster are called *light-harvesting* or *antenna molecules.* They function to absorb light energy, which they transmit at a very high rate to the single reaction center where the photochemical act occurs (Figure 23-11).

The thylakoid membranes of plant chloroplasts have two different kinds of photosystems, each with its own set of light-harvesting chlorophyll and carotenoid molecules and photochemical reaction center. *Photosystem I,* which is maximally excited by light at longer wavelengths, has a high ratio of chlorophyll *a* to chlorophyll *b*. *Photosystem II,* which is maximally activated by light below 680 nm, contains relatively more chlorophyll *b* and may also contain chlorophyll *c*. The thylakoid membranes of a single spinach chloroplast have many hundreds of each kind of photosystem. These two photosystems have different functions, as we shall see. However, we can make an important generalization at this point. *All oxygen-evolving photosynthetic cells, i.e., those of higher plants and the cyanobacteria, contain both photosystems I and II, whereas all other species of photosynthetic bacteria, which do not evolve oxygen, contain only photosystem I.*

Figure 23-11
Schematic diagram of the surface of a photosystem in the thylakoid membrane. It contains a patchlike mosaic of several hundred chlorophyll and carotenoid antenna molecules oriented in the membrane. An exciton absorbed by an antenna molecule quickly migrates via the pigment molecules to the reaction center, P700; its path is shown by the colored arrows. Although all the antenna molecules can absorb light, only the reaction-center molecule can convert the exciton energy into electron flow.

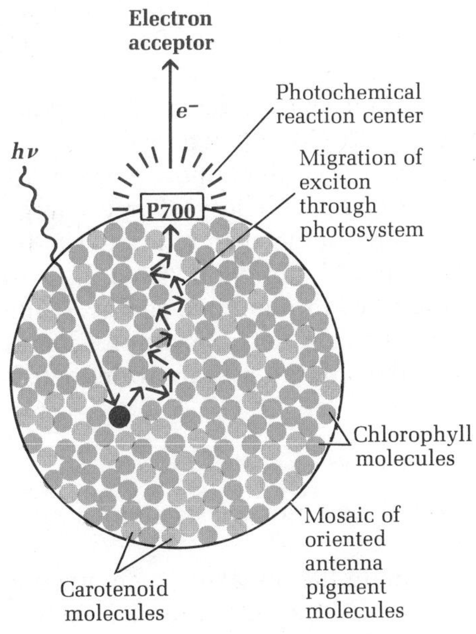

Electron Flow Is Induced by Illumination of Chloroplasts

Exactly how does the absorption of light by the pigment molecules of the thylakoid membranes bring about a chemical change resulting in the conversion of light energy into chemical energy?

The key to this question came from a discovery made in 1937 at the University of Cambridge by Robert Hill, a pioneer in photosynthesis research. He found that when leaf extracts containing chloroplasts were supplemented with a nonbiological hydrogen-acceptor molecule and then illuminated, evolution of oxygen and simultaneous reduction of the hydrogen acceptor took place, according to the equation

$$2H_2O + 2A \xrightarrow{\text{light}} 2AH_2 + O_2 \qquad (2)$$

in which A is the artificial hydrogen acceptor and AH_2 its reduced form. One of the nonbiological hydrogen acceptors Hill used was the dye *2,6-dichlorophenolindophenol,* which in its oxidized form (A) is blue and in its reduced form (AH_2) is colorless. When the leaf extract supplemented with the dye was illuminated, the blue dye became colorless and oxygen was evolved. In the dark neither oxygen evolution nor dye reduction took place. This was the first specific clue to how absorbed light energy is converted into chemical energy: it causes electrons to flow from H_2O to an electron-acceptor molecule. Moreover, Hill found that carbon dioxide was not required for this reaction, nor was it reduced to a stable form under these conditions. He therefore concluded that oxygen evolution can be dissociated from carbon dioxide reduction. The reaction summarized in equation (2) is known as the *Hill reaction* and the artificial acceptor A as a *Hill reagent*.

A search then began to identify the naturally occurring, biologically active counterpart of the Hill reagent: the electron acceptor, present in chloroplasts, that normally accepts hydrogen atoms from water during illumination. Several years later it was found that the coenzyme $NADP^+$ is the natural biological electron acceptor in chloroplasts, according to the equation

$$2H_2O + 2NADP^+ \xrightarrow{\text{light}} 2NADPH + 2H^+ + O_2 \qquad (3)$$

We must note a very important characteristic of this reaction: electrons flow from water to $NADP^+$, whereas in mitochondrial respiration electrons flow in the opposite direction, from NADH or NADPH to oxygen, with loss of free energy (page 475). Because electron flow induced in chloroplasts by light is in the reverse or "uphill" direction, from H_2O to $NADP^+$, it cannot occur without the input of free energy. Quite simply, the energy required to push electrons uphill comes from the light absorbed when the chloroplasts are illuminated.

Captured Light Energy Causes Uphill Flow of Electrons

How can light energy captured by chloroplasts induce electrons to flow "uphill"?

When a chlorophyll molecule in the thylakoid membrane is excited by light, the energy level of an electron in its structure is "boosted" by an amount equivalent to the energy of the absorbed light and the chlorophyll becomes excited. The packet of excitation energy (the exciton) now migrates rapidly through the cluster of light-harvesting pigment molecules to the reaction center of the photosystem, where it causes an electron to acquire a large amount of energy. This "hot" electron is expelled from the reaction center and is accepted by the first member of a chain of electron carriers. As a result, the first electron carrier of this chain becomes reduced, i.e., gains an electron, whereas the reaction center has become oxidized, i.e., has lost an electron. The reaction center in this oxidized condition is said to have an *electron hole*. The energy-rich electron, which has a very high reducing "pressure," now passes from the first electron acceptor downhill along a chain of electron-carrier molecules to $NADP^+$, reducing it to NADPH (Figure 23-12). Clearly the standard reduction potential of the photochemical reaction center must have a very negative value in order to pass electrons *downhill* to $NADP^+$, since the $NADP^+$–NADPH redox couple has a rather negative standard potential, −0.32 V (page 474).

Two questions now arise. How is the electron hole left in the reaction center filled again? And how can we explain the formation of O_2 from water? To answer these questions we must look at the overall scheme of photosynthetic electron flow.

Figure 23-12
The cooperation of photosystems I and II. This zigzag scheme (the Z scheme) shows the pathway of electron flow from H_2O *(lower right) to* $NADP^+$ *(upper left) in noncyclic plant photosynthesis. It also shows the energy relationships. To raise the energy of electrons derived from water to the energy level required to reduce* $NADP^+$ *to NADPH, each electron must be "boosted" twice (heavy colored arrows) by photons absorbed in photosystems I and II. One quantum or photon is required per electron boosted in each photosystem. After each boosting step, the high-energy electrons flow "downhill" (heavy gray arrows) via the routes shown. Photophosphorylation of ADP to yield ATP is coupled to electron flow in the central or connecting electron-transport chain leading from photosystem II to photosystem I (see page 660). The dashed black arrow from P430 to cytochrome b is the alternative or shunt pathway taken by electrons in cyclic electron flow and phosphorylation (see text and Figure 23-14). In cyclic electron flow only photosystem I is involved; electrons return via the shunt pathway to photosystem I, instead of reducing* $NADP^+$ *to NADPH.*

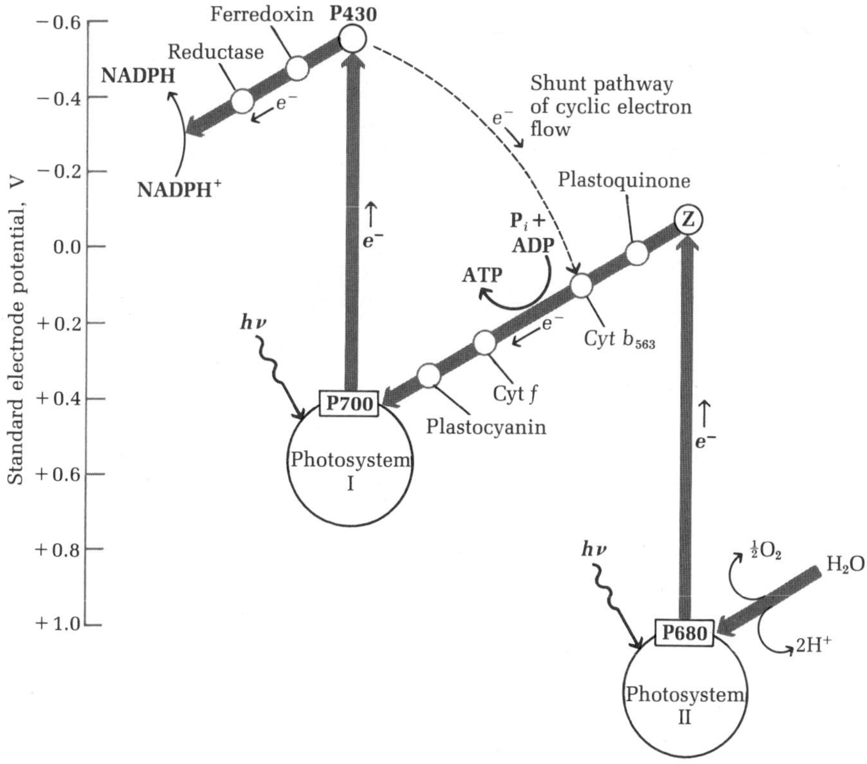

Photosystems I and II Cooperate to Carry Electrons from H_2O to $NADP^+$

The set of light-harvesting or antenna pigments and its reaction center that donates energy-rich electrons for the reduction of $NADP^+$ is *photosystem I*, which is maximally excited by light of wavelength 700 nm. It has been found, however, that chloroplasts must be illuminated not only at 700 nm but also at lower wavelengths, for example at 600 nm, for maximal rates of oxygen evolution. If chloroplasts are illuminated only at 600 nm and not at 700 nm, there is a large drop in oxygen evolution, called *red drop*, since 700 nm is at the red end of the spectrum. These observations have led to the conclusion that there are two photosystems absorbing light at different peak wavelengths and that they function together in the oxygen-evolving light reactions of plant photosynthesis. The diagram in Figure 23-12, often called the *Z scheme*, outlines the pathway of electron flow between the two photosystems, as well as the energy relationships in the light reactions.

Let us look first at the pathway of electron flow driven by light. When light quanta are absorbed by photosystem I, energy-rich electrons are expelled from the reaction center and flow down a chain of electron carriers to $NADP^+$ to reduce it to NADPH. This process leaves an empty electron hole in photosystem I. This hole is in turn filled by an electron expelled by illumination of photosystem II, which arrives via a connecting chain of electron carriers. But this leaves an electron hole in photosystem II. This hole is filled in turn by electrons arising from H_2O. The water molecule is split to yield (1) electrons, which are donated to the holes in photosystem II; (2) H^+ ions, which are released to the medium; and (3) molecular oxygen, which is released into the gas phase. The equation for water cleavage is

$$2H_2O \longrightarrow 4H^+ + 4e^- + O_2$$

The Z scheme thus describes the complete route by which electrons flow from H_2O to $NADP^+$ according to the equation

$$2H_2O + 2NADP^+ \xrightarrow{\text{light}} O_2 + 2NADPH + 2H^+$$

For each electron flowing from H_2O to $NADP^+$, two light quanta are absorbed, one by each photosystem. To form one molecule of O_2, which requires flow of four electrons from two H_2O to two $NADP^+$, a total of eight quanta must be absorbed, four by each photosystem.

The Z Scheme Shows the Energy Profile of Photosynthetic Electron Transport

Figure 23-12 shows not only the pathway of electrons from H_2O to $NADP^+$ but also the energy relationships involved. The vertical axis in this diagram is in energy units. The electron-transfer processes occurring in the upward direction (colored arrows) require input of light energy, and processes in which electrons are shown moving downward (gray arrows) proceed with a decrease in free energy. Absorption of a light quantum by photosystem I boosts an electron from a relatively energy-poor state to an energy-rich state, causing the excited photochemical reaction center I to become an extremely potent reducing agent, so that electrons flow downhill from it to $NADP^+$, reducing it to NADPH. The high-energy electrons that flow downhill from photosystem II to photosystem I in turn acquired their energy from light quanta absorbed by photosystem II. To refill the electron hole left in photosystem II, which is now an extremely potent oxidizing agent or electron acceptor, electrons flow downhill from H_2O. The Z scheme thus describes the pathway by which electrons flow from H_2O, which has a very positive standard reduction potential (+0.82 V) to reduce $NADP^+$ to NADPH, which has a very negative standard reduction potential (−0.32 V). The free energy needed to move one electron from H_2O to $NADP^+$ is provided by the two light quanta absorbed, one by each of the two photosystems.

Several Electron Carriers Participate in Photosynthetic Electron Transport

When the reaction center of photosystem I, a complex of a chlorophyll *a* molecule with a specific protein, is excited by light quanta received from the antenna molecules, there is a decrease in the light absorption of the chloroplasts at 700 nm. For this reason the reaction center of photosystem I is usually designated *P700* (Figure 23-12). The first electron-carrier molecule in the chain from P700 to $NADP^+$ is believed to be an *iron-sulfur protein,* designated *P430*. The next electron carrier is *ferredoxin,* also an iron-sulfur protein but different from P430. Spinach ferredoxin, which has been isolated and crystallized, has a molecular weight of about 10,700 and contains two iron atoms bound to two acid-labile sulfur atoms. The iron atoms in P430 and ferredoxin transfer electrons via one-electron Fe(II)–Fe(III) valence changes.

The third electron carrier is a flavoprotein called *ferredoxin-NADP oxidoreductase*. It transfers electrons from reduced ferredoxin (Fd_{red}) to $NADP^+$, reducing the latter to NADPH:

$$2Fd_{red}^{2+} + 2H^+ + NADP^+ \longrightarrow 2Fd_{ox}^{3+} + NADPH + H^+$$

Next we have the *connecting chain* of electron carriers that leads electrons "downhill" from the excited reaction center of photosystem II to the empty holes in photosystem I (Figure 23-12). The oxidized reaction center of photosystem II absorbs at 680 nm (hence, P680). Very little is known of its chemical nature; presumably it resembles P700 of photosystem I in being a chlorophyll-protein complex. The first electron carrier in the

Figure 23-13
Plastoquinone A, the most abundant plastoquinone in plants and algae. Other plastoquinones differ in the length of the side chain and the nature of the substituents in the quinone ring. Plastoquinol is the reduced form of plastoquinone.

Plastoquinone A

Plastoquinol A

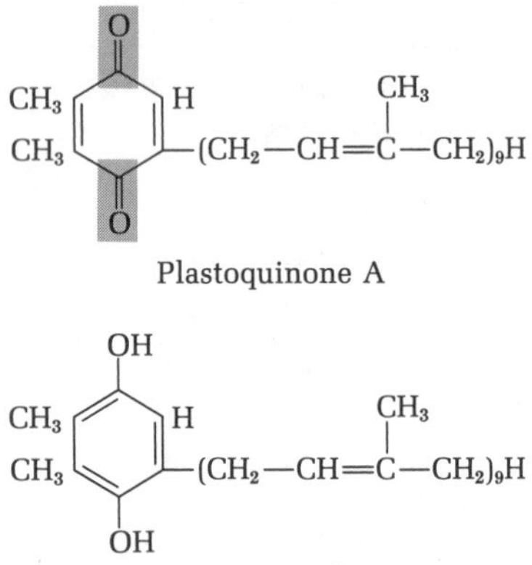

chain is not well understood either; it is usually designated Z. The reduced form of Z passes electrons downhill to plastoquinone or PQ (Figure 23-13), a fat-soluble quinone with a long isoprenoid side chain that resembles ubiquinone of the mitochondrial respiratory chain. The reduced form of plastoquinone then donates electrons in turn to a *b*-type cytochrome, called *cytochrome* b_{563}, which in turn passes electrons to cytochrome *f* (Latin *frons*, "leaf"). Cytochrome *f* resembles *cytochrome c* of mitochondria. Cytochrome *f* in turn passes electrons to *plastocyanin*, a blue copper protein. The copper atom of this protein is the actual carrier of the electrons, since it is capable of undergoing Cu(I)–Cu(II) cycles. Plastocyanin is the immediate donor of electrons to the empty holes in P700 of photosystem I.

The electron holes left in the reaction center P680 of photosystem II are refilled by electrons removed from H_2O by a little-understood Mn^{2+}-containing enzyme complex called *H_2O dehydrogenase*.

Now we must make a very important point. We speak of the entire set of reactions in Figure 23-12 as the light reactions of photosynthesis. This definition is useful because it clearly separates the energy-producing phase of photosynthesis from the dark reactions, in which CO_2 is reduced to form glucose. However, the term "light reactions" is not entirely accurate. Actually, the only points in the "light reactions" that really require light are the two steps in which the two photochemical reaction centers are excited (Figure 23-12). All the other steps in photosynthetic electron-transport actually can occur in the dark, once the electrons have been boosted by absorption of light energy. We thus must use the term "light reactions" with full understanding of the nature of the processes involved.

Phosphorylation of ADP Is Coupled to Photosynthetic Electron Transport

We have now seen how one of the two energy-rich products formed in the light reactions, namely, NADPH, is generated by photosynthetic electron transport from H_2O to $NADP^+$. What about the other energy-rich product, ATP?

In 1954 Daniel Arnon and his colleagues at the University of California in Berkeley discovered that ATP is generated from ADP and phosphate during photosynthetic electron transport in illuminated spinach chloroplasts. Simultaneously and independently, Albert Frankel at the University of Minnesota made a similar observation when he illuminated membranous pigment-containing structures called *chromatophores*, derived from photosynthetic bacteria. Both groups concluded that some of the light energy captured by the photosynthetic systems of these organisms is transformed into the phosphate-bond energy of ATP. This process is called *photosynthetic phosphorylation* or *photophosphorylation*, to distinguish it from oxidative phosphorylation in respiring mitochondria.

Recall (page 484) that oxidative phosphorylation of ADP to ATP in mitochondria occurs at the expense of the free energy released as high-energy electrons flow *downhill* along the electron-transport chain from substrates to oxygen. In a similar way, photophosphorylation of ADP to ATP is coupled to the energy released as high-energy electrons flow down the photosynthetic electron-transport chain from excited photosystem II to the electron holes in photosystem I. Most evidence suggests that one ATP is formed per pair of electrons passing down the connecting chain (Figure 23-12), but some researchers believe that two ATPs are formed.

Chloroplasts Also Promote Cyclic Electron Flow and Cyclic Photophosphorylation

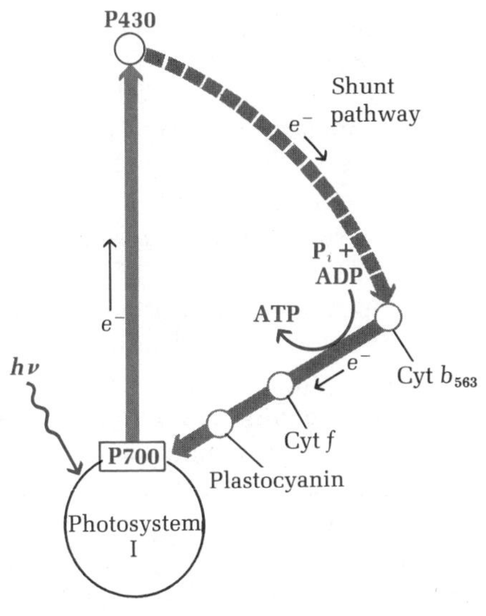

Figure 23-14
Pathway of electrons in cyclic photophosphorylation. Only photosystem I is involved. The absorbed light energy is used only for generating ATP. Compare with Figure 23-12.

Another type of light-induced electron flow that can take place in chloroplasts is called *cyclic electron flow,* to differentiate it from the normally unidirectional or *noncyclic* electron flow that proceeds from H_2O to $NADP^+$. Cyclic electron flow involves only photosystem I (Figures 23-12 and 23-14). It is called cyclic because the electron boosted to the first electron acceptor P430 by illumination of photosystem I, instead of passing to $NADP^+$, flows back into the electron hole of photosystem I by a shunt or bypass pathway. This shunt involves some of the electron carriers of the chain between photosystems I and II, including the segment that contains the phosphorylation step. Thus illumination of photosystem I can cause electrons to cycle continuously out of the reaction center of photosystem I and back into it, each electron being propelled around the cycle by the energy yielded by absorption of one light quantum. During cyclic electron flow there is no net formation of NADPH, nor is there any oxygen evolution. However, cyclic electron flow is accompanied by the phosphorylation of ADP to yield ATP, referred to as *cyclic photophosphorylation* (Figure 23-14). The overall reaction equation for cyclic electron flow and photophosphorylation is simply

$$P_i + ADP + \textbf{light energy} \longrightarrow ATP + H_2O$$

Cyclic electron flow and photophosphorylation are believed to occur when the plant cell is already amply supplied with reducing power in the form of NADPH but requires additional ATP for other metabolic needs. Little is known about the regulation of the cyclic pathway, however.

Photosynthetic Phosphorylation Resembles Oxidative Phosphorylation

Photosynthetic electron transport and photophosphorylation in chloroplasts have many similarities with electron transport and oxidative phosphorylation in mitochondria: (1) the reaction centers, electron carriers, and ATP-forming enzymes are

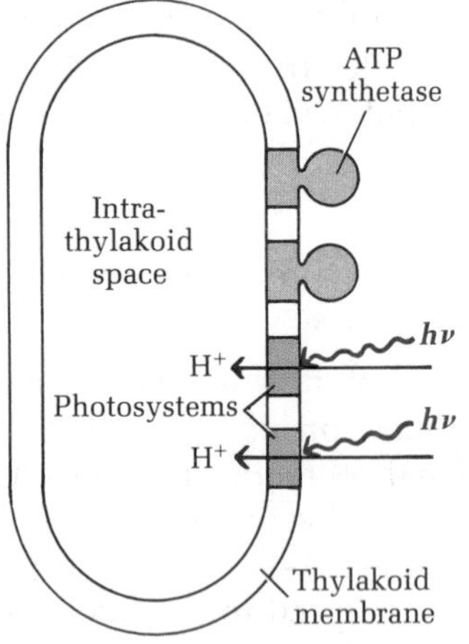

Figure 23-15
The thylakoid membrane has sidedness. The photosystems and the electron-transport chain are oriented in such a way that they pump H^+ into the thylakoids. The ATP synthetase "knobs" (CF_1) are on the outer surface.

located in the thylakoid membrane; (2) photophosphorylation requires intact thylakoid membranes; (3) the thylakoid membrane is impermeable to H^+; (4) photophosphorylation can be uncoupled from electron flow by reagents that promote the passage of H^+ through the thylakoid membrane; (5) photophosphorylation can be blocked by *oligomycin* and similar agents that inhibit the formation of ATP from ADP and phosphate by ATP synthetase of mitochondria (page 485); and (6) ATP is formed by knoblike enzyme molecules on the outer surface of the thylakoid membranes, which are very similar in structure and function to the F_1ATPase of mitochondria and thus are often designated CF_1 (C for chloroplast).

Like the inner mitochondrial membrane (page 470), the thylakoid membrane has an asymmetric molecular organization (Figure 23-15). The electron-transferring molecules in the connecting chain between photosystem II and photosystem I are oriented in the thylakoid membrane in such a way that electron flow results in the net movement of H^+ ions across the membrane, from the *outside* of the thylakoid membrane to the inner compartment. Thus photoinduced electron flow generates a gradient of H^+ ions across the thylakoid membrane, so that the inside of the thylakoid vesicles becomes more acid than the outside. All these properties are consistent with the chemiosmotic hypothesis originally proposed for oxidative phosphorylation and since applied to photosynthetic phosphorylation. The diagram in Figure 23-16 shows the flow of H^+, driven by light energy, from the stroma to the interior of the thylakoid and out again via the ATP synthetase.

In 1966 André Jagendorf performed an important experiment which proved that an alkaline-outside pH gradient across the thylakoid membrane can furnish the driving force to generate ATP. He soaked chloroplasts in the dark in a buffer at pH 4, which slowly penetrated into the inner compartment of the thylakoids, lowering their internal pH. He added ADP and phosphate to the dark suspension of chloroplasts and then suddenly raised the outer medium to pH 8 by adding an alkaline buffer, momentarily creating a large pH gradient across the membrane. As the pH gradient decreased, owing to the passage of H^+ out of the thylakoids into the medium, ATP was generated from ADP and phosphate. Since the formation of ATP occurred in the dark, this experiment showed that a pH gradient across the membrane is an energy-rich state that can act as the medium for transferring electron-transport energy to the ATP synthetase for generating ATP, as the chemiosmotic hypothesis proposes.

Figure 23-16
The photosynthetic H^+ cycle. Photosynthetic electron transport "pumps" H^+ into the thylakoids to form an acid-inside H^+ gradient. The return of H^+ ions through the oriented ATP synthetase molecule drives the energy-requiring synthesis of ATP from ADP and phosphate.

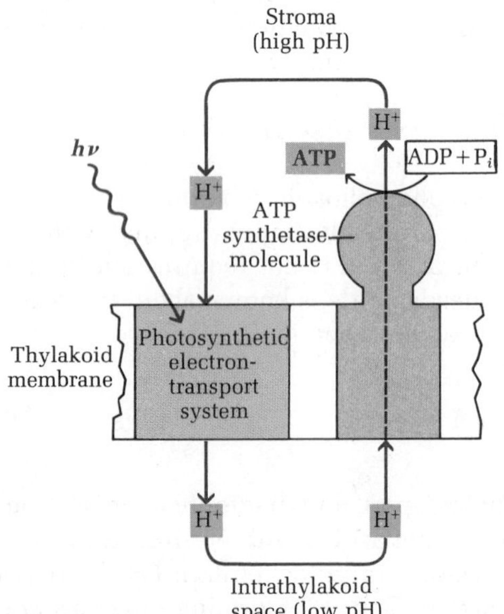

The Overall Equation of Plant Photosynthesis

The standard-free-energy change for the synthesis of glucose from CO_2 and H_2O by the reaction

$$6CO_2 + 6H_2O \longrightarrow C_6H_{12}O_6 + 6O_2$$

is +686 kcal/mol (recall that oxidation of glucose by the reverse of this equation proceeds by a *decrease* of 686 kcal/mol). Now let us compare this energy requirement with the energy yielded by the light reactions of plant photosynthesis. Recall that two light quanta must be absorbed, one by each photosystem, to cause flow of one electron from H_2O to $NADP^+$. To generate one molecule of O_2, four electrons must be transferred, which requires eight quanta. To evolve six molecules of O_2, as in the equation above, 6(8) = 48 light quanta must be absorbed and used. Since the energy of a "mole" of light quanta may range from 72 kcal at 400 nm to about 41 kcal at 700 nm (Table 23-1), anywhere from 48(41) = 1968 to 48(72) = 3456 kcal, depending upon the wavelength of the absorbed light, is required by green cells under standard conditions to make 1 mol of glucose "costing" 686 kcal.

Photosynthetic Formation of Hexoses Involves Net Reduction of Carbon Dioxide

Now let us see how photosynthetic organisms generate glucose and other carbohydrates from CO_2 and H_2O at the expense of energy furnished by the ATP and NADPH generated by photosynthetic electron transport. Here we see a distinct difference between photosynthetic and heterotrophic organisms. Green plants and photosynthetic bacteria can use carbon dioxide as the sole source of all the carbon atoms required not only for biosynthesis of cellulose and starch but also of the lipids and proteins and many other organic components of plant cells. In contrast, animals and heterotrophic organisms in general are unable to bring about the net reduction of CO_2 to form "new" glucose in any significant amounts. Although we have seen that CO_2 can be taken up by animal tissues, for example, by the acetyl-CoA carboxylase reaction during fatty acid synthesis.

$$\text{Acetyl-CoA} + CO_2 + \text{ATP} + H_2O \longrightarrow \text{malonyl-CoA} + \text{ADP} + P_i$$

the CO_2 molecule incorporated into malonyl-CoA is lost again in a subsequent reaction step (page 595). Similarly, the CO_2 taken up by pyruvate carboxylase during gluconeogenesis in animal tissues (page 564) or by carbamoyl phosphate synthetase I during urea formation (page 551) is lost to the body in later steps. Clearly, plants and other photosynthetic organisms must have a distinctive metabolic route for the use of CO_2 as sole carbon source for the net synthesis of glucose. This route does not require light per se, since it occurs in the dark phase of photosynthesis.

Carbon Dioxide Is Fixed into Phosphoglycerate

An important clue to the nature of the CO_2 fixation mechanisms in photosynthetic organisms first came from the work of Melvin Calvin and his associates at the University of California

$^{1}CH_2—O—PO_3^{2-}$
C=O
H—C—OH
H—C—OH
$^{2}CH_2—O—PO_3^{2-}$
Ribulose 1,5-diphosphate

CO_2, H_2O → ribulose diphosphate carboxylase

$CH_2—O—PO_3^{2-}$
H—C—OH
COO^-
3-Phosphoglycerate

+

COO^-
H—C—OH
$CH_2—O—PO_3^{2-}$
3-Phosphoglycerate

Figure 23-17
Carbon dioxide fixation in the ribulose diphosphate carboxylase reaction. The CO_2 fixed appears as the carboxyl group of one of the 3-phosphoglycerate molecules formed as products.

at Berkeley in the late 1940s. They illuminated a suspension of green algae in the presence of radioactive carbon dioxide ($^{14}CO_2$) for only a few seconds and then quickly killed the cells, extracted them, and with the help of chromatographic methods searched for the metabolites in which the labeled carbon appeared earliest. The first compound that became labeled was *3-phosphoglycerate*, an intermediate of glycolysis (page 410). On degrading the 3-phosphoglycerate molecule they found that the carbon isotope was predominantly located in the carboxyl carbon atom. This was a very significant finding, because this carbon atom does not become labeled rapidly in animal tissues in the presence of radioactive CO_2. These experiments strongly suggested that 3-phosphoglycerate is an early intermediate in photosynthesis, a view supported by the fact that 3-phosphoglycerate is readily converted into glucose in plant extracts.

Further research led to the identification of an enzyme in green leaf extracts that catalyzes incorporation of $^{14}CO_2$ into organic form. This enzyme, *ribulose diphosphate carboxylase,* catalyzes the covalent insertion of CO_2 and simultaneous cleavage of the 5-carbon sugar *ribulose 1,5-diphosphate* to form two molecules of 3-phosphoglycerate, one of which bears the isotopic carbon introduced as CO_2 in its carboxyl group (Figure 23-17). This enzyme, which is not present in animal tissues, has a very complex structure (Figure 23-18), a molecular weight of 550,000, and is located on the outer surface of the thylakoid membranes. It makes up about 15 percent of the total chloroplast protein. Ribulose diphosphate carboxylase is the most abundant enzyme in the biosphere. It is the key enzyme in biomass production from CO_2 in the plant world.

The 3-phosphoglycerate formed by this enzyme can then be converted into glucose 6-phosphate by reversal of the glycolytic reactions and the fructose diphosphatase "bypass" reaction (page 565), exactly as in animal tissues. This sequence of reactions is schematized in Figure 23-19.

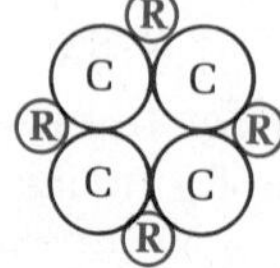

Figure 23-18
Subunit structure of ribulose diphosphate carboxylase (top view). A second set of identical subunits lies directly underneath the set shown. The molecule thus contains eight large or catalytic (C) subunits and eight small or regulatory (R) subunits.

Glucose Is Made from CO_2 by the Calvin Cycle

The reactions in Figures 23-17 and 23-19 account for the *net* conversion of CO_2 into only one of the six carbon atoms of glucose. How, then, can the other five carbons of glucose arise from CO_2? To answer this question, Calvin and his colleagues proposed a complex cyclic mechanism for the total biosynthesis of all six carbon atoms of glucose from carbon dioxide (Figure 23-20). In the Calvin cycle, one molecule of ribulose 1,5-diphosphate is consumed for each molecule of CO_2 fixed, but at the end of the cycle a molecule of ribulose diphosphate is regenerated. This is the same chemical trick used in the citric acid cycle and in the urea cycle, both of which end by regenerating a compound required in the first step, oxaloacetate (page 441) and ornithine (page 550), respectively. Figure 23-21

Figure 23-19
Because the two molecules of 3-phosphoglycerate formed cannot be distinguished from each other by enzymes, the isotopic carbon of the $^{14}CO_2$ introduced will be found in both carbons 3 and 4 of the glucose ultimately formed.

Ribulose diphosphate		3-Phosphoglycerate		Glucose formed
^1C		^3C		^1C
^2C		^2C		^2C
^3C	$\xrightarrow{^{14}CO_2}$	^1C	$\longrightarrow$	^3C
^4C		^1C		^4C
^5C		^2C		^5C
		^3C		^6C

shows the individual reactions of the cyclic Calvin pathway in the form of balanced equations. It uses seven of the steps [reactions (2) to (8)] of the gluconeogenesis pathway of animal tissues (Chapter 20). The only difference is that in photosynthesis NADPH rather than NADH is the reductant of glyceraldehyde 3-phosphate. The remaining reactions (Figure 23-21) are catalyzed by six additional enzymes. An instructive way of writing the overall equation of this complex cycle is

$$\text{6 Ribulose 1,5-diphosphate} + 6CO_2 + 18ATP + 12H_2O + 12NADPH + 12H^+ \longrightarrow \text{6 ribulose 1,5-diphosphate} + \text{glucose} + 18P_i + 18ADP + 12NADP^+$$

Ribulose 1,5-diphosphate is written on both sides of the equation to show that it is a necessary component regenerated at the

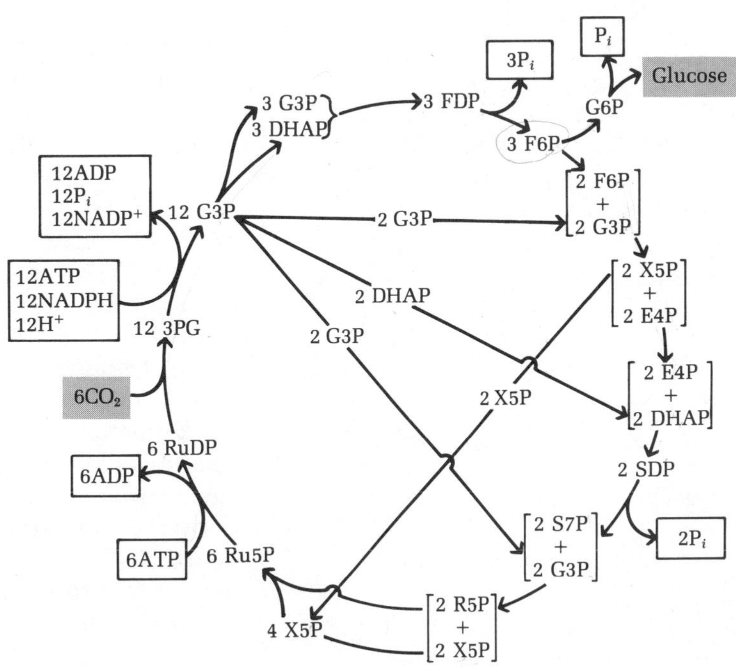

Figure 23-20
The Calvin cycle for conversion of CO_2 into D-glucose during photosynthesis. The entering CO_2 and glucose formed as end product are in color. Other inputs and outputs are in boxes. See Figure 23-21 for balanced equations.

Key:
- 3PG = 3-phosphoglycerate
- G3P = glyceraldehyde 3-phosphate
- DHAP = dihydroxyacetone phosphate
- FDP = fructose 1,6-diphosphate
- F6P = fructose 6-phosphate
- G6P = glucose 6-phosphate
- E4P = erythrose 4-phosphate
- X5P = xylulose 5-phosphate
- SDP = sedoheptulose 1,7-diphosphate
- S7P = sedoheptulose 7-phosphate
- R5P = ribose 5-phosphate
- Ru5P = ribulose 5-phosphate
- RuDP = ribulose 1,5-diphosphate

$6CO_2$ + 6 ribulose 1,5-diphosphate + $6H_2O \longrightarrow$ 12 3-phosphoglycerate (1)

12 3-Phosphoglycerate + 12ATP $\longrightarrow$ 12 3-phosphoglyceroyl phosphate + 12ADP (2)

12 3-Phosphoglyceroyl phosphate + 12NADPH + $12H^+ \longrightarrow$
12 glyceraldehyde 3-phosphate + $12NADP^+$ + $12P_i$ (3)

5 Glyceraldehyde 3-phosphate $\longrightarrow$ 5 dihydroxyacetone phosphate (4)

3 Glyceraldehyde 3-phosphate + 3 dihydroxyacetone phosphate $\longrightarrow$ 3 fructose 1,6-diphosphate (5)

3 Fructose 1,6-diphosphate + $3H_2O \longrightarrow$ 3 fructose 6-phosphate + $3P_i$ (6)

Fructose 6-phosphate $\longrightarrow$ glucose 6-phosphate (7)

Glucose 6-phosphate + $H_2O \longrightarrow$ $\boxed{\text{glucose}}$ + P_i (8)

2 Fructose 6-phosphate + 2 glyceraldehyde 3-phosphate $\xrightarrow{\text{transketolase}}$
2 xylulose 5-phosphate + 2 erythrose 4-phosphate (9)

2 Erythrose 4-phosphate + 2 dihydroxyacetone phosphate $\xrightarrow{\text{aldolase}}$ 2 sedoheptulose 1,7-diphosphate (10)

2 Sedoheptulose 1,7-diphosphate + $2H_2O$ $\xrightarrow{\text{phosphatase}}$ 2 sedoheptulose 7-phosphate + $2P_i$ (11)

2 Sedoheptulose 7-phosphate + 2 glyceraldehyde 3-phosphate $\xrightarrow{\text{transketolase}}$
2 ribose 5-phosphate + 2 xylulose 5-phosphate (12)

2 Ribose 5-phosphate $\xrightarrow{\text{isomerase}}$ 2 ribulose 5-phosphate (13)

4 Xylulose 5-phosphate $\xrightarrow{\text{epimerase}}$ 4 ribulose 5-phosphate (14)

6 Ribulose 5-phosphate + 6ATP $\xrightarrow{\text{phosphoribulokinase}}$ 6 ribulose 1,5-diphosphate + 6ADP (15)

Sum: $6CO_2 + 18ATP + 12H_2O + 12NADPH + 12H^+ \longrightarrow C_6H_{12}O_6 + 18P_i + 18ADP + 12NADP^+$

Figure 23-21
Balanced reaction equations for the conversion of CO_2 into glucose (in box) by the Calvin cycle. Reactions (1) to (8) generate D*-glucose, whereas reactions (9) to (15) regenerate ribulose 1,5-diphosphate.*

end of each cycle. After canceling out the ribulose diphosphate on both sides, the *net* reaction equation of the Calvin cycle is

$$6CO_2 + 18ATP + 12H_2O + 12NADPH + 12H^+ \longrightarrow C_6H_{12}O_6 + 18P_i + 18ADP + 12NADP^+$$

To synthesize each molecule of glucose from 6 molecules of CO_2, 18 ATPs and 12 NADPHs are consumed, which are replenished at the expense of the light reactions of photosynthesis.

The sequential reactions of the Calvin cycle can now be examined (Figure 23-21). Reactions (1) to (8) describe the formation of glucose from CO_2 and ribulose 1,5-diphosphate. Reactions (9) to (15) are concerned with the regeneration of ribulose 1,5-diphosphate, required to begin another round of the Calvin cycle. Reaction (9) is catalyzed by the enzyme *transketolase*, which contains thiamine pyrophosphate as its prosthetic group and requires Mg^{2+}. Transketolase catalyzes reversible transfer of a ketol ($CH_2OH—CO—$) group from a ketose phosphate, in this case fructose 6-phosphate,

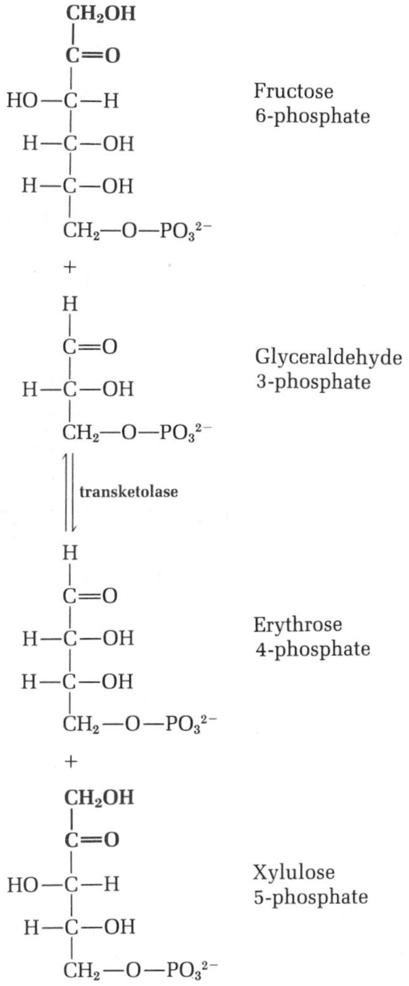

Figure 23-22
The transketolase reaction. This enzyme, which requires thiamine pyrophosphate and Mg^{2+}, transfers a ketol group (color) reversibly from a ketose phosphate to an aldose phosphate.

$$\begin{array}{c} CH_2OH \\ | \\ C{=}O \\ | \\ HO{-}C{-}H \\ | \\ H{-}C{-}OH \\ | \\ H{-}C{-}OH \\ | \\ CH_2{-}O{-}PO_3^{2-} \end{array} \quad \text{Fructose 6-phosphate}$$

$$+$$

$$\begin{array}{c} H \\ | \\ C{=}O \\ | \\ H{-}C{-}OH \\ | \\ CH_2{-}O{-}PO_3^{2-} \end{array} \quad \text{Glyceraldehyde 3-phosphate}$$

$$\Big\Updownarrow \text{ transketolase}$$

$$\begin{array}{c} H \\ | \\ C{=}O \\ | \\ H{-}C{-}OH \\ | \\ H{-}C{-}OH \\ | \\ CH_2{-}O{-}PO_3^{2-} \end{array} \quad \text{Erythrose 4-phosphate}$$

$$+$$

$$\begin{array}{c} CH_2OH \\ | \\ C{=}O \\ | \\ HO{-}C{-}H \\ | \\ H{-}C{-}OH \\ | \\ CH_2{-}O{-}PO_3^{2-} \end{array} \quad \text{Xylulose 5-phosphate}$$

$$\begin{array}{c} CH_2{-}O{-}PO_3^{2-} \\ | \\ C{=}O \\ | \\ CH_2OH \end{array} \quad \text{Dihydroxyacetone phosphate}$$

$$+$$

$$\begin{array}{c} H \\ | \\ C{=}O \\ | \\ H{-}C{-}OH \\ | \\ H{-}C{-}OH \\ | \\ CH_2{-}O{-}PO_3^{2-} \end{array} \quad \text{Erythrose 4-phosphate}$$

$$\Big\Updownarrow \text{ aldolase}$$

$$\begin{array}{c} CH_2{-}O{-}PO_3^{2-} \\ | \\ C{=}O \\ | \\ HO{-}C{-}H \\ | \\ H{-}C{-}OH \\ | \\ H{-}C{-}OH \\ | \\ H{-}C{-}OH \\ | \\ CH_2{-}O{-}PO_3^{2-} \end{array} \quad \text{Sedoheptulose 1,7-diphosphate}$$

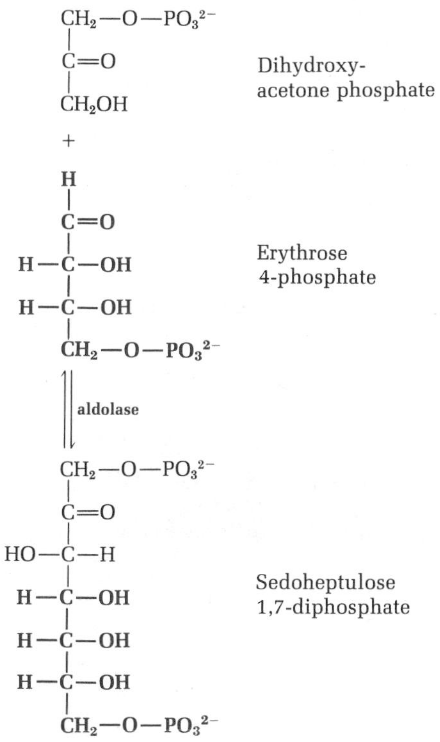

Figure 23-23
The formation of sedoheptulose 1,7-diphosphate by aldolase, which catalyzes the condensation of various aldehydes with dihydroxyacetone phosphate.

to an aldose phosphate, in this case glyceraldehyde 3-phosphate (Figure 23-22). Reaction (10) is promoted by aldolase, which catalyzes reversible condensation of an aldehyde, in this case, erythrose 4-phosphate, with dihydroxyacetone phosphate, yielding the 7-carbon *sedoheptulose 1,7-diphosphate* (Figure 23-23). After the latter loses its 1-phosphate group, it undergoes another transketolase reaction to yield two different pentose phosphates, which are ultimately converted into ribulose 1,5-diphosphate.

All the reactions in Figure 23-21 except the first, catalyzed by ribulose diphosphate carboxylase, also take place in animal tissues. For lack of ribulose diphosphate carboxylase animals cannot carry out net conversion of CO_2 into glucose. Plants that use ribulose diphosphate carboxylase as the first CO_2 fixation step are called C_3 plants, because the CO_2 is incorporated into a 3-carbon compound.

Glucose Is the Precursor of the Plant Carbohydrates Sucrose, Starch, and Cellulose

Glucose 6-phosphate generated during photosynthesis is the precursor of three carbohydrates characteristic of plants, sucrose, starch, and cellulose, which are not made by animals.

Sucrose is made by transfer of a D-glucose residue from UDP-glucose (page 418) to D-fructose 6-phosphate to form *sucrose 6-phosphate,* which is then hydrolyzed by a phosphatase to yield sucrose, as shown in Figure 23-24. Sucrose is the major transport sugar in plants (Figure 23-25). It is generated in the leaves during photosynthesis, secreted in high concentration into the *sieve tubes,* the "capillaries" of leaves. Sucrose may have been selected for this role because its unusual linkage, which joins the anomeric 1 carbon of D-glucose and the anomeric 2 carbon of D-fructose, is not hydrolyzed by amylases or other common carbohydrate-cleaving enzymes.

Cellulose, the major extracellular structural polymer of most plants (page 289), also is made from D-glucose in plants. The

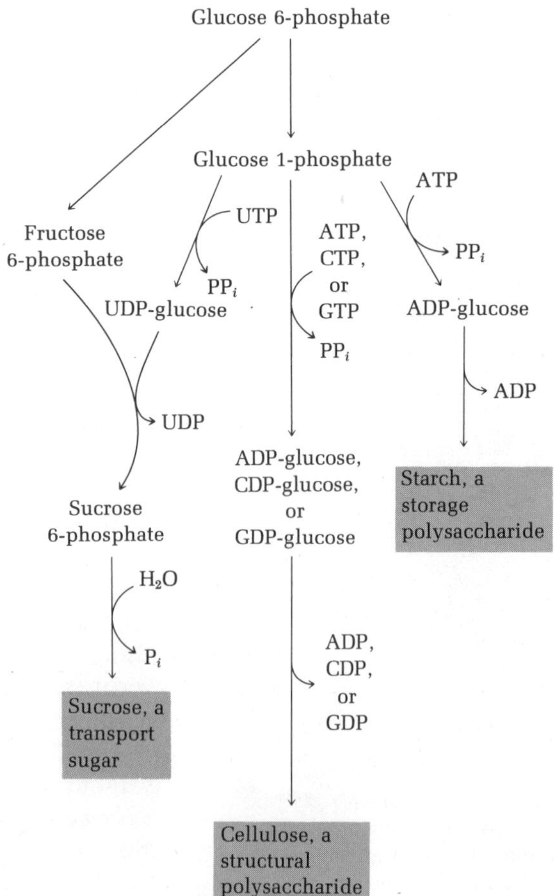

Figure 23-24
Synthesis of characteristic plant carbohydrates from glucose 6-phosphate generated by photosynthesis.

Figure 23-25
Function of sucrose in drawing water into sieve tubes by osmosis, thus providing the driving force for the flow of sap to the roots.

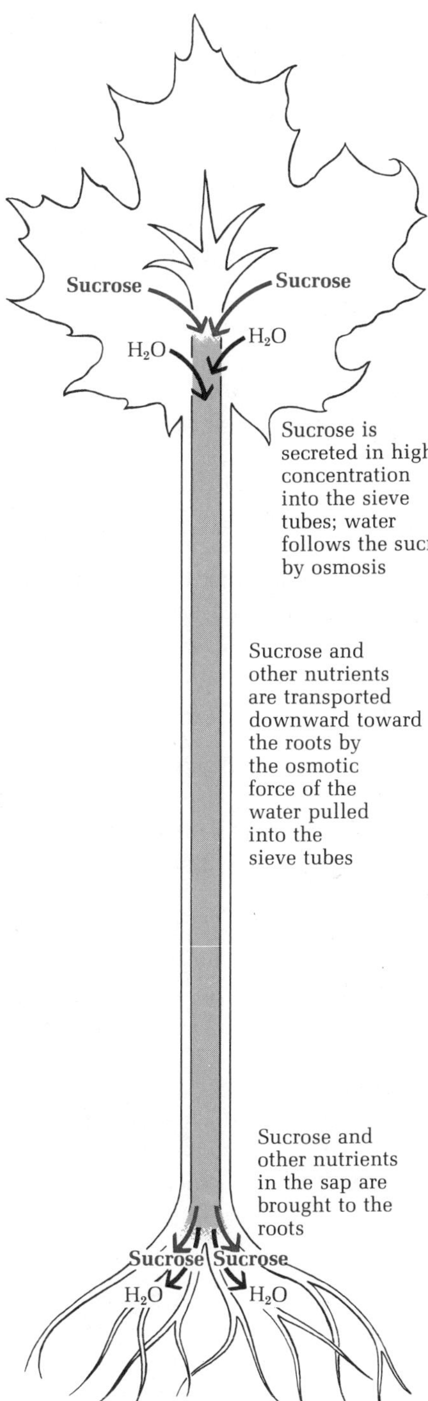

immediate precursor of the glucose units in cellulose, which has $\beta(1 \rightarrow 4)$ linkages between successive D-glucose units, is either ADP-glucose, CDP-glucose, or GDP-glucose, depending upon the plant species. These nucleoside diphosphate glucose compounds are analogous in structure and function to UDP-glucose (page 572), the precursor of glycogen in animal cells. Here is another example of the role of different nucleotides in channeling metabolic intermediates into specific biosynthetic pathways (page 389).

Starch, which has $\alpha(1 \rightarrow 4)$ linkages in its main chains (page 287), is made in a similar way from ADP-glucose in most plants (Figure 23-24).

Regulation of the Dark Reactions

The rate-limiting step in the dark reactions is the fixation of CO_2 by the ribulose diphosphate carboxylase reaction to form 3-phosphoglycerate. This enzyme, which has 16 subunits, is an allosteric enzyme stimulated by three different changes that result from illumination of chloroplasts:

1. It is stimulated by an increase in pH. When chloroplasts are illuminated, H^+ ions are transported from the stroma into the thylakoids, resulting in an increase in the stroma pH, which stimulates the carboxylase, located on the outer surface of the thylakoid membrane.
2. It is stimulated by Mg^{2+}, which enters the stroma as H^+ ions leave when chloroplasts are illuminated.
3. It is stimulated by NADPH, which is generated by photosystem I during illumination.

Thus, although CO_2 fixation by ribulose diphosphate carboxylase is a dark reaction, it is stimulated indirectly by illumination of chloroplasts. In a similar indirect way certain other enzymes involved in the Calvin cycle, as well as the ATP synthetase, are also promoted by illumination of chloroplasts.

Tropical Plants Employ the C_4 or Hatch-Slack Pathway

Most plants in the tropics, as well as temperate-zone crop plants native to the tropics, such as corn, sugar cane, and sorghum, fix CO_2 by a route called the Hatch-Slack or *C_4 pathway,* but it must be made clear at the outset that both the C_3 and C_4 plants ultimately use the C_3 pathway as described above and summarized in Figure 23-21. The important difference is that in the C_4 plants, the C_3 pathway is preceded by additional steps in which there is a preliminary fixation of CO_2 into a compound with four carbon atoms before the CO_2 is incorporated into phosphoglycerate (Figure 23-26). Let us see how the C_4 pathway functions.

Figure 23-26
C_4 plants first incorporate CO_2 into a C_4 compound, in a set of two reactions that precede the C_3 pathway.

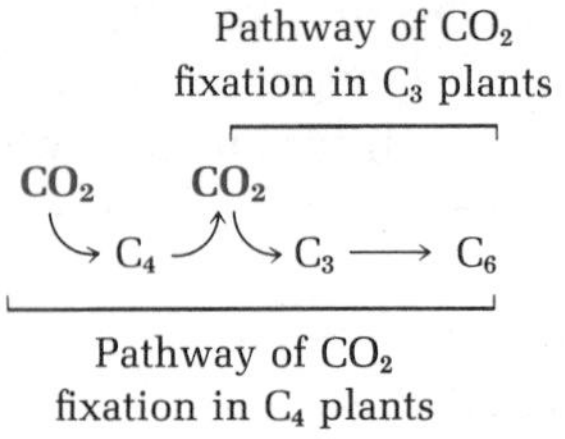

Pathway of CO_2 fixation in C_3 plants

$$CO_2 \searrow C_4 \nearrow CO_2 \searrow C_3 \longrightarrow C_6$$

Pathway of CO_2 fixation in C_4 plants

In the 1960s two Australian plant biochemists, M. D. Hatch and C. R. Slack, discovered that in plants of tropical origin the earliest intermediate in which radioactive CO_2 is fixed is *oxaloacetate*, a 4-carbon compound. This reaction, which occurs in the mesophyll cells of the leaves (Figure 23-27), is catalyzed by *phosphoenolpyruvate carboxylase*

$$\text{Phosphoenolpyruvate} + CO_2 \longrightarrow \text{oxaloacetate} + P_i$$

This enzyme does not occur in animal tissues and is not to be confused with *phosphoenolpyruvate carboxykinase* (page 565), which catalyzes the reaction

$$\text{Phosphoenolpyruvate} + \text{GDP} + CO_2 \rightleftharpoons \text{oxaloacetate} + \text{GTP}$$

during gluconeogenesis in animals.

The oxaloacetate formed in the mesophyll cells is then reduced to malate at the expense of NADPH

$$\text{Oxaloacetate} + \text{NADPH} + H^+ \longrightarrow \text{malate} + \text{NADP}^+$$

Now we come to a crucial point in the C_4 cycle. The malate formed in the mesophyll cells, which contains the fixed CO_2, is now transferred into the neighboring *bundle-sheath* cells via special junctions or channels between these cells. In the bundle-sheath cells the malate is now decarboxylated to yield pyruvate and CO_2 by the action of *malate enzyme*

$$\text{Malate} + \text{NADP}^+ \longrightarrow \text{pyruvate} + CO_2 + \text{NADPH} + H^+$$

The free CO_2 formed in the bundle-sheath cells is the same CO_2 molecule that was originally fixed into oxaloacetate in the mesophyll cells.

In the bundle-sheath cells the CO_2 arising from the decarboxylation of malate is now fixed again, this time by ribulose diphosphate carboxylase, in exactly the same reaction that occurs in C_3 plants, leading to incorporation of CO_2 into the carboxyl carbon of 3-phosphoglycerate. The pyruvate formed by the decarboxylation of malate in the bundle-sheath cells now is transferred back to the mesophyll cells, where it is converted into phosphoenolpyruvate by an unusual enzymatic reaction, catalyzed by the enzyme *pyruvate-phosphate dikinase*

$$\text{Pyruvate} + P_i + \text{ATP} \longrightarrow \text{phosphoenolpyruvate} + \text{AMP} + PP_i$$

This enzyme is called a dikinase because *two* different molecules are simultaneously phosphorylated by one molecule of ATP: pyruvate is phosphorylated to phosphoenolpyruvate, and phosphate is phosphorylated to pyrophosphate. The pyrophosphate formed is subsequently hydrolyzed to phosphate; so two high-energy phosphate groups of ATP are used. This reaction thus regenerates phosphoenolpyruvate, which is now ready to fix another molecule of CO_2 in the mesophyll cell.

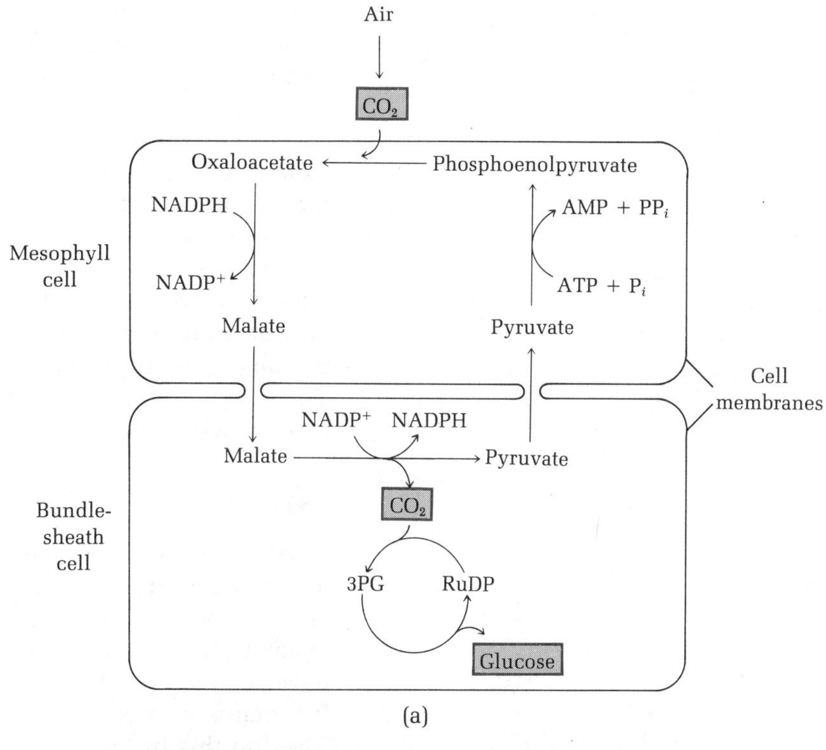

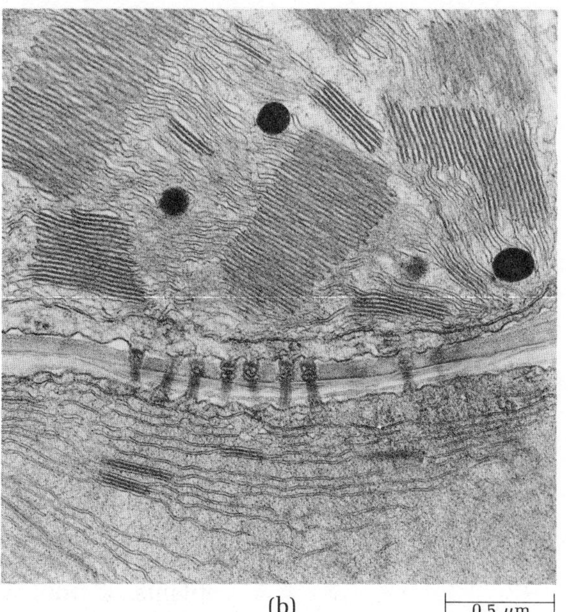

Figure 23-27
(a) The Hatch-Slack pathway of CO_2 fixation, via a 4-carbon intermediate. This pathway prevails in plants of tropical origin. (b) An electron micrograph showing connected mesophyll (above) and bundle-sheath (below) cells.

Once CO_2 is fixed into 3-phosphoglycerate in the bundle-sheath cells, after the preliminary fixation into malate in the mesophyll cells, all the other reactions of the C_3 or Calvin cycle take place exactly as described in Figures 23-20 and 23-21. Thus in C_4 plants the mesophyll cells carry out CO_2 fixation by the C_4 pathway, but glucose biosynthesis occurs by the C_3 pathway in the bundle-sheath cells.

There is a second important point about the pathway of CO_2 fixation in C_4 plants: it has a larger energy cost than in C_3 plants. For each molecule of CO_2 fixed in the C_4 pathway, a molecule of phosphoenolpyruvate must be regenerated at the expense of two high-energy phosphate groups of ATP, as shown above. The C_4 plants thus need five ATPs altogether to fix one molecule of CO_2, whereas the C_3 plants need only three.

The C_4 Pathway Serves to Concentrate CO_2

What can possibly be gained by fixing CO_2 in one kind of cell only to unfix it and refix it in the other, especially since this pathway requires extra energy? Fundamental studies of the biochemistry and histology of CO_2 fixation in tropical plants have revealed what may be the purpose of the C_4 cycle. Tropical plants must avoid excessive loss of water by evaporation. They do this by closing the *stomata*, the "breathing tubes" of the leaves. But this also reduces the flow of CO_2 from the atmosphere into the bundle-sheath cells. Consequently the CO_2 concentration in the bundle-sheath cells is relatively low, so that the ribulose diphosphate carboxylase would be kept from acting at its maximal possible rate. But the phosphoenolpyruvate carboxylase reaction in the mesophyll cells has a much higher affinity for CO_2 and thus can fix CO_2 more efficiently. This reaction serves to fix and concentrate CO_2 in the form of malate. Release of CO_2 from malate in the bundle-sheath cells yields a sufficiently high concentration of CO_2 so that ribulose diphosphate carboxylase can function near its maximal rate.

Paradoxically, although the Hatch-Slack pathway requires five high-energy phosphate groups per molecule of CO_2 fixed, compared with only three in the C_3 plants, C_4 plants of tropical origin grow faster and make more biomass per unit leaf area than the C_3 plants of temperate zones. Unfortunately for gardeners, crabgrass and many other weeds are among the C_4 plants of tropical origin that are very efficient in converting light energy into biomass.

Let us examine the basis of this paradox.

Photorespiration Limits the Efficiency of C_3 Plants

At night green leaf cells of both C_3 and C_4 plants carry out respiration and phosphorylation in their mitochondria at the expense of substrates generated by photosynthesis in preceding light periods. A question now arises: Do leaf cells also respire in the light, during active photosynthesis, or is mitochondrial respiration turned off? Careful measurements of the rates of oxygen and carbon dioxide exchanges have shown that C_3 plants do indeed respire in the light and consume some oxygen while they are carrying out oxygen-forming photosynthesis. However, this is not entirely mitochondrial respiration, since it is only partially inhibited by cyanide, an inhibitor of mitochondrial cytochrome oxidase (page 483). The cyanide-insensitive respiration observed in C_3 plants in the light is called *photorespiration*.

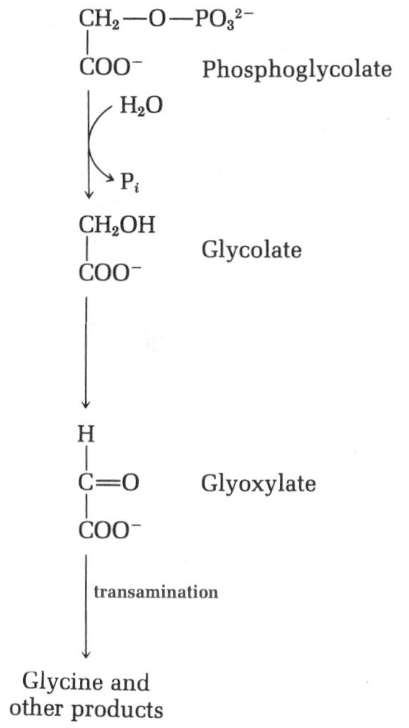

Photorespiration appears to be a wasteful process. First, it diverts some of the reducing power generated in the light reactions from the biosynthesis of glucose into the reduction of oxygen. Second, unlike mitochondrial respiration, photorespiration is not accompanied by oxidative phosphorylation. Thus photorespiration wastes a good part of the solar energy trapped by the light reactions. There is a third important point: photorespiration is very active in the C_3 plants but nearly absent in the C_4 plants of tropical origin.

The major substrate oxidized by photorespiration in C_3 plants is *glycolic acid* (Figure 23-28). Glycolate is oxidized in the peroxisomes of leaf cells to glyoxylate, which is converted into glycine and other products. Glycolate arises in plant cells in a most unusual reaction. It is formed by the oxidative breakdown of ribulose 1,5-diphosphate by ribulose diphosphate carboxylase, *the very same enzyme that fixes CO_2 into phosphoglycerate.* How can this come about?

Ribulose diphosphate carboxylase can promote the reaction of ribulose diphosphate with *either* CO_2 or O_2. When the CO_2 concentration is low and that of O_2 is relatively high, the O_2 molecule not only competes with CO_2 but can also replace it. The result of this curious reaction is that ribulose diphosphate in C_3 plants becomes *oxygenated* rather than carboxylated. The products of oxygenation are *phosphoglycolate* and *3-phosphoglycerate* (Figure 23-29), instead of the two molecules of 3-phosphoglycerate normally formed in the carboxylation. The phosphoglycolate formed in the oxygenation reaction then undergoes enzymatic hydrolysis to yield free glycolate, which is the substrate oxidized during photorespiration.

In the C_4 plants, on the other hand, the CO_2/O_2 ratio in the bundle-sheath cells always remains relatively high, because of the preceding C_4 step, and this favors the carboxylation of ribulose 1,5-diphosphate. Moreover, the closing of the stomata in the leaves of C_4 plants not only avoids loss of water but also limits entry of atmospheric oxygen.

$CH_2—O—PO_3^{2-}$ | COO^- Phosphoglycolate

↓ H_2O → P_i

CH_2OH | COO^- Glycolate

↓

H | C=O | COO^- Glyoxylate

↓ transamination

Glycine and other products

Figure 23-28
Hydrolysis of phosphoglycolate yields glycolate, the substrate of photorespiration. It is oxidized to glyoxylate, CO_2, and other products.

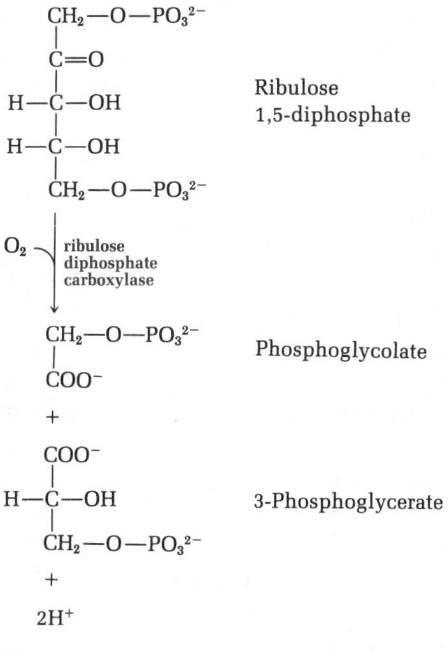

$CH_2—O—PO_3^{2-}$ | C=O | H—C—OH | H—C—OH | $CH_2—O—PO_3^{2-}$ Ribulose 1,5-diphosphate

O_2 ↓ ribulose diphosphate carboxylase

$CH_2—O—PO_3^{2-}$ | COO^- Phosphoglycolate

\+

COO^- | H—C—OH | $CH_2—O—PO_3^{2-}$ 3-Phosphoglycerate

\+

$2H^+$

Figure 23-29
Oxygenation of ribulose 1,5-phosphate. In this reaction oxygen replaces the normal substrate CO_2 to yield phosphoglycolate instead of a second molecule of 3-phosphoglycerate.

Photorespiration Is an Important Problem in Temperate-Zone Agriculture

Although the precise biological function of photorespiration is still not understood, it is important in temperate-zone agriculture because of the lower efficiency of C_3 plants in generating biomass. In a field of C_3 crop plants during the course of a hot windless day, the concentration of CO_2 in the air over the plants decreases from its normal level of about 0.03 percent to as low as 0.005 percent because of the rapid use of CO_2 for photosynthesis. This change decreases the CO_2/O_2 ratio in the air to which the plants are exposed, allowing O_2 to compete more effectively with CO_2 in the ribulose diphosphate carboxylase reaction, with the result that CO_2 fixation is decreased and the wasteful photorespiration process is increased. Photorespiration may inhibit net biomass formation by C_3 plants as much as 50 percent.

Efforts are under way to inhibit photorespiration in C_3 plants, in order to increase their efficiency in fixing CO_2 and thus their biomass production. One approach is to inhibit the enzymes oxidizing glycolate. Another is to breed C_3 plants with lower rates of photorespiration. But the increasing level of CO_2 in the air, caused by combustion of fossil fuels (page 334), may solve the problem for us!

Salt-Loving Bacteria Use Light Energy to Make ATP

The halophilic ("salt-loving") bacterium *Halobacterium halobium* conserves energy derived from absorbed sunlight by an entirely different principle from that employed by true photosynthetic organisms. These unusual bacteria live only in brine ponds and salt lakes (Great Salt Lake and the Dead Sea, for example), where the high salt concentration results from water loss by evaporation; indeed, they cannot live in NaCl concentrations lower than 3 *M*. These bacteria are aerobes and normally use oxygen to oxidize organic fuel molecules. However, the solubility of oxygen is so low in brine ponds, in which the NaCl concentration may exceed 4 *M*, that the halobacteria must sometimes call on another source of energy, namely sunlight. The plasma membranes surrounding the cells of *H. halobium* contain patches of light-absorbing pigments, called *purple patches*. These patches are made up of closely packed molecules of the protein *bacteriorhodopsin* (MW 26,000), which contains a molecule of bound *retinal* or *vitamin A aldehyde* (page 266) as prosthetic group. When the cells are illuminated, the excited bacteriorhodopsin molecules undergo transient bleaching. As the light-excited bacteriorhodopsin molecules in the membrane revert to their initial ground state, the energy released is harnessed to translocate H^+ ions from the inside to the outside of the cells, to form an acid-outside pH gradient across the cell membrane. Because its concentration is higher on the outside, H^+ ions tend to diffuse back into the cell through an ATP-forming enzyme in the membrane, similar to the ATP synthetase of mitochondria and of chloroplasts. As H^+ ions pass through the bacterial ATPase, they supply the energy required to make ATP from ADP and phosphate (Figure 23-30). Thus the halobacteria can conserve light

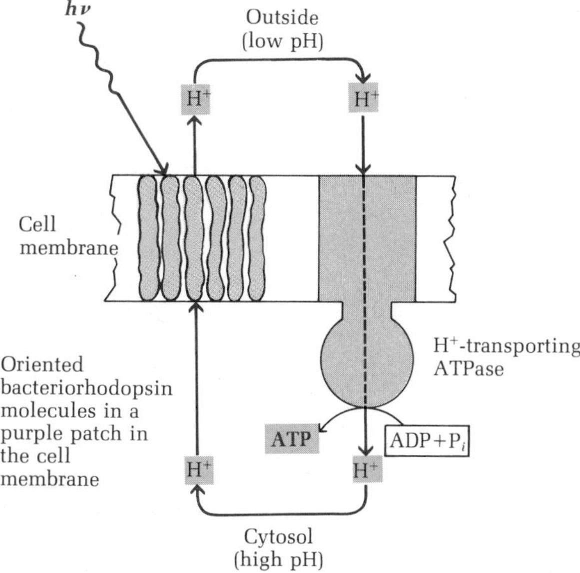

Figure 23-30
In the halobacteria, bacteriorhodopsin molecules oriented in the cell membrane serve as outward-directed light-driven H^+ pumps. The H^+ gradient formed drives the synthesis of ATP by an ATP synthetase.

energy in the form of ATP, to supplement their ability to carry out oxidative phosphorylation when oxygen is available. However, the halobacteria do not evolve oxygen, nor do they carry out photoreduction of $NADP^+$. Bacteriorhodopsin, a relatively small protein molecule, is the simplest light-driven H^+ pump known (Figure 23-31). Its molecular structure and mechanism of action are expected to yield important new insights into light-dependent energy transduction and into the action of different types of H^+ pumps functioning in respiration and photosynthesis.

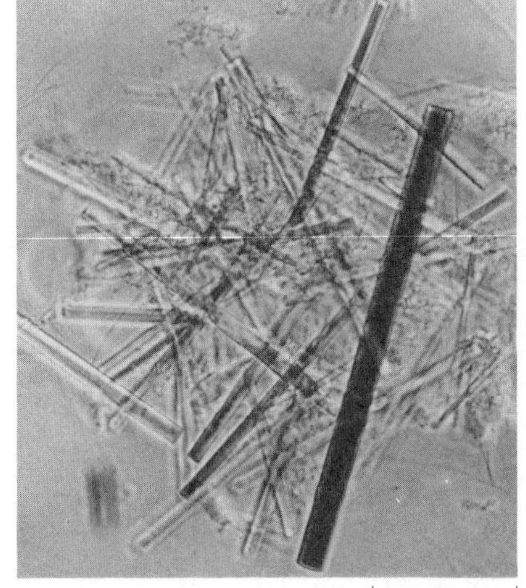

Figure 23-31
Crystals of bacteriorhodopsin.

Photosynthetic Organisms Are Models for the Design of Solar Energy Cells

Because the photosynthetic mechanisms of chloroplasts are so efficient in transforming solar energy into the chemical energy of ATP and NADPH, much research is being carried out to simulate these processes with simpler artificial molecular systems, with the hope of harnessing the inexhaustible flow of solar energy constantly striking the surface of the earth. Present solar-energy cells use very expensive solid-state materials, such as crystalline silicon, as receptors of light energy, and they are not as efficient as chloroplasts. If we can fully understand the molecular and subatomic principles by which chlorophyll and bacteriorhodopsin function as efficient traps for light energy, and if we can further understand how electric charges and H^+ can be separated across membranes to create energy-rich electrochemical gradients, we may be in a position to simulate these processes efficiently with cheap materials. Fundamental biochemical and biophysical research on the energy-transducing membranes of photosynthetic organisms is therefore not only important for an understanding of nature but may also have long-range practical implications in agriculture, energy production, and the protection of our atmosphere and environment from pollution by CO_2, H_2SO_4, and other by-products of the combustion of fossil fuels.

Summary

In the light reactions of green plant photosynthesis, absorbed light energy causes electrons to flow from H_2O to $NADP^+$, reducing it to form NADPH; oxygen is thereby evolved from the H_2O. ATP is the other product of the light reactions. In the dark reactions ATP and NADPH are used to reduce CO_2 to form glucose. In green plant cells photosynthesis takes place in chloroplasts. The light reactions are localized in the thylakoids, flattened membranous vesicles in the chloroplasts. Photosynthetic plant cells contain two major types of light-capturing pigments, chlorophylls and carotenoids, arranged into two kinds of photosystems. Each photosystem has a set of light-harvesting or antenna pigments and a reaction center which uses light energy to bring about electron transfer to a chain of electron-carrying molecules. Photosystem I is excited by longer wavelengths of light; it causes electrons to reduce $NADP^+$ to NADPH. Photosystem II is activated by shorter wavelengths; it removes electrons from H_2O and causes oxygen evolution. Excitation of photosystem I leads to reduction of $NADP^+$ via ferredoxin and ferredoxin-NADP oxidoreductase. The electrons required to fill the electron holes left in photosystem I come from excited photosystem II. They are transferred by a central electron-transport chain to which photosynthetic phosphorylation is coupled. The electrons required to fill the electron holes in photosystem II, which has a highly oxidizing potential, come from H_2O. The energy required for ATP synthesis comes from an H^+ gradient across the membrane generated by downhill electron flow. Eight light quanta are required to cause evolution of a molecule of oxygen and the production of two NADPHs and two ATPs.

In the dark reactions CO_2 is fixed into the carbon backbone of glucose by a reaction with ribulose 1,5-diphosphate to yield two molecules of 3-phosphoglycerate, which are converted into glucose via the Calvin cycle at the expense of 18 ATPs and 12 NADPHs, generated in the light reactions. The Calvin cycle consists of interconnected reactions of the pentose phosphate and glycolytic pathways. In C_4 plants CO_2 is first fixed in mesophyll cells into malate, which is passed into bundle-sheath cells; there CO_2 is released in high concentrations for the ribulose diphosphate carboxylase reaction, which is then followed by the C_3 pathway. Photorespiration wastes photosynthetic energy in C_3 plants by oxidizing glycolate, a product of the oxygenation of ribulose 1,5-diphosphate.

When illuminated, bacteriorhodopsin of the cell membrane of halophilic bacteria causes translocation of H^+ out of the cells. The H^+ gradient so formed is used by the cells to generate ATP.

References

Light in Biology

HENDRICKS, S. B.: "How Light Interacts with Living Matter," *Sci. Am.*, **219**:174–186, September (1968).

The Photoreactions

GOVINDJEE and R. GOVINDJEE: "The Primary Events in Photosynthesis," *Sci. Am.*, **231:**8–82, December (1974).

Photosynthetic Electron Transport and Phosphorylation

BLANKENSHIP, R. E., and W. W. PARSON: "The Photochemical Electron Transfer Reactions of Photosynthetic Bacteria and Plants," *Annu. Rev. Biochem.*, **44:**635–653 (1978).

HINKLE, P. C., and R. E. McCARTY: "How Cells Make ATP," *Sci. Am.*, **238:**104–123 (1978).

MILLER, K. R.: "The Photosynthetic Membrane," *Sci. Am.*, **241:** 102–113, October (1979).

The Path of Carbon

BASSHAM, J. A.: "The Path of Carbon in Photosynthesis," *Sci. Am.*, **206:** 88–100, June (1962).

ZELITCH, I.: "Pathways of Carbon Fixation in Green Plants," *Annu. Rev. Biochem.*, **44:**123–145 (1975).

Photorespiration

BJORKMAN, O., and J. BERRY: "High-Efficiency Photosynthesis," *Sci. Am.*, **229:**80–93, October (1973).

OSMOND, C. B.: "Photorespiration and Photoinhibition. Implications for the Bioenergetics of Photosynthesis," *Biochem. Biophys. Acta* **639:**77–156 (1981).

Bacteriorhodopsin

SINGH, K., and S. R. CAPLAN: "The Purple Membrane and Solar Energy Conversion," *Trends Biochem. Sci.*, **5:**62–64, March (1980).

STOECKENIUS, W.: "The Purple Membrane of Salt-Loving Bacteria," *Sci. Am.*, **234:**38–46, June (1976).

Books

GOVINDJEE (ed.): *Bioenergetics of Photosynthesis,* Academic, New York, 1975.

GREGORY, R. P. F.: *The Biochemistry of Photosynthesis,* 2d ed., Wiley, New York, 1977.

STUMPF, P. K., and E. E. CONN (eds.): *The Biochemistry of Plants,* 8 vols., Academic, New York, 1980–1981. Comprehensive encyclopedia. Volume 8 covers photosynthesis.

Problems

1. *Phases of Photosynthesis* When a suspension of green algae is illuminated in the absence of carbon dioxide and then incubated with $^{14}CO_2$ in the dark, $^{14}CO_2$ is converted into ^{14}C-glucose for a brief time. What is the significance of this observation with regard to the two phases of photosynthesis? Why does the conversion of $^{14}CO_2$ into ^{14}C-glucose stop after a brief time?

2. *Photochemical Efficiency of Light at Different Wavelengths* The rate of photosynthesis, measured by O_2 production, is higher when a green plant is illuminated with 680-nm light than with 700-nm light. However, illumination by a combination of 680- and 700-nm light gives a higher rate of photosynthesis than light of either wavelength alone. Explain.

3. *Role of H_2S in Some Photosynthetic Bacteria* Illuminated purple sulfur bacteria carry out photosynthesis in the presence of H_2O and $^{14}CO_2$ but only if H_2S is added and oxygen is absent. During the course of photosynthesis, measured by formation of ^{14}C-labeled glucose, H_2S is converted into elemental sulfur, but no oxygen is evolved. What is the role of the conversion of H_2S into sulfur? Why is no oxygen evolved?

4. *Boosting the Reducing Power of Photosystem I by Light Absorption* When photosystem I absorbs red light at 700 nm, the reduction potential of P700 changes from +0.4 to −0.6 V. What fraction of the absorbed light is trapped in the form of reducing power?

5. *Mode of Action of the Herbicide DCMU* When chloroplasts are treated with 3-(3,4-dichlorophenyl)-1,1-dimethylurea (DCMU, or Diuron), a potent herbicide, oxygen evolution and photophosphorylation cease. Oxygen evolution but *not* photophosphorylation can be restored by the addition of an external electron acceptor, i.e., a Hill reagent. How does this herbicide act as a weed killer? Suggest a location for the inhibitory site of this herbicide in Figure 23-12. Explain.

6. *Bioenergetics of Photophosphorylation* The steady-state concentrations of ATP, ADP, and P_i in isolated spinach chloroplasts under full illumination at pH 7.0 are 120, 6, and 700 μM, respectively.
   (a) What is the free-energy requirement for the synthesis of 1 mol of ATP under these conditions?
   (b) The energy for ATP synthesis is furnished by light-induced electron transport in the chloroplasts. What is the minimum voltage drop necessary during the transport of a pair of electrons to synthesize ATP under these conditions?

Problem 7a

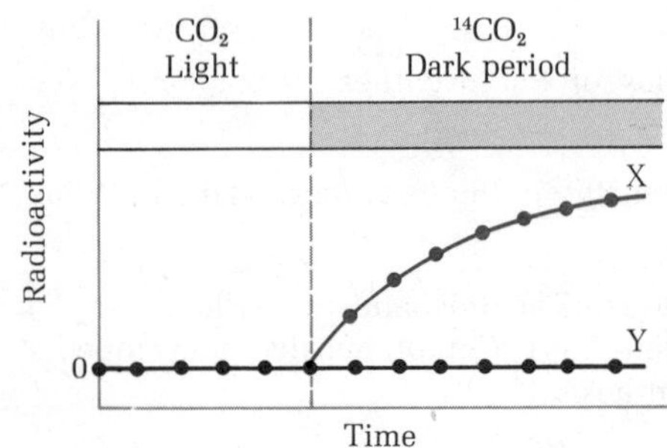

7. *Identification of Key Intermediates in the Dark Reactions of Photosynthesis* Calvin and his colleagues used the unicellular green algae *Chlorella* to study the dark reactions of photosynthesis. In their experiments $^{14}CO_2$ was incubated with illuminated suspensions of algae under different conditions. They followed the time course of the appearance of ^{14}C in two compounds X and Y under two sets of conditions.
   (a) Illuminated *Chlorella* were grown on unlabeled CO_2; the lights were turned off, and $^{14}CO_2$ was added. Under these conditions X was the first compound found in the algae to become labeled with ^{14}C. Compound Y was unlabeled.

Problem 7b

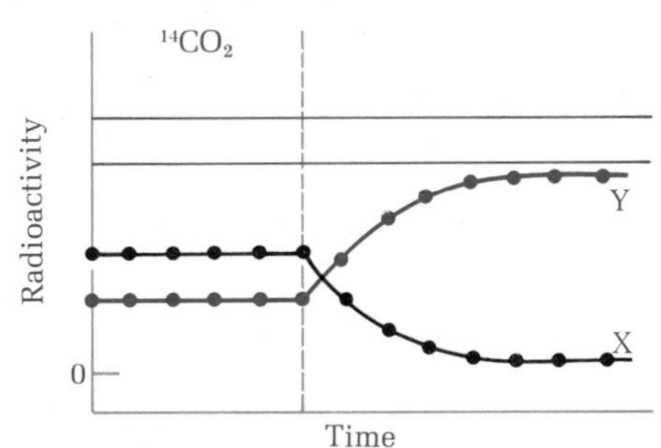

(b) Illuminated *Chlorella* cells were grown on radioactive $^{14}CO_2$. Illumination was continued until all the $^{14}CO_2$ had disappeared (vertical dashed line). Under these conditions compound X became labeled quickly but lost its radioactivity with time. On the other hand, compound Y became more radioactive with time.

Suggest the identities of X and Y based on your understanding of the Calvin cycle.

8. *Regulation of Ribulose-1,5-Diphosphate Carboxylase by pH* The K_M value for CO_2 of ribulose-1,5-diphosphate carboxylase decreases significantly when the pH of the medium is increased. What is the effect of this decrease on the rate of CO_2 fixation in the ribulose diphosphate carboxylation reaction? How does this property serve to regulate the rate of photosynthesis during illumination of the plant? What role does this regulation process play in the plant during the dark hours?

9. *Pathway of CO_2 Fixation in Maize* If a maize plant is illuminated in the presence of gaseous $^{14}CO_2$, after about 1 s over 90 percent of all the radioactivity incorporated in the leaves is found in the C-4 atoms of malate, aspartate, and oxaloacetate. Only after 60 s does ^{14}C appear in the C-1 atom of 3-phosphoglycerate. Explain.

10. *Chemistry of Malate Enzyme: Variation on a Theme* Malate enzyme, found in the bundle-sheath cells of C_4 plants, carries out a reaction which has an analogous counterpart in the citric acid cycle. What is the analogous reaction? Explain.

11. *Absence of Photosystem II in Mesophyll Cells* Mesophyll cells in tropical grasses contain photosystem I but not photosystem II. In contrast, bundle-sheath cells of the same plant contain both photosystems I and II. Are these observations consistent with the role of the Hatch-Slack pathway? Explain.

Problem 12

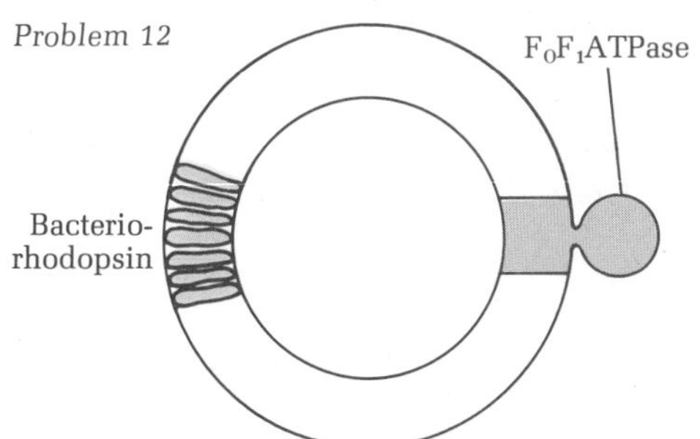

12. *Reconstitution Experiments: ATP-Synthesizing Vesicles* W. Stoeckenius and E. Racker have reported some interesting reconstitution experiments in which inverted synthetic phospholipid vesicles have been prepared containing bacteriorhodopsin molecules from *Halobacterium halobium* and the F_0F_1ATPase from beef heart mitochondria (see illustration). When illuminated, these vesicles synthesized ATP from ADP and P_i. However, illumination in the presence of dinitrophenol abolished ATP production. Explain these experiments in terms of the chemiosmotic hypothesis.

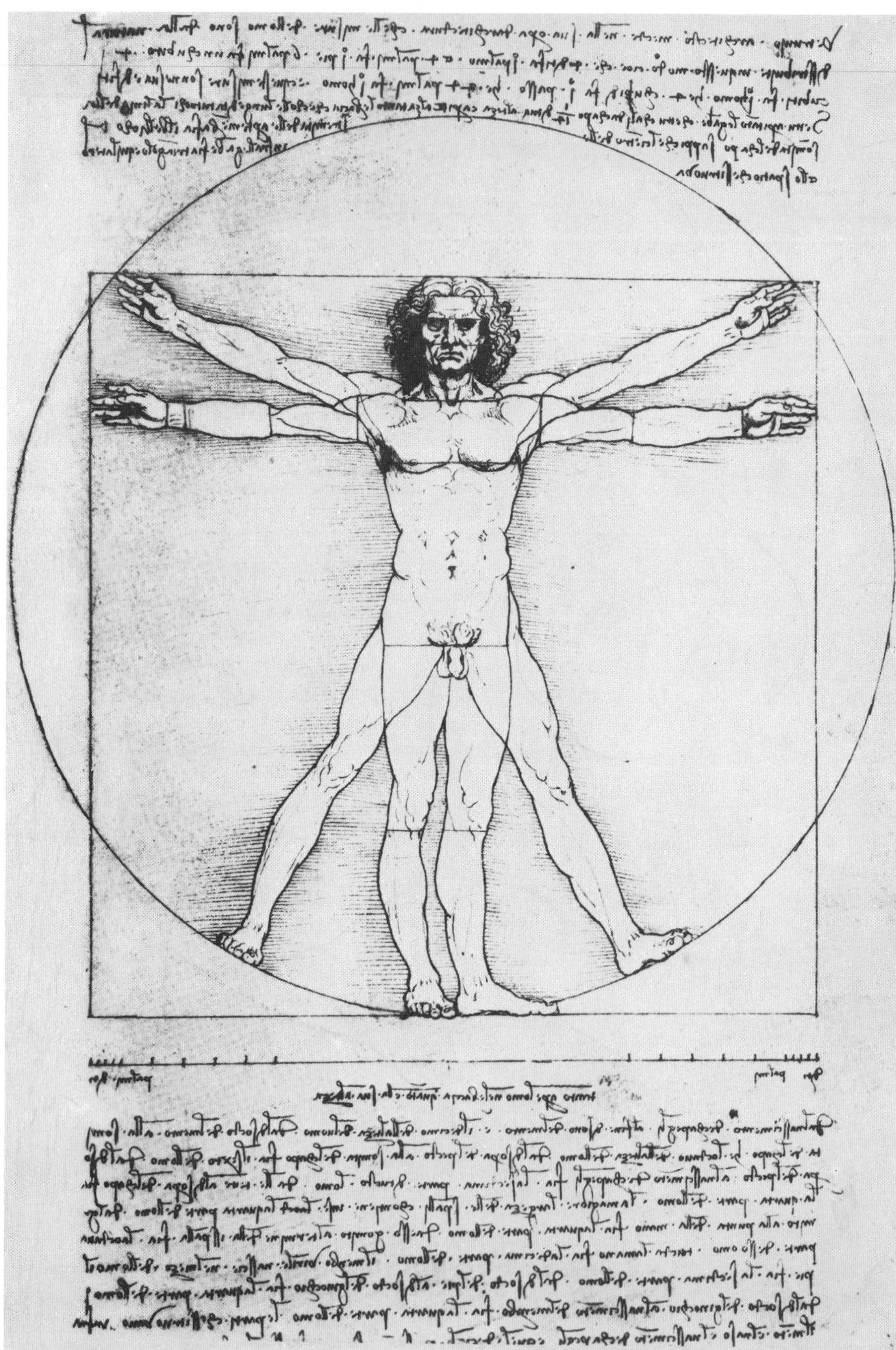

PART III

Some Aspects of Human Biochemistry

We have seen in preceding chapters that there is an essential continuity in different forms of life, in cell structure and its molecular components, cell bioenergetics, and in most of the metabolic pathways. Indeed, this continuity has enabled biochemists to select certain specific organisms, for reasons of availability, convenience, or some special biological advantage, for detailed study of fundamental processes that occur in all living forms. For example, yeast cells and muscle extracts have taught us much about glycolysis and *E. coli* cells have revealed basic principles of molecular genetics. Moreover, much important biochemical information on regulatory mechanisms is coming from such unusual organisms as the slime mold *Dictyostelium discoideum* and the South African clawed toad *Xenopus laevis*. Quite literally, as far as fundamental biochemical processes are concerned, there seems to be little difference between cabbages and kings.

But now we shall examine one specific form of life to determine some of the ways its particular biochemical make-up adapts it for its activities and its niche in the biosphere. We might examine a bacterium, especially since we are just beginning to appreciate that bacteria are far more complex organisms than has been thought. Or we might study the special biochemistry of plant life, particularly because of the growing disparity between agricultural yield and world food needs. But as humans we want most of all to know more about the biochemistry of *Homo sapiens*, the biochemical characteristics and functions of the different tissues and organs, how they coordinate their metabolic activities, and in what ways biochemical abnormalities affect health.

The next three chapters develop some of the basic biochemical characteristics of the human organism, which are at the same time applicable to many other forms of higher animal life. In particular, we shall see how the metabolism of the different tissues and organs is coordinated, how hormones regulate the complex responses of the human organism to external and

Manuscript illustration by Leonardo da Vinci: "The Proportions of Man." There are some 10^{13} cells in the adult human body, each consisting of a set of biomolecules in specific proportion, each with a characteristic ultrastructure. Cells are grouped into tissues, tissues into organs, and organs into organ systems, their biochemical activities marvelously coordinated into an organism that not only is and does, but also thinks and creates.

internal changes, and, not least, the proper nutrition of the human organism, one of the most important contributions made by biochemistry to human welfare. These three chapters can only be stepping-stones, since human biochemistry has much broader, more complex, and more profound aspects that cannot be developed in this book. Indeed, the ultimate goals in the study of human biochemistry are still far in the future, since we know only a very small fraction of what there is to know. The ultimate biochemical understanding of man's own nature must come when we examine those higher functions of the human brain that set off *Homo sapiens* from all other organisms: how memory is stored, how brain cells interpret sensory images, how logic and thought are developed, and finally how human behavior, intelligence, and creativity arise.

But let us begin with the stepping-stones.

CHAPTER 24

Digestion, Transport, and the Integration of Metabolism

This chapter is the first of three surveying the biochemical aspects of metabolism and its regulation in the human body. We begin by examining the biochemistry of digestion and absorption in the gastrointestinal tract, the distribution of nutrients to the various organs, and the metabolic cooperation between the different tissues. We shall also survey the biochemical mechanisms by which oxygen is brought to the tissues and carbon dioxide and other end products of metabolism are eliminated.

One of the most important practical applications of biochemistry is in the diagnosis of diseases involving metabolic disturbances. We shall see how biochemical measurements are useful in medicine, taking as an example the metabolic disease diabetes mellitus.

Foods Are Enzymatically Digested to Prepare Them for Absorption

During digestion in the gastrointestinal tract of mammals the three major nutrients (carbohydrates, lipids, and proteins) undergo enzymatic hydrolysis into their building-block components. This is necessary for their utilization, since the cells lining the intestine are able to absorb into the bloodstream only relatively small molecules. For example, polysaccharides and even disaccharides must be completely hydrolyzed into their constituent monosaccharides by digestive enzymes before they can be absorbed. Similarly, proteins and lipids must also be hydrolyzed into their building-block components.

Figure 24-1 is an orienting "map" of the human digestive system. Although digestion begins in the mouth and stomach, the final stages of digestion of all major food components and absorption of their building-block components into the blood take place in the small intestine. It is anatomically well adapted for these functions, since it presents a very large surface area through which absorption can take place. Not only is the small intestine long (12 to 14 ft) but its inner surface has many folds

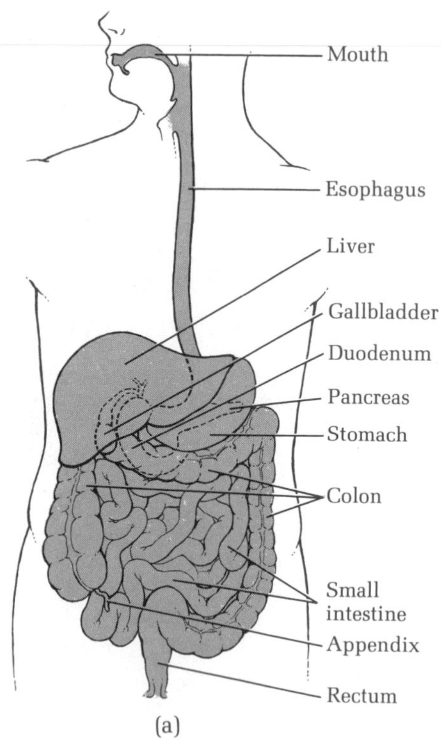

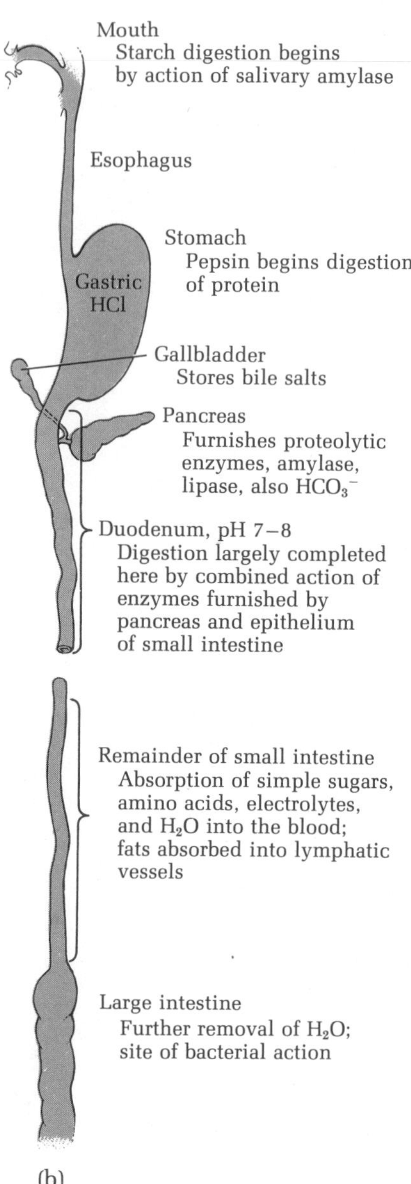

Figure 24-1
(a) The human gastrointestinal tract. (b) Flowsheet of digestion and absorption.

with numerous fingerlike protrusions, the villi. Each villus in turn is coated with many epithelial cells, having numerous microvilli (Figure 24-2). The villi provide a very large surface area through which the end products of digestion can be rapidly transported through the epithelial cells and into the blood capillaries and lymph vessels inside. The surface area of the human small intestine is about 180 m^2, a little less than the playing area of a tennis court.

The microvilli contain bundles of actin microfilaments (page 163) which connect with a web of myosin filaments at the base of the microvilli. This filament system imparts a wavelike motion to the microvilli, providing local stirring to enhance the absorption of the digested nutrients.

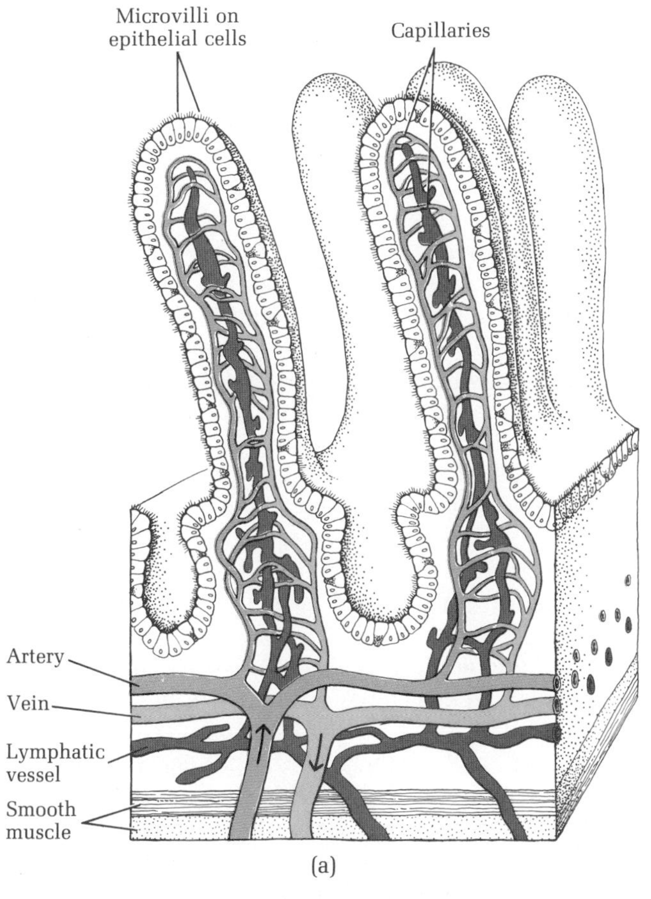

Figure 24-2
(a) *A group of villi of the small intestine, showing the large surface area available for absorption of digestion products. Amino acids, sugars, and salts are absorbed into the blood capillaries, whereas triacylglycerols enter the central lymphatic vessels. Each epithelial cell has many microvilli.* (b) *to* (d) *Three views of microvilli.* (b) *Scanning electron micrograph.* (c) *and* (d) *Transmission electron micrographs of longitudinal and cross sections, showing the internal microfilaments, which impart wavelike motion.*

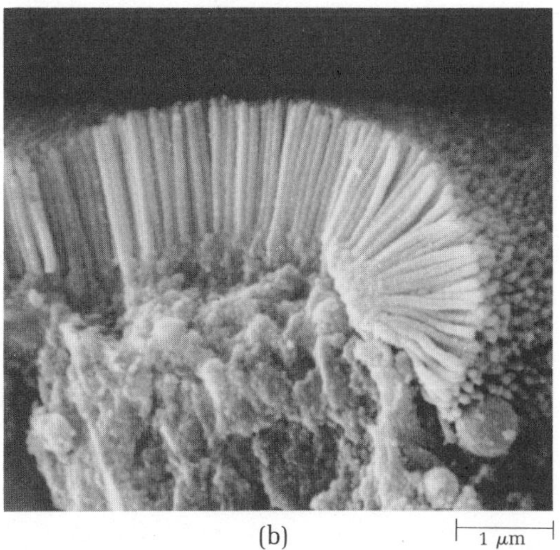

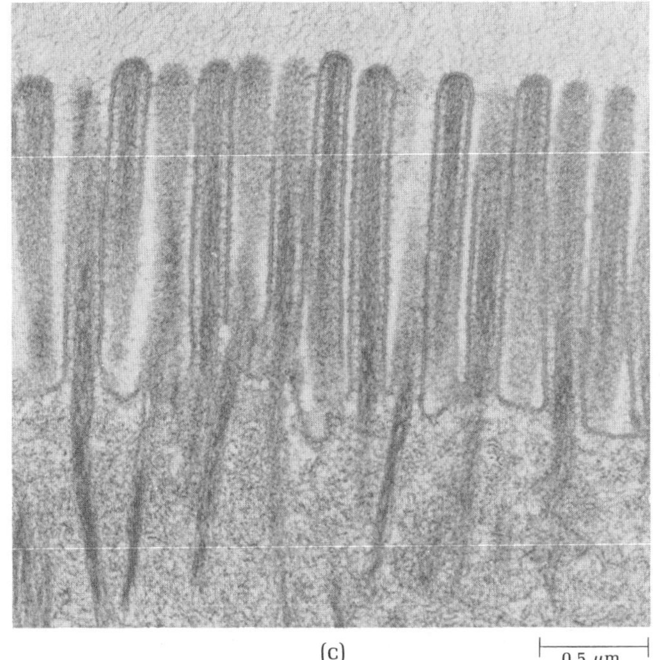

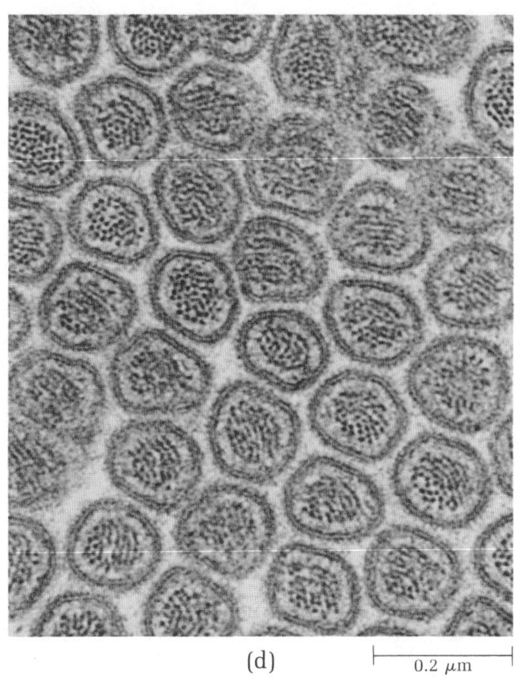

Digestion of Carbohydrates

The most abundant carbohydrates ingested by human beings are the polysaccharides starch and cellulose, furnished by plant foods, and glycogen, provided by foods of animal origin. Starch and glycogen are completely hydrolyzed by enzyme action in the gastrointestinal tract to yield their building blocks, free D-glucose. This process begins in the mouth during chewing, through the action of *amylase* secreted by the salivary glands. Salivary amylase hydrolyzes many of the $\alpha(1 \rightarrow 4)$ glycosidic linkages of starch and glycogen to yield a mixture of maltose, glucose, and oligosaccharides. Soda crackers gradually taste sweeter as we chew them because the starch they contain, which has no taste, is enzymatically hydrolyzed to yield sugars. The digestion of starch, glycogen, and other digestible polysaccharides to yield D-glucose is continued and completed in the small intestine, largely by the action of *pancreatic amylase*, made by the pancreas and secreted via the pancreatic duct into the upper portion of the small intestine. This segment of the small intestine, in which most of its digestive activity occurs, is called the *duodenum*.

Cellulose cannot be enzymatically digested and used by most mammals, for lack of enzymes capable of hydrolyzing the $\beta(1 \rightarrow 4)$ linkages between the successive D-glucose residues of cellulose (page 291). Nevertheless, undigested cellulose residues of plant foods provide bulk or fiber (also called "roughage") in the diet and are desirable for proper motility of the intestine. Cellulose can be digested by ruminant animals, but only indirectly. The rumen bacteria hydrolyze cellulose to yield D-glucose, which they ferment to yield lactate, acetate, and propionate, absorbed into the blood. Lactate and propionate are converted by the liver into blood sugar in ruminants (page 569).

Disaccharides are hydrolyzed by enzymes located in the outer border of the epithelial cells lining the small intestine. Sucrose or cane sugar is hydrolyzed to D-glucose and D-fructose by *sucrase*, also called *invertase*; lactose is hydrolyzed to D-glucose and D-galactose by *lactase*, also called *β-galactosidase*; and maltose is hydrolyzed by *maltase*, yielding two molecules of D-glucose. Recall (page 285) that most adults of Asian and African races show *lactose intolerance*, due to a loss of lactase activity in the small intestine after infancy and childhood. In lactose-intolerant individuals the undigested lactose remains in the intestine, where some of it undergoes fermentation by intestinal microorganisms. Diarrhea and formation of intestinal gases are the consequences.

In the epithelial cells lining the small intestine, D-fructose, D-galactose, and D-mannose are converted in part into D-glucose (page 417). The resulting mixture of simple hexoses is absorbed into the epithelial cells lining the small intestine and brought via the blood to the liver.

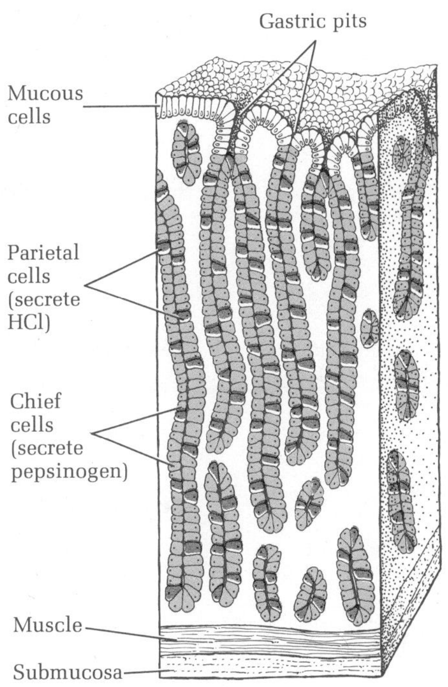

Figure 24-3
Gastric glands in the stomach lining. The parietal cells secrete HCl in response to the hormone gastrin, formed by epithelial cells when protein enters the stomach. The chief cells secrete pepsinogen.

Digestion of Proteins

Ingested proteins are enzymatically hydrolyzed into their constituent amino acids in the gastrointestinal tract. When protein enters the stomach, it stimulates the secretion of the hormone gastrin, which in turn stimulates the secretion of hydrochloric acid by the parietal cells of the gastric glands (Figure 24-3) and pepsinogen by the chief cells. The gastric juice has a pH between 1.5 and 2.5. The acidity of gastric juice acts as an antiseptic and kills most bacteria and other cells. In addition, it causes globular proteins to undergo denaturation or unfolding at this low pH, rendering their internal peptide bonds more accessible to enzymatic hydrolysis. Pepsinogen (MW 40,000), an inactive precursor or zymogen (page 229), is converted into active pepsin in the gastric juice by the enzymatic action of pepsin itself, an example of autocatalysis. In this process (Figure 24-4) 42 amino acid residues are removed from the amino-terminal end of the polypeptide chain of pepsinogen as a mixture of small peptides. The rest of the pepsinogen molecule, which remains intact, is enzymatically active pepsin (MW 33,000). In the stomach, pepsin hydrolyzes those peptide bonds of ingested proteins involving the aromatic amino acids tyrosine, phenylalanine, and tryptophan, among others (Table 24-1), thus cleaving long polypeptide chains into a mixture of smaller peptides.

As the acid stomach contents pass into the small intestine, the low pH triggers the secretion of the hormone secretin into

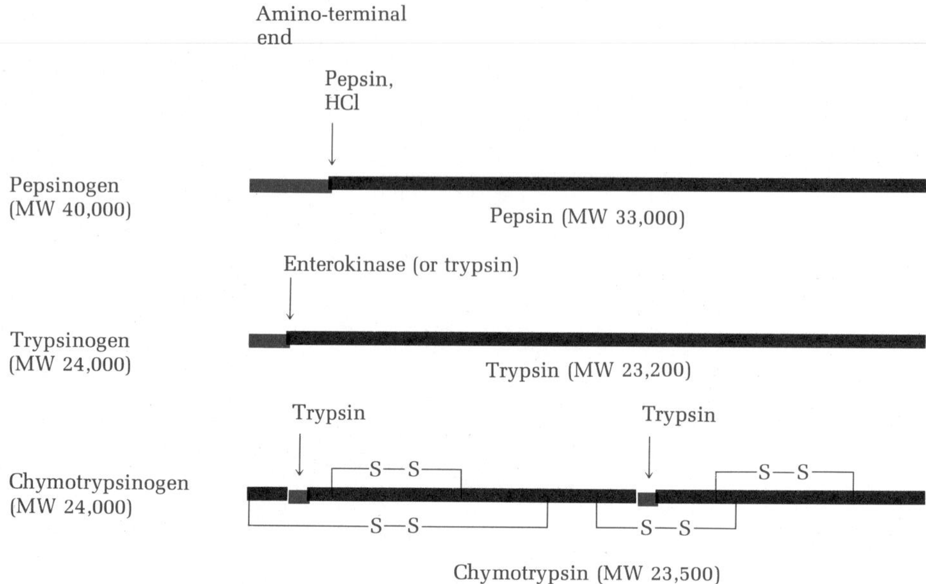

Figure 24-4
Activation of the zymogens of pepsin, trypsin, and chymotrypsin. The diagrams show the sites of proteolytic attack in the zymogens that set free the active enzymes (color). Segments of the polypeptide chains of the zymogens that are removed or excised are in black. Note that chymotrypsin has three polypeptide chains, held together covalently by two disulfide linkages and noncovalently by hydrogen bonds and hydrophobic interactions (page 229).

the blood. Secretin stimulates the pancreas to secrete bicarbonate into the small intestine to neutralize the gastric HCl. The pH then rises abruptly from between pH 1.5 and 2.5 to about pH 7. In the small intestine the digestion of proteins continues. The entry of amino acids into the duodenum releases the hormone cholecystokinin (page 746), which stimulates secretion of several pancreatic enzymes, whose optimum pHs are near 7. Three of these, trypsin, chymotrypsin, and carboxypeptidase, are made by the exocrine cells of the pancreas (Figure 24-5) as their respective enzymatically inactive zymogens, trypsinogen, chymotrypsinogen, and procarboxypeptidase. Synthesis of these enzymes as inactive precursors protects the exocrine cells from destructive proteolytic attack. After trypsinogen enters the small intestine, it is converted into its active form trypsin (Figure 24-4) by enterokinase, a specialized proteolytic enzyme secreted by intestinal cells. Once some free trypsin has been formed, it also can catalyze the conversion of trypsinogen into trypsin. The formation of free trypsin is brought about by removal of a hexapeptide from the amino-terminal end of the trypsinogen chain. As we have seen (page 131), trypsin hydrolyzes those peptide bonds whose carbonyl groups are contributed by lysine and arginine residues (Table 24-1).

Chymotrypsinogen has a single polypeptide chain with a number of intrachain disulfide bonds. When it reaches the small intestine, it is converted into chymotrypsin by trypsin, which cleaves the single long polypeptide chain of chymotrypsinogen at two points by excision of dipeptides (Figure 24-4). The three segments formed from the original chymotrypsinogen chain are still held together, however, by disulfide

Table 24-1 Enzymes Involved in Protein Digestion and Their Peptide Bond Specificity

Pepsin	Tyr, Phe, Trp; also Leu, Glu, Gln
Trypsin	Lys, Arg
Chymotrypsin	Tyr, Phe, Trp
Carboxypeptidase	Successive carboxy-terminal residues
Aminopeptidase	Successive amino-terminal residues (except proline)

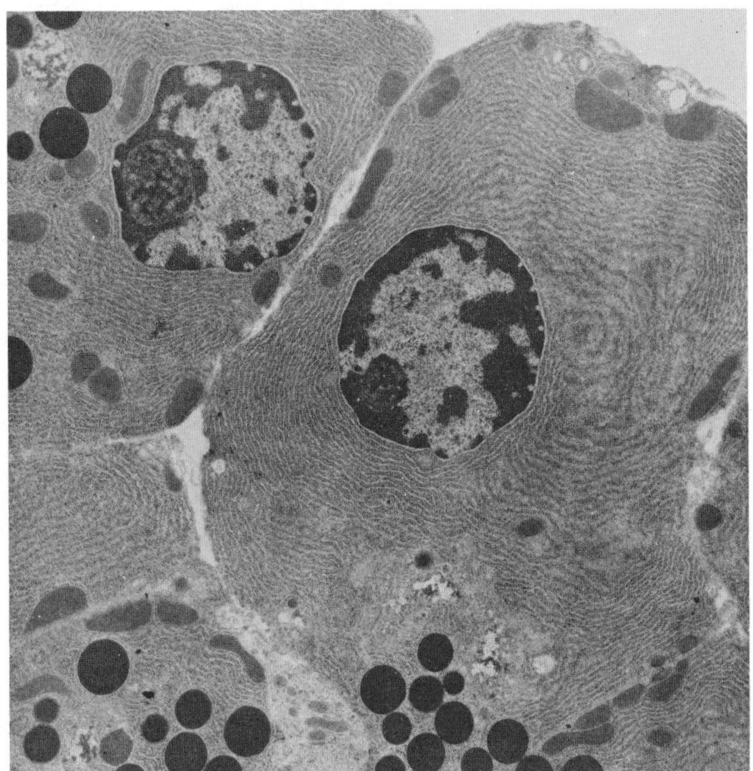

Figure 24-5
Adjacent exocrine cells of the pancreas. The cytoplasm is completely filled with rough-surfaced endoplasmic reticulum. The ribosomes on the latter synthesize the polypeptide chains of the zymogens of many digestive enzymes. The zymogens are concentrated in condensing vacuoles, ultimately forming mature zymogen granules. On stimulation of the cell the plasma membrane fuses with the membrane around the zymogen granules, the dense dark spheres at the bottom. They are released into the lumen of the collecting duct (the light region in lower left area) by exocytosis. The collecting ducts ultimately lead to the pancreatic duct and thence to the small intestine.

cross-linkages (page 229). Chymotrypsin hydrolyzes those peptide bonds involving *phenylalanine, tyrosine,* and *tryptophan* residues (Table 24-1). Trypsin and chymotrypsin thus hydrolyze into smaller peptides the polypeptides resulting from the action of pepsin in the stomach. This stage of protein digestion is accomplished very efficiently because pepsin, trypsin, and chymotrypsin have different amino acid specificities in hydrolyzing polypeptide chains.

Degradation of the short peptides in the small intestine is now completed by other peptidases. The first is *carboxypeptidase,* a zinc-containing enzyme (page 271), which the pancreas makes as its inactive zymogen *procarboxypeptidase.* Carboxypeptidase removes successive carboxyl-terminal residues from peptides. The small intestine also secretes an *aminopeptidase,* which can hydrolyze off successive amino-terminal residues from short peptides (Table 24-1). By the sequential action of these proteolytic enzymes and peptidases, ingested proteins are ultimately hydrolyzed to yield a mixture of free amino acids, which are then transported across the epithelial cells lining the small intestine. The free amino acids enter the blood capillaries in the villi and are transported to the liver.

Not all proteins are completely digested by human beings (Chapter 26). Most animal proteins are almost completely hydrolyzed into amino acids, but some fibrous proteins, such as

Prolipase (inactive)

↓ colipase

Lipase-colipase complex (active)

(a)

$$
\begin{array}{l}
H_2C-O-\underset{\substack{\| \\ O}}{C}-R_1 \\
| \\
HC-O-\underset{\substack{\| \\ O}}{C}-R_2 \quad \text{Triacylglycerol} \\
| \\
H_2C-O-\underset{\substack{\| \\ O}}{C}-R_3
\end{array}
$$

↓ lipase-colipase complex, bile salts, Na^+

$$
\begin{array}{l}
H_2C-OH \\
| \\
HC-O-\underset{\substack{\| \\ O}}{C}-R_2 \quad \text{2-Monoacylglycerol} \\
| \\
H_2C-OH \\
+ \\
R_1COONa \\
+ \\
R_3COONa
\end{array}
$$

Sodium soaps of fatty acids, emulsified with bile salts

(b)

Figure 24-6
(a) Activation of lipase and (b) its action on triacylglycerols. (a) Prolipase, secreted by the pancreas, becomes active in the small intestine. Colipase is a small protein (MW 10,000) that binds to lipase and stabilizes the enzyme. (b) Triacylglycerols are hydrolyzed by lipase to yield 2-monoacylglycerol and the 1- and 3-acyl groups as fatty acid soaps. The reaction is aided by the bile salts, which emulsify the fatty acid soaps.

keratin, are only partially digested. Many proteins of plant foods, such as cereal grains, are incompletely digested because the protein part of grains or seeds is surrounded by nondigestible cellulose husks.

Celiac disease is a rare condition in which the intestinal enzymes are unable to digest certain water-insoluble proteins of wheat, particularly *gliadin*, which is injurious to the cells lining the small intestine. Wheat products must therefore be avoided by such patients. Another disease involving abnormal activity of the proteolytic enzymes of the digestive tract is *acute pancreatitis*. In this condition, caused by obstruction of the normal pathway of secretion of pancreatic juice into the intestine, the zymogens of the proteolytic enzymes are converted into their catalytically active forms prematurely, *inside* the pancreatic cells. As a result these powerful enzymes attack the pancreatic tissue itself, causing a painful and serious destruction of the organ, which can be fatal. Normally the pancreatic zymogens are not activated until they reach the small intestine. The pancreas protects itself against self-digestion in another way: it makes a specific *trypsin inhibitor*, itself a protein. Since free trypsin can activate not only trypsinogen and chymotrypsinogen but also two other digestive zymogens, procarboxypeptidase and *proelastase*, trypsin inhibitor effectively prevents premature production of free proteolytic enzymes in pancreatic cells.

Digestion of Lipids

The digestion of triacylglycerols begins in the small intestine, into which the zymogen *prolipase* is secreted by the pancreas. There it is converted into active lipase, which, in the presence of *bile salts* (see below) and a special protein called *colipase*, binds to droplets of triacylglycerols and catalyzes the hydrolytic removal of one or both of the outer fatty acid residues to yield a mixture of free fatty acids as their Na^+ or K^+ soaps and 2-monoacylglycerols (Figure 24-6). A small fraction of the triacylglycerols remains unhydrolyzed.

The fatty acid soaps and the uncleaved acylglycerols are emulsified into fine droplets by *peristalsis*, the churning action of the intestine, aided by the detergent effect of the bile salts and the monoacylglycerols, which are amphipathic molecules. The fatty acids and monoacylglycerols in these droplets are absorbed by intestinal cells, where they are largely reassembled into triacylglycerols (page 595). The triacylglycerols do not pass into the blood capillaries but into the *lacteals*, small lymph vessels in the villi. The lymph draining the small intestine, called *chyle*, has a milky appearance after a fat-rich meal, due to the suspended *chylomicrons*, droplets of highly emulsified triacylglycerols about 1 μm in diameter. Chylomicrons have a hydrophilic coat of phospholipids and a special protein, which function to keep the chylomicrons suspended. The chylomicrons pass from the thoracic duct into the subclavian vein

Figure 24-7
Schematic drawing of the elements in fat digestion and absorption. Note that bile salts cycle from the liver to the small intestine and are then reabsorbed through the lymph vessels and portal vein and returned to the liver. Some of the bile salts are usually lost in the feces in each cycle.

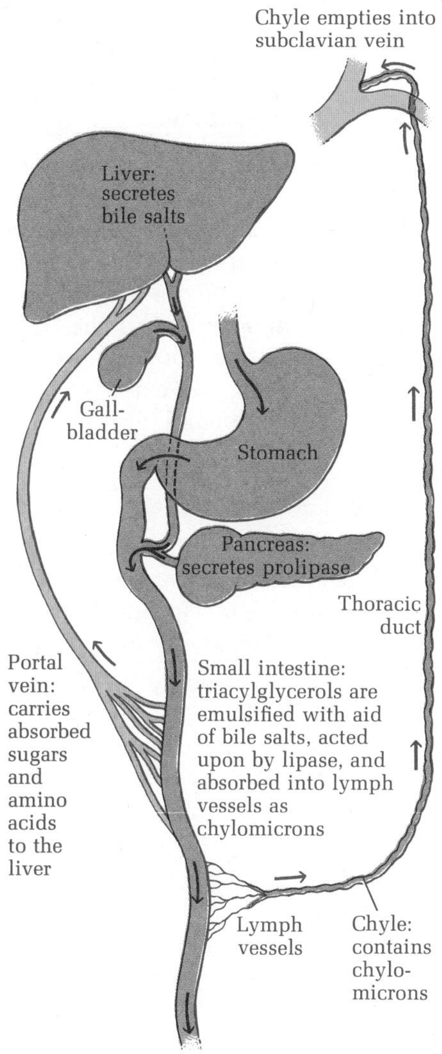

(Figure 24-7). After a meal rich in lipids the blood plasma itself becomes opalescent from the high concentration of chylomicrons, but the opalescence disappears in 1 or 2 h, as the triacylglycerols are removed from the blood, especially by the adipose tissues.

The emulsification and digestion of lipids in the small intestine is facilitated by the bile salts. The major human bile salts are sodium glycocholate and sodium taurocholate, derivatives of cholic acid (Figure 24-8), the most abundant of four major human bile acids. The bile salts are powerful emulsifying agents secreted by the liver into the bile, which empties into the upper portion of the small intestine. After the fatty acids and monoacylglycerols of the emulsified fat droplets have been absorbed in the lower small intestine, the bile salts aiding this process are also reabsorbed. They return to the liver, to be used over again. The bile salts thus constantly cycle between the liver and small intestine (Figure 24-7).

The bile salts are extremely important in absorption of all fat-soluble nutrients, not only triacylglycerols. When the formation or secretion of bile salts is deficient, which happens in some diseases, undigested and unabsorbed fats appear in the stool. In such conditions the fat-soluble vitamins A, D, E, and K (page 264) are not completely absorbed, and nutritional deficiency of vitamin A can result.

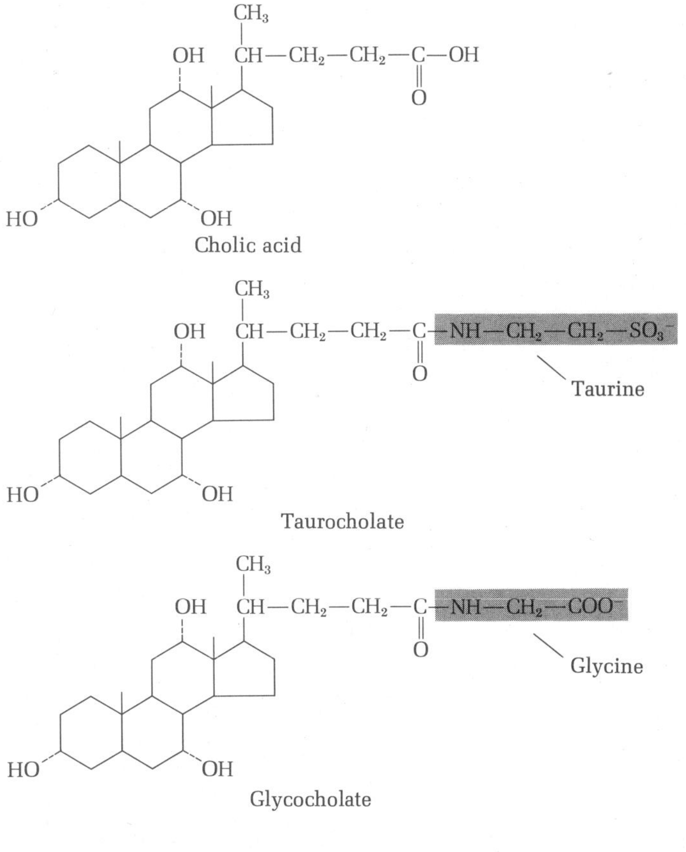

Figure 24-8
Cholic acid and its conjugated forms taurocholate and glycocholate. Because of their amphipathic qualities they are excellent detergents and emulsifiers. The glycine and taurine groups (color) are hydrophilic, while the steroid nucleus is hydrophobic.

The Liver Processes and Distributes Nutrients

After being absorbed from the intestinal tract the incoming nutrients, except for a large portion of the triacylglycerols, pass directly to the liver, the major nutrient distribution center in vertebrates. Here the incoming sugars, amino acids, and some of the lipids are processed and distributed to the other organs and tissues. Let us now see how the metabolic pathways of the major nutrients are integrated in the liver.

In the Liver Sugars Follow Five Metabolic Routes

In the liver much of the incoming free D-glucose is phosphorylated by ATP to yield glucose 6-phosphate. D-Fructose, D-galactose, and D-mannose, absorbed from the small intestine, are also converted into D-glucose 6-phosphate by enzymatic pathways examined earlier (page 417). D-Glucose 6-phosphate is at the crossroads of carbohydrate metabolism in the liver. There are five major metabolic routes that may be taken by glucose 6-phosphate in the liver, depending upon the minute-by-minute and hour-by-hour supply and demand (Figure 24-9).

Conversion into Blood Glucose

Glucose 6-phosphate is dephosphorylated by glucose 6-phosphatase to yield free D-glucose, which passes into the systemic blood to be transported to other tissues. This pathway has first call on glucose 6-phosphate, since the blood glucose concentration must be kept sufficiently high to provide adequate energy for the brain and other tissues.

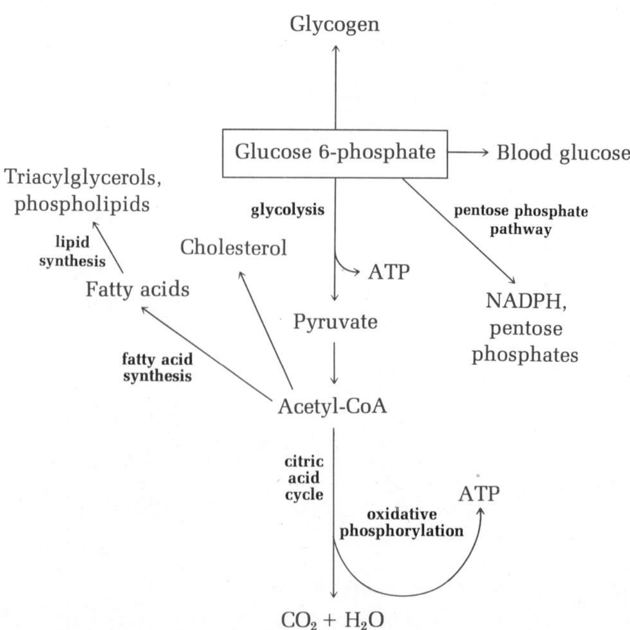

Figure 24-9
Metabolic pathways for glucose 6-phosphate in the liver. Here, as well as in Figures 24-10 and 24-11, biosynthetic pathways lead upward, degradative pathways downward, and distribution to other organs horizontally.

Conversion into Glycogen

Glucose 6-phosphate not immediately needed to form blood glucose is converted into liver glycogen by the sequential action of *phosphoglucomutase* and *glycogen synthase* (page 572).

Conversion into Fatty Acids and Cholesterol

Excess glucose 6-phosphate not used to make blood glucose or liver glycogen is degraded via glycolysis and pyruvate dehydrogenase into acetyl-CoA, which is converted into malonyl-CoA, and thence into fatty acids (page 595). These are used to form triacylglycerols and phospholipids (pages 596 and 600), which are in part exported to other tissues, carried there by plasma lipoproteins. Some acetyl-CoA is also used by the liver to make cholesterol (page 607).

Oxidative Degradation to CO_2

Acetyl-CoA, yielded from glucose 6-phosphate via glycolysis and decarboxylation of pyruvate, may be oxidized via the citric acid cycle. The ensuing electron transport and oxidative phosphorylation yield energy in the form of ATP. Normally, however, fatty acids are the major oxidative fuel for the citric acid cycle in the liver.

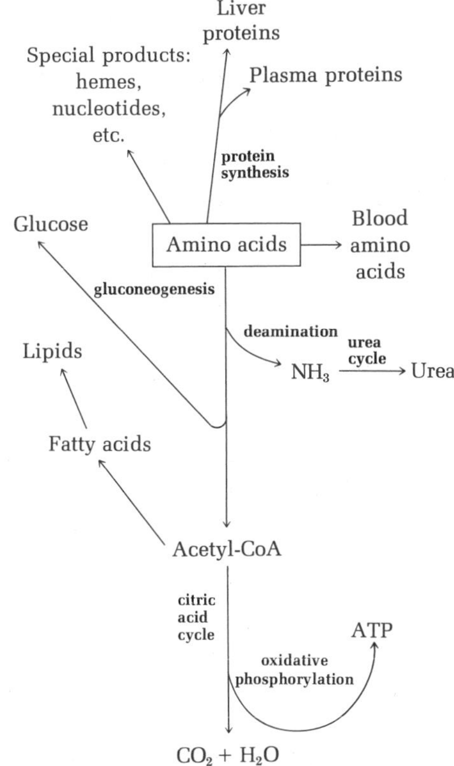

Figure 24-10
Metabolic pathways of amino acids in the liver.

Degradation via the Pentose Phosphate Pathway

Glucose 6-phosphate is the substrate for the *pentose phosphate pathway*, yielding (1) reducing power in the form of NADPH, needed in the reduction steps in the biosynthesis of fatty acids and cholesterol (page 592), and (2) D-ribose 5-phosphate, a precursor in nucleotide biosynthesis (page 457).

Through the action of various regulatory enzymes and through hormonal regulation (Chapter 25), the liver directs the flow of glucose residues into these different pathways according to the prevailing supply-and-demand economy of the organism.

Amino Acids Also Have Five Metabolic Routes

The amino acids entering the liver, following absorption from the intestinal tract, also have several important metabolic routes (Figure 24-10).

Transport to Other Tissues

Amino acids may pass into the systemic blood and thus to other organs, to be used as building blocks in the biosynthesis of tissue proteins (Chapter 29).

Biosynthesis of Liver Proteins and Plasma Proteins

The liver constantly renews its own intrinsic proteins, which have a very high turnover rate, with an average half-life of only a few days. The liver is also the site of biosynthesis of most of the plasma proteins of the blood.

Deamination and Degradation

Amino acids not needed for protein biosynthesis in the liver or elsewhere are deaminated and degraded to yield acetyl-CoA and citric acid cycle intermediates (page 636). Citric acid cycle intermediates so formed may be converted into glucose and glycogen via the gluconeogenesis pathway (page 562). Acetyl-CoA may be oxidized via the citric acid cycle for ATP energy, or it may be converted into lipids for storage, described before. The ammonia released on degradation of amino acids is converted by the liver into the excretory product urea by the urea cycle (page 549).

Participation in the Glucose-Alanine Cycle

The liver also participates in the metabolism of amino acids arriving intermittently from the peripheral tissues. Some hours after each meal there is a period when alanine is carried by the blood from the muscles to the liver, where it is deaminated and the resulting pyruvate converted into blood glucose via gluconeogenesis (page 547). The glucose returns to the skeletal muscles to replenish muscle glycogen stores. One purpose of this cyclic process, the *glucose-alanine cycle*, is to smooth out fluctuations in the blood glucose level in the periods between meals. The blood is adequately supplied with glucose just after the digestion and absorption of dietary carbohydrate and subsequently by the conversion of some of the liver glycogen into blood glucose. But in the period preceding the next meal, there is some degradation of muscle protein to amino acids, which donate their amino groups by transamination to pyruvate, the product of glycolysis, to yield alanine. The alanine thus transports both pyruvate and NH_3 to the liver. There alanine is deaminated, the pyruvate converted into blood glucose, and the NH_3 converted into urea for excretion. The amino acid deficit incurred in the muscles is made up after the next meal from incoming dietary amino acids.

Conversion into Nucleotides and Other Products

Amino acids are precursors in the biosynthesis of the purine and pyrimidine bases of the nucleotides (page 627) and in the synthesis of specialized products such as porphyrins (page 625), hormones (page 625), and other nitrogenous compounds. These relationships are summarized in Figure 24-10.

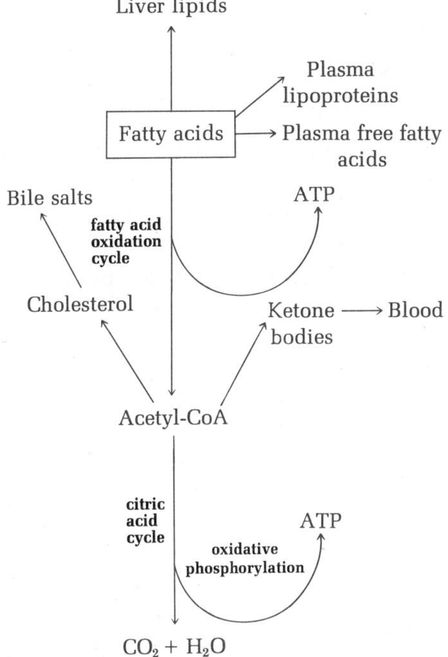

Figure 24-11
Metabolism of fatty acids in the liver.

Lipids Follow Five Pathways

The fatty acid components of the lipids entering the liver also have several different pathways, as outlined in Figure 24-11.

Oxidation to CO_2 with ATP Production

Free fatty acids may be activated and oxidized to yield acetyl-CoA and ATP (page 521). The acetyl-CoA is oxidized via the citric acid cycle to yield ATP by oxidative phosphorylation. Fatty acids are the major oxidative fuel in the liver.

Formation of Ketone Bodies

Excess acetyl-CoA released on oxidation of fatty acids and not required by the liver is converted into the ketone bodies, acetoacetate and D-β-hydroxybutyrate, which are circulated via the blood to peripheral tissues, to be used there as fuel for the citric acid cycle (page 524). The ketone bodies may be regarded as a transport form of acetyl groups. They can supply a significant fraction of the energy of some peripheral tissues, up to one-third in the case of the heart.

Biosynthesis of Cholesterol and Bile Salts

Some of the acetyl-CoA derived from fatty acids (and from glucose) will be used as the major precursor for the biosynthesis of cholesterol, which in turn is the precursor of the bile salts, which are essential for the digestion and absorption of lipids (pages 610 and 691).

Biosynthesis of Plasma Lipoproteins

Fatty acids are also used as precursors for the synthesis of the lipid portion of the plasma lipoproteins, which carry lipids to adipose or fat tissue for storage as triacylglycerols.

Formation of Plasma Free Fatty Acids

Free fatty acids become bound to serum albumin and are carried via the blood to the heart and skeletal muscles, which absorb and oxidize free fatty acids as a major fuel.

The liver therefore has an extraordinary flexibility and range in its metabolism. It is well adapted to serve as the body's distribution center, to export nutrients in the correct proportions to the other organs, to smooth out fluctuations in metabolism caused by the intermittent way food is taken, and to process excess amino groups into urea and other products to be disposed of by the kidneys.

In addition to processing and distribution of carbohydrates, fat, and amino acids, the liver is also active in the enzymatic detoxification of foreign organic compounds, such as drugs, food additives, preservatives, and other possibly harmful agents with no food value. Detoxification usually involves the enzymatic hydroxylation of relatively insoluble organic compounds to make them more soluble for further breakdown and excretion.

Each Organ Has Specialized Metabolic Functions

Nearly all cells of vertebrates are equipped with the appropriate enzymes to catalyze the central pathways of metabolism, particularly those yielding energy in the form of ATP, those required for replacement of the organisms' glycogen and lipid stores and for maintenance of their proteins and nucleic acids. However, over and above the basic "housekeeping" metabolism that occurs in all cells, the various organs show characteristic differences in the way they share in various body functions and how they use their ATP energy. We have seen that the liver plays a central processing and distributing role and furnishes all the other organs and tissues with a proper mix of nutrients via the blood. We shall now examine the metabolic characteristics of other major organs and tissues and see how they use their ATP energy.

Skeletal Muscle Uses ATP for Intermittent Mechanical Work

The skeletal muscle mass accounts for over 50 percent of the total oxygen consumption in the resting human being and up to 90 percent during very active muscular work. The metabolism of skeletal muscle is primarily specialized to generate ATP as the immediate source of energy for contraction and relaxation. Moreover, skeletal muscle is adapted to do its mechanical work in an intermittent fashion, on demand. Sometimes skeletal muscles must deliver an enormous amount of work in a very short time, as in a 100-meter sprint.

Skeletal muscles can use glucose, free fatty acids, or ketone bodies as fuel, depending on the degree of activity. In *resting muscle* the basic fuels are free fatty acids and ketone bodies, carried from the liver via the blood. These are oxidized and degraded to yield acetyl-CoA, which enters the citric acid cycle for oxidation to CO_2. The ensuing transport of electrons to oxygen provides the energy for the oxidative phosphorylation of ADP to ATP. *Moderately active muscles* use blood glucose in addition to fatty acids and ketone bodies. The glucose is phosphorylated and degraded by the glycolytic sequence to pyruvate, which is then oxidized via acetyl-CoA by the citric acid cycle. However, in *maximally active muscles,* the demand for ATP for contraction is so great that the blood flow cannot provide oxygen and fuels fast enough. Under these conditions, the stored muscle glycogen is called upon. The muscle glycogen is broken down to lactate by anaerobic glycolysis, with a yield of two ATPs per glucose unit degraded (page 413). Anaerobic glycolysis thus provides extra ATP energy, supplementing the basal ATP production resulting from the aerobic oxidation of other fuels via the citric acid cycle. The use of blood glucose and muscle glycogen as emergency fuel for muscular activity is greatly enhanced by the secretion of adrenaline, which stimulates formation of blood glucose from glycogen in the liver and the breakdown of glycogen to lactate in muscle tissue (Chapter 25). Because skeletal muscle contains no glucose 6-phosphatase, its glycogen is completely dedicated to providing energy via glycolytic breakdown.

Since skeletal muscles lack a large store of glycogen, there is an upper limit to the amount of glycolytic energy available during an all-out sprint. Moreover, the accumulation of lactic acid, the consequent decrease in pH, and the rise in temperature occurring in maximally active muscles causes them to become less efficient. During the recovery period after such a maximal burst of muscular activity an athelete continues to breathe hard for some time. The extra oxygen consumed is used to oxidize pyruvate, lactate, and other fuels to regenerate ATP and phosphocreatine in the muscles. Simultaneously, in the liver, blood glucose is rebuilt from blood lactate by gluconeogenesis and conveyed to the muscles to replenish their glycogen (page 562). The extra oxygen consumed, the "oxygen debt," thus restores the normal metabolic state of the body by a complex series of adjustments in which the muscles and liver cooperate metabolically (Figure 24-12).

Skeletal muscles are adapted in another way to provide maximum ATP for emergencies. They contain considerable amounts of *phosphocreatine* (page 383), which can rapidly replenish the terminal phosphate groups of ATP after they have been expended in contraction, by the creatine kinase reaction

$$\text{Phosphocreatine} + \text{ADP} \overset{Mg^{2+}}{\rightleftharpoons} \text{creatine} + \text{ATP}$$

which proceeds predominantly to the right during periods of active contraction and glycolysis. Phosphocreatine is resynthesized from creatine at the expense of ATP during recovery.

Skeletal muscle requires ATP not only to bring about sliding of the actin filaments along the myosin or thick filaments (page 381) but also to relax again. Muscle contraction is initiated by a motor nerve impulse, which is transmitted to the transverse tu-

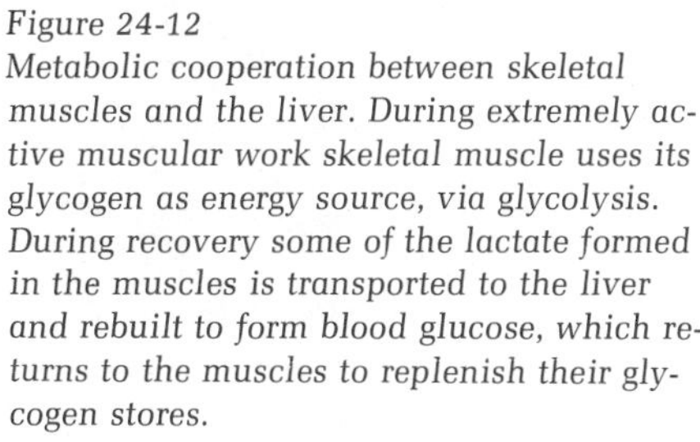

Figure 24-12
Metabolic cooperation between skeletal muscles and the liver. During extremely active muscular work skeletal muscle uses its glycogen as energy source, via glycolysis. During recovery some of the lactate formed in the muscles is transported to the liver and rebuilt to form blood glucose, which returns to the muscles to replenish their glycogen stores.

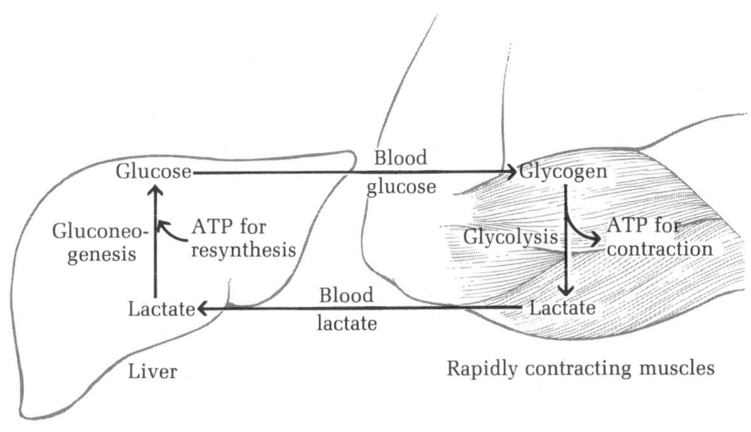

bules and sarcoplasmic reticulum, from which Ca^{2+} is released into the sarcoplasm. The Ca^{2+} is bound to *troponin*, a regulatory protein that translates this signal into sliding of the actin filaments at the expense of ATP energy. When the motor nerve impulses cease, the Ca^{2+} in the sarcoplasm must be sequestered again to relax the muscle. The Ca^{2+} is transported back into the sarcoplasmic reticulum by a Ca^{2+}-transporting membrane ATPase. Two Ca^{2+} ions are transported inward for each molecule of ATP hydrolyzed. Almost as much ATP is required for relaxing skeletal muscle as for supporting its contraction.

Heart Muscle Must Work Constantly and Rhythmically

Heart muscle also contains myosin and actin filaments but differs from skeletal muscle in that it is continuously active in a regular rhythm of contraction and relaxation. Although the heart must sometimes work harder and faster than normally, e.g., when the body's demand for oxygen increases or when the heart is stimulated by adrenaline (Chapter 25), it does not have the very large range of work output shown by skeletal muscle. Moreover, the heart has a completely aerobic metabolism at all times, in contrast to skeletal muscle, which can function anaerobically for short periods. Mitochondria are much more profuse in heart muscle than in skeletal muscles; they make up almost half the volume of the cells (Figure 24-13). As fuel the heart uses a mixture of glucose, free fatty acids, and ke-

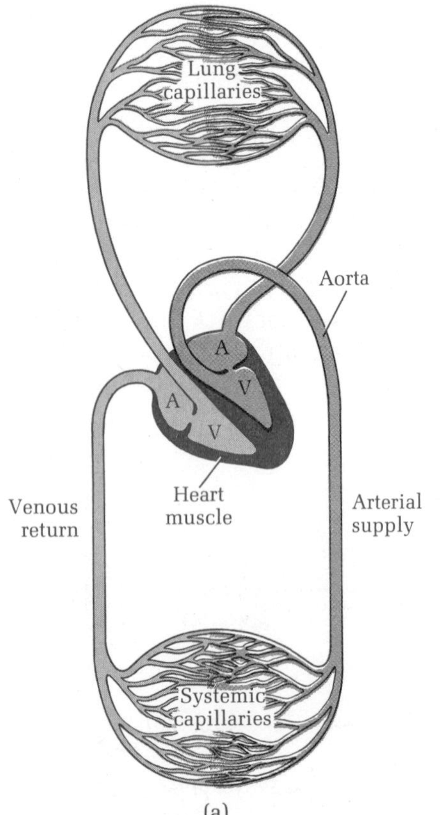

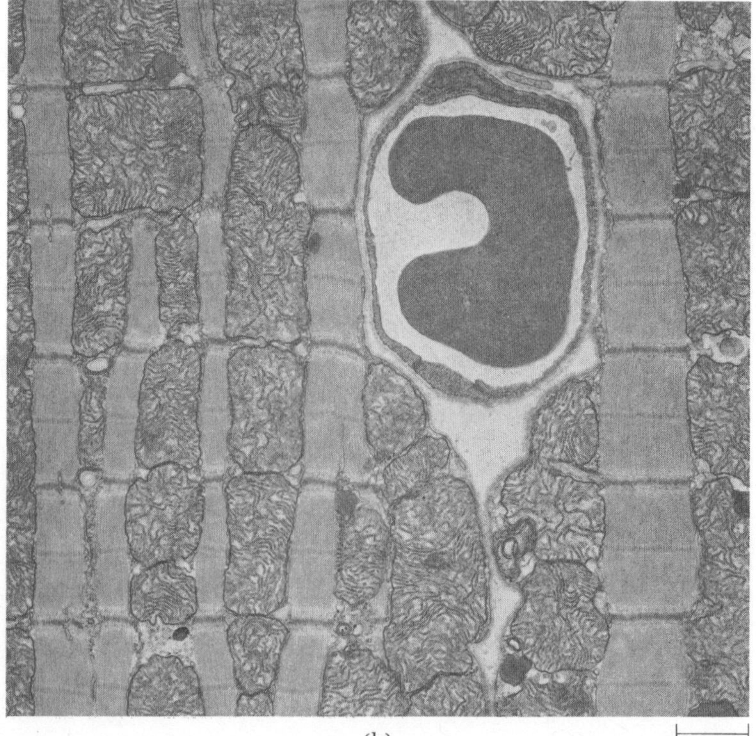

Figure 24-13
(a) Outline of the pumping circuit of the cardiovascular system. A = atrium, V = ventricle, the two types of chambers in the heart. The human heart pumps blood at a rate of 5 quarts per minute, or 75 gallons per hour, or 18 million barrels in a 70-year lifetime. (b) Electron micrograph of heart muscle, showing the profuse mitochondria. In the center is the lumen of a blood capillary and a single red blood cell.

tone bodies arriving from the blood. These fuels are oxidized via the citric acid cycle to deliver the energy required to generate ATP by oxidative phosphorylation. Like skeletal muscle, heart muscle does not store lipids or glycogen in large amounts. Small amounts of reserve energy are stored in the form of phosphocreatine. Individual beats or contractions of the heart muscle are initiated or triggered by nerve impulses, which cause release of free Ca^{2+} into the cytosol bathing the myofibrils; conversely, relaxation is brought about by the ATP-dependent reuptake of Ca^{2+} by the sarcoplasmic reticulum. Because the heart is normally aerobic and obtains nearly all its energy from oxidative phosphorylation, the failure of oxygen to reach a portion of the heart muscle when blood vessels are blocked by lipid deposits (page 610) can cause this region of the heart muscle to die, a process known as *myocardial infarction*.

The Brain Uses Energy for Transmission of Impulses

The metabolism of the brain is remarkable in several respects. First, the brain of adult mammals normally uses only glucose as fuel. Second, the brain has a very active respiratory metabolism; it uses almost 20 percent of the total oxygen consumed by a resting human adult. Moreover, the use of oxygen by the brain is fairly constant in rate and does not change significantly during active thought or sleep. Because the brain contains very little glycogen, it depends on the incoming glucose from the blood on a minute-to-minute basis. If the blood glucose should fall significantly below certain critical levels for even short periods of time, severe and sometimes irreversible changes in brain function may occur. For this reason brain surgery can be carried out only if the brain is constantly supplied with blood glucose.

Although the brain cannot directly use free fatty acids or lipids from the blood as fuels, it can use blood β-hydroxybutyrate formed from fatty acids in the liver. The capacity of the brain to oxidize β-hydroxybutyrate via acetyl-CoA (page 525) becomes important during prolonged fasting or starvation, after essentially all the liver glycogen has been depleted, since it allows the brain to use body fat as a source of energy. Stores of body fat are much larger than the stores of glycogen, which are used up in only a few days. The utilization of β-hydroxybutyrate by the brain during starvation also spares muscle proteins, which, via gluconeogenesis, are the ultimate source of glucose for the brain during starvation.

Glucose is used by the brain via the glycolytic sequence and the citric acid cycle; the ATP supply of the brain is almost entirely generated via catabolism of glucose. ATP energy is required to support the capacity of nerve cells (neurons) to maintain an electric potential across the plasma membrane, particularly that surrounding their long processes, the *axons* and *dendrites*, which form the "transmission lines" of the ner-

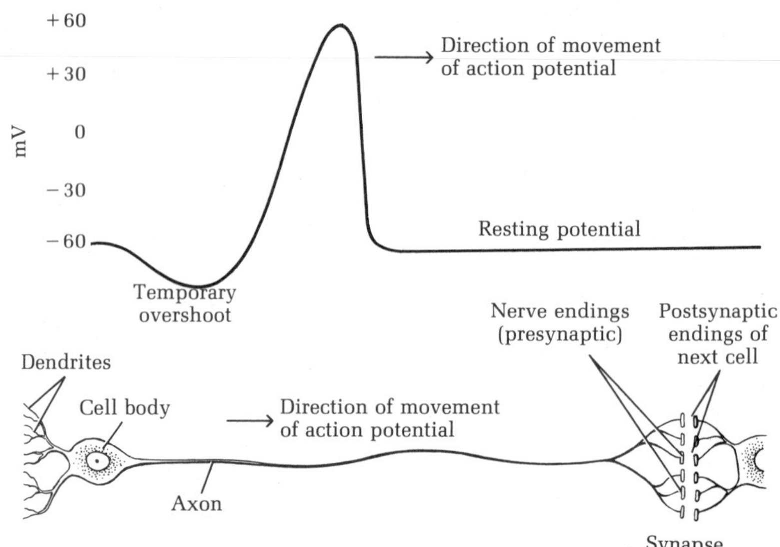

Figure 24-14
A neuron and the action potential. Impulses received by the dendrites are transmitted as a wavelike action potential along the axon to the next neuron. The resting potential is normally −60 mV (negative inside). The reversal of the sign of the potential results from a rapid transient influx of Na^+ *from the extracellular space due to selective opening of* Na^+ *gates. The resting potential is restored by the action of the* Na^+K^+*-transporting ATPase of the axonal membrane.*

vous system. A wavelike change in the membrane's electrical properties, called the *action potential,* is the means of transmitting nerve impulses along neurons. The plasma membrane Na^+K^+ATPase (page 386) requires a steady supply of ATP energy to pump K^+ into and Na^+ out of the axons (Figure 24-14). Three Na^+ ions are transported out and two K^+ ions in for each molecule of ATP hydrolyzed. Because of this imbalance of electric charges, the Na^+K^+ATPase generates an electric potential difference across the axonal membrane, which is normally positive outside.

ATP is also used in large amounts by the brain to synthesize the neurotransmitter substances that carry impulses from one neuron to the next across *synapses,* the junctions between successive nerve cells. Many different neurotransmitter and inhibitor substances are known; each is specific for certain types of neurons or for certain regions of the brain. Among them are the amino acids *glutamate, glutamine, aspartate, glycine,* and *γ-aminobutyrate.* In other synapses certain other amino acid derivatives, peptides, or *acetylcholine* function as transmitters or inhibitors. Neurotransmitters are stored in special vesicles in the presynaptic nerve ending (Figure 24-15). In response to the action potential arriving via the axonal membrane, the contents of some of these vesicles are released into the synaptic cleft and become bound to specific receptor sites in the sensitive ending of the postsynaptic neuron, stimulating it to transmit the impulse. After stimulating the postsynaptic neuron, the neurotransmitter released into the cleft must be quickly destroyed by enzymes or reabsorbed back into the presynaptic terminal so that the synapse is ready to transmit another impulse. Acetylcholine, the transmitter substance in some nerve circuits and in neuromuscular junctions, is enzymatically inactivated in the synaptic cleft by the enzyme *acetylcholine esterase,* which hydrolyzes it to acetate and free choline (page 221).

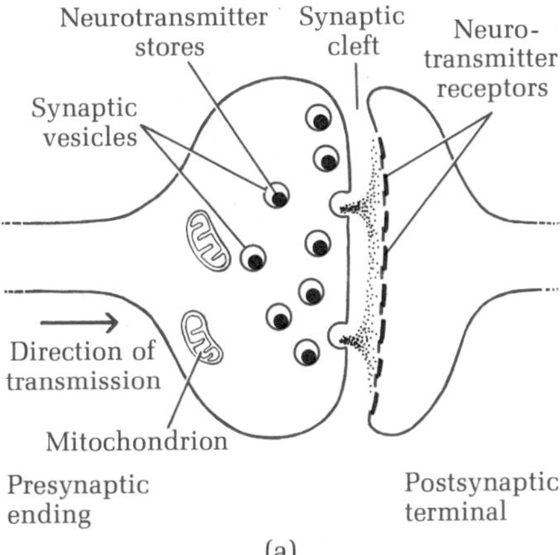

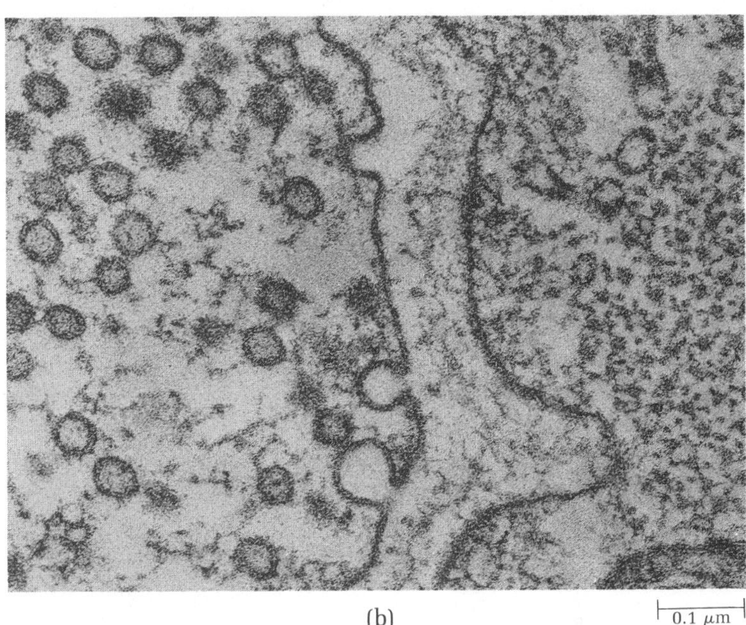

Figure 24-15
(a) Drawing of a synapse, showing its major functional components. The mitochondria provide ATP for concentrating the neurotransmitter in the secretory vesicles, as well as the energy for reabsorbing the neurotransmitter from the synaptic cleft. (b) An electron micrograph showing the release of neurotransmitters from the secretory vesicles into a neuromuscular junction.

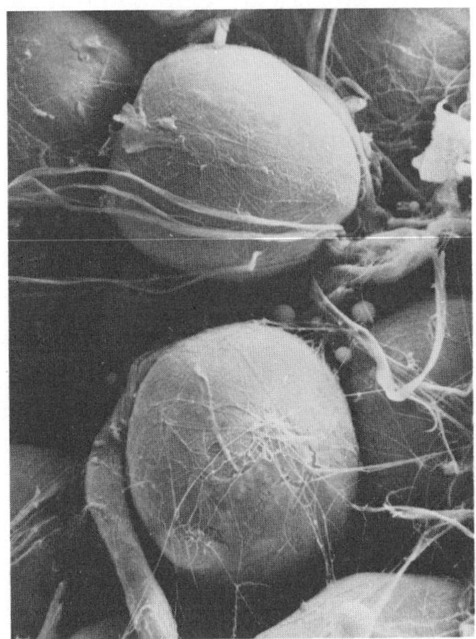

Figure 24-16
Scanning electron micrograph of adipocytes. Capillaries and collagen fibers form a supporting network around adipocytes in fat tissues. Nearly all the volume of the adipocytes is filled with fat droplets, which are very active metabolically.

Adipose Tissue Has an Active Metabolism

Adipose tissue, which consists of *adipocytes* or fat cells (Figure 24-16), is amorphous and widely distributed in the body: under the skin, around the deep blood vessels, and in the abdominal cavity. The total amount of adipose tissue in a young adult male of average weight is about 20 kg, almost equal to the total muscle mass. Approximately 65 percent of the weight of adipose tissue consists of stored triacylglycerols. Although fat tissue may give the impression of being inert, it is metabolically very active. It responds quickly to metabolic and hormonal stimuli and takes part in an active interplay with the liver, skeletal muscles, and the heart.

Like other cell types in the body, fat cells have an active glycolytic metabolism, they use the citric acid cycle to oxidize pyruvate and fatty acids, and they carry out oxidative phosphorylation. During periods of high carbohydrate intake, adipose tissue actively converts glucose via pyruvate and acetyl-CoA into fatty acids, from which triacylglycerols are made and stored as large fat globules (Figure 24-16). In this process they require NADPH as a reducing agent; it is generated by the pentose phosphate cycle and also by malate enzyme (page 593).

Adipocytes also actively store triacylglycerols arriving from the intestinal tract in the form of chylomicrons (pages 316 and 690), particularly after meals rich in fat. Chylomicrons arriving at adipose tissue are acted upon by *lipoprotein lipase,* which is located in cells of the blood capillaries. This enzyme hydrolyzes one or more fatty acids from the triacylglycerols of chylomicrons. The free fatty acids released by lipoprotein lipase are then absorbed into the adipocytes and are enzymatically converted into triacylglycerols for storage. Lipoprotein lipase also occurs in capillaries of skeletal muscles and the heart, which also absorb the free fatty acids from chylomicrons but oxidize them as fuel. As their triacylglycerols are degraded by lipoprotein lipase, the chylomicrons in the blood shrink in size but retain their phospholipids, cholesterol esters, and proteins. The resulting structures, called chylomicron remnants, are removed from the blood by the liver. Triacylglycerols already stored in adipocytes are not attacked by lipoprotein lipase, which is located in cells of the blood vessels. Rather, they are hydrolyzed by lipases within the adipocytes to release free fatty acids, which may be delivered into the blood where they are bound by serum albumin.

Each molecule of serum albumin can bind two molecules of long-chain fatty acid very tightly and one or two more loosely. Because of its very high concentration in the blood plasma, serum albumin is the major carrier of fatty acids in the blood. They are carried to skeletal muscles and the heart, which use most of the free fatty acids bound to serum albumin.

The release of fatty acids from adipocytes is greatly accelerated by the hormone *adrenaline* (Chapter 25), which is bound to receptors on the cell surface and stimulates the conversion of the inactive form of adipocyte lipase into its active form through a phosphorylation reaction. On the other hand, when insulin is bound to the cell surface of adipocytes, it cancels the effect of adrenaline and decreases the action of adipocyte lipase.

Some people have a genetic deficiency in lipoprotein lipase of adipocytes. As a consequence chylomicrons persist in the bloodstream for long periods after fat-rich meals. The triacylglycerols that cannot be properly used for lack of lipoprotein lipase are deposited in yellow, lipid-filled swellings under the skin. In other human genetic diseases the metabolism of one type or another of plasma lipoprotein is defective. There appears to be a greater incidence of atherosclerosis and coronary infarction among persons with excessively high blood triacylglycerol concentrations.

Human beings, as well as many animals, particularly those which hibernate, have a specialized type of adipose tissue called *brown fat* (Figure 24-17). It is particularly prominent in newborn infants, where it is present in the neck, upper chest, and upper back. Brown fat owes its color to numerous mitochondria, which in turn are rich in cytochromes (page 491). Brown fat is specialized to generate heat rather than ATP during the oxidation of fatty acids. The inner membranes of the mitochondria in brown fat have specific H^+-transporting pores, which are under regulation. These allow H^+, ejected during electron transport (page 489), to flow back into the respiring mitochondria, thus causing futile cycling of H^+, which generates heat instead of ATP (page 492). When thermal energy is not needed, the H^+ pores are closed, allowing the brown fat mitochondria to make ATP again.

Figure 24-17
Location of brown fat (color) in the neck and back in adults.

The Kidneys Use ATP to Do Osmotic Work

The kidneys have a very active respiratory metabolism and considerable metabolic flexibility. They can use blood glucose, ketone bodies, free fatty acids, and amino acids as sources of fuel, degrading them ultimately via the citric acid cycle to make ATP energy available by oxidative phosphorylation. Most of this energy is used to bring about the formation of urine, which takes place in a two-stage process. In the first stage, the blood plasma is filtered through microscopic structures called *glomeruli* in the cortex or outer layer of the kidney (Figure 24-18). The glomeruli allow all components of the blood plasma, except the proteins and their ligands, to pass into the long *renal tubules*, which are lined with epithelial cells capable of ATP-dependent active transport of certain ions and metabolites between the tubule contents and the blood in the capillaries surrounding the tubules. As the plasma filtrate passes down these tubules, water is reabsorbed back into blood capillaries surrounding the tubules. As a result the glomerular filtrate undergoes concentration, as well as changes in its composition, as it proceeds down the tubules. Each milliliter of the final bladder urine is formed by the concentration of 50 to 100 mL of the glomerular filtrate. The posterior pituitary hormone vasopressin (Chapter 25) promotes the reabsorption of H_2O from the tubules. The solute composition of normal human urine is shown in Table 24-2.

Table 24-2 Major Components of Human Urine†

Component	Grams per 24 h	Approximate urine/plasma concentration ratio
Glucose	<0.05	<0.05
Amino acids	0.80	1.0
Ammonia	0.80	100
Urea	25	70
Creatinine	1.5	70
Uric acid	0.7	20
H^+	pH 5–8	Up to 300
Na^+	3.0	1.0
K^+	1.7	15
Ca^{2+}	0.2	5
Mg^{2+}	0.15	2
Cl^-	6.3	1.5
HPO_4^{2-}	1.2 g P	25
SO_4^{2-}	1.4 g S	50
HCO_3^-	0–3	0–2

† The 24-h volume and composition of urine vary widely depending on fluid intake and diet. The data shown are for an average 24-h specimen of total volume 1200 mL.

Some components of urine, particularly glucose, are normally present in lower concentration in the urine than in the blood. Components of this group are reabsorbed from the glo-

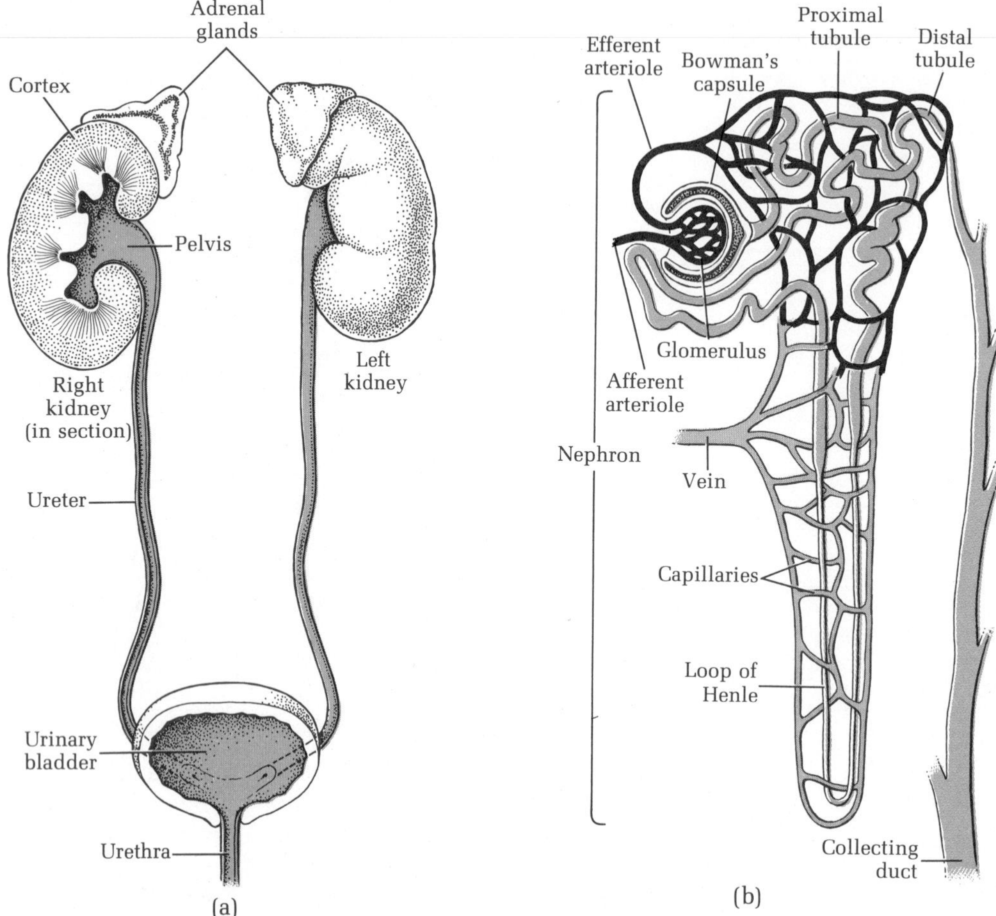

Figure 24-18
(a) The kidney contains many functional units called nephrons. Urine collected from the nephrons passes into the pelvis of the kidney and thence into the bladder via the ureters. (b) Schematic drawing of a nephron. Blood plasma is filtered by the glomerulus. The filtrate is caught by Bowman's capsule and passes down the long renal tubule, which is lined with epithelial cells. The tubule urine is ultimately concentrated by removal of water into the surrounding blood capillaries. Some substances, e.g., glucose, are reabsorbed back into the blood, and others are secreted into the urine, in both cases against concentration gradients. Such active-transport mechanisms in the renal tubule cells require input of large amounts of ATP energy.

merular filtrate back into the blood, against a concentration gradient, by the action of ATP-dependent membrane transport systems. A second group of solutes, including NH_4^+, K^+, and phosphate, occurs in relatively high concentration in urine compared with blood; these solutes are actively transported from the blood into the tubules, also against a concentration gradient. A third group of substances, including urea and creatinine, the end product of phosphocreatine degradation, is not reabsorbed and gradually increases in concentration as the urine passes down the tubules. Na^+ is a special case. It is reabsorbed from the glomerular filtrate into the blood by active transport in the first portion of the tubule, but some of it later passes back into the urine again by secondary exchange with other cations.

The transport of Na^+ and K^+ is especially important in the kidney, which must preserve the proper concentrations of these vital cations in the body by conserving Na^+ and secreting K^+. Virtually all mammalian cells contain a relatively high concentration of K^+ and a low concentration of Na^+, whereas the blood plasma and most other extracellular fluids have a high concentration of Na^+ and low K^+ (Figure 24-19).

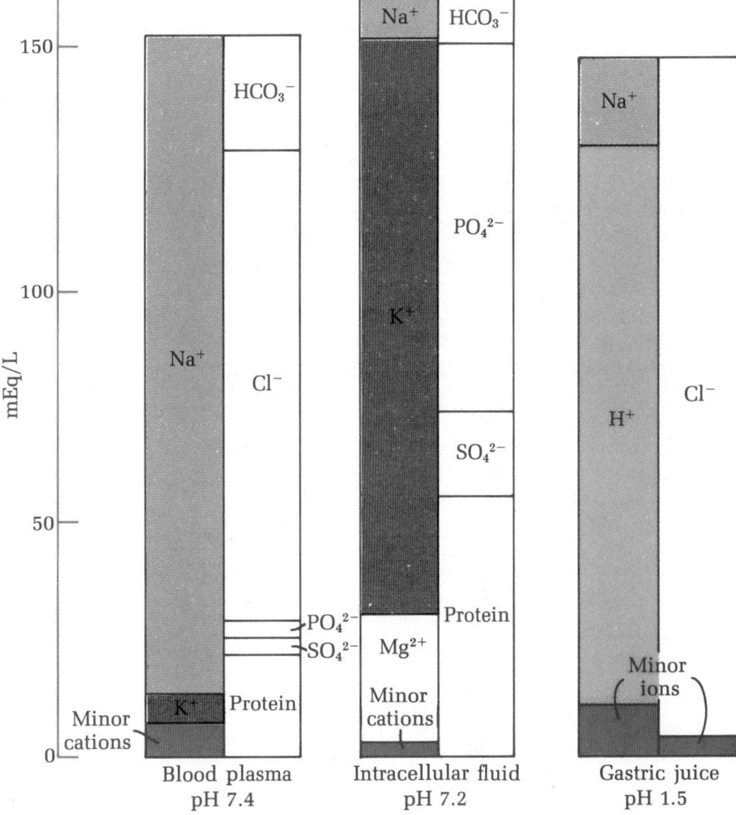

Figure 24-19
The electrolyte composition of blood plasma, compared with that of intracellular fluid and gastric juice. The left portion of each bar graph shows the cation composition; the right portion the anion composition. The dark gray zones represent the sums of the minor components. Note the large differences in Na^+ and K^+ content between blood plasma and intracellular fluid. These gradients are maintained by the Na^+K^+ATPase of the plasma membrane in nearly all cells in the body.

Similarly, note the large gradient of H^+ ions between gastric juice and blood plasma, from which gastric juice is made. This gradient results from the action of an H^+-transporting ATPase in the parietal cells of the stomach.

The plasma membranes of most cells contain Na^+K^+ATPase (page 386), which carries K^+ into cells and simultaneously transports Na^+ out. This energy-dependent process is coupled to hydrolysis of cytosolic ATP to ADP and phosphate. The Na^+K^+-ATPase of the tubule cells functions in such a way as to allow constant loss of K^+ in the urine, whereas loss of Na^+ can be kept to very low levels even if very little Na^+ is taken in (see also Chapter 26).

Through the action of the Na^+K^+-transporting ATPase, as well as other energy-requiring membrane transport systems for glucose and amino acids, the urine is so formed that those substances whose concentration in the blood must be lowered are excreted and those substances required to maintain the composition of the blood constant are reabsorbed from the tubules. Over three-fourths of the ATP generated by respiration in the kidneys is used in the formation of urine by such active membrane transport processes.

Blood Is a Very Complex Fluid

The blood is the vehicle for metabolic communication between the organs of the body. It transports nutrients from the small intestine to the liver and other organs and transports waste products to the kidneys for excretion. The blood is also the vehicle for transport of oxygen from the lungs to the tissues and for the transport of CO_2 generated during the respiratory metabolism of the tissues to the lungs for excretion. Moreover, hormones are transported from the endocrine glands via the blood to their specific target organs, in their function as chemical messengers (Chapter 25). The blood is very complex in its chemical composition since it carries a large number of nutrients, metabolites, waste products, and inorganic ions, making possible the coordinated interplay and integration of metabolism in the various organs of higher animals.

The blood in the vascular system of an adult human being has a volume of about 5 to 6 L. Almost one-half its volume is occupied by cells, which consist largely of red blood cells (erythrocytes), much smaller numbers of white blood cells (leukocytes), and blood platelets (Figure 24-20). The liquid portion is the blood plasma, which is 90 percent H_2O and 10 percent dissolved matter. Over 70 percent of the plasma solids is con-

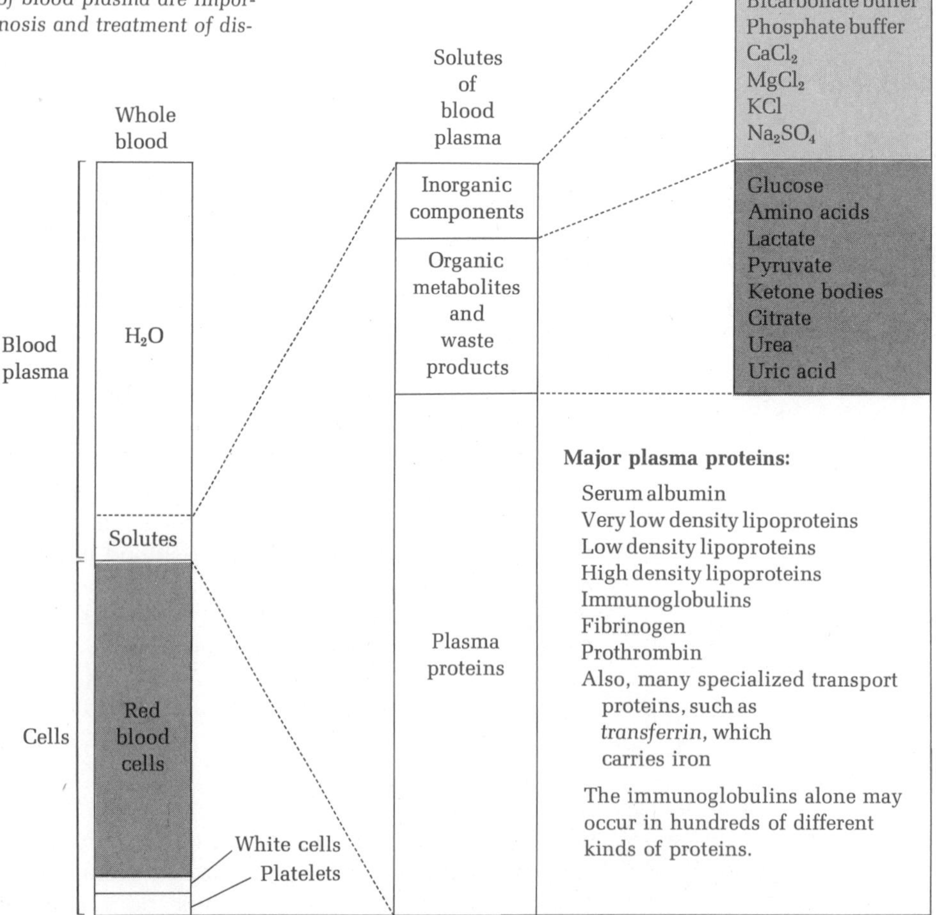

Figure 24-20
The composition of blood. Whole blood is separated into blood plasma and cells by centrifugation. The blood plasma contains about 10 percent dissolved solids, of which about 70 percent consists of plasma proteins, about 10 percent inorganic salts, and about 20 percent of small organic molecules. The major components of each fraction are given at the right. The quantitative composition of the inorganic components of blood plasma is given in Figure 24-19, plasma proteins in Table 24-3, and the nonprotein organic components in Table 24-4. Blood plasma also contains almost 700 mg of lipids per 100 mL, which are bound to α- and β-globulins (Table 24-3). Blood contains many other substances, often in trace amounts, including other metabolites, hormones, vitamins, trace elements, and bile pigments. Measurement of the concentration of components of blood plasma are important in the diagnosis and treatment of disease.

Table 24-3 Major Plasma Protein Fractions†

Protein	Amount, mg/100 mL	MW	Function
Serum albumin	3500–4500	66,000	Regulation of blood volume; transport of fatty acids
α_1-Globulins	300–600	40,000–60,000	Transport of lipids, thyroxine, adrenal cortical hormones
α_2-Globulins	400–900	100,000–400,000	Transport of lipids, copper
β-Globulins	600–1100	110,000–120,000	Transport of lipids, iron, hemes; antibody activity
γ-Globulins	700–1500	150,000–200,000	Most of the circulating antibodies
Fibrinogen	3000	340,000	Precursor of fibrin of blood clots
Prothrombin	100	69,000	Precursor of thrombin, required for blood clotting

† The total plasma protein content is 7000–7500 mg per 100 mL. Many different proteins are found in blood plasma; only the major classes are listed.

Table 24-4 Concentration of Major Nonprotein Organic Components of Blood Plasma

Component	Normal range, mg/100 mL
Nitrogenous	
Urea	20–30
Amino acids	35–65
Uric acid	2–6
Creatinine	1–2
Carbohydrates	
Glucose	70–90
Fructose	6–8
Organic acids	
Ketone bodies	1–4
Lactate	8–17
Pyruvate	0.4–2.5
Citrate	1.5–3.0
Lipids (entirely protein-bound, to α- and β-globulins)	
Total lipids	300–700
Triacylglycerols	80–240
Cholesterol and its esters	130–260
Phospholipids	160–300

tributed by the *plasma proteins* (Table 24-3). About 20 percent consists of organic metabolites, which are passing between various organs, and the waste products urea and uric acid, which pass to the kidneys to be excreted into the urine (Table 24-4). The remaining 10 percent of plasma solids consists of inorganic salts. Figure 24-20 summarizes the distribution of the major solutes in normal human blood plasma.

Some of the blood components normally vary somewhat in concentration, depending on the nature of the nutrient intake and when they are measured. The blood glucose level, for example, reaches a maximum directly after a meal, especially if it is rich in sugar, and then may decline below the mean level several hours later. Similarly, the concentration of chylomicrons in the blood varies during the periods between meals. The concentrations of the various components of the blood plasma are maintained at characteristic levels by various regulatory systems (Chapter 25).

Large Volumes of Oxygen Are Transported by the Blood

A normal adult male at complete rest requires about 375 L/d of pure oxygen gas, equivalent to the oxygen content of about 1900 L of air. If he works at a sedentary occupation, his oxygen requirement might then be at least twice as great. A trained athlete in an all-out running or swimming race may consume oxygen at a rate 10 times greater than the resting rate. Table

Table 24-5 Relative Amounts of Oxygen Consumed by the Major Organs in an Adult Human Male

	Rest	Light work	Heavy work
Skeletal muscles	0.30	2.05	6.95
Abdominal organs	0.25	0.24	0.24
Heart	0.11	0.23	0.40
Kidneys	0.07	0.06	0.07
Brain	0.20	0.20	0.20
Skin	0.02	0.06	0.08
Other	0.05	0.06	0.06
Total	1.00	2.90	8.00

24-5 shows the relative amounts of oxygen consumed by the major organs of the adult male at rest and during heavy muscular work.

As we have seen (pages 185 to 194), most of the oxygen carried by the blood is transported by the hemoglobin of the red blood cells. These cells constitute a major "tissue" of the human body, since they have a total volume of almost 3 L in an adult male and altogether weigh about the same as the liver. Erythrocytes are very small degenerate cells that contain no nuclei, mitochondria, or any other intracellular organelles. They do not use oxygen in their own metabolism. The relatively small amounts of ATP they require are supplied entirely by the glycolysis of blood glucose to yield lactate. The major function of red blood cells is to transport O_2 from the lungs to the tissues and to assist in the transport of CO_2 from the tissues to the lungs. They contain about 35 percent by weight of hemoglobin, which represents about 90 percent of the total cell protein.

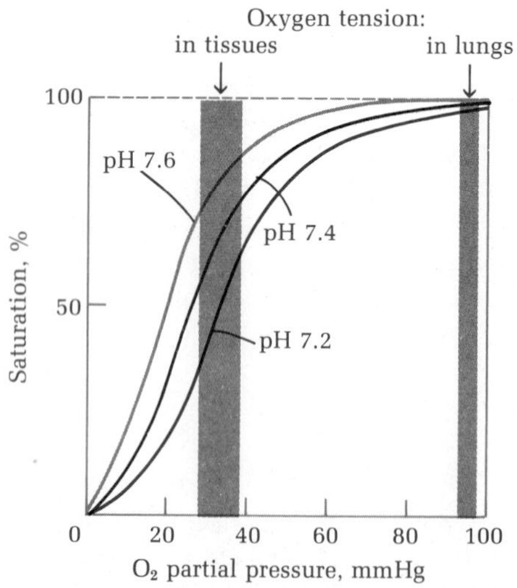

Figure 24-21
Effect of pH on the oxygen saturation curve of hemoglobin. At the low pH of the tissues, oxygen is more readily released, whereas at the higher pH in the lungs oxygen is more readily taken up.

Hemoglobin Is the Carrier of Oxygen

Hemoglobin contains two α and two β chains, as well as four heme groups, one bound to each polypeptide chain (pages 181 to 184). Each of the heme groups can bind one molecule of molecular oxygen reversibly. Because of the large amount of hemoglobin in the red cells, 100 mL of whole mammalian blood, when fully oxygenated, can carry about 21 mL of gaseous oxygen. The amount of oxygen bound by hemoglobin depends upon four factors: (1) the partial pressure of O_2, (2) the pH, (3) the concentration of 2,3-diphosphoglycerate, and (4) the CO_2 concentration (pages 185 to 194). Figure 24-21 shows oxygen saturation curves for hemoglobin. The sigmoid shape of these curves indicates that binding the first molecule of oxygen increases the affinity of the remaining subunits of hemoglobin for binding additional oxygen molecules. As the partial pressure of oxygen is increased further, a plateau is approached at which each of the hemoglobin molecules is saturated and contains the limit of four molecules of oxygen. The reversible binding of oxygen by hemoglobin is accompanied by release of protons, according to the approximate equation

$$HHb^+ + O_2 \rightleftharpoons HbO_2 + H^+$$

Thus increasing the pH will pull the equilibrium to the right and cause hemoglobin to bind more oxygen at a given partial pressure; conversely, decreasing the pH will decrease the amount of oxygen bound.

In the lungs, where the partial pressure of oxygen is high (90 to 100 mmHg) and the pH also relatively high (up to 7.6), hemoglobin will tend to become almost maximally saturated with oxygen (Figure 24-21). On the other hand, in the capillaries in the interior of the peripheral tissues, where the oxygen tension is low (25 to 40 mmHg) and the pH also relatively low (about 7.2 to 7.3), the release of some of the bound oxygen to the respiring tissue mass is favored. In the venous blood leaving the tissue the hemoglobin is only about 65 percent saturated. Therefore hemoglobin cycles between about 65 and 97 percent saturation with oxygen in its repeated circuits between the lungs and the peripheral tissues.

An important regulator of the degree of oxygenation of hemoglobin is 2,3-diphosphoglycerate (DPG). The higher the concentration of DPG in the cell, the lower the affinity of hemoglobin for oxygen. When the delivery of oxygen to the tissues is chronically limited, as in people who have a deficiency of red cells or who live in high altitudes, the concentration of DPG in the cells is higher than in normal individuals at sea level. This biochemical adjustment allows the hemoglobin to release its bound oxygen more readily to the tissues, to compensate for the decreased oxygenation of hemoglobin in the lungs.

Measurements of the oxygen content of the blood are important in the diagnosis and management of diseases in which there is an impairment in oxygen transport. Among these are severe *anemias,* in which either the number of red blood cells or the content of hemoglobin per cell is decreased; *asthma,* in which the blood may be insufficiently oxygenated due to constriction of the bronchioles; or *heart failure,* in which blood is pumped at an insufficient rate to keep the tissues well supplied with oxygen.

Red Blood Cells Also Transport CO_2

The blood also transports carbon dioxide from the tissues, where it is formed as the end product of the oxidation of fuels, to the lungs, where it is eliminated in the expired air. Venous blood leaving the tissues contains the equivalent of about 60 mL of CO_2 gas per 100 mL of blood, whereas arterial blood on leaving the lungs contains only about 50 mL CO_2 per 100 mL. About two-thirds of the total blood CO_2 is present in the plasma and about one-third in the red blood cells. However, nearly all the CO_2 of the blood must pass into and out of the red blood cells during CO_2 transport from the tissues to the lungs. In both the plasma and the red blood cells the total CO_2 is present in two forms, as dissolved CO_2 and as bicarbonate (HCO_3^-). Since dissolved CO_2 can be reversibly hydrated to yield carbonic acid (H_2CO_3), the mixture of H_2CO_3 and HCO_3^- in the blood constitutes a buffer system (page 85), in which H_2CO_3 is the proton donor and the HCO_3^- ion the proton acceptor. The H_2CO_3–HCO_3^- system is the major buffer of blood plasma.

The following sequence of events occurs in CO_2 transport from the tissues to the lungs (Figure 24-22). Dissolved CO_2, a product of the citric acid cycle oxidations as well as other enzymatic decarboxylation reactions, diffuses out of the tissues into the blood plasma and then into the erythrocytes. In the erythrocytes the CO_2 is rapidly hydrated to yield free carbonic acid in the reversible reaction

$$CO_2 + H_2O \rightleftharpoons H_2CO_3$$

In the absence of a catalyst this reaction is relatively slow and does not proceed fast enough to keep pace with the production of CO_2 by the respiring tissues. However, erythrocytes contain *carbonic anhydrase*, an extremely active enzyme that greatly accelerates the reaction. Once H_2CO_3 is formed, it spontaneously ionizes to yield bicarbonate

$$H_2CO_3 \rightleftharpoons H^+ + HCO_3^-$$

The HCO_3^- so formed then passes out of the red blood cells into the blood plasma, in exchange for chloride ions (Cl^-). The H^+ resulting from the ionization of the carbonic acid (H_2CO_3) in the erythrocyte promotes the loss of oxygen from oxyhemoglobin, in the reverse of the reaction described earlier

$$H^+ + HbO_2 \rightleftharpoons HHb^+ + O_2$$

Thus the H^+ formed as the result of the uptake of CO_2 and its conversion into HCO_3^- in the red blood cells promotes unloading of oxygen by hemoglobin as the blood passes through the peripheral tissues.

When the CO_2-rich venous blood returns to the lungs, the reverse of this cycle takes place. The binding of oxygen by hemoglobin in the lung capillaries causes H^+ to be formed

$$HHb^+ + O_2 \rightleftharpoons H^+ + HbO_2$$

The H^+ so formed now promotes the formation of carbonic acid from HCO_3^- in the erythrocyte

$$H^+ + HCO_3^- \rightleftharpoons H_2CO_3$$

and the H_2CO_3 formed is then dehydrated by carbonic anhydrase to yield dissolved CO_2

$$H_2CO_3 \rightleftharpoons H_2O + CO_2$$

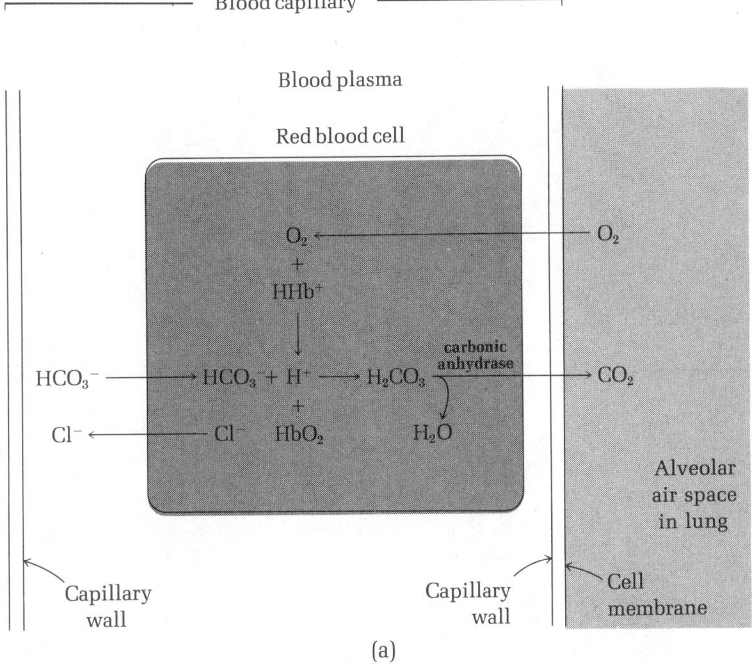

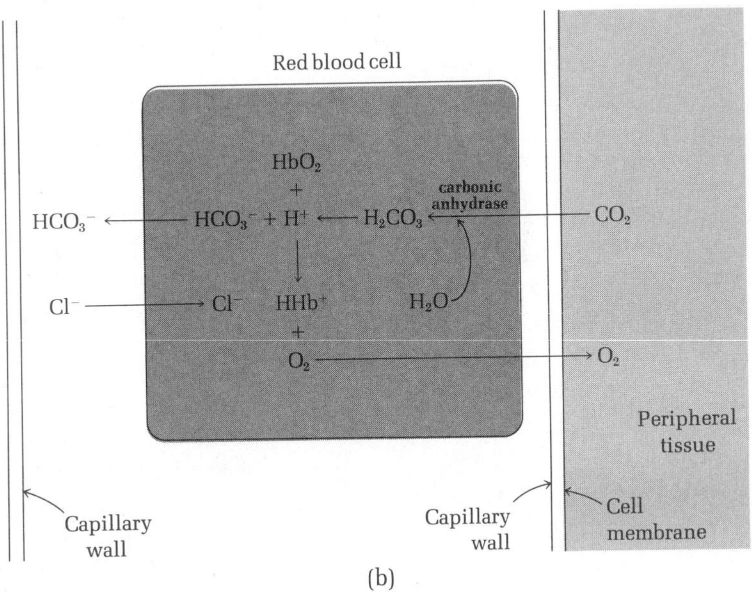

Figure 24-22
Coordination of oxygen and carbon dioxide transport in the red blood cell. (a) Oxygenation of hemoglobin in the lung results in release of H^+, which combines with HCO_3^- to give H_2CO_3. The H_2CO_3 is then dehydrated by carbonic anhydrase to give dissolved CO_2, which diffuses into the blood plasma and thence into the air space of the lungs, to be exhaled. (b) Uptake of dissolved CO_2 by the red blood cells in the peripheral tissues requires hydration by carbonic anhydrase to give H_2CO_3, which loses an H^+ to become HCO_3^-. The H^+ so released helps shift the hemoglobin equilibrium in the direction of O_2 release to the tissue.

O_2 and CO_2, being soluble in lipids, pass through cell membranes readily without the need of membrane transport systems. However, the exchange of Cl^- and HCO_3^- across the membrane of a red blood cell is promoted by an anion transport system.

The dissolved CO_2 now passes out of the erythrocytes through the blood plasma and capillaries and escapes into the air space via the vast surface area of the lung tissue (Figure 24-23). In this way the transport of oxygen and transport of CO_2 promote each other through the action of hemoglobin, which is beautifully adapted for these special transport functions.

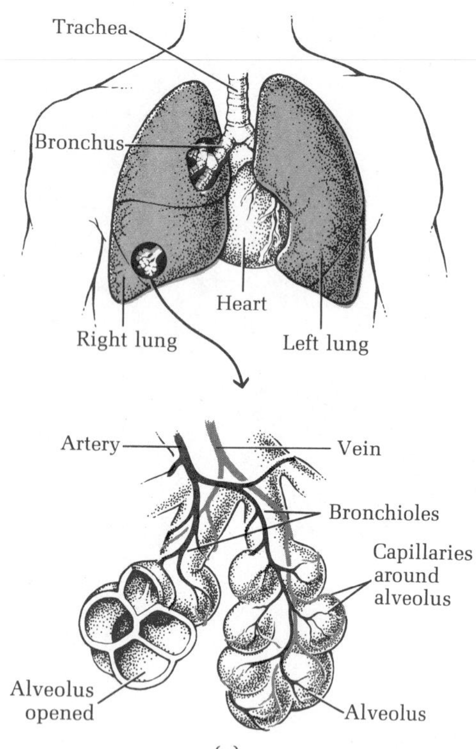

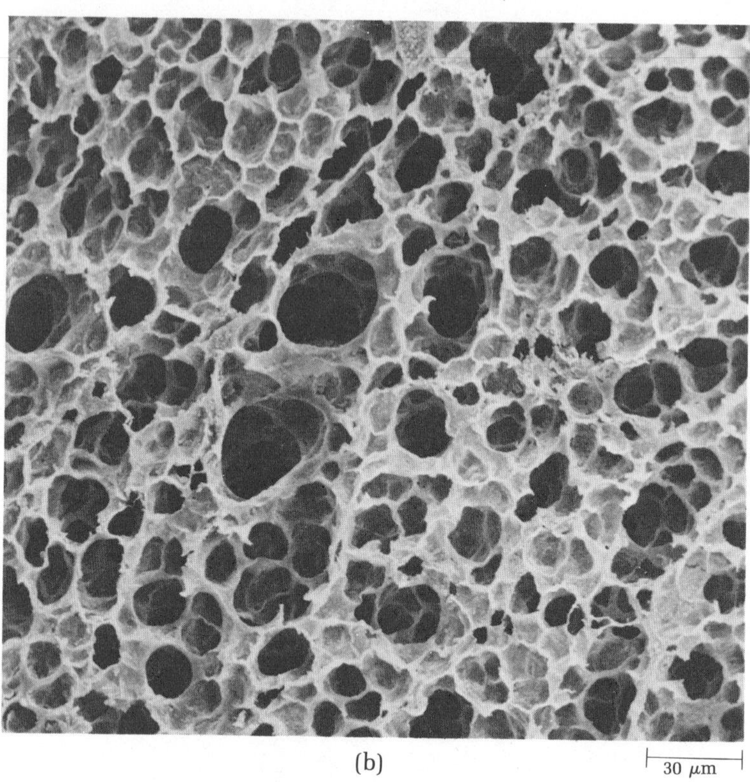

Figure 24-23
The alveoli or air sacs of the lung (a) *present a large surface area for exchange of* O_2 *and* CO_2 *between the alveolar air and the blood capillaries, as is shown in the electron micrograph* (b).

Diagnosing and Treating Diabetes Mellitus Relies on Biochemical Measurements

Biochemical measurements of specific components of the blood and urine serve as important indicators of metabolic state and are used in diagnosing disease and regulating treatment. A cardinal example is *diabetes mellitus,* produced by a deficiency in the secretion or action of the pancreatic hormone insulin, which in turn produces profound abnormalities of metabolism. Diabetes mellitus ranks third as the cause of death in the United States. It is relatively common: nearly 5 percent of the United States population shows some degree of abnormality in glucose metabolism indicative of diabetes or a tendency toward it. Diabetes mellitus is really a group of diseases in which the regulatory activity of insulin may be defective in different ways. Moreover, several other hormones can influence the metabolism of glucose. Diabetes has a genetic component; in addition it is also suspected that viral infection may play a role in its development. There are two major classes of diabetes: *juvenile-onset* and *adult-onset.* In the former, the disease begins early in life and quickly becomes severe. The latter is slow to develop, milder, and often goes unrecognized. Juvenile-onset diabetes requires insulin therapy and careful, life-long control of the balance between glucose intake and insulin dose. Biochemical measurements on the blood and urine

Table 24-6 Significant Changes in Blood and Urine Components in Uncontrolled Diabetes Mellitus

↑ indicates increases, ↓ indicates decreases

Urine	
Glucose	↑
Ketone bodies	↑
pH	↓
Na^+	↑
NH_4^+	↑
Urea	↑
Volume	↑
Blood	
Glucose	↑
Ketone bodies	↑
Urea	↑
pH	↓
Total CO_2 (sum of CO_2 + HCO_3^-)	↓

are essential in the diagnosis and treatment of diabetes, which causes profound changes in metabolism (Table 24-6).

Characteristic symptoms of diabetes are excessive thirst and frequent urination (*polyuria*), leading to intake of large volumes of water (*polydipsia*). These changes are due to the excretion of large amounts of glucose in the urine, a condition known as *glucosuria*. The term diabetes mellitus means "excessive excretion of sweet urine." In severe uncontrolled diabetes mellitus the amount of glucose in the urine may exceed 100 g per 24 h, whereas in normal individuals only traces of glucose are excreted. The large volume of the urine in diabetes reflects the kidney's need to excrete a certain amount of water along with the glucose since the kidney's capacity to concentrate the solutes in urine has an upper limit. Measurement of the amount of glucose in the urine secreted in a 24-h period is one of the diagnostic tests for diabetes.

More revealing is the blood glucose level and how it responds to intake of glucose. Diabetics usually have an abnormally high concentration of glucose in the blood, a condition called *hyperglycemia*. In very severe uncontrolled diabetes the blood glucose level may be enormously elevated, to as high as 100 m*M*, or 25 times higher than the normal value of about 4 m*M*. In mild diabetes the blood sugar level may not be much higher than in a normal person. A more sensitive diagnostic criterion is provided by the *glucose-tolerance test*. After a night without food, the patient drinks a test dose of 100 g of glucose dissolved in a glass of water. The blood glucose concentration is measured before and at 30-min intervals after the test dose of glucose. A normal person will assimilate the glucose readily, the blood glucose rising to no more than about 9 or 10 m*M*, because the increased concentration of blood glucose evokes secretion of insulin by the pancreas, which in turn increases the rate of uptake of glucose by the tissues. Normal people show little or no tendency for glucose to appear in the urine during such a test (Figure 24-24).

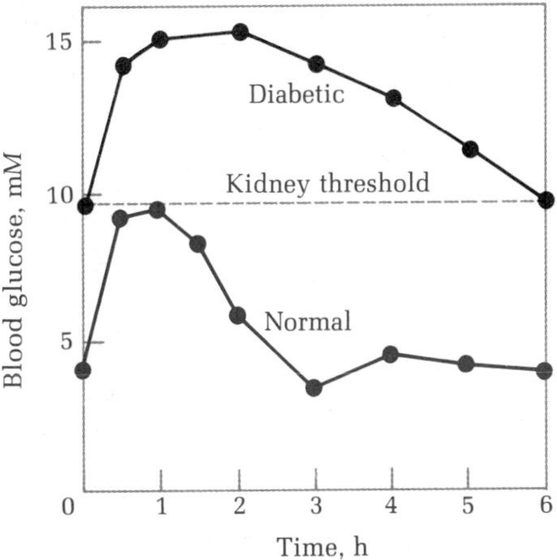

Figure 24-24
Glucose-tolerance curves for a normal subject and a diabetic. After the test dose of glucose the blood glucose level of the normal subject may rise to twice its normal level but quickly declines as insulin secretion is stimulated by the increase in blood glucose. In fact, there is usually an overshoot, a transient decrease in blood glucose below the normal level, due to a lag before the effect of the increased insulin wears off and the glucose returns to its normal level. In the diabetic, however, the blood glucose level is already very high, in this case about at the kidney threshold. After the test dose of glucose the blood level increases and remains at a high level because of the deficient secretion of insulin, before it slowly returns to the initial level. At the peak the diabetic subject will excrete considerable glucose into the urine.

However, diabetic patients, who may already have a high fasting blood glucose level, will show a marked deficiency in assimilating the test dose of glucose. The blood glucose level will increase far above the kidney threshold, which is about 10 m*M*, causing glucose to appear in the urine; furthermore, the blood glucose level may remain above normal for several hours (Figure 24-24). The failure of the blood glucose level to subside at a normal rate indicates that the secretion of insulin in response to the rise in blood glucose is defective.

Hyperglycemia and glycosuria reflect another profound change in metabolism in *diabetes mellitus:* the almost complete cessation of the conversion of excess glucose into fatty acids for storage as triacylglycerols. Severe diabetics may lose weight even if the caloric content of the diet is high. Quite simply, instead of being stored as fat, excess glucose is excreted in the urine.

Ketosis Occurs in Diabetes

Another characteristic metabolic change in diabetes is excessive but incomplete oxidation of fatty acids in the liver, with the result that the ketone bodies acetoacetate and β-hydroxybutyrate are overproduced and cannot be used by the peripheral tissues as fast as they are made by the liver. In addition to β-hydroxybutyrate and acetoacetate, the blood of diabetics also contains acetone, which results from the spontaneous decarboxylation of acetoacetate

$$\underset{\text{Acetoacetate}}{CH_3{-}\underset{\underset{O}{\|}}{C}{-}CH_2{-}COO^-} + H_2O \longrightarrow \underset{\text{Acetone}}{CH_3{-}\underset{\underset{O}{\|}}{C}{-}CH_3} + HCO_3^-$$

Acetone is very volatile and is present in the breath of diabetics, to which it gives a characteristic sweet "organic" odor. A diabetic in coma is occasionally mistaken for a drunk because of the odor of acetone on the breath. The overproduction of ketone bodies, called *ketosis*, results in their appearance in greatly increased concentrations in the blood (*ketonemia*) and urine (*ketonuria*).

Urea Excretion Is Elevated in Diabetics

Another characteristic of severe diabetes is an elevated excretion of urea, the major nitrogenous waste product of the oxidative degradation of amino acids (pages 549 to 554). The amount of urea excreted per day is a measure of the total amount of amino acids broken down oxidatively, which in turn reflects the balance between the protein intake and the normal daily breakdown of body protein. The concentration of urea in the blood in diabetes may be as high as 25 m*M*, or about 5 times higher than the normal value of about 5 m*M*.

The excessive oxidative breakdown of amino acids by diabetics reflects a greatly increased rate of gluconeogenesis from amino acids. In the absence of insulin the liver tends to pour glucose into the blood. As a result, liver glycogen stores are very low and all available amino acids capable of yielding carbon for gluconeogenesis are degraded in order to form more blood glucose. Thus the measurement of urea in the blood and urine gives valuable information about the metabolic state of diabetic patients.

Acidosis Accompanies Severe Diabetes

Very striking and quite serious is the large decrease in blood pH in severe uncontrolled diabetes, which may fall as low as pH 6.8, compared with the normal value of about pH 7.4. While this increase in acidity may seem very small on an absolute basis, it is indicative of a very profound change in the acid-base balance of the body. The increased acidity is due to the extensive formation of the ketone bodies in the liver and their delivery into the blood. When one molecule of a triacylglycerol, a neutral molecule, is oxidized by the diabetic liver, at least 12 H^+ are produced in the form of β-hydroxybutyric and acetoacetic acids. The body compensates for the continuous production of these acids by decreasing the concentration of H_2CO_3, the proton donor or acid species of the bicarbonate buffer system (page 85). This is accomplished by an increase in the rate of excretion of CO_2 by the lungs (recall that H_2CO_3 dissociates reversibly to yield dissolved CO_2 and H_2O). The loss of CO_2 via the lungs will thus tend to bring the ratio of the proton-acceptor (HCO_3^-) and proton-donor (H_2CO_3) species of the bicarbonate buffer pair back to the proper ratio to keep the blood pH near its normal value of 7.4. However, in severe diabetes so much CO_2 is "blown off" by the lungs to compensate for the acidosis produced by the excess ketone bodies that the *sum* of the concentrations of HCO_3^- and H_2CO_3 becomes very low, greatly decreasing the buffering capacity of the blood, a serious complication.

These biochemical abnormalities in a severely diabetic individual can be life-threatening, particularly the changes in acid-base balance. Administration of insulin to correct the endocrine deficiency and administration of $NaHCO_3$ to correct the loss of both the Na^+ and the bicarbonate buffer capacity can bring the entire body chemistry back into near-normal balance within 12 to 24 h. To follow the course of such treatment, measurements of blood glucose, blood pH, or blood CO_2 are made frequently.

Summary

Starch and other polysaccharides are partially hydrolyzed by salivary amylase in the mouth. Digestion of polysaccharides and disaccharides is completed in the small intestine by the action of pancreatic amylase and lactase, sucrase, and maltase in the intestinal epithelial cells. Proteins are digested by the successive action of pepsin in the acid contents of the stomach, followed by the action of trypsin and chymotrypsin at pH 7 to 8 in the small intestine. Short peptides are then hydrolyzed to amino acids by carboxypeptidase and aminopeptidase. Triacylglycerols are digested by pancreatic lipase to 2-monoacylglycerols and free fatty acids, which are emulsified with the help of bile salts and absorbed. Pepsin, trypsin, chymotrypsin, carboxypeptidase, and lipase are secreted into the gastrointestinal tract as inactive zymogens.

The liver is the central distributing and processing organ for nutrients. Glucose 6-phosphate, the key intermediate in carbohydrate metabolism, may be converted into glycogen, into blood glucose, or into fatty acids via acetyl-CoA. It may undergo degradation via glycolysis and the citric acid cycle to yield ATP energy or via the pentose phosphate cycle to yield pentoses and NADPH. Amino acids may be built up into liver and plasma proteins, or they may be converted into glucose and glycogen by gluconeogenesis. The ammonia formed by their deamination is converted into urea by the urea cycle. Fatty acids may be converted by the liver into triacylglycerols, cholesterol, and into plasma lipoproteins for transport and storage in adipose tissue. They may also be oxidized to yield ATP energy and to form the ketone bodies, to be circulated to other tissues.

Skeletal muscle is specialized to produce ATP for contraction and relaxation. In very severe muscular work, glycogen is the ultimate fuel and is converted into lactate. During recovery the lactate is rebuilt to form liver glycogen and glucose again. The brain uses only glucose and β-hydroxybutyrate as fuels, the latter being important in fasting or starvation. The brain uses most of its ATP energy for active transport of Na^+ and K^+ and in the maintenance of the action potential of neuronal membranes.

Red blood cells participate in transport of both oxygen and CO_2 between the lungs and peripheral tissues. The loading and unloading of oxygen is promoted by the relatively high pH in the lungs and by the lower pH in the tissues; it is also helped by the sigmoid relationship between O_2 pressure and percent saturation. CO_2 transport in the red blood cells is linked to oxygen transport by means of compensating H^+ and Cl^- ion shifts.

Biochemical methods are intensively used in diagnosis and in following treatment of diseases, particularly diabetes mellitus, in which there are characteristic changes in blood glucose, urea, ketone bodies, and pH.

References

Books

MONTGOMERY, R., R. L. DRYER, T. W. CONWAY, and A. A. SPECTOR: *Biochemistry: A Case-Oriented Approach,* 3d ed., Mosby, St. Louis, 1980. The biochemistry of human health and disease, developed from consideration of patient case histories.

WHITE, A., P. HANDLER, E. L. SMITH, R. L. HILL, and I. R. LEHMAN: *Principles of Biochemistry,* 6th ed., McGraw-Hill, New York, 1978. The metabolism of the different organs, the components of blood and urine, and their variations in health and disease are developed in much useful detail.

Articles

AXELROD, J.: "Neurotransmitters," *Sci. Am.*, **230**:58–71, June (1974).

DAVENPORT, H. W.: "Why the Stomach Does Not Digest Itself," *Sci. Am.*, **226**:86–93, January (1972).

KAPPAS, A., and A. P. ALVARES: "How the Liver Metabolizes Foreign Substances," *Sci. Am.*, **232**:22–31, June (1975).

KEYNES, R. D.: "Ion Channels in the Nerve-Cell Membrane," *Sci. Am.*, **240**:126–135, March (1979).

NOTKINS, A. L.: "The Causes of Diabetes," *Sci. Am.*, **241**:62–73, November (1979).

PERUTZ, M. F.: "Hemoglobin Structure and Respiratory Transport," *Sci. Am.*, **239**:92–125, December (1978).

Problems

1. *Synthesis of Gastric Juice* Gastric juice (pH 1.5) is produced by "pumping" protons from blood plasma (pH 7.4) into the stomach. Calculate the amount of free energy required to concentrate the H^+ in 1 L of gastric juice. Under cellular conditions, how many moles of ATP must be hydrolyzed to provide this amount of free energy? Recall that ΔG_p of ATP hydrolysis under cellular conditions is approximately 14 kcal/mol.

2. *pH Adjustment of Gastric Juice* The protein-digesting enzymes trypsin, chymotrypsin, and carboxypeptidase have optimal activity in the pH range 7–8 in the small intestine. The substrates for these enzymes, on the other hand, arrive from the stomach in gastric juice of pH 1.5–2.5. How is the pH of gastric juice adjusted to the pH optimum of the enzymes of the small intestine?

3. *Digestibility of Casein and Keratin* Bovine β-casein (MW 23,600), one member of a family of proteins found in cow's milk, lacks cysteine and cystine residues. In addition, this protein has little tertiary structure, and its native conformation resembles a random coil. In contrast, α-keratins (proteins found in hair, feathers, nails, etc.) are rich in cysteine and cystine residues. Furthermore, these proteins have a high degree of secondary and tertiary structure. Explain how the properties of these two groups of proteins affect their digestibility. For example, why is milk an excellent dietary amino acid source in the growing cat while its own fur is indigestible ("hairballs" can lead to blockage of the intestine)?

4. *Prevention of Pepsin-Catalyzed Digestion of Chief Cells* Pepsinogen, the zymogen precursor of pepsin, is activated by the removal of a 42–amino acid segment from the amino terminal end. The activation process is usually catalyzed by pepsin itself, although pepsinogen shows some catalytic activity at pH values less than 5. In addition, the 42–amino acid segment removed during activation binds tightly to the active site of pepsin at pH values above 2 but weakly at pH values below 2.
   (a) How is the activation of pepsinogen initiated in the stomach?
   (b) How do the properties of pepsin and pepsinogen described above serve to protect the cells of the intestinal mucosa from self-digestion?

5. *Fate of Digestive Enzymes* During digestion of a meal rich in proteins large amounts of trypsin, chymotrypsin, and carboxypeptidase are secreted into the gastrointestinal tract by the pancreas. Although premature release of these enzymes in their active forms within the pancreas can cause severe damage to this organ, these enzymes do not damage the epithelial cells of the small intestine during normal digestion of a protein-rich meal. Moreover, tests of the contents of the lower part of the small intestine during digestion show that only traces of these enzymes are present, and active pepsin cannot even be detected. Suggest explanations for these observations.

6. *Efficiency of Carboxypeptidases* Carboxypeptidases secreted by the pancreas are much more efficient at protein hydrolysis in the later stages of the digestion of a meal than the initial stages. Explain.

7. *Milk Intolerance in Orientals and Blacks* The consumption of milk by oriental and black adults is frequently followed by abdominal distention, cramps, pain, and diarrhea. In fact, these symptoms will occur within 1 to 4 hours after drinking as little as one glass of milk. These symptoms occur equally with fresh and powdered milk. What substance in milk is responsible for these symptoms? How does this substance cause the observed symptoms?

8. *Causes of Steatorrhea* The clinical symptoms of steatorrhea, characterized by excessive lipid in the stools, can be caused by either insufficient bile secretion or the absence of pancreatic secretion. Why should these conditions lead to excessive lipid in the stools? How can these two causes be differentiated by analysis of the stool specimen of the patient. Explain.

9. *Alanine and Glutamine in the Blood* The blood plasma contains all the amino acids required for the synthesis of body proteins, but they are not present in equal concentrations. Two amino acids, alanine and glutamine, are present in much higher concentrations in normal human blood plasma than any of the other amino acids. Suggest possible reasons for their abundance.

10. *Action of Ouabain on Kidney Tissue* Ouabain (pronounced wah-bane) is a toxic glycoside that specifically inhibits Na^+K^+-ATPase activity of animal tissues but does not inhibit any other known enzyme. However, when ouabain is added in graded concentrations to thin slices of living kidney tissue, it inhibits oxygen consumption up to a maximum of 66 percent. Explain the basis of this observation. What does it tell us about the use of respiratory energy by kidney tissue?

11. *Is Glycolysis Necessary for Contraction of Skeletal Muscle?* The alkylating agent iodoacetate (ICH_2COO^-) is a potent inhibitor of glyceraldehyde 3-phosphate dehydrogenase, since it combines with its essential —SH groups. When skeletal muscle strips are treated with a mixture of potassium iodoacetate and antimycin A (Chapter 17), they continue to contract on repeated electric stimulation, but no lactate is formed. Explain what has happened. What is the energy source for contraction under these circumstances?

12. *ATP and Phosphocreatine as Sources of Energy for Muscle* In contracting skeletal muscle, the concentration of phosphocreatine drops while the concentration of ATP remains fairly constant. Explain how this happens. In a classical experiment Robert Davies found that if the muscle is first treated with fluoro-2,4-dinitrobenzene, the concentration of ATP in the muscle declines rapidly, whereas the concentration of phosphocreatine remains unchanged during a series of contractions. Suggest an explanation.

13. *Metabolism of Glutamate in the Brain* Glutamate in the blood flowing into the brain is transformed into glutamine, which appears in the blood leaving the brain. What is the purpose of this metabolic conversion? How does it take place? Actually, the brain can generate more glutamine than can be made from the glutamate entering in the blood. How does this extra glutamine arise?

14. *Absence of Glycerol Kinase in Adipose Tissue* Glycerol 3-phosphate is a key intermediate in the biosynthesis of triacylglycerols. Adipose cells, which are specialized for the synthesis and degradation of triacylglycerols, cannot directly use glycerol due to the lack of glycerol kinase, which catalyzes the reaction

$$\text{Glycerol} + \text{ATP} \longrightarrow \text{glycerol 3-phosphate} + \text{ADP}$$

How does adipose tissue obtain the glycerol 3-phosphate necessary for triacylglycerol synthesis? Explain.

15. *Edema Associated with Nephrotic Syndrome* Patients with nephrosis excrete large quantities of serum albumin in their urine. In fact, in such patients the serum albumin level in the blood may drop to 1.0 g per 100 mL, compared with a normal value of 3.5 to 4.5 g per 100 mL (Table 24-3). Patients with this condition develop severe edema (swelling) of the extremities due to the accumulation of fluid in the extracellular spaces. Explain the origin of these symptoms.

16. *Hyperglycemia in Patients with Acute Pancreatitis* Patients with acute pancreatitis are treated by withholding protein from the diet and by intravenous administration of glucose-saline solution. What is the biochemical basis for these measures? Patients with acute pancreatitis undergoing treatment commonly experience hyperglycemia. Why?

17. *Oxygen Consumption during Exercise* A sedentary adult consumes about 0.05 L of oxygen during a 10 s period. A sprinter, running a 100 m race, consumes about 1 L of oxygen during the same time period. After finishing the race, the sprinter will continue to breathe at an elevated but declining rate for some minutes, consuming an extra 4 L above the sedentary amount. Why do the oxygen needs increase dramatically during the sprint? Why do the oxygen demands remain high after the sprint is completed?

18. *The Effect of a Milk Diet on the pH of Urine* On a diet consisting entirely of whole milk the pH of the urine will tend to be lower (pH $\sim$ 5) than with a normal mixed diet. Yet whole milk has a pH of 7.5. How can you explain this paradox, given the gross composition of milk, which contains about 3.5 percent lactose,

3.5 percent butter fat, and about 3.5 percent casein, a phosphoprotein with many residues of serine phosphate?

19. *Respiratory Acidosis* A 22-year-old male was brought to the emergency room of a local hospital and diagnosed as having an overdose of a narcotic drug. Upon arrival his breathing was very shallow and irregular. His condition deteriorated, and he was placed on a mechanical respirator. An arterial blood sample indicated that his blood pH was 7.18. He was then given 12 g of sodium bicarbonate intravenously.
    (a) Suggest why a narcotic overdose might lead to respiratory acidosis.
    (b) How does mechanical venting of the lungs affect the blood pH of the patient? How does the mechanical venting affect the ability of blood to transport oxygen to the tissues?
    (c) Explain the purpose of administering bicarbonate.

20. *Thiamine Deficiency and Brain Function* Patients with thiamine deficiency display a number of characteristic neurological signs: loss of reflexes, anxiety states, and mental confusion. Suggest a reason why thiamine deficiency is manifested in brain function.

21. *Adrenal Cortical Hyperactivity and Hyperkaluria* Hyperactivity of the human adrenal cortex caused by cortical tumors leads to excessive Na^+ reabsorption in the renal tubes. A patient with this condition also excretes excessive amounts of K^+ in the urine. Why is excessive Na^+ reabsorption accompanied by excessive K^+ excretion?

22. *The Toxic Effect of Drinking Seawater* The human kidney is a marvelous device that can regulate the removal of Na^+ ions from the blood by producing urine containing Na^+ in concentrations as high as 340 mM. Seawater, however, contains Na^+ at twice the concentration that can be produced by the kidneys of a healthy adult. When a person's sole fluid intake consists of seawater, sodium chloride accumulates in the extracellular fluid (the fluid surrounding the cells of the body) but not in the intracellular fluid. Prolonged ingestion of seawater leads to death because of damage to cells in the brain. How can drinking seawater over a prolonged period cause injury to cells?

CHAPTER 25

Hormones

We shall now examine the hormones and how they regulate the harmonious interplay of different tissues and organs.

The word *hormone* is derived from a Greek verb meaning "to stir up or excite." A hormone is a chemical messenger, secreted in trace amounts by one type of tissue and carried by the blood to a target tissue elsewhere in the body to stimulate a specific biochemical or physiological activity. *Endocrinology*, the branch of biomedical science dealing with hormones and their action, is one of the most exciting fields of biochemistry because new insights have recently come from a number of major discoveries. Moreover, because alterations in hormone action can lead to disease, endocrinology has also become one of the most directly useful fields of biochemistry.

Many hormones are known and more are being discovered. Hormones control not only different aspects of metabolism but also many other functions: cell and tissue growth, heart rate, blood pressure, kidney function, motility of the gastrointestinal tract, secretion of digestive enzymes, lactation, and the reproductive system. We shall not examine them all; indeed, the action of most hormones is still unknown in biochemical terms. Instead we shall focus on the biochemistry of those hormones which regulate the mainstreams of metabolism, including adrenaline, insulin, glucagon, thyroxine, and the adrenocortical hormones.

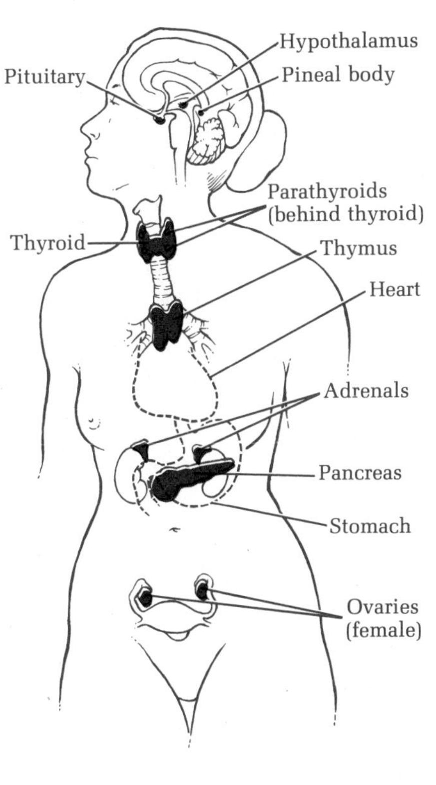

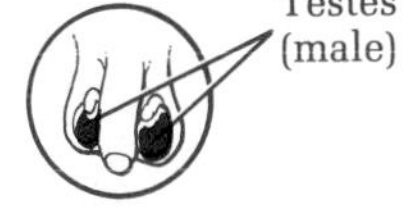

Figure 25-1
The major endocrine glands.

Hormones Function in a Complex Interrelated Hierarchy

First let us examine the major endocrine systems of the body and some of their functional interrelationships. Figure 25-1 shows the anatomical location of the major endocrine glands important in the regulation of metabolism. The word endocrine ("to secrete within") means that the secretions of such glands are internal, i.e., into the blood. In addition to the more familiar endocrine glands, such as the thyroid and pituitary, there are

Figure 25-2
The major endocrine systems and their target tissues. Signals arising from the nervous system are passed via a series of relays to the ultimate target tissue. In addition to the systems shown the thymus and pineal glands, as well as groups of cells in the gastrointestinal tract, also secrete hormones. FSH designates follicle-stimulating hormone; LH, luteinizing hormone.

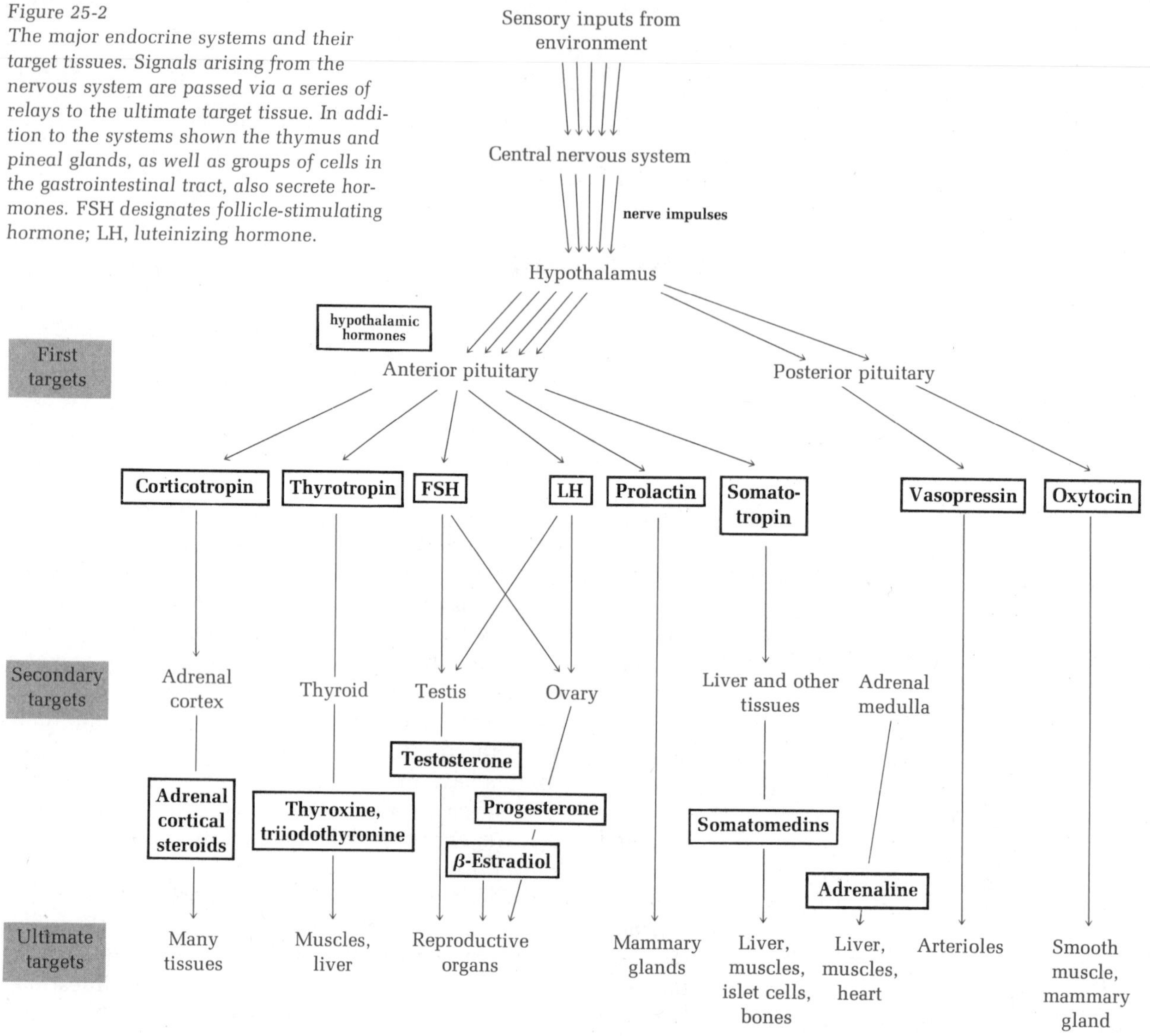

Sensory inputs from environment
Central nervous system
nerve impulses
Hypothalamus
hypothalamic hormones
First targets
Anterior pituitary
Posterior pituitary
Corticotropin
Thyrotropin
FSH
LH
Prolactin
Somatotropin
Vasopressin
Oxytocin
Secondary targets
Adrenal cortex
Thyroid
Testis
Ovary
Liver and other tissues
Adrenal medulla
Testosterone
Adrenal cortical steroids
Thyroxine, triiodothyronine
Progesterone
β-Estradiol
Somatomedins
Adrenaline
Ultimate targets
Many tissues
Muscles, liver
Reproductive organs
Mammary glands
Liver, muscles, islet cells, bones
Liver, muscles, heart
Arterioles
Smooth muscle, mammary gland

many other tissues, such as the pineal, the thymus, and numerous groups of cells in the gastrointestinal tract that secrete a variety of different hormones.

Figure 25-2 is a schematic master plan of the regulatory relationship between the endocrine glands and their *target tissues.* The *hypothalamus,* a specialized portion of the brain, is the coordination center of the endocrine system; it receives and integrates messages from the central nervous system. In response to these messages the hypothalamus produces a number of *hypothalamic regulatory hormones,* which are sent to the *anterior pituitary gland,* located just below the hypothalamus. Each hypothalamic hormone regulates the secretion of a specific hormone by the anterior portion of the pituitary. Some hypothalamic hormones stimulate the pituitary to secrete a given hormone; others are inhibitory. Once it is stimulated, the pitu-

Figure 25-3
Feedback regulation of the secretion of the thyroid hormones thyroxine and triiodothyronine. When their blood concentration increases, they inhibit secretion of thyrotropin-releasing hormone (TRH) by the hypothalamus and of thyrotropin by the pituitary. Somatostatin, another hypothalamic hormone, also inhibits TRH secretion.

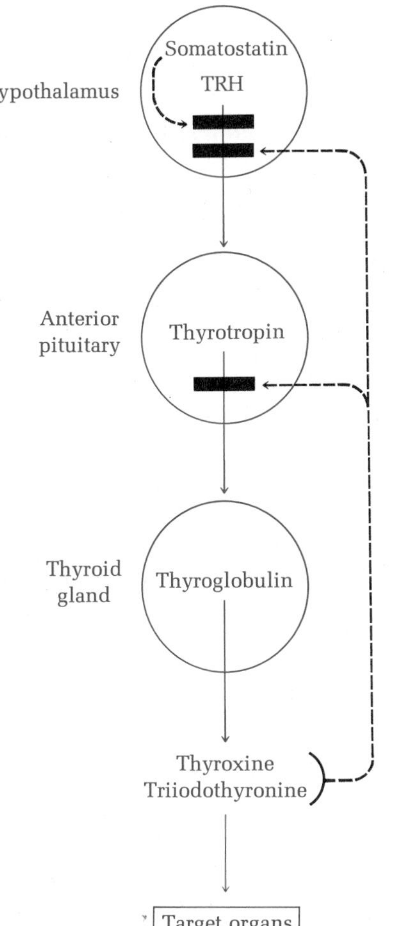

itary secretes hormones into the blood to be carried to the next rank of endocrine glands, which includes the *adrenal cortex*, the endocrine cells of the *pancreas*, the *thyroid gland*, and the *ovary* and *testis*. These glands in turn are stimulated to secrete their specific hormones, which are carried by the blood to *hormone receptors* on or in the cells of the ultimate target tissues. There is one more link in this relay system. In the cells of the target tissue yet another molecular signaling agent, an *intracellular messenger*, carries the message from the hormone receptor to the specific cell structure or enzyme that is the ultimate target. Thus each endocrine system resembles a set of relays, designed to carry messages from the nervous system to a specific effector molecule in its target cells.

The endocrine system is also modulated by interconnecting feedback controls. Figure 25-3 shows by an example how such controls function. The hypothalamus can send out thyrotropin-releasing hormone (TRH) to the anterior pituitary, causing it to release more thyrotropin; this in turn stimulates the thyroid gland to release the thyroid hormones *thyroxine* and *triiodothyronine*, which are sent to their target tissues. However, the circulating thyroid hormones in the blood also act as feedback inhibitors of the secretion of TRH by the hypothalamus and of thyrotropin by the pituitary. Moreover, *somatostatin*, secreted by the hypothalamus (and also by the pancreas), can inhibit secretion of TRH. Hence the secretion or action of one hormone may be strongly influenced or regulated by other hormones. Thus a very complex regulatory network controls the activities of the different endocrine systems.

Some General Characteristics of Hormones

Some generalizations regarding the structure and function of hormones may be made.

There Are Three Classes of Hormones: Peptides, Amines, and Steroids

The *peptide hormones* (Table 25-1), which may have from 3 to over 200 amino acid residues, include all the hormones of the hypothalamus and pituitary, as well as insulin and glucagon of the pancreas. The *amine hormones*, small, water-soluble compounds containing amino groups, include adrenaline of the adrenal medulla and the thyroid hormones. The *steroid hormones*, which are fat-soluble, include the adrenal cortical hormones, *androgens* (male sex hormones), and *estrogens* (female hormones).

Some Polypeptide Hormones Are Made as Inactive Precursors

Several polypeptide hormones, including insulin and glucagon, are synthesized by their parent endocrine cells as inactive precursors called *prohormones*. Such inactive precursors

Table 25-1 Classes of Hormones and Examples

	Secreted by
Peptide hormones	
Thyrotropin-releasing hormone	Hypothalamus
Corticotropin	Anterior pituitary
Vasopressin	Posterior pituitary
Insulin	Pancreas
Glucagon	Pancreas
Amine hormones	
Adrenaline	Adrenal medulla
Thyroxine	Thyroid
Steroid hormones	
Cortisol	Adrenal cortex
β-Estradiol	Ovary
Testosterone	Testis
Progesterone	Corpus luteum

contain longer polypeptide chains than the active hormone itself. An example is *proinsulin*, a polypeptide chain of about 80 amino acid residues, which is converted into active insulin (51 residues) by enzymatic removal of part of the chain. Prohormones are stored in an inactive form in the endocrine cell, often in secretory granules, ready to be quickly converted into their active forms by enzymatic action when the cell receives the appropriate signal.

Hormones Function in Very Low Concentrations and Most Are Short-Lived

Hormones occur in very low resting concentrations in the blood, in the *micromolar* (10^{-6} *M*) to the *picomolar* (10^{-12} *M*) range, which may be contrasted with the normal concentration of glucose, which is in the millimolar range, about 4×10^{-3} *M*. For this reason hormones have been very difficult to isolate, identify, and measure accurately. The exceedingly sensitive technique of radioimmunoassay, to be described, has revolutionized hormone research by making possible quantitative and specific measurement of many hormones in minute concentrations.

When the secretion of a given hormone is stimulated, its concentration in the blood rises, sometimes by orders of magnitude. On cessation of secretion, the hormone concentration quickly returns to the resting level. Hormones have a short lifetime in the blood, often only minutes. Once their presence is no longer required, hormones are quickly inactivated by enzyme action.

Some Hormones Act Immediately; Others Act Slowly

Some hormones yield immediate physiological or biochemical responses. Seconds after adrenaline is secreted into the bloodstream, the liver responds by pouring glucose into the blood. On the other hand, the thyroid hormones or the estrogens yield maximal response in their target tissues only after hours or even days. As we shall see, such differences in response time correspond to a difference in mode of action.

Hormones Bind to Specific Receptors on or in Target Cells

The first step in the action of a hormone is its binding to a specific molecule or set of molecules, called the *hormone receptor*, which is located on the cell surface or in the cytosol of the target cell. Such a receptor has a very high specificity and affinity for its hormone molecule. The receptors for the *water-soluble* peptide and amine hormones, which do not penetrate cell membranes readily, are located on the outer surface of the target cells. The receptors for the *lipid-soluble* steroid hormones, which readily pass through the plasma membrane of their target cells, are specific proteins located in the cell cytosol.

Hormones May Have Intracellular "Second Messengers"

Once the hormone receptor on or in the target cell is occupied by the hormone molecule, the receptor undergoes a characteristic change which forms or sets free an intracellular messenger molecule, often called the second messenger. This messenger conveys the signal from the hormone receptor to some enzyme or molecular system in the cell that carries out the instructions brought by the hormone. The intracellular messenger either regulates a specific enzymatic reaction or causes an inactive gene or set of genes to be expressed.

Table 25-2 Some Hypothalamic Hormones

Adrenocorticotropin-releasing hormone
Thyrotropin-releasing hormone
Somatotropin-releasing hormone
Somatostatin
Prolactin-releasing hormone
Prolactin-inhibitory hormone
FSH- and LH-releasing hormone

The Hormones of the Hypothalamus and Pituitary Gland Are Peptides

The hormones secreted by the hypothalamus (Table 25-2) are relatively short peptides having from 3 to 15 amino acid residues. The structures of two such hormones are shown in Figure 25-4. These hormones have been isolated and identified only after years of painstaking research. Of all the known hormones, the hypothalamic factors are secreted in the smallest amounts. For example, only 1 mg of the thyrotropin-releasing hormone could be isolated from over 4 tons of hypothalamic tissue ob-

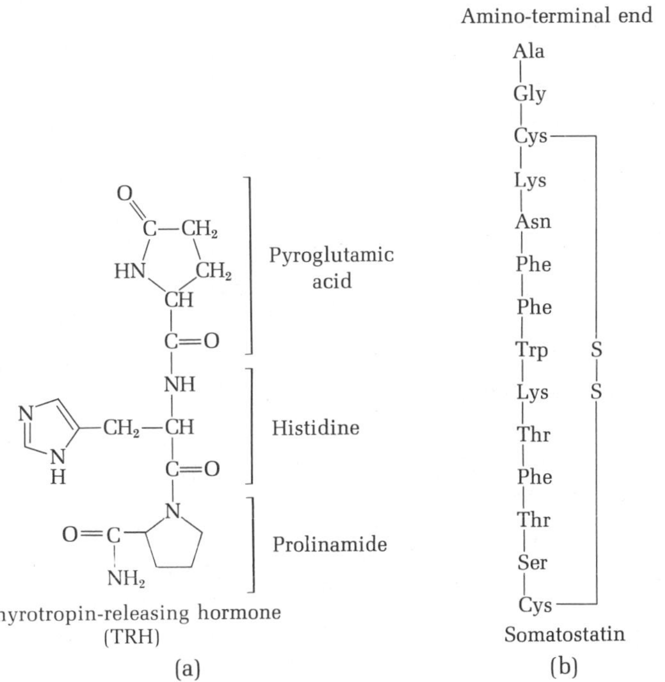

Figure 25-4
Two hormones secreted by the hypothalamus. (a) Thyrotropin-releasing hormone (TRH) promotes release of thyrotropin from the anterior pituitary gland. Thyrotropin in turn stimulates the thyroid gland to secrete thyroxine and triiodothyronine. (b) Sheep somatostatin, which inhibits release of somatotropin from the anterior pituitary.

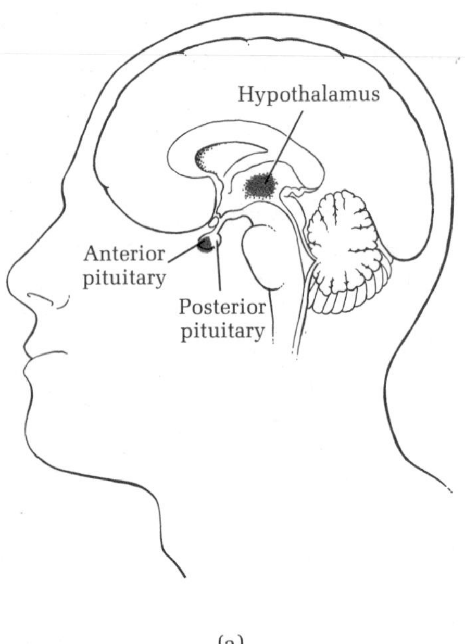

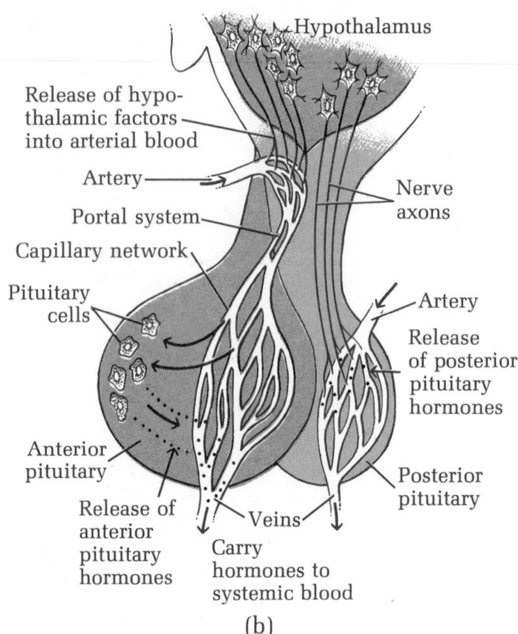

Figure 25-5
(a) *Location of the hypothalamus and pituitary gland.* (b) *Details of the hypothalamus-pituitary system. Signals coming from connecting neurons stimulate the hypothalamus to secrete hormones destined for the anterior pituitary into a special blood vessel, which carries the hormones directly to a capillary network in the anterior pituitary. The hormones of the latter are then released into the blood and passed into the general circulation. Posterior pituitary hormones are made in nerve cells arising in the hypothalamus and are conveyed by their axons to the posterior pituitary, where they are released into the blood.*

Table 25-3 Pituitary Hormones

	MW
Anterior pituitary hormone	
Corticotropin	4,500
Thyrotropin	28,000
Somatotropin	21,500
Follicle-stimulating hormone	34,000
Luteinizing hormone	28,500
Prolactin	23,500
Lipotropin	11,800
Posterior pituitary	
Vasopressin	1,070
Oxytocin	1,070

tained from slaughterhouse animals. Roger Guillemin of San Diego and Andrew Schally of New Orleans were honored with a Nobel prize in 1977 for being the first to succeed in isolating hypothalamic hormones and proving their chemical structure. They shared the award with Rosalind Yalow, who perfected the extremely sensitive radioimmunoassay method (Box 25-1) for measurement of hormones, without which much of the recent progress in hormone research would have been impossible.

The hypothalamic hormones do not enter the general circulation but are conveyed directly to the nearby pituitary gland through special vessels (Figure 25-5). The pituitary gland has two parts, with different embryological origins, the *anterior* and *posterior* lobes. The anterior pituitary produces several different hormones, all relatively long polypeptides (Table 25-3). They are called *tropic hormones* or *tropins*, since they have an affinity for the next rank of endocrine glands and stimulate them. Thus *corticotropin* stimulates the adrenal cortex and *thyrotropin* stimulates the thyroid gland. The major anterior pituitary hormones have been isolated and their amino acid sequences determined. Some of the anterior pituitary hormones, such as corticotropin, are used medically to replace or augment the natural secretion of the hormone when it is deficient.

The posterior pituitary releases two characteristic hormones, *oxytocin* and *vasopressin*, peptides of nine residues that are formed from longer-chain precursors (Figure 25-6). Formation of these hormones begins in cells of the hypothalamus. Oxytocin (Greek, "rapid birth") acts on certain smooth muscles, especially in the uterus. It is used in obstetrics to induce labor and to stimulate lactation. Vasopressin increases blood pressure and increases water reabsorption in the kidney.

Box 25-1 Radioimmunoassay of polypeptide hormones

Animals form antibodies to specific antigens, foreign proteins that gain entrance into the body. The antibody protein, which appears in the blood serum, can bind the antigen molecule very tightly but reversibly. Each antibody is very specific and can bind only the antigen that evoked its formation. The specificity and affinity of antibodies for their antigens were exploited by Rosalind Yalow and her colleagues to measure extremely low concentrations of polypeptide hormones in the blood and tissues. The hormone to be determined is used as the antigen (Ag) and is injected into guinea pigs. After a series of such injections, a high concentration of the antibody to the injected hormone appears in the blood plasma. The antibody (Ab), which can be isolated from the serum, is then mixed with a known amount of radioactively labeled hormone ($\overset{*}{\text{Ag}}$) to form the antigen-antibody complex ($\overset{*}{\text{Ag}}$-Ab), in a reversible reaction whose equilibrium is far to the right:

$$\overset{*}{\text{Ag}} + \text{Ab} \rightleftharpoons \overset{*}{\text{Ag}}\text{-Ab}$$

To determine the concentration of this hormone in a sample of blood serum from a patient, the sample is mixed with a known amount of the radioactive $\overset{*}{\text{Ag}}$-Ab complex. The antibody cannot distinguish between the labeled hormone already bound to it and the unlabeled hormone (Ag) in the blood sample. The latter will therefore compete with $\overset{*}{\text{Ag}}$ for binding to Ab and will displace some of the labeled hormone from the $\overset{*}{\text{Ag}}$-Ab complex, as shown in the equation

$$\text{Ag} + \overset{*}{\text{Ag}}\text{-Ab} \rightleftharpoons \overset{*}{\text{Ag}} + \text{Ag-Ab}$$

After the antibody comes into full equilibrium with the labeled and unlabeled hormones, the amount of radioactive antigen displaced from the $\overset{*}{\text{Ag}}$-Ab complex by the unlabeled hormone in the test sample is measured, from which the concentration of the latter in the blood sample is calculated. For example, if the amount of the hormone in the sample to be tested is exactly equal to the amount of radioactive hormone already bound to the antibody, exactly one-half the radioactive hormone will be displaced from the antibody by the unlabeled hormone. The higher the concentration of the unlabeled hormone in the serum to be analyzed, the greater the amount of radioactive hormone that is displaced from the $\overset{*}{\text{Ag}}$-Ab complex.

Radioimmunoassays are not only highly specific but extremely sensitive. In Yalow's Nobel prize citation the extraordinary sensitivity of radioimmunoassay was compared to detection of a half-cube of sugar dissolved in a very large lake.

Figure 25-6
Hormones of the posterior pituitary gland. The carboxyl-terminal residues indicated by —Gly—NH_2 (color) are glycinamide (—NH—CH_2—CO—NH_2) residues.

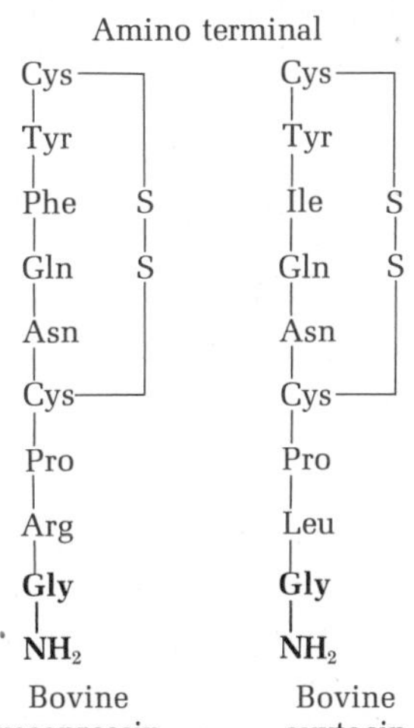

Deficient secretion of vasopressin results in <u>diabetes insipidus</u> (excessive excretion of insipid, i.e., tasteless urine), in which the urine is very dilute and its volume large, up to 10 L/d. This condition is to be distinguished from <u>diabetes mellitus</u>, also an endocrine disease. We shall return to the action of these and other hypothalamic and pituitary hormones later.

Now we shall examine those hormones controlling the metabolism of the major nutrients. These are the best-understood hormones with regard to their mode of action, and they are important in the treatment of a number of endocrine diseases affecting metabolism.

The Adrenal Medulla Secretes the Amine Hormones Adrenaline and Noradrenaline

<u>*Adrenaline*</u> is the best understood of all the hormones. The mechanism of its action, which is well known, has served as a model for research on other hormones. The target tissues of adrenaline are the liver and skeletal muscles, as well as the heart and vascular system.

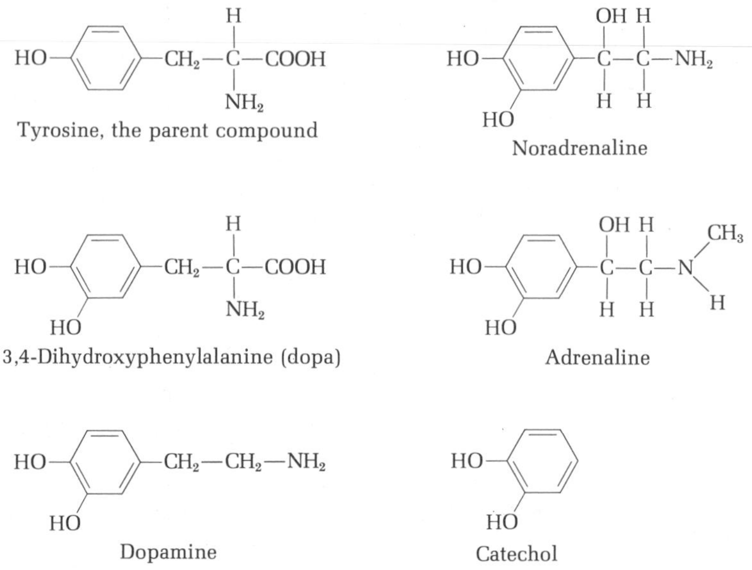

Figure 25-7
The catecholamine hormones. They are formed from the amino acid tyrosine and are derivatives of catechol or 1,2-dihydroxybenzene, shown at bottom right. The abbreviation dopa is derived from the German name of this compound, dioxyphenylalanine.

Tyrosine, the parent compound

Noradrenaline

3,4-Dihydroxyphenylalanine (dopa)

Adrenaline

Dopamine

Catechol

Adrenaline (also called epinephrine) and *noradrenaline* (norepinephrine) are closely related hormones, made and secreted by the *medulla,* or inner portion, of the adrenal glands, which are located just above the kidneys (Figure 25-1). The adrenal medulla is essentially part of the nervous system, from which it receives signals. Both adrenaline and noradrenaline are water-soluble amines derived from tyrosine via 3,4-dihydroxyphenylalanine (*dopa*), as shown in Figure 25-7. Another intermediate in this conversion, *3,4-dihydroxyphenylethylamine,* usually called *dopamine,* is also a hormonelike substance. Adrenaline, noradrenaline, and dopamine are called *catecholamines* since they may be regarded as derivatives of *catechol* or *1,2-dihydroxybenzene* (Figure 25-7). The catecholamines are also made in the brain and nervous system, where they function as neurotransmitter substances. In Parkinson's disease the formation of dopamine in the brain is defective, but the symptoms can be relieved by administration of dopa, the precursor of dopamine.

Adrenaline is stored in *chromaffin vesicles* in cells of the adrenal medulla. These membrane-surrounded structures, about 0.1 μm in diameter (Figure 25-8), contain about 20 percent adrenaline and about 4 percent ATP. Nerve impulses reaching the medulla cause exocytosis of adrenaline from these granules into the surrounding extracellular fluid and thence into the blood. Normally the adrenaline level in blood is only about 0.06 μg/L, or about 10^{-10} M, but sensory stimuli that alarm the animal and galvanize it into readiness to fight or flee cause the concentration of adrenaline in the blood to increase almost a thousandfold in seconds or minutes. Adrenaline prepares the animal for emergency in several ways. It increases the heart rate, the heart output, and the blood pressure, preparing the cardiovascular system for emergency action; it stimulates the breakdown of liver glycogen into blood glucose, the fuel for anaerobic muscular work; and it promotes the anaerobic breakdown of the glycogen of skeletal muscle into lactate via glycolysis, thus stimulating glycolytic ATP formation. These properties also make adrenaline one of the most valuable drugs used in medicine, particularly in life-threatening situations when the cardiovascular system is in acute collapse. It also relaxes the smooth muscle surrounding the bronchioles of the lungs, relieving the symptoms of acute asthma.

Today we have a rather complete picture of the biochemical mechanism by which adrenaline stimulates the breakdown of glycogen, revealed in classical investigations by the American biochemist Earl W. Sutherland, Jr., and his colleagues.

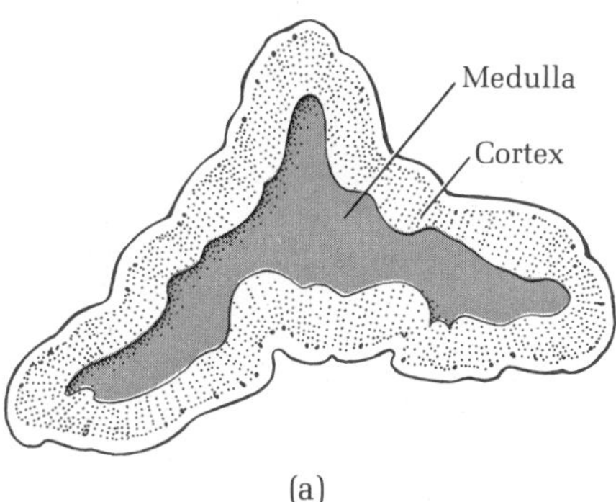

(a)

(b)

Figure 25-8
(a) Drawing of the adrenal gland. The medulla secretes adrenaline and noradrenaline; the cortex secretes adrenal cortical steroid hormones. (b) Freeze-fracture replica of chromaffin granules in the cytoplasm of the cells of the rat adrenal medulla. These granules contain large amounts of adrenaline and ATP. When the medulla is stimulated by incoming nerve signals, the contents of the granules are extruded out of the cell by exocytosis.

Adrenaline Stimulates the Formation of Cyclic Adenylate

In the early 1950s Sutherland found that when it is added to slices of intact liver tissue suspended in a buffered medium, adrenaline accelerates the breakdown of liver glycogen and causes free glucose to appear in the medium. When he extracted the adrenaline-treated liver tissue and measured the enzymes concerned in the conversion of glycogen to glucose, he found that the activity of glycogen phosphorylase was greatly increased over its normal level in untreated liver slices. He concluded that the activity of glycogen phosphorylase, which promotes breakdown of glycogen to glucose 1-phosphate, is the rate-limiting step in the breakdown of glycogen and that it is accelerated when adrenaline stimulates the liver. However, when Sutherland added adrenaline to isolated, purified preparations of glycogen phosphorylase, no increase in its activity occurred, indicating that the stimulating effect of adrenaline on phosphorylase is indirect and depends on factors contributed by the intact liver cell.

On the other hand, when adrenaline was added to homogenized liver tissue, supplemented with ATP and Mg^{2+}, the activity of glycogen phosphorylase in the homogenate was greatly stimulated. This effect was found to require two stages. In the first, adrenaline, acting on the membranous fraction of the liver homogenate, causes it to form a soluble heat-stable stimulatory factor in a reaction requiring ATP and Mg^{2+}. In the second stage, which also requires ATP, the heat-stable factor promotes the conversion of phosphorylase *b* (page 422), in the soluble portion of the liver homogenate into its active form phosphorylase *a*. The heat-stable stimulatory factor formed in the first stage, which normally occurs in only minute amounts in cells, was ultimately isolated and crystallized. It was found to contain adenine, ribose, and phosphate in the ratio 1:1:1, suggesting that it arose from the ATP required for its formation. In 1960 it was identified as *3′,5′-cyclic adenylic acid*, a derivative of adenylic acid never before observed in biological material. In cyclic adenylate the single phosphate group is esterified

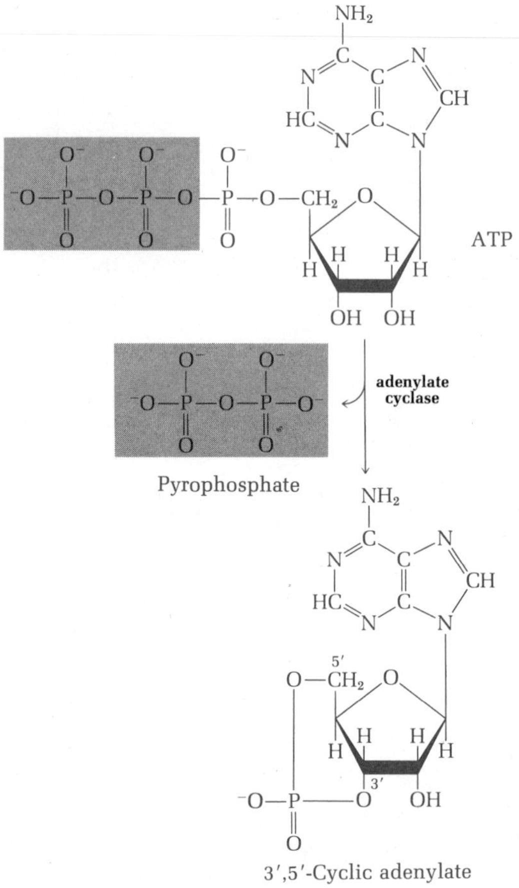

Figure 25-9
Enzymatic formation of 3′,5′-cyclic adenylate (cyclic AMP) from ATP. The elimination of pyrophosphate results in closure of a six-membered ring involving the α-phosphate group. The phosphate group is thus esterified to the 3′- and 5′-hydroxyl groups of the ribose.

to two hydroxyl groups of the ribose, the 3′ and 5′ groups. It is thus a cyclic phosphodiester (Figure 25-9). Addition of very small amounts of cyclic adenylate (cyclic AMP or cAMP) to soluble liver extracts was found to promote the formation of phosphorylase *a* from the less active phosphorylase *b* in the presence of ATP.

Further research by Sutherland and his colleagues revealed that adrenaline greatly stimulates a Mg^{2+}-dependent enzymatic reaction of ATP in the plasma-membrane fraction of liver cells. In this reaction ATP is split to yield cyclic AMP, with loss of inorganic pyrophosphate

$$\text{ATP} \xrightarrow{Mg^{2+}} 3',5'\text{-cyclic AMP} + PP_i$$

The enzyme catalyzing this reaction, adenylate cyclase, is found in many animal tissues. It is tightly bound to the inner surface of the plasma membrane and is not easily extracted in soluble form. Thus the binding of adrenaline (the first messenger) to its receptor sites on the cell surface promotes the formation within the cell of cyclic AMP, the second messenger, which in turn promotes activation of glycogen phosphorylase.

Cyclic AMP Stimulates Protein Kinase Activity

How does cyclic AMP stimulate the conversion of inactive phosphorylase *b* into phosphorylase *a*? We will recall (page 574) that this conversion is promoted by the enzyme *phosphorylase kinase*, which catalyzes the transfer of the terminal phosphate group of two ATPs to the hydroxyl groups of two specific serine residues in phosphorylase *b*, thus converting the latter into phosphorylase *a*

$$2\text{ATP} + \text{phosphorylase } b \longrightarrow 2\text{ADP} + \text{phosphorylase } a$$

Cyclic AMP does not affect phosphorylase kinase itself. However, phosphorylase kinase also exists in active and less active forms (page 422). The inactive form of phosphorylase kinase is converted into its active form by phosphorylation, also at the expense of ATP.

Now we come to the regulatory connection between cyclic AMP and the activity of glycogen phosphorylase. The missing link is an enzyme called *protein kinase*, which also occurs in active and inactive forms. The active form catalyzes the phosphorylation of inactive phosphorylase kinase by ATP to yield the active, phosphorylated form, in a reaction in which ATP is the phosphate-group donor and Ca^{2+} is required as an activator

$$\underset{\text{Inactive form}}{\text{Phosphorylase kinase}} + \text{ATP} \xrightarrow{\text{active protein kinase, } Ca^{2+}} \underset{\text{Active form}}{\text{phosphorylase kinase—Ⓟ}} + \text{ADP}$$

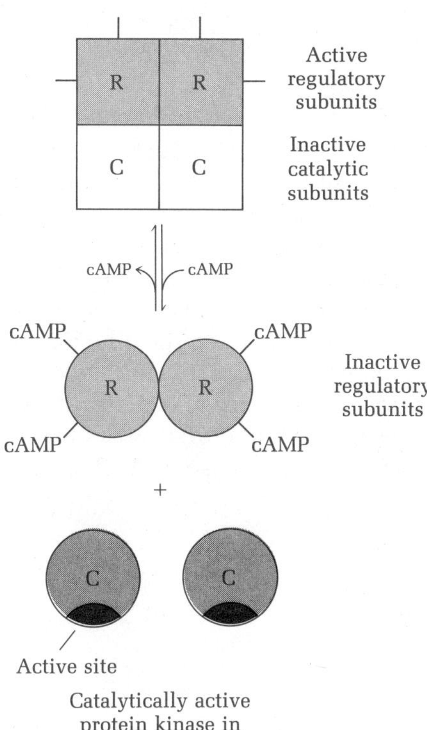

Figure 25-10
Activation of inactive protein kinase by cyclic adenylate (cAMP), which binds to the two regulatory subunits (R) and sets free the two active catalytic subunits (C) of protein kinase.

Phosphorylase kinase is a very large protein, over 1 million in molecular weight. It has 16 subunits, each containing a specific serine residue that becomes phosphorylated by ATP through the action of activated protein kinase.

Protein kinase, the key enzyme linking cyclic AMP to the activation of phosphorylase, is an allosteric enzyme. Its inactive form contains two *catalytic subunits* (C) and two *regulatory subunits* (R) (Figure 25-10). When the subunits are associated together as a complex, designated C_2R_2, the enzyme is inactive. However, cyclic AMP is the allosteric stimulator of protein kinase. When four molecules of cyclic AMP bind to their sites on the two regulatory subunits, the inactive C_2R_2 complex dissociates to yield the free catalytic subunits in enzymatically active form and the R_2–$cAMP_4$ complex, in which cyclic AMP remains bound. Cyclic AMP therefore removes the inhibition of protein kinase activity imposed by the bound regulatory subunit (Figure 25-10).

Cyclic AMP functions as an intracellular messenger in the action of many hormones besides adrenaline (Table 25-4). When protein kinase is activated by cyclic AMP, it can phosphorylate a number of other important enzymes in different kinds of target cells.

Table 25-4 Hormones Using Cyclic AMP as a Second Messenger

Adrenaline
Corticotropin
Lipotropin
Parathyroid hormone
Thyroid-stimulating hormone
Vasopressin

The Stimulation of Glycogen Breakdown by Adrenaline Occurs by an Amplification Cascade

Now let us assemble the various elements described above and trace the chain of events when adrenaline stimulates the breakdown of glycogen to blood glucose in the liver (Figure 25-11). Adrenaline arrives at the surface of the liver cell, where it binds to the specific <u>*adrenaline receptor*</u>. The binding of adrenaline, which never actually enters the cell, causes a change in the receptor protein. The change is somehow transmitted through the membrane and "turns on" adenylate cyclase, bound at the inner surface of the cell membrane. The active form of adenylate cyclase now converts ATP into cyclic AMP, the second messenger, which quickly attains a peak concentration of about 10^{-6} *M* in the cytosol. The cyclic AMP in turn binds to the regulatory subunits of protein kinase, causing its catalytic portion to be set free in its active form. The activated protein kinase then catalyzes phosphorylation of the inactive or dephospho form of phosphorylase kinase, at the expense of ATP, to yield its active or phospho form. Active phosphorylase kinase, which requires Ca^{2+} for activity, then catalyzes the phosphorylation of the relatively inactive phosphorylase *b* at the expense of ATP to yield

Figure 25-11
Adrenaline triggers an amplification cascade in liver cells. Binding of a few molecules of adrenaline to its specific receptors on the cell surface initiates a series of enzymatic reactions that result in release of a very large amount of glucose into the blood.

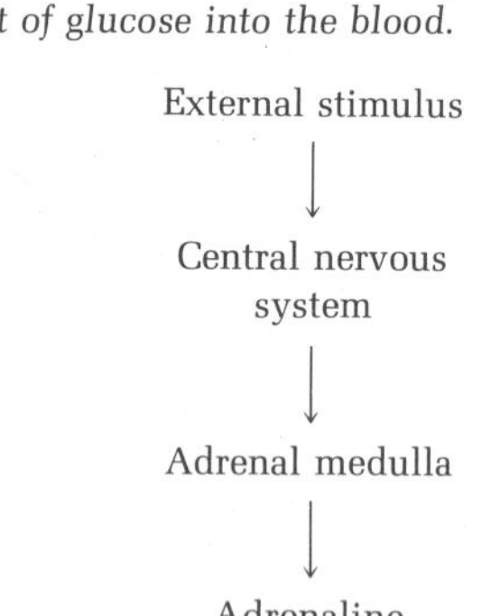

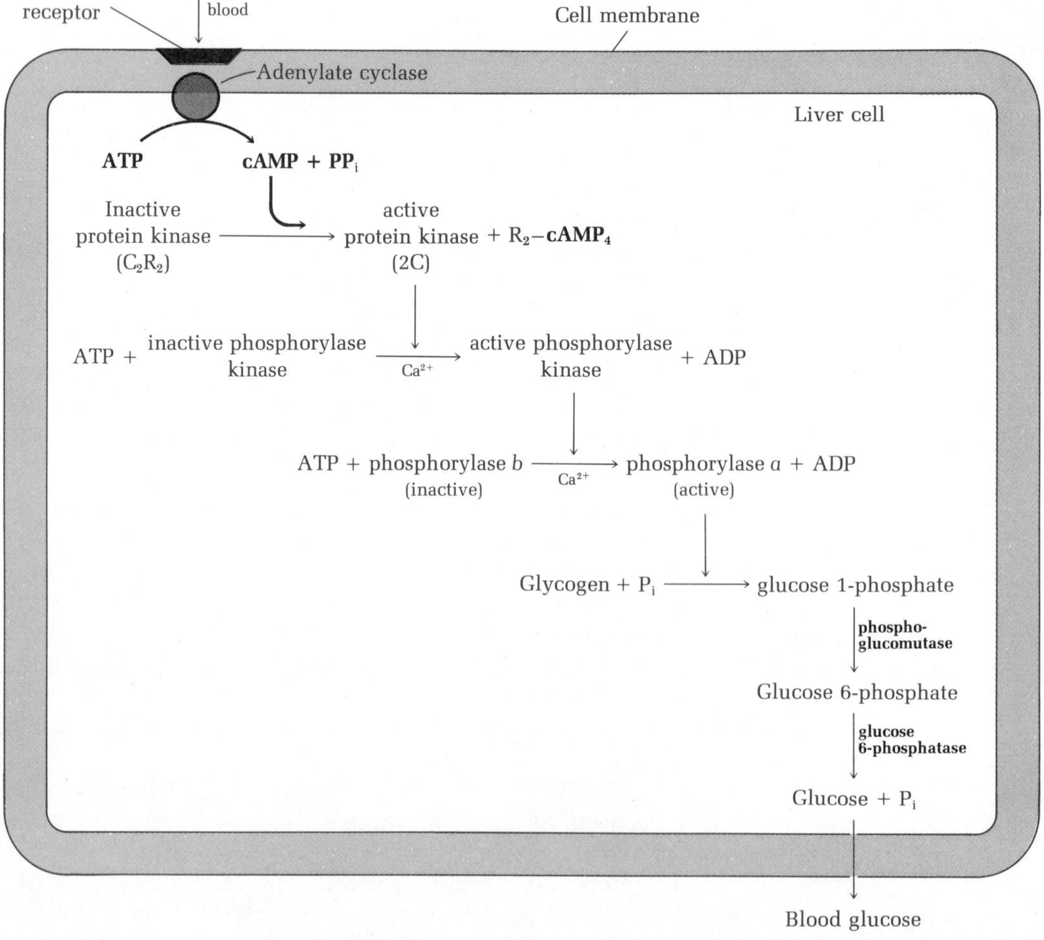

active phosphorylase *a*. The latter in turn catalyzes breakdown of glycogen at a very high rate to yield glucose 1-phosphate, from which glucose 6-phosphate and then free blood glucose are formed (Figure 25-11). Although there are many steps in this sequence of events, glycogen phosphorylase can reach peak activity in a few minutes after adrenaline is bound to the liver cell.

The sequence of steps in Figure 25-11 may be regarded as a cascade of enzymes acting upon enzymes. Each enzyme in the cascade promotes the activation of many molecules of the next enzyme. In this way there is a large and rapid amplification of the incoming signal, estimated to be some 25-million-fold. Thus binding of a few adrenaline molecules to the adrenaline receptors in the liver quickly results in the release of many grams of glucose into the blood.

The cascade shown in Figure 25-11, up to the point of glucose 6-phosphate, is identical in the liver and skeletal muscles. Since muscles contain no glucose 6-phosphatase, they form no blood glucose. Instead, the increased formation of glucose 6-phosphate in muscle leads to a greatly increased rate of glycolysis to lactate, making ATP available for contraction. More recent research has shown that adrenaline also stimulates glycogen breakdown in the liver by a second amplification pathway, parallel to that described in Figure 25-11. The alternative pathway, which predominates in some conditions, involves Ca^{2+} as an intracellular messenger.

The cascade shown in Figure 25-11 is set into operation in the liver not only by adrenaline but also by the pancreatic hormone glucagon, as we shall see.

Adrenaline Also Inhibits Glycogen Synthesis

Adrenaline not only stimulates glycogen breakdown but at the same time inhibits synthesis of glycogen from glucose in the liver. Thus it has a reciprocal action that results in making glucose maximally available in the blood. Figure 25-12 shows that the binding of adrenaline to the liver cell and the subsequent formation of cyclic AMP promote the phosphorylation by protein kinase of the active, or *dephospho*, form of *glycogen synthase* to its phosphorylated, inactive form (page 574). Inhibition of glycogen synthase is thus brought about by a chain of events triggered by the same stimulus that causes acceleration of glycogen breakdown to yield blood glucose. In this way all available glycogen, glucose 6-phosphate, and other precursors are directed into the production of free blood glucose, thus providing the muscles with a maximal supply of fuel, ready to face the emergency.

Besides its action on the liver, adrenaline promotes glycogen breakdown in skeletal muscle and the heart, also by stimulating muscle phosphorylase via cyclic AMP. Because the muscles and heart lack glucose 6-phosphatase, the product of glycogen breakdown in those organs is not blood glucose but

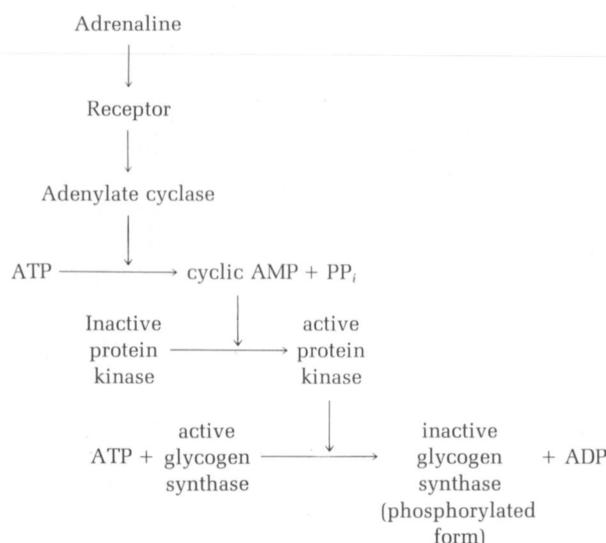

Figure 25-12
Inhibition of glycogen synthesis by adrenaline. Protein kinase, activated by cyclic AMP, promotes phosphorylation of glycogen synthase to yield its less active form (see page 574).

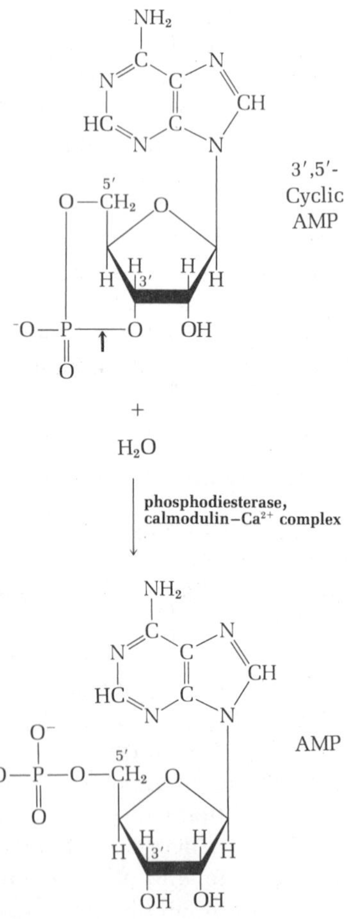

Figure 25-13
Action of phosphodiesterase on cyclic AMP. In many tissues phosphodiesterase is stimulated by Ca^{2+}. This effect is indirect, initiated by the binding of Ca^{2+} to the regulatory protein calmodulin (see facing page). The subsequent binding of the Ca^{2+}–calmodulin complex activates phosphodiesterase.

lactate, formed from glucose 6-phosphate by glycolysis. In muscles glycogen breakdown is stimulated, thus accelerating glycolysis and ATP formation to make possible a rapid increase in muscular activity.

Phosphodiesterase Inactivates Cyclic Adenylate

As long as an emergency persists and adrenaline is actively secreted into the blood by the adrenal medulla, the liver adenylate cyclase system remains fully activated. Thus cyclic AMP is maintained at a relatively high concentration in the target cells, yielding a high rate of glycogen breakdown. Once the emergency is over, however, adrenaline secretion stops and its level in the blood rapidly falls due to its enzymatic destruction in the liver. Once the adrenaline receptors become empty, adenylate cyclase reverts to its inactive form and cyclic AMP is no longer formed. The remaining cyclic AMP in the cell is destroyed by phosphodiesterase (Figure 25-13), an enzyme that catalyzes the hydrolysis of the 3′ phosphate bond of cyclic AMP to yield free 5′-adenylate (5′-AMP)

$$3',5'\text{-Cyclic AMP} + H_2O \xrightarrow{\text{phosphodiesterase}} 5'\text{-adenylate}$$

As the cyclic AMP in the cytosol decreases, the bound cyclic AMP is released from the regulatory subunit of protein kinase. As a consequence the regulatory subunit recombines with the catalytic subunit of protein kinase, converting it back into its inactive form. The phosphorylated form of phosphorylase kinase then undergoes dephosphorylation, as does phosphorylase *a* itself, by the action of phosphorylase phosphatase. In this way the glycogenolytic system relaxes into its normal resting state. Simultaneously, glycogen synthase is reactivated again by its dephosphorylation.

Figure 25-14
Calmodulin is the mediator in many Ca^{2+}-stimulated enzymatic reactions and membrane transport systems.

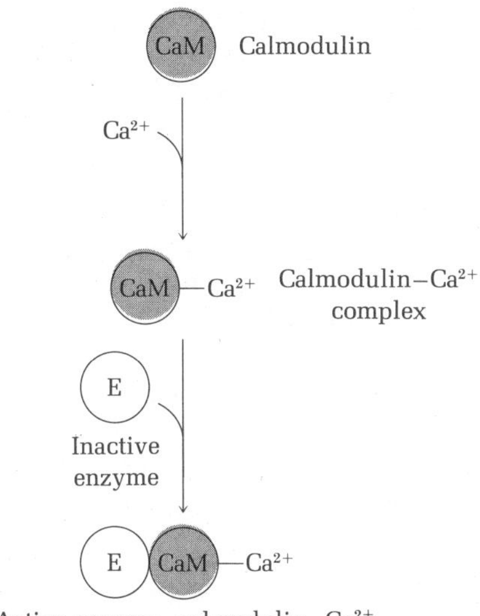

Phosphodiesterase is characteristically inhibited by caffeine and theophylline, alkaloids present in small amounts in coffee and tea, respectively. These drugs prolong or intensify the activity of adrenaline by decreasing the rate of breakdown of cyclic AMP. Phosphodiesterase in some tissues is stimulated by Ca^{2+}. This effect is mediated by the binding of Ca^{2+} to a specific Ca^{2+}-binding protein called calmodulin. The Ca^{2+}–calmodulin complex binds to phosphodiesterase and stimulates it. Calmodulin is a newly discovered Ca^{2+}-binding protein that is widely distributed in all forms of animal life. Its amino acid sequence is nearly identical in all animal species. Calmodulin is thus one of the most ancient and most highly conserved animal proteins, in an evolutionary sense. Many cell functions are regulated by the concentration of Ca^{2+} in the cytosol; for this reason Ca^{2+}, like cyclic AMP, is also an important second messenger. Calmodulin transmits the message carried by an increase in Ca^{2+} in the cytosol by forming a Ca^{2+}–calmodulin complex, which then binds to the specific Ca^{2+}-regulated protein, stimulating its activity (Figure 25-14).

The Pancreas Secretes Several Hormones Regulating Metabolic Pathways

We now examine another way in which carbohydrate metabolism, as well as the metabolism of amino acids and lipids, is controlled by a group of hormones secreted by the pancreas, whose endocrine activity is in turn regulated by the anterior pituitary. The pancreas has two major biochemical functions. One is the biosynthesis of several enzymes, such as trypsin, chymotrypsin, and carboxypeptidase, that are secreted into the intestine to carry out digestion of food. This set of functions is carried out by the exocrine cells ("secreting outwardly," i.e., into the pancreatic duct) (Figure 25-15). The other major func-

Figure 25-15
The endocrine tissue of the pancreas. In addition to the exocrine or acinar cells, which secrete digestive enzymes as their zymogens (Chapter 24), the pancreas contains endocrine tissue in the form of the islets of Langerhans. The islets contain several different types of cells, each of which secretes a specific polypeptide hormone.

(a) Drawing of the pancreas. (b) Schematic drawing of an islet showing the different cell types listed in Table 25-5. (c) Electron micrograph of a portion of an islet of Langerhans of human pancreas. It shows an A cell, a B cell, and a D cell.

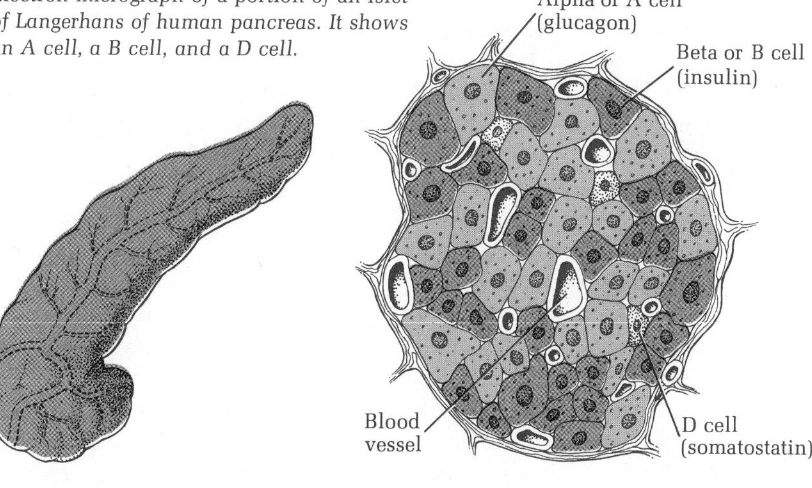

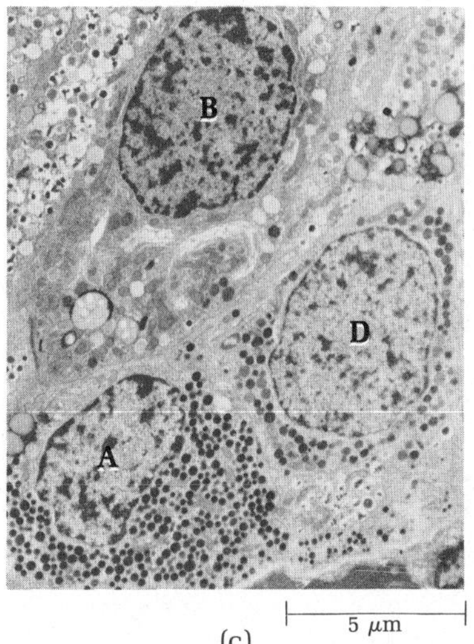

	A chain			B chain
	Gly			Phe
	Ile			Val
	Val			Asn
	Glu			Gln
5	Gln		5	His
	Cys (S—S to Cys 11)			Leu
	Cys	—S—S—		Cys
	Ala			Gly
	Ser			Ser
10	Val		10	His
	Cys (S—S to Cys 6)			Leu
	Ser			Val
	Leu			Glu
	Tyr			Ala
15	Gln		15	Leu
	Leu			Tyr
	Glu			Leu
	Asn			Val
	Tyr			Cys
20	Cys	—S—S— (to B Cys 19)	20	Gly
	Asn			Glu
				Arg
				Gly
				Phe
	Carboxyl-terminal ends		25	Phe
				Tyr
				Thr
				Pro
				Lys
			30	Ala

Figure 25-16
The amino acid sequence of the A and B chains of bovine insulin.

tion of the pancreas is the biosynthesis of insulin and several other polypeptide hormones that regulate the metabolism of glucose and other major nutrients. This function is carried out by the *endocrine* tissue of the pancreas, clusters of specialized cells called the *islets of Langerhans*. The islets contain several different types of related cells, each type forming a single kind of pancreatic hormone (Figure 25-15 and Table 25-5). The hormones secreted by the islets are *glucagon*, made by the A cells; *insulin*, made by the B cells; *somatostatin*, made by the D cells; and *pancreatic polypeptide*, a nondistinctive name for a recently discovered hormone made by the F cells of the islet tissue. Although all these hormones have pronounced effects on metabolism, particularly on that of carbohydrates, insulin is central to their action. Despite the great medical importance of insulin in the treatment of diabetes mellitus, we do not yet know precisely how it performs its life-preserving action. The quest to understand the mode of action of insulin is one of the most intense research efforts in biochemistry and medicine today.

Insulin Is the Hypoglycemic Hormone

In the late nineteenth century it was found that surgical removal of the pancreas of dogs causes a condition closely resembling human diabetes mellitus, discussed in Chapter 24. In such animals, as in people with diabetes, the blood glucose level is abnormally high, a condition called *hyperglycemia*. So much glucose is excreted in the urine that it tastes sweet, a condition called *glycosuria*. (Diabetes mellitus and diabetes insipidus, both of which cause excretion of large volumes of urine, were once differentiated by tasting the urine.) Attempts to cure the depancreatized dogs by feeding them raw pancreatic tissue obtained from normal animals failed, but injection of extracts of normal pancreas into depancreatized dogs alleviated the diabetic symptoms. After many unsuccessful attempts the active factor present in pancreatic extracts was finally isolated in pure form in 1922. It was named *insulin* ("islet substance") since the islet tissue had been recognized to be the source of the hormone. In a short time insulin came into use in the treatment of human diabetes. Since then it has become one of the most important therapeutic agents known to medicine: it has saved or prolonged countless lives.

Insulin, a small protein of molecular weight 5700 (page 125), consists of two polypeptide chains, A and B, joined to each other by two disulfide cross-links (Figure 25-16). Recall that it was the first protein for which the amino acid sequence was deduced (page 134). The insulin used in medicine is isolated from the pancreatic tissue of slaughterhouse animals.

Insulin is synthesized by the B or β cells of the pancreas in inactive precursor forms. The immediate precursor of insulin is *proinsulin*, a single-chain polypeptide with 78 to 86 residues, depending upon the species (Figure 25-17). Proinsulin from

Table 25-5 Pancreatic Hormones

Source	Hormone
A cells	Glucagon
B cells	Insulin
D cells	Somatostatin
F cells	Pancreatic polypeptide

bovine pancreas has 81 residues and three intrachain disulfide cross-links. It is stored in granules inside the B cells of the islet tissue until the signal comes for its secretion. At that time proinsulin is converted into active insulin by the action of specific peptidases, which cleave two peptide bonds in the proinsulin chain, removing a middle segment. Two amino acid residues are then removed from the ends of the middle segment by peptidase action (Figure 25-17) to yield *C-peptide*. The two terminal segments of the original proinsulin chain become the A and B chains of insulin, held together by the two disulfide cross-linkages.

Proinsulin itself is made from an earlier precursor, *preproinsulin*, which has 23 more amino acid residues at the amino-terminal end of proinsulin (Figure 25-17). This amino-terminal sequence is removed by a peptidase to form proinsulin. The extra amino acid sequence at the amino-terminal end of proinsulin is a genetically determined "leader" or "signaling" sequence, which directs the newly made proinsulin to its specific destination in the cell, in this case the proinsulin secretory vesicles. Such signal sequences are coded into many proteins during their synthesis, as we shall see in Chapter 29.

Insulin Secretion Is Regulated Primarily by Blood Glucose

Insulin is secreted from the B cells of the islets into the blood by a complex process; it requires Ca^{2+} and its last step is the extrusion into the blood of the contents of secretion granules in which insulin and C-peptide are formed. The rate of insulin secretion is determined primarily by the concentration of glucose in the blood. When the blood sugar level rises, insulin is secreted at an increased rate. The increased insulin level accelerates the entrance of glucose from the blood into the liver and muscles, where it is largely converted into glycogen. This causes the blood glucose concentration to fall to its normal

Figure 25-17
Formation of insulin. Preproinsulin (the entire structure shown below) is the ultimate precursor of insulin. It is converted into proinsulin by enzymatic removal of 23 residues from the amino-terminal end. Proinsulin is then converted into insulin (color) by peptidase action at the two points shown by arrows. A dipeptide is then removed from each end of the intervening peptide to yield C-peptide.

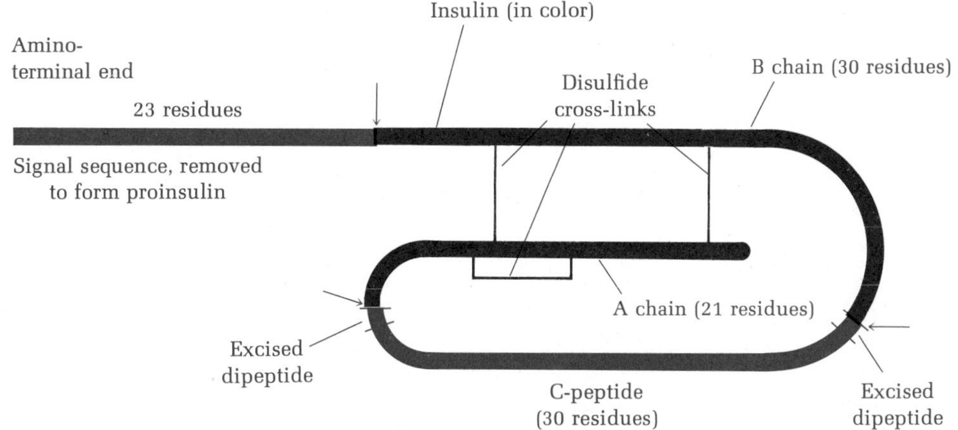

level, which causes insulin secretion in turn to slow down to the normal rate. Thus there is a closely adjusted feedback relationship between the rate of insulin secretion and the blood glucose concentration.

The Second Messenger of Insulin Is Still Unknown

Insulin receptors have been detected on the surfaces of liver and skeletal muscle cells and also in fat cells or adipocytes. Insulin receptors have been extracted from cell membranes and purified. The isolated insulin receptor is a specific glycoprotein which binds insulin very tightly. The number of insulin receptors on cell surfaces changes depending on metabolic conditions; moreover, they have a rather high turnover rate. Despite intensive efforts for many years, the cytoplasmic second messenger released when insulin is bound to its receptors on the cell surface has not yet been identified. However, it is known that intracellular Ca^{2+} is important in triggering the action of insulin; another clue is that the pyruvate dehydrogenase activity of mitochondria (page 437) is strongly promoted when liver cells are exposed to insulin. Identification of the intracellular messenger for insulin is one of the most important objectives in biochemical research, since it is certain to be fundamental to our understanding of diabetes and its treatment.

Figure 25-18
Structure of alloxan, a derivative of pyrimidine that produces experimental diabetes in animals by destroying islet cells.

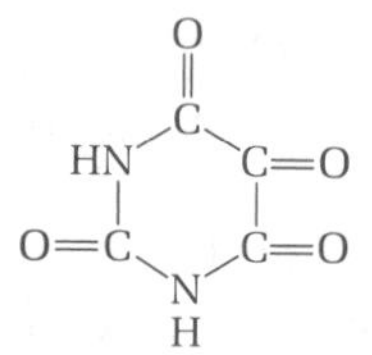

```
        O
        ‖
        C
   HN⁄     ⁀C=O
    |        |
 O=C        C=O
      ⁀N⁄
       H
```

Insulin Influences Many Other Aspects of Metabolism

Some of the other metabolic effects of insulin deficiency have been described in Chapter 24. Human diabetics or animals made diabetic by removal of the pancreas or by destruction of islet tissue with the drug alloxan (Figure 25-18) fail to synthesize fatty acids and lipids from glucose. They oxidize fatty acids at higher-than-normal rates, causing overproduction of the ketone bodies, which accumulate in the tissues, blood, and urine, a condition called ketosis. In diabetic animals there is also failure of circulating amino acids of the blood to enter the peripheral tissues from the blood at the normal rate, causing a decreased rate of protein biosynthesis. Instead amino acids are deaminated in the liver and their carbon chains converted via gluconeogenesis (page 562) into blood glucose. These metabolic aberrations (Table 25-6) are brought back into normal balance by administration of insulin.

All the complex metabolic effects caused by insulin deficiency have been interpreted as an effort by the diabetic organism to convert all available nutrients into blood glucose. The tissues are desperately hungry for glucose, and the liver is just as desperately making it, only to have much of the blood glucose "spill" into the urine. According to this view of the metabolic defect in diabetes, the tissues are unable to absorb glucose from the blood at its normal concentration of about 4.5 mM and require a much higher concentration to absorb it efficiently.

Table 25-6 Metabolic Effects of Insulin Deficiency (Diabetes Mellitus)
Accelerated glycogenolysis in liver
Increased gluconeogenesis
Decreased entrance of glucose into peripheral tissues
Hyperglycemia
Glucosuria
Accelerated oxidation of fatty acids in the liver
Overproduction of ketone bodies
Ketonuria
Decreased fatty acid synthesis
Decreased protein synthesis in peripheral tissues
Increased urea formation and excretion

However, when the blood glucose exceeds about 10 mM, the <u>kidney threshold</u> for glucose, the excess glucose is excreted in the urine, resulting in a large loss of glucose from the body.

Other serious abnormalities in diabetes mellitus are not cured by insulin. In particular, there is defective biosynthesis of the basement membrane of blood capillaries, leading to diseased blood vessels in the heart, the kidneys, the extremities, and the retina. Blindness and renal failure are late complications of diabetes. These conditions are believed to result because administration of insulin once a day by injection cannot simulate exactly the minute-by-minute, hour-by-hour fluctuations in the secretion of insulin into the blood by healthy islet tissue, in response to variations in the blood glucose level. Automated devices are therefore being developed to inject insulin into the circulation of diabetics at a rate determined continuously by the blood glucose concentration, to simulate more exactly the secretory action of normal islet tissue in the body.

Amino-terminal end

Proglucagon:
His
Ser
Gln
Gly
Thr
Phe
Thr
Ser
Asp
10 Tyr
Ser
Lys
Tyr
Leu
Asp
Ser
Arg
Arg
Ala
20 Gln
Asp
Phe
Val
Gln
Tyr
Leu
Met
Asn
Thr (His through Thr: Glucagon)
30 **Lys**
Arg
Asn
Asn
Lys
Asn
Ile
Ala

Figure 25-19
The structures of bovine proglucagon and glucagon. Glucagon has 29 amino acid residues and a molecular weight of 3500. It is formed by enzymatic removal of eight amino acids (color) from the carboxyl end of proglucagon.

Glucagon Is the Hyperglycemic Pancreatic Hormone

Glucagon, also a polypeptide hormone, is secreted by the A cells of the pancreatic islets and by related cells in the gastrointestinal tract. It is a single-chain polypeptide of 29 amino acid residues and molecular weight 3500 (Figure 25-19). Recent research indicates that glucagon, like insulin, has two inactive biosynthetic precursors, <u>proglucagon</u> and <u>preproglucagon</u>. The latter has a polypeptide signaling extension at the amino terminal end, removed in two stages to yield proglucagon.

Glucagon promotes an *increase* in blood glucose concentration; its action is therefore opposed to that of insulin (Table 25-6). The hyperglycemic effect of glucagon is brought about by two different actions. First, glucagon promotes breakdown of liver glycogen to yield blood glucose, by a mechanism similar to that of adrenaline. The plasma membrane of liver cells contains specific surface receptors for glucagon. When they are occupied by the hormone, adenylate cyclase in the plasma membrane is activated, setting into motion an amplification cascade resembling that promoted by adrenaline (Figure 25-11). Second, glucagon, unlike adrenaline, *inhibits* the breakdown of glucose to lactate by glycolysis. This effect is caused by an indirect inhibition of the liver-type or L isozyme of <u>pyruvate kinase</u> (page 424) in the glycolytic sequence. Glucagon differs from adrenaline also in that it has a much longer duration of action and does not stimulate the heart rate or raise blood pressure.

Somatostatin Inhibits Secretion of Insulin and Glucagon

<u>*Somatostatin*</u>, also a polypeptide hormone (Figure 25-20), was first discovered in extracts of the hypothalamus, where it acts as an inhibitor of the secretion of somatotropin and other hormones by the anterior pituitary (next section). Somatostatin is

Figure 25-20
The structure of sheep somatostatin. It contains an intrachain disulfide bond. Somatostatin is secreted by the hypothalamus, the pancreatic islets, and by certain cells in the intestine.

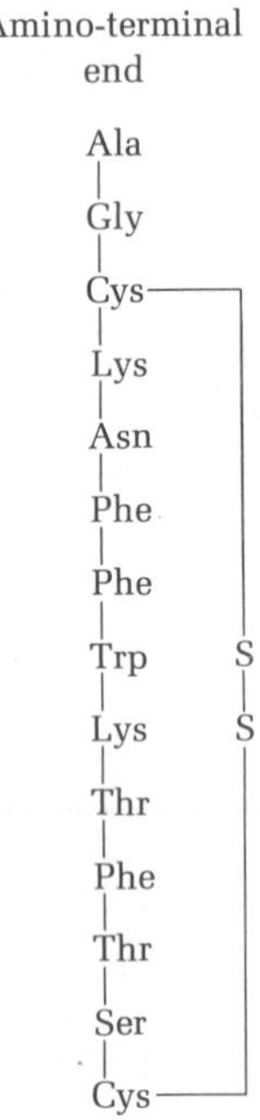

also made by the D cells of the pancreatic islets and by related cells in the gastrointestinal tract. The somatostatin formed in the pancreas influences secretion of insulin and glucagon in a complex manner. Somatostatin is useful in the treatment of some types of diabetes mellitus.

Somatotropin Also Affects the Action of Insulin

Somatotropin, or *anterior pituitary growth hormone*, was first recognized through its capacity to promote the growth of the skeleton and increase body weight in young animals. Its deficiency in people results in *dwarfism* (Figure 25-21). When it is secreted in excess, it results in *gigantism* and *acromegaly*, in which the growth of the hands, feet, and especially the facial bones is greatly accentuated, resulting in a massive "lantern jaw" and heavy brows. Somatotropin also has profound effects on carbohydrate metabolism. Administration of excess somatotropin to animals causes *pituitary diabetes*, brought about because somatotropin inhibits insulin secretion. Somatotropin is a polypeptide of 191 residues (MW 21,000). As indicated above, its secretion is inhibited by somatostatin of the hypothalamus.

Figure 25-22 shows the complex interrelationships in the hormonal regulation of glucose metabolism. Hormones secreted by five different endocrine glands, including the adrenal cortex (below), affect glucose metabolism.

Figure 25-21
"General" Tom Thumb, who probably was deficient in somatotropin secretion, with P. T. Barnum, the circus impresario.

Adrenal Cortical Hormones Are Steroids

The *adrenal cortical hormones*, the *androgens*, and the *estrogens* are the major lipid-soluble steroid hormones. We consider the adrenal cortical hormones first since they also influence carbohydrate metabolism, among other activities. The secretion of the adrenal cortical hormones is controlled by the hypothalamus. In response to stress situations, the hypothalamus secretes *corticotropin-releasing hormone*, which is sent to the anterior pituitary and stimulates it to release *corticotropin* into the blood. Corticotropin is a polypeptide of 39 amino acids (page 135). Its concentration in the blood is normally between 10^{-11} and 10^{-12} *M*. Since its half-life is only 10 min, corticotropin is continuously made and destroyed in a steady state. Corticotropin binds to receptors on the surface of cells of the adrenal cortex, stimulating them to produce their characteristic steroid hormones. Over 30 steroids are made by the adrenal cortex. Collectively called *corticoids*, they can be grouped into three main classes. The *glucocorticoids*, of which *cortisol* is the most important (Figure 25-23), oppose some of the actions of insulin. Cortisol promotes gluconeogenesis from amino acids and glycogen deposition in the liver, increases blood glucose, and decreases peripheral utilization of glucose. It also stimulates utilization of fatty acids and ketogenesis. The glucocorticoids also have remarkable anti-inflammatory and

Figure 25-22
Hormonal control of the blood glucose level.

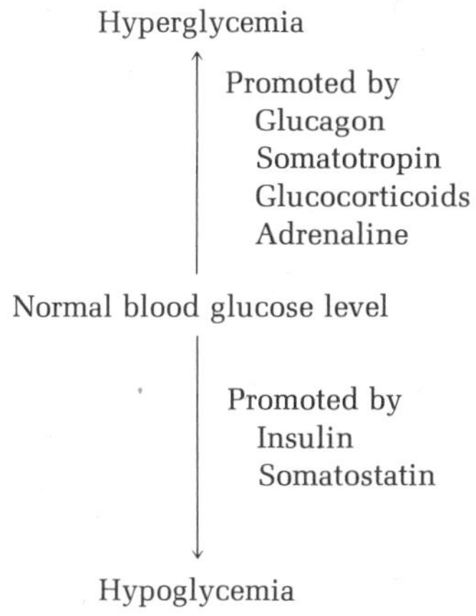

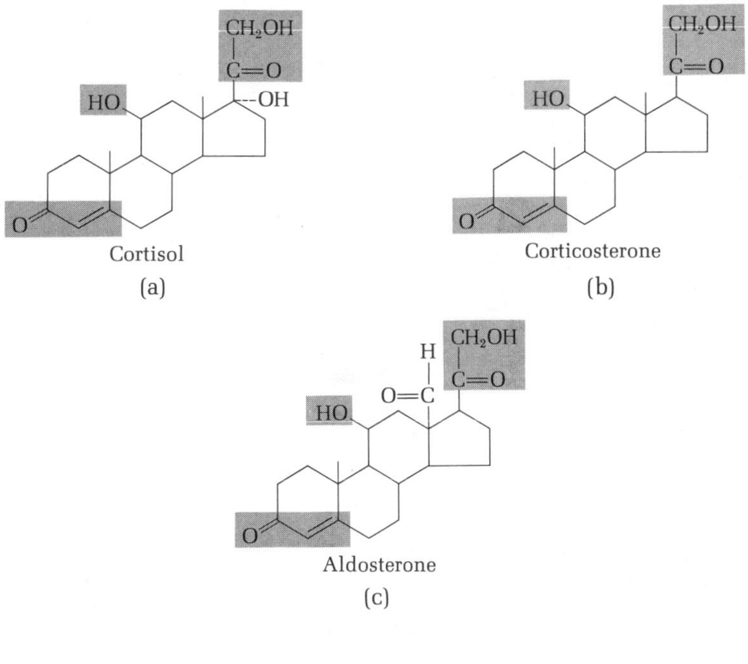

antiallergenic action. Excessive secretion of glucocorticoids is the cause of *Cushing's disease*, in which there is fatigue and loss of muscle mass because of excess conversion of amino acids into glucose and redistribution of body fat, causing the condition of "moon face."

The second class of adrenal cortical steroids comprises the *mineralocorticoids*, which promote the retention of Na^+ and the loss of K^+ by the kidneys; through this action these hormones maintain water and salt balance in the body. The major corticosteroid of this group is *aldosterone* (Figure 25-23). This hormone also has weak glucocorticoid activity; conversely, cortisol has weak mineralocorticoid activity.

The third class of adrenal cortical steroids has properties midway between those of the glucocorticoids and the mineralocorticoids. The major hormone of this class is *corticosterone*. The three classes of corticosteroids have some common denominators in their structures, as seen in Figure 25-23.

The adrenal cortical hormones are lipid-soluble and readily pass through cell membranes of target tissues into the cytoplasm, where they combine with specific *intracellular* receptor proteins. Once the adrenal hormones are bound by these receptors, the hormone-receptor complexes, which may be regarded as intracellular messengers, move into the cell nucleus. There they regulate the transcription of certain genes and thus promote the biosynthesis of specific enzymes and proteins responsible for producing the characteristic end effects of these hormones.

Deficiency in the secretion of adrenal cortical hormones is the cause of *Addison's disease*, in which there is fatigue, weakness, skin pigmentation, and craving for salt. Such patients are very sensitive to stress and infections.

Figure 25-23
The three important adrenal cortical steroids. The shaded portions are common structures. Although all three hormones have both glucocorticoid and mineralocorticoid activities, the glucocorticoid action is predominant in cortisol and the mineralocorticoid action is predominant in aldosterone. Corticosterone has both activities in equal measure. (a) Cortisol, a glucocorticoid. It promotes gluconeogenesis from amino acids and has anti-inflammatory activity. (b) Corticosterone has both glucocorticoid and mineralocorticoid activity. (c) Aldosterone, a mineralocorticoid, promotes Na^+ retention and K^+ loss by the kidneys.

CH_2OH, C=O, HO, OH, O
Cortisol
(a)

CH_2OH, C=O, HO, O
Corticosterone
(b)

CH_2OH, C=O, H, O=C, HO, O
Aldosterone
(c)

The Thyroid Hormones Control Metabolic Rate

The thyroid gland secretes two characteristic hormones, *L-thyroxine* and *L-triiodothyronine* (Figure 25-24). Like adrenaline, these two hormones, abbreviated T4 and T3, respectively, are derivatives of tyrosine. The thyroid hormones are made in response to signals received by the hypothalamus, which secretes *thyrotropin-releasing hormone*, as well as an inhibitor of thyrotropin release. The thyrotropin-releasing hormone stimulates the anterior pituitary to secrete thyrotropin into the blood. Thyrotropin binds to its receptors on the cells of the thyroid gland, stimulating them to produce the thyroid hormones. Thyroxine and triiodothyronine are synthesized in a series of enzymatic reactions beginning with the iodination of L-tyrosine residues in *thyroglobulin*, a glycoprotein (MW 650,000), converting them into *L-monoiodotyrosine* residues (Figure 25-24). The iodine required, an essential trace element (Chapter 26), comes to the thyroid via the blood. The iodine is avidly bound and collected by the thyroglobulin, also called "colloid protein," and then used to iodinate tyrosine residues of thyroglobulin. Subsequent steps in thyroid hormone synthesis occur while the iodotyrosine residues remain attached to thyroglobulin. Nearly all the newly formed thyroxine and triiodothyronine remains bound to thyroglobulin in the thyroid vesicles until they are released into the blood in free form by the action of proteolytic enzymes. It is generally believed that thyroxine is the precursor or prohormone of triiodothyronine, the latter being more active in stimulating metabolism.

The thyroid hormones are then carried by the blood to their targets. Most tissues are stimulated by the thyroid hormones, with the notable exception of the adult brain and some of the

Figure 25-24
The thyroid hormones are shown in (a) and (b). Monoiodotyrosine, the first iodinated precursor of the thyroid hormones, is shown in (c).

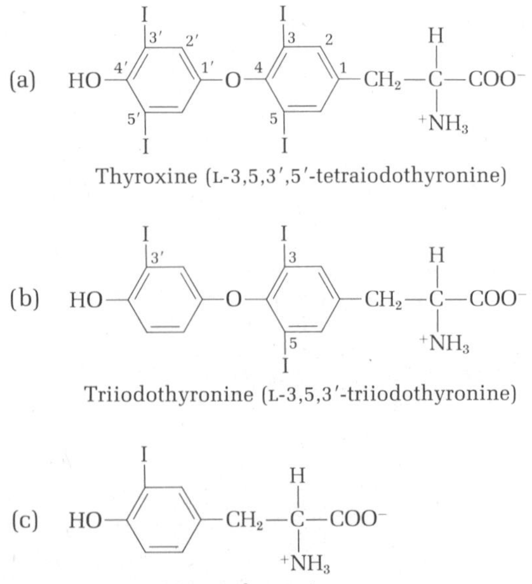

(a) Thyroxine (L-3,5,3′,5′-tetraiodothyronine)

(b) Triiodothyronine (L-3,5,3′-triiodothyronine)

(c) L-3-Monoiodotyrosine

reproductive tissues. The thyroid hormones are especially active in stimulating the metabolism of the liver and muscles. They are bound by specific receptor proteins, which in turn convey thyroxine into the cell nucleus. By interaction of the thyroxine-receptor complexes with specific genes the target cells are induced to synthesize greatly increased amounts of certain enzymes and enzyme systems. The major result is the stimulation of the *basal metabolic rate* of the animal.

The basal metabolic rate (BMR) is a measure of the rate of oxygen consumption of an individual at complete rest, 12 h after a meal, in relation to the surface area of the body. It is expressed in terms of the percentage deviation from the BMR of normal individuals, i.e., as percent above or below the norm for individuals of the same sex, weight, and height. BMR measurements are made in the diagnosis of diseases affecting thyroid function. When thyroid hormone secretion is excessive, the condition is called *hyperthyroidism*. Hyperthyroid individuals have an elevated basal metabolic rate. They burn their foods up at a higher rate than normal individuals, produce more body heat, and tend to be hyperactive. The condition of deficient thyroid hormone secretion, *hypothyroidism*, is characterized by a lowered basal metabolic rate. Hypothyroid patients burn fuels more slowly, have a lower rate of heat production, and tend to be sluggish. Nutritional deficiency of iodine results in the condition of *colloid goiter* (Chapter 26), in which the thyroid gland becomes greatly enlarged due to the abnormally large amount of colloid protein that is made to trap the minimal amounts of circulating iodine.

Just exactly how thyroid hormones regulate the rate of aerobic metabolism is still a mystery, despite much research. Although the mitochondria are the site of organized respiratory activity and ATP formation in cells, it is not entirely clear how the thyroid hormones influence mitochondrial activity. Thyroid hormones also accelerate the maturation and development of certain tissues. For example, the metamorphosis of tadpoles into adult frogs is stimulated by thyroid hormones.

The Sex Hormones Are Steroids

Like the adrenal cortical hormones, the *androgens* (male sex hormones) and the *estrogens* (female sex hormones) are also steroids. The adrenal cortex, testis, and ovary have a common embryological origin. This relationship is shown in another way. Androgens are made not only by the testis but also, in lesser amounts, by the adrenal cortex and the ovary. Similarly, estrogens are made not only by the ovary but also by the adrenal cortex and the testis. In principle, maleness and femaleness are a matter of the balance between the secretion of androgens and estrogens. All the steroid hormones are ultimately made from a single precursor, *cholesterol*, which in turn is made from acetyl-CoA (page 608). The most important sex or gonadal steroids are shown in Figure 25-25.

Figure 25-25
The three major sex hormones. (a) *Testosterone, the major androgen or male sex hormone.* (b) *β-Estradiol, the major estrogen or female sex hormone, made from testosterone.* (c) *Progesterone, the hormone of the corpus luteum. It is the precursor of both testosterone and of β-estradiol in the female.*

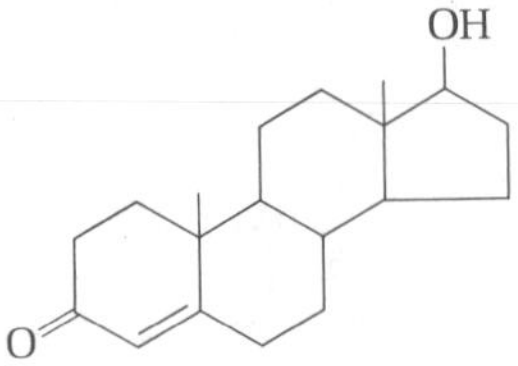

Testosterone
(a)

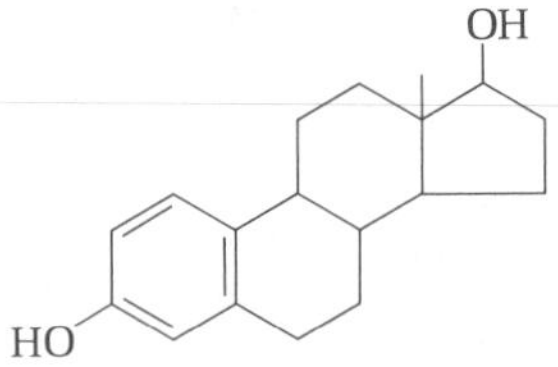

β-Estradiol
(b)

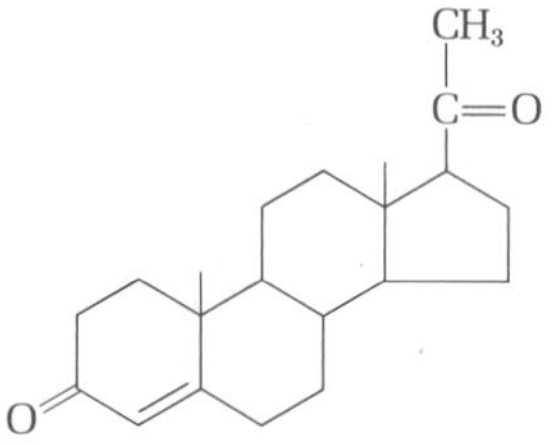

Progesterone
(c)

Androgens stimulate the growth, maturation, and maintenance of the male reproductive system and accessory sex tissues; estrogens regulate the activities of the female reproductive system. However, androgens and estrogens also have significant effects on most of the other nonreproductive tissues of the body. For example, androgens stimulate the growth of skeletal muscles. Androgens and certain androgen derivatives are also called *anabolic steroids*. They are often used by weight lifters, football players, and wrestlers to increase muscle mass and strength. In most sports they confer no demonstrable advantage. Anabolic steroids are also known to be used by some female athletes, probably with advantage; however, they produce other masculinizing effects as well.

A third sex hormone, *progesterone* (Figure 25-25), is a precursor in the biosynthesis of adrenal cortical steroids and of the sex hormones. It also functions to stimulate implantation of the ovum.

Estrogen Action on Target Cells Is Becoming Understood

The principal estrogen secreted by the human female is *β-estradiol*, made by the ovary, from, of all things, *testosterone*, the principal *male* sex hormone (Figure 25-26). Eve was made from more than Adam's rib! The specific intracellular receptors for β-estradiol are located in its primary target tissues, the uterus and mammary glands. The estrogen receptor, also called *estrophilin I*, has a molecular weight of about 200,000. On binding of the estrogen molecule it undergoes a molecular change to yield *estrophilin II*, which may be regarded as the second messenger in estrogen action. Estrophilin II enters the

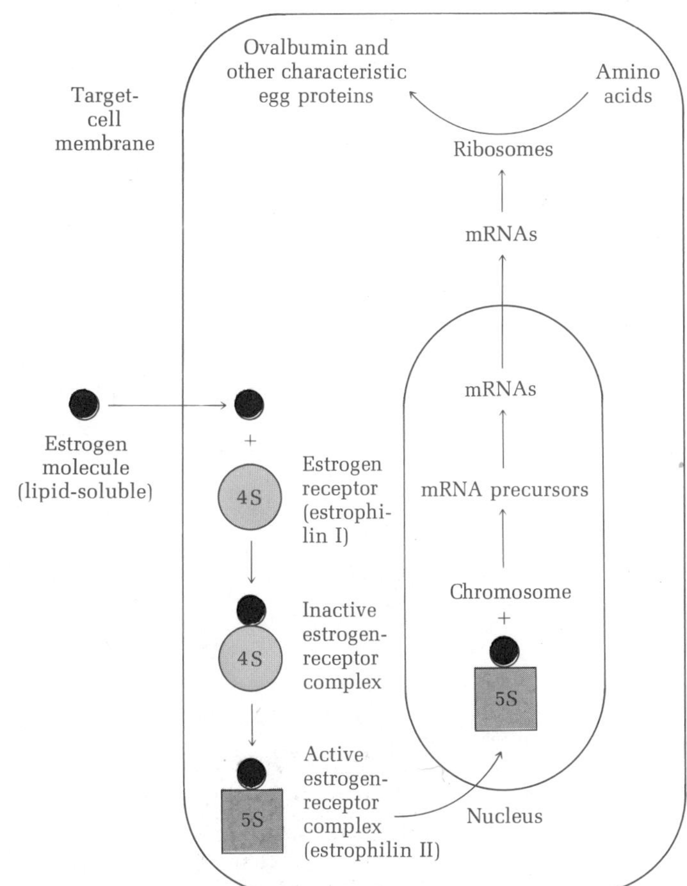

Figure 25-26
Schematic representation of estrogen action on target cells of the hen oviduct. The lipid-soluble estrogen passes through the cell membrane and is bound by the estrogen receptor, a protein with a sedimentation coefficient of 4S. The estrogen-receptor complex now undergoes an alteration into its active 5S form. The latter is the second messenger and enters the nucleus, where it interacts with specific portions of the chromatin, causing certain genes to be transcribed into messenger RNAs. These leave the nucleus and serve as templates on the ribosomes, where they code for synthesis of certain proteins characteristic of the stimulated oviduct, such as ovalbumin.

cell nucleus, where it interacts with the chromatin to instruct certain genes to produce specific proteins characteristic of the stimulated sex accessory tissues (Figure 25-26). For example, estradiol administered to chicks causes greatly increased rates of synthesis of characteristic egg proteins by the oviduct, particularly ovalbumin and ovovitellin. Thus estradiol prepares the ovary for egg formation.

The estrogen receptors in the human mammary gland decrease during the development and growth of breast cancer. Measurement of the amount of estrogen receptors in a small sample of mammary tissue is a diagnostic aid in recognizing the stage of the disease and for designing treatment. Once recognized, early breast cancer can sometimes be checked by changing the balance of androgens and estrogens in the body.

Many Other Hormones Are Known

Many other hormones have been discovered, but relatively little is known of their mode of action in biochemical terms. Some are listed in Table 25-7, together with their major functions. Especially noteworthy is the anterior pituitary hormone

Table 25-7 Other Important Polypeptide Hormones

Hormone and its source	MW	Effect
Parathyroid hormone (parathyroid)	9,500	Mobilizes Ca^{2+} from bone; controls its retention by the kidney
Calcitonin (thyroid)	3,600	Inhibits Ca^{2+} loss from bones
Thymosin (thymus)	12,500	Promotes growth of lymphoid tissue
Gastrin (stomach)	2,000	Stimulates gastric secretion
Secretin (small intestine)	3,500	Promotes secretion of pancreatic juice
Cholecystokinin (small intestine)	4,200	Promotes secretion of digestive enzymes

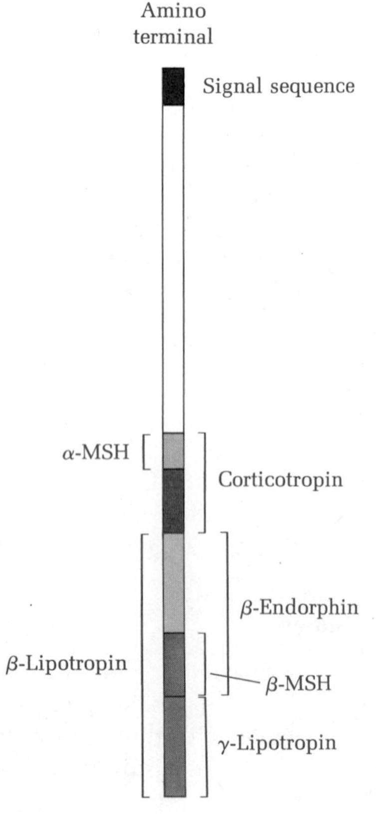

Figure 25-27
Proopiocortin, the precursor of several anterior pituitary hormones. It has about 260 amino acid residues. The polypeptide chain is cleaved at the points shown to yield several hormones: two kinds of melanocyte-stimulating hormones (α- and β-MSH), β- and γ-lipotropins, corticotropin, and β-endorphin.

precursor proopiocortin (Figure 25-27). This large polypeptide of about 260 amino acid residues was given this name because it is a precursor of both opiatelike and cortical-stimulating hormones. Actually, by peptidase cleavage at certain points proopiocortin is the precursor of several hormones (Figure 25-27), including two lipotropins, which stimulate utilization of lipids as fuel, and two kinds of endorphins, hormones with analgesic (pain-relieving) properties, similar to those of morphine and other opiate drugs. Endorphins are the "body's own opium." The pentapeptide comprising the first five residues of the amino-terminal end of the endorphin is an enkephalin ("in the brain"), which binds to the opiate receptors in the brain and has very potent morphinelike activity. These recent discoveries are leading to new insights regarding the biochemistry of pain and the mechanism of addiction to opiate drugs.

Prostaglandins and Thromboxanes Modulate the Action of Some Hormones

The prostaglandins (Figure 25-28), a family of lipid-soluble organic acids containing 5-carbon rings, are derived from essential fatty acids (page 595) via the unsaturated fatty acid arachidonic acid. They are regulators of hormone action and were named prostaglandins because they were first discovered in the secretion of the prostate gland and thought to regulate the activity of male reproductive tissues. Actually prostaglandins are made by, and function in, virtually all organs. They have many different kinds of physiological effects, and some are in use as therapeutic agents. Some prostaglandins stimulate and enhance the action of adenylate cyclase. Aspirin, a valuable drug in relieving the pain of arthritis, is a potent inhibitor of prostaglandin synthetase, an enzyme functioning in the biosynthesis of prostaglandins from arachidonic acid. Thus it appears that certain prostaglandins may function in regulating the perception of pain. Thromboxanes (Figure 25-28) are labile

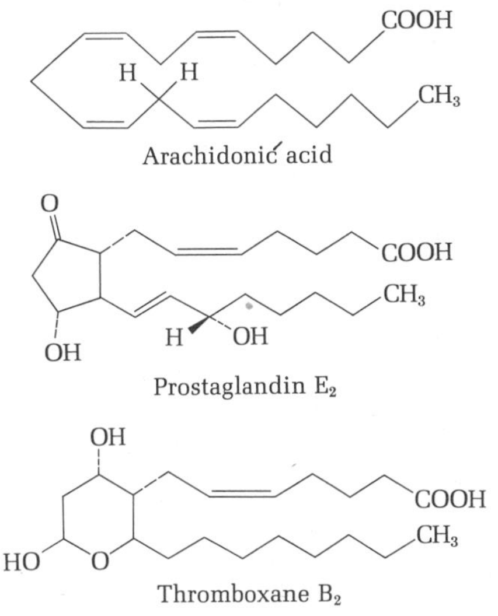

Figure 25-28
A typical prostaglandin (PGE_2) and thromboxane (B_2). They are derived from arachidonic acid (page 595).

Arachidonic acid

Prostaglandin E_2

Thromboxane B_2

transformation products of certain prostaglandins, believed to function in regulating the activities of blood platelets and other cells.

Ca^{2+} and the cytoplasmic Ca^{2+}-binding protein *calmodulin* are also believed to function as modulators or even as intracellular messengers for some hormones. Free Ca^{2+} is present in the cytosol in very low concentration, less than 10^{-6} *M*. Its concentration increases during the action of certain hormones. Ca^{2+} modulates many intracellular activities, probably through binding to calmodulin, which stimulates phosphodiesterase and various protein kinases involved in carbohydrate metabolism, muscle contraction, and intracellular membrane transport mechanisms.

Summary

Hormones are chemical messengers secreted by certain tissues into the blood, serving to regulate the activity of certain other tissues. They act in a hierarchy of functions. Nerve impulses received by the hypothalamus cause it to send specific hormones to the pituitary gland, stimulating (or inhibiting) the release of different tropic hormones. The anterior pituitary hormones in turn can stimulate other endocrine glands to secrete their characteristic hormones, which in turn stimulate specific target tissues.

Adrenaline, one of three catecholamine hormones made by the adrenal medulla from tyrosine, helps prepare the body to "fight or flee" by mobilizing blood glucose from glycogen and other precursors. Adrenaline binds to specific receptors on the outer surface of liver and muscle cells, causing activation of adenylate cyclase on the inner surface to convert ATP into cyclic adenylate (cAMP). The cAMP then binds to the regulatory subunit of protein kinase, causing it to dissociate from the catalytic subunit, which is thus activated. Protein kinase phosphorylates inactive phosphorylase kinase, which, through a subsequent step, stimulates glycogen phosphorylase. Cyclic adenylate is destroyed by phosphodiesterase, which is activated by Ca^{2+} and calmodulin, a regulatory Ca^{2+}-binding protein, and inhibited by theophylline.

Insulin, one of three major pancreatic hormones, is secreted by the B cells of the islets of Langerhans. In excess it causes lowered blood sugar levels by promoting entry of glucose into the tissues. Its deficiency causes diabetes mellitus, characterized by hyperglycemia, glycosuria, failure of fatty acid synthesis, and excessive fatty acid oxidation and ketone body formation. Insulin is bound to specific insulin receptors on the surface of many tissues, but its intracellular action is not yet known. Glucagon, secreted by the A cells, opposes the action of insulin by causing glycogen breakdown to blood glucose. Somatostatin, another pancreatic hormone, regulates insulin secretion.

The adrenal cortex secretes glucocorticoids, such as cortisol, which stimulates gluconeogenesis and also inhibits inflammatory responses; mineralocorticoids, principally aldosterone, which promotes secretion of Na^+; and other steroids of intermediate action, such as corticosterone. The adrenal cortical steroids, as well as the estrogens and androgens, pass through the plasma membranes of their target cells and bind to receptors in the cytosol, which then act in the nucleus to cause certain genes to be expressed.

References

General

WHITE, A., P. HANDLER, E. L. SMITH, R. L. HILL, and I. R. LEHMAN: *Principles of Biochemistry*, 6th ed., McGraw-Hill, New York, 1978. Chapters 41 to 48 give a comprehensive account of the hormones and their action, with many references.

Special Topics

BRADSHAW, R. A., and W. A. FRAZIER: "Hormone Receptors as Regulators of Hormone Action," *Curr. Top. Cell Regul.*, **12**:1–35 (1977).

GUILLEMIN, R.: "Endorphins: Brain Peptides That Act Like Opiates," *New Engl. J. Med.*, **296**:226–228 (1977).

GUILLEMIN, R.: "Peptides in the Brain: The New Endocrinology of the Neuron," *Science*, **202**:390–402 (1978).

GREENGARD, P.: "Phosphorylated Proteins as Physiological Effectors," *Science*, **199**:146–152 (1978).

JENSEN, E. V.: "Estrogen Receptors in Human Cancers," *JAMA*, **238**:59 (1977).

MEANS, A. R., and J. R. DEDHAM: "Calmodulin: An Intracellular Calcium Receptor," *Nature*, **285**:73–77 (1980).

NOTKINS, A. L.: "The Causes of Diabetes," *Sci. Am.*, **241**:62–73, November (1979).

O'MALLEY, B. W., and W. T. SCHRADER: "The Receptors of Steroid Hormones," *Sci. Am.*, **234**:32 (1976).

PASTAN, I.: "Cyclic AMP," *Sci. Am.*, **227**:97–105, August (1972).

RENOLD, A. E., D. H. MINTZ, W. A. MULLER, and G. F. CAHILL, JR.: "Diabetes Mellitus," in J. B. Stanbury, J. B. Wyngaarden, and D. S. Fredrickson (eds.), *The Metabolic Basis of Inherited Disease*, 4th ed., McGraw-Hill, New York, 1978. Detailed account of various forms of diabetes mellitus.

RYAN, A. J.: "Anabolic Steroids Are Fool's Gold," *Fed. Proc.*, **40:** 2682–2688 (1981). Interesting discussion of the use of steroids by athletes.

SNYDER, S. H.: "Opiate Receptors and Internal Opiates," *Sci. Am.*, **236:** 44–56 March (1977).

STEINER, D. F.: "Insulin Today," *Diabetes*, **26:**322–340 (1977).

SUTHERLAND, E. W.: "Studies on the Mechanism of Hormone Action," *Science*, **177:**401–408 (1972). Account of the classic experiments on cyclic AMP.

YALOW, R. S.: "Radioimmunoassay: A Probe for the Fine Structure of Biologic Systems," *Science*, **200:**1236–1246 (1978).

Problems

1. *Significance of Hormone Concentration* Under normal conditions, the human adrenal medulla secretes adrenaline ($C_9H_{13}NO_3$) at a rate to maintain a concentration of 10^{-10} *M* in the circulating blood. To appreciate what that concentration means, calculate the diameter in meters that you would need for a round swimming pool 2 m deep (~6 ft) in order to dissolve 1 g (about 1 teaspoon) of adrenaline to its physiological concentration in the blood.

2. *Regulation of Hormone Levels in the Blood* The lifetime of most hormones in the blood is relatively short. For example, if radioactively labeled insulin is injected into an animal, within 30 min half the hormone will have disappeared from the blood. What is the importance of the relatively rapid inactivation of circulating hormones? In view of this rapid inactivation how can the circulating hormone level be kept constant under normal conditions? In what ways can the organism make possible such rapid changes in the level of circulating hormones?

3. *Water-Soluble vs. Lipid-Soluble Hormones* Based on their physical properties, hormones fall into one of two categories: those that are very soluble in water but relatively insoluble in lipids, e.g., adrenaline, and those that are relatively insoluble in water but highly soluble in lipids, e.g., steroid hormones. In their role as regulators of cellular activity, most water-soluble hormones do not penetrate into the interior of the cells of target tissues. The lipid-soluble hormones, on the other hand, do penetrate into their target cells and ultimately act in the nucleus. What is the basis for the correlation between solubility, the location of receptors, and the mode of action of the two classes of hormones?

4. *Hormone Experiments in Cell-Free Systems* In the 1950s, Earl Sutherland and his colleagues carried out pioneering experiments to elucidate the mechanism of action of adrenaline and glucagon.

In the light of our current understanding of hormone action (see text) interpret *each* of their experiments as described below. Identify the components and indicate the signifcance of the results.

(a) The addition of adrenaline to a homogenate or broken-cell preparation of normal liver resulted in an increase in the activity of glycogen phosphorylase. However, if the homogenate was first centrifuged at a high speed and adrenaline or glucagon added to the clear supernatant fraction, no increase in its phosphorylase activity was observed.

(b) When the particulate fraction sedimented from a liver homogenate by centrifugation was separated and treated with adrenaline, a new substance was produced. This substance was isolated and purified. Unlike adrenaline, this substance activated glycogen phosphorylase when added to the supernatant fraction of the homogenate.

(c) The substance formed by the particulate fraction was heat-stable; i.e., heat treatment did not prevent its capacity to activate phosphorylase. (*Hint:* Would this be the case if the substance were a protein?) The substance was identical to a compound obtained when pure ATP was treated with barium hydroxide.

5. *Effect of Dibutyryl Cyclic AMP vs. Cyclic AMP on Intact Cells* The physiological effects of the hormone adrenaline should in principle be mimicked by the addition of cyclic AMP to the target cells. In actual practice, the addition of cyclic AMP to intact target cells elicits only a minimal physiological response. Why? However, when the structurally related derivative dibutyryl cyclic AMP

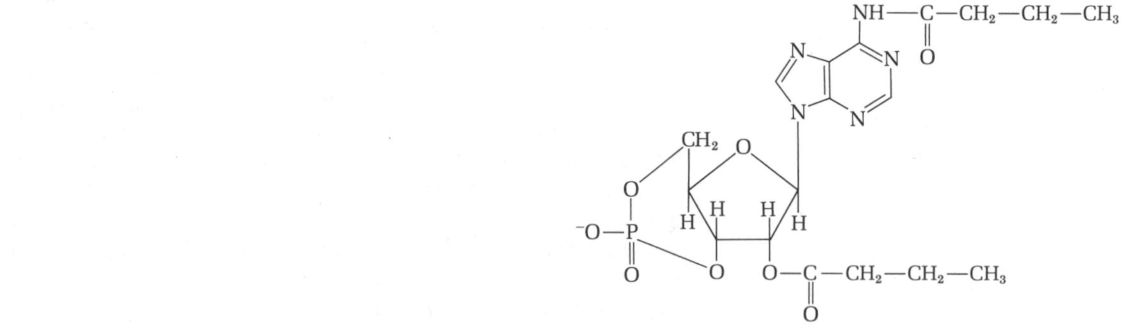

$NH-C(=O)-CH_2-CH_2-CH_3$

$^{-}O-P(=O)$ ... $O-C(=O)-CH_2-CH_2-CH_3$

is added to intact cells, the expected physiological responses can readily be seen. Explain the basis for the difference in response of cells to these two substances. Dibutyryl cyclic AMP is a widely used derivative in studies of cyclic AMP function.

6. *The Effect of Cholera Toxin on Adenylate Cyclase* The gram-negative bacterium *Vibrio cholerae* produces a protein, cholera toxin (MW 90,000), responsible for the characteristic symptoms of cholera, namely, extensive loss of body water and Na^+ through continuous, debilitating diarrhea. If body fluids and Na^+ are not replaced, severe dehydration will occur; untreated, the disease is often fatal. If the cholera toxin gains access to the intestinal tract of human beings, it binds tightly to specific sites in the plasma membrane of the epithelial cells lining the small intestine,

causing adenylate cyclase to undergo activation which persists for hours or days. What is the effect of cholera toxin on the level of cyclic AMP in the intestinal cells? Based on the information above, can you suggest how cyclic AMP normally functions in the cells of the intestinal mucosa? Suggest a possible treatment for cholera.

7. *Different Metabolism in Muscle vs. Liver during a "Fight or Flee" Situation* During a "fight or flee" situation, the release of adrenaline promotes glycogen breakdown in the liver, heart, and skeletal muscle. The product of glycogen breakdown in the liver is glucose. In contrast, the skeletal muscle glycogen is broken down by glycolysis.
   (a) Why are different products of glycogen breakdown observed in the two different tissues?
   (b) What is the advantage to the organism during a "fight or flee" condition of having these specific glycogen breakdown routes?

8. *Excessive Amounts of Insulin Secretion: Hyperinsulinism* Certain malignant tumors of the pancreas cause excessive production of insulin by the B cells. Such patients exhibit shaking and trembling, weakness and fatigue, sweating, and hunger. If this condition is prolonged, brain damage occurs. What is the effect of hyperinsulinism on the metabolism of carbohydrate, amino acids, and lipids by the liver? What are the causes of the observed symptoms? Suggest why this condition, if prolonged, leads to brain damage.

9. *Thermogenesis Caused by Thyroid Hormones* Thyroid hormones are intimately involved in regulating the basal metabolic rate. Liver tissue of animals given excess thyroxine shows an increased rate of O_2 consumption and increased heat output (thermogenesis), but the ATP concentration in the tissue is normal. Different explanations have been offered for the thermogenic effect of thyroxine. One is that excess thyroid hormone causes uncoupling of oxidative phosphorylation in mitochondria. How could such an effect account for the observations? Another explanation suggests that the thermogenesis is due to an increased rate of ATP utilization by the thyroid-stimulated tissue. Is this a reasonable explanation? Why?

10. *Ovariectomy in the Treatment of Mammary Carcinoma* One of the treatments of breast carcinoma is to perform an ovariectomy, i.e., surgical removal of the ovaries. Explain the biochemical basis of this treatment. An additional treatment is the administration of male sex hormones to a female subject, which has the effect of inhibiting the physiological responses brought about by the naturally occurring sex hormones.

11. *Endorphin Synthesis in the Adrenal Gland* In addition to manufacture of catecholamine hormones the adrenal medulla also makes some endorphins, sometimes called the "brain's own opiates." Can you provide a possible explanation for the synthesis of endorphins by both the brain and the adrenal medulla?

12. *Function of Prohormones* What are the possible advantages in the synthesis of hormones as prohormones or preprohormones?

13. *Action of Aminophylline* Aminophylline, a purine derivative resembling theophylline of tea, is often administered to patients with acute asthma together with adrenaline. What is the purpose and biochemical basis for this treatment?

14. *Calmodulin* When calmodulin isolated from clam muscles is added to phosphodiesterase isolated from rat liver, there is no effect on the low rate of the hydrolysis of cyclic AMP to AMP. But when Ca^{2+} is added to the system, phosphodiesterase activity is greatly accelerated. What biochemical information is revealed by these observations?

CHAPTER 26

Human Nutrition

Table 26-1 Nutrients Probably Required by Human Beings†

Energy sources	Essential fatty acids
Carbohydrates	Linoleic acid
Fats	Linolenic acid
Proteins	Mineral elements
Essential amino acids	Arsenic
Arginine (in adults)	Calcium
Histidine	Chlorine
Isoleucine	Chromium
Leucine	Copper
Lysine	Fluorine
Methionine	Iodine
Phenylalanine	Iron
Threonine	Magnesium
Tryptophan	Manganese
Valine	Nickel
Vitamins	Molybdenum
Thiamine	Phosphorus
Riboflavin	Potassium
Nicotinamide	Selenium
Pyridoxine	Silicon
Pantothenic acid	Sodium
Folic acid	Tin
Biotin	Vanadium
Vitamin B_{12}	Zinc
Ascorbic acid	
Vitamins A, D, E, and K	

† Water, although not a nutrient in a strict sense, is of course required to replace that lost in the urine, breath, and sweat.

Now that we have examined the coordination of metabolism in the human body, let us see how this complex organism is sustained and kept in metabolic health by its intake of nutrients.

Today's scientific understanding of human nutrition is one of the most important contributions of biochemistry; it has saved countless human lives or improved their quality. Until relatively recent times pellagra, beriberi, and rickets were endemic in many parts of the world. Today these diseases need no longer occur since we have the knowledge to prevent them. Yet even today, with our highly developed knowledge of nutrition, more than one-eighth of the world's population is undernourished. And, paradoxically, many people in the most affluent countries are malnourished, not through deficiency of food but through overconsumption and other imbalances. One of the most important missions of biochemistry is to inform people of the scientific basis of nutrition and to dispel irrational beliefs in food folklore, food faddism, and diet quackery.

There Are Five Basic Components in an Adequate Diet

There are five classes of nutrients that contribute to an adequate diet (Table 26-1); each plays a special role.

Carbohydrates

Carbohydrates are the most abundant bulk nutrients and the major source of biological energy through their oxidation in the tissues. They also furnish organic precursors for the biosynthesis of many cell components.

Fats

Triacylglycerols from animal and plant sources rank close behind carbohydrates as major sources of energy. They also are important carbon sources for biosynthesis of cholesterol and other steroids. Moreover, plant triacylglycerols provide *essential fatty acids.*

Table 26-2 Daily Dietary Allowances Recommended by the Food and Nutrition Board, National Academy of Sciences–National Research Council† (Revised 1980)

					Fat-soluble vitamins			Water-soluble vitamins						
	Age, years	Weight, kg	Height, cm	Protein, g	Vitamin A, µg‡	Vitamin D_3, µg	Vitamin E, mg	Vitamin C, mg	Thiamine, mg	Riboflavin, mg	Nicotinamide, mg	Vitamin B_6, mg	Folic acid, µg	Vitamin B_{12}, µg
Infants	0.0–0.5	6	60	kg × 2.2	420	10	3	35	0.3	0.4	6	0.3	30	0.5
	0.5–1.0	9	71	kg × 2.0	400	10	4	35	0.5	0.6	8	0.6	45	1.5
Children	1–3	13	90	23	400	10	5	45	0.7	0.8	9	0.9	100	2.0
	4–6	20	112	30	500	10	6	45	0.9	1.0	11	1.3	200	2.5
	7–10	28	132	34	700	10	7	45	1.2	1.4	16	1.6	300	3.0
Males	11–14	45	157	45	1000	10	8	50	1.4	1.6	18	1.8	400	3.0
	15–18	66	176	56	1000	10	10	60	1.4	1.7	18	2.0	400	3.0
	19–22	70	177	56	1000	7.5	10	60	1.5	1.7	19	2.2	400	3.0
	23–50	70	178	56	1000	5	10	60	1.4	1.6	18	2.2	400	3.0
	51+	70	178	56	1000	5	10	60	1.2	1.4	16	2.2	400	3.0
Females	11–14	46	157	46	800	10	8	50	1.1	1.3	15	1.8	400	3.0
	15–18	55	163	46	800	10	8	60	1.1	1.3	14	2.0	400	3.0
	19–22	55	163	44	800	7.5	8	60	1.1	1.3	14	2.0	400	3.0
	23–50	55	163	44	800	5	8	60	1.0	1.2	13	2.0	400	3.0
	51+	55	163	44	800	5	8	60	1.0	1.2	13	2.0	400	3.0
Pregnant				+30	+200	+5	+2	+20	+0.4	+0.3	+2	+0.6	+400	+1.0
Lactating				+20	+400	+5	+3	+40	+0.5	+0.5	+5	+0.5	+100	+1.0

† The allowances are intended to provide for individual variations among most normal people as they live in the United States under usual environmental stresses. Diets should be based on a variety of common foods in order to provide other nutrients for which human requirements have been less well defined.

‡ Retinol equivalents; 1 retinol equivalent = 1 µg retinol or 6 µg β-carotene.

§ An ordinary diet cannot fulfill the iron requirements of a pregnant or lactating woman, so a supplement of 30 to 60 mg of iron is recommended.

Proteins

Proteins have three major roles in nutrition. They provide both essential and nonessential amino acids as building blocks for protein biosynthesis, not only for the growth of infants and children but also for the constant replacement and turnover of body proteins in adults. Amino acids are also precursors of hormones, porphyrins, and many other biomolecules. Oxidation of the carbon skeletons of amino acids also furnishes a minor but significant fraction of the total daily energy requirement.

Carbohydrates, fats, and proteins are bulk nutrients or *macronutrients*. Together they are consumed in hundreds of grams per day, depending upon body weight, age, and sex.

Vitamins

The vitamins, classified into water-soluble and fat-soluble groups, are organic micronutrients, required in only milligram or microgram quantities per day. They serve as essential components of specific coenzymes or enzymes participating in metabolism and other specialized activities.

Minerals					
Ca, mg	P, mg	Mg, mg	Fe, mg	Zn, mg	I, μg
360	240	50	10	3	40
540	360	70	15	5	50
800	800	150	15	10	70
800	800	200	10	10	90
800	800	250	10	10	120
1200	1200	350	18	15	150
1200	1200	400	18	15	150
800	800	350	10	15	150
800	800	350	10	15	150
800	800	350	10	15	150
1200	1200	300	18	15	150
1200	1200	300	18	15	150
800	800	300	18	15	150
800	800	300	18	15	150
800	800	300	10	15	150
+400	+400	+150	S§	+5	+25
+400	+400	+150	S§	+10	+50

Minerals and Trace Elements

The required inorganic nutrients can be grouped into two classes. Calcium, phosphorus, and magnesium are required in relatively large amounts per day, whereas iron, iodine, zinc, copper, and many others are needed in only milligram or microgram quantities. The inorganic elements have many functions: as components of bones and teeth, as electrolytes in maintaining water balance in the vascular system and tissues, and in the prosthetic groups of enzymes, among others.

Over 40 different substances are indispensable in human nutrition and must be obtained from food (Table 26-1). They include 10 amino acids, 13 vitamins, 20 or more mineral elements, usually in the form of a soluble salt, and one or more polyunsaturated fatty acids. We may also include *fiber*, consisting largely of cellulose and other nondigestible cell-wall polymers of plant origin. Although fiber is nondigestible and plays no metabolic role, it helps maintain proper motility of the intestinal tract.

The Food and Nutrition Board of the National Academy of Sciences–National Research Council has developed a table of *Recommended Daily Allowances* (RDAs) of various nutrients for optimum nutrition of infants, children, adults, and pregnant women in the United States population (Table 26-2). These allowances are not the *minimum* daily requirements but are intended to provide an ample safety margin.

Energy Is Provided by Oxidation of the Bulk Organic Nutrients

The first requisite of an adequate diet is a source of energy, provided by oxidation of the three bulk nutrients: carbohydrate, fat, and protein. The unit of energy used in calculating energy requirements is the *kilocalorie* (kcal), also called the *nutritional calorie* (abbreviated Cal, with a capital C), the amount of energy which in the form of heat is required to raise the temperature of 1.0 kg of water by 1 degree, from 15 to 16°C. Kilocalories are the same units used in calculating standard-free-energy changes of metabolic reactions (page 368).

Table 26-3 shows the Food and Nutrition Board's recommended daily allowance of energy in kilocalories for different age groups. College-age males require, on the average, about 2900 kcal/d, females about 2100 kcal/d. Infants, children, and older people generally require less energy. We may compare these values with the *basal* caloric requirement, i.e., the amount of energy needed by the body at complete rest, 12 h after a meal, as defined in Chapter 25 (page 743). For college-age males the basal requirement is approximately 1800 kcal/d and for females about 1300 kcal/d. The excess of the recommended daily allowance over the basal rate is of course required to carry out physical work. Table 26-4 shows the relative energy requirements for different kinds and rates of physical activity.

Table 26-3 Recommended Daily Energy Allowances (Food and Nutrition Board, National Academy of Sciences–National Research Council, 1980)†

	Age, years	Weight, kg	Energy, kcal
Infants	0.0–0.5	6	650
	0.5–1.0	9	970
Children	1–3	13	1300
	4–6	20	1700
	7–10	28	2400
Males	11–14	45	2700
	15–18	66	2800
	19–22	70	2900
	23–50	70	2700
	51+	70	2400
Females	11–14	46	2200
	15–18	55	2100
	19–22	55	2100
	23–50	55	2000
	51+	55	1800
Pregnant			Add 300
Lactating			Add 500

† The daily energy requirements listed are midpoints of ranges extending at least 15 percent above and below the values listed.

Table 26-4 Energy Needs for Various Activities

Activity	Men, kcal/kg·h	Women, kcal/kg·h
Very light		
Seated and standing activities, painting, auto and truck driving, laboratory work, typing	1.5	1.3
Light		
Walking (2.5–3 mi/h), carpentry, shopping, restaurant trades, washing clothes, golf	2.9	2.6
Moderate		
Walking fast (3.5–4 mi/h), jogging, weeding, hoeing, cycling, tennis, dancing, volleyball	4.3	4.1
Heavy		
Walking with load uphill, sawing wood, pick-and-shovel labor, swimming, climbing, football	8.4	8.0

The amount of energy released by the oxidation of carbohydrates, fats, and proteins has been determined by burning weighed samples in an atmosphere of oxygen in a bomb calorimeter and measuring the total amount of heat produced (Figure 26-1). Pure carbohydrates yield on the average about 4.2 kcal/g, fats about 9.5 kcal/g, and proteins about 4.3 kcal/g (Table 26-5). The caloric value of specific foods, such as bread, potatoes, meat, fruit, and so on, can also be determined by combustion in a calorimeter. Another way is to determine the weight of carbohydrate, fat, and protein in a given sample of the food by chemical analysis; the weight of each is then multiplied by the caloric equivalents shown in Table 26-5.

When oxidized in the body, foods that are completely

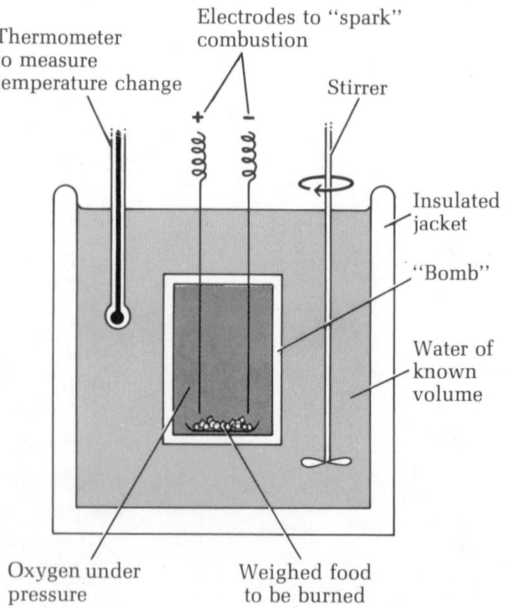

Figure 26-1
Principle of the bomb calorimeter for measuring caloric value of foods. A weighed amount of the food is ignited electrically inside a pressure-resistant bomb filled with excess oxygen under pressure. Combustion of the food causes an increase in the temperature of the known weight of water in the jacket. Since 1.0 kcal is required to raise the temperature of 1 kg of H_2O through 1.0°C, the heat evolved by combustion of the food is easily calculated.

Very large calorimeters have been used to measure the heat evolved by human subjects in an enclosed chamber supplied with oxygen and provided with an arrangement to absorb CO_2.

Table 26-5 Caloric Equivalent of Major Nutrients†

	Energy equivalent, kcal/g
Carbohydrates	4.2
Fats	9.5
Proteins	4.3

† Since different carbohydrates, fats, and proteins vary somewhat in chemical composition, these values represent averages.

digested and absorbed ultimately yield an amount of heat equal to the heat released when they are oxidized in a calorimeter. This equality has been confirmed in studies of human subjects placed in a very large *human calorimeter*. Because the human body at all times conforms to the laws of thermodynamics, there is no way some "magic" diet can circumvent such an energy accounting. Calories are calories.

Now let us examine the nutritional characteristics of two of the three major energy-supplying nutrients, carbohydrates and fats.

Carbohydrates Are the Major Source of Energy

Carbohydrates per se are not essential in the human diet, but because carbohydrate-rich foods are abundant and cheap compared with fats and protein, they naturally form a major part of the diet in most of the world. In the four-fifths of the world's population that relies mainly on plant food, carbohydrates provide at least 70 percent and often up to 90 percent of the total caloric intake. But in affluent countries, where meat and dairy products are consumed in relatively large amounts, carbohydrates furnish only about 45 percent of the total daily caloric intake. About 400 g of carbohydrate is consumed per day by males of college age in the United States.

In affluent countries over 40 percent of the dietary carbohydrate is furnished by sucrose and other refined sugars, mostly glucose and fructose, and the remainder by the polysaccharide starch. In less affluent countries, starch furnishes nearly all the carbohydrate intake, and very little sucrose is consumed. Two hundred years ago, at the beginning of the Industrial Revolution, the average sugar consumption in England was only about 5 g/d per person; today it is over 200 g. A similar change has taken place in the United States (Figure 26-2). As each country develops and becomes affluent, its sucrose consumption rises. One reason is that sucrose is the most convenient and often the cheapest source of carbohydrate in these countries. In November 1981 the retail price of table sugar in the United States

Figure 26-2
The changing United States diet. (a) *Diet in 1910;* (b) *current diet; and* (c) *a recently recommended long-term goal, showing percentage of total calories provided by each nutrient.*

Carbohydrate
Protein
Fat

56%
Sugar 18%
Starch 38%
32%
12%

(a)

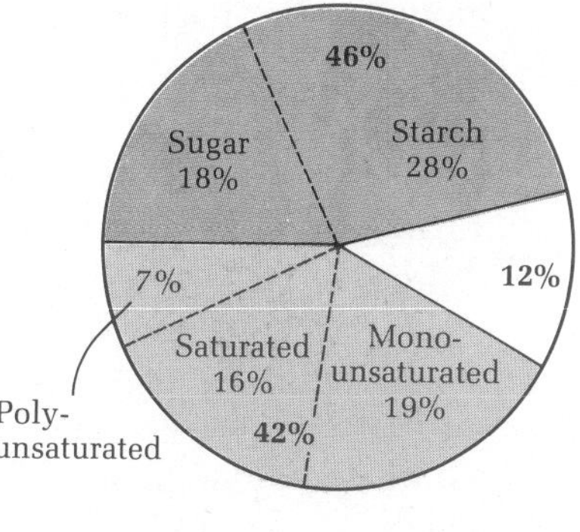

(b)

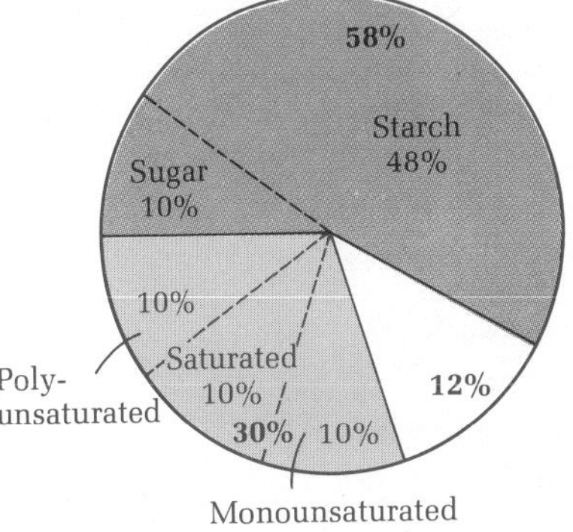

(c)

was 34 cents per pound, equivalent to 1880 kcal, or over 60 percent of the daily caloric requirement for a college-age male. Less land is needed to provide calories from sugar cane or beets than from potatoes or cereals. In fact, sugar cane is perhaps the most productive crop plant known (page 672). Here we may have a conflict between agricultural economics and good nutrition, since sucrose and other sugars are deleterious to dental health (page 782).

Sweet foods are often consumed for pleasure; indeed, sweets are irresistible to some individuals. The craving for sweet foods may reflect a primitive drive for oral satisfaction originating during infancy. (Human milk contains over twice as much sugar as cow's milk.) Many animals also prefer sweet foods, but some species are indifferent or even avoid them.

Noncaloric Sweeteners Are Widely Used

The artificial sweetener saccharin (Figure 26-3) has been used for many years as a noncaloric dietary aid in diabetes and obesity, with no convincing evidence that it is harmful to human life. However, in 1969 it was found to cause cancer when given in extremely large amounts to rats, and its incorporation into "diet" drinks or prepared foods became a matter of some controversy. Since the usefulness of saccharin as a sweetener has been deemed to outweigh the very low risk, it is still used as a sweetener in diet drinks. Sodium cyclamate, another synthetic noncaloric sweetener (Figure 26-3), much more potent as a carcinogenic agent in animals, has been banned from prepared foods.

Many efforts are underway to find new, nontoxic sweeteners. One compound that has received much study is aspartame (Figure 26-3). Since it is the methyl ester of a dipeptide of two amino acids that normally occur in proteins, it is believed to be nontoxic. Aspartame has been ruled acceptable by the Food and Drug Administration for use as a sugar substitute in some commercial food products. Another candidate is monellin, a protein (MW 11,000) from the African serendipity berry; it is 2000 times as sweet as sucrose per unit weight (Table 26-6). Its sweetness is due to the specific three-dimensional conformation of its polypeptide chain. When monellin is heated or otherwise denatured, it loses its sweetness.

Fats Provide Calories and Essential Fatty Acids

Triacylglycerols constitute about 98 percent of total dietary lipids; the remaining 2 percent consists of phospholipids and cholesterol and its esters. At room temperature triacylglycerols of animal origin, which contain a relatively large proportion of saturated fatty acids, are usually solid; those of plant origin, which contain a greater fraction of unsaturated fatty acids, are usually liquid. On oxidation in the tissues triacylglycerols of both types provide over twice as much energy per gram as car-

Figure 26-3
Some noncaloric sweeteners. See Table 26-6 for their relative sweetness. Saccharin tastes bitter to some people, presumably because of a genetic difference in taste perception.

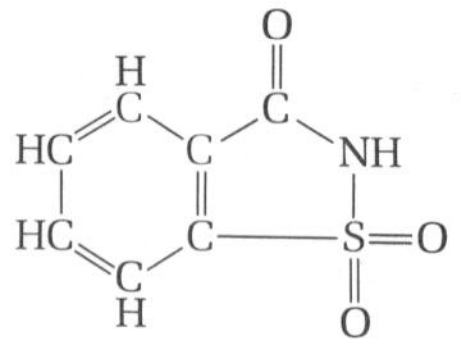

Saccharin

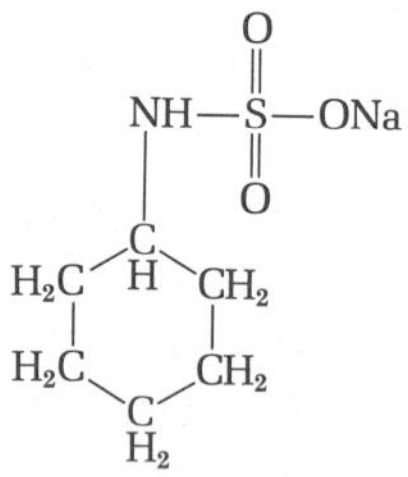

Sodium cyclamate

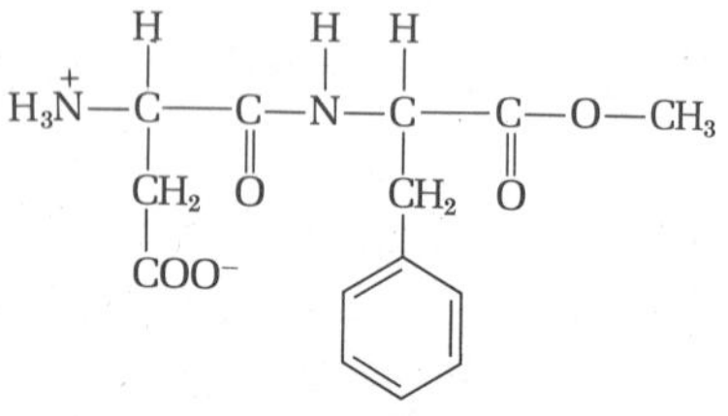

Aspartame
(L-aspartyl-L-phenylalanine methyl ester)

Table 26-6 Sweetness of Some Sugars and Noncaloric Sweeteners, Relative to Sucrose

Sucrose	1.0
Glucose	0.5
Fructose	1.7
Lactose	0.2
Saccharin	400
Sodium cyclamate	30
Aspartame	180
Monellin	2000

Figure 26-4
Essential fatty acids. Mammals lack the enzymes required to introduce double bonds beyond Δ^9 and therefore must obtain linoleic and linolenic acids from plant sources. They are required as precursors of other polyunsaturated fatty acids in the tissues, particularly arachidonic acid and other 20-carbon polyunsaturated acids, which are, in turn, precursors of prostaglandins. Deficiency of essential fatty acids may be the cause of eczema in infants.

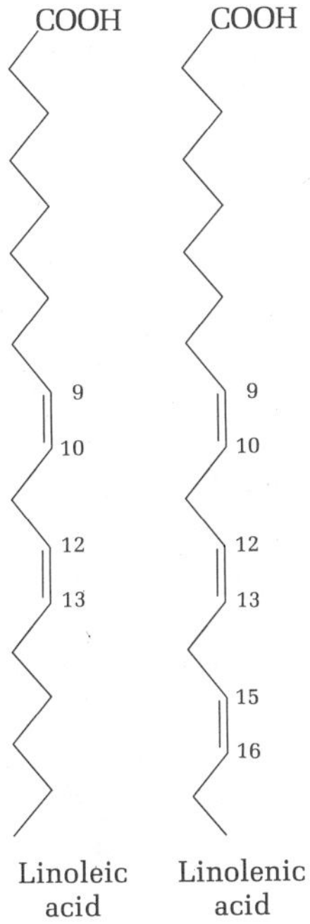

bohydrates (Table 26-5). Since fats tend to remain in the stomach longer than carbohydrates and are digested more slowly, they also have greater satiety value than carbohydrates.

Experimental animals are unable to synthesize *linoleic* and *linolenic acids* (page 595), and thus require them in the diet. However, deficiency of these essential fatty acids probably occurs only rarely in people, simply because these fatty acids are abundantly present in many plant foods, in fish, and in fowl. Meat and dairy products have much lower amounts. Linoleic acid (Figure 26-4) is needed as the precursor of *arachidonic acid* (page 595), which in turn is the precursor of *prostaglandins* and *thromboxanes* (page 746).

Along with the large amount of refined sugar in the diet in affluent countries, there is also a large proportion of fat, particularly animal fat (Figure 26-2). This is suspected to be one possible factor in the increased incidence of atherosclerosis, heart disease, and strokes in highly developed countries. In atherosclerosis abnormal deposits of lipids form on the inner lining of arterial blood vessels, reducing blood flow. Coronary artery disease and strokes occur when such lipid deposits obstruct a blood vessel in the heart or brain, causing the tissue served by the vessel to die for lack of oxygen or fuels (Figure 26-5).

Animal fats contain two components believed to predispose to atherosclerosis, *saturated fatty acids* and *cholesterol*, although the statistical evidence supporting this view is questioned by some authorities. Most animal fats, e.g., those from meat, milk, and eggs, are relatively rich in saturated fatty acids but contain a rather low content of polyunsaturated acids (Table 26-7); two exceptions are chicken fat and fish fat. Plant fats, on the other hand, are very rich in polyunsaturated fatty acids. While saturated and unsaturated fats have nearly equal caloric value, consumption of a diet rich in saturated animal fats but poor in polyunsaturated fats tends to *decrease* the concentration of *high-density lipoproteins* and *increase* the concentra-

Figure 26-5
Atherosclerosis, the progressive blocking of a small artery by accumulating lipid deposits. The photos show cross sections of (a) a normal artery, (b) an artery with lipid deposits forming, (c) an artery with hardened deposits, and (d) an artery blocked completely by a blood clot.

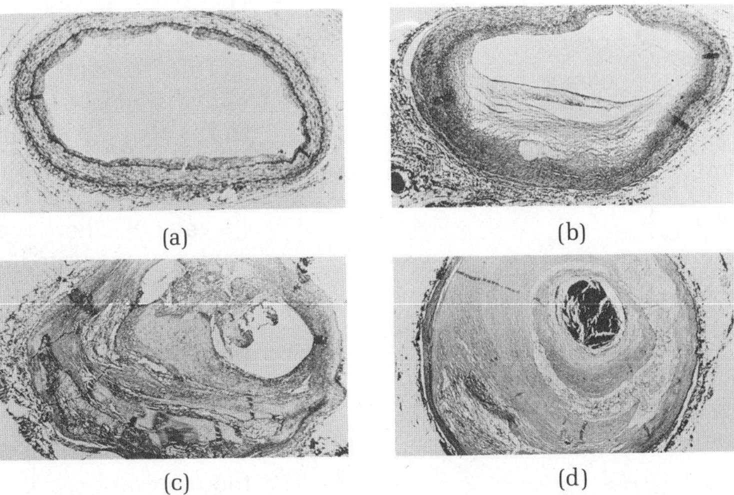

Table 26-7 Fatty Acid Composition of Typical Animal and Plant Fats

	Percentage of total fatty acids		
	Saturated	Monounsaturated	Polyunsaturated
Butterfat	60	36	4
Pork fat	59	39	2
Beef fat	53	44	2
Chicken fat	39	44	21
Corn oil	15	31	53
Soy bean oil	14	24	53
Soft margarine	23	22	52

tion of *low-density lipoproteins* (page 316) and of total cholesterol in the blood, in many but not all individuals. There is a positive statistical correlation between the incidence of coronary heart disease and low levels of high-density plasma lipoproteins and also high levels of low-density lipoproteins and total cholesterol. It is believed, therefore, that saturated animal fats, those contributed by meat, eggs, milk, butter, and cheese, should be replaced in part by plant fats rich in polyunsaturated fatty acids. Use of margarine rather than butter is also considered desirable, since margarine is made by partial hydrogenation of plant oils (page 307). Although hydrogenation increases the degree of saturation of these oils, it can be controlled to produce "soft" margarine, preferred nutritionally to "hard" margarine, since it still contains a large fraction of polyunsaturated fats (Table 26-7).

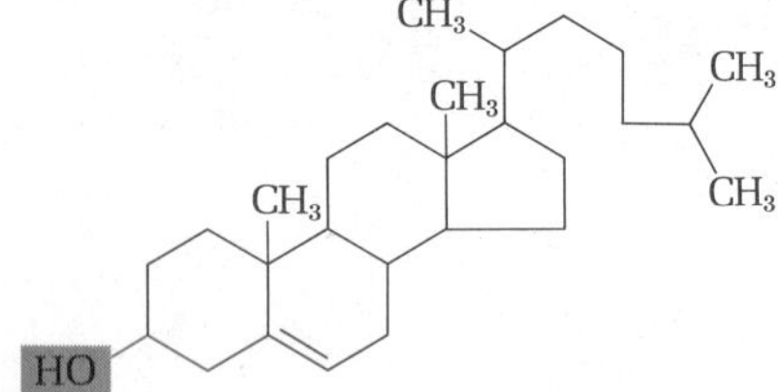

Figure 26-6
Cholesterol. In cholesterol esters the hydroxyl group (color) is esterified with long-chain fatty acids.

Cholesterol in the diet also appears to affect the proportions of the lipoproteins in the blood in some individuals. Cholesterol (Figure 26-6) is present in significant amounts in animal products, particularly egg yolk, butterfat, and meat, but is absent from plant foods. The average American diet contains about 600 to 800 mg of cholesterol per day, most of which comes from egg yolks. Cholesterol is also made in the body, starting from acetyl-CoA (page 608). It is lost from the body by conversion into bile salts (page 610), which are only slowly lost in the feces. Although a diet high in cholesterol will increase the blood cholesterol level, dietary cholesterol is at the same time inhibitory to the biosynthesis of cholesterol in the tissues (page 609). There is a delicately regulated balance between the amount of ingested cholesterol, the amount of cholesterol synthesized in the body, and the amount excreted. Diets low in cholesterol and in which saturated fats are in part replaced by polyunsaturated fats are often prescribed for people who have suffered coronary artery disease. However, there is a strong genetic factor in the incidence of coronary artery disease, which is also influenced by smoking and by hypertension. It is therefore not certain that all individuals will necessarily ben-

Table 26-8 Annual per Capita Consumption of Alcoholic Beverages in the United States, 1978†

Beverage	Volume, gal	Ethanol, vol %	Energy equivalent, kcal
Hard liquor	2.9	50	33,000
Beer	23.4	5	25,000
Wine	1.9	12	4,900
Total			62,900

† The caloric intake as ethanol per capita by the drinking-age population is about 240 kcal/d. Since the per capita caloric intake of all foods in the same population is believed to total about 3000 kcal/d, ethanol contributes about 8 percent on the average. In some European countries, particularly France, average ethanol consumption is much higher. Alcohol provides over half the caloric intake in heavy drinkers and alcoholics.

efit from a diet in which the animal fat and cholesterol content is reduced. Atherosclerosis is a disease with complex origins and individuals differ markedly in their susceptibility. It is undoubtedly influenced by the diet, but probably the best prescription is to be born with the right genes.

Alcohol Also Contributes Calories

In affluent countries the consumption of alcoholic beverages has greatly increased in the past 25 years, to the point where alcohol is making a significant contribution to the average caloric intake, which in nonalcoholic adults may approach 12 percent (Table 26-8). Alcohol, i.e., ethanol, has a high energy content, yielding about 7.1 kcal/g on oxidation; this value lies between those of carbohydrate and of fat. Moreover, its energy is biologically available in the form of ATP through well-known metabolic pathways. Ethanol is oxidized to acetaldehyde in the liver by the action of cytosolic *alcohol dehydrogenase*, which uses NAD^+ as acceptor

$$\underset{\text{Ethanol}}{CH_3CH_2OH} + NAD^+ \rightleftharpoons \underset{\text{Acetaldehyde}}{CH_3{-}\overset{\displaystyle O}{\overset{\|}{C}}{-}H} + NADH + H^+$$

Acetaldehyde is then oxidized to acetate by a mitochondrial enzyme, *aldehyde dehydrogenase*, also NAD-linked

$$CH_3{-}\overset{\displaystyle O}{\overset{\|}{C}}{-}H + NAD^+ + H_2O \rightleftharpoons CH_3COOH + NADH + H^+$$

The two NADHs formed in these reactions ultimately donate their reducing equivalents to the mitochondrial respiratory chain. The subsequent electron transport to oxygen results in the formation of 2(3) = 6 molecules of ATP from ADP and P_i.

Acetate formed from ethanol is then activated in the liver by short-chain acyl-CoA synthetase to form acetyl-CoA

$$CH_3COOH + CoA\text{-}SH + ATP \rightleftharpoons CH_3{-}\underset{\underset{O}{\|}}{C}{-}S{-}CoA + AMP + PP_i$$

The acetyl-CoA so produced is ultimately oxidized via the citric acid cycle. Alcohol is often spoken of as containing "empty calories." This term is misleading since it suggests that calories in alcohol are not used by the body. What this term is *intended* to mean is that liquor, wine, and beer contain insignificant amounts of vitamins and minerals.

Quite apart from the deleterious social and economic effects of alcoholism, alcohol has a number of disadvantages as a source of food or energy. First, calories consumed as ethanol in excess over the daily caloric requirement are converted into fat via acetyl-CoA. Ethanol cannot be converted into glucose or glycogen in the body. Second, acute ethanol consumption brings on hypoglycemia in many individuals, since it inhibits gluconeogenesis from lactate and amino acids. Third, ethanol is a very expensive source of calories. A six-pack of beer (six 12-oz cans), which contains about 1130 kcal, costs about 20 times as much in the United States as 10 oz of sugar, its energy equivalent.

Obesity Is the Result of Caloric Overnutrition

Obesity, which increases the risk of cardiovascular diseases, hypertension, and diabetes, is quite simply the result of caloric intake in excess of body needs. It usually begins in childhood or adolescence, and the longer it is allowed to persist, the less likely that it can be controlled. Sensible dietary and exercise habits begun in early adult life are the most assured way to control obesity. Some people tend to transform a larger fraction of a given caloric intake into body fat than others, perhaps because of decreased capacity to generate heat by futile cycling (page 571) or by the thermogenic action of brown fat (page 492). It is important to understand that obesity is not necessarily the result of overconsumption of fat per se; it can result from excess caloric intake from any source, whether fat, carbohydrate, or protein.

One pound of body fat is the equivalent of about 3500 kcal. From this fact and the data in Table 26-3 we can make some estimates of the amount of overeating required to produce a given degree of overweight. The accumulation of 25 lb of excess body fat by a 28-year-old male is the consequence of the overconsumption of 25(3500) = 87,500 kcal. This accumulation could result from consumption of 150 excess calories per day, only 5.6 percent above the average requirement of 2700 kcal, for a period of 583 days, a little over 19 months. Usually, however, excess weight is gained over periods of several years. For example, if our subject had acquired his 25 excess pounds of body

Table 26-9 A Weight-Watcher's List of Caloric Values of Some Energy-Rich Foods

Food	Portion	kcal
Bacon	2 slices	96
Pizza	5-in cut	184
Baked potato	1	93
French fries	10 medium	156
Potato chips	10 medium	114
Rice	1 cup	210
Spaghetti	1 cup	178
White bread	1 slice	62
Butter	1 pat	50
Mayonnaise	1 tbsp	93
Peanut butter	1 tbsp	87
Sweet roll	1	174
Chocolate cake, frosted	1-in cut	407
Cupcake, plain	1	204
Doughnut, plain	1	125
Cookies	3	240
Apple pie	$\frac{1}{7}$ cut	343
Milk chocolate	1 oz	147
Sugar	1 tbsp	46
Ice cream	1 scoop	147
Peanuts	1 oz	160
Beer	1 12-oz can	188
Soft drink	1 12-oz can	90

fat over a period of 5 years, it would be the result of excess consumption of about 48 kcal/d, or only 1.8 percent above his real needs. These data show how responsive body weight can be to even a slight excess (or a deficiency) of calories over long periods.

Now let us see how long it would take to remove the excess weight by restricting calorie intake. To lose 25 lb of body fat our overweight subject would have to limit himself to 2400 kcal/d for 42 weeks, or 2100 kcal/d for 21 weeks, or 1800 kcal/d for 14 weeks. Higher rates of weight loss require rather drastic curtailment of daily food intake, to the point that work efficiency may be impaired. Clearly, to lose 25 excess pounds of fat quickly requires near-heroic effort. Table 26-9 shows the caloric values of some foods that are often consumed in excess. From these one can determine which foods and how much of each would have to be subtracted from a normal daily diet to achieve different rates of weight loss. Data in Table 26-4 show how much daily physical exercise is required to lose a given amount of weight. The best way to accomplish weight balance is by a combination of regular exercise and regular eating habits. It is far easier to prevent weight gain than to lose weight once gained.

"Crash" diets give the illusion of weight loss, but much of the lost weight is body water. They are usually ineffective over the long run; almost invariably they are followed by a return to the original calorie-excessive diet. "Formula" diets, which specify precisely the kind and amount of foods that may be consumed per day, have the advantage that they require no calculation of calories from food tables, but they limit food choices too severely for longer-term use and may result in a deficiency of certain vitamins or minerals.

Proteins Are Required for Their Amino Acid Content

Proteins per se are not required in the human diet. It is their content of certain amino acids that is essential in nutrition (Table 26-10). Adults require nine essential amino acids in amounts that range from about 0.5 g/d (tryptophan) to about 2 g/d (leucine, phenylalanine). Infants and growing children require 10 amino acids, the extra one being *arginine*. Although arginine is normally made by the liver as a step in the synthesis of urea (page 552), children cannot make arginine fast enough to support both urea synthesis and the synthesis of body proteins.

The recommended allowance of protein for college-age males is about 54 g/d, but this assumes that the diet contains a variety of different animal and plant proteins. From Table 26-10 we see that of the 54 g at least 12 must be in the form of the essential amino acids, the remainder as nonessential amino acids. The nutritional *value* or *quality* of a given protein depends upon two factors: (1) its content of the essential amino acids and (2) its digestibility. Proteins differ considerably in

Table 26-10 Daily Requirement of Essential Amino Acids (for College-Age Males)

Amino acid	g
Arginine	0†
Histidine	Unknown‡
Isoleucine	1.30
Leucine	2.02
Lysine	1.50
Methionine	2.02
Phenylalanine	2.02
Threonine	0.91
Tryptophan	0.46
Valine	1.50

† Required by infants and growing children.

‡ Essential but precise requirement not yet established.

the relative proportion of amino acids they contain (page 125). Some proteins contain a complete set of essential amino acids in the proper proportions; others may be deficient in one or more essential amino acids. Plant proteins, particularly those of wheat and other grains, are not completely hydrolyzed during digestion because the protein-rich portions of the grains are surrounded by protective husks of cellulose and other polysaccharides that are not hydrolyzed by intestinal enzymes. Since only free amino acids can be absorbed from the intestine, not all the amino acid content of most plant foods is actually biologically available.

The nutritional quality of proteins can be determined in two ways. The protein is completely hydrolyzed and its amino acid composition measured and compared with that of egg or milk protein as a standard. This *chemical score* of a protein indicates the *potential* value of the protein. A more accurate measure is the *biological value* of the protein, a factor that is inversely proportional to the amount of a given protein source that must be consumed to keep an adult human subject or experimental animal in *nitrogen balance*, the condition in which the intake of protein nitrogen exactly balances the loss of nitrogen in the urine and feces. If a given protein provides all the essential amino acids in the proper proportions and all are released in free form and absorbed, it will have a biological value of 100, much higher than a protein that is complete in amino acid content but is incompletely digested or one that is completely digested but is low in one or more essential amino acids. According to this test, if a given protein is totally deficient in only one essential amino acid, it will have zero biological value. If a protein has a very low biological value, a very large amount must be consumed in order to provide the minimal requirement of whatever essential amino acid it provides in least amount. Each of the other amino acids in such a protein would be consumed in excess of the amount needed for synthesis of body protein. The excess would be deaminated in the liver and converted into glycogen or fat or burned as fuel.

Table 26-11 shows the relative quality of the proteins of some common foods. Animal proteins, for example, those of milk, beefsteak, and eggs, rank high in both chemical score and biological value. At the other end of the scale the proteins of corn or whole wheat bread have a low chemical score because they are deficient in one or more essential amino acids. However, they have an even lower biological value, since they are incompletely digested. Thus a much larger amount of these plant proteins must be consumed to provide the minimum daily requirement of all the amino acids. This is not to say that whole wheat bread is an undesirable food, but it means that if one lived on whole wheat bread alone, a very large amount of it would be required to provide the minimum daily requirement of the essential amino acids. For example, since a slice of whole wheat bread contains less than 2 g of protein, and since this protein has less than one-third the quality of good protein,

Table 26-11 The Quality of Proteins in Some Foods

Source	Chemical score	Biological value
Human milk	100	95
Beefsteak	98	93
Whole egg	100	87
Cow's milk	95	81
Corn	49	36
Polished rice	67	63
Whole wheat bread	47	30

some 73 slices of whole wheat bread would have to be consumed by a 21-year-old male to meet the recommended daily allowance of protein, about 56 g. Moreover, foods of plant origin as they are usually prepared generally contain less protein per unit dry weight than foods of animal origin.

All the essential amino acids must be available simultaneously for body protein synthesis. When rats are fed a synthetic diet containing all the essential amino acids but one, and the missing one is fed 3 h after the rest, the rats will not grow. Amino acids cannot be stored.

Certain Plant Proteins Can Supplement Each Other in the Diet

Although plant proteins are in general nutritionally inferior to those from animal sources, certain combinations of plant proteins can provide a completely adequate and balanced mixture of essential amino acids. For example, corn proteins are low in lysine but contain adequate tryptophan, whereas bean proteins contain adequate lysine but are low in tryptophan. Neither is a "good" protein. A mixture of corn and beans, however, provides a nutritionally adequate supply of essential amino acids. This mixture, called *succotash*, was instinctively "discovered" by New World Indians. Having only beans for breakfast and only corn for dinner would obviously cancel the usefulness of this combination of vegetable proteins. Oriental peoples have similarly learned to combine certain plant foods to provide a nutritionally complete mixture of amino acids; one such combination is rice and soybeans. In Central and South America, where protein undernutrition has been common, an international nutrition board has promoted the consumption of Incaparina, a nutritionally complete mixture of relatively cheap native plant proteins, principally corn, sorghum, and cottonseed meal. Although each has low biological value, in combined form these proteins have almost the same biological value as milk protein.

Marasmus and Kwashiorkor Are World Health Problems

In some parts of the world there is an increasing mismatch between the rate of population growth and the rate of food production, particularly in the Southern Hemisphere. At any given time famine is endemic somewhere in the world and is made worse by poor crop years and by war. It is estimated that 500 million people live on the edge of starvation and that 12,000 people die of hunger every day. In India alone it is estimated that 1 million children die of undernutrition annually. Two forms of child undernutrition, often occurring together, are mainly responsible, *marasmus* and *kwashiorkor* (Figure 26-7).

Marasmus (from Greek, "to waste") is the term applied to chronic deficiency of calories in children; it is caloric starvation. Marasmus occurs in famine areas when infants are

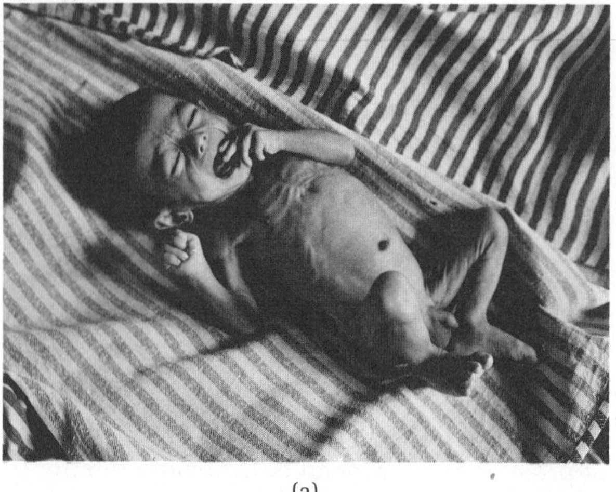

(a)

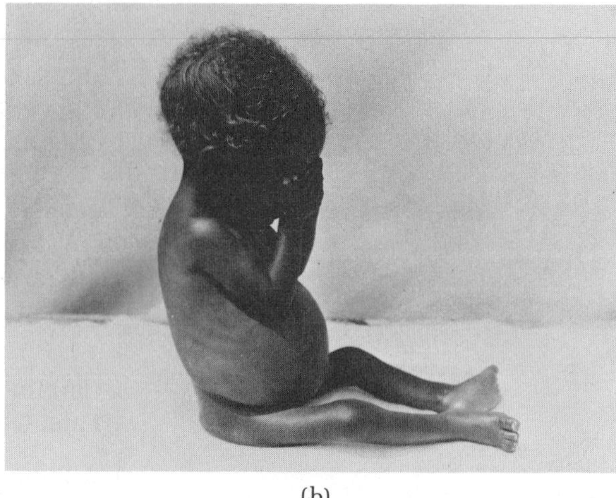

(b)

Figure 26-7
(a) *Marasmus, severe caloric deficiency, in an Indonesian infant. It is usually complicated by deficiency of several vitamins and minerals.* (b) *An Angolan child, a victim of kwashiorkor. The deficiency of dietary protein causes some organs to become watery and bloated, due to inadequate synthesis of serum albumin, required to maintain water balance.*

weaned from breast milk and given inadequate bottle feedings of thin watery gruels of native cereals or other plant foods, usually deficient in both calories and protein. This is an especially crucial stage in growth of infants and children, when both calories and protein are needed in plenty. Marasmus is characterized by arrested growth, extreme muscle wasting, weakness, and anemia; it is usually complicated by multiple deficiencies of vitamins and minerals. Calorie deficiency in early childhood, even if ultimately alleviated with an ample diet, leaves a permanent deficit in body growth. The mortality rate in calorie-deficient children in famine areas is very high; up to half do not survive until 5 years of age. Marasmus is one end of a spectrum of famine-induced nutritional deficiency diseases. The other is protein deficiency. In famine there is usually a deficiency of both calories and protein, but in different proportions; the combined deficiency is called *protein-calorie malnutrition.*

In many parts of the world little animal protein is available. Since the protein content of plant foods is generally low and plant proteins are also low in quality, a serious shortage of "good" protein often occurs in areas where population growth is especially great. Chronic protein deficiency in children is called *kwashiorkor,* an African word that means "weaning disease." Native children are nursed for relatively long periods, but when they are weaned, usually to make way for the next baby, they receive insufficient protein. The growth of protein-deficient children is retarded, they become anemic, and the tissues become watery and bloated because of the low serum protein levels, which upset the normal distribution of water between the tissues and the blood. Moreover, the liver, kidneys, and pancreas undergo severe degeneration. The mortality rate of kwashiorkor is very high. Even if children survive a long period of protein lack, there is a permanent physiological deficit. More important is evidence that protein deprivation in early childhood also leaves a deficit in learning and

Table 26-12 Vitamins Required in Human Nutrition

Deficiencies relatively common
Thiamine
Niacin
Ascorbic acid
Riboflavin
Folic acid
Deficiencies rare in the United States
Pantothenic acid
Pyridoxine
Biotin
Vitamin B_{12}
Vitamin A
Vitamin D
Vitamin E
Vitamin K

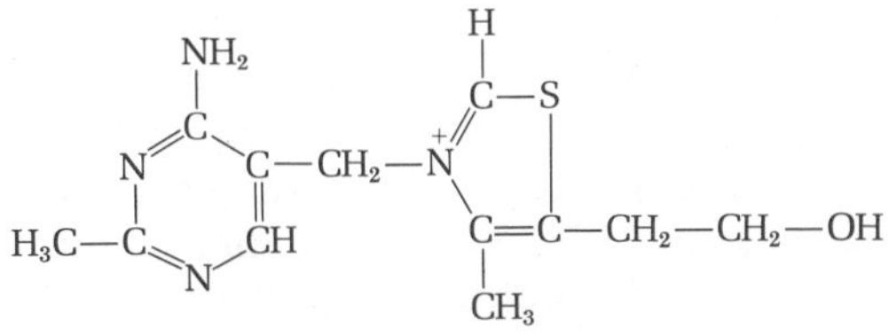

other aspects of mental ability. The deficit is especially severe if protein deprivation has occurred in two or three successive generations. Kwashiorkor, although first named in Africa, is nearly worldwide in incidence, wherever proteins are in short supply.

Deficiencies of Some Vitamins Can Be Life-Threatening

Now let us examine the role of vitamins in nutrition. In Chapter 10 we have already considered the structure and coenzyme functions of the vitamins. Here we are concerned with the requirements of these vitamins in human nutrition and the consequences of their deficiency.

Today we are reasonably certain that all the vitamins required in human and rat nutrition have been identified. They can be divided into two classes, depending on the extent to which their deficiency constitutes a significant health problem (Table 26-12). Deficiencies of thiamine, niacin, riboflavin, folic acid, and ascorbic acid are life-threatening in many parts of the world. Even in highly developed countries, marginal deficiencies of these vitamins are relatively common. However, spontaneous human deficiencies of pantothenic acid, pyridoxine, biotin, vitamin B_{12}, and vitamins A, D, E, and K are extremely rare in affluent countries.

The requirements of individuals for any given vitamin may vary considerably depending on the adequacy of the rest of the diet, on the activity of intestinal microorganisms, and on genetic and racial factors. For example, the requirement for nicotinamide depends very much on protein nutrition, particularly on the intake of tryptophan, which can be converted into nicotinamide; furthermore, the requirement for pyridoxine increases with the protein content of the diet. On the other hand, biotin, pantothenic acid, and vitamin B_{12} are made in sufficient quantities by intestinal bacteria to cover ordinary human needs. Only under very unusual dietary circumstances are deficiencies of these vitamins observed.

Most of the water-soluble vitamins must be obtained from the diet regularly since they are excreted or destroyed by enzyme action during normal metabolic turnover. Consumption of huge amounts of water-soluble vitamins in "mega-vitamin" pills simply is followed by excretion of the excess beyond daily needs, since most water-soluble vitamins cannot be stored. Excessive consumption of the fat-soluble vitamins A and D can, however, lead to toxic effects.

NH$_2$ H C—S N C—CH$_2$—N$^+$ H$_3$C—C CH C=C—CH$_2$—CH$_2$—OH N CH$_3$

Thiamine

Coenzyme form: Thiamine pyrophosphate

Function: Coenzyme for pyruvate dehydrogenase, α-ketoglutarate dehydrogenase, and transketolase

Recommended daily allowance (21-year-old males): 1.5 mg

Deficiency syndrome: Beriberi

Figure 26-8
Structure, function, and requirement of thiamine (vitamin B_1). See also Figure 10-2.

Thiamine Deficiency Is Still a Nutritional Problem

Deficiency of thiamine (Figure 26-8) in the human leads to beriberi ("I cannot"), a neurological disorder that was endemic in the rice-eating countries of the Orient in the nineteenth and early twentieth centuries (page 252). This disease was unknown until the invention of rice-polishing machinery in the early 1800s. For many years beriberi was thought to be an infectious disease. The first real clue that it was a nutritional disorder came from observations made by a Dutch physician, C. Eijkman, in what is now Indonesia. He reported in 1897 that chickens fed left-over white rice that had been cooked for human consumption developed a nerve disease resembling human beriberi, but when the chickens were fed either whole rice or the rice polishings or husks, they did not develop this condition. This report provided the clue that led physicians of the Japanese navy to feed unpolished rice to sailors suffering from beriberi, with dramatic benefit. Today we know that the husks of rice, which are removed in the polishing process, contain most of the thiamine. Severe beriberi is no longer the endemic scourge it once was, but it still is a medical problem among the undernourished in the rice-eating countries of Asia, Indonesia, and the Philippines, as well as in Africa, where there is increasing consumption of refined white flour.

In beriberi there is muscle weakness and wasting, lack of coordination, peripheral neuritis, mental confusion, apathy, low heart rate, and heart enlargement. Edema or swelling of the extremities also may occur. Heart failure is the usual cause of death. Beriberi is particularly severe in infants of nursing mothers who consume diets deficient in thiamine. In beriberi the concentration of pyruvate in the blood is greatly elevated, a finding consistent with the requirement of thiamine pyrophosphate as the coenzyme for the pyruvate dehydrogenase complex (page 437). Administration of thiamine to deficient animals or children with beriberi results in almost miraculous recovery, within hours (Figure 26-9).

Few foods are exceptionally rich in thiamine. The best sources are lean meat, beans, nuts, whole-grain cereals, and fish. Although the need for thiamine has been known for many years, a significant fraction of the American population today is thought to have a marginal intake of thiamine. The recommended daily allowance of thiamine is from 1.0 to 1.5 mg/d for adults (Figure 26-8; Table 26-2), but a great many Americans have an intake of less than 1 mg/d. For this reason, white bread, which is very low in thiamine, is usually enriched with thiamine. White flour, breakfast foods, noodles, spaghetti, and soda crackers are also inherently low in thiamine and are often enriched or fortified with it.

Because alcoholics obtain a large fraction of their caloric requirement from alcoholic beverages, which contain no thiamine, they are likely to become thiamine-deficient. A particularly severe thiamine deficiency in alcoholics is Wernicke-Korsakoff syndrome, characterized by neurological disorders, psychosis, and loss of memory. This disease, which is irreversible and usually requires hospitalization, is not caused by alcohol per se but by a combination of thiamine deficiency and a possible defect in the thiamine-dependent enzyme *transketolase* (page 666). It has been proposed that the permanent mental impairment of Wernicke-Korsakoff disease, as well as

(a)

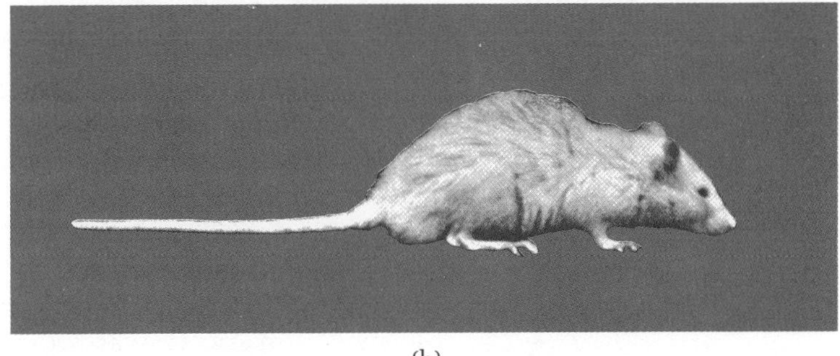
(b)

Figure 26-9
(a) An albino rat reared on a diet essentially free of thiamine. The deficiency affects nerve cells, leading to polyneuritis and paralysis. The rat is unable to coordinate its movements. (b) The dramatic recovery of the same rat after only 24 hours on a diet containing a normal level of thiamine.

other symptoms of thiamine deficiency in alcoholics, might be avoided by fortifying liquor, wine, and beer with thiamine, but no success has attended efforts to carry out such a plan, which some believe would encourage consumption of alcohol.

Nicotinamide and Tryptophan Nutrition Are Related

Deficiency of nicotinamide (Figure 26-10, see also page 255) in people leads to *pellagra* (Italian, "rough skin"), first observed in Europe, but endemic in most regions where corn is a staple food and the diet is low in meat or fish. Today, pellagra is still seen in very poor rural areas and in prisons and mental institutions where inmates are given marginal food. Pellagra may be recognized by the three d's: dermatitis, diarrhea, and dementia.

Corn, i.e., maize, has two biochemical peculiarities that lead to nicotinamide deficiency when it is a major dietary component and meat is simultaneously lacking. Corn contains considerable nicotinamide, but it is present in a bound form which cannot be absorbed by the intestine and thus is unavailable. When corn is treated with dilute bases, however, its bound nicotinamide is released in free form and can then be absorbed. Long before this was discovered by biochemists, Indians in Mexico and Central America somehow learned to soak cornmeal in lime water (dilute calcium hydroxide), thus releasing free thiamine, before baking the meal into tortillas.

The other special feature of corn relevant to pellagra is that corn proteins contain very little tryptophan. Actually, human

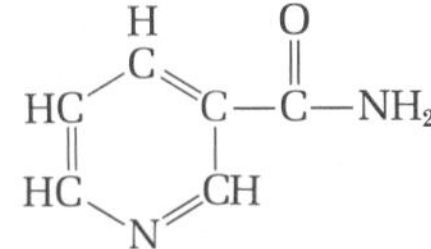

Nicotinamide (niacinamide)

Coenzyme forms: Nicotinamide adenine dinucleotide and its phosphate (NAD and NADP)

Function: Coenzymes for many dehydrogenases

Recommended daily allowance: 19 mg

Deficiency syndrome: Pellagra

Figure 26-10
Structure, function, and requirement of nicotinamide, also known as niacinamide.

beings and most animals can make nicotinamide from tryptophan (page 538), but when the diet is low in tryptophan, it is used almost entirely for protein biosynthesis, leaving little or none available as a precursor of nicotinamide. Some 60 mg of dietary tryptophan is equivalent to 1 mg of nicotinamide.

Lean meat, peas, beans, nuts, and fish are the best sources of nicotinamide, but milk and eggs contain very little. Refined grains are also poor sources. White bread, breakfast cereals, and other corn, wheat, and rice products are commonly fortified with added nicotinamide.

Many Foods Are Low in Ascorbic Acid

For centuries it has been known that the disease scurvy, once very common, particularly among seamen and explorers subsisting on dried foods, is alleviated by ingestion of certain plant tissues or juices. One of the very first accounts of the disease was recorded in the journals of Jacques Cartier, who explored Newfoundland and the St. Lawrence river in the mid-1500s (Box 26-1), but two more centuries were to pass before the cure of scurvy was proved by systematic experiments. In the 1750s James Lind fed a group of sailors suffering from scurvy six different diets. Only one of these diets cured the scurvy: it contained the juice of lemons. Lind found that a variety of fresh greens, vegetables, or fruits, but especially citrus, could cure scurvy. Although Lind recommended that lemon juice be included in the diet of sailors, it took another half century before this advice was accepted by the British Admiralty. Deaths from scurvy still occurred among Arctic and Antarctic explorers as late as the early twentieth century. The antiscurvy vitamin was finally isolated from lemon juice in 1932 and synthesized shortly later. It was given the name ascorbic acid (Figure 26-11).

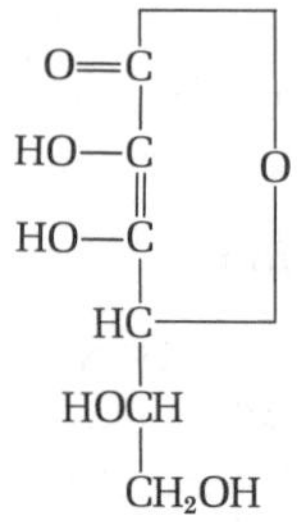

O=C
HO—C
HO—C
HC
HOCH
CH_2OH

Ascorbic acid (vitamin C)

Active form: Unknown
Function: Cofactor in some hydroxylation reactions
Recommended daily allowance: 60 mg
Deficiency syndrome: Scurvy

Figure 26-11
Structure, function, and requirement of ascorbic acid.

While outright scurvy in adults is uncommon today, many people do not have an optimum daily intake of ascorbic acid, particularly nursing mothers. Ascorbic acid is not as widely distributed as other vitamins, and it is also an unstable substance that is destroyed by heat under alkaline conditions or by exposure to oxygen in the presence of iron or copper ions, which catalyze its oxidation to inactive products. Meat, eggs, and fish are relatively poor sources of ascorbic acid. Milk that has been pasteurized, i.e., heated to about 80 to 85°C, contains very little ascorbic acid. Moreover, dried cereal grains are virtually devoid of ascorbic acid. People who do not regularly consume fresh fruit and vegetables, particularly elderly people living alone, are prone to subclinical deficiency of ascorbic acid.

The recommended daily allowance of ascorbic acid in the United States (Figure 26-11; Table 26-2) is about 60 mg for adults, but this may be in great excess of the minimal requirement. In the United Kingdom only 20 mg/d is considered sufficient for adults and only 10 mg/d appears to be sufficient to

Box 26-1 One of the first recorded accounts of the cure of scurvy, among the men of Jacques Cartier's Expedition to Newfoundland, 1535†

"Some did lose all their strength. . . . Others also had all their skins spotted with spots of blood of a purple color: then did it ascend up to their ankles, knees, thighs, shoulders, arms, and necks. Their mouths became stinking, their gums so rotten, that all the flesh did fall off, even to the roots of the teeth, which did almost all fall out. Our Captain, considering our estate, and how sickness was increased and hot amongst us, one day went forth from the fort, and walking upon the ice saw a troop of those countrymen coming from Stadacona, among which was Domagaia, who not ten or twelve days afore, had been very sick with that disease, and had his knees swollen as big as a child of two years old, all his sinews shrunk together, his teeth spoiled, his gums rotten and stinking. Our Captain, seeing him whole and sound, was marvelous glad, hoping to understand how he had healed himself, to the end he might ease and help his own men. As soon as they came near he asked Domagaia how he had healed himself: he answered that he had taken the juice and sap of the leaves of a certain tree, and had with these healed himself. Then our Captain asked him if any were to be had thereabout. . . . Domagaia straight sent two women to fetch some of it, who brought ten or twelve branches of it, and then showed the way how to use it and that is, to take the bark and leaves and boil them together, then to drink of said decoction every other day. . . . The tree in their language is called Ameda or Hannedew, this thought to be the sassafras tree. Our Captain presently caused some of that drink to be made for his men, but there was none durst taste of it, except one or two, who ventured the drinking of it: the others seeing did the same, and presently recovered their health and were delivered of their sickness, and with this drink were clean healed. After this medicine was found and proved to be true, there was such a strife about it, who should be the first to take it, that they were ready to kill one another. A tree as big as any oak in France was spoiled and lopped bare, and occupied all for 5 or 6 days, and it wrought so well, that if all the physicians in Montpelier or Louvain had been there with all the drugs of Alexandria, they would not have alone so used in one year, as that tree did in six days, for it did so prevail, that as many as used of it, by the grace of God recovered their health."

† From Hakluyt's *Principal Navigators,* 1600; reprinted in S. Davidson et al. (eds.): *Human Nutrition and Dietetics,* 6th ed., Churchill Livingstone, Edinburgh, 1975.

prevent scurvy. Citrus fruit, tomatoes, pineapple, cabbage, and green vegetables are the best sources of ascorbic acid, but most fresh fruits and vegetables are good sources. As pointed out above, the leaves and bark of many shrubs and trees are rich in ascorbic acid (Box 26-1).

Some nutritionists feel that the body should be kept fully "saturated" with a relatively large intake of ascorbic acid. One view is that the human body may have been genetically adapted to a much larger dietary intake of ascorbic acid in the hunter-gatherer era, when uncooked plant and animal tissues furnished much of the food intake. It has been claimed that massive doses of ascorbic acid, as high as several grams a day, should be taken for optimum health and, particularly, for prevention of the common cold. Well-controlled clinical tests have shown no statistically significant effect of such massive doses on the incidence of colds, although it is possible that some individuals may benefit significantly. High doses of ascorbic acid do not seem to be toxic; the excess intake above the amount required to saturate the body is excreted as such or in the form of various oxidized products.

Figure 26-12
Structure, function, and requirement of riboflavin.

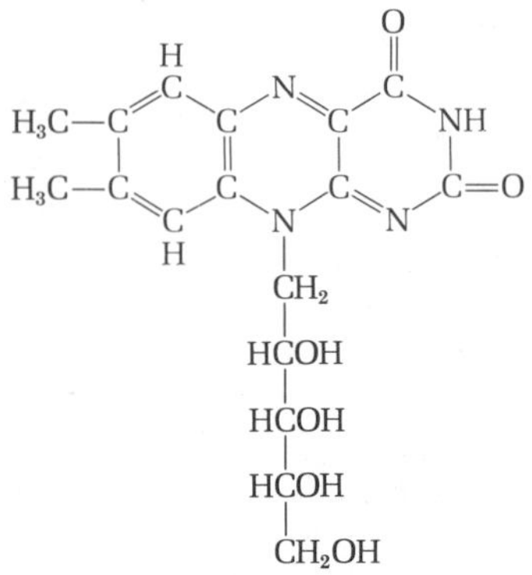

Riboflavin (vitamin B_2)

Coenzyme forms: Flavin mononucleotide and flavin adenine dinucleotide

Function: Coenzyme in oxidation-reduction reactions

Recommended daily allowance: 1.7 mg

Deficiency syndrome: Skin and mucous membrane disorders

The form of scurvy seen most often today, *infantile scurvy*, occurs in babies who are bottle-fed with pasteurized milk or reconstituted dried milk and who receive no supplementation with ascorbic acid. Such infants suffer from spontaneous hemorrhages under the skin, are easily bruised, and have very painful limbs.

Marginal Riboflavin Deficiency Is Also Common

Marginal deficiency of riboflavin (Figure 26-12) is relatively common in most parts of the world, including the United States. It has not received much attention because it is almost never life-threatening. It is most frequent in pregnancy, in growing children, or in those undergoing physiological stress. Riboflavin deficiency is characteristically manifested in the form of soreness and cracking of the lips and corners of the mouth, as well as an oily dermatitis of the face. Anemia is also observed. Riboflavin deficiency often occurs together with other deficiency diseases, particularly pellagra.

The best sources of riboflavin are milk, liver, eggs, meat, and yellow vegetables. Since cereals and bread contain very little riboflavin, they are usually fortified with the vitamin.

The Most Common Vitamin Deficiency Is of Folic Acid

On a worldwide basis, deficiency of folic acid (Figure 26-13) is believed to be the most common form of vitamin undernutrition. It is particularly prevalent in underdeveloped countries in the tropics, where most of the population is believed to have at least marginal folic acid deficiency. In the United States many indigents and elderly people suffer from deficiency of folic acid, manifested as anemia, weight loss, and weakness. Pregnant women and infants are also particularly vulnerable. Folic acid deficiency is a major feature of *tropical sprue*, in which there is a general deficiency in absorption of many nutrients from the small intestine.

Folic acid is widely distributed in leafy green vegetables, liver, yeast, and meat but is destroyed by cooking and exposure to reducing agents.

Figure 26-13
Structure, function, and requirement of folic acid.

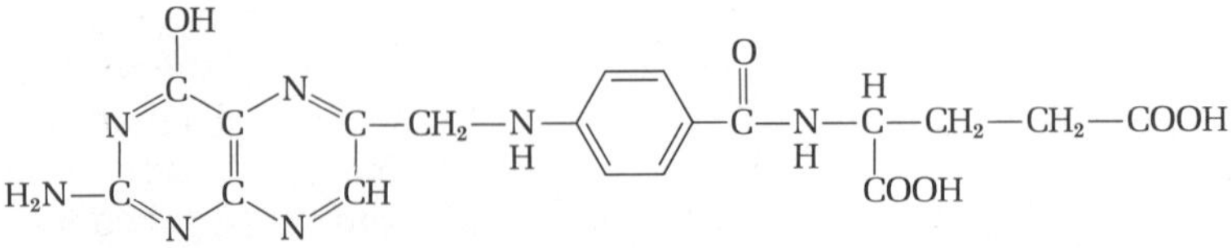

Folic acid

Active form: Tetrahydrofolate

Function: Coenzyme in enzymatic reactions involving 1-carbon-group transfer

Recommended daily allowance: 400 μg

Deficiency syndrome: Anemia

Figure 26-14
Structure, function, and requirement of pyridoxine.

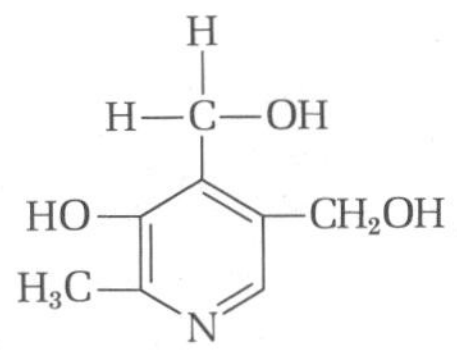

Pyridoxine (vitamin B_6)

Coenzyme form: Pyridoxal phosphate
Function: Coenzyme in transaminations and other reactions of amino acids
Recommended daily allowance: 2.2 mg

Figure 26-15
Structure, function, and requirement of biotin.

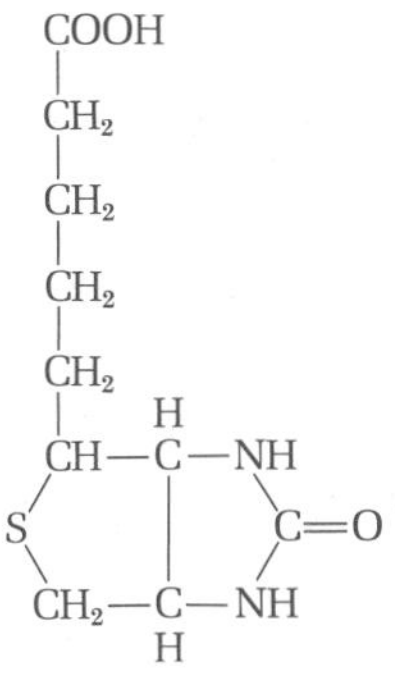

Biotin

Active form: Biocytin
Function: Prosthetic group of pyruvate carboxylase and other CO_2-transferring enzymes
Probable daily requirement: 150 μg

Figure 26-16
Structure, function, and requirement of pantothenic acid.

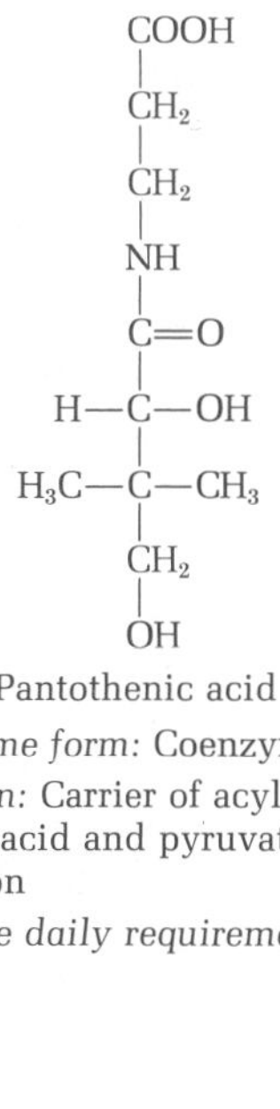

Pantothenic acid

Coenzyme form: Coenzyme A
Function: Carrier of acyl groups in fatty acid and pyruvate oxidation
Probable daily requirement: 5 to 10 mg

Human Deficiencies of Pyridoxine, Biotin, and Pantothenic Acid Are Rare

Pyridoxine, biotin, and pantothenic acid (Figures 26-14 to 26-16) are present in ample amounts in many foods. Under natural conditions human deficiencies of these vitamins are very rare, but they have been produced in human volunteers or detected in people following unusual diets. A classic example is a case of biotin deficiency discovered in a Boston hospital in a patient who had consumed nothing but raw eggs and wine over a period of many months. Although raw eggs are rich in proteins and most vitamins and minerals, they contain a protein, *avidin*, which binds biotin and prevents it from being absorbed by the intestine. If the eggs had been cooked, this deficiency would not have occurred, since avidin after heat denaturation no longer binds biotin.

Deficiencies of these three vitamins probably occur in alcoholics. A tendency to form calcium oxalate kidney stones is characteristic of marginal pyridoxine deficiency in some Oriental races. Vitamin B_6 deficiency has also been observed in tuberculosis patients treated with the drug *isoniazid*, which inactivates pyridoxal phosphate, the coenzyme form of pyridoxine.

True Dietary Deficiency of Vitamin B_{12} Is Very Rare

Vitamin B_{12} (Figure 26-17) is the anti–pernicious anemia vitamin (page 262). Neither plants nor animals can make vitamin B_{12}; it is formed only by certain bacteria. The bacteria in the human intestinal tract can make enough vitamin B_{12} for the normal daily requirement. Vitamin B_{12} is also made in large amounts by the rich population of bacteria in the rumen of

Figure 26-17
Vitamin B_{12} (cobalamine).

Structure: See page 263; shown is a stereo model of the backbone structure of vitamin B_{12}

Coenzyme form: Deoxyadenosylcobalamine (coenzyme B_{12})

Function: Coenzyme in conversion of methylmalonyl-CoA to succinyl-CoA (page 522), and a few other reactions

Recommended daily allowance: 3.0 μg

Deficiency syndrome: Pernicious anemia

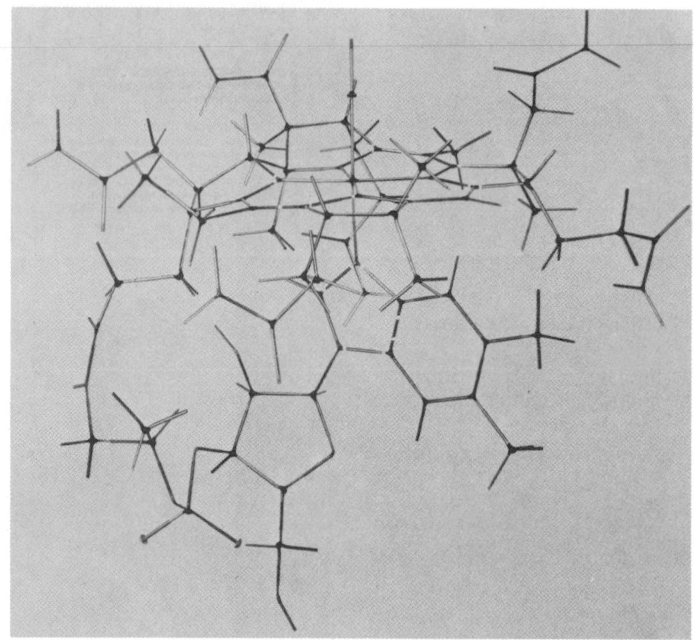

ruminant animals and in the <u>cecum</u> of other herbivorous species, such as the rabbit. The cecum is an appendage to the intestine, of which the human appendix is a vestige. Rabbits obtain their requirement of vitamin B_{12} and certain other vitamins by periodically eating some of their own feces.

Pernicious anemia is a serious disease in which there is a deficiency in red cells and hemoglobin formation and severe impairment of the central nervous system. It is not due to a dietary deficiency of vitamin B_{12} but to a failure to absorb the vitamin from the intestine because of deficient secretion of a glycoprotein in the stomach. This glycoprotein, called the <u>intrinsic factor</u>, is required for absorption of vitamin B_{12}. Pernicious anemia is treated by injection of vitamin B_{12} or by feeding large amounts of the vitamin, which can overcome the defect in absorption. Because the human liver can store sufficient vitamin B_{12} to last for several years, a true dietary deficiency is extremely rare. People who are strict vegetarians for long periods obtain their vitamin B_{12} requirement from either intestinal bacteria or from bacteria ingested with plant foods.

Figure 26-18
Vitamin A.

```
              H2
              C
         H2C     CH2
  CH3     |       |
      C         C—CH3
  CH3     C  =  C
              |
             HC
             ||
             CH
              |
             C—CH3
             ||
             HC
              |
             CH
             ||
             HC
              |
             C—CH3
             ||
             HC
              |
             CH2OH
```

Vitamin A_1 (retinol)

Active form: Unknown

Function: Intermediate in visual cycle; normal tissue development

Recommended daily allowance: 1.0 mg

Deficiency syndrome: Night blindness, susceptibility to infections

Vitamin A Deficiency Has Multiple Effects

The nutritional diseases <u>xerophthalmia</u> ("dry eyes") and <u>keratomalacia</u> (excessive keratin formation in the skin and the cornea of the eye) are common in Southeast Asia, Central and South America, and parts of Africa but rare in the United States. These conditions are due to lack of vitamin A (Figure 26-18) or carotene, the precursor of vitamin A. Carotene is abundant in yellow plants, such as carrots and sweet potatoes.

(a)

(b)

(c)

Figure 26-19
Night blindness in vitamin A deficiency. (a) *Both the normal person and the vitamin A–deficient one see the headlights of an approaching car.* (b) *After the car has passed, the normal person sees a wide stretch of road.* (c) *The vitamin A–deficient person can barely see a few feet ahead and cannot see the road sign at all.*

Children in areas where protein-calorie malnutrition is endemic are especially prone to vitamin A deficiency. In the United States it occurs in people with intestinal or pancreatic diseases in which there is defective fat absorption; such people fail to absorb dietary vitamin A or carotene, which are fat-soluble. In its early stages xerophthalmia is manifested as *night blindness*, owing to deficient synthesis of the visual pigment *rhodopsin*, which contains as its active group *retinal*, for which vitamin A is the precursor (page 265) (Figure 26-18). In vitamin A deficiency there is increased susceptibility of all tissues to infections.

Vitamin A is stored in the liver (page 265) in amounts that are sufficient for many months. For example, a single 30-μg dose of vitamin A is sufficient to protect children for a 6-month period; the adult human liver can store more than 300 μg. The livers of cold-water fish and Arctic mammals are very rich in preformed vitamin A. However, vitamin A in great excess (20 to 30 times the RDA) is toxic, with many painful symptoms. Arctic explorers have died after eating polar bear liver, which contains enormous amounts of vitamin A. Vitamin A toxicity has been frequently observed in persons consuming very large quantities of vitamin pills.

Figure 26-20
Vitamin D.

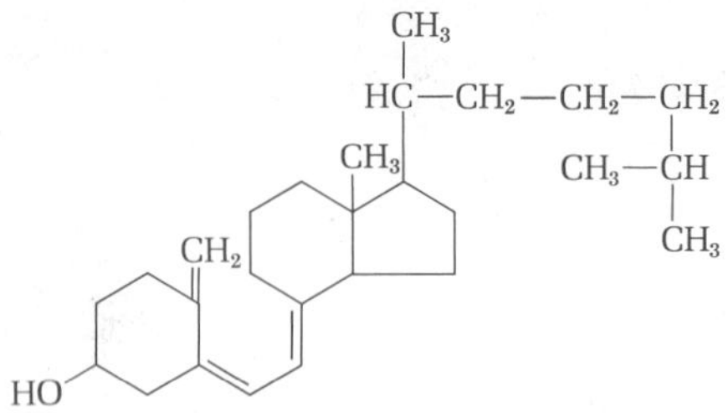

Vitamin D_3 (cholecalciferol)

Active form: 1,25-dihydroxy-cholecalciferol

Function: Active form is a hormone controlling Ca and phosphate metabolism

Recommended daily allowance: 10 μg

Deficiency syndrome: Rickets; osteomalacia

Vitamin D Deficiency Causes Rickets and Osteomalacia

Vitamin D deficiency has become quite rare in the United States, although it was once prevalent in many northern countries (page 267). In the childhood disease *rickets* (and its adult form *osteomalacia*) the bones are soft and malformed because they do not retain calcium. This condition is not caused by a lack of calcium in the diet but by lack of a hormone, *1,25-dihydroxycholecalciferol* (page 268), whose normal precursor is *vitamin D_3* or *cholecalciferol*. Normally vitamin D is not required in the diet so long as there is ample exposure of the skin to sunlight, which causes the formation of cholecalciferol or vitamin D_3 (Figure 26-20) from 7-dehydrocholesterol in a photochemical reaction (page 268). However, when the skin is not exposed more or less regularly to sunlight, vitamin D must be provided in the diet (page 267). Half an hour of direct sunlight on the cheeks of a baby each day is sufficient to generate the minimal daily requirement of vitamin D. Although Eskimos are exposed to very little sunlight in the Arctic winter, they obtain ample vitamin D from fish (Figure 26-21).

Babies and young children with rickets seem well fed and fat but lack muscle tone and are slow to walk. Deformations of bones appear in the skull, chest, spine, and particularly as characteristic bowlegs or knock-knees. Early treatment with vitamin D can reverse mild deformities, but in long-standing rickets the deformities are permanent.

Osteomalacia is the adult form of rickets. Child-bearing women on marginal diets are particularly susceptible to vitamin D deficiency, since the developing fetus has first call on Ca^{2+} derived from the mother's skeleton.

When dietary supplements of vitamin D are required, the usual form is irradiated *ergosterol;* this yeast sterol is readily converted into *ergocalciferol* (page 267), which has potent vitamin D activity. The almost complete disappearance of rickets in the United States has been the result of enriching cow's milk with irradiated ergosterol. The amount of vitamin D required by adults in the absence of sunlight is about 10 μg/d. Doses exceeding 1.5 mg/d are very toxic.

Figure 26-21
Vitamin D deficiency usually results from lack of ultraviolet radiation in sunlight, needed to convert 7-dehydrocholesterol into vitamin D_3. It has been postulated that the ancestors of Homo sapiens *arose in the tropics and were dark-skinned. As they migrated to the north, the ultraviolet-screening effect of dark skin pigments was unfavorable for synthesis of vitamin D in the skin. Genetic selection for light skin color allowed northern peoples to absorb more ultraviolet radiation. Such selection did not occur among Eskimos, who consume a fish diet rich in vitamin D.*

Human Deficiencies of Vitamin E or K Are Very Rare

In the United States the average diet contains more than ample amounts of vitamins E and K to satisfy minimal requirements; spontaneous deficiencies are rarely seen. In addition, vitamin K is synthesized by intestinal bacteria. Because vitamins E and K are fat-soluble, their absorption from the small intestine may fail in conditions in which there is defective lipid absorption, particularly when the secretion of bile salts is impaired. Vitamin E (Figure 26-22) protects membrane lipids against oxidative destruction of polyunsaturated fatty acids. About 10 to 30 mg of α-tocopherol is recommended as the daily intake. Massive doses of tocopherol are not toxic, but there is no evidence

Figure 26-22
Vitamin E.

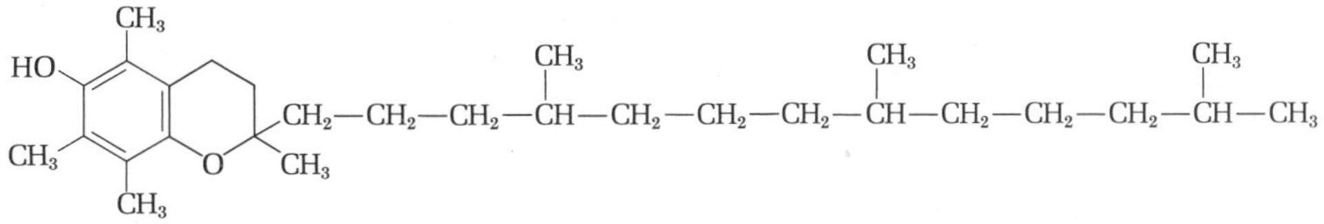

Vitamin E (α-tocopherol)

Active form: Unknown
Function: Not precisely known, but it protects against damage of membranes by oxygen
Recommended daily allowance: 10 mg

Figure 26-23
Vitamin K_1.

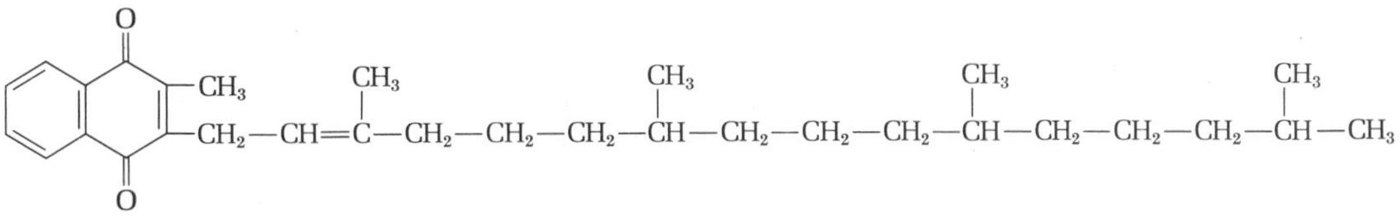

Vitamin K_1

Active form: Unknown
Function: Coenzyme in carboxylation of glutamyl residues of prothrombin and other proteins
Probable daily requirement: 1 mg

that such doses improve the complexion or cure infertility, as has been claimed by some enthusiasts.

Vitamin K (Figure 26-23) is given to newborn infants and to patients before and after surgery on the gallbladder or liver, in order to assure that the prothrombin levels in the blood will be normal. Recall (page 269) that in vitamin K deficiency there is deficient enzymatic carboxylation of certain glutamate residues in prothrombin and other proteins participating in blood clotting. Very large doses of vitamin K are toxic.

Many Elements Are Required in Human Nutrition

In addition to the six "foundation" elements carbon, hydrogen, nitrogen, oxygen, sulfur, and phosphorus, of which carbohydrates, fats, proteins, and nucleic acids are composed, many other elements are known to be required in the nutrition of different experimental animals and humans (Table 26-13). Human nutritional deficiencies of some of the trace elements have not yet been observed, presumably because of their ubiquitous occurrence in most foods or in drinking water. It is highly likely, however, that all the elements proved to be essential for the rat or chick are also required by humans. It is also probable that other elements in addition to those listed in Table 26-13 will be found nutritionally essential in the future, as better

Table 26-13 Elements Required in Human Nutrition

Bulk elements
Calcium
Chlorine
Magnesium
Phosphorus
Potassium
Sodium
Trace elements
Copper
Fluorine
Iodine
Iron
Manganese
Molybdenum
Selenium
Zinc
Other trace elements known to be essential in animals and very likely essential for humans
Arsenic
Chromium
Nickel
Silicon
Tin
Vanadium

experimental methods for demonstrating deficiencies are developed and as increased knowledge of enzyme active groups becomes available.

It is not possible to determine whether a given element is essential in nutrition simply by carrying out a microanalysis of the tissues of an experimental animal. Animal and human tissues invariably contain trace amounts of some elements that are not intrinsic to biological structure or function but are present by accident, carried into the body as contaminants with food. An example is mercury, accumulated in toxic amounts by tuna and other fish from industrial pollutants dumped into the sea.

Here we briefly consider those elements whose deficiency in the human diet is relatively common and is known to impair health. The elements required in human nutrition can be grouped into *bulk* and *trace* groups (Table 26-13). The bulk elements are required in excess of 100 mg/d; the trace elements are required in amounts no greater than a few milligrams per day. The nutritionally important elements are biologically available only as salts or other chemically combined, soluble forms.

Calcium and Phosphorus Are Essential for Development of Bones and Teeth

The adult human body contains more than a kilogram of calcium, nearly all of which is in bones and teeth, together with phosphate, as the insoluble crystalline mineral *calcium hydroxyapatite*. Calcium also plays an important role in all cells as an intracellular regulator or messenger (page 747); it helps regulate the activity of skeletal muscle, the heart, and many other tissues. Calcium is very abundant in foods, particularly in milk and cheese and also in cereal grains, legumes, nuts, and vegetables. Nevertheless, it is of great importance in human nutrition because of the very high requirement for calcium during infancy and childhood, when the skeleton is actively growing, and during pregnancy and lactation. Moreover, the absorption of calcium from the intestine is regulated by many complex factors, including pH, the ratio of calcium to phosphorus in the diet, the presence of fatty acids and also certain plant acids, and most of all vitamin D, which regulates calcium absorption. Because of these factors, only a fraction of the calcium in ingested food is actually absorbed.

Calcium nutrition is rendered complex because of the fact that the bones constitute a very large and labile reservoir of calcium that can be drawn upon when calcium is low in the diet. Pregnant or lactating women on a calcium-deficient diet withdraw calcium from the skeletal reservoir for the growth of the fetus or for production of milk, which is rich in calcium. The calcium in bones is not permanently laid down; much of it is constantly undergoing turnover. About 700 to 800 mg of cal-

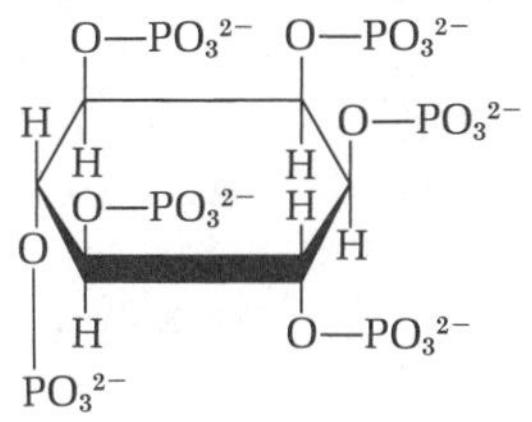

Figure 26-24
The phytate anion. Phytate can bind Ca^{2+}, Mg^{2+}, and Zn^{2+} very tightly at the multiple phosphate groups, preventing these essential metal ions from being absorbed. Phytates are found only in plant foods, especially grains.

cium may leave and enter the skeletal mass per day. Short-term needs for calcium in other parts of the body are thus readily met from this large reservoir.

The recommended daily allowance of calcium for adults (Table 26-2) is 800 mg/d. More, 1200 mg/d, is recommended for women during pregnancy and lactation, and for teenagers. The calcium present in cereal grains is not readily absorbed because much of it is tightly bound to inositol hexaphosphate, also called phytate (Figure 26-24), forming phytin, a calcium-magnesium salt of phytate. Phytate also binds other required elements, particularly zinc, as we shall see.

Phosphorus also is very abundant in the body, not only in bones but also in the nucleic acids, in the nucleotide coenzymes, and in the ATP-ADP-phosphate energy-transferring system in cells. Phosphorus is so widespread in natural foods that outright deficiency is unknown. However, not all the phosphorus of the food is absorbed readily, since its absorption is determined by many of the factors that determine calcium absorption, particularly the state of vitamin D nutrition.

Marginal Magnesium Deficiency Is Relatively Common

The body contains about 25 g of magnesium, most of which is present in the bones. All cells contain rather high concentrations of magnesium (5–10 mM). Mg^{2+} ions play a very important role in the action of many enzymes, particularly those of glycolysis and many ATP-dependent reactions. Although most foods contain considerable magnesium (it is especially abundant in the chlorophyll of green leafy vegetables), there is increasing evidence that the intake of magnesium is marginal in the United States diet, particularly in the elderly and poor. Alcoholics are very prone to magnesium deficiency, which also occurs in protein-calorie deficiency. The recommended magnesium intake is 350 mg/d for adult males.

Sodium and Potassium Intake Are Important in Prevention and Treatment of Hypertension

Sodium and potassium are present in ample amounts in most foods, and outright deficiencies of these elements are rare. The problem with these elements is one of nutritional imbalance, rather than undernutrition. Na^+, which is the principal extracellular cation, and K^+, the principal intracellular cation, are very important in regulation of water and electrolyte balance and of acid-base balance in the body (page 741). They in turn are subject to regulation by the mineralocorticoid hormones of the adrenal cortex (page 741).

The actual nutritional requirement for sodium is only about 1 g/d, but the average intake in the United States is about 5 g/d. As with sugar, the human consumption of salt has increased greatly in modern times. Moreover, many people appear to have a deep craving for salt and may consume 10 g of sodium per day. Continuous overconsumption of NaCl not only hastens the onset of high blood pressure or *hypertension* in later life but also seriously worsens this condition, which has a large genetic component and to which black people seem more prone. People with hypertension are advised to limit sodium chloride intake.

About 4 g of potassium is ingested per day in the average United States diet, but excess potassium loss from the body, as in diabetes and diarrhea, or loss in the urine after use of diuretic drugs employed in treatment of hypertension, can lead to serious potassium depletion. Tomato juice, citrus fruit, and bananas are especially rich in potassium.

Iron and Copper Are Needed for Heme Protein Synthesis

Iron deficiency is one of the most common nutritional disorders in the North American population. It is most frequent in children, in adolescent girls, and women in their reproductive years. Iron is absorbed only in its ferrous form (Fe^{2+}); its absorption and excretion are relatively slow and controlled by many complex factors. Only a small fraction of the iron present in most foods is actually absorbed. Moreover, the iron present in different foods varies a great deal in its availability for absorption. Iron in meat is absorbed best, but that in cereal grains is poorly absorbed. Milk contains very little iron.

Iron is required for the synthesis of the iron-porphyrin proteins hemoglobin, myoglobin, cytochromes, and cytochrome oxidase (Figure 26-25). It is carried in the blood bound to the plasma protein *transferrin*, and in the tissues it is stored in the form of *ferritin*, an iron-protein containing ferric hydroxide and ferric phosphate. The liver, spleen, and bone marrow contain much ferritin. Iron is not excreted in the urine but is lost from the body via the bile and feces and in menstrual blood. Because the rate of loss of iron from the body is doubled or tripled during menstruation, women require larger amounts of iron than men. Although bread and certain other grain products are fortified with iron, this has not solved the problem of iron deficiency, since weight watching by girls and women usually results in the elimination of bread from the diet. Deficiency of iron leads to *iron-deficiency anemia*, in which the number of red blood cells is normal but the amount of hemoglobin in the cells is relatively low.

Copper is essential in the diet because it is involved in the proper utilization of iron and especially for the synthesis of cytochrome oxidase, which contains both iron and copper. Copper is also necessary for proper development of connective tissues and blood vessels. About 2.5 to 5.0 mg/d is required. It is abundant in organ meats, seafood, vegetables, and nuts but absent from dairy products.

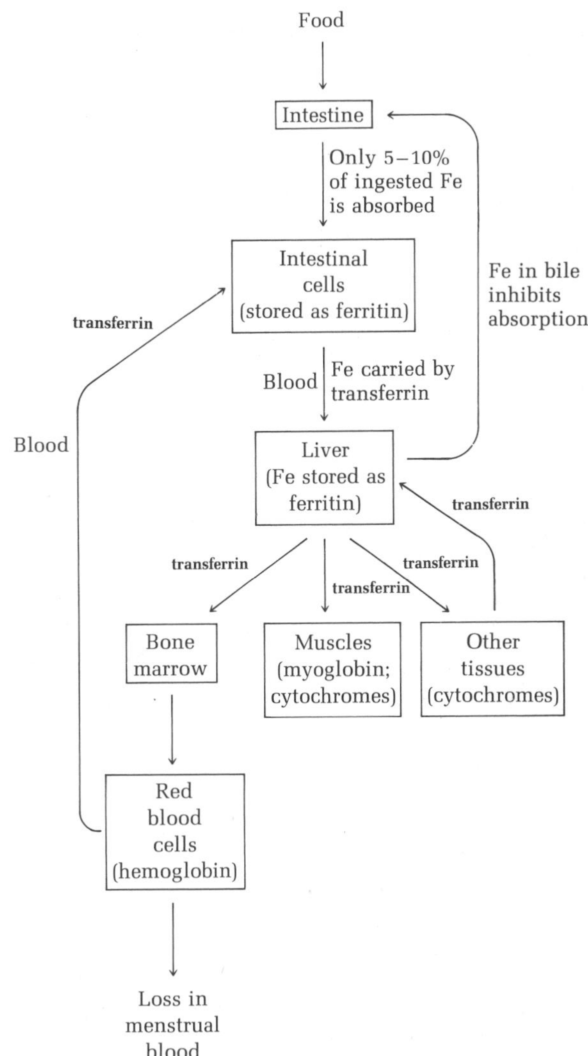

Figure 26-25
The absorption, storage, and utilization of iron in the synthesis of cytochromes, myoglobin, and hemoglobin. Once it is absorbed from the intestine, iron is always bound to proteins, in the form of transferrin during transport by the blood and in the form of ferritin for storage in cells. Iron is excreted only very slowly. When the capacity of ferritin to store iron is exceeded, iron is accumulated as insoluble hemosiderin granules within mitochondria of some tissues.

Goiter Is the Result of Iodine Deficiency

In some inland areas, e.g., the mountainous regions of Europe, central Africa, and the American Midwest near the Great Lakes, many people suffer from simple goiter, enlargement of the thyroid gland (Figure 26-26), due largely to a deficiency of iodine in the soil in areas far from the sea. Goiter in the United States has become much less common since frozen seafood is widely available; nevertheless, it is still a medical problem.

The thyroid gland collects iodine from the blood and uses it to make the thyroid hormones (page 742). In iodine deficiency the thyroid gland undergoes compensatory enlargement in order to extract iodine from the blood more efficiently. In long-standing iodine deficiency, the goiterous thyroid may enlarge to enormous size, weighing several pounds, and can become a threat to life. Severe iodine deficiency of mothers also

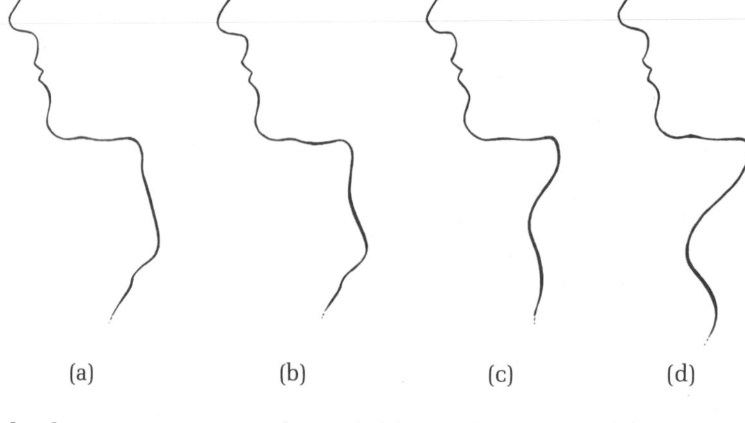

Figure 26-26
Neck profiles showing stages in the development of iodine-deficiency goiter. (a) Normal, (b) early, (c) moderate, and (d) severe goiter.

leads to <u>cretinism</u> in their children, characterized by mental retardation, slow body development, dwarfism, and a characteristic facial structure.

Goiter is easily preventable. Although iodized salt (0.5 g of potassium iodide per kilogram of NaCl) is widely available in food stores and very effective in controlling goiter, it is not always chosen by those who need it. Unfortunately, iodized table salt is not available in many states.

Dental Caries Is an Important Nutritional Problem

Dental caries is considered to be the most widespread disease condition in the United States; it is also the most widespread disorder in which nutrition is a factor. In some regions up to 90 percent of the population shows the results of dental caries: missing teeth, fillings, and restorations. The incidence of dental caries correlates positively with the consumption of free sugars and may be regarded as a disease of affluent populations.

However, there is another nutritional factor involved in the incidence of dental caries. Comprehensive statistical studies have been carried out on the incidence of dental caries in communities in which fluoride was added in controlled amounts to reservoir water, compared with similar communities in which the water supply contained no fluoride (Figure 26-27). Such studies show that fluoride in a concentration of 1 ppm (part per million) greatly diminishes the incidence of dental caries. Today over 9000 communities in the United States, serving about half of the United States population, have fluoride-treated city water supplies.

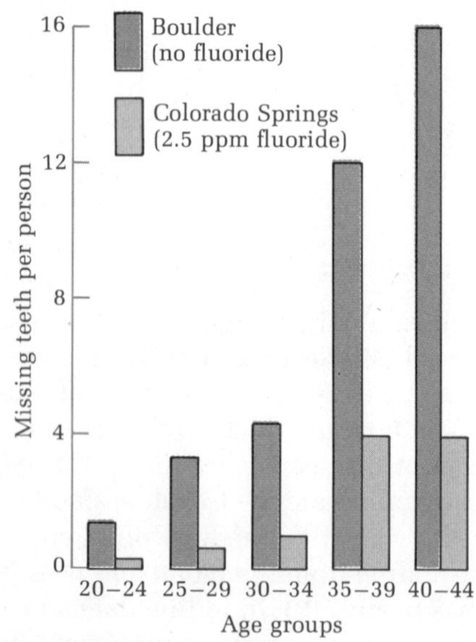

Figure 26-27
Lasting beneficial effects of fluoride in two comparable communities in Colorado. At age 40 to 44 the total tooth loss was 16 per person in Boulder but only 4 per person in Colorado Springs.

Unfortunately, some citizen groups have vigorously opposed fluoridation of city water supplies. Some oppose it on the basis that fluoride is a drug or medication, some feel that their freedom of choice or their individual rights are violated, and some are pure food extremists. It is a fact that *excess* fluoride in drinking water is deleterious and can cause <u>fluorosis</u>, manifested as mottled teeth, a condition that occurs in children in certain regions of the United States because of a

Figure 26-28
Effect of fluoride on hydroxyapatite.

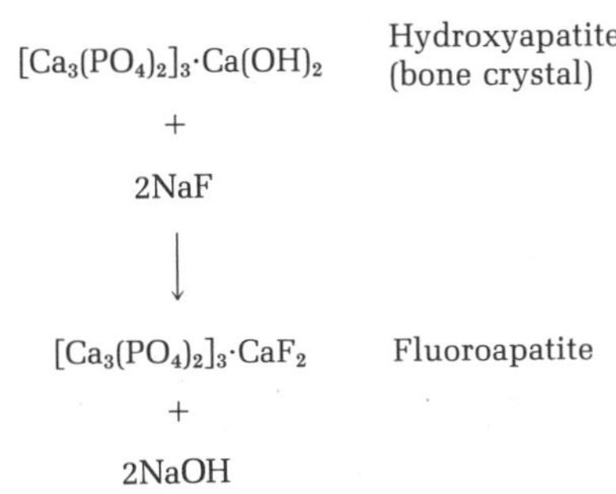

very high content of fluoride in the local soil and water. But fluorine is not a drug. It is an essential element required in the diet for proper formation of bones and teeth. Fluoride becomes incorporated into *hydroxyapatite,* the crystalline mineral of bones and teeth, to form *fluoroapatite* (Figure 26-28). Although only a small fraction of bone crystal consists of fluoroapatite, its presence causes hydroxyapatite crystals to become larger, harder, and more resistant against attack by acids. Because fluoride is not always available in sufficient amounts in natural or refined foods, fluoridation of drinking water, which can be done safely, is especially important for growing children, since adequate intake of fluoride beginning in childhood has very beneficial long-term effects (Figure 26-27).

Zinc and Several Other Microelements Are Essential in the Diet

Over 80 enzymes are known to require zinc as part of their prosthetic groups. These include alcohol dehydrogenase, carbonic anhydrase, DNA and RNA polymerases, and carboxypeptidase. Zinc is found in high concentrations in the prostate gland, sperm cells, and the eyes, where it presumably plays important but still unknown functions. Endemic zinc deficiency, which has been discovered in some village populations in Iran, leads to small stature, anemia, low serum albumin, and retarded development of the reproductive system. It is caused by consumption of a traditional local staple of diet, unleavened bread made from partially refined flour. Such bread contains much *phytate* (Figure 26-24), which binds zinc very tightly, thus preventing its absorption by the intestine. An unusual feature of zinc deficiency in rare individuals is an alteration in taste and smell perception. Some odors are enormously distorted, so that normal food or cooking smells are perceived as foul and disgusting, leading to refusal of food and consequent weight loss. At least 15 mg/d of zinc is required by adults; pregnant and lactating women require more (Table 26-2). Zinc is abundant in meat, eggs, seafood, milk, and liver but rather low in fruit and green vegetables.

Rats, chickens, and other small animals raised under "ultraclean" conditions in all-plastic cages devoid of trace metals and given only triple-distilled water, filtered air, and highly purified diets require several other trace elements for proper growth and function. Presumably they are also required by human beings. Among these are *tin, nickel, vanadium, chromium,* and *silicon,* which is needed for proper growth of bone and connective tissue. *Cobalt* is a vital part of the vitamin B_{12} molecule and is therefore required by the microorganisms that make this vitamin. Cobalt is required by some animals, probably because it is needed for synthesis of vitamin B_{12} by microorganisms of the gastrointestinal tract. Also required are *selenium,* an essential component of the enzyme *glutathione peroxidase,* and *molybdenum,* part of the prosthetic group of the enzymes *xanthine oxidase* and *aldehyde oxidase*. Selenium in excess is very toxic. In some areas of the western United States and New Zealand excess selenium in the soil and consequently in the vegetation is the cause of "blind staggers" in horses and *alkali disease* in cattle. Copper, cobalt, zinc, manganese, and nickel, although essential in the diet, are also toxic in excess and are health hazards in the mining and metallurgical industries.

A Balanced Diet Must Have Variety

There is no single perfect food that provides all nutritional needs for everyone. The 40 different required nutrients occur in very different proportions in different foods. A useful, simple guide for an adequate diet is given by the Basic Four food groups (Table 26-14). This table gives the kinds of foods that should be eaten *daily from each group* in order to provide a reasonably balanced diet. The choices from each group should be varied from day to day, to avoid consumption of one or a few foods to the exclusion of most others in that group. The best guarantee of adequate nutrition is a highly varied diet, together with an appropriate balance of calories and proteins in relation to individual needs, considering height, weight, and the degree of physical activity.

Table 26-14 The Four Basic Food Groups

A guide to an adequate daily intake of required nutrients; variety within each food group is essential

Milk group
Two glasses of milk or servings of cheese, cottage cheese, ice cream, or other dairy products
Meat group
Two servings of meat, fish, poultry or eggs; peas, beans, or nuts are alternatives
Vegetable and fruit group
Four servings of green or yellow vegetables, tomatoes, citrus fruit
Bread and cereal group
Four servings of whole grains or fortified cereal products

People have always had irrational obsessions about food, however, endowing some with near-miraculous powers or imposing strict taboos on others. Food choices are often set by religious, cultural, or ethnic considerations. Furthermore, food folklore has it that some foods are good for the brain, some for the hair, and others for fertility. Even today, with our highly developed scientific knowledge of nutrition, food superstitions and fads are not only prevalent but sometimes even fashionable, as in the case of "organic," "macrobiotic," or Zen diets. Food fallacies are also common. For example, athletes are often advised to consume nothing but protein-rich steak and eggs for breakfast on the morning of a contest. However, proteins cannot be stored and are not as useful as energy sources as carbohydrates. It would make more biochemical sense to prepare for a contest or race by stuffing with carbohydrates, to build up liver and muscle glycogen stores, since glycogen is the ultimate fuel during intense physical activity.

Labeling Food Products Protects Consumers

Consumer-protection laws require that the labels of many packaged food products provide a nutritional analysis. Together with unit costs, i.e., price per pound or other unit, such labeling allows consumers to make quantitative judgments of different brands of a given type of product. Nutritional labeling is especially informative in the case of breakfast foods, which are heavily promoted. Table 26-15 shows the nutritional informa-

Table 26-15 The Nutritional Information Printed on the Box of a Popular American Breakfast Cereal

Nutrition information, per serving, 1 oz (28.4 g)		
	Alone	With ½ cup vitamin D–fortified whole milk
Calories	110	180
Protein	2 g	6 g
Carbohydrate	26 g	31 g
Fat	0 g	4 g
Sodium	195 mg	255 mg
Percent of U.S. recommended daily allowances		
	1 oz alone	With ½ cup whole milk
Protein	2	10
Vitamin A	25	30
Vitamin C	25	25
Thiamine	25	30
Riboflavin	25	35
Nicotinamide	25	25
Calcium	<2	15
Iron	10	10
Vitamin D	10	25
Vitamin B_6	25	25
Folic acid	25	25
Phosphorus	2	15
Magnesium	<2	4
Zinc	<2	4
Copper	2	2
Carbohydrate information		
	1 oz alone	With ½ cup whole milk
Starch and related carbohydrates	13 g	13 g
Sucrose and other sugars	13 g	18 g
Total carbohydrates	26 g	31 g

Ingredients: Milled corn, sugar, salt, malt flavoring, ascorbic acid, vitamin A palmitate, niacinamide, reduced iron, pyridoxine hydrochloride, riboflavin, thiamine hydrochloride, folic acid, and vitamin D_3. BHA added to preserve freshness.

tion taken from the label of a well-advertised children's breakfast food, typical of many rich in free sugar. Note that this cereal contains as its only natural ingredient milled corn; all the other ingredients represent flavoring, preservatives, and several vitamins and minerals. A 1-oz serving (28.4 g) contains 26 g of total carbohydrates, 2 g protein, and no fat; it is therefore 93 percent carbohydrate. The carbohydrate consists of equal amounts of starch and of sugars; the free sugar content is therefore about 47 percent, much higher than recommended for a good diet (see Figure 26-2) and unfavorable with respect to the problem of dental caries in children. The milk added to a serving of this cereal contributes 67 percent of the protein and almost 40 percent of the calories. This cereal is enriched or fortified with various vitamins, as well as with iron and zinc. Also listed is the sodium content, of concern to those with hypertension.

Sugar in the form of such a sugar-rich breakfast food is very expensive. In November 1981 the cereal described in Table 26-15 retailed for $2.03 per pound. For comparison, ordinary table sugar at that time was $0.34 per pound. Thus the sugar in 1 lb of this cereal, if purchased as sugar, would cost only 16 cents.

References

For General Reference

DAVIDSON, S., R. PASSMORE, J. F. BROCK, and A. S. TRUSWELL (eds.): *Human Nutrition and Dietetics,* 6th ed., Churchill Livingstone, Edinburgh, 1975.

Human Nutrition: Readings from Scientific American, Freeman, San Francisco, 1978. An extremely interesting collection of articles (some listed below).

"The Nutrition of Man," special issue of the *British Medical Journal,* vol. 37 (1981). More specialized, up-to-date articles.

UNDERWOOD, E. J.: *Trace Elements in Human and Animal Nutrition,* Academic Press, New York, 1977.

Special Topics in Nutrition

CHISHOLM, J. J., JR.: "Lead Poisoning," *Sci. Am.,* **224:**15–23, February (1971).

FEENSTROM, W. D., and R. J. WURTMAN: "Nutrition and the Brain," *Sci. Am.,* **230:**84–91, February (1974).

"Food and Agriculture," *Sci. Am.,* **235,** September (1976). The entire issue is devoted to articles on this topic.

GOLDWATER, L. J.: "Mercury in the Environment," *Sci. Am.,* **224:** 15–21, May (1971).

HARPSTEAD, D. D.: "High-Lysine Corn," *Sci. Am.,* **225**:34–42, August (1971).

JANICK, J., C. H. NOLLER, and C. L. RHYKERD: "The Cycles of Plant and Animal Nutrition," *Sci. Am.*, **235**:74–86, September (1976).

KERMODE, G. O.: "Food Additives," *Sci. Am.*, **226**:15–21, March (1972).

KRETCHMER, N.: "Lactose and Lactase," *Sci. Am.*, **227**:70–78, October (1972).

LIEBER, C. S.: "The Metabolism of Alcohol," *Sci. Am.*, **234**:25–33, March (1976).

LOOMIS, W. F.: "Rickets," *Sci. Am.*, **223**:76–91, December (1970).

MAYER, J.: "The Dimensions of Human Hunger," *Sci. Am.*, **235**:40–49, September (1976).

"Nutrition Misinformation and Food Faddism," *Nutrition Reviews*, Special Supplement, July 1974.

Problems

1. *Purpose of Recommended Dietary Allowances* Why are *recommended* daily dietary allowances set forth by the Food and Nutrition Board (Table 26-2) different from the *minimal* daily requirements?

2. *ATP Generated by Glucose vs. Fatty Acid* Calculate the number of moles of ATP generated by oxidative phosphorylation per gram of glucose and per gram of palmitic acid, under standard conditions. Compare your results to the amount of heat released on combustion in a calorimeter per gram of glucose and per gram of palmitic acid (Table 26-5).

3. *Weight Loss during Fasting* When a crash diet is first begun, the initial weight loss observed is largely due to loss of body water. Why? When fasting is continued, weight loss per day is less than in the initial period. Can you account for this?

4. *Nutrients Tending toward Obesity* Excessive intake of calories over long periods above actual needs can cause obesity. Which type of diet is the more likely to lead to overconsumption and obesity, one rich in sugar or one rich in fat? Give your reasons.

5. *Calorimetry of Foods* A 9.5-g sample of bran flakes is oxidized completely to CO_2 and H_2O by ignition in a bomb calorimeter. The 2500 g of water in the water jacket increases in temperature from 15 to 27°C.
   (a) Calculate the caloric value of the bran flakes in kilocalories per gram.
   (b) If the initial moisture content of the bran flakes was 25 percent, what is the caloric value of the solids in bran flakes?
   (c) From these data and other considerations is carbohydrate, fat, or protein the major component of the bran flakes? Explain your answer.
   (d) When the bran flakes are consumed by a person, will they yield the same amount of energy in the body as in the calorimeter? If not, why?

6. *Caloric Balance* A young woman, a college senior, finds that she has gained weight over the past year. Her diet provides 2400

kcal/d. Standard tables correlating height, body build, age, and weight indicate she is about 20 percent overweight. The average daily intake that would maintain her weight constant at a "normal" level is 2100 kcal. From data in this chapter recommend six different foods and their amounts that might be avoided daily in order to maintain her weight at the "normal" level.

7. *Nitrogen Balance and Dietary Protein* A healthy adolescent person living on a diet containing a balanced array of essential amino acids exhibits *positive* nitrogen balance; i.e., the total daily amount of ingested nitrogen is larger than the total daily amount of excreted nitrogen. In contrast, an adult living on a diet deficient in one (or more) of the essential amino acids exhibits *negative* nitrogen balance; i.e., the total daily amount of ingested nitrogen is smaller than the total daily amount of excreted nitrogen. Explain.

8. *Experimental Determination of Amino Acid Requirements* Describe the design of an experiment to determine the minimum daily requirement of the amino acid phenylalanine for the albino rat.

9. *Initial Weight Loss Associated with Kwashiorkor Treatment* Young children suffering from kwashiorkor actually *lose* weight initially when placed on an adequate diet. Explain.

10. *Nutrition and Kidney Disease* Patients with limited kidney function are unable to excrete waste products at an adequate rate. They must undergo regular periods of dialysis, in which their blood is dialyzed through a membrane to remove waste products such as urea and uric acid. In addition, their diets are controlled with respect to the amount and kind of protein intake. Explain. On such a restricted diet, which would be a better protein source for the patient, eggs or corn? Why?

11. *Vitamin B_6 Requirements and Diet* The dietary requirement for vitamin B_6 increases when a person is placed on a high-protein diet. Give possible reasons for this observation.

12. *Nutritional Studies and Intestinal Bacteria* One of the complications in human nutrition studies is the uncertainty associated with the effect of the test diet on the intestinal bacteria. Why is this a concern?

13. *Requirement for a Specific Food* Some people maintain that milk is the perfect food and must be included in the diet of all people for proper nutrition. Is this statement true? Provide a biochemical basis for your answer.

14. *Food for Mountaineers* Suppose you had to design a food supply to be carried by a mountaineer planning a 48-h ascent of a peak in the Himalayas.
    (a) What would you consider to be the essential characteristics and requirements of a diet for this purpose?
    (b) What specific types of foods would you include?
    (c) Which foods would you regard as unnecessary?
    (d) Which vitamins and minerals would you add as supplements to the diet? Give reasons for your answers.

15. *Alcohol as a Precursor of Fat and Carbohydrate* Alcohol is readily converted into triacylglycerols but cannot be converted into glucose or glycogen. Why?

16. *Caloric Content of Beer* A college senior who has maintained his weight constant on a total caloric intake of 2900 kcal/d acquires the habit of consuming one 12-oz can of beer per day in addition. If all other factors, such as exercise, remain constant, how much new body fat will he accumulate over a 3-year period? (His actual weight gain will be more, since deposition of fat in adipose tissue requires an increase in the volume of blood and extracellular fluid.)

17. *Meat as a Source of Calories* In a few parts of the world where the relevant natural resources are available, meat is consumed in large amounts, often at all meals. Obesity can occur in meat eaters if their meat intake exceeds their caloric needs.
    (a) By what metabolic routes can meat, which is rich in protein, lead to deposition of triacylglycerols?
    (b) What other metabolic changes would be evoked by such a diet?

```
                                mRNA START ──→                    A START
CCGTCAGGATTGACACCCTCCCAATTGTATGTTTTCATGCCTCCAAATCTTGGAGGCTTTTTTATGGTTCGTTCTTATTACCCTTCTGAA
TGTCACGCTGATTATTTTGACTTTGAGCGTATCGAGGCTCTTAAACCTGCTATTGAGGCTTGTGGCATTTCTACTCTTTCTCAATCCCCA
ATGCTTGGCTTCCATAAGCAGATGGATAACCGCATCAAGCTCTTGGAAGAGATTCTGTCTTTTCGTATGCAGGGCGTTGAGTTCGATAAT
GGTGATATGTATGTTGACGGCCATAAGGCTGCTTCTGACGTTCGTGATGAGTTTGTATCTGTTACTGAGAAGTTAATGGATGAATTGGCA
                |──────── REGION OF ORIGIN OF DNA REPLICATION ────────|
CAATGCTACAATGTGCTCCCCCAACTTGATATTAATAACACTATAGACCACCGCCCCGAAGGGGACGAAAAATGGTTTTTAGAGAACGAG
AAGACGGTTACGCAGTTTTGCCGCAAGCTGGCTGCTGAACGCCCTCTTAAGGATATTCGCGATGAGTATAATTACCCCAAAAAGAAAGGT
ATTAAGGATGAGTGTTCAAGATTGCTGGAGGCCTCCACTAAGATATCGCGTAGAGGCTTTGCTATTCAGCGTTTGATGAATGCAATGCGA
CAGGCTCATGCTGATGGTTGGTTTATCGTTTTTGACACTCTCACGTTGGCTGACGACCGATTAGAGGCGTTTTATGATAATCCCAATGCT
TTGCGTGACTATTTTCGTGATATTGGTCGTATGGTTCTTGCTGCCGAGGGTCGCAAGGCTAATGATTCACACGCCGACTGCTATCAGTAT
TTTTGTGTGCCTGAGTATGGTACAGCTAATGGCCGTCTTCATTTCCATGCGGTGCACTTTATGCGGACACTTCCTACAGGTAGCGTTGAC
                                                                          mRNA START ──→
CCTAATTTTGGTCGTCGGATACGCAATCGCCGCCAGTTAAATAGCTTGCAAAATACGTGGCCTTATGGTTACAGTATGCCCATCGCAGTT
CGCTACACGCAGGACGCTTTTTCACGTTCTGGTTGGTTGTGGCCTGTTGATGCTAAAGGTGAGCCGCTTAAAGCTACCAGTTATATGGCT
                                                                              B START
GTTGGTTTCTATGTGGCTAAATACGTTAACAAAAAGTCAGATATGGACCTTGCTGCTAAAGGTCTAGGAGCTAAAGAATGGAACAACTCA
CTAAAAACCAAGCTGTCGCTACTTCCCAAGAAGCTGTTCAGAATCAGAATGAGCCGCAACTTCGGGATGAAAATGCTCACAATGACAAAT
CTGTCCACGGAGTGCTTAATCCAACTTACCAAGCTGGGTTACGACGCGACGCCGTTCAACCAGATATTGAAGCAGAACGCAAAAAGAGAG
ATGAGATTGAGGCTGGGAAAAGTTACTGTAGCCGACGTTTTGGCGGCGCAACCTGTGACGACAAATCTGCTCAAATTTATGCGCGCTTCG
                                                                             B END
ATAAAAATGATTGGCGTATCCAACCTGCAGAGTTTTATCGCTTCCATGACGCAGAAGTTAACACTTTCGGATATTTCTGATGAGTCGAAA
                                                                       C START  A END
AATTATCTTGATAAAGCAGGAATTACTACTGCTTGTTTACGAATTAAATCGAAGTGGACTGCTGGCGGAAAATGAGAAAATTCGACCTAT
CCTTGCGCAGCTCGAGAAGCTCTTACTTTGCGACCTTTCGCCATCAACTAACGATTCTGTCAAAAACTGACGCGTTGGATGAGGAGAAGT
GGCTTAATATGCTTGGCACGTTCGTCAAGGACTGGTTTAGATATGAGTCACATTTTGTTCATGGTAGAGATTCTCTTGTTGACATTTTAA
                 mRNA START ──→                           D START  C END
AAGAGCGTGGATTACTATCTGAGTCCGATGCTGTTCAACCACTAATAGGTAAGAAATCATGAGTCAAGTTACTGAACAATCCGTACGTTT
CCAGACCGCTTTGGCCTCTATTAAGCTCATTCAGGCTTCTGCCGTTTTGGATTTAACCGAAGATGATTTCGATTTTCTGACGAGTAACAA
                                                          E START
AGTTTGGATTGCTACTGACCGCTCTCGTGCTCGTCGCTGCGTTGAGGCTTGCGTTTATGGTACGCTGGACTTTGTAGGATACCCTCGCTT
TCCTGCTCCTGTTGAGTTTATTGCTGCCGTCATTGCTTATTATGTTCATCCCGTCAACATTCAAACGGCCTGTCTCATCATGGAAGGCGC
TGAATTTACGGAAAACATTATTAATGGCGTCGAGCGTCCGGTTAAAGCCGCTGAATTGTTCGCGTTTACCTTGCGTGTACGCGCAGGAAA
                                                            E END  D END  J START
CACTGACGTTCTTACTGACGCAGAAGAAAACGTGCGTCAAAAATTACGTGCGGAAGGAGTGATGTAATGTCTAAAGGTAAAAAACGTTCT
GGCGCTCGCCCTGGTCGTCCGCAGCCGTTGCGAGGTACTAAAGGCAAGCGTAAAGGCGCTCGTCTTTGGTATGTAGGTGGTCAACAATTT
J END                                  F START
TAATTGCAGGGGCTTCGGCCCCTTACTTGAGGATAAATTATGTCTAATATTCAAACTGGCGCCGAGCGTATGCCGCATGACCTTTCCCAT
CTTGGCTTCCTTGCTGGTCAGATTGGTCGTCTTATTACCATTTCAACTACTCCGGTTATCGCTGGCGACTCCTTCGAGATGGACGCCGTT
GGCGCTCTCCGTCTTTCTCCATTGCGTCGTGGCCTTGCTATTGACTCTACTGTAGACATTTTTACTTTTTATGTCCCTCATCGTCACGTT
TATGGTGAACAGTGGATTAAGTTCATGAAGGATGGTGTTAATGCCACTCCTCTCCCGACTGTTAACACTACTGGTTATATTGACCATGCC
GCTTTTCTTGGCACGATTAACCCTGATACCAATAAAATCCCTAAGCATTTGTTTCAGGGTTATTTGAATATCTATAACAACTATTTTAAA
GCGCCGTGGATGCCTGACCGTACCGAGGCTAACCCTAATGAGCTTAATCAAGATGATGCTCGTTATGGTTTCCGTTGCTGCCATCTCAAA
AACATTTGGACTGCTCCGCTTCCTCCTGAGACTGAGCTTTCTCGCCAAATGACGACTTCTACCACATCTATTGACATTATGGGTCTGCAA
GCTGCTTATGCTAATTTGCATACTGACCAAGAACGTGATTACTTCATGCAGCGTTACCATGATGTTATTTCTTCATTTGGAGGTAAAACC
TCATATGACGCTGACAACCGTCCTTTACTTGTCATGCGCTCTAATCTCTGGGCATCTGGCTATGATGTTGATGGAACTGACCAAACGTCG
TTAGGCCAGTTTTCTGGTCGTGTTCAACAGACCTATAAACATTCTGTGCCGCGTTTCTTTGTTCCTGAGCATGGCACTATGTTTACTCTT
GCGCTTGTTCGTTTTCCGCCTACTGCGACTAAAGAGATTCAGTACCTTAACGCTAAAGGTGCTTTGACTTATACCGATATTGCTGGCGAC
CCTGTTTTGTATGGCAACTTGCCGCCGCGTGAAATTTCTATGAAGGATGTTTTCCGTTCTGGTGATTCGTCTAAGAAGTTTAAGATTGCT
GAGGGTCAGTGGTATCGTTATGCGCCTTCGTATGTTTCTCCTGCTTATCACCTTCTTGAAGGCTTCCCATTCATTCAGGAACCGCCTTCT
GGTGATTTGCAAGAACGCGTACTTATTCGCAACCATGATTATGACCAGTGTTTCAGTCGTTCAGTTGTTGCAGTGGATAGTCTTACCTCA
                                                         F END
TGTGACGTTTATCGCAATCTGCCGACCACTCGCGATTCAATCATGACTTCGTGATAAAAGATTGAGTGTGAGGTTATAACCGAAGCGGTA
                                                                                G START
AAAATTTTAATTTTTGCCGCTGAGGGGTTGACCAAGCGAAGCGCGGTAGGTTTTCTGCTTAGGAGTTTAATCATGTTTCAGACTTTTATT
TCTCGCCACAATTCAAACTTTTTTTCTGATAAGCTGGTTCTCACTTCTGTTACTCCAGCTTCTTCGGCACCTGTTTTACAGACACCTAAA
GCTACATCGTCAACGTTATATTTTGATAGTTTGACGGTTAATGCTGGTAATGGTGGTTTTCTTCATTGCATTCAGATGGATACATCTGTC
AACGCCGCTAATCAGGTTGTTTCAGTTGGTGCTGATATTGCTTTTGATGCCGACCCTAAATTTTTTGCCTGTTTGGTTCGCTTTGAGTCT
TCTTCGGTTCCGACTACCCTCCCGACTGCCTATGATGTTTATCCTTTGGATGGTCGCCATGATGGTGGTTATTATACCGTCAAGGACTGT
GTGACTATTGACGTCCTTCCCCGTACGCCGGGCAATAACGTCTACGTTGGTTTCATGGTTTGGTCTAACTTTACCGCTACTAAATGCCGC
                                                               G END        H START
GGATTGGTTTCGCTGAATCAGGTTATTAAAGAGATTATTTGTCTCCAGCCACTTAAGTGAGGTGATTTATGTTTGGTGCTATTGCTGGCG
GTATTGCTTCTGCTCTTGCTGGTGGCGCCATGTCTAAATTGTTTGGAGGCGGTCAAAAAGCCGCCTCCGGTGGCATTCAAGGTGATGTGC
TTGCTACCGATAACAATACTGTAGGCATGGGTGATGCTGGTATTAAATCTGCCATTCAAGGCTCTAATGTTCCTAACCCTGATGAGGCCG
CCCCTAGTTTTGTTTCTGGTGCTATGGCTAAAGCTGGTAAAGGACTTCTTGAAGGTACGTTGCAGGCTGGCACTTCTGCCGTTTCTGATA
AGTTGCTTGATTTGGTTGGACTTGGTGGCAAGTCTGCCGCTGATAAAGGAAAGGATACTCGTGATTATCTTGCTGCTGCATTTCCTGAGC
TTAATGCTTGGGAGCGTGCTGGTGCTGATGCTTCCTCTGCTGGTATGGTTGACGCCGGATTTGAGAATCAAAAAGAGCTTACTAAAATGC
AACTGGACAATCAGAAAGAGATTGCCGAGATGCAAAATGAGACTCAAAAAGAGATTGCTGGCATTCAGTCGGCGACTTCACGCCAGAATA
CGAAAGACCAGGTATATGCACAAAATGAGATGCTTGCTTATCAACAGAAGGAGTCTACTGCTCGCGTTGCGTCTATTATGGAAAACACCA
ATCTTTCCAAGCAACAGCAGGTTTCCGAGATTATGCGCCAAATGCTTACTCAAGCTCAAACGGCTGGTCAGTATTTTACCAATGACCAAA
TCAAAGAAATGACTCGCAAGGTTAGTGCTGAGGTTGACTTAGTTCATCAGCAAACGCAGAATCAGCGGTATGGCTCTTCTCATATTGGCG
CTACTGCAAAGGATATTTCTAATGTCGTCACTGATGCTGCTTCTGGTGTGGTTGATATTTTTCATGGTATTGATAAAGCTGTTGCCGATA
                                                           H END
CTTGGAACAATTTCTGGAAAGACGGTAAAGCTGATGGTATTGGCTCTAATTTGTCTAGGAAATAA
```

PART IV

Molecular Transmission of Genetic Information

The nucleotide sequence of the DNA chromosome of the small bacterial virus φX174, as it was worked out in 1977 by Frederick Sanger and his colleagues in Cambridge, England. This achievement marked the beginning of a new era in genetic biochemistry and gained a second Nobel prize for Sanger, who was also the first to have deduced the complete amino acid sequence of a protein, insulin, twenty-five years earlier. After this sequence was published, it was found to be incomplete. Eleven additional nucleotides were found, making a total of 5386.

An important by-product of this effort was the discovery of overlapping genes and genes-within-genes in the φX174 chromosome, revealed by painstaking comparison of its base sequence with the amino acid sequence of the proteins for which it codes. The φX174 DNA contains nine genes, designated A–J. Their "start" and "end" codons are indicated in boxes. Note that gene E lies within gene D. The screened boxes indicate ribosome-recognition sites.

In the last part of this book we shall consider biochemical questions raised by the genetic continuity and the evolution of living organisms. What is the molecular nature of the genetic material? How is genetic information transmitted with such fidelity? How is it ultimately translated into the amino acid sequence of protein molecules?

Modern biochemical research on gene structure and function has brought to biology an intellectual revolution comparable to that evoked over a hundred years ago by Darwin's theory on the origin of species. Virtually every aspect of biology and medicine has been profoundly influenced by this new knowledge. Biochemical genetics has brought penetrating new insight into some of the most fundamental problems in the structure and function of cells. Moreover, it has led to a more comprehensive conceptual framework for the science of biochemistry.

Today's knowledge of the molecular aspects of genetics has arisen from the convergence of three different disciplines: *genetics*, *biochemistry*, and *molecular physics*. The contributions of these three fields are epitomized by the event that opened the modern era of genetic biochemistry. In 1953 James Watson and Francis Crick postulated the double-helical structure of DNA. Their hypothesis, which has been verified by many kinds of experiments, represented the confluence of genetic theory, which contributed the concept of coding by genes, of physics, which made possible determination of molecular structure by x-ray analysis, and of biochemical research, which revealed the chemical composition of DNA, particularly the pairing of nucleotides. The Watson-Crick hypothesis not only accounted for the structure of the DNA molecule but also showed how it could be replicated with precision. It soon led to the *central dogma* of molecular genetics (Figure 27-1), which defines three major steps in the processing of genetic information. The first is *replication*, the copying of parent DNA to form daughter DNA molecules having nucleo-

The central dogma of molecular genetics, showing the flow of genetic information via the three fundamental processes of replication, transcription, and translation. Later we shall see that the central dogma had to be modified.

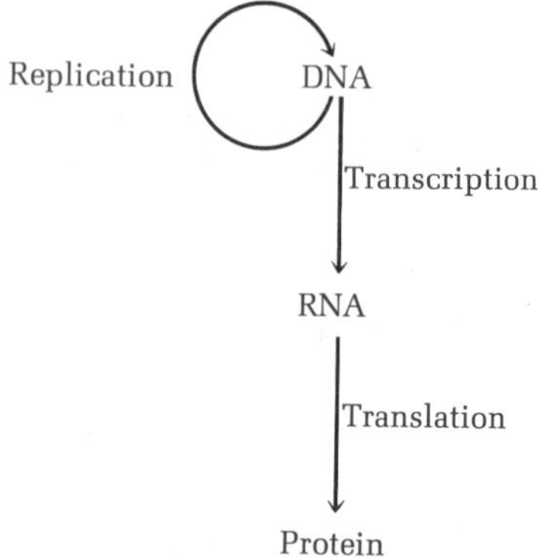

tide sequences identical to those of the parent DNA. The second step is *transcription*, the process in which parts of the genetic message in DNA are rewritten in the form of ribonucleic acid (RNA). The third step is *translation*, in which the genetic message coded by RNA is translated by the ribosomes into the 20-letter alphabet of protein structure.

In the following chapters we shall examine the successive steps in this sequence. The new element we will meet is the concept of molecular information storage and transmission, which we have touched upon only lightly in earlier sections of this book. First, we will examine the nature, dimensions, and conformation of the functional units of the genetic material of cells and viruses: chromosomes and genes. Then we shall examine the pathways and mechanisms of the extraordinarily complex enzyme systems responsible for replication and transcription of DNA. Here we will see that biosynthesis of information-rich DNA and RNA molecules requires dozens of different enzymes and specialized proteins, compared to the few enzymes required to make a noninformational macromolecule such as glycogen. Then we will examine the mechanism of protein biosynthesis, the most complex of all known biosynthetic pathways, which requires participation of well over 200 different enzymes and other specialized macromolecules, all needed to decipher and translate the symbols of the genetic code into the three-dimensional structure of proteins.

In the final chapter we will see that chromosomes and genes are not stable, inert structures. They may undergo mutation, sometimes to produce serious defects in the biological function of a protein, sometimes to produce a superior protein. Genes or sets of genes often undergo exchange and recombination to produce new sets of characteristics in progeny organisms. Moreover, pieces of genes are often exchanged and recombined, making possible, for example, the marvelously effective immune system that protects vertebrates from invasion by microbes and helps to preserve their species identity.

Our understanding of this area of biochemistry is growing at a dazzling pace. Scarcely a month goes by without some major advance or discovery. The "cracking" of the genetic code in the early 1960s has been followed by a seemingly endless succession of spectacular discoveries and achievements. Among these are determination of the nucleotide sequences of many genes, the laboratory synthesis of genes, the splicing of genes into new combinations, the incorporation of genes into cells of other species, and the use of such genetically altered "producer" cells to make many new proteins that are useful in different ways. Indeed, another new era in genetic biochemistry has begun, one that will surely affect many aspects of human health and endeavor in the years to come.

CHAPTER 27

DNA: The Structure of Chromosomes and Genes

Before we begin study of DNA as a repository of genetic information, it will be useful to consider again the nature of information. We have already seen that information represents order and is the opposite of entropy, which reflects disorder or randomness (Box 14-1, page 364); information has also been called "negative entropy." Thus we see that information is related to energy. In fact, it is possible to measure or quantitate information and relate it to entropy units and free energy, but this requires rather complex considerations of probability and statistics.

Today, in the era of electronic calculators and computers we are all aware of their labor-saving usefulness in storing, processing, and retrieval of information. In the language of digital computers, the unit of information is the *bit* (abbreviated from *binary digit*), which represents the amount of information needed to make a correct choice between two alternative possibilities. The information required to make two successive choices, each between two possibilities, is given by two bits; similarly, the information required to select the correct card from 16 cards, by a sequence of binary choices, would be four bits. In this way the digital computer can convert information into a series of binary choices and thus keep inventories, prepare payrolls, and even record symphonies.

But the amount and kind of information inherent in a single human cell is still far beyond the present programming art of the digital computer and still beyond the ability of the biochemist to provide the facts and relationships to be digitalized. The 20 amino acids from which all proteins are made are not simply 20 coding units, because any given amino acid may have different meanings in a protein. Serine, for example, in one "meaning" can furnish a polar hydroxyl group capable of hydrogen bonding, or an essential group in the active site of an enzyme (as in trypsin), or an essential group in a regulatory site (as in glycogen phosphorylase), or as a carrier of phosphate groups (as in the milk protein casein). To convert into digital language the four-unit language of DNA and the 20-unit language of proteins, in which the units may have multiple meanings, is not yet possible.

Perhaps a simpler way to illustrate the vast amount of information inherent in DNA is to return to the base sequence of the small virus ϕX174 on page 790, which has 5386 base pairs, requiring one page of very fine print. If the base sequence of the *E. coli* chromosome, which has 4 million base pairs, were printed, it would require about 740 pages of this book. If the entire base sequence of the 46 chromosomes of a human cell were printed in the same way, it would require over 820,000 pages, equivalent to 820 volumes of the size of this book. But all this print would be useless without full knowledge of the coding and programming principles involved in transcription, translation, and regulation of gene expression, which might require a great many additional volumes of information.

Let us now begin to examine the structure of DNA, the evidence that it stores genetic information, and the nature of the major functional units of the genetic material—chromosomes and genes.

DNA and RNA Perform Different Functions

First, for orientation, let us briefly survey the nature, function, and intracellular location of the major types of nucleic acids. DNAs are extremely long molecules, consisting of many thousands of deoxyribonucleotides of four different kinds, joined in a sequence that is characteristic for each organism. They are usually double-stranded. The chromosome of prokaryotic cells is a single large DNA molecule, tightly bunched into a *nuclear zone* or *nucleoid*. Recall that prokaryotes do not possess a membrane surrounding the genetic material (page 18).

Eukaryotic cells contain many DNA molecules, each generally much larger than the single DNA molecule in prokaryotes. The DNA molecules in eukaryotes are combined with proteins and organized into *chromatin* fibers in the nucleus, which is surrounded by a complex double-membrane system. DNA functions to store the complete genetic information required to specify the structure of all the proteins and RNAs of each species of organism, to program in time and space the orderly biosynthesis of cell and tissue components, to determine the activities of an organism throughout its life cycle, and to determine the individuality of a given organism.

Viruses (page 37) also contain nucleic acids as their genetic material; some contain DNA and some RNA (Table 27-1). Viral nucleic acids, which are relatively small compared with the DNAs of bacteria, code for the characteristic proteins found in viral particles and also for certain enzymes required for replication of the virus in its host cell.

Ribonucleic acids consist of long strings of *ribonucleotides*. Although they are much shorter than DNAs, they are much more abundant in most cells. In both prokaryotic and eu-

Table 27-1 Some Well-Known Viruses†

Bacterial viruses (bacteriophages)
- With DNA as chromosomes
  - ϕX174
  - λ (lambda)
  - T2
  - T4
- With RNA as chromosomes
  - f2
  - MS2
  - R17
  - Qβ

Animal viruses
- DNA
  - Simian virus 40 (SV40)
  - Mouse polyoma
  - Rabbit papilloma
  - Herpes simplex (human)
  - Adenovirus (human)
- RNA
  - Rous sarcoma (fowl)
  - Poliomyelitis
  - Influenza
  - Reovirus (human)

Plant viruses (RNA)
- Tobacco mosaic
- Tomato bushy stunt

† All the bacteriophages listed have *E. coli* as host.

karyotic cells the three major classes of RNA are *messenger RNA* (mRNA), *ribosomal RNA* (rRNA), and *transfer RNA* (tRNA). Each consists of a single strand of ribonucleotides, and each has a characteristic molecular weight, nucleotide sequence, and biological function (Table 27-2). *Messenger RNA* functions as the template used by ribosomes for the translation of the genetic information into the amino acid sequence of proteins. The nucleotide sequence of mRNA is complementary to the genetic message contained in a specific segment of the template strand of DNA. A single eukaryotic cell may contain over 10^4 different mRNA molecules, each coding for one or more different polypeptide chains.

Transfer RNAs also consist of a single strand of ribonucleotides but in a highly folded conformation. They have between 70 and 95 ribonucleotides, corresponding to molecular weights between 23,000 and 30,000. Each of the 20 amino acids found in proteins has one or more corresponding tRNAs to bind it, carry it to the ribosomes, and serve as an "adapter" for translating the genetic code words of the mRNA into the amino acid sequence of proteins. Each tRNA contains a specific trinucleotide sequence, called its *anticodon,* which is complementary to a *codon,* the trinucleotide sequence of mRNA that codes for one specific amino acid.

Ribosomal RNAs (rRNAs) are the major components of ribosomes and constitute up to 65 percent of their weight. There are three kinds of rRNA in prokaryotic ribosomes (Table 27-2), whereas eukaryotic ribosomes contain four kinds, as we shall see in Chapter 29. Ribosomal RNAs play an important role in the structure and biosynthetic function of ribosomes.

In eukaryotic cells, there are two additional kinds of RNAs: *heterogeneous nuclear RNAs* (hnRNAs), which are nuclear precursors of mRNAs, and *small nuclear RNAs* (snRNAs), which participate in RNA processing.

Now let us examine the structure of nucleic acids in more detail, beginning with their building blocks, the *nucleotides.*

Table 27-2 Properties of E. coli RNAs

Type	S†	MW	Approximate number of nucleotide residues	Percentage of total cell RNA
mRNA	6–25	25,000–1,000,000	75–300	~2
tRNA	~4	23,000–30,000	73–93	16
rRNA	5	~35,000	~100	82
	16	~550,000	~1500	
	23	~1,100,000	~3100	

† S stands for the svedberg, a unit denoting the rate of sedimentation of a macromolecule in a centrifugal field. It is related to molecular size. It was named after The Svedberg, the Swedish physical chemist who invented the ultracentrifuge.

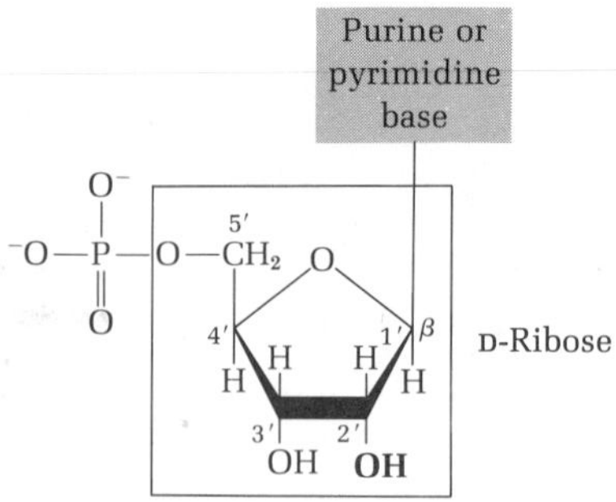

Figure 27-1
The general structure of nucleotides. The structure shown is that of a ribonucleotide. In deoxyribonucleotides the —OH group in color is replaced with —H. The carbon atoms of the pentose are numbered 1′, 2′, etc., to distinguish them from the numbering of atoms in the base.

Figure 27-2
The principal pyrimidine and purine bases of DNA and RNA.

	DNA	RNA
Purines	Adenine	Adenine
	Guanine	Guanine
Pyrimidines	Cytosine	Cytosine
	Thymine	Uracil

The Nucleotide Units of DNA and RNA Have Characteristic Bases and Pentoses

Recall from Chapter 3 that nucleotides contain three characteristic components: (1) a nitrogenous base, (2) a pentose, and (3) phosphoric acid, linked as shown in Figure 27-1, so that the base is covalently joined in an *N*-glycosyl linkage to carbon atom 1′ of the pentose and the phosphoric acid is esterified to carbon 5′. The nitrogenous bases are derivatives of two parent heterocyclic compounds, *pyrimidine* and *purine* (Figure 27-2). DNA contains two principal pyrimidine bases, *cytosine* (C) and *thymine* (T), and two principal purine bases, *adenine* (A) and *guanine* (G). RNAs also contain two principal pyrimidines, *cytosine* (C) and *uracil* (U), and two principal purine bases, *adenine* (A) and *guanine* (G). The single important difference in the bases of DNA and RNA is that thymine is a principal pyrimidine in DNAs but does not occur often in RNAs; conversely, uracil is a principal pyrimidine in RNAs but occurs only rarely in DNAs. The pyrimidine and purine bases are nearly flat molecules that are relatively insoluble in water (Figure 27-3).

Two kinds of pentoses are found in nucleic acids. The recurring deoxyribonucleotide units of DNA contain *2′-deoxy-D-ribose,* and the ribonucleotide units of RNA contain *D-ribose.* Both pentoses occur in nucleotides in their β-furanose forms.

Figure 27-4 gives the structures and names of the four principal *deoxyribonucleotides* (deoxyribonucleoside 5′-monophosphates) functioning as the structural and coding units of DNAs and the four principal *ribonucleotides* (ribonucleoside 5′-monophosphates), the coding units of RNAs. Specific long sequences of A, T, G, and C bases in DNA encode the genetic information. In addition to nucleotides bearing the four principal or *major* bases, DNA also contains some *minor* bases. Usually the minor bases are methylated forms of the principal bases, but in some viral DNAs certain bases may be hydroxymethylated or glucosylated. Such altered or unusual bases in DNA molecules are in many cases specific signals that play important roles in programming or protecting the genetic information. Minor bases are also found in RNAs, especially in tRNAs.

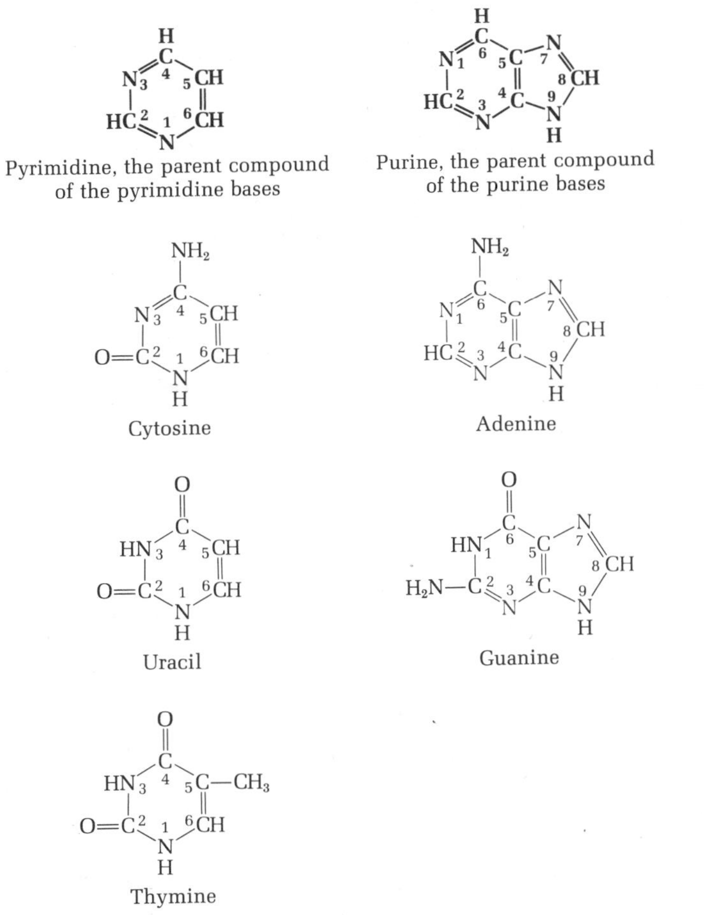

Figure 27-3
The structures of the parent compounds pyrimidine and purine, as well as the principal pyrimidine and purine bases of nucleic acids.

Pyrimidine, the parent compound of the pyrimidine bases

Purine, the parent compound of the purine bases

Cytosine

Adenine

Uracil

Guanine

Thymine

Phosphodiester Bonds Link the Successive Nucleotides of Nucleic Acids

The successive nucleotides of both DNA and RNA are covalently linked to each other through phosphate-group "bridges." The 5′-hydroxyl group of the pentose of one nucleotide unit is joined to the 3′-hydroxyl group of the pentose of the next nucleotide by a *phosphodiester linkage* (Figure 27-5). Thus the covalent backbones of nucleic acids consist of alternating phosphate and pentose groups, whereas the characteristic bases may be regarded as side groups joined to the backbone at regular intervals. Also note that the backbones of both DNA and RNA are highly polar, since the phosphate groups are acidic and have negative charges at the pH of the cell. On the other hand, the purine and pyrimidine bases, which are relatively insoluble in water, are hydrophobic. We also note that DNA and RNA strands have a specific *polarity,* or direction, because all the internucleotide phosphodiester linkages have the same orientation along the chain (Figure 27-5). Because of this polarity, each linear nucleic acid strand has a *5′ end* and a *3′ end*.

Figure 27-4
(a) *The deoxyribonucleotide units of DNA in free form at pH 7.0. They are usually symbolized as* A, G, T, *and* C *(or, less often,* dA, dG, dT, *and* dC*) when they are part of the DNA molecule. In free form these nucleotides are commonly abbreviated* dAMP, dGMP, dTMP, *and* dCMP. (b) *The ribonucleotide units of RNAs. In both cases the abbreviations assume that the phosphate group is at the* 5′ *position.*

(a)

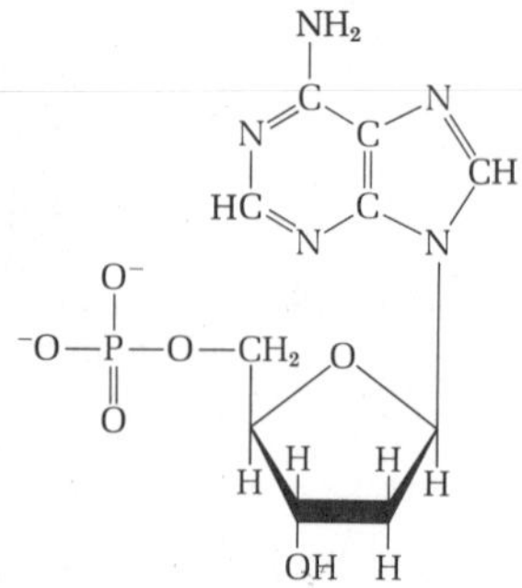

Deoxyadenylate, deoxyadenosine 5′-monophosphate
Symbols: A, dAMP

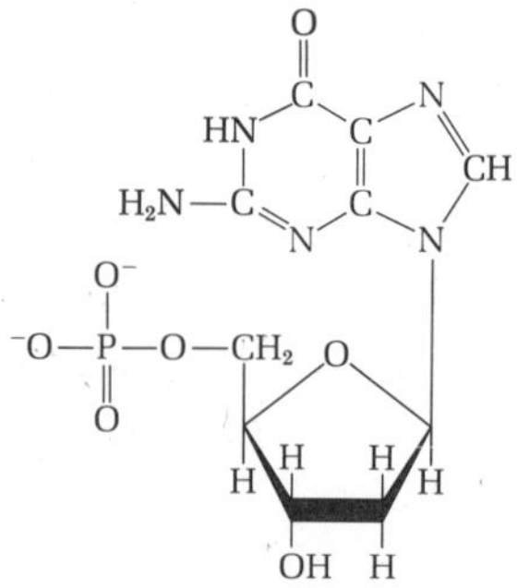

Deoxyguanylate, deoxyguanosine 5′-monophosphate
Symbols: G, dGMP

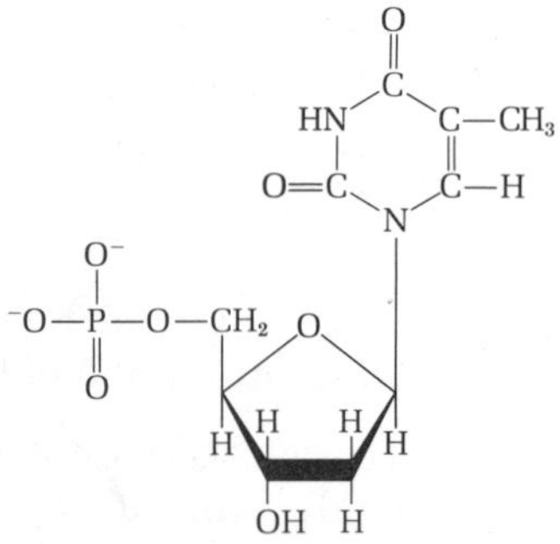

Deoxythymidylate, deoxythymidine 5′-monophosphate
Symbols: T, dTMP

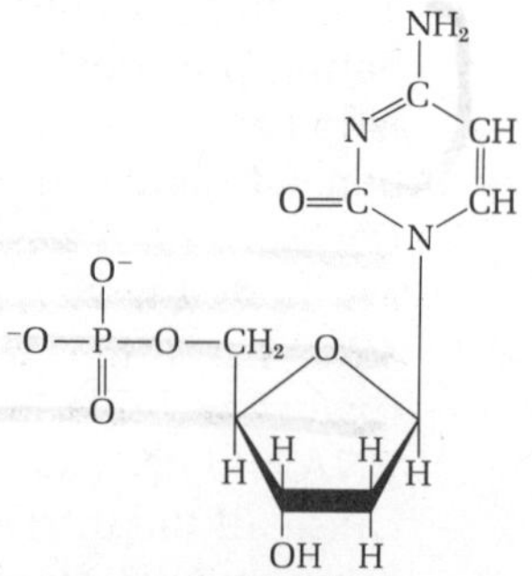

Deoxycytidylate, deoxycytidine 5′-monophosphate
Symbols: C, dCMP

(b)

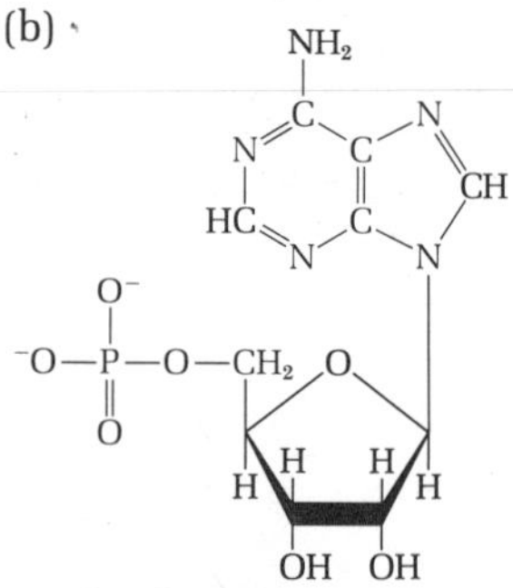

Adenylate, adenosine 5′-monophosphate
Symbols: A, AMP

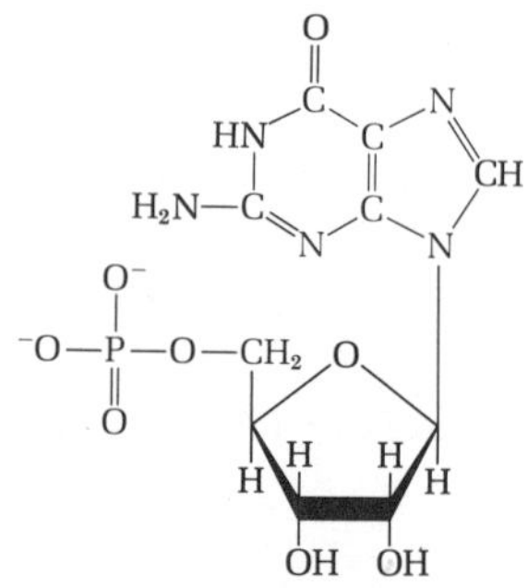

Guanylate, guanosine 5′-monophosphate
Symbols: G, GMP

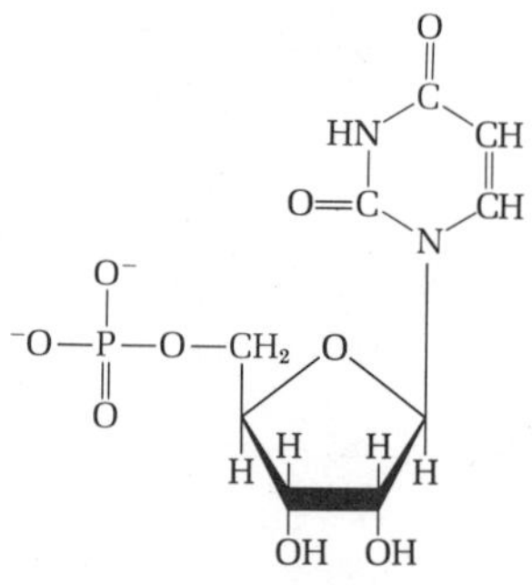

Uridylate, uridine 5′-monophosphate
Symbols: U, UMP

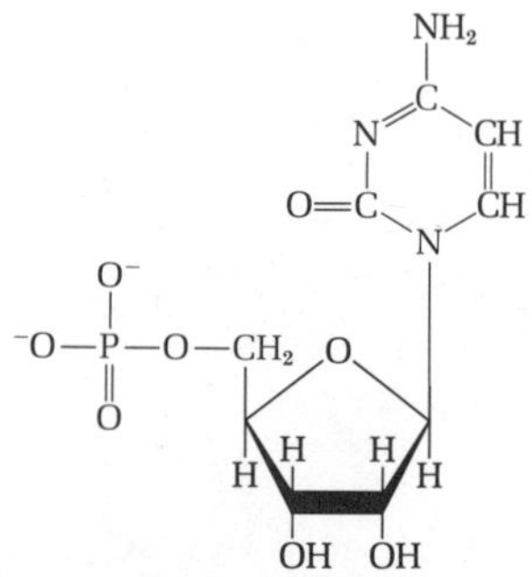

Cytidylate, cytidine 5′-monophosphate
Symbols: C, CMP

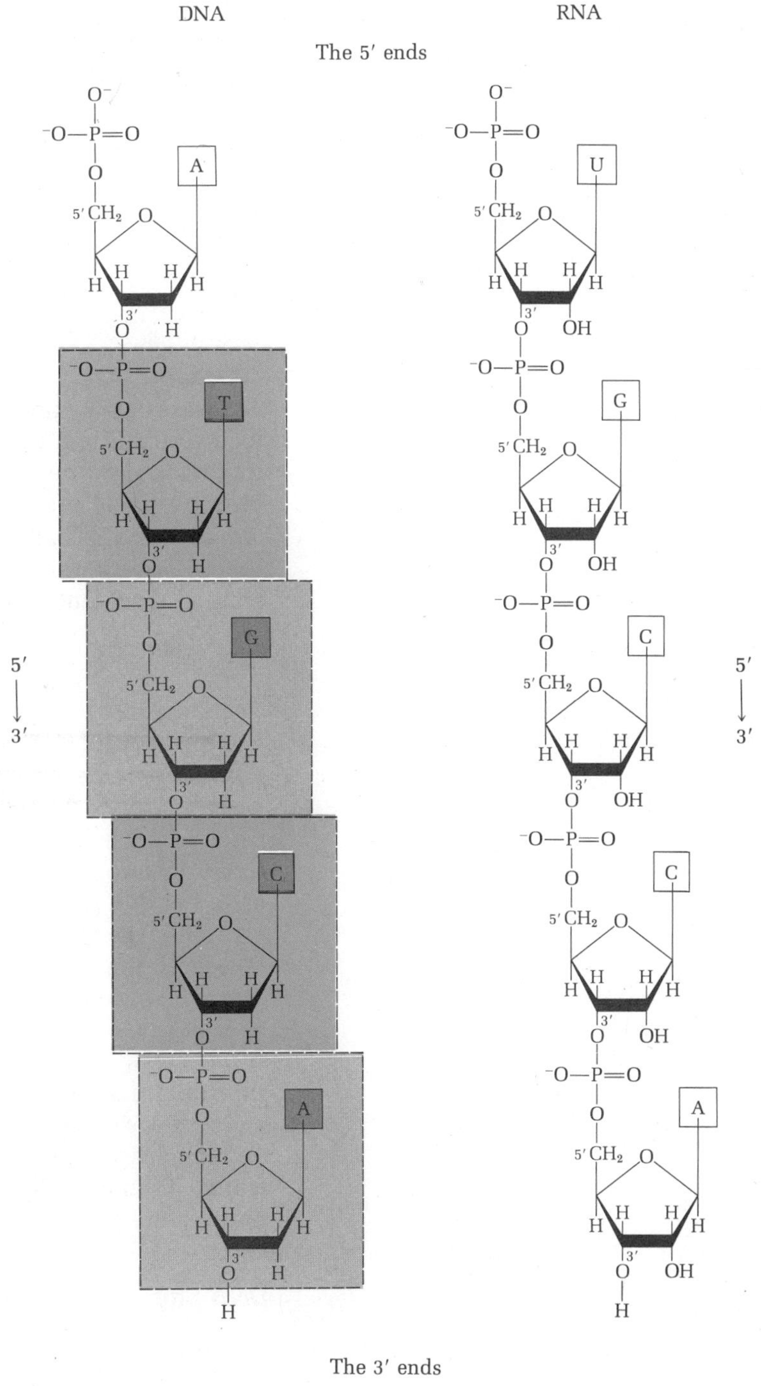

Figure 27-5
The covalent backbone structures of DNA and RNA, showing the phosphodiester bridges linking the successive nucleotide units (shaded in the DNA). Note that the backbone of alternating pentose and phosphate groups of both DNA and RNA is highly polar, whereas the bases are nonpolar and hydrophobic.

DNA

RNA

The 5′ ends

5′ → 3′

5′ → 3′

The 3′ ends

The nucleotide sequences of nucleic acids can be represented schematically, as illustrated by a segment of DNA having five nucleotide units

A T G C A

(P) (3′) (5′) (P) (3′) (5′) (P) (3′) (5′) (P) (3′) (5′) (P) (3′) (5′) OH

5′ ⟶ 3′

The bases are symbolized by A, T, G, and C, each deoxyribose by a vertical line, and the phosphate groups by (P). The small numerals give the positions in the deoxyribose units to which the phosphate groups are attached. The structure of a single strand of DNA is always written with the 5′ end at the left and the 3′ end at the right, i.e., in the 5′ → 3′ direction. Two simpler representations of the above pentadeoxyribonucleotide are

pApTpGpCpA and pATGCA

The internucleotide linkages of both DNA and RNA can be chemically cleaved by hydrolysis. They can also be enzymatically hydrolyzed by enzymes called *nucleases*. Some nucleases can hydrolyze linkages between two adjacent nucleotides at *internal* positions in the DNA or RNA strand; such nucleases are called *endonucleases*. Another class of nucleases can hydrolyze only the *terminal* nucleotide linkage, some at the 5′ and others at the 3′ end; these are *exonucleases*. *Deoxyribonucleases*, specific for hydrolysis of certain internucleotide linkages of DNAs, and *ribonucleases*, specific for RNAs, are found in all cells. They are also secreted by the pancreas into the intestinal tract, where they participate in the hydrolysis of nucleic acids during digestion. We shall see that different types of endonucleases are important biochemical tools for the controlled fragmentation of DNA and RNA into smaller pieces, preparatory to determination of their nucleotide sequences.

DNA Stores Genetic Information

The history of DNA begins with Friedrich Miescher, a Swiss biologist, who carried out the first systematic chemical studies of cell nuclei. In 1868 Miescher isolated a phosphorus-containing substance he called *nuclein* from the nuclei of pus cells obtained from discarded surgical bandages. (Pus cells are white blood cells, or leukocytes.) Miescher found nuclein to consist of an acidic portion, which we know today as DNA, and a basic portion, contributed by protein. He later found a similar substance in the heads of salmon sperm cells. Although he separated the nucleic acid fraction and studied its properties, the covalent structure of DNA as shown in Figure 27-5 did not become known with certainty until the late 1940s.

Although Miescher and many others following him suspected that nuclein or nucleic acid had something to do with cell inheritance, the first direct evidence that DNA is the bearer of genetic information came in 1943, through a discovery made by Oswald T. Avery, Colin MacLeod, and Maclyn McCarty at the Rockefeller Institute. They found that DNA extracted from a virulent (disease-causing) strain of the bacterium *Streptococcus pneumoniae*, also known as pneumococcus, permanently transformed a nonvirulent strain of this organism into a virulent form (Figure 27-6). Avery and his colleagues con-

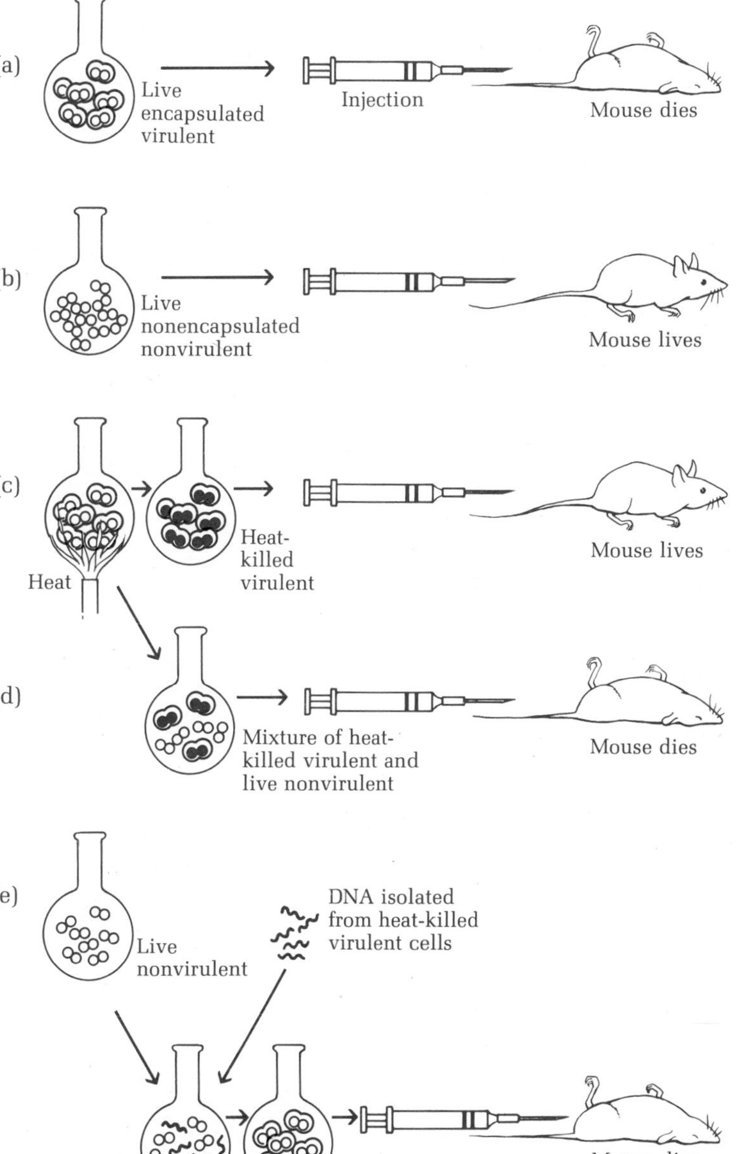

Figure 27-6
The Avery-MacLeod-McCarty experiment. When injected into mice, (a) the encapsulated strain of pneumococcus is lethal, while (b) the nonencapsulated strain is harmless, as is (c) the heat-killed encapsulated strain. (d) Earlier research by Frederick Griffith, a bacteriologist, had shown that adding heat-killed virulent cells, which alone are harmless to mice, to live nonvirulent pneumococci permanently transformed them into virulent encapsulated cells. He concluded that a transforming factor was present in the heat-killed virulent cells, which gained entrance into live nonvirulent cells and rendered them virulent and encapsulated.

Avery and his colleagues identified the Griffith transforming factor as DNA. (e) They extracted the DNA from heat-killed virulent pneumococci, removed the protein from the DNA as completely as possible, and added this DNA to nonvirulent cells. The nonvirulent pneumococci were permanently converted into a virulent strain. The DNA evidently gained entrance into the nonvirulent cells, and the genes for virulence and capsule formation became incorporated into the chromosomes of the nonvirulent cells.

cluded that the DNA extracted from the virulent strain carried the heritable genetic message for virulence and that it became permanently incorporated into the DNA of the recipient non-virulent cells. At first not everyone accepted these conclusions. Some critics suggested that traces of protein impurities present in the DNA may have been the actual carrier of the genetic information in Avery's experiments. This possibility was soon eliminated by the finding that treatment of the DNA with proteolytic enzymes did not destroy the transforming activity but deoxyribonucleases did.

Later a second important experiment provided independent evidence that DNA carries genetic information. In 1952 Alfred D. Hershey and Martha Chase showed by means of isotope tracer experiments that when bacteriophage T2 infects its host cell, *E. coli*, it is the DNA of the T2 viral particle, and not its protein part, that actually enters the host cell and furnishes the genetic information for the replication of the virus (Figure 27-7).

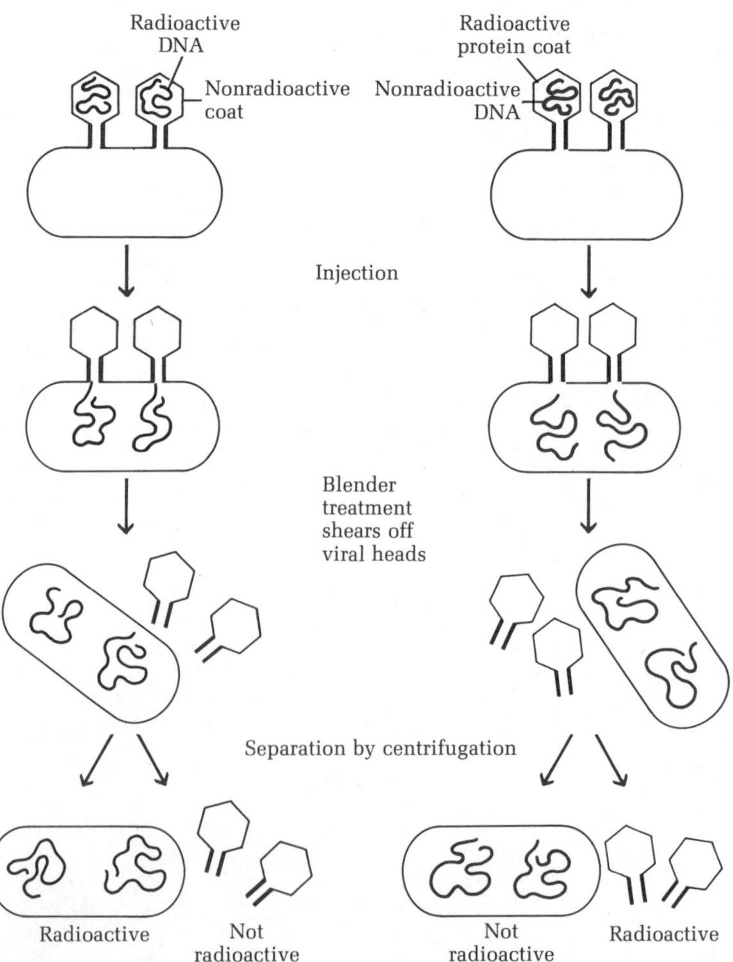

Figure 27-7
Summary of the Hershey-Chase experiment. Two batches of isotopically labeled bacteriophage particles were prepared. One was labeled with ^{32}P in the phosphate groups of the phage DNA and the other with ^{35}S in the sulfur-containing amino acids of the protein coat of the phage. Both types of labeled phage were then added to suspensions of unlabeled bacteria. The two batches of phage-infected cells were then agitated in a blender. The cells infected with the ^{32}P-labeled virus particles were found to contain ^{32}P, indicating that the labeled viral DNA was injected. The separated virus ghosts (empty protein coats) contained no radioactivity. The cells infected with ^{35}S-labeled viral particles were found to have no radioactivity after blender treatment, but the viral ghosts released by the blender contained the ^{35}S. Since in both cases progeny virus particles were found, the genetic message for their replication was introduced by the viral DNA, not by the viral protein.

From these important early experiments and many other lines of evidence developed since, it is now completely certain that DNA is the chromosomal material bearing the genetic information of living cells.

DNAs of Different Species Have Distinctive Base Compositions

A most important clue to the structure of DNA came from a discovery made by Erwin Chargaff and his colleagues at Columbia University in the late 1940s. They found that the four bases occur in different ratios in the DNAs of different organisms and that the bases have a numerical relationship to each other. Table 27-3 shows the relative amounts of the four bases (A, T, G, and C) of DNA specimens isolated from representative species. Such data, collected from DNAs of a great many different species, led Chargaff and subsequent investigators to the following conclusions:

1. DNA specimens isolated from different tissues of the same species have the same base composition.
2. The base composition of DNA varies from one species to another.
3. The base composition of DNA in a given species does not change with age of the organism, its nutritional state, or changes in its environment.

Table 27-3 Base Equivalences in DNAs. Note that the base composition in humans and the bacterium *S. aureus* is almost the same. Base composition, although fixed in each species, is not distinctively different for all species.

	Base composition, mol %				Base ratio		
Organism	A	G	C	T	A/T	G/C	Purines/Pyrimidines
Homo sapiens	30.9	19.9	19.8	29.4	1.05	1.00	1.04
Sheep	29.3	21.4	21.0	28.3	1.03	1.02	1.03
Hen	28.8	20.5	21.5	29.3	1.02	0.95	0.97
Turtle	29.7	22.0	21.3	27.9	1.05	1.03	1.00
Salmon	29.7	20.8	20.4	29.1	1.02	1.02	1.02
Sea urchin	32.8	17.7	17.3	32.1	1.02	1.02	1.02
Locust	29.3	20.5	20.7	29.3	1.00	1.00	1.00
Wheat germ	27.3	22.7	22.8	27.1	1.01	1.00	1.00
Yeast	31.3	18.7	17.1	32.9	0.95	1.09	1.00
E. coli	24.7	26.0	25.7	23.6	1.04	1.01	1.03
Staphylococcus aureus	30.8	21.0	19.0	29.2	1.05	1.11	1.07
Phage T7	26.0	24.0	24.0	26.0	1.00	1.00	1.00
Phage λ	21.3	28.6	27.2	22.9	0.92	1.05	0.79
Phage φX174 (replicative)	26.3	22.3	22.3	26.4	1.00	1.00	1.00

4. The number of adenine residues in all DNAs, regardless of the species, is equal to the number of thymine residues (that is, A = T), and the number of guanine residues is always equal to the number of cytosine residues (G = C). From these relationships it follows that the sum of the purine residues equals the sum of the pyrimidine residues; that is, A + G = T + C.

These quantitative relationships, confirmed by many subsequent researchers, not only became important in establishing the three-dimensional structure of DNA but also yielded clues to how genetic information is encoded in DNA and passed from one generation to the next.

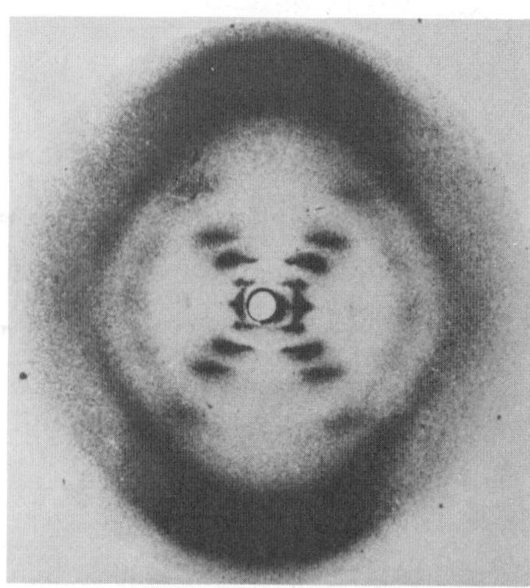

Figure 27-8
The x-ray diffraction pattern of DNA. The spots forming a cross in the center denote a helical structure. The heavy bands at the top and bottom correspond to the recurring bases.

Watson and Crick Postulated a Double-Helix Model of DNA Structure

We have seen how the powerful method of x-ray diffraction analysis revealed the structures of fibrous and globular proteins (Chapters 7 and 8). X-ray studies of the structure of DNA fibers, made by Rosalind Franklin and Maurice Wilkins, yielded a characteristic diffraction pattern (Figure 27-8). From this pattern it was deduced that DNA fibers have two periodicities along their long axis, a major one of 0.34 nm and a secondary one of 3.4 nm. The problem was to formulate a three-dimensional model of the DNA molecule that could account not only for these periodicities but also for the specific A = T and G = C base equivalences discovered by Chargaff.

Figure 27-9
Watson and Crick with one of their models of DNA, in a photograph taken in 1953.

In 1953 James Watson, an American geneticist, and Francis Crick, a British physicist (shown in Figure 27-9), working at the University of Cambridge, postulated a three-dimensional model of DNA structure that accounted for both the x-ray data and the characteristic base pairing in DNA (Figure 27-10). It consists of two helical DNA chains coiled around the same axis to form a right-handed double helix. In the helix the two chains or strands are *antiparallel;* i.e., their 5′,3′ internucleotide phosphodiester bridges run in opposite directions. Their hydrophilic backbones, which consist of alternating deoxyribose and negatively charged phosphate groups, are on the outside of the double helix, facing the surrounding water. The hydrophobic purine and pyrimidine bases of both strands are stacked inside the double helix, so that the nearly planar base molecules are very close together and perpendicular to the long axis of the double helix. The spatial relationship between these strands creates a *major groove* and a *minor groove* between the two strands. The bases of one strand are paired in the same planes with the bases of the other strand. Only certain base pairs fit precisely inside this structure. The allowed pairs are always a purine and a pyrimidine, specifically the pairs A-T and G-C, which are the base pairs found by Chargaff to occur in exact equivalence in DNA specimens (Table 27-3). Moreover,

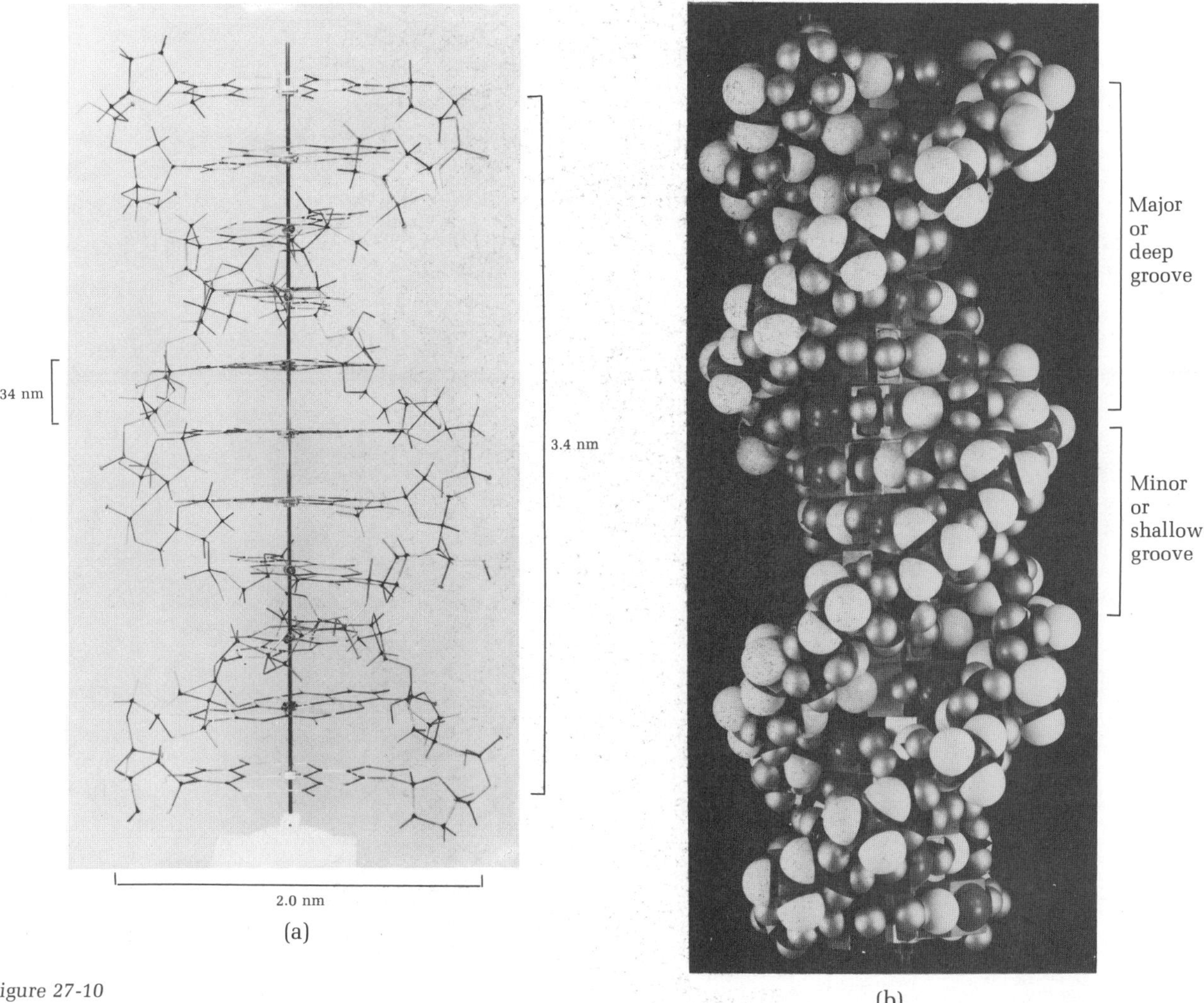

Figure 27-10
The Watson-Crick model for the structure of DNA. (a) *Wire model of the backbone.* (b) *Space-filling model.*

the bases of each pair are close enough for their edges to hydrogen-bond to each other. How hydrogen bonds form between adenine and thymine and between guanine and cytosine is shown in Figure 27-11. It is very important to note that *three* hydrogen bonds can form between G and C, symbolized G≡C, but only *two* can form between A and T, symbolized A=T. Other pairings of bases do not fit the double-helix structure. A base pair made up of two purines (A and G) would be too large to fit inside a helix having these dimensions, and the bases in a C-T pair would be too far apart to form stable hydrogen bonds. Furthermore, A cannot hydrogen bond with C or G with T and still fit within the helix.

To account for the periodicities observed by the x-ray analysis, Watson and Crick showed with the use of molecular models that the vertically stacked bases inside the double helix

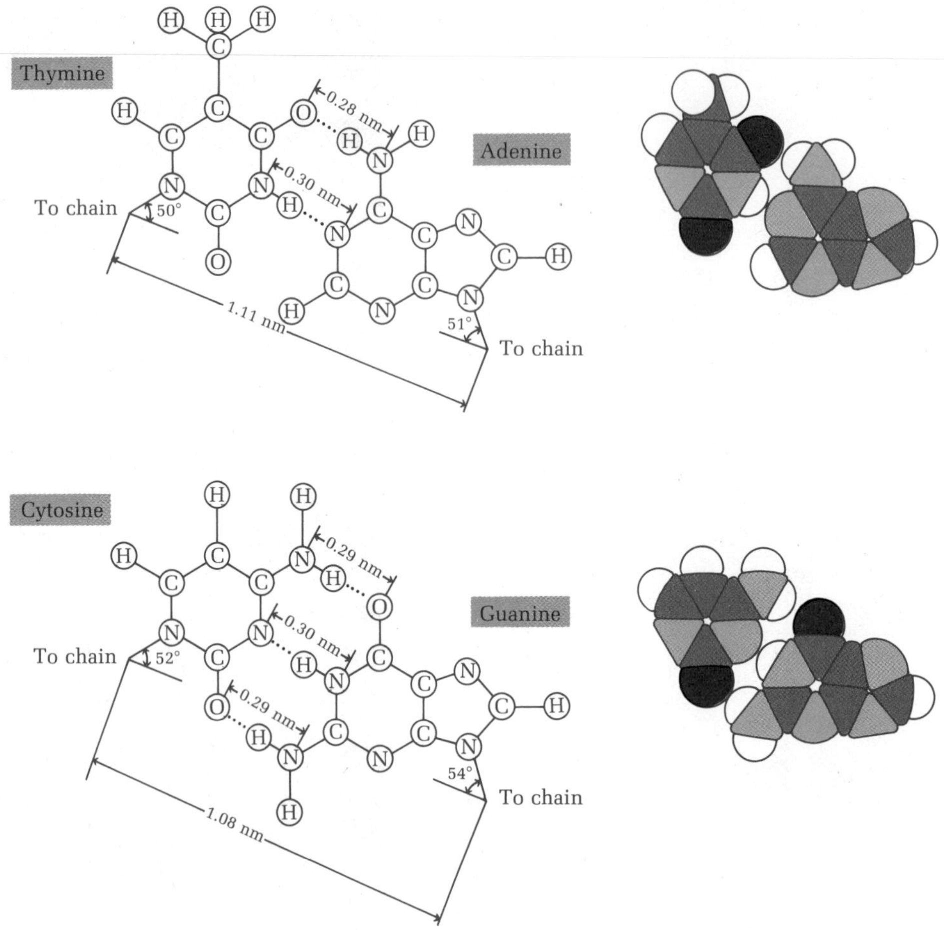

Figure 27-11
Scale drawings and space-filling models of the hydrogen-bonded base pairs adenine-thymine and guanine-cytosine. The former has two hydrogen bonds, the latter three. Guanine-cytosine pairs are slightly denser than adenine-thymine pairs.

would be 0.34 nm apart. The models also showed that the secondary repeat distance of about 3.4 nm could be accounted for, since there are about 10 nucleotide residues in each complete turn of the double helix (Figure 27-10). The double helix is about 2 nm thick. It is very important to note that the two antiparallel polynucleotide chains of double-helical DNA are *not* identical in either base sequence or composition, as can be seen in Figure 27-12. Instead they are *complementary* to each other. Wherever adenine appears in one chain, thymine is found in the other; similarly, wherever guanine is found in one chain, cytosine is found in the other.

The DNA double helix, or *duplex*, as it is often called, is held together by two sets of forces: (1) hydrogen bonding between complementary base pairs (Figure 27-11) and (2) hydrophobic interactions, which cause the stacked bases to be largely hidden within the double helix, shielded from water, and the highly polar backbones on the outside, exposed to water. The hydrophobic interactions make the major contribution to the stability of the double helix, as they also do for the tertiary

Figure 27-12
Schematic drawing of complementary antiparallel strands of DNA proposed by Watson and Crick. Note that base-paired antiparallel strands differ in base composition. They also differ in sequence when each chain is read in the 5′ → 3′ direction. Note the base equivalences: A = T and G = C.

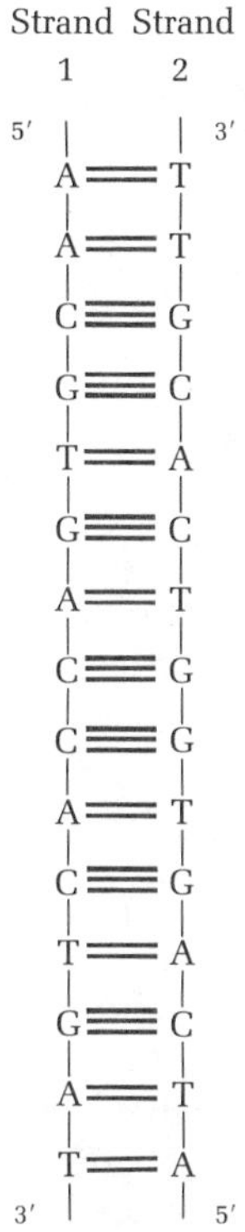

Strand 1 Strand 2

5′ 3′
A=T
A=T
C≡G
G≡C
T=A
G≡C
A=T
C≡G
C≡G
A=T
C≡G
T=A
G≡C
A=T
T=A
3′ 5′

Base compositions
Strand 1: A_5 T_3 G_3 C_4
Strand 2: A_3 T_5 G_4 C_3

structure of globular proteins (page 179). Note that all the phosphate groups in the polar backbones of the double helix are ionized and negatively charged at pH 7; thus DNA is strongly acidic.

Much evidence, both chemical and biological, indicates that the double-helix model of DNA structure is essentially correct. Now let us see how this structure allows for accurate replication of genetic information.

The Base Sequence of DNA Constitutes a Template

DNAs are very long molecules containing specific sequences of the four principal bases A, T, G, and C, which are the symbols for coding the genetic information. We speak of the sequence of bases in DNA as providing a template for DNA replication. However, it is important to understand why we need templates for the precise replication, transcription, and translation of genetic information.

In the biosynthesis of the noninformational macromolecule glycogen, which has only a single type of recurring unit, D-glucose, the identity and purity of the end product is guaranteed by the active site of glycogen synthase (page 572). This enzyme is substrate-specific: its active site can accept only a UDP-glucose molecule (page 572) and the nonreducing end of the glycogen chain that is to be lengthened. In principle, the active site (of this and all enzymes, for that matter) may be regarded as a template (the word means a pattern or mold), since there is a complementary "fit" between the substrate molecule(s) and the active site.

However, the specific sequence of the coding units in DNA, RNA, and polypeptides cannot be provided by an enzyme active site alone. Active sites of enzymes are relatively small and can bind only one or a few building-block molecules at a time in the proper positions to specify that they will be assembled in the proper sequence. However, nucleic acid molecules are so large, with thousands or millions of nucleotide units, that enzyme active sites are simply too small to specify the complete sequence in which the nucleotide units are to be assembled. One strand of DNA must therefore serve as the template for the base sequence of its complementary partner.

The double-helical structure of DNA leads to another aspect of the Watson-Crick hypothesis, namely, a way in which genetic information can be accurately replicated (Figure 27-13). Since the two strands of double-helical DNA are structurally complementary to each other, they contain complementary information in their base sequences. Replication of DNA during cell division was postulated by Watson and Crick to begin by separation of the two strands, each becoming the template specifying the base sequence of a new complementary strand made by the replicating enzymes. The fidelity of replication of each DNA strand, they postulated, would be guaranteed

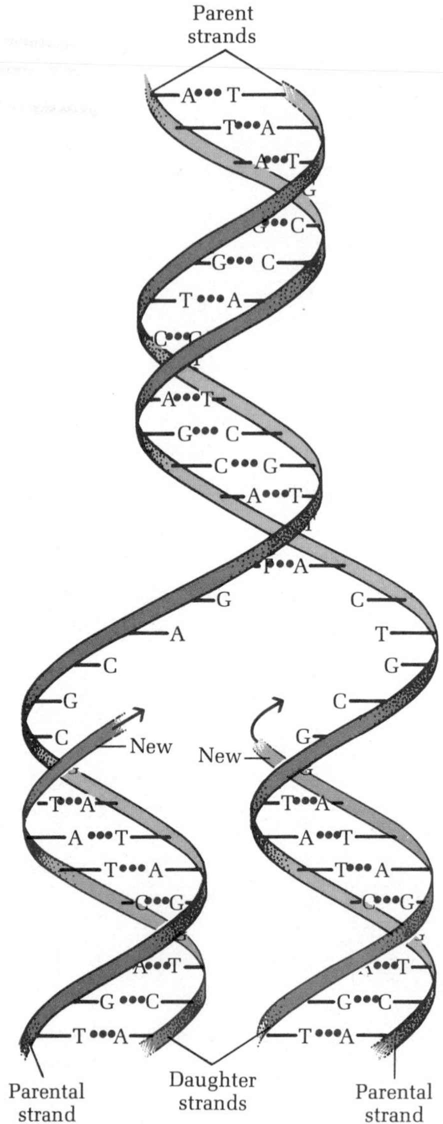

Figure 27-13
Replication of DNA as suggested by Watson and Crick. The complementary parent strands become separated, and each forms the template for biosynthesis of a complementary daughter strand (color).

by the fit and stability of the complementary base pairs A=T and G≡C in the two daughter duplexes, each of which would contain one strand from the parent DNA and a new strand complementary to the parent strand. It was further proposed that each newly formed daughter double helix would pass intact into a daughter cell. In Chapter 28 we shall see how this hypothesis was verified experimentally.

Double-Helical DNA May Undergo Denaturation or Unwinding

Now let us examine some chemical and physical properties of DNA in the light of its double-helix structure. Solutions of carefully isolated DNA are highly viscous at pH 7.0 and room tem-

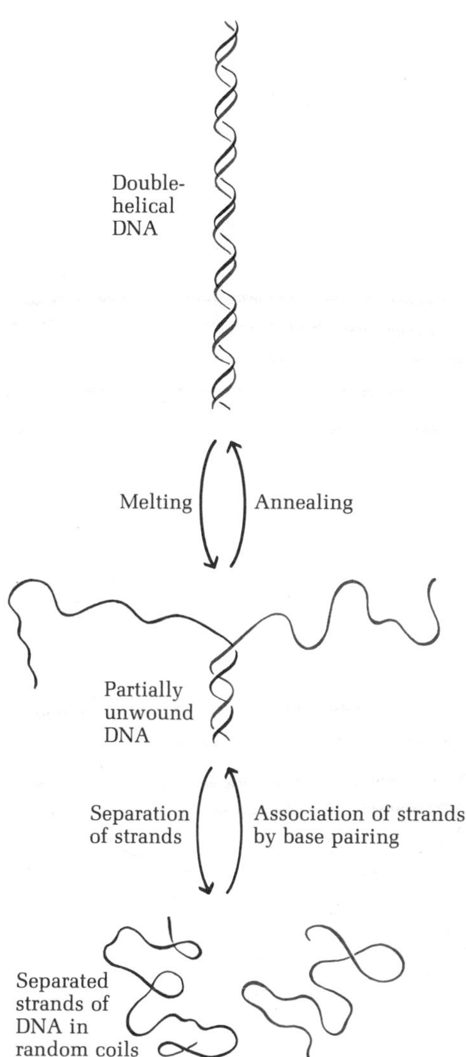

Figure 27-14
Stages in the reversible denaturation of DNA *and its reannealing.*

perature (20 to 25°C). When such a solution is subjected to extremes of pH or to temperatures above 80 to 90°C, its viscosity decreases sharply, indicating that the DNA undergoes a physical change. We have already seen that heat and extremes of pH cause denaturation or unfolding of globular proteins (page 140). In the same way high temperatures or extremes of pH cause denaturation or unfolding of double-helical DNA, by disrupting the hydrogen bonds between the paired bases and the hydrophobic interactions that hold the stacked bases together. As a result, the double helix unwinds into random, disordered coils until, finally, the two strands separate completely from each other. During denaturation of DNA, also called *melting*, no covalent bonds in the backbone are broken (Figure 27-14).

Denaturation of homogeneous DNA is readily reversible as long as a double-helical segment of a dozen or more residues unites the two strands, with base pairs in register. When the temperature or pH is brought back into the biological range, the unwound segments of the two strands will spontaneously rewind or *anneal* to yield the intact duplex (Figure 27-14). However, if the two strands are completely separated, renaturation occurs in two stages. The first stage is relatively slow, since the two strands must first "find" each other by random collisions and form a short segment of complementary double helix. The second stage is much faster, as the remaining bases successively come into register and form base pairs. The two strands then "zipper" themselves together to form the double helix again.

DNA Strands from Two Different Species Can Form DNA-DNA Hybrids

If duplex DNAs isolated from human cells and from mouse cells are heated separately, in order to denature them, and then are mixed and kept at 65°C for many hours, most of the mouse DNA strands will reanneal with complementary mouse DNA strands to form mouse duplex DNA; similarly, most of the human DNA strands will reanneal with complementary human DNA strands. In addition, however, a few of the mouse DNA single strands will associate with human DNA single strands to yield *hybrid duplexes,* in which segments of the mouse DNA strand will form base-paired regions with segments of a human DNA strand. Only limited portions of mouse DNA and human DNA will base-pair in this manner; the remainder of the bases will be unable to pair. Formation of hybrid duplexes can occur only if there is some sequence similarity in the DNA of two different species. The more closely two species are related, the more extensively will their DNAs hybridize with each other. For example, DNA of the human hybridizes much more extensively with mouse DNA than it does with DNA from yeast.

Figure 27-15
Principle of the hybridization test. Two DNAs from different species are heated to denature them completely, with separation of strands. When they are mixed and slowly cooled, complementary DNAs of each species will find each other and reanneal to form normal duplexes. But if the two DNAs have significant sequence homology, they will tend to form partial duplexes or hybrids with each other. The greater the sequence homology between two DNAs, the greater the number of hybrids formed. Hybrid formation can be measured by different procedures, e.g., chromatography or density-gradient centrifugation. Usually one of the DNAs is labeled with a radioactive isotope to simplify the measurements.

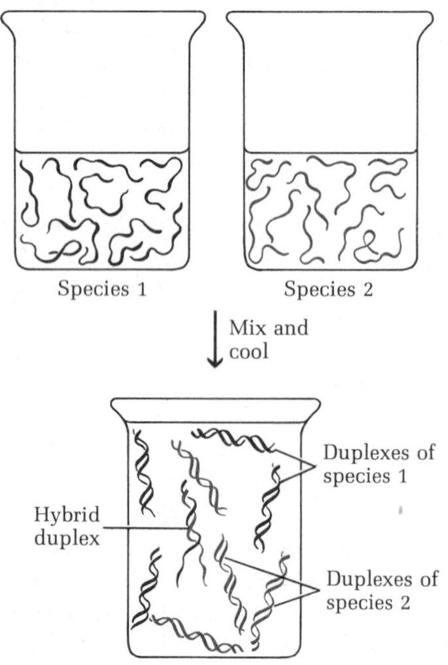

Hybridization tests (Figure 27-15) are very powerful tools for the study of many aspects of genetic biochemistry. Not only can they determine how closely two species are related but they can also test the relationship between a given DNA and an RNA, through DNA-RNA hybridization. Hybridization has also been used to isolate and purify genes and RNAs.

Some Physical Properties of Double-Helical DNAs Reflect the Proportion of G≡C versus A=T Pairs

Viral or bacterial DNA molecules denature rather precisely at characteristic temperatures when their solutions are slowly heated (Figure 27-16). The transition from the native duplex DNA into the unwound randomly coiled denatured form can be detected by an increase in the absorption of ultraviolet light or a decrease in the viscosity of the DNA solution. Each species of DNA has a characteristic denaturation temperature or "melting point." The higher its content of G≡C base pairs, the higher the melting point of the DNA. This is because G≡C base pairs are more stable and require more heat energy to dissociate than A=T base pairs, in part because G≡C pairs have three hydrogen bonds, whereas A=T pairs have only two. Careful determination of the melting point of a DNA specimen, under fixed conditions of pH and ionic strength, can therefore yield an estimate of its base composition.

A second physical property of DNA determined by the relative proportion of G≡C versus A=T base pairs is its *buoyant density*. A DNA specimen having a relatively high content of G≡C base pairs has a slightly greater density than a DNA with a high content of A=T pairs. DNA specimens are centrifuged at high speeds in a concentrated solution of cesium chloride (CsCl), which has a specific gravity in the same range as that of DNAs. The CsCl forms a density gradient in the centrifuge tube, with the highest density at the bottom of the tube. The DNA in the CsCl solution comes to equilibrium at a point in the tube where it will remain stationary, i.e., buoyant. It will neither rise nor settle, since its density will be equal to the density of the CsCl solution at that point. With this method, described more fully in Chapter 28, it is possible to separate DNA molecules of differing G≡C contents from each other since they will have different buoyant densities. From its buoyant density, the proportions of G≡C and A=T pairs in a given DNA can be calculated.

Native DNA Molecules Are Very Fragile

Early measurements indicated that DNAs had molecular weights of 10 million or less, equivalent to about 15,000 base pairs. But with improved methods for isolation of native DNAs their molecular weights were found to be much higher. Today we know that native DNA molecules, such as those from *E. coli* cells, are so large that they are not readily isolated in intact

Figure 27-16
The denaturation or melting curve of two DNA specimens. The temperature at the midpoint of the transition (T_m) is the melting point. T_m depends on pH and salt concentration and must be specified.

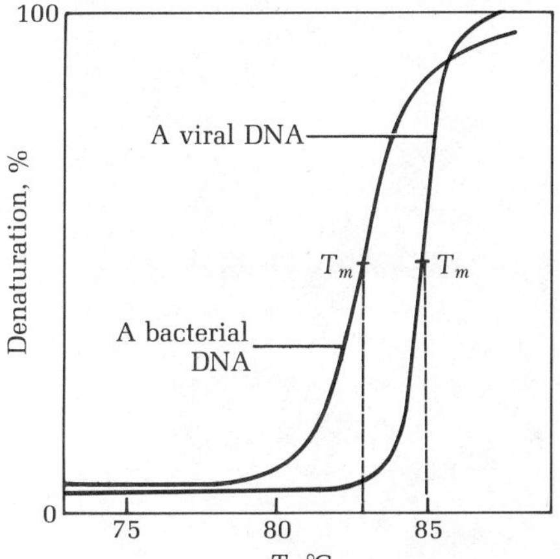

form, since they are easily broken by mechanical shear forces. Mere stirring or pipetting of solutions of native DNA can cause fragmentation of these molecules into many smaller pieces. With careful handling, it is possible to obtain the DNAs of large DNA viruses in intact form and to determine their molecular weights by physical methods. Single DNA molecules can also be visualized directly with the electron microscope, permitting their length to be measured. Such studies have shown that DNAs are molecules having an exact size and composition; they are not merely mixtures of polymers of varying size. Let us now examine the molecular size and other characteristics of the DNAs of viruses, bacteria, and eukaryotic cells.

Viral DNA Molecules Are Relatively Small

Table 27-4 lists the particle weights of some DNA viruses, the molecular weights of their DNAs, the number of base pairs they contain, and an approximation of their length in nanometers. From the molecular weight of a double-stranded viral DNA it is possible to calculate its *contour length*, since each nucleotide pair has an average molecular weight of about 650 and there is one nucleotide pair for every 0.34 nm of the duplex, as we have seen. A typical small DNA virus is *bacteriophage λ* (lambda) of *E. coli*. In its intracellular or replicative form its DNA is a double helix whose 5′ and 3′ ends are covalently joined to make a circle (in the sense of an endless belt, rather than a perfect round). Double-stranded λ DNA has a molecular weight of about 32 million, contains about 48,000 base pairs, and has a contour length of about 17.2 μm (Figure 27-17). The DNAs of many other DNA viruses are circular duplexes. In some viruses, however, e.g., *bacteriophage T2*, the DNA is a linear double-stranded molecule, i.e., it has two ends; and in others, e.g., *bacteriophage φX174*, the DNA is a single-stranded circle. During the replication cycle linear DNAs often become circular and all single-stranded viral DNAs become double-stranded. Such specific types of DNA that appear only during viral replication are called *replicative forms*.

Table 27-4 The DNAs of Some Bacterial Viruses

Virus	Viral particle weight, 10^6 daltons	Long dimension of particle, nm	DNA MW, millions	Approximate number of base pairs
φX174 (duplex form)	6	15	3.4	5,386†
T7	38	6	25	40,000
λ (lambda)	50	20	32	48,000
T2, T4	220	18	120	182,000

† Since the complete base sequence of the φX174 DNA is known, this figure is the exact number of base pairs.

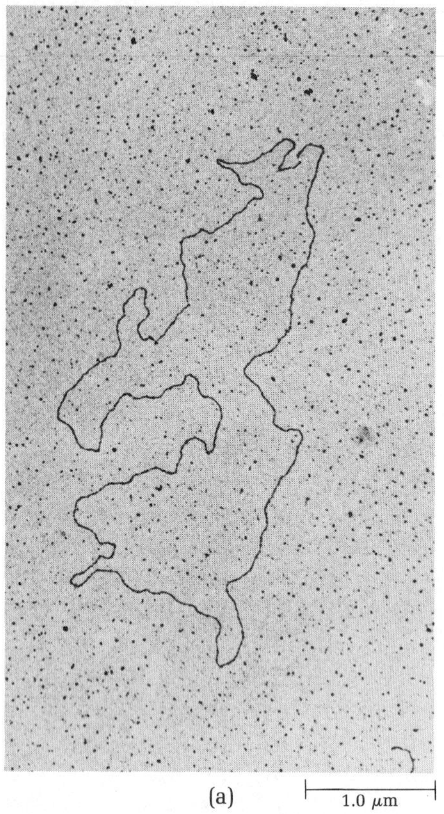

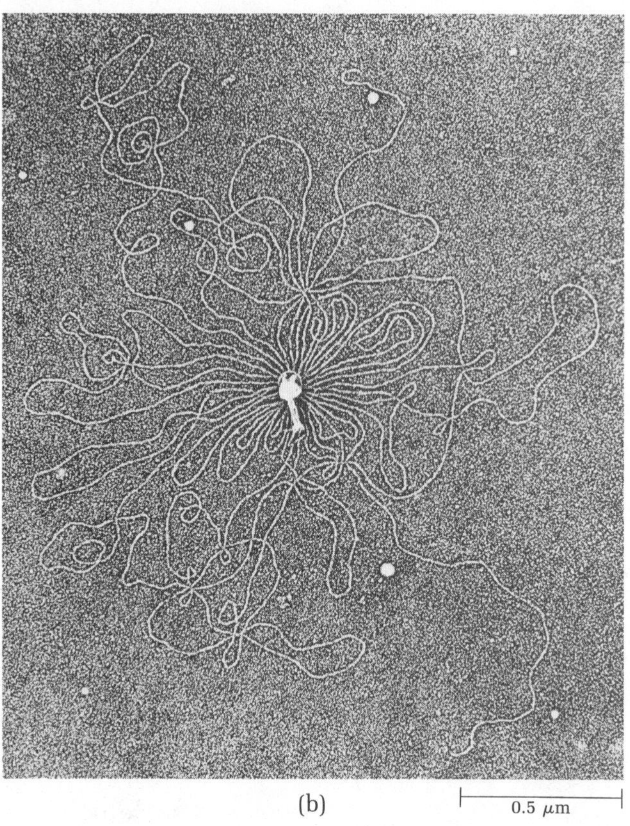

Figure 27-17
Electron micrographs of two bacteriophage DNAs. (a) *DNA of bacteriophage lambda* (λ). *Its molecular weight is 32 million, and its length is 17.2 μm.* (b) *Electron micrograph of bacteriophage T2 surrounded by its single linear molecule of DNA. The DNA was released by lysing the bacteriophage in distilled water and allowing the DNA to spread on the water surface.*

Another important point about viral DNAs: their contour lengths are much greater than the long dimensions of the viral particles in which they are found, as can be seen in the case of T2 (Figure 27-17; Table 27-4). Clearly, the DNA of viruses must be very compactly bunched to fit inside the viral particle.

In RNA viruses the RNA serves as chromosome. Viral RNAs are generally relatively small, contain few genes, and are usually single-stranded. All plant viruses contain RNA.

The Chromosomes of Prokaryotic Cells Are Single, Very Large DNA Molecules

Prokaryotic cells contain much more DNA than the DNA viruses. For example, a single *E. coli* cell contains almost 200 times as much DNA as a bacteriophage λ particle. Genetic experiments, as well as direct microscopy, have shown that the DNA in an *E. coli* cell is a single, very large molecule. It is a covalently closed double-stranded circle with a molecular weight of about 2600 million. It contains about 4 million base pairs and has a contour length of about 1400 μm = 1.4 mm, some 700 times the length of an *E. coli* cell (2 μm). Here again we see that the DNA molecule must be very tightly bunched or coiled, since it is contained entirely within the nuclear zone (page 20) of *E. coli* cells. The DNA of bacterial cells appears to be attached at one or more points to the inner surface of the cell membrane.

Circular DNAs Are Supercoiled

Figure 27-18
Supercoiling of a circular DNA by negative twists, opposite to the direction of twist of the double helix. Supercoiling can be abolished by nicking one strand, yielding a relaxed circular DNA.

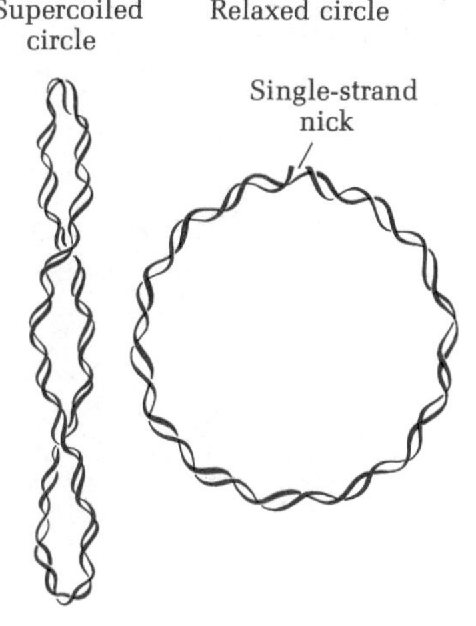

Circular viral DNAs, when carefully isolated, are supercoiled or *supertwisted;* i.e., they appear as though the double helix had been partly untwisted before the ends of the strands were joined into a circle. This reverse twist imparts a torque to the circular DNA molecule so that it twists on itself (Figure 27-18). If such a supercoiled DNA, in which extra energy is stored, is exposed to an endonuclease that can "nick" one strand, the torsion caused by the reverse twist is relieved and the circular DNA assumes its normal low-energy, *relaxed* state (Figure 27-18). Supertwisted viral DNAs are more compact than relaxed circles.

Examination of carefully isolated DNA from *E. coli* has shown that it has a great many loops held together by proteins. Each of these loops is in turn supertwisted (Figure 27-19). Looping and supercoiling help make possible packaging of very large circular DNA molecules into small volumes without a surrounding membrane. Linear DNAs cannot be supercoiled unless both ends are anchored. Supercoiling is important for many processes involving DNA. Many proteins or enzymes will not bind to DNA unless it is supercoiled. Enzymes known as *topoisomerases* can regulate the degree of supercoiling by introducing or removing supertwists.

Figure 27-19
Electron micrograph of the chromosome of a single E. coli *cell, still intact and supercoiled, as determined from greatly enlarged prints. At the left center are fragments of the cell membrane.*

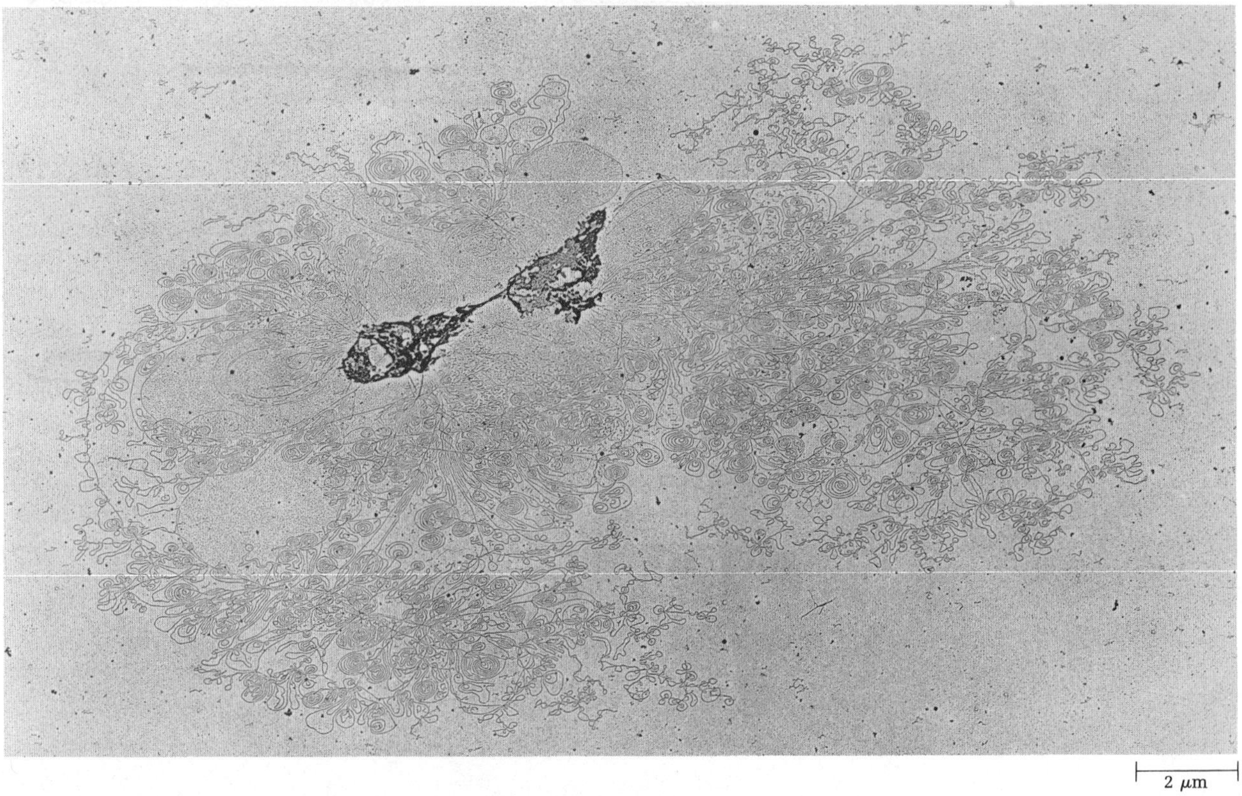

2 μm

(a) 0.25 μm

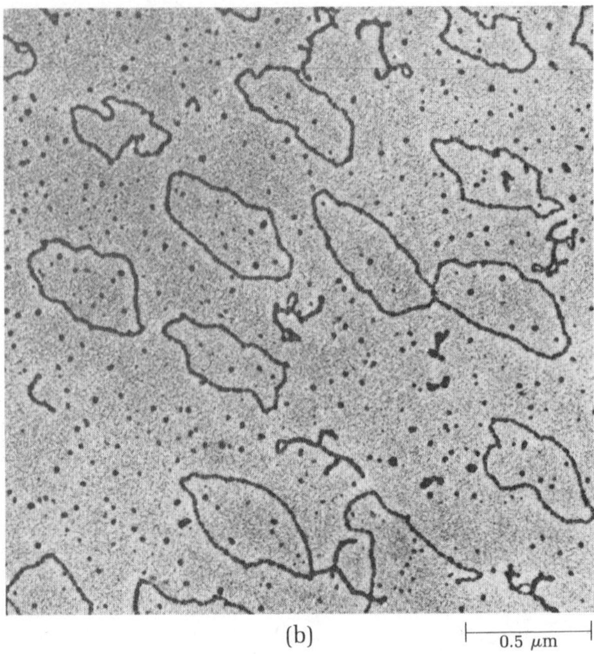

(b) 0.5 μm

Figure 27-20
Electron micrographs of plasmids from two bacterial species. (a) pSC101, *a plasmid of* E. coli *that confers resistance to tetracycline. (b) Plasmids from* Neisseria gonorrhoeae, *the bacterium that causes gonorrhea. Most are relaxed. The irregularly twisted supercoiled plasmids (such as that in the precise center) show how effectively circular DNA can be condensed or bunched by negative twists.*

Some Bacteria Also Contain DNA in the Form of Plasmids

In addition to the very large circular DNA chromosome located in the nuclear zone most species of bacteria contain one or more small circular DNA molecules that are free in the cell cytoplasm. These extrachromosomal elements are called *plasmids* (Figure 27-20). Most plasmids are very small and contain only a few genes, compared with the bacterial chromosome, which contains thousands of genes. In some cells, however, plasmids may be quite large. Plasmids carry genetic information and undergo replication to yield daughter plasmids, which pass into the daughter cells when the cell divides. Ordinarily plasmids appear to maintain a separate life, detached from the chromosomal DNA through many cell divisions. However, sometimes plasmids integrate into the chromosomal DNA and may leave again in a coordinated manner.

Some plasmids carry genes that specify the resistance of the host bacterium to antibiotics such as tetracycline and streptomycin. Bacterial cells containing such plasmids are resistant to these antibiotics. These cells may survive in the human body after treatment of a bacterial infection with an antibiotic, which causes death of only antibiotic-sensitive cells. The antibiotic-resistant cells then can multiply and produce an infection that may be uncontrollable by antibiotics. For this reason antibiotics should not be used indiscriminately to treat infections, without some knowledge that the causative organism is sensitive to the antibiotic used. Plasmids also may pass from an antibiotic-resistant cell to an antibiotic-sensitive cell of the same or another species, thus rendering the latter resistant.

Table 27-5 Normal Chromosome Number in Different Species

Prokaryotic organisms	
Bacteria	1
Eukaryotic organisms	
Drosophila	8
Red clover	14
Garden peas	14
Honeybee	16
Corn	20
Frog	26
Hydra	30
Fox	34
Cat	38
Mouse	40
Rat	42
Rabbit	44
Human being	46
Chicken	78

Another important aspect of plasmids is that they can be isolated quite easily from bacterial cells. New genes from other species can be inserted into isolated plasmids and the modified plasmid then can be put back into its normal host cell. Such a plasmid, containing a foreign gene, will be replicated and transcribed and may also cause the host cell to make the proteins coded by the artificially introduced gene, even though it is not part of the normal genome of the cell. Later we shall see how such *recombinant DNAs* are made, transcribed, and translated into potentially useful products.

Eukaryotic Cells Contain Much More DNA than Prokaryotes

Now we take a big jump—from prokaryotic cells to the much more complex eukaryotic cells. Eukaryotes contain far more DNA than prokaryotes. An individual cell of a slime mold, one of the lowest eukaryotes, has over 10 times the DNA content of an *E. coli* cell. Cells of *Drosophila,* the fruit fly used in classical genetic studies, have over 25 times as much DNA as *E. coli* cells. And the cells of human beings and many other mammals have about 600 times as much DNA as *E. coli.*

The total contour length of all the DNA in a *single* human cell is about 2 m, compared with 1.4 mm for *E. coli* DNA. Since there are about 10^{13} cells in the adult human body, the total length of all the DNA in the body would be about 2×10^{13} m or 2×10^{10} km. Compare this with the circumference of the earth (25,000 mi = 40,000 km) or the distance between the earth and the sun (90,000,000 mi = 1.44×10^8 km).

Microscopic observation of nuclei in dividing eukaryotic cells has shown that the genetic material is subdivided into chromosomes, their number depending upon the species of organism (Table 27-5). Human cells, for example, have 46 chromosomes. It is now certain that each chromosome of a eukaryotic cell, such as that shown in Figure 27-21, contains a single, very large duplex DNA molecule, which may be from 4 to 100 times larger than those in *E. coli* cells. For example, the

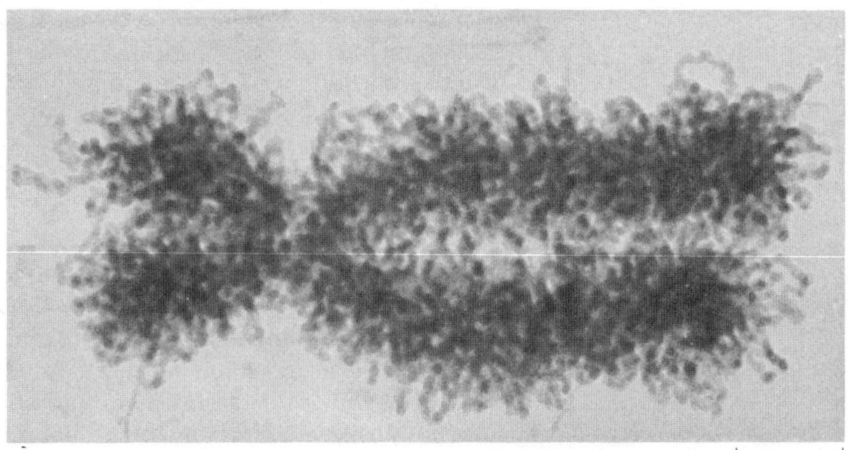

Figure 27-21
Electron micrograph of human chromosome 12.

DNA of one of the smaller human chromosomes has a contour length of about 30 mm, almost 15 times longer than the DNA of *E. coli*. The DNA molecules in the 46 human chromosomes of human cells are not all identical in size; they vary over a 25-fold range. Eukaryotic DNAs are linear, not circular. Each chromosome in eukaryotes carries a unique set of genes. All the genes of a cell comprise the *genome*.

A typical human cell, such as a liver cell, is about 25 μm in diameter. Its nucleus is about 5 μm in diameter and holds 46 chromosomes, containing altogether 2 m of DNA. The "packaging" of DNA in eukaryotic chromosomes is very different from that in prokaryotes, as we shall now see.

Eukaryotic Chromosomes Consist of Chromatin Fibers

We have used the term chromosome to refer to the nucleic acid molecule that is the repository of the genetic information of a virus, a prokaryote, or a eukaryotic cell. But the word chromosome ("colored body") was originally used in another sense, to refer to the densely staining bodies in eukaryotic nuclei that can be visualized by the light microscope after the cells are stained with a dye. Eukaryotic chromosomes, in the original sense of the word, appear as sharply defined bodies in the eukaryotic nucleus only in the period just before and during *mitosis*, the process of nuclear division in somatic cells (Figure 27-22). In resting, nondividing eukaryotic cells, the chromosomal material, called *chromatin*, is amorphous and appears to be rather randomly dispersed throughout the nucleus. But when the cells prepare to divide, the chromatin condenses and assembles itself into a species-specific number of well-defined chromosomes (Figure 27-22).

Chromatin has been isolated and analyzed. It consists of very thin fibers that contain about 60 percent protein, about 35 percent DNA, and perhaps 5 percent RNA (page 25). Within a chromosome the chromatin fibers are folded and looped into many bundles (Figure 27-21). The DNA in the chromatin is very tightly associated with proteins called *histones*, which function to package and order the DNA into structural units called *nucleosomes*. Also found in chromatin are some nonhistone proteins. In contrast, the chromosomes of bacteria contain no histones but do have small amounts of proteins that facilitate the looping and condensation of their DNA.

Histones Are Small Basic Proteins

Histones are found in the chromatin of all somatic eukaryotic cells but never in prokaryotes. They have molecular weights between 11,000 and 21,000. Histones are very rich in the basic amino acids arginine and lysine, which together make up about one-fourth of the amino acid residues. Because the R groups of their arginine and lysine residues are protonated at pH 7 and thus have a positive charge, histones combine with the nega-

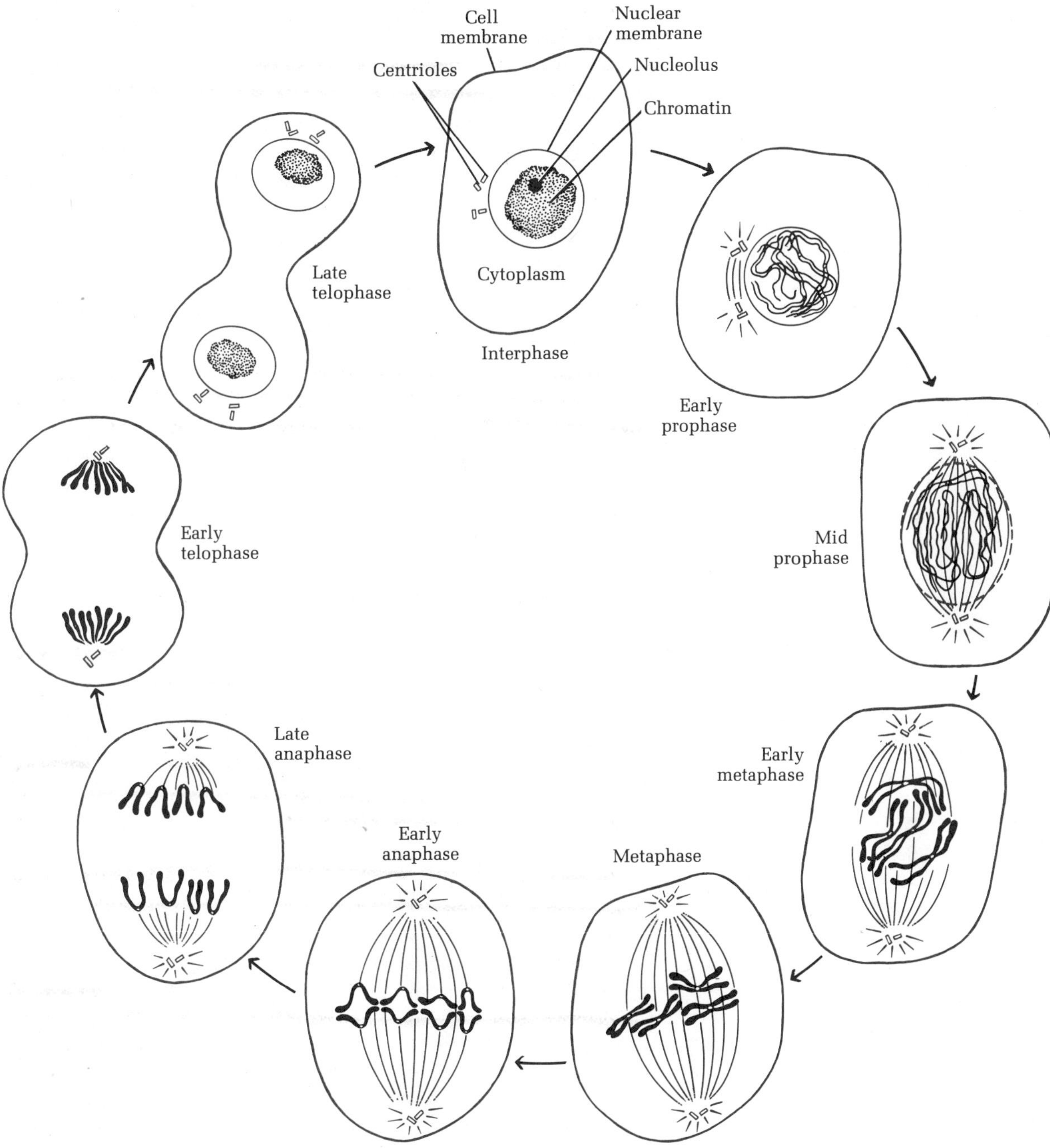

Figure 27-22
Stages in mitosis. Note the randomly dispersed chromatin in interphase (between divisions). As the cell prepares for division, the chromatin becomes organized into visible chromosomes. The paired daughter chromosomes are then separated in anaphase. The chromatin becomes randomized again in the daughter cells at late telophase, just before they separate.

Table 27-6 Histones

Histone	MW	Lysine residues, %	Arginine residues, %
H1	21,000	29	1.5
H2A	14,500	11	9.5
H3B	13,700	16	6.5
H3	15,300	10	13.5
H4	11,300	11	14

tively charged double-helical DNA to form DNA-histone complexes, which are held together by electrostatic attraction.

The five major classes of histones found in all eukaryotic cells differ in molecular weight and amino acid composition (Table 27-6). Histone H1 is *lysine-rich* (29 percent lysine), histones H2A and H2B contain large amounts of both lysine and arginine, the former predominating, whereas histones H3 and H4 contain somewhat more arginine than lysine and are referred to as *arginine-rich*. Two of the histones, H3 and H4, are nearly identical in amino acid sequence in all forms of eukaryotic life and thus perform the same function in all eukaryotic cells. Histones H1, H2A, and H2B show a lesser degree of sequence homology in different eukaryotic species.

Each of the histones can exist in different forms because the R groups of some of their amino acids can be enzymatically modified by methylation, phosphorylation, or acetylation. Such modifications of histone R groups may change their net electric charge and other properties. For example, acetylation of the ϵ-amino groups of lysine residues results in elimination of their positive charges.

DNA-Histone Complexes Form Beadlike Nucleosomes

Chromatin fibers resemble a string of beads (Figure 27-23). The repeating beadlike structures in these fibers are *nucleosomes*. They consist of a segment of duplex DNA containing about 200 base pairs, wound twice around a set of histone molecules to make a bead having a diameter of about 10 to 11 nm. There are eight histone molecules in each nucleosome core, two each of histones H2A, H2B, H3, and H4. The DNA is wound around the outside of the nucleosome core.

Between individual nucleosome cores is a connecting or spacer link of DNA to which is bound histone H1. These linking segments differ in length, from 20 to 120 nucleotide pairs, depending on the species and cell type. In human chromatin fibers they are about 50 nucleotide pairs long. Nucleosomes are structural units of chromatin, largely serving a packaging function. In addition to the net shortening of a DNA strand produced by winding segments of it around the histones, additional shortening and packaging of eukaryotic DNA is brought about by an orderly arrangement of the nucleosomes (Figure 27-23). The chromatin is also held in place in the nucleus by nonhistone proteins, which form a scaffolding or nuclear matrix.

Eukaryotic Cells Also Contain Cytoplasmic DNA

In addition to the DNA found in the nucleus of eukaryotic cells, very small amounts of DNA, differing in base composition from nuclear DNA, are present in the cytoplasm, localized in the mitochondria. Chloroplasts of photosynthetic cells also contain DNA. Usually less than 0.1 percent of all the cell DNA

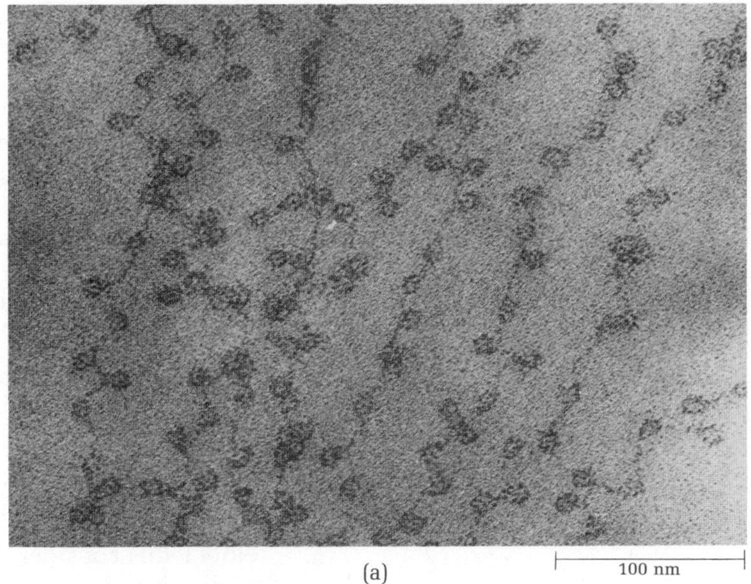

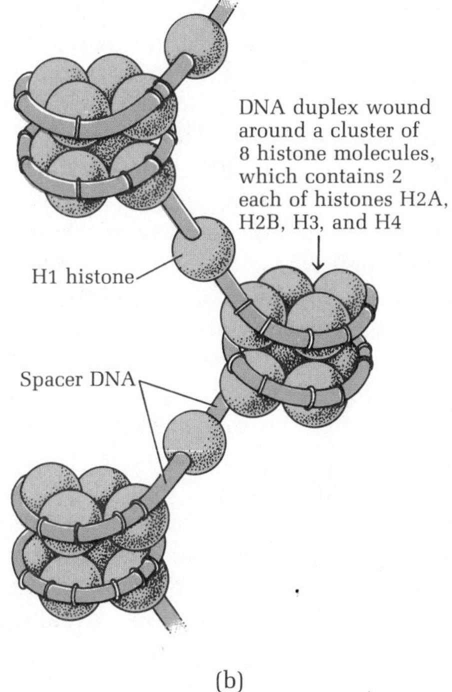

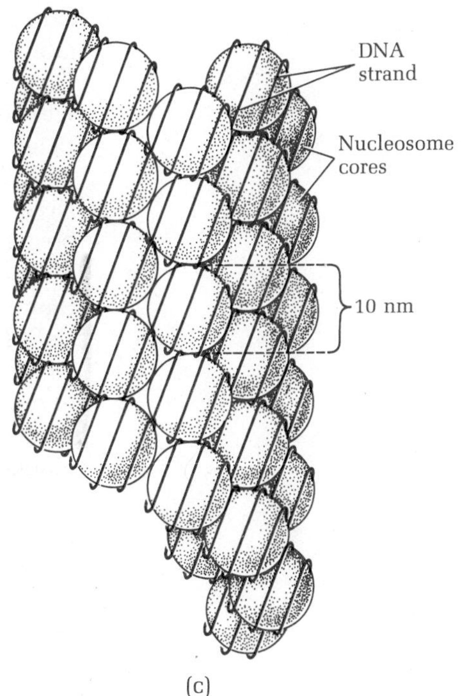

Figure 27-23
Nucleosomes. (a) Electron micrograph of extended chromatin fibers, showing the bead-like nucleosomes. (b) Schematic drawing of an extended segment of a chromatin fiber, showing the structure of the nucleosomes. (c) Schematic drawing showing that nucleosomes and the spacers are wound into a close-packed array.

is present in these organelles in resting somatic cells, but in fertilized and dividing egg cells, where the mitochondria are much more numerous, the amount of mitochondrial DNA is very much larger. Mitochondrial DNA (mDNA) is a very small molecule compared with the nuclear chromosomes. It has a molecular weight of only about 10 million in animal cells and occurs as a circular duplex. Chloroplast DNA molecules are considerably larger than those of mitochondria. The DNAs of these organelles are not associated with histones.

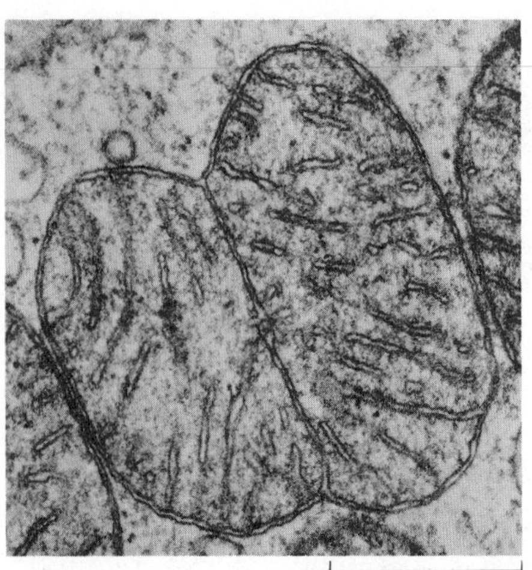

Figure 27-24
A dividing mitochondrion from the fat body of an insect.

The origin of mitochondrial and chloroplast DNAs has been the subject of much speculation. One view is that they are vestiges of the chromosomes of ancient bacteria that invaded the cytoplasm of host cells and became the precursors of these organelles. Mitochondrial DNA codes for the mitochondrial tRNAs and rRNAs and for a few mitochondrial proteins. Since over 95 percent of mitochondrial proteins are encoded by nuclear DNA, the reason for the occurrence of DNA in mitochondria and chloroplasts is one of the mysteries of cell genetics. Mitochondria and chloroplasts undergo a division process during division of the host cell (Figure 27-24). Before and during division of mitochondria their DNA is replicated and the daughters pass into the daughter mitochondria.

Genes Are Segments of DNA That Code for Polypeptide Chains and RNAs

Now let us consider the major functional subdivisions of DNA molecules, the genes. A gene is defined in the classical biological sense as a portion of a chromosome that determines or specifies a single character or *phenotype*, for example, eye color. (The word phenotype means "visible appearance.") But today we have a molecular definition, first proposed by George Beadle and Edward Tatum in 1940. They exposed spores of the mold *Neurospora crassa* to x-rays and other agents that damage DNA and thus cause mutations. Some mutants were found to be deficient in one or another specific enzyme, resulting in the failure of a metabolic pathway (page 349). This observation led them to conclude that a gene is a segment of the genetic material that determines or codes for one enzyme: the *one gene–one enzyme* hypothesis. Later this concept was made more general to mean *one gene–one protein,* since some genes code for proteins that are not enzymes. But today we have a more precise biochemical definition.

Recall that many proteins have multiple polypeptide chains (page 180). In some multichain proteins, all the polypeptide chains are identical, in which case they can all be encoded by the same gene. But others have two or more different kinds of polypeptide chains, each with a distinctive amino acid sequence. Hemoglobin, for example, has two kinds of polypeptide chains, α and β chains, which differ in length and in amino acid sequence. Today we know that the α and β chains are encoded by two different genes. Thus the gene-protein relationship is more accurately described by the phrase "one gene–one polypeptide."

However, not all genes are ultimately expressed in the form of polypeptide chains. Some genes code for the different kinds of transfer RNAs, and other genes code for the different kinds of ribosomal RNAs. Genes that code for polypeptides and RNAs are known as *structural genes:* they determine the structure of some final gene product, such as an enzyme or a stable RNA. DNA also contains other segments or sequences that have a purely regulatory function. Some of these regulatory segments constitute signals denoting the beginning and end of structural genes; others participate in turning the transcription of structural genes on or off. Thus chromosomes contain *structural genes* and *regulatory sequences*.

There Are Many Genes in a Single Chromosome

How many genes are there in a single chromosome? We can give an approximate answer to this question in the case of *E coli*. It is estimated that the single *E. coli* chromosome has in excess of 3000 genes, perhaps as many as 5000. A direct count of all the different polypeptides present in *E. coli* cells, following their separation by two-dimensional electrophoresis, has been attempted. Some 1100 different spots, each corresponding to a different polypeptide chain, were found (Figure 27-25). This is a minimum estimate, however, since this method is not able to separate all the polypeptides and is not sensitive enough to reveal the presence of proteins occurring in only a few copies per cell.

The sequential arrangement of many of the genes in the chromosomes of viruses and bacteria has been "mapped" by various genetic procedures. A partial genetic map of *E. coli* in Figure 27-26 shows the relative location of some of its genes in the circular DNA molecule.

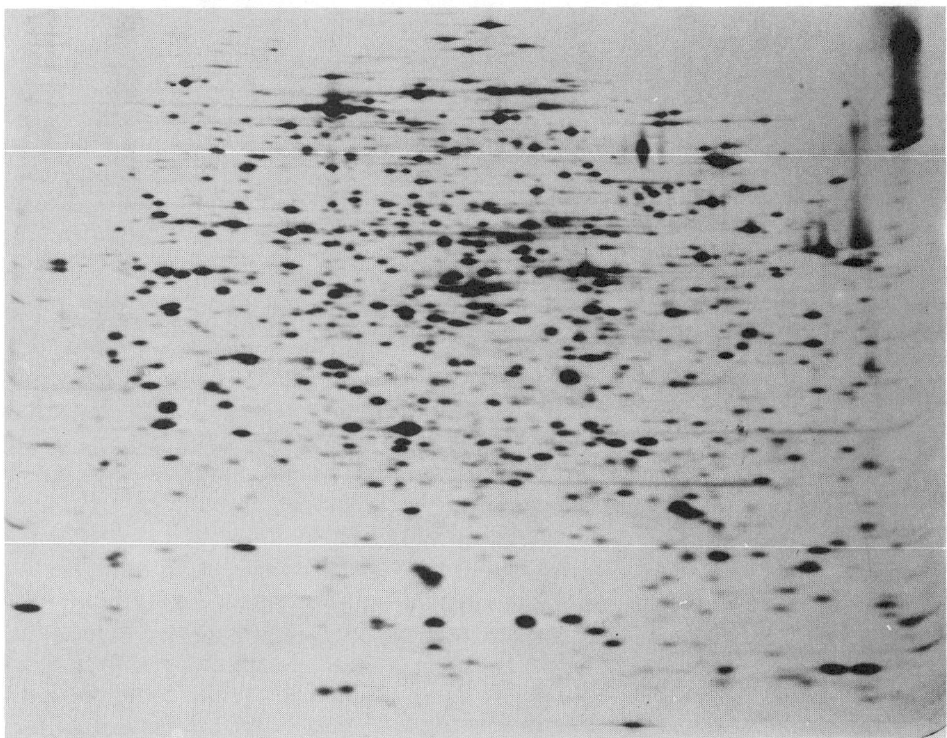

Figure 27-25
Fractionation of E. coli *proteins. A two-dimensional chromatogram of polypeptide chains present in an extract of* E. coli. *Over 1100 distinct polypeptides can be detected in this "map." Presumably many more polypeptides are present but are not detectable. Attempts are being made to separate and count all the polypeptides in different types of human cells.*

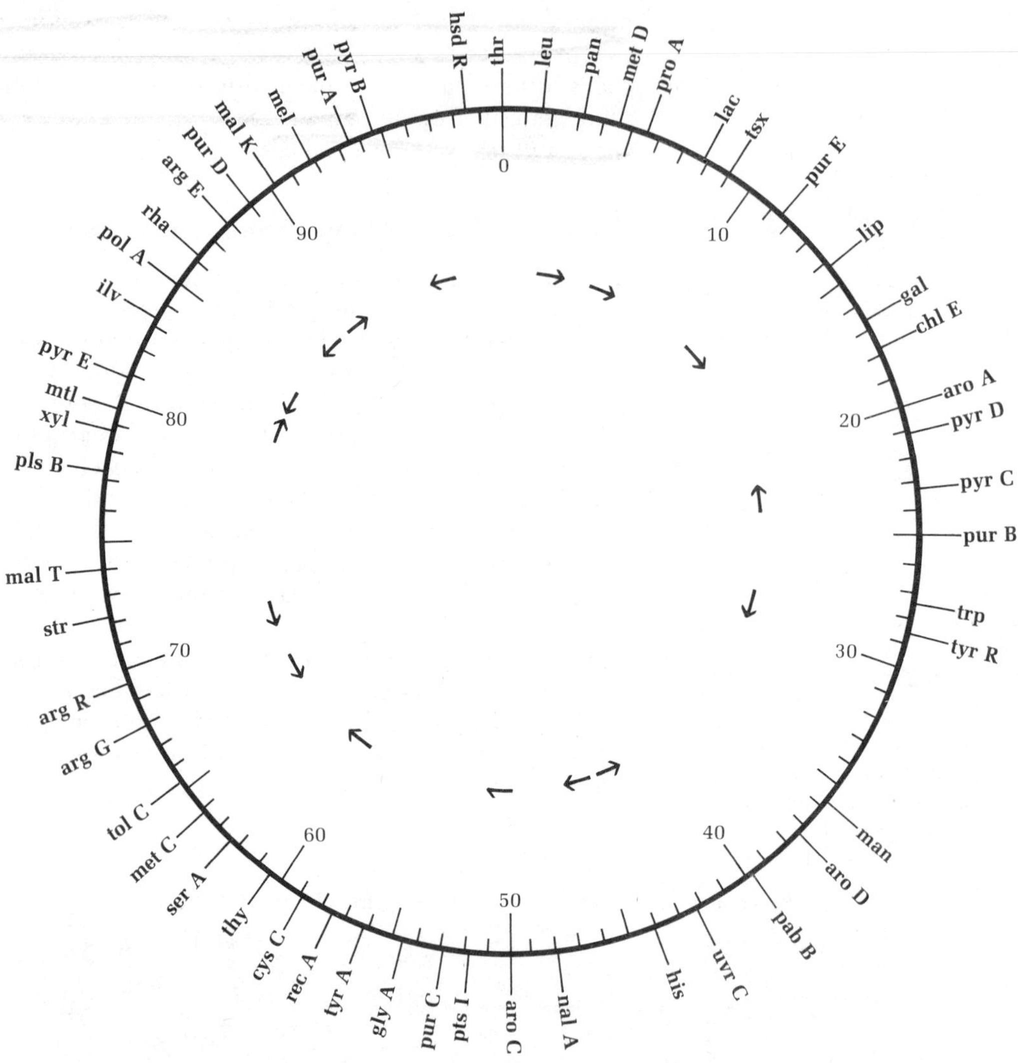

Figure 27-26
The circular reference map of the chromosome of E. coli K12. Around the outer periphery are the symbols of 52 genes whose relative loci in the chromosome are known with great accuracy and thus serve as landmarks in mapping of other genes. The numbers inside the map correspond to the number of minutes required for the male chromosome to move into the female cell during the sexual conjugation of E. coli. The zero point is the gene thr. The direction of entry of the male chromosome is given by the arrows. Of the 3000 genes of E. coli, about 1000 have been mapped to date.

How Large Are Genes?

We can make an approximation of the size of genes. We have seen (page 812) that the duplex DNA molecule *E. coli* contains about 4 million nucleotide pairs. If there are 3000 genes in *E. coli*, each gene, on the average, would consist of a length of DNA containing $(4 \times 10^6)/3000 \approx 1300$ nucleotide pairs. This is probably an overestimate, since it does not allow for the occurrence of signals, spacers, or other regions of unknown function in the DNA.

We can approximate the size of genes more directly. Today we know that each amino acid of a polypeptide chain is coded by a sequence of three consecutive nucleotides in a single strand of prokaryotic DNA (Figure 27-27). Since there are no signals for "commas" in the genetic code, the coding triplets of DNA are arranged sequentially, corresponding to the sequence

Figure 27-27
Colinearity of the nucleotide sequences of DNA, mRNA, and the amino acid sequence of polypeptide chains. The triplets of nucleotide units in DNA determine the sequence of amino acids in proteins through the intermediary formation of mRNA, which has nucleotide triplets (codons) complementary to those of a DNA template strand.

DNA (The complementary strand)	DNA (The coding strand)	RNA	Polypeptide
5′	**3′**	5′	Amino-terminal end
C	**G**	C	
G	**C**	G	Arginine
T	**A**	U	
G	**C**	G	
G	**C**	G	Glycine
A	**T**	A	
T	**A**	U	
A	**T**	A	Tyrosine
C	**G**	C	
A	**T**	A	
C	**G**	C	Threonine
T	**A**	U	
T	**A**	U	
T	**A**	U	Phenylalanine
T	**A**	U	
G	**C**	G	
C	**G**	C	Alanine
C	**G**	C	
G	**C**	G	
T	**A**	U	Valine
T	**A**	U	
T	**A**	U	
C	**G**	C	Serine
T	**A**	U	
3′	**5′**	3′	Carboxyl-terminal end

of amino acids in the polypeptide for which it codes. Figure 27-27 shows the principle of the coding relationships between DNA, RNA, and proteins. Since a single polypeptide chain may have anywhere from about 50 up to 2000 or more amino acid residues in a specific sequence (page 126), a gene coding for the biosynthesis of a polypeptide chain would have, correspondingly, at least 150 to 6000 or more nucleotide units. If the average polypeptide chain has about 350 residues, this would correspond to 1050 nucleotide pairs. Since there are about 4 million nucleotide pairs in *E. coli* DNA, they could account for $(4 \times 10^6)/1050 \approx 3800$ genes.

The fact that the base pairs in duplex DNA are about 0.34 nm apart (page 804) allows us to calculate the length of the gene for an average polypeptide: $0.34(1050) = 357$ nm $= 0.36$ μm. Since each nucleotide pair has a molecular weight of about 650, the molecular weight of this gene would be about $650(1050) = 680{,}000$.

The genes that code for transfer RNAs are much smaller than those for polypeptides, since each nucleotide unit of a transfer RNA is coded by a single nucleotide unit of DNA.

Bacterial DNA Is Protected by Restriction-Modification Systems

It has long been known that among the millions of the standard bases A, T, G, and C in a bacterial DNA some have extra methyl groups. The biological significance of these methylated bases has become apparent through some important discoveries which have had a major impact on genetics and genetic biochemistry. Each species of bacteria has a characteristic pattern of methylated bases in its DNA, which distinguishes it from the DNAs of other species. If DNA from some other species gains entry into a living bacterial cell, it will be recognized as "foreign" by its lack of the species-specific methylation pattern characteristic of that cell's own DNA. The foreign DNA will then be destroyed by a specific nuclease that cleaves both its strands at or near the location that lacks the methylation pattern possessed by the host cell's DNA. Thus foreign DNAs are *restricted;* they are destroyed by the distinctive nucleases elaborated by each bacterial species.

Two enzymes, closely related in their specificity, protect the DNA of a given bacterial species: (1) a *modification methylase* and (2) a *restriction endonuclease.* The modification methylase is responsible for producing a species-characteristic methylation pattern in a specific short base sequence in the host cell's own DNA. This methylated sequence will occur many times in the host-cell DNA. The methyl groups in these sequences remain intact for the lifetime of the cell. The corresponding restriction endonuclease, on the other hand, cleaves both strands of any other DNA that has this particular base sequence if it is *unmethylated.* An example is the restriction endonuclease *Hin* dII of the bacterium *Hemophilus influenzae* (each restriction endonuclease has a symbol). This enzyme cleaves, at the points shown by arrows, both strands of any DNA having the specific base sequence:

```
              Site of
              cleavage
                 ↓
5′  –G–T–Py–Pu–A–C–  3′
             •
3′  –C–A–Pu–Py–T–G–  5′
                 ↑
```

but it will *not* cleave the same sequence if the bases indicated by colored asterisks are methylated:

```
                   *
5′  –G–T–Py–Pu–A–C–  3′
             •
3′  –C–A–Pu–Py–T–G–  5′
       *
```

It is important to note that this short DNA segment, methylated or unmethylated, has internal symmetry about a central point shown in color. If this segment is rotated 180° in the plane of the paper around the central point, it will read exactly the same as before rotation. This type of symmetry is called *twofold rotational symmetry*. Most modification-restriction sequences identified to date have twofold symmetry. The *Hin* dII restriction endonuclease cleaves both strands at the center of this segment in any DNA in which this sequence is unmethylated. Once the foreign DNA has been cleaved in this way, it cannot be repaired and thus cannot be replicated. It is degraded to mononucleotides by other cell nucleases.

Table 27-7 shows the specific sequences that are attacked by typical restriction endonucleases from different bacterial species. Each restriction enzyme has a symbol, and each sequence it recognizes has twofold symmetry. Some restriction endonucleases yield flush or blunt ends, as in the *Hin* dII example above, but others, such as the *Eco* RI endonuclease of *E. coli*, leave staggered ends. In the latter type the two overlapping ends are called *cohesive ends* or *sticky ends*, since they can base-pair with each other.

Sequences of six nucleotides having twofold symmetry have been calculated to occur by random chance in any DNA, regardless of species, about once every 4000 nucleotide pairs. Since the DNA molecules of different bacteria have millions of nucleotide pairs, it is highly probable that any foreign bacterial DNA will be cleaved at least once by any given restriction endonuclease, but the host cell protects its own DNA by methylating one or more of the bases in its restriction sequence; the methylated sequence does not bind to the restriction endonuclease and thus cannot be cleaved.

Over 150 different restriction endonucleases have been discovered in different species of bacteria. Some bacteria have more than one set of corresponding modification methylases and restriction endonucleases. DNA viruses of bacteria, however, have learned several ways to penetrate the restriction de-

Table 27-7 Specificity of Some Restriction Endonucleases†

Producing flush ends

Hin dII

↓ *

5′ –G–T–Py–Pu–A–C– 3′

•

3′ –C–A–Pu–Py–T–G– 5′

* ↑

Hpa I

↓

5′ –G–T–T–A–A–C– 3′

•

3′ –C–A–A–T–T–G– 5′

↑

Producing staggered ends

Eco RI

↓ *

5′ –G–A–A–T–T–C– 3′

•

3′ –C–T–T–A–A–G– 5′

* ↑

Eco RII

↓ *

5′ –N–C–C–N–G–G–N– 3′

•

3′ –N–G–G–N–C–C–N– 5′

* ↑

Hin dIII

*↓

5′ –A–A–G–C–T–T– 3′

•

3′ –T–T–C–G–A–A– 5′

↑*

† The colored dot indicates the axis of twofold rotational symmetry, and the colored arrows indicate the site of cleavage. The colored asterisks show the methylation sites (where known) in the parent organism, which is *Hemophilus influenzae* for *Hin* dII and dIII, *E. coli* for *Eco* RI and *Eco* RII, and *Hemophilus parainfluenzae* for *Hpa* I. Pu = purine, Py = pyrimidine, and N = A or T.

fenses of their host cells. Some viral DNAs contain different kinds of base modifications that enable them to escape destruction by restriction endonucleases of the host cells they invade. The modifying groups in such viral DNAs are methyl, hydroxymethyl, and glucosyl groups. Other viruses have evolved DNA sequences that do not contain sequences recognized by some restriction endonucleases.

The restriction endonucleases have become extremely important for another reason: they are very useful tools, since they can reproducibly cleave both strands of DNAs but only at specific points. This property has opened a new era in the biochemistry of genes. Restriction endonucleases, many of which are now commercially available, have made possible the systematic dissection and mapping of chromosomes and are essential tools in determination of the base sequence of DNAs. Together with other recent developments, restriction endonucleases have made it possible to splice or recombine genes from one organism into the genome of another (Chapter 30). The Nobel prize in Physiology or Medicine in 1978 was awarded to Werner Arber of Switzerland, and Hamilton Smith and Daniel Nathans of the United States, the discoverers of DNA restriction, the nature of the action of the restriction endonucleases, and their usefulness in dissection of genes.

Eukaryotic DNA Contains Base Sequences That Are Repeated Many Times

Prokaryotes usually have only one copy of DNA per cell, and in nearly all cases each DNA molecule contains only one copy of any given gene. Apart from regulatory and signaling sequences, there is relatively little silent or nontranslated DNA in prokaryotes. Moreover, each gene is precisely colinear with the amino acid sequence (or RNA sequence) for which it codes (Figure 27-27).

The organization of genes in eukaryotic DNA is structurally and functionally much more complex. Tests made of the extent to which segments of mouse DNA occur in multiple copies had a surprising outcome. About 10 percent of mouse DNA consists of short lengths of less than 10 base pairs that are repeated millions of times. These are called *highly repetitive* segments. Another 20 percent of mouse DNA was found to occur in lengths that are repeated at least 1000 times, designated *moderately repetitive*. The remainder, some 70 percent of the DNA, consists of unique, i.e., nonrepeated segments, as well as segments that are repeated only a few times. The most highly repeated sequences are called *satellite DNA* and they are believed to be nontranslated segments.

Such experiments have been extended to many other species. It now appears probable that all eukaryotic chromosomes contain repetitive DNA, but prokaryotes generally do not. The amounts of highly and moderately repetitive sequences vary from one eukaryotic species to another.

A Few Eukaryotic Genes Occur in Multiple Copies

Some genes occur in many copies, at least at certain points in the cell life cycle. The most conspicuous example is the set of genes that codes for the four ribosomal RNAs, which occur in many copies in normal somatic cells. The egg cells of amphibia can amplify the genes for three of the four rRNAs even further. This is a necessity since the egg cells, once fertilized, undergo very rapid growth and division, for which they require many ribosomes to make all the cell proteins. The genes coding for the histones also occur in multiple copies, up to a thousand in different eukaryotic species. Early embryos must make histones very rapidly during this period of rapid growth. The genes for feather keratins also occur in multiple copies in the genome of the chicken.

One might expect that the genes for other proteins occurring in very large amounts in certain cells and tissues of eukaryotes, e.g., hemoglobin, serum albumin, collagen, or egg albumin, would also occur in multiple copies, but this is not generally true. Most eukaryotic genes occur in only single copies or in a very small number.

Eukaryotic DNA Has Many Palindromes

Eukaryotic DNA has another characteristic property. It contains numerous (perhaps thousands) of palindromes. A palindrome (Greek, meaning "to run back again") is a word, phrase, or sentence that is spelled identically reading forward or backward. Two examples are

Able was I ere I saw Elba.

Madam, in Eden I'm Adam.

The term is applied to specific regions in eukaryotic DNA in which there are inverted repetitions of base sequence with twofold symmetry, resembling the short sequences attacked by restriction endonucleases. An example is given in Figure 27-28.

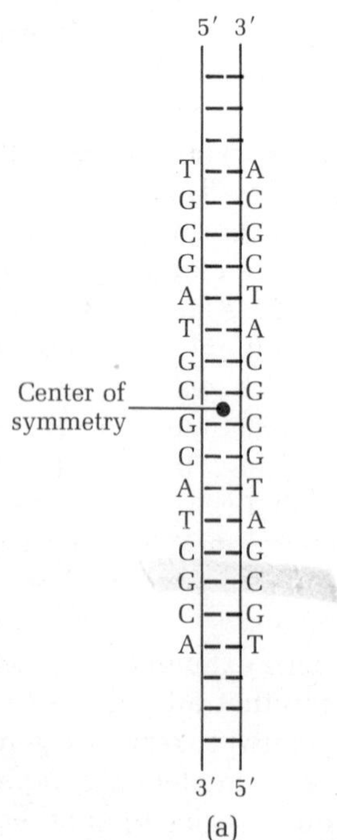

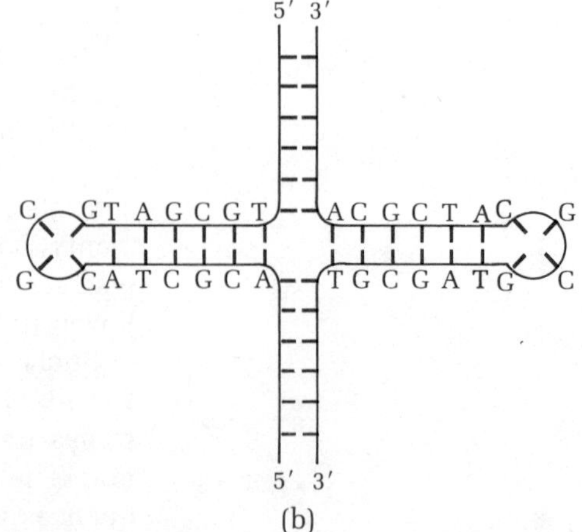

Figure 27-28
(a) *A palindrome, also called an inverted repetition, showing its twofold rotational symmetry about the central point.* (b) *A cruciform (crosslike) structure produced when the palindrome base-pairs within each strand, rather than between the strands. Some palindromes in native eukaryotic DNA contain dozens or hundreds of bases in inverted sequences.*

Many palindromes involve up to a thousand base pairs. The shorter ones are probably special signals, as is the case in restriction sequences. Long palindromes have the potential to form *cruciform* (crosslike) loops by base pairing within each strand (Figure 27-28). The function of long palindromes is unknown.

Many Eukaryotic Genes Contain Intervening Nontranscribed Sequences (Introns)

Many if not most eukaryotic genes have a distinctive and very puzzling structural feature: their base sequences contain one or more intervening segments of DNA that do not code for the amino acid sequence of the polypeptide product. These nontranslated inserts interrupt the otherwise precisely colinear relationship between the nucleotide sequence of the gene and the amino acid sequence of the polypeptide it encodes (see Figure 27-29). Such nontranslated DNA segments in genes are called *intervening sequences* or *introns,* whereas the coding segments of the gene are called *exons.* A well-known example is the gene coding for the single polypeptide chain of the egg protein *ovalbumin.* As can be seen in Figure 27-29, this gene has six introns, which split the ovalbumin gene into seven exons. Moreover, the introns of this particular gene are much longer than the exons; altogether the introns make up 85 percent of the total length of DNA containing this gene (Figure 27-29). With few exceptions, *all eukaryotic genes that have been examined to date appear to contain introns, which vary in number, position, and in the fraction of the total length of the gene they occupy.* For example, the serum albumin gene contains 6 introns, the protein conalbumin of the chicken egg contains 17 introns, and a collagen gene has been found to have over 50 introns. The genes for histones are an exception; they appear to have no introns.

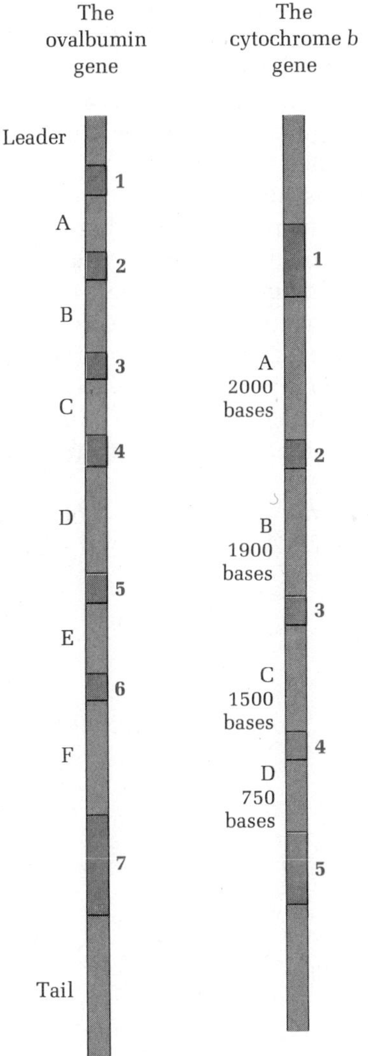

Figure 27-29
Intervening sequences (introns) in two eukaryotic genes. The gene for ovalbumin has six introns (gray, A to F) and thus is split into seven pieces or exons (color, 1 to 7). The gene for cytochrome b has four introns (gray, A to D) and five exons (color, 1 to 5). In both cases the introns contain most of the DNA. The number of bases in the introns of the cytochrome b gene are shown.

The function of the introns is not entirely clear, and they are the subject of much speculation. One idea is that they contain regulatory signals. Another is that introns separate genes into exchangeable units ("minigenes") that can be recombined to make new genes during the evolution of species. Whatever the function of introns and exons, they pose some problems in the transcription of genes (Chapter 28).

The Base Sequences of Some DNAs Have Been Determined

In 1977 the base sequence of an entire DNA molecule, that of bacteriophage ϕX174 (see page 790), was determined. This extraordinary achievement signaled the opening of a new era in the biochemistry of genes and chromosomes. Since then the base sequences of a number of different genes have been determined, and it now appears possible, in principle, to determine the base sequence of any DNA.

Before 1977 the base sequences of many transfer RNAs and several small mRNAs had been determined. Robert Holley and his colleagues were the first to determine the base sequence of a nucleic acid, a yeast tRNA for alanine. This accomplishment, completed in 1965, required several years of work. Although tRNAs have fewer than 100 nucleotide units, they contain many unusual modified bases, which had to be identified. However, the sequencing of DNAs presented difficulties of a different kind. We have seen that the average gene in *E. coli* has about 1200 nucleotide pairs and the entire DNA molecule of bacteriophage ϕX174 has over 5000. In the past no procedures were available for selectively cleaving DNA at a specific nucleotide, let us say, at all A residues. Even if such a method were available, a very large number of small fragments would result, which would be exceedingly difficult to separate. Moreover, even if they could be separated and sequenced, ordering these fragments into the correct overall sequence would be virtually impossible.

Figure 27-30
Simian virus 40 produces cancers in hamsters and other small animals. It is one of the smallest cancer-causing viruses. It has an icosahedral (20-sided) protein coat.

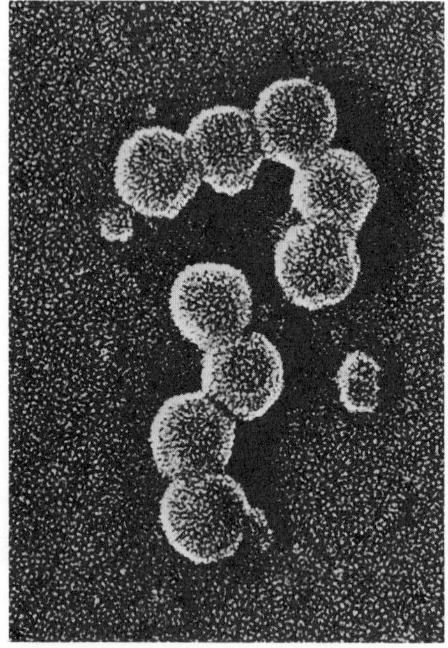

Three major developments made a breakthrough possible. The first was the discovery of restriction endonucleases, which cleave DNA molecules only at a relatively few specific points. Use of two or more different restriction endonucleases (Table 27-7) has made possible the fragmentation of DNA molecules in different ways to yield sequence overlaps, just as the use of two different proteolytic enzymes, such as trypsin and chymotrypsin, allowed fragmentation of polypeptide chains in different ways to establish amino acid sequence overlaps (page 133). As an example, the DNA of simian virus 40 (SV40), an animal virus (Figure 27-30) that can transform some cells into a malignant state, has been cleaved at different specific points with restriction endonucleases to yield fragments useful in determining the position of individual genes. Figure 27-31 shows cleavage maps of SV40 DNA obtained with three restriction endonucleases.

The second major advance was the refinement of electrophoretic methods for separating DNA fragments on the basis of

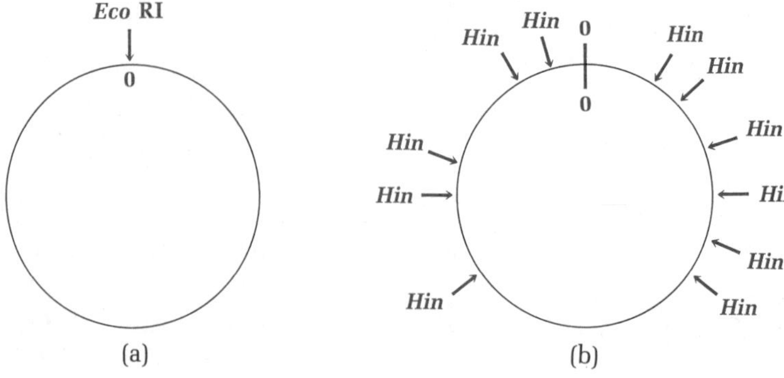

Figure 27-31
Points of cleavage of the circular SV40 DNA molecule produced by three different restriction endonucleases, Eco RI, Hin, and Hpa, each of which recognizes and catalyzes a double-strand cleavage of a specific restriction sequence. (a) Eco RI cleaves at only one point in the circular SV40 genome, causing its conversion into a linear form. This is the arbitrary zero point, to which the cleavage points of other endonucleases are referred. (b) Hin (a mixture of Hin dII and Hin dIII) cleaves at 11 points, thus yielding 12 fragments. (c) Hpa cleaves at only 4 points, yielding 5 fragments. These were the first endonucleases used by Daniel Nathans and his colleagues in mapping the genes in SV40 DNA. The cleavage points of many other endonucleases on SV40 DNA are now known.

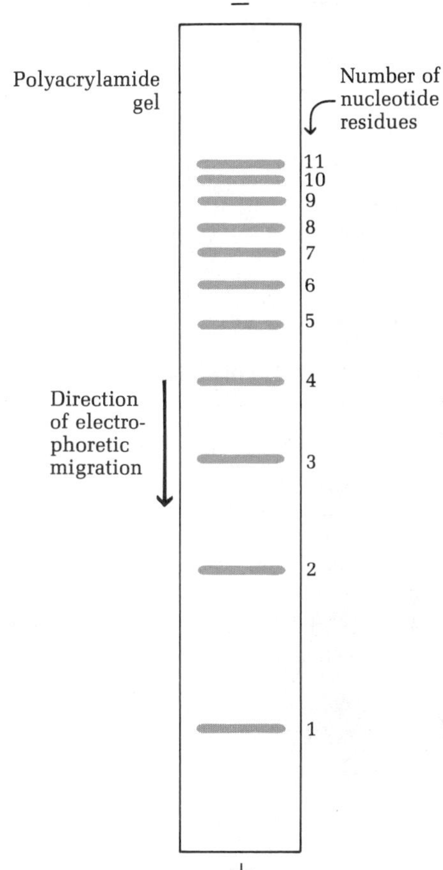

Figure 27-32
Electrophoretic separation of oligonucleotides on the basis of chain length. It is carried out on gel blocks of polyacrylamide. The shortest nucleotide travels fastest toward the positive electrode. By varying the porosity of the gel polymer the procedure can be adapted to separate much longer oligonucleotides up to 200 or more residues, even if they differ by only one residue.

the *number* of nucleotide units they contain. These methods are so sensitive that DNA fragments up to 200 nucleotides long can be separated from each other, even if they differ by only one nucleotide (Figure 27-32).

The third advance was the technique of DNA cloning (Chapter 30), which made possible the preparation of relatively large amounts of pure genes as starting material for sequencing. Two basic approaches for sequencing DNAs have been developed, each having a number of variations. Frederick Sanger, the first to determine the amino acid sequence of a protein, namely, insulin (page 134), also became the first to deduce the base sequence of a DNA molecule, that of the phage ϕX174, reported in 1977. Sanger and his colleagues developed the *chain-termination method*, also referred to as the "plus-minus" system, a very elegant procedure. Independently, Alan Maxam and Walter Gilbert in the United States developed a somewhat different procedure, called the *chemical method*. Both procedures begin with fragments of DNA obtained by cleavage of the starting DNA with restriction endonucleases. Box 27-1 shows the working principle of the Maxam and Gilbert sequencing method.

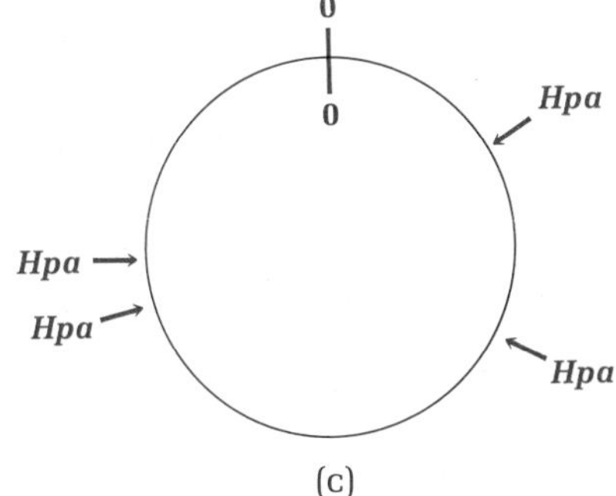

Box 27-1 The sequencing of a short fragment of DNA by the chemical method of Maxam and Gilbert

The following description omits some details in order to focus on the basic principle of the method. Assume that we have a fragment of DNA obtained by the action of a restriction endonuclease and that it has 10 residues with the sequence

5′ end 3′ end
G-A-T-C-A-G-C-T-A-G

The first step is the attachment of a radioactive label to the residue at the 5′ end, which we indicate by the colored shading of the G residue at the 5′ end

G-A-T-C-A-G-C-T-A-G

The 5′-labeled oligonucleotide is now divided into four portions. The first portion is treated by a chemical procedure that cleaves the oligonucleotide into smaller pieces by randomly deleting C residues. The labeled oligonucleotide shown above, when cleaved by the C-specific method, can result in the following mixture

G-A-T-C-A-G
T-A-G
G-A-T
A-G-C-T-A-G
A-G

In this set of fragments the labeled 5′-terminal residue is indicated by color. Note that two of the fragments are labeled and thus include the 5′ end of the original oligonucleotide, whereas T-A-G, A-G-C-T-A-G, and A-G are unlabeled and thus have lost the 5′ end. We are concerned only with the labeled fragments.

The second portion of the starting labeled oligonucleotide is treated by another chemical procedure that deletes only G residues, to yield another set of labeled fragments, as shown in Figure 1; the unlabeled fragments are not shown. This process is now repeated with the third portion of the starting labeled oligonucleotide, which is fragmented by a procedure excising only A residues. Similarly, the fourth portion is cleaved by excising only T residues. We now have four different mixtures of labeled fragments, produced by four different chemical procedures (Figure 1).

Each of the four mixtures of fragments is now subjected to an electrophoretic separation procedure in a gel block, under conditions in which the nucleotide fragments are separated according to the *number* of nucleotide residues they contain, regardless of the identity of their bases; in this procedure the smallest fragments will move at the highest rate. The precise position of each of the labeled fragments in the gel support is then determined by autoradiography on a photographic film. The positions of the labeled fragments resulting from each type of cleavage are shown in Figure 1. The positions of the unlabeled fragments are not detected by this process; in any case they are not needed for the sequence analysis.

The results of the first fragmentation, in which C residues were excised, are shown schematically in Figure 1, compared with the calibrated electrophoretic pattern corresponding to nucleotides having from 1 to 10 residues. Two labeled fragments were found in the first electrophoretic pattern. These of course contained the 5′ end of the starting polynucleotide. They were found in positions in the gel block corresponding to oligonucleotides having three and six residues. Both these labeled fragments originally had a C as the *next* residue toward the 3′ end, because the procedure that was used to fragment the starting oligonucleotide excised only C residues. From this information we now know that residues 4 and 7, starting from the 5′ end of the original oligonucleotide, must have been C residues.

Exactly the same procedure is now repeated with the three other sets of fragments obtained when the original oligonucleotide was cleaved by methods specific for deleting G, A, and T residues, respectively. The results are tallied in Figure 1. We see that the labeled fragments obtained by deletion of G residues moved at rates indicating that they had five and nine nucleotide units; thus, residues 6 and 10 of the original oligonucleotide must have been G. The third set of fragments, obtained on excising A's, showed that residues 2, 5, and 9 were A, and the fourth set obtained by excising T's tells us that residues 3 and 8 were T. At the bottom of Figure 1 is shown the base sequence of the original oligonucleotide deduced from this simple procedure and logic. The base sequence of oligonucleotides having 200 or more residues can be determined in this manner, often in less than 2 days.

To deduce the sequence of an entire DNA molecule it is fragmented by a restriction endonuclease. Each of the resulting fragments is then sequenced separately in the manner shown in Figure 1. A second sample of the original DNA is then cleaved, but at other points, by a different restriction endonuclease, to obtain a second set of fragments. After all the fragments of the second set have been sequenced, they can provide the overlaps required to assemble the first set of fragments into their proper order, thus giving the base sequence of the original native DNA. Sometimes a third or fourth set of fragments must be analyzed to resolve portions of the sequence unresolved by the first two cleavages.

Figure 1

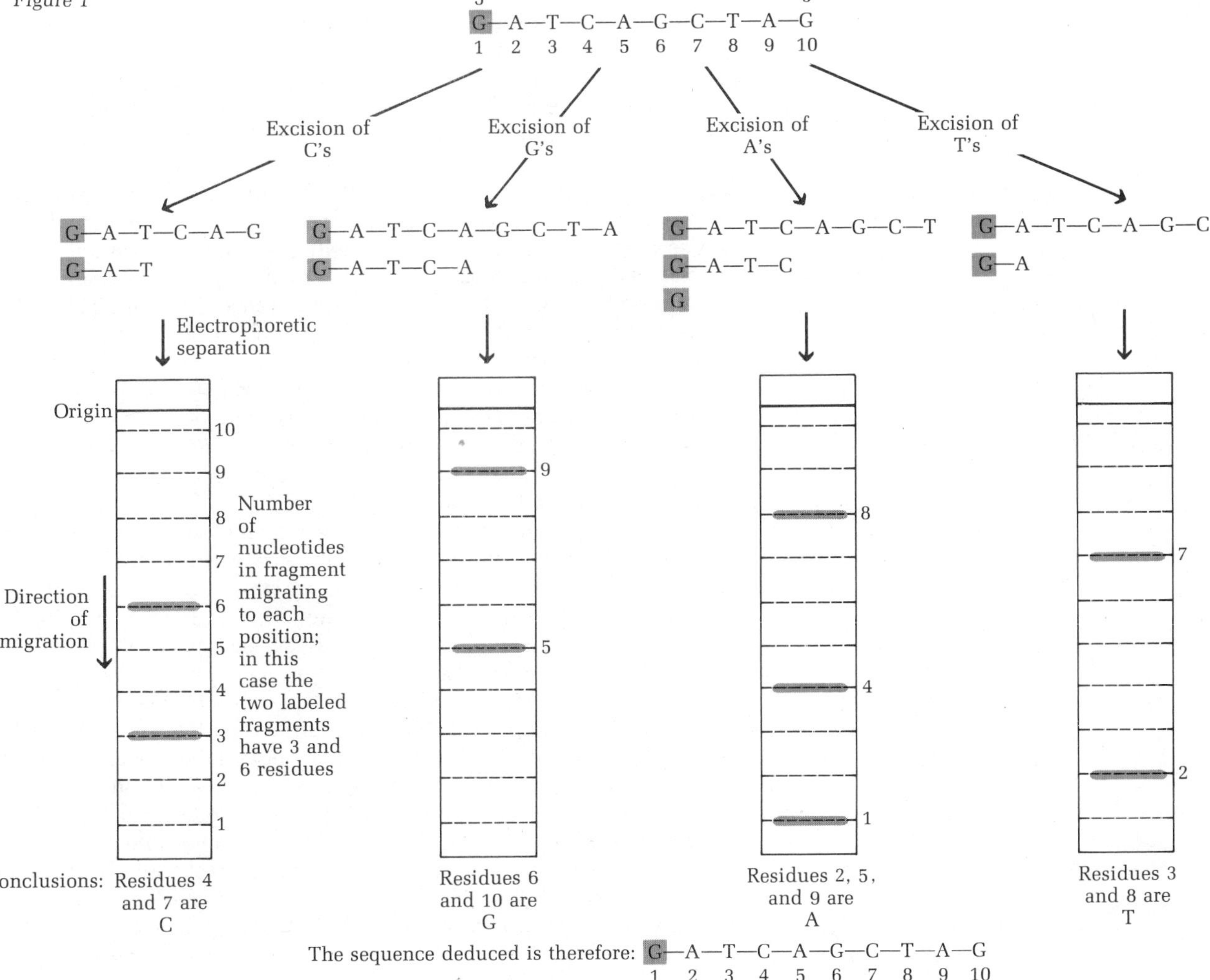

Summary

Several lines of evidence show that DNA bears genetic information. The Avery-McLeod-MacCarty experiment showed that DNA isolated from one strain of a bacterium can enter and transform the cells of another strain, endowing it with some of the heritable characteristics of the donor. The Hershey-Chase experiment showed that the DNA of a bacterial virus, but not its protein coat, carries the genetic message for replication of the virus in its host cell. All somatic cells of a given species of organism contain DNA having the same base composition, which is not modified by diet or environmental circumstances. Although the base composition of DNA varies characteristically from one species to another, in all double-stranded DNAs the number of adenine residues always equals the number of thymine residues; similarly, guanine residues equal the cytosine residues.

From x-ray analysis of DNA fibers and the base equivalences in DNA, Watson and Crick postulated that native DNA consists of two antiparallel chains in a double-helical arrangement. The complementary bases A-T and G-C are paired by hydrogen bonding within the helix, and the hydrophilic sugar-phosphate backbone is located on the outside. The base pairs are closely stacked perpendicular to the long axis, 0.34 nm apart; there are about 10 base pairs in each complete turn of the double helix. The complementary nature of the double helix suggests a mechanism for accurate replication of the two strands.

Native DNA undergoes reversible unwinding and separation of strands on heating or at extremes of pH. Because G≡C base pairs are more stable than A═T pairs, the melting point of DNAs rich in G≡C pairs is higher than that of DNAs rich in A═T pairs. A denatured single-strand DNA from one species can form a hybrid duplex with a denatured DNA strand from another species having some homology of base sequence. The degree to which such hybrids form is the basis for testing possible relationships between different species and for testing homology between DNAs and RNAs.

In different DNA viruses of bacteria the single molecule of DNA may be a circular or linear duplex; some viral DNAs, such as that of ϕX174, are single-stranded. Viral DNAs are supercoiled, which aids in packaging the DNA in the interior of the virion. In bacteria the single chromosome is a much larger covalently closed circular duplex. The DNA of bacteria is folded into many loops, each independently supercoiled. Eukaryotic cells have multiple chromosomes, each a single, very large linear DNA, 4 to 100 times larger than the single chromosome of prokaryotes. Eukaryotic DNA is wound around a set of basic proteins, the histones, at regular intervals to form nucleosomes.

Structural genes are segments of DNA that code for polypeptide chains, tRNAs, and ribosomal RNAs. Viral DNAs have relatively few genes, but *E. coli* probably has 3000 or more, many of which have been mapped at specific points on the circular chromosome. Many bacterial species protect their own DNA by methylating specific bases at distinctive locations, through the action of modification methylases, and destroying foreign DNAs lacking such identifying methyl groups, by the action of restriction endonucleases. Eukaryotic DNA has many highly repetitive short sequences, an intermediate number of longer, moderately repetitive sequences, which have been proposed to play a regulatory role, and a number of unique, single-copy segments that appear to be structural genes. Eukaryotic genes contain intervening nontranslated nucleotide sequences called introns. The nucleotide sequences of some genes and viral DNAs have been determined by newly developed procedures.

References

General

AYALA, F., and J. KIGER: *Modern Genetics*, Benjamin-Cummings, Menlo Park, Calif., 1980. Excellent description of fundamentals.

KORNBERG, A.: *DNA Replication*, Freeman, San Francisco, 1980. Comprehensive, authoritative, and encyclopedic.

Packaging

BAUER, W. R., F. H. C. CRICK, and J. H. WHITE, "Supercoiled DNA," *Sci. Am.*, **243**:118–133, July (1980).

Plasmids

NOVICK, R. P.: "Plasmids," *Sci. Am.*, **243**:102–127, December (1980).

Restriction-Modification Enzymes

NATHANS, D.: "Restriction Endonucleases, Simian Virus 40, and the New Genetics," *Science*, **206**:903–909 (1979).

SMITH, H.: "Nucleotide Sequence Specificity of Restriction Endonucleases," *Science*, **205**:455–462 (1979).

Isolation and Synthesis of Genes

KHORANA, H. G.: "Total Synthesis of a Gene," *Science*, **203**:614–625 (1979).

Repetitive Sequences

BRITTEN, R. J., and D. E. KOHNE: "Repeated Segments of DNA," *Sci. Am.*, **222**:24–31, April (1970).

DAVIDSON, E. H., and R. J. BRITTEN: "Possible Role of Repetitive Sequences," *Science*, **204**:1052–1059 (1979).

Intervening Sequences

CATTERALL, J. F., and colleagues: "The Chick Ovomucoid Gene Contains at Least Six Intervening Sequences," *Science*, **204**:264–271 (1979).

CRICK, F.: "Split Genes and RNA Splicing," *Science*, **204**:264–271 (1979).

Chromatin and Nucleosomes

KORNBERG, R. D., and A. KLUG: "The Nucleosome," *Sci. Am.*, **244**: 52–78, February (1981).

OLINS, D. E., and A. L. OLINS: "Nucleosomes: The Quantum Beyond DNA," *Am. Sci.*, **66**:704–711 (1978).

Nucleotide Sequence of DNAs

FIDDES, J. C.: "The Nucleotide Sequence of a Viral DNA," *Sci. Am.*, **237**:55–67, December (1977). Excellent account of how the nucleotide sequence of ϕX174 DNA was unraveled.

MAXAM, A. M., and W. GILBERT: "A New Method for Sequencing DNA," *Proc. Natl. Acad. Sci. US*, **74**:560–564 (1977).

REDDY, V. B., and colleagues: "The Genome of Simian Virus 40," *Science*, **200**:494–502 (1978).

SANGER, F.: "Determination of Nucleotide Sequences in DNA," *Biosci. Repts.*, **1**:3–18 (1981).

Historical

CAIRNS, J., G. STENT, and J. D. WATSON (eds.): *Phage and the Origins of Molecular Biology*, Cold Spring Harbor Laboratory, New York, 1966. The beginnings of molecular genetics and some of the people who helped bring the new era to flower.

JUDSON, H.: *The Eighth Day of Creation*, Simon & Schuster, New York, 1979.

OLBY, R.: *The Path to the Double Helix*, University of Washington Press, Seattle, 1974.

WATSON, J. D.: *The Double Helix*, Atheneum, 1968. A personal account of the human aspects of the discovery.

Problems

1. *Base Pairing in DNA* In samples of DNA isolated from two unidentified species of bacteria, adenine makes up 32 and 17 percent, respectively, of the total bases. What relative proportions of adenine, guanine, thymine, and cytosine would you expect to find in the two DNA samples? What assumptions have you made? One of these bacteria was isolated from a hot spring (64°C). Which DNA came from this organism, a thermophilic bacterium? What is the basis for your answer?

2. *Base Sequence of Complementary DNA Strands* Write the base sequence of the complementary strand of double helical DNA in which one strand has the sequence (5′)ATGCCGTATGCATTC(3′).

3. *DNA of the Human Body* Calculate the weight in grams of a double-helical DNA molecule stretching from the earth to the moon (~200,000 mi). The DNA double helix weighs about 1×10^{-18} g per 1000 nucleotide pairs. There are 1.6×10^{12} nm per mile and each base pair extends 0.34 nm. For an interesting comparison, your body contains about 0.5 g of DNA!

4. *How Long Is the Ribonuclease Gene?* What is the minimum number of nucleotide pairs in the gene for pancreatic ribonuclease (124 amino acids long)? Why might the number of nucleotide pairs be much larger than your answer? What is the uncertainty due to?

5. *Packaging of DNA in a Virus* The DNA of bacteriophage T2 has a molecular weight of 130 million. The head of T2 phage is about 100 nm long. Assuming the molecular weight of a base pair is 660, calculate the length of T2 DNA and compare it with the length of the T2 head. Your answer will show the necessity of very compact packaging of DNA in viruses.

6. *Packaging of DNA in Eukaryotic Cells* Compare the length of the DNA in a single nucleosome with the diameter of a nucleosome, 10–11 nm. Then compare the length of the total DNA in a human cell with the diameter of its nucleus, about 2 μm. In which structure is the packaging of DNA more compact?

7. *A Palindrome* Would the palindrome of Figure 27-28 be likely to assume the illustrated cruciform structure spontaneously in isolated pure DNA? Explain. What might your answer be if this palindrome occurred in a chromosome in an intact cell? Why?

8. *The DNA of Phage M13* Bacteriophage M13 DNA has the following base composition: A, 23 percent; T, 36 percent; G, 21 percent; C, 20 percent. What does this information tell us about the DNA of this phage?

9. *Separation of DNAs in a Density Gradient* Sedimentation in alkaline sucrose density gradients can separate DNAs according to their size and shape and also cause DNA to denature. This technique can be used to differentiate linear from circular DNA and to determine the relative size of DNA fragments. Replicative form II (RFII) of ϕX174 DNA is a double-stranded circle with a nick in one strand.
   (a) What types and lengths of molecules should be distinguishable upon sedimentation of RFII DNA in an alkaline sucrose density gradient? ϕX174 has 5386 base pairs.
   (b) What types and lengths of molecules will be found if the RFII DNA is first treated with a restriction enzyme that cleaves RFII only once?

10. *Sequencing of Bases in DNA* In DNA sequencing by the chemical method, why must the DNA be labeled at only one end, rather than being labeled uniformly throughout the molecule?

11. *Base Composition of ϕX174 DNA* Bacteriophage ϕX174 DNA occurs in two forms, single-stranded in the isolated virion and double-stranded during viral replication in the host cell. Would you expect them to have the same base composition? Give your reasons.

12. *Size of Eukaryotic Genes* An enzyme present in rat liver has a polypeptide chain of 192 amino residues. It is coded for by a gene having 1440 base pairs. Explain the relationship between the number of amino acid residues in this enzyme and the number of nucleotide pairs in its gene.

13. *Species Differences in DNAs* If two tubes containing DNA specimens carefully isolated from *E. coli* and the sea urchin (Table 27-3) accidentally lost their labels, how would you determine which was which?

14. *"Eyes" in Partially Denatured DNA* A sample of double-helical linear DNA carefully isolated from a species of crustacean was spread on a grid at 20°C and examined with the electron microscope. Another sample was kept at 60°C for 30 min and then

examined with the electron microscope. The following represent the images seen

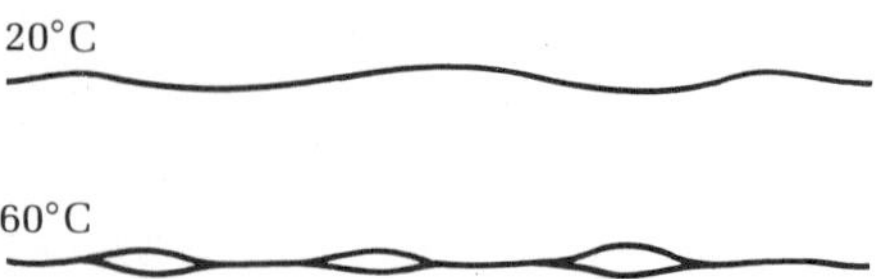

What is your interpretation of this outcome? How can this phenomenon give useful information?

15. *DNA Hybridization* From what you know of the structure of homologous proteins, why would you expect DNA strands from different vertebrate species to hybridize with each other to some extent?

16. *Action of a Restriction Endonuclease* A closed circular viral DNA is treated with a restriction endonuclease. This DNA has a single restriction site with the structure

5′–A–T–G–C–T–A–G–C–A–T–3′
3′–T–A–C–G–A–T–C–G–T–A–5′

(a) Indicate with a dot the probable center of the restriction site.
(b) How do you know that the point you have chosen is the center? What are its characteristics?
(c) After the restriction enzyme has cleaved both strands, the mixture is heated to destroy the enzyme and then slowly cooled. The viral DNA, on examination under the electron microscope, is circular. How do you account for this?
(d) When the final solution in part (c) is made alkaline with 0.1 *M* NaOH, it is found to contain only linear, single strands of DNA. How do you account for this finding?

17. *RNA Viruses: Can Genes Consist of RNA?* RNA viruses of *E. coli* contain no DNA, only RNA, which serves as the viral chromosome. This implies that these viruses have genes of RNA rather than DNA. Does this constitute a violation of the central dogma of molecular genetics? Give your reasons.

CHAPTER 28

Replication and Transcription of DNA

Now that we have examined the structure of DNA and the nature of chromosomes and genes, let us see how DNA is replicated to yield daughter molecules and transcribed to yield complementary messenger RNAs.

The enzymes and other proteins participating in replication and transcription of DNA are among the most remarkable biological catalysts known. Not only can they form these enormously long macromolecules from mononucleotide precursors at the expense of phosphate-group energy, but they can also transmit genetic information from the template strand to the new strand with extraordinary fidelity. In addition, these enzymes must solve complex mechanical problems, since the parental duplex DNA must be unwound in advance of the replication enzymes, so that they can gain access to the information encoded in the sequence of bases within the duplex. Moreover, in eukaryotic cells the replication system must cope with the complex three-dimensional organization of chromatin and the nucleosomes.

The transcription enzymes also have extraordinary properties. Not only can they make a large assortment of different RNAs, but they must begin and end their action at specific points on chromosomes in response to various regulatory signals, so that only certain genes are transcribed at certain times in the life cycle of the cell. DNA and RNA polymerases and the other proteins that help carry out the replication and transcription of DNA are thus vitally important instruments in the perpetuation of genetic information.

DNA Undergoes Semiconservative Replication

The Watson-Crick hypothesis proposed that each strand of double-helical DNA is used as a template for the replication of complementary daughter strands. In this way two daughter duplex DNA molecules identical to the parent DNA would be formed, each containing one intact strand from the parental

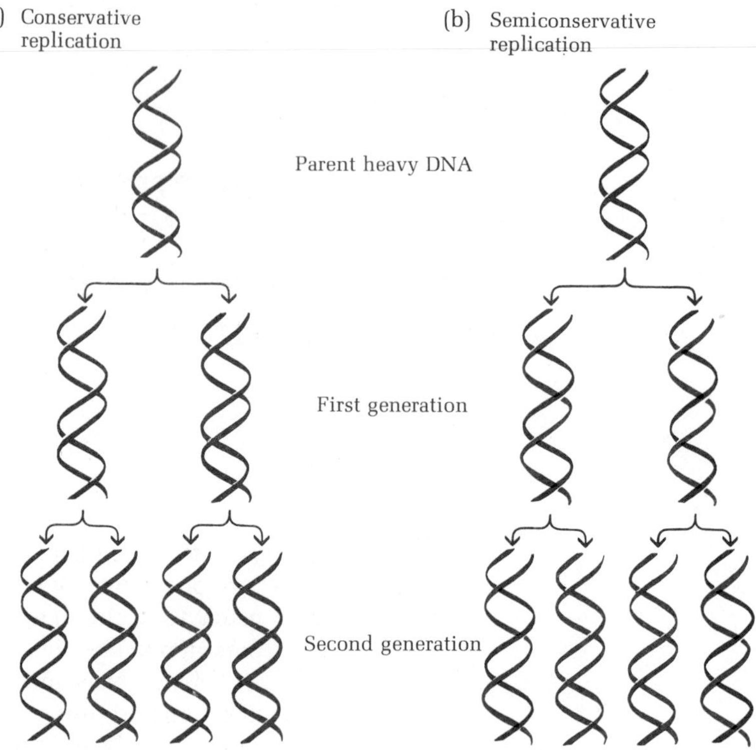

Figure 28-1
Principle of the Meselson-Stahl experiment to distinguish between two possible mechanisms of replication of heavy or [^{15}N]DNA (black) when replication takes place in a light or ^{14}N medium. Light strands are in color. (a) Conservative replication. If this occurs, each of the two heavy strands of parent DNA will be replicated to yield a DNA duplex containing two light strands and the original heavy duplex DNA. Continuation of conservative replication will yield in the next generation one heavy DNA and three light DNAs but no hybrid DNAs. (b) Semiconservative replication. This results in formation of two daughter duplexes, each containing one parent heavy strand and one light strand. The next generation will yield two hybrid DNAs and two light DNAs.

DNA. This hypothesis was verified in ingeniously designed experiments carried out by Matthew Meselson and Franklin Stahl in 1957. Figure 28-1 shows the principle of their experiments. They grew *E. coli* cells for several generations in a medium in which the ammonium chloride (NH_4Cl) used as sole nitrogen source contained ^{15}N, the "heavy" isotope of nitrogen, instead of the common, abundant isotope ^{14}N. Thus all the nitrogenous components of the cells grown in this medium, including the bases in their DNA, became highly enriched in ^{15}N. The DNA isolated from such cells has a density about 1 percent greater than that of normal [^{14}N]DNA. Although this is only a small difference, a mixture of heavy (^{15}N) and light (^{14}N) DNAs in a concentrated cesium chloride solution can be separated by centrifugation. Cesium chloride is used because its aqueous solution has a specific gravity close to that of DNAs. When such a CsCl solution is centrifuged for a long time at a high speed, it attains an equilibrium in which the CsCl forms a continuous density gradient. Because of the sedimentation force, the concentration of CsCl at the bottom of the tube is higher and the solution is therefore denser than at the top. A DNA specimen dissolved in the CsCl will come to an equilibrium position in the tube at which its density will be equal to that of the CsCl solution. Since the [^{15}N]DNA is slightly denser than [^{14}N]DNA, it reaches a lower equilibrium position in the CsCl gradient than the [^{14}N]DNA (Figure 28-2).

Meselson and Stahl transferred *E. coli* cells grown on the ^{15}N medium, in which all the DNA strands were heavy, to a fresh

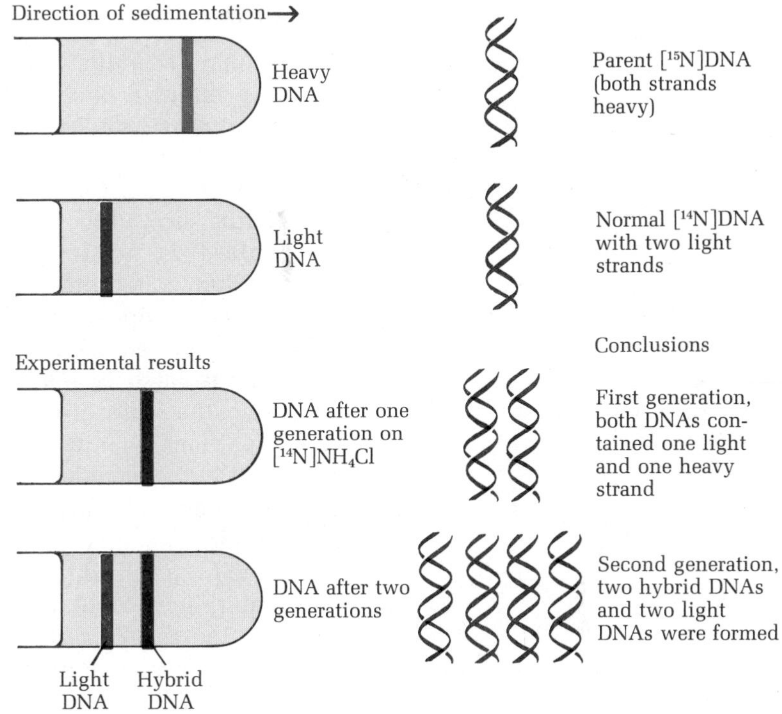

Figure 28-2
Results of the Meselson-Stahl experiment. The heavy or [^{15}N]DNA comes to equilibrium at a lower position than light or [^{14}N]DNA in the CsCl density gradient. Hybrid DNA equilibrates in an intermediate position. Test of the daughter DNAs in the first and second generations showed that DNA replication is semiconservative.

medium in which the NH_4Cl contained the normal ^{14}N isotope. They allowed these cells to grow in the ^{14}N medium until they had just doubled in number. The DNA was then isolated from the cells and its density analyzed by the sedimentation procedure described above. The DNA formed but a single band in the CsCl gradient, midway in density between normal "light" DNA containing ^{14}N and the "heavy" DNA of cells grown exclusively on ^{15}N (Figure 28-2). This is exactly the result expected if the double-helical DNA of the daughter cells contained one new ^{14}N strand and one old ^{15}N strand from the parent DNA (Figure 28-2).

When the cells were allowed to double in number again in the ^{14}N medium, the DNA isolated exhibited *two* bands, one having a density equal to that of normal light DNA and the other having the density of the hybrid DNA observed after the first cell doubling. Meselson and Stahl thus came to the conclusion that each daughter DNA duplex in the two generations of cells contained one parental strand and one newly made strand, in precise agreement with the Watson-Crick hypothesis. This type of replication is called *semiconservative,* since only one parental strand is conserved in each daughter DNA (Figures 28-1 and 28-2). Their observations clearly excluded *conservative replication,* in which one daughter DNA retains both original strands and the other daughter DNA duplex has the two new strands. They also excluded a *dispersive* mechanism in which each daughter DNA strand contains short segments of both the parental and new DNA spliced together at random.

Circular DNA Is Replicated Bidirectionally

We have seen that the DNA of bacteria and many DNA viruses is a circular double helix. This discovery raised questions about how circular DNA is replicated. Is the circular DNA first cleaved to yield a linear molecule, or can the circle be replicated as such? Important experiments carried out by John Cairns showed that the DNA of intact *E. coli* cells is replicated while it is in circular form. He allowed *E. coli* to grow in a medium containing thymidine labeled with the radioactive hydrogen isotope *tritium* (^3H). In this way the DNA in these cells was made radioactive. When the DNA of these cells was carefully isolated in relaxed form and spread on a photographic plate, the radioactive thymidine residues caused the formation of a "track" of silver grains on the exposed plate, an image of the DNA molecule. From these tracks, Cairns concluded that the intact chromosome was a giant circle, conforming to the circular genetic map of *E. coli* DNA, but radioactive DNA isolated from cells during replication showed an extra radioactive loop (Figure 28-3). Cairns concluded that the loop in the DNA was the result of the formation of two radioactive daughter strands, each complementary to a parent strand, as the replication "fork" moved around the parental DNA circle.

At first, this image was taken to mean that replication begins at a fixed origin or initiation point on the parental DNA and

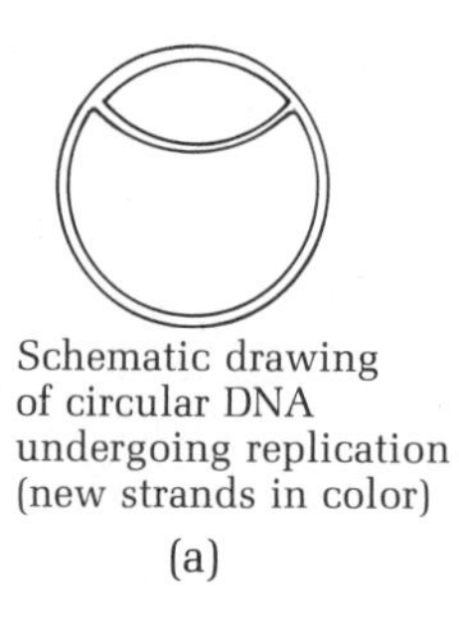

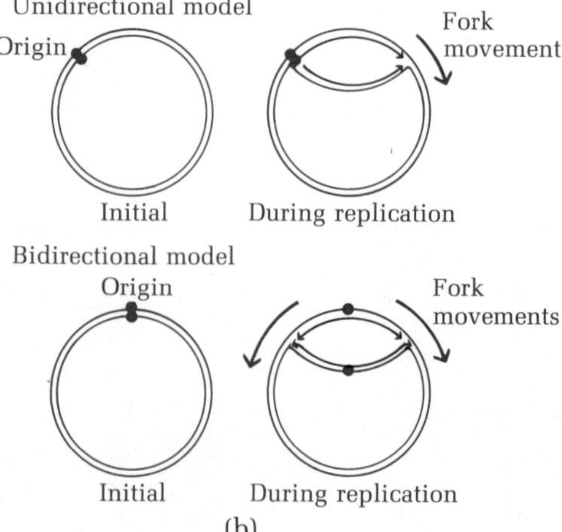

Figure 28-3
Replication of the E. coli chromosome. (a) Drawing of the track of the tritium-labeled chromosome of E. coli during replication. (b) Schematic interpretations of replication with the new daughter strands in color. In the unidirectional model only one replication fork proceeds from the origin. In the bidirectional model two forks begin from the origin and move in opposite directions until they meet. Chromosomes in E. coli and other bacteria, as well as those of many DNA viruses, are replicated bidirectionally.

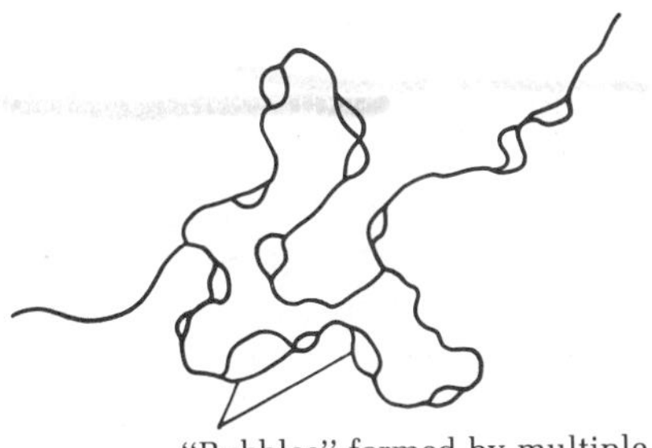

(a)

Parent DNA | Early stage in replication | Later stage | Daughter duplexes

Origin

(b)

Figure 28-4
Replication of a eukaryotic chromosome. (a) *Drawing of a replicating segment of DNA of* Drosophila melanogaster *eggs, showing "bubbles" or " eyes" formed by multiple points of bidirectional replication.* (b) *Bidirectional replication begins simultaneously at thousands of origins. It proceeds until the daughter strands* (color) *are complete. The new duplexes then separate; each contains a parent strand* (black) *and a daughter strand* (color).

that a single replication fork moves in one direction around the entire circular DNA molecule (Figure 28-3). More recent experiments on the chromosome from *E. coli* and from viruses, however, have led to the conclusion that replication usually occurs in a *bidirectional* manner; i.e., there are *two* replication forks. Both begin at the origin and proceed away from it in both directions simultaneously, until they meet (Figure 28-3). At this point the two completed daughter duplex circles separate, each containing an old and a new strand.

Chromosomes contain an *origin,* where the replication forks begin. This is a nucleotide sequence of 100 to 200 base pairs, without which a DNA cannot be replicated. The origin is recognized by specific cell proteins which initiate the replication cycle at this point. It is this initiation process at the origin which is under cellular regulation.

From the rate of movement of an *E. coli* replication fork, new DNA is made at the rate of over 45,000 nucleotide residues per fork per minute at 37°C. Since each full turn of the double helix has about 10 base pairs (page 806), the rate of unwinding of the parent DNA at a replication fork in *E. coli* cells is over 4500 r/min, which is faster than the rotation of the engine of an automobile going 70 mi/h. We can imagine that this very rapid unwinding and replication process might create some mechanical problems because of the double-helical nature of native DNA molecules. Later we shall see how cells deal with this problem.

Eukaryotic DNAs Have Many Origins of Replication

Clearly, the replication of eukaryotic DNA, which is organized into nucleosomes in the chromatin fibers, must be much more complex than replication of bacterial chromosomes. Nevertheless, radioautographic experiments, such as those of Cairns on *E. coli* replication, have also been carried out to determine how DNA replication proceeds in eukaryotes. Such experiments have shown that the eukaryotic DNA also is replicated in a bidirectional manner, but the replication forks move very slowly, less than a tenth of the rate of those in *E. coli.* Because eukaryotic chromosomes are so large, almost 2 months would be required for their replication if there were only one pair of replication forks per chromosome. The answer to this problem came with the discovery that replication of a eukaryotic DNA begins at many origins simultaneously, perhaps over a thousand. From each origin two replication forks move simultaneously in opposite directions (Figure 28-4). In this way the replication of entire eukaryotic chromosomes can be completed in less time than replication of a bacterial chromosome. Since a eukaryotic cell has multiple chromosomes, all of them must be replicated simultaneously; thus many thousands of replication forks are at work simultaneously in the nucleus of a eukaryotic cell.

Sometimes DNA Is Replicated by a Rolling-Circle Process

Some viral DNAs are replicated by a unidirectional *rolling-circle mechanism,* one version of which is shown in Figure 28-5. One of the two strands of the circular parent DNA is first cleaved by an enzyme. New nucleotide units are then added to the 3′ end of the broken strand. The growth of the new strand around the circular template continuously displaces the 5′ tail of the broken strand from the rolling template circle. As growth of the new strand proceeds further, the separated 5′ tail now becomes a linear template for synthesis of a new complementary strand. The 5′ tail, now a duplex, is elongated until a daughter DNA strand complementary to one circumference of the circular template has been formed. The duplex tail can be cleaved away by an enzyme and another length of new DNA is begun on the 5′ end of the DNA strand. In this way many complementary copies of a single circular DNA strand can be rolled off the

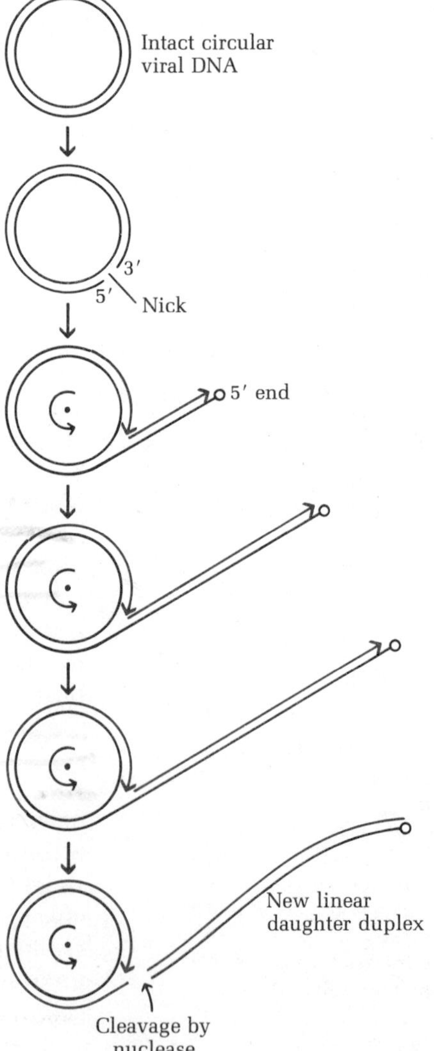

Figure 28-5
Rolling-circle replication, characteristic of some viral DNAs. A nick is formed in one strand, and new nucleotide units are added to the 3′ end. As the nicked parent strand is elongated in a manner complementary to the other parent strand, its loose 5′ end is displaced. Replication of the loose strand then begins as the tail continually peels away from the circle. Once a complete new daughter strand has been made, complementary to the tail, the new linear duplex can be cleaved away by a nuclease, yielding a linear viral DNA. The other new daughter strand now repeats this process: its 3′ end undergoes elongation and its 5′ end peels off and becomes the template for a daughter strand. Thus many new linear duplexes can be formed from the rolling-circle template. Rolling-circle replication also occurs during the synthesis of tandem repeated genes for ribosomal RNA in eukaryotes. In this case the new genes are not cleaved away but are retained in one unbroken strand. See pages 826 and 862.

template circle. This rolling-circle process also occurs in oocytes to amplify the action of the genes for ribosomal RNAs, by making many copies of them in tandem, so that a great many ribosomal RNAs can be made simultaneously. Oocytes must make a very large number of ribosomes quickly for fast synthesis of cell proteins during the rapid growth of the early embryo.

Bacterial Extracts Contain DNA Polymerase

Now that we have seen the gross features of DNA replication, let us examine the enzymatic mechanisms by which DNA is replicated. These were first opened to direct biochemical investigation by the important research of Arthur Kornberg and his colleagues, beginning in 1956. They incubated extracts of *E. coli* cells with mixtures of ^{32}P-labeled dATP, dTTP, dGTP, and dCTP, in which the ^{32}P was in the α-phosphate group (Figure 28-6). They found that a very small amount of new DNA was formed containing the ^{32}P isotope in its phosphate groups. The enzyme catalyzing this reaction, now called DNA polymerase I, was ultimately purified and its properties examined in detail. It was found to catalyze the addition of successive deoxyribonucleotide units to the end of a DNA strand, with simultaneous release of inorganic pyrophosphate, which contained the β- and γ-phosphate groups of each incoming deoxyribonucleoside 5′-triphosphate. The reaction equation in its simplest form is

Figure 28-6
Deoxyribonucleoside 5′-triphosphate labeled with ^{32}P in the α position.

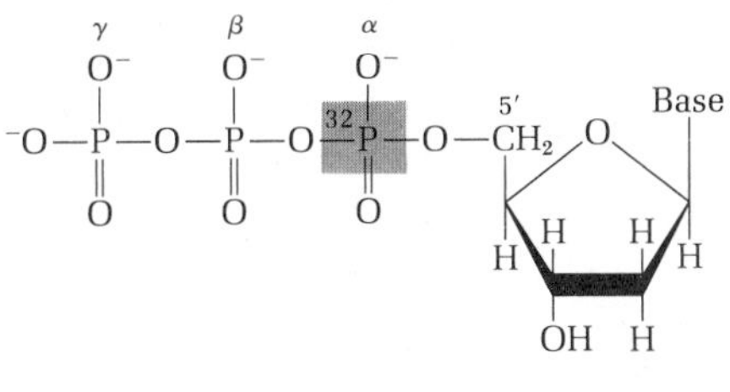

$$\underset{\text{DNA}}{(\text{dNMP})_n} + \text{dNTP} \rightleftharpoons \underset{\text{Lengthened DNA}}{(\text{dNMP})_{n+1}} + \text{PP}_i$$

where dNMP and dNTP denote deoxyribonucleoside 5′-phosphate and 5′-triphosphate, respectively. If any one of the four precursors was omitted, no new DNA was formed. Thus net DNA synthesis occurs only when all four nucleotide precursors are available. The 5′-triphosphates of the four deoxyribonucleosides cannot be replaced by the corresponding 5′-diphosphates or 5′-monophosphates, nor will the enzyme act upon ribonucleoside 5′-triphosphates. DNA polymerase requires Mg^{2+} for its action and also contains tightly bound Zn^{2+} in its active site.

DNA polymerase catalyzes the covalent attachment of new deoxyribonucleotide units by their α-phosphate groups to the free 3′-hydroxyl end of a preexisting DNA chain; the direction of synthesis of the DNA strand is thus from the 5′ end toward the 3′ end (Figure 28-7). The energy for the formation of each new phosphodiester backbone linkage is yielded by the cleavage of the pyrophosphate bond between the α- and β-phosphate groups of the precursor deoxyribonucleoside 5′-triphosphates. The pyrophosphate so formed is ultimately cleaved to phosphate, which can pull the reaction further in the direction of

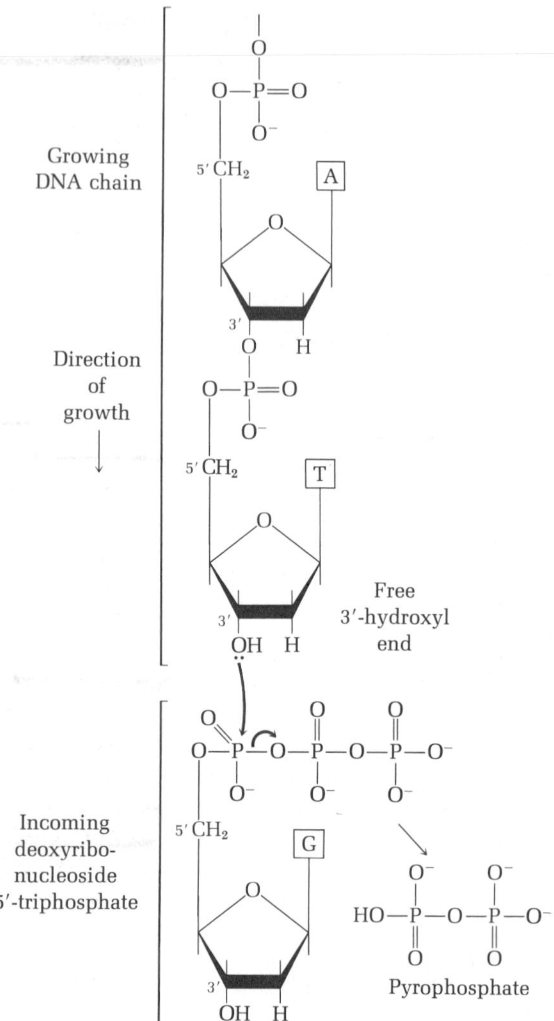

Figure 28-7
Elongation of a DNA chain by DNA polymerase. By the nucleophilic attack of the free 3′ hydroxyl group on the α phosphorus atom of the incoming deoxyribonucleoside triphosphate, the new internucleotide linkage is formed with the departure of free pyrophosphate. The template DNA strand is not shown.

completion. The very significant observation was made that the DNA polymerase reaction occurs only if there is some preformed double-helical DNA already present in the system.

Preformed DNA Is Necessary for the Action of DNA Polymerase

Kornberg and his colleagues studied the requirement of preformed DNA for the action of DNA polymerase and found that it plays two crucial roles in the polymerase reaction: as a *primer* and as a *template.*

One Strand of the Preformed DNA Serves as the Primer

DNA polymerase adds nucleotides sequentially to the 3′ end of one strand, the primer. Therefore, the synthesis of the new DNA strand is in the 5′ → 3′ direction. The DNA polymerase cannot start a new DNA strand by itself; it can only add to a

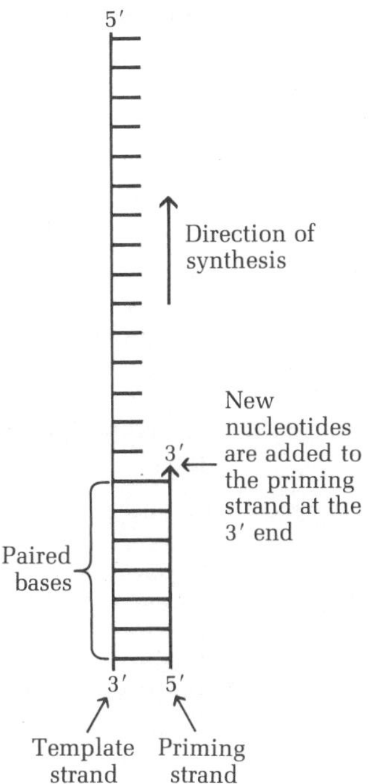

Figure 28-8
Structure of preformed duplex DNA required for the action of DNA polymerase. A primer strand is needed to which new nucleotide units are added. A single unpaired strand is also required to act as template.

preexisting strand and even then can add new nucleotide residues only if a template strand is available.

The Other Strand of the Preformed DNA Serves as a Template

Nucleotides are added to the primer strand in accordance with the base sequence of the template strand, following Watson-Crick base pairing. Wherever a thymine residue occurs in the template strand, an adenine residue is added to the daughter strand, and vice versa. Similarly, wherever a guanine residue is present in the template strand, a cytosine residue is inserted in the daughter strand, and vice versa. The product of DNA polymerase is thus a base-paired duplex. Figure 28-8 shows a schematic representation of the requirement of preformed DNA for the action of DNA polymerase.

Because of the requirement for both a primer strand and a free template strand, DNA polymerase alone cannot replicate an intact native chromosome, whether in the form of a double-strand circle, a single-strand circle, or an intact linear DNA duplex with all bases paired. These observations on the template and primer requirements of DNA polymerase raised many fundamental questions regarding the initiation and elongation of DNA strands.

DNA Replication Requires Many Enzymes and Protein Factors

Although these early discoveries of Kornberg and his colleagues opened the door to the direct study of DNA replication, we still do not have a complete picture of all the details of the replication process, even for DNA viruses, which have but a single small chromosome. Today, through the efforts of Kornberg and many other investigators, we know that replication requires not just a single DNA polymerase but perhaps 20 or more different enzymes and proteins, each performing a specific task in the complex process by which the parent DNA is replicated. Replication takes place in many sequential steps, which include recognition of the origin, unwinding of the parent duplex, holding the template strands apart, initiation of the new daughter strands, elongation of the daughters, rewinding, and the termination of replication. These all proceed at very high rates, and the replication process is extremely accurate. The entire complex of 20 or more replication enzymes and factors is variously called the *DNA replicase system* or the *replisome*. We shall now outline the major steps in replication as we know them today.

There Are Three DNA Polymerases in *E. coli*

E. coli cells contain three different DNA polymerases, designated I, II, and III (Table 28-1). DNA polymerase I, the most

Table 28-1 DNA Polymerases of *E. coli*†

	I	II	III complex
Activity catalyzed			
5′ → 3′ polymerization	+	+	+
5′ → 3′ exonuclease	+	−	+
3′ → 5′ exonuclease	+	+	+
Molecular weight	109,000	120,000	400,000
Molecules per cell	400		10
Activity (nucleotides per minute per molecule at 37°C)	600	30	9,000

† From A. Kornberg, *DNA Replication*, Freeman, San Francisco, 1980, p. 168.

abundant, is the enzyme described above. Although Kornberg has shown that this enzyme can replicate the entire DNA molecule of the small ϕX174 virus in the test tube to yield a biologically active daughter DNA, we know that DNA polymerase I is not the *principal* enzyme that normally carries out DNA elongation. DNA polymerase I does participate in replication but in a specialized manner, as we shall see later.

DNA polymerase III is the enzyme mainly responsible for DNA chain elongation in intact *E. coli*. It functions in the form of a large complex of molecular weight about 550,000 called *DNA polymerase III holoenzyme* (Figure 28-9). The DNA polymerase III system contains Zn^{2+} and requires Mg^{2+}. It also requires both a template and a primer strand and thus cannot initiate replication. Like DNA polymerase I, it also elongates DNA in the 5′ → 3′ direction by adding new units to the 3′ end of the primer strand. The holoenzyme has a number of subunits. Subunit β, or copolymerase III, is required to recognize and bind to a primer strand on the parental DNA. Once the holoenzyme is bound to the correct initiation position, copolymerase III is released in free form. Elongation of the daughter DNA strand is then carried out by the DNA polymerase III complex. Both DNA polymerases I and III have *three* enzymatic activities. In addition to polymerase activity, these DNA polymerases can also act as 5′ → 3′ exonucleases and as 3′ → 5′ exonucleases. They can therefore hydrolyze away terminal nucleotides from either end of a strand of DNA. Later we shall see the purpose of the exonuclease activities of DNA polymerases. The function of DNA polymerase II is not yet known.

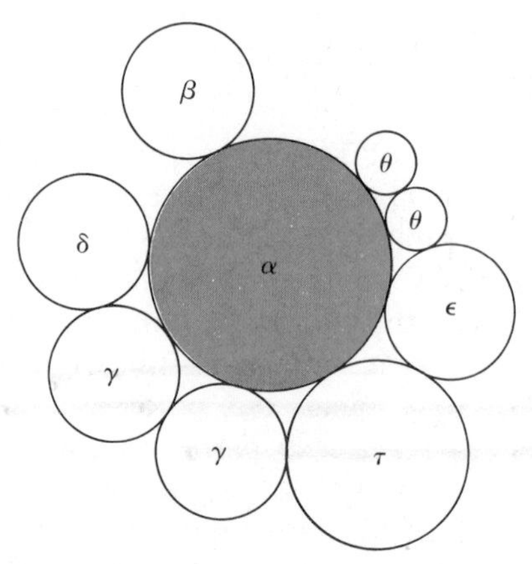

Figure 28-9
The DNA polymerase III holoenzyme. It consists of the polymerase itself (subunit α) plus a number of other subunits.

Now we shall examine some of the problems that arose en route to our present understanding of DNA replication.

Simultaneous Replication of Both Strands of DNA Poses a Problem

The fact that the DNA polymerases can add new units to DNA only on the 3′ end raises an important question. We have seen

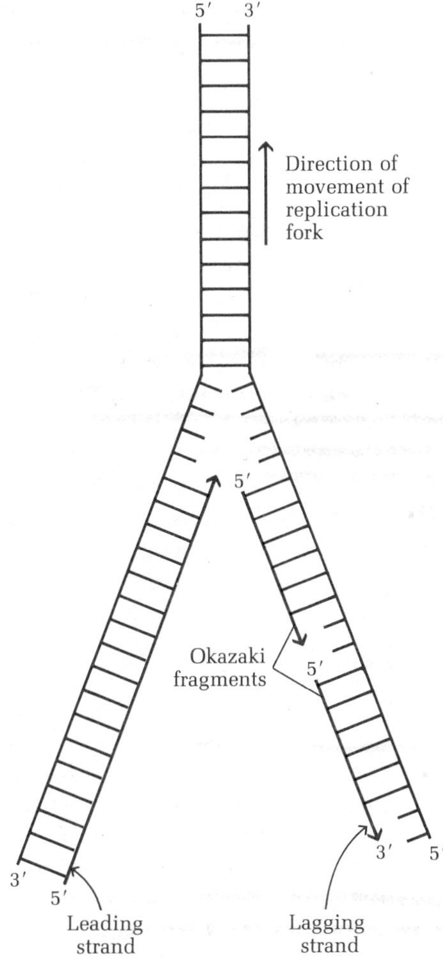

Figure 28-10
Discontinuous replication of one strand of DNA in short pieces. The strand that is continuously replicated (in the direction taken by the replication fork) is the leading strand. The other strand is replicated discontinuously in short pieces (Okazaki fragments), in a direction opposite to the direction of movement of the replication fork. The Okazaki fragments are then spliced together by DNA ligase, to become the lagging strand. In prokaryotes the Okazaki fragments are about 1000 to 2000 nucleotides long. In animal cells they are 150 to 200 nucleotides long.

that when the chromosome of *E. coli* is replicated, both strands appear to be replicated simultaneously at the replication fork. We recall, however, that the two strands in duplex DNA are antiparallel. This means that the growing daughter strands are also antiparallel. One growing strand would have its 3′ end at the fork, the other would have its 5′ end at the fork. Since DNA polymerase can add nucleotides to the 3′ end but not to the 5′ end, it would appear that DNA polymerase can extend only one of the two growing strands in the direction of movement of the fork. Questions arise. Are there *two* classes of DNA polymerases, one adding new residues only to the 3′ end and the other adding residues at the 5′ end? Or is there a type of DNA polymerase that can replicate at both the 5′ and 3′ ends? Or is one strand replicated by enzymes moving in a direction opposite to the direction of movement of the replication fork?

The Discovery of Okazaki Fragments Solved the Problem

The answer to these questions came from important observations made by Reiji Okazaki in Japan. He discovered that during replication of DNA in *E. coli* and other bacteria much newly formed DNA could be detected in the form of small pieces. These pieces of DNA, now called *Okazaki fragments,* are about 1000 to 2000 bases long. Okazaki postulated that these fragments represent short lengths of DNA that are replicated in a discontinuous manner and then are spliced together. This discovery ultimately led to the conclusion that one strand of the DNA is replicated *continuously* in the 5′ → 3′ direction, moving in the same direction as the replication fork; this is called the *leading* strand. The other strand is made *discontinuously* in short pieces, also by adding new units to the 3′ end, but in the direction *opposite to* the movement of the replication fork. The Okazaki fragments are then enzymatically spliced together to make the second daughter strand, which is called the *lagging* strand (Figure 28-10). Okazaki fragments are also formed in animal cells, but they are much shorter, less than 200 nucleotides.

Synthesis of Okazaki Fragments Requires an RNA Primer

Now things become more complicated, with an unexpected turn. After the discovery of the Okazaki fragments a new problem arose. Since DNA polymerase cannot initiate a new strand, how is each Okazaki fragment initiated? Unexpectedly it was found that the formation of Okazaki fragments in cell extracts required the presence not only of dATP, dGTP, dCTP, and dTTP but also a mixture of the ribonucleoside 5′-triphosphates (ATP, GTP, CTP, and UTP). These and other observations suggested that RNA synthesis is somehow required for DNA synthesis.

This was ultimately found to be true. The synthesis of Okazaki fragments requires as a primer a short length of RNA, com-

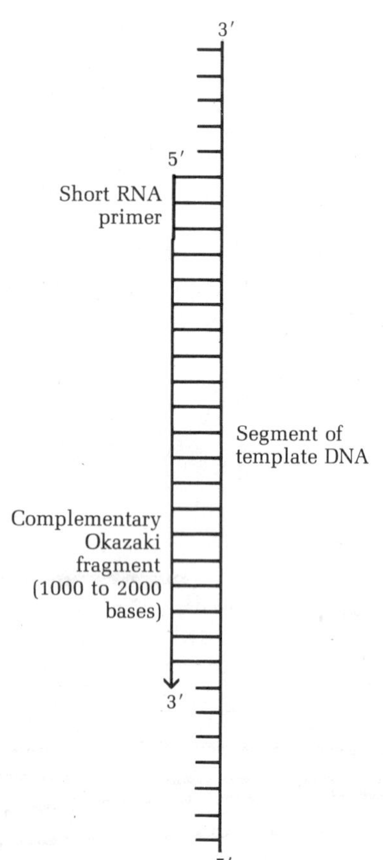

Figure 28-11
Priming of the synthesis of each Okazaki fragment by a short length of complementary RNA made by primase, which requires no primer itself.

plementary to the DNA template strand. The RNA is made in the 5′ → 3′ direction from ATP, UTP, CTP, and GTP precursors by an enzyme called primase. To the 3′ end of this short single-strand RNA primer are now added successive deoxyribonucleotide units complementary to the template DNA strand. As shown in Figure 28-11, the RNA primer usually has only a few ribonucleotide residues, to which are then added, by the action of DNA polymerase III, from 1000 to 2000 deoxyribonucleotide units, to complete the Okazaki fragment. The base sequence of the newly made Okazaki fragment is of course complementary to the sequence of bases in the template strand. The RNA primer is then removed, nucleotide by nucleotide, by the 5′ → 3′ exonuclease activity of DNA polymerase I. As each ribonucleotide unit is removed, it is replaced with complementary deoxyribonucleotides by DNA polymerase I acting in its polymerase mode and using the 3′ end of the preceding Okazaki fragment as a primer. However, DNA polymerase I cannot make the final covalent splice that joins the Okazaki fragment to the growing DNA strand; another enzyme is required.

The Okazaki Fragments Are Spliced by DNA Ligase

The new Okazaki fragment is joined to the lagging strand by the enzyme *DNA ligase,* which can form a phosphodiester bond between the 3′-hydroxyl group at the end of the DNA undergoing elongation and the 5′-phosphate group of the newly made Okazaki fragment. The formation of this new bond requires input of energy, which is provided by the coupled hydrolysis of the pyrophosphate bond of NAD^+ (in bacteria) or ATP (in animal cells) (Figure 28-12). The DNA ligase reaction, which is rather complex in its mechanism, functions most efficiently when the two DNA fragments to be joined are completely base-paired with the complementary template strand of DNA.

DNA ligase has other splicing functions, as we shall see later (Chapter 30).

Replication Requires Physical Separation of the Parent Duplex DNA

We have seen that circular duplex DNA is replicated in both directions simultaneously, so that the two replication forks travel around the circular chromosome toward each other. If we now recall that the DNA double helix is a tightly wound structure and that the coding bases are *within* the helix, it is clear that the two strands of the parent DNA must be separated for at least a short segment so that the DNA replicating enzymes can "read" the base sequence of the template. Although bacterial DNA is already negatively supercoiled, i.e., slightly untwisted, further unwinding is required as the replication fork moves forward.

Figure 28-12
Splicing an Okazaki fragment to the lagging strand by DNA ligase. The two nucleotides to be joined must be base-paired to the template strand. The reaction actually takes place in a number of steps.

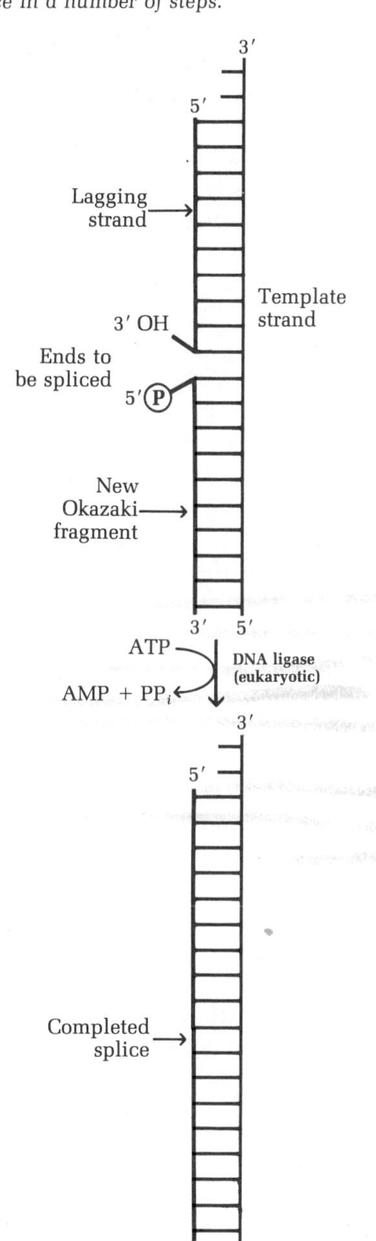

Figure 28-13
Summary of the major steps in DNA replication. There is some uncertainty regarding the exact location of DNA gyrase action.

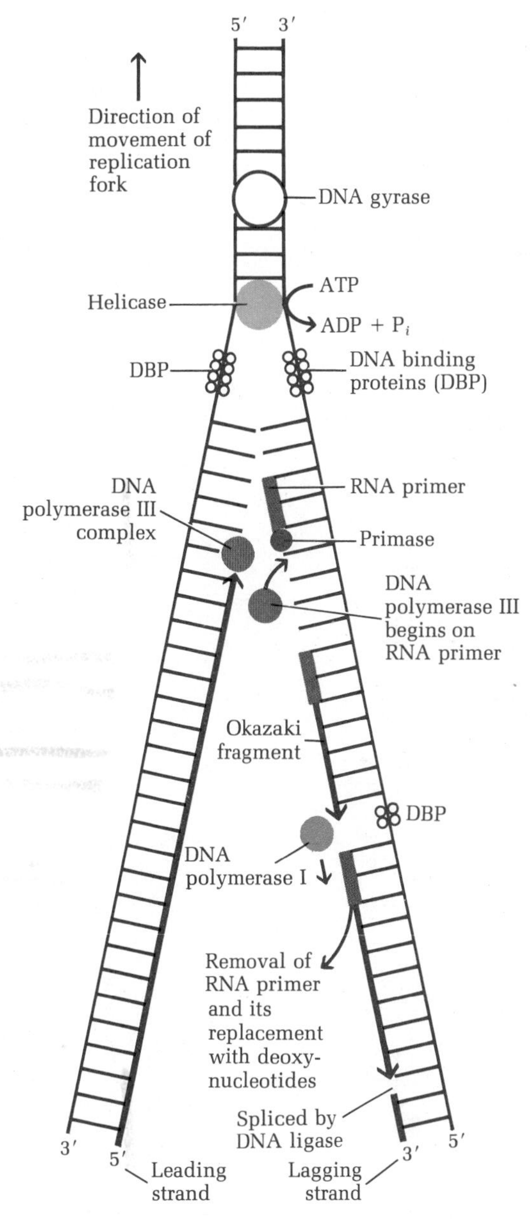

Unwinding the double helix and keeping the two strands apart so that they can be replicated is made possible by several other specialized proteins (Figure 28-13). Enzymes known as *helicases* unwind short segments of DNA just ahead of the replicating fork. The unwinding of DNA requires energy to separate each base pair, yielded by the hydrolysis of two molecules of ATP to ADP and phosphate. As soon as a short sequence has been unwound, several molecules of a *DNA binding protein*

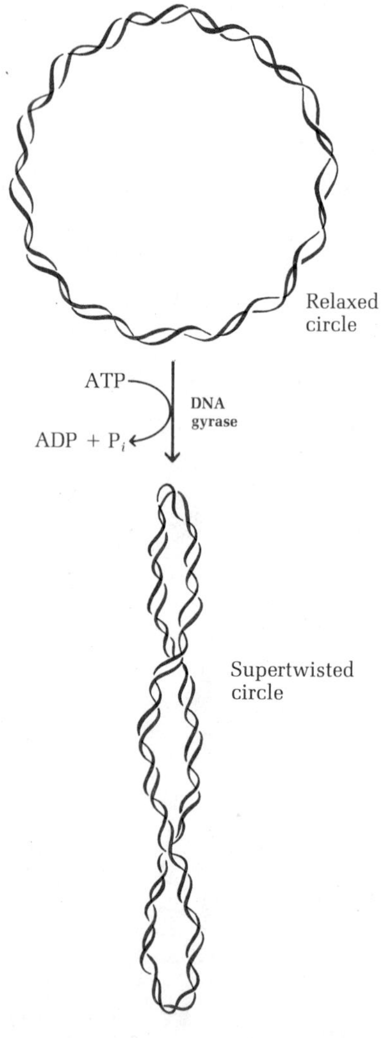

(DBP) bind tightly to each of the separated strands, preventing them from coming together into base pairing again. Their base sequences are thus exposed to the replication system. DNA polymerase can directly extend the leading strand by adding nucleotides to the 3′ end of the leading strand. Other specific proteins make possible the access of primase to the template of the lagging strand. Primase can then bind to the lagging strand and create the RNA primers for the Okazaki fragments. The unwinding of DNA is one of the most interesting aspects of DNA replication, but it is a very complex process. Much still needs to be done before we fully understand how this mechanical and biochemical problem has been solved by living cells. Figure 28-13 summarizes the major proteins required in DNA replication.

The rapid unwinding of the parental strands during replication (4500 r/min) creates another problem. Without some special "swivel" mechanism, the entire chromosome ahead of the replication fork would be forced to rotate at this speed. To counteract this problem it is thought that the cell introduces a "swivel" into the DNA, probably just ahead of the fork so that only a short sequence must rotate at this high speed. This is brought about by a transient break in one strand of the DNA, which is very quickly and accurately respliced after one or more revolutions. The transient break-and-splice is introduced by enzymes known as *topoisomerases*. In prokaryotes, the topoisomerase is known as *DNA gyrase*. Not only does this enzyme permit swiveling of the DNA, but it actively twists it in the direction which favors unwinding of the template strands at the fork. Thus gyrase helps helicase to unwind the DNA for replication. The twisting of the DNA by gyrase, which is coupled to the hydrolysis of ATP to ADP and P_i, also results in the supercoiling of the chromosome. Gyrase keeps all circular DNAs in bacterial cells in a supercoiled form (Figure 28-14).

As the replicase system completes each splice in the new lagging strand, it moves on along the DNA being replicated. The two new strands, together with their base-paired template strands, form two daughter duplexes automatically, each containing a new and a parental strand. No energy is required for formation of the new helixes, nor is a "winding" enzyme necessary.

Relaxed circle

ATP

DNA gyrase

ADP + P_i

Supertwisted circle

Figure 28-14
Generation of negative supertwists by DNA gyrase. The energy required is furnished by ATP hydrolysis. The enzyme nicks one strand and reseals it after the intact strand has been passed through the gap or nick.

DNA Polymerases Can Proofread and Correct Errors

It has been determined that the replication of DNA in *E. coli* proceeds with no more than about 1 error in every 10^9 to 10^{10} nucleotides. Since there are about 4.5×10^6 base pairs in the *E. coli* chromosome, only one wrong nucleotide would be inserted per 10,000 cells undergoing one division. It was long thought that this high degree of fidelity in replication of genetic information was entirely imparted by the precision of Watson-Crick base pairing between the template strand and the new daughter strand, but more recent analyses of the problem indicate that if fidelity of replication depended solely on the accuracy of base pairing, the error frequency would be much higher, about one wrong nucleotide for every 10^4 to 10^5 residues. Some other factor or factors must therefore operate in order to account for the much lower error frequency of replication in vivo.

More detailed research on the properties of highly purified DNA polymerases has yielded at least part of the answer. DNA polymerases I and III, we recall, have three different enzymatic activities. We have seen how the enzyme functions as a DNA polymerase and also how it can remove nucleotide residues from the 5′ end of a DNA segment. However, the 3′ exonuclease activity of DNA polymerases I and III was very puzzling, since it meant that these enzymes can also move backwards, taking off 3′-nucleotide residues in the direction opposite to the direction in which it functions as a polymerase. The 3′ exonuclease activity of DNA polymerases I and III is a device for proofreading the newly made DNA strand and for correcting errors made by the polymerase activity. As Figure 28-15 shows, if a "wrong" nucleotide is inserted by the DNA polymerase, the enzyme can recognize its failure to form a correct base pair with the corresponding nucleotide in the template. It then "backs up" and hydrolyzes off the wrong nucleotide from the 3′ end of the chain. The polymerase then proceeds to add the correct nucleotide as it resumes its normal progress in the 5′ → 3′ direction. Thus addition of each nucleotide is checked as the replication fork moves along the template strand. The proofreading function of the DNA polymerase is very efficient and contributes a factor of at least 10^4 in guaranteeing fidelity of replication. The total error rate is the product of the error rates of the polymerase and the proofreading functions, to give no more than one error in about 10^9 to 10^{10} residues.

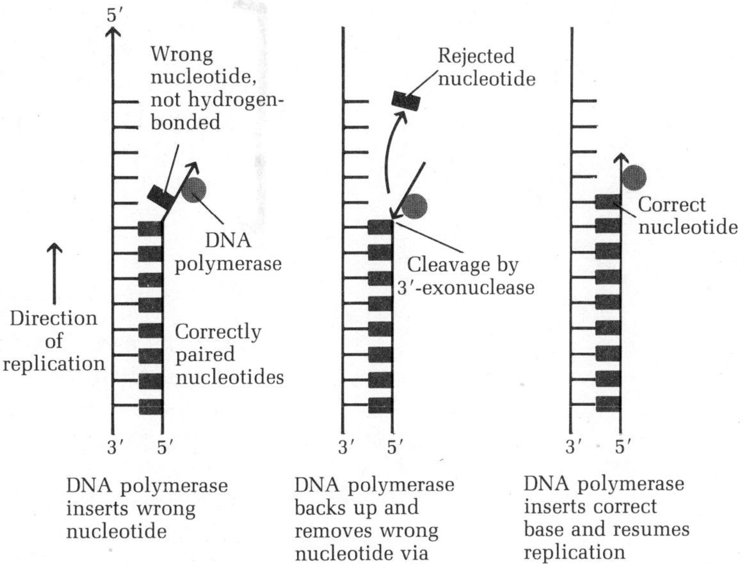

Figure 28-15
Error correction by the 3′ exonuclease activity of DNA polymerase.

It is much more important that replication proceed with high fidelity than transcription or translation. Frequent errors in replication would place the identity of a species or its viability at high risk. Errors in transcription or translation can be tolerated much more readily, since they affect the development of an RNA or a protein in only one cell, without altering the entire genetic lineage of the species. Proofreading by DNA polymerases may be only one of several ways for ensuring fidelity. In fact, the extreme complexity of the replication process, with its many participating proteins, may be required to ensure fidelity. It is interesting that some eukaryotic DNA polymerases do not carry out proofreading. Presumably eukaryotes have other mechanisms for ensuring accuracy of replication.

Replication in Eukaryotic Cells Is Very Complex

The replication of eukaryotic DNA must obviously be much more complex than replication in *E. coli,* since eukaryotic cells have many chromosomes, which are replicated simultaneously, and they are much larger than in prokaryotes. Moreover, short segments of the DNA are wound around the beadlike nucleosomes, which in turn are tightly packed into helical arrays (page 819). Replication of eukaryotic DNA thus involves a number of mechanical or geometrical changes, in addition to the untwisting of the DNA to expose its base sequence to the action of the replicase system.

Although we know very little yet about the specific steps and all the enzymatic factors required for these preliminary events in eukaryotic cells, there is much evidence that the basic enzymatic processes in the replication of eukaryotic DNA, once it is "opened" and made available, are similar to those in prokaryotes. Eukaryotic nuclei contain DNA polymerases, DNA ligase, and various unwinding enzymes and proteins. Eukaryotic DNA is also replicated via Okazaki fragments, which are much shorter than those in bacteria, although they are also initiated with RNA primers. In mammalian cells it has been calculated that the rate of movement of replication forks is only about 60 bases per second, or less than 1 μm/min. However, as we have seen (Figure 28-4), a thousand or more replication forks function simultaneously in eukaryotic chromosomes.

Genes Are Transcribed to Yield RNAs

Now let us turn to the next major stage in the flow of genetic information: *transcription* of the genetic information of DNA into the form of RNA. In this process an RNA strand having a base sequence complementary to one of the DNA strands is synthesized by an enzyme system. Transcription must be carried out faithfully if the cell is to have proteins with their normal genetically determined amino acid sequences. Three kinds of RNA are the products of transcription. The first is *messenger RNA* (mRNA), sent to the ribosomes for the purpose of coding the amino acid sequence of one or more polypeptides specified by a gene or set of genes in the chromosomes. Some 90 to 95 percent of the *E. coli* chromosome codes for messenger RNAs. The remainder codes for *transfer RNAs* and *ribosomal RNAs*, and also for regulatory sequences, leaders, spacers, and tails.

There is an important difference between replication and transcription. During replication the entire chromosome is copied to yield daughter DNAs identical to the parent DNA. But in the transcription process not all of the cell DNA is necessarily transcribed. Usually only individual genes or groups of genes are transcribed. Thus the transcription of DNA is selective, turned on by specific regulatory sequences indicating the beginning and end of the segments of DNA to be transcribed.

Messenger RNAs Code for Polypeptide Chains

That a form of RNA carries the genetic message for protein biosynthesis was first suggested by the fact that in eukaryotes DNA is almost entirely confined to the nucleus, whereas protein synthesis takes place largely on the ribosomes in the cytoplasm. Some macromolecule other than DNA must therefore carry the genetic message from the nucleus to the ribosomes. RNA was the logical candidate, since it is found in both the nucleus and cytoplasm. It was also observed that onset of protein synthesis in cells is accompanied by an increase in the RNA content of the cytoplasm and an increase in its rate of turnover. These and other observations led Francis Crick to propose, as part of the central dogma of molecular genetics, that RNA serves to carry genetic information from DNA to the process of protein biosynthesis in the ribosome. Later, in 1961, Francois Jacob and Jaques Monod proposed the name *messenger RNA* for that portion of the total cell RNA carrying the genetic information from DNA to the ribosomes, where the messengers provide the templates for the biosynthesis of polypeptide chains with specific amino acid sequences.

At any given time in growing *E. coli* cells there is an exceedingly complex mixture of hundreds of mRNAs, each coding for one or more polypeptide chains. For a long time it was virtually impossible to isolate a single type of mRNA molecule from this complex mixture, but today, by virtue of greatly improved methods, we have rather complete knowledge of mRNA structure.

Messenger RNAs are single-stranded molecules that vary greatly in length. In prokaryotes a single mRNA molecule may code for a single polypeptide or for two or more polypeptide chains. If it carries the code for only one polypeptide, the mRNA is *monogenic* or *monocistronic;* if it codes for two or more different polypeptides, the mRNA is *polygenic* or *polycistronic*. The minimum length of an mRNA is set by the length of the polypeptide chain for which it codes. For example, a poly-

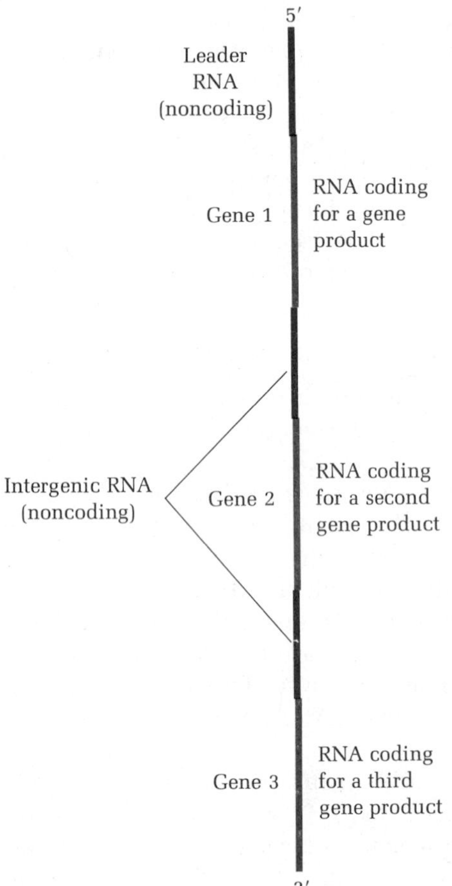

Figure 28-16
Schematic diagram of a polygenic mRNA of prokaryotic cells, in this case a transcript of three genes. Intergenic or spacer RNA separates the gene transcripts.

peptide chain of 100 amino acid residues requires an RNA coding sequence of at least 300 nucleotides, since each amino acid is coded by a nucleotide triplet (page 822).

Today, however, we know that mRNAs formed in bacterial transcription are always somewhat longer than needed for the polypeptide(s) which they code, since they have a noncoding polynucleotide "leader" at the 5′ end. Such leaders may be anywhere from 25 to 150 bases long. Polygenic mRNAs may also contain nontranslated *intergenic* or *spacer* regions, which separate the segments coding the individual polypeptide chains and appear to help regulate the rate of translation. Polygenic mRNAs usually code for two or more different polypeptide chains that function together, e.g., two or more enzymes participating in the same metabolic sequence, such as the biosynthesis of an amino acid. Figure 28-16 summarizes the general structure of prokaryotic mRNAs.

Messenger RNA Is Made by a DNA-Directed RNA Polymerase

The discovery of DNA polymerase and its dependence on a DNA template naturally led to a search for enzymes that can make an RNA strand complementary to a DNA template. In 1959 such an enzyme, capable of forming an RNA polymer from ribonucleoside 5′-triphosphates, was isolated from bacterial extracts almost simultaneously by four different groups of American biochemists working independently. This enzyme, called *DNA-directed RNA polymerase,* is similar in some ways to DNA polymerase. RNA polymerase requires all four ribonucleoside 5′-triphosphates (ATP, GTP, UTP, and CTP) as precursors of the nucleotide units of RNA, as well as Mg^{2+}. The purified enzyme also contains zinc as an essential part of its active group. RNA polymerase elongates an RNA strand by adding ribonucleotide units to the 3′-hydroxyl end of the RNA chain and thus builds RNA chains in the 5′ → 3′ direction. Inorganic pyrophosphate is formed from the β- and γ-phosphate groups of the nucleoside 5′-triphosphate precursors. The reaction equation is

$$\underset{\text{RNA}}{n(\text{NMP})_n} + \underset{\substack{\text{Ribonucleoside}\\ \text{5}'\text{-triphosphate}}}{\text{NTP}} \rightleftharpoons \underset{\substack{\text{Elongated}\\ \text{RNA}}}{(\text{NMP})_{n+1}} + \text{PP}_i$$

RNA polymerase requires preformed DNA for activity. The enzyme is most active with a natural double-stranded DNA as template, which may come from the same or some other organism. Only one of the two DNA strands is transcribed. In the newly formed RNA, uracil (U) residues are inserted in the positions specified by adenine residues in the DNA template; adenine and uracil form complementary base pairs. Adenine residues are inserted at the RNA positions specified by thymine residues of the DNA. Guanine and cytosine residues in DNA specify cytosine and guanine, respectively, in the new RNA strand. Analysis of the base composition and sequence of the new RNA strand showed it to have the polarity opposite that of the template strand and a base sequence complementary to that of the template. Although RNA polymerase does not require a primer strand, it cannot function until it binds to a specific initiation signal on the DNA template strand. The enzyme then starts the new RNA, at its 5′ end, usually with a GTP or ATP residue, whose 5′-triphosphate group, designated ppp, is not cleaved to PP_i, but remains intact throughout transcription. During transcription the new RNA strand base-pairs temporarily with the template DNA strand to form a short length of hybrid DNA-RNA double helix, which is essential to assure correct readout of the DNA strand. The hybrid duplex has only a transitory existence, since the RNA "peels off" shortly after its formation (Figure 28-17).

In *E. coli* there is but a single DNA-directed RNA polymerase that can make not only mRNAs but also tRNAs and rRNAs. It is a large (MW 500,000) and complex enzyme, containing five polypeptide subunits: two α chains, one β, one β', and one σ (sigma) chain. The holoenzyme form of the polymerase has a structure symbolized as α_2, β, β', σ. The first step in transcrip-

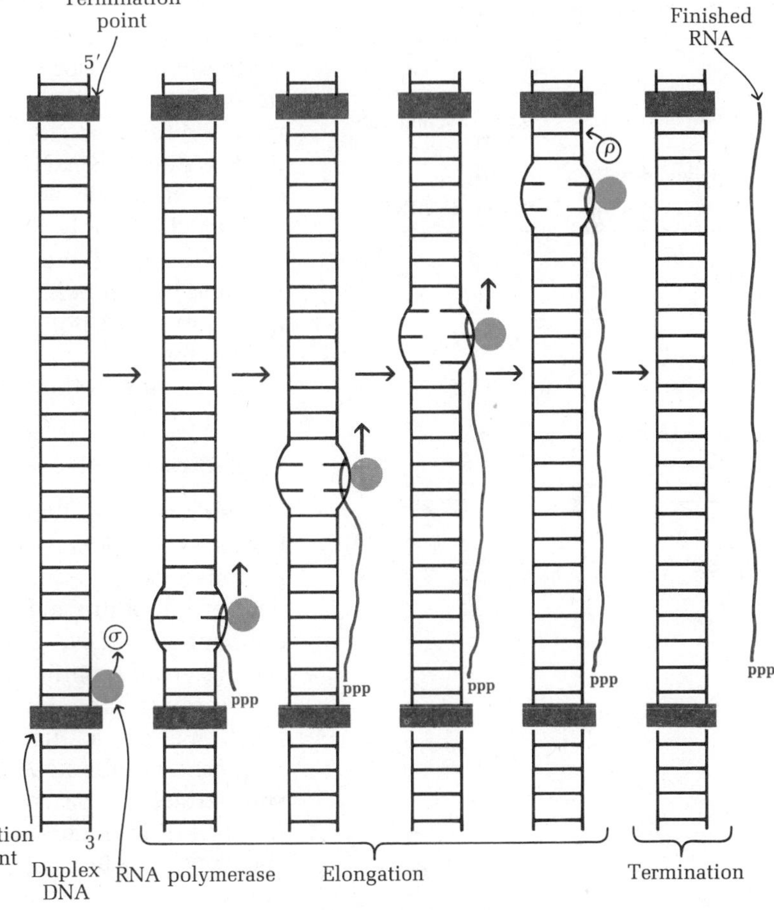

Figure 28-17
Summary of steps in transcription. RNA polymerase in prokaryotes must first enter DNA at the promoter site (Chapter 29). It then moves to the initiation sequence and begins to transcribe at the correct nucleotide, building the RNA in the 5′ to 3′ direction. This diagram does not show the actual length of the DNA-RNA hybrid, which is about a dozen nucleotides. After initiation the σ subunit departs. Termination of the chain requires the ρ subunit.

tion is the binding of the holoenzyme to a specific site in the DNA, called the promoter site, which consists of a short sequence recognized by the RNA polymerase. Different promoters have somewhat different sequences, which probably determine how efficiently a particular gene is transcribed. Once the RNA polymerase is correctly positioned at the promoter site and has made a few phosphodiester bonds, the σ subunit dissociates from the holoenzyme. The RNA is then elongated, step by step, by the remaining "core enzyme." The end of the gene or genes being transcribed is signaled by a specific termination sequence in the DNA template. To terminate transcription and bring about release of the RNA polymerase from the DNA, another specific protein, designated ρ (rho) is required. In the synthesis of RNAs, we therefore have three phases, initiation, chain elongation, and termination (Figure 28-17).

Many investigations have been directed toward identifying the DNA nucleotide sequences at which RNA polymerase starts and stops transcription, but much still remains to be learned about these important signals.

Eukaryotic Cell Nuclei Have Three RNA Polymerases

Table 28-2 Products of Eukaryotic DNA-Directed RNA polymerases

Enzyme	Product
RNA polymerase I	5.8S, 18S, and 28S rRNAs
RNA polymerase II	mRNA
RNA polymerase III	tRNAs; 5S rRNA

In eukaryotic cells there are three nuclear RNA polymerases, designated I, II, and III (Table 28-2). RNA polymerase I is localized in the nucleolus and is concerned largely with biosynthesis of ribosomal RNA, whereas RNA polymerases II and III are found in the chromatin and nucleoplasm. RNA polymerase II functions in mRNA synthesis and RNA polymerase III in tRNA and 5S rRNA synthesis. Mitochondria, like bacteria, contain a single RNA polymerase. The eukaryotic DNA-directed RNA polymerases elongate RNA chains in the same way as the *E. coli* enzyme, but they differ in their subunit structure and their regulatory units. This is to be expected, since the eukaryotic RNA polymerases must transcribe from DNA that is tightly bound in nucleosomes and associated with other nuclear proteins in the chromatin.

DNA-Directed RNA Polymerase Can Be Selectively Inhibited

The elongation of RNA chains by RNA polymerase in both prokaryotes and eukaryotes is specifically and characteristically inhibited by the antibiotic actinomycin D (Figure 28-18). The flat portion of this molecule intercalates (inserts itself) into the double-helical DNA between successive G≡C base pairs, producing a deformation in the DNA template. This local alteration prevents the movement of the polymerase along the template. In effect, actinomycin D jams the zipper. Because actinomycin D inhibits RNA elongation in intact cells, as well as in cell extracts, it has become a very useful diagnostic tool for identification of cell processes that depend upon RNA synthesis. Another intercalating inhibitor is acridine, also a flat molecule.

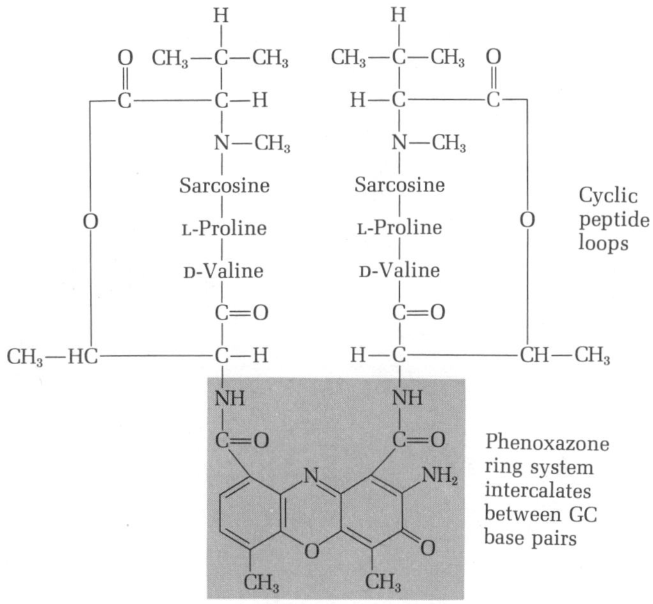

Figure 28-18
Structure of actinomycin D, an inhibitor of DNA transcription. The shaded portion intercalates between two successive GC base pairs in duplex DNA. The two cyclic peptide structures of the actinomycin D molecule bind to the minor groove of the double helix. Sarcosine is N-methylglycine. The linkages between sarcosine, L-proline, and D-valine are peptide bonds.

Another important antibiotic inhibitor of RNA synthesis is *rifampicin,* which binds to the β subunit of RNA polymerase of prokaryotes and prevents the initiation of RNA synthesis. However, rifampicin does not inhibit RNA synthesis in eukaryotes. A characteristic inhibitor of RNA synthesis in animal cells is *α-amanitin,* a toxic principle of the poisonous mushroom *Amanita phalloides.* It blocks messenger RNA synthesis by the RNA polymerase II of eukaryotic cell nuclei but does not affect RNA synthesis in prokaryotes. It is clever of this mushroom to develop a substance that is poisonous to mRNA formation in other eukaryotic organisms but is evidently harmless to its own transcription mechanism.

RNA Transcripts Are Processed Further

RNA transcripts made by RNA polymerases normally undergo further enzymatic alteration, called *posttranscriptional processing,* before they become active in their different roles. Ribosomal RNAs and transfer RNAs are made in the form of longer precursors, which are then modified and cleaved to yield the finished products. Eukaryotic mRNA transcripts are also processed, but prokaryotic mRNAs are not.

Ribosomal RNAs of both prokaryotic and eukaryotic cells are made from longer precursors called *preribosomal RNAs.* In prokaryotes 16S and 23S ribosomal RNAs (Chapter 29) may arise from a single long 30S RNA precursor, which has a molecular weight of about 2 million. This is methylated at specific bases and cleaved to yield 17S and 25S intermediate RNAs, which are then processed by "trimming" of residues by nucleases to yield the 16S and 23S ribosomal RNAs characteristic of prokaryotes (Figure 28-19). The 5S ribosomal RNA arises separately from the 3′ end of the 30S precursor.

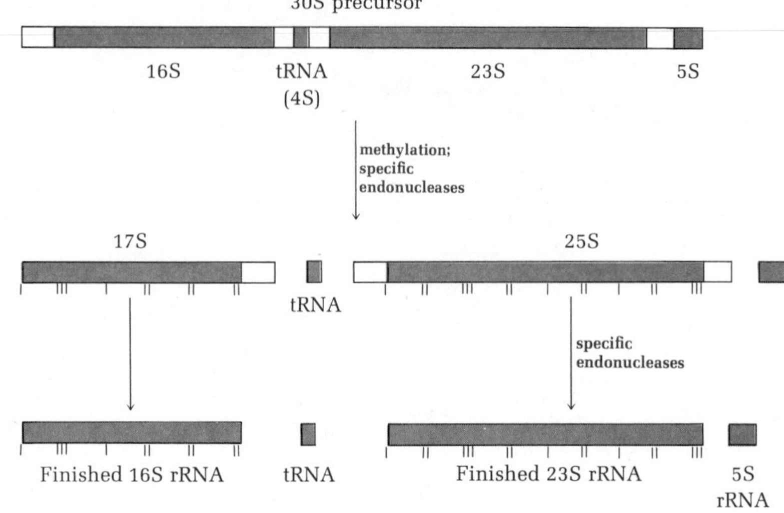

Figure 28-19
Processing of ribosomal RNA transcript in prokaryotes. The final 16S and 23S rRNA products are made from a longer 30S RNA precursor by the action of specific nucleases. Prior to cleavage of 30S RNA, it is methylated at specific bases (short colored lines). From the midsection, a single tRNA is formed.

In eukaryotes the 18S and 28S ribosomal RNAs are made in a series of steps from a large 45S preribosomal RNA. Processing of 45S RNA takes place in the nucleolus. The 45S precursor is first modified by the methylation of over 100 of its 14,000 nucleotides, mostly on the 2′-hydroxyl groups of their ribose units. As is shown in Figure 28-20, the methylated 45S RNA then undergoes a series of enzymatic cleavages which ultimately yield the 18S, 28S, and 5.8S ribosomal RNAs characteristic of eukaryotic ribosomes. The 5S ribosomal RNA of eukaryotes is made separately.

Transfer RNAs are also made from longer RNA precursors by enzymatic removal of extra nucleotide units from the 5′ and 3′ ends. In some cases two or more different tRNAs arise by enzymatic cleavage of a single long RNA precursor molecule. As we shall see (page 898), there are at least 32 different tRNAs, possibly many more.

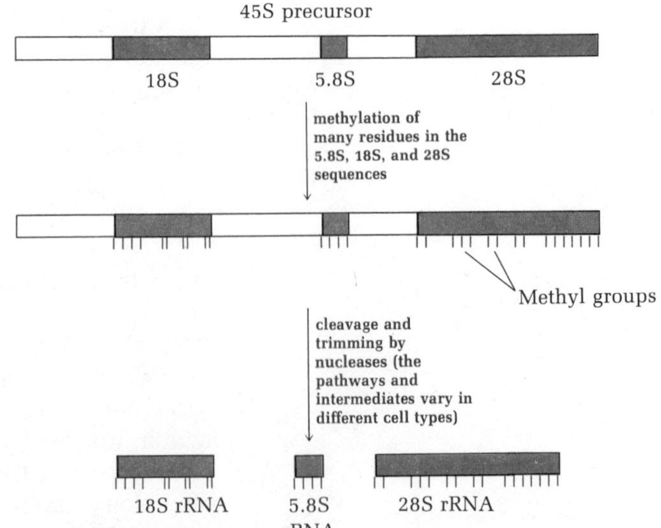

Figure 28-20
Processing of the transcripts for eukaryotic ribosomal RNAs. The methylation occurring in the first step involves the 2′-hydroxyl groups of the ribose units of those segments that appear in the final products. The 5S rRNA arises separately.

In addition to removal of leaders and tails, tRNA precursors undergo two other types of posttranscriptional processing. First, some tRNAs undergo attachment of the 3′ terminal trinucleotide sequence –C–C–A (3′); other tRNAs already contain this 3′ terminal sequence as transcribed. We shall see that the terminal A residue at the 3′ end is the part of the tRNA molecule to which its specific amino acid is covalently attached, to prepare the amino acid for addition to the growing polypeptide chain on the ribosome. Second, some of the bases in tRNAs are characteristically modified by methylation, some by deamination, and some by reduction. All tRNAs have characteristic modified bases at certain positions, as we shall see (Chapter 29).

Heterogeneous Nuclear RNAs Are Precursors of Eukaryotic Messenger RNAs

Processing of the precursors of messenger RNAs in eukaryotes is a very complex process. There are three distinctive structural features of eukaryotic messenger RNAs, as they occur in the cytoplasm. First, eukaryotic mRNAs are generally monogenic, whereas many prokaryotic mRNAs are polygenic. Second, at the 3′ end most eukaryotic mRNAs have a "tail" of 100 to 200 successive A residues, called the *poly-A tail*. The tail is made separately, starting from ATP, by *polyadenylate polymerase*, an enzyme that functions in much the same way as RNA polymerase. It catalyzes the reaction

$$n\text{ATP} \longrightarrow \underset{\text{Poly A}}{(\text{AMP})_n} + n\text{PP}_i$$

This enzyme requires no template but does require the mRNA as a primer. The third distinctive feature of most eukaryotic mRNAs is the *5′ cap*, a residue of *7-methylguanosine* linked to the 5′ terminal residue of the mRNA in a most unusual way, i.e., through a triphosphate linkage (Figure 28-21). The functions of the 5′ cap and the 3′ poly-A tail are not known with certainty. The 5′ cap may participate in the binding of the mRNA to the ribosome to initiate translation (page 883). It is also possible that the 5′ cap and poly-A tail protect the messenger RNA from enzymatic destruction.

Intron RNA Must Be Removed from mRNA Precursors

Messenger RNAs in the eukaryotic nucleus must undergo further processing to remove *intron RNA* (page 827). The nucleus contains a special type of RNA that "turns over" very rapidly during protein synthesis, called *heterogeneous nuclear RNA (hnRNA)*, which consists of a mixture of very long RNA molecules. Although it was long suspected that the hnRNAs are the precursors of cytoplasmic messenger RNAs, there appeared to be some puzzling discrepancies between the structure of hnRNAs and the finished cytoplasmic mRNAs. For one thing,

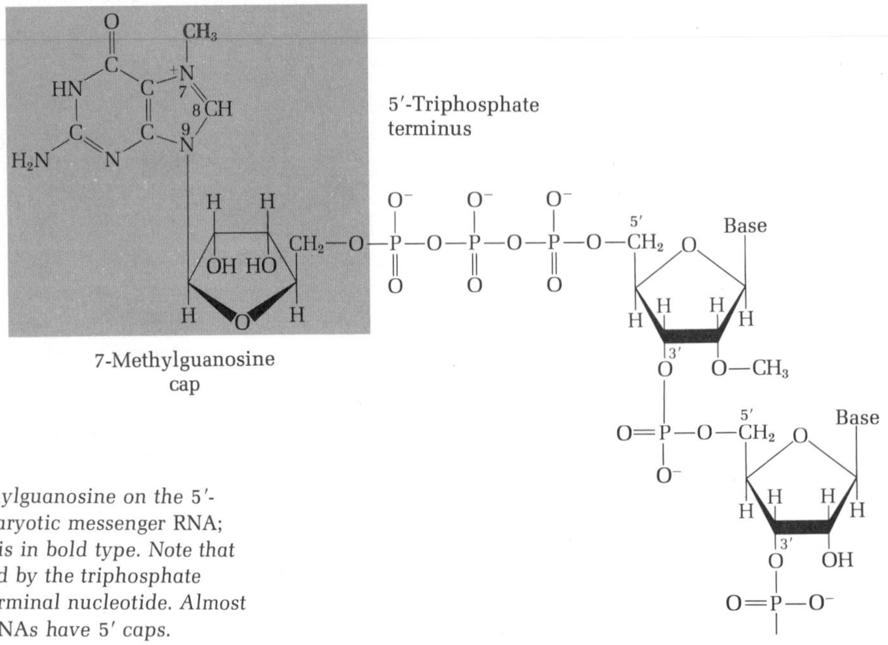

Figure 28-21
The cap of 7-methylguanosine on the 5′-terminus of a eukaryotic messenger RNA; the methyl group is in bold type. Note that the cap is attached by the triphosphate bridge to the 5′-terminal nucleotide. Almost all eukaryotic mRNAs have 5′ caps.

hnRNAs are much longer than mRNAs; moreover, the diversity of nucleotide sequences in hnRNAs is greater than in mRNAs. This long-standing mystery has been resolved by the discovery of the nontranslated intervening sequences or introns in eukaryotic genes. Recall that introns are often much longer than exons, the coding portions of the genes (page 827). After the discovery of introns in DNA the question naturally arose whether they are also transcribed in a colinear fashion, along with the exons, to yield a very long mRNA precursor complementary to, and colinear with, both the exons and introns in the DNA, or whether the RNA polymerase "jumps" over the introns and transcribes only the exons.

This question has been answered. RNA polymerase of eukaryotes transcribes both the exons and introns, precisely in the sequence they occur in the gene, to form a very long precursor of RNA that contains blocks of nontranslated RNA complementary to the intron base sequences. These mRNA precursors are found only in the nucleus and represent a large part of the heterogeneous nuclear RNA. Moreover, their occurrence also accounts for the anomalous sequence relationships between hnRNAs and corresponding cytoplasmic mRNAs. Once this mystery was solved, another arose.

How is the nuclear mRNA precursor, which contains blocks of intron RNA, converted into the finished cytoplasmic mRNA, which obviously must not contain the RNA blocks corresponding to introns? If the mRNA precursor were simply cleaved at each point where an exon ends and an intron begins or where an intron ends and an exon begins, the nucleus would end up with many segments of the precursor mRNA, some coding for a portion of a polypeptide chain and some that are nontranslated. How can the separate coding pieces be assembled in the correct order and spliced to form the final mRNA?

Small Nuclear RNAs Aid in Removing Intron RNA

Detailed answers to these questions are emerging. Much evidence suggests that the processing of mRNA precursors to remove the nontranslated introns takes place in such a way that the successive exons or coding segments of the mRNA are never physically separated from each other. The successive exons are spliced together very precisely through the help of another class of RNA molecules present in the nucleus, called *small nuclear RNAs (snRNAs)*. The function of these short nuclear RNAs, which have about 100 nucleotide units, had been a puzzle, but it turns out their base sequence is complementary to the sequences at the ends of each intron. By base pairing between the snRNA and the ends of the looped intron, the sequences of the two exons are brought into correct juxtaposition for enzymatic splicing of the exons and the removal of the intron separating them. The snRNA therefore functions as a temporary template to hold the ends of the two exons together for splicing at the correct point (Figure 28-22). A splicing error of only a single nucleotide would change the reading frame downstream of the error, resulting in synthesis of a defective protein molecule from the mRNA.

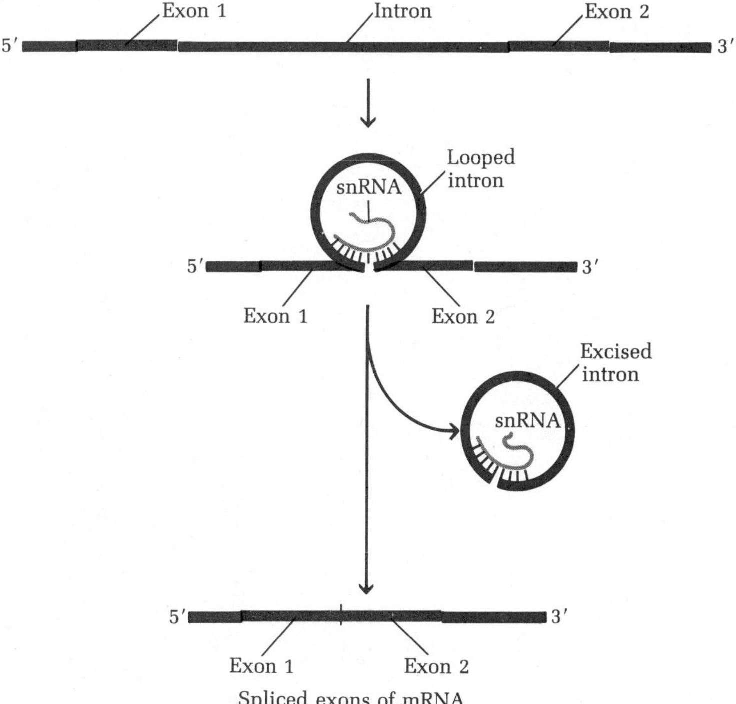

Figure 28-22
Function of small nuclear RNA (snRNA) in splicing exons and excising introns. The ends of the intron base-pair with a segment of snRNA. The splicing reaction is accompanied by excision of the intron.

After all the introns have been removed in this manner, which completes the processing of the mRNA precursor, the finished mRNA leaves the nucleus. In order to do so, the mRNA first binds to two special proteins which appear to guide the mRNA through the pores in the nuclear envelope (page 25) into the cytoplasm. These pores, which are surrounded by a complex arrangement of protein molecules, appear to allow only the completely processed and "finished" mRNAs to leave the nucleus. The bits and pieces of RNA that are left over after processing and trimming are degraded by nucleases. The resulting nucleoside 5′-phosphates are rephosphorylated by ATP to the triphosphates and used over again for the synthesis of RNAs in the nucleus.

The Transcription Process Can Be Visualized

In oocytes of the South African clawed toad *Xenopus*, there are many copies of the genes coding for the synthesis of ribosomal RNA. These genes must be transcribed very rapidly to provide for manufacture of the ribosomes needed for the rapid growth and division of the early embryo following fertilization. Electron microscopy of such cells reveals many thin, looped *core fibers*, about 20 nm in diameter; at periodic intervals they are coated with extended, hairlike *radial fibrils*, which gradually increase in length (Figure 28-23). The core fiber extending continuously through these structures consists of DNA associated with protein. The lateral hairlike fibrils, of which about 100 occur in each segment, are growing strands of RNA. The successive DNA segments coated with RNA fibrils are repeated genes for ribosomal RNA arranged in a sequence; all are un-

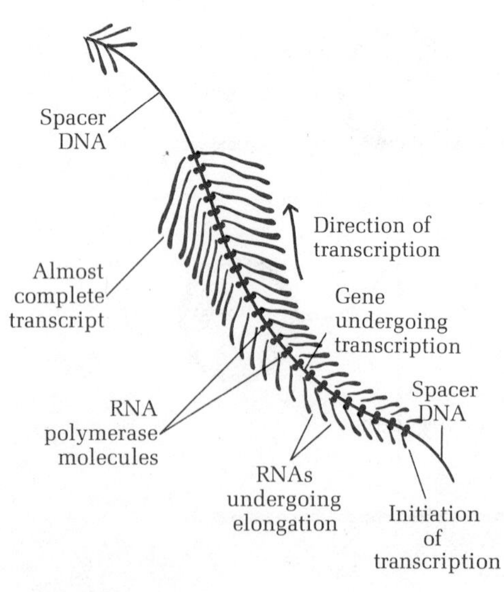

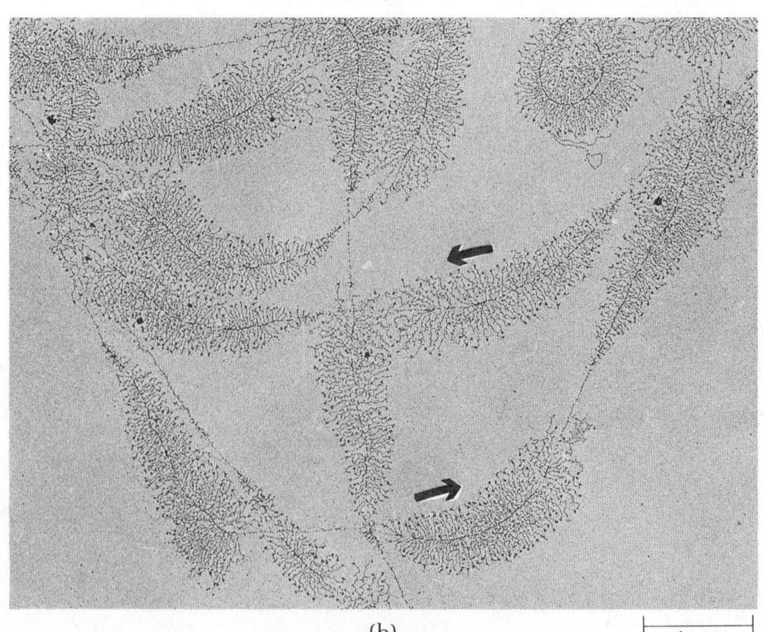

Figure 28-23
(a) *Schematic interpretation of the transcription process. The new single-strand* RNA *molecules appear much shorter than the* DNA *from which they were transcribed, owing to formation of secondary structure.* (b) *Electron micrograph of ribosomal genes of* Xenopus *undergoing transcription by many* RNA *polymerase molecules simultaneously.*

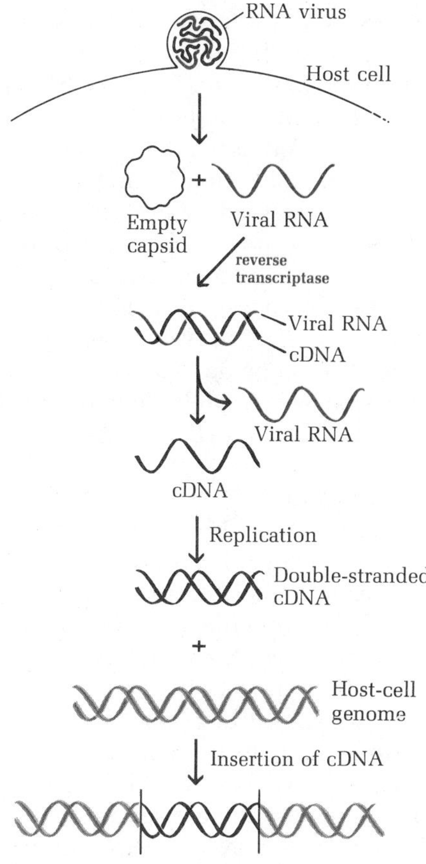

Figure 28-24
Action of reverse transcriptase in forming complementary DNA from single-strand viral RNA in an animal cell. The new cDNA may be inserted into the host-cell genome.

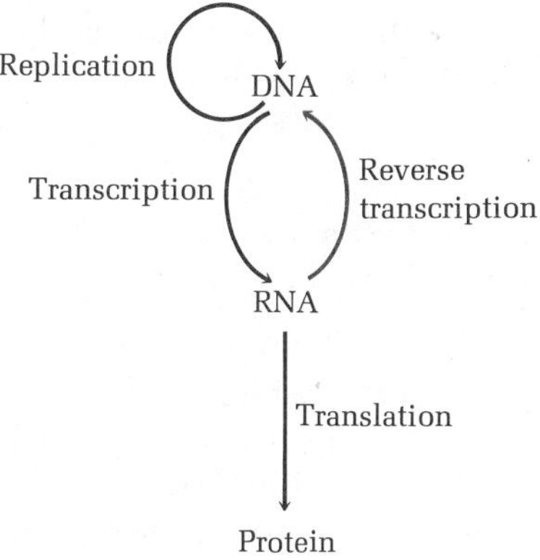

Figure 28-25
Extension of the central dogma of molecular genetics to encompass the flow of genetic information from RNA to DNA required by the discovery of reverse transcriptases.

dergoing active transcription. Many RNA chains are transcribed from a single gene at the same time, each being formed by an RNA polymerase molecule moving from one end of the gene to the other. Simultaneously each RNA chain grows in length. The dark granules on the DNA, at the base of the RNA fibrils, are molecules of DNA-directed RNA polymerase. The RNA polymerase molecules are very close together, moving along the gene in "bumper-to-bumper" fashion. The DNA segment undergoing transcription is 2 to 3 μm long, about the length of DNA required to code for the 45S precursor of ribosomal RNAs of eukaryotes. The segments of the core fiber that are not transcribed are spacer DNA.

DNA Is Transcribed from Some Viral RNAs by Reverse Transcriptase

Certain cancer-causing RNA viruses of animal tissues, such as the *Rous sarcoma virus* of fowl, contain within the viral particle a unique RNA-directed DNA polymerase, often called *reverse transcriptase*. After such a virus gains entrance into a host cell, the viral polymerase can catalyze the enzymatic synthesis of a DNA complementary to the viral RNA. The DNA so formed, which contains genes for causing cancer, often becomes incorporated into the genome of the host eukaryotic cell, where it may remain dormant, i.e., unexpressed, through many generations (Figure 28-24). Under some conditions such dormant viral genes become activated and cause the virus to replicate; under other conditions they may cause the cell to undergo transformation into a cancer cell.

The reverse transcriptases, whose existence in RNA tumor viruses was predicted by Howard Temin of the University of Wisconsin in 1962, were ultimately demonstrated to occur in such viruses by Temin and independently by David Baltimore of the Massachusetts Institute of Technology in 1970. Their discovery aroused much attention, particularly because it constituted molecular proof that genetic information can sometimes flow "backward," i.e., from RNA to DNA. It also provided a mechanism for the incorporation into the host-cell genome of cancer genes carried in the form of RNA by RNA viruses. Because of this discovery the central dogma of molecular biology has had to be restated, as shown in Figure 28-25. The RNA viruses containing reverse transcriptases are also known as *retroviruses* (*retro-* is from Latin, "turning back").

The reverse transcriptases provided an explanation for a long-standing question: How can the genetic message of cancer-causing RNA viruses become incorporated into the DNA of the host cell? There is increasing evidence that the DNA of many species of animals contains genes originating from RNA viruses, even though the animals from which such DNA was isolated had not themselves been exposed to such viruses. These observations suggest that the genes of some RNA viruses had been transcribed into DNA and become incorporated into the chromosomes of the ancestors of these animals, possibly early in biological evolution of these species. Thus they were passed along from generation to generation by replication of the entire DNA complement of the cell, including the cancer genes originally introduced as viral RNA. Indeed, one theory of the origin of cancer holds that in our chromosomes we all carry dormant, unexpressed cancer genes that gained entry into the genomes of our ancestors in the form of RNA viruses, perhaps thousands or millions of years ago. This theory holds further that such cancer genes (oncogenes) are not usually transcribed but that when they are activated, perhaps by exposure of the genome to carcinogenic agents, the oncogenes are transcribed and translated, to yield gene products causing transformation of normal human cells into malignant cells.

Viral reverse transcriptases contain Zn^{2+}, as do all DNA and RNA polymerases. They are most active with the RNA of their own type of virus but can be used to make DNA complementary to a variety of RNAs. The reverse transcriptases do require a primer; they make the new DNA strand in the $5' \rightarrow 3'$ direction and in many other respects resemble DNA polymerases.

Reverse transcriptases, like the restriction endonucleases (page 823), have become very important biochemical tools in the study of DNA-RNA relationships and in cloning DNA, as we shall see. Reverse transcriptase makes possible the laboratory synthesis of a DNA complementary in base sequence to an RNA template, whether it is an mRNA, tRNA, or rRNA. A synthetic DNA prepared in this manner is called a *complementary DNA (cDNA)*. With reverse transcriptase it is possible, for example, to make a synthetic gene, i.e., a cDNA, for one of the hemoglobin polypeptide chains from its mRNA. The mRNAs for the hemoglobin chains are readily isolated from immature red blood cells. In this and many other cases where the natural gene for a eukaryotic polypeptide cannot readily be isolated but its mRNA is available, it is possible to make a synthetic gene from the mRNA by the action of reverse transcriptase. Later we shall see how cDNAs are used to clone recombinant DNA (Chapter 30).

Some Viral RNAs Are Replicated by an RNA-Directed RNA Polymerase

Some *E. coli* bacteriophages contain RNA as their chromosomes rather than DNA. These viruses, which include the f2, MS2, R17, and Qβ phages (Table 27-1), have become important tools in the study of mRNA structure and function. The RNAs of these viruses, which function as mRNAs for the synthesis of viral proteins, are replicated in the host cell by the action of enzymes called *RNA-directed RNA polymerases* or *RNA replicases*. These enzymes are not normally present in the host *E. coli* cell but are produced by it in response to the viral RNA.

RNA replicase isolated from *E. coli* cells infected with the Qβ virus catalyzes the formation of an RNA complementary to the viral RNA from the ribonucleoside 5′-triphosphates. The reaction equation is similar to that of DNA-directed RNA polymerases

$$\underset{}{\text{NTP}} + \underset{\text{RNA}}{(\text{NMP})_n} \xrightarrow{\substack{\text{viral}\\ \text{RNA}\\ \text{template}}} \underset{\substack{\text{Lengthened}\\ \text{RNA}}}{(\text{NMP})_{n+1}} + \text{PP}_i$$

Synthesis of the new RNA strand proceeds in the 5′ → 3′ direction. RNA replicase requires RNA as template and will not function with DNA, but unlike the DNA and RNA polymerases, the RNA replicases are template-specific. Thus the Qβ RNA replicase can use as template only the RNA of the Qβ virus; the RNAs of the host cell are not replicated. This fact obviously explains how RNA viruses are preferentially replicated in the host cell, which contains many other types of RNA.

Purified Qβ RNA replicase thus can bring about the net synthesis of new, biologically active Qβ RNA molecules. Starting with the infectious Qβ (+) strand of RNA as template, the purified replicase can make a complementary Qβ RNA (−) strand, which can then be used as the template in another incubation with the enzyme to yield synthetic, fully infectious Qβ RNA identical with the starting (+) strand. These properties thus require that the central dogma of molecular genetics be modified even further, as shown in Figure 28-26.

The complete nucleotide sequence of the RNA chromosome of the R17 virus has been worked out, as well as the amino acid sequences of the three viral coat proteins for which the RNA codes. This achievement has revealed important information on the genetic code and on the specific signals for initiation and termination of polypeptide chain synthesis (Chapter 29).

Figure 28-26
Further extension of the central dogma, to include replication of RNA.

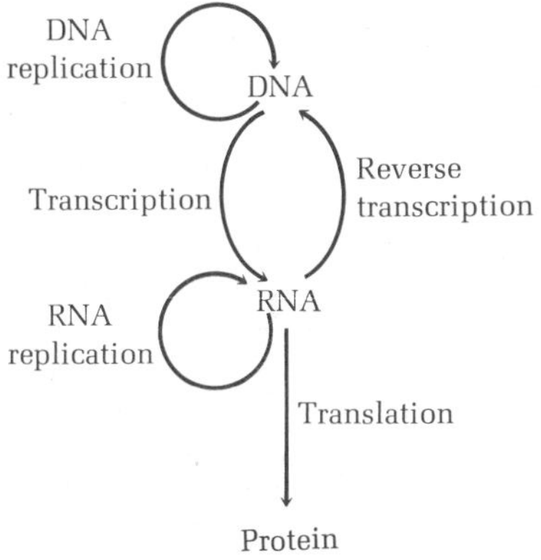

Polynucleotide Phosphorylase Makes Random RNA-Like Polymers

In 1955 Marianne Grunberg-Manago and Severo Ochoa discovered polynucleotide phosphorylase, the first enzyme found to synthesize long-chain polynucleotides. (Shortly thereafter Arthur Kornberg reported the discovery of DNA polymerase.) Polynucleotide phosphorylase, which apparently occurs only in bacteria, catalyzes the reaction

$$\text{NDP} + (\text{NMP})_n \rightleftharpoons (\text{NMP})_{n+1} + \text{P}_i$$

where NDP is a nucleoside 5′-diphosphate, $(\text{NMP})_{n+1}$ is a lengthened polynucleotide, and phosphate is the leaving group. The enzyme requires the 5′-diphosphates of ribonucleosides and cannot act on the homologous 5′-triphosphates or on deoxyribonucleoside 5′-diphosphates. Mg^{2+} is required for its action. The RNA-like polymer formed by polynucleotide phosphorylase contains 3′,5′-phosphodiester linkages, which can be hydrolyzed by ribonuclease. The reaction is readily reversible and can be pushed in the direction of breakdown of the polyribonucleotide by increasing the phosphate concentration.

The polynucleotide phosphorylase reaction does not use a template and does not form a polymer having a specific base sequence. It does require a priming strand of RNA, which merely furnishes a free 3′-hydroxyl terminus to which additional residues can be added. The reaction proceeds as well with one species of monomer as with all four. The base composition of the polymer formed by the enzyme reflects the relative concentrations of the 5′-diphosphate precursors in the medium. It is therefore unlikely that polynucleotide phosphorylase normally functions to make RNA; it appears instead to participate in the degradation of RNAs.

Polynucleotide phosphorylase is a very valuable tool, since it can be used for the laboratory preparation of many different kinds of RNA-like polymers with differing sequences and frequencies of bases. Such synthetic RNA polymers made it possible to deduce the genetic code words for the amino acids, as we shall see (Chapter 29).

Summary

The DNA of *E. coli* is replicated semiconservatively, so that the daughter double helixes each contain one parental strand and one newly made strand. The circular bacterial chromosome is replicated bidirectionally starting from a single origin point. Some viral DNAs are replicated by a rolling-circle model.

DNA polymerase I of *E. coli* catalyzes synthesis of DNA from the four deoxyribonucleoside 5′-triphosphates in the presence of Mg^{2+}, with elimination of pyrophosphate. Chain growth is in the $5' \rightarrow 3'$ direction. The reaction requires preexisting DNA to provide both a template strand and a primer strand. The enzyme makes a strand of DNA complementary to the template strand; the polarity of the new strand is opposite to that of the template strand. *E. coli* contains three DNA polymerases. DNA polymerase III is the primary replication enzyme, whereas DNA polymerase I functions in an auxiliary capacity in replication. One strand of DNA is replicated continuously in the $5' \rightarrow 3'$ direction (the leading strand), whereas the other strand, the lagging strand, is replicated in short pieces called Okazaki fragments. Okazaki fragments, which have up to 2000 nucleotides in prokaryotes, are made in the direction opposite that of the movement of the replication fork. Each Okazaki fragment is initiated by formation of a short primer segment of complementary RNA, made by primase. DNA is then made on the 3′ end of the RNA primer by DNA polymerase III. Next the RNA primer is excised and replaced by complementary DNA, which is then spliced to the lagging strand by DNA ligase. Replication also requires helicase and DNA-binding proteins, to unwind and hold the template chains apart prior to action of DNA polymerase. In addition, swiveling of DNA by gyrase favors unwinding of the strand. After replication, DNA gyrase is also required to produce supercoiling. DNA polymerase

I has 3′ → 5′ exonuclease and 5′ → 3′ exonuclease activities as well. The former serves in a proofreading capacity by eliminating unpaired nucleotides, the latter in removing RNA primers from Okazaki fragments and in repair of DNA. DNA polymerase III also has 3′ → 5′ and 5′ → 3′ exonuclease activities.

Transcription is catalyzed by DNA-directed RNA polymerase, a complex enzyme that makes RNA complementary to one strand of duplex DNA, starting from ribonucleoside 5′-triphosphates. Prokaryotic RNA polymerase requires a special subunit σ (sigma) for recognizing the promoter region on DNA, the signal for the initiation of RNA. Many RNA chains can be transcribed simultaneously from a single gene. Ribosomal RNAs and transfer RNAs are made from longer precursor RNAs, which are then trimmed by nucleases and further modified enzymatically to yield the finished products. In eukaryotes messenger RNAs are formed from longer precursors known as heterogeneous nuclear RNAs. They are subsequently modified by attachment of a long poly-A tail at the 3′ end and a methylguanosine cap at the 5′ end. Introns are removed with the aid of small nuclear RNAs.

RNA-directed DNA polymerases, also called reverse transcriptases, are made in animal cells infected by cancer-causing RNA viruses. This enzyme transcribes the RNA viral chromosome into a complementary DNA. In this way cancer genes (oncogenes) may be incorporated into the genome of animal cells.

RNA-directed RNA replicases are found in bacterial cells infected by certain RNA viruses. They are template-specific for the viral RNA. Polynucleotide phosphorylase can reversibly form RNA-like polymers from ribonucleoside 5′-diphosphates. Although it can add or remove ribonucleotides at the 3′-hydroxyl end of the polymer, it normally acts to degrade RNA.

References

Replication

ALBERTS, B., and R. STERNGLANZ: "Recent Excitement in the DNA Replication Problem," *Nature*, **269:**655–661 (1977). Excellent survey of the complexities of replication, especially supercoiling, unwinding, and fidelity.

KORNBERG, A.: "Aspects of DNA Replication," *Cold Spring Harbor Symp. Quant. Biol.*, **43:**1–9 (1979). Survey of the present status and problems. This volume has many valuable articles.

KORNBERG, A.: *DNA Replication*, Freeman, San Francisco, Calif., 1980. The latest word and complete references on replication.

DNA Ligase

LEHMAN, I. R.: "DNA Ligase: Structure, Mechanism, Function," *Science*, **186:**790–797 (1974).

Transcription

CHAMBERLIN, M. J.: "RNA Polymerase: An Overview," in R. Losick and M. J. Chamberlin (eds.), *RNA Polymerase*, pp. 17–67, Cold Spring Harbor Laboratory, New York, 1976.

MILLER, O. L., JR.: "The Visualization of Genes in Action," *Sci. Am.*, **228**:34–42, March (1973).

PEDERSON, T.: "Messenger RNA Biosynthesis and Nuclear Structure," *Am. Sci.*, **69**(1):76–84 (1981).

Reverse Transcriptase

TEMIN, H.: "RNA-Directed DNA Synthesis," *Sci. Am.*, **226**:24–33, January (1972).

Processing of RNA

ABELSON, J.: "RNA Processing and the Intervening Sequence Problem," *Annu. Rev. Biochem.*, **48**:1035–1069 (1979).

Problems

1. *Conclusions from the Meselson-Stahl Experiment* The Meselson-Stahl experiment proved that DNA undergoes semiconservative replication in *E. coli.* In the "dispersive" model of DNA replication, the parent DNA strands are cleaved into pieces of random size and are then joined with pieces of the newly replicated DNA to yield daughter duplexes in which both strands contain random segments of both heavy and light DNA. Explain how the Meselson-Stahl experiment ruled out such a model.

2. *The Cairns Experiment*
   (a) Why did Cairns use radioactive thymidine to follow the course of DNA replication?
   (b) Would radioactive adenosine or guanosine have been equally useful?
   (c) Show the enzymatic pathway by which radioactive thymidine is incorporated into *E. coli* DNA.

3. *Number of Turns in the E. coli Chromosome* How many turns of the *E. coli* chromosome must be unwound during its replication?

4. *Replication Time in E. coli*
   (a) From the data in this chapter, how long would it take to replicate the *E. coli* chromosome at 37°C, if two replication forks start from the origin?
   (b) Under some conditions *E. coli* cells can grow and divide in 20 min. Can you suggest how this is possible?

5. *Replication Forks in E. coli and Human Cells*
   (a) Calculate the time required for the gene of ribonuclease (104 residues) to be replicated in *E. coli* if a replication fork moves 750 base pairs per second.
   (b) A replication fork in a human cell moves only one-tenth as fast as it does in *E. coli.* What additional information do you need to calculate the minimum rate of replication of a human gene for a protein having 104 amino acid residues?

6. *Base Pairing during Replication and Transcription*
   (a) Write the base sequence of a segment of DNA replicated by DNA polymerase from the following DNA template, keeping

in mind that base sequences of DNA and RNA are written in the 5′ → 3′ direction:

(5′) AGCTTGCAACGTTGCATTAG (3′)

(b) Now write the base sequence of a segment of messenger RNA transcribed by RNA polymerase from the newly replicated DNA strand of part (a).

7. *Base Composition of an RNA Transcript* A strand of DNA having 10^5 nucleotide residues with the base composition A, 21 percent; G, 29 percent; C, 29 percent; and T, 21 percent is replicated by DNA polymerase to yield a complementary strand. The resulting duplex DNA is then used as a template by RNA polymerase, which transcribes the new strand to yield an RNA strand having the same number of residues.
   (a) Determine the base composition of the RNA formed.
   (b) Suppose the RNA polymerase stopped after transcribing only 2000 residues of the new DNA strand. What might the base composition of the new short RNA be?

8. *Base Composition of DNAs Made from Single-Stranded Templates* Determine the base composition you might expect in the total DNA synthesized by DNA polymerase on templates provided by an equimolar mixture of the two complementary strands of circular ϕX174 DNA (i.e., the replicative form) if the base composition of one strand is A, 24.7 percent; G, 24.1 percent; C, 18.5 percent; and T, 32.7 percent. What assumption is necessary to answer this problem?

9. *Hybridization of DNA with mRNAs* DNA will hybridize with mRNAs transcribed from it. How do you explain that no more than 50 percent of the entire *E. coli* DNA can hybridize with all known *E. coli* mRNAs?

10. *Okazaki Fragments*
    (a) In the replication of the *E. coli* chromosome, about how many Okazaki fragments would be formed?
    (b) What factors guarantee that the numerous Okazaki fragments are assembled in the correct order in the new DNA?

11. *Leading and Lagging Strands* Make a list and compare the precursors and enzymes needed to make the leading versus lagging strands during DNA replication.

12. *Fidelity of Replication of DNA*
    (a) What factors participate in ensuring the fidelity of replication during the synthesis of the leading strand of the new DNA?
    (b) Would you expect the lagging strand to be made with the same fidelity as the leading strand? Give your reasons.

13. *Initiation of Replication* The DNA replicase system requires both a template strand and a primer strand; moreover, it cannot replicate an intact circular DNA, except under special circumstances.
    (a) Of what biological importance is this property?
    (b) What are the special circumstances in which the DNA replicase system can replicate an intact circular DNA?

14. *The Differences between RNA Polymerase and Polynucleotide Phosphorylase* RNA polymerase requires nucleoside 5′-triphosphates as precursors during transcription and cannot act on nucleoside 5′-diphosphates. Polynucleotide phosphorylase, on the other hand, requires nucleoside 5′-diphosphates but does not act upon the 5′-triphosphates.
    (a) Give possible reason(s) for the difference between the precursor requirements of the two enzymes.
    (b) In relation to your answer to (a), what other differences between the two enzymes are relevant?

15. *Error Correction by RNA Polymerases* DNA polymerases are capable of editing and error correction. However, RNA polymerases do not appear to have this capacity. Since a single base error in either replication or transcription can lead to an error in protein synthesis, can you give a possible biological explanation for this striking difference?

CHAPTER 29

Protein Synthesis and Its Regulation

The mechanism of biosynthesis of proteins, with all their variety of biological activities and their species specificity, has been one of the greatest challenges in the history of biochemistry. Yet for many years even very simple questions about protein synthesis could not be answered. For example, are proteins made one residue at a time or are they made by assembling many short, prefabricated peptides? Or are all proteins in the cell made from a single, long precursor polypeptide, followed by specific alterations of their R groups? Indeed, until the early 1950s it was not even certain that proteins were specific chemical compounds of definite molecular weight, amino acid composition, and sequence.

Today we know a great deal about the overall process of protein synthesis, but perhaps only a fraction of what there is yet to learn. By all odds, protein synthesis is the most complex of biosynthetic mechanisms, requiring a very large number of enzymes and other specific macromolecules. In eukaryotic cells protein synthesis requires participation of over 70 different ribosomal proteins; 20 or more enzymes needed to activate the amino acid precursors; a dozen or more auxiliary enzymes and other specific protein factors for initiation, elongation, and termination of polypeptides; and possibly 100 or more additional enzymes required for the final processing of different kinds of proteins. To these we must add 70 or more kinds of transfer and ribosomal RNAs. Thus almost 300 different macromolecules must cooperate to synthesize polypeptides. Moreover, many of these macromolecules are organized into the complex three-dimensional structure of the ribosomes to carry out stepwise translocation of the mRNA as the polypeptide is assembled.

Despite this great complexity, proteins can be made at exceedingly high rates. For example, only about 5 seconds are required for an *E. coli* ribosome to make a complete polypeptide chain of 100 residues. Moreover, we shall see that the synthesis of the thousands of different proteins in each cell is tightly regulated, so that only the required number of molecules of each is made under any given set of metabolic circumstances.

Early Discoveries Set the Stage

Three major advances in the 1950s set the stage for our present knowledge of protein biosynthesis. In the early 1950s Paul Zamecnik and his colleagues at the Massachusetts General Hospital asked: Where in the cell are proteins synthesized? To seek an answer they injected radioactive amino acids into rats. At different time intervals after the injection, the liver was removed, homogenized, and fractionated by centrifugation (page 355). The subcellular fractions were then examined for the presence of radioactive protein. When hours or days were allowed to elapse after injection of the labeled amino acid, *all* the intracellular fractions contained labeled proteins. However, if the liver was removed and fractionated only minutes after the injection of the labeled amino acid, newly labeled protein was found only in a fraction containing small ribonucleoprotein particles. These particles, earlier discovered by electron microscopy in animal tissues (page 32), were thus identified as the site of protein synthesis from amino acids; later they were named *ribosomes* (Figure 29-1).

The second advance was made by Mahlon Hoagland and Zamecnik, when they found that incubation of amino acids with ATP and the cytosol fraction of liver cells "activated" them. In this enzymatic process amino acids were attached to a special form of heat-stable soluble RNA, later called *transfer RNA*.

The third major advance occurred when Francis Crick asked: How is the genetic information coded in the four-letter language of nucleic acids translated into the 20-letter language of proteins? Crick reasoned that transfer RNA must serve the role of an *adapter*, so that one part of the tRNA molecule can bind a specific amino acid and some other part of the tRNA can recognize a short nucleotide sequence in the messenger RNA coding for that amino acid (Figure 29-2).

These developments soon led to recognition of the major stages in protein biosynthesis and ultimately to the elucidation of the genetic code words for the amino acids.

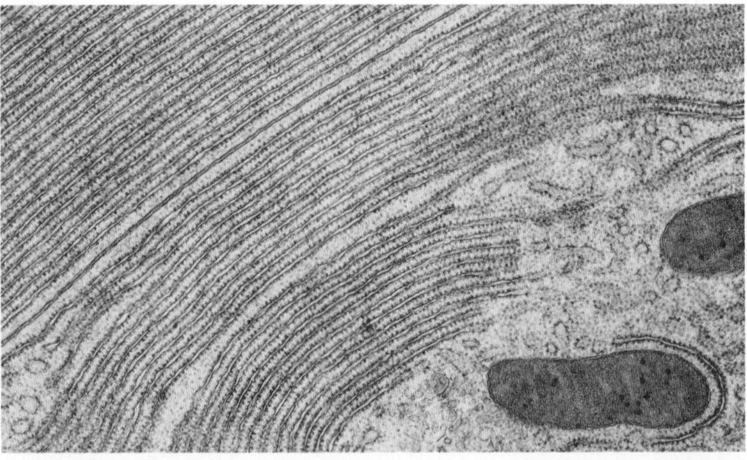

Figure 29-1
Portion of a pancreatic cell, showing ribosomes attached to the outer surface of the endoplasmic reticulum.

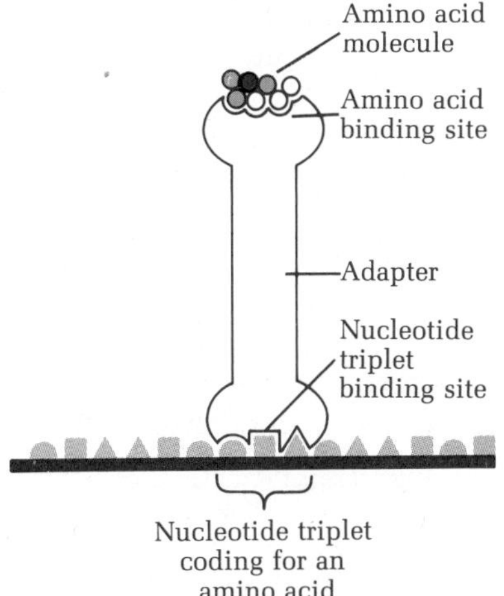

Figure 29-2
Crick's hypothesis of the adapter function of transfer RNA. Today we know that the amino acid is covalently bound, whereas a specific nucleotide triplet at another point in the transfer RNA molecule recognizes its coding triplet in the mRNA through hydrogen bonding of complementary bases.

Protein Synthesis Occurs in Five Major Stages

Today we know that there are five major stages in protein synthesis, each requiring a number of components. Table 29-1 shows the components required in *E. coli* and other prokaryotes; protein synthesis in eukaryotic cells follows the same pattern, but there are some differences in details. An outline follows:

Stage 1: Activation of Amino Acids

In this stage, which takes place in the cytosol, not on the ribosome, each of the 20 amino acids is covalently attached to a specific transfer RNA at the expense of ATP energy. These reactions are catalyzed by a group of Mg^{2+}-dependent activating enzymes, each specific for one amino acid and for a corresponding tRNA.

Stage 2: Initiation of the Polypeptide Chain

Next, the messenger RNA bearing the code for the polypeptide to be made is bound to the smaller subunit of a ribosome, followed by the initiating amino acid, attached to its tRNA to form

Table 29-1 Components Required in the Five Major Stages in Polypeptide Synthesis in *E. coli*

Stage	Necessary components
1. Activation of the amino acids	20 amino acids 20 aminoacyl-tRNA synthetases 20 or more transfer RNAs ATP Mg^{2+}
2. Initiation of the polypeptide chain	Messenger RNA *N*-Formylmethionyl-tRNA Initiation codon in mRNA (AUG) 30S ribosomal subunit 50S ribosomal subunit GTP Mg^{2+} Initiation factors (IF-1, IF-2, IF-3)
3. Elongation	Functional 70S ribosome (initiation complex) Aminoacyl-tRNAs specified by codons Mg^{2+} Elongation factors (Tu, Ts, and G) GTP Peptidyl transferase
4. Termination	ATP Termination codon in mRNA Polypeptide releasing factors (R_1, R_2, and S)
5. Folding and processing	Specific enzymes and cofactors for removal of initiating residues and signaling leaders, for modification of terminal residues, attachment of enzyme prosthetic groups, and covalent modification of specific amino acid R groups by attachment of phosphate, methyl, carboxyl, or carbohydrate groups

an *initiation complex*. The tRNA of the initiating amino acid base-pairs with a specific nucleotide triplet or codon on the mRNA that signals the beginning of the polypeptide chain. This process, which requires guanosine triphosphate (GTP), is promoted by three specific cytosolic proteins called *initiation factors*.

Stage 3: Elongation

The polypeptide chain is now lengthened by covalent attachment of successive amino acid units, each carried to the ribosome and put into its proper position by its corresponding tRNA, which is base-paired to its corresponding codon in the messenger RNA. Elongation is promoted by cytosolic proteins called *elongation factors*. The energy required for binding each incoming aminoacyl-tRNA and for the movement of the ribosome along the messenger RNA by one codon comes from the hydrolysis of two molecules of GTP for each residue added to the growing polypeptide.

Stage 4: Termination and Release

The completion of the polypeptide chain, which is signaled by a *termination codon* in the mRNA, is followed by its release from the ribosome, promoted by *releasing factors*.

Stage 5: Folding and Processing

In order to achieve its native biologically active form (page 169) the polypeptide must undergo folding into its proper three-dimensional conformation. Before or after folding the new polypeptide may undergo processing by enzymatic action to remove initiating amino acids, to introduce phosphate, methyl, carboxyl, or other groups into certain amino acid residues, or to attach oligosaccharides or prosthetic groups.

Now let us examine each stage in more detail.

Transfer RNAs Are Required for the Activation of Amino Acids

To understand how transfer RNAs can serve as adapters in translating the language of nucleic acids into the language of proteins, we must first examine their structure. Transfer RNAs are relatively small single-strand molecules. In bacteria and in the extramitochondrial cytosol of eukaryotes tRNAs have between 73 and 93 nucleotide residues, corresponding to molecular weights between 24,000 and 31,000. (Mitochondria contain distinctive tRNAs that are somewhat smaller.) There is at least one kind of tRNA for each amino acid; some amino acids have two or more specific tRNAs. At least 32 tRNAs are required to recognize all the amino acid codons (page 899), but some cells possess many more than 32.

Many tRNAs have been isolated in homogeneous form. In 1965, after several years of work, Robert W. Holley and his colleagues at Cornell University worked out the complete nucleotide sequence of alanine transfer RNA from yeast. This, the very first nucleic acid to be sequenced in its entirety, was found to contain 76 nucleotide residues, 10 of which are modified forms. Its complete base sequence is shown in Figure 29-3. Since then the base sequences of dozens of other tRNAs for different amino acids and from various species have been worked out and have revealed many common denominators of structure. Eight or more of the nucleotide residues of all tRNAs have unusual modified bases, many of which are methylated derivatives of the principal bases. Most tRNAs have a guanylic acid (pG) residue at the 5′ end, and all have the trinucleotide sequence –C–C–A(3′) at the 3′ end. All tRNAs, if written in a form in which there is maximum intrachain base pairing through the allowed pairs A–U, G–C, and G–U, form a "cloverleaf" with four arms; the longer tRNAs have a short fifth or extra arm (Figures 29-3 and 29-4). Two of these arms participate in the adapter function of tRNAs. The <u>amino acid</u> or <u>AA</u>

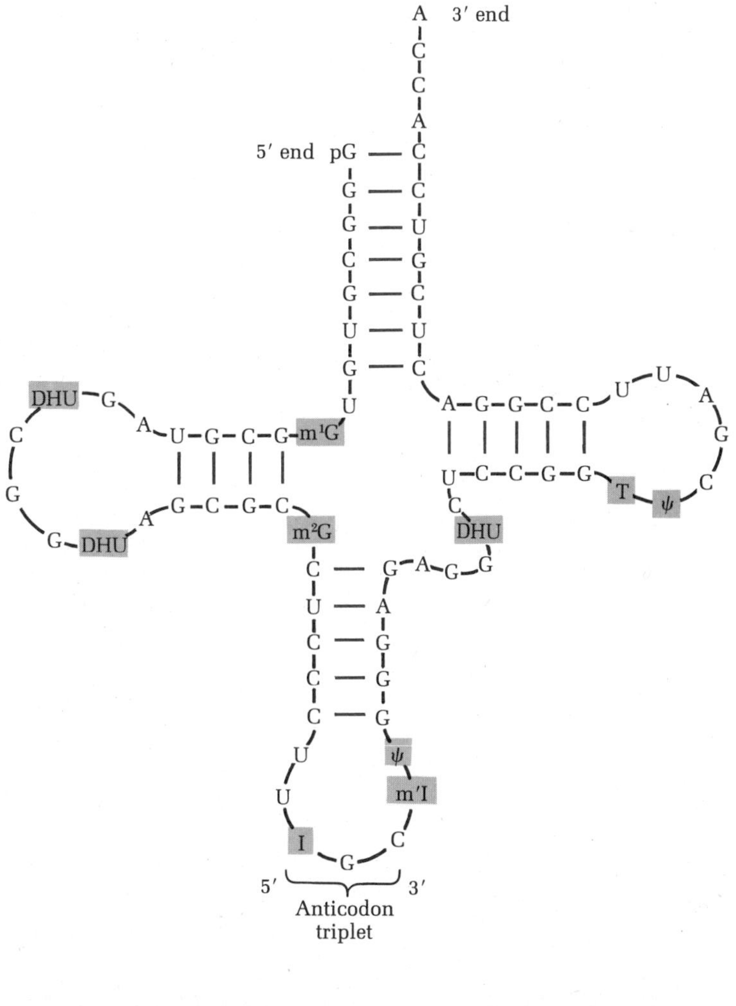

Figure 29-3
The nucleotide sequence of yeast alanine tRNA as deduced by Holley and his colleagues, written in the cloverleaf conformation. In addition to A, G, U, and C, the following symbols are used for the modified nucleosides: ψ = pseudouridine, I = inosine, T = ribothymidine, DHU = 5,6-dihydrouridine, m^1I = 1-methylinosine, m^1G = 1-methylguanosine, m^2G = N^2-dimethylguanosine. The symbols for the modified nucleosides are in color. The colored lines between the parallel sections indicate base pairs. The anticodon is capable of recognizing the codons for alanine in the mRNA. Other features of tRNA structure are given in the text and in Figure 29-4. In RNAs G can base-pair with U as well as with C, although the G–U pair is not as stable as the Watson-Crick pair G–C.

3′ end
5′ end
5′
3′
Anticodon triplet

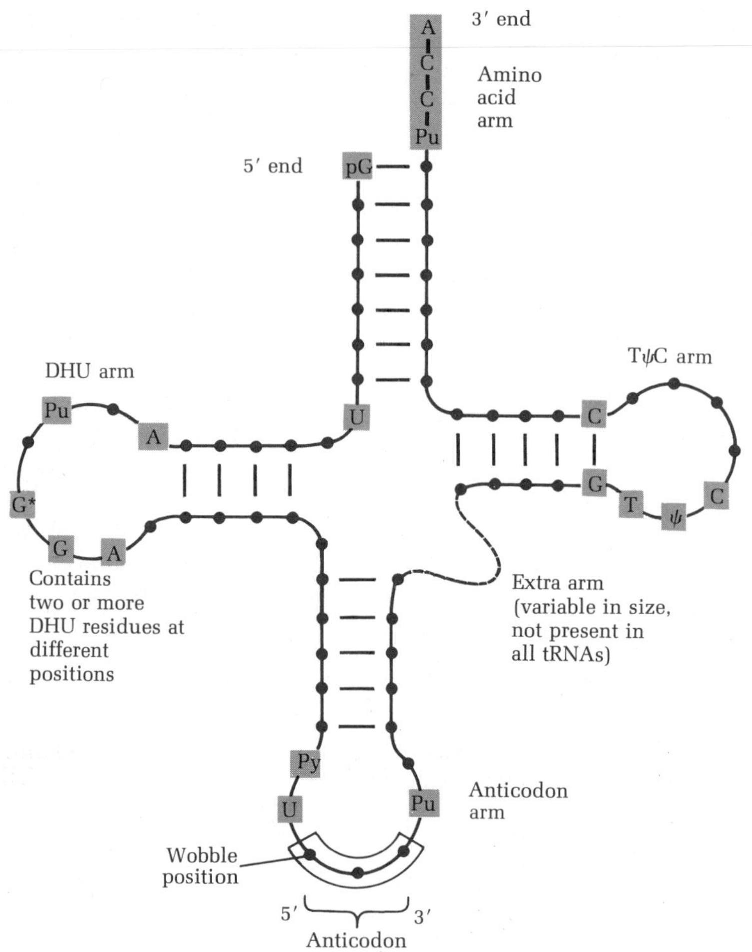

Figure 29-4
General structure of all tRNAs. When they are written with maximum intrachain base pairing, all tRNAs show the cloverleaf structure. The large dots on the backbone are nucleoside residues, and the colored lines represent base pairings. Characteristic and/or invariant residues common to all tRNAs are in color. tRNAs differ in length, from 73 to 93 nucleotides. Extra nucleotides occur in the extra arm or in the DHU arm. At the end of the anticodon arm is the anticodon loop, which always contains seven unpaired nucleotides. The DHU arm contains up to three DHU residues, depending upon the tRNA. In some tRNAs, the DHU arm has only three hydrogen-bonded base pairs.

Key:
Pu = *purine nucleoside*
Py = *pyrimidine nucleoside*
ψ = *pseudouridine*
G* = *guanosine or 2′-O-methylguanosine*
T = *ribothymidine*
DHU = *dihydrouridine*

arm carries a specific amino acid esterified by its carboxyl group to the 2′- or 3′-hydroxyl group of the A residue at the 3′ end of the tRNA. The *anticodon arm* contains the *anticodon,* the specific triplet of nucleotides that is complementary to, and can base-pair in an antiparallel manner with, its corresponding codon triplet in the messenger RNA. Each tRNA has a characteristic anticodon triplet. The other major arms are the *DHU* or *dihydrouridine arm,* which contains the unusual nucleoside *dihydrouridine,* and the *TψC arm,* which contains the nucleoside ribothymidine (T), not usually present in RNAs, and the nucleoside *pseudouridine* (ψ), which has an unusual carbon-carbon bond between the base and pentose (Figure 29-5).

Yeast phenylalanine tRNA has been crystallized and subjected to structure analysis by x-ray diffraction, which has verified that there is maximal intrachain base pairing through hydrogen bonding. However, the actual three-dimensional shape of a tRNA looks more like a twisted L than a cloverleaf (Figure 29-6). In addition to the hydrogen bonds between base pairs, tRNAs also have other kinds of hydrogen bonds that participate in maintaining their tertiary structure. Because base pairing in

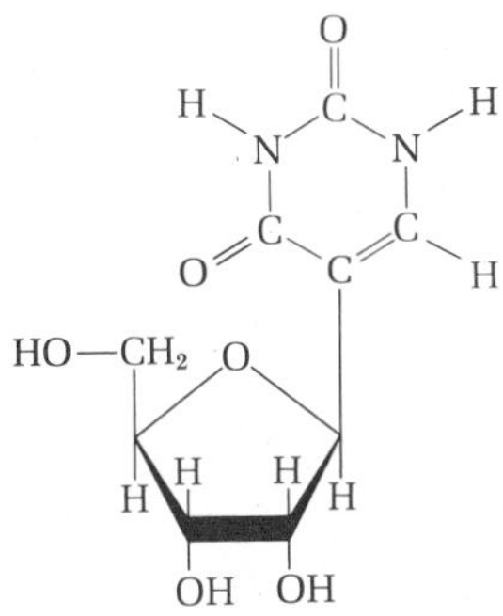

Pseudouridine

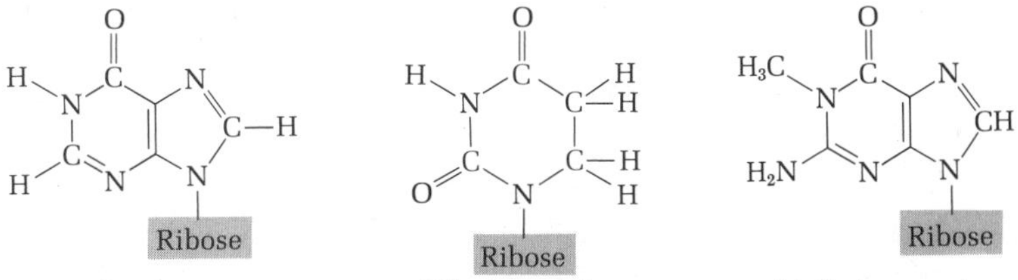

Inosine

Dihydrouridine

1-Methylguanosine

Figure 29-5
Some unusual or modified nucleosides present in tRNAs.

RNAs is not as strict as in DNA, the base-paired segments of tRNAs are somewhat irregular in form and not as rigid and rodlike as the DNA double helix; in fact, the tRNA molecule has considerable flexibility.

Now let us see how its specific amino acid is enzymatically attached to a tRNA molecule.

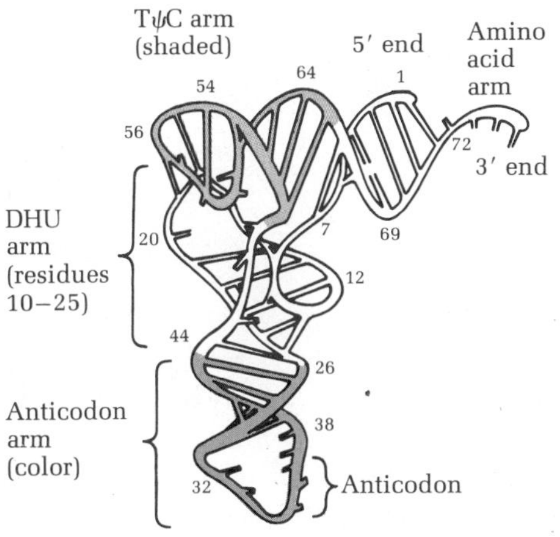

Figure 29-6
The three-dimensional conformation of yeast phenylalanine tRNA *deduced from x-ray diffraction analysis at* 0.3-nm *resolution. It resembles an inverted* L. *[Adapted from S. H. Kim and colleagues,* Science, **185:** 436 (1974).]

Aminoacyl-tRNA Synthetases Attach the Correct Amino Acids to Their tRNAs

In the first stage of protein biosynthesis, which takes place in the cell cytosol, the 20 different amino acids are esterified to their corresponding transfer RNAs by 20 different activating enzymes, called *aminoacyl-tRNA synthetases*, each of which is specific for one amino acid and a corresponding tRNA. Nearly all the aminoacyl-tRNA synthetases of *E. coli* have been isolated, and a number have been crystallized. The overall reaction they catalyze is

$$\text{Amino acid} + \text{tRNA} + \text{ATP} \xrightleftharpoons{Mg^{2+}} \text{aminoacyl-tRNA} + \text{AMP} + \text{PP}_i$$

The activation reaction occurs in two separate steps on the enzyme catalytic site. In the first step an enzyme-bound intermediate, *aminoacyl adenylate* (Figure 29-7), is formed by reaction of ATP and the amino acid at the active site. In this reaction the carboxyl group of the amino acid is bound in anhydride linkage with the 5′-phosphate group of the AMP, with displacement of pyrophosphate

$$\text{Amino acid} + \text{ATP} + \text{E} \rightleftharpoons \text{E–[aminoacyl adenylate]} + \text{PP}_i$$

In the second step the aminoacyl group is transferred from enzyme-bound aminoacyl adenylate to its corresponding specific transfer RNA

$$\text{E–[aminoacyl adenylate]} + \text{tRNA} \rightleftharpoons \text{aminoacyl-tRNA} + \text{adenylate} + \text{E}$$

The aminoacyl group

Adenine

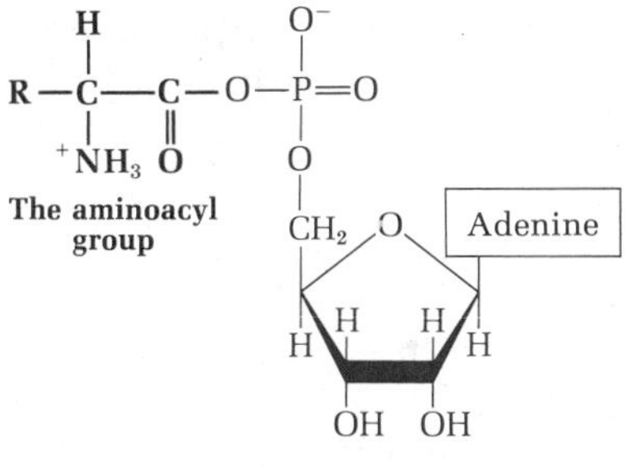

Figure 29-7
*General structure of aminoacyl adenylates, formed on the active site of aminoacyl-*tRNA *synthetases.*

In the latter step the aminoacyl group is transferred to either the free 2′- or 3′-hydroxyl group of the terminal A residue of the tRNA molecule (Figure 29-8); however, once attached it can "jump" back and forth between the 2′- and 3′-hydroxyl groups.

Figure 29-8
General structure of aminoacyl-tRNAs. The aminoacyl group, shown on the 3′ position of the terminal A *(adenylate) residue, is shaded.* R *represents the* R *group of the amino acid. The aminoacyl group can switch between positions 2′ and 3′ on the ribose.*

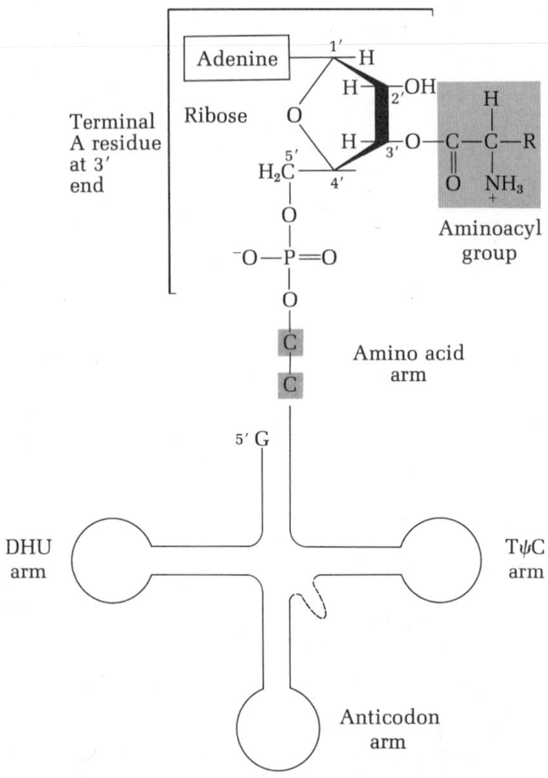

The ester linkage between the amino acid and the tRNA has high-energy characteristics; its $\Delta G^{\circ\prime}$ of hydrolysis is approximately −7.0 kcal/mol. The inorganic pyrophosphate formed in the activation reaction undergoes hydrolysis to orthophosphate by pyrophosphatase (page 386). Thus *two* high-energy phosphate bonds may ultimately be expended for each amino acid molecule activated, rendering the overall reaction for amino acid activation essentially irreversible

$$\text{Amino acid} + \text{tRNA} + \text{ATP} \xrightarrow{Mg^{2+}} \text{aminoacyl-tRNA} + \text{AMP} + 2P_i \qquad \Delta G^{\circ\prime} = -7.0 \text{ kcal/mol}$$

The aminoacyl-tRNA synthetases are very specific for both the tRNA and its corresponding amino acid. If the wrong amino acid is joined to a tRNA to form a mismatched aminoacyl-tRNA, an incorrect amino acid residue will be inserted into the polypeptide chain. However, some of the aminoacyl-tRNA synthetases are "smart" enzymes; like DNA polymerases, they are capable of proofreading and correcting their own errors. For example, because the R groups of the amino acids valine and isoleucine resemble each other structurally (the only difference is that isoleucine has one more —CH_2— group) it might be expected that valine would often be incorporated into polypeptides in place of isoleucine. However, the error frequency in isoleucine incorporation is no greater than for other amino acids, about 1 per 3000–4000 residues, because the isoleucyl-tRNA synthetase can proofread and prevent such an error. It recognizes when it has formed the wrong aminoacyl adenylate and corrects the error by hydrolyzing the valyl-AMP while it is still on the active site

$$\text{E–[valyl-AMP]} + H_2O \longrightarrow \text{valine} + \text{AMP} + \text{E}$$

The isoleucyl-tRNA synthetase then starts over again and correctly forms the isoleucyl-AMP intermediate, which in turn leads to the formation of the correct isoleucyl-tRNA^{Ile}

$$\text{Isoleucine} + \text{ATP} \rightleftharpoons \text{isoleucyl-AMP} + \text{PP}_i$$

$$\text{E–[isoleucyl-AMP]} + \text{tRNA}^{\text{Ile}} \rightleftharpoons \text{isoleucyl-tRNA}^{\text{Ile}} + \text{AMP} + \text{E}$$

Since the R group of valine is slightly smaller than that of isoleucine it appears that valyl-AMP fits the hydrolytic site of the isoleucine-tRNA synthetase, but isoleucyl-AMP does not. Aminoacyl-tRNA synthetases appear to have four specific sites that participate in recognition, catalysis, and correction of errors: one for the amino acid, one for the tRNA, another for the ATP, and yet another for the entry of H_2O for hydrolysis of incorrect aminoacyl-AMPs.

Transfer RNA Is an Adapter

Once an amino acid has been esterified to its corresponding tRNA, it makes no contribution to the specificity of the aminoacyl-tRNA, since the aminoacyl group itself is not recognized by either the ribosome or the mRNA template. The specificity of the aminoacyl-tRNA is furnished by the structure of the tRNA portion alone. This was conclusively proved in clever experiments in which enzymatically formed cysteinyl-tRNA^{Cys} was isolated and then chemically converted into alanyl-tRNA^{Cys}. This hybrid aminoacyl-tRNA, which carries alanine but contains the anticodon for cysteine, was then incubated with a cell-free system capable of protein synthesis. The newly synthesized polypeptide was found to contain alanine in positions that should have been occupied by cysteine residues. This experiment thus provided proof for Crick's adapter hypothesis.

Polypeptide Chains Originate at the Amino-Terminal End

One more question arises. Does polypeptide chain growth begin from the amino-terminal or from the carboxyl-terminal end? The answer came from isotope tracer experiments. Radioactive leucine was incubated with reticulocytes, immature red blood cells that actively synthesize hemoglobin. This amino acid was chosen because it occurs frequently along both the α- and β-globin chains. Samples of completed α chains were isolated from the reticulocytes at various time intervals after addition of radioactive leucine, and the distribution of radioactivity along the α chain was determined. Nearly all the leucine resi-

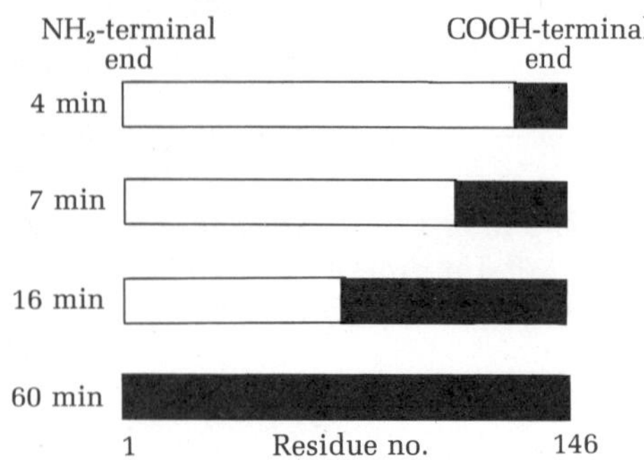

Figure 29-9
Proof that polypeptide chains grow by addition of new amino acid residues to the carboxyl end. The colored zones show the portions of hemoglobin chains containing radioactive leucine residues at different time intervals after addition of radioactive leucine. At 4 min only a few residues at the carboxyl end of α-globin were labeled. On longer times of incubation with labeled leucine, successively longer segments of the polypeptide chain contained labeled leucine residues, always in a block at the carboxyl end of the chain. Thus the polypeptide chain grows by successive addition of amino acids at the carboxyl end.

dues were radioactive in those globin chains isolated after 60 min of incubation, but completed chains that were isolated only a few minutes after radioactive leucine was added contained radioactive leucine residues only at the carboxyl-terminal end (Figure 29-9). From these observations it was concluded that polypeptide chains are begun at the amino-terminal end and are elongated by addition of residues to the carboxyl-terminal end.

We shall now see how the amino-terminal residue initiates the biosynthesis of polypeptides.

N-Formylmethionine Is the Initiating Amino Acid in Prokaryotes and Methionine in Eukaryotes

In *E. coli* and all other prokaryotes the starting amino acid residue at the amino-terminal end is always *N-formylmethionine* (Figure 29-10). It enters as *N*-formylmethionyl-$tRNA^{fMet}$ (symbolized as fMet-$tRNA^{fMet}$), which is formed in two successive reactions. First, methionine is attached to a special initiating methionine tRNA, designated $tRNA^{fMet}$, by the methionine aminoacyl-tRNA synthetase

$$\text{Methionine} + \text{tRNA}^{\text{fMet}} + \text{ATP} \longrightarrow \text{methionyl-tRNA}^{\text{fMet}} + \text{AMP} + \text{PP}_i$$

In the second reaction a formyl group is transferred to the amino group of the methionyl residue from its donor, *N^{10}-formyltetrahydrofolate* (page 261), by a specific transformylase enzyme

$$N^{10}\text{-Formyltetrahydrofolate} + \text{Met-tRNA}^{\text{fMet}} \longrightarrow \text{tetrahydrofolate} + \text{fMet-tRNA}^{\text{fMet}}$$

```
    H  COO⁻
    |   |
H—C—N—C—H
  ‖     |
  O    CH2
        |
       CH2
        |
        S
        |
       CH3
```

Figure 29-10
N-Formylmethionine, the initiating amino acid in all prokaryotes. The N-formyl group is in color.

This enzyme cannot formylate free methionine. There are two species of tRNA specific for methionine. One is designated $tRNA^{Met}$ and the other $tRNA^{fMet}$. Both can accept methionine groups in the activation reaction, but only methionyl-$tRNA^{fMet}$ can accept the formyl group to become the initiating amino acid. The other species, methionyl-$tRNA^{Met}$, is used to insert methionine in *interior* positions in the polypeptide chain. Blocking of the amino group of methionine by the *N*-formyl group not only prevents it from entering into interior positions but also allows fMet-$tRNA^{fMet}$ to be bound at a specific initiation site on the ribosome which does not accept Met-$tRNA^{Met}$ or any other free aminoacyl-tRNA.

In eukaryotic cells, on the other hand, all polypeptides synthesized by extramitochondrial ribosomes begin with a *methionine* residue, donated by a special initiating methionyl-tRNA. Polypeptides synthesized by the ribosomes in the mitochondria and chloroplasts of eukaryotic cells begin with *N*-formylmethionine. This and other similarities in the protein-synthesizing machinery of these organelles and bacteria have been taken to support the view (pages 26 and 493) that mitochondria and chloroplasts may have originated from bacteria during the early evolution of eukaryotic cells. However, in certain other respects there are significant differences between bacterial and mitochondrial transcription and translation mechanisms.

We are now left with a puzzle. There is only *one* codon for methionine, namely (5′)AUG(3′). How can this single codon serve to identify both the *starting* N-formylmethionine residue (or methionine in the case of eukaryotes) and those methionine residues which occur in interior positions in polypeptide chains? The answer will come later when we examine the *initiation step*, but first we must examine the structure of ribosomes.

Ribosomes Are Molecular Machines for Making Polypeptide Chains

In *E. coli* cells there are 15,000 or more ribosomes, and they make up almost a quarter of the dry weight of the cell. Prokaryotic ribosomes contain about 65 percent ribosomal RNA and about 35 percent protein. They have a particle weight of about 2.8 million daltons, a diameter of about 18 nm, and a sedimentation coefficient of 70S.

Prokaryotic ribosomes consist of two subunits of unequal size (Figure 29-11), the larger having a sedimentation coefficient of 50S and the smaller of 30S, corresponding to particle weights of 1.8 million and 1.0 million daltons, respectively.

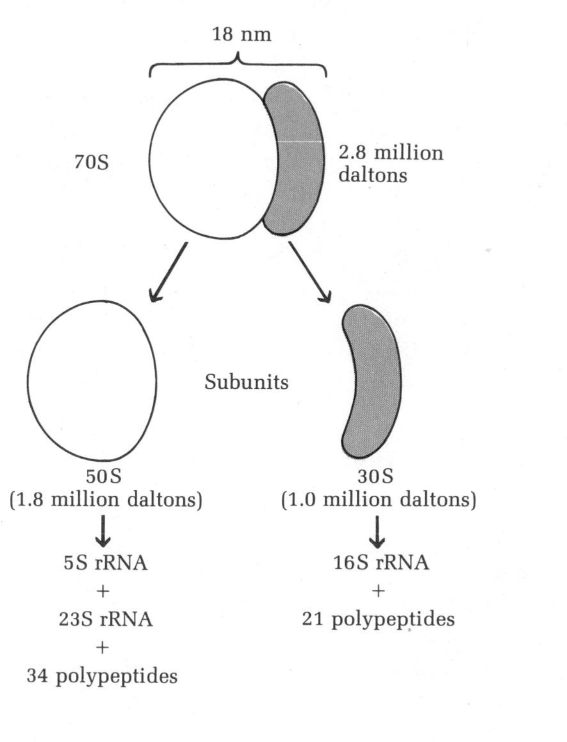

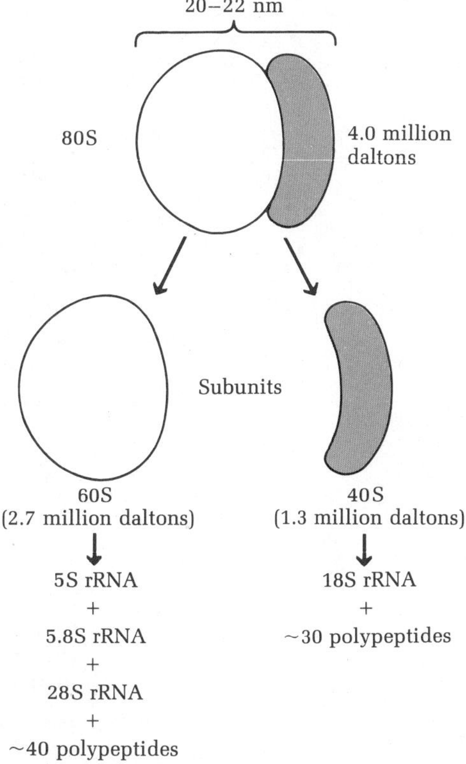

Figure 29-11
Composition of (a) prokaryotic and (b) eukaryotic ribosomes (cytoplasmic). In the mitochondria and chloroplasts of eukaryotes the ribosomes resemble those of prokaryotic cells.

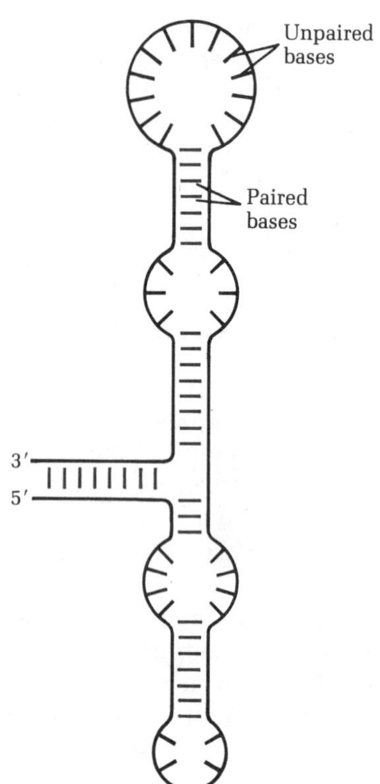

Figure 29-12
Schematic drawing of one possible model for the secondary folded structure of the 5S rRNA of prokaryotes, assuming maximal intrachain base pairing (color).

The 50S subunit contains one molecule of 23S rRNA (about 3200 nucleotides), one molecule of 5S rRNA (about 120 nucleotides), and 34 proteins. The 30S subunit contains one molecule of 16S rRNA (1600 nucleotides) and 21 proteins. The proteins of the subunits are designated by numbers. Those in the large 50S subunit are numbered L1 to L34 (L for large) and those in the smaller subunit S1 to S21 (S for small). All the proteins of *E. coli* ribosomes have been isolated and many have been sequenced; they are markedly different from each other. Their molecular weights range from about 6000 to about 75,000.

The sequences of nucleotides in the single-strand ribosomal RNAs of *E. coli* have also been determined. Each of the three rRNAs has a specific three-dimensional conformation conferred by intrachain base pairing. Figure 29-12 shows a postulated representation of 5S rRNA in a maximally base-paired conformation. The ribosomal RNAs appear to serve as framework structures, to which the polypeptide subunits are bound in a very specific ordered arrangement. When the 21 different polypeptides and the 16S rRNA of the 30S subunit are isolated in pure form and then mixed in the proper specific order at the correct temperature, they spontaneously reassemble to form 30S subunits identical in structure and activity to native 30S subunits. Similarly, the 50S subunit can assemble itself from its 34 polypeptides and its 5S and 23S rRNAs, providing the 30S subunit is also present. Every one of the 55 proteins in the prokaryotic ribosome is believed to play a specific role in the synthesis of polypeptides, either as an enzyme or as a "helper" in the overall process. However, the specific function of only a few of the ribosomal proteins is known.

Although ribosomes are usually drawn with the 30S subunit represented as a "cap" set on a 50S sphere (as in Figure 29-11), they are not symmetrical structures. In fact, the two subunits have surprisingly irregular shapes. Figure 29-13 shows the three-dimensional structure of the 30S and 50S subunits of *E. coli* ribosomes, deduced from x-ray analysis and electron microscopy. The two oddly shaped subunits fit together in such a way that a cleft is formed through which the mRNA passes as the ribosome moves along it during the translation process and from which the newly formed polypeptide chain emerges.

The Cytoplasmic Ribosomes of Eukaryotes Are Larger and More Complex

The ribosomes in the extramitochondrial portions of eukaryotic cells are substantially larger than prokaryotic ribosomes (Figure 29-11). They have a diameter of about 21 nm, a sedimentation coefficient of about 80S and a particle weight of about 4 million daltons. They also have two subunits, which vary in size in different species, but on the average are 60S and 40S. The small subunit contains an 18S rRNA, and the large subunit

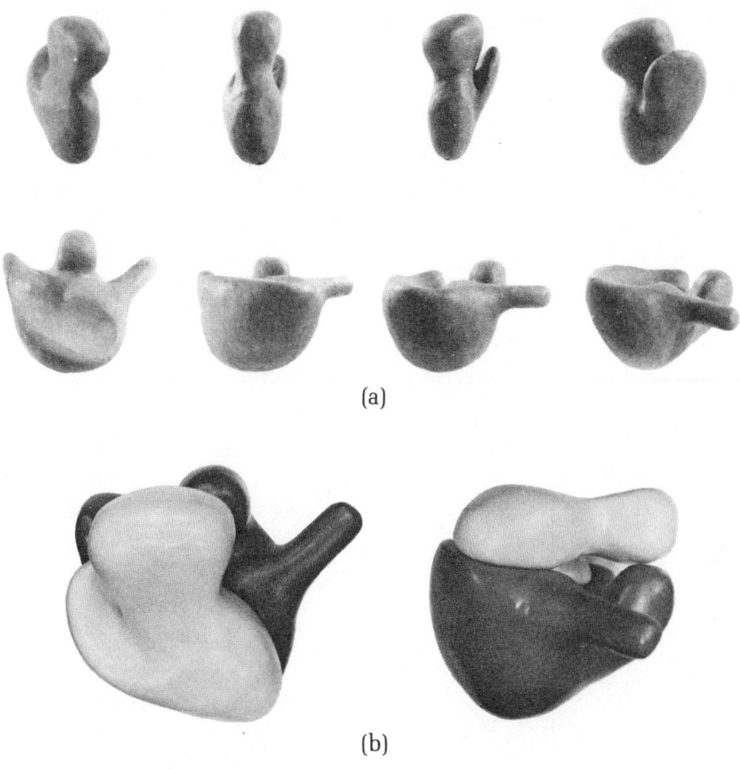

Figure 29-13
E. coli ribosomal subunits have odd shapes, revealed by x-ray analysis and electron microscopy. (a) Several views of models of the 30S (top) and 50S (below) subunits. (b) Two views of complete 70S ribosomes. The 30S subunit is white, the 50S black.

contains 5S, 5.8S, and 28S rRNAs. Altogether eukaryotic ribosomes contain over 70 different proteins. The ribosomal RNAs and most of the proteins of eukaryotic ribosomes have also been isolated.

The two subunits of ribosomes are not permanently attached to each other. As we shall see, ribosomes must dissociate into their two subunits each time a new polypeptide chain is begun.

Initiation of a Polypeptide Has Several Steps

The initiation of a polypeptide chain in prokaryotes requires (1) the 30S subunit, which contains 16S ribosomal RNA, (2) the mRNA coding for the polypeptide to be made, (3) the initiating N-formylmethionyl-$tRNA^{fMet}$, (4) a set of three proteins called *initiation factors* (IF-1, IF-2, and IF-3), and (5) GTP (Table 29-1).

The formation of the *initiation complex* takes place in three steps. In the first, the 30S ribosomal subunit binds *initiation factor 3* (IF-3), which prevents the 30S and 50S subunits from recombining. Binding of the mRNA to the 30S subunit then takes place, in such a way that the initiating codon in the mRNA [which is (5′)AUG(3′)] binds to a special location on the 30S subunit (Figure 29-14). The initiating codon AUG is guided to the correct position on the 30S subunit by a special *initiating signal* in the mRNA, located to the 5′ side of the AUG codon. This signal contains predominantly A and G residues,

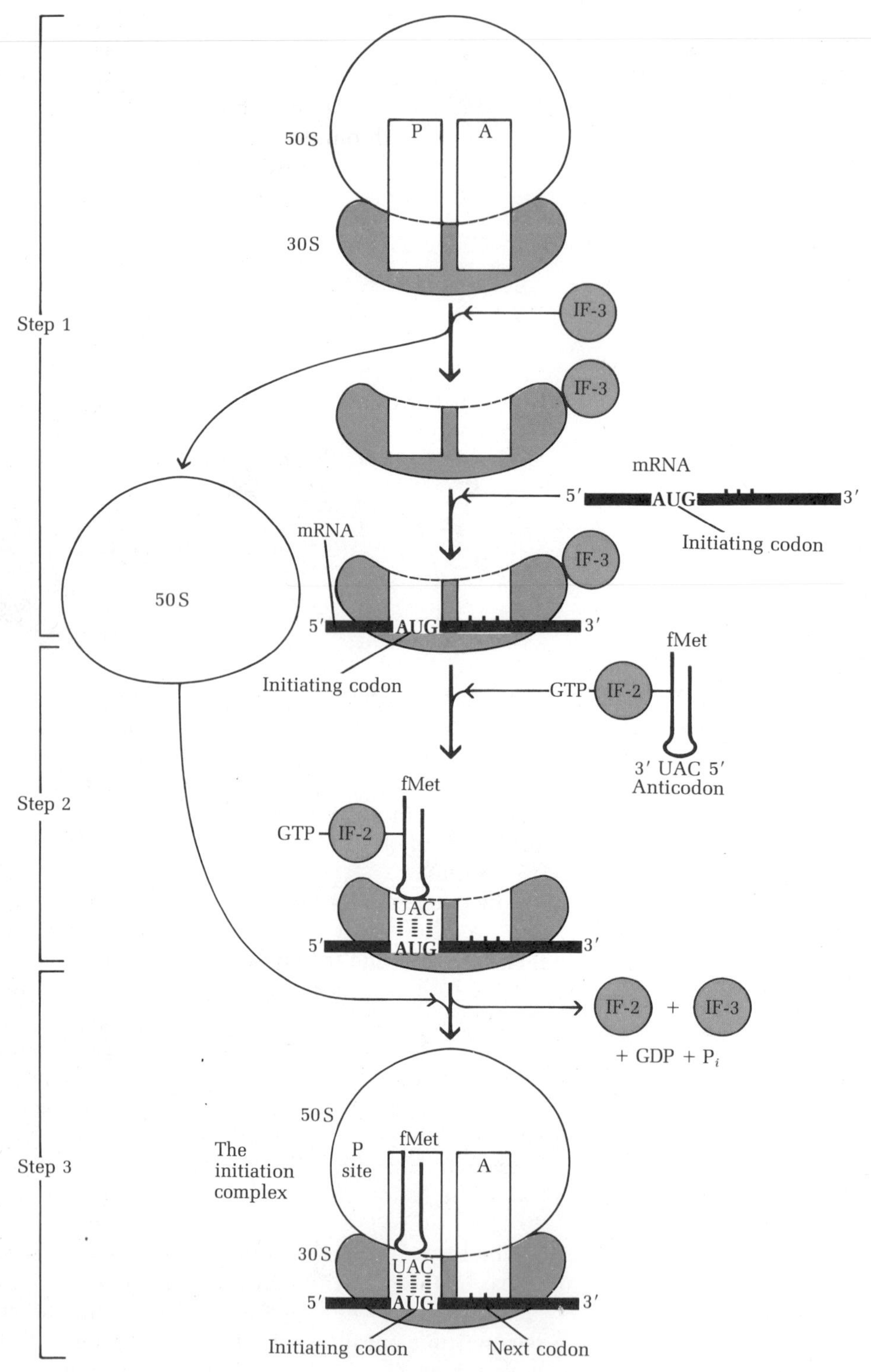

Step 1
50 S
P
A
30 S
IF-3
mRNA
5′
AUG
3′
Initiating codon
50 S
Step 2
fMet
GTP
IF-2
3′ UAC 5′
Anticodon
UAC
Step 3
IF-2 + IF-3
+ GDP + P_i
The initiation complex
P site
Next codon

Figure 29-14
Formation of the initiation complex in three steps at the expense of the hydrolysis of GTP *to* GDP *and* P_i.

IF-1, IF-2, *and* IF-3 *are initiation factors.* P *designates the peptidyl site,* A *the aminoacyl site. The role of* IF-1 *is not fully known.*

usually six to eight in number. This sequence is recognized by, and base-pairs with, a complementary sequence in the 16S ribosomal RNA of the 30S subunit, which positions the mRNA correctly for initiation of translation. Since there is only one codon for methionine, which codes for both the initiating and interior methionine residues, the initiation signal for the 5′ side of AUG identifies the site where fMet-tRNA^{fMet} is to be bound. Interior AUG codons are specific for Met-tRNA^{Met} and cannot bind fMet-tRNA^{fMet}.

In the second step of the initiation process (Figure 29-14), the complex of the 30S subunit, IF-3, and mRNA now forms a still larger complex by binding initiation protein IF-2, which already contains bound GTP and the initiating *N*-formylmethionyl-tRNA^{fMet}, which is correctly placed on the initiation codon.

This large complex now combines in the third step of initiation, with the 50S ribosomal subunit; simultaneously, the GTP molecule bound to IF-2 is hydrolyzed to GDP and phosphate, which are set free. IF-3 and IF-2 also depart from the ribosome. We now have a functional 70S ribosome, called the *initiation complex,* containing the mRNA and the initiating *N*-formylmethionyl-tRNA^{fMet}. The correct binding of the *N*-formylmethionyl-tRNA^{fMet} in the complete 70S initiation complex is assured by two points of recognition and attachment. In the first the anticodon triplet of the initiating aminoacyl-tRNA base-pairs in an antiparallel manner with the codon triplet AUG in the mRNA. The second point of attachment of the initiating aminoacyl-tRNA is to the ribosomal P site. Ribosomes have two sites for binding aminoacyl-tRNAs, the *aminoacyl* or *A site* and the *peptidyl* or *P site*. Each is a composite of specific portions of the 50S and 30S subunits. The initiating fMet-tRNA can bind only to the P site (Figure 29-14), but it is the exception: all other new incoming aminoacyl-tRNAs bind to the A site, whereas the P site is the site from which the "empty" tRNAs leave and to which the growing peptidyl-tRNA becomes bound.

The initiation complex is now ready for the elongation steps.

Elongation of the Polypeptide Chain Is a Repetitive Process

Three steps take place in the addition of each amino acid residue to the growing polypeptide chain. This cycle is repeated as many times as there are residues to be added. For elongation to occur we need (1) the initiation complex described above, (2) the next aminoacyl-tRNA, specified by the next coding triplet in the mRNA, (3) a set of three soluble proteins of the cytosol called *elongation factors,* EF-Tu, EF-Ts, and EF-G, and (4) GTP. The elongation factors are often designated simply as Tu, Ts, and G.

In the first step in the elongation cycle (Figure 29-15) the next aminoacyl-tRNA is first bound to a complex of elongation factor Tu containing a molecule of bound GTP. The resulting

Figure 29-15
First step in elongation, the binding of the second aminoacyl-tRNA. It enters bound to elongation factor Tu, which also contains bound GTP. Binding of the second aminoacyl-tRNA is accompanied by hydrolysis of the bound GTP. The bound GDP so formed is rephosphorylated by GTP, in a reaction catalyzed by elongation factor Ts. The nucleotides for the anticodon and codon for subsequent amino acids are symbolized by small open circles.

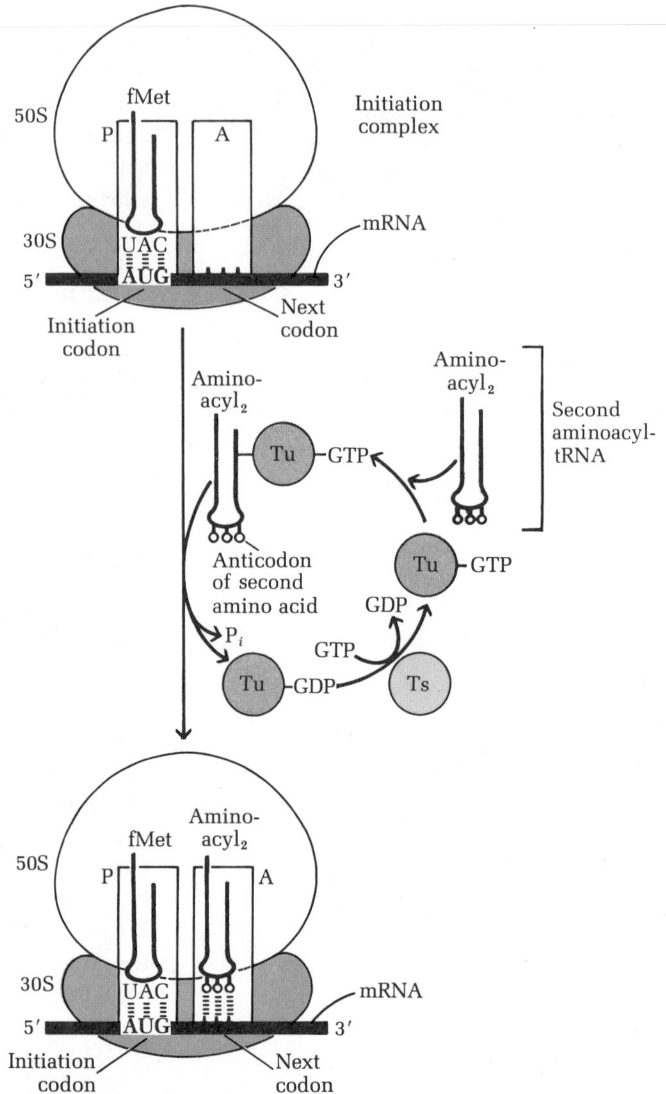

aminoacyl-tRNA-Tu-GTP complex is now bound to the 70S initiation complex; simultaneously, the GTP is hydrolyzed and a Tu-GDP complex is released from the 70S ribosome. The Tu-GTP complex is regenerated from the Tu-GDP complex by the Ts factor and GTP.

The new aminoacyl-tRNA is now bound to the *aminoacyl* or *A site* on the ribosome. This occurs through antiparallel base pairing between the anticodon of the new aminoacyl-tRNA and the corresponding codon in the messenger RNA (we shall examine the different codon and anticodon sequences later). However, the anticodon-codon interaction is not enough to ensure that the correct aminoacyl-tRNA has been bound. The identity of the latter is checked by a second binding interaction within the A site, between another part of the tRNA molecule and the ribosomal RNA. Only if both binding interactions are correct will the next step occur.

In the second step in the elongation cycle the new peptide bond is formed between the amino acids whose tRNAs are located on the A and P sites on the ribosome. This step occurs by the transfer of the initiating *N*-formylmethionine acyl group from its tRNA to the amino group of the new amino acid that has just entered the A site. This step is catalyzed by *peptidyl transferase,* a specific ribosomal protein in the 50S subunit (Figure 29-16). As the result of this reaction a dipeptidyl-tRNA is formed on the A site and the now "empty" initiating $tRNA^{fMet}$ remains bound to the P site.

In the third step of the elongation cycle the ribosome moves along the mRNA, toward its 3′ end, by a distance of one codon (three bases). Since the dipeptidyl-tRNA is still attached to the second codon of the mRNA, the movement of the ribosome shifts the dipeptidyl-tRNA from the A site to the P site, which

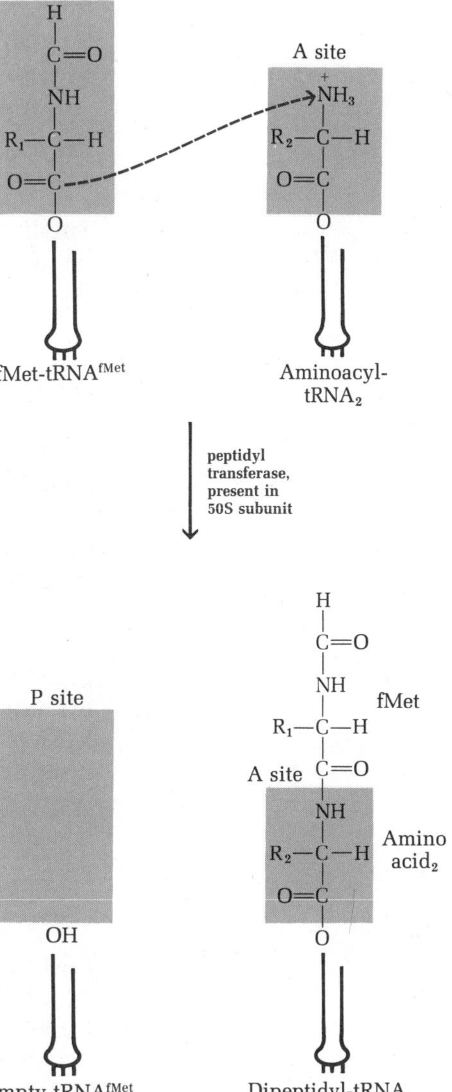

Figure 29-16
Formation of the first peptide bond. The N-formylmethionyl group is transferred to the amino group of the second aminoacyl-tRNA to make a dipeptidyl-tRNA on the A site.

causes the release of the preceding tRNA, which is empty, from the P site back into the cytosol. Now the third codon of the mRNA is on the A site and the second codon on the P site. This shift of the ribosome along the mRNA is called the *translocation* step; it requires elongation factor G (also called the *translocase*), as well as the simultaneous hydrolysis of another molecule of GTP (Figure 29-17). A change in the three-dimensional conformation of the entire ribosome is believed to take place in this step, in order to move the ribosome to the next codon toward the 3′ end of the mRNA. The hydrolysis of GTP provides the energy for the translocation.

The ribosome, with its attached dipeptidyl-tRNA and mRNA, is now ready for another elongation cycle to attach the third amino acid residue, which proceeds in precisely the same way as the addition of the second. For each amino acid residue added to the chain, two GTPs are hydrolyzed to GDP and P_i. As the ribosome moves from codon to codon along the mRNA toward its 3′ end, adding an amino acid residue each time, the polypeptide chain always remains attached to the tRNA of the last amino acid to be inserted.

Termination of Polypeptide Synthesis Requires a Special Signal

Finally, the ribosome adds the last amino acid, which completes the polypeptide coded by the mRNA. The termination of the polypeptide is signaled by one of three termination triplets in the mRNA, immediately following the last amino acid codon. The termination triplets, UAA, UAG, and UGA, do not code for any amino acids. They are called *nonsense triplets* because they were first found to result from single-base mutations of some of the amino acid codons, resulting in *nonsense mutants* of *E. coli*, in which certain polypeptide chains are prematurely terminated. These nonsense mutants, arbitrarily named *amber*, *ochre*, and *opal*, ultimately made possible identification of UAA, UAG, and UGA as termination codons.

Once the ribosome reaches a termination codon, three *termination* or *releasing factors*, the proteins R_1, R_2, and S, participate in causing (1) the hydrolytic cleavage of the polypeptide from its terminal tRNA and its release in free form, (2) the release of the last tRNA, now empty, from the P site, and (3) the dissociation of the 70S ribosome into its 30S and 50S subunits, ready to start a new polypeptide chain.

Energy Is Required to Guarantee Fidelity of Protein Synthesis

Two high-energy phosphate groups are used in the enzymatic formation of each aminoacyl-tRNA from its free amino acid, as we have seen (page 878). Additional ATPs may be used in this step for correction of errors detected by the hydrolytic action of the aminoacyl-tRNA synthetases. One molecule of GTP is

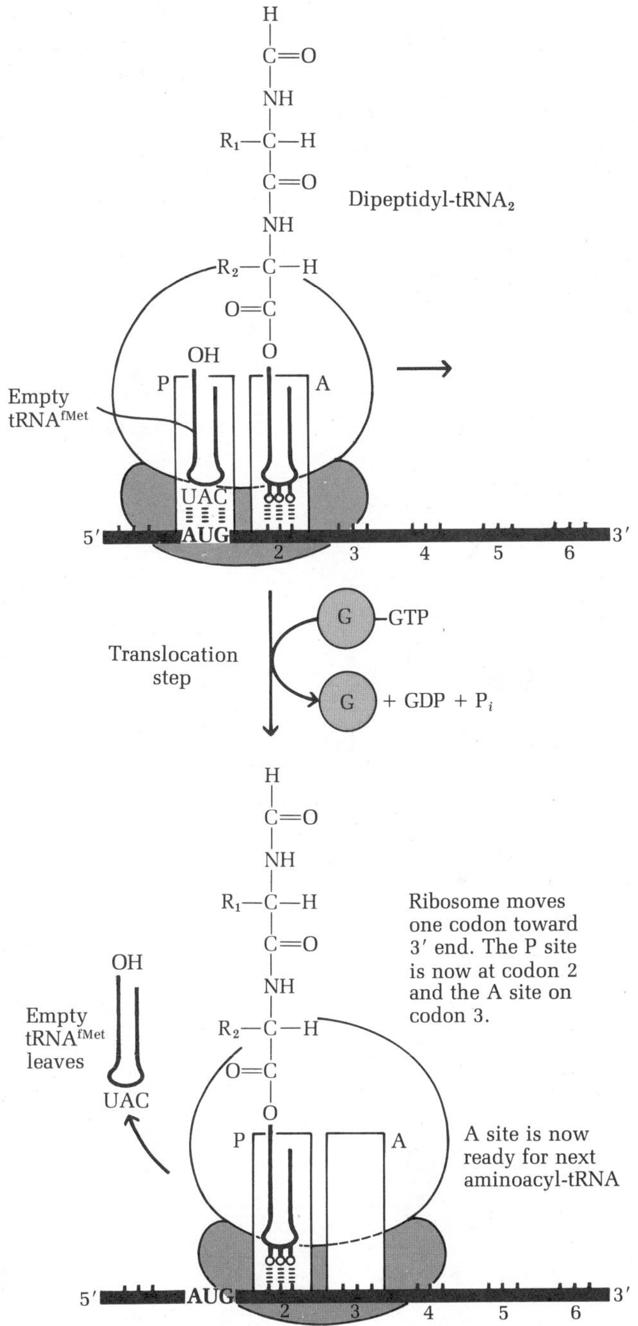

Figure 29-17
The translocation step. The ribosome moves one codon toward the 3′ end of mRNA, at the expense of the energy provided by hydrolysis of GTP bound to elongation factor G. The dipeptidyl-tRNA₂ is now on the P site, leaving the A site open for the incoming aminoacyl-tRNA₃.

cleaved to GDP and phosphate during the first elongation step, and another GTP is hydrolyzed in the translocation step. Therefore a total of at least four high-energy bonds is ultimately required for the formation of each peptide bond of the completed polypeptide chain. This represents an exceedingly large thermodynamic "push" in the direction of synthesis, since at least 7.3(4) = 29.2 kcal of phosphate-group energy is required to generate a peptide bond whose standard free energy of hydrolysis is only about −5.0 kcal. The net free-energy requirement for peptide-bond synthesis is thus −24.2 kcal/mol. Although this large energy expenditure may appear wasteful, it is one of the important factors making possible nearly perfect fidelity in the biological translation of the genetic message of mRNA into the amino acid sequence of proteins.

Polyribosomes Make Possible Rapid Translation of a Single Messenger

When ribosomes are carefully isolated from tissues that are very active in protein synthesis, such as the pancreas, they are often arranged in clusters containing several or many ribosomes, sometimes 80 or more. Such clusters, called polyribosomes or polysomes, have been examined with the electron microscope and studied chemically. Since they can be fragmented into single individual ribosomes by the action of ribonuclease, polyribosomes are held together by a strand of RNA. A connecting fiber between adjacent ribosomes can in fact be seen in electron micrographs (Figure 29-18). The connecting strand of RNA is actually messenger RNA, which is being translated by many ri-

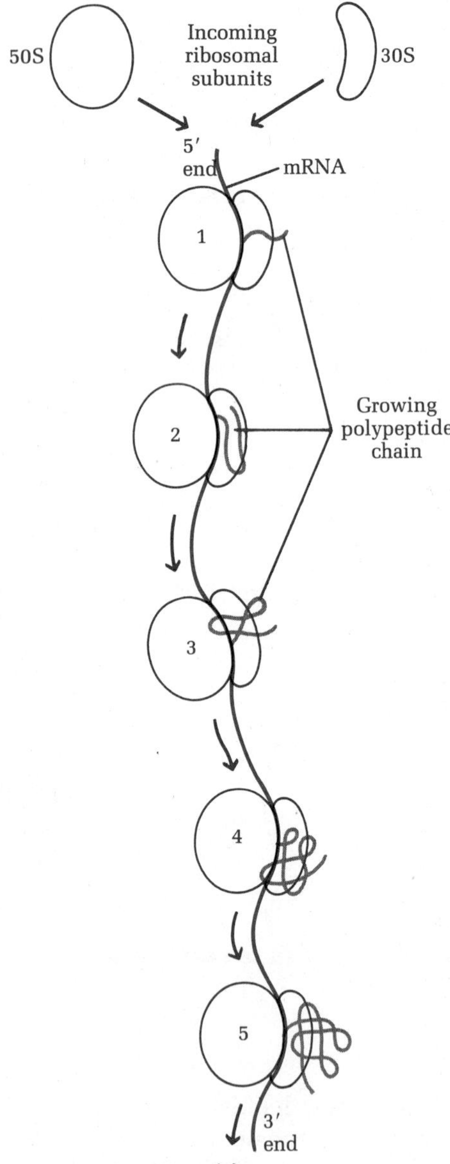

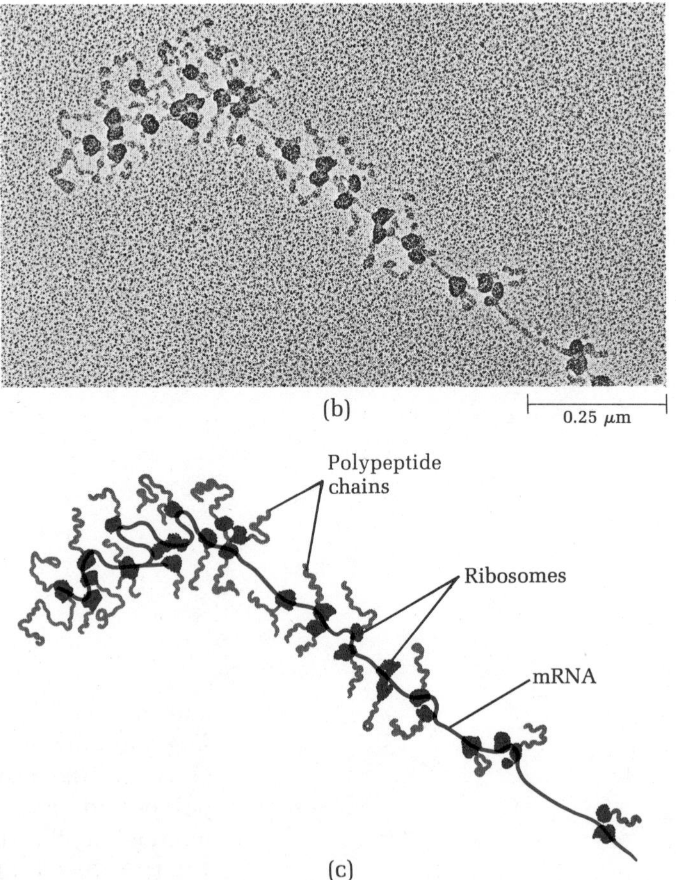

Figure 29-18
A polyribosome. (a) Five ribosomes are reading an mRNA molecule simultaneously, moving from its 5′ end to the 3′ end. (b) Electron micrograph of a polyribosome from the silkworm gland. The mRNA is being translated by many ribosomes simultaneously. The polypeptide chains become longer as the ribosomes move toward the 3′ end of this mRNA, which codes for silk fibroin. (c) Diagram identifying the components in (b).

Figure 29-19
The coupling of transcription and translation in bacteria. The mRNA is translated by ribosomes while it is still being transcribed from DNA by RNA polymerase. This is possible because the mRNA in bacteria does not need to be transported from the nucleus to the cytoplasm.

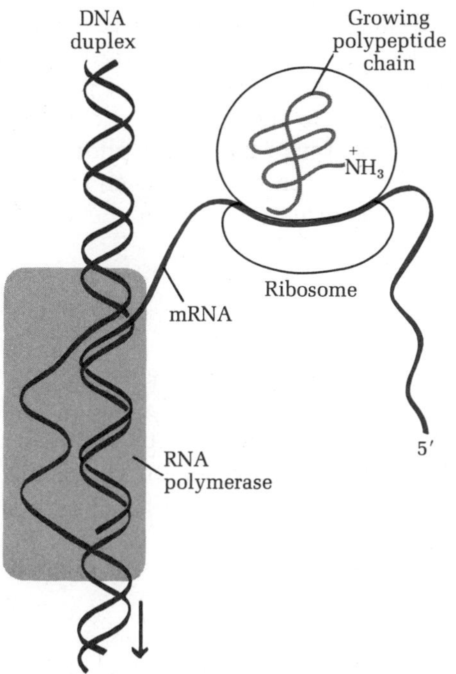

bosomes simultaneously (Figure 29-18), spaced rather closely together. Thus simultaneous translation of a single mRNA by many ribosomes greatly increases the efficient use of the mRNA template.

In bacteria there is a very tight coupling between transcription and translation. As shown in Figure 29-19, ribosomes begin translating messenger RNA while the latter is still being made by the DNA-directed RNA polymerase. Another feature of protein synthesis in bacteria is that mRNA molecules have a very short life, only a few minutes; they are quickly degraded by nucleases. For this reason the mRNA for a given protein or set of proteins must be made continuously and used as template with maximum efficiency, in order to maintain constant rates of protein synthesis. The short lifetime of mRNAs in prokaryotes makes possible rapid turnoff of the synthesis of a protein when it is no longer needed, as we shall see later.

Polypeptide Chains Undergo Folding and Processing

As we have seen in Chapters 7 and 8, a protein is not biologically active until it is in its native folded conformation, which is determined by its amino acid sequence. At some point during or after its synthesis, the polypeptide chain spontaneously assumes its native conformation (page 179). In this way the linear or one-dimensional genetic message brought by the messenger RNA is converted into the specific three-dimensional structure of the newly synthesized polypeptide. However, the newly made polypeptide chain often does not attain its final biologically active conformation until it has been subjected to *processing* or *covalent modification*. Such changes are called *posttranslational modifications*. Several kinds of processing may take place, depending upon the protein.

Amino-Terminal and Carboxyl-Terminal Modifications

All polypeptides are begun with a residue of N-formylmethionine in prokaryotes and with methionine in eukaryotes (page 880). However, the formyl group, the initiating methionine residue, and often one or more succeeding residues may be removed by the action of specific enzymes and thus may not appear in the finished proteins.

In some proteins the amino group of the amino-terminal residue is acetylated after transcription; in others the carboxyl-terminal residues may be modified.

Loss of Signaling Sequences

As we shall see, some proteins are made with an extra polypeptide sequence of 15 to 30 residues at the amino-terminal end, to direct the protein to its ultimate destination in the cell. Such *signaling sequences* are ultimately removed by specific peptidases.

Phosphorylation of Hydroxy Amino Acids

The hydroxyl groups of certain serine, threonine, and tyrosine residues of some proteins are enzymatically phosphorylated by ATP to yield phosphoserine, phosphothreonine, and phosphotyrosine residues; the phosphate groups thus add negative charges to these polypeptides. The milk protein casein has many phosphoserine groups, which function to bind Ca^{2+}. Since both Ca^{2+} and phosphate, as well as amino acids, are required by suckling young, casein of milk provides three essential nutrients. The phosphorylation of the hydroxyl group of specific serine residues is required to activate certain enzymes, such as glycogen phosphorylase (page 239). Phosphorylation of specific tyrosine residues of some proteins has been found to be an important step in the transformation of normal cells into cancer cells.

Carboxylation Reactions

Extra carboxyl groups may be added to aspartic and glutamic acid residues of some proteins. For example, the blood-clotting protein prothrombin contains a number of γ-carboxyglutamyl residues (Chapter 24) in its amino-terminal region, introduced by a vitamin K–requiring enzyme. These groups bind Ca^{2+}, required to initiate the clotting mechanism.

Methylation of R Groups

In some proteins certain lysine residues are methylated enzymatically. Monomethyl- and dimethyllysine residues are present in some muscle proteins and in cytochrome *c*. In other proteins the carboxylate groups of some glutamate residues undergo methylation, which removes their negative charges.

Attachment of Carbohydrate Side Chains

The carbohydrate side chains of glycoproteins are attached covalently during or after the synthesis of the polypeptide chain. In some glycoproteins the carbohydrate side chain is attached enzymatically to asparagine residues, in others to serine or threonine residues. Many proteins that function extracellularly, as well as the "lubricating" proteoglycans coating mucous membranes, contain oligosaccharide side chains.

Addition of Prosthetic Groups

Many enzymes contain covalently bound prosthetic groups necessary for their activity; these are also attached to the polypeptide chain after it leaves the ribosome. Two examples are the covalently bound biotin molecule in acetyl-CoA carboxylase (page 586) and the heme group of cytochrome *c* (page 480).

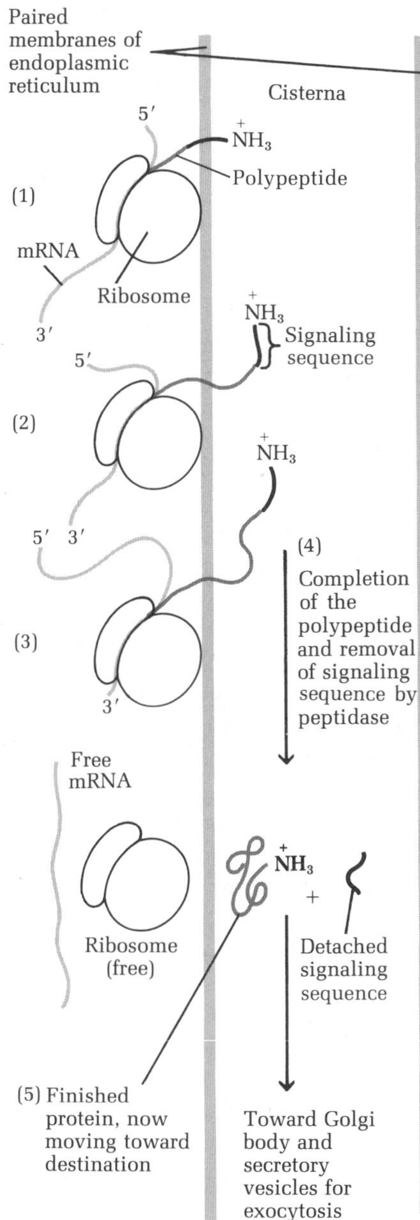

Figure 29-20
Steps in the synthesis of a protein to be exported from the cell (1 to 5), showing the formation and fate of the signaling sequence, a polypeptide leader at the amino-terminal end of many proteins made by ribosomes on rough-surfaced endoplasmic reticulum. Signaling sequences help guide the new polypeptide chain through the membrane into the cisterna. The signaling sequence is removed by a peptidase during or after the entry of the polypeptide into the cisterna.

Formation of Disulfide Cross-Links

Many proteins to be exported from eukaryotic cells, after undergoing spontaneous folding into their native conformations, are covalently cross-linked by the enzymatic formation of the disulfide groups of cystine residues within one polypeptide chain or between two chains (pages 134 and 135). The cross-links formed in this way help to protect the native folded conformation of the protein molecule from denaturation.

Newly Made Proteins Are Often Directed to Their Destinations

Some newly made proteins will simply be delivered into the cell cytosol, some will be delivered to different cell organelles, some will be secreted to the exterior of the cell, and some will be inserted into one cell membrane or another, there to function as transport proteins or membrane enzymes. It is therefore important that a newly synthesized protein find its way to its correct site in the cell. How are newly made proteins earmarked and dispatched to their destination?

Many proteins contain specific polypeptide "leaders" on their amino-terminal ends that function as signals for directing them to their proper destinations. These *signaling sequences* have aptly been compared to the zip code of a postal address. Exported proteins made by ribosomes on the rough endoplasmic reticulum of pancreatic cells, such as trypsinogen and procarboxypeptidase (page 689), contain polypeptide leaders or sequences at their amino-terminal ends. Such signaling sequences have from 15 to 30 or more amino acid residues, many of which have hydrophobic R groups (Figure 29-20). Signaling leaders, being located at the amino-terminal end, are formed first during the synthesis of proteins that are to be exported. Such leaders are recognized by specific receptor sites on the outer surface of the endoplasmic reticulum, even before the protein has been completed by the ribosome. The hydrophobic, lipid-soluble part of the leader sequence penetrates through the membrane into the cisterna of the endoplasmic reticulum, followed by the growing polypeptide chain. The signaling leader is removed by the action of a special peptidase in the cisterna. The finished protein then makes its way to the Golgi body, becomes encapsulated in a secretory vesicle, and finally is secreted to the exterior of the cell. Many other exported proteins functioning outside of cells, such as the blood plasma proteins, polypeptide hormones, antibodies, and mucoproteins, may be guided to their destinations in the same way.

Bacteria also exhibit a rather similar process. *E. coli* cells have an outer membrane consisting of lipids and proteins. The latter are synthesized by ribosomes bound to the inner surface of the inner membrane. Signaling sequences on the amino-terminal ends of these proteins guide them through the inner membrane and through the wall to the appropriate sites in the outer membrane where the polypeptide is incorporated. Bacterial mutants allow study of the genetics of the signaling sequences and of the peptidases that ultimately remove them.

Protein Synthesis Is Inhibited by Many Different Antibiotics

Protein synthesis is characteristically inhibited by many different antibiotics. This is of some interest from the biological point of view, since antibiotics may be regarded as "chemical warfare" agents, made by some species of microorganisms but extremely toxic for others. Some antibiotics have become valuable tools in the study of protein synthesis, since nearly every step in this process has been found to be specifically inhibited by one antibiotic or another.

One of the most important inhibitory antibiotics is *puromycin,* made by the mold *Streptomyces alboniger.* Puromycin has a structure very similar to that of the 3′ end of an aminoacyl-tRNA (Figure 29-21). It acts by interrupting peptide-chain elongation by virtue of its ability to replace an entering aminoacyl-tRNA, thus causing formation of a *peptidyl-puromycin.* No new amino acid residues can be added to a peptidyl-puromycin; as a consequence it is discharged from the ribosome, thus terminating synthesis of the polypeptide.

Tetracyclines, another class of antibiotics, inhibit protein synthesis by blocking the A site on the ribosome, rendering it incapable of binding aminoacyl-tRNAs. *Chloramphenicol* inhibits protein synthesis by prokaryotic (and mitochondrial)

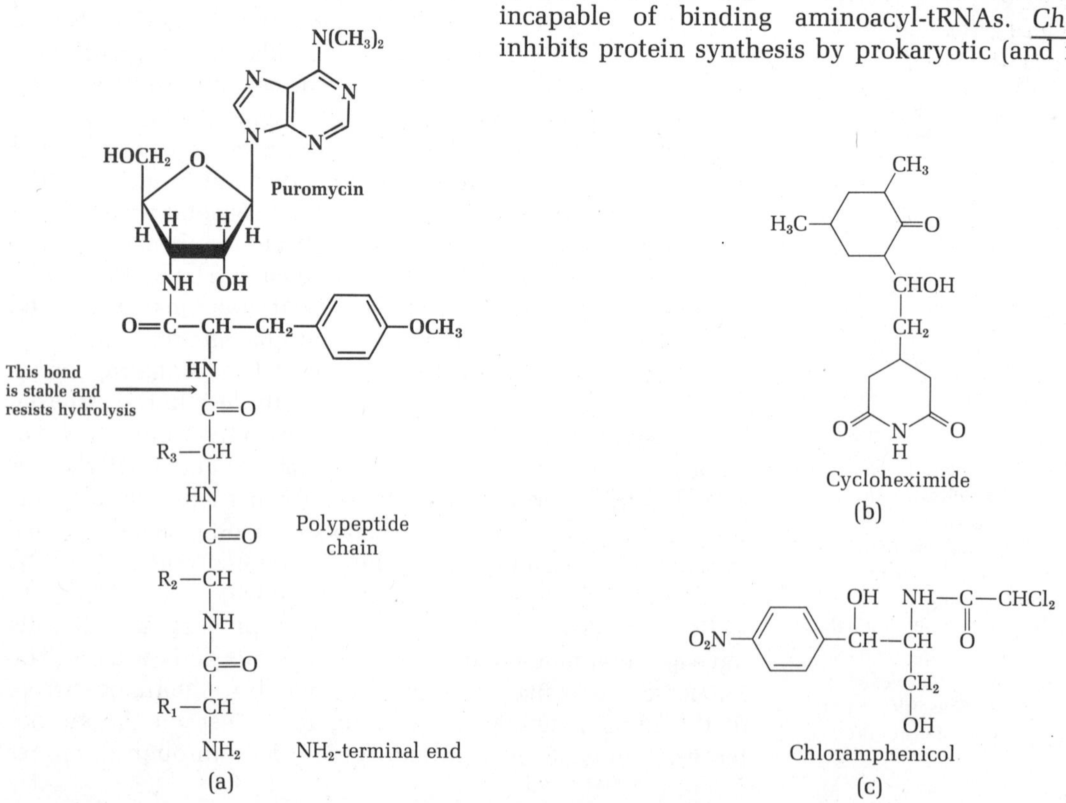

Figure 29-21
Antibiotic inhibitors of protein biosynthesis. (a) Puromycin (color) resembles the terminal aminoacyl adenylate end of an aminoacyl-tRNA and reacts with the carboxyl end of the growing peptidyl-tRNA to yield a peptidyl-puromycin complex. However, because the bond between the puromycin and peptide is stable to enzymatic cleavage, the peptide cannot be elongated further, and the peptidyl-puromycin is released from the ribosome. (b) Cycloheximide inhibits protein synthesis by 80S ribosomes of eukaryotic cells but does not inhibit protein synthesis in prokaryotes or in mitochondria. (c) Chloramphenicol inhibits protein synthesis by prokaryotic and mitochondrial ribosomes but not by 80S ribosomes of eukaryotes.

ribosomes but does not affect extramitochondrial protein synthesis in eukaryotes. Conversely, *cycloheximide* inhibits protein synthesis by 80S eukaryotic ribosomes but not by 70S prokaryotic or mitochondrial ribosomes. Note again the similarity between mitochondria and prokaryotes. *Streptomycin* causes misreading of the genetic code, and *tunicamycin* prevents attachment of oligosaccharide side chains to certain glycoproteins.

Other inhibitors of protein synthesis include *diphtheria toxin*, which inactivates an elongation factor, and *ricin*, an extremely toxic protein of the castor bean, which inactivates the 60S subunit of eukaryotic ribosomes.

The Genetic Code Has Been Solved

We shall now consider in more detail how the four-letter language of DNA is translated into the 20-letter language of proteins. It had long appeared likely that at least three nucleotide residues of DNA were required to code for each amino acid, since the four code letters of DNA (A, T, G, and C) in groups of two can yield only $4^2 = 16$ different combinations, not sufficient to code for 20 amino acids. But if the 4 bases are taken in groups of 3, they can yield $4^3 = 64$ different combinations. Early genetic experiments conclusively proved not only that the genetic code words for amino acids are triplets of nucleotides but also that there is no punctuation between codons for successive amino acid residues. But the ultimate question still loomed: What are the three-letter code words for the different amino acids? How could they be identified experimentally?

In 1961 Marshall Nirenberg and Heinrich Matthaei reported an observation that provided the breakthrough. They incubated the synthetic polyribonucleotide *polyuridylic acid* with an *E. coli* extract, GTP, and a mixture of the 20 amino acids in 20 different tubes. In each tube a different amino acid was radioactively labeled. Since the polyuridylic acid (designated poly U) can be regarded as an artificial messenger RNA containing many successive UUU triplets, it should promote the synthesis of a radioactive polypeptide from only one of the 20 different amino acids—that coded by the triplet UUU. A radioactive polypeptide was formed in only one of the 20 tubes, that containing radioactive phenylalanine. It was found that the radioactive polypeptide was polyphenylalanine; i.e., it contained only phenylalanine residues. Nirenberg and Matthaei therefore concluded that the triplet UUU codes for phenylalanine. Very soon, with the same approach, it was found that the synthetic messenger polycytidylic acid (poly C) codes for formation of a polypeptide containing only proline residues (polyproline) and polyadenylic acid (poly A) codes for polylysine. Thus the triplet CCC must code for proline and the triplet AAA for lysine.

The synthetic polynucleotides used in such experiments were made by the action of *polynucleotide phosphorylase* (page 865), which readily forms RNA-like polymers starting from ADP, UDP, CDP, and GDP. This enzyme requires no template and makes polymers whose base composition simply reflects the starting nucleoside 5′-diphosphates present in the medium and their relative concentrations. If polynuclectide phosphorylase is presented with uridine diphosphate, it makes only poly U. If it is presented with a mixture of two parts of ADP and one of GDP, it will make a polymer in which about two thirds of the residues are A and one third G. Such a random polymer is likely to have many triplets of AAA, lesser numbers of AAG, AGA, and GAA triplets, relatively few AGG, GGA, and GAG triplets, and very few GGG triplets. With the use of different artificial messenger RNAs made by polynucleotide phosphorylase from different starting mixtures of ADP, GDP, UDP, and CDP, the bases present in the triplets coding for all the amino acids were soon identified. However, these experiments could not reveal the sequence of the bases in each coding triplet, i.e., how the codons are "spelled."

In 1964 Nirenberg and Philip Leder made another breakthrough, which led to a solution. They found that isolated *E. coli* ribosomes will bind a specific aminoacyl-tRNA in the absence of GTP if at the same time the corresponding synthetic polynucleotide messenger is present. For example, ribosomes incubated with poly U and phenylalanyl-$tRNA^{Phe}$ will bind both, but if the ribosomes are incubated with poly U and some other aminoacyl-tRNA, the latter will not be bound because its anticodon will not recognize the UUU triplets in poly U. The shortest polynucleotide that could promote specific binding of phenylalanyl-$tRNA^{Phe}$ was the trinucleotide UUU. By use of simple trinucleotides of known sequence it was possible to determine the sequence of bases in the codons specifying the binding of different aminoacyl-tRNAs. With this and other approaches base sequences in all the triplet code words for each of the amino acids were soon established. These code words have been checked in many different ways. The complete codon "dictionary" for the amino acids is given in Figure 29-22. The cracking of the genetic code has been regarded as the greatest scientific discovery of the 1960s.

The Genetic Code Has Some Interesting Characteristics

First, we must note again that no punctuation or signal is required to indicate the end of one codon and the beginning of the next. The reading frame must therefore be correctly set at the beginning of the readout of an mRNA molecule and then moved sequentially from one triplet to the next. If the initial reading frame is off by one or two bases, or if the ribosome accidentally skips a nucleotide in the mRNA, all the subsequent codons will be out of register and will lead to formation of a missense protein with a garbled amino acid sequence.

Figure 29-22
The "dictionary" of amino acid code words as they occur in messenger RNAs. The codons are written in the 5′ → 3′ direction. The first and second bases are in black type; the third base of each codon, which is in colored type, is less specific than the first two. The three termination codons are shaded in gray, and the initiation codon AUG is shaded in color. Note that all the amino acids except methionine and tryptophan have more than one codon.

The amino acid code words, as they occur in DNA, will be complementary to the mRNA code words, but will be antiparallel and will contain T residues at positions complementary to A and A residues at positions complementary to U. For example, the mRNA and DNA codons for methionine would be

mRNA (5′) AUG (3′)
DNA (3′) TAC (5′)

Ordinarily, codons and anticodons are written in the 5′ → 3′ direction, from left to right.

The first letter of the codons (5′ end)	The second letter of the codons: U		C		A		G	
U	UU**U**	Phe	UC**U**	Ser	UA**U**	Tyr	UG**U**	Cys
U	UU**C**	Phe	UC**C**	Ser	UA**C**	Tyr	UG**C**	Cys
U	UU**A**	Leu	UC**A**	Ser	UA**A**	End	UG**A**	End
U	UU**G**	Leu	UC**G**	Ser	UA**G**	End	UG**G**	Trp
C	CU**U**	Leu	CC**U**	Pro	CA**U**	His	CG**U**	Arg
C	CU**C**	Leu	CC**C**	Pro	CA**C**	His	CG**C**	Arg
C	CU**A**	Leu	CC**A**	Pro	CA**A**	Gln	CG**A**	Arg
C	CU**G**	Leu	CC**G**	Pro	CA**G**	Gln	CG**G**	Arg
A	AU**U**	Ile	AC**U**	Thr	AA**U**	Asn	AG**U**	Ser
A	AU**C**	Ile	AC**C**	Thr	AA**C**	Asn	AG**C**	Ser
A	AU**A**	Ile	AC**A**	Thr	AA**A**	Lys	AG**A**	Arg
A	AU**G**	Met	AC**G**	Thr	AA**G**	Lys	AG**G**	Arg
G	GU**U**	Val	GC**U**	Ala	GA**U**	Asp	GG**U**	Gly
G	GU**C**	Val	GC**C**	Ala	GA**C**	Asp	GG**C**	Gly
G	GU**A**	Val	GC**A**	Ala	GA**A**	Glu	GG**A**	Gly
G	GU**G**	Val	GC**G**	Ala	GA**G**	Glu	GG**G**	Gly

Second, we note that 3 of the 64 possible nucleotide triplets (UAG, UAA, and UGA) do not code for any known amino acids (Figure 29-22); they are the nonsense codons that normally signal termination of polypeptide chains. The codon AUG is not only the initiation codon in both prokaryotes and eukaryotes but also codes for methionine residues in internal positions of polypeptides.

The third major point is that the amino acid code words (Figure 29-22) are identical in all the species that have been tested, including human beings, *E. coli*, the tobacco plant, and amphibia, among many other species, as well as in viral genomes. Thus it would appear that all species of plants and animals had a common evolutionary precursor with a single genetic code that has been completely preserved throughout the course of biological evolution. It has therefore been widely accepted that the genetic code is universal. But a surprise has arisen. It has recently been discovered that in the synthesis of proteins by mitochondria, using mitochondrial ribosomes, tRNAs, and mRNAs, some of the amino acid codons employed are not identical with those in the standard code-word dictionary (Figure 29-22). For example, human mitochondria use AUA as a codon for methionine rather than isoleucine. And they also use UGA, normally a termination codon, to code for tryptophan. The meaning of these observations is a tantalizing puzzle. It has set off an intensive search for other significant differences in the genetic biochemistry of mitochondria, which may yield important clues to the origin of mitochondria and of eukaryotic cells.

Table 29-2 Degeneracy of the Amino Acid Code

Amino acid	Number of codons
Ala	4
Arg	6
Asn	2
Asp	2
Cys	2
Gln	2
Glu	2
Gly	4
His	2
Ile	3
Leu	6
Lys	2
Met	1
Phe	2
Pro	4
Ser	6
Thr	4
Trp	1
Tyr	2
Val	4

Perhaps the most striking feature of the genetic code is that it is *degenerate,* a mathematical term indicating that a given amino acid may have more than one specific codon (Table 29-2). Only methionine and tryptophan have single codons. Degenerate should not be taken to mean imperfect, for there is no codon that specifies more than one amino acid. The degeneracy of the codons is not uniform. Thus, leucine and serine have six codons, glycine and alanine have four, and glutamic acid, tyrosine, and histidine have two.

In most cases where an amino acid has multiple codons, the difference between the codons is in the third base, i.e., that at the 3′ end (see Figure 29-22). For example, alanine is coded by the triplets GCU, GCC, GCA, and GCG. Thus, the first two bases GC are common to all the codons for alanine. In fact, the codons for nearly all of the amino acids consist of triplets which can be symbolized by XY^{A}_{G} or XY^{U}_{C}. Evidently the first two letters of each codon are the primary determinants of its specificity; the third position, i.e., the nucleotide at the 3′ end of the codon, is less specific. Let us examine this matter more closely.

Wobble Allows Some tRNAs to Recognize More Than One Codon

One might expect that the anticodon triplet of a given tRNA would recognize only one codon triplet through Watson-Crick base pairing, so that there would be a different tRNA for each codon of an amino acid. However, the number of different tRNAs for each amino acid is not the same as the number of its codons. Moreover, some of the tRNAs contain the nucleoside *inosine* (designated I), which contains the base *hypoxanthine,* the product arising when the 6-amino group of adenine is removed by hydrolysis. Molecular models show that I can form hydrogen bonds with three different bases, U, C, and A, but these pairings are rather weak compared with the strong hydrogen bonds between the Watson-Crick base pairs G–C and A–U. For example, one of the arginine tRNAs has the anticodon (5′)I–C–G(3′), which can recognize three different arginine codons, (5′)C–G–A, (5′)C–G–U, and (5′)C–G–C. The first two bases of these codons are identical (C–G) and form strong Watson-Crick base pairs (color) with the corresponding bases of the anticodon:

Anticodon	(3′) G—C—I (5′)	(3′)G—C—I	(3′)G—C—I
Codon	(5′) C—G—A(3′)	(5′)C—G—U	(5′)C—G—C

However, the third bases of these arginine codons (A, U, and C) form rather weak hydrogen bonds (black) with the I residue of the anticodon. Examination of these and other codon-anticodon pairings led Francis Crick to conclude that the third base of most codons pairs rather loosely with the corresponding base of their anticodons; to use his picturesque word, the third bases of such codons "wobble." Crick proposed a set of four relationships called the *wobble hypothesis:*

1. The first two bases of a codon always form strong Watson-Crick base pairs with the corresponding bases of the anticodon and confer most of the coding specificity.
2. The first base of some anticodons (reading in the $5' \rightarrow 3'$ direction) allows them to read more than one codon for a given amino acid. When the first base of the anticodon is C or A, it can read only one codon; when it is U or G, it can read two different codons. When the wobble base of an anticodon is I, or certain other modified bases, it can read three different codons. Thus I in the first position of an anticodon permits it to recognize the maximum number of codons for any given amino acid. These relationships are summarized in Table 29-3.
3. For a given amino acid, codons that differ in either of the first two bases require different tRNAs.
4. A minimum of 32 tRNAs is required to translate all 61 different codons for the amino acids (recent research indicates that in mitochondria a smaller number of tRNAs suffices to translate all the amino acid codons).

Table 29-3 The Base in the Wobble (5′) Position of the Anticodon of a tRNA Determines How Many Codons of a Given Amino Acid It Can Recognize

In the following, X and Y denote complementary bases capable of strong Watson-Crick base pairing with each other. The bases in the wobble or 3′ position of the codons and 5′ position of the anticodons are in color.

1. When C or A is in the 5′ position of an anticodon the tRNA can recognize only one codon, which must contain G or U, respectively, in its 3′ position. C and A form Watson-Crick base pairs with G and U, respectively.

Anticodon	(3′) X-Y-**C** (5′)	(3′) X-Y-**A** (5′)
Codon	(5′) Y-X-**G** (3′)	(5′) Y-X-**U** (3′)

2. When either U or G is in the 5′ position of an anticodon the tRNA can recognize 2 different codons. One of them forms a weak or wobbly base pair at its 3′ position, the other forms a strong Watson-Crick pair.

Anticodon	(3′) X-Y-**U** (5′)	(3′) X-Y-**G** (5′)
Codon	(5′) Y-X-**A** (strong) **G** (wobbles)	(5′) Y-X-**C** (strong) **U** (wobbles)

3. When I or some other modified base is in the 5′ position of an anticodon the tRNA can recognize 3 different codons, all of which form a wobbly base pair at the 3′ position.

Anticodon	(3′) X-Y- **I** (5′)
Codon	**A** (5′) Y-X- **U** (all pairings wobble) **C**

What can the reason be for this unexpected complexity of codon-anticodon interactions? In brief, it is believed that the first two bases of codons confer most of the codon-anticodon specificity. The wobble or third base also contributes to specificity but because it pairs only loosely with its corresponding base it permits rapid dissociation of tRNAs from their codons in the mRNA during protein synthesis. If all three bases of codons engaged in strong Watson-Crick pairing with the corresponding bases of their anticodons, the anticodon-codon association would be so tight that tRNAs would dissociate only slowly from their codons, severely limiting the rate of protein synthesis. Biochemical evolution has therefore optimized most of the codon-anticodon interactions for *both* accuracy and speed.

Viral DNAs Sometimes Contain Genes within Genes or Overlapping Genes

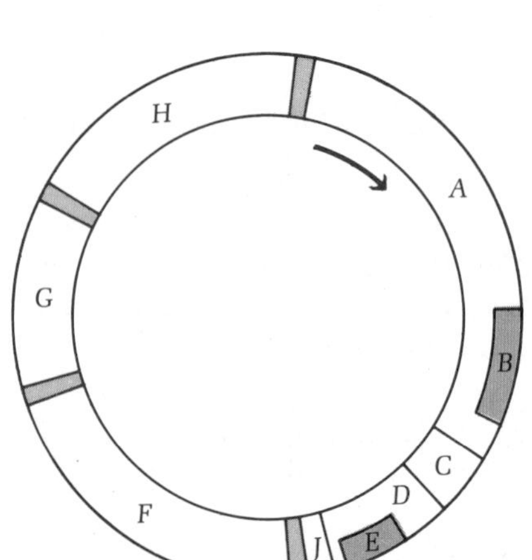

Figure 29-23
Genes within genes. The DNA of ϕX174 *contains nine genes (A to J). Because the proteins coded by the DNA contain more amino acid residues than correspond to the 5386 nucleotides present in the DNA, it was deduced that some of the DNA must code for more than one gene. Comparison of the nucleotide sequence of the DNA with the amino acid sequences of the proteins coded by it revealed that gene B (color) lies within the sequence of gene A but uses a different reading frame. Similarly, gene E (color) lies within gene D and also uses a different reading frame (see Figure 29-24). The gray segments of the DNA are untranslated spacer regions.*

In addition to genes B and E nestled in genes A and D, the DNA of ϕX174 *contains short overlaps between other genes. In five cases the initiation signal for one gene overlaps the termination signal of the preceding gene; in these overlaps the signals utilize different reading frames.*

It has long been a fundamental principle of molecular biology that the nucleotide sequence of a gene is precisely colinear with the sequence of its messenger RNA and with that of the polypeptide for which it codes. But we have seen that many eukaryotic genes have intervening nontranslated nucleotide sequences, or introns, which interrupt the precise colinear relationship between the gene and its polypeptide (pages 827, 860).

Another surprising discovery raises further questions about the principle of colinearity. In a number of viruses one and the same DNA base sequence codes for two different proteins, which employ two different reading frames for their codons. The discovery of such "genes within genes" arose from the observation that the DNA of the bacteriophage ϕX174, which contains 5386 nucleotide residues, is not long enough to code for the nine different proteins that are known to be the products of the ϕX174 DNA genome. Once the entire nucleotide sequence of the ϕX174 chromosome was worked out by Sanger and his colleagues (page 790), they scanned it carefully and compared it with the amino acid sequences of the proteins coded by the ϕX174 genes. They found several overlapping gene sequences in the ϕX174 DNA. Figure 29-23 shows that genes B and E are nestled within A and D, respectively, and that there are five cases in which the start codon of one gene overlaps the termination signal of the other gene. Figure 29-24 shows how genes D and E share a segment of DNA but use different reading frames; this is also the case for genes A and B. The sum of all the nestled and overlapping sequences accounts completely for the deficit in the number of nucleotides in ϕX174 DNA, compared with the number of amino acid residues in the nine proteins for which it codes (see also page 790).

This discovery was quickly followed by similar observations on other viral DNAs, including those of phage λ, the cancer-

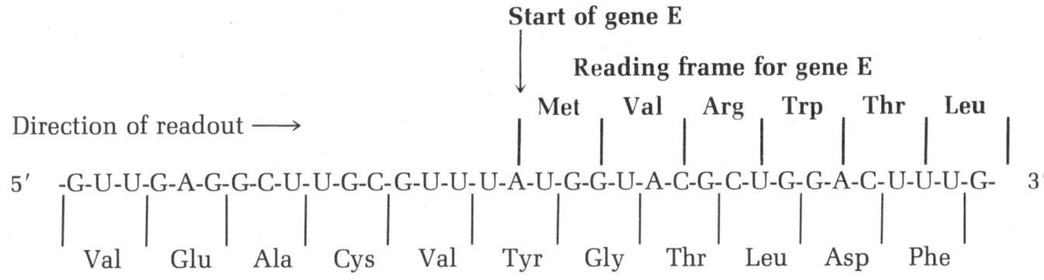

Figure 29-24
Portion of the nucleotide sequence of the mRNA *transcript of gene* D *of* ϕX174 DNA, *showing how gene* E, *which is nestled within gene* D, *is coded by a different reading frame* (color) *than that used by gene* D (black).

causing simian virus 40 (SV40), and phage G4, a close relative of ϕX174. G4 is remarkable in that at least one codon is shared by *three* different genes. It has been suggested that overlapping genes or genes within genes may be found only in viruses. The fixed size of the viral capsid requires economical use of a limited amount of DNA to code for the proteins needed to ensure the ability of the viral particle to infect its host cell and be replicated by it.

Protein Synthesis Is Regulated

Living cells have accurately programmed mechanisms for regulating the synthesis of their different proteins, so that each cell has the proper number of copies of each protein to carry out its metabolic activities smoothly and with maximum efficiency. We have seen that *E. coli* contains the genes for over 3000 different proteins. However, the 3000 *E. coli* proteins do not occur in the same number of copies per cell. The actual number of copies per protein differs over a wide range; moreover, some are fixed in number, and others may vary. A single *E. coli* cell contains about 15,000 ribosomes; thus, each of the 50 or more ribosomal proteins is present in about 15,000 copies. The enzymes of the glycolytic pathway also appear to be fixed in number and occur in a great many copies per cell. On the other hand, there are only about five copies of the enzyme β-galactosidase in an *E. coli* cell, but, as we shall see, the number may increase dramatically in response to changes in the availability of certain nutrients in the environment. Regulation of enzyme synthesis provides each type of cell with the proper ensemble of enzymes required in its basic housekeeping activities. It also allows bacteria to economize in the use of amino acids for the synthesis of enzymes that are required in only small amounts or used only occasionally.

In higher organisms, regulation of protein biosynthesis is much more complex. Although every cell of a vertebrate contains the entire genome of the organism, only a fraction of its structural genes are expressed in any given cell type. Nearly all the cells of higher animals contain the basic ensembles of enzymes required for the central metabolic pathways. However, different cell types, such as muscle, brain, and liver, have characteristically different structures and biological functions, each dependent upon distinctive sets of specialized proteins. Thus, skeletal muscle cells contain a great abundance of oriented myosin and actin filaments (page 380), but these proteins are present in only small amounts in the liver. Similarly, the brain contains the enzymes required to make many different neurotransmitter substances, but liver cells do not. On the other hand, the liver of mammals contains all the enzymes required to make urea, but no other tissue has this capacity (page 549). Moreover, the biosynthesis of different sets of specialized proteins during the orderly differentiation and growth of higher organisms must also be accurately programmed with respect to the time and sequence of their appearance. We know relatively little as yet of the regulation of gene expression in the multiple chromosomes of eukaryotic organisms. However, today we have considerable information on the regulation of protein synthesis in prokaryotes, aspects of which we shall now examine.

Bacteria Contain Constitutive and Induced Enzymes

Constitutive enzymes are those present in bacterial cells in constant amounts, regardless of the metabolic state of the organism. Examples of constitutive enzymes are those participating in the central pathways of catabolism, such as the glycolytic sequence. *Induced* enzymes, on the other hand, vary in their concentration in cells. An induced enzyme is normally present in only trace amounts in a bacterial cell, but its concentration can quickly increase a thousandfold or more when its substrate is present in the medium, particularly when the substrate is the only carbon source of the cell. Under these conditions the induced enzymes may be required to transport the substrate into the cell and convert it into a metabolite that can be used by the cell. Such an inducible enzyme is *β-galactosidase,* which catalyzes the hydrolysis of lactose to yield D-glucose and D-galactose, the first step in the utilization of lactose (page 419). Normally, *E coli* does not utilize lactose if plenty of glucose is available, since each cell contains only about five copies of β-galactosidase. However, if *E. coli* cells are placed in a culture medium containing lactose as the only source of energy and carbon, within 1 or 2 min they begin to synthesize β-galactosidase in large amounts, to yield over 1000 copies per cell. The induced β-galactosidase can hydrolyze lactose to yield D-glucose and D-galactose, which can then be used as fuel and carbon sources. If the lactose-induced *E. coli* cells are now transferred to a fresh medium containing glucose but no lactose, further synthesis of β-galactosidase is immediately halted. Thus we see that enzyme induction is an economical process; inducible enzymes are made only when they are needed. An agent capable of inducing the synthesis of an enzyme or group of enzymes is called an *inducer* or *inducing agent.*

When a β-galactoside such as lactose is added to *E. coli* cells in the absence of glucose, not only β-galactosidase but also two related proteins, *β-galactoside permease* and *A protein*, are made in large numbers. The permease is a membrane protein that promotes the transport of β-galactosides from the medium into the cell. The function of the A protein is not entirely clear, but it may be related to metabolic utilization of galactosides. When a group of related enzymes or proteins is induced by a single inducing agent, as in this case, the process is known as *coordinate induction*. Today we know that *E. coli* and other bacteria have the capacity to make many different enzymes or sets of related enzymes, in response to different specific inducing agents. This capacity enables bacteria to adapt themselves quickly and economically to use many different possible nutrients that may become available in their environment.

Prokaryotes Also Can Repress Protein Synthesis

Another important type of change in the concentration of an enzyme in bacteria, seemingly opposite to that observed in enzyme induction, is *enzyme repression*. When *E. coli* cells are grown on a medium containing an ammonium salt as the sole nitrogen source, they must make all their nitrogenous components from NH_4^+ and a carbon source. Such cells obviously contain all the enzyme systems needed for synthesis of the 20 different amino acids. However, if only one of the amino acids, say histidine, is added to the culture medium, the entire set of enzymes required for making histidine from ammonia and a carbon source will no longer be made by the cells, but the rest of the enzyme systems required to make the other 19 amino acids will remain active. Since the cell population will continue to grow and divide, the specific activity of the remaining histidine-synthesizing enzymes will decline. Turning off the synthesis of the histidine-synthesizing enzymes evoked by the addition of histidine is called *enzyme repression*. Repression, like induction, is a reflection of the principle of cell economy: when the enzymes required for histidine biosynthesis are no longer required, they are no longer made. Most cases of enzyme repression involve enzymes participating in biosynthesis, particularly of the amino acids. Repression of an entire set of enzymes catalyzing a sequence of biosynthetic reactions by the presence of its end product, as in the case of the histidine enzymes, is called *coordinate* or *end-product repression*.

The Operon Hypothesis

The molecular and genetic relationships between enzyme induction and enzyme repression have been clarified by the genetic research of François Jacob and Jacques Monod at the Pasteur Institute in Paris. Their classical work on the induction of β-galactosidase activity in *E. coli* cells led them to propose the *operon hypothesis* for the genetic control of protein synthesis in prokaryotes, which has since been completely verified by direct biochemical experiments. The type of regulation of protein synthesis provided by the operon concept is spoken of as *transcriptional control,* since the control is exerted primarily on the rate of transcription of genes into their corresponding mRNAs. There is another general way in which protein synthesis is regulated, namely, by *translational control,* i.e., control of the rate of synthesis of the polypeptide chain from its mRNA template. Transcriptional control appears to be the primary mechanism for the regulation of gene expression in bacteria. Translational control, which is not well understood, appears to be a secondary mechanism in bacteria but very important in eukaryotes. In addition, there are other mechanisms that make possible the fine-tuning of the rate of protein synthesis.

From their experiments Jacob and Monod proposed that the three structural genes *z, y,* and *a,* coding for the synthesis of β-galactosidase, β-galactoside permease, and A protein, respectively, all of which can be induced by lactose, are located adjacent to each other in the *E. coli* chromosome (Figure 29-25). They further proposed that the DNA contains another site near these genes, an inhibitory site *i*, capable of inhibiting the transcription of the three structural genes, *z, y,* and *a.* The *i* region was proposed to be a *regulatory gene* (page 821) coding for the amino acid sequence of a regulatory protein called a *repressor.* When the *i* gene is transcribed to form its corresponding mRNA, the latter diffuses to the ribosomes and there acts as the template for the synthesis of the repressor. The repressor protein has a binding site for another specific segment on the DNA called the *operator* (Figure 29-25). Binding of the repressor protein to the operator site in DNA was postulated to prevent or repress the transcription by RNA polymerase of the three structural genes *z, y,* and *a* coding for the three enzymes induced by β-galactosides, thus repressing their synthesis for lack of the mRNA template.

To account for the action of the inducer, when glucose is not available but lactose is, Jacob and Monod proposed that it combines with a second specific binding site on the repressor protein, the *inducer site,* to form an *inducer-repressor complex.* The binding of the inducer causes the repressor to be released from the operator site on the DNA by decreasing its affinity for the latter. Once the repressor-inducer complex is released, the structural genes for β-galactosidase and the other two proteins become available for transcription by RNA polymerase to yield the corresponding mRNA. The three proteins are then synthesized from the mRNA template on the ribosomes, enabling the cells to use lactose as a carbon and energy source.

Now, let us suppose that the cells are taken out of the lactose medium, washed, and placed in a medium containing, instead of lactose, D-glucose, which the cells are always able to use. Because the concentration of lactose in the cell will now become

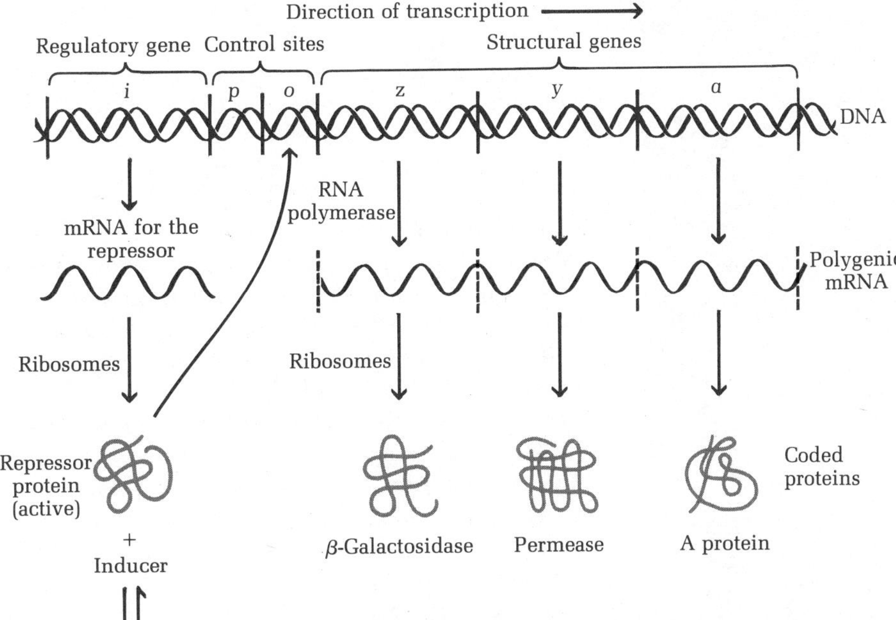

Figure 29-25
Schematic diagram of the lac *operon. The three* lac *structural genes* z, y, *and* a *are adjacent to each other. They are preceded by two control sites,* p *(for promoter) and* o *(for operator). The drawing is not in scale; the* p *and* o *regions are very short compared with the genes. A regulatory gene* i *codes for the repressor protein. The repressor protein has two binding sites, one for the operator and the other for the inducer. The active form of the repressor protein can bind to the operator, thus preventing the binding of* RNA *polymerase and subsequent transcription of the structural genes* z, y, *and* a. *Under these conditions β-galactosidase and the other two proteins are not made by the cells. However, when lactose is available and glucose is simultaneously lacking, the inducer binds to the repressor, converting it into an inactive form that will not bind to the operator. In this case* RNA *polymerase can bind to the* p *site, move through the* o *site, and begin to transcribe the three structural genes into polygenic* mRNA, *which codes for the synthesis of the three* lac *proteins on the ribosomes. The function of the* p *site is given in more detail in Figure 29-27.*

Lactose itself is not an inducer of the lac *operon, but it is converted into its isomer allolactose, the actual inducer.*

vanishingly low, the inducer bound to the repressor protein will dissociate away, causing the repressor molecule to revert to its active form, so that it now binds with high affinity to the operator site. As a consequence, the structural genes for β-galactosidase and the two other proteins can no longer be transcribed, and for lack of their mRNA these proteins can no longer be made. Thus a repressor protein, through its capacity to bind either the inducer or the operator reversibly but not both simultaneously, can account for both induction and repression of galactosidase synthesis.

The three structural genes *z*, *y*, and *a*, together with their operator *o*, were designated by Jacob and Monod as an operon, specifically, the *lac* operon (Figure 29-25). An operon thus consists of a group of functionally related structural genes, which can be turned on or off coordinately, plus their operator. Many operons have been identified in *E. coli*, *Salmonella typhimurium*, and other bacteria (Table 29-4). One of the most complex operons is that for histidine biosynthesis: it has nine structural genes coding for the set of enzymes required in the

Table 29-4 Some Operons in Bacteria

Operon	Number of enzyme proteins	Function
lac	3	Hydrolysis and transport of β-galactosides
his	9	Synthesis of histidine
leu	4	Conversion of α-ketoisovalerate to leucine
ara	4	Transport and utilization of arabinose

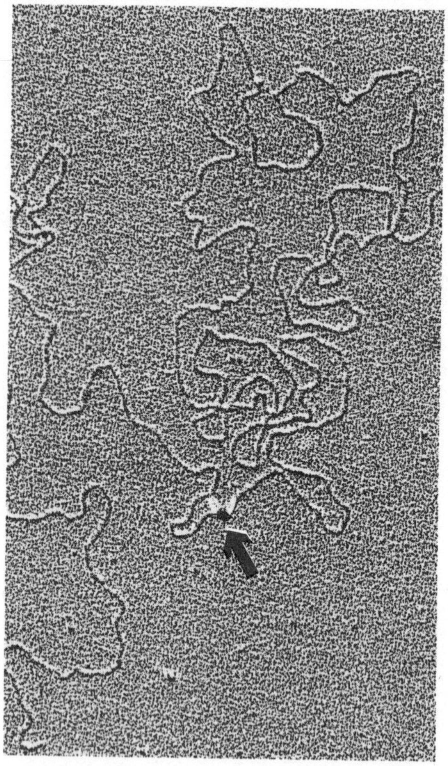

0.2 μm

Figure 29-26
Electron micrograph of a segment of E. coli *DNA, showing a* lac *repressor molecule (arrow) bound to the* lac *operator.*

biosynthesis of histidine. The regulatory gene of the *his* operon codes for a repressor protein that binds to the *his* operator, thus preventing transcription of the entire set of nine proteins when ample histidine is present in the medium.

Repressor Molecules Have Been Isolated

In 1967 Walter Gilbert and Benno Müller-Hill succeeded in isolating the postulated *lac* repressor. They showed that it is indeed a protein and that it has separate binding sites for the inducer and for a specific site on *E. coli* DNA. When its inducer site is unoccupied, the *lac* repressor protein binds very tightly to a specific site on the DNA; but when the inducer site is occupied, the repressor is released from the operator. The isolation of the *lac* repressor was very difficult since a normal *E. coli* cell contains only about 10 copies.

The *lac* repressor protein has a molecular weight of about 150,000. It has an extraordinarily high affinity for its specific locus in *E. coli* DNA in the absence of inducer; half-maximal binding of the repressor to the operator site occurred at a repressor concentration of only 10^{-13} *M*. Figure 29-26 shows an electron micrograph of the *lac* repressor protein bound to the operator region of *E. coli* DNA.

Operons Also Have a Promoter Site

We have seen that when glucose is absent and lactose is present in the medium, the inducer binds to the repressor, releasing the latter from the operator, thus allowing transcription of the *lac* genes and subsequent synthesis of the *lac* proteins. Now suppose that *both* glucose and lactose are present in the medium. Under these conditions *E. coli* uses only glucose and ignores the lactose; moreover, the cells no longer make the *lac* proteins. The repression of the *lac* proteins by glucose is called *catabolite repression*. *E. coli* cells can sense whether glucose is available by another regulatory mechanism that cooperates with the *lac* repressor and operator for controlling the synthesis of the *lac* enzymes.

In addition to the *i* gene and the operator or *o* site, there is another distinctive control site, the *promoter* or *p* site, which lies between the *i* and *o* sites (Figure 29-27). The promoter in turn contains two specific regions, each with a distinctive function. Next to the operator is the *RNA polymerase entry site*, to which RNA polymerase first becomes bound. The other portion of the promoter is the specific binding site for another regulatory protein, the *catabolite activator protein* (CAP). The CAP site controls the polymerase site. When a complex of CAP with cyclic AMP forms, which occurs when glucose is lacking, it is bound to the CAP site and allows the entry site to bind RNA polymerase. If lactose is present, the *o* site will be open, since the operator cannot bind the repressor-inducer complex. Under these conditions the RNA polymerase moves from its entry site

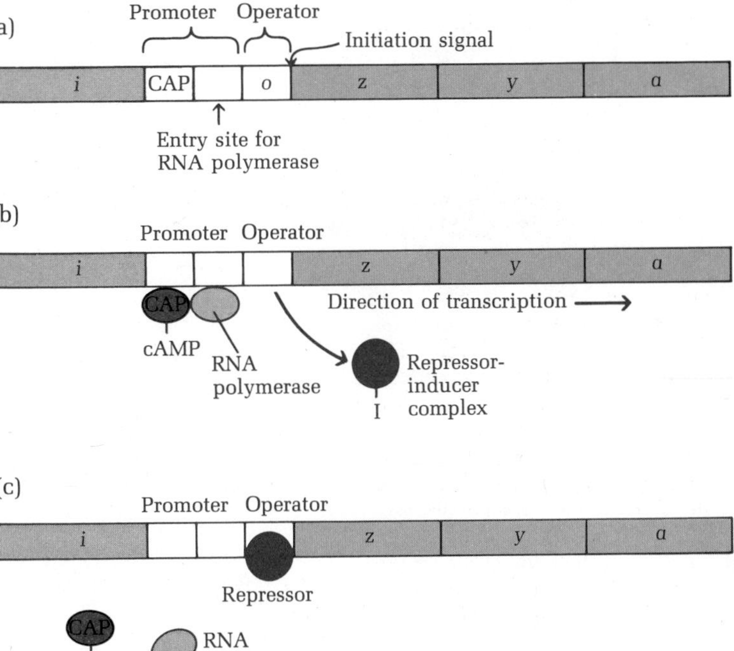

Figure 29-27
(a) The control regions for the lac *operon. The CAP site in the promoter can bind the CAP–cyclic AMP complex, but not CAP alone. The RNA polymerase can bind to its entry site only when the CAP site is occupied. The repressor can bind to the operator only in the absence of inducer.*

(b) The three structural genes z, y, *and* a *of the* lac *operon are transcribed when lactose is available to the cells but glucose is not. In this case the operator site is not occupied by the repressor, and the CAP–cyclic AMP complex binds to the promoter, allowing RNA polymerase to bind to its entry site, move "downstream" to the initiation codon, and begin to transcribe the three structural genes.*

(c) When ample glucose is available in the medium, cyclic AMP is not formed and the CAP protein is unable to bind to the promoter. Under these conditions RNA polymerase cannot enter and the lac *genes are not transcribed.*

through the operator and begins to transcribe the three *lac* genes. On the other hand, when ample glucose is available the concentration of cyclic AMP is very low and the CAP–cyclic AMP complex cannot form. Since the CAP site can make the entry site available to RNA polymerase only if it binds the CAP–cyclic AMP complex, RNA polymerase cannot enter and the *lac* genes cannot be transcribed. Thus transcription of the three *lac* genes is possible only when glucose is unavailable; the *lac* operon is therefore subject to both positive and negative control by the *p* and *o* control sites, respectively.

How does CAP sense whether glucose is available? The CAP molecule has two binding sites, one for the CAP site in the promoter DNA, as indicated above, and the other for cyclic AMP. Recall that cyclic AMP is the second or intracellular messenger in the action of a number of hormones in vertebrate cells (Chapter 25). In *E. coli* cyclic AMP also serves as a messenger, but for a different purpose: it signals whether or not glucose is available as a cell fuel. *E. coli* cells contain the enzyme adenylate cyclase (page 730), which can make cyclic AMP from ATP; they also contain a phosphodiesterase, which can hydrolyze cyclic AMP and thus inactivate it. When glucose concentration is high and adequate for the cell's needs, the concentration of cyclic AMP in the cell is very low (Figure 29-27). However, when the glucose concentration becomes low, the cyclic AMP level increases, due to increased adenylate cyclase activity and decreased phosphodiesterase activity (page 734). The cyclic AMP so formed binds to the CAP protein, and the CAP–cyclic AMP complex in turn binds to the CAP site of the promoter. Only when the CAP site is occupied in this manner can the

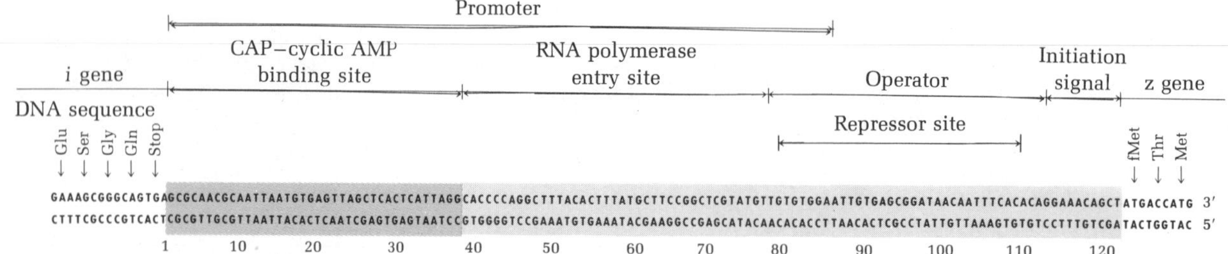

Figure 29-28
The structure of the promoter-operator region of the lac operon in E. coli. The base sequence of both DNA strands is shown, from the last 15 bases of the i or regulatory gene through the first 9 bases of the z gene. The promoter overlaps the operator, as shown. The binding site for the CAP–cyclic AMP complex is about 38 bases long, and the RNA polymerase entry site is about 40 bases long. The binding site for the lac repressor in the operator is about 28 base pairs long and shows twofold symmetry.

RNA polymerase bind to the initiation site and start transcribing the *lac* genes into mRNA (Figure 29-27). Cyclic AMP in bacteria thus has been called a "hunger signal."

Most of the base sequence of the *lac* operon of *E. coli* has been determined. The complete base sequence of the operator site and the promoter site is known. The promoter has altogether 85 base pairs. The CAP site contains about 38 and the RNA polymerase entry site about 40 (Figure 29-28).

In addition to modulation of operon activity by operators and promoters, bacteria possess a variety of other mechanisms that regulate protein synthesis. Some permit gradual attenuation or slowing down of the rate of synthesis of enzymes, such as those involved in making histidine. Others are responsive to the concentration of ammonia or other nitrogen sources and thus allow bacteria to adjust their protein economy to lean times. Bacteria thus have exquisitely sensitive mechanisms for regulation of enzyme synthesis, enabling them to optimize their metabolism according to the principle of maximum economy.

Summary

Amino acids are first activated for protein synthesis by specific aminoacyl-tRNA synthetases in the cytosol. They catalyze the formation of the aminoacyl esters of homologous transfer RNAs, with simultaneous cleavage of ATP to AMP and pyrophosphate. Transfer RNAs have from 73 to 93 nucleotide units, several of which have modified bases. They have an amino acid arm with the terminal sequence (3′)A–C–C–, to which an amino acid is esterified, an anticodon arm, a TψC arm, and a DHU arm; some tRNAs have a fifth or extra arm. The anticodon nucleotide triplet of tRNA is responsible for the specificity of interaction of the aminoacyl-tRNA with the complementary codon triplet on the mRNA. The growth of polypeptide chains on ribosomes begins with the NH_2-terminal amino acid and proceeds by successive additions of new residues to the COOH-terminal end. Prokaryotes contain 70S ribosomes, with a large 50S subunit and a small 30S subunit. Ribosomes of eukaryotes are significantly larger and contain more proteins than do prokaryotic ribosomes.

In bacteria the initiating NH_2-terminal residue in all proteins is *N*-formylmethionyl-tRNA. It forms a complex with initiation factor IF-2, the 30S ribosomal subunit, mRNA, and GTP; this complex combines with the 50S subunit to form the initiation complex, with simultaneous cleavage of GTP to GDP and dissociation of IF-2. In the subsequent elongation steps, GTP and three elongation factors are required for binding the incoming aminoacyl-tRNA to the aminoacyl binding site on the ribosome. In the peptidyl transferase reaction the fMet residue is transferred to the amino group of the incoming aminoacyl-tRNA. The elongated peptidyl-tRNA is then translocated from the aminoacyl site to the peptidyl site, a process requiring hydrolysis of GTP. After many such elongation cycles termination of the polypeptide chain takes place with the aid of releasing factors. Polyribosomes consist of mRNA molecules to which are attached several or many ribosomes, each independently reading the mRNA and forming a protein. At least four high-energy phosphate bonds are required to generate each peptide bond, presumably to guarantee fidelity of translation.

The codons for the amino acids consist of specific nucleotide triplets. The base sequences of the codons were deduced from experiments with the aid of synthetic messenger RNAs of known composition and sequence. The amino acid code has multiple code words for nearly all the amino acids. The third position in each codon is much less specific than the first and second and is said to wobble. The standard genetic code words are probably universal in all species, although some mitochondrial codons differ. The initiating amino acid *N*-formylmethionine is coded by AUG but requires an A- and G-rich initiation signal at the 5′ side of AUG. The nonsense triplets UAA, UAG, and UGA code for no amino acids but are signals for chain termination. In some viral DNAs two different proteins may be coded by the same nucleotide sequence but transcribed with different reading frames.

Protein synthesis in prokaryotes is regulated primarily at the level of the transcription of DNA to yield mRNA. Transcription of a set of metabolically related genes is regulated by binding of a specific protein, the repressor, to an operator site in the DNA. The operator and the set of related genes together consitute an operon. The transcription of such a set of genes can be induced by a specific nutrient, such as lactose, which can bind to the repressor to cause its release from the operator, thus allowing transcription of the genes coding for the proteins needed by the cell to use lactose as a carbon and energy source. Some operons also have a promoter site; it contains a regulatory site, the CAP site, for binding a complex of the catabolite activator protein with cyclic AMP. This complex, formed when no glucose is available to the cell, allows entry of RNA polymerase to the initiation site for transcription of the genes for lactose catabolism.

References

Activation of Amino Acids

RICH, A., and S. H. KIM: "The Three-Dimensional Structure of Transfer RNA," *Sci. Am.*, **238**:52–62, January (1978).

SCHIMMEL, P. R.: "Understanding the Recognition of Transfer RNAs by Aminoacyl Transfer RNA Synthetases," *Adv. Enzymol.*, **49**: 187–222 (1979).

Ribosomes

LAKE, JAMES: "The Ribosome," *Sci. Am.*, **245**:84–97, August (1981).

WITTMAN, H. G.: "Structure and Function of *E. coli* Ribosomes," *Fed. Proc.*, **36**:2025–2080 (1977).

Steps in Initiation, Elongation, and Termination

WEISSBACH, H., and S. PESTKA (eds.): *Molecular Mechanisms of Protein Biosynthesis*, Academic, New York, 1977. Review articles on different aspects of protein synthesis.

The Genetic Code

BARRELL, B. G., G. M. AIR, and C. A. HUTCHINSON, III: "Overlapping Genes in Bacteriophage ϕX174," *Nature*, **264**:34–40 (1976).

CRICK, F. H. C.: "The Genetic Code III," *Sci. Am.*, **215**:55–62, October (1966).

FIDDES, J. C.: "The Nucleotide Sequence of a Viral DNA," *Sci. Am.*, **237**:54–67, December (1977).

HALL, B. D.: "Mitochondria Spring Surprises," *Nature*, **282**:129–130 (1979). Exceptions to the universality of the genetic code.

NIRENBERG, M.: "The Genetic Code II," *Sci. Am.*, **208**:80–94, March (1963). Description of the original experiments.

Regulation of Gene Expression

BROWN, D. D.: "Gene Expression in Eukaryotes," *Science*, **211**:667–674 (1981).

LEWIN, B.: *Gene Regulation II*, 2d ed., Wiley, New York, 1980. A comprehensive volume including much information on eukaryotic regulation.

MANIATIS, T., and M. PTASHNE: "A DNA Operator-Repressor System," *Sci. Am.*, **234**:64–76, January (1976).

O'MALLEY, B. W., et al.: "The Ovalbumin Gene: Organization, Structure, Transcription, and Regulation," *Recent Progr. Horm. Res.*, **35**:1–42 (1979).

PASTAN, I.: "Cyclic AMP," *Sci. Am.*, **227**:97–105, August (1972).

Special Topics

PALADE, G.: "Intracellular Aspects of the Process of Protein Synthesis," *Science*, **189**:347–357 (1975).

YARUS, M.: "Accuracy of Translation," *Progr. Nucleic Acid Res.*, **23**:195–225 (1979).

Problems

1. *Messenger RNA Translation* Predict the amino acid sequences of peptides formed by ribosomes in response to the following messengers, assuming that the initial codon begins with the first base on the left.
   (a) GGUCAGUCGCUCCUGAUU
   (b) UUGGAUGCGCCAUAAUUUGCU
   (c) CAUGAUGCCUGUUGCUAC
   (d) AUGGACGAA

2. *Can the Base Sequence of Its Messenger RNA Be Predicted from the Amino Acid Sequence of a Polypeptide?* A given sequence of bases in a messenger RNA will code for one and only one sequence of amino acids in a polypeptide, if the reading frame is specified. From a given sequence of amino acid residues in a protein such as cytochrome *c*, can we predict the base sequence of the unique messenger RNA that coded for it? Give reasons for your answer.

3. *How Many mRNAs Can Specify One Amino Acid Sequence?* To illustrate further the point made in the preceding problem, write all the possible mRNA sequences that can code for the simple tripeptide segment Leu·Met·Tyr. Your answer will give you some idea as to the number of possible mRNAs that can code for one polypeptide.

4. *Coding of a Polypeptide by Duplex DNA* The transcribed strand of a sample of double-helical DNA contains the sequence

   (5′)CTTAACACCCCTGACTTCGCGCCGTCG

   (a) What is the sequence of mRNA that can be transcribed from this strand?
   (b) What amino acid sequence could be coded by this sequence, starting from the 5′ end?
   (c) Suppose the other strand of this DNA sample is transcribed and translated. Will the resulting amino acid sequence be the same as in (b)? Explain the biological significance of your answers to (b) and (c).

5. *Methionine Has Only One Codon* Methionine is one of the two amino acids having only one codon. Yet the single codon for methionine can specify both the initiating residue as well as interior methionine residues of polypeptides synthesized by *E. coli*. Explain exactly how this is possible.

6. *Synthetic Messenger RNAs* How would you make a polyribonucleotide that would serve as a messenger RNA coding predominantly for many phenylalanine residues and a small number of leucine and serine residues? What other amino acids would be coded for by this polyribonucleotide but in much smaller amounts?

7. *The Direct Energy Cost of Protein Biosynthesis* Determine the minimum energy cost, in terms of high-energy phosphate groups, required for the biosynthesis of the β-globin chain of hemoglobin (146 residues), starting from a pool of all necessary amino acids, ATP, and GTP. Compare your answer with the direct energy cost of the biosynthesis of a linear glycogen chain of 146 glucose residues in $\alpha(1 \rightarrow 4)$ linkage, starting from a pool of glucose and ATP. From your data what is the *extra* energy cost of imparting the genetic information inherent in the β-globin molecule?

8. *Indirect Costs of Protein Synthesis* In addition to the direct energy cost for the synthesis of a protein, as developed in the preceding question, there are indirect energy costs, those required for the cell to make the necessary biocatalysts for protein synthesis. Contrast the relative magnitude of the indirect costs to a eukaryotic cell for the biosynthesis of linear $\alpha(1 \rightarrow 4)$ glycogen chains versus the indirect costs for the biosynthesis of polypeptides.

9. *Predicting Anticodons from Codons* Most amino acids have more than one codon, more than one tRNA, and more than one anticodon. Write all possible anticodons for the four codons for glycine, (5′)GGU(3′), GGC, GGA, and GGG.
   (a) From your answer, which of the positions in the anticodons are primary determinants of their codon specificity in the case of glycine?
   (b) Which of these anticodon-codon pairings will have a wobbly base pair?
   (c) In which of the anticodon-codon pairings will all three positions exhibit strong Watson-Crick hydrogen bonding?
   (d) Which of the anticodon-codon pairings will be least likely to be used in biological protein synthesis? Why?

10. *An Unusual tRNA* Recently a tRNA whose anticodon recognizes and binds to a tetranucleotide sequence of mRNA was discovered. Predict its effect on the amino acid sequence of the polypeptide in whose synthesis this unusual tRNA participated.

11. *The Effect of Single-Base Changes on Amino Acid Sequence* Much important confirmatory evidence on the amino acid code words has come from the nature of single-residue changes in the amino acid sequence of mutant proteins. Which of the following single-residue amino acid replacements would be consistent with the genetic code? Which cannot be the result of single-base mutations? Why?
   (a) Phe, Leu
   (b) Lys, Ala
   (c) Ala, Thr
   (d) Phe, Lys
   (e) Ile, Leu
   (f) His, Glu
   (g) Pro, Ser

12. *The Basis of the Sickle-Cell Mutation* In sickle-cell hemoglobin there is a valine residue at position 6 of the β-globin chain, instead of the glutamic acid residue found in this position in normal hemoglobin A. Can you predict what change took place in the DNA codon for glutamic acid to account for its replacement by valine?

CHAPTER 30

More about Genes: Repair, Mutation, Recombination, and Cloning

Chromosomes are by no means inert, stable structures holding genetic information in dead storage. They are constantly undergoing changes of different kinds. Some of these changes are accidental and are readily repaired. For example, DNA may undergo occasional single-strand breaks during replication, when the duplex must be parted and unwound at a high rate. These breaks are repaired by the action of DNA polymerase I and DNA ligase. Similarly, an incorrectly inserted nucleotide that fails to base-pair properly can be excised and replaced by the proofreading activity of DNA polymerases (page 851). In this chapter we shall see that DNA may also undergo changes induced by environmental agents. When such a change escapes repair by cellular mechanisms, a permanent, heritable mutation results.

Chromosomes also undergo other kinds of alterations and rearrangements that are a normal part of their biological function. Genetic recombination also occurs following the union of egg and sperm cells of eukaryotes, to yield progeny with new combinations of genes. Moreover, genes and parts of genes may be transposed and spliced into other locations in chromosomes. Genes may also be exchanged and recombined when viruses infect cells.

Despite the continuous damage, repair, exchange, transposition, and splicing of genes, the species identity of living organisms is transferred from generation to generation with extraordinary fidelity. This is made possible by the remarkable ability of certain enzymes to protect or restore the specific base sequences of each organism's chromosomes. Discovery of these enzymes has also made possible the laboratory construction of new kinds of DNA molecules, in which, for example, genes from one species have been combined with genes of another. These achievements have brought a new era of genetic biochemistry and a new field, genetic engineering.

Now let us examine some of the different kinds of DNA alterations.

DNA Is Constantly Subject to Damage

Virtually all forms of life are exposed to energy-rich radiations capable of causing chemical changes in DNA. *Ultraviolet radiation* (wavelength 200 to 400 nm), which makes up a significant portion of the solar spectrum, can cause chemical changes in the DNA of bacteria and the skin cells of human beings. Absorption of ultraviolet (UV) light can bring a purine or pyrimidine base into an excited state that may lead to covalent changes in its structure. Another form of radiant energy, *ionizing radiation*, can cause expulsion of one or more electrons from a biomolecule, thus forming an extremely unstable ion or free radical. Such highly reactive substances can cause abnormal chemical changes in DNA. There is a constant field of ionizing radiations around us in the form of *cosmic rays*, which can penetrate deep into the earth, as well as those emitted from natural radioactive elements, such as radium, plutonium, ^{14}C, and tritium (^{3}H). *X-rays* used in medical or dental examinations, as well as in radiation therapy of cancer and other diseases, are another form of ionizing radiation. Other possible sources are radioactive fallout from atom-bomb tests and radioactive waste that may escape from nuclear power reactors. Ultraviolet and ionizing radiations are responsible for about 10 percent of all DNA damage caused by nonbiological agents. Fortunately, most of this damage is quickly repaired by cells, by means of specific enzymatic mechanisms.

Figure 30-1
Formation of a thymine dimer by ultraviolet irradiation. Two new carbon-carbon bonds (color) are formed between two successive thymine residues in a DNA strand. (a) The structure of the dimer is shown in two dimensions but is better visualized as a three-dimensional projection, as in (b). The four-membered ring created by the two new bonds is shaded.

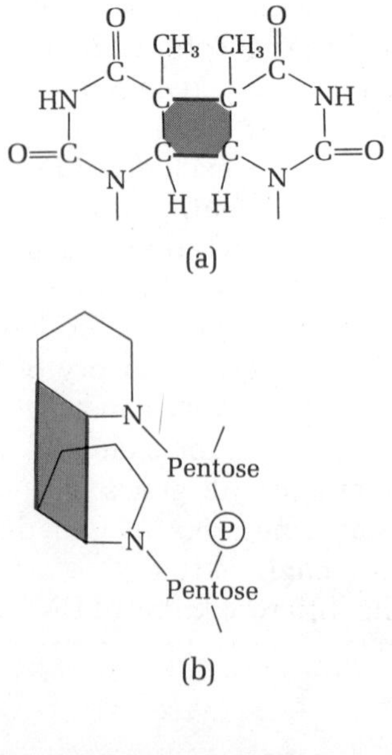

O O CH₃ CH₃ C C HN C C NH O=C C C C=O N N H H

(a)

N Pentose P N Pentose

(b)

Ultraviolet Damage Can Be Excised and Repaired

Exposure of bacteria to ultraviolet light can cause two successive pyrimidine residues in a strand of DNA, most often two adjacent thymine residues, to become covalently joined to form a *dimeric* base (Figure 30-1). If not removed and repaired, such a *thymine dimer* prevents DNA polymerase from replicating that strand beyond the point of the damage. Thymine dimers are excised and the gap patched by the sequential action of four enzyme activities (Figure 30-2). The first, called *ultraviolet-* or *UV-endonuclease*, cleaves the damaged strand at the 5′ side of the thymine dimer. In the second step DNA polymerase I adds the correct deoxyribonucleotides to the open 3′ end of the damaged strand to make a short length of DNA complementary to the template strand. During this process the strand containing the thymine dimer peels away. In the third step an endonuclease excises the defective section. In the last step the new patch of correctly base-paired DNA is spliced into the strand by DNA ligase (Figure 30-2).

Pyrimidine dimers may be formed and repaired not only in ultraviolet-irradiated bacteria but also in skin cells of people exposed to unfiltered sunlight. However, in the rare human disease *xeroderma pigmentosum* the enzymatic repair mechanism for ultraviolet damage is genetically defective. In this condition the skin is extremely sensitive to sunlight. It becomes very dry and thick, the skin cells proliferate abnormally, and skin cancer

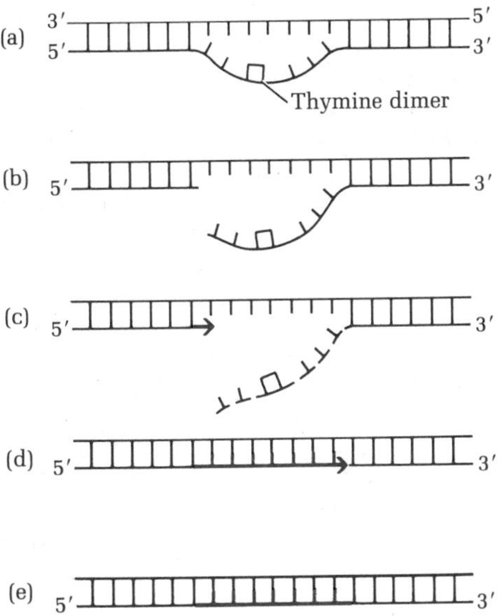

Figure 30-2
Repair of a thymine dimer (a). A special UV-endonuclease nicks the strand at the 5′ side of the dimer (b). DNA polymerase I begins to patch the strand, and 5′ → 3′ endonuclease removes the thymine dimer and some adjacent nucleotides (c). The new patch is completed (d) and is then spliced into the chain by DNA ligase (e).

almost invariably develops. The disease is often fatal unless the skin is carefully shielded from sunlight. Biochemical and genetic research has revealed that in the most common form of xeroderma the defective enzyme is the UV-endonuclease required to nick the defective strand at the 5′ side of the pyrimidine dimer. For lack of this single enzyme exposure of the skin to sunlight can have disastrous consequences.

The ultraviolet portion of sunlight, which is responsible for suntanning, is filtered out by glass and also by protective "sun-block" preparations, which contain ultraviolet-absorbing chemical compounds.

Figure 30-3
Spontaneous decomposition of a cytosine residue of DNA to yield uracil can cause a mutation unless repaired.

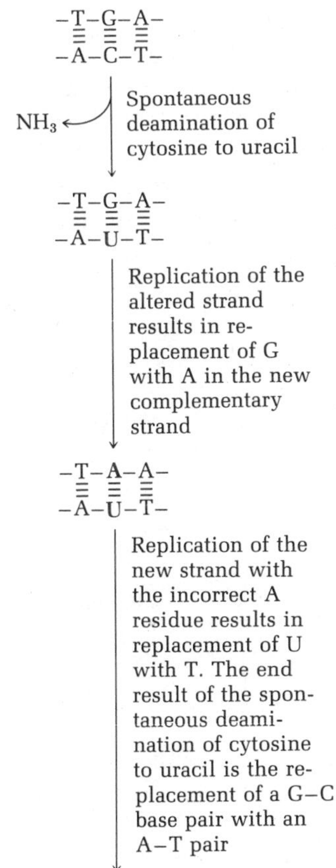

Spontaneous Deamination of Cytosine to Uracil Can Be Repaired

DNA also may undergo alteration caused by the chemical instability of the base cytosine in aqueous systems. Cytosine residues very slowly undergo spontaneous loss of their amino groups by hydrolysis to form uracil residues, which are not normally present in DNA (Figure 30-3). When a DNA strand containing a uracil residue is replicated, the uracil cannot form strong hydrogen bonds with a guanine (G) residue, the normal partner of cytosine. Instead the uracil will tend to pair with an adenine residue. When the new DNA strand containing a wrong A residue is in turn replicated, it will of course specify a T residue in its complementary strand. The result will be a new daughter duplex DNA that will contain an A–T base pair rather than the G–C pair specified by the original undamaged parental DNA.

This type of damage is repaired in a novel way (Figure 30-4). A special repair enzyme, *uracil–DNA glycosidase*, removes by hydrolysis the wrong uracil base from the damaged strand. The remaining deoxyribose phosphate residue, which now lacks a

Figure 30-4
Repair of the spontaneous conversion of cytosine to uracil. The uracil can be removed enzymatically and the "empty" deoxyribose phosphate replaced with the correct deoxycytidine phosphate residue.

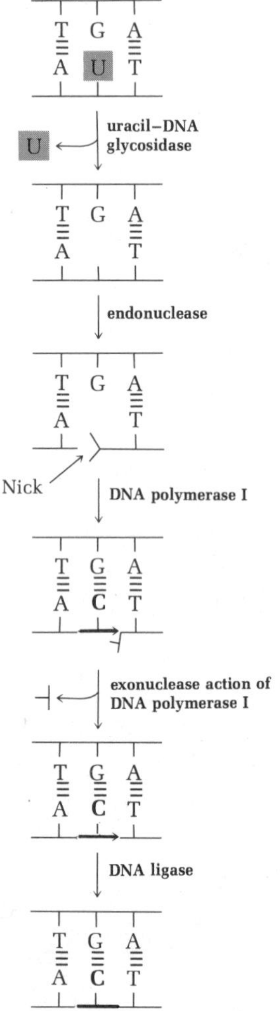

base, is then split at the 5′ side of the phosphodiester bond by DNA polymerase I, which proceeds to insert the correct cytidine phosphate unit on the now open 3′ end of the damaged strand, to base-pair with the G residue in the undamaged strand. The strand is then covalently spliced by DNA ligase to complete the repair.

The glycosidase that removes the uracil base from DNA is of necessity very specific; if it were not, it would cause much more damage than it could repair. Fortunately, this enzyme does not remove uracil residues from RNA, nor does it remove thymine residues from DNA. This observation suggests a reason for the long-puzzling fact that DNA contains thymine instead of uracil. If DNA normally contained both uracil and cytosine residues, there would be no way to distinguish between normal, preformed uracil residues and uracil residues formed by spontaneous decomposition of cytosine. It has therefore been argued that DNA normally contains thymine residues, which are stable, but no uracil, so that a repair system can recognize and delete uracil residues that arise by spontaneous hydrolysis of cytosine.

Damage by External Chemical Agents Can Also Be Repaired

DNA may also be damaged by reactive chemicals introduced into the environment as products of industrial activity. Such products may not necessarily be injurious per se but may be metabolized by cells into forms that are. There are three major classes of such reactive chemical agents: (1) deaminating agents, particularly nitrous acid (HNO_2) or compounds that can be metabolized to nitrous acid or nitrites, (2) alkylating agents, and (3) compounds that can simulate or mimic the normal bases present in DNA (Figure 30-5).

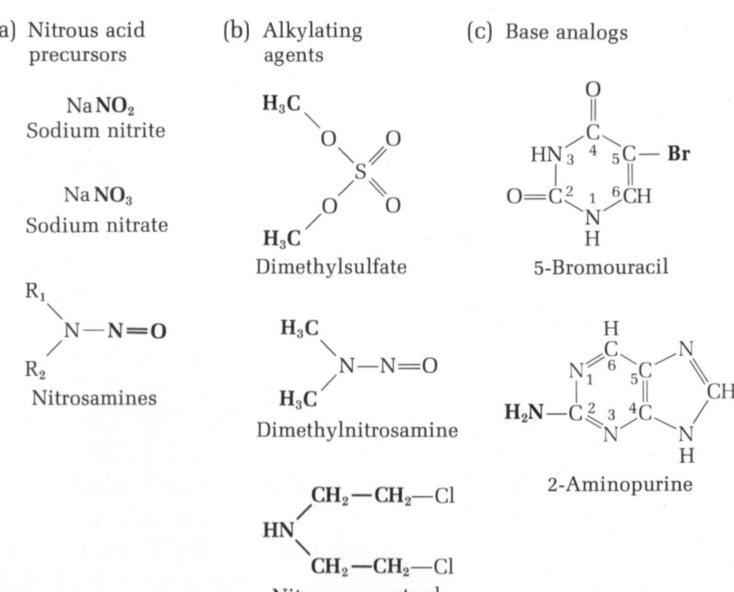

Figure 30-5
Some chemical agents that may alter the structure of purine or pyrimidine bases of DNA. Because such agents can cause permanent, heritable mutations if their effects escape repair, they are called mutagens. The most active deaminating agent is nitrous acid, which can be formed from various precursors (a). Alkylating agents (b) act by transferring an alkyl group to a susceptible oxygen or nitrogen atom of a base, thus altering its hydrogen-bonding characteristics. Base analogs (c) cause mutations by replacing a normal base during synthesis of the DNA, thus leading to incorrect base pairing. The toxic or abnormal groups are in color.

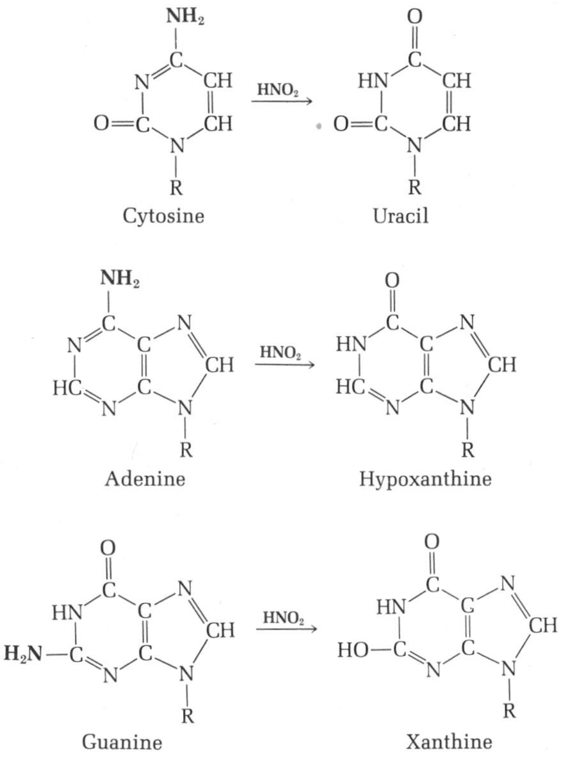

Figure 30-6
Action of HNO_2 in promoting deamination of base residues in DNA. The amino group removed is in color. R is the remainder of the deoxyribonucleotide. Removal of the amino groups alters the hydrogen-bonding properties of the bases.

<u>*Nitrous acid*</u>, which is formed from organic precursors, such as nitrosamines, and from nitrite and nitrate salts, is a very potent reagent capable of removing the amino groups from the bases <u>*cytosine*</u>, <u>*adenine*</u>, and <u>*guanine*</u> (Figure 30-6). Nitrous acid greatly accelerates the deamination of cytosine to yield uracil, which we have examined above. In a similar way nitrous acid deaminates adenine to yield <u>*hypoxanthine*</u> and guanine to yield <u>*xanthine*</u> (Figure 30-6). Hypoxanthine and xanthine residues so formed can be recognized and removed by specific enzymes, followed by the sequential action of DNA polymerase I and DNA ligase, much as was shown in Figure 30-4 for the removal of uracil residues. Nitrates and nitrites are used as preservatives in sandwich meat, frankfurters, and other meat products, but there is some disagreement whether they are injurious to people when used in this way. Other precursors of nitrous acid are widely used in industry.

Alkylating agents can alter certain bases of DNA. For example, the highly reactive chemical dimethylsulfate (Figure 30-5) can methylate a guanine residue to yield <u>*O-methylguanine*</u> (Figure 30-7), which is unable to base-pair with cytosine, the normal partner of guanine. Both bacteria and animal tissues contain enzymes that can specifically remove O-methylguanine and replace it with the normal base guanine, again by a cut, patch, and splice mechanism, similar to that shown in Figure 30-4.

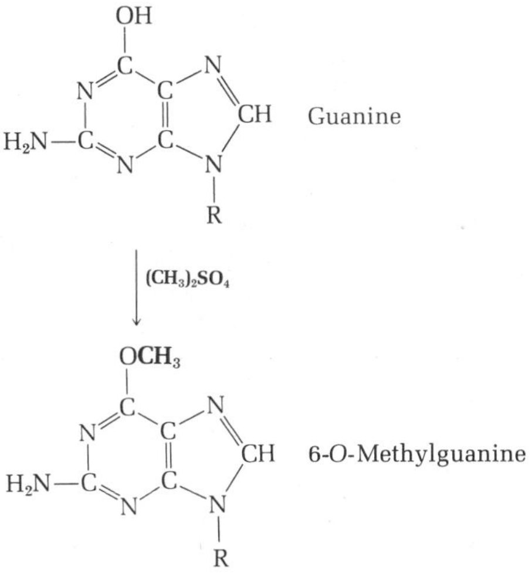

Figure 30-7
Methylation of guanine (enolic form) by an active methylating agent. Methylation alters the hydrogen-bonding specificity of guanine, leading to faulty base pairing.

Many other types of DNA repair mechanisms have been discovered, but these examples suffice to show how specific repair enzymes help preserve the identity of each species' chromosomes. Although cells contain numerous enzymes capable of repairing damage to DNA, they have no enzymes that can repair damaged RNA. The sequence integrity of DNA is vital for the preservation of a whole species, whereas the sequence integrity of RNAs is vital only for a specific cell in which an RNA molecule happens to contain an error or has been damaged.

Alteration of a Single Base Pair Causes Point Mutations

Although DNA proofreading and repair mechanisms are exceedingly effective, inevitably some errors in replication will remain uncorrected or some damage in DNA will go unrepaired, resulting in a change that may become perpetuated in a heritable manner in the genome of the organism. Such permanent changes transmitted by replication are called *mutations*.

Mutations caused by the replacement of a single base with an incorrect one are called *substitution* mutants. Such a mutation will have the effect of changing only one codon in the affected gene. Thus it may or may not result in the replacement of a single amino acid by another in the sequence of the polypeptide encoded by this gene. Examples of such single-point substitutions and their consequences are shown in Table 30-1. Often replacement of one amino acid by another will cause no significant change in the biological properties of the protein translation product; these are called *silent mutations*. In other cases, however, the replaced amino acid residue may be very critical and leave the protein totally unable to perform its normal biological function. Such mutations are often lethal to the cell. For example, a specific serine residue is a vital part of the active site in the serine group of enzymes (page 222). A point mutation in the codon for this serine residue, resulting in its replacement by some other amino acid (Table 30-1), would cause a total loss of activity of a serine enzyme encoded by the mutated gene. If this enzyme is in a central metabolic pathway, such a mutation may be lethal.

It is also possible that a single amino acid change will result in a protein product still capable of functioning but with different characteristics, so that it is not as efficient as the normal protein product. For example, if the mutant protein is an enzyme, it may have a higher K_M, a lower V_{max}, or both. Mutants that form such an altered but still partially functional protein are called *leaky mutants* (Table 30-1). Sometimes a mutation may leave the polypeptide gene product *better* able to perform its function under the special environmental circumstances in which the organism finds itself. Such mutations may result in progeny better equipped in the struggle for the survival of the fittest. Indeed, a series of desirable mutations may lead to the evolution of a new species.

Table 30-1 Effects of Some Hypothetical Single-Base Mutations on the Biological Activity of the Resulting Protein Products

Mutation		Wild-type (unmutated) DNA triplet)	Mutated triplet
A single-base substitution causing no change in the amino acid sequence; a silent mutation	DNA template RNA codon Amino acid	(3′)–GGT–(5′) (5′)–CCA–(3′) –Pro–	–GGA– –CCU– –**Pro**–
A single-base mutation resulting in an amino acid change that may not alter the biological activity of the protein because the amino acid replacement is in a noncritical position and also resembles the normal amino acid; also a silent mutation		(3′)–TAA–(5′) (5′)–AUU–(3′) –Ile–	–GAA– –CUU– –**Leu**–
A lethal single-base mutation in which a serine residue essential for enzyme activity is replaced by phenylalanine to give an enzymatically inactive product		(3′)–AGA–(5′) (5′)–UCU–(3′) –Ser–	–AAA– –UUU– –**Phe**–
A leaky mutation in which the amino acid change results in a protein that retains at least some of its normal activity		(3′)–CGT–(5′) (5′)–GCA–(3′) –Ala–	–CCT– –GGA– –**Gly**–
A hypothetical beneficial mutation, in which the amino acid replacement yields a protein with improved biological activity, giving the mutated organism an advantage; it is not possible to predict advantageous amino acid replacements		(3′)–TTC–(5′) (5′)–AAG–(3′) –Lys–	–TCC– –AGG– –**Arg**–

Substitution of one base for another accounts for only a minority of permanent mutations in bacteria. More numerous and more lethal are *insertion* and *deletion* mutations.

Insertion or Deletion of Nucleotides Causes Frame-Shift Mutations

When a mutation is caused by the insertion or deletion of one base pair in a gene, a more extensive type of genetic damage may result. The consequence of such a mutation is the disruption of the normal colinearity between the codons in the DNA and the amino acid sequence of the polypeptide for which it codes. The disruption will begin at the site of the gain or loss of a base, because there will be a shift in the reading frame of the DNA. As a result the polypeptide product will have the correct amino acid sequence up to the point of the mutation but will have a garbled, totally different amino acid sequence from that point on (Figure 30-8). Frame-shift mutations often generate an internal termination codon that results in a prematurely released, shortened polypeptide chain. The great majority of

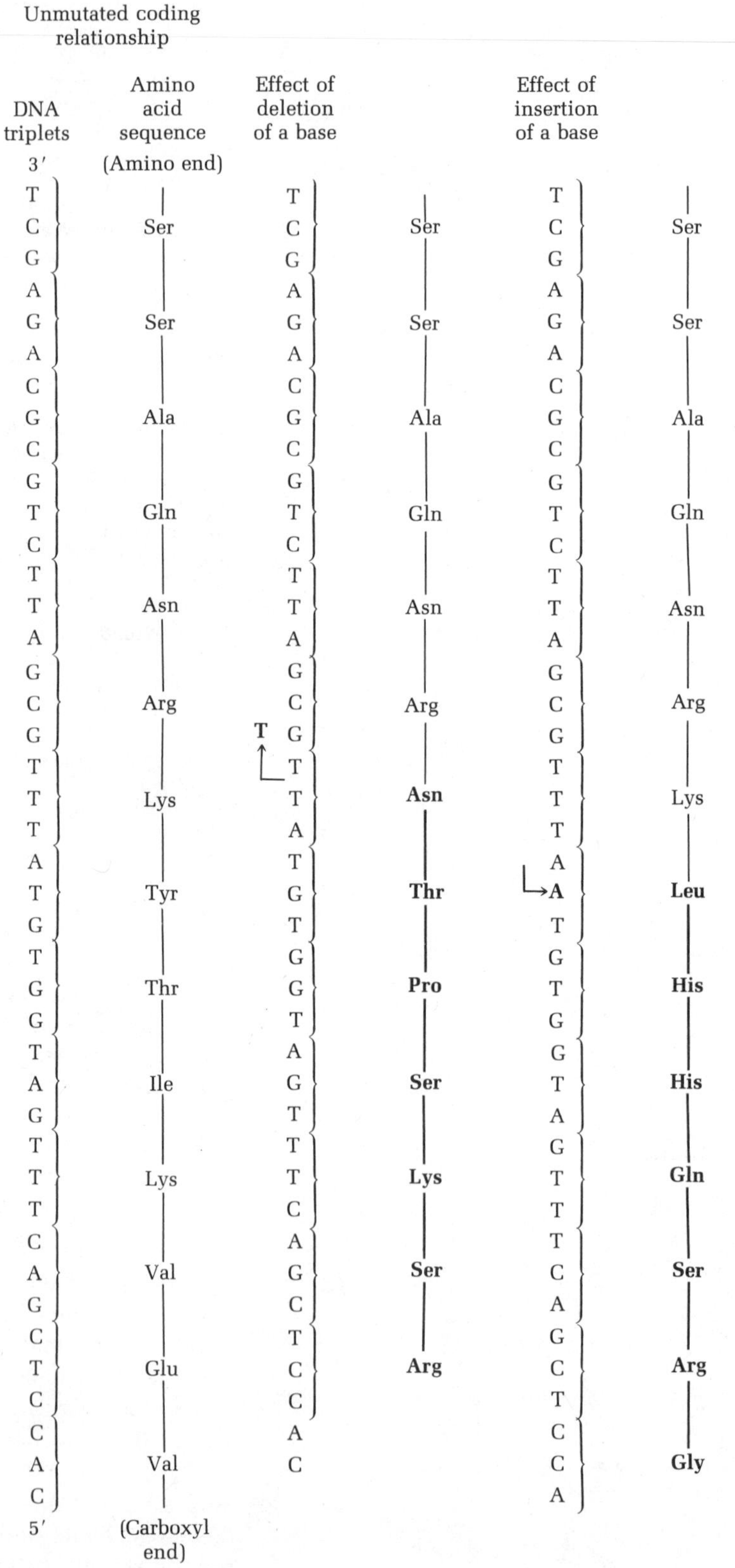

Figure 30-8
Frame-shift mutations caused by a deletion or an insertion of a base (colored arrows). Beginning with the codon in which a base has been lost or gained, the amino acid sequence will be completely garbled (color). Most frame-shift mutations are lethal.

Unmutated coding relationship: DNA triplets	Unmutated coding relationship: Amino acid sequence	Effect of deletion of a base		Effect of insertion of a base	
3′	(Amino end)				
TCG	Ser	TCG	Ser	TCG	Ser
AGA	Ser	AGA	Ser	AGA	Ser
CGC	Ala	CGC	Ala	CGC	Ala
GTC	Gln	GTC	Gln	GTC	Gln
TTA	Asn	TTA	Asn	TTA	Asn
GCG	Arg	GCG (T ↑ deleted)	Arg	GCG	Arg
TTT	Lys	TTA	**Asn**	TTT	Lys
ATG	Tyr	TGT	**Thr**	A→AT (A inserted)	**Leu**
TGG	Thr	GGT	**Pro**	GTG	**His**
TAG	Ile	AGT	**Ser**	GTA	**His**
TTT	Lys	TTC	**Lys**	GTT	**Gln**
CAG	Val	AGC	**Ser**	TCA	**Ser**
CTC	Glu	TCC	**Arg**	GCT	**Arg**
CAC	Val	AC		CCA	**Gly**
5′	(Carboxyl end)				

single-base frame-shift mutations result in biologically inactive gene products.

Sometimes one frame-shift mutation can be canceled out by a second one. If a frame-shift mutation of a given gene is caused by the loss of one base pair, a second mutation in the same gene, in which there is a gain of one base pair downstream of the first, can restore the normal reading frame beyond the point of the second mutation. The second mutation in this case suppresses the effect of the first and is therefore called a *suppressor mutation*. Sometimes, although this is very rare, a group of three successive bases is lost or gained. In this case the polypeptide product formed will have one missing (or one extra) amino acid at a given point, but the rest of the amino acid sequence will be correct. Such mutations are usually harmless. Frame-shift mutations, suppressor mutations, and three-base mutations were historically important in establishing that the genetic code consists of nucleotide triplets.

Frame-shift mutations can be induced by certain large, flat basic molecules that resemble a normal base or base pair. Such molecules tend to *intercalate* (insert themselves) between two adjacent base pairs, in effect adding an extra base to the DNA. When this altered strand is replicated, an extra base may be incorporated into the daughter strand through faulty base pairing with the intercalated molecule. *Acridine* is an example of such an intercalating mutagen (Figure 30-9).

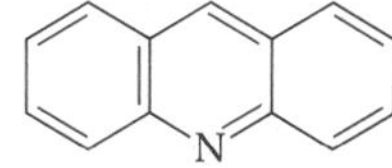

Figure 30-9
Acridine, a mutagenic agent producing frame-shift mutations. The acridine molecule is flat, and its ring structure has some resemblance to a purine base. Acridine inserts itself (intercalates) between two successive base pairs of DNA, separating them somewhat. When the DNA is replicated, an extra base pair is added into the new chain, opposite the acridine.

Mutations Are Random, Rare Events in Individuals

Mutations in real life are very rare events as far as individual organisms are concerned. The probability that a mutation will occur in the lifetime of one *E. coli* cell is only about 1 chance in 10^9. For a human cell the probability is greater, perhaps 1 chance in 10^5; this was calculated from the natural incidence of *hemophilia*, a genetic defect in the blood-clotting mechanism leading to prolonged bleeding before clotting occurs. Hemophilia was one of the first genetic diseases of human beings to be recognized. A classic case is represented in the family history of Queen Victoria of England. Hemophilia has been traced through three generations of her descendants, in the royal families of England, Prussia, Spain, Greece, and Russia. Although some mutations in human DNA are silent, harmless, or favorable, and thus cause no problems, many result in genetic defects that may impair normal human activities or functions. Mutations of some 2500 different human genes have been detected to date, many impairing function or ultimately lethal. Presumably all the other human genes that undergo mutations have not yet been detected. This is why we can expect the number of recognized human genetic diseases to increase as methods capable of detecting the results of mutation become available. Genetic diseases pose an especially great challenge to biochemistry and medicine for their recognition and treatment.

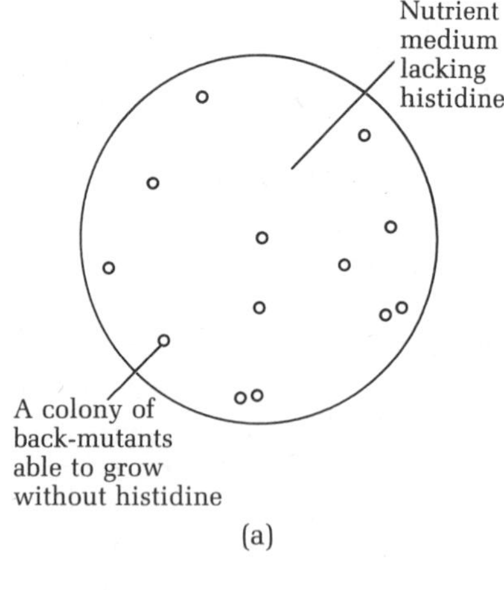

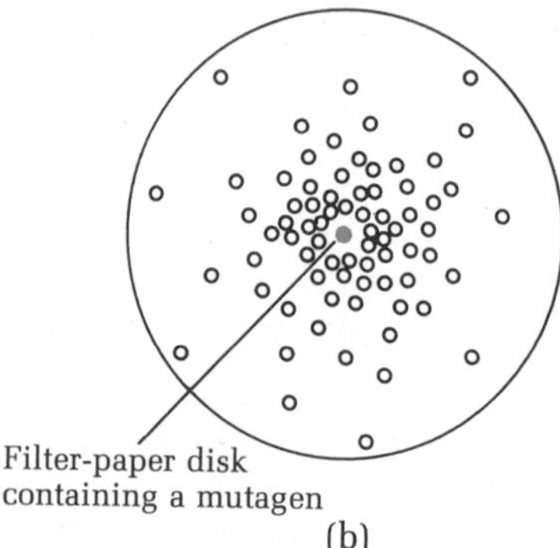

Figure 30-10
The Ames test for carcinogens, based on their mutagenicity. (a) *The few small colonies of back-mutants of histidine-less* Salmonella typhimurium *grown on a medium free of histidine are the result of spontaneous back-mutations.* (b) *To an identical nutrient plate inoculated with an equal number of cells has been added a disk of filter paper containing a mutagen (color), which greatly increases the rate of back-mutation and hence the number of colonies. In the clear area around the spot of mutagen, the concentration is so high it is lethal to the cells. As the mutagen diffuses outward, away from the central spot, it is diluted and promotes back-mutation at concentrations that are less lethal. Mutagens are compared on the basis of the increased mutation rate they produce.*

Many Mutagenic Agents Are Also Carcinogenic

Statistical evidence strongly indicates that continued exposure of human beings to certain chemical agents, especially in the workplace, results in increased incidence of specific kinds of cancer. For example, workers in chemical industries using or producing naphthylamines have a much higher incidence of bladder cancer than the general population. Indeed, it has been estimated that up to 90 percent of human cancers may be caused by exposure to noxious physical or chemical agents capable of transforming a normal cell into a malignant one. For this reason there is considerable public concern over possible carcinogenic effects of industrial chemicals, food additives, automobile exhaust gases, dyes, flavors, medicines, cosmetics, and other chemicals to which we are constantly exposed. Much effort is being expended in the assessment of the possible carcinogenicity of such agents. However, this is a difficult problem, since it is estimated that at least 50,000 different chemical compounds are currently used in industry to prepare or process the many different products with which we have frequent contact. And every year 1000 or more new chemicals are brought into use. Moreover, a thorough animal test of one new chemical for toxicity and carcinogenicity may require 2 or 3 years and may cost as much as $100,000. Thus it seems almost impossible to carry out animal tests of the potential carcinogenic effects of all the many chemical agents to which human beings are increasingly exposed.

A simple bacterial test of carcinogenicity has been developed by Bruce Ames and his colleagues at the University of California. It is very cheap and can be done in less than a day. The test was developed on the assumption that carcinogenic agents are also mutagens. It employs a histidine-requiring mutant of the common bacterium *Salmonella typhimurium,* which cannot make histidine because an enzyme in the pathway of histidine biosynthesis is genetically defective. Occasionally, however, the histidine-requiring mutant will undergo a spontaneous back-mutation, so that it regains its normal capacity to make histidine from its normal precursors. Such back-mutations are easily detected since they will grow on a medium containing ammonia as nitrogen source but no histidine (Figure 30-10). Because the rate of back-mutation is markedly increased by mutagens, it is possible to compare the relative mutagenicity of different substances, such as those suspected to be carcinogens. The histidine-less nutrient medium used in these tests is supplemented with an extract of rat liver, which provides enzymes of the endoplasmic reticulum capable of hydroxylating or otherwise converting many foreign organic compounds into their ultimate carcinogenic forms.

Over 300 chemical compounds definitely known from animal studies to be carcinogens have been tested by the Ames method; over 90 percent were mutagenic in the bacterial test. The very high degree of correlation between carcinogenicity and mutagenicity therefore indicates that the bacterial test is a valid predictor of carcinogenicity. Almost 3000 other chemicals have been subjected to the Ames test, including representative industrial chemicals, food preservatives, pesticides, flavoring agents, synthetic polymers and monomers, and cosmetics. Many such products have been found to be mutagenic. It was found, for example, that 90 percent of the hair dyes once sold in the United States, used by tens of millions of people, were mutagenic. The cosmetic industry has since changed the formulas of these preparations to render them nonmutagenic. Several different variations and adaptations of the Ames test are now in wide use to screen chemical products for possible carcinogenicity.

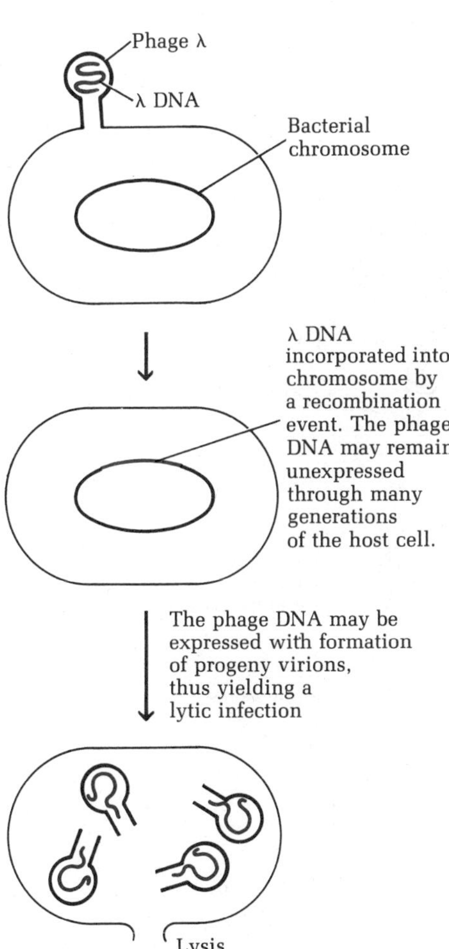

Figure 30-11
Incorporation of the DNA of phage λ into the E. coli chromosome in a nonexpressed form, which may be replicated over many generations. As a result of some triggering event the viral genome may be expressed, with the formation of λ phage particles and subsequent lysis of the cells.

Genes Often Undergo Recombination

Up to now we have considered changes in genes that are spontaneous, accidental, or caused by environmental agents. Let us now examine alterations in genes and chromosomes that are normal events in cells.

The normal biological exchange or addition of genes from different sources to form an altered chromosome, which can then be replicated, transcribed, and translated, is called *genetic recombination*. It occurs in a number of different biological situations. We are already familiar with one type of genetic recombination, the *transformation* of bacteria by exogenous DNA, as in the classical Avery-McLeod-McCarty experiment (Figure 27-6, page 801). We recall that DNA from a virulent strain of pneumococcus can gain entry into cells of a nonvirulent strain and permanently transform the latter into a virulent strain. The gene for virulence present in the DNA of the donor cell evidently was permanently incorporated or recombined into the genome of the recipient cell. Such transformation of bacterial cells by gene recombination can occur not only in the laboratory, but also under natural conditions.

Another natural genetic recombination process is the phenomenon of *lysogeny*. The DNA of certain kinds of phages, when they infect a bacterial cell, may become covalently incorporated into the circular host-cell chromosome, instead of leading to the immediate propagation of daughter viral particles and subsequent lysis of the cells, as is usually the case in viral infections. Once the viral genome has been incorporated into the host chromosome, it can be replicated for many generations without being expressed in the form of progeny viral particles. However, at a later time some particular event may trigger the expression of these dormant viral genes, causing the formation of viral progeny and the lysis of the host cell (Figure 30-11). Phages that can incorporate their DNA into the host-cell chromosome in such a nonexpressed form are called *temperate* or *lysogenic* phages. Phage λ is the best-known temperate phage; its incorporation into the *E. coli* chromosome has been studied in great detail. This type of genetic recombination is believed to occur in human infections with the *herpes simplex* virus, which causes cold sores, as well as sores and ulcers of the genitals. The DNA of herpes simplex virus can become incorporated into the genome of human cells and there remain dormant until some event triggers translation into infectious viral particles.

Another type of genetic recombination takes place in *transduction* (Figure 30-12). When a bacterial cell becomes infected with certain DNA phages, a small portion of the host-cell chromosome may become covalently attached to the viral DNA, undergo replication with the viral DNA, and thus become incor-

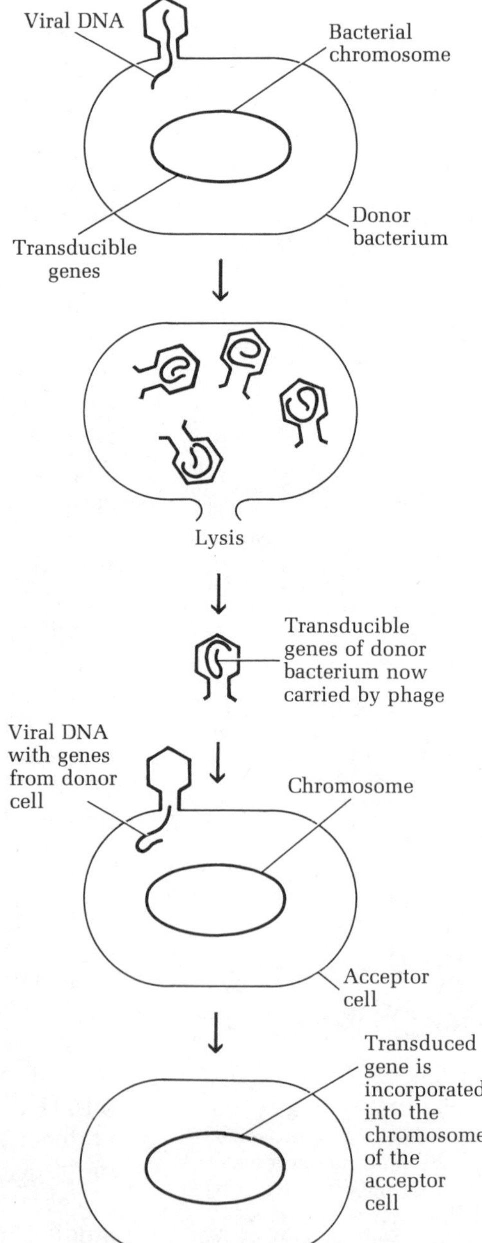

Figure 30-12
Genetic recombination during viral transduction of bacterial genes into a recipient cell.

porated into the DNA of the progeny viral particles. When these particles infect another cell, the viral DNA carries a portion of the first cell's chromosome into the chromosome of the second cell. Transduction, which means "leading over" or "carrying over," is a natural process that has been exploited in the laboratory to map bacterial chromosomes.

Bacterial conjugation is another example of genetic recombination. Bacteria normally reproduce asexually, by simple growth and division. Some species of bacteria occasionally undergo *sexual conjugation*. In this process part or all of one strand of the chromosome of the donor cell, designated the F^+ or (+) cell because it carries a sex factor F, is transferred into a recipient cell of the same species, the (−) cell, which lacks F, (Figure 30-13). As a consequence the recipient cell now acquires some new genes which are combined into its chromosome.

In eukaryotic organisms genetic recombination occurs by the sexual union of egg and sperm cells, in which both parental chromosomes contribute certain genes to the new daughter chromosomes appearing in the progeny cells (Figure 30-14). In this process the chromosomes of both the sperm and the egg cells undergo cleavage at homologous points, and pieces of the chromosomes from the two parent cells are then exchanged and spliced together to yield new combinations of genes, with

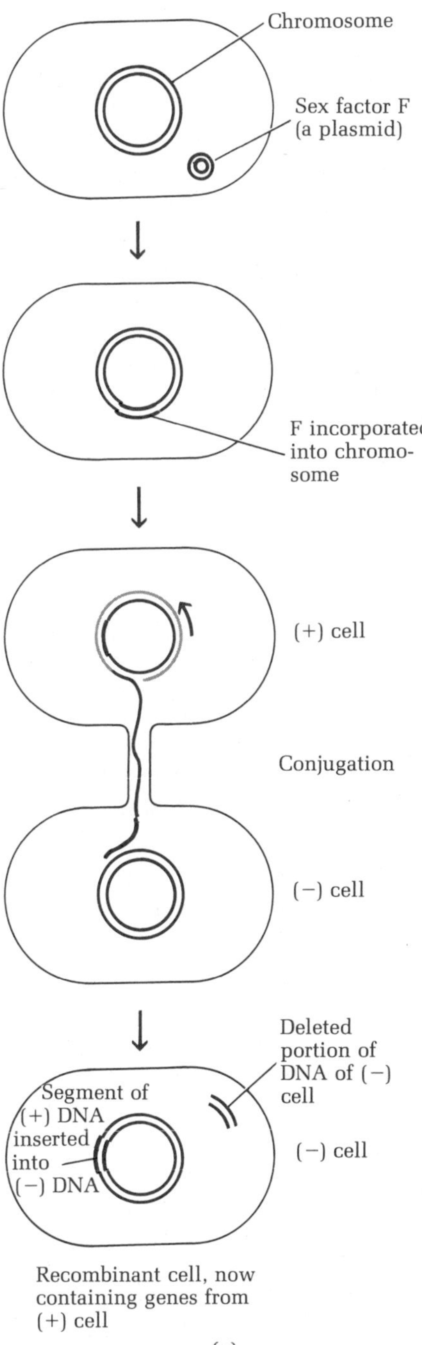

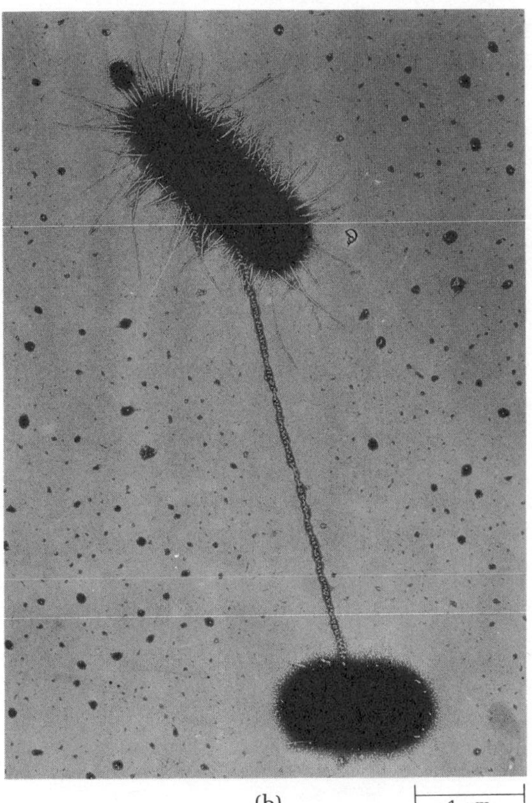

Figure 30-13
(a) Transfer and recombination of genes during bacterial conjugation. The DNA of the (+) cell is replicated by the rolling-circle process, and the resulting single strand containing F is introduced into the (−) cell. (b) Electron micrograph of conjugating E. coli cells. The (+) cell at the top of the micrograph is connected to a (−) cell by a single long tubelike structure.

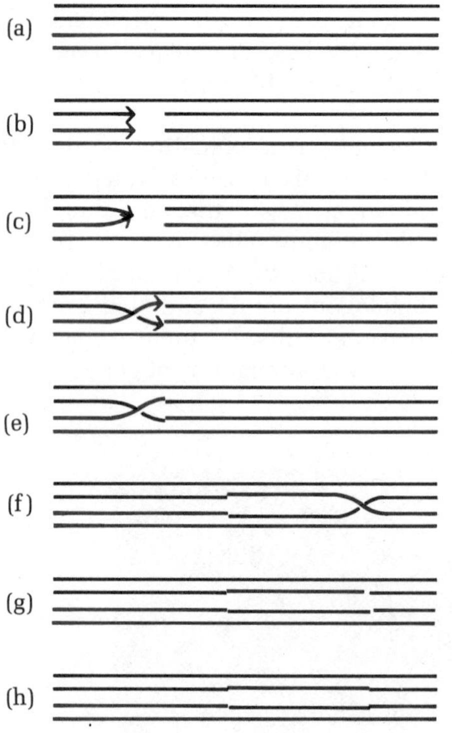

Figure 30-14
Schematic diagram of the "single-strand switch" model for recombination of genes between two homologous parental DNAs (black and color), shown in (a). One strand of each DNA is broken (b), switched over to the other DNA (c, d), and joined to the opposite switched strand (e). The exchange of strands between DNAs proceeds along the chromosome (f), and at a specific point the switched strands are again broken (g) and resealed (h), completing the exchange and recombination of the genes.

the result that the progeny now contain an assortment of phenotypic characteristics derived from both parents. Such naturally occurring cutting, reassembly, and splicing of genes and sets of genes during sexual conjugation of eukaryotes occur with great precision, without disturbing the reading frame or signals in the DNA sequence.

Segments of Chromosomes Often Undergo Transposition

Chromosomes may undergo other kinds of biological alteration. Both genetic and biochemical evidence has shown that a gene or set of genes of a prokaryotic or eukaryotic chromosome often leaves its original position and undergoes transposition to some other site in the genome. Such mobile genetic elements are called transposable elements or transposons. The capacity of transposons to be inserted into different locations in DNA is conferred by short extensions at each of their ends called insertion sequences (Figure 30-15). The transposon and its two insertion sequences, which contain inverted base sequences, may be inserted into different sites in chromosomes and plasmids by a specific enzyme system that recognizes the insertion sequence and splices the transposon into its new site. In this way a gene or set of genes may be transposed from one site to another in the same chromosome, or from a plasmid or phage into a bacterial chromosome, or from a plasmid to a phage.

Antibody Variety Is the Result of Transposition and Recombination Events

One of the most extraordinary types of genetic recombination occurs in the formation of the genes that code for the synthesis of different antibodies in vertebrates. Antibodies or immune globulins are proteins made by specific lymphocytes or immunocytes of vertebrates in response to the introduction of a foreign macromolecule, the antigen, into the body (page 138). Each type of antigen can bind to and stimulate the growth and division of one specific type of immunocyte, to yield a unique

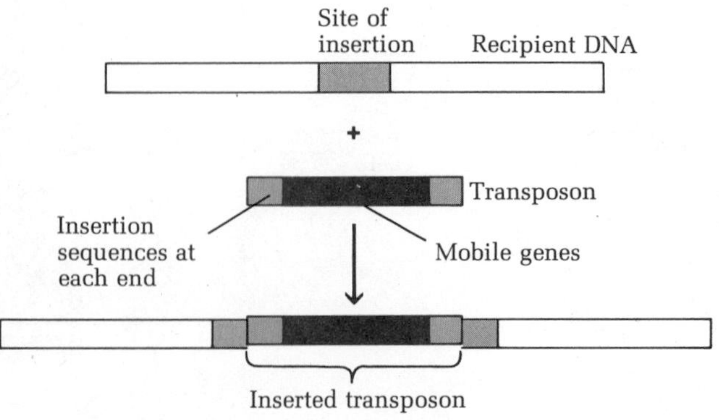

Figure 30-15
Insertion of a transposon into recipient DNA. At each end of the transposon are inverted insertion sequences.

line or *clone* of cells, all identical. The cells of this clone make only one type of immune globulin, which is specific for binding only that antigen which induced their proliferation. Binding of the antigen to its antibody yields an antigen-antibody complex in which the antigen is usually made biologically inactive. What is extraordinary is that the human body can form literally millions of different kinds of antibodies, each capable of binding just one of the millions of different antigens to which the organism may become exposed. Antibodies can be elicited not only against most proteins of other animal species, bacteria, viruses, parasites, and plants, but also against almost any macromolecule, even if it is not a naturally occurring substance.

In order to manufacture different antibody proteins, each capable of combining with just one of many millions of different possible antigens, one might think that the immunocytes would have to possess, correspondingly, millions of different antibody genes, each coding for only one kind of antibody. This has long seemed improbable. There simply is not enough DNA in the nucleus of human cells to code for millions of different antibodies, in addition to the many thousands of normal body proteins characteristic of the structure, metabolism, and individuality of a human organism.

The answer to this puzzle is coming from studies of the structure of antibody molecules and the genes that code for them. Figure 30-16 shows the structure of an antibody molecule, which has a molecular weight of about 160,000. It contains two *heavy* or long polypeptide chains (446 amino acid residues) and two *light* or short chains (214 residues). The chains are attached to each other by disulfide cross-linkages; in addition, both chains have intrachain —S—S— cross-links. Each heavy and each light chain contains a region of invariant amino acid sequence characteristic of each species, called the *constant* or C region. Each chain also has a *variable* or V region in which the amino acid sequence appears to be different for

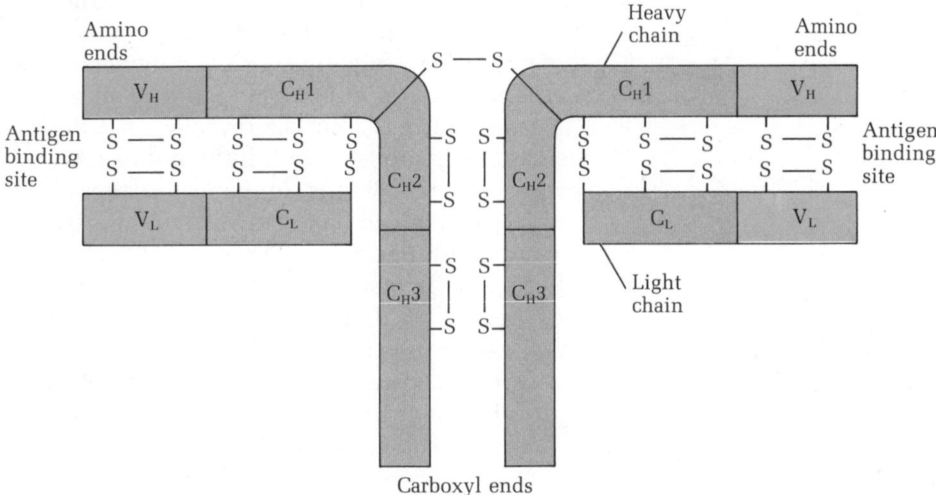

Figure 30-16
Structure of an antibody molecule. It has two light (L) and two heavy (H) chains, each of which has a variable (V) region (color) and a constant (C) region (gray). The constant regions of the heavy chains have three distinct domains, C_H1, C_H2, and C_H3. The molecule has numerous disulfide cross-bridges, as well as carbohydrate groups attached to the heavy chains.

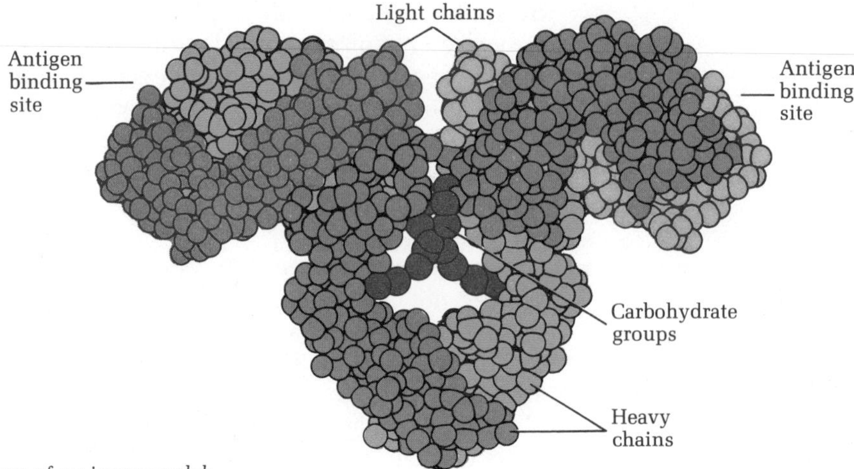

Figure 30-17
Model of the structure of an immune globulin, deduced from x-ray diffraction analysis.

each specific antibody. The constant region of the heavy chains contains three domains in which the amino acid sequences are similar. The antibody molecule has two *antigen binding sites*, each located in the cleft or pocket between the ends of the variable regions of a heavy and light chain. A molecular model of an antibody is shown in Figure 30-17.

The presence of both constant and variable regions long ago suggested that the DNA coding for the light chains may have originated by the splicing of two genes, one for the variable and one for the constant region. This was proved by Susumu Tonegawa and his colleagues, who found that the genes for the constant and variable portions of one specific type of light chain are very close together in the DNA of immunocytes producing this type of light chain but very far apart in the DNA of other lines of immunocytes unable to make this type of light chain. They concluded that immunocytes can select segments of DNA coding for the variable portion of an antibody chain from different and widely separated parts of the cell genome and transpose them to a position next to the gene for the constant region of a given light chain. After the constant and variable sequences are spliced, RNA polymerase then can make a single mRNA molecule to code for the entire light chain.

Subsequent research has revealed that the DNA coding for the variable regions of both light and heavy chains consists of several different kinds of genes, scattered throughout the genome, which can be transposed and assembled into various sequential arrangements (Figure 30-18). The DNA for the variable regions is made from perhaps 400 different *variable* (V) genes, about 12 *diversity* (D) genes, and 4 *joining* (J) genes, which can be assembled in different combinations to make the DNA for 20,000 or more variable regions. These in turn undergo additional alterations in base sequence and are spliced to different constant-region DNAs to yield millions of different

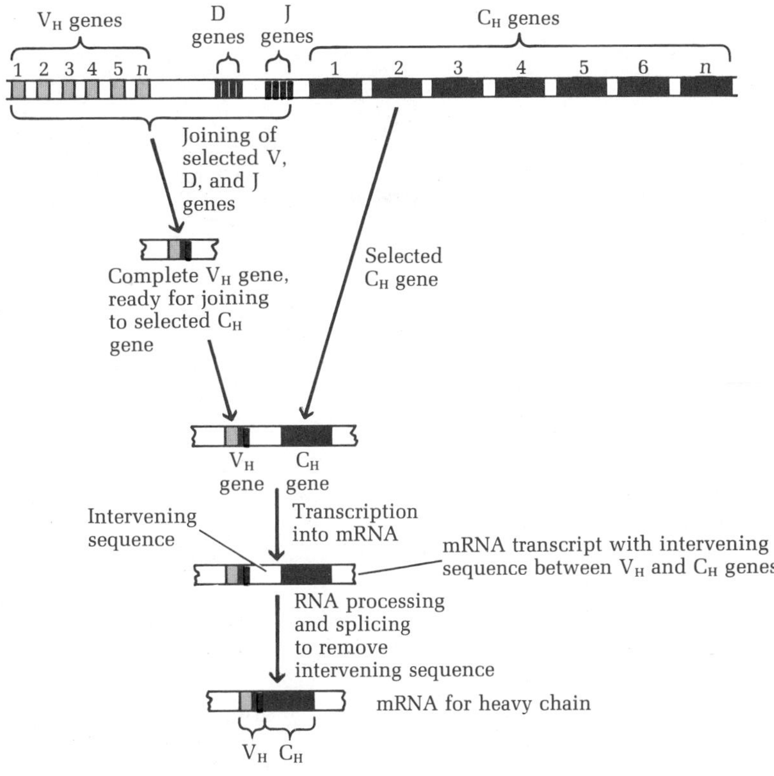

Figure 30-18
Schematic diagram showing how a gene for an antibody heavy chain may be assembled by transposition of V, D, and J genes from different portions of the genome to form a complete V_H gene that joins to a C_H gene. The mRNA transcript is then processed to remove the spacer, yielding the final mRNA for the heavy-chain gene.

antibody genes. While many questions remain to be answered, transposition and recombination of genes or parts of genes is a very active and precise process by which immunocytes can produce antibodies against almost any macromolecule.

Genes from Different Organisms Can Be Artificially Recombined

We have examined several different kinds of natural genetic recombination processes that are normal biological events in different kinds of cells. Genes or sets of genes can also be recombined in the test tube to produce new combinations that do *not* occur biologically. For example, it is possible to isolate genes specifying two different proteins from two different species of organism and splice them together to form a new combination. Such artificial recombinant DNAs are extremely useful tools in genetic research. They can also be turned to practical use, as we shall see. The development of methods to isolate and splice genes into new combinations was a major biochemical advance that initiated a new era in genetic research.

The discovery of restriction endonucleases gave the first clue that showed how genes can be recombined in the laboratory. Suppose we wish to splice together two duplex DNAs, each arising from a different species (Figure 30-19). Each DNA

Figure 30-19
Use of a specific restriction endonuclease producing staggered ends to cleave two different DNAs (DNA_1 and DNA_2) and prepare the resulting fragments for recombination. Since many restriction endonucleases cleave only specific restriction sites having twofold rotational symmetry, the cohesive ends formed will base-pair with the ends of any other DNA cleaved by the same endonuclease. In the annealing step DNA_1 and DNA_2 can be reformed again from their respective pieces, but new recombinant DNAs will be formed when pieces of DNA_1 and DNA_2 anneal together, as shown. Covalent splicing is carried out by DNA ligase.

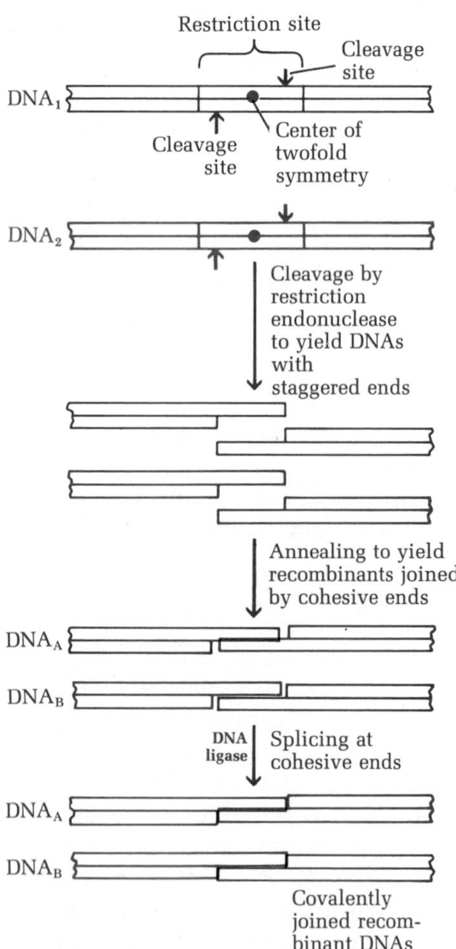

is separately treated with the same restriction endonuclease, one that yields a staggered two-strand cut (page 824). Assume that each of the two DNAs has only one restriction site. The staggered ends of the two DNAs will therefore be complementary in sequence. If the two cut DNAs are now mixed, heated, and slowly cooled, their cohesive ends will base-pair to produce a new kind of noncovalently joined recombinant DNA, as shown in Figure 30-19. If these DNAs, held together by their cohesive ends, are now treated with DNA ligase and the required energy source, a new covalently joined recombinant DNA results.

Another key enzyme that is widely useful in splicing DNAs together is *terminal transferase,* which can add many successive deoxyribonucleotide residues to the 3′ end of DNA strands. This enzyme is nonspecific and can use dATP, dTTP, dGTP, or dCTP as precursors. Since it requires no template, terminal transferase can generate 3′ tails of a single type of residue. Poly G tails can thus be built on the 3′ ends of the two strands of a duplex DNA and poly C tails on the 3′ ends of the other DNA. Since these tails are complementary, they will enable the two DNAs to be joined by base pairing of their cohesive ends (Figure 30-20), which are then spliced by the action of DNA ligase.

By using these and other enzymes many DNAs of different kinds and species have been spliced together. In one of the earliest successes the gene for a ribosomal RNA obtained from the clawed toad *Xenopus laevis* was spliced into an *E. coli* plasmid. In another early experiment the DNA of the monkey virus SV40 (page 828) was spliced to the DNA of phage λ, thus joining the chromosome of an animal virus with that of a bacterial virus. Since these pioneering experiments were carried out, hundreds of different kinds of artifical recombinant DNAs have been made in the laboratory.

Plasmids and Phage Lambda Are Vectors for Introducing Foreign Genes into Bacteria

The next step in developing recombinant DNA technology was to find the means of introducing foreign genes into a host cell. Plasmids and phage λ DNA have become the most widely used carriers or *vectors* to introduce foreign genes into the *E. coli* genome. Plasmids (page 814) are small circular duplex DNAs found in the cytoplasm of most bacterial species. Each plasmid may contain anywhere from 2000 to 100,000 bases. The smaller ones may occur in 20 or more copies in a single cell, the larger ones in only 1 or 2 copies. Each plasmid contains several or sometimes many genes, which are replicated, transcribed, and translated independently of, but simultaneously with, the chromosomal genes. They are readily isolated and separated from bacterial chromosomes, from which they differ in size, base composition, and density. Plasmids have two remarkable properties useful in genetic manipulation. They can pass from one

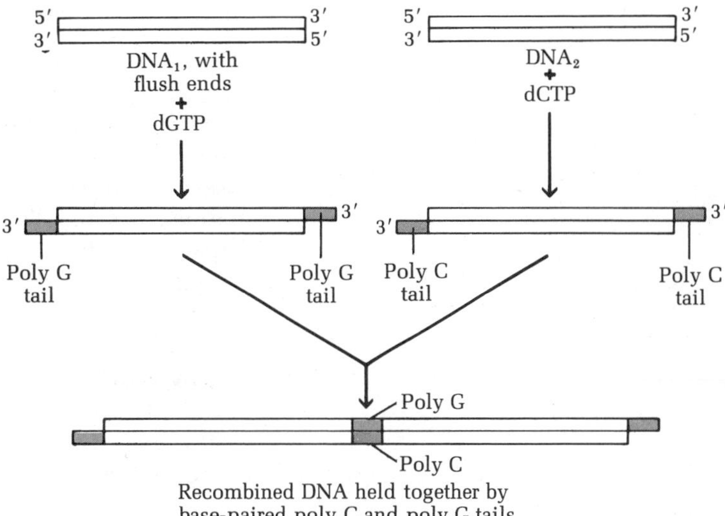

Figure 30-20
Use of terminal transferase to provide complementary cohesive ends.

cell to another and, indeed, from one species of bacteria to another. For example, *Salmonella typhimurium* cells can acquire permanent resistance to certain antibiotics, such as penicillin, when they are mixed with a strain of *E. coli* cells that is resistant to penicillin. The gene for penicillin resistance, called an R factor, is present in a plasmid of *E. coli,* which can be transmitted from *E. coli* to *S. typhimurium* cells. Secondly, foreign genes can be spliced into plasmids quite easily and may then be carried as "passengers" into *E. coli* cells and become part of the host-cell genome.

The DNA of λ phage also can be used to carry a foreign gene into *E. coli.* When the recombinant λ DNA, with its "passenger" gene, is mixed with the coat protein of λ virus, infectious phage particles will be assembled if the size of the recombinant DNA is not greatly different from that of natural λ DNA. This is the preferred way of introducing a foreign gene into *E. coli,* since λ phage is highly efficient in injecting its DNA into the host cell, whereas plasmids only occasionally succeed in entering an intact *E. coli* cell. Since λ is a temperate phage (page 923), its DNA can become incorporated into the *E. coli* chromosome, together with the foreign gene it carries. In this case the λ DNA and the foreign gene will be replicated at each cell-division cycle.

Let us see in more detail how genes are isolated, introduced into host cells, cloned, and caused to be translated into various gene products. The word clone is derived from the Greek *klon,* a slip or cutting used to propagate a plant. It is used in two senses. *Cell cloning* is the formation of a group of genetically identical cells, all arising from a single cell, as in the case of a line of immunocytes programmed to produce a single type of antibody. *Molecular* or *gene cloning* is the formation of many identical gene copies replicated from a single gene introduced into a host cell.

Isolation of Genes and the Preparation of cDNAs

Although a number of genes have been isolated directly from fragments of viral and bacterial chromosomes, the isolation of specific genes from fragmented eukaryotic chromosomes is still rather difficult and time-consuming. There are two general approaches for obtaining a specific gene to be recombined and cloned. In the "shotgun" approach the entire cell DNA is treated with a restriction endonuclease, one that makes staggered ends. The resulting pieces of DNA are then spliced into *E. coli* plasmids "opened" with the same restriction endonuclease. The product is an exceedingly complex mixture of perhaps thousands of different recombinant plasmids, of which only one may contain the desired gene. Screening procedures have been developed to select the plasmid carrying this gene. One such screening procedure is described on page 936.

The second procedure for obtaining a gene is to construct its cDNA (complementary DNA) from its mRNA. Although, as we recall, most cells contain mixtures of many different mRNAs, which are difficult to separate, sometimes pure mRNAs for one specific protein can be isolated from cell types that make predominantly one kind of protein. For example, the mRNAs for the α- and β-polypeptide chains of hemoglobin can be isolated from reticulocytes, immature red blood cells, in which 90 percent of the protein being synthesized is hemoglobin. Similarly, the mRNA for proinsulin can be isolated from B cells of the islet tissue of human pancreas (page 736). A more general way of obtaining the mRNA for a specific gene is to lyse the cells, collect the polyribosomes by centrifugation, and treat them with an antibody specific for the protein whose gene is being sought. Among the many polyribosomes in the population there will be some synthesizing this particular protein from its mRNA template. In such polyribosomes this protein will be present in various stages of completion (page 890). The specific antibody will combine only with the complete or nearly complete protein, still attached to the polyribosomes making it from its mRNA template. The specific antibody-polyribosome complex will be precipitated and thus separated from the mixture of polyribosomes. The specific mRNA from which the protein was made can then be extracted from the antibody-precipitated polyribosomes and isolated in nearly pure form, free of other mRNAs, by chromatographic methods.

The specific mRNA for the protein whose gene is being sought is now used as the template for the enzymatic synthesis of its complementary DNA (cDNA) with reverse transcriptase (page 863) as catalyst. It is first necessary, however, to construct the primer DNA required by reverse transcriptase. We recall (page 859) that mRNAs contain a poly A tail at the 3′ end. To the mRNA is added poly T, which base-pairs with the poly A tail of the mRNA (Figure 30-21). This serves as the primer for reverse transcriptase, which now transcribes the mRNA to make a complementary strand of cDNA from a mixture of

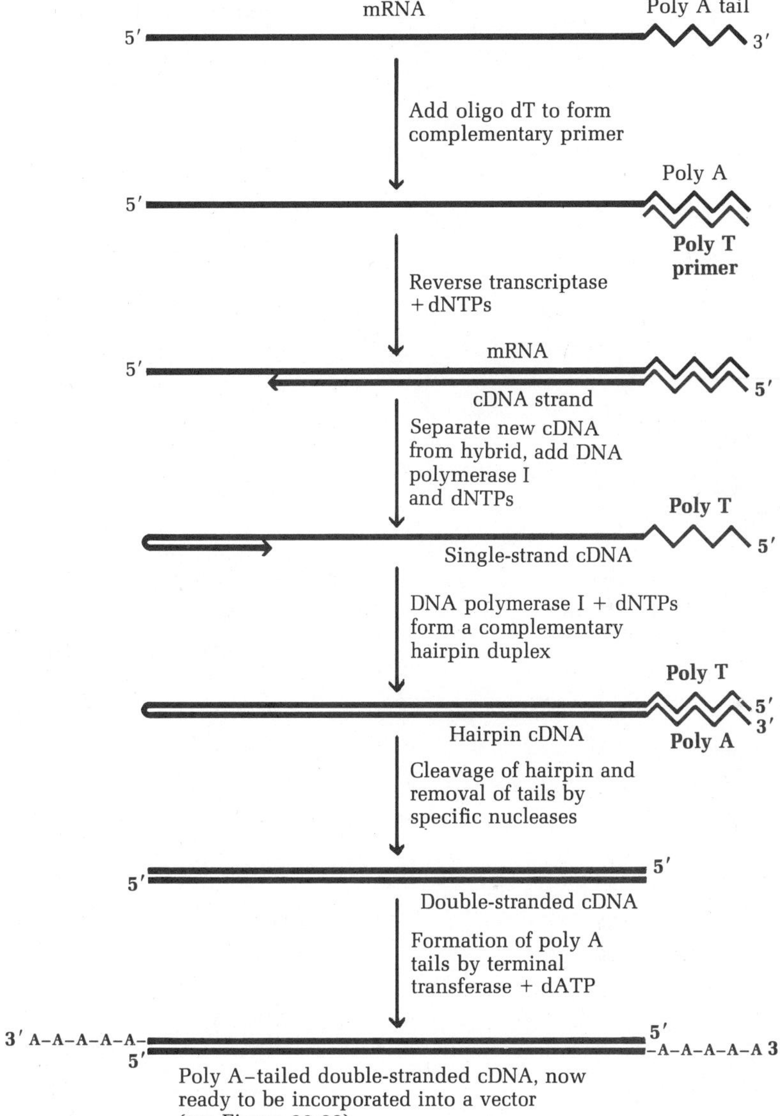

Figure 30-21
Construction of a double-stranded cDNA *from an* mRNA. *Once an* mRNA *for a given gene has been isolated, it can be transcribed to form complementary* DNA *by reverse transcriptase. This enzyme requires a* DNA *primer, furnished by poly* T, *complementary to the 3′ poly* A *tail of the* mRNA. *The new* cDNA *strand, which is usually isotopically labeled from radioactive* dNTP *precursors, is separated and used as both primer and template by* DNA *polymerase* I *(page 845) to make a hairpin duplex* cDNA. *The hairpin is cleaved and trimmed, and the resulting* cDNA *duplex with flush ends is tailed with poly* A *(Figure 30-20). The tailed* cDNA *may now be spliced into the vector (Figure 30-22). Radioactive* cDNA *is useful in the recovery of the corresponding natural gene from fragmented chromosomes.*

dATP, dTTP, dGTP, and dCTP. The mRNA is then removed from the mRNA–cDNA hybrid and the single-strand cDNA is now replicated by DNA polymerase I to yield a "hairpin" double-stranded DNA. The hairpin is cleaved (Figure 30-21), leaving as product a synthetic double-stranded cDNA, specific for the protein whose gene is being sought. In order to simplify later procedures the cDNA is usually made radioactive by using ^{32}P-labeled deoxyribonucleoside 5′-triphosphates as precursors in the reverse transcriptase reaction.

We now have a synthetic cDNA that can specify the amino acid sequence of a given protein. Note, however, that if this synthetic cDNA has been obtained from a eukaryotic mRNA, it is not identical with the natural gene for this protein because it does not contain the introns or intervening sequences, nor the "start" and "stop" signals, characteristic of the genes for most proteins in eukaryotes.

Construction of the Gene-Bearing Vector

The cDNA prepared as above is now inserted into a plasmid or viral vector. If the cDNA is to be incorporated into a plasmid, the cDNA must be given appropriate "tails" or cohesive ends (Figure 30-21). This is best done by adding to the opposite 3′ ends of the two strands of the duplex cDNA a series of repeating deoxyribonucleotide residues of a single type, for example, A residues. They are added by the action of terminal transferase. With dATP as precursor, a poly A tail of perhaps 50 or 100 residues is added to the two 3′ ends of the duplex cDNA. The newly tailed cDNA is now ready to be incorporated into the plasmid vector (Figure 30-21).

The plasmid is now opened at a single point to yield its linear form by the action of a restriction endonuclease that produces flush ends (page 824). The 3′ ends of the linear plasmid are now given tails complementary to the tails built on the cDNA to be inserted. Since poly A tails were added to the cDNA, the plasmid would be provided with 3′ tails of poly T. The tailed linear plasmid and the tailed cDNA are now simply mixed and allowed to undergo base pairing. Among the products will be an enlarged circular plasmid now containing the new gene, bound into the plasmid only through base pairing of their sticky ends. The ends can now be covalently joined by the action of DNA ligase (page 848) to make a covalently closed circular plasmid containing the new gene (Figure 30-22).

Insertion of the "Loaded" Plasmids into the *E. coli* Chromosome

The recombinant plasmids are now mixed with the *E. coli* cells to be transformed. Only a few plasmids will succeed in entering and becoming incorporated into the chromosome, but the efficiency of plasmid entry can be increased by addition of Ca^{2+}. To detect those *E. coli* cells which contain the recombined genes, the simplest way is to grow the cells on a plate of nutrient medium, extract the DNA of each cell colony, and determine which colony contains DNA that hybridizes with the radioactive cDNA originally prepared from the isolated mRNA. The colony or clone of *E. coli* cells containing the recombined plasmid is now grown for many generations on a large scale in a new culture medium, increasing the number of recombinant plasmids by a factor of 10^{12} or more.

Such a recombinant DNA, carrying unrelated genes from two different species, is called a *DNA chimera* or *chimeric DNA*. In Homer's *Iliad*, Chimera was a mythological creature with the head of a lion, the body of a goat, and the tail of a snake.

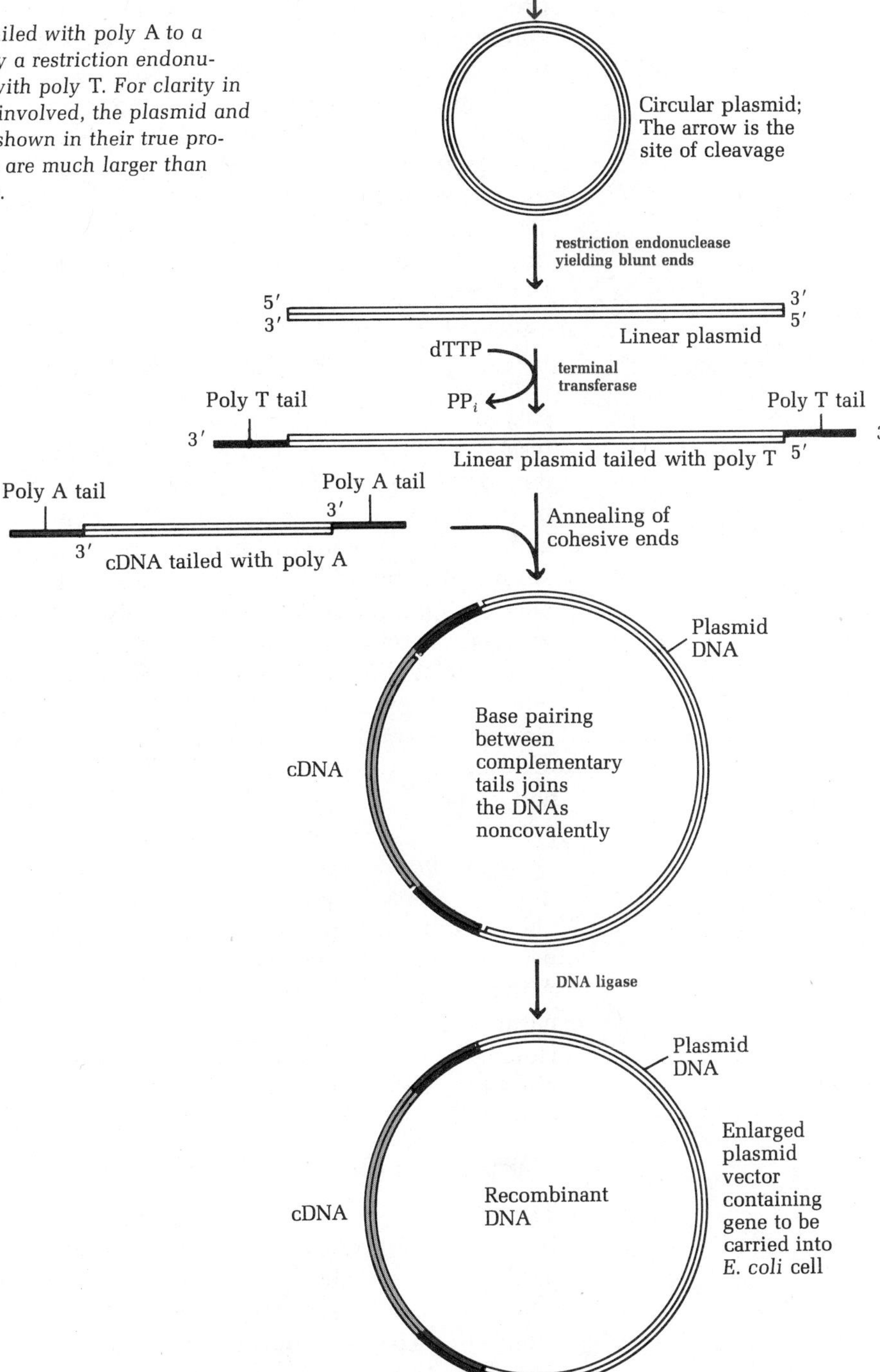

Figure 30-22
Joining of cDNA tailed with poly A to a plasmid cleaved by a restriction endonuclease and tailed with poly T. For clarity in showing the steps involved, the plasmid and the cDNA are not shown in their true proportions. Plasmids are much larger than single-gene cDNAs.

Cloned cDNAs Can Be Used to Recover the Corresponding Natural Gene

The cloned cDNA recovered as a recombinant plasmid, in the manner just described, is not identical with the native gene for that protein since it does not contain the introns that are normally present in many, if not most, eukaryotic genes coding for proteins (page 827). Recall that when mRNAs are formed in a colinear fashion from such eukaryotic genes during normal biological transcription, the mRNA segments complementary to the introns in the DNA are deleted (page 859) before translation of the mRNA by ribosomes. The cloned cDNA can now be used to "fish out" the *real* native gene for this protein from the thousands of genes present in the genome of the parent cell. The entire DNA of the organism in question is extracted and digested by a restriction endonuclease, and the many resulting DNA fragments are inserted into an appropriate vector, usually λ phage DNA, by methods described above. The recombinant λ DNA is then packaged into viral particles by adding coat protein, and the resulting altered λ phage particles carrying the passenger gene are allowed to infect normal *E. coli* cells. From lysates of these cells many different viral DNAs can be obtained, which will contain most if not all of the genes present in the donor organism. From this "gene library" the natural gene for the protein in question can be isolated, simply and specifically, by adding the ^{32}P-labeled cDNA for the protein, heating the mixture, and allowing it to cool. Hybrid duplexes will form between the cDNA strands and the strands of the natural gene. Although the natural gene will have many intron sequences, its exons will recognize and hybridize with the cDNA strands. The radioactive hybrids so formed can then be isolated and the natural gene for the protein in question recovered and cloned via a plasmid of λ phage vector, as already described.

Since the base sequence of DNAs can now be carried out quickly and accurately (page 827), the base sequence of a cDNA obtained by cloning will give the amino acid sequence of the protein for which it codes, as well as the sequence of its mRNA. Although the cDNA cannot give the full base sequence of the natural gene for a eukaryotic protein, the natural gene can be isolated and cloned with the help of the labeled cDNA originally obtained from its mRNA.

Expression of Cloned Genes Is Enhanced by a Promoter

The transcription of many genes and the subsequent translation of their mRNAs into specific protein products requires the action of a promoter and operator, in order to signal the point where RNA polymerase binds and begins to transcribe (page 906); otherwise it cannot be expected that a foreign gene inserted into a recipient cell will necessarily be transcribed and translated. To ensure that an incorporated gene is transcribed it is often necessary to insert the gene in question into a portion of a plasmid or λ DNA "downstream" of its promoter-operator region. For example, the gene can be inserted into a segment of *E. coli* DNA containing the *lac* operon, but at a point downstream of the *lac* promoter-operator region (page 906). When *E. coli* cells containing such a recombinant DNA are placed in a lactose medium containing no glucose, the inserted gene will be transcribed and translated, along with the *lac* DNA lying between the operator and the inserted gene.

This approach was taken in the cloning of the gene for *somatostatin*, a 14-residue hypothalamic peptide hormone that regulates secretion of insulin, glucagon, and growth hormone. The gene for somatostatin was in this case synthesized chemically and joined to the end of the β-galactosidase gene. The two connected genes were recombined into an *E. coli* plasmid and the resulting passenger-bearing plasmid introduced into *E. coli* cells. As a consequence the cells made large amounts of a *hybrid protein*, in which β-galactosidase and somatostatin were covalently joined. Such a hybrid pair of proteins, in this case, one native to *E. coli* and the other to a vertebrate, is called a *chimeric protein*. The β-galactosidase–somatostatin chimera could then be cleaved at the peptide bond joining the two proteins to yield free somatostatin having biological activity.

Many Genes Have Been Cloned in Different Host Cells

With variations of the approaches described above many genes have been recombined and cloned in *E. coli* cells. In many cases the new genes have been expressed in the form of specific proteins or mRNAs. In one type of cloning, extra copies of an *E. coli* gene have been recombined into normal *E. coli* cells. Such cells accordingly produce many extra copies of the product of that gene. In another type of recombination, genes obtained from various eukaryotic species have been successfully cloned in *E. coli* cells. Among these are the genes for insulin, pituitary growth hormone (somatotropin), α- and β-globins, ribosomal RNAs, somatostatin, and ovalbumin. The reverse type of recombination has also been carried out: bacterial genes have been inserted into the genome of some eukaryotic cells. For example, the *E. coli* gene for β-galactosidase (page 902) has been incorporated into mouse cells grown in culture. An especially interesting case is the incorporation of the *E. coli* gene coding for the enzyme *hypoxanthine-guanine phosphoribosyltransferase* (page 635) into cultured human connective tissue cells taken from a patient with Lesch-Nyhan disease (page 635), in which this enzyme is genetically defective. The inserted *E. coli* gene was expressed by the formation of the bacterial enzyme. Thus a genetic defect in human cells could be corrected in vitro by incorporation of the corresponding bacterial gene.

Eukaryotic genes have also been recombined and expressed in cells of other eukaryotic species. For example, the rabbit gene for the α chain of hemoglobin has been incorporated and expressed in cultured mouse cells. Recombination of a foreign gene into eukaryotic cells may not always lead to its transcription and translation into active proteins. The regulation of gene expression in eukaryotes is not yet well understood (page 902); for this reason a complete rationale for expressing recombined genes in eukaryotes is still under development at the time of writing.

Recombinant DNA and Gene Cloning Open New Lines of Genetic Research

Although the possible practical uses of recombinant DNA receive much attention in the popular press, the capacity to clone genes has opened new approaches to the solution of many fundamental problems in molecular genetics that would otherwise be most difficult to solve. The way is now open to isolate and prepare in quantity virtually any gene for study of its nucleotide sequence and the sequences of the mRNAs or protein which it codes. More direct mapping of genes in chromosomes is also possible. An especially important objective is the isolation, cloning, and identification of various signaling and regulatory segments of DNA, such as *i* genes, promoters, and operators. The base sequences and function of the introns in eukaryotic structural genes can also be studied. Gene amplification, the synthesis of many copies of a single gene, which occurs early in embryonic development, is another process whose nature is being examined by cloning methods, as is the identification of the various genes of vertebrates that are spliced together to form the DNA coding for the light and heavy chains of antibodies. Also rendered more approachable is the identification of the regulatory mechanisms that bring about repression and derepression of specific genes in eukaryotic chromosomes. In all these problems recombinant DNA techniques and cloning methods have and will play a most important role.

Recombinant DNA Research May Have Many Practical Applications

Cloning of recombinant genes and their expression in the form of protein products by *E. coli* or yeast cells, which can be grown in enormous amounts, may make possible the commercial production of many practically useful proteins that are otherwise very difficult to obtain in large amounts. These prospects have given rise to the field of *genetic engineering*.

A simple example is the bacterial enzyme DNA ligase, which is so useful in research on gene biochemistry that it has become an article of commerce. It is produced in large quantities by *E. coli* cells into which many extra copies of the DNA ligase gene have been inserted. Such cells produce the enzyme in amounts several hundredfold larger than normal *E. coli* cells. Many other enzymes that are useful in industry can be made in very high yields by cloning extra gene copies in *E. coli* or yeast cells. Another project is to insert the genes capable of oxidizing petroleum hydrocarbons into common harmless bacteria, to be utilized in cleaning up oil spills.

The genes for some proteins required in medicine have been cloned. Insulin, needed for the treatment of diabetes, is now obtained from the pancreas of slaughterhouse animals. Although enough insulin is available from this source to cover present needs, the increasing incidence of diabetes mellitus, which affects over 5 percent of the United States population, may some day outstrip available supplies. Moreover, the insulin of slaughterhouse animals is not identical in amino acid sequence with human insulin and is either ineffective or not tolerated by some people. The synthesis of human insulin by *E. coli* cloned with the human insulin gene has been achieved. Synthetic human insulin produced in this way is already being used in treating diabetes. Similarly, the production of human pituitary growth hormone (somatotropin), never before available for medical use, may be possible using recombinant DNA approaches. Growth hormone from slaughterhouse animals is ineffective in treating dwarfism in human beings, presumably because it differs in amino acid sequence from the human hormone.

Various proteins useful in agriculture may also be produced in large amounts in host cells carrying appropriate recombined genes. For example, a new and highly effective vaccine has been produced by genetic engineering methods against hoof-and-mouth disease of cattle, sheep, and swine. This disease, which is viral, is otherwise incurable and renders stock animals useless for food. Because this disease is endemic in the Third World, it has an immense economic and human cost. The protein antigen of this virus has been cloned and is being used to prepare vaccine against the disease. Another particularly important objective is the incorporation of the genes for the enzymes and other proteins in nitrogen fixation into the genomes of crop plants that do not normally fix nitrogen. While practical success in this effort still seems far distant, it could have a profound effect upon world agriculture.

Interferon Genes Have Been Cloned

Interferons are proteins that are secreted by certain cells of vertebrates when they are infected by a virus. They bind to the plasma membrane of uninfected cells and render them immune to infection by the same or some other virus. The existence of such substances was first inferred from the medical observation that patients already suffering from one viral disease do not simultaneously contract a second viral disease, suggesting that the first viral infection interferes with the development of the second. When interferon was first discovered in the 1950s, it was hoped that it might be useful in treatment of viral diseases, which for the most part are not susceptible to treatment by drugs or antibiotics. Such diseases include the common cold, influenza, poliomyelitis, chicken pox, herpes, viral hepatitis, and many others. Most of all, it appeared possible that interferon might be used to treat some kinds of human cancer, which have long been suspected to be caused by viruses. However, for many years interferon remained mysterious, since it is formed in only minute amounts by infected cells, leaving little hope that enough could be isolated to determine its structure and to test its biological activity.

With the development of powerful methods of isolating and identifying trace amounts of proteins and their genes, interferons have since been found to be glycoproteins having about 160 amino acid residues. Each vertebrate species can produce at least three different kinds of interferon during viral infections, one by fibroblast cells of connective tissues, another by white blood cells, and a third type by T lymphocytes (page 138). On binding to the membrane of healthy cells interferons elicit the formation of specific enzymes that can destroy viral mRNAs and inactivate an initiation factor for ribosomal protein synthesis, thus preventing the expression of viral genes by the host cell.

In 1980, the gene for human white blood cell interferon was isolated and recombined into a plasmid of *E. coli* by research groups in Switzerland and the United States. Efforts are under way to produce human interferons on a large scale from interferon genes cloned in *E. coli* cells. While it is still uncertain whether interferons will be effective in the treatment of cancer, at the least these new advances may lead to a new era of understanding of human viral diseases.

Summary

Both prokaryotic and eukaryotic cells contain enzyme systems capable of correcting replication errors and various forms of damage to DNA caused by hydrolysis or external mutagenic agents, such as ultraviolet and ionizing radiations, as well as deaminating and alkylating agents. DNA damage that is uncorrected by such mechanisms results in heritable mutations, which may be lethal, leaky, silent, or even beneficial, depending upon the site and nature of the damage. One type of mutation causes insertion of an incorrect base in DNA, leading to a change in a single codon and in most cases an amino acid replacement in the protein product. In the other major type, called frame-shift mutations, loss or gain of one or more nucleotides occurs. Such mutations result in a shift of the reading frame of codons and thus yield protein products in which the amino acid sequence is completely garbled beyond the point of gain or loss of a nucleotide. Mutations are usually random, but their rates can be greatly accelerated by mutagenic agents, which include deaminating reagents, alkylating agents, and other substances capable of altering the structure of a base. Nearly all carcinogenic compounds are mutagens and can be screened with simple bacterial growth tests.

Genes and sets of genes normally undergo recombination during such biological processes as bacterial transformation, viral transduction, bacterial conjugation, and exchange of genes during the sexual conjugation of eukaryotic germ cells. Genes and sets of genes may also undergo transposition from one location to another in the same or a different chromosome. For example, antibody proteins, which can be formed by plasma cells or immunocytes of vertebrates against millions of different foreign macromolecules, are coded for by DNA sequences generated by the transposition and reshuffling of a relatively small number of different kinds of genes. Such sequences on transcription yield the mRNAs for the millions of different antibody proteins, each capable of binding specifically to one and only one type of antigen molecule.

New combinations of genes can also be constructed artificially in the laboratory with the help of such enzymes as restriction endonucleases, DNA ligase, and terminal transferase. To insert a foreign gene into the genome of *E. coli* cells the gene is first spliced into a carrier or vector, either an *E. coli* plasmid or the DNA of phage λ. The recombinant DNA so formed can then enter the *E. coli* cell, become covalently incorporated into its chromosome, and undergo replication. If the new gene contained in the recombinant DNA also has appropriate signaling sequences to indicate initiation and termination points, it will also be transcribed and translated into its protein product. Many genes from animal cells have been introduced into bacteria; bacterial genes have also been introduced into eukaryotic cells. Many medically useful human proteins, such as insulin, growth hormone, and interferons, can be made by bacteria into which the appropriate genes have been introduced.

References

Repair, Mutation, and Carcinogens

AMES, B. N.: "Identifying Environmental Chemicals Causing Mutations and Cancer," *Science,* **204:**587–593 (1979).

CAIRNS, J.: "The Cancer Problem," *Sci. Am.,* **233:**64–78, November (1975). A thoughtful analysis by a molecular biologist.

DEVORET, R.: "Bacterial Tests for Potential Carcinogens," *Sci. Am.,* **241:**40–49, August (1979).

DRAKE, J. W.: *The Molecular Basis of Mutation,* Holden-Day, San Francisco, 1970.

HANAWALT, P., et al.: "DNA Repair in Bacteria and Mammalian Cells," *Annu. Rev. Biochem.,* **48:**783–836 (1979). A detailed review of current knowledge.

Natural Recombination Events

AYALA, F. J., and J. A. KIGER, JR.: *Modern Genetics,* Benjamin/Cummings, Menlo Park, Calif., 1980. Bacterial transformation, viral transduction, and bacterial conjugation are described.

CAMPBELL, A. M.: "How Viruses Insert Their DNA into the DNA of the Host Cell," *Sci. Am.,* **235:**102–113, December (1976).

COHEN, S. N., and J. A. SHAPIRO: "Transposable Genetic Elements," *Sci. Am.,* **242:**40–49, February (1980).

Antibodies

HOOD, L. E., I. L. WEISSMAN, and W. B. WOOD: *Immunology,* Benjamin/Cummings, Menlo Park, Calif., 1978.

MILSTEIN, C.: "Monoclonal Antibodies," *Sci. Am.,* **243:**66–74, October (1980). A technique for producing antibodies of a single kind in large amounts in a cell culture rather than in an animal.

DNA Cloning and Genetic Engineering

ABELSON, J., and E. BUTZ (eds.): "Recombinant DNA," *Science,* vol. 209 (1980). This entire issue is devoted to a series of important research papers and a valuable glossary of terms.

ANDERSON, W. B., and E. G. DIACUMAKOS: "Genetic Engineering in Mammalian Cells," *Sci. Am.,* **245**:106–121, July (1981).

DEROBERTIS, E. M., and J. B. GURDON: "Gene Transplantation and the Analysis of Development," *Sci. Am.,* **241**:74–82, December (1979).

GROBSTEIN, CLIFFORD: "The Recombinant-DNA Debate," *Sci. Am.,* **237**:22–33, July (1977).

Industrial Microbiology and the Advent of Genetic Engineering, Freeman, San Francisco, 1981. A collection of articles that appear in the September 1981 issue of *Scientific American,* which is devoted to genetic engineering and the use of microorganisms in industry, with special attention to genetic alteration and cloning.

SETLOW, J. K., and A. HOLLAENDER (eds.): *Genetic Engineering: Principles and Methods,* Plenum, New York, 1979. A collection of papers by experts.

WATSON, JAMES D., and JOHN TOOZE: *The DNA Story: A Documentary History of Gene Cloning,* Freeman, San Francisco, 1981.

Interferons

DERYNCK, R., J. CONTENT, E. DECLERCQ, G. VOLCKAERT, J. TAVERNIER, R. DEVOS, and W. FIERS: "Isolation and Structure of a Human Fibroblast Interferon Gene," *Nature,* **285**:542–547 (1980).

Problems

1. *DNA Repair Mechanisms*
   (a) Why would you expect that UV-endonuclease cleaves at the 5′ rather than the 3′ side of a thymine dimer?
   (b) Why is it unlikely that enzymes will be found that repair damaged mRNA as efficiently as damaged DNA is repaired?
   (c) From what you know about DNA repair mechanisms, what is the probable consequence of damage to a gene by x-rays, which often cause double-strand breaks?

2. *Effects of a Single-Base Replacement* Enumerate the possible effects of a mutation caused by a single-base replacement on a segment of eukaryotic DNA coding for an enzyme.

3. *Recombination of Restriction Endonuclease Fragments* Suppose you have two double-stranded DNA molecules, A and B. Strand 1 of DNA_A has the sequence

   (5′)ATATGAATTCAATT(3′)

   and strand 1 of DNA_B has the sequence

   (5′)CGCGGAATTCCCGG(3′)

   Each DNA is cleaved by the restriction endonuclease *Eco* RI, the fragments of both are mixed, allowed to recombine, and are then covalently spliced by DNA ligase.

   Write the sequences of all the recombined DNAs. Indicate which are the new recombinants.

4. *UAA as Signal* The nucleotide residues of a prokaryotic mRNA are numbered so that those coding for the AUG of the initiating methionine of the protein product (140 amino acid residues) are numbered 1, 2, and 3. What is the effect on the product of having the triplet (5′) UAA at nucleotide numbers 330–332, 334–336, and 338–340? What would your answer be in the case of a eukaryotic gene?

5. *Construction of a Restriction Map of a Virus* The DNA isolated from a virus infecting animal cells shows the usual base equivalences and is resistant to all known exonucleases.
   (a) What do these facts tell us about its three-dimensional structure?
   (b) Cleavage with restriction endonuclease *Eco* RI yields a single kind of duplex DNA of molecular weight 3.4×10^6. What does this outcome tell about its structure?
   (c) Cleavage of the original viral DNA with restriction endonuclease *Hpa* I from *Hemophilus parainfluenzae* produces three fragments that are separable by gel electrophoresis and have molecular weights 1.4×10^6, 0.7×10^6, and 1.3×10^6. When the original viral DNA is subjected to the *combined* action of *Eco* RI and *Hpa* I, four fragments are produced. The smallest has a molecular weight of 0.6×10^6; the largest, 1.3×10^6. The fragments are lettered by size, A being the largest. What are the molecular weights of the two other fragments produced by the combined action of *Eco* RI and *Hpa* I?
   (d) After *partial* or incomplete action of both enzymes, the electrophoretic band corresponding to the fragment of molecular weight 1.3×10^6 was isolated and subjected to *complete* cleavage by *Hpa* I. This treatment produced three bands, one that migrated at a rate corresponding to a molecular weight of 1.3×10^6, and bands corresponding to two smaller fragments. Construct a possible cleavage map of the fragments A to D of the viral DNA and indicate the cleavage sites.
   (e) The DNA from a mutant strain of the same animal virus yielded only three fragments upon complete cleavage by the two restriction enzymes. Their molecular weights were 2.1×10^6, 0.6×10^6, and 0.7×10^6. What is the simplest explanation of the mutation that results in this new cleavage pattern?

6. *Functional Domains in Proteins* There is evidence that proteins of eukaryotes are composed of functional domains, i.e., independent regions each specifying one aspect of the protein's function or properties. It has been postulated that there might be an evolutionary advantage in such domains, since the segments of DNA coding for each domain may be able to undergo reassortment, i.e., exchange and recombination to yield new genes that specify proteins with new combinations of functions and properties. What aspect of eukaryotic transcription might aid in the origin of such new genes and why?

7. *Cloning Human Interferon* Recombinant DNA technology can make possible greatly increased production of scarce proteins. Interferons—antiviral and potential anticancer agents—can be isolated from white blood cells, but the yield is only about 1 μg/L in human blood, a precious commodity. Genes for interferons can be cloned and expressed in bacteria, which can be grown easily in hundreds or thousands of liters.

(a) Calculate how much interferon could be generated by cloning an interferon gene in bacteria, assuming that there are 10^9 bacterial cells per mL in a fully grown culture and that each cell contains 10^{-1} picogram of protein, of which 5 percent is interferon.
(b) If the interferon has a molecular weight of 30,000, how many molecules of interferon are produced per cell? (*Hint:* Avogadro's number, the number of molecules in one mole of a compound, is 6.02×10^{23}.)
(c) What weight of cloned human interferon could theoretically be obtained from 100 L of culture medium assuming the conditions in part (b)?

8. *Recombining Circular DNAs* Suppose you have two circular duplex DNAs, one large and the other small, that are to be combined into a single circular duplex. You have available a restriction endonuclease that can cleave each DNA once to produce staggered ends.
(a) How would you proceed to combine the large and small DNAs into a single larger circle?
(b) Indicate the structure of the major by-products of recombination.

APPENDIXES

APPENDIX A Common Abbreviations in Biochemical Research Literature

A	Adenine or adenosine
ACP	Acyl carrier protein
ACTH	Adrenocorticotropic hormone
Acyl-CoA (acyl-S-CoA)	Acyl derivatives of coenzyme A
AMP, ADP, ATP	Adenosine 5′-mono-, -di-, -triphosphate
cAMP	3′,5′-Cyclic AMP
dAMP, dGMP, dADP, etc.	Deoxyadenosine 5′-monophosphate, deoxyguanosine 5′-monophosphate, deoxyadenosine 5′-diphosphate, etc.
Ala	Alanine
Arg	Arginine
Asn	Asparagine
Asp	Aspartic acid
ATPase	Adenosine triphosphatase
C	Cytosine or cytidine
CAP	Catabolite activator protein
cDNA	Complementary DNA
CMP, CDP, CTP	Cytidine 5′-mono-, -di-, -triphosphate
CoA (CoA-SH)	Coenzyme A
CoQ	Coenzyme Q (ubiquinone)
Cys	Cysteine
dATP, dGTP, etc.	2′-Deoxyadenosine 5′-triphosphate, 2′-deoxyguanosine 5′-triphosphate, etc.
DFP (DIFP)	Diisopropylphosphofluoridate
DNA	Deoxyribonucleic acid
DNase	Deoxyribonuclease
DNP	2,4-Dinitrophenol
DOPA	Dihydroxyphenylalanine
EC (followed by numbers)	Enzyme Commission (numbers indicate formal classification of an enzyme)
EDTA	Ethylenediaminetetraacetic acid
ETP	Electron-transfer particle (from mitochondrial membrane)
FA	Fatty acid
FAD, $FADH_2$	Flavin adenine dinucleotide and its reduced form
FCCP	Carbonylcyanide *p*-trifluoromethoxyphenylhydrazone
Fd	Ferredoxin
FDNB (DNFB)	1-Fluoro-2,4-dinitrobenzene
FDP	Fructose 1,6-diphosphate
FFA	Free fatty acid
FH_2, FH_4 (THFA)	Dihydro- and tetrahydrofolic acid
fMet	*N*-Formylmethionine
FMN, $FMNH_2$	Flavin mononucleotide and its reduced form
FP	Flavoprotein
$\Delta G^{\circ\prime}$	Standard-free-energy change
ΔG	Free-energy change
ΔG_p	Free-energy change of ATP hydrolysis under nonstandard conditions
G	Guanine or guanosine
Gal	D-Galactose
GalNAc	*N*-Acetyl-D-galactosamine
GDH	Glutamate dehydrogenase
GH	Growth hormone
Glc	D-Glucose
GlcNAc	*N*-acetyl-D-glucosamine
Gln	Glutamine
Glu	Glutamic acid
Gly	Glycine
cGMP	3′,5′-Cyclic GMP
GMP, GDP, GTP	Guanosine 5′-mono-, -di-, -triphosphate
G3P	Glyceraldehyde 3-phosphate
G6P	Glucose 6-phosphate
GSH, GSSG	Glutathione and its oxidized form
Hb, HbO_2, HbCO, MetHb	Hemoglobin, oxyhemoglobin, carbon monoxide hemoglobin, methemoglobin
His	Histidine
hnRNA	Heterogeneous nuclear RNA
Hyp	Hydroxyproline
I	Inosine
Ile	Isoleucine
IMP, IDP, ITP	Inosine 5′-mono-, -di-, -triphosphate
αKG	α-Ketoglutarate
LDH	Lactate dehydrogenase
Leu	Leucine
LH	Luteinizing hormone
Lys	Lysine

Mb, MbO_2	Myoglobin, oxymyoglobin
MDH	Malate dehydrogenase
MtDNA	Mitochondrial DNA
Met	Methionine
MSH	Melanocyte-stimulating hormone
NAD^+, NADH	Nicotinamide adenine dinucleotide and its reduced form
$NADP^+$, NADPH	Nicotinamide adenine dinucleotide phosphate and its reduced form
NMN^+, NMNH	Nicotinamide mononucleotide and its reduced form
OAA	Oxaloacetate
P_i	Inorganic orthophosphate
PAB or PABA	p-Aminobenzoic acid
PEP	Phosphoenolpyruvate
3PG	3-Phosphoglycerate
PGA	Pteroylglutamic acid (folic acid)
PGP	3-Phosphoglyceroyl phosphate
Phe	Phenylalanine
PP_i	Inorganic pyrophosphate
PQ	Plastoquinone
Pro	Proline
PRPP	5-Phosphoribosyl 1-pyrophosphate
Q	Coenzyme Q (ubiquinone)
Rib	D-Ribose
RNA	Ribonucleic acid
hnRNA	Heterogeneous nuclear RNA
mRNA	Messenger RNA
rRNA	Ribosomal RNA
snRNA	Small nuclear RNA
tRNA	Transfer RNA
RNase	Ribonuclease
RQ	Respiratory quotient
Ser	Serine
T	Thymine or thymidine
TH	Thyrotropic hormone
Thr	Threonine
TMP, TDP, TTP	Thymidine 5′-mono-, -di-, -triphosphate
TMV	Tobacco mosaic virus
TPP	Thiamine pyrophosphate
Trp	Tryptophan
Tyr	Tyrosine
U	Uracil or uridine
UDP-gal	Uridine diphosphate galactose
UDP-glc	Uridine diphosphate glucose
UMP, UDP, UTP	Uridine 5′-mono-, -di-, -triphosphate
UV	Ultraviolet light
Val	Valine

APPENDIX B Unit Abbreviations, Prefixes, Constants, and Conversion Factors

A	Ampere	kJ	Kilojoule	mol	Mole
Å	Angstrom	K	Kelvin	MW	Molecular weight
atm	Atmosphere	K'	Equilibrium constant	mV	Millivolt
C	Coulomb	L	Liter	N	Normal concentration
cal	Calorie	ln	Logarithm to the base e	nm	Nanometer
Ci	Curie	log	Logarithm to the base 10	pH	$-$Log H^+ concentration
cm	Centimeter	μm	Micrometer	pK'	$-$Log K'
cpm	Counts per minute	μmol	Micromole	r	Revolution
d	Day	M	Molar concentration	s	Second
dm	Decimeter	m	Molal concentration	S	Svedberg unit
dpm	Disintegrations per minute	m	Meter	STP	Standard temperature and pressure
$\mathcal{F}$	Faraday	mg	Milligram	V	Volt
g	Gram	min	Minute		
h	Hour	mL	Milliliter		
J	Joule	mm	Millimeter		
kcal	Kilocalorie	mmHg	Millimeters of mercury pressure		

Some Prefixes Used in the International System of Units

Multiple	*Prefix*	*Abbreviation*
10^6	mega	M
10^3	kilo	k
10^{-1}	deci	d
10^{-2}	centi	c
10^{-3}	milli	m
10^{-6}	micro	μ
10^{-9}	nano	n
10^{-12}	pico	p

Some Conversion Factors

Length	1 cm = 10 mm = 10^4 μm = 10^7 nm = 10^8 Å = 0.394 in, 1 in = 2.54 cm
Mass	1 g = 10^{-3} kg = 10^3 mg = 10^6 μg = 3.53×10^{-2} oz, 1 oz = 28.3 g
Temperature	°C = $\frac{5}{9}$(°F − 32), K = °C + 273
Energy	1 J = 10^7 erg = 0.239 cal, 1 cal = 4.184 J
Pressure	1 torr = 1 mmHg = 1.32×10^{-3} atm, 1 atm = 760 torr

Some Physical Constants

	Symbol or abbreviation	*Value*
Avogadro's number	N	6.02×10^{23} mol
Curie	Ci	3.70×10^{10} disintegrations s^{-1}
Atomic mass unit (dalton)	amu	1.661×10^{-24} g
Faraday constant	$\mathcal{F}$	23,062 cal V^{-1} mol^{-1} 96,485 C mol^{-1}
Gas constant	R	1.987 cal mol^{-1} K^{-1} 8.314 J mol^{-1} K^{-1}
Planck's constant	h	1.584×10^{-34} cal·s 6.626×10^{-34} J·s

Mathematical Constants

$\pi = 3.1416$

$e = 2.718$

$\ln (\log_e) x = 2.303 \log_{10} x$

APPENDIX C **International Atomic Weights**

Element	Symbol	Atomic number	Atomic weight	Element	Symbol	Atomic number	Atomic weight
Aluminum	Al	13	26.98	Neodymium	Nd	60	144.27
Antimony	Sb	51	121.75	Neon	Ne	10	20.18
Argon	Ar	18	39.95	Nickel	Ni	28	58.71
Arsenic	As	33	74.92	Niobium	Nb	41	92.91
Barium	Ba	56	137.34	Nitrogen	N	7	14.01
Beryllium	Be	4	9.01	Osmium	Os	76	190.2
Bismuth	Bi	83	208.98	Oxygen	O	8	16.00
Boron	B	5	10.81	Palladium	Pd	46	106.4
Bromine	Br	35	79.90	Phosphorus	P	15	30.97
Cadmium	Cd	48	112.40	Platinum	Pt	78	195.09
Calcium	Ca	20	40.08	Potassium	K	19	39.09
Carbon	C	6	12.01	Praseodymium	Pr	59	140.91
Cerium	Ce	58	140.12	Protactinium	Pa	91	231.04
Cesium	Cs	55	132.91	Radium	Ra	88	226.03
Chlorine	Cl	17	35.45	Radon	Rn	86	222
Chromium	Cr	24	52.00	Rhenium	Re	75	186.2
Cobalt	Co	27	58.93	Rhodium	Rh	45	102.91
Copper	Cu	29	63.55	Rubidium	Rb	37	85.47
Dysprosium	Dy	66	162.50	Ruthenium	Ru	44	101.07
Erbium	Er	68	167.26	Samarium	Sm	62	150.43
Europium	Eu	63	151.96	Scandium	Sc	21	44.96
Fluorine	F	9	19.00	Selenium	Se	34	78.96
Gadolinium	Gd	64	157.25	Silicon	Si	14	28.09
Gallium	Ga	31	69.72	Silver	Ag	47	107.87
Germanium	Ge	32	72.59	Sodium	Na	11	23.00
Gold	Au	79	196.97	Strontium	Sr	38	87.62
Hafnium	Hf	72	178.49	Sulfur	S	16	32.06
Helium	He	2	4.00	Tantalum	Ta	73	180.95
Holmium	Ho	67	164.93	Tellurium	Te	52	127.60
Hydrogen	H	1	1.008	Terbium	Tb	65	158.93
Indium	In	49	114.82	Thallium	Tl	81	204.39
Iodine	I	53	126.91	Thorium	Th	90	232.04
Iridium	Ir	77	192.22	Thulium	Tm	69	169.93
Iron	Fe	26	55.84	Tin	Sn	50	118.69
Krypton	Kr	36	83.80	Titanium	Ti	22	47.90
Lanthanum	La	57	138.91	Tungsten	W	74	183.85
Lead	Pb	82	207.21	Uranium	U	92	238.03
Lithium	Li	3	6.94	Vanadium	V	23	50.94
Lutecium	Lu	71	174.97	Xenon	Xe	54	131.30
Magnesium	Mg	12	24.31	Ytterbium	Yb	70	173.04
Manganese	Mn	25	54.94	Yttrium	Y	39	88.91
Mercury	Hg	80	200.59	Zinc	Zn	30	65.38
Molybdenum	Mo	42	95.94	Zirconium	Zr	40	91.22

APPENDIX D Logarithms

N	0	1	2	3	4	5	6	7	8	9
10	0000	0043	0086	0128	0170	0212	0253	0294	0334	0374
11	0414	0453	0492	0531	0569	0607	0645	0682	0719	0755
12	0792	0828	0864	0899	0934	0969	1004	1038	1072	1106
13	1139	1173	1206	1239	1271	1303	1335	1367	1399	1430
14	1461	1492	1523	1553	1584	1614	1644	1673	1703	1732
15	1761	1790	1818	1847	1875	1903	1931	1959	1987	2014
16	2041	2068	2095	2122	2148	2175	2201	2227	2253	2279
17	2304	2330	2355	2380	2405	2430	2455	2480	2504	2529
18	2553	2577	2601	2625	2648	2672	2695	2718	2742	2765
19	2788	2810	2833	2856	2878	2900	2923	2945	2967	2989
20	3010	3032	3054	3075	3096	3118	3139	3160	3181	3201
21	3222	3243	3263	3284	3304	3324	3345	3365	3385	3404
22	3424	3444	3464	3483	3502	3522	3541	3560	3579	3598
23	3617	3636	3655	3674	3692	3711	3729	3747	3766	3784
24	3802	3820	3838	3856	3874	3892	3909	3927	3945	3962
25	3979	3997	4014	4031	4048	4065	4082	4099	4116	4133
26	4150	4166	4183	4200	4216	4232	4249	4265	4281	4298
27	4314	4330	4346	4362	4378	4393	4409	4425	4440	4456
28	4472	4487	4502	4518	4533	4548	4564	4579	4594	4609
29	4624	4639	4654	4669	4683	4698	4713	4728	4742	4757
30	4771	4786	4800	4814	4829	4843	4857	4871	4886	4900
31	4914	4928	4942	4955	4969	4983	4997	5011	5024	5038
32	5051	5065	5079	5092	5105	5119	5132	5145	5159	5172
33	5185	5198	5211	5224	5237	5250	5263	5276	5289	5302
34	5315	5328	5340	5353	5366	5378	5391	5403	5416	5428
35	5441	5453	5465	5478	5490	5502	5514	5527	5539	5551
36	5563	5575	5587	5599	5611	5623	5635	5647	5658	5670
37	5682	5694	5705	5717	5729	5740	5752	5763	5775	5786
38	5798	5809	5821	5832	5843	5855	5866	5877	5888	5899
39	5911	5922	5933	5944	5955	5966	5977	5988	5999	6010
40	6021	6031	6042	6053	6064	6075	6085	6096	6107	6117
41	6128	6138	6149	6160	6170	6180	6191	6201	6212	6222
42	6232	6243	6253	6263	6274	6284	6294	6304	6314	6325
43	6335	6345	6355	6365	6375	6385	6395	6405	6415	6425
44	6435	6444	6454	6464	6474	6484	6493	6503	6513	6522
45	6532	6542	6551	6561	6571	6580	6590	6599	6609	6618
46	6628	6637	6646	6656	6665	6675	6684	6693	6702	6712
47	6721	6730	6739	6749	6758	6767	6776	6785	6794	6803
48	6812	6821	6830	6839	6848	6857	6866	6875	6884	6893
49	6902	6911	6920	6928	6937	6946	6955	6964	6972	6981
50	6990	6998	7007	7016	7024	7033	7042	7050	7059	7067
51	7076	7084	7093	7101	7110	7118	7126	7135	7143	7152
52	7160	7168	7177	7185	7193	7202	7210	7218	7226	7235
53	7243	7251	7259	7267	7275	7284	7292	7300	7308	7316
54	7324	7332	7340	7348	7356	7364	7372	7380	7388	7396
N	0	1	2	3	4	5	6	7	8	9

N	0	1	2	3	4	5	6	7	8	9
55	7404	7412	7419	7427	7435	7443	7451	7459	7466	7474
56	7482	7490	7497	7505	7513	7520	7528	7536	7543	7551
57	7559	7566	7574	7582	7589	7597	7604	7612	7619	7627
58	7634	7642	7649	7657	7664	7672	7679	7686	7694	7701
59	7709	7716	7723	7731	7738	7745	7752	7760	7767	7774
60	7782	7789	7796	7803	7810	7818	7825	7832	7839	7846
61	7853	7860	7868	7875	7882	7889	7896	7903	7910	7917
62	7924	7931	7938	7945	7952	7959	7966	7973	7980	7987
63	7993	8000	8007	8014	8021	8028	8035	8041	8048	8055
64	8062	8069	8075	8082	8089	8096	8102	8109	8116	8122
65	8129	8136	8142	8149	8156	8162	8169	8176	8182	8189
66	8195	8202	8209	8215	8222	8228	8235	8241	8248	8254
67	8261	8267	8274	8280	8287	8293	8299	8306	8312	8319
68	8325	8331	8338	8344	8351	8357	8363	8370	8376	8382
69	8388	8395	8401	8407	8414	8420	8426	8432	8439	8445
70	8451	8457	8463	8470	8476	8482	8488	8494	8500	8506
71	8513	8519	8525	8531	8537	8543	8549	8555	8561	8567
72	8573	8579	8585	8591	8597	8603	8609	8615	8621	8627
73	8633	8639	8645	8651	8657	8663	8669	8675	8681	8686
74	8692	8698	8704	8710	8716	8722	8727	8733	8739	8745
75	8751	8756	8762	8768	8774	8779	8785	8791	8797	8802
76	8808	8814	8820	8825	8831	8837	8842	8848	8854	8859
77	8865	8871	8876	8882	8887	8893	8899	8904	8910	8915
78	8921	8927	8932	8938	8943	8949	8954	8960	8965	8971
79	8976	8982	8987	8993	8998	9004	9009	9015	9020	9025
80	9031	9036	9042	9047	9053	9058	9063	9069	9074	9079
81	9085	9090	9096	9101	9106	9112	9117	9122	9128	9133
82	9138	9143	9149	9154	9159	9165	9170	9175	9180	9186
83	9191	9196	9201	9206	9212	9217	9222	9227	9232	9238
84	9243	9248	9253	9258	9263	9269	9274	9279	9284	9289
85	9294	9299	9304	9309	9315	9320	9325	9330	9335	9340
86	9345	9350	9355	9360	9365	9370	9375	9380	9385	9390
87	9395	9400	9405	9410	9415	9420	9425	9430	9435	9440
88	9445	9450	9455	9460	9465	9469	9474	9479	9484	9489
89	9494	9499	9504	9509	9513	9518	9523	9528	9533	9538
90	9542	9547	9552	9557	9562	9566	9571	9576	9581	9586
91	9590	9595	9600	9605	9609	9614	9619	9624	9628	9633
92	9638	9643	9647	9652	9657	9661	9666	9671	9675	9680
93	9685	9689	9694	9699	9703	9708	9713	9717	9722	9727
94	9731	9736	9741	9745	9750	9754	9759	9763	9768	9773
95	9777	9782	9786	9791	9795	9800	9805	9809	9814	9818
96	9823	9827	9832	9836	9841	9845	9850	9854	9859	9863
97	9868	9872	9877	9881	9886	9890	9894	9899	9903	9908
98	9912	9917	9921	9926	9930	9934	9939	9943	9948	9952
99	9956	9961	9965	9969	9974	9978	9983	9987	9991	9996
N	0	1	2	3	4	5	6	7	8	9

APPENDIX E Answers to Problems

Chapter 2

1 (*a*) 625 cells (*b*) 1×10^5 mitochondria (*c*) 2×10^{10} molecules
2 (*a*) 1.1×10^4 molecules (*b*) 1×10^{-4} *M*
3 (*a*) 1×10^{-12} g (1 picogram) (*b*) 5.9% (*c*) 4.6%
4 (*a*) 1.3 mm; DNA is 650 times longer than the cell and thus must be tightly coiled up (*b*) 3156 proteins
5 (*a*) The metabolic rate is limited by diffusion, which is limited by surface area. (*b*) 12×10^6 m^{-1} for bacterium; 4×10^4 m^{-1} for ameba; ratio = 300 (*c*) Surface-to-volume of human being is 19 m^{-1}; ratio of surface-to-volume for a bacterium to surface-to-volume for a human is $(1.2 \times 10^6)/1$. This answer will vary with assumptions made about body dimensions.
6 (*a*) 7850 (*b*) 3.14×10^{-10} m^2 (*c*) 2.72×10^{-9} m^2 (*d*) 765% improvement in surface-to-volume ratio

Chapter 3

1 The vitamins from the two sources are identical and the body cannot distinguish the source.
2 (*a*) H_2N—CH$_2$—CH$_2$—OH: Amine, Hydroxyl
(*b*) H—C(H)(OH)—C(H)(OH)—C(H)(OH)—H: Hydroxyls
(*c*) Phospho; Carboxylate
(*d*) Carboxyl; Amine; Hydroxyl; Methyl
(*e*) Carboxylate; Amido; Hydroxyl; Methyl; Hydroxyl
(*f*) Aldehyde; Amine; Hydroxyl; Hydroxyls

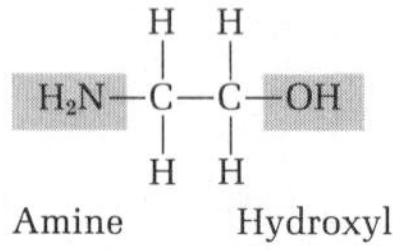

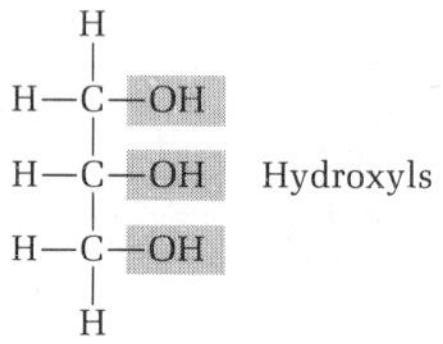

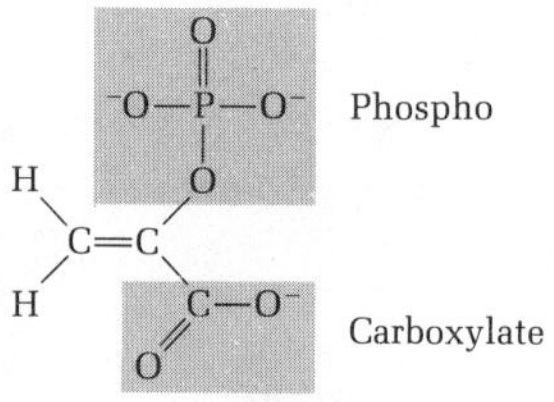

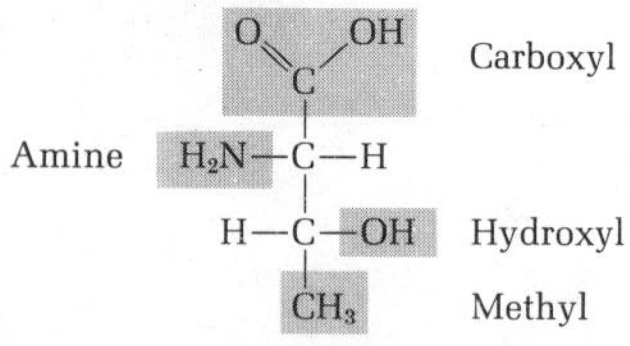

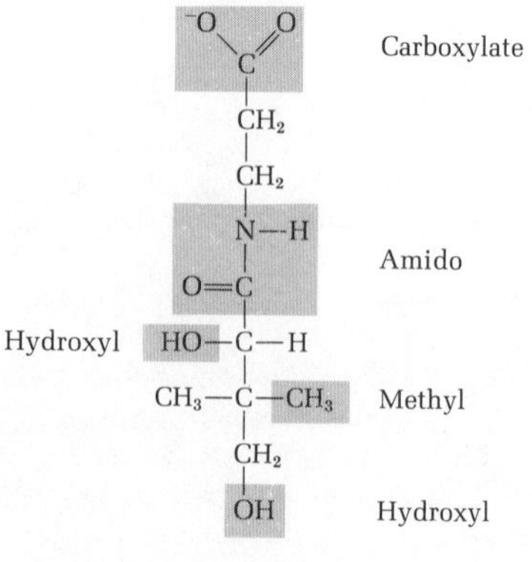

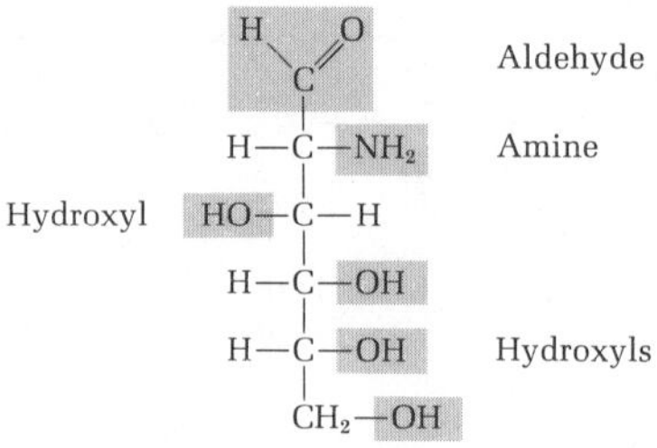

3 HO, HO (ring)—C*(OH)(H)—CH_2—N(H)—C(H)(CH_3)—CH_3

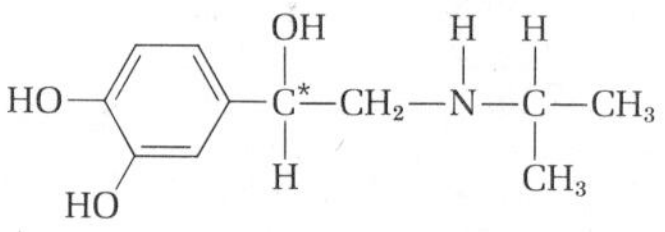

The two enantiomers have different interactions with a chiral biological "receptor" (a protein).
4 Dexedrine consists of a single enantiomer, whereas Benzedrine consists of a racemic mixture.
5 (*a*) 3 phosphoric acid molecules, α-D-ribose, adenine (*b*) choline, phosphoric acid, glycerol, oleic acid, palmitic acid (*c*) tyrosine, 2 glycine molecules, phenylalanine, methionine
6 (*a*) CH_2O; $C_3H_6O_3$
(*b*) Structures 1–12

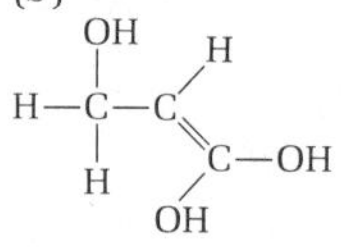

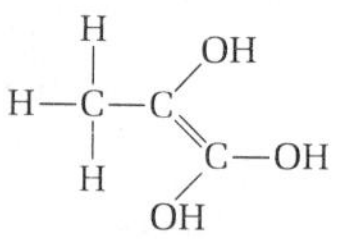

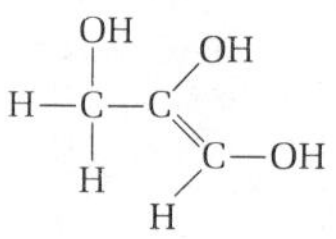

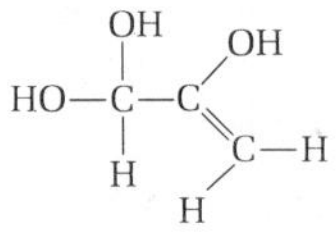

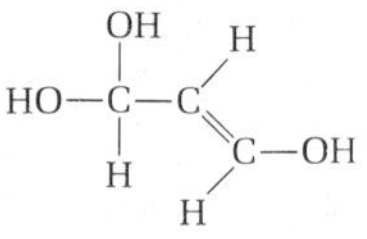

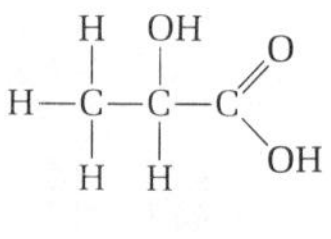

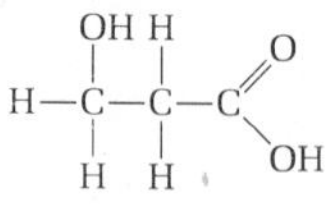

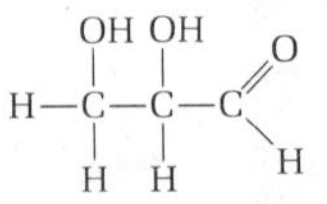

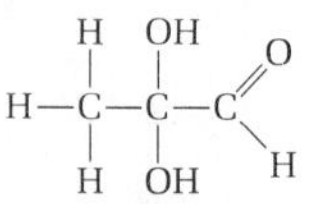

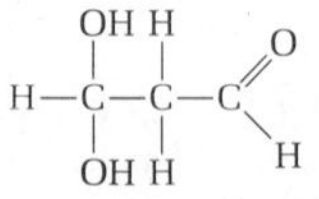

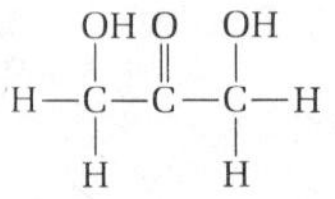

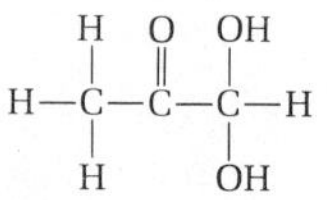

(*c*) that X contains a chiral center; eliminates all but **6** and **8** (*d*) that X contains an acid functional group; eliminates **8**; structure **6** is consistent with all data (*e*) structure **6**; we cannot distinguish between two possible enantiomers

Chapter 4

1 9.6 molal ≈ 9.6 *M*
2 3.35 mL
3 1.1
4 1.7×10^{-9} mol
5 For the equilibrium $HA \rightleftharpoons H^+ + A^-$, the corresponding Henderson-Hasselbalch equation is $pK' = pH + \log [A^-]/[HA]$. When the acid (HA) is half-ionized, $[HA] = [A^-]$. Hence $[A^-]/[HA] = 1$, $\log 1 = 0$, and $pK' = pH$.
6 (*a*) In a zone centered about pH 9.3 (*b*) $\frac{2}{3}$ (*c*) 10^{-2} L (*d*) $pH - pK' = -2$
7 (*a*) 0.1 *M* HCl (*b*) 0.1 *M* NaOH (*c*) 0.1 *M* NaOH
8 (*d*)
9 Stomach
10 $NaH_2PO_4 \cdot H_2O$, 5.80 g; Na_2HPO_4, 8.23 g
11 (*a*) The blood pH is controlled by the carbon dioxide–bicarbonate buffer system, as shown in the net equation

$$CO_2 + H_2O \rightleftharpoons H^+ + HCO_3^-$$

During *hypoventilation* the concentration of CO_2 increases in the lungs and arterial blood, driving the above equilibrium to the right and raising the hydrogen ion concentration; i.e., the pH is lowered. (*b*) During *hyperventilation* the concentration of CO_2 is lowered in the lungs and arterial blood. This drives the equilibrium to the left, which requires the consumption of hydrogen ions. Thus the concentration of hydrogen ions is reduced; i.e., the pH is raised from the normal 7.4 value. (*c*) Lactic acid is a moderately strong acid (pK′ 3.86) that completely dissociates under physiological conditions

$$CH_3CHOHCOOH \rightleftharpoons CH_3CHOHCOO^- + H^+$$

which lowers the pH of the blood and muscle tissue. Hyperventilation is useful because it removes hydrogen ions [part (*b*)], raising the pH of the blood and tissues in anticipation of the acid buildup.

Chapter 5

1 +17.9 deg · mL dm^{-1} g^{-1}; does not indicate whether citrulline is a D- or L-amino acid
2 Determine the absolute configuration of the α carbon and compare to D- and L-glyceraldehyde.
3 (1) Glycine, (b); (2) alanine, (f); (3) valine, (f); (4) serine, (a); (5) proline, (h); (6) phenylalanine, (e); (7) tryptophan, (e); (8) tyrosine, (j); (9) aspartate, (i); (10) glutamate, (i); (11) methionine, (d); (12) cysteine, (k); (13) histidine, (g); (14) arginine, (l); (15) lysine, (c); (16) asparagine, (m)
4 (*a*) I (*b*) II (*c*) IV (*d*) II (*e*) IV (*f*) II and IV (*g*) III (*h*) III (*i*) II (*j*) V (*k*) III (*l*) IV (*m*) V (*n*) II (*o*) III (*p*) IV (*q*) V (*r*) I, III, and V (*s*) V
5 (*b*) 1 part in 10^7 (one ten millionth)
6 (*a*)

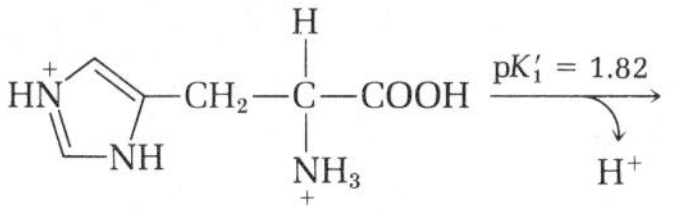

1: HN^+(imidazolium)–CH_2–C(H)(NH_3^+)–COOH $\xrightarrow{pK_1' = 1.82}$ H^+

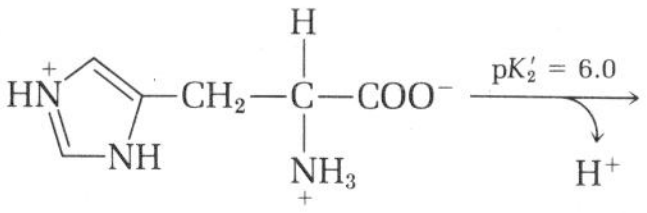

2: HN^+(imidazolium)–CH_2–C(H)(NH_3^+)–COO^- $\xrightarrow{pK_2' = 6.0}$ H^+

3: N(imidazole)–CH_2–C(H)(NH_3^+)–COO^- $\xrightarrow{pK_3' = 9.17}$ H^+

4: N(imidazole)–CH_2–C(H)(NH_2)–COO^-

(*b*)

pH	Structure	Net charge	Migrates toward
1	**1**	+2	cathode (−)
4	**2**	+1	cathode (−)
8	**3**	0	does not migrate
12	**4**	−1	anode (+)

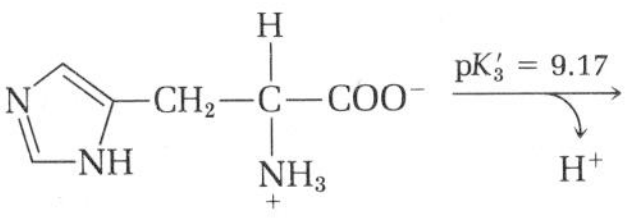

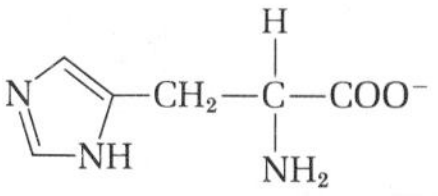

7 0.879 L of 0.1 *M* glycine and 0.121 L of 0.1 *M* glycine hydrochloride
8 (*a*) Towards anode: Glu (*b*) towards cathode: Lys, Arg, and His (*c*) remains at origin: Gly and Ala
9 (*a*) Asp (*b*) Met (*c*) Glu (*d*) Gly (*e*) Ser
10 (*a*) 27 (*b*) 6
11 (*a*) 2 (*b*) 4
(*c*)

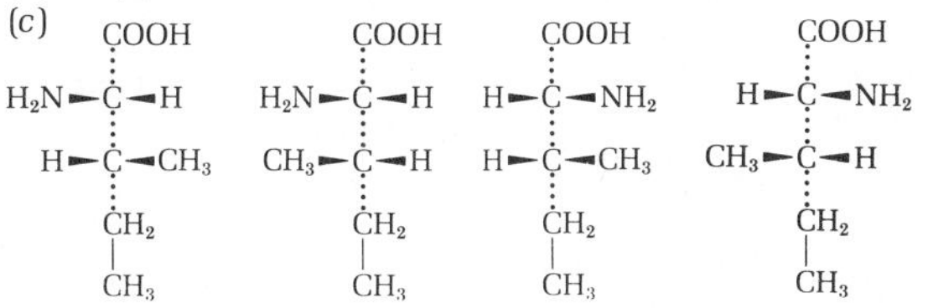

COOH	COOH	COOH	COOH
H_2N–C–H	H_2N–C–H	H–C–NH_2	H–C–NH_2
H–C–CH_3	CH_3–C–H	H–C–CH_3	CH_3–C–H
CH_2	CH_2	CH_2	CH_2
CH_3	CH_3	CH_3	CH_3

(*d*) 2*S*,3*R*, 2*S*,3*S*, 2*R*,3*R*, and 2*R*,3*S*, respectively
12 (*a*)

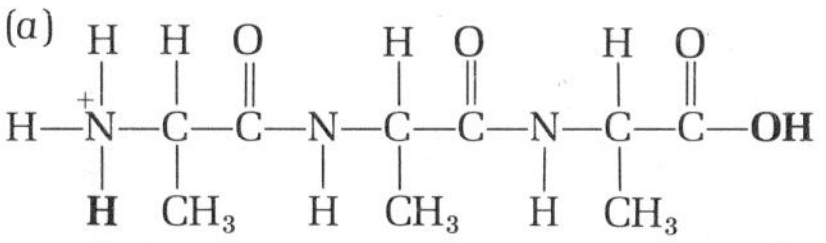

H_3N^+–CH(CH_3)–C(=O)–NH–CH(CH_3)–C(=O)–NH–CH(CH_3)–C(=O)–**OH**

$pK_2' = 8.03$ $pK_1' = 3.39$

(*b*) The ionization of the first proton in both Ala and Ala oligopeptide results in the formation of the zwitterionic conjugate base. This equilibrium is driven to the right due to the formation of a favorable charge-charge interaction between the carboxylate anion and the protonated amino group. Since the protonated amino group is closer to the carboxylate anion in Ala than Ala oligopeptide, the equilibrium for the former is more strongly driven, as shown by the lower pK_1' value.
(*c*) The ionization of the second proton in both Ala and Ala oligopeptide destroys the favorable charge-charge interaction. Since this favorable charge-charge interaction is closer together in Ala than Ala oligopeptide, it is more difficult to remove the second proton in Ala than Ala oligopeptide and consequently the pK_2' of Ala is higher than the pK_2' of Ala oligopeptide.

Chapter 6

1 3500 molecules

2 (a) 32,100 g/mol (b) 2

3 1200; 12,200 g/mol

4 +2; +1; 0; −2; pI = 7.8

5 $-COO^-$; Asp and Glu

6 Lys, His, Arg; electrostatic attraction between negatively charged phosphate residues in DNA and positively charged basic residues in histones

7 (a) $(Glu)_{20}$ (b) $(Lys-Ala)_3$ (c) $(Asn-Ser-His)_5$ (d) $(Asn-Ser-His)_5$

8 Phe⋮Val–Asn–Glu–His–Leu–$CySO_3H$–Gly–Ser–
(C)

His–Leu–Val–Glu–Ala–Leu–Tyr|Leu–Val–$CySO_3H$–
(C)

Gly–Glu–Arg|Gly–Phe⋮Phe|Tyr|Thr–Pro–Lys|Arg,
(T) (C) (C) (C) (T)

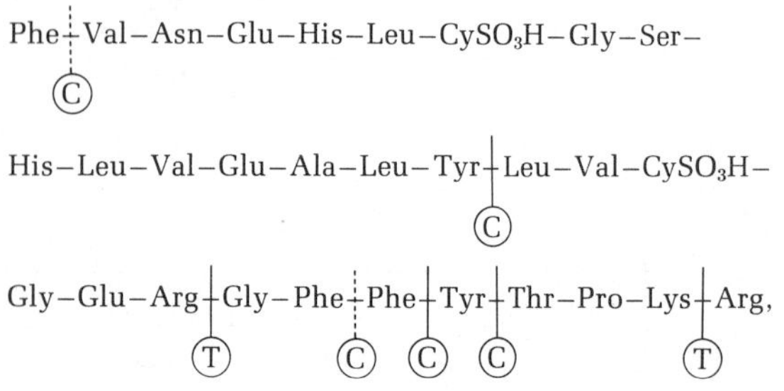

where dotted lines indicate cleavage points with low yields

9 Tyr–Gly–Gly–Phe–Leu

10 (a) (1) Anode; (2) cathode; (3) cathode; (4) anode (b) pH 7–9

11 (a) High concentrations of added salts remove the water of hydration from protein molecules, and thus reduce their solubility. (b) Choose a concentration of $(NH_4)_2SO_4$ that will precipitate protein *A* while protein *B* remains in solution. Collect precipitated protein *A* by centrifugation.

12 Wash the column with a large excess of free ligand to displace the polymer-attached ligand.

13 Orn → Leu → Phe → Pro → Val → Orn → Leu → Phe → Pro → Val → Orn (cyclic)

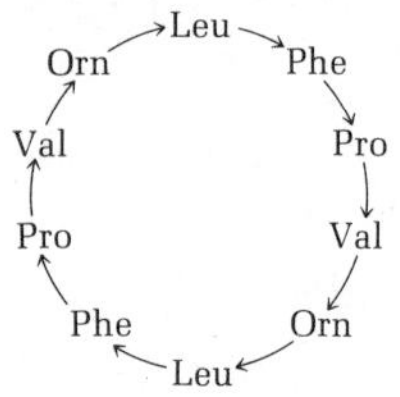

Chapter 7

1 (a) Short bonds are strong, have a high bond order, and are multiple rather than single bonds. The C—N bond is midway between a single and a double bond. (b) The structure of the peptide bond is represented by two resonance structures. (c) Rotation about the peptide bond is difficult at physiological temperatures.

2 The principal structural units in the wool fiber polypeptides are successive turns of the α-helix, which are spaced at 0.54-nm intervals. Stretching and steaming the fiber yields an extended polypeptide chain where the distances between the R groups in the β conformation are about 0.70 nm.

3 About 40 peptide bonds per second

4 Repulsion among the negatively charged carboxylate groups of polyglutamic acid at pH 7 leads to an unfolding. Similarly, repulsion among the positively charged ammonium groups of polylysine at pH 7 also leads to unfolding.

5 The disulfide bridges in cystine residues serve to cross-link protein chains, increasing their stiffness, mechanical strength, and hardness.

6 Wool shrinks due to conversion of a polypeptide chain in the extended conformation (β-pleated sheet) to an α-helical conformation.

7 Cystine residues prevent the complete unfolding of the protein.

8 (a) $-S-S- + 2\ HSCH_2CH_2OH$ (excess) $\rightleftharpoons$ $-SH\ \ HS- + S-CH_2CH_2OH$ / $S-CH_2CH_2OH$

(b) Air oxidation of cysteine to cystine residues

9 The sheets are packed Gly against Gly and Ala/Ser against Ala/Ser.

10 30 amino acids; 89%

11 The observation that ^{14}C-hydroxyproline is not incorporated into collagen argues against route (1) but is consistent with route (2).

12 The invasiveness of the bacterium is due to its ability to destroy the connective-tissue barrier of the host through the secretion of the collagenase enzyme. Bacteria do not contain collagen.

Chapter 8

1 At positions 7 and 19; at positions 13 and 24

2 External surface: Asp, Gln, Lys; interior: Leu, Val; either: Ser

3 In most cases, the functional three-dimensional folding (native structure) represents the most stable conformation of the protein. Consequently, although proteins are synthesized as linear polymers, they spontaneously fold into the proper three-dimensional conformation. Support for this thesis can be found in the classic work of Anfinsen on ribonuclease (see Figure 8-8).

4 (a) Only combinations found in the native structure give functional activity. (b) The native structure is determined by the primary sequence. (c) The native structure of insulin is not the most stable conformation.

5 (a) Comparing the mol of derivative and mol of Val gives the number of amino termini and thus the number of polypeptide chains. (b) 4

6 (a) 16,400 g/mol (b) that there are 4

7 (a) 3.2×10^{-11} g (b) 300 million (c) 90 μm^3 (d) 0.55 (e) volume of array = 94 μm^3 (f) Since the above calculation showed that hemoglobin molecules were touching their neighbors and filled the entire red blood cell, if the interaction among neighbors were to change, the cell shape must also change. In sickle-cell anemia the hemoglobin prefers to pack in long parallel filaments rather than in the cubic array. Consequently, the red blood cell also becomes elongated in the direction of the filaments.

8 (a) 7.8×10^{-4} g O_2 per kg tissue (b) 1.3×10^{-2} g O_2 per kg tissue; 17/1 (c) 7.8%

9 (*a*) Hemoglobin F (*b*) They assure that oxygen will flow from maternal blood to fetal blood. (*c*) DPG reduces the oxygen affinity. The observation that the saturation curve for hemoglobin A displays a larger shift upon DPG binding than the saturation curve for hemoglobin F suggests that hemoglobin A binds DPG more tightly than hemoglobin F.

10 (*a*) It cleaves the peptide bonds on the carboxyl side of Lys and Arg residues. (*b*) $G_{Philadelphia}$ (*c*) electrophoresis of intact α chain

11 Electrophoresis at pH 7

Chapter 9

1 Enzyme activity (sugar to starch conversion) is destroyed by heat denaturation.

2 2.4×10^{-6} M

3 9.5×10^{8} y

4 (*a*) 15.5 nm; 18.8 nm (*b*) three-dimensional folding of the enzyme brings these amino acids into close proximity (*c*) The protein serves as "scaffolding" to keep the catalytic groups in a precise orientation.

5 Determine the K_M value; measure the initial rate (rate of NADH disappearance with time measured spectrophotometrically) at several known enzyme concentrations; make a plot of the initial rates at increasing concentrations of enzyme.

6 $V_{max} \sim 140$ μmol/L·min; $K_M \sim 1 \times 10^{5}$ M

7 They had probably isolated the same form of the enzyme. The value of V_{max} is dependent on the enzyme concentration. To resolve the discrepancy they need to determine the turnover number of each enzyme sample.

8 (*a*) 1.7×10^{-3} M (*b*) 0.33; 0.67; 0.91

9 $K_M = 2.2$ mM; $V_{max} = 0.51$ mg/min

10 2.0×10^{7} min^{-1}

11 29,000; we need to assume that each enzyme molecule contains only one titratable sulfhydryl group

12 The enzyme-substrate complex is more stable than the isolated enzyme and substrate.

13 Measure the total acid phosphatase activity in the presence and absence of tartrate ion.

14 The observation that acetazolamine decreases the V_{max} of the enzyme but leaves K_M unchanged indicates that the inhibitor acts in a noncompetitive manner.

15 Ethanol competes with methanol for the active site of alcohol dehydrogenase.

16 Glu 35: protonated; Asp 52: deprotonated

Chapter 10

1 (*a*) Either nicotinic acid is essential to the biosynthesis of Trp, or nicotinic acid can be synthesized from Trp. (*b*) Corn is low in Trp.

2 Thiamine deficiency

3 The rate of lactic acid production is dependent on the amount of riboflavin in the culture medium.

4 The pyridoxine is converted to pyridoxal phosphate, a prosthetic group which serves a central role in transamination reactions.

5 (*a*) Bacterial growth (*b*) Avidin binds free biotin and prevents bacterial growth. (*c*) It protects the developing embryo from destructive bacterial growth during the incubation period.

6 Supplementing the growth medium with thymidine circumvents the need to carry out the folic-acid-requiring biosynthesis of tetrahydrofolate.

7 Vitamin B_{12} deficiency in bacterial flora

8 The high solubility of the B-complex vitamins leads to their rapid excretion.

9 The mature liver stores vitamin A.

10 Vitamin D_3; damaged kidneys prevent the complete hydroxylation of vitamin D_3 to its biologically active form

11 (*a*) It could act as an inhibitor in vitamin K–dependent enzymatic reactions. (*b*) They develop abnormal bleeding due to apparent vitamin K deficiency. (*c*) The vitamin K antagonist will lower the concentration of blood coagulation factors.

12 Vitamin B_{12} deficiency

13 (*a*) Phytic acid binds zinc and prevents intestinal absorption. (*b*) The yeast destroys the phytic acid.

Chapter 11

1 (*a*)

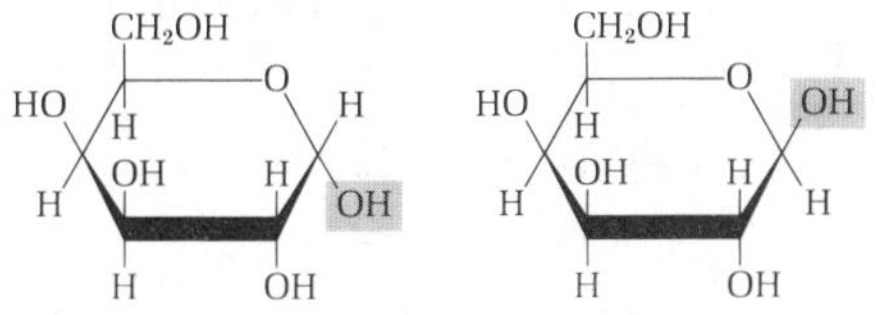

α-D-Galactose β-D-Galactose

(*b*) A freshly prepared solution of α-D-galactose undergoes mutarotation to yield an equilibrium mixture of α- and β-D-galactose. Mutarotation of either pure α- or β-D-galactose will yield the same mixture of forms.
(*c*) 72% of the β form and 28% of the α form

2 (*a*) Measure the change in optical rotation with time. (*b*) The optical rotation of the mixture is negative (inverted) relative to the sucrose solution. (*c*) 0.63 of sucrose is hydrolyzed; final composition = 0.77 glucose and fructose, 0.23 sucrose.

3 Prepare a slurry of sucrose and water for the core; add a small amount of invertase; immediately coat with chocolate.

4 (*a*)

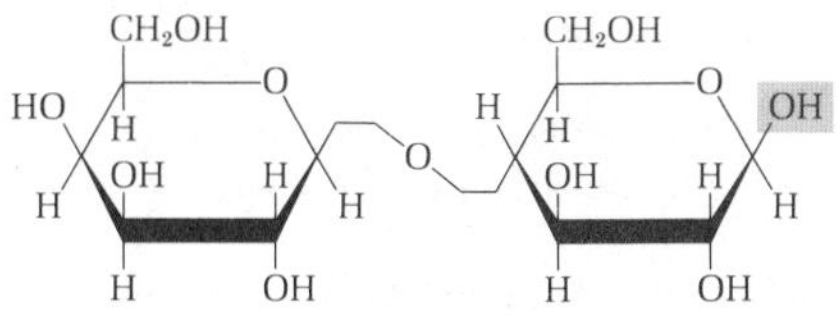

β anomer

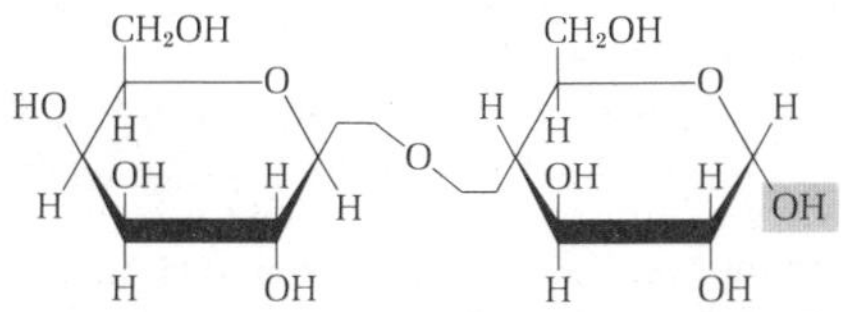

α anomer

(*b*) The hydrolysis of either α- or β-lactose will yield a mixture consisting of α- and β-D-glucose and α- and β-D-galactose.

5 Sucrose is not a reducing sugar.

6 7840 residues/s

7 Native cellulose consists of glucose units linked via a β(1 → 4) glycosidic bond. The β linkage between glucose units forces the polymer chain into an extended conformation (see Figure 11-16). A series of these parallel chains will form intermolecular hydrogen bonds and

aggregate into long, tough, insoluble fibers. Glycogen also consists of glucose units but they are linked together via $\alpha(1 \rightarrow 4)$ linkages. The α linkage between the glucose units causes a bend in the chain and prevents the formation of long fibers. In addition, glycogen is highly branched (Figure 11-15). These structural features cause glycogen to be highly hydrated, since many of the hydroxyl groups are exposed to water. Hence, glycogen can be extracted as a dispersion in hot water.

The physical properties of these two polymers are well suited for their biological roles. Cellulose serves as a structural material in plants, which is consistent with its side-by-side aggregation into insoluble fibrils. Glycogen serves as a stored fuel in animals. The highly hydrated and exposed glycogen granules can be hydrolyzed rapidly by glycogen phosphorylase to release glucose 1-phosphate. Since this enzyme only acts on the nonreducing termini, extensive branching produces many sites on which glycogen phosphorylase can act.

8 10.8 s

9 (a) The branch points yield 2,3-dimethylglucose, while the other residues yield 2,3,6-trimethylglucose.
(b) 3.74%

10 D-Glucopyranosyl-$(1 \rightarrow 1)$D-glucopyranoside

Chapter 12

1 The number of cis double bonds. Each cis double bond causes a bend in the hydrocarbon tails, making it more difficult to pack the tails in the crystal lattice.

2 Unsaturated fats are susceptible to oxidation by molecular oxygen

3 Phosphatidylcholine emulsifies fat droplets.

4 (a) The sodium salts of palmitic and stearic acids, plus glycerol (b) under mild hydrolysis, the sodium salts of palmitic and oleic acids plus glycerol-3-phosphorylcholine; under vigorous conditions, the sodium salts of palmitic, oleic, and phosphoric acids, plus glycerol and choline

5 (a) 0 (b) 0 (c) −1

6 It prevents water loss.

7 63

8 Hydrophobic units: (a) 2 fatty acids; (b), (c), (d) 1 fatty acid and the hydrocarbon chain of sphingosine; (e) the hydrocarbon backbone
Hydrophilic units: (a) phosphoethanolamine; (b) phosphocholine; (c) D-galactose; (d) several sugar molecules; (e) alcohol group (—OH)

9 (a) Lipids that form bilayers are amphipathic molecules; i.e., they contain a hydrophilic and a hydrophobic unit. In order to minimize the hydrophobic area that is exposed to the water surface, they form two-dimensional sheets with the hydrophilic unit exposed to water and the hydrophobic unit buried in the interior of the sheet. Furthermore, to avoid exposing the hydrophobic edges of the sheet to water, lipid bilayers close upon themselves. Similarly, if the sheet is perforated, the hole will seal because the membrane is semifluid. (b) These properties have important biological consequences, namely, that these sheets form closed membrane surfaces as cells or compartments within cells (organelles).

10 They must pass through the nonpolar environment of the inner core of the membrane.

11 Sodium dodecyl sulfate and sodium cholate solubilize the hydrophobic portions of integral membranes by acting as soaps or detergents (Figure 12-2).

12 (a) Sugars are hydrophilic. (b) Glycoproteins cannot tumble.

13 (a) Integral membrane proteins must exist in a fluid environment to assure their functional conformation.
(b) The increased level of unsaturated acids lowers the melting point of the membrane.

Chapter 13

1 (a) Glyceraldehyde 3-phosphate + P_i + NAD^+ $\rightarrow$
3-phosphoglyceroyl phosphate + NADH + H^+
3-Phosphoglyceroyl phosphate + ADP $\rightarrow$
3-phosphoglycerate + ATP
3-Phosphoglycerate $\rightarrow$ 2-phosphoglycerate
2-Phosphoglycerate $\rightarrow$ phosphoenolpyruvate + H_2O
Phosphoenolpyruvate + ADP $\rightarrow$ pyruvate + ATP
Pyruvate $\rightarrow$ CO_2 + acetaldehyde
Acetaldehyde + NADH + H^+ $\rightarrow$ ethanol + NAD^+
(b) Glyceraldehyde 3-phosphate + P_i + 2ADP $\rightarrow$
ethanol + CO_2 + 2ATP + H_2O
(c)

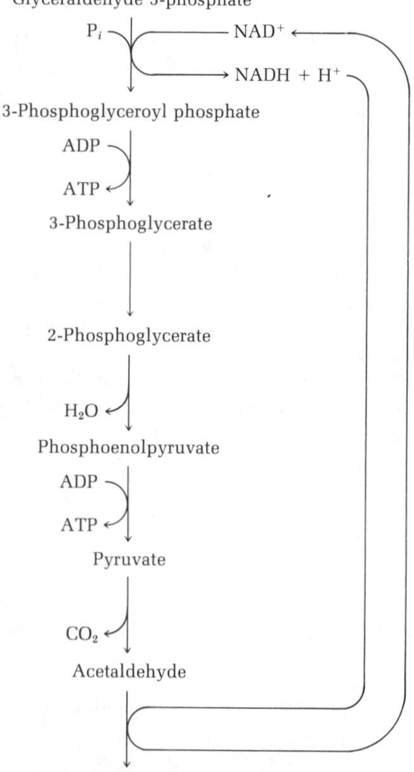

2 (a) Oxalate + formyl-CoA $\rightarrow$ formate + oxalyl-CoA
Oxalyl-CoA + H^+ $\rightarrow$ CO_2 + formyl-CoA
Formate + NAD^+ $\rightarrow$ CO_2 + NADH
(b) Oxalate + H^+ + NAD^+ $\rightarrow$ $2CO_2$ + NADH

3 ^{14}C-glucose is broken down to smaller units which are subsequently used for the biosynthesis of histidine. Unlabeled histidine acts as a feedback inhibitor and shuts down the pathway leading to histidine biosynthesis.
4 (*a*) By the level of β-galactosidase (*b*) enzyme turnover (*c*) The induction process is highly specific.
5 (*a*) Glucose + 2ATP $\rightarrow$ fructose 1,6-diphosphate + 2ADP
(*b*) Fructose 1,6-diphosphate + $2H_2O \rightarrow$ glucose + $2P_i$
(*c*) The distinguishing differences are that the catabolic route consumes two molecules of ATP while the anabolic route consumes two molecules of water. Hence, the two routes are not reverses of each other. (*d*) two thermodynamically favored phosphate transfers from ATP to glucose (*e*) The glucose–glucose 6-phosphate interconversion in the two routes cannot be catalyzed by the same enzyme because different reactions are being carried out, as shown by the net reaction. In contrast, the glucose 6-phosphate–fructose 6-phosphate interconversion is carried out by the same enzyme.
6 (*a*) $5 \times 10^{-4}\ M$ (*b*) 3.9×10^6 cpm
7 $1.1 \times 10^{-4}\ M$

Chapter 14

1 (*a*) -1.13 kcal/mol (*b*) $+1.80$ kcal/mol (*c*) -3.27 kcal/mol
2 (*a*) 267 *M* (*b*) 622 *M* (*c*) 0.28
3 9.7
4 (*a*) $3.75 \times 10^{-3}\ M^{-1}$; $8.64 \times 10^{-8}\ M$; no (*b*) 13.9 *M*; no (*c*) $\Delta G^{\circ\prime} = -4.0$ kcal/mol; $K'_{eq} = 873$; glucose concentration $1.12 \times 10^{-7}\ M$; yes (*d*) no (*e*) By directly transferring the phosphate group from ATP to glucose, the phosphate-group-transfer potential ("tendency" or "pressure") of ATP is utilized without generating high concentrations of intermediates. The essential part of this transfer is, of course, the enzymatic catalysis.
5 (*a*) -3.0 kcal (*b*) -3.5 kcal
6 -2.4 kcal
7 11.0 kcal/mol
8 (*a*) 11.0 kcal/mol (*b*) 46 kg; 68% (*c*) ATP is synthesized as it is needed, and its concentration is maintained at a steady state.
9 (*a*) 1.1 s
(*b*) Phosphocreatine + ADP $\rightarrow$ creatine + ATP
(*c*) ATP synthesis via the catabolism of glucose, amino acids, and fatty acids
10 (*a*) $+0.2$ kcal/mol (*b*) Pyrophosphatase catalyzes the hydrolysis of pyrophosphate and drives the net reaction towards the synthesis of acetyl-CoA.

Chapter 15

1 Glucose + 2ATP $\rightarrow$ 2 glyceraldehyde 3-phosphate + 2ADP + $2H^+$; $\Delta G^{\circ\prime} = +0.56$ kcal/mol
2 Glyceraldehyde 3-phosphate + 2ADP + P_i + $H^+ \rightarrow$ lactate + 2ATP + H_2O; $\Delta G^{\circ\prime} = -15.0$ kcal/mol
3 Fructose + 2ADP + $2P_i \rightarrow$ 2 lactate + 2ATP + $2H_2O$
4 (*a*) $^{14}CH_3$—CH_2—OH (*b*) 3-^{14}C glucose or 4-^{14}C glucose
5 The K_M value of hexokinase ($K_M = 0.1$ m*M*) is 100 times lower than the K_M value of glucokinase ($K_M = 10.0$ m*M*). At a normal blood glucose concentration of 5 m*M*, hexokinase is fully saturated and operating at full capacity, while glucokinase is only partially saturated. Unless glucose 6-phosphate demands in muscle are high (as in strenuous activity), the increased concentration of glucose 6-phosphate shuts down hexokinase. Thus, glucose utilization by muscle occurs even when blood glucose levels are below normal, but is shut down when there is little demand for glucose 6-phosphate. Unlike hexokinase, glucokinase is not inhibited by glucose 6-phosphate, an important property which assures glucose utilization by the liver even when the demand for glucose 6-phosphate is minimal (e.g., in glycogen biosynthesis). As blood glucose levels return to normal, glucokinase shuts down and glucose utilization by the liver is shut down.
6 No; lactate dehydrogenase is required to recycle the NADH formed during the oxidation of glyceraldehyde 3-phosphate.
7 (*a*) 3-Phosphoglycerate is the product. (*b*) In the presence of arsenate there is no net ATP synthesis under anaerobic conditions.
8 (*a*) The stoichiometry of alcoholic fermentation requires 2 mol of P_i per mol of glucose. (*b*) Reduction of acetaldehyde to ethanol is required to recycle NADH. During fermentation, fructose 1,6-diphosphate accumulates in order to recycle the adenine nucleotide pool.
(*c*) There is no net ATP synthesis in the presence of arsenate.
9 Glycerol + $2NAD^+$ + ADP + $P_i \rightarrow$ pyruvate + 2NADH + ATP + $2H^+$
10 (*a*) Q = 0.029 (*b*) $K'_{eq} = 316$ (*c*) The physiological reaction is not at equilibrium; phosphofructokinase is regulated.
11 (*a*) There are two binding sites for ATP: a catalytic site and a regulatory site. (*b*) Glycolytic flux is reduced when ATP is plentiful. (*c*) The graph indicates that the addition of ADP suppresses the inhibition of ATP. Since the pool of adenylate phosphates is fairly constant, consumption of ATP leads to an increase in ADP levels. The data indicate that the activity of phosphofructokinase may be regulated by the ATP/ADP ratio.
12 (*a*) Glycogen phosphorylase catalyzes the conversion of stored glycogen to glucose 1-phosphate. Glucose 1-phosphate is a precursor of glucose 6-phosphate, an intermediate in glycolysis. During strenuous activity, skeletal muscle requires large quantities of glucose 6-phosphate. The liver, on the other hand, uses the breakdown of glycogen to maintain a steady level of blood glucose between meals. (*b*) In actively working muscle, where ATP flux requirements are very high, glucose 1-phosphate needs to be produced rapidly, requiring a high V_{max} value.
13 Case A: (f), (3); case B: (c), (3); case C: (a), (4); case D: (d), (6)
14 With galactokinase deficiency, galactose accumulates. With galactose 1-phosphate uridylyltransferase deficiency, galactose 1-phosphate accumulates. The latter is more toxic.

Chapter 16

1 (a) *Citrate synthetase:*
Acetyl-CoA + oxaloacetate + $H_2O \rightarrow$ citrate + CoA-SH + H^+
Aconitase: Citrate $\rightarrow$ isocitrate
Isocitrate dehydrogenase:
Isocitrate + $NAD^+ \rightarrow \alpha$-ketoglutarate + CO_2 + NADH
α-Ketoglutarate dehydrogenase:
α-Ketoglutarate + NAD^+ + CoA-SH $\rightarrow$ succinyl-CoA + CO_2 + NADH
Succinyl-CoA synthetase:
Succinyl-CoA + P_i + GDP $\rightarrow$ succinate + GTP + CoA-SH
Succinate dehydrogenase:
Succinate + FAD $\rightarrow$ fumarate + $FADH_2$
Fumarase: Fumarate + $H_2O \rightarrow$ malate
Malate dehydrogenase:
Malate + $NAD^+ \rightarrow$ oxaloacetate + NADH + H^+
(b), (c): Step 1: CoA, condensation; step 2: isomerization; step 3: NAD^+, oxidation, decarboxylation; step 4: NAD^+, CoA, thiamine pyrophosphate, oxidation, decarboxylation; step 5: CoA, phosphorylation; step 6: FAD, oxidation; step 7: hydration; step 8: NAD^+, oxidation
(d) Acetyl-CoA + $3NAD^+$ + FAD + GDP + P_i + $2H_2O \rightarrow 2CO_2$ + 3NADH + $FADH_2$ + GTP + $2H^+$ + CoA

2 (a) Oxidation; methanol $\rightarrow$ formaldehyde + H—H
(b) Oxidation; formaldehyde $\rightarrow$ formic acid + H—H
(c) Reduction; CO_2 + H—H $\rightarrow$ formic acid
(d) Reduction; glyceric acid + H—H $\rightarrow$ glyceraldehyde
(e) Oxidation; glycerol $\rightarrow$ dihydroxyacetone + H—H
(f) Oxidation; $2H_2O$ + toluene $\rightarrow$ benzoic acid + 3 H—H
(g) Oxidation; succinate $\rightarrow$ fumarate + H—H
(h) Oxidation; pyruvic acid + $H_2O \rightarrow$ acetic acid + H—H + CO_2

3 (a) Ethanol + $NAD^+ \rightarrow$ acetaldehyde + NADH + H^+
(b) 3-Phosphoglyceroyl phosphate + NADH + $H^+ \rightarrow$ glyceraldehyde 3-phosphate + NAD^+ + HPO_4^{2-}
(c) Pyruvate + $H^+ \rightarrow$ acetaldehyde + CO_2
(d) Pyruvate + $NAD^+ \rightarrow$ acetate + NADH + H^+ + CO_2
(e) Oxaloacetate + NADH + $H^+ \rightarrow$ malate + NAD^+
(f) Acetoacetate + $H^+ \rightarrow$ acetone + CO_2

4 (a) Oxygen consumption is a measure of the activity of the first two stages of cell respiration, glycolysis and the citric acid cycle. The addition of oxaloacetate or malate stimulates the citric acid cycle and thus stimulates respiration. (b) The added oxaloacetate or malate serves a catalytic role in the citric acid cycle, since it is regenerated in the latter part of the cycle.

5 (a) 6.0×10^{-6} (b) 1.2×10^{-8} *M* (c) 30 molecules

6 (a) $^-OOCCH_2CH_2COO^-$ (succinate) (b) Malonate is a competitive inhibitor of succinate dehydrogenase. (c) A block in the citric acid cycle stops NADH formation, which stops electron transport, which stops respiration. (d) by a large excess of succinate

7 (a) Add uniformly labeled ^{14}C-glucose and look for the release of $^{14}CO_2$. (b) equally distributed in the 2 and 3 positions of the oxaloacetate (c) an infinite number

8 (a) 1 position (b) 3 position (c) 3 position (d) methyl group (e) equally distributed in the —CH_2— groups (f) 4 position (g) equally distributed in the —CH_2— groups

9 No; carboxylation of pyruvate

10 (a) Inhibition of aconitase (b) fluorocitrate; competes with citrate; by a large excess of citrate (c) Citrate and fluorocitrate are inhibitors of phosphofructokinase. (d) The catabolic processes to ATP production are all shut down.

11 2 pyruvate + ATP + $2NAD^+$ + $H_2O \rightarrow$ α-ketoglutarate + CO_2 + ADP + P_i + 2NADH + $3H^+$

12 2 acetyl-CoA + $2NAD^+$ + FAD + $H_2O \rightarrow$ oxaloacetate + 2NADH + $4H^+$ + 2CoA + $FADH_2$

13 The ratio of $^{14}CO_2$ production from 1-^{14}C-glucose to 6-^{14}C-glucose is 2/1.

Chapter 17

1 (1): (a), (d) NADH; (b), (e) E—FMN; (c) $NADH/NAD^+$ and E—$FMNH_2$/E—FMN
(2): (a), (d) E—$FMNH_2$; (b), (e) Fe^{3+}; (c) E—$FMNH_2$/E—FMN and Fe^{2+}/Fe^{3+}
(3): (a), (d) Fe^{2+}; (b), (e) Q; (c) Fe^{2+}/Fe^{3+} and Q/QH_2

2 (a) $NAD^+/NADH$ (b) pyruvate (c) in the direction of lactate formation (d) −6.0 kcal (e) 2.58×10^4

3 Ferredoxin-reducing substrate $\rightarrow$ ferredoxin $\rightarrow$ cytochrome $b_6 \rightarrow$ cytochrome $f \rightarrow$ plastocyanin; the first and last steps are unlikely to yield ATP under standard conditions.

4 (a) 39 (b) 12 (c) 20 (d) 76

5 (a) 1.14 V (b) −52.6 kcal (c) about 7

6 The oxidation of succinate by FAD is favored by the negative standard free energy change ($\Delta G^{\circ\prime} = -0.9$ kcal), whereas oxidation by NAD^+ is opposed by the large positive value of $\Delta G^{\circ\prime} = +16.1$ kcal.

7 (a) Cyanide blocks the reduction of oxygen catalyzed by cytochrome aa_3

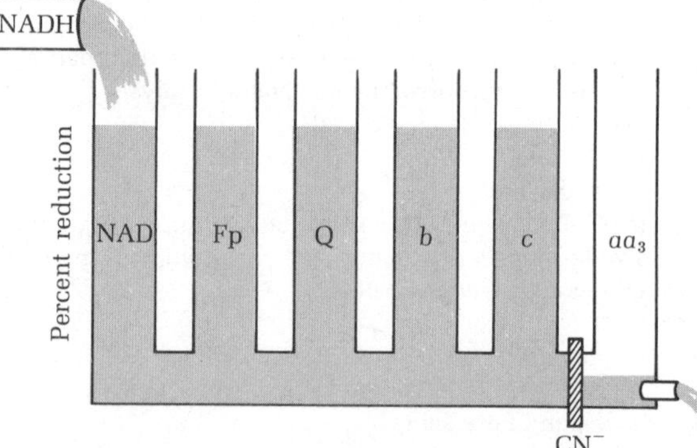

(b) In the absence of oxygen, the reduced carriers do not get reoxidized

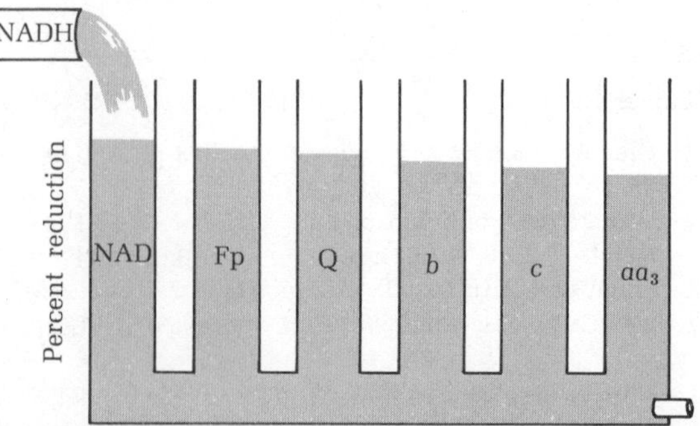

(c) In the absence of NADH, all the carriers get oxidized by O_2

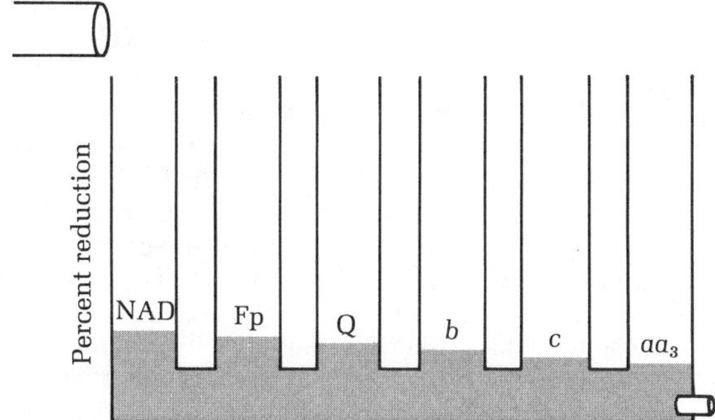

(d) See Figure 17-14(a).

8 (a) The inhibition of NADH dehydrogenase by rotenone decreases the rate of electron transport, which in turn decreases the rate of ATP production. If the reduced ATP production rate is unable to meet the needs of the organism, it dies. (b) Antimycin A strongly inhibits the oxidation of ubiquinol in the electron transport chain. This reduces the rate of electron transport leading to the consequences described in (a). (c) Antimycin

9 (a) In the presence of an uncoupler, the level of electron-transport activity necessary to meet ATP demands increases, and consequently the P/O ratio decreases. (b) The P/O ratio decreases. (c) Increased electron-transport activity in the presence of an uncoupler requires the degradation of additional fuel. (d) It could act as an uncoupler.

10 (a) The formation of ATP is inhibited. (b) Formation of ATP is tightly coupled to electron transport; 2,4-dinitrophenol is an uncoupler of oxidative phosphorylation. (c) oligomycin

11 (c) Oligomycin inhibits both; atractyloside will not affect electron transport or ATP synthesis in inverted vesicles.

12 (a) Brown fat generates heat to keep the infant warm. (b) uncoupling oxidative phosphorylation; increased ATP utilization; fewer than 3 phosphorylation sites

13 (a) Glycolysis will go into an anaerobic phase. (b) Oxygen consumption will cease. (c) Lactate formation should increase. (d) ATP synthesis will shut down.

14 (a) NADH was then recycled via electron transport. (b) Oxidative phosphorylation is more efficient. (c) The high mass-action ratio of the ATP system inhibits phosphofructokinase.

15 (a) Muscle contraction (b) It stimulated glycolysis and respiration. (c) See Figure 17-29.

16 (a) External medium: 4.0×10^{-8} M; matrix: 2.0×10^{-8} M (b) 2/1 (c) 21 (d) no (e) transmembrane potential

Chapter 18

1 In the fatty acid portion

2 (a) 9.5×10^4 kcal (b) 48 d (c) 0.5 lb/d

3 The first step in fatty acid oxidation is analogous to the conversion of succinate to fumarate; the second step is analogous to the conversion of fumarate to malate; the third step is analogous to the conversion of malate to oxaloacetate.

4 (a) $R—COOH + ATP \rightarrow \text{acyl-AMP} + PP_i$
Acyl-AMP + CoA-SH $\rightarrow$ R—CO—S—CoA + AMP
(b) Irreversible hydrolysis of pyrophosphate to inorganic phosphate by cellular pyrophosphatase

5 Yes; the removed tritium appears as tritiated water.

6 0.13 gal of water/lb of tripalmitin

7 Complete oxidation of the hydrocarbon to carbon dioxide and water

8 (a) Phenylacetic acid; MW = 136 (b) even

9 Since the mitochondrial pool of coenzyme A is small, the coenzyme A pool must be recycled via the formation of ketone bodies in order to maintain the β-oxidation pathway necessary for energy.

10 (a) Carbohydrate deprivation would lead to a citric acid cycle with low activity. (b) odd-numbered

11 (a)

$$CH_3(CH_2)_{12}—\overset{\overset{\large O}{\|}}{C}—S—CoA$$

$$6CoA\text{-}SH + 6FAD + 6NAD^+ + 6H_2O \rightarrow 7\ \text{acetyl-CoA} + 6FADH_2 + 6NADH + 6H^+$$

(b)

$$CH_3(CH_2)_{16}COOH + ATP + 9CoA\text{-}SH + 8FAD + 8NAD^+ + 9H_2O \rightarrow 9\ \text{acetyl-CoA} + AMP + 2P_i + 8FADH_2 + 8NADH + 10H^+$$

(c)

$$CH_3—\underset{\underset{\large H}{|}}{\overset{\overset{\large OH}{|}}{C}}—CH_2COOH + NAD^+ + ATP + 2CoA\text{-}SH + H_2O \rightarrow 2\ \text{acetyl-CoA} + NADH + AMP + 2P_i + 3H^+$$

12 (a)

$$CH_3—{}^{14}\overset{\overset{\large O}{\|}}{C}—S—CoA$$

(b)

$$^-OOC—CH_2—\underset{\underset{\large COO^-}{|}}{\overset{\overset{\large OH}{|}}{C}}—CH_2—{}^{14}\overset{\overset{\large O}{\|}}{C}—O^-$$

(c) No label

13

$$CH_3—\underset{\underset{\large H}{|}}{\overset{\overset{\large OH}{|}}{C}}—CH_2—\overset{\overset{\large O}{\|}}{C}—OH + \tfrac{9}{2}O_2 + 25ADP + 25P_i + 25H^+ \longrightarrow 4CO_2 + 29H_2O + 25ATP$$

Chapter 19

1 (a) $^-OOC—CH_2—\overset{\overset{\large O}{\|}}{C}—COO^-$ Oxaloacetate

(b) $^-OOC—CH_2—CH_2—\overset{\overset{\large O}{\|}}{C}—COO^-$ α-Ketoglutarate

(c) $CH_3—\overset{\overset{\large O}{\|}}{C}—COO^-$ Pyruvate

(d) $H—C_6H_4—CH_2—\overset{\overset{\large O}{\|}}{C}—COO^-$ Phenylpyruvate

2 The described procedure is a coupled reaction assay where the product of the slow transamination (pyruvate) is rapidly consumed in the subsequent indicator reaction catalyzed by lactate dehydrogenase. The indicator reaction is monitored by observing the loss of the characteristic yellow color of NADH (due to absorbance at 340 nm) with a spectrophotometer.

3 No; the nitrogen in alanine can be transferred to oxaloacetate via transamination to form aspartate.

4 (*a*) Phenylalanine 4-monooxygenase; a low-phenylalanine diet (*b*) The normal route of Phe metabolism via hydroxylation to tyrosine is blocked, and Phe accumulates. (*c*) Phe is transformed to phenylpyruvate by transamination, and to phenyllactate by reduction. (*d*) Tyrosine is a precursor for the synthesis of melanin, a pigment that serves as the coloring matter for hair.

5 Catabolism of the carbon skeletons of valine, methionine, and isoleucine will be impaired.

6 17 ATP produced per lactate, and 15 ATP produced per alanine when nitrogen removal is included

7 (*a*) $^{15}NH_2—CO—^{15}NH_2$
(*b*) $^-OOC—CH_2—CH_2—^{14}COO^-$
(*c*) $R—NH—C(=^{15}NH)—^{15}NH_2$
(*d*) $R—NH—C(=O)—^{15}NH_2$
(*e*)
(*f*) $^-OOC—CH(^{15}\overset{+}{N}H_3)—CH_2—^{14}COO^-$ and

$$^-OO^{14}C—CH(^{15}\overset{+}{N}H_3)—CH_2—COO^-$$

8 (*a*) Isoleucine → **II** → **IV** → **I** → **V** → **III** → acetyl-CoA + propionyl-CoA (*b*) Step 1: transamination; step 2: oxidative decarboxylation analogous to the oxidative decarboxylation of pyruvate to acetyl-CoA; step 3: oxidation analogous to dehydrogenation of succinate; step 4: hydration analogous to hydration of fumarate to malate; step 5: oxidation analogous to dehydrogenation of malate to oxaloacetate; step 6: thiolysis (reverse aldol condensation) analogous to thiolase reaction.

9 (*a*) Fasting results in low blood glucose levels; subsequent administration of the experimental diet leads to rapid catabolism of glucogenic amino acids. (*b*) Oxidative deamination causes the rise in ammonia levels; the absence of Arg (an intermediate in the urea cycle) prevents the conversion of ammonia to urea; Arg is not synthesized in sufficient quantities in the cat to meet the needs imposed by the stress of the experiment. (*c*) Orn is converted to Arg by the urea cycle.

10 $O_2 + 2\text{ glutamate} + CO_2 + 2ADP + 2P_i \rightarrow 2\ \alpha\text{-ketoglutarate} + 3H_2O + 2ATP + \text{urea}$

Chapter 20

1 No. The transformation of 2 molecules of pyruvate to one molecule of glucose requires the input of energy (4ATP + 2GTP) and reducing power (2NADH). This energy and reducing power is obtained through the citric acid cycle/oxidative phosphorylation pathway by the catabolism of amino acids, fatty acids, or other carbohydrate.

2 (*a*) No initial ^{14}C incorporation (*b*) 3,4-^{14}C-glucose

3 Pyruvate carboxylase is a mitochondrial enzyme. The ^{14}C-oxaloacetate formed is mixed with the oxaloacetate pool used in the tricarboxylic acid cycle. Consequently, ^{14}C-oxaloacetate is equilibrated with the tricarboxylic acid cycle intermediates to form a mixture of 1-^{14}C- and 4-^{14}C-oxaloacetate. Oxaloacetate labeled with ^{14}C in the 1 position leads to the formation of 3,4-^{14}C-glucose (see Problem 2).

4 Phosphofructokinase: activated by AMP and inhibited by ATP, regulates glycolysis; fructose diphosphatase: activated by ATP and inhibited by AMP, regulates gluconeogenesis

5 (*a*), (*b*), and (*d*) are glucogenic; (*c*) and (*e*) aren't.

6 (*a*) The rapid increase in glycolysis; the rise in pyruvate and NADH results in a rise in lactate (*b*) Lactate is transformed to glucose. (*c*) The equilibrium for the lactate dehydrogenase reaction is in favor of lactate formation.

7 If the catabolic and anabolic pathways of glucose are operating simultaneously, ATP will be consumed without the net synthesis of glucose. These cyclic processes are referred to as futile cycles.

8 The observation that glycogen synthase has the lowest measured activity of the enzymes in glycogen synthesis suggests that this enzyme step is the bottleneck in the flow of metabolites, and is thus a regulatory point. This is confirmed by the observation that the stimulation of glycogen synthesis by the activation of the regulatory enzyme leads to a decrease in the concentration of the intermediates before the regulation point, and an increase in the concentration of metabolites after the regulation point. Activation of glycogen synthase increases the flow of metabolites through this point. This has the effect of lowering the steady-state concentration of UDP-glucose and increasing the steady-state concentrations of UDP.

9 Storage consumes 1 ATP per glucose molecule; the overall efficiency of storage is 97.4%.

10 UDP-glucose pyrophosphorylase

11 The diversion of glucose and its precursor oxaloacetate to milk production under conditions of extensive fatty acid catabolism results in ketosis. Ruminants can readily transform propionate to succinyl-CoA (Figure 20-7) and thus into oxaloacetate to avert the ketosis.

12 The presence of UDP-galactose pyrophosphorylase provides a means to catabolize galactose by the following route

$$\text{Galactose} + \text{ATP} \xrightarrow{\text{galactokinase}} \text{galactose-1-phosphate} + \text{ADP}$$

$$\text{UTP} + \text{galactose-1-phosphate} \xrightarrow{\text{(new enzyme)}} \text{UDP-galactose} + PP_i$$

$$\text{UDP-galactose} \xrightarrow{\text{galactose-4-epimerase}} \text{UDP-glucose}$$

$$\text{Net: Galactose} + \text{ATP} + \text{UTP} \longrightarrow \text{UDP-glucose} + \text{ADP} + PP_i$$

UDP-glucose is subsequently utilized to synthesize glycogen; or it can be hydrolyzed to UMP and glucose-1-phosphate.

anabolic route requires the hydrolysis of 3 additional ATPs per turn. (*b*) the differences summarized in Table 21-1

Chapter 21

1 CO_2 is a reactant in the acetyl-CoA carboxylase reaction; incubation with $^{14}CO_2$ will not lead to ^{14}C incorporation into palmitate.

2 (*a*) Uniformly labeled ^{14}C-acetyl-CoA is transformed to ^{14}C-malonyl-CoA; these precursors are transformed to uniformly ^{14}C-labeled palmitate. (*b*) If a trace of uniformly labeled ^{14}C-acetyl-CoA in the presence of a large excess of unlabeled malonyl-CoA is introduced, the metabolic pool of malonyl-CoA does not become labeled with ^{14}C; hence, only 15,16-^{14}C-palmitate is formed.

3 8 acetyl-CoA (mitochondrial) + 15ATP + 14NADPH + $9H_2O$ → palmitate + 8CoA-SH + 15ADP + $15P_i$ + $14NADP^+$ + $2H^+$

4 (*a*) 3 deuteriums per palmitate; all 3 located on C-16 (*b*) 7 deuteriums per palmitate;

$$CH_3{-}CH_2{-}(\overset{\displaystyle H}{\underset{\displaystyle D}{C}}{-}CH_2)_6{-}\overset{\displaystyle H}{\underset{\displaystyle D}{C}}{-}COO^-$$

5 Acetyl-CoA (mitochondrial) + NADH + $NADP^+$ + 2ATP + $2H_2O$ → acetyl-CoA (cytosolic) + NAD^+ + NADPH + 2ADP + $2P_i$ + $2H^+$

6 The rate-limiting step in the biosynthesis of fatty acids is the carboxylation of acetyl-CoA catalyzed by acetyl-CoA carboxylase. High levels of citrate and isocitrate indicate that the conditions for fatty acid synthesis are favorable, due to an active citric acid cycle with the formation of a plentiful supply of ATP, reduced pyridine nucleotides, and acetyl-CoA. Consequently, citrate stimulates (increases the V_{max} of) the rate-limiting enzymatic step in fatty acid biosynthesis. Furthermore, since citrate binds more tightly to the filamentous form of the enzyme (the active form), the presence of citrate drives the protomer-filamentous equilibrium in the direction of the active form. In contrast, palmitoyl-CoA (the end product from fatty acid biosynthesis) drives the equilibrium in the direction of the inactive form. Hence, when the end product of fatty acid biosynthesis builds up, fatty acid biosynthesis is slowed down.

7 3 palmitic acid + glycerol + 7ATP + $4H_2O$ → tripalmitin + 7ADP + $7P_i$ + $7H^+$

8 Dihydroxyacetone phosphate + NADH + palmitic acid + oleic acid + 3ATP + CTP + choline + $4H_2O$ → phosphatidylcholine + NAD^+ + 2AMP + ADP + H^+ + CMP + $5P_i$; 7 ATP per phosphatidylcholine

9 β-Sitosterol mimics some of the regulatory actions of cholesterol, e.g., enzyme inhibition during cholesterol synthesis, absorption of cholesterol, inhibition of enzyme synthesis.

10 $^{14}CH_3{-}CH_2{-}(^{14}CH_2{-}CH_2)_6{-}^{14}CH_2{-}COO^-$; Ala can serve as a palmitate precursor

11 (*a*) They are not simply the reverse of each other; the

Chapter 22

1 If phenylalanine hydroxylase is defective, the biosynthetic route to Tyr is blocked, and Tyr must be obtained from the diet.

2 Glucose + $2CO_2$ + $2NH_4^+$ → 2 aspartate + $4H^+$ + $2H_2O$

3 The ammonia acceptor, 5-phosphoribosyl-1-pyrophosphate, will accumulate.

4 Bacterial mutants that are unable to synthesize Gly, Asp, or Glu would require the addition of the purines adenine and guanine. Furthermore, uridine and cytosine would be required for Asp and Glu auxotrophs, but no additional pyrimidines would be required for Gly auxotrophs.

5 (*a*) The rapid division of cells such as cancer cells depends on the rapid synthesis of DNA. Since DNA synthesis is limited by the availability of deoxythymidylate, a block in its synthesis caused by the irreversible inhibition of thymidylate synthesis via F-dUMP reduces the rate of cell division and thus tissue growth. (*b*) Tetrahydrofolate is converted to N^5,N^{10}-methylenetetrahydrofolate during the biosynthesis of glycine from serine, i.e.,

Serine + tetrahydrofolate ⟶ glycine + N^5,N^{10}-methylenetetrahydrofolate + H_2O

The inhibition of dihydrofolate reductase by methotrexate prevents the recycling of dihydrofolate back to N^5,N^{10}-methylenetetrahydrofolate. Consequently, as cells become starved in the N^5,N^{10}-methylenetetrahydrofolate necessary for dTMP synthesis, DNA synthesis is reduced, the rate of cell division is reduced, and tissue growth is reduced. Although the drug also affects normal cells, it is not as critical because normal cells grow more slowly anyway.

6 Organisms do not store nucleotides to be used as fuel and do not completely degrade them, but rather hydrolyze them to their bases and recover the bases in salvage pathways. The low ratio of carbon to nitrogen makes nucleotides poor sources of energy.

7 (*a*) As shown in Figure 22-7, *p*-aminobenzoic acid is a component of N^5,N^{10}-methylenetetrahydrofolate, the cofactor involved in the transfer of one-carbon units. (*b*) Sulfanilamide is a structural analog of *p*-aminobenzoic acid. In the presence of sulfanilamide, bacteria are unable to synthesize tetrahydrofolate, a cofactor necessary for the transformation of 5′-phosphoribosyl-4-carboxamide-5-aminoimidazole to 5′-phosphoribosyl-4-carboxamide-N^5-formylaminoimidazole by the addition of —CHO, and the precursor accumulates. (*c*) The addition of excess *p*-aminobenzoic acid reverses the growth inhibition and ribonucleotide accumulation because *p*-aminobenzoic acid and sulfanilamide compete for the same active site in the enzyme (competitive inhibition) involved in tetrahydrofolate biosynthesis. Such competitive inhibition (Chapter 9) is removed by the addition of excess substrate.

8 Treatment with allopurinol has two biochemical consequences. First, the conversion of hypoxanthine to uric acid is inhibited, causing the accumulation of hypoxanthine, which is more readily excreted (more soluble). This alleviates the clinical problems associated with AMP degradation. Second, the conversion of guanine to uric acid is also inhibited, causing the accumulation of xanthine, which is, unfortunately, even less soluble than uric acid. This is the source of xanthine stones. Since the amount of GMP degradation is low relative to AMP degradation, the incidence of xanthine stones is lower than in untreated cases of gout because the main route to xanthine from hypoxanthine has been inhibited by allopurinol.

9 The bacteria in the root nodules maintain a symbiotic relationship with the plant; namely, the plant supplies ATP and reducing power while the bacteria supply ammonium ion by reducing atmospheric nitrogen. This reduction requires large quantities of ATP.

10 O H N * * * O N * COO⁻ H

Chapter 23

1 These observations suggest that ATP and NADPH are generated in the light; conversion stops as the supply of NADPH and ATP becomes exhausted.

2 For the maximum photosynthetic rate, photosystem I (which absorbs at 700 nm) and photosystem II (which absorbs at 680 nm) must be operating simultaneously.

3 Purple sulfur bacteria use H_2S as the hydrogen donor in photosynthesis. No oxygen is evolved because photosystem II is absent.

4 0.57

5 DCMU blocks electron transport between photosystem II and the first ATP production site.

6 (*a*) +13.4 kcal (*b*) 0.29 V

7 X is 3-phosphoglycerate and Y is ribulose 1,5-diphosphate.

8 Lowering the K_M for CO_2 when the pH of the medium is increased serves to activate ribulose-1,5-diphosphate carboxylase and thus increase the rate of CO_2 fixation. Illumination increases the pH of the medium. At night, the absence of illumination decreases the activity of ribulose-1,5-diphosphate carboxylase and lowers the rate of photorespiration.

9 In maize, carbon dioxide is fixed by the Hatch-Slack pathway, in which phosphoenolpyruvate is carboxylated rapidly to oxaloacetate (some of which undergoes transamination into aspartate) and reduced to malate. Only after subsequent decarboxylation does the CO_2 enter the Calvin cycle.

10 The enzyme reaction analogous to malate enzyme in the Hatch-Slack pathway is isocitrate dehydrogenase in the citric acid cycle.

11 Since photosystem I can synthesize ATP by cyclic photophosphorylation, only photosystem I is required in the mesophyll cell to operate the Hatch-Slack pathway.

12 Illumination will cause protons to be pumped into the phospholipid vesicles by bacteriorhodopsin. This proton gradient is then used by the mitochondrial F_0F_1ATPase to synthesize ATP.

Chapter 24

1 8 kcal; 0.60 mol

2 Pancreatic secretions contain high concentrations of HCO_3^-, which adjusts the pH to the optimal range when mixed with gastric secretions.

3 Since β-casein has little structure and the native conformation resembles a random coil, all sections of the protein chain are readily accessible to hydrolysis by pepsin and pancreatic enzymes. In contrast, since α-keratins have a high degree of secondary and tertiary structure, with extensive S—S cross-linking, their denaturation at the low pH of the stomach is limited, access to the peptide bonds by digestive enzymes is difficult, and hydrolysis is slow.

4 (*a*) Activation of pepsinogen occurs spontaneously in the acidic environment of the stomach. The resulting pepsin serves to catalyze the activation of the remaining pepsinogen molecules. (*b*) Pepsinogen is produced and stored in the chief cells as an inactive precursor in a medium that is slightly alkaline. If the fragment should be removed from pepsinogen, the fragment binds tightly to the active site of pepsin, acting as an inhibitor.

5 Proteolytic enzymes secreted by the pancreas do not damage the epithelial cells of the small intestine because: (a) the concentration of dietary protein is higher than that of protein on the surface of epithelial cells; (b) pancreatic enzymes undergo self-proteolysis after digestion is complete; (c) secretion of pancreatic enzymes is regulated by hormones; i.e., secretion only occurs when a protein-rich meal is present in the stomach; and (d) the epithelial cells of the small intestine secrete a mucous coating that protects the surface of the cells.

6 As digestion proceeds, proteins are cleaved into smaller polypeptides by the pancreatic endopeptidases. Consequently, in the later stages of digestion, the number of carboxyl terminals available for carboxypeptidase increases the carboxypeptidase efficiency.

7 The substance responsible is lactose. The ability to secrete lactase, the enzyme that hydrolyzes lactose, decreases with age. Unabsorbed lactose passes through the small intestine to the large intestine, where it is fermented by intestinal bacteria.

8 If lipids in the stool contain unhydrolyzed triacylglycerols, insufficient pancreatic secretion, which contains lipase, is suggested as a probable cause of the steatorrhea. If lipids in the stool contain hydrolyzed triacylglycerols, insufficient bile secretion, which contains bile salts necessary for absorption, is suggested as the probable cause.

9 The higher concentrations of alanine and glutamine reflect the importance of the glucose-alanine cycle and the transport of ammonia via glutamine.

10 About two-thirds of the ATP synthesized by the kidney is used for ion pumping.

11 Antimycin A and iodoacetate inhibit electron transport and glycolysis, respectively. The energy source for contraction is the phosphocreatine reservoir.

12 Steady-state levels of ATP are maintained by phosphate transfer from phosphocreatine. Fluoro-2,4-dinitrobenzene inhibits creatine kinase.

13 Ammonia is very toxic to nervous tissue, especially in the brain. Excess ammonia is removed by the transformation of glutamate to glutamine, which travels to the liver and is subsequently transformed to urea. The additional glutamate arises from the transformation of glucose to α-ketoglutarate. Additional ammonia is removed by transamination of α-ketoglutarate to glutamate.

14 $$\text{Glucose} \xrightarrow{\text{glycolysis}} \text{dihydroxyacetone phosphate}$$

$$\xrightarrow[\text{NAD}^+]{\text{NADH} + \text{H}^+} \text{glycerol 3-phosphate} \longrightarrow \text{triacylglycerols}$$

15 The amount of circulating albumin is related to the total plasma volume. The loss of albumin in nephrotic patients leads to an osmotic-pressure difference between the blood plasma and the extracellular space, and hence results in a net movement of water into the extracellular space.

16 Pancreatic secretion is governed by the type and amount of food ingested, especially protein. The reduction of fluid, caloric, and electrolyte intake is compensated by the intravenous administration of glucose-saline. Irritated pancreatic tissue has a reduced ability to release insulin; hence hyperglycemia results.

17 Increased muscular activity increases the necessary ATP flux, which is met by increased oxygen consumption. During the sprint, muscle transforms some glycogen to lactate. After the sprint, lactate is transported to the liver where it is converted back to glucose and glycogen. This process requires ATP and thus requires oxygen above the amount necessary in the resting state.

18 The excretion of phosphate (PO_4^{3-}) must be accompanied by the excretion of cationic counterions (Na^+ or H^+) in order to maintain electrical neutrality in the urine. To minimize Na^+ loss during phosphate secretion, the distal tubule exchanges sodium for a proton. This increases the concentration of protons in the urine.

19 (*a*) The shallow and irregular breathing in a narcotic overdose causes the accumulation of CO_2 in the lungs. The increase in CO_2 concentration drives the following equilibrium toward the right, lowering the pH of blood plasma

$$CO_2 + H_2O \rightleftharpoons H_2CO_3 \rightleftharpoons H^+ + HCO_3^-$$

(*b*) Mechanical venting lowers the CO_2 concentration and drives the equilibrium to the left. The resulting higher pH of the blood plasma in the lungs increases the affinity of hemoglobin for oxygen. (*c*) It increases the pH and increases the buffer capacity of blood plasma.

20 Glucose is the primary fuel of the brain. A key reaction in the catabolism of glucose is the thiamine pyrophosphate–dependent oxidative decarboxylation of pyruvate to acetyl-CoA. Hence thiamine deficiency reduces glucose utilization by the brain.

21 Active transport of Na^+ during reabsorption in the renal tubes is carried out by the ATP-dependent Na^+K^+ATPase. In this coupled process the reabsorption of three Na^+ ions requires the simultaneous secretion of two K^+ ions.

22 Drinking large amounts of seawater causes the total electrolyte concentration in the extracellular fluid to be much higher than the total electrolyte concentration in the intracellular fluid. This concentration gradient produces a net transport of water from the interior of the cell to the extracellular compartment. As the cell becomes depleted of water, the delicate internal organelles of the cell (e.g., mitochondria) shrink, leading to irreversible damage.

Chapter 25

1 1.87×10^2 m (approximately the length of two football fields)

2 Rapid inactivation provides a rapid means to change hormone concentrations. A constant concentration of insulin is maintained by having equal rates of synthesis and degradation. Other means of varying hormone concentration include changes in the rate of release from storage, transport, and conversion from prohormone to active hormone.

3 Because of their low solubility in a lipid environment, water-soluble hormones cannot penetrate the cell membrane. Rather, they bind to a receptor on the surface of the cell. In the case of adrenaline, this receptor is an enzyme that catalyzes the formation of a large amount of second messenger (cyclic AMP) on the *interior* of the cell. In contrast, lipid-soluble hormones can readily penetrate the hydrophobic core of the cell membrane. Once inside the cell they can act on receptors directly.

4 (*a*) Since adenylate cyclase is a membrane-bound protein, centrifugation of the preparation sediments the adenylate cyclase activity into the particulate fraction. (*b*) Adrenaline stimulates the production of cyclic AMP, a soluble substance that stimulates glycogen phosphorylase. (*c*) Cyclic AMP is the heat-stable substance. Cyclic AMP can be prepared by treating ATP with barium hydroxide.

5 Unlike cyclic AMP, dibutyryl cyclic AMP passes through the cellular membrane.

6 Cholera toxin increases the level of cyclic AMP in intestinal epithelial cells. The observations suggest that cyclic AMP regulates Na^+ permeability. Treatment consists of replacing lost body fluids and electrolytes.

7 (*a*) Heart and skeletal muscle lack the enzyme glucose-6-phosphate phosphatase. Consequently any glucose 6-phosphate that is produced enters the glycolytic pathway, and under oxygen-deficient conditions is converted into lactate. (*b*) The charged nature of phosphorylated intermediates prevents their escape from the cell, because the membrane is not permeable to charged species. In a "fight or flee" situation the concentration of glycolytic precursors in muscle tissue needs to be high in preparation for muscular activity. By maintaining phosphorylated precursors, their escape from muscle tissue is prevented. The liver, on the other hand, provides the glucose necessary to keep up the blood glucose level. Consequently, it is desirable that glucose can readily move from the liver cells to the blood stream in a "fight or flee" situation, by glucose-6-phosphate phosphatase–catalyzed dephosphorylation.

8 Excessive secretion of insulin by the pancreas promotes excessive utilization of blood glucose by the liver, leading to hypoglycemia. In addition, high insulin levels also shut down amino and fatty acid catabolism. Thus, patients have little circulating fuel available for ATP requirements. If prolonged, hyperinsulinism leads to brain damage because glucose is the main source of fuel for the brain.

9 The observation that the incubation of liver tissue with thyroxine increases respiration and heat output while ATP concentrations remain constant is consistent with the thesis that thyroxine acts as an uncoupler of oxidative phosphorylation. Uncouplers act to lower the P/O ratio of tissues, and thus the tissue must increase respiration to meet the normal ATP demands. The observed thermogenesis could also be due to the increased rate of ATP utilization by the thyroid-stimulated tissue. In such tissue the increased ATP demands are met by increased oxidative phosphorylation (respiration), and thermogenesis results. Despite much research, the details of how thyroid hormones regulate the rate of aerobic metabolism remain a mystery.

10 The female hormone estrogen serves to promote the development of female secondary sex characteristics. One such characteristic is the growth and development of breast tissue. Since the carcinoma occurs in the breast tissue, any effect that reduces the growth rate of breast tissue will slow down the growth of the carcinoma. Reduced growth rate of breast tissue can be achieved by lowering the estrogen concentration by ovariectomy. Sometimes the treatment includes administration of an antagonist (a hormone that opposes the effect of estrogen—e.g., testosterone) to estrogen.

11 The adrenal glands, which are located just above the kidneys, are essentially part of the nervous system, from which they receive signals. Consequently it is not unexpected that the synthesis of endorphins occurs by both the brain and the adrenal medulla.

12 Several of the polypeptide hormones, namely insulin and glucagon, are synthesized as inactive prohormones that contain larger polypeptide chains than the active hormones themselves. One advantage of the prohormone is that it can be stored in quantity in secretory granules since it is inactive. Activation can be achieved rapidly by enzymatic cleavage in response to an appropriate signal.

13 Adrenaline is administered to patients suffering from acute asthma because it relaxes the smooth muscle surrounding the bronchioles of the lungs. The administration of adrenaline stimulates the formation of cyclic AMP in target cells. Cyclic AMP is destroyed, however, by phosphodiesterase-catalyzed hydrolysis. Since phosphodiesterase is inhibited by the purine derivatives caffeine, theophylline, and aminophylline, the administration of these drugs prolongs and intensifies the activity of adrenaline by decreasing the rate of breakdown of cyclic AMP.

14 These observations indicate that the activity of phosphodiesterase is stimulated by Ca^{2+} mediated by calmodulin. This is consistent with the known properties of calmodulin as a Ca^{2+}-binding protein. Binding of the calmodulin–Ca^{2+} complex to phosphodiesterase stimulates its activity toward cyclic AMP hydrolysis.

Chapter 26

1 The recommended levels contain an appreciable margin of nutritional safety above the minimal requirements.

2 0.21 ATP/g glucose, 0.50 ATP/g palmitic acid; 4.2 kcal/g glucose, 9.5 kcal/g palmitic acid; comparing the two sets of results

$$\frac{0.50\ \text{ATP/g}}{0.21\ \text{ATP/g}} = 2.4 \qquad \frac{9.5\ \text{kcal/g}}{4.2\ \text{kcal/g}} = 2.3$$

indicates that the ratio of the number of ATPs produced is the same as the ratio of the amount of heat released.

3 Within several hours after a fast is begun, blood glucose levels start to drop as glucose is absorbed by the tissues. To maintain the level of the blood glucose, humans catabolize glucogenic amino acids. This process requires the excretion of nitrogen in the form of urea, accompanied by large amounts of water. Within a few days of continued fasting, the body switches from amino acid catabolism to fatty acid catabolism to meet its energy needs, and hence reduces daily water loss substantially.

4 Obesity is the result of caloric intake in excess of body needs. On a per-gram-of-diet basis, a diet rich in fat is more likely to lead to obesity than a diet rich in sugar, because the former contains a higher caloric content per gram. However, most obesity problems result from the ingestion of too many total calories. Since sweet foods are consumed for pleasure, the prevalent tendency for obese people is a diet too rich in sugar.

5 (*a*) 3.16 kcal/g (*b*) 4.21 kcal/g (*c*) Based on the information, bran flakes consist of a mixture of protein and carbohydrate. Elemental analysis would show that little nitrogen is present, so the principal component must be carbohydrate. (*d*) Only when the food is completely digested and absorbed will the heat released when oxidized in the body be the same as the heat released when oxidized in the calorimeter.

6 3.2 baked potatoes; 19.2 medium french fries; 26.3 medium potato chips; 4.8 slices of white bread; 6 pats of butter; 1.6 cans of beer

7 If one of the essential amino acids is deficient in the diet, protein synthesis will proceed until the supply is exhausted. To replenish the supply, additional intake of the unbalanced mixture is needed or body protein will be broken down. This process will yield "extra" amino acids that will subsequently be catabolized, causing the nitrogen to be excreted and resulting in a negative nitrogen balance.

8 The rat is placed on a carefully controlled diet consisting of all the amino acids except Phe. To keep the total nitrogen intake constant over the set of experiments, an appropriate amount of urea is added. Since the rat is deficient in one amino acid, it will show a negative nitrogen balance; i.e., the amount of dietary nitrogen will be smaller than the amount of excreted nitrogen (see Problem 7). Phe is then added to the diet and the nitrogen balance remeasured. The minimal daily requirement is the amount of Phe necessary to establish a normal (or slightly positive) nitrogen balance.

9 Young children suffering from kwashiorkor show distended bellies due to retained water in the intracellular spaces; this results from inadequate levels of serum albumin in the blood plasma. When an adequate protein

diet is provided, serum albumin levels return to normal, osmotic-pressure differences are reversed, and water loss results. Thus, the initial weight loss is due to water.

10 Patients with limited kidney function must undergo regular dialysis to remove toxic wastes, principally urea and uric acid, from the blood. To minimize the need for this procedure, nitrogen excretion should be minimized. This is accomplished through a diet that is nutritionally balanced in the amount and distribution of amino acids, i.e., proteins with a biological value near 100. Since egg contains all the essential amino acids and has a higher biological value than corn, it is more nutritionally balanced than corn and would be the better food for the patient.

11 Vitamin B_6, pyridoxine, is used in the synthesis of the coenzyme pyridoxal phosphate. This enzyme is required by transaminase enzymes, which catalyze the first step in amino acid catabolism.

12 Gastrointestinal bacteria supply a significant fraction of necessary vitamins (bacteria synthesize them) to their host. A diet that reduces the intestinal population could lead to a vitamin deficiency.

13 From a nutritional standpoint, people need a balanced diet to meet their biochemical requirements for energy, and to provide building blocks and cofactors for biosynthesis. These requirements can be met by dietary intake from a large variety of sources; there is no evidence that a specific food (e.g., milk) is required for proper nutrition.

14 (*a*) An adequate supply of fuel to carry out strenuous work and maintain body temperature (*b*) foods that build up the stores of glycogen; e.g., easily digestible carbohydrate (honey, dried fruit, chocolate, pancakes, etc.) (*c*) proteins and fats (*d*) possibly Na^+ and K^+

15 Mammals, including humans, have no available route for converting two-carbon units (acetyl-CoA) to the three-carbon units (pyruvate) necessary for gluconeogenesis. Hence, although ethanol is transformed into acetyl-CoA, it cannot be converted into glucose.

16 47.8 lb

17 (*a*) The hydrolysis of protein in the small intestine leads to a mixture of amino acids. These amino acids fall into two categories: glucogenic (catabolism of the carbon skeleton leads to the formation of pyruvate) and ketogenic (catabolism of the carbon skeleton leads to the formation of acetyl-CoA). Since pyruvate is transformed into acetyl-CoA, the carbon skeletons of all amino acids can lead to the formation of fatty acids and thus the deposition of triacylglycerols. (*b*) The catabolism to amino acids is accompanied by the excretion of nitrogen in the form of urea. Large volumes of water are needed to produce dilute urine.

Chapter 27

1 One DNA contains 32% A, 32% T, 18% G, and 18% C; the other, 17% A, 17% T, 33% G, and 33% C. This conclusion assumes that both were double-stranded. The DNA containing 33% G and 33% C must have come from the thermophilic bacterium, since this DNA would be much more stable to heat than the other. G–C base pairs have three hydrogen bonds and thus are associated more strongly than A–T base pairs, which have only two.

2 (5′)GAATGCATACGGCAT(3′)

3 0.94 mg

4 372 base pairs; the actual length is probably much greater because nearly all eukaryotic genes contain introns, which may be longer than the exons; most eukaryotic genes also code for a leader or signal sequence in their protein products.

5 200,000 base pairs; the total T2 DNA is 68,000 nm, whereas the T2 head is only 100 nm.

6 The ratio of DNA length to diameter is 6.8 for the nucleosome and 10^6 for the nucleus. The packaging is much more compact in the nucleus.

7 No, because it would not be maximally base-paired. In an intact cell, there would be DNA segments in which the negative charges in the backbone were neutralized by histones, allowing the DNA more flexibility to assume a cruciform structure.

8 Since neither the A and T residues nor the G and C residues are paired, the M13 DNA is single-stranded.

9 (*a*) A single-stranded linear DNA with the contour length of the nicked strand of RFII, and a single-stranded circular DNA, i.e., the intact strand of RFII (*b*) There will be three linear single-stranded DNAs. The one arising from the un-nicked strand will have 5386 bases; the other two, from the nicked strand, will be smaller.

10 If there is no distinctive label to mark the 5′ end of the DNA, there will be no reference point for assignment of the base sequence.

11 No; the two strands of double-helical DNA differ in base composition (see Figure 27-12).

12 The exons of this gene contain 3 × 192 = 576 base pairs. The 864 remaining base pairs are contributed by introns and possibly a leader or signal sequence.

13 Centrifugation of the two DNAs in a cesium chloride density gradient would show that the *E. coli* DNA, which has a higher content of G–C pairs, would be denser and would reach equilibrium at a lower point in the gradient than the sea urchin DNA (see Table 27-3).

14 The "eyes" are due to unfolding of local segments of duplex DNA. These segments are rich in A–T base pairs, which are less stable to heat than G–C pairs. This procedure is useful in detecting differences in base composition along the length of DNA duplexes.

15 Because homologous proteins from different species show amino acid sequence homology

16 (*a*)

—A—T—G—C—T—A—G—C—A—T—
•
—T—A—C—G—A—T—C—G—T—A—

(*b*) This is the center of twofold rotational symmetry; 180° rotation about this point yields identical base sequences. (*c*) The restriction endonuclease evidently produced a staggered cut, leaving the cohesive ends still adhering to each other. The DNA therefore remained a circle. (*d*) The alkaline medium caused unwinding of the circular duplex DNA and separation of the cohesive ends, to yield two linear single strands.

17 No; the central dogma says that genetic information flows from DNA to RNA to protein. The RNA chromosome can transfer the information to protein without involving DNA.

Chapter 28

1 If dispersive replication had taken place at random, the density of the first-generation DNAs would have been the same as actually observed, a single band midway between heavy and light DNA. But in the second generation all the DNAs would have had the same density and would have appeared as a single band, midway between that observed in the first generation and that of light DNA.

2 (*a*) Since thymidine residues are present in DNA but not in RNAs, only DNA would become labeled and the experiment could be conducted without interference from labeled RNAs. (*b*) Neither guanosine nor adenosine would be useful, since they are present in both DNA and all RNAs. (*c*) Thymidine is phosphorylated by ATP in steps to yield thymidine 5′-triphosphate

Thymidine + ATP ⟶ thymidine 5′-phosphate + ADP
TMP + ATP ⟶ TDP + ADP
TDP + ATP ⟶ TTP + ADP

3 400,000 turns

4 (*a*) 44 min (*b*) One possibility is that each *E. coli* chromosome is replicated by four forks starting from two origins.

5 (*a*) 0.42 s (*b*) the number of base pairs in the introns, and the leader or signal sequence

6 (*a*) (5′)CTAATGCAACGTTGCAAGCT(3′) (*b*) (5′)AGCUUGCAACGUUGCAUUAG(3′)

7 (*a*) A, 21%; U, 21%; G, 29%; C, 29% (*b*) It may or may not be the same as in part (*a*).

8 New DNA made from the given template would have A, 32.7%; G, 18.5%; C, 24.1%, and T, 24.7%. New DNA made from the template complementary to the given one would have A, 24.7%; G, 24.1%; C, 18.5%; and T, 32.7%. The base composition of the two new DNAs together would be A, 28.7%; G, 21.3%; C, 21.3%; and T, 28.7%. It is assumed that the two template strands are replicated completely.

9 RNAs are transcribed from only one strand of duplex DNA.

10 (*a*) About 2000 (*b*) Okazaki fragments are formed by DNA polymerase III from an RNA primer and a DNA template. Since the Okazaki fragments in *E. coli* are about 2000 bases long, they are firmly bound to the template strand by base pairing. Each fragment is quickly joined to the lagging strand by the successive action of DNA polymerase I and DNA ligase, thus preserving the correct order of the fragments. Therefore a mixed pool of different Okazaki fragments, detached from their template, does not form during normal replication.

11 Leading strand
Precursors: dATP, dGTP, dCTP, dTTP
Enzymes: DNA gyrase, helicase, DNA-binding proteins, DNA polymerase III, inorganic pyrophosphatase
Cofactors: Zn^{2+}, Mg^{2+}
Lagging strand
Precursors: ATP, GTP, CTP, UTP, dATP, dGTP, dCTP, dTTP
Enzymes: DNA gyrase, helicase, DNA-binding proteins, primase, DNA polymerase III, DNA polymerase I, DNA ligase, pyrophosphatase
Cofactors: Zn^{2+}, Mg^{2+}, NAD^+

12 (*a*) (1) Watson-Crick base pairing between the template and leading strand (2) hydrophobic stabilization of stacked base pairs (3) enzymatic hydrolysis of pyrophosphate formed in the DNA polymerase reaction, ensuring that the reaction goes to near-completion at each step (4) proofreading and removal of wrongly inserted nucleotides by the 3′ exonuclease activity of DNA polymerase III
(*b*) Because the four factors ensuring fidelity of replication are operative in the leading and the lagging strands, it is expected that the lagging strand would probably be made with the same fidelity. However, because the number of distinct chemical operations involved in making the lagging strand is greater than that for the leading strand, there would seem to be greater opportunity for errors to arise in replication of the lagging strand.

13 (*a*) It ensures that foreign intact circular DNAs cannot be replicated in, e.g., *E. coli* cells, unless they possess an origin identical to that of *E. coli*. (*b*) DNA replicase of *E. coli* must replicate an *E. coli* chromosome beginning at its origin, which has a distinctive base sequence.

14 (*a*) The requirement of NTPs by RNA polymerase would enable the resulting pyrophosphate to be cleaved by pyrophosphatase, thus pulling the reaction further to completion and ensuring high fidelity. Polynucleotide phosphorylase requires NDPs, and thus the leaving group is phosphate. Since phosphate concentration in cells is relatively high, polynucleotide phosphorylase is more likely to function in reverse, i.e., in degradation of RNAs. (*b*) RNA polymerase requires all four NTPs and a template, whereas polynucleotide phosphorylase does not require all four NDPs or a template, factors which further support its function in degrading RNAs.

15 A single base error in DNA replication, if not corrected, would cause one of the two daughter cells, as well as all its descendants, to have a mutated chromosome. A single base error by RNA polymerase in a cell would not affect the DNA chromosome; it would lead to formation of some defective copies of one protein, but since mRNAs turn over rapidly, most copies of this protein would be normal. The progeny of this cell would be normal.

Chapter 29

1 (*a*) Gly–Gln–Ser–Leu–Leu–Ile (*b*) Leu–Asp–Ala–Pro (*c*) His–Asp–Ala–Cys–Cys–Tyr (*d*) Met–Asp–Glu in eukaryotes; fMet–Asp–Glu in prokaryotes

2 Because nearly all the amino acids have more than one codon (e.g., leucine has six), any given polypeptide can be coded for by a large number of different base sequences, if all the codons for each amino acid are employed in different combinations.

3 UUAAUGUAU, UUGAUGUAU, CUUAUGUAU, CUCAUGUAU, CUAAUGUAU, CUGAUGUAU, UUAAUGUAC, UUGAUGUAC, CUUAUGUAC, CUCAUGUAC, CUAAUGUAC, CUGAUGUAC

4 (a) (5′)CGACGGCGCGAAGUCAGGGGUGUUAAG(3′)
(b) Arg–Arg–Arg–Glu–Val–Arg–Gly–Val–Lys
(c) No, because the complementary antiparallel strands in double-helical DNA do not have the same base sequence in the 5′ → 3′ direction. RNA is transcribed from only one specific strand of duplex DNA. The RNA polymerase must therefore recognize and bind to the correct strand.

5 There are two tRNAs for methionine; one is the initiating $tRNA^{fMet}$, and the other is $tRNA^{Met}$, which can carry methionine to interior positions in a polypeptide. $tRNA^{fMet}$ reacts with methionine to yield methionine-$tRNA^{fMet}$, promoted by methionine aminoacyl-tRNA synthetase. The amino group of its methionine residue is then formylated by N^{10}-formyltetrahydrofolate to yield N-formylmethionine-$tRNA^{fMet}$. Free methionine or methionine-$tRNA^{Met}$ cannot be formylated. Only N-formylmethionine-$tRNA^{fMet}$ can bind to the initiating methionine codon AUG in the mRNA, because it can also recognize a special initiation signal containing six or more A and G residues upstream of AUG. This signal cannot be recognized by methionine-$tRNA^{Met}$. AUG in interior positions can bind and incorporate only methionine-$tRNA^{Met}$.

6 Polynucleotide phosphorylase would be allowed to act on a mixture of UDP and CDP in which UDP would have, say, five times the concentration of CDP. The result would be an RNA-like polymer in which there would be many UUU triplets (coding for Phe) and a small number of UUC (Phe), UCU (Ser), and CUU (Leu). Also coded, but in much smaller amounts, would be UCC (also Ser), CCU (Pro), and CUC (Leu).

7 At least 583 high-energy phosphate groups, depending upon the number of errors that are detected and corrected by the aminoacyl-tRNA synthetases. Correction of each error requires 2 high-energy phosphate groups. For glycogen synthesis, 292 high-energy phosphate groups. The extra energy cost for β-globin synthesis is 291 high-energy phosphate groups, or about twice the cost of glycogen synthesis.

8 At least 20 activating enzymes, 70 ribosomal proteins, 4 ribosomal RNAs, 20 or more tRNAs, an mRNA, and 10 or more auxiliary enzymes must be made by the eukaryotic cell in order to synthesize a protein from amino acids. In contrast, the synthesis of an $\alpha(1 \rightarrow 4)$ chain of glycogen from glucose requires only 4 or 5 enzymes.

9

Glycine codons	Recognized by anticodons	
(5′)GGU	(5′)ACC and	(5′)GCC and ICC
GGC	GCC	ICC
GGA	UCC	ICC
GGG	CCC	UCC

(a) The base pairs at the 5′ end and middle positions (b) The anticodons (5′)GCC, ICC, and UCC will have a wobbly base pair with their codons. (c) in the pairings involving the anticodons (5′)ACC, GCC, UCC, and CCC (d) those with the anticodons in part (c), since the tRNAs containing them will be released at a lower rate than the other tRNAs for Gly, because of the strong binding of all three anticodon bases

10 It would cause a frame-shift error in translation, beginning at the site of binding of the unusual aminoacyl-tRNA.

11 (a), (c), (e), and (g); mutations (b), (d), and (f) cannot be the result of single-base mutations; (b) and (f) would require substitutions of two bases, and (d) would require substitutions of all three bases.

12 The two DNA codons for Glu are (5′)TTC and (5′)CTC, and the four DNA codons for Val are (5′)TAC, (5′)CAC, (5′)AAC, and (5′)GAC. A single-base change in (5′)TTC (Glu) to (5′)TAC (Val) or in (5′)CTC (Glu) to (5′)CAC (Val) could account for the replacement in sickle-cell hemoglobin. Much less likely are two-base changes from (5′)TTC to (5′)CAC, (5′)AAC, and (5′)GAC; and from (5′)CTC to (5′)TAC, (5′)AAC, and (5′)GAC.

Chapter 30

1 (a) The lesion must be repaired by DNA polymerase I, which acts only in the 5′ → 3′ direction. (b) mRNA is single-stranded; hence there is no template strand to correct from. (c) Double-strand breaks caused by x-rays cannot be repaired, since the broken ends are unlikely to rejoin properly.

2 (1) No change, if the mutation yields another codon for the same amino acid (2) no change, if the mutation occurs in a functionless region within an intron (3) an amino acid substitution yielding an improved, unchanged, less active, or inactive protein (4) shortening of the protein by mutation of a codon to a termination signal, or lengthening by mutation of a termination signal to an amino acid codon

3 The products resulting will be as follows, with the cohesive ends formed by *Eco* RI indicated.

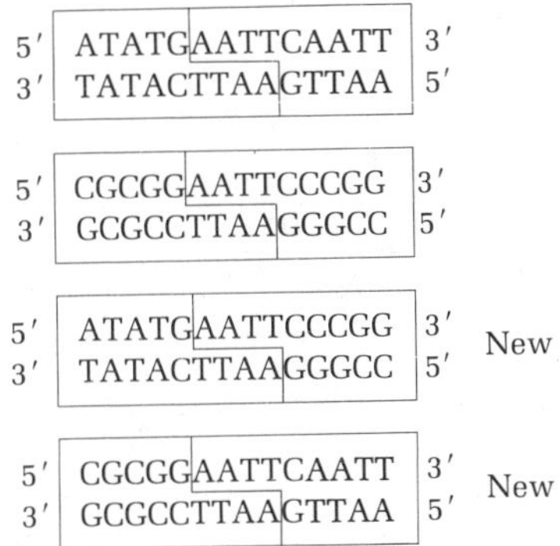

4 UAA will cause termination of the polypeptide product only at position 334–336 of a prokaryotic mRNA. UAA will not cause termination at positions 330–332 or 338–340 because they are not in the correct reading frame. In a eukaryotic mRNA, termination will occur at 334–336 if there are no intervening sequences before that point. If UAA occurs within an intervening sequence, there will probably be no effect. If it is in an exon but not the first one, the effect on the protein product will depend on the number of nucleotides removed in the preceding introns.

5 (*a*) It's a double- or single-stranded circle. (*b*) It's a double-stranded circle, since *Eco* RI would not cut a single-stranded DNA. (*c*) A = 1.3×10^6; B = 0.8×10^6; C = 0.7×10^6; D = 0.6×10^6 (*d*) Fragments C and D are adjacent. From part (*c*), we know that fragments B and D are adjacent.

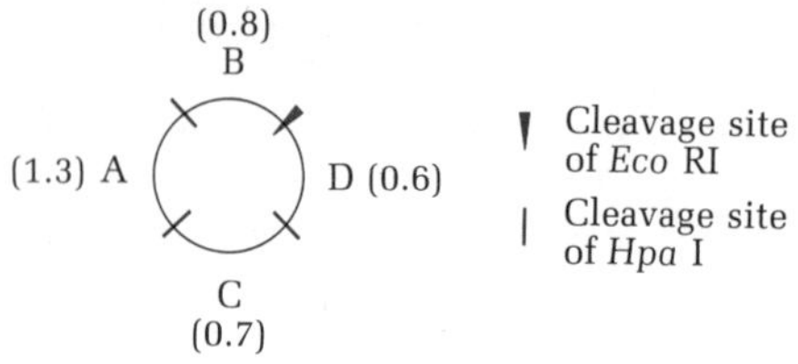

(0.8)
B

(1.3) A D (0.6)

C
(0.7)

Cleavage site of *Eco* RI

Cleavage site of *Hpa* I

(*e*) The restriction site at the border between fragments A and B was destroyed.

6 Exons may code for functional domains of proteins. A specific exon coding one domain of a given protein may be combined with an exon for a different domain of another protein, to form a gene for a new protein having the domains of both proteins.

7 (*a*) 5 mg/L (*b*) 10^5 molecules (*c*) 0.5 g

8 (*a*) Cleave the two DNAs with the restriction endonuclease, mix and warm them to inactivate the endonuclease and to separate the cohesive ends at the cleavage points. Allow the mixture to cool to form the noncovalent recombinants and splice them covalently with DNA ligase. (*b*) In addition to the desired circular duplex in which the original circular DNAs are combined, the by-products will include the original small and large circular DNAs and two other circular DNAs, one a recombinant of two large and the other of two small DNAs. In addition, there will be linear recombinants made up of the two large DNAs, the two small DNAs, and two kinds of linear recombinants of small plus large DNAs in two different sequences.

APPENDIX F Glossary

Absolute configuration: The configuration of four different substituent groups around an asymmetric carbon atom, in relation to D- and L-glyceraldehyde.

Absorption: Transport of the products of digestion from the intestinal tract into the blood.

Acceptor control: The regulation of the rate of respiration by the availability of ADP as phosphate acceptor.

Acid mucopolysaccharides: Acidic polysaccharides found in mucous secretions and in the intercellular space of higher animals.

Acidosis: A metabolic condition in which the capacity of the body to buffer H^+ is diminished; usually accompanied by decreased blood pH.

Actin: A protein constituting the thin filaments of muscle, but present in most other animal cells.

Activation energy: The amount of energy (in kcal) required to bring all the molecules in 1 mol of a reacting substance to the transition state.

Active site: The region of an enzyme surface that binds the substrate molecule and transforms it.

Active transport: Energy-requiring transport of a solute across a membrane in the direction of increasing concentration.

Activity: The true thermodynamic activity or potential of a substance, as distinguished from its molar concentration.

Activity coefficient: The factor by which the concentration of a solute must be multiplied to give its true thermodynamic activity.

Adenosine diphosphate: See ADP.

Adenosine triphosphate: See ATP.

Adipose tissue: Specialized connective tissue that functions to store large amounts of triacylglycerols.

ADP (adenosine diphosphate): A ribonucleoside 5′-diphosphate serving as phosphate-group acceptor in the cell energy cycle.

Aerobe: An organism that lives in air and utilizes oxygen.

Aerobic: Exposed to air or oxygen.

Aldose: A simple sugar in which the carbonyl carbon atom is at one end of the carbon chain.

Alkaloids: Nitrogen-containing organic compounds of plant origin; they are often basic and have intense biological activity.

Alkalosis: A metabolic condition in which the capacity of a body to buffer OH^- is diminished; usually accompanied by an increase in blood pH.

Allosteric enzymes: Regulatory enzymes whose catalytic activity is modulated by the noncovalent binding of a specific metabolite at a site other than the catalytic site.

Allosteric site: The specific "other" site on the surface of an allosteric enzyme molecule to which the modulator or effector molecule is bound.

α helix: See Helix, α.

Ames test: A simple bacterial test for carcinogens, based on the assumption that they are mutagens.

Amino acid activation: ATP-dependent enzymatic esterification of the carboxyl group of an amino acid with the 3′-hydroxyl group of its corresponding transfer RNA.

Amino acids: α-Amino-substituted carboxylic acids, the building blocks of proteins.

Aminoacyl-tRNA: An aminoacyl ester of a tRNA.

Aminoacyl-tRNA synthetase: An enzyme that catalyzes synthesis of an aminoacyl-tRNA at the expense of ATP energy.

Aminotransferases: Enzymes that catalyze transfer of amino groups from one metabolite to another; also called *transaminases*.

Amniocentesis: A procedure for withdrawing amniotic fluid for diagnosis of genetic abnormalities.

Amphibolic pathway: A metabolic pathway used in both catabolism and anabolism.

Amphipathic compound: A compound whose molecules contain both polar and nonpolar domains.

Amphoteric compound: A compound capable of both donating and accepting protons, and thus able to serve as either an acid or a base.

Anabolism: The phase of intermediary metabolism concerned with the energy-requiring biosynthesis of cell components from smaller precursor molecules.

Anaerobe: An organism that lives without oxygen.

Anaplerotic reaction: An enzyme-catalyzed reaction that can replenish the supply of intermediates of the citric acid cycle.

Angstrom (Å): A unit of length (10^{-8} cm) used to indicate molecular dimensions.

Anion-exchange resin: A polymeric resin with fixed cationic groups used in chromatographic separation of anions.

Anomers: Two stereoisomers of a given sugar that differ only in the configuration about the carbonyl (anomeric) carbon atom.

Antibiotic: One of many different organic compounds that are formed and secreted by various species of microorganisms and plants, are toxic to other species, and presumably have a defensive function.

Antibody: A defense protein synthesized by the immune system of a higher organism; it binds specifically to the foreign molecule (antigen) that induced its synthesis.

Anticodon: A specific sequence of three nucleotides in a transfer RNA, complementary to a codon for an amino acid in a messenger RNA.

Antigen: A molecule capable of eliciting synthesis of a specific antibody in vertebrates.

Antiparallel: Opposite in polarity or orientation (said of two linear polymers).

Asymmetric carbon atom: A carbon atom that is covalently bonded to four different groups and that may occupy two different tetrahedral configurations.

ATP (adenosine triphosphate): A ribonucleoside 5′-triphosphate functioning as a phosphate-group donor in the cell energy cycle.

ATPase: An enzyme that hydrolyzes ATP to yield ADP and phosphate; usually coupled to some process requiring energy.

ATP synthetase: An enzyme complex that forms ATP from ADP and phosphate during oxidative phosphorylation in the inner mitochondrial membrane.

Autotroph: An organism that can live on very simple carbon and nitrogen sources, such as carbon dioxide and ammonia.

Auxotrophic mutant (auxotroph): A mutant defective in the synthesis of a given biomolecule, which must thus be supplied for its normal growth.

Avogadro's number: The number of molecules in a gram molecular weight of any compound (6.023×10^{23}).

Back-mutation: A mutation that causes a mutant gene to regain its wild-type base sequence.

Bacteriophage: A virus capable of replicating in a bacterial cell.

Basal metabolic rate: The rate of oxygen consumption of a body at complete rest, long after a meal.

Base pair: Two nucleotides that are in different nucleic acid chains and whose bases pair by hydrogen bonding; e.g., A with T or U, and G with C.

β conformation: See Conformation, β.

β oxidation: See Oxidation, β.

Bilayer: A double layer of oriented amphipathic lipid molecules; the hydrocarbon tails face inward to form a continuous nonpolar phase.

Bile salts: Amphipathic steroid derivatives with detergent properties, participating in digestion and absorption of lipids.

Biomolecule: An organic compound normally present as an essential component of living organisms.

Biosphere: All the living matter on or in the earth, the seas, and the atmosphere.

Bond energy: The energy required to break a bond.

Buffer: A system capable of resisting changes in pH, consisting of a conjugate acid-base pair in which the ratio of proton acceptor to proton donor is near unity.

Building-block molecule: A biomolecule serving as a structural unit of a biological macromolecule; e.g., an amino acid, a sugar, or a fatty acid.

Calorie: The amount of heat required to raise the temperature of 1.0 g of water from 14.5 to 15.5°C.

CAP: See Catabolite activator protein.

Capsid: The protein coat of a virion or virus particle.

Carbohydrate: A polyhydroxylic aldehyde or ketone.

Carcinogen: A cancer-causing chemical agent.

Carotenoids: Lipid-soluble photosynthetic pigments formed of isoprene units.

Catabolism: The phase of metabolism involved in the energy-yielding degradation of nutrient molecules.

Catabolite activator protein (CAP): A specific regulatory protein that controls initiation of transcription of the genes of enzymes required for a cell to use some other nutrient when glucose is lacking.

Catalytic site: On an enzyme molecule, the site involved in the catalytic process.

Catecholamines: Hormones, such as adrenaline, that are amino derivatives of catechol.

Cation-exchange resin: An insoluble polymer with fixed negative charges; used in chromatographic separation of cationic substances.

cDNA: See Complementary DNA.

Central dogma: The organizing principle of genetic chemistry: genetic information flows from DNA to RNA to protein.

Chemiosmotic coupling: Coupling of ATP synthesis to electron transport via an electrochemical H^+ gradient across a membrane.

Chemotaxis: Sensing of and movement toward (or away from) a specific chemical agent by cells.

Chimeric DNA: Recombinant DNA containing genes from two different species.

Chimeric protein: Covalently joined proteins from different species; made in response to chimeric DNA.

Chiral compound: A compound that contains an asymmetric center and is capable of occurring in two nonsuperimposable mirror images.

Chlorophylls: Green pigments functioning as receptors of light energy in photosynthesis; they consist of magnesium-porphyrin complexes.

Chloroplasts: Chlorophyll-containing photosynthetic organelles in eukaryotic cells.

Chromatin: A filamentous complex of DNA, histones, and other proteins constituting eukaryotic chromosomes.

Chromatography: A process by which complex mixtures of molecules may be separated by many repeated partitionings between a flowing phase and a stationary phase.

Chromosome: A single large DNA molecule containing many genes and functioning to store and transmit genetic information.

Chylomicron: A blood plasma component consisting of a large droplet of triacylglycerols stabilized by a coat of protein and phospholipid.

Citric acid cycle: A cyclic system of enzymatic reactions for the oxidation of acetyl residues to CO_2, in which formation of citrate is the first step.

Clones: The descendants of a single cell.

Codon: A sequence of three adjacent nucleotides in a nucleic acid that codes for a specific amino acid.

Coenzyme: An organic cofactor required for the action of certain enzymes; often contains a vitamin as a component.

Coenzyme A: A pantothenic acid–containing coenzyme serving as an acyl-group carrier in certain enzymatic reactions.

Cofactor: A small-molecular-weight, heat-stable inorganic or organic substance required for the action of an enzyme.

Cohesive end: A single-strand loose end of duplex DNA complementary to a single-strand end of the same or another DNA molecule of the same polarity.

Colinearity: The linear correspondence between the sequence of amino acids in a polypeptide and the corresponding nucleotide sequence in the DNA and mRNA coding for it.

Colligative properties: Properties of solutions that depend on the number of solute particles per unit volume; e.g., freezing-point depression.

Common intermediate: A chemical compound common to two chemical reactions, as a product of one and a reactant in the other.

Competitive inhibition: A type of enzyme inhibition reversed by increasing the substrate concentration.

Complementary DNA (cDNA): A DNA that is usually made by reverse transcriptase and is complementary to a given messenger RNA; used in DNA cloning.

Configuration: The arrangement in space of substituent groups around an asymmetric carbon atom.

Conformation: The three-dimensional shape or form of a macromolecule.

Conformation, β: An extended, zigzag arrangement of a polypeptide chain.

Conjugate acid-base pair: A proton donor and its deprotonated species; e.g., acetic acid–acetate.

Conjugated protein: A protein containing a metal or an organic prosthetic group, or both.

Conjugation: The process by which DNA is transferred from an F^+ bacterium to an F^- bacterium.

Constitutive enzymes: Enzymes of central metabolic pathways; always present in normal cells.

Coordinate induction: Induction of a set of related enzymes by a single inducer.

Coordinate repression: Repression of a set of related enzymes.

Corticosteroids: Steroid hormones formed by the adrenal cortex.

Coupled reactions: Two chemical reactions that have a common intermediate and thus a means of energy transfer from one to the other.

Covalent bond: A chemical bond involving sharing of electron pairs.

Crossover point: In a multienzyme system, a point at which inhibition causes accumulation of the preceding intermediates and decreased concentration of the subsequent products.

Cyclic AMP (cyclic adenylate): The second messenger within cells; its formation by adenylate cyclase is stimulated by certain hormones.

Cyclic electron flow: In chloroplasts, the light-induced flow of electrons originating from and returning to photosystem I.

Cytochromes: Heme proteins serving as electron carriers in respiration and photosynthesis.

Cytoplasm: The portion of a cell's contents outside the nucleus or nucleoid.

Cytoskeleton: The filamentous skeleton in cytoplasm.

Cytosol: The continuous aqueous phase of cytoplasm, with its dissolved solutes.

Dalton: The weight of a single hydrogen atom (1.66×10^{-24} g).

Dark reactions: In photosynthetic cells, the light-independent enzymatic reactions concerned in the synthesis of glucose from CO_2, ATP, and NADPH.

Deamination: The enzymatic removal of amino groups from amino acids.

Degenerate code: A code in which a single element in one language is specified by more than one element in a second language.

Dehydrogenases: Enzymes catalyzing the removal of pairs of hydrogen atoms from their substrates.

Deletion mutation: Mutation resulting from deletion of one or more nucleotides from a gene.

Denaturation: Partial or complete unfolding of the specific native conformation of the polypeptide chain(s) of proteins.

Denatured protein: A protein that has lost its native conformation by exposure to a destabilizing agent such as heat.

Deoxyribonucleic acid: See DNA.

Deoxyribonucleotides: Nucleotides containing 2-deoxy-D-ribose as the pentose component.

Dextrorotatory isomer: A stereoisomer that rotates the plane of plane-polarized light to the right.

Diabetes mellitus: A metabolic disease due to insulin deficiency, characterized by a failure in glucose transport from the blood into cells at normal glucose concentrations.

Dialysis: Removal of small molecules from a solution of a macromolecule, by allowing them to diffuse through a semipermeable membrane into water.

Diastereoisomers: A pair of stereoisomers about a second asymmetric carbon atom for each of the two isomers of the first asymmetric carbon.

Differential centrifugation: Separation of cell organelles, etc. by their different rates of sedimentation in a centrifugal field.

Differentiation: Specialization of cell structure and function during embryonic growth and development.

Diffusion: The tendency for molecules to move in the direction of a lower concentration.

Digestion: Enzymatic hydrolysis of major nutrients in the gastrointestinal system to yield their building-block components.

Dipole: A molecule having both positive and negative charges.

Diprotic acid: An acid capable of dissociating two protons.

Disaccharides: Carbohydrates consisting of two covalently joined monosaccharide units.

Dissociation constant: An equilibrium constant for the dissociation of a compound into its components; e.g., dissociation of an acid to yield its anion and a proton.

Disulfide bridge: A covalent cross-link formed between two polypeptide chains by a cystine residue.

DNA (deoxyribonucleic acid): A polynucleotide having a specific sequence of deoxyribonucleotide units and serving as the carrier of genetic information.

DNA chimera: A DNA containing genes from two different species.

DNA ligase: An enzyme that creates a phosphodiester bond between the 3′ end of one DNA segment and the 5′ end of another, while they are base-paired to a template strand.

DNA polymerase: An enzyme catalyzing template-dependent synthesis of DNA from its deoxyribonucleoside 5′-triphosphate precursors.

DNA replicase system: The entire complex of enzymes and specialized proteins required in biological DNA replication.

Double helix: The natural coiled conformation of two complementary, antiparallel DNA chains.

E_0': See Standard reduction potential.

***E. coli* (*Escherichia coli*):** A common aerobic bacterium found in the small intestine of vertebrates.

Effector (modulator): A metabolite which, when bound to the allosteric site of a regulatory enzyme, alters its kinetic characteristics.

Electrochemical gradient: The sum of the gradients of mass and electric charge for an ion across a membrane.

Electron acceptor: A substance acting to receive electrons in an oxidation-reduction reaction.

Electron carrier: A protein, such as a flavoprotein or a cytochrome, that can gain and lose electrons reversibly and function to transfer electrons from organic nutrients toward oxygen.

Electron donor: A donor of electrons in an oxidation-reduction reaction.

Electron transport: Movement of electrons from substrates to oxygen; promoted by the respiratory chain.

Electrophoresis: Transport of charged solutes in response to an electrical field; often used to separate mixtures of ions.

Elongation factors: Specific proteins required in the elongation of polypeptide chains by ribosomes.

Eluate: The effluent from a chromatographic column.

Enantiomers: Isomers that are nonsuperimposable mirror images of each other.

Endergonic reaction: A chemical reaction with a positive standard-free-energy change; an "uphill" reaction.

Endocrine glands: Groups of cells specialized to make and secrete hormones into the blood, for regulation of other types of cells.

Endonuclease: An enzyme capable of hydrolyzing interior phosphodiester bonds of a nucleic acid, at points other than the terminal bonds.

Endoplasmic reticulum: An extensive system of double membranes in the cytoplasm of eukaryotic cells; it encloses secretory channels and is often coated with ribosomes.

End-product (feedback) inhibition: Inhibition of an allosteric enzyme at the beginning of a metabolic sequence, by the end product of the sequence.

Energy charge: The fractional degree to which the ATP-ADP-AMP system is filled with high-energy phosphate groups.

Energy coupling: Transfer of energy from one process to another.

Enthalpy: The heat content of a system.

Entropy: The state of randomness or disorder.

Enzyme: A protein specialized to catalyze a specific metabolic reaction.

Enzyme repression: Inhibition of enzyme synthesis; caused by the availability of the product of that enzyme.

Epimerase: An enzyme capable of the reversible interconversion of two epimers.

Epimers: Two stereoisomers differing in configuration at one asymmetric center in a compound having two or more asymmetric centers.

Equilibrium: The state of a system in which no further change is occurring and in which its free energy is at a minimum.

Equilibrium constant: A constant characteristic for each chemical reaction; relates the specific concentrations of all reactants and products at equilibrium at a given temperature and pressure.

Escherichia coli: See *E. coli.*

Essential amino acids: Amino acids that cannot be synthesized by humans and other vertebrates, and must be obtained from the diet.

Essential fatty acids: The group of polyunsaturated fatty acids of plants required in the diet of mammals.

Eukaryotes: The family of cells having a membrane-surrounded nucleus, multiple chromosomes, and internal organelles.

Excited state: An energy-rich state of an atom or molecule; produced by the absorption of light energy.

Exergonic reaction: A chemical reaction with a negative standard-free-energy change; a "downhill" reaction.

Exon: The segment of a eukaryotic gene that is transcribed into mRNA; it codes for a specific domain of a protein.

Exonuclease: An enzyme that hydrolyzes only a terminal phosphodiester bond of a nucleic acid.

Facultative cells: Cells that can live either in the presence or absence of oxygen.

FAD (flavin adenine dinucleotide): The coenzyme of some oxidation-reduction enzymes; it contains riboflavin.

Fatty acid: A long-chain aliphatic acid found in natural fats and oils.

Feedback inhibition: See End-product inhibition.

Fermentation: Energy-yielding anaerobic breakdown of a nutrient molecule, such as glucose.

Fibrous proteins: Insoluble proteins that serve in a protective or structural role, and in which the polypeptide chain is extended or coiled along one dimension.

First law of thermodynamics: The law stating that in all processes, the total energy of the universe remains constant.

Flagellum: A cell appendage used in propulsion.

Flavin-linked dehydrogenases: Dehydrogenases requiring one of the riboflavin coenzymes, FMN or FAD.

Flavin nucleotides: Nucleotide coenzymes (FMN and FAD) containing riboflavin.

Flavoprotein: An enzyme containing a flavin nucleotide as a prosthetic group.

Fluorescence: Emission of light by excited molecules as they revert to the ground state.

FMN (flavin mononucleotide): Riboflavin phosphate, a coenzyme of certain oxido-reduction enzymes.

Frame shift: A mutation that is caused by insertion or deletion of one or more paired nucleotides, and whose effect is to change the reading frame of codons during protein synthesis, thus yielding a garbled amino acid sequence beginning at the mutated codon.

Free energy: The component of the total energy of a system that can do work at constant temperature and pressure.

Furanose: A sugar containing the five-membered furan ring.

Futile cycle: An enzyme-catalyzed set of cyclic reactions that results in release of thermal energy by the hydrolysis of ATP.

$\Delta G^{\circ\prime}$: See Standard-free-energy change.

Gel filtration: A chromatographic procedure for separation of a mixture of molecules on the basis of size, through the capacity of porous polymers to exclude solutes above a certain size.

Gene: A chromosomal segment that codes for a single polypeptide chain or RNA molecule.

Gene library: A random collection of DNA fragments that includes all the genetic information of a given species; sometimes called a shotgun collection.

Gene splicing: The enzymatic attachment of one gene or part of a gene to another; also removal of introns and splicing of exons during mRNA synthesis.

Genetic code: The set of triplet code words in DNA coding for the amino acids of proteins.

Genetic information: The hereditary information contained in a sequence of nucleotide bases in chromosomal DNA or RNA.

Genetic map: A diagram showing the relative sequence and position of specific genes along a chromosome molecule.

Genome: All the genes of an organism or individual.

Germ cell: A type of animal cell formed early in embryogenesis that may multiply by mitosis or produce, by meiosis, cells that develop into either eggs or sperm.

Globular protein: A soluble protein in which the polypeptide chain is tightly folded in three dimensions to yield a globular shape.

Glucogenic amino acids: Amino acids whose carbon chains can be metabolically converted into glucose or glycogen.

Gluconeogenesis: The biosynthesis of new carbohydrate from noncarbohydrate precursors.

Glycolipid: A lipid containing a carbohydrate group.

Glycolysis: A form of fermentation in which glucose is broken down into two molecules of pyruvate.

Glycoprotein: A protein containing at least one carbohydrate group.

Glyoxylate cycle: A variant of the citric acid cycle; present in bacteria and some plant cells, for net conversion of acetate into succinate and, eventually, new carbohydrate.

Glyoxysome: A membranous vesicle containing certain enzymes of the glyoxylate cycle.

Golgi body: A complex membranous organelle of eukaryotic cells; functions in secretion to the exterior and formation of the plasma membrane.

Gram molecular weight: The weight in grams of a compound numerically equal to its molecular weight; the weight of 1 mol.

Ground state: The normal, stable form of an atom or molecule; distinguished from the excited state.

Ground substance: A jellylike hydrated polysaccharide complex in the space between the cells of animal tissues.

Half-life: The time required for the disappearance or decay of one-half of a given component.

Heat of vaporization: The number of calories required to convert 1 g of a liquid into the vapor state at the same temperature.

Helix, α: A coiled, helical conformation of the polypeptide chain with maximal intrachain hydrogen bonding; found in α-keratins.

Heme: The iron-porphyrin prosthetic group of heme proteins.

Heme protein: A protein containing a heme as a prosthetic group.

Hemoglobin: A heme protein of the red blood cell functioning in O_2 transport.

Henderson-Hasselbalch equation: An equation relating the pH, the pK', and the ratio of proton-acceptor (A^-) to proton-donor (HA) species.

Heterotroph: An organism that requires complex nutrient molecules, such as glucose, as a source of energy and carbon.

Heterotropic enzyme: An allosteric enzyme requiring a modulator other than its substrate.

Hexose: A simple sugar having a linear backbone chain of six carbon atoms.

High-energy compound: A compound that undergoes a large decrease in free energy upon hydrolysis under standard conditions.

Hill reaction: The oxygen evolution and photoreduction of an artificial electron acceptor by a chloroplast preparation in the absence of carbon dioxide.

Histones: The group of five basic proteins associated with the chromosomes of all eukaryotic cells.

Homologous protein: A protein having identical functions and similar properties in different species; e.g., the hemoglobins.

Homotropic enzyme: An allosteric enzyme that utilizes its substrate as a modulator.

Hormone: A chemical substance that is synthesized in trace amounts by an endocrine tissue and that acts as a messenger to regulate the function of another tissue or organ.

Hormone receptor: A specific hormone-binding site on the cell surface or within the cell.

Hydrogen bond: A weak electrostatic attraction between one electronegative atom and a hydrogen atom covalently linked to a second electronegative atom.

Hydrolysis: Cleavage of a molecule into two or more smaller molecules by reaction with water.

Hydronium ion: The hydrated hydrogen ion (H_3O^+).

Hydrophilic: "Water-loving"; said of polar or charged molecules or groups that associate with H_2O.

Hydrophobic: "Water-hating"; said of nonpolar molecules or groups that are insoluble in water.

Hydrophobic interactions: The association of nonpolar groups with each other in aqueous systems because of the tendency of the surrounding water molecules to seek their most stable state.

Hyperchromic effect: The large increase in light absorption at 260 nm occurring as a double-helical DNA is melted.

Immune response: The capacity of a vertebrate to generate antibodies to an antigen, a macromolecule foreign to it.

Immunoglobulin: An antibody protein generated to a specific antigen.

Induced enzyme: An enzyme that is not made by the cell (i.e., is repressed) unless it is induced by its substrate or a closely related compound.

Induced fit: A change in the shape of an enzyme to conform to the structure of its substrate.

Inducer: A molecule capable of inducing the synthesis of a given enzyme; usually the enzyme's substrate.

Informational molecules: Molecules containing information in the form of specific sequences of different building blocks; they include proteins and nucleic acids.

Initiation codon: AUG. It codes for the first amino acid in a polypeptide sequence, which is N-formylmethionine in prokaryotes and methionine in eukaryotes.

Initiation complex: A complex of a ribosome with an mRNA and the initiating Met-$tRNA^{Met}$ or fMet-$tRNA^{fMet}$, ready for the elongation steps.

Initiation factors: Specific proteins required to initiate synthesis of a polypeptide by ribosomes.

Insertion mutation: A mutation caused by insertion of an extra base or a mutagen between two successive bases in DNA.

Insertion sequence: Specific base sequences at either end of a transposable segment of DNA.

Intercalating mutagen: A mutagen that inserts itself between two successive nucleotides and causes a frame-shift mutation.

Interferon: A protein made by virus-infected cells of vertebrates; it prevents infection by a second kind of virus.

Intermediary metabolism: In cells, the enzyme-catalyzed reactions that extract chemical energy from nutrient molecules and utilize it to synthesize and assemble cell components.

Intron: An intervening sequence in a gene; it is transcribed but excised before the gene is translated.

In vitro: "In glass," i.e., in the test tube.

In vivo: "In life," i.e., in the cell or organism.

Ion-exchange resin: A polymeric resin that contains fixed charged groups and is used in chromatographic columns to separate ionic compounds.

Ionizing radiation: A type of radiation, such as x-rays, that causes loss of electrons from some organic molecules, thus making them more reactive.

Ion product of water: $K_w = [H^+][OH^-] = 1 \times 10^{-14}$ at 25°C.

Irreversible process: A process in which the entropy of the universe increases.

Isoelectric pH: The pH at which a solute has no net electrical charge.

Isomerase: An enzyme catalyzing transformation of a compound into its positional isomer.

Isoprene: The hydrocarbon 2-methyl-1,3-butadiene, a recurring structural unit of the terpenoid biomolecules.

Isothermal process: A process occurring at constant temperature.

Isotopes: Stable or radioactive forms of an element that differ in atomic weight but are otherwise chemically identical with the naturally abundant form of the element; used as tracers.

Isozymes (isoenzymes): Multiple forms of an enzyme that differ from each other in their substrate affinity, in their maximum activity, or in regulatory properties.

K_M: See Michaelis constant.

Keratins: Insoluble protective or structural proteins consisting of parallel polypeptide chains in α-helical or β conformations.

Ketogenic amino acids: Amino acids whose carbon skeletons can be precursors of the ketone bodies.

Ketone bodies: Acetoacetate, D-β-hydroxybutyrate, and acetone, products of partial oxidation of fatty acids.

Ketose: A simple monosaccharide having its carbonyl group at other than a terminal position.

Ketosis: A condition in which the ketone body concentration of the blood, tissues, and urine is abnormally high.

Kinase: An enzyme catalyzing phosphorylation of an acceptor molecule by ATP.

Law of mass action: The law stating that the rate of any given chemical reaction is proportional to the product of the activities (or concentrations) of the reactants.

Leaky mutant: A mutant in which the mutated gene product, such as an enzyme, still possesses a fraction of its normal biological activity.

Lethal mutation: Mutation of a gene to yield a totally defective gene product unable to sustain the life of the organism.

Levorotatory isomer: An isomer of an optically active compound; rotates the plane of plane-polarized light to the left.

Ligand: A molecule or ion that is bound to a protein.

Light reactions: The reactions of photosynthesis that require light and cannot occur in the dark.

Lineweaver-Burk equation: An algebraic transformation of the Michaelis-Menten equation, allowing more accurate determination of V_{max} and K_M.

Lipid: A water-insoluble biomolecule having greasy or oily character.

Lipoic acid: A vitamin for some microorganisms; it serves as an intermediate carrier of hydrogen atoms and acyl groups in α-ketoacid dehydrogenases.

Lipoprotein: A conjugated protein containing a lipid or a group of lipids.

Lithosphere: The inorganic (mineral) portion of the earth's surface.

Low-energy phosphate compound: A phosphorylated compound with a relatively low standard free energy of hydrolysis.

Lysogeny: One of two outcomes of the infection of a host cell by a temperate phage. It occurs when the phage genome becomes repressed and is replicated as part of the host DNA; infrequently it may be induced and the phage particles so produced cause the host cell to lyse.

Lysosome: A membrane-surrounded organelle in the cytoplasm of eukaryotic cells; it contains many hydrolytic enzymes.

Macromolecule: A molecule having a molecular weight in the range of a few thousand to many millions.

Membrane transport: Facilitated transport of a solute across a membrane, usually by a specific membrane protein.

Messenger RNA (mRNA): A class of RNA molecules, each of which is complementary to one strand of cell DNA and serves to carry the genetic message from the chromosome to the ribosomes.

Metabolic turnover: The constant, steady-state metabolic replacement of cell components.

Metabolism: The entire set of enzyme-catalyzed transformations of organic nutrient molecules in living cells.

Metabolite: A chemical intermediate in the enzyme-catalyzed reactions of metabolism.

Metalloenzyme: An enzyme having a metal ion as its prosthetic group.

Micelle: An association of a number of amphipathic molecules in water into a structure in which their nonpolar portions are in the interior and the polar portions on the exterior exposed to water.

Michaelis constant (K_M): The substrate concentration at which an enzyme shows one-half its maximum velocity.

Michaelis-Menten equation: An equation relating the velocity and the substrate concentration of an enzyme.

Microbodies: Cytoplasmic membrane-surrounded vesicles containing peroxide-forming and -destroying enzymes.

Microfilaments: Very thin filaments found in the cytoplasm of cells.

Microsomes: Membranous vesicles formed by fragmentation of the endoplasmic reticulum of eukaryotic cells and recovered by differential centrifugation.

Microtrabecular network: A complex network of very thin cytoplasmic filaments; visible only by high-voltage electron microscopy.

Microtubules: Thin tubules assembled from two types of globular protein subunits; present in cilia, flagella, and other contractile or motile structures.

Mitochondria: Membrane-surrounded organelles in the cytoplasm of eukaryotic cells; they contain the enzyme systems required in the citric acid cycle, electron transport, and oxidative phosphorylation.

Mitosis: Replication of chromosomes in somatic cells of eukaryotes.

Mixed-function oxygenases: Enzymes catalyzing simultaneous oxidation of two substrates by oxygen, one of which is usually NADPH or NADH.

Molal (m) solution: One mole of a solute dissolved in 1000 g of water.

Molar (M) solution: One mole dissolved in water and made up to 1000 mL.

Mole: One gram molecular weight of a compound.

Monolayer: A single layer of oriented lipid molecules.

Monoprotic acid: An acid capable of losing only one proton.

Monosaccharides: Carbohydrates consisting of single sugar units.

mRNA: See Messenger RNA.

Mucopolysaccharides: Acidic polysaccharides found in mucous secretions and in the intercellular space of higher animals.

Mucoproteins: Conjugated proteins containing an acid mucopolysaccharide. Also called proteoglycans.

Multienzyme system: A sequence of related enzymes participating in a given metabolic pathway.

Mutagenic agent (mutagen): A chemical agent capable of producing a genetic mutation.

Mutarotation: The change in specific rotation of a pyranose or furanose sugar or glycoside accompanying equilibration of its α and β forms.

Mutase: An enzyme catalyzing transposition of a functional group in the substrate.

Mutation: A heritable change in a chromosome.

Myofibril: A unit of thick and thin filaments of muscle fibers.

Myosin: A muscle protein, the major component of the thick filaments of the contractile system.

NAD, NADP (nicotinamide adenine dinucleotide, nicotinamide adenine dinucleotide phosphate): Nicotinamide-containing coenzymes functioning as carriers of hydrogen atoms and electrons in some oxidation-reduction reactions.

Native conformation: The biologically active conformation of a protein molecule.

Neurospora crassa: A common bread mold used in genetic analysis of metabolic pathways.

Neurotransmitter: A small-molecular-weight compound (usually containing nitrogen) secreted by a neuron from its terminal and bound by the next neuron; serves to transmit a nerve impulse.

Neutral fats: The trivial name of fatty acid esters of the three hydroxyl groups of glycerol; usually called triacylglycerols.

Nicotinamide adenine dinucleotide, nicotinamide adenine dinucleotide phosphate: See NAD, NADP.

Ninhydrin reaction: A color reaction given by amino acids and peptides on heating with ninhydrin; it is widely used for their detection and estimation.

Nitrogenase system: A system of enzymes capable of reducing atmospheric nitrogen to ammonia in the presence of ATP.

Nitrogen cycle: The cycling of various forms of biologically available nitrogen through the plant, animal, and microbial worlds, as well as the atmosphere and geosphere.

Nitrogen fixation: Conversion of atmospheric nitrogen (N_2) into a soluble biologically available form by nitrogen-fixing organisms.

Noncompetitive inhibition: A type of enzyme inhibition not reversed by increasing the substrate concentration.

Noncyclic electron flow: The light-induced flow of electrons from water to $NADP^+$ in oxygen-evolving photosynthesis; it involves both photosystems I and II.

Nonessential amino acids: Amino acids of proteins that can be made by humans and other vertebrates from simpler precursors, and are thus not required in the diet.

Nonheme-iron proteins: Proteins containing iron but no porphyrin groups.

Nonpolar group: A hydrophobic group, usually hydrocarbon in nature.

Nonsense codon: A codon that does not specify an amino acid, but indicates the termination of a polypeptide chain.

Nonsense mutation: A mutation that results in the premature termination of a polypeptide chain.

Nuclear body (nucleoid): The nuclear zone in a prokaryotic cell; it contains the chromosome but has no surrounding membrane.

Nuclease: An enzyme capable of hydrolyzing internucleotide linkages of a nucleic acid.

Nucleic acids: Biologically occurring polynucleotides in which the nucleotide residues are linked in a specific sequence by phosphodiester bonds.

Nucleolus: A densely staining structure found in the nucleus of eukaryotic cells; involved in rRNA synthesis and ribosome formation.

Nucleophilic group: An electron-rich group with a strong tendency to donate electrons to an electron-deficient nucleus.

Nucleoside: A compound consisting of a purine or pyrimidine base covalently linked to a pentose.

Nucleoside diphosphate sugar: A coenzyme-like carrier of a sugar molecule functioning in enzymatic synthesis of polysaccharides and sugar derivatives.

Nucleoside diphosphokinase: An enzyme catalyzing transfer of the terminal phosphate of a nucleoside 5′-triphosphate to a nucleoside 5′-monophosphate.

Nucleotide: A nucleoside phosphorylated at one of its pentose hydroxyl groups.

Nucleus: In eukaryotes, a membrane-enclosed organelle that contains chromosomes.

Oligomeric protein: A protein having two or more polypeptide chains.

Oligosaccharide: Several monosaccharide groups joined by glycosidic bonds.

Open system: A system that exchanges matter and energy with its surroundings.

Operator: A region of DNA that interacts with a repressor protein to control the expression of a gene or group of genes.

Operon: A unit of genetic expression consisting of one or more related genes and the operator and promoter sequences that regulate their transcription.

Optical activity: The capacity of a substance to rotate the plane of plane-polarized light.

Optimum pH: The pH at which an enzyme shows maximum catalytic activity.

Organelles: Membrane-surrounded structures found in eukaryotic cells; they contain enzymes and other components required for specialized cell functions.

Orthophosphate cleavage: Enzymatic cleavage of ATP to yield ADP and phosphate, usually coupled to an energy-requiring process or reaction.

Osmosis: Bulk flow of water through a semipermeable membrane into another aqueous phase containing a solute in a higher concentration.

Osmotic pressure: Pressure generated by the osmotic flow of water through a membrane into an aqueous phase containing a solute in higher concentration.

Oxidation: The loss of electrons from a compound.

Oxidation, β: Oxidative degradation of fatty acids into acetyl-CoA by successive oxidations at the β carbon atom.

Oxidation-reduction reaction: A reaction in which electrons are transferred from a donor to an acceptor molecule.

Oxidative phosphorylation: The enzymatic phosphorylation of ADP to ATP coupled to electron transport from a substrate to molecular oxygen.

Oxidizing agent (oxidant): The acceptor of electrons in an oxidation-reduction reaction.

Oxygenase: An enzyme catalyzing a reaction in which oxygen is introduced into an acceptor molecule.

Oxygen debt: The extra oxygen (above the normal resting level) consumed in the recovery period after strenuous physical exertion.

Palindrome: A segment of duplex DNA in which the base sequences of the two strands exhibit two-fold rotational symmetry about a center.

Partition coefficient: A constant that expresses the ratio in which a given solute will be partitioned or distributed between two given immiscible liquids at equilibrium.

Pathogenic: Disease-causing.

Pentose: A simple sugar whose backbone contains five carbon atoms.

Pentose phosphate pathway: The pathway of oxidation of glucose 6-phosphate to yield pentose phosphates.

Peptidase: An enzyme hydrolyzing a peptide bond.

Peptide: Two or more amino acids covalently joined by peptide bonds.

Peptide bond: A substituted amide linkage between the α-amino group of one amino acid and the α-carboxyl group of another.

Peptide mapping (fingerprinting): The characteristic two-dimensional pattern (on paper) of a mixture of peptides resulting from partial hydrolysis of a protein.

Peroxisomes: See Microbodies.

pH: The negative logarithm of the hydrogen ion concentration of an aqueous solution.

Phenotype: The observable characteristics of an organism.

Phosphagen: In excitable tissues, an energy-storing compound containing a high-energy phosphate group, usually in enzymatic equilibrium with the terminal phosphate of ATP.

Phosphate-group energy: The decrease in free energy as 1 mol of a phosphorylated compound at 1.0 *M* concentration undergoes hydrolysis to equilibrium at pH 7.0 and 25°C.

Phosphodiester: A molecule that contains two alcohols esterified to one molecule of phosphoric acid, which thus serves as a bridge between them.

Phosphogluconate pathway: An oxidative pathway beginning with glucose 6-phosphate and leading via 6-phosphogluconate to NADPH, pentoses, and other products.

Phospholipid: A lipid containing one or more phosphate groups.

Phosphorolysis: Enzymatic cleavage of a compound by interaction with phosphate; analogous to hydrolysis.

Phosphorylation: Formation of a phosphate derivative of a biomolecule, usually by enzymatic transfer of a phosphate group from ATP.

Phosphorylation potential (ΔG_p): The actual free-energy change of ATP hydrolysis under a given set of conditions.

Photon: The ultimate unit of light energy.

Photoreduction: Light-induced reduction of an electron acceptor in photosynthetic cells.

Photorespiration: Oxygen consumption occurring in illuminated temperate-zone plants, largely due to oxidation of phosphoglycolate.

Photosynthesis: The utilization of light energy to form carbohydrates from CO_2 and a reducing agent.

Photosynthetic phosphorylation (photophosphorylation): The enzymatic formation of ATP from ADP coupled to light-dependent transport of electrons in photosynthetic organisms.

Photosystem: A functional set of light-absorbing pigments and its reaction center in photosynthetic cells.

Pili: Bacterial appendages employed in the transfer of genes from one cell to another during conjugation.

p*K*′: The negative logarithm of an equilibrium constant.

Plasma membrane: The membrane immediately surrounding the cytoplasm of a cell.

Plasma proteins: The proteins present in blood plasma.

Plasmid: An extrachromosomal, independently replicating small circular DNA molecule.

Plastid: In plants, a self-replicating organelle; it may differentiate into a chloroplast.

Pleated sheet: The side-by-side, hydrogen-bonded arrangement of polypeptide chains in the extended β conformation.

Polar group: A hydrophilic or water-loving group.

Polarimeter: An instrument for determining the degree of rotation of plane-polarized light by a solution.

Polarity: In genetic biochemistry, the distinction between the 5′ and 3′ ends of nucleic acids.

Polynucleotide: A covalently linked sequence of nucleotides in which the 3′ position of the pentose of one nucleotide is joined by a phosphodiester group to the 5′ position of the pentose of the next.

Polypeptide: A long chain of amino acids linked by peptide bonds.

Polyribosome (polysome): A complex of a messenger RNA molecule and two or more ribosomes.

Polysaccharides: Linear or branched macromolecules composed of many monosaccharide units linked by glycosidic bonds.

Polysome: See Polyribosome.

Porphyrins: Complex nitrogenous compounds containing four substituted pyrroles covalently joined into a ring, often complexed with a central metal atom.

Posttranslational modification: Enzymatic processing of a polypeptide chain after its translation from its mRNA.

Primary structure of proteins: The covalent backbone structure of a protein, including its amino acid sequence and its inter- and intrachain disulfide bridges.

Prochiral molecule: A symmetrical molecule that may react asymmetrically with an enzyme having an asymmetric active site.

Prokaryotes: Simple unicellular organisms (e.g., bacteria and blue-green algae) with a single chromosome, no nuclear membrane, and no membrane-bound organelles.

Promoter: A DNA sequence at which RNA polymerase may bind, leading to initiation of transcription.

Prostaglandins: A class of lipid-soluble, hormone-like regulatory molecules derived from arachidonic and other polyunsaturated fatty acids.

Prosthetic group: A heat-stable metal ion or an organic group (other than an amino acid) that is bound to a protein and serves as its active group.

Protein: A macromolecule composed of one or more polypeptide chains, each possessing a characteristic amino acid sequence and molecular weight.

Protein kinases: Enzymes capable of phosphorylating certain amino acid residues in specific proteins.

Proteoglycan: A hybrid macromolecule consisting of an oligo- or polysaccharide joined to a polypeptide; the polysaccharide is the major component.

Proteolytic enzyme: An enzyme catalyzing hydrolysis of proteins or peptides.

Proton acceptor: An anionic compound capable of accepting a proton from a proton donor.

Proton donor: The donor of a proton in an acid-base reaction—i.e., an acid.

Protoplasm: A general term referring to the entire contents of a living cell.

Purine: A basic nitrogenous heterocyclic compound found in nucleotides and nucleic acids; it contains fused pyrimidine and imidazole rings.

Puromycin: An antibiotic that inhibits polypeptide synthesis by being incorporated into a polypeptide chain, causing its premature termination.

Pyranose: A simple sugar containing the six-membered pyrane ring.

Pyridine-linked dehydrogenases: Dehydrogenases requiring as a coenzyme either one of the pyridine coenzymes NAD or NADP.

Pyridine nucleotide: A nucleotide coenzyme containing the pyridine derivative nicotinamide; i.e., NAD or NADP.

Pyridoxal phosphate: A coenzyme containing the vitamin pyridoxine and functioning in reactions involving amino-group transfer.

Pyrimidine: A nitrogenous heterocyclic base serving as a component of a nucleotide or nucleic acid.

Pyrophosphatase: An enzyme hydrolyzing inorganic pyrophosphate to yield two molecules of (ortho)phosphate.

Pyrophosphate cleavage: Enzymatic cleavage of ATP to yield AMP and pyrophosphate, energetically coupled to formation of a covalent bond joining two biomolecules.

Pyrophosphorylases: Enzymes catalyzing the formation of nucleoside diphosphate sugars and pyrophosphate from a sugar phosphate and a nucleoside 5′-triphosphate.

Quantum: The ultimate unit of light energy.

Quaternary structure: The three-dimensional structure of an oligomeric protein; particularly the manner in which the subunit chains fit together.

R group: The distinctive side chain of an α-amino acid.

Racemate: An equimolar mixture of the D and L stereoisomers of an optically active compound.

Radioactive isotope: An isotopic form of an element with an unstable nucleus that stabilizes itself by emitting ionizing radiation.

Radioimmunoassay: Sensitive quantitative determination of trace amounts of a hormone (or some other biomolecule) by its capacity to displace the radioactive form of the hormone from combination with its specific antibody.

Recombinant DNA: DNA formed by the joining of genes into new combinations.

Recombination: The joining of genes, sets of genes, or parts of genes into new combinations, either biologically or through laboratory manipulation.

Redox couple: An electron donor and its corresponding oxidized form.

Reducing agent (reductant): An electron donor in an oxidation-reduction reaction.

Reducing equivalent: A general or neutral term for an electron or an electron equivalent in the form of a hydrogen atom or a hydride ion.

Reduction: The gain of electrons by a compound.

Regulatory enzyme: An enzyme having a regulatory function through its capacity to undergo a change in catalytic activity by noncovalent or covalent binding of a specific modulating metabolite.

Regulatory gene: A gene whose product is involved in the regulation of the expression of another gene: e.g., the gene coding for a repressor protein.

Regulatory sequence: A DNA sequence involved in regulating the expression of a gene; e.g., a promoter or operator.

Releasing factors (termination factors): Protein factors of the cytosol; required in releasing a completed polypeptide chain from the ribosome.

Renaturation: Refolding of an unfolded (i.e., denatured) globular protein.

Replication: Synthesis of a daughter duplex DNA molecule identical to the parental duplex DNA.

Repressible enzyme: An enzyme whose synthesis is inhibited when its reaction product is readily available to a bacterial cell.

Repressor: The protein that binds to the regulatory sequence or operator for a gene and blocks its transcription.

Resonance hybrid: A stable form of an organic molecule caused by "sinking" of certain electrons into a lower energy state, thus forming a hybrid between two or more canonical forms.

Respiration: The oxidative breakdown and release of energy from nutrient molecules by reaction with oxygen.

Respiratory chain: The electron-transport chain, a sequence of electron-carrying proteins that transfer electrons from substrates to molecular oxygen in aerobic cells.

Respiratory-chain phosphorylation: Oxidative phosphorylation; phosphorylation of ADP coupled to electron transport between a substrate and oxygen.

Restriction endonuclease: A site-specific endodeoxyribonuclease causing cleavage of both strands of DNA at points signaled by two-fold symmetry of base sequence about a center; important tool in genetic engineering.

Retrovirus: RNA virus containing a reverse transcriptase, i.e., an RNA-directed DNA polymerase.

Reverse transcriptase: An RNA-directed DNA polymerase formed in retroviruses; capable of making DNA complementary to an RNA.

Reversible process: A process that proceeds with no change in entropy.

Ribonuclease: A nuclease capable of hydrolyzing certain internucleotide linkages of RNA.

Ribonucleic Acid: See RNA.

Ribonucleotide: Nucleotide containing D-ribose as its pentose component.

Ribosomal RNA (rRNA): A class of RNA molecules serving as components of ribosomes.

Ribosome: A macromolecular assembly of rRNAs and proteins, about 20 nm in diameter, which serves as the site of protein synthesis.

RNA (ribonucleic acid): A polyribonucleotide of a specific sequence linked by successive 3′,5′-phosphodiester linkages.

RNA polymerase: An enzyme that catalyzes the formation of RNA from ribonucleoside 5′-triphosphates, using a strand of DNA or RNA as a template.

rRNA: See Ribosomal RNA.

Saponification: Alkaline hydrolysis of triacylglycerols to yield fatty acids as soaps.

Sarcomere: A functional and structural unit of the muscle contractile system.

Satellite DNA: Highly repeated, nontranslated segments of DNA in eukaryotic chromosomes.

Saturated fatty acid: A fatty acid containing a fully saturated alkyl chain.

Secondary structure: The residue-by-residue conformation of the backbone of a polypeptide chain.

Second law of thermodynamics: The law stating that in any chemical or physical process, the entropy of the universe tends to increase.

Sedimentation coefficient: A physical constant specifying the rate of sedimentation of a particle in a centrifugal field under specified conditions.

Sickle-cell anemia: A human disease characterized by defective hemoglobin molecules and due to a homozygous allele coding for the beta chain of hemoglobin.

Sickle-cell trait: A human condition recognized by the sickling of red blood cells when exposed to low oxygen tension; heterozygous for the allele responsible for sickle-cell anemia.

Signaling sequence: A polypeptide 5′-leader sequence that signals the destination of a newly synthesized protein; it guides the protein through a specific membrane.

Silent mutation: A mutation in a gene that causes no detectable change in the biological characteristics of the gene product.

Simple protein: A protein yielding only amino acids on hydrolysis.

Somatic cells: All body cells except the gametes and the cells from which they develop.

Specific activity: The number of μmol of a substrate transformed by an enzyme preparation per min per mg protein at 25°C.

Specific heat: The amount of heat in cal required to raise the temperature of 1 g of a substance 1°C.

Specific rotation: The rotation in degrees of the plane of plane-polarized light (D-line of sodium) for an optically active compound at 25°C at a specified concentration and light path.

Spontaneous process: A process accompanied by an increase in entropy in the universe.

Standard-free-energy change ($\Delta G°$): The gain or loss of free energy in cal as 1 mol of reactants in the standard state is converted into 1 mol of products under standard conditions of temperature, pressure, concentration, and pH.

Standard reduction potential (E_0'): The electromotive force exhibited at an electrode by 1 *M* concentrations of a reducing agent and its oxidized form at 25°C at pH 7.0; a measure of the relative tendency of the reducing agent to lose electrons.

Standard state: The most stable form of a pure substance at 1.0 atm pressure and 25°C (298 K). For reactions occurring in solution, the standard state of a solute is a 1.0 *m* solution.

Steady state: A nonequilibrium state of a system through which matter is flowing and in which all components remain in constant concentration.

Stereoisomers: Isomers that are nonsuperimposable mirror images of each other.

Steroids: A class of lipids containing the cyclopentanophenanthrene ring system.

Structural gene: A gene coding for the structure of a protein.

Substitution mutation: A mutation caused by the replacement of one base by another.

Substrate: The specific compound acted upon by an enzyme.

Substrate-level phosphorylation: Phosphorylation of ADP or some other nucleoside 5′-diphosphate coupled to a dehydrogenation of an organic substrate, independent of the electron-transport chain.

Suppressor mutation: A mutation that totally or partially restores a function lost by a primary mutation; located at a site in the gene different from the site of the primary mutation.

Svedberg (S): A unit of measure of the rate at which a particle sediments in a centrifugal field.

System: An isolated collection of matter. All other matter in the universe apart from the system is called the surroundings.

Temperate phage: A phage whose DNA may be incorporated into the host-cell genome without being expressed. In contrast, a virulent phage destroys the host cell.

Template: A macromolecular mold or pattern for the synthesis of an informational macromolecule.

Terminal transferase: An enzyme capable of adding nucleotide residues of a single kind to the 3′ end of DNA chains.

Termination codons: The three codons UAA, UAG, and UGA, which signal termination of a polypeptide chain.

Termination factors: See Releasing factors.

Termination sequence: A DNA sequence that appears at the end of a transcriptional unit and signals the end of transcription.

Terpene: An organic hydrocarbon or hydrocarbon derivative constructed from recurring isoprene units.

Tertiary structure (of a protein): The three-dimensional conformation of the polypeptide chain of a globular protein in its native folded state.

Tetrahydrofolic acid: The reduced, active coenzyme form of the vitamin folic acid.

Thioester: An ester of a carboxylic acid with a thiol or mercaptan.

Thymine dimer: A covalently joined dimer of two adjacent thymine residues in DNA; induced by absorption of ultraviolet light.

Titration curve: A plot of the pH versus the equivalents of base added during titration of an acid.

Tocopherols: Forms of vitamin E.

Topoisomerases: Enzymes capable of positive or negative supercoiling of duplex DNA circles.

Toxins: Proteins that are elaborated by some organisms and are toxic to some other species.

Trace element: A chemical element required by an organism in only a trace amount.

Transaminases: Enzymes catalyzing transfer of amino groups from α-amino to α-keto acids; also called aminotransferases.

Transamination: Enzymatic transfer of an amino group from an α-amino acid to an α-keto acid.

Transcription: The enzymatic process whereby the genetic information contained in one strand of DNA is used to specify a complementary sequence of bases in an mRNA chain.

Transcriptional control: The regulation of protein synthesis by regulation of the formation of its mRNA.

Transduction: Transfer of genetic material from one cell to another by means of a viral vector.

Transfer RNA (tRNA): A class of RNA molecules (MW 25,000–30,000), each of which combines covalently with a specific amino acid as the first step in protein synthesis.

Transformation: Introduction of an exogenous DNA into a cell, causing it to acquire a new phenotype.

Transition state: An activated form of a molecule, in which the molecule is capable of undergoing a chemical reaction.

Translation: The process in which the genetic information present in an mRNA molecule directs the sequence of amino acids during protein synthesis.

Translational control: Regulation of protein synthesis by the rate of its translation on the ribosome.

Translocase: A membrane transport system. The term is also applied to an enzyme causing a movement, such as the movement of a ribosome along an mRNA.

Transpiration: Passage of water from the roots of a plant to the atmosphere via the vascular system and leaves.

Transposable element (transposon): A segment of DNA that can be moved from one position in the genome to another.

Transposition: Movement of a gene or set of genes from one site in the genome to another.

Triacylglycerol: An ester of glycerol with three molecules of fatty acid; also called a neutral fat.

Tricarboxylic acid cycle: See Citric acid cycle.

tRNA: See Transfer RNA.

Tropic hormone (tropin): A peptide hormone that stimulates its target gland to secret *its* hormone; e.g., thyrotropin of the pituitary stimulates secretion of thyroxine by the thyroid.

Turnover number: The number of times an enzyme molecule transforms a substrate molecule per min under conditions giving maximal activity.

Ultraviolet radiation: Electromagnetic radiation in the region 200–400 nm.

Uncoupling agent: A substance that uncouples phosphorylation of ADP from electron transport; e.g., 2,4-dinitrophenol.

Unsaturated fatty acid: A fatty acid containing one or more double bonds.

Urea cycle: A metabolic pathway in the liver; promotes the synthesis of urea from amino groups and CO_2.

UV endonuclease: An endonuclease capable of cleaving a DNA chain at the 5′ side of a thymine dimer.

V_{max}: The maximum velocity of an enzymatic reaction.

Vector: A DNA molecule known to replicate autonomously in a host cell, to which a segment of DNA may be spliced in order to bring about its replication; e.g., a plasmid or a temperate-phage DNA.

Virion: A virus particle.

Virus: A self-replicating, infectious, nucleic acid–protein complex that requires an intact host cell for its replication and that contains a chromosome of either DNA or RNA.

Vitamin: A trace organic substance required in the diet of some species; most vitamins function as components of certain coenzymes.

Wobble: The relatively loose base-pairing between the base at the 3′ end of a codon and the complementary base at the 5′ end of the anticodon.

X-ray crystallography: Use of x-ray diffraction by crystals of a compound to determine its three-dimensional structure.

Zwitterion: A dipolar ion, one with spatially separated positive and negative charges.

Zymogen: An inactive precursor of an enzyme; e.g., pepsinogen.

Illustration Acknowledgments

Cover, Robert Langridge, Computer Graphics Laboratory, University of California, San Francisco; © Regents of the University of California

Facing page 1, Part I Opener, © California Institute of Technology and the Carnegie Institution of Washington

1-1 (a) Michael A. Walsh; (b) Treat Davidson, National Audubon Society Collection/PR; (c) Languepin, Rapho/PR; (d) Camera Pix, Rapho/PR

1-9 L. A. MacHattie, *J. Mol. Biol.*, **11**:648 (1965)

Page 14 Micrograph by Karen L. Anderson, supplied by K. R. Porter

2-2 E. S. Barghoorn, *Science*, **152**: 758–763 (1966); © by AAAS

2-3 D. Greenwood

2-4 Micrographs by A. Ryter

2-5 Micrograph by Norma J. Lang, *J. Phycol.*, **1**:127–143 (1965)

2-7 G. Decker

2-8 (a) U. Goodenough; (b) D. Branton

2-10 Micrograph by K. R. Porter

2-12 (a) K. R. Porter; (b) D. Fawcett

2-13 (a) C. J. Flickinger, *J. Cell Biol.*, **49**:221 (1971); (b) D. Fawcett, from William Bloom and D. W. Fawcett, *A Textbook of Histology*, 10th ed., Saunders, Philadelphia, 1975

2-14 (a) D. Fawcett; (b) G. Decker

2-15 (a) E. Lazarides, *J. Cell Biol.*, **65**:549 (1975); (b) K. Weber and U. Groeschel-Stewart, *Proc. Nat. Acad. Sci., U.S.*, **171**:4561 (1974)

2-17 K. R. Porter

2-19 D. Fawcett

2-20 J. C. Wanson

2-21 Micrograph by Michael A. Walsh

2-22 Micrograph by C. Arntzen

2-23 R. D. Preston

2-25 (a), (b) L. D. Simon; (c) K. Allen; (d), (e) J. D. Almeida and A. F. Howatson, *J. Cell Biol.*, **16**:616 (1963)

Page 44 D. P. Wilson, Eric and David Hosking

3-9 (a) L. M. Beidler; (b) R. D. Preston

3-14 (a) H. Ris

3-15 S. Jonasson, from R. Anderson et al., *Science*, **148**:1179–1190 (1965); © by AAAS

4-6 After S. E. Luria, S. J. Gould, and S. Singer, *A View of Life*, Benjamin/Cummings, Menlo Park, Calif., 1981

4-14 Larry Pringle, Photo Researchers

5-1 Jack Dermid

6-2 (a) E. Margoliash; (b) M. Kunitz

6-14 (b) Adapted from M. O. Dayhoff, C. M. Park, and P. J. McLaughlin, p. 8, in M. O. Dayhoff (ed.), *Atlas of Protein Sequences and Structure*, vol. 5, National Biomedical Research Foundation, Washington, D.C., 1972

7-3 John Mais

7-6 (b) From L. Pauling and R. B. Corey, *Proc. Int. Wool Text. Res. Conf.*, **B**:249 (1955)

7-12 J. Gross

7-14 Adapted from R. E. Dickerson and I. Geis, *The Structure and Action of Proteins*, Benjamin/Cummings, Menlo Park, Calif., 1969

8-4 I. Geis

8-5 I. Geis

8-6 After Dickerson and Geis, *op. cit.*

8-7 I. Geis

8-10 I. Geis, adapted from R. E. Dickerson and Irving Geis, *Proteins: Structure, Function, and Evolution*, 2d ed., Benjamin/Cummings, Menlo Park, Calif., 1982

8-11 (b) Adapted from M. F. Perutz, "The Hemoglobin Molecule"; © 1964 by Scientific American

8-13 From C. R. Cantor and P. R. Schimmel, *Biophysical Chemistry*, Part I, Freeman, San Francisco, 1980; originally supplied by Dr. Thomas Steitz, Department of Molecular Biophysics, Yale University

8-15 Margaret Clark

Page 193, Box 8-1, Figure 1(b), I. Geis

8-20 (a) Patricia Farnsworth; (b) Margaret Clark; (c) Johanna Dobler

8-21 After Cantor and Schimmel, *op. cit.*

8-24 After A. C. Allison, "Sickle Cells and Evolution"; © 1956 by Scientific American

9-1 M. Kunitz

9-13 Adapted from A. Dafforn and D. E. Koshland, Jr., *Biochem. Biophys. Res. Commun.*, **52**:780 (1973)

Page 228, Box 9-4A, Figure 1, I. Geis

Page 229, Box 9-4B, Figure 2, I. Geis

Pages 230–231, Box 9-4C, Figures 1–8, I. Geis, based on I. Geis, in M. Stroud, "A Family of Protein-Cutting Proteins," *Sci. Am.*, **231**:244–288, January (1974); © 1974 by Stroud, Dickerson, and Geis

Page 232, Box 9-4D, Figures 1 and 2, I. Geis

Page 238, Box 9-5, Figure 1, from Cantor and Schimmel, *op. cit.*, originally from H. L. Monaco, J. L. Crawford, and W. N. Lipscomb. *Proc. Nat. Acad. Sci., U.S.* **75**:5276 (1978)

9-24 Ylla, Rapho-Guillemette/PR

11-14 (a) Myron C. Ledbetter; (b) D. Fawcett

11-16 (c), (d) I. Geis

11-17 Eric V. Gravé

11-19 (a) R. D. Preston; (b) Grant Heilman

11-21 D. Greenwood

11-23 S. Ito, *Fed. Proc.*, **23**:12 (1969)

Page 302, A. D. Greenwood

12-5 (a) D. Fawcett, from D. W. Fawcett, *The Cell, Its Organelles and Inclusions*, Saunders, Philadelphia, 1967; (b) K. R. Porter

12-6 Russ Kinne, Photo Researchers

Page 320, Box 12-1, Figure 1, J. D. Robertson; Figure 2, H. Latta, W. Johnson, and T. Stanley, *J. Ultrastruct, Res.*, **51**:354 (1975); Figure 4, P. Pinto da Silva and D. Branton, *J. Cell Biol.*, **45**:598 (1970)

12-18 Modified from S. J. Singer and G. L. Nicolson, *Science*, **175**:720–731 (1972); © by AAAS

Page 330, Part II Opener, Prepared by D. E. Nicholson, University of Leeds, published by Koch-Light Laboratories, Ltd., Colnbrook, England

13-3 R. D. Estes

13-9 Alfred Sussman

13-23 G. Decker

14-11 John Dominis, LIFE Magazine; © 1967, Time, Inc.

Page 388, Box 14-3, Figure 1, Grant Heilman

16-5 L. J. Reed, p. 213, in P. D. Boyer (ed.), *The Enzymes*, vol. 1, 3d ed.; © 1970 by Academic Press

16-10 *Biochem. Soc. Bull.*, **2**(2):5 (1980)

17-15 Adapted from E. Racker, *Essays in Biochemistry*, vol. 6, pp. 1–22, Academic, New York, 1970

17-16 (a) D. E. Smith; (c), (d) M. Amzel and P. Pedersen

17-28 C. R. Hackenbrock

19-26 Bernard G. Silberstein, Photo Researchers

Page 560 P. Drochmans

Page 582 Cedric S. Raine, from P. Morell (ed.), *Myelin*, Plenum, New York, 1977

21-14 N. Mrosovsky

21-22 (a) Herbert A. Fischler, Isaac Albert Research Institute of the Kingsbrook Jewish Medical Center; (b) John S. O'Brien

22-7 The Nitragin Co.

Page 644 T. E. Adams

23-2 (c) Eric V. Gravé, Photo Researchers; (d) Walter Dawn, Photo Researchers; (e) Kent Cambridge Scientific Instruments; (f), (g) D. P. Wilson, Eric and David Hosking

23-6 (a) A. D. Greenwood; (c) D. Branton

23-27 Ray F. Evert

23-31 D. Oesterhelt

Page 680, Part III Opener, The Bettmann Archive

24-2 (b)–(d) From Richard D. Kessel and Randy Kardon, *Tissues and Organs: A Text-Atlas of Scanning Electron Microscopy*, Freeman, San Francisco, 1979

24-5 Kessel and Kardon, *op. cit.*

24-13 (b) K. R. Porter

24-15 (b) John Heuser

24-16 Kessel and Kardon, *op. cit.*

24-23 (b) K. R. Porter

25-8 (b) Daniel Friend

25-15 (c) Marie H. Greider and Paul E. Lacy

25-21 The Bettmann Archive

26-5 American Heart Association

26-7 (a) Bernard Wolff, Photo Researchers; (b) F.A.O.

26-9 From Bernard L. Oser (ed.), *Hawk's Physiological Chemistry*, 14th ed., p. 601, McGraw-Hill, New York, 1965; originally from Smith and Munsell, USDA Circ. No. 84, 1929

26-17 Dreiding Stereo-Models, Büchi Laboratory Techniques, Ltd.

26-19 The Upjohn Co.

26-21 L. L. Rue, III, Photo Researchers

Page 790, Part IV Opener, From John Fiddes, "Nucleotide Sequence of a Viral DNA," *Sci. Am.*, **236**:64–65, December (1977)

27-6 After Derry D. Koob and William E. Boggs, *The Nature of Life*, Addison-Wesley, Reading, Mass., 1972

27-8 J. Watson, from James D. Watson, *The Double Helix*, Atheneum, New York, 1968

27-9 *Ibid.*

27-10 (a) Dreiding Stereo-Models, Büchi Laboratory Techniques, Ltd.; (b) M. H. F. Wilkins, King's College Biophysics Department, London

27-11 Redrawn from L. Pauling and R. B. Corey, *Arch. Biochem. Biophys.*, **65**:164 (1956)

27-13 From James D. Watson, *Molecular Biology of the Gene*, 3d ed., Benjamin/Cummings, Menlo Park, Calif., 1976

27-14 Adapted from A. Worcel and E. Burgi, *J. Mol. Biol.*, **71**:143 (1972)

27-16 Redrawn from P. Doty, p. 8, in D. J. Bell and J. K. Grant (eds.), *Biochem. Soc. Symp.*, vol. 21, Cambridge University Press, London, 1962

27-17 (a) R. Davis; (b) A. K. Kleinschmidt, D. Land, D. Jacherts, and R. K. Zahn, *Biochem. Biophys. Acta*, **61**:857–864 (1962)

27-19 H. Delius and A. Worcel, *J. Mol. Biol.*, **82**:108 (1974)

27-20 (a) Stanley N. Cohen; (b) S. Palchaudhuri, M. R. J. Salton, and E. Bell, *Infect. Immun.*, **11**:1141 (1975)

27-21 E. J. DuPraw, *DNA and Chromosomes*, Holt, New York, 1970

27-23 (a) Ada L. Olins, *Amer. Scientist*, **66**(6):704–711, November-December (1978); (b), (c) adapted from F. J. Ayala and J. A. Kiger, Jr., *Modern Genetics*, Benjamin/Cummings, Menlo Park, Calif., 1980

27-24 W. J. Larsen, *J. Cell Biol.*, **47**:353 (1970)

27-25 P. H. O'Farrell, *J. Biol. Chem.*, **250**:4007–4021 (1975)

27-26 Redrawn from Barbara J. Bachman, K. Brooks Low, and Austin L. Taylor, *Bacteriol. Rev.*, pp. 116–117, March (1976)

27-30 Jack D. Griffith

28-4 (a) Redrawn from H. J. Kriegstein and D. S. Hogness, *Proc. Nat. Acad. Sci., U.S.*, **71**:136 (1974)

28-9 From A. Kornberg, *DNA Replication*, Freeman, San Francisco, 1980

28-13 Adapted from *ibid.*

28-23 (b) O. L. Miller, Jr., and Barbara R. Beatty, *J. Cell Physiol.*, **74**(Supplement):225–232 (1969)

29-1 K. R. Porter

29-6 Adapted from S. H. Kim et al., *Science*, **185**:436 (1974), © by AAAS

29-13 James A. Lake, *J. Mol. Biol.*, **105**:131–159 (1976)

29-18 (b) Steven L. McKnight and O. L. Miller, Jr., Department of Biology, University of Virginia

29-26 R. Abermann

29-28 From R. C. Dickson, J. Abelson, W. M. Barnes, and W. S. Reznikoff, *Science*, **187**:32 (1975); © by AAAS

30-13 (b) Judith Carnahan and Charles C. Brinton, Jr.

30-17 Redrawn from image supplied by Richard J. Feldmann, National Institutes of Health

Index

Configurational and positional designations in chemical names (for example, 1,2-, α,β-, D-, L-, p-, *cis*-, *erythro*-) are disregarded in alphabetizing; the same or similar forms used as adjectives (α carbon atom, cis form) are not disregarded. Greek letters used as adjectives are alphabetized as if they were spelled out. Word division by space or hyphen is ignored; for example, alkyl ether acylglycerols *follows* alkylation, and cell wall *follows* cellubiose.

Page numbers in boldface type indicate where a structural formula is given.

A band, 381
abdominal organs, oxygen consumption of, 708
abiotic synthesis, 61
absolute configuration, 97
absorption, of digestion products, 684, 685
acceptor control, 499
accessory pigments, 654
A cell, 735–737
acetaldehyde, in fermentation, 426, 427
acetate, from glucose, 570
  in metabolism, 351, 352
acetate thiokinase, 465
acetazolamide, 247
acetic acid, oxidation, 449
  p*K*, 79, 80, 105, 106
  titration curve, 80, 82
  as weak acid, 78
acetic acid–acetate couple, as buffer, 83
acetic acid bacteria, 427
acetoacetate, activation, 525
  in amino acid oxidation, 539, 544
  in diabetes mellitus, 714
  formation, 524
acetoacetate decarboxylase, 525
acetoacetic acid, **51**, 524
acetoacetyl-CoA, **524**
  in amino acid oxidation, 536–540, 545
  in cholesterol synthesis, 608, 609
acetoacetyl-S-ACP, 589–591
α-aceto-α-hydroxybutyrate, **622**
acetolactate synthase, 622
acetone, 524, 525
  in diabetes mellitus, 714
  from photosynthesis, 648
  physical properties, 67
acetylcholine, 90, 221, **222**
  as neurotransmitter, 700
acetylcholinesterase, 222, 700
  inhibition, 221, 222
acetyl-CoA (acetyl–coenzyme A), as acetyl-group carrier, 257
  from alcohol oxidation, 762
  from amino acid oxidation, 536–538, 558, 694
  in cholesterol synthesis, 607–609
  in citric acid cycle, 398, 441, 444, 445, 450, 515, 526, 693
  in control of carbohydrate metabolism, 501
  in control of gluconeogenesis, 567
  in control of glycolysis, 567
  from fatty acid oxidation, 515–521, 695
  in fatty acid synthesis, 526, 584–589, 592, 595
  in glyoxylate cycle, 455, 456
  in ketone body metabolism, 524–526
  in metabolism, 338, 339, 526, 568, 585, 693
  mitochondrial transport of, 585, 586
  from pyruvate, 398, 437–441
  in pyruvate carboxylase regulation, 454, 455, 563, 564
  in pyruvate dehydrogenase regulation, 449–452
  in pyruvate kinase regulation, 421, 424
  in respiration, 435, 436
  standard free energy of hydrolysis, 396
acetyl-CoA acetyltransferase (*see also* thiolase), 516, 517
acetyl-CoA carboxylase, 586, 587, 613, 663, 892
  regulation, 595
acetyl-coenzyme A, *see* acetyl-CoA
acetylene, in nitrogenase reaction, 639
N-acetyl-D-galactosamine, **314**
  in chondroitin, 296, 297
  in dermatan, 296, 297
  in glycolipids, 312, 314
N-acetyl-D-glucosamine, **292**, 293
  in chitin, 292
  in heparin, 297
  in hyaluronic acid, 296
  in lactose synthesis, 576
  in peptidoglycans, 293
N-acetylglutamate, 552
acetyl-group shuttle, 585, 586
N-acetylhexosaminidase, genetic defect in, 604, 606
acetyllipoic acid, **438**
N-acetylmuramic acid, **293**
N-acetylneuraminic acid, **314**
N-acetyltyrosine methyl ester, **91**
acid mucopolysaccharides, 296
acidosis, 78, 87, 276, 581, 715
acid phosphatase, 246
acid rain, 88
acids, 50, 78
  dissociation constants, 79
  in rainwater, 88
ackee plant, 523
aconitase, 445
  active site, 451
  in glyoxylate cycle, 455
cis-aconitate, **442**
  in citric acid cycle, 441, 442, 445, 451
ACP, *see* acyl carrier protein
ACP-acetyltransferase, 588, 589
ACP-acyltransferase, 591
ACP-malonyltransferase, 588, 589
acridine, 856, **921**
acromegaly, 740
actin, 29, 123, 124, 163, 380–383
  electron micrograph, 29
  fibrous, 163, 380, 381
  globular, 380, 381
  in membranes, 323, 324
  in microvilli, 684
  molecular weight, 163
actinomycin D, 856, **857**
action potential, 700
activation energy, 211
active site, 209, 221, 226, 227, 244, 807
active transport (*see also* transport systems), 384–386
  in kidney, 703–705
acute pancreatitis, 690, 719
acyl carrier protein (ACP), 584, 587–591
  molecular weight, 587
acyl-CoA dehydrogenase, 484, 515–517
acyl-CoA synthases, 512, 528, 571, 762
acyl phosphates, 408, 431
adapter hypothesis, 872, 873
Addison's disease, 741
adductor muscle, 382
adenine, **57**, 796, 797
  deamination, 917
  in DNA, 803–806
  in RNA, 854
  salvage, 635
adenine nucleotides, synthesis, 627–630
  regulation of, 630, 631
adenine nucleotide translocase, 495, 496
adenine ribonucleotides, 373
adenosine, catabolism, 634
adenosine 5′-diphosphate (ADP), 9, **343**, 372
  as allosteric effector, 346
  cell concentrations, 373
  in chloroplasts, 678
  in control of carbohydrate metabolism, 500–502
  in control of citric acid cycle, 452, 453
  in control of glycolysis, 406, 567
  in control of respiration, 498, 499
  in glutamate dehydrogenase regulation, 535
  heart muscle concentration, 433
  in isocitrate dehydrogenase regulation, 445, 453
  magnesium complexes of, 373
  in metabolism, 343, 344, 346
  mitochondrial transport of, 491, 495, 496
  in phosphofructokinase regulation, 567
  reduction to dADP, 633
  resonance hybrids, 375
  standard free energy of hydrolysis, 374
adenosine 5′-monophosphate (AMP), **372**, 798
  as allosteric effector, 346
  from ATP, 387
  cell concentrations, 373
  in control of carbohydrate metabolism, 500, 501
  in control of gluconeogenesis, 563, 566, 567
  in control of glycolysis, 406
  in control of purine nucleotide synthesis, 630
  deamination, 546, 547
  degradation, 634
  in glutamine synthetase regulation, 623
  in phosphofructokinase regulation, 423, 424, 567
  in phosphorylase regulation, 422, 574
  standard free energy of hydrolysis, 374
  synthesis, 628–630
    regulation of, 630, 631
adenosine triphosphatase, *see* ATPase
adenosine 5′-triphosphate (ATP), **9**, 26, 65, 343, 372, 390
  in active transport, 384–386, 703, 704
  in *S*-adenosylmethionine synthesis, 618
  in adenylate cyclase reaction, 730, 732
  from alcohol oxidation, 761
  as allosteric inhibitor, 346
  from amino acid oxidation, 545
  in aspartate transcarbamoylase regulation, 632, 633
  in bioluminescence, 388
  in biosynthesis, 562
  in brain cell metabolism, 699, 700
  in carbon dioxide fixation, 672
  cell concentrations, 373
  in chloroplasts, 678
  from citric acid cycle, 447, 448
  in contractile processes, 32
  in control of carbohydrate metabolism, 500–502
  in control of gluconeogenesis, 566
  in control of glycolysis, 406, 421, 453

in dikinase reaction, 670
in DNA ligase reaction, 848
in DNA replication, 849, 850
in energy cycle, 371, 372
in energy transfer, 371–392
in enzyme phosphorylation, 731
in fatty acid activation, 387, 512, 513
from fatty acid oxidation, 515–519
in fatty acid synthesis, 586, 587, 592
free energy of hydrolysis, 377
in futile cycles, 571, 572
in gluconeogenesis, 564–566, 569
from glucose oxidation, 497, 498
in glycogen synthase regulation, 574, 575
in glycolysis, 399, 400, 402, 408, 410, 413, 414, 417, 418, 431
in heart muscle, 433, 699
in hexokinase reaction, 394, 395, 403
hydrolysis, 485
for heat production, 572
ionization, 373, 375
in kidney, 705
magnesium complexes of, 373
in metabolism, 337, 343, 344, 346
mitochondrial transport of, 491, 495, 496
in mRNA synthesis, 859
in muscle, 396, 401, 718
in muscle activity, 344, 380–384, 388, 389, 401, 569, 696–698
in nerve cell transmission, 699–701
in nitrogenase activity, 638, 639
in nitrogen fixation, 643
in nucleoside diphosphokinase reactions, 389
orthophosphate cleavage of, 387
as phosphate-group carrier, 376–380
γ-phosphate-group turnover, 391
in phosphofructokinase reaction, 406
in phosphofructokinase regulation, 423, 424, 433, 567
in phosphoglyceride synthesis, 600, 602
in phosphorylase regulation, 422, 574, 575
from photosynthesis, 645, 649, 652, 657, 660–662, 666
in propionyl-CoA carboxylation, 522
in protein phosphorylation, 892
in protein synthesis, 873, 877, 878, 880, 888, 889
pyrophosphate cleavage of, 386–388, 396, 554
in pyruvate carboxylase reaction, 454
in pyruvate dehydrogenase regulation, 450–452
in pyruvate kinase regulation, 568
standard free energy of hydrolysis, 374, 375, 377, 380
synthesis (*see also* oxidative phosphorylation), 435, 436, 446, 467–469, 475, 484–494, 510
in triacylglycerol synthesis, 596
turnover, 467
in urea cycle, 552, 554
S-adenosylhomocysteine, **602,** 619
in cysteine synthesis, 619
S-adenosylmethionine (SAM), **602,** 618
in creatine synthesis, 625
in cysteine synthesis, 619
in phosphatidylcholine synthesis, 600, 602, 603
adenovirus 38, 795
electron micrograph, 39
adenylate, *see* adenosine 5′-monophosphate
adenylate cyclase, 347, 732
adrenaline stimulation of, 730–732
effects of cholera toxin on, 750
glucagon stimulation of, 739
in operon function, 907
regulation, 734, 746
adenylate kinase, 387–389, 629
molecular weight, 180
adenylic acid, *see* adenosine 5′-monophosphate
adipocytes, 308, 318
electron micrograph, 700
adipose tissue, as energy store, 528
in metabolism, 701, 702
pentose phosphate pathway in, 456
triacylglycerol synthesis, 719
ADP, *see* adenosine 5′-diphosphate
ADP-glucose, in cellulose synthesis, 668, 669
in starch synthesis, 574, 668, 669
adrenal cortex, 135, 722, 723, 729, 743
hormone synthesis in, 504
hyperactivity of, 720
pentose phosphate pathway in, 456
adrenal cortical hormones, 740, 741
in control of triacylglycerol synthesis, 598
synthesis, 504
adrenal cortical steroids, 722
adrenal glands, 704, 721, 729
adrenaline, 346, 347, 540, 722, 724, 727, **728,** 729–734, 750
in adenylate cyclase stimulation, 730, 732
in blood, 728, 749
in control of glycogen metabolism, 575
in control of lipid metabolism, 702
in glycogen phosphorylase regulation, 422, 423, 729–733
in glycogen synthase regulation, 733, 734
in muscular activity, 697
physiological effects of, 728, 729
receptor sites, 326, 346, 347, 732
storage, 728, 729
adrenal medulla, 722, 727–729
hormones, 752
adrenocorticotropin (*see also* corticotropin), amino acid sequence, 135
adrenocorticotropin-releasing hormone, 725
aerobes, 334
aerobic glycolysis, 414, 435
affinity chromatography, 145
alanine, 49, 56, **97,** 98, 101, 102
acid-base properties, 104
in amino-group transport, 694
in ammonia transport, 546, 547, 694
asymmetry, 51, 52
in blood, 718
as buffer, 106
chemical properties, 51
dimensions, 17
in glucose synthesis, 568
in glutamine synthetase regulation, 623
isoelectric point, 106
optical activity, 97, 148
oxidative degradation, 536, 537
p*K*, 105, 107
in proteins, 156, 158, 160, 173
in pyruvate kinase regulation, 568
specific rotation, 97
synthesis, 618
titration curve, 105
in transaminase reaction, 532
alanine transaminase, 532, 534, 535, 547, 557
serum levels, in myocardial infarction, 535
alanine tRNA, 875
albinism, 241, 349, 542
alcohol, in diet, 761, 762
inhibition of gluconeogenesis, 570, 571
alcohol dehydrogenase, 126, 209, 247, 271, 426, 427, 761
alcoholic beverages, consumption of, 761, 762, 768
alcoholic fermentation, 398, 401, 426–428, 432
alcoholism, 762
alcohols, 50
solubility, 70
aldehyde dehydrogenase, 761
aldehyde oxidase, 784
aldehydes, 50
solubility, 70
aldofuranose, 282
aldohexose, 280, 282
aldolase, 404, 406, 407, 667
aldolase reaction, standard-free-energy change, 406
aldonolactonase, 458
aldopentoses, 280
aldopyranoses, 282
aldoses, 278, 279, 282
aldosterone, **741**
aldotetroses, 280
aldotrioses, 280
algae (*see also* cyanobacteria), cell walls, 293
photosynthesis in, 303, 644, 646
alkali disease, 784
alkaline phosphatase, disulfide cross-linkages, 203
pH activity curve, 87
alkaloids, 735
alkalosis, 78
alkaptonuria, 241, 349, 540, 542
alkylating agents, as mutagens, 916, 917
allantoin, **635**
from purine catabolism, 635
from purine nucleotide catabolism, 634
allergy, 35
alligators, anaerobic glycolysis in, 401
allopurinol, **636**
xanthine oxidase inhibition, 643
D-allose, **280**
allosteric enzymes, 233–238, 346
allosteric inhibition, in amino acid synthesis, 622–624
allosteric site, 234
alloxan, **738**
alpha cell, 735
α_1-globulins, 707
α_2-globulins, 707
α helix, 151–153, 155, 162, 175
in cytochrome *c*, 175
disruption, 155, 156
in myoglobin, 172, 175
unfolding, 165
α particles (in glycogen), 560
D-altrose, **280**
aluminum, as trace element, 273
alveolus, 712
Amanita phalloides, 857
α-amanitin, 114, 857
amber mutant, 888
Ames, B., 922
Ames test, 922, 923
amides, 50
amidophosphoribosyltransferase, 628
amination, 532
amine hormones, 723, 724, 727, 728
amines, 50
amino acid analyzer, 110, 137
amino acid auxotrophs, 642
amino acid code words, 897
amino acid residues, 112
amino acids, 6, 56, 61, 95–111
abbreviations, 96
abiotic synthesis, 61
absorption, 685, 689, 693
acid-base properties, 103–108
activation, 873, 877, 888
ammonia production, 545

in biosynthesis, 625
in blood, 707
as buffers, 105, 108
catabolism, 694
regulation of, 347
charged, 100
chemical properties, 50, 51, 110
chromatography, 109
classification, 100
in diet, 763–765
dipolar nature, 103
electrophoresis, 108
essential, 158, 335, 555, 616, 753, 754, 763–765
synthesis, 621, 622
genetic code words, 897, 898
glucogenic, 545, 568
ketogenic, 545
hydrophobic, 102
in lipid synthesis, 597
in metabolism, 338, 339, 754
metabolism, genetic defects in, 349–351, 540–542
metabolism, in hibernation, 598
as neurotransmitters, 700
ninhydrin reaction, 110
in nitrogen cycle, 335, 637
nonessential, 616
synthesis, 616–621
nonpolar, 100
in nucleotide synthesis, 627, 628, 694
optical activity, 96
oxidative degradation, 531–545
in diabetes mellitus, 531, 715
in peptides, 112
p*K* values, 107
polar, 100
from protein digestion, 689
in proteins, 121, 125, 154, 172, 173, 764
in protein synthesis, 693, 694, 872–874, 877
in pyruvate kinase regulation, 425
racemization, 100
R groups, 100
salvage, 531
separation, 108
specific rotation, 97
stereoisomers, 152
synthesis, 615–625
from citric acid cycle intermediates, 453
enzyme repression, 358, 624, 625, 903
feedback inhibition, 622–624
titration curves, 104
transamination reactions, 258, 532–534
in urine, 703
x-ray analysis, 150
aminoacyl adenylate, 877
hydrolysis, 879
aminoacyl-tRNA, 874, 877–879
in protein synthesis, 884–888
standard free energy of hydrolysis, 878
trinucleotide binding experiments, 896
aminoacyl-tRNA synthetases, 877–880
p-aminobenzoic acid, in bacterial growth, 643
in folic acid synthesis, **262**
amino-group transfer, 532–534
α-amino-β-ketoadipic acid, 625, 626
δ-aminolevulinic acid, **626**
in porphyrin synthesis, 626
aminopeptidase, 688
specificity, 689
aminophylline, 752
2-aminopurine, as mutagen, 916
aminotransferases (*see also* transaminases), 259
ammonia, from amino acids, 338, 339, 534, 535
in amino acid synthesis, 535
excretion, 548, 549
in fertilizers, 639
in nitrogen cycle, 335, 336, 636–638
toxicity, 545, 555, 559
transport in blood, 546, 547
in urine, 703
ammonium, p*K*, 79
as weak acid, 81, 82
ammonium nitrate, as fertilizer, 638
ammonotelic animals, 548, 549
amniocentesis, 542, 604
amoeba, 32
AMP, *see* adenosine 5′-monophosphate
amphibia, glycolysis, 569
nitrogen excretion, 549
amphibolic pathways, 342, 453
amphipathic compounds, 71, 305, 311
ampholytes, 104
amphoteric compounds, 104
amylase, 284, 288, 686
α-amylase, 288, 289
in brewing, 428
β-amylase, 289
turnover number, 220
amylopectin, 287–289, 300
amylose, 289, 290
α-amylose, 287, 288
amytal, 483
Anabaena azollae, 21
anabolic pathways, 561, 562
anabolic steroids, 744
anabolism (*see also* biosynthesis), 337, 338, 340–346, 359
regulation, 342, 346
anaerobes, 334, 335
facultative, 335
strict, 335
anaphase, 817
anaplerotic reactions, 454, 455, 564
androgens, 740, 743, 744
anemia, 260, 262, 457, 709, 772, 780
pernicious, 523, 558, 774
Anfinsen, C., 135, 178
angstrom units, 17
animal viruses, 795
anion-exchange resin, 109
anomers, 282, 299, 300
antenna molecules, 655
anterior pituitary gland, 135, 726
anterior pituitary hormones, 740, 745, 746
antibiotics, 145, 294, 489
inhibition of protein synthesis, 894, 895
peptide, 114
resistance to, 814
antibodies (*see also* immunoglobulins), 102, 123, 138, 707, 926, 927
genes for, 927–929, 938
in isolation of mRNA, 932
molecular weights, 927
in radioimmunoassay, 727
structure, 927, 928
anticancer drugs, 642, 643
anticodon, 795, 875, 876, 898, 912
antidepressant drugs, 65
antifreeze, 90
"antifreeze" proteins, 124, 294
antigen-antibody complex, 138, 927
antigen binding sites, 928
antigens, 138, 926, 927
in radioimmunoassay, 727
antimycin A, 483, 508
antiserum, 138
aorta, 698
apoenzyme, 209
appendix, 684
A protein, 903–905
D-arabinose, **280**
arachidic acid, 305
arachidonic acid, 305, 746, **747**, 759
synthesis, 595
ara operon, 905
Arber, W., 825
arginase, 209, 271, 550, 553, 554
optimum pH, 219
arginine, **56,** 101, 103
in creatine synthesis, 625
deficiency, 559
in diet, 763
in enzyme action, 244
in glucose synthesis, 568
in histones, 816, 818
oxidative degradation, 536, 543
p*K* values, 107
in proteins, 153, 173
specific rotation, 97
synthesis, 622, 763
genetic defects, 350, 351
in urea cycle, 550–554
argininosuccinate, **553**
in urea cycle, 551–553
argininosuccinate lyase, 542, 552, 553
argininosuccinate synthetase, 552, 553
argininosuccinic acidemia, 542
Arnon, D., 660
arsenate, 431, 432
arsenic, as trace element, 270
arterial blood, carbon dioxide levels, 709
arteries, elastic tissue, 160
lipid deposits, 759
ascorbic acid (*see also* vitamin C), 252, **264,** 308, 457, 770
deficiency, 770–772
in diet, 767, 770, 771
recommended daily allowance, 770
synthesis, 458, 459
asparaginase, 544
asparagine, **56,** 96, 101, 102
in glucose synthesis, 568
oxidative degradation, 536, 544
in proteins, 153, 173
synthesis, 618
asparagine synthetase, 618
aspartame, **758**
aspartase, specificity, 220
aspartate (*see also* aspartic acid), in amino acid synthesis, 623, 624
in glucose synthesis, 568
oxidative degradation, 536, 544
in purine nucleotide synthesis, 628
in pyrimidine nucleotide synthesis, 631
synthesis, 618
transaminase reaction, 532
in urea cycle, 552, 553
aspartate aminotransferase, 216
aspartate semialdehyde, 624
aspartate transaminase, 532, 534, 535
serum levels, in myocardial infarction, 535
aspartate transaminase reaction, equilibrium constant, 394
aspartate transcarbamoylase, 181, 631, 632
K_M, 633
molecular weight, 180
regulation, 237, 238, 632, 633
tertiary structure, 238
aspartic acid (*see also* aspartate), **56,** 101, 102
in enzymes, 227, 229–231, 247
p*K* values, 107
in proteins, 173, 892
aspartyl phosphate, 624
aspirin, **92,** 746
Astbury, W., 150, 165
asthma, 709, 729, 752
asymmetric compounds, 51

atherogenesis, 610
atherosclerosis, 316, 702, 759, 761
Atkinson, D., 500
atmospheric pollution, 676
ATP, *see* adenosine 5′-triphosphate
ATPase (adenosine triphosphatase), bacterial, 674, 675
  in brown fat, 491, 702
  Ca^{2+}-transporting, 382, 386, 698
  in cilia, 383
  H^+-transporting, 385, 386, 705
  in muscle, 163, 380–382
  myosin, 163
  Na^+K^+-transporting, 322, 385, 386
    inhibition, 718
    in kidney, 705
    in nerve cell membrane, 700
  in sarcoplasmic reticulum, 382, 386
ATP cycle, 372
ATP:glucose phosphotransferase (*see also* hexokinase), 210
ATP synthetase, 470, 484–487, 493
  bacterial, 674, 675
  of chloroplasts, 662
  electron micrograph, 486
ATP system, in control of carbohydrate metabolism, 500
  in control of respiration, 499
  mass-action ratio, 499, 500, 502
atractyloside, 495
atrium, 698
autocatalysis, 687
autotrophic cells, 333–335
autoxidation, 307, 308
auxotrophic mutants, 350, 351, 642
Avery, O. T., 801, 802
Avery-MacLeod-McCarty experiment, 801, 802, 923
avidin, 259, 773
axons, 30, 325, 582, 699, 700
azaserine, **642**
Azobacter, 637

bacilli, 19
Bacillus brevis, 145
back-mutation, 922
bacteria (*see also* prokaryotes), 17
  amino acid synthesis, 616
    regulation, 623–625
  antibiotic-resistant, 814
  ATP synthesis, 492, 493
  auxotrophic, 642
  cell walls, 293, 294
    proteins, 893, 894
  chemotaxis, 22
  conjugation, 925
  denitrifying, 335
  dimensions, 17
  DNA, 11, 810, 812, 813, 823–825
  electron transport, 493
  enzyme localization, 353
  flagella, 21, 353, 494, 495
  gram-negative, 294
  gram-positive, 294
  halophilic, 266, 674, 675
  lactic acid fermentation, 398
  membranes, 329, 353, 493
  nitrifying, 335, 336, 636
  nitrogen-fixing, 335, 336, 637, 640
  nucleoid, 353
  pathogenic, 18, 262, 296
  phage infection, 923–925
  photosynthetic, 18, 21, 334, 646, 648
  plasmids, 814, 815
  protein synthesis, regulation, 901–908
  rate of metabolism, 43
  ribosomes, 353
  root-nodule, 637, 638, 643
  rumen, 570, 686, 773, 774
  soil, 636, 637
  structure, 17–22
  thermophilic, 834
  toxins, 123
  transformation, 923
  viruses, *see* bacteriophages
bacteriophage f2, 795
  replication, 864
bacteriophage G4, overlapping genes, 901
bacteriophage λ, 38, 795, 811, 923
  DNA, 803, 811
    electron micrograph, 812
    as gene vector, 931, 936
  overlapping genes, 900
  particle weight, 811
bacteriophage M13, DNA, 835
bacteriophage MS2, 38, 795
  replication, 864
bacteriophage ϕX174, 37, 795, 811
  DNA, 811
    base composition, 803
    base sequence, 790, 794, 827
  overlapping genes, 791, 900, 901
  particle weight, 811
  replicative form, 835
bacteriophage Qβ, 795
  replication, 864, 865
bacteriophage R17, 795, 865
  replication, 864
bacteriophages, 38, 795
  DNA, 802, 803, 811, 812
  gene transposition, 926
  replication, 38
  temperate (lysogenic), 923
  transformation of host cells, 924
bacteriophage T2, 795, 802, 811
  DNA, 811, 834
    electron micrograph, 812
  particle weight, 811
bacteriophage T4, 38, 795, 811
  DNA, 811
  electron micrograph, 39
  particle weight, 811
bacteriophage T7, DNA, 811
    composition, 803
  particle weight, 811
bacteriorhodopsin, 167, 266, 674, 675, 679
  crystals, 675
ball-and-stick models, 48
Baltimore, D., 863
barley, in brewing, 428
Barnum, P. T., 740
basal metabolic rate (BMR), 743
  thyroid hormone regulation, 743, 751
base analogs, as mutagens, 916
base pairs, in DNA, 804–807, 810
  in RNA, 882
  in tRNA, 875
bases, 78
bases (purine and pyrimidine), 796, 797
  abiotic synthesis, 61
  methylation, 823
Basic Four food groups, 784
B cell, 735–737
Beadle, G., 820
beans, as food source, 765
bears, hibernation, 598
beef, as food source, 764
beef fat, fatty acid composition, 307, 760
beer, 427, 428
  consumption, 761, 762
Benesch, Reinhold, 192
Benesch, Ruth, 192
Benzedrine, **65**
benzene, physical properties, 67
benzoic acid, solubility, **91**
beriberi, 250, 252, 440, 767, 768
β cells, 735, 736
β conformation, 155, 162, 167, 175
β-globulins, 707
β oxidation, 511–519
β particles (glycogen), 560
bicarbonate, in blood plasma, 85–87, 554, 705, 710, 711
    in diabetes mellitus, 715
  buffer systems, 85–87, 710, 711
  in intracellular fluid, 705
  membrane transport, 711
  pK, 79
  in red blood cells, 710, 711
  in urea cycle, 552, 554
  in urine, 703
bilayers, lipid, 317, 328, 329
    in membranes, 321, 322
bile, 691
  secretion, deficiency in, 718
bile pigments, 627
bile salts, 610, 695
  in lipid digestion, 690, 691
bilirubin, **627**
biochemical engineering, 607
biocytin, 209, 252, 259, **260**, 773
bioelements, 45
bioenergetics, 361
biological evolution, 62
biological value, of protein, 764
bioluminescence, 388
biomolecules, 1, 45–62
biosphere, energy flow, 336
biosynthesis (*see also* anabolism), 338, 340, 345, 561, 562
  ATP in, 344
  NADPH in, 344
  regulation, 561
biotin, 252, 259, **260**, 454, 455, 773
  in acetyl-CoA carboxylase, 586, 587
  daily requirement, 773
  deficiency, 767, 773
  in propionyl-CoA carboxylation, 522
  in pyruvate carboxylase, 259, 260, 454, 455
biotinyllysine, **259**
birds, uric acid excretion, 548, 549, 555, 627
bit (binary digit), 793
blacktongue, 251
"blind staggers," 784
Bloch, K., 607, 609
blood, adrenaline levels, 728, 749
  alanine level, 718
  ammonia levels, 545
  bicarbonate buffer system, 85–87, 554, 710, 711
    in diabetes mellitus, 715
  carbon dioxide transport, 189, 706, 708–711
  composition, 706
  free fatty acids, 695
  glucose levels, 405, 420, 692, 694, 707, 713, 737, 738
    in diabetes mellitus, 284, 405, 713, 714, 738, 739
    regulation of, 422, 423, 741
  glutamine levels, 546, 718
  hormone levels, 706, 707, 723, 724, 727, 749
  ketone body levels, 525, 707
    in diabetes mellitus, 525, 713, 714
  in metabolism, 705–711
  oxygen transport, 185, 706–711
  serum albumin levels, 719
  solutes, 72
blood clotting, 269, 707, 777, 892, 921
  heparin inhibition, 297

blood-group antigens, 323
blood plasma, 185, 706
  cholesterol level, 700, 707
  chylomicrons, 691
  composition, 705–707
  kidney filtration, 703
  lactate levels, 580, 707
  lipoproteins, 315, 316, 760
  pH, 87, 705, 719
    in diabetes mellitus, 715
    regulation of, 93, 554
  potassium level, 385, 704, 705
  sodium level, 385, 704, 705
blood serum, creatine kinase levels, 535
  hormone levels, 727
  lactate dehydrogenase levels, 535
  transaminase levels, 535
blood types, 323–325
Blow, D. M., 229
blue-green algae (*see also* cyanobacteria), 18, 335
B lymphocyte, 139
boat conformation, 283
body fat, 762, 763
Bohr, C., 189
Bohr effect, 189, 192
bombardier beetle, 481
bomb calorimeter, 756
Bombyx mori, 167
bond energy, 69
bond rotation, 48
bone, 157, 160, 162
  calcium store, 778, 779
bone mineral, 269
bony fishes, ammonia excretion, 548
boron, as trace element, 273
botulinus toxin, 123
Bowman's capsule, 704
bradykinin, 114
brain, adenine nucleotides in, 373
  ammonia toxicity, 545, 546
  glucose requirement, 562
  metabolic turnover of components, 353
  in metabolism, 699, 700
  opiate receptors, 746
  oxygen consumption, 708
  phosphate concentration, 373
  phosphocreatine concentration, 373
  sphingolipid metabolism, defects in, 606
  in Tay-Sachs disease, electron micrograph, 605
branched-chain α-ketoacid dehydrogenase, 542
branching enzyme, genetic defect, 576
breast cancer, 745, 751
brewer's yeast, 427
brewing, 427, 428
5-bromouracil, as mutagen, 916
bronchioles, 712
bronchus, 712
brown fat, 491, 509, 702, 703, 762
  mitochondria, 491, 509, 702
Buchanan, J., 628
Buchner, E., 208, 348
buffers, 81–86, 91, 92, 106
  amino acids, 108
  bicarbonate, 85–87
  biological, 83
  phosphate, 83, 85
  standard, 92
building-block molecules, 6, 55, 56, 339
bulk element, 269
bundle-sheath cells, 670–672, 679
  electron micrograph, 671
buoyant density, 810
trans-Δ^2-butenoyl-S-ACP, 590, **591,** 592
butter, 307
  fatty acid content, 307, 760
n-butylmalonate, inhibition of transport systems, 509
butyrate, in blood, 523
  from glucose, 570
butyroyl-CoA, oxidation, 523
butyryl-S-ACP, 591, 592

caffeine, 636, 735
Cairns, J., 840, 868
calcitonin, 746
calcium, absorption, 778, 779
  in blood clotting, 269, 892
  in bone, 778, 779
  calmodulin binding, 734, 735, 747, 752
  in diet, 778, 779
  in enzyme regulation, 747
  as essential element, 269
  in hormone action, 738, 747
  intracellular concentration, 747
  as intracellular messenger, 733, 735
  metabolism of, 776
    regulation, 268
  in milk, 892
  mitochondrial transport, 491, 492, 494
  in muscle activity, 27, 381, 382, 698, 699
  in phosphodiesterase activity, 734, 735
  in phosphorylase activity, 422, 731, 732
  in pyruvate dehydrogenase regulation, 451, 452
  recommended daily allowance, 755, 779
  role in metabolism, 778
  in urine, 703
calcium aluminum silicate, 272
calcium hydroxyapatite, 778
calmodulin, 734, 735, 747, 752
calorie, 367, 368
calorie deficiency, 765, 766
calorimeter, 756, 757
Calvin, M., 663, 664, 678
Calvin cycle, 664–667, 672
camel, triacylglycerol storage, 598
  urea production, 554
cancer, 261, 864, 892, 922, 923
  artificial sweeteners and, 758
  breast, 745, 751
  cholesterol synthesis in, 610
  drug therapy, 642, 643
  glycolysis in, 502
  interferon and, 940, 941
  regulatory behavior in, 502
  skin, 914, 915
  viruses, 828, 863, 864, 939
cane sugar, *see* sucrose
capping, 322
CAP protein, 906–908
capsid, 37
carbaminohemoglobin, 189
N-carbamoylaspartate, **631,** 632
carbamoyl phosphate, 551, **552,** 554
  in glutamine synthetase regulation, 623
  as high-energy compound, 552
  in pyrimidine nucleotide synthesis, 631, 632
carbamoyl phosphate synthetase I (mitochondrial), 551, 552, 632, 663
carbamoyl phosphate synthetase II (cytosolic), 552, 632
carbamoyl phosphate synthetase reaction, standard-free-energy change, 552
carbohydrates (*see also* monosaccharides, disaccharides, polysaccharides), 277–297
  catabolism, regulation, 500–502
  classification, 277
  in diet, 753, 757
  digestion, 686
  as energy source, 755–758, 784
  in glycoproteins, 892
  in lipid synthesis, 597
  metabolism, genetic defects, 433, 434, 581
    futile cycles, 571
    hormonal regulation, 575, 576
  synthesis, 465, 562–577
carbon, asymmetric, 51
  bonding characteristics, 46
  double bonds, 47, 48
  isotopes, 351, 352, 358
  in living organisms, 46
  single bonds, 46, 48
  van der Waals radius, 49
carbon cycle, 333–335
carbon dioxide, in atmosphere, 88, 334, 674
  in blood, 554, 706, 708–711
  as carbon source, 333, 334, 663
  from catabolism, 338, 339
  from citric acid cycle, 441, 445, 446, 448
  excretion, in diabetes mellitus, 715
  in fatty acid synthesis, 584, 586, 587, 589, 590, 592
  fixation (*see also* photosynthesis), 663, 664, 667, 674
    by C_3 plants, 667, 669
    by Hatch-Slack pathway, 669–672
  in gluconeogenesis, 564, 565, 567, 570, 571, 579
  in lungs, 92
  in photorespiration, 673, 674
  in photosynthesis, 645, 646, 649, 656, 663–666, 674
  in propionyl-CoA carboxylation, 522
  in pyruvate carboxylation, 454, 455
  from respiration, 86, 435, 436
  in ribulose diphosphate carboxylase reaction, 673, 674
  transport in blood, 189, 706, 708–711
  in urea cycle, 551, 552, 554
1-carbon-group transfer, 621
carbonic acid, p*K*, 79
  in red blood cells, 710, 711
carbonic anhydrase, 209, 271, 710, 711
  inhibition, 247
  K_M, 216
  molecular weight, 246
  turnover number, 220, 246
carbon monoxide, cytochrome oxidase inhibition, 483
  cytochrome P-450 reaction, 503
  hemoglobin inhibition, 171
carbon monoxide poisoning, 171
γ-carboxyglutamic acid, 103, 892
  in blood clotting, 269
carboxylation reactions, biotin requirement, 259
carboxypeptidase, 130, 271, 735
  active site, 244
  in digestion, 688, 718
  optimum pH, 717
  specificity, 688, 689
  tertiary structure, 175
carcinogens, 864, 922, 923
cardiolipin, **311**
  synthesis, 603
cardiovascular disease, 762
cardiovascular system, 698
carnitine, 251, **514**
  in fatty acid oxidation, 513, 514, 526
carnitine acyltransferase I, 513, 514
  regulation, 526
carnitine acyltransferase II, 514
β-carotene, **265,** 654
  deficiency, 774
  in diet, 775
  in photosynthesis, 654
carotenoids, 265, 610, 654
  in photosystems, 655

Cartier, J., 770, 771
cartilage, 157, 160
casein, 123, 126
  calcium binding, 892
β-casein, 717
catabolic repression, 906, 907
catabolism, 337–347, 359, 561
  energy release, 371
  NADPH production, 345
  regulation, 342, 346
catabolite activator protein (CAP), 906–908
catalase, 29, 209, 270, 481
  K_M, 216
  optimum pH, 219
catalysis, acid-base, 226
  covalent, 226
catalysts, 208, 211, 212
catalytic site, *see* active site
catecholamines, 728
catechols, 92, 503, **728**
cation-exchange resin, 109
cattle, cellulose digestion, 292, 570
  gluconeogenesis, 570
CDP-choline, in phosphatidylcholine synthesis, 600, 602
CDP-ethanolamine, **601**
  in phosphatidylethanolamine synthesis, 600–602
  synthesis, 600, 601
CDP-glucose, in cellulose synthesis, 668, 669
cecum, 774
celiac disease, 690
cell cloning, 931
cell coat, 34, 295
cell-free extracts, 348
cell membrane, 29, 33
  bacterial, 21
  enzyme activities, 353, 354
  lipids, 315, 603
cellobiose, 284
cell-recognition sites, 34, 325
cells, aerobic, 334
  anaerobic, 334, 335
  autotrophic, 333–335
  dimensions, 16
  division, 22, 25, 30, 383, 816, 817
  energy status, 500
  eukaryotic, 18, 22–37, 352–354
  evolutionary origins, 60
  fractionation, 354–356
  glycoproteins, 295
  heterotrophic, 334, 335, 366, 371, 645
  localization of metabolic pathways, 353–356
  prokaryotic, 17–22, 33, 352, 353
  structure, 15, 58, 647
  surface area, 43
  water, 67
cell surface, 34
cellulase, 291
cellulose, 37, 54, 278, 289–292, 300
  bacterial hydrolysis, 570, 686, 773, 774
  digestion, 275, 291, 292, 686
  electron micrograph, 54, 293
  synthesis, 465, 668, 669
cell walls, 292–294, 353
  formation, 28
  plants, 37, 292
  proteins, synthesis, 893, 894
central dogma, 792, 853, 863, 865
centrifugation, differential, 355, 356
centriole, 25, 817
cerebrosides, 312, 313
cesium chloride density gradient, 810, 838
CF_1 ATPase, 662
Chaetomorpha, 293
chair conformation, 283
Chargaff, E., 803, 804
Chase, M., 802
chemical coupling hypothesis, 487, 488
chemical energy, 331, 333, 343, 344, 366
chemical evolution, 59–62
chemical reactions, equilibrium constant, 366–368
  standard-free-energy change, 366–371
chemical score, of protein, 764
chemiosmotic hypothesis, 487–493, 509, 510
  in photophosphorylation, 662
chemotaxis, 22, 325
chicken fat, fatty acid composition, 760
chief cells, 687
chimeric DNA, 934
chimeric protein, 937
chiral center, 52
chiral compounds, 52
  amino acids as, 96
chitin, 292
Chlamydomonas, 24, 27, 647
chloramphenicol, **894**, 895
Chlorella, 678, 679
chloride, in blood, 705
  in gastric juice, 705
  membrane transport, 711
  in urine, 703
chloroform, physical properties, 67
chlorophyll, 36, 610
  bacterial, 21
  light absorption, 651–657
  in photosynthesis, 649–656, 659
  synthesis, 625
chlorophyll *a*, **653**, 654, 659
  absorption spectrum, 655
  in photosystems, 655
chlorophyll *b*, **653**, 654
  in photosystems, 655
chlorophyll *c*, in photosystems, 655
chlorophylls, 653
  in photosystems, 655
chloroplasts, 4, 35, 303, 649, 650
  ADP content, 678
  ATP content, 678
  ATP synthesis (*see also* photophosphorylation), 492–494
  dimensions, 17
  division, 820
  DNA, 818–820
  electron micrograph, 36, 650
  evolutionary origins, 36, 820, 881
  inner membrane, 649, 650
    H^+-ATPase, 386
  isolation, 354, 651
  membranes, 325, 494, 649, 650, 662
  phosphate content, 678
  photosynthesis [*see also* photosynthesis (*main entry*)], 649
  in plankton, 647
  protein synthesis, 880, 881
  starch granules, 288
  structure, 650
cholecalciferol (*see also* vitamin D_3), **267**, 776
cholecystokinin, 746
cholera toxin, 750
cholesterol, **315**
  arterial deposits, 610
  in blood, 700, 707
  in diet, 759, 760
  in hormone synthesis, 743
  in lipoproteins, 316
  in membranes, 319
  metabolic turnover, 353
  metabolism, 760
  synthesis, 583, 607, 608, 693, 695
    regulation, 609, 610
cholesterol esters, 315
  in membranes, 319
cholic acid, **691**
choline, **57**, 222, 311
  as vitamin, 603
choline kinase, 602
chondrocytes, 157
chondroitin, 296, 297
chromaffin vesicles, 728, 729
chromatin, 25, 58, 794, 816, 818, 856
  electron micrograph, 819
  in mitosis, 817
chromatography, 128
  affinity, 145
  ion-exchange, 109
  of peptides, 131
chromatophores, 660
chromium, as trace element, 270, 273, 783
chromosome number, 815
chromosomes, 25, 815, 816, 913
  in genetic recombination, 923–926
  gene transposition, 926
  mapping, 825, 938
  in mitosis, 816, 817
  origin, 841
  prokaryotic, 794, 812, 813
  replication, 840, 841
chyle, 690, 691
chylomicrons, 316, 318, 690, 691, 701, 702, 707
chymotrypsin, 113, 134, 735
  active site, 227, 229, 230, 231
  crystals, 208
  in digestion, 688
  inhibition, 222
  K_M, 216
  molecular weight, 126, 688
  optimum pH, 717
  reaction mechanism, 227, 230, 231
  specificity, 131, 220, 221, 688, 689
  tertiary structure, 175, 229
chymotrypsinogen, 125, 229
  activation, 688, 690
  amino acid composition, 125
  isoelectric point, 128
  molecular weight, 688
  secretion, 688
cilia, 30–32, 123, 163, 383
  ATPase activity, 383
  electron micrograph, 32
circular DNA, 811, 813
  replication, 840–843
cis-aconitate, **442**
  in citric acid cycle, 441, 442, 445, 451
cisternae, 27
cis-trans isomers, 148
citrate, **442**
  in acetyl-CoA carboxylase regulation, 587, 595
  binding to aconitase, 451
  in blood, 707
  in citrate synthase regulation, 452, 453
  in citric acid cycle, 441, 442, 444, 445, 449–451
  in control of carbohydrate metabolism, 501, 502
  in glyoxylate cycle, 455, 456
  mitochondrial transport, 496, 586
  in phosphofructokinase regulation, 421, 423, 424, 453, 567
  as prochiral compound, 451
citrate-cleavage enzyme, 586
citrate lyase, 586
citrate synthase, 257, 444, 445, 448, 450
  in acetyl-group shuttle, 585
  in glyoxylate cycle, 455
  regulation, 444, 452, 453
citrate synthase reaction, standard-free-energy change, 444
citric acid cycle, 257, 398, 435–437, 444–449, 461

in amino acid oxidation, 536, 537, 544, 545
as amphibolic pathway, 453
anaplerotic reactions, 564
enzymes, localization, 444, 470
experimental investigation, 441–444, 449–451, 464
in fatty acid oxidation, 515, 519, 521
fluoroacetate inhibition, 465
intermediates, in amino acid synthesis, 453, 616
concentrations, 454
in glucose synthesis, 568
replenishment, 454, 455
malonate block, 443
in metabolism, 339, 342, 693–695
regulation, 444, 445, 449, 452, 453, 500–502
in skeletal muscle, 696
and urea cycle, 553
citrulline, 116, **550**
in urea cycle, 550–552
citryl-CoA, 444
Clausius, R., 364
clones, 931
of immunocytes, 927, 931
Clostridium perfringens, 168
CMP, *see* cytidine 5′-monophosphate
CoA, *see* coenzyme A
cobalamine (*see also* vitamin B_{12}), 774
in amino acid catabolism, 558
cobalt, nutritional requirement in cattle, 275
as trace element, 270, 783
in vitamin B_{12}, 263, 272
cocci, 19
cocoonase, 223
codon, 795, 895
codon-anticodon binding, 898–900
coelacanth, anaerobic metabolism, 401
coenzyme A (CoA), 209, 252, 256, **257,** 773
in acetate activation, 396
as acyl-group carrier, 257
in citrate synthase reaction, 257
cytosolic, 514
in fatty acid oxidation, 387, 512–514, 517, 518, 588
in α-ketoglutarate dehydrogenase complex, 446
mitochondrial, 514
in pyruvate dehydrogenase reaction, 257, 438–440
coenzyme B_{12} (*see also* 5′-deoxyadenosylcobalamine), 209, 262, 263, 522, 523, 570, 571
coenzyme Q (*see also* ubiquinone), 479
as electron carrier, 476
coenzymes, 33, 209, 249, 252–264
cofactors, 209
colinearity, 823, 827, 860, 900
colipase, 690
collagen, 103, 123, 124, 157–160, 162
in adipose tissue, 700
amino acid composition, 158, 168
cross-linkage formation, 271
electron micrograph, 157
gene for, 827
helical structure, 160
hydroxyproline content, 158, 264
synthesis, 168
x-ray diffraction analysis, 159
colloid goiter, 743
colloid protein, 742, 743
colon, 684
competitive inhibition, 223–225
complementary DNA (cDNA), 863, 864, 932, 933
cloned, 935, 936
in gene isolation, 936
incorporation into plasmid, 934, 935
conalbumin, gene for, 827
concanavalin A, 324
concerted inhibition, 623, 624
configuration, 147
conformation, 147, 148
conformational coupling hypothesis, 487
conformational formulas, 283
conjugate acid-base pair, 78, 81, 471, 472
conjugated proteins, 125, 126, 315
conjugate redox pairs, 471–473
standard reduction potentials, 474
connective tissue, 157
acid mucopolysaccharides, 296
constitutive enzymes, 902
contractile proteins, 123, 163, 380, 381
cooperative interactions, 187
coordinate induction, 903
coordinate repression, 903
copepod, 644
copolymerase III, 846
copper, in cytochrome oxidase, 481, 780
in diet, 780
recommended daily allowance, 780
as enzyme cofactor, 209
in plastocyanin, 660
in respiratory chain, 476
as trace element, 270, 271
Corey, R., 150, 165
corn, as food source, 764, 765, 769, 770
cornea, 158, 160
Cornforth, J., 609
corn oil, fatty acid composition, 760
corn syrup, 286
coronary artery disease, 760, 761
coronary infarction (*see also* myocardial infarction), 316
corpus luteum, 724, 744
cortical steroids (corticoids), 740, 741
corticosterone, **741**
corticotropin, 114, 123, 135, 722, 726, 731, 746
amino acid sequence, 135
in blood, 740
molecular weight, 726
corticotropin-releasing hormone, 740
cortisol, 740, **741**
metabolic effects of, 740
cosmic rays, DNA damage, 914
cotton, 291, 300
coumarin, **276**
covalent bonding, 46, 58, 69
C_4 pathway, *see* Hatch-Slack pathway
C-peptide, 737
C_3 plants, 667
photorespiration, 672–674
C_4 plants, 672–674
creatine, 384, **625**
synthesis, 625
creatine kinase, 384, 389, 625, 697
serum levels, in myocardial infarction, 535
creatine phosphate, *see* phosphocreatine
creatinine, in blood, 707
in urine, 703, 704
cretinism, 782
Crick, F., 129, 804, 805, 807, 808, 853, 872, 873, 898, 899
cristae, 25, 470, 485, 491
critical micelle concentration, 328
crop rotation, 638
cross-feeding, 351
crossover point, 425, 483
crystal lattice, ice, 68
sodium chloride, 70
CTP, *see* cytidine 5′-triphosphate
CTP synthetase, 631
Cu^{+}–Cu^{2+} redox pair, 471
curie, 360
Cushing's disease, 741
cuticle, 149
cyanate, 200
cyanide, 483
cyanobacteria, 18, 36, 334, 335
nitrogen fixation, 637, 648
photosynthesis, 646, 648
cyanocobalamine (*see also* vitamin B_{12}), **263**
cyanogen bromide, 131, 132
cyclamate, **758**
3′,5′-cyclic adenylic acid (*see also* cyclic AMP), 729, **730**
cyclic AMP, 347, **730**, 731, 732, 750
cell concentration, 732
in control of *lac* operon, 907, 908
protein kinase regulation, 731, 732, 734
cycloheximide, **894**, 895
cystathionine, **619**
in cysteine synthesis, 619, 620
cystathionine γ-lyase, 619, 620
cystathionine β-synthase, 241, 542, 619, 620
cysteic acid, 134
cysteine, **56,** 101, 102
in enzymes, 223, 588–593
in glucose synthesis, 568
oxidative degradation, 536, 537
pK values, 107
in proteins, 173, 717
synthesis, 618–620
cystine, **102**
cleavage, 134
in proteins, 134, 153, 156, 166, 178, 717, 893
cytidine 5′-monophosphate (CMP), **798**
synthesis, 631, 632
cytidine 5′-triphosphate (CTP), **390**
in aspartate transcarbamoylase regulation, 632, 633
in glutamine synthetase regulation, 623
as high-energy compound, 389
in phosphoglyceride synthesis, 600–602
synthesis, 631, 632
cytidylic acid, **798**
synthesis, 631, 632
cytochrome *a*, in electron-transport chain, 476
standard reduction potential, 474, 475
cytochrome a_3, in electron-transport chain, 476
standard reduction potential, 474
cytochrome aa_3, 481
in electron-transport chain, 467–469, 476, 480, 482
cytochrome *b*, in electron-transport chain, 468, 476, 480
gene for, 827
standard reduction potential, 474, 475
cytochrome b_6, standard reduction potential, 507
cytochrome b_{562}, 476
cytochrome b_{563}, 657, 660, 661
cytochrome b_{566}, 476
cytochrome *c*, 173, 270, 480, 481
absorption spectrum, 480
amino acid composition, 125
amino acid sequence, 137, 481
crystals, 125
in electron-transport chain, 468, 476, 480–483
heme group, 892
invariant residues, 136, 173, 174
isoelectric point, 128
molecular weight, 173, 480
species variations, 136, 137
standard reduction potential, 474, 475
tertiary structure, 174
cytochrome c_1, standard reduction potential, 474, 475
cytochrome *f*, 657, 660, 661
standard reduction potential, 507

cytochrome oxidase, 209, 270, 271, 467, 469, 482, 502, 503
molecular weight, 180
cytochrome P-450, 270, 503, 504
cytochromes, 476, 480, 481
absorption spectra, 480
synthesis, 625
cytoplasm, 15, 16, 22, 303
cytoplasmic membrane, 16
cytosine, **57**, 796, 797
deamination, 915, 917
in DNA, 803–806
in RNA, 854
cytoskeleton, 30, 31
cytosol, 16, 22, 23
enzyme activities, 353, 354, 356

dansyl chloride, **130**
dark reactions, 649, 651, 660, 663–669
regulation, 669
Darwin, C., 62
Davies, R., 718
D cell, 735–737, 739
deaminating agents, as mutagens, 916, 917
debranching enzyme [*see also* $\alpha(1 \rightarrow 6)$ glucosidase], genetic defect, 576
dehydroascorbic acid, **264**
7-dehydrocholesterol, **267**
in vitamin D_3 synthesis, 776
dehydrogenases, 256, 462, 467–469
cytosolic, 477
in fatty acid oxidation, 515, 516
flavin-linked, 478–480, 484
in metabolism, 344
mitochondrial, 469, 477
NAD-linked, 271, 471, 476, 477, 484
NADP-linked, 271, 476, 477
dehydrogenation reactions, 461
deletion mutations, 919, 920
denaturation, DNA, 808–810
enzymes, 177, 246
proteins, 140, 166, 169, 177
dendrites, 699, 700
denitrification, 637
denitrifying bacteria, 335
density gradient centrifugation, 810, 835, 838
dental caries, 782, 783, 786
dentine, 100
2′-deoxyadenosine 5′-diphosphate, **633**
synthesis, 632, 633
2′-deoxyadenosine 5′-monophosphate, **798**
2′-deoxyadenosine 5′-triphosphate (dATP), 389, **390**
5′-deoxyadenosylcobalamine (*see also* coenzyme B_{12}), 209, 252, 263, 774
in methylmalonyl-CoA mutase reaction, 522, 523, 570
deoxyadenylate, **798**
2′-deoxycytidine 5′-monophosphate, **798**
2′-deoxycytidine 5′-triphosphate (dCTP), 389
deoxycytidylate, **798**
2′-deoxyguanosine 5′-diphosphate, synthesis, 633
2′-deoxyguanosine 5′-monophosphate, **798**
2′-deoxyguanosine 5′-triphosphate (dGTP), 389
deoxyguanylate, **798**
deoxyhemoglobin, 182, 190
deoxyhemoglobin S, 195
association, 198
deoxynucleoside 5′-triphosphates, 391
deoxyribonucleases, 800
deoxyribonucleic acid (DNA), 6, 11, 16, 42, 54, 55, 129, 794
in antibody coding, 928, 929
bacterial, 11, 810, 812, 813, 823–825
base composition, 802, 803
base deamination, 915, 917
base pairs, 804–807, 870
base sequence, 827, 828
determination, 825, 828–831
buoyant density, 810
as carrier of genetic information, 800–802
chemical damage, 916, 917
chimeric, 934
in chloroplasts, 818–820
in chromatin, 816, 818
in chromosomes, 815, 816
circular, 811, 813
replication, 840–843
cleavage maps, 828
cloning, 829, 864
cohesive (sticky) ends, 824, 825
in recombinant DNA formation, 930
complementary, 863, 864, 932, 933
denaturation, 808–810
density gradient centrifugation, 810, 835, 838
double helix, 804–806
unwinding, 848–850
E. coli., 20, 42, 803, 812, 813, 906
electron micrograph, 11
elongation, 844–846
enzymatic hydrolysis, 800
eukaryotic, 794, 815, 816, 825–827
replication, 841, 852
flush (blunt) ends, 824, 825
fragments, 811
separation, 829–831
in human body, 834
hybridization, 809, 810
introns, 827, 860, 938
melting, 809–811
methylated bases, 717, 823, 824
minor bases, 796
mitochondrial, 26, 818–820
molecular weights, 810
in mRNA synthesis, *see* DNA, transcription
mutation (*see also* genetic mutations), 915, 916, 918–923
nontranslated, 827
in nucleosomes, 818, 819
nucleotide units, 796–800
palindromes, 826, 827
plasmids, 814, 815
as primer, 844–846
prokaryotic, 19, 794, 812, 813, 825
promoter site, 855, 856
radiation damage, 914
recombinant, 815, 929, 930, 938
regulatory sequences, 821
renaturation, 809
repair, 913–918
repeated sequences, 825
replication, *see* DNA replication
replicative form, 811
restriction, 823–825
restriction sequences, 827, 828
satellite, 825
sequence homologies, 810
shotgunning, 932
spacer, 818, 819, 862, 863
species variations, 802, 803, 809
splicing, 930
structural complementarity, 12
supercoils, 813
of temperate phage, 923, 931
as template, 807, 844–846, 854, 855
termination sequence, 856
three-dimensional structure, 804–806
transcription, 792, 837, 852–863, 891
electron micrograph, 862
errors in, 852
inhibition, 856, 857
twofold symmetry, 824, 825
uracil residues, 915, 916
viral (*see also* DNA viruses), 38, 795, 796, 801, 802, 811, 813, 900
replication, 842, 843
restriction, 824, 825
x-ray diffraction analysis, 804
deoxyribonucleotides, 796–798
in DNA synthesis, 843, 847, 848
synthesis, 632–634
2-deoxy-D-ribose, **57**, 279
in DNA, 796
2′-deoxythymidine 5′-monophosphate, **798**
synthesis, 642
2′-deoxythymidine 5′-triphosphate (dTTP), 389, **399**
deoxythymidylate, **262**, 798
synthesis, 633
2′-deoxyuridine 5′-monophosphate (dUMP), methylation, 633, 642
deoxyuridylate, **262**
dermatan, 296, 297
dermatitis, 595
desmosine, **103**, 161
in elastin, 161
detergent, 328, 329
detoxification, 457, 458, 504, 695, 696
Dexedrine, **65**
dextrins, 288
dextrorotation, 96
diabetes insipidus, 736
diabetes mellitus, 78, 87, 284, 712–715, 736, 762
acetone in blood, 525
adult-onset, 712
amino acid oxidation, 531, 715
blood glucose, 284, 405, 713, 714, 738, 739
blood ketone bodies, 525, 713, 714
blood pH, 713, 715
blood urea, 713–715
carbon dioxide excretion, 715
diagnosis, 713
fatty acid metabolism, 530, 714
glucokinase deficiency, 405, 421
glucose synthesis, 568
incidence of, 938, 939
juvenile-onset, 712
ketone body formation, 545, 598, 714
lipid metabolism, 597, 598
metabolic aberrations, 738, 739
treatment of (*see also* insulin therapy), 712, 715
urea formation, 568
urine, glucose in, 713, 714
ketone bodies in, 713, 714
pH, 713
production, 713, 736
sodium in, 713
urea in, 713–715
1,2-diacylglycerol, 596, **597**
in phosphoglyceride synthesis, 600–602
in triacylglycerol synthesis, 596, 597, 600
diacylglycerol 3-phosphate, 596
dialysis, 127
diastereoisomers, 99
diatoms, 24
photosynthesis, 646, 647
O-(2-diazoacetyl)-L-serine, *see* azaserine
dibutyryl cyclic AMP, **750**
dicarboxylate-transport system, 496, 497, 509, 586
dicarboxylic acids, in citric acid cycle, 442
2,6-dichlorophenolindophenol, 656
3-(3,4-dichlorophenyl)-1,1-dimethylurea (DCMU), 678
dicoumarol, **276**

dicyclohexylcarbodiimide (DCCD), 508, 509
diet (*see also* nutrition), 753
balanced, 784
calorie-deficient, 765, 766
cholesterol in, 760, 761
fats in, 759, 760
mineral elements in, 777, 778
protein in, 763–765
deficiency, 766, 767
in United States, 757
vitamins in, *see* vitamins
weight-loss, 763, 787
differential centrifugation, 355, 356
differentiation, 902
digestion, 683–685
of carbohydrates, 686
of lipids, 690, 691
of proteins, 687–690
dihydrofolate reductase, 261
inhibition, 643
dihydrogen phosphate, p*K*, 79, 82
titration curve, 82
dihydrolipoic acid, **438**
dihydrolipoyl dehydrogenase, 438–440
dihydrolipoyl transacetylase, 438–440
molecular weight, 439
dihydroorotase, 631, 632
L-dihydroorotate, **631**, 632
dihydroorotate oxidase, 631
dihydrosphingosine, 312
5,6-dihydrouridine, in tRNA, 875, 876, **877**
dihydroxyacetone, 278, **279**
dihydroxyacetone phosphate, **404**
in Calvin cycle, 665–667
in gluconeogenesis, 567
in glycolysis, 404, 406, 407, 415, 417
dihydroxyacid dehydratase, 622
1,2-dihydroxybenzene (*see also* catechol), **728**
1,25-dihydroxycholecalciferol, 252, 267, **268**
deficiency, 776
α,β-dihydroxy-β-methylvalerate, **622**
3,4-dihydroxyphenylalanine (dopa), **728**
3,4-dihydroxyphenylethylamine (dopamine), **728**
diisopropylfluorophosphate, 221, **222**, 227
enzyme inhibition, 417
2,3-dimethylglucose, **300**
N^2-dimethylguanosine, in tRNA, 875
dimethylnitrosamine, as mutagen, 916
dimethylsulfate, as mutagen, 916, 917
dimethyl sulfide, in brewing, 428
2,4-dinitrophenol, **488**
as uncoupler, 488, 508
dinoflagellates, photosynthesis, 646, 647
dioxygenases, 502, 503
dipeptide, 111
2,3-diphosphoglycerate, **192**
effect on hemoglobin, 192, 205, 235, 708, 709
diphtheria toxin, 895
diplococci, 19
dipolar ions, 103
disaccharides, 278, 284–287
digestion, 686
hydrolysis, 419
dissociation constant, 78
distal tubule, 704
disulfide cross-linkages, 179, 202
in antibodies, 927
in chymotrypsin, 229
cleavage, 167, 177
formation, 893
in keratins, 153, 156
in ribonuclease, 176
diversity genes, 928, 929
diving mammals, anaerobic glycolysis, 401
DNA, *see* deoxyribonucleic acid
DNA binding protein (DBP), 849, 850
DNA chimera, 934
DNA-directed RNA polymerase, *see* RNA polymerase
DNA gyrase, 849, 850
DNA ligase, 847–849
in DNA repair, 914–917
in formation of gene vector, 934, 935
from gene cloning, 938
in recombinant DNA formation, 930
DNA polymerase, RNA-directed, *see* reverse transcriptase
DNA polymerase I, 843–846
in complementary DNA synthesis, 933
in DNA repair, 914–917
exonuclease activity, 848, 851
molecular weight, 846
DNA polymerase II, 846
DNA polymerase III, 846, 848–850
exonuclease activity, 851
holoenzyme, 846
molecular weight, 846
DNA polymerases, 209, 271
DNA replicase system, 845, 850, 869
DNA replication, 50, 792, 807, 808, 837–853
continuous, 847
discontinuous, 847
errors in, 850–852
helix unwinding, 848, 849
labeling experiments, 838–840, 843
lagging strand, 847, 849, 850
leading strand, 847, 849, 850
Okazaki fragments, 847–849, 852
proofreading, 851, 852
rate of, 841, 852
rolling-circle mechanism, 842, 843, 925
semiconservative, 838, 839
DNA viruses, 38, 795, 796, 802, 811, 813, 824, 825, 900
dopa, **728**
dopamine, **728**
dormouse, hibernation, 598
double-displacement reactions, 218, 219
double helix, in DNA, 804–806
DNA-RNA, 855
in replication, 807, 808
unwinding, 808, 809, 848–850
Drosophila, DNA, 815
DNA replication, 841
drugs, detoxification, 457, 458, 504, 695, 696
overdose, 719
dTMP, *see* 2′-deoxythymidine 5′-monophosphate
dTTP, *see* 2′-deoxythymidine 5′-triphosphate
duodenum, 684, 686, 688
dwarfism, 740, 939
dynein, 123, 383

earth's crust, composition, 46
E. coli., *see Escherichia coli*
edema, 719
Edman, P., 131
Edman degradation, 131, 132, 135
effectors (*see also* modulators), 234
egg, 275
as food source, 764
egg albumin, denaturation, 140
isoelectric point, 128
egg cells, DNA, 819
mutiple genes, 826
in sexual conjugation, 925
eicosatrienoic acid, synthesis, 595
Eijkman, C., 768
einstein, 651
elastase, inhibition, 222
elastic tissue, 160
elastin, 103, 123, 157, 160–162
amino acid content, 160
cross-linkage formation, 271
helical structure, 160
electric dipole, 68
electrochemical gradient, 489
electrode, 472, 473
electromotive force, 472
electron carriers, 476
absorption spectra, 480, 483
mitochondrial location, 476
reduction state, 508
of respiratory chain, 476
standard reduction potentials, 474, 475, 482, 507
electron carrier Z, 657, 660
electron hole, 657
electron microscopy, membranes, 320
scanning, 320
electron transfer, 270, 471, 472
free-energy change, 474, 475
electron-transferring flavoprotein, 516, 517
electron transport, 435, 436, 467–484, 487–494, 507
in bacteria, 493
coupling to ATP synthesis (*see also* oxidative phosphorylation), 484, 487–493, 508
in flagella rotation, 495
heat generation, 491
inhibitors, 483, 508
photosynthetic, 661, 662
producing electrochemical gradient, 489
producing proton gradient, 488–491
in submitochondrial vesicles, 485
electron-transport chain, 468, 469
complexes, 482
crossover points, 483
energy-conserving sites, 484
in fatty acid oxidation, 517
free-energy change, 475, 484
mitochondrial location, 470
photosynthetic, 657
sequence, 492, 493
determination, 483
electrophoresis, 108
DNA fragment separation, 829, 830
isozyme separation, 240
peptide separation, 131
protein separation, 128
electrostatic attraction, 68
in proteins, 179
elimination reactions, 369
elongation factors, 874, 885, 886
Elvehjem, C., 251
emulsification, 690, 691
enantiomers (*see also* stereoisomers), 51
amino acids, 96
endocrine cells, 735
secretory vesicles, 724
endocrine glands, 721, 722
endocrine system, 721–723
endocrinology, 721
endonucleases, 800
in DNA repair, 914–916
restriction, 823–825, 828–831, 929, 930, 932, 934, 935
endoplasmic reticulum, 27
cytochromes, 503
electron micrograph, 872
enzyme activities, 354
fatty acid elongation, 594
glucose 6-phosphatase activity, 566
isolation, 355
membrane composition, 319, 603
in muscle cells, *see* sarcoplasmic reticulum
phosphoglyceride synthesis, 600, 603
protein export, 893

rough-surfaced, 23, 689
electron micrograph, 27
in protein synthesis, 893
smooth, 27
endorphins, 746, 751
end-product inhibition (*see also* feedback inhibition), 233–235, 346
end-product repression, 903
energy, 361–371
chemical, 366, 371, 383
conservation of, 362
free, 362–367
heat, 362–366
interconversions, 361, 383
mechanical, 381, 383
energy charge, 500, 510
energy cycle, 344, 371
enkephalin, **66**, 114, 144, 746
enolase, 408, 411
molecular weight, 411
enolase reaction, standard-free-energy change, 411
enoyl-ACP reductase, 588, 590–592
cis-Δ^3-enoyl-CoA, **520**, 521
trans-Δ^2-enoyl-CoA, 515–517, 520, **521**
enoyl-CoA hydratase, 516, 517, 520, 521
enoyl-CoA isomerase, 520, 521
enterokinase, 688
enthalpy, 363–365
entropy, 7, 362–365
enzyme cofactors, 209, 249, 270
enzyme induction, 347, 902, 903
enzyme inhibition, 221–225
competitive, 223–225
irreversible, 221, 222, 246
noncompetitive, 224, 225
reversible, 221, 223
enzyme modifier, 577
enzyme repression, 903
enzymes, 8, 122, 169, 207–242, 337
active site, 50, 209, 221, 226, 227, 244, 807
allosteric, 233–238, 346
blood serum levels, 535
as catalysts, 211, 212, 225, 226, 230, 231
cell concentrations, 347
classification, 210
coenzyme requirement (*see also* coenzymes), 209
cofactor requirement, 209, 249, 270
conformational changes, 179, 226, 232–234
constitutive, 902
covalent modification, 233, 237–239, 422, 450, 451
denaturation, 177, 246
double-reciprocal plots, 217, 225
feedback inhibition, 233–235, 346
gene cloning, 938
genetic defects, 241, 242, 349
glycolytic, 403
heterotropic, 235, 237
homotropic, 235, 236
hormonal regulation, 346, 347
inactivation, 209
induced fit, 194, 226, 232, 233
isolation, 145, 208, 348
K_M, 212–217, 225, 245, 246
kinetics, 214, 215, 217, 218
inhibition, 225
in membranes, 325
metal ion requirement, 209
molecular weights, 209
optimum pH, 87, 218
ping-pong reactions, 218, 259, 533
prosthetic groups, 125, 209, 270, 892
as proteins, 208, 209
proteolytic, in brewing, 428
in digestion, 688
quantitative assay, 218, 219
reaction mechanisms, 230, 231
regulation, 342
crossover point, 425
regulatory, 185, 207, 233–239, 241
regulatory site, 234
repressible, 625
serine-class, 223, 227, 229–231, 238, 417, 422, 574, 892, 918, 919
specific activity, 219
specificity, 100, 220
substrate-saturation curves, 212–214, 236, 237
synthesis, 892
regulation, 347, 625, 901–908
thiol groups, 409, 588–590
turnover number, 220, 246
V_{max}, 212–217, 225, 236, 245, 246
x-ray analysis, 226–233
enzyme-substrate complex, 212–215, 226, 228, 230, 232
Eobacterium isolatum, 18
epimerase, 520
epimers, 281
epinephrine (*see also* adrenaline), 728
epithelial cells, kidney, 703
small intestine, 684–686
stomach, 687
equilibrium constant, 73, 74, 366, 367
equilibrium reactions, 426
ergocalciferol, **267**, 776
ergosterol, 267, 776
erythrocyte (*see also* red blood cells), 185, 186
adenine nucleotide concentrations, 373, 377
phosphate concentration, 373
D-erythrose, **280**
erythrose 4-phosphate, **667**
in Calvin cycle, 665–667
Escherichia coli, 5, 20–22, 42
adenine nucleotide concentrations, 373
amino acid synthesis, 358–360, 616
auxotrophic, 642
cell wall, 21, 294
chromosome, 822
clones, 934
conjugation, 822
electron micrograph, 925
dimensions, 16
division, 19, 20
DNA, 20, 42, 803, 812, 813
electron micrograph, 813, 906
molecular weight, 812
DNA replication, 838–841, 843, 847
electron micrograph, 20
F factor, 925
foreign gene introduction, 930, 931
β-galactosidase, 142, 359, 901–905
in gene cloning, 936, 937
genes, 828
genetic map, 821, 822
glutamine metabolism, 623
growth, 358, 359
isotopic labeling experiments, 358, 360
lac operon, 905–908, 936
lactose metabolism, regulation, 359
lipids, 53
messenger RNA, 795, 853
molecular components, 53
mutants, 351
nucleic acids, 53
operons, 905
phosphate concentration, 373
plasmid, electron micrograph, 814
polysaccharides, 53
proteins, 53, 821, 901
protein synthesis, 179, 871, 873, 880
regulation, 901
purine nucleotide synthesis, 630
pyruvate dehydrogenase complex, electron micrograph, 440
restriction endonucleases, 824, 825
ribosomes, 881–883
RNA, 795
RNA polymerase, 855, 856
transformation, 934
esophagus, 684
essential amino acids, 158, 335, 555, 616, 753, 754, 763–765
essential fatty acids, 595, 753, 759
essential minerals, 753, 755
esters, 50
β-estradiol, 722, **744**, 745
estrogen receptors, 744, 745
estrogens, 724, 740, 743–745
estrophilin I, 744, 745
estrophilin II, 744, 745
ethane, conformations, 148
ethanol, in diet, 761, 762
from fermentation, 398, 401, 426–428, 432
gluconeogenesis inhibition, 570, 571
physical properties, 67
from pyruvate, 398
ethanolamine, **64**, 311
ethanolamine kinase, 600
ethanolamine phosphotransferase, 601
ethers, 50
ethylene, from nitrogenase reaction, 639
ethylene glycol, 90
Euglena, 647
electron micrograph, 303
euglenids, photosynthesis, 646
eukaryotes, 18, 22–37, 352–354
cell division, 25
cell structure, 22
DNA, 794, 815, 816, 825–827
DNA replication, 841, 852
evolutionary origins, 494
genes, 826, 827
recombination, 925, 937
messenger RNA, 859
synthesis, 857
protein synthesis, 871, 880
regulation, 902
ribosomes, 881, 882
RNA polymerase, 856
evolution, biological, 62
chemical, 59
evolutionary maps, 136, 138
excited state, 651, 652
exciton, 651, 657
exocrine cells, 688, 689, 735
exocytosis, 28, 603, 689, 728, 893
exons, DNA, 827, 936
messenger RNA, 860, 861
exonucleases, 800, 846
exoskeletons, 292

F_0, 485
F_1, 485–487
Fabry's disease, 606
F actin, 380, 381
facultative cells, 335
FAD, *see* flavin adenine dinucleotide
FAD–$FADH_2$ couple, 461
faraday, 475
farnesyl pyrophosphate, **608**
in cholesterol synthesis, 608
fasting, 308, 597, 699, 787
amino acid oxidation, 531
cholesterol synthesis, 610
ketone body metabolism, 525, 526

fats (*see also* triacylglycerols), in adipose tissue, 701
brown, 491, 509, 702, 703, 762
in diet, 753, 757
as energy source, 755–759
as insulator, 309, 598, 599
oxidation, heat generation, 491
storage, 301, 318, 598, 701
fat cells (*see also* adipocytes), 308, 318
fat droplets, 32
F_1ATPase, 485
molecular weight, 180
fatty acid elongation systems, 594
fatty acids, 57, 303–305
absorption, 690, 691
activation, 387, 512–514, 520
deficiency, 759
in diet, 760
digestion, 690
essential, 595, 753, 759
melting points, 328
in membrane lipids, 329
metabolism, 338, 339, 695, 696
micelle formation, 305
mitochondrial transport of, 512–514, 520, 526
odd number, oxidation, 521–523
oxidation, 342, 452, 511–527, 584, 693, 695
in diabetes mellitus, 530
inhibition, 523
regulation, 514
in skeletal muscle, 696
in phosphoglycerides, 310, 311
in pyruvate kinase regulation, 421, 424
saturated, 304, 305
in diet, 759
oxidation, 511–519
in phosphoglycerides, 311
in triacylglycerols, 307, 758
in waxes, 309
soaps, 71, 305, 690
synthesis, 342, 526, 583–595, 693
in diabetes mellitus, 714
regulation, 587, 595
transport, 707
from triacylglycerols, 701, 702
in triacylglycerols, 306
unsaturated, 304, 305, 515
in diet, 759
oxidation, 520, 521
in phosphoglycerides, 311
synthesis, 594, 595
in triacylglycerols, 307, 758
in waxes, 309
fatty acid synthase system, 584, 587–593
molecular weight, 587
fatty acyl adenylate, 513, **514**, 528
fatty acyl–carnitine, 513, **514**
fatty acyl–CoA, 387, 512, **513,** 514
in citrate synthase regulation, 453
dehydrogenation, 515
oxidation, 526
inhibition, 523
in phospholipid synthesis, 526, 595, 596
standard free energy of hydrolysis, 513
in triacylglycerol synthesis, 526, 595, 596
fatty acyl–CoA dehydrogenase, *see* acyl-CoA dehydrogenase
fatty acyl–CoA oxygenase, 594
fatty acyl–CoA synthetase, 596
F cells, 736, 737
feathers, 162
electron micrograph, 149
feedback inhibition, 233–235, 346
in amino acid synthesis, 617, 622–624
concerted, 624
sequential, 624
Fe^{2+}–Fe^{3+} redox couple, 471
fermentation, 208, 358, 398, 426–428, 432
lactic acid, 398, 427
ferredoxin, 270
molecular weight, 659
in nitrogenase reaction, 638, 639
in photosynthesis, 657, 659
standard reduction potential, 507
ferredoxin-NADP oxidoreductase, 659
ferredoxin-reducing substrate, standard reduction potential, 507
ferric ion, as electron acceptor, 471
ferritin, 123, 126, 780, 781
ferrous ion, as electron donor, 471
fertilizers, 638–640
fetal hemoglobin, 184, 205
F factor, 925
F_0F_1ATPase, 470, 484–487, 679
in ATP synthesis, 495, 509, 510
in brown fat, 491
in chemiosmotic hypothesis, 490, 491
electron micrograph, 486
fiber, in diet, 755
fibrin, 269, 707
fibrinogen, 123, 269, 706, 707
fibroblasts, 29, 157, 940
electron micrograph, 31
fibroin, 123, 124, 150, 155, 167
fibronectin, 295, 324
fibrous actin, 163, 380, 381
fibrous proteins, 124, 147–163
digestion, 689, 690
x-ray analysis, 150
fingerprinting, *see* peptide mapping
firefly luminescence, 388
first law of thermodynamics, 362, 646
first messenger, 730
Fischer, E., 208
fish, "antifreeze" proteins, 294
Fiske, C., 372
flagella, 30–32, 123, 163, 353, 383
bacterial, 21, 353, 494, 495
E. coli, 20
plankton organisms, 647
rotation, 491, 494, 495
flavin adenine dinucleotide (FAD), 209, 252, **254**
in fatty acid oxidation, 515–519
in α-ketoglutarate complex, 446
in pyruvate dehydrogenase complex, 438–440
in succinate dehydrogenase, 447, 507
flavin adenine dinucleotide, reduced form ($FADH_2$), 447, 448
flavin dehydrogenases, 254, 272, 478–480
flavin mononucleotide (FMN), 252, **254,** 478
as electron carrier, 476
in NADH dehydrogenase reaction, 478, 479
flavin mononucleotide, reduced form ($FMNH_2$), 478
flavin nucleotides, 254, 772
in monooxygenase reactions, 503
flavoproteins, 126, 254, 271, 479
in electron transport, 517
flickering clusters, 69
flight muscle, 382
fluid mosaic model, 321, 322
fluorescence, 651, 652
fluoride, in diet, 783
in drinking water, 782, 783
enolase inhibition, 411
fluorine, as trace element, 270
fluoroacetate, 465
fluoroacetyl-CoA, 465
fluoroapatite, 783
1-fluoro-2,4-dinitrobenzene, **111,** 113, 114, 130, 718
fluorophosphate, enolase inhibition, 411
fluorosis, 782
fluorouracil, **642**
as anticancer drug, 642, 643
FMN, *see* flavin mononucleotide
folic acid, 252, 260, **261,** 275, 621, 772
deficiency, 772
in diet, 767, 772
recommended daily allowance, 754, 772
Folkers, K., 263
follicle-stimulating hormone (FSH), 722, 726
molecular weight, 726
food groups, 784
foods, caloric values, 756, 757, 763
nutritional labeling, 784
formaldehyde, 247
formic acid, pK, 79
formimino-group transfer, 538
formyl-group transfer, 538
N-formylmethionine, **880**
in protein synthesis, 880, 881, 891
N-formylmethionyl-$tRNA^{fMet}$, 880, 883–885, 887
N^{10}-formyltetrahydrofolate, 880
fossil, 18
fossil fuels, 334, 646, 674, 676
frame-shift mutations, 919–921
Frankel, A., 660
Franklin, R., 804
free energy, 336, 343, 362–367
released in catabolism, 337
free-energy change (ΔG), 363, 365, 369, 370, 376, 377
free fatty acids, 512, 695, 701, 702
freeze-clamping, 425, 432
freeze-fracture method, 320
frogs, metamorphosis, 743
α-D-fructofuranose, **283**
β-D-fructofuranose, **283**
fructokinase, 415, 417
D-fructose, 57, **279,** 281
in blood, 707
from digestion, 686
glycolysis, 415, 417, 430
Haworth projection formulas, 283
optical activity, 280
in semen, 430
specific rotation, 280
from sucrose, 417, 419
sweetness, 286, 758
fructose diphosphatase, 571
genetic defect, 576
in gluconeogenesis, 565–567
molecular weight, 566
fructose diphosphatase reaction, standard-free-energy change, 565
fructose 1,6-diphosphate, **404**
in Calvin cycle, 665, 666
in gluconeogenesis, 565, 567
in glycolysis, 404, 406
heart muscle concentration, 433
in phosphofructokinase regulation, 423, 424
fructose diphosphate aldolase, 404, 406, 407
fructose 1-phosphate, **417**
in fructose metabolism, 415, 417
fructose 6-phosphate, **404**
in Calvin cycle, 665–667
in gluconeogenesis, 565, 567
in glycolysis, 404, 405, 415
heart muscle concentration, 433
in mannose catabolism, 419
standard free energy of hydrolysis, 374
in sucrose synthesis, 668
fructose phosphate aldolase, 415, 417
fucose, 325
fumarase, 446, 447

molecular weight, 447
optimum pH, 219
fumarase reaction, standard-free-energy change, 369, 394
fumarate (*see also* fumaric acid), **442**
in amino acid oxidation, 536, 537, 540, 544
in citric acid cycle, 441, 442, 446, 447
in urea cycle, 552, 553
fumarate hydratase (*see also* fumarase), 447
fumaric acid (*see also* fumarate), **148**
fumarylacetoacetase, 540
4-fumarylacetoacetate, **540**
functional groups, 50
Funk, K., 250
furan, 282, **283**
furanoses, 282, 283
futile cycles, 571, 572, 702, 762

ΔG, *see* free-energy change
$\Delta G°$, $\Delta G°'$, *see* standard-free-energy change
G actin, 163, 380, 381
galactocerebroside, 312, **313**
galactokinase, 418
deficiency, 418, 419, 434
D-galactose, **280,** 281
from digestion, 686
in glucose formation, 418, 419
in glycolipids, 312, 314
glycolysis, 415, 418
from lactose, 419
optical rotation, 299
synthesis, 419
galactosemia, 241, 418–420, 434, 581
galactose 1-phosphate, in galactose catabolism, 418, 419
galactose 1-phosphate uridylyltransferase, 241
genetic defect, 434, 581
β-galactosidase, 142, 359, 686, 901–905
gene cloning, 937
genetic defect, 606
induction, 903–905
K_M, 216
repression, 904, 905
turnover number, 220
β-galactosidase reaction, standard-free-energy change, 369, 394
β-galactoside permease, 903–905
galactosylceramide β-galactosylhydrolase, 606
galactosyltransferase, 576, 577
gallbladder, 684
γ-globulins, 126, 707
isoelectric point, 128
molecular weight, 126
ganglioside G_{M2}, **605**
ganglioside G_{M3}, 605
gangliosides, 312, 314
degradation, genetic defect, 604–606
lysosomal deposits, electron micrograph, 605
in membranes, 324, 326
gangliosidosis, 606
gargoylism, 606
gas gangrene, 168
gastric glands, 687
gastric juice, 90, 385, 386, 687, 717
composition, 705
pH, 687, 705
gastrin, 687
molecular weight, 746
gastrointestinal tract, 683, 684, 686, 687
hormone secretion by, 722
Gaucher's disease, 606
GDP, *see* guanosine 5′-diphosphate
GDP-glucose, in cellulose synthesis, 668, 669
gelatin, 158
gel filtration, 128
gene amplification, 938
gene cloning (*see also* gene recombination), 931, 936–939
gene library, 936
genes, 820–822
base sequence, 827, 828
determination, 828, 829
cancer-causing, 863, 864
cloning, 936–939
colinearity, 825
dissection, 825
eukaryotic, 826, 827, 937
isolation, 932, 936, 938
mapping, 828, 938
multiple copies, 826
mutations (*see also* genetic mutations), 921
overlapping, 900, 901
recombination (*see also* genetic recombination), 825, 913, 923–940
regulatory, 904–908, 938
repression, 625
size, 822, 823
splicing, 825, 929, 930
structural, 820, 904–908
synthetic, 864
transcription, 855, 856, 862
hormone regulation, 741, 745
transducible, 924
transposition, 926
in antibody synthesis, 928
genetic code, 895–900
degeneracy, 898
experimental investigation, 895, 896
universality, 897
wobble hypothesis, 898–900
genetic counseling, 604
genetic diseases, 241, 542, 921
of amino acid metabolism, 540–542
of glycogen metabolism, 576, 606
heart disease, 760, 761
of lipid metabolism, 604–607, 702
lysosomal, 606
of pentose phosphate pathway, 457
of porphyrin synthesis, 626, 627
of purine nucleotide synthesis, 635, 636
of urea cycle, 554, 555
genetic engineering, 938
genetic mutation, 12, 198, 349–351
beneficial, 918, 919
deletion, 919, 920
frame-shift, 919–921
insertion, 919, 920
lethal, 918–920
point, 918, 919
silent, 918, 919
spontaneous, 921
suppression, 921
genetic recombination, 825, 913, 923–926, 937
in antibody formation, 926–929
experimental, 929–940
gene transfer, 640
gene vectors, 934–936
genome, 816
geometrical isomers, 148
geranyl pyrophosphate, **608**
in cholesterol synthesis, 608
germ cells, 926
gigantism, 740
Gilbert, W., 829, 906
glass electrode, 77, 92
glaucoma, 247
gliadin, 123, 690
globins, 135, 181
gene cloning, 937
globular actin, 163, 380, 381
globular proteins, 124, 147, 169–200
denaturation, 169, 177, 687
solubility, 154
tertiary structure, 173, 178
x-ray analysis, 170
globulins, *see* α_1-, α_2-, β-, γ-globulins
glomerulus, 703, 704
glucagon, 114, 739, 750
in control of glycogen metabolism, 575
in control of glycolysis, 739
in control of triacylglycerol synthesis, 598
in endocrine system, 722
hyperglycemic effect, 739
molecular weight, 739
receptor sites, 326
secretion, regulation, 739, 740
synthesis, 735, 736
glucocerebrosidase, 606
glucocerebrosides, 312
glucocorticoids, 740, 741
glucogenic amino acids, 545, 568
glucokinase, 395, 404, 405, 420
deficiency, 405, 421
K_M, 431
gluconeogenesis, 547, 562–571, 579, 694
ATP requirement, 569
inhibition, 570, 571
in recovery period, 569
regulation, 563, 564, 567, 568, 571, 572
α-D-glucopyranose, **282,** 283
β-D-glucopyranose, **282,** 283
D-glucosamine, **64**
glucosamine 6-phosphate, in glutamine synthetase regulation, 623
D-glucose, **51,** 278–280
in ascorbic acid synthesis, 458, 459
asymmetry, 52
blood levels, 405, 420, 692, 694, 707, 713, 737, 738
in diabetes mellitus, 284, 405, 713, 714, 738, 739
regulation, 422, 423, 741
in brain metabolism, 699
from carbohydrate digestion, 686, 694
catabolism, 465, 466
from cellulose hydrolysis, 570
conformations, 283
dimensions, 17
epimers, 281
excretion, in diabetes mellitus, 713, 714
in fatty acid synthesis, 526
fermentation, 348, 398, 426–428
functional groups, 51
from galactose, 418, 419
from glucose 6-phosphate, 692
from glycogen, 732, 733
in glycogen synthesis, 572–574
in glycolipids, 312, 314
in glycolysis, 34, 403–405, 415
Haworth projection formulas, 283
in hexokinase reaction, 403–405, 431
kidney threshold, 739
in lactose metabolism, 419, 577
from maltose, 419
metabolism of, 359
mutarotation, 282
optical activity, 280
oxidation, 397, 398, 400, 437
ATP production, 497, 498
energetics, 364, 365
in skeletal muscle, 696
standard-free-energy change, 437
from photosynthesis, 648, 649, 663–666
in regulation of β-galactosidase synthesis, 902–907
ring structure, 281–283
specific rotation, 280, 282
from sucrose, 419
sweetness, 286, 758

synthesis, *see* gluconeogenesis
in treatment of hypoglycemia, 571
in urine, 703
glucose-alanine cycle, 547, 694
glucose 1,6-diphosphate, **416**
in phosphoglucomutase reaction, 416, 417
glucose isomerase, 286
glucose 6-phosphatase, 209, 423, 566, 571, 692
genetic deficiency, 576
pH-activity profile, 218
glucose 6-phosphatase reaction, standard-free-energy change, 369, 394, 566
glucose 1-phosphate, **416**
in galactose catabolism, 418, 419
in glycogen metabolism, 415–417, 572
in glycolysis, 415–417
standard free energy of hydrolysis, 374
glucose 6-phosphate, **404**
in Calvin cycle, 665, 666
cell concentration, 394, 395
in cellulose synthesis, 668
in gluconeogenesis, 566, 567
glucose formation, 692
in glycogen synthesis, 572, 693
regulation of, 575
in glycolysis, 403–406, 415
in hexokinase inhibition, 405, 420
in lipid synthesis, 693
as low-energy phosphate compound, 379, 380
in metabolism, 692
in pentose phosphate pathway, 457, 693
from photosynthesis, 664
standard free energy of hydrolysis, 374, 375
in starch synthesis, 668
in sucrose synthesis, 668
glucose 6-phosphate dehydrogenase, 457, 476, 593
deficiency, 457
glucose 1-phosphate uridylyltransferase, 572
glucose tolerance test, 713
$\alpha(1 \rightarrow 6)$-glucosidase, 288, 289, 416
genetic defect, 576
glucosuria, 713, 714
D-glucuronate, **296,** 457–459
in chondroitin, 296, 297
in heparin, 458
in hyaluronic acid, 296, 458
glucuronate reductase, 458
glutamate (*see also* glutamic acid), in amino acid metabolism, 532–535, 547, 616, 618
in ammonia excretion, 549
in glucose synthesis, 568
oxidative deamination, 534, 535
oxidative degradation, 536, 543
synthesis, 616
in transaminase reactions, 532
glutamate dehydrogenase, 476, 478, 534, 535
in amino acid metabolism, 547, 616
in ammonia excretion, 545, 546, 549, 551
molecular weight, 126, 180, 535
regulation, 535
glutamate kinase dehydrogenase, 617
glutamate-oxaloacetate transaminase (GOT), *see* aspartate transaminase
glutamate-pyruvate transaminase (GPT), *see* alanine transaminase
glutamate semialdehyde, 543, 617
glutamic acid (*see also* glutamate), **56,** 96, 101, 102
in enzyme action, 244, 247
isoelectric point, 108
p*K* values, 107
in proteins, 152, 173, 892
specific rotation, 97
titration curve, 107
glutaminase, in ammonia excretion, 548, 549
glutamine, **56,** 101, 102
in ammonia transport, 546, 548, 549, 617
in blood, 546, 718
in glucose synthesis, 568
oxidative degradation, 536, 543
p*K* values, 107
in proteins, 173
in purine nucleotide synthesis, 628
in pyrimidine nucleotide synthesis, 631
synthesis, 616, 617
glutamine synthetase, 546, 616, 617
in ammonia excretion, 549
molecular weight, 180
quaternary structure, 185
regulation, 239, 623
glutamyl 5-phosphate, **546,** 616, 617
glutaryl-CoA, **539**
in amino acid oxidation, 539
glutathione, **272**
glutathione peroxidase, 209, 272, 783
glutenin, 166
glycans (*see also* polysaccharides), 287
glyceraldehyde, 278, **279**
optical activity, 279, 280
phosphorylation, 417
stereoisomers, 97
glyceraldehyde 3-phosphate, **404**
in Calvin cycle, 665–667
in gluconeogenesis, 567
in glycolysis, 400, 402, 404, 406–408, 415, 416
glyceraldehyde 3-phosphate dehydrogenase, 135, 408–410, 477, 497
iodoacetate inhibition, 409, 410, 718
molecular weight, 410
reaction mechanism, 409
glyceraldehyde 3-phosphate dehydrogenase reaction, standard-free-energy change, 408
glycerol, 56, **57**
metabolism, 432
in phosphoglycerides, 310
in triacylglycerols, 306
glycerol kinase, 596, 719
in phosphate-group transfer, 379
glycerol 1-phosphate, standard free energy of hydrolysis, 374
glycerol 3-phosphate, **597**
formation, 596
as low-energy phosphate compound, 379, 380
in phospholipid synthesis, 596, 600
in triacylglycerol synthesis, 596, 600
glycerol phosphate acyltransferase, 597
glycerol phosphate dehydrogenase, 596
glycerol phosphate shuttle, 497
glycine, **56,** 96, 101, 102
as buffer, 91
in creatine synthesis, 625
in glucose synthesis, 568
in glutamine synthetase regulation, 623
from glyoxylate, 673
oxidative degradation, 536–538
p*K* values, 107
in porphyrin synthesis, 625, 626
in proteins, 156, 158, 160, 173
in purine nucleotide synthesis, 628
synthesis, 620, 621
titration curve, 118
glycine synthase, 538, 621
glycocalyx, 295, 320
glycogen, 32, 278, 287, 300
branching, 574
digestion, 686
glycolysis, 401, 414–417
regulation, 421–423, 574–576
granules, 32, 354
electron micrograph, 288, 560
metabolic turnover, 353
metabolism, genetic defects, 576
regulation, 574–576
in muscle, 401, 696, 697
storage, 288, 308, 420, 583
synthesis, 562, 569, 572–574, 693, 807
regulation, 581
glycogen-branching enzyme, 574
glycogenolysis, 729–733, 739
glycogen phosphorylase, 289, 414–416, 574, 892
genetic defect, 576
hormonal regulation, 346, 347, 422, 423, 575, 576, 729–733
in liver, 433
in muscle, 433
regulation, 238, 239, 421, 422, 426, 574, 575
V_{max}, 433
glycogen phosphorylase reaction, standard-free-energy change, 415
glycogen storage diseases, 576, 606
glycogen synthase, 572–574, 693, 807
genetic defect, 576
hormonal regulation, 575, 576, 733, 734
regulation, 574, 575
glycogen synthase *a*, 574, 575
glycogen synthase *b*, 574, 575
glycogen synthase *D*, 575
glycogen synthase *I*, 575
glycolic acid, **673**
in photorespiration, 673
glycolipids, 295, 312
in membranes, 319, 320, 322, 324
glycolysis, 397–428, 437, 441, 563, 564, 566
aerobic, 414, 435
anaerobic, 398, 399, 401, 412–414
ATP in, 399, 400, 402
in cancer cells, 502
energy conservation, 378, 379, 400, 402, 408
glucagon stimulation, 739
intermediates, assay, 425
isotope tracer technique, 465, 466
NADH production, 412–414, 496, 497
oxidoreduction reactions, 408–410, 412–414
phosphorylated intermediates, 403
regulation, 405, 406, 420–426, 449, 453, 500–502, 567, 571, 572
crossover point, 425
hormonal, 422, 423
in skeletal muscle, 696, 697
standard-free-energy change, 399, 400, 437
glycolytic enzymes, in gluconeogenesis, 403, 562
glycophorin, 295, 323, 324
glycoproteins, 125, 126, 157, 278, 294, 295
digestion, 419
in membranes, 320, 322–325, 329
synthesis, 577, 892
inhibition, 895
glycosaminoglycans, 296
glycosidic bond, 284
glycosphingolipids, 312
glycosuria, 736
glycosyl-$(4 \rightarrow 6)$-transferase, 574
glyoxylate, 455, **456**
in photorespiration, 673
glyoxylate cycle, 455, 456
glyoxysomes, 456
GMP, *see* guanosine 5′-monophosphate
goiter, 781, 782
Golgi, C., 28

Golgi body, 23, 28
  electron micrograph, 28
  enzyme activities, 354
  membrane composition, 319, 603
  in protein export, 893
Gonionemus murbachii, 44
gout, 636, 643
gramicidin S, 115, 489
gram stain, 294
grana, 649, 650
Greenberg, G. R., 628
green sulfur bacteria, 646, 648
Griffith, F., 801
ground state, 651, 652
ground substance, 15, 30, 157, 296
growth hormone, *see* pituitary growth hormone
Grunberg-Manago, M., 865
GTP, *see* guanosine 5′-triphosphate
guanidinoacetate, **625**
guanido group, 51
guanine, **57**, 796, 797
  catabolism, 634
  deamination, 917
  in DNA, 803–806
  methylation, 917
  in RNA, 854
  salvage, 635
guanine nucleotides, synthesis, 627–630
  regulation, 630, 631
guano, 555, 556, 638
guanosine, catabolism, 634
guanosine 5′-diphosphate (GDP), in substrate-level phosphorylation, 446
guanosine 5′-monophosphate (GMP), **798**
  in control of purine nucleotide synthesis, 630
  degradation, 634
  synthesis, 628, 629
    regulation, 630, 631
guanosine 5′-triphosphate (GTP), **390**
  from citric acid cycle, 446–448
  in glucose synthesis, 563, 565, 566
  in glutamate dehydrogenase regulation, 535
  as high-energy compound, 389
  from phosphoenolpyruvate carboxykinase reaction, 455
  in protein synthesis, 874, 883–886, 888, 889
  from substrate-level phosphorylation, 446
  from succinyl-CoA synthetase reaction, 446, 447
guanylate (*see also* guanosine 5′-monophosphate), **798**
guanylic acid (*see also* guanosine 5′-monophosphate), 629, 630
Guillemin, R., 726
L-gulonate, **458**, 459
L-gulonolactone, **458**, 459
gulonolactone oxidase, 458, 459
D-gulose, **280**
guttapercha, 610

Haber process, 639
hair, 149, 154–156, 162, 165
  permanent waving, 156
half-cell, 472
halobacteria, 266, 674, 675
Halobacterium halobium, 167, 674
Harden, A., 432
Hatch, M. D., 670
Hatch-Slack pathway, 669–672
H^+–ATPase, 385, 386, 675, 705
Haworth projection formulas, 280, 283
heart attack (*see also* myocardial infarction), 610
heart disease, 759–761
heart failure (*see also* myocardial infarction), 709
heart muscle, 382, 698, 699
  adrenaline stimulation, 733
  electron micrograph, 698
  metabolism, 698, 699
  mitochondria, 470, 698
    electron micrograph, 510
  oxygen consumption, 708
heat energy, 343, 362–366
  from electron transport, 491
  from futile cycles, 571, 572, 702, 762
heat of vaporization, 67
helicases, 849, 850
helix, in collagen, 160
  in elastin, 160
hemaglutinins, 324, 325
heme, 170, **171**, 173, 270, 271
  in cytochromes, 174, 480, 481
  degradation, 627
  in enzymes, 270
  in hemoglobin, 171, 181
  in myoglobin, 170–172
  oxygen binding by, 708, 709
  synthesis, 625, 626
heme enzymes, 271
heme proteins, 175, 480
hemiacetals, 282
hemiketals, 282
hemoglobin, 58, 122, 123, 126, 135, 180, 181, 270, 272, 708, 711
  amino acid sequence, 135
  Bohr effect, 189, 192
  buffering power, 108
  carbon dioxide transport, 188
  carbon monoxide binding, 171
  conformational changes, 179, 190
  dimensions, 17
  2,3-diphosphoglycerate binding, 192, 205, 235, 708, 709
  evolution, 184
  fetal, 184, 205
  genes for, 820
  heme group, 171
    degradation, 627
    synthesis, 625, 626
  invariant residues, 183, 197
  isoelectric point, 128
  messenger RNA for, 932
  molecular weight, 126, 180, 204
  mutations, 197–199, 206
  oxygen binding by, 190–194, 205, 708, 709
  oxygen saturation curve, 187, 189, 193, 236
    pH effect on, 708
  oxygen transport, 185, 188, 708–711
  peptide mapping, 131, 196
  quaternary structure, 181, 191
  in red blood cells, 204
  regulation of, 236, 237
  sickle-cell, 194–200, 912
  species variations, 183
  x-ray analysis, 181
hemoglobin A, 195
  peptide mapping, 196
hemoglobin C, 199, 206
hemoglobin D_{Punjab}, 199
hemoglobin E, 199
hemoglobin F, 205
hemoglobin $G_{Honolulu}$, 199
hemoglobin $G_{Philadelphia}$, 199
hemoglobin $G_{San\ Jose}$, 199
hemoglobin I, 199
hemoglobin M_{Boston}, 199
hemoglobin $M_{Milwaukee}$, 199
hemoglobin $M_{Saskatoon}$, 199
hemoglobin Norfolk, 199
hemoglobin $O_{Indonesia}$, 199
hemoglobin S, 195, 204, 206
  peptide mapping, 196
hemoglobin Zurich, 199
hemolysis, 457
hemophilia, 921
Hemophilus influenzae, DNA, 11
  restriction endonuclease, 823, 824
hemoprotein, 126
hemosiderin granules, 781
hen, DNA composition, 803
Henderson-Hasselbalch equation, 83, 84, 106
Henri, V., 212
Henseleit, K., 443, 549, 550
heparin, 297
hepatocytes (*see also* liver cells), 22
heptoses, 278
herpes simplex, 795
  DNA, 923, 924
Hershey, A. D., 802
Hershey-Chase experiment, 802
heterogeneous nuclear RNA (hnRNA), 796, 859, 860
heteropolysaccharides, 287, 296
heterotrophs, 334, 335, 336, 371, 645
  aerobic, 334
  anaerobic, 334, 335
  facultative, 335
heterotropic enzymes, 235, 237
hexane, physical properties, 67
hexokinase, 184, 209, 210, 217, 403–405, 415, 566, 571
  conformational changes, 185, 232, 405
  denaturation, 246
  in glycogen synthesis, 572
  isozymes, 405
  K_M, 216, 431
  molecular weight, 126, 180, 184
  in phosphate-group transfer, 379
  quaternary structure, 184, 232
  regulation, 405, 420, 421, 426, 449
  specificity, 405, 417, 419
hexokinase reaction, standard-free-energy change, 403
hexosaminidase A, 241, 604
hexoses, 278
hibernating animals, futile cycles, 572
  triacylglycerol oxidation, 598, 599
high-density lipoproteins, 706, 759, 760
high-energy phosphate bonds, 375, 376
high-energy phosphate compounds, 374–380
  in glycolysis, 408, 410, 411
  super, 374, 378–380, 408
Hill, R., 656
Hill reaction, 656
Hill reagent, 656
his operon, 905, 906
histidine, **56**, 101, 103
  as buffer, 108
  in enzymes, 223, 227, 229–231, 446
  in glucose synthesis, 568
  in glutamine synthetase regulation, 623
  oxidative degradation, 536, 543
  p*K* values, 107
  in proteins, 173
  specific rotation, 97
  synthesis, 621, 622
    enzyme repression, 903, 905
  titration curve, 107
histocompatibility sites, 326
histohematins, 480
histones, 143, 816–819
  genes for, 826, 827
  molecular weights, 818
hnRNA, *see* heterogeneous nuclear RNA

Hoagland, M., 872
H_2O dehydrogenase, 660
Holley, R. W., 828, 875
holoenzyme, 209
homocysteine, **619**
  in cysteine synthesis, 619, 620
homocystinuria, 241, 542
homogenizer, 355
homogentisate, **540**
  excretion, 542
homogentisate 1,2-dioxygenase, 241, 540, 542
homologous proteins, 137, 139, 173, 177, 183
homopolysaccharides, 287
Homo sapiens, 681, 682
  DNA composition, 803
homoserine, in amino acid synthesis, 624
homotropic enzymes, 235, 236
hoof-and-mouth disease, 939
Hooker, Sir Joseph, 62
hormone-receptor sites, 34, 314, 325, 326, 346, 347, 723, 724, 741, 750
hormones, 34, 50, 123, 268, 346, 347, 721–747
  amine, 723, 724, 727, 728
  in blood, 706, 707, 723, 724, 727, 749
  classification, 724
  isolation, 724, 725
  lipid-soluble, 749
  peptide, 114, 723–727
  in radioimmunoassay, 726, 727
  regulation of blood glucose, 422, 423, 741
  secretion, 722, 723
    regulation, 723
  solubility, 724, 725, 749
  steroid, 723, 724, 740–745
  structure, 723–728
  water-soluble, 749
H^+-transporting ATPase, 385, 386, 675, 705
human cells, chromosomes, 815, 816
  electron micrograph, 815
human nutrition, *see* diet, nutrition
Hurler's syndrome, 606, 607
hyaluronic acid, 287, 296
hyaluronidase, 296
hybrid duplexes, in DNA, 809
hybridization, 809, 810
hybrid protein, 937
hydride ion transfer, 408, 409, 471, 472, 477, 478
hydrocarbons, 50
hydrochloric acid, in stomach, 687
hydrogen, isotopes, 352
  in living organisms, 46
  van der Waals radius, 49
hydrogenation reactions, 461
hydrogen atom transfer, 471, 472
hydrogen bonds, 58, 68–70
  in α helix, 152, 153
  in β conformation, 155
  in biological systems, 69
  in DNA structure, 69, 70, 805, 806, 809
  in enzyme-substrate complex, 228
  interchain, 155, 162
  intrachain, 155, 162
  in proteins, 69, 70, 160, 178
  in tRNA, 876
  in water, 68
hydrogen cyanide, 61
hydrogen electrode, 472, 473
hydrogen peroxide, 29
  in catalase reaction, 481
  toxicity, 481
hydrogen sulfide, in cytochrome oxidase inhibition, 483
  in photosynthesis, 648
hydrolases, 210
hydrolysis reactions, standard-free-energy changes, 369
hydronium ion, 74
hydrophilic groups, 71
hydrophobic groups, 71
hydrophobic interactions, 71
  in DNA structure, 806, 809
  in proteins, 179
3-hydroxyacyl-ACP dehydratase, 588, 590, 591
D-3-hydroxyacyl-CoA, **521**
L-3-hydroxyacyl-CoA, **516,** 517, 520, 521
3-hydroxyacyl-CoA dehydrogenase, 477, 516, 517
3-hydroxyacyl-CoA epimerase, 521
hydroxyapatite, 269, 783
D-β-hydroxybutyrate, **524**
  from acetoacetate, 525
  in brain metabolism, 699
  in diabetes mellitus, 714
  oxidation, 525
β-hydroxybutyrate dehydrogenase, 478, 525
D-3-hydroxybutyryl-ACP, 590, 591
hydroxylases, 503
5-hydroxylysine, **103,** 158
β-hydroxy-β-methylglutaryl-CoA, **524**
  in cholesterol synthesis, 608, 609
  in ketone body formation, 524
hydroxymethylglutaryl-CoA lyase, 524
hydroxymethylglutaryl-CoA reductase, 608–610
hydroxymethylglutaryl-CoA synthase, 524, 609
hydroxymethyl-group transfer, 538
4-hydroxyphenylpyruvate, **540**
4-hydroxyphenylpyruvate dioxygenase, 540
4-hydroxyproline, **103,** 158
  in collagen, 158, 264
  formation, 168, 264
hypercholesterolemia, 614
hyperglycemia, 713, 714, 719, 736
hyperinsulinism, 751
hyperkaluria, 720
hypertension, 760, 762, 780
hyperthyroidism, 743
hypervalinemia, 542
hyperventilation, 93
hypoglycemia, 523, 570, 571
  in ethanol consumption, 762
hypoglycin, 523
hypothalamic hormones, 725–727
hypothalamic regulatory hormones, 722
hypothalamus, 721–726, 740, 742
  hormones, 725–727
hypothyroidism, 743
hypoventilation, 93
hypoxanthine, **635**
  in AMP degradation, 634
  in DNA, 917
  in purine catabolism, 635
  salvage, 635
  in tRNA, 898
hypoxanthine-guanine phosphoribosyltransferase, 635
  from gene cloning, 937
hypoxia, 192
H zone, 381

I band, 381
ice, 68
D-idose, **280**
D-iduronate, in dermatan, 296, 297
  in heparin, 297
α-L-iduronidase, 606
i gene (*see also* regulatory gene), 908, 938
imidazole group, 51
immune globulins, *see* immunoglobulins
immune response, 138
immunity, 138
immunocytes, 926–929
immunoglobulins (*see also* antibodies), 102, 123, 138, 706, 926–928
  isolation, 145
Incaparina, 765
indicator dyes, 77
indoleacetate, **539**
induced enzymes, 347, 902, 903
induced fit, 194, 226, 232, 233
inducer (inducing agent), 902–907
inducer-repressor complex, 904–907
influenza, 795
information, 793, 794
informational macromolecules, 55
information theory, 365
Ingenhousz, J., 646
Ingram, V., 196
initiation codon, 883, 897
initiation complex, 874, 883–885
initiation factors, 874, 883–885
initiation signal, 883
inorganic pyrophosphatase (*see also* pyrophosphatase), 513
inosine, **877**
  in AMP degradation, 634
  in tRNA, 875, 877, 898, 899
inosine monophosphate (IMP) (*see also* inosinic acid), 635
inosinic acid, **629,** 635
  in purine nucleotide synthesis, 628–630
  synthesis, 628, 629
inositol, 251, 311
inositol hexaphosphate (phytate), 193, **779**
insect flight muscle, cytochromes, 480
insects, futile cycles, 572
insertion mutations, 919, 920
insertion sequences, 926
insulin, 102, 114, 123, 134, 271, 736–739
  amino acid sequence, 129, 736
    analysis, 134
  in blood, 749
  in control of lipid metabolism, 702
  in control of triacylglycerol synthesis, 597
  deficiency (*see also* diabetes mellitus), 738
  disulfide cross-linkages, 203
  in endocrine system, 722
  from gene cloning, 937, 939
  glucokinase stimulation, 420, 421
  molecular weight, 126, 736
  precursor, 723
  receptor sites, 325, 738
  secretion, 713, 714
    excess, 751
    regulation, 737–740
  species variations, 134
  synthesis, 735–737
insulin therapy, 712, 715, 736, 739, 939
integral proteins, 319, 321, 323, 329
intercellular cement, 296
interferon, 934, 939–941, 944
intermediary metabolism, 337
interphase, 817
intervening sequences, *see* introns
intestine, cells, 43
intracellular fluid, composition, 704, 705
intracellular messenger, 731, 741
intrinsic factor, for vitamin B_{12} absorption, 774
introns, DNA, 827, 860, 935, 936, 938
  RNA, 859–861, 936
invariant residues, 137
invertase (*see also* sucrase), 286, 299
invert sugar, 299
iodine, deficiency, 743, 781
  in diet, 782

isotopes, 352
recommended daily allowance, 755
in thyroid, 742
as trace element, 270
iodoacetamide, **223**
iodoacetate, 718
glyceraldehyde 3-phosphate dehydrogenase inhibition, 409, 410, 718
ion-exchange chromatography, 109, 129
ionization, 74
ionization constant, 78
ionizing radiation, DNA damage, 914
ionophores, 489
ion product, 75
iron, absorption, 780, 781
in cytochrome *c*, 174
deficiency, 780
in diet, 754, 780, 781
in enzymes, 209, 270
in heme, 171, 173, 270
in hemoglobin, 181, 457
isotopes, 352
in nitrogenase, 638, 639
recommended daily allowance, 755
transport in blood, 780, 781
storage, 780, 781
as trace element, 270
iron-deficiency anemia, 780
iron-porphyrin, *see* heme
iron-sulfur centers, 479
in aconitase, 445
in NADH dehydrogenase, 479, 506
in respiratory chain, 476
in succinate dehydrogenase, 447
in xanthine oxidase, 634
iron-sulfur enzymes, 270
iron-sulfur proteins, 659
irreversible reactions, 426
islets of Langerhans, 735, 736
alloxan effects, 738
isocitratase, *see* isocitrate lyase
isocitrate, **442**
in citric acid cycle, 441, 442, 445
in glyoxylate cycle, 455, 456
mitochondrial transport of, 496
isocitrate dehydrogenase, 445, 477
molecular weight, 180
NAD-linked, 445
regulation, 453
NADP-linked, 445
regulation, 452, 453
isocitrate dehydrogenase reaction, standard-free-energy change, 445
isocitrate lyase, 455, 456, 465
isodesmosine, 161
isoelectric pH, 106
isoelectric point, 106
isoenzymes, *see* isozymes
isoleucine, **56,** 98, 101, 102
daily requirement, 763
in glucose synthesis, 568
as inhibitor, 234
oxidative degradation, 523, 536, 544, 558
in proteins, 173
specific rotation, 97
synthesis, 622, 623
regulation, 624
isomerases, 210, 520
isoniazid, 773
Δ^3-isopentenyl pyrophosphate, **608**
in cholesterol synthesis, 608, 609
in isoprenoid synthesis, 610
isoprene, 264, **265**
isoprene unit, in cholesterol, 609
in ubiquinone, 479
isoprenoid compounds, 264, 610
isopropanol, in photosynthesis, 648
isoproterenol, 64
isotopes, 352
radioactive, 352, 360
isotopic labeling experiments, 351, 352, 358, 360
in cholesterol synthesis, 607
in citric acid cycle, 449
in DNA base-sequence determination, 830
in glycolysis, 465, 466
in Hershey-Chase experiment, 802
in phosphogluconate pathway, 465, 466
in photosynthesis, 664
in protein synthesis, 872, 879, 880, 895
radioimmunoassay, 727
isozymes, 240, 241
in amino acid synthesis, 623
of hexokinase, 405
of lactate dehydrogenase, 240, 413
of malate dehydrogenase, 240
of pyruvate kinase, 424
Itano, H., 196

Jacob, F., 853, 903–905
Jagendorf, A., 662
jaundice, 627
joining genes, 928, 929
Joule, J., 367
joule, 367, 368

$K_{0.5}$, 236
K_I, 224
K_M, Michaelis constant, 212–217, 225, 245, 246
kala azar, 318
Kaplan, N., 256
Keilin, D., 480
Kendrew, J., 170
Kennedy, E. P., 444, 602
keratins, 95, 123, 149–157
x-ray analysis, 150
α-keratins, 124, 149–156, 162
amino acid content, 153
in hair, 154, 165
secondary structure, 151, 717
solubility, 154
tertiary structure, 717
β-keratins, 150, 155
amino acid content, 156
keratomalacia, 774
α-keto acid dehydrogenase, 544
α-keto acids, amination, 555
in amino acid synthesis, 616
oxidation, 536
from transaminase reactions, 532–534
3-ketoacyl-ACP reductase, 588, 590, 591
3-ketoacyl-ACP synthase, 588, 589
3-ketoacyl-CoA, **516,** 517, 518
3-ketoacyl-CoA transferase, 525, 540
α-ketobutyrate, **622**
in amino acid synthesis, 624
ketogenic amino acids, 545
α-ketoglutarate, **442**
in amino acid oxidation, 536, 537, 543
in amino acid synthesis, 453, 465
in citric acid cycle, 441, 442, 445, 446, 450, 451
in glucose synthesis, 568
in monooxygenase reactions, 503
oxidative decarboxylation, 446
in transaminase reactions, 259, 532
α-ketoglutarate dehydrogenase complex, 254, 445, 446, 477
localization, 444
regulation, 452, 453
ketohexoses, 283
ketol-acid reductoisomerase, 622
α-keto-β-methylvalerate, **622**
ketone bodies, 524, 526
from amino acids, 545
in blood, 525, 707
in diabetes mellitus, 525, 530, 545, 598, 713, 714
formation, 524, 525, 695
regulation, 526
oxidation, 524–526, 695
in skeletal muscle, 696
ketonemia, 714
ketones, 50
solubility, 70
ketonuria, 714
ketoses, 278, 281, 283
ketosis, 525, 581, 714, 738
kidney disease, 788
kidneys, 548
gluconeogenesis, 562
in metabolism, 703–705
Na^+K^+–ATPase, 386, 718
oxygen consumption, 708
regulation of blood sodium, 720
kilocalorie, 755
kinases, 377–380
King, C. G., 264
King George III, 627
Knoop, F., 511, 512
Kögl, F., 259
Kornberg, A., 843–846, 865
Krabbe's disease, 606
Krebs, H., 441–443, 549, 550
Krebs cycle (*see also* citric acid cycle), 444
kwashiorkor, 766, 788

lac operon, 905–908, 936
base sequence, 908
lac repressor, 906–908
electron micrograph, 906
molecular weight, 906
α-lactalbumin, 577
lactalbumin-galactosyltransferase complex, 577
lactase, 285, 419, 420, 686
lactate (*see also* lactic acid), **408**
from anaerobic glycolysis, 399–402, 408
in blood, 580, 707
from fermentation, 398
from glucose, in ruminants, 570
in glucose synthesis, 569, 570
in glycolysis, 408, 412–414, 437, 569
in muscle, 401, 569, 696, 697
in photosynthesis, 648
from pyruvate, 398
lactate dehydrogenase, 408, 412, 413, 426, 431, 477
isozymes, 240, 413
molecular weight, 180
quantitative assay, 244
quaternary structure, 185
serum levels, in myocardial infarction, 535
lactate dehydrogenase reaction, standard-free-energy change, 413
lactate fermentation, 398, 427
lacteals, 690
lactic acid (*see also* lactate), **51**
pK, 79
lactic acid bacteria, 427
amino acid synthesis, 616
lactic acid fermentation, 398, 427
Lactobacillus casei, 274, 275
β-lactoglobulin, isoelectric point, 128
lactonase, 457
lactose, 284, **285,** 286
anomers, 299
hydrolysis, 418–420, 902
metabolism, 359

regulation of β-galactosidase synthesis, 902–907
sweetness, 286
synthesis, 419, 576, 577, 581
lactose intolerance, 285, 420, 686
lactose synthetase, 577
lactose tolerance test, 434
lanosterol, **608**
in cholesterol synthesis, 608, 609
large intestine, 684
Larinus maculatus, 301
laruic acid, 305, 518
lauroyl-CoA, 518
law of mass action, 73
leaf, absorption spectrum, 655
leaf cell, 35
leaky mutation, 918, 919
leather, 158
lectins, 324, 325
Leder, P., 896
legumes, 335
nitrogen fixation, 637, 638
Lehninger, A., 444, 512
leishmaniasis, 318
Leloir, L., 419, 574
Lesch, M., 635
Lesch-Nyhan syndrome, 635, 636, 937
lethal mutations, 918–920
leucine, **56**, 101, 102
daily requirement, 763
oxidative degradation, 536, 538, 539
pK values, 107
in proteins, 153, 173
transaminase reaction, 532
leucine enkephalin, 144
leucine transaminase, 532
leucocytes, 706, 940
leucoplasts, 36
leu operon, 905
levorotation, 96
ligaments, 157, 160, 162
ligand, 187
ligases, 210
light energy, 651, 652
light-harvesting molecules, 655, 658
light reactions, 649, 651–662
energy production, 663
lignin, 292
lignoceric acid, 305
limit dextrin, 288
Lind, J., 770
Lineweaver-Burk equation, 217, 225
linoleic acid, 305, **759**
as essential fatty acid, 759
melting point, 328
oxidation, 520
synthesis, 595
linolenic acid, 305, **759**
as essential fatty acid, 759
melting point, 328
synthesis, 595
α-linolenic acid, synthesis, 595
γ-linolenic acid, synthesis, 595
linoleyl-CoA, 520
linseed oil, 308
lipases, 307, 512, 690, 702
lipid bilayers, 317, 328, 329
in membranes, 320–322
lipids, 54, 303–319
absorption, 690, 691
in blood, 707
deposits, in atherosclerosis, 759
digestion, 683, 690, 691
hydrolysis, 328, 604
lysosomal hydrolysis, 28
in membranes, 303, 310–312, 314, 315, 317–319, 321, 322, 328, 603
metabolism, 695, 696
genetic defects, 604–607, 702
micelle formation, 71
nonsaponifiable, 315
polar, 310, 317, 318
saponifiable, 307, 315
in stool, 718
storage, 303
synthesis, 27, 583–603, 701
transport, 316, 707
Lipmann, F., 256, 372, 443
lipoic acid, 251, **438**
in α-ketoglutarate dehydrogenase complex, 446
in pyruvate dehydrogenase complex, 438–440
β_1-lipoprotein, 123, 126
lipoprotein lipase, 701, 702
lipoproteins, 122, 125, 126, 315, 316, 760
liposomes, 317, 318, 329
lipotropin, 731, 746
molecular weight, 726
lipoyllysyl group, 438–440
Lipscomb, W., 238
liquid scintillation counting, 360
liver, 684, 691
adrenaline stimulation, 732, 733, 751
amino acid metabolism, 693, 694
in ammonia metabolism, 546, 547
damage to, 535
in detoxification, 457, 458, 504, 695, 696
in digestion, 692
lipids, 695, 696
sugars, 692–694
glucagon stimulation, 739
glucokinase activity, 405, 431
gluconeogenesis, 562
glucose metabolism, 692, 693
glycogen content, 288, 420, 576
glycogen phosphorylase activity, 423, 433
glycogen synthesis, 400, 401, 572–574
ketone body formation, 524–526
lipid metabolism, 695, 696
metabolic turnover of components, 353
in metabolism, 695–697
mitochondria, 25, 470
electron micrographs, 499
pentose phosphate pathway, 456
protein synthesis, 694
urea cycle, 546–555, 694
vitamin A storage, 775
liver cells, 22
adenine nucleotide concentrations, 373
compartmentation, 354
dimensions, 17
DNA, 816
electron micrograph, 23
fat droplets, 308
phosphate concentration, 373
surface, electron micrograph, 34
lizards, uric acid excretion, 548, 555
locust, DNA composition, 803
Lohmann, K., 372
loop of Henle, 704
low-density lipoproteins, 706, 760
low-energy phosphate compounds, 374, 378–380
luciferase, 388
luciferin, **388**
luciferyl adenylate, **388**
lungs, carbon dioxide in, 86, 87, 710, 711
in cardiovascular system, 698
electron micrograph, 712
oxygen in, 187, 188, 709
luteinizing hormone (LH), 722
molecular weight, 726
luteinizing-releasing hormone, 725
lyases, 210
lymphoblast, 139
lymphocytes, 123, 138
antibody formation, 926
lymph vessels, 685, 690, 691
Lynen, F., 512, 609
lysine, **56**, 101, 103
daily requirement, 763
in glucose synthesis, 568
in histones, 816, 818
isoelectric point, 107
methylation, 892
oxidative degradation, 536, 538, 539
pK values, 107
in proteins, 153, 160, 173, 892
specific rotation, 97
synthesis, 622
regulation, 623, 624
lysogeny, 923
lysosomal diseases, 606
lysosomes, 23, 28, 606
electron micrograph, 29
enzyme activities, 354, 606
genetic defects, 604–606
ganglioside deposits, electron micrograph, 605
isolation, 355
lysozyme, 175
active site, 175
disulfide cross-linkages, 203
enzyme-substrate complex, 228
isoelectric point, 128
molecular weight, 126
pH activity curve, 247
tertiary structure, 175
lysyl oxidase, 271
D-lyxose, **280**

McCarty, M., 801
McCollum, E., 265
McElroy, W., 388
MacLeod, C., 801
macrofibril, 154
macromolecules, 5, 54, 58
catabolism, 338
informational, 54
synthesis, 338
macronutrients, 250, 754
magnesium, in chlorophylls, 653, 654
deficiency, 779
in diet, 779
as enzyme cofactor, 209, 403, 779
in glycolytic enzyme activity, 403
in intracellular fluid, 705, 779
in pyruvate dehydrogenase regulation, 451
recommended daily allowance, 755, 779
in urine, 703
malaria, 198
malate, **442**
in citric acid cycle, 441, 442, 446–448
in gluconeogenesis, 564
in glyoxylate cycle, 455, 456
in Hatch-Slack pathway, 670–672
mitochondrial concentration, 464
mitochondrial transport, 496, 586
stimulation of oxygen consumption, 463
malate-aspartate shuttle, 496, 497
malate dehydrogenase, 446–448, 477
in acetyl-group shuttle, 586
in gluconeogenesis, 564
in glyoxylate cycle, 455
isozymes, 240
in NADH shuttle system, 496, 497
NAD requirement, 256
malate dehydrogenase reaction, standard-free-energy change, 448, 463
malate enzyme, 670, 679
NADPH production, 593

malate synthase, 455, 456, 465
malathion, 222
maleic acid, **148**
4-maleylacetoacetate, **540**
maleylacetoacetate isomerase, 540
malnutrition, 766
malonate, **224**
  succinate dehydrogenase inhibition, 224, 442, 443, 447, 584
malonyl-CoA, **584**
  in control of fatty acid oxidation, 526
  in fatty acid synthesis, 584–589, 592–595
malt, 428
maltase, 282, 419
  in brewing, 428
malting, 428
maltose, 284, **285**, 286
  hydrolysis, 419
mammary gland, estrogen receptors, 744, 745
  lactose synthesis, 419, 577
  pentose phosphate pathway, 456
manganese, as enzyme cofactor, 209, 412
  as trace element, 270, 271
D-mannose, 57, **280**, 281
  from digestion, 686
  glycolysis, 415, 419
mannose 6-phosphate, 415, 419
α-mannosidase, 606
mannosidosis, 606
maple syrup urine disease, 542, 544
marasmus, 765, 766
margarine, fatty acid composition, 760
Matthaei, H., 895
Maxam, A., 829
Mayer, R., 646
melanin, 349
melanocyte-stimulating hormones, 746
melting curve, DNA, 811
membranes, 58, 318–326
  ammonium permeability, 545
  asymmetry, 329
  cell, enzyme activities, 353, 354
  chloroplast, 325, 494, 649, 650, 662
  composition, 318, 319
  concentration gradients, 385
  electric potential, 325, 700
  electron microscopy of, 320
  enzymes in, 325
  evolutionary origins, 60
  fluid-mosaic model, 321, 322
  glycolipids, 319, 320, 322, 324
  glycoproteins, 320, 322–325, 329
  lipids, 303, 310–322, 328, 603
    oxidation, 481, 776
    synthesis, 600–603
    turnover, 604
  mitochondrial, 318, 319, 325, 469, 470, 485, 495, 603
    calcium transport, 492
    proton translocation, 492, 493
    transport systems, 469, 470
  nerve cells, 699–701
  permeability, 319
  phospholipids, synthesis, 583
  proteins, 318, 319, 321–324
  recognition sites, 325, 326
  reconstitution, 485
  sidedness, 322
  transport, *see* transport systems
menaquinone, **269**
Menten, M., 212
mercaptoethanol, 167, 178
Meselson, M., 838, 839
Meselson-Stahl experiment, 838, 839, 868
mesophyll cells, 670–672, 679
  electron micrograph, 671
messenger RNA (mRNA), 795, 852–854, 859
  for antibodies, 928, 929
  base sequence, 828
  5′ cap, 859, 860
  in complementary DNA synthesis, 932, 933
  coding of amino acids, 896
  codon-anticodon binding, 898–900
  in *E. coli*, 795, 853
  eukaryotic, 795, 859
  exons, 860, 861
  in gene synthesis, 864
  initiation codon, 883, 897
  intergenic (spacer) regions, 854
  introns, 859–861, 936
  isolation, 932
  leader, 854
  lifetime of, 891
  molecular weights, 795
  monogenic (monocistronic), 853
  poly-A tail, 859
  polygenic (polycistronic), 853, 854
  in polyribosomes, 890, 891
  posttranscriptional processing, 857–861, 936
  precursors, 859, 860
  prokaryotic, 857, 859, 891
  in protein synthesis, 853, 872–874, 876, 882–891
  size, 853, 854
  synthesis (*see also* DNA, transcription), 854–856
    inhibition, 856, 857
    regulation, 904, 905
  synthetic, 895, 896
  termination codon, 874, 888, 897, 919
  translation, 792, 795, 891
  transport to cytoplasm, 861, 862
metabolic intermediates, 337
metabolic pathways, 331, 333, 337
  regulation, 426
metabolic turnover, 352, 353
metabolism, 333, 337, 342, 343, 345, 348
  compartmentation, 342, 353
  experimental investigation, 348–352
  rate of, 352
  regulation, 345–347
  secondary, 347
metabolites, 33, 337
  assay, 425, 432
metal ions, as cofactors, 209
metalloenzymes, 270, 638
metalloproteins, 125, 126
metaphase, 817
methane, 46
  physical properties, 67
methanol poisoning, 247
methemoglobin, 272
methionine, **56**, 101, 102
  in S-adenosylmethionine synthesis, 603
  in creatine synthesis, 625
  in cysteine synthesis, 618
  daily requirement, 763
  in glucose synthesis, 568
  oxidative degradation, 523, 536, 544, 570
  in proteins, 173
  in protein synthesis, 880, 881, 891
  synthesis, 622
    regulation, 623, 624
methionine adenosyltransferase, 619
methionine enkephalin, **65**
methionyl-tRNA, 880, 885
methotrexate, **643**
  as anticancer drug, 643
methylaspartate mutase, 264
2-methyl butadiene, 264, **265**
methylcobalamine, 264
methylene cyclopropylacetate, **523**
methylene cyclopropylacetyl-CoA, **523**
methylene-group transfer, 621
N^5,N^{10}-methylenetetrahydrofolate, 262, 538, **621**
  in dTMP synthesis, 633, 634, 642
methyl-group transfer, 538, 618, 619
  by *S*-adenosylmethionine, 602
O-methylguanine, 917
1-methylguanosine, in tRNA, 875, **877**
7-methylguanosine, in mRNA, 859
1-methylinosine, in tRNA, 875
6-*N*-methyllysine, **103**
methylmalonic acid, excretion, 276, 523
methylmalonic acidemia, 523
methylmalonyl-CoA, **522**
  in amino acid degradation, 523, 544
  from fatty acid oxidation, 522, 523
  in glucose synthesis, 570, 571
methylmalonyl-CoA mutase, 522, 523, 570, 571
  genetic defect, 523
methylmalonyl-CoA racemase, 571
methylmalonyl epimerase, 522
mevalonate 5-phosphate, **609**
mevalonate 5-pyrophosphate, **609**
mevalonic acid, **608**
  in cholesterol synthesis, 608, 609
micelles, 71, 305, 317, 328
Michaelis, L., 212
Michaelis-Menten constant, *see* K_M
Michaelis-Menten equation, 214–215, 217, 245
  for enzyme inhibition, 223–225
microbody (*see also* peroxisome), enzyme activities, 354
  isolation, 355
microfibril, 154
microfilaments, 29
micronutrients, 250, 754
microorganisms, mutant, 349, 350
microsome fraction, 355
microtrabecular network, 30
microtubules, 30–32, 123, 124, 163
  in cilia, 383
  in flagella, 383
  in mitosis, 25
microvilli, 33, 43, 295, 684
  electron micrograph, 685
Miescher, F., 800, 801
milk, 758
  calcium content, 778
  as food source, 717, 764
milk intolerance, 718
milk sugar, *see* lactose
Miller, S., 61
mineralocorticoids, 741
minerals, deficiency, 766
  in diet, 777, 778
  essential, 249, 753, 755
Minot, G., 262
missense protein, 896
Mitchell, P., 487, 492
mitochondria, 23, 25, 303
  acetyl-group shuttle, 585, 586
  ammonia permeability, 545
  in brown fat, 491, 509, 702
  calcium transport, 491, 492, 494
  citric acid cycle enzymes, 444
  coupled, 508
  cristae, 25, 470, 485, 491
  dehydrogenase, 477
  dicarboxylate-transport system, 496, 497, 509, 564, 586
  dimensions, 17
  division, electron micrograph, 820

DNA, 818–820
electron micrograph, 26, 499
electron-transport chain, 469, 470
enzyme activities, 354, 356
evolutionary origins, 26, 494, 820, 881, 897
fatty acid oxidation, 512–523, 526
fatty acid transport, 512–514, 520, 526
in gluconeogenesis, 563, 564
of heart muscle, 470, 698
electron micrograph, 510
inner membrane, 325, 469, 470
composition, 319, 328
conformational changes, 487
in energy coupling, 487–493
F_0F_1ATPase, 484, 485
H^+–ATPase, 386
in oxidative phosphorylation, 488
permeability, 477, 488, 489, 491–493, 495
pH gradient across, 489
proton pump, 509
transport systems, 495–497
vesicles, 485, 509
isolation, 355
matrix, 469, 470
membrane lipids, 603
outer membrane, 470
fatty acid activation, 512
oxygen consumption, 464, 499, 502
in plants, 649
proteins, metabolic turnover, 353
protein synthesis in, 880, 894, 895, 897
ribosomes in, 26, 881
shuttle systems, 496
sonication, 485
structure, 25, 470
tricarboxylate-transport system, 586
urea cycle, 551, 552
mitosis, 22, 25, 30, 383, 816, 817
mixed-function oxidases, 503, 594, 618
modification methylase, 823, 824
modulators, 234–237
molecular cloning, 931
molecular logic of the living state, 5, 45, 562
molecular sieve, 128
molybdenum, as enzyme cofactor, 209
in nitrogenase, 638, 639
as trace element, 270, 272, 783
in xanthine oxidase, 634
monellin, 124, 758
2-monoacylglycerols, 690
Monod, J., 853, 903–905
monohydrogen phosphate, pK, 79
L-monoiodotyrosine, **742**
monolayer, 317
monooxygenases, 502, 503, 618
monosaccharides, 278–284
aldoses, 278, 279, 282
anomers, 282
as chiral molecules, 280
conformations, 283
epimers, 281
hemiacetal formation, 282
hemiketal formation, 282, 283
ketoses, 278, 281, 283
optical activity, 279–281
projection formulas, 280
pyranoses, 282
reducing properties, 284
ring structure, 281–283
Moore, S., 135
mouse polyoma virus, 795
mouth, in digestion, 684
mRNA, *see* messenger RNA
mucin, 296
mucous cells, 687
Müller-Hill, B., 906
multicellular organisms, 16
multienzyme systems, 233, 333, 337, 340
regulation, 346
murein, 293, 294
Murphy, W., 262
muscle (*see also* heart muscle, skeletal muscle), adductor, 382
adenine nucleotide concentrations, 373
ammonia production, 546, 547
ATP content, 373, 384
ATP production, 569
flight, 382
glycogen content, 300, 569
glycogen phosphorylase activity, 422, 423, 433
glycolysis, 398–401, 431, 569, 696
glycolytic intermediates, assay, 425
hexokinase, 405, 431
metabolic turnover of components, 353
oxygen in, 188, 401, 441
phosphate concentration, 373
phosphocreatine content, 373, 384
recovery period, 401, 569, 697
red, 401
relaxation, 382, 396
smooth, 382
white, 401, 431
muscle contraction, 27, 29, 163, 380–384, 697
ATP requirement, 344, 380–384, 388, 389, 401, 569, 696–698
calcium in, 27, 381, 382, 698, 699
mutagens, 349, 916
detection, 922, 923
intercalating, 921
mutants (*see also* genetic mutations), 349–351, 820
auxotrophic, 350, 351
mutarotation, 282
mutase, 411
mutations, *see* genetic mutations
Mycoplasma pneumoniae, 42
mycoplasmas, 17, 42
myelin membranes, composition, 318, 582
myelin sheath, 312, 313, 582
myocardial infarction, 316, 535, 610, 699, 702
serum transaminase level in, 534
myofibril, 381
myoglobin, 123, 170, 183, 188, 270
carbon monoxide binding, 171
dimensions, 17
evolutionary origins, 184
heme, synthesis, 626
invariant residues, 173
isoelectric point, 128
molecular weight, 126, 170
oxygen saturation curve, 187, 236
oxygen storage, 205
secondary structure, 172
species variations, 173
tertiary structure, 171
x-ray analysis, 170, 171
myoinositol hexaphosphate, *see* phytate, phytic acid
myosin, 30, 103, 123, 124, 163, 380–383
ATPase activity, 163, 381, 382
dimensions, 17
electron micrograph, 29
in microvilli, 684
molecular weight, 163
in muscle contraction, 381, 382, 697, 698
myristic acid, 305, 518
myristoyl-CoA, 518
NAD, *see* nicotinamide adenine dinucleotide
NADH, *see* nicotinamide adenine dinucleotide, reduced form
NADH dehydrogenase, 478, 479, 482, 496, 506, 517
in electron-transport chain, 468, 476, 478
inhibition, 508
standard reduction potential, 474, 475
NADH/NAD^+ ratio, in cytosol, 593
NADH shuttle, 496, 497
NADH-ubiquinone reductase, 479, 482
NAD-linked dehydrogenases, 271, 471, 476, 477, 484
NAD^+/NADH couple, 461
standard reduction potential, 474, 475
NADP, *see* nicotinamide adenine dinucleotide phosphate
NADPH, *see* nicotinamide adenine dinucleotide phosphate, reduced form
NADPH/NAD^+ ratio, in cytosol, 593
NADP-linked dehydrogenases, 271, 476, 477
$NADP^+$/NADPH couple, standard reduction potential, 474, 657
Na^+K^+–ATPase, 322, 385, 386
inhibition, 717, 718
in kidney, 705
in nerve cell membrane, 700
naphthylamine, as carcinogen, 922
narcotic overdose, 719
Nathans, D., 825, 828
native conformation, 148
native proteins, 140
negative-contrast staining, 320
Neisseria gonorrhoeae, plasmids, 814
nephron, 704
nephrosis, 719
nerve cells, 30, 33, 221, 222, 315, 325, 699–701
nerve fibers, 582
nerve impulses, 700
in heart contractions, 699
neuromuscular junction, electron micrograph, 701
neurons, *see* nerve cells
Neurospora crassa, 349, 350
mutants, 349–351, 820
neurotransmitter, 221, 315, 700, 728
niacin, 256
in diet, 767
nickel as enzyme cofactor, 209
as trace element, 270, 273, 783
Nicolson, G., 321
nicotinamide (niacinamide), **251**, 255, 769
deficiency, 769, 770
in diet, 767, 769, 770
in NAD, 477
in pyruvate dehydrogenase complex, 438
recommended daily allowance, 754, 769
nicotinamide adenine dinucleotide (NAD), 209, 252, **255**, 256, 409, 477
in alcohol oxidation, 761
in dehydrogenase reactions, 476, 477
in DNA ligase reaction, 848
in electron transport, 436, 476, 478
in fatty acid oxidation, 516–519
in glutamate dehydrogenase reaction, 534, 535
in glyceraldehyde phosphate dehydrogenase reaction, 408, 409
in glycolysis, 408–410, 412–414
in hydride ion transfer, 408, 409, 471, 472, 477, 478
in α-ketoglutarate dehydrogenase complex, 446
in oxidation reactions, 462

in pyruvate dehydrogenase complex, 438–440
nicotinamide adenine dinucleotide, reduced form (NADH), **256**
in citrate synthase regulation, 452, 453
from citric acid cycle, 448
in control of carbohydrate metabolism, 501
in electron transport, 414, 468, 469
in fermentation, 427
in gluconeogenesis, 566, 567
in glycolysis, 412–414, 496, 497
regulation, 453
in isocitrate dehydrogenase regulation, 453
mitochondrial shuttles, 496, 497
in monooxygenase reactions, 503
from NADPH, 478
oxidation, 414, 496–498
in pyruvate dehydrogenase regulation, 451, 452
in reduction reactions, 462
nicotinamide adenine dinucleotide phosphate (NADP), 250, 252, **255**, 256, 345, 477
in dehydrogenase reactions, 476, 477
in glutamate dehydrogenase reaction, 535
in hydride ion transfer, 477, 478
in metabolism, 337, 345
in photosynthesis, 656–659
nicotinamide adenine dinucleotide phosphate, reduced form (NADPH), 345
in biosynthesis, 562, 593
in deoxyribonucleotide synthesis, 633
in fatty acid synthesis, 456, 584, 590–593
in Hatch-Slack pathway, 670, 671
in isocitrate dehydrogenase regulation, 453
in metabolism, 338, 344, 345
in mixed-function oxygenation, 594
in monooxygenase reactions, 503
in nitrogenase reaction, 638, 639
from pentose phosphate pathway, 456, 467, 593
in steroid synthesis, 456
from photosynthesis, 645, 649, 652, 656–659, 665, 666
in red blood cells, 457
nicotinamide coenzymes, 462
nicotinic acid, 252, **255**, 274, 538, 539
deficiency, *see* pellagra
Niemann-Pick disease, 604, 606
night blindness, 266, 775
ninhydrin, 109, **111**
ninhydrin reaction, 110, 111
Nirenberg, M., 895, 896
nitrate, as mutagen, 916, 917
in nitrogen cycle, 335, 636, 637
nitrate reductase, 209, 637
nitric acid, in rainwater, 88
nitrification, 636, 637
nitrifying bacteria, 335, 336, 636
nitrite, as mutagen, 916, 917
in nitrogen cycle, 335, 636, 637
nitrogen, in atmosphere, 335, 636, 637
excretion, 548–555
as ammonia, 549
as urea, 548, 549
as uric acid, 548, 549, 555
isotopes, 352
in living organisms, 46
in metabolism, 764
sources of, 334, 335
van der Waals radius, 49
nitrogenase system, 638–640
nitrogen balance, 764, 788
nitrogen cycle, 335, 336, 636–640
nitrogen fixation, 335, 336, 636–640, 648
nitrogen-fixing enzymes, gene cloning, 939
nitrogen mustard, as mutagen, 916
nitrosamines, as mutagens, 916
nitrous acid, as mutagen, 916, 917
Nocardia, 529
noncompetitive inhibition, 224, 225
nonheme iron, in NADH dehydrogenase, 479
noninformational macromolecules, 807
nonsense codons, 897
nonsense mutations, 888
noradrenaline, 540, **728**
norepinephrine (*see also* noradrenaline), 728
Northrop, J., 208
nuclear body (*see also* nucleoid), 18
nuclear envelope, 18, 24
nuclear pores, 24, 862
nucleases, 780, 800
in posttranscriptional processing, 858
nucleic acid (*see also* deoxyribonucleic acid, ribonucleic acid), 54, 794, 795
amphipathic nature, 71
digestion, 800
nucleotide units, 796–800
synthesis, 457, 807
viral, 37
nuclein, 800
nucleoid (*see also* nuclear body), 353
nucleolus, 23, 24
enzyme activities, 354
ribosomal RNA processing, 858
nucleoside diphosphate kinase, 389, 447, 630
nucleoside diphosphate sugars, in biosynthesis, 574, 669
nucleoside monophosphate kinase, 630
nucleoside 5′-triphosphates (*see also* nucleotides), 389, 391
nucleosomes, 816, 818, 819, 852
electron micrograph, 819
5′-nucleotidase, 634
nucleotides, 6, 55, 373, 796–800
synthesis, 615–636, 694
nucleus, 16, 23, 24, 58
enzyme activities, 354
isolation, 355
membrane composition, 319
nutrient proteins, 123
nutrients, bulk, 754
essential, 753
inorganic, 755
recommended daily allowances, 754, 755
nutrition (*see also* diet), bulk elements, 788
energy requirements, 753, 755, 756
trace elements, 778
nutritional calorie (Cal), 755
nutritional deficiency diseases, 249, 252, 255, 259, 260, 264, 266–269, 274–276, 440, 766
Nyhan, W., 635

obesity, 488, 762, 787
Ochoa, S., 865
ochre mutant, 888
Ogston, A., 451
Okazaki, R., 847
Okazaki fragments, 847–849, 852
oleic acid, **57**, 65, 71, 304, 305
melting point, 328
oxidation, 520
synthesis, 594, 595
oleyl-carnitine, 520
oleyl-CoA, **520**
oligomeric proteins, 126, 180, 203
cooperative interactions, 187
oligomycin, 485, 509, 662
oligonucleotides, separation of, 829, 830
oligosaccharides, 278
in cell coats, 295
in membranes, 322
olive oil, fatty acid content, 307
oncogenes, 864
oocytes, DNA replication, 843
opal mutant, 888
Oparin, A. I., 60
operator, 904–908, 936, 938
operon, 905–908
operon hypothesis, 903–905
opiates, 144, 746, 751
opiate receptors, 746
opioids, 144
opsin, 266
optical activity, 96
monosaccharides, 279–281
optical isomers, 148
amino acids, 96
RS system, 99
optimum pH, 87, 218
organelles, 17, 58
organic compounds, 47
chiral, 52
evolutionary origins, 60
functional groups, 50
origins of living matter, 59
ornithine, **550**
in urea cycle, 550–554
ornithine transcarbamoylase, 552
orotate, **631**, 632
orotate phosphoribosyltransferase, 631
orotidylate, **631**
in pyrimidine nucleotide synthesis, 631
orotidylate decarboxylase, 631
osmosis, 72
osmotic pressure, 72
osteomalacia, 776
ouabain, 718
ovalbumin, 123
gene for, 827
gene cloning, 937
synthesis, hormonal regulation, 745
ovariectomy, 751
ovaries, 721–723, 743
overlapping genes, 791, 900, 901
oviduct, 745
ovovitellin, synthesis, hormonal regulation, 745
ovum, fertilization, 296
oxaloacetate, **442**
in acetyl-group shuttle, 586
in amino acid oxidation, 536, 544
in amino acid synthesis, 453
in carbon dioxide fixation, 670
catabolism, in bacteria, 358
in citric acid cycle, 441, 442, 444–446, 448–451, 464
control of, 452, 453
in control of acetyl-CoA metabolism, 526
in glucose synthesis, 563–565, 567, 568
in glyoxylate cycle, 455, 456
in malate-aspartate shuttle, 496, 497
from phosphoenolpyruvate carboxylation, 455
from pyruvate carboxylation, 454, 455
stimulation of oxygen consumption, 463
succinate dehydrogenase inhibition, 224
synthesis from acetyl-CoA, 456, 465
oxidant, 471
oxidation reactions, standard-free-energy changes, 369
oxidation-reduction reactions, 461, 462, 471
oxidative phosphorylation, 435, 436, 446, 467–469, 484–494, 510
bacterial, 675
chemical coupling hypothesis, 487, 488

chemiosmotic hypothesis, 487–493, 509, 510
conformational-coupling hypothesis, 487
in fatty acid oxidation, 517–519
membrane transport systems in, 495, 496
mitochondrial location, 470
oligomycin inhibition, 485
regulation, 498–502
in submitochondrial vesicles, 485, 509
uncoupling, 488, 490, 508
oxidizing agent, 471
oxidoreductases, 210
oxidoreductions, 471, 472
in glycolysis, 408–410, 412–414
oxygen, arterial pressure, 187
consumption in brain, 699
consumption in muscle, 697
during exercise, 719
during rest, 719
cytochrome oxidase reaction, 480, 481
as electron acceptor, 469, 472, 473, 481
isotopes, 352
in living organisms, 46
nitrogenase inhibition, 638
in oxygenase reactions, 502–504, 594
in Pasteur effect, 509, 510
in photorespiration, 673, 674
from photosynthesis, 645, 646, 648, 649, 655–658
in respiration, 499
transport in blood, 185, 706–711
transport by hemoglobin, 189
van der Waals radius, 49
venous pressure, 187
oxygenases, 502–504
oxygen cycle, 334, 335
oxygen debt, 401, 569, 697
oxygen saturation curves, 187, 189, 193, 236
oxygen-water redox couple, standard reduction potential, 474, 475
oxyhemoglobin, 182, 191
oxyluciferin, 388
oxytocin, 114, 722, 726, 727
molecular weight, 726

P430, 657, 659, 661
P680, 657, 659
P700, 655, 657, 659, 661, 678
palindromes, 826, 827
palmitic acid, 56, **57**, 305
elongation, 594
oxidation, 515, 516, 593, 594
standard-free-energy change, 519
synthesis, 584, 592–594
palmitoleic acid, 305
synthesis, 594, 595
palmitoyl-ACP, 592
palmitoyl carnitine, **514**
palmitoyl-CoA, **513**
in control of fatty acid synthesis, 595
oxidation, 516, 518, 519
standard free energy of hydrolysis, 387
pancreas, 684, 686, 688, 690, 691, 721
electron micrograph, 735
endocrine activity, 735, 736
endocrine cells, 723, 735
enzyme synthesis, 735
exocrine cells, 735
insulin production, 736
tumors, 751
pancreatic amylase, 686
pancreatic cell, electron micrograph, 872
pancreatic hormones, 737
pancreatic juice, 690
pancreatic polypeptide, 736
pantothenic acid, 252, 256, **257**, 587, 773
daily requirement, 773
deficiency, 767, 773
in pyruvate dehydrogenase complex, 438
paper electrophoresis, 108
parasites, 18
viral, 37
parathyroid gland, 721
parathyroid hormone, 124, 731
molecular weight, 746
parietal cells, 385, 386, 687, 705
Parkinson's disease, 728
Pasteur, L., 208, 427, 509
Pasteur effect, 509, 510
pathogenic bacteria, 262, 296
pathogenic viruses, 39
Pauling, L., 150, 165, 196
pellagra, 251, 255, 274, 769, 770
penicillin, 294
penicillin resistance, 931
pentapeptides, 112
pentose phosphate pathway, 456, 457
defect in, 457
NADPH production, 593
pentoses, 278
in nucleic acids, 796
pepsin, 134
in digestion, 687
isoelectric point, 128, 143
molecular weight, 687
optimum pH, 219
pH-activity curve, 87, 218
specificity, 131, 687, 688
pepsinogen, activation, 687, 717
molecular weight, 688
secretion, 687
peptidases, 688, 689
in digestion, 688
peptide bond, 111, 150, 165
in α helix, 152
formation, 887, 890
freedom of rotation, 151
hydrolysis, 113, 131
peptide hormones, 114, 723–727
peptide map, 131, 196
peptides, 94, 111–115
amino acid sequence analysis, 132
antibiotic, 145
biological activity, 114
chemical reactions, 113
ionization, 112
overlaps, 133
separation, 112, 131
x-ray analysis, 150
peptidoglycan, 293, 294
peptidyl puromycin, **894**
peptidyl transferase, 887
peptidyl-tRNA, 887–889
performic acid, 134
peripheral proteins, 319, 321
peristalsis, 690
permanent waving, 156
pernicious anemia, 523, 558, 774
peroxidase, 209, 270
peroxisome (*see also* microbody), 23, 29, 673
electron micrograph, 29
perspective formulas, 97
Perutz, M., 181
petroleum, as microbial food source, 529, 938
pH, measurement, 77
phages, *see* bacteriophages
phenol, detoxification, 458
phenol glucosiduronide, **458**
phenotype, 820
phenylalanine, **56**, 101, 102
daily requirement, 763
in glucose synthesis, 568
metabolism, genetic defects, 349, 540–542, 557, 558
oxidative degradation, 536, 538–541, 544
in proteins, 173
specific rotation, 97
synthesis, 621
in tyrosine synthesis, 618
in urine, 558
phenylalanine hydroxylase (*see also* phenylalanine monooxygenase), genetic defects, 642
phenylalanine 4-monooxygenase (*see also* phenylalanine hydroxylase), 241, 540–542, 618
phenyl isothiocyanate, 131
phenylketonuria (PKU), 241, 349, 503, 540–542
phenyllactate, in urine, 558
phenylpyruvate, **541**
in urine, 557, 558
pheoporphyrin, 653
Phillips, D. C., 228
pH meter, 77, 92
phosphagens, 384
phosphatase, specificity, 220
phosphate, in blood plasma, 705
cell concentrations, 373
in chloroplasts, 678
in intracellular fluid, 705
in metabolism, 268, 343
mitochondrial transport of, 491, 495
resonance hybrids, 375
in urine, 703
phosphate buffer system, 83, 85
phosphate carrier, 495
phosphate-group transfer, 376–380
phosphatidate phosphatase, 596, 597
phosphatidic acid, 310, 311, **597**
in lipid synthesis, 596, 597, 600
phosphatidylcholine, 65, **310**, 311, 328
dimensions, 17
hydrolysis, 604
in membranes, 322
synthesis, 595, 600, 602
phosphatidylethanolamine, **310**, 311
in membranes, 322
synthesis, 595, 600, 601
phosphatidylinositol, **311**
synthesis, 603
phosphatidylserine, **310**, 311
in membranes, 322
synthesis, 603
phosphoarginine, 384
phosphocholine, **601**, 602
phosphocreatine, **384**
cell concentration, 373
in heart, 699
in muscle, 396, 397, 718
as phosphate-group reserve, 380, 383, 384, 389, 395
standard free energy of hydrolysis, 374, 383
phosphodiesterase, 734, 735
in *E. coli*, 907
phosphodiester linkage, 799, 800
phosphoenolpyruvate, **379**
carboxylation, 455
in glucose synthesis, 456, 563–565, 567
in glycolysis, 408, 411, 412
in Hatch-Slack pathway, 670, 671
as high-energy compound, 378–380, 411
standard free energy of hydrolysis, 374, 411
phosphoenolpyruvate carboxykinase, 455, 456
genetic defect, 576
in gluconeogenesis, 565
phosphoenolpyruvate carboxylase, 670, 672

phosphoethanolamine, 600, **601**
phosphoethanolamine cytidylyltransferase, 601
phosphofructokinase, 404, 406, 565, 571
  genetic defect, 576
  regulation, 406, 421, 423–426, 433, 453, 501, 502, 567
phosphofructokinase reaction, equilibrium constant, 394
  standard-free-energy change, 406
phosphoglucoisomerase, 404–406
phosphoglucoisomerase reaction, standard-free-energy change, 405
phosphoglucomutase, 414, 416–419, 693
  in glycogen synthesis, 572
  in glycolysis, 415–417
  inhibition, 222
  turnover number, 220
phosphoglucomutase reaction, standard-free-energy change, 368, 416
6-phosphogluconate, **457**
6-phosphogluconate dehydrogenase, 457
phosphogluconate pathway, 456, 457
  isotopic labeling experiments, 465, 466
6-phosphoglucono-δ-lactone, **457**
2-phosphoglycerate, **408**
  in gluconeogenesis, 567
  in glycolysis, 408, 411
3-phosphoglycerate, **408**
  in gluconeogenesis, 567
  in glycolysis, 408, 410
  in photosynthesis, 664–666
  from ribulose diphosphate carboxylase reaction, 673
  in serine synthesis, 620
phosphoglycerate dehydrogenase, 620
phosphoglycerate kinase, 378, 408, 410
phosphoglycerate kinase reaction, standard-free-energy change, 410
phosphoglycerate mutase, 408, 411
phosphoglycerate mutase reaction, standard-free-energy change, 411
phosphoglycerides (*see also* phospholipids), 304, 310–312
  bilayers, 317, 318
  in lipoproteins, 316
  in membranes, 318
  metabolic turnover, 353
  micelles, 317
  monolayers, 317
  stereoisomers, 311
  synthesis, 600–603
3-phosphoglyceroyl phosphate, **408**
  in gluconeogenesis, 567
  in glycolysis, 408–410
  as high-energy phosphate compound, 378–380
  standard free energy of hydrolysis, 374, 408
phosphoglycolate, **673**
3-phosphohydroxypyruvate, **620**
phospholipase A_1, 604
phospholipase A_2, 604
phospholipase C, 604
phospholipase D, 604
phospholipases, 312, 604
phospholipids (*see also* phosphoglycerides), 310–312, 316
  amphipathic nature, 71
  bilayers, 317
  in blood, 707
  in lipoproteins, 316
  liposomes, 317, 318
  in membranes, 319, 322
  metabolic turnover, 353
  synthesis, 583, 693
  vesicles, ATP-synthesizing, 679
phosphomannoisomerase, 415, 419
4′-phosphopantetheine, **587**
  in ACP, 587–593
phosphopentose isomerase, 457
phosphoprotein, 126
phosphoprotein phosphatase, 574, 575
3-phospho-5-pyrophosphomevalonate, **608**, 609
5-phospho-β-D-ribosylamine, **628**, 630
5′-phosphoribosyl-4-carboxamide-5-aminoimidazole, **643**
5-phosphoribosyl-1-pyrophosphate (PRPP), **628**
  in purine nucleotide synthesis, 628, 630, 635, 636
  in pyrimidine nucleotide synthesis, 631, 632
phosphoric acid, 57
  p*K*, 79
phosphorolysis, 415
phosphorus, absorption, 779
  in diet, 779
  as essential element, 270
  isotopes, 352, 391
  in metabolism, 779
  recommended daily allowance, 755
  van der Waals radius, 49
phosphorylase, *see* glycogen phosphorylase
phosphorylase *a*, 238, 239, 421, 422, 574, 575, 729, 734
  molecular weight, 421
phosphorylase *a* phosphatase, 238, 239, 422, 574, 734
phosphorylase *b*, 238, 239, 421, 422, 574, 575
  activation, 729–732
phosphorylase *b* kinase, 239, 422, 423, 574, 731, 732, 734
phosphorylation, *see* oxidative phosphorylation, photophosphorylation, substrate-level phosphorylation
phosphorylation potential, 376
3-phosphoserine, **620**, 892
phosphoserine phosphatase, 620
phosphoserine transaminase, 620
phosphothreonine, 892
phosphotyrosine, 892
photochemical action spectrum, 655
photochemical reaction center, 655, 657
photon, 651
photophosphorylation, 492–494, 660–662
  chemiosmotic hypothesis, 662
  cyclic, 661
photorespiration, 672–674
photosynthesis, 36, 645–676
  accessory pigments, 654
  bacterial, 18, 21, 334, 646, 648
  Calvin cycle, 664–667, 672
  dark reactions, 649, 651, 660, 663–669
  electron carriers, 652, 657, 659
  electron transport, 656–658, 661, 662
    cyclic, 657, 661
    noncyclic, 657–659
  hydrogen donors, 648, 649
  light reactions, 649, 651–662
  oxygen production, 645, 646, 648, 649, 655–657, 658
  photosystems, 654, 655
  Z scheme, 657, 658
photosynthetic algae, 644
photosynthetic cell, 4, 334, 335, 645, 647
  accessory pigments, 654
  chlorophylls, 654
photosynthetic phosphorylation, *see* photophosphorylation
photosystem I, 655, 657–659, 679
  cyclic electron flow, 661
photosystem II, 655, 657, 658, 679
photosystems, 654, 655
pH scale, 76
phylloquinone (*see also* vitamin K_1), **269**
phytate (*see also* phytic acid), 193, **779**, 783
phytic acid (*see also* phytate), 276
phytin, 779
phytohemogglutinins, 324
phytol, 610
  in chlorophyll *a*, 653
phytoplankton, 648
phytosterols, 614
pili, *E. coli*, 20
  in sexual conjugation, 925
pineal body, 721
ping-pong reactions, 218, 259, 533
pituitary diabetes, 740
pituitary gland, 721–723, 726
pituitary growth hormone, 124, 598
pituitary hormones, 725–727
p*K*, amino acids, 105–108
p*K*′, 79, 91
plankton, 309
  photosynthesis, 647
plant cell, chloroplasts, 4, 35, 303, 649, 650
  electron micrograph, 35
  membranes, 315
  mitochondria, 649
  structure, 35
plant cell walls, 292
  electron micrograph, 37
plants, amino acid synthesis, 616
  ammonia utilization, 637
  electron-transport chain, 507
  glyoxylate cycle, 456, 465
  lectins, 324, 325
  nitrogen fixation, 335, 336, 637–640
  photorespiration, 672–674
  proteins, as food source, 764, 765
  respiration, 672, 673
  sieve tubes, 668, 669
  triacylglycerol synthesis, 583
  tropical, 669, 670, 672
  viruses, 38, 794
  waxes, 309
plasma, *see* blood plasma
plasma cell, 139
plasma lipoproteins, 315, 316, 760
  synthesis, 695
plasma membrane, 16, 23, 324
  composition, 318, 319, 322
  electron carriers, 493
  electron micrographs, 320
  enzyme activities, 354
  H^+–ATPase, 386
  lipids, 603
  Na^+K^+–ATPase, 385, 386, 705
plasma proteins, 694, 706, 707
plasmid pSC101, electron micrograph, 814
plasmids, 814, 815, 925, 930, 931
  in gene isolation, 932
  gene transposition, 926
  as gene vectors, 930, 931, 934, 935
  recombinant, 934, 935
plastids, 35
plastocyanin, 657, 660, 661
  standard reduction potential, 507
plastoquinol A, **660**
plastoquinone, in photosynthesis, 657, 660
plastoquinone A, **660**
pleated sheet, 155, 167
pneumococcus, DNA, 801, 802
pneumonia, 42
pOH, 77
point mutations, 918, 919
poison ivy, 92
poison oak, 92
polar compounds, solubility, 70
polarimeter, 51
polar liquids, 68

poliomyelitis virus, 38, 795
  dimensions, 17
polyacrylamide gel, 829
polyadenylate polymerase, 859
polydipsia, 713
polyneuritis, 274, 440, 769
polynucleotide phosphorylase, 865, 866, 869, 895, 896
polynucleotides, 895, 896
polyoma virus, 794
polypeptides (*see also* proteins), 112
  amino acid sequence determination, 129–134
  folding, 179
  hydrolysis, 131
polyprenols, 610
polyribosomes, 21, 890, 891
  electron micrograph, 890
  isolation of mRNA from, 932
polysaccharides (*see also* carbohydrates), 54, 278, 287–297
  digestion, 683, 686
  lysosomal hydrolysis, 28
  storage, 287
  structural, 289–294
polysomes, *see* polyribosomes
polyuria, 713
P/O ratio, 508
pork fat, fatty acid composition, 760
porphobilinogen, 625, **626**, 627
porphyrias, 627
porphyrins, 653
  synthesis, 625, 626
    genetic defects, 626, 627
portal vein, 691
posterior pituitary gland, 726
posterior pituitary hormones, 726
postsynaptic neuron, 700, 701
posttranscriptional processing, 857–861, 936
potassium, in blood, 385, 704, 705
  in diet, 779, 780
  as enzyme cofactor, 209, 412
  in intracellular fluid, 704, 705
  isotopes, 352
  mitochondrial membrane permeability, 488, 489
  transport in kidney, 704, 705
  in urine, 703
potassium cyanate, 200
precipitin reaction, 138
preproglucagon, 739
preproinsulin, 737
preribosomal RNA, 857, 858
presynaptic nerve ending, 700, 701
Priestley, J., 646
primaquine, 457
primary structure, 121, 140, 153
primase, 848–850
procarboxypeptidase, activation, 690
  export from cell, 893
  secretion, 688, 689
prochiral compounds, 451
proelastase, activation, 690
progesterone, 722, **744**
proglucagon, 739
prohormones, 724
proinsulin, 724, 736, 737
  messenger RNA, 932
projection formulas, 97, 280
prokaryotes, 17–22, 33, 352, 353
  chromosomes, 794, 812, 813
  DNA, 19, 794, 812, 813, 825
  mRNA, 859, 891
    synthesis, 857
  photosynthetic, 646
  protein synthesis, 873, 880
  ribosomes, 881–883
prolactin, 722
  molecular weight, 726
prolactin-inhibitory hormone, 725
prolactin-releasing hormone, 725
proline, **56**, 101, 102
  in glucose synthesis, 568
  ninhydrin reaction, 110
  oxidative degradation, 536, 543
  in proteins, 153, 158, 173
  specific rotation, 97
  synthesis, 617
prolipase, 690, 691
promoter, 856, 905–908, 936, 938
proopiocortin, 746
propanol, physical properties, 67
prophase, 817
propionate, **571**
  from fatty acid oxidation, 570
  from glucose, in ruminants, 570
  in glucose synthesis, 570, 571
propionic acid (*see also* propionate), 521
  p*K*, 79
propionyl-CoA, **522**
  in amino acid degradation, 544, 558
  from fatty acid oxidation, 521, 522
  in glucose synthesis, 570, 571
  oxidation, 522
propionyl-CoA carboxylase, 522, 571
prostaglandin E_2, **747**
prostaglandins, 746, 747, 759
  synthesis, 595
prostaglandin synthetase, 746
prostate gland, 246, 746
prosthetic group, 125, 209, 270, 874, 892
protein-calorie malnutrition, 766
protein kinase, 574, 575, 731, 732
  in glycogen synthase regulation, 733, 734
  regulation of, 731, 732, 734
proteins, 6, 54, 95, 121–141
  α-helical structure, 151–153, 155, 162, 175
  amino acid analysis, 108, 113
  amino acid content, 764
  amino acid sequence, 121, 122, 135–138, 177, 178
    determination of, 129–134
  amino-terminal residue, 130
  amphipathic nature, 71
  "antifreeze," 124, 294
  in blood (*see also* plasma proteins), 705
  carboxyl-terminal residue, 130
  in cell membrane, 16
  chimeric, 937
  in chromatin, 816
  classification, 123, 124
  conjugated, 125, 126, 315
  contractile, 123, 163, 380, 381
  covalent modification, 891–893
  dansyl derivatives, 130
  deficiency, 766, 767
  denaturation, 140, 166, 169, 177
  in diet, 757, 763–765
  digestion, 683, 687–690, 717, 764
  2,4-dinitrophenyl derivative, 130
  disulfide cross-linkages, 102, 179, 202, 893
  as energy source, 755–757
  enzymatic hydrolysis, 687
  evolutionary origins, 60
  export, 893
  extracellular, 294, 892
  fibrous, 124, 147–163
  filamentous, 28, 30, 162
  folding, 179, 873, 891
  functional domains, 943
  functions, 122–124
  genetic coding, 822, 823
  globular, 124, 147, 154, 169–200
  half-life, 352
  homologous, 137, 139, 173, 177, 183
  hydrolysis, 124, 131
  in intracellular fluid, 705
  invariant residues, 177
  isoelectric points, 128
  in lipoproteins, 316
  lysosomal hydrolysis, 28
  in membranes, 318, 319, 321–324
  metabolic turnover, 353
  molecular weights, 126
  mutant, 198, 918
  native, 140
  nutrient, 123, 754
  oligomeric, 126, 180, 203
  phosphorylation, 892
  primary structure (*see also* amino acid sequence), 121, 140, 153
  purification, 127, 145
  quaternary structure, 181
  recommended daily allowance, 754, 763
  regulatory, 123, 193
  renaturation, 177
  ribosomal, 881–883
  salting out, 144
  secondary structure, 153, 162, 177
  secretion, 893
  separation, 127
  signaling sequences, 737, 891, 893, 894
  simple, 125
  solubility, 73, 154
  storage, 123
  structural, 123
  tertiary structure, 12, 141, 147, 170, 177, 178
  transport, 122
  viral, 38
  x-ray analysis, 147, 150, 170, 176
protein synthesis, 27, 32, 340, 871–901
  amino acid activation, 873, 877
  amino acid modification, 874
  in chloroplasts, 880
  direction of chain growth, 879, 880
  elongation, 873, 874, 879, 880, 885–888
    inhibition, 894, 895
  energy requirement, 888–890
  eukaryotic, inhibition, 894, 895
  folding, 873, 891
  GTP in, 874, 883–886, 888, 889
  inhibition, 894, 895
  initiation, 873, 874, 880, 881, 883
  isotopic labeling experiments, 872, 879, 880, 895
  in mitochondria, 880, 897
  mRNA in, 853, 872–874, 876, 882–891
  posttranslational modifications, 891–893
  prokaryotic, 873, 880
    inhibition, 894, 895
  rate of, 179
  regulation, 901–908
  termination, 873, 874, 888
  translocation, 888, 889
proteoglycans, 123, 157, 278, 296, 297, 892
  degradation, genetic defects, 606
proteolytic enzymes, in brewing, 428
  in digestion, 688
prothrombin, 103, 269, 706, 707, 777, 892
protofibril, 154
proton acceptors, 78
proton ATPase, *see* H^+–ATPase
proton donors, 78
protonophores, 488
proton pump, bacterial, 674
  mitochondrial, 509
protons, mitochondrial membrane permeability, 488, 489
proton "turbine," 495
protoporphyrin, 171

protoporphyrin IX, **626**
synthesis, 626
protozoa, 24
proximal tubule, of kidney, 704
Pseudomonas, 529
oxalate catabolism, 359
pseudouridine, **877**
in tRNA, 875, 876
pseudovitamin, 438
pteridine, 261
pteroylglutamic acid (*see also* folic acid), 260
pulse-chase experiments, 430, 431
purine, 796, **797**
purine bases, 796, 797
in DNA, 803, 804
salvage pathways, 634–636
synthesis, 694
uric acid formation, 555
purine nucleotides, 627–630
degradation, 634
synthesis, genetic defects, 635, 636
regulation, 630, 631
salvage pathways, 634–636
puromycin, **894**
purple sulfur bacteria, 646, 678
pus cells, 800
pyran, **282**
pyranoses, 282
pyrenoid, 647
pyridine nucleotides, in dehydrogenase reactions, 476, 477
pyridine nucleotide transhydrogenase, 478
pyridoxal, **258**
pyridoxal phosphate, 209, 252, **258**, 533, 773
in transaminase reactions, 258, 532–534
pyridoxamine, **258**
pyridoxamine phosphate, **258**, 533
pyridoxine (*see also* vitamin B_6), 252, **258**, 275, 532, 773
deficiency, 767, 773
in diet, 773
recommended daily allowance, 773
pyrimidine, 796, **797**
pyrimidine bases, 796, 797
in DNA, 803, 804
dimer formation, 914
synthesis, 694
pyrimidine dimers, 914
pyrimidine nucleotides, synthesis, 631, 632
regulation, 632
pyrocatechase, 503
pyrophosphatase, 387, 396, 572
pyrophosphate, from amino acid activation reaction, 877, 878
from DNA polymerase reaction, 843
from fatty acid activation, 512, 513
from RNA polymerase reaction, 854
standard free energy of hydrolysis, 513
Δ'-pyrroline 5-carboxylate, **617**
pyrroline carboxylate reductase, 617
pyruvate, **408**
in amino acid oxidation, 536, 537
in blood, 547, 707
carboxylation, 454, 455
in gluconeogenesis, 547, 562–568, 694
from glycolysis, 397, 408, 412, 414
in Hatch-Slack pathway, 670, 671
in lactate dehydrogenase regulation, 413
mitochondrial transport, 496
oxidation, 398, 441, 442
inhibition, 442, 443
oxidative decarboxylation, 438–440
pyruvate carboxylase, 454, 455, 564, 567, 663
biotin requirement, 259, 260
genetic defect, 576
molecular weight, 454
pyruvate carboxylase reaction, standard-free-energy change, 454
pyruvate decarboxylase, 426, 427
thiamine pyrophosphate requirement, 253
pyruvate dehydrogenase, 438–440
pyruvate dehydrogenase complex, 181, 254, 257, 437–441, 446, 477
electron micrograph, 440
insulin stimulation, 738
localization, 444
particle weight, 438
regulation, 441, 449–452, 567
thiamine pyrophosphate requirement, 437, 768
pyruvate dehydrogenase kinase, 450–452
pyruvate dehydrogenase phosphate, 451, 452
pyruvate dehydrogenase phosphate phosphatase, 451, 452
pyruvate dehydrogenase reaction, standard-free-energy change, 437
pyruvate kinase, 209, 379, 408, 412, 564
isozymes, 424
molecular weight, 412
regulation, 421, 423–426, 567, 568, 739
pyruvate kinase reaction, standard-free-energy change, 412, 564
pyruvate phosphate dikinase, 670

quantum, 651
quaternary structure, 181
Queen Victoria, 921
quinone, 479

rabbit papilloma virus, 795
racemate, 98
racemic mixture, 98
Racker, E., 485, 679
radioactive isotopes (*see also* isotopic labeling experiments), 351, 352, 360
radioactive waste, 914
radioimmunoassay, 726, 727
rearrangement reactions, standard-free-energy changes, 369
receptor proteins, isolation, 145
receptor sites, 34, 314, 325, 326, 346, 347, 723, 724, 741, 750
recognition sites (*see also* receptor sites), 325, 326
recombinant DNA, 815, 929, 930, 938
gene vectors, 930, 931
rectum, 684
red blood cells (*see also* erythrocytes), 185, 706
agglutination, 324, 325
carbon dioxide transport, 708–711
electron micrographs, 186, 320, 698
hemoglobin in, 204, 780
hemolysis, 457
membrane, 322–324
Na^+K^+–ATPase, 384, 386
oxygen transport, 708–711
pentose phosphate pathway, 457
potassium in, 322, 385
in sickle-cell anemia, 195
sodium in, 322, 385
type, 324, 325
red drop, 658
redox couple, 473
standard reduction potential, 473
redox potential, 473
redox reactions, 461, 462, 471
reducing agent, 471
reducing equivalent, 472
reducing sugars, 284
reductant, 471
reduction potential, 472, 473
Reed, L., 438
reference half-cell, 472, 473
regulatory enzymes, 185, 207, 233–239, 241
regulatory gene, 904–908, 938
regulatory proteins, 123, 193
regulatory site, 234
releasing factors, 874, 888
renal osteodystrophy, 276
renal rickets, 276
renal tubules, 703, 704
renaturation, DNA, 809
proteins, 177
reovirus, 795
replication, *see* DNA replication
replication fork, 840, 841, 847, 852, 868
replisome, 845
repressor, 123, 124, 904–908
isolation, 906
resilin, 124
resonance hybrids, 375
respiration, 435, 467, 468
control of, 441, 498, 499
malonate inhibition, 464
resting, 498, 499
respiratory chain (*see also* electron-transport chain), 435, 469
energy-conserving sites, 469, 476
free-energy changes, 475
mitochondrial location, 469
respiratory-chain phosphorylation, *see* oxidative phosphorylation
restriction endonuclease, 823–825
in DNA sequence determination, 828–831
Eco RI, 824, 825, 828
in gene isolation, 932
in gene vector formation, 934, 935
Hin, 828
Hin dII, 824, 825
Hpa, 828
in recombinant DNA formation, 929, 930
reticulocytes, 186
hemoglobin synthesis, 879, 880
reticuloendothelial system, 318
retina, visual cycle, 266
retinal, 167, 252, 266, 674, 775
11-*cis*-retinal, **266**
all-*trans*-retinal, **266**
retinol (*see also* vitamin A_1), **265**
retroviruses, 863
reverse transcriptase, 863, 864
in complementary DNA synthesis, 932, 933
in genetic engineering, 864
reversible reactions, 73
R groups, 96
rhodopsin, 266
deficiency, 775
riboflavin, 252, **254**, 274, 772
deficiency, 772
in diet, 767, 772
in FMN, 479
in pyruvate dehydrogenase complex, 438
recommended daily allowance, 754, 772
ribonuclease, 123, 125, 143, 175, 209
amino acid sequence, 135
denaturation, 267, 277
disulfide cross-linkages, 176, 203
molecular weight, 126
optimum pH, 219
renaturation, 177
tertiary structure, 175, 176
ribonucleic acid (RNA) (*see also* messenger RNA, ribosomal RNA, transfer RNA), 54, 795
in chromatin, 816
damage to, 918

enzymatic hydrolysis, 800
heterogeneous nuclear, 796, 859, 860
hybridization, 810
minor bases, 796
mitochondrial, 26
nucleotide units, 796–800
as primer in DNA replication, 847–849
small nuclear, 796
synthesis (*see also* DNA, transcription), 391
synthetic, 866
translation, *see* messenger RNA, translation
errors in, 852
viral (*see also* RNA viruses), 795, 812, 836, 863–865
transcription, 863–865
ribonucleoside 5′-diphosphates, in polynucleotide phosphorylase reaction, 865
ribonucleotide reductase, 633
ribonucleotides, 796, 797
in mRNA synthesis, 854
in RNA primer synthesis, 848
synthesis, 627–632
D-ribose, **57**, 279, 280
in RNA, 796
ribose 5-phosphate, **457**
in Calvin cycle, 665, 666
from pentose phosphate pathway, 456, 457
in purine nucleotide synthesis, 628
in pyrimidine nucleotide synthesis, 631
ribose phosphate diphosphotransferase, 628
ribosomal RNA (rRNA), 795
in *E. coli*, 795
gene cloning, 937
genes for, 820, 826
methylation, 858
molecular weights, 795
in protein synthesis, 885, 886
in ribosomes, 881–883
sedimentation coefficients, 881–883
structure, 882
synthesis, 843, 856–858, 862
ribosomes, 16, 27, 32, 58, 353, 354
aminoacyl site, 884–889
blocking of, 894
bacterial, 21
dimensions, 17
electron micrographs, 59, 872
eukaryotic, 881, 882
inactivation, 895
isolation, 355
mitochondrial, 26, 881
particle weights, 881, 882
peptidyl site, 884–889
prokaryotic, 881–883
proteins, 881–883
in protein synthesis, 872, 883–888, 890, 891
reassembly, 882
RNA (*see also* ribosomal RNA), 881–883
sedimentation coefficients, 881, 882
structure, 59
synthesis, 843
ribothymidine, in tRNA, 875, 876
D-ribulose, **281**
ribulose 1,5-diphosphate, **664**
in Calvin cycle, 665, 666
glycolate formation, 673
in photosynthesis, 664, 665
ribulose diphosphate carboxylase, 664, 667, 670, 672
molecular weight, 664
with oxygen as substrate, 673
regulation, 669, 679
ribulose 5-phosphate, **457**
in Calvin Cycle, 665, 666
in pentose phosphate pathway, 457
rice, as food source, 764, 765, 768
ricin, 123, 324, 895
Rickes, E., 263
rickets, 267, 776
rifampicin, 857
RNA, *see* ribonucleic acid
RNA-directed DNA polymerase (*see also* reverse transcriptase), 863
RNA-directed RNA polymerase, 864, 865
RNA polymerase, 180, 181, 271, 854–856, 860, 863, 869, 870, 891
electron micrograph, 862
inhibition, 856, 857
interaction with operon, 906–908
molecular weight, 180.
regulation, 904, 905
rRNA synthesis, 857–859
tRNA synthesis, 857–859
RNA polymerase I, 856
RNA polymerase II, 856
RNA polymerase III, 856
RNA replicases, 864, 865
RNA viruses, 38, 795, 812, 836, 863–865
Roberts, R. B., 358
rod cells, 266
rotation of plane-polarized light, 51
rotenone, 483, 508
rous sarcoma virus, 795, 863
rRNA, *see* ribosomal RNA
RS system, 99
rubber, 610
rumen, 275, 292, 570
microorganisms, 554
rumen bacteria, cellulose digestion, 686
vitamin B_{12} synthesis, 773, 774
ruminants, gluconeogenesis, 569, 570
glucose fermentation, 570
propionic acid formation, 521
urea production, 554

saccharin, 286, **287**, 758
salivary glands, 686
salmon, DNA composition, 803
Salmonella typhimurium, in Ames test, 922
histidine-requiring mutant, 922
operons, 905
salt (sodium chloride), in diet, 779
iodized, 782
salting out, 144
salts, absorption, 685
solubility, 70, 73
Sanger, F., 129, 130, 134, 791, 829, 900
saponification, 307
sarcolemma, 381
sarcomere, 380, 381
sarcoplasm, 382
sarcoplasmic reticulum, 382
Ca^{2+}–ATPase, 382, 386, 698
in heart, 699
in muscle contraction, 698
in skeletal muscle, 697, 698
sarcosine, 857
satellite DNA, 825
scanning electron microscopy, 320
Schally, A., 726
Schiff base, 533, 534
Schwann cell, 582
SCK test, 535
scurvy, 262, 770–772
sea, in chemical evolution, 62
seals, glycolysis, 569
sea urchin, DNA composition, 803
seawater, 46
toxicity, 720
secondary structure, 153, 162, 177
second law of thermodynamics, 331, 362–365
second messenger (*see also* intracellular messenger), 725, 730
secretin, 687, 688, 746
secretory vesicles, 28, 603, 893
D-sedoheptulose, 281
sedoheptulose 1,7-diphosphate, 667
in Calvin cycle, 665–667
sedoheptulose 7-phosphate, in Calvin cycle, 665, 666
selenium, as enzyme cofactor, 209
as trace element, 270, 272, 783, 784
selenocysteine, 272
sequenator, 135, 137
sequence homology, 137
serine, **56**, 101, 102, 794
in cysteine synthesis, 618, 620
in enzymes, 222, 223, 227, 229–231, 238, 417, 422, 574, 892, 918, 919
in glucose synthesis, 568
in glycine metabolism, 538, 620
oxidative degradation, 536–538
in phosphoglycerides, 311
p*K* values, 107
in proteins, 153, 173, 892
specific rotation, 97
synthesis, 620
serine-class enzymes, *see* enzymes, serine-class
serine hydroxymethyltransferase, 620
serotonin, 538, **539**
serum albumin, 123, 135, 142, 170, 706
in blood, 707, 719
dimensions, 170
excretion, 719
fatty acid transport, 512, 701, 702
gene for, 827
isoelectric point, 128
molecular weight, 126, 170
solubility, 154
sex hormones, 743–745
sexual conjugation, 925
gene recombination, 925, 926
SGOT test, 535
SGPT test, 535
sheep, DNA composition, 803
milk production, 581
Shemin, D., 625
Shorb, M., 262
shuttle systems, 496, 497
sialic acid, **314**
sickle-cell anemia, 58, 194–200, 542
sickle-cell hemoglobin, 194–200, 912
sickle-cell trait, 195
sickled red blood cells, electron micrograph, 195
sieve tubes, 668, 669
signaling sequences, 737, 891, 893, 894
silent mutations, 918, 919
silica, 272
in diatoms, 647
silicon, as trace element, 270, 272, 783
silk, 150, 155, 162, 166, 167
simian virus 40 (SV40), 38, 795
DNA, cleavage maps, 828
electron micrograph, 828
overlapping genes, 901
simple proteins, 125
Singer, J., 321
single-displacement reactions, 218
β-sitosterol, **614**
skeletal muscle, 162, 163, 382, 696
adrenaline stimulation, 733, 734, 751
ATP in, 718
contractile elements (*see also* muscle contraction), 29
glycolysis, 401, 696, 697

metabolism, 696
oxygen consumption, 708
phosphocreatine content, 718
relaxation, 382, 697, 698
skin, cancer, 914, 915
oxygen consumption, 708
UV damage, 914
Slack, C. R., 670
slime molds, 24
DNA, 815
small intestine, 683–686, 688, 690
small nuclear RNA (snRNA), 796, 861
smallpox virus, 38
Smith, E. L., 262
Smith, H., 825
smooth muscle, 382
snakes, nitrogen excretion, 548, 555
snake venom, 123
snRNA, *see* small nuclear RNA
soaps, 71, 305, 307
sodium, in blood, 385, 704, 705
in diet, 779, 780
in intracellular fluid, 704, 705
isotopes, 352
kidney transport, 704, 705
mitochondrial membrane permeability, 489
recommended daily allowance, 779
in urine, 703
sodium cholate, 329
sodium cyclamate, **758**
sodium dodecyl sulfate, 328, 329
sodium glycocholate, **691**
sodium oleate, **71**
sodium taurocholate, **691**
soil bacteria, 636, 637
solar energy, 8, 334, 336
in photosynthesis, 645, 646, 649
solar-energy cell, 675
solar light, 651
solutions, 72, 75, 76
somatic cells, 816
somatostatin, 722, 723, 725, 739, 740
gene cloning, 937
synthesis, 735, 736
somatotropin, 123, 722, 740
gene cloning, 939
molecular weight, 726, 740
secretion, 739, 740
somatotropin-releasing hormone, 725
soy beans, fatty acid composition, 760
space-filling models, 48, 49
spacer DNA, 818, 819, 862, 863
spark-discharge apparatus, 61
specific activity, enzymes, 219
isotopically labeled compounds, 360
specific rotation, 97
spectrin, 323, 324
spermaceti oil, 599
spermaceti organ, 599
spermatozoa, fructose metabolism, 430
sperm cell, 32, 925
sperm whale, 599
sphingolipids, 304, 312–314
metabolism, genetic defects, 604, 606
synthesis, 583
sphingomyelin, 312, **313**
degradation, genetic defect, 604, 606
in membranes, 322
sphingomyelinase, 604, 606
sphingosine, **312**
spinach cells, chloroplasts, electron micrograph, 650
photosystems, 654
electron-transport chain, 507
spinach ferredoxin, 659
spindle apparatus, 25, 30, 383
spirillae, 19
Spirogyra, 644
squalene, **608**, 610
in cholesterol synthesis, 608, 609
Stahl, F., 838, 839
standard-free-energy change (ΔG°, $\Delta G^{\circ\prime}$), 366–371, 376, 377
standard free energy of hydrolysis, 374, 375
standard oxidation-reduction potential, (E_0'), 472, 473
standard potential, 473
standard reduction potential, 473, 474, 506
staphylococci, 19
Staphylococcus aureus, cell wall, 293, 294
DNA composition, 803
starch, 54, 278, 287
in diet, 757
digestion, 686
glycolysis, 414, 416
granules, electron micrographs, 54, 288, 671
storage, 288
synthesis, 574, 668, 669
starch phosphorylase, 414, 416
starvation, 765
stearic acid, **304**, 305
melting point, 328
synthesis, 594, 595
steatorrhea, 718
Stein, W., 135
Steitz, T. A., 232
stereoisomers, 51–53, 97, 148
absolute configuration, 97
amino acids, 96
classification, 97
monosaccharides, 279–281
phosphoglycerides, 311
RS system, 99
stereospecificity, 53
steroid hormones, 610, 723, 724, 740–745
steroids, 315
hydroxylation, 504
sterol carrier protein, 610
sterols, 304, 315
stigmasterol, 315
Stoeckenius, W., 679
stomach, 684, 687, 691
stomata, 672, 673
storage granules, 353
storage proteins, 123
streptococci, 19
Streptococcus faecalis, 275
Streptococcus pneumoniae, DNA, 801, 802
Streptomyces, 483
streptomycin, 895
resistance, 814
strict anaerobes, 335
stroke, 315, 610, 759
stroma, 36, 650, 651
structural complementarity, 12
structural genes, 820, 904–908
structural proteins, 123
Subbarow, Y., 372
subclavian vein, 690, 691
substrate-level phosphorylation, 410–412, 446
succinate, **442**
in amino acid synthesis, 453
in citric acid cycle, 441, 442, 445, 446
in glyoxylate cycle, 455, 456
mitochondrial transport of, 496
succinate dehydrogenase, 126, 446, 447, 482, 484, 517
FAD requirement, 254
inhibition, 223, 442, 443, 447, 584
iron-sulfur centers, 479
molecular weight, 447
succinate-ubiquinone reductase, 482
succinyl-CoA, **446**
in amino acid oxidation, 536, 537, 544
in citric acid cycle, 441, 445, 446
control of, 452, 453
from fatty acids, 522, 523
in glucose synthesis, 568, 570, 571
as high-energy compound, 446
in ketone body metabolism, 525
in porphyrin synthesis, 625, 626
from propionyl-CoA, 522
standard free energy of hydrolysis, 446
succinyl-CoA synthetase, 446
succotash, 765
sucrase (*see also* invertase), 286, 299, 419, 680
sucrose, 57, 278, 284, **285**, 286
in diet, 757, 758
hydrolysis, 299, 417, 419
as sweetener, 758
synthesis, 668
sucrose density gradient, 835
sucrose 6-phosphate, 668
sugar, in diet, 757, 786
sugar cane, 758
sugars (*see also* monosaccharides, disaccharides), absorption, 685, 686
in glycolipids, 312
in metabolism, 338, 339
solubility, 70
sweetness, 286, 758
UDP derivatives, 458
sulfa drugs, 643
sulfanilamide, 262, **643**
sulfate, in blood, 705
in intracellular fluid, 705
in urine, 703
sulfhydryl groups, in fatty acid synthase, 588–590
sulfonic acid, in ion-exchange resins, 109
sulfur, isotopes, 352, 360
from photosynthesis, 648
van der Waals radius, 49
sulfuric acid, in rainwater, 88
Sumner, J., 208
superoxide dismutase, 481
superoxide radical, 481
suppressor mutation, 921
supramolecular assemblies, 37, 58
Surtsey, volcanic eruption, 60
Sutherland, E. W., Jr., 729, 730, 750
Suttie, J., 269
Svedberg, T., 795
svedberg, 795
symbiosis, 335
symbiotic nitrogen fixation, 637, 640
synapse, 222, 700, 701
synaptic cleft, 700, 701
synaptic vesicles, 700, 701
synovial fluid, 296
Szent-Györgyi, A., 442, 463

D-talose, **280**
target cells, 724
target tissues, 722, 723
Tatum, E., 820
Tay-Sachs disease, 28, 241, 542, 604–606
telophase, 817
Temin, H., 863
temperate phage, DNA, 923
as gene vectors, 931
tendons, 157, 160, 162
terminal transferase, 930, 931, 934, 935
termination codons, 874, 888, 897, 919
termination factors, 888
termites, 291
tertiary structure, 12, 141, 147, 170, 177, 178, 891
testes, 721–723, 743

water strider, 88
Watson, J. D., 129, 804, 805, 807, 808
Watson-Crick base pairs, 845, 850, 898–900
Watson-Crick hypothesis, 837
Waugh, W. A., 264
waxes, 304, 309
weak acids, 78
  as buffers, 82
  titration, 79
weak bases, 78
Wernicke-Korsakoff syndrome, 768
wheat, as food source, 764
wheat germ, DNA composition, 803
wheat proteins, 690
white blood cells, 706, 940
Wilkins, M., 804
Williams, Robert R., 252
Williams, Roger, 256
Willstätter, R., 208
wine, 427, 761, 762
wobble hypothesis, 898–900
wood, 291, 293
Woodward, R., 607
wool, 165, 166
Woolley, D. W., 251
wort, 428

xanthine, **635**
  in DNA, 917
  in purine catabolism, 635
  in purine nucleotide catabolism, 634
  in urine, 643
xanthine oxidase, 272, 634, 635, 784
  inhibition, 636
xanthophyll, 654
Xenopus, gene transcription, electron micrograph, 862
xeroderma pigmentosum, 914, 915
xerophthalmia, 266, 275, 774, 775
x-ray analysis, 150
  of enzymes, 226–233
  of myoglobin, 170
  of proteins, 150, 181
x-rays, as mutagen, 349, 914
D-xylose, **280**
xylulose 5-phosphate, in Calvin cycle, 665–667

Yalow, R., 726
yeast, 24, 208, 334
  aerobic glycolysis, 428
  in brewing, 428
  DNA composition, 803
  fermentation, 348, 358, 398, 426–428
  ribosomes, 59
Young, W., 432

Zamecnik, P., 872
zinc, in aldolase activity, 406
  deficiency, 276, 783
  in diet, 783
  as enzyme cofactor, 209, 783
  recommended daily allowance, 755, 783
  as trace element, 251, 270, 271
Z lines, 381
zwitterion, 103
zymogen granules, 689
zymogens, 229, 687–690
  secretion, 689